CLINICAL CASES

SPOTLIGHT FIGURES

HUMAN ANATOMY

NINTH EDITION

FREDERIC H. MARTINI, PH.D.
UNIVERSITY OF HAWAII AT MANOA

ROBERT B. TALLITSCH, PH.D.
AUGUSTANA COLLEGE (IL)

JUDI L. NATH, PH.D.
LOURDES UNIVERSITY

William C. Ober, M.D., *Art Coordinator and Illustrator*

Claire E. Ober, R.N., *Illustrator*

Kathleen Welch, M.D., *Clinical Consultant*

Ruth Anne O'Keefe, M.D., *Clinical Consultant*

Ralph T. Hutchings, *Biomedical Photographer*

330 Hudson Street, NY NY 10013

Editor-in-Chief: Serina Beauparlant
Courseware Portfolio Manager: Cheryl Cechvala
Content Producer: Caroline Ayres
Managing Producer: Nancy Tabor
Courseware Director, Content Development: Barbara Yien
Courseware Sr. Analysts: Suzanne Olivier, Alice Fugate
Development Editor: Molly Ward
Courseware Editorial Assistant: Kimberly Twardochleb
Director of Digital Product Development: Lauren Fogel
Executive Content Producer: Laura Tommasi
Rich Media Content Producer: Patrice Fabel, Lauren Chen
Content Developer, A&P and Microbiology: Cheryl Chi
Senior Mastering Media Producer: Katie Foley
Associate Mastering Media Producer: Kristin Sanchez
Full-Service Vendor: Cenveo Publisher Services
Full-Service Vendor Project Manager: Norine Strang
Copyeditor: Joanna Dinsmore
Compositor: Cenveo Publisher Services
Contributing Illustrators: Imagineeringart.com; Anita Impagliazzo
Art Coordinator:: Wynne Auyeung,
Design Manager: Marilyn Perry
Interior Designer: Elise Lansdon
Cover Designer: Elise Lansdon
Rights & Permissions Project Manager: Kathleen Zander, Jason Perkins
Rights & Permissions Management: Cenveo Publisher Services
Photo Researcher: Kristin Piljay
Manufacturing Buyer: Stacey Weinberger
Executive Marketing Manager: Allison Rona
Cover Illustration Credit: Sebastian Kaulitzki/Science Photo Library/Getty

Credits and acknowledgments borrowed from other sources and reproduced, with permission, in this textbook appear on the appropriate page within the text or on page 826.

Library of Congress Cataloging-in-Publication Data
Names: Martini, Frederic, author. | Tallitsch, Robert B., author. | Nath, Judi Lindsley, author.
Title: Human anatomy / Frederic H. Martini, Ph.D., University of Hawaii at Manoa, Robert B. Tallitsch, Ph.D., Augustana College, Rock Island, IL, Judi L. Nath, Ph.D., Lourdes University, Sylvania, OH; with William C. Ober, M.D., art coordinator and illustrator; Claire E. Ober, R.N., illustrator; Kathleen Welch, M.D., clinical consultant: Ruth Anne O'Keefe, clinical consultant; Ralph T. Hutchings, biomedical photographer.
Description: Ninth edition. | Glenview, IL: Pearson [2018]
Identifiers: LCCN 2016033415 | ISBN 9780134320762 (alk. paper)
Subjects: LCSH: Human anatomy. | Human anatomy—Atlases.
Classification: LCC QM23.2 .M356 2018 | DDC 612--dc23 LC record available at https://lccn.loc.gov/2016033415

4 18

ISBN 10: 0-13-432076-X; ISBN 13: 978-0-13-432076-2 (Student edition)
ISBN 10: 0-13-429229-4; ISBN 13: 978-0-13-429229-8 (Exam copy)
ISBN 10: 0-13-442494-8; ISBN 13: 978-0-13-4424941 (Books a la Carte edition)

Author and Illustration Team

Frederic (Ric) Martini
Author

Dr. Martini received his Ph.D. from Cornell University in comparative and functional anatomy for work on the pathophysiology of stress. In addition to professional publications that include journal articles and contributed chapters, technical reports, and magazine articles, he is the lead author of ten undergraduate texts on anatomy and physiology or anatomy. Dr. Martini is currently affiliated with the University of Hawaii at Manoa and has a long-standing bond with the Shoals Marine Laboratory, a joint venture between Cornell University and the University of New Hampshire. He has been active in the Human Anatomy and Physiology Society (HAPS) for over 24 years and was a member of the committee that established the course curriculum guidelines for A&P. He is now a President Emeritus of HAPS after serving as President-Elect, President, and Past-President over 2005–2007. Dr. Martini is also a member of the American Physiological Society, the American Association of Anatomists, the Society for Integrative and Comparative Biology, the Australia/New Zealand Association of Clinical Anatomists, the Hawaii Academy of Science, the American Association for the Advancement of Science, and the International Society of Vertebrate Morphologists.

Robert B. Tallitsch
Author

Dr. Tallitsch received his Ph.D. in physiology with an anatomy minor from the University of Wisconsin–Madison at the ripe old age of 24. Dr. Tallitsch has been on the biology faculty at Augustana College (Illinois) since 1975. His teaching responsibilities include human anatomy, neuroanatomy, histology, and cadaver dissection. He is also a member of the Asian Studies faculty at Augustana College, teaching a course in traditional Chinese medicine. Dr. Tallitsch has been designated as one of the "unofficial teachers of the year" by the graduating seniors at Augustana 19 out of the last 20 years. Dr. Tallitsch is a member of the American Association of Anatomists, where he serves as a Career Development Mentor; the American Association of Clinical Anatomists; and the Human Anatomy and Physiology Society. In addition to his teaching responsibilities, Dr. Tallitsch currently serves as a reviewer for the Problem-Based Learning Clearinghouse and has served as a visiting faculty member at the Beijing University of Chinese Medicine and Pharmacology (Beijing, PRC), the Foreign Languages Faculty at Central China Normal University (Wuhan, PRC), and in the Biology Department at Central China Normal University (Wuhan, PRC).

Judi L. Nath
Author

Dr. Judi Nath is a biology professor and the writer-in-residence at Lourdes University, where she teaches at both the undergraduate and graduate levels. Primary courses include anatomy, physiology, pathophysiology, medical terminology, and science writing. She received her Bachelor's and Master's degrees from Bowling Green State University, which included study abroad at the University of Salzburg in Austria. Her doctoral work focused on autoimmunity, and she completed her Ph.D. from the University of Toledo. Dr. Nath is devoted to her students and strives to convey the intricacies of science in captivating ways that are meaningful, interactive, and exciting. She has won the Faculty Excellence Award—an accolade recognizing effective teaching, scholarship, and community service—multiple times and in 2013 was named as an Ohio Memorable Educator. She is active in many professional organizations, notably the Human Anatomy and Physiology Society (HAPS), where she has served several terms on the board of directors. Dr. Nath is a coauthor of *Fundamentals of Anatomy & Physiology, Visual Anatomy & Physiology, Visual Essentials of Anatomy & Physiology,* and *Anatomy & Physiology* (published by Pearson); and she is the sole author of *Using Medical Terminology* and *Stedman's Medical Terminology* (published by Wolters Kluwer). Her favorite charities are those that have significantly affected her life, including the local Humane Society, the Cystic Fibrosis Foundation, and the ALS Association. In 2015, she and her husband established the Nath Science Scholarship at Lourdes University to assist students pursuing science-based careers. When not working, days are filled with family life, bicycling, and hanging with the dogs.

William C. Ober

Art Coordinator and Illustrator

Dr. Ober received his undergraduate degree from Washington and Lee University and his M.D. from the University of Virginia. He also studied in the Department of Art as Applied to Medicine at Johns Hopkins University. After graduation, Dr. Ober completed a residency in Family Practice and later was on the faculty at the University of Virginia in the Department of Family Medicine and in the Department of Sports Medicine. He also served as Chief of Medicine of Martha Jefferson Hospital in Charlottesville, VA. He is currently a Visiting Professor of Biology at Washington and Lee University, where he has taught several courses and led student trips to the Galápagos Islands. He was on the Core Faculty at Shoals Marine Laboratory for 24 years, where he taught Biological Illustration every summer. Dr. Ober has collaborated with Dr. Martini on all of his textbooks in every edition.

Claire E. Ober

Illustrator

Claire E. Ober, R.N., B.A., practiced family, pediatric, and obstetric nursing before turning to medical illustration as a full-time career. She returned to school at Mary Baldwin College, where she received her degree with distinction in studio art. Following a five-year apprenticeship, she has worked as Dr. Ober's partner at Medical & Scientific Illustration since 1986. She was on the Core Faculty at Shoals Marine Laboratory and co-taught the Biological Illustration course with Dr. Ober for 24 years. The textbooks illustrated by Medical & Scientific Illustration have won numerous design and illustration awards.

Kathleen Welch

Clinical Consultant

Dr. Welch received her B.A. from the University of Wisconsin–Madison and her M.D. from the University of Washington in Seattle, and did her residency in Family Practice at the University of North Carolina in Chapel Hill. Participating in the Seattle WWAMI regional medical education program, she studied in Fairbanks, Anchorage, and Juneau, Alaska, with time in Boise, Idaho, and Anacortes, Washington, as well. For two years, she served as Director of Maternal and Child Health at the LBJ Tropical Medical Center in American Samoa and subsequently was a member of the Department of Family Practice at the Kaiser Permanente Clinic in Lahaina, Hawaii, and on the staff at Maui Memorial Hospital. She was in private practice from 1987 until her retirement in 2012. Dr. Welch is a Fellow of the American Academy of Family Practice and a member of the Hawaii Medical Association, the Maui County Medical Association, and the Human Anatomy and Physiology Society (HAPS). With Dr. Martini, she has coauthored both a textbook on anatomy and physiology and the *A&P Applications Manual*. She and Dr. Martini were married in 1979, and they have one son.

Ruth Anne O'Keefe

Clinical Consultant

Dr. O'Keefe did her undergraduate studies at Marquette University, attended graduate school at the University of Wisconsin, and received her M.D. from George Washington University. She was the first woman to study orthopedic surgery at The Ohio State University. She did fellowship training in trauma surgery at Loma Linda University. She has always been passionate about global health and has done orthopedic surgery in high-need areas around the world, taking her own surgical teams to places such as the Dominican Republic, Honduras, Peru, Burkina Faso, and New Zealand. She serves on the board of Global Health Partnerships, a group that partners with a clinic serving 50,000 very poor people in rural Kenya. Dr. O'Keefe has enjoyed teaching at all levels and at all the universities in places where she has lived. She now lives in Albuquerque with her Sweet Ed. She is mother of four, grandmother of thirteen, and foster grandmother to many.

Ralph T. Hutchings

Biomedical Photographer

Mr. Hutchings was associated with The Royal College of Surgeons of England for 20 years. An engineer by training, he has focused for years on photographing the structure of the human body. The result has been a series of color atlases, including the *Color Atlas of Human Anatomy,* the *Color Atlas of Surface Anatomy,* and *The Human Skeleton* (all published by Mosby-Yearbook Publishing). For his anatomical portrayal of the human body, the International Photographers Association chose Mr. Hutchings as the best photographer of humans in the twentieth century. He lives in North London, where he tries to balance the demands of his photographic assignments with his hobbies of early motor cars and airplanes.

Preface

Welcome to the ninth edition of *Human Anatomy*! This edition marks a significant change to the author team with the retirement of Michael Timmons and the addition of a fine colleague and excellent writer, Judi Nath.

We have made significant changes to *every* chapter of the text. As a result, this book—which was already highly visual—is now even more visual and engaging. These changes will enhance students' understanding of the chapters and the intricacies of the human body. Our new and revised visuals will promote student involvement with the figures.

In addition, the author team has revised the chapter narratives to be even more "student friendly" with a lively writing style. We have repositioned figure callouts and tried to place all graphics on the same two-page spread with their anatomical descriptions.

New to the Ninth Edition

Our goal is to build on the strengths of previous editions while meeting the needs of today's students. The author team has paid significant attention to the latest research on the science of teaching and learning. Our reading of this research has informed the revision of both the art program and text narratives in this edition. As a result, we believe this edition will prove even more effective for attracting students' attention, enhancing their understanding, and promoting their retention of anatomical concepts.

- **EVERY ILLUSTRATION** has been revised, either partially or totally.
- **EVERY CHAPTER** has been extensively rewritten to
    - Engage students with an informal, friendly approach
    - Reposition figure callouts for easy reading and understanding
    - Place figures in a logical design that is both attractive and effective
    - Place figures as close to their anatomical descriptions as possible
    - Increase the number of bullet lists and numbered lists to better facilitate student learning
    - Use standardized terminology of the latest editions of *Terminologia Anatomica*, *Terminologia Histologica*, *Terminologia Embryologica*, and *Stedman's Medical Dictionary*

- **NEW Chapter Opener Clinical Cases** have been added to every chapter. These clinical cases increase student interest in the topics and vividly demonstrate the importance of anatomical concepts in the health professions. In addition, all of the existing Clinical Notes features, found within the chapters, have been updated or replaced to reflect current topics and the latest research.
- **NEW Tips & Tools** boxes are concise, catchy memory devices to help students easily remember anatomical facts and concepts.
- **NEW Key Points** boxes give students a quick summary of the material discussed in the upcoming section of the chapter.
- **Improved text-art integration** throughout enhances the readability of figures with the text.
- **NEW MasteringA&P features** include the following:
    - Ready to Go Teaching Modules, created by teachers for teachers, are organized around eight of the toughest topics in human anatomy. They provide suggestions to instructors on which assets in MasteringA&P can best be used before, during, and after class to effectively teach the topic.
    - A Coaching Activity for the new Spotlight Figure in Chapter 17 on the sympathetic nervous system.
    - Revised and updated Dynamic Study Modules.

Chapter-by-Chapter Revisions

In addition to a significant rewriting of every chapter within the text, as outlined above, the following changes have been made in each chapter of the ninth edition of *Human Anatomy*:

1 Foundations: An Introduction to Anatomy

- Nine illustrations either are new or have been significantly revised.
- All Clinical Notes within this chapter have been revised.
- One new Tips & Tools box was added to this chapter.
- The section dealing with sectional anatomy was extensively revised to better facilitate student learning

2 Foundations: The Cell

- Eight illustrations either are new or have been significantly revised.
- The sections dealing with the plasma membrane, cellular cytoskeleton and intercellular attachments were reorganized and revised to better facilitate student learning.

3 Foundations: Tissues and Early Embryology

- Sixteen illustrations either are new or have been significantly revised.
- Four new Tips & Tools boxes were added to this chapter.

4 The Integumentary System

- Nine illustrations either are new or have been significantly revised.
- One new Tips & Tools box was added to this chapter.

5 The Skeletal System: Osseous Tissue and Bone Structure

- Nine illustrations either are new or have been significantly revised.
- The sections dealing with blood and nerve supply to bones and factors regulating growth were reorganized and revised to better facilitate student learning.

6 The Skeletal System: Axial Division

- Twenty-nine illustrations either are new or have been significantly revised.
- Two new Tips & Tools boxes were added to this chapter.

7 The Skeletal System: Appendicular Division

- Twenty-four illustrations either are new or have been significantly revised.
- One new Tips & Tools box was added to this chapter.

8 The Skeletal System: Joints

- The chapter title has been changed from Articulations to Joints.
- Twenty-one illustrations either are new or have been significantly revised.
- The sections dealing with darthroses (freely movable synovial joints) and the elbow and radio-ulnar joints were reorganized and revised.

9 The Muscular System: Skeletal Muscle Tissue and Muscle Organization

- Thirteen illustrations either are new or have been significantly revised.
- All sections dealing with the microanatomy and the physiology of skeletal muscle contraction were extensively revised.
- One new Tips & Tools box was added to this chapter.

10 The Muscular System: Axial Musculature

- Fourteen illustrations either are new or have been significantly revised.
- The organization of the sections dealing with muscles of the vertebral column and muscles of the perineum and the pelvic diaphragm was changed to better facilitate student learning.
- One new Tips & Tools box was added to this chapter.

11 The Muscular System: Appendicular Musculature

- Thirty illustrations either are new or have been significantly revised.
- Five new Tips & Tools boxes were added to this chapter.

12 Surface Anatomy and Cross-Sectional Anatomy

- Eighteen illustrations either are new or have been significantly revised.
- Four Clinical Note illustrations have been added to this chapter.

13 The Nervous System: Nervous Tissue

- The chapter title has been changed from Neural Tissue to Nervous Tissue.
- Sixteen illustrations either are new or have been significantly revised.
- The section dealing with synaptic transmission was reorganized and revised to better facilitate student learning.

14 The Nervous System: The Spinal Cord and Spinal Nerves

- Seventeen illustrations either are new or have been significantly revised.
- The sections dealing with the spinal meninges and the peripheral distribution of spinal nerves were reorganized and revised to better facilitate student learning.

15 The Nervous System: Sensory and Motor Tracts of the Spinal Cord

- Seven illustrations either are new or have been significantly revised.
- The entire chapter was significantly revised to better facilitate student learning.

16 The Nervous System: The Brain and Cranial Nerves

- Thirty-four illustrations either are new or have been significantly revised.
- One new Tips & Tools box was added to this chapter.

17 The Nervous System: Autonomic Nervous System

- Eleven illustrations either are new or have been significantly revised.
- All material describing the anatomy of the sympathetic nervous system was revised to better facilitate student learning
- A new Spotlight Figure on the sympathetic nervous system has been added.
- New material was added to clarify the anatomy of the sympathetic ganglia

18 The Nervous System: General and Special Senses

- Twenty-eight illustrations either are new or have been significantly revised.
- All sections dealing with the physiology of the general and special senses were extensively revised.

19 The Endocrine System

- Eleven illustrations either are new or have been significantly revised.
- All sections dealing with the physiology of the endocrine glands were extensively revised.
- All material describing the anatomy of the pituitary gland was reorganized and revised to better facilitate student learning.

20 The Cardiovascular System: Blood

- Eight illustrations either are new or have been significantly revised.
- One new Tips & Tools box was added to this chapter.

21 The Cardiovascular System: The Heart

- Twelve illustrations either are new or have been significantly revised.
- All material describing the anatomy of the pericardium and the surface anatomy of the heart were revised to better facilitate student learning.

22 The Cardiovascular System: Vessels and Circulation

- Twenty-six illustrations either are new or have been significantly revised.
- One new Tips & Tools box was added to this chapter.

23 The Lymphatic System

- Seventeen illustrations either are new or have been significantly revised.
- All sections dealing with the development and immunological functions of the lymphatic cells, lymphatic vessels, and lymph nodes were extensively revised.

24 The Respiratory System

- Eighteen illustrations either are new or have been significantly revised.
- The organization of several sections was changed to better facilitate student learning.

25 The Digestive System

- Twenty-three illustrations either are new or have been significantly revised.

26 The Urinary System

- Thirteen illustrations either are new or have been significantly revised.
- All sections dealing with the anatomy of the nephron were revised to better facilitate student learning
- All sections dealing with the physiology of the urinary system were extensively revised.

27 The Reproductive System

- Twenty-two illustrations either are new or have been significantly revised.
- All sections dealing with the physiology of the male and female reproductive systems were extensively revised

28 The Reproductive System: Embryology and Human Development

- All of the Embryology Summaries have been revised.

Acknowledgments

Once again, the creative talents and patience brought to this project by our artist team, William Ober, M.D., Claire E. Ober, R.N., and Anita Impagliazzo, M.F.A., are inspiring and valuable beyond expression. Bill, Claire, and Anita worked intimately and tirelessly with us, imparting a unity of vision to the book while making each illustration clear and beautiful. Their superb art program is greatly enhanced by the incomparable bone and cadaver photographs of Ralph T. Hutchings, formerly of The Royal College of Surgeons of England. In addition, Dr. Pietro Motta, Professor of Anatomy, University of Roma, La Sapienza, provided several superb SEM images for use in the text. Thanks also to Dr. Ruth Anne O'Keefe for her excellent work on the clinical material, and to Colonel (ret) Michael Yard of Indiana University – Purdue University Indianapolis, for his additional feedback on clinical cases and notes. We are grateful to Elise Lansdon of Elise Lansdon Design for her excellent work on the design of the ninth edition of *Human Anatomy.*

Special thanks also goes to our new Portfolio Manager, Cheryl Cechvala, who came in the midst of revisions and supported us to the end. Content Producer, Caroline Ayres, guided us through all the stages from development to pages. This text wouldn't be what it is today without their valuable expertise and help.

We would like to acknowledge the many users and reviewers whose advice, comments, and collective wisdom helped shape this text into its final form. Their passion for the subject, their concern for accuracy and method of presentation, and their experience with students of widely varying abilities and backgrounds have made the revision process interesting and educating.

Reviewers

Jeffrey Blodig, *Johnson County Community College*
Lisa Brinn, *Florida International University*
Diep Burbridge, *Long Beach City College*
Anne Burrows, *Duquesne University*
Annamaria Crescimanno, *Golden West College*
Kimberly Dudzik, *Cuyamaca Community College*
Leticia Gallardo, *West Valley College*
Patricia Mansfield, *Santa Ana College*
Julie Porterfield, *Tulsa Community College*
Kimberly Ritterhoff, *University of Dayton*
Divya Sharma, *Triton College*
Deborah Shelley, *Fresno City College*
Michael Yard, *Indiana University – Purdue University Indianapolis*

We are also indebted to the Pearson staff, whose efforts were vital to the creation of this edition. A special note of thanks and appreciation goes to the editorial staff at Pearson. Thanks also to Barbara Yien, Courseware Director, Courseware Analysts Alice Fugate and Molly Ward, and Kimberly Twardochleb, Editorial Coordinator. We express thanks to Patrice Fabel and Lauren Chen for their work on the media programs that support *Human Anatomy,* especially MasteringA&P and Practice Anatomy Lab™ (PAL™). Thanks also to Norine Strang for her role in the production of the text.

We are very grateful to Adam Jaworski, Vice President, and Serina Beauparlant, Editor in Chief, for their continued enthusiasm and support of this project. We appreciate the contributions of Derek Perrigo, Senior Anatomy and Physiology Specialist, and Allison Rona, Executive Marketing Manager, who keep their fingers on the pulse of the market and help us meet the needs of our customers. Thanks also to the remarkable and tireless Pearson Science sales reps.

We are also grateful that the contributions of all the aforementioned people have led to this text receiving the following awards: the Association of Medical Illustrators Award, the Text and Academic Authors Award, the New York International Book Fair Award, the 35th Annual Bookbuilders West Award, and the 2010 Text and Academic Authors Association "Texty" Textbook Excellence Award.

Finally, we would like to thank our families for their love, patience, and support during the revision process. We could not have accomplished this without the help of our spouses—Kitty, Mary, and Mike.

In an effort to improve future editions, we ask that readers with pertinent information, suggestions, or comments concerning the organization or content of this textbook send their remarks to Robert Tallitsch directly, by the email address below, or care of Publisher, Applied Sciences, Pearson Benjamin Cummings, 1301 Sansome Street, San Francisco, CA 94111.

Frederic H. Martini
Robert B. Tallitsch
(RobertTallitsch@augustana.edu)
Judi L. Nath

Get Ready for a Whole New Human Anatomy Experience

Celebrated author Judi Nath (*Fundamentals of Anatomy & Physiology* and *Visual Anatomy & Physiology*) brings a fresh voice and a clear, engaging writing style to the **Ninth Edition** of **Human Anatomy**. The Ninth Edition continues the Martini legacy of a visually stunning presentation with exceptionally clear photographs, detailed illustrations, and captivating clinical content.

NEW! Ready-to-Go Teaching Modules help instructors find the best assets to use before, during, and after class to teach the toughest topics in Human Anatomy. Created by teachers for teachers, these curated sets of teaching tools save you time by highlighting the most effective and engaging animations, videos, quizzing, coaching and active learning activities from MasteringA&P.

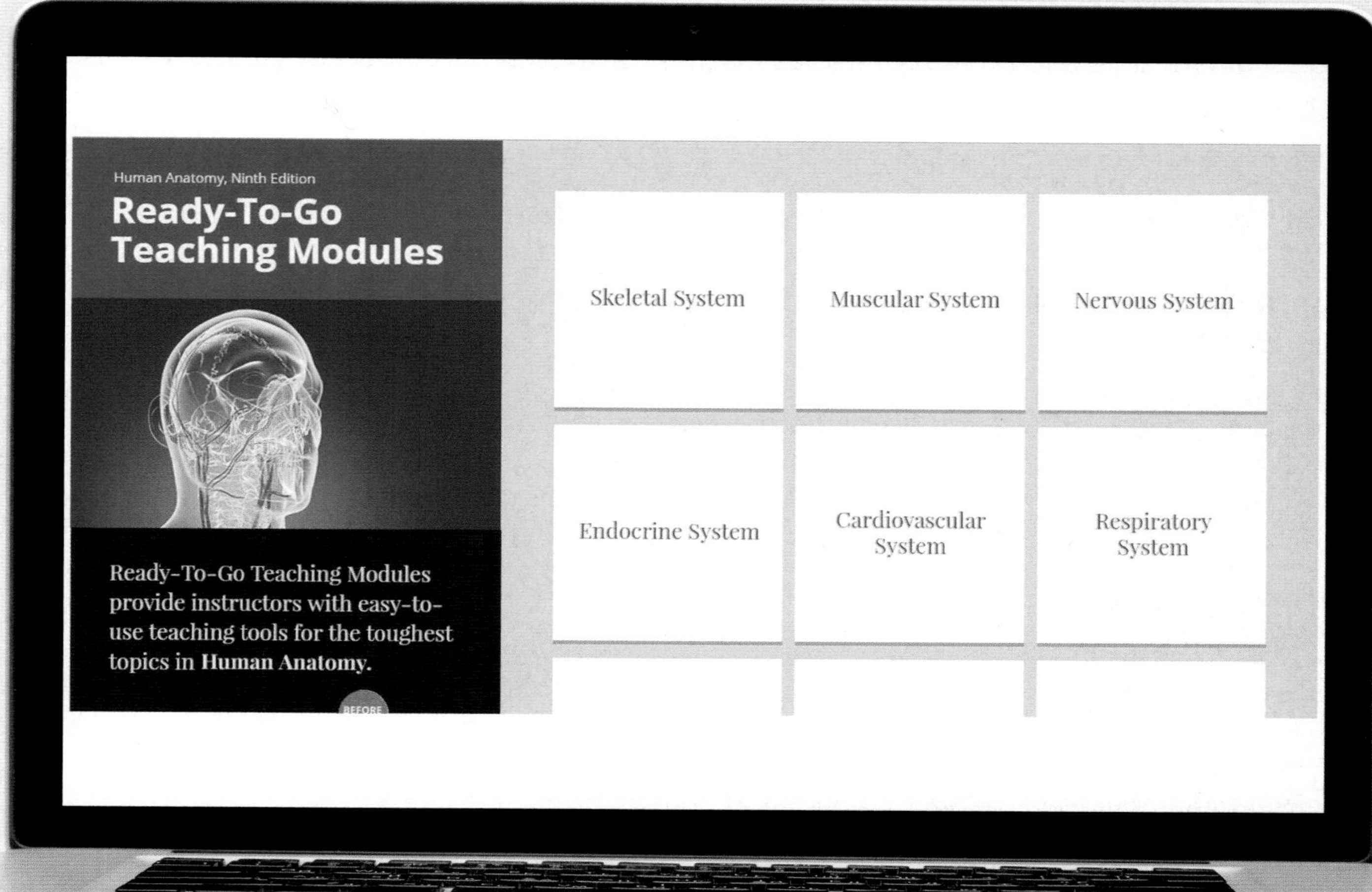

Prepare for the Classroom

New Study Tools
throughout each chapter help students understand and navigate the content.

NEW! Summary Boxes
at the beginning of each section outline the key points from that reading.

▶ **KEY POINT** Dermal ridges form friction ridges, ensuring a secure grip on objects. Dermal ridges also form fingerprints, a unique genetic identifier of an individual.

▶ **KEY POINT** The position of the wrist affects the functioning of the hand. Many muscles of the forearm, therefore, affect the actions of the wrist because (1) all of the muscles that flex or extend the wrist originate on the humerus, radius, and/or ulna and (2) many muscles that flex or extend the fingers originate on the radius and/or ulna.

TIPS & TOOLS

Remembering the names of the epidermal layers of thick skin

A mnemonic to help you remember the names of the epidermal layers of thick skin, from deep to superficial, is "Brent Spiner gained Lieutenant Commander" (basale, spinosum, granulosum, lucidum, corneum).

NEW! Tips & Tools
offer advice on how to approach some of the toughest topics.

TIPS & TOOLS

Here is a simple trick to remember the four anterior superficial forearm muscles originating from the medial epicondyle of the humerus. Hold both arms out, palms touching. Then slide your right hand proximally until your palm reaches your elbow with your fingers pointing toward your wrist. With each finger representing one of the four muscles, think PFPF: **P**ronator teres (index finger), **F**lexor carpi radialis (middle finger), **P**almaris longus (ring finger), and **F**lexor carpi ulnaris (little finger).

and Future Careers

NEW! Clinical Cases help motivate students for their future careers. Each chapter opens with a story-based Clinical Case related to the chapter content and ends with a Clinical Case Wrap-Up.

CLINICAL CASE

A Neuroanatomist's Stroke of Insight

Dr. Jill Taylor, a neuroanatomist, is 37 and at the top of her field. One morning she develops a throbbing headache behind her left eye. She then notices that her thoughts and movements are slowing down. Soon she realizes her right arm is paralyzed, and she is barely able to call for help. When she arrives at the hospital, she cannot walk, talk, read, write, or recall anything. She feels her spirit surrender and braces for death.

Dr. Taylor awakes later that day, shocked to be alive. She still cannot speak or understand speech, or recognize or use numbers. She can, however, appreciate the irony of her situation: a neuroscientist (scientist who studies the brain) witnessing her very own brain emergency, an evolving cerebrovascular accident (CVA) or stroke. Doctors perform open brain surgery to remove a large blood clot that was pressing on the left side of her brain near her language area.

Will Dr. ... rn to the Clinical Case Wrap-Up on p. 448.

CLINICAL CASE | WRAP-UP

A Neuroanatomist's Stroke of Insight

While her stroke affected the left side of Dr. Taylor's brain, the right side continued functioning. Because language and thoughts are typically controlled in the left hemisphere (the dominant hemisphere of a right-handed person), Dr. Taylor "sat in an absolutely silent mind" for the first month. Since the center for mathematical calculation is situated in the left hemisphere, she had to learn to use numbers all over again. And because the primary motor cortex governing the right side of the body resides in the precentral gyrus of the left hemisphere, she had to learn to use her right arm again. Full recovery took 8 years.

The stroke destroyed some brain cells, but others were able to form new neuronal connections. Neuroplasticity, this ability of nerve cells to make new connections, allows the brain to reorganize itself after injury.

Dr. Taylor wants anatomy students to know two things. First, "if you study the brain, you will never be bored." Second, "if you treat stroke patients like they will recover, they are more likely to recover." She has written a best-selling memoir about her experience, *My Stroke of Insight: A Brain Scientist's Personal Journey.*

1. How would you know, based on signs and symptoms, which side of Dr. Taylor's brain was injured by the stroke?
2. What is neuroplasticity, and why was it important in Dr. Taylor's recovery?

See the blue Answers tab at the back of the book.

CLINICAL NOTE

Skin Cancer

SKIN CANCER, the abnormal growth of skin cells, is often caused by exposure to UV radiation, primarily sunlight.

Basal cell carcinoma originates in the stratum basale. This is the most common skin cancer and the slowest growing, and it most often arises in areas that receive UV exposure. Although basal cell carcinomas almost never metastasize, they should be treated quickly to prevent local spread.

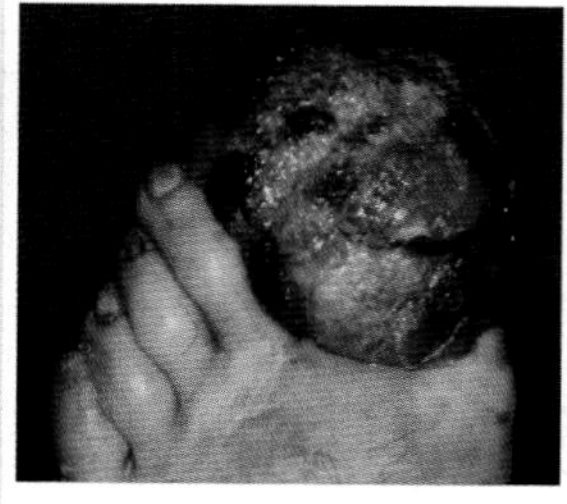

Squamous cell carcinoma

Squamous cell carcinoma, the second most common skin cancer, is an uncontrolled growth of abnormal squamous cells in the epidermis. They most often occur in UV-exposed areas of skin, but tobacco can also be a trigger. They can metastasize to tissues, bones, and nearby lymph nodes, and they often cause local disfigurement.

Malignant melanoma develops in melanocytes in the basal layer. These cancerous melanocytes multiply rapidly and metastasize to distant sites. Malignant melanomas cause the most deaths from skin cancer.

Clinical Terms end every chapter with a list of relevant clinical terms and definitions.

Related Clinical Terms

aphasia: A neurological condition caused by damage to the portions of the brain that are responsible for language.

ataxia: Loss of muscle coordination in the arms or legs due to cerebellar dysfunction.

chronic traumatic encephalopathy (CTE): A traumatic brain injury resulting from repeated sports-related head trauma.

concussion: A mild traumatic brain injury that may be accompanied by a period of unconsciousness.

dementia: A chronic or persistent disorder of the mental processes caused by brain disease or injury and marked by memory disorders, personality changes, and impaired reasoning.

epidural hematoma: The accumulation of blood between the inner table of the skull and the dura mater.

hydrocephalus: A condition marked by an excessive accumulation of cerebrospinal fluid within the brain ventricles.

microcephaly: A birth defect in which the head circumference is much smaller than expected for the age and sex of the child.

Parkinson's disease: A neurological disorder resulting from a degeneration of the dopaminergic neurons in the substantia nigra.

Clinical Notes appear within every chapter, expand upon topics just discussed, and present diseases and pathologies along with their relationship to normal function.

Continuous Learning
Before, During, and After Class

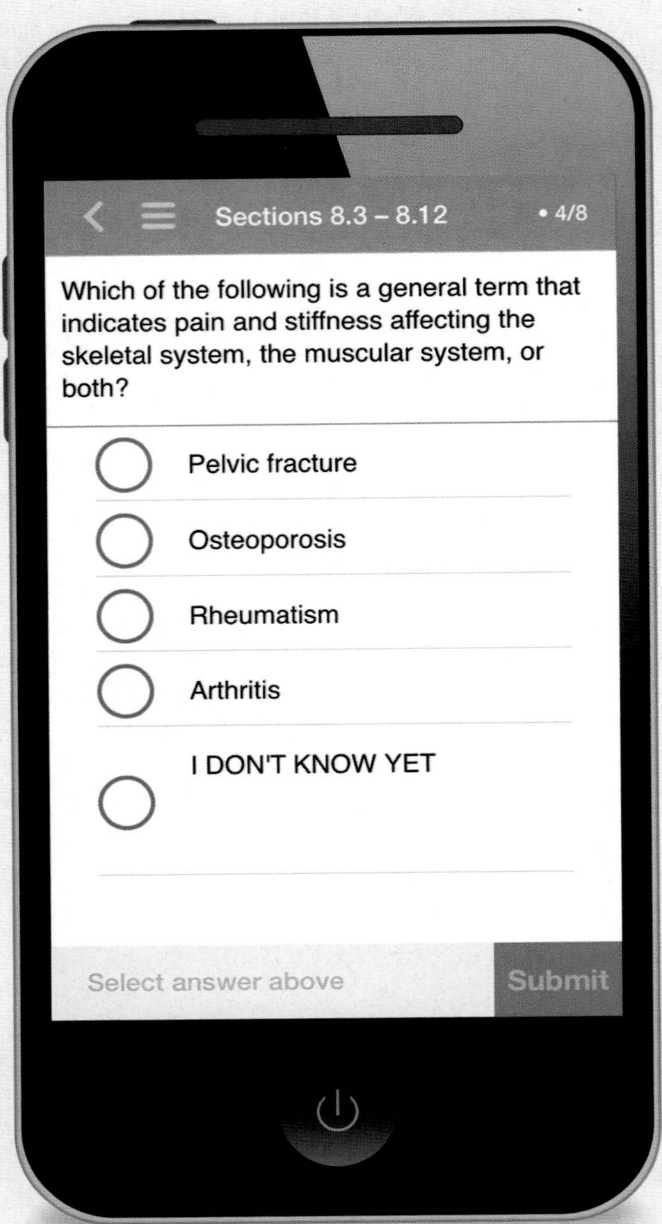

Sections 8.3 – 8.12
• 4/8
Which of the following is a general term that indicates pain and stiffness affecting the skeletal system, the muscular system, or both?
Pelvic fracture
Osteoporosis
Rheumatism
Arthritis
I DON'T KNOW YET
Select answer above
Submit

Dynamic Study Modules
enable students to study more effectively on their own. With the Dynamic Study Modules mobile app, students can quickly access and learn the concepts they need to be more successful on quizzes and exams.

NEW! Instructors can now select which questions to assign to students.

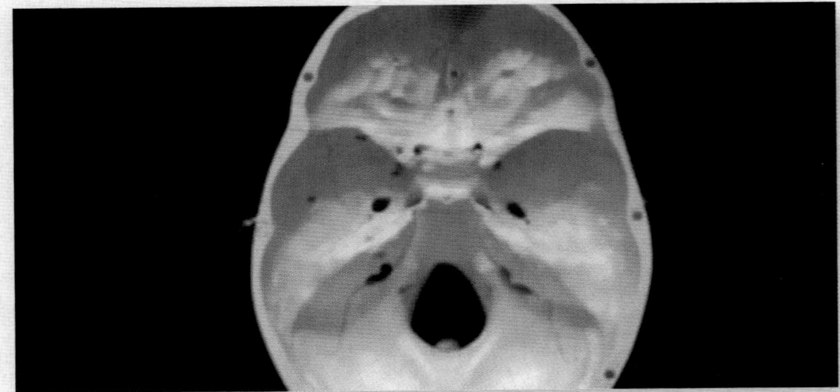

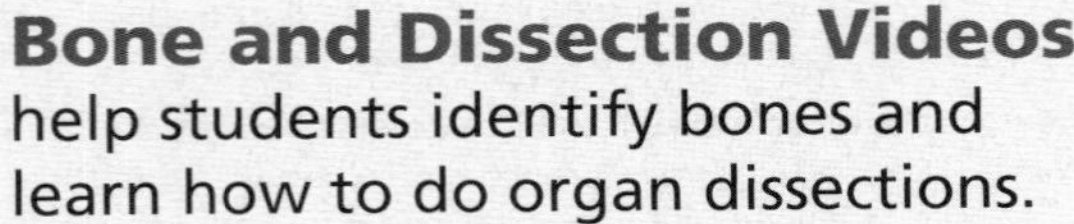

Bone and Dissection Videos
help students identify bones and learn how to do organ dissections.

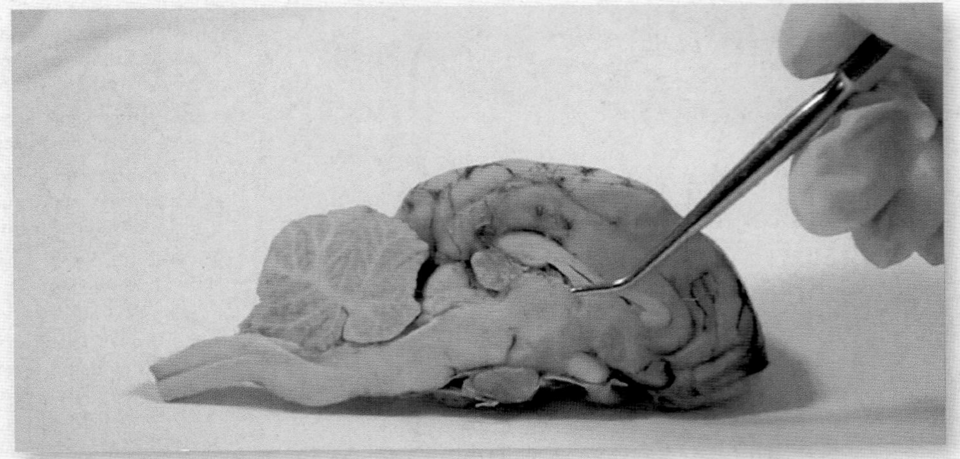

with MasteringA&P™

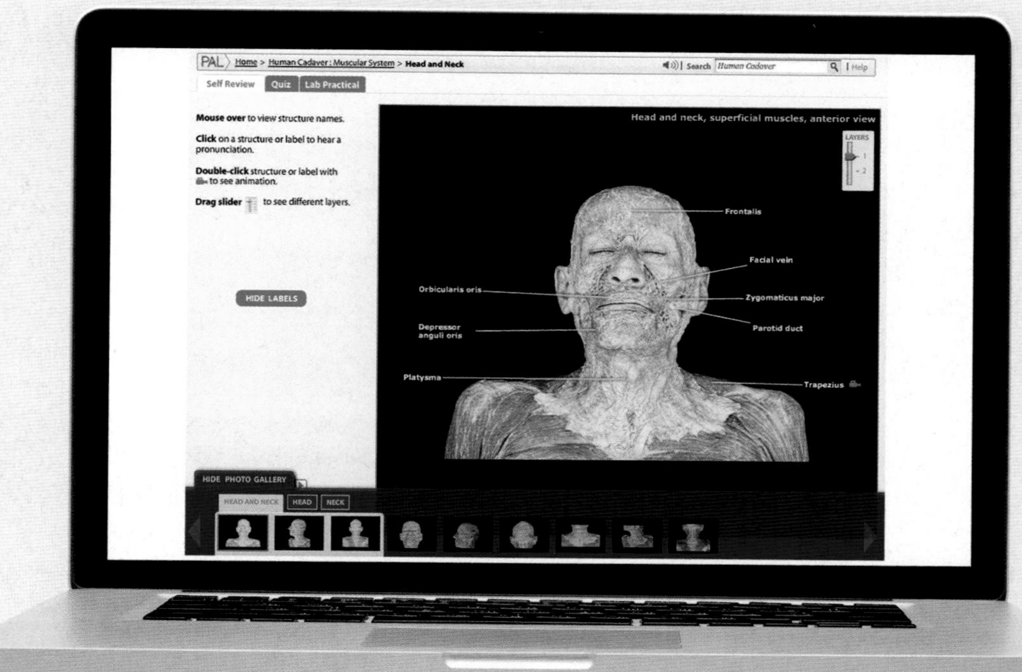

Practice Anatomy Lab (PAL™ 3.0) is a virtual anatomy study and practice tool that gives students 24/7 access to the most widely used lab specimens, including the human cadaver, anatomical models, histology, cat, and fetal pig. PAL 3.0 is easy to use and includes built-in audio pronunciations, rotatable bones, and simulated fill-in-the-blank lab practical exams.

Additional assignable MasteringA&P activities include:

- Bone & Dissection Coaching Activities
- A&P Flix Activities for Anatomy Topics
- Spotlight Figure Coaching Activities
- Clinical Case Activities
- PAL Assessments
- Art-Labeling Questions
- And More!

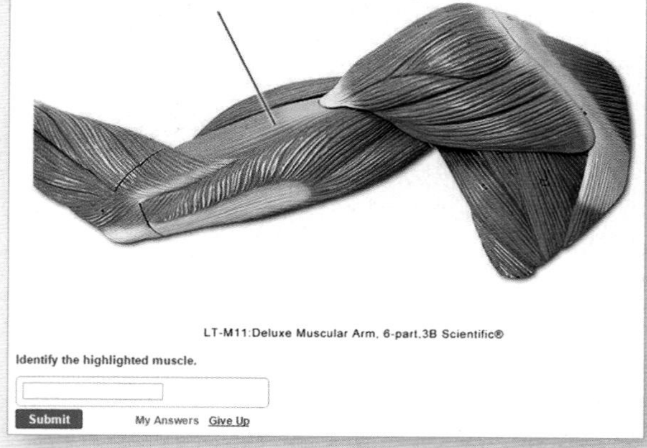

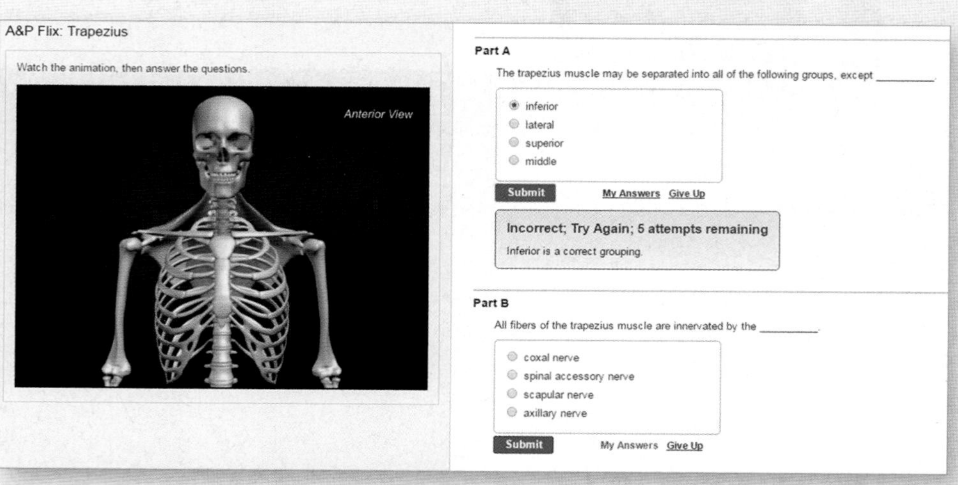

MasteringA&P™

Learning Catalytics is a "bring your own device" (laptop, smartphone, or tablet) engagement, assessment, and classroom intelligence system. Students use their device to respond to open-ended questions and then discuss answers in groups based on their responses.

"My students are so busy and engaged answering Learning Catalytics questions during lecture that they don't have time for Facebook."
—Declan De Paor, Old Dominion University

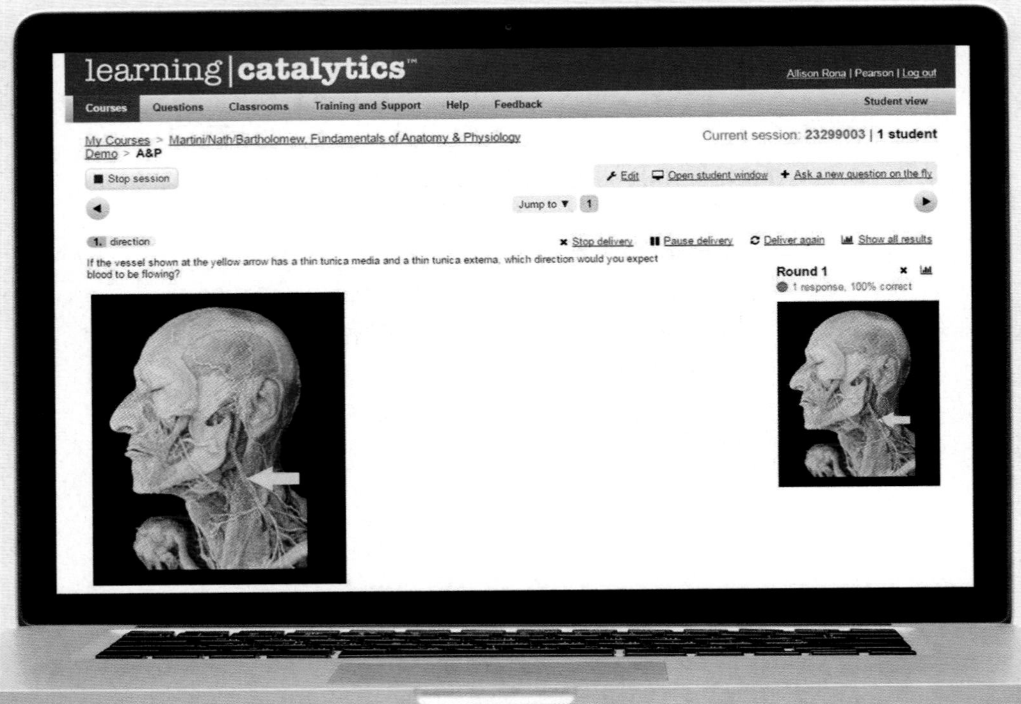

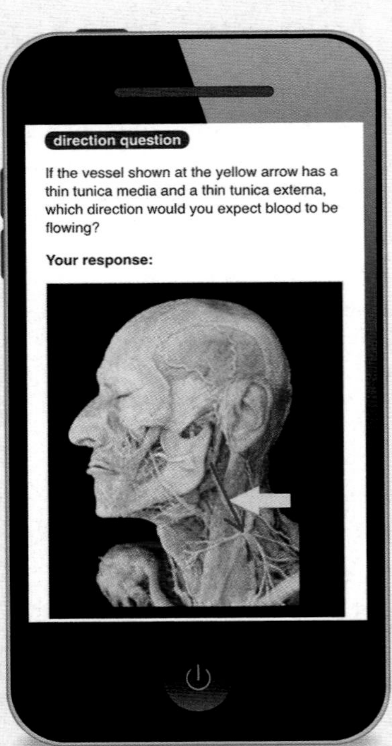

Access the Complete Textbook on or offline with eText 2.0

New! The **Ninth Edition** is available in Pearson's fully-accessible eText 2.0 platform.*

NEW! The eText 2.0 mobile app offers offline access and can be downloaded for most iOS and Android phones and tablets from the iTunes or Google Play stores.

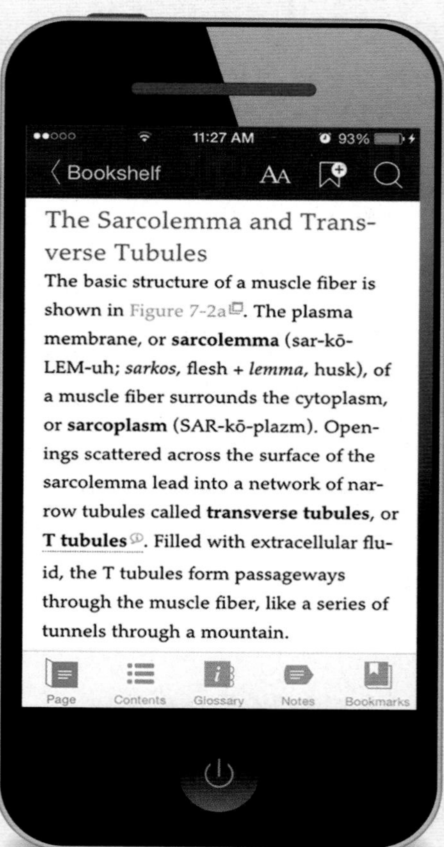

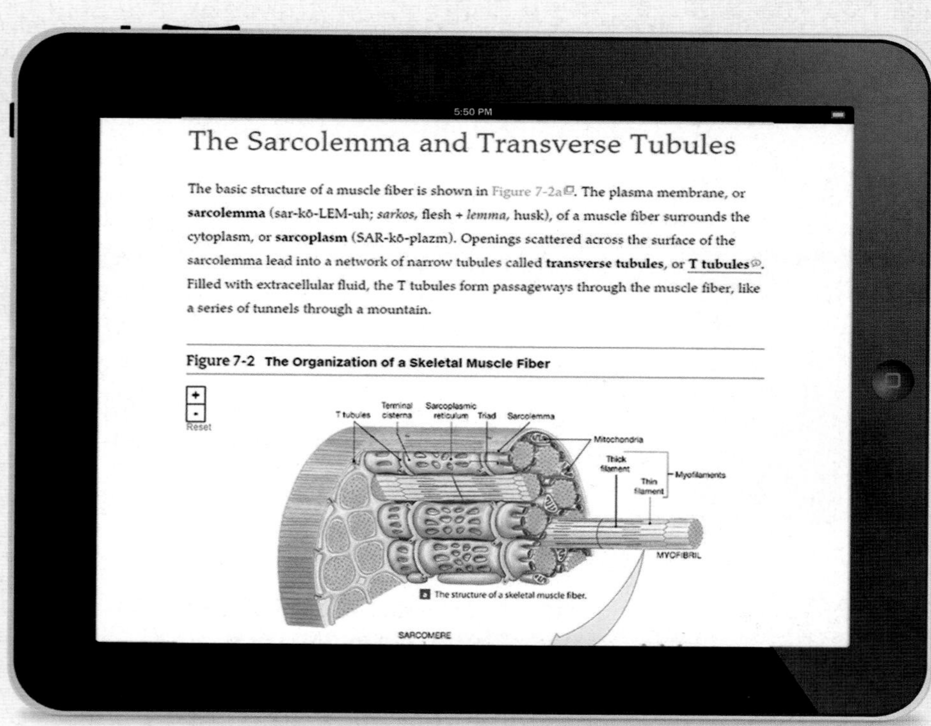

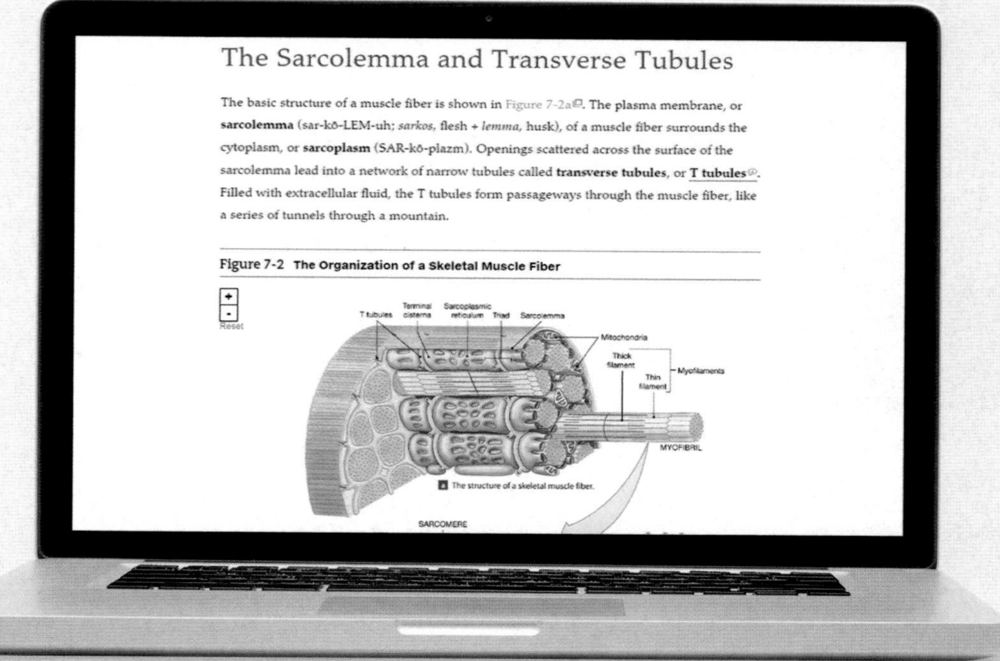

Powerful interactive and customization functions include instructor and student note-taking, highlighting, bookmarking, search, and links to glossary terms.

*The eText 2.0 edition will be live for Fall 2017 classes.

Instructor and Student Support

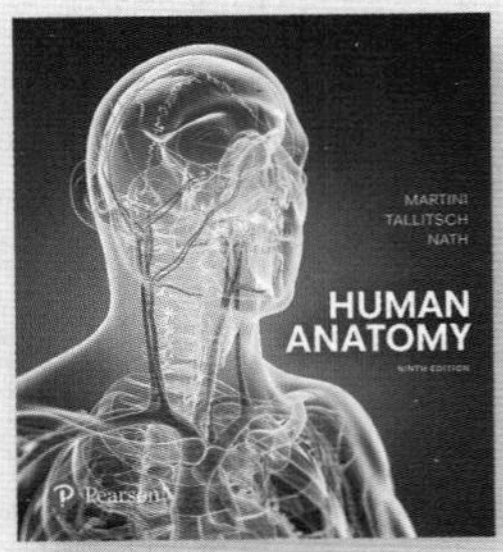

A complete package of instructor resources includes:

- Customizable PowerPoint slides
- All figures from the book in JPEG format
- A&P Flix 3D movie-quality animations on tough topics
- Test Bank
- And more!

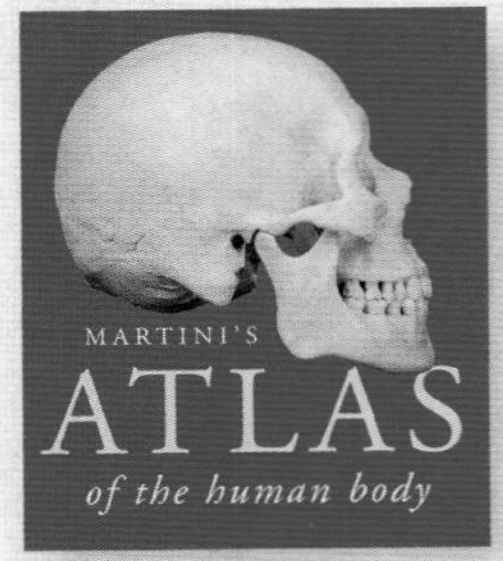

Martini's Atlas of the Human Body by Frederic H. Martini
978-0-321-94072-8 / 0-321-94072-5
The Atlas offers an abundant collection of anatomy photographs, radiology scans, and embryology summaries, helping students visualize structures and become familiar with the types of images seen in a clinical setting. Free when packaged with the textbook.

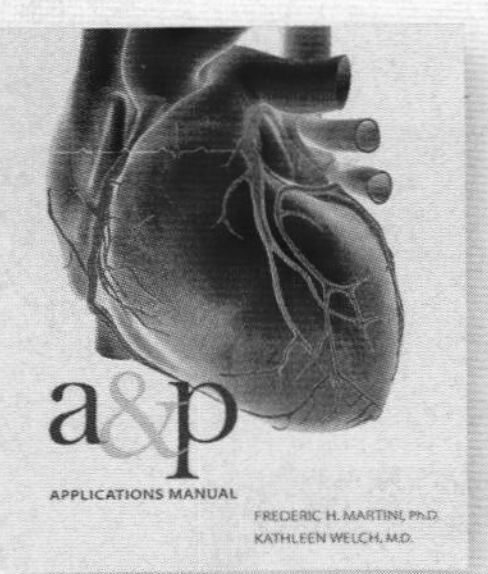

A&P Applications Manual by Frederic H. Martini and Kathleen Welch
978-0-321-94973-8 / 0-321-94973-0
This manual contains extensive discussions on clinical topics and disorders to help students apply the concepts of anatomy and physiology to daily life and their future health professions. Free when packaged with the textbook.

Get Ready for A&P by Lori K. Garrett
978-0-321-81336-7 / 0-321-81336-7
This book and online component were created to help students be better prepared for their A&P course. Features include pre-tests, guided explanations followed by interactive quizzes and exercises, and end-of-chapter cumulative tests. Also available in the Study Area of MasteringA&P. Free when packaged with the textbook.

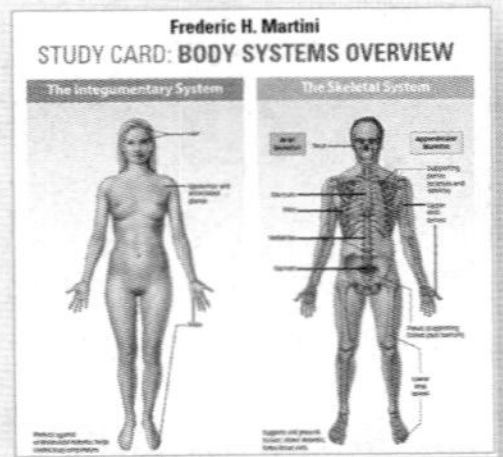

Study Card for Martini: Body Systems Overview
978-0-321-92930-3 / 0-321-92930-6
A six-panel laminated card showing all body systems and their organs and functions. Free when packaged with the textbook

Contents

8 | THE SKELETAL SYSTEM
Joints 203

9 | THE MUSCULAR SYSTEM
Skeletal Muscle Tissue and Muscle Organization 235

1

Foundations

An Introduction to Anatomy

Learning Outcomes

These Learning Outcomes correspond by number to this chapter's sections and indicate what you should be able to do after completing the chapter.

1.1 Define the limits of microscopic anatomy and compare and contrast cytology and histology. p. 2

1.2 Compare and contrast the various ways to approach gross anatomy. p. 2

1.3 Define the various subspecialties of anatomy. p. 2

1.4 Explain the major levels of organization in a living organism. p. 5

1.5 Identify the organ systems of the human body and compare and contrast their functions. p. 7

1.6 Understand and correctly apply descriptive anatomical and directional terminology. p. 14

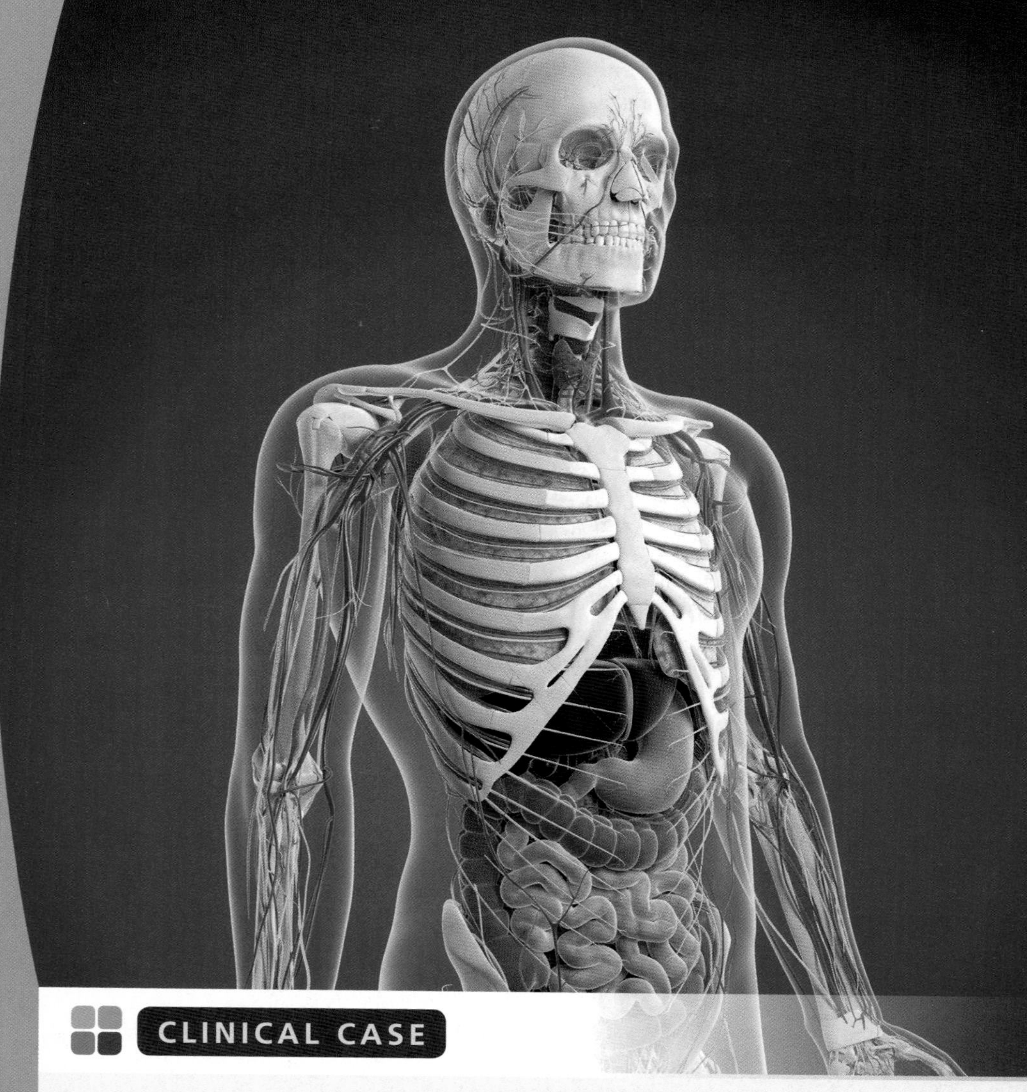

CLINICAL CASE

Using Anatomy to Save a Life

Zach, a 20-year-old college sophomore, is late for his anatomy class, so he decides to ride his bike to class instead of walking. As he enters an intersection, he is struck by a speeding pickup truck. The impact throws him 50 feet, and he lands on his head and left side.

Emergency medical technicians (EMTs) arrive within minutes. They roll the unconscious Zach onto his back for initial assessment. He has an obvious open skull fracture (bone break with pierced skin), open fractures of his left upper and lower extremities, and multiple rib fractures on his left side, and he exhibits rapid, shallow breathing. Assuming he has neck and back injuries, the EMTs splint him carefully for transport to the nearest Level I (highest designation) trauma center.

En route, an EMT calls the triage nurse in the emergency room (who assigns medical priority) and reports that he is arriving with a young male trauma victim with an Injury Severity Score (ISS) of 57. The nurse tells him to immediately report to the trauma room and sounds the alert for the trauma team.

With an ISS of 57, what are Zach's chances of survival? To find out, turn to the Clinical Case Wrap-Up on p. 26.

1

WE ALL USE our knowledge of human anatomy in our daily lives: We remember specific anatomical features to identify friends and family, and we observe changes in body movements and facial expressions for clues to what others are thinking. **Anatomy** is the study of the external and internal structures of the body and the physical relationships between body parts. In practical terms, anatomy is the careful observation of the human body.

Anatomical information provides clues about probable functions. **Physiology** is the study of the function of bodily structures, and we explain physiological mechanisms in terms of the underlying anatomy. *All specific physiological functions are performed by specific anatomical structures.* For instance, functions of the nasal cavity include filtering, warming, and humidifying inhaled air. The shapes of the bones projecting into the nasal cavity cause turbulence in the inhaled air. As the air swirls, it contacts the moist lining of the nasal cavity, which warms and humidifies the air, and any suspended particles stick to the moist surfaces. In this way, the air is conditioned and filtered before it reaches the lungs.

This text discusses the anatomical structures and functions that make human life possible. Our goals are to help you

1. develop a three-dimensional understanding of anatomical relationships,
2. prepare for more advanced courses in anatomy, physiology, and related subjects, and
3. make informed decisions about your personal health.

1.1 | Microscopic Anatomy

▶ **KEY POINT** Microscopic anatomy—the study of structures too small to be seen by the naked eye—includes the specialties of cytology and histology.

Microscopic anatomy is the study of structures that cannot be seen without magnification. The boundaries of microscopic anatomy are established by the limits of the equipment used (**Figure 1.1**). A simple hand lens shows details that barely escape the naked eye, while an electron microscope shows structural details that are more than a million times smaller. As we proceed through the text, we will consider details at various size levels.

Microscopic anatomy is subdivided into two specialties that consider features within a characteristic range of sizes:

- **Cytology** (sī-TOL-ō-jē) analyzes the internal structure of **cells**, the smallest units of life. Living cells are composed of complex chemicals in various combinations, and our lives depend on the chemical processes occurring in the trillions of cells that form our body.
- **Histology** (his-TOL-ō-jē) takes a broader perspective and examines **tissues**, groups of specialized cells and cell products that work together and perform specific functions. The human body has four basic tissue types: epithelial tissue, connective tissue, muscle tissue, and neural tissue (which will be described in Chapter 3).

Tissues combine to form organs such as the heart, kidney, liver, and brain. An **organ** is an anatomical structure that has multiple functions. Many tissues and most organs are examined easily without a microscope, and at this point we cross the boundary from microscopic anatomy into gross anatomy.

1.1 CONCEPT CHECK ✔

1 Histologists study what structures?

2 Define an organ.

See the blue Answers tab at the back of the book.

1.2 | Gross Anatomy

▶ **KEY POINT** We study gross anatomy—the study of structures visible to the naked eye—by examining surface anatomy, regional anatomy, or systemic anatomy.

Gross anatomy (*macroscopic anatomy*) is the study of structures and features that are visible to the unaided (naked) eye**.** There are several ways to approach gross anatomy:

- **Surface anatomy** is the study of general anatomical form, or **morphology**, and how superficial (surface) anatomical markings relate to deeper anatomical structures.
- **Regional anatomy** is the study of the superficial and internal features in a specific area of the body, such as the head, neck, or trunk. Advanced courses in anatomy often stress a regional approach because it emphasizes the relationships among structures.
- **Systemic anatomy** is the study of anatomy based upon the body's organ systems. An **organ system** is a group of organs that function together to produce coordinated effects. For example, the heart, blood, and blood vessels form the cardiovascular system, which distributes oxygen and nutrients throughout the body. There are 11 organ systems in the human body, which we will introduce later in the chapter. Introductory anatomy texts, including this one, usually use a systemic approach to organize information about important structural and functional patterns.

1.2 CONCEPT CHECK ✔

3 How does the work of a gross anatomist differ from that of a histologist?

4 What is an organ system, and how does it apply to systemic anatomy?

See the blue Answers tab at the back of the book.

1.3 | Other Types of Anatomical Studies

▶ **KEY POINT** Other anatomical specialties that are important in the understanding of the human body are developmental, comparative, clinical, surgical, radiographic, and cross-sectional anatomy.

Other anatomical specialties you will read about in this text include the following:

- **Developmental anatomy** studies the changes in form that take place between conception and physical maturity. Because it considers anatomical structures with a broad range of sizes (from a single cell to an adult human), developmental anatomy involves both microscopic and gross anatomy. Developmental anatomy is important in medicine because many structural abnormalities result from errors that occur during development. The most extensive structural changes occur during the first two months of development; **embryology** (em-brē-OL-ō-jē) is the study of these early developmental processes.
- **Comparative anatomy** studies the anatomical organization of different types of animals. Observed similarities may reflect evolutionary relationships. For example, humans, chickens, and salmon are all called vertebrates because they share a combination of anatomical features not found in any other group of animals, including a spinal column composed of individual structures called vertebrae (**Figure 1.2a**). Comparative anatomy uses the techniques of gross, microscopic, and developmental anatomy.

Figure 1.1 The Study of Anatomy at Different Scales. The amount of detail recognized depends on the method of study and the degree of magnification.

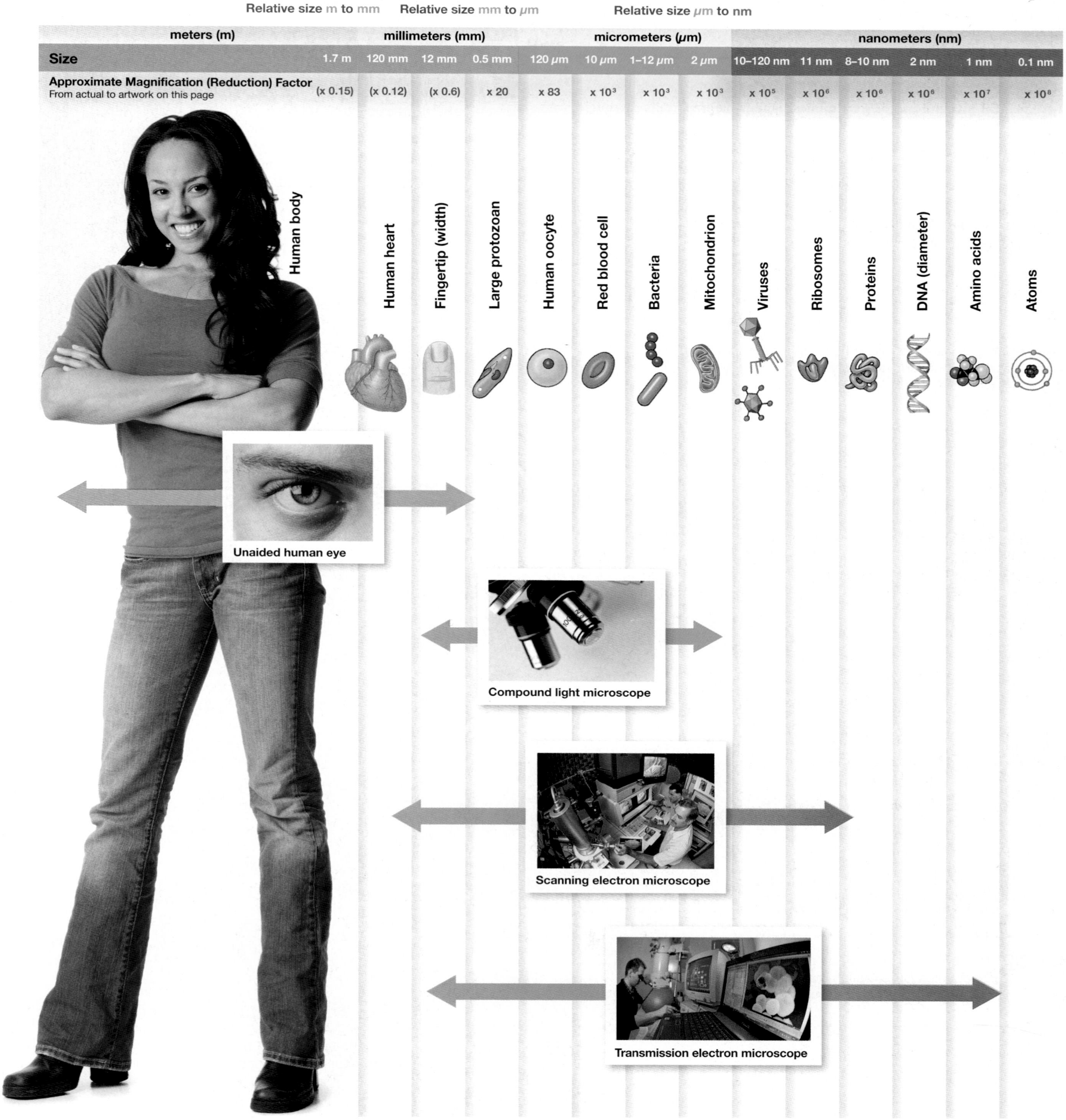

Relative size m to mm | Relative size mm to μm | Relative size μm to nm

	Human body	Human heart	Fingertip (width)	Large protozoan	Human oocyte	Red blood cell	Bacteria	Mitochondrion	Viruses	Ribosomes	Proteins	DNA (diameter)	Amino acids	Atoms
	meters (m)	millimeters (mm)			micrometers (μm)				nanometers (nm)					
Size	1.7 m	120 mm	12 mm	0.5 mm	120 μm	10 μm	1–12 μm	2 μm	10–120 nm	11 nm	8–10 nm	2 nm	1 nm	0.1 nm
Approximate Magnification (Reduction) Factor From actual to artwork on this page	(x 0.15)	(x 0.12)	(x 0.6)	x 20	x 83	x 10^3	x 10^3	x 10^3	x 10^5	x 10^6	x 10^6	x 10^6	x 10^7	x 10^8

Unaided human eye

Compound light microscope

Scanning electron microscope

Transmission electron microscope

Research shows that related animals typically go through similar developmental stages **(Figure 1.2b,c)**.

Several other gross anatomical specialties are important in medical diagnosis:

- **Clinical anatomy** focuses on anatomical features that may undergo recognizable pathological changes during illness.
- **Surgical anatomy** studies anatomical landmarks important for surgical procedures.
- **Radiographic anatomy** utilizes x-rays, ultrasound scans, or other specialized procedures performed on an intact body to visualize and study anatomical structures.
- **Cross-sectional anatomy** has emerged due to advances in radiographic anatomy, such as computerized tomography (CT) and spiral CT scans.

The Diagnosis of Disease

Homeostasis is the maintenance of a relatively constant internal environment suitable for the survival of cells, tissues, and organs. It is achieved by a system of control mechanisms activated by negative feedback. **Disease** is the failure to maintain homeostatic conditions. The disease process may affect any aspect of physiology from the cellular to the organismic level. The body's defenses can overcome some diseases, but others require medical intervention.

Figure 1.2 Comparative Anatomy. Humans are classified as vertebrates, a group that also includes animals as different in appearance as salmon and chickens.

Basic Vertebrate Body Plan

Dorsal, hollow nerve cord forms **brain** and **spinal cord**

Notochord, a stiffened rod below spinal cord, is usually replaced by vertebrae

Muscular tail extends beyond exit of digestive tract

Digestive tract

Mouth

Heart

Anus

Braincase of cartilage or bone surrounds the brain

Pharyngeal (gill) arches may persist or be modified to form other structures in adult

Ventral body cavity contains thoracic and abdominopelvic organs

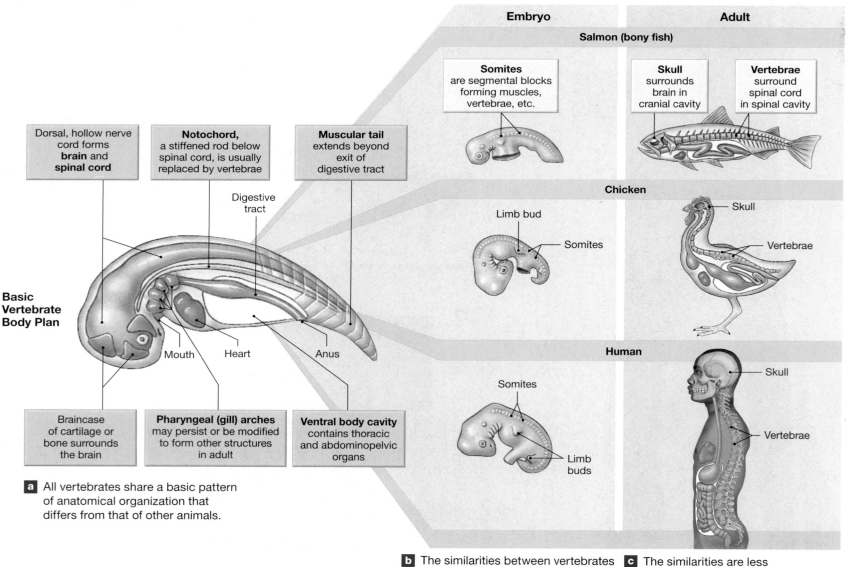

a All vertebrates share a basic pattern of anatomical organization that differs from that of other animals.

Embryo

Adult

Salmon (bony fish)

Somites are segmental blocks forming muscles, vertebrae, etc.

Skull surrounds brain in cranial cavity

Vertebrae surround spinal cord in spinal cavity

Chicken

Limb bud

Somites

Skull

Vertebrae

Human

Somites

Limb buds

Skull

Vertebrae

b The similarities between vertebrates are most apparent when comparing embryos at comparable stages of development.

c The similarities are less obvious when comparing adult vertebrates.

1.3 CONCEPT CHECK

5 How does surgical anatomy differ from clinical anatomy?

6 Cross-sectional anatomy is a subspecialty of which anatomical specialty?

See the blue Answers tab at the back of the book.

1.4 Levels of Organization

▶ **KEY POINT** The levels of structural organization in the human body range from the chemical/molecular level (the simplest level) to the entire organism (the most complex level).

Our study of the human body begins at the chemical, or molecular, level of organization. The human body consists of more than a dozen different elements, but four of them (hydrogen, oxygen, carbon, and nitrogen) account for more than 99 percent of the total number of atoms **(Figure 1.3a)**. At the chemical level, atoms interact to form three-dimensional molecules with distinctive properties. The major classes of molecules in the human body are indicated in **Figure 1.3b**.

The next level of organization, the cellular level, includes cells, the smallest living units in the body **(Figure 1.4)**. Cells contain internal structures called organelles. Cells and their organelles are made of complex chemicals. (Cell structure and the function of the major organelles found within cells are presented in Chapter 2.) As shown in **Figure 1.4**, chemical interactions produce complex proteins within a muscle cell in the heart. Muscle cells are unusual because they can contract powerfully, shortening along their longitudinal axis.

Heart muscle cells are connected to form a distinctive muscle tissue, an example of the tissue level of organization. Layers of muscle tissue form most of the wall of the heart, a hollow, three-dimensional organ. We are now at the organ level of organization **(Figure 1.4)**.

Normal functioning of the heart depends on interrelated events at the chemical, cellular, tissue, and organ levels of organization. Coordinated contractions in the muscle cells of cardiac muscle tissue produce a heartbeat. When that beat occurs, the internal anatomy of the organ enables it to function as a pump. With each contraction, the heart pushes blood into the vascular system, a network of blood vessels. Together, the heart, blood, and vascular system form an organ system: the cardiovascular system (CVS).

Each level of organization is dependent on the others. Damage at any level may affect the entire system. A chemical change in heart muscle cells may cause abnormal contractions or even stop the heartbeat. Physical damage to muscle tissue, such as a chest wound, can make the heart ineffective even when most of the heart muscle cells are intact. An inherited abnormality in heart structure can make it an ineffective pump even if muscle cells and tissues are normal.

Note that anything affecting a system ultimately affects all the components of that system. For example, damage to a major blood vessel somewhere else in the body can cause the heart to lose the ability to pump blood effectively. If the heart cannot pump and blood cannot flow, oxygen and nutrients cannot be distributed to tissues. In a very short time, the tissue breaks down as heart muscle cells die from oxygen and nutrient starvation.

Of course, the changes that occur when the heart is not pumping effectively are not limited to the cardiovascular system; all the cells, tissues, and organs in the body will be damaged. This observation brings us to the highest level of organization, an organism—in this case, a human. The organism level reflects the interactions among organ systems **(Figure 1.4)**. All are vital; every system must be working properly and in harmony with every other system, or survival will be impossible.

When all systems are functioning normally, the characteristics of the internal environment are relatively stable at all levels. This tendency toward stability, called **homeostasis** (hō-mē-ō-STĀ-sis; *homeo*, unchanging, + *stasis*, standing), is maintained by physiological processes.

Figure 1.3 Composition of the Body at the Chemical Level of Organization.

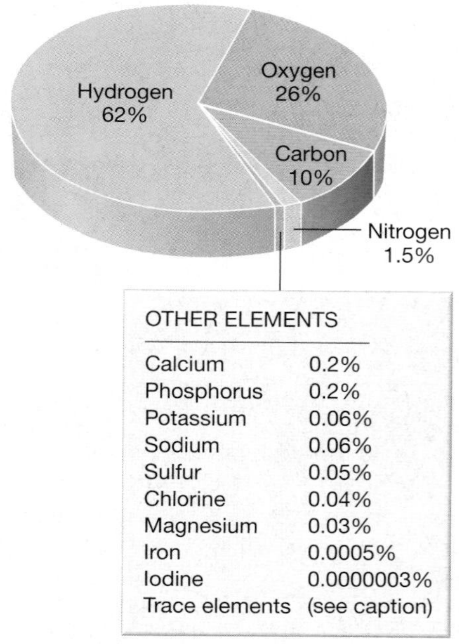

OTHER ELEMENTS	
Calcium	0.2%
Phosphorus	0.2%
Potassium	0.06%
Sodium	0.06%
Sulfur	0.05%
Chlorine	0.04%
Magnesium	0.03%
Iron	0.0005%
Iodine	0.0000003%
Trace elements	(see caption)

a Elemental composition of the body. Trace elements include silicon, fluorine, copper, manganese, zinc, selenium, cobalt, molybdenum, cadmium, chromium, tin, aluminum, and boron.

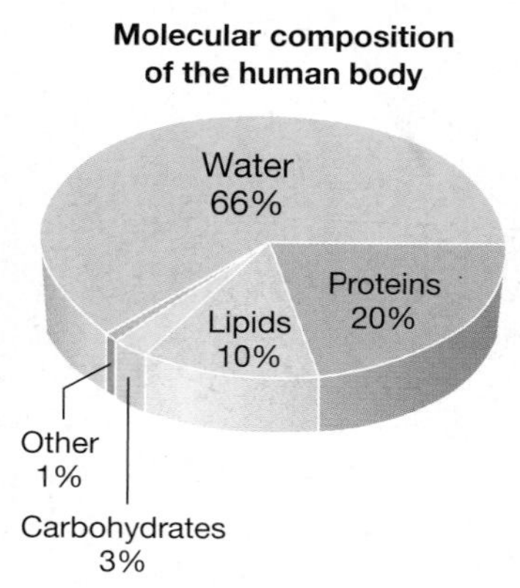

b Molecular composition of the body.

Figure 1.4 Levels of Organization

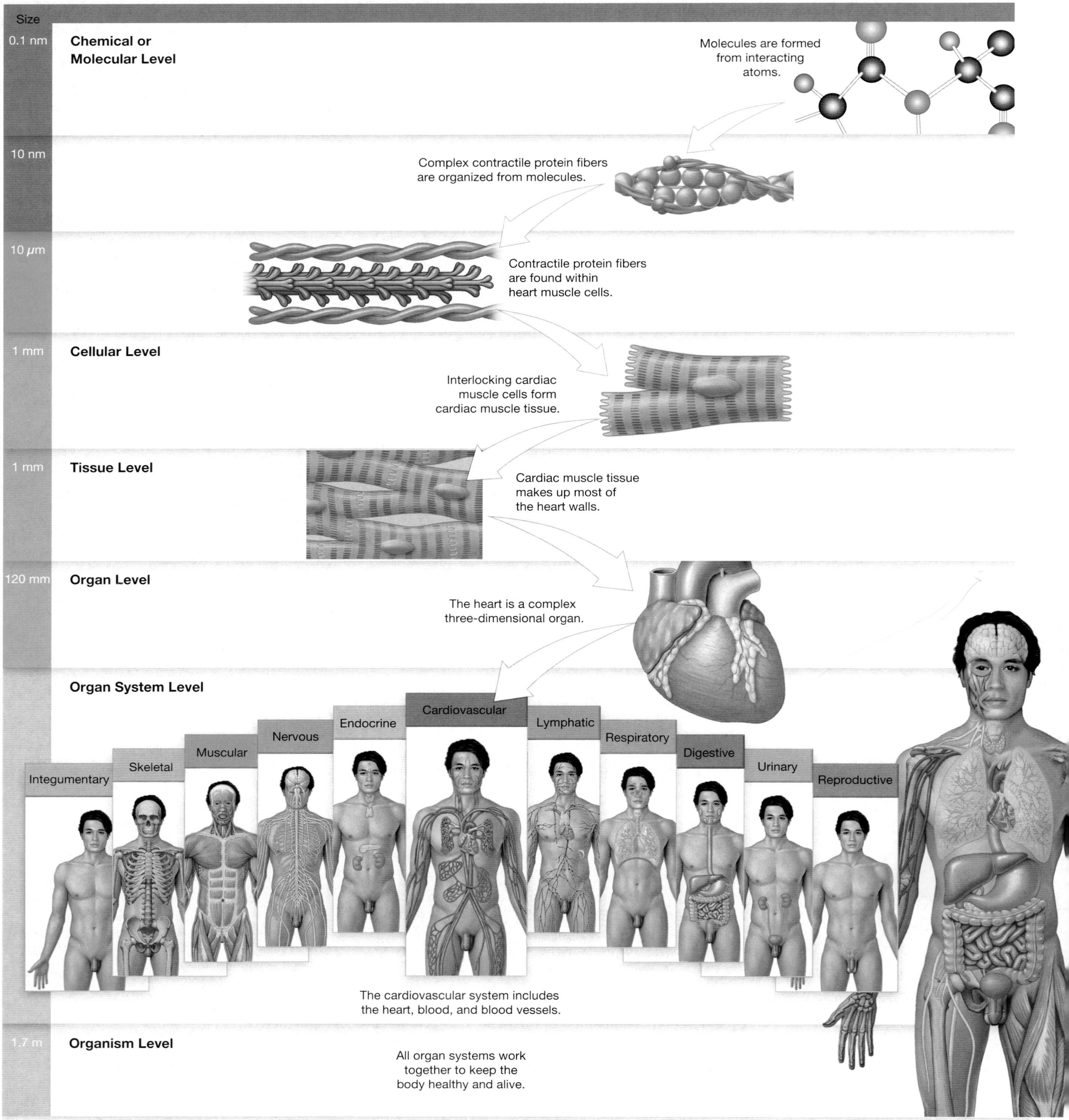

Size

0.1 nm **Chemical or Molecular Level**

Molecules are formed from interacting atoms.

10 nm

Complex contractile protein fibers are organized from molecules.

10 μm

Contractile protein fibers are found within heart muscle cells.

1 mm **Cellular Level**

Interlocking cardiac muscle cells form cardiac muscle tissue.

1 mm **Tissue Level**

Cardiac muscle tissue makes up most of the heart walls.

120 mm **Organ Level**

The heart is a complex three-dimensional organ.

Organ System Level

Integumentary
Skeletal
Muscular
Nervous
Endocrine
Cardiovascular
Lymphatic
Respiratory
Digestive
Urinary
Reproductive

The cardiovascular system includes the heart, blood, and blood vessels.

1.7 m **Organism Level**

All organ systems work together to keep the body healthy and alive.

1.4 CONCEPT CHECK

7 Cyanosis is a medical condition in which a person's lips and fingertips turn blue due to the inadequate delivery of oxygen to tissues. If a patient is exhibiting cyanosis, why should the patient's heart be examined *in addition to* the patient's lungs?

See the blue Answers tab at the back of the book.

CLINICAL NOTE

Disease, Pathology, and Diagnosis

Pathology is the study of disease. Diseases produce **signs** (objective evidence that the health provider can detect, such as fever or limited motion) and **symptoms** (subjective indications that the patient perceives, such as pain or fatigue). A **diagnosis** is an identification of the nature of an illness based on its signs and symptoms.

The World Health Organization (WHO) developed the International Classification of Diseases (ICD) as an international diagnostic standard. The ICD is important for health management and epidemiology (the study of disease occurrence, distribution, and cause). The current ICD-10 contains 69,823 diseases.

1.5 An Introduction to Organ Systems

▶ **KEY POINT** The 11 organ systems of the human body enable us to carry out vital life functions such as responsiveness, growth and differentiation, reproduction, movement, and metabolism and excretion.

Figure 1.5 summarizes the functions of the 11 organ systems of the human body. **Figure 1.6** details the components and primary functions of each organ system. Like all living organisms, humans share vital characteristics and processes:

- **Responsiveness:** The ability of an organism to respond to changes in its immediate environment is termed **responsiveness**. Examples include you jerking your hand away from a hot stove, your dog barking at approaching strangers, and amoebas gliding toward potential prey. Organisms also make longer-lasting responses as they adjust to their environments. For example, as winter approaches, an animal grows a heavier coat or migrates to a warmer climate. Adaptability is the capacity to make longer-lasting adjustments.
- **Growth and Differentiation:** Over a lifetime, organisms grow larger, increasing in size by increasing the size or number of their cells. In multicellular organisms, the individual cells become specialized to perform particular functions. This specialization is called **differentiation**. Growth and differentiation in cells and organisms produce changes in form and function. For example, the anatomical proportions and physiological capabilities of an adult human are quite different from those of an infant.
- **Reproduction:** Organisms reproduce, creating subsequent generations of their own kind, whether unicellular or multicellular.
- **Movement:** Organisms produce movement, which may be internal (transporting food, blood, or other materials inside the body) or external (moving through the environment).

Figure 1.5 An Introduction to Organ Systems. An overview of the 11 organ systems and their major functions.

ORGAN SYSTEM	MAJOR FUNCTIONS
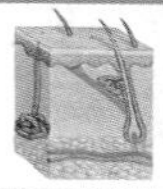 Integumentary	Protects against environmental hazards; controls temperature
 Skeletal	Supports and protects soft tissues; stores minerals; forms blood
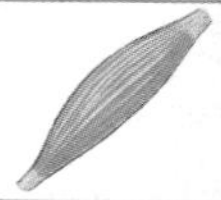 Muscular	Provides movement and support; generates heat
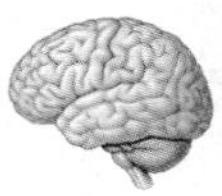 Nervous	Directs immediate responses to stimuli, usually by coordinating the activities of other organ systems
 Endocrine	Directs long-term changes in the activities of other organ systems
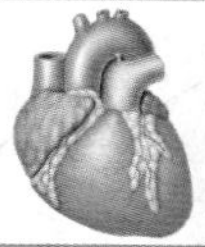 Cardiovascular	Distributes cells and dissolved materials, including nutrients, wastes, and gases
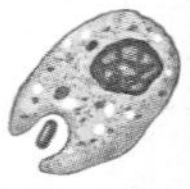 Lymphatic	Defends against infection and disease
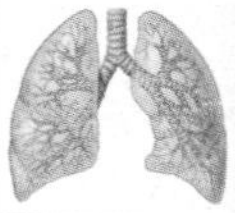 Respiratory	Delivers air to sites where gas exchange occurs between the air and circulating blood
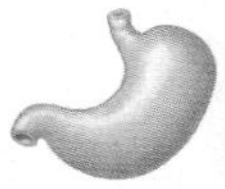 Digestive	Processes and digests food; absorbs nutrients; stores energy reserves
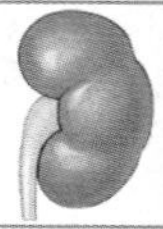 Urinary	Eliminates excess water, salts, and wastes; controls pH; regulates blood pressure
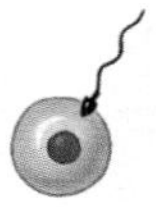 Reproductive	Produces sex cells and hormones

- **Metabolism and Excretion:** Organisms rely on chemical reactions to provide energy for responsiveness, growth, reproduction, and movement. They also synthesize complex chemicals, such as proteins. The term **metabolism** refers to *all* the chemical operations under way in the body. Types of metabolic reactions include **catabolism** (the *breakdown* of complex molecules into simple ones) and **anabolism** (the *synthesis* of complex molecules from simple ones). Normal metabolic operations require the **absorption** (*taking in*) of materials from the environment. To generate energy efficiently, cells require various nutrients, as well as oxygen, an atmospheric gas. The term **respiration** refers to cells' absorption, transport, and use of oxygen. Metabolic operations generate potentially harmful wastes that must be removed through the process of **excretion**.

1

Figure 1.6 The Organ Systems of the Body

The Integumentary System

Protects against environmental hazards; helps control body temperature

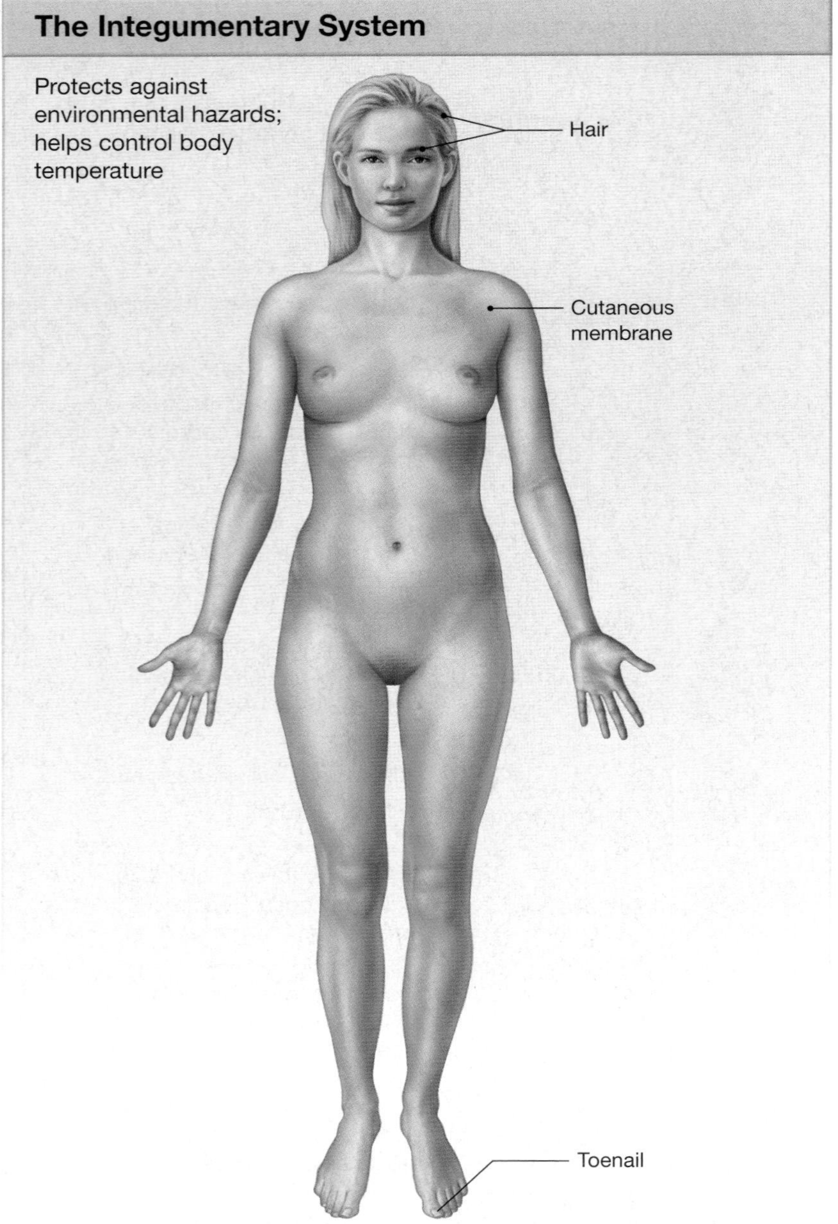

Organ/Componenent	Primary Functions
Skin (Cutaneous Membrane)	
Epidermis	Covers surface; protects deeper tissues
Dermis	Nourishes epidermis; provides strength; contains glands
Hair Follicles	Produce hair; innervation provides sensation
Hairs	Provide protection for head
Sebaceous glands	Secrete lipid coating that lubricates hair shaft and epidermis
Sweat Glands	Produce perspiration for evaporative cooling
Nails	Protect and stiffen distal tips of digits
Sensory Receptors	Provide sensations of touch, pressure, temperature, and pain
Subcutaneous Layer	Stores lipids; attaches skin to deeper structures

The Skeletal System

Supports and protects tissues; stores minerals; forms blood cells

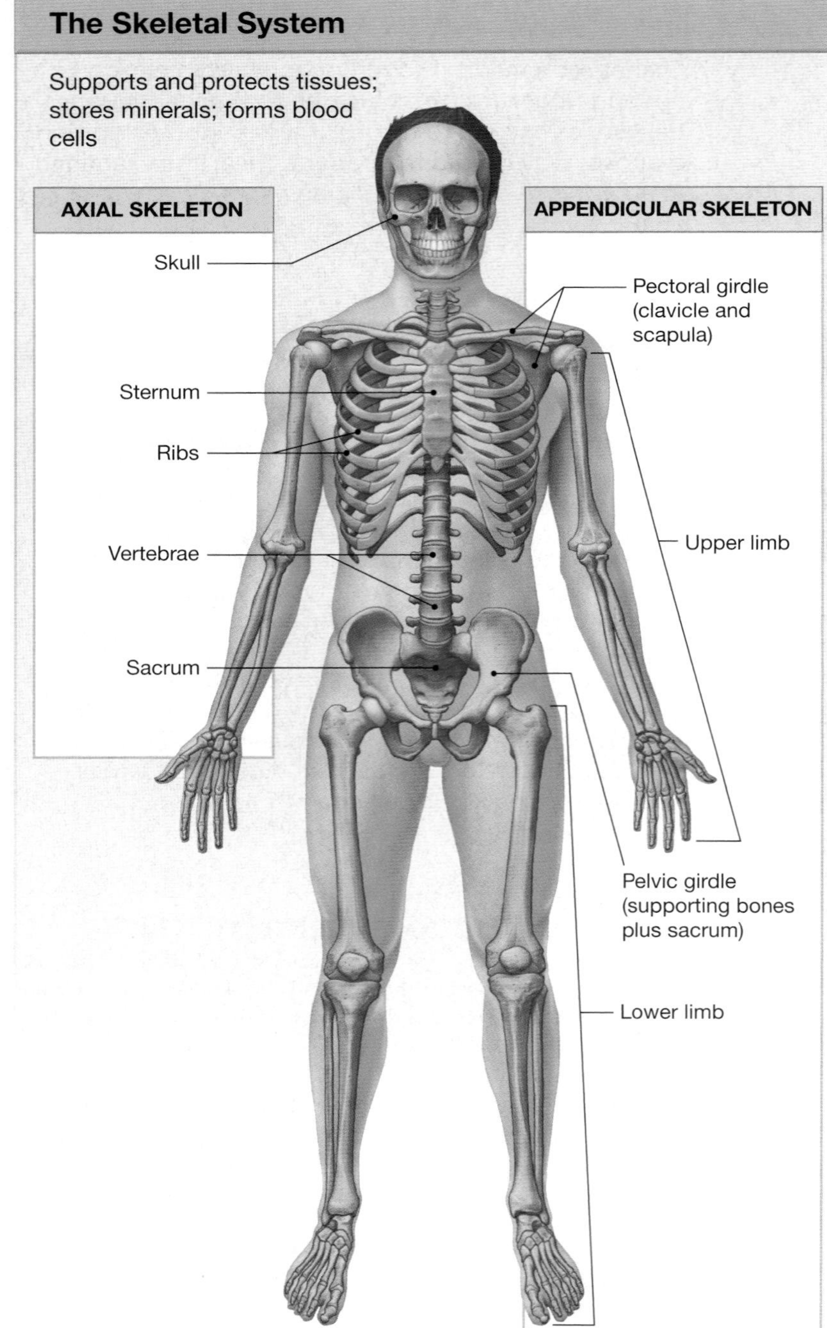

Organ/Componenent	Primary Functions
Bones, Cartilages, and Joints	Support and protect soft tissues; bones store minerals
Axial skeleton (skull, vertebrae, sacrum, coccyx, sternum, supporting cartilages and ligaments)	Protects brain, spinal cord, sense organs, and soft tissues of thoracic cavity; supports the body weight over lower limbs
Appendicular skeleton (limbs and supporting bones and ligaments)	Provides internal support and positioning of the limbs; supports and moves axial skeleton
Ligaments	Connect bone to bone, bone to cartilage, or cartilage to cartilage
Bone Marrow	Primary site of blood cell production (red bone marrow); storage of energy reserves in fat cells (yellow bone marrow)

The Muscular System

Allows for locomotion; provides support; produces heat

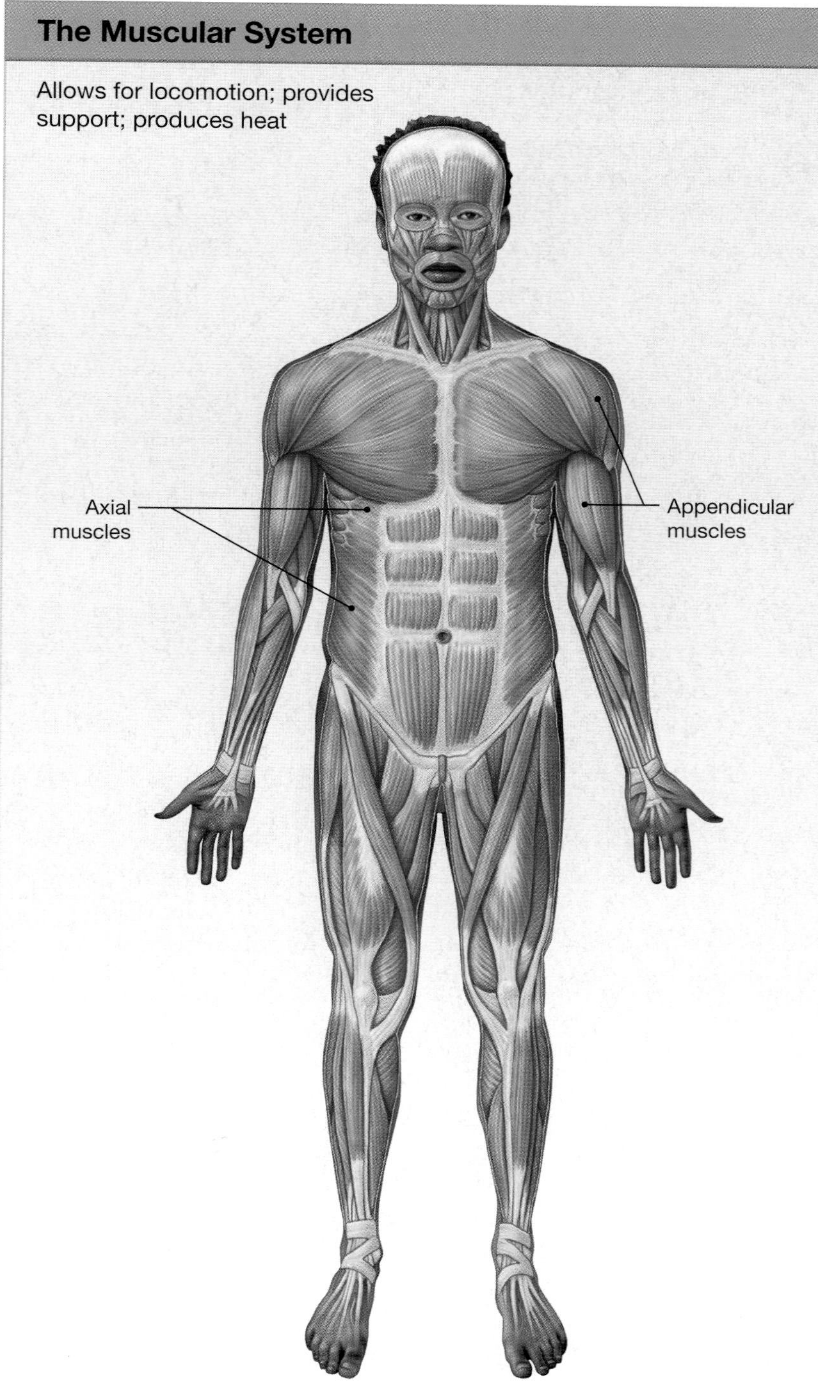

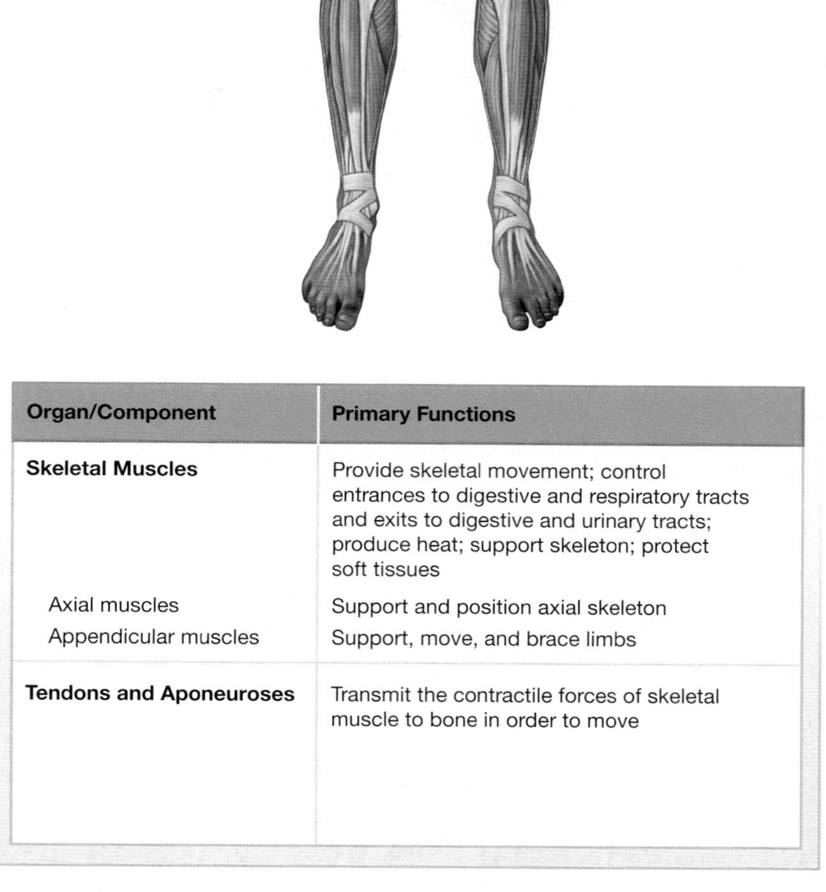

Organ/Component	Primary Functions
Skeletal Muscles	Provide skeletal movement; control entrances to digestive and respiratory tracts and exits to digestive and urinary tracts; produce heat; support skeleton; protect soft tissues
Axial muscles	Support and position axial skeleton
Appendicular muscles	Support, move, and brace limbs
Tendons and Aponeuroses	Transmit the contractile forces of skeletal muscle to bone in order to move

The Nervous System

Directs immediate responses to stimuli, usually by coordinating the activities of other organ systems

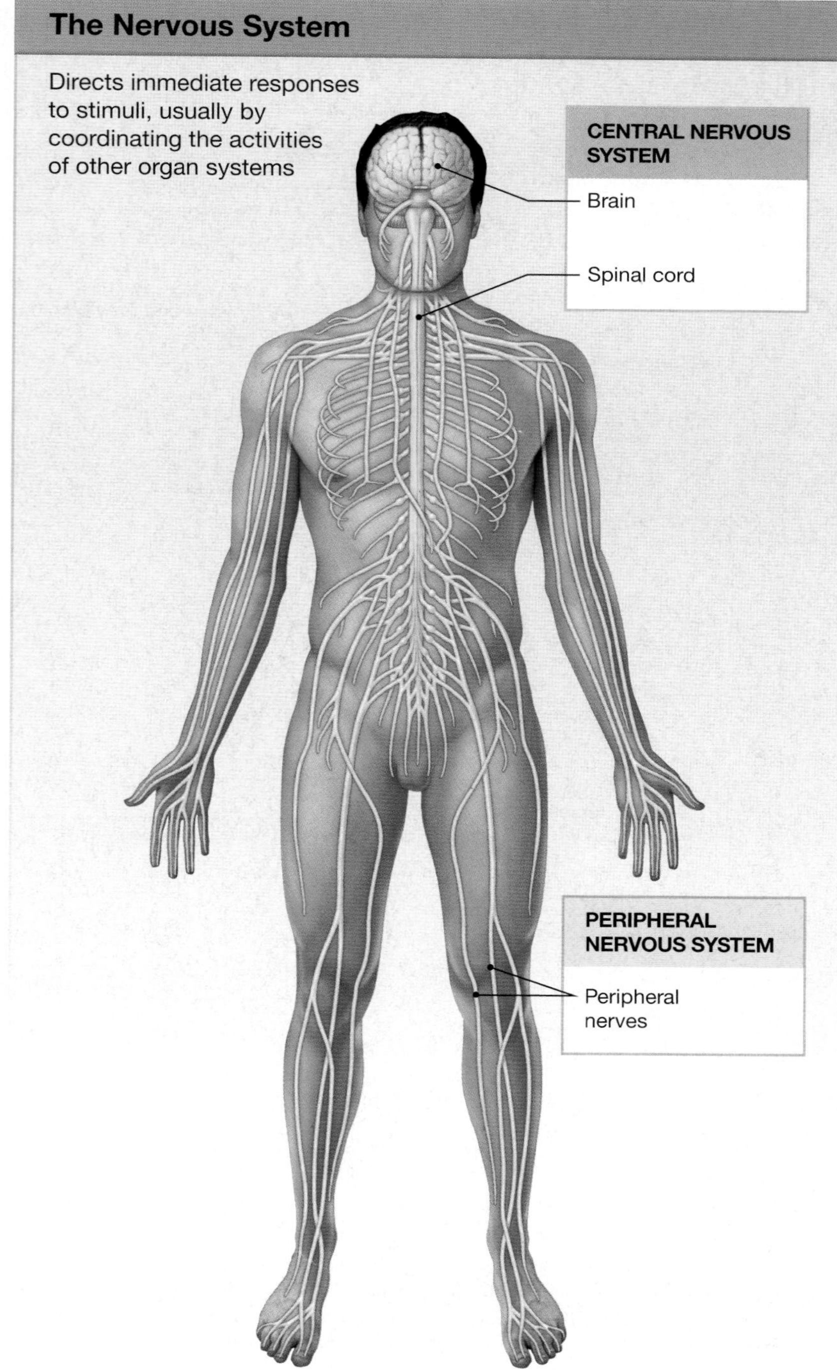

Organ/Component	Primary Functions
Central Nervous System (CNS)	Control center for nervous system; processes information; short-term control over activities of other systems
Brain	Performs complex integrative functions; controls both voluntary and autonomic activities
Spinal cord	Relays information to and from brain; performs less-complex integrative activities
Special senses	Provide sensory input to the brain relating to sight, hearing, smell, taste, and equilibrium
Peripheral Nervous System (PNS)	Links CNS with other systems and with sense organs

Figure 1.6 *(continued)*

1

The Endocrine System

Directs long-term changes in activities of other organ systems

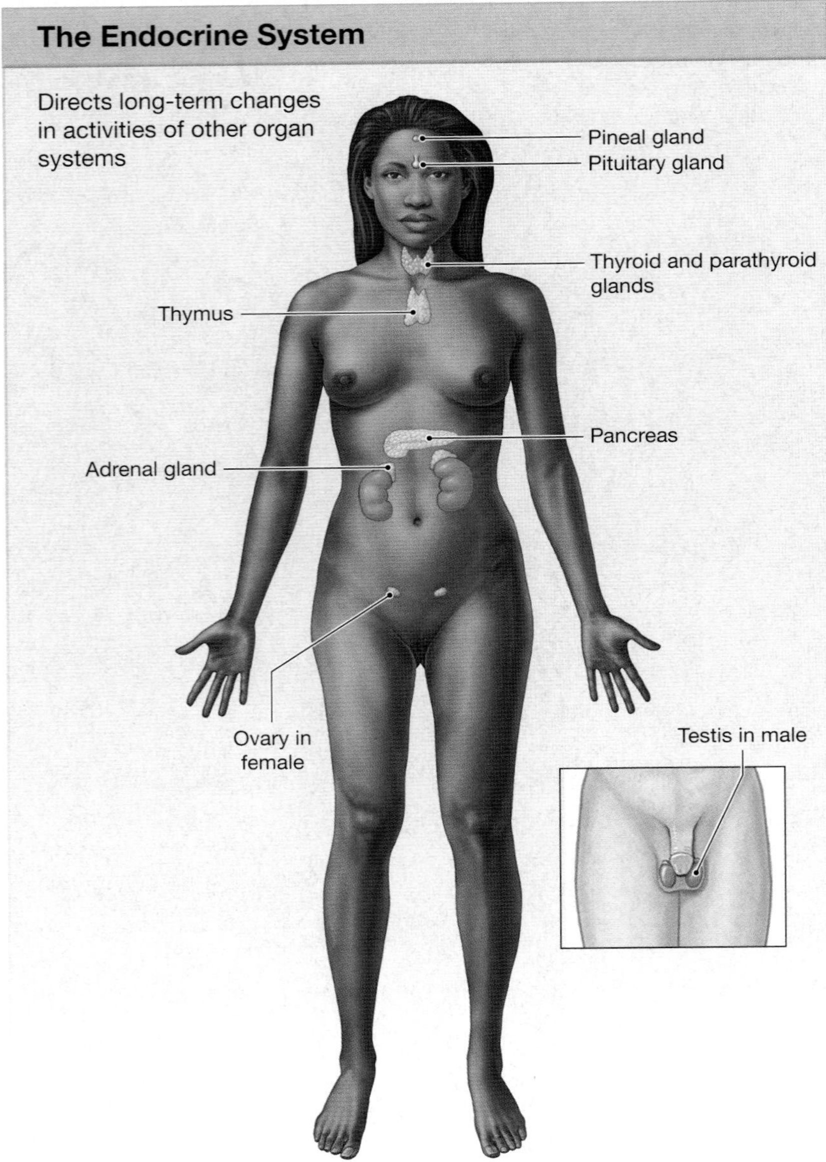

Organ/Component	Primary Functions
Pineal Gland	May control timing of reproduction and set day-night rhythms
Pituitary Gland	Controls other endocrine glands; regulates growth and fluid balance
Thyroid Gland	Controls tissue metabolic rate; regulates calcium levels
Parathyroid Glands	Regulate calcium levels (with thyroid)
Thymus	Controls maturation of lymphocytes
Adrenal Glands	Regulate water balance, tissue metabolism, and cardiovascular and respiratory activity
Kidneys	Control red blood cell production and elevate blood pressure
Pancreas	Regulates blood glucose levels
Gonads	
Testes	Support male sexual characteristics and reproductive functions
Ovaries	Support female sexual characteristics and reproductive functions

The Cardiovascular System

Transports cells and dissolved materials, including nutrients, wastes, and gases

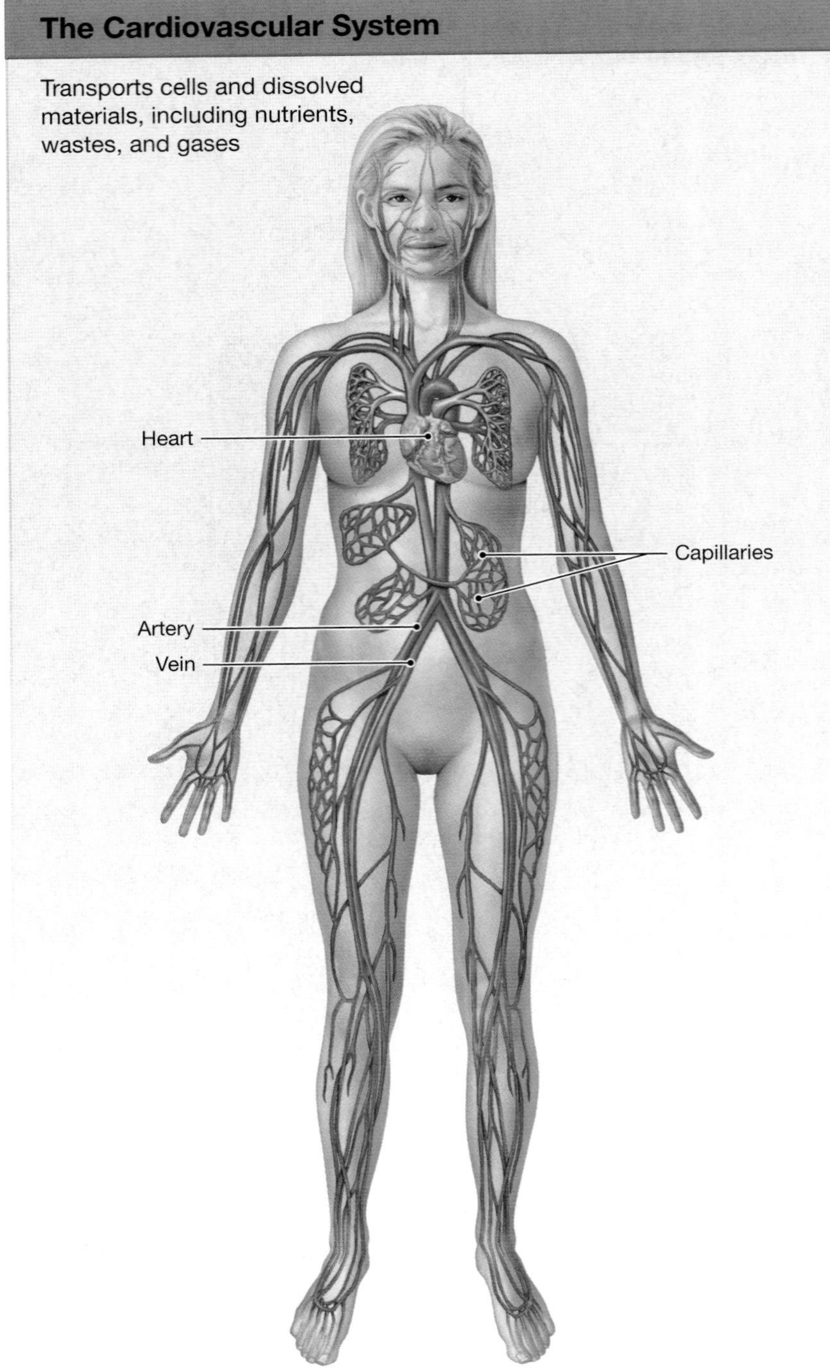

Organ/Component	Primary Functions
Heart	Propels blood; maintains blood pressure
Blood Vessels	Distribute blood around the body
Arteries	Carry blood from the heart to capillaries
Capillaries	Permit diffusion between blood and interstitial fluids
Veins	Return blood from capillaries to the heart
Blood	Transports oxygen, carbon dioxide, and blood cells; delivers nutrients and hormones; removes wastes; assists in temperature regulation and defense against disease

The Lymphatic System

Defends against infection and disease; returns tissue fluid to the bloodstream

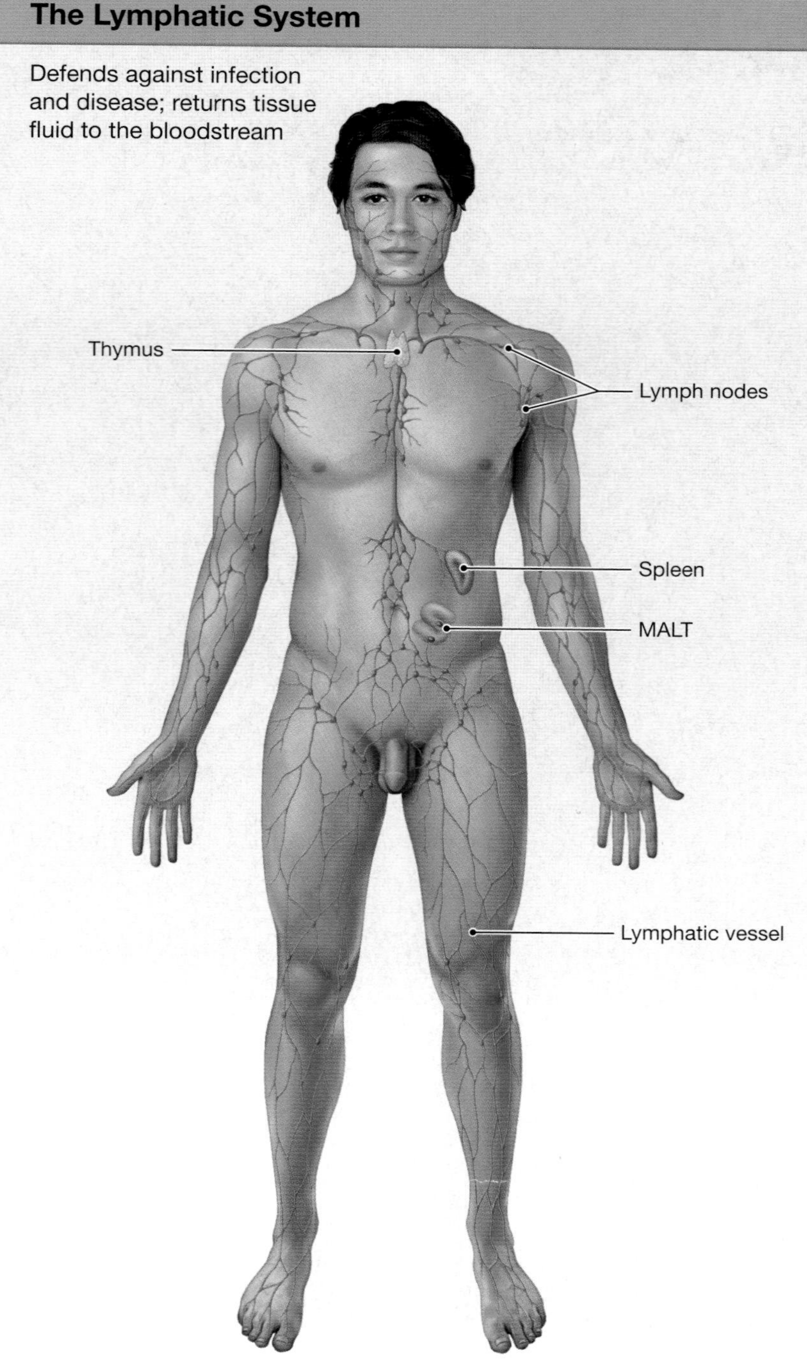

Organ/Component	Primary Functions
Lymphatic Vessels	Carry lymph (fluid with cells and proteins) and lymphocytes from peripheral tissues to veins of the cardiovascular system
Lymph Nodes	Monitor the composition of lymph; engulf pathogens; stimulate immune response
Spleen	Monitors circulating blood; engulfs pathogens and recycles red blood cells; stimulates immune response
Thymus	Controls development and maintenance of one class of lymphocytes (T cells)

The Respiratory System

Delivers air to sites where gas exchange can occur between the air and circulating blood; produces sound

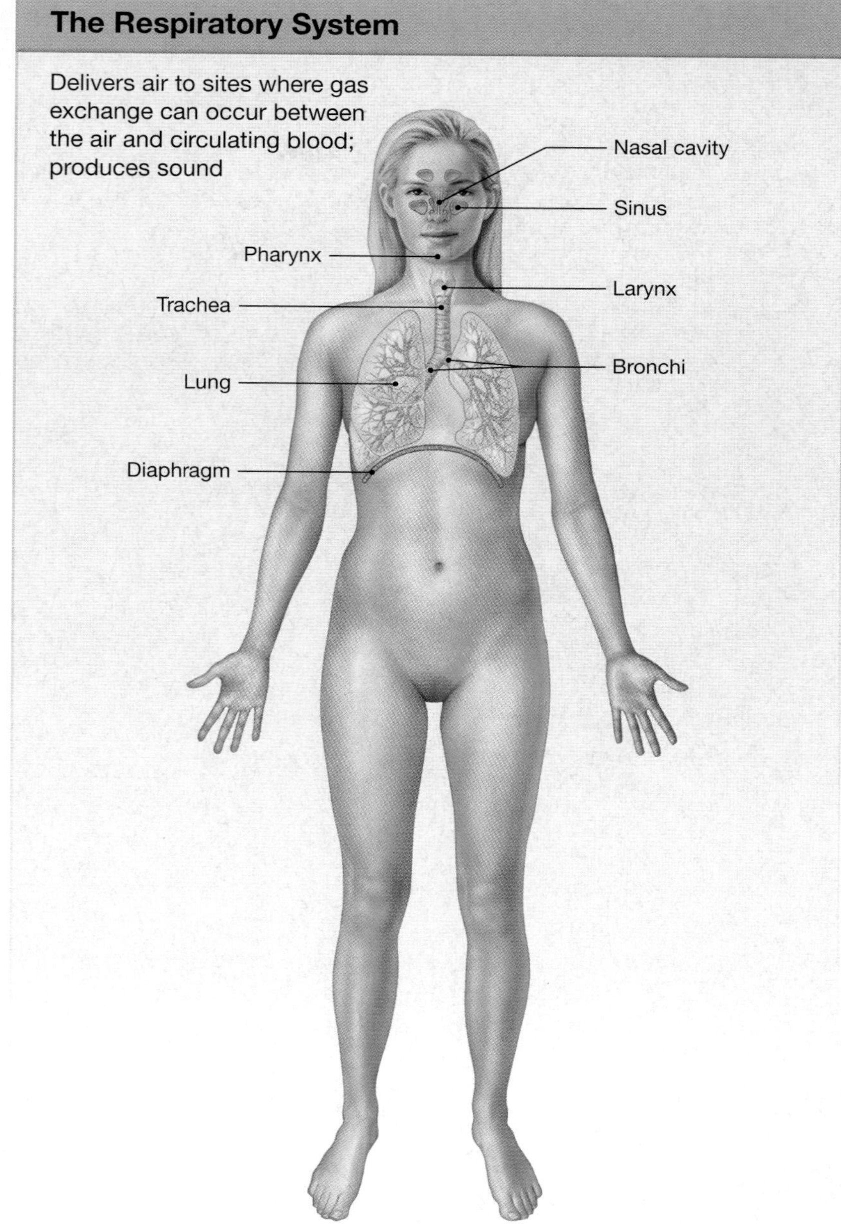

Organ/Component	Primary Functions
Nasal Cavities and Paranasal Sinuses	Filter, warm, humidify air; detect smells
Pharynx	Conducts air to larynx, a chamber shared with the digestive tract
Larynx	Protects opening to trachea and contains vocal cords
Trachea	Filters air, traps particles in mucus, conducts air to lungs; cartilages keep airway open
Bronchi	Same functions as trachea; diameter decreases as branching occurs
Lungs	Responsible for air movement during movement of ribs and diaphragm; include airways and alveoli
Alveoli	Blind pockets at the end of the smallest branches of the bronchioles; sites of gas exchange between air and blood

Figure 1.6 *(continued)*

The Digestive System

Processes food and absorbs nutrients

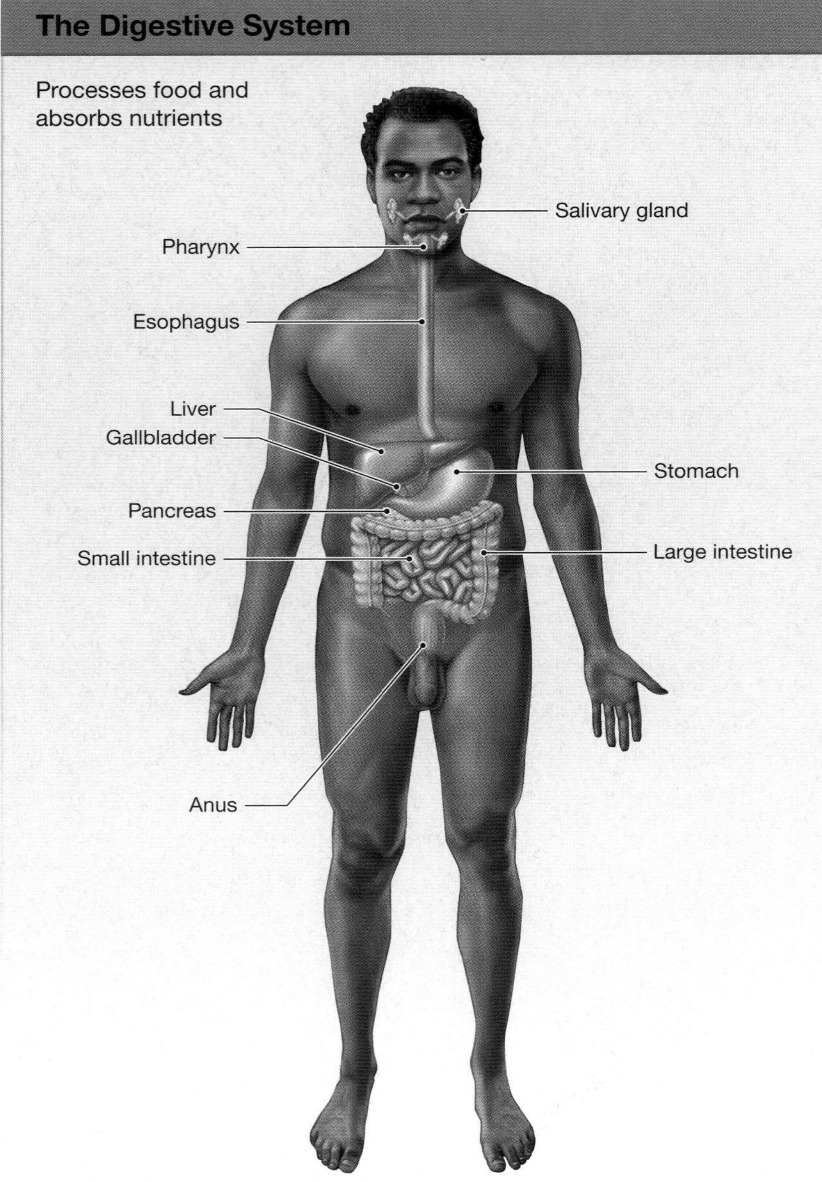

Organ/Component	Primary Functions
Oral Cavity	Receptacle for food; works with associated structures (teeth, tongue) to break up food and pass food and liquids to pharynx
Salivary Glands	Provide buffers and lubrication; produce enzymes that begin digestion
Pharynx	Conducts solid food and liquids to esophagus; chamber shared with respiratory tract
Esophagus	Delivers food to stomach
Stomach	Secretes acids and enzymes
Small Intestine	Secretes digestive enzymes, buffers, and hormones; absorbs nutrients
Liver	Secretes bile; regulates nutrient composition of blood
Gallbladder	Stores and concentrates bile for release into small intestine
Pancreas	Secretes digestive enzymes and buffers; contains endocrine cells
Large Intestine	Removes water from fecal material; stores wastes

The Urinary System

Eliminates excess water, salts, and wastes

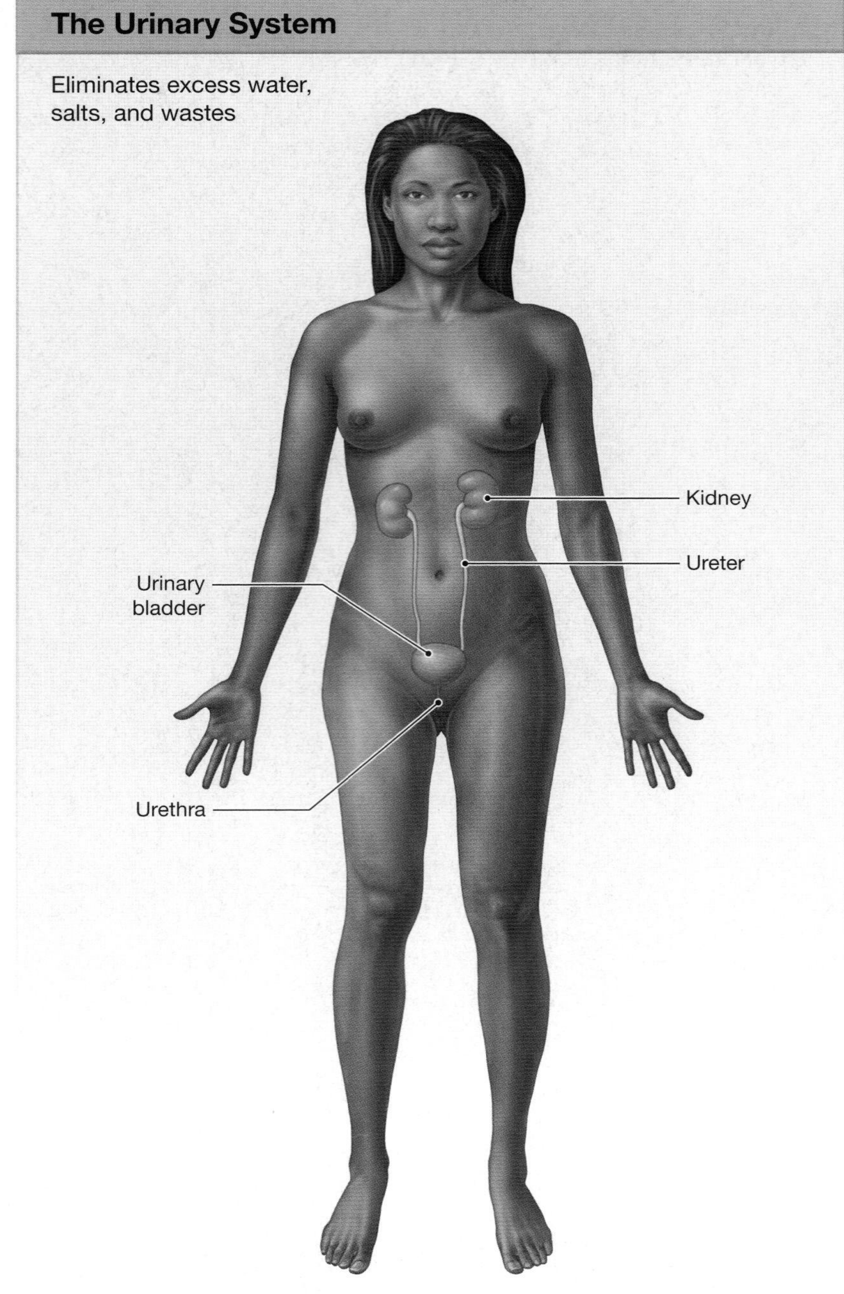

Organ/Component	Primary Functions
Kidneys	Form and concentrate urine; regulate blood pH, ion concentrations, blood pressure; perform endocrine functions
Ureters	Conduct urine from kidneys to urinary bladder
Urinary Bladder	Stores urine for eventual elimination
Urethra	Conducts urine to exterior

The Male Reproductive System

Produces sex cells and hormones

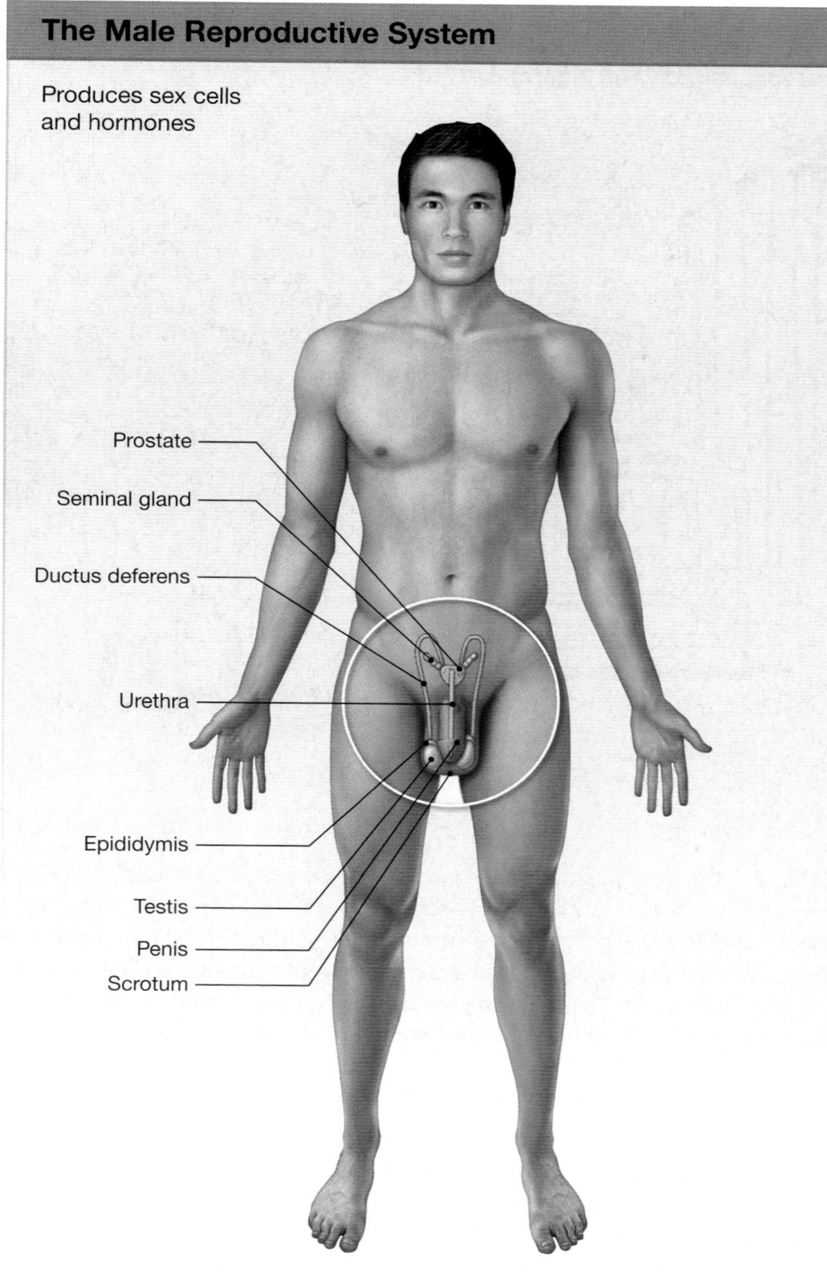

Organ/Component	Primary Functions
Testes	Produce sperm and hormones
Accessory Organs	
Epididymis	Site of sperm maturation
Ductus deferens	Conducts sperm from the epididymis and merges with the duct of the seminal gland
Seminal glands	Secrete fluid that makes up much of the volume of semen
Prostate	Secretes fluid and enzymes
Urethra	Conducts semen to exterior
External Genitalia	
Penis	Contains erectile tissue; deposits sperm in vagina of female; produces pleasurable sensations during sexual activities
Scrotum	Surrounds the testes and controls their temperature

The Female Reproductive System

Produces sex cells and hormones; supports embryonic and fetal development from fertilization to birth

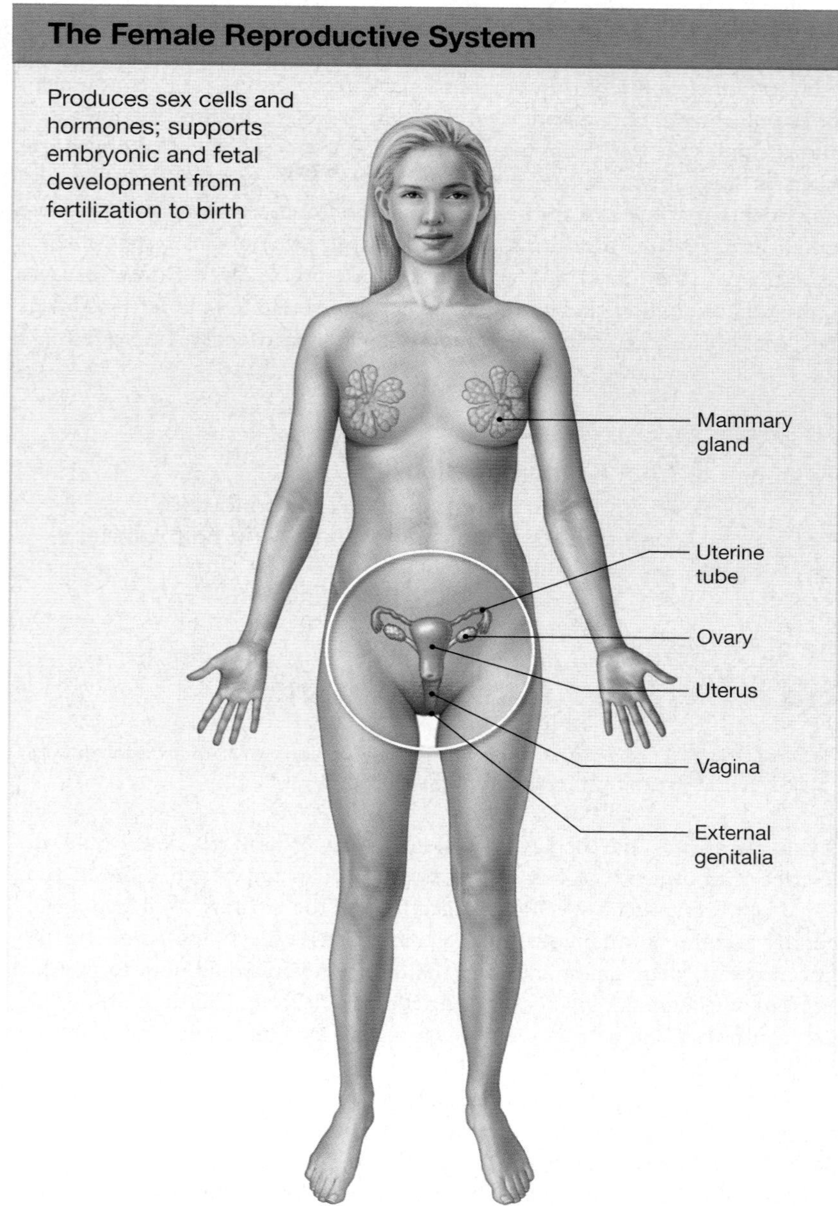

Organ/Component	Primary Functions
Ovaries	Produce oocytes and hormones
Uterine Tubes	Deliver oocyte or embryo to uterus; normal site of fertilization
Uterus	Site of embryonic and fetal development; site of exchange between maternal and fetal bloodstreams
Vagina	Site of sperm deposition; birth canal during delivery; provides passageway for fluids during menstruation
External Genitalia	
Clitoris	Contains erectile tissue; provides pleasurable sensations during sexual activities
Labia	Contain glands that lubricate entrance to vagina
Mammary Glands	Produce milk that nourishes newborn infant

For very small organisms, absorption, respiration, and excretion involve the movement of materials across exposed surfaces. But creatures larger than a few millimeters seldom absorb nutrients directly from their environment. For example, we cannot absorb steaks, apples, or ice cream directly—our bodies must first alter the foods' chemical structure. That processing, called **digestion**, occurs in specialized areas where complex foods are broken down into simpler components that are absorbed easily. Respiration and excretion are also more complicated for large organisms, so we have specialized organs responsible for gas exchange (the lungs) and waste excretion (the kidneys). Finally, because absorption, respiration, and excretion are performed in different portions of the body, we have an internal transportation system, or **cardiovascular system**.

1.5 CONCEPT CHECK

8 What is differentiation?

9 Which organ system includes the following components: sweat glands, nails, and hair follicles?

See the blue Answers tab at the back of the book.

1.6 The Language of Anatomy

▶ **KEY POINT** Learning the specialized terminology of human anatomy will make it easier to understand anatomical concepts.

If you discovered and then fully explored a new continent, how would you describe it to others in a way that everyone would understand? One method would be to construct a specific, detailed map of the territory. Your map would identify prominent landmarks, such as mountains, valleys, and rivers; the distance between these landmarks; and the direction you would need to travel to get from one landmark to another using compass bearings (north, south, northeast, southwest, and so on). With such a map, anyone could find a specific location on that continent.

Early anatomists faced a similar challenge when trying to communicate their findings. Identifying a particular location on the human body proved to be difficult. Stating that a bump is "on the back," for instance, does not specify its exact location. So anatomists created maps of the human body. The landmarks are the prominent anatomical structures, and distances are measured in centimeters or inches.

Anatomy uses a special language that you must learn at the start. It will take some time and effort, but it is absolutely essential if you want to avoid a situation like that shown in **Figure 1.7**.

New anatomical terms are introduced as technology advances, but many older words and phrases persist. Latin and Greek words form the basis for an impressive number of anatomical terms. Many Latin names assigned to specific structures 2000 years ago are still in use today.(For more information, see the Appendix "Foreign Word Roots, Prefixes, Suffixes, and Combining Forms.")

Some anatomical structures and clinical conditions were named after the discoverer or, in the case of diseases, after the most famous victim. Over time, most of these commemorative names, or *eponyms*, have been replaced by more descriptive terms. (For information about commemorative names still being used today, see the Appendix "Eponyms in Common Use.")

Superficial Anatomy

▶ **KEY POINT** Learning anatomical landmarks, regions, and directions will help you create "mental maps" of internal structures.

Figure 1.7 The Importance of Precise Vocabulary. Would you want to be this patient? []

Except for the skin, hair, and nails (which are parts of the integumentary system), you cannot see any of the organ systems from the body surface. To understand structures that are deep to (internal to) the integument, it is important to create your own mental maps based on the illustrations and discussions throughout this text. The following sections discuss anatomical landmarks and regions that will help you create these mental maps.

Anatomical Landmarks

Figure 1.8 presents important anatomical landmarks. Become familiar with both the anatomical term (for instance, *nasus*) and its adjective form (*nasal*). Learning these terms will help you remember the location of a particular structure as well as its name. For example, the term **brachium** refers to the arm, and in later chapters you will learn about the brachialis muscle and the brachial artery, both of which are located in the arm.

Standard anatomical illustrations show a human figure in the **anatomical position**: standing with legs together, feet flat on the floor, with hands at the sides and palms facing forward. **Figure 1.8a** shows the anatomical position from the anterior (front) view, and **Figure 1.8b** shows it from the posterior (back) view. The anatomical position is the standard by which the language of anatomy is communicated. *Therefore, unless otherwise noted, all the descriptions in this text refer to the body in the anatomical position.* A person lying down in the anatomical position is said to be **supine** (sū -PĪN) when lying face up and **prone** when lying face down.

TIPS & TOOLS

Remembering Supine Position

When you are in the supine position, you can hold a bowl of soup in the palm of your hand or on your navel without spilling it.

Figure 1.8 Anatomical Landmarks. Anatomical terms are shown in boldface type, common names are in plain type, and anatomical adjectives are in parentheses.

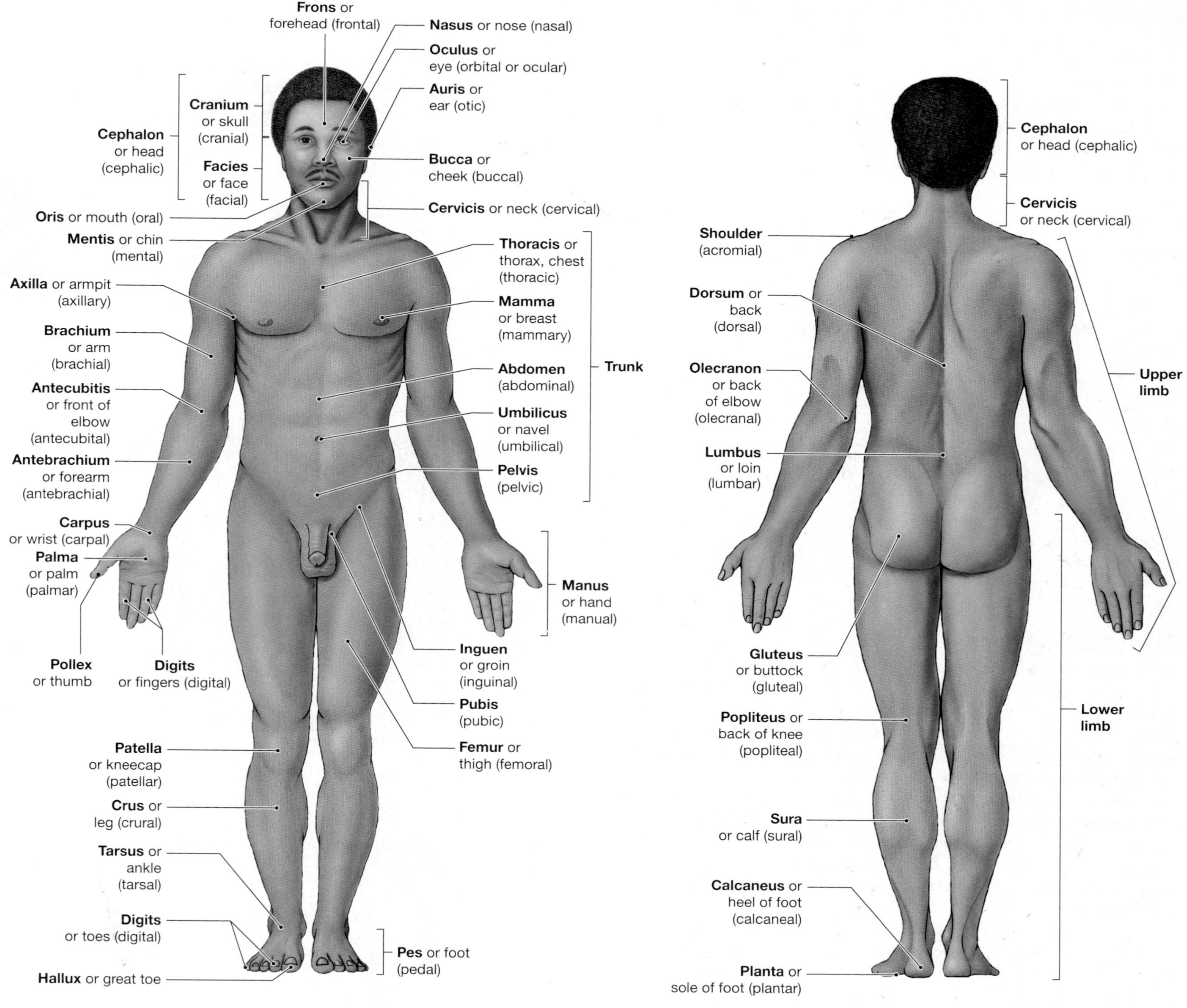

Table 1.1 | Regions of the Human Body*

Anatomical Name	Anatomical Region	Area Indicated
Cephalon	Cephalic	Head
Cervicis	Cervical	Neck
Thoracis	Thoracic	Chest
Brachium	Brachial	Segment of the upper limb closest to the trunk; the arm
Antebrachium	Antebrachial	Forearm
Carpus	Carpal	Wrist
Manus	Manual	Hand
Abdomen	Abdominal	Abdomen
Pelvis	Pelvic	Pelvis (in general)
Pubis	Pubic	Anterior pelvis
Inguen	Inguinal	Groin (crease between thigh and trunk)
Lumbus	Lumbar	Lower back
Gluteus	Gluteal	Buttock
Femur	Femoral	Thigh
Patella	Patellar	Kneecap
Crus	Crural	Leg, from knee to ankle
Sura	Sural	Calf
Tarsus	Tarsal	Ankle
Pes	Pedal	Foot
Planta	Plantar	Sole region of foot

* See **Figure 1.8**.

Anatomical Regions

Table 1.1 summarizes the major regions of the body, and **Figure 1.9** labels these regions (additional regions and anatomical landmarks are noted in **Figure 1.8**). Anatomists and clinicians use special terminology to describe specific areas of the abdominal and pelvic regions. There are two different methods in use. In the first, the abdominopelvic surface is divided into four sections, called the **abdominopelvic quadrants**, using a pair of imaginary lines (one horizontal and one vertical) that intersect at the umbilicus (navel) **(Figure 1.9a)**. This simple method is useful for describing pain and injuries. Knowing the location of an ache or pain helps a clinician determine the possible cause; for example, tenderness in the right lower quadrant (RLQ) is a symptom of appendicitis, whereas tenderness in the right upper quadrant (RUQ) may indicate gallbladder or liver problems.

In the second method, nine **abdominopelvic regions** are used to more precisely describe the location and orientation of internal organs **(Figure 1.9b)**. **Figure 1.9c** shows the relationship between abdominopelvic quadrants, regions, and internal organs.

Anatomical Directions

Figure 1.10 shows the principal directional terms used in anatomy and examples of their use. There are many different directional terms, and some can be used interchangeably. *As you learn these terms, it is important to remember that all anatomical directions use the anatomical position as the standard point of reference. When following anatomical descriptions, it is useful to remember that the terms left and right refer to the left and right sides of the subject, not the observer.* You should also note that although some reference terms are equivalent, such as posterior and dorsal and anterior and ventral, anatomical descriptions do not mix the terms of the opposing pairs. For example, a discussion would reference either posterior versus anterior or dorsal versus ventral; it would not reference posterior versus ventral.

Figure 1.9 Abdominopelvic Quadrants and Regions. The abdominopelvic surface is separated into sections to identify anatomical landmarks more clearly and to define the location of contained organs more precisely.

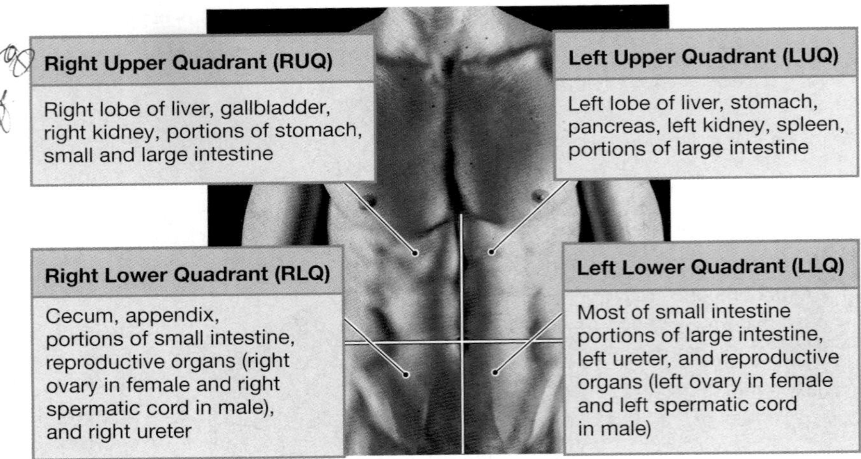

a Abdominopelvic quadrants divide the area into four sections.

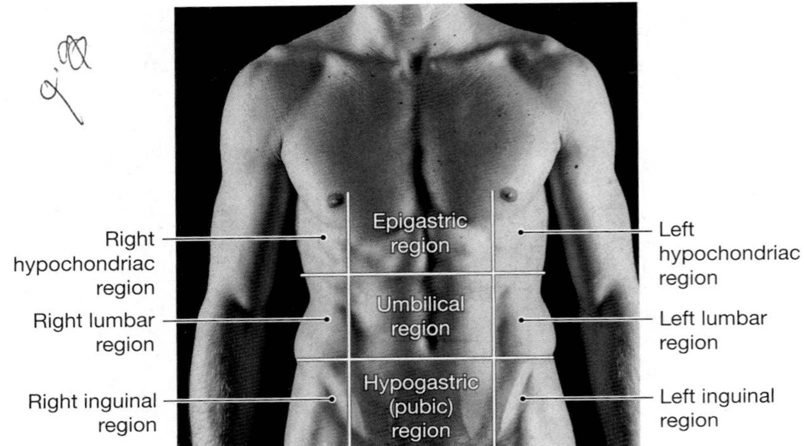

b More precise anatomical descriptions are provided by reference to the appropriate abdominopelvic region.

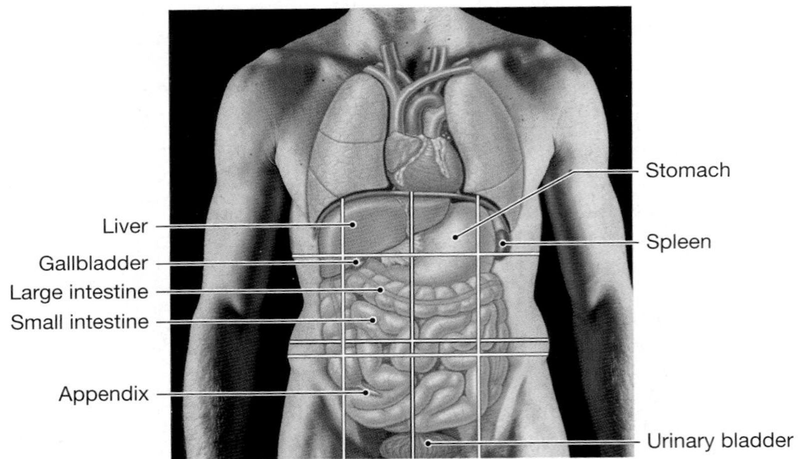

c Dividing the abdominal/pelvic area into quadrants or regions is useful because there is a known relationship between superficial anatomical landmarks and the underlying organs.

Figure 1.10 Directional References. The arrows indicate important directional references used in this text.

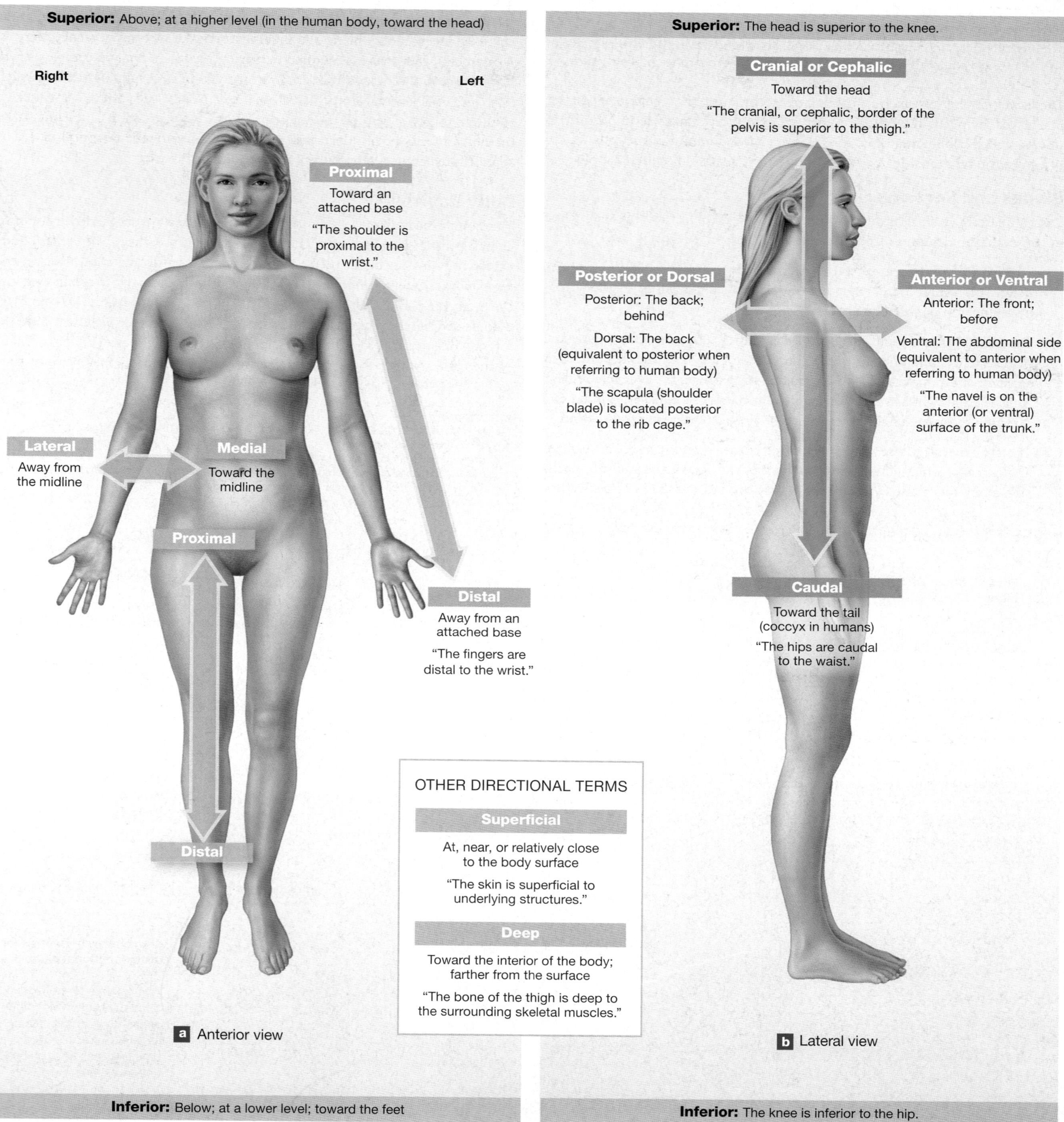

Superior: Above; at a higher level (in the human body, toward the head)
Right
Left
Proximal
Toward an attached base
"The shoulder is proximal to the wrist."
Lateral
Away from the midline
Medial
Toward the midline
Proximal
Distal
Away from an attached base
"The fingers are distal to the wrist."
Distal
a Anterior view
Inferior: Below; at a lower level; toward the feet
Superior: The head is superior to the knee.
Cranial or Cephalic
Toward the head
"The cranial, or cephalic, border of the pelvis is superior to the thigh."
Posterior or Dorsal
Posterior: The back; behind
Dorsal: The back (equivalent to posterior when referring to human body)
"The scapula (shoulder blade) is located posterior to the rib cage."
Anterior or Ventral
Anterior: The front; before
Ventral: The abdominal side (equivalent to anterior when referring to human body)
"The navel is on the anterior (or ventral) surface of the trunk."
Caudal
Toward the tail (coccyx in humans)
"The hips are caudal to the waist."
OTHER DIRECTIONAL TERMS
Superficial
At, near, or relatively close to the body surface
"The skin is superficial to underlying structures."
Deep
Toward the interior of the body; farther from the surface
"The bone of the thigh is deep to the surrounding skeletal muscles."
b Lateral view
Inferior: The knee is inferior to the hip.

Sectional Anatomy

1

▶ **KEY POINT** The word anatomy comes from a Greek word meaning "to cut apart." To fully understand anatomy, you must understand how the plane of section—how something is cut apart—changes the appearance of a structure.

The development of electronic imaging techniques that enable us to see inside the living body without resorting to surgery makes it important to understand sectional anatomy. A sectional view is sometimes the only way to illustrate the relationships between the parts of a three-dimensional object.

Planes and Sections

You can describe a slice through a three-dimensional object by referencing one of three **sectional planes**: frontal, sagittal, or transverse (**Figure 1.11**).

- The **frontal plane**, or *coronal plane*, parallels the longitudinal axis of the body. The frontal plane extends from side to side, dividing the body into **anterior** and **posterior** sections.
- The **sagittal plane** also parallels the longitudinal axis of the body. The sagittal plane extends from anterior to posterior, dividing the body into left and right sections. A section passing along the midline that divides the body into roughly equal left and right halves is a **midsagittal section**, or a *median sagittal section*. A section that runs parallel to the midsagittal line is a **parasagittal section**.
- The **transverse plane**, or *horizontal* or *cross-sectional plane*, lies at right angles to the longitudinal axis of the part of the body being studied. A division along this plane is a **transverse section**, or *horizontal* or *cross section*.

Each sectional plane gives a different perspective on the structure of the body. When combined with observations of external anatomy, they create a reasonably complete picture. You could develop an even more complete picture by choosing one sectional plane and making a series of sections at small intervals. This process, called **serial reconstruction**, allows us to analyze complex structures. **Figure 1.12** shows the serial reconstruction of a simple bent tube, such as a piece of elbow macaroni. This procedure can show the path of a small blood vessel or follow a loop of the intestine. Serial reconstruction is an important method for studying histological structure and analyzing the images produced by sophisticated clinical procedures (see the Clinical Note on pp. 20–21).

Body Cavities

The human body is not a solid object; many organs are suspended in internal chambers called **body cavities*** that protect and cushion them. The **ventral body cavity** contains organs of the respiratory, cardiovascular, digestive, urinary, and reproductive systems (**Figure 1.13**). The ventral body cavity is subdivided into the thoracic cavity and the abdominopelvic cavity; the **diaphragm** (DĪ-a-fram) is a dome-shaped sheet of skeletal muscle that separates them (**Figure 1.13**). The internal organs that project into these cavities are called **viscera** (VIS-er-a). Many of the organs within these cavities change size and shape as they perform their

*In the human adult, the thoracic, abdominal, and pelvic cavities share a common embryological origin. The term *dorsal body cavity* is sometimes used to refer to the internal chamber of the skull and the space enclosed by the vertebral arches. These chambers, which are defined by bony structures, are anatomically and embryologically distinct from true body cavities, and the term dorsal body cavity is not encountered in either clinical anatomy or comparative anatomy.

Figure 1.11 Sectional Planes. The three primary planes of section are frontal, sagittal, and transverse.

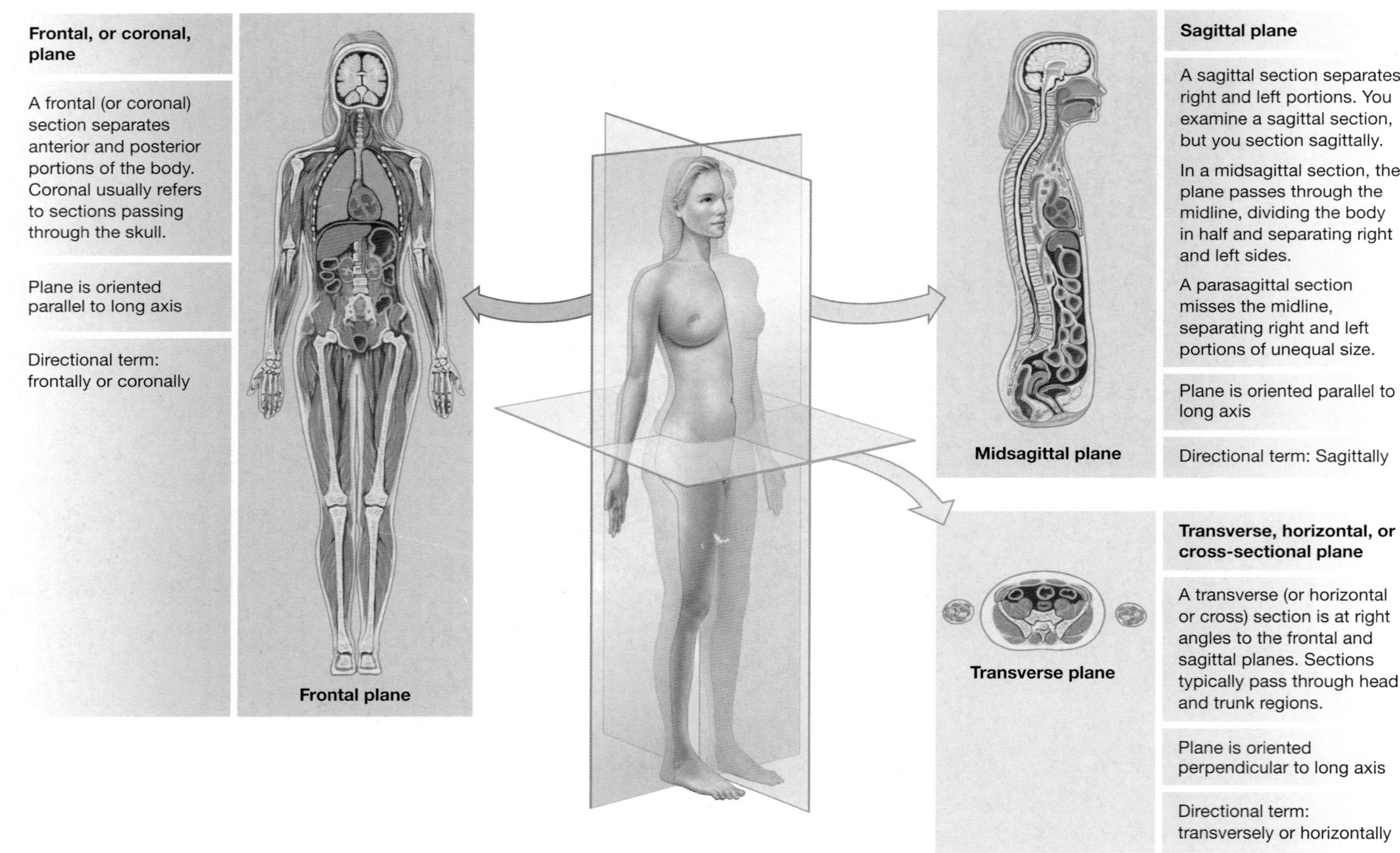

Frontal, or coronal, plane
A frontal (or coronal) section separates anterior and posterior portions of the body. Coronal usually refers to sections passing through the skull.
Plane is oriented parallel to long axis
Directional term: frontally or coronally
Frontal plane
Sagittal plane
A sagittal section separates right and left portions. You examine a sagittal section, but you section sagittally.
In a midsagittal section, the plane passes through the midline, dividing the body in half and separating right and left sides.
A parasagittal section misses the midline, separating right and left portions of unequal size.
Plane is oriented parallel to long axis
Directional term: Sagittally
Midsagittal plane
Transverse, horizontal, or cross-sectional plane
A transverse (or horizontal or cross) section is at right angles to the frontal and sagittal planes. Sections typically pass through head and trunk regions.
Plane is oriented perpendicular to long axis
Directional term: transversely or horizontally
Transverse plane

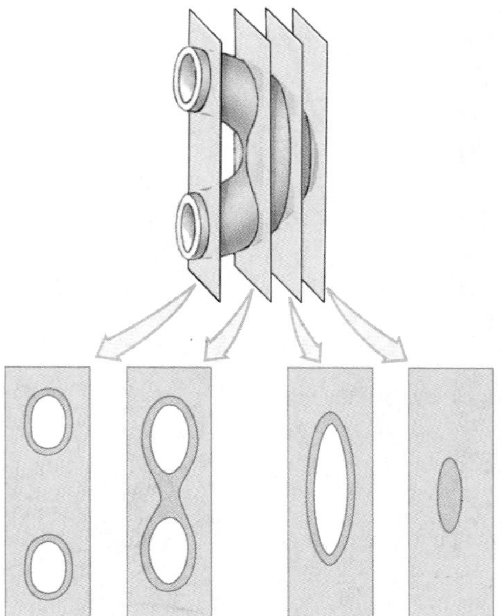

Figure 1.12 Sectional Planes and Visualization. This diagram shows the serial reconstruction of a bent tube (like a piece of elbow macaroni). Notice how the sectional views change as the plane approaches the curve. Keep the effects of sectioning in mind when looking at slides under the microscope. Sectional views of internal organs, such as those taken via a CT or MRI scan (see pp. 20–21), can vary widely. For example, although it is a simple tube, the small intestine can look like a pair of tubes, a dumbbell, an oval, or a solid, depending on where the section was taken.

functions. For example, the stomach swells at each meal, and the heart contracts and expands with each beat. These organs project into moist internal chambers that allow expansion and limited movement, but prevent friction.

The Thoracic Cavity The **thoracic cavity** contains organs of the respiratory, cardiovascular, and lymphatic systems, as well as the thymus and inferior portions of the esophagus. The muscles and bones of the chest wall and the diaphragm form the boundaries of the thoracic cavity. The thoracic cavity is subdivided into the left and right pleural cavities, which are separated by the mediastinum (Look ahead to **Figure 1.13a–c**).

Each **pleural cavity** contains a lung. A shiny, slippery serous membrane called a **pleura** (plūr-ah) lines each pleural cavity and reduces friction as the lung expands and recoils during breathing. The **visceral pleura** covers the outer surfaces of each lung, and the **parietal pleura** covers the opposing mediastinal surface and the inner body wall.

The **mediastinum** (mē-dē-as-TĪ-num) is connective tissue that surrounds, stabilizes, and supports the esophagus, trachea, thymus, and major blood vessels that originate or end at the heart. The mediastinum also contains the **pericardial cavity**, a small chamber that surrounds the heart. The serous membrane covering the heart is called the **pericardium** (*peri*, around, + *kardia*, heart). To visualize the relationship between the heart and pericardial cavity, think of a fist pushing into a balloon (**Figure 1.13d**): The wrist corresponds to the base (attached portion) of the heart, and the balloon corresponds to the pericardium.

The pericardium is composed of two parts: an outer sac of tough, fibrous connective tissue termed the **parietal layer of the serous pericardium** an inner **visceral layer of the serous pericardium**. During each beat, the heart changes in size and shape. The pericardial cavity permits these changes, and the slippery pericardial lining prevents friction between the heart and adjacent structures in the mediastinum.

The Abdominopelvic Cavity The **abdominopelvic cavity** is divided into (1) a superior abdominal cavity, (2) an inferior pelvic cavity, and (3) an internal chamber called the **peritoneal (**per-i-tō-NĒ-al**) cavity** (**Figure 1.13a,c**).

The peritoneal cavity is lined by a serous membrane called the **peritoneum** (per-i-tō-NĒ-um). The **parietal peritoneum** lines the body wall. A narrow, fluid-filled space separates the parietal peritoneum from the **visceral peritoneum**, which covers the enclosed organs. Double sheets of peritoneum, called **mesenteries** (MES-en-ter-ēs), suspend organs such as the stomach, small intestine, and portions of the large intestine within the peritoneal cavity. Mesenteries provide blood supply, support, lubrication, and stability while permitting limited movement.

The **abdominal cavity** extends from the inferior surface of the diaphragm to an imaginary plane extending from the inferior surface of the lowest spinal vertebra to the anterior and superior margins of the pelvic girdle (**Figure 1.13a,c**). The abdominal cavity contains the liver, stomach, spleen, kidneys, pancreas, and small intestine, and most of the large intestine. (Refer to **Figure 1.9a,c** on page 16 for the positions of many of these organs.) These organs project partially or completely into the peritoneal cavity, much as the heart and lungs project into the pericardial and pleural cavities, respectively.

The inferior portion of the abdominopelvic cavity is the **pelvic cavity** (**Figure 1.13a,c**). The pelvic cavity is enclosed by the bones of the pelvis and contains the last segments of the large intestine, the urinary bladder, and various reproductive organs: The pelvic cavity of a female contains the ovaries, uterine tubes, and uterus; in a male, it contains the prostate and seminal glands. The inferior portion of the peritoneal cavity extends into the pelvic cavity. The peritoneum covers the uterine tubes, the ovaries, and the superior portion of the uterus in females, as well as the superior portion of the urinary bladder in both males and females.

The Clinical Note on pp. 20–21 summarizes modern methods of visualizing anatomical structures in living people. A true understanding of anatomy involves integrating the information provided by sectional images, interpretive artwork based on sections and dissections, and direct observation. It is up to you to integrate these views and develop your ability to observe and visualize anatomical structures. Remember that every structure you encounter has a specific function. The goal of anatomy isn't simply to identify structural details, but to understand the three-dimensional relationships between bodily structures and how those structures interact to perform the varied functions of the human body.

CLINICAL NOTE

Pericarditis and Peritonitis

The suffix *–itis* means "inflammation." Thus, pericarditis means inflammation of the pericardium. Pericarditis can be caused by any disease-causing agent or trauma, and it can severely restrict the function of the heart. Peritonitis is an inflammation of the peritoneum (the serous membrane lining the abdomen). It may be due to bacterial infection, liver failure, kidney failure, or many other causes. Peritonitis affects all the organs within the peritoneal cavity.

1.6 CONCEPT CHECK

10 You fall and break your antebrachium. What part of your body is affected?

11 What is the anatomical name for each of the following areas: groin, buttock, and hand?

12 What type of section would separate the two eyes?

13 What is the general function of the mesenteries?

14 If a surgeon makes an incision just inferior to the diaphragm, which body cavity will be opened?

See the blue Answers tab at the back of the book.

CLINICAL NOTE

Clinical Anatomy and Technology

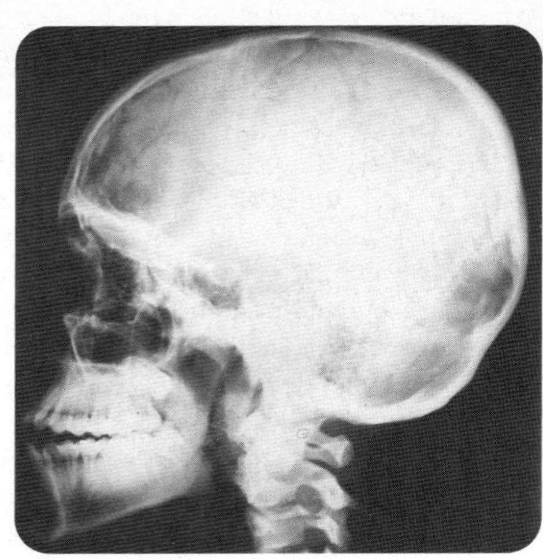

X-ray

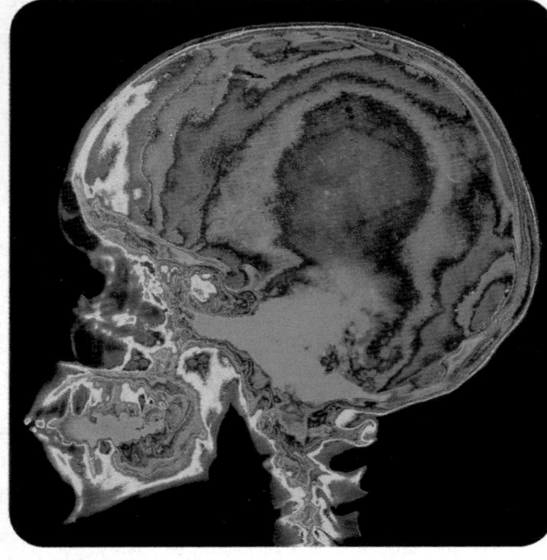

Color-enhanced x-ray

Stomach

Small intestine

Barium-contrast x-ray

Radiological procedures

include various noninvasive techniques that use radioisotopes, radiation, and magnetic fields to produce images of internal structures. Physicians who specialize in the performance and analysis of these diagnostic images are called radiologists. Radiological procedures can provide detailed information about internal systems and structures.

X-rays are a form of high-energy radiation that can penetrate living tissues. In the most familiar procedure, a beam of x-rays travels through the body and strikes a photographic plate. Not all of the projected x-rays arrive at the film; some are absorbed or deflected as they pass through the body. The resistance to x-ray penetration is called radiodensity. In the human body, the order of increasing radiodensity is as follows: air, fat, liver, blood, muscle, bone. The result is an image with radiodense tissues, such as bone, appearing white, while less dense tissues are seen in shades of gray to black.

A **barium-contrast x-ray** of the upper digestive tract. Barium is very radiodense, and the contours of the gastric and intestinal lining can be seen outlined against the white of the barium solution.

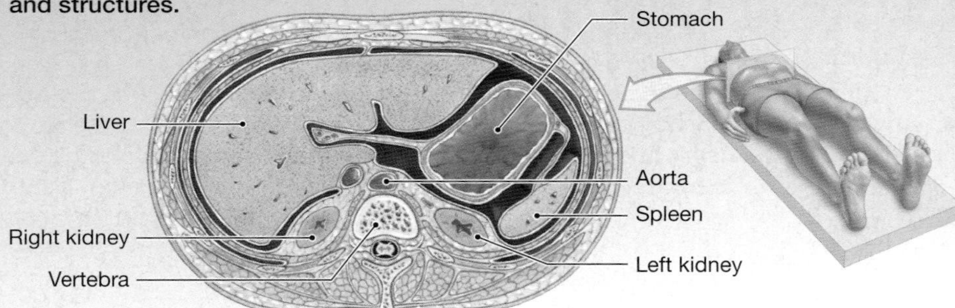

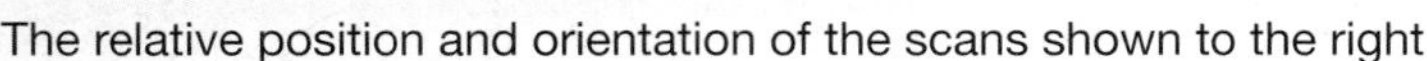

The relative position and orientation of the scans shown to the right.

Note that when anatomical diagrams or scans present cross-sectional views, the sections are presented as though the observer were standing at the feet of a person in the supine position and looking toward the head of the subject.

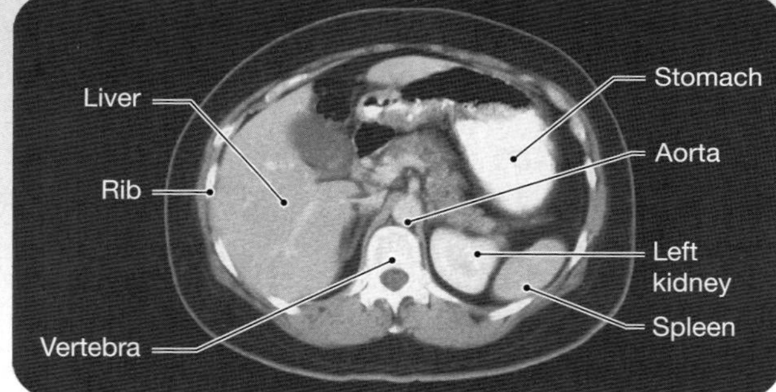

CT scan of the abdomen

CT scans, formerly called CAT (computerized axial tomography), use a single x-ray source rotating around the body. The x-ray beam strikes a sensor monitored by a computer. The source completes one revolution around the body every few seconds; it then moves a short distance and repeats the process. By comparing the information obtained at each point in the rotation, the computer reconstructs the three-dimensional structure of the body. The result is usually displayed as a sectional view in black and white, but it can be colorized.

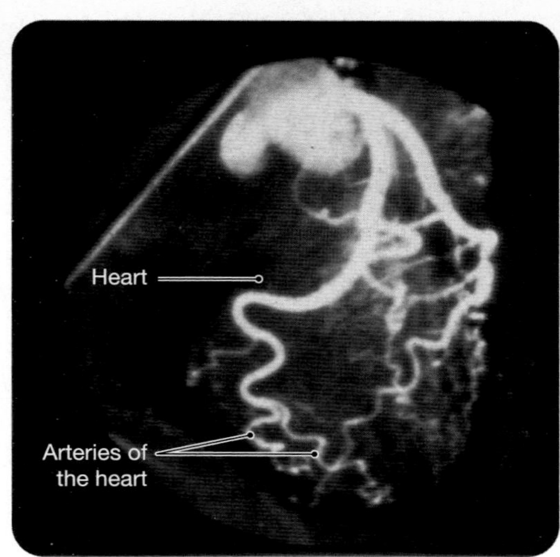

Digital subtraction angiography

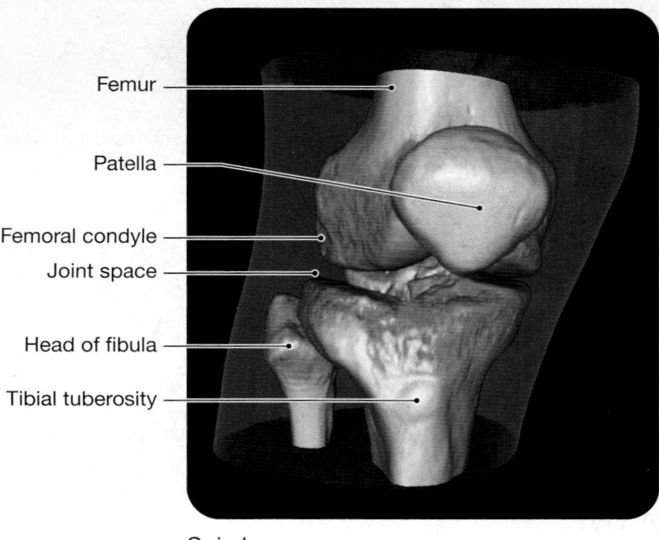

Spiral scan

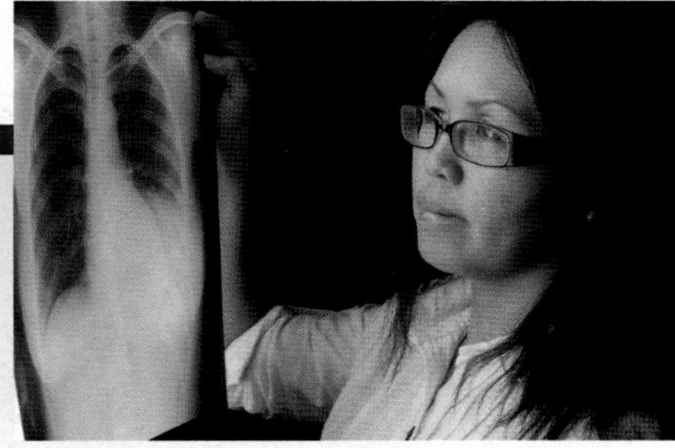

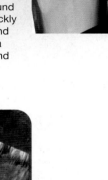

Digital subtraction angiography (DSA) is used to monitor blood flow through specific organs, such as the brain, heart, lungs, and kidneys. X-rays are taken before and after radiopaque dye is administered, and a computer "subtracts" details common to both images. The result is a high-contrast image showing the distribution of the dye.

A **spiral CT scan** (also termed a helical CT scan) is a new form of three-dimensional imaging technology that is becoming increasingly important in clinical settings. With a spiral CT scan the patient is placed on a platform that advances at a steady pace through the scanner while the imaging source, usually x-rays, rotates continuously around the patient. Because the x-ray detector gathers data quickly and continuously, a higher quality image is generated, and the patient is exposed to less radiation as compared to a standard CT scanner, which collects data more slowly and only one slice of the body at a time.

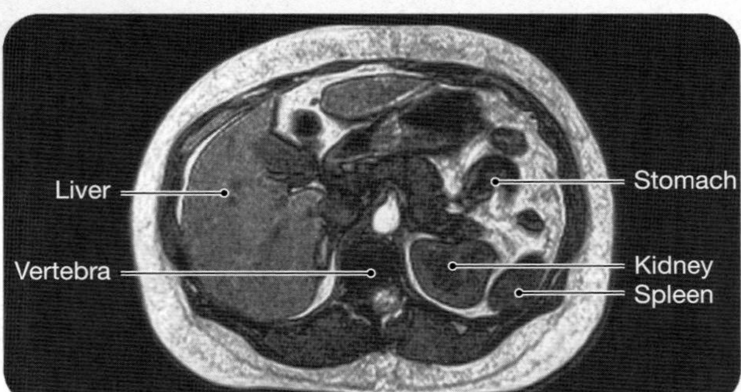

MRI scan of the abdomen

An **MRI (magnetic resonance imaging)** scan surrounds part or all of the body with a magnetic field about 3000 times as strong as that of Earth. This field affects protons within atomic nuclei throughout the body, which line up along the magnetic lines of force like compass needles in Earth's magnetic field. When struck by a radio wave of the proper frequency, a proton will absorb energy. When the pulse ends, that energy is released, and the energy source of the radiation is detected by the MRI computers.

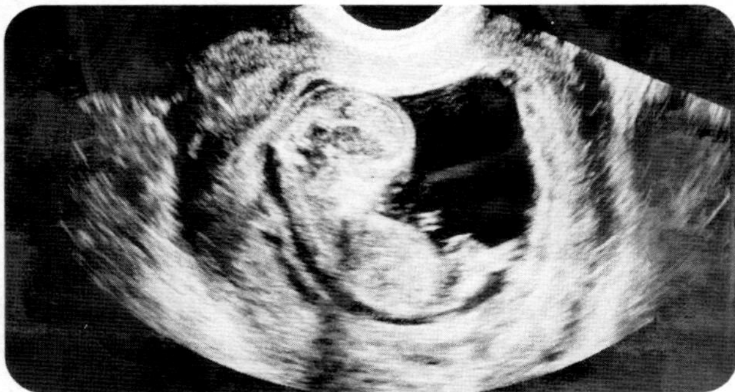
Ultrasound scan of the abdomen

In **ultrasound** procedures, a small transmitter contacting the skin broadcasts a brief, narrow burst of high-frequency sound and then picks up the echoes. The sound waves are reflected by internal structures, and a picture, or echogram, can be assembled from the pattern of echoes. These images lack the clarity of other procedures, but no adverse effects have been attributed to the sound waves, and fetal development can be monitored without a significant risk of birth defects.

1

Figure 1.13 **Body Cavities.** Relationships, contents, and selected functions of the subdivisions of the thoracic and abdominopelvic body cavities.

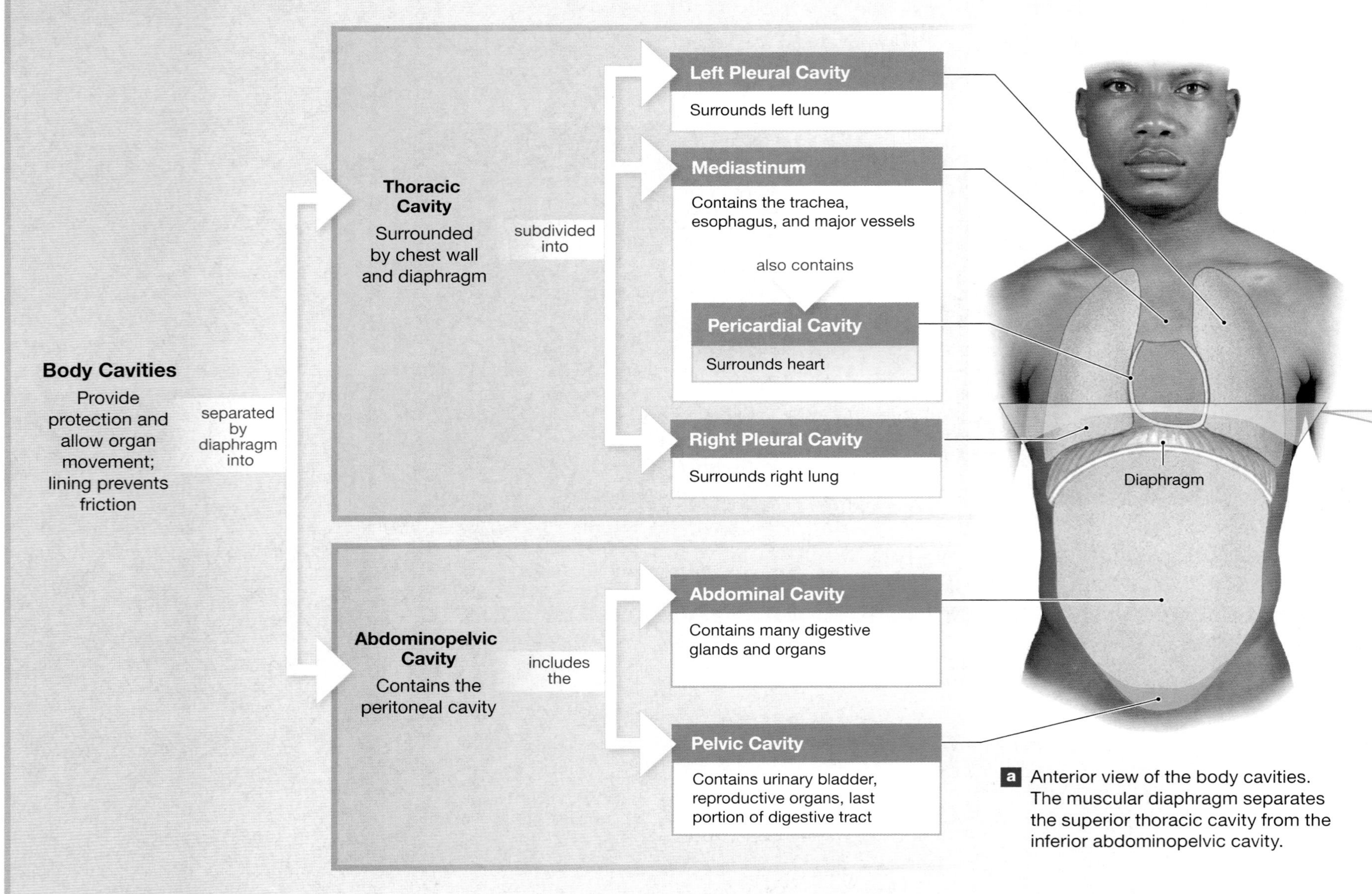

a Anterior view of the body cavities. The muscular diaphragm separates the superior thoracic cavity from the inferior abdominopelvic cavity.

Study Outline

Introduction p. 2

- **Anatomy** is the study of internal and external structures and the physical relationships between body parts. Specific anatomical structures perform specific functions.

1.1 Microscopic Anatomy p. 2

- The boundaries of **microscopic anatomy** are established by the limits of the equipment used. **Cytology** is the study of the internal structure of individual **cells**, the smallest units of life. **Histology** examines **tissues**, groups of cells that work together to perform specific functions. Specific arrangements of tissues form an **organ**, an anatomical unit with multiple functions. A group of organs that function together forms an **organ system**. *(See Figure 1.1.)*

1.2 Gross Anatomy p. 2

- **Gross** (macroscopic) **anatomy** considers features visible without a microscope. It includes **surface anatomy** (general form and superficial markings), **regional anatomy** (superficial and internal features in a specific area of the body), and **systemic anatomy** (structure of major organ systems).

1.3 Other Types of Anatomical Studies p. 2

- **Developmental anatomy** examines the changes in form that occur between conception and physical maturity. **Embryology** studies the processes that occur during the first two months of development.
- **Comparative anatomy** considers the similarities and relationships in anatomical organization of different animals. *(See Figure 1.2.)*

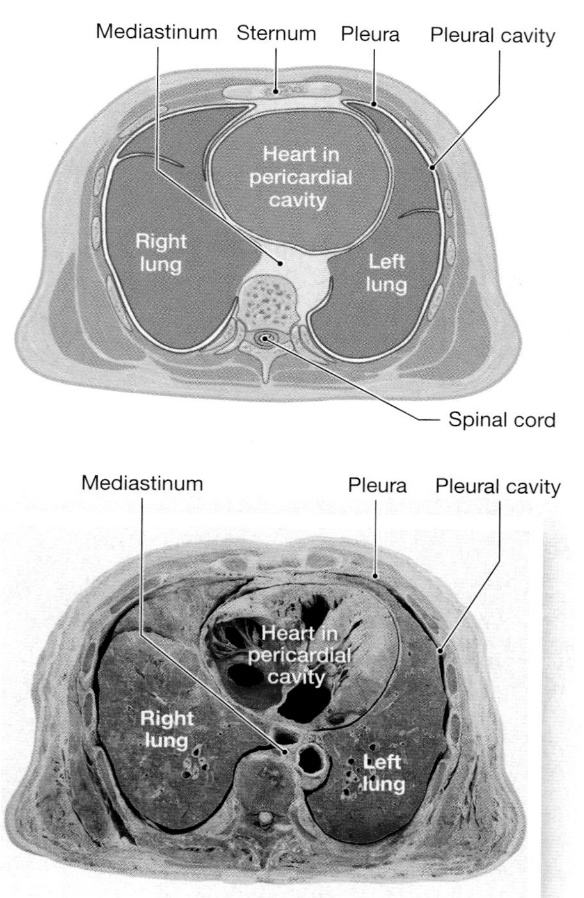

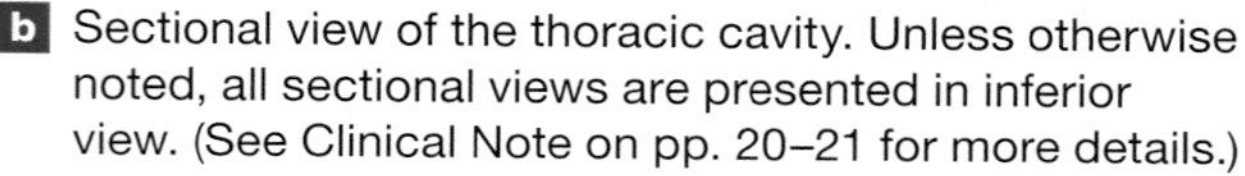

b Sectional view of the thoracic cavity. Unless otherwise noted, all sectional views are presented in inferior view. (See Clinical Note on pp. 20–21 for more details.)

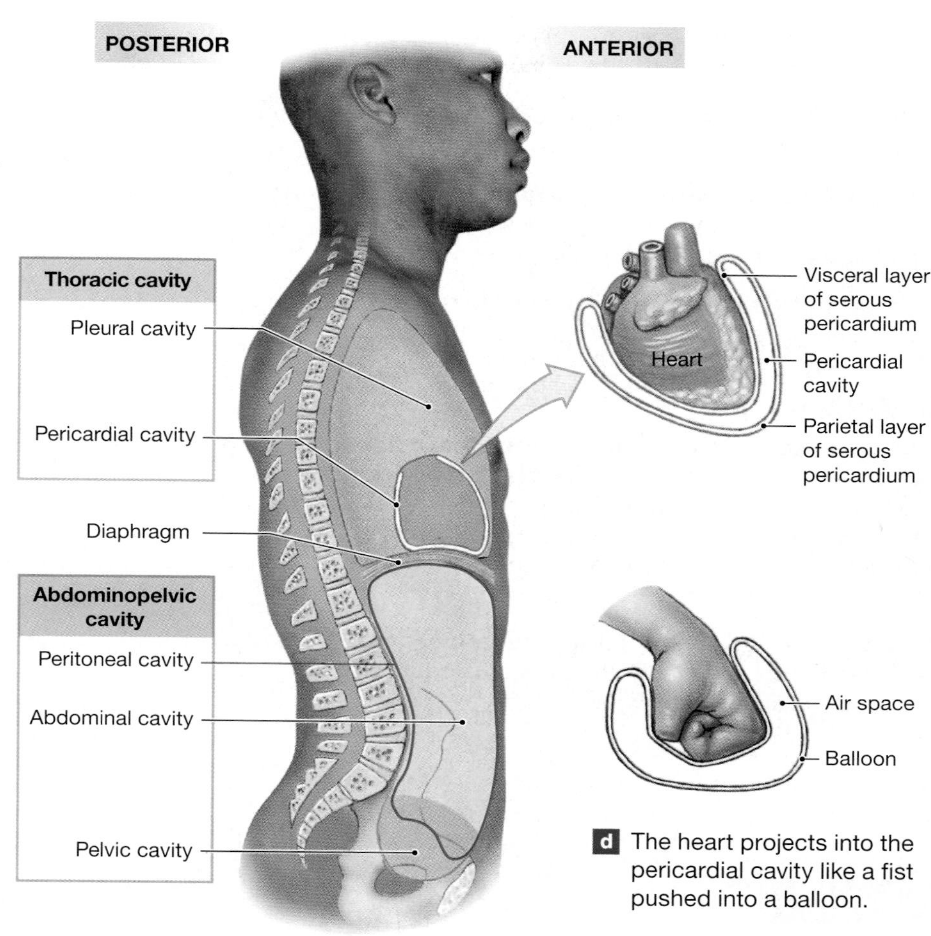

c Lateral view of the subdivisions of the body cavities.

d The heart projects into the pericardial cavity like a fist pushed into a balloon.

- Anatomical specialties important to clinical practice include **clinical anatomy** (anatomical features that undergo characteristic changes during illness), **surgical anatomy** (landmarks important for surgical procedures), **radiographic anatomy** (anatomical structures that are visualized by specialized procedures performed on an intact body), and **cross-sectional anatomy**. *(See Clinical Note on pp. 20–21.)*

1.4 | Levels of Organization p. 5

- Anatomical structures are arranged in a series of interacting levels of organization ranging from the chemical/molecular level, through cell/tissue levels, to the organ/organ system/organism levels. *(See Figures 1.3 and 1.4.)*
- When the body's internal environment is relatively stable, this is called **homeostasis**.

1.5 | An Introduction to Organ Systems p. 7

- All living organisms share a set of vital properties and processes: **responsiveness** to changes in their environment, growth and **differentiation**, reproduction, movement, and **metabolism** and **excretion**. Organisms absorb and consume oxygen during **respiration** and discharge waste products during excretion. **Digestion** breaks down complex foods for use by the body. The **cardiovascular system** forms an internal transportation system between areas of the body. *(See Figures 1.5 and 1.6.)*
- The 11 organ systems of the human body perform these vital functions. *(See Figure 1.5.)*

1.6 | The Language of Anatomy p. 14

- Anatomy usess a specialized language. *(See Figures 1.7 to 1.13.)*

Superficial Anatomy p. 14

- Standard anatomical illustrations show the body in the **anatomical position**. *(See Figures 1.8 and 1.10.)*
- A person lying down in the anatomical position may be **supine** (face up) or **prone** (face down).
- Specific terms identify specific anatomical regions. *(See Figure 1.8 and Table 1.1.)*
- **Abdominopelvic quadrants** and **abdominopelvic regions** represent two different approaches to describing locations in the abdominal and pubic areas of the body. *(See Figure 1.9.)*
- Specific directional terms are used to indicate relative location on the body. *(See Figure 1.10.)*

Sectional Anatomy p. 18

- There are three **sectional planes: frontal plane** or *coronal plane* (anterior versus posterior), **sagittal plane** (right versus left sides), and **transverse plane** (superior versus inferior). These sectional planes and related reference terms describe relationships between the parts of the three-dimensional human body. *(See Figure 1.11.)*
- **Serial reconstruction** is an important technique for studying histological structure and analyzing images produced by radiological procedures. *(See Figure 1.12.)*
- **Body cavities** protect delicate organs and permit changes in the size and shape of visceral organs.
- The **diaphragm** separates the superior **thoracic cavity** from the inferior **abdominopelvic cavity**. *(See Figure 1.13.)*
- The **abdominal cavity** extends from the inferior surface of the diaphragm to an imaginary line drawn from the inferior surface of the most inferior spinal vertebra to the anterior and superior margin of the pelvic girdle. Inferior to this imaginary line is the **pelvic cavity**. *(See Figure 1.13.)*
- The thoracic and abdominopelvic cavities contain narrow, fluid-filled spaces lined by a serous membrane. The thoracic cavity contains two **pleural cavities** (each surrounding a lung) separated by the **mediastinum**. *(See Figure 1.13.)*
- The mediastinum contains the thymus, trachea, esophagus, blood vessels, and the **pericardial cavity**, which surrounds the heart. The membrane lining the pleural cavities is called the **pleura**; the membrane lining the pericardial cavity is called the **serous pericardium**. *(See Figure 1.13.)*
- The **abdominopelvic cavity** contains the **peritoneal cavity**, which is lined by the **peritoneum**. Many digestive organs are supported and stabilized by **mesenteries**. *(See Figure 1.13.)*
- Important **radiological procedures**, which can provide detailed information about internal systems, include **x-rays**, **CT scans**, **MRI**, and **ultrasound**. Physicians who perform and analyze these procedures are called radiologists. *(See Clinical Note on pp. 20–21.)*

Chapter Review

For answers, see the blue Answers tab at the back of the book.

Level 1 Reviewing Facts and Terms

Match each numbered item with the most closely related lettered item.

1. supine .... ☐
2. cytology .... ☐
3. homeostasis .... ☐
4. lumbar .... ☐
5. prone .... ☐
6. metabolism .... ☐
7. histology .... ☐

a. study of tissues
b. face down
c. all chemical activity in body
d. study of cells
e. face up
f. constant internal environment
g. lower back

8. Label the planes on the diagram below.

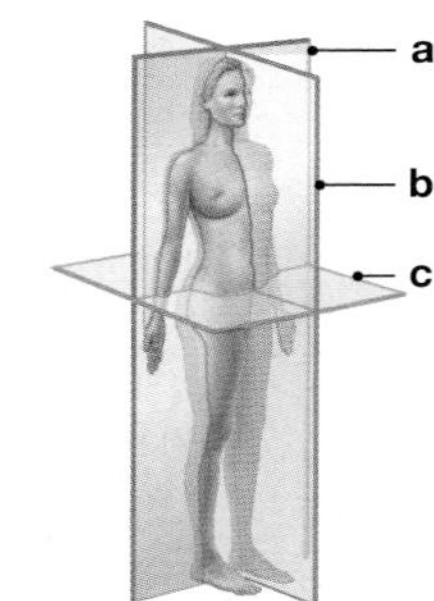

(a) ________
(b) ________
(c) ________

9. Label the abdominal, pleural, pelvic, and pericardial cavities on the diagram below.

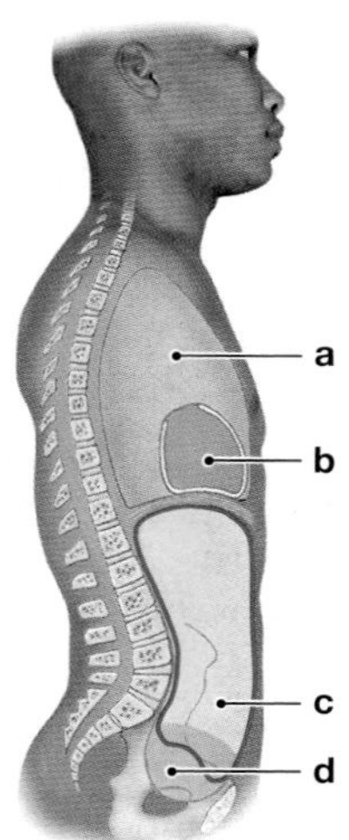

(a) ________
(b) ________
(c) ________
(d) ________

10. The major function of the _____ system is the internal transport of nutrients, wastes, and gases.
(a) digestive
(b) cardiovascular
(c) respiratory
(d) urinary

11. Which of the following includes only structures enclosed within the mediastinum?
(a) lungs, esophagus, heart
(b) heart, trachea, lungs
(c) esophagus, trachea, thymus
(d) pharynx, thymus, major vessels

12. The primary site of blood cell production is within the
(a) cardiovascular system.
(b) skeletal system.
(c) integumentary system.
(d) lymphatic system.

13. Which of the following regions corresponds to the arm?
(a) cervical
(b) brachial
(c) femoral
(d) pedal

Level 2 Reviewing Concepts

1. From the following selections, demonstrate your understanding of anatomical terminology by selecting the directional terms equivalent to *ventral*, *posterior*, *superior*, and *inferior* in the correct sequence.
(a) anterior, dorsal, cephalic, caudal
(b) dorsal, anterior, caudal, cephalic
(c) caudal, cephalic, anterior, posterior
(d) cephalic, caudal, posterior, anterior

2. Explain the properties and processes that are associated with all living things.

3. Using the following diagram, illustrate the relationships of proximal and distal and medial and lateral.

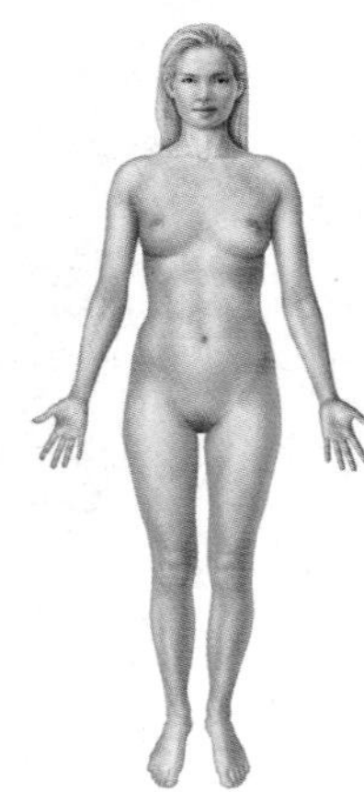

4. The body system that performs crisis management by directing rapid, short-term, and very specific responses is the
(a) lymphatic system.
(b) nervous system.
(c) cardiovascular system.
(d) endocrine system.

5. Applying the concept of sectional planes, how could you divide the body so that the face remains intact?
(a) sagittal section
(b) coronal section
(c) midsagittal section
(d) none of the above

6. Analyze why large organisms must have a vascular system.

Level 3 Critical Thinking

1. Defend the following statement: A disruption in normal cellular division within the red bone marrow supports the view that all levels of organization within an organism are interdependent.

2. A child born with a severe cleft palate may require surgery to repair the nasal cavity and reconstruct the roof of the mouth. Which body systems are affected by the cleft palate? Studies of other mammals that develop cleft palates have helped us understand the origins and treatment of such problems. Which anatomical specialties are involved in identifying and correcting a cleft palate?

1

CLINICAL CASE WRAP-UP

Using Anatomy to Save a Life

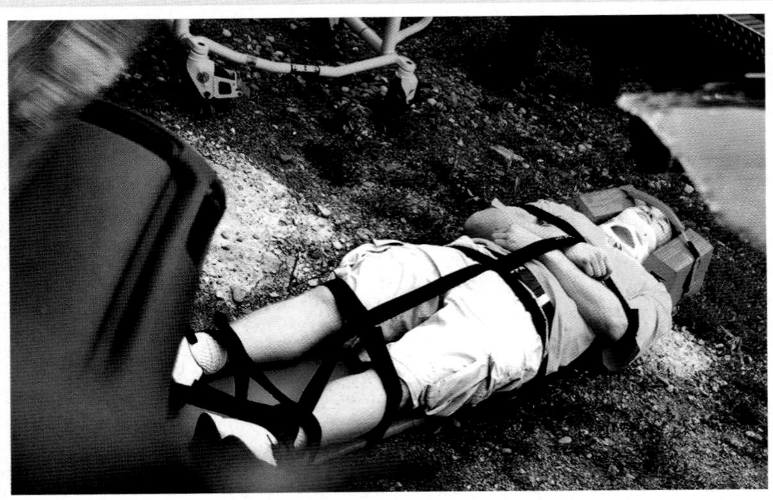

The Injury Severity Score (ISS) is based on the concept of *regional* anatomy. It is a tool used for triaging (prioritizing) injured patients, determining what mode of transport is best, and predicting outcomes. The ISS correlates with morbidity (injury or illness) and mortality (death).

To calculate ISS, six body regions are defined: (1) head/neck, (2) face, (3) chest, (4) abdomen, (5) appendicular skeleton (see Chapter 7), and (6) external (skin). Each region is given a score ranging from 0 (no apparent injury) to 6 (not compatible with life). The scores for the three most severely injured regions are squared and added together to produce the ISS score. A score of 15 is considered multiple trauma (polytrauma). The highest possible score is 75 $(5^2 + 5^2 + 5^2 = 25 + 25 + 25)$. Note that a score of 6 in any region is not possible because it indicates death.

Because of his open skull fracture, the EMT gives Zach a score of 5 in the head/neck category. Multiple rib fractures with breathing difficulty rate him a score of 4 in the chest region. Due to the open fractures of his upper and lower extremities, his appendicular skeleton score is 4. The sum of the squares of these three highest regions is 57 $(5^2 + 4^2 + 4^2)$, placing Zach in the critical category.

Zach's survival now depends on many factors, but his diagnosis and treatment began with an EMT's assessment of regional anatomy.

1. If Zach has a fracture of his hip and an open fracture of his ankle, which injury is proximal and which is distal?
2. If Zach sustains a laceration to his spleen, what are the abdominopelvic regions and quadrant of that injury?

See the blue Answers tab at the back of the book.

Related Clinical Terms

acute: A disease of short duration but typically severe.

chemotherapy: The treatment of disease by the use of chemical substances, especially the treatment of cancer by cytotoxic and other drugs.

chronic: An illness persisting for a long time or constantly recurring. Often contrasted with acute.

epidemiology: The branch of science that deals with the incidence, distribution, and possible control of diseases and other factors relating to health.

etiology: The science and study of the cause of diseases.

idiopathic: Denoting any disease or condition of unknown cause.

morbidity: The state of being diseased or unhealthy, or the incidence of disease in a population.

pathophysiology: The functional changes that accompany a particular syndrome or disease.

syndrome: A condition characterized by a group of associated symptoms.

2 Foundations

The Cell

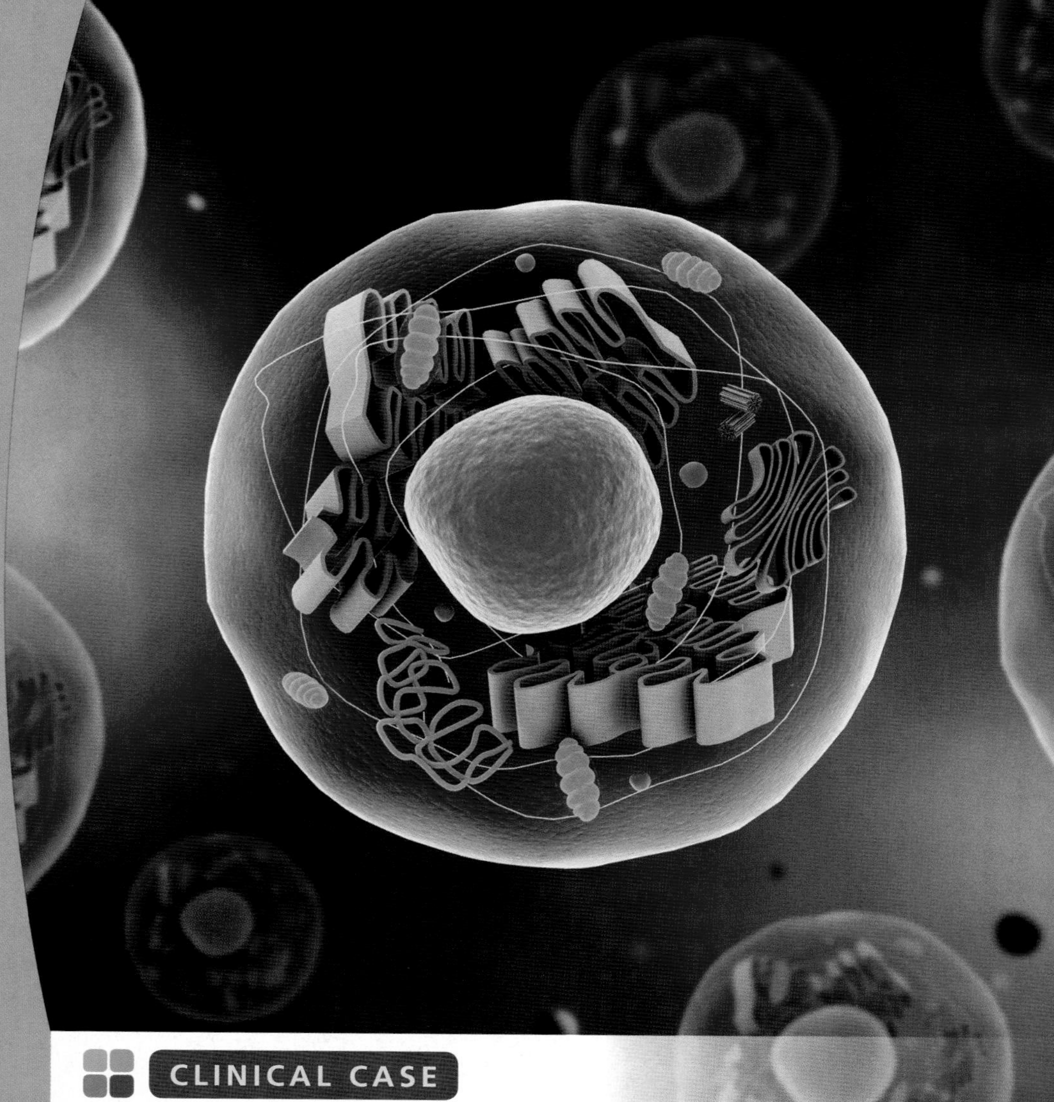

Learning Outcomes

These Learning Outcomes correspond by number to this chapter's sections and indicate what you should be able to do after completing the chapter.

2.1 Identify and explain the functions of the plasma membrane, cytoplasm, and nonmembranous and membranous organelles. p. 28

2.2 Explain how cells can be interconnected to maintain structural stability in body tissues. p. 42

2.3 Summarize the life cycle of a cell and how cells divide by the process of mitosis. p. 42

CLINICAL CASE

Inheritance from Mom

Jessica has always been weak and uncoordinated. As a child, she had some developmental delays, was always shorter than her schoolmates, and had hearing and visual impairments. She suffered her first seizure (an episode of symptoms resulting from irregular electrical activity of brain cells) at age 8. While pregnant with her daughter at age 32, she developed gestational diabetes (high blood sugar during pregnancy).

Jessica's mother, now deceased, as well as Jessica's sisters experienced similar physical ailments. Her older sister's symptoms were less severe than Jessica's, while her younger sister's condition was much worse.

When Jessica's daughter began to show similar symptoms, she sought help from a genetics (inheritance) counselor. She learned that mitochondria (the tiny "power factories" of cells) have a small amount of DNA (genetic material) that can mutate and fail to function properly. All cytoplasmic organelles (intracellular structures), including mitochondria and their mtDNA (mitochondrial DNA), are inherited via the oocyte (egg) of the mother. Jessica's family's symptoms were indicative of mitochondrial disease. Jessica's mother must have had a mtDNA mutation causing this disease, and Jessica and both of her sisters had inherited the mutation, and thus the disease, from their mother. All of Jessica's children would inherit it as well.

After some frantic research, Jessica learned about a procedure that might allow her to have additional children without mitochondrial disease.

Can Jessica bear a healthy child? To find out, turn to the Clinical Case Wrap-Up on p. 48.

2

WALK through a building supply store, and you will see many individual items—bricks, floor tiles, lumber. Each item by itself is of limited use, but put them together and you can build a functional unit such as a house. The human body is also made up of many individual components, called cells. Much as individual bricks and lumber collectively form a house, individual cells work together to build tissues, organs, and organ systems, as discussed in Chapter 1.

According to the **cell theory**, cells are the fundamental units of all living things. The cell theory has three basic concepts:

- Cells are the structural "building blocks" of all plants and animals.
- Cells are produced by the division of preexisting cells.
- Cells are the smallest structural units that perform all vital functions.

The human body contains trillions of cells. All of our activities, from thinking to running, result from the combined and coordinated responses of millions or even billions of cells. Yet each individual cell remains unaware of its role in the "big picture"—it is simply responding to changes in its local environment. Because cells form all the structures in the body and perform all vital functions, our exploration of the human body must begin with basic cell biology.

Our bodies contain two types of cells: sex cells and somatic cells. **Sex cells** (*germ cells* or *reproductive cells*) are either the sperm of males or the oocytes of females. **Somatic cells** (*soma*, body) are all the other cells in the body. We will discuss sex cells in Chapter 27, The Reproductive System. In this chapter we will focus on somatic cells.

2.1 | Cellular Anatomy

▶ **KEY POINT** A cell consists of cytoplasm—the cytosol and organelles—enclosed in a plasma membrane.

The "typical" cell is like the "average" person—we can only describe it in general terms because enormous individual variations occur. Our typical cell shares features with most cells of the body without being identical to any of them, because the components within a cell vary based on each cell's function. **Figure 2.1** shows a typical cell, and **Table 2.1** summarizes the structures and functions of its parts.

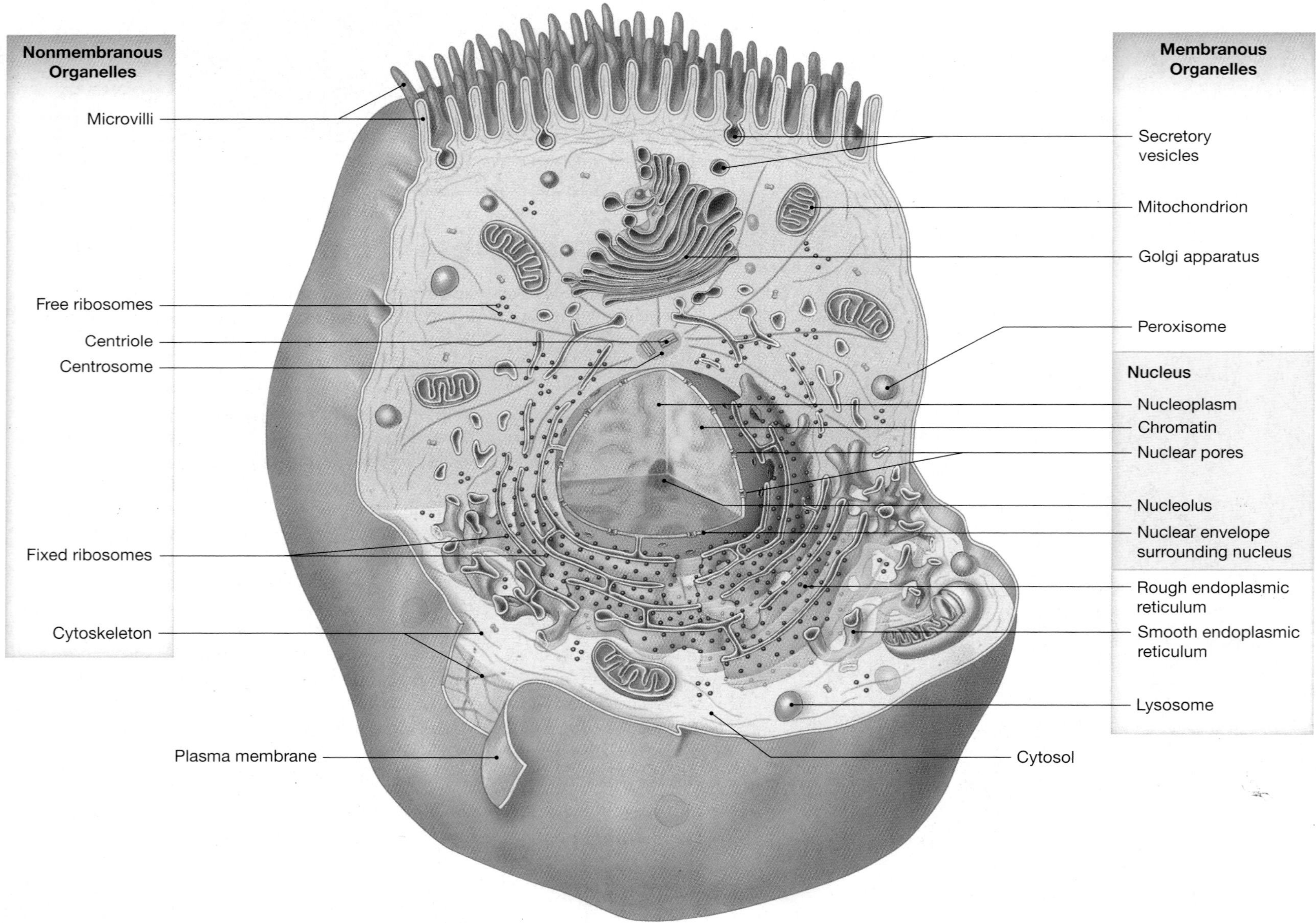

Figure 2.1 Anatomy of a Typical Cell. See Table 2.1 for a summary of the functions of the various cell structures.

Table 2.1 | **Anatomy of a Representative Cell**

Appearance	Cell Structure	Composition	Function
Plasma Membrane and Cytosol			
	Plasma membrane	Lipid bilayer, containing phospholipids, steroids, proteins, and carbohydrates	Isolates; protects; senses; supports; controls entrance/exit of materials
	Cytosol	Fluid component of cytoplasm; may contain inclusions of insoluble materials	Distributes materials by diffusion; stores glycogen, pigments, and other materials
Nonmembranous Organelles			
	Cytoskeleton Microtubule Microfilament	Proteins organized in fine filaments or slender tubes	Strengthens and supports; moves cellular structures and materials
	Microvilli	Membrane extensions containing microfilaments	Increase surface area to facilitate absorption of extracellular materials
	Centrosome Centrioles	Cytoplasm containing two centrioles, at right angles; each centriole is composed of nine microtubule triplets	Essential for movement of chromosomes during cell division; organizes microtubules in cytoskeleton
	Cilia	Membrane extensions containing microtubule doublets in a 9 + 2 array	Move materials over cell surface
	Ribosomes	RNA + proteins; attached ribosomes bound to rough endoplasmic reticulum; free ribosomes scattered in cytoplasm	Synthesize proteins
Membranous Organelles			
	Mitochondria	Double membrane, with inner membrane folds (cristae) enclosing metabolic enzymes	Produce 95 percent of the ATP required by cell
	Nucleus Nuclear envelope Nuclear pore	Nucleoplasm containing nucleotides, enzymes, nucleoproteins, and chromatin; surrounded by a double-layered membrane (nuclear envelope) containing nuclear pores	Controls metabolism; stores and processes genetic information; controls protein synthesis
	Nucleolus	Dense region in nucleoplasm containing DNA and RNA	Site of rRNA synthesis and assembly of ribosomal subunits
	Endoplasmic reticulum (ER)	Network of membranous channels extending throughout the cytoplasm	Synthesizes secretory products; intracellular storage and transport
	Rough ER	Has ribosomes bound to membranes	Modifies and packages newly synthesized proteins
	Smooth ER	Lacks attached ribosomes	Synthesizes lipids, steroids, and carbohydrates; stores calcium ions
	Golgi apparatus	Stacks of flattened membranes (cisternae) containing chambers	Stores, alters, and packages secretory products and lysosomal enzymes
	Lysosome	Vesicles containing digestive enzymes	Removes damaged organelles or pathogens
	Peroxisome	Vesicles containing degradative enzymes	Catabolizes fats and other organic compounds; neutralizes toxic compounds generated in the process

Figure 2.2 **A Flowchart for the Study of Cell Structure.** Cytoplasm is composed of the cytosol and organelles. Organelles are classified as either nonmembranous organelles or membranous organelles.

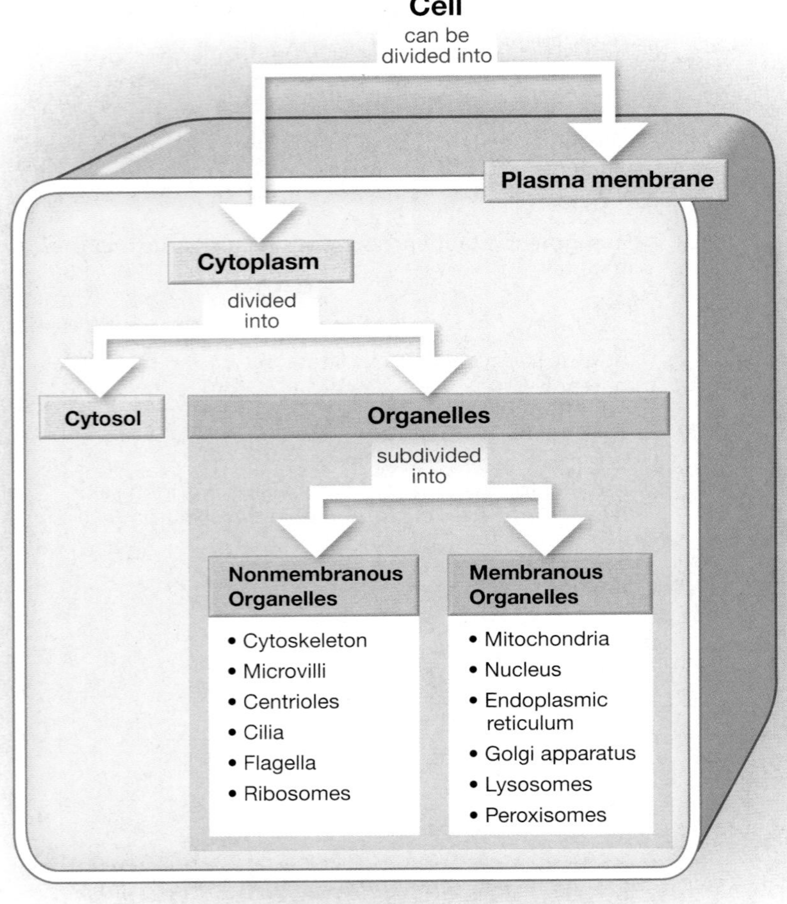

Figure 2.2 outlines the organization of a cell, and thus this chapter. Our cells float in a watery medium known as the **extracellular fluid**. A plasma membrane separates the cell contents, or cytoplasm, from the extracellular fluid. The cytoplasm can be subdivided into a fluid, called the cytosol, and intracellular structures collectively known as organelles (or-ga-NELZ, "little organs").

The Plasma Membrane

▶ **KEY POINT** The functions of the plasma membrane relate to its crucial characteristic of selective permeability.

The outer boundary of a cell is the **plasma membrane**, also termed the **cell membrane** or *plasmalemma*. It is extremely thin and delicate, ranging from 6 to 10 nm (1 nm = 0.001 μm) in thickness. The plasma membrane has four important functions:

- Physical isolation: The lipid bilayer of the plasma membrane forms a physical barrier separating the inside of the cell from the surrounding extracellular fluid.
- Regulation of exchange with the environment: The plasma membrane controls the entry of ions and nutrients, the elimination of wastes, and the release of secretory products.
- Sensitivity: The plasma membrane is the first part of the cell affected by changes in the extracellular fluid. It also contains a variety of receptors that allow the cell to recognize and respond to specific molecules in its environment and to communicate with other cells. Any alteration in the plasma membrane affects all cellular activities.
- Cell-to-cell communication, adhesion, and structural support: Specialized connections between adjacent plasma membranes or between plasma membranes and extracellular materials allow cells to communicate with and attach to each other and give tissues a stable structure.

Components of the Plasma Membrane

The plasma membrane is composed of complex chemical compounds, including phospholipids, proteins, glycolipids, and sterols such as cholesterol (Figure 2.3a). The chemical composition of the plasma membrane varies from cell to cell depending on the function of that cell.

Phospholipids The plasma membrane is called a **phospholipid bilayer** because its phospholipids form two distinct layers. In each layer the phospholipid molecules are arranged so that their heads are at the cell surface and their tails are on the inside (Figure 2.3b). Dissolved ions and water-soluble compounds cannot cross the lipid portion of the plasma membrane because the lipid tails will not associate with water molecules. This feature makes the membrane very effective in isolating the cytoplasm from the surrounding fluid environment. This isolation is important because the composition of the cytoplasm is very different from that of the extracellular fluid, and that difference must be maintained.

Proteins There are two general types of membrane proteins. **Peripheral proteins** are attached to either the inner or outer membrane surface, depending on their function. **Integral proteins** (also called *transmembrane proteins*) are embedded in the membrane. Most integral proteins are quite long and folded and, therefore, span the entire width of the membrane one or more times. Some integral proteins form **channels** that let water molecules, ions, and small water-soluble substances into or out of the cell. Communication between the interior and exterior of the cell occurs through these channels. Some of the channels are **gated channels** that can open or close to regulate the passage of materials. Other integral proteins function in cell-to-cell recognition or as catalysts or receptor sites. Because of the membrane's fluidity, integral proteins move within the membrane like ice cubes drifting in a bowl of punch.

Glycolipids The inner and outer surfaces of the plasma membrane differ in protein and lipid composition. The carbohydrate (*glyco-*) component of the glycolipids and glycoproteins that extend away from the outer surface of the plasma membrane form a viscous, superficial coating known as the **glycocalyx** (*calyx*, cup). Some of these molecules function as receptors. When a membrane receptor binds to a specific molecule in the extracellular fluid, the membrane receptor triggers a change in cellular activity. For example, cytoplasmic enzymes on the inner surface of the plasma membrane are bound to integral proteins, and events on the membrane surface affect the activities of these enzymes.

Sterols Sterols stabilize the membrane structure and maintain its fluidity. The most common sterol in the plasma membrane of human cells is cholesterol. The composition of the plasma membrane changes over time through the removal and replacement of membrane components.

Membrane Permeability: Passive and Active Processes

The **permeability** of a membrane determines its effectiveness as a barrier. The greater its permeability, the easier it is for substances to cross the membrane. If nothing crosses a membrane, it is **impermeable**. If all substances cross without difficulty, it is **freely permeable**.

Figure 2.3 **The Plasma Membrane**

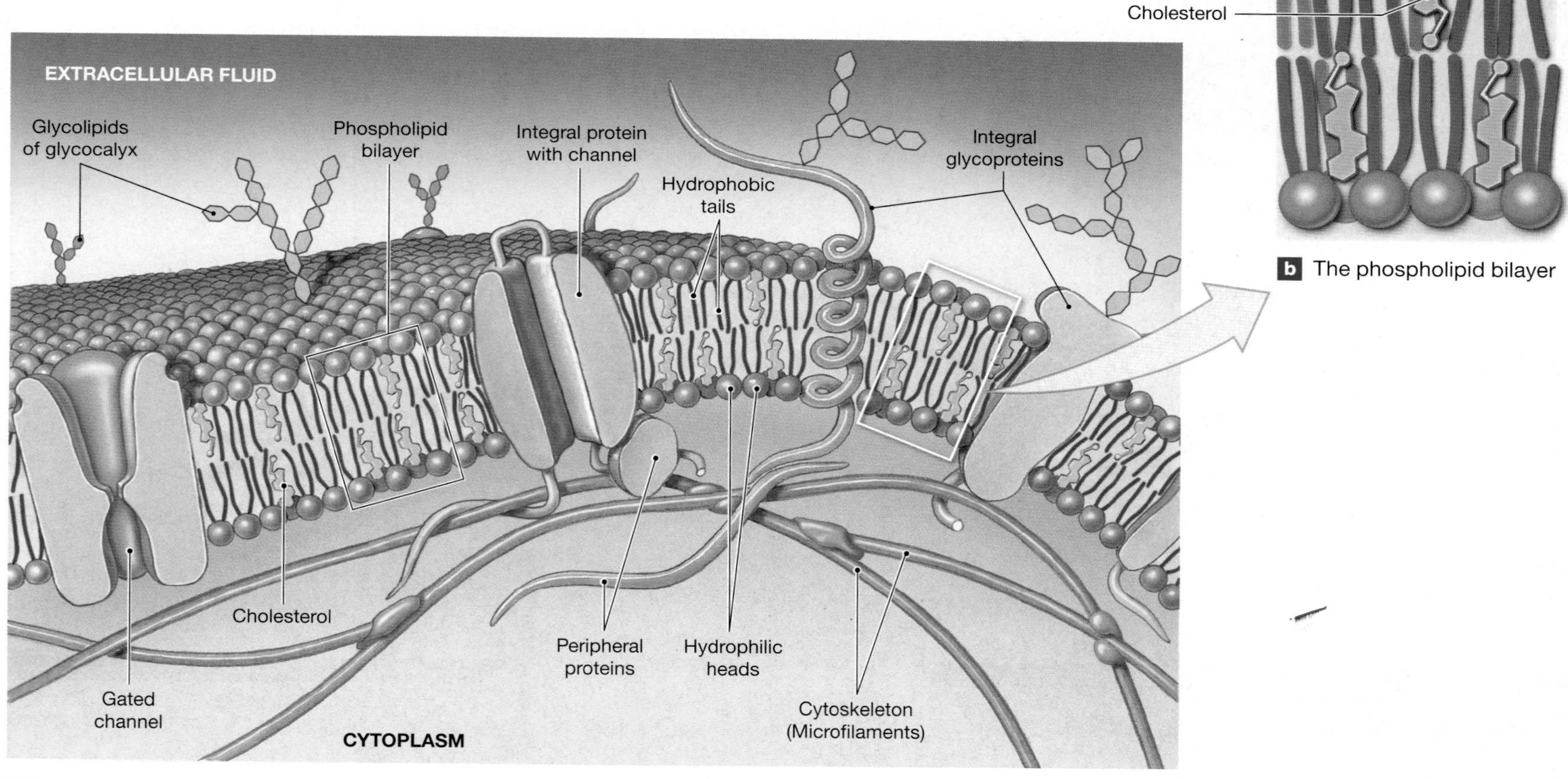

a The plasma membrane

b The phospholipid bilayer

Plasma membranes fall somewhere in between and thus are **selectively permeable**. *A selectively permeable membrane permits some substances to cross freely but restricts others from crossing.* Differences in permeability may be due to a substance's size, electrical charge, molecular shape, solubility, or any combination of these factors.

The permeability of the plasma membrane varies depending on the organization and characteristics of the lipids and proteins within it. A substance may pass through the membrane by a passive or active process. **Spotlight Figure 2.4** explains three passive processes, **diffusion**, **osmosis**, and **facilitated diffusion**, and three active processes, **active transport**, **endocytosis**, and **exocytosis**.

Extensions of the Plasma Membrane: Microvilli

Microvilli (singular, *microvillus*) are small, finger-shaped projections of the plasma membrane. They are found in cells that absorb materials from the extracellular fluid, such as in the small intestine and kidneys (look ahead to **Figure 2.5a,b**, p. 34).

Microvilli promote absorption by increasing the surface area exposed to the extracellular environment. A network of microfilaments stiffens each microvillus and anchors it to the terminal web, a dense supporting network within the underlying cytoskeleton. The cytoskeleton is the cell's internal framework of filaments and fibers. Interactions between these microfilaments and the cytoskeleton produce a waving or bending action. Movements of the microvilli circulate fluid close to the plasma membrane, bringing dissolved substances into contact with receptors on the membrane surface.

The Cytoplasm

▶ **KEY POINT** Inside the cell, the cytoplasm consists of the cytosol and various organelles.

Cytoplasm is the general term for all the material found inside the cell. The cytoplasm contains more proteins than the extracellular fluid; proteins account for 15–30 percent of the weight of the cell. The cytoplasm has two major components:

1. **Cytosol**, or *intracellular fluid*, contains dissolved nutrients, ions, soluble and insoluble proteins, and wastes. The plasma membrane separates the cytosol from the surrounding extracellular fluid.
2. **Organelles** (or-ga-NELZ) are intracellular structures that have specific functions.

The Cytosol

The composition of the cytosol differs from that of extracellular fluid in three important ways:

1. The cytosol contains a high concentration of potassium ions, while extracellular fluid contains a high concentration of sodium ions. The numbers of positive and negative ions are not in balance across the plasma membrane. Outside the plasma membrane there is a net excess of positive charges, and inside there is a net excess of negative charges. This separation of unlike charges creates a membrane potential, like a miniature battery. (The significance of the membrane potential will be discussed in Chapter 13.)

SPOTLIGHT

FIGURE 2.4

Membrane Permeability: Passive and Active Processes

Materials cross the plasma membrane passively (without using cellular energy) or actively (using cellular energy).

Passive Processes

Do not require or utilize cellular energy

Diffusion

Diffusion is the movement of molecules from an area of higher concentration to an area of lower concentration. The difference between the high and low concentrations is a **concentration gradient**. In diffusion, molecules move down a concentration gradient until the gradient is eliminated.

Factors Affecting Rate:
Membrane permeability; concentration gradient; size, charge, and lipid solubility of the diffusing molecules; presence of membrane channel proteins; temperature

Substances Involved (all cells):
Gases, small inorganic ions and molecules, lipid-soluble materials

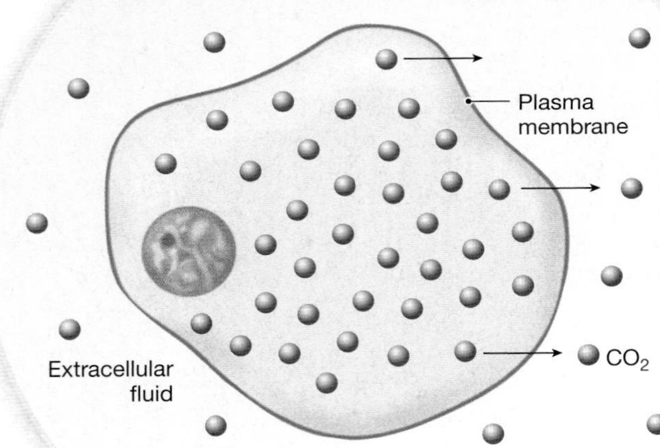

Example:
When the concentration of CO_2 inside a cell is greater than outside the cell, CO_2 diffuses out of the cell and into the extracellular fluid.

Osmosis

Osmosis is the diffusion of water molecules (rather than solutes) across a selectively permeable membrane. Note that water molecules diffusing toward an area of lower *water* concentration are moving toward an area of higher *solute* concentration. Because solute concentrations can easily be determined, they are used to determine the direction and force of osmotic water movement.

Factors Affecting Rate:
Concentration gradient; opposing pressure

Substances Involved:
Water only

Water

Solute

Example:
If the solute concentration outside a cell is greater than inside the cell, water molecules will move across the plasma membrane into the extracellular fluid.

Facilitated diffusion

In **facilitated diffusion**, solutes are passively transported across a plasma membrane by a carrier protein. As in simple diffusion, the direction of movement follows the concentration gradient.

Factors Affecting Rate:
Concentration gradient; size, charge, and solubility of the solutes; temperature; availability of carrier proteins

Substances Involved (all cells):
Glucose and amino acids

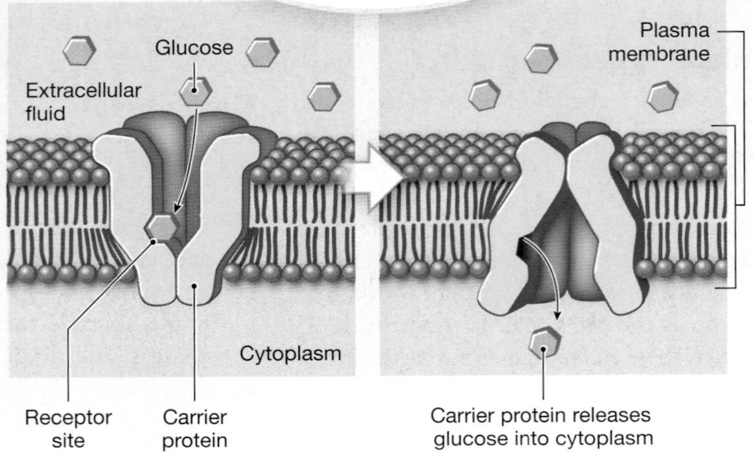

Example:
Nutrients that are insoluble in lipids or too large to fit through membrane channels may be transported across the plasma membrane by carrier proteins. Many carrier proteins move a specific substance in one direction only, either into or out of the cell, after first binding the substance at a specific **receptor site**.

Active Processes

Require ATP or other energy sources

Active transport

Using **active transport**, carrier proteins can move specific substances across the plasma membrane despite an opposing concentration gradient. Carrier proteins that move one solute in one direction and another solute in the opposite direction are called **exchange pumps**.

Factors Affecting Rate:
Availability of carrier proteins, substrate, and ATP

Substances Involved:
Na^+, K^+, Ca^{2+}, Mg^{2+}; other solutes in special cases

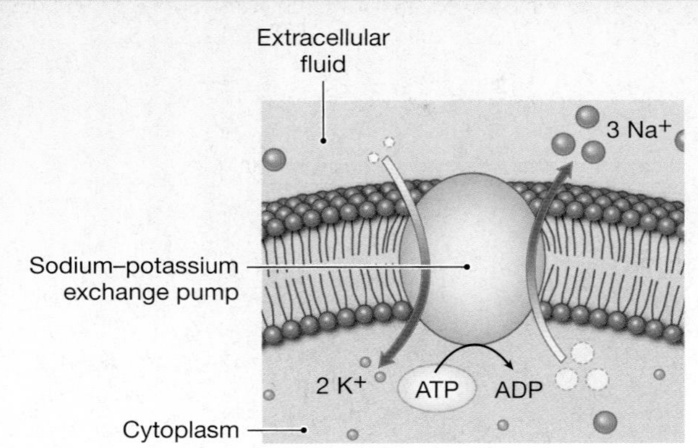

Example:
One of the most common examples of active transport is the **sodium–potassium exchange pump**. For each molecule of ATP consumed, three sodium ions are ejected from the cell and two potassium ions are reclaimed from the extracellular fluid.

Endocytosis

Endocytosis is the packaging of extracellular materials into a **vesicle** (a membrane-bound sac) for importation into the cell.

Pinocytosis

In **pinocytosis**, vesicles form at the plasma membrane and bring extracellular fluid and small molecules into the cell. This process is often called "cell drinking."

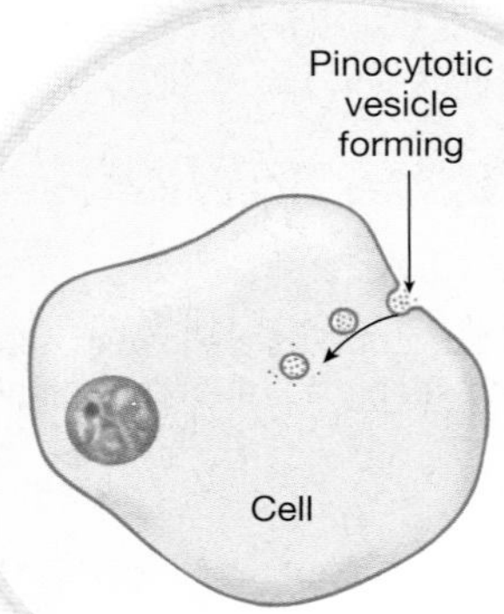

Example:
Water and small molecules within a vesicle may enter the cytoplasm through carrier-mediated transport or diffusion.

Factors Affecting Rate:
Stimulus and mechanism not understood

Substances Involved:
Extracellular fluid and its associated solutes

Phagocytosis

In **phagocytosis**, vesicles form at the plasma membrane to bring solid particles into the cell. This process is often called "cell eating."

Factors Affecting Rate:
Presence and abundance of extracellular pathogens or debris

Substances Involved:
Bacteria, viruses, cell debris, and other foreign material

Example:
Large particles are brought into the cell when cytoplasmic extensions (called **pseudopodia**) engulf the particle and form a phagocytic vesicle.

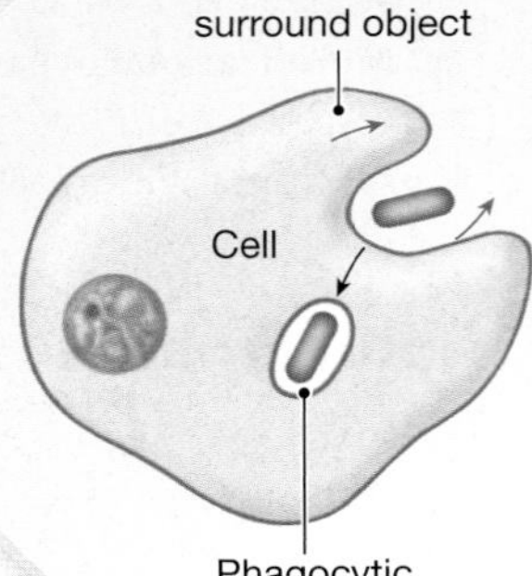

Receptor-mediated endocytosis

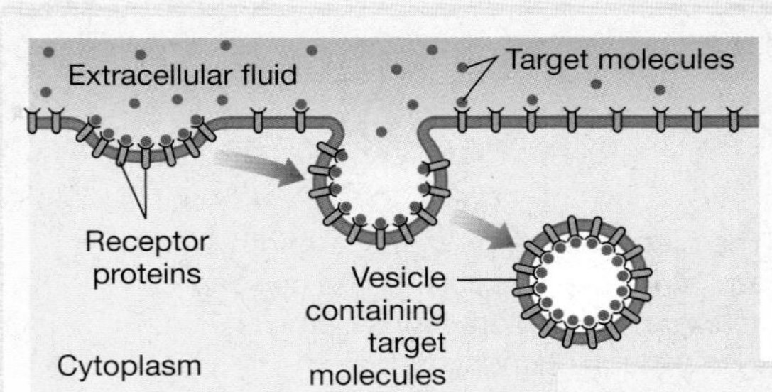

Example:
Each cell has specific sensitivities to extracellular materials, depending on the kind of receptor proteins present in the plasma membrane.

In **receptor-mediated endocytosis**, target molecules bind to specific receptor proteins on the membrane surface, triggering vesicle formation.

Factors Affecting Rate:
Number of receptors on the plasma membrane and the concentration of target molecules (called **ligands**)

Substances Involved:
Many examples, including cholesterol and iron ions

Exocytosis

Exocytosis is the release of fluids and/or solids from cells when intracellular vesicles fuse with the plasma membrane.

Factors Affecting Rate:
Stimulus and mechanism incompletely understood; requires ATP and calcium ions

Substances Involved:
Fluid and cellular wastes; secretory products from some cells

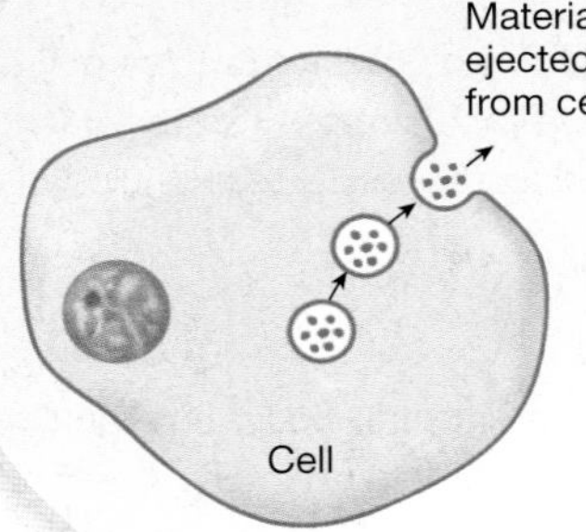

Example:
Cellular wastes that accumulate in vesicles are ejected from the cell.

Figure 2.5 The Cytoskeleton

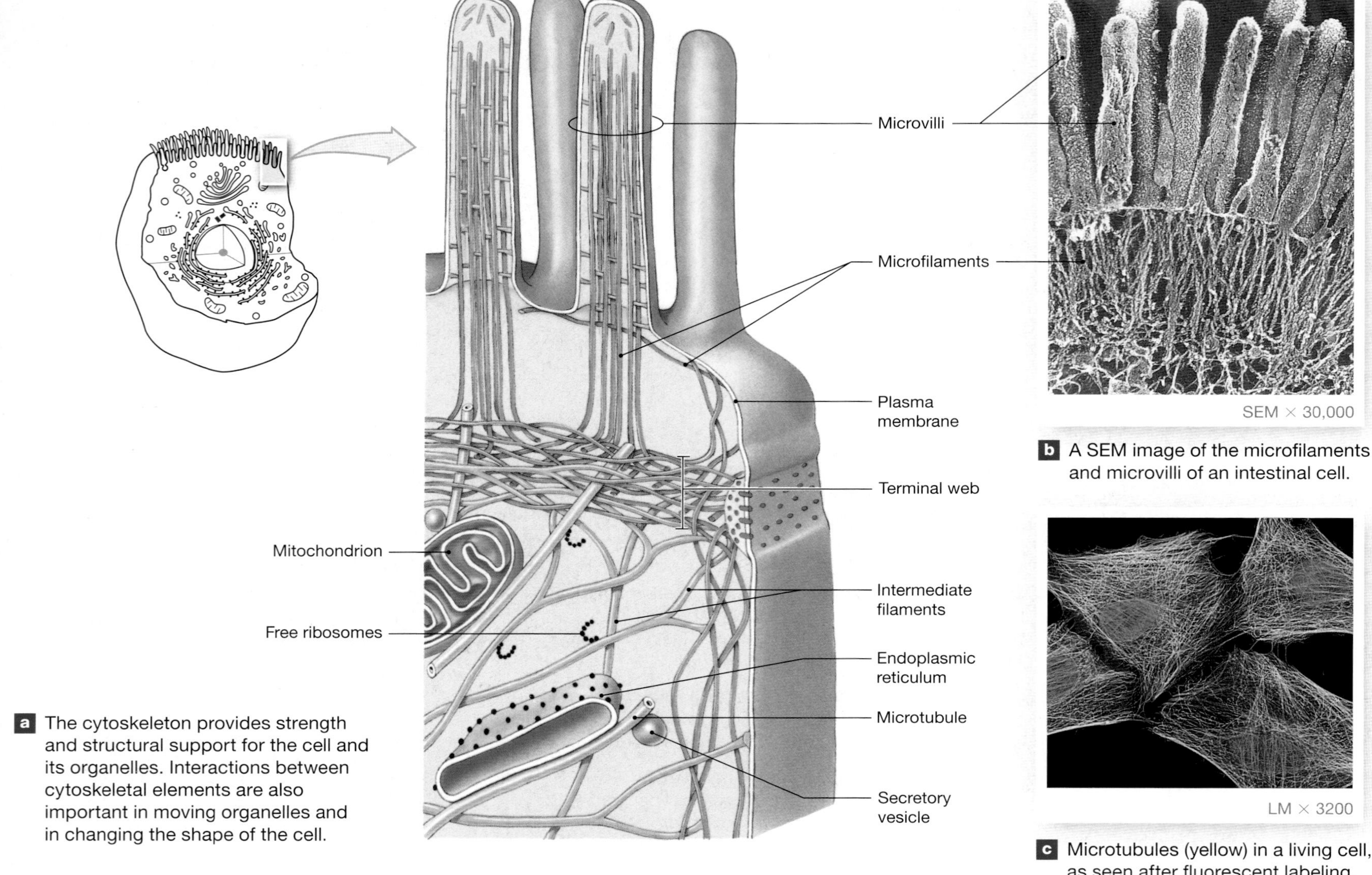

a The cytoskeleton provides strength and structural support for the cell and its organelles. Interactions between cytoskeletal elements are also important in moving organelles and in changing the shape of the cell.

b A SEM image of the microfilaments and microvilli of an intestinal cell.

c Microtubules (yellow) in a living cell, as seen after fluorescent labeling.

2. The cytosol contains a relatively high concentration of dissolved and suspended proteins. Many of these proteins are enzymes that regulate metabolic operations, while others are associated with various organelles. These proteins give the cytosol a consistency ranging from that of thin maple syrup to that of almost-set gelatin, depending on the cytosol's composition.
3. The cytosol contains small quantities of carbohydrates and large amounts of amino acids and lipids. The cell breaks down the carbohydrates for energy, uses the amino acids to manufacture proteins, and uses the lipids to maintain plasma membranes and provide energy when carbohydrates are unavailable.

The cytosol also contains masses of insoluble substances known as **inclusions**, or *inclusion bodies*. The most common inclusions are stored nutrients, for example, glycogen granules in liver or skeletal muscle cells and lipid droplets in fat cells.

Organelles

Organelles are found in all cells of the human body, but their types and numbers differ depending on the cell type. Each organelle has specific functions that are essential to normal cell structure, maintenance, and/or metabolism. Cellular organelles can be divided into two broad categories (Table 2.1, p. 29):

- **Nonmembranous organelles** are always in contact with the cytosol.
- **Membranous organelles** are surrounded by membranes that isolate their contents from the cytosol, just as the plasma membrane isolates the cytosol from the extracellular fluid.

Nonmembranous Organelles

▶ **KEY POINT** Nonmembranous organelles—the cytoskeleton, centrioles, cilia, flagella, and ribosomes—lack a membrane.

Nonmembranous organelles include the cytoskeleton, centrioles, cilia, flagella, and ribosomes.

The Cytoskeleton

Examination of a cell with the electron microscope demonstrates a dense, seemingly disorganized mat of filaments. These filaments are often grouped into bundles, forming a framework that makes the cytoplasm strong and flexible. This internal framework of fibers is the **cytoskeleton**. The cytoskeleton has four major components: microfilaments, intermediate filaments, thick filaments, and microtubules.

Microfilaments **Microfilaments** are slender strands composed mainly of the protein **actin**. In most cells, microfilaments are scattered throughout the cytosol and form a dense network deep to the plasma membrane. Figure 2.5a,b shows the superficial layers of microfilaments in a cell of the small intestine.

Microfilaments have two major functions:

1. Microfilaments anchor the cytoskeleton to integral proteins of the plasma membrane. This stabilizes the position of the membrane proteins, strengthens the cell, and attaches the plasma membrane to the underlying cytoplasm.
2. Actin microfilaments interact with other microfilaments or larger structures composed of the protein myosin. This interaction allows part of the cell to move or changes the shape of the entire cell.

Intermediate Filaments The composition of **intermediate filaments**, which are defined mainly by their size, varies depending on cell type. Intermediate filaments stabilize organelles, transport materials within the cytoplasm, or provide strength. For example, specialized intermediate filaments in nerves called **neurofilaments** support the axons, which are cellular processes that may be up to a meter in length.

Thick Filaments Filaments composed of **myosin** protein subunits are termed **thick filaments** because of their large size. Thick filaments are abundant in muscle cells, where they interact with actin filaments to produce powerful contractions.

Microtubules All cells possess **microtubules**, small, hollow tubes built from the protein **tubulin**. Figures 2.5a,c and 2.6 show microtubules in the cytoplasm of various cells. A microtubule forms temporarily from a collection of tubulin molecules, but then disassembles into individual tubulin molecules once again. The microtubular array is centered in a region called the centrosome. Microtubules radiate outward from the centrosome into the edge of the cell.

Microtubules have a variety of functions:

- Microtubules form the main components of the cytoskeleton, giving the cell strength and rigidity and anchoring major organelles.
- As microtubules assemble and disassemble, they change the shape of the cell and may help the cell move.
- Microtubules attach to organelles and other intracellular materials and move them around within the cell.
- During cell division, microtubules form the spindle apparatus that distributes the duplicated chromosomes (genetic material) to opposite ends of the dividing cell. We will discuss this process later in the chapter.
- Microtubules form structural components of organelles such as centrioles, cilia, and flagella. Although these organelles are associated with the plasma membrane, they do not have their own enclosing membrane; therefore, they are nonmembranous organelles.

The cytoskeleton as a whole incorporates microfilaments, intermediate filaments, and microtubules into a network that extends throughout the cytoplasm. The organizational details are as yet poorly understood because the network is delicate and difficult to study in an intact state.

Figure 2.6 Centrioles and Cilia

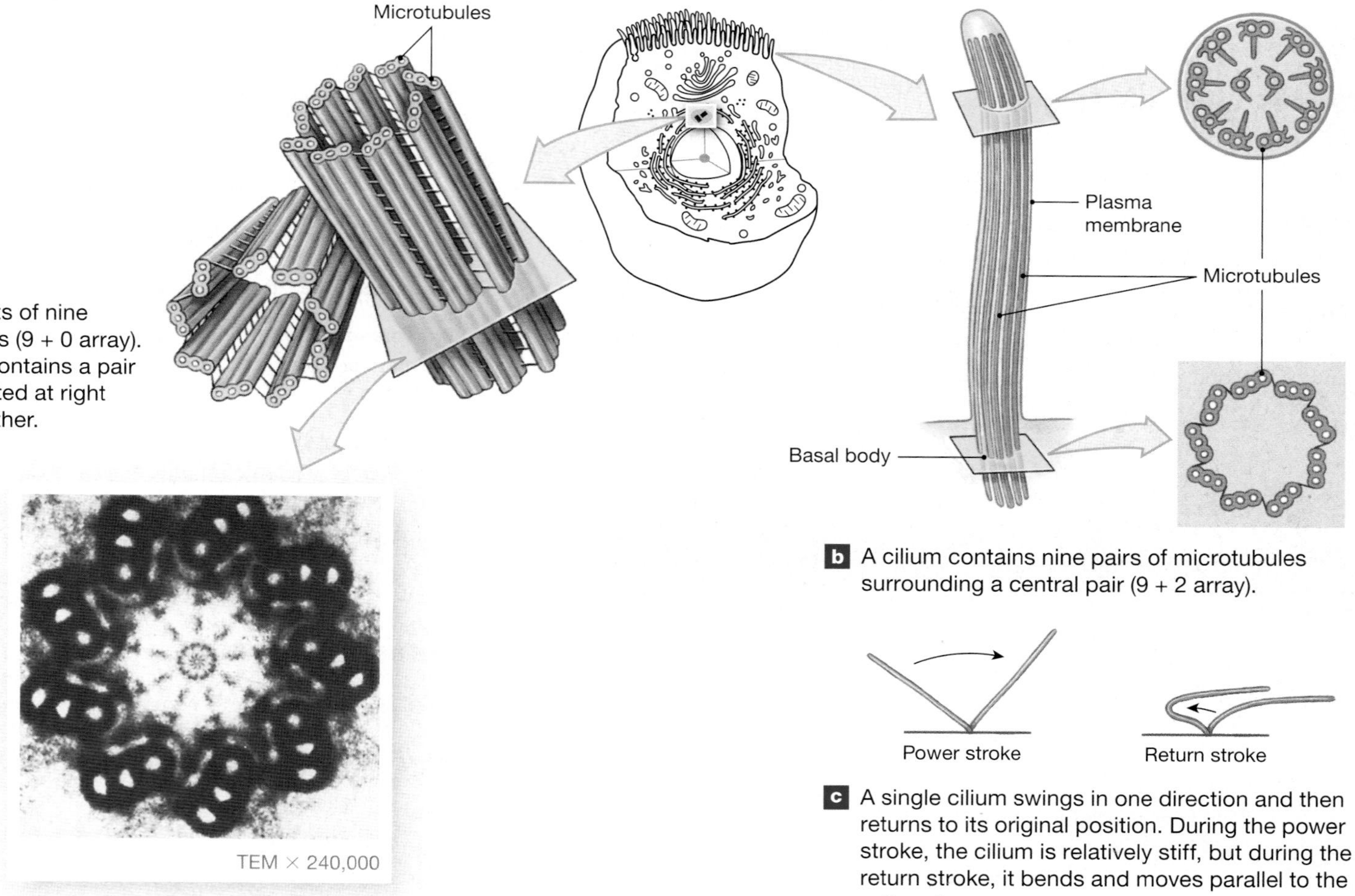

Microtubules

a A centriole consists of nine microtubule triplets (9 + 0 array). The centrosome contains a pair of centrioles oriented at right angles to one another.

TEM × 240,000

Plasma membrane

Microtubules

Basal body

b A cilium contains nine pairs of microtubules surrounding a central pair (9 + 2 array).

Power stroke

Return stroke

c A single cilium swings in one direction and then returns to its original position. During the power stroke, the cilium is relatively stiff, but during the return stroke, it bends and moves parallel to the cell surface.

2

CLINICAL NOTE

Bodybuilding

Skeletal muscle cells lack centrioles and do not divide. They do, however, hypertrophy, or grow larger. The bodybuilder shown in these before and after photos has the same number of individual muscle cells, but with exercise, each cell has grown larger.

a Before b After

Centrioles, Cilia, and Flagella

The cytoskeleton contains groups of microtubules that function individually as centrioles, cilia, and flagella (Table 2.2).

Centrioles A **centriole** is a cylindrical structure composed of short microtubules (Figure 2.6a). Within a centriole there are nine groups of microtubules; each group is composed of a triplet of microtubules. Because there are no central microtubules in the centriole, this is called a 9 + 0 array. The first number (9) indicates the number of peripheral groups of microtubules in the ring, and the second number (zero) indicates the number of microtubules at the center of the ring.

Table 2.2 | Characteristics of Centrioles, Cilia, and Flagella

Structure	Microtubule Organization	Location	Function
Centrioles	Nine groups of microtubule triplets form a short cylinder	In centrosome near nucleus	Organize microtubules in the spindle to move chromosomes during cell division
Cilia	Nine groups of long microtubule doublets form a cylinder around a central pair	At cell surface	Propel fluids or solids across cell surface
Flagella	Same as Cilia	At cell surface	Propel sperm cells through fluid

Cells that do not divide, such as mature red blood cells and skeletal muscle cells, lack centrioles. Cells that can divide contain a pair of centrioles arranged at right angles to each other. The **centrosome** is a clear region of cytoplasm that contains this pair of centrioles. It directs the organization of the microtubules of the cytoskeleton.

Cilia A **cilium** (plural, *cilia*) is composed of nine groups of microtubule doublets surrounding a central pair of microtubules, an arrangement known as a 9 + 2 array (Figure 2.6b). Cilia are anchored to a compact **basal body** located just beneath the cell surface. The structure of the basal body resembles that of a centriole. The plasma membrane covers the exposed portion of the cilium completely.

Cilia "beat" rhythmically to move fluids or secretions across the cell surface (Figure 2.6c). For example, cilia lining the respiratory tract beat in a coordinated manner to move sticky mucus and trapped dust particles toward the throat and away from delicate respiratory surfaces. If the cilia are damaged due to smoking or a metabolic problem, the irritants stay in the lungs and chronic respiratory infections develop.

Flagella Flagella (fla-JEL-ah; singular, *flagellum*, "whip") resemble cilia but are much longer. A flagellum moves a cell through the surrounding fluid, rather than moving the fluid past a stationary cell. The sperm cell is the only cell in the human body that has a flagellum; it moves the cell along the female reproductive tract. If a man's sperm flagella are paralyzed or abnormal, the man is sterile because immobile sperm cannot reach and fertilize an oocyte (a female sex cell).

Ribosomes

Ribosomes are too small to be seen with a light microscope. In an electron micrograph, they appear as dense granules about 25 nm in diameter (Figure 2.7a). Ribosomes are found in all cells, but their number varies depending on the type of cell and its activities. These nonmembranous organelles are intracellular factories that manufacture proteins using information provided by the DNA (the carrier of genetic information) of the nucleus. Each ribosome, which is composed of about 60 percent RNA (a macromolecule) and 40 percent protein, consists of two subunits that interlock as protein synthesis begins (Figure 2.7b). When protein synthesis is complete, the subunits separate.

There are two major types of ribosomes (Figure 2.7a):

- **Free ribosomes** are scattered throughout the cytosol.
- **Attached ribosomes** are temporarily attached to the endoplasmic reticulum, a membranous organelle.

Free and attached ribosomes are identical in structure and function; they differ only in location and the proteins they produce. Proteins manufactured by free ribosomes enter the cytosol. Proteins manufactured by attached ribosomes enter the lumen, or internal cavity, of the endoplasmic reticulum, where they are modified and packaged for export. We will detail these processes later in this chapter.

Membranous Organelles

▶ **KEY POINT** Each membranous organelle—the mitochondrion, nucleus, endoplasmic reticulum, Golgi apparatus, lysosome, and peroxisome—is completely surrounded by a membrane that isolates its contents from the cytosol.

Membranous organelles are surrounded by a phospholipid bilayer membrance that is similar in structure to the plasma membrane. The isolation provided by its membrane allows a membranous organelle to manufacture or store substances that could harm the cytoplasm. There are six types of **membranous organelles**:

Figure 2.7 Ribosomes. These small, dense structures are involved in protein synthesis.

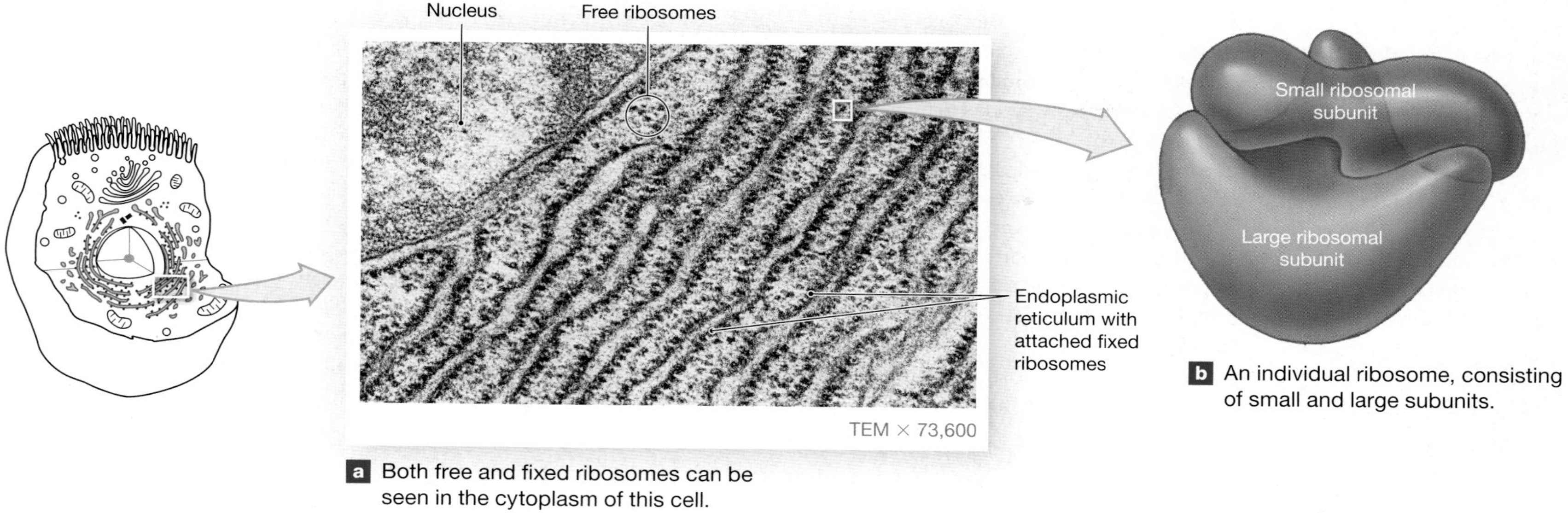

a Both free and fixed ribosomes can be seen in the cytoplasm of this cell.

b An individual ribosome, consisting of small and large subunits.

mitochondria, nucleus, endoplasmic reticulum, Golgi apparatus, lysosomes, and peroxisomes (see **Table 2.1** on p. 29).

Mitochondria

Mitochondria (mī-tō-KON-drē-ah; singular, *mitochondrion*; *mitos*, thread, + *chondrion*, small granules) produce ATP by breaking down organic molecules in a series of reactions that consume oxygen (O_2) and generate carbon dioxide (CO_2). This series of catabolic chemical reactions within the mitochondria produces about 95 percent of the energy needed to keep a cell alive. Mitochondria have an unusual double-layered membrane **(Figure 2.8)**. An outer membrane surrounds the entire organelle, and a second, inner membrane contains numerous folds, called **cristae**. These cristae surround the fluid contents, or **matrix**, and provide a large surface area for the organization of the enzymes that produce most of the ATP. Mitochondria have various shapes, ranging from long and slender to short and fat. They control their own maintenance, growth, and reproduction.

The number of mitochondria in a cell varies depending on the cell's energy demands. Red blood cells lack mitochondria—they obtain energy in other ways—but skeletal muscle cells typically contain as many as 300 mitochondria. Muscle cells have high rates of energy consumption, and over time the mitochondria respond to increased energy demands by reproducing. The increased numbers of mitochondria can provide energy faster and in greater amounts, thereby improving muscular performance.

The Nucleus

The **nucleus** is the control center for cellular operations. A single nucleus stores all the information needed to control the synthesis of the approximately 100,000 different proteins in the human body. The nucleus determines the

Figure 2.8 Mitochondria. The three-dimensional organization of a mitochondrion, and a color-enhanced TEM showing a typical mitochondrion in section.

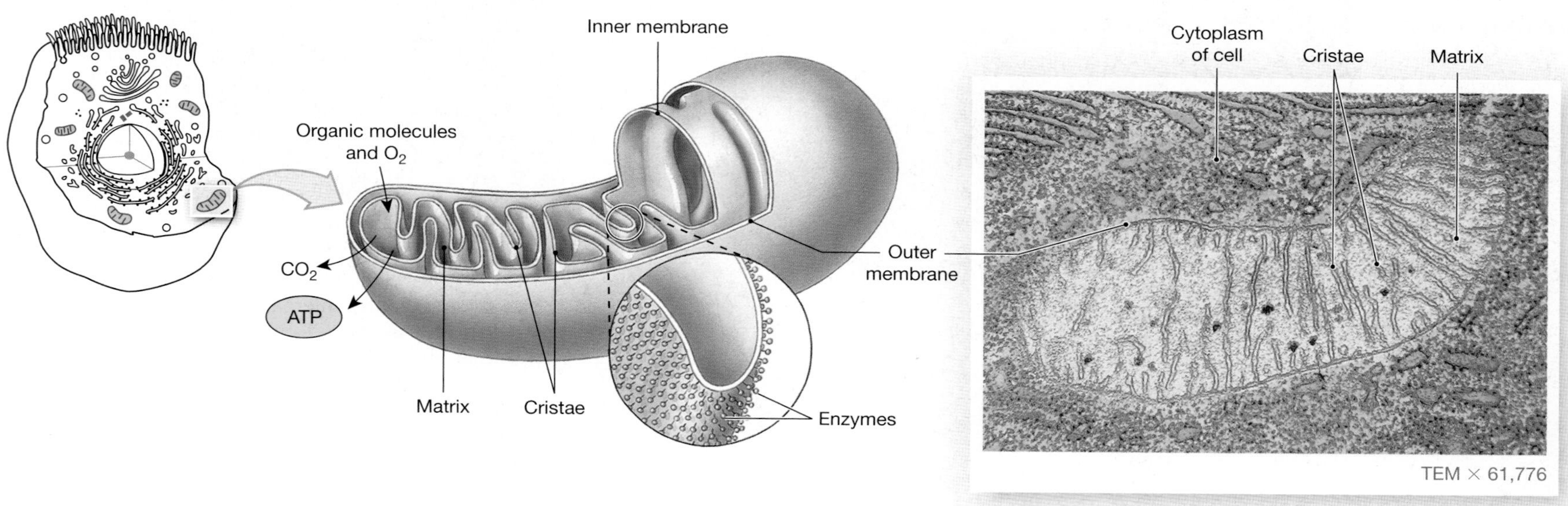

2

structural and functional characteristics of the cell by controlling which proteins are synthesized, when they are synthesized, and in what amounts.

Most cells contain a single nucleus, but there are exceptions. For example, skeletal muscle cells are **multinucleate** (*multi*–, many) because they have many nuclei, whereas mature red blood cells are **anucleate** (*a*–, without) because they lack a nucleus. A cell without a nucleus will survive only three to four months.

Figure 2.9 details the structure of a typical nucleus. Surrounding the nucleus and separating it from the cytosol is the **nuclear envelope**, a double-layered membrane that encloses a narrow **perinuclear space** (*peri*–, around). The inner membrane contains proteins that are binding sites for chromosomes, while the composition of the external membrane is similar to that of the endoplasmic reticulum. At several locations, the nuclear envelope is connected to the rough endoplasmic reticulum (see **Figure 2.1**, p. 29).

The nucleus directs processes taking place in the cytosol and also receives information about conditions and activities in the cytosol. Chemical communication between the nucleus and cytosol occurs through **nuclear pores**, a complex of proteins that regulates the movement of macromolecules into and out of the nucleus. These pores, which account for about 10 percent of the surface of the nucleus, permit the movement of water, ions, and small molecules but regulate the passage of large proteins, RNA, and DNA.

The **nucleoplasm** is a jelly-like fluid substance within the nucleus. The nucleoplasm contains a network of fine filaments, the **nuclear matrix**, which provides structural support and may help regulate genetic activity. Ions, enzymes, proteins, small amounts of RNA and DNA are also suspended within the nucleoplasm. The DNA strands form complex structures known as chromosomes (*chroma*, color). Each **chromosome** contains DNA strands bound to special proteins called **histones**. The nucleus of each of your somatic cells contains 23 pairs of chromosomes; one member of each pair came from your mother and one from your father. **Figure 2.10** diagrams a typical chromosome.

At intervals, the DNA strands wind around the histones, forming a complex called a **nucleosome**. The entire chain of nucleosomes may coil around

Figure 2.9 The Nucleus. The nucleus is the control center for cellular activities.

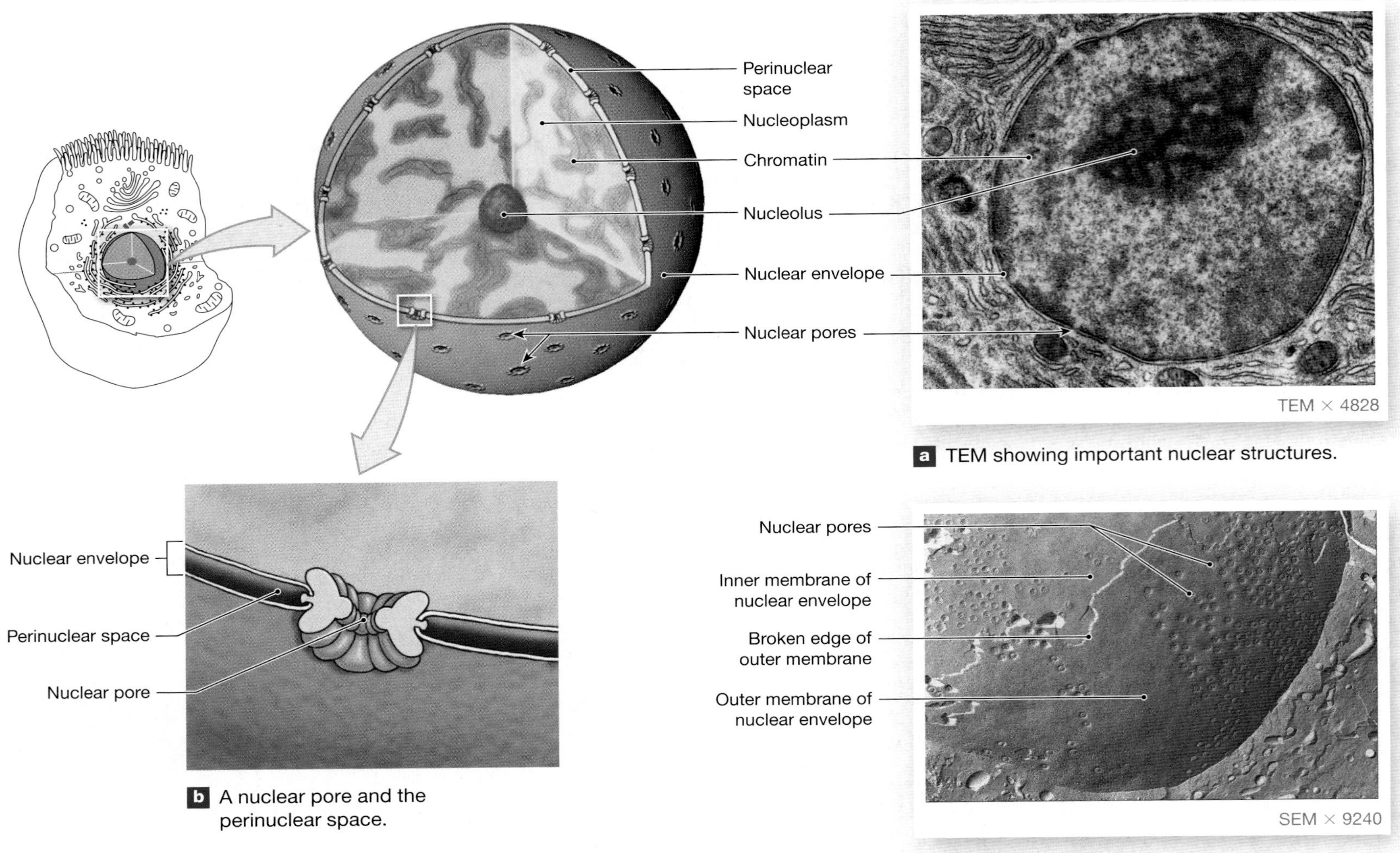

a TEM showing important nuclear structures.

b A nuclear pore and the perinuclear space.

c The cell seen in this SEM was frozen and then broken apart so that internal structures could be seen. This technique, called freeze-fracture, provides a unique perspective on the internal organization of cells. The nuclear envelope and nuclear pores are visible; the fracturing process broke away part of the outer membrane of the nuclear envelope, and the cut edge of the nucleus can be seen.

Figure 2.10 **Chromosome Structure.** DNA strands coil around histones to form nucleosomes. Nucleosomes form coils that may be very tight or rather loose.

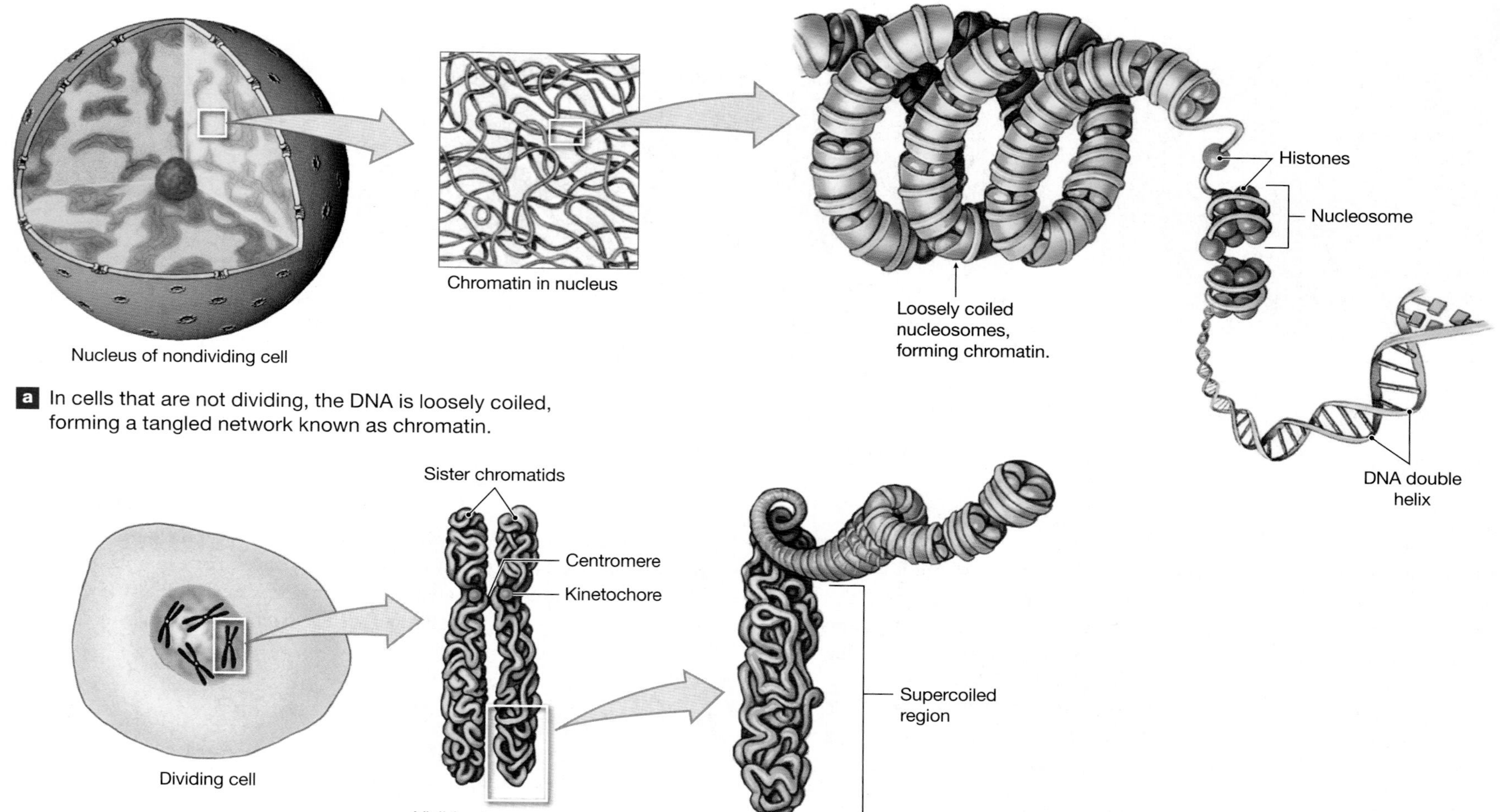
Nucleus of nondividing cell
Chromatin in nucleus
Loosely coiled nucleosomes, forming chromatin.
Histones
Nucleosome
DNA double helix
a In cells that are not dividing, the DNA is loosely coiled, forming a tangled network known as chromatin.
Sister chromatids
Centromere
Kinetochore
Dividing cell
Visible chromosome
Supercoiled region
b When the coiling becomes tighter, as it does in preparation for cell division, the DNA becomes visible as distinct structures called chromosomes. Chromosomes are composed of two sister chromatids that attach at a single point, the centromere. Kinetochores are the region of the centromere where spindle fibers attach during mitosis.

other histones. The degree of coiling determines whether the chromosome is long and thin or short and fat. In dividing cells, chromosomes are tightly coiled and clearly visible in light or electron micrographs. In cells that are not dividing, the chromosomes are loosely coiled, forming a tangle of fine filaments known as **chromatin** (KRŌ-ma-tin). Each chromosome has some tightly coiled regions, and only these areas stain clearly. This makes the nucleus look clumped and grainy.

The chromosomes also have direct control over the synthesis of RNA. Most nuclei contain one to four dark-staining areas called **nucleoli** (nū-KLĀ-ō-lī; singular, *nucleolus*). Nucleoli are nuclear organelles that synthesize the components of ribosomes. A nucleolus contains histones and enzymes as well as RNA, and it forms around a chromosomal region containing the genetic instructions for producing ribosomal proteins and RNA. Nucleoli are most prominent in cells that manufacture many proteins, such as liver cells and muscle cells, because these cells need large numbers of ribosomes.

The Endoplasmic Reticulum

The **endoplasmic reticulum** (en-dō-PLAZ-mik re-TIK-ū-lum), or **ER**, is a network of intracellular membranes forming hollow tubes, flattened sheets, and rounded chambers (**Figure 2.11**). The chambers are called **cisternae** (sis-TUR-nē; singular, *cisterna*, a reservoir for water).

The endoplasmic reticulum has four major functions:

1. **Synthesis:** The ER contains specialized regions that manufacture lipids, proteins, and carbohydrates. The cisternae of the ER store these manufactured products.
2. **Storage:** The ER stores synthesized molecules and substances absorbed from the cytosol without affecting other cellular operations.
3. **Transport:** Substances travel from place to place within the cell inside the endoplasmic reticulum.
4. **Detoxification:** The ER absorbs drugs and toxins and neutralizes them with enzymes.

The ER thus functions as a combination workshop, storage area, and shipping depot. It is where many newly synthesized proteins undergo chemical modification and are packaged for export to their next destination, the Golgi apparatus.

There are two distinct types of endoplasmic reticulum: rough and smooth. The outer surface of the **rough endoplasmic reticulum (RER)** contains

Figure 2.11 The Endoplasmic Reticulum. This organelle is a network of intracellular membranes. The figure shows the three-dimensional relationships between the nucleus and the rough and smooth endoplasmic reticulum.

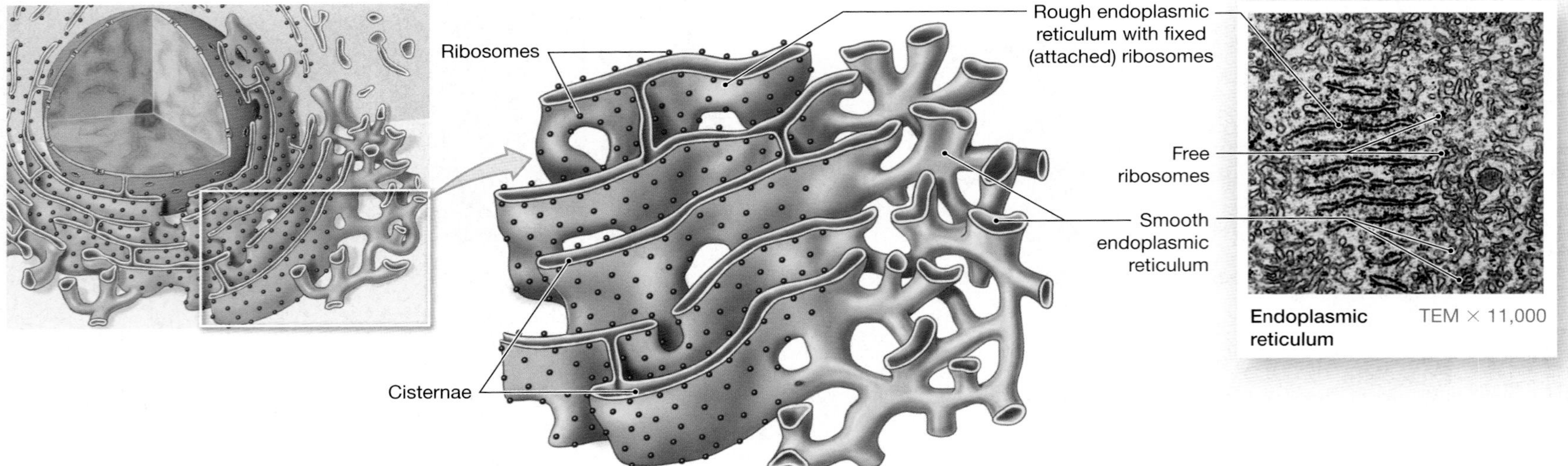

attached ribosomes. Ribosomes synthesize proteins using instructions provided by a strand of RNA. As the polypeptide chains grow, they enter the cisternae of the endoplasmic reticulum, where they are further modified. The proteins and glycoproteins produced by the RER are packaged into small membrane sacs that pinch off the edges or surfaces of the ER. These **transport vesicles** deliver the proteins to another organelle called the Golgi apparatus.

The **smooth endoplasmic reticulum (SER)** lacks ribosomes. The SER has a variety of functions that center on synthesizing lipids, steroids, and carbohydrates; storing calcium ions; and removing and inactivating toxins.

Figure 2.12 TEM of the Golgi Apparatus

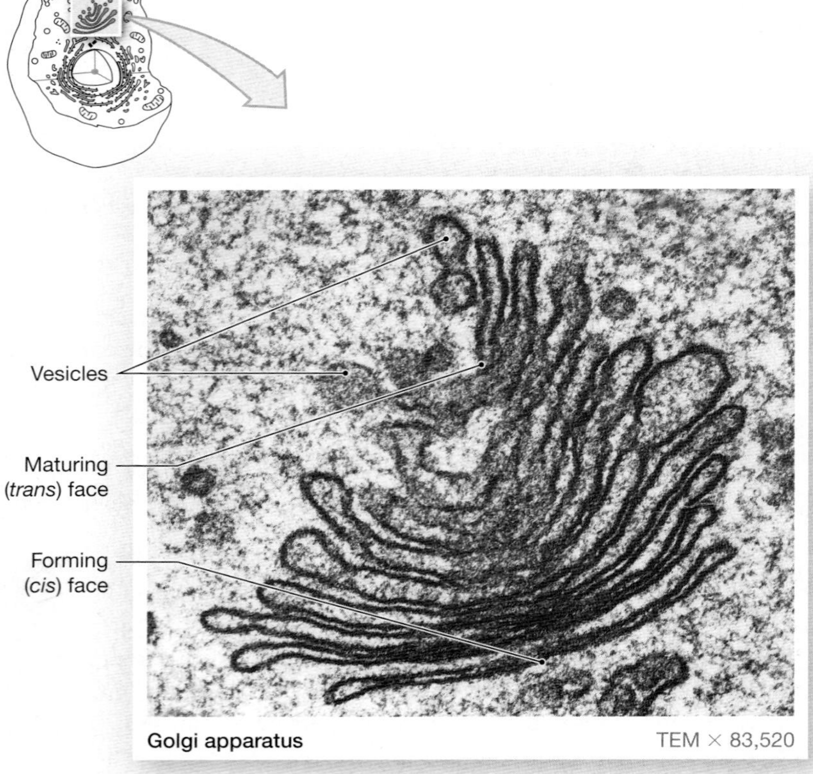

The amount of endoplasmic reticulum and the proportion of RER to SER vary depending on the type of cell and its activities. For example, pancreatic cells manufacturing digestive enzymes contain extensive RER and little SER. The situation is reversed in cells synthesizing steroid hormones in reproductive organs.

The Golgi Apparatus

The **Golgi** (GŌL-jē) **apparatus**, or *Golgi complex*, consists of flattened membrane discs called cisternae. A typical Golgi apparatus consists of five to six cisternae (**Figure 2.12**). Actively secreting cells have larger and more numerous cisternae than resting cells. The most actively secreting cells contain several sets of cisternae, each resembling a stack of dinner plates. Most often these stacks lie near the nucleus of the cell.

The major functions of the Golgi apparatus are to:

- Package enzymes for use in the cytosol
- Renew or modify the plasma membrane
- Synthesize and package secretions

Spotlight Figure 2.13 illustrates these functions.

Lysosomes

Most of the vesicles produced at the Golgi apparatus never leave the cytoplasm. The most important of these are **lysosomes** (LĪ-sō-sōms; *lyso-*, dissolution, + *soma*, body), vesicles filled with digestive enzymes formed by the rough endoplasmic reticulum and then packaged within the lysosomes by the Golgi apparatus. **Primary lysosomes** contain inactive enzymes. **Secondary lysosomes**, *which contain activated enzymes*, form when a primary lysosome fuses with another membrane-bound vesicle.

The functions of lysosomes include defending against disease. Using the process of endocytosis, cells remove bacteria, fluids, and organic debris from their surroundings and isolate them within vesicles. Primary lysosomes fuse with these vesicles, forming secondary lysosomes, and the digestive enzymes within the secondary lysosomes break down the contents. Reusable substances such as sugars or amino acids are released into the cytosol, and the remaining wastes

SPOTLIGHT

FIGURE 2.13

Functions of the Golgi Apparatus

The flattened membrane discs, or cisternae, of the **Golgi apparatus** communicate with the **endoplasmic reticulum (ER)** and with the cell surface by the formation, movement, and fusion of vesicles.

Three major functions of the Golgi apparatus

Packaging of Enzymes for Use in the Cytosol

Among the vesicles that are packaged by the Golgi and remain in the cytoplasm are **lysosomes**, which are filled with enzymes that break down engulfed foreign material or pathogens as well as damaged membranous organelles.

Renewal or Modification of the Plasma Membrane

As the Golgi apparatus loses membrane through the generation of vesicles at the **maturing** (or *trans*) **face**, it gains membrane by the fusion of **transport vesicles** at the **forming** (or *cis*) **face**. When Golgi-generated vesicles fuse with the plasma membrane, it adds to the surface area of the cell, balancing the membrane loss that occurs during endocytosis. Over time, this process can change the sensitivity and functions of the cell. In an actively secreting cell the change can be rapid, and the entire plasma membrane may be replaced every hour.

Synthesis and Packaging of Secretions

At the maturing face, vesicles form that carry materials away from the Golgi apparatus. Vesicles containing material that will be secreted from the cell are called **secretory vesicles**. When a secretory vesicle fuses with the plasma membrane, its contents are released into the extracellular fluid; this process is known as **exocytosis**. The events in the synthesis and packaging of secretions are described on the right side of this figure.

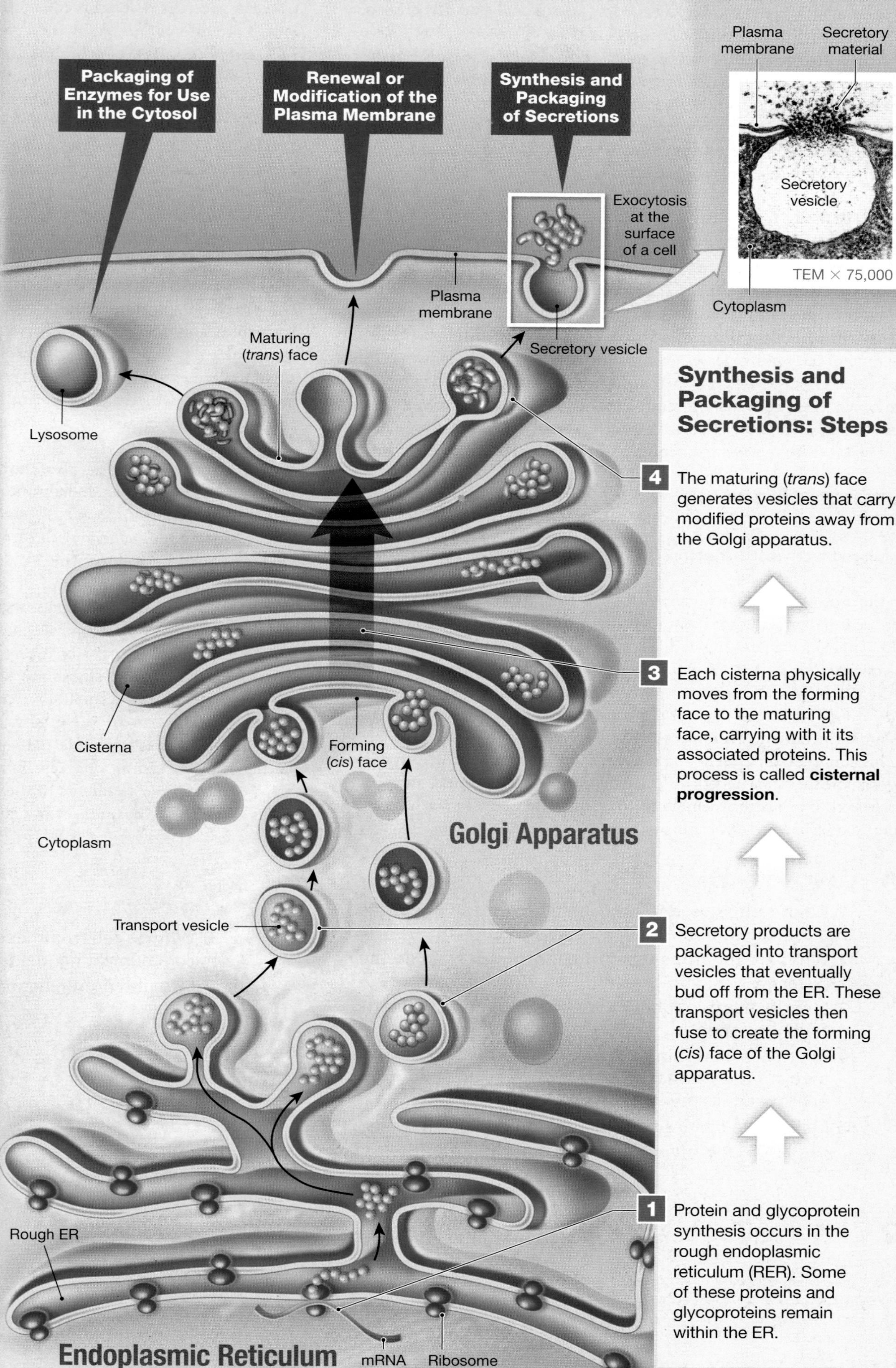

Synthesis and Packaging of Secretions: Steps

4 The maturing (*trans*) face generates vesicles that carry modified proteins away from the Golgi apparatus.

3 Each cisterna physically moves from the forming face to the maturing face, carrying with it its associated proteins. This process is called **cisternal progression**.

2 Secretory products are packaged into transport vesicles that eventually bud off from the ER. These transport vesicles then fuse to create the forming (*cis*) face of the Golgi apparatus.

1 Protein and glycoprotein synthesis occurs in the rough endoplasmic reticulum (RER). Some of these proteins and glycoproteins remain within the ER.

are eliminated by exocytosis. In this way, the cell not only protects itself against pathogens (disease-causing organisms) but also obtains valuable nutrients.

Lysosomes also perform essential cleanup and recycling activities inside the cell. For example, when muscle cells are inactive, lysosomes gradually break down their contractile proteins—this process accounts for the decreased muscle mass that occurs with aging or disuse. If the muscle cells become active once again, the destruction ceases. In a damaged or dead cell, lysosomes disintegrate and release active enzymes into the cytosol. These enzymes rapidly destroy the proteins and organelles of the cell, a process called **autolysis** (aw-TOL-i-sis; *auto-*, self). Because the breakdown of lysosomal membranes can destroy a cell, lysosomes have been called cellular "suicide packets."

Peroxisomes

Peroxisomes are smaller than lysosomes and carry a different group of enzymes. Peroxisome enzymes are formed by free ribosomes and are inserted into the membranes of pre-existing peroxisomes. Therefore, new peroxisomes result from the cell recycling older, pre-existing peroxisomes that no longer contain active enzymes.

Enzymes within peroxisomes perform a wide variety of cellular functions. One group of enzymes, oxidases, breaks down organic compounds into hydrogen peroxide (H_2O_2), which is toxic to cells. Catalase, another enzyme within peroxisomes, then converts hydrogen peroxide to water and oxygen. Peroxisomes also absorb and break down fatty acids. Peroxisomes are most abundant in liver cells, which remove and neutralize toxins absorbed in the digestive tract.

Membrane Flow

▶ **KEY POINT** Membrane flow is the cellular mechanism that changes the anatomical and functional characteristics of the plasma membrane.

With the exception of mitochondria, all the membranous organelles in the cell either are interconnected or communicate through the movement of vesicles. The RER and SER are continuous and connected to the nuclear envelope. Transport vesicles connect the ER with the Golgi apparatus, and secretory vesicles link the Golgi apparatus with the plasma membrane. Finally, vesicles forming at the exposed surface of the cell remove and recycle segments of the plasma membrane (see **Spotlight Figure 2.13** on p. 41).

This continual movement and exchange, called **membrane flow**, allows a cell to change the characteristics of the lipids, receptors, channels, anchors, and enzymes within its plasma membrane. Membrane flow enables the cell to grow, mature, and respond to environmental stimuli.

2.1 CONCEPT CHECK

1. What term is used to describe the permeability of the plasma membrane?
2. What are the three types of endocytosis? How do they differ?
3. Identify the two subdivisions of the cytoplasm and the functions of each.
4. Microscopic examination of a cell reveals that it has many mitochondria. What do you conclude about the energy requirements of this cell?
5. Cells in the ovaries and testes contain large amounts of smooth endoplasmic reticulum (SER). Why?

See the blue Answers tab at the back of the book.

2.2 Intercellular Attachments

▶ **KEY POINT** Specialized attachments between cells or between cells and underlying structures involve cell adhesion molecules and cell junctions.

Many cells form permanent or temporary attachments to other cells or extracellular materials (**Figure 2.14**). Cellular connections may involve extensive areas of opposing plasma membranes, or they may be concentrated at specialized attachment sites. Large areas of opposing plasma membranes are interconnected by transmembrane proteins called **cell adhesion molecules (CAMs)**, which bond to each other and to extracellular materials. For example, CAMs on the attached base of an epithelium bind the basal surface (where the epithelium attaches to underlying tissues) to the underlying basement membrane. The membranes of adjacent cells may also be bonded by a thin layer of proteoglycans that contain polysaccharide derivatives known as glycosaminoglycans, most notably **hyaluronan** (*hyaluronic acid*).

Cell junctions are specialized areas of the plasma membrane that attach a cell to another cell or to extracellular materials. There are three common types of cell junctions: (1) gap junctions, (2) tight junctions, and (3) desmosomes (**Figure 2.14a**). At a **gap junction**, two cells are held together by two interlocking transmembrane proteins called connexons (**Figure 2.14b**). Each connexon is composed of six connexin proteins that form a cylinder with a central pore. Two aligned connexons form a narrow passageway that lets small molecules and ions pass from cell to cell. Gap junctions are common in epithelial cells, where the movement of ions helps coordinate functions such as the beating of cilia. They are also common in cardiac muscle and smooth muscle tissues.

At a **tight junction** (also known as an *occluding junction*), the lipid portions of two plasma membranes are tightly bound together by interlocking membrane proteins (**Figure 2.14c**). Inferior to the tight junctions, a continuous *adhesion belt* forms a band that encircles cells and binds them to their neighbors. The bands are attached to the microfilaments of the terminal web. This kind of attachment is so tight that tight junctions largely prevent water and solutes from passing between the cells. At a **desmosome**, CAMs and proteoglycans link the opposing plasma membranes. Desmosomes are very strong and can resist stretching and twisting. There are two types of desmosomes: spot desmosomes and hemidesmosomes. **Spot desmosomes** are small discs connected to bands of intermediate filaments (**Figure 2.14d**). The intermediate filaments stabilize the cell shape. **Hemidesmosomes** resemble half of a spot desmosome (**Figure 2.14e**). Rather than attaching one cell to another, a hemidesmosome attaches a cell to extracellular filaments in the basement membrane. This attachment helps stabilize the position of the epithelial cell and anchors it to underlying tissues.

2.2 CONCEPT CHECK

6. Why do cells that need to coordinate functions have communicating junctions between them?
7. Describe tight junctions.

See the blue Answers tab at the back of the book.

2.3 The Cell Life Cycle

▶ **KEY POINT** Somatic cell division duplicates genetic material (DNA replication) and distributes one copy to each of the two daughter cells (mitosis).

Between fertilization and physical maturity, a human increases in complexity from a single cell to roughly 75 trillion cells. This amazing increase occurs through **cell division**, a form of cellular reproduction. The division of a single cell produces a pair of daughter cells, each half the size of the original. Thus, two new cells have replaced the original one.

Figure 2.14 **Cell Attachments.**

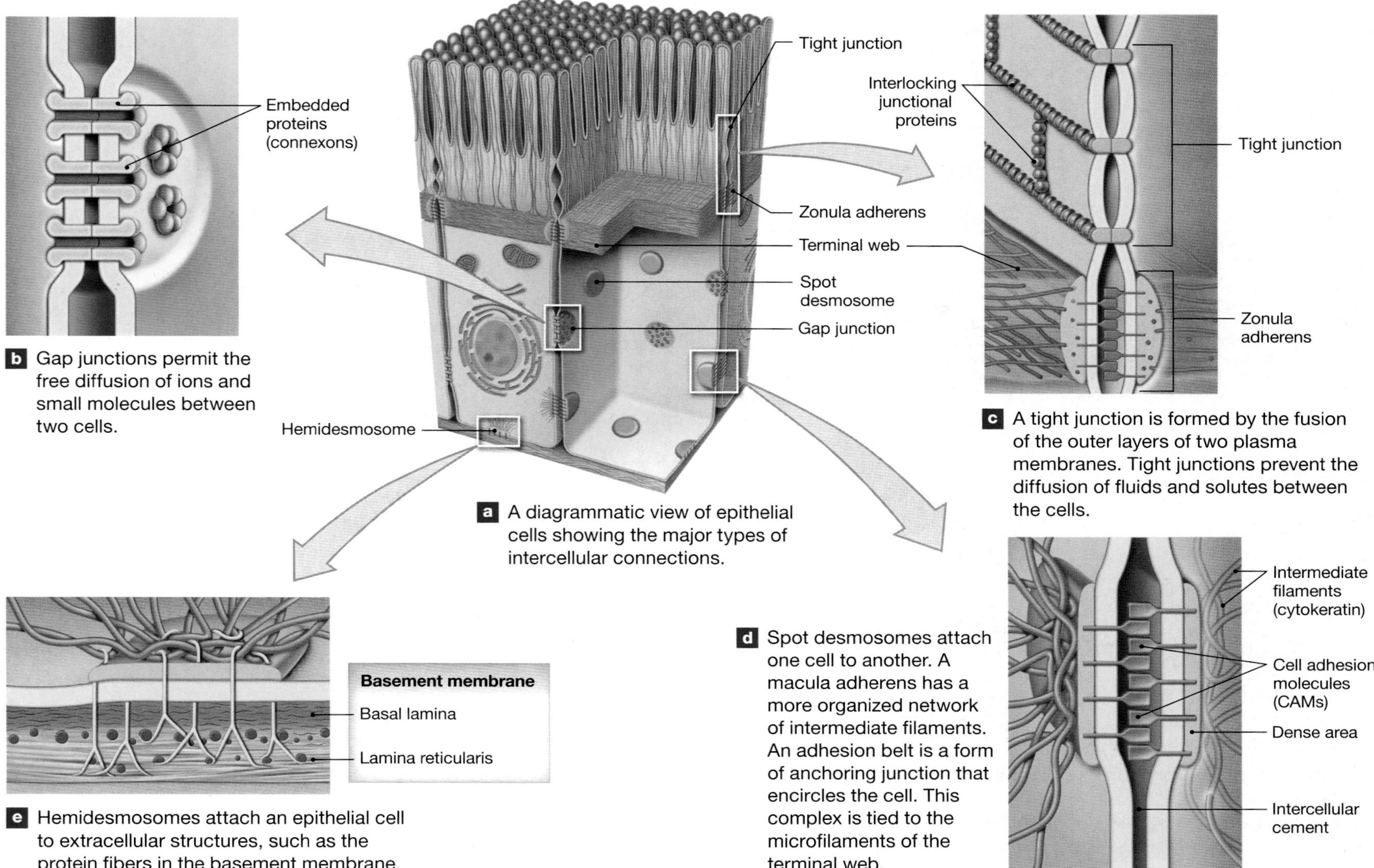

a A diagrammatic view of epithelial cells showing the major types of intercellular connections.

b Gap junctions permit the free diffusion of ions and small molecules between two cells.

c A tight junction is formed by the fusion of the outer layers of two plasma membranes. Tight junctions prevent the diffusion of fluids and solutes between the cells.

d Spot desmosomes attach one cell to another. A macula adherens has a more organized network of intermediate filaments. An adhesion belt is a form of anchoring junction that encircles the cell. This complex is tied to the microfilaments of the terminal web.

e Hemidesmosomes attach an epithelial cell to extracellular structures, such as the protein fibers in the basement membrane.

Even after development ends, cell division continues to be essential to survival. Although cells are highly adaptable, they are damaged by physical wear and tear, toxic chemicals, temperature changes, and other environmental hazards. The life span of a cell ranges from hours to decades, depending on the type of cell and the environmental stresses involved. A typical cell does not live nearly as long as a typical person, so cells must divide to maintain cell populations.

There are two important steps in cell division: DNA replication and mitosis. DNA replication is the accurate duplication of the cell's genetic material, and **mitosis** (mī-TŌ-sis) is the distribution of one copy of the genetic information to each of the two new daughter cells. Mitosis occurs during the division of somatic cells, all the cells in the body other than the sex cells. (Recall that sex cells give rise to sperm or oocytes. A distinct process, meiosis (mī-Ō-sis), produces sex cells. We will describe meiosis in Chapter 27.)

Next, let's look at the steps involved in the division of somatic cells.

Interphase

▶ **KEY POINT** During interphase, the cell grows, duplicates organelles, and replicates DNA as it prepares for the next mitosis.

Figure 2.15 summarizes the life cycle of a typical somatic cell. A somatic cell spends most of its functional life in **interphase**. During interphase the cell performs all of its normal functions and, if necessary, prepares for division. Stages of interphase include G_0, G_1, S, and G_2.

An interphase cell in the **G_0 phase** is performing all normal cell functions and is not preparing for division. Some mature cells, such as skeletal muscle cells and most neurons, remain in G_0 indefinitely and never undergo mitosis. In contrast, stem cells, which divide repeatedly with very brief interphase periods, never enter G_0.

In the **G_1 phase**, the cell manufactures enough mitochondria, cytoskeletal elements, endoplasmic reticula, ribosomes, Golgi membranes, and cytosol to make two functional cells. In cells dividing at top speed, G_1 may last as little as 8–12 hours. Such cells pour all their energy into mitosis, and all other activities cease. If G_1 lasts for days, weeks, or months, preparation for mitosis occurs as the cells perform their normal functions.

After completing the activities of G_1, the cell enters the **S phase**. Over the next 6–8 hours, the cell duplicates its chromosomes. This involves DNA replication and the synthesis of histones (proteins in chromosomes) and other proteins in the nucleus.

DNA Replication

The DNA strands in the nucleus remain intact throughout the life of a cell. DNA synthesis, or **DNA replication**, occurs in cells preparing to undergo mitosis or meiosis. The goal of replication is to copy the genetic

Figure 2.15 The Cell Life Cycle. The cell cycle is divided into interphase, comprising the G_1, S, and G_2 stages, and the M phase, which includes mitosis and cytokinesis. The result is the production of two identical daughter cells.

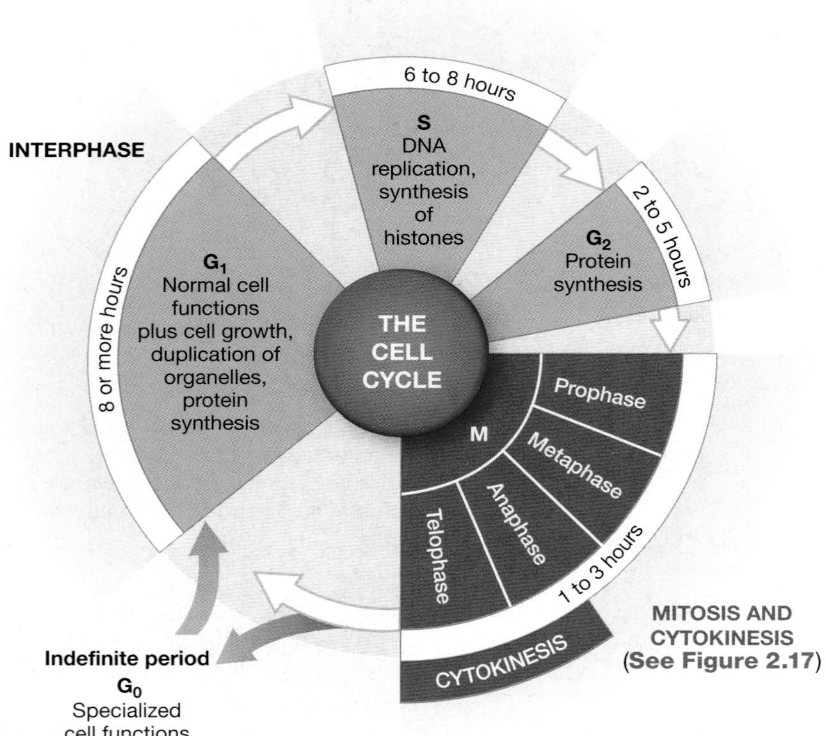

information in the nucleus so that one set of chromosomes is distributed to each of the two cells produced. Several different enzymes are needed for the process.

Each DNA molecule consists of a pair of nucleotide strands held together by hydrogen bonds between complementary nitrogen bases. **Figure 2.16** diagrams the process of DNA replication. DNA replication starts when the weak bonds between the nitrogenous bases are broken and the strands unwind, forming two segments—segment 1 and segment 2. As they unwind, molecules of the enzyme **DNA polymerase** bind to the exposed nitrogenous bases and allow the attachment of complementary DNA nucleotides suspended in the nucleoplasm.

Many molecules of DNA polymerase work simultaneously along different portions of each DNA strand. This produces short complementary nucleotide chains that are then linked together by enzymes called **ligases** (LĪ-gās-ez; *liga*, to tie). The final result is a pair of identical DNA molecules.

Once DNA replication is completed, there is a brief (2–5 hours) **G_2 phase** for last-minute protein synthesis. The cell then enters the **M phase**, and mitosis begins (see **Figure 2.15**). The M phase of the cell cycle includes mitosis and cytokinesis, which are detailed in **Spotlight Figure 2.17**.

Figure 2.16 DNA Replication. In DNA replication the original paired strands unwind, and DNA polymerase begins attaching complementary DNA nucleotides along each strand. This process produces two identical copies of the original DNA molecule.

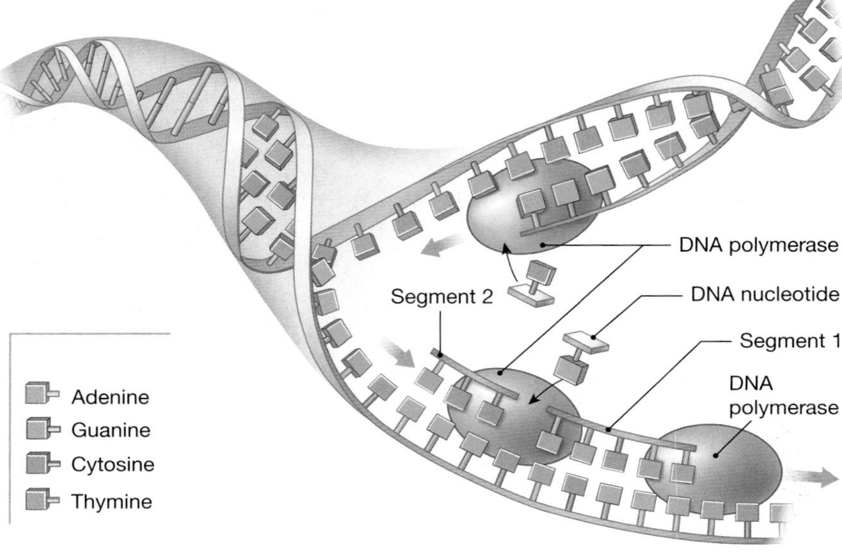

Mitosis

▶ **KEY POINT** Mitosis is the process by which a cell divides its nucleus and cytoplasm. As a result of mitosis two daughter cells are formed.

Mitosis consists of four stages: **prophase**, **metaphase**, **anaphase**, and **telophase** **(Spotlight Figure 2.17)**. The transitions from stage to stage are seamless.

We can estimate the frequency of cell division by the number of cells in mitosis at any given time. We use the term **mitotic rate** to discuss rates of cell division. The longer the life span of a cell type, the slower its mitotic rate. Long-lived cells, such as muscle cells and neurons, either never divide or do so only under special circumstances. Other cells, such as those lining the digestive tract, survive only days or even hours because they are constantly under attack by chemicals, pathogens, and abrasion. Special cells called **stem cells** maintain these cell populations through repeated cycles of cell division.

2.3 CONCEPT CHECK

8 What is cell division?

9 Prior to cell division, mitosis must occur. What is mitosis?

10 List, in order of appearance, the stages of interphase and mitosis and the events that occur in each.

See the blue Answers tab at the back of the book.

SPOTLIGHT

FIGURE 2.17
Mitosis

Mitosis is the process that delivers replicated DNA to two daughter cells. It consists of four stages, but the transitions from stage to stage are seamless.

Interphase

During **interphase**, the DNA strands are loosely coiled and chromosomes cannot be seen.

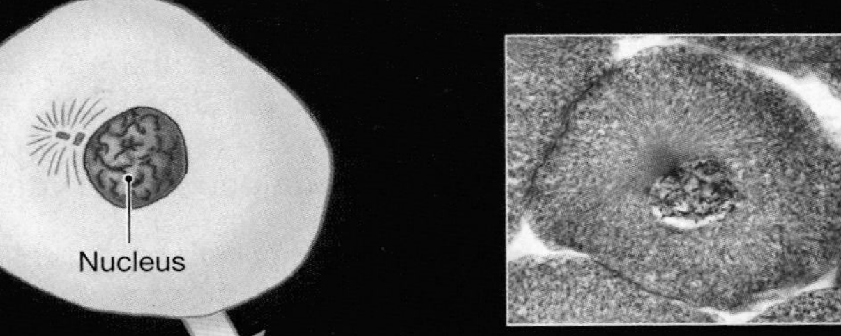

Prophase

Early prophase

Prophase (PRŌ-fāz; *pro-*, before) begins only after DNA replication has been completed. During prophase the DNA strands coil so tightly that the duplicated chromosomes become visible as single structures. During early prophase, the two pairs of centrioles, which are connected to an array of microtubules called **spindle fibers**, move apart from each other. Smaller microtubules called astral rays radiate into the surrounding cytoplasm.

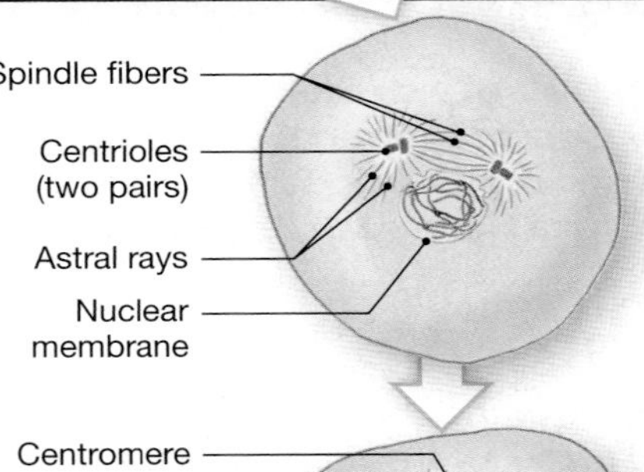

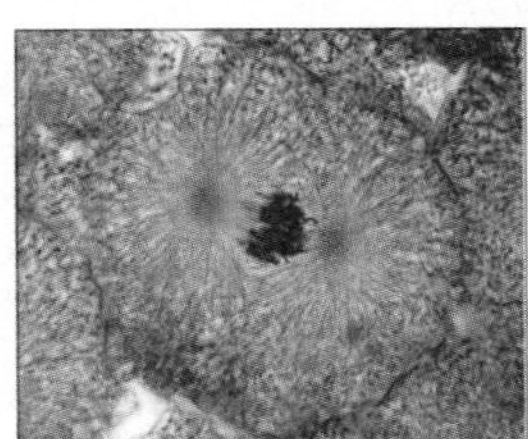

Late prophase

As the chromosomes finish their coiling, the nuclear membrane and nucleoli disintegrate. The two copies of each chromosome are called **chromatids** (KRŌ-ma-tids), and at this stage they are connected at a single point, the **centromere** (SEN-trō-mēr). The spindle fibers now extend among the chromosomes. Some of the spindle fibers bind to the centromeres; these fibers are called **chromosomal microtubules**.

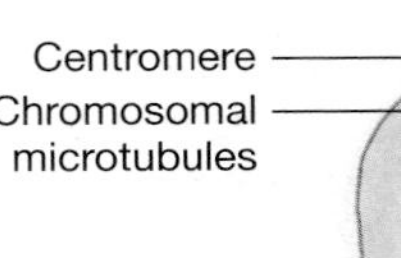
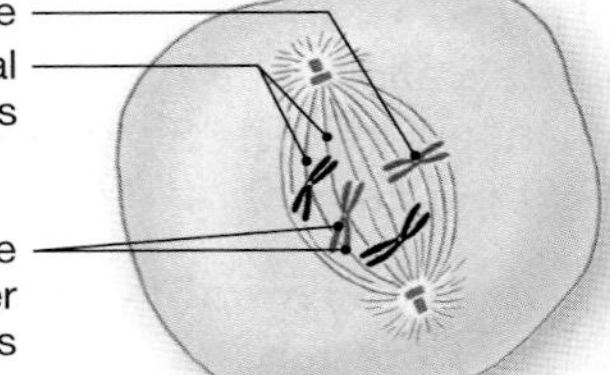

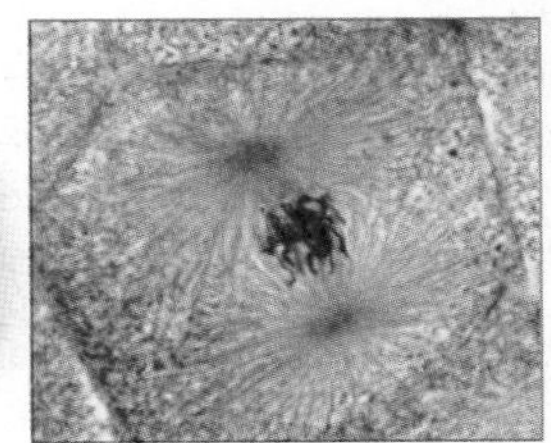

Metaphase

In **metaphase** (MET-a-fāz; *meta-*, after), the centromeres move to a narrow central zone called the **metaphase plate**.

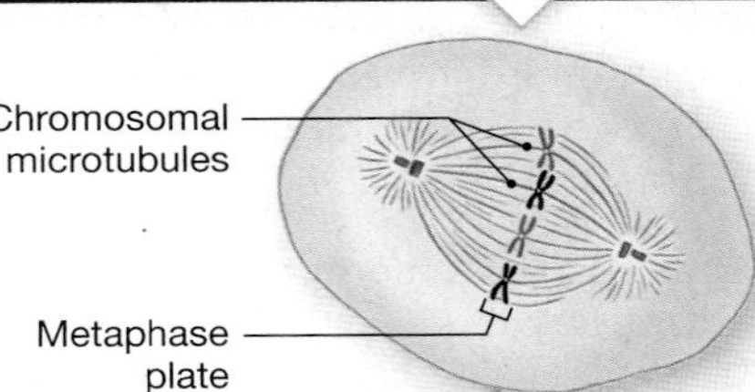

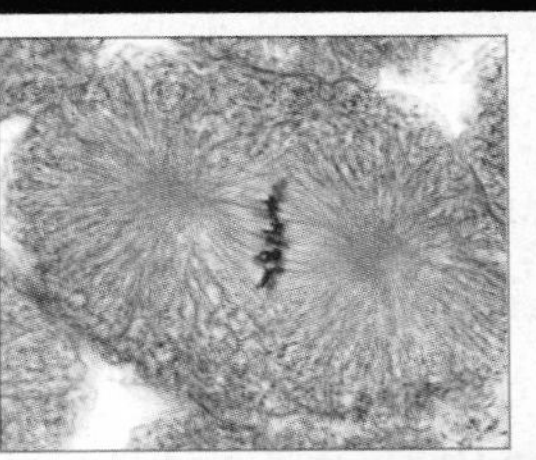

Anaphase

Anaphase (AN-a-fāz; *ana-*, apart) begins when the centromere of each chromatid pair splits apart. The two chromatids, now called **daughter chromosomes**, are pulled toward opposite ends of the cell by the chromosomal microtubules. Anaphase ends as the daughter chromosomes arrive near the centrioles at opposite ends of the dividing cell.

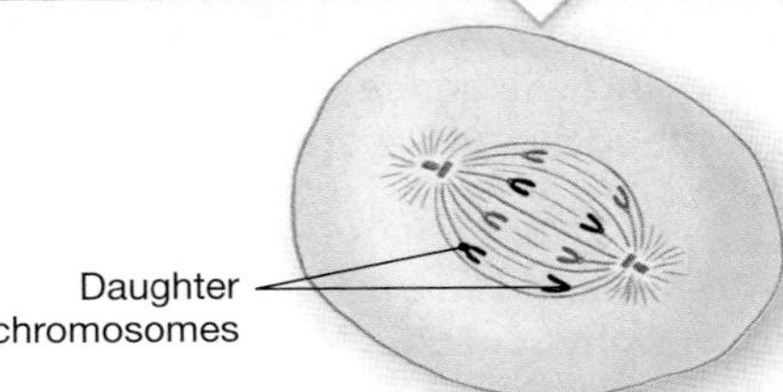

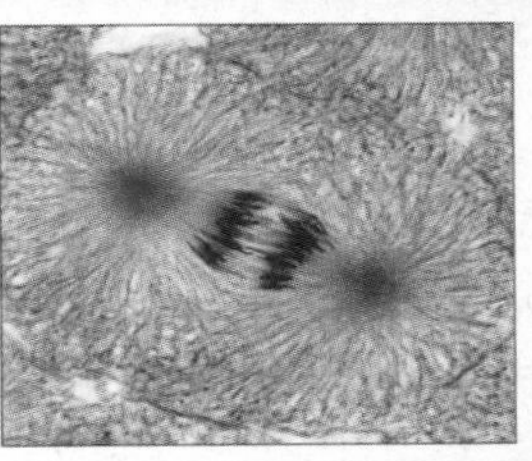

Telophase

Telophase (TĒL-ō-fāz; *telo-*, end) is the reverse of prophase, because the cell prepares to return to the interphase state. The nuclear membranes form and the nuclei enlarge as the chromosomes gradually uncoil. Once the chromosomes uncoil and are no longer visible, nucleoli reappear and the nuclei resemble those of interphase cells.

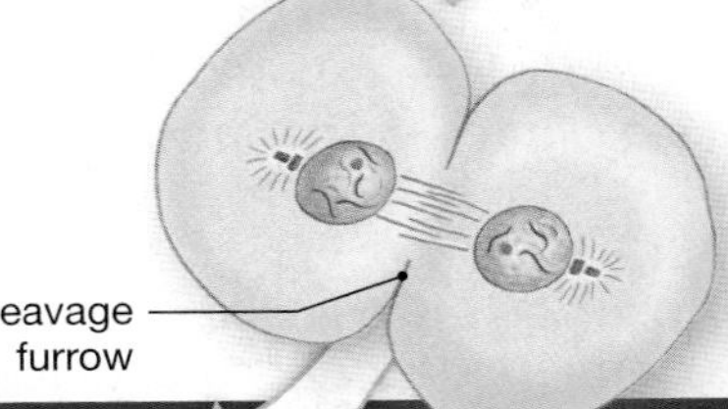

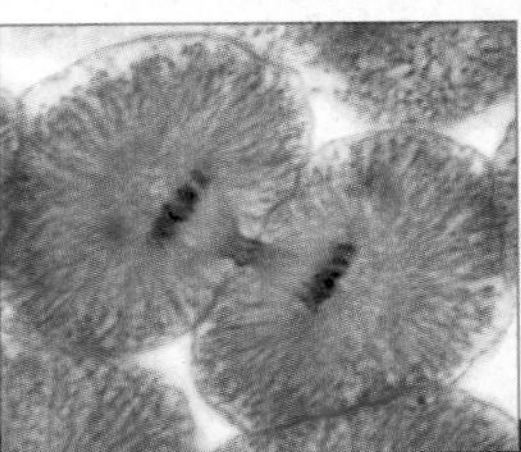

Cytokinesis

Telophase is the end of mitosis proper, but before cell division is completed the cytoplasm of the original cell must be divided between two daughter cells. This separation process, called **cytokinesis** (*cyto-*, cell, + *kinesis*, motion), begins when the daughter chromosomes near the ends of the spindle apparatus. The cytoplasm then constricts along the plane of the metaphase plate, forming a **cleavage furrow** that deepens until the two daughter cells separate. This event is the end of cell division and the beginning of the next interphase period.

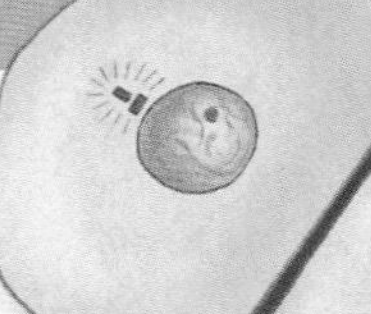
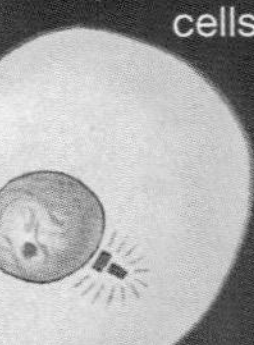

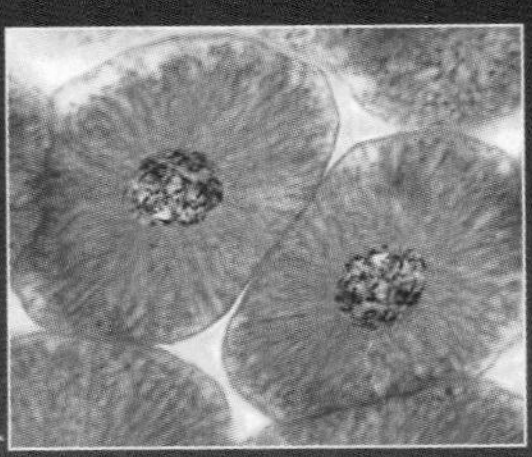

2

Study Outline

Introduction p. 28

- All living things are composed of cells. The cell theory incorporates several basic concepts: (1) Cells are the building blocks of all plants and animals; (2) cells are produced by the division of pre-existing cells; (3) cells are the smallest units that perform all vital functions.
- The body contains two types of cells: sex cells (germ cells or reproductive cells) and somatic cells (all other body cells).

2.1 | Cellular Anatomy p. 28

- A cell is surrounded by a thin layer of **extracellular fluid**. The cell's outer boundary, the **plasma membrane**, is a **phospholipid bilayer** containing proteins and cholesterol. Table 2.1 summarizes the anatomy of a typical cell. *(See Figures 2.1, 2.3, 2.5, Spotlight Figure 2.4, and Table 2.1.)*

The Plasma Membrane p. 30

- **Integral proteins** are embedded in the phospholipid bilayer of the membrane, while **peripheral proteins** are attached to the membrane but can separate from it. **Channels** allow water and ions to move across the membrane; some channels are **gated channels** because they can open or close. *(See Figure 2.3 and Spotlight Figure 2.4.)*
- Plasma membranes are **selectively permeable**; that is, they permit the free passage of some materials.
- Passive transport mechanisms include diffusion, osmosis, and facilitated diffusion. **Diffusion** is the net movement of material from an area where its concentration is high to an area where its concentration is lower. Diffusion occurs until the **concentration gradient** is eliminated. *(See Spotlight Figure 2.4.)*
- **Osmosis** is the diffusion of water across a membrane in response to differences in water concentration.
- **Facilitated diffusion** requires **carrier proteins**.
- **Active transport** mechanisms consume ATP and are independent of concentration gradients. They include active transport, endocytosis, and exocytosis. Some **ion pumps** are **exchange pumps**. *(See Spotlight Figure 2.4.)*
- **Endocytosis** is movement into a cell; it is an active process that occurs in one of three forms: **pinocytosis** (cell drinking), **phagocytosis** (cell eating), or **receptor-mediated endocytosis** (selective movement). *(See Spotlight Figure 2.4.)*
- **Microvilli** are small, fingerlike projections of the plasma membrane that increase the surface area exposed to the extracellular environment. *(See Figure 2.5 and Table 2.1.)*

The Cytoplasm p. 31

- The **cytoplasm** contains **cytosol**, an intracellular fluid that surrounds structures that perform specific functions, called **organelles**. *(See Figures 2.1 and 2.2 and Table 2.1.)*

Nonmembranous Organelles p. 34

- **Nonmembranous organelles** are not enclosed in membranes and are always in contact with the cytosol. They include the cytoskeleton, microvilli, centrioles, cilia, flagella, and ribosomes. *(See Figures 2.5 and 2.6 and Table 2.1.)*
- The **cytoskeleton** is an internal protein network that gives the cytoplasm strength and flexibility. It has four components: **microfilaments**, **intermediate filaments**, **thick filaments**, and **microtubules**. *(See Figure 2.5 and Table 2.1.)*
- **Centrioles** are small, microtubule-containing cylinders that direct the movement of chromosomes during cell division. *(See Figure 2.6 and Table 2.2.)*
- **Cilia**, anchored by a **basal body**, are microtubules containing hair-like projections from the cell surface that beat rhythmically to move fluids or secretions across the cell surface. *(See Figure 2.6 and Table 2.2.)*
- A whip-like **flagellum** moves a cell through surrounding fluid. Table 2.2 compares centrioles, cilia, and flagella. *(See Table 2.2.)*
- **Ribosomes** are intracellular factories consisting of small and large subunits; together they manufacture proteins. Two types of ribosomes, **free** (within the cytosol) and **fixed** (bound to the endoplasmic reticulum), are found in cells. *(See Figure 2.7 and Table 2.1.)*

Membranous Organelles p. 36

- **Membranous organelles** are surrounded by lipid membranes that isolate them from the cytosol. They include the mitochondria, nucleus, endoplasmic reticulum (rough and smooth), Golgi apparatus, lysosomes, and peroxisomes.
- **Mitochondria** produce 95 percent of the ATP within a typical cell. *(See Figure 2.8 and Table 2.1.)*
- The **nucleus** is the control center for cellular operations. It is surrounded by a **nuclear envelope**, through which it communicates with the cytosol through **nuclear pores**. The nucleus contains 23 pairs of **chromosomes**. *(See Figures 2.9 and 2.10 and Table 2.1.)*
- The **endoplasmic reticulum (ER)** is a network of intracellular membranes involved in synthesis, storage, transport, and detoxification. The ER forms hollow tubes, flattened sheets, and rounded chambers termed **cisternae**. There are two types of ER: rough and smooth. **Rough endoplasmic reticulum (RER)** has attached ribosomes; **smooth endoplasmic reticulum (SER)** does not. *(See Figure 2.11 and Table 2.1.)*
- The **Golgi apparatus** packages materials for **lysosomes**, **peroxisomes**, **secretory vesicles**, and membrane segments that are incorporated into the plasma membrane. Secretory products are discharged from the cell through the process of **exocytosis**. *(See Figure 2.12 and Spotlight Figures 2.4 and 2.13.)*
- **Lysosomes** are vesicles filled with digestive enzymes. The process of endocytosis is important for ridding the cell of bacteria and debris. The endocytic vesicle fuses with a lysosome, which digests its contents. *(See Spotlight Figure 2.13.)*
- **Peroxisomes** carry enzymes used to break down organic molecules and neutralize toxins.

Membrane Flow p. 42

- **Membrane flow** is the continuous movement of membrane components among the nuclear envelope, Golgi apparatus, endoplasmic reticulum, vesicles, and plasma membrane.

2.2 | Intercellular Attachments p. 42

- Cells can attach to other cells or to extracellular protein fibers by means of **cell adhesion molecules (CAMSs)** or at specialized attachment sites called **cell junctions**. The three major types of cell junctions are **gap junctions**, **tight junctions**, and **desmosomes**. *(See Figure 2.14.)*

- There are two types of desmosomes: **spot desmosomes** (small discs connected to intermediate filaments) and **hemidesmosomes** (resembling half of a spot desmosome). (*See Figure 2.14.*).

2.3 | The Cell Life Cycle p. 42

- **Cell division** is the reproduction of cells. In a dividing cell, interphase alternates with **mitosis**. *(See Spotlight Figure 2.17.)*

Interphase p. 43

- Most somatic cells spend most of their time in **interphase**, a time of growth, organelle duplication, and DNA replication. *(See Spotlight Figure 2.17.)*

Mitosis p. 44

- **Mitosis** refers to the nuclear division of somatic cells.
- Mitosis proceeds in four distinct, contiguous stages: **prophase**, **metaphase**, **anaphase**, and **telophase**. *(See Spotlight Figure 2.17.)*
- During **cytokinesis**, the cytoplasm is divided between the two daughter cells and cell division ends.
- In general, the longer a cell's life span, the slower its **mitotic rate**. **Stem cells** undergo frequent mitosis to replace other, more specialized cells.

Chapter Review

For answers, see the blue Answers tab at the back of the book.

Level 1 Reviewing Facts and Terms

Match each numbered item with the most closely related lettered item.

1. ribosomes .... ☐
2. lysosomes .... ☐
3. integral proteins .... ☐
4. Golgi apparatus .... ☐
5. endocytosis .... ☐
6. cytoskeleton .... ☐
7. tight junction .... ☐
8. nucleus .... ☐
9. S phase .... ☐

(a) DNA replication
(b) flattened membrane discs, packaging
(c) adjacent plasma membranes bound by bands of interlocking proteins
(d) packaging of materials for import into cell
(e) RNA and protein; protein synthesis
(f) control center; stores genetic information
(g) cell vesicles with digestive enzymes
(h) embedded in the plasma membrane
(i) internal protein framework in cytoplasm

10. All of the following membrane transport mechanisms are passive processes except
(a) facilitated diffusion.
(b) vesicular transport.
(c) diffusion.
(d) osmosis.

11. The viscous, superficial coating on the outer surface of the plasma membrane is the
(a) phospholipid bilayer.
(b) gated channel network.
(c) glycocalyx.
(d) plasma membrane.

12. Compared to the intracellular fluid, the extracellular fluid contains
(a) equivalent amounts of sodium ions.
(b) a consistently higher concentration of potassium ions.
(c) many more enzymes.
(d) a lower concentration of dissolved proteins.

13. Label the following organelles on the diagram below.
- free ribosomes
- mitochondrion
- Golgi apparatus
- nucleolus
- peroxisome
- rough endoplasmic reticulum
- lysosome

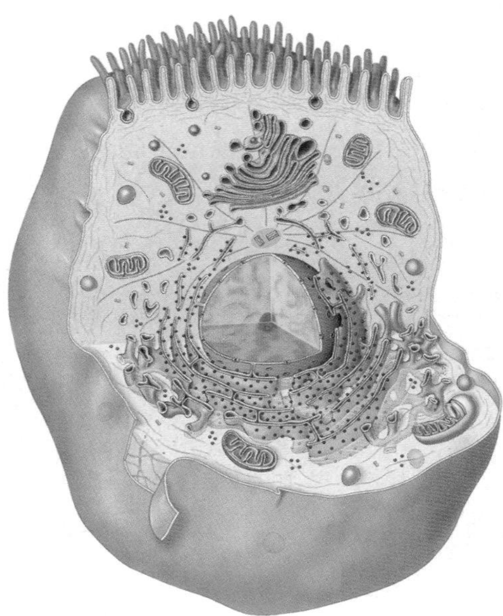

14. Membrane flow provides a mechanism for
(a) continual change in the characteristics of membranes.
(b) increase in the size of the cell.
(c) response of the cell to a specific environmental stimulus.
(d) all of the above.

15. If a cell lacks mitochondria, the direct result will be that it cannot
(a) manufacture proteins.
(b) produce substantial amounts of ATP.
(c) package proteins manufactured by the attached ribosomes.
(d) reproduce itself.

16. Three major functions of the endoplasmic reticulum are
(a) hydrolysis, diffusion, and osmosis.
(b) detoxification, packaging, and modification.
(c) synthesis, storage, and transport.
(d) pinocytosis, phagocytosis, and storage.

17. A selectively permeable plasma membrane
(a) permits only water-soluble materials to enter or leave the cell freely.
(b) prohibits entry of all materials into the cell at certain times.
(c) permits the free passage of some materials but restricts passage of others.
(d) allows materials to enter or leave the cell only using active processes.

18. The presence of invading pathogens in the extracellular fluid stimulates immune cells to engage the mechanism of
(a) pinocytosis.
(b) phagocytosis.
(c) receptor-mediated pinocytosis.
(d) bulk transport.

Level 2 Reviewing Concepts

1. Identify the advantages a cell has because its nucleus is enclosed within a membrane.
2. Identify the three basic concepts that make up the cell theory.
3. Identify the three passive processes by which substances get into and out of cells.
4. Compare and contrast facilitated diffusion and active transport.
5. Analyze the three major factors that determine whether a substance can diffuse across a plasma membrane.
6. Define an organelle and then compare and contrast the two broad categories of organelles.

7. Name the stages of mitosis shown in this figure.

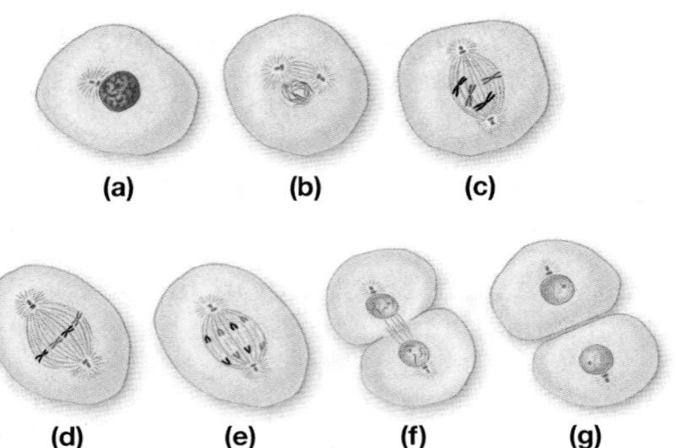

a. ________
b. ________
c. ________
d. ________
e. ________
f. ________
g. ________

8. Identify the four general functions of the plasma membrane.

Level 3 Critical Thinking

1. When skin that is damaged by sunburn "peels," large amounts of epidermal cells are often shed simultaneously. Hypothesize why the shedding occurs in this manner.
2. Experimental evidence demonstrates that in the transport of a certain molecule, the molecule moves against its concentration gradient, and cellular energy is required for the transport to occur. Justify what type of transport process is at work based on this experimental evidence.
3. Solutions A and B are separated by a selectively permeable barrier. Over time, the level of fluid on side A increases. Which solution initially had the higher concentration of solute?

CLINICAL CASE WRAP-UP

Inheritance from Mom

Mitochondria produce 95 percent of the energy that cells need for life. Cells that use the most energy, including cardiac muscle, skeletal muscle, and brain cells, have the most mitochondria. Inadequate cell energy due to mutant mitochondria can cause poor growth, weakness, and seizures.

When a sperm fertilizes an oocyte (see Chapter 28), all of the cytoplasmic organelles come from the mother's oocyte. Therefore, all of a child's mitochondrial DNA (mtDNA) comes from the mother and will be affected by mutated mtDNA.

Genetic engineering uses biotechnology to manipulate inheritance. If the nucleus from one of Jessica's abnormal oocytes, containing normal nuclear DNA (nDNA), is transferred to a surrogate's enucleated normal oocyte (cell without a nucleus), they would form one normal oocyte containing Jessica's nuclear DNA—a genetically engineered cure for this type of mitochondrial disease.

Researchers have successfully performed this procedure, called a spindle transfer or spindle-chromosomal complex transfer, in monkeys. It is currently being tested and is showing much promise in preventing the transmission of mitochondrial disorders.

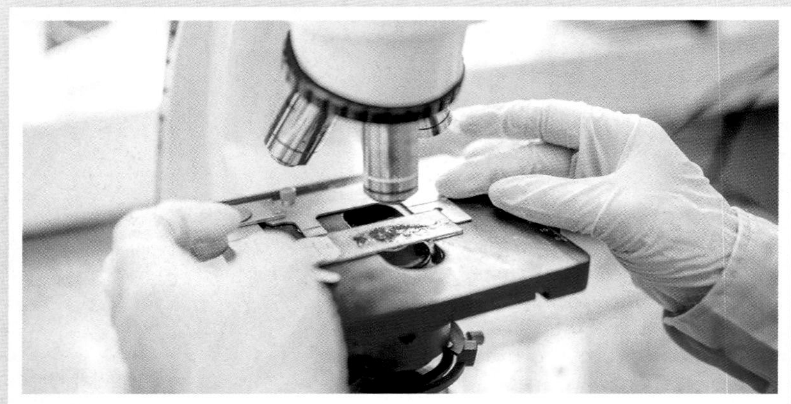

1. Are Jessica's red blood cells affected by her mitochondrial disease?
2. Why was an enucleated surrogate cell used?

See the blue Answers tab at the back of the book.

Related Clinical Terms

adhesions: Restrictive fibrous connections that can result from surgery, infection, or other injuries to serous membranes.

anaplasia: An irreversible change in the size and shape of tissue cells.

dysplasia: A reversible change in the normal shape, size, and organization of tissue cells.

hyperplasia: An increase in the number of normal cells (not tumor formation) in a tissue or organ, thus enlarging that tissue or organ.

hypertrophy: The enlargement of an organ or tissue due to an increase in the size of its cells.

necrosis: Death of one or more cells in an organ or tissue due to disease, injury, or inadequate blood supply.

3 Foundations

Tissues and Early Embryology

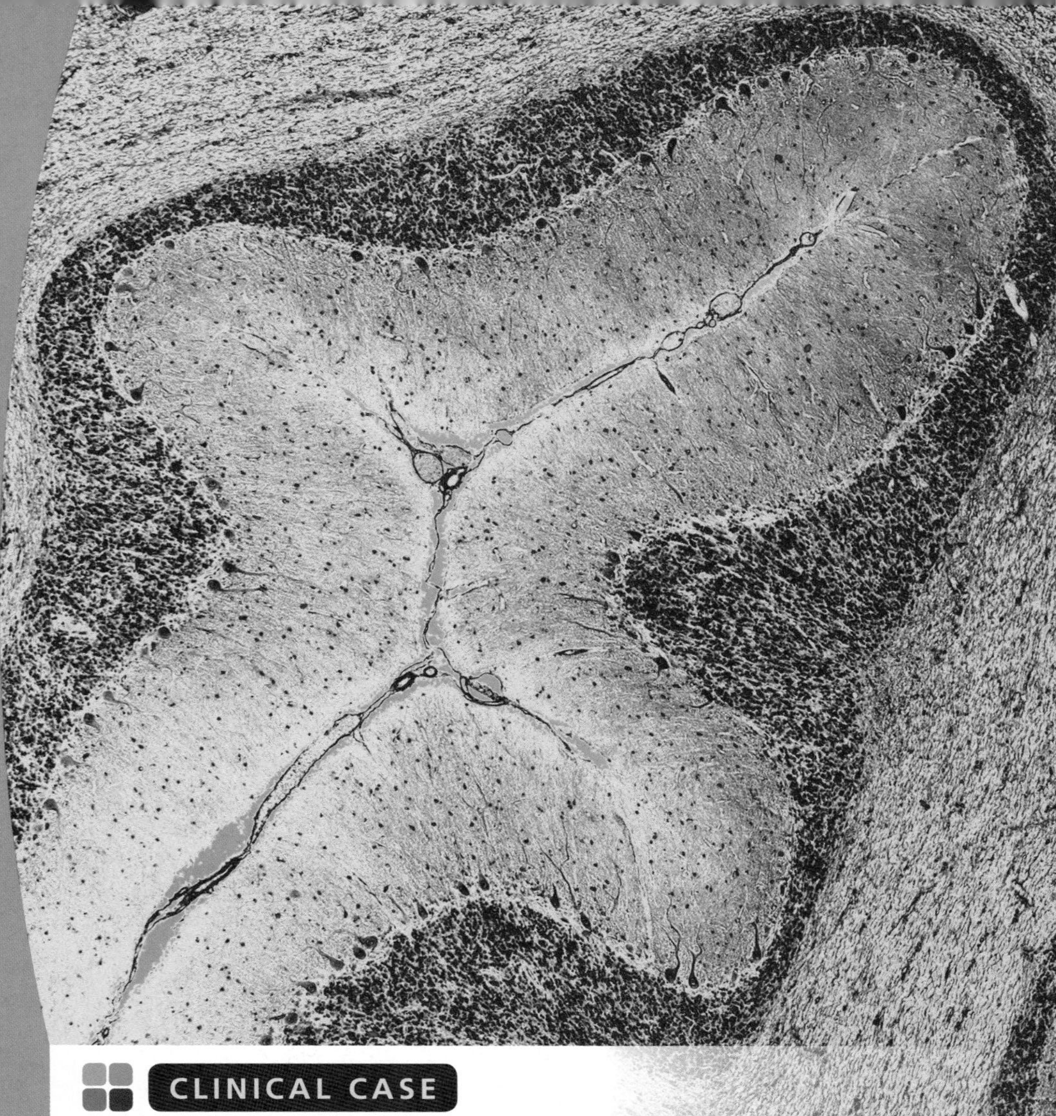

Learning Outcomes

These Learning Outcomes correspond by number to this chapter's sections and indicate what you should be able to do after completing the chapter.

3.1 List the functions of epithelial tissues, the criteria used to classify epithelial tissue, the functions of each type of epithelial tissue, and at least one location for each type of epithelial tissue. p. 50

3.2 List the three categories of connective tissue, their functions, and the cell types that would be found within each category. p. 59

3.3 Compare and contrast the characteristics of mucous, serous, cutaneous, and synovial membranes. p. 70

3.4 Summarize how connective tissues establish the framework of the body. p. 73

3.5 Compare and contrast the three forms of muscle tissue in terms of structure, function, and location. p. 74

3.6 Differentiate between neurons and neuroglia and discuss the functions of each. p. 76

3.7 Describe how nutrition and aging affect tissues. p. 76

3.8 List and explain the key embryological steps in the formation of epithelial and connective tissues. p. 78

CLINICAL CASE

The Tallest in the School

Everyone expects Elijah to be a basketball star because he's so tall. Actually, he isn't that good at the sport. For one thing, he has long, delicate fingers that have trouble gripping the ball. He is also "double-jointed," has very flat feet, and has scoliosis (spinal curvature), which cause him discomfort during exercise, and he's embarrassed to take his shirt off because his chest sinks in. Furthermore, he has trouble with his vision. But because Elijah's father died suddenly when in his early 30s and money is tight for his family, Elijah feels that he must try out for his high school team with the goal of earning a college basketball scholarship.

At tryouts, the coach asks the players to run around the gym a dozen times. On the last round, Elijah feels nauseated and dizzy, and then everything goes dark.

Elijah has collapsed due to a sudden cardiac arrhythmia (abnormal heart rhythm). The coach, thanks to his recent Red Cross training, uses the AED (automated external defibrillator) mounted on the gym wall to re-start a functional rhythm in Elijah's heart. Elijah is then rushed to the emergency room.

Can Elijah be saved? To find out, turn to the Clinical Case Wrap-Up on p. 85.

3

A BIG CORPORATION is a lot like a living organism, although it depends on its employees, rather than cells, to ensure its survival. Keeping the corporation in business may require thousands of employees, and their duties vary—no one employee can do everything. Instead, groups of employees form departments, with each department performing specialized functions.

Similarly, the functions of an organism's body are diverse, and no single cell contains the metabolic machinery and organelles to perform every function. Instead, cells combine to form **tissues**, groups of specialized cells and cell products that work together to perform specific functions. Histology is the study of tissues (as noted in Chapter 1). A solid understanding of an organ's histology is essential for understanding that organ's functional anatomy.

In your study of tissues, be aware that tissue samples usually undergo considerable manipulation before microscopic examination. For example, the photomicrographs in this chapter show tissue samples that were removed, preserved in a fixative solution, and embedded in a medium that made thin sectioning possible. The orientation of the embedded tissue and the knife blade determine the plane of section (the direction in which the tissue is cut). By changing the sectional plane, we can obtain useful information about the three-dimensional anatomy of a structure (see **Figure 1.11**, ↩ p. 18). However, the appearance of a tissue in histological preparations will change markedly depending on the plane of section, as indicated in **Figure 1.12**, ↩ p. 19. Even within a single plane of section, the internal organization of a cell or tissue will vary as the level of section changes. As you review the micrographs throughout this text, keep these limitations in mind.

There are four **primary tissue types**: epithelial tissue, connective tissue, muscle tissue, and nervous (or neural) tissue (**Figure 3.1**). This chapter discusses the characteristics of each major tissue type, focusing on the relationships between cellular organization and tissue function

3.1 | Epithelial Tissue

▶ **KEY POINT** Epithelial tissue is composed of tightly bound, avascular sheets of cells that cover exposed surfaces and line internal cavities and passageways.

Epithelia (ep-i-THĒ-lē-a; singular, *epithelium*) are sheets of cells that cover every exposed body surface and line any internal cavities and passageways. Epithelial tissue includes epithelia and glands.

The surface of the skin is a good example of epithelia covering an exposed surface, but epithelia also line the digestive, respiratory, reproductive, and urinary tracts—passageways that communicate with the outside world. Other internal epithelia line the chest cavity; the fluid-filled chambers in the brain, eye, and inner ear; and the inner surfaces of the blood vessels and the heart.

Important characteristics of epithelia include the following:

- Cellularity: Epithelia are composed almost entirely of cells bound tightly together by specialized junctions. ↩ p. 42 There is little or no intercellular space between these cells.
- Polarity: An epithelium generally* has both an exposed **apical surface** (apical aspect) that faces the exterior of the body or an internal space and an attached **basal surface** (basal aspect), where it attaches to adjacent tissues. These two surfaces differ in plasma membrane structure and function. Because of the differing functions of the apical, basal, and lateral surfaces, the organelles and other cytoplasmic structures within epithelial cells are distributed unevenly between the exposed and attached surfaces, a property known as **polarity**. Polarity occurs whether the epithelium contains a single layer or multiple layers of cells.
- Attachment: The basal surface of a typical epithelium is bound to a thin basement membrane, a complex structure produced by the epithelium and cells of the underlying connective tissue.
- Avascularity: Epithelia do not contain blood vessels and are therefore **avascular** (ā-VAS-kū-ler; *a*, without, + *vas*, vessel). Because of this, epithelial cells must obtain nutrients by diffusion or absorption across their apical or basal surfaces.
- Sheets or layers: All epithelial tissue is composed of a sheet of cells one or more layers thick.
- Regeneration: Stem cells located within the epithelium divide to continually replace surface epithelial cells that are damaged or lost.

Functions of Epithelial Tissue

▶ **KEY POINT** Epithelial tissues protect surfaces, control permeability, provide sensation, or produce secretions.

Epithelia are specialized to perform several essential functions:

- Protect surfaces: Epithelia protect exposed and internal surfaces from abrasion, dehydration, and destruction by chemical or biological agents.
- Control permeability: Any substance that enters or leaves the body has to cross an epithelium. Some epithelia are relatively impermeable, whereas others are permeable to substances as large as proteins. Many epithelia contain the molecular "machinery" needed for selective absorption or secretion. The epithelial barrier can be regulated and modified in response to various stimuli. For example, hormones can affect the transport of ions and nutrients through epithelial cells. Even physical stress can alter the structure and properties of epithelia—think of the calluses that form on your hands and feet.
- Provide sensation: Sensory nerves innervate most epithelia. Specialized epithelial cells can detect changes in the environment and convey information about such changes to the nervous system. For example, touch receptors in the deepest epithelial layers of the skin respond to pressure by stimulating adjacent sensory nerves. **Neuroepithelia** are specialized sensory epithelia found in special sense organs that provide our sensations of smell, taste, sight, balance, and hearing.
- Produce specialized secretions: **Gland cells** are epithelial cells that produce **secretions**, substances produced in and discharged from a cell. **Unicellular glands** are individual gland cells scattered among other cell types in an epithelium. In **glandular epithelia**, most or all of the epithelial cells produce secretions.

Specializations of Epithelial Cells

▶ **KEY POINT** The apical and lateral surfaces of epithelial cells are specialized in a variety of ways to perform cellular functions.

Many epithelial cells are specialized for secreting substances, moving fluids over the epithelial surface, or moving fluids through the epithelium itself. These cells usually show a definite polarity along the axis that extends from the apical surface, where the cell is exposed to an internal or external environment, to the **basolateral surfaces**, where the epithelium contacts the basement membrane and neighboring epithelial cells. This polarity means that intracellular organelles are unevenly distributed, and the apical and basolateral plasma membranes differ in terms of their associated proteins and functions. The actual arrangement of organelles varies depending on cell function (**Figure 3.2**).

* In special situations, epithelial cells may lack a free surface. These **epithelioid cells** are found in most endocrine glands.

Figure 3.1 An Orientation to the Tissues of the Body. An overview of the levels of organization in the body and an introduction to some of the functions of the four tissue types.

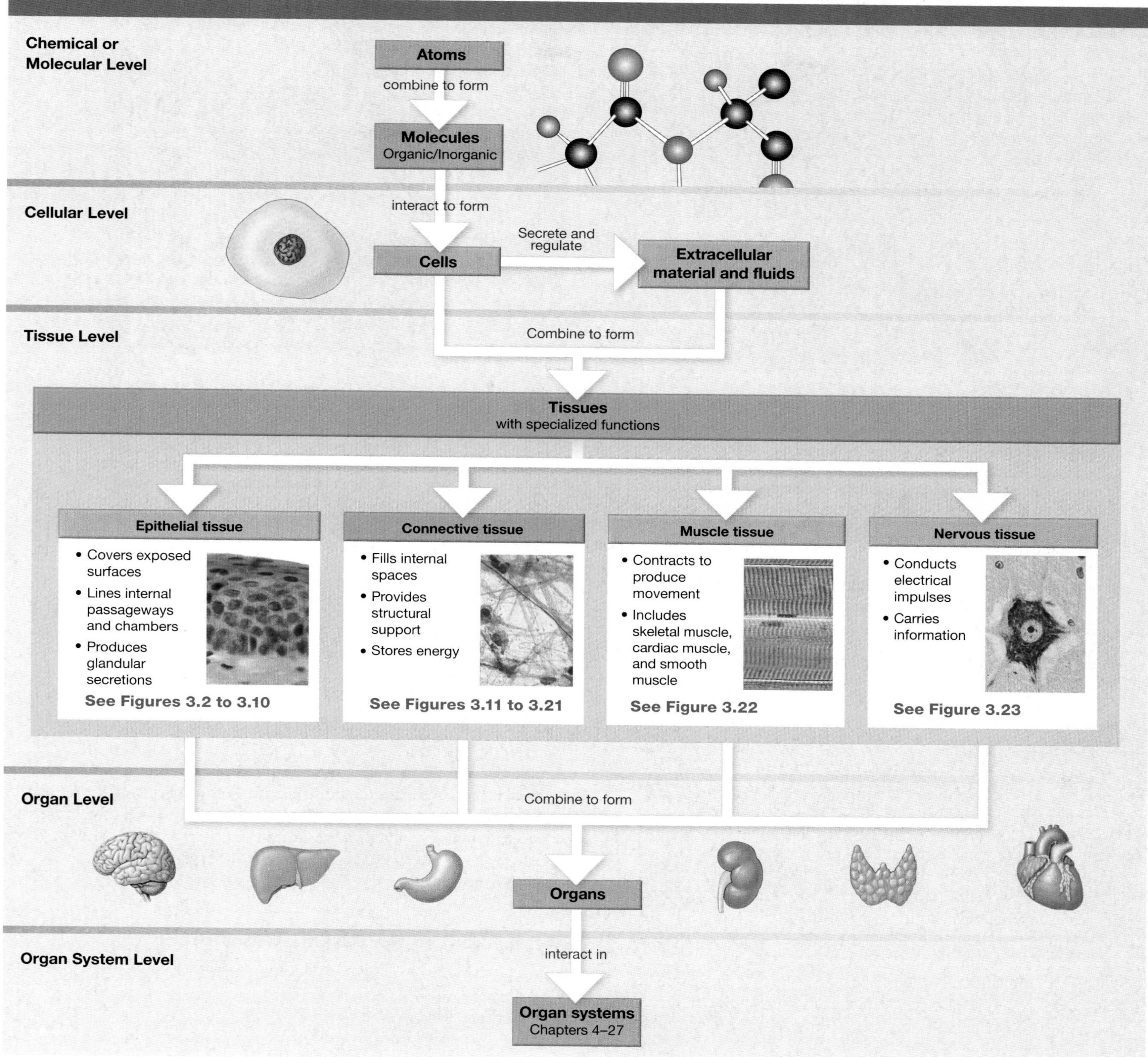

Figure 3.2 **Polarity of Epithelial Cells**

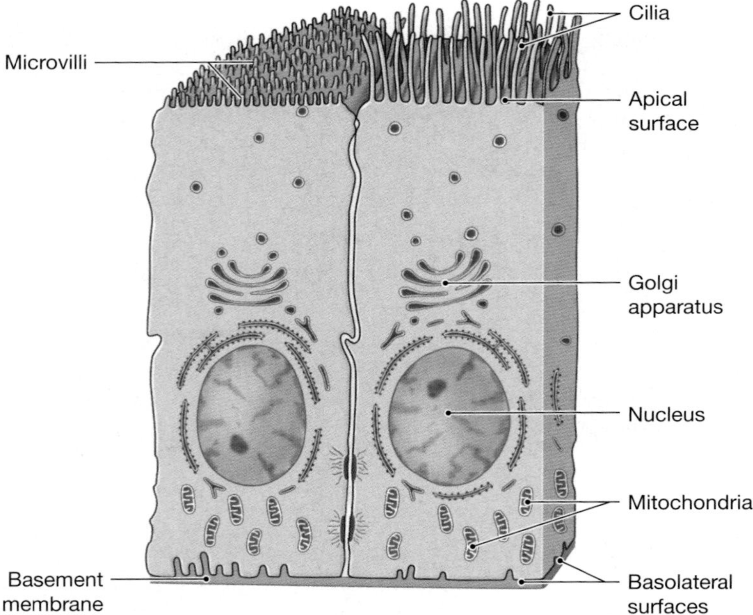

a Many epithelial cells differ in internal organization along an axis between the apical surface and the basement membrane. The apical surface often has microvilli; less often, it may have cilia or (very rarely) stereocilia. A single cell typically has only one type of process; cilia and microvilli are shown together to highlight their relative proportions. Tight junctions prevent movement of pathogens or diffusion of dissolved materials between the cells. Folds of plasma membrane near the base of the cell increase the surface area exposed to the basement membrane. Mitochondria are typically concentrated at the basolateral region, probably to provide energy for the cell's transport activities.

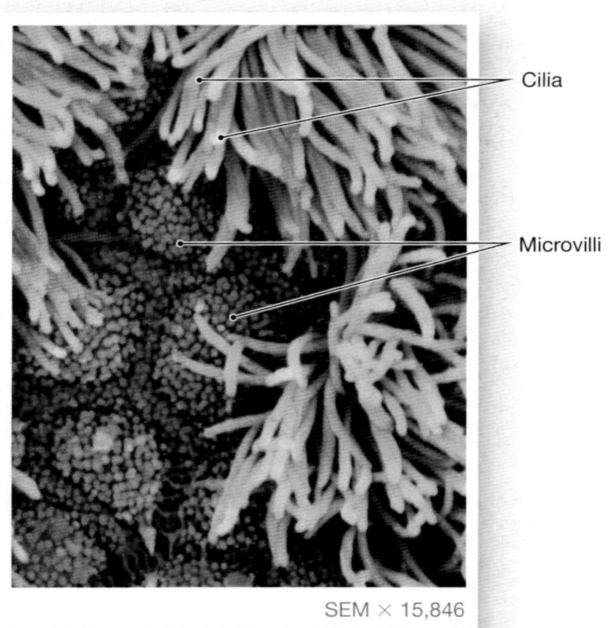

b An SEM showing the surface of the epithelium that lines most of the respiratory tract. The small, bristly areas are microvilli found on the exposed surfaces of mucus-producing cells that are scattered among the ciliated epithelial cells.

Most epithelial cells have microvilli, small, finger-shaped projections, on their exposed apical surfaces. Microvilli are especially abundant on epithelial surfaces where absorption and secretion occur, such as in the digestive and urinary tracts. ⮌ **p. 31** A cell with microvilli has at least 20 times more surface area for absorption and secretion than one without microvilli. **Stereocilia** (not shown in the figure) are very long microvilli (up to 200 μm); unlike cilia, they cannot move. Stereocilia are found along portions of the male reproductive tract and on receptor cells of the inner ear.

Recall that cilia are microtubule structures that move together to propel substances over an epithelial surface. ⮌ **p. 36** A **ciliated epithelium** has cilia on its apical surface that usually beat together **(Figure 3.2)**. For example, the ciliated epithelium lining the respiratory tract beat in a coordinated fashion to move mucus from the lungs toward the throat. The mucus traps particles and pathogens (disease-causing organisms) and carries them away from the lungs.

Maintaining the Integrity of the Epithelium

▶ **KEY POINT** Three factors maintain the physical integrity of an epithelium: intercellular connections, attachment to the basement membrane, and epithelial maintenance and renewal.

Three factors interact to keep epithelial tissues intact and healthy: (1) intercellular connections, (2) attachment to the basement membrane, and (3) epithelial maintenance and renewal.

Intercellular Connections

A variety of cell junctions bind epithelial cells together. ⮌ **p. 42** Extensive infolding of adjacent cell membranes interlocks the cells and increases the surface area of the cell junctions **(Figure 3.3a,c)**. These connections may prevent chemicals or pathogens from entering the cells. The combination of cell junctions and physical interlocking, along with CAMs (cell adhesion molecules) and intercellular cement, gives the epithelium strength and stability **(Figure 3.3b)**.

Attachment to the Basement Membrane

The basal surface of a typical epithelium attaches to the **basement membrane**, a complex structure produced by the epithelium and cells of the underlying connective tissue.

The superficial portion of the basement membrane consists of the **basal lamina**, which is secreted by epithelial cells. This is a region dominated by glycoproteins, proteoglycans, and a network of fine microfilaments. The basal lamina has numerous functions, one of which is restricting the movement of proteins and other large molecules from the underlying connective tissue into the epithelium. Deep to the basal lamina is the second layer of the basement membrane, the **reticular lamina** (*lamina reticularis*), which is secreted by the underlying connective tissue. The reticular lamina contains bundles of coarse protein fibers that anchor the basement membrane to the underlying connective tissue.

Figure 3.3 **Epithelia and Basement Membrane.** The integrity of an epithelium depends on connections between adjacent epithelial cells and their attachment to the underlying basement membrane.

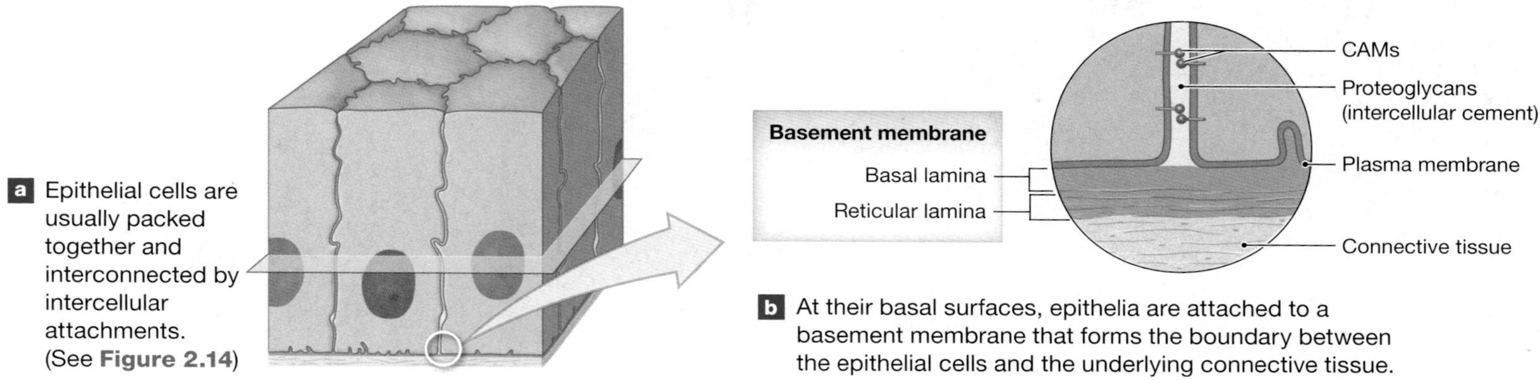

a Epithelial cells are usually packed together and interconnected by intercellular attachments. (See Figure 2.14)
Basement membrane
Basal lamina
Reticular lamina
CAMs
Proteoglycans (intercellular cement)
Plasma membrane
Connective tissue
b At their basal surfaces, epithelia are attached to a basement membrane that forms the boundary between the epithelial cells and the underlying connective tissue.

Epithelial Maintenance and Renewal

Exposure to enzymes, toxic chemicals, pathogens, and abrasion damages epithelial cells, so epithelial tissues must continually repair and renew themselves. The faster epithelial cells die, the faster they have to be replaced. Under severe conditions, such as those inside the small intestine, an epithelial cell may survive for only a day or two. The only way the epithelium can maintain itself is through continual division of **stem cells**, which are usually found close to the basal lamina.

Classification of Epithelia

▶ **KEY POINT** Epithelial tissue is classified according to the number of cell layers and the shape of the cells at the exposed surface.

We classify epithelial tissue based on its layers and the shape of the epithelial cells.

- There are two types of layering: simple and stratified.
- There are three cell shapes: squamous, cuboidal, and columnar.

A **simple epithelium** has only one layer of cells covering its basement membrane. All the cells in a simple epithelium have the same polarity, so the nuclei form a row at the same distance from the basement membrane. A single layer of cells cannot provide much protection, so simple epithelia are thin and fragile. They are found only in protected areas inside the body, such as in the thoracic, abdominal, and pelvic cavities, the chambers of the heart, and all blood vessels.

Simple epithelia are also found where secretion, absorption, or filtration occurs, such as the lining of the intestines and the gas exchange surfaces of the lungs. In these locations, their thin single layer provides an advantage by decreasing the diffusion distance and the time needed for materials to pass through the epithelial barrier.

A **stratified epithelium** has two or more layers of cells. The height and shape of the cells in stratified epithelium differ from layer to layer, but we use only the shape of the most superficial cells in epithelium classification. Their multiple layers of cells make stratified epithelia thicker and stronger than simple epithelia; correspondingly, they occur in areas where mechanical or chemical stresses are severe, such as the surface of the skin and the lining of the oral cavity.

Whether an epithelium is simple or stratified, it must regenerate and replace its cells over time via the division of stem cells, which are located at or near the basement membrane. In a simple epithelium these stem cells are a part of the exposed epithelial surface, while in a stratified epithelium the stem cells are found within the deeper layers of the epithelium.

By combining the number of layers (simple or stratified) with the three possible cell shapes (squamous, cuboidal, or columnar), we can describe almost every epithelium in the body.

Squamous Epithelia

In a **squamous epithelium** (SKWĀ-mus; *squama*, plate or scale), the cells are thin, flat, and irregular in shape, like puzzle pieces (**Figure 3.4**). In a side view the nucleus occupies the thickest portion of each cell and looks flattened, like the cell as a whole. From the surface, the cells look like fried eggs laid side by side.

A **simple squamous epithelium** is the most delicate epithelium in the body (**Figure 3.4a**). It is found in protected regions where diffusion or other forms of transport take place or where a slick, slippery surface reduces friction. Examples include the respiratory exchange surfaces (alveoli) of the lungs, the serous membranes lining the thoracic, abdominal, and pelvic cavities, and the inner surfaces of the circulatory system.

Some simple squamous epithelia line chambers and passageways that do not communicate with the outside world. The simple squamous epithelium that lines body cavities is termed **mesothelium** (mez-ō-THĒ-lē-um; *mesos*, middle). The pleura lining the thoracic cavity and the peritoneum lining the abdominal cavity both contain a superficial mesothelium. The simple squamous epithelium lining the heart and blood vessels is termed **endothelium** (en-dō-THĒ-lē-um).

In a **stratified squamous epithelium**, the cells form a series of layers (**Figure 3.4b**); correspondingly, it occurs where mechanical stresses are severe. For instance, stratified squamous epithelium protects the surface of the skin and the lining of the oral cavity, throat, esophagus, rectum, vagina, and anus.

Stratified squamous epithelium may be keratinized or nonkeratinized. On surfaces where mechanical stress and dehydration are potential problems, the apical layers of epithelial cells are packed with filaments of the protein **keratin**. The superficial layers of this **keratinized** stratified squamous epithelium are tough and water resistant. Keratinized stratified squamous epithelia are found in the hair shafts and palmar skin.

A **nonkeratinized**, or *mucosal*, stratified squamous epithelium also resists abrasion but must be kept moist, or it will dry out and deteriorate. Nonkeratinized stratified squamous epithelia occur in the oral cavity, pharynx, esophagus, rectum, anus, and vagina.

Figure 3.4 **Histology of Squamous Epithelia**

Simple Squamous Epithelium

LOCATIONS: Mesothelia lining pleural, pericardial, and peritoneal body cavities; endothelia lining heart and blood vessels; portions of kidney tubules (thin sections of nephron loops); inner lining of cornea; alveoli of lungs

FUNCTIONS: Reduces friction; controls vessel permeability; performs absorption and secretion

Connective tissue

Lining of peritoneal cavity

Cytoplasm

Nucleus

LM × 238

a A superficial view of the simple squamous epithelium (mesothelium) lining the peritoneal cavity

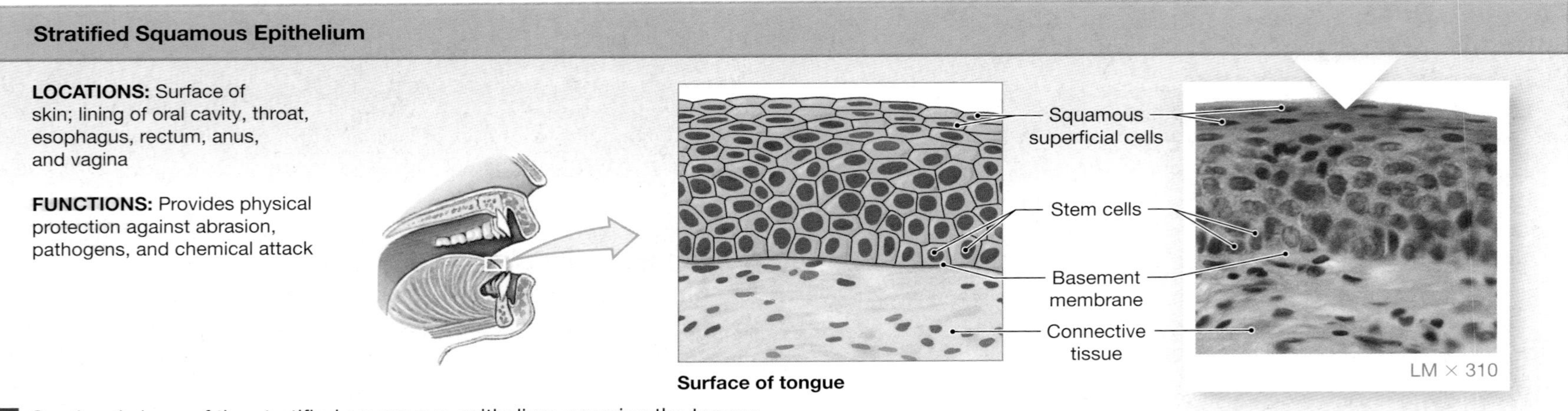
Stratified Squamous Epithelium
LOCATIONS: Surface of skin; lining of oral cavity, throat, esophagus, rectum, anus, and vagina
FUNCTIONS: Provides physical protection against abrasion, pathogens, and chemical attack
Squamous superficial cells
Stem cells
Basement membrane
Connective tissue
Surface of tongue
LM × 310

b Sectional views of the stratified squamous epithelium covering the tongue

Cuboidal Epithelia

The cells of a **cuboidal epithelium** look like little hexagonal boxes and appear square in typical sectional views. Each nucleus is centrally located.

Simple cuboidal epithelia provide limited protection and are found in regions where secretion or absorption takes place. The kidney tubules are lined with simple cuboidal epithelium **(Figure 3.5a)**.

Stratified cuboidal epithelia are rare; they are found lining the ducts of sweat glands and mammary glands **(Figure 3.5b)**.

Columnar Epithelia

The cells of a **columnar epithelium** are also hexagonal in cross section. However, unlike cuboidal cells, their height is much greater than their width **(Figure 3.6)**. The nuclei are typically located within the basal portion of the cell.

A **simple columnar epithelium** is found in areas where absorption or secretion occurs, such as the lining of the stomach, intestinal tract, uterine tubes, and many excretory ducts **(Figure 3.6a)**. Columnar epithelia provide slightly more protection than simple cuboidal epithelia.

Stratified columnar epithelia are rare. They occur in the pharynx, urethra, and anus and in a few large excretory ducts. The epithelium may have two or more layers **(Figure 3.6b)**. If it has more than two layers, only the superficial cells are columnar in shape.

Pseudostratified and Transitional Epithelia

Two specialized epithelia line the respiratory system and the hollow conducting organs of the urinary system.

Portions of the respiratory tract contain **pseudostratified ciliated columnar epithelium** (*pseudo*, fake), a specialized columnar epithelium that includes a mixture of cell types **(Figure 3.7a)**. Because the cells' nuclei are located at varying distances from the surface, the epithelium looks stratified. However, all cells rest on the basement membrane, so it is actually a simple epithelium. The surface epithelial cells possess cilia. Pseudostratified ciliated columnar epithelium lines most of the nasal cavity, trachea, and bronchi and also portions of the male reproductive tract.

Transitional epithelium lines the renal pelvis, ureters, and urinary bladder **(Figure 3.7b)**. Transitional epithelium is a stratified epithelium that can stretch without damaging the epithelial cells. In an empty urinary bladder, transitional epithelium seems to have many layers, and its outermost cells are rounded or balloon-shaped cuboidal cells. As the bladder fills and stretches it, transitional epithelium resembles a stratified, nonkeratinized epithelium with two or three layers.

Glandular Epithelia

▶ **KEY POINT** We classify glandular epithelia based on the glands' secretions, structure, and mechanism of secretion.

Figure 3.5 Histology of Cuboidal Epithelia

Simple Cuboidal Epithelium

LOCATIONS: Glands; ducts; portions of kidney tubules; thyroid gland

FUNCTIONS: Limited protection; secretion; absorption

Kidney tubule

Connective tissue

Nucleus

Cuboidal cells

Basement membrane

LM × 1400

a A section through the simple cuboidal epithelium lining a kidney tubule. The diagrammatic view emphasizes structural details that classify the epithelium as cuboidal.

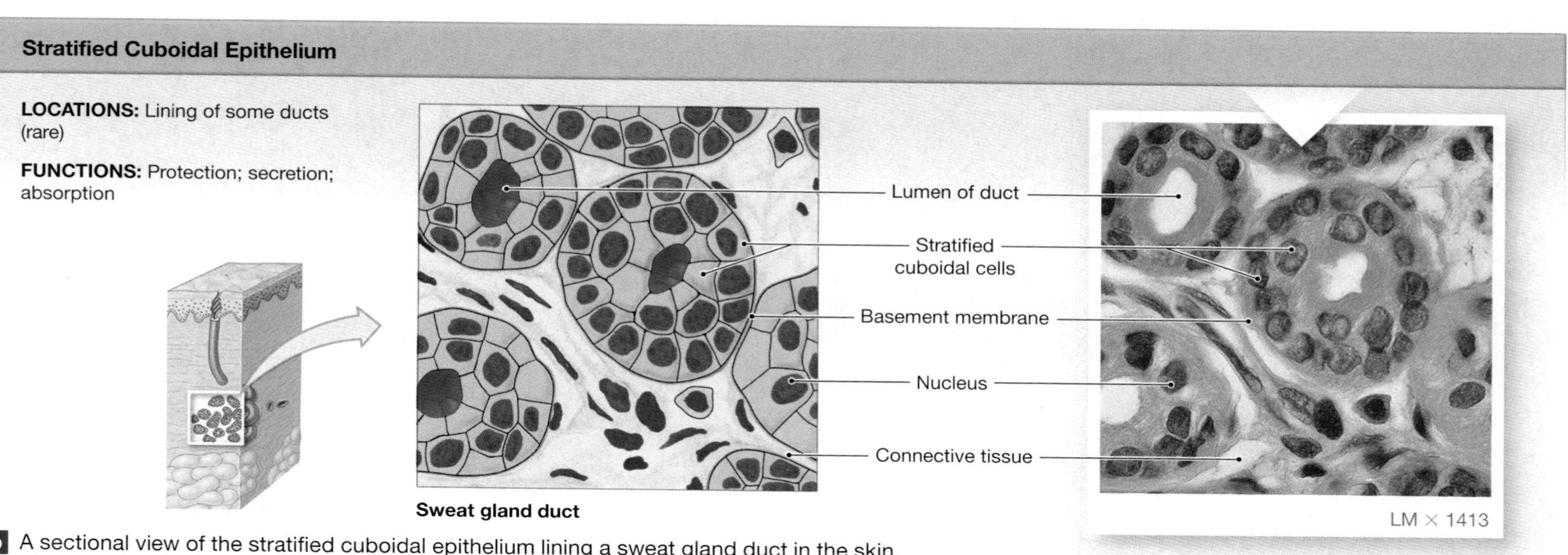

b A sectional view of the stratified cuboidal epithelium lining a sweat gland duct in the skin.

TIPS & TOOLS

Identifying Stratified Squamous and Transitional Epithelia

Stratified squamous epithelium (nonkeratinized type) and transitional epithelium look quite similar. Keep the following differences in mind.

In stratified squamous epithelium **(Figure 3.4b)**:

- There are many layers.
- The basal cells may look cuboidal or columnar.
- The surface layer is *always* composed of flattened cells.
- The number and extent of epithelial layers, and the shape of the most superficial cells, are relatively constant.

In transitional epithelium **(Figure 3.7b)**:

- There are fewer layers than in stratified squamous epithelium.
- The most superficial cells may be balloon- or dome-shaped (in a relaxed organ) or flattened (in a stretched organ).
- The number and extent of epithelial layers, and the shape of the most superficial layer, vary widely within the section.

Many epithelia contain gland cells that produce secretions. We classify these glandular epithelia based on the (1) type of secretion they release, (2) structure of the gland, and (3) method of secretion. Glandular epithelia may be exocrine or endocrine.

Type of Secretions

Exocrine glands (*exo–*, outside) release their secretions onto an epithelial surface through epithelial **ducts**. These ducts may release the secretion unaltered, or they may alter it by a variety of mechanisms, including reabsorption, secretion, or countertransport. Examples of exocrine secretions include enzymes entering the digestive tract, perspiration on the skin, and milk produced by mammary glands.

There are three types of exocrine glands, based on the secretions they produce:

- **Serous glands** secrete a watery solution that usually contains enzymes, such as the salivary amylase in saliva.
- **Mucous glands** secrete glycoproteins called **mucins** (MŪ-sins) that absorb water to form a slippery *mucus*, such as the mucus in saliva.
- **Mixed exocrine glands** contain more than one type of gland cell and may produce both serous and mucous secretions.

Figure 3.6 Histology of Columnar Epithelia

Simple Columnar Epithelium

LOCATIONS: Lining of stomach, intestine, gallbladder, uterine tubes, and collecting ducts of kidneys

FUNCTIONS: Protection; secretion; absorption

Microvilli

Cytoplasm

Nucleus

Basement membrane

Loose connective tissue

Intestinal lining

LM × 350

a A sectional view of the simple columnar epithelium in the intestinal lining. In the diagrammatic sketch, note the relationship between the height and width of each cell; the relative size, shape, and location of nuclei; and the distance between adjacent nuclei. Contrast these observations with the characteristics of simple cuboidal epithelia.

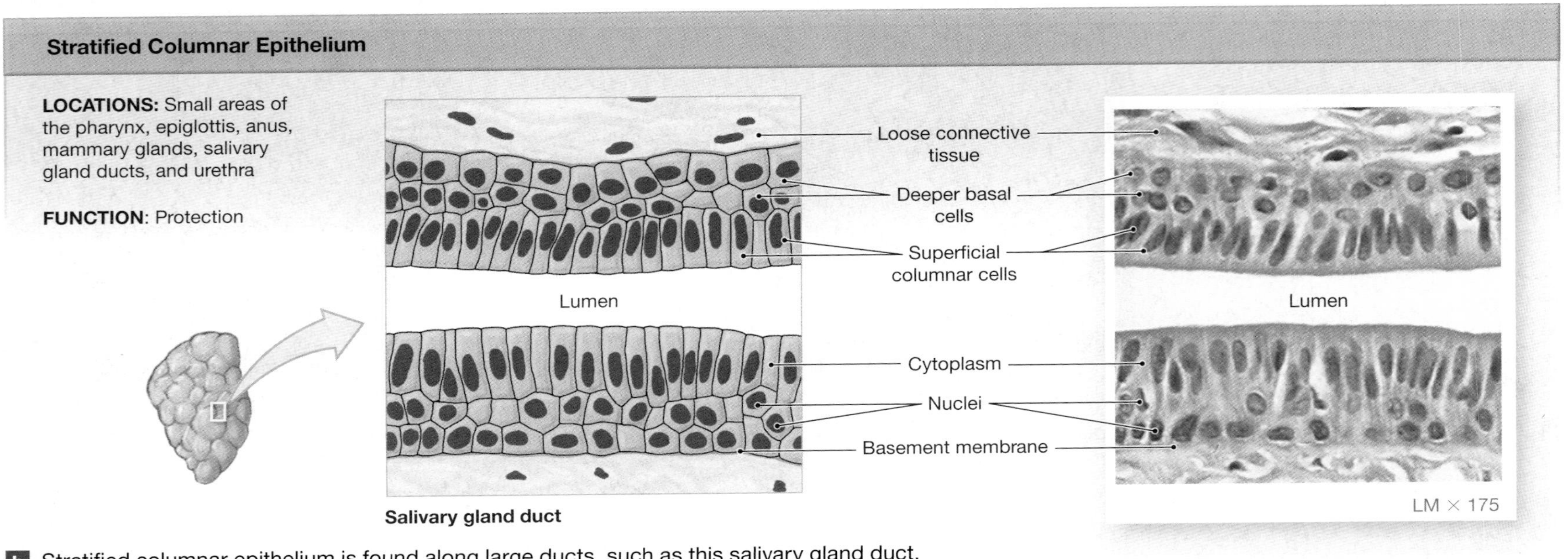

b Stratified columnar epithelium is found along large ducts, such as this salivary gland duct. Note the overall height of the epithelium and the location and orientation of the nuclei.

Endocrine glands (*endo-*, inside) are ductless glands that release their secretions by exocytosis directly into the interstitial fluid surrounding the cell. These secretions, called **hormones**, diffuse into the blood for distribution to other regions of the body. Hormones regulate or coordinate the activities of other tissues, organs, and organ systems. (We will discuss endocrine cells and hormones in Chapter 19.)

Structure of the Gland

In epithelia containing scattered gland cells, the individual secretory cells are called **unicellular glands**. Unicellular exocrine glands secrete mucins. There are two types of unicellular glands, **mucous cells** and **goblet cells**. For example, the epithelium of some salivary glands contains mucous cells. The columnar epithelium of the small and large intestines and the pseudostratified ciliated epithelium that lines the trachea contain goblet cells.

Multicellular glands include glandular epithelia and clusters of gland cells that produce exocrine or endocrine secretions. The simplest multicellular exocrine gland is a **secretory sheet**, in which glandular cells dominate the epithelium and release their secretions into an inner compartment **(Figure 3.8a)**. For instance, the mucus-secreting cells that line the stomach form a secretory sheet that protects the stomach from acids and enzymes.

Other multicellular glands occur in pockets set back from the epithelial surface. For example, the submandibular salivary gland is a multicellular exocrine gland that produces mucus and digestive enzymes **(Figure 3.8b)**. These glands have two epithelial components: a glandular portion that produces the secretion and a duct that carries the secretion to the epithelial surface.

Figure 3.7 **Histology of Pseudostratified Ciliated Columnar and Transitional Epithelia**

Pseudostratified Ciliated Columnar Epithelium

LOCATIONS: Lining of nasal cavity, trachea, and bronchi; portions of male reproductive tract

FUNCTIONS: Protection; secretion

Trachea
Cilia
Cytoplasm
Nuclei
Basement membrane
Loose connective tissue
LM × 350

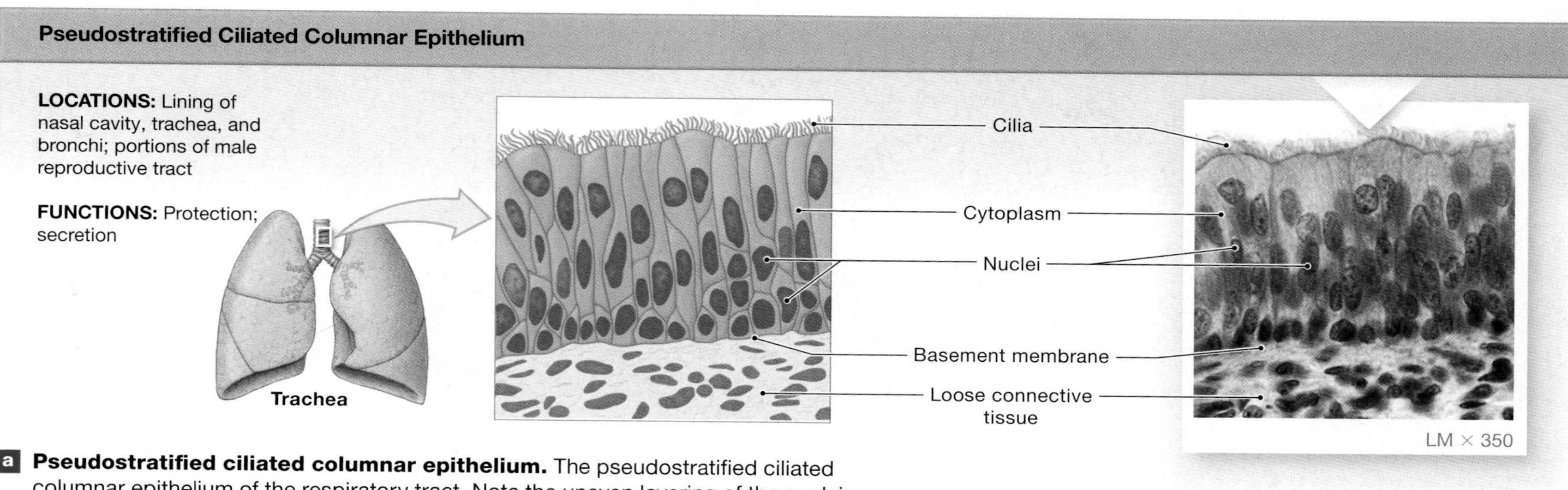

a **Pseudostratified ciliated columnar epithelium.** The pseudostratified ciliated columnar epithelium of the respiratory tract. Note the uneven layering of the nuclei.

Transitional Epithelium

LOCATIONS: Urinary bladder; renal pelvis; ureters

FUNCTIONS: Permits expansion and recoil after stretching

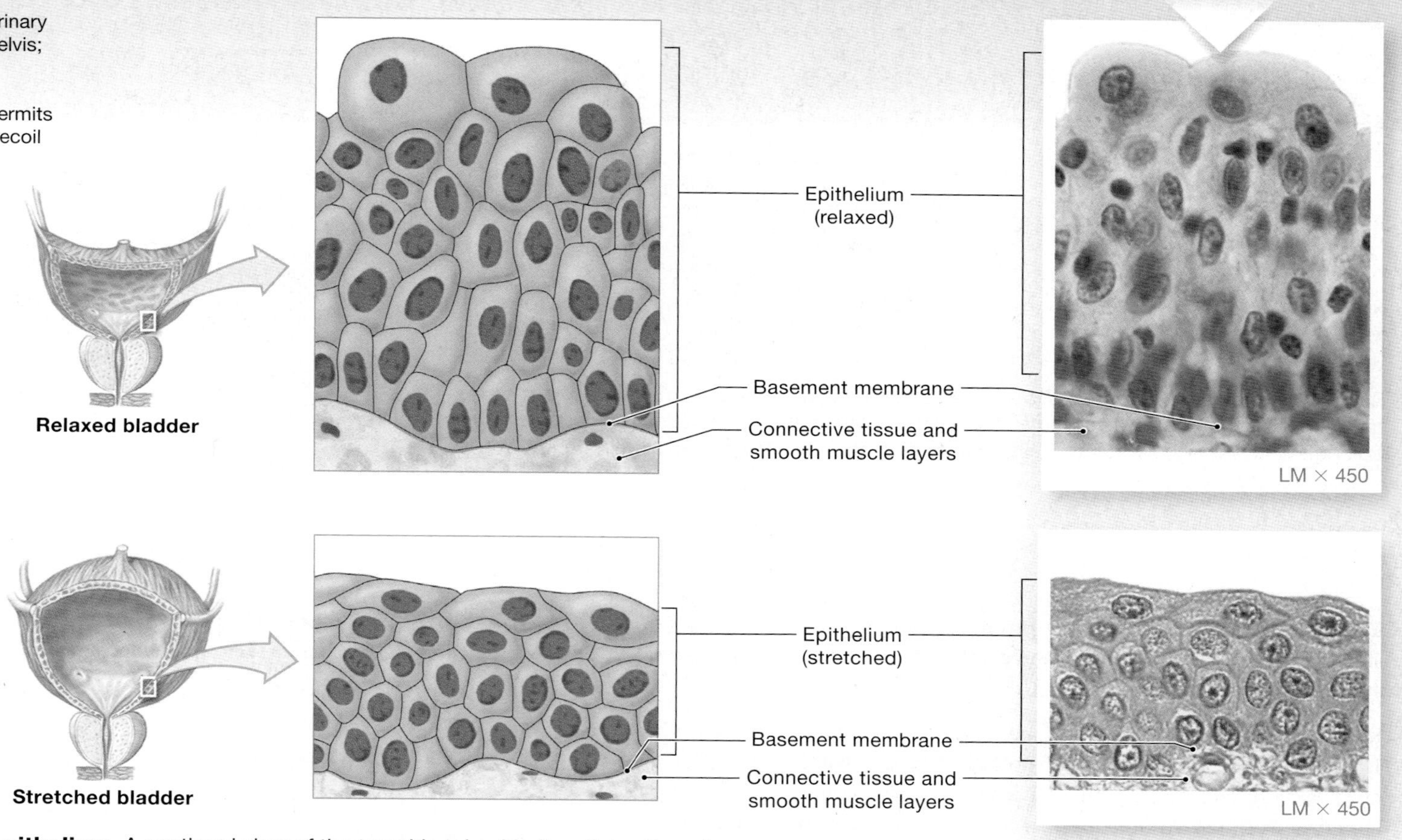

Relaxed bladder
Epithelium (relaxed)
Basement membrane
Connective tissue and smooth muscle layers
LM × 450
Stretched bladder
Epithelium (stretched)
Basement membrane
Connective tissue and smooth muscle layers
LM × 450

b **Transitional epithelium.** A sectional view of the transitional epithelium lining the urinary bladder. The cells from an empty bladder are in the relaxed state, while those lining a full urinary bladder show the effects of stretching on the arrangement of cells in the epithelium.

Figure 3.8 **Histology of Mucous and Mixed Glandular Epithelia**

Secretory Sheet

Columnar mucous epithelium

LM × 250

a The interior of the stomach is lined by a secretory sheet whose secretions protect the walls from acids and enzymes. (The acids and enzymes are produced by glands that discharge their secretions onto the mucous epithelial surface.)

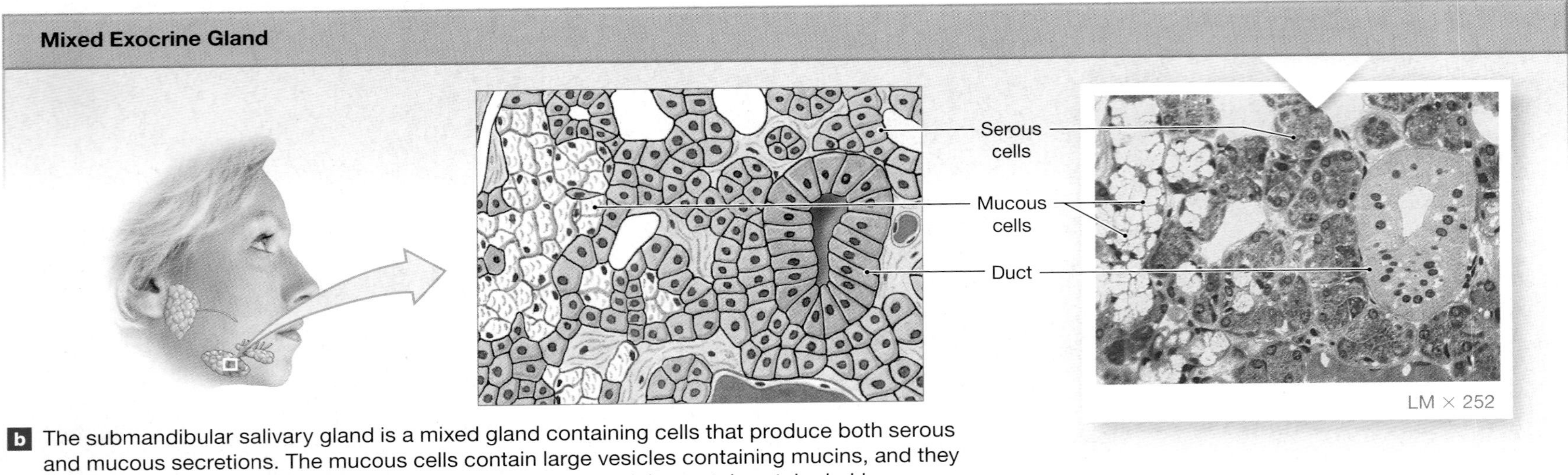

b The submandibular salivary gland is a mixed gland containing cells that produce both serous and mucous secretions. The mucous cells contain large vesicles containing mucins, and they look pale and foamy. The serous cells secrete enzymes, and the proteins stain darkly.

Two characteristics describe the organization of a multicellular gland: (1) the shape of the secretory portion of the gland and (2) the branching pattern of the duct **(Figure 3.9)**.

- In a **tubular gland**, the cells making up the gland are arranged in a tube. If the gland's cells form a blind pocket, it is an **alveolar** (al-VĒ-ō-lar; *alveolus*, sac), or **acinar** (AS-i-nar; *acinus*, chamber), gland. A gland that combines both tubular and alveolar arrangements is a **tubuloalveolar** or **tubuloacinar gland**.
- The duct of a **simple** exocrine gland does not branch. The duct of a **compound** exocrine gland branches repeatedly. Each glandular area may have its own duct; in the case of branched glands, several glands share a common duct.

Method of Secretion

A glandular epithelial cell may use one of three methods to release its secretions: **eccrine secretion** (also termed *merocrine secretion*), **apocrine secretion** (AP-ō-krin; *apo-*, off), or **holocrine secretion** (HOL-ō-krin; *holos*, entire). **Spotlight Figure 3.10** explains these three mechanisms.

3.1 CONCEPT CHECK

1. You look at a tissue under a microscope and see a simple squamous epithelium. Can it be a sample of the skin surface? Why or why not?
2. Why is epithelium regeneration necessary in a gland that releases its product by holocrine secretion?
3. Ceruminous glands of the external acoustic meatus of the ear release their products by apocrine secretion. What occurs in this mode of secretion?
4. What functions are associated with a simple columnar epithelium?

See the blue Answers tab at the back of the book.

Figure 3.9 **A Structural Classification of Simple and Compound Exocrine Glands**

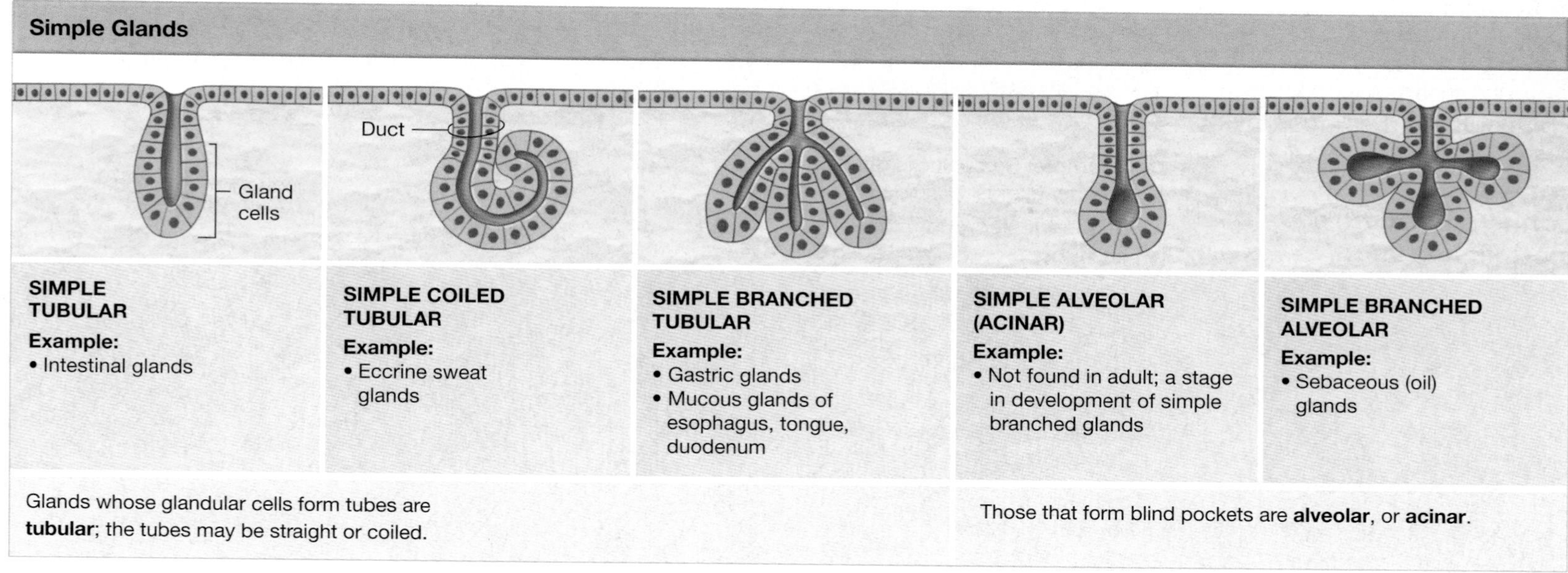

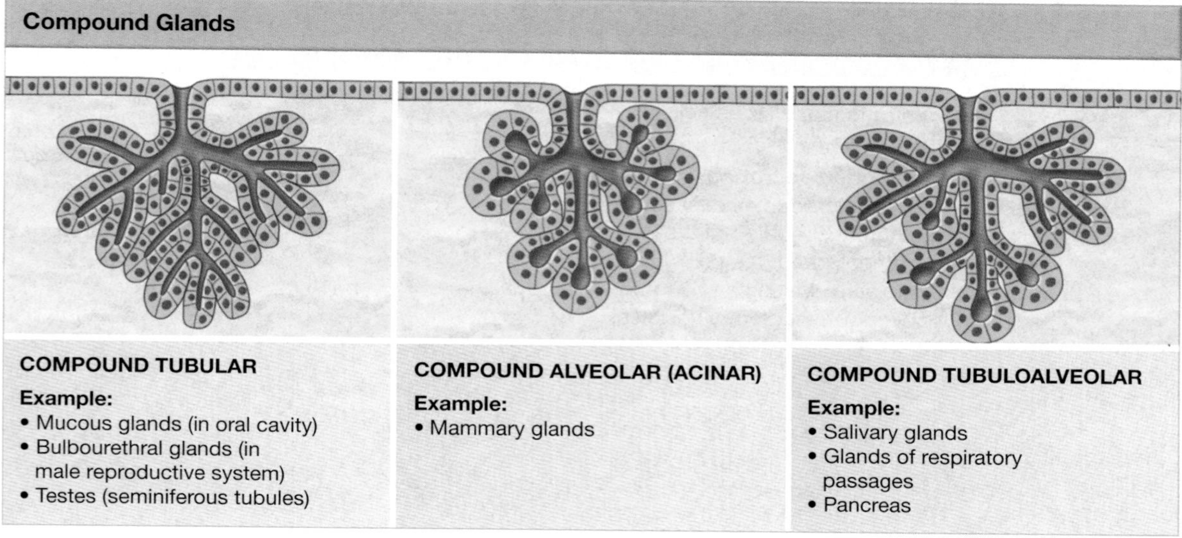

3.2 | Connective Tissues

▶ **KEY POINT** All connective tissues have three basic components—specialized cells, extracellular protein fibers, and ground substance—but perform a wide variety of functions.

Connective tissues are found throughout the body. Unlike epithelial tissues, connective tissues are never exposed to the environment outside the body. All connective tissues have three basic components: (1) specialized cells, (2) extracellular protein fibers, and (3) a fluid known as the **ground substance**. The extracellular fibers and ground substance form the **matrix** that surrounds the cells. *Although epithelial tissue consists almost entirely of cells, connective tissue consists mostly of extracellular matrix.*

Connective tissues do far more than just connect body parts. Connective tissues:

- Establish a structural framework for the body
- Transport fluids and dissolved materials from one region of the body to another
- Protect delicate organs
- Support, surround, and interconnect other tissue types
- Store energy, especially in the form of lipids
- Defend the body from invasion by microorganisms

Although connective tissues have multiple functions, no single category of connective tissue performs all of these functions.

Classification of Connective Tissues

▶ **KEY POINT** There are three categories of connective tissue: connective tissue proper, fluid connective tissue, and supporting connective tissue.

Figure 3.11 introduces the three main categories of connective tissue:

1. Connective tissue proper is composed of many types of cells and extracellular fibers in a syrupy ground substance. These connective tissues differ in the number of cell types they contain and the properties and proportions of fibers and ground substance. Adipose (fat) tissue, ligaments, and tendons differ greatly, but all three are connective tissue proper.

SPOTLIGHT

FIGURE 3.10

Mechanisms of Glandular Secretion

A glandular epithelial cell may use one of three methods to release its secretions: **eccrine secretion**, **apocrine secretion**, or **holocrine secretion**.

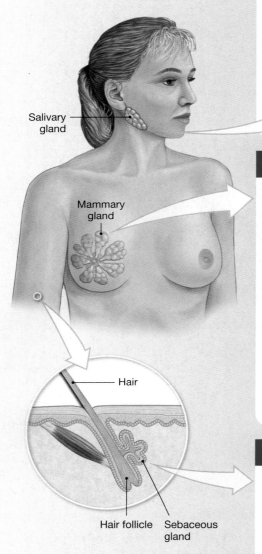

Salivary gland

Mammary gland

Hair

Hair follicle

Sebaceous gland

Eccrine secretion

Example: Serous cells of the salivary glands

In **eccrine secretion**, the secretory product, packaged into secretory vesicles, is released through exocytosis onto the surface of the cell. This is the most common mode of secretion. An example of this type of secretion is the release of saliva from serous cells in the salivary gland, or mucins from goblet cells in the intestine.

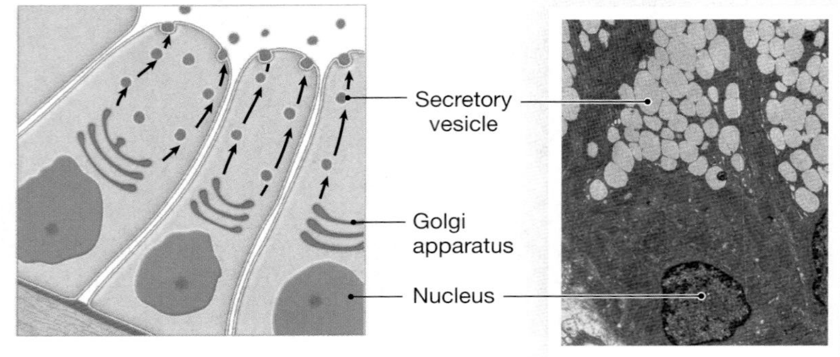

Apocrine secretion

Example: Lactiferous cells of the mammary glands

In **apocrine secretion**, the secretory product is released during the shedding of the apical portion of the cell's cytoplasm, which has become packed with secretory vesicles. The gland cells then undergo regrowth and produce additional secretory vesicles.

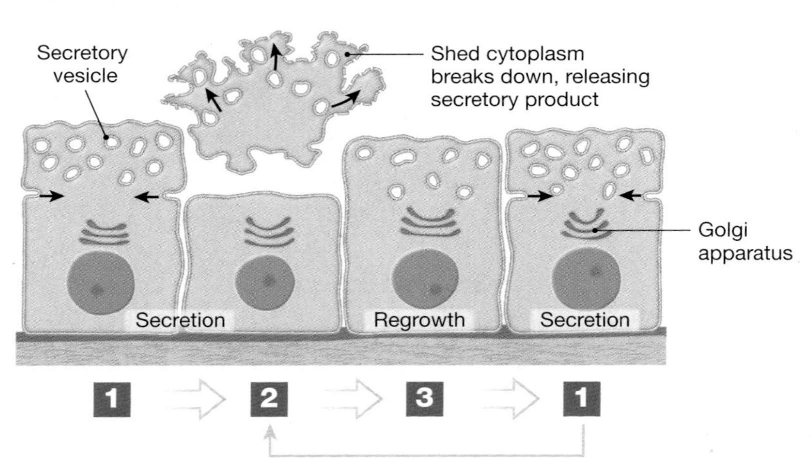

Holocrine secretion

Example: Sebaceous gland cells

Holocrine secretion destroys the gland cell. During holocrine secretion, the entire cell becomes packed with secretory products and then bursts apart. The secretion is released and the cell dies. Further secretion depends on gland cells being replaced by the division of stem cells found deeper within the epithelium.

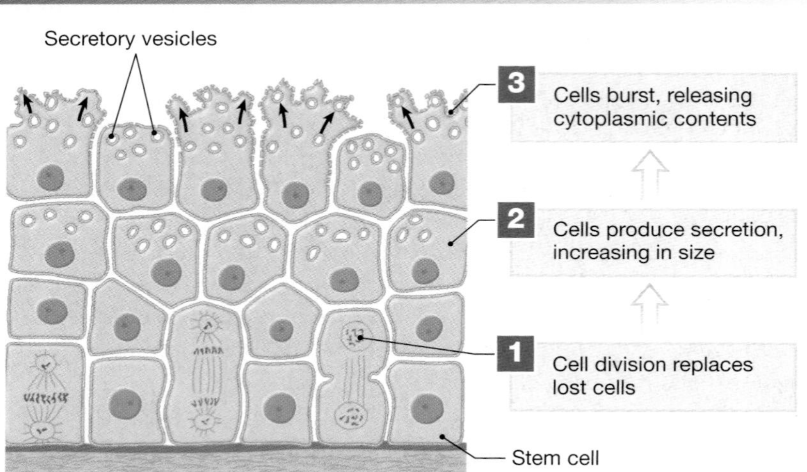

Figure 3.11 **Classification of Connective Tissues**

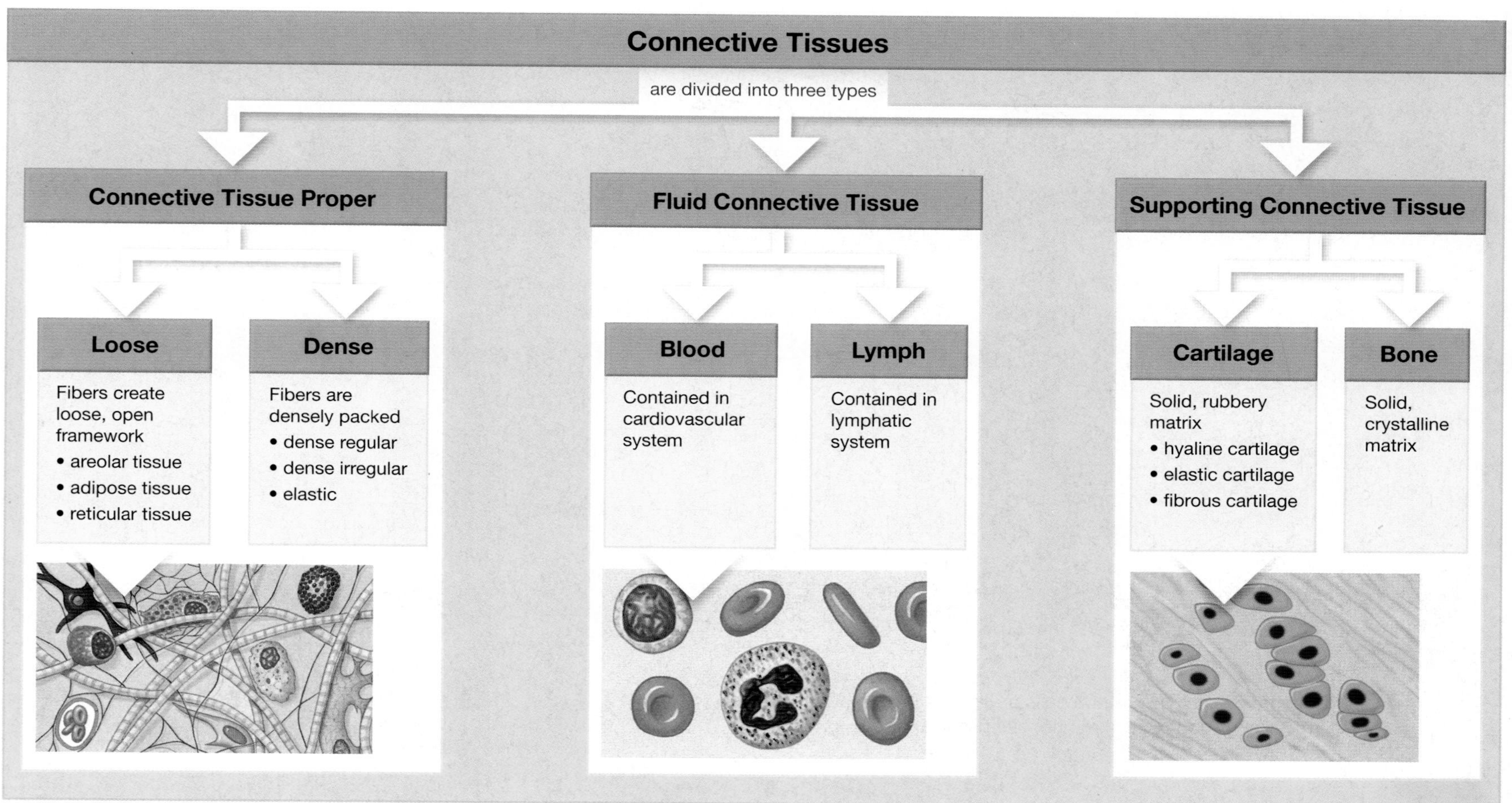

2 Fluid connective tissues have a distinctive population of cells suspended in a watery matrix containing dissolved proteins. There are two types of fluid connective tissues: blood and lymph.

3 Supporting connective tissues have a smaller cell population than connective tissue proper and a matrix of closely packed fibers. There are two types of supporting connective tissues: cartilage and bone. The matrix of cartilage is a gel whose characteristics vary depending on the dominant fiber type. The matrix of bone is **calcified** because it contains mineral deposits, primarily calcium salts, that give the bone strength and rigidity.

Connective Tissue Proper

▶ **KEY POINT** There are two types of connective tissue proper—loose connective tissues and dense connective tissues, based on their relative proportions of cells, fibers, and ground substance.

Connective tissue proper contains extracellular fibers and a viscous (syrupy) ground substance. It has two classes of cells, fixed cells and wandering cells. The number of cells at any given moment varies depending on local conditions. Refer to **Figure 3.12** and **Table 3.1** as we describe connective tissue proper.

Cells of Connective Tissue Proper

Fixed Cells **Fixed cells** are stationary and are involved with local maintenance, repair, and energy storage. The fixed cells of connective tissue proper include mesenchymal cells, fibroblasts, fibrocytes, fixed macrophages, adipocytes, and, in a few locations, melanocytes.

- **Mesenchymal cells** (or *mesenchymal stem cells*) (MES-en-kī-mul) are present in many connective tissues. These cells respond to local injury or infection by dividing to produce daughter cells that differentiate into fibroblasts, macrophages, or other connective tissue cells.
- **Fibroblasts** (FĪ-brō-blasts) are one of the two most abundant fixed cells in connective tissue proper and are the only cells always present. These slender, star-shaped cells produce all connective tissue fibers. Each fibroblast manufactures and secretes protein subunits that interact to form large extracellular fibers. In addition, fibroblasts secrete hyaluronan, which makes the ground substance viscous.
- **Fibrocytes** (FĪ-brō-sīts) (or *activated fibroblasts*) differentiate from fibroblasts and are the second most abundant fixed cell in connective tissue proper. These star-shaped cells maintain the connective tissue fibers of connective tissue proper. Their cytoplasm stains poorly, so only the nucleus is visible in a standard histological preparation. If connective tissue is injured, fibrocytes have the ability to differentiate back into fibroblasts that help repair the damaged tissue.
- **Fixed macrophages** (MAK-rō-fā-jez; *phagein*, to eat) (also termed *resident* or *resting macrophages*) are large, amoeboid cells scattered among the connective tissue fibers. They engulf damaged cells, dead cells, and pathogens that enter the tissue. Although not abundant, they play an important role in mobilizing the body's defenses. When stimulated, they release chemicals that attract wandering cells involved in the body's defense mechanisms.
- **Adipocytes** (AD-i-pō-sīts) (*fat cells* or *adipose cells*) contain a single lipid droplet that occupies almost the entire cell, squeezing the nucleus and other organelles to one side. The number of adipocytes varies according to the type of connective tissue, the region of the body, and the individual.

Figure 3.12 Histology of the Cells and Fibers of Connective Tissue Proper

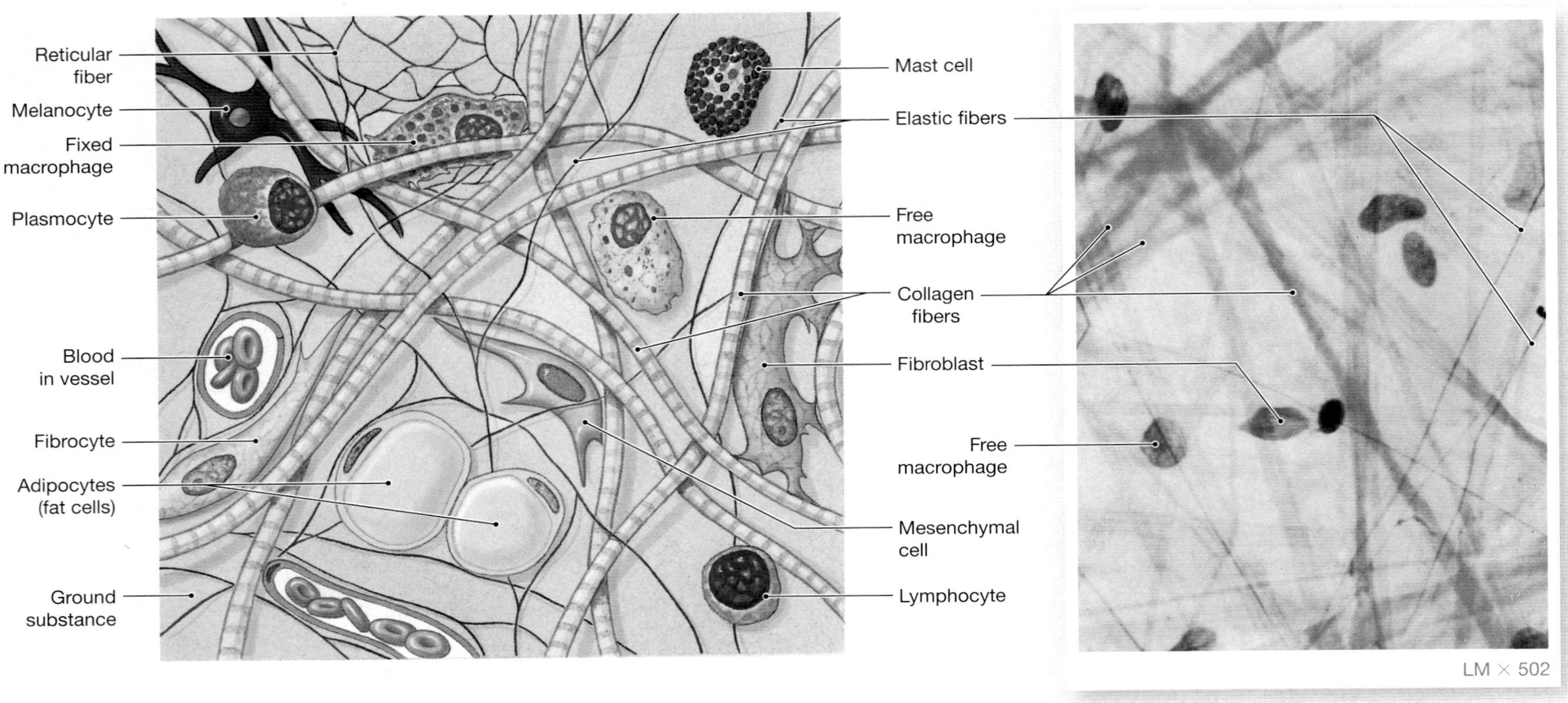

a Diagrammatic view of the cells and fibers in areolar tissue, the most common type of connective tissue proper

b A light micrograph showing the areolar tissue that supports the mesothelium lining the peritoneum

- **Melanocytes** (MEL-an-ō-sīts or me-LAN-ō-sīts) synthesize and store a brown pigment, **melanin** (MEL-a-nin), which gives this tissue a dark color. Melanocytes are common in the epithelium of the skin and in the underlying connective tissue (the dermis), where they help determine skin color. Melanocytes are also abundant in connective tissues of the eyes and dermis of the skin, although the number present differs by body region and among individuals.

Wandering Cells **Wandering cells** help defend and repair damaged tissues. The wandering cells of connective tissue proper include free macrophages, mast cells, lymphocytes, plasma cells, neutrophils, and eosinophils.

Table 3.1 | Functions of Fixed Cells and Wandering Cells

Cell Type	Function
FIXED CELLS	
Fibroblasts	Produce connective tissue fibers
Fibrocytes	Maintain connective tissue fibers and matrix
Fixed macrophages	Phagocytize pathogens and damaged cells
Adipocytes	Store lipid reserves
Mesenchymal cells	Connective tissue stem cells that can differentiate into other cell types
Melanocytes	Synthesize melanin
WANDERING CELLS	
Free macrophages	Mobile/traveling phagocytic cells (derived from monocytes of the blood)
Mast cells	Stimulate local inflammation
Lymphocytes	Participate in immune response
Neutrophils and eosinophils	Mobilize during infection or tissue injury

- **Free macrophages** (also called *wandering macrophages, elicited macrophages*, or *histocytes*) are relatively large phagocytic cells that wander rapidly through the connective tissues of the body. When circulating within the blood, these cells are called **monocytes**. Fixed macrophages in a tissue provide a "frontline" defense that is reinforced by the arrival of free macrophages and other specialized cells.
- **Mast cells** are small, mobile connective tissue cells that are found near blood vessels. The cytoplasm of a mast cell is filled with secretory granules of **histamine** (HIS-ta-mēn) and **heparin** (HEP-a-rin). These chemicals are released after injury or infection and stimulate local inflammation.
- **Lymphocytes** (LIM-fō-sīts), like free macrophages, migrate throughout the body. Lymphocytes multiply wherever tissue damage occurs, and some then develop into **plasma cells** (*plasmocytes*). Plasma cells produce **antibodies**, proteins that help defend the body against disease.
- **Neutrophils** and **eosinophils** are phagocytic blood cells that are smaller than monocytes. Small numbers of these cells migrate through connective tissues. When an infection or injury occurs, chemicals released by macrophages and mast cells attract neutrophils and eosinophils in large numbers.

Fibers of Connective Tissue Proper

Connective tissue proper contains three types of fibers: collagen, reticular, and elastic fibers. Fibroblasts produce all three types of fibers by synthesizing and secreting protein subunits that combine or cluster within the matrix. Fibrocytes maintain these connective tissue fibers.

Collagen Fibers The strongest and most common fibers in connective tissue proper, **collagen fibers** are long, straight, and unbranched **(Figure 3.12)**.

Each collagen fiber consists of three fibrous protein subunits wound together like the strands of a rope. Also like a rope, a collagen fiber is flexible but very strong when pulled from either end. This kind of applied force is called **tension**, and the ability to resist tension is called **tensile strength**.

Tendons and ligaments consist almost entirely of collagen fibers. **Tendons** connect skeletal muscles to bones, and **ligaments** (LIG-a-ments) connect bone to bone, bone to cartilage, or cartilage to cartilage **(Figure 3.14a,b)**. The parallel alignment of collagen fibers in tendons and ligaments allows them to withstand tremendous forces. In fact, sudden, severe muscle contractions and skeletal movements are more likely to snap a bone than a tendon or ligament.

Reticular Fibers **Reticular fibers** (*reticulum*, network) contain the same protein subunits as collagen fibers, but the subunits interact in a different way. Reticular fibers are thinner than collagen fibers, and they form a branching, interwoven framework that is tough but flexible. These fibers are abundant in organs such as the spleen and liver, where they create a complex three-dimensional network that supports the functional cells of these organs **(Figures 3.12a** and **3.13c)**. Because they form a mesh within the organs, reticular fibers resist forces applied from many different directions and stabilize the organ's cells, blood vessels, and nerves despite the pull of gravity.

Elastic Fibers Branching and wavy, **elastic fibers** contain the protein **elastin**. After stretching up to 150 percent of their resting length, they recoil to their original dimensions.

Ground Substance of Connective Tissue Proper

A solution called **ground substance** surrounds the cellular and fibrous components of connective tissue proper **(Figure 3.12a)**. Ground substance in connective tissue proper is clear, colorless, and similar in consistency to maple syrup. It contains hyaluronan and a mixture of other proteoglycans and glycoproteins that interact to determine its consistency.

Loose Connective Tissues

Loose connective tissues are the "packing material" of the body. These tissues fill spaces between organs, provide cushioning, and support epithelia. Loose connective tissues also surround and support blood vessels and nerves, store lipids, and provide a route for the diffusion of materials. There are three types of loose connective tissues: areolar, adipose, and reticular.

Areolar Tissue The least specialized connective tissue in the adult body, **areolar tissue** (*areola*, a little space) contains all the cells and fibers found in connective tissue proper **(Figure 3.13a)**. Areolar tissue has an open framework, and ground substance accounts for most of its volume. The ground substance cushions shocks, and because the fibers within the ground substance are loosely organized, areolar tissue can be distorted without damage. The presence of elastic fibers makes it fairly resilient, so this tissue returns to its original shape after external pressure is relieved.

A layer of areolar tissue separates the skin from deeper structures. In addition to providing padding, the elastic properties of this layer allow a considerable amount of independent movement. Thus, pinching the skin of your arm does not affect the underlying muscle. Conversely, contractions of the underlying arm muscles do not pull against your skin—as the muscle bulges, the areolar tissue stretches. Because this tissue has an extensive circulatory supply, drugs injected into the areolar tissue layer under the skin are quickly absorbed into the bloodstream.

In addition to delivering oxygen and nutrients and removing carbon dioxide and waste products, the capillaries (tiny blood vessels) in areolar tissue carry wandering cells to and from the tissue. When epithelial tissue covers a layer of areolar tissue, the fibrocytes are responsible for maintaining the dense layer of the basement membrane. The epithelial cells rely on diffusion across the basement membrane, and the capillaries in the underlying connective tissue provide the necessary oxygen and nutrients.

Adipose Tissue Adipocytes are found in almost all areolar connective tissues. Adipocytes can become so abundant that any resemblance to normal areolar connective tissue disappears: They become immobile, are surrounded by a basal lamina, and cluster together like tightly packed grapes. The tissue is then called **adipose tissue**. In areolar connective tissue, most of the tissue volume consists of intercellular fluids and fibers. In adipose tissue, most of the tissue volume consists of adipocytes **(Figure 3.13b)**.

There are two types of adipose tissue: white fat and brown fat. **White fat**, which is more common in adults, has a pale, yellow-white color. Its adipocytes (termed **white adipose cells**) contain a single large lipid droplet and are therefore called *unilocular adipose cells* (*uni*, one, + *locular*, chamber). White adipose tissue cushions shocks, insulates the body to slow heat loss through the skin, and serves as padding or filler around structures. White adipose tissue is found under the skin of the groin, sides, buttocks, and breasts. It also surrounds the kidneys and fills the bony sockets behind the eyes and areas of loose connective tissue in the pericardial and abdominal cavities.

CLINICAL NOTE

Liposuction

Liposuction is a popular surgical procedure for reducing subcutaneous (under the skin) adipose tissue. Although many patients expect liposuction to be a quick, safe way to lose weight, there is no scientific evidence that liposuction provides any health benefits. Liposuction does not alter any obesity-related diseases, the procedure has significant risks, and the adipose tissue that has been removed can replace itself.

Several methods are used for liposuction, but in all, the subcutaneous adipose tissue is broken up and suctioned away. All the blood vessels and nerves that traverse this subcutaneous space are destroyed. Bleeding can be significant, and the skin can be left permanently numb. The procedure itself or a postoperative infection can destroy the overlying skin. Complications of liposuction are worse in people who smoke, have diabetes, or are older, due to reduced skin elasticity.

Anyone thinking about such a drastic disruption of normal anatomy might consider first watching a video of the procedure on a reputable medical website.

Figure 3.13 Histology of Loose Connective Tissues. This is the "packing material" of the body, filling spaces between other structures.

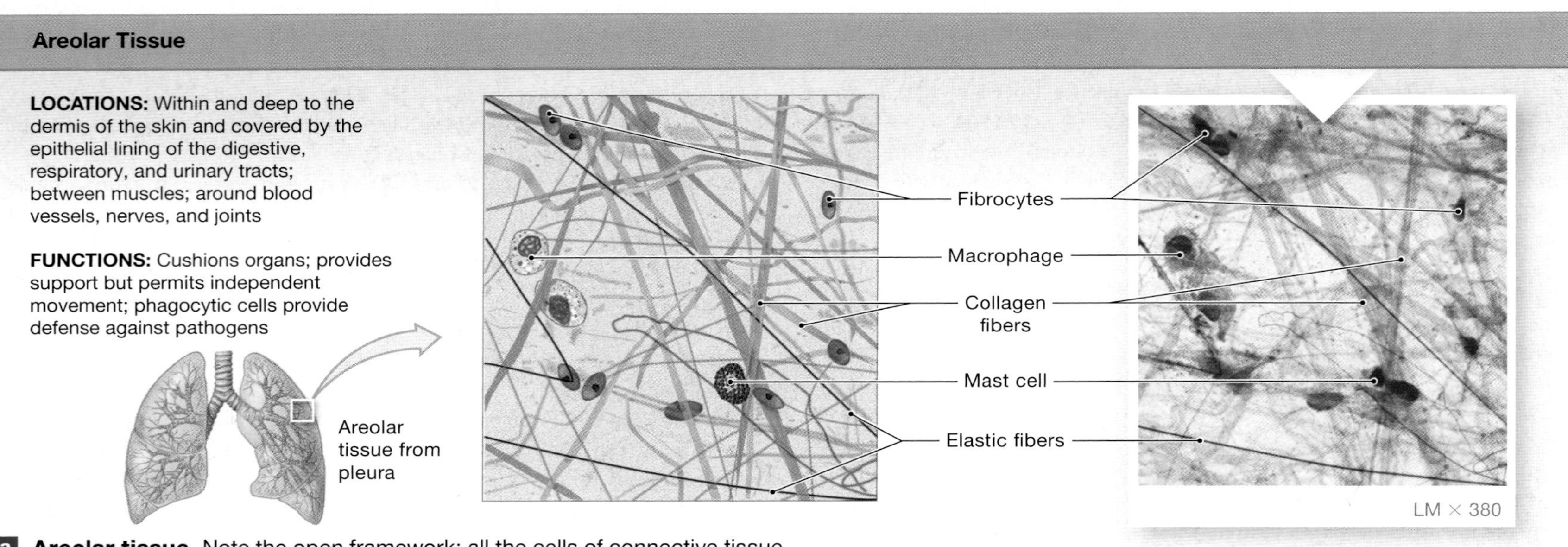

a Areolar tissue. Note the open framework; all the cells of connective tissue proper are found in areolar tissue.

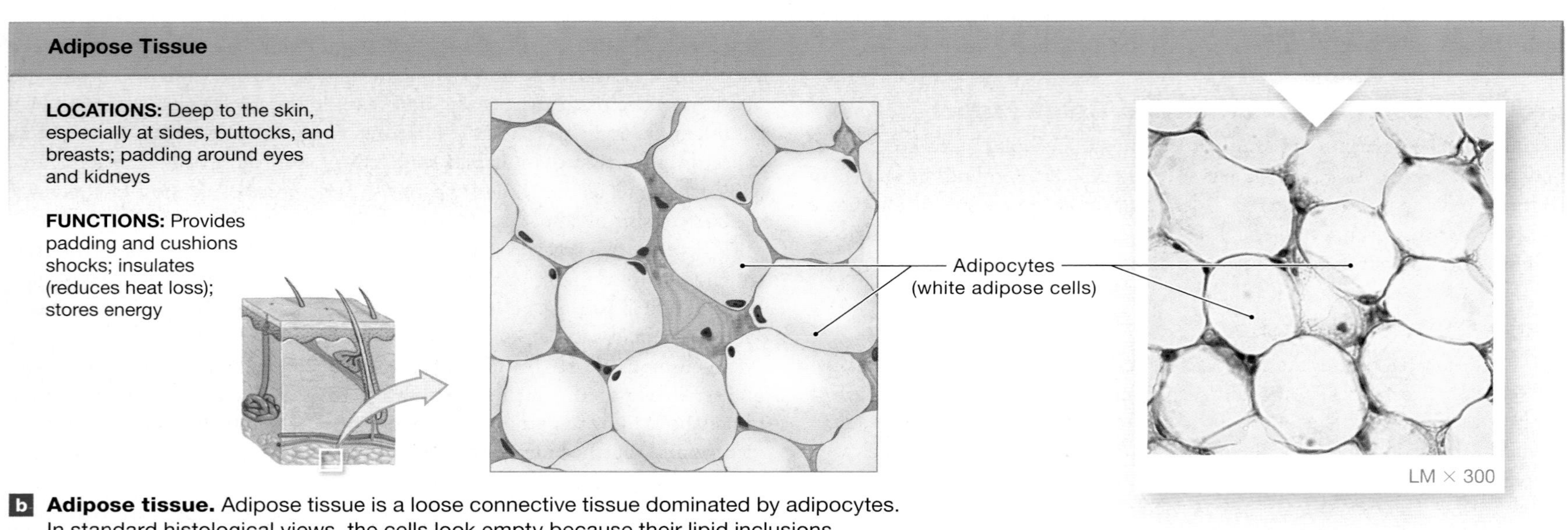

b Adipose tissue. Adipose tissue is a loose connective tissue dominated by adipocytes. In standard histological views, the cells look empty because their lipid inclusions dissolve during slide preparation.

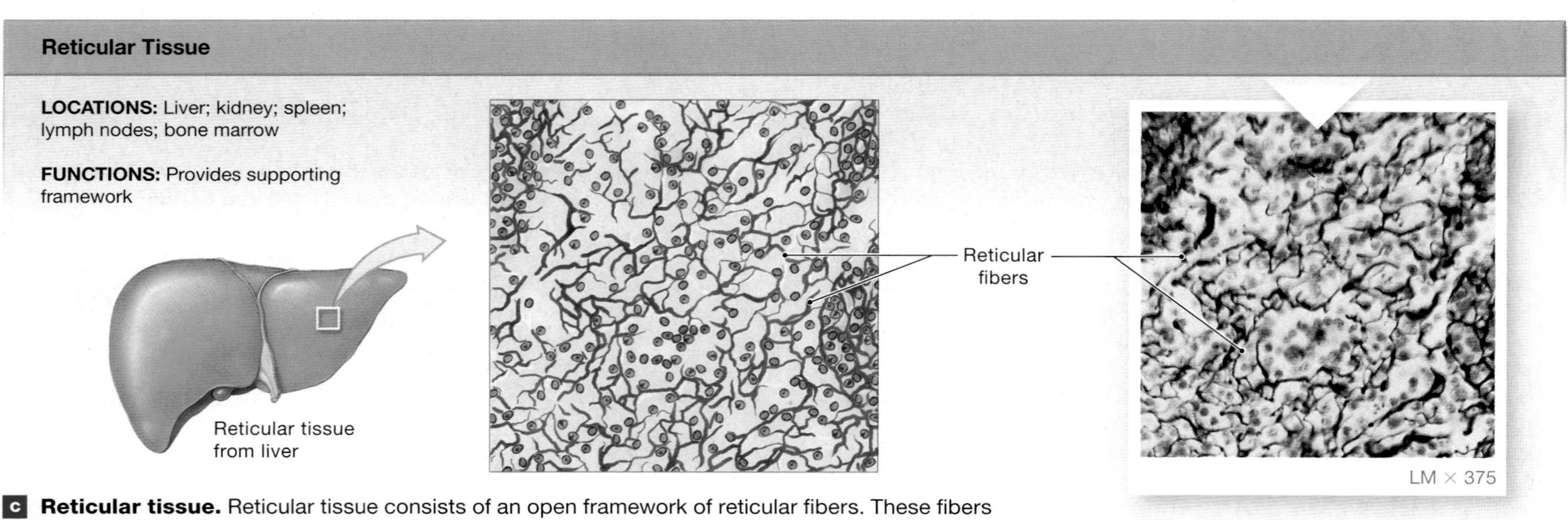

c Reticular tissue. Reticular tissue consists of an open framework of reticular fibers. These fibers are usually very difficult to see because of the large numbers of cells organized around them.

Brown fat is more abundant in infants and children than in adults. Fat is stored in numerous cytoplasmic vacuoles in **brown adipose cells** (*multilocular adipose cells*). This tissue is highly vascularized, and the individual cells contain numerous mitochondria, giving the tissue a deep, rich color.

Brown fat, found between the shoulder blades, around the neck, and possibly elsewhere in the upper body of newborn children, is biochemically active and is important in temperature regulation of newborns and young children. At birth, an infant's temperature-regulating mechanisms are not fully functional, and brown fat provides a mechanism for raising body temperature rapidly. Sympathetic fibers of the autonomic nervous system stimulate brown fat cells and accelerate their breakdown. The energy released from this chemical reaction radiates into the surrounding tissues as heat, which is distributed throughout the body. In this way, an infant can accelerate metabolic heat generation by 100 percent very quickly. With increasing age and size, body temperature becomes more stable, so the importance of brown fat declines. Adults have little if any brown fat.

Reticular Tissue Connective tissue consisting of reticular fibers, macrophages, fibroblasts, and fibrocytes is called **reticular tissue** (**Figure 3.13c**). The fibers of reticular tissue form the supporting connective tissue of the liver, spleen, lymph nodes, and bone marrow.

Dense Connective Tissues

Most of the volume of **dense connective tissues** consists of fibers. Dense connective tissues are also called *collagenous* (ko-LAJ-in-us) *tissues* because collagen fibers are the dominant fiber type. There are two types of dense connective tissue: dense regular connective tissue and dense irregular connective tissue.

Dense Regular Connective Tissue In **dense regular connective tissue** the collagen fibers are packed tightly and aligned parallel to applied forces. Four examples are tendons, aponeuroses, elastic tissue, and ligaments.

- **Tendons** are cords of dense regular connective tissue that attach skeletal muscles to bones and cartilage (**Figure 3.14a**). The collagen fibers run along the longitudinal axis of the tendon and transfer the pull of the contracting muscle to the bone or cartilage. Large numbers of fibrocytes are found between the collagen fibers.
- **Aponeuroses** (ap-ō-nū-RŌ-sēz) are collagenous sheets or ribbons that resemble flat, broad tendons. Aponeuroses often cover the surface of a muscle and help attach superficial muscles to another muscle or structure.
- **Elastic tissue** contains large numbers of elastic fibers, making it springy and resilient (**Figure 3.14b**). This ability to stretch and rebound allows it to tolerate expansion and contraction. Elastic tissue is found deep to transitional epithelia (**Figure 3.7b**, p. 57); it is also found in the walls of blood vessels and respiratory passageways.
- **Ligaments** resemble tendons, but they connect cartilage to cartilage, bone to cartilage, or bone to bone. Ligaments contain significant numbers of elastic fibers as well as collagen fibers, and they can tolerate a modest amount of stretching. **Elastic ligaments** have an even higher proportion of elastic fibers (**Figure 3.14b**). Although uncommon elsewhere, elastic ligaments along the vertebral column are very important in stabilizing the vertebrae.

Dense Irregular Connective Tissue The fibers in **dense irregular connective tissue** form an interwoven meshwork and do not show any consistent pattern (**Figure 3.14c**). This tissue strengthens and supports areas subjected to stresses from many directions. Except at joints, dense irregular connective tissue forms a sheath around cartilage (the perichondrium) and bone (the periosteum). It also forms the thick fibrous **capsule** that surrounds many internal organs, such as the dermis, liver, kidneys, and spleen, and encloses the cavities of joints.

TIPS & TOOLS

Identifying Tendons and Elastic Ligaments

It can be difficult to distinguish between tendons and ligaments since both contain fibrocytes and closely packed connective tissue fibers, and both occur as sheets, bands, and cordlike structures. Here's how to tell the difference.

In tendons (**Figure 3.14a**):

- There are relatively few fibrocytes.
- Fibrocytes are located between bundles of collagen fibers.
- Fibrocytes tend to be elongated.

In ligaments (**Figure 3.14b**):

- Fibrocytes are more numerous than in tendons.
- Fibrocytes are found among bundles of collagen fibers.
- Fibrocytes tend to be less elongated in shape.

Fluid Connective Tissues

▶ **KEY POINT** Two types of fluid connective tissue, blood and lymph, consist of cells within a liquid matrix.

Blood and lymph are **fluid connective tissues** that contain distinctive collections of cells in a fluid matrix called plasma. (We will discuss blood and lymph in detail in Chapters 20 and 23.)

Blood contains red blood cells, white blood cells, and platelets (**Figure 3.15**):

- **Red blood cells** (erythrocytes) (e-RITH-rō-sītz; *erythros*, red) account for almost half the volume of blood. Red blood cells transport oxygen and carbon dioxide in the blood.
- **White blood cells** (leukocytes) (LŪ-kō-sīts; *leuko*, white) include neutrophils, eosinophils, basophils, lymphocytes, and monocytes. White blood cells help protect the body against infection and disease.
- **Platelets** (*thrombocytes*), tiny membrane-enclosed packets of cytoplasm, contain enzymes and special proteins. Platelets function in the clotting response that seals breaks in blood vessel walls.

The extracellular fluid of fluid connective tissue includes three major subdivisions: plasma, interstitial fluid, and lymph. Plasma is normally confined to the blood vessels, and contractions of the heart keep it in motion. In tissues, filtration moves water and small solutes out of capillaries and into the interstitial fluid, which bathes the body's cells. The major difference between plasma and interstitial fluid is that plasma contains a large number of suspended proteins.

Lymph forms as interstitial fluid and then enters lymphatic vessels, small passageways that return it to the cardiovascular system. Along the way, cells of the immune system monitor the composition of the lymph and respond to signs of injury or infection. The number of cells in lymph varies, but ordinarily 99 percent of them are lymphocytes. The rest are primarily phagocytic macrophages, eosinophils, and neutrophils.

Figure 3.14 Histology of Dense Connective Tissues

Dense Regular Connective Tissue

LOCATIONS: Between skeletal muscles and skeleton (tendons and aponeuroses); between bones or stabilizing positions of internal organs (ligaments); covering skeletal muscles; deep fasciae

FUNCTIONS: Provides firm attachment; conducts pull of muscles; reduces friction between muscles; stabilizes relative positions of bones

Collagen fibers

Fibrocyte nuclei

LM × 440

a Tendon. The dense regular connective tissue in a tendon consists of densely packed, parallel bundles of collagen fibers. The fibrocyte nuclei are seen flattened between the bundles. Most ligaments resemble tendons in their histological organization.

Elastic Tissue

LOCATIONS: Between vertebrae of the spinal column (ligamentum flavum and ligamentum nuchae); ligaments supporting penis; ligaments supporting transitional epithelia; in blood vessel walls

FUNCTIONS: Stabilizes positions of vertebrae and penis; cushions shocks; permits expansion and contraction of organs

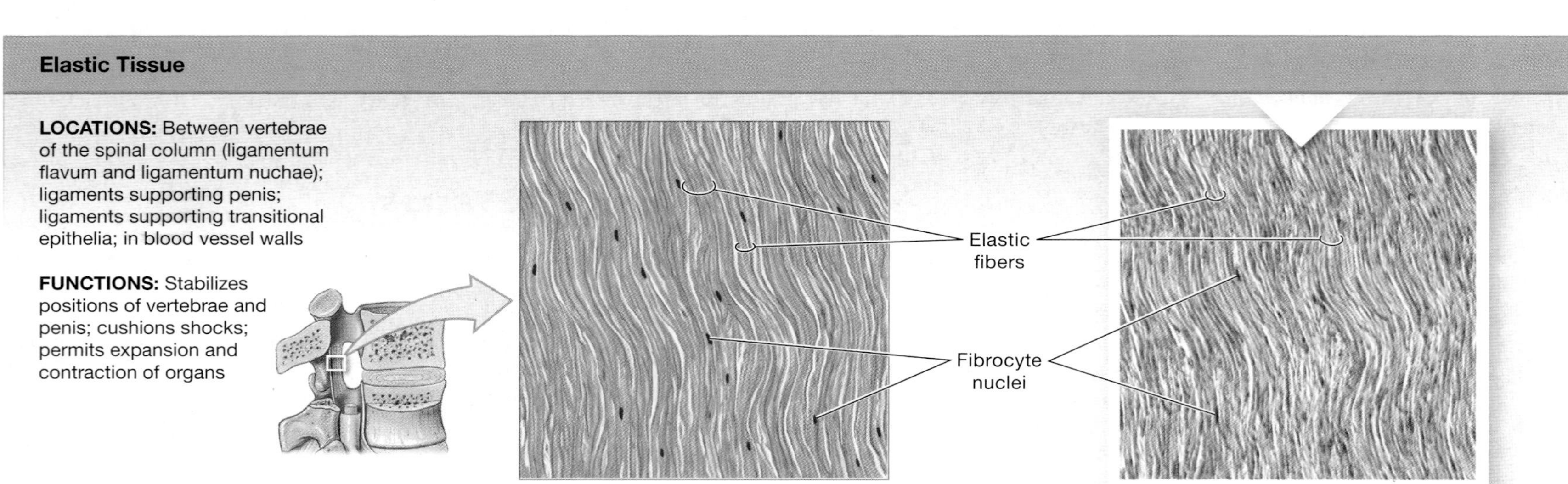

b Elastic Ligament. Elastic ligaments extend between the vertebrae of the spinal column. The bundles of elastic fibers are fatter than the collagen fiber bundles of a tendon or typical ligament.

Dense Irregular Connective Tissue

LOCATIONS: Capsules of visceral organs; periostea and perichondria; nerve and muscle sheaths; dermis

FUNCTIONS: Provides strength to resist forces applied from many directions; helps prevent overexpansion of organs, such as the urinary bladder

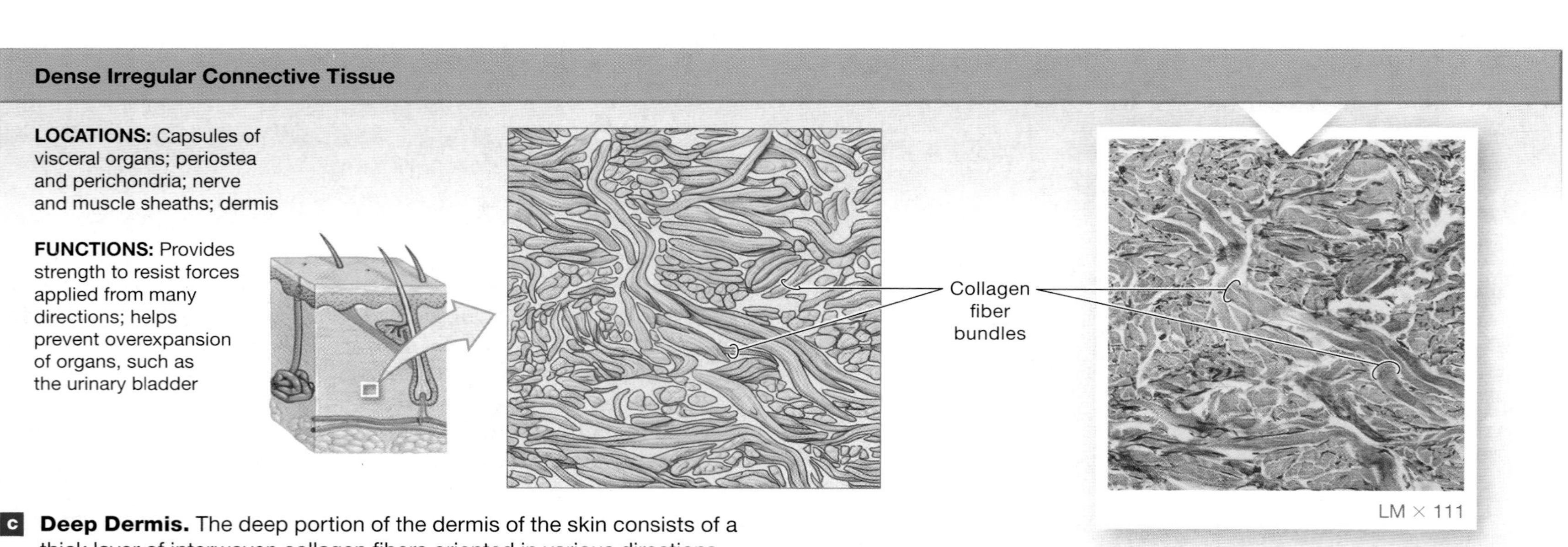

c Deep Dermis. The deep portion of the dermis of the skin consists of a thick layer of interwoven collagen fibers oriented in various directions.

Figure 3.15 Blood Cells and Platelets

Red Blood Cells

Red blood cells, or erythrocytes, are responsible for the transport of oxygen (and, to a lesser degree, of carbon dioxide) in the blood.

Red blood cells account for about half the volume of whole blood and give blood its color.

White Blood Cells

White blood cells, or leukocytes, defend the body from infection and disease.

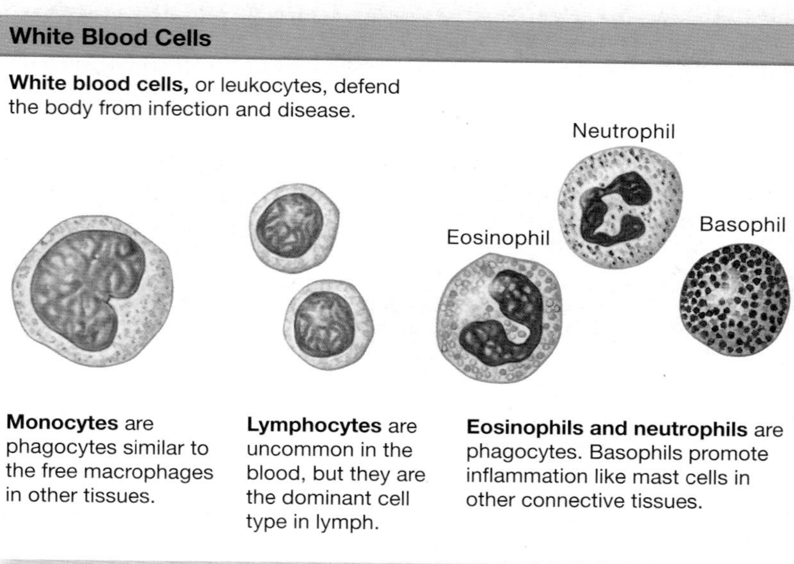

Monocytes are phagocytes similar to the free macrophages in other tissues.

Lymphocytes are uncommon in the blood, but they are the dominant cell type in lymph.

Eosinophils and neutrophils are phagocytes. Basophils promote inflammation like mast cells in other connective tissues.

Platelets

Platelets, or thrombocytes, are membrane-enclosed packets of cytoplasm.

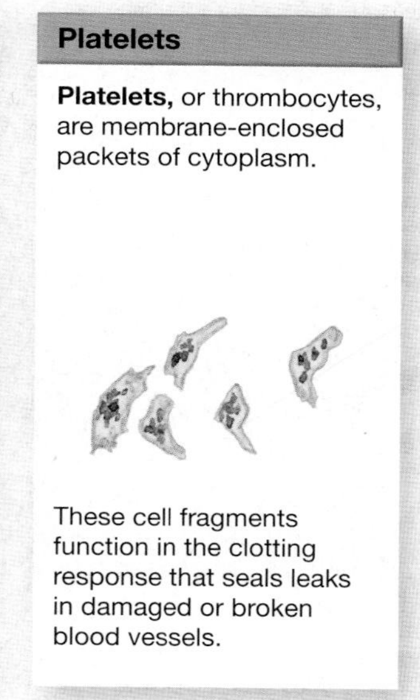

These cell fragments function in the clotting response that seals leaks in damaged or broken blood vessels.

Supporting Connective Tissues

▶ **KEY POINT** Cartilage and bone are the two types of supporting connective tissue.

Cartilage and bone are **supporting connective tissues** that provide a strong framework to support the rest of the body. In these connective tissues, the matrix contains numerous fibers and, in some cases, deposits of insoluble calcium salts.

Cartilage

The matrix of **cartilage** is a firm gel that contains complex polysaccharides called **chondroitin sulfates** (kon-DRŌ-i-tin; *chondros*, cartilage). The chondroitin sulfates form complexes with proteins, forming proteoglycans. Cartilage cells, or **chondrocytes** (KON-drō-sīts), are the only cells within the cartilage matrix. Chondrocytes live in small chambers known as **lacunae** (la-KOO-nē; *lacus*, pool).

The physical properties of cartilage depend on the composition of its matrix. Collagen fibers provide cartilage with its tensile strength, and extracellular fibers and ground substance give cartilage its flexibility and resilience.

Cartilage is avascular because chondrocytes produce a chemical that discourages the formation of blood vessels. All nutrients and waste products must diffuse through the matrix. A fibrous **perichondrium** (per-i-KON-drē-um; *peri*, around) usually separates cartilage from the surrounding tissues **(Figure 3.16a)**. The perichondrium contains two distinct layers: an outer **fibrous layer** of dense irregular connective tissue and an inner **cellular layer**. The fibrous layer provides mechanical support and protection and attaches the cartilage to other structures. The cellular layer is important for the growth and maintenance of the cartilage.

Cartilage grows by two mechanisms, appositional growth and interstitial growth **(Figure 3.16b,c)**:

- In **appositional growth**, stem cells of the inner layer of the perichondrium undergo repeated cycles of division. The innermost cells differentiate into chondroblasts, which begin producing cartilage matrix. After they are completely surrounded by matrix, the chondroblasts differentiate into chondrocytes. Appositional growth gradually increases the dimensions of the cartilage by adding to its surface.
- In **interstitial growth**, chondrocytes within the cartilage matrix divide and their daughter cells produce additional matrix. This cycle enlarges the cartilage from within, much like a balloon inflating.

Neither appositional nor interstitial growth occurs in adult cartilage. Most cartilage cannot repair itself after a severe injury.

Types of Cartilage There are three types of cartilage: hyaline, elastic, and fibrous.

- In **hyaline cartilage** (HĪ-a-lin; *hyalos*, glass), the most common type, the matrix contains closely packed collagen fibers. Although it is tough and somewhat flexible, this is the weakest type of cartilage. Because the collagen fibers of the matrix do not stain well, they are not apparent in light microscopy **(Figure 3.17a)**. Examples of hyaline cartilage in the adult body include (1) the connections between the ribs and the sternum, (2) supporting cartilages along the conducting passageways of the respiratory tract, and (3) articular cartilages covering opposing bone surfaces within synovial joints, such as the shoulder.
- **Elastic cartilage** contains numerous elastic fibers that make it extremely flexible. Elastic cartilage, along with other structures, forms the external flap (auricle) of the external ear **(Figure 3.17b)**, the epiglottis, the auditory canal (the airway to the middle ear), and the cuneiform cartilages of the larynx. Although the cartilage at the tip of the nose is very flexible, scientists disagree about whether it is elastic cartilage because the elastic fibers are less abundant than in the auricle or epiglottis.
- **Fibrous cartilage**, or *fibrocartilage*, has little ground substance and may lack a perichondrium, and the matrix is dominated by collagen fibers **(Figure 3.17c)**. Fibrocartilaginous pads lie in areas of high stress, such as the intervertebral discs between the spinal vertebrae, between the pubic bones of the pelvis, and around or within a few joints and tendons. Fibrous cartilage resists compression, absorbs shocks, and prevents damaging bone-to-bone contact. The collagen fibers within fibrous cartilage follow the stress lines encountered at that particular location and therefore are more regularly arranged than those of hyaline or elastic cartilage. Cartilages heal slowly and poorly, and damaged fibrous cartilage in joints can interfere with normal movement.

Figure 3.16 **The Formation and Growth of Cartilage**

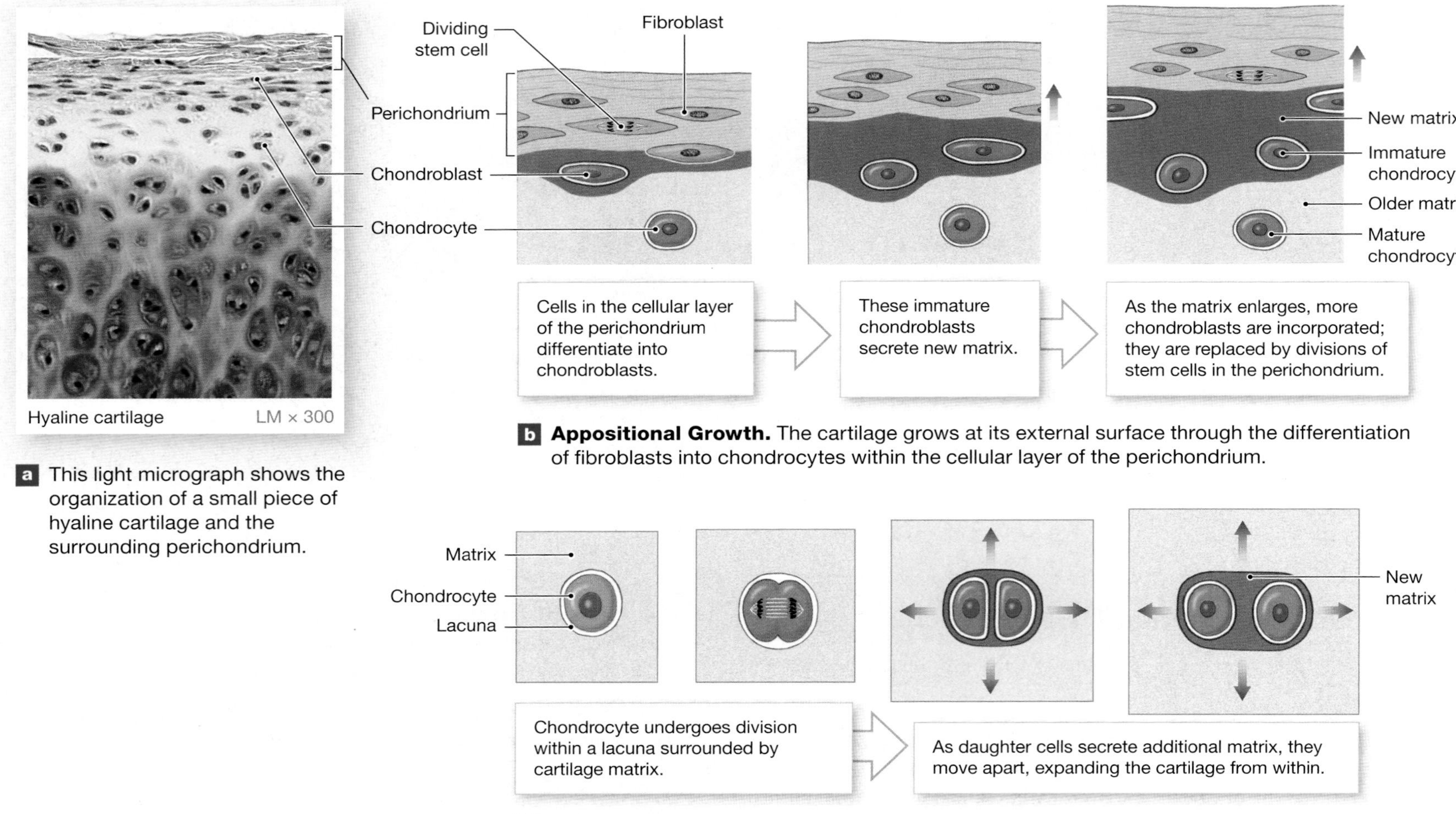
Hyaline cartilage
LM × 300
Dividing stem cell
Fibroblast
Perichondrium
Chondroblast
Chondrocyte
New matrix
Immature chondrocyte
Older matrix
Mature chondrocyte
Cells in the cellular layer of the perichondrium differentiate into chondroblasts.
These immature chondroblasts secrete new matrix.
As the matrix enlarges, more chondroblasts are incorporated; they are replaced by divisions of stem cells in the perichondrium.
Matrix
Chondrocyte
Lacuna
New matrix
Chondrocyte undergoes division within a lacuna surrounded by cartilage matrix.
As daughter cells secrete additional matrix, they move apart, expanding the cartilage from within.

a This light micrograph shows the organization of a small piece of hyaline cartilage and the surrounding perichondrium.

b **Appositional Growth.** The cartilage grows at its external surface through the differentiation of fibroblasts into chondrocytes within the cellular layer of the perichondrium.

c **Interstitial Growth.** The cartilage expands from within as chondrocytes in the matrix divide, grow, and produce new matrix.

TIPS & TOOLS

Identifying Hyaline and Elastic Cartilage

It's not easy to distinguish hyaline cartilage from fibrous cartilage. To make matters more confusing, the section you are examining may not show a perichondrium because of the plane of section, even if one is present. Here's how to tell them apart.

In hyaline cartilage **(Figure 3.17a)**:

- The matrix is homogeneous in appearance.
- There is usually a perichondrium.
- Lacunae are randomly arranged.

In fibrous cartilage **(Figure 3.17c)**:

- Collagenous fibers are visible within the matrix.
- There is no perichondrium.
- Lacunae are widely spaced and regularly arranged.

Bone

Bone is the second type of supporting connective tissue. There are significant differences between cartilage and **bone**, or *osseous tissue* (OS-ē-us; *os*, bone) **(Table 3.2)**.(We will discuss the histology of bone in detail in Chapter 5.) Approximately one-third of the matrix of bone consists of collagen fibers. The rest is a mixture of calcium salts—primarily calcium phosphate with lesser amounts of calcium carbonate.

This combination gives bone remarkable properties. By themselves, calcium salts are strong but rather brittle. Collagen fibers are weaker, but relatively flexible. In bone, the minerals are organized around the collagen fibers. This results in a strong, somewhat flexible combination that is very resistant to shattering. In its overall properties, bone can compete with the strongest steel-reinforced concrete.

Figure 3.18 shows the general organization of bone. Lacunae contain **osteocytes** (OS-tē-ō-sīts) and are often organized around blood vessels that branch through the bony matrix. Because diffusion cannot occur through the calcium salts, osteocytes communicate with blood vessels and with one another through slender cytoplasmic extensions termed filapodia. These extensions run through **canaliculi** (kan-a-LIK-ū-lī; "little canals"), long slender passages in the matrix. Canaliculi form a branching network for the exchange of materials between blood vessels and osteocytes.

There are two types of bone. **Compact bone** contains blood vessels trapped within the matrix, and **spongy bone** does not.

Except at joints, all bone surfaces are covered by a **periosteum** (per-ē-OS-tē-um) composed of an outer fibrous layer and an inner cellular layer. The periosteum helps attach a bone to surrounding tissues and to tendons and ligaments. The cellular layer functions in bone growth and in repairs after an injury.

Figure 3.17 Histology of the Three Types of Cartilage. Cartilage is a supporting connective tissue with a firm, gelatinous matrix.

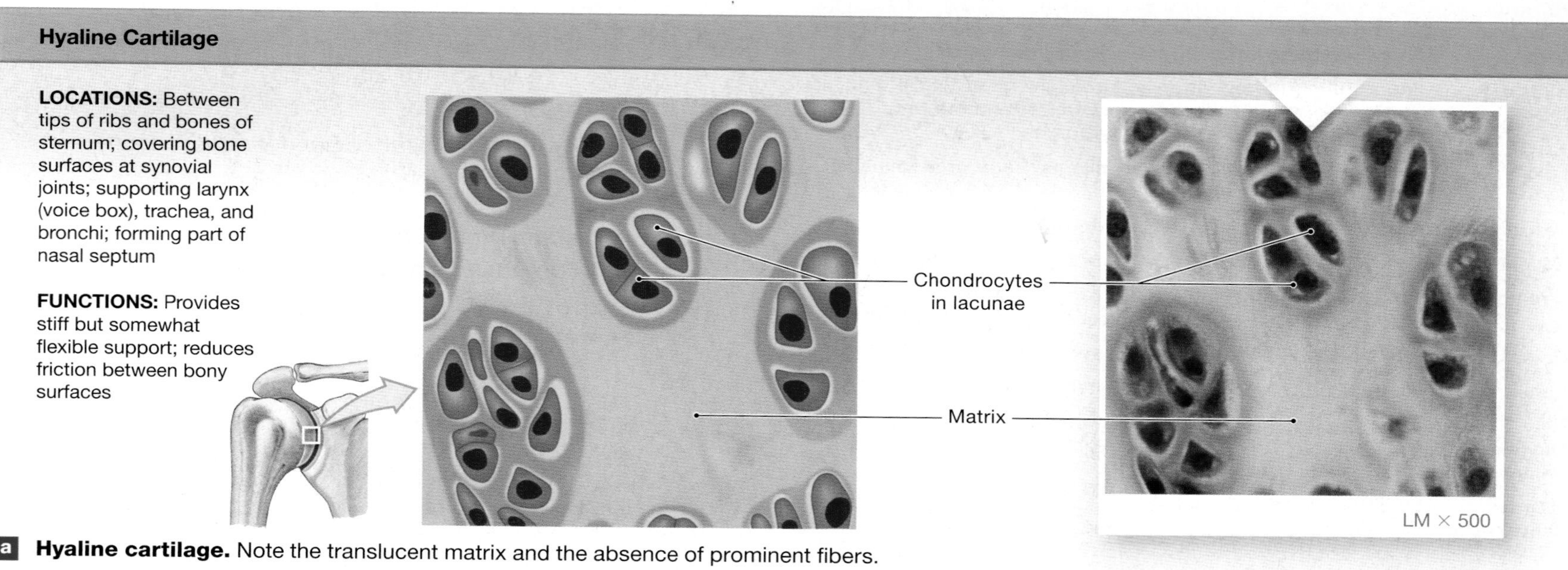
Hyaline Cartilage
LOCATIONS: Between tips of ribs and bones of sternum; covering bone surfaces at synovial joints; supporting larynx (voice box), trachea, and bronchi; forming part of nasal septum
FUNCTIONS: Provides stiff but somewhat flexible support; reduces friction between bony surfaces
Chondrocytes in lacunae
Matrix
LM × 500

a Hyaline cartilage. Note the translucent matrix and the absence of prominent fibers.

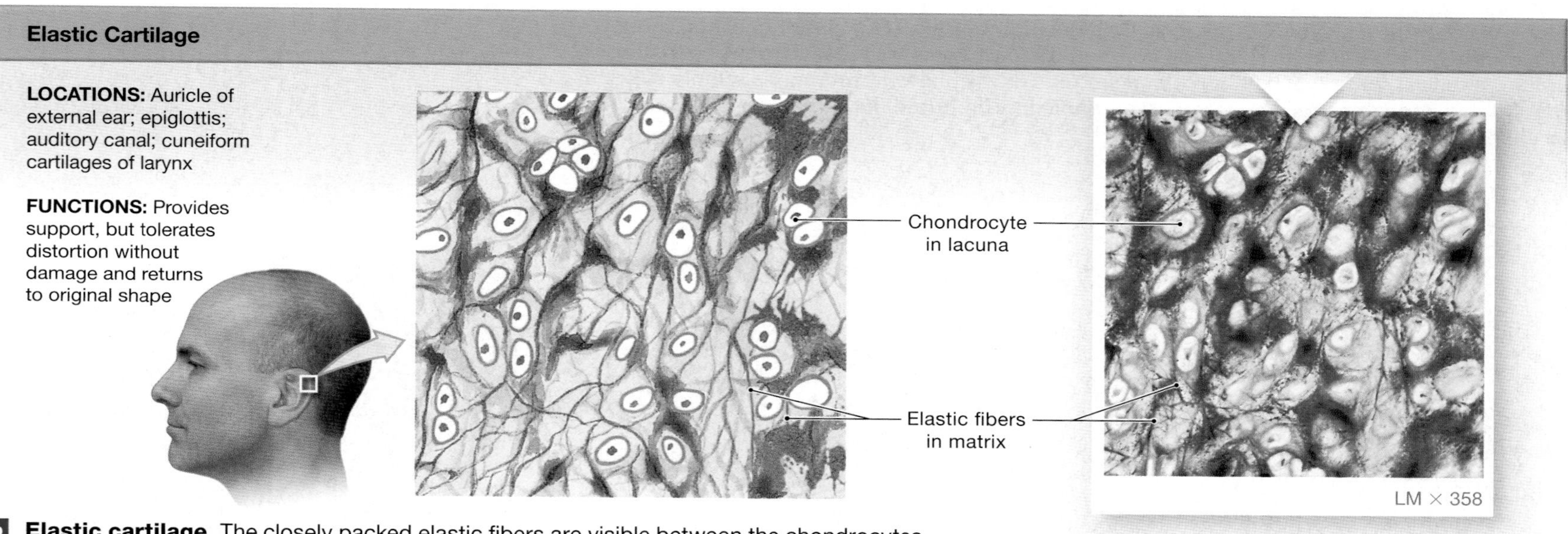
Elastic Cartilage
LOCATIONS: Auricle of external ear; epiglottis; auditory canal; cuneiform cartilages of larynx
FUNCTIONS: Provides support, but tolerates distortion without damage and returns to original shape
Chondrocyte in lacuna
Elastic fibers in matrix
LM × 358

b Elastic cartilage. The closely packed elastic fibers are visible between the chondrocytes.

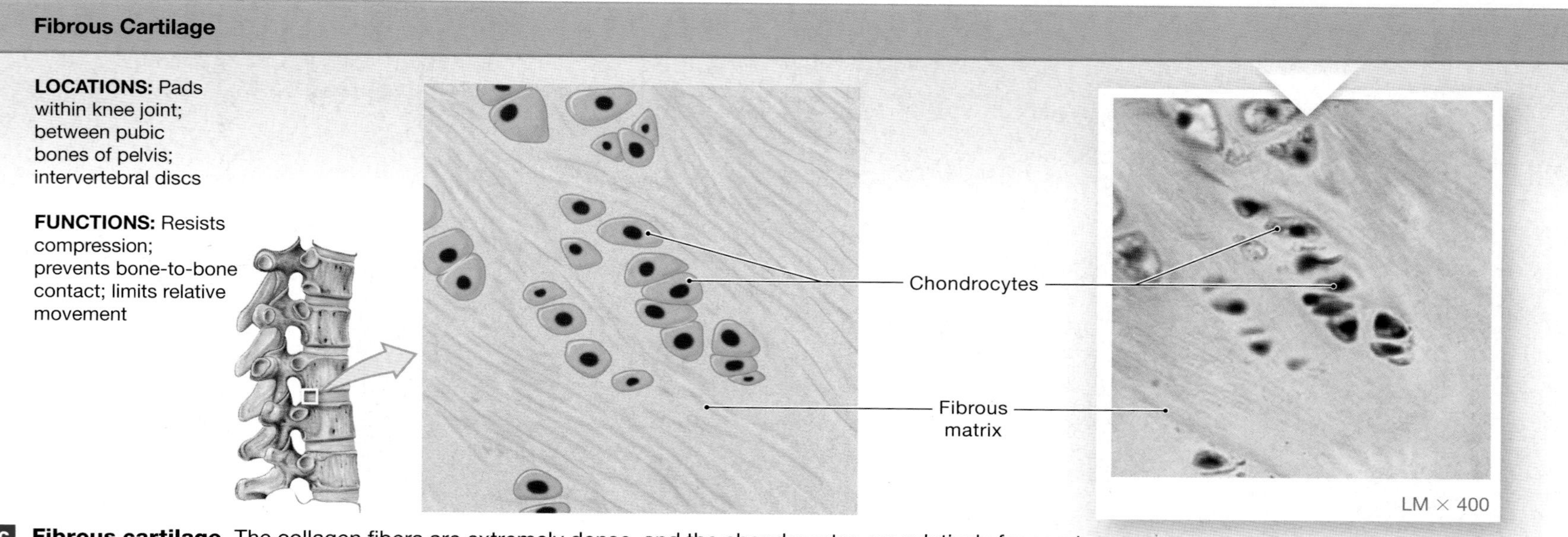
Fibrous Cartilage
LOCATIONS: Pads within knee joint; between pubic bones of pelvis; intervertebral discs
FUNCTIONS: Resists compression; prevents bone-to-bone contact; limits relative movement
Chondrocytes
Fibrous matrix
LM × 400

c Fibrous cartilage. The collagen fibers are extremely dense, and the chondrocytes are relatively far apart.

Table 3.2 A Comparison of the Structural Features of Cartilage and Bone

Feature	Cartilage	Bone
Cells	Chondrocytes in lacunae	Osteocytes in lacunae
Matrix	Chondroitin sulfates with proteins, forming hydrated proteoglycans	Insoluble crystals of calcium phosphate and calcium carbonate
Fibers	Collagen, elastic, reticular fibers (proportions vary)	Collagen fibers predominate
Vascularity	Avascular	Extensive
Covering	Perichondrium, two layers	Periosteum, two layers
Strength	Limited: bends easily but hard to break	Strong: resists distortion until breaking point is reached
Growth	Interstitial and appositional	Appositional only
Repair capabilities	Limited ability	Extensive ability
Oxygen demands	Low	High
Nutrient delivery	By diffusion through matrix	By diffusion through cytoplasm and fluid in canaliculi

Unlike cartilage, bone undergoes extensive remodeling on a regular basis and can repair itself completely, even after severe damage. Bone also responds to the stresses placed on it. It grows thicker and stronger with exercise, but becomes thin and brittle with inactivity.

3.2 CONCEPT CHECK

5 Identify the three basic components of all connective tissues.

6 What is a major difference between connective tissue proper and supporting connective tissue?

7 What are the two general classes of cells in connective tissue proper? What cells are found in each class?

See the blue Answers tab at the back of the book.

3.3 Membranes

▶ **KEY POINT** Epithelia and connective tissues combine to form membranes, which cover and protect other structures and tissues.

Figure 3.18 Anatomy and Histological Organization of Bone. Bone is a supporting connective tissue with a hardened matrix. The osteocytes in compact bone are usually organized in groups around a central space that contains blood vessels. For the photomicrograph, a sample of bone was ground thin enough to become transparent. Bone dust produced during the grinding filled the lacunae, making them appear dark.

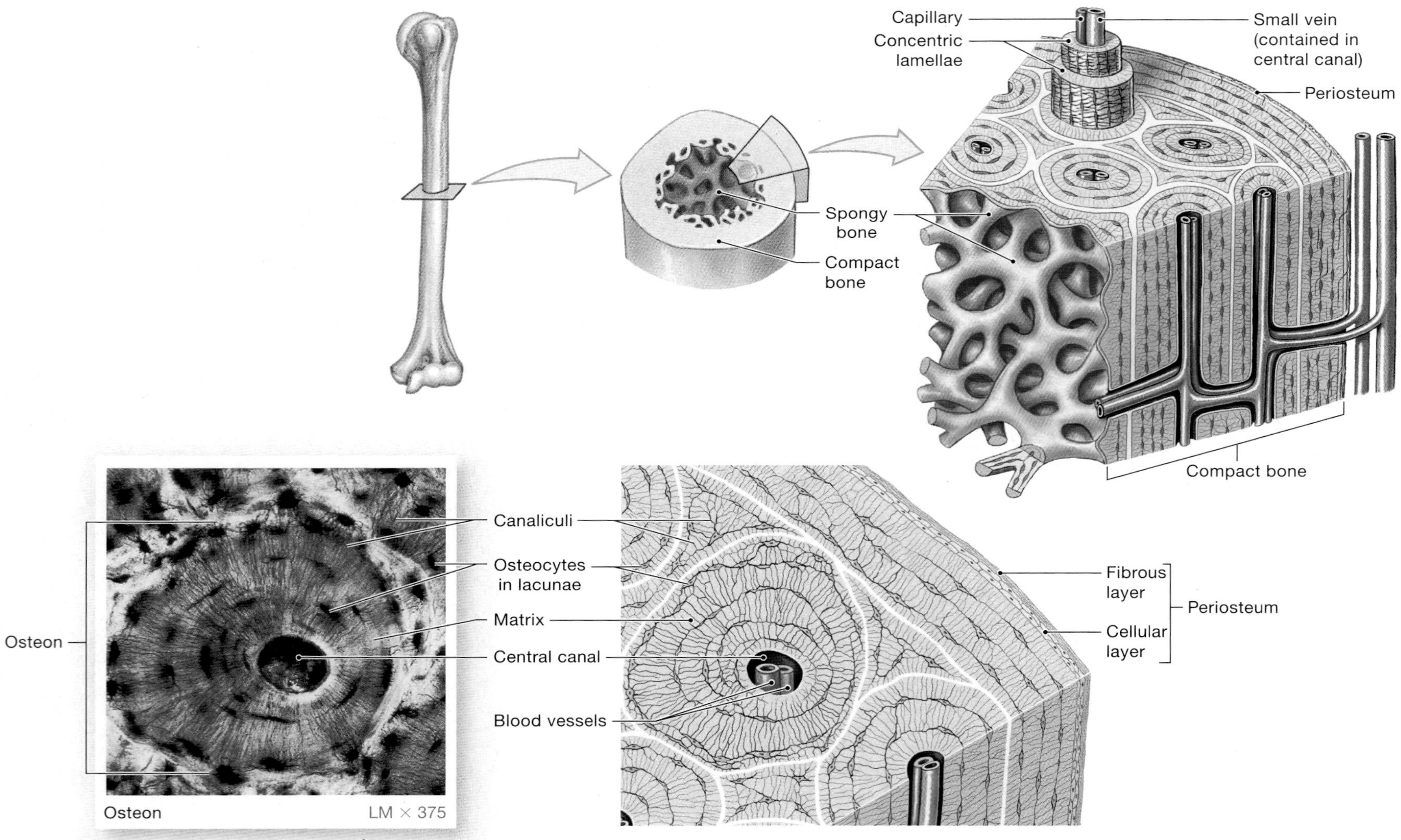

Capillary
Concentric lamellae
Small vein (contained in central canal)
Periosteum
Spongy bone
Compact bone
Compact bone
Osteon
Canaliculi
Osteocytes in lacunae
Matrix
Central canal
Blood vessels
Fibrous layer
Cellular layer
Periosteum
Osteon LM × 375

A **membrane** consists of an epithelial sheet with an underlying connective tissue layer. There are four types of epithelial membranes: mucous, serous, cutaneous, and synovial.

Mucous Membranes

▶ **KEY POINT** Mucous membranes are moist and line passageways that open to the exterior of the body.

A **mucous membrane**, or *mucosa* (mū-KŌ-sa; plural, *mucosae*), forms a barrier that resists the entry of pathogens (**Figure 3.19a**). The epithelial surfaces of the mucosa are moist because they are lubricated by mucus or other glandular secretions or by fluids such as urine or semen. The areolar tissue component of a mucous membrane is called the **lamina propria** (PRŌ-prē-a). The lamina propria forms a bridge that connects the epithelium to underlying structures. It also supports blood vessels and nerves that supply the epithelium.

Examples of mucous membranes are (1) the simple columnar epithelium of the digestive tract, (2) the stratified squamous epithelium of the oral cavity, and (3) the transitional epithelium found in most of the urinary tract. We will discuss specific mucous membranes in greater detail in later chapters.

Serous Membranes

▶ **KEY POINT** Serous membranes line body cavities that lack openings to the exterior; they minimize friction between opposing surfaces.

A **serous membrane** consists of a mesothelium (⮌ p. 53) supported by a thin layer of areolar connective tissue rich in blood and lymphatic vessels (**Figure 3.19b**). The three types of serous membranes are the pleura, peritoneum, and pericardium.

- The **pleura** lines the pleural cavities and covers the lungs
- The **peritoneum** lines the peritoneal cavity and covers the surfaces of the enclosed organs
- The **pericardium** lines the pericardial cavity and covers the heart. ⮌ p. 19

Figure 3.19 Membranes. Membranes are composed of epithelia and connective tissues, which cover and protect other tissues and structures.

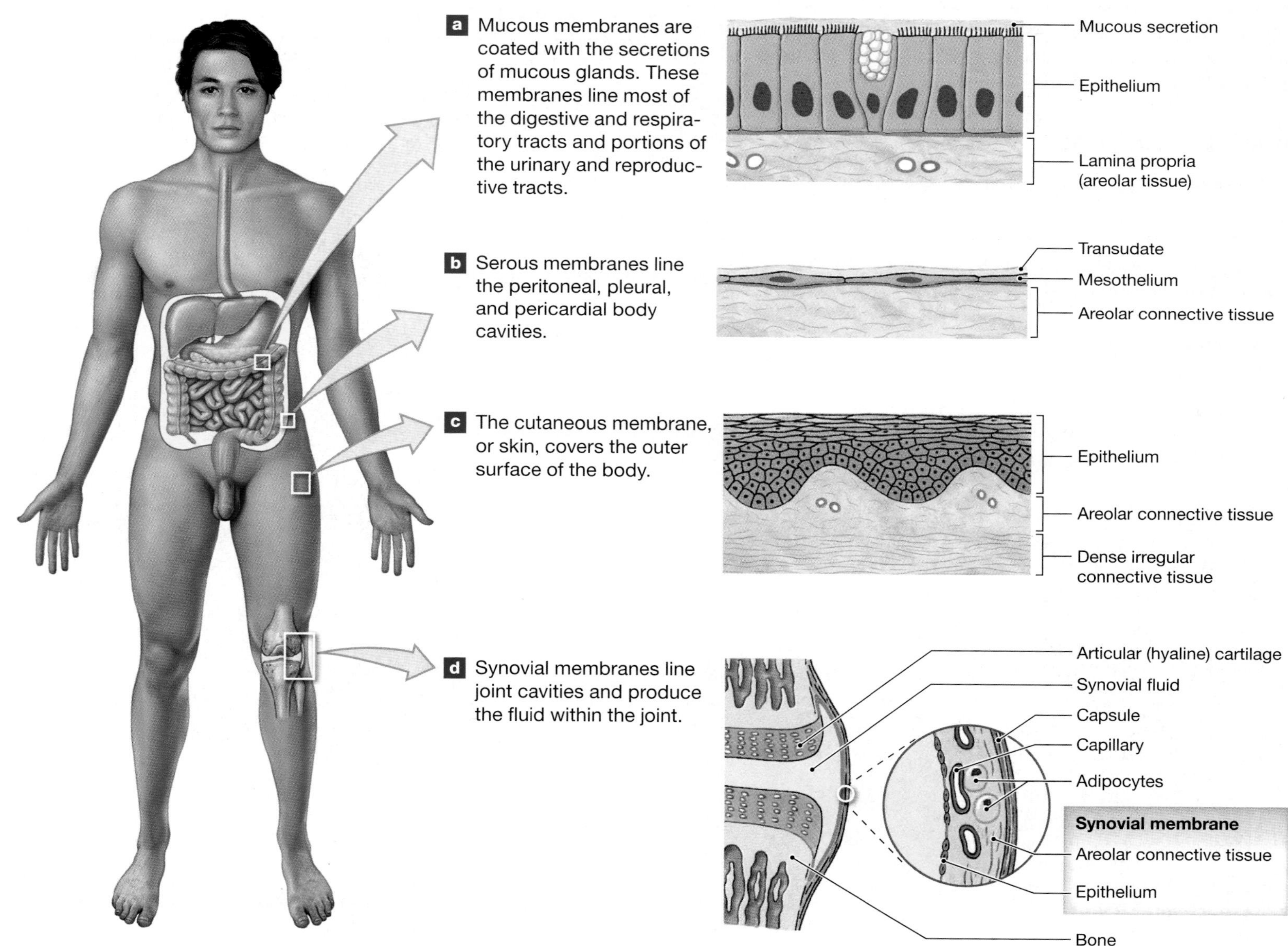
a Mucous membranes are coated with the secretions of mucous glands. These membranes line most of the digestive and respiratory tracts and portions of the urinary and reproductive tracts.
Mucous secretion
Epithelium
Lamina propria (areolar tissue)
b Serous membranes line the peritoneal, pleural, and pericardial body cavities.
Transudate
Mesothelium
Areolar connective tissue
c The cutaneous membrane, or skin, covers the outer surface of the body.
Epithelium
Areolar connective tissue
Dense irregular connective tissue
d Synovial membranes line joint cavities and produce the fluid within the joint.
Articular (hyaline) cartilage
Synovial fluid
Capsule
Capillary
Adipocytes
Synovial membrane
Areolar connective tissue
Epithelium
Bone

Serous membranes are very thin and attach firmly to the body wall and the organs they cover. When you look at an organ, such as the heart or stomach, you are seeing its tissues through a transparent serous membrane.

The parietal and visceral portions of a serous membrane are in close contact at all times. The membrane's primary function is to minimize friction between these two surfaces. Because the mesothelial cells are very thin, serous membranes are relatively permeable, and tissue fluids diffuse onto the exposed surface, keeping it moist and slippery.

The fluid formed on the surfaces of a serous membrane is called a **transudate** (TRANS-ū-dāt; *trans-*, across). Three examples of transudates are the pleural fluid, peritoneal fluid, and pericardial fluid. In normal, healthy individuals, the total volume of transudate at any given time is extremely small, just enough to prevent friction between the walls of the cavities and the surfaces of internal organs. After an injury or in certain diseases, the volume of transudate may increase dramatically, complicating existing medical problems or producing new ones.

The Cutaneous Membrane (Skin)

▶ **KEY POINT** The cutaneous membrane covers the body surface.

The **cutaneous membrane**, or the *skin*, covers the surface of the body. It consists of a keratinized stratified squamous epithelium and an underlying layer of areolar connective tissue that is reinforced by a layer of dense connective tissue **(Figure 3.19c)**. Unlike serous and mucous membranes, the cutaneous membrane is thick, relatively waterproof, and usually dry. The skin is the first line of defense against environmental pathogens. (We will discuss the skin in detail in Chapter 4.)

Synovial Membranes

▶ **KEY POINT** Synovial membranes line and lubricate joint cavities.

Bones contact one another at joints, or articulations. Joints that permit significant movement are surrounded by a fibrous capsule and contain a joint cavity lined by a **synovial membrane**. This synovial (si-NŌ-vē-al) membrane consists of areolar tissue covered by an incomplete superficial layer of squamous or cuboidal cells **(Figure 3.19d)**.

Although called an epithelial tissue, the synovial membrane lining a joint cavity develops within connective tissue. Synovial membrane differs from other epithelial tissues in three respects:

1. It has no basement membrane.
2. Its cellular layer is incomplete, with gaps between adjacent cells.
3. Its "epithelial cells" derive from macrophages and fibroblasts of the adjacent connective tissue.

Some of the lining cells are phagocytic, and others are secretory. The phagocytic cells remove cellular debris or pathogens that could disrupt joint function. The secretory cells regulate the composition of the **synovial fluid** within the joint cavity. Synovial fluid lubricates the cartilages in the joint, distributes oxygen and nutrients, and cushions shocks.

Embryonic Connective Tissues

▶ **KEY POINT** Mesenchyme and mucoid connective tissue are two forms of embryonic connective tissues.

Mesenchyme is the first connective tissue to appear in the developing embryo. Mesenchyme contains star-shaped cells separated by a matrix that contains very fine protein filaments. This connective tissue gives rise to all other connective tissues, including fluid connective tissues, cartilage, and bone **(Figure 3.20a)**. **Mucoid connective tissue** (*gelatinous connective tissue* or *Wharton's jelly*) is a loose connective tissue found in many regions of the embryo, including the umbilical cord **(Figure 3.20b)**.

Neither of these embryonic connective tissues is found in the adult. However, many adult connective tissues contain scattered mesenchymal (*stem*) cells that assist in repairs after the connective tissue has been injured or damaged.

3.3 CONCEPT CHECK

8 Identify the four types of membranes in the body and list their functions.

See the blue Answers tab at the back of the book.

Figure 3.20 Histology of Embryonic Connective Tissues. These connective tissue types give rise to all other connective tissue types.

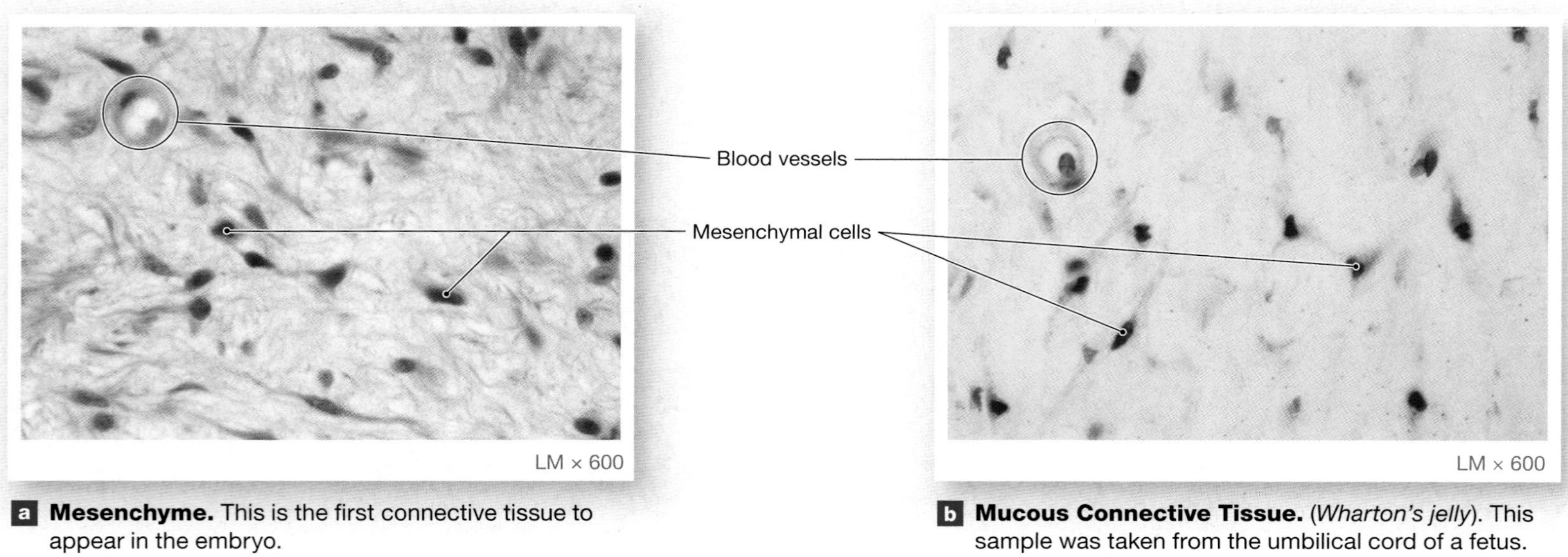

a Mesenchyme. This is the first connective tissue to appear in the embryo.

b Mucous Connective Tissue. (*Wharton's jelly*). This sample was taken from the umbilical cord of a fetus.

CLINICAL NOTE

Do Chondroitin and Glucosamine Help Arthritis Pain?

Chondroitin and glucosamine are two substances found in articular hyaline cartilage. Loss of this articular cartilage results in arthritis. Many people believe that taking supplements containing chondroitin and/or glucosamine will make their aching joints feel better. When chondroitin and glucosamine reach the stomach and intestines, they are broken down (catabolized) into their amino acids and are not absorbed as intact molecules.

Many studies have compared the effects of glucosamine and chondroitin with a placebo (an inert substance that has no active ingredients). Studies that are sponsored by the supplement industry appear more favorable than independent studies, but there is little overall evidence to prove that either supplement works better than a placebo.

3.4 The Connective Tissue Framework of the Body

▶ **KEY POINT** Connective tissues provide the internal framework of the body.

Layers of connective tissue connect the organs within the body cavities with the rest of the body. These layers provide strength and stability, maintain the positions of internal organs, and provide a route for the distribution of blood vessels, lymphatic vessels, and nerves.

Fascia (FASH-ē-a; plural, *fasciae*) is a layer or sheet of connective tissue that you can see on gross dissection. There are three types of fascia: superficial, deep, and subserous (**Figure 3.21**):

- The **superficial fascia**, or **subcutaneous layer** (*sub*, below, + *cutis*, skin), is also termed the *hypodermis* (*hypo*, below, + *derma*, skin). This layer of loose connective tissue separates the skin from underlying tissues and organs. It provides insulation and padding and lets the skin and underlying structures move independently.
- The **deep fascia** consists of dense regular connective tissue. Its fiber organization resembles that of plywood because all of the connective tissue fibers in a layer run in the same direction, but the orientation of the fibers changes from one layer to another. This structure helps the deep fascia resist forces from many different directions. Structures that connect to the deep fascia include the tough capsules that surround organs (including the organs in the thoracic and peritoneal cavities), the perichondrium around cartilages, the periosteum around bones, and the connective tissue sheaths of muscle. The deep fascia of the neck and limbs, the **intermuscular fascia**, passes between groups of muscles and divide the muscles into compartments that differ functionally and developmentally. These dense connective tissue components are interwoven. For example, the deep fascia around a muscle blends into the tendon, whose fibers blend into the periosteum. This arrangement creates a strong, fibrous network for the body and ties structural elements together.
- The **subserous fascia** is a layer of loose connective tissue that lies between the deep fascia and the serous membranes that line body cavities. The subserous fascia separates the serous membranes from the deep fascia, preventing the movements of muscles and muscular organs from severely distorting the delicate lining.

To illustrate the strength of fascia, fascia can hold surgical sutures, but the muscle, areolar, and adipose tissues that fascia encloses cannot.

Figure 3.21 The Fasciae. The anatomical relationship of connective tissue elements in the body.

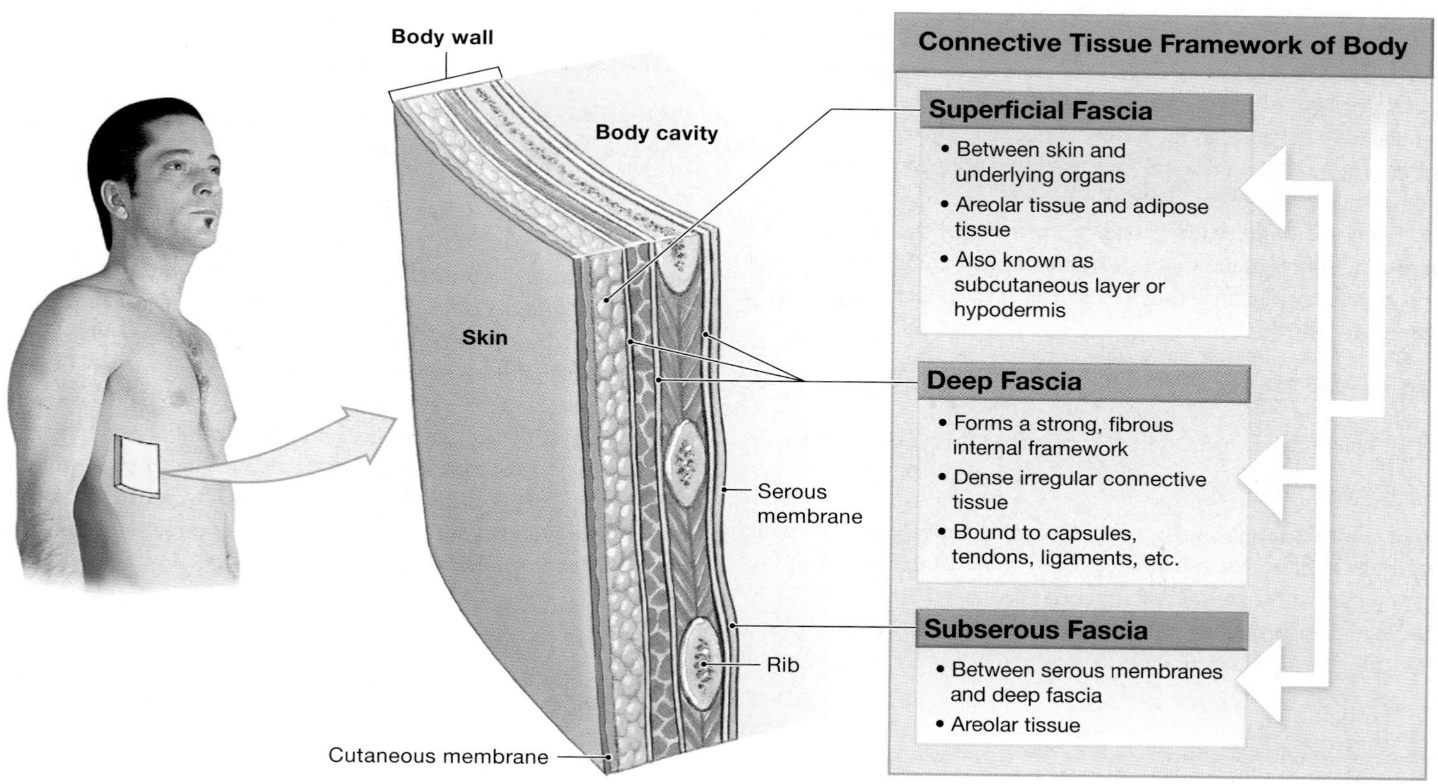

Body wall
Body cavity
Skin
Serous membrane
Rib
Cutaneous membrane
Connective Tissue Framework of Body
Superficial Fascia
• Between skin and underlying organs
• Areolar tissue and adipose tissue
• Also known as subcutaneous layer or hypodermis
Deep Fascia
• Forms a strong, fibrous internal framework
• Dense irregular connective tissue
• Bound to capsules, tendons, ligaments, etc.
Subserous Fascia
• Between serous membranes and deep fascia
• Areolar tissue

3.4 CONCEPT CHECK

9 Give another name for the superficial fascia. What does it do?

10 Which layer of fascia lies between the deep fascia and the serous membranes, and what is its function?

See the blue Answers tab at the back of the book.

3.5 Muscle Tissue

▶ **KEY POINT** The three types of muscle tissue—skeletal, cardiac, and smooth—are specialized for contraction.

Muscle tissue is capable of powerful contractions that shorten its cells along the longitudinal axis **(Figure 3.22)**. A muscle cell possesses organelles and properties distinct from those of other cells. Because it is so different from "typical" cells, the cytoplasm of a muscle cell is called **sarcoplasm**, and its plasma membrane is called a **sarcolemma**.

Our bodies have three types of muscle tissue*: skeletal, cardiac, and smooth. The contraction mechanism is similar in all three, but their internal organization is different. Here we will focus on general characteristics; we will discuss each muscle type in detail in Chapters 9, 21, and 25.

Skeletal Muscle Tissue

▶ **KEY POINT** Skeletal muscle is composed of slender multinucleate cells that form striated voluntary muscle.

Skeletal muscle tissue contains very large cells. Individual skeletal muscle cells, called **muscle fibers**, are relatively long and slender, and some may be a foot (0.3 m) or more in length. Each cell is **multinucleate**, containing hundreds of nuclei lying just deep to the sarcolemma **(Figure 3.22a)**. Skeletal muscle fibers cannot divide, but new muscle fibers can be produced through the division of **myosatellite cells** (also termed *satellite cells*), a type of stem cell that persists in adult skeletal muscle tissue. As a result, skeletal muscle tissue can at least partially repair itself after an injury.

Skeletal muscle fibers contain actin and myosin contractile filaments arranged in parallel within organized functional groups. As a result, skeletal muscle fibers look banded, or *striated* **(Figure 3.22a)**. Normally, skeletal muscle fibers will not contract unless stimulated by nerves, and the nervous system provides voluntary control over their activities. Thus, skeletal muscle is called **striated voluntary muscle**.

Areolar connective tissue binds skeletal muscle tissue together. The collagen and elastic fibers surrounding each skeletal muscle cell and group of cells blend into those of a tendon or aponeurosis that conducts the force of contraction—usually to a bone of the skeleton. When the muscle tissue contracts, it pulls on the bone, and the bone moves.

Cardiac Muscle Tissue

▶ **KEY POINT** Cardiac muscle tissue, found only in the heart, is striated involuntary muscle.

Cardiac muscle tissue is found in the heart. A **cardiac muscle cell** is smaller than a skeletal muscle fiber and has one centrally placed nucleus. Its prominent striations resemble those of skeletal muscle **(Figure 3.22b)**.

Cardiac muscle cells form extensive connections with one another at specialized regions called **intercalated discs**. As a result, cardiac muscle tissue consists of a branching network of interconnected muscle cells. The intercalated discs help channel the forces of contraction, and gap junctions within the intercalated discs help coordinate the activities of individual cardiac muscle cells. Like skeletal muscle fibers, cardiac muscle cells cannot divide, and because this tissue lacks myosatellite cells, damaged cardiac muscle tissue cannot regenerate.

Cardiac muscle cells do not rely on the nervous system to start a contraction. Instead, specialized cardiac muscle cells called **pacemaker cells** establish a regular rate of contraction. Although the nervous system can alter the rate of pacemaker activity, it does not provide voluntary control over individual cardiac muscle cells. Therefore, cardiac muscle is **striated involuntary muscle**.

Smooth Muscle Tissue

▶ **KEY POINT** Smooth muscle tissue, found in the walls of blood vessels and various organs, is nonstriated involuntary muscle.

Smooth muscle tissue is found in the walls of blood vessels, around hollow organs such as the urinary bladder, and in layers around the respiratory, circulatory, digestive, and reproductive tracts. A smooth muscle cell is small, with tapering ends, and contains a single, centrally located oval nucleus **(Figure 3.22c)**. Because smooth muscle cells can divide, smooth muscle can regenerate after an injury.

The actin and myosin filaments in smooth muscle cells are organized differently from those of skeletal and cardiac muscle, and as a result there are no striations; *it is the only nonstriated muscle tissue*. Some smooth muscle cells contract on their own, through the action of pacemaker cells, while others contract when stimulated by the nervous system. However, the nervous system does not provide voluntary control over those contractions, so smooth muscle is **nonstriated involuntary muscle**.

TIPS & TOOLS

Identifying Skeletal, Smooth, and Cardiac Muscle

When viewing skeletal, smooth, and cardiac muscle, it is important to keep in mind several distinguishing characteristics. Listed below are some of the more obvious characteristics that may be used to help identify the muscle type.

Muscle (Section Type)	Nucleus	Shape; Size Variation
Skeletal muscle (cross section)	Multiple nuclei; peripherally located	Rounded cells; minimal size variation
Skeletal muscle (longitudinal section)	Multiple nuclei; peripherally located	Rounded cells; minimal size variation
Smooth muscle (cross section)	Single nucleus; centrally located	Circular cells; considerable size variation
Smooth muscle (longitudinal section)	Single nucleus; centrally located	Spindle-shaped cells; considerable size variation
Cardiac muscle (cross section)	Single nucleus; centrally located	Rounded cells; moderate size variation
Cardiac muscle (longitudinal section)	Single nucleus; centrally located	Branched cells; moderate size variation

3.5 CONCEPT CHECK

11 Muscle tissue is specialized to accomplish what function?

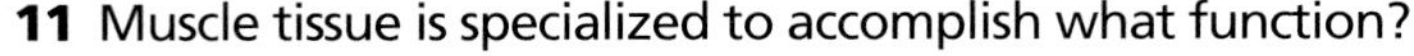

12 How do skeletal, cardiac, and smooth muscle cells differ?

See the blue Answers tab at the back of the book.

* *Terminologia Histologica: International Terms for Human Cytology and Histology* (2008) splits striated muscle tissue into noncardiac striated muscle tissue and cardiac striated muscle tissue. We will use skeletal muscle tissue and cardiac muscle tissue respectively..

Figure 3.22 Histology of Muscle Tissue

3

Skeletal Muscle Tissue

Cells are long, cylindrical, striated, and multinucleate.

LOCATIONS: Combined with connective tissues and neural tissue in skeletal muscles

FUNCTIONS: Moves or stabilizes the position of the skeleton; guards entrances and exits to the digestive, respiratory, and urinary tracts; generates heat; protects internal organs

Striations

Nuclei

Muscle fiber

LM × 180

a Skeletal Muscle Fibers. Note the large fiber size, prominent banding pattern, multiple nuclei, and unbranched arrangement.

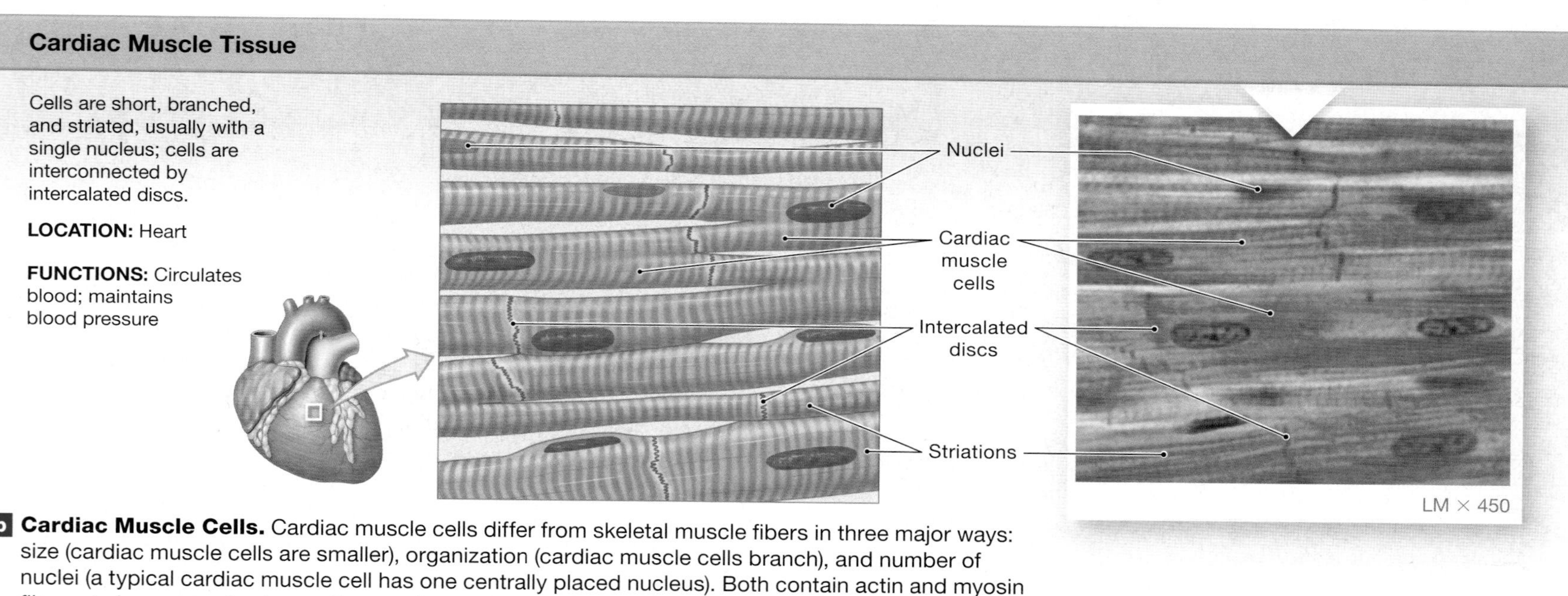

Cardiac Muscle Tissue
Cells are short, branched, and striated, usually with a single nucleus; cells are interconnected by intercalated discs.
LOCATION: Heart
FUNCTIONS: Circulates blood; maintains blood pressure
Nuclei
Cardiac muscle cells
Intercalated discs
Striations
LM × 450

b Cardiac Muscle Cells. Cardiac muscle cells differ from skeletal muscle fibers in three major ways: size (cardiac muscle cells are smaller), organization (cardiac muscle cells branch), and number of nuclei (a typical cardiac muscle cell has one centrally placed nucleus). Both contain actin and myosin filaments in an organized array that produces the striations seen in both types of muscle cell.

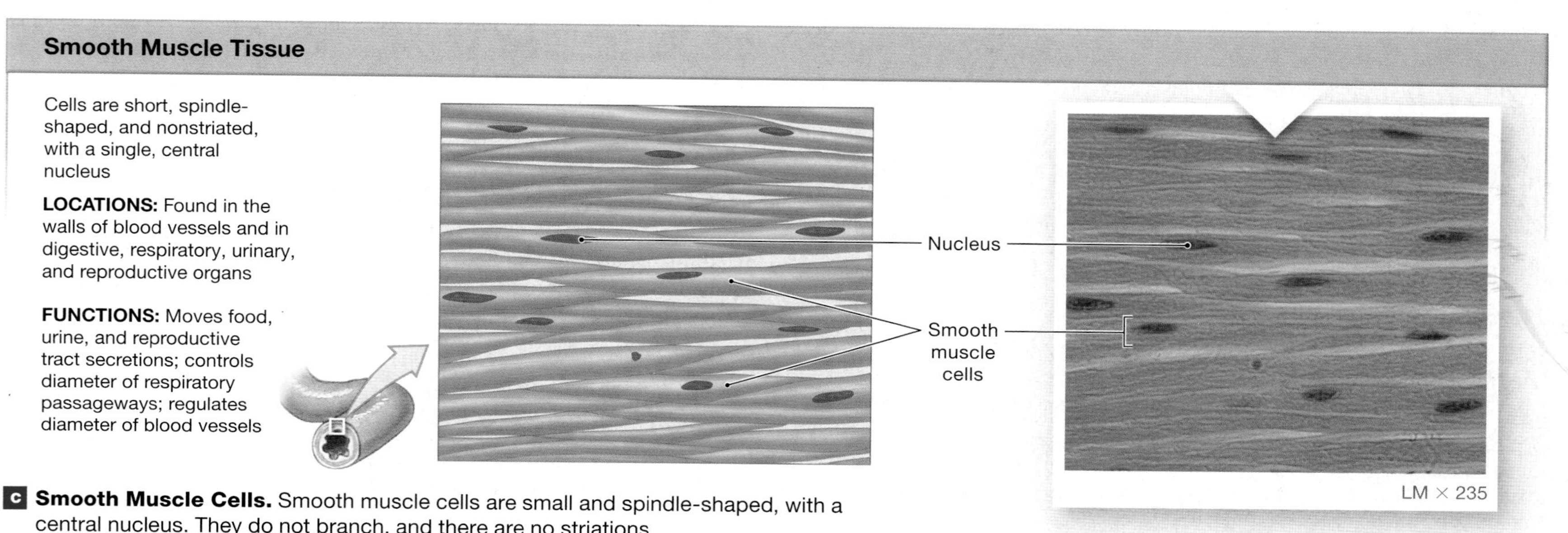

Smooth Muscle Tissue
Cells are short, spindle-shaped, and nonstriated, with a single, central nucleus
LOCATIONS: Found in the walls of blood vessels and in digestive, respiratory, urinary, and reproductive organs
FUNCTIONS: Moves food, urine, and reproductive tract secretions; controls diameter of respiratory passageways; regulates diameter of blood vessels
Nucleus
Smooth muscle cells
LM × 235

c Smooth Muscle Cells. Smooth muscle cells are small and spindle-shaped, with a central nucleus. They do not branch, and there are no striations.

Figure 3.23 Histology of Nervous Tissue. Diagrammatic and histological views of a representative neuron. Neurons conduct electrical impulses over relatively long distances within the body.

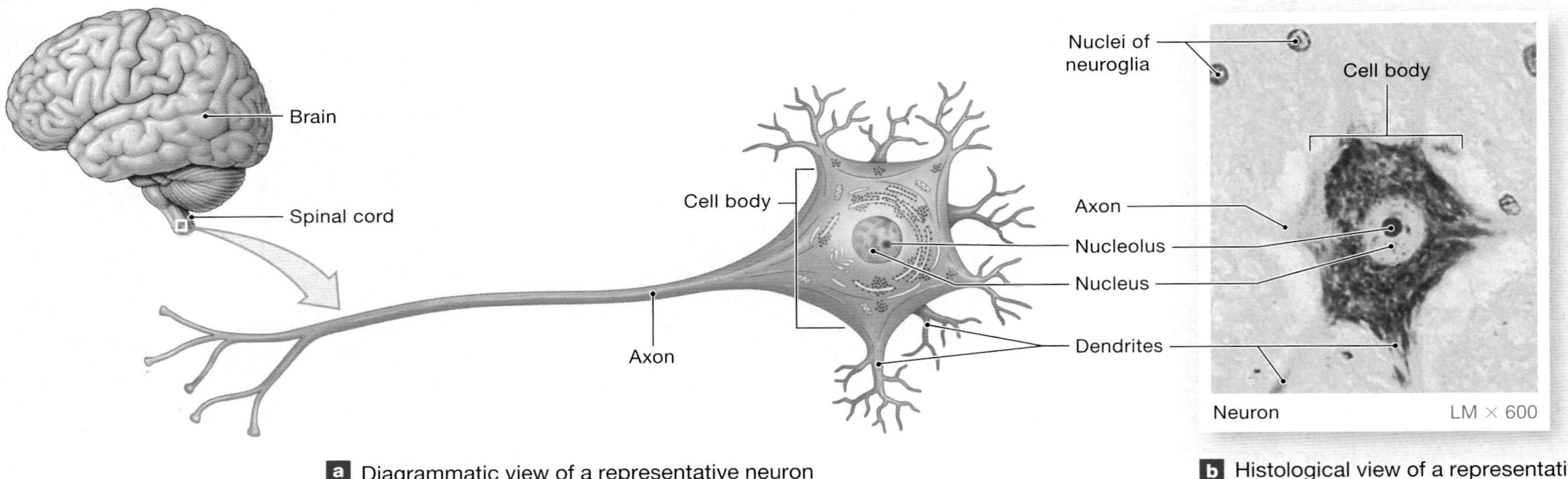

a Diagrammatic view of a representative neuron

b Histological view of a representative neuron

3.6 Nervous Tissue

▶ **KEY POINT** Nervous tissue is specialized to conduct electrical impulses from one region of the body to another.

Most of the **nervous tissue** in the body (roughly 96 percent) is concentrated in the brain and spinal cord, the control centers for the nervous system. Nervous tissue contains two basic types of cells: neurons and neuroglia.

- **Neurons** (NŪ-ronz; *neuro*, nerve), or *nerve cells*, transmit electrical impulses along their plasma membrane. All of the functions of the nervous system involve changes in the pattern and frequency of the impulses carried by individual neurons.
- **Neuroglia** (nū-ROG-lē-a; *glia*, glue) is a general term for several different kinds of supporting cells. Neuroglia have various functions, such as supporting nervous tissue, regulating the composition of the interstitial fluid, and providing nutrients to neurons.

Neurons are the longest cells in the body, with many reaching a meter in length. Most neurons are incapable of dividing under normal circumstances, and they have a very limited ability to repair themselves after injury. A neuron has a **cell body**, or *soma*, that contains a large, prominent nucleus **(Figure 3.23)**. Typically, the cell body is attached to several branching processes, called **dendrites** (DEN-drīts; *dendron*, tree), and a single **axon**. Dendrites receive incoming messages; axons conduct outgoing messages. It is the length of the axon that can make a neuron so long; because axons are very slender, they are also called **nerve fibers**. In Chapter 13 we will discuss nervous tissue in more detail.

3.6 CONCEPT CHECK

13 Nervous tissue is specialized to accomplish what function?

14 What two types of cells are found in nervous tissue, and what are their functions?

See the blue Answers tab at the back of the book.

3.7 Tissues and Aging

▶ **KEY POINT** Tissue structure, repair, and maintenance undergo normal changes with aging.

As tissues age, repair and maintenance activities become less efficient, and a combination of hormonal changes and alterations in lifestyle affect the structure and composition of many tissues. Epithelial tissues get thinner, and connective tissues become more fragile. We bruise easily, and bones become brittle. Joint pains and broken bones are common complaints. Because cardiac muscle cells and neurons cannot be replaced, over time, cumulative losses from relatively minor damage can contribute to major health problems such as cardiovascular disease and impaired mental function.

In future chapters we will discuss the effects of aging on specific organs and systems. Some of these changes are genetically programmed. For example, the chondrocytes of older individuals produce a slightly different form of proteoglycan than those of younger people. The difference probably accounts for the observed changes in the thickness and elasticity of cartilage. In other cases, tissue degeneration may be temporarily slowed or even reversed.

The age-related reduction in bone strength in women, a condition called **osteoporosis**, is often caused by a combination of inactivity, low dietary calcium levels, and a reduction in circulating estrogens (female sex hormones). Exercise and calcium supplements, sometimes combined with hormone replacement therapies, can usually maintain normal bone structure for many years. (The risks versus potential benefits of hormone replacement therapies must be carefully evaluated on an individual basis.)

In this chapter we have introduced the four basic types of tissue found in the human body. In combination, these tissues form all the organs and systems that we will discuss in future chapters and in the Embryology Summary at the end of this chapter.

3.7 CONCEPT CHECK

15 What general changes occur within the tissues of the body as a person ages?

See the blue Answers tab at the back of the book.

CLINICAL NOTE

Cell Division, Tumor Formation, and Cancer

When cells divide and grow at an abnormal rate, they form a **tumor,** or **neoplasm.** If the tumor is **benign,** the cells remain within a connective tissue capsule. If the cells no longer respond to feedback mechanisms and spread to surrounding or distant tissues, the tumor is **malignant** and is known as **cancer.** Malignant cells that spread to distant tissues through the lymphatic system or blood form **metastatic tumors.**

Cancer cells lose their resemblance to normal cells, both in appearance and in function. They grow and multiply rapidly at the expense of normal tissues, often causing progressive weight loss.

Oncologists are physicians who specialize in treating cancer. Cancers are classified in two ways: by the tissue of origin (histological type) and by the primary site of the tumor (Table 3.3).

Table 3.3 | Cancer Classification by Tissue of Origin

Tissue of Origin	Category
Epithelial Tissue	Carcinomas
Epithelial tissue of organs or glands	Adenocarcinomas
Squamous epithelium	Squamous cell carcinomas
Connective Tissue	Sarcomas
Fibrous connective tissue	Fibrosarcomas
Adipose tissue	Liposarcomas
Supporting connective tissue	Chondrosarcomas and osteosarcomas
Fluid Connective Tissue	
Plasma cells of bone marrow	Myelomas
Blood	Leukemias
Lymph	Lymphomas
Muscle Tissue	
Skeletal muscle	Rhabdomyosarcomas
Smooth muscle	Leiomyosarcomas
Neurogenic Connective Tissue Found in the Brain	
Neural interstitial tissue	Gliomas

The Development of Cancer. Diagram of abnormal cell divisions leading to the formation of a tumor. Blood vessels grow into the tumor, and tumor cells invade the blood vessels to travel throughout the body.

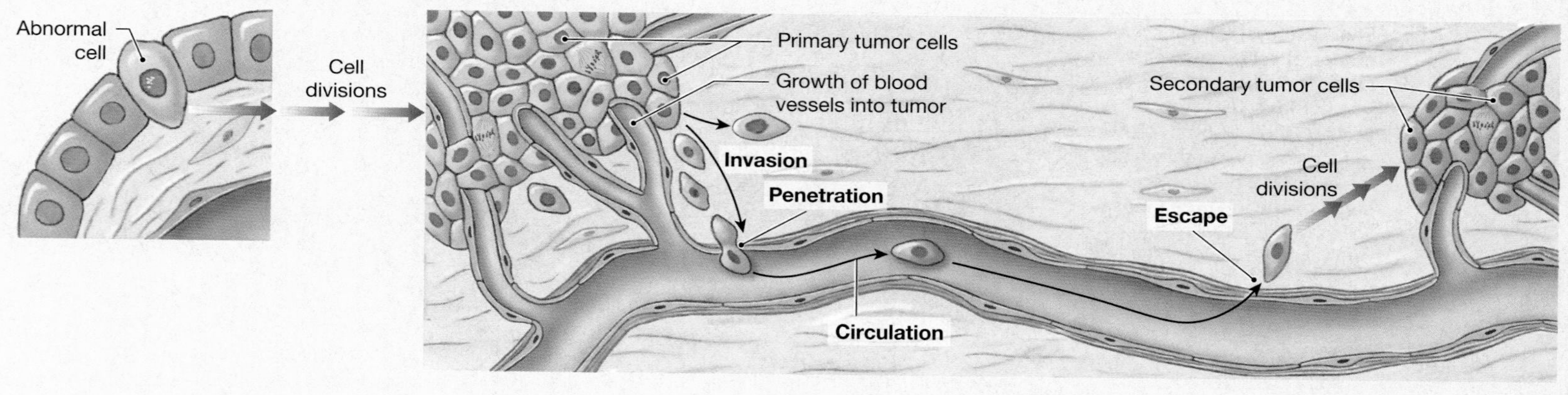

3.8 | Summary of Early Embryology
The Formation of Tissues

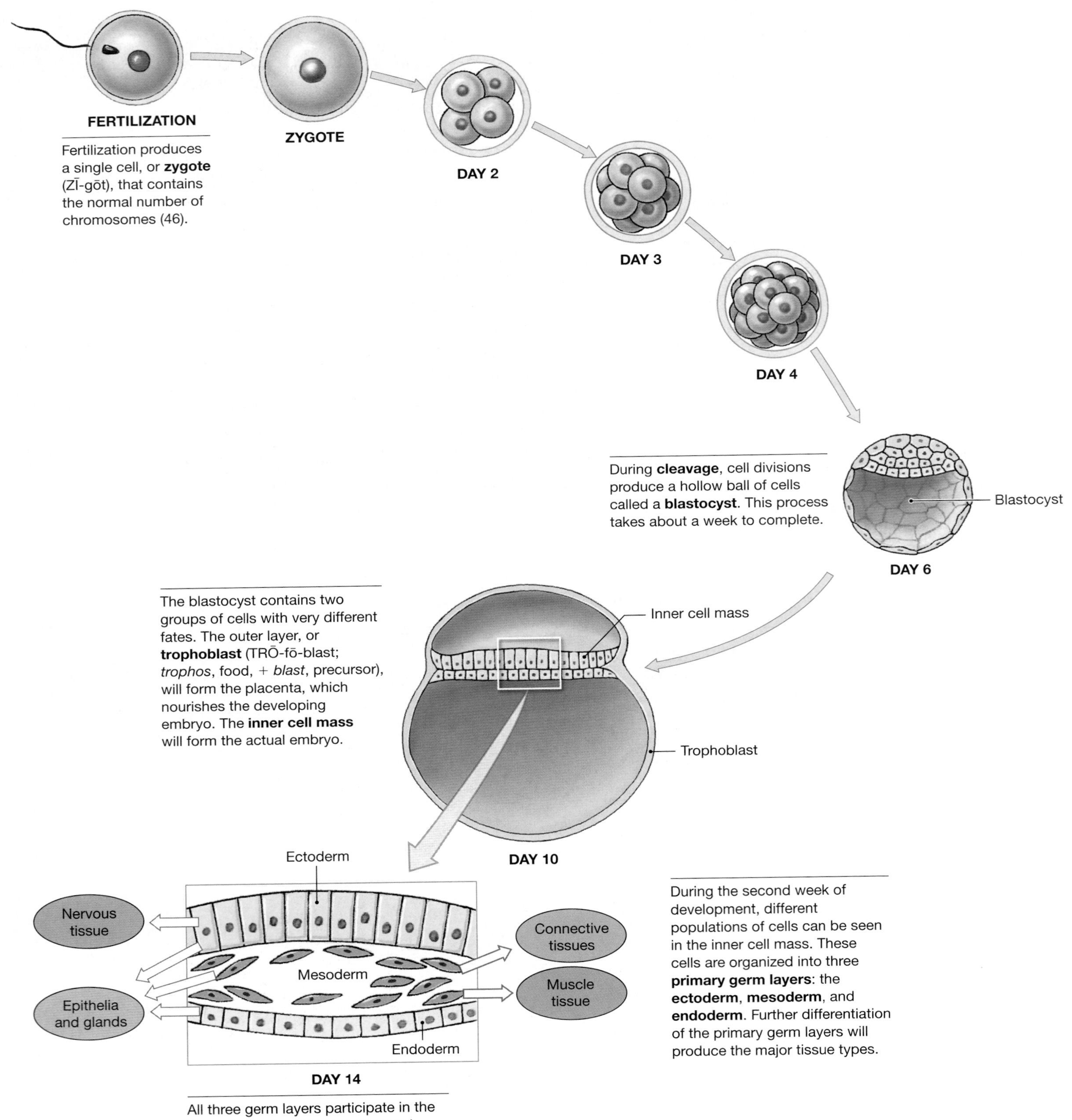

FERTILIZATION
Fertilization produces a single cell, or **zygote** (ZĪ-gōt), that contains the normal number of chromosomes (46).
ZYGOTE
DAY 2
DAY 3
DAY 4
During **cleavage**, cell divisions produce a hollow ball of cells called a **blastocyst**. This process takes about a week to complete.
Blastocyst
DAY 6
The blastocyst contains two groups of cells with very different fates. The outer layer, or **trophoblast** (TRŌ-fō-blast; *trophos*, food, + *blast*, precursor), will form the placenta, which nourishes the developing embryo. The **inner cell mass** will form the actual embryo.
Inner cell mass
Trophoblast
DAY 10
Ectoderm
Nervous tissue
Mesoderm
Epithelia and glands
Connective tissues
Muscle tissue
Endoderm
DAY 14
During the second week of development, different populations of cells can be seen in the inner cell mass. These cells are organized into three **primary germ layers**: the **ectoderm**, **mesoderm**, and **endoderm**. Further differentiation of the primary germ layers will produce the major tissue types.
All three germ layers participate in the formation of functional organs and organ systems. Their interactions will be detailed in later Embryology Summaries dealing with specific systems.

The Development of Epithelia

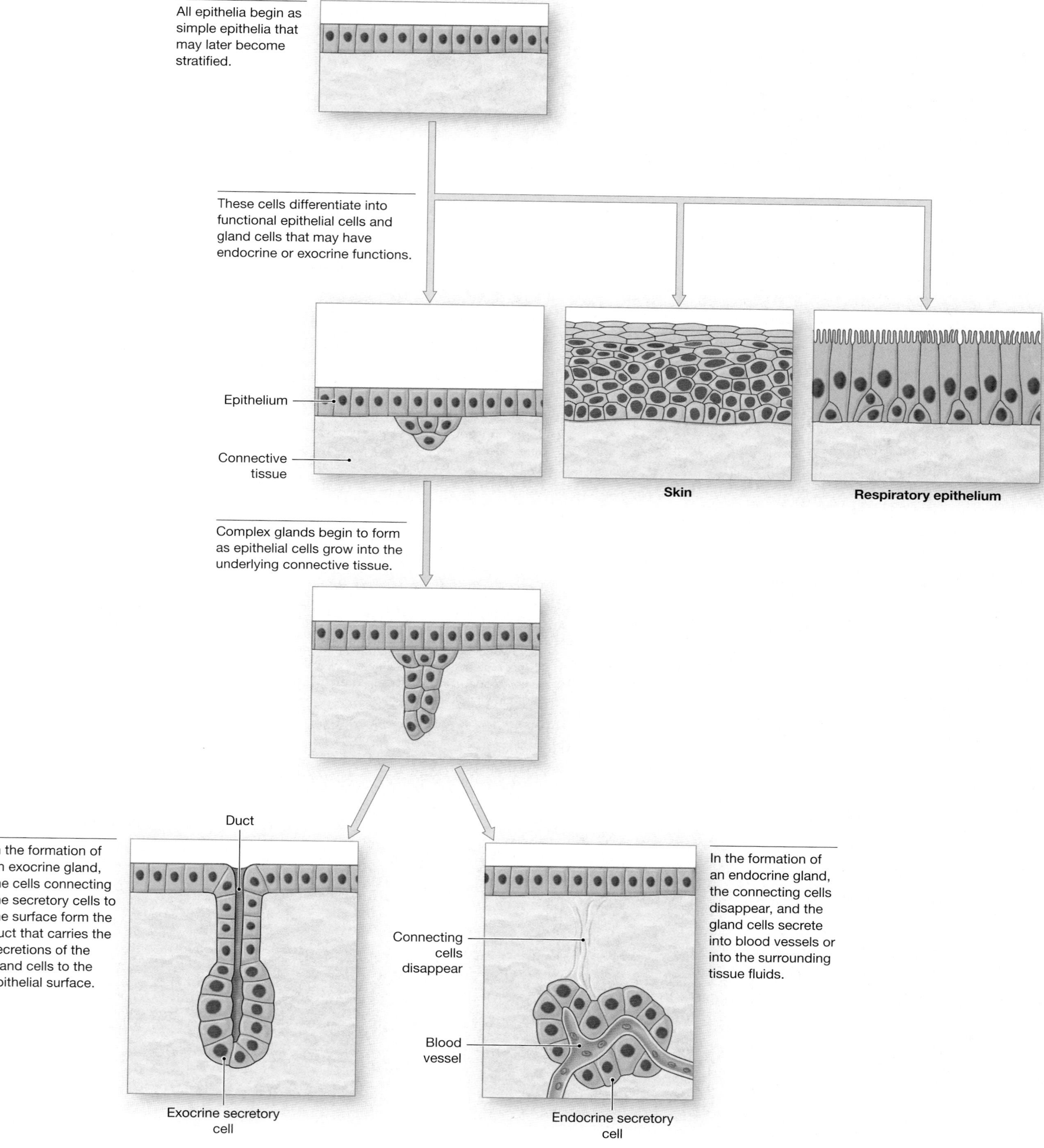

All epithelia begin as simple epithelia that may later become stratified.
These cells differentiate into functional epithelial cells and gland cells that may have endocrine or exocrine functions.
Epithelium
Connective tissue
Skin
Respiratory epithelium
Complex glands begin to form as epithelial cells grow into the underlying connective tissue.
Duct
In the formation of an exocrine gland, the cells connecting the secretory cells to the surface form the duct that carries the secretions of the gland cells to the epithelial surface.
Exocrine secretory cell
Connecting cells disappear
Blood vessel
Endocrine secretory cell
In the formation of an endocrine gland, the connecting cells disappear, and the gland cells secrete into blood vessels or into the surrounding tissue fluids.

Origins of Connective Tissues

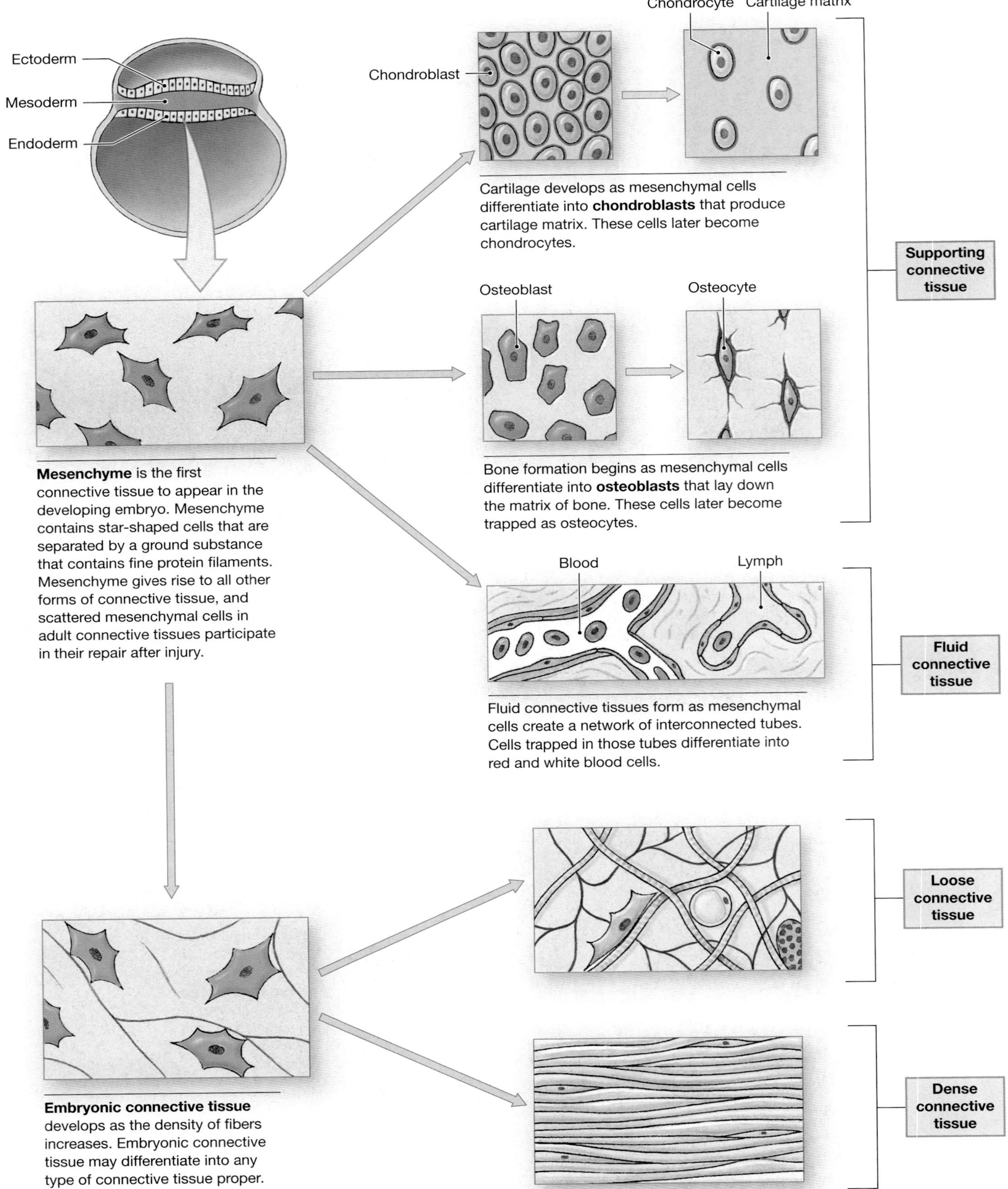

Cartilage develops as mesenchymal cells differentiate into **chondroblasts** that produce cartilage matrix. These cells later become chondrocytes.

Bone formation begins as mesenchymal cells differentiate into **osteoblasts** that lay down the matrix of bone. These cells later become trapped as osteocytes.

Mesenchyme is the first connective tissue to appear in the developing embryo. Mesenchyme contains star-shaped cells that are separated by a ground substance that contains fine protein filaments. Mesenchyme gives rise to all other forms of connective tissue, and scattered mesenchymal cells in adult connective tissues participate in their repair after injury.

Fluid connective tissues form as mesenchymal cells create a network of interconnected tubes. Cells trapped in those tubes differentiate into red and white blood cells.

Embryonic connective tissue develops as the density of fibers increases. Embryonic connective tissue may differentiate into any type of connective tissue proper.

The Development of Organ Systems

Many different organ systems show similar patterns of organization. For example, the digestive, respiratory, urinary, and reproductive systems each include passageways lined by epithelia and surrounded by layers of smooth muscle. These patterns are the result of developmental processes that take place in the first two months of embryonic life.

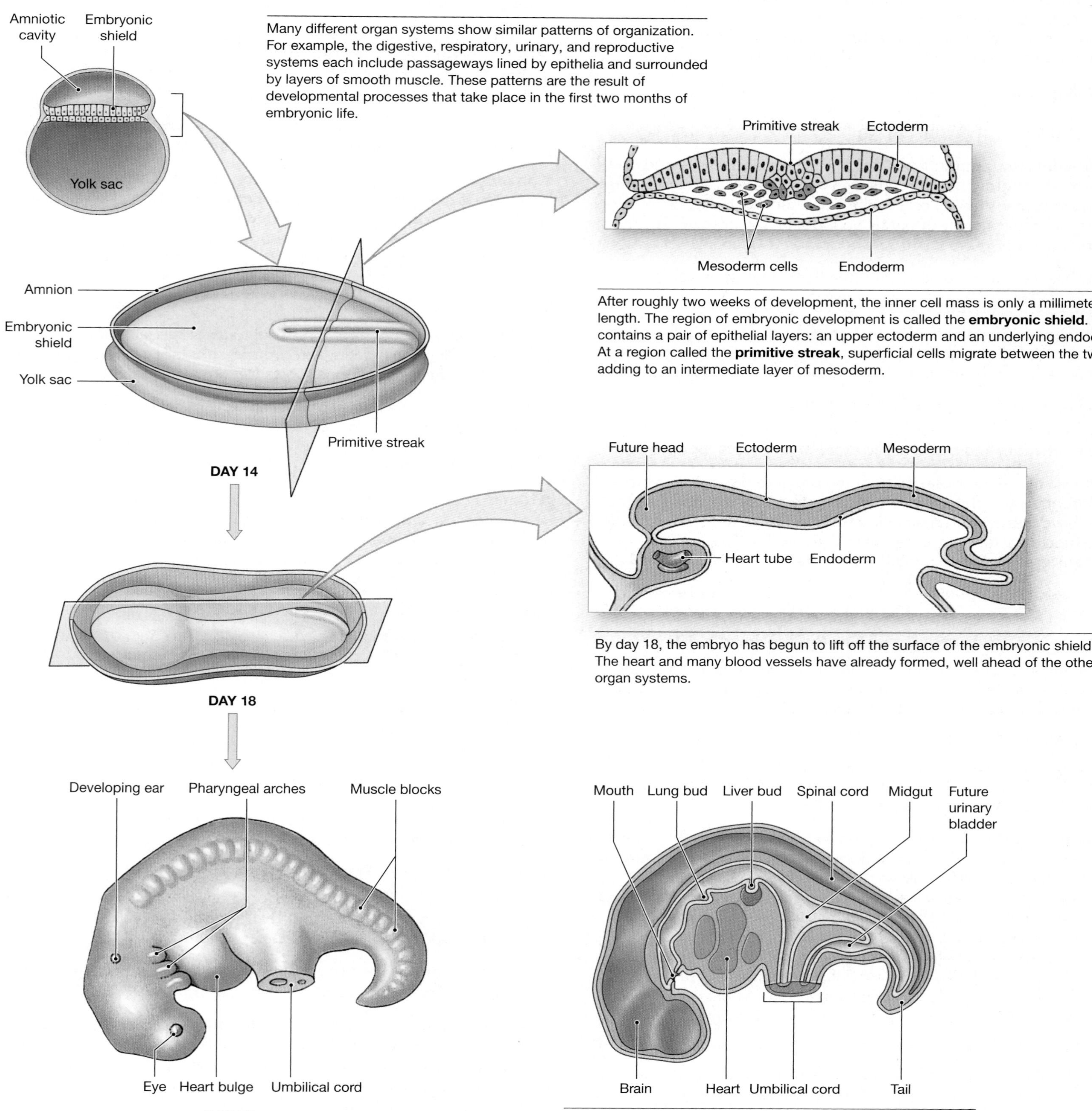

After roughly two weeks of development, the inner cell mass is only a millimeter in length. The region of embryonic development is called the **embryonic shield**. It contains a pair of epithelial layers: an upper ectoderm and an underlying endoderm. At a region called the **primitive streak**, superficial cells migrate between the two, adding to an intermediate layer of mesoderm.

By day 18, the embryo has begun to lift off the surface of the embryonic shield. The heart and many blood vessels have already formed, well ahead of the other organ systems.

After one month, all major organ systems have begun to develop.

3

Study Outline

Introduction p. 50

- **Tissues** are collections of specialized cells and cell products that are organized to perform a relatively limited number of functions. There are four **primary tissue types**: epithelial tissue, connective tissue, muscle tissue, and nervous tissue. *(See Figure 3.1.)*
- Histology is the study of tissues.

3.1 Epithelial Tissue p. 50

- **Epithelial tissues** include *epithelia*, which cover surfaces, and *glands*, which are secretory structures derived from epithelia. An **epithelium** is an **avascular** sheet of cells that forms a surface, lining, or covering. Epithelia consist mainly of tightly bound cells, rather than extracellular materials. *(See Figures 3.2 to 3.9 and Spotlight Figure 3.10.)*
- Epithelial cells are replaced continually through stem cell activity.

Functions of Epithelial Tissue p. 50

- Epithelia provide physical protection, control permeability, provide sensation, and produce specialized **secretions**. **Gland cells** are epithelial cells (or cells derived from them) that produce secretions.

Specializations of Epithelial Cells p. 50

- Epithelial cells are specialized to maintain the physical integrity of the epithelium and perform secretory or transport functions.
- Epithelia may show **polarity** from the **basal** to the **apical surface**; cells connect to neighbor cells on their lateral surfaces; some epithelial cells have microvilli on their apical surfaces. There are often structural and functional differences between the apical surface and the **basolateral surfaces** of individual epithelial cells. *(See Figure 3.2.)*
- The coordinated beating of the cilia on a **ciliated epithelium** moves materials across the epithelial surface. *(See Figure 3.2.)*

Maintaining the Integrity of the Epithelium p. 52

- All epithelial tissue rests on an underlying **basement membrane**. In areas exposed to extreme chemical or mechanical stresses, divisions by **stem cells** replace the short-lived epithelial cells. *(See Figure 3.3a.)*

Classification of Epithelia p. 53

- Epithelia are classified on the basis of the number of cell layers in the epithelium and the shape of the exposed cells at the surface of the epithelium. *(See Figures 3.4 to 3.7.)*
- A **simple epithelium** has a single layer of cells covering the basement membrane. A **stratified epithelium** has several layers. In a **squamous epithelium** the surface cells are thin and flat; in a **cuboidal epithelium** the cells resemble short hexagonal boxes; in a **columnar epithelium** the cells are also hexagonal, but they are relatively tall and slender. **Pseudostratified columnar epithelium** contains columnar cells, some of which possess cilia, and mucous (secreting) cells that appear stratified but are not. A **transitional epithelium** is characterized by a mixture of what appears to be both cuboidal and squamous cells arranged to permit stretching. *(See Figures 3.4 to 3.7.)*

Glandular Epithelia p. 54

- We classify glands by the type of secretion produced, the structure of the gland, and their mode of secretion. *(See Figures 3.8/3.9 and Spotlight Figure 3.10.)*
- **Exocrine** secretions are discharged through **ducts** onto the skin or an epithelial surface that communicates with the exterior. **Endocrine** secretions, known as **hormones**, are released by gland cells into the interstitial fluid surrounding the cell.
- Exocrine glands are classified as **serous** (producing a watery solution usually containing enzymes), **mucous** (producing a viscous, sticky mucus), or **mixed** (producing both types of secretions.)
- In epithelia that contain scattered gland cells, the individual secretory cells are called **unicellular glands**. **Multicellular glands** are glandular epithelia or aggregations of gland cells that produce exocrine or endocrine secretions. *(See Figures 3.8/3.9.)*
- A glandular epithelial cell may release its secretions through a eccrine, apocrine, or holocrine mechanism. *(See Spotlight Figure 3.10.)*
- In **eccrine secretion**, the most common method of secretion, the product is released by exocytosis. **Apocrine secretion** involves the loss of both secretory product and some cytoplasm. Unlike the other two methods, **holocrine secretion** destroys the cell, which becomes packed with secretory product before bursting. *(See Spotlight Figure 3.10.)*

3.2 Connective Tissues p. 59

- All connective tissues have three components: specialized cells, extracellular protein fibers, and **ground substance**. The combination of protein fibers and ground substance forms the **matrix** of the tissue.
- Whereas epithelia consist almost entirely of cells, the extracellular matrix accounts for most of the volume of a connective tissue. Therefore, connective tissues, with the exception of adipose tissue, are identified by the characteristics of the extracellular matrix.
- Connective tissue is an internal tissue with many important functions, including establishing a structural framework; transporting fluids and dissolved materials; protecting delicate organs; supporting, surrounding, and interconnecting tissues; storing energy reserves; and defending the body from microorganisms.

Classification of Connective Tissues p. 59

- **Connective tissue proper** refers to all connective tissues that contain varied cell populations and fiber types suspended in a viscous ground substance. *(See Figure 3.11.)*
- **Fluid connective tissues** have a distinctive population of cells suspended in a watery ground substance containing dissolved proteins. Blood and lymph are examples of fluid connective tissues. *(See Figure 3.11.)*
- **Supporting connective tissues** have a less diverse cell population than connective tissue proper. Additionally, they have a dense matrix that contains closely packed fibers. The two types of supporting connective tissues are cartilage and bone. *(See Figure 3.11.)*

Connective Tissue Proper p. 61

- Connective tissue proper is composed of extracellular fibers, a viscous ground substance, and two categories of cells: **fixed cells** and **wandering cells**. *(See Figure 3.12 and Table 3.1.)*
- There are three types of fibers in connective tissue: **collagen fibers**, **reticular fibers**, and **elastic fibers**. *(See Figures 3.12/3.13/3.14.)*
- Connective tissue proper includes **loose** and **dense connective tissues**. There are three types of loose connective tissues: **areolar tissue**, **adipose tissue**, and **reticular tissue**. Most of the volume of loose connective tissue is ground substance, a viscous fluid that cushions shocks. Most of the volume in dense connective tissue consists of extracellular protein fibers. There are two types of dense connective tissue: **dense regular connective tissue**, in which fibers are parallel and aligned along lines of stress, and **dense irregular connective tissue**, in which fibers form an interwoven meshwork. *(See Figures 3.13/3.14.)*

Fluid Connective Tissues p. 65

- **Blood** and **lymph** are examples of fluid connective tissues, each with a distinctive collection of cells in a watery matrix. Both blood and lymph contain cells and many different types of dissolved proteins. *(See Figure 3.15.)*
- Extracellular fluid includes the plasma of blood; the interstitial fluid within other connective tissues and other tissue types; and lymph, which is confined to vessels of the lymphatic system.

Supporting Connective Tissues p. 67

- Cartilage and bone are called supporting connective tissues because they support the rest of the body. *(See Figures 3.16/3.17.)*
- The matrix of **cartilage** is a firm gel that contains **chondroitin sulfates**. It is produced by immature cells called chondroblasts and maintained by mature cells called **chondrocytes**. A fibrous covering called the **perichondrium** separates cartilage from surrounding tissues. Cartilage grows by two different mechanisms, **appositional growth** (growth at the surface) and **interstitial growth** (growth from within). *(See Figure 3.16).*
- There are three types of cartilage: **hyaline cartilage**, **elastic cartilage**, and **fibrous cartilage**. *(See Figure 3.17 and Table 3.2.)*
- **Bone** (*osseous tissue*) has a matrix consisting of collagen fibers and calcium salts, giving it unique properties. *(See Figure 3.18.)*
- **Osteocytes** in **lacunae** depend on diffusion through intercellular connections or **canaliculi** for nutrient intake. *(See Figure 3.18 and Table 3.2.)*
- All bone surfaces except those inside joint cavities are covered by a **periosteum** that has fibrous and cellular layers. The periosteum helps attach the bone to surrounding tissues, tendons, and ligaments, and it helps repair bone after an injury.

3.3 Membranes p. 70

- **Membranes** form a barrier or interface. Epithelia and connective tissues combine to form membranes that cover and protect other structures and tissues. There are four types of membranes: mucous, serous, cutaneous, and synovial. *(See Figure 3.19.)*

Mucous Membranes p. 71

- **Mucous membranes** line passageways that communicate with the exterior, such as the digestive and respiratory tracts. These surfaces are usually moistened by mucous secretions. They contain areolar tissue called the **lamina propria**. *(See Figure 3.19a.)*

Serous Membranes p. 71

- **Serous membranes** line internal cavities and are delicate, moist, and very permeable. Examples include the **pleural**, **peritoneal**, and **pericardial** membranes. Each serous membrane forms a fluid called a **transudate**. *(See Figure 3.19b.)*

The Cutaneous Membrane (Skin) p. 72

- The **cutaneous membrane**, or *skin*, covers the body surface. Unlike other membranes, it is relatively thick, waterproof, and usually dry. *(See Figure 3.19c.)*

Synovial Membranes p. 72

- The **synovial membrane**, located within the cavity of synovial joints, produces **synovial fluid** that fills joint cavities. Synovial fluid helps lubricate the joint and promotes smooth movement. *(See Figure 3.19d.)*

Embryonic Connective Tissues p. 72

- All connective tissues are derived from embryonic **mesenchyme**. *(See Figure 3.20.)*

3.4 The Connective Tissue Framework of the Body p. 73

- All organ systems are interconnected by a network of connective tissue proper that includes the **superficial fascia** (the **subcutaneous layer** or *hypodermis*, separating the skin from underlying tissues and organs), the **deep fascia** (dense connective tissue), and the **subserous fascia** (the layer between the deep fascia and the serous membranes that line body cavities). *(See Figure 3.21.)*

3.5 Muscle Tissue p. 74

- **Muscle tissue** consists primarily of cells that are specialized for contraction. There are three different types of muscle tissue: skeletal muscle, cardiac muscle, and smooth muscle. *(See Figure 3.22.)*

Skeletal Muscle Tissue p. 74

- **Skeletal muscle tissue** contains large, cylindrical **muscle fibers** interconnected by collagen and elastic fibers. Skeletal muscle fibers have striations due to the organization of their contractile proteins. Because we can control the contraction of skeletal muscle fibers through the nervous system, skeletal muscle is classified as **striated voluntary muscle**. **Myosatellite cells** divide to produce new muscle fibers. *(See Figure 3.22a.)*

Cardiac Muscle Tissue p. 74

- **Cardiac muscle tissue** is found only in the heart. It is composed of unicellular, branched short cells. The nervous system does not provide voluntary control over cardiac muscle cells; thus, cardiac muscle is **striated involuntary muscle**. *(See Figure 3.22b.)*

Smooth Muscle Tissue p. 74

- **Smooth muscle tissue** is composed of short, tapered cells containing a single nucleus. It is found in the walls of blood vessels, around hollow organs, and in layers around various tracts. It is **nonstriated involuntary muscle**. Smooth muscle cells can divide and therefore regenerate after injury. *(See Figure 3.22c.)*

3.6 Nervous Tissue p. 76

- **Nervous tissue** is specialized to conduct electrical impulses from one area of the body to another.
- Nervous tissue consists of two cell types: neurons and neuroglia. **Neurons** transmit information as electrical impulses. There are different kinds of **neuroglia**; among their other functions, these cells provide a supporting framework for nervous tissue and play a role in providing nutrients to neurons. *(See Figure 3.23.)*
- Neurons have a **cell body**, or *soma*, that contains a large prominent nucleus. Various branching processes termed **dendrites** and a single **axon** or **nerve fiber** extend from the cell body. Dendrites receive incoming messages; axons conduct messages toward other cells. *(See Figure 3.23.)*

3.7 Tissues and Aging p. 76

- Tissues change with age. Repair and maintenance grow less efficient, and the structure and chemical composition of many tissues are altered.

3.8 Summary of Early Embryology p. 78

- Early embryology includes formation of tissues, development of epithelia, development of connective tissues, and development of organ systems. *(See Embryology Summary illustrations on pp. 78–81.)*

Chapter Review

For answers, see the blue Answers tab at the back of the book.

3

Level 1 Reviewing Facts and Terms

Match each numbered item with the most closely related lettered item.

1. skeletal muscle .... [ ]
2. mast cell .... [ ]
3. avascular .... [ ]
4. transitional .... [ ]
5. goblet cell .... [ ]
6. collagen .... [ ]
7. cartilage .... [ ]
8. simple epithelium .... [ ]
9. ground substance .... [ ]
10. holocrine secretion .... [ ]

a. all epithelia
b. single cell layer
c. urinary bladder
d. cell destroyed
e. connective tissue component
f. unicellular, exocrine gland
g. tendon
h. wandering cell
i. lacunae
j. striated

11. Label the following photomicrographs with the proper identifying terms.
   - skeletal muscle
   - smooth muscle
   - transitional epithelium
   - tendon
   - hyaline cartilage

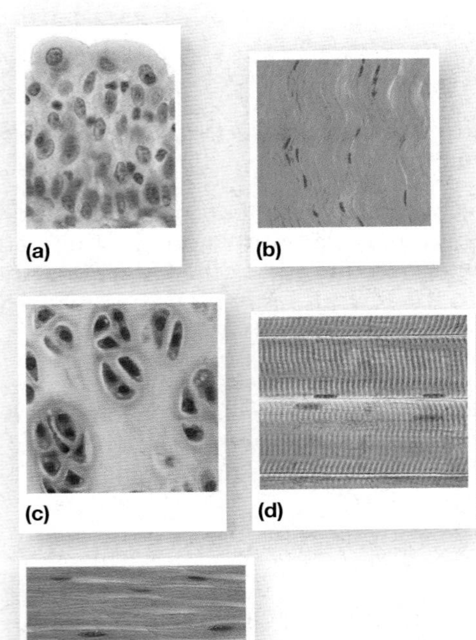

(a) __________
(b) __________
(c) __________
(d) __________
(e) __________

12. Which of the following refers to the dense connective tissue that binds the capsules that surround many organs?
(a) superficial fascia
(b) hypodermis
(c) deep fascia
(d) subserous fascia

13. The reduction of friction between the parietal and visceral surfaces of an internal cavity is the function of
(a) cutaneous membranes.
(b) mucous membranes.
(c) serous membranes.
(d) synovial membranes.

14. Which of the following is not a characteristic of epithelial cells?
(a) They may consist of a single or multiple cell layers.
(b) They always have a free surface exposed to the external environment or some inner chamber or passageway.
(c) They are avascular.
(d) They consist of a few cells but have a large amount of extracellular material.

15. Functions of connective tissue include all of the following except
(a) establishing a structural framework for the body.
(b) transporting fluids and dissolved materials.
(c) storing energy.
(d) providing sensation.

16. What type of supporting tissue is found in the auricle of the ear and the tip of the nose?
(a) bone
(b) fibrous cartilage
(c) elastic cartilage
(d) hyaline cartilage

17. An epithelium is connected to underlying connective tissue by
(a) a basement membrane.
(b) canaliculi.
(c) stereocilia.
(d) proteoglycans.

18. Which of the following are wandering cells found in connective tissue proper?
(a) fixed macrophages
(b) mesenchymal cells and adipocytes
(c) fibroblasts and melanocytes
(d) eosinophils, neutrophils, and mast cells

Level 2 Reviewing Concepts

1. Compare and contrast the role of a tissue in the body to that of a single cell.
2. Compare and contrast the functions of a tendon and a ligament.
3. Compare and contrast exocrine and endocrine secretions.
4. Analyze the significance of the cilia on the respiratory epithelium.
5. Compare a tendon to an aponeurosis.
6. A layer of glycoproteins and a network of fine protein filaments that perform limited functions together act as a barrier that restricts the movement of proteins and other large molecules from the connective tissue to epithelium. This description illustrates the structure and function of
(a) interfacial canals.
(b) the reticular lamina.
(c) the basement membrane.
(d) areolar tissue.
7. Some connective tissue cells respond to an injury or infection by dividing to produce daughter cells that differentiate into other cell types. This statement illustrates the behavior of which of the following cell types?
(a) mast cells
(b) fibroblasts
(c) plasma cells
(d) mesenchymal cells
8. Analyze why pinching the skin usually does not distort or damage the underlying muscles.
9. Identify what stem cells are and analyze their functions.

Level 3 Critical Thinking

1. Analysis of a glandular secretion indicates that it contains some DNA, RNA, and membrane components such as phospholipids. What mode of secretion is this? Explain your answer.
2. Smoking destroys the cilia found on many cells of the respiratory epithelium. Formulate a hypothesis as to why this contributes to a "smoker's cough."
3. Assess why cardiac muscle ischemia (inadequate blood supply) is more life-threatening than skeletal muscle ischemia.

CLINICAL CASE WRAP-UP

The Tallest in the School

Elijah exhibits the classic signs and symptoms of Marfan syndrome, an inherited disorder of connective tissue. A genetic mutation, inherited from his father, has caused abnormal elastic fibers in the connective tissues throughout his body. The skeletal system shows the most visible signs of Marfan syndrome, including Elijah's height, slender build, and disproportionately long, slender limbs, fingers, and toes. The weakened connective tissues making up his ligaments (which connect bone to bone) and tendons (which connect muscle to bone) result in "double-jointedness," flat feet, pectus excavatum (sunken anterior chest), and scoliosis. Even the connective tissue holding the lenses in his eyes (see Chapter 18) is too weak to do the job efficiently.

The most serious complication of Elijah's Marfan syndrome involves his cardiovascular system. The weakened connective tissue of his aorta (the large artery exiting the heart) and the aortic valve "stretched out" (forming an aneurysm) during tryouts, resulting in sudden cardiac arrhythmia and Elijah's collapse. Such a cardiac arrhythmia may have been what caused his father's premature death.

At the hospital, a cardiologist (heart doctor) prescribes medications to slow Elijah's heart rate. He also discusses with Elijah and his family the possibility of surgical replacement of Elijah's aorta and aortic valve. With proper exercise and follow-up, Elijah is expected to lead a good, long life.

1. Describe how Marfan syndrome develops.
2. What other types of connective tissue would be affected by abnormal elastin?

See the blue Answers tab at the back of the book.

Related Clinical Terms

adhesions: Restrictive fibrous connections that can result from surgery, infection, or other injuries to serous membranes.

angiogenesis: The growth of new blood vessels.

ascites (a-SĪ-tēz)**:** An accumulation of peritoneal fluid that creates characteristic abdominal swelling.

effusion: The accumulation of fluid in body cavities.

oncologists (ong-KOL-ō-jists)**:** Physicians who specialize in identifying and treating cancers.

pathologists (pa-THOL-ō-jists)**:** Physicians who specialize in the diagnosis of diseases, primarily from an examination of body fluids, tissue samples, and other anatomical clues.

pericarditis: An inflammation of the pericardium.

peritonitis: An inflammation of the peritoneum.

pleuritis (pleurisy)**:** An inflammation of the lining of the pleural cavities.

4

The Integumentary System

Learning Outcomes

These Learning Outcomes correspond by number to this chapter's sections and indicate what you should be able to do after completing the chapter.

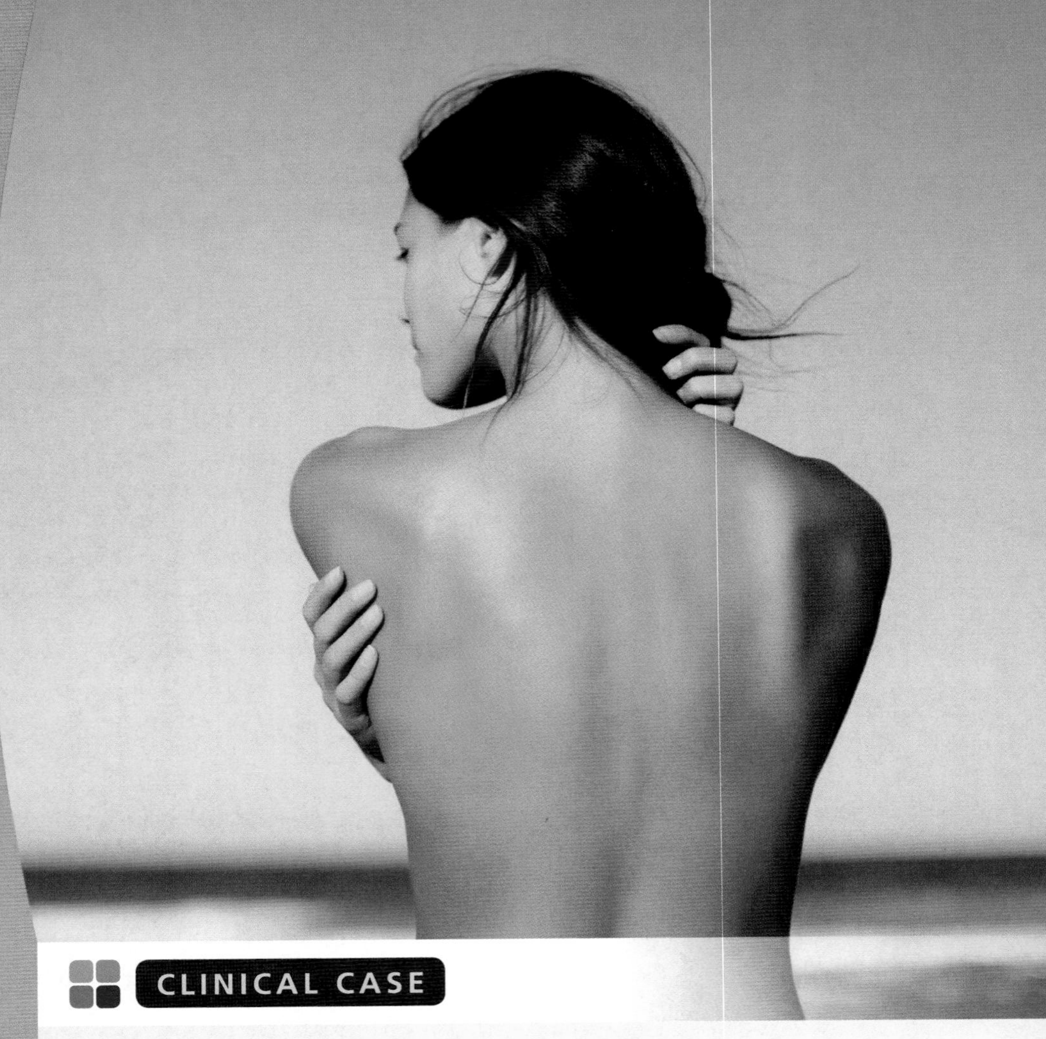

CLINICAL CASE

Flesh-Eating Bacteria

Martin, a 52-year-old with diabetes, has to regularly give himself insulin injections to control his blood sugar level. He hates the shots, and often he doesn't bother to first clean his skin with alcohol. One day while Martin is at work, a few hours after his last injection, his thigh begins throbbing with pain, and he feels feverish and nauseated. He heads to a bathroom, barely able to walk.

In the bathroom, Martin steps into a stall and pulls down his pants. His right thigh is extremely tender, red, and swollen, and when he touches his thigh it makes a crinkling sound, as if there are gas bubbles beneath the skin. Suddenly, he feels dizzy and faints.

A coworker hears the thump as Martin hits the floor. He calls 911, and within minutes the EMTs are there. Martin's blood pressure is so low they are barely able to measure it; his pulse is rapid and weak. Because Martin's pants are down, they can see the inflammation in his thigh. They start intravenous (IV) fluids and transport him to the closest emergency room.

Will Martin survive? To find out, turn to the Clinical Case Wrap-Up on p. 106.

THE **INTEGUMENTARY SYSTEM,** or *integument*, is composed of the skin (cutaneous membrane) and its derivatives: hair, glands, and nails **(Figure 4.1)**. The integument is probably our most closely watched yet underappreciated organ system. We devote a lot of time to improving the appearance of the integument and associated structures: washing our faces, styling our hair, and trimming our nails, for instance.

The skin has more than a cosmetic role, however. Many people use the general appearance of the skin to estimate other people's overall health and age. Indeed, the skin mirrors the general health of other systems, and clinicians use its appearance to detect underlying disease; for example, liver disease results in changes to skin color. Your skin protects you from the surrounding environment; its receptors tell you a lot about the outside world; and it helps regulate your body temperature. You will learn several more important functions as we examine the functional anatomy of the integumentary system in this chapter.

4.1 Structure and Function of the Integumentary System

▶ **KEY POINT** The integument, which has numerous functions, has two major components: the cutaneous membrane (skin) and the accessory structures (hair, nails, and several types of exocrine glands).

The integument covers the entire body surface, including the anterior surface of the eyes and the tympanic membrane (eardrum) at the boundary between the external and middle ear. At the nostrils and lips, anus, urethral opening, and vaginal opening, the integument turns inward, meeting the mucous membranes lining the respiratory, digestive, urinary, and reproductive tracts, respectively. The transition at these sites is seamless, and the epithelial defenses remain intact and functional.

The integument contains all four primary tissue types:

- An epithelium covers its surface.
- Underlying connective tissues make it strong and resilient.
- Smooth muscle tissue within the integument controls the diameters of the blood vessels and adjusts the positions of the hairs that project above the body surface.
- Nervous tissue controls these smooth muscles and monitors sensory receptors that provide the sensations of touch, pressure, temperature, and pain.

The integument has numerous functions, including physical protection, regulation of body temperature, excretion, secretion, nutrition (through vitamin D_3 synthesis), sensation, and immunity. **Figure 4.1** outlines its functional organization.

Figure 4.1 Functional Organization of the Integumentary System. Flowchart showing the relationships among the components of the integumentary system.

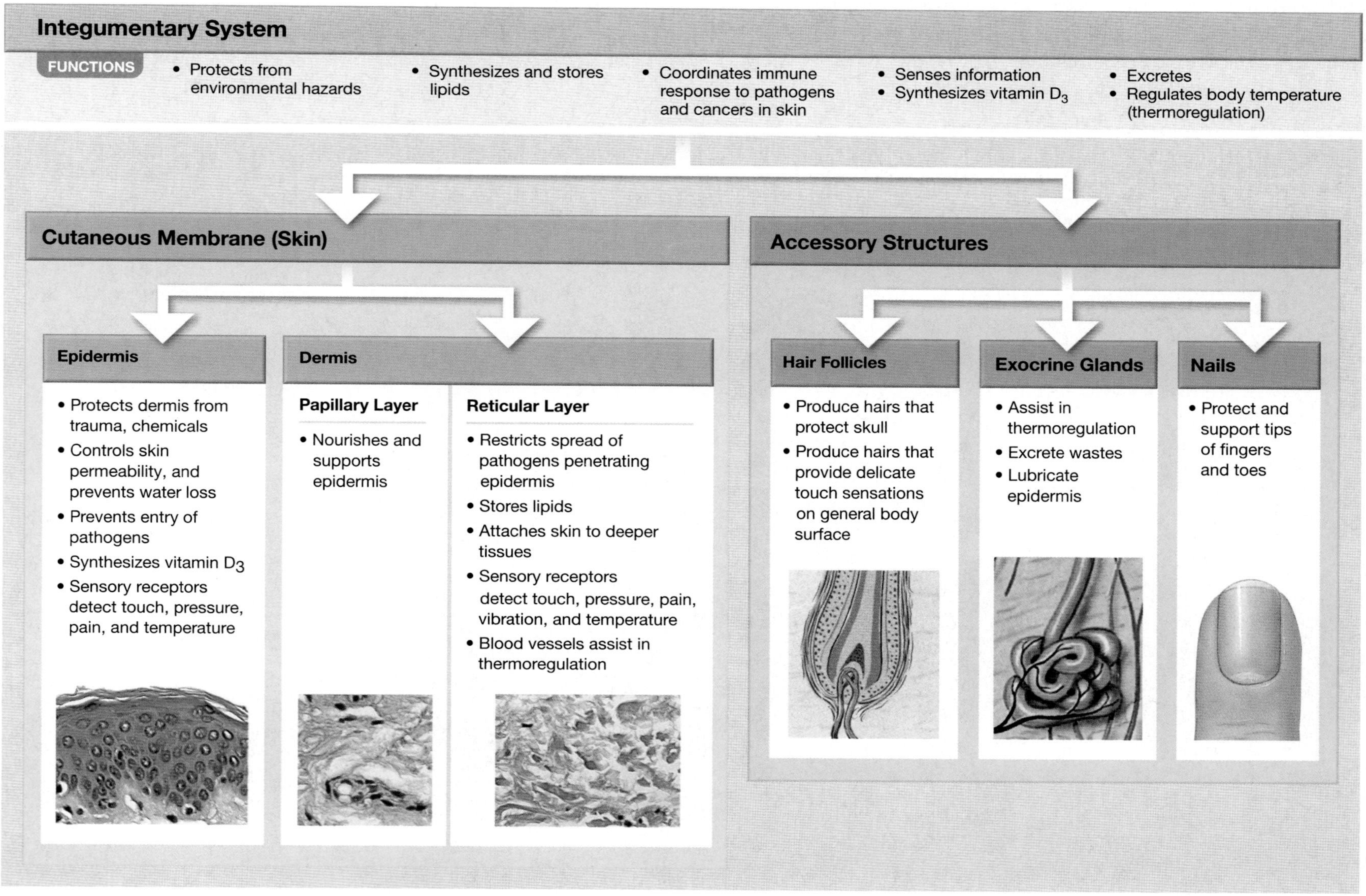

Integumentary System
FUNCTIONS
• Protects from environmental hazards
• Synthesizes and stores lipids
• Coordinates immune response to pathogens and cancers in skin
• Senses information
• Synthesizes vitamin D_3
• Excretes
• Regulates body temperature (thermoregulation)
Cutaneous Membrane (Skin)
Accessory Structures
Epidermis
• Protects dermis from trauma, chemicals
• Controls skin permeability, and prevents water loss
• Prevents entry of pathogens
• Synthesizes vitamin D_3
• Sensory receptors detect touch, pressure, pain, and temperature
Dermis
Papillary Layer
• Nourishes and supports epidermis
Reticular Layer
• Restricts spread of pathogens penetrating epidermis
• Stores lipids
• Attaches skin to deeper tissues
• Sensory receptors detect touch, pressure, pain, vibration, and temperature
• Blood vessels assist in thermoregulation
Hair Follicles
• Produce hairs that protect skull
• Produce hairs that provide delicate touch sensations on general body surface
Exocrine Glands
• Assist in thermoregulation
• Excrete wastes
• Lubricate epidermis
Nails
• Protect and support tips of fingers and toes

4

The integumentary system has two major components: the cutaneous membrane (skin) and the accessory structures. The **skin** is the largest organ of the body. It consists of the **epidermis** (*epi-*, above, + *derma*, skin)—the superficial epithelium—and the underlying connective tissues of the **dermis**. Deep to the dermis, the loose connective tissue of the subcutaneous layer, also known as the *hypodermis* or *superficial fascia*, separates the integument from the deep fascia around other organs, such as muscles and bones. ↺ p. 73 Although it is not usually considered part of the integument, we will consider the subcutaneous layer in this chapter because of its extensive interconnections with the dermis.

The **accessory structures** include hair, nails, and a variety of multicellular exocrine glands. These structures are located in the dermis and protrude through the epidermis to the surface.

4.1 ✔ CONCEPT CHECK

1 Explain how each of the four different tissue types is a component of the integumentary system.

See the blue Answers tab at the back of the book.

4.2 | The Epidermis

▶ **KEY POINT** The epidermis, the most superficial layer of the skin, is composed of a stratified squamous epithelium.

The epidermis of the skin consists of a stratified squamous epithelium (**Figure 4.2**). There are four cell types in the epidermis: keratinocytes, melanocytes, Merkel cells, and Langerhans cells. **Keratinocytes** (ke-RAT-i-NŌ-sīts) are the most numerous cells within the epidermis. Melanocytes are the pigment-producing cells in the epidermis. Merkel cells have a role in detecting sensation, and Langerhans cells (also termed *dendritic cells*) are wandering phagocytic cells that are important in the body's immune response. Melanocytes, Merkel cells, and Langerhans cells are scattered among keratinocytes.

Layers of the Epidermis

▶ **KEY POINT** From deep to superficial, the epidermal layers of thick skin are the stratum basale, stratum spinosum, stratum granulosum, stratum lucidum, and stratum corneum.

The epidermis of thick skin has five layers. Beginning at the basal lamina and traveling superficially toward the epithelial surface, we find the stratum basale, stratum spinosum, stratum granulosum, stratum lucidum, and stratum corneum. Refer to **Figure 4.3** as we describe the layers in a section of thick skin.

Stratum Basale

The deepest epidermal layer is the **stratum basale** (STRĀ-tum BASA-le), or *stratum germinativum* (jer-mi-na-TĒ-vum). This single layer of cells is firmly attached to the basal lamina, which separates the epidermis from the loose connective tissue of the adjacent dermis. Large stem cells, termed **basal cells**, dominate the stratum basale. As basal cells undergo mitosis, new keratinocytes are formed and move into the more superficial layers of the epidermis. This upward migration of cells replaces more superficial keratinocytes that are shed at the epithelial surface.

The brown tones of the skin result from the pigment-producing cells called **melanocytes**. ↺ p. 62 Melanocytes are scattered among the basal cells

Figure 4.2 Major Components of the Integumentary System. The epidermis is a keratinized stratified squamous epithelium that overlies the dermis, a connective tissue region containing glands, hair follicles, and sensory receptors. Underlying the dermis is the subcutaneous layer, which contains fat and blood vessels supplying the dermis. (Nails are not shown; see **Figure 4.15**.)

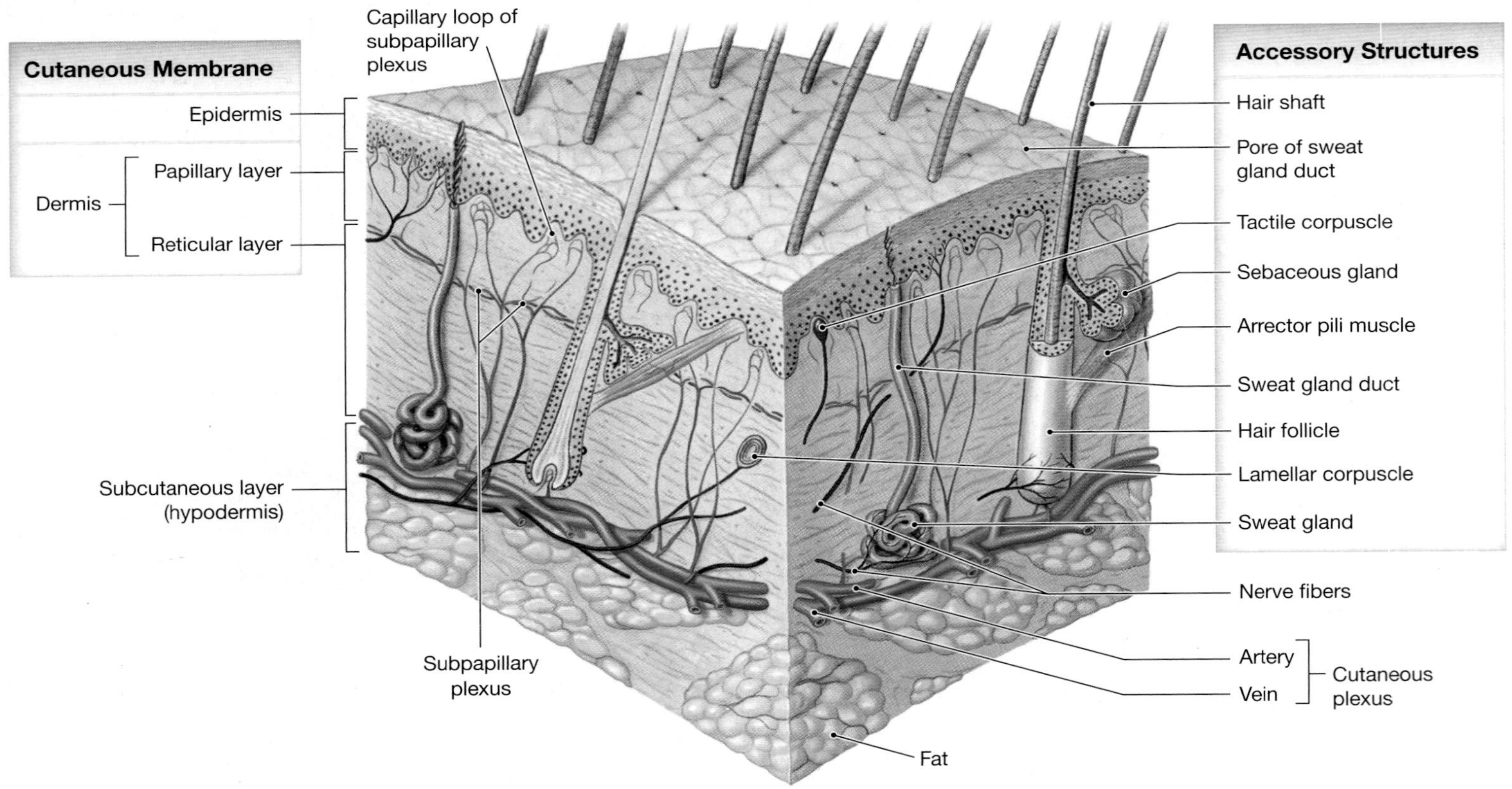

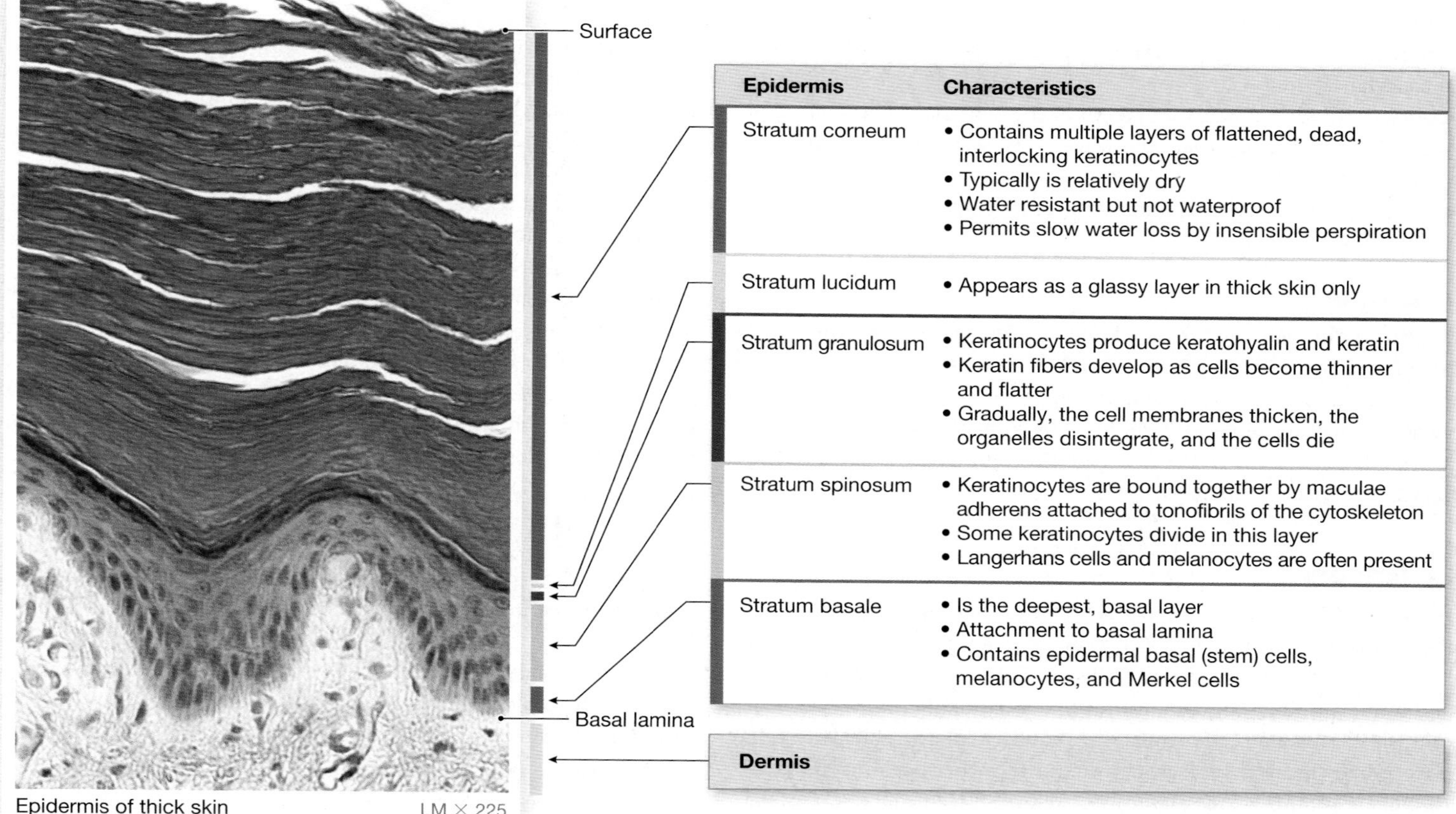

Epidermis	Characteristics
Stratum corneum	• Contains multiple layers of flattened, dead, interlocking keratinocytes • Typically is relatively dry • Water resistant but not waterproof • Permits slow water loss by insensible perspiration
Stratum lucidum	• Appears as a glassy layer in thick skin only
Stratum granulosum	• Keratinocytes produce keratohyalin and keratin • Keratin fibers develop as cells become thinner and flatter • Gradually, the cell membranes thicken, the organelles disintegrate, and the cells die
Stratum spinosum	• Keratinocytes are bound together by maculae adherens attached to tonofibrils of the cytoskeleton • Some keratinocytes divide in this layer • Langerhans cells and melanocytes are often present
Stratum basale	• Is the deepest, basal layer • Attachment to basal lamina • Contains epidermal basal (stem) cells, melanocytes, and Merkel cells
Dermis	

Figure 4.3 The Structure and Layers of the Epidermis. The light micrograph shows the five major stratified layers of epidermal cells in thick skin.

of the stratum basale. They have numerous cytoplasmic processes that inject **melanin**—a black, yellow-brown, or brown pigment—into the basal cells in this layer and into the keratinocytes of more superficial layers. The ratio of melanocytes to stem cells ranges between 1:4 and 1:20 depending on the region examined. Melanocytes are most abundant in the cheeks, forehead, nipples, and genital region.

Differences in skin color result from varying levels of melanocyte activity, not varying numbers of melanocytes. **Albinism** is an inherited disorder characterized by deficient melanin production; individuals with this condition have a normal distribution of melanocytes, but the cells cannot produce melanin. It affects approximately one person in 10,000.

Skin surfaces that lack hair contain specialized epithelial cells known as **Merkel cells**. These cells are found among the cells of the stratum basale and are most abundant in skin where sensory perception is most acute, such as fingertips and lips. Merkel cells are sensitive to touch and, when compressed, release chemicals that stimulate sensory nerve endings, providing information about objects touching the skin. (There are many other kinds of touch receptors, but they are located in the dermis and will be introduced in later sections.) Chapter 18 will describe all the integumentary receptors.

Stratum Spinosum

Each time a basal cell divides, one of the daughter cells is pushed into the next, more superficial layer, the **stratum spinosum**. The stratum spinosum is several cells thick. Each keratinocyte in the stratum spinosum contains bundles of protein filaments that extend from one side of the cell to the other. These bundles, called **tonofibrils**, begin and end at a desmosome (*macula adherens*) that connects the keratinocyte to its neighbors. ⮌ p. 42 The tonofibrils act as cross braces, strengthening and supporting the cell junctions. This interlocking network of desmosomes and tonofibrils ties all the cells in the stratum spinosum together. Standard histological procedures, used to prepare tissue for microscopic examination, shrink the cytoplasm but leave its tonofibrils and desmosomes intact. This makes the cells look like miniature pincushions, which is why early histologists called this stratum the "spiny layer."

The deepest cells within the stratum spinosum are mitotically active and continue to divide, making the epithelium thicker. Melanocytes are common in this layer, as are **Langerhans cells** (also termed *dendritic cells*), although you cannot see Langerhans cells in standard histological preparations. Langerhans cells, which account for 3–8 percent of the cells in the epidermis, are most common in the superficial portion of the stratum spinosum. These cells play an important role in triggering an immune response against epidermal cancer cells and pathogens that have penetrated the superficial layers of the epidermis.

Stratum Granulosum

Superficial to the stratum spinosum is the **stratum granulosum** (*granular layer*). This is the most superficial layer of the epidermis in which all the cells still possess a nucleus.

The stratum granulosum consists of keratinocytes that have moved out of the stratum spinosum. By the time cells reach this layer, they have begun to manufacture large quantities of the proteins **keratohyalin** (ker-a-tō-HĪ-a-lin) and **keratin** (KER-a-tin; *keros*, horn). Keratohyalin accumulates in electron-dense keratohyalin granules. These granules form an intracellular matrix that surrounds the keratin filaments. Cells of this layer also contain membrane-bound granules that release their contents by exocytosis, which forms sheets of a lipid-rich substance that begins to coat the cells of the stratum granulosum. In more superficial layers, this substance forms a complete water-resistant layer around the cells that protects the epidermis, but also prevents the diffusion of nutrients and wastes into and out of the cells. As a result, cells in the more superficial layers of the epidermis die.

Environmental factors often influence the rate at which keratinocytes synthesize keratohyalin and keratin. Increased friction against the skin, for example, stimulates increased synthesis, thickening the skin and forming a **callus** (also termed a *clavus*).

In humans, keratin forms the basic structural component of hair and nails. It is a very versatile material, however, and it also forms the claws of dogs and cats, the horns of cattle and rhinos, the feathers of birds, the scales of snakes, the baleen of whales, and a variety of other interesting epidermal structures.

Stratum Lucidum

In the thick skin of the palms of the hands and soles of the feet, there is a glassy **stratum lucidum** (*clear layer*) superficial to the stratum granulosum of the epidermis. The cells in this layer lack organelles and nuclei, are flattened and densely packed, and are filled with keratin filaments that are oriented parallel to the surface of the skin. The cells of the stratum lucidum do not stain well in standard histological preparations.

Stratum Corneum

The **stratum corneum** (KŌR-nē-um; *cornu*, horn) is the most superficial layer of both thick and thin skin. It consists of numerous layers of flattened, *dead* cells that possess a thickened plasma membrane. These dehydrated cells lack organelles and a nucleus, but still contain many keratin filaments. Because the interconnections established in the stratum spinosum remain intact, the cells of this layer are usually shed in large groups or sheets, rather than individually.

An epithelium containing large amounts of keratin is termed a **keratinized** (KER-a-ti-nīzd), or *cornified* (KŌR-ni-fīd; *cornu*, horn, + *facere*, to make), epithelium. Normally, the stratum corneum is relatively dry, which makes the surface unsuitable for the growth of many microorganisms. Maintenance of this barrier involves coating the surface with the secretions of sebaceous and sweat glands (discussed in a later section). The process of **keratinization** occurs everywhere on exposed skin surfaces except over the anterior surface of the eyes.

Although the stratum corneum is water resistant, *it is not waterproof*. Water from the interstitial fluids slowly penetrates the surface and evaporates into the surrounding air. This process, called insensible perspiration, accounts for a loss of roughly 500 ml (about 1 pint) of water per day.

It takes 15–30 days for a cell to move superficially from the stratum basale to the stratum corneum. The dead cells in the exposed stratum corneum layer usually remain for two weeks before they are shed or washed away. Thus, the deeper portions of the epithelium—and all underlying tissues—are always protected by a barrier composed of dead, durable, and expendable cells.

We can appreciate the protective role of the skin when an injury destroys large areas of it. For example, a second-degree (partial-thickness) or third-degree (full-thickness) burn can lead to absorption of toxic substances, loss of excess fluids, and infection at the site of the burn—medical problems that result from the loss of the skin.

Thin and Thick Skin

▶ **KEY POINT** The terms thin and thick skin refer to the relative thickness of the epidermis, not to the thickness of the integument as a whole.

Most of the body is covered by **thin skin**, which has only four layers because the stratum lucidum is typically absent. In thin skin, the epidermis is a mere 0.08 mm thick, and the stratum corneum is only a few cell layers deep **(Figure 4.4a,b)**.

Thick skin, found only on the palms of the hands and soles of the feet, contains all five layers and may be covered by 30 or more layers of keratinized cells. As a result, the epidermis in these locations is up to six times thicker than the epidermis covering the general body surface **(Figure 4.4c)**.

Figure 4.4 Thin and Thick Skin. The epidermis is a stratified squamous epithelium that varies in thickness.

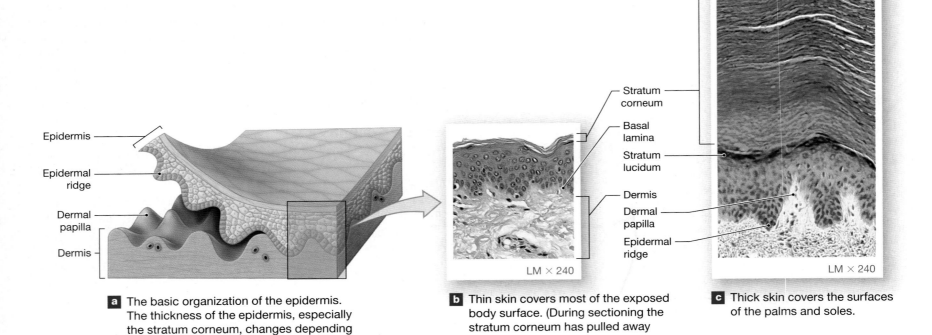

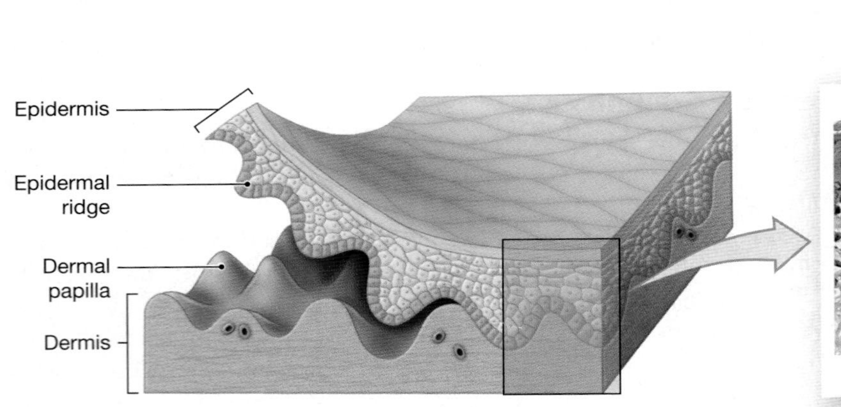

a The basic organization of the epidermis. The thickness of the epidermis, especially the stratum corneum, changes depending on the location sampled.

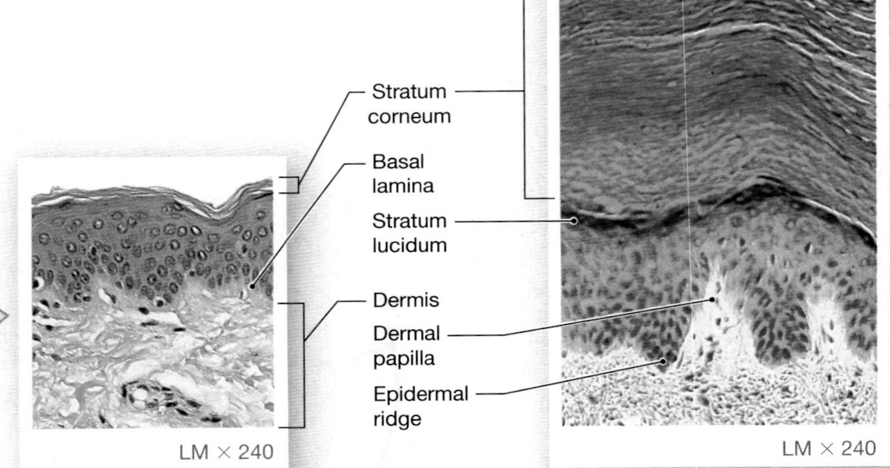

b Thin skin covers most of the exposed body surface. (During sectioning the stratum corneum has pulled away from the rest of the epidermis.)

c Thick skin covers the surfaces of the palms and soles.

TIPS & TOOLS

Remembering the names of the epidermal layers of thick skin

A mnemonic to help you remember the names of the epidermal layers of thick skin, from deep to superficial, is "Brent Spiner gained Lieutenant Commander" (basale, spinosum, granulosum, lucidum, corneum).

Dermal Ridges

▶ **KEY POINT** Dermal ridges form friction ridges, ensuring a secure grip on objects. Dermal ridges also form fingerprints, a unique genetic identifier of an individual.

The stratum basale of the epidermis forms **dermal ridges** (also known as *friction ridges*) that extend into the dermis, increasing the area of contact between the two regions. Projections from the dermis toward the epidermis, called **dermal papillae** (singular, *papilla*; "nipple-shaped mound"), extend between adjacent ridges (**Figure 4.4a,c**).

The contours of the skin surface follow the ridge patterns, which vary from small conical pegs (in thin skin) to the complex whorls seen on the thick skin of the palms and soles. Ridges on the palms and soles increase the surface area of the skin and promote friction, ensuring a secure grip.

Ridge shapes are genetically determined: Those of each person are unique and do not change during a lifetime. Ridge patterns on the fingertips (**Figure 4.5**) can therefore identify individuals, and criminal investigators have used fingerprints for this purpose for over a century.

Figure 4.5 The Epidermal Ridges of Thick Skin. Fingerprints reveal the pattern of epidermal ridges in thick skin. This scanning electron micrograph shows the ridges on a fingertip.

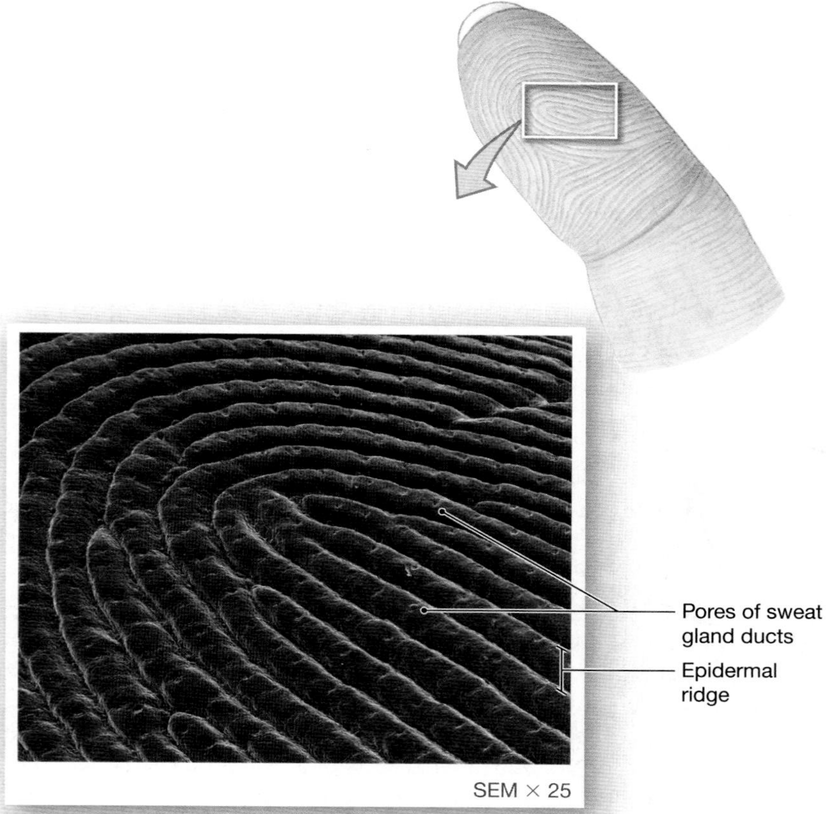

Skin Color

▶ **KEY POINT** Factors affecting skin color include dermal blood supply and the amounts of epidermal pigments that are present.

The color of the epidermis is due to a combination of factors: circulation in the dermis and variable quantities of two epidermal pigments, carotene and melanin.

Dermal Blood Supply

Blood contains red blood cells that carry the protein hemoglobin. When hemoglobin is bound to oxygen, it has a bright red color, giving blood vessels in the dermis a reddish tint that is most visible in fair-skinned people. When those vessels dilate, as during inflammation, the red tones become much more pronounced.

When circulation in the dermis is temporarily reduced, the skin becomes relatively pale. A frightened Caucasian may "turn white" because of a sudden drop in blood flow to the skin.

During a sustained reduction in oxygen content in the blood, hemoglobin becomes a darker red. Seen from the surface, the skin takes on a bluish coloration called **cyanosis** (sī-a-NŌ-sis; *kyanos*, blue). In people of any skin color, cyanosis is most apparent in areas of thin skin, such as the lips or beneath the nails. It can be a response to extreme cold or the result of circulatory or respiratory disorders, such as heart failure or severe asthma.

Epidermal Pigments

Two pigments determine skin color: carotene and melanin.

Carotene (KAR-ō-ten) is an orange-yellow pigment also found in various orange- or yellow-colored vegetables, such as carrots, corn, and squashes. It can be converted to vitamin A, which is required for epithelial maintenance and the synthesis of photoreceptor pigments in the eye. Carotene normally accumulates in the subcutaneous fat as well as inside keratinocytes, becoming especially evident in the dehydrated cells of the stratum corneum.

Melanin (MEL-a-nin) is produced and stored in melanocytes of thin skin, and the amount produced is genetically determined (**Figure 4.6**). The black, yellow-brown, or brown melanin forms in intracellular vesicles called **melanosomes**. These vesicles, which are transferred to keratinocytes, color the keratinocytes temporarily, until lysosomes destroy the melanosomes. As a result, the cells in more superficial layers of the epidermis have less melanin, and therefore are lighter in color, than cells within the deeper layers of the epithelium. In light-skinned individuals, melanosome transfer occurs in the stratum basale and stratum spinosum, and the cells of more superficial layers lose their pigmentation. In dark-skinned individuals, the melanosomes are larger and the transfer may occur in the stratum granulosum as well, making the pigmentation darker and more persistent.

Little or no melanin is produced in the thick skin of the palms of the hands and soles of the feet. Melanin produced in thick skin is difficult to see because of the thickness of the stratum corneum.

Melanin pigments help protect the underlying dermis and also prevent skin damage by absorbing ultraviolet (UV) radiation in sunlight. A little ultraviolet radiation is necessary because the skin requires it to form vitamin D.* The small intestine needs vitamin D to absorb calcium and phosphorus; inadequate vitamin D impairs bone maintenance and growth. However, too much UV radiation may damage chromosomes and

*Specifically, vitamin D_3, or *cholecalciferol*, which undergoes further modification in the liver and kidneys before circulating as the active hormone *calcitriol*.

Figure 4.6 Melanocytes. The micrograph and accompanying drawing indicate the location and orientation of melanocytes in the stratum basale of a dark-skinned person.

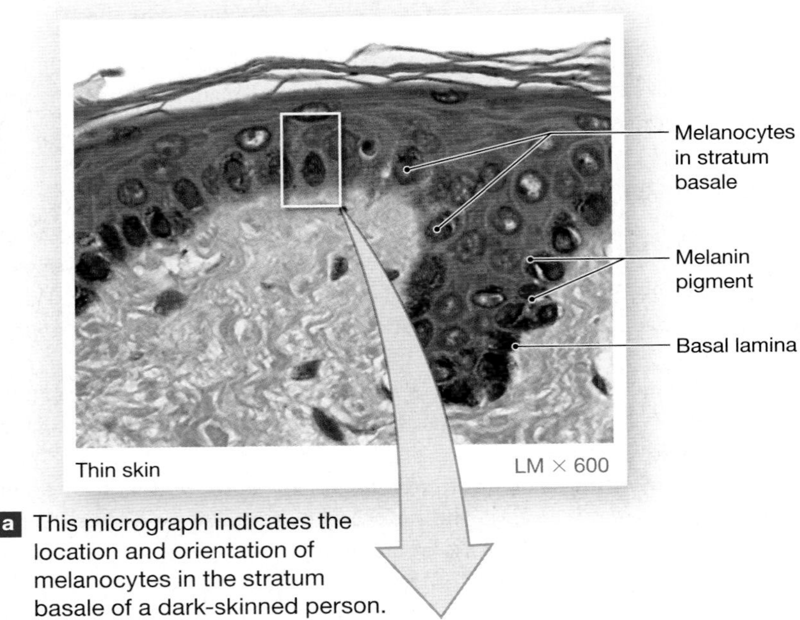

a This micrograph indicates the location and orientation of melanocytes in the stratum basale of a dark-skinned person.

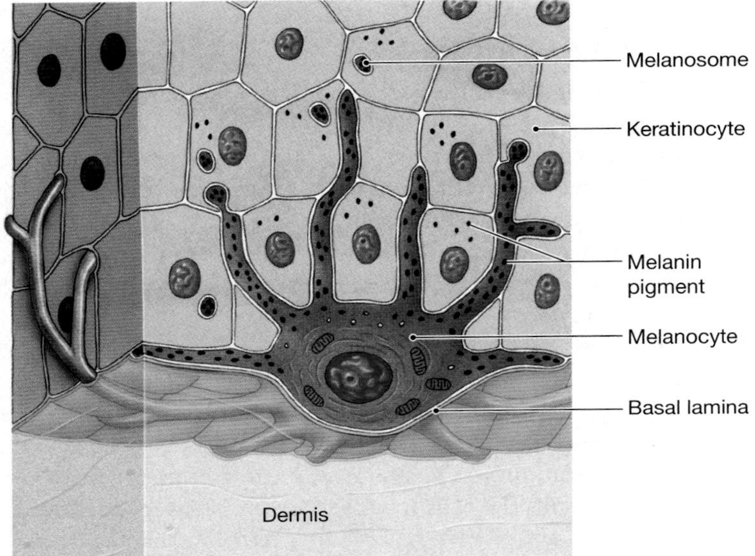

b Melanocytes produce and store melanin.

cause widespread tissue damage similar to that caused by mild to moderate burns. Within each keratinocyte, the melanosomes are most abundant around the cell's nucleus, helping absorb the UV radiation before it can damage the nuclear DNA.

Melanocytes respond to UV exposure by synthesizing and transferring more melanin. The skin "tans," but this response is not quick enough to prevent a sunburn on the first day at the beach; it takes about 10 days. Anyone can get a sunburn, but dark-skinned individuals have greater initial protection against the effects of UV radiation. Repeated UV exposure sufficient to stimulate tanning will result in long-term skin damage in the dermis and epidermis. In the dermis, damage to fibrocytes causes abnormal connective tissue structure and premature wrinkling. In the epidermis, chromosomal damage in basal cells or melanocytes can lead to skin cancer.

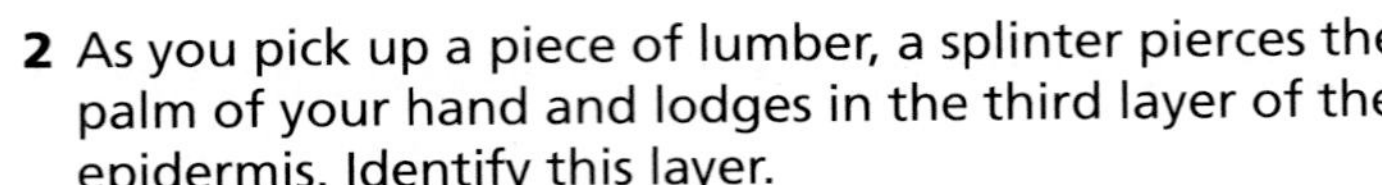

4.2 ✔ CONCEPT CHECK

2 As you pick up a piece of lumber, a splinter pierces the palm of your hand and lodges in the third layer of the epidermis. Identify this layer.

3 What is keratinization? What are the stages of this process?

4 Identify the sources of the color of the epidermis.

5 Describe the relationship between epidermal ridges and dermal papillae.

See the blue Answers tab at the back of the book.

4.3 The Dermis

▶ **KEY POINT** The dermis, which is immediately deep to the epidermis, consists of loose connective tissue above a dense network of connective tissue fibers.

Deep to the epidermis is the dermis (**Figure 4.2**, p. 88). The dermis has two major parts: a superficial papillary layer and a deeper reticular layer.

Dermal Organization

▶ **KEY POINT** The two layers of the dermis are the superficial papillary layer and the deeper reticular layer.

The **papillary layer**, the superficial layer of the dermis, consists of loose connective tissue (**Figure 4.7a**). This region is specialized to provide mechanical attachment for the more superficial epidermis. It also contains capillaries and axons of neurons. Capillaries are branching blood vessels that supply the epidermis, and axons of sensory neurons monitor epidermal receptors. The papillary layer gets its name from the dermal papillae that project between the epidermal ridges (**Figure 4.4**, p. 90).

The deeper **reticular layer** consists of fibers in an interwoven meshwork of dense irregular connective tissue that surrounds blood vessels, hair follicles, nerves, sweat glands, and sebaceous glands (**Figure 4.7b**). Its name comes from the interwoven arrangement of collagen fiber bundles in this region (*reticulum*, a little net). Some of the collagen fibers in the reticular layer extend into the papillary layer, tying the two layers together; therefore, the boundary between these layers is not clear. Collagen fibers of the reticular layer also extend into the deeper subcutaneous layer (**Figure 4.7c**). The arrangement of these connective tissue fibers in the reticular layer is responsible for the strength, toughness and elasticity of the skin.

Wrinkles, Stretch Marks, and Tension Lines

The interwoven collagen fibers of the reticular layer provide considerable tension, and the extensive array of elastic fibers enables the dermis to stretch and recoil repeatedly during normal movements. Age, hormones, and the destructive effects of ultraviolet radiation reduce the thickness and flexibility of the dermis, producing wrinkles and sagging skin.

Tretinoin (Retin-A) is a derivative of vitamin A that can be applied to the skin as a cream or gel. This drug was originally developed to treat acne, but it also increases blood flow to the dermis and stimulates dermal repairs. As a result, the rate of wrinkle formation decreases, and existing wrinkles become smaller. The degree of improvement varies from individual to individual.

Extensive distortion of the dermis over the abdomen, such as during pregnancy or after a substantial weight gain, often exceeds the elastic capabilities of the skin. Elastic and collagen fibers break, and although the skin stretches, it does not recoil to its original size after delivery or a rigorous diet. The skin then wrinkles and creases, creating a network of **stretch marks**.

Figure 4.7 **The Structure of the Dermis and the Subcutaneous Layer.** The dermis is a connective tissue layer deep to the epidermis; the subcutaneous layer (superficial fascia) is a connective tissue layer deep to the dermis.

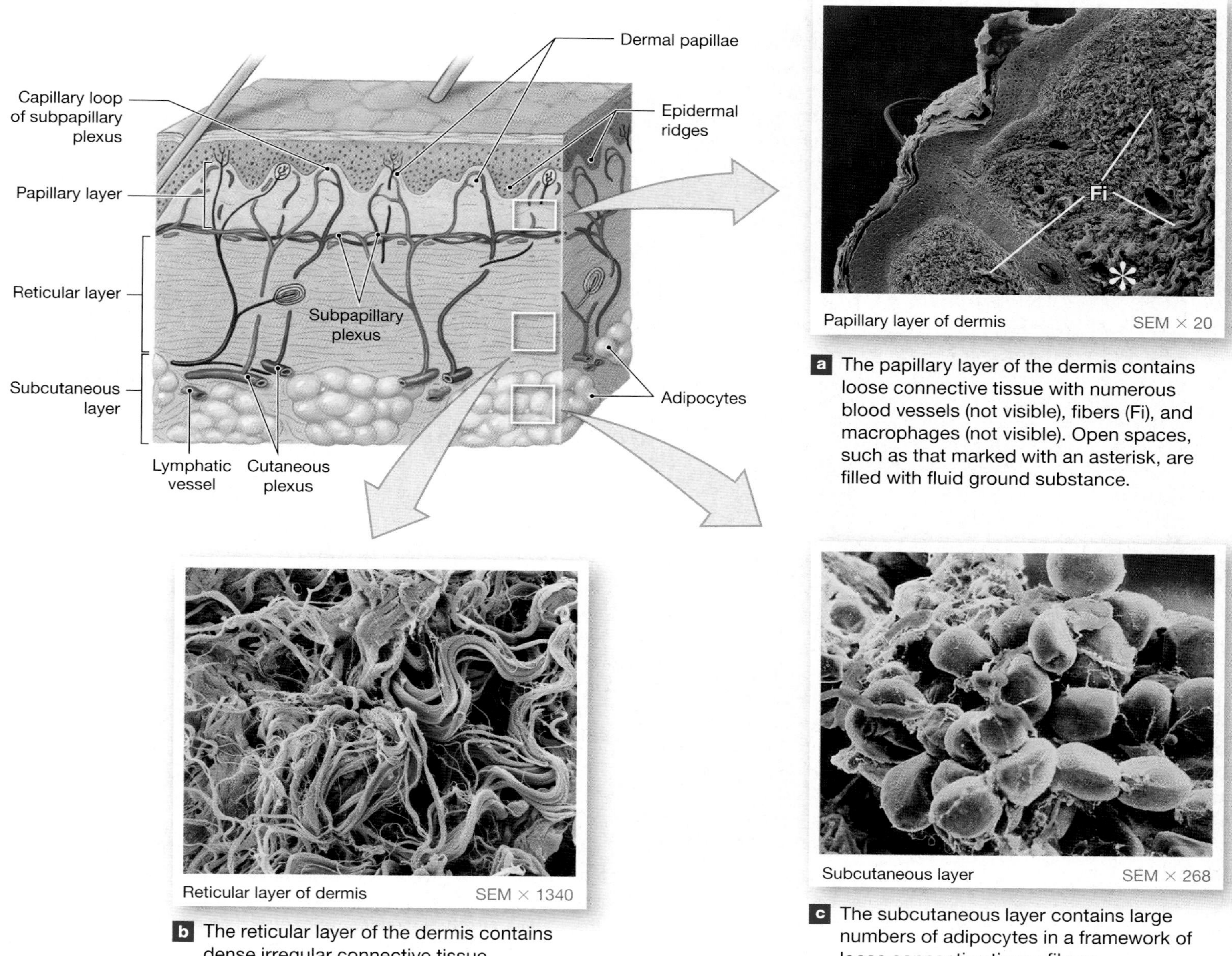

a The papillary layer of the dermis contains loose connective tissue with numerous blood vessels (not visible), fibers (Fi), and macrophages (not visible). Open spaces, such as that marked with an asterisk, are filled with fluid ground substance.

b The reticular layer of the dermis contains dense irregular connective tissue.

c The subcutaneous layer contains large numbers of adipocytes in a framework of loose connective tissue fibers.

At any one location, most of the collagen and elastic fibers are arranged in parallel bundles. The orientation of these bundles depends on the stress placed on the skin during normal movement; the bundles are aligned to resist the applied forces. The resulting pattern of fiber bundles establishes the **tension lines** (also termed *cleavage lines*) of the skin (**Figure 4.8**). Tension lines are clinically significant because a cut parallel to a tension line will usually remain closed, whereas a cut at right angles to a tension line will pull open as cut elastic fibers recoil. When possible, surgeons choose their incision patterns accordingly, since an incision parallel to the tension lines will heal fastest with minimal scarring.

Other Dermal Components

▶ **KEY POINT** The dermis contains connective tissue cells, connective tissue fibers, accessory structures, blood vessels, lymphatic vessels, and nerves.

In addition to extracellular protein fibers, the dermis contains all the cells of connective tissue proper. Accessory organs of epidermal origin, such as hair follicles and sweat glands, extend into the dermis (**Figure 4.9**). In addition, the reticular and papillary layers of the dermis contain networks of blood vessels, lymph vessels, and nerve fibers (**Figure 4.2**, p. 88).

Blood Supply to the Skin

Arteries and veins supplying the skin form an interconnected network at the junction between the reticular layer of the dermis and the subcutaneous layer. This network is called the **cutaneous plexus** (**Figure 4.2**). Branches of the arteries supply the adipose tissue of the subcutaneous layer as well as more superficial tissues of the skin. As small arteries travel toward the epidermis, branches supply the hair follicles, sweat glands, and other structures in the dermis.

4

CLINICAL NOTE

Skin Cancer

Skin cancer, the abnormal growth of skin cells, is often caused by exposure to UV radiation, primarily sunlight.

Basal cell carcinoma originates in the stratum basale. This is the most common skin cancer and the slowest growing, and it most often arises in areas that receive UV exposure. Although basal cell carcinomas almost never metastasize, they should be treated quickly to prevent local spread.

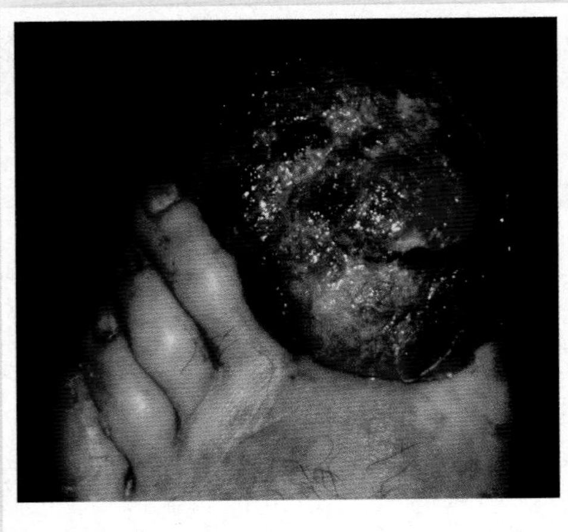

Squamous cell carcinoma

Squamous cell carcinoma, the second most common skin cancer, is an uncontrolled growth of abnormal squamous cells in the epidermis. They most often occur in UV-exposed areas of skin, but tobacco can also be a trigger. They can metastasize to tissues, bones, and nearby lymph nodes, and they often cause local disfigurement.

Malignant melanoma develops in melanocytes in the basal layer. These cancerous melanocytes multiply rapidly and metastasize to distant sites. Malignant melanomas cause the most deaths from skin cancer.

Upon reaching the papillary layer, these small arteries enter another branching network, the **subpapillary plexus**. From there, capillary loops follow the contours of the epidermal–dermal boundary (**Figure 4.7a**). These capillaries empty into a network of delicate veins that rejoin the papillary plexus. From there, larger veins drain into a network of veins in the deeper cutaneous plexus.

Circulation to the skin must be tightly regulated. Skin plays a key role in **thermoregulation**, the control of body temperature. When body temperature increases, increased circulation to the skin permits the loss of excess heat. When body temperature decreases, reduced circulation to the skin promotes retention of body heat.

Total blood volume in the body is relatively constant. Thus, increased blood flow to the skin means a decreased blood flow to some other organ(s). The nervous, cardiovascular, and endocrine systems interact to regulate blood flow to the skin while maintaining adequate blood flow to other organs and systems.

Nerve Supply to the Skin

Nerve fibers in the skin control blood flow, adjust gland secretion rates, and monitor sensory receptors in the dermis and the deeper layers of the epidermis. We have already noted the presence of Merkel cells in the deeper layers of the epidermis. These cells are touch receptors monitored by sensory nerve endings known as **tactile discs**. The epidermis also contains sensory nerves that are believed to respond to pain and temperature. The dermis contains similar receptors as well as other, more specialized receptors. In Chapter 18 we will discuss receptors that are sensitive to light touch, stretch, deep pressure, and vibration.

Figure 4.8 Tension Lines of the Skin. Tension lines reflect the orientation of collagen fiber bundles in the dermis of the skin.

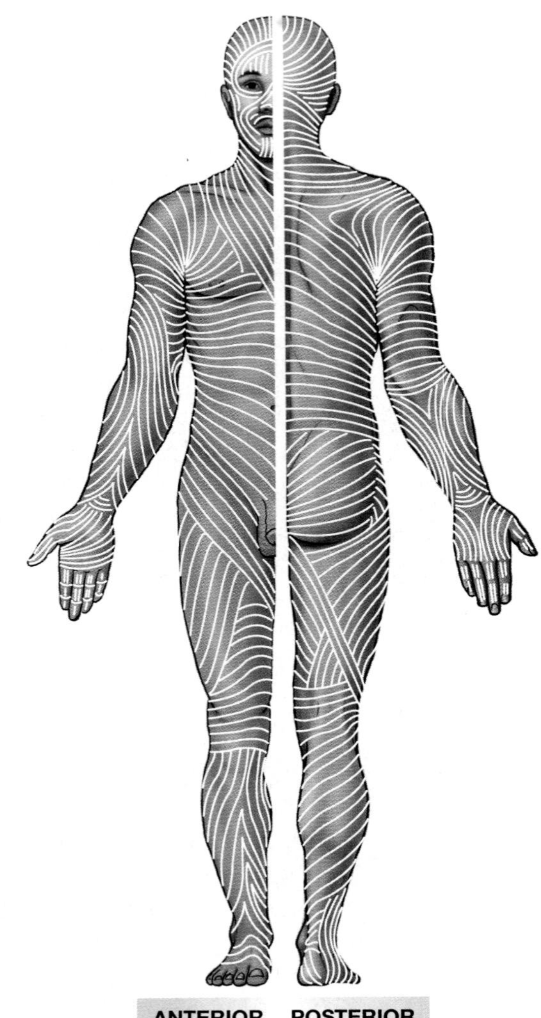

4.3 CONCEPT CHECK

6 What is the relationship between the collagen bundles of the dermis and the tension lines of the skin?

7 Explain the anatomy of the cutaneous plexus.

See the blue Answers tab at the back of the book.

4.4 | The Subcutaneous Layer

▶ **KEY POINT** The subcutaneous layer, composed of loose connective tissue with many adipocytes, stabilizes the integument.

The connective tissue fibers of the reticular layer are extensively interwoven with those of the subcutaneous layer. The boundary between the two layers is usually indistinct (**Figure 4.2**, p. 88). Although the subcutaneous layer is not

Figure 4.9 **Accessory Structures of the Skin.**

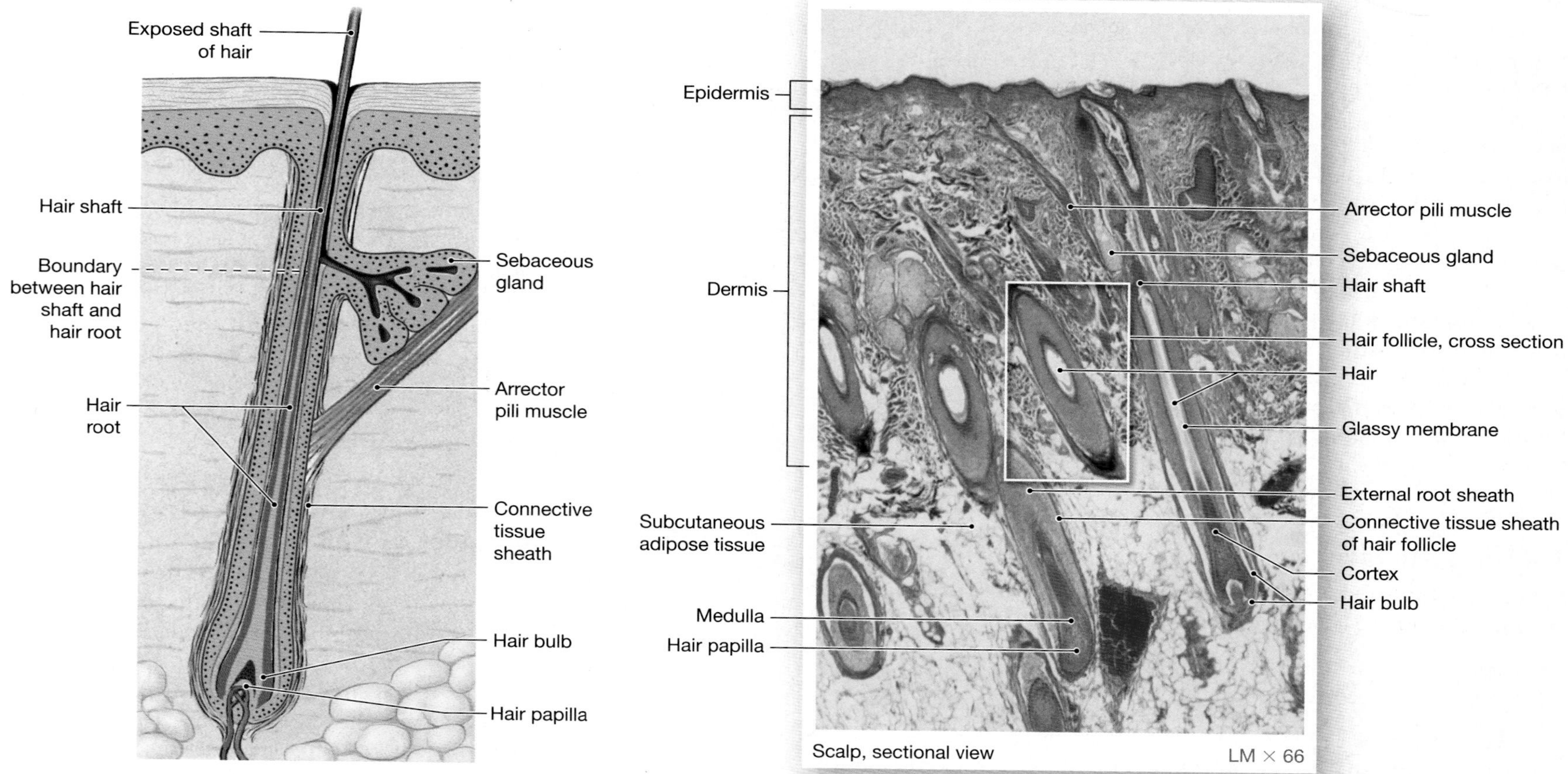

a A diagrammatic view of a single hair follicle.

Scalp, sectional view LM × 66

b A light micrograph showing the sectional appearance of the skin of the scalp. Note the abundance of hair follicles and the way they extend into the dermis and hypodermis.

usually considered part of the integument, it is important in stabilizing the position of the skin in relation to underlying tissues, such as skeletal muscles or other organs, while still permitting independent movement.

The subcutaneous layer consists of loose connective tissue with abundant adipocytes (**Figure 4.7c**). Infants and small children usually have extensive "baby fat," which helps reduce heat loss. Subcutaneous fat also serves as an energy reserve and shock absorber for the rough-and-tumble activities of our early years.

As we grow, the distribution of subcutaneous fat changes. Men accumulate subcutaneous fat in the neck, upper arms, lower back, and over the buttocks; women store it primarily in the breasts, buttocks, hips, and thighs. In adults of either sex, the subcutaneous layer of the backs of the hands and the upper surfaces of the feet contain few adipocytes, whereas distressing amounts of adipose tissue can accumulate in the abdominal region, producing a prominent "pot belly."

The subcutaneous layer is quite elastic. Only its superficial region contains large arteries and veins; the remaining areas contain a limited number of capillaries and no vital organs. This last characteristic makes subcutaneous injection a useful method for administering drugs. The familiar term hypodermic needle refers to the region (the hypodermis) targeted for injection.

4.4 CONCEPT CHECK

8 What is one function of the subcutaneous layer?

See the blue Answers tab at the back of the book.

4.5 Accessory Structures

▶ **KEY POINT** The accessory structures of the skin include hair follicles, sebaceous glands, sweat glands, and nails.

Recall that, in addition to the skin, the integument consists of several accessory structures: hair follicles, sebaceous glands, sweat glands, and nails (**Figure 4.2**, p. 88). During embryological development, these structures form through invagination (infolding) of the epidermis.

Hair Follicles and Hair

▶ **KEY POINT** Hair and hair follicles develop from the epidermis, extend deep into the dermis and subcutaneous layer, and protect the body in a variety of ways.

Hairs project beyond the surface of the skin almost everywhere except the sides and soles of the feet, the palms of the hands, the sides of the fingers and toes, the lips, and portions of the external genitalia.* There are about 5 million hairs on the human body, and 98 percent of them are on the general body surface, not the head. Hairs are nonliving structures that form in organs called **hair follicles**.

*External genitalia include the glans penis and prepuce in the male and the clitoris, labia minora, and inner surfaces of the labia majora in the female.

Hair Production

Hair follicles extend deep into the dermis, often projecting into the underlying subcutaneous layer. The epithelium at the follicle base surrounds a small **hair papilla**, a peg of connective tissue containing capillaries and nerves. The **hair bulb** consists of epithelial cells that surround the papilla.

Hair production involves a specialized keratinization process. The **hair matrix** is the epithelial layer involved in hair production. When the superficial basal cells divide, they produce daughter cells that are pushed toward the surface as part of the developing hair.

Most hairs have an inner medulla and an outer cortex. The **medulla** contains relatively soft and flexible **soft keratin**. Matrix cells closer to the edge of the developing hair form the relatively hard **cortex** (**Figures 4.9b** and **4.10**). The cortex contains **hard keratin**, which gives hair its stiffness. A single layer of dead, keratinized cells at the outer surface of the hair overlap and form the **cuticle** that coats the hair.

The **hair root** anchors the hair into the skin. The root begins at the hair bulb and extends distally to the point where the internal organization of the hair is complete, about halfway to the skin surface. The **hair shaft** extends from this halfway point to the skin surface, where we see the exposed hair tip. The size, shape, and color of the hair shaft are highly variable.

Follicle Structure

Beginning at the hair cuticle and moving outward, the cells of the follicle walls are organized into the following three concentric layers (**Figure 4.10a**):

- **Internal root sheath:** This layer is produced by cells at the periphery of the hair matrix.
- **External root sheath:** This layer includes all the cell layers found in the superficial epidermis. However, where the external root sheath joins the hair matrix, all the cells resemble those of the stratum basale.
- **Glassy membrane:** This layer is a thickened, specialized basal lamina.

Functions of Hair

The 5 million hairs on the human body have important functions. The roughly 100,000 hairs on the head protect the scalp from ultraviolet light and bumps to the head and insulate the skull. The hairs guarding the entrances to our nostrils and external auditory canals help block foreign particles and insects, and eyelashes perform a similar function for the surface of the eye. A **root hair plexus** of sensory nerves surrounds the base of each hair follicle (**Figure 4.10a**). As a result, we can feel the movement of even a single hair. This sensitivity gives an early-warning system that may help prevent injury. For example, you may be able to swat a mosquito before it reaches your skin.

A ribbon of smooth muscle, the **arrector pili** (a-REK-tor PĪ-lī) **muscle**, extends from the papillary layer of the dermis to the connective tissue sheath surrounding the hair follicle (**Figures 4.9** and **4.10a**). When stimulated, the arrector pili muscle pulls on the follicle and raises the hair. Contraction may be due to an emotional state, such as fear or rage, or to cold temperatures that produce characteristic "goose bumps." In a furry mammal, this action thickens the insulating coat, rather like putting on an extra sweater. Although we do not receive any comparable insulating benefits, the reflex persists.

Types of Hairs

Hairs first appear after about three months of embryonic development. These hairs, collectively known as **lanugo** (la-NŪ-gō), are extremely fine and unpigmented. Most lanugo hairs are shed before birth.

The two types of hairs in the adult integument are vellus hairs and terminal hairs:

- **Vellus hairs** are the fine "peach fuzz" hairs found over much of the body surface.
- **Terminal hairs** are heavy, more deeply pigmented, and sometimes curly. The hairs on your head, including your eyebrows and eyelashes, are terminal hairs.

The hair structure we described earlier in the chapter was that of terminal hairs. Vellus hair is similar, though it does not have a distinct medulla. Hair follicles may alter the structure of the hairs they produce in response to circulating hormones.

Hair Color

Variations in hair color reflect differences in hair structure and in the pigment produced by melanocytes at the papilla. These characteristics are genetically determined, although hormonal or environmental factors may influence the condition of your hair. Whether your hair is black or brown depends on the density of melanin in your cortex. Red hair results from a biochemically distinct form of melanin.

As pigment production decreases with age, hair color lightens toward gray. White hair results from the combination of a lack of pigment and the presence of air bubbles within the medulla of the hair shaft. Because the hair itself is dead and inert, changes in coloration are gradual; your hair can't "turn white overnight," as some horror stories suggest.

Growth and Replacement of Hair

A hair in the scalp grows for two to five years, at a rate of around 0.33 mm/day (about 1/64 inch). Variations in hair growth rate and the duration of the **hair growth cycle** account for individual differences in uncut hair length (**Figure 4.11**).

While a hair is growing, its root is firmly attached to the matrix of the follicle. At the end of the growth cycle, the follicle becomes inactive, and the hair is termed a **club hair**. The follicle shrinks, and over time the connections between the hair matrix and the root of the club hair break down. When another growth cycle begins, the follicle produces a new hair, and the old club hair is pushed toward the surface.

A healthy adult loses about 50–100 head hairs each day, but several factors may affect this rate. Sustained losses of more than 100 hairs per day usually indicate that something is wrong. Temporary increases in hair loss can result from drugs, dietary factors, radiation, high fever, stress, or hormonal factors related to pregnancy.

In males, changes in the level of the sex hormones circulating in the blood can affect the scalp, causing a shift from terminal hair to vellus hair production. This result of this alteration is called **male pattern baldness**.

Glands in the Skin

▶**KEY POINT** Sebaceous glands and sweat glands are the two types of exocrine glands found within the skin.

Glands are another accessory structure of the integument. The skin contains two types of exocrine glands: sebaceous glands and sweat (*sudoriferous*) glands. **Figure 4.12** summarizes their functional classification.

Sebaceous Glands

Sebaceous (se-BĀ-shus) **glands**, or *oil glands*, produce a waxy, oily lipid that coats hair shafts and the epidermis (**Figure 4.13**). The gland cells manufacture large quantities of lipids as they mature, and the lipid product is released through holocrine secretion, a process that involves the rupture of the secretory cell. ⮌ p. 60 The ducts are short, and several sebaceous glands may open into a single follicle. Depending on whether the glands share a common duct, they may be simple alveolar glands (each gland has its own duct) or simple branched alveolar glands (several glands empty into a single duct). ⮌ pp. 58–59

Figure 4.10 Hair Follicles. Hairs originate in hair follicles, which are complex organs.

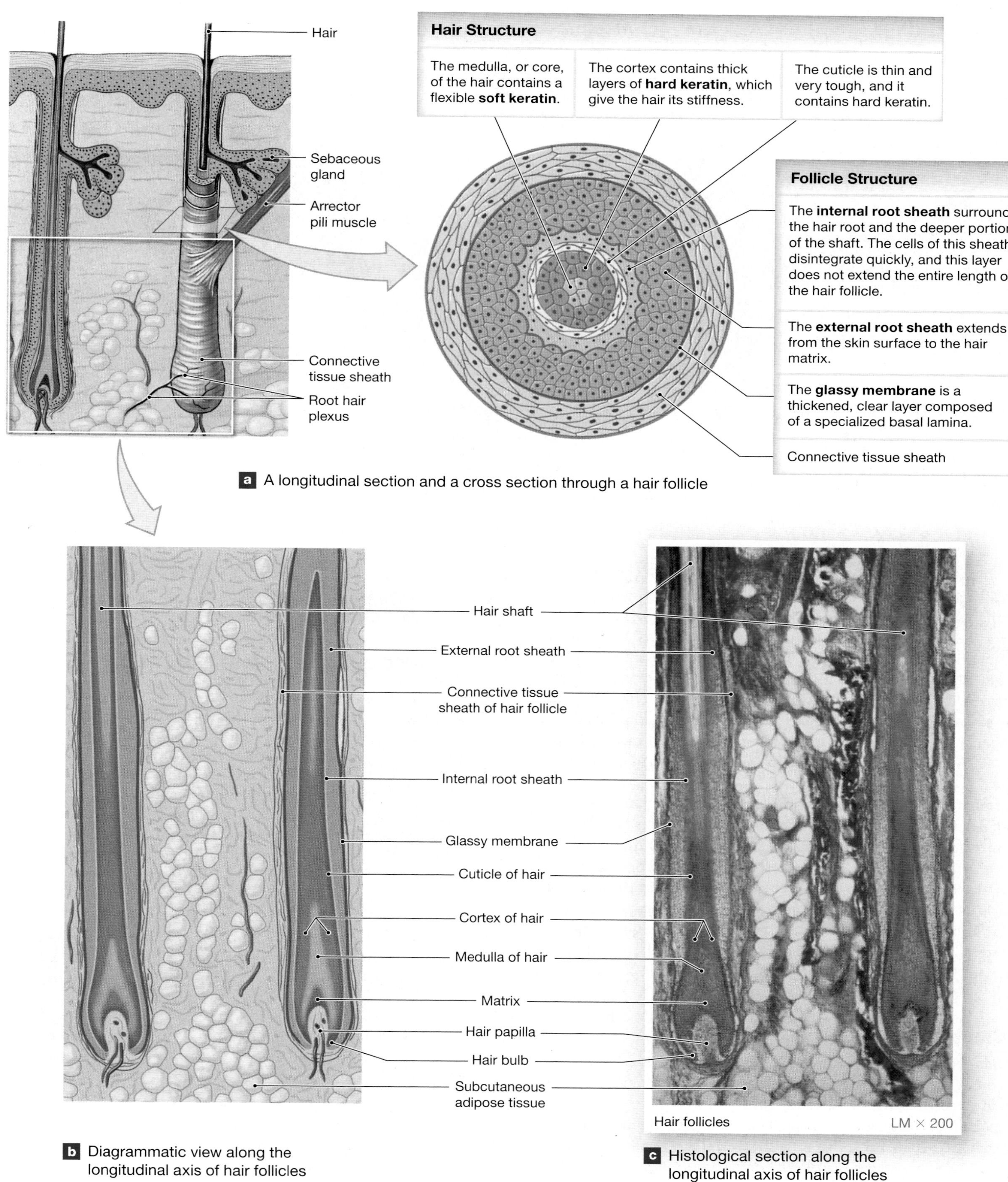

Hair
Sebaceous gland
Arrector pili muscle
Connective tissue sheath
Root hair plexus
Hair Structure
The medulla, or core, of the hair contains a flexible **soft keratin**.
The cortex contains thick layers of **hard keratin**, which give the hair its stiffness.
The cuticle is thin and very tough, and it contains hard keratin.
Follicle Structure
The **internal root sheath** surrounds the hair root and the deeper portion of the shaft. The cells of this sheath disintegrate quickly, and this layer does not extend the entire length of the hair follicle.
The **external root sheath** extends from the skin surface to the hair matrix.
The **glassy membrane** is a thickened, clear layer composed of a specialized basal lamina.
Connective tissue sheath
Hair shaft
External root sheath
Connective tissue sheath of hair follicle
Internal root sheath
Glassy membrane
Cuticle of hair
Cortex of hair
Medulla of hair
Matrix
Hair papilla
Hair bulb
Subcutaneous adipose tissue
Hair follicles
LM × 200

a A longitudinal section and a cross section through a hair follicle

b Diagrammatic view along the longitudinal axis of hair follicles

c Histological section along the longitudinal axis of hair follicles

Figure 4.11 **The Hair Growth Cycle.** Each hair follicle goes through growth cycles involving active and resting stages.

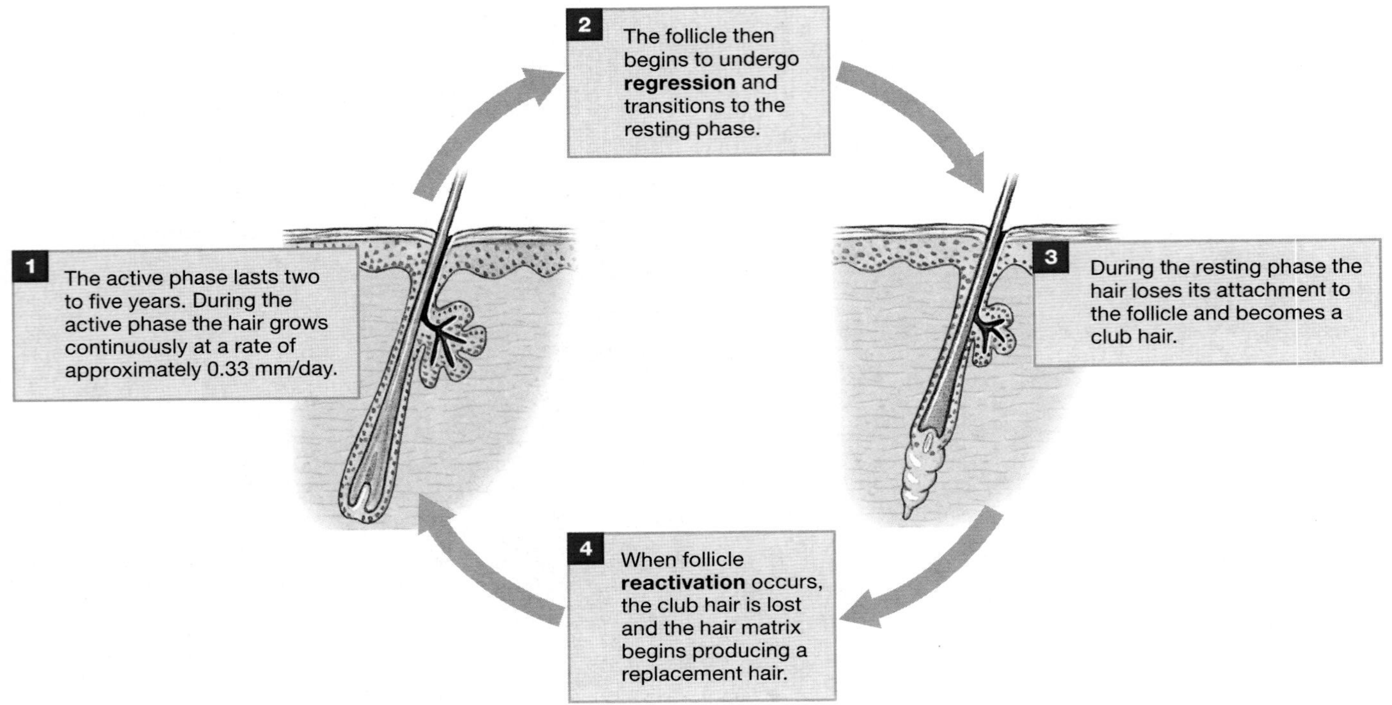

1 The active phase lasts two to five years. During the active phase the hair grows continuously at a rate of approximately 0.33 mm/day.
2 The follicle then begins to undergo regression and transitions to the resting phase.
3 During the resting phase the hair loses its attachment to the follicle and becomes a club hair.
4 When follicle reactivation occurs, the club hair is lost and the hair matrix begins producing a replacement hair.

The lipids released by sebaceous gland cells enter the open passageway, or lumen, of the gland. The arrector pili muscle that elevates the hair contracts and squeezes the sebaceous gland, forcing its waxy secretions, called **sebum** (SĒ-bum), into the follicle and onto the surface of the skin. Keratin is a tough protein, but dead, keratinized cells become dry and brittle once exposed to the environment. Sebum lubricates and protects the keratin of the hair shaft, conditions the surrounding skin, and inhibits the growth of bacteria. Shampooing removes this natural oily coating, and excessive washing can make hairs stiff and brittle. Conditioners reduce structural damage by rehydrating and coating the hair shaft.

Sebaceous follicles are large sebaceous glands that communicate directly with the epidermis. These follicles, which never produce hairs, are found on the integument covering the face, back, chest, nipples, and external genitalia. Although sebum has bactericidal (bacteria-killing) properties, under some conditions bacteria can invade sebaceous glands or follicles. The presence of bacteria in glands or follicles can produce local inflammation known as **folliculitis** (fo-lik-ū-LĪ-tis). If the duct of the gland becomes blocked, a distinctive abscess called a **furuncle** (FŪ-rung-kel), or "boil," develops. The usual treatment for a furuncle is to cut it open, or "lance" it, so that normal drainage and healing can occur.

Sweat Glands

Sweat glands produce a watery solution and perform other special functions. The skin contains two groups of sweat glands: apocrine sweat glands and eccrine sweat glands (**Figures 4.12** and **4.14**).

Both apocrine and eccrine sweat glands contain **myoepithelial cells** (*myo*-, muscle), specialized epithelial cells located between the gland cells and the underlying basal lamina. Myoepithelial cell contractions squeeze the gland and discharge the accumulated secretions. The autonomic nervous system and circulating hormones control the secretory activities of the gland cells and the contractions of myoepithelial cells.

CLINICAL NOTE

Acne and Seborrheic Dermatitis

Sebaceous glands and follicles are sensitive to the changes in sex hormones that occur during puberty. Sebaceous ducts of the face, back, and chest can become blocked, allowing secretions and dead skin cells to accumulate. This is a fertile environment for inflammation and bacterial infection, known as **acne**.

Seborrheic dermatitis is an inflammatory condition of abnormally active sebaceous glands, usually involving the scalp, eyebrows, or beard. The affected area becomes red with epidermal scaling. In adults; this condition is known as dandruff; in children, it is commonly known as cradle cap.

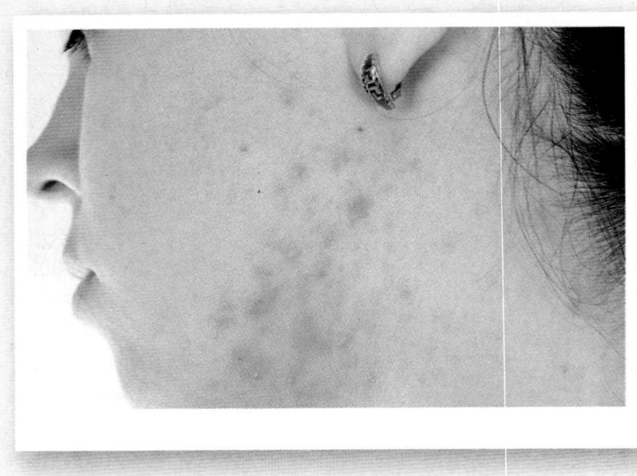

Figure 4.12 A Classification of Exocrine Glands in the Skin. Relationship of sebaceous glands and sweat glands and some characteristics and functions of their secretions.

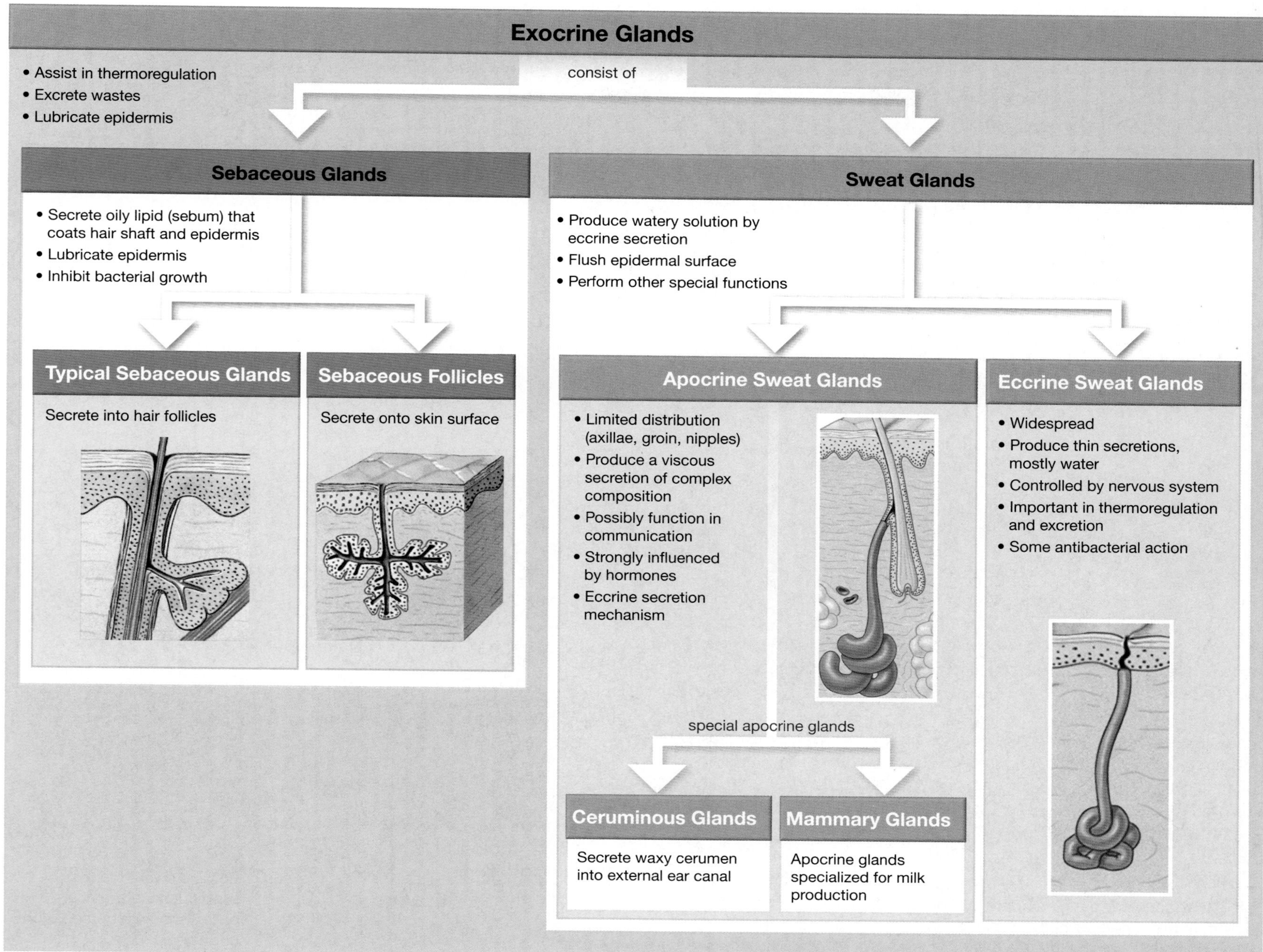

Apocrine Sweat Glands Sweat glands that release their secretions into hair follicles in the axillae (*armpits*), around the areolae (*nipples*), and in the groin are **apocrine sweat glands** (**Figures 4.12** and **4.14a**). They were originally named apocrine because scientists thought their cells used an apocrine method of secretion. ⮌ p. 60 Although we now know that they release their products through eccrine secretion, the name has not changed.

Apocrine sweat glands are coiled tubular glands that produce a viscous, cloudy, and potentially odorous secretion. They begin secreting at puberty. This sweat is a nutrient source for bacteria, whose growth and breakdown intensify the odor, which may be masked by deodorants. Apocrine gland secretions may also contain **pheromones**, chemicals that communicate information to other individuals at a subconscious level. The apocrine secretions of mature women have been shown to alter the menstrual timing of other women. The significance of these pheromones, and the role of apocrine secretions in males, remains unknown.

Eccrine Sweat Glands Far more numerous and widely distributed than apocrine sweat glands are **eccrine** (EK-rin) **sweat glands**, also known as *merocrine sweat glands* (**Figures 4.12** and **4.14b**). The adult integument contains around 3 million eccrine glands. They are smaller than apocrine sweat glands, and they do not extend as far into the dermis. Palms and soles have the highest numbers; the palm of your hand has about 500 glands per square centimeter (3000 glands per square inch). Eccrine sweat glands are coiled tubular glands that discharge their secretions directly onto the surface of the skin.

Eccrine glands produce a clear secretion called **sweat**, or **sensible perspiration**. Sweat is 99 percent water, but it does contain electrolytes (mainly sodium chloride, which gives sweat its salty taste), metabolites, and wastes. The functions of eccrine sweat gland activity include:

Figure 4.13 Sebaceous Glands and Follicles. The structure of sebaceous glands and sebaceous follicles in the skin.

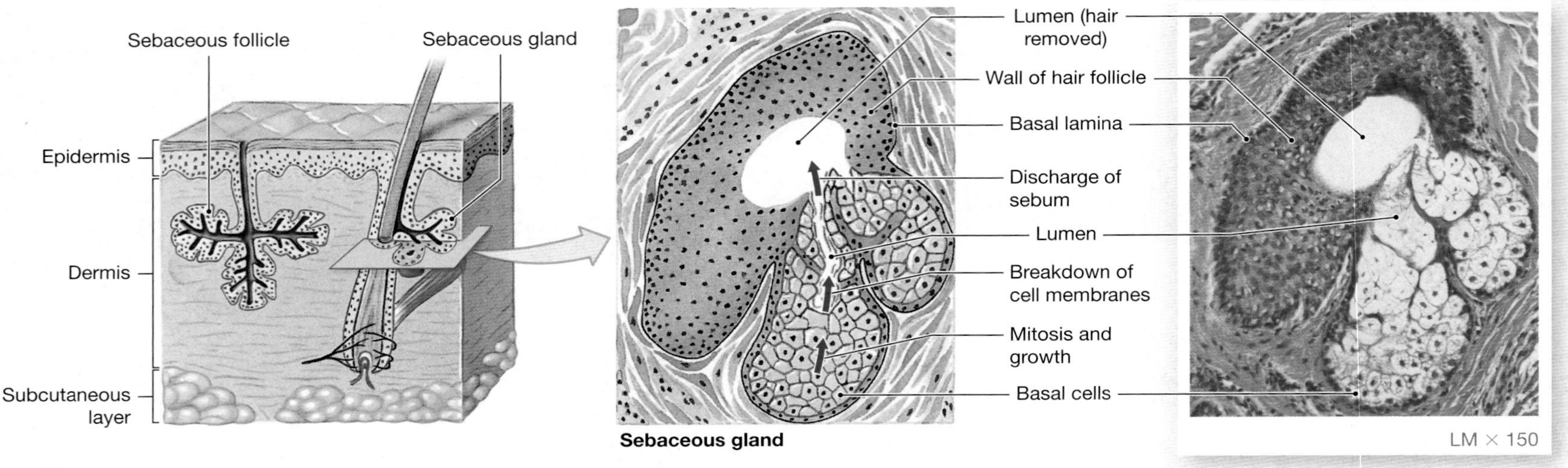

Figure 4.14 Sweat Glands.

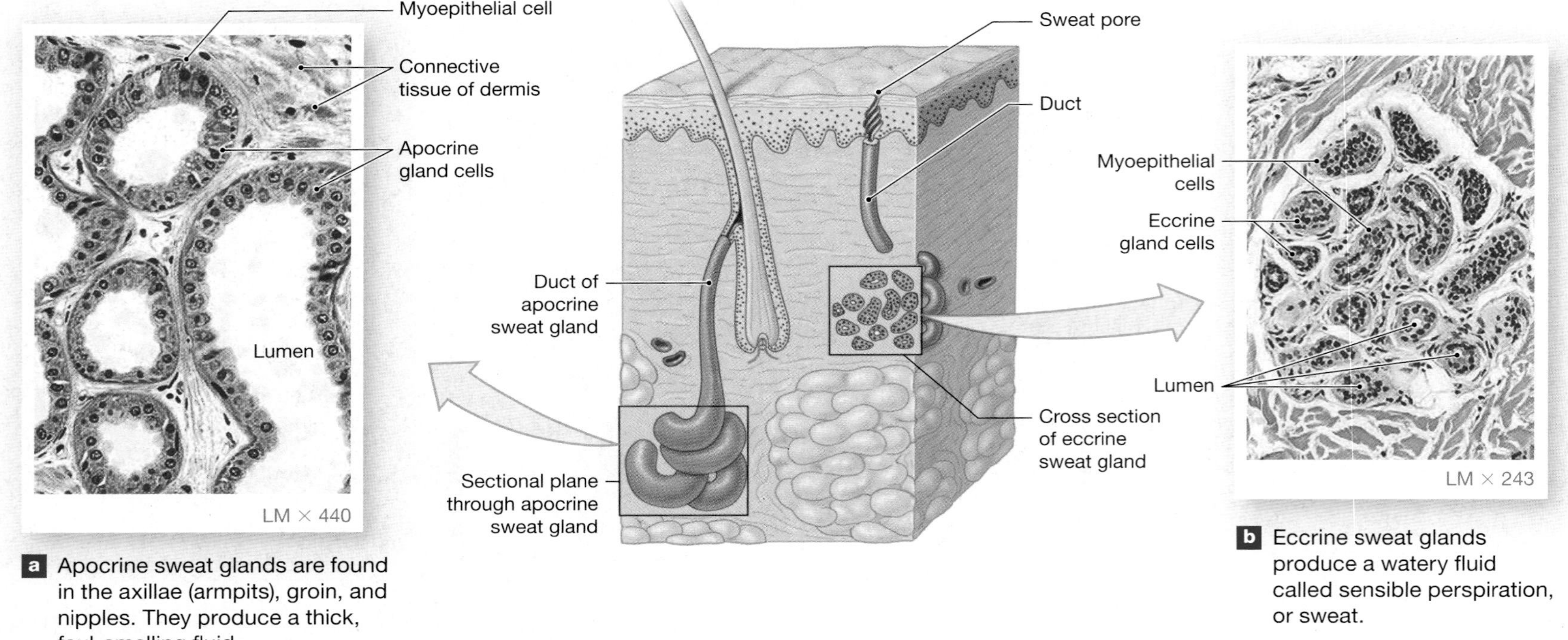

a Apocrine sweat glands are found in the axillae (armpits), groin, and nipples. They produce a thick, foul-smelling fluid.

b Eccrine sweat glands produce a watery fluid called sensible perspiration, or sweat.

- Thermoregulation. Sweat cools the surface of the skin and decreases body temperature. This cooling is the primary function of sensible perspiration, and the nervous system and hormonal mechanisms regulate the degree of secretory activity. When all the eccrine sweat glands are working at maximum, the rate of perspiration may exceed a gallon per hour, and dangerous fluid and electrolyte losses can occur. For this reason athletes participating in endurance sports must pause frequently to drink fluids.
- Excretion. Eccrine sweat gland secretion can also provide a significant excretory route for water and electrolytes, as well as for a number of prescription and nonprescription drugs.
- Protection. Eccrine sweat gland secretion protects us from environmental hazards by diluting harmful chemicals and discouraging the growth of microorganisms.

Control of Glandular Secretions

The autonomic nervous system can turn sebaceous glands and apocrine sweat glands on or off, but no regional control is possible. This means that when one sebaceous gland is activated, so are all the other sebaceous glands in the body. Eccrine sweat glands are much more precisely controlled; the amount of secretion and the area of the body involved can vary independently. For example, when you are nervously awaiting an anatomy exam, your palms may begin to sweat.

Other Integumentary Glands

Sebaceous glands and eccrine sweat glands are found over most of the body surface. Apocrine sweat glands are found in relatively restricted areas. The skin also contains a variety of specialized sweat glands that are restricted to specific locations. We will discuss many of them in later chapters, so we will note just two important examples here.

CLINICAL NOTE

Repairing Injuries to the Skin

The skin can regenerate effectively, even after considerable damage, because stem cells persist in both epithelial and connective tissues. The speed and effectiveness of skin repair vary with the wound's type and location and the person's age and health. Regeneration of skin after an injury involves four overlapping stages.

Step 1: Inflammation Phase

When damage to skin extends through the epidermis and into the dermis, bleeding occurs. Immune cells in the area, including mast cells, trigger an inflammatory response.

Step 2: Migration Phase

A blood clot, or **scab**, forms over the wound, blocking additional microorganisms from entry. Most of the clot consists of an insoluble network of fibrin, a fibrous protein that forms from blood proteins during the clotting response. The color of the clot reflects the presence of trapped red blood cells. Cells of the stratum basale divide rapidly and migrate along the edges of the wound to replace the missing epidermal cells. Simultaneously, macrophages patrol the damaged dermis, phagocytizing debris and pathogens.

If the wound involves a large area of thin skin, dermal repairs must be under way before epithelial cells can cover the surface. Divisions of fibroblasts and mesenchymal cells produce mobile cells that invade the deeper areas of the injury. Endothelial cells of damaged blood vessels also begin to divide, and capillaries follow the fibroblasts, enhancing circulation. The combination of blood clot, fibroblasts, and capillary network is called **granulation tissue**.

Step 3: Proliferation Phase

Over time, deeper portions of the clot dissolve and the number of capillaries decreases. Fibroblasts form collagen fibers and ground substance. Migrating epithelial cells surround the clot, forming new epidermis beneath the scab.

Step 4: Scarring Phase

After the scab is shed, the new epidermis is thin and the new dermis contains an abnormally large number of collagen fibers and relatively few blood vessels. Severely damaged hair follicles, sebaceous and sweat glands, muscle cells and nerves are seldom restored; they, too, are replaced by fibrous tissue. This rather inflexible, fibrous, noncellular **scar tissue** is the final result of the healing process.

After Healing

Scar tissue formation is highly variable. For example, surgical procedures on a fetus do not leave scars. In some adults, particularly those with dark skin, scar tissue formation may continue, producing a thickened scar called a **keloid**.

1 Bleeding occurs at the site of injury immediately after the injury, and mast cells in the region trigger an inflammatory response.

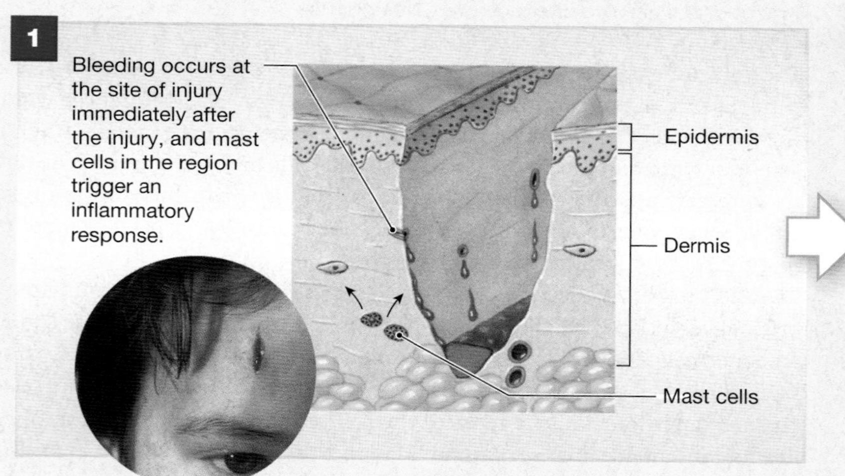

2 After several hours, a scab has formed and cells of the stratum basale are migrating along the edges of the wound. Phagocytic cells are removing debris, and more of these cells are arriving via the enhanced circulation in the area. Clotting around the edges of the affected area partially isolates the region.

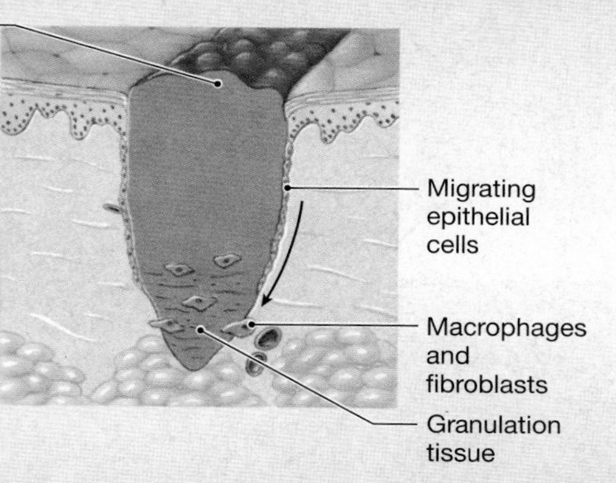

3 One week after the injury, the scab has been undermined by epidermal cells migrating over the meshwork produced by fibroblast activity. Phagocytic activity around the site has almost ended, and the fibrin clot is disintegrating.

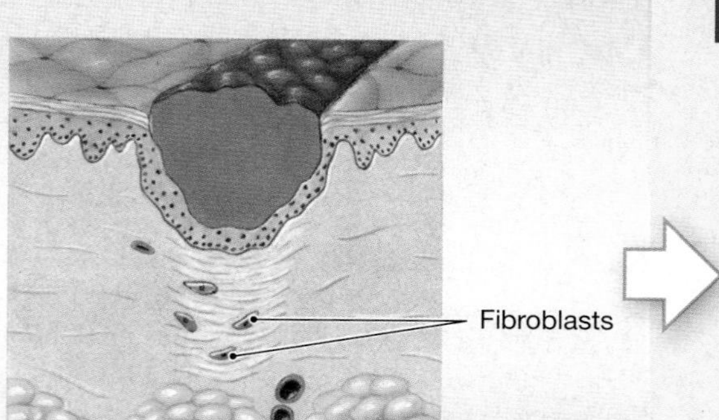

4 After several weeks, the scab has been shed, and the epidermis is complete. A shallow depression marks the injury site, but fibroblasts in the dermis continue to create scar tissue that will gradually elevate the overlying epidermis.

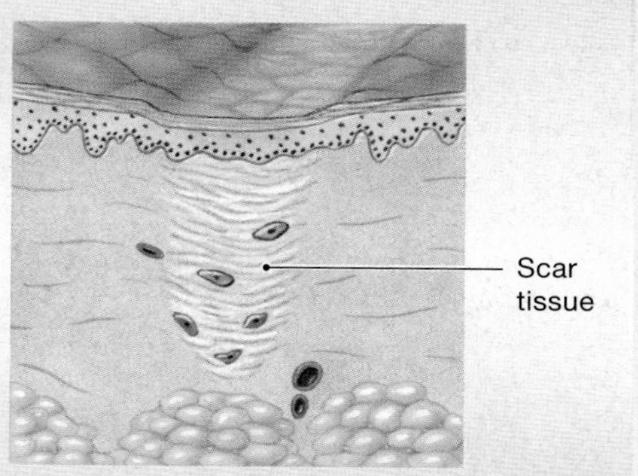

1. The **mammary glands** of the breasts are anatomically related to apocrine sweat glands. A complex interaction between sexual and pituitary hormones controls their development and secretion. We will discuss mammary gland structure and function in Chapter 27.
2. **Ceruminous** (se-RŪ-mi-nus) **glands** are modified sweat glands in the external auditory canal of the ear. They have a larger lumen than eccrine sweat glands do, and their gland cells contain pigment granules and lipid droplets not found in other sweat glands. Their secretions combine with those of nearby sebaceous glands, forming a mixture called **cerumen**, or "earwax." Earwax, together with tiny hairs along the ear canal, helps trap foreign particles and small insects, preventing them from reaching the tympanic membrane (eardrum).

Nails

▶ **KEY POINT** Nails, composed of keratinized epithelial cells arranged into plates, protect our fingers and toes.

Nails form on the dorsal surfaces of the tips of the toes and fingers. They protect the exposed tips and help limit distortion when they are subjected to mechanical stress—for example, when we run or grasp objects.

Figure 4.15 shows the structure of a nail. The **nail body** covers the **nail bed**. Nail production occurs at the **nail root**, an epithelial fold not visible from the surface. The deepest portion of the nail root lies very close to the periosteum of the bone of the fingertip.

Figure 4.15 Structure of a Nail. These drawings illustrate the prominent structures of a typical fingernail.

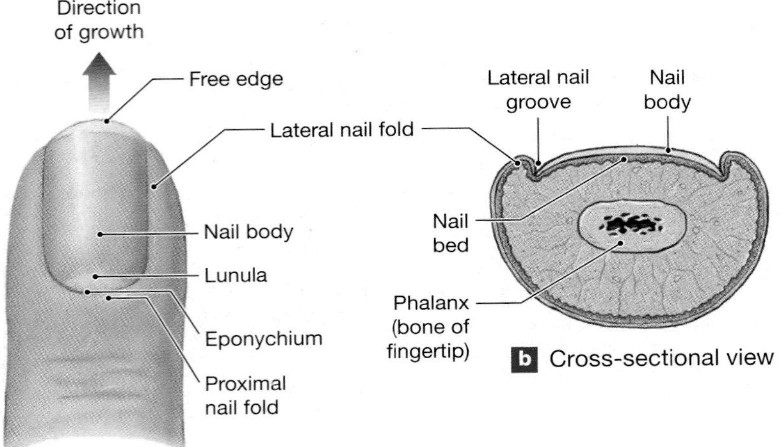

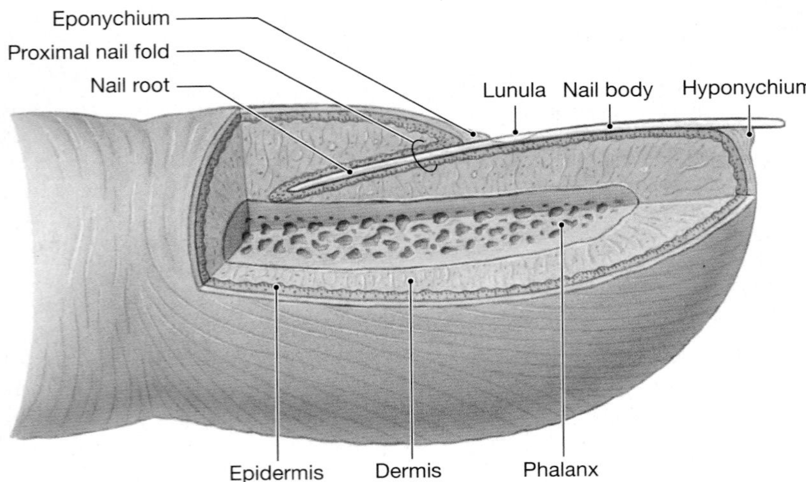

The nail body is recessed beneath the level of the surrounding epithelium, and it is bordered by **lateral nail grooves** and **lateral nail folds**. A portion of the stratum corneum of the nail fold extends over the exposed nail nearest the root, forming the **eponychium** (ep-ō-NIK-ē-um; *epi-*, over, + *onyx*, nail), or *cuticle*. The **proximal nail fold** is the portion of skin nearest the eponychium. Underlying blood vessels give the nail its characteristic pink color, but near the root these vessels may be obscured, leaving a pale crescent known as the **lunula** (LŪ-nū-la; *luna*, moon). The distal free edge of the nail body extends over a thickened stratum corneum, the **hyponychium** (hī-pō-NIK-ē-um).

Changes in the shape, structure, or appearance of the nails are clinically significant. For example, the nails may turn yellow in patients who have chronic respiratory disorders, thyroid gland disorders, or AIDS. Nails may become pitted and distorted in psoriasis and concave in some blood disorders.

4.5 ✔ CONCEPT CHECK

9 What are the functions of hair?
10 What are the two types of hair?
11 Compare and contrast apocrine and eccrine sweat glands.
12 List the main structures of a typical fingernail.

See the blue Answers tab at the back of the book.

4.6 Local Control of Integumentary Function

▶ **KEY POINT** Skin responds to environmental influences without the involvement of the nervous or endocrine systems.

The integumentary system displays a significant degree of functional independence. It responds directly and automatically to local influences without the involvement of the nervous or endocrine systems. For example, when the skin is subjected to mechanical stresses, stem cells in the stratum basale divide more rapidly and the epithelium thickens. That is why calluses form on your palms when you do a lot of work with your hands. We can see a more dramatic display of local regulation when the skin is injured.

After severe damage, the repair process does not return the integument to its original condition. (See the Clinical Note on ⮌ p. 101.) The injury site contains an abnormal density of collagen fibers and relatively few blood vessels. Damaged hair follicles, sebaceous glands, sweat glands, muscle cells, and nerves are seldom repaired, and fibrous tissue replaces them. The formation of this rather inflexible, fibrous, noncellular **scar tissue** is a practical limit to the healing process.

4.6 ✔ CONCEPT CHECK

13 List one example of local control of the integumentary system.

See the blue Answers tab at the back of the book.

4.7 Aging and the Integumentary System

▶ **KEY POINT** Aging affects all the components of the integumentary system.

Figure 4.16 summarizes the effects of aging on the integument.

- The epidermis thins as basal cell activity decreases, making older people more prone to injury and skin infections.
- The number of Langerhans cells decreases to around 50 percent of levels seen at maturity (approximately age 21). This decrease may make the immune system less sensitive and further encourage skin damage and infection.

Figure 4.16 The Skin during the Aging Process. Characteristic changes, causes, and effects in the skin during aging.

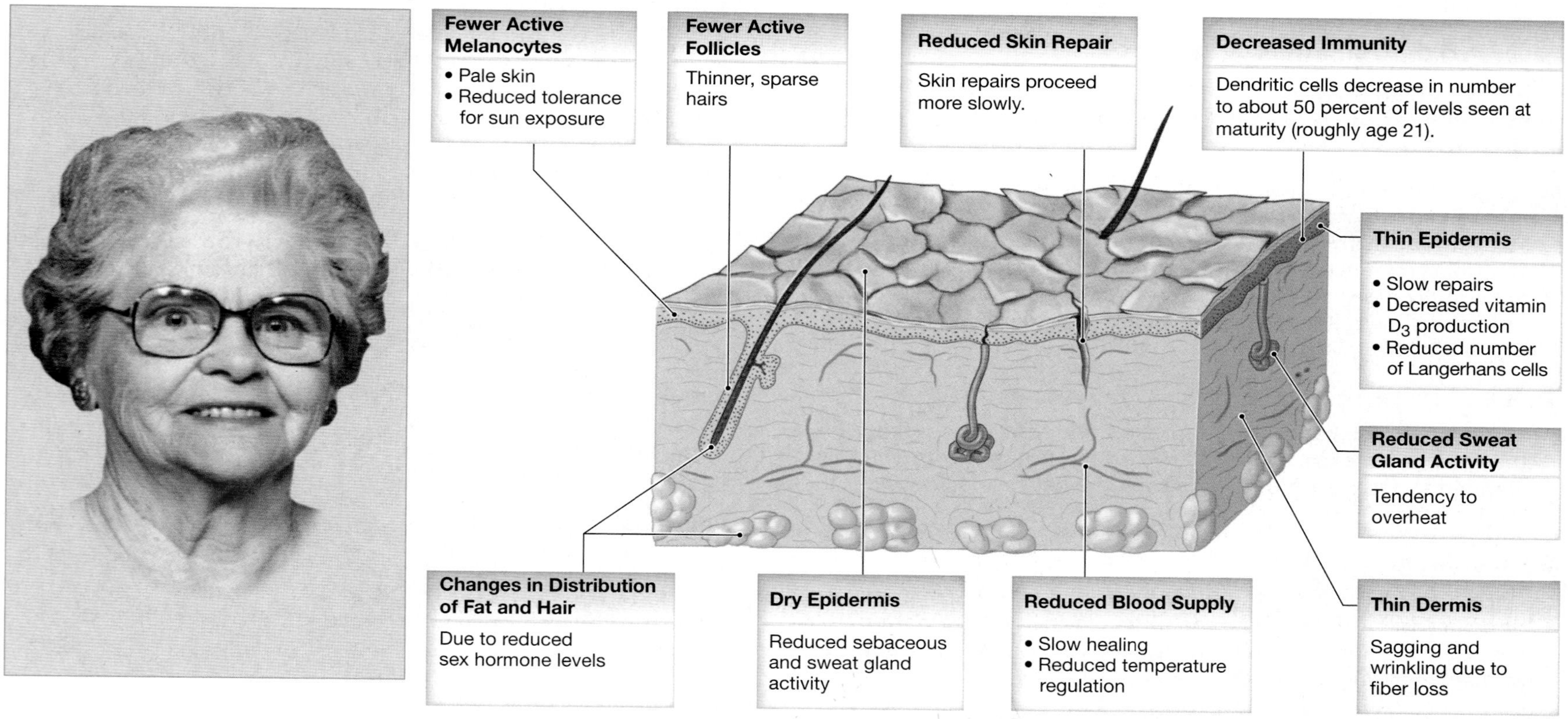

Fewer Active Melanocytes
- Pale skin
- Reduced tolerance for sun exposure

Fewer Active Follicles
Thinner, sparse hairs

Reduced Skin Repair
Skin repairs proceed more slowly.

Decreased Immunity
Dendritic cells decrease in number to about 50 percent of levels seen at maturity (roughly age 21).

Thin Epidermis
- Slow repairs
- Decreased vitamin D_3 production
- Reduced number of Langerhans cells

Reduced Sweat Gland Activity
Tendency to overheat

Changes in Distribution of Fat and Hair
Due to reduced sex hormone levels

Dry Epidermis
Reduced sebaceous and sweat gland activity

Reduced Blood Supply
- Slow healing
- Reduced temperature regulation

Thin Dermis
Sagging and wrinkling due to fiber loss

- Vitamin D_3 production decreases by around 75 percent. The result can be muscle weakness and weaker bones.
- Melanocyte activity declines, and in Caucasians the skin becomes very pale. With less melanin in the skin, older people are more sensitive to sun exposure and more likely to experience sunburn.
- Glandular activity decreases. The skin becomes dry and often scaly because sebum production decreases; eccrine sweat glands are also less active. With impaired perspiration, older people cannot lose heat as efficiently as younger people can. Thus, the elderly are at greater risk of overheating in warm environments.
- The blood supply to the dermis decreases at the same time that sweat glands become less active. This combination lessens the elderly person's ability to lose body heat. For this reason, overexertion or warm temperatures (such as a sauna or hot tub) can cause dangerously high body temperatures.
- Hair follicles stop functioning or produce thinner, finer hairs. With decreased melanocyte activity, these hairs are gray or white.
- The dermis and its elastic fiber network become thinner. The integument becomes weaker and less resilient; sagging and wrinkling occur. These effects are most noticeable in areas exposed to the sun.
- Secondary sexual characteristics in hair and body fat distribution begin to fade as the result of changes in levels of sex hormones.
- Skin repairs proceed more slowly, and recurring infections may result. It takes three to four weeks to repair a blister site at age 21, but six to eight weeks at age 65–75.

4.7 CONCEPT CHECK

14 List three examples of how the integumentary system changes as we get older.

See the blue Answers tab at the back of the book.

Study Outline

Introduction p. 87

- The **integumentary system**, or integument, protects us from the surrounding environment. Its receptors tell us about the outside world, and it helps regulate body temperature.

4.1 Structure and Function of the Integumentary System p. 87

- The integumentary system consists of the cutaneous membrane or **skin**, which includes the superficial **epidermis**, the deeper **dermis**, and the **accessory structures**, including hair follicles, nails, and exocrine glands. The subcutaneous layer is deep to the cutaneous membrane. *(See Figures 4.1 and 4.2.)*

4.2 The Epidermis p. 88

- The epidermis is a stratified squamous epithelium.
- There are four cell types in the epidermis: **keratinocytes**, the most abundant epithelial cells; **melanocytes**, pigment-producing cells; **Merkel cells**, involved in detecting sensation; and **Langerhans cells**,

which are phagocytic cells of the immune system. Melanocytes, Merkel cells, and Langerhans cells are scattered among the keratinocytes.

Layers of the Epidermis p. 88

- Division of basal cells in the **stratum basale** produces new keratinocytes, which replace more superficial cells. *(See Figure 4.3.)*
- As new committed epidermal cells differentiate, they pass through the **stratum spinosum**, the **stratum granulosum**, the **stratum lucidum** (of thick skin), and the **stratum corneum**. The keratinocytes move toward the surface, and through the process of keratinization, the cells accumulate large amounts of **keratin**. Ultimately, the cells are shed or lost at the epidermal surface. *(See Figure 4.3.)*

Thin and Thick Skin p. 90

- **Thin skin**, which has four layers of keratinocytes, covers most of the body; **thick skin**, which has five layers, covers only the heavily abraded surfaces, such as the palms of the hands and the soles of the feet. *(See Figure 4.4.)*

Dermal Ridges p. 91

- **Dermal ridges**, such as those on the palms and soles, improve our gripping ability and increase the skin's sensitivity. Their pattern is determined genetically. The ridges interlock with **dermal papillae** of the underlying dermis. *(See Figures 4.4 and 4.5.)*

Skin Color p. 91

- The color of the epidermis depends on a combination of factors: the dermal blood supply and variable quantities of two pigments: **carotene** and **melanin**. Melanin helps protect the skin from the damaging effects of excessive ultraviolet radiation. *(See Figure 4.6.)*

4.3 | The Dermis p. 92

Dermal Organization p. 92

- Two layers compose the dermis: the superficial **papillary layer** and the deeper **reticular layer**. *(See Figures 4.2, 4.4, and 4.7–4.9.)*
- The papillary layer derives its name from its association with the dermal papilla. It contains blood vessels, lymphatics, and sensory nerves. This layer supports and nourishes the overlying epidermis. *(See Figures 4.4 and 4.7.)*
- The reticular layer consists of a meshwork of collagen and elastic fibers oriented in all directions to resist tension in the skin. *(See Figures 4.7 and 4.8.)*

Other Dermal Components p. 93

- An extensive blood supply to the skin includes the **cutaneous** and **papillary plexuses**. The papillary layer contains abundant capillaries that drain into the veins of these plexuses. *(See Figure 4.2.)*

4.4 | The Subcutaneous Layer p. 94

- Although not part of the integument, the subcutaneous layer (*hypodermis* or *superficial fascia*) stabilizes the skin's position against underlying organs and tissues yet permits limited independent movement. *(See Figures 4.2 and 4.7.)*

4.5 | Accessory Structures p. 95

Hair Follicles and Hair p. 95

- **Hairs** originate in complex organs called **hair follicles**, which extend into the dermis. Each hair has a **bulb**, **root**, and **shaft**. Hair production involves a special keratinization of the epithelial cells of the **hair matrix**. At the center of the matrix, the cells form a soft core, or **medulla**; cells at the edge of the hair form a hard **cortex**. The **cuticle** is a hard layer of dead, keratinized cells that coats the hair. *(See Figures 4.9 and 4.10.)*
- The lumen of the follicle is lined by an **internal root sheath** produced by the hair matrix. An **external root sheath** surrounds the internal root sheath, between the skin surface and hair matrix. The **glassy membrane** is the thickened basal lamina external to the external root sheath; a dense connective tissue layer surrounds it. *(See Figure 4.9.)*
- A **root hair plexus** of sensory nerves surrounds the base of each hair follicle and detects the movement of the shaft. Contraction of the **arrector pili muscle** elevates the hair by pulling on the follicle. *(See Figures 4.9 and 4.10.)*
- **Vellus hairs** ("peach fuzz") and heavy **terminal hairs** make up the hair population on our adult bodies.
- Hairs grow and are shed according to the **hair growth cycle**. A single hair grows for two to five years and is subsequently shed. *(See Figure 4.11.)*

Glands in the Skin p. 96

- **Sebaceous** (*oil*) **glands** discharge a waxy, oily secretion (**sebum**) into hair follicles. **Sebaceous follicles** are large sebaceous glands that produce no hair; they communicate directly with the epidermis. *(See Figures 4.12 and 4.13.)*
- **Apocrine sweat glands** produce an odorous secretion. The more numerous **eccrine sweat glands**, or *merocrine sweat glands*, produce a thin, watery secretion known as **sensible perspiration**, or **sweat**. *(See Figures 4.12 and 4.14.)*
- The **mammary glands** of the breast resemble larger and more complex apocrine sweat glands. Active mammary glands secrete milk. **Ceruminous glands** in the ear canal are modified sweat glands that produce waxy **cerumen**.

Nails p. 102

- The **nails** protect the exposed tips of the fingers and toes and help limit their distortion when they are subjected to mechanical stress.
- The **nail body** covers the **nail bed**, with nail production occurring at the **nail root**. The **eponychium** (*cuticle*) is formed by a fold of the stratum corneum called the **nail fold**, which extends from the nail root to the exposed nail. *(See Figure 4.15.)*

4.6 | Local Control of Integumentary Function p. 102

- The skin can regenerate effectively even after considerable damage, such as severe cuts or moderate burns.
- Severe damage to the dermis and accessory glands cannot be completely repaired, and fibrous **scar tissue** remains at the injury site. *(See Clinical Note on p. 101.)*

4.7 | Aging and the Integumentary System p. 102

- Aging affects all layers and accessory structures of the integumentary system. *(See Figure 4.16.)*

Chapter Review

For answers, see the blue Answers tab at the back of the book.

Level 1 Reviewing Facts and Terms

Match each numbered item with the most closely related lettered item.

1. subcutaneous layer .... [ ]
2. dermis .... [ ]
3. stem cell .... [ ]
4. keratinized/cornified .... [ ]
5. melanocytes .... [ ]
6. epidermis .... [ ]
7. sebaceous gland .... [ ]
8. sweat gland .... [ ]
9. scar tissue .... [ ]

a. fibrous, noncellular
b. holocrine; oily secretion
c. pigment cells
d. stratum basale
e. superficial fascia
f. papillary layer
g. stratum corneum
h. stratified squamous epithelium
i. eccrine; clear secretion

10. Label the following structures on the accompanying diagram of the integumentary system:
- dermis
- epidermis
- (sweat gland)
- lamellar corpuscle
- sebaceous gland
- arrector pili muscle

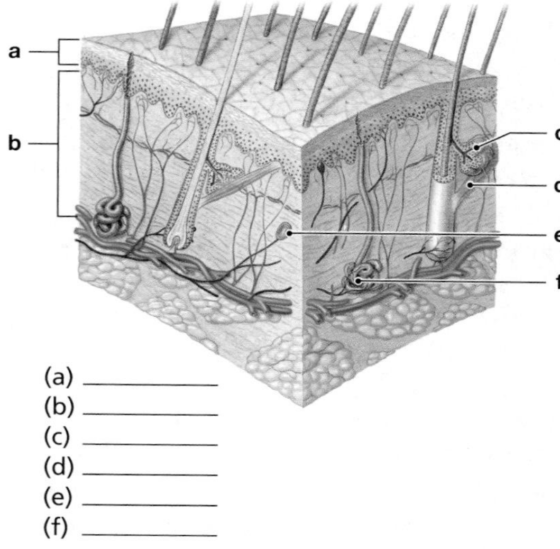

(a) ________
(b) ________
(c) ________
(d) ________
(e) ________
(f) ________

11. The effects of aging on the skin include
(a) a decrease in sebaceous gland activity.
(b) increased production of vitamin D_3.
(c) thickening of the epidermis.
(d) an increased blood supply to the dermis.

12. Skin color is the result of
(a) the dermal blood supply.
(b) pigment composition.
(c) pigment concentration.
(d) all of the above.

13. Label the following structures on the accompanying figure:
- eponychium
- phalanx
- hyponychium
- nail root

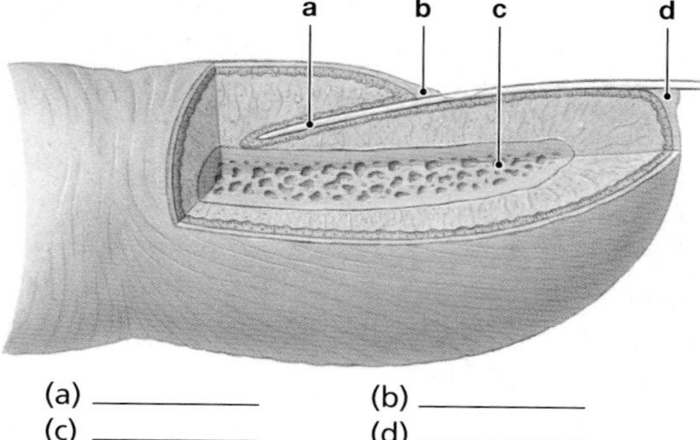

(a) ________ (b) ________
(c) ________ (d) ________

14. The layer of the skin that contains both interwoven bundles of collagen fibers and the protein elastin, and is responsible for the strength of the skin, is the
(a) papillary layer. (b) reticular layer.
(c) epidermal layer. (d) hypodermal layer.

15. The layer of the epidermis that contains cells undergoing division is the
(a) stratum corneum.
(b) stratum basale.
(c) stratum granulosum.
(d) stratum lucidum.

16. All of the following are effects of aging *except*
(a) the thinning of the epidermis of the skin.
(b) an increase in the number of Langerhans cells.
(c) a decrease in melanocyte activity.
(d) a decrease in glandular activity.

17. Each of the following is a function of the integumentary system *except*
(a) protection of underlying tissue.
(b) excretion.
(c) synthesis of vitamin C.
(d) thermoregulation.

18. Carotene is
(a) an orange-yellow pigment that accumulates inside epidermal cells.
(b) another name for melanin.
(c) deposited in stratum granulosum cells to protect the epidermis.
(d) a pigment that gives the characteristic color to hemoglobin.

19. Which statement best describes a hair root?
(a) It extends from the hair bulb to the point where the internal organization of the hair is complete.
(b) It is the nonliving portion of the hair.
(c) It encompasses all of the hair deep to the surface of the skin.
(d) It includes all of the structures of the hair follicle.

Level 2 Reviewing Concepts

1. Explain why fair-skinned individuals have greater need to shield themselves from the sun than do dark-skinned individuals.
2. Explain how and why calluses form.
3. Weight lifters often have stretch marks. Explain why, utilizing sound anatomical logic.
4. Describe how the protein keratin affects the appearance and function of the integument.
5. List the characteristics that make the subcutaneous layer a region commonly used for hypodermic injections.
6. Explain how washing the skin and applying deodorant reduce the odor of apocrine sweat glands.
7. Explain what is happening to a person who is cyanotic, and what body structures would show this condition most easily.
8. Explain why elderly people are less able to adapt to temperature extremes.

Level 3 Critical Thinking

1. You are about to undergo surgery. Explain why you want your physician to have an excellent understanding of the tension lines of the skin.
2. Defend the following statement: Many medications can be administered transdermally by applying patches that contain the medication to the surface of the skin. These patches can be attached anywhere on the skin except the palms of the hands and the soles of the feet.

4

CLINICAL CASE | WRAP-UP

Flesh-Eating Bacteria

The surgeon evaluating Martin notes his high fever, low blood pressure, and rapid pulse, indicating shock. The swelling and subcutaneous air in his right thigh indicate the presence of gas-forming bacteria. Her diagnosis is necrotizing fasciitis, a serious, often fatal necrosis (tissue death) of the deep fascia caused by bacteria, primarily *Streptococcus*, the same microbe that causes strep throat. The bacteria spread along fascial planes and destroy tissue with their toxins; thus, they are popularly known as "flesh-eating bacteria." The surgical team takes Martin directly to the operating room.

Upon incising Martin's thigh, the surgeon suctions a large amount of "dishwater"-appearing fluid that contains gas. The subcutaneous tissue appears gray and necrotic. She pushes her finger along the plane between the subcutaneous layer and the dermis and notes that the integument is no longer firmly attached to the deep fascia around the muscles of the anterior thigh. She sees that the cutaneous plexus of blood vessels is thrombosed (clotted), with no effective blood flow to the overlying dermis, and assumes the subpapillary plexus is similarly thrombosed. She debrides (removes dead tissue from) the area and hopes that further debridements, and eventual skin grafting, will save Martin's life and limb.

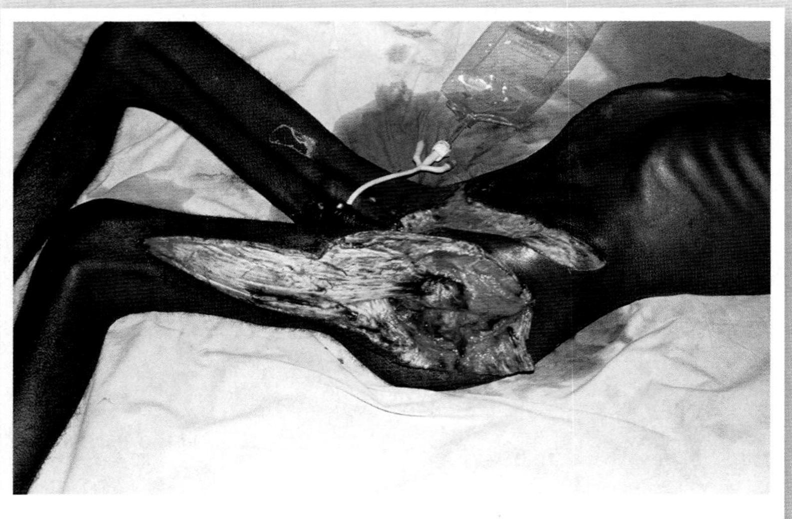

1. How did the bacteria causing the necrotizing fasciitis get into the loose connective tissue of the subcutaneous layer?
2. With occlusion (blockage due to clotting) of the deep cutaneous plexus and the superficial subpapillary plexus of blood vessels, what will happen to the skin in this area?

See the blue Answers tab at the back of the book.

Related Clinical Terms

cold sore: A lesion that typically occurs in or around the mouth and is caused by a dormant herpes simplex virus that may be reactivated by factors such as stress, fever, or sunburn. Also called *fever blister*.

comedo: The primary sign of acne consisting of an enlarged pore filled with skin debris, bacteria, and sebum (oil); the medical term for a blackhead.

dermatology: The branch of medicine concerned with the diagnosis, treatment, and prevention of diseases of the skin, hair, and nails.

eczema: Rash characterized by inflamed, itchy, dry, scaly, or irritated skin.

frostbite: Injury to body tissues caused by exposure to below-freezing temperatures, typically affecting the nose, fingers, or toes and sometimes resulting in gangrene.

gangrene: A term that describes dead or dying body tissue that occurs when the local blood supply to the tissue either is lost or is inadequate to keep the tissue alive.

impetigo: An infection of the surface of the skin, caused by staphylococcus ("staph") and streptococcus ("strep") bacteria.

nevus: A benign pigmented spot on the skin such as a mole.

onycholysis: A nail disorder characterized by a spontaneous separation of the nail bed starting at the distal free margin and progressing proximally.

pallor: An unhealthy pale appearance.

porphyria: A rare hereditary disease in which the blood pigment hemoglobin is abnormally metabolized. Porphyrins are excreted in the urine, which becomes dark; other symptoms include mental disturbances and extreme sensitivity of the skin to light.

rosacea: A condition in which certain facial blood vessels enlarge, giving the cheeks and nose a flushed appearance.

scleroderma: An idiopathic chronic autoimmune disease characterized by hardening and contraction of the skin and connective tissue, either locally or throughout the body.

tinea: A skin infection caused by a fungus; also called ringworm.

urticaria: Skin condition characterized by red, itchy, raised areas that appear in varying shapes and sizes; commonly called hives.

5

The Skeletal System

Osseous Tissue and Bone Structure

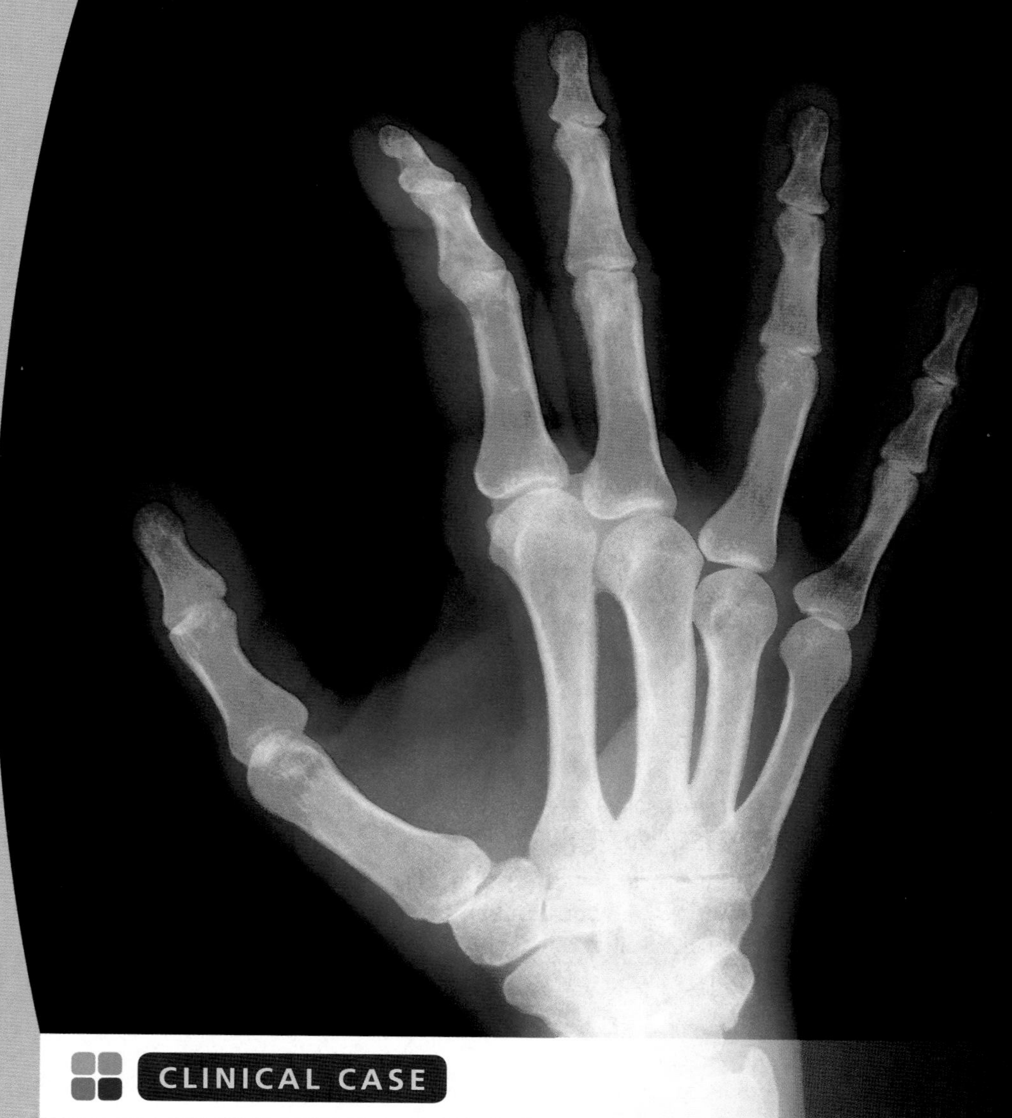

Learning Outcomes

These Learning Outcomes correspond by number to this chapter's sections and indicate what you should be able to do after completing the chapter.

- **5.1** Compare and contrast the structure and function of the various cell types found within developing and mature bone, how these cells contribute to the formation of compact bone and cancellous bone, and how these cells contribute to the structure and function of the periosteum and endosteum. p. 108
- **5.2** Compare and contrast the processes involved in the formation of bone and the growth of bone, and explain the factors involved in the regulation of these processes. p. 113
- **5.3** Describe the different types of fractures and outline how fractures heal. p. 118
- **5.4** Classify bones according to their shapes and give one or more examples for each type. p. 120
- **5.5** Explain how the normal functioning, growth, remodeling, and repair of the skeletal system is integrated with other systems of the body. p. 125

CLINICAL CASE

Pushing Beyond Her Limits

Emily, a university freshman, had always wanted to run competitively during her university career. Now, only four weeks into the season, she is changing her mind.

During the second week of training Emily began to feel vague discomfort at the junction of the middle and distal thirds of her right leg. In the third week she started limping after practice. Now the pain starts soon after she begins running, and she limps the rest of the day.

Emily goes to the student health center for an x-ray of her right leg. While waiting for the doctor, she palpates her leg and finds a very tender spot right where the pain is—the anterior surface of the tibia directly over the crest (anterior margin). It even looks a little red and swollen there.

"Emily, I cannot see anything wrong on these x-rays," says the radiologist. "But something is causing your pain, so I want you to have an MRI of that leg."

What could be causing Emily's pain? To find out, turn to the Clinical Case Wrap-Up on p. 130.

THE SKELETAL SYSTEM includes the bones of the skeleton and the cartilages, ligaments, and other connective tissues that stabilize or interconnect them. Bones are the organs of the skeletal system, and they do more than serve as racks that muscles hang from; they support our weight and work together with muscles, producing controlled, precise movements. Without a framework of bones to connect to, contracting muscles would just get shorter and fatter. Our muscles must pull against the skeleton to make us sit, stand, walk, or run. The skeleton has many other vital functions; some may be unfamiliar to you, so we begin this chapter by summarizing the major functions of the skeletal system.

- Support: The skeletal system provides structural support for the entire body. Individual bones or groups of bones provide a framework for the attachment of soft tissues and organs.
- Mineral storage: The calcium salts of bone are a valuable mineral reserve that maintains normal concentrations of calcium and phosphate ions in body fluids. Calcium is the most abundant mineral in the human body. A typical human body contains 1–2 kg (2.2–4.4 lb) of calcium, with more than 98 percent of it in the bones of the skeleton.
- Blood cell production: Red blood cells, white blood cells, and platelets are produced in red marrow, which fills the internal cavities of many bones. The role of bone marrow in blood cell formation will be described in later chapters on the cardiovascular and lymphatic systems (Chapters 20 and 23).
- Protection: Delicate tissues and organs are surrounded by skeletal elements. The ribs protect the heart and lungs, the skull encloses the brain, the vertebrae shield the spinal cord, and the pelvis cradles delicate digestive and reproductive organs.
- Leverage: Many bones of the skeleton function as levers. They change the magnitude and direction of the forces generated by skeletal muscles. The movements produced range from the delicate motion of a fingertip to powerful changes in the position of the entire body.

This chapter describes the structure, development, and growth of bone. The two chapters that follow organize bones into two divisions: the **axial skeleton** (the bones of the skull, vertebral column, sternum, and ribs) and the **appendicular skeleton** (the bones of the limbs and the associated bones that connect the limbs to the trunk at the shoulders and pelvis). The fourth and final chapter in this group examines articulations or joints. structures where the bones meet and may move with respect to each other.

The bones of the skeleton are complex, dynamic organs that contain osseous tissue, other connective tissues, smooth muscle tissue, and neural tissue. We now consider the internal organization of a typical bone.

5.1 Structure and Function of Bone

▶ **KEY POINT** Bone is a specialized form of connective tissue—a supporting connective tissue. Like all connective tissues, it is composed of specialized cells, protein fibers, and an extracellular matrix.

Bone tissue, or **osseous tissue**, is one of the supporting connective tissues. (Review the sections on dense connective tissues, cartilage, and bone at this time. ↺ **pp. 65, 67–70**) Like other connective tissues, osseous tissue contains specialized cells and an extracellular matrix of protein fibers and a ground substance. The matrix of bone tissue is solid and sturdy because of the deposition of calcium salts around the protein fibers.

Osseous tissue is separated from surrounding tissues by a fibrous periosteum. When osseous tissue surrounds another tissue, the inner bony surfaces are lined by a cellular endosteum.

The Histological Organization of Mature Bone

▶ **KEY POINT** Mature bone is composed of four types of highly specialized cells and an extracellular matrix of calcium, phosphate, and connective tissue fibers.

The basic organization of bone tissue was introduced in Chapter 3. We now take a closer look at the organization of the matrix and cells of bone.

The Matrix of Bone

Calcium phosphate, $Ca_3(PO_4)_2$, accounts for almost two-thirds of the weight of bone. It interacts with calcium hydroxide $[Ca(OH)_2]$ to form crystals of hydroxyapatite (hī-DROK-sē-ap-a-tīt)$[Ca_{10}(PO_4)_6(OH)_2]$. As these crystals form, they incorporate other calcium salts, such as calcium carbonate, and ions such as sodium, magnesium, and fluoride. These inorganic components enable bone to resist compression. Roughly one-third of the weight of bone is from collagen fibers and other noncollagenous proteins, which give bone considerable tensile strength. Osteocytes and other cell types account for only 2 percent of the mass of a typical bone.

Calcium phosphate crystals are very strong, but relatively inflexible. They withstand compression, but the crystals shatter when exposed to bending, twisting, or sudden impacts. Collagen fibers are tough and flexible. They easily tolerate stretching, twisting, and bending but, when compressed, they simply bend out of the way. In bone, the collagen fibers and other noncollagenous proteins provide an organic framework for the formation of mineral crystals. The hydroxyapatite crystals form small plates that lie alongside these ground substance proteins. The result is a protein–crystal combination with properties intermediate between those of collagen and those of pure mineral crystals.

The Cells of Mature Bone

Bone contains four cell types: osteoblasts, osteocytes, osteoprogenitor cells, and osteoclasts **(Figure 5.1a)**.

Osteocytes Mature bone cells are **osteocytes** (*osteon*, bone). They maintain and monitor the protein and mineral content of the surrounding matrix. The minerals in the matrix are continually recycled. Each osteocyte directs the release of calcium from bone into blood and the deposition of calcium salts into the surrounding matrix. Osteocytes occupy small chambers, called **lacunae**, that are sandwiched between layers of calcified matrix. These matrix layers are called **lamellae** (la-MEL-lē; singular, *lamella*; "thin plate") **(Figure 5.1b–d)**. Channels called **canaliculi** (kan-a-LIK-ū-lī; "little canals") radiate through the matrix from lacuna to lacuna and toward free surfaces and adjacent blood vessels. The canaliculi connect adjacent lacunae and bring the processes of neighboring osteocytes into close contact. Tight junctions interconnect these processes and provide a route for the diffusion of nutrients and waste products from one osteocyte to another across gap junctions.

Osteoblasts Cuboidal cells found in a single layer on the inner or outer surfaces of a bone are called **osteoblasts** (OS-tē-ō-blasts; *blast*, precursor). These cells secrete the organic components of the bone matrix. This material, called **osteoid** (OS-tē-oyd), later becomes mineralized through a complicated, multistep mechanism. Osteoblasts are responsible for the production of new bone—a process called **osteogenesis** (os-tē-ō-JEN-e-sis; *gennan*, to produce). It is thought that osteoblasts respond to a variety of different stimuli, including mechanical and hormonal, to initiate osteogenesis. If an osteoblast becomes surrounded by matrix, it differentiates into an osteocyte.

Osteoprogenitor Cells Bone tissue also contains small numbers of stem cells termed **osteoprogenitor cells** (os-tē-ō-prō-JEN-i-tor; *progenitor*, ancestor). Osteoprogenitor cells differentiate from mesenchyme and are found in numerous locations, including the innermost layer of the periosteum and the endosteum lining the medullary cavities. Osteoprogenitor cells divide to produce daughter cells that differentiate into osteoblasts. The ability to produce additional osteoblasts becomes extremely important after a bone is cracked or broken. We will consider the repair process further in a later section.

Figure 5.1 Histological Structure of a Typical Bone. Osseous tissue contains specialized cells and a dense extracellular matrix containing calcium salts.

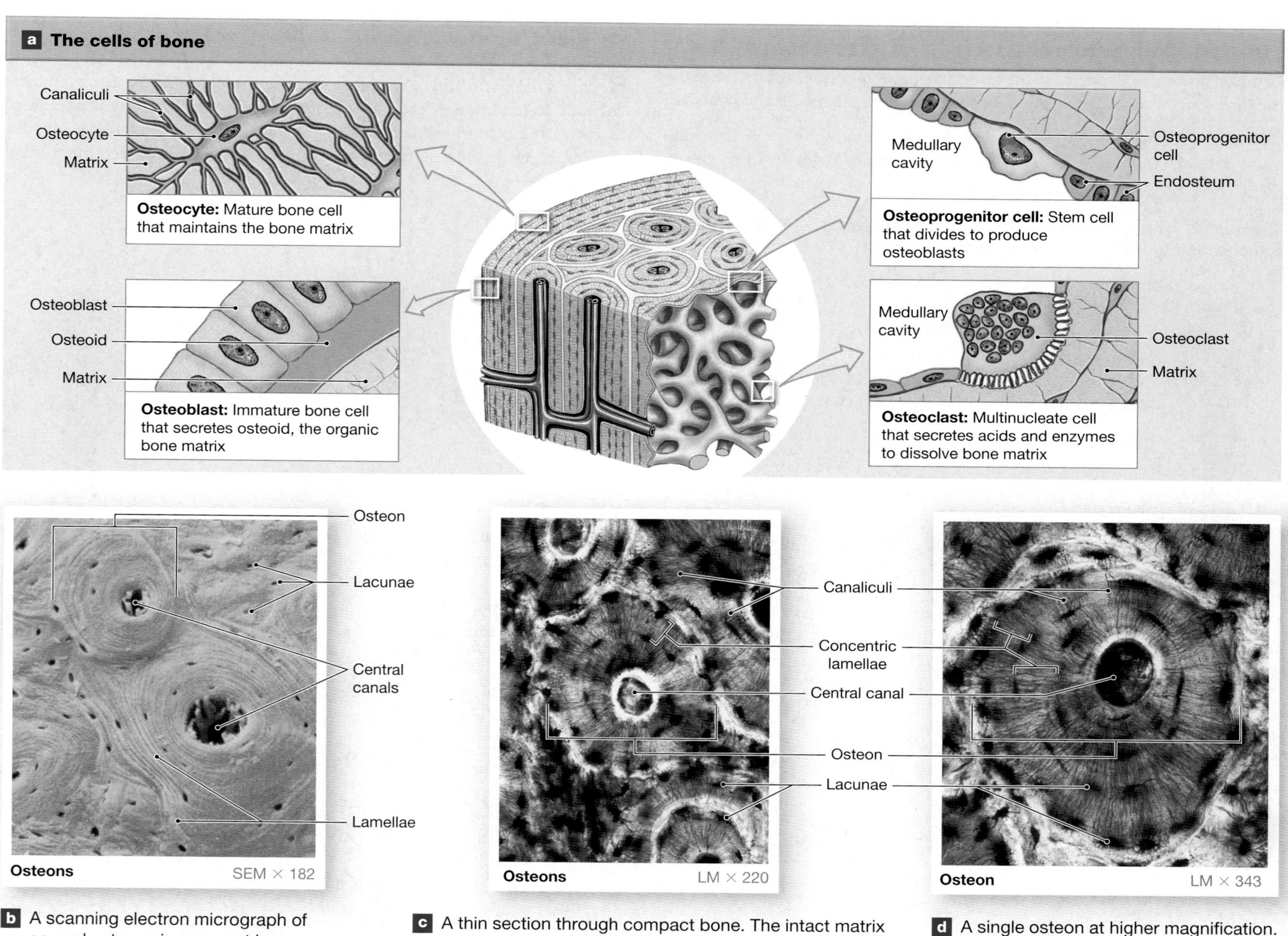

b A scanning electron micrograph of several osteons in compact bone.

c A thin section through compact bone. The intact matrix making up the lamellae and central canal is white, while lacunae and canaliculi appear black in this section.

d A single osteon at higher magnification. The central canal appears black on this section.

Osteoclasts Large, multinucleate cells found at sites where bone is being removed are termed **osteoclasts** (OS-tē-ō-klasts). They are derived from the same stem cells that produce monocytes and neutrophils. ⮌ **pp. 62, 65** They secrete acids through a process involving the exocytosis of lysosomes. The acids dissolve the bony matrix and release amino acids and the stored calcium and phosphate. This erosion process, called **osteolysis** (os-tē-OL-ī-sis), increases the calcium and phosphate concentrations in body fluids. Osteoclasts are always removing matrix and releasing minerals, and osteoblasts are always producing matrix that quickly binds minerals. The balance between the activities of osteoblasts and osteoclasts is very important; when osteoclasts remove calcium salts faster than osteoblasts deposit them, bones become weaker. When osteoblasts are more active than osteoclasts, bones become stronger and more massive.

New research indicates that osteoclasts may also be involved in osteoblast differentiation, immune system activation, and the proliferation of tumor cells in bone.

Compact and Spongy Bone

▶ **KEY POINT** There are two forms of adult bone: compact bone and spongy bone. Because of their different structure, compact bone is heavier and better at resisting forces that occur parallel to the bone. Spongy bone is lighter and better at resisting forces that occur from more than one direction at the same time.

There are two types of osseous tissue: compact bone and spongy bone. **Compact bone** is relatively dense and solid, whereas **spongy bone**, also termed *trabecular* (tra-BEK-ū-lar) *bone* or *cancellous bone*, forms an open network of struts and plates. Both are found in typical bones of the skeleton, such as the humerus, the proximal bone of the upper limb, and the femur, the proximal bone of the lower limb. Compact bone forms the walls, and an internal layer of spongy bone surrounds the **medullary** (*marrow*) **cavity** (**Figure 5.2a**). The medullary cavity contains **bone marrow**, a loose connective tissue that is dominated by either adipocytes (**yellow marrow**) or a mixture of mature and immature red and white blood cells and the stem cells that produce them (**red marrow**). Yellow marrow, often found in the medullary cavity of the shaft, is an important energy reserve. Extensive areas of red marrow, such as in the spongy bone of the femur, are important sites of blood cell formation.

Structural Differences between Compact and Spongy Bone

Compact and spongy bone have the same matrix composition, but they differ in the three-dimensional arrangement of the osteocytes, canaliculi, and lamellae.

Compact Bone The functional unit of mature compact bone is the cylindrical **osteon** (OS-tē-on), or *Haversian system* **(Figure 5.1b-d)**. In an osteon the osteocytes are arranged in circular layers around a **central canal** (or *Haversian canal*), which contains the blood vessels that supply the osteon. Central canals run parallel to the surface of the bone **(Figure 5.2b,c)**. Other passageways known as **perforating canals**, or *Volkmann's canals*, extend perpendicular to the surface **(Figure 5.2b)**. Blood vessels in the perforating canals deliver blood to osteons deeper in the bone and service the medullary cavity. The cylindrical **concentric lamellae** of each osteon are arranged parallel to the long axis of the bone. These lamellae form a series of concentric rings, resembling a "bull's-eye" target, around the central canal **(Figure 5.2b,c)**. The collagen fibers spiral

5

Figure 5.2 The Internal Organization in Representative Bones. The structural relationship of compact and spongy bone in representative bones.

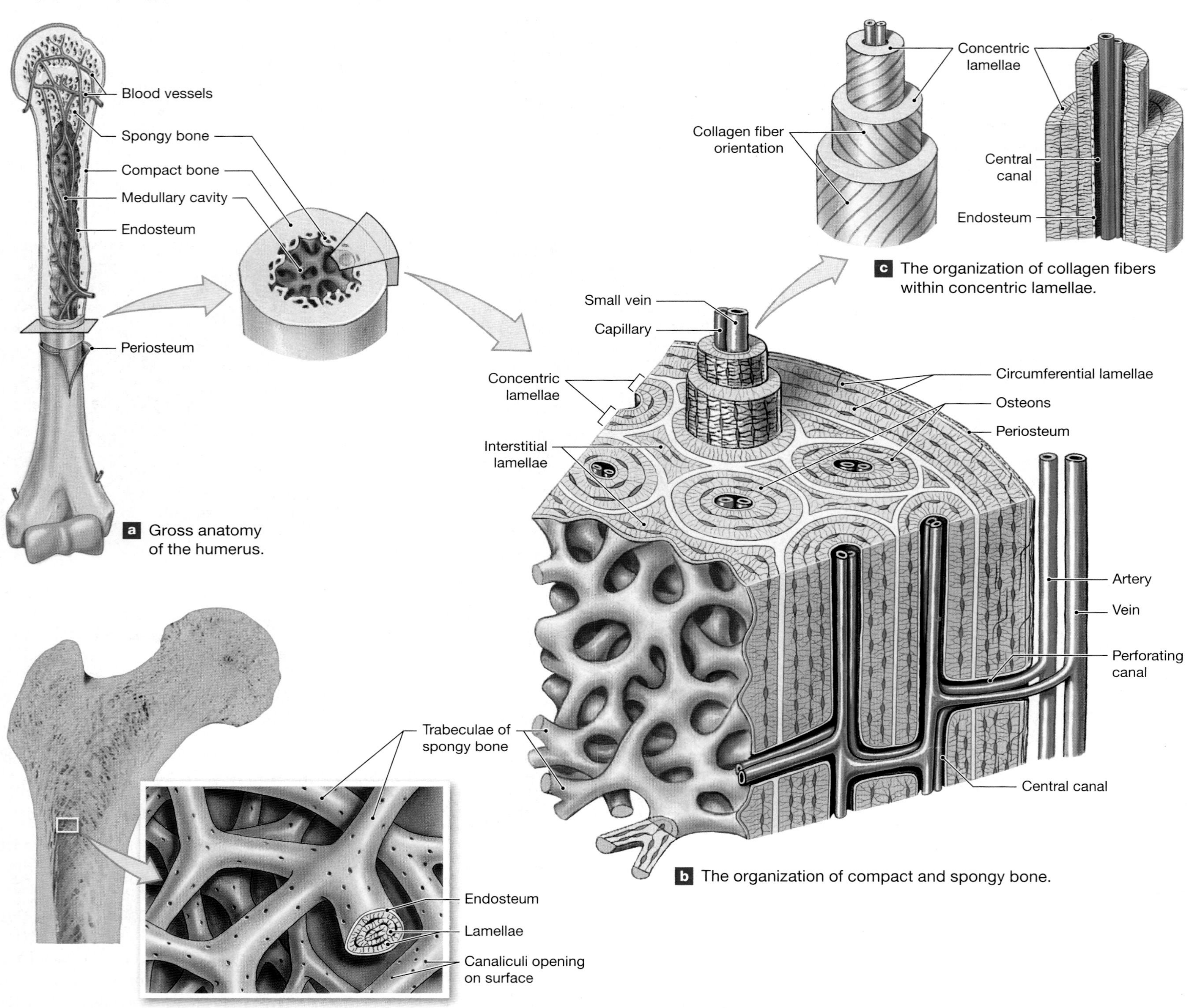

along the length of each lamella, and differences in the direction of spiraling within adjacent lamellae strengthen the osteon. Canaliculi form an interconnecting network with the osteocytes within their lacunae. This enables the passage of nutrients and wastes to and from the blood vessels within the central canal. **Interstitial lamellae** fill in the spaces between the osteons in compact bone. Depending on their location, these lamellae either have been produced during the growth of the bone or may represent what is left of osteons that have been recycled by osteoclasts during bone repair or remodeling. **Circumferential lamellae**, a third type of lamellae found in bone, occur at the external and internal surfaces of the bone. In a bone such as the humerus or femur, the circumferential lamellae form the outer and inner surfaces of the shaft **(Figure 5.2b)**.

Spongy Bone The major difference between compact and spongy bone is the arrangement of spongy bone into parallel struts or thick, branching plates called **trabeculae** (tra-BEK-ū-lē; "a little beam") (also termed *spicules*). Numerous interconnecting spaces occur between the trabeculae in spongy bone. Spongy bone possesses lamellae, and if the trabeculae are thick enough, osteons will be present.

In terms of the associated cells and the structure and composition of the lamellae, spongy bone is no different from compact bone. Spongy bone forms an open framework **(Figure 5.2d)**, and as a result it is much lighter than compact bone. However, the branching trabeculae give spongy bone considerable strength despite its relatively light weight. *Thus, the presence of spongy bone reduces the weight of the skeleton and makes it easier for muscles to move the bones.* Spongy bone is found wherever bones are not stressed heavily or where stresses arrive from many directions.

Functional Differences between Compact and Spongy Bone

A layer of compact bone covers the surface of all bones. The thickness of this layer varies from region to region and from one bone to another, but compact bone is thickest where stresses arrive from a limited number of directions. This superficial layer of compact bone is in turn covered by the **periosteum**, a connective tissue wrapping that is connected to the deep fascia **(Figure 5.2a,d)**. The periosteum is complete everywhere except within a joint, where the edges or ends of two bones contact one another.

Figure 5.3a shows the general anatomy of the femur, the proximal bone of the lower limb. Compact bone surrounds the medullary cavity. The bone

Figure 5.3 Anatomy of a Representative Bone

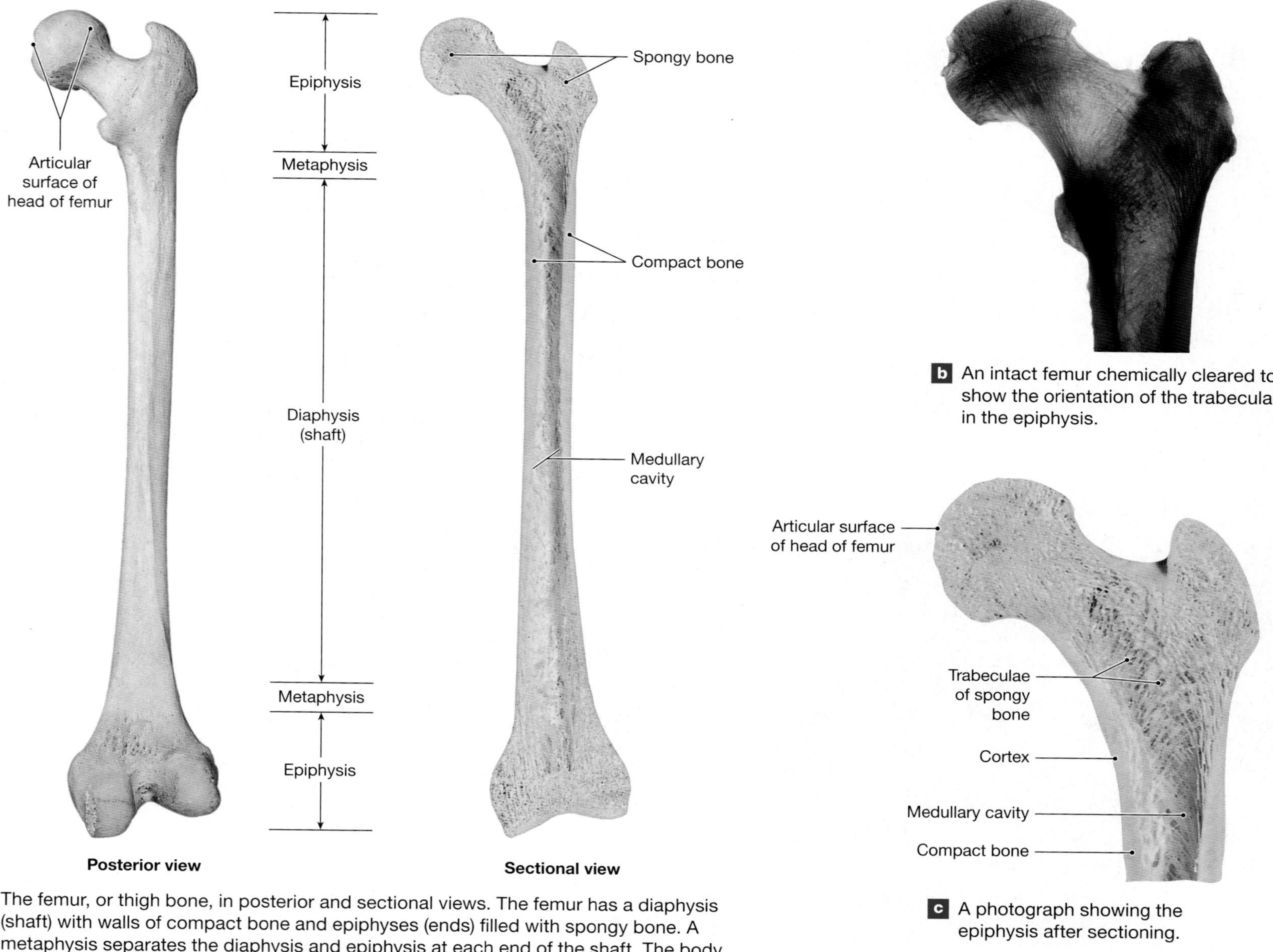

a The femur, or thigh bone, in posterior and sectional views. The femur has a diaphysis (shaft) with walls of compact bone and epiphyses (ends) filled with spongy bone. A metaphysis separates the diaphysis and epiphysis at each end of the shaft. The body weight is transferred to the femur at the hip joint. Because the hip joint is off center relative to the axis of the shaft, the body weight is distributed along the bone so that the medial portion of the shaft is compressed and the lateral portion is stretched.

b An intact femur chemically cleared to show the orientation of the trabeculae in the epiphysis.

c A photograph showing the epiphysis after sectioning.

has two ends, or **epiphyses** (e-PIF-i-sēs; singular, *epiphysis*; *epi*, above, + *physis*, growth), separated by a tubular **diaphysis** (dī-AF-i-sis; "a growing between"), or *shaft*. The diaphysis is connected to the epiphysis at a narrow zone known as the **metaphysis** (me-TAF-i-sis). The shaft of compact bone transfers stresses from one epiphysis to another. For example, when you are standing, the shaft of the femur transfers your body weight from your hip to your knee. The osteons within the shaft are parallel to its long axis, and as a result the femur is very strong when stressed along that axis. Imagine a single osteon as a drinking straw with very thick walls. When you push the ends of a straw together, it seems quite strong. However, when you hold the ends and push the side of the straw, it breaks easily. Similarly, a long bone does not bend when forces are applied to either end, but an impact to the side of the shaft can easily cause a break, or **fracture**.

Spongy bone is not as massive as compact bone, but it is much more capable of resisting stresses applied from many different directions. The epiphyses of the femur are filled with spongy bone, and the alignment of the trabeculae within the proximal epiphysis is shown in **Figure 5.3b,c**. The trabeculae are oriented along the stress lines, but with extensive cross-bracing. At the proximal epiphysis, the trabeculae transfer forces from the hip across the metaphysis to the femoral shaft; at the distal epiphysis, the trabeculae transfer the forces across the knee joint to the leg. In addition to reducing weight and handling stress from many directions, the open trabecular framework supports and protects the cells of the bone marrow.

The Periosteum and Endosteum

▶ **KEY POINT** Most bones are covered by a strong and highly sensitive connective tissue called the periosteum. The endosteum, which is an incomplete layer of connective tissue, lines the hollow, central marrow cavity of bones. Both layers are involved in the repair and remodeling of bone.

The outer surface of a bone is usually covered by a **periosteum** (**Figure 5.4a**). The periosteum isolates and protects the bone from surrounding tissues,

Figure 5.4 Anatomy and Histology of the Periosteum and Endosteum. Diagrammatic representation of periosteum and endosteum locations and their association with other bone structures; histology section shows both periosteum and endosteum.

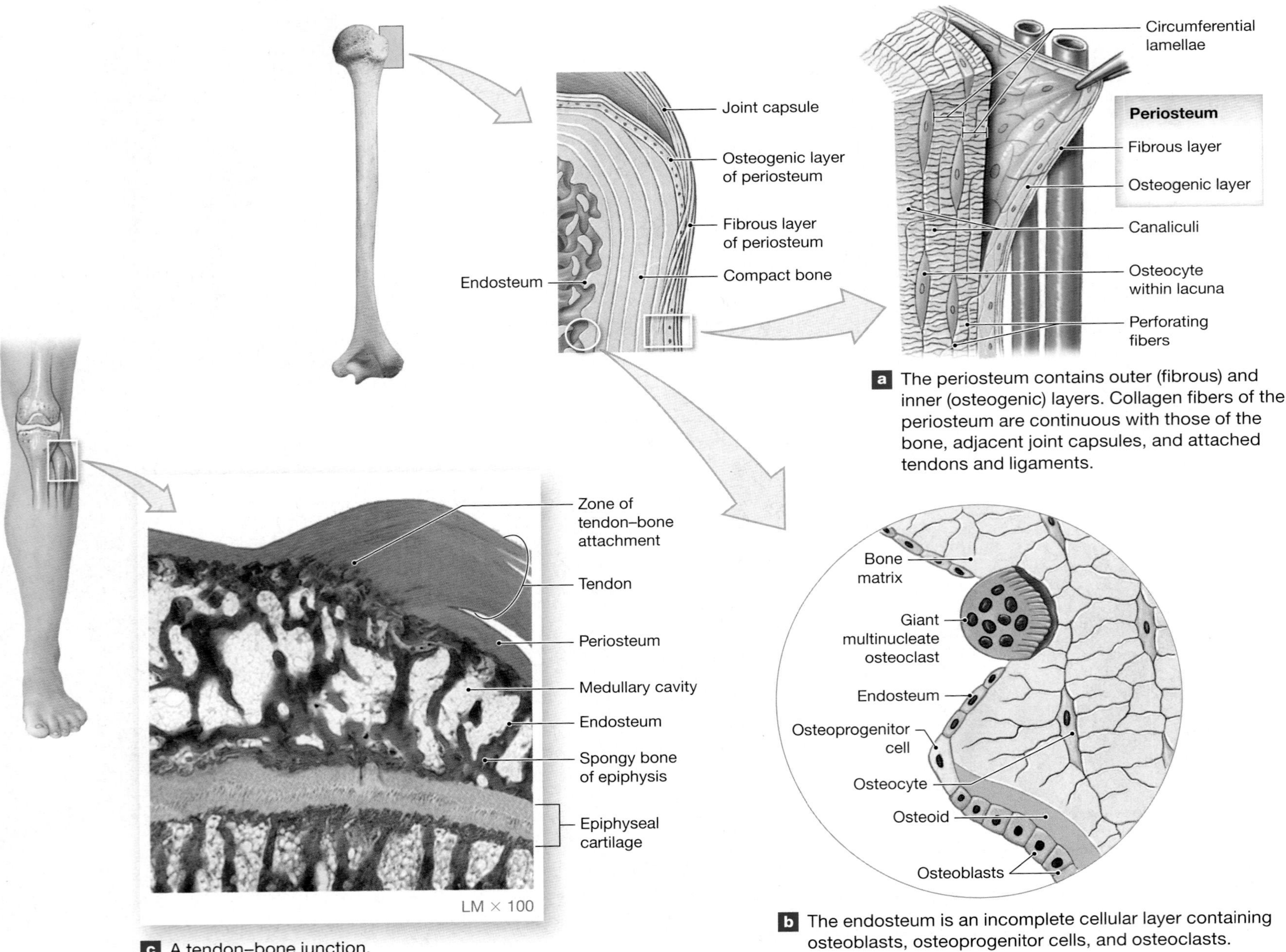

a The periosteum contains outer (fibrous) and inner (osteogenic) layers. Collagen fibers of the periosteum are continuous with those of the bone, adjacent joint capsules, and attached tendons and ligaments.

b The endosteum is an incomplete cellular layer containing osteoblasts, osteoprogenitor cells, and osteoclasts.

c A tendon–bone junction.

provides a route and a place of attachment for circulatory and nervous supply, actively participates in bone growth and repair, and attaches the bone to the connective tissue network of the deep fascia. A periosteum does not surround sesamoid bones, nor is it present where tendons, ligaments, or joint capsules attach or where bone surfaces are covered by articular cartilages.

The periosteum consists of an outer **fibrous layer** of dense fibrous connective tissue and an inner **osteogenic layer** containing osteoprogenitor (*stem*) cells. When a bone is not undergoing growth or repair, few osteoprogenitor cells are visible within the cellular layer.

Near joints, the periosteum becomes continuous with the connective tissue network that surrounds and helps stabilize the joint. At a fluid-filled (synovial) joint, the periosteum is continuous with the joint capsule that encloses the joint complex. The fibers of the periosteum are also interwoven with those of the tendons attached to the bone (**Figure 5.4c**). As the bone grows, these tendon fibers are cemented into the superficial lamellae by osteoblasts from the cellular layer of the periosteum. The collagen fibers incorporated into bone tissue from tendons and from the superficial periosteum are called **perforating fibers** or *Sharpey's fibers* (**Figure 5.4a**). The cementing process makes the tendon fibers part of the general structure of the bone, providing a much stronger bond than would otherwise be possible. An extremely powerful pull on a tendon or ligament will break the bone rather than snap the collagen fibers at the bone surface.

Inside the bone, a cellular **endosteum** lines the medullary (*marrow*) cavity (**Figure 5.4b**). This layer is only one cell thick and is an incomplete layer. The endosteum contains osteoprogenitor cells and covers the trabeculae of spongy bone. It also lines the inner surfaces of the central canals and perforating canals. The endosteum is active during the growth of bone and whenever repair or remodeling is under way.

5.1 CONCEPT CHECK

1. How would the strength of a bone be affected if the ratio of collagen to calcium salts (hydroxyapatite) increased?
2. A sample of bone shows concentric lamellae surrounding a central canal. Is the sample from the cortex or the medullary cavity of a long bone?
3. If the activity of osteoclasts exceeds the activity of osteoblasts in a bone, how is the mass of the bone affected?
4. If a poison selectively destroyed the osteoprogenitor cells in bone tissue, what future, normal process may be impeded?

See the blue Answers tab at the back of the book.

5.2 Bone Development and Growth

▶ **KEY POINT** Bone grows by two processes: intramembranous ossification and endochondral ossification.

The process of bone development and growth is carefully regulated, and a breakdown in regulation affects all body systems. In this section we discuss the process of osteogenesis (bone formation) and bone growth.

From fertilization to about eight weeks of age, an embryo's skeletal elements are composed of either mesenchyme or hyaline cartilage. The bony skeleton begins to form at eight weeks. During subsequent development, the bones increase tremendously in size.

Bone growth continues through adolescence, and portions of the skeleton usually do not stop growing until age 25. The growth of the skeleton determines the size and proportions of the body.

During embryonic development, bone replaces both mesenchyme and cartilage. This process of replacing other tissues with bone is **ossification**. **Calcification** refers to the deposition of calcium salts within a tissue. *Any tissue can be calcified, but only ossification forms bone.*

There are two major forms of ossification:

- In **intramembranous** (in-tra-MEM-bra-nus) **ossification**, bone develops from mesenchyme or fibrous connective tissue. Intramembranous ossification forms bones such as the clavicle, mandible, and flat bones of the face and skull.
- In **endochondral** (en-dō-KON-dral; *endo*, *inside*, + *chondros*, cartilage) **ossification**, bone replaces an existing cartilage model. The bones of the limbs and other bones that bear weight, such as the vertebral column, develop by endochondral ossification.

Intramembranous Ossification

▶ **KEY POINT** During intramembranous ossification, embryonic mesenchyme condenses into a thick "membrane-like" layer that is replaced by bone.

Intramembranous ossification occurs within embryonic mesenchymal tissue before cartilage develops. **Figure 5.5a** shows skull bones forming through intramembranous ossification in the head of a 10-week-old fetus. **Spotlight Figure 5.6** explains the process of intramembranous ossification, also called *dermal ossification.*

Endochondral Ossification

▶ **KEY POINT** During endochondral ossification, embryonic mesenchyme forms a cartilage model of a developing bone that is gradually replaced by bone.

Endochondral ossification occurs within a hyaline cartilage model in which bone replaces cartilage. **Figure 5.5** shows the appendicular skeleton forming by endochondral ossification in 10- and 16-week-old fetuses. **Spotlight Figure 5.7** illustrates the process of endochondral ossification.

Epiphyseal Closure

At maturity, an event called **epiphyseal closure** stops bone growth (**Spotlight Figure 5.7**). X-rays can often detect the former location of the epiphyseal cartilage as a distinct **epiphyseal line** that remains after epiphyseal growth has ended (look forward to **Figure 5.8** on page 118).

Increasing the Diameter of a Developing Bone

The diameter of a bone enlarges through growth at the outer surface of the bone. In this process of **appositional growth**, stem cells of the inner layer of the periosteum differentiate into osteoblasts and add bone matrix to the surface. This adds layers of circumferential lamellae to the superficial surface of the bone. Over time, the deeper lamellae are recycled and replaced with the osteons typical of compact bone.

Blood vessels and collagen fibers of the periosteum can and do become enclosed within the matrix. Where this occurs, the process of appositional bone growth is somewhat more complex (look forward to **Figure 5.8** on page 119).

While osteoblasts add bone to the outer surface, osteoclasts reabsorb (remove) bone matrix at the inner surface. As a result, the medullary cavity gradually enlarges as the bone increases in diameter.

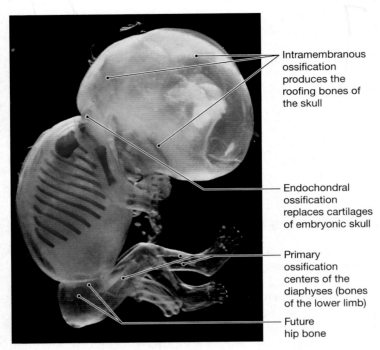

Figure 5.5 Fetal Intramembranous and Endochondral Ossification. These 10- and 16-week human fetuses have been specially stained (with alizarin red) and cleared to show developing skeletal elements.

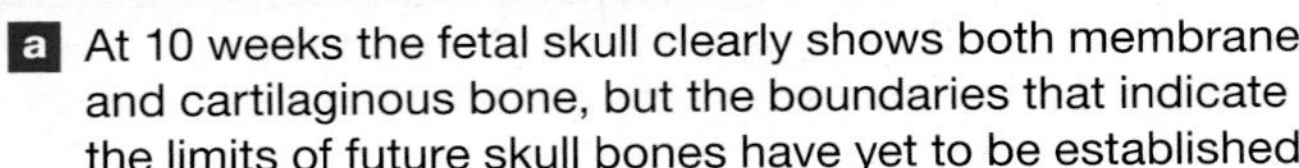

a At 10 weeks the fetal skull clearly shows both membrane and cartilaginous bone, but the boundaries that indicate the limits of future skull bones have yet to be established.

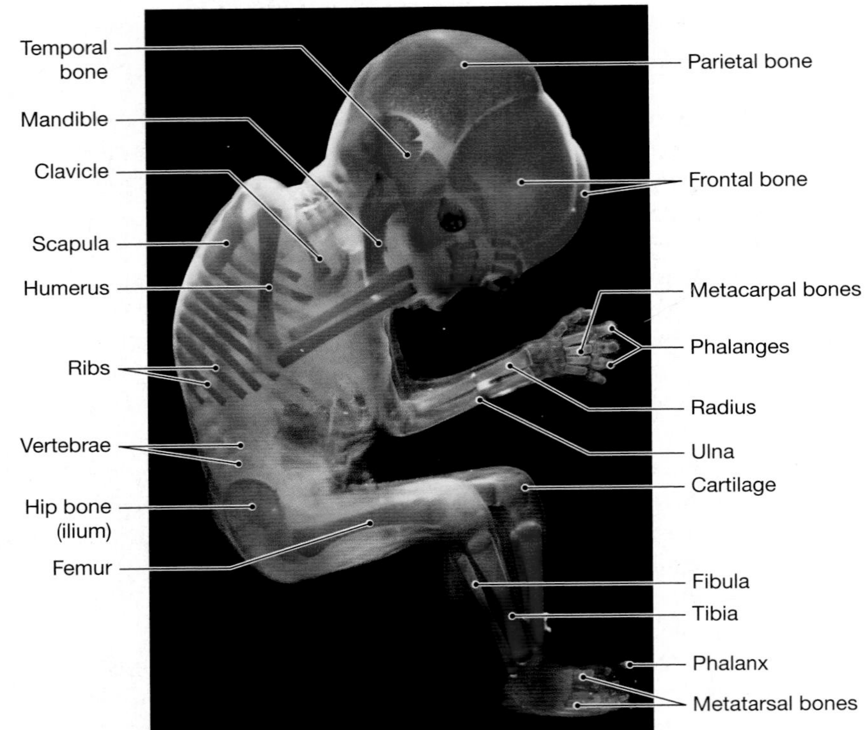

b At 16 weeks the fetal skull shows the irregular margins of the future skull bones. Most elements of the appendicular skeleton form through endochondral ossification. Note the appearance of the wrist and ankle bones at 16 weeks versus at 10 weeks.

Blood and Nerve Supply to Bones

▶ **KEY POINT** In certain locations, four sets of blood vessels nourish adult bones.

Like any living tissue, bones need nourishment. Osseous tissue is very vascular, and the bones of the skeleton have an extensive blood supply. In a long bone such as the humerus, four major sets of blood vessels develop (look forward to **Figure 5.10** on page 120):

1. **The nutrient artery** and **nutrient vein:** These vessels form as blood vessels invade the cartilage model at the start of endochondral ossification. There is usually only one nutrient artery and one nutrient vein entering the diaphysis through a **nutrient foramen**. A foramen (fō-RĀ-men; plural, *foramina*) is an opening in a bone. However, a few bones, including the femur, have two or more nutrient arteries. These vessels penetrate the shaft to reach the medullary cavity. The nutrient artery divides into ascending and descending branches, which approach the epiphyses. These vessels then re-enter the compact bone through perforating canals and extend along the central canals to supply the osteons of the compact bone (**Figure 5.2b**, p. 110).
2. **Metaphyseal vessels**: These vessels supply blood to the inner (diaphyseal) surface of each epiphyseal cartilage, where bone is replacing cartilage.
3. **Epiphyseal vessels**: The epiphyseal ends of long bones contain numerous smaller foramina. Epiphyseal vessels enter the bone through these foramina to supply the osseous tissue and medullary cavities of the epiphyses.
4. **Periosteal vessels**: Blood vessels from the periosteum are incorporated into the developing bone surface (look forward to **Figure 5.9** on page 119). These vessels provide blood to the superficial osteons of the shaft. During endochondral ossification, branches of periosteal vessels enter the epiphyses, bringing blood to the secondary ossification centers.

After the epiphyses close, all of these blood vessels become extensively interconnected (look forward to **Figure 5.10** on page 120). The periosteum also contains an extensive network of lymphatic vessels and sensory nerves. The lymphatic vessels collect lymph (fluid derived from the interstitial fluid) from branches that enter the bone and reach individual osteons through perforating canals. The sensory nerves penetrate the compact bone with the nutrient artery to innervate the endosteum, medullary cavity, and epiphyses. Because of this rich sensory innervation, injuries to bones are usually very painful.

Factors Regulating Bone Growth

▶ **KEY POINT** Vitamins A, C, and D_3, calcium and other ions, and many hormones have significant effects on bone development and maintenance in children and adults.

Normal bone growth depends on a combination of factors, including nutrition and the effects of hormones:

- Minerals. Normal bone growth cannot occur without a constant dietary source of calcium and phosphate salts, as well as other ions such as magnesium, citrate, carbonate, and sodium.

SPOTLIGHT

FIGURE 5.6

Intramembranous Ossification

Intramembranous ossification, also called *dermal ossification*, typically starts during the eighth week of embryonic development. This type of ossification occurs in the deeper layers of the dermis, and the bones that result are called **membrane bones**, or *dermal bones*.

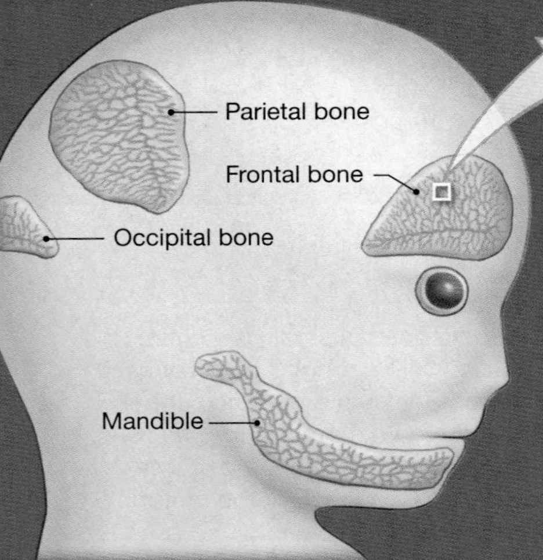

Examples of dermal bones include the roofing bones of the skull (the frontal, parietal, and occipital bones) and the mandible (lower jaw).

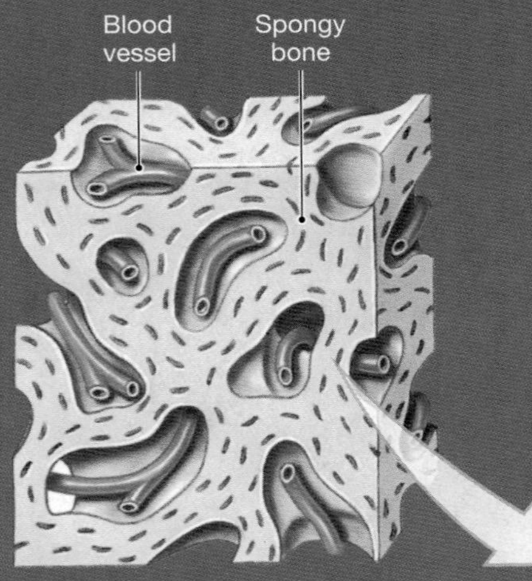

This is a three-dimensional view of **spongy bone** (also termed *cancellous* or *trabecular* bone). Areas of spongy bone may later be removed, creating **medullary cavities**. Through remodeling by osteoclasts and osteoblasts, spongy bone formed in this way can be converted to **compact bone** seen in the mature bones of the skull.

1 Differentiation of Osteoblasts within Mesenchyme

Mesenchymal tissue becomes highly vascularized, and the mesenchymal cells aggregate, enlarge, and then differentiate into **osteoblasts**. The osteoblasts then cluster together and start to secrete the organic components of the matrix. The resulting **osteoid** then becomes mineralized through the crystallization of calcium salts. The location where ossification begins is called an **ossification center**.

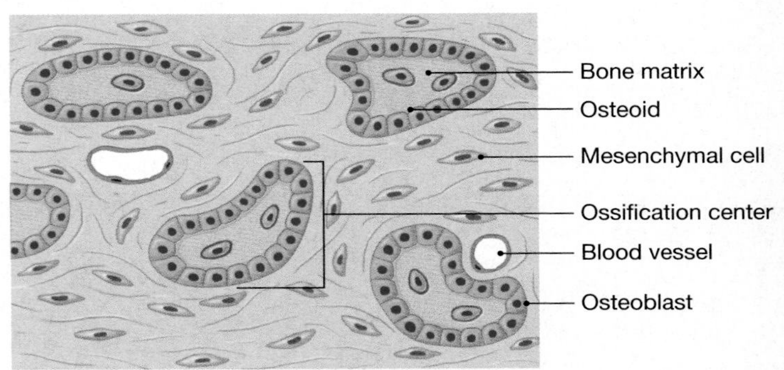

2 Formation of Bony Spicules

As ossification proceeds, osteoblasts that become surrounded by osteoid differentiate into **osteocytes**. These cells will remain trapped within tiny spaces known as **lacunae** (singular, *lacuna*). The developing bone grows outward from the ossification center in small struts called **spicules**. Although osteoblasts are still being trapped in the expanding bone, mesenchymal cell divisions continue to produce additional osteoblasts.

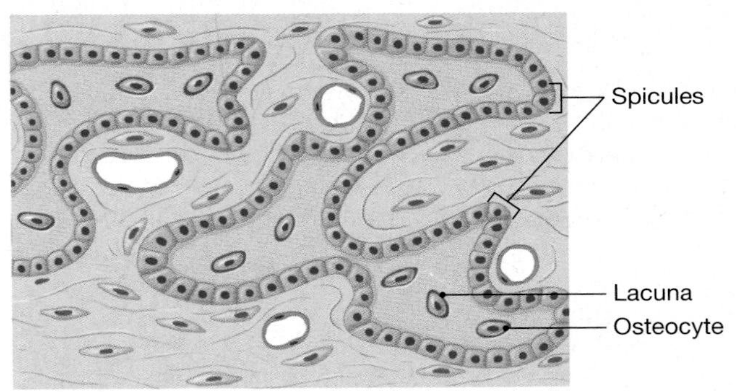

3 Entrapment of Blood Vessels

Bone growth is an active process, and osteoblasts require oxygen and a reliable supply of nutrients. The rate of bone growth accelerates as blood vessels branch within the region and grow between the spicules. As spicules interconnect, they trap blood vessels within the bone.

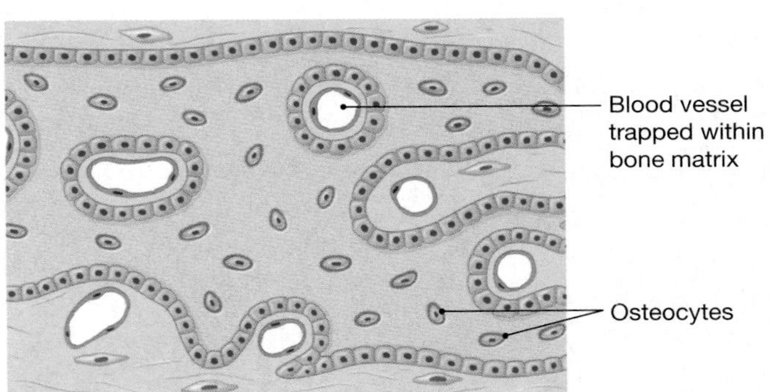

4 Formation of Spongy Bone

Continued deposition of bone by osteoblasts creates a bony plate that is perforated by blood vessels. As adjacent plates fuse together, the bone structure becomes increasingly complex.

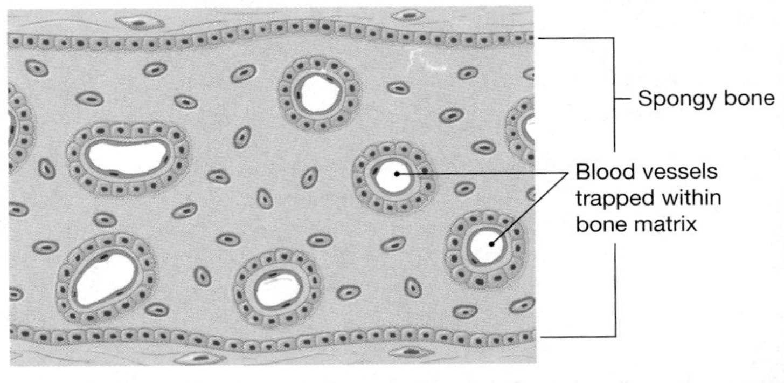

SPOTLIGHT

FIGURE 5.7
Endochondral Ossification

Endochondral ossification begins with the formation of a hyaline cartilage model. The bones of the limbs form in this way. By the time an embryo is six weeks old, the proximal bones of the limbs, the humerus (upper limb) and femur (lower limb), have formed, but they are composed entirely of cartilage. These cartilage models continue to grow by expansion of the cartilage matrix **(interstitial growth)** and by the production of more cartilage at the outer surface **(appositional growth)**.

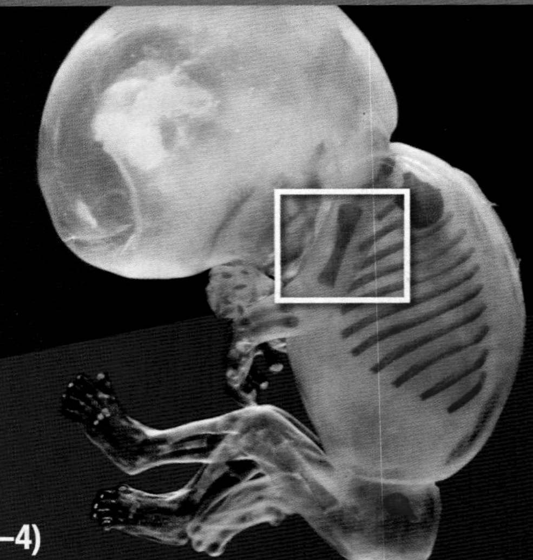

Initiation of Ossification in the Developing Bone (Steps 1–4)

1

As the cartilage enlarges, chondrocytes near the center of the shaft increase greatly in size, and the surrounding matrix begins to calcify. Deprived of nutrients, these chondrocytes die and disintegrate, leaving cavities within the cartilage.

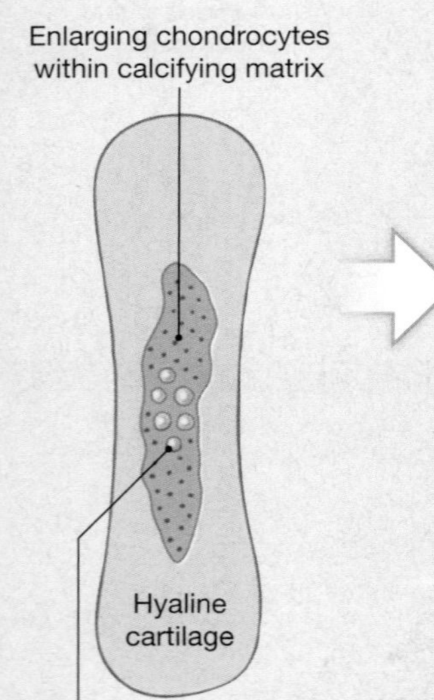

2

Blood vessels grow around the edges of the cartilage, and the cells of the perichondrium begin differentiating into osteoblasts. The perichondrium has now been converted into a periosteum, and the inner **osteogenic** (os-tē-ō-JEN-ik) **layer** soon produces a bone collar, a thin layer of compact bone around the shaft of the cartilage.

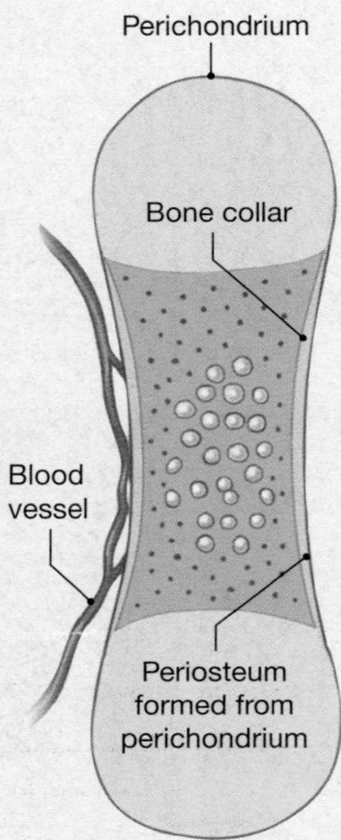

3

While these changes are under way, the blood supply to the periosteum increases, and capillaries and osteoblasts migrate into the heart of the cartilage, invading the spaces left by the disintegrating chondrocytes. The calcified cartilaginous matrix then breaks down, and osteoblasts replace it with spongy bone. Bone development proceeds from this **primary ossification center** in the shaft, toward both ends of the cartilaginous model.

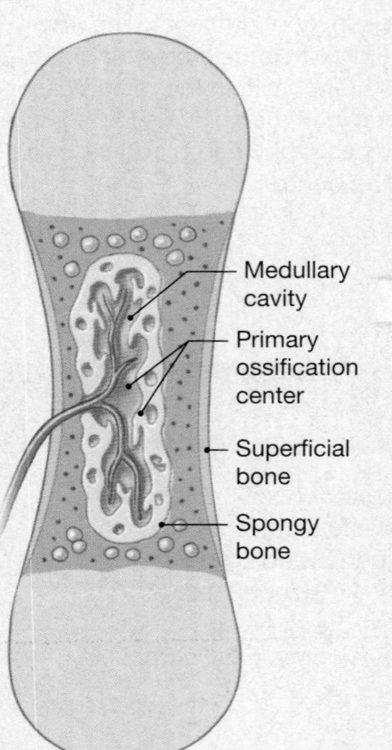

4

While the diameter is small, the entire shaft is filled with spongy bone, but as it enlarges, osteoclasts erode the central portion and create a **medullary** (*marrow*) **cavity**. The bone of the shaft becomes thicker, and the cartilage of the metaphysis is invaded by osteoblasts that produce columns of bone. Further growth involves two distinct processes: an increase in length and an enlargement in diameter **(Figure 5.9)**.

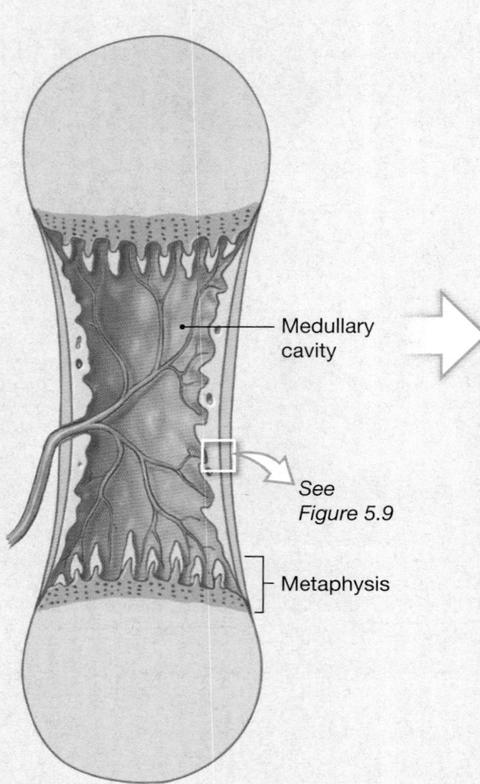

Increasing the Length of the Developing Bone (Steps 5–7)

During the initial stages of osteogenesis, osteoblasts move away from the primary ossification center toward the epiphyses. But they do not manage to complete the ossification of the model immediately, because the cartilages of the epiphyses continue to grow. The situation is like a pair of joggers, one in front of the other. As long as they are running at the same speed, they can run for miles without colliding. In this case, the osteoblasts and the epiphysis are both "running away" from the primary ossification center. As a result, the osteoblasts never catch up with the epiphysis, although the skeletal element continues to grow longer and longer.

5

Capillaries and osteoblasts then migrate into the centers of the epiphyses, creating **secondary ossification centers**. The time of appearance of secondary ossification centers varies from one bone to another and from individual to individual. Secondary ossification centers may be present at birth in both ends of the humerus (arm), femur (thigh), and tibia (leg), but the epiphyses of some other bones remain cartilaginous through childhood.

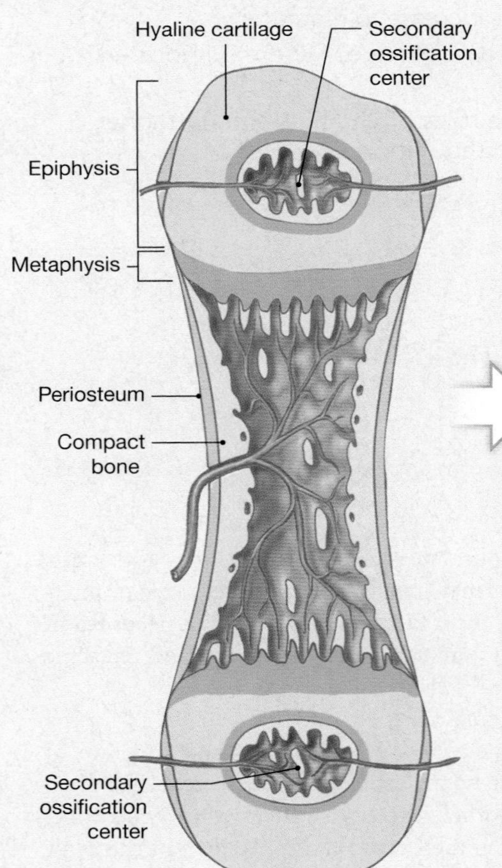

6

The epiphyses eventually become filled with spongy bone. The epiphysis and diaphysis are now separated by a narrow **epiphyseal cartilage**, or *epiphyseal plate*, within the metaphysis. Osteoblasts invade the shaft side of the epiphyseal cartilage, replacing the cartilage with bone, at the same rate that the epiphyseal cartilage enlarges through interstitial growth. This enlargement pushes the epiphysis away from the diaphysis, and the length of the bone increases.

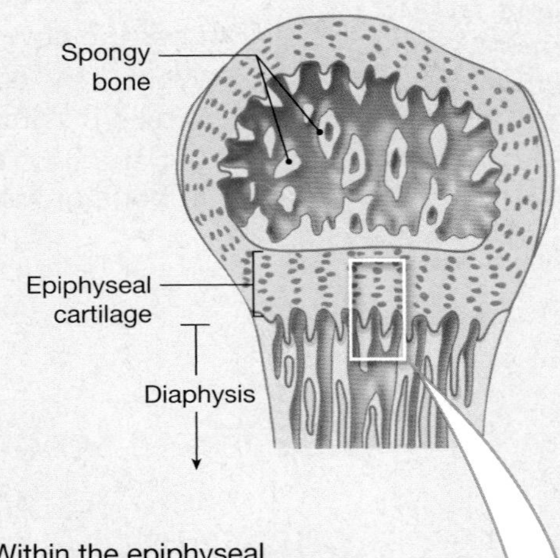

Within the epiphyseal cartilage, the chondrocytes are organized into zones.

Chondrocytes at the epiphyseal side of the cartilage continue to divide and enlarge.

Chondrocytes degenerate at the diaphyseal side.

Osteoblasts migrate upward from the diaphysis, and the degenerating cartilage is gradually replaced by bone.

7

At maturity, the rate of epiphyseal cartilage enlargement slows and the rate of osteoblast activity accelerates. As a result, the epiphyseal cartilage gets narrower and narrower, until it ultimately disappears. This event is called **epiphyseal closure**. The former location of the epiphyseal cartilage becomes a distinct **epiphyseal line** that remains after epiphyseal growth has ended.

A thin cap of the original cartilage model remains exposed to the joint cavity as the **articular cartilage**. This cartilage prevents damaging bone-to-bone contact within the joint.

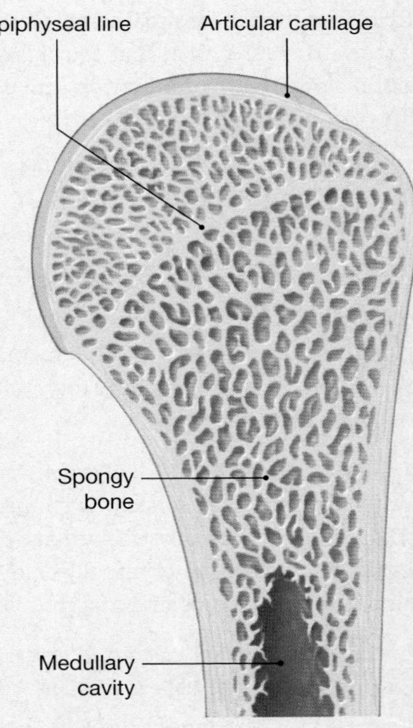

Figure 5.8 Epiphyseal Cartilages and Lines. The epiphyseal cartilage is the location of long bone growth in length prior to maturity; the epiphyseal line marks the former location of the epiphyseal cartilage after growth has ended.

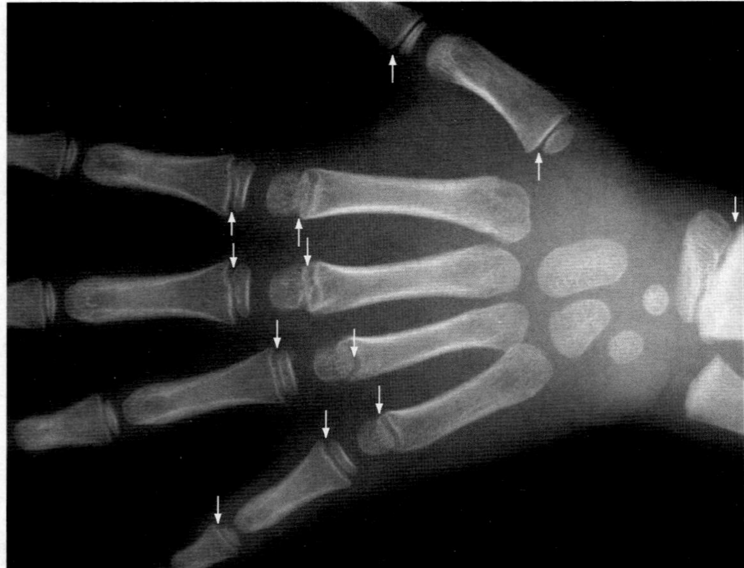

a X-ray of the hand of a young child. The arrows indicate the locations of the epiphyseal cartilages.

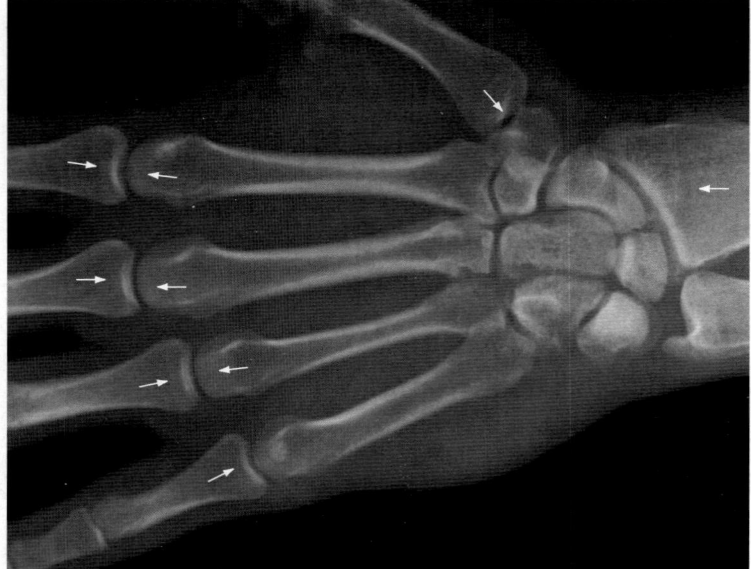

b X-ray of the hand of an adult. The arrows indicate the locations of epiphyseal lines.

- Vitamins. Vitamins A and C are essential for normal bone growth and remodeling. Vitamin A stimulates osteoblast activity, and vitamin C is required for enzymatic reactions in collagen synthesis and osteoblast differentiation.
- Calcitriol and vitamin D_3. The hormone calcitriol is essential for normal calcium and phosphate ion absorption into the blood. Calcitriol is synthesized in the kidneys from a related steroid, cholecalciferol (vitamin D_3), which may be produced in the skin in the presence of UV radiation or obtained from the diet.
- Parathyroid hormone. The parathyroid glands release parathyroid hormone, which stimulates osteoclasts and osteoblasts, increases the rate of calcium absorption along the small intestine, and decreases the rate of calcium loss in urine.
- Calcitonin. The thyroid glands of children and pregnant women secrete the hormone calcitonin (kal-si-TŌ-nin), which inhibits osteoclasts and increases the rate of calcium loss in the urine. Calcitonin is of uncertain significance in the healthy nonpregnant adult.
- Growth hormone and thyroxine. Growth hormone, produced by the pituitary gland, and thyroxine, from the thyroid gland, stimulate bone growth. In proper balance, these hormones maintain normal activity at the epiphyseal cartilages until puberty.
- Sex hormones. At puberty, bone growth accelerates dramatically. The sex hormones estrogen and testosterone stimulate osteoblasts to produce bone faster than the rate of epiphyseal cartilage expansion. Over time, the epiphyseal cartilages narrow and eventually ossify, or "close." The continued production of sex hormones is essential for maintaining bone mass in adults.

The timing of epiphyseal cartilage closure varies from bone to bone and individual to individual. The toes may ossify completely by age 11, while portions of the pelvis or wrist may continue to enlarge until age 25. Differences in male and female sex hormones account for the variation between the sexes and for related variations in body size and proportions.

5.2 CONCEPT CHECK

5. How can x-rays of the femur be used to determine whether a person has reached full height?
6. Briefly describe the major steps in the process of intramembranous ossification.
7. Describe how bones increase in diameter.
8. What is the epiphyseal cartilage? Where is it located? Why is it significant?
9. List and describe the roles of vitamins and hormones in bone growth and regulation.

See the blue Answers tab at the back of the book.

5.3 Bone Maintenance, Remodeling, and Repair

▶ **KEY POINT** The human skeleton constantly maintains, remodels, and repairs itself as needed.

Bone grows when osteoblasts produce more bone matrix than osteoclasts remove. Bone remodeling and repair may involve changing the shape or internal structure of a bone or changing the total amount of minerals deposited in the skeleton. In the adult, osteocytes are continually removing and replacing the surrounding calcium salts.

Osteoblasts and osteoclasts remain active throughout life. In young adults, osteoblast and osteoclast activity are in balance, and the rate of bone formation equals the rate of bone reabsorption. As osteoblasts form one osteon, osteoclasts destroy another osteon elsewhere within the bone. The rate of mineral turnover is high—each year almost one-fifth of the adult skeleton is demolished and then rebuilt or replaced. Every part of every bone

Figure 5.9 Appositional Bone Growth

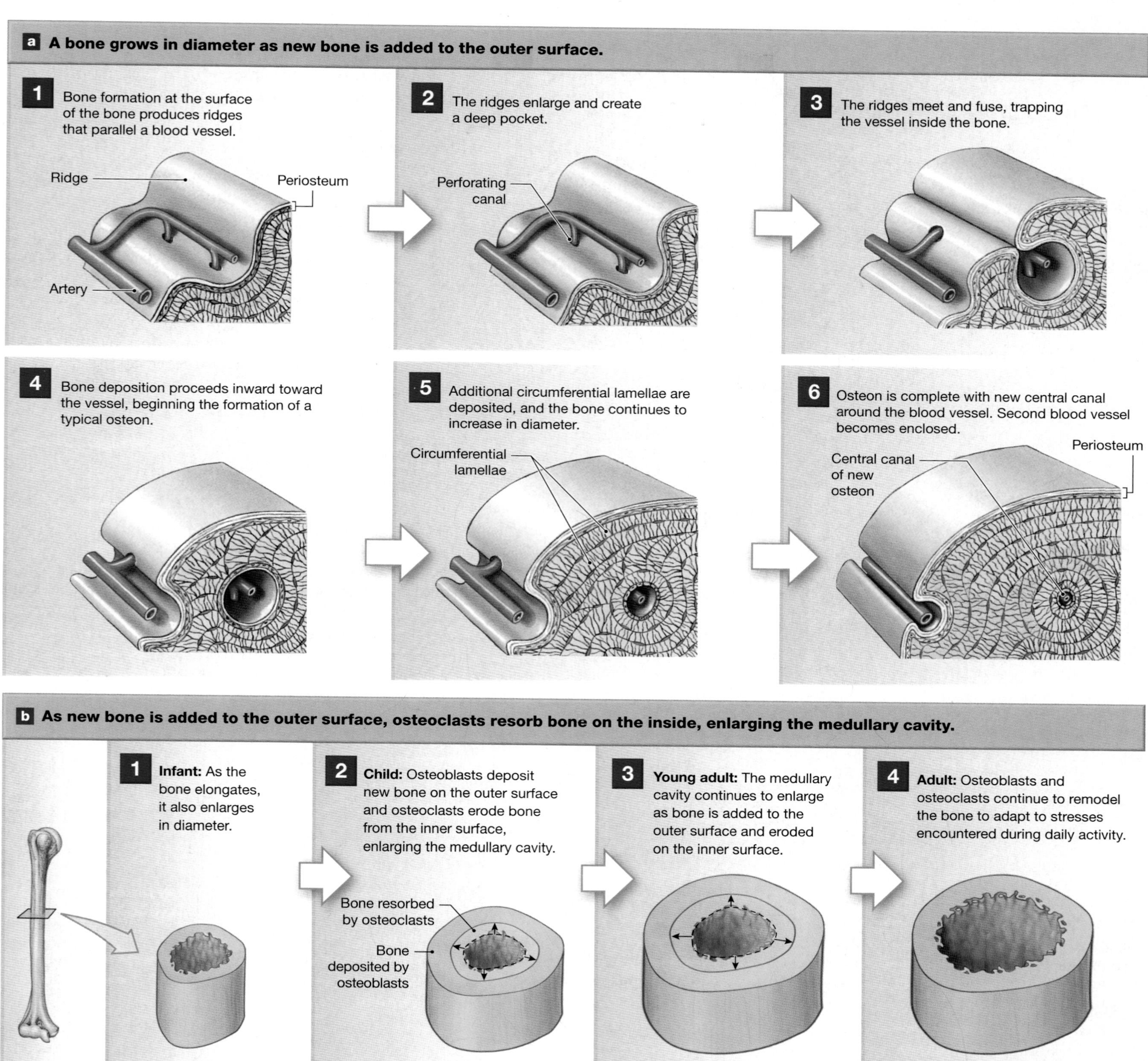

may not be affected, as there are regional and even local differences in the rate of turnover. For example, the spongy bone in the head of the femur may be replaced two or three times each year, whereas the compact bone along the shaft remains largely untouched.

a A bone grows in diameter as new bone is added to the outer surface.
1 Bone formation at the surface of the bone produces ridges that parallel a blood vessel.
Ridge
Periosteum
Artery
2 The ridges enlarge and create a deep pocket.
Perforating canal
3 The ridges meet and fuse, trapping the vessel inside the bone.
4 Bone deposition proceeds inward toward the vessel, beginning the formation of a typical osteon.
5 Additional circumferential lamellae are deposited, and the bone continues to increase in diameter.
Circumferential lamellae
6 Osteon is complete with new central canal around the blood vessel. Second blood vessel becomes enclosed.
Central canal of new osteon
Periosteum
b As new bone is added to the outer surface, osteoclasts resorb bone on the inside, enlarging the medullary cavity.
1 Infant: As the bone elongates, it also enlarges in diameter.
2 Child: Osteoblasts deposit new bone on the outer surface and osteoclasts erode bone from the inner surface, enlarging the medullary cavity.
Bone resorbed by osteoclasts
Bone deposited by osteoblasts
3 Young adult: The medullary cavity continues to enlarge as bone is added to the outer surface and eroded on the inner surface.
4 Adult: Osteoblasts and osteoclasts continue to remodel the bone to adapt to stresses encountered during daily activity.

This high turnover rate continues into old age, but in older people, osteoclast activity outpaces osteoblast activity. As a result, bone reabsorption exceeds bone deposition, and the skeleton gradually gets weaker.

Remodeling of Bone

▶ **KEY POINT** Bone is adaptable; the demands placed on the skeleton determine its maintenance and remodeling.

Although bone is hard and dense, its shape changes in response to environmental conditions. Bone remodeling involves the simultaneous processes of adding

Figure 5.10 Circulatory Supply to a Mature Bone. Arrangement and association of blood vessels supplying the humerus.

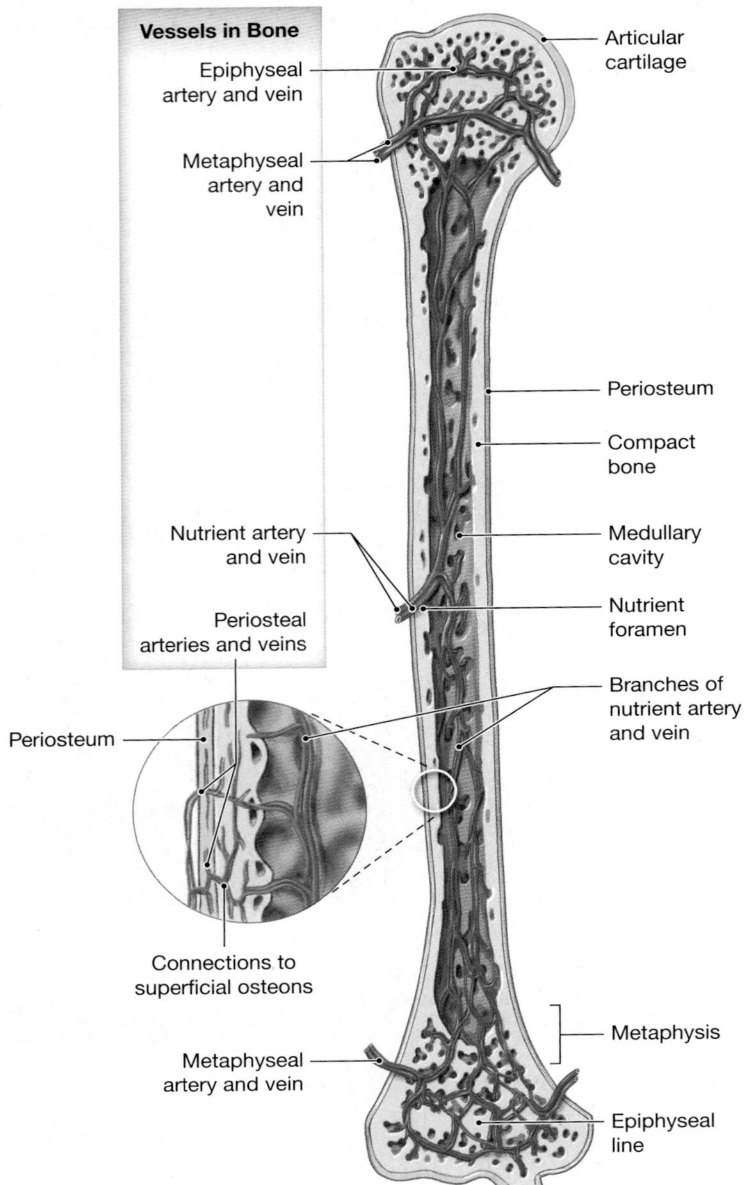

new bone and removing previously formed bone. For example, if you have dental braces, remodeling is occurring in your jaw. Old bone is resorbed and new bone is deposited, altering the shape of your tooth sockets to accommodate the new position of your teeth. If you lift weights, you are placing new and additional stresses on your skeleton. Your skeleton responds by remodeling at the sites of muscular and tendon attachment.

Bones adapt to stress by altering the turnover and recycling of minerals. Osteoblast sensitivity to electrical events may be the mechanism controlling the internal organization and structure of bone. Whenever a bone is stressed, its mineral crystals generate minute electrical fields. Apparently, these electrical fields attract osteoblasts, and once in the area they begin to produce bone. (Clinicians sometimes use electrical fields to stimulate the repair of severe fractures.)

Because bones are adaptable, their shapes and surface features reflect the forces applied to them. For example, bumps and ridges on the surface of a bone mark the sites where tendons attach. If muscles become more powerful, the corresponding bumps and ridges enlarge to withstand the increased forces. Heavily stressed bones become thicker and stronger, while bones not subjected to ordinary stresses become thin and brittle. Regular exercise is important as a stimulus that maintains normal bone structure, especially in growing children, postmenopausal women, and elderly men.

Degenerative changes in the skeleton occur after even brief periods of inactivity. For example, using a crutch while wearing a cast takes weight off the injured limb. After a few weeks, the unstressed bones lose up to a third of their mass. However, the bones rebuild when normal loading resumes.

Injury and Repair

▶ **KEY POINT** Bone repair occurs in a series of steps involving all four bone cell types.

Despite its mineral strength, bone may crack or even break if subjected to extreme loads, sudden impacts, or stresses from unusual directions. The damage produced constitutes a **fracture**. Even a severe fracture may heal, provided the blood supply and the cellular components of the endosteum and periosteum survive (look forward to Clinical Note on pp. 122–123).

The repaired bone will be slightly thicker and stronger than the original bone. Under comparable stress, a second fracture will usually occur at a different site.

Aging and the Skeletal System

▶ **KEY POINT** As we age, bones become thinner and weaker due to a variety of factors, including decreased osteoblast activity.

The bones of the skeleton become thinner and weaker as a normal part of the aging process. Inadequate ossification is called **osteopenia** (os-tē-ō-PĒ-nē-a; *penia*, lacking), and we all become slightly osteopenic as we age. This reduction in bone mass begins between ages 30 and 40. Osteoblast activity decreases while osteoclast activity continues at previous levels. Once the reduction begins, women lose about 8 percent of their skeletal mass every 10 years; the skeletons of men deteriorate at the slower rate of about 3 percent over the same time period. All parts of the skeleton are not equally affected. Epiphyses, vertebrae, and the jaws lose more than their fair share, resulting in fragile limbs, reduced height, and the loss of teeth.

A significant percentage of older women and a smaller proportion of older men suffer from **osteoporosis** (os-tē-ō-pō-RŌ-sis; *porosus*, porous). This condition is characterized by decreased bone mass and microstructural changes that compromise normal function and increase the risk of fractures (look forward to Clinical Note: Osteoporosis on p. 125).

5.3 CONCEPT CHECK

10 What differences would you expect to see in the bones of an athlete before and after extensive training to increase muscle mass?

11 What major difference might you expect to find when comparing bone growth in a 15-year-old and a 30-year-old?

See the blue Answers tab at the back of the book.

5.4 Anatomy of Skeletal Elements

▶ **KEY POINT** We classify bones into seven categories according to their shapes. Bone markings are a useful way to identify bones and determine the positions of other tissues.

The human skeleton has 206 major bones. We group these bones into anatomical categories based on their shapes and identify them based on their bone markings (surface features).

Classification of Bone Shapes

Figure 5.11 describes the anatomical classification of seven categories based on bone shape: **sutural** (*Wormian*), **pneumatized**, **short**, **irregular**, **flat**, **long**, and **sesamoid bones**.

Bone Markings

Most bones have a pattern of surface markings. Elevations or projections form where tendons and ligaments attach and where adjacent bones form joints. Depressions, grooves, and tunnels indicate sites where blood vessels and nerves lie alongside or penetrate the bone.

Figure 5.12 illustrates important examples of these **bone markings**, or *surface features*. Specific anatomical terms describe the various features. Bone markings are useful in identifying a bone. They also provide landmarks that help us determine the position of the soft tissue components of other systems. Indeed, forensic anthropologists can often determine the age, size, sex, and general appearance of an individual on the basis of skeletal remains. (We will discuss this topic further in Chapter 6.)

5

Figure 5.11 Shapes of Bones

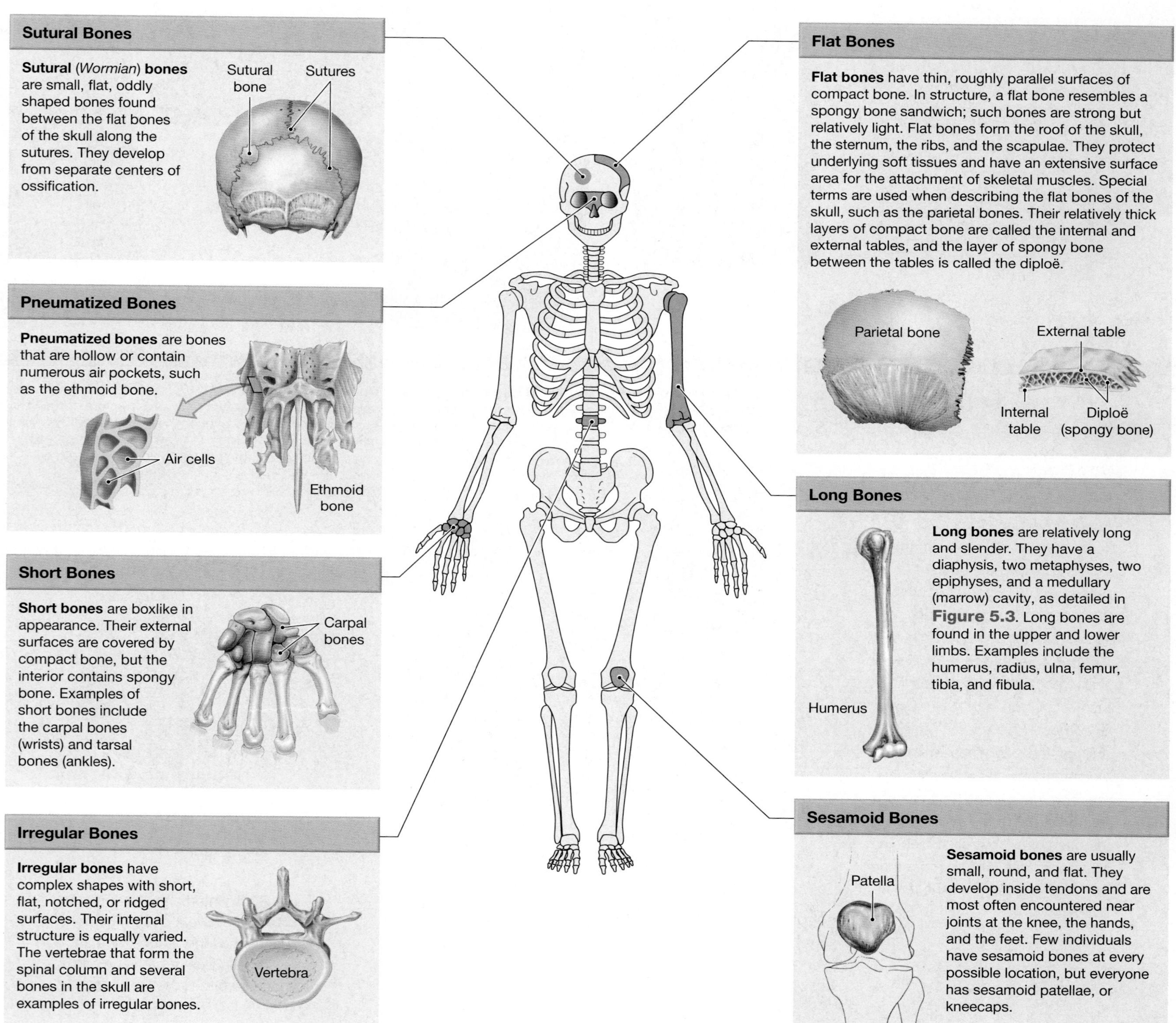

Sutural Bones
Sutural (Wormian) bones are small, flat, oddly shaped bones found between the flat bones of the skull along the sutures. They develop from separate centers of ossification.
Sutural bone
Sutures
Pneumatized Bones
Pneumatized bones are bones that are hollow or contain numerous air pockets, such as the ethmoid bone.
Air cells
Ethmoid bone
Short Bones
Short bones are boxlike in appearance. Their external surfaces are covered by compact bone, but the interior contains spongy bone. Examples of short bones include the carpal bones (wrists) and tarsal bones (ankles).
Carpal bones
Irregular Bones
Irregular bones have complex shapes with short, flat, notched, or ridged surfaces. Their internal structure is equally varied. The vertebrae that form the spinal column and several bones in the skull are examples of irregular bones.
Vertebra
Flat Bones
Flat bones have thin, roughly parallel surfaces of compact bone. In structure, a flat bone resembles a spongy bone sandwich; such bones are strong but relatively light. Flat bones form the roof of the skull, the sternum, the ribs, and the scapulae. They protect underlying soft tissues and have an extensive surface area for the attachment of skeletal muscles. Special terms are used when describing the flat bones of the skull, such as the parietal bones. Their relatively thick layers of compact bone are called the internal and external tables, and the layer of spongy bone between the tables is called the diploë.
Parietal bone
External table
Internal table
Diploë (spongy bone)
Long Bones
Long bones are relatively long and slender. They have a diaphysis, two metaphyses, two epiphyses, and a medullary (marrow) cavity, as detailed in Figure 5.3. Long bones are found in the upper and lower limbs. Examples include the humerus, radius, ulna, femur, tibia, and fibula.
Humerus
Sesamoid Bones
Sesamoid bones are usually small, round, and flat. They develop inside tendons and are most often encountered near joints at the knee, the hands, and the feet. Few individuals have sesamoid bones at every possible location, but everyone has sesamoid patellae, or kneecaps.
Patella

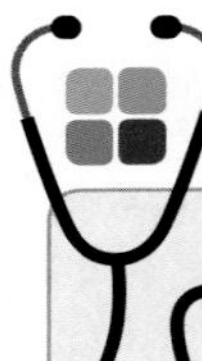

CLINICAL NOTE

Fractures and Their Repair

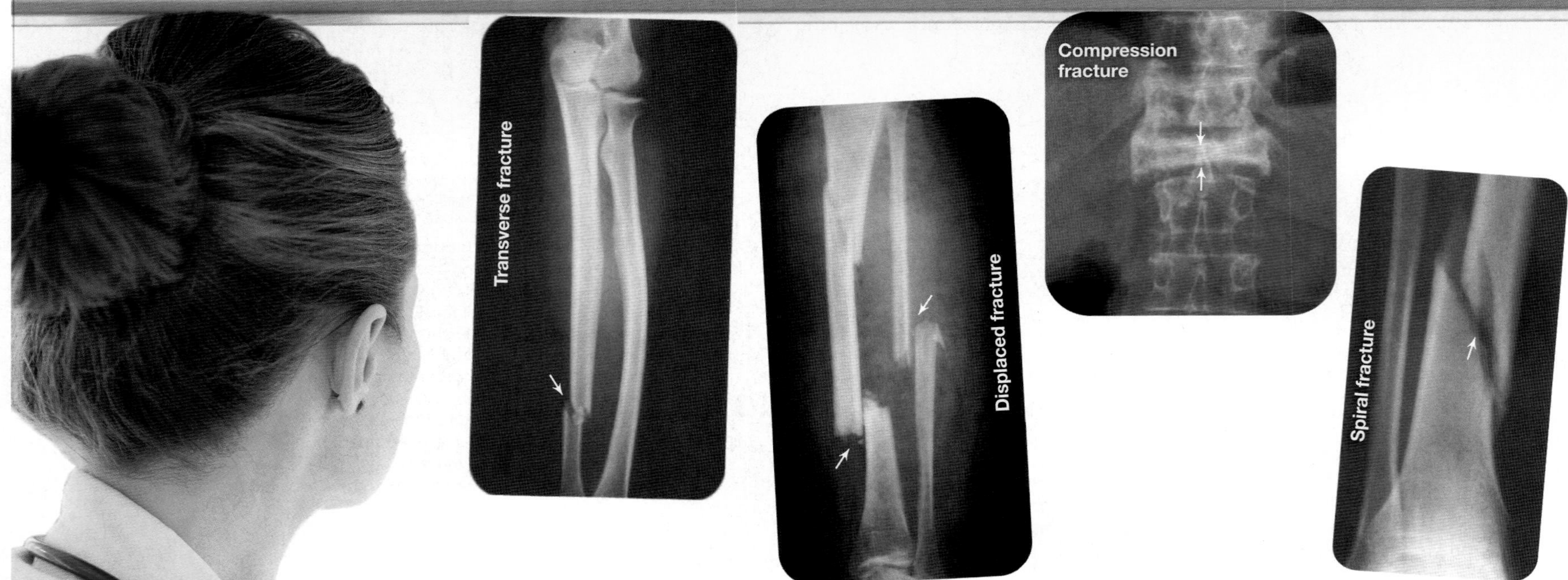

Types of Fractures

Fractures are named according to their external appearance, their location, and the nature of the crack or break in the bone. Important types of fractures are illustrated here by representative x-rays. The broadest general categories are closed fractures and open fractures. **Closed**, or *simple*, fractures are completely internal. They can be seen only on x-rays because they do not involve a break in the skin. **Open**, or *compound*, fractures project through the skin. These fractures, which are obvious on inspection, are more dangerous than closed fractures due to the possibility of infection or uncontrolled bleeding. Many fractures fall into more than one category because the terms overlap.

Transverse fractures, such as this fracture of the ulna, break a bone shaft across its long axis.

Displaced fractures produce new and abnormal bone arrangements; **nondisplaced fractures** retain the normal alignment of the bones or fragments.

Compression fractures occur in vertebrae subjected to stress, such as when you fall on your tailbone. They can also be caused by tumors or occur spontaneously in brittle bone.

Spiral fractures, such as this fracture of the tibia, are produced by twisting stresses that spread along the length of the bone.

Repair of a fracture

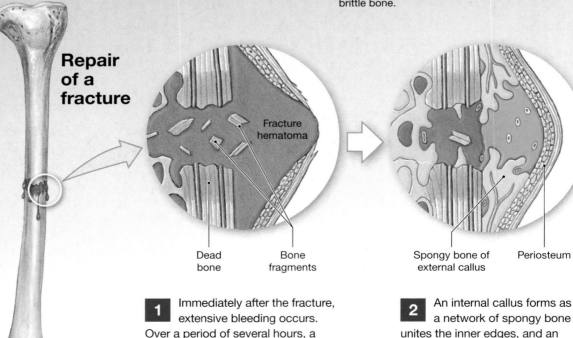

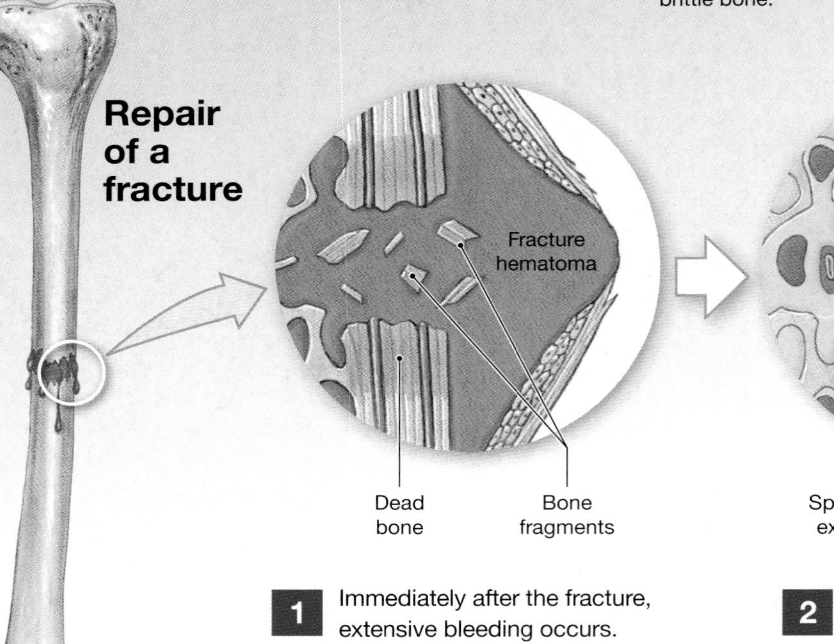

1 Immediately after the fracture, extensive bleeding occurs. Over a period of several hours, a large blood clot, or fracture hematoma, develops.

2 An internal callus forms as a network of spongy bone unites the inner edges, and an external callus of cartilage and bone stabilizes the outer edges.

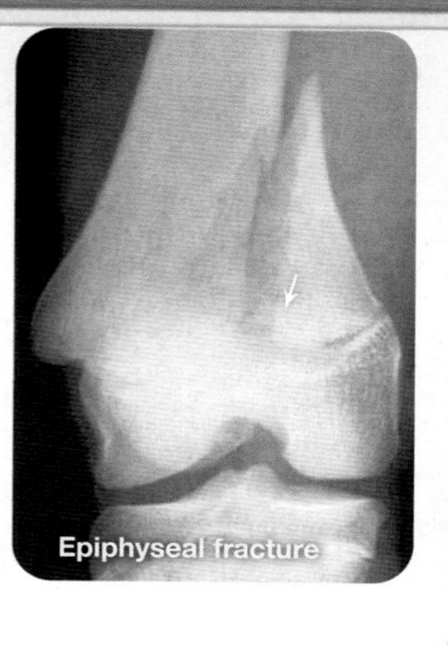

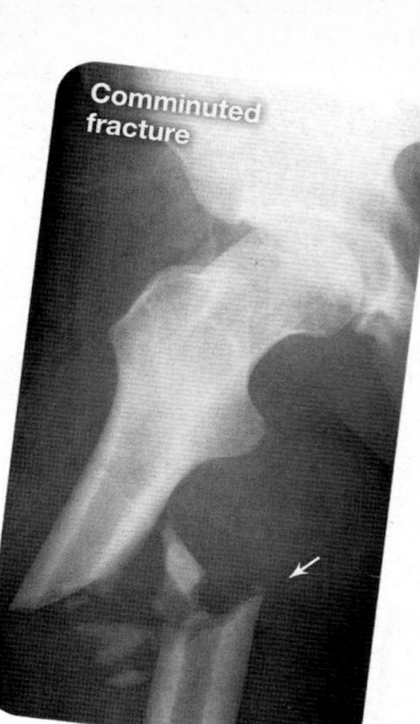

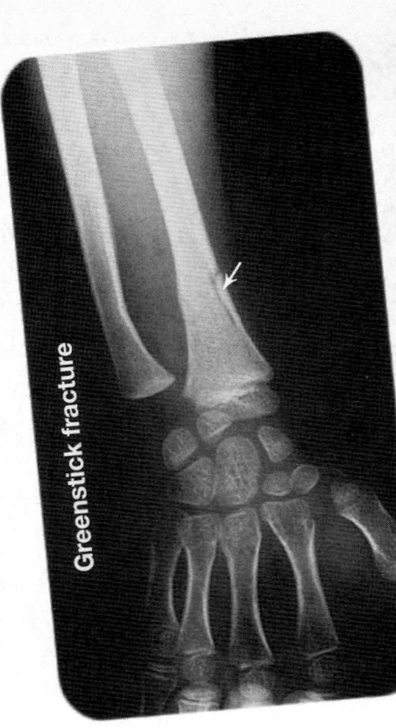

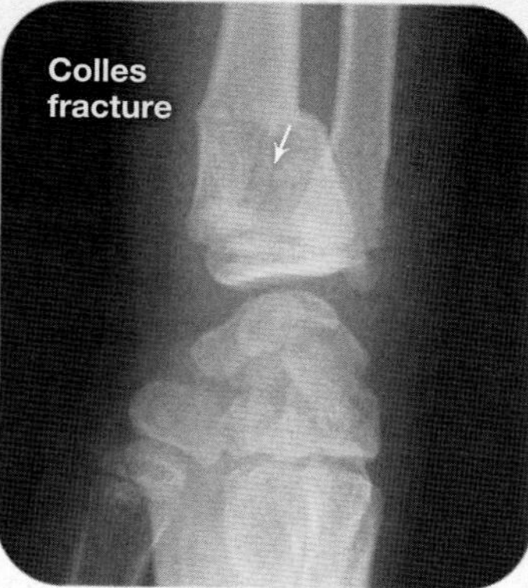

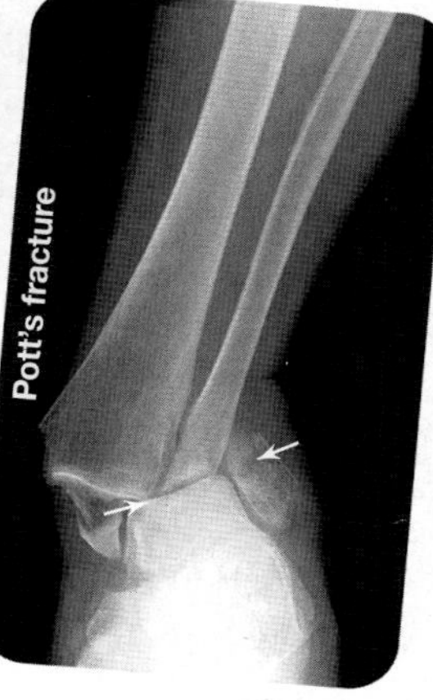

5

Epiphyseal fractures, such as this fracture of the femur, tend to occur where the bone matrix is undergoing calcification and chondrocytes are dying. A simple transverse fracture along this line generally heals well as long as it is lined up perfectly. Unless carefully treated, fractures between the epiphysis and the epiphyseal cartilage can permanently distort growth.

Comminuted fractures, such as this fracture of the femur, shatter the affected area into a multitude of bony fragments.

In a **greenstick fracture**, such as this fracture of the radius, only one side of the shaft is broken, and the other is bent. This type of fracture generally occurs in children, whose long bones have yet to ossify fully.

A **Colles fracture**, a break in the distal portion of the radius, is typically the result of reaching out to cushion a fall.

A **Pott's fracture**, also called a *bimalleolar fracture*, occurs at the ankle and affects both the medial malleolus of the tibia and the lateral malleolus of the fibula.

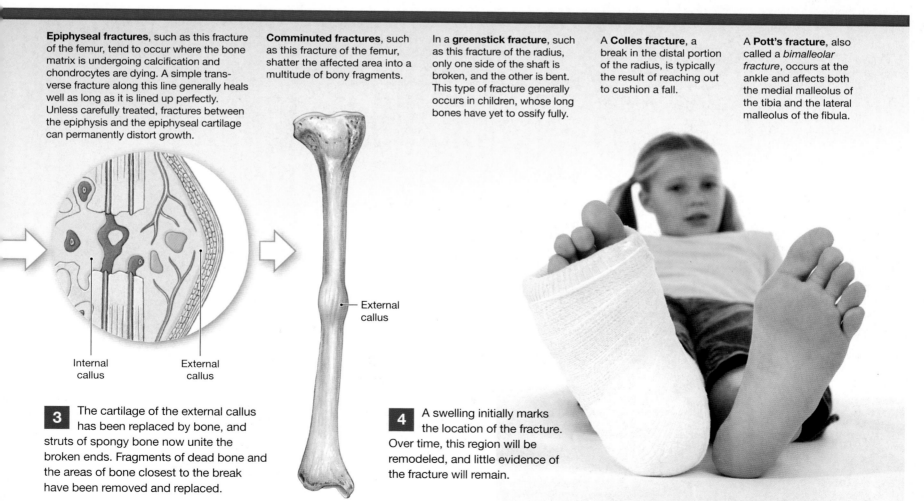

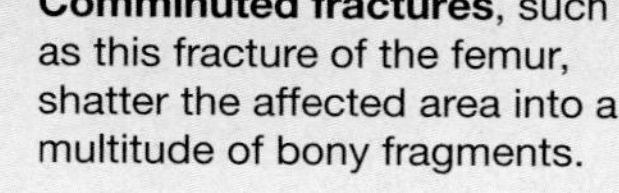

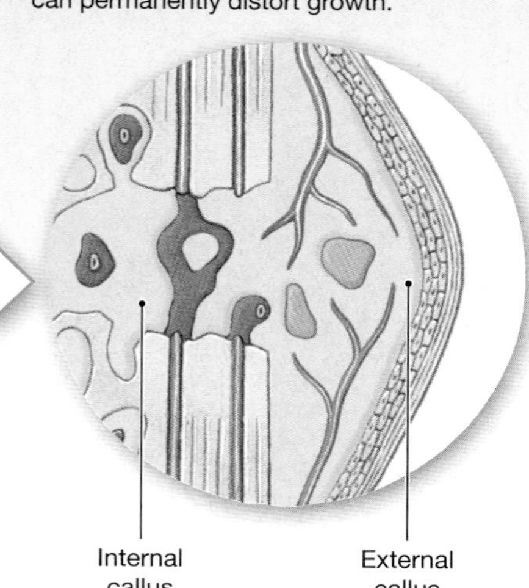

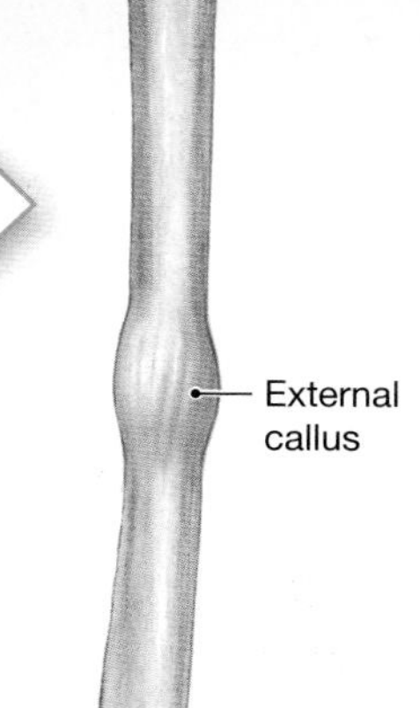

3 The cartilage of the external callus has been replaced by bone, and struts of spongy bone now unite the broken ends. Fragments of dead bone and the areas of bone closest to the break have been removed and replaced.

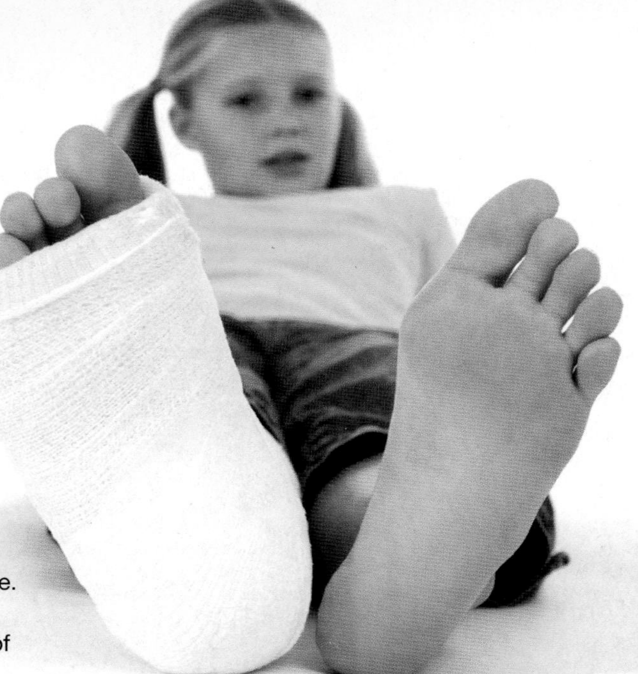

4 A swelling initially marks the location of the fracture. Over time, this region will be remodeled, and little evidence of the fracture will remain.

Figure 5.12 Examples of Bone Markings (Surface Features). Bone markings provide distinct and characteristic landmarks for orientation and identification of bones and associated structures.

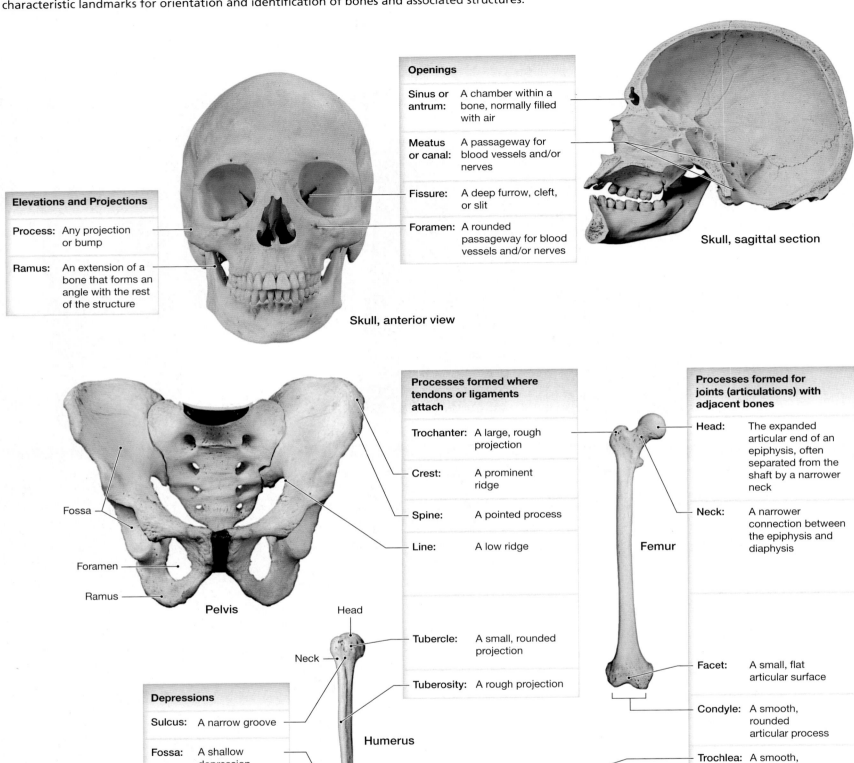

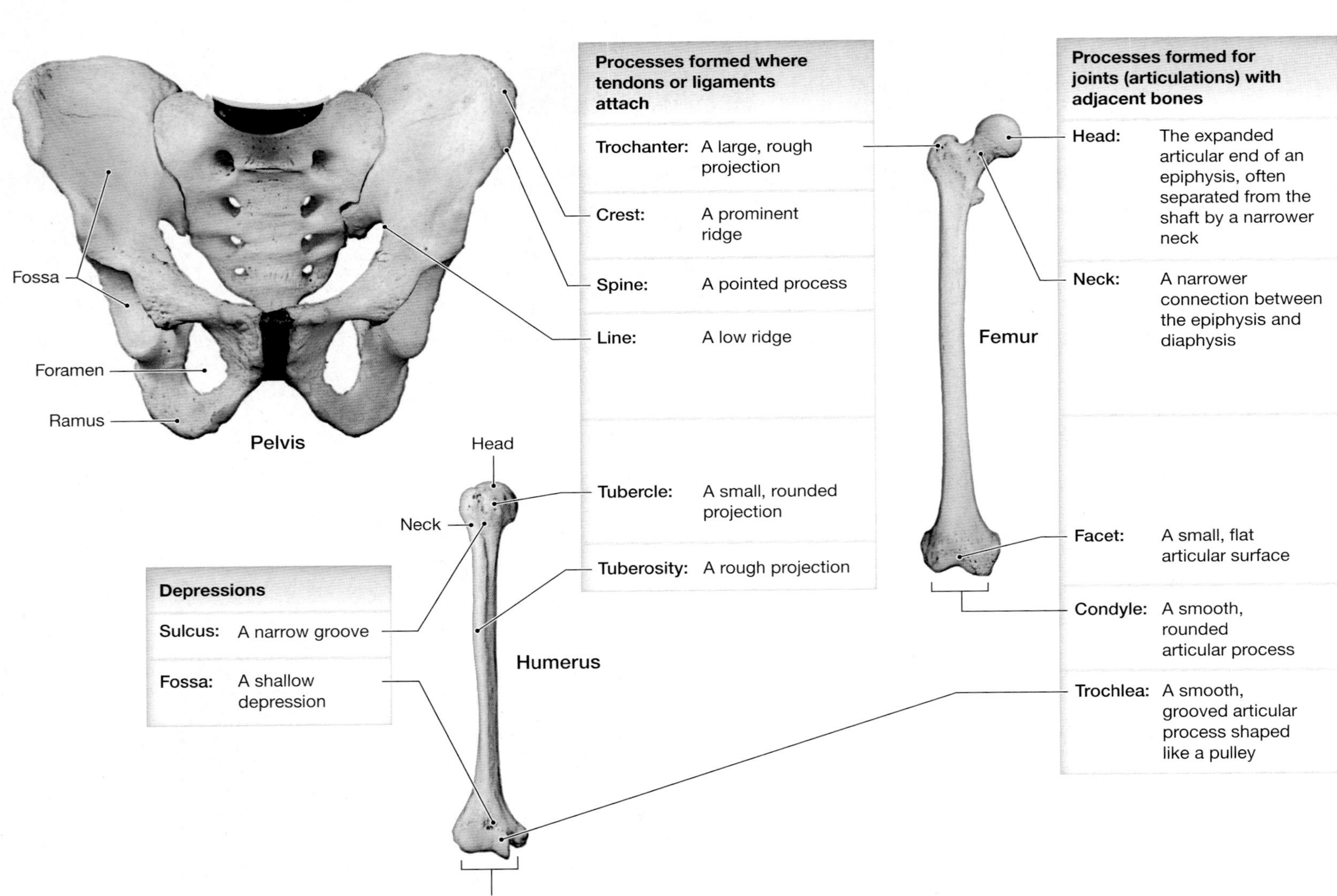

Openings
Sinus or antrum: A chamber within a bone, normally filled with air
Meatus or canal: A passageway for blood vessels and/or nerves
Fissure: A deep furrow, cleft, or slit
Foramen: A rounded passageway for blood vessels and/or nerves
Skull, sagittal section
Elevations and Projections
Process: Any projection or bump
Ramus: An extension of a bone that forms an angle with the rest of the structure
Skull, anterior view
Processes formed where tendons or ligaments attach
Trochanter: A large, rough projection
Crest: A prominent ridge
Spine: A pointed process
Line: A low ridge
Tubercle: A small, rounded projection
Tuberosity: A rough projection
Processes formed for joints (articulations) with adjacent bones
Head: The expanded articular end of an epiphysis, often separated from the shaft by a narrower neck
Neck: A narrower connection between the epiphysis and diaphysis
Facet: A small, flat articular surface
Condyle: A smooth, rounded articular process
Trochlea: A smooth, grooved articular process shaped like a pulley
Fossa
Foramen
Ramus
Pelvis
Femur
Head
Neck
Humerus
Condyle
Depressions
Sulcus: A narrow groove
Fossa: A shallow depression

CLINICAL NOTE

Osteoporosis

In **osteoporosis**, bones become fragile due to decreased or insufficient bone mass and thus become porous and more likely to fracture. We reach peak bone density in our early 20s and it decreases as we age. Inadequate calcium intake in childhood reduces peak bone density and increases the risk of osteoporosis. The distinction between the "normal" osteopenia (decreased calcification) of aging and the clinical condition of osteoporosis is a matter of degree.

Current projections indicate there will be more than 14 million osteoporotic Americans by 2020, most of them elderly women. The increase in incidence after menopause has been linked to decreased estrogen (female sex hormone) production. Men have heavier skeletons and produce testosterone (male sex hormone) throughout life, so they are less likely to develop osteoporosis.

Osteoporosis can also develop as a secondary effect of some cancers. Cancers of the bone, breast, or other tissues may release a chemical known as **osteoclast-activating factor**. This compound increases both the number and activity of osteoclasts and may produce severe osteoporosis.

The excessive fragility of osteoporotic bones commonly leads to fractures, particularly of the vertebrae, wrist, and hip. Supplemental estrogen, increased dietary intake of calcium and vitamin D_3, exercise to stimulate osteoblastic activity, and calcitonin (a hormone produced in the thyroid) are recommended to slow the progression of osteoporosis. Drugs called bisphosphonates can help by inhibiting osteoclastic activity.

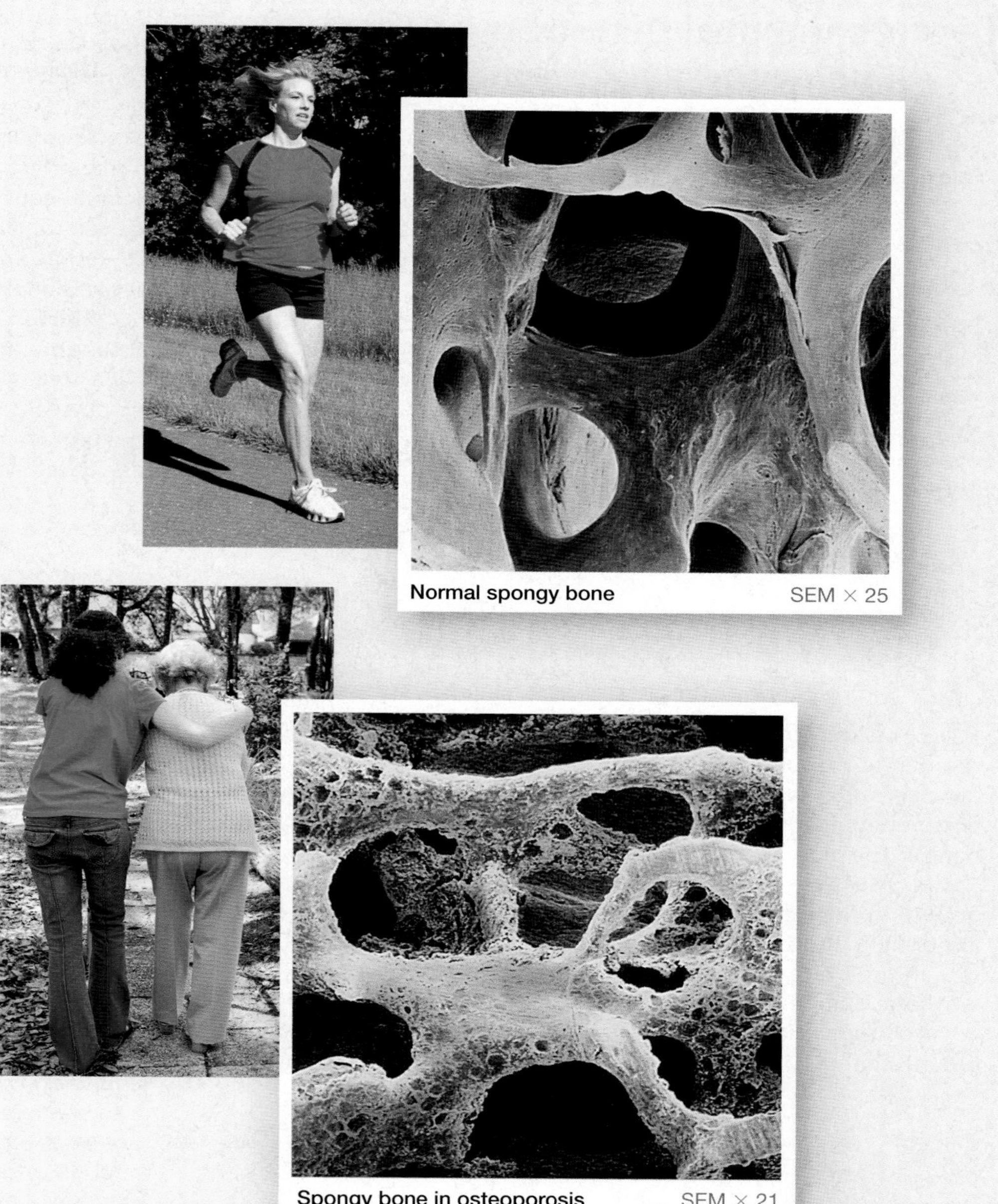

Normal spongy bone SEM × 25

Spongy bone in osteoporosis SEM × 21

5.4 CONCEPT CHECK

12 Why is a working knowledge of bone markings important in a clinical setting?

13 What is the primary difference between sesamoid and irregular bones?

14 Where would you look for sutural bones in a skeleton?

See the blue Answers tab at the back of the book.

5.5 | Integration with Other Systems

▶ **KEY POINT** The skeletal system is dynamic and interacts with other body systems in a variety of interesting ways.

Although bones may seem inert, you should now realize that they are dynamic structures. The entire skeletal system is intimately associated with other systems. Bones attach to the muscular system and have extensive connections with the cardiovascular and lymphatic systems. Bones are largely controlled by the endocrine system. The digestive and excretory systems play important roles in providing the calcium and phosphate minerals needed for bone growth. In return, the skeleton provides a reserve of calcium, phosphate, and other minerals that can compensate for changes in the dietary supply of these ions.

5.5 CONCEPT CHECK

15 How would the bones of an individual who is eating a diet low in calcium be affected?

See the blue Answers tab at the back of the book.

CLINICAL NOTE

Congenital Disorders of the Skeleton

Excessive secretion (hypersecretion) of growth hormone (GH) can result in two disorders: **gigantism** and **acromegaly**. Pre-puberty hypersecretion results in gigantism (abnormally long bones), and post-puberty hypersecretion results in acromegaly (abnormally thick bones). The most common cause is a pituitary tumor.

In **pituitary growth failure**, inadequate production of growth hormone leads to reduced epiphyseal cartilage activity and abnormally short bones. This condition is becoming increasingly rare in the United States because children can be treated with synthetic human GH.

Achondroplasia (a-kon-drō-PLĀ-sē-uh) is a disorder of bone growth that results from abnormal epiphyseal activity. The child's epiphyseal cartilages grow unusually slowly and are replaced by bone early in life. As a result, the person develops short, stocky limbs. Although other skeletal abnormalities occur, the trunk is normal in size, and sexual and mental development are unaffected.

Achondroplasia results from an abnormal gene on chromosome 4 that affects a fibroblast growth factor. Most cases result from spontaneous mutations. If both parents have achondroplasia, 25 percent of their children will be unaffected, 50 percent will have the condition, and 25 percent will inherit two abnormal genes, leading to severe abnormalities and early death.

Marfan syndrome is a genetic disorder that affects connective tissue structure. Individuals with Marfan syndrome are very tall and have long, slender limbs due to excessive cartilage formation at the epiphyseal cartilages. An abnormal gene on chromosome 15 that affects the protein fibrillin is responsible. The skeletal effects are striking, but associated weaknesses in blood vessel walls are more dangerous.

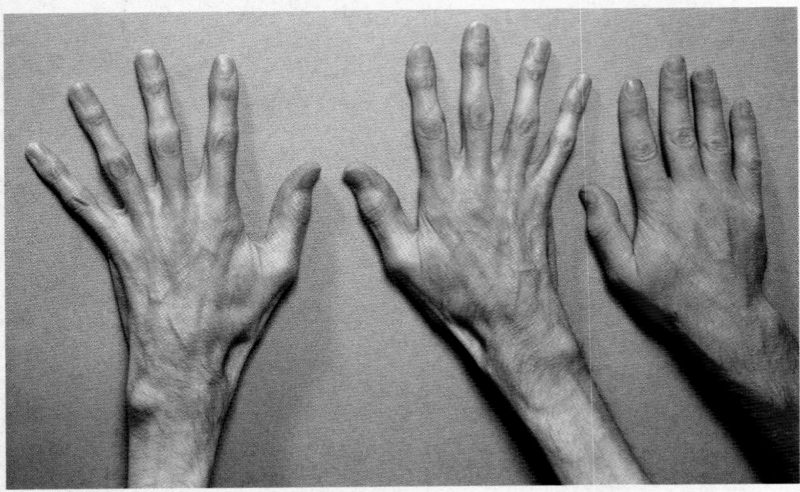

Osteomalacia (os-tē-ō-ma-LĀ-shē-uh; *malakia*, softness), also called *adult rickets*, is characterized by a gradual softening and bending of the bones as a result of poor mineralization. The bones appear normal, but are weak and flexible because the bone matrix cannot accumulate enough calcium salts. **Rickets**, a form of osteomalacia that affects children, generally results from a vitamin D_3 deficiency caused by inadequate skin exposure to sunlight or an inadequate dietary supply of the vitamin. The bones of children with rickets are so poorly mineralized they bend laterally, producing a bowlegged appearance.

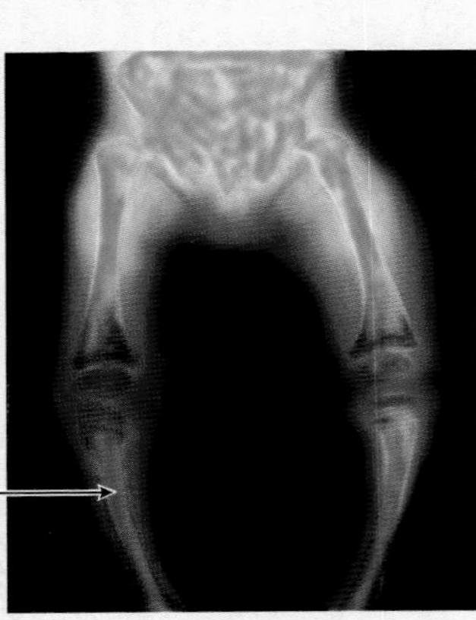

Examination of the Skeletal System

Clinicians use relatively sophisticated equipment to view the skeleton. However, a clinician's most important tools are a careful medical history and physical examination. Information that the patient can provide includes:

- Onset, location, quality, and severity of the pain or stiffness
- What makes the pain better or worse
- Associated signs and symptoms
- What the patient thinks is causing it

A physical examination can be supplemented with diagnostic procedures and laboratory tests, summarized in Table 5.1.

Table 5.1 | Examples of Tests Used in the Diagnosis of Bone and Joint Disorders

Diagnostic Procedure	Method and Result	Representative Uses
X-ray	A beam of radiation that passes through the body and strikes a photographic film, forming an image of body structures	Detects fractures, tumors, dislocations, reduction in bone density (osteopenia), and bone infections (osteomyelitis)
Bone scan	A nuclear imaging test, which uses a radioactive tracer that demarcates "hot spots" of increased bone turnover (osteoblastic activity). "Cold spots" that lack tracer may indicate cancer (such as multiple myeloma) or impaired blood supply to the bone.	Detects occult fractures that are healing, osteomyelitis, areas of metastatic cancer to bone, arthritis, and diseases of abnormal bone metabolism (such as Paget's disease)
Arthrocentesis	Insertion of a needle into a joint to aspirate synovial fluid	Detects abnormalities in synovial fluid, including bleeding, infection, inflammation, gout, and pseudogout
MRI (magnetic resonance imaging)	Uses powerful magnets and radio waves that produce 3-D images, including soft tissue detail, without radiation	Detects infections, areas of inflammation, fractures, and tumors including their exact dimensions
DEXA (dual-energy X-ray absorptiometry) or bone densitometry	Enhanced form of low-dose x-ray technology that measures bone loss	Quantifies and monitors bone density loss and predicts risk of bone fractures

Study Outline

Introduction p. 108

- The skeletal system includes the bones of the skeleton and the cartilages, ligaments, and other connective tissues that stabilize or interconnect bones. Its functions include structural support, storage of minerals and lipids, blood cell production, protection of delicate tissues and organs, and leverage.

5.1 | Structure and Function of Bone p. 108

- **Osseous (bone) tissue** is a supporting connective tissue with specialized cells and a solid, extracellular matrix of protein fibers and a ground substance.

The Histological Organization of Mature Bone p. 108

- Bone matrix consists largely of crystals of hydroxyapatite, accounting for almost two-thirds of the weight of bone. The remaining third is dominated by collagen fibers and small amounts of other calcium salts; bone cells and other cell types contribute only about 2 percent to the volume of bone tissue.
- **Osteocytes** are mature bone cells that are completely surrounded by hard bone matrix. Osteocytes reside in spaces termed **lacunae**. Osteocytes in lacunae are interconnected by small, hollow channels called **canaliculi**. **Lamellae** are layers of calcified matrix. *(See Figure 5.1.)*
- **Osteoblasts** are bone-forming cells. By the process of **osteogenesis**, osteoblasts synthesize **osteoid**, the matrix of bone prior to its calcification. *(See Figure 5.1.)*
- **Osteoprogenitor cells** are mesenchymal cells that play a role in the repair of bone fractures. *(See Figure 5.1.)*
- **Osteoclasts** are large, multinucleate cells that help dissolve the bony matrix through the process of **osteolysis**. They are important in the regulation of calcium and phosphate concentrations in body fluids. *(See Figure 5.1.)*

Compact and Spongy Bone p. 109

- There are two types of bone: **compact** (*dense*) **bone**, and **spongy** (*trabecular* or *cancellous*) **bone**. Compact and spongy bone have the same matrix composition, but they differ in the three-dimensional arrangement of osteocytes, canaliculi, and lamellae. *(See Figures 5.1 and 5.2.)*
- The basic functional unit of compact bone is the **osteon**, or *Haversian system*. Osteocytes in an osteon are arranged in concentric layers around a **central canal**. *(See Figures 5.1b-d and 5.2.)*
- Spongy bone contains struts or plates called **trabeculae**, often in an open network. *(See Figure 5.2.)*
- Compact bone covers bone surfaces. It is thickest where stresses come from a limited range of directions. Spongy bone is located internally in bones. It is found where stresses are few or come from many different directions. *(See Figure 5.3.)*

The Periosteum and Endosteum p. 112

- A bone is covered externally by a two-layered **periosteum** (outer fibrous, inner cellular) and lined internally by a cellular **endosteum**. *(See Figure 5.4.)*

5.2 | Bone Development and Growth p. 113

- **Ossification** replaces other tissue with bone; **calcification** deposits calcium salts within a tissue.

Intramembranous Ossification p. 113

- **Intramembranous ossification** (*dermal ossification*) begins when osteoblasts differentiate within a mesenchymal or fibrous connective tissue. This process can ultimately produce spongy or compact bone. Such ossification begins at an **ossification center**. *(See Spotlight Figure 5.6.)*

5

Endochondral Ossification p. 113

- **Endochondral ossification** begins with the formation of a cartilaginous model. Osseous tissue gradually replaces this hyaline cartilage model. *(See Spotlight Figure 5.7.)*
- The length of a developing bone increases at the epiphyseal cartilage, which separates the epiphysis from the diaphysis. Here, new cartilage is added at the epiphyseal side, while osseous tissue replaces older cartilage at the diaphyseal side. The time of closure of the epiphyseal cartilage differs among bones and among individuals. *(See Figure 5.8.)*
- The diameter of a bone enlarges through **appositional growth** at the outer surface. *(See Figure 5.9.)*

Blood and Nerve Supply to Bones p. 114

- A typical bone formed through endochondral ossification has several sets of vessels: the **nutrient artery**, **nutrient vein**, **metaphyseal vessels**, **epiphyseal vessels**, and **periosteal vessels**. Lymphatic vessels are distributed in the periosteum and enter the osteons through the nutrient and perforating canals. *(See Spotlight Figure 5.7 and Figure 5.10.)*
- Sensory nerve endings branch throughout the periosteum, and sensory nerves penetrate the cortex with the nutrient artery to innervate the endosteum, medullary cavity, and epiphyses.

Factors Regulating Bone Growth p. 114

- Normal osteogenesis requires a continual and reliable source of minerals, vitamins, and hormones.
- Parathyroid hormone, secreted by the parathyroid glands, stimulates osteoclast and osteoblast activity. In contrast, calcitonin, secreted by the thyroid gland, inhibits osteoclast activity and increases calcium loss in the urine. These hormones control the rate of mineral deposition in the skeleton and regulate the calcium ion concentrations in body fluids.
- Growth hormone, thyroxine, and sex hormones stimulate bone growth by increasing osteoblast activity.
- There are differences between individual bones and between individuals with respect to the timing of epiphyseal cartilage closure.

5.3 | Bone Maintenance, Remodeling, and Repair p. 118

- The turnover rate for bone is high. Each year almost one-fifth of the adult skeleton is broken down and then rebuilt or replaced.

Remodeling of Bone p. 119

- Bone remodeling involves the simultaneous processes of adding new bone and removing previously formed bone.
- Mineral turnover and recycling allow bone to adapt to new stresses.

Injury and Repair p. 120

- A **fracture** is a crack or break in a bone. Healing of a fracture can usually occur if portions of the blood supply, endosteum, and periosteum remain intact. For a classification of fracture types, see the Clinical Note on pp. 122–123.

Aging and the Skeletal System p. 120

- The bones of the skeleton become thinner and relatively weaker as a normal part of the aging process. **Osteopenia** usually develops to some degree, but in some cases this process progresses to **osteoporosis** and the bones become dangerously weak and brittle.

5.4 | Anatomy of Skeletal Elements p. 120

Classification of Bone Shapes p. 121

- The seven categories of bones are based on anatomical classification: **sutural** (*Wormian*), **pneumatized**, **short**, **irregular**, **flat**, **long**, and **sesamoid bones**. *(See Figure 5.11.)*

Bone Markings p. 121

- **Bone markings** (*surface features*) identify specific elevations, depressions, and openings of bones. *(See Figure 5.12.)*

5.5 | Integration with Other Systems p. 125

- The skeletal system is anatomically and physiologically linked to other body systems and represents a reservoir for calcium, phosphate, and other minerals.

Chapter Review

For answers, see the blue Answers tab at the back of the book.

Level 1 Reviewing Facts and Terms

1. Label the following structures on the accompanying diagram of a long bone.
   - diaphysis
   - articular surfaces of the proximal end of the bone
   - epiphysis
   - articular surfaces of the distal end of the bone

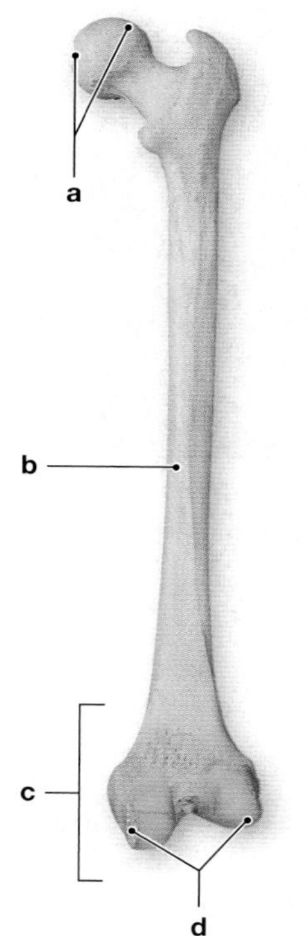

(a) __________ (c) __________
(b) __________ (d) __________

2. Spongy bone is formed of
   (a) osteons.
   (b) struts and plates.
   (c) concentric lamellae.
   (d) spicules only.
3. The basic functional unit of mature compact bone is the
   (a) osteon.
   (b) canaliculus.
   (c) lamella.
   (d) central canal.
4. Endochondral ossification begins with the formation of
   (a) a fibrous connective tissue model.
   (b) a hyaline cartilage model.
   (c) a membrane model.
   (d) a calcified model.
5. When sex hormone production increases, bone production
   (a) slows down.
   (b) increases.
   (c) both increases and decreases.
   (d) is not affected.
6. The presence of an epiphyseal line indicates that
   (a) epiphyseal growth has ended.
   (b) epiphyseal growth is just beginning.
   (c) growth in bone diameter is just beginning.
   (d) the bone is fractured at that location.
7. The inadequate ossification that occurs with aging is called
   (a) osteopenia.
   (b) osteomyelitis.
   (c) osteitis.
   (d) osteoporosis.
8. The process by which the diameter of a developing bone enlarges is
   (a) appositional growth at the outer surface.
   (b) interstitial growth within the matrix.
   (c) lamellar growth.
   (d) Haversian growth.
9. The sternum is an example of a(n)
   (a) flat bone.
   (b) long bone.
   (c) irregular bone.
   (d) sesamoid bone.
10. A small, rough projection of a bone is termed a
    (a) ramus.
    (b) tuberosity.
    (c) trochanter.
    (s) spine.

Level 2 Reviewing Concepts

1. How would decreasing the proportion of organic molecules to inorganic components in the bony matrix affect the physical characteristics of bone?
   (a) The bone would be less flexible.
   (b) The bone would be stronger.
   (c) The bone would be more brittle.
   (d) The bone would be more flexible.
2. Which of the following could cause premature closure of the epiphyseal cartilages?
   (a) increased levels of sex hormones
   (b) high levels of vitamin D_3
   (c) too little parathyroid hormone
   (d) an excess of growth hormone
3. Identify the factors that determine the type of ossification that occurs in a specific bone.
4. Identify the events that signal the end of long bone elongation.
5. Compare and contrast the advantages of spongy bone and compact bone in an area such as the expanded ends of long bones.
6. Identify the steps involved in the process by which a bone grows in diameter.
7. Explain why a healed area of bone is less likely to fracture in the same place again from similar stresses.
8. Explain why a diet that consists mostly of junk foods will hinder the healing of a fractured bone.
9. Identify the properties that are used to distinguish a sesamoid bone from a sutural bone.
10. Compare and contrast the processes of ossification and calcification.

Level 3 Critical Thinking

1. A small child falls off a bicycle and breaks an arm. The bone is set correctly and heals well. After the cast is removed, an enlarged bony bump remains at the region of the fracture. After several months this enlargement disappears, and the arm is essentially normal in appearance. What happened during this healing process?
2. Most young children who break a bone in their upper or lower limbs experience a greenstick fracture. This type of fracture is rare in an adult. What is the reason for this difference?

 CLINICAL CASE WRAP-UP

Pushing Beyond Limits

Emily has a stress fracture of her tibia where the cross-sectional diameter is thinnest. She first sustained a minor injury from the repetitive stress of running, and only the periosteum tore and bled. Then the circumferential lamellae of the anterior cortex of her tibia gave way with a tiny horizontal crack. As she continued to run, adjacent osteons broke and their central veins bled, causing further damage. Because the periosteum is loaded with sensory nerve endings, her stress fracture caused increasing pain.

This tiny fracture, involving only the anterior cortex of the tibia, was not yet visible on an x-ray. However, the MRI revealed the stress fracture and the swelling within the periosteum, cortex, and adjacent endosteum.

Emily is relieved to hear that she only needs to wear a walking boot and restrict her running for six weeks. By then, an MRI will show the fracture healing, and perhaps she will be able to increase her activity.

1. What do you think would happen if Emily ignored the pain and continued to run on this cortical stress fracture?
2. If the stress fracture continued across the entire anterior cortex, where else could osteoprogenitor cells be recruited for healing?

See the blue Answers tab at the back of the book.

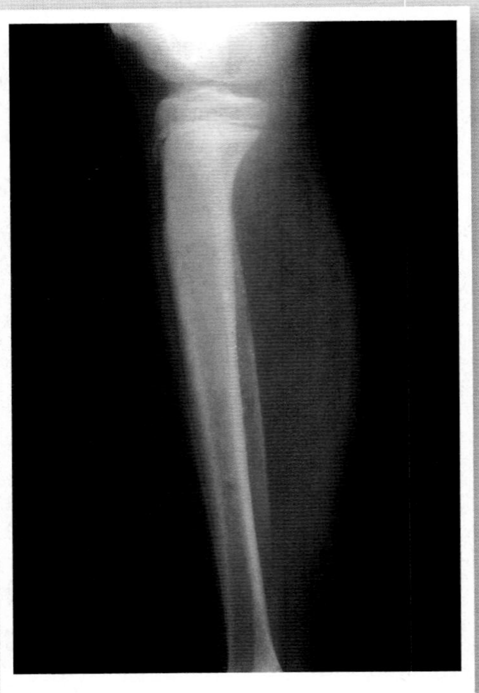

5

Related Clinical Terms

bone marrow transplant: Transferring healthy bone marrow stem cells from one person into another to replace bone marrow that either is dysfunctional or has been destroyed by chemotherapy or radiation.

bone mineral density test (BMD): A test to predict the risk of bone fractures by measuring how much calcium and other types of minerals are present in the patient's bones.

closed reduction: The correction of a bone fracture by manipulation without incision into the skin.

open reduction: The correction of a bone fracture by making an incision into the skin and rejoining the fractured bone parts, often by mechanical means such as a rod, plate, or screw.

orthopedics: The branch of medicine dealing with the correction of deformities of bones or muscles.

osteogenesis imperfecta (OI): An inherited (genetic) disorder characterized by extreme fragility of the bones; also called brittle bone disease.

osteomyelitis: An acute or chronic bone infection.

osteopetrosis: A rare hereditary bone disorder in which the bones become overly dense; it presents in one of three forms: osteopetrosis tarda, osteopetrosis congenita, or "marble bone" disease.

osteosarcoma: A type of cancer that starts in the bones; also called osteogenic sarcoma.

Paget's disease: A chronic disorder that can result in enlarged and misshapen bones due to abnormal bone destruction and regrowth.

traction: The application of a sustained pull on a limb or muscle in order to maintain the position of a fractured bone until healing occurs or to correct a deformity.

6

The Skeletal System

Axial Division

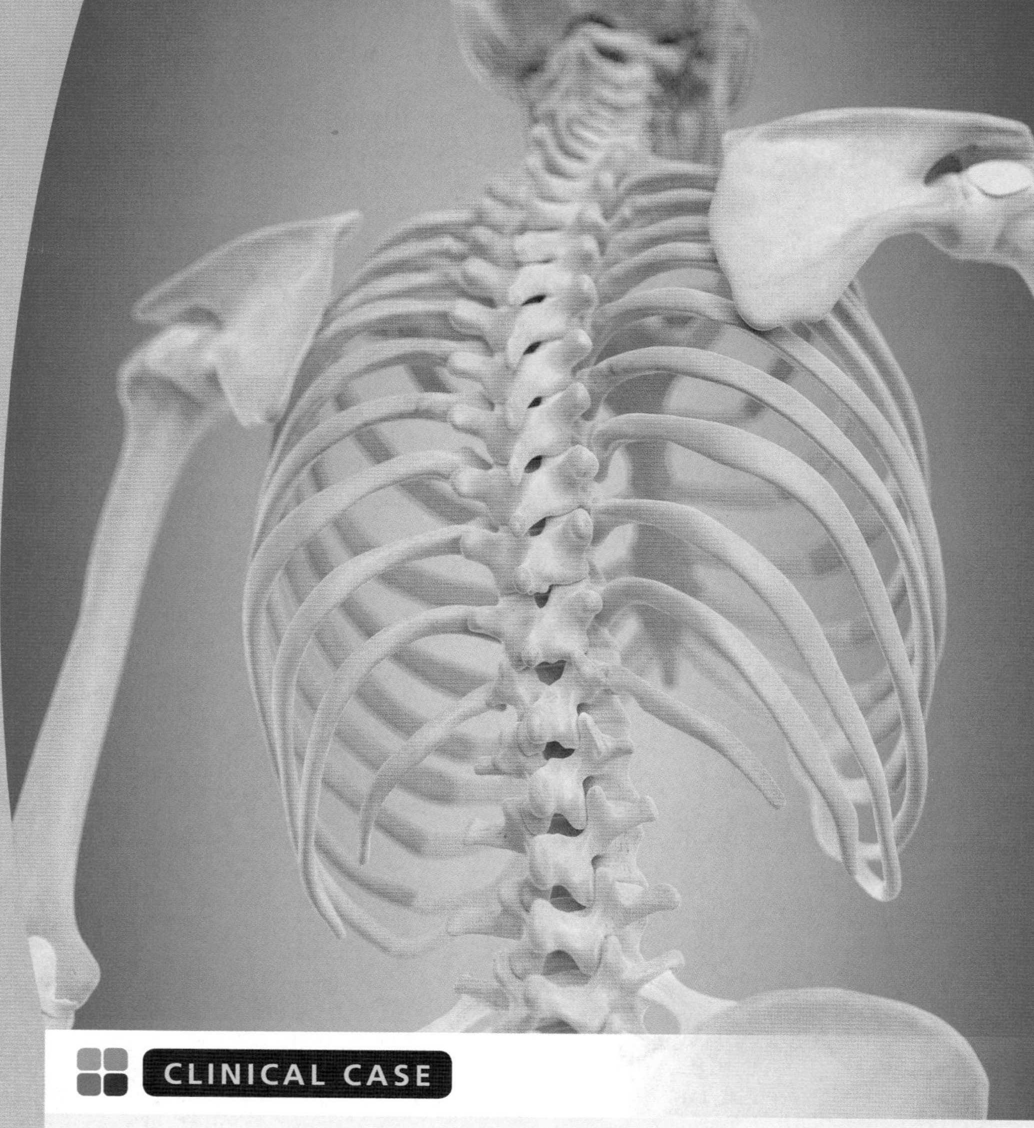

Learning Outcomes

These Learning Outcomes correspond by number to this chapter's sections and indicate what you should be able to do after completing the chapter.

6.1 List the names of the bones that constitute the skull and the associated skull bones. p. 133

6.2 Compare and contrast the sutures of the skull. p. 140

6.3 List and describe the bones of the cranium. p. 140

6.4 Identify and describe the bones of the face. p. 150

6.5 Identify and list the functions of the bones of the orbital and nasal complexes. p. 154

6.6 Compare and contrast structural differences among the skulls of infants, children, and adults. p. 156

6.7 Compare and contrast the vertebral groups and describe the structural and functional differences among them. p. 158

6.8 Explain the significance of the articulations of the thoracic vertebrae, ribs, and sternum. p. 167

CLINICAL CASE

The Last Lap

By 2001, Dale Earnhardt was one of the most beloved NASCAR drivers of all time. At age 49 he had won seven NASCAR championships. In the final lap of the 2001 Daytona 500, Earnhardt's car clipped another car, spun counterclockwise, was broadsided by a third car, and smashed into the wall at 175 mph.

Earnhardt was the only NASCAR driver still wearing an open-faced helmet. In addition, his team had altered the seatbelts for his comfort, and Earnhardt did not wear a head and neck support (HANS) device, a semi-hard collar harnessed to the driver's upper body and tethered to the helmet, designed to prevent the driver's head from snapping forward in a high-speed collision.

Earnhardt was unresponsive, not breathing, and pulseless from the moment he was pulled from his car. Medics initiated CPR immediately and transported him to the ER where, despite vigorous resuscitation efforts, he died.

The autopsy stated that Earnhardt had a displaced fracture of the left ankle, abrasions on the right side of his chin (from contact with the steering wheel), fractures of left ribs 2 to 9, and a contusion of the left side of the head. The fatal injury was indicated by the abundant blood in each external acoustic meatus.

What do you think was the fatal injury? How did the blood in the ears indicate this injury?

To find out, turn to the Clinical Case Wrap-Up on p. 171.

THE BASIC FEATURES of the human skeleton have been shaped by evolution, but because no two people have exactly the same combination of age, diet, activity pattern, and hormone levels, the bones of each individual are unique. As discussed in Chapter 5, bones are continually remodeled and reshaped. Your skeleton changes throughout your lifetime; examples include the proportional changes at puberty and the gradual osteoporosis of aging. This chapter gives other examples of the dynamic nature of the human skeleton, such as the changes in the shape of the vertebral column during the transition from crawling to walking.

The skeletal system, composed of 206 separate bones and a number of associated cartilages, is divided into the axial and appendicular skeletons (**Figure 6.1**). The **axial skeleton** consists of the bones of the skull, thorax, and vertebral column. These structures form the longitudinal axis of the body. There are 80 bones in the axial skeleton, roughly 40 percent of the bones in the human body. The remaining 126 bones of the human skeleton make up the

Figure 6.1 The Axial Skeleton

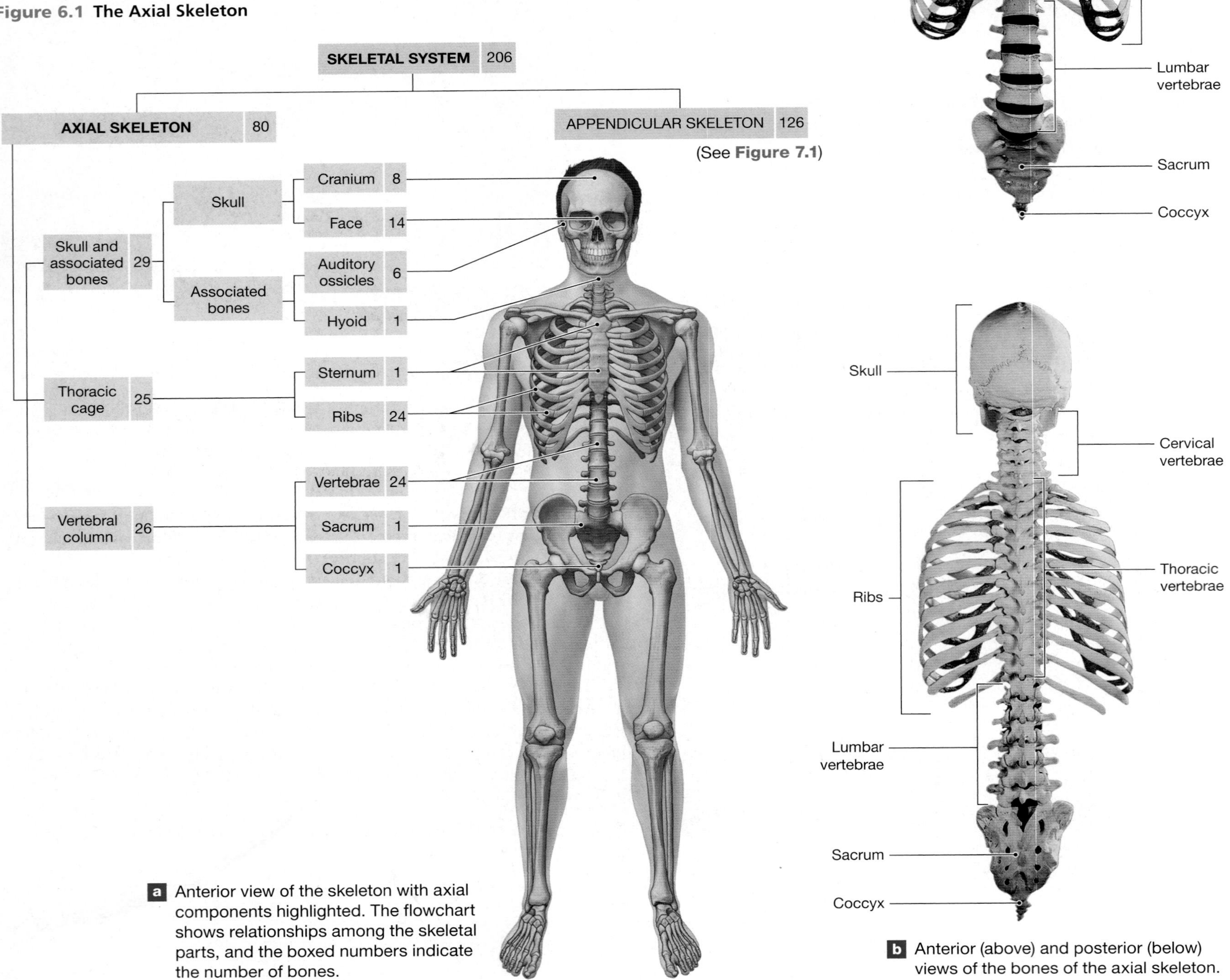

a Anterior view of the skeleton with axial components highlighted. The flowchart shows relationships among the skeletal parts, and the boxed numbers indicate the number of bones.

b Anterior (above) and posterior (below) views of the bones of the axial skeleton.

appendicular skeleton (discussed in Chapter 7). This division includes the bones of the limbs and the pectoral and pelvic girdles that attach the limbs to the trunk.

The axial components, shown in yellow and blue in **Figure 6.1**, include:

- the skull (22 bones),
- bones associated with the skull (6 auditory ossicles and 1 hyoid bone),
- the vertebral column (24 vertebrae, 1 sacrum, and 1 coccyx), and
- the thoracic cage (24 ribs and 1 sternum).

The axial skeleton has several functions:

- forms a framework that supports and protects organs in the thoracic, abdominal, and pelvic body cavities;
- houses special sense organs for taste, smell, hearing, balance, and sight;
- provides areas for the attachment of muscles that adjust the positions of the head, neck, and trunk; performs respiratory movements; and stabilizes or positions structures of the appendicular skeleton; and
- produces blood cells using the red bone marrow in portions of the vertebrae, sternum, and ribs.

This chapter describes the structural anatomy of the axial skeleton, and we begin with the skull. Before proceeding, you will find it helpful to review the directional references included in **Figure 1.10**, p. 17.

6.1 | The Skull and Associated Bones

▶ **KEY POINT** The skull is composed of the cranial and facial bones. The cranial bones enclose the cranial cavity, which encloses the brain. The facial bones protect and support the eyes and the entrances to the respiratory and digestive systems.

The bones of the head are collectively known as the **skull**. The skull contains 22 bones: 8 form the cranium and 14 are associated with the face **(Figures 6.2 to 6.5)**.

Figure 6.2 Cranial and Facial Subdivisions of the Skull. The skull can be divided into the cranial and the facial divisions. The palatine bones and the inferior nasal conchae of the facial division are not visible from this perspective.

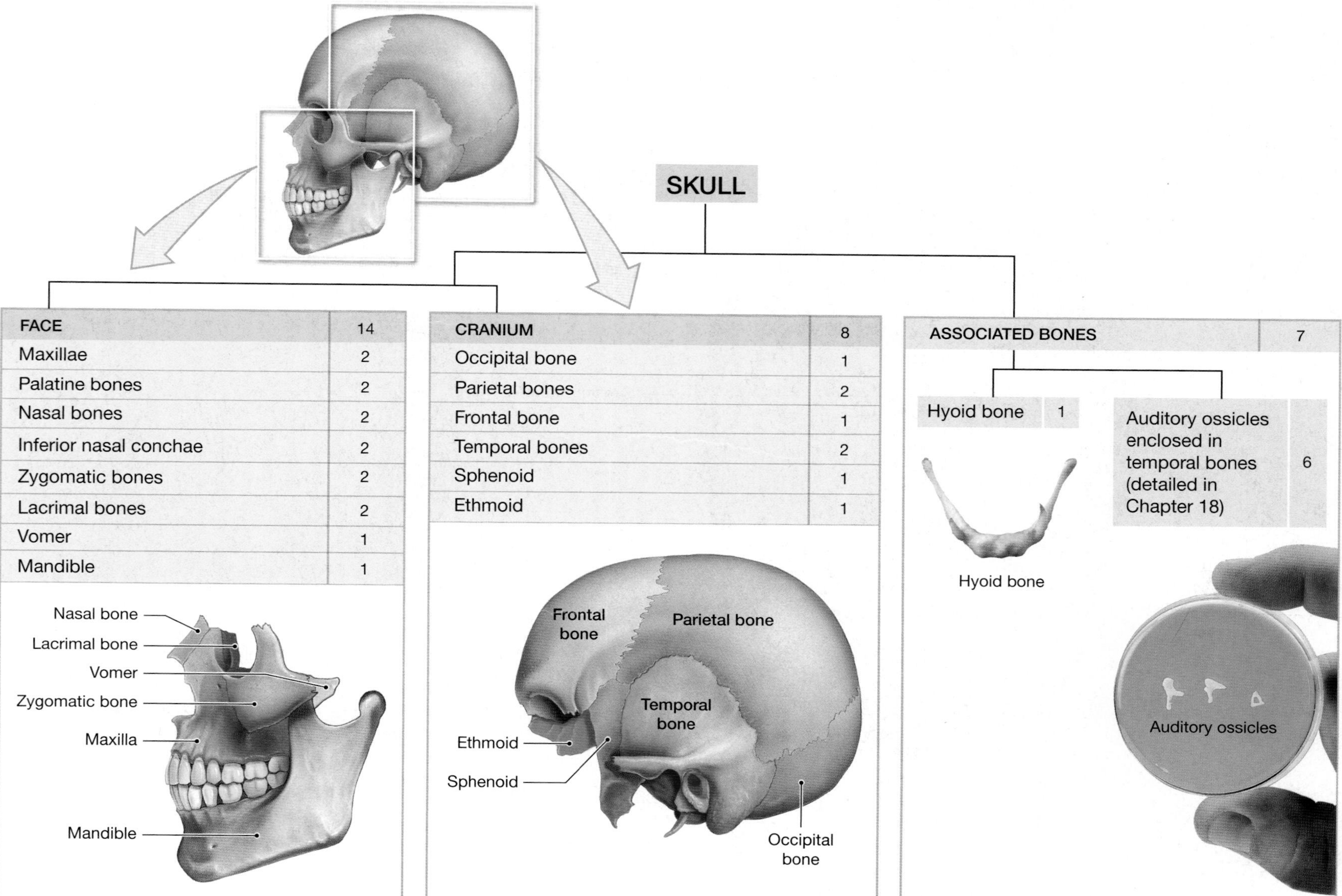

FACE	14
Maxillae	2
Palatine bones	2
Nasal bones	2
Inferior nasal conchae	2
Zygomatic bones	2
Lacrimal bones	2
Vomer	1
Mandible	1

CRANIUM	8
Occipital bone	1
Parietal bones	2
Frontal bone	1
Temporal bones	2
Sphenoid	1
Ethmoid	1

ASSOCIATED BONES	7
Hyoid bone	1
Auditory ossicles enclosed in temporal bones (detailed in Chapter 18)	6

Figure 6.3 The Adult Skull

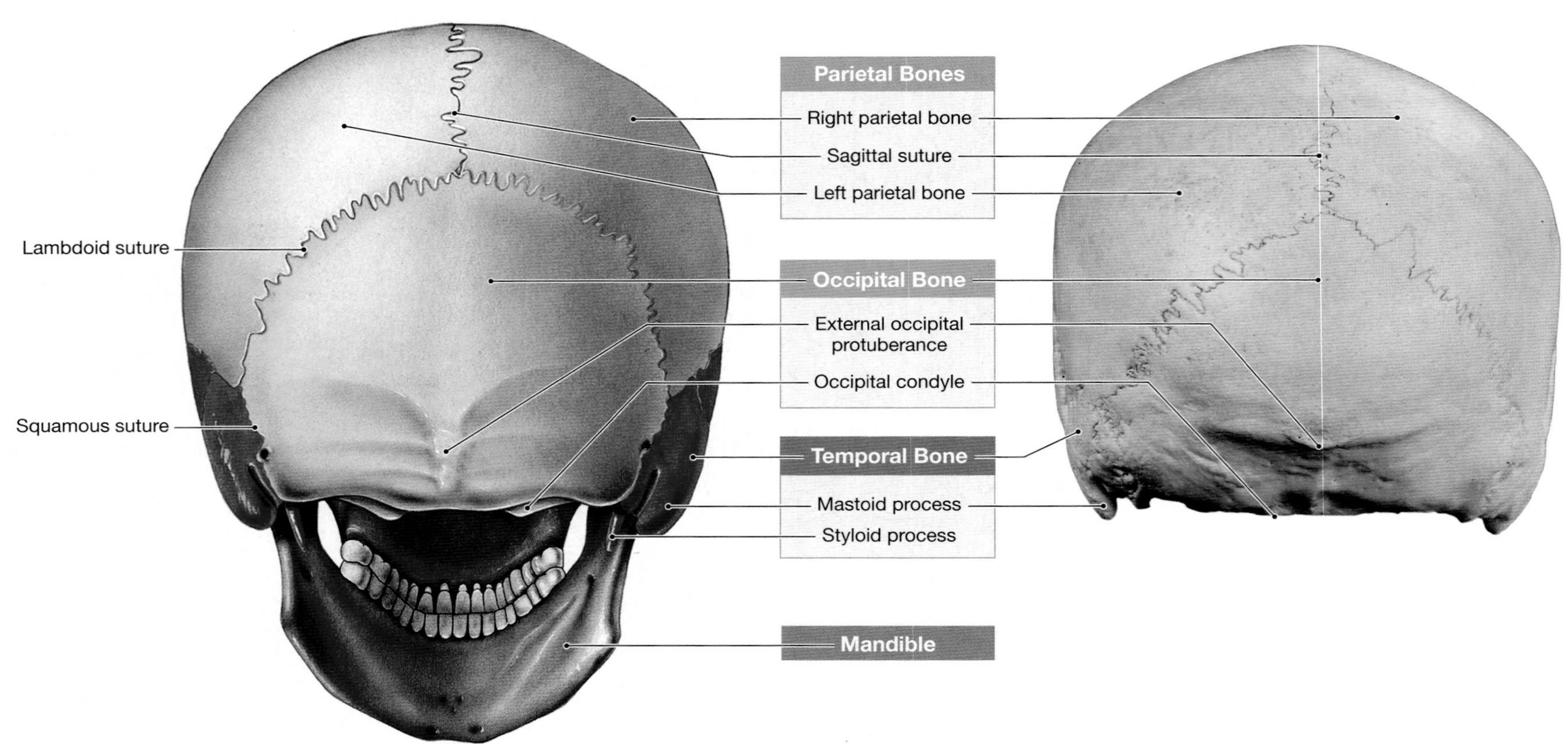

a Posterior view of the adult skull

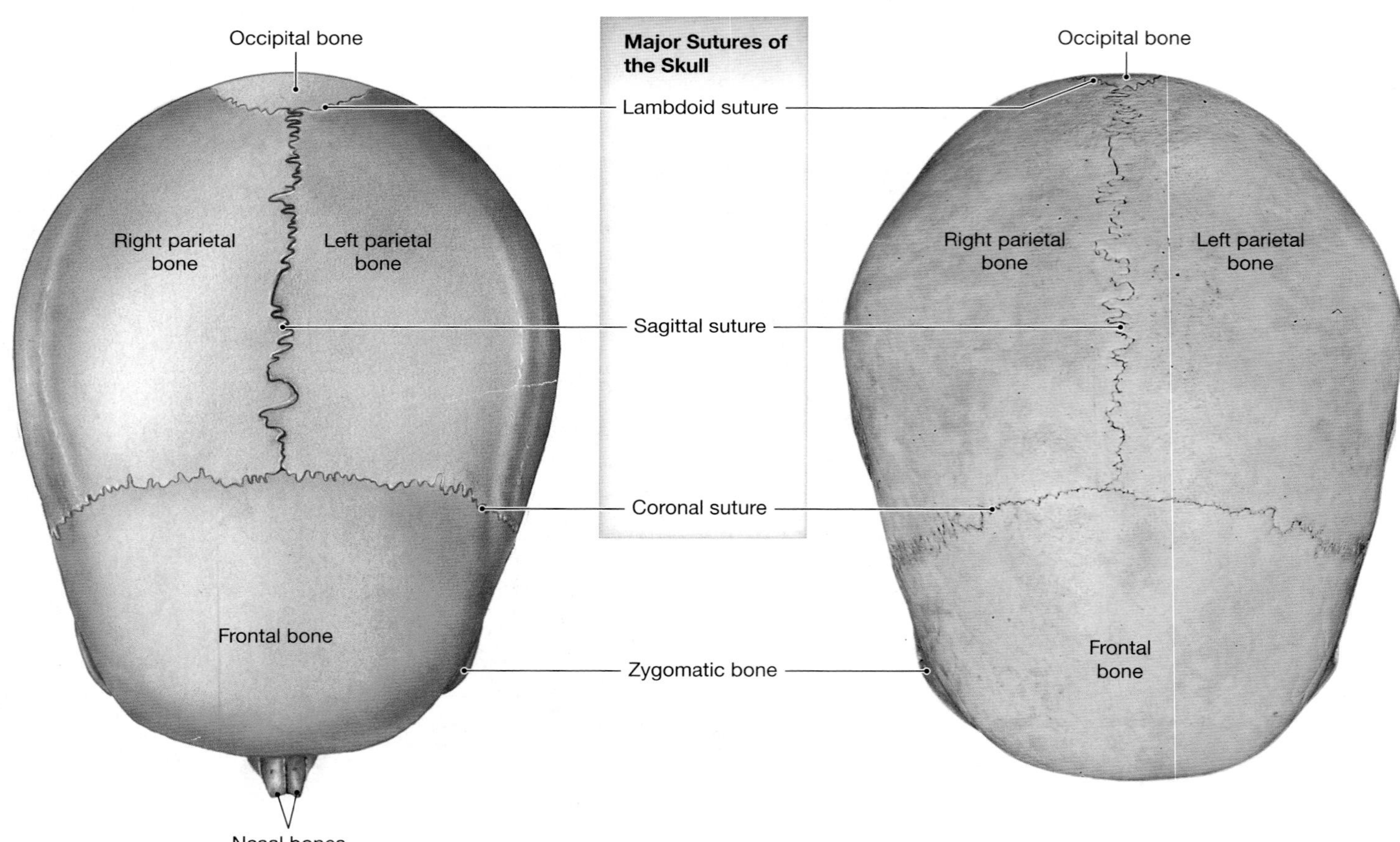

b Superior view of the adult skull

Figure 6.3 (*continued*)

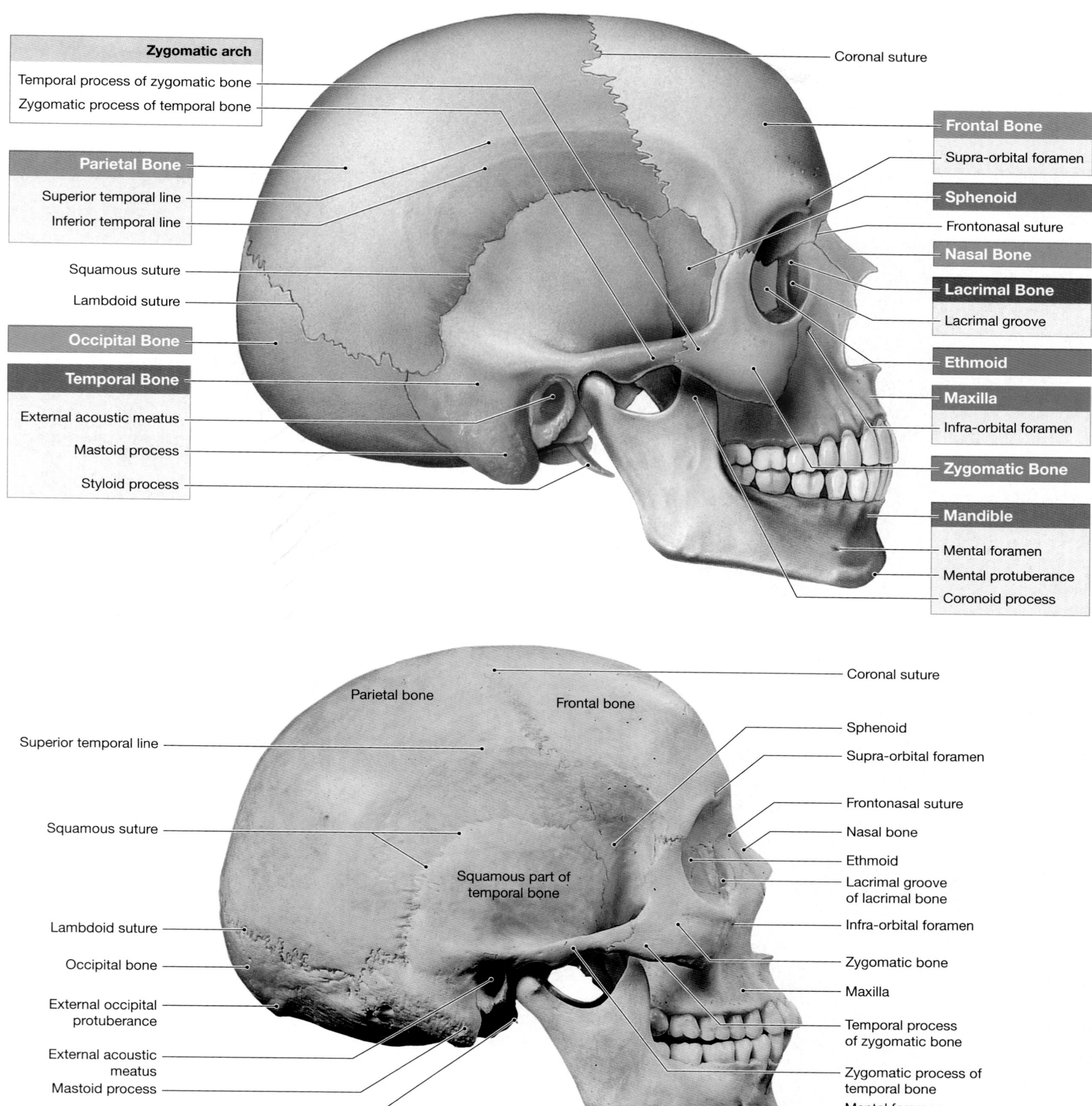

Zygomatic arch
Temporal process of zygomatic bone
Zygomatic process of temporal bone
Parietal Bone
Superior temporal line
Inferior temporal line
Squamous suture
Lambdoid suture
Occipital Bone
Temporal Bone
External acoustic meatus
Mastoid process
Styloid process
Coronal suture
Frontal Bone
Supra-orbital foramen
Sphenoid
Frontonasal suture
Nasal Bone
Lacrimal Bone
Lacrimal groove
Ethmoid
Maxilla
Infra-orbital foramen
Zygomatic Bone
Mandible
Mental foramen
Mental protuberance
Coronoid process
Parietal bone
Frontal bone
Coronal suture
Superior temporal line
Sphenoid
Supra-orbital foramen
Squamous suture
Frontonasal suture
Nasal bone
Ethmoid
Squamous part of temporal bone
Lacrimal groove of lacrimal bone
Lambdoid suture
Infra-orbital foramen
Occipital bone
Zygomatic bone
Maxilla
External occipital protuberance
Temporal process of zygomatic bone
External acoustic meatus
Mastoid process
Zygomatic process of temporal bone
Mental foramen
Styloid process
Mental protuberance

c Lateral view of the adult skull

Figure 6.3 ***(continued)***

Parietal bone
Coronal suture
Frontonasal suture
Temporal bone
Ethmoid
Palatine bone
Lacrimal bone
Temporal process of zygomatic bone
Nasal bone
Middle nasal concha
Perpendicular plate of ethmoid
Inferior nasal concha
Vomer
Mental protuberance

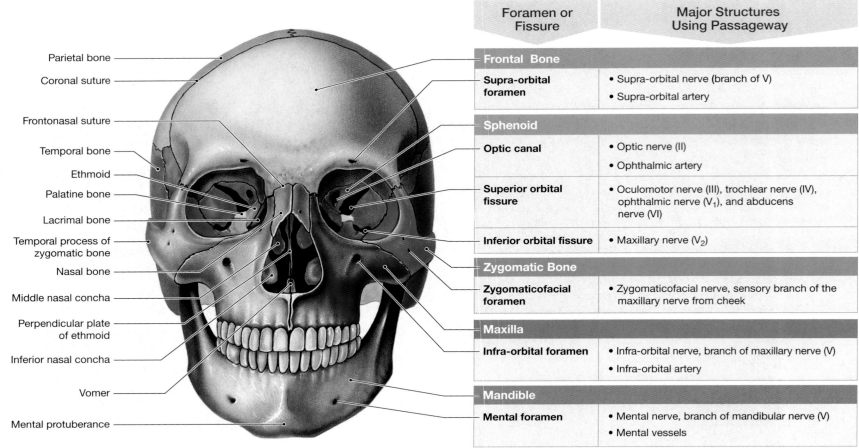

Foramen or Fissure	Major Structures Using Passageway
Frontal Bone	
Supra-orbital foramen	• Supra-orbital nerve (branch of V) • Supra-orbital artery
Sphenoid	
Optic canal	• Optic nerve (II) • Ophthalmic artery
Superior orbital fissure	• Oculomotor nerve (III), trochlear nerve (IV), ophthalmic nerve (V_1), and abducens nerve (VI)
Inferior orbital fissure	• Maxillary nerve (V_2)
Zygomatic Bone	
Zygomaticofacial foramen	• Zygomaticofacial nerve, sensory branch of the maxillary nerve from cheek
Maxilla	
Infra-orbital foramen	• Infra-orbital nerve, branch of maxillary nerve (V) • Infra-orbital artery
Mandible	
Mental foramen	• Mental nerve, branch of mandibular nerve (V) • Mental vessels

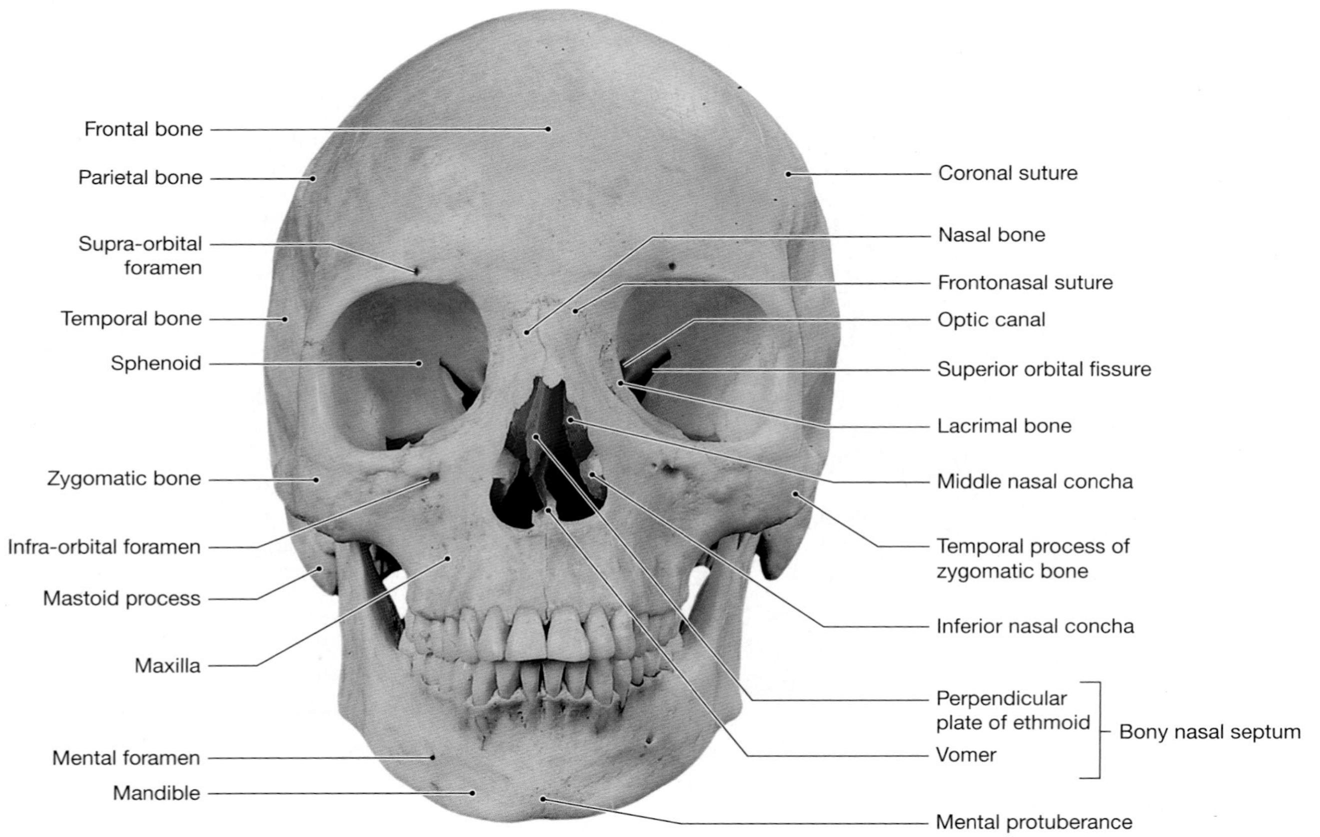

d Anterior view of the adult skull

Figure 6.3 ***(continued)***

Frontal bone
Palatine process of maxilla
Zygomatic bone
Internal naris
Lesser palatine foramen
Zygomatic arch
Plates of pterygoid process
Vomer
Styloid process
Mandibular fossa
Temporal squama
Occipital condyle
Mastoid process
Occipitomastoid suture
Condylar fossa
Lambdoid suture
External occipital protuberance
Superior nuchal line

Foramen or Fissure	Major Structures Using Passageway
Maxilla	
Incisive fossa	• Nasopalatine nerves • Small arteries to the palate
Palatine Bone	
Greater palatine foramen	• Anterior palatine nerve
Sphenoid	
Foramen ovale	• Mandibular nerve (V_3)
Foramen lacerum (with temporal and occipital bones)	• Internal carotid artery after leaving carotid canal • Auditory tube
Temporal Bone	
External acoustic meatus	• Air in meatus conducts sound to eardrum
Carotid canal	• Internal carotid artery
Stylomastoid foramen	• Facial nerve (V)
Occipital Bone	
Foramen magnum	• Medulla oblongata (most caudal portion of brain) • Accessory nerve (XI) • Vertebral arteries
Jugular foramen (with temporal bone)	• Glossopharyngeal, vagus, and accessory nerves (IX, X, XI)

6

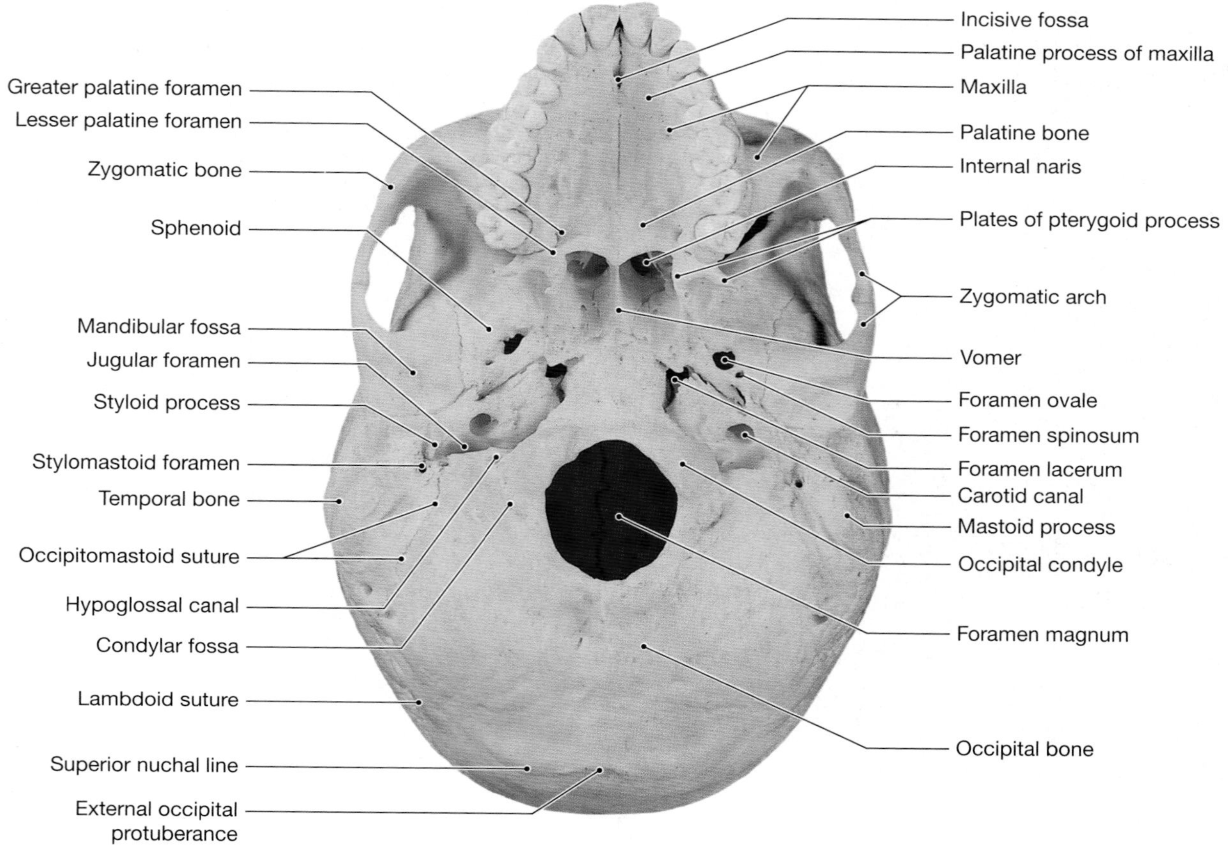

e Inferior view, mandible removed

Figure 6.4 Sectional Anatomy of the Skull, Part I. Horizontal section: A superior view showing major landmarks in the floor of the cranial cavity.

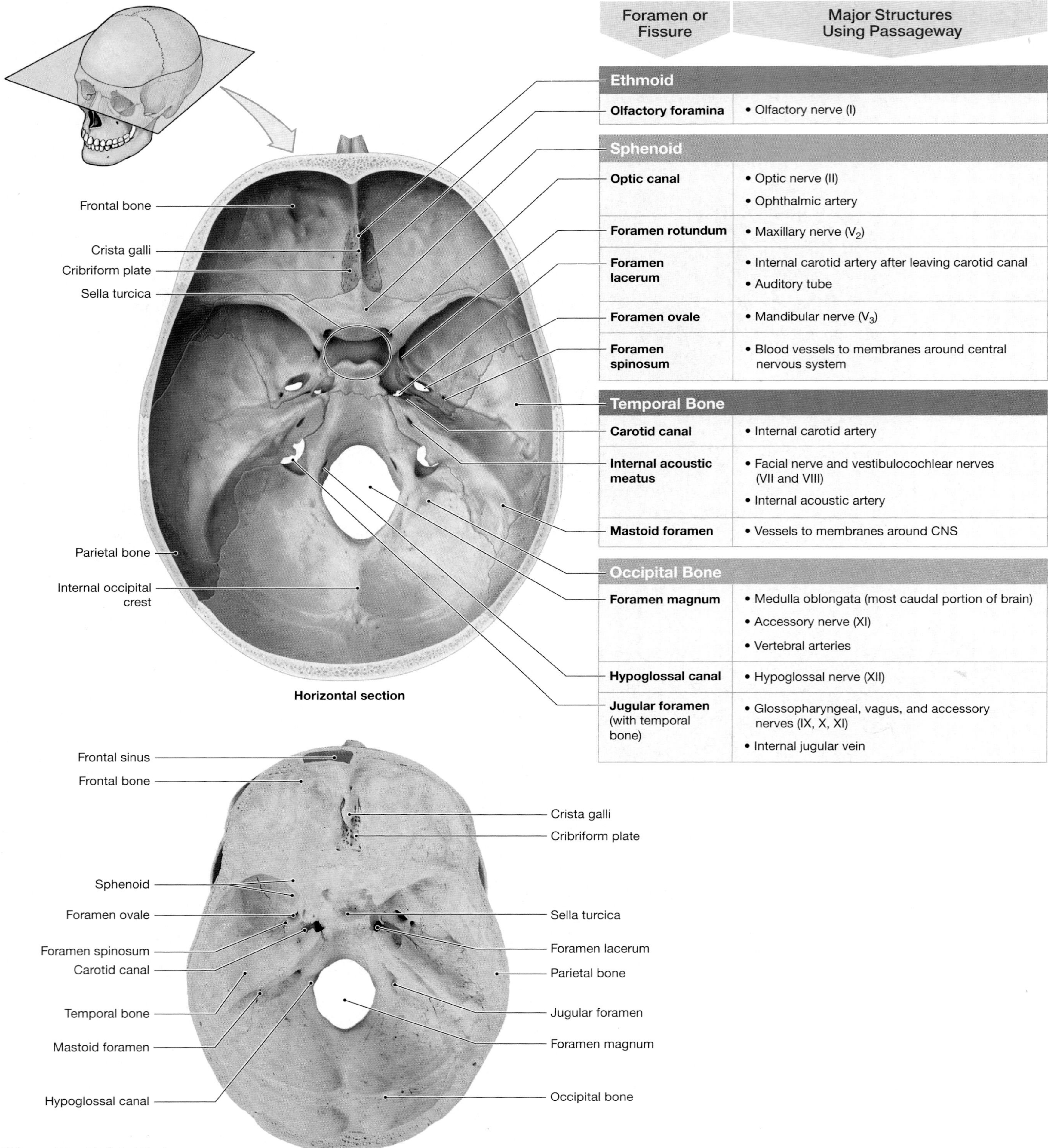

Foramen or Fissure	Major Structures Using Passageway
Ethmoid	
Olfactory foramina	• Olfactory nerve (I)
Sphenoid	
Optic canal	• Optic nerve (II) • Ophthalmic artery
Foramen rotundum	• Maxillary nerve (V_2)
Foramen lacerum	• Internal carotid artery after leaving carotid canal • Auditory tube
Foramen ovale	• Mandibular nerve (V_3)
Foramen spinosum	• Blood vessels to membranes around central nervous system
Temporal Bone	
Carotid canal	• Internal carotid artery
Internal acoustic meatus	• Facial nerve and vestibulocochlear nerves (VII and VIII) • Internal acoustic artery
Mastoid foramen	• Vessels to membranes around CNS
Occipital Bone	
Foramen magnum	• Medulla oblongata (most caudal portion of brain) • Accessory nerve (XI) • Vertebral arteries
Hypoglossal canal	• Hypoglossal nerve (XII)
Jugular foramen (with temporal bone)	• Glossopharyngeal, vagus, and accessory nerves (IX, X, XI) • Internal jugular vein

Figure 6.5 **Sectional Anatomy of the Skull, Part II.** Sagittal section: A medial view of the right half of the skull. Because the bony nasal septum is intact, the right nasal cavity cannot be seen.

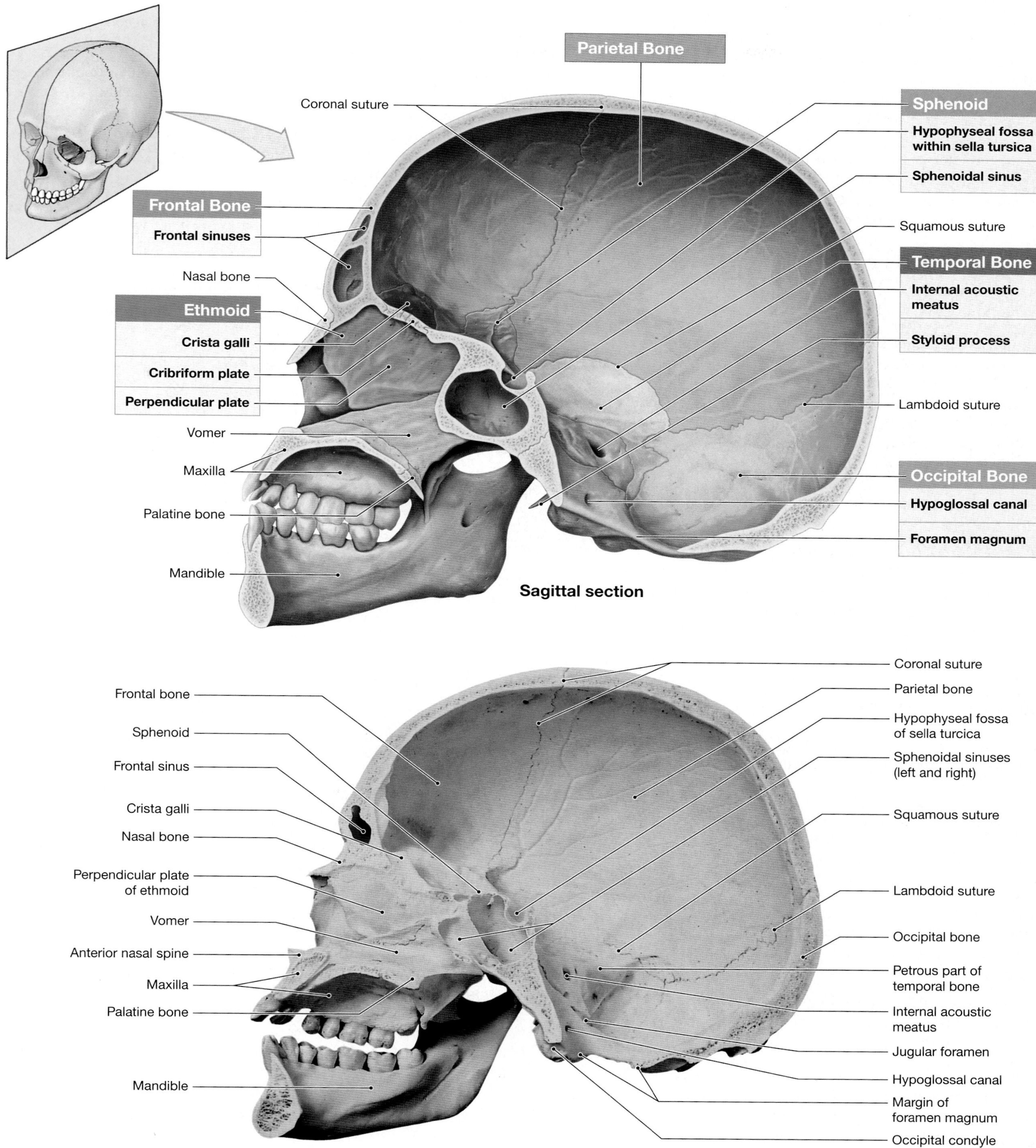

6

The **cranium**, which surrounds and protects the brain, consists of the occipital, parietal, frontal, temporal, sphenoid, and ethmoid bones. These cranial bones enclose the **cranial cavity**, a fluid-filled chamber that cushions and supports the brain. Blood vessels, nerves, and membranes that stabilize the position of the brain attach to the inner surface of the cranium. Its outer surface provides an extensive area for attachment of muscles that move the eyes, jaw, and head. A specialized joint between the occipital bone and the first spinal vertebra both stabilizes the positions of the cranium and vertebral column and allows a wide range of head movements.

If the cranium is the house where the brain resides, the *facial complex* is the front porch. The superficial **facial bones**—the maxillae, palatine, nasal, zygomatic, lacrimal, vomer, and mandible (**Figure 6.2**, p. 133)—provide areas for the attachment of muscles that control facial expressions and assist in manipulating food.

TIPS & TOOLS

- The following mnemonic will help you remember the bones of the face: "Virgil Can Not Make My Pet Zebra Laugh!" **V**omer, **C**onchae, **N**asal, **M**axilla, **M**andible, **P**alatine, **Z**ygomatic, **L**acrimal
- The following mnemonic will help you remember the bones of the cranium: "Old People From Timbuktu Eat Spiders." **O**ccipital, **P**arietal, **F**rontal, **T**emporal, **E**thmoid, **S**phenoid

6.1 CONCEPT CHECK

1 What are the associated bones of the skull, and how many are there?

2 What is the function of the cranial cavity, and what are the anatomical names of the bones that make up the cranial cavity?

See the blue Answers tab at the back of the book.

6.2 | Sutures of the Skull

▶ **KEY POINT** The bones of the skull join at immovable joints called sutures. Five major sutures of the skull are the lambdoid, sagittal, coronal, squamous, and frontonasal sutures.

At a **suture**, dense fibrous connective tissue joins the bones firmly together. At this time, you need to know only the lambdoid, sagittal, coronal, squamous, and frontonasal sutures (**Figure 6.3**, p. 134).

- Lambdoid (lam-DOYD) suture. The **lambdoid suture** arches across the posterior surface of the skull joining the occipital bone to the parietal bones. Often there are one or more **sutural bones** (*Wormian bones*) along this suture; they range from the size of a grain of sand to as large as a quarter.
- Sagittal suture. Posteriorly, the **sagittal suture** begins at the superior midline of the lambdoid suture and extends anteriorly between the parietal bones to the coronal suture.
- Coronal suture. Anteriorly, the sagittal suture ends at the coronal suture. The **coronal suture** crosses the superior surface of the skull, joining the anterior frontal bone to the posterior parietal bones. Together, the occipital, parietal, and frontal bones form the **calvaria** (kal-VAR-ē-a), also called the *cranial vault*.
- Squamous suture. On each side of the skull a **squamous suture** marks the boundary between the temporal bone and the parietal bone of that side.
- Frontonasal suture. The **frontonasal suture** is the boundary between the superior aspects of the two nasal bones and the frontal bone.

Many superficial bumps and ridges in the axial skeleton are associated with the skeletal muscles that are described in Chapter 10; learning the names now will help you master the muscular material in Chapter 10. Use **Table 6.1** (look ahead to p. 142) as a reference for foramina and fissures of the skull and **Table 6.2** (look ahead to p. 143) as a reference for surface features of the skull. These references will be especially important in later chapters dealing with the nervous and cardiovascular systems.

6.2 CONCEPT CHECK

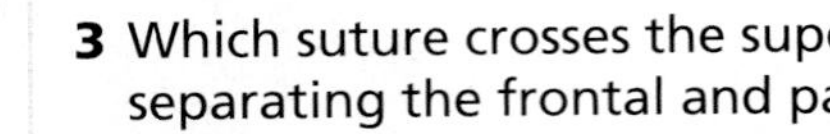

3 Which suture crosses the superior aspect of the skull, separating the frontal and parietal bones?

4 Which suture separates the occipital bone from the parietal bones?

See the blue Answers tab at the back of the book.

6.3 | Bones of the Cranium

▶ **KEY POINT** The eight bones of the cranium determine the shape of the head and protect the brain and the special sense organs related to vision, smell, hearing, and balance.

We now examine each of the bones of the cranium. **Figures 6.3, 6.4**, and **6.5** (pp. 134–139) show the adult skull in superficial and sectional views. (*Refer to Chapter 12*, **Figure 12.1b**, *to identify these anatomical structures from the body surface.*) As we proceed, refer to these figures to develop a three-dimensional perspective on the individual bones.

Occipital Bone

▶ **KEY POINT** The shape of the occipital bone makes the posterior surface of the skull round. The occipital bone is seen from the posterior, lateral, and inferior aspects of the cranium.

The **occipital bone** forms part of the posterior, lateral, and inferior surfaces of the cranium. The inferior surface of the occipital bone contains a large, circular opening, the **foramen magnum** which connects the cranial cavity to the vertebral canal enclosed by the vertebral column. At the adjacent **occipital condyles**, the skull forms two joints with the first cervical vertebra. The posterior, external surface of the occipital bone has a number of prominent ridges. The **external occipital crest** extends posteriorly from the foramen magnum, ending in a small midline bump called the **external occipital protuberance**. Two horizontal ridges, the **inferior** and **superior nuchal** (NU-kal) **lines**, cross the crest. These lines are the attachments for muscles and ligaments that stabilize the joints between the first vertebra and the skull at the occipital condyles; they help balance the weight of the head over the cervical vertebrae of the neck. The occipital bone forms part of the wall of the large **jugular foramen**, a passageway for vital arteries, veins, and nerves. The internal jugular vein passes through this foramen to drain venous blood from the brain. The **hypoglossal canals** begin at the lateral base of each occipital condyle, just superior to the condyles (**Figure 6.6a-c, e**). The hypoglossal nerves, cranial nerves that control the tongue muscles, pass through these canals.

Figure 6.6 **The Occipital and Parietal Bones**

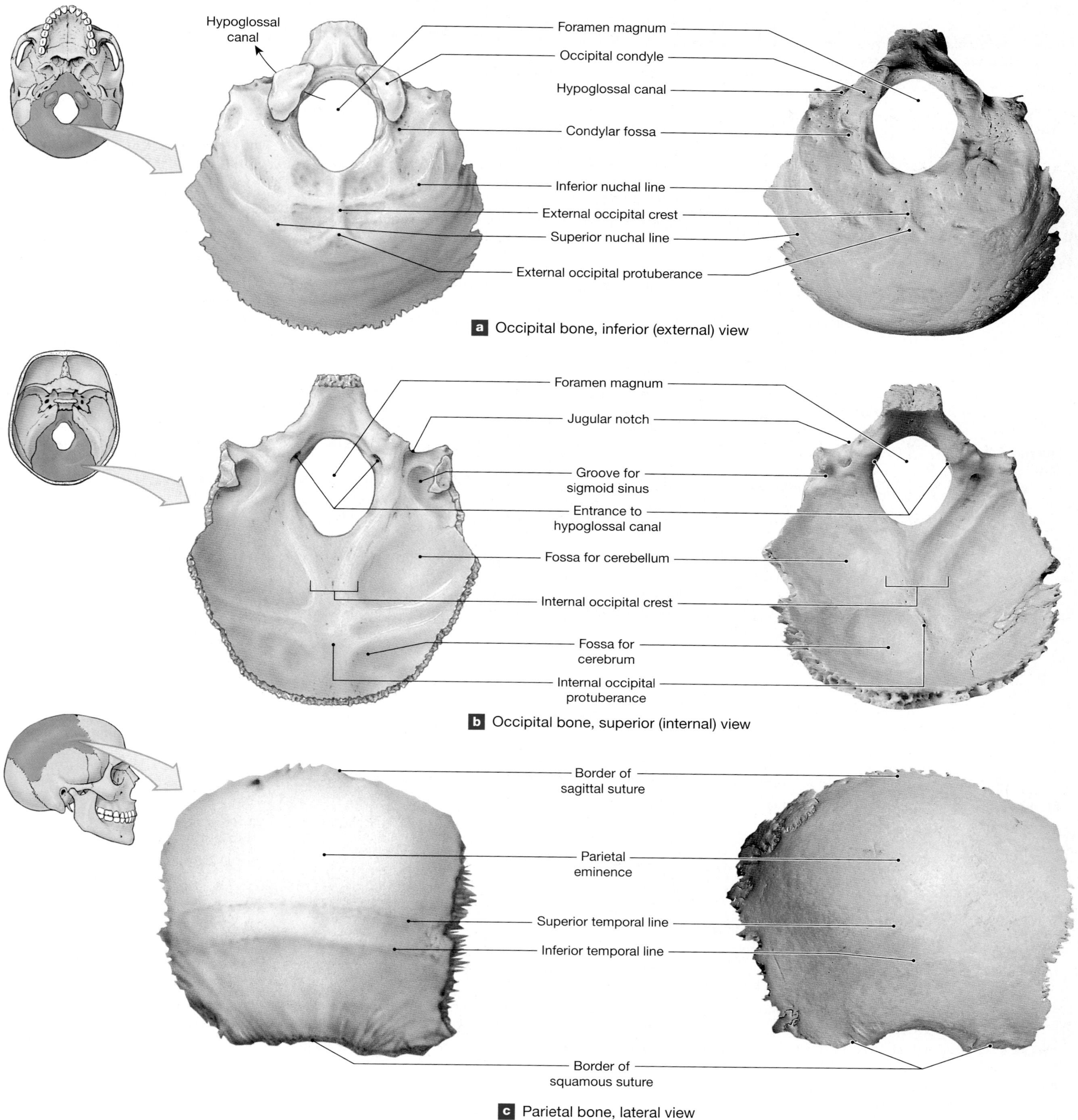
Hypoglossal canal
Foramen magnum
Occipital condyle
Hypoglossal canal
Condylar fossa
Inferior nuchal line
External occipital crest
Superior nuchal line
External occipital protuberance
a Occipital bone, inferior (external) view
Foramen magnum
Jugular notch
Groove for sigmoid sinus
Entrance to hypoglossal canal
Fossa for cerebellum
Internal occipital crest
Fossa for cerebrum
Internal occipital protuberance
b Occipital bone, superior (internal) view
Border of sagittal suture
Parietal eminence
Superior temporal line
Inferior temporal line
Border of squamous suture
c Parietal bone, lateral view

6

Table 6.1 | **A Key to the Foramina and Fissures of the Skull**

Bone	Foramen/Fissure	Major Structures Using Passageway: Neural Tissue	Major Structures Using Passageway: Vessels and Other Structures
Occipital Bone	Foramen magnum	Medulla oblongata (most caudal portion of the brain) and accessory nerve (XI)* controlling several muscles of the back, pharynx, and larynx	Vertebral arteries to brain and supporting membranes around the central nervous system (CNS)
	Hypoglossal canal	Hypoglossal nerve (XII) provides motor control to muscles of the tongue	
With temporal bone	Jugular foramen	Glossopharyngeal nerve (IX), vagus nerve (X), accessory nerve (XI). Nerve IX provides taste sensation; X is important for visceral functions; XI innervates important muscles of the back and neck	Internal jugular vein (important vein returning blood from brain to heart)
Frontal Bone	Supra-orbital foramen (or notch)	Supra-orbital nerve, sensory branch of the ophthalmic nerve, innervating the eyebrow, eyelid, and frontal sinus	Supra-orbital artery delivers blood to same region
Temporal Bone	Mastoid foramen		Vessels to membranes around CNS
	Stylomastoid foramen	Facial nerve (VII) provides motor control of facial muscles	
	Carotid canal		Internal carotid artery (major arterial supply to the brain)
	External acoustic meatus		Air conducts sound to eardrum.
	Internal acoustic meatus	Vestibulocochlear nerve (VIII) from sense organs for hearing and balance. Facial nerve (VII) enters here and exits at stylomastoid foramen	Internal acoustic artery to internal ear
Sphenoid	Optic canal	Optic nerve (II) brings information from the eye to the brain	Ophthalmic artery brings blood into orbit
	Superior orbital fissure	Oculomotor nerve (III), trochlear nerve (IV), ophthalmic nerve (V_1), abducens nerve (VI). Ophthalmic nerve provides sensory information about eye and orbit; other nerves control muscles that move the eye	Ophthalmic vein returns blood from orbit
	Foramen rotundum	Maxillary nerve (V_2) provides sensation from the face	
	Foramen ovale	Mandibular nerve (V_3) controls the muscles that move the lower jaw and provides sensory information from that area	
With temporal and occipital bones	Foramen spinosum		Vessels to membranes around CNS; internal carotid artery leaves carotid canal, passes along superior margin of foramen lacerum
	Foramen lacerum		Vessels to membranes around CNS; internal carotid artery leaves carotid canal, passes along superior margin of foramen lacerum
With maxillae	Inferior orbital fissure	Maxillary nerve (V_2) (See *foramen rotundum* of the sphenoid)	
Ethmoid	Cribriform foramina	Olfactory nerve (I) provides sense of smell	
Maxilla	Infra-orbital foramen	Infra-orbital nerve, maxillary nerve (V_2) from the inferior orbital fissure to face	Infra-orbital artery with the same distribution
	Incisive canals	Nasopalatine nerve	Small arteries to the palatal surface
Zygomatic Bone	Zygomaticofacial foramen	Zygomaticofacial nerve, sensory branch of maxillary nerve to cheek	
Lacrimal Bone	Lacrimal groove, nasolacrimal canal (with maxilla)		Tear duct drains into the nasal cavity
Mandible	Mental foramen	Mental nerve, sensory nerve branch of the mandibular nerve; provides sensation from the chin and lower lip	Mental vessels to chin and lower lip
	Mandibular foramen	Inferior alveolar nerve, sensory branch of the mandibular nerve; provides sensation from the gums, teeth	Inferior alveolar vessels supply the same region

* We are using the classical definition of cranial nerves based on the nerve's anatomical structure as it leaves the brain stem.

Table 6.2 | Surface Features of the Skull

		Surface Features			
Region/Bone	**Articulates with**	**Structures**	**Functions**	**Foramina**	**Functions**
CRANIUM (8)					
Occipital bone (1) **(Figure 6.6)**	Parietal bone, temporal bone, sphenoid	**External:** Occipital condyles	Articulate with first cervical vertebra	Jugular foramen (with temporal)	Carries blood from smaller veins in the cranial cavity
		Occipital crest, external occipital protuberance, and inferior and superior nuchal lines	Attachment of muscles and ligaments that move the head and stabilize the atlanto-occipital joint	Hypoglossal canal	Passageway for hypoglossal nerve that controls tongue muscles
		Internal: Internal occipital crest	Attachment of membranes that stabilize position of the brain		
Parietal bones (2) **(Figure 6.6)**	Occipital, frontal, temporal bones, sphenoid	**External:** Superior and inferior temporal lines	Attachment of major jaw-closing muscle		
		Parietal eminence	Attachment of scalp to skull		
Frontal bone (1) **(Figure 6.7)**	Parietal, nasal, zygomatic bones, sphenoid, ethmoid, maxillae	**External:** Frontal suture	Marks fusion of frontal bones in development	Supra-orbital foramina	Passageways for sensory branch of ophthalmic nerve and supra-orbital artery to the eyebrow and eyelid
		Squamous part	Attachment of scalp muscles		
		Supra-orbital margin	Protects eye		
		Lacrimal fossae	Recesses containing the lacrimal glands		
		Frontal sinuses	Lighten bone and produce mucous secretions		
		Frontal crest	Attachment of stabilizing membranes (meninges) within the cranium		
Temporal bones (2) **(Figure 6.8)**	Occipital, parietal, frontal, zygomatic bones, sphenoid and mandible; enclose auditory ossicles and suspend hyoid bone by stylohyoid ligaments	**External:** Squamous part: Squama	Attachment of jaw muscles	**External:** Carotid canal	Entryway for carotid artery bringing blood to the brain
		Mandibular fossa and articular tubercle	Form articulation with mandible	Stylomastoid foramen	Exit for nerve that controls facial muscles
		Zygomatic process	Articulates with zygomatic bone	Jugular foramen (with occipital bone)	Carries blood from smaller veins in the cranial cavity
		Petrous part: Mastoid process	Attachment of muscles that extend or rotate head	External acoustic meatus	Entrance and passage to tympanum
		Styloid process	Attachment of stylohyoid ligament and muscles attached to hyoid bone	Mastoid foramen	Passage for blood vessels to membranes of brain
		Internal: Mastoid cells	Lighten mastoid process	**External:** Foramen lacerum between temporal and occipital bones	Passage for nerves and arteries to the inner surface of the cranium
		Petrous part	Protects middle and internal ear	**Internal:** Auditory tube	Connects air space of middle ear with pharynx
				Internal acoustic meatus	Passage for blood vessels and nerves to the internal ear and stylomastoid foramen

Table 6.2 | **Surface Features of the Skull (*continued*)**

		Surface Features			
Region/Bone	**Articulates with**	**Structures**	**Functions**	**Foramina**	**Functions**
Sphenoid (1) (Figure 6.9)	Occipital, frontal, temporal, parietal, zygomatic, palatine bones, maxillae, ethmoid, and vomer	**Internal:** Sella turcica	Protects pituitary gland	Optic canal	Passage of optic nerve
		Anterior and posterior clinoid processes, optic groove	Protects pituitary gland and optic nerve	Superior orbital fissure	Entrance for nerves that control eye movements
		External: Pterygoid processes and spines	Attachment of jaw muscles	Foramen rotundum	Passage for sensory nerves from face
				Foramen ovale	Passage for nerves that control jaw movement
				Foramen spinosum	Passage of vessels to membranes around brain
Ethmoid (1) (Figure 6.10)	Frontal, nasal, palatine, lacrimal bones, sphenoid, maxillae, and vomer	Crista galli	Attachment of membranes that stabilize position of brain	Cribriform foramina	Passage of olfactory nerves
		Ethmoidal labyrinth	Lightens bone, site of mucus production		
		Superior and middle conchae	Create turbulent airflow		
		Perpendicular plate	Separates nasal cavities (with vomer and nasal cartilage)		
FACE (14)					
Maxillae (2) (Figure 6.12)	Frontal, zygomatic, palatine, lacrimal bones, sphenoid, ethmoid, and inferior nasal concha	Orbital margin	Protects eye	Inferior orbital fissure and infra-orbital foramen	Exit for nerves entering skull at foramen rotundum
		Palatine process	Forms most of the bony palate	Greater and lesser palatine foramina	Passage of sensory nerves from face
		Maxillary sinus	Lightens bone, secretes mucus	Nasolacrimal canal (with lacrimal bone)	Drains tears from lacrimal sac to nasal cavity
		Alveolar process	Surrounds articulations with teeth		
Palatine bones (2) (Figure 6.13)	Inferior nasal conchae, sphenoid, maxillae, and vomer		Contribute to bony palate and orbit		
Nasal bones (2) (Figures 6.3c,d and 6.15)	Frontal bone, ethmoid, maxillae		Support bridge of nose		
Vomer (1) (Figures 6.3d,e, 6.5, 6.16)	Ethmoid, maxillae, palatine bones		Forms inferior and posterior part of nasal septum		
Inferior nasal conchae (2) (Figures 6.3d and 6.16)	Maxillae and palatine bones		Create turbulent airflow		
Zygomatic bones (2) (Figures 6.3c,d and 6.15)	Frontal and temporal bones, sphenoid, maxillae	Temporal process	With zygomatic process of temporal, complete zygomatic arch for attachment of jaw muscles		
Lacrimal bones (2) (Figures 6.3c,d and 6.15)	Ethmoid, frontal bones, maxillae, inferior nasal conchae			Nasolacrimal groove	Contains lacrimal sac

Table 6.2 | **Surface Features of the Skull (*continued*)**

		Surface Features			
Region/Bone	**Articulates with**	**Structures**	**Functions**	**Foramina**	**Functions**
Mandible (1) (Figure 6.14)	Temporal bones	Ramus		Mandibular foramen	Passage for sensory nerve from teeth and gums
		Condylar process	Articulates with temporal bone		
		Coronoid process	Attachment of temporalis muscle from parietal surface	Mental foramen	Passage for sensory nerve from chin and lips
		Alveolar part	Protects articulations with teeth		
		Mylohyoid line	Attachment of muscle supporting floor of mouth		
		Submandibular fossa	Protects submandibular salivary gland		
ASSOCIATED BONES (7)					
Hyoid bone (1) (Figure 6.17)	Suspended by ligaments from styloid process of temporal bone; connected by ligaments to larynx	Greater horns	Attachment of tongue muscles and ligaments to larynx		
		Lesser horns	Attachment of stylohyoid ligaments		
Auditory ossicles (6) (Figure 6.2)	Three are enclosed by the petrous part of each temporal bone		Conduct sound vibrations from tympanic membrane to fluid-filled chambers of internal ear		

Inside the skull, the hypoglossal canals begin on the internal surface of the occipital bone near the foramen magnum **(Figure 6.6b)**. Note the concave internal surface of the occipital bone, which closely follows the contours of the brain. The grooves mark the paths of major vessels, and the ridges are the attachment sites of membranes (the meninges) that stabilize the position of the brain.

Parietal Bones

▶ **KEY POINT** The two parietal bones form most of the superior and lateral surfaces of the cranium.

The paired **parietal** (pa-RĪ-e-tal) **bones** form the major part of the calvaria **(Figure 6.3b,c)**. The external surface of each parietal bone bears a pair of low ridges, the **superior** and **inferior temporal lines (Figure 6.6c)**. These lines indicate the attachment of the large temporalis muscle, which closes the mouth. The smooth parietal surface superior to these lines is called the **parietal eminence**. The internal surfaces of the parietal bones have impressions of the cranial veins and arteries that branch inside the cranium **(Figure 6.5)**.

Frontal Bone

▶ **KEY POINT** The frontal bone forms the forehead and the superior portions of the orbit of the eyes.

Figure 6.3b–d shows the **frontal bone**. During development, the cranial bones form through the fusion of separate ossification centers, and the fusions are not yet complete at birth. At this time two frontal bones join along the **frontal** (*metopic*) **suture**. Although the suture usually disappears by age 8 with the fusion of the bones, the frontal bones of some adults retain some sign of this suture line.

The coronal suture, or what remains of it, runs down the center of the **frontal eminence** of the frontal bone **(Figure 6.7a)**. The anterior surface of the frontal bone is the **squamous part**, or forehead. The lateral surfaces contain the anterior continuations of the **superior temporal lines (Figures 6.3c and 6.7a)**. The frontal part of the frontal bone ends at the **supra-orbital margins**, which are the superior limits of the orbits. Above the supra-orbital margins are thickened ridges, the **superciliary arches**, which support the eyebrows. At the center of each margin is a single **supra-orbital foramen**, or *supra-orbital notch*.

The **orbital part** of the frontal bone forms the roof of each orbit. The inferior surface of the orbital part is smooth and contains small openings for blood vessels and nerves going to or from structures in the orbit. This region is called the orbital surface of the frontal bone. The shallow **lacrimal fossa** houses the lacrimal gland, which forms tears that lubricate the surface of the eye **(Figure 6.7b)**.

The internal surface of the frontal bone conforms to the shape of the anterior portion of the brain **(Figures 6.4 and 6.7c)**. This region has a prominent **frontal crest** for the attachment of membranes that prevent the delicate brain tissues from touching the cranial bones.

The **frontal sinuses** vary in size and usually develop after age 6, but some people never develop them at all. The frontal sinuses and other sinuses are described in a later section **(Figures 6.5 and 6.7b)**.

Temporal Bones

▶ **KEY POINT** The two temporal bones, on the lateral sides of the skull, form part of the cranium and a portion of the "cheekbones."

The paired **temporal bones** form part of the lateral and inferior walls of the cranium and the zygomatic arches of the cheek. They are easily seen on the

Figure 6.7 The Frontal Bone

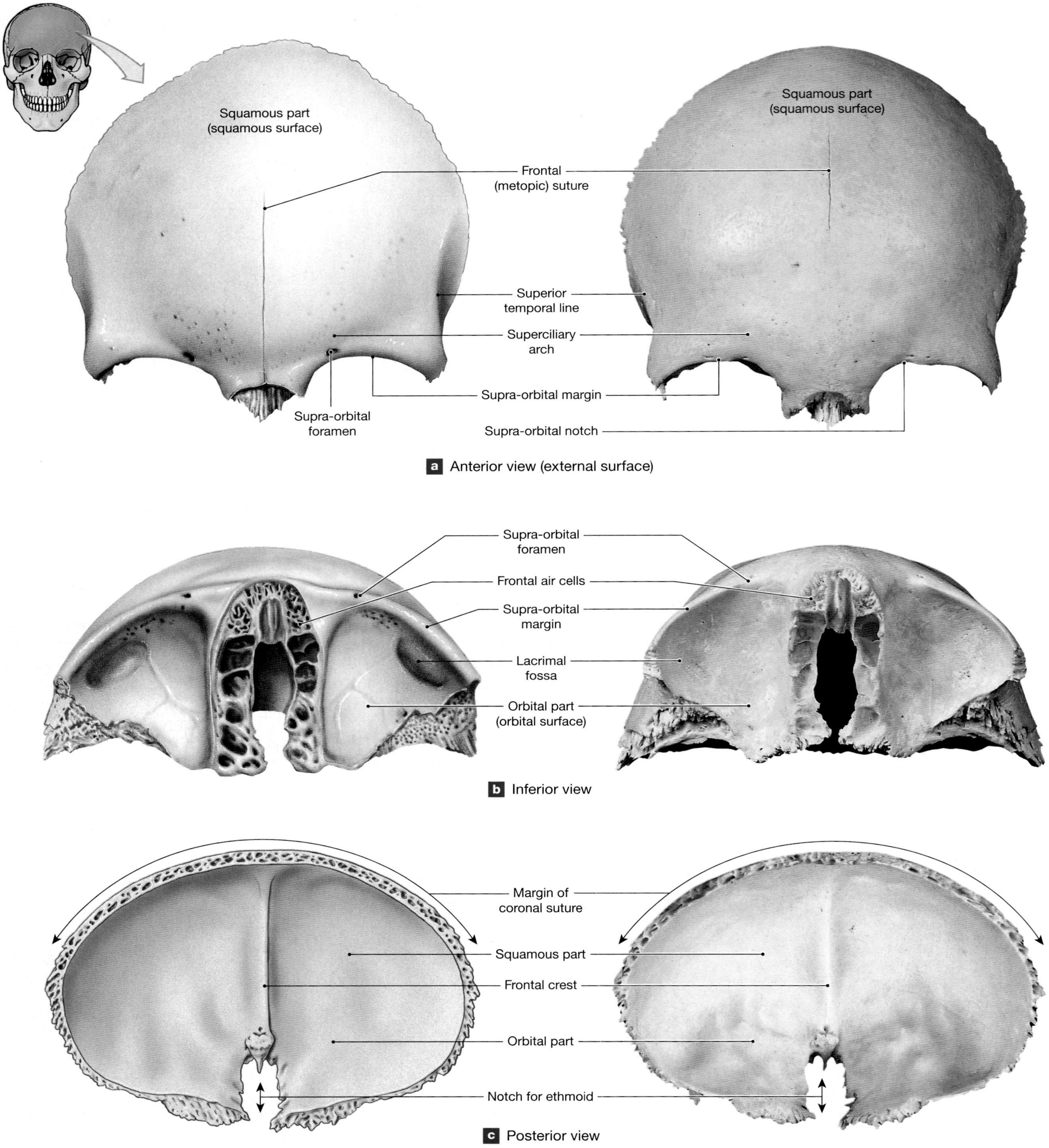

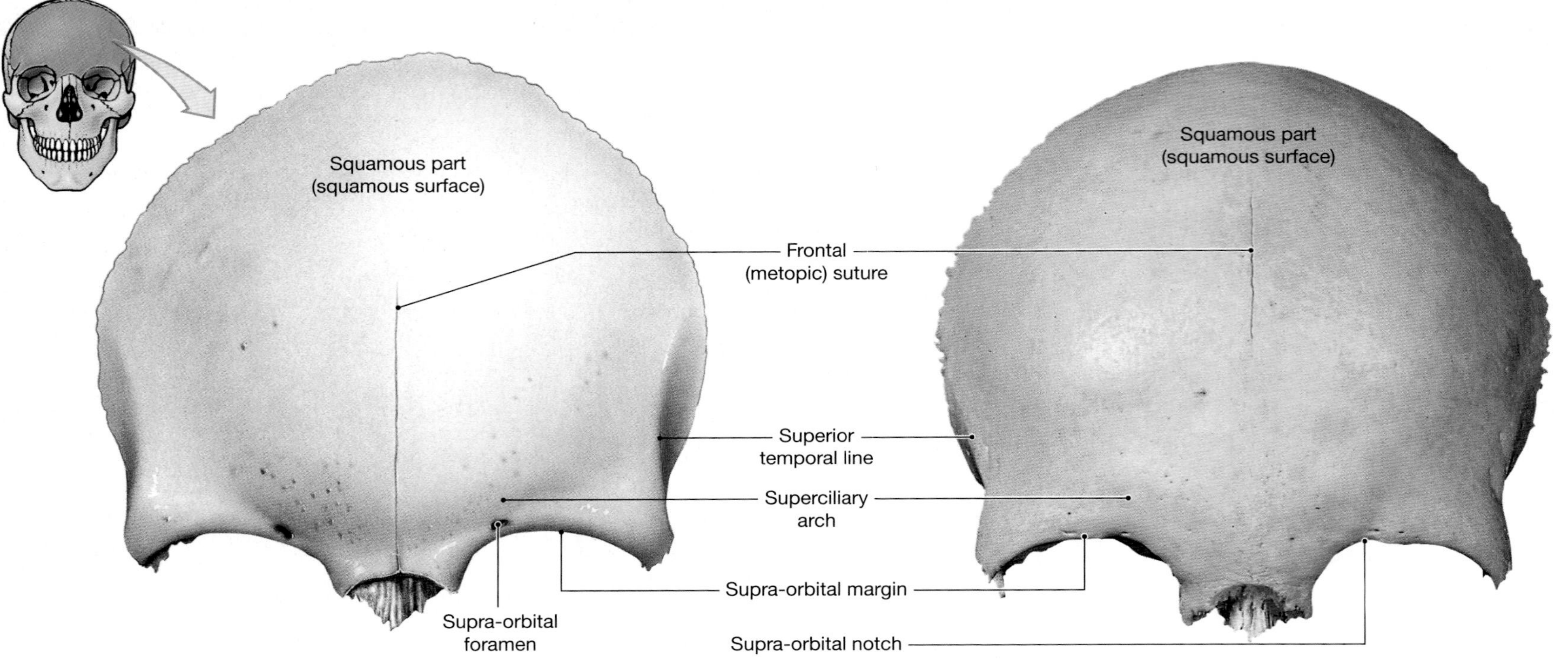

a Anterior view (external surface)

Supra-orbital foramen

Frontal air cells

Supra-orbital margin

Lacrimal fossa

Orbital part (orbital surface)

b Inferior view

Margin of coronal suture

Squamous part

Frontal crest

Orbital part

Notch for ethmoid

c Posterior view

lateral sides of the cranium. They also form the only articulations with the mandible and protect the sense organs of the internal ear. In addition, the parts of the temporal bone inferior to each parietal bone form a broad area for the attachment of muscles that close the jaws and move the head. The temporal bones articulate with the zygomatic, parietal, and occipital bones and with the sphenoid and mandible. Each temporal bone has squamous, tympanic, and petrous parts.

The **squamous part** of the temporal bone is the lateral surface bordering the squamous suture (**Figure 6.8a,d**). The prominent **zygomatic process** forms the inferior margin of the squamous part. The zygomatic process curves laterally and anteriorly to meet the **temporal process** of the zygomatic bone. Together, these processes form the **zygomatic arch**, or *cheekbone*. Inferior to the base of the zygomatic process, the temporal bone articulates with the mandible. A depression called the **mandibular fossa** and an elevated **articular tubercle** mark this site (**Figure 6.8a,c**).

Immediately posterior and lateral to the mandibular fossa is the **tympanic part** of the temporal bone (**Figure 6.8b**). This region surrounds the entrance to the **external acoustic meatus**. This passageway ends at the delicate **tympanic membrane**, or *eardrum*. (The tympanic membrane disintegrates during preparation of a dried skull and thus cannot be seen.)

The **petrous part** (*petrous*, stone) of the temporal bone surrounds and protects the sense organs of hearing and balance. On the lateral surface, the bulge just posterior and inferior to the external acoustic meatus is the **mastoid process** (**Figure 6.8a–d**). (*Refer to Chapter 12,* **Figure 12.1b,c**, *to identify this structure from the body surface.*) This process is an attachment site for muscles that rotate or extend the head. The mastoid process contains numerous interconnected mastoid sinuses, termed mastoid cells. Mastoiditis, an inflammation of the mastoid cells, results when infection from the respiratory tract spreads to this area. Several other landmarks on the petrous part of the temporal bone are seen on its inferior surface. Near the base of the mastoid process, the **mastoid foramen**

Figure 6.8 The Temporal Bone. Major anatomical landmarks are shown on a right temporal bone.

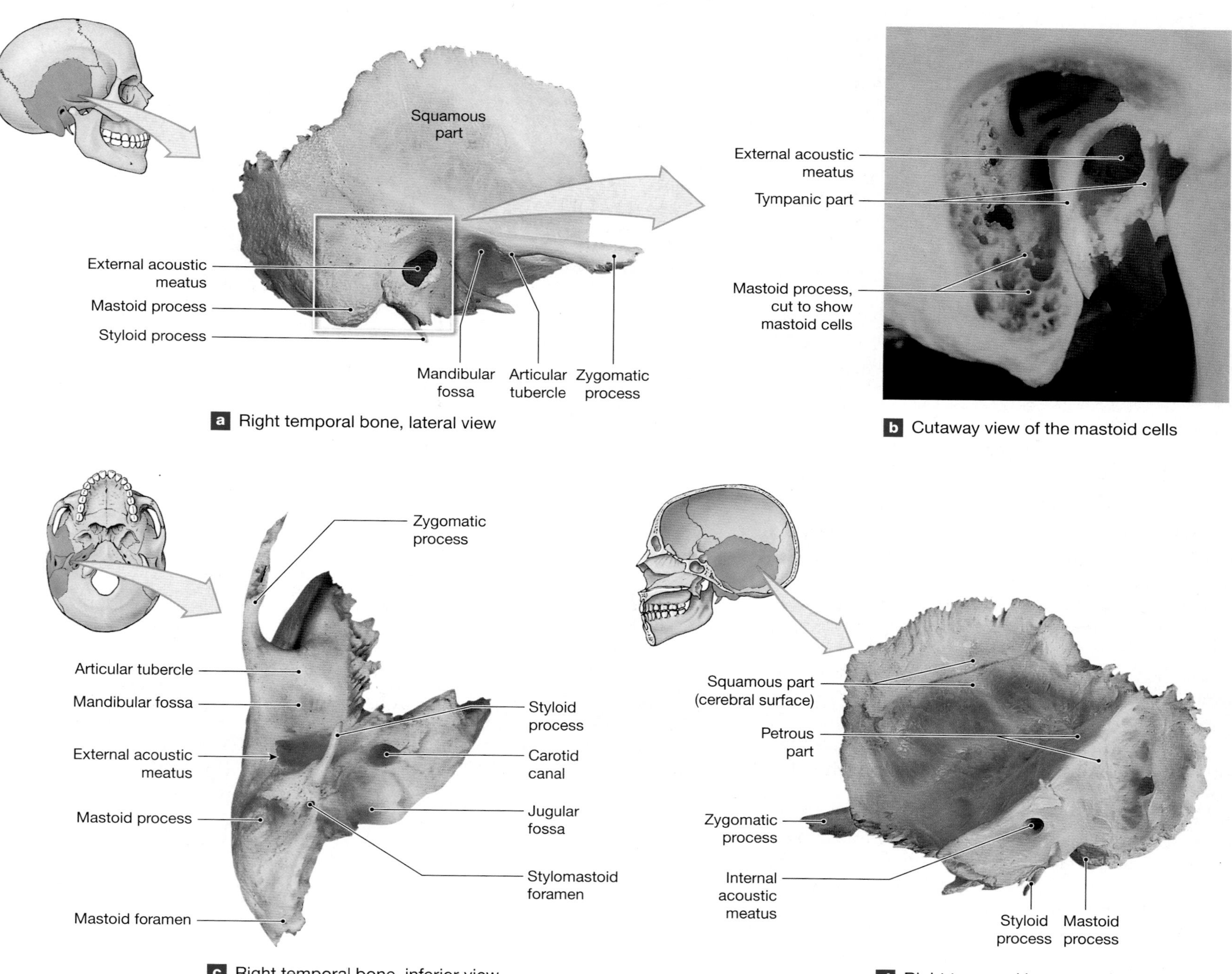

penetrates the temporal bone. Blood vessels travel through this passageway and reach the membranes surrounding the brain. Ligaments supporting the hyoid bone attach to the sharp **styloid process** (STĪ-loyd; *stylos*, pillar), as do muscles of the tongue, pharynx, and larynx. The **stylomastoid foramen** lies posterior to the base of the styloid process. The facial nerve passes through this foramen to control the facial muscles. Medially, the **jugular fossa** and **jugular foramen** are found at the junction of the temporal and occipital bones (**Figure 6.8c**). Anterior and slightly medial to the jugular foramen is the entrance to the **carotid canal**. The internal carotid artery, a major artery supplying blood to the brain, penetrates the skull through this passageway. Anterior and medial to the carotid canal, a jagged slit, the **foramen lacerum** (LA-se-rum; *lacerare*, to tear), extends between the occipital and temporal bones. In a living person, this space contains hyaline cartilage and small arteries supplying the inner surface of the cranium.

Lateral and anterior to the carotid canal, the temporal bone articulates with the sphenoid. A small musculotubal canal begins at that articulation and ends inside the temporal bone (**Figure 6.8c**). This canal surrounds the **auditory tube**. The auditory tube (or *eustachian* [yū-STĀ-shan] *tube*, or *pharyngotympanic tube*) is an air-filled passageway that begins at the pharynx and ends at the **tympanic cavity**, or *middle ear*, a chamber inside the temporal bone. The tympanic cavity contains the **auditory ossicles**, or *ear bones*. These tiny bones transfer sound vibrations from the tympanic membrane (eardrum) toward receptors in the internal ear, which provides the sense of hearing.

The petrous part dominates the medial surface of the temporal bone (**Figure 6.8d**). The **internal acoustic meatus** carries blood vessels and nerves to the internal ear and carries the facial nerve to the stylomastoid foramen. The medial surface of the temporal bone features grooves that indicate the location of blood vessels that pass along the internal surface of the cranium. The sharp ridge on the internal surface of the petrous part is the attachment point for a membrane that helps stabilize the brain's position within the skull.

Sphenoid

▶ **KEY POINT** The sphenoid has a complex shape and contacts every other bone of the cranium.

The **sphenoid**, or *sphenoidal bone*, extends from one side to the other across the floor of the cranium. Although it is relatively large, much of the sphenoid lies deep to several superficial bones. The sphenoid unites all the cranial and facial bones: It articulates with the frontal, occipital, parietal, ethmoid, and temporal bones of the cranium and with the palatine bones, zygomatic bones, maxillae, and vomer of the facial complex (**Figure 6.9**). The sphenoid acts as a cross-brace that strengthens the sides of the skull.

The general shape of the sphenoid resembles a giant bat with its wings extended (**Figures 6.9a**). The **body** is the central portion of the bone. A central depression between the wings cradles the pituitary gland below the brain. This depression is the **hypophysial** (hī-pō-FIZ-ē-al) **fossa**, and its bony enclosure is the **sella turcica** (TUR-si-ka) ("Turkish saddle"). A rider facing forward would grasp the **anterior clinoid** (KLĪ-noyd) **processes** on either side. The anterior clinoid processes are posterior projections of the **lesser wings** of the sphenoid. The **tuberculum sellae** forms the anterior border of the sella turcica; the **dorsum sellae** forms the posterior border. A **posterior clinoid process** extends laterally on each side of the dorsum sellae. The lesser wings are triangular, and the superior surfaces support the frontal lobe of the brain. The inferior surfaces form part of the orbit of the eye and the superior part of the **superior orbital fissure**, a passageway for blood vessels and cranial nerves of the eye.

The transverse groove crossing the floor of the cranium anterior to the saddle, above the level of the seat, is the **optic groove**. At each end of this groove is an **optic canal**. The optic nerves that carry visual information from the eyes to the brain travel through these canals. On each side of the sella turcica, the **foramen rotundum**, the **foramen ovale** (ō-VAH-lē), and the **foramen spinosum** penetrate the **greater wings** of the sphenoid. These passages carry blood vessels and cranial nerves to structures of the orbit, face, and jaws. Posterior and lateral to these foramina, the greater wings end at a sharp

Figure 6.9 The Sphenoid. Views of the sphenoid showing major anatomical landmarks.

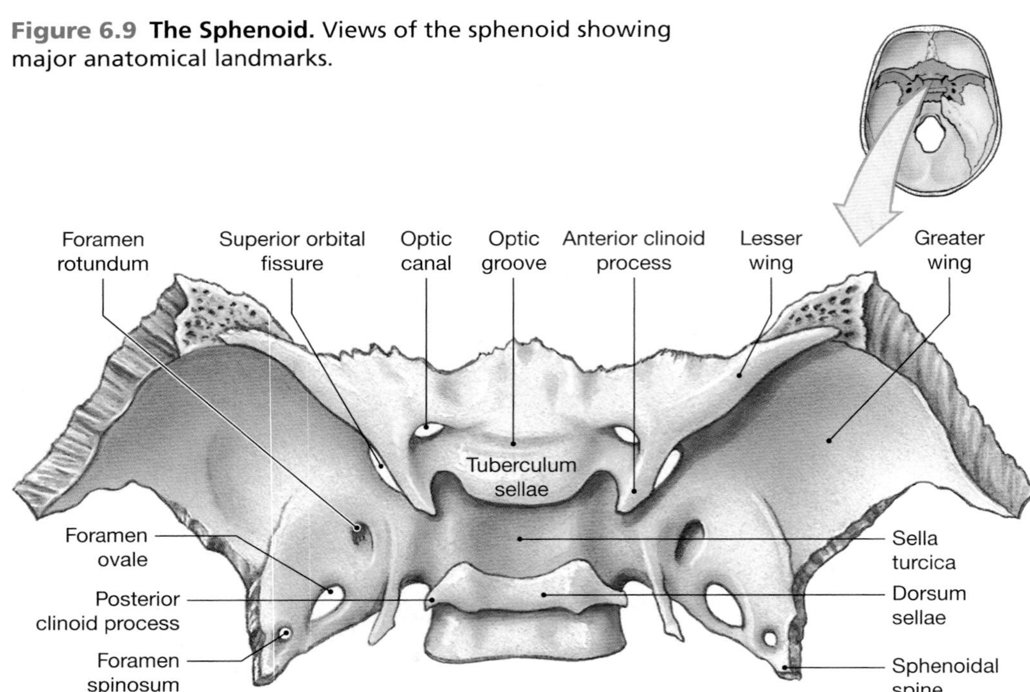

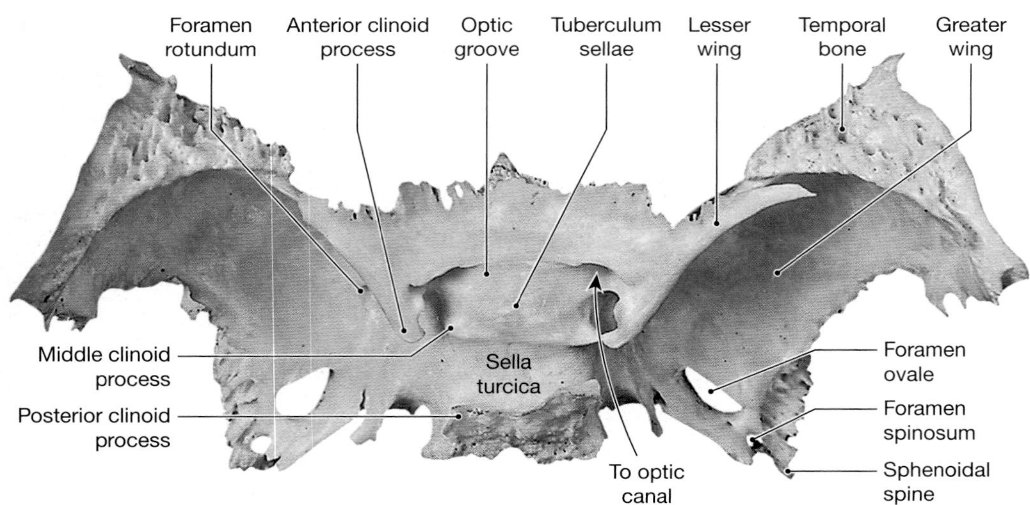

a Superior surface

sphenoidal spine. The superior orbital fissures and the left and right foramen rotundum are also seen in an anterior view **(Figure 6.9b)**.

The **pterygoid processes** (TER-i-goyd; *pterygion*, wing) of the sphenoid are vertical projections beginning at the boundary between the greater and lesser wings. Each process forms a pair of plates, the lateral and medial pterygoid plates, which are important attachment sites for the muscles that move the lower jaw and soft palate. At the base of each pterygoid process, the **pterygoid canal** is a route for a small nerve and an artery supplying the soft palate and adjacent structures.

Ethmoid

▶ KEY POINT The ethmoid, like the sphenoid, has a highly complex shape. The ethmoid makes up part of the nasal cavity and nasal septum and forms part of the orbit of the eye.

Figure 6.9 ***(continued)***

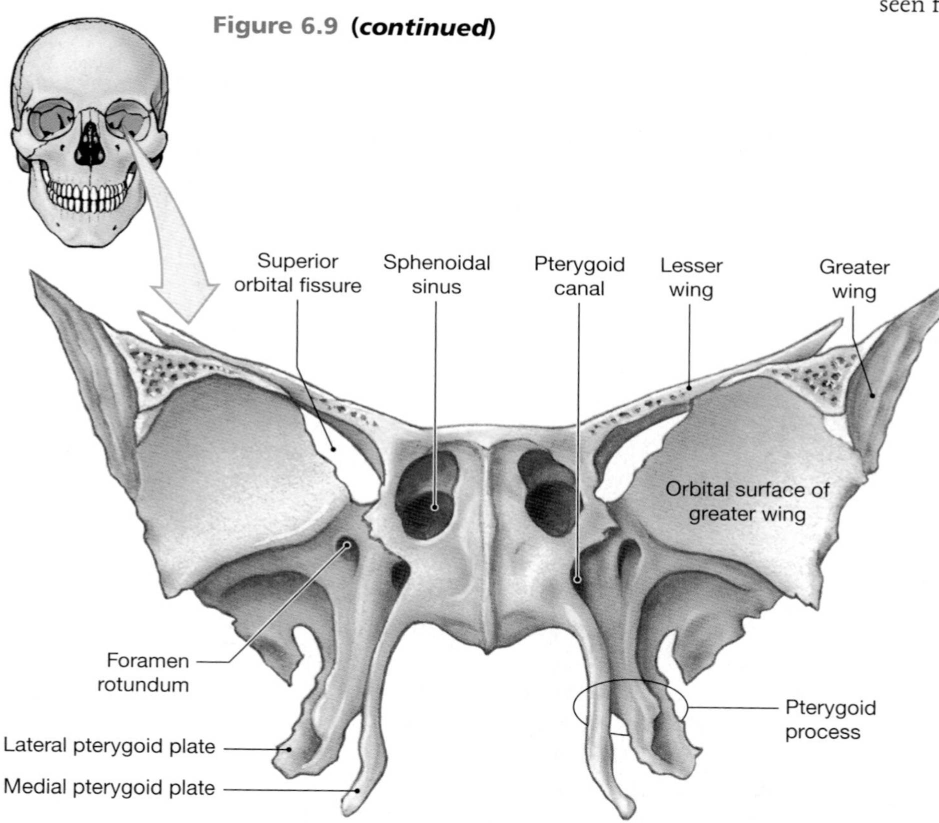

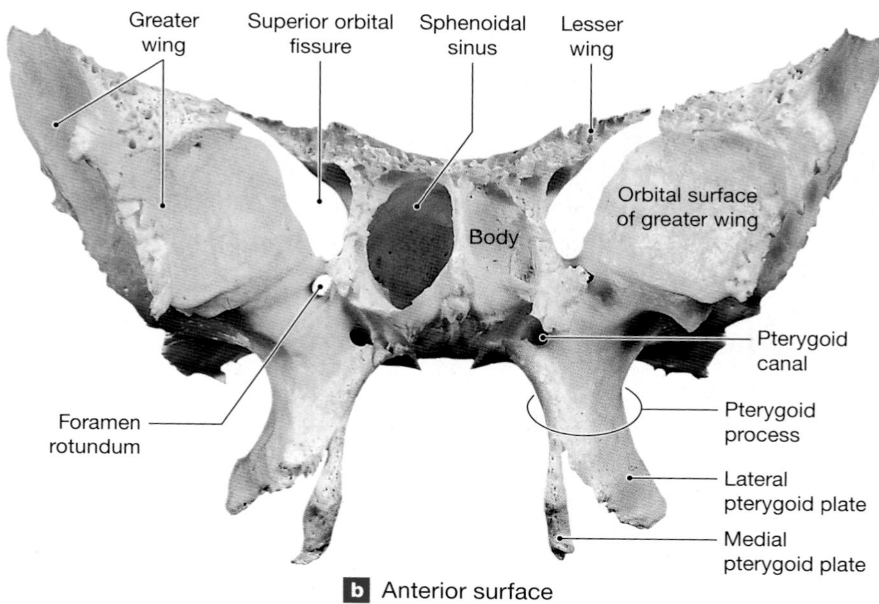

b Anterior surface

The **ethmoid**, or *ethmoidal bone*, is an irregularly shaped bone that forms part of the orbital wall, the anteromedial floor of the cranium the roof of the nasal cavity, and part of the nasal septum. The ethmoid has three parts: the cribriform plate, the ethmoidal labyrinth, and the perpendicular plate **(Figure 6.10)**.

The superior surface of the ethmoid **(Figure 6.10a)** contains the **cribriform plate**, an area perforated by the cribriform foramina. Branches of the olfactory nerves, which provide the sense of smell, pass through these openings. A prominent ridge, the **crista galli** (*crista*, crest, + *gallus*, chicken; "cockscomb"), separates the right and left sides of the cribriform plate. The falx cerebri, a membrane that stabilizes the brain's position within the skull, attaches to this bony ridge.

The ethmoidal labyrinth, dominated by the **superior nasal conchae** (KON-kē; singular, *concha*; "snail shell") and the **middle nasal conchae**, is seen from the anterior and posterior surfaces of the ethmoid **(Figure 6.10b,c)**. The **ethmoidal labyrinth** is an interconnected network of ethmoidal cells. These ethmoidal cells are continuous with the ethmoidal cells of the inferior portion of the frontal bone. The ethmoidal cells also open into the nasal cavity on each side. Mucus from these cells flushes the surfaces of the nasal cavities.

The nasal conchae are thin scrolls of bone that project into the nasal cavity on either side of the perpendicular plate. The nasal conchae break up the airflow, creating swirls and eddies that slow air movement, adding extra time for warming, humidifying, and removing dust before the air reaches more delicate portions of the respiratory tract.

The **perpendicular plate** is a thin plate of bone that projects downward from the crista galli of the ethmoid and forms part of the nasal septum. The nasal septum is a partition that also includes the vomer and a piece of hyaline cartilage. Olfactory receptors are located in the epithelium covering the inferior surfaces of the cribriform plate, the medial surfaces of the superior nasal conchae, and the superior portion of the perpendicular plate.

Cranial Fossae

▶ KEY POINT The skull interior contains depressions known as the cranial fossae. These fossae consist of three parts, and each supports part of the brain.

The contours of the cranium closely follow the shape of the brain. When you view the cranium from anterior to posterior, you see that the floor of the cranium is not horizontal; it descends in two steps **(Figure 6.11a,b)**. The cranial floor at each level forms a curving depression known as a **cranial fossa**. The frontal bone, the ethmoid, and the lesser wings of the sphenoid form the **anterior cranial fossa**, which cradles the frontal lobes of the cerebral hemispheres. The **middle cranial fossa** extends from the **posterior nasal apertures** to the petrous parts of the temporal bones. The sphenoid, temporal, and parietal bones form this fossa, which cradles the temporal lobes of the cerebral hemispheres, the diencephalon, and the anterior portion of the brain stem (mesencephalon). The more inferior **posterior cranial fossa** extends from the petrous parts of the temporal bones to the posterior skull surface. The posterior fossa is formed primarily by the occipital bone, with contributions from the temporal and parietal bones. The posterior cranial fossa supports the occipital lobes of the cerebral hemispheres, the cerebellum, and the posterior brain stem (pons and medulla oblongata).

Figure 6.10 The Ethmoid. Views of the ethmoid showing major anatomical landmarks.

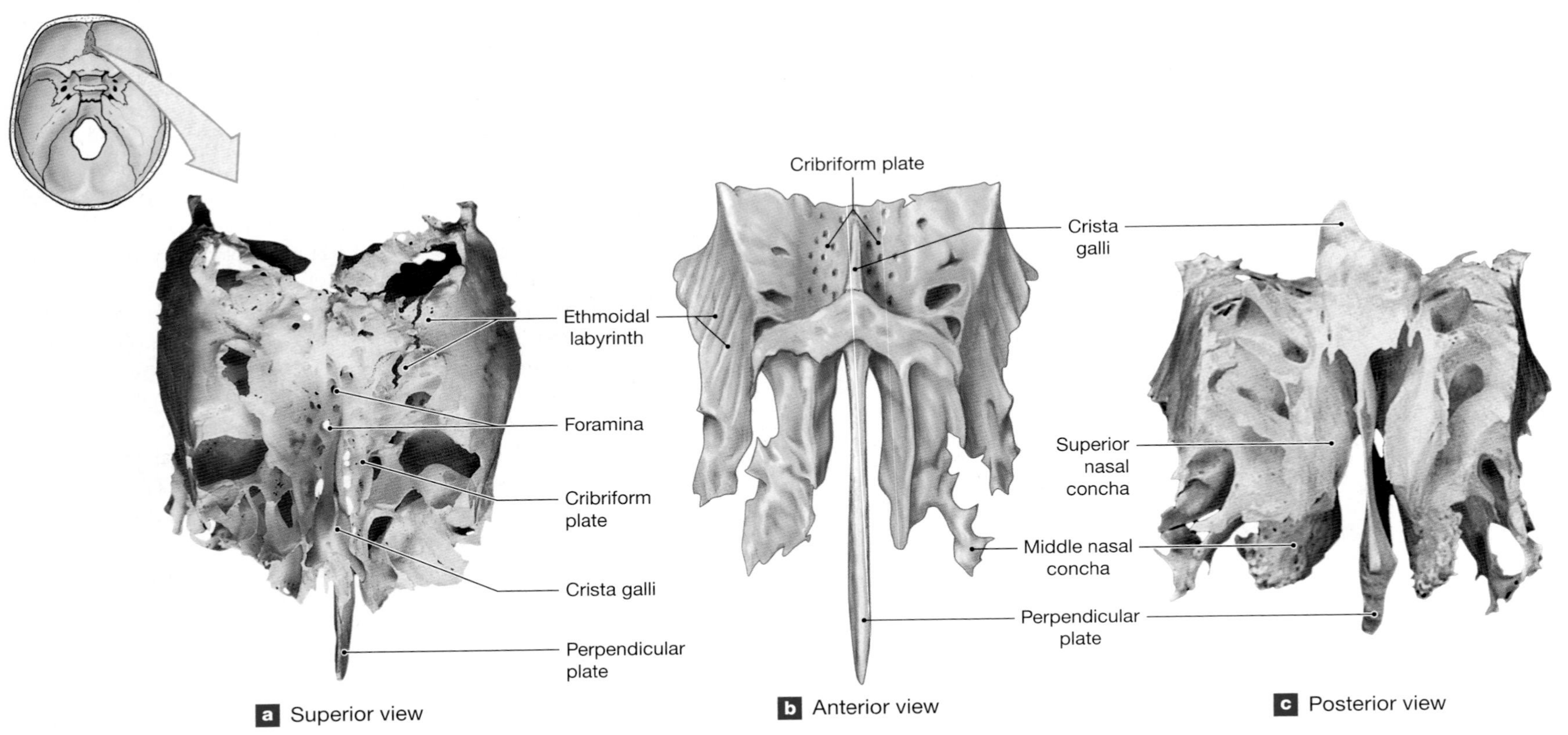

6

6.3 CONCEPT CHECK

5 The internal jugular veins are important blood vessels of the head. Through what opening do these blood vessels pass?

6 What bone contains the depression called the sella turcica? What is located in this depression?

7 Which of the five senses would be affected if the cribriform plate of the ethmoid failed to form?

See the blue Answers tab at the back of the book.

6.4 | Bones of the Face

▶ **KEY POINT** Six pairs of bones and two singular bones make up the bony structure of the face.

The facial bones are the paired maxillae, palatine bones, nasal bones, inferior nasal conchae, zygomatic bones, and lacrimal bones and the single vomer and mandible. (*Refer to Chapter 12,* **Figure 12.1**, *to identify these structures from the body surface and to* **Figure 12.9** *to visualize the maxilla in a cross section of the body at the level of* C_2.)

The Maxillae

▶ **KEY POINT** The paired maxillae form the upper jaw and most of the roof of the oral cavity. The maxillae are the largest bones of the face and feature cavities (maxillary sinuses) that are lined with a mucous membrane.

The **maxillae** (singular, *maxilla*), or *maxillary bones*, articulate with all other facial bones except the mandible (**Figure 6.3d**, p. 136). The **orbital surface** protects the eye and other structures in the orbit **(Figure 6.12a)**. The **frontal process** of each maxilla articulates with the frontal bone of the cranium and with a nasal bone. The **alveolar processes** of the maxillae contain the upper teeth. An elongated **inferior orbital fissure** within each orbit lies between the maxillae and the sphenoid. The **infra-orbital foramen** penetrates the orbital rim and carries a major facial sensory nerve. In the orbit, this nerve runs along the **infra-orbital groove (Figure 6.15)** before passing through the inferior orbital fissure and the foramen rotundum to reach the brain stem.

The large **maxillary sinuses** are seen in a medial view and in a horizontal section **(Figure 6.12b,c)**. These are the largest sinuses in the skull; they lighten the portion of the maxillae superior to the teeth and produce mucus that flushes the inferior surfaces of the nasal cavities. The sectional view also shows the extent of the **palatine processes** that form most of the bony palate, or bony roof, of the mouth. The **incisive fossa** on the inferior midline of the palatine process marks the openings of the incisive canals **(Figure 6.3e)**, which contain small arteries and nerves.

The Palatine Bones

▶ **KEY POINT** The two palatine bones form the most posterior part of the oral cavity roof. These small bones can be seen from the inferior view of the skull.

The **palatine bones** are small, L-shaped bones **(Figure 6.13)**. The horizontal plates articulate with the maxillae to form the posterior portions of the bony palate **(Figure 6.12c)**. On its inferior surface, a greater palatine groove lies between the palatine bone and the maxilla on each side. The left and right

Figure 6.11 The Cranial Fossae. Cranial fossae are curved depressions in the cranium.

Optic groove
Crista galli
Frontal sinus
Nasal conchae (superior, middle, and inferior)
Sphenopalatine foramen
Sphenoidal sinus
Anterior cranial fossa
Middle cranial fossa
Posterior cranial fossa
Sella turcica
Hypoglossal canal
Internal acoustic meatus
Jugular foramen

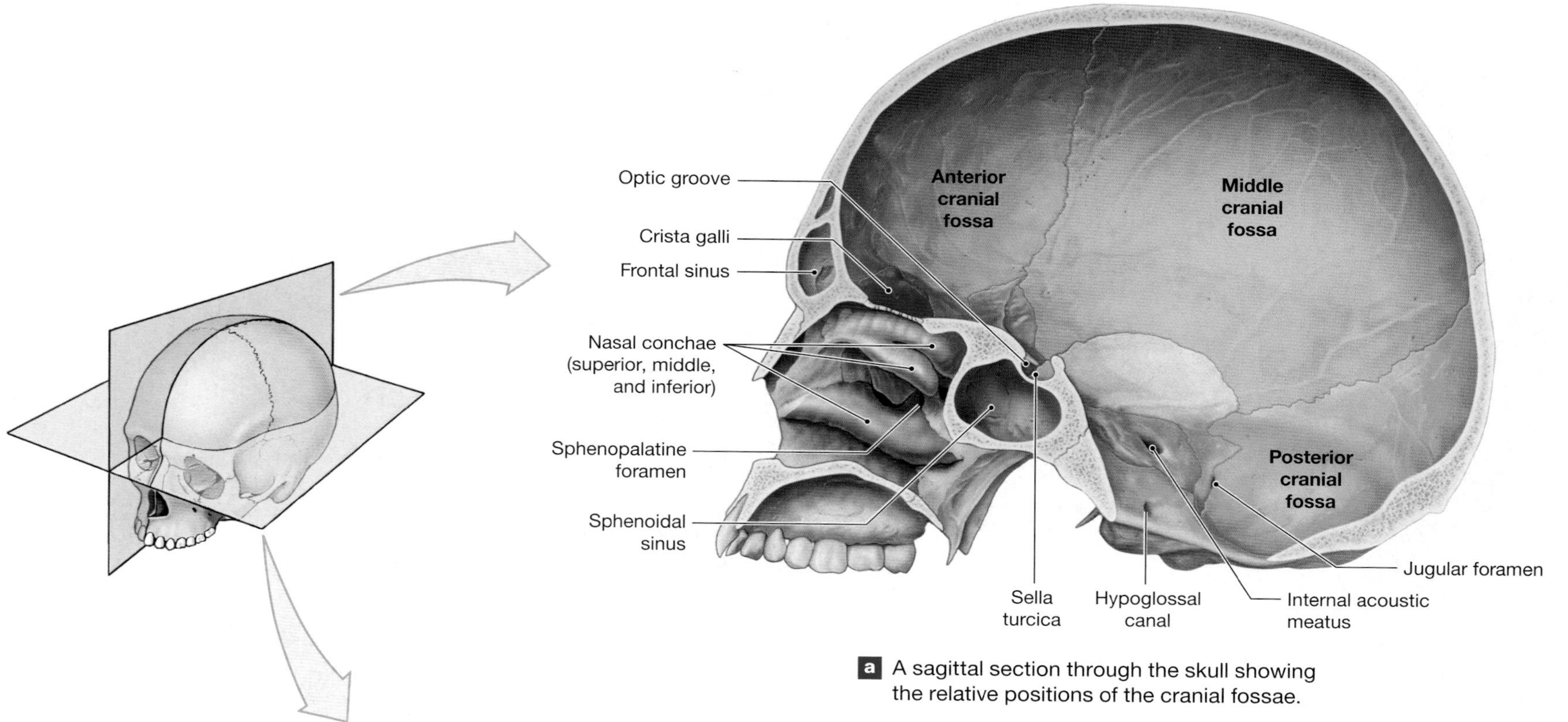

a A sagittal section through the skull showing the relative positions of the cranial fossae.

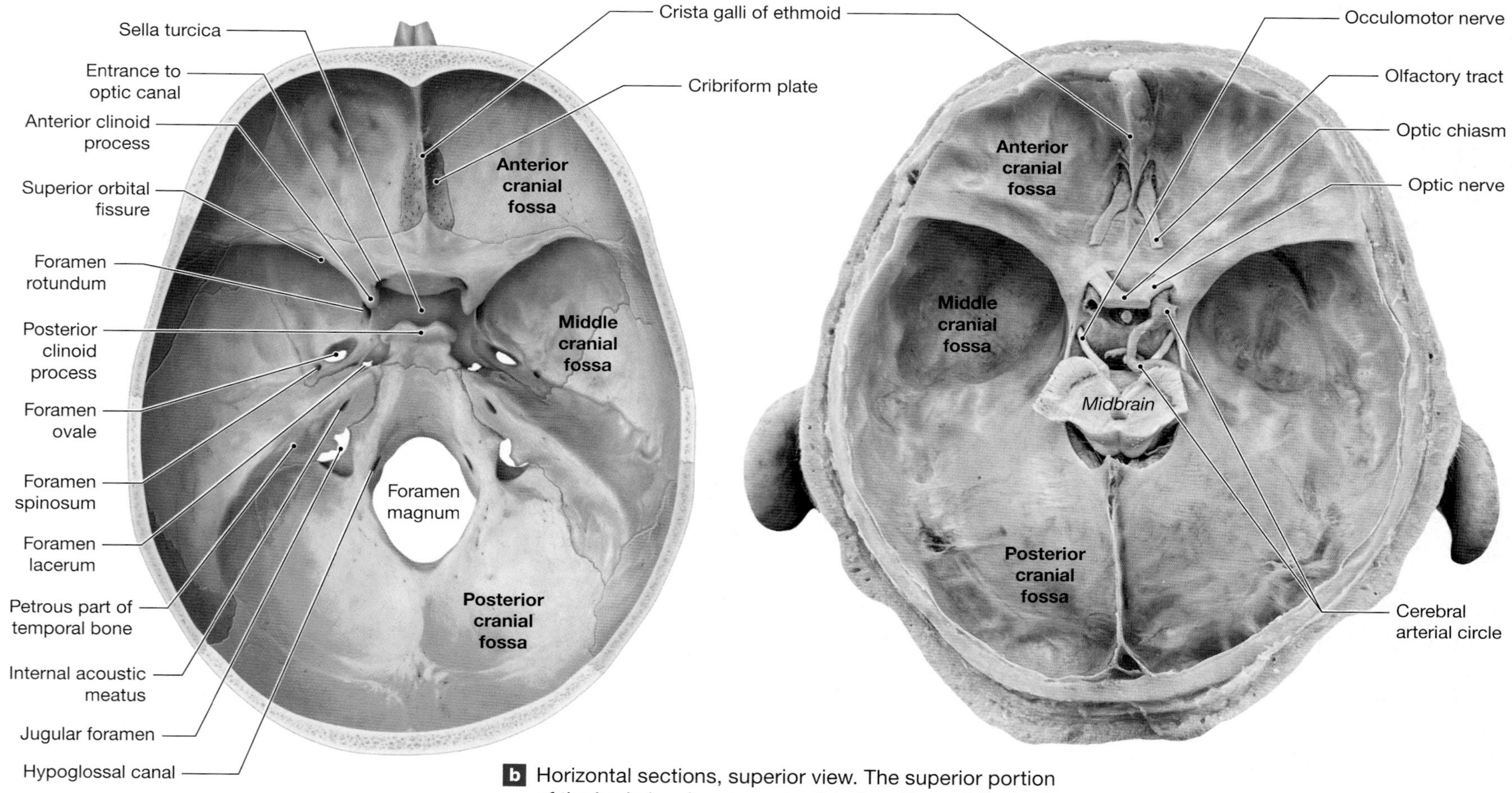

Sella turcica
Entrance to optic canal
Anterior clinoid process
Superior orbital fissure
Foramen rotundum
Posterior clinoid process
Foramen ovale
Foramen spinosum
Foramen lacerum
Petrous part of temporal bone
Internal acoustic meatus
Jugular foramen
Hypoglossal canal
Crista galli of ethmoid
Cribriform plate
Anterior cranial fossa
Middle cranial fossa
Foramen magnum
Posterior cranial fossa
Occulomotor nerve
Olfactory tract
Optic chiasm
Optic nerve
Anterior cranial fossa
Middle cranial fossa
Midbrain
Posterior cranial fossa
Cerebral arterial circle

b Horizontal sections, superior view. The superior portion of the brain has been removed, but portions of the brain stem and associated nerves and blood vessels remain.

Figure 6.12 **The Maxillae.** Views of the right maxilla showing major anatomical landmarks.

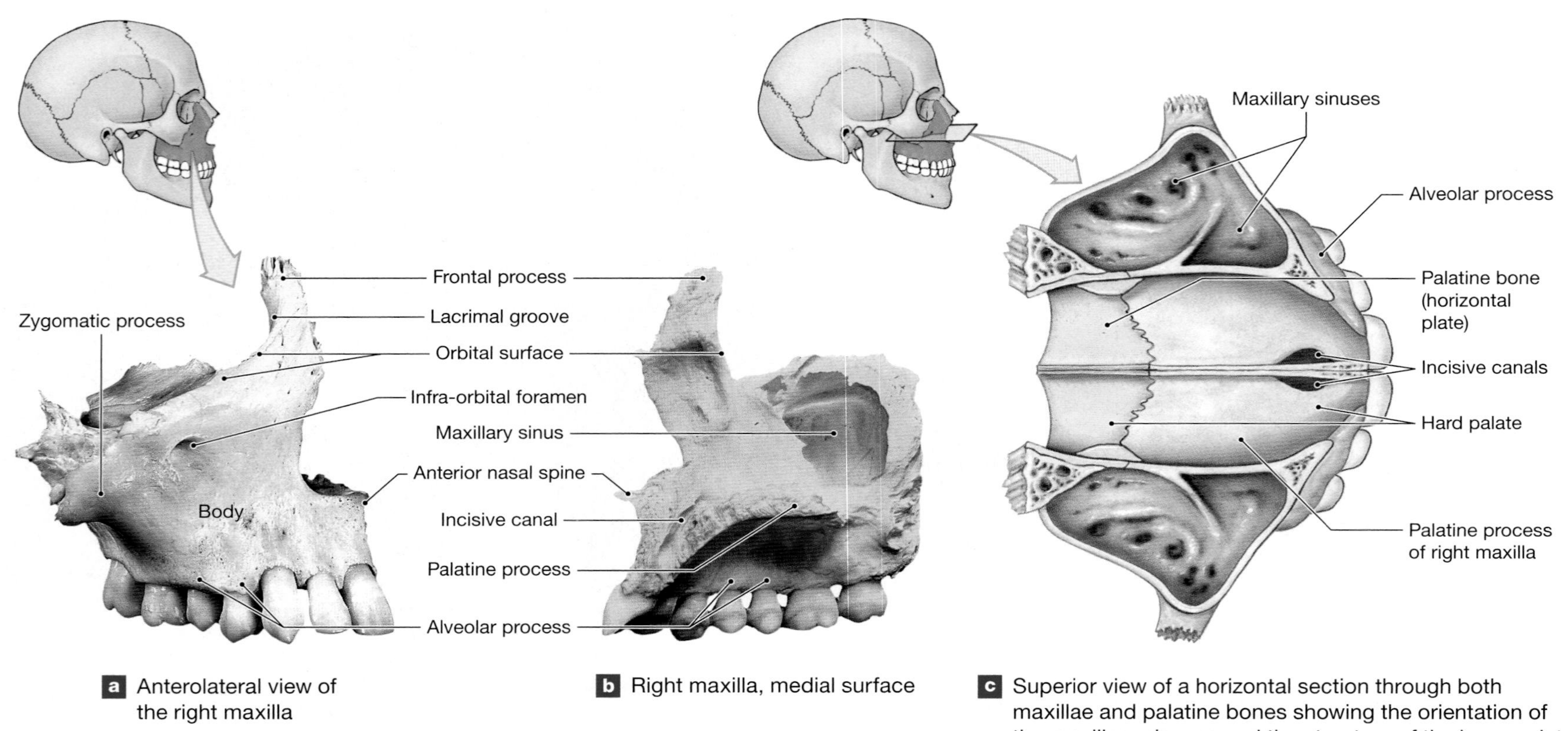

a Anterolateral view of the right maxilla

b Right maxilla, medial surface

c Superior view of a horizontal section through both maxillae and palatine bones showing the orientation of the maxillary sinuses and the structure of the bony palate

Figure 6.13 **The Palatine Bones.** Views of the palatine bones showing major anatomical landmarks.

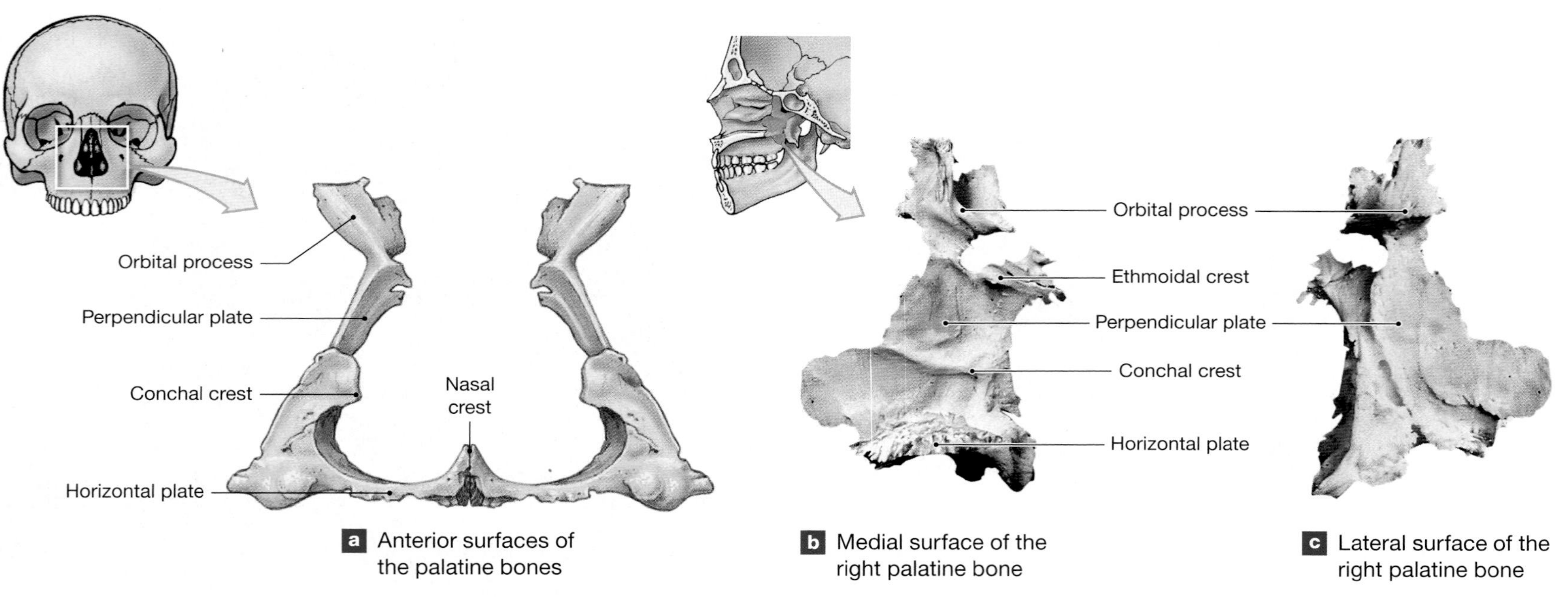

a Anterior surfaces of the palatine bones

b Medial surface of the right palatine bone

c Lateral surface of the right palatine bone

palatine bones join at the nasal crest, a ridge that articulates with the vomer. The perpendicular plate of the palatine bone forms the vertical portion of the "L," which articulates with the maxillae, sphenoid, ethmoid, and inferior nasal concha. The medial surface of the perpendicular plate has two ridges: the conchal crest, which articulates with the inferior nasal concha, and the ethmoidal crest, which articulates with the middle nasal concha of the ethmoid. The orbital process, a part of the perpendicular plate of the palatine bone, forms a small portion of the posterior floor of the orbit **(Figure 6.15)**.

The Nasal Bones

▶ **KEY POINT** The nasal bones are small, paired bones that form the bony portion of the nose.

The paired **nasal bones** articulate with the frontal bone at the frontonasal suture at the midline of the face **(Figure 6.3c,d)**. The bridge of your sunglasses rests on the nasal bones. Cartilages attached to the inferior margins of the nasal bones support the flexible portion of the nose, which extends to the nostrils, or nasal openings. The lateral edge of each nasal bone articulates with the frontal process of a maxilla (**Figures 6.3c** and **6.15**).

The Inferior Nasal Conchae

▶ **KEY POINT** The inferior nasal conchae are small bones within the nasal cavity. Their shape causes air to swirl when you breathe through your nose, which warms, moistens, and cleanses the air.

The **inferior nasal conchae** are paired scroll-like bones that resemble the superior and middle conchae of the ethmoid. One inferior concha is found on each side of the nasal septum, attached to the lateral wall of the nasal cavity (**Figures 6.3d** and **6.16**). They have the same functions as the conchae of the ethmoid.

The Zygomatic Bones

▶ **KEY POINT** The zygomatic bones form the lateral part of the orbit of the eye and the anterior portion of the "cheekbones."

As noted earlier, the temporal process of the **zygomatic bone** articulates with the zygomatic process of the temporal bone to form the zygomatic arch (**Figure 6.3c,d**). A **zygomaticofacial foramen** on the anterior surface of each zygomatic bone carries a sensory nerve innervating the cheek. The zygomatic bone also forms the lateral rim of the orbit and contributes to the inferior orbital wall (**Figure 6.15**).

The Lacrimal Bones

▶ **KEY POINT** The lacrimal bones are small, paired bones that form the medial portion of the orbit of the eye. Your tear ducts pass through these bones.

The paired **lacrimal bones** (*lacrima*, tear) are the smallest bones in the skull. Each lacrimal bone is found in the medial portion of the orbit, where it articulates with the frontal bone, maxilla, and ethmoid. A shallow depression, the **lacrimal groove**, or *lacrimal sulcus*, leads to the **nasolacrimal canal**, a narrow passageway formed by the lacrimal bone and the maxilla. This canal encloses the tear duct as it passes toward the nasal cavity (**Figures 6.3c,d** and **6.15**).

The Vomer

▶ **KEY POINT** The vomer is a singular bone of the face that forms the inferior portion of the nasal septum.

The **vomer** (**Figure 6.5**, p. 139) is located on the floor of the nasal cavity and articulates with both the maxillae and palatine bones along the midline. The vertical portion of the vomer is thin. Its curving superior surface articulates with the sphenoid and the perpendicular plate of the ethmoid, forming a bony **nasal septum** (*septum*, wall) that separates the right and left nasal cavities (**Figure 6.3d,e**). Anteriorly, the vomer supports a cartilaginous extension of the nasal septum that continues into the fleshy portion of the nose and separates the nostrils.

The Mandible

▶ **KEY POINT** The mandible forms the lower jaw.

The **mandible** is subdivided into the horizontal **body** and the ascending **ramus** of the mandible on either side (**Figures 6.3c,d** and **6.14**). The mandibular body supports the teeth. Each ramus meets the body at the **angle** of the mandible. The **condylar processes** extend to the smooth articular surface of the **head** of the mandible. The head articulates with the mandibular fossae of the temporal bone at the temporomandibular joint (TMJ). This joint is quite mobile, as evidenced by jaw movements during chewing or talking. The disadvantage of such mobility is that forceful forward thrusting or side-to-side movements of the mandible can easily dislocate the jaw.

The temporalis muscle, one of the strongest muscles involved with closing the mouth, inserts onto the mandible at the **coronoid** (kor-ō-noyd) **process**. Anteriorly, the **mental foramina** (*mentalis*, chin) penetrate the body on each

Figure 6.14 **The Mandible.** Views of the mandible showing major anatomical landmarks.

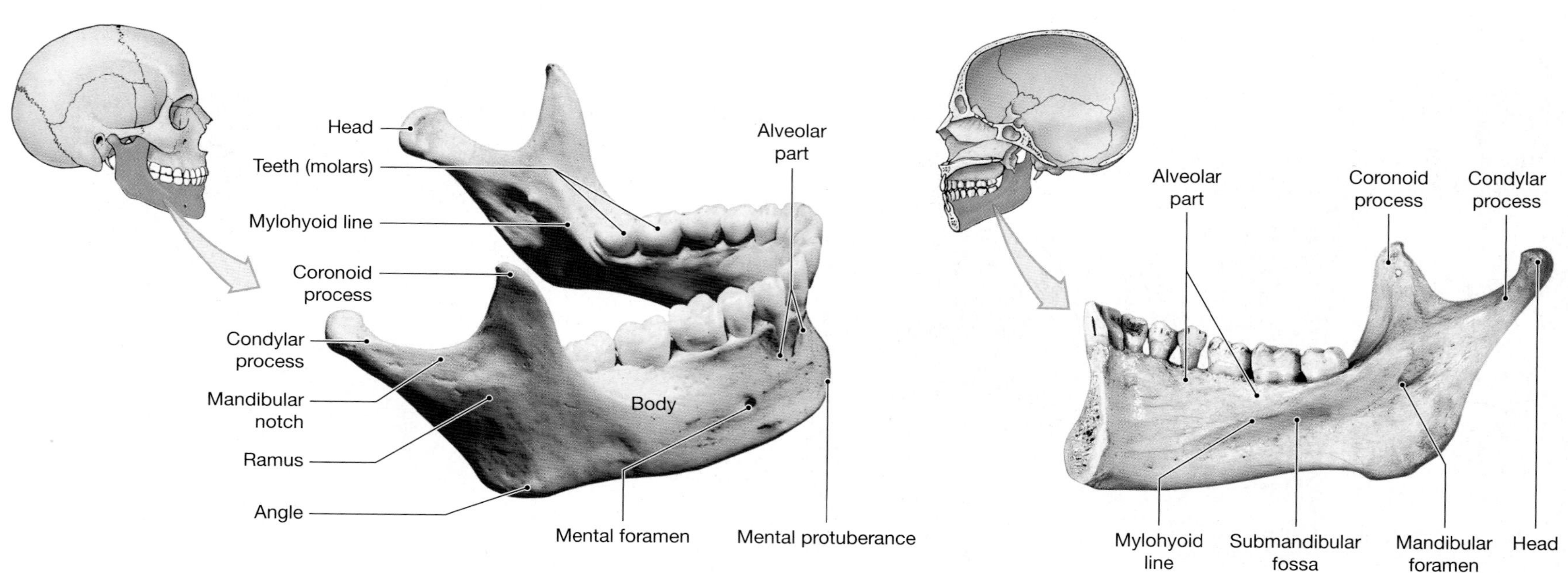

side of the chin. Nerves pass through these foramina, carrying sensory information from the chin and the lower lips back to the brain. The **mandibular notch** is the depression between the condylar and coronoid processes.

The **alveolar part** of the mandible is a thickened area that surrounds and supports the roots of the lower teeth (**Figure 6.14b**). The **mylohyoid line** on the medial aspect of the body of the mandible marks the origin of the mylohyoid muscle that supports the floor of the mouth and tongue. The submandibular salivary gland lies in the **submandibular fossa**, a depression located inferior to the mylohyoid line. Near the posterior, superior end of the mylohyoid line, a prominent **mandibular foramen** leads into the **mandibular canal**. This canal is a passageway for blood vessels and nerves that service the lower teeth. The nerve that uses this passage carries sensory information from the teeth and gums. Dentists anesthetize this nerve before working on the lower teeth.

6.4 CONCEPT CHECK

8 What are the names and functions of the facial bones?

See the blue Answers tab at the back of the book.

6.5 The Orbits, Nasal Complex and the Hyoid Bone

▶ **KEY POINT** Several facial bones articulate with cranial bones to form the orbit surrounding each eye and the nasal complex surrounding the nasal cavities.

The Orbits

▶ **KEY POINT** Seven bones form the orbits of the eye.

The **orbits** are the bony recesses that enclose and protect the eyes (**Figure 6.15**). In addition to the eye, each orbit also houses a lacrimal gland, adipose tissue, muscles that move the eye, blood vessels, and nerves. The frontal bone forms the roof, and the maxilla forms most of the floor of the orbit. Proceeding from medial to lateral, the first portion of the wall of the orbit is formed by the maxilla, the lacrimal bone, and the lateral mass of the ethmoid, which articulates with the sphenoid and a small process of the palatine bone. The sphenoid forms most of the posterior orbital wall. Several prominent foramina and fissures penetrate the sphenoid or lie between the sphenoid and maxilla. Laterally, the sphenoid and maxilla articulate with the zygomatic bone, which forms the lateral wall and rim of the orbit.

The Nasal Complex

▶ **KEY POINT** Seven bones form the nasal complex.

The **nasal complex** consists of the bones and cartilage that enclose the nasal cavities and the paranasal sinuses. The nasal complex extends from the **nostrils** (**Figure 24.3**) to the **posterior nasal apertures** (**Figure 6.3e**).

The frontal bone, sphenoid, and ethmoid form the superior wall of the nasal cavities. The perpendicular plate of the ethmoid and the vomer form the bony portion of the nasal septum (**Figures 6.5**, p. 139, and **6.16**). The lateral walls are formed by the maxillae, the lacrimal bones, the ethmoid, and the inferior nasal conchae. The maxillae and nasal bones support the bridge of the nose. The soft tissues of the nose enclose the anterior extensions of the nasal cavities. Cartilaginous extensions of the bridge of the nose and the nasal septum support the nasal cavities.

The Paranasal Sinuses

The frontal bone, sphenoid, ethmoid, and maxillae contain the **paranasal sinuses**, air-filled chambers that open into the nasal cavities. **Figure 6.16** shows the location of the **frontal**, **sphenoidal**, and **maxillary sinuses** and the **ethmoidal cells**. These sinuses lighten the skull, produce mucus, and resonate when a person produces any sound. The mucus is released into the nasal cavities, and the ciliated epithelium moves the mucus back toward the throat, where it is swallowed. Incoming air is humidified and warmed as it flows across this carpet of mucus. Foreign particulate matter, such as dust and microorganisms, becomes trapped in this sticky mucus and is also swallowed. This mechanism protects the delicate gas exchange surfaces of the lungs.

Figure 6.15 The Orbital Complex. The structure of the orbital complex on the right side. Seven bones form the bony orbit, which encloses and protects the right eye.

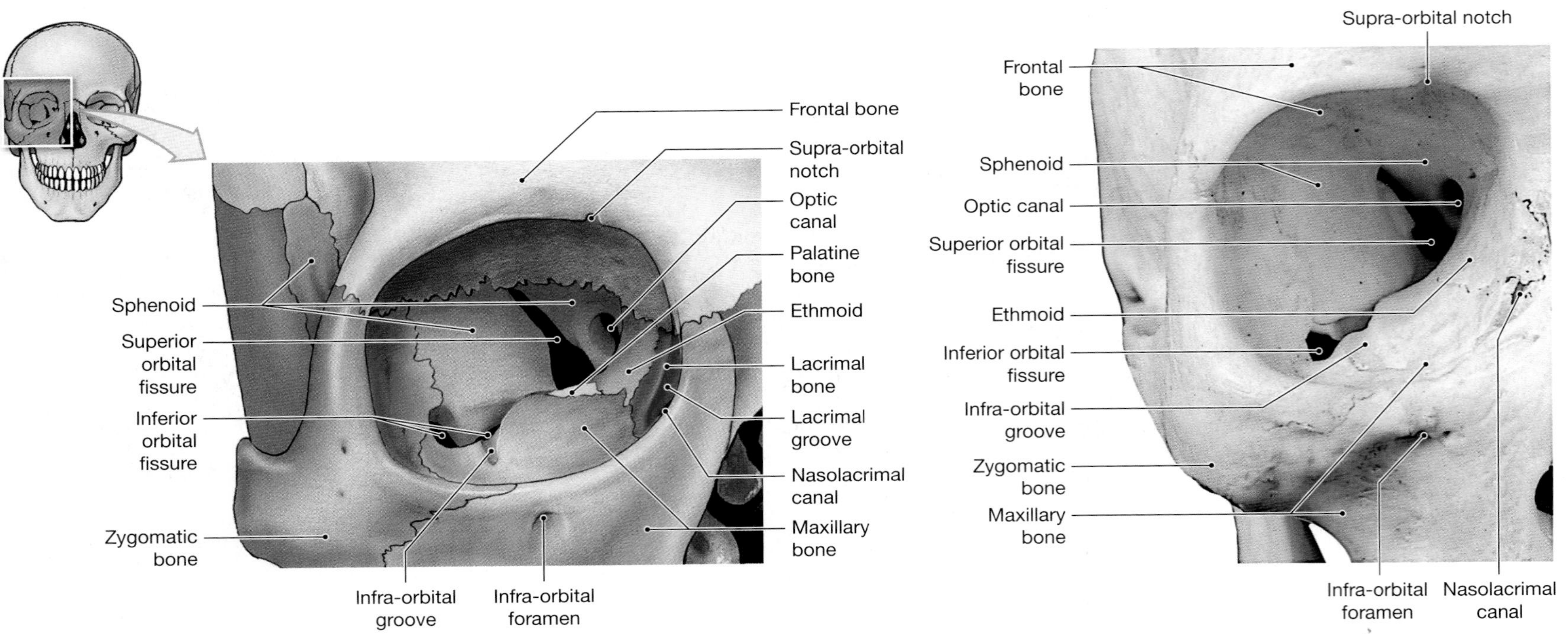

Figure 6.16 The Nasal Complex and Paranasal Sinuses. Sections through the skull and head showing relationships among the bones of the nasal complex and the positions of the paranasal sinuses.

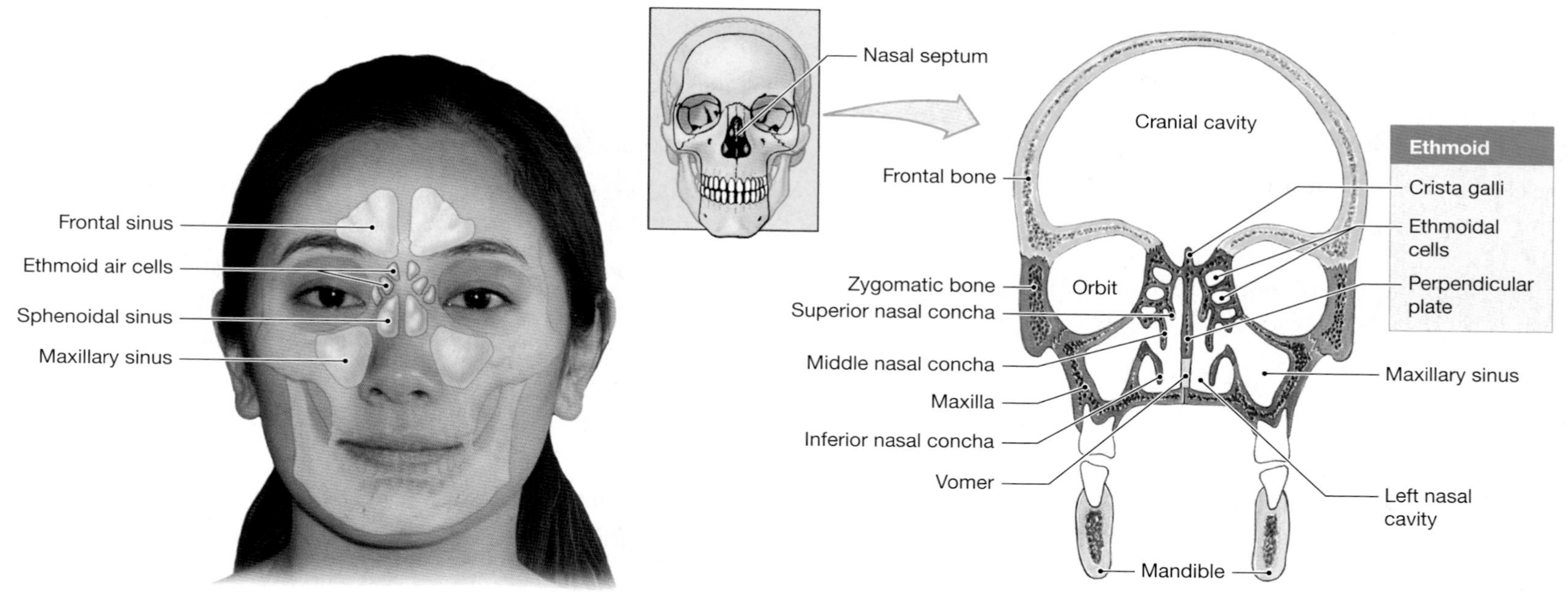

a Locations of the paranasal sinuses

b A diagrammatic frontal section showing the positions of the paranasal sinuses

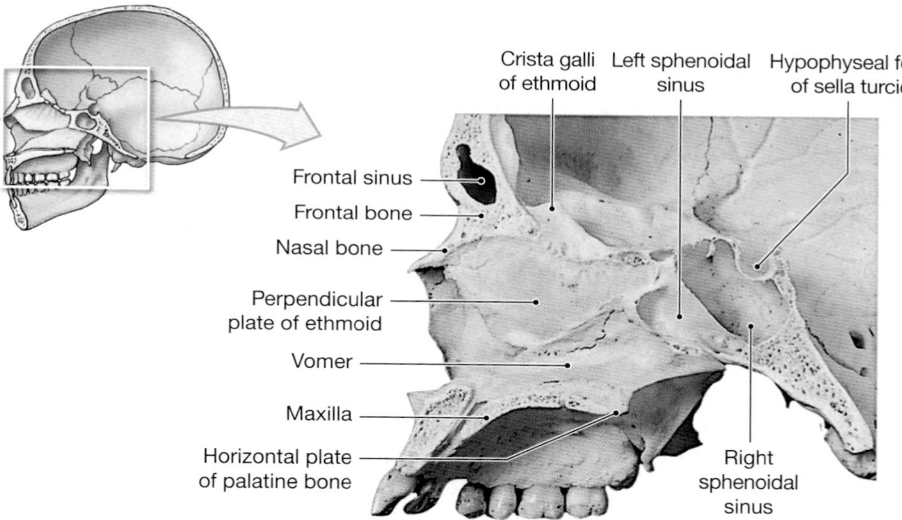

c Sagittal section with the nasal septum in place

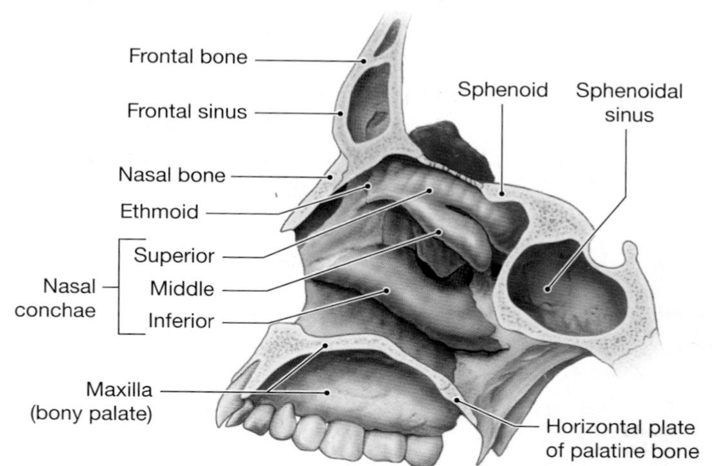

d Diagrammatic sagittal section with the nasal septum removed to show major features of the wall of the right nasal cavity

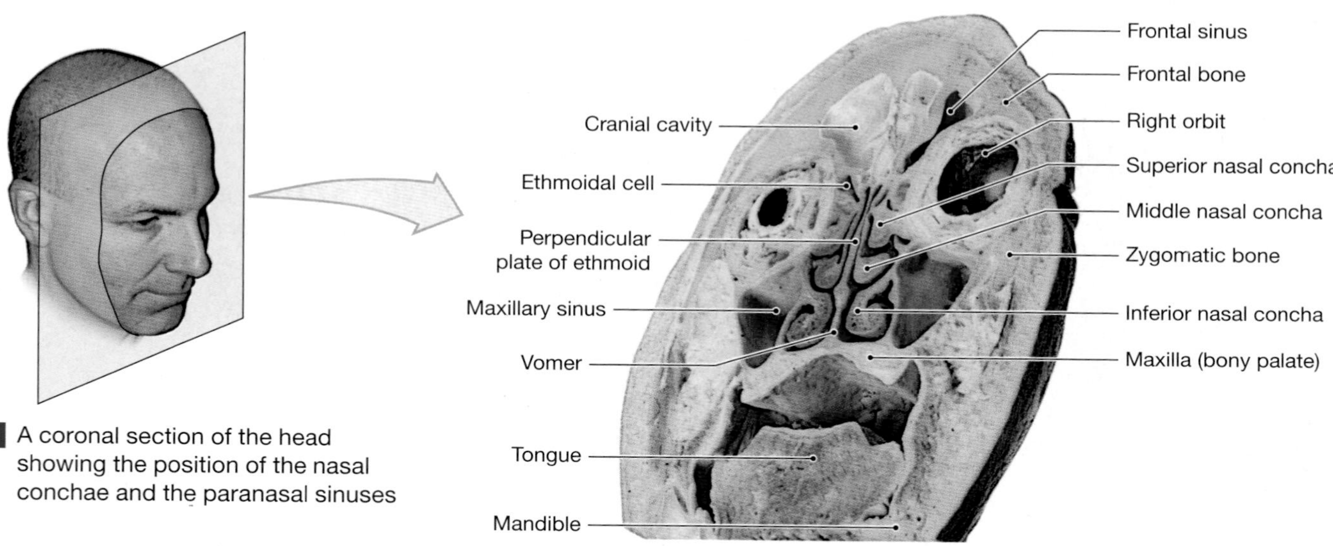

e A coronal section of the head showing the position of the nasal conchae and the paranasal sinuses

The Hyoid Bone

▶ **KEY POINT** The hyoid bone is the attachment site for several muscles of the tongue and larynx. The hyoid is the only bone of the skull that is not attached to another bone.

The **hyoid bone** lies inferior to the skull, suspended by the **stylohyoid ligaments** (**Figure 6.17**). The body of the hyoid serves as a base for several muscles that move the tongue and larynx. Because muscles and ligaments form the only connections between the hyoid and other skeletal structures, the entire complex is quite mobile. The larger processes on the hyoid are the **greater horns**, which support the larynx and serve as the base for muscles that move the tongue. The **lesser horns** are connected to the stylohyoid ligaments. The hyoid and larynx hang beneath the skull from these ligaments as a swing hangs from a tree limb.

6.5 **CONCEPT CHECK**

9 What are the functions of the paranasal sinuses?

10 Which bones form the orbital complex?

See the blue Answers tab at the back of the book.

6.6 The Skulls of Infants, Children, and Adults

▶ **KEY POINT** Bones of the cranium, which are formed by intramembranous ossification, are not completely ossified at birth. These areas of incomplete ossification are called fontanelles, the "soft spots" of an infant's skull.

Many different ossification centers are involved in forming the skull. As development proceeds, these centers fuse, reducing the number of individual bones. However, fusion is still not complete at birth; there are two frontal bones, four occipital bones, and several sphenoid and temporal elements.

The skull organizes around the developing brain. As birth approaches, the brain enlarges rapidly. Although the bones of the skull are also growing, they fail to keep pace, and at birth areas of fibrous connective tissue connect the cranial bones. These connections are flexible, and the skull can be distorted without damage during delivery, which eases the passage of the infant through the birth canal. The largest fibrous regions between the cranial bones are known as **fontanelles** (fon-tah-NELS; sometimes spelled *fontanels*) (**Figure 6.18**):

- The **anterior fontanelle** is the largest. It lies at the junction of the frontal, sagittal, and coronal sutures.
- The **posterior fontanelle** is at the junction between the lambdoid and sagittal sutures.
- The **sphenoidal fontanelles** are at the junctions between the squamous sutures and the coronal suture.
- The **mastoid fontanelles** are at the junctions between the squamous sutures and the lambdoid suture.

The skulls of infants and adults differ in the shape and structure of cranial elements, and this difference accounts for variations in both proportion and size. The most significant growth in the skull occurs before age 5, at which point the brain stops enlarging and the cranial sutures develop. As a result, when compared with the skull as a whole, the cranium of a young child is relatively larger than that of an adult.

6.6 **CONCEPT CHECK**

11 What are fontanelles?

12 What are the names and locations of the four fontanelles of the fetal skull?

See the blue Answers tab at the back of the book.

Figure 6.17 **The Hyoid Bone**

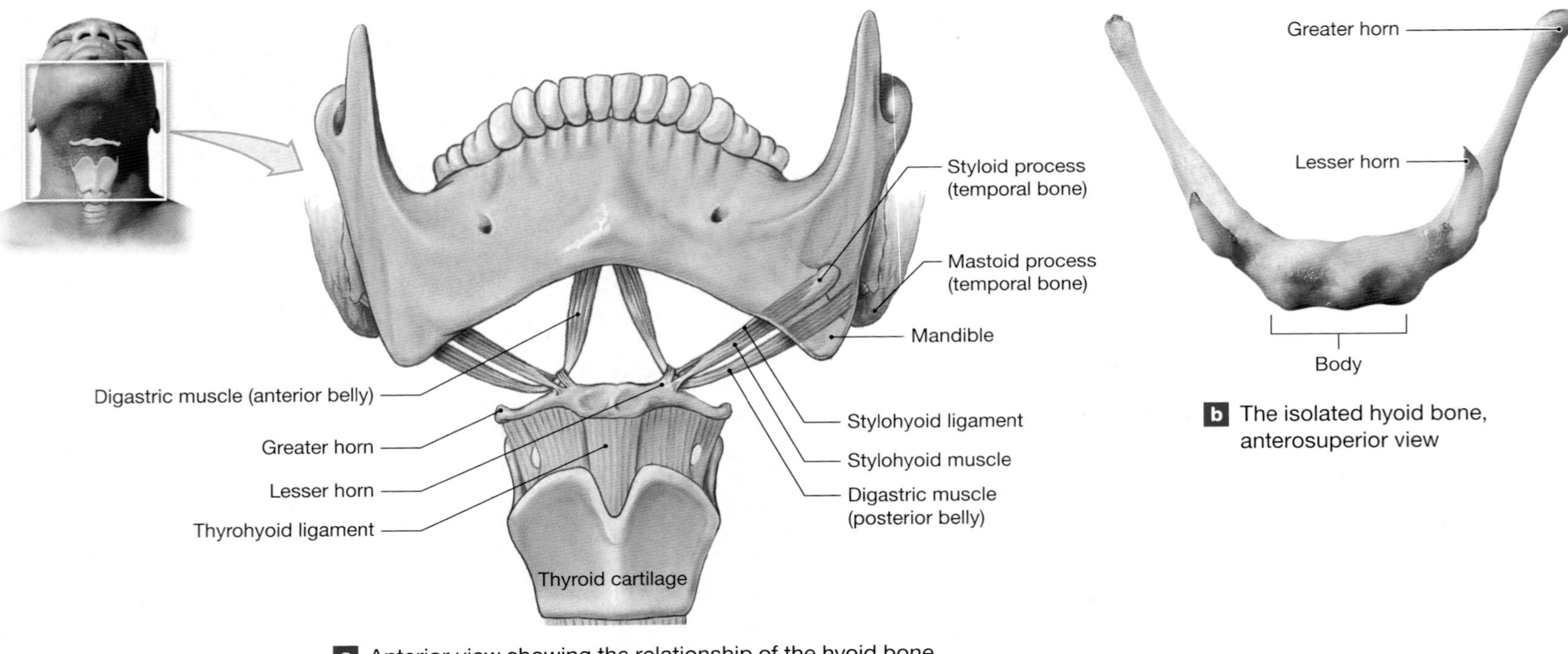

a Anterior view showing the relationship of the hyoid bone to the skull, larynx, and selected skeletal muscles

b The isolated hyoid bone, anterosuperior view

Figure 6.18 The Skull of an Infant. The flat bones in the infant skull are separated by fontanelles, which allow for cranial expansion and the distortion of the skull during birth. By about age 5 these areas will disappear, and skull growth will be completed.

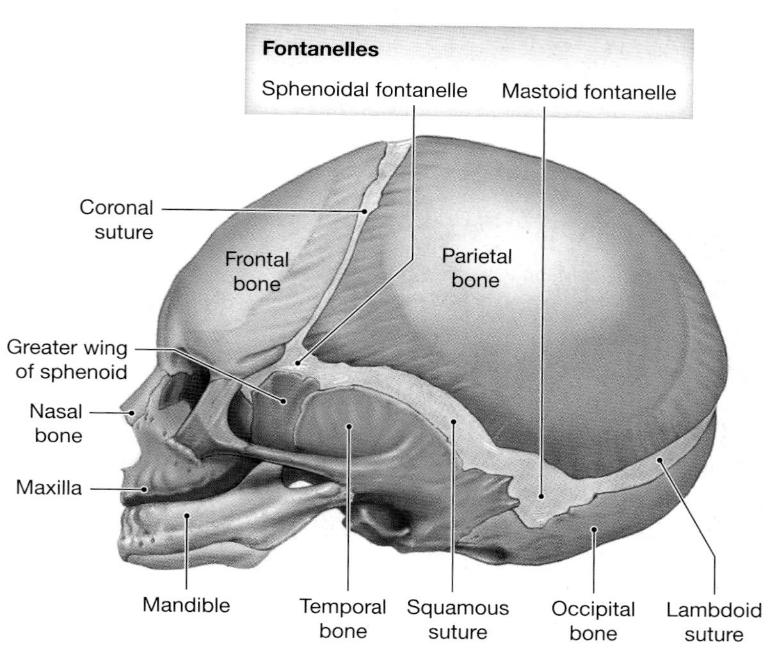

a Lateral view

Sagittal suture
Parietal bone
Anterior fontanelle
Coronal suture
Frontal suture
Frontal bone
Frontal suture

b Anterior/superior view

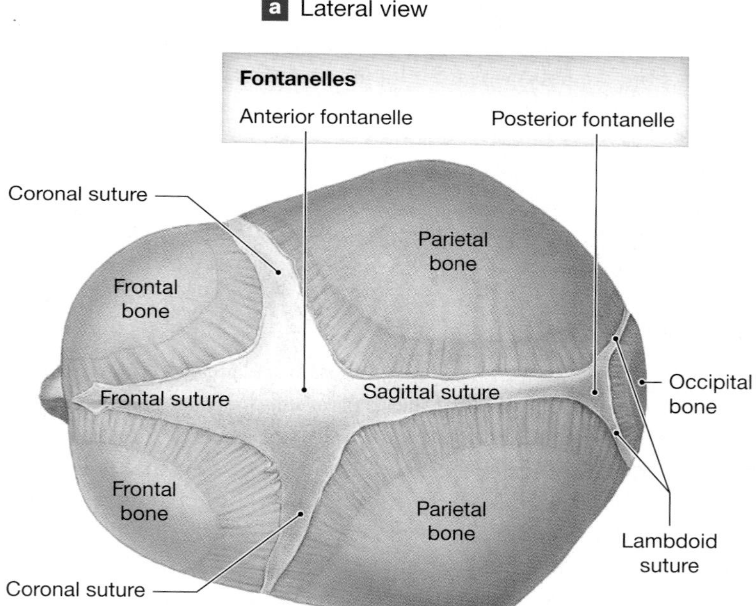

c Superior view

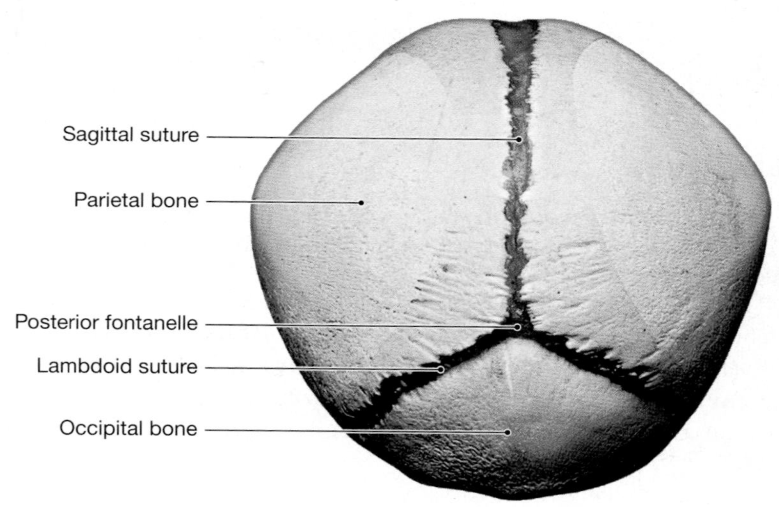

d Posterior view

CLINICAL NOTE

Coccygeal Fractures

During vaginal delivery, a baby's head, face, or shoulder can fracture or dislocate a woman's coccyx, causing a **coccygeal fracture**. When a coccygeal fracture occurs due to a difficult vaginal delivery, the distal segment tilts backward. When a coccygeal fracture occurs due to a fall on the buttocks, the distal segment of the coccyx tilts forward.

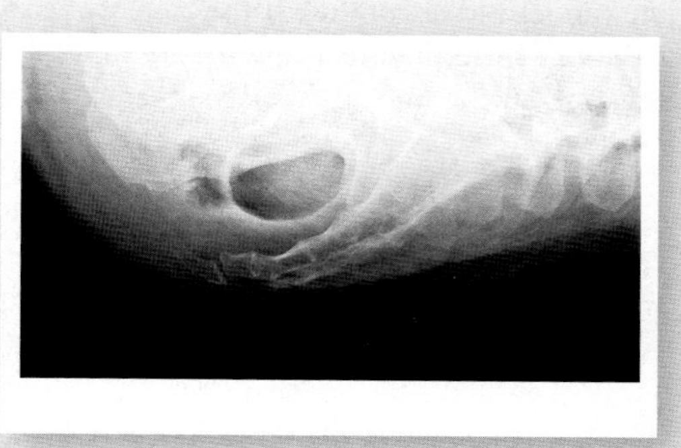

6.7 | The Vertebral Column

▶ **KEY POINT** The vertebral column is composed of 7 cervical vertebrae, 12 thoracic vertebrae, 5 lumbar vertebrae, 5 sacral vertebrae, and 3 to 5 coccygeal vertebrae. The sacral vertebrae fuse to form the sacrum, and the coccygeal vertebrae fuse to form the coccyx.

The rest of the axial skeleton is subdivided into the vertebral column and rib cage. The vertebrae support the weight of the head, neck, and trunk and transfer that weight to the appendicular skeleton of the lower limbs. The vertebrae also protect the spinal cord, provide a passageway for spinal nerves that begin or end at the spinal cord, and maintain an upright body position during sitting or standing.

The **vertebral column** is divided into the cervical, thoracic, lumbar, sacral, and coccygeal regions **(Figure 6.19)**. Each region has different functions; as a result, vertebrae within each region have anatomical specializations that allow for these functional differences. In addition, the vertebrae located at the junction between two regions of the vertebral column share some anatomical characteristics of the region above and the region below.

Seven cervical vertebrae make up the neck and extend inferiorly to the trunk. The first cervical vertebra forms a pair of joints, or articulations, with the occipital condyles of the skull. The seventh cervical vertebra articulates with the first thoracic vertebra. Twelve thoracic vertebrae form the midback region, and each forms joints with one or more pairs of ribs. The 12th thoracic vertebra articulates with the first lumbar vertebra. Five lumbar vertebrae form the lower back; the fifth articulates with the sacrum, which in turn articulates with the coccyx. During development, the sacrum starts as a group of five vertebrae, and the coccyx (KOK-siks), or "tailbone," begins as three to five small individual vertebrae. Fusion of the vertebrae of the sacrum is usually complete by age 25. Fusion of the distal coccygeal vertebrae occurs at a variable pace, and complete fusion may not occur at all in some adults. The total length of the vertebral column of an adult averages 71 cm (28 in.).

Spinal Curves

▶ **KEY POINT** The four curves of the vertebral column establish the body's center of gravity, balance the head at the top of the vertebral column, and permit upright walking and running.

The vertebral column is not a straight and rigid structure. A side view of the adult vertebral column shows four **spinal curves**: the **cervical curve**, **thoracic curve**, **lumbar curve**, and **sacral curve**. The development of the spinal curves from fetus to newborn, child, and adult is illustrated in **Figure 16.9**.

When you are standing, the weight of your body is transmitted through the vertebral column to the hips and the lower limbs. Most of the body weight lies in front of the vertebral column. The spinal curves bring the body weight in line with the body axis and its center of gravity. Consider what people do automatically when they stand holding a heavy object: To avoid falling forward, they exaggerate the lumbar curve, bringing the weight and center of gravity closer to the body axis. This posture can lead to discomfort at the base of the spinal column. Similarly, women in the last three months of pregnancy often develop chronic back pain from the changes in the lumbar curve that adjust for the increasing weight of the fetus.

Vertebral Anatomy

▶ **KEY POINT** The vertebrae enclose and protect the spinal cord. Structures on the vertebrae feature processes for muscle attachments and for joint formation between adjacent vertebrae or between vertebrae and ribs.

Each **vertebra** (pleural, *vertebrae*) has a common structural plan **(Figure 6.20)**. Anteriorly, each vertebra has a relatively thick, spherical to oval body. A vertebral arch extends posteriorly from the body of the vertebra. Various processes for muscle attachment or for rib articulation extend from the vertebral arch. Paired articulating processes on the superior and inferior surfaces project from the vertebral arch.

The Vertebral Body

The **vertebral body** transfers weight along the axis of the vertebral column **(Figure 6.20e)**. Each vertebra articulates with neighboring vertebrae; the bodies are interconnected by ligaments and separated by pads of fibrous cartilage, the **intervertebral discs**.

The Vertebral Arch

The **vertebral arch** forms the lateral and posterior margins of the **vertebral foramen** that surrounds a portion of the spinal cord **(Figure 6.20a,c)**. The vertebral foramen has a floor (the posterior surface of the body), walls (the pedicles), and a roof (the lamina). The **pedicles** (PED-i-kels) arise along the posterolateral (posterior and lateral) margins of the body. The **laminae** (LAM-i-nē; singular, *lamina*; "thin layer") extend dorsomedially (toward the back and middle) to complete the roof. From the fusion of the laminae, a **spinous process** projects posteriorly and often caudally from the midline. These processes can be seen and felt through the skin of the back. **Transverse processes** project laterally or dorsolaterally on both sides from the point where the laminae join the pedicles. These processes are sites of muscle attachment, and they may also articulate with the ribs.

The Articular Processes

The **articular processes** are seen at the junction between the pedicles and laminae. There is a **superior articular process** and an **inferior articular process** on each side of the vertebra **(Figure 6.20)**.

Vertebral Articulation

The inferior articular processes of one vertebra articulate with the superior articular processes of the more caudal vertebra. Each articular process has a polished surface called an **articular facet**. The superior processes have articular facets on their posterior surfaces, whereas the inferior processes articulate along their anterior surfaces.

The vertebral foramina of the vertebral column together form the **vertebral canal**, a space that encloses the spinal cord. However, the spinal cord is not completely encased in bone. The intervertebral discs separate the vertebral bodies, and between the pedicles of adjacent vertebrae are gaps called **intervertebral foramina (Figure 6.20)**. Nerves running to and from the spinal cord pass through these intervertebral foramina.

Vertebral Regions

▶ **KEY POINT** All vertebrae are fundamentally alike, sharing common characteristics and structures. Various minor differences exist among vertebrae from the five regions of the vertebral column. These differences determine the vertebral regions' basic functions.

Figure 6.19 The Vertebral Column. Lateral views of the vertebral column.

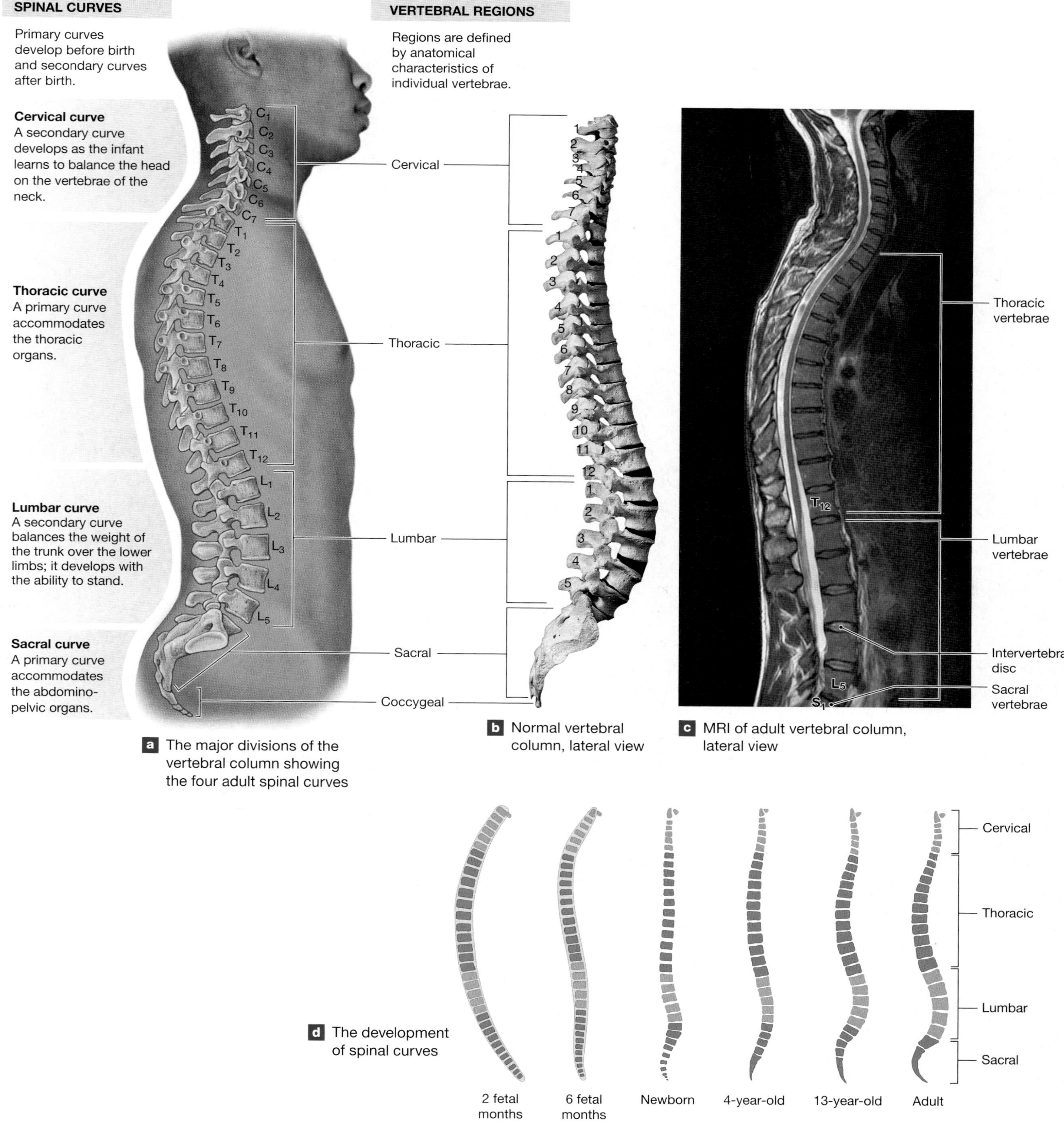

a The major divisions of the vertebral column showing the four adult spinal curves

b Normal vertebral column, lateral view

c MRI of adult vertebral column, lateral view

d The development of spinal curves

Figure 6.20 Vertebral Anatomy. The anatomy of a typical vertebra and the arrangement of articulations between vertebrae.

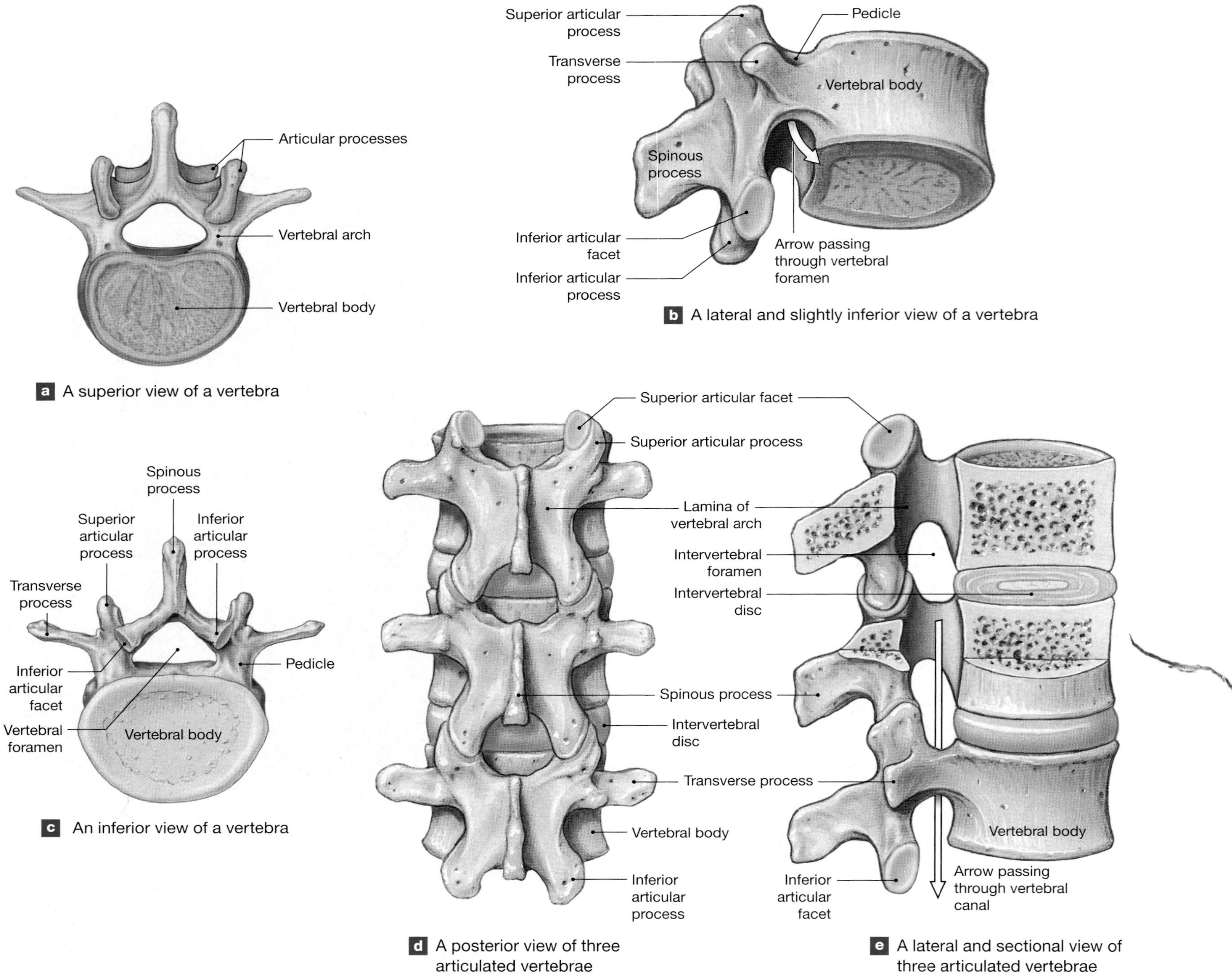

When referring to vertebrae, a capital letter indicates the vertebral region, and a subscript number indicates the specific vertebra, starting with the cervical vertebra closest to the skull. For example, C_3 refers to the third cervical vertebra; L_4 is the fourth lumbar vertebra **(Figure 6.19a)**.

Although each vertebra bears characteristic markings and articulations, focus on the general characteristics of each region and how the regional variations determine the vertebral group's basic function. **Table 6.3** compares typical vertebrae from each region of the vertebral column.

Cervical Vertebrae

The seven **cervical vertebrae** are the smallest of the vertebrae **(Figure 6.21)**. They extend from the occipital bone of the skull to the thorax. As you will see, the first, second, and seventh cervical vertebrae possess unique characteristics and are termed atypical cervical vertebrae, whereas the third through the sixth display similar characteristics and are termed typical cervical vertebrae. The body of a cervical vertebra is small compared with the size of the triangular vertebral foramen. At the cervical level, the spinal cord contains

Figure 6.21 Cervical Vertebrae. These are the smallest and most superior vertebrae.

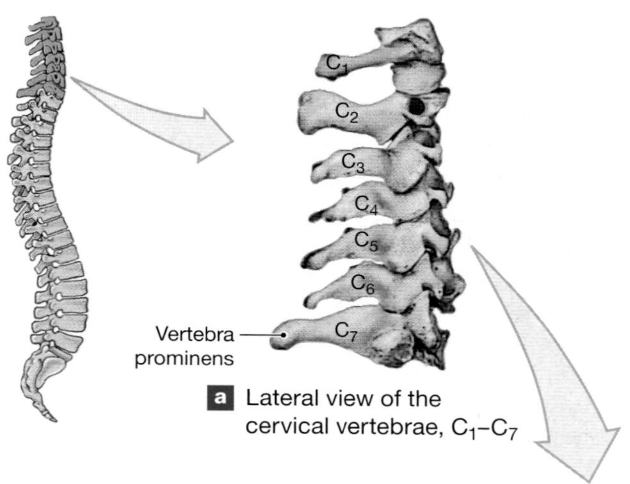

a Lateral view of the cervical vertebrae, C_1–C_7

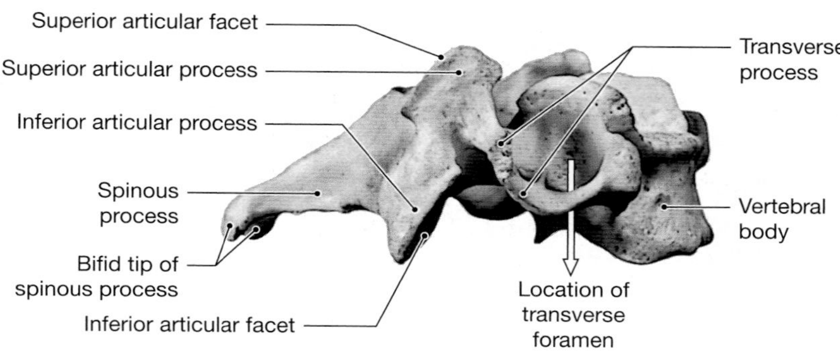

b Lateral view of a typical (C_3–C_6) cervical vertebra

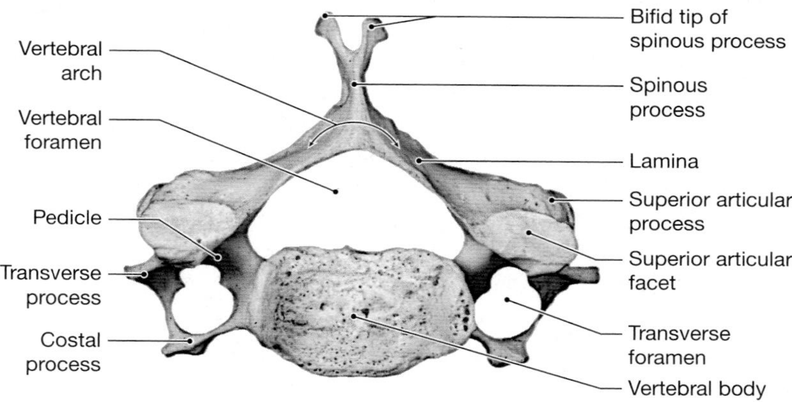

c Superior view of the vertebra in (b). Note the characteristic features listed in Table 6.3.

most of the nerves that connect the brain to the rest of the body. As you continue caudally along the vertebral canal, the diameter of the spinal cord decreases, and so does the diameter of the vertebral foramina. Cervical vertebrae support only the weight of the head, so the vertebral bodies can be relatively small and light. Continuing caudally along the vertebral column, the vertebrae support more and more weight, and the vertebral bodies gradually enlarge.

In a typical cervical vertebra (C_3–C_6), the spinous process is stumpy and shorter than the diameter of the vertebral foramen. The tip of each process other than C_7 has a prominent notch. A notched spinous process is described as **bifid** (BĪ-fid; *bifidus*, split in two parts). Laterally, the transverse processes are fused to the **costal processes** that originate near the ventrolateral portion of the vertebral body. *Costal* refers to a rib, and these processes represent the fused remnants of cervical ribs. The costal and transverse processes encircle the round **transverse foramina**. These foramina protect the vertebral arteries and vertebral veins, important blood vessels supplying the brain.

CLINICAL NOTE

Spina Bifida

Spina bifida is a congenital neural tube defect in which part of the spinal cord is exposed through a gap in the vertebral arches. During the third week of embryonic development, the vertebral arches form around the developing spinal cord, known as the *neural tube*. If this neural tube fails to form completely, the fetal spinal column does not close completely. Spinal cord development may also be incomplete, causing paralysis of the legs. Consuming enough folic acid before and during pregnancy prevents most neural tube defects. Spina bifida affects 1,500 babies every year in the United States.

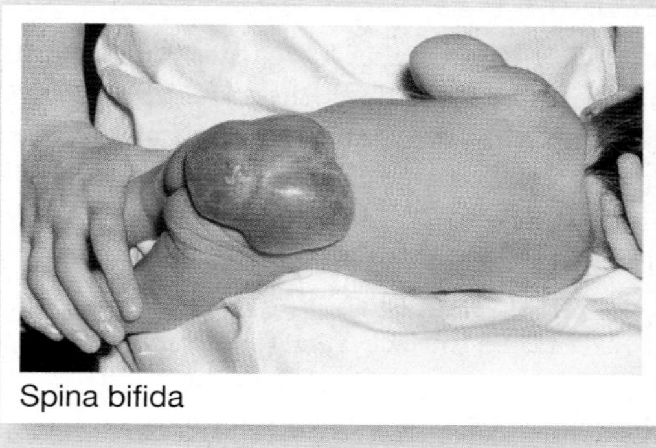

Spina bifida

Table 6.3 | Regional Differences in Vertebral Structure and Function

Type Number	Vertebral Body	Vertebral Foramen	Spinous Process	Transverse Process	Functions
Cervical vertebrae (7) (Figure 6.21)	Small; oval; curved faces	Large	Long; split; tip points inferiorly	Has transverse foramen	Support skull; stabilize relative positions of brain and spinal cord; allow controlled head movement
Thoracic vertebrae (12) (Figure 6.23)	Medium; heart-shaped; flat faces; facets for rib articulations	Smaller	Long; slender; not split; tip points inferiorly	All but two (T_{11}, T_{12}) have facet for rib articulation	Support weight of head, neck, upper limbs, organs of thoracic cavity; articulate with ribs to allow changes in volume of thoracic cage
Lumbar vertebrae (5) (Figure 6.24)	Massive; oval; flat faces	Smallest	Blunt; broad tip points posteriorly	Short; no articular facets or transverse foramen	Support weight of head, neck, upper limbs, organs of thoracic and abdominal cavities

When cervical vertebrae C_3–C_7 articulate, their interlocking vertebral bodies permit a relatively greater degree of flexibility than do those of other regions. **Table 6.3** summarizes the features of cervical vertebrae.

The Atlas (C_1)

Articulating with the occipital condyles of the skull, the **atlas** (C_1) holds up the head (**Figure 6.22a,b**). It is named after Atlas, a figure in Greek mythology who held up the world. The articulation between the occipital condyles and the atlas is a joint that permits nodding (as when indicating "yes") but prevents twisting. The atlas differs from the other vertebrae in several important ways: (1) lacking a body; (2) possessing semicircular **anterior** and **posterior vertebral arches**, each containing **anterior** and **posterior tubercles**; (3) containing oval **superior articular facets** and round **inferior articular facets**; and (4) having the largest vertebral foramen of any vertebra. These modifications provide more free space for the spinal cord, which prevents the cord from damage during the wide range of movements possible in this region. The atlas articulates with the second cervical vertebra, the axis.

The Axis (C_2)

During development, the body of the atlas fuses to the body of the second cervical vertebra, called the **axis** (C_2) (**Figure 6.22c,d**). (*Refer to Chapter 12,* (**Figure 12.9**, *to visualize this structure in a cross section of the body.*) This fusion creates the prominent **dens** of the axis. Therefore, there is no intervertebral disc between the atlas and the axis. A transverse ligament binds the dens to the inner surface of the atlas, allowing for rotation of the atlas and skull, permitting the head to turn from side to side (as when indicating "no"; **Figure 6.22e,f**). Important muscles controlling the position of the head and neck attach to the spinous process of the axis.

In a child, the fusion between the dens and axis is incomplete; impacts or even severe shaking dislocates the dens and severely damages the spinal cord. In an adult, a hit to the base of the skull can be equally dangerous because dislocating the axis–atlas joint can force the dens into the base of the brain, with fatal results.

Vertebra Prominens (C_7)

The transition from one vertebral region to another is not abrupt, and the last vertebra of one region usually resembles the first vertebra of the next. The **vertebra prominens** (C_7) has the most prominent spinous process. Its long, slender spinous process ends in a broad tubercle. You can feel C_7 beneath the skin at the base of the neck. This vertebra, shown in **Figures 6.21a** and **6.23a**, is the junction between the cervical curve, which arches anteriorly, and the thoracic curve, which arches posteriorly. The transverse processes on C_7 are large, providing additional surface area for muscle attachment, and the transverse foramina are smaller or absent. A large elastic ligament, the **ligamentum nuchae** (lig-a-MEN-tum NU-kē; *nucha*, nape), begins at the vertebra prominens and extends cranially to attach to the spinous processes of the other cervical vertebrae and to the external occipital crest of the skull. When the head is upright, this ligament acts like a bow string, maintaining the cervical curvature without muscular effort. If the neck is bent forward, the elasticity in this ligament helps return the head to an upright position.

The head is relatively massive, and it sits atop the cervical vertebrae like a soup bowl on the tip of a finger. With this arrangement, small muscles can produce significant effects by tipping the balance one way or another. But if the body suddenly changes position, as in a fall or during rapid acceleration (a jet taking off) or deceleration (a car crash), the balancing muscles are not strong enough to stabilize the head. A dangerous partial or complete dislocation of the cervical vertebrae can result, injuring muscles and ligaments and potentially injuring the spinal cord. The term **whiplash** describes such an injury because the movement of the head resembles the cracking of a whip.

Thoracic Vertebrae

There are 12 **thoracic vertebrae**. A typical thoracic vertebra has a distinctive heart-shaped body that is larger than that of a cervical vertebra (**Figure 6.23**). The round vertebral foramen is smaller than that of a cervical vertebra, and the long, slender spinous process projects posterocaudally. The spinous processes of T_{10}, T_{11}, and T_{12} resemble those of the lumbar vertebrae more and more as the transition between the thoracic and lumbar curves approaches. Because of the amount of weight supported by the lower thoracic and lumbar vertebrae, it is difficult to stabilize the transition between the thoracic and lumbar curves. As a result, compression fracture dislocations after a hard fall often involve the last thoracic and first two lumbar vertebrae.

Each thoracic vertebra articulates with ribs along the vertebral body along the dorsolateral surfaces. The location and structure of the articulations vary from vertebra to vertebra (**Figure 6.23b,c**). Thoracic vertebrae T_1 to T_8 have *both* **superior** *and* **inferior costal facets**, as they articulate with two pairs of ribs. Vertebrae T_9 to T_{12} have only a *single* costal facet on each side, as they articulate with only one pair of ribs.

The transverse processes of vertebrae T_1 to T_{10} are thick, and their anterolateral surfaces contain **transverse costal facets** for articulation with the rib tubercles. As a result, ribs 1 through 10 contact their vertebrae at *two* points: at a costal facet and at a transverse costal facet, limiting the mobility of the thoracic vertebrae. **Table 6.3**, p. 161, summarizes the features of the thoracic vertebrae.

Lumbar Vertebrae

The **lumbar vertebrae** are the largest vertebrae. The body of a lumbar vertebra is thicker than that of a thoracic vertebra, and the superior and inferior surfaces are oval rather than heart-shaped (**Figure 6.24**). There are no articular facets for ribs on either the body or the transverse processes, and the vertebral foramen is triangular. The transverse processes are slender and project dorsolaterally, and the stumpy spinous processes project posteriorly.

The lumbar vertebrae bear the most weight. For this reason, a compression injury to the vertebrae or intervertebral discs occurs most often in this region. The most common injury is a tear or rupture in the connective tissues of the intervertebral disc, a condition known as a herniated disc. The massive spinous processes of the lumbar vertebrae provide surface area for the attachment of lower back muscles that reinforce or adjust the lumbar curvature. **Table 6.3**, p. 161, summarizes the characteristics of lumbar vertebrae.

The Sacrum

The **sacrum** consists of the fused components of five sacral vertebrae (**Figure 6.25**). These vertebrae begin fusing shortly after puberty and are completely fused between ages 25 and 30. When this fusion is complete, prominent transverse lines mark the former boundaries of individual vertebrae. The sacrum protects reproductive, digestive, and excretory organs and, by paired articulations, attaches the axial skeleton to the pelvic girdle of the appendicular skeleton. The broad surface area of the sacrum provides an extensive area for muscle attachment, especially those muscles responsible for movement of the thigh. (*Refer to Chapter 12,* **Figure 12.14**, *to visualize this structure in a cross section of the body at the level of* L_5.)

The sacrum is curved, with a convex posterior surface (**Figure 6.25a**). The narrow, caudal portion is the sacral **apex**, whereas the broad superior surface forms the **base**. The **sacral promontory**, a prominent bulge at the anterior tip of the base, is an important landmark in females during pelvic examinations and during labor and delivery. The **superior articular processes** form synovial joints with the last lumbar vertebra. The **sacral canal** begins between the superior articular processes and extends caudally the length of the sacrum. Nerves and membranes that line the vertebral canal and the spinal cord continue into the sacral canal.

Figure 6.22 Atlas and Axis. Unique anatomical characteristics of vertebrae C_1 (atlas) and C_2 (axis).

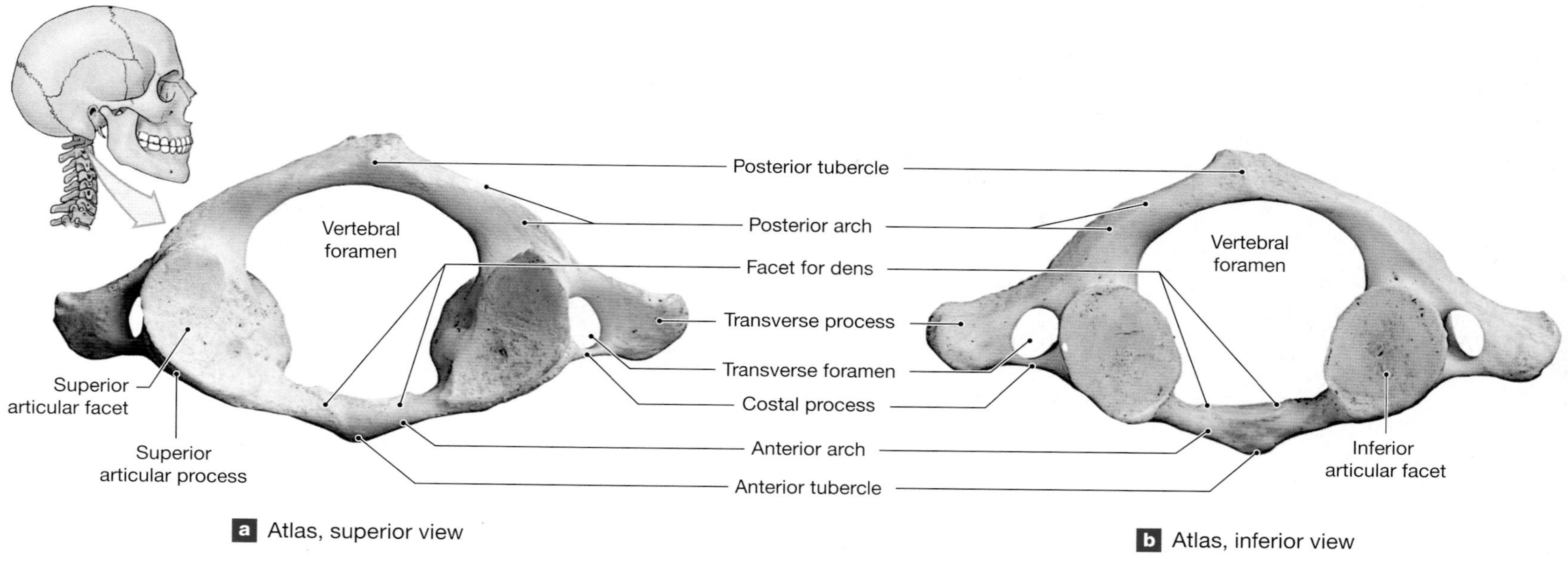

a Atlas, superior view

b Atlas, inferior view

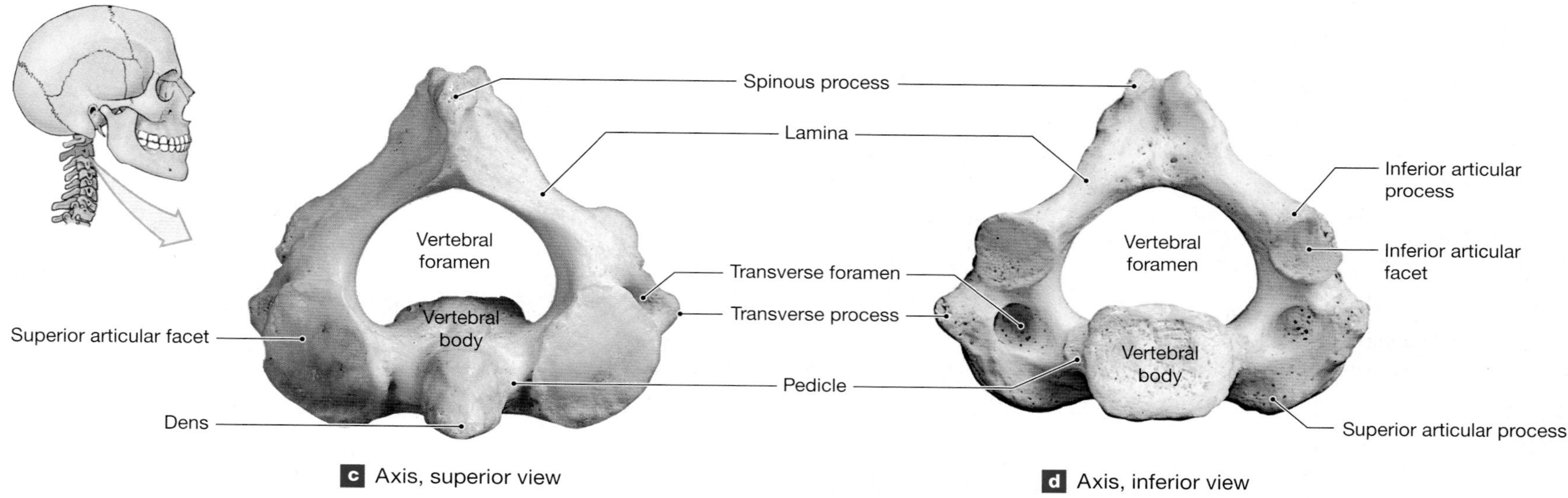

c Axis, superior view

d Axis, inferior view

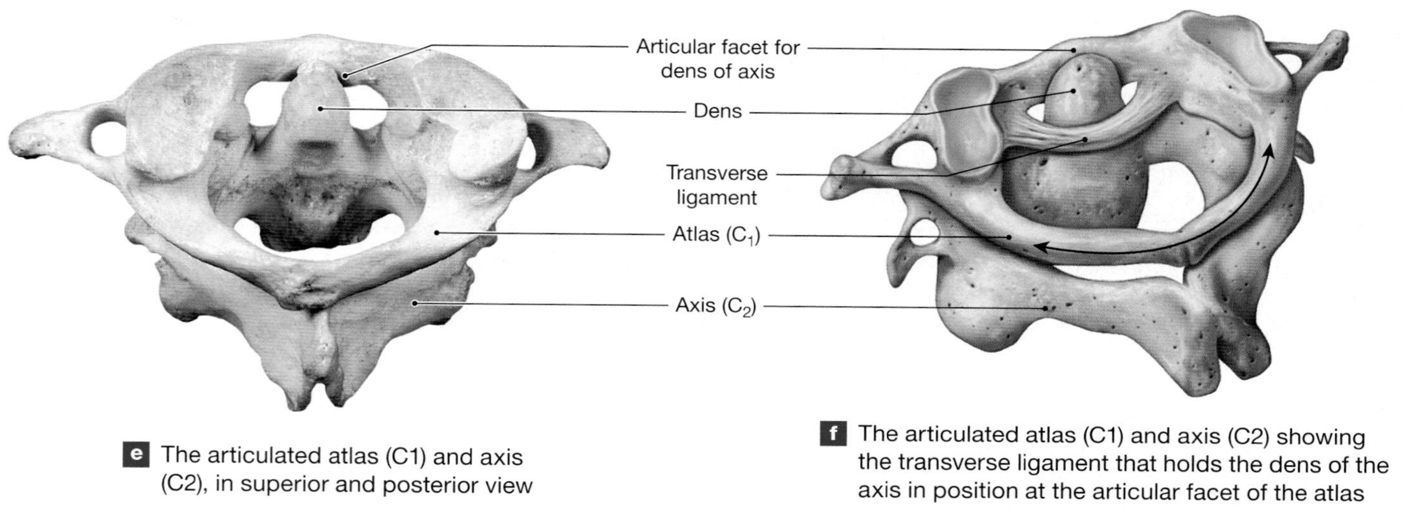

e The articulated atlas (C1) and axis (C2), in superior and posterior view

f The articulated atlas (C1) and axis (C2) showing the transverse ligament that holds the dens of the axis in position at the articular facet of the atlas

Figure 6.23 **Thoracic Vertebrae.** The body of each thoracic vertebra articulates with ribs. Note the characteristic features listed in Table 6.3.

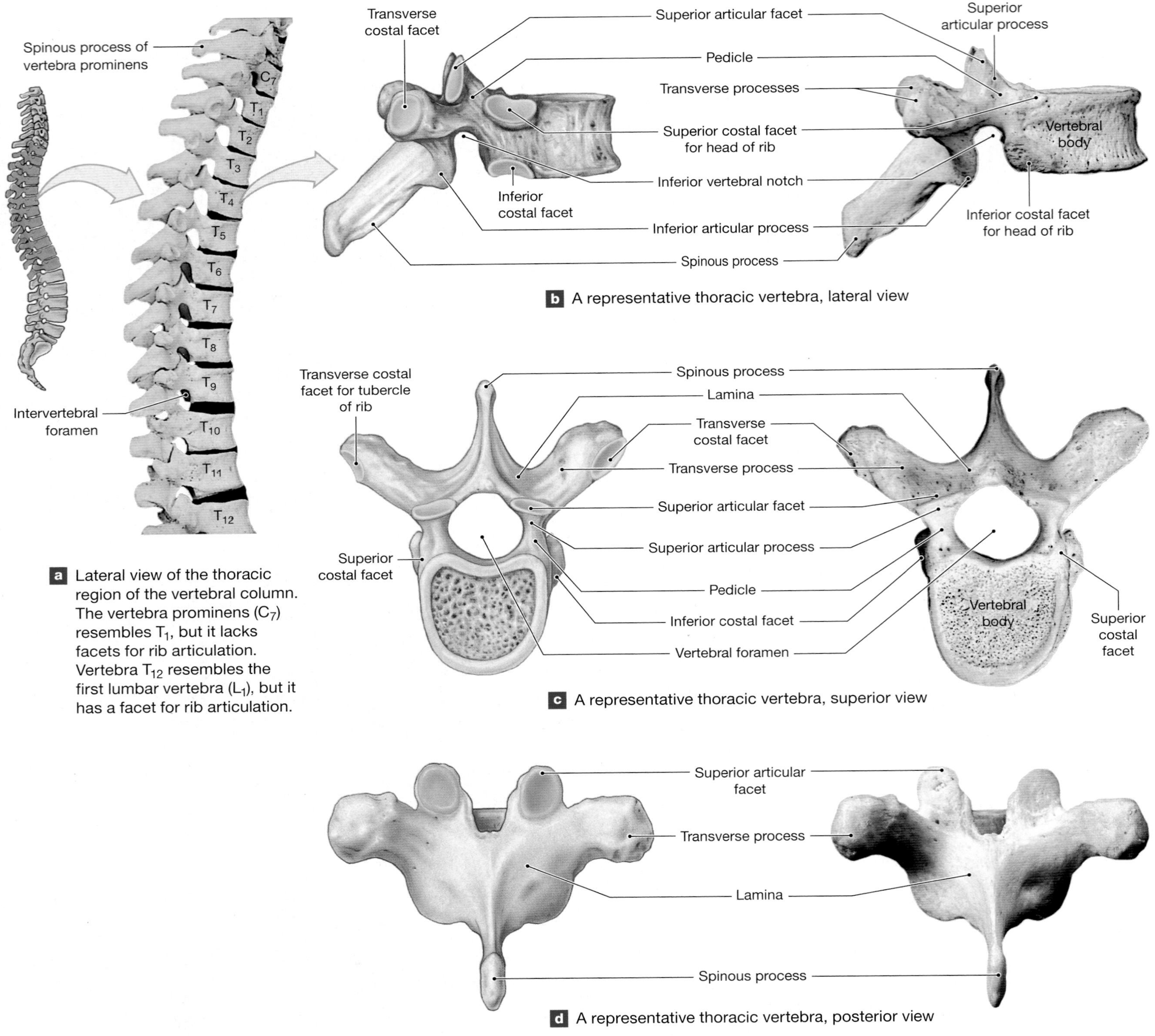

a Lateral view of the thoracic region of the vertebral column. The vertebra prominens (C_7) resembles T_1, but it lacks facets for rib articulation. Vertebra T_{12} resembles the first lumbar vertebra (L_1), but it has a facet for rib articulation.

b A representative thoracic vertebra, lateral view

c A representative thoracic vertebra, superior view

d A representative thoracic vertebra, posterior view

The spinous processes of the five fused sacral vertebrae form a series of elevations along the **median sacral crest**. The laminae of the fifth sacral vertebra do not contact one another at the midline, and they form the **sacral cornua** (singular *cornu*; "horn"). These ridges establish the margins of the **sacral hiatus** (hī-Ā-tus), the end of the sacral canal. In life, this opening is covered by connective tissues. On both sides of the median sacral crest are the **sacral foramina**. The intervertebral foramina, now enclosed by the fused sacral bones, open into these passageways. A broad sacral *wing*, or **ala**, extends laterally from each **lateral sacral crest**. The median and lateral sacral crests provide surface area for the attachment of lower back and hip muscles.

Figure 6.24 **Lumbar Vertebrae.** The lumbar vertebrae are the largest vertebrae and bear the most weight.

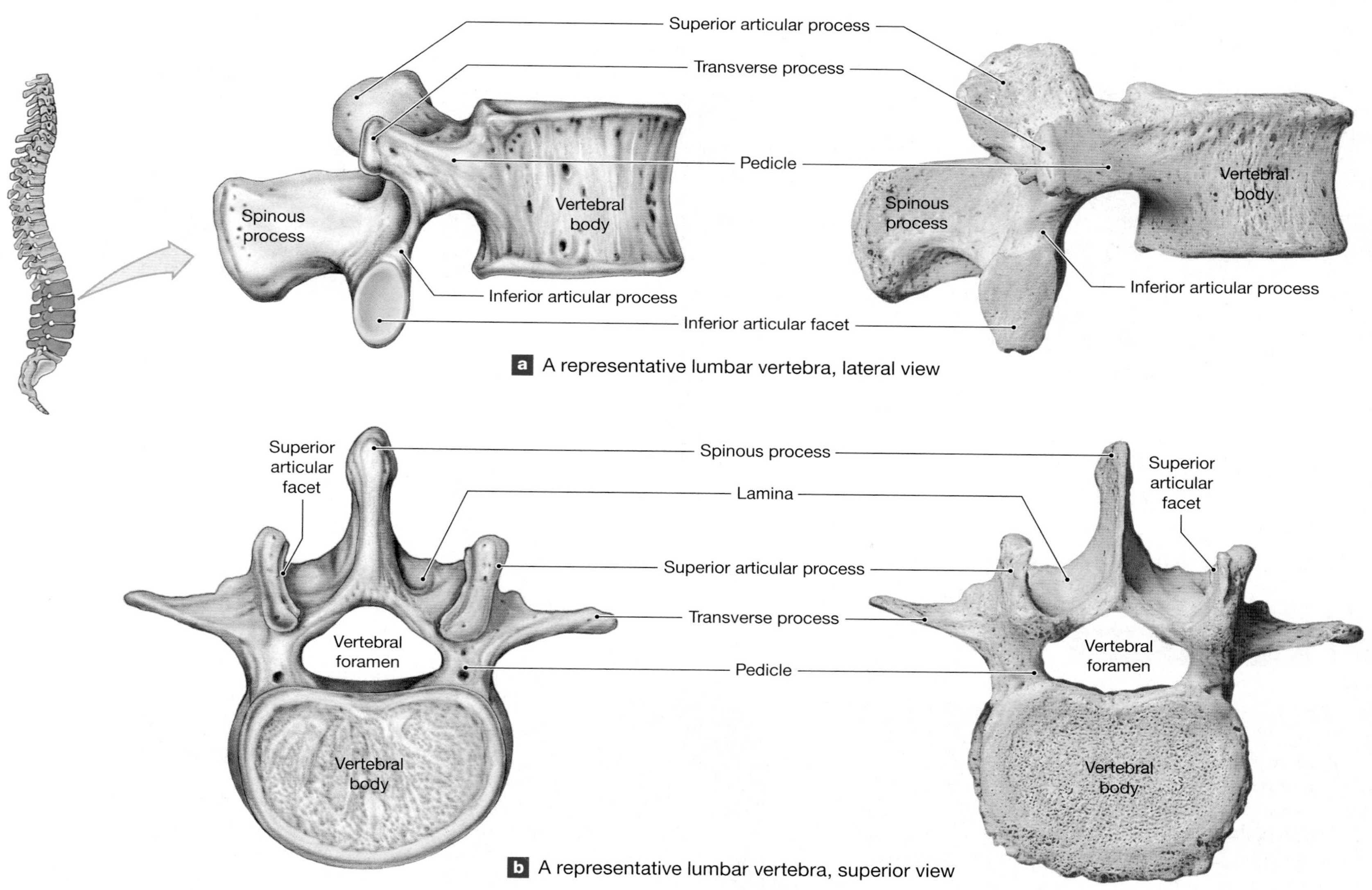

Viewed laterally, the sacral curve is more apparent **(Figure 6.25b)**. The curvature is greater in males than in females (see **Table 7.1**). Laterally, the **auricular surface** of the sacrum articulates with the pelvic girdle at the **sacro-iliac joint**. Posterior to the auricular surface is a roughened area, the **sacral tuberosity**, which marks the attachment of a ligament that stabilizes the sacro-iliac joint. The anterior surface, or **pelvic surface**, of the sacrum is concave **(Figure 6.25c)**. At the apex, a flattened area articulates with the coccyx. The wedge-like shape of the mature sacrum provides a strong foundation for transferring the body's weight from the axial skeleton to the pelvic girdle.

The Coccyx

The small **coccyx** consists of three to five (most often four) coccygeal vertebrae that have begun fusing by age 26 **(Figure 6.25)**. The coccyx is an attachment site for a number of ligaments and for a muscle that constricts the anal opening. The first two coccygeal vertebrae have transverse processes and unfused vertebral arches. The prominent laminae of the first coccygeal vertebra are known as the **coccygeal cornua**; they curve to meet the cornua of the sacrum. The coccygeal vertebrae do not complete their fusion until late in adulthood. In males, the adult coccyx points anteriorly, whereas in females, it points inferiorly. In very elderly people, the coccyx may fuse with the sacrum.

6.7 CONCEPT CHECK

13 Joe suffered a hairline fracture (fracture without separation of the fragments) at the base of the dens. What bone is fractured, and where would you find it?

14 What are the five vertebral regions? What are the identifying features of each region?

15 List the spinal curves in order from superior to inferior.

See the blue Answers tab at the back of the book.

Figure 6.25 The Sacrum and Coccyx. Fused vertebrae form the adult sacrum and coccyx.

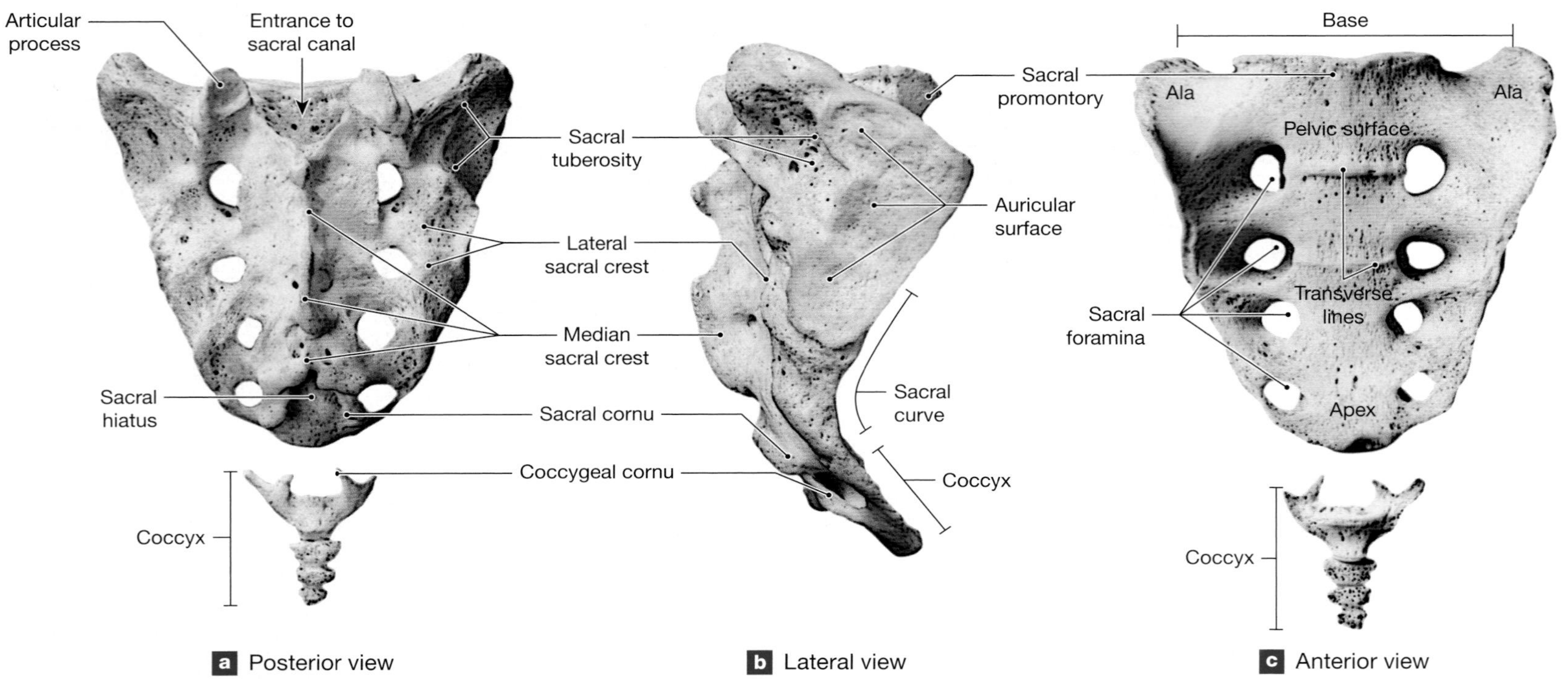

CLINICAL NOTE

Kyphosis, Lordosis, and Scoliosis

The vertebral column, with its many bones and joints, must move, balance, and support the trunk and head.

Kyphosis (kī-FŌ-sis) is an exaggeration of the normal posterior curvature of the thoracic spine, producing a "round back" deformity. This can occur at any age, but is more common in elderly individuals. It is often related to compression fractures of the anterior vertebral bodies or poor posture. Severe cases can cause pain and disfigurement.

Lordosis (lōr-DŌ-sis), or "swayback," is an abnormal anterior curvature of the lumbar spine. Lordosis may result from abdominal wall obesity, pregnancy, or weakness of the trunk muscles.

Scoliosis (skō-lē-Ō-sis) is an abnormal lateral curvature of the spine. Scoliosis may result from abnormal growth of one or more vertebral bodies. It may also be the result of muscular imbalance from muscular dystrophy or cerebral palsy. It is common for scoliosis to develop in adolescent girls during their "growth spurt." The cause of this adolescent scoliosis is often idiopathic, or unknown.

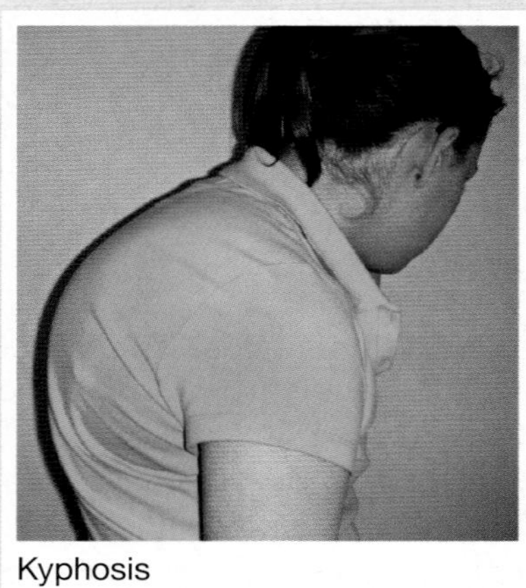
Kyphosis

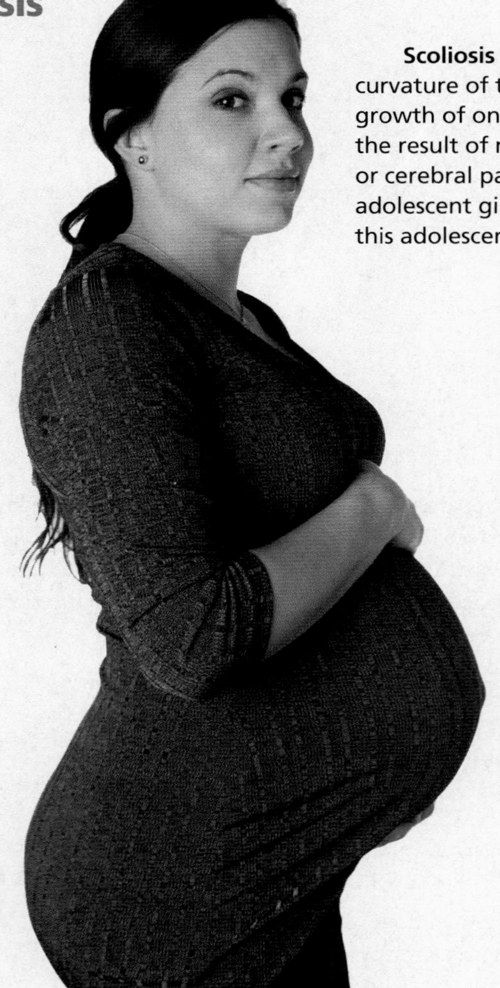

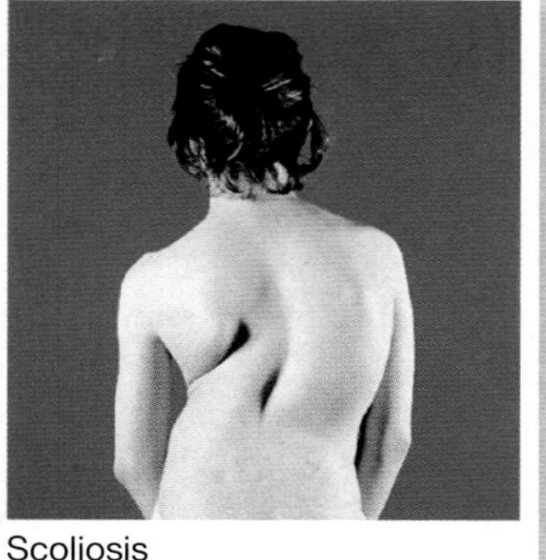
Scoliosis

6.8 The Thoracic Cage

▶ **KEY POINT** The skeleton of the thoracic cage (chest) is composed of the ribs, costal cartilages, and sternum.

The **thoracic cage (Figure 6.26a,c)** serves two functions:

- Protects the heart, lungs, thymus, and other structures in the thoracic cavity and
- Serves as an attachment point for muscles involved with (1) breathing, (2) maintaining the position of the vertebral column, and (3) movements of the pectoral girdle (clavicles and scapulae) and upper limbs.

The Ribs

▶ **KEY POINT** Ribs are elongated, curved, flattened bones that form the borders of the rib cage. All ribs attach posteriorly to or between thoracic vertebrae. Only the first seven ribs attach to the sternum anteriorly.

There are 12 pairs of **ribs (Figure 6.26)**. The upper first seven pairs are **true ribs**, or *vertebrosternal ribs*. Anteriorly, the true ribs are connected to the sternum by separate cartilages, the **costal cartilages**. Beginning with the first rib, the vertebrosternal ribs gradually increase in length and in the radius of curvature.

Ribs 8–10 are called **false ribs** or *vertebrochondral ribs*, because they do not attach directly to the sternum. The costal cartilages of ribs 8–10 fuse together

Figure 6.26 The Thoracic Cage

Jugular notch
T_1
Clavicular articulation
Sternum
Manubrium
Body
Xiphoid process
Costal cartilages
Floating ribs (ribs 11–12)
Vertebrochondral ribs (ribs 8–10)
True ribs (ribs 1–7)
False ribs (ribs 8–12)

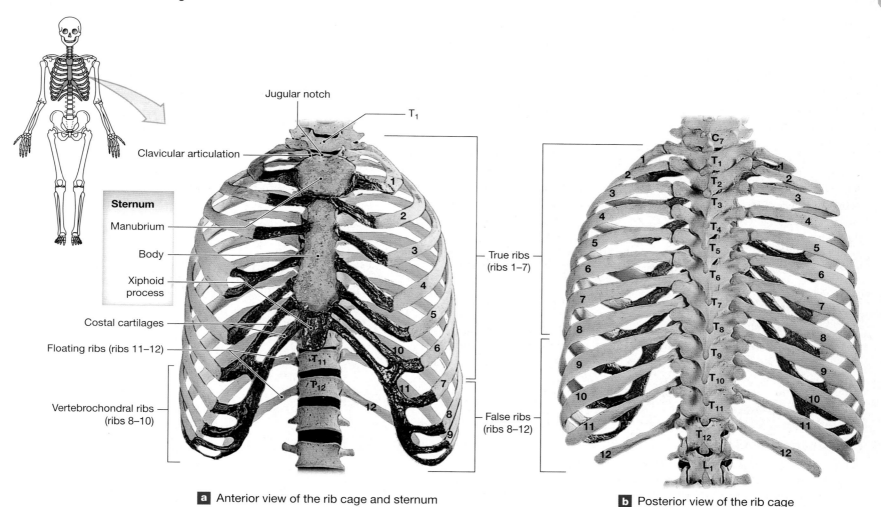

a Anterior view of the rib cage and sternum

b Posterior view of the rib cage

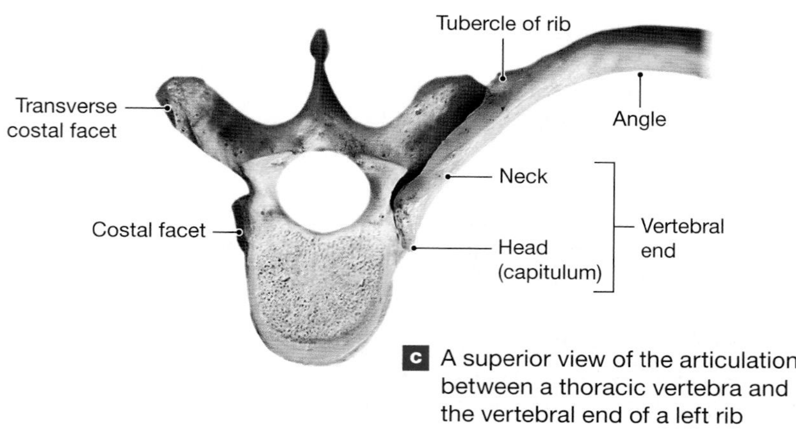

c A superior view of the articulation between a thoracic vertebra and the vertebral end of a left rib

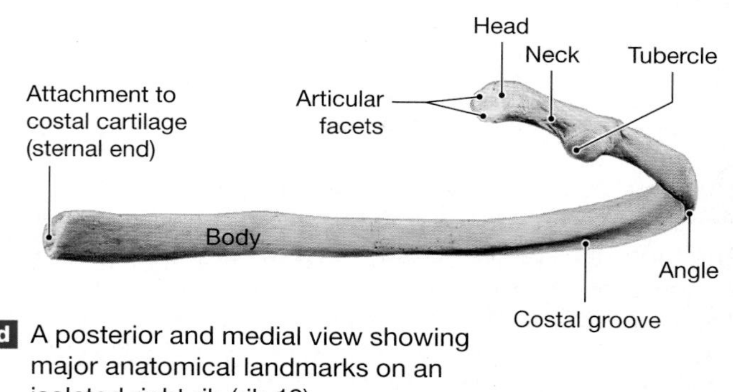

d A posterior and medial view showing major anatomical landmarks on an isolated right rib (rib 10) before reaching the sternum (**Figure 6.26a**). The last two pairs of ribs are called **floating ribs** because they have no connection with the sternum.

Figure 6.26c shows the superior surface of the vertebral end of a rib. The **head** of each rib articulates with the body of a thoracic vertebra or between adjacent vertebral bodies. After a short **neck**, the **tubercle** projects posteriorly. The inferior portion of the tubercle contains an articular facet that contacts the transverse process of thoracic vertebrae T_1 to T_{11}. When the rib articulates between adjacent vertebrae, the **interarticular crest** divides the articular surface into **superior** and **inferior articular facets** (**Figure 6.26c,d**). Ribs 1 through 10 originate at costal facets on the bodies of vertebrae T_1 to T_{10}. The articular facets articulate with the transverse costal facets of their respective vertebrae. Ribs 11 and 12 originate at costal facets on T_{11} and T_{12}. These ribs do not have tubercular facets and do not articulate with transverse processes. The difference in rib orientation and articulation with the vertebral column can be seen by comparing **Figures 6.23**, p. 164, and **6.26c,d**.

The bend, or **angle**, of the rib indicates where the tubular **body**, or *shaft*, begins curving toward the sternum. The internal rib surface is concave, and the **costal groove** on the inferior border marks the path of the intercostal (*inter*, between, + *costal*, ribs) nerves and blood vessels. The superficial surface is convex and is the attachment site for muscles of the pectoral girdle and trunk. The intercostal muscles that move the ribs are attached to the superior and inferior surfaces.

With their complex musculature, dual articulations at the vertebrae, and flexible connection to the sternum, the ribs are quite mobile. Note how the ribs curve away from the vertebral column to angle downward. Functionally, a typical rib acts as if it were the handle on a bucket, lying just below the horizontal plane. Pushing it down forces it inward; pulling it up swings it outward. In addition, because of the curvature of the ribs, the same movements change the position of the sternum. Depressing the ribs moves the sternum posteriorly (inward), whereas elevating the ribs moves it anteriorly (outward). As a result, rib movements affect both the width and the depth of the thoracic cage, increasing or decreasing its volume accordingly.

The Sternum

▶ **KEY POINT** The sternum has three parts and is found on the anterior wall of the thorax. It forms joints with the clavicle (collarbone) and the upper seven pairs of ribs.

The adult **sternum** is a flat bone in the anterior midline of the thoracic wall (**Figure 6.26a**). (*Refer to Chapter 12,* **Figures 12.2a** *and* **12.3**, *to identify this anatomical structure from the body surface.*) There are three parts to the sternum:

- The **manubrium** (ma-NU-brē-um), which articulates with the clavicles (or *collarbones*) of the appendicular skeleton and the costal cartilages of the first pair of ribs. The manubrium is the widest and most superior portion of the sternum. The **jugular notch** is the shallow indentation on the superior surface of the manubrium. It is located between the clavicular articulations.
- The **body** attaches to the inferior surface of the manubrium and extends caudally along the midline. Individual costal cartilages from rib pairs 2–7 are attached to this portion of the sternum. The rib pairs 8–10 are also attached to the body, but by a single pair of cartilages shared with rib pair 7.
- The **xiphoid** (ZĪ-foyd) **process**, the smallest part of the sternum, is attached to the inferior surface of the body. The muscular diaphragm (muscle used in breathing) and the rectus abdominis muscle attach to the xiphoid process.

Ossification of the sternum begins in 6 to 10 different ossification centers, and fusion is completed by age 25 or later. Before age 25, the sternal body consists of four separate bones. Their boundaries are detected as a series of transverse lines crossing the adult sternum. The xiphoid process is the last component of the sternum to undergo ossification and fusion. Its connection to the body of the sternum can be broken by an impact or strong pressure, creating a spear of bone that can severely damage the liver. To reduce the chances of that happening, strong emphasis is placed on the proper positioning of the hand during cardiopulmonary resuscitation (CPR) training.

6.8 CONCEPT CHECK ✔

16 Improper administration of CPR could fracture which bone?

17 What structures are found within the costal groove?

See the blue Answers tab at the back of the book.

Study Outline

Introduction p. 132

- The skeletal system consists of the axial skeleton and the appendicular skeleton. The **axial skeleton** can be subdivided into the **skull** and associated bones (the **auditory ossicles** and **hyoid bone**), the **vertebral column**, and the **thoracic cage**, composed of the **ribs** and **sternum**. *(See Figure 6.1.)*
- The **appendicular skeleton** includes the **pectoral** and **pelvic girdles**, which support and attach the upper and lower limbs to the trunk. *(See Figure 6.1.)*

6.1 The Skull and Associated Bones p. 133

- The **skull** consists of the cranium and the bones of the face. Skull bones protect the brain and guard entrances to the respiratory and digestive systems. Eight skull bones form the **cranium**, which encloses the **cranial cavity**, a division of the posterior body cavity. The **facial bones** protect and support the entrances to the respiratory and digestive systems. *(See Figures 6.2–6.5, 12.1, and 12.9 and Tables 6.1 and 6.2.)*

6.2 Sutures of the Skull p. 140

- Prominent superficial landmarks on the skull include the **lambdoid**, **sagittal**, **coronal**, **squamous**, and **frontonasal sutures**. **Sutures** are immovable joints that form boundaries between skull bones. *(See Figure 6.3a–d and Tables 6.1 and 6.2.)*

6.3 Bones of the Cranium p. 140

- For articulations of cranial bones with other cranial bones and/or facial bones, *see Table 6.2*.

Occipital Bone p. 140

- The **occipital bone** forms part of the base of the skull. It surrounds the **foramen magnum** and forms part of the wall of the **jugular foramen**. *(See Figures 6.3a–c,e and 6.6a,b.)*

Parietal Bones p. 145

- The **parietal bones** form part of the superior and lateral surfaces of the cranium. *(See Figure 6.3b,c, 6.5, and 6.6c.)*

Frontal Bone p. 145

- The **frontal bone** forms the forehead and roof of the orbits. *(See Figures 6.3b–d, 6.5, and 6.7.)*

Temporal Bones p. 145

- The **temporal bone** forms part of the wall of the jugular foramen and houses the **carotid canal**. The thick **petrous part** of the temporal bone houses the tympanic cavity containing the **auditory ossicles**. The auditory ossicles transfer sound vibrations from the tympanic membrane to a fluid-filled chamber in the internal ear. *(See Figures 6.3c–e and 6.8.)*

Sphenoid p. 148

- The **sphenoid** contributes to the floor of the cranium and is a bridge between the cranial and facial bones. Optic nerves pass through the **optic canal** in the sphenoid to reach the brain. **Pterygoid processes** form plates that serve as sites for attachment of muscles that move the mandible and soft palate. *(See Figures 6.3c–e, 6.4, and 6.9.)*

Ethmoid p. 149

- The **ethmoid** is an irregularly shaped bone that forms part of the orbital wall and the roof of the nasal cavity. The **cribriform plate** of the ethmoid contains perforations for olfactory nerves. The **perpendicular plate** forms part of the nasal septum. *(See Figures 6.3d, 6.4, 6.5, and 6.10.)*

Cranial Fossae p. 149

- **Cranial fossae** are curving depressions in the cranial floor that closely follow the shape of the brain. The frontal bone, the ethmoid, and the **lesser wings** of the sphenoid form the **anterior cranial fossa**. The sphenoid, temporal, and parietal bones form the **middle cranial fossa.** The **posterior cranial fossa** is formed primarily by the occipital bone, with contributions from the temporal and parietal bones. *(See Figure 6.11.)*

6.4 | Bones of the Face p. 150

- For articulations of facial bones with other facial bones and/or cranial bones, *see Table 6.2.*

The Maxillae p. 150

- The left and right **maxillae**, or maxillary bones, are the largest facial bones and form the upper jaw. *(See Figures 6.3d and 6.12.)*

The Palatine Bones p. 150

- The **palatine bones** are small, L-shaped bones that form the posterior portions of the bony palate and contribute to the floor of the orbit. *(See Figures 6.3e and 6.13.)*

The Nasal Bones p. 152

- The paired **nasal bones** articulate with the frontal bone at the midline and articulate with cartilages that form the superior borders of the **nostrils**. *(See Figures 6.3c,d, 6.15, and 6.16.)*

The Inferior Nasal Conchae p. 153

- One **inferior nasal concha** is located on each side of the nasal septum, attached to the lateral wall of the nasal cavity. They increase the epithelial surface area and create turbulence in the inspired air. The superior and middle conchae of the ethmoid also create turbulence in the inspired air. *(See Figures 6.3d and 6.16.)*

The Zygomatic Bones p. 153

- The **temporal process** of the **zygomatic bone** articulates with the **zygomatic process** of the temporal bone to form the **zygomatic arch** (*cheekbone*). *(See Figures 6.3c,d and 6.15.)*

The Lacrimal Bones p. 153

- The paired **lacrimal bones** are the smallest bones in the skull. They are located in the medial portion of each orbit. Each lacrimal bone forms a **lacrimal groove** with the adjacent maxilla, and this groove leads to a **nasolacrimal canal** that delivers tears to the nasal cavity. *(See Figures 6.3c,d and 6.15.)*

The Vomer p. 153

- The **vomer** forms the inferior portion of the **nasal septum**. It is based on the floor of the nasal cavity and articulates with both the maxillae and the palatines along the midline. *(See Figures 6.3d,e, 6.5, and 6.16.)*

The Mandible p. 153

- The **mandible** is the entire lower jaw. It articulates with the temporal bone at the temporomandibular joint (TMJ). *(See Figures 6.3c,d and 6.14.)*

6

6.5 | The Orbits, Nasal Complex and the Hyoid Bone p. 154

The Orbits p. 154

- Seven bones form each **orbit**, a bony recess that contain an eye: frontal, lacrimal, palatine, and zygomatic bones and the ethmoid, sphenoid, and maxillae. *(See Figure 6.15.)*

The Nasal Complex p. 154

- The **nasal complex** includes the bones and cartilage that enclose the nasal cavities and the **paranasal sinuses**. Paranasal sinuses are hollow airways that interconnect with the nasal passages. Large paranasal sinuses are present in the frontal bone and the sphenoid, ethmoid, and maxillae. *(See Figures 6.5 and 6.16.)*

The Hyoid Bone p. 156

- The **hyoid bone**, suspended by **stylohyoid ligaments**, consists of a **body**, the **greater horns**, and the **lesser horns**. The hyoid bone serves as a base for several muscles concerned with movements of the tongue and larynx. *(See Figure 6.17.)*

6.6 | The Skulls of Infants, Children, and Adults p. 156

- Fibrous connections at **fontanelles** permit the skulls of infants and children to continue growing. *(See Figure 6.18.)*

6.7 | The Vertebral Column p. 158

- The adult **vertebral column** consists of 26 bones (24 individual **vertebrae**, the **sacrum**, and the **coccyx**). There are 7 **cervical vertebrae** (the first articulates with the occipital bone), 12 **thoracic vertebrae** (which articulate with the ribs), and 5 **lumbar vertebrae** (the fifth articulates with the sacrum). The sacrum and coccyx consist of fused vertebrae. *(See Figures 6.19–6.25.)*

Spinal Curves p. 158

- The spinal column has four **spinal curves**: the **thoracic** and **sacral curves** are called primary curves; the **lumbar** and **cervical curves** are known as secondary **curves**. *(See Figure 6.19.)*

Vertebral Anatomy p. 158

- A typical **vertebra** has a thick, supporting **body**; it has a **vertebral arch** formed by walls (**pedicles**) and a roof (**lamina**) that provide a space for the spinal cord; and it articulates with other vertebrae at the **superior** and **inferior articular processes**. *(See Figure 6.20.)*
- Adjacent vertebrae are separated by **intervertebral discs**. Spaces between successive pedicles form the **intervertebral**

foramina, through which nerves pass to and from the spinal cord. (See Figure 6.20.)

Vertebral Regions p. 158

- **Cervical vertebrae** are distinguished by the shape of the vertebral body, the relative size of the vertebral foramen, the presence of **costal processes** with **transverse foramina**, and **bifid spinous processes**. *(See Figures 6.19, 6.21, and 6.22 and Table 6.3.)*
- **Thoracic vertebrae** have distinctive heart-shaped bodies, long, slender spinous processes, and articulations for the ribs. *(See Figures 6.19 and 6.23.)*
- The **lumbar vertebrae** are the largest and least mobile; they are subjected to the greatest strains. *(See Figures 6.19 and 6.24.)*
- The **sacrum** protects reproductive, digestive, and excretory organs. It has an **auricular surface** for articulation with the pelvic girdle. The sacrum articulates with the fused elements of the **coccyx**. *(See Figures 6.25 and 12.14.)*

6.8 The Thoracic Cage p. 167

- The skeleton of the **thoracic cage** consists of the thoracic vertebrae, ribs, costal cartilages, and sternum. The ribs and sternum form the rib cage. *(See Figure 6.26a,c.)*

The Ribs p. 167

- **Ribs** 1–7 are **true**, or *vertebrosternal*, **ribs**. Ribs 8–10 are called **false**, or *vertebrochondral*, **ribs**. The last two pairs of ribs are **floating ribs**. The vertebral end of a typical rib articulates with the vertebral column at the **head**. After a short **neck**, the **tubercle** projects posteriorly. A bend, or **angle**, of the rib indicates the site where the tubular **body**, or shaft, begins curving toward the sternum. A prominent, inferior **costal groove** marks the path of nerves and blood vessels. *(See Figures 6.23 and 6.26.)*

The Sternum p. 168

- The **sternum** consists of a **manubrium**, a **body**, and a **xiphoid process**. *(See Figures 6.26a, 12.2a, and 12.3.)*

Chapter Review

For answers, see the blue Answers tab at the back of the book.

Level 1 Reviewing Facts and Terms

Match each numbered item with the most closely related lettered item.

1. suture ______
2. foramen magnum ______
3. mastoid process ______
4. optic canal ______
5. crista galli ______
6. condylar process ______
7. transverse foramen ______
8. costal facets ______
9. manubrium ______
10. upper jaw ______

a. mandible
b. boundary between skull bones
c. maxillae
d. cervical vertebrae
e. occipital bone
f. sternum
g. thoracic vertebrae
h. temporal bone
i. ethmoid
j. sphenoid

11. Which of the following is/are true of the ethmoid?
    (a) It contains the crista galli.
    (b) It contains the cribriform plate.
    (c) It serves as the anterior attachment of the falx cerebri.
    (d) all of the above
12. Which statement applies to the sella turcica?
    (a) It supports and protects the pituitary gland.
    (b) It is bounded directly laterally by the foramen spinosum.
    (c) It does not develop until after birth.
    (d) It permits passage of the optic nerves.
13. Label the following structures on the diagram of the skull below.
    - frontal bone
    - temporal process of the zygomatic bone
    - lambdoid suture
    - sphenoid
    - styloid process

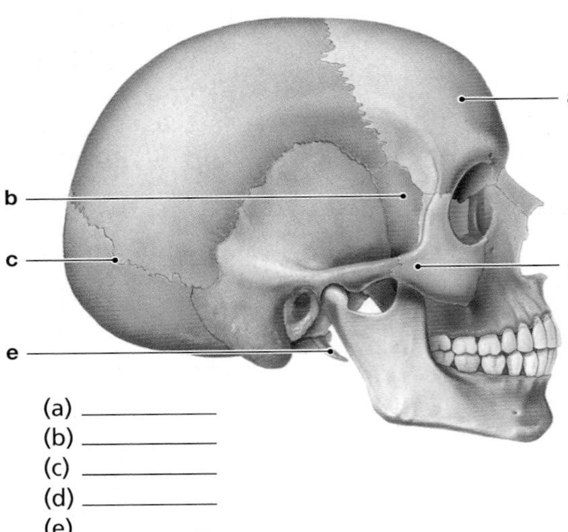

(a) ______
(b) ______
(c) ______
(d) ______
(e) ______

14. The portion of the sternum that articulates with the clavicles is the
    (a) manubrium.
    (b) body.
    (c) xiphoid process.
    (d) angle.
15. The role of fontanelles is to
    (a) allow for compression of the skull during childbirth.
    (b) serve as ossification centers for the facial bones.
    (c) serve as the final bony plates of the skull.
    (d) lighten the weight of the skull bones.
16. Label the following structures on the following diagram of a superior view of a vertebra.
    - pedicle
    - transverse process
    - lamina
    - spinous process

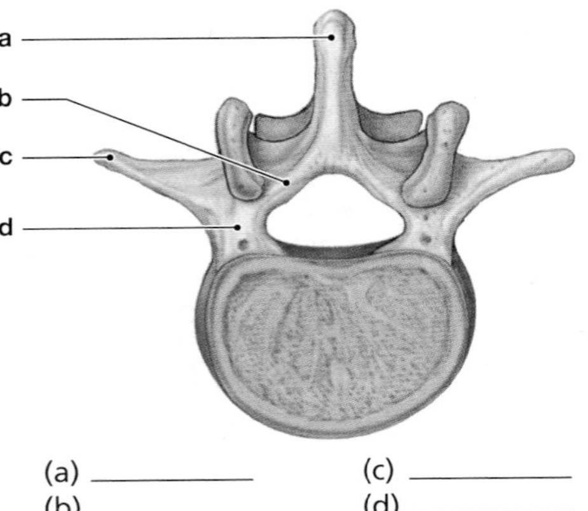

(a) ______ (b) ______ (c) ______ (d) ______

17. The sacrum
    (a) provides protection for reproductive, digestive, and excretory organs.
    (b) bears the most weight in the vertebral column.
    (c) articulates with the pectoral girdle.
    (d) is composed of vertebrae that are completely fused by puberty.
18. The prominent groove along the inferior border of the internal rib surface
    (a) provides an attachment for intercostal muscles.
    (b) is called the costal groove.
    (c) marks the path of nerves and blood vessels.
    (d) Both b and c are correct.

19. The lower jaw articulates with the temporal bone at the
    (a) mandibular fossa.
    (b) mastoid process.
    (c) anterior clinoid process.
    (d) cribriform plate.
20. The hyoid bone
    (a) serves as a base of attachment for muscles that move the tongue.
    (b) is part of the mandible.
    (c) is located inferior to the larynx.
    (d) articulates with the maxillae.
21. The vertebral structure that has a pedicle and a lamina, and from which the spinous process projects, is the
    (a) centrum.
    (b) transverse process.
    (c) inferior articular process.
    (d) vertebral arch.

Level 2 Reviewing Concepts

1. As you move inferiorly from the atlas, you will note that free space for the spinal cord is greatest at C_1. What function does this increased space serve?
2. What is the relationship between the pituitary gland and the sphenoid bone?
3. The secondary curves of the vertebral column, which develop several months after birth, shift the trunk weight over the legs. What does this shifting of weight help accomplish?
4. Describe the relationship between the ligamentum nuchae and the axial skeleton with respect to holding the head in the upright position.
5. Discuss factors that can cause increased mucus production by the mucous membranes of the paranasal sinuses.
6. Why are the largest vertebral bodies found in the lumbar region?
7. What is the relationship between the temporal bone and the ear?
8. What is the purpose of the many small openings in the cribriform plate of the ethmoid bone?

Level 3 Critical Thinking

1. Symptoms of the common cold or the flu include an ache in the teeth in the maxillae, even though there is nothing wrong with them, and a heavy feeling in the front of the head. What anatomical response to the infection causes these unpleasant sensations?
2. A model is said to be very photogenic and is often complimented for her high cheekbones and large eyes. Do these features have an anatomical basis, or could they be explained in another manner?

MasteringA&P™

Access more chapter study tools online in the Study Area:

- Chapter Quizzes, Chapter Practice Test, Clinical Cases, and more!
- Practice Anatomy Lab (PAL) PAL™
- A&P Flix for anatomy topics A&PFlix™

6

CLINICAL CASE WRAP-UP

The Last Lap

Earnhardt suffered a basilar skull fracture, which involves the base of the cranium. His was a ring pattern fracture, forming a half ring around the foramen magnum of the occipital bone. This extended through both temporal bones, which disrupted the internal and external acoustic meatuses, tore the tympanic membranes, and formed an abnormal pathway for blood to exit both ears. His entire cranial cavity was disconnected from his body. The trauma did massive damage to the brain—causing immediate death.

After Earnhardt's death, NASCAR made the HANS a mandatory piece of safety equipment, preventing many injuries and deaths when drivers hit the wall at high speed.

1. Aside from the occipital and temporal bones, what other bones could a basilar skull fracture involve?
2. If the carotid artery within each carotid canal and the jugular vein within each jugular foramen were disrupted by a fracture, which bone would be involved?

See the blue Answers tab at the back of the book.

Related Clinical Terms

craniotomy: The surgical removal of a section of bone (bone flap) from the skull for the purpose of operating on the underlying tissues.

deviated nasal septum: A bent nasal septum (cartilaginous structure dividing the left and right nasal cavities) that slows or prevents sinus drainage.

laminectomy: A surgical operation to remove the posterior vertebral arch on a vertebra, usually to give access to the spinal cord or to relieve pressure on nerves.

spina bifida: A condition resulting from the failure of the vertebral laminae to unite during development; commonly associated with developmental abnormalities of the brain and spinal cord.

spinal fusion: A surgical procedure that stabilizes the spine by joining together (fusing) two or more vertebrae using bone grafts, metal rods, or screws.

7 The Skeletal System

Appendicular Division

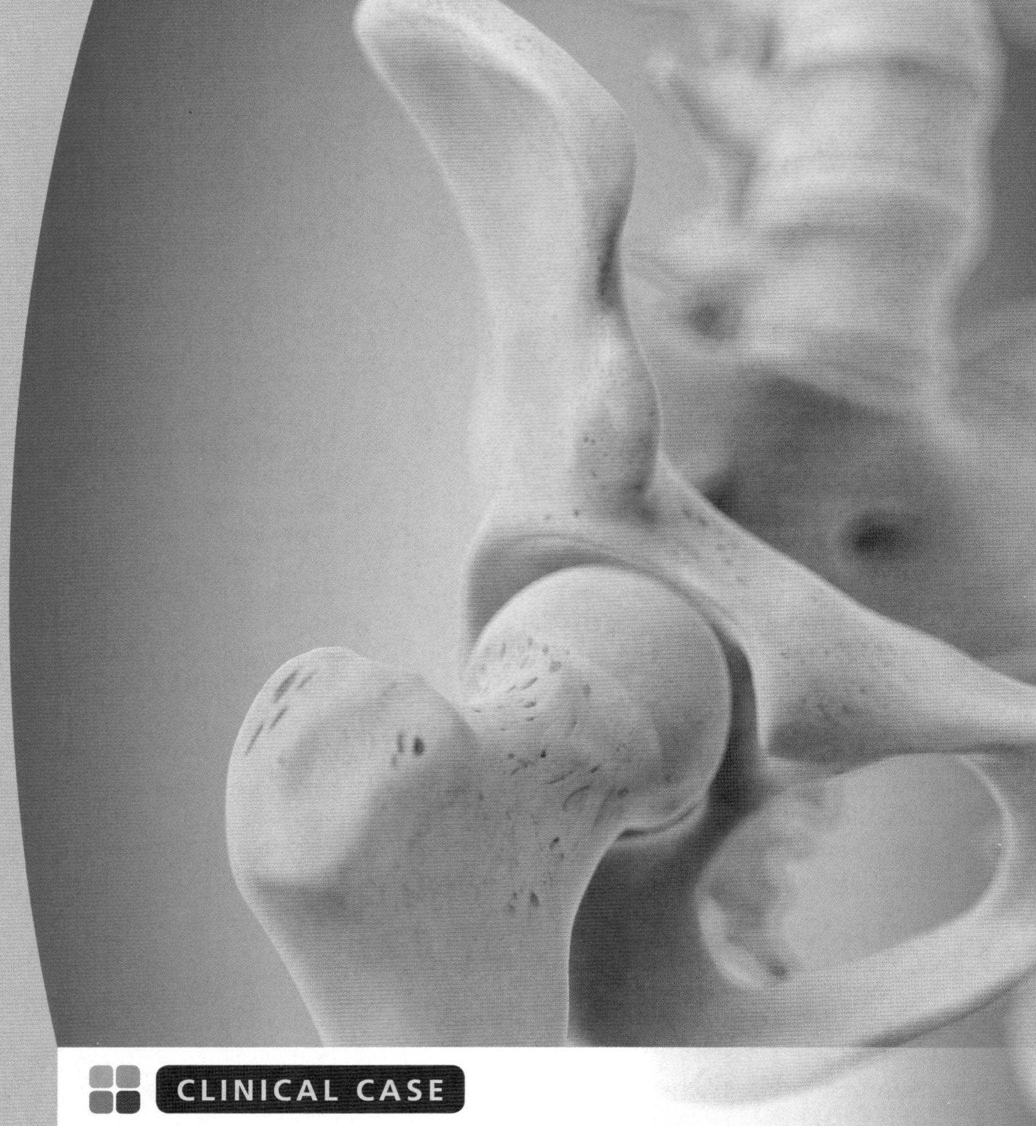

Learning Outcomes

These Learning Outcomes correspond by number to this chapter's sections and indicate what you should be able to do after completing the chapter.

- **7.1** Identify the bones that form the pectoral girdle and upper limb and their prominent surface features. p. 174
- **7.2** Identify the bones that form the pelvic girdle and lower limb and their prominent surface features. p. 184
- **7.3** Explain how studying the skeleton can reveal important information about an individual. p. 197

CLINICAL CASE

Double Jeopardy

Monique, a 28-year-old receptionist, is 7 months pregnant. As she drives to her ultrasound appointment, safely buckled in, a pickup truck runs a red light at 50 mph and broadsides Monique's car. Monique's head breaks the driver's-side door window, and she loses consciousness. The crushed door pins her to the car seat.

The emergency response team who arrives at the scene uses the Jaws of Life to cut Monique out of the car, straps a cervical collar on her, administers oxygen by mask, and transports her to the nearest trauma center.

As soon as Monique regains consciousness, she cries, "My baby!" She feels intense pain in the entire left side of her body, particularly in her hip. She cannot move her left lower limb, and she cannot feel any movement of her fetus. An orthopedic surgeon and an obstetrician at her bedside explain the next steps in her care. "Fetal monitoring shows your baby's heart rate is normal and steady, Monique. First we'll get an MRI of your pelvis, then we'll head to the operating room."

What has happened to Monique and to her fetus? What plans do the doctors have to treat them both?

To find out, turn to the Clinical Case Wrap-Up on p. 202.

MAKE A LIST of the things you've done today, and you'll see that your appendicular skeleton plays a major role in your life. Your axial skeleton protects and supports internal organs and participates in vital functions, such as breathing. But your appendicular skeleton gives you control over your environment, changes your position in space, and gives you mobility.

The bones of the upper and lower limbs and the supporting elements, called **girdles**, make up the **appendicular skeleton** (**Figure 7.1**). This chapter describes the bones of the appendicular skeleton, emphasizing surface features that have functional importance and highlighting the interactions between the skeletal system and other systems. For example, many of the anatomical features noted in this chapter are attachment sites for skeletal muscles or openings for nerves and blood vessels that supply the bones or other organs of the body.

There are direct anatomical connections between the skeletal and muscular systems. The deep fascia surrounding a skeletal muscle is continuous with its tendon, which continues into the periosteum and becomes part of the bone matrix at its attachment site. ⮌ **pp. 73–74, 110–114** Muscles and bones are also linked physiologically, because muscle contractions occur only when the extracellular calcium concentration stays within narrow limits. The skeleton contains most of the body's calcium, and these reserves are vital to calcium homeostasis.

Figure 7.1 The Appendicular Skeleton. A flowchart showing the relationship of the components of the appendicular skeleton: pectoral and pelvic girdles, and upper and lower limbs.

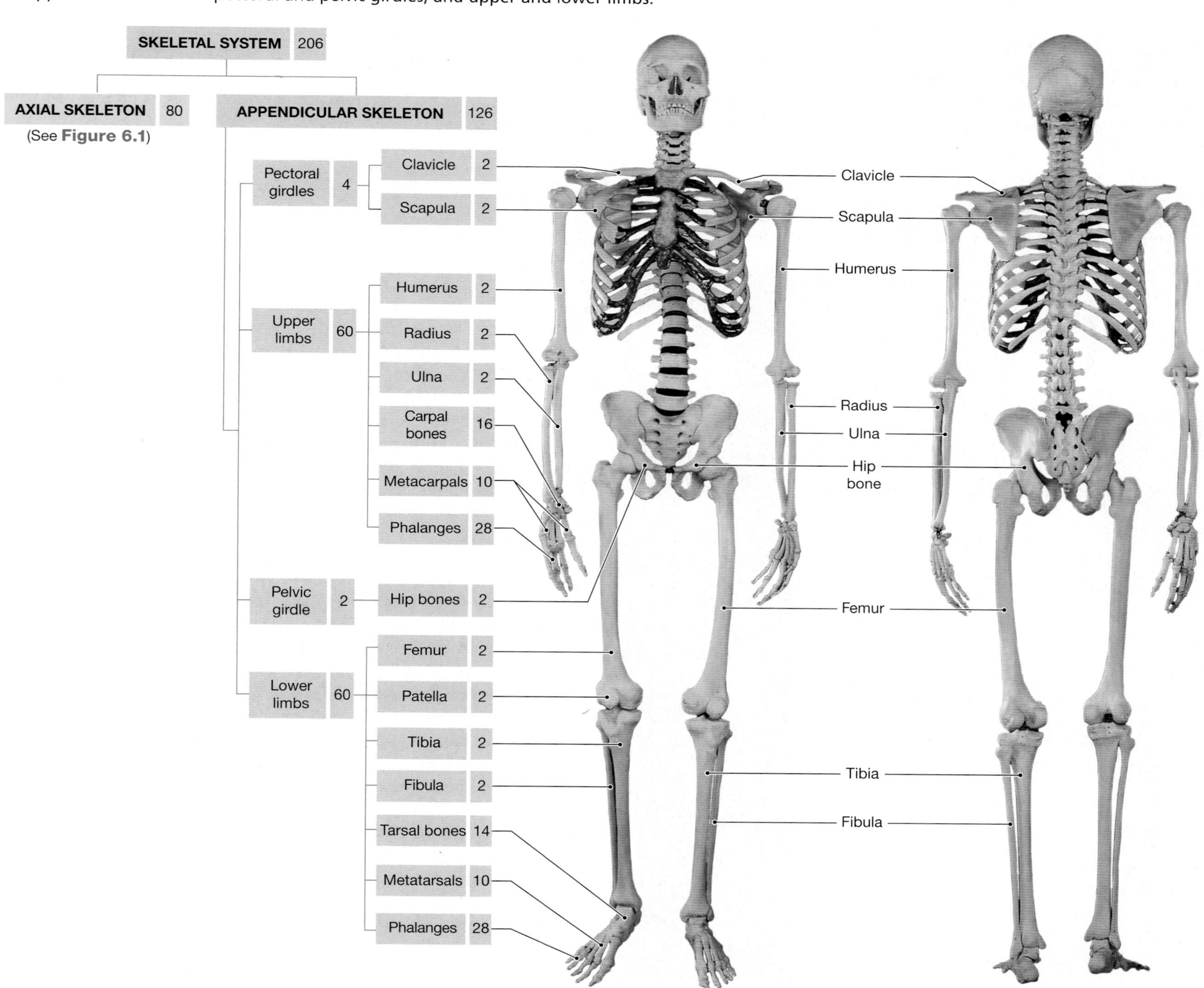

a Anterior view of the skeleton with appendicular components indicated. The flowchart shows relationships among the skeletal parts, and the boxed numbers indicate the number of bones.

b Posterior view of the skeleton.

7.1 | The Pectoral Girdle and Upper Limb

▶ **KEY POINT** Each pectoral girdle and upper limb consists of 32 bones: clavicle, scapula, humerus, radius, ulna, 8 carpal bones of the wrist, and 5 metacarpal and 14 phalanges of the hand. The most important function of the upper limb is to position the hand correctly so you can complete your desired task.

The skeleton of the upper limb consists of the humerus of the arm, the ulna and radius of the forearm, the carpal bones of the wrist, and the metacarpals and phalanges of the hand. Each arm articulates (forms a joint) with the trunk at the **pectoral girdle**, or *shoulder girdle* (**Figure 7.2**). The S-shaped clavicle articulates with the manubrium of the sternum in the *only* direct bony connection between the pectoral girdle and the axial skeleton. Skeletal muscles support and position the scapulae, which have no direct bony or ligamentous connections to the thoracic cage. As a result, the shoulders are extremely mobile, but not very strong.

The Pectoral Girdle

▶ **KEY POINT** The position of the clavicle allows increased movement of the upper limb. The clavicle forms joints with the manubrium of the sternum and the acromion of the scapula. The scapula is a triangular bone lying on the posterior surface of the thorax.

Figure 7.2 The Pectoral Girdle and Upper Limb. Each upper limb articulates with the axial skeleton at the trunk through the pectoral girdle.

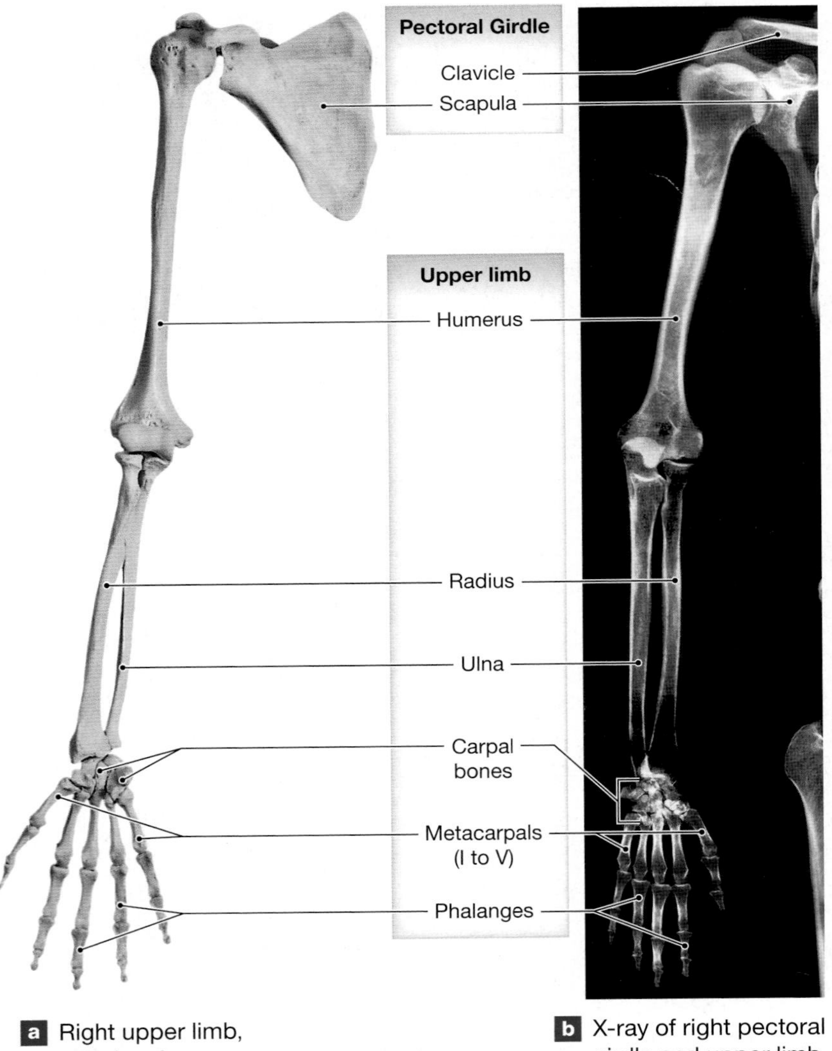

a Right upper limb, anterior view

b X-ray of right pectoral girdle and upper limb, posterior view

Movements of the clavicle and scapula position the shoulder joint, provide a base for arm movement, and help maximize the range of motion. Skeletal muscles between the pectoral girdle and axial skeleton stabilize the joint and help move the upper limb. The surfaces of the scapula and clavicle are extremely important sites for muscle attachment. Where major muscles attach, they leave their marks, creating bony ridges, tubercles, and tuberosities. Other bone markings, such as grooves or foramina, indicate the position of nerves or blood vessels that control the muscles and nourish the muscles and bones.

The Clavicle

The **clavicle** (KLAV-i-kel) connects the pectoral girdle and the axial skeleton and transfers some of the weight of the upper limb to the axial skeleton (**Figure 7.3**). *(Refer to Chapter 12,* **Figure 12.2a**, *to identify these anatomical structures from the body surface and* **Figure 12.10** *to visualize this structure in a cross section of the body at the level of* T_2.*)* Each clavicle originates at the superolateral border of the manubrium of the sternum, lateral to the jugular notch (see **Figure 6.26a,** ↺ p. 167 and **Figure 7.4**). From the **sternal end**, the clavicle curves in an S shape laterally and posteriorly until it articulates with the acromion of the scapula. The **acromial end** of the clavicle is broader and flatter than the sternal end.

The smooth, superior surface of the clavicle lies just deep to the skin. Lines and tubercles indicating the attachment sites for muscles and ligaments mark the rough inferior surface of the acromial end. The **conoid tubercle** is on the inferior surface at the acromial end, and the **costal tuberosity** is at the sternal end. These are attachment sites for ligaments of the shoulder.

You can explore the interaction between scapulae and clavicles. Place your fingers in the jugular notch of your sternum and locate the clavicle on either side. As you move your shoulders, you will feel the clavicles change their positions. Because the clavicles are so close to the skin, you can trace one laterally until it articulates with the scapula. The clavicle's position at the sternoclavicular joint limits shoulder movements, as shown in **Figure 7.4** (The structure of the sternoclavicular joint will be described in Chapter 8.)

Fractures of the medial portion of the clavicle are common because a fall on the palm of an outstretched hand produces compressive forces that are conducted to the clavicle and its joint with the manubrium of the sternum. Most clavicular fractures heal fairly quickly with a simple clavicle strap.

The Scapula

The **body** of the **scapula** (SKAP-ū-lah) forms a broad triangle, and its many surface markings are attachment sites for muscles, tendons, and ligaments (**Figure 7.5a,c,d,f**). *(Refer to Chapter 12,* **Figures 12.2b** *and* **12.10**, *to identify this structure from the body surface and to visualize this structure in a cross section of the body at the level of* T_2.*)* The three sides of the scapular triangle are the **superior border**; the **medial**, or *vertebral*, **border**; and the **lateral**, or *axillary*, **border**. Muscles that position the scapula attach along these edges. The corners of the scapular triangle are called the **superior angle**, the **inferior angle**, and the **lateral angle**. The lateral angle forms a broad process that supports the cup-shaped **glenoid cavity**. At the glenoid cavity, the scapula articulates with the proximal end of the humerus (the bone of the arm) to form the **glenohumeral joint**, or *shoulder joint*. The **neck** separates the body of the scapula from the lateral angle. The smooth, concave **subscapular fossa** forms the anterior surface of the scapula.

Two large scapular processes extend over the superior margin of the glenoid cavity, superior to the head of the humerus. The **coracoid** (KOR-a-koyd; *korakodes*, like a crow's beak) **process** is the smaller, anterior projection. This process projects anteriorly and laterally and serves as the origin (more proximal and more immovable muscle attachment) for the short head of the biceps brachii muscle, a muscle on the anterior surface of the arm. The **suprascapular notch** is an indentation medial to the base of the coracoid process.

Figure 7.3 The Clavicle. The clavicle is the only direct connection between the pectoral girdle and the axial skeleton.

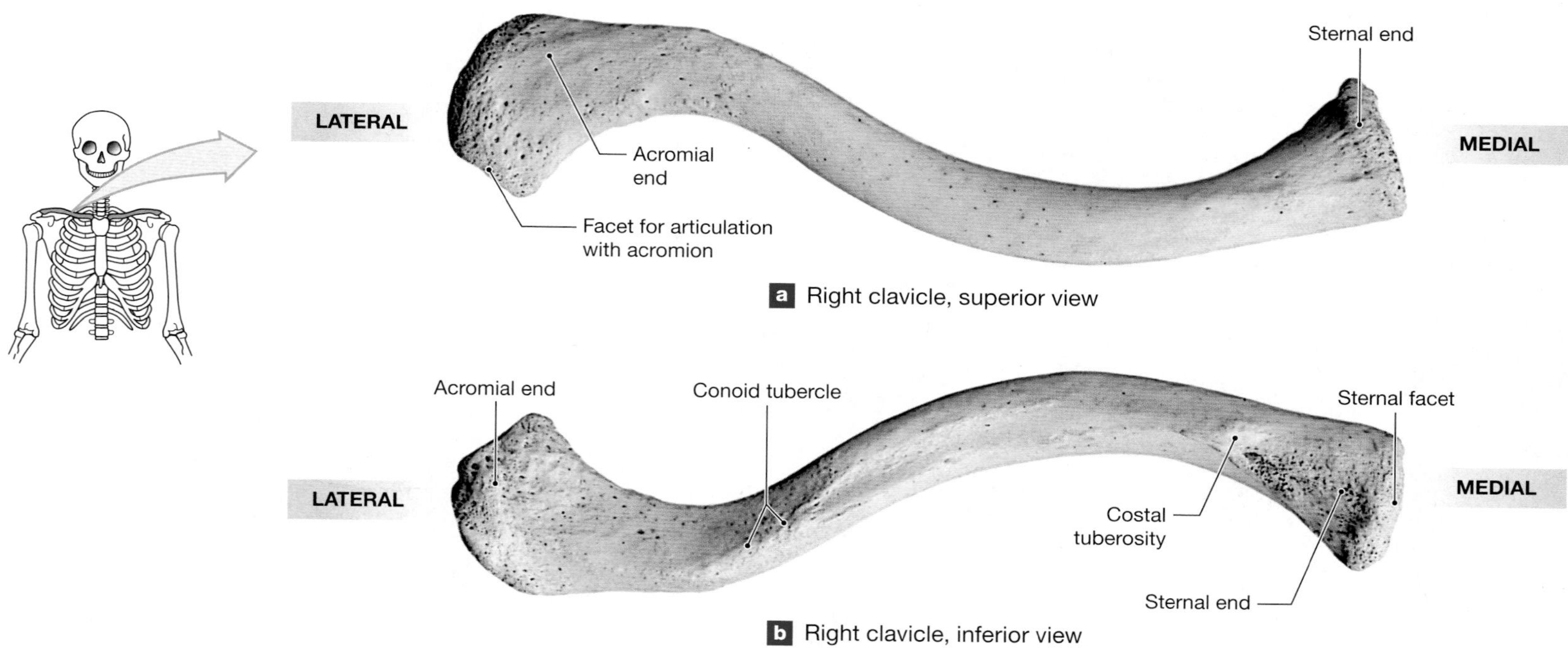

a Right clavicle, superior view

b Right clavicle, inferior view

Figure 7.4 Mobility of the Pectoral Girdle. Diagrammatic representation of normal movements of the pectoral girdle.

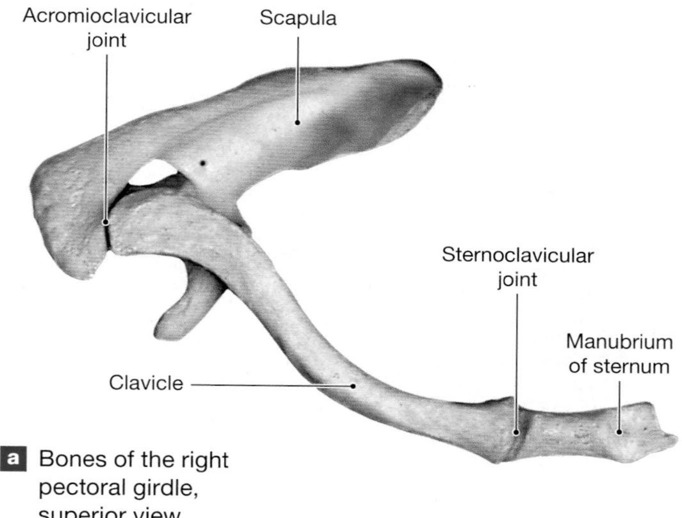

a Bones of the right pectoral girdle, superior view.

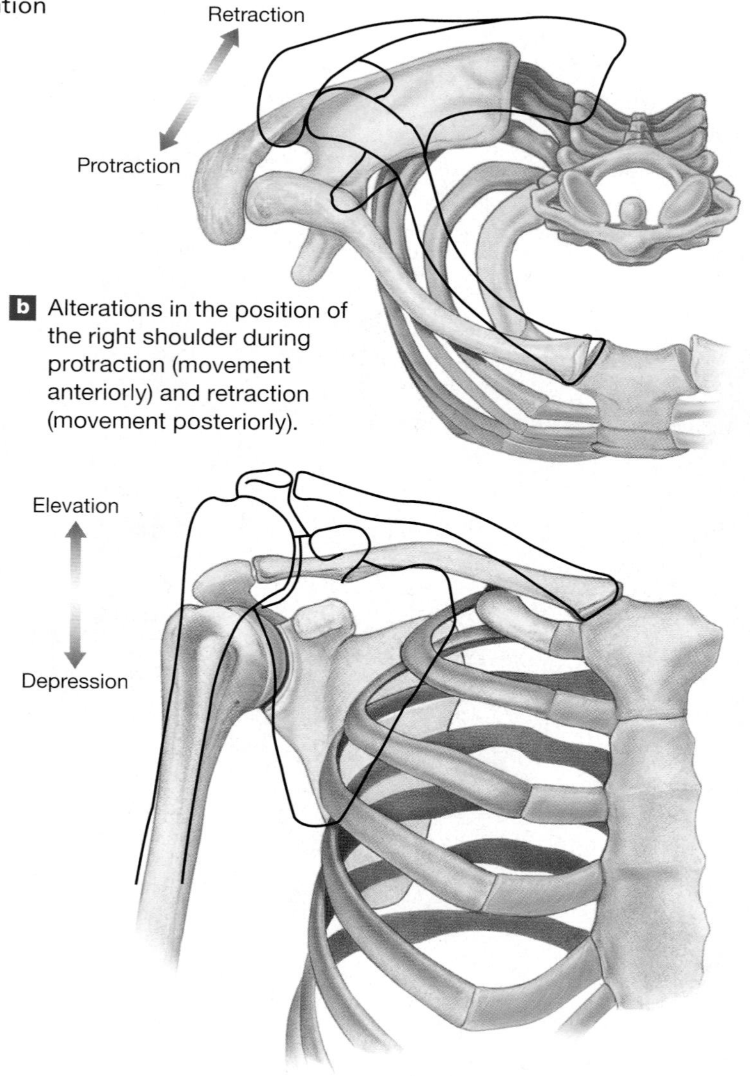

b Alterations in the position of the right shoulder during protraction (movement anteriorly) and retraction (movement posteriorly).

c Alterations in the position of the right shoulder during elevation (superior movement) and depression (inferior movement). Note that the clavicle is responsible for limiting the range of motion.

Figure 7.5 The Scapula. The scapula, which is part of the pectoral girdle, articulates with the upper limb.

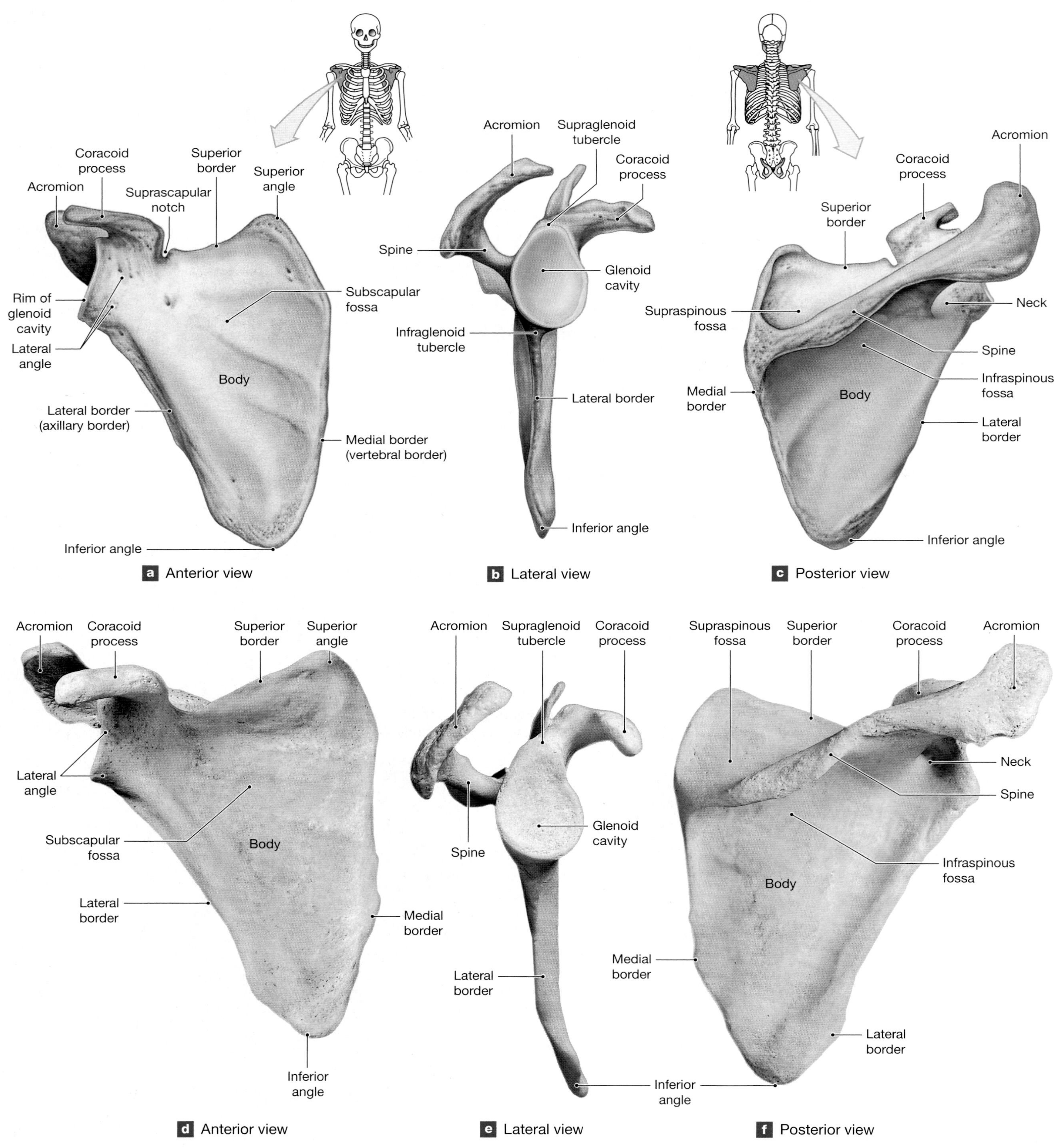

The larger process is the **acromion** (a-KRŌ-mē-on; *akron*, tip, + *omos*, shoulder). The acromion projects anteriorly at a 90° angle from the lateral end of the scapular spine and serves as an attachment point for part of the trapezius of the back. When you run your fingers along the superior surface of your shoulder joint, you will feel this process. The acromion articulates with the clavicle at the **acromioclavicular joint** (Figure 7.4a). Both the coracoid process and the acromion are attached to ligaments and tendons associated with the shoulder joint. The surface markings of the scapula are attachment sites for muscles that position the shoulder and arm. For example, the **supraglenoid tubercle** is the origin (less movable muscle attachment site) of the long head of the biceps brachii. The **infraglenoid tubercle** is the origin of the long head of the triceps brachii, a prominent muscle on the posterior surface of the arm. The **scapular spine** crosses the scapular body, ending at the medial border. The scapular spine divides the dorsal body of the scapula into two regions. The area superior to the spine is the **supraspinous fossa** (*supra*, above), an attachment for the supraspinatus. The region inferior to the spine is the **infraspinous fossa** (*infra*, beneath), an attachment for the infraspinatus. The surfaces of the scapular spine separate these muscles, and the prominent posterior ridge of the scapular spine is an attachment site for the deltoid and trapezius.

The Upper Limb

▶ **KEY POINT** The bones of each upper limb are the humerus, ulna, radius, carpal bones of the wrist, and metacarpals and phalanges of the hand.

The Humerus

The proximal bone of the upper limb is the **humerus**. The superior, medial portion of the proximal epiphysis is smooth and round. This is the **head** of the humerus, which articulates with the glenoid cavity of the scapula. The **greater tubercle** is on the lateral edge of the epiphysis (Figure 7.6a,b). The greater tubercle forms the lateral margin of the shoulder; you can find it by feeling for a bump a few centimeters anterior and inferior to the tip of the acromion. The greater tubercle has three smooth, flat impressions that are sites for three muscles that originate on the scapula: The supraspinatus inserts onto the uppermost impression, the infraspinatus onto the middle, and the teres minor onto the lowermost. The **lesser tubercle** is on the anterior and medial surface of the epiphysis. The lesser tubercle is the insertion point of another scapular muscle, the subscapularis. The **intertubercular sulcus**, or *intertubercular groove*, separates the lesser tubercle and greater tubercles. A tendon of the biceps brachii runs along this sulcus from its origin at the supraglenoid tubercle of the scapula. The **anatomical neck** is a constriction between the tubercles and the head of the humerus. It marks the extent of the joint capsule of the shoulder joint. Distal to the tubercles, the narrow **surgical neck** marks the metaphysis of the growing bone. The name reflects the fact that fractures typically occur at this site.

The proximal **shaft**, or *body*, of the humerus is round in cross section. The **deltoid tuberosity** is an elevation that runs halfway along its lateral length. The deltoid tuberosity is named after the deltoid that attaches to it. On the anterior surface of the shaft, the intertubercular sulcus continues alongside the deltoid tuberosity.

The articular **condyle** is on the distal end of the humerus (Figure 7.6a,c). A low ridge crosses the condyle, dividing it into two joint surfaces. The **trochlea** (*trochlea*, pulley) is the spool-shaped medial portion that articulates with the ulna, the medial bone of the forearm. The trochlea extends from the base of the **coronoid fossa** (KOR-ō-noyd; *corona*, crown) on the anterior surface to the **olecranon fossa** on the posterior surface (Figure 7.6a,d). Projections from the ulna enter these fossae as the elbow flexes (bends) or extends (straightens). The rounded **capitulum** forms the lateral surface of the condyle. The capitulum articulates with the head of the radius, the lateral bone of the forearm. The shallow **radial fossa** is superior to the capitulum. A small part of the radial head enters the radial fossa as the forearm flexes at the elbow.

On the posterior surface the **radial groove** runs alongside the posterior margin of the deltoid tuberosity (Figure 7.6d). The radial nerve, a large nerve that provides sensory information from the back of the hand and controls the large muscles that extend the elbow, runs in this groove. The radial groove ends at the inferior margin of the deltoid tuberosity, where the nerve turns toward the anterior surface of the arm. At the distal end of the humerus the shaft expands to form a broad triangle. The **medial** and **lateral epicondyles**, processes that provide additional surface area for muscle attachment, project to either side of the distal humerus at the elbow joint (Figure 7.6a,c,d). The ulnar nerve crosses the posterior surface of the medial epicondyle. Bumping the posteromedial side of the elbow joint strikes this nerve and produces a temporary numbness and paralysis of muscles on the anterior surface of the forearm. This causes an odd sensation, so the area is sometimes called the *funny bone*.

The Ulna

The **ulna** and radius are parallel bones that support the forearm (Figure 7.2). In the anatomical position, the ulna lies medial to the radius (Figure 7.7a). The **olecranon** (ō-LEK-ra-non), or *olecranon process*, of the ulna forms the point of the elbow (Figure 7.7b). This process is the superior and posterior portion of the proximal epiphysis. On the ulna's anterior surface, the **trochlear notch** (or *semilunar notch*) articulates with the trochlea of the humerus (Figure 7.7c–e). The olecranon forms the superior lip of the trochlear notch, and the **coronoid process** forms the inferior lip. When the elbow is extended, the olecranon projects into the olecranon fossa on the posterior surface of the humerus. When the elbow is flexed, the coronoid process projects into the coronoid fossa on the anterior humeral surface. Lateral to the coronoid process, a smooth **radial notch** articulates with the head of the radius at the proximal radio-ulnar joint (Figure 7.7d,e).

The shaft of the ulna is triangular in cross section. A fibrous sheet, the **interosseous membrane**, connects the lateral margin of the ulna to the medial margin of the radius and provides additional surface area for muscle attachment (Figure 7.7a,d). Distally, the ulnar shaft narrows before ending at a disc-shaped **ulnar head**. The posterior margin of the ulnar head has a short **ulnar styloid process** (*styloid*, long and pointed). A triangular articular cartilage attaches to the styloid process, separating the ulnar head from the bones of the wrist. The distal radio-ulnar joint lies near the lateral border of the ulnar head (Figure 7.7f).

The elbow joint is a stable, two-part joint that functions like a hinge (Figure 7.7b,c). The stability of the elbow comes from the joints between the trochlea of the humerus and the trochlear notch of the ulna. This is the humero-ulnar joint. The capitulum of the humerus and the flat superior surface of the head of the radius form the other part of the elbow joint, the humeroradial joint. (We will discuss the structure of the elbow joint in Chapter 8.)

The Radius

The **radius** is the lateral bone of the forearm (Figure 7.7). The disc-shaped **head** of the radius articulates with the capitulum of the humerus. A narrow neck extends from the radial head to the **radial tuberosity**, which is the attachment site of the biceps brachii. The shaft of the radius curves along its length, and the distal end is considerably larger than the distal end of the ulna. Because the articular cartilage and an articulating disc separate the ulna from the wrist, only the distal end of the radius forms the wrist joint. The **radial styloid process** on the lateral surface of the distal end helps stabilize the wrist.

Figure 7.6 The Humerus

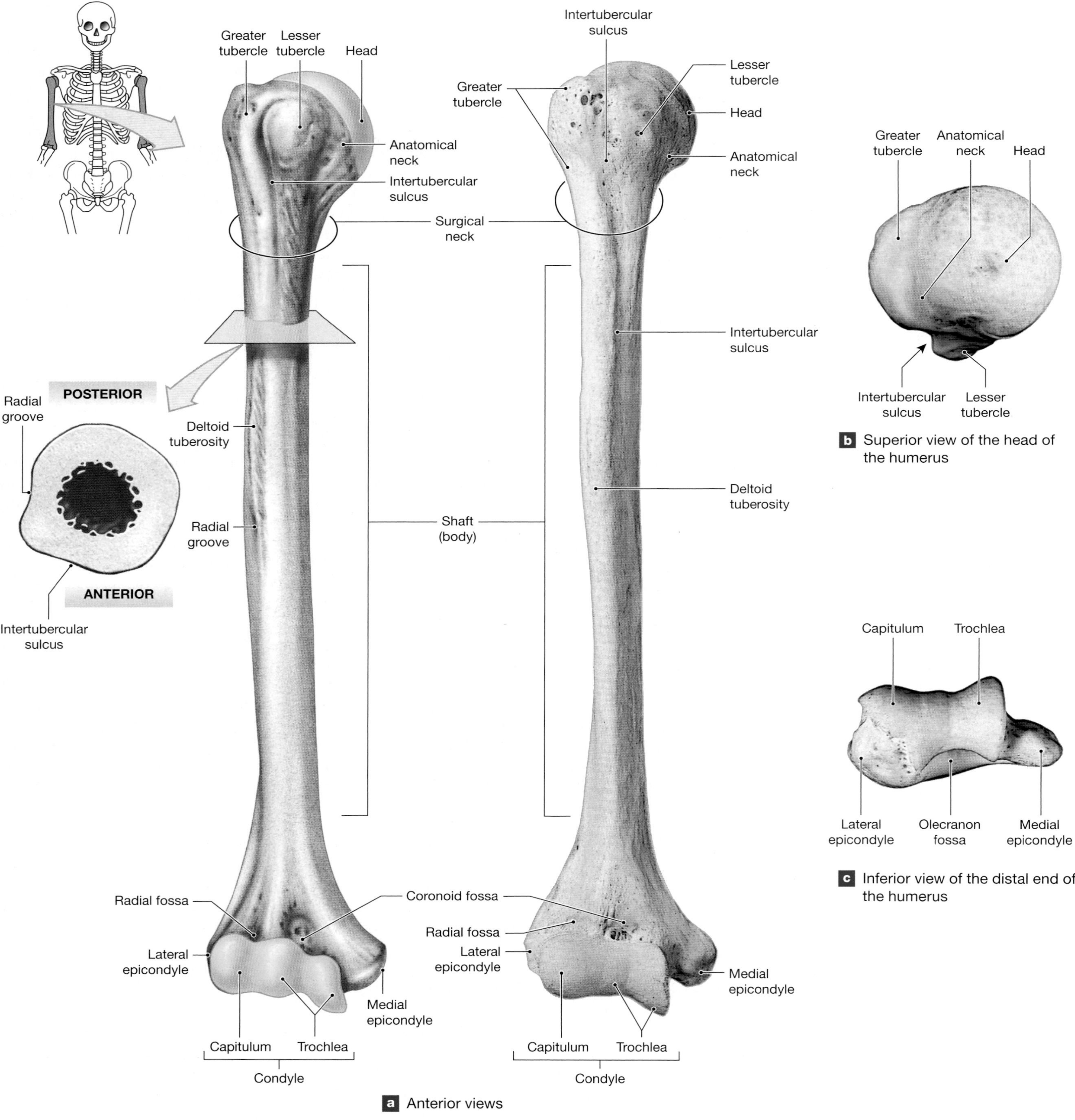

a Anterior views

b Superior view of the head of the humerus

c Inferior view of the distal end of the humerus

Figure 7.6 ***(continued)***

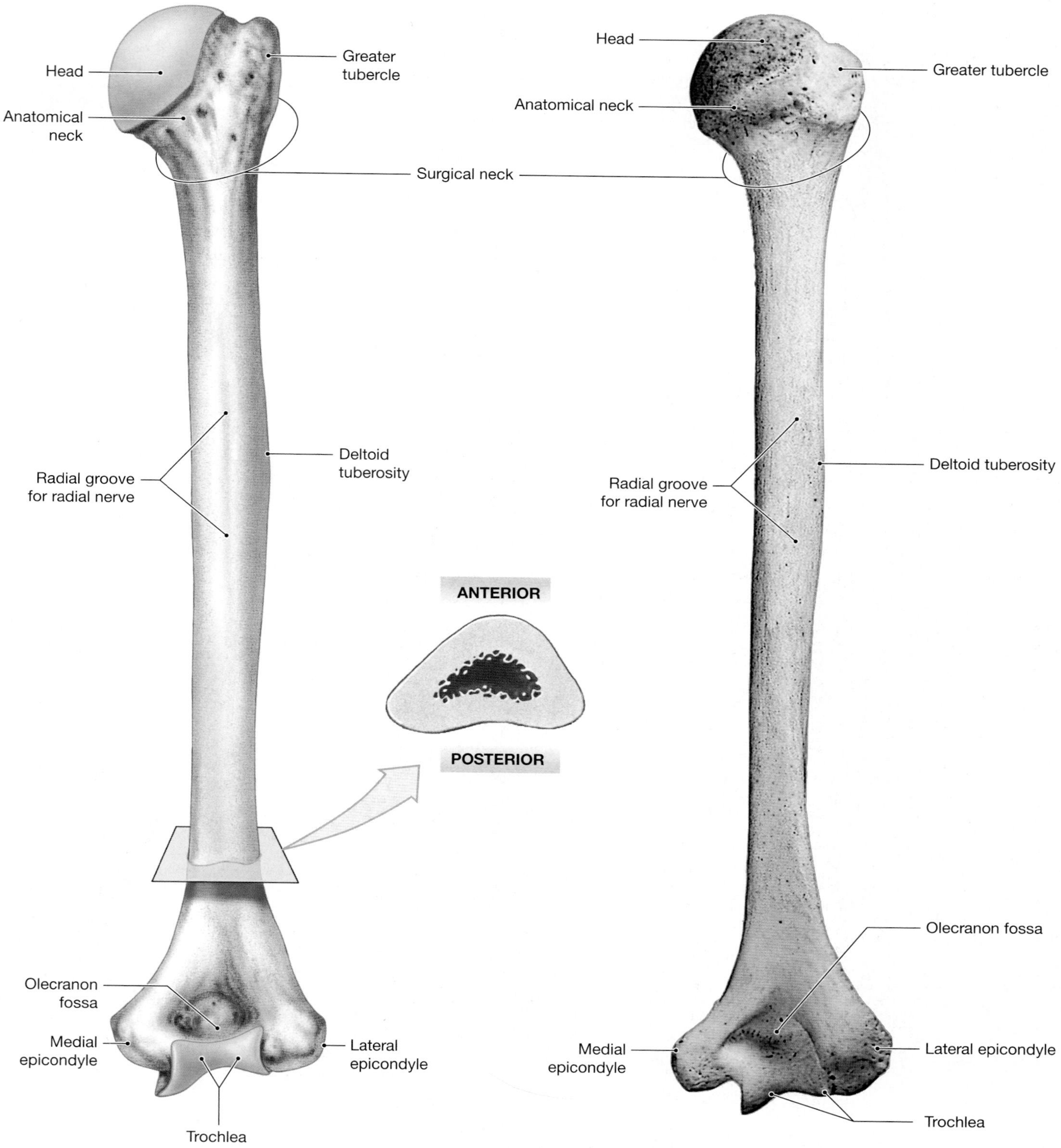

Head
Greater tubercle
Anatomical neck
Surgical neck
Radial groove for radial nerve
Deltoid tuberosity
ANTERIOR
POSTERIOR
Olecranon fossa
Medial epicondyle
Lateral epicondyle
Trochlea
Head
Greater tubercle
Anatomical neck
Radial groove for radial nerve
Deltoid tuberosity
Olecranon fossa
Medial epicondyle
Lateral epicondyle
Trochlea

d Posterior views

Figure 7.7 The Radius and Ulna. The radius and ulna are the bones of the forearm.

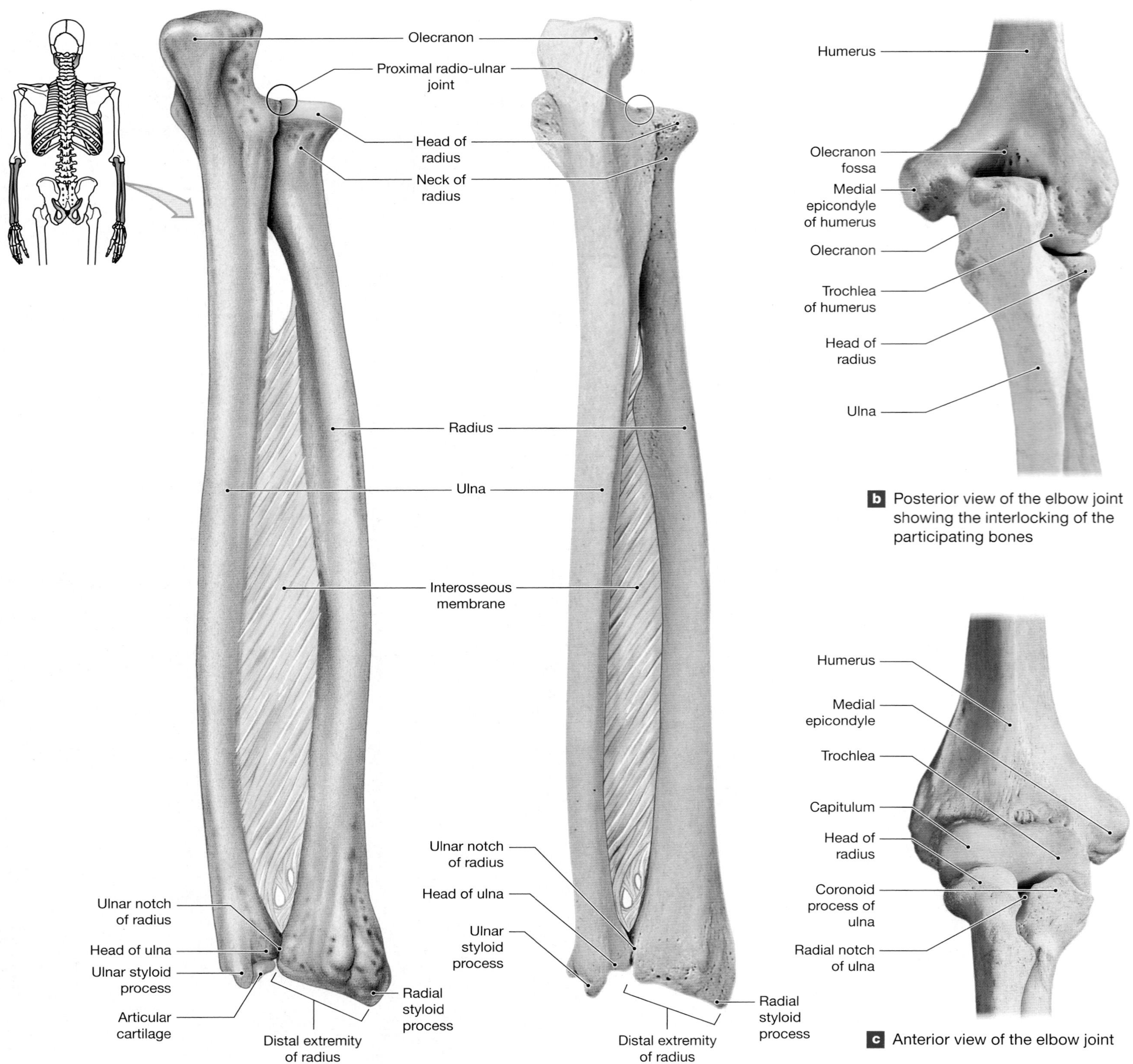

a Posterior view of the right radius and ulna

b Posterior view of the elbow joint showing the interlocking of the participating bones

c Anterior view of the elbow joint

Figure 7.7 ***(continued)***

Olecranon
Trochlear notch
Coronoid process
Radial notch of ulna
Head of radius
Head of radius
Neck of radius
Ulnar tuberosity
Radial tuberosity
Ulna
Radius
Interosseous membrane
Attachment surfaces for interosseous membrane
Ulnar notch of radius
Distal radio-ulnar joint
Head of ulna
Ulnar notch of radius
Ulnar styloid process
Radial styloid process
Carpal articular surface
Radial styloid process
Carpal articular surface

d Anterior view of the radius and ulna

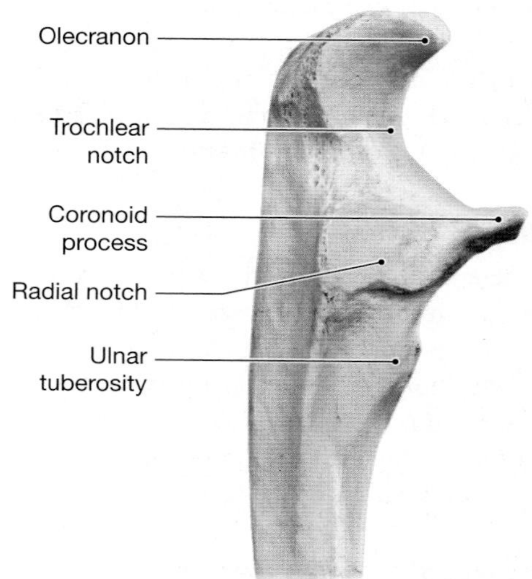

e Lateral view of the proximal end of the ulna

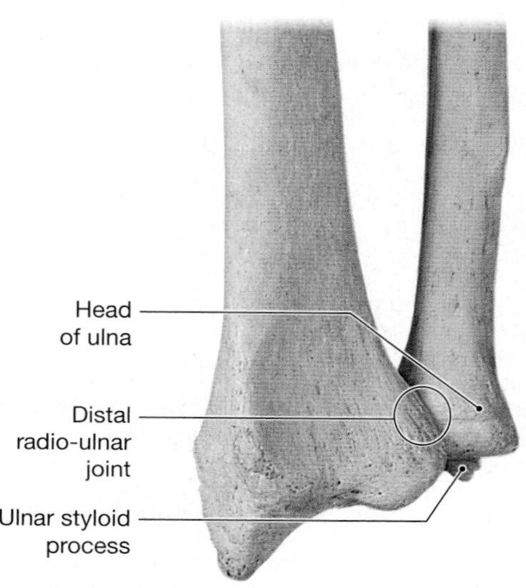

f Anterior view of the distal ends of the radius and ulna and the distal radio-ulnar joint

The medial surface of the distal end of the radius forms a joint with the ulnar head at the **ulnar notch of the radius**, forming the distal radio-ulnar joint. The proximal radio-ulnar joint allows medial or lateral rotation of the radial head. When medial rotation occurs at the proximal radio-ulnar joint, the ulnar notch of the radius rolls across the rounded surface of the ulnar head. Medial rotation at the radio-ulnar joints in turn rotates the wrist and hand medially, from the anatomical position. This rotational movement is called **pronation**. The reverse movement, which involves lateral rotation at the radio-ulnar joints, is called **supination**.

7

The Carpal Bones

The eight **carpal bones** form the **wrist**, or *carpus*. The bones form two rows, with four **proximal carpal bones** and four **distal carpal bones**. The proximal carpal bones are the scaphoid, lunate, triquetrum (tri-KWEĒ-trum), and pisiform (PIS-i-form). The distal carpal bones are the trapezium, trapezoid, capitate, and hamate (**Figure 7.8**). The joints between the carpal bones permit limited sliding and twisting movements. Ligaments interconnect the carpal bones and help stabilize the wrist.

The Proximal Carpal Bones

- The **scaphoid** is the proximal carpal bone on the lateral border of the wrist adjacent to the styloid process of the radius.
- The comma-shaped **lunate** (*luna*, moon) lies medial to the scaphoid. Like the scaphoid, the lunate articulates with the radius.
- The **triquetrum** (*triangular bone*) is medial to the lunate. It has the shape of a small pyramid. The triquetrum articulates with the cartilage that separates the ulnar head from the wrist.
- The small, pea-shaped **pisiform** lies anterior to the triquetrum and extends farther medially than any other carpal bone in the proximal or distal rows.

The Distal Carpal Bones

- The **trapezium** is the lateral bone of the distal row. It forms a proximal joint with the scaphoid.
- The wedge-shaped **trapezoid** lies medial to the trapezium. It is the smallest distal carpal bone. Like the trapezium, it has a proximal joint with the scaphoid.
- The **capitate** is the largest carpal bone. It sits between the trapezoid and the hamate.
- The hook-shaped **hamate** (*hamatum*, hooked) is the medial distal carpal bone.

TIPS & TOOLS

The following mnemonic will help you remember the names of the carpal bones, proceeding lateral to medial; the first four are proximal, the last four distal: "Sam Likes To Push The Toy Car Hard." **S**caphoid, **L**unate, **T**riquetrum, **P**isiform; **T**rapezium, **T**rapezoid, **C**apitate, **H**amate

The Metacarpals and Phalanges

Five **metacarpal** (met-a-KAR-pal) **bones** articulate with the distal carpal bones and support the palm of the hand (**Figure 7.8b,c**). Roman numerals I–V identify the metacarpals beginning with the lateral metacarpal (thumb). Each metacarpal looks like a miniature long bone, possessing a wide, concave, proximal base, a small body, and a distal head. Distally, the metacarpals articulate with the **phalanges** (fa-LAN-jēz; singular, *phalanx*), or finger bones. There are 14 phalanges in each hand. The thumb, or **pollex** (POL-eks), has two phalanges (proximal phalanx and distal phalanx), and each of the fingers has three phalanges (proximal, middle, and distal).

Figure 7.8 The Bones of the Wrist and Hand. Carpal bones form the wrist; metacarpals and phalanges form the hand.

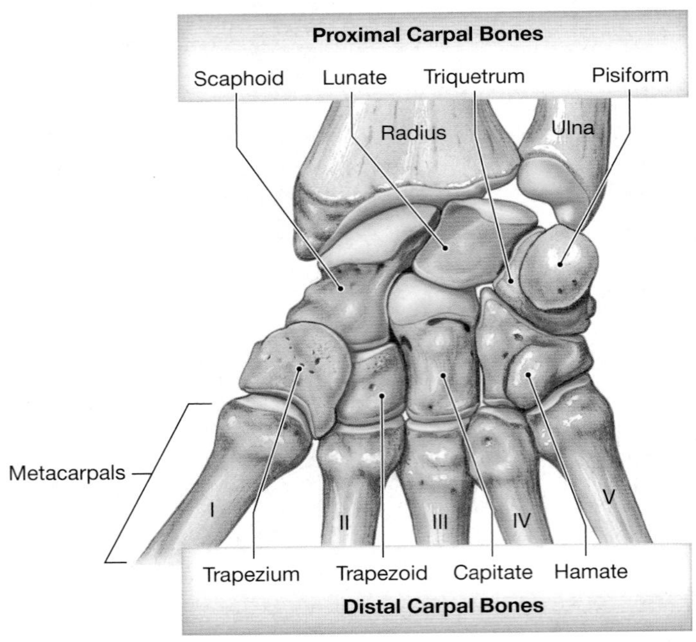

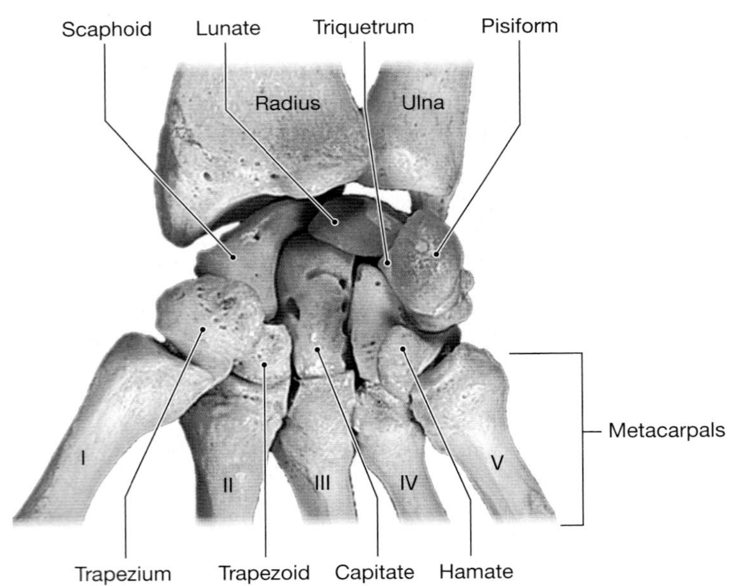

a Anterior (palmar) view of the bones of the right wrist

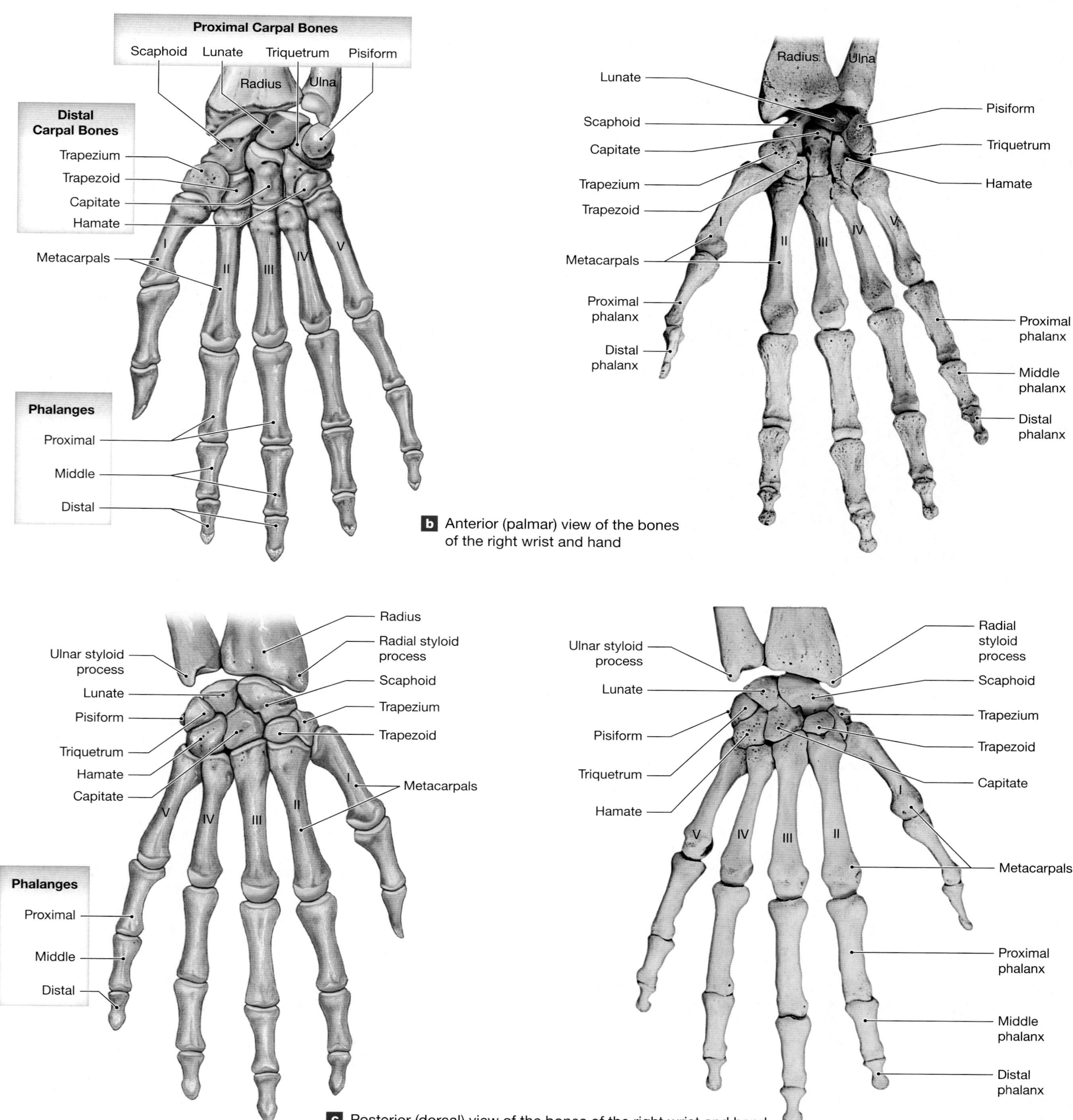

b Anterior (palmar) view of the bones of the right wrist and hand

c Posterior (dorsal) view of the bones of the right wrist and hand

7

CLINICAL NOTE

Scaphoid Fractures

THE SCAPHOID is the most frequently fractured carpal bone, usually resulting from a fall onto an outstretched hand. The fracture pattern most often is transverse at the "waist" of the bone. Because the scaphoid functionally crosses the proximal and distal carpal rows, the proximal scaphoid fragment stays with the proximal carpal row, and the distal fragment breaks off and dorsiflexes with the distal carpal row when forced dorsiflexion occurs during a fall on the outstretched hand.

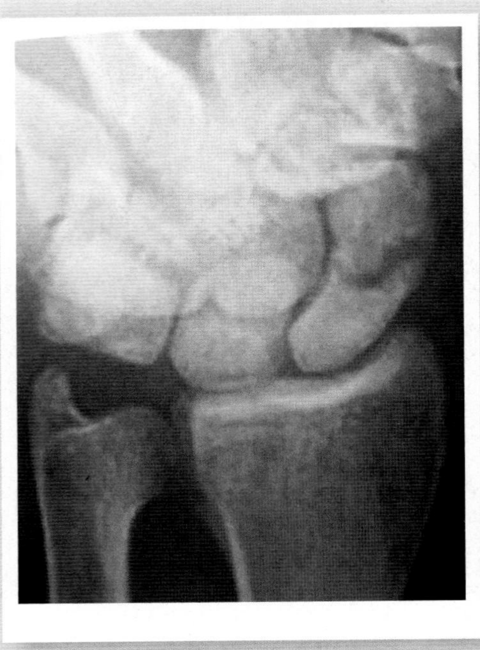

7

7.1 **CONCEPT CHECK**

1 Why does a broken clavicle affect the mobility of the scapula?
2 Which forearm bone is lateral in the anatomical position?
3 What is the function of the olecranon?
4 Which bone is the only direct connection between the pectoral girdle and the axial skeleton?

See the blue Answers tab at the back of the book.

7.2 | The Pelvic Girdle and Lower Limb

▶ **KEY POINT** The pelvic girdle transmits force from the lower limbs to the axial skeleton, supports the body while standing, allows walking, and protects the organs of the pelvic cavity. The 30 bones of the lower limb are the femur, patella, tibia, fibula, 7 tarsal bones, 5 metatarsals and 14 phalanges.

The pelvic girdle consists of two **hip bones**, also called *pelvic bones* or *coxal bones*. The bones of the **pelvic girdle** support and protect the lower viscera, including the reproductive organs, and the developing fetus in females. The pelvic girdle is more massive than the pectoral girdle because of the stresses of weight bearing and walking and running. The pelvis is a composite structure made up of the hip bones of the appendicular skeleton and the sacrum and coccyx of the axial skeleton. The skeleton of each lower limb includes the femur (thigh), patella (kneecap), tibia and fibula (leg), and bones of the ankle (tarsal bones) and foot (metatarsals and phalanges) (**Figure 7.9**). In anatomical terms, *thigh* refers to the proximal part of the limb and *leg* to the distal part.

Figure 7.9 The Pelvic Girdle and Lower Limb. Each lower limb articulates with the axial skeleton at the trunk through the pelvic girdle.

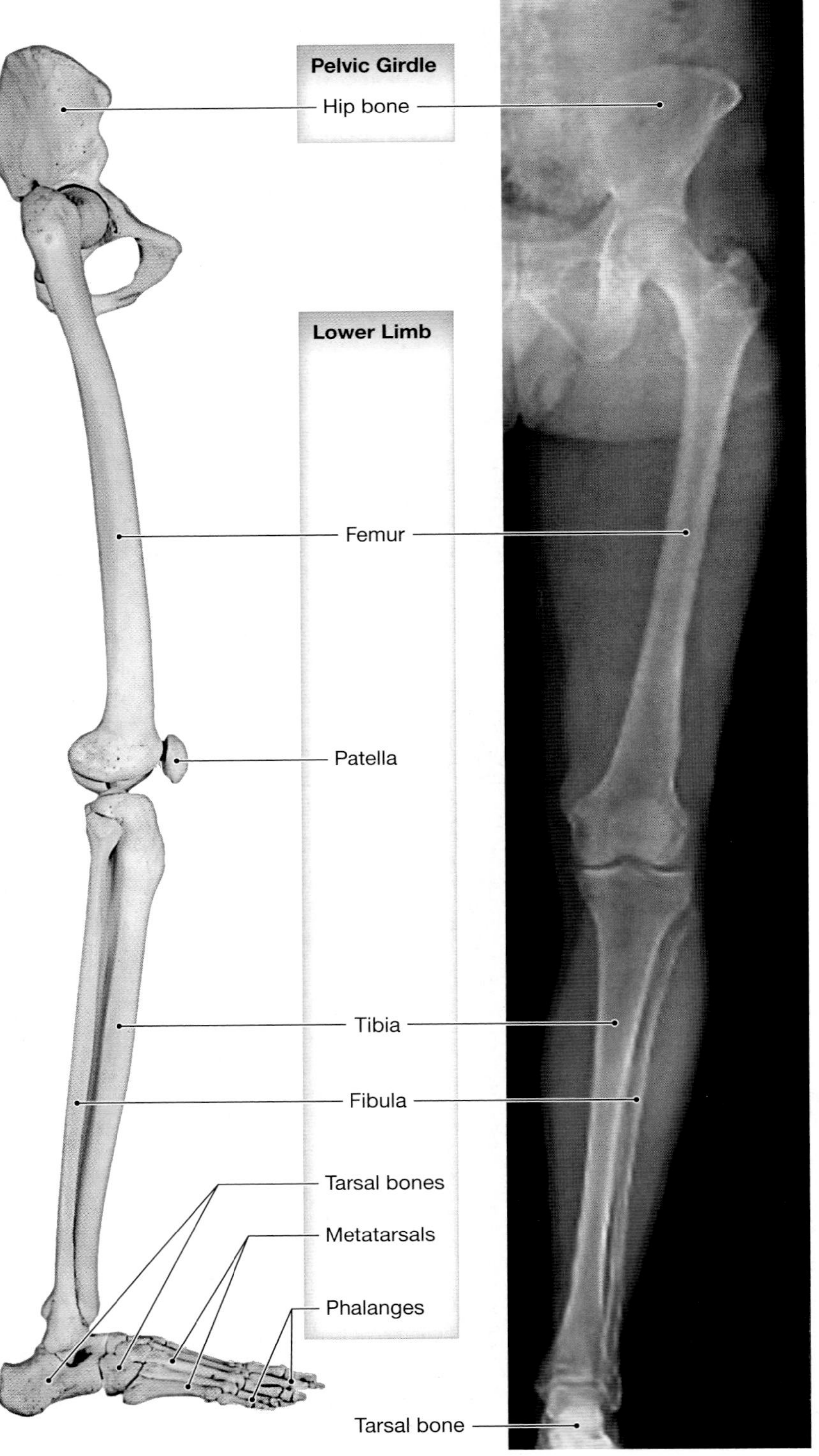

a Right lower limb, lateral view

b X-ray, pelvic girdle and lower limb, anterior/posterior projection

The Pelvic Girdle

▶ **KEY POINT** The hip bones protect the organs of the pelvic cavity and are attachment sites for the large muscles of the buttocks and thighs of the lower limbs.

Each hip bone of the adult pelvic girdle consists of three fused bones: the ilium (IL-ē-um), ischium (IS-kē-um), and pubis (PŪ-bis) **(Figures 7.10, 7.11)**. At birth, hyaline cartilage separates the three bones. Growth and fusion of the three bones into a single hip bone are usually complete by age 25. The sacro-iliac joint between a hip bone and the auricular surfaces of the sacrum occurs at the posterior and medial aspect of the ilium. A pad of fibrous cartilage at the pubic symphysis connects the anterior and medial portions of the hip bones. The **acetabulum** (as-e-TAB-ū-lum; *acetabulum*, vinegar cup) is a concave socket on the lateral surface of the hip bone. The head of the femur articulates with the acetabulum.

The acetabulum is inferior and anterior to the center of the hip bones **(Figure 7.10a)**. A ridge of bone forms the lateral and superior margins of the acetabulum. The anterior and inferior portion of the ridge is incomplete, leaving a gap called the **acetabular notch**. The space enclosed by the walls of the acetabulum is the **acetabular fossa**. The smooth, cup-shaped **lunate surface** articulates with the head of the femur.

The Hip Bones

The ilium, ischium, and pubis meet inside the acetabular fossa, as if it were a pie sliced into three pieces **(Figure 7.10a)**. The **ilium** (plural, *ilia*) is the largest of the bones, forming two-fifths of the acetabular surface. Superior to the acetabulum, the broad, curved, lateral surface of the ilium is an extensive area for the attachment of muscles, tendons, and ligaments **(Figure 7.10a)**. The **anterior**, **posterior**, and **inferior gluteal lines** are the attachment sites for the gluteal muscles that move the femur. The ala (*wing*) of the ilium begins superior to the **arcuate** (AR-kū-āt) **line** **(Figure 7.10b)**. The anterior border of the wing is the **anterior inferior iliac spine**, superior to the **inferior iliac notch**, and continues anteriorly to the **anterior superior iliac spine**. The **iliac crest** is a ridge for the attachments of several ligaments and muscles. *(Refer to Chapter 12,* **Figures 12.3** *and* **12.14**, *to identify these anatomical structures from the body surface and in a cross section of the body at the level of* L_5.*)* The iliac crest ends at the **posterior superior iliac spine**. Inferior to the spine, the posterior border of the ilium continues inferiorly to the rounded **posterior inferior iliac spine**, which is superior to the **greater sciatic** (sī-AT-ik) **notch**. The sciatic nerve passes through the sciatic notch as it travels into the lower limb.

At the superior and posterior margin of the acetabulum, the ilium fuses with the **ischium**. The ischium is the strongest of the hip bones. Posterior to the acetabulum, the prominent **ischial spine** is superior to the **lesser sciatic notch**. The rest of the ischium forms a sturdy process that turns medially and inferiorly. A roughened projection called the **ischial tuberosity** forms the posterolateral border of the ischium. The ischial tuberosities support your weight when you are sitting. The narrow **ischial ramus** continues toward its anterior fusion with the **pubis**.

Continuing inferiorly, the ischial ramus fuses with the **inferior pubic ramus**. Anteriorly, the inferior pubic ramus begins at the **pubic tubercle**, where it meets the **superior pubic ramus**. The anterior, superior surface of the superior pubic ramus has a roughened ridge, the **pubic crest**, which extends laterally from the pubic tubercle. The pubic and ischial rami encircle the **obturator** (OB-tū-rā-tor) **foramen**. This space is closed by a sheet of collagenous connective tissue fibers that provide a firm base for the attachment of hip muscles. The superior pubic ramus originates at the anterior margin of the acetabulum. Inside the acetabulum, the pubis contacts the ilium and ischium.

Figures 7.10b and **7.11a** show additional features visible on the medial and anterior surfaces of the right hip bone:

- The concave medial surface of the **iliac fossa** supports the abdominal organs and provides surface area for muscle attachment. The arcuate line marks the inferior border of the iliac fossa.
- The anterior and medial surfaces of the pubis contain a roughened area where the pubis articulates with the pubis of the opposite side. At this joint, termed the **pubic symphysis**, a pad of fibrous cartilage connects the two pubic bones.
- The **pectineal** (pek-TIN-ē-al) **line** begins near the pubic symphysis and extends diagonally across the pubis to merge with the arcuate line, which continues toward the **auricular surface** of the ilium. The auricular surfaces of the ilium and sacrum unite to form the sacro-iliac joint. Ligaments at the **iliac tuberosity** stabilize this joint.
- On the medial surface of the superior pubic ramus lies the **obturator groove**. The obturator blood vessels and nerves are within this groove.

7

The Pelvis

Figure 7.11 shows anterior and posterior views of the **pelvis**, a ring of bone that consists of four individual bones: the two hip bones, the sacrum, and the coccyx. The hip bones form the anterior and lateral parts of the pelvis; the sacrum and coccyx form the posterior part. A network of ligaments increases the stability of the pelvis by connecting the lateral borders of the sacrum with the iliac crest, the ischial tuberosity, the ischial spine, and the iliopectineal line. Additional ligaments bind the ilia to the posterior lumbar vertebrae.

The pelvis is subdivided into the **greater** (*false*) **pelvis** and the **lesser** (*true*) **pelvis** **(Figure 7.12)**. The greater pelvis consists of the expanded, bladelike portions of each ilium superior to the iliopectineal line. The greater pelvis protects organs within the inferior portion of the abdominal cavity. Structures inferior to the iliopectineal line form the lesser pelvis, which is the boundary of the pelvic cavity. ⮌ **pp. 19** The structures of the lesser pelvis are the inferior portions of each ilium, both pubic bones, the ischia, the sacrum, and the coccyx. In a medial view, the superior limit of the lesser pelvis is a line that extends from either side of the base of the sacrum, along the iliopectineal lines to the superior margin of the pubic symphysis **(Figure 7.12b)**. The bony edge of the lesser pelvis is called the **pelvic brim**. The space enclosed by the pelvic brim is the **pelvic inlet**.

The borders of the **pelvic outlet** are the coccyx, the ischial tuberosities, and the inferior border of the pubic symphysis **(Figure 7.12a–c)**. (The pubic symphysis is the fibrous cartilage joint between the pubic bones.) In life, the region of the pelvic outlet is called the perineum (per-i-NĒ-um). Pelvic muscles form the floor of the pelvic cavity and support the enclosed organs. (These muscles are described in Chapter 10.)

The shape of the female pelvis differs from that of the male pelvis **(Figure 7.12d)**. Some of these differences result from variations in body size and muscle mass. Because women are typically less muscular than men, the adult female pelvis is usually smoother and lighter, and the markings where muscles or ligaments attach are less prominent. Other differences are adaptations for childbearing, as explained in **Figure 7.12d**.

The Lower Limb

▶ **KEY POINT** The lower limb is constructed on the same basic plan as the upper limb, but its functional anatomy is different. The lower limb transmits force from the ground to the axial skeleton and positions the lower limb so that movement from one place to another can occur.

The skeleton of the lower limb consists of the femur, patella (*kneecap*), tibia, fibula, tarsal bones of the ankle, and metatarsals and phalanges of the foot **(Figure 7.9)**.

Figure 7.10 The Pelvic Girdle. The pelvic girdle consists of the two hip bones. Each hip bone forms as a result of the fusion of an ilium, an ischium, and a pubis.

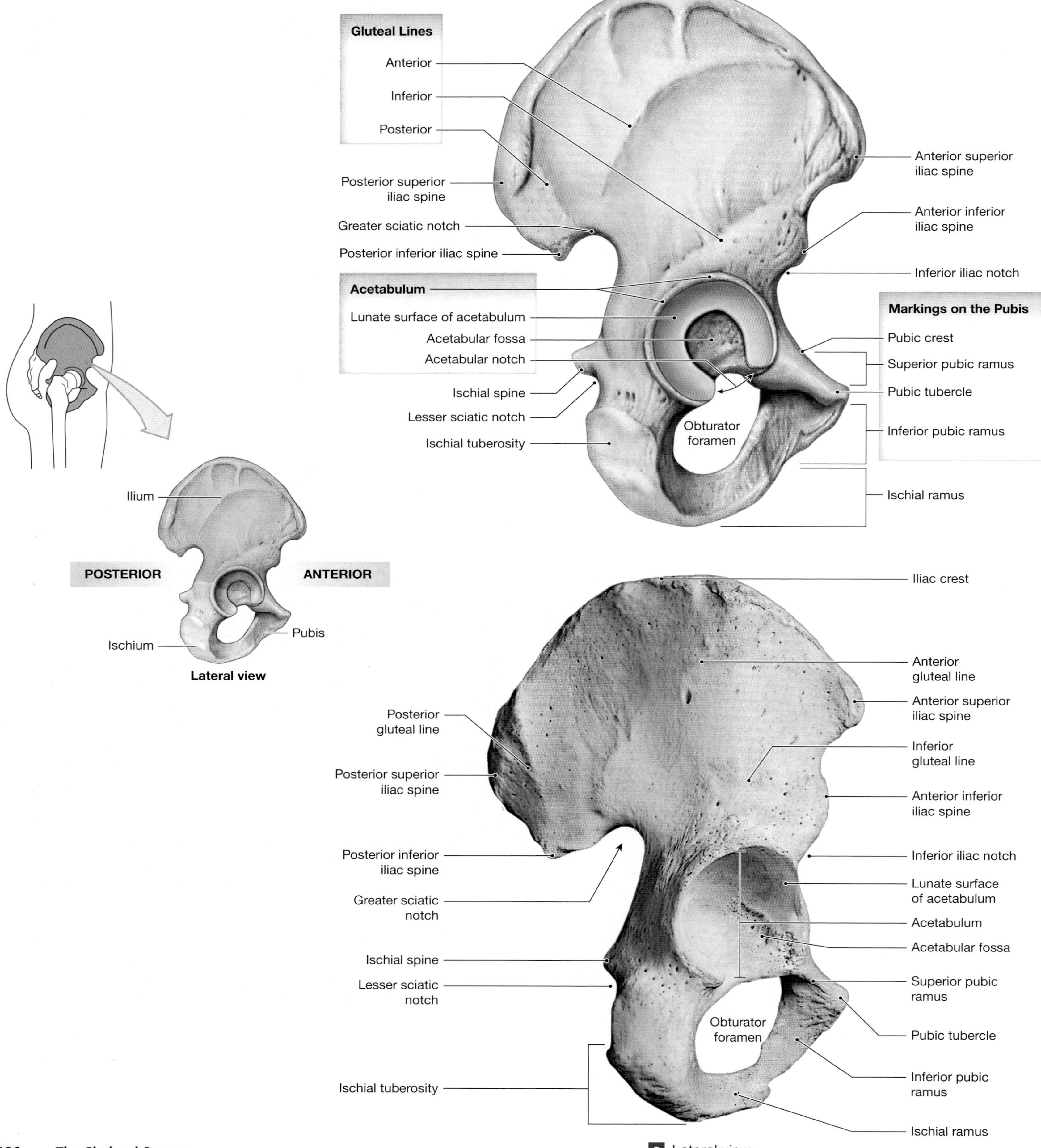

a Lateral view

Figure 7.10 ***(continued)***

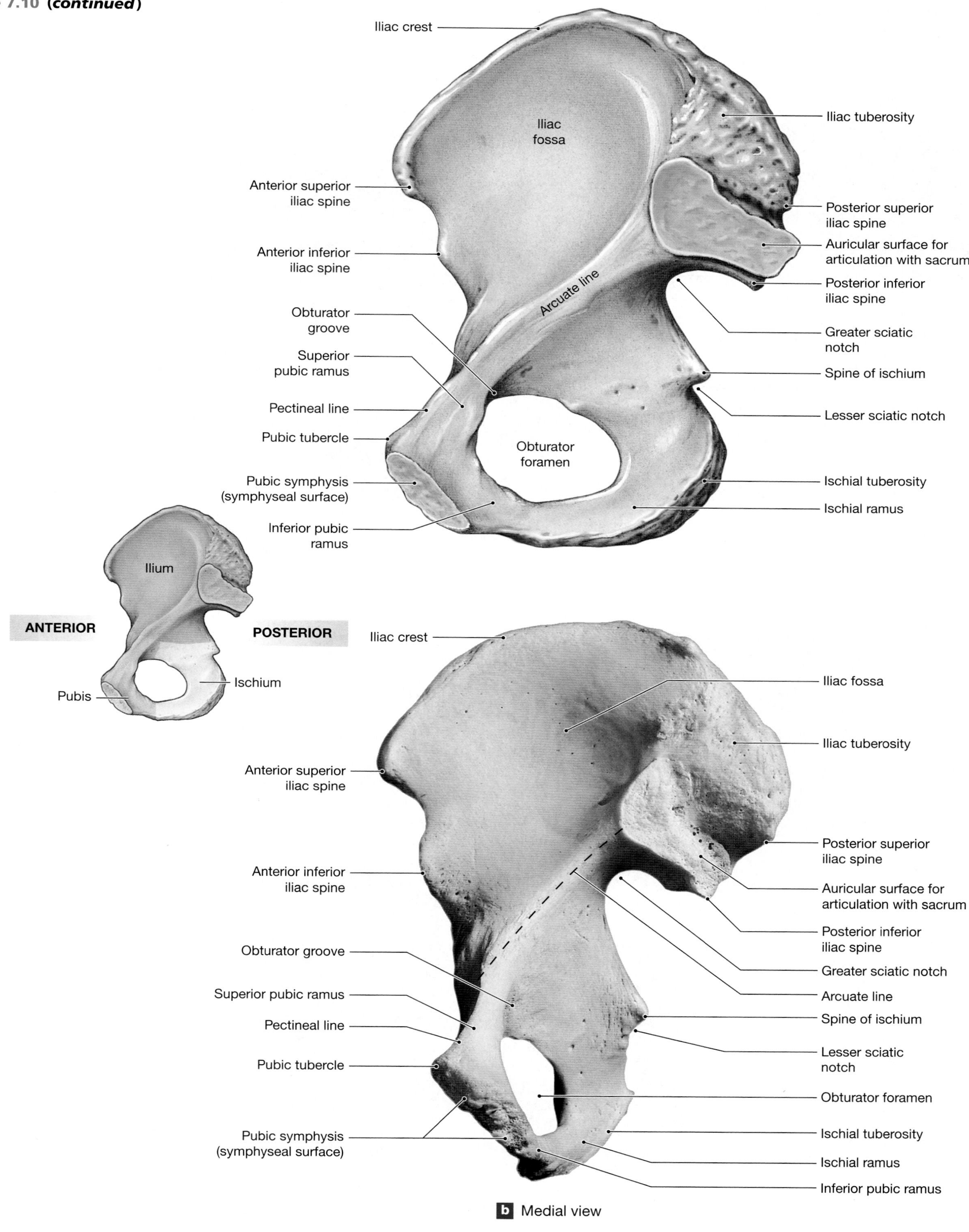

b Medial view

Figure 7.11 The Pelvis. The pelvis consists of two hip bones, the sacrum, and the coccyx.

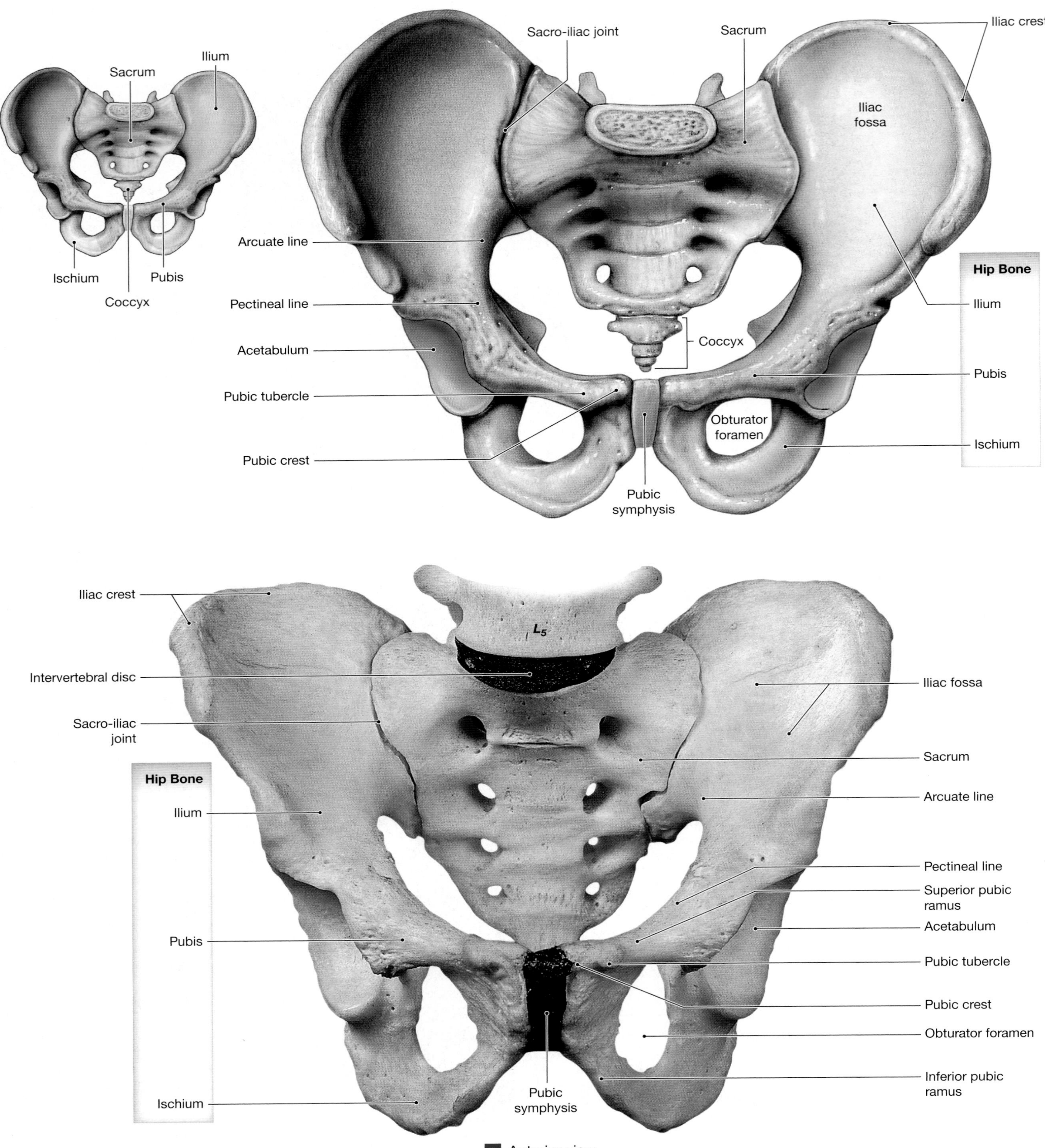

a Anterior view

Figure 7.11 ***(continued)***

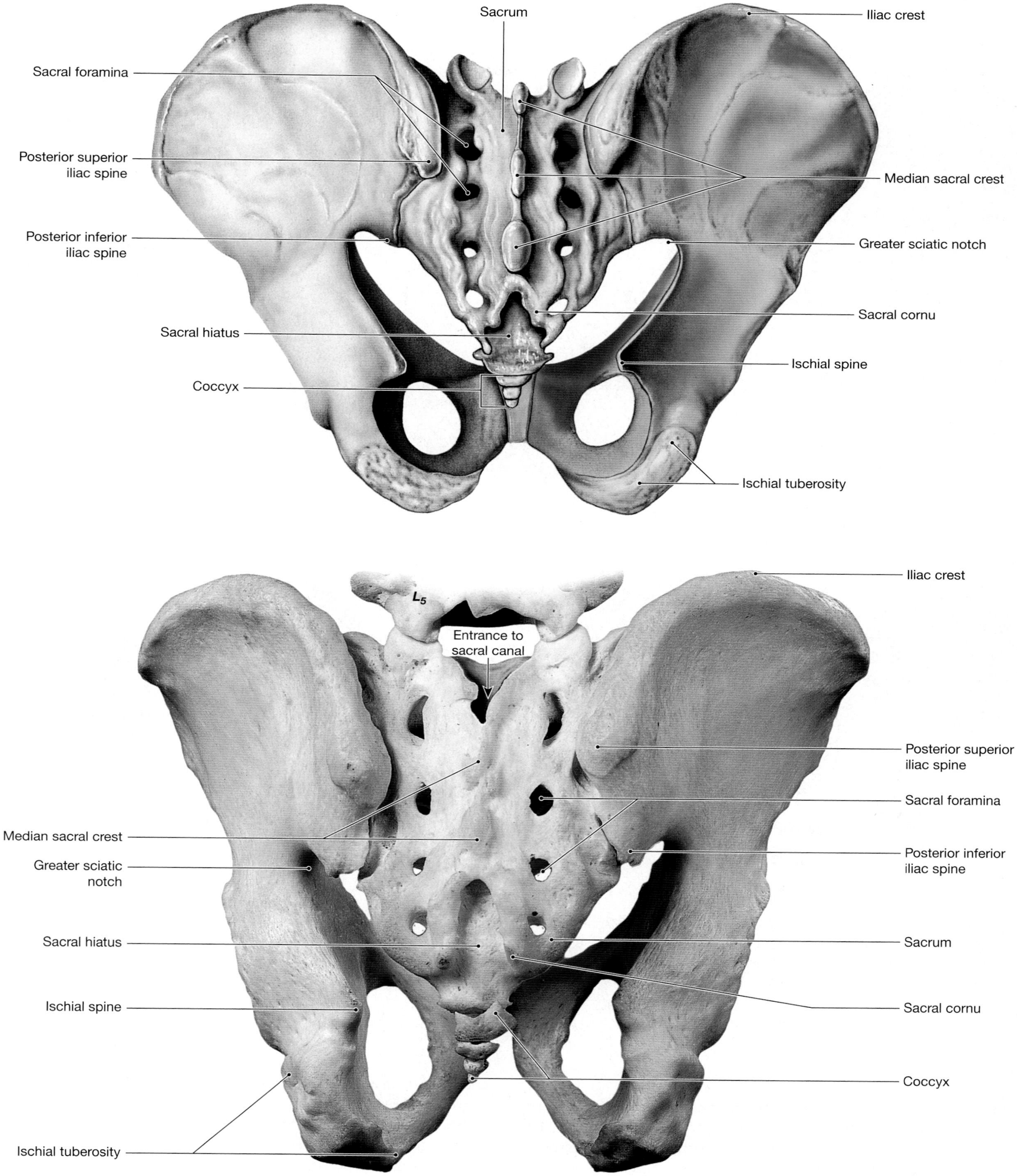

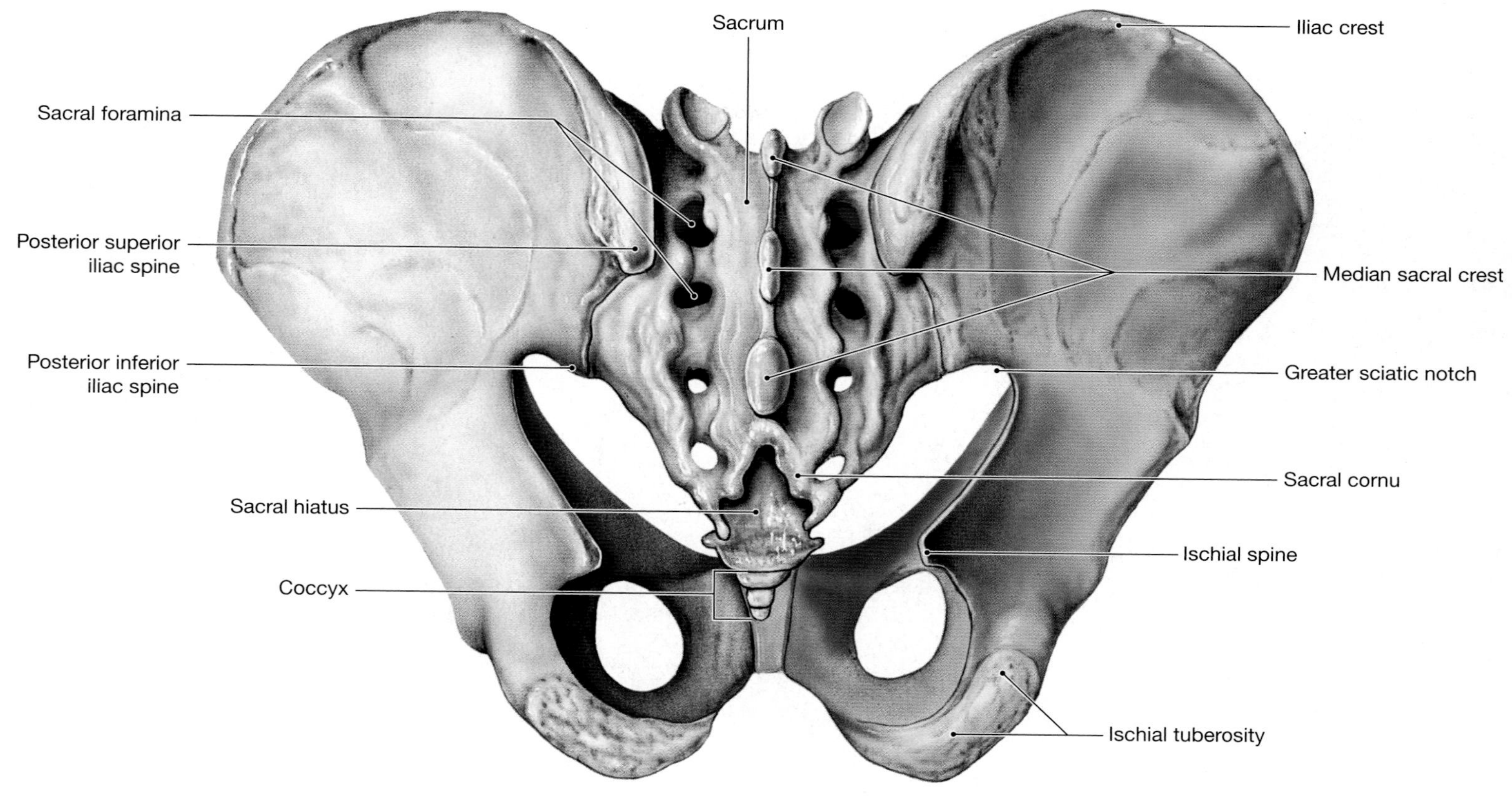

Sacrum
Iliac crest
Sacral foramina
Posterior superior iliac spine
Median sacral crest
Posterior inferior iliac spine
Greater sciatic notch
Sacral cornu
Sacral hiatus
Ischial spine
Coccyx
Ischial tuberosity
Iliac crest
L5
Entrance to sacral canal
Posterior superior iliac spine
Sacral foramina
Median sacral crest
Posterior inferior iliac spine
Greater sciatic notch
Sacral hiatus
Sacrum
Sacral cornu
Ischial spine
Coccyx
Ischial tuberosity

b Posterior view

Figure 7.12 Divisions of the Pelvis and Anatomical Differences in the Male and Female Pelvis. A pelvis is subdivided into the lesser (true) and greater (false) pelvis.

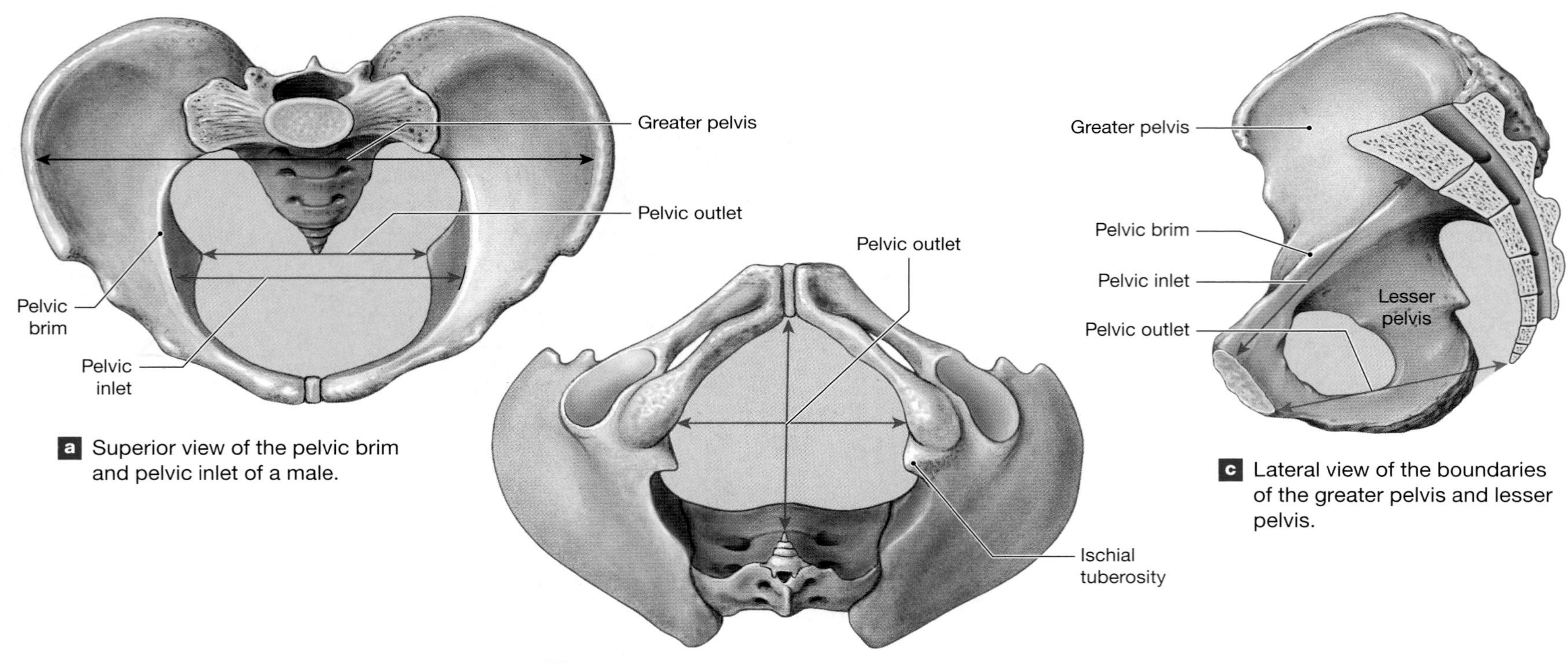

b Inferior view of the limits of the pelvic outlet.

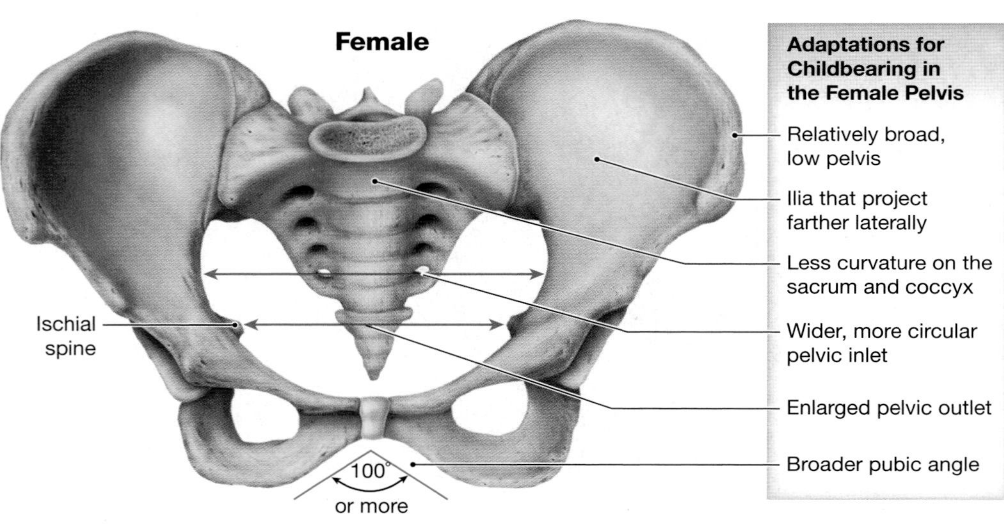

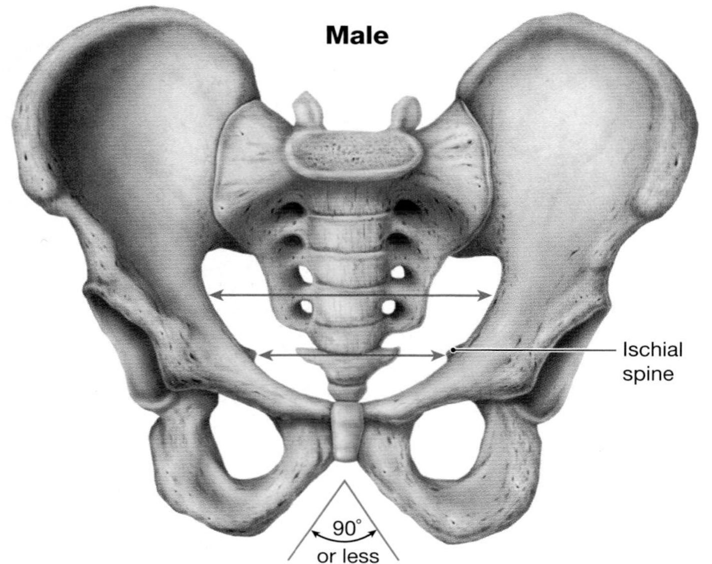

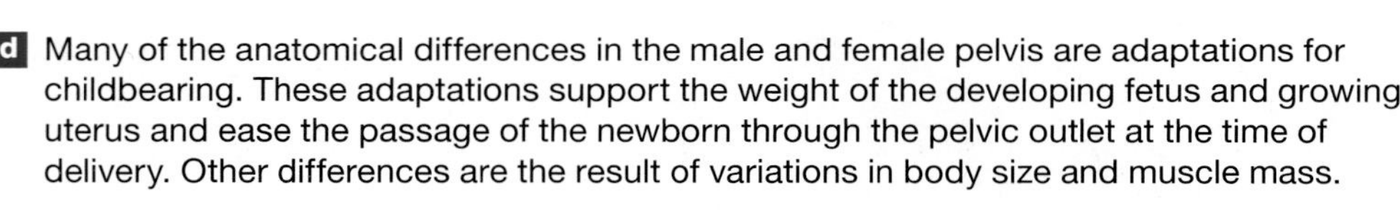

d Many of the anatomical differences in the male and female pelvis are adaptations for childbearing. These adaptations support the weight of the developing fetus and growing uterus and ease the passage of the newborn through the pelvic outlet at the time of delivery. Other differences are the result of variations in body size and muscle mass.

The Femur

The **femur** (**Figure 7.13**) is the longest and heaviest bone in the body. Distally, the femur articulates with the patella and the tibia of the leg at the knee joint. Proximally, the rounded **head** of the femur articulates with the pelvis at the acetabulum (**Figure 7.9**). A stabilizing ligament (the ligament of the head) attaches to the femoral head at a depression, the **fovea** (**Figure 7.13b**). Distal to the head, the **neck** joins the shaft of the femur. The shaft is strong and massive and curves along its length (**Figure 7.13a,d**). This lateral curve helps with weight bearing and balance and becomes greatly exaggerated if the skeleton weakens. A bowlegged stance is characteristic of rickets, a metabolic disorder discussed in Chapter 5. ⮌ **p. 126**

The **greater trochanter** (trō-KAN-ter) projects laterally from the junction of the neck and shaft of the femur. The **lesser trochanter** is on the posteromedial surface of the femur. Both trochanters develop where large

Figure 7.13 The Femur

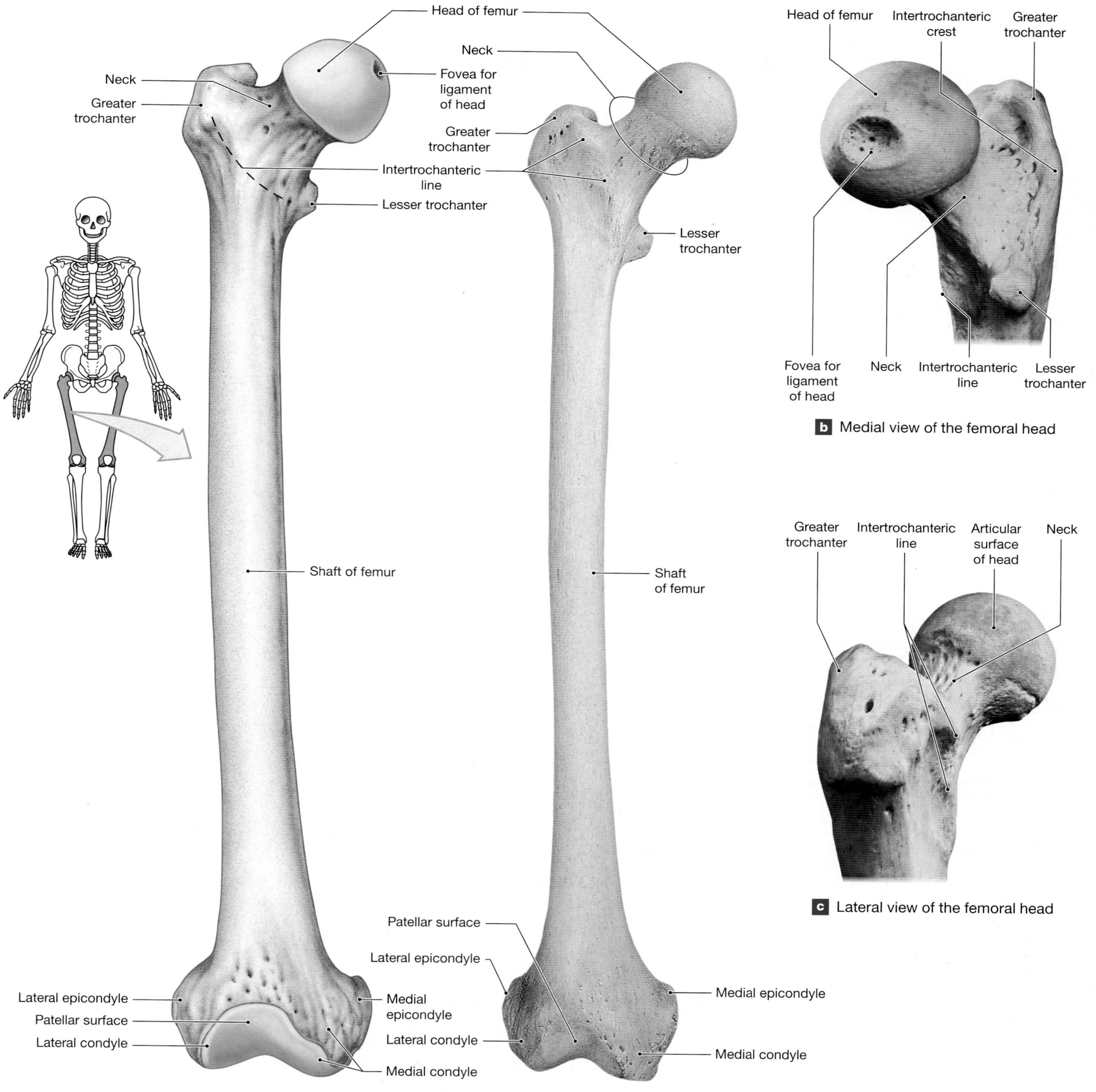

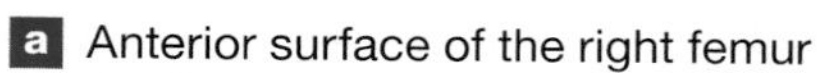

a Anterior surface of the right femur

b Medial view of the femoral head

c Lateral view of the femoral head

Figure 7.13 (*continued*)

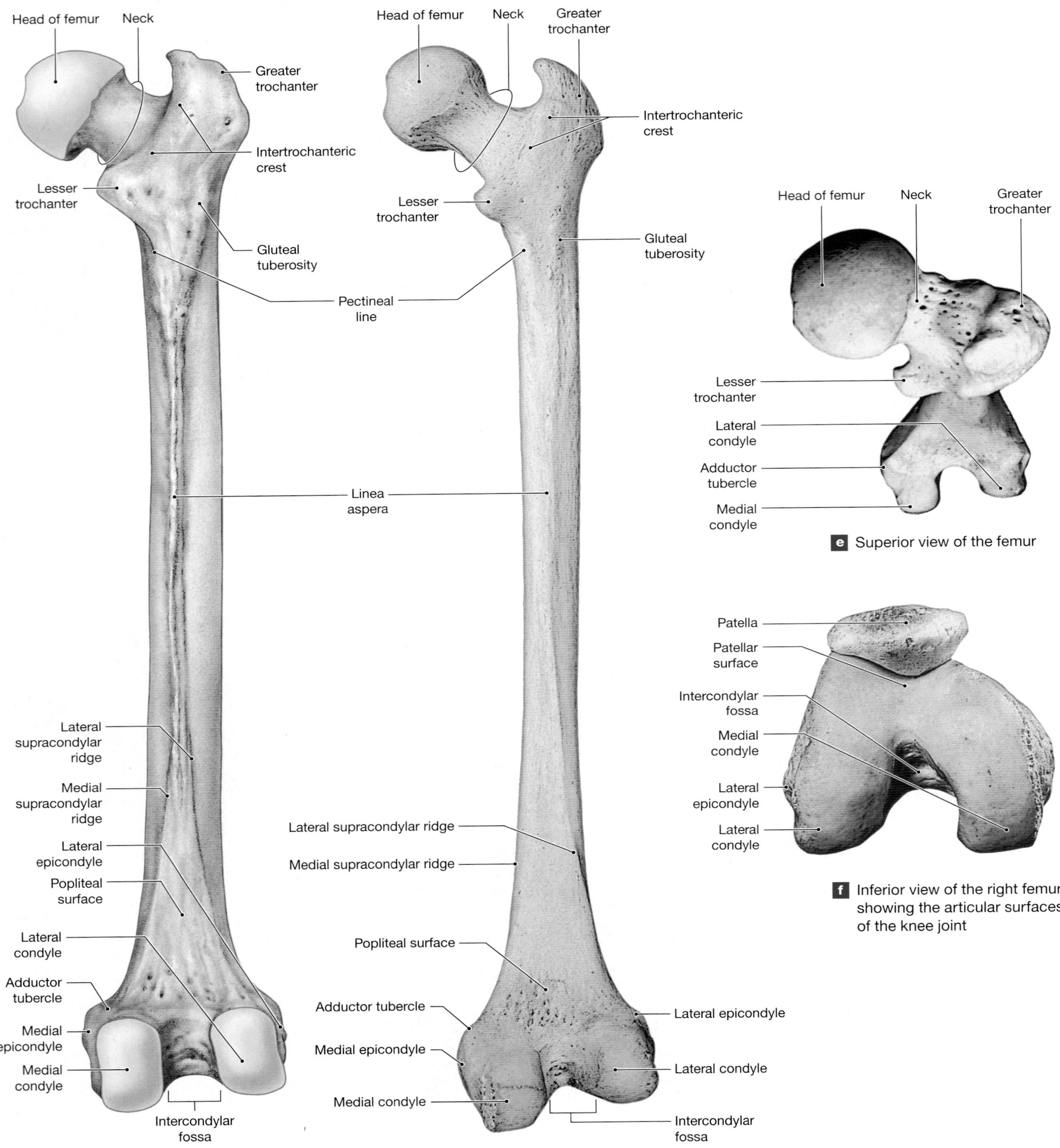

d Posterior surface of the right femur

e Superior view of the femur

f Inferior view of the right femur showing the articular surfaces of the knee joint tendons attach to the femur. On the anterior surface of the femur, the raised **intertrochanteric** (in-ter-trō-kan-TER-ik) **line** marks the distal edge of the capsule of the hip joint **(Figure 7.13a,c)**. This line continues around to the posterior surface, passing inferior to the trochanters as the **intertrochanteric crest (Figure 7.13b,d)**. Inferior to the intertrochanteric crest, the medial **pectineal line** and the lateral **gluteal tuberosity** are the attachment sites for the pectineus and the gluteus maximus respectively. A prominent elevation called the **linea aspera** (*aspera*, rough) runs along the center of the posterior surface of the femoral shaft. This ridge is the attachment site of other powerful adductor hip muscles. Distally, the linea aspera divides into a **medial** and a **lateral supracondylar ridge**, forming a flattened triangular area called the **popliteal** (pop-LITe-al) **surface**. The medial supracondylar ridge ends in a raised, rough projection, the **adductor tubercle**, which is located on the **medial epicondyle**. The lateral supracondylar ridge ends at the **lateral epicondyle**. The smooth, rounded **medial** and **lateral condyles** are distal to the epicondyles. The condyles continue from the posterior, inferior surface of the femur to the anterior surface, but the intercondylar fossa does not. As a result, the smooth articular surfaces merge, producing an articular surface with elevated lateral borders. The patella (*kneecap*) glides over this **patellar surface (Figure 7.13a,f)**. On the posterior surface, the two condyles are separated by a deep **intercondylar fossa**.

The Patella

The **patella** (pa-TEL-a) is a large sesamoid bone that forms within the tendon of the quadriceps femoris, a group of anterior thigh muscles that extends the knee. *(Refer to Chapter 12,* **Figure 12.7a**, *to identify this anatomical structure from the body surface.)* The patella strengthens the quadriceps tendon, protects the anterior surface of the knee joint, and serves as an anatomical pulley that increases the contraction force of the quadriceps femoris. The triangular patella has a rough, convex anterior surface, a broad superior **base**, and a pointed inferior **apex (Figure 7.14a)**. The roughened surface and apex are attachment sites for the quadriceps tendon and the patellar ligament, respectively. The patellar ligament extends from the apex of the patella to the tibia. The posterior patellar surface has a **medial facet** and a **lateral facet**, which articulate with the medial and lateral condyles of the femur **(Figure 7.14b)**.

The Tibia

The **tibia** (TIB-ē-a) is the large medial bone of the leg **(Figure 7.15)**. The medial and lateral condyles of the femur articulate with the **medial** and **lateral condyles** of the proximal end of the tibia. The lateral condyle is larger and has a facet for articulating with the fibula at the superior tibiofibular joint. The **intercondylar eminence** is an elevation that separates the medial and lateral condyles of the tibia **(Figure 7.15b,d)**. There are two **tubercles** (**medial** and **lateral**) on the intercondylar eminence. The anterior surface of the tibia near the condyles has a prominent, rough **tibial tuberosity** that you can easily feel beneath the skin of the leg. This tuberosity is the attachment site for the tough patellar ligament.

The **anterior margin**, or *border*, of the tibia is another feature you can feel under the skin of the leg. It is a ridge that begins at the tibial tuberosity and extends distally along the anterior tibial surface. The lateral margin of the shaft is the **interosseous border**. A collagenous sheet of connective tissue extends from the lateral margin of the tibia to the medial margin of the fibula. Distally, the tibia narrows, and the medial border ends in a large process called the **medial malleolus** (ma-LĒ-ō-lus; *malleolus*, hammer). *(Refer to Chapter 12,* **Figure 12.7**, *to identify this anatomical structure from the body surface.)* The inferior surface of the tibia forms a hinge joint with the talus, the proximal bone of the ankle **(Figure 7.15c)**. Here, the tibia, having received the weight of the body from the femur at the knee, passes that weight across the ankle joint to the foot. The medial malleolus supports the ankle joint and prevents the tibia from sliding laterally across the talus. On the posterior surface of the tibia, the **soleal line**, or *popliteal line*, is the attachment site for several leg muscles, including the popliteus and the soleus **(Figure 7.15d)**.

The Fibula

The slender **fibula** (FIB-ū-la) parallels the lateral border of the tibia **(Figure 7.15)**. The **head** of the fibula, or *fibular head*, articulates at the lateral margin of the tibia on the inferior and posterior surface of the lateral tibial condyle. The **interosseous membrane of the leg** (*crural interosseous membrane*) attaches the medial border of the thin shaft to the tibia. A sectional view through the shafts of the tibia and fibula shows the locations of the tibial and fibular interosseous borders and the fibrous interosseous membrane that extends between them **(Figure 7.15e)**. This membrane stabilizes the positions of the two bones and provides additional surface area for muscle attachment.

Figure 7.14 The Patella. This sesamoid bone forms within the tendon of the quadriceps femoris.

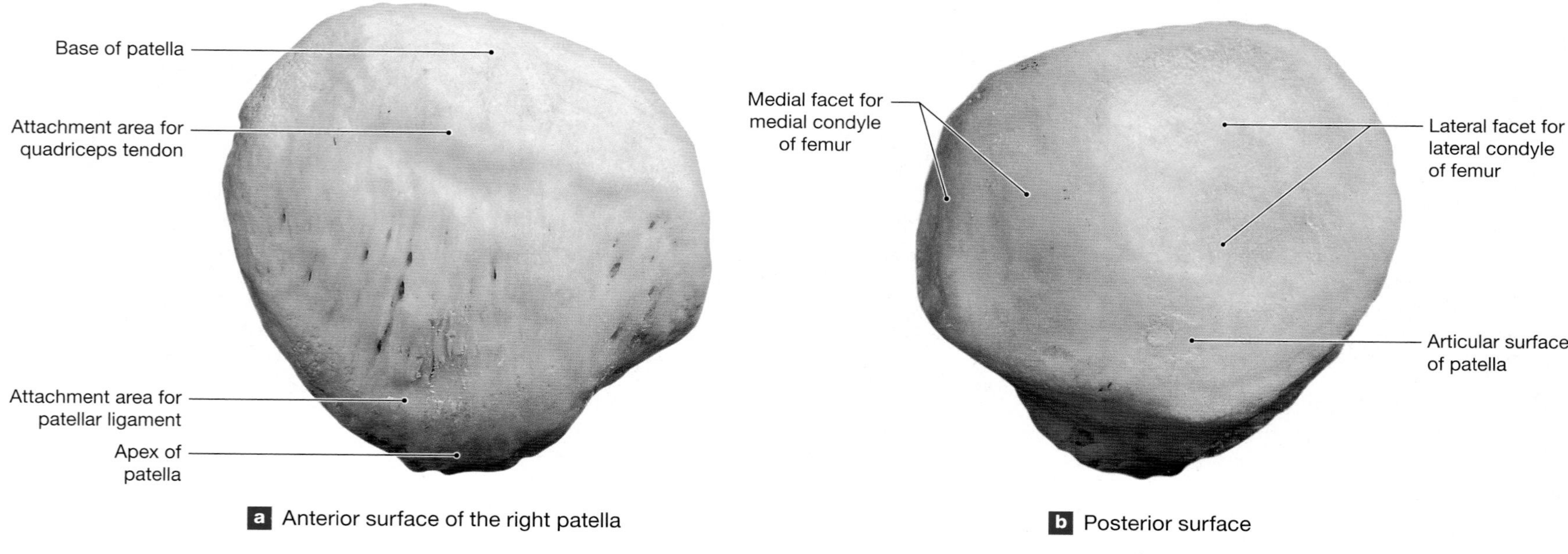

a Anterior surface of the right patella

b Posterior surface

Figure 7.15 **The Tibia and Fibula**

7

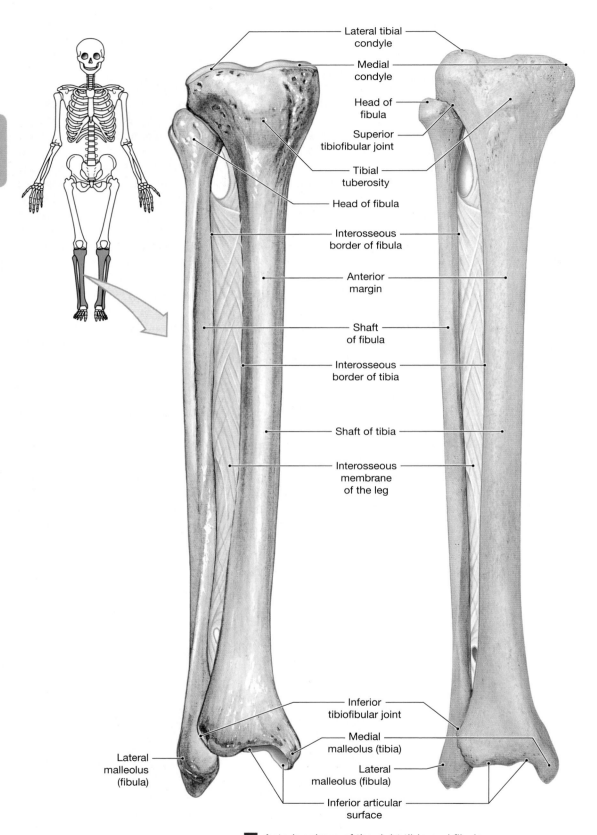

Lateral tibial condyle
Medial condyle
Head of fibula
Superior tibiofibular joint
Tibial tuberosity
Head of fibula
Interosseous border of fibula
Anterior margin
Shaft of fibula
Interosseous border of tibia
Shaft of tibia
Interosseous membrane of the leg
Inferior tibiofibular joint
Medial malleolus (tibia)
Lateral malleolus (fibula)
Lateral malleolus (fibula)
Inferior articular surface

a Anterior views of the right tibia and fibula

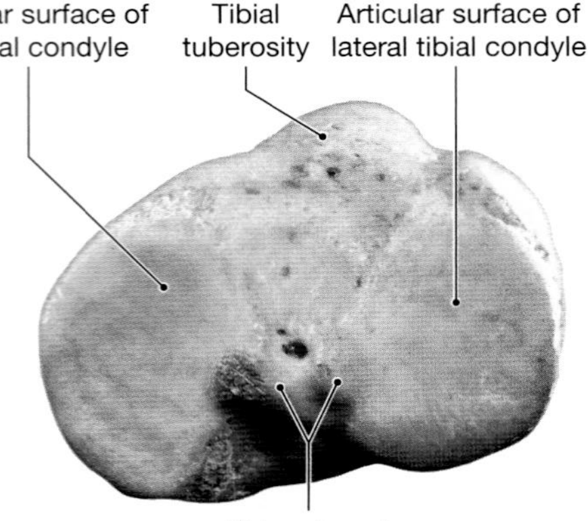

b Superior view of the proximal end of the tibia showing the articular surface

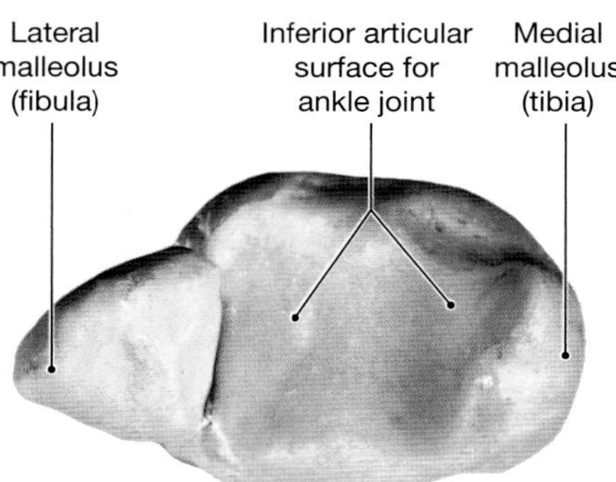

c Inferior view of the distal surfaces of the tibia and fibula showing the surfaces of the ankle joint

Figure 7.15 **(continued)**

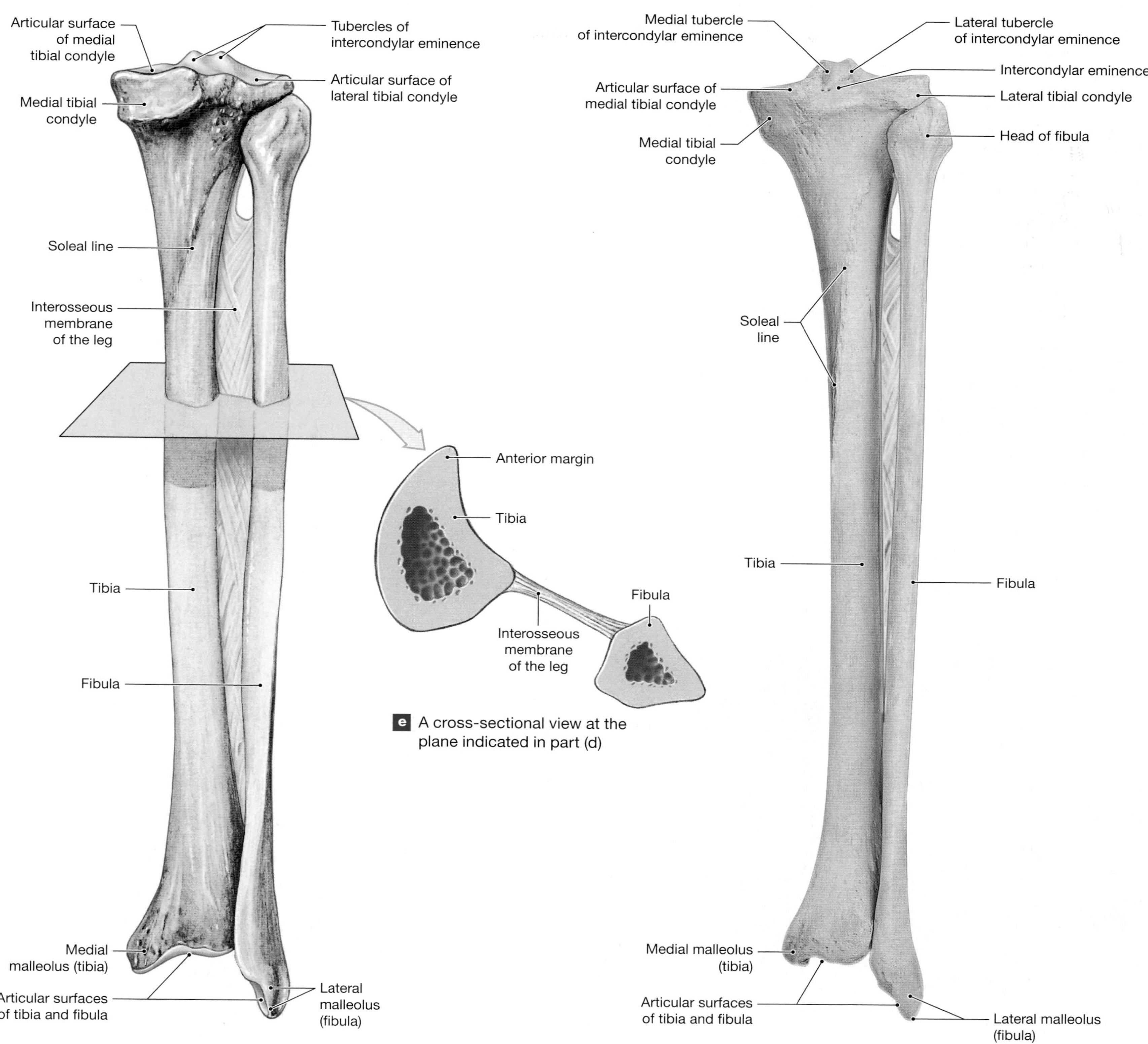

e A cross-sectional view at the plane indicated in part (d)

d Posterior views of the right tibia and fibula

The fibula is not part of the knee joint and does not transfer weight to the ankle and foot. However, it is an important site for muscle attachment. The distal fibular process, termed the **lateral malleolus**, also gives stability to the ankle joint by preventing the tibia from sliding medially across the surface of the talus. *(Refer to Chapter 12, **Figure 12.7**, to identify this anatomical structure from the body surface.)*

The Tarsal Bones

The **ankle**, or **tarsus**, contains seven **tarsal bones**: the talus, calcaneus, cuboid, navicular, and three cuneiform bones **(Figure 7.16)**.

7

- The **talus** is the second largest bone in the foot. It transfers the weight of the body from the tibia anteriorly, toward the toes. The most important distal tibial joint is between the talus and the tibia. This involves the smooth superior surface of the **trochlea** of the talus. The trochlea has lateral and medial extensions that articulate with the lateral malleolus of the fibula and the medial malleolus of the tibia. Ligaments attach the lateral surfaces of the talus to the tibia and fibula, further stabilizing the ankle joint.
- The **calcaneus** (kal-KĀ-nē-us), or *heel bone*, is the largest tarsal bone and is easily palpated. When you are standing normally, the pelvis transmits your weight to the ground as follows: pelvis → femur → tibia → calcaneus → ground. The posterior surface of the calcaneus is a rough, knob-shaped projection. This is the attachment site for the calcaneal tendon (*Achilles tendon*) that comes from the strong calf muscles. These muscles raise the heel and lift the sole of the foot off the ground, as when standing on tiptoe. The superior and anterior surfaces of the calcaneus have smooth facets for articulation with other tarsal bones. *(Refer to Chapter 12, **Figure 12.7**, to identify this anatomical structure from the body surface.)*
- The **cuboid** articulates with the anterolateral surface of the calcaneus.
- The **navicular**, located on the medial side of the ankle, articulates with the anterior surface of the talus. The distal surface of the navicular articulates with the three cuneiform bones.
- The three **cuneiform bones** are wedge-shaped bones arranged in a row, with joints between them, located anterior to the navicular. They are named according to their position: **medial cuneiform**, **intermediate** (or *middle*) **cuneiform**, and **lateral cuneiform bones**. Proximally, the cuneiform bones articulate with the anterior surface of the navicular. The lateral cuneiform bone also articulates with the medial surface of the cuboid. The distal surfaces of the cuboid and the cuneiform bones articulate with the metatarsals of the foot.

The Metatarsals and Phalanges

The **metatarsals** are five long bones between the instep and the toes that form the distal portion (or *metatarsus*) of the foot **(Figure 7.16)**. The metatarsals are identified with Roman numerals I–V, going from medial (great toe) to

Figure 7.16 Bones of the Ankle and Foot

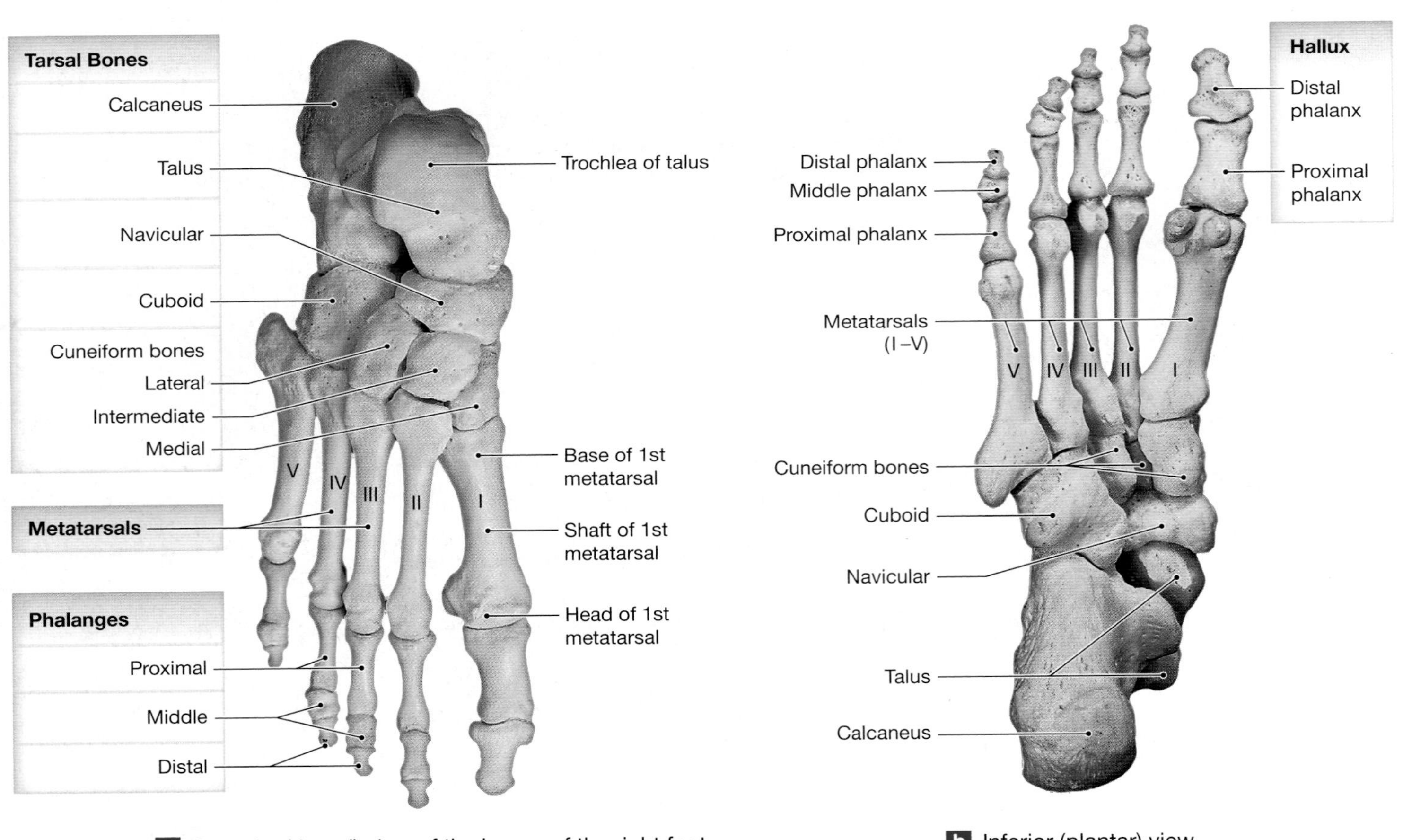

a Superior (dorsal) view of the bones of the right foot

b Inferior (plantar) view

lateral. Proximally, the first three metatarsals articulate with the three cuneiform bones, and the last two articulate with the cuboid. Distally, each metatarsal articulates with a different proximal phalanx. The metatarsals help support the weight of the body during standing, walking, and running.

The 14 **phalanges**, or *toe bones*, have the same anatomical organization as the phalanges of the fingers. The **hallux**, or *great toe*, has two phalanges (proximal phalanx and distal phalanx), and the other four toes have three phalanges each (proximal, middle, and distal).

Arches of the Foot The arches of the foot support and transfer the weight of the body and adapt to walking or running on uneven surfaces. In order to do this, the arches are flexible and function as levers.

Weight transfer occurs along the **longitudinal arch** of the foot (**Figure 7.16d**). Ligaments and tendons maintain this arch by "tying" the calcaneus to the distal portions of the metatarsals. The lateral, or *calcaneal*, part of the longitudinal arch carries most of the weight of the body while standing. This part of the arch has less curvature than the medial, or *talar*, part of the longitudinal arch. The medial part is more elastic than the lateral part of the longitudinal arch. Therefore, the medial, plantar (*sole*) surface remains elevated, and the muscles, nerves, and blood vessels supplying the inferior surface of the foot are not squeezed between the metatarsals and the ground. This elasticity also absorbs the shocks that occur with sudden shifts in weight load. For example, the stresses involved with running or ballet dancing are cushioned by the elasticity of the medial part of the longitudinal arch. The change in the degree of curvature from the medial to the lateral borders of the foot is the **transverse arch**.

When you stand normally, your body weight is distributed evenly between the calcaneus and the distal ends of the metatarsals. The amount of weight transferred forward depends on the position of the foot and the placement of your body weight. During dorsiflexion of the foot, as when "digging in the heels," all of the body weight rests on the calcaneus. During plantar flexion, as when "standing on tiptoe," the talus and calcaneus transfer the weight to the metatarsals and phalanges through more anterior tarsal bones.

7.2  **CONCEPT CHECK**

5 What three bones make up the hip bone?

6 The fibula does not form part of the knee joint, nor does it bend, but if it fractures, walking is difficult. Why?

7 Ten-year-old Mark jumps off the back steps of his house, lands on his right heel, and breaks his foot. What foot bone is most likely broken?

8 Compared to males, how is the female pelvis adapted for childbirth?

9 Where does the weight of the body rest during dorsiflexion? During plantar flexion?

See the blue Answers tab at the back of the book.

7

7.3 Individual Variation in the Skeletal System

▶ **KEY POINT** Bones and teeth form the most lasting record of a person's life. With proper examination and knowledge of anatomy, a pathologist or crime scene investigator can determine the size, age, sex, and genetic heritage of a person from skeletal remains.

A detailed study of the axial and appendicular divisions of a human skeleton can reveal important information about the person. For example, there are characteristic genetic differences in portions of the skeleton, and the development of various ridges and general bone mass can permit an estimation of muscular development. The condition of the teeth or the presence of healed fractures can provide information about the person's medical history. Specific measurements (see **Tables 7.1** and **7.2**) can help determine or closely estimate sex and age.

Figure 7.16 ***(continued)***

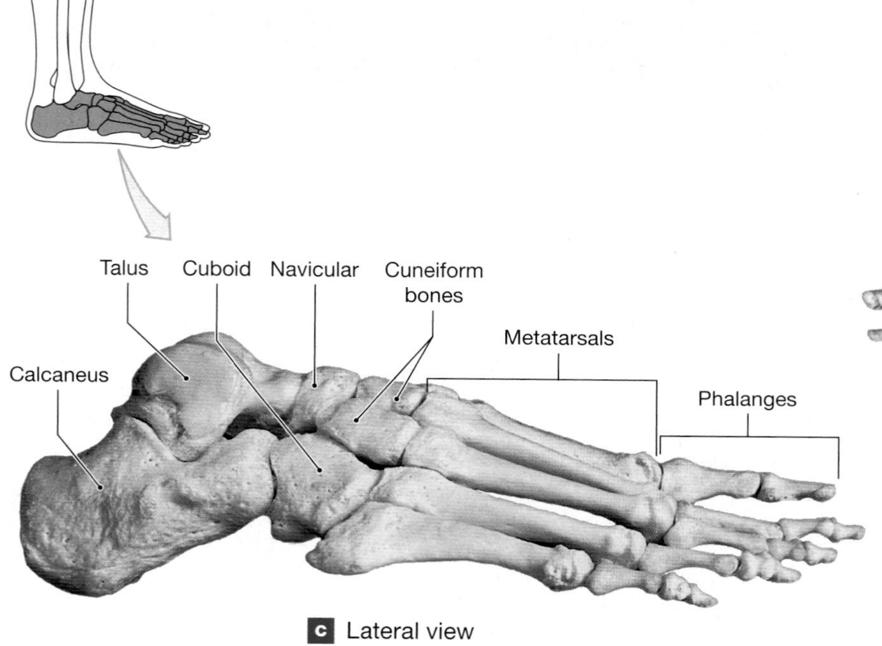

c Lateral view

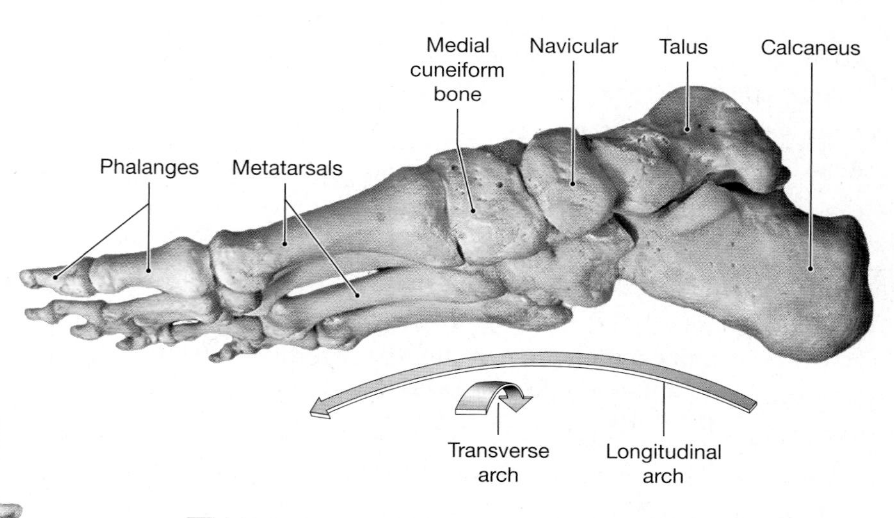

d Medial view of the positions of the tarsal bones and the orientation of the transverse and longitudinal arches. The orientation of the tarsal bones transfer the weight of the body to the heel and the plantar surfaces of the foot

Table 7.1 identifies characteristic differences between the skeletons of males and females, but not every skeleton shows every feature in classic detail. Many differences, including markings on the skull, cranial size, and general skeletal features, reflect differences in average body size, muscle mass, and muscular strength. **Table 7.2** summarizes age-related skeletal changes. These changes begin at age 3 months and continue throughout life. For example, fusion of the epiphyseal cartilages begins at about age 3, whereas degenerative changes in the skeletal system as a whole, such as reduced mineral content in the bony matrix, do not begin until age 30–45.

7

Table 7.1 | Sexual Differences in the Adult Human Skeleton

Region/Feature	Male	Female
SKULL		
General appearance	Heavier; rougher surface	Lighter; smoother surface
Forehead	More sloping; presence of supra-orbital ridges	More vertical; absence of supra-orbital ridges
Sinuses	Larger	Smaller
Cranium	About 10% larger (average)	About 10% smaller
Mandible	Larger, more robust	Lighter, smaller
Teeth	Larger	Smaller
PELVIS		
General appearance	Narrow; robust; heavier; rougher surface	Broad; light; smoother surface
Pelvic inlet	Heart shaped	Oval to round
Iliac fossa	Deeper	Shallower
Ilium	More vertical; extends farther superiorly	Less vertical; less extension superiorly to the sacro-iliac joint
Angle inferior to pubic symphysis	Less than 90°	100° or more
Acetabulum	Directed laterally	Faces slightly anteriorly and laterally
Obturator foramen	Oval	Triangular
Ischial spine	Points medially	Points posteriorly
Sacrum	Long, narrow triangle with pronounced curvature	Broad, short triangle with less curvature
Coccyx	Points anteriorly	Points inferiorly
OTHER SKELETAL ELEMENTS		
Bone weight	Heavier	Lighter
Bone markings	More prominent	Less prominent

7.3 CONCEPT CHECK

10 List at least three regions or features of the skeleton that can help an investigator determine whether an unknown skeleton was that of a male or female individual.

See the blue Answers tab at the back of the book.

CLINICAL NOTE

Hip Dysplasia

Developmental Dysplasia of the Hip (DDH), formerly known as congenital dislocation of the hip, refers to the dysplastic (malformed) growth of the acetabulum. At birth, the ilium, ischium, and pubis are cartilaginous and malleable. As these bones grow, they ossify and eventually fuse to form the acetabulum in response to the presence of the femoral head. If the femoral head is not seated well in the acetabulum, the socket becomes shallow and malformed. If the femoral head is completely dislocated, a "false" acetabulum will form on the outer ilium, superior and lateral to the true acetabulum. Even subtle degrees of developmental dysplasia of the hip can cause early arthritis with disability.

Factors that seem to contribute to DDH include genetics (increased incidence in Native Americans, decreased incidence in Africans), female sex (susceptibility to hormonal laxity during birth), firstborn child (tighter uterus and birth canal), and breech position (buttocks-first in the birth canal).

Newborn assessments include hip joint screening exams to assess the stability of the femoral head in the developing socket.

Diapering an infant with the hips widely abducted (frog-leg position) is often enough to keep the femoral head within the developing acetabulum.

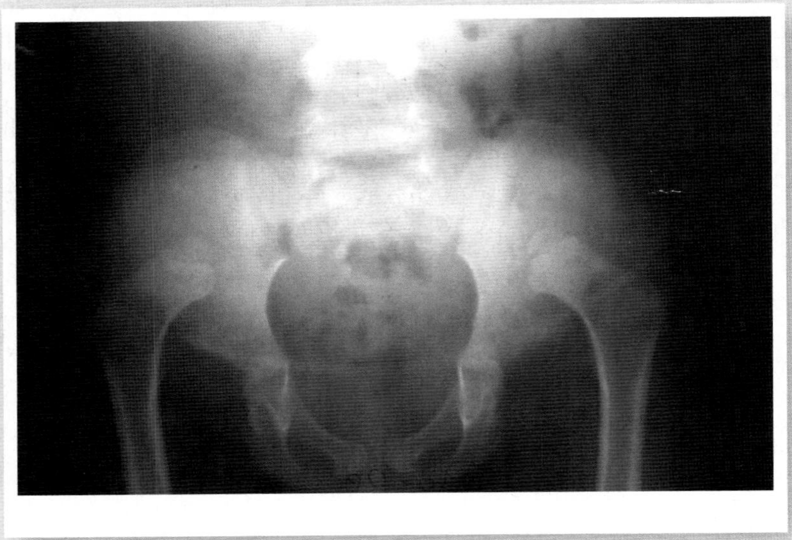

EMBRYOLOGY SUMMARY

For a summary of the development of the appendicular skeleton, see Chapter 28 (Embryology and Human Development).

Table 7.2 | **Age-Related Changes in the Skeleton**

Region/Structure	Event(s)	Age (Years)
GENERAL SKELETON		
Bony matrix	Reduction in mineral content	Begins at age 30–45; values differ for males versus females between ages 45 and 65; similar reductions occur in both sexes after age 65
Markings	Reduction in size, roughness	Gradual reduction with increasing age and decreasing muscular strength and mass
SKULL		
Fontanelles	Closure	Completed by age 2
Frontal suture	Fusion	2–8
Occipital bone	Fusion of ossification centers	1–6
Styloid process	Fusion with temporal bone	12–16
Hyoid bone	Complete ossification and fusion	25–30 or later
Teeth	Loss of "baby teeth"; appearance of permanent teeth; eruption of permanent molars	Detailed in Chapter 25 (Digestive System)
Mandible	Loss of teeth; reduction in bone mass; change in angle at mandibular notch	Accelerates in later years (age 60)
VERTEBRAL COLUMN		
Curvature	Appearance of major curves	3 months–10 years
Intervertebral discs	Reduction in size, percentage contribution to height	Accelerates in later years (age 60)
LONG BONES		
Epiphyseal cartilages	Fusion	Ranges vary according to specific bone under discussion, but general analysis permits determination of approximate age (3–7, 15–22, etc.)
PECTORAL AND PELVIC GIRDLES		
Epiphyseal cartilages	Fusion	Overlapping ranges are somewhat narrower than for long bones, including 14–16, 16–18, 22–25

7

CLINICAL NOTE

Women and Sports Injuries

Title IX became law in 1972, entitling women to equal participation in sports in institutions of public learning. Since then we have learned a lot about female sports injuries.

Women's skeletons are lighter, less dense, and more delicate than men's. Due to hormonal differences, women also have greater flexibility and less strength than men. Unfortunately, these traits make women more vulnerable to head and neck injuries, including concussions.

Women have less upper body strength and weaker shoulder girdles, leading to a higher incidence of shoulder injuries, including separations and dislocations, compared with men.

Because of the shape of the female pelvis, women's hips are wider, causing an increased genu valgum, or "knock-knee" deformity. This, combined with decreased quadriceps muscle mass and strength, leads to patellofemoral pain syndromes. The female patella tends to track laterally, out of its femoral groove, causing anterior knee pain, or "miserable malalignment" syndrome. Female ligaments, particularly the anterior cruciate ligament (ACL), are smaller and more prone to injury than their male counterparts.

Because the cross-sectional diameter of the female tibia and metatarsals is smaller than the male lower extremity, and the bones are less dense, women are more prone to stress fractures of the lower extremities.

A solution to women's increased vulnerability to sports injuries is greater emphasis on training for strength, stability, balance, and coordination.

Study Outline

Introduction p. 173

- The **appendicular skeleton** includes the bones of the upper and lower limbs and the pectoral and pelvic **girdles** that support the limbs and connect them to the trunk. *(See Figure 7.1.)*

7.1 | The Pectoral Girdle and Upper Limb p. 174

7

- Each upper limb articulates with the trunk through the **pectoral girdle**, or *shoulder girdle*, which consists of the **clavicle** and the **scapula**. *(See Figures 7.2 to 7.5.)*

The Pectoral Girdle p. 174

- The clavicle and scapula position the shoulder joint, help move the upper limb, and provide a base for muscle attachment. *(See Figures 7.3–7.5, 12.2, and 12.10.)*
- The **clavicle** is an S-shaped bone that extends between the manubrium of the sternum and the **acromion** of the scapula. This bone provides the only direct bony connection between the pectoral girdle and the axial skeleton.
- The **scapula** articulates with the round head of the humerus at the **glenoid cavity** of the scapula, the **glenohumeral joint** (*shoulder joint*). Two scapular processes, the **coracoid** and the **acromion**, are attached to ligaments and tendons associated with the shoulder joint. The acromion articulates with the clavicle at the **acromioclavicular joint**. The acromion is continuous with the **scapular spine**, which crosses the posterior surface of the scapular body. *(See Figures 7.5, 12.2b, and 12.10.)*

The Upper Limb p. 177

- The **humerus** articulates with the glenoid cavity of the scapula. The joint capsule of the shoulder attaches distally to the humerus at its **anatomical neck**. Two prominent tubercles, the **greater tubercle** and **lesser tubercle**, are important sites for muscle attachment. Other prominent surface features include the **deltoid tuberosity**, site of deltoid attachment; the articular **condyle**, divided into two articular regions, the **trochlea** (medial) and **capitulum** (lateral); the **radial groove**, marking the path of the radial nerve; and the **medial** and **lateral epicondyles** for other muscle attachment. *(See Figures 7.2, 7.6, and 7.7.)*
- Distally, the humerus articulates with the ulna (at the trochlea) and the radius (at the capitulum). The trochlea extends from the **coronoid fossa** to the **olecranon fossa**. *(See Figure 7.6.)*
- The **ulna** and **radius** are the parallel bones of the forearm. The **olecranon** of the ulna enters the olecranon fossa of the humerus during straightening (extension) of the elbow joint. The **coronoid process** of the ulna enters the coronoid fossa during bending (flexion) of the elbow joint. *(See Figures 7.2 and 7.7.)*
- The **carpal bones** of the wrist form two rows, **proximal** and **distal**. From lateral to medial, the proximal row consists of the **scaphoid**, **lunate**, **triquetrum**, and **pisiform**. From lateral to medial, the distal row consists of the **trapezium**, **trapezoid**, **capitate**, and **hamate**. *(See Figure 7.8.)*
- Five **metacarpals** articulate with the distal carpal bones. Distally, the metacarpals articulate with the phalanges. Four of the fingers contain three **phalanges**; the **pollex** (thumb) has only two. *(See Figure 7.8.)*

7.2 | The Pelvic Girdle and Lower Limb p. 184

The Pelvic Girdle p. 185

- The pelvic girdle consists of two **hip bones**, also called *coxal bones* or *pelvic bones*; each hip bone forms through the fusion of three bones—an ilium, an ischium, and a pubis. *(See Figures 7.9 and 7.10.)*
- The **ilium** is the largest of the hip bones. Inside the **acetabulum** (the fossa on the lateral surface of the hip bone that accommodates the **head** of the femur) the ilium fuses to the **ischium** (posteriorly) and to the **pubis** (anteriorly). The **pubic symphysis** limits movement between the pubic bones of the left and right hip bones. *(See Figures 7.10–7.12, 12.3, and 12.14.)*
- The **pelvis** consists of the two hip bones, the sacrum, and coccyx. It may be subdivided into the **greater** (*false*) **pelvis** and the **lesser** (*true*) **pelvis**. The lesser pelvis encloses the pelvic cavity. *(See Figures 7.11 and 7.12.)*

The Lower Limb p. 185

- The **femur** is the longest bone in the body. At its rounded **head**, it articulates with the pelvis at the acetabulum; distally, its **medial** and **lateral condyles** articulate with the tibia at the knee joint. The **greater** and **lesser trochanters** are projections near the head where large tendons attach to the femur. *(See Figures 7.9 and 7.13.)*
- The **patella** (kneecap) is a large sesamoid bone that forms within the tendon of the quadriceps femoris. The patellar ligament extends from the patella to the **tibial tuberosity**. *(See Figures 7.13f, 7.14, and 12.7a.)*
- The **tibia** is the large medial bone of the leg. The prominent rough surface markings of the tibia include the **tibial tuberosity**, the **anterior margin**, the **interosseous border**, and the **medial malleolus**. The medial malleolus is a large process that gives medial support for the ankle joint. *(See Figures 7.15 and 12.7.)*
- The **fibula** is the slender leg bone lateral to the tibia. The **head** articulates with the tibia inferior to the knee, inferior and slightly posterior to the lateral tibial condyle. A fibular process, the **lateral malleolus**, stabilizes the ankle joint by preventing medial movement of the tibia across the talus. *(See Figures 7.15 and 12.7.)*
- The **tarsus**, or ankle, includes seven **tarsal bones**; only the smooth superior surface of the **trochlea** of the **talus** articulates with the tibia and fibula. The lateral and medial extensions of the tarsus articulate with the lateral and medial malleoli of the fibula and tibia, respectively. When standing, most of the body weight transfers to the **calcaneus**; the rest passes to the **metatarsals**. *(See Figure 7.16.)*
- The basic organizational pattern of the **metatarsals** and **phalanges** of the foot is the same as that of the metacarpals and phalanges of the hand. *(See Figure 7.16.)*
- Weight transfer occurs along the **longitudinal arch** and **transverse arch** of the foot. *(See Figure 7.16.)*

7.3 Individual Variation in the Skeletal System p. 197

- Studying a human skeleton can reveal important information such as sex, genetic heritage, medical history, body size, muscle mass, and age. *(See Tables 7.1 and 7.2.)*
- A number of age-related changes and events take place in the skeletal system. These changes begin at about age 3 months and continue throughout life. *(See Tables 7.1 and 7.2.)*

Chapter Review

For answers, see the blue Answers tab at the back of the book.

Level 1 Reviewing Facts and Terms

Match each numbered item with the most closely related lettered item. Use letters for answers in the spaces provided.

1. shoulder ____
2. hip ____
3. scapula ____
4. trochlea ____
5. ulnar notch ____
6. one hip bone ____
7. greater trochanter ____
8. medial malleolus ____
9. heel bone ____
10. toes ____

(a) tibia
(b) pectoral girdle
(c) radius
(d) phalanges
(e) pelvic girdle
(f) femur
(g) infraspinous fossa
(h) calcaneus
(i) ilium
(j) humerus

11. Label the following structures on the lateral view of the scapula below.
    - acromion
    - glenoid cavity
    - coracoid process
    - spine of scapula

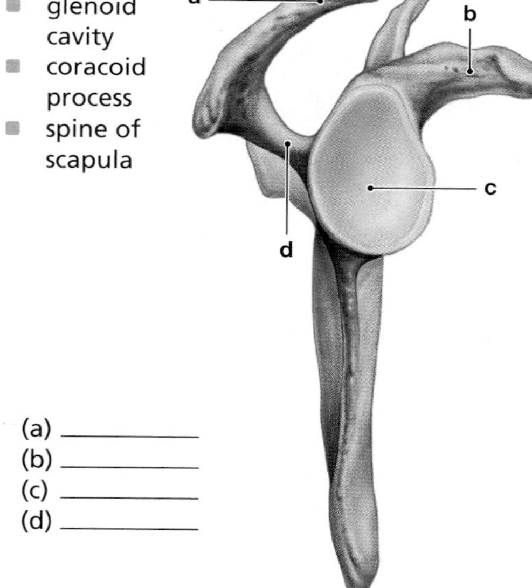

(a) ________
(b) ________
(c) ________
(d) ________

12. The broad, relatively flat portion of the clavicle that articulates with the scapula is the
    (a) sternal end.
    (b) conoid tubercle.
    (c) acromial end.
    (d) costal tuberosity.

13. Label the following structures on the lateral view of the proximal end of the ulna below.
    - ulnar tuberosity
    - olecranon
    - radial notch
    - coronoid process

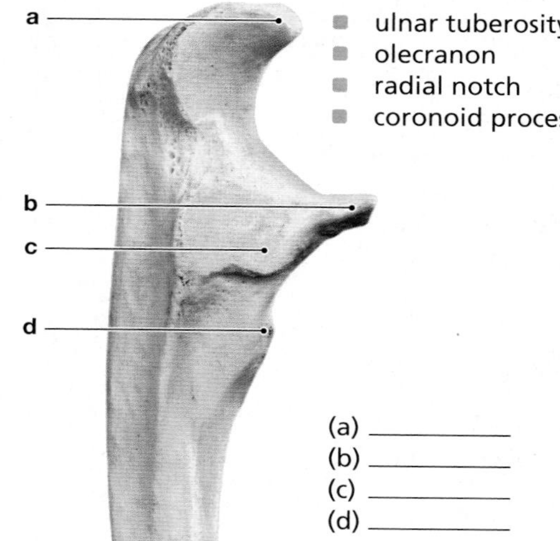

(a) ________
(b) ________
(c) ________
(d) ________

14. The protuberance that you can feel on the lateral side of the ankle is the
    (a) lateral malleolus.
    (b) lateral condyle.
    (c) tibial tuberosity.
    (d) lateral epicondyle.

15. Structural characteristics of the pelvic girdle that allow it to bear the body's weight include
    (a) heavy bones.
    (b) stable joints.
    (c) limited range of movement.
    (d) all of the above at some joints.

16. Which of the following is not a carpal bone?
    (a) scaphoid (b) hamate
    (c) cuboid (d) triquetrum

17. Label the following structures on the lateral view of the pelvis below.
    - posterior inferior iliac spine
    - anterior superior iliac spine
    - superior pubic ramus
    - anterior gluteal line

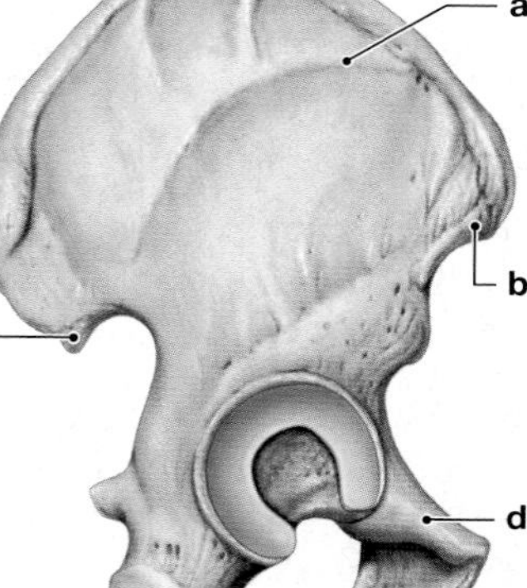

(a) ________
(b) ________
(c) ________
(d) ________

18. The ________ of the radius assists in stabilizing the wrist joint.
    (a) olecranon
    (b) coronoid process
    (c) styloid process
    (d) radial tuberosity

Level 2 Reviewing Concepts

1. What pieces of information are helpful in determining the age of a skeleton?
2. What is the importance of maintaining the correct amount of curvature of the longitudinal arch of the foot?
3. Why are fractures of the clavicle so common?
4. Why is the tibia, but not the fibula, involved in weight transfer to the ankle and foot?
5. What is the function of the olecranon of the ulna?
6. How is body weight passed to the metatarsals?

Level 3 Critical Thinking

1. Why would a person who has osteoporosis be more likely to suffer a broken hip than a broken shoulder?
2. How would a forensic scientist decide whether a partial skeleton found in the forest is that of a male or a female?

7

CLINICAL CASE WRAP-UP

Double Jeopardy

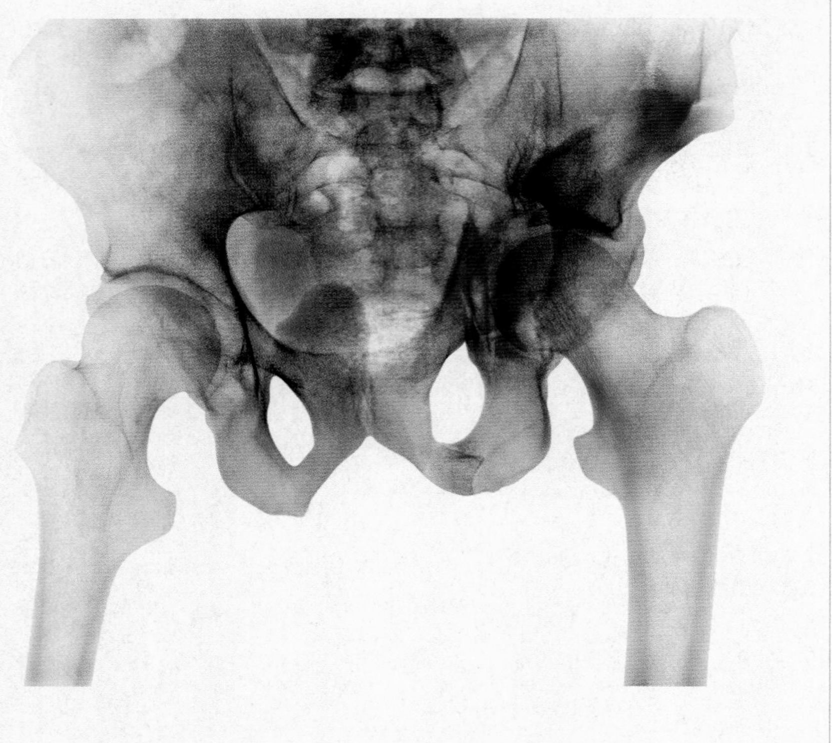

Monique has suffered a transverse fracture of her left acetabulum. Look closely, and you can see the fetal head engaged within the pelvis.

The impact to the greater trochanter of the femur pushed the femoral head medially into the acetabulum, fracturing both the anterior and posterior segments of the acetabulum. This horizontal fracture pattern follows the fusion lines of the ilium with the pubis anteriorly and with the ischium posteriorly. The superior surface of the acetabulum, including two-fifths of the joint surface, is still attached to the ilium. The inferior surface of the acetabulum, including the pubis anteriorly and the ischium posteriorly, has been pushed medially. The femoral head is pushed medially, toward the head of the fetus, as well.

In the operating room, the obstetrician monitors the fetus while the orthopedist performs an internal fixation of the anterior and posterior segments of the acetabulum, restoring the lunate surface of the joint. Because this fracture takes at least eight weeks to heal, a healthy baby boy is born by Cesarean section seven weeks later. Monique's baby shows no signs of visible trauma.

1. Is Monique's acetabular fracture located in the greater (false) pelvis or the lesser (true) pelvis?
2. Using anatomical terms, describe the structures involved in the acetabular fracture.

See the blue Answers tab at the back of the book.

Related Clinical Terms

bone graft: A surgical procedure that transplants bone tissue to repair and rebuild diseased or damaged bone.

genu valgum: Deformity in which the knees angle medially and touch one another while standing; commonly called "knock-knee."

pelvimetry: Measurement of the dimensions of the female pelvis.

8

The Skeletal System

Joints

Learning Outcomes

These Learning Outcomes correspond by number to this chapter's sections and indicate what you should be able to do after completing the chapter.

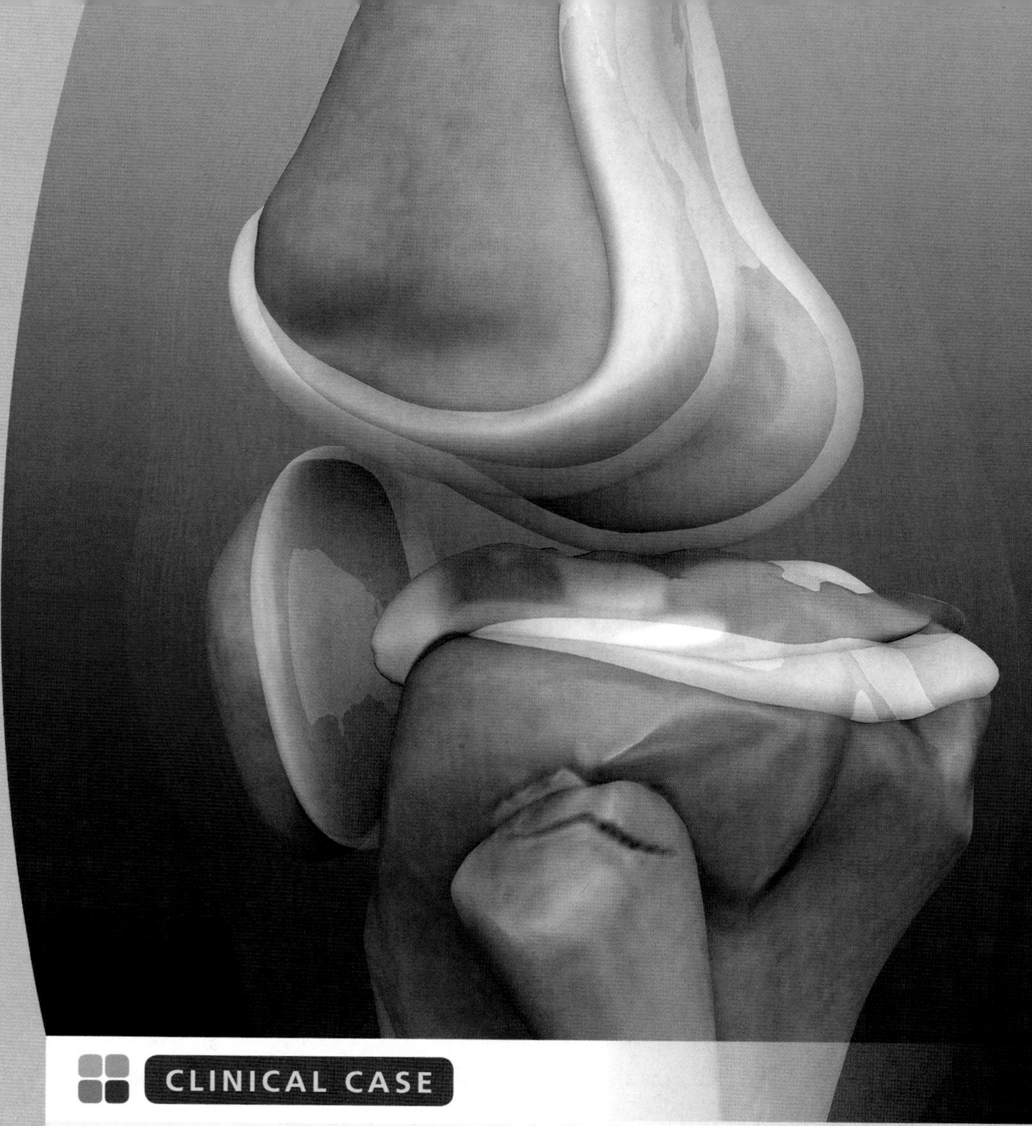

CLINICAL CASE

Why Does My Knee Hurt So Much?

Molly is a 21-year-old college pole vaulter; she also competes in the long jump, triple jump, and hurdles. In all of these events, she uses the left limb as her "take-off leg." Molly trains by pole vaulting twice a week, strength training daily, and running three miles every other night. Early in the season, Molly experiences a sharp pain on the medial side of her left knee while running. When palpating the knee, she hears "clicking." She also reports a feeling she describes as "catching on something" when she flexes her leg at the knee. After experiencing this pain for a few days, she decides to go see her physician.

What could be causing Molly's knee symptoms? To find out, turn to the Clinical Case Wrap-Up on p. 234.

WE DEPEND ON our bones for support, but support without mobility would make us immovable statues. Body movements cannot exceed the stresses that compact and spongy bone can handle. For example, you cannot bend the shaft of the humerus or femur; such movements are restricted to joints. **Joints**, or **articulations** (ar-tik-ū-LĀ-shuns), are junctions between two or more bones. The bones may be in direct contact or separated by fibrous tissue, cartilage, or fluid. Each joint has a normal range of motion, and bony surfaces, cartilages, ligaments, tendons, and muscles work together to keep movement within this normal range. The anatomy of a joint determines its function and range of motion. Some joints interlock and prohibit movement completely, and other joints permit either slight movement or extensive movement. *Immovable and slightly movable joints are more common in the axial skeleton, whereas freely movable joints are more common in the appendicular skeleton.*

8.1 | Joint Classification

▶ **KEY POINT** A common way to classify joints is by the joint's anatomy and range of motion. That classification method produces three types of joints: immovable (synarthrosis), slightly movable (amphiarthrosis), and freely movable (diarthrosis).

There is no simple, single way to classify joints. Anatomists classify joints based on either the histological structure of the joint (fibrous, cartilaginous, bony, or synovial) or the range of motion at the joint, as outlined in **Table 8.1**. In the classification method based on the range of motion, there are three categories of joints.

1. An immovable joint is termed a **synarthrosis** (sin-ar-THRŌ-sis; *syn-*, together, + *arthro-*, joint).
2. A slightly movable joint is an **amphiarthrosis** (am-fē-ar-THRŌ-sis; *amphi-*, on both sides).
3. A freely movable joint is a **diarthrosis** (dī-ar-THRŌ-sis; *dia-*, through).

Subdivisions within these categories are based on the histological structural differences of the joints. Synarthrotic and amphiarthrotic joints are classified as fibrous or cartilaginous, and diarthrotic joints are subdivided according to the range of motion. Here, we will use the classification method based on the range of motion rather than the histological structure.

Synarthroses (Immovable Joints)

▶ **KEY POINT** A synarthrosis is a joint held together by dense, irregularly arranged connective tissue. Synarthroses allow little or no movement.

At a synarthrosis, the bony edges are close together and may even interlock. A **suture** (*sutura*, a sewing together) is a fibrous synarthrotic joint found only between the bones of the skull. The edges of these bones are bound together at sutures by the **sutural ligament**. This fibrous connective tissue is the unossified remains of the embryonic mesenchyme in which the bones developed. ⮌ **pp. 113–115** A synarthrosis allows forces to be spread easily from one bone to another with little or no joint movement, thereby decreasing the chance of injury. A **gomphosis** (gom-FŌ-sis; *gomphos*, bolt) is a specialized fibrous synarthrosis that binds each tooth to its bony socket. This fibrous connection is the **periodontal ligament** (per-ē-ō-DON-tal; *peri-*, around, + *odous*, tooth).

In a growing bone, the diaphysis and epiphyseal ends are bound together by an epiphyseal cartilage, which is an example of a cartilaginous synarthrosis. This connection is called a **synchondrosis** (sin-kon-DRŌ-sis; *syn-*, together, + *chondros*, cartilage). Sometimes two separate bones fuse, and the boundary between them disappears. The result is a **synostosis** (sin-os-TŌ-sis), a totally rigid, immovable joint.

Amphiarthroses (Slightly Movable Joints)

▶ **KEY POINT** In an amphiarthrosis, the bones of the joint are held together by fibrous cartilage, hyaline cartilage, or fibrous connective tissue (a ligament). These joints allow limited movement.

An amphiarthrosis permits limited movement, and the bones are farther apart than in a synarthrosis. In a fibrous amphiarthrosis the bones are connected by collagen fibers, and the bones in a cartilaginous amphiarthrosis are connected by fibrous cartilage.

In a **syndesmosis** (sin-dez-MŌ-sis; *desmos*, band), a ligament connects the bones and limits movement at the joint. The distal joint between the tibia and fibula and the interosseous membrane between the radius and ulna are two examples of a syndesmosis. At a **symphysis** the bones are separated by a pad of fibrous cartilage. The joints between adjacent vertebral bodies (the intervertebral discs) and the anterior connection between the two pubic bones (the pubic symphysis) are examples of this type of joint.

Diarthroses (Freely Movable Synovial Joints)

▶ **KEY POINT** Diarthroses contain a fluid-filled cavity between the bones of the joint. Because these joints have a synovial membrane and contain synovial fluid, they are called synovial joints. Diarthroses are specialized for movement and have seven components.

A diarthrosis, or **synovial** (si-NŌ-vē-al) **joint**, is specialized for movement and permits a wide range of motion. The bony surfaces within a synovial joint are covered by **articular cartilages** and therefore are not in direct contact with one another. These cartilages act as shock absorbers and help reduce friction within the joint. Articular cartilage resembles hyaline cartilage elsewhere in the body, but it has no perichondrium. In addition, the matrix contains more fluid than typical hyaline cartilage. Synovial joints are found at the ends of the long bones of the upper and lower limbs.

Figure 8.1 shows the structure of a typical synovial joint. All synovial joints have the same basic components: (1) a joint capsule, (2) articular cartilages, (3) a joint cavity filled with synovial fluid, (4) a synovial membrane lining the joint capsule, (5) accessory structures, (6) sensory nerves, and (7) blood vessels that supply the exterior and interior of the joint.

Synovial Fluid

A synovial joint is surrounded by a **joint capsule**, or **articular capsule**. The joint capsule has an outer layer of thick, dense, regularly arranged connective tissue and an inner **synovial membrane** that lines the joint cavity. The synovial membrane stops at the edges of the articular cartilages. ⮌ **p. 72** Synovial membranes produce **synovial fluid** that fills the joint cavity. Synovial fluid serves three functions:

1. Lubrication and friction reduction: A thin layer of synovial fluid covers the inner surface of the joint capsule and the exposed surfaces of the articular cartilages. This layer lubricates and reduces friction within the joint. In particular, the substances hyaluronan and lubricin in synovial fluid significantly reduce friction between the cartilage surfaces within the joint.
2. Nutrient distribution: The total amount of synovial fluid in any joint is normally less than 3 mL. Whenever the joint moves, synovial fluid circulates within the cavity and articular cartilages to nourish the tissues, distribute dissolved gases, and remove wastes. When the joint compresses the cartilage, synovial fluid and the dissolved gases and waste are forced out of the articular cartilages. When the joint allows the articular cartilage to expand, the fluid and all of the dissolved gases and nutrients are pulled back into the cartilages.

Table 8.1 | **Function and Structural Classification of Joints**

Functional Category	Structural Category and Type		Description and Examples
SYNARTHROSIS (NO MOVEMENT)			
At a synarthrosis, the bony edges are close together and may even interlock. These extremely strong joints are located where movement between the bones must be prevented.	Fibrous	Suture	A suture is a synarthrotic joint located only between the bones of the skull. The edges of the bones interlock; dense fibrous connective tissue binds the bones together at the suture.
		Gomphosis	A gomphosis is a synarthrosis that binds the teeth to bony sockets in the maxillae and mandible. The fibrous connection between a tooth and its socket is a periodontal ligament.
	Cartilaginous	Synchondrosis	A synchondrosis is a rigid, cartilaginous bridge between two articulating bones. The cartilaginous connection between the ends of the first pair of vertebrosternal ribs and the sternum is a synchondrosis.
	Bony fusion	Synostosis	A synostosis is a totally rigid, immovable joint created when two bones fuse and the boundary between them disappears. The frontal suture of the frontal bone and the epiphyseal lines of mature long bones are synostoses.
AMPHIARTHROSIS (LITTLE MOVEMENT)			
An amphiarthrosis permits more movement than a synarthrosis, but is much stronger than a freely movable joint. Collagen fibers or cartilage connect the articulating bones.	Fibrous	Syndesmosis	At a syndesmosis, a ligament connects the bones. One example is the distal articulation between the tibia and fibula.
	Cartilaginous	Symphysis	At a symphysis, a wedge or pad of fibrous cartilage separates the articulating bones. The articulation between the two pubic bones (the pubic symphysis) is an example.
DIARTHROSIS (FREE MOVEMENT)			
		Synovial	Diarthroses, or synovial joints, permit a wider range of motion than do other types of joints. They are typically located at the ends of long bones, such as those of the upper and lower limbs.
	■ **Monaxial** (movement in one plane)		The elbow and ankle are monaxial joints.
	■ **Biaxial** (movement in two planes)		The ribs and wrist are biaxial joints.
	■ **Triaxial** (movement in three planes)		The shoulder and hip are triaxial joints.

3. Shock absorption: Synovial fluid cushions joints that are subjected to compression. For example, the hip, knee, and ankle joints are compressed during walking, and they are severely compressed during jogging or running. When the pressure suddenly increases, the synovial fluid absorbs the shock and distributes it evenly across the articular surfaces.

Accessory Structures

Synovial joints have a variety of accessory structures, including cartilages, fat pads, menisci, ligaments, tendons, and bursae **(Figure 8.1)**.

Cartilages and Fat Pads In complex joints such as the knee, these accessory structures lie between the articular surfaces and modify the shapes of the joint surfaces:

- **Menisci** (me-NIS-kē; singular, *meniscus*) are crescent-shaped pads of fibrous cartilage that often subdivide a synovial cavity. These structures channel the flow of synovial fluid, allow for variations in the shapes of the articular surfaces, or restrict movements at the joint.
- **Fat pads** are found around the periphery of the joint and are lightly covered by the synovial membrane. Fat pads protect the articular cartilages and serve as a sort of "packing material" for the joint as a whole. When the bones move, fat pads also fill spaces created as the joint cavity changes shape.

Ligaments The joint capsule surrounding the entire joint is continuous with the periostea of the articulating bones. Three **accessory ligaments** support, strengthen, and reinforce synovial joints. **Capsular ligaments** are

Figure 8.1 Structure of a Synovial Joint. Synovial joints are diarthrotic joints that permit a wide range of motion.

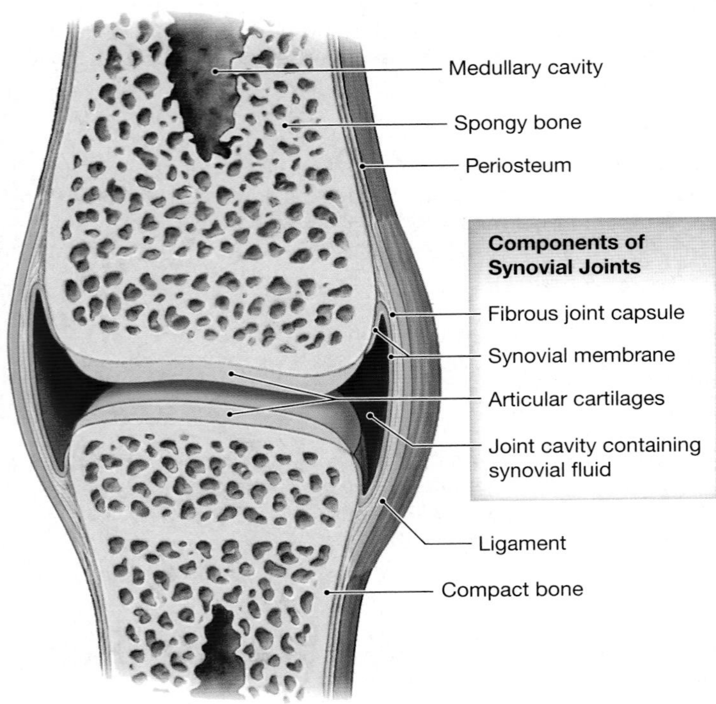

a Synovial joint, sagittal section

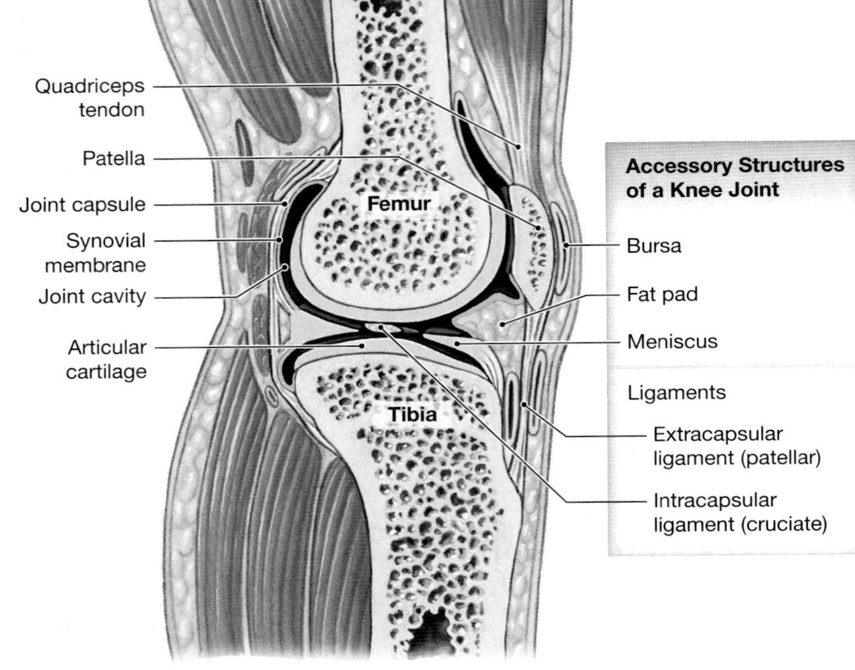

b Knee joint, sagittal section

thickenings of the joint capsule itself. **Extracapsular ligaments** are found outside the joint capsule, while **intracapsular ligaments** are inside the joint capsule **(Figure 8.1b)**.

Tendons Tendons are not a part of the joint itself, but are accessory structures that pass across or around it **(Figure 8.1b)**. Normal muscle tone keeps tendons taut, and they often limit the joint's range of motion. In some joints, tendons are an integral part of the joint capsule, providing significant strength to the capsule.

Bursae **Bursae** (singular, *bursa*) are small, fluid-filled pockets in connective tissue that reduce friction and act as shock absorbers **(Figure 8.1b)**. They are filled with synovial fluid and lined by a synovial membrane. Bursae are often connected to the joint cavity, but they may be completely separate from it. Bursae form where a tendon or ligament rubs against other tissues and are found around most synovial joints, such as the shoulder joint. **Synovial tendon sheaths** are tubular bursae that surround tendons where they pass across bony surfaces. Bursae also appear beneath the skin covering a bone or within other connective tissues exposed to friction or pressure.

Strength versus Mobility

A joint cannot be both highly mobile *and* very strong. *The greater the range of motion of a joint, the weaker it is.* A synarthrosis, the strongest type of joint, permits no movement, whereas a diarthrosis, with its wide range of motion, is weak and easily damaged. Several factors combine to limit mobility and reduce the chance of injury:

- accessory ligaments and the collagen fibers of the joint capsule;
- the shapes of the articulating surfaces preventing movement in specific directions;

CLINICAL NOTE

Dislocation of a Synovial Joint

A **dislocation**, or **luxation**, is a painful displacement of articulating bones that temporarily deforms and immobilizes the joint. It can damage the joint surface or stretch or tear the joint capsule and supporting ligaments and tendons. The more mobile the joint, the greater the chance of dislocation. Shoulders and fingers are the most frequently dislocated. A **subluxation** is an incomplete dislocation.

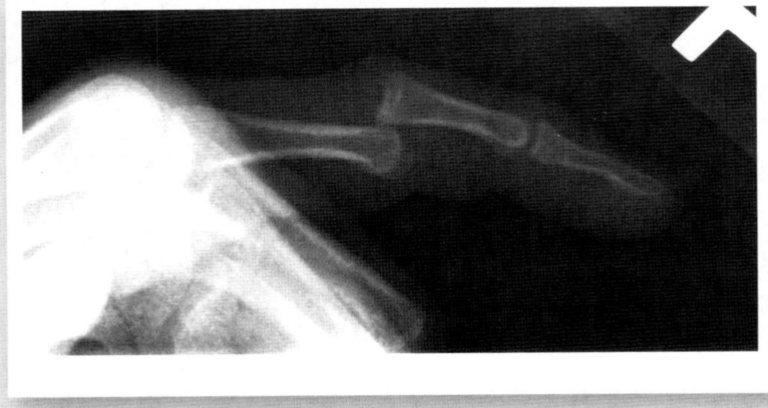

- other bones, bony processes, skeletal muscles, or fat pads around the joint; and
- tension in tendons attached to the articulating bones. When a skeletal muscle contracts and pulls on a tendon, it either causes or opposes movement in a specific direction.

8.1 CONCEPT CHECK

1 Distinguish between the types of movement at a synarthrosis and an amphiarthrosis.
2 List two functions of synovial fluid.

See the blue Answers tab at the back of the book.

8.2 | Articular Form and Function

▶ **KEY POINT** To *understand* human movement you must become aware of the relationship between structure and function at each joint. To *describe* human movement you need a frame of reference that permits accurate and precise communication.

Describing Dynamic Motion and the Structural Classification of Synovial Joints

▶ **KEY POINT** Synovial joints move by linear, angular, or rotational motions. All of these movements occur along one or more axes of rotation: the superior-inferior axis, lateral-medial axis, or anterior-posterior axis. Synovial joints are classified by a system that describes joints as familiar objects or shapes.

Synovial joints are freely movable diarthrotic joints that are classified according to their anatomical and functional properties. **SpotLight Figure 8.2** describes

- the types of movement that occur at a typical synovial joint, using a simplified model;
- the axes of motion around which all joint movements occur; and
- how synovial joints are classified according to the type and range of movement at that joint.

Types of Movements

▶ **KEY POINT** The types of movements at joints are linear motion, angular motion, circumduction, rotation, and a series of special movements that are unique to a limited number of specialized joints.

Anatomists use descriptive terms to illustrate movements at a synovial joint. To understand the types of movements at a joint you must (1) think about the motion of the particular joint movement (angular, rotation, or special), (2) remember that the terms for joint movements occur in pairs, and (3) remember that all joint movements, unless otherwise indicated, are described with reference to a figure in the anatomical position. ↺ pp. 14–16

Angular Motion

Examples of angular motion include abduction, adduction, flexion, and extension **(Figure 8.3)**.

- **Abduction** (*ab*-, away from) is movement away from the longitudinal axis of the body in the frontal plane. For example, swinging the upper limb away from the side is abduction of the limb. The opposite motion—moving it back to center is called **adduction** (*ad*-, toward). Spreading the fingers or toes apart abducts them, because they move away from a central finger or toe. Bringing them together is adduction. *Abduction and adduction always refer to movements of the appendicular skeleton* **(Figure 8.3a,c)**.
- **Flexion** (FLEK-shun) is movement in the anterior-posterior plane that decreases the angle between the bones of the joint. **Extension** occurs in the same plane, but it increases the angle between the bones of the joint **(Figure 8.3b)**. Flexion at the elbow or hip swings the limbs anteriorly, whereas extension moves them posteriorly. Flexion at the wrist moves the palm forward, and extension moves it back. When you bring your head toward your chest, you flex the intervertebral articulations of the neck. When you bend down to touch your toes, you flex the intervertebral joints of the vertebral column. Extension is a movement in the same plane as flexion, but in the opposite direction. Extension may return the limb to or beyond the anatomical position. **Hyperextension** is a term applied to any movement in which a limb is extended beyond its normal limits, resulting in joint damage. Ligaments, bony processes, or surrounding soft tissues usually prevent hyperextension.
- A special type of angular motion, **circumduction** involves moving the arm in a circle, as when drawing a large circle on a chalkboard in one continuous motion **(Figure 8.3d)**.

Rotation

Rotation of the head involves either **left rotation** or **right rotation**, as in shaking the head "no." In movements of the limbs, when the anterior surface of the limb rotates *inward*, toward the anterior surface of the body, it is termed **internal rotation**, or **medial rotation**. When it turns *outward*, it is **external rotation**, or **lateral rotation**. These rotational movements are illustrated in **Figure 8.4**.

The joints between the radius and ulna permit the distal end of the radius to rotate from the anatomical position across the anterior surface of the ulna. This motion, called **pronation** (prō-NĀ-shun), moves the wrist and hand from the palm-facing-front position to the palm-facing-back position. The opposing movement, which turns the palm forward, is **supination** (sū-pi-NĀ-shun).

Special Movements

Special terms apply to specific joints or unusual types of movement **(Figure 8.5)**.

- **Eversion** (ē-VER-zhun; *e*-, out, + *everto*, to overturn) is a motion of the foot that turns the sole outward **(Figure 8.5a)**. The opposite movement, turning the sole inward, is called **inversion** (*inverto*, to turn upside down).
- **Dorsiflexion** and **plantar flexion** (*plantar*, sole of the foot) also refer to movements of the foot **(Figure 8.5b)**. Dorsiflexion elevates the distal portion of the foot and the toes, as in "digging in your heels." Plantar flexion elevates the heel and the proximal portion of the foot, as when standing on tiptoe.
- **Lateral flexion** occurs when the vertebral column bends to the side. This movement is most pronounced in the cervical and thoracic regions **(Figure 8.5c)**. Lateral flexion to the left is counteracted by lateral flexion to the right.
- **Protraction** is moving a part of the body anteriorly in the horizontal plane. **Retraction** is the reverse movement **(Figure 8.5d)**. You protract your jaw when you grasp your upper lip with your lower teeth, and you retract your jaw when you return it to its normal position.
- **Opposition** is a special movement of the thumb that produces pad-to-pad contact of the thumb with the palm or any other finger. Flexion of the fifth metacarpophalangeal joint can assist this movement. **Reposition** is the opposite movement that returns the thumb and fingers to their normal position **(Figure 8.5e)**.

8

SPOTLIGHT

FIGURE 8.2

Joint Motion

A Simple Model of Joint Movement

Take a pencil as your model and stand it upright on the surface of a desk. The pencil represents a bone, and the desk is an articular surface. A lot of twisting, pushing, and pulling will demonstrate that there are only three ways to move the pencil.

Axes of Motion

Movements at a joint occur along specific axes of motion. An axis of motion is an imaginary plane along which joint movement occurs. There are three possible axes of motion, just as there are three dimensions in the world around us. A joint that permits movements in all three axes, such as the shoulder joint illustrated below, is described as **triaxial**. A joint that permits movement along two axes is termed **biaxial**, whereas a joint that permits movement in only one axis is termed **monaxial**.

Moving the Point

Linear motion

Gliding is an example of **linear motion**. The pencil remains vertical, but the pencil tip moves away from its original position.

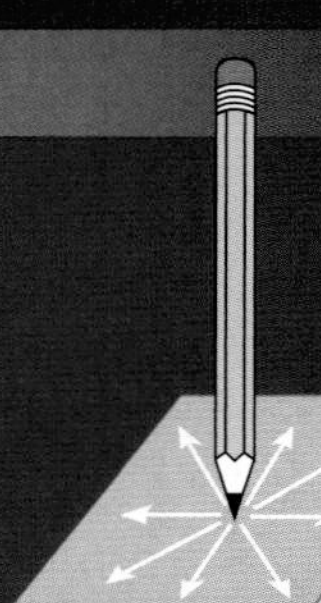

Changing the Shaft Angle

Angular motion

During **angular motion** the pencil tip remains stationary, but the angle between the shaft and the surface changes.

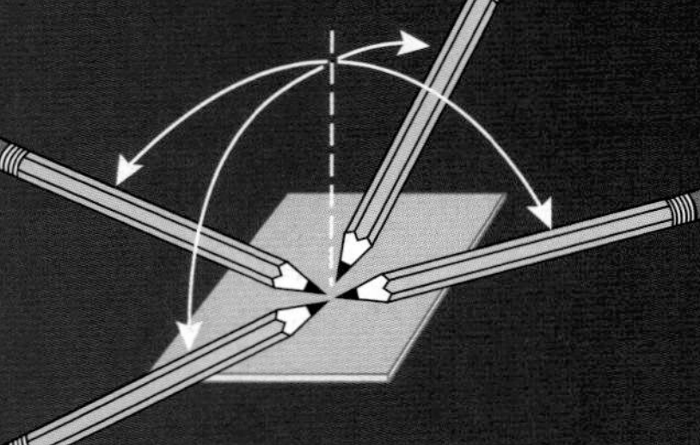

Circumduction

Circumduction is a special type of angular motion. In circumduction the tip of the pencil remains stationary while the pencil shaft moves in a circle.

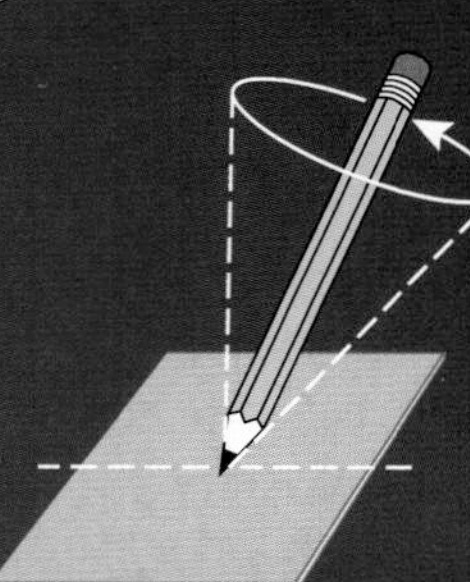

Rotating the Shaft

Rotation

In **rotation** the pencil tip remains in position and the angle of the shaft does not change, but the shaft spins around its longitudinal axis.

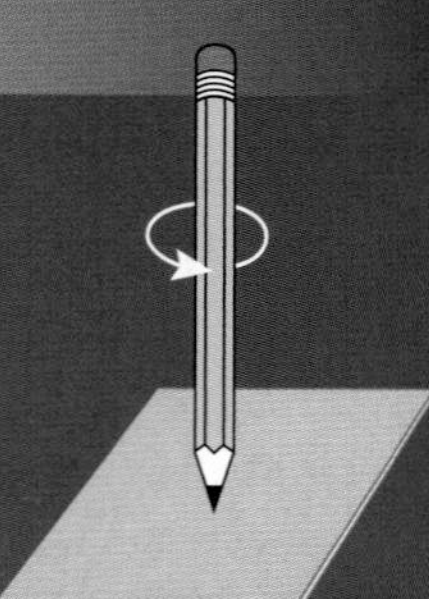

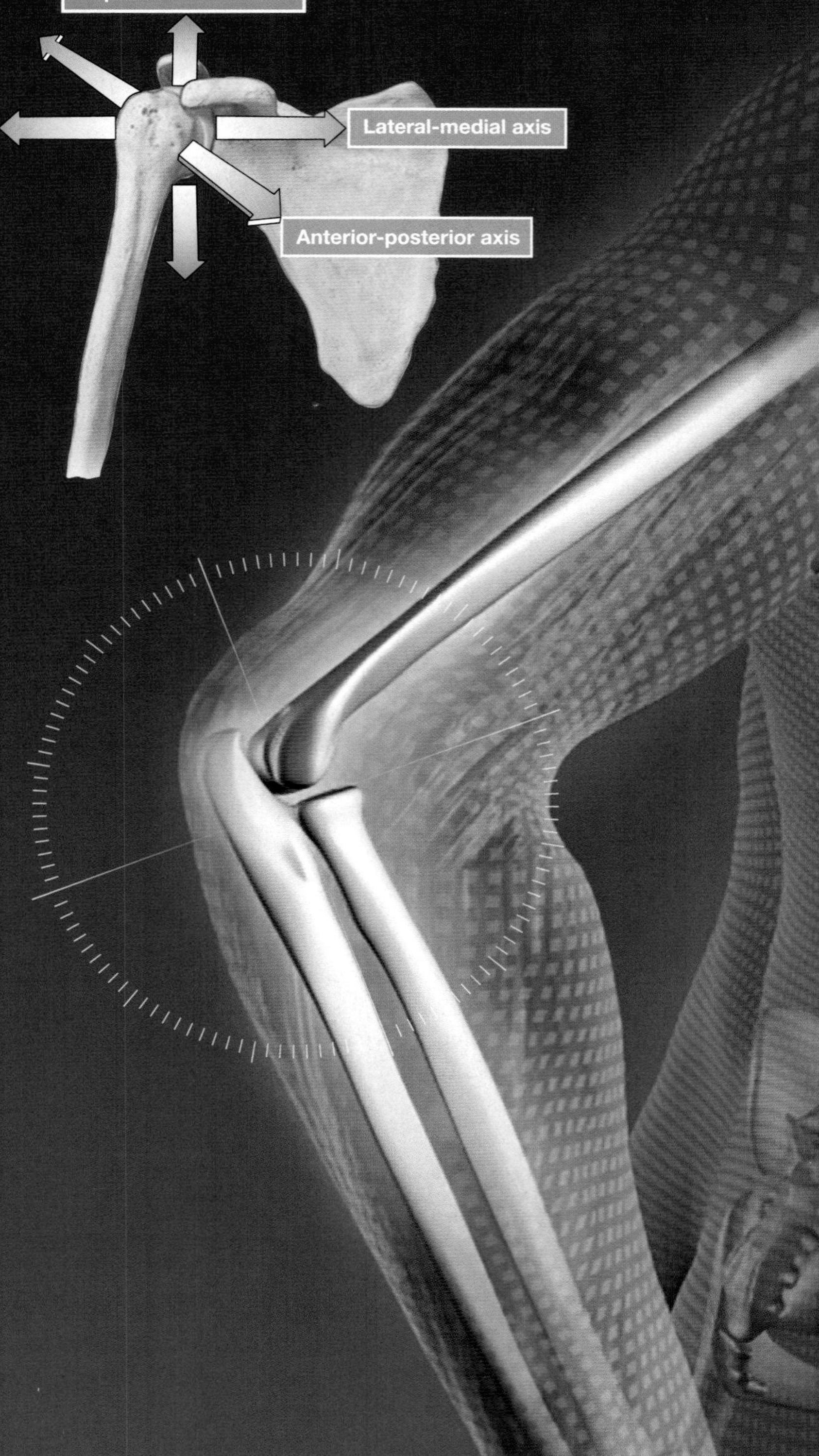

Classification of Synovial Joints

Synovial joints are freely movable diarthrotic joints, and they are are divided into six types by the range of movement permitted. Synovial joints are described as **plane**, **pivot**, **saddle, hinge**, **condylar**, or **ball-and-socket** on the basis of the shapes of the articulating surfaces, which in turn determine the joint movement.

Plane joint

Plane joints, or *gliding joints*, have flattened or slightly curved surfaces that slide across one another, but the amount of movement is very slight.

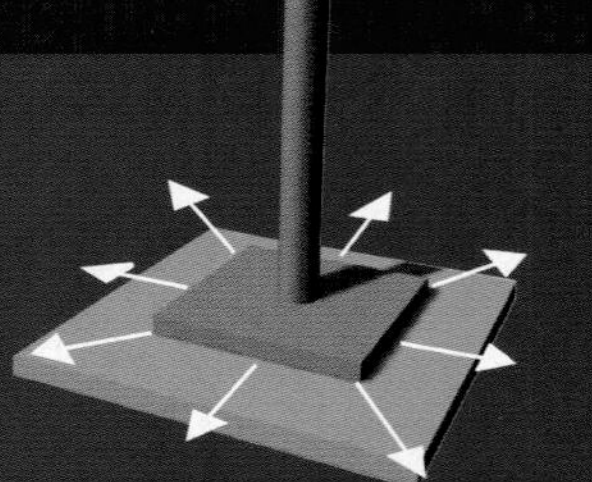

Description: Monaxial
Movement: Slight linear motion

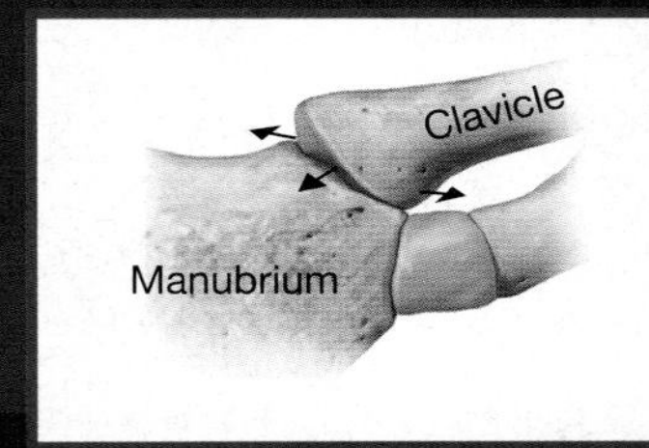

Examples:
- Sternoclavicular and acromioclavicular joints
- Intercarpal and intertarsal joints
- Vertebrocostal joints
- Sacro-iliac joints

Pivot joint

Pivot joints permit rotation only.

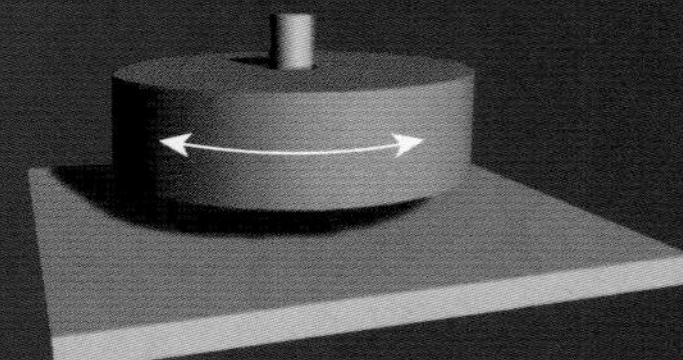

Description: Monaxial
Movement: Rotation

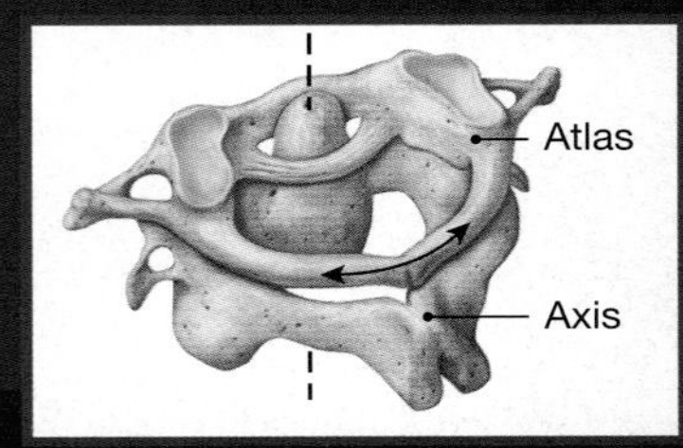

Examples:
- Atlanto-axial joint
- Proximal radio-ulnar joint

Saddle joint

Saddle joints have complex articular faces and fit together like a rider in a saddle. Each face is concave along one axis and convex along the other.

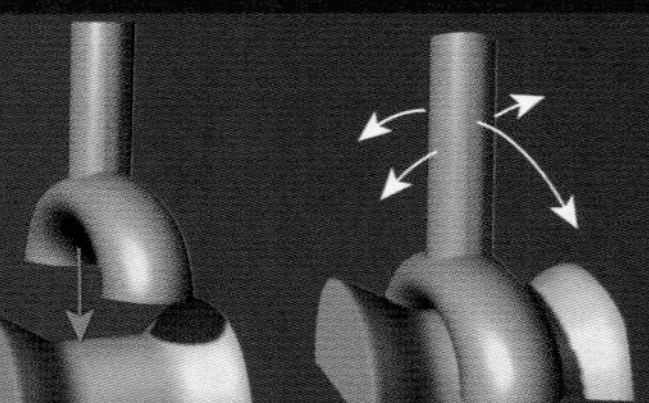

Description: Biaxial
Movement: Angular motion

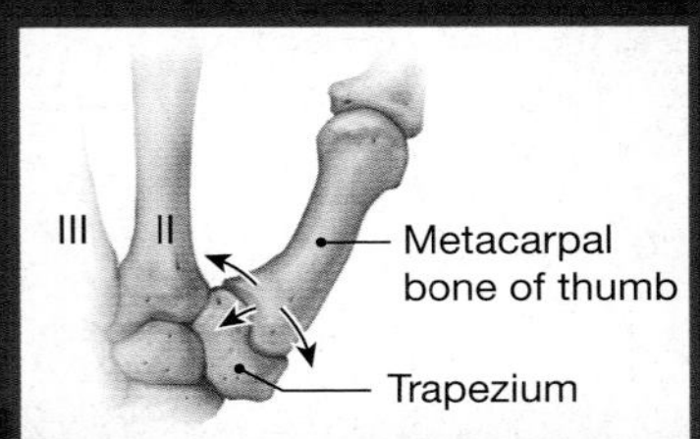

Example:
- First carpometacarpal joint

Hinge joint

Hinge joints permit angular motion in a single plane, like the opening and closing of a door.

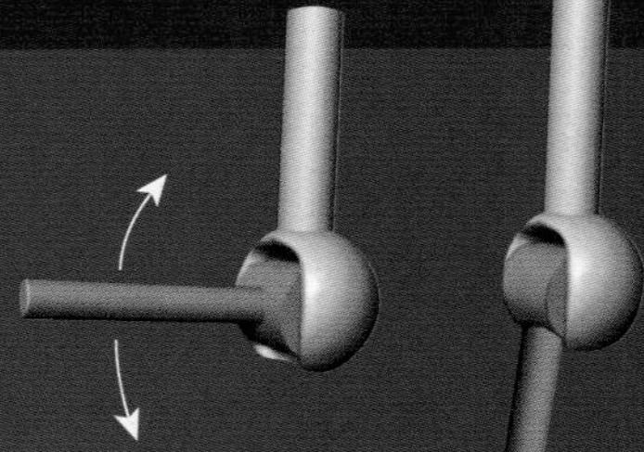

Description: Monaxial
Movement: Angular motion

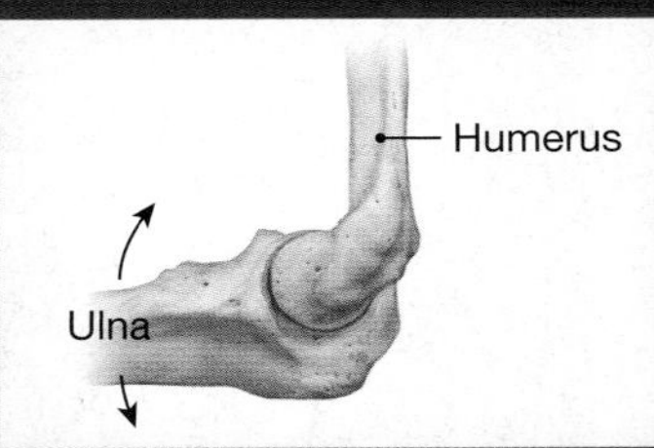

Examples:
- Elbow joint
- Knee joint
- Ankle joint
- Interphalangeal joint

Condylar joint

Condylar joints, or *ellipsoidal joints*, have an oval articular face nestled within a depression on the opposing surfaces

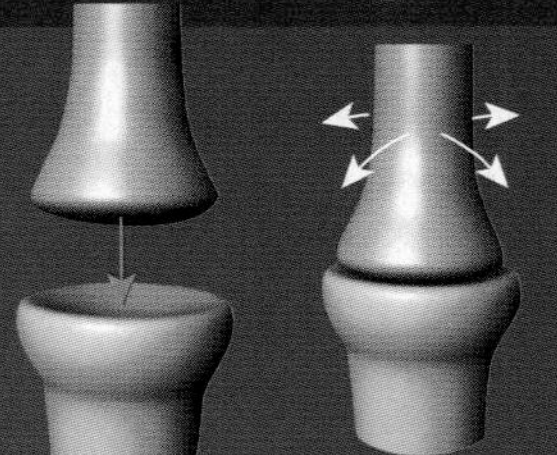

Description: Biaxial
Movement: Angular motion

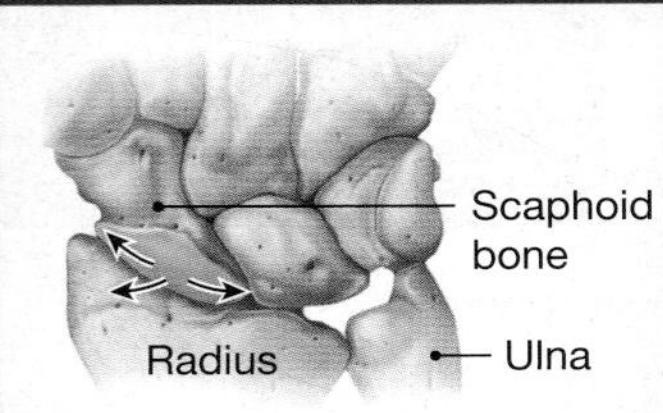

Examples:
- Metacarpophalangeal joints 2–5
- Radiocarpal joint
- Metatarsophalangeal joints

Ball-and-socket joint

In a ball-and-socket joint, the round head of one bone rests within a cup-shaped depression in another.

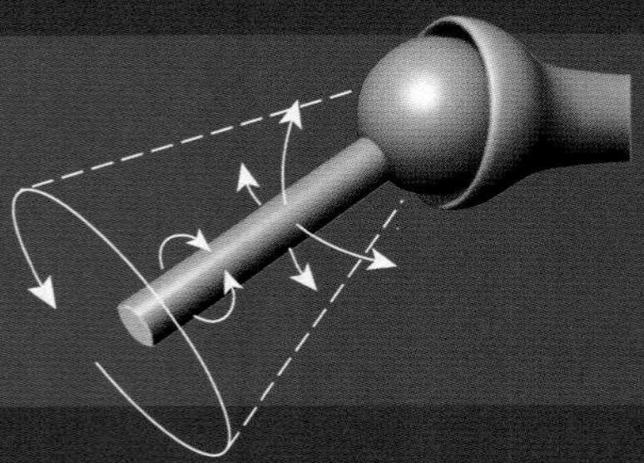

Description: Triaxial
Movement: Angular motion, circumduction, and rotation

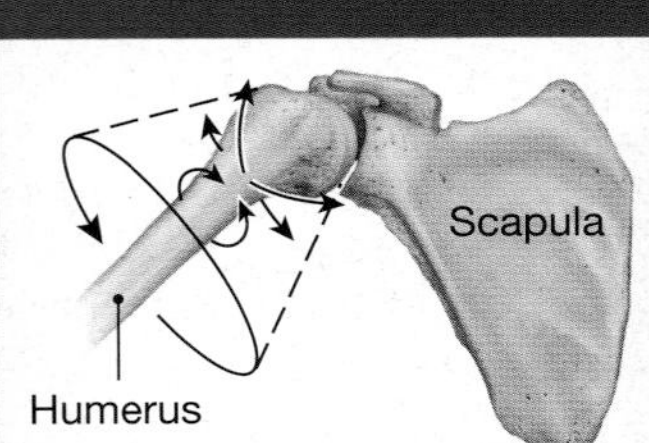

Examples:
- Shoulder joint
- Hip joint

Figure 8.3 **Angular Movements.** Examples of movements that change the angle between the shaft and the articular surface. The red dots indicate the locations of the joints involved in the illustrated movement.

8

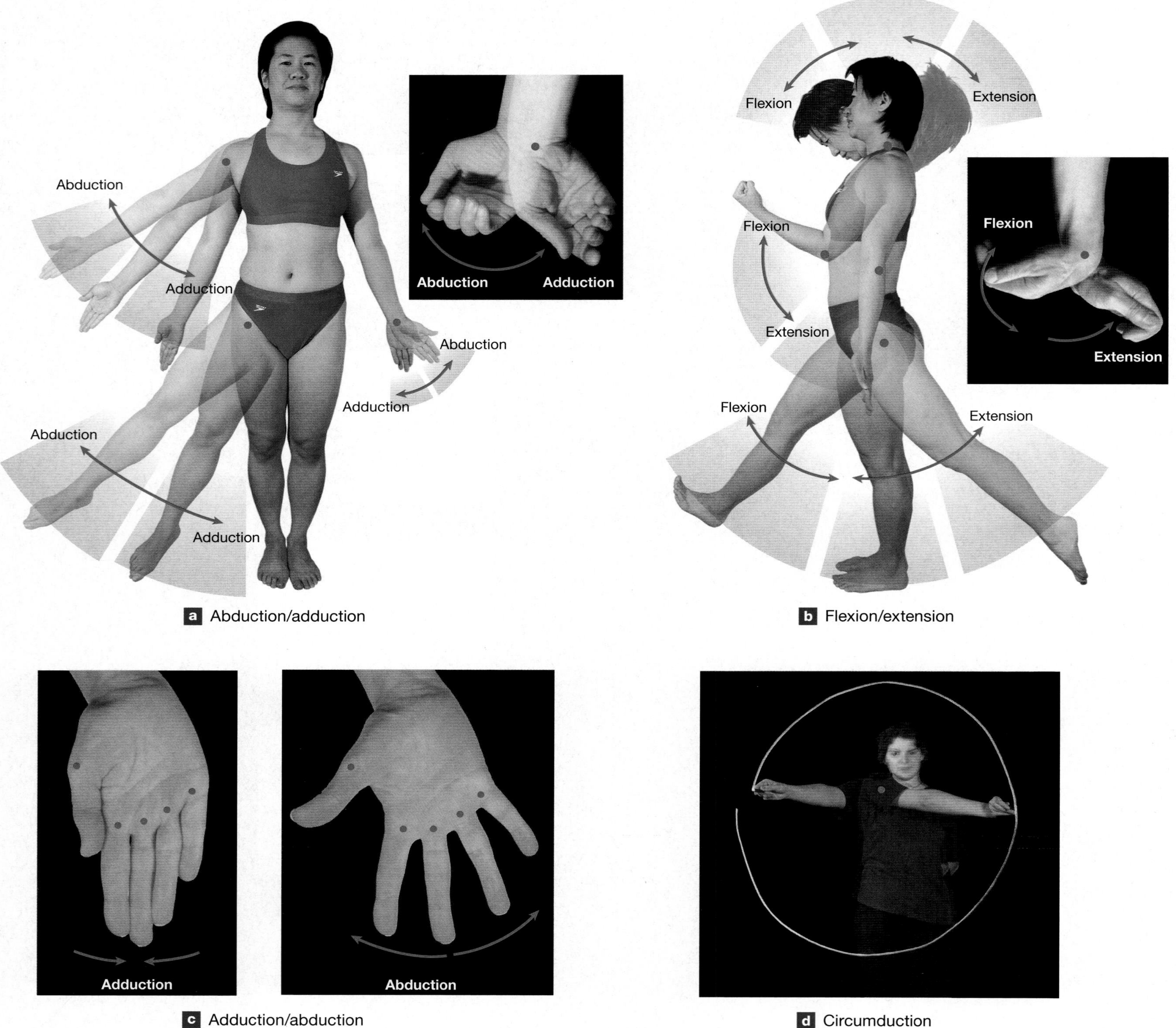

a Abduction/adduction

b Flexion/extension

c Adduction/abduction

d Circumduction

- **Elevation** and **depression** occur when a structure moves in a superior or inferior direction. You depress your mandible when you open your mouth and elevate it as you close it (**Figure 8.5f**). Another familiar elevation occurs when you shrug your shoulders.

In the remainder of this chapter we discuss several joints of the axial skeleton: (1) the temporomandibular joint (TMJ), between the mandible and the temporal bone, (2) the intervertebral joints between adjacent vertebrae, and (3) the sternoclavicular joint between the clavicle and the sternum. We then examine the synovial joints of the appendicular skeleton. The shoulder has great mobility, the elbow has great strength, and the wrist makes fine adjustments in the movement of the palm and fingers. The functional requirements of the joints in the lower limb differ from those of the upper limb. Hip, knee, and ankle joints must transfer body weight to the ground, and during running, jumping, or twisting movements, the applied forces are much greater than the weight.

Figure 8.4 Rotational Movements. Examples of motion in which the shaft of the bone rotates.

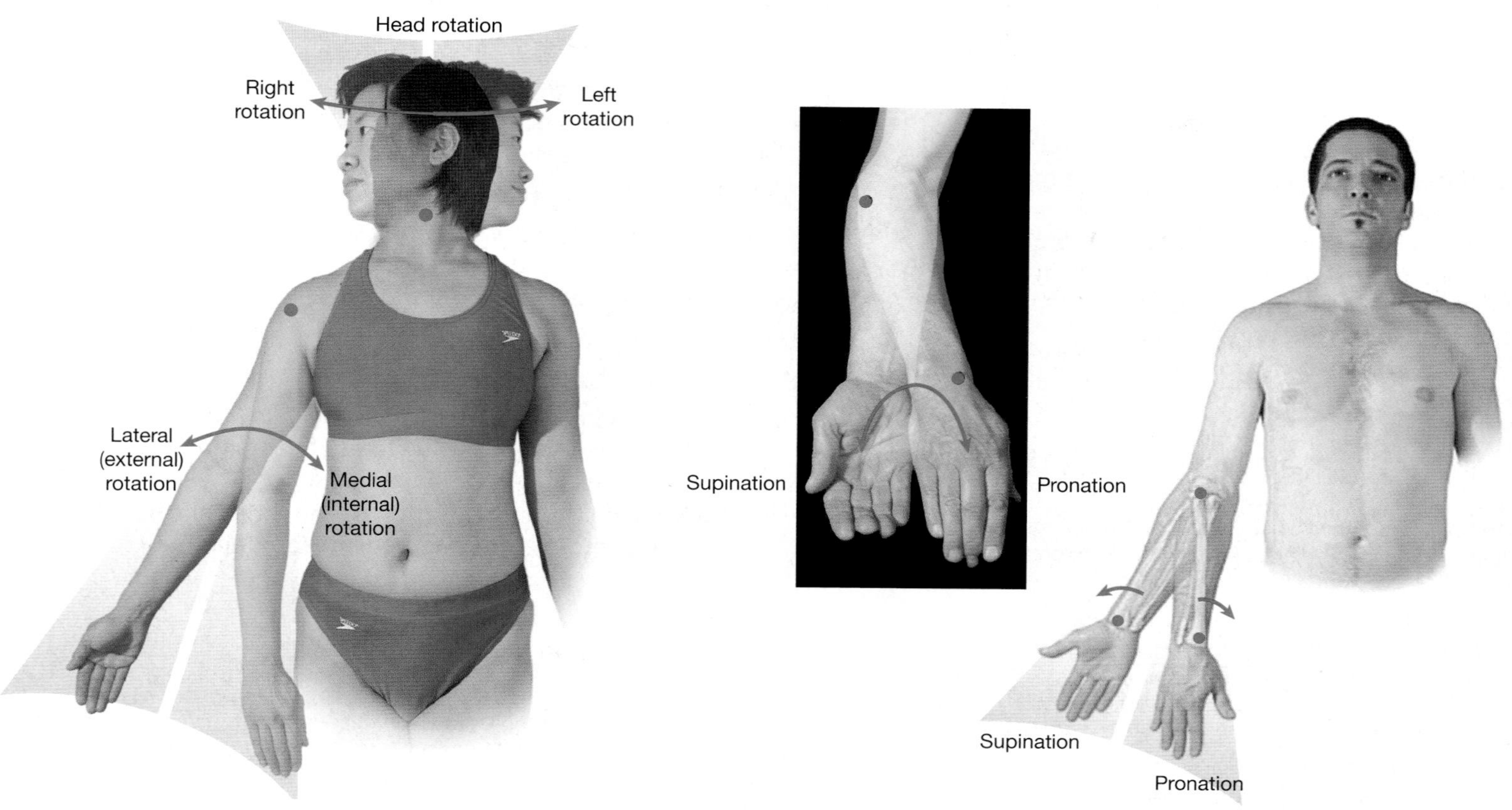

Figure 8.5 Special Movements. Examples of special terms used to describe movement at specific joints or unique directions of movement.

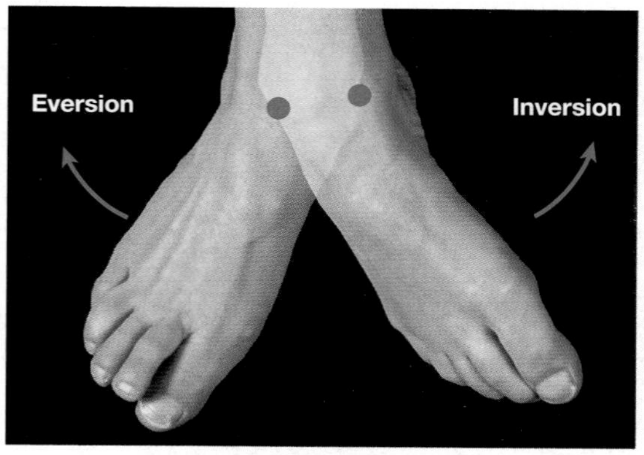

Eversion

Inversion

a Eversion/inversion

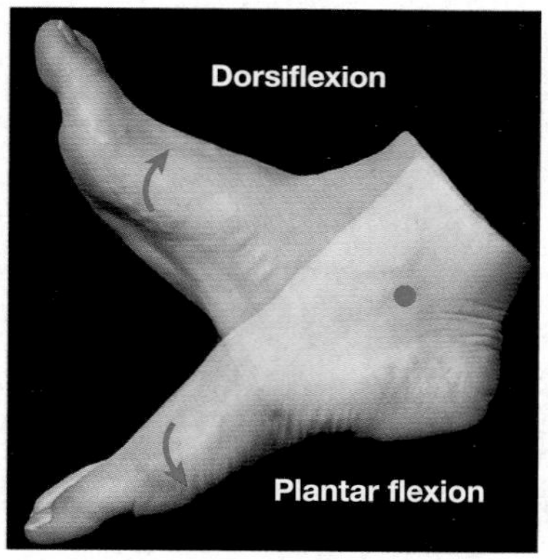

Dorsiflexion

Plantar flexion

b Dorsiflexion/plantar flexion

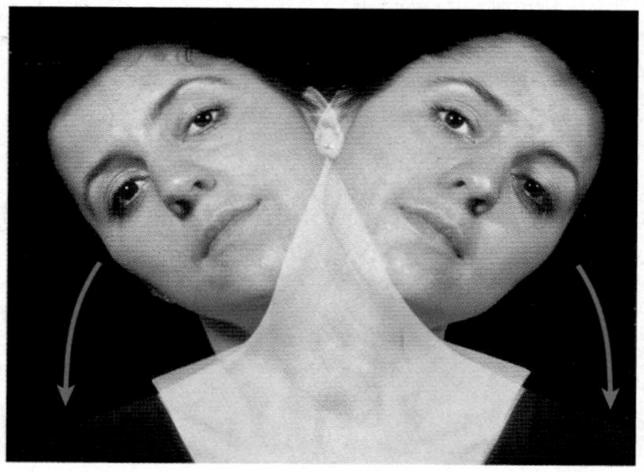

c Lateral flexion

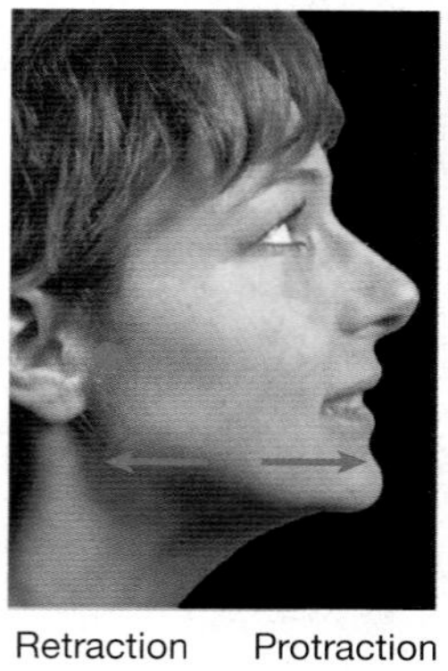

Retraction Protraction

d Retraction/protraction

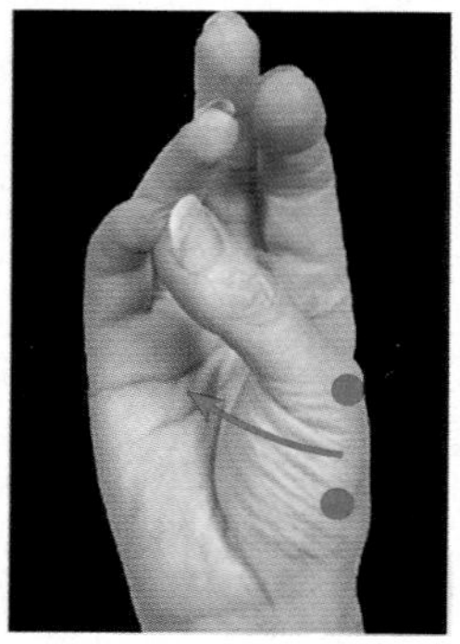

e Opposition

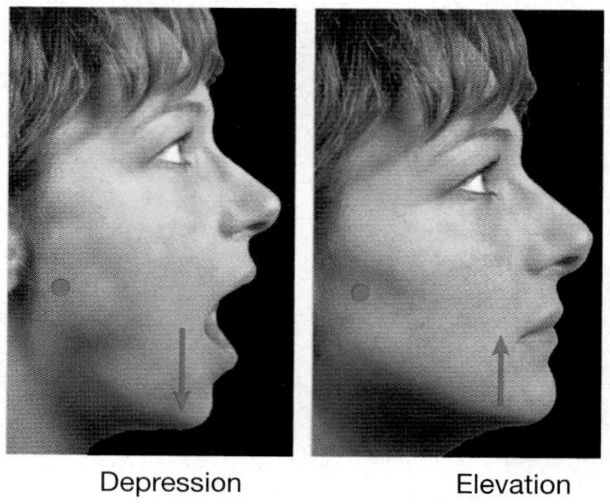

Depression Elevation

f Depression/elevation

Although the remainder of this chapter considers only some joints, Tables 8.2, 8.3, and 8.4 summarize information about most of the joints in the body.

8.2 CONCEPT CHECK

3 What classifications of synovial joints are monaxial? Biaxial? Triaxial?

4 Give the proper term for each of the following types of motion: (a) moving the humerus away from the midline of the body; (b) turning the palms so that they face forward; (c) bending the elbow.

See the blue Answers tab at the back of the book.

8.3 The Temporomandibular Joint

▶ **KEY POINT** The temporomandibular joint is a synovial joint between the mandible and temporal bone. This hinge joint allows elevation, depression, protraction, retraction, and slight right and left movements.

The **temporomandibular joint (TMJ)** (Figure 8.6) is a triaxial joint between the mandibular fossa of the temporal bone and the condylar process of the mandible. The temporomandibular joint is unique because (1) the articulating surfaces on the temporal bone and mandible are covered with fibrous cartilage rather than hyaline cartilage, and (2) a thick disc of fibrous cartilage separates the bones of the joint. This articular disc extends horizontally and divides the joint cavity into two separate chambers. Therefore, the temporomandibular joint is really two synovial joints: one between the temporal bone and the articular disc and the other between the articular disc and the mandible.

The articular capsule surrounding this joint complex is poorly defined. The portion of the capsule superior to the neck of the condyle is loose, whereas the portion of the capsule inferior to the cartilage disc is tight. Although the capsule's structure permits a wide range of motion, because the joint is rather unstable, forceful lateral or anterior movements of the mandible can partially or completely dislocate the joint.

The lateral portion of the articular capsule is called the **lateral ligament**. There are also two extracapsular ligaments:

- The **stylomandibular ligament** extends from the styloid process to the posterior margin of the angle of the mandibular ramus.
- The **sphenomandibular ligament** extends from the sphenoidal spine to the medial surface of the mandibular ramus. Its insertion covers the posterior portion of the mylohyoid line.

The temporomandibular joint is primarily a hinge joint, but the loose capsule and flat articular surfaces also permit small side-to-side gliding movements. These movements allow you to chew.

8.3 CONCEPT CHECK

5 How do the unique structure of the joint capsule and the relatively flat articular surfaces of the temporomandibular joint affect the function of this joint?

6 How does the orientation of the articular disc affect the anatomy of the temporomandibular joint?

See the blue Answers tab at the back of the book.

8.4 Intervertebral Joints

▶ **KEY POINT** There are three joints between adjacent vertebrae: one median joint and two lateral joints. The median joint involves the intervertebral disc, and the lateral joints involve the vertebral facets.

All vertebrae from C_2 to S_1 articulate with symphysis joints between the vertebral bodies and synovial joints between the articulating facets. Figure 8.7 illustrates the structure of the intervertebral joints.

Zygapophysial Joints

▶ **KEY POINT** Zygapophysial joints are diarthrotic synovial joints that exhibit gliding motion.

Figure 8.6 The Temporomandibular Joint. This hinge joint forms between the condylar process of the mandible and the mandibular fossa of the temporal bone.

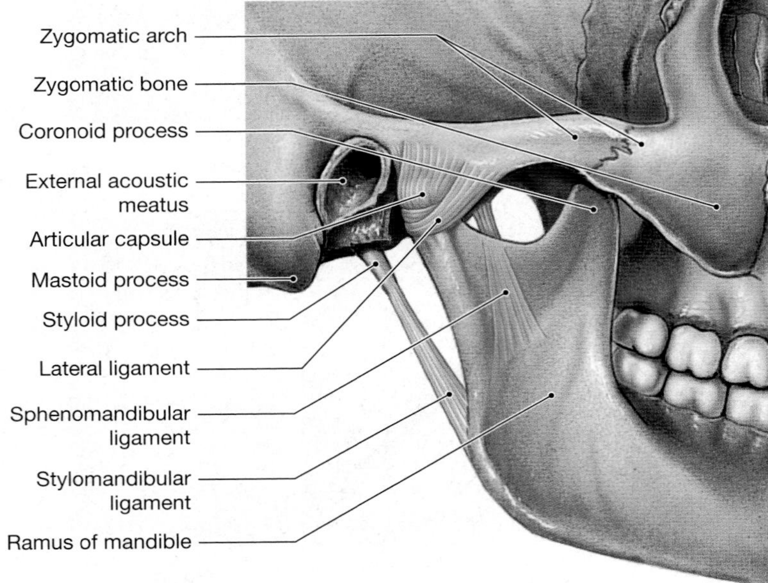

a Lateral view of the right temporomandibular joint

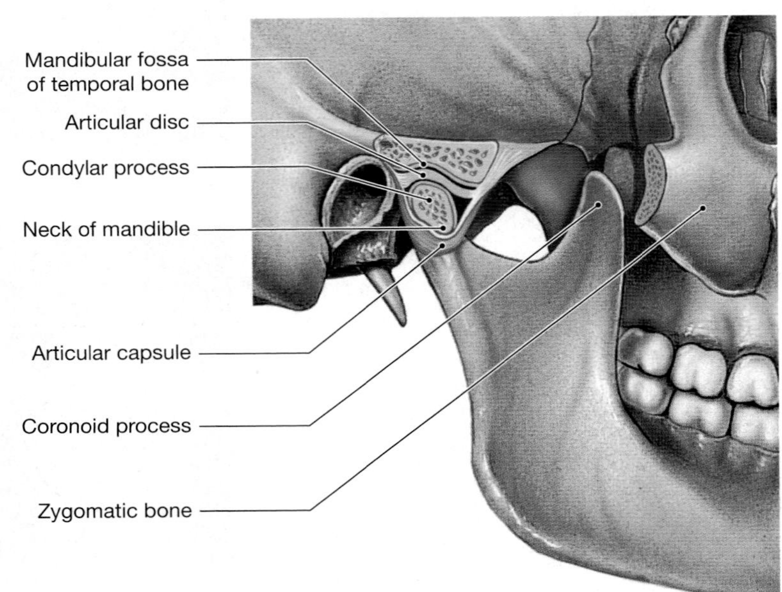

b Sectional view of the temporomandibular joint

Figure 8.7 **Intervertebral Joints.** Adjacent vertebrae articulate at their superior and inferior articular processes; their bodies are separated by intervertebral discs.

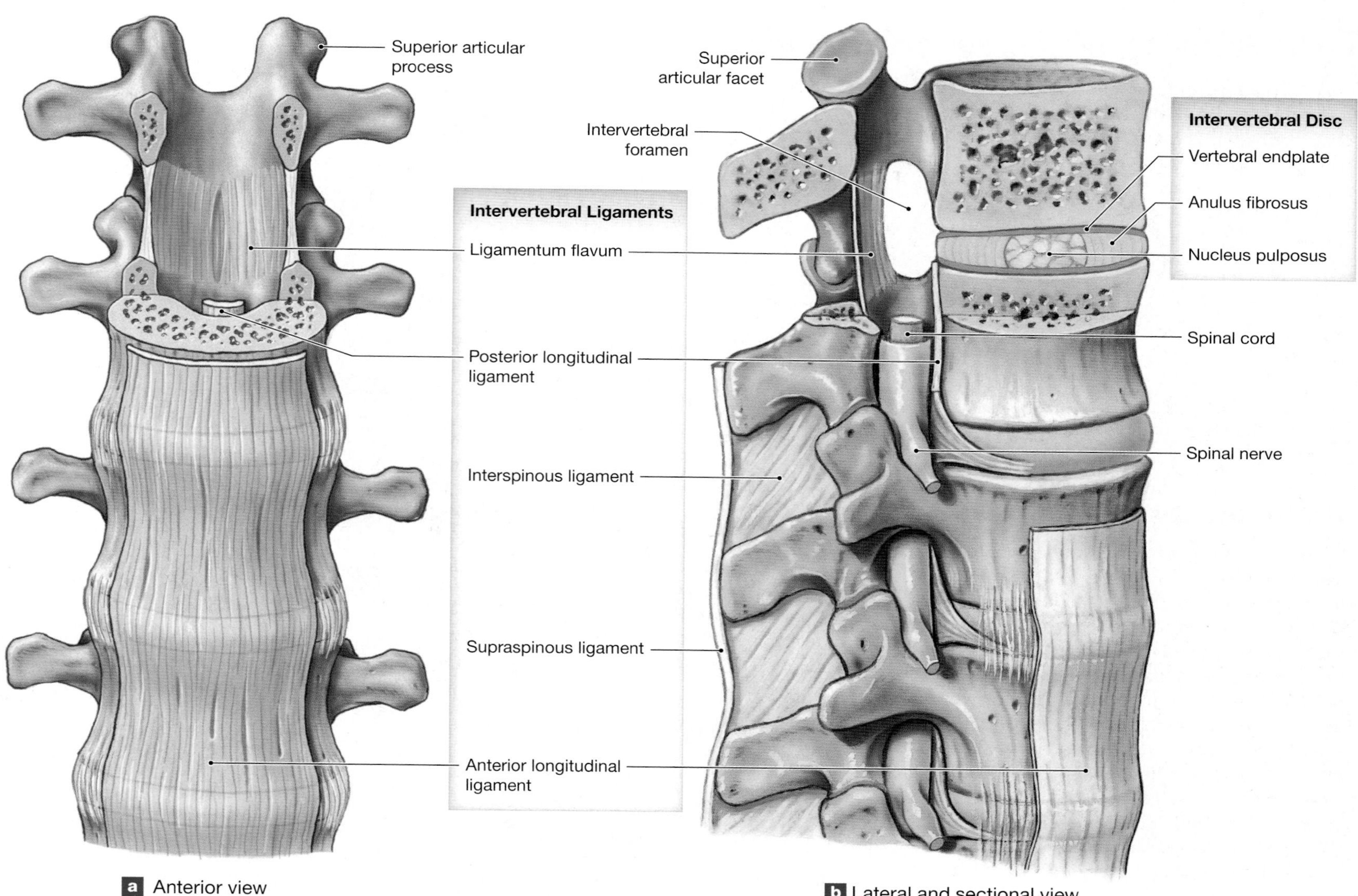

The **zygapophysial** zī-ga-pō-FIZ-ē-al **joints** (*facet joints*) are the synovial joints found between the superior and inferior articulating facets of adjacent vertebrae (**Figures 8.7** and **6.20**, **p. 160**). The articulating surfaces of these plane joints are covered with hyaline cartilage, and the size, structure, and complexity of the zygapophysial joints vary from region to region within the vertebral column. These joints permit small movements associated with flexion and extension, lateral flexion, and rotation of the vertebral column.

The Intervertebral Discs

▶ **KEY POINT** Intervertebral discs, made of fibrous cartilage, separate adjacent vertebrae and transmit forces between adjacent vertebrae. An intervertebral disc is composed of an outer fibrous part (anulus fibrosus) and an inner gelatinous mass (nucleus pulposus). A vertebral endplate covers the surface of the intervertebral disc.

The joints between adjacent vertebral bodies from C_2 to the sacrum are symphysis joints. In these joints, pads of fibrous cartilage called **intervertebral discs** separate and cushion the adjacent vertebral bodies. Intervertebral discs are not found either in the sacrum and coccyx, where vertebrae have fused, or between the first and second cervical vertebrae.

The intervertebral discs have two functions: (1) to separate individual vertebrae and (2) to transmit the load from one vertebra to another. Each intervertebral disc has a tough outer layer of fibrous cartilage called the **anulus fibrosus** (AN-ū-lus fī-BRŌ-sus) and an inner **nucleus pulposus** (pul-PŌ-sus). The nucleus pulposus is a soft, elastic, gelatinous core, composed primarily of water (about 75 percent) with scattered reticular and elastic fibers. The nucleus pulposus enables the disc to act as a shock absorber. The superior and inferior surfaces of the disc are almost completely covered by thin **vertebral endplates**. These vertebral endplates are made of hyaline and fibrous cartilage. They attach tightly to the anulus fibrosus of the intervertebral disc and weakly to the adjacent vertebrae (**Figure 8.7**). The vertebral attachments help stabilize the position of the intervertebral disc. The intervertebral ligaments, discussed in the next section, provide additional reinforcement to the disc.

Movements of the vertebral column compress the nucleus pulposus and move it in the opposite direction. This movement permits smooth gliding movements by each vertebra while still maintaining the alignment of all the vertebrae. The discs contribute significantly to an individual's height, accounting for one-quarter of the length of the vertebral column above the sacrum. As we age, the water content of the nucleus pulposus within each disc decreases, and fibrous cartilage replaces the gelatinous core. The discs gradually become less effective as a cushion, and the risk of vertebral injury increases. As the discs lose water, the vertebral column shortens, which accounts for the characteristic decrease in height with advanced age.

CLINICAL NOTE

Problems with Intervertebral Discs

With aging and use, intervertebral discs lose their water content and elasticity, stiffen, and often become brittle. Discs shrink with age, contributing to loss of height.

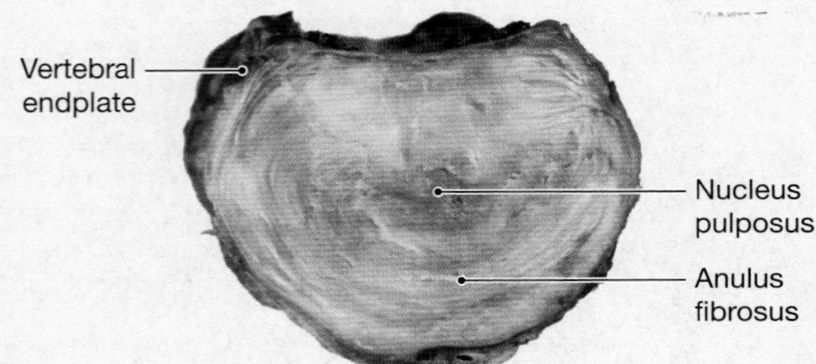

The superior surface of an isolated normal intervertebral disc

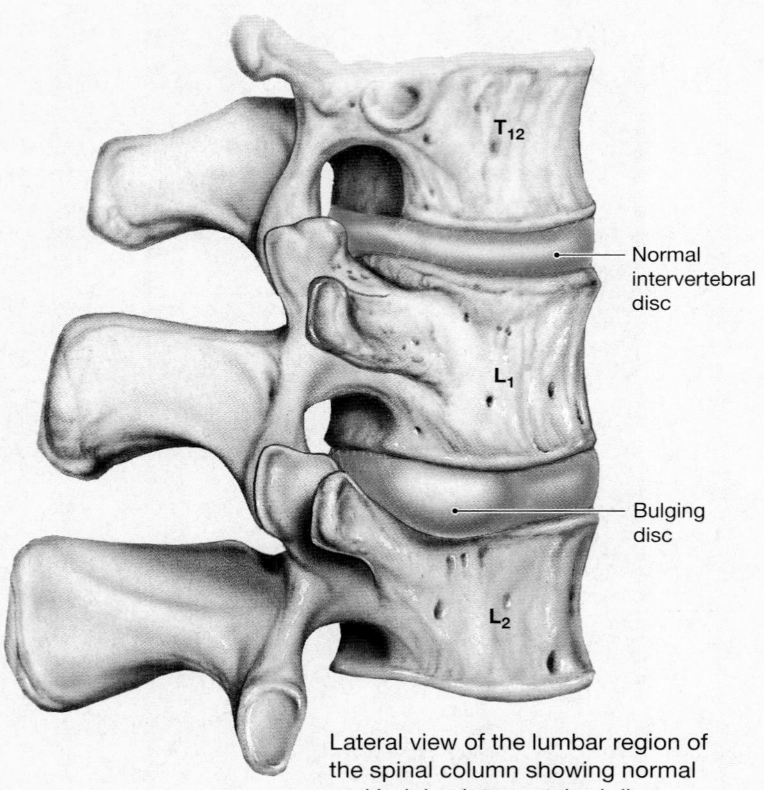

Lateral view of the lumbar region of the spinal column showing normal and bulging intervertebral discs

Herniated Disc

With age and repetitive stress, the annulus fibrosus can rupture, causing the nucleus pulposus to ooze into the vertebral canal or intervertebral foramen. This condition is called a **herniated disc**. Within the vertebral canal, the protruding nucleus pulposus can compress the spinal nerves as they pass through their tight intervertebral foramen. This compression can cause severe pain to radiate along the course of the involved nerve. Herniated discs occur most often in areas of the vertebral column subject to the greatest stress and where the greatest spinal motion takes place: between C_5 and C_6 and C_6 and C_7 in the cervical spine, and between L_4 and L_5 and L_5 and S_1 in the lumbar spine. **Sciata**, pain in the lower back and hip radiating down the posterior thigh into the leg, results when a herniated lumbar disc compresses the sciatic nerve roots.

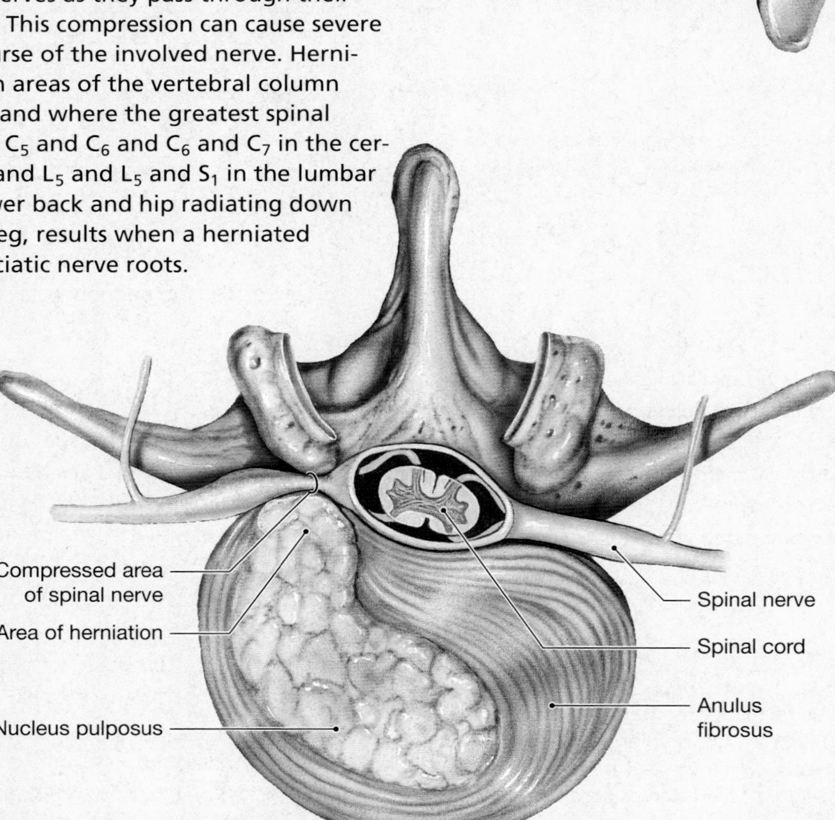

A sectional view through a herniated disc showing displacement of the nucleus pulposus and its effect on the spinal cord and adjacent nerves

Bulging Disc

Intervertebral disc disease (IVDD) is a painful condition associated with disc degeneration that affects the spinal nerves. Causes include a bulging disc or herniated disc. If the posterior longitudinal ligaments weaken, as often occurs with age, the compressed nucleus pulposus may distort the anulus fibrosus, forcing it partway into the vertebral canal. This condition, seen here in lateral view, is called a **bulging disc.**

Intervertebral Ligaments

▶ **KEY POINT** Numerous ligaments attached to the bodies and processes of all vertebrae bind them together and stabilize the vertebral column. Ligaments interconnecting adjacent vertebrae are the anterior longitudinal ligament, the posterior longitudinal ligament, the ligamentum flavum, the interspinous ligament, and the supraspinous ligament.

- The strong collagenous fibers of the **anterior longitudinal ligament** connect the intervertebral discs and anterior surfaces of each vertebral body.
- The **posterior longitudinal ligament** parallels the anterior longitudinal ligament but passes across the posterior surfaces of each intervertebral disc and vertebral body.
- The **ligamenta flava** are paired ligaments that connect the laminae of adjacent vertebrae within the vertebral arch.
- The **interspinous ligament** connects the spinous processes of adjacent vertebrae.
- The **supraspinous ligament**, which is often missing, interconnects the tips of the spinous processes from C_7 to L_3 or L_4. The **ligamentum nuchae** is a supraspinous ligament that extends from C_7 to the base of the skull.

The posterior longitudinal ligament, ligamenta flava, and ligamentum nuchae between adjacent vertebrae limit flexion of the vertebral column. The anterior longitudinal ligament limits extension, and the interspinous ligaments limit rotation and lateral flexion **(Figure 8.7)**.

Vertebral Movements

▶ **KEY POINT** The vertebral column is capable of flexion, extension, rotation, and lateral flexion. The size of the intervertebral discs determines the *amount* of movement at any segment of the vertebral column; the facets of the zygapophysial joints determine the *direction* of movement.

CLINICAL NOTE

Ankylosing Spondylitis

Ankylosing spondylitis is an inflammatory arthritis that affects the joints of the axial skeleton. Contrary to most inflammatory arthropathies, men are affected more than women. This disease can cause complete fusion (synostosis) of the sacroiliac joints and all the zygapophysial joints as well as the pubic symphysis, or intervertebral joints between vertebral bodies.

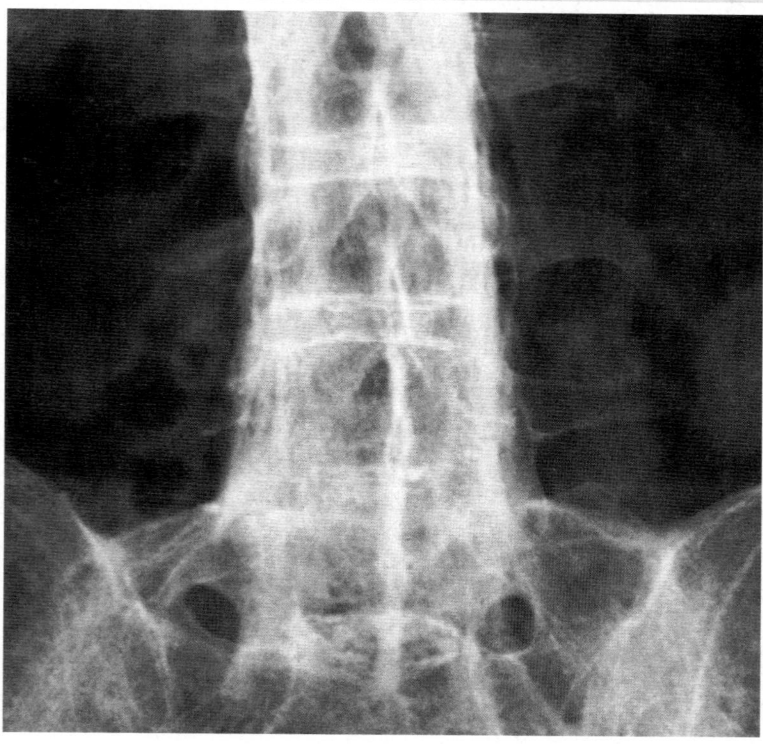

At right, spine from the sixth century showing ankylosing spondylitis, on display at the Landesmuseum Württemberg, Stuttgart, Germany

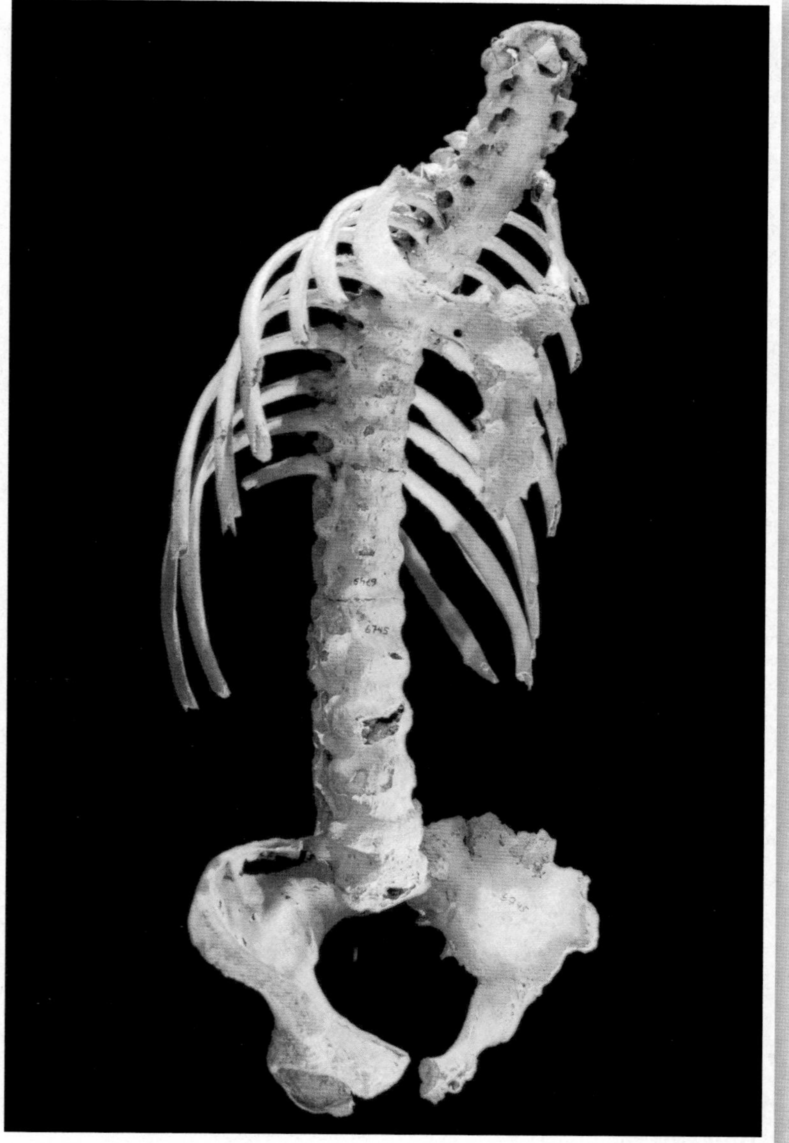

8

The following movements of the vertebral column are possible: (1) **flexion**, bending forward; (2) **extension**, bending backward; (3) **lateral flexion**, bending to the side; and (4) **rotation**, or twisting.

Table 8.2 summarizes information related to joints and movements of the axial skeleton.

8.4 CONCEPT CHECK

7 Name the ligaments that stabilize the vertebral column.

8 Where are the symphysis joints in the vertebral column, and how do they differ from the structure of the zygapophysial joints?

See the blue Answers tab at the back of the book.

8

8.5 The Shoulder Complex

▶ **KEY POINT** The shoulder complex has only one attachment to the axial skeleton (the clavicle to the sternum), and the entire complex must move as one to allow for maximum range of motion at the shoulder. Proper movement of the shoulder complex positions the hand for a wide variety of functions.

The **shoulder complex** includes the clavicle, scapula, and humerus and their associated joints (sternoclavicular and shoulder) and supporting structures. The shoulder complex links the upper limb to the thorax. In addition, proper movement of the sternoclavicular joint is essential for maximum range of motion of the shoulder (*glenohumeral*) joint.

The Sternoclavicular Joint

▶ **KEY POINT** The sternoclavicular joint is the *master joint* of the shoulder complex because it determines the position of the scapula, which is critical for allowing maximum range of motion of the shoulder joint. Movement of the arm at the shoulder depends on the action of other joints of the shoulder complex, particularly the sternoclavicular joint.

The **sternoclavicular joint** is a synovial joint between the medial end of the clavicle and the manubrium of the sternum. This joint, which is the only joint between the upper limb and the axial skeleton, anchors the scapula to the axial skeleton. Movement at the sternoclavicular joint changes the position of the scapula on the thoracic wall, helping the shoulder joint achieve maximum range of motion.

As at the temporomandibular joint, an articular disc divides the sternoclavicular joint into two synovial cavities (**Figure 8.8**). The articular capsule provides stability and limits movement. Two accessory ligaments, the **anterior sternoclavicular ligament** and the **posterior sternoclavicular ligament**, reinforce the joint capsule. There are also two extracapsular ligaments:

- The **interclavicular ligament** connects the clavicles and reinforces the superior portions of the adjacent articular capsules. This ligament attaches firmly to the superior border of the manubrium and prevents dislocation when the shoulder is depressed.
- The broad **costoclavicular ligament** extends from the costal tuberosity of the clavicle to the superior and medial borders of the first rib and the first costal cartilage. This ligament prevents dislocation when the shoulder is elevated.

Table 8.2 | Joints of the Axial Skeleton

Element	Joint	Type of Articulation	Movements
SKULL			
Cranial and facial bones of skull	Various	Synarthroses (suture or synostosis)	None
Maxillae/teeth	Alveolar	Synarthrosis (gomphosis)	None
Mandible/teeth	Alveolar	Synarthrosis (gomphosis)	None
Temporal bone/mandible	Temporomandibular	Combined plane and hinge diarthrosis	Elevation/depression, lateral gliding, limited protraction/retraction
VERTEBRAL COLUMN			
Occipital bone/atlas	Atlanto-occipital	Condylar diarthrosis	Flexion/extension
Atlas/axis	Atlanto-axial	Pivot diarthrosis	Rotation
Other vertebral elements	Intervertebral (between vertebral bodies)	Amphiarthrosis (symphysis)	Slight movement
	Intervertebral (between articular processes)	Plane diarthrosis	Slight rotation and flexion/extension
Thoracic vertebrae/ribs	Vertebrocostal	Plane diarthrosis	Elevation/depression
Rib/costal cartilage	Costochondral	Synchondrosis	None
Costal cartilage/sternum	Sternocostal	Synchondrosis (rib 1)	None
	Sternocostal	Plane diarthrosis (ribs 2–7)	Slight gliding movement
L_5/sacrum	Between body of L_5 and sacral body	Amphiarthrosis (symphysis)	Slight movement
	Between inferior articular processes of L_5 and articular processes of sacrum	Plane diarthrosis	Slight flexion/extension
Sacrum/hip	Sacro-iliac	Plane diarthrosis	Slight gliding movement
Sacrum/coccyx	Sacrococcygeal	Plane diarthrosis (may become fused)	Slight movement
Coccygeal bones		Synarthrosis (synostosis)	None

Figure 8.8 The Sternoclavicular Joint. An anterior view of the thorax showing the bones and ligaments of the sternoclavicular joint. This joint is classified as a stable, heavily reinforced plane diarthrosis.

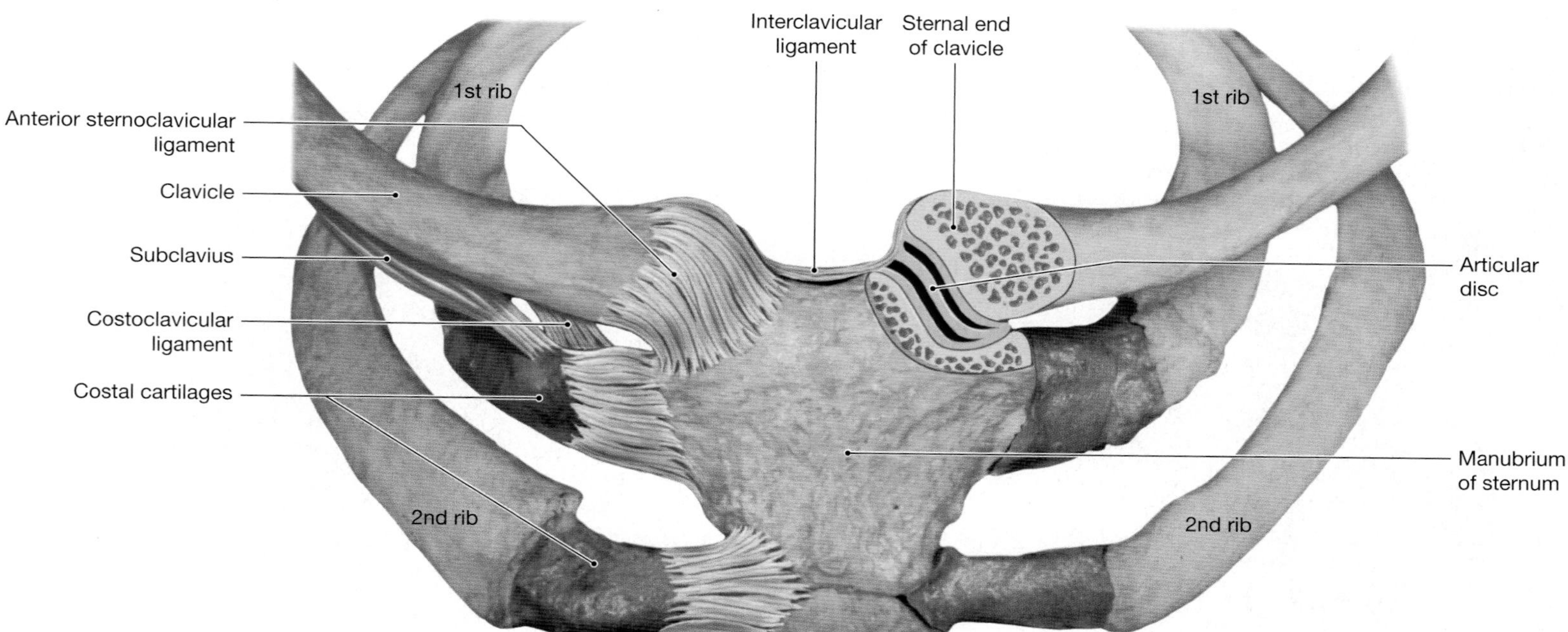

The sternoclavicular joint is a plane joint, but the capsular fibers allow slight rotation and circumduction of the clavicle.

The Shoulder Joint

▶ **KEY POINT** The shoulder (or glenohumeral) joint is a triaxial, ball-and-socket synovial joint. It is located between the glenoid cavity of the scapula and the head of the humerus. The shoulder joint has the greatest range of motion of any joint and is therefore one of the most easily damaged joints of the body.

The **shoulder joint**, or **glenohumeral joint**, is a loose and shallow joint that permits the greatest range of motion of any joint in the body. The shape of the joint, and the accompanying wide range of motion, allows us to position the hand to do a wide variety of functions. The fact that the shoulder joint is also the most frequently dislocated joint clearly demonstrates that strength and stability must be sacrificed to obtain mobility.

This joint is a ball-and-socket type between the head of the humerus and the glenoid cavity of the scapula **(Figure 8.9)**. *(Refer to Chapter 12,* **Figure 12.10**, *to visualize this structure in a cross section of the body at the level of* T_2*.)* The **glenoid labrum** (*labrum*, lip or edge), a ring of dense, irregular fibrous cartilage, covers the edge of the glenoid cavity **(Figure 8.9c,d)**. The glenoid labrum attaches to the margin of the glenoid cavity and both enlarges and deepens the cavity. It is also an attachment site for the glenohumeral ligaments and the long head of the biceps brachii muscle, a flexor of the shoulder and elbow.

The articular capsule extends from the scapular neck to the humerus. It is an oversized capsule that is weakest at its inferior surface. When the upper limb is in the anatomical (neutral) position, the capsule is tight superiorly and loose inferiorly and anteriorly. The capsule's construction contributes to the wide range of motion of the shoulder joint. The bones of the pectoral girdle stabilize the superior surface of the joint somewhat, but ligaments and surrounding skeletal muscles and their associated tendons provide most of the stability at this joint.

Ligaments

The major ligaments that help stabilize the shoulder joint are shown in **Figure 8.9a–c**.

- The capsule surrounding the shoulder joint is thin. Areas of localized thickening of the anterior capsule surface are known as the **glenohumeral ligaments**. These ligaments help stabilize the shoulder joint only when the humerus approaches or exceeds maximum normal motion.
- The **coracohumeral ligament** originates at the base of the coracoid process and inserts on the head of the humerus. This ligament strengthens the superior part of the articular capsule and supports the weight of the upper limb.
- The **coraco-acromial ligament** spans the gap between the coracoid process and the acromion, just superior to the capsule. This ligament provides additional support to the superior surface of the capsule.
- The strong **acromioclavicular ligament** attaches the acromion to the clavicle and restricts movement of the clavicle at the acromial end. A shoulder separation is a relatively common injury involving partial or complete dislocation of the acromioclavicular joint.
- The **coracoclavicular ligaments** attach the clavicle to the coracoid process and limit motion between the clavicle and scapula.
- The **transverse humeral ligament** extends between the greater and lesser tubercles and holds the tendon of the long head of the biceps brachii in the intertubercular groove of the humerus.

Skeletal Muscles and Tendons

Muscles that move the humerus stabilize the shoulder joint more than all the ligaments and capsular fibers combined. Muscles originating on the axial skeleton, scapula, and clavicle cover the anterior, superior, and posterior surfaces of the capsule. Tendons passing across the joint reinforce the anterior and superior portions of the capsule. Tendons of specific appendicular muscles support the shoulder and limit its movement. These muscles, collectively called the **rotator cuff** (discussed in Chapter 11), are a frequent site of sports injury.

Bursae

As they do at other joints, bursae at the shoulder reduce friction where large muscles and tendons pass across the joint capsule. The **subacromial bursa** and the

Figure 8.9 The Glenohumeral Joint. This ball-and-socket joint connects the humerus and the scapula.

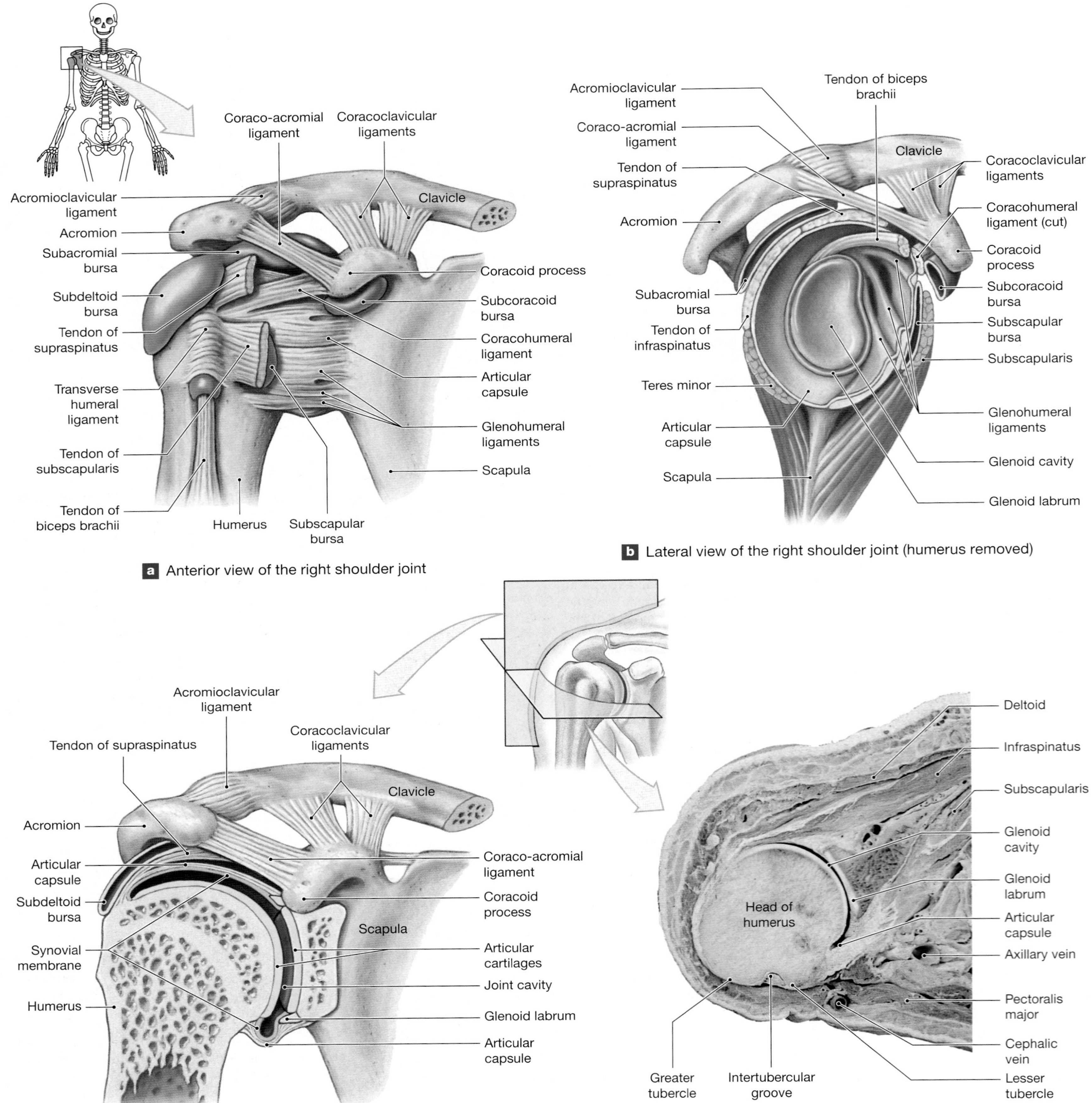

CLINICAL NOTE

Shoulder Injuries

In a shoulder dislocation, the humerus dislocates from the glenoid cavity. The glenoid labrum, shoulder capsule, glenohumeral ligaments, and coracohumeral ligament may be damaged, stretched, or torn.

In a shoulder **separation**, the clavicle separates from the acromion. The acromioclavicular ligament and capsule and the coracoclavicular ligaments stretch or tear.

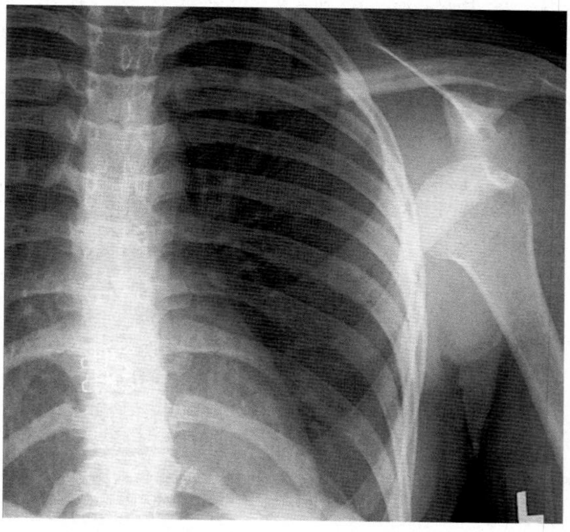

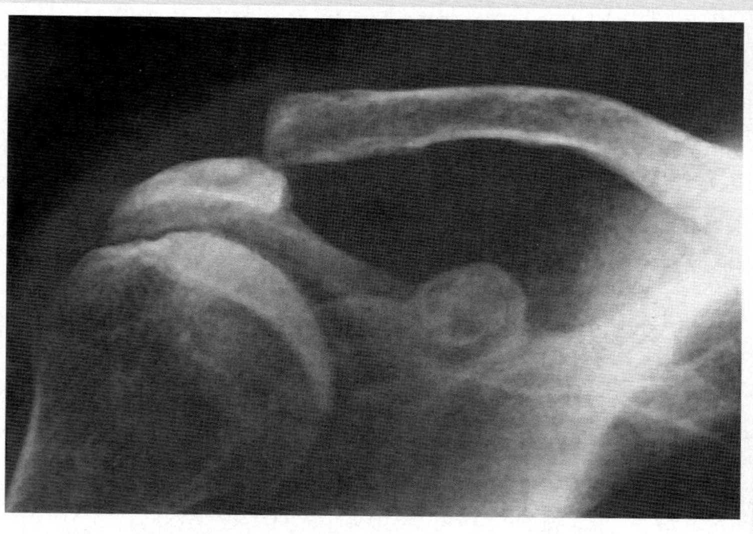

subcoracoid bursa prevent contact between the acromion and coracoid process and the capsule **(Figure 8.9a,b)**. The **subdeltoid bursa** and the **subscapular bursa** lie between large muscles and the capsular wall **(Figure 8.9a–c)**. Inflammation of one or more of these bursae restricts motion and produces the painful signs and symptoms of **bursitis**.

8.5 CONCEPT CHECK

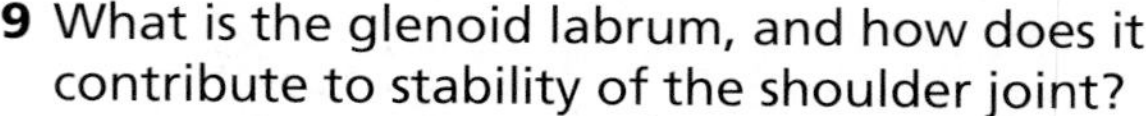

9 What is the glenoid labrum, and how does it contribute to stability of the shoulder joint?

10 Who would be more likely to develop inflammation of the subscapular bursa—a tennis player or a jogger? Why?

See the blue Answers tab at the back of the book.

8.6 | The Elbow and Radio-ulnar Joints

▶**KEY POINT** The movements of the elbow and proximal and distal radio-ulnar joints serve to position the hand. The elbow is a uniaxial, synovial hinge joint that allows flexion and extension. The proximal and distal radio-ulnar joints are uniaxial syndesmoses that work together and allow rotation of the forearm.

The Elbow Joint

▶**KEY POINT** The elbow joint is unusual because there are three joints (humero-ulnar, humeroradial, and the proximal radio-ulnar) within the joint capsule. The term "elbow joint" typically refers to the humero-ulnar and humeroradial joints.

The **elbow joint** is composed of the joints between (1) the humerus and the ulna, and (2) the humerus and the radius. The joints between the humerus and the radius and ulna enable flexion and extension of the elbow. These movements, combined with the radio-ulnar joints discussed below, allow for positioning of the hand to perform numerous activities, such as feeding, grooming, or defense, simply by changing the position of the hand with respect to the trunk.

The largest and strongest joint at the elbow is the **humero-ulnar joint**, where the trochlea of the humerus articulates with the trochlear notch of the ulna. At the smaller **humeroradial joint**, which is lateral to the humero-ulnar joint, the capitulum of the humerus articulates with the head of the radius **(Figure 8.10)**.

The elbow joint is a very stable joint because (1) the bony surfaces of the humerus and ulna interlock to prevent lateral movement and rotation, (2) the articular capsule is very thick, and (3) strong ligaments reinforce the capsule. The **ulnar collateral ligament** stabilizes the medial surface of the joint. This ligament extends from the medial epicondyle of the humerus anteriorly to the coronoid processes of the ulna and posteriorly to the olecranon **(Figure 8.10a,b)**. The **radial collateral ligament** stabilizes the lateral surface of the joint. It extends between the lateral epicondyle of the humerus and the **annular ligament** that binds the proximal radial head to the ulna **(Figure 8.10e)**.

Despite the strength of the capsule and ligaments, severe impact or unusual stress can damage the elbow joint. For example, the repetitive, high-velocity motions involved in throwing can injure the ulnar collateral ligament. (The common term for the surgical procedure to repair the ulnar collateral ligament is *Tommy John surgery*—named for a pitcher for the Los Angeles Dodgers baseball team who was one of the first professional athletes to undergo the procedure, in 1974.) Less violent stresses can dislocate or otherwise injure the elbow, especially if epiphyseal growth is not complete. For example, parents in a hurry may drag a child along behind them, exerting an upward, twisting pull on the elbow joint that results in a partial dislocation known as "nursemaid's elbow."

Figure 8.10 The Elbow Joint. The elbow joint is a complex hinge joint formed between the humerus and the ulna and radius. All views are of the right elbow joint.

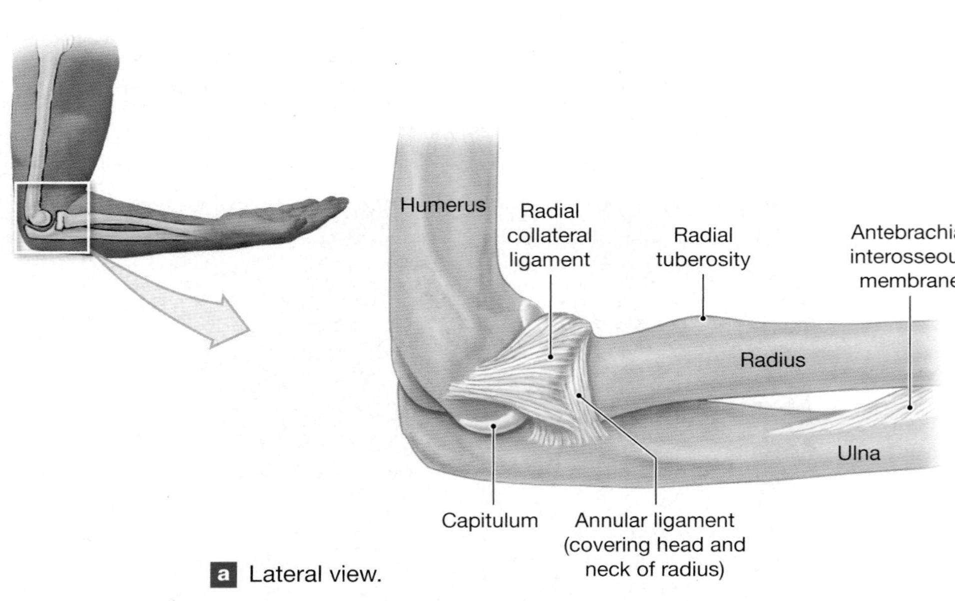

a Lateral view.

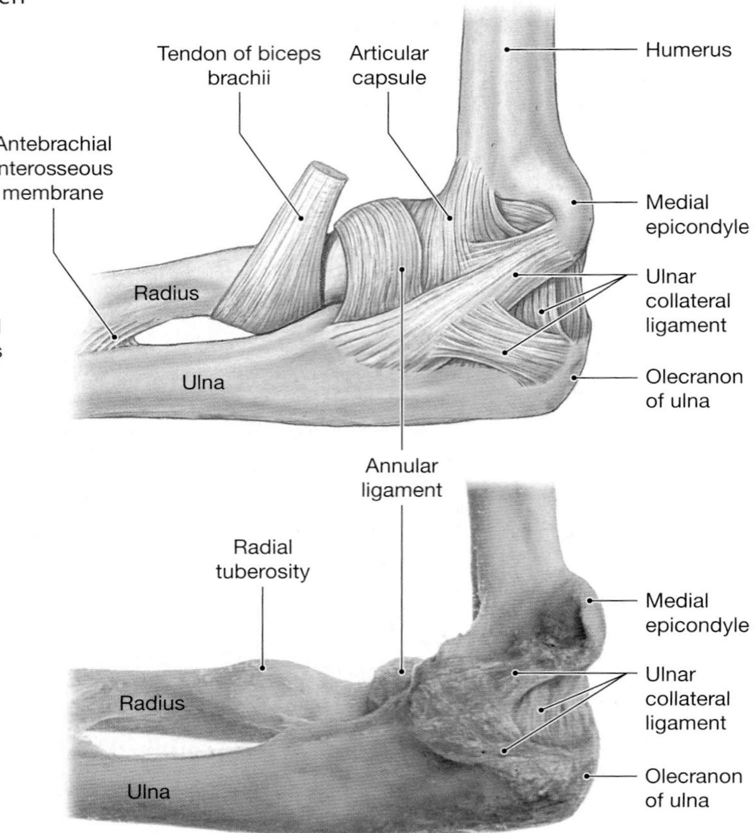

b Medial view. The radius is shown pronated; note the position of the biceps brachii tendon, which inserts on the radial tuberosity.

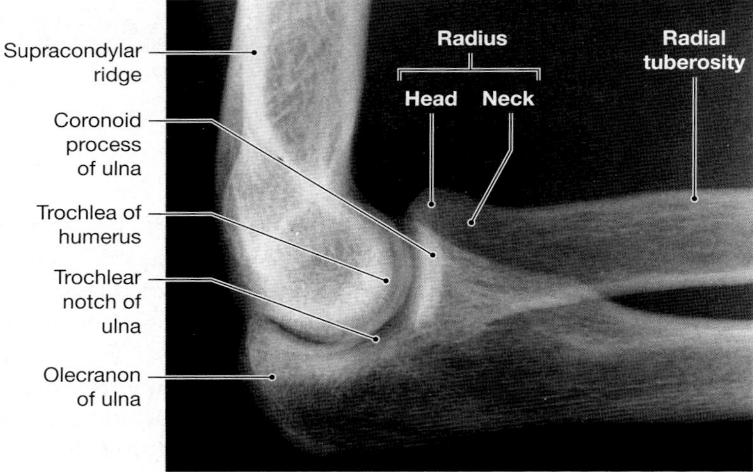

c X-ray.

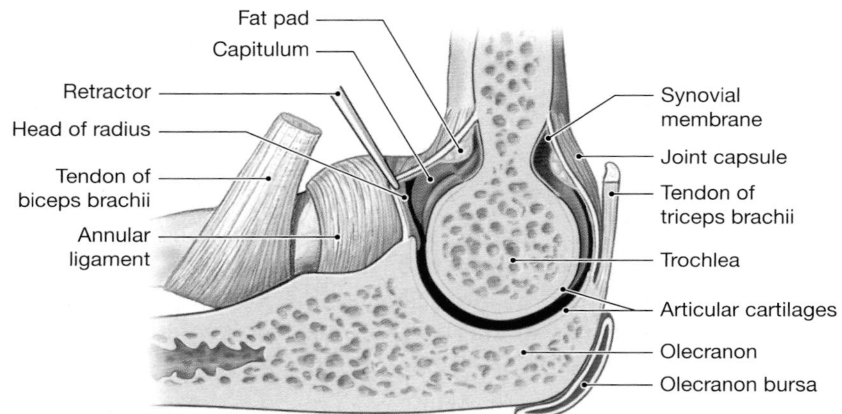

d Sagittal view of the elbow. The radius is pronated.

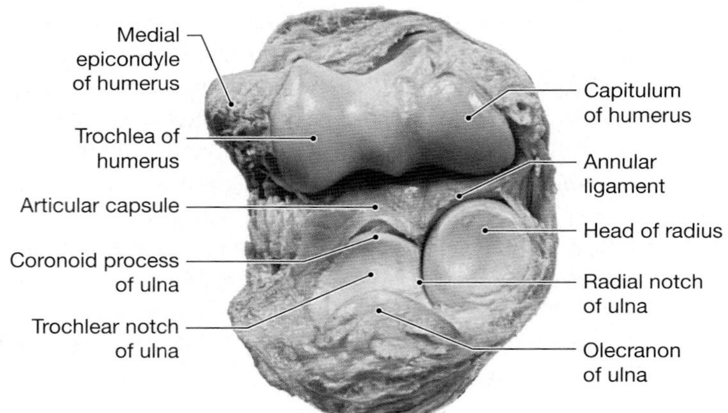

e A posterior view; the posterior portion of the capsule has been cut and the joint cavity opened to show the opposing surfaces.

The Radio-ulnar Joints

▶ **KEY POINT** The antebrachial interosseous membrane anatomically links the proximal and distal radio-ulnar joints; motion at one joint is always accompanied by motion at the other joint. The anatomy of the radio-ulnar joints allows pronation (medial rotation) and supination (lateral rotation) of the forearm.

At the **proximal radio-ulnar joint**, the head of the radius articulates with the radial notch of the ulna. The **anular ligament** and the **quadrate ligament** hold the head of the radius in place (**Figure 8.11a**). The articulating surfaces of the **distal radio-ulnar joint** are the ulnar notch of the radius, the radial notch of the ulna, and the articular disc. A series of radio-ulnar ligaments and the antebrachial interosseous membrane hold these articulating surfaces together (**Figure 8.11b**).

Muscles that insert on the radius control pronation and supination at the radio-ulnar joints. The largest is the biceps brachii, which covers the anterior surface of the arm. Its tendon is attached to the radius at the radial tuberosity, and muscle contraction here both flexes the elbow and supinates the forearm. (The muscles that move the elbow and radio-ulnar joints are explained in detail in Chapter 11.)

8.6 CONCEPT CHECK

11 What bones make up the elbow joint, and how do they articulate?

12 Your roommate complains of elbow pain while he is pronating and supinating his forearm. Is the pain really originating in the elbow joint? How would you explain this to your roommate?

See the blue Answers tab at the back of the book.

Figure 8.11 The Radio-ulnar Joints

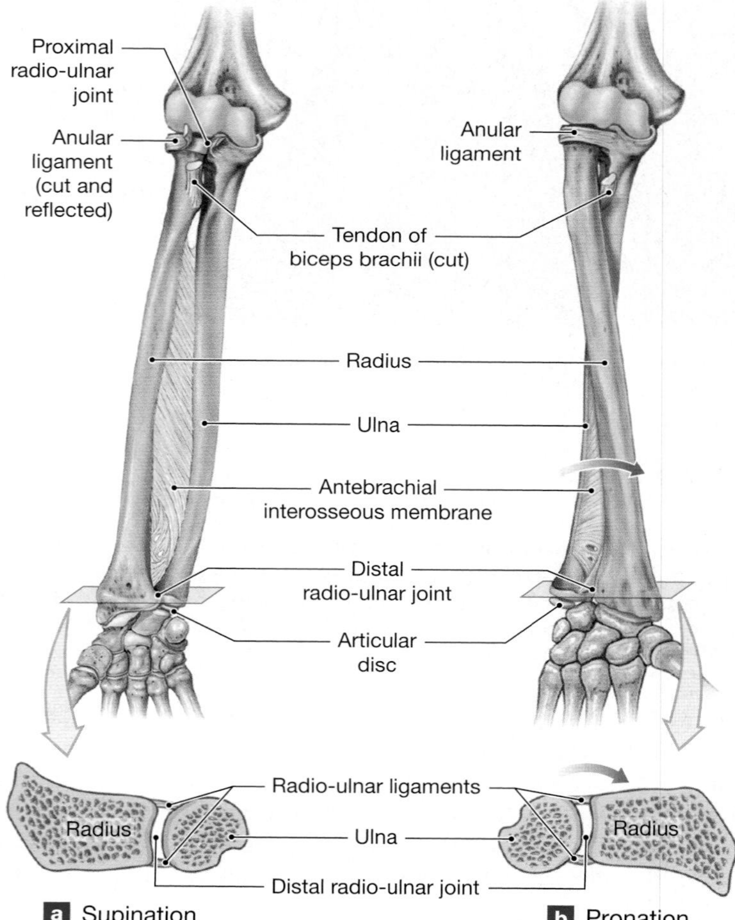

8.7 The Joints of the Wrist and Hand

▶ **KEY POINT** The joints of the wrist and hand are numerous and complex. Wrist movement relative to the forearm occurs at only two points: the radio-ulnar joint and the intercarpal joints (the joints between the proximal and distal rows of carpal bones). The joints between the individual carpal bones of the wrist do not play a role in moving the hand relative to the forearm. Other joints of the hand are between the (1) distal carpal row of carpal bones and the metacarpals, (2) metacarpal bones and proximal phalanges, (3) proximal and middle phalanges, and (4) middle and distal phalanges.

The Joints of the Wrist

▶ **KEY POINT** Unlike the shoulder joint, the elbow joint, or the radio-ulnar joint, the wrist joint plays little or no role in positioning the hand. The position of the wrist, however, does affect the function of the tendons of the muscles that move the fingers.

The **wrist joint** consists of the **radiocarpal joint** and the **intercarpal joints** (**Figure 8.12**). The radiocarpal joint involves the distal articulating surface of the radius and three of the four proximal carpal bones: the scaphoid, lunate, and triquetrum. The radiocarpal joint is a condylar joint that allows flexion/extension, adduction/abduction, and circumduction. The intercarpal joints are plane joints that allow sliding and slight twisting movements.

Wrist Stability

Not all carpal surfaces form intercarpal joints. Those surfaces that do not form a joint are roughened for the attachment of ligaments and for the passage of tendons. A tough connective tissue capsule, reinforced by broad ligaments, surrounds the wrist and stabilizes the positions of the individual carpal bones (**Figure 8.12b,c**). Four major ligaments stabilize the wrist:

- The **palmar radiocarpal ligament** connects the distal radius to the anterior surfaces of the scaphoid, lunate, and triquetrum.
- The **dorsal radiocarpal ligament** connects the distal radius to the posterior surfaces of the scaphoid, lunate, and triquetrum.
- The **ulnar collateral ligament** extends from the ulnar styloid process to the medial surface of the triquetrum.
- The **radial collateral ligament** extends from the radial styloid process to the lateral surface of the scaphoid.

In addition to these prominent ligaments, **intercarpal ligaments** connect the carpal bones, and **digitocarpal ligaments** attach the distal carpal bones to the metacarpal bones (**Figure 8.12c**). Tendons pass across the anterior and posterior surfaces of the wrist joint and provide additional reinforcement. Tendons of muscles that flex the wrist and finger joints pass over the anterior surface of the wrist joint superficial to the ligaments of the wrist joint, whereas tendons of muscles that extend the wrist and finger joints pass across the posterior surface. A pair of broad transverse ligaments arch across the anterior and posterior surfaces of the wrist superficial to these tendons, holding the tendons in position.

The Joints of the Hand

▶ **KEY POINT** The anatomy of the hand permits balance, stability, and control—in particular, the fine control of a finger grip.

The carpal bones articulate with the metacarpal bones of the palm (**Figure 8.12a**). The first metacarpal bone has a saddle joint at the wrist, the **carpometacarpal joint** of the thumb (**Figure 8.12b,d**). All other carpal/metacarpal joints are

Figure 8.12 The Joints of the Wrist and Hand

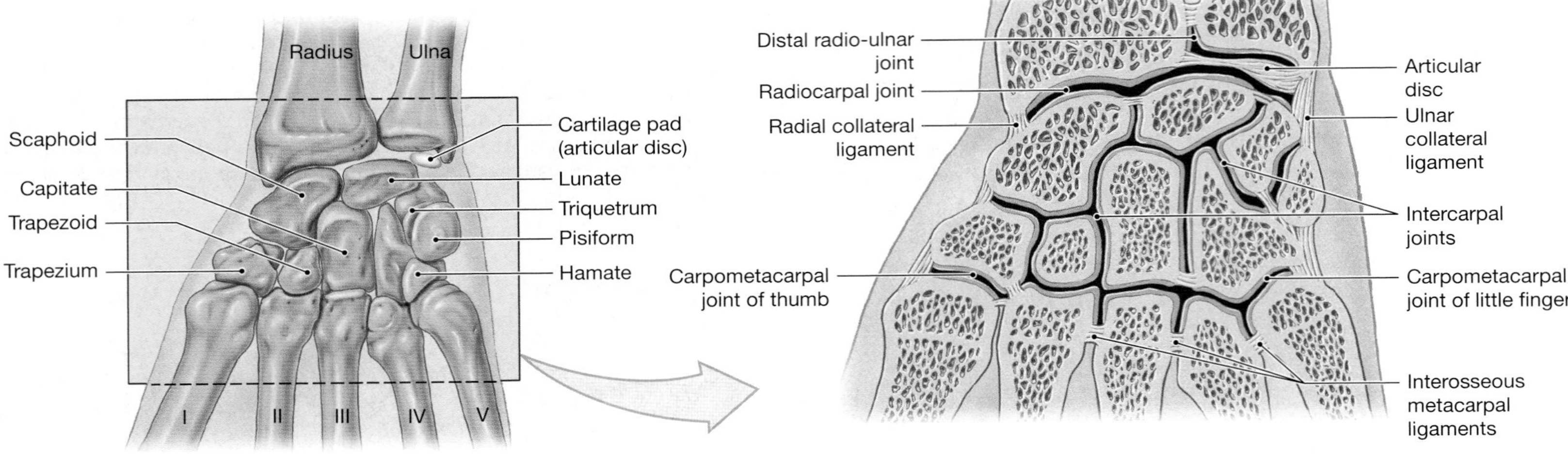

a Anterior view of the right wrist identifying the structures of the wrist joint

b Sectional view through the wrist showing the radiocarpal, intercarpal, and carpometacarpal joints

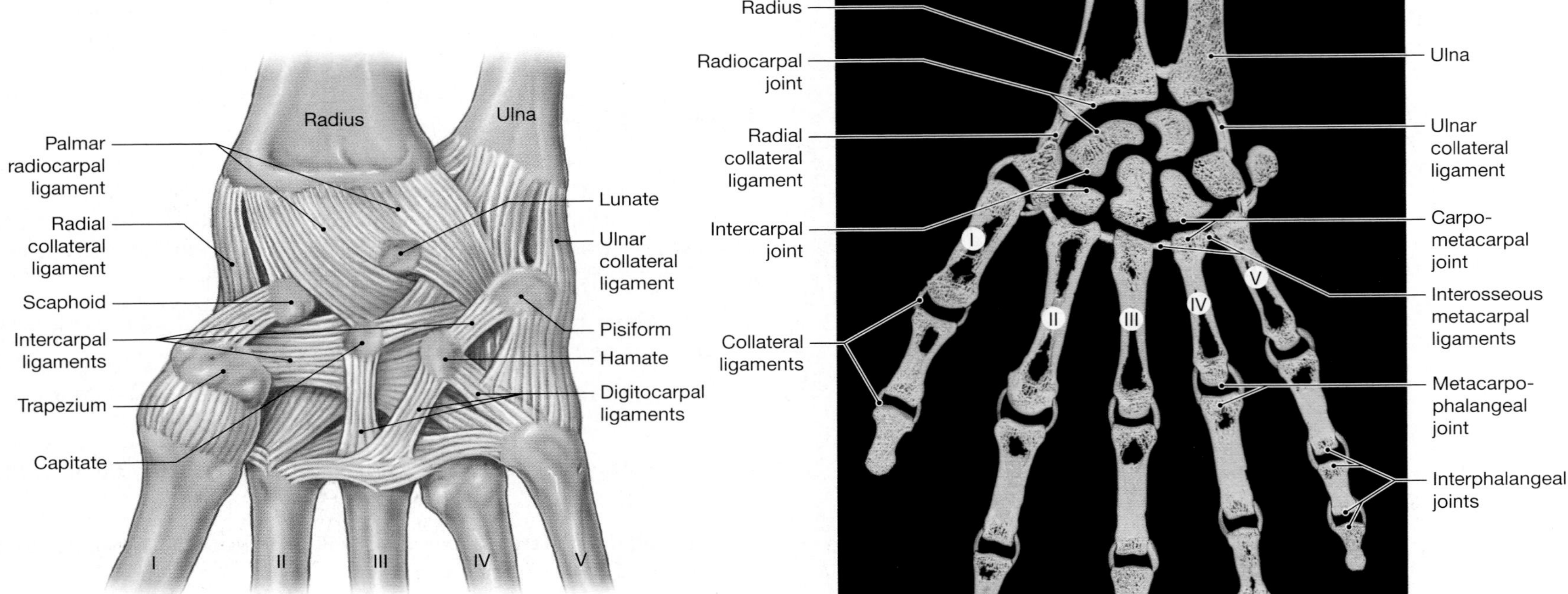

c Stabilizing ligaments on the anterior (palmar) surface of the wrist

d Sectional view of the bones that form the wrist and hand

plane joints. An **intercarpal joint** is formed by carpal/carpal articulations. The joints between the metacarpal bones and the proximal phalanges (**metacarpophalangeal joints**) are condylar, permitting flexion/extension and adduction/abduction. The **interphalangeal joints** are hinge joints that allow flexion and extension (**Figure 8.12d**).

Table 8.3 summarizes the characteristics of the joints of the upper limb.

8.7 CONCEPT CHECK

13 What six structures contribute to the stability of the wrist?

14 What types of movements are permitted between the metacarpophalangeal joints?

See the blue Answers tab at the back of the book.

8.8 The Hip Joint

▶ **KEY POINT** The hip joint, between the head of the femur and the acetabulum of the hip, is the strongest synovial joint of the appendicular skeleton. This joint is a triaxial, synovial ball-and-socket joint that transmits forces from the femur to the pelvis and positions the femur for movement of the body.

The **hip joint** is a ball-and-socket joint (**Figure 8.13**). A pad of fibrous cartilage, the **acetabular labrum**, covers the articular surface of the acetabulum and extends like a horseshoe along the sides of the **acetabular notch** (**Figure 8.13a,c**). A fat pad encased in a synovial membrane covers the center of the acetabulum. This fibrous cartilage pad acts as a shock absorber, and the adipose tissue stretches and cushions without causing damage.

Figure 8.13 The Hip Joint. Views of the hip joint and supporting ligaments.

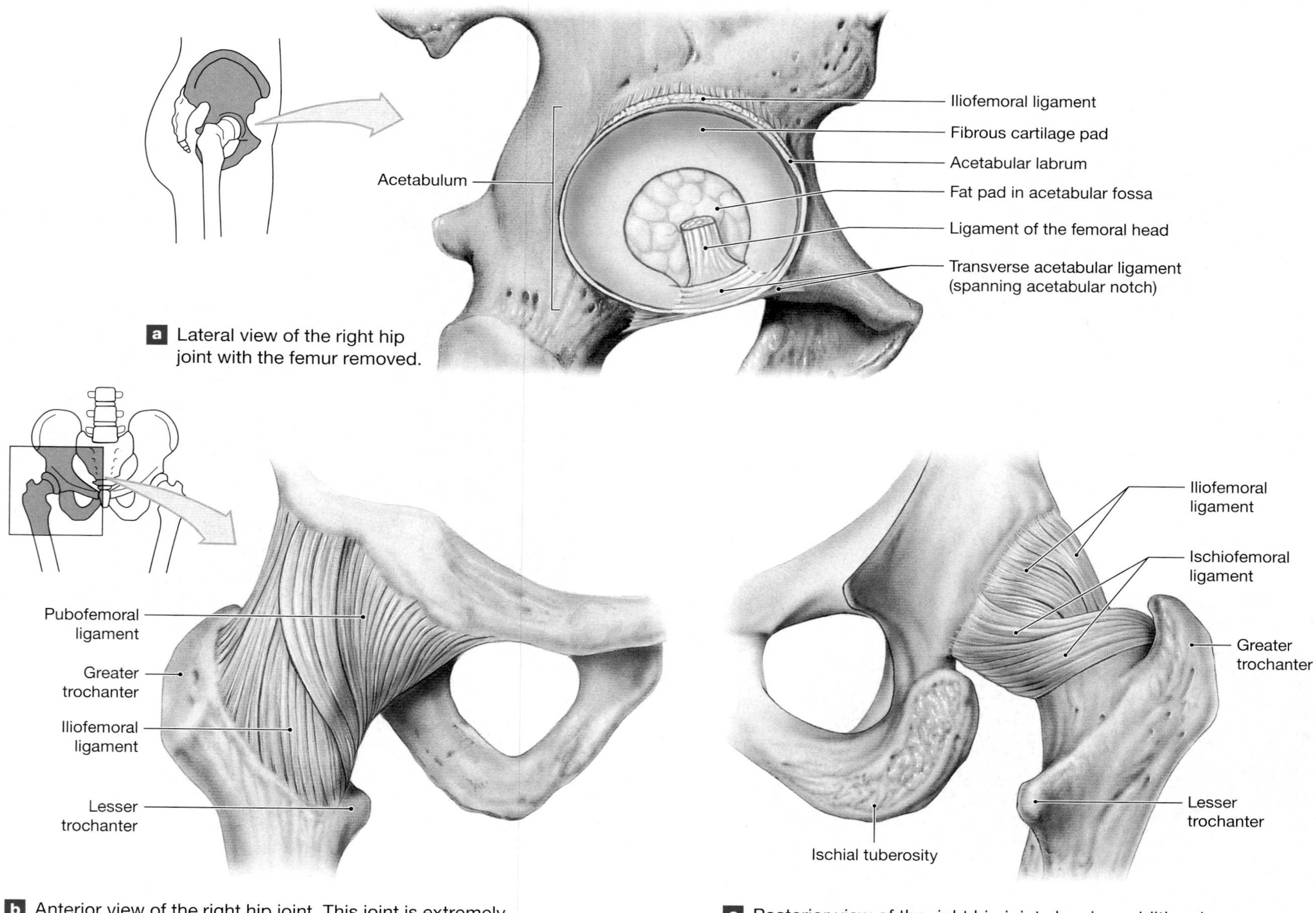

a Lateral view of the right hip joint with the femur removed.

b Anterior view of the right hip joint. This joint is extremely strong and stable, mostly because of the massive capsule.

c Posterior view of the right hip joint showing additional ligaments that strengthen the capsule.

The Articular Capsule

▶ **KEY POINT** The capsule of the hip joint is strong and dense and provides stability for the hip joint. The capsule is composed of strong circular fibers that surround the neck of the femur and three capsular ligaments that twist as they pass from the pelvis to the femur.

The articular capsule of the hip joint is dense, strong, and deep **(Figure 8.13b,c)**. Unlike the capsule of the shoulder joint, the capsule of the hip joint contributes extensively to joint stability. The capsule extends from the lateral and inferior surfaces of the hip bones to the intertrochanteric line and intertrochanteric crest of the femur. This arrangement encloses the femoral head and neck of the femur, preventing the head from moving away from the acetabulum. The acetabular labrum also increases the depth of the acetabulum and, therefore, the stability of the hip joint **(Figure 8.13a,c)**.

Hip Stabilization

▶ **KEY POINT** The hip joint has four reinforcing capsular ligaments: two located anteriorly and two posteriorly. These ligaments are so strong that the neck of the femur will fracture before the capsule tears or the hip dislocates.

Four broad ligaments strengthen the articular capsule **(Figure 8.13b,c)**. Three of them are thickenings of the capsule: the **iliofemoral**, **pubofemoral**, and **ischiofemoral ligaments**. The **transverse acetabular ligament** crosses the acetabular notch and completes the inferior border of the acetabular fossa. A fifth ligament, the **ligament of the femoral head** originates along the transverse acetabular ligament and attaches to the center of the femoral head **(Figures 8.13a** and **8.14)**. This ligament tightens when the thigh is flexed and externally rotated. Surrounding muscles also stabilize the hip joint. Although flexion, extension, adduction, abduction, and rotation occur at the

Figure 8.14 Articular Structure of the Hip Joint. Coronal sectional views of the hip joint.

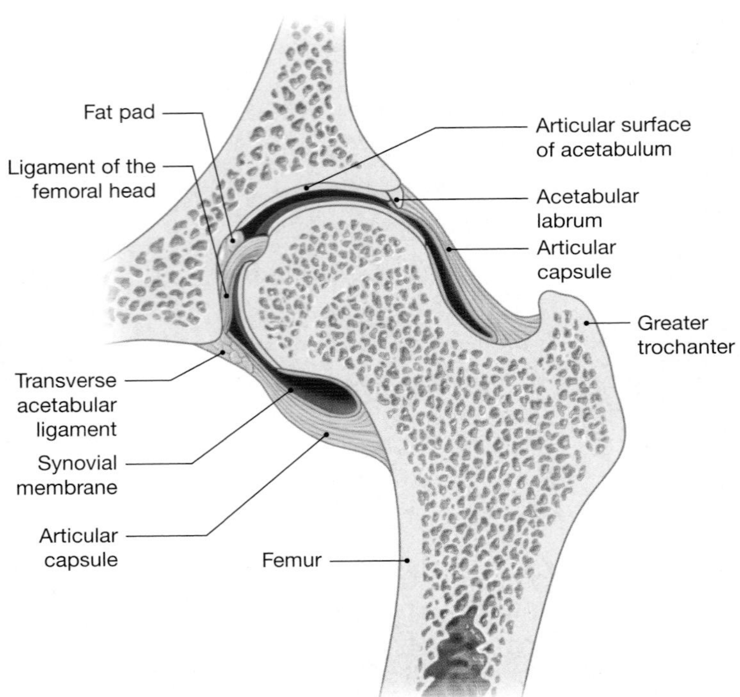

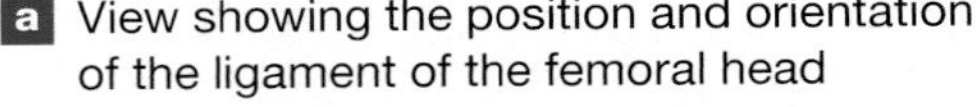
a View showing the position and orientation of the ligament of the femoral head

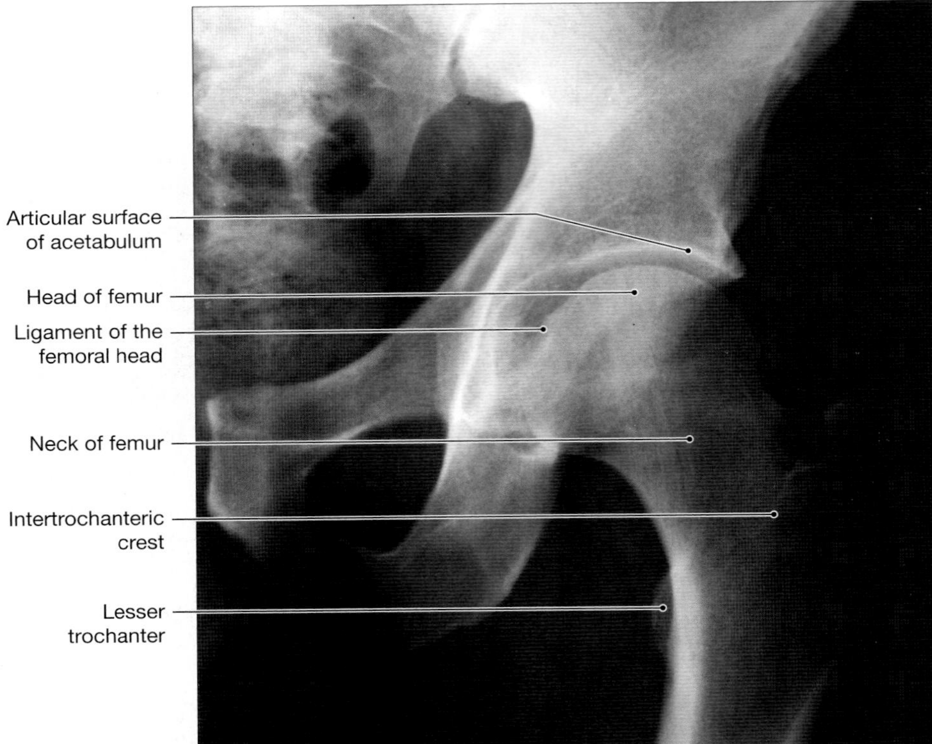

b X-ray of right hip joint, anterior/posterior view

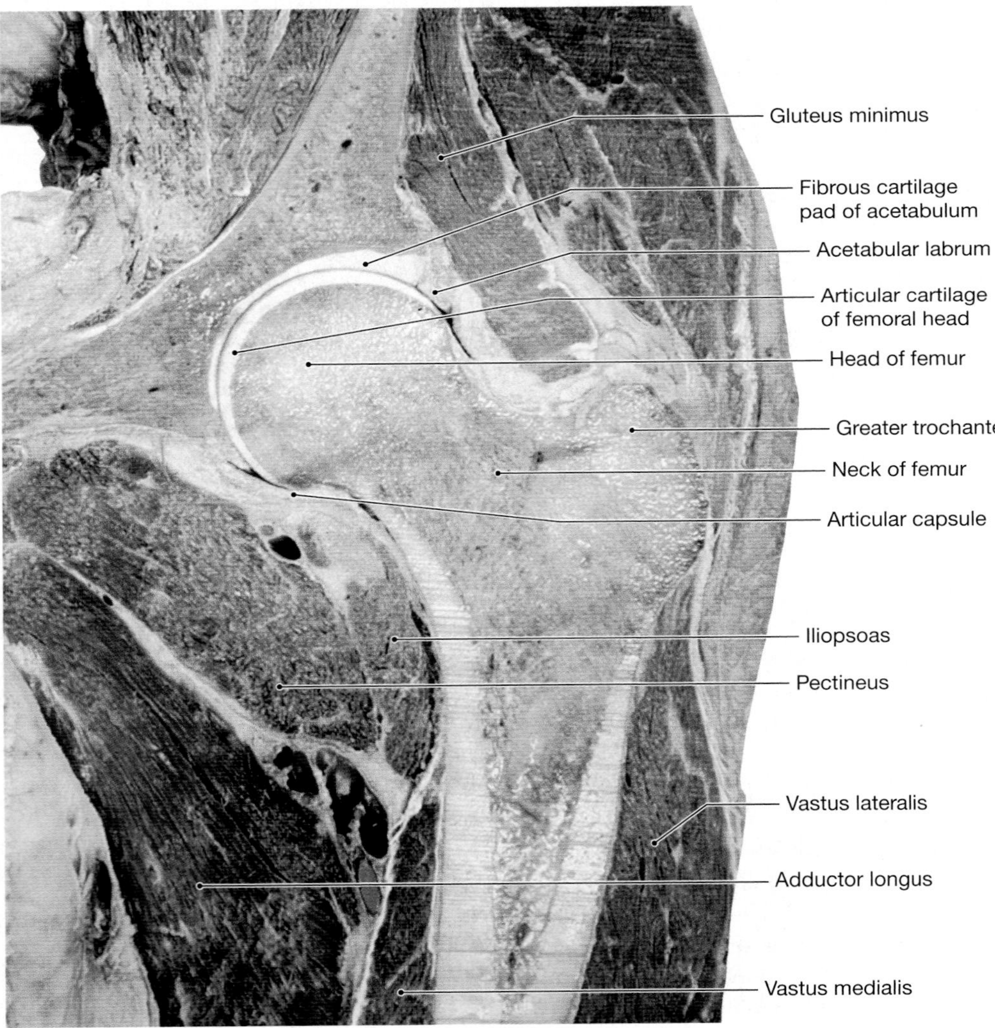

c Coronal section through the hip

hip, flexion is the most important normal movement. The combination of ligaments, capsular fibers, the depth of the acetabulum, and the surrounding muscles helps to limit the range of motion of these movements. The almost complete bony socket enclosing the head of the femur, the strong articular capsule, the stout supporting ligaments, and the dense muscular padding make this an extremely stable joint. Because of this stability, fractures of the femoral neck or between the trochanters are more common than hip dislocations.

8.8 CONCEPT CHECK

15 Explain why dislocations of the hip are rare.

See the blue Answers tab at the back of the book.

8.9 The Knee Joint

▶ **KEY POINT** The knee joint has two articulations within one capsule: the joint between the tibia and the femur and the joint between the femur and the patella. The tibiofemoral joint is a synovial hinge joint. This biaxial joint allows flexion, extension, and slight rotational movements.

The knee joint, along with the hip and ankle joints, supports the body's weight during a variety of activities, such as standing, walking, and running. However, the knee must provide this support while (1) permitting the largest range of motion of any joint of the lower limb, (2) lacking the large muscle mass that

Table 8.3 | **Joints of the Pectoral Girdle and Upper Limb**

Element	Joint	Type of Articulation	Movements
Sternum/clavicle	Sternoclavicular	Plane diarthrosis (a double "plane joint," with two joint cavities separated by an articular cartilage)	Protraction/retraction, depression/elevation, slight rotation
Scapula/clavicle	Acromioclavicular	Plane diarthrosis	Slight gliding movement
Scapula/humerus	Glenohumeral (shoulder)	Ball-and-socket diarthrosis	Flexion/extension, adduction/abduction, circumduction, rotation
Humerus/ulna and humerus/radius	Elbow (humeroulnar and humeroradial)	Hinge diarthrosis	Flexion/extension
Radius/ulna	Proximal radio-ulnar	Pivot diarthrosis	Rotation
	Distal radio-ulnar	Pivot diarthrosis	Pronation/supination
Radius/carpal bones	Radiocarpal	Condylar diarthrosis	Flexion/extension, adduction/abduction, circumduction
Carpal bone/carpal bone	Intercarpal	Plane diarthrosis	Slight gliding movement
Carpal bone/metacarpal bone I	Carpometacarpal of thumb	Saddle diarthrosis	Flexion/extension, adduction/abduction, circumduction, opposition
Carpal bones/metacarpal bones II–V	Carpometacarpal	Plane diarthrosis	Slight flexion/extension, adduction/abduction
Metacarpal bones/ phalanges	Metacarpophalangeal	Condylar diarthrosis	Flexion/extension, adduction/abduction, circumduction
Phalanx/phalanx	Interphalangeal	Hinge diarthrosis	Flexion/extension

supports and strengthens the hip, and (3) lacking the strong ligaments that support the ankle joint.

Although the knee functions as a hinge joint, it is more complex than the elbow. Because the rounded femoral condyles roll and glide across the superior surface of the tibia, the points of contact are constantly changing. The knee is less stable than other hinge joints, and a small amount of rotation occurs in addition to flexion and extension. The knee is composed of two joints within a single joint capsule: a joint between the tibia and femur (the tibiofemoral joint) and one between the patella and the patellar surface of the femur (the patellofemoral joint).

The Articular Capsule

▶ **KEY POINT** The capsule of the knee joint is large and extends from the distal femur to the proximal tibia, and includes the patella. Note that the joint between the tibia and fibula is *not* located within the knee capsule. The knee capsule and its reinforcing ligaments and tendons are important for limiting the motion of the knee.

The capsule and synovial cavity of the knee joint are complex structures (**Figure 8.15**). A pair of fibrous cartilage pads, the **medial** and **lateral menisci**, lie between the femoral and tibial surfaces (**Figure 8.16b,c**). The menisci (1) cushion, (2) conform to the shape of the articulating surfaces as the femur changes position, (3) increase the surface area of the tibiofemoral joint, and (4) provide lateral stability to the joint. Prominent **fat pads** at the margins of the joint help the bursae reduce friction between the patella and other tissues (**Figure 8.15a,b,d**).

Supporting Ligaments

▶ **KEY POINT** The ligaments of the knee (1) prevent hyperextension of the knee, (2) prevent excessive adduction and abduction of the tibia, (3) prevent anterior or posterior dislocation of the tibia, (4) prevent excessive medial and lateral rotation of the tibia, and (5) help control locking and unlocking of the knee.

Seven major ligaments stabilize the knee joint.

- The **quadriceps tendon** from the muscles that extend the knee passes over the anterior surface of the joint (**Figure 8.15a,d**). The patella is embedded within this tendon, and the **patellar ligament** continues to its attachment on the anterior surface of the tibia. The quadriceps tendon and the patellar ligament support the anterior surface of the knee joint (**Figure 8.15b**), where there is no continuous capsule.

The remaining supporting ligaments are grouped as either extracapsular ligaments or intracapsular ligaments, depending on the location of the ligament with respect to the articular capsule. The extracapsular ligaments are discussed below:

- The **tibial collateral ligament** (*medial collateral ligament*) reinforces the medial surface of the knee joint, and the **fibular collateral ligament** (*lateral collateral ligament*) reinforces the lateral surface (**Figures 8.15a** and **8.16**). These ligaments tighten only at full extension, and in this position they stabilize the joint.
- Two superficial **popliteal ligaments** extend between the femur and the heads of the tibia and fibula (**Figure 8.16**). These ligaments reinforce the back of the knee joint.

The intracapsular ligaments are the following:

- The **anterior cruciate ligament (ACL)** and **posterior cruciate ligament (PCL)** (*cruciatuss*, cross) attach the intercondylar area of the tibia to the condyles of the femur. *Anterior and posterior refer to their sites of origin on the tibia*, and they cross one another as they proceed to their destinations on the femur (**Figure 8.16b,c**). These ligaments limit the anterior and posterior movement of the femur and maintain the alignment of the femoral and tibial condyles.

Locking of the Knee

▶ **KEY POINT** Involuntary lateral rotation of the tibia during the final stages of knee extension plays a large role in stabilizing the extended knee and reducing the work of the knee extensors during prolonged standing. This process, however, also contributes to ACL injuries.

The knee joint normally "locks" in the extended position. As the knee approaches full extension, a slight lateral rotation of the tibia tightens the

Figure 8.15 **The Knee Joint, Part I**

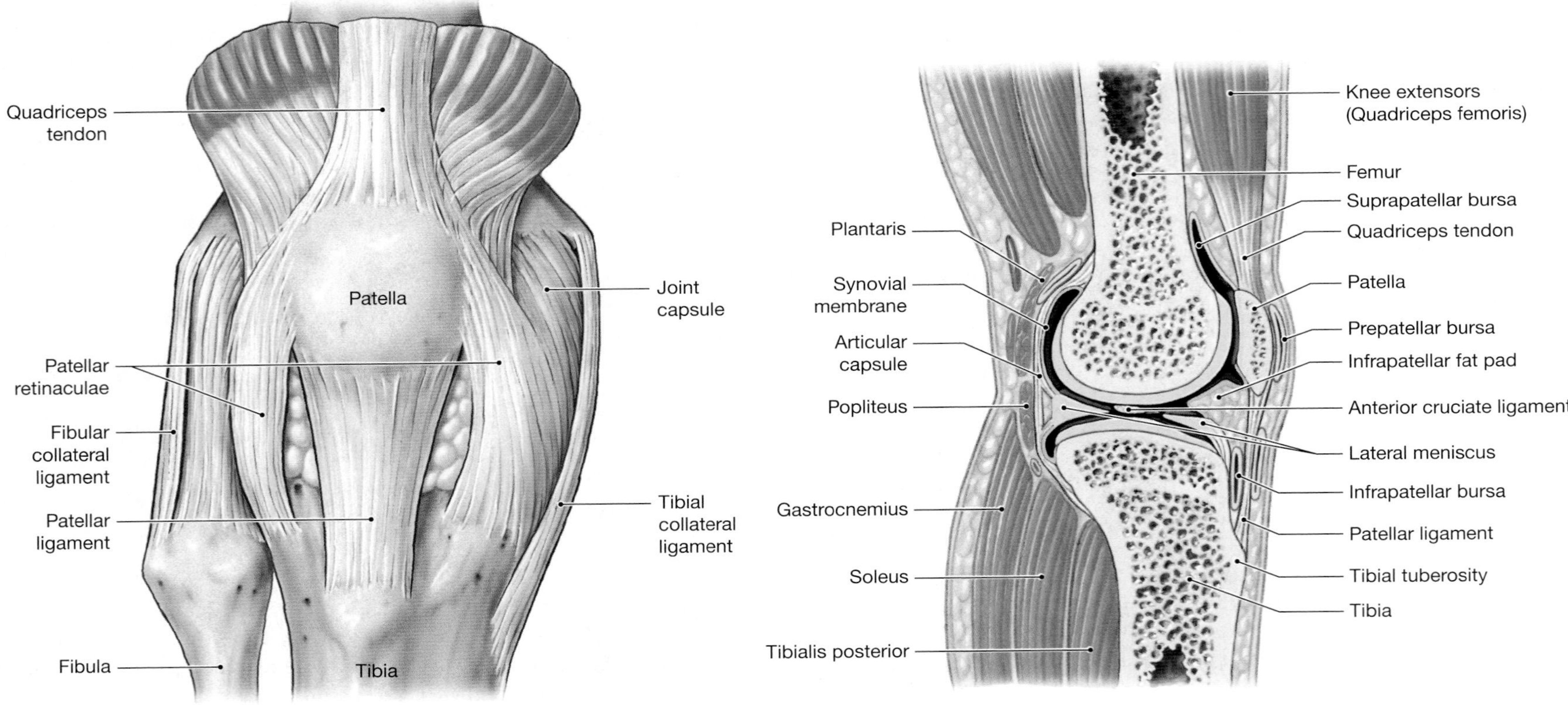

a Anterior view of extended right knee

b Parasagittal section of extended right knee

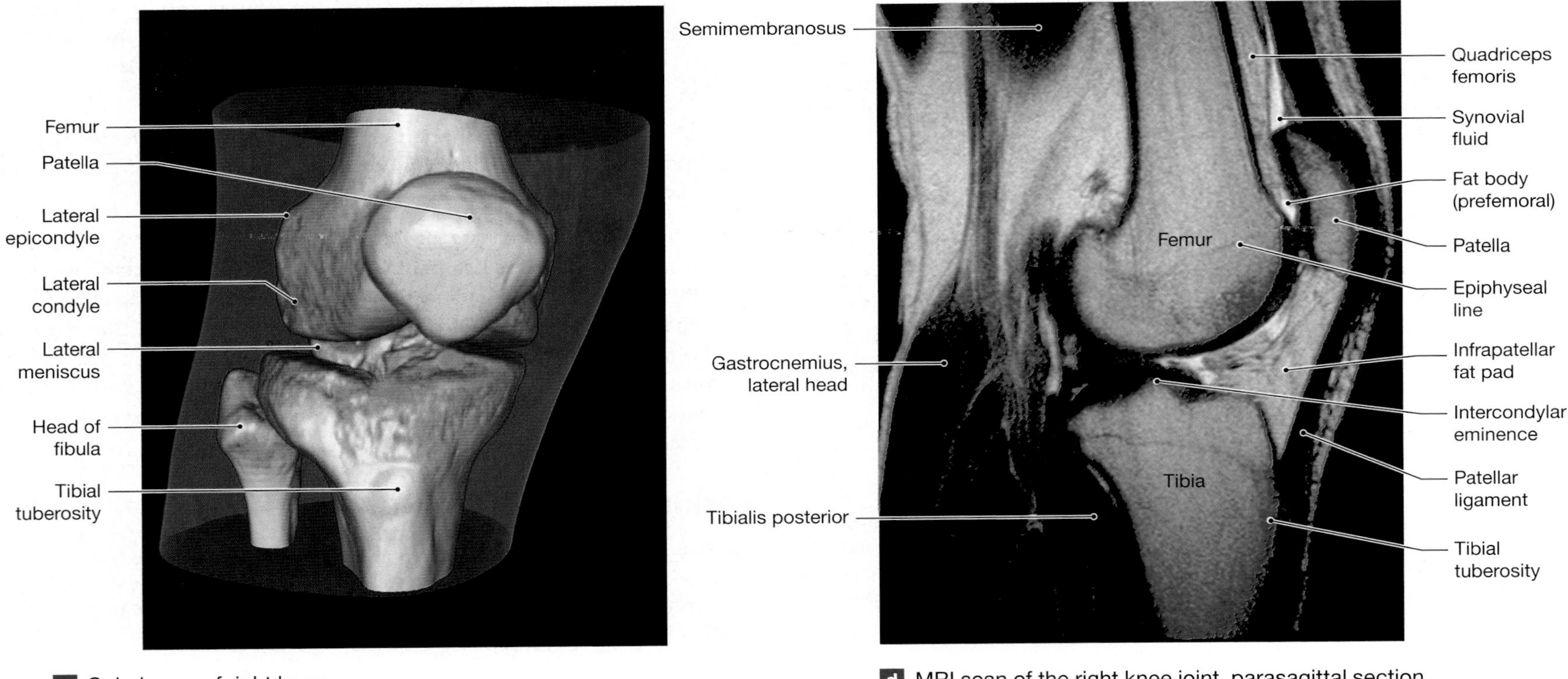

c Spiral scan of right knee

d MRI scan of the right knee joint, parasagittal section, lateral to medial sequence

Figure 8.16 **The Knee Joint, Part II**

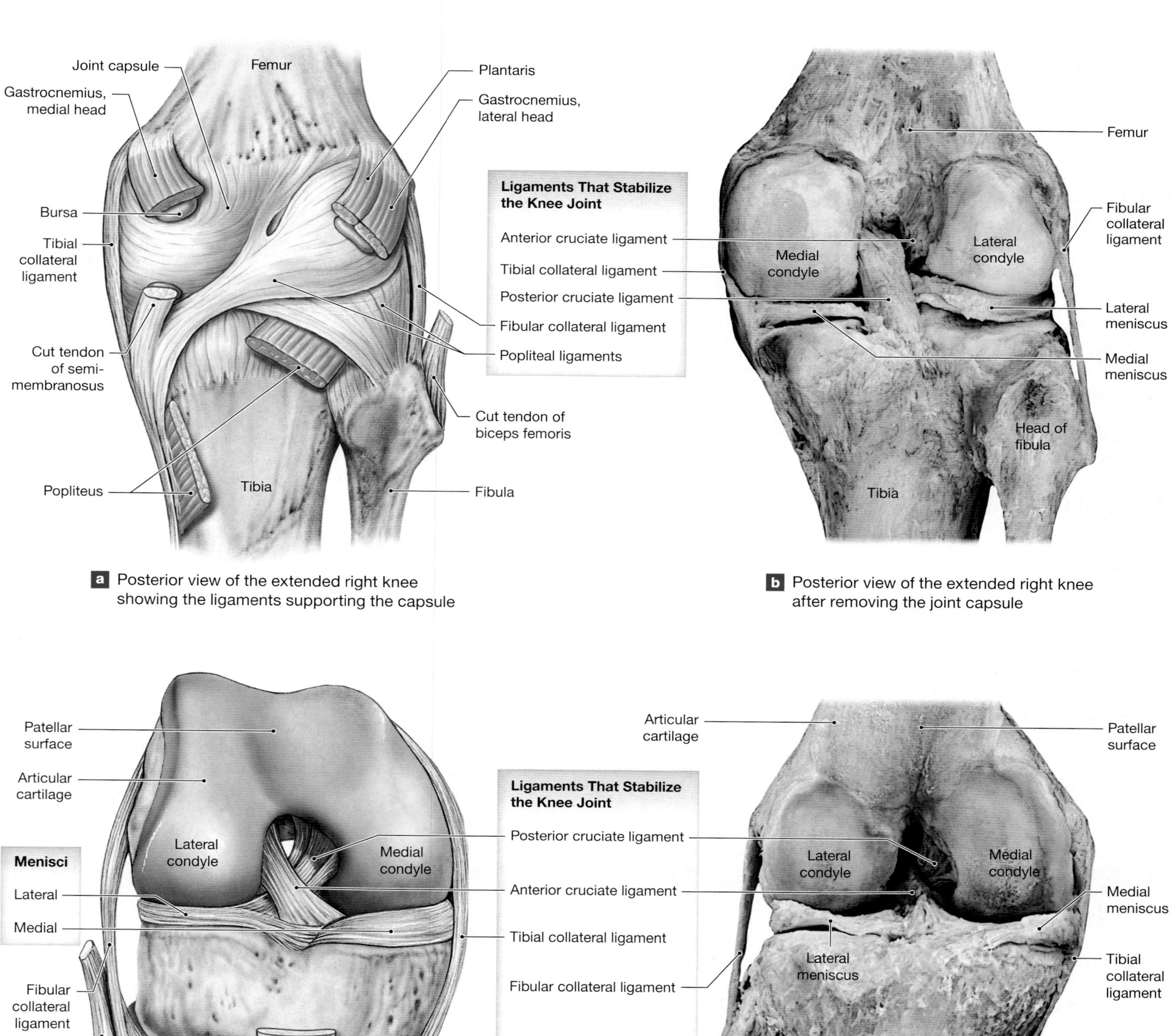

a Posterior view of the extended right knee showing the ligaments supporting the capsule

b Posterior view of the extended right knee after removing the joint capsule

c Anterior views of the right knee at full flexion after removing the joint capsule, patella, and associated ligaments anterior cruciate ligament, causes the tibial tubercles to become "stuck" in the intercondylar notch of the femur, and jams the menisci between the tibia and femur. This mechanism stabilizes the knee joint and permits standing for long periods without using (and tiring) the extensor muscles. Knee flexion is preceded by unlocking the knee—a reversal of the locking motions, which loosens the ACL and slightly rotates the tibia medially. Table 8.4 summarizes information about the joints of the lower limb.

8.9 CONCEPT CHECK

16 What are the cruciate ligaments and what is their role in maintaining stability of the knee joint?

See the blue Answers tab at the back of the book.

8.10 The Joints of the Ankle and Foot

▶ **KEY POINT** The joints between the tibia, fibula, and talus are commonly referred to as the ankle. The joints of the foot lend both flexibility and stability to the foot so that it can successfully support the weight of the body.

The Ankle Joint

▶ **KEY POINT** The ankle is a synovial joint that dorsiflexes and plantar flexes, so it is a uniaxial joint.

The **ankle joint**, or **talocrural joint**, is a hinge joint formed by joints of the tibia, the fibula, and the talus (**Figures 8.17** and **8.18**). The ankle joint permits limited dorsiflexion and plantar flexion.

The weight-bearing joint of the ankle is the **tibiotalar joint**, the joint between the distal articular surface of the tibia and the trochlea of the talus. Normal functioning of the tibiotalar joint depends on medial and lateral stability at this joint. Three joints provide this stability: (1) the proximal tibiofibular joint, (2) the distal tibiofibular joint, and (3) the fibulotalar joint.

The **proximal tibiofibular joint** is a plane joint between the posterolateral surface of the tibia and the head of the fibula. The **distal tibiofibular joint** is a fibrous syndesmosis between the distal facets of the tibia and fibula. The joint formed between the lateral malleolus of the fibula and the lateral articular surface of the talus is the **fibulotalar joint**. Ligaments on the tibia and fibula hold these two bones in place, limiting movement at the two tibiofibular joints and the fibulotalar joint. Medial and lateral stability of the ankle results from maintaining the correct amount of movement at these joints.

The anterior and posterior portions of the joint capsule are thin, but the lateral and medial surfaces are strong and reinforced by ligaments (**Figure 8.18b–d**). The major ligaments are the medial **deltoid ligament** and the three **lateral ligaments**. The malleoli, supported by these ligaments and bound together by the **tibiofibular ligaments**, prevent the ankle bones from sliding from side to side.

The Joints of the Foot

▶ **KEY POINT** The foot must be both stable *and* flexible. It must be stable to support the weight of the body above and absorb shock from contacting the ground below. At the same time, it must be flexible enough to walk or run on uneven surfaces. The foot must also be strong enough to propel the body forward during walking and running. The foot can do this because of its joint anatomy.

Four groups of synovial joints are found in the foot (**Figures 8.17** and **8.18**):

1. The **intertarsal joints** are plane joints that permit limited sliding and twisting movements. The joints between the tarsal bones are similar to those between the carpal bones of the wrist.
2. The **tarsometatarsal joints** are plane joints that also allow limited sliding and twisting movements. The first three metatarsal bones articulate with the medial, intermediate, and lateral cuneiform bones. The fourth and fifth metatarsal bones articulate with the cuboid.
3. The **metatarsophalangeal joints** are condylar joints that permit flexion/extension and adduction/abduction. Joints between the metatarsal bones and phalanges resemble those between the metacarpal bones and phalanges of the hand. The first metatarsophalangeal joint is condylar, rather

Table 8.4 Joints of the Pelvic Girdle and Lower Limb

Element	Joint	Type of Articulation	Movements
Sacrum/hip bones	Sacro-iliac	Plane diarthrosis	Gliding movements
Pubic bone/pubic bone	Pubic symphysis	Amphiarthrosis	None*
Hip bones/femur	Hip	Ball-and-socket diarthrosis	Flexion/extension, adduction/abduction, circumduction, rotation
Femur/tibia	Knee	Complex, functions as hinge	Flexion/extension, limited rotation
Tibia/fibula	Tibiofibular (proximal)	Plane diarthrosis	Slight gliding movements
	Tibiofibular (distal)	Plane diarthrosis and amphiarthrotic syndesmosis	Slight gliding movements
Tibia and fibula with talus	Ankle, or talocrural	Hinge diarthrosis	Dorsiflexion/plantar flexion
Tarsal bone to tarsal bone	Intertarsal	Plane diarthrosis	Slight gliding movements
Tarsal bones to metatarsal bones	Tarsometatarsal	Plane diarthrosis	Slight gliding movements
Metatarsal bones to phalanges	Metatarsophalangeal	Condylar diarthrosis	Flexion/extension, adduction/abduction
Phalanx/phalanx	Interphalangeal	Hinge diarthrosis	Flexion/extension

*During pregnancy, hormones weaken the symphysis and permit movement important to childbirth (see Chapter 28).

Figure 8.17 The Joints of the Ankle and Foot, Part I

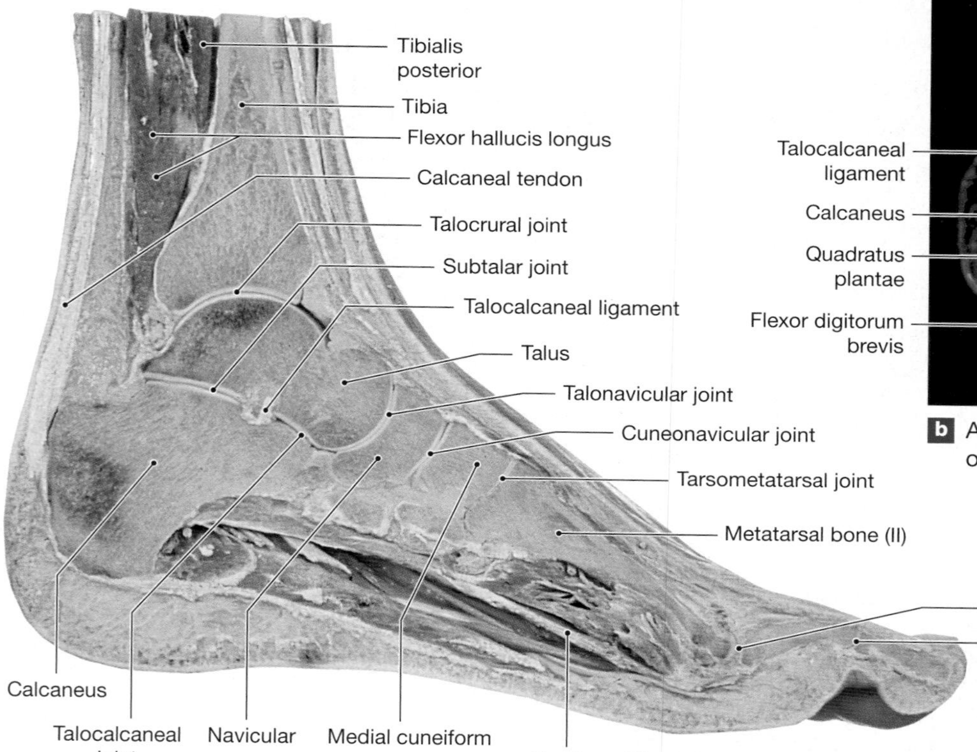

Tibialis posterior
Tibia
Flexor hallucis longus
Calcaneal tendon
Talocrural joint
Subtalar joint
Talocalcaneal ligament
Talus
Talonavicular joint
Cuneonavicular joint
Tarsometatarsal joint
Metatarsal bone (II)
Metatarsophalangeal joint
Interphalangeal joint
Calcaneus
Talocalcaneal joint
Navicular
Medial cuneiform
Tendon of flexor digitorum brevis

a Longitudinal section of the left foot identifying major joints and associated structures

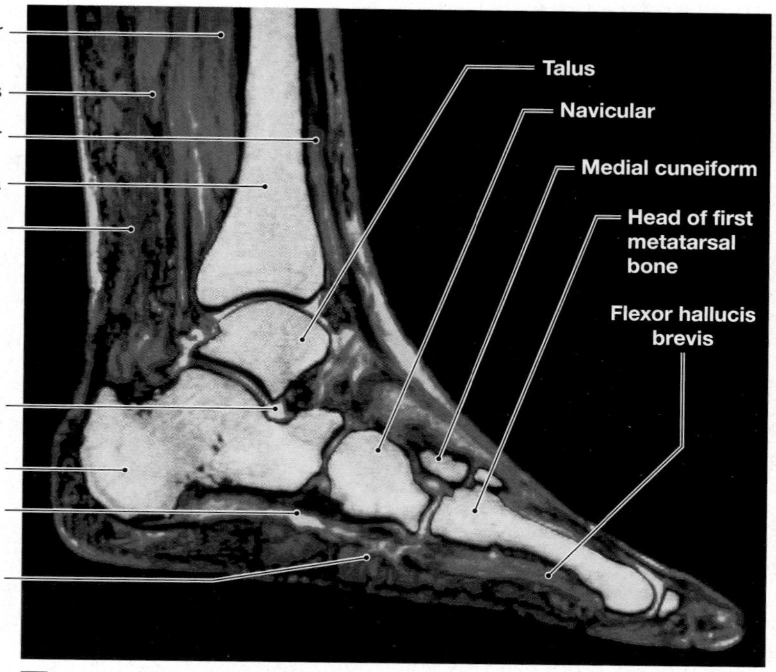

Tibialis posterior
Flexor hallucis longus
Tendon of tibialis anterior
Tibia
Calcaneal tendon
Talocalcaneal ligament
Calcaneus
Quadratus plantae
Flexor digitorum brevis
Talus
Navicular
Medial cuneiform
Head of first metatarsal bone
Flexor hallucis brevis

b A corresponding MRI scan of the left ankle and proximal portion of the foot

CLINICAL NOTE

Ankle Injuries

Ankle injuries can involve the traversing muscles and tendons (strains), the stabilizing ligaments (sprains), the bones of the ankle (fractures), or a combination of tissues. Ankles are injured when they are twisted out of their normal position, usually with weight-bearing activity. Inversion injuries are most common.

Excessive eversion of the ankle can result in a **Pott's fracture**, also called a *bimalleolar fracture*. This excessive eversion puts significant strain on the strong medial ligaments of the ankle, causing fractures of the medial malleolus of the distal tibia and the lateral malleolus of the distal fibula.

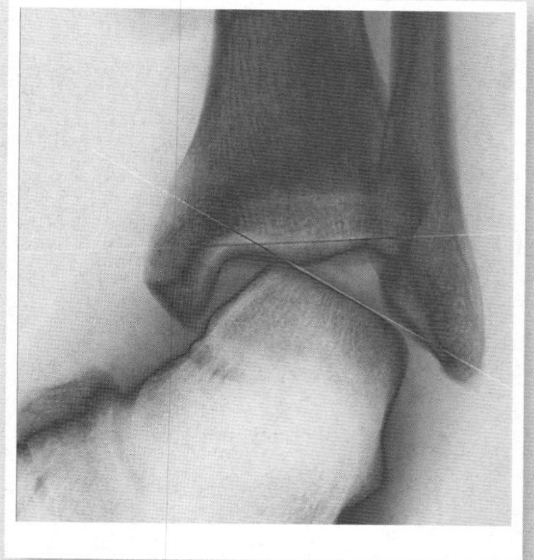

Ligamentous injury with lateral ankle instability

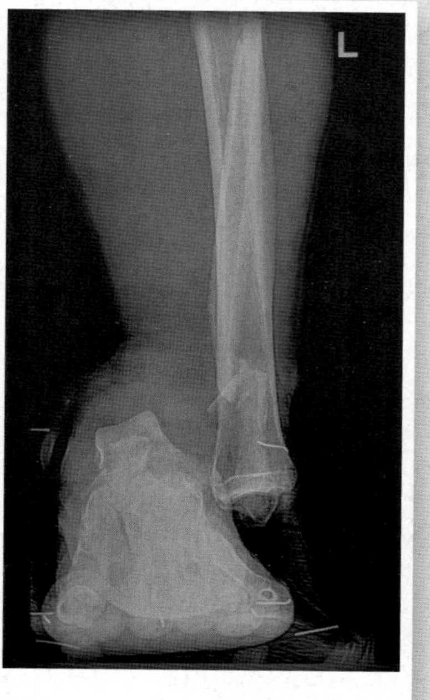

Bony injury with tibial and fibular fractures

Figure 8.18 The Joints of the Ankle and Foot, Part II

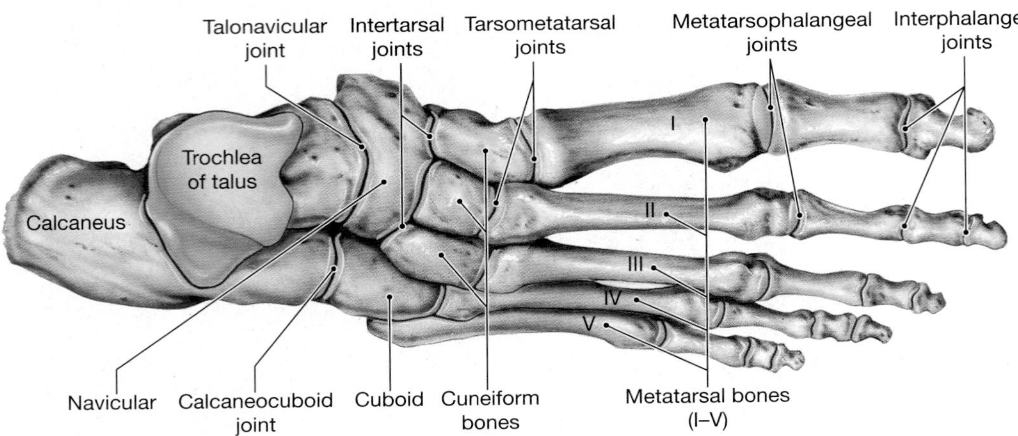

a Superior view of bones and joints of the right foot.

Tibia
Fibula
Talocrural (ankle) joint
Medial malleolus
Talus
Lateral malleolus
Deltoid ligament
Calcaneus
Talocalcaneal ligament
Cuboid
Calcaneocuboid joint

b Posterior view of a coronal section through the right ankle after plantar flexion. Note the placement of the medial and lateral malleoli.

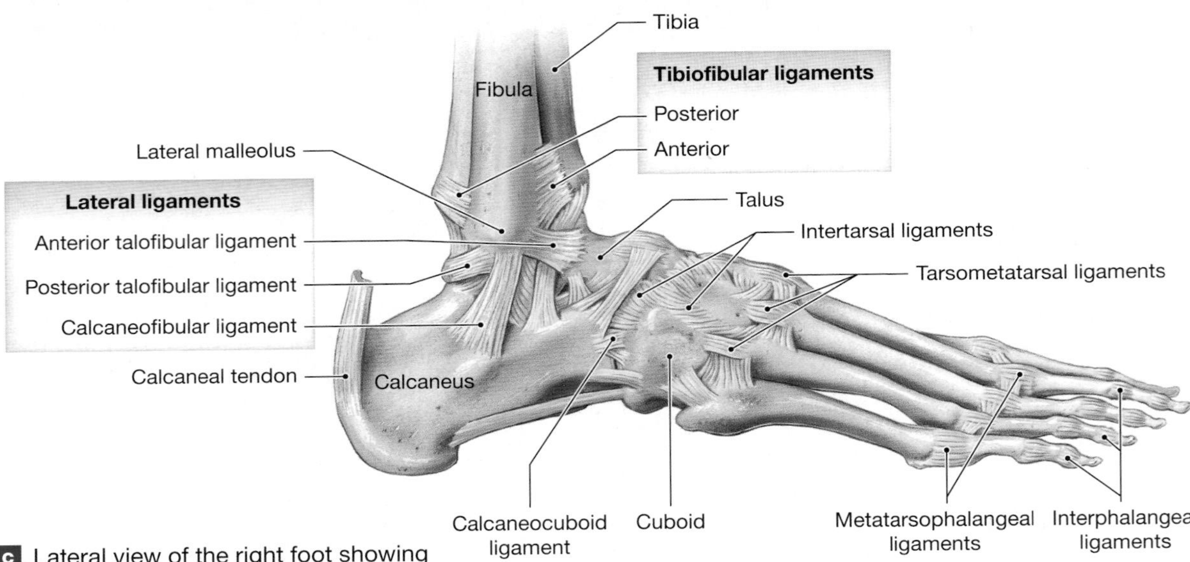

c Lateral view of the right foot showing ligaments that stabilize the ankle joint.

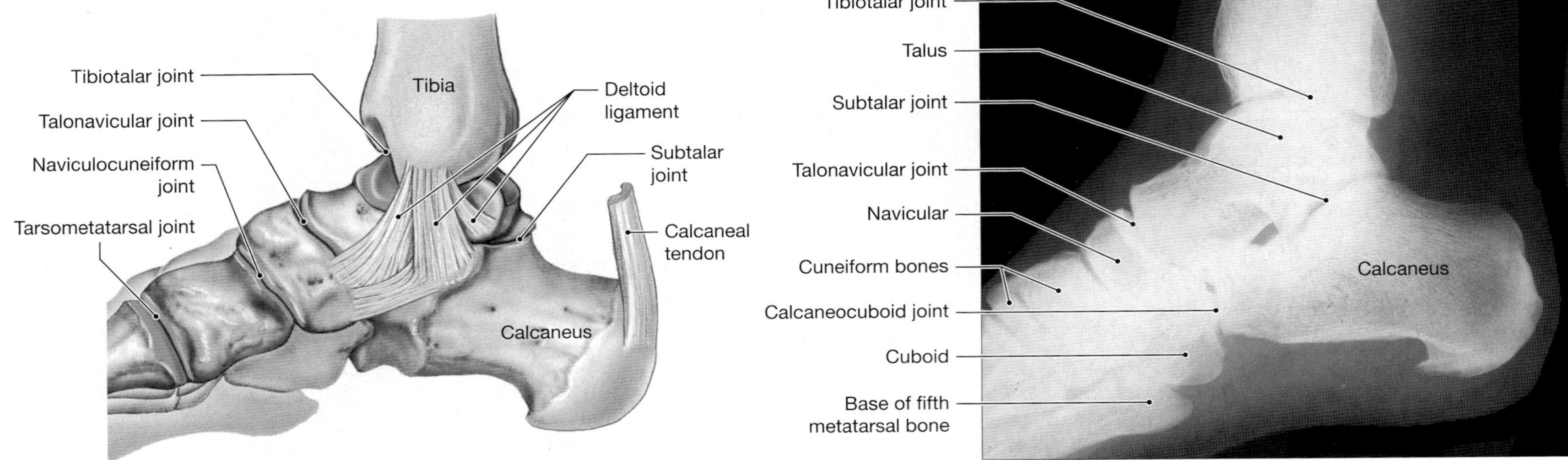

d Medial view of the right ankle showing the medial ligaments.

e X-ray of right ankle, medial/lateral projection.

than saddle-shaped like the first metacarpophalangeal joint of the hand; therefore, the great toe is not as flexible as the thumb. A pair of sesamoid bones often forms in the tendons that cross the inferior surface of this joint, and their presence further restricts movement.

4 The **interphalangeal joints** are hinge joints that permit flexion and extension.

8.10 CONCEPT CHECK

17 What is the primary weight-bearing joint in the ankle?

18 What joints provide medial and lateral stability in the ankle?

See the blue Answers tab at the back of the book.

8.11 Aging and Joints

▶ **KEY POINT** Aging affects bones and joints in a variety of ways, many of which result in a decreased range of motion at joints and an increased risk of fractures. Regular exercise, particularly weight-bearing and flexibility exercises, can slow the effects of aging and help maintain the normal functioning of bones and joints.

We subject our joints to heavy wear and tear throughout our lifetime, so it is no surprise that problems with joint function are common, especially in older people. **Rheumatism** (RŪ-ma-tizm) is a general term for the pain and stiffness affecting joints or other parts of the the skeletal system, the muscular system, or both. **Arthritis** (ar-THRĪ-tis), inflammation of the joints, is a type of rheumatic disease that affects synovial joints. Arthritis always damages the articular cartilages, but the specific causes vary. For example, arthritis can result from bacterial or viral infection, injury to the joint, metabolic problems, or severe physical stresses.

With age, bone mass decreases and bones become weaker, so the risk of fractures increases. If osteoporosis develops, bones weaken to the point where fractures occur in response to stresses that could easily be tolerated by normal bones.

Bone healing proceeds very slowly. In the case of hip fractures, the powerful muscles that surround the hip joint often prevent proper alignment of the bone fragments. Fractures at the greater or lesser trochanter generally heal well if the joint can be stabilized; steel frames, pins, screws, or some combination of these devices may be used to preserve alignment and permit healing.

Although hip fractures are most common among those over 60, in recent years the incidence of hip fractures has increased dramatically among young, healthy professional athletes.

8.11 CONCEPT CHECK

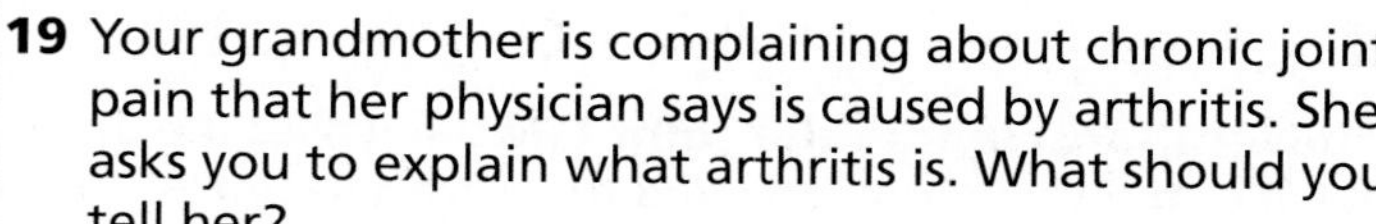

19 Your grandmother is complaining about chronic joint pain that her physician says is caused by arthritis. She asks you to explain what arthritis is. What should you tell her?

See the blue Answers tab at the back of the book.

Study Outline

Introduction p. 204

- **Joints (articulations)** occur wherever two bones contact one another. The function of a joint depends on its anatomy. Joints may permit (1) no movement, (2) slight movement, or (3) extensive movement.

8.1 Joint Classification p. 204

- Three categories of joints are based on range of movement. *Immovable joints* are **synaroses**, *slightly movable joints* are **amphiarthroses**, and *freely movable joints* are **diarthroses**. Joints may be classified by function or by structure. *(See Table 8.1.)*

Synarthroses (Immovable Joints) p. 204

- In a **synarthrosis**, bony edges are close together and may interlock. Examples of synarthroses include a **suture** between skull bones, a **gomphosis** between teeth and jaws, a **synchondrosis** between bone and cartilage in an epiphyseal plate, and a **synostosis**, where two bones fuse and the boundary between them disappears.

Amphiarthroses (Slightly Movable Joints) p. 204

- Very limited movements are permitted in an **amphiarthrosis**. Examples of amphiarthroses are a **syndesmosis**, where collagen fibers connect bones of the leg, and a **symphysis**, where bones are separated by a pad of fibrous cartilage.

Diarthroses (Freely Movable Joints) p. 204

- A wide range of movement is permitted at a **diarthrosis**, or **synovial joint**. These joints possess seven common characteristics: a **joint capsule, articular cartilages,** a fluid-filled synovial cavity, a **synovial membrane**, accessory **capsular ligaments**, sensory nerves**,** and blood vessels that supply the synovial membrane. The articular cartilages are lubricated by **synovial fluid**. Other accessory structures can include **menisci, fat pads, ligaments**, tendons, **bursae**, and **tendon sheaths**. *(See Figure 8.1.)*
- A joint cannot have both great strength and great mobility at the same time. The stronger the joint, the less mobile it is; the more mobile the joint, the weaker it is.

8.2 Articular Form and Function p. 207

Describing Dynamic Motion and the Structural Classification of Synovial Joints p. 207

- Synovial joints are freely movable diarthrotic joints that can be classified according to their anatomical functional properties. *(See Spotlight Figure 8.2.)*

Types of Movements p. 207

- Examples of angular motion include abduction, adduction, flexion, and extension. All movements are described with reference to a figure in the anatomical position. *(See Figure 8.3.)*

- **Abduction** is the movement away from the longitudinal axis of the body in the frontal plane. *(See Figure 8.3a,c.)* **Adduction** is the movement toward the longitudinal axis of the body in the frontal plane.
- **Flexion** is a movement in the anterior-posterior plane that reduces the angle between the articulating elements. **Extension** is a movement in the same plane as flexion, but in the opposite direction. *(See Figure 8.3b.)*
- **Hyperextension** is a term applied to any movement in which a limb is extended beyond its normal limits.
- A special type of angular motion is **circumduction**, movement of an extremity in a circular direction. *(See Figure 8.3d.)*
- Rotation of the head may involve **left rotation** or **right rotation**. Rotation of the limbs may involve **internal rotation** (or **medial rotation**) and **external rotation** (or **lateral rotation**). *(See Figure 8.4.)*
- Special terms apply to specific joints or unusual types of movements. These include **eversion**, **inversion**, **dorsiflexion**, **plantar flexion**, **lateral flexion**, **protraction**, **retraction**, **opposition**, **reposition**, **elevation**, and **depression**. *(See Figure 8.5.)*

8.3 | The Temporomandibular Joint p. 212

- The **temporomandibular joint (TMJ)** involves the mandibular fossa of the temporal bone and the condylar process of the mandible. This joint has a thick pad of fibrous cartilage, the articular disc. Supporting structures include the **lateral ligament**, the **stylomandibular ligament**, and the **sphenomandibular ligament**. This relatively loose hinge joint permits small amounts of gliding and rotation. *(See Figure 8.6.)*

8.4 | Intervertebral Joints p. 212

Zygapophysial Joints p. 212

- The **zygapophysial joints** are plane joints that are formed by the superior and inferior articular processes of adjacent vertebrae. The bodies of adjacent vertebrae form symphyseal joints.

The Intervertebral Discs p. 213

- The vertebrae are separated by **intervertebral discs** containing an inner soft, elastic gelatinous core, the **nucleus pulposus**, and an outer layer of fibrous cartilage, the **anulus fibrosus**. *(See Figure 8.7 and Clinical Note on p. 214.)*

Intervertebral Ligaments p. 215

- Numerous ligaments bind together the bodies and processes of all vertebrae. *(See Figure 8.7.)*

Vertebral Movements p. 215

- The joints of the vertebral column permit anterior **flexion** and posterior **extension**, **lateral flexion**, and **rotation**.
- Joints of the axial skeleton are summarized in *Table 8.2*.

8.5 | The Shoulder Complex p. 216

- The **shoulder complex** is composed of the clavicle, scapula, and humerus and their associated joints and supporting structures. These bones link the upper limb to the thorax.

The Sternoclavicular Joint p. 216

- The **sternoclavicular joint** is a plane joint that lies between the sternal end of each clavicle and the manubrium of the sternum. An articular disc separates the opposing surfaces. The **anterior** and **posterior sternoclavicular ligaments**, along with the **interclavicular** and **costoclavicular ligaments**, reinforce the joint capsule. *(See Figure 8.8.)*

The Shoulder Joint p. 217

- The **shoulder joint**, or **glenohumeral joint**, formed by the glenoid fossa and the head of the humerus, is a loose, shallow joint that permits the greatest range of motion of any joint in the body. It is a ball-and-socket diarthrosis. Strength and stability are sacrificed to gain mobility.
- The ligaments and surrounding muscles and tendons provide strength and stability. The shoulder has a large number of bursae that reduce friction as large muscles and tendons pass across the joint capsule. *(See Figures 8.9 and 12.10 and Clinical Note on p. 219.)*

8.6 | The Elbow and Radio-ulnar Joints p. 219

The Elbow Joint p. 219

- The **elbow joint** is composed of the joints between (1) the humerus and the ulna, (2) the humerus and the radius, (3) between the humerus and the ulna (**humero-ulnar joint**) and (4) between the humerus and the radius (**humeroradial joint**).
- **Radial** and **ulnar collateral ligaments** and **anular ligaments** stabilize this joint. *(See Figure 8.10.)*

The Radio-ulnar Joints p. 221

- The **proximal radio-ulnar** and **distal radio-ulnar joints** allow for supination and pronation of the forearm. The head of the radius is held in place by the **anular ligament**, whereas the distal radio-ulnar articulating surfaces are held in place by a series of radio-ulnar ligaments and the antebrachial interosseous membrane. *(See Figure 8.11.)*

8.7 | The Joints of the Wrist and Hand p. 221

The Joints of the Wrist p. 221

- The **wrist joint** is formed by the **radiocarpal joint** and the **intercarpal joints**. The radiocarpal joint is a condylar joint that involves the distal articular surface of the radius and three proximal carpal bones (scaphoid, lunate, and triquetrum). The radiocarpal joint permits flexion/extension, adduction/abduction, and circumduction. A connective tissue capsule and broad ligaments stabilize the positions of the individual carpal bones. The intercarpal joints are plane joints. *(See Figure 8.12.)*

The Joints of the Hand p. 221

- Five types of diarthrotic joints are found in the hand: (1) carpal bone/carpal bone (**intercarpal joints**), plane diarthrosis; (2) carpal bone/first metacarpal bone (**carpometacarpal joint** of the thumb), saddle diarthrosis, permitting flexion/extension, adduction/abduction, circumduction, opposition; (3) carpal bones/metacarpal bones II–V (**carpometacarpal joints**), plane diarthrosis, permitting slight flexion/extension and adduction/abduction; (4) metacarpal bone/phalanx (**metacarpophalangeal joints**), condylar diarthrosis, permitting flexion/extension, adduction/abduction, and circumduction; and (5) phalanx/phalanx (**interphalangeal joints**), hinge diarthrosis, permitting flexion/extension. *(See Figure 8.12 and Table 8.3.)*

8.8 | The Hip Joint p. 222

- The **hip joint** is a ball-and-socket diarthrosis between the head of the femur and the acetabulum. The joint permits flexion/extension, adduction/abduction, circumduction, and rotation. *(See Figures 8.13 and 8.14.)*

The Articular Capsule p. 223

- The articular capsule of the hip joint is reinforced and stabilized by four broad ligaments: the **iliofemoral**, **pubofemoral**, **ischiofemoral**, and **transverse acetabular ligaments**. *(See Figure 8.13.)*

Hip Stabilization p. 223

- The **ligament of the femoral head** helps stabilize the hip joint. *(See Figures 8.13 and 8.14.)*

8.9 | The Knee Joint p. 224

- The knee joint functions as a hinge joint, but is more complex than standard hinge joints such as the elbow. Structurally, the knee is composed of two joints: (1) one between the tibia and femur and (2) one between the patella and the patellar surface of the femur. The joint permits flexion/extension and limited rotation. *(See Figures 8.15 and 8.16, and Table 8.4.)*

The Articular Capsule p. 225

- The articular capsule of the knee is not a single unified capsule with a common synovial cavity. It contains (1) fibrous cartilage pads, called the **medial** and **lateral menisci**, and (2) **fat pads**. *(See Figures 8.15 and 8.16.)*

Supporting Ligaments p. 225

- Seven major ligaments bind and stabilize the knee joint: the **patellar**, **tibial collateral**, **fibular collateral**, **popliteal** (two), and **anterior** and **posterior cruciate ligaments** (**ACL** and **PCL**, respectively). *(See Figures 8.15 and 8.16.)*

Locking of the Knee p. 225

- The knee joint typically "locks" in the extended position, thereby stabilizing the knee. At full extension, a slight lateral rotation of the tibia tightens the anterior cruciate ligament and jams the meniscus between the tibia and femur, thereby "locking" the knee.

8.10 | The Joints of the Ankle and Foot p. 228

The Ankle Joint p. 228

- The **ankle joint**, or **talocrural joint**, is a hinge joint formed by the inferior surface of the tibia, the lateral malleolus of the fibula, and the trochlea of the talus. The primary joint is the **tibiotalar joint**. The tibia and fibula are bound together by anterior and posterior **tibiofibular ligaments**. With these stabilizing ligaments holding the bones together, the medial and lateral malleoli prevent lateral or medial sliding of the tibia across the trochlear surface. The ankle joint permits dorsiflexion/plantar flexion. The medial **deltoid ligament** and three **lateral ligaments** further stabilize the ankle joint. *(See Figures 8.17 and 8.18.)*

The Joints of the Foot p. 228

- Four types of diarthrotic joints are in the foot: (1) tarsal bone/tarsal bone (**intertarsal joints**, named after the participating bone), plane diarthrosis; (2) tarsal bone/metatarsal bone (**tarsometatarsal joints**), plane diarthrosis; (3) metatarsal bone/phalanx (**metatarsophalangeal joints)**, condylar diarthrosis, permitting flexion/extension and adduction/abduction; and (4) phalanx/phalanx (**interphalangeal joints**), hinge diarthrosis, permitting flexion/extension. *(See Figures 8.17 and 8.18, Table 8.4, and Clinical Note on p. 229.)*

8.11 | Aging and Joints p. 231

- Problems with joint function are relatively common, especially in older people. **Rheumatism** is a general term for pain and stiffness affecting joints or other parts of the skeletal system, the muscular system, or both; several major forms exist. **Arthritis** is inflammation of the joints and is a type of rheumatic disease that affects synovial joints. Both conditions become increasingly common with age.

Chapter Review

For answers, see the blue Answers tab at the back of the book.

Level 1 Reviewing Facts and Terms

Match each numbered item with the most closely related lettered item.

1. no movement ______
2. synovial ______
3. increased angle ______
4. bursae ______
5. palm facing anteriorly ______
6. digging in heels ______
7. fibrous cartilage ______
8. carpus ______
9. menisci ______

(a) wrist joint
(b) dorsiflexion
(c) fluid-filled pockets
(d) diarthrosis
(e) knee
(f) intervertebral discs
(g) supination
(h) extension
(i) synarthrosis

10. The function of a bursa is to
    (a) reduce friction between a bone and a tendon.
    (b) absorb shock.
    (c) smooth the surface outline of a joint.
    (d) both a and b.
11. Which of the following is *not* a function of synovial fluid?
    (a) absorb shocks
    (b) increase osmotic pressure within joint
    (c) lubricate the joint
    (d) provide nutrients
12. Match each diagram to the right to the following terms:
    - plane joint
    - hinge joint
    - pivot joint
    - saddle joint
    - condylar joint
    - ball-and-socket joint

(a) __________ (d) __________
(b) __________ (e) __________
(c) __________ (f) __________

8

13. Which of the following ligaments is not associated with the hip joint?
    (a) iliofemoral ligament
    (b) pubofemoral ligament
    (c) ligament of the femoral head
    (d) ligamenta flava

14. Label the following structures on the diagram of a frontal section through the right shoulder joint (anterior view).
    - acromioclavicular ligament
    - subdeltoid bursa
    - coraco-acromial ligament
    - coracoid process

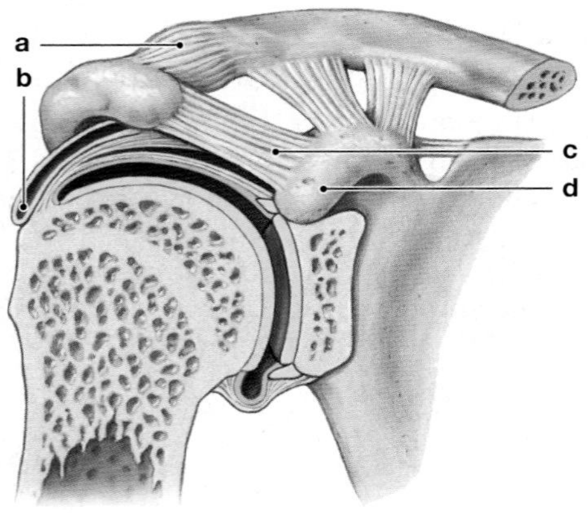

(a) __________
(b) __________
(c) __________
(d) __________

15. A twisting motion of the foot that turns the sole inward is
    (a) dorsiflexion.
    (b) eversion.
    (c) inversion.
    (d) protraction.

16. The ligaments that limit the anterior and posterior movement of the femur and maintain the alignment of the femoral and tibial condyles are the _______ ligaments.
    (a) cruciate
    (b) fibular collateral
    (c) patellar
    (d) tibial collateral

Level 2 Reviewing Concepts

1. Compare and contrast the strength and stability of a joint with respect to the amount of mobility in the joint.
2. How does the classification of a joint change when an epiphysis fuses at the ends of a long bone?
3. How do the malleoli of the tibia and fibula function to retain the correct positioning of the tibiotalar joint?
4. How do articular cartilages differ from other cartilages in the body?
5. What factors limit the range of motion of a mobile diarthrosis?
6. What role do capsular ligaments play in a complex synovial joint? Use the humero-ulnar joint to illustrate your answer.
7. What common mechanism holds together immovable joints such as skull sutures and the gomphoses, holding teeth in their alveoli?
8. How can pronation be distinguished from circumduction of a skeletal element?
9. What would you tell your grandfather about his decrease in height as he grows older?

Level 3 Critical Thinking

1. When a person involved in an automobile accident suffers from "whiplash," what structures have been affected? What movements could be responsible for this injury?
2. Almost all football knee injuries occur when the player has the foot "planted" and extended rather than flexed. What anatomical facts would account for that?

CLINICAL CASE WRAP-UP

Why Does My Knee Hurt So Much?

Molly is suffering from tenosynovitis of the tendon of one of the hamstring muscles—the semimembranosus. Tenosynovitis is an inflammation of the lining of the synovial tendon sheath surrounding a tendon. This inflammation causes all of Molly's symptoms. Her key signs and symptoms—"clicking" and the feeling that something is "catching"—are indicators that the tendon is rolling over the femoral condyle of the knee joint when she flexes and extends her leg.

Physical therapists emphasize the difference between tendonitis and tenosynovitis. The key symptom of tendonitis is pain at the point of tendon attachment to the bone, whereas that of tenosynovitis is pain along the tendon for an inch or more.

1. What is a synovial tendon sheath?
2. The tendon of the semimembranosus muscle helps form what medial structure of the knee joint?

See the blue Answers tab at the back of the book.

Related Clinical Terms

Bouchard's nodes: Bony enlargements on the proximal interphalangeal joints due to osteoarthritis.

chondromalacia: Softening of cartilage as a result of strenuous activity or an overuse injury.

Heberden's nodes: Bony overgrowths on the distal interphalangeal joints due to osteoarthritis that cause the patient to have knobby fingers.

joint mice:–Small fibrous, cartilaginous, or bony loose bodies in the synovial cavity of a joint.

pannus:–Granulation tissue (combination of fibrous connective tissue and capillaries), forming within a synovial membrane, which releases cartilage-destroying enzymes.

tophi: Deposits of uric acid crystals often found around joints and usually associated with gout.

9

The Muscular System

Skeletal Muscle Tissue and Muscle Organization

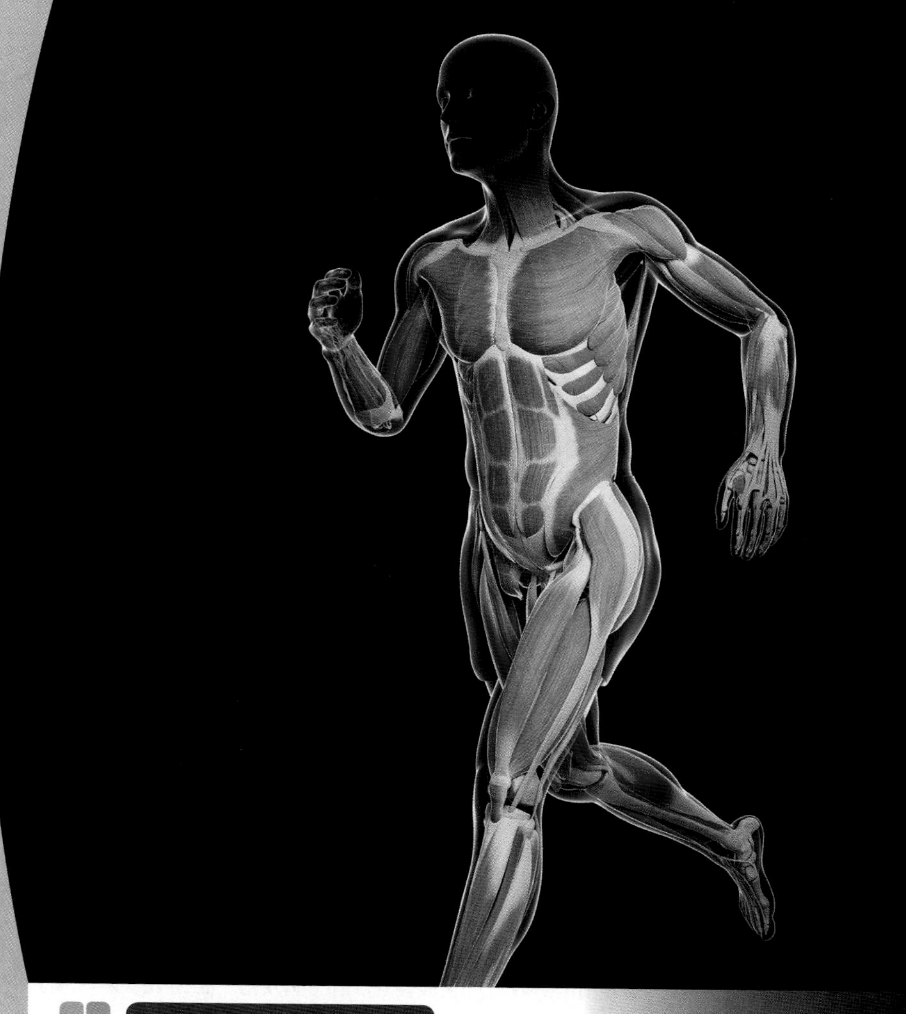

Learning Outcomes

These Learning Outcomes correspond by number to this chapter's sections and indicate what you should be able to do after completing the chapter.

- **9.1** List five functions of skeletal muscle. p. 236
- **9.2** Compare and contrast the gross and microscopic anatomy of a skeletal muscle. p. 236
- **9.3** List and explain the steps involved in the contraction and relaxation of a skeletal muscle fiber. p. 243
- **9.4** Define a motor unit and explain how a motor unit is controlled by a single motor neuron. p. 247
- **9.5** Compare and contrast fast, slow, and intermediate skeletal muscle fibers. p. 248
- **9.6** Describe the arrangement of muscle fibers in parallel, convergent, pennate, and circular skeletal muscles. p. 249
- **9.7** Explain how a term may indicate the action of a muscle, the specific region of the body where a muscle might be found, or the structural characteristics of that muscle. p. 251
- **9.8** Explain how levers and pulleys are used to describe how a skeletal muscle produces movement. p. 253
- **9.9** Describe the effects of aging on skeletal muscles. p. 253

CLINICAL CASE

A Case of Asymmetrical Development

Abdul, a 26-year-old Afghani refugee, just moved to Michigan from a Pakistani refugee camp. Abdul's host family brings him to their physician for a physical examination and immunizations. This is the first time in his life that Abdul has seen a doctor.

Abdul's medical history includes an early childhood illness followed by a long recovery. Afterward, he was unable to walk for over a year, and he continues to have significant weakness in his left lower limb. When he was a young teen, Abdul broke his left tibia, and it took a long time to heal. As he grew, the musculature of his left pelvis, thigh, and leg became thinner and weaker than on his right side. In addition, his left lower limb is shorter than the right. His gait is described as a "short leg limp." Because of the difference in the length of his lower limbs, Abdul has scoliosis of his vertebral column. He has no reflexes in his left lower limb, but he has normal sensation.

What could have caused the asymmetrical muscular development that has persisted since Abdul's early childhood? To find out, turn to the Clinical Case Wrap-Up on p. 258.

IT IS HARD TO IMAGINE what life would be like without muscle tissue. We couldn't sit, stand, walk, speak, or grasp objects. Blood would not circulate because the heart couldn't propel it through the vessels. The lungs couldn't empty and fill, nor could food move along the digestive tract. This is not to say that all life depends on muscle tissue. There are large organisms that get by very nicely without it—we call them plants.

Many of *our* physiological processes, however, and virtually all our dynamic interactions with the environment, involve muscle tissue. Muscle tissue is one of the four primary tissue types. There are three types of muscle tissue: **skeletal muscle**, **cardiac muscle**, and **smooth muscle**. ⮌ pp. 74–75 Skeletal muscle tissue moves the body by pulling on bones of the skeleton, making it possible for us to walk, dance, or play a musical instrument. Cardiac muscle tissue pushes blood through the blood vessels of the cardiovascular system; smooth muscle tissue pushes fluid and solids along the digestive tract and performs varied functions in other systems. These muscle tissues share four properties:

9

1. Excitability: The ability to respond to stimulation. For example, skeletal muscles respond to stimulation by the nervous system, and some smooth muscles respond to circulating hormones.
2. Contractility: The ability to shorten actively and exert a pull or tension that is harnessed by connective tissues.
3. Extensibility: The ability to contract over a range of resting lengths. For example, a smooth muscle cell can be stretched to several times its original length and still contract when stimulated.
4. Elasticity: The ability of a muscle to return to its original length after a contraction.

This chapter focuses on skeletal muscle tissue. Cardiac muscle tissue is discussed along with heart anatomy in Chapter 21. Smooth muscle tissue is discussed in context of the digestive system in Chapter 25.

Skeletal muscles are organs composed mainly of skeletal muscle, but they also contain parts of all four tissue types. The **muscular system** of the human body has more than 700 skeletal muscles and includes all the skeletal muscles that are under voluntary control. This chapter discusses the functions, gross anatomy, microanatomy, and organization of skeletal muscles, as well as muscle terminology. The muscular system is also discussed in the next two chapters:

- Chapter 10 discusses the gross anatomy of the axial musculature—skeletal muscles associated with the axial skeleton.
- Chapter 11 discusses the gross anatomy of the appendicular musculature—skeletal muscles associated with the appendicular skeleton.

9.1 | Functions of Skeletal Muscle

▶ **KEY POINT** Skeletal muscle has five functions, all of which are essential for normal functioning of the human body.

Skeletal muscles are contractile organs directly or indirectly attached to bones of the skeleton. Skeletal muscles have the following functions:

- Produce skeletal movement: Muscle contractions pull on tendons and move the bones of the skeleton. The effects range from simple motions, such as extending the arm, to the highly coordinated movements of swimming, skiing, or texting.
- Maintain posture and body position: Skeletal muscle contraction maintains body posture. Without constant muscular contraction, we could not sit upright without collapsing or stand without falling over.
- Support soft tissues: The abdominal wall and the floor of the pelvic cavity contain layers of skeletal muscle. These muscles support the visceral organs and protect internal tissues from injury.
- Regulate the entry and exit of material: Skeletal muscles encircle the openings, or **orifices**, of the digestive and urinary tracts. These muscles provide voluntary control over swallowing, defecation, and urination.
- Maintain body temperature: Muscle contractions require energy, and some of that energy is converted to heat. This heat released by contracting muscles helps maintain the body's normal temperature.

9.1 CONCEPT CHECK ✔

1 What is shivering, and why do you do it when you are cold?

See the blue Answers tab at the back of the book.

9.2 | Anatomy of Skeletal Muscles

▶ **KEY POINT** To understand how skeletal muscle moves parts of the body, you must understand the gross and microscopic anatomy of skeletal muscles.

Gross Anatomy

▶ **KEY POINT** Skeletal muscle is an organ of the musculoskeletal system. The role of skeletal muscle is to to create a pulling force and enable body movement.

Our study of the gross anatomy of muscle begins with a description of the connective tissues that attach skeletal muscles to other structures.

Connective Tissue

Each skeletal muscle has three concentric layers, or wrappings, of connective tissue: an outer epimysium, a central perimysium, and an inner endomysium (**Figure 9.1**).

- The **epimysium** (ep-i-MIS-ē-um; *epi-*, on, + *mys*, muscle) is a layer of dense irregular connective tissue surrounding the entire skeletal muscle. The epimysium separates the muscle from surrounding tissues and organs and is connected to the deep fascia. ⮌ pp. 65–66, 73
- The connective tissue fibers of the **perimysium** (per-i-MIS-ē-um; *peri-*, around) divide the muscle into internal compartments. Each compartment contains a bundle of muscle fibers called a **fascicle** (FAS-i-kul; *fasciculus*, bundle). The perimysium contains collagen and elastic fibers, and numerous blood vessels and nerves supply each fascicle.
- The **endomysium** (en-dō-MIS-ē-um; *endo-*, inside, + *mys*, muscle) surrounds each skeletal muscle fiber (individual skeletal muscle cell), binds each muscle fiber to its neighbor, and supports the capillaries that supply the individual fiber. The endomysium consists of a delicate network of reticular fibers. Scattered **myosatellite cells** that lie between the endomysium and the muscle fibers are stem cells that repair damaged muscle tissue.

Tendons and Aponeuroses At each end of the muscle, the collagen fibers of the epimysium, perimysium, and endomysium come together and form a **tendon** that attaches the muscle to bone, cartilage, skin, or another

Figure 9.1 Structural Organization of Skeletal Muscle. A skeletal muscle consists of bundles of muscle fibers (fascicles) enclosed within a connective tissue sheath, the epimysium. Each fascicle is then ensheathed by the perimysium, and within each fascicle the individual muscle fibers are surrounded by the endomysium. Each muscle fiber has many nuclei as well as mitochondria and other organelles seen here and in Figure 9.3.

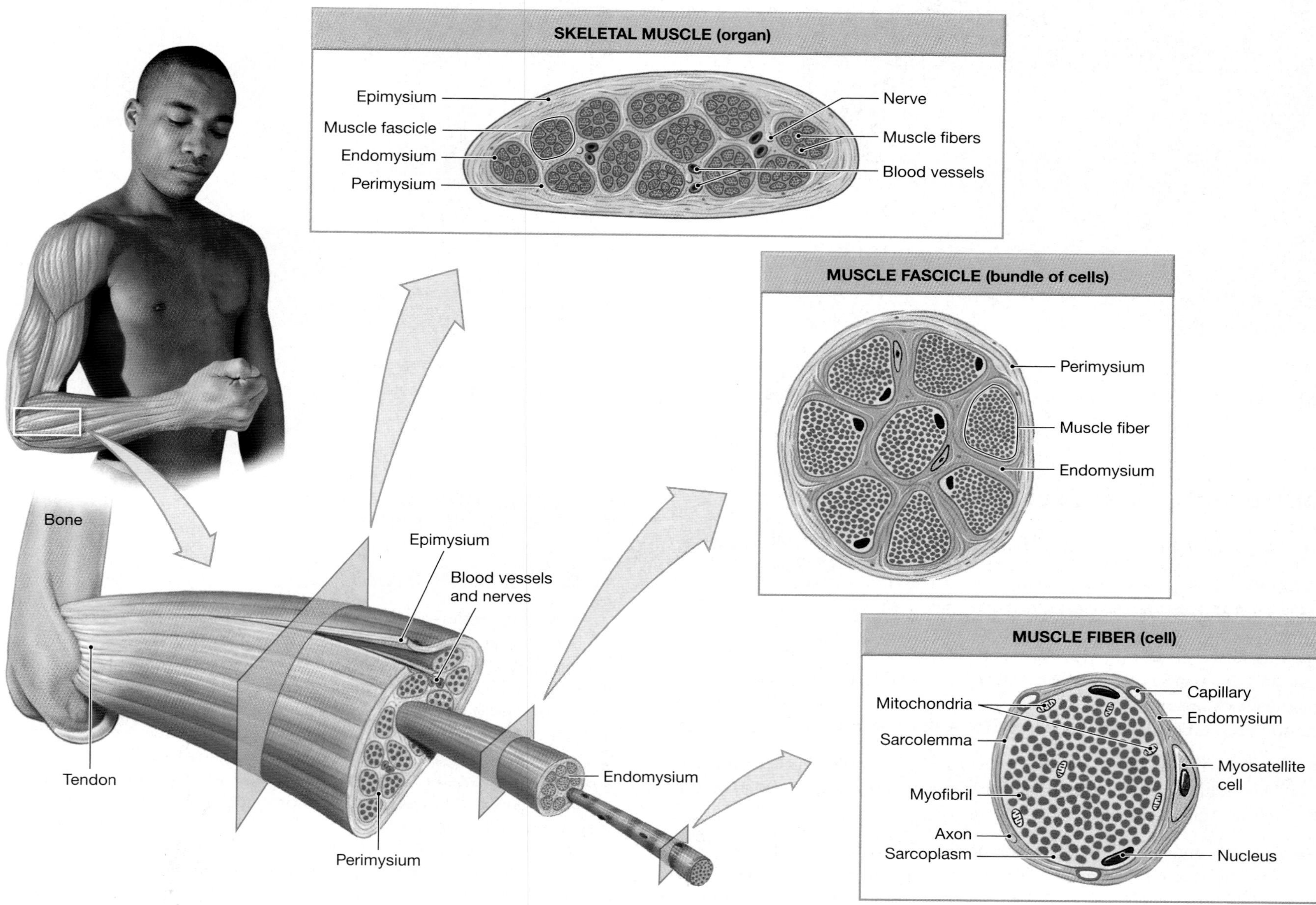

muscle. Tendons that form thick, flattened sheets are called **aponeuroses**. (The anatomy of tendons and aponeuroses was discussed in Chapter 3. ⮌ p. 65)

The collagen fibers of the tendon are continuous with the periosteum and matrix of bone to which it attaches, making an extremely strong bond, so that any contraction of the muscle pulls on the attached bone.

Nerves and Blood Vessels

Nerves and blood vessels supplying the muscle fibers lie within the connective tissues of the epimysium, perimysium, and endomysium. Skeletal muscles are also called *voluntary muscles* because their contractions are consciously controlled. The nervous system provides this control. Nerves, which are bundles of axons, enter the epimysium, branch through the perimysium, and enter the endomysium to attach to individual muscle fibers. Chemical communication between a neuron and a skeletal muscle fiber occurs at a site called the **neuromuscular junction (NMJ)** (Figure 9.2). This is the area made up of an axon terminal of a neuron, a specialized region of the plasma membrane called the motor end plate, a narrow space in between called the synaptic cleft, and the membrane of the muscle fiber. Each muscle fiber has one neuromuscular junction, usually located midway along its length. At the NMJ, the axon terminal of the neuron attaches to the motor end plate of the skeletal muscle fiber. The motor end plate is a specialized area where the axon of a motor neuron establishes synaptic contact with a skeletal muscle fiber.

Muscle contraction requires tremendous quantities of energy, and the blood vessels deliver the oxygen and nutrients needed to produce energy in the form of ATP in skeletal muscles. These blood vessels often enter the epimysium alongside the nerves and follow the same branching pattern through the perimysium. Within the endomysium the arteries supply a large capillary network around each muscle fiber. These capillaries are coiled rather than straight, so they can withstand changes in the length of the muscle fiber.

Figure 9.2 **Skeletal Muscle Innervation.** Each skeletal muscle fiber is stimulated by a nerve fiber at a neuromuscular junction.

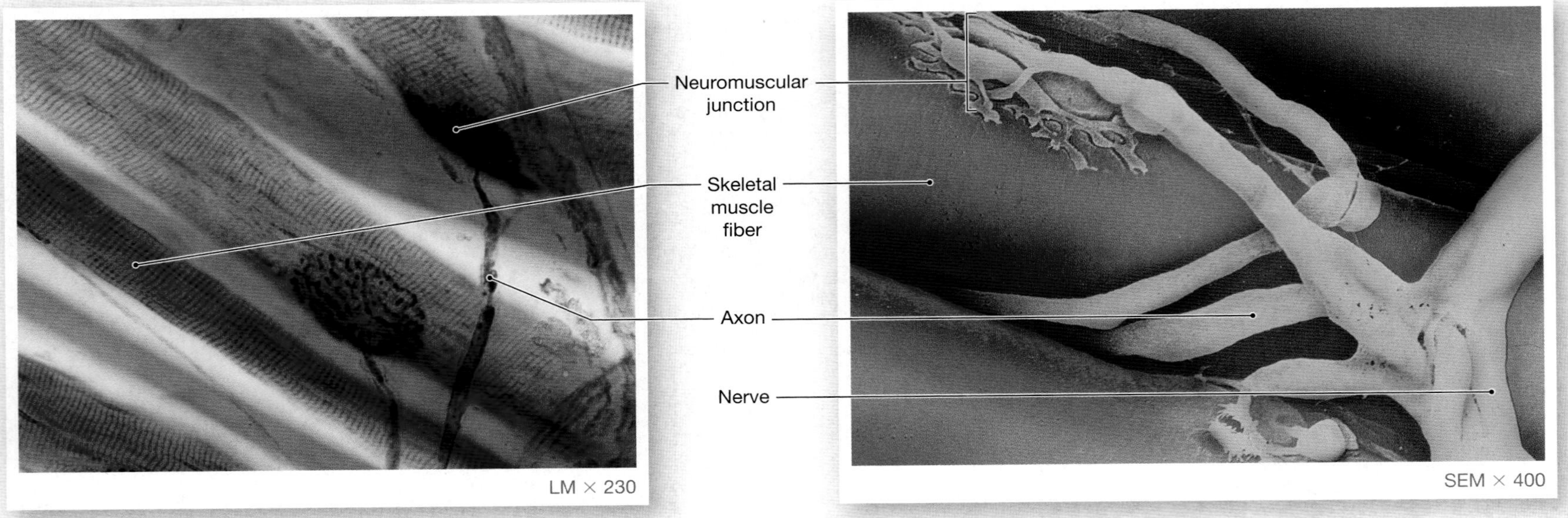

a A neuromuscular junction seen on a muscle fiber of this fascicle

b Colorized SEM of a neuromuscular junction

9

Microanatomy of Skeletal Muscle Fibers

▶ **KEY POINT** Like other cells within the body, skeletal muscle fibers are enclosed by a plasma membrane and contain a large number of organelles. However, the names given to these structures differ from those of other cells. That is, the word part *sarco–*, meaning flesh, is often part of the term.

In a skeletal muscle cell, the plasma membrane is called the **sarcolemma** (sar-kō-LEM-a; *sarkos*, flesh, + *lemma*, husk). Within the sarcolemma is the cytoplasm, which in a muscle cell is called the **sarcoplasm** (SAR-kō-plazm). Skeletal muscle fibers differ in several ways from the "typical" cell described in Chapter 2:

- Skeletal muscle fibers are large compared with the cells of other tissues. A fiber in a leg muscle could have a diameter of 100 μm and a length equal to that of the entire muscle (30–40 cm, or 12–16 in.).
- Skeletal muscle fibers are multinucleate (containing more than one nucleus). During development, groups of embryonic cells called **myoblasts** fuse to form individual skeletal muscle fibers (**Figure 9.3a**). Each skeletal muscle fiber contains hundreds of nuclei deep to the sarcolemma (**Figure 9.3b,c**). This characteristic distinguishes skeletal muscle fibers from cardiac and smooth muscle fibers. Some myoblasts do not fuse with developing muscle fibers, but remain in adult skeletal muscle tissue as stem cells called **myosatellite cells** (**Figures 9.1** and **9.3a**). When a skeletal muscle is injured, these stem cells differentiate and assist in repairing and regenerating the muscle.
- Deep indentations in the sarcolemmal surface form a network of narrow tubules called **transverse tubules**, or **T tubules**, that extend into the sarcoplasm. The sarcolemma and these T tubules conduct electrical impulses, called action potentials, to stimulate muscle fiber contraction.

Myofibrils and Myofilaments

▶ **KEY POINT** Skeletal muscle cells feature intracellular myofibrils that extend the length of the cell. The organization of the myofibrils and their molecular components gives skeletal muscle a striated (striped) appearance when viewed through a microscope.

CLINICAL NOTE

Fibromyalgia

Fibromyalgia (*–algia*, pain) is a disorder characterized by widespread, chronic (more than three months) musculoskeletal pain usually accompanied by fatigue and disturbances in sleep, memory, and mood. Anxiety, depression, tension headaches, temporomandibular joint disorders, and irritable bowel syndrome often accompany fibromyalgia. "Fibro fog" is a descriptive term for the cognitive difficulties that may accompany this disorder. Women are affected more frequently than men. Evidence suggests that environmental factors, physical or emotional stress, and genes play a role in its development. The American College of Rheumatology has described nine paired trigger points (sites of reproducible, significant tenderness) that indicate fibromyalgia. All laboratory and radiographic studies are normal in fibromyalgia, differentiating it from other rheumatologic diseases.

The sarcoplasm of a skeletal muscle fiber contains hundreds to thousands of fine cylindrical fibers called **myofibrils**. Each myofibril is 1–2 μm in diameter and as long as the entire cell (**Figure 9.3c,d**). The active shortening of myofibrils is responsible for skeletal muscle fiber contraction.

Myofibrils are made of protein filaments called **myofilaments**. There are two types: thin filaments and thick filaments. Each contains several different types of proteins, each with a specific function during muscle cell contraction and relaxation. For example, **actin** and **myosin** are the contractile proteins in thin filaments and thick filaments, respectively. Other proteins include the regulatory proteins tropomyosin and troponin, while titin and nebulin are the accessory proteins.

Surrounding each myofibril is the **sarcoplasmic reticulum (SR)**, a membrane complex similar to the smooth endoplasmic reticulum of other cells (**Figure 9.3d**). The SR is the storage and release site of calcium ions and plays an essential role in controlling individual myofibril contraction. On each side of a T (*transverse*) tubule, the tubules of the SR enlarge, fuse, and form

Figure 9.3 The Formation and Structure of a Skeletal Muscle Fiber

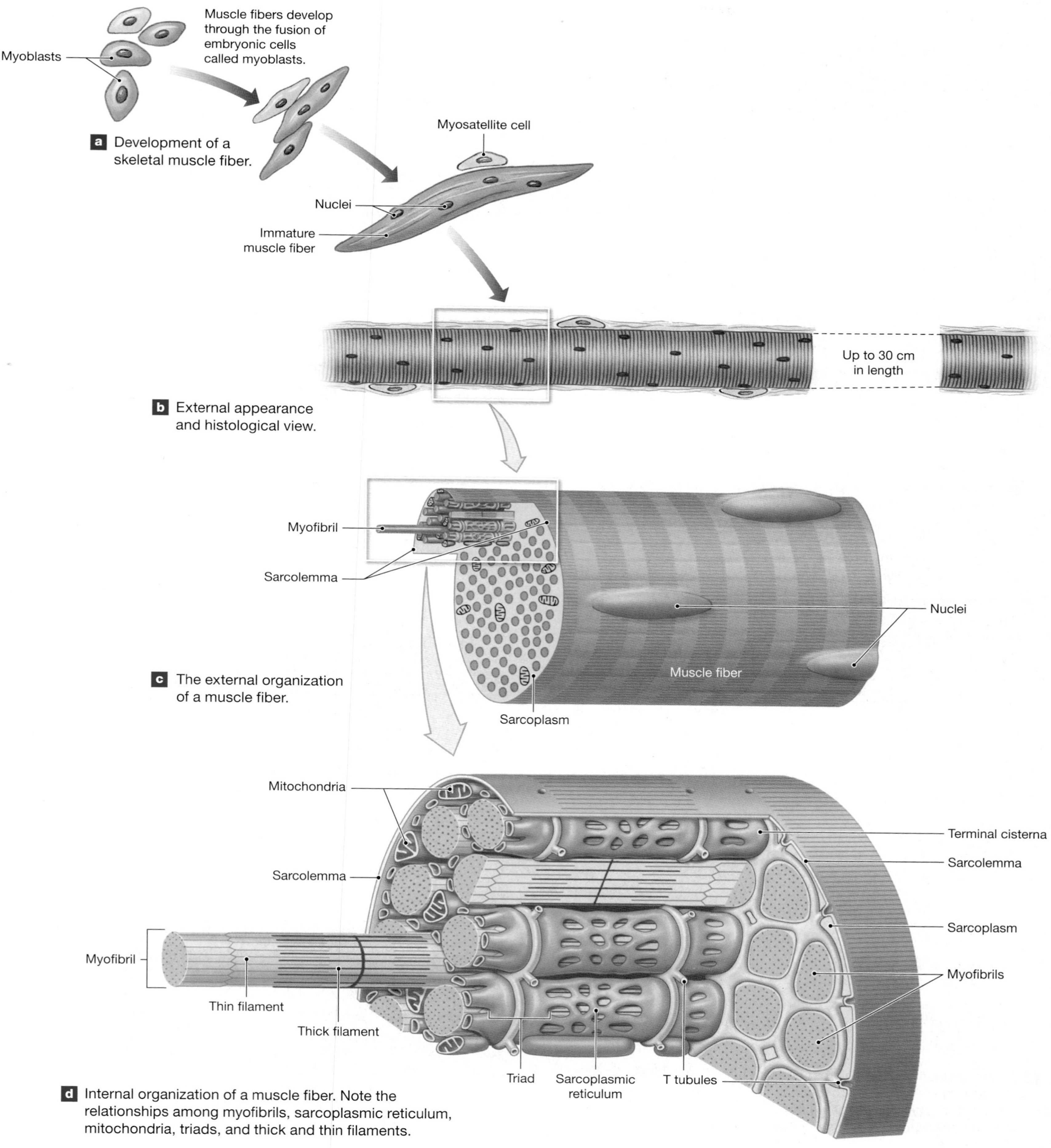

a Development of a skeletal muscle fiber.

b External appearance and histological view.

c The external organization of a muscle fiber.

d Internal organization of a muscle fiber. Note the relationships among myofibrils, sarcoplasmic reticulum, mitochondria, triads, and thick and thin filaments.

9

Figure 9.4 Sarcomere Structure

Sarcomere
Myofibril
Thin filament
Thick filament
Actinin filaments
Attachment of titin
Titin filament
Z line
I band
M line
H band
Zone of overlap

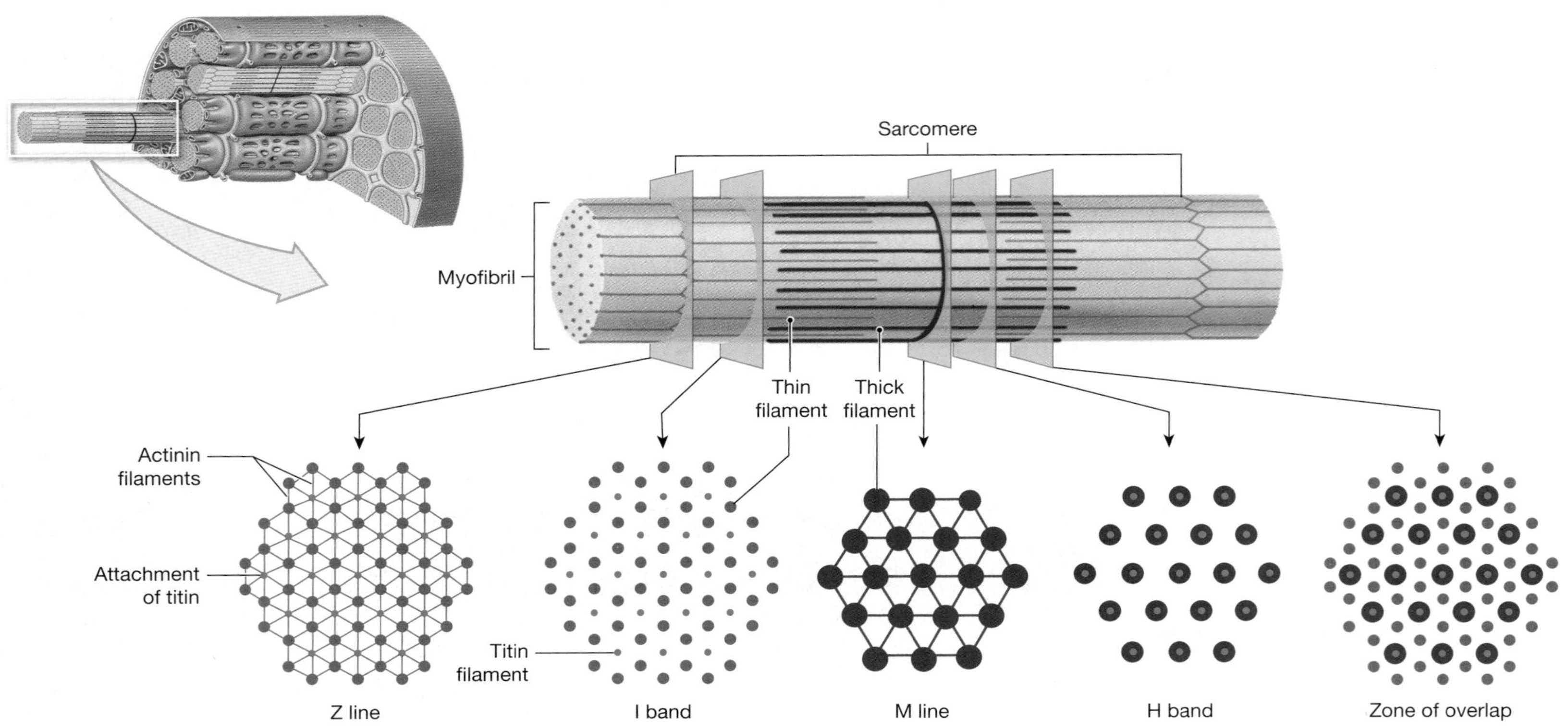

a The arrangement of thin and thick filaments within a sarcomere and cross-sectional views of each region of the sarcomere

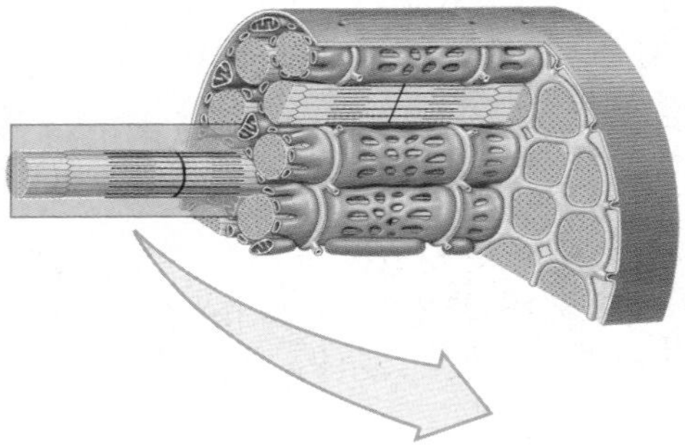

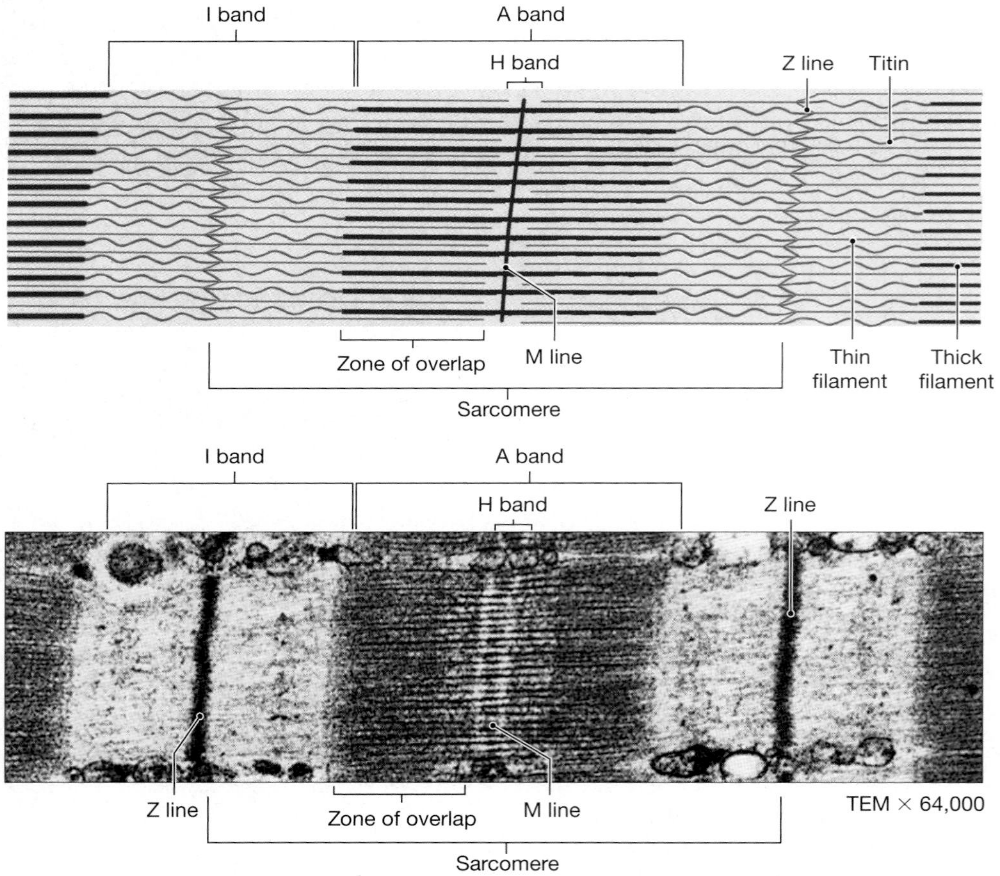

I band
A band
H band
Z line
Titin
Zone of overlap
M line
Thin filament
Thick filament
Sarcomere
I band
A band
H band
Z line
Z line
Zone of overlap
M line
Sarcomere
TEM × 64,000

b A transmission electron microscope image of a sarcomere in the gastrocnemius of the calf and a diagram showing the various components of this sarcomere expanded chambers called **terminal cisternae**. The combination of a pair of terminal cisternae plus a T tubule is known as a **triad**. The membranes of the triad are in close contact and tightly bound together, but there is no direct connection between them **(Figure 9.3c,d)**.

Mitochondria and glycogen granules are scattered among the myofibrils. Mitochondrial activity and the chemical breakdown of glycogen provide the ATP needed to power muscular contractions. A skeletal muscle fiber has hundreds of mitochondria, more than other cells in the body.

Sarcomere Organization

Myofibrils are organized in repeating units called **sarcomeres** (SAR-kō-mērz; *sarkos*, flesh, + *meros*, part). Sarcomeres are the smallest functional units of muscle fibers. Differences in the size, density, and distribution of the thin and thick filaments give the sarcomere a banded appearance (see **Figures 9.2** and **9.3**).

Figure 9.4 shows the structure of an individual sarcomere. The dark bands are called **A bands** and the light bands are called **I bands**. These names are derived from the terms *anisotropic* (A band) and *isotropic* (I band), which refer to their appearance when viewed using polarized light microscopy. The thick filaments are at the center of each sarcomere, in the A band. The A band contains the M line, H band, and zone of overlap.

The **M line** is the center of the A band; the M stands for *middle*. Proteins of the M line connect the central portion of each filament to the neighboring thick filaments. M lines stabilize the positions of the thick filaments.

The **H band** is the lighter region on each side of the M line. The H band contains thick filaments, but no thin filaments.

The **zone of overlap** is the dark region where thin filaments are found between the thick filaments. Here, three thick filaments surround each thin filament, and six thin filaments surround each thick filament. Two tubules encircle each sarcomere, and the triads containing them are found in the zones of overlap. As a result, calcium ions released by the SR enter the area where thin and thick filaments interact.

The I band is the region of the sarcomere that contains thin filaments but no thick filaments. The I band extends from the A band of one sarcomere to the A band of the next sarcomere. **Z lines**, or *Z discs*, bisect the I bands and mark the boundary between adjacent sarcomeres. Z lines are made up of proteins called **actinin**, which interconnect thin filaments of adjacent sarcomeres. Strands of the elastic protein **titin** extend from the tips of the thick filaments to the attachment sites at the Z line. **Figure 9.5** reviews the levels of organization we have discussed so far.

Thin Filaments Each **thin filament** is a twisted strand 5–6 nm in diameter and 1 μm long **(Figure 9.6a,b)**. A single thin filament contains four proteins: F-actin, nebulin, tropomyosin, and troponin. Filamentous actin, or **F-actin**, is a twisted strand composed of two rows of 300–400 globular molecules of **G-actin**. G-actin is the globular (G) subunit of the actin molecule. A slender strand of the protein **nebulin** extends along the F-actin strand in the cleft between the rows of G-actin molecules. Nebulin holds the F-actin strand together. Each molecule of G-actin contains an **active site** where myosin in the thick filaments can bind.

A thin filament also contains the regulatory proteins **tropomyosin** (trō-pō-MĪ-ō-sin) and **troponin** (TRŌ-pō-nin; *trope*, turning). Tropomyosin molecules form a long chain that covers the active sites on G-actin, preventing actin-myosin interaction. Troponin holds the tropomyosin strand in place. Before a contraction can begin, the troponin molecules must change position, moving the tropomyosin molecules and exposing the active sites. (This mechanism is discussed in a later section.)

Thick Filaments Each **thick filament** is 10–12 nm in diameter and 1.6 μm long and is composed of a bundle of **myosin** molecules, each made up of a pair of myosin subunits twisted around one another **(Figure 9.6c)**. The long tail is bound to the other myosin molecules in the thick filament. The free head, with two globular protein subunits, projects outward toward the nearest thin filament. When the myosin heads interact with thin filaments during a contraction they are known as **cross-bridges**.

Figure 9.5 Levels of Functional Organization in a Skeletal Muscle Fiber

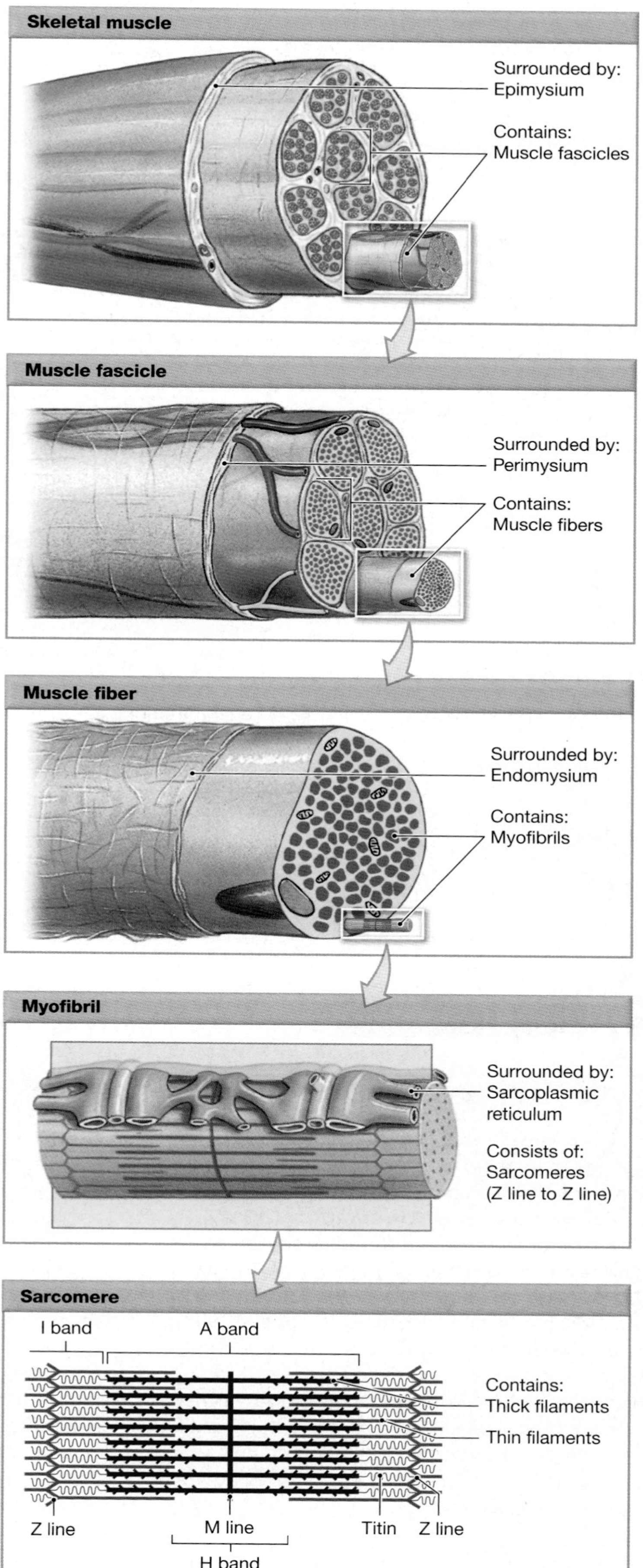

Figure 9.6 Thin and Thick Filaments. Myofilaments are bundles of thin and thick filament proteins.

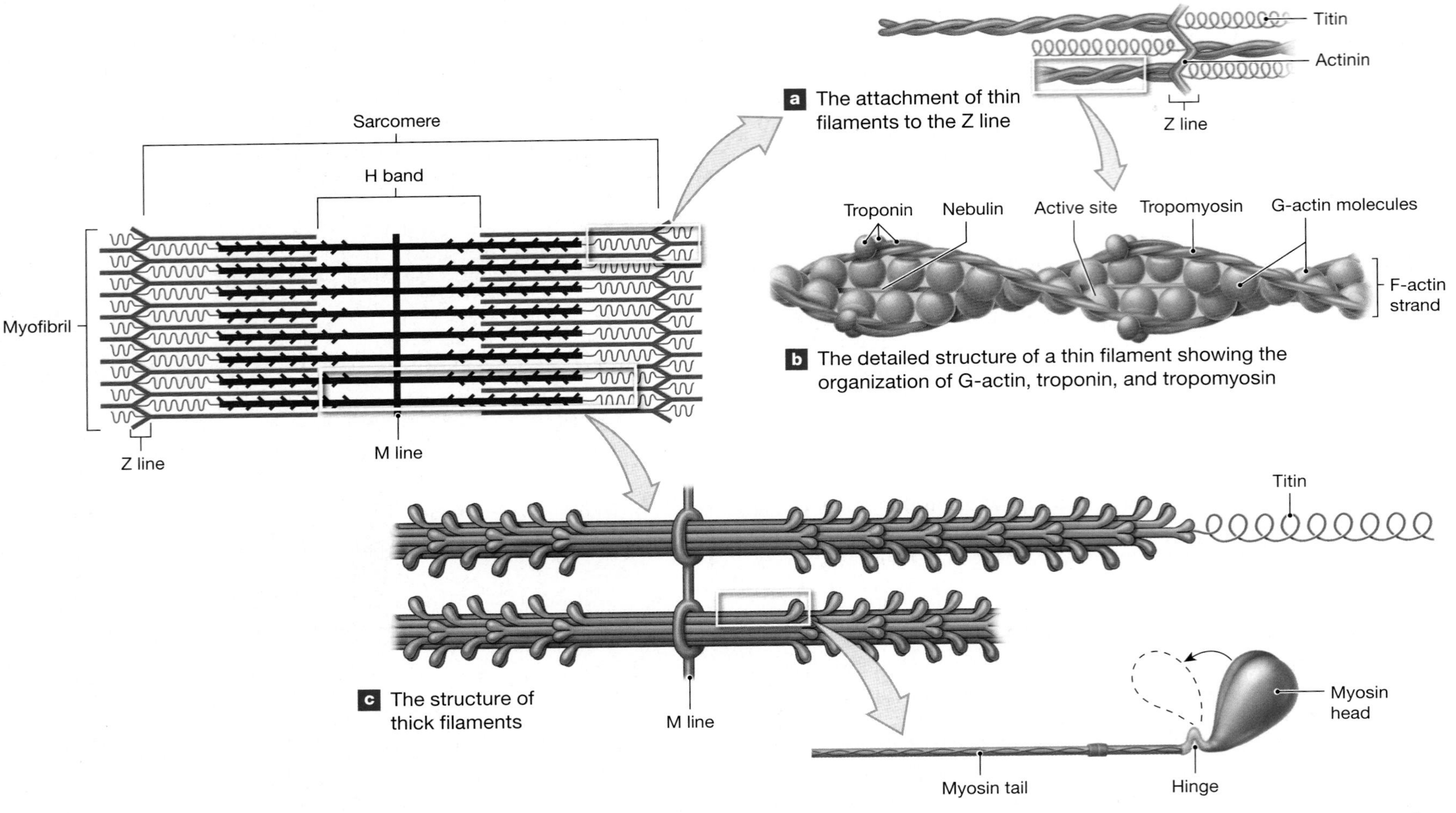

Titin
Actinin
Z line

a The attachment of thin filaments to the Z line

Sarcomere
H band
Myofibril
Z line
M line

Troponin
Nebulin
Active site
Tropomyosin
G-actin molecules
F-actin strand

b The detailed structure of a thin filament showing the organization of G-actin, troponin, and tropomyosin

Titin
M line

c The structure of thick filaments

Myosin head
Myosin tail
Hinge

d A single myosin molecule detailing the structure and movement of the myosin head after cross-bridge binding occurs

CLINICAL NOTE

Delayed-Onset Muscle Soreness

You have probably experienced muscle soreness the day *after* a rigorous physical workout. This common phenomenon is known as **delayed-onset muscle soreness (DOMS)**. This type of soreness is different from the acute soreness that develops during activity. The pain of DOMS begins within hours of the activity, peaks at 24–72 hours, and eventually disappears within five to seven days. DOMS occurs most frequently in the first part of an athletic season and can affect both elite and novice athletes. Any activity placing unaccustomed loads on muscles, particularly activities involving eccentric (lengthening) contractions, such as walking or running downhill, can cause DOMS.

During DOMS, levels of creatine phosphokinase (CPK) and myoglobin are elevated in the blood, indicating damage to muscle sarcolemmae. DOMS hampers athletic performance by causing a reduction in joint range of motion, muscle tenderness to touch and pressure, muscle swelling, and reversible loss of muscle power.

Microscopic ruptures at the Z line of the sarcomeres within muscle fibers is thought to contribute to DOMS. In addition, micro trauma to the sarcoplasmic reticulum causes calcium normally stored in the sarcoplasmic reticulum to accumulate in the damaged muscle cell sarcoplasm. This additional calcium may activate enzymes that break down sarcomeres.

Thick filaments have a core of titin (**Figures 9.4a** and **9.6c**). On both sides of the M line, a strand of titin extends the length of the filament and attaches at the Z line. In the resting sarcomere, the titin strands are completely relaxed; they become tense only when some external force stretches the sarcomere. When the sarcomere stretches, the titin strands maintain the normal alignment of the thick and thin filaments. When the tension is removed, the titin fibers help return the sarcomere to its normal resting length.

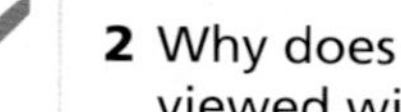

9.2 CONCEPT CHECK

2 Why does skeletal muscle appear striated when viewed with a microscope?

3 What two proteins help regulate the interaction between actin and myosin?

See the blue Answers tab at the back of the book.

9.3 Muscle Contraction

▶ **KEY POINT** The explanation for how skeletal muscle contracts is called the sliding filament theory because the actin and myosin filaments within the sarcomeres slide along each other during contraction. When the sarcomeres of a myofibril shorten, the muscle fiber shortens. When enough muscle fibers shorten, the muscle shortens, exerting a pulling force on the muscle's attachments. If this force is strong enough, movement occurs.

A contracting muscle fiber exerts a pull, or **tension**, and shortens in length. Muscle fiber contraction results from interactions between the thick and thin filaments in each sarcomere. The mechanism for muscle contraction is explained by the sliding filament theory. The trigger for a contraction is the presence of calcium ions (Ca^{2+}), and the contraction itself requires ATP.

The Sliding Filament Theory

▶ **KEY POINT** The sliding filament theory explains the process of skeletal muscle contraction. Contraction of skeletal muscle is seen at both the gross anatomy and microscopic level. At the gross anatomy level, skeletal muscle contraction is obvious when you move a body part. At the microscopic level, it is seen by a change in the size of the sarcomere, the organization of the A band, I band, and H zone, and the distance between the Z lines.

Observation of contracting muscle fibers reveals several physical changes: (1) the H band and I band get smaller, (2) the zone of overlap gets larger, and (3) the Z lines move closer together. However, the width of the A band remains constant throughout the contraction. The **sliding filament theory** explains the physical changes occurring between thick and thin filaments during contraction and is illustrated in **Spotlight Figure 9.7**.

Neural Control of Muscle Fiber Contraction

▶ **KEY POINT** The central nervous system controls skeletal muscle contraction by sending a message to the muscle by a motor neuron. The electrical message that the motor neuron carries to the skeletal muscle cell sets off a series of steps, resulting in skeletal muscle cell contraction.

The sequence of events in the process of skeletal muscle contraction is as follows:

1. Chemicals released by the motor neuron at the neuromuscular junction alter the membrane potential of the sarcolemma. This change sweeps across the surface of the sarcolemma and into the T tubules.
2. The change in the membrane potential of the T tubules triggers the SR to release calcium ions. This release initiates the contraction.

Each skeletal muscle fiber is controlled by a motor neuron whose nucleus is located in the central nervous system in either the brain or the spinal cord. The axon reaches the neuromuscular junction of the muscle fiber (**Figures 9.2, p. 238**, and **9.8**). The expanded tip of the axon at the neuromuscular junction is called the **axon terminal**. The cytoplasm of the axon terminal contains mitochondria and small secretory vesicles, called **synaptic vesicles**. These vesicles contain molecules of the neurotransmitter **acetylcholine** (as-e-til-KŌ-lēn) **(ACh)**. A neurotransmitter is a chemical released by a neuron that communicates with another cell. That communication causes a change in the membrane potential of that cell. A narrow space called the **synaptic cleft** separates the axon terminal from the motor end plate of the skeletal muscle fiber. The synaptic cleft contains the enzyme **acetylcholinesterase (AChE)**, which breaks down molecules of ACh.

An action potential is a sudden change in the membrane potential that travels the length of an axon. The stimulus for ACh release is the arrival of an action potential. When an action potential arrives at the axon terminal, ACh is released into the synaptic cleft. The ACh diffuses across the synaptic cleft and binds to receptor sites on the motor end plate, generating an action potential in the sarcolemma and into each T tubule. Action potentials continue to be generated, one after another, until ACh is removed from the synaptic cleft. This removal occurs in two ways: ACh diffuses away from the synapse or is broken down by AChE.

9

Muscle Contraction: A Summary

Figure 9.9 summarizes the entire sequence of events in muscle contraction from neural activation to relaxation. Key steps are as follows:

1. At the neuromuscular junction ACh released by the axon terminal binds to receptors on the sarcolemma.
2. The resulting change in the membrane potential of the muscle fiber leads to the production of an action potential that spreads across its entire surface and along the T tubules.

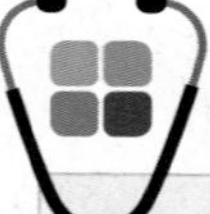

CLINICAL NOTE

Rigor Mortis

When death occurs, circulation ceases and the skeletal muscles are deprived of oxygen and nutrients. Within a few hours, the skeletal muscle fibers run out of ATP, and the sarcoplasmic reticulum becomes unable to remove calcium ions from the sarcoplasm. Calcium ions diffusing into the sarcoplasm from the extracellular fluid or leaking out of the sarcoplasmic reticulum then trigger a sustained contraction. Without ATP, the cross-bridges cannot detach from the myosin active sites, and the muscles lock in the contracted position. All skeletal muscles are involved, beginning in the face and head and progressing caudally. Depending on ambient temperature, this physical state, called **rigor mortis** (*rigor*, stiffness, + *mortis*, death), begins within a few hours after death and peaks at 12 hours. The rigidity lasts until the lysosomal enzymes released by autolysis break down the myofilaments, typically 48–60 hours after death.

SPOTLIGHT

FIGURE 9.7

Sliding Filament Theory

The **tension** produced by a contracting skeletal muscle fiber results from the interaction between the thick and thin filaments within sarcomeres. The mechanism of skeletal muscle contraction is explained by the **sliding filament theory**.

Resting Sarcomere

A resting sarcomere showing the locations of the I band, A band, H band, M line, and Z lines.

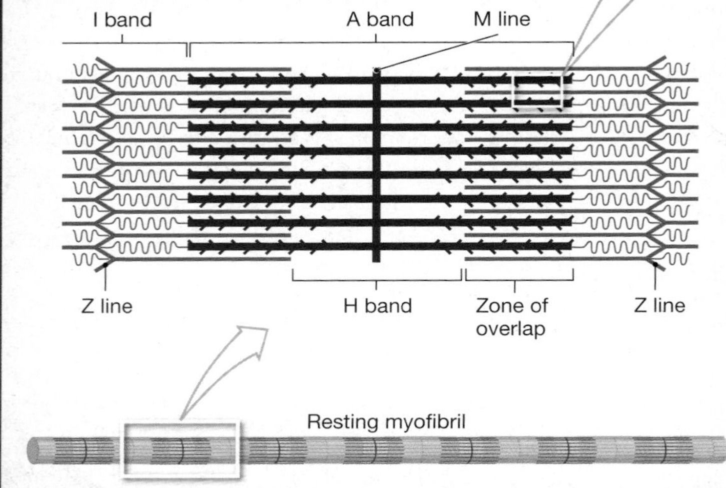

Contracted Sarcomere

After repeated cycles of "bind, pivot, detach, and reactivate," the entire muscle completes its contraction.

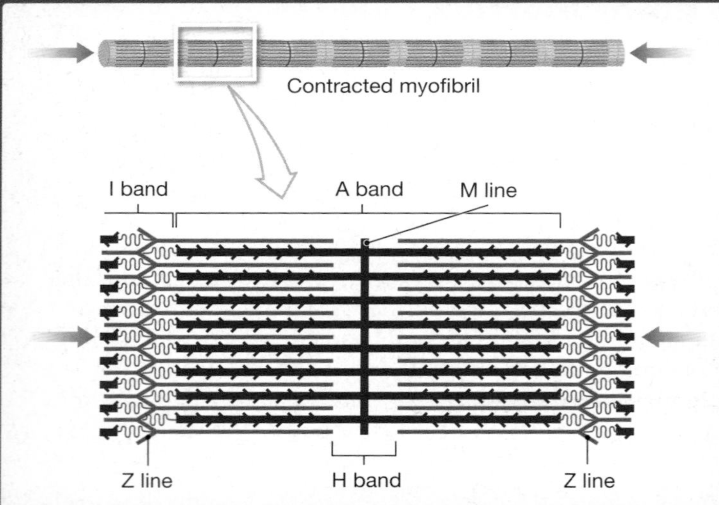

In a contracting sarcomere the A band stays the same width, but the Z lines move closer together and the H band and the I bands get smaller.

1 Contraction Cycle Begins

The contraction cycle involves a series of interrelated steps. The cycle begins with electrical events in the sarcolemma that trigger the release of calcium ions from the terminal cisternae of the sarcoplasmic reticulum (SR). These calcium ions enter the zone of overlap.

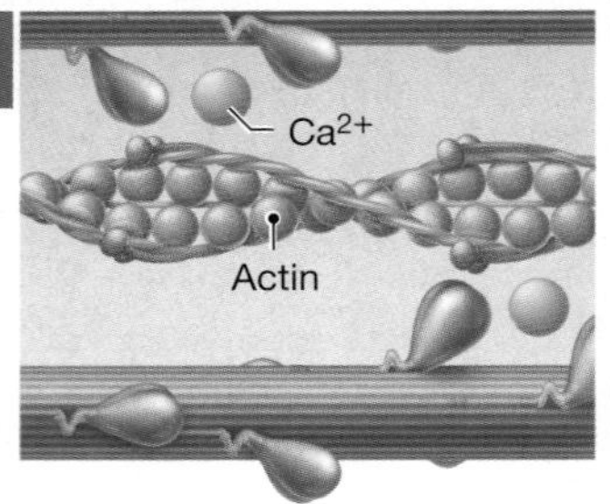

2 Active-Site Exposure

Calcium ions bind to troponin in the troponin–tropomyosin complex. The troponin molecule then changes position, rolling the tropomyosin molecule away from the active sites on actin.

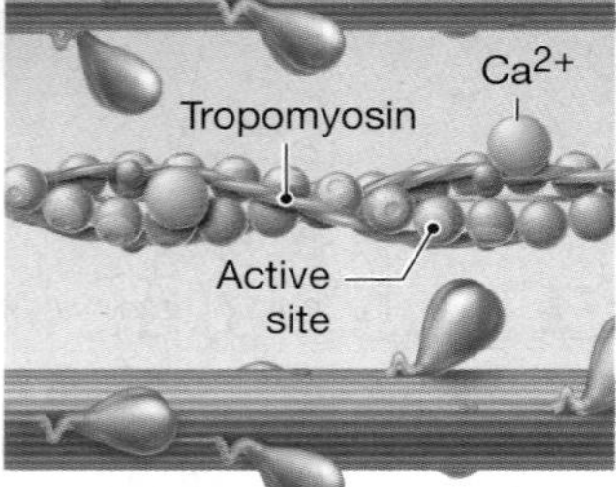

3 Cross-Bridge Formation

Once the active sites are exposed, the myosin heads of adjacent thick filaments bind to them, forming cross-bridges.

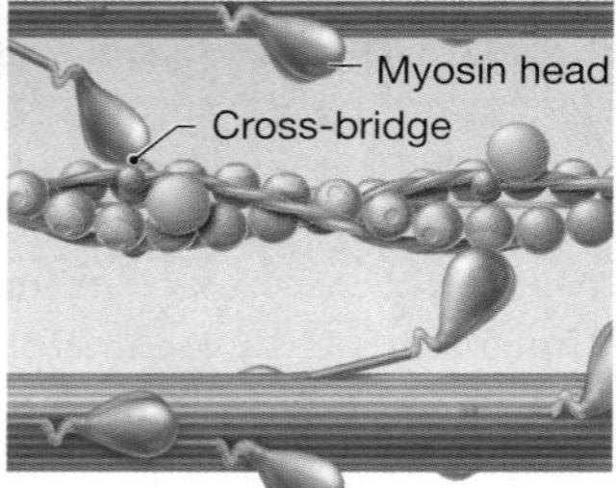

4 Myosin Head Pivoting

After cross-bridge formation, energy is released as the myosin heads pivot toward the M line. This action is called the power stroke.

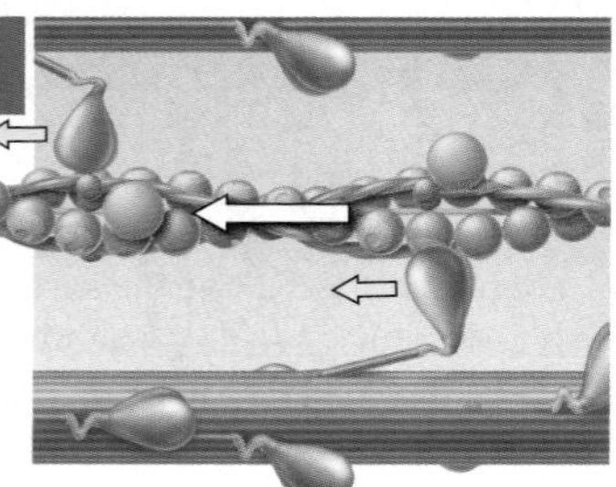

5 Cross-Bridge Detachment

ATP binds to the myosin heads, breaking the cross-bridges between the myosin heads and the actin molecules. The active site is now exposed and able to form another cross-bridge.

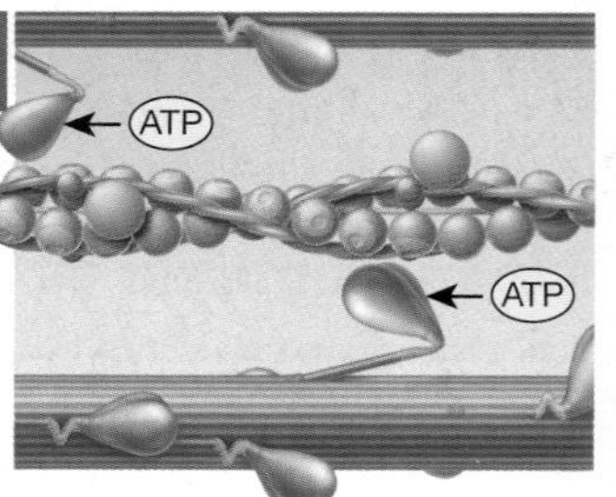

6 Myosin Reactivation

ATP provides the energy to reactivate the myosin heads and return them to their original positions. The entire process is repeated as long as calcium ion concentrations remain elevated and ATP reserves are sufficient.

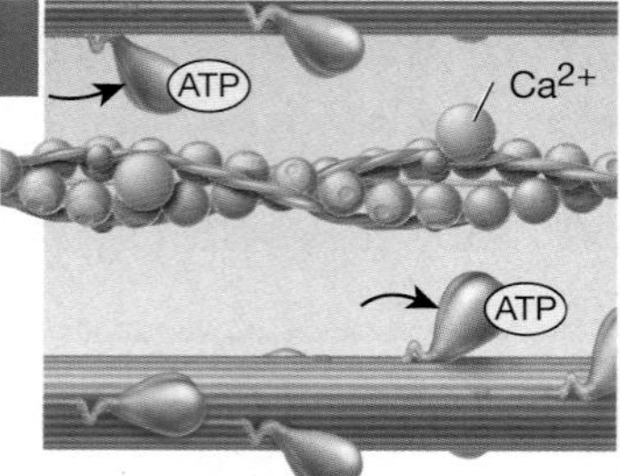

Effect of Sarcomere Length on Tension Production

When many people are pulling on a rope, the amount of tension produced is proportional to the number of people involved. In a muscle fiber, the amount of tension generated during a contraction depends on the number of cross-bridge interactions occurring in the sarcomeres of the myofibrils. The number of cross-bridges is determined by the degree of overlap between thick and thin filaments. Only myosin heads within the zone of overlap can bind to active sites on the actin molecule to produce tension. The tension produced by the muscle fiber is related directly to the structure of an individual sarcomere. If sarcomeres are too short or too long, the tension of a contraction is reduced.

Maximum tension is produced when the zone of overlap is large, but the thin filaments do not extend across the sarcomere's center.

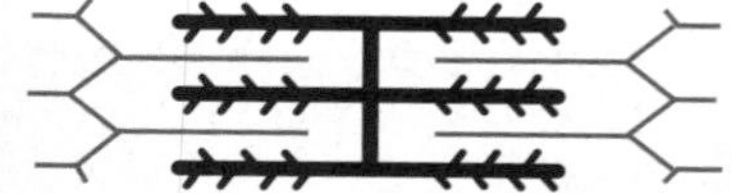

At short resting lengths, thin filaments extending across the center of the sarcomere interfere with the normal orientation of thick and thin filaments, decreasing tension.

An increase in sarcomere length reduces the tension produced by reducing the size of the zone of overlap and the number of potential cross-bridge interactions.

Tension production falls to zero when the thick filaments are pressed against the Z lines and the sarcomere cannot shorten further.

Tension (percent of maximum)

100, 80, 60, 40, 20, 0

1.2 μm | 1.6 μm | 2.6 μm | 3.6 μm

Normal range

Decreased length | Increased sarcomere length

Optimal resting sarcomere length: The normal range of sarcomere lengths in the body is 75 to 130 percent of the optimal length.

When the zone of overlap is reduced to zero, thin and thick filaments cannot interact at all. Under these conditions, the muscle fiber cannot produce any active tension, and a contraction cannot occur. Such extreme stretching of a muscle fiber is normally prevented by the titin filaments in the muscle fiber (which tie the thick filaments to the Z lines) and by the surrounding connective tissues (which limit the degree of muscle stretch).

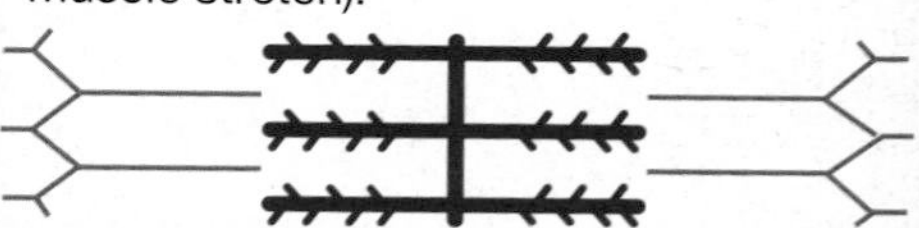

Figure 9.8 The Neuromuscular Junction

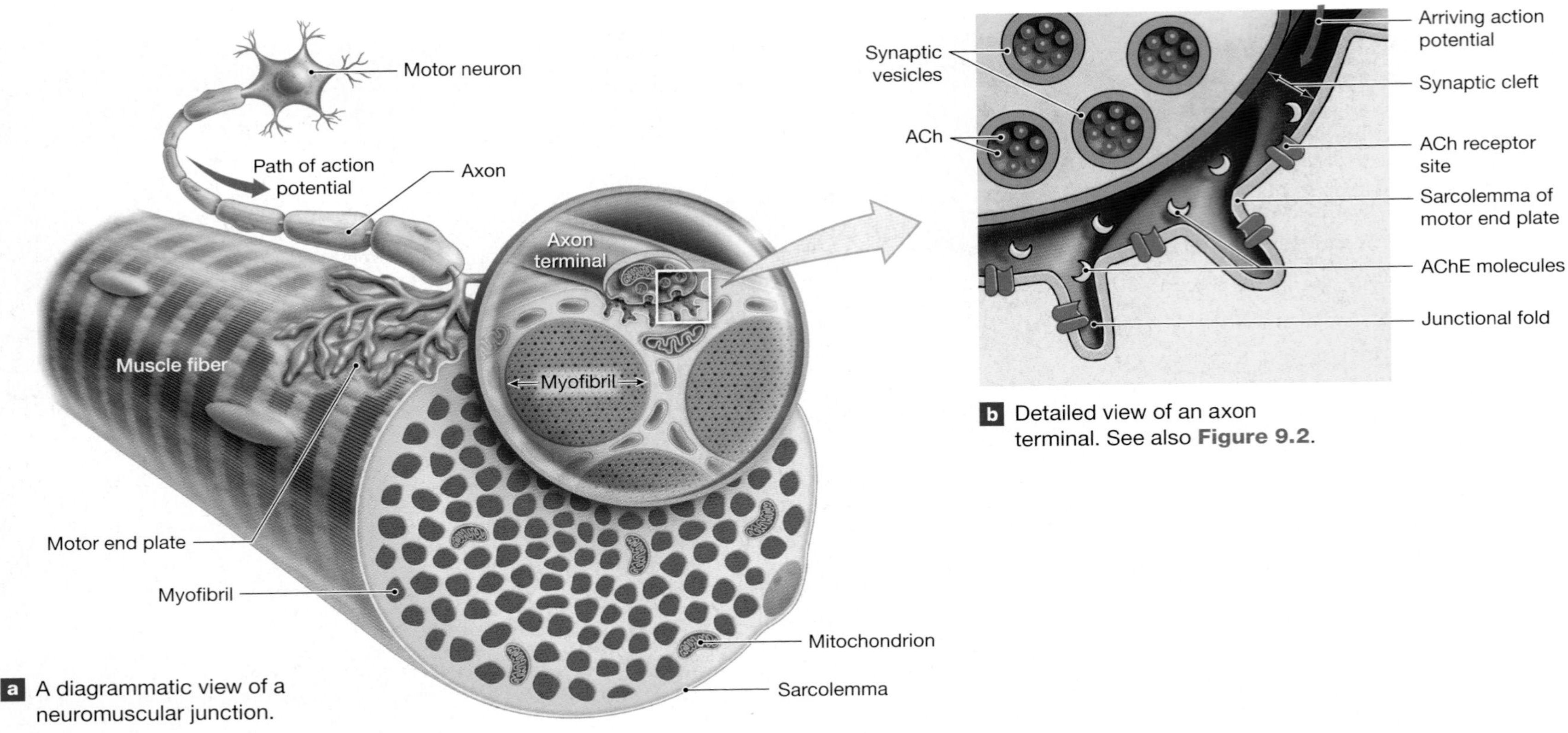

a A diagrammatic view of a neuromuscular junction.

b Detailed view of an axon terminal. See also **Figure 9.2**.

Figure 9.9 The Events in Muscle Contraction. A summary of the sequence of events in a muscle contraction.

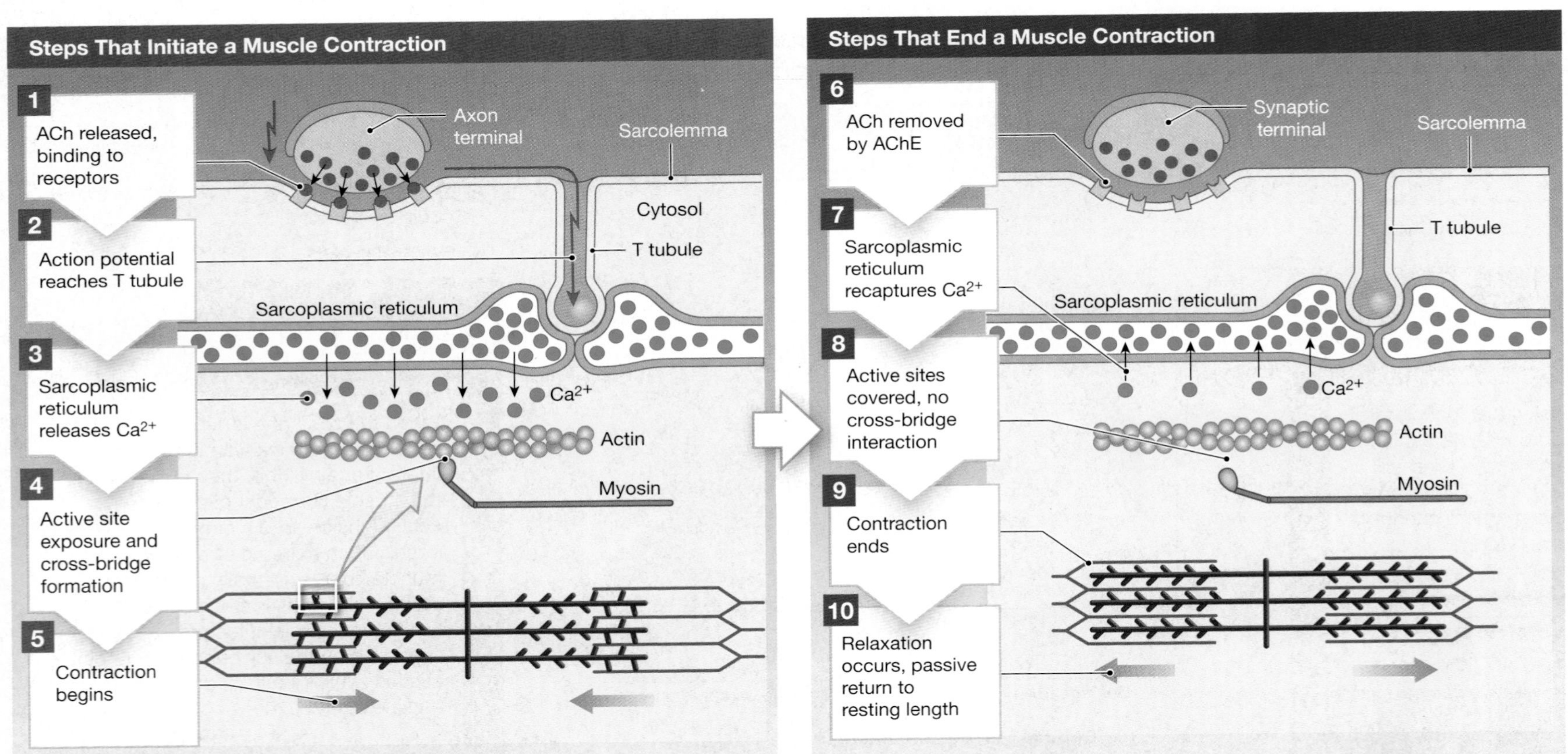

3 The sarcoplasmic reticulum (SR) releases stored calcium ions, increasing the calcium concentration in the sarcoplasm and around the sarcomeres.

4 Calcium ions bind to troponin, producing a change in the orientation of the troponin-tropomyosin complex that exposes active sites on the thin (actin) filaments. Myosin cross-bridges form when myosin heads bind to active sites.

5 Repeated cycles of cross-bridge binding, pivoting, and detachment occur, powered by the hydrolysis of ATP. These events produce filament sliding, and the muscle fiber shortens.

This process continues for a brief period, until:

6 Action potential generation stops as ACh diffuses out of the synapse or is broken down by AChE.

7 The sarcoplasmic reticulum (SR) reabsorbs calcium ions, and the concentration of calcium ions in the sarcoplasm decreases.

8 When calcium ion concentrations near normal resting levels, the troponin-tropomyosin complex returns to its normal position. This change covers the active sites and prevents further cross-bridge interaction.

9 Without cross-bridge interactions, further sliding does not take place, and the contraction ends.

10 Muscle relaxation occurs, and the muscle fiber returns passively to resting length.

9.3 CONCEPT CHECK

4 What happens to the widths of the A bands and I bands during a muscle contraction?

5 How do terminal cisternae and T tubules interact to cause a skeletal muscle contraction?

See the blue Answers tab at the back of the book.

9.4 | Motor Units and Muscle Control

▶ **KEY POINT** All motor units are similar in their anatomical characteristics; they are differentiated by the number of muscle fibers controlled by a single motor unit. Smaller motor units control the contraction of a small number of muscle fibers and, therefore, create smaller and more precise muscular contractions.

A **motor unit** is a single motor neuron and all of the muscle fibers it controls. Some motor neurons control a single muscle fiber, but most control hundreds. The smaller the size of a motor unit, the finer the control of movement will be. In the eye, where precise muscular control is critical, a motor neuron may control only two or three muscle fibers. We have less precise control over power-generating muscles, such as our leg muscles, where a single motor neuron may control up to 2000 muscle fibers.

A skeletal muscle contracts when its motor units are stimulated. Two factors determine the amount of tension produced: (1) the frequency of stimulation and (2) the number of motor units involved. A single, momentary contraction is called a **muscle twitch**—the response to a single stimulus. As the rate of stimulation increases, tension production rises to a peak and plateaus at a maximum level. Most muscle contractions occur in a sequence of stimulus, contraction, relaxation.

Each muscle fiber contracts completely or does not contract at all. This is called the **all or none principle**. All the fibers in a motor unit contract at the same time, and the amount of force exerted by the muscle depends on how many motor units contract. By varying the number of motor units contracting at any one time, the nervous system precisely controls the amount of force that a muscle generates.

The decision to move stimulates specific groups of motor neurons. The stimulated neurons do not respond simultaneously, and over time, the number of activated motor units gradually increases. **Figure 9.10** shows how the muscle fibers of different motor units are arranged. Because of this arrangement, the direction of pull exerted on the tendon doesn't change, but the total amount of force generated changes as more motor units are stimulated. The smooth but steady increase in muscular tension produced by increasing the number of active motor units is called **recruitment**.

Peak tension occurs when all the motor units in the muscle are contracting at the maximum rate of stimulation. However, these powerful contractions cannot last long because the individual muscle fibers soon use up their energy reserves. To delay the onset of fatigue during periods of sustained contraction, motor units are activated on a rotating basis. Therefore, some motor units are resting and recovering while others are actively contracting.

Muscle Tone

▶ **KEY POINT** At rest, skeletal muscle exhibits a low level of contraction independent of the voluntary control of the central nervous system.

Even when a muscle is resting, some motor units are always active. Their contractions do not produce enough tension to cause movement, but they do tense the muscle. This resting tension in a skeletal muscle is called **muscle tone**. Motor units are randomly stimulated, so there is a constant tension in the attached tendon as some motor units contract and others relax. Resting muscle tone stabilizes the position of bones and joints. For example, in muscles involved with balance and posture, enough motor units are stimulated to produce the tension needed to maintain body position. Specialized muscle cells called **muscle spindles** are monitored by sensory nerves that control the muscle tone in the surrounding muscle tissue. This sensory organ is sensitive to passive stretching of the muscles in which it is enclosed.

Muscle Hypertrophy

▶ **KEY POINT** Muscle cells increase the amount of intracellular contractile proteins in response to exercise. This increase in cell size is called hypertrophy.

Figure 9.10 The Arrangement of Motor Units in a Skeletal Muscle

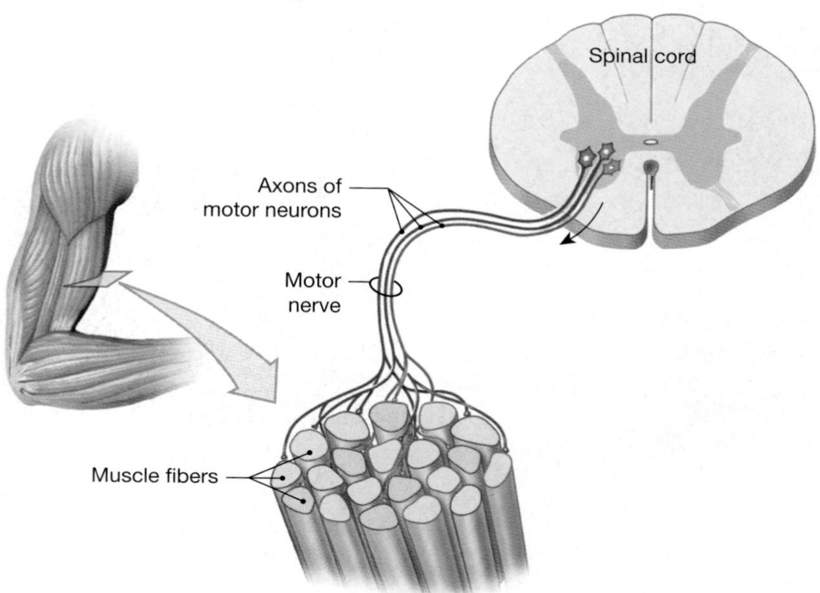

Exercise increases the activity of muscle spindles and may enhance muscle tone. As a result of repeated, exhaustive stimulation, muscle fibers develop a greater number of myofibrils and mitochondria, a higher concentration of glycolytic enzymes, and larger glycogen reserves. The net effect is an enlargement, or **hypertrophy** (hī-PER-trō-fē), of the stimulated muscle. Hypertrophy occurs in muscles that have been repeatedly stimulated to produce near-maximal tension; the intracellular changes that occur increase the amount of tension produced when these muscles contract. The muscles of a champion weight lifter or bodybuilder are an excellent example of muscular hypertrophy.

Muscle Atrophy

▶ **KEY POINT** The reverse of hypertrophy is atrophy, a degenerative process that occurs when skeletal muscle cells are not stimulated for a prolonged period of time.

9

When a skeletal muscle is not regularly stimulated by a motor neuron, it loses muscle tone and mass. The muscle becomes flaccid and the muscle fibers become smaller and weaker due to a loss of intracellular contractile proteins. This reduction in muscle size, tone, and power is called **atrophy**. People who are paralyzed by spinal injuries or other damage to the nervous system gradually lose muscle tone and size in the affected areas. Even a temporary reduction in muscle use leads to muscular atrophy; this loss of tone and size is clearly visible when you compare limb muscles before and after being in a cast. Muscle atrophy is initially reversible, but dying muscle fibers are not replaced, and in extreme atrophy the functional losses are permanent. That is why physical therapy is crucial in cases in which people are temporarily unable to move normally. Direct electrical stimulation by an external device can substitute for nerve stimulation and prevent or decrease muscle atrophy.

9.4 CONCEPT CHECK

6 Why do some motor units control only a few muscle fibers, whereas others control many fibers?

7 You are watching a movie with your roommate. In one scene, a person is shot with a dart dipped in a chemical that blocks the action of acetylcholine. Your roommate does not understand why the person then falls down. How do you explain this to your roommate?

See the blue Answers tab at the back of the book.

9.5 | Types of Skeletal Muscle Fibers

▶ **KEY POINT** There are three types of muscle fibers in the human body: slow (red), fast (white), and intermediate fibers. Slow fibers are slow to contract and slow to fatigue. Fast fibers contract and fatigue quickly. Intermediate fibers have characteristics between those of slow and fast fibers. (Red and white fibers are analogous to the dark and white meat, respectively, in chicken and turkey.)

Skeletal muscles are organized for a variety of actions. The types of fibers making up a muscle will, in part, determine its action. There are three major types of skeletal muscle fibers in the body: fast, slow, and intermediate (**Figure 9.11**). The fiber types differ in how they obtain ATP to fuel their contractions.

Fast fibers, or *white fibers*, have (1) a large diameter, (2) densely packed myofibrils, (3) large glycogen reserves, and (4) relatively few mitochondria. Most of the skeletal muscle fibers in the body are called fast fibers because they contract quickly following stimulation. The tension produced by a muscle fiber is directly proportional to the number of myofibrils, so fast-fiber

Figure 9.11 Types of Skeletal Muscle Fibers. Fast fibers produce rapid contractions, and slow fibers produce slower, but extended contractions.

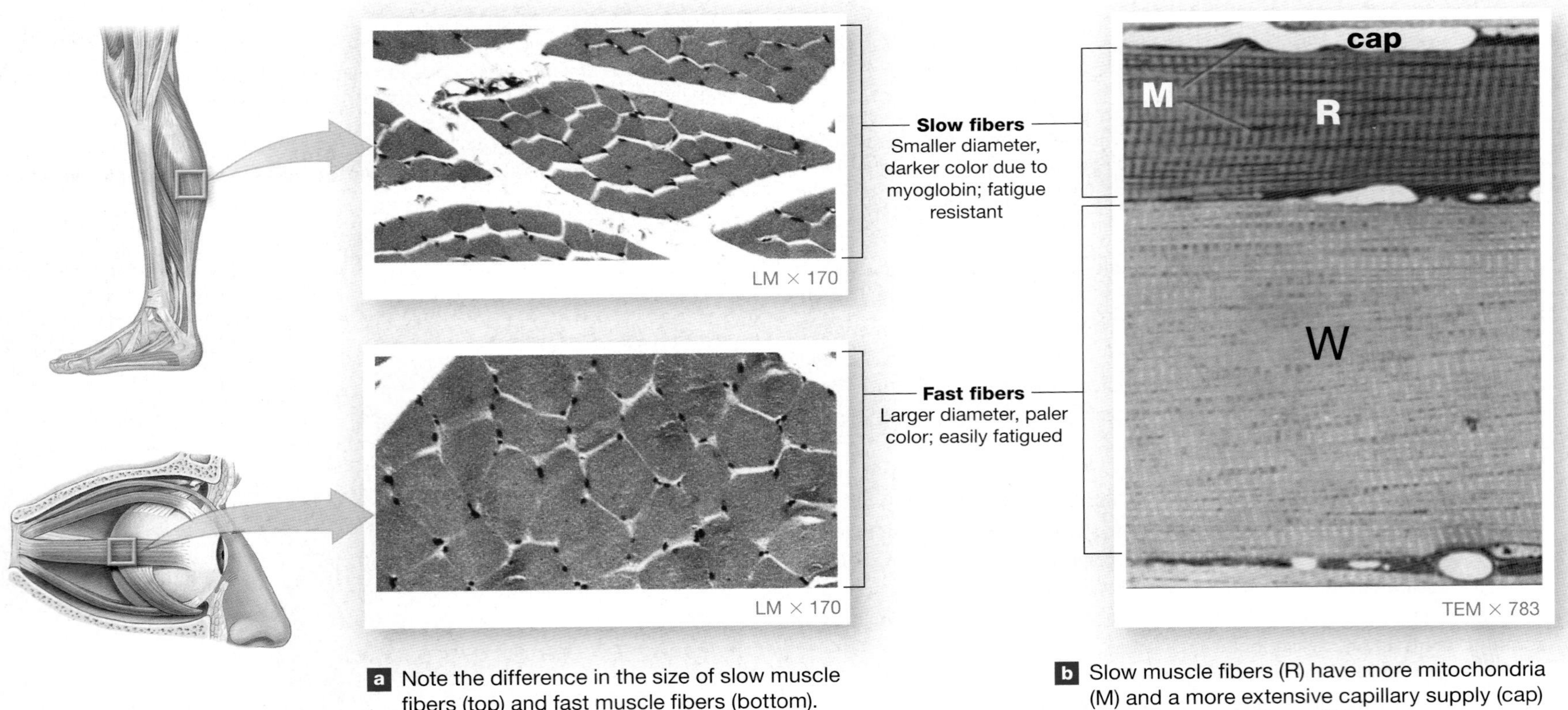

a Note the difference in the size of slow muscle fibers (top) and fast muscle fibers (bottom).

b Slow muscle fibers (R) have more mitochondria (M) and a more extensive capillary supply (cap) than fast muscle fibers (W).

muscles produce powerful contractions. However, these contractions use large amounts of ATP, and their mitochondria are unable to meet the demand. As a result, their contractions are supported primarily by **anaerobic metabolism** (glycolysis). **Glycolysis** does not require oxygen and converts stored glycogen to lactic acid. Fast fibers fatigue rapidly because their glycogen reserves are limited and because lactic acid builds up and the resulting acidic pH interferes with the contraction mechanism.

Slow fibers, or *red fibers*, have a smaller diameter than fast fibers and take three times as long to contract after stimulation. Slow fibers are specialized to continue contracting for extended periods of time, long after a fast muscle fatigues. They fatigue slowly because their mitochondria continue producing ATP throughout the contraction process. Mitochondria absorb oxygen and generate ATP by a pathway called **aerobic metabolism**. The oxygen required comes from two sources:

1. Skeletal muscles containing slow muscle fibers have a larger network of capillaries than muscles dominated by fast muscle fibers. This means that there is greater blood flow to the muscle, and the red blood cells can deliver more oxygen to the active muscle fibers.
2. Slow fibers are red because they contain the red pigment **myoglobin** (MĪ-ō-glō-bin). This globular protein, like hemoglobin (the oxygen-binding pigment found in red blood cells), binds oxygen molecules. As a result, slow muscle fibers contain large oxygen reserves that are mobilized during a contraction.

Slow muscles also contain a larger number of mitochondria than fast muscles. Whereas fast muscle fibers must rely on their glycogen reserves during peak levels of activity, the mitochondria in slow muscle fibers can break down carbohydrates, lipids, or even proteins. Therefore, slow muscles can continue to contract for extended periods; for example, the leg muscles of marathon runners are dominated by slow muscle fibers.

Intermediate fibers have properties between those of fast and slow fibers. For example, intermediate fibers contract faster than slow fibers but slower than fast fibers.

The properties of the various types of skeletal muscles are detailed in Table 9.1.

Distribution of Fast, Slow, and Intermediate Fibers

▶ **KEY POINT** Muscle motor units are composed of the same type of muscle fibers, meaning that a motor unit is composed of all slow, all fast, or all intermediate fibers. The ratio of slow, fast, and intermediate fibers is genetically determined. Intermediate fibers can convert to slow or fast fibers in response to exercise and training.

The percentage of fast, slow, and intermediate muscle fibers varies from one skeletal muscle to another. Most muscles contain a mixture of fiber types, although all the fibers within one motor unit are of the same type. However, there are no slow fibers in muscles of the eye and hand, where swift but brief contractions are required. Many back and calf muscles are dominated by slow fibers; these muscles contract almost continually to maintain an upright posture.

The percentage of fast versus slow fibers in each muscle is genetically determined, and there are significant individual differences. These variations have an effect on endurance. A person with more slow muscle fibers in a particular muscle will be better able to perform repeated contractions under aerobic conditions. For example, marathon runners with high proportions of slow muscle fibers in their leg muscles outperform those with more fast muscle fibers. For brief periods of intense activity, such as a sprint or a weight-lifting event, the person with the higher percentage of fast muscle fibers will have the advantage.

The characteristics of the muscle fibers change with physical conditioning. Repeated intense workouts promote the enlargement of fast muscle fibers and muscular hypertrophy. Training for endurance events, such as cross-country or marathon running, increases the proportion of intermediate fibers in the active muscles. This occurs through the gradual conversion of fast fibers to intermediate fibers.

9.5 CONCEPT CHECK ✔

8 Why does a sprinter experience muscle fatigue after a few minutes, whereas a marathon runner can run for hours?

See the blue Answers tab at the back of the book.

9.6 Organization of Skeletal Muscle Fibers

▶ **KEY POINT** Skeletal muscles of the human body are arranged into one of four patterns: parallel, convergent, pennate, and circular. These patterns are determined by the shape or arrangement of the fibers relative to the direction of pull.

Most skeletal muscle fibers contract at comparable rates and shorten to the same degree. However, variations in microscopic and macroscopic organization affect the power, range, and speed of movement produced when a muscle contracts.

Table 9.1 | Properties of Skeletal Muscle Fiber Types

Property	Slow	Intermediate	Fast
Cross-sectional diameter	Small	Intermediate	Large
Tension	Low	Intermediate	High
Contraction speed	Slow	Fast	Fast
Fatigue resistance	High	Intermediate	Low
Color	Red	Pink	White
Myoglobin content	High	Low	Low
Capillary supply	Dense	Intermediate	Scarce
Mitochondria	Many	Intermediate	Few
Glycolytic enzyme concentration in sarcoplasm	Low	High	High
Substrates used to generate ATP during contraction	Lipids, carbohydrates, amino acids (aerobic)	Primarily carbohydrates (anaerobic)	Carbohydrates (anaerobic)
Alternative names	Type I, S (slow), red, SO (slow oxidizing), slow-twitch oxidative	Type II-A, FR (fast resistant), fast-twitch oxidative	Type II-B, FF (fast fatigue), white, fast-twitch glycolytic

Muscles are classified according to the shape or arrangement of their fibers relative to the direction of pull. The muscle fibers of each fascicle lie parallel to one another, but the organization of the fascicles within a muscle and their relationship to the muscle's tendon, varies from one muscle to another. Based on the pattern of fascicle arrangement, we can classify skeletal muscles as (1) parallel muscles, (2) convergent muscles, (3) pennate muscles, or (4) circular muscles. **Figure 9.12** illustrates the fascicle organization of skeletal muscle fibers.

Parallel Muscles

▶ **KEY POINT** In a parallel muscle, the fascicles are parallel to the longitudinal axis of the muscle. Most of the skeletal muscles in the body are parallel muscles.

In a **parallel muscle**, the individual fibers may run the entire length of the muscle, as in the biceps brachii of the arm **(Figure 9.12a)**, or they may be interrupted by transverse pieces of connective tissue, as in the rectus abdominis of the anterior surface of the abdomen **(Figure 9.12b)**. Other parallel muscles have a twisted or spiral arrangement. The supinator of the forearm is an example of this arrangement; the muscle wraps around the proximal portion of the radius, allowing supination of the hand **(Figure 9.12c)**.

The anatomical characteristics of a parallel muscle resemble those of an individual muscle fiber. For example, the biceps brachii has a tendon that extends from the free tip to a movable bone of the skeleton and a central **body** (or belly) **(Figure 9.12a)**. When this muscle contracts, it shortens and the

Figure 9.12 Skeletal Muscle Fiber Organization. There are four different arrangements of muscle fiber patterns: parallel (a, b, c), convergent (d), pennate (e, f, g), and circular (h).

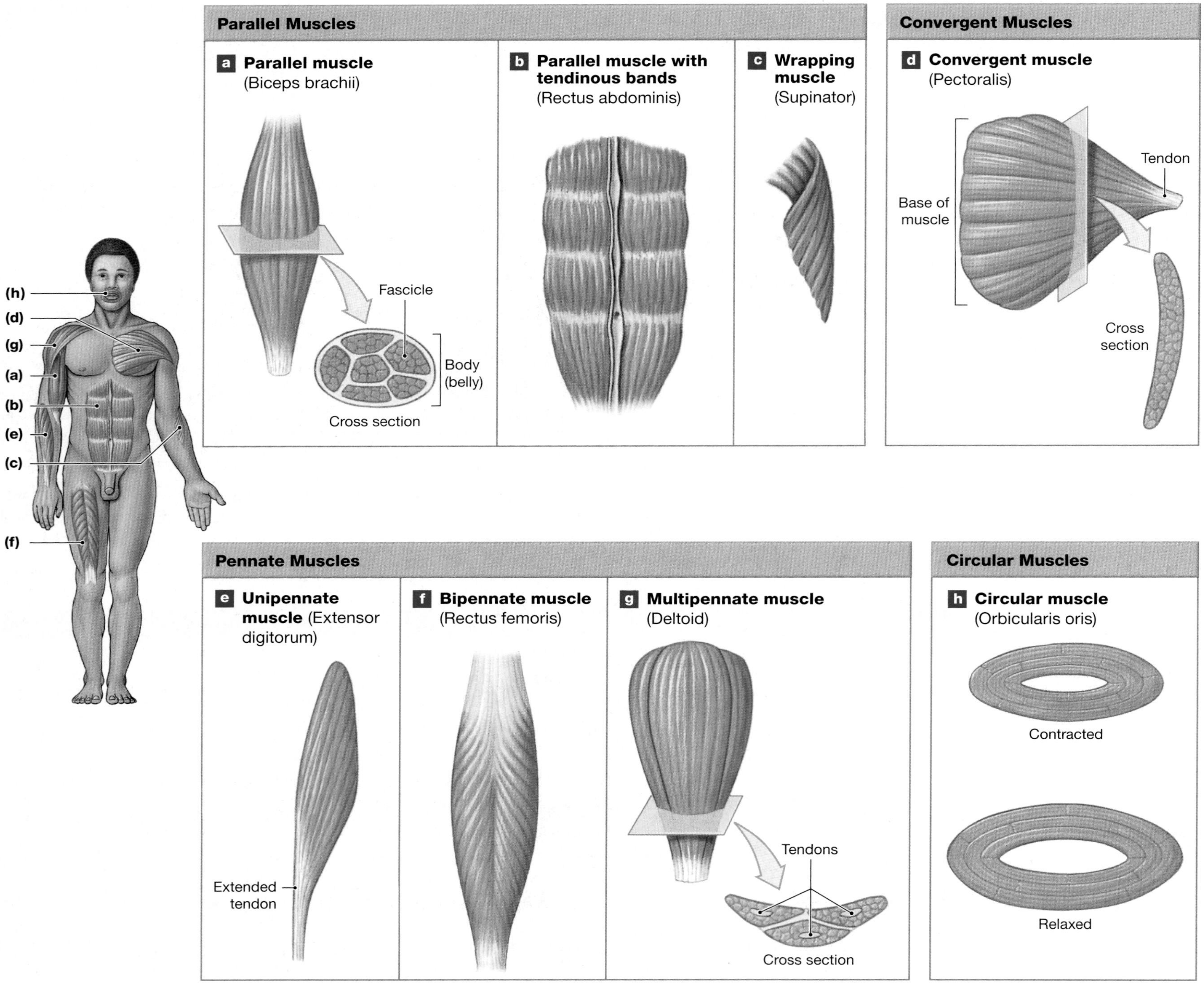

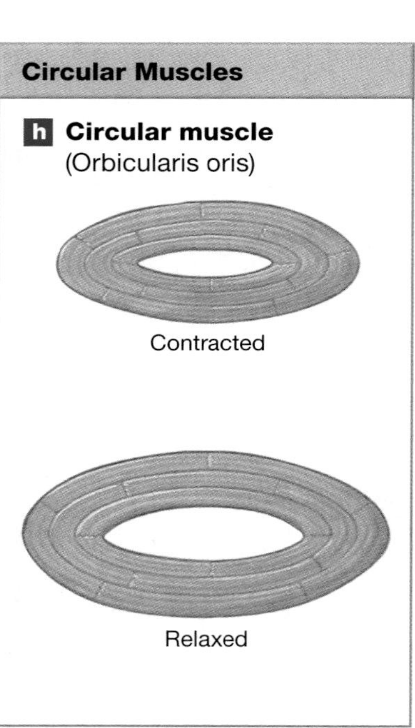

body increases in diameter. The bulge of the contracting biceps is seen on the anterior surface of the arm when the elbow flexes.

Because the muscle fibers are parallel to the long axis of the muscle, when they contract together, the entire muscle shortens by the same amount. The tension the muscle develops during this contraction depends on the total number of myofibrils it contains.

Convergent Muscles

▶ **KEY POINT** The fibers in convergent muscles are spread out, like a fan or a broad triangle, with a tendon at the tip.

In a **convergent muscle**, the muscle fibers cover a broad area, but all the fibers come together at a common attachment site **(Figure 9.12d)**. They may pull on a tendon, a tendinous sheet, or a slender band of collagen fibers known as a **raphe** (RĀ-fē; "seam"). The prominent pectoralis muscles of the chest have this shape. A convergent muscle is versatile: The direction of pull in a convergent muscle can be changed by stimulating only one group of muscle cells at any one time. However, when they all contract at once, they do not pull as hard on the tendon as a parallel muscle of the same size because the muscle fibers on opposite sides of the tendon pull in different directions rather than all pulling in the same direction.

Pennate Muscles

▶ **KEY POINT** The fibers in a pennate muscle are arranged like a feather. The muscle fibers sit at an angle and attach to a tendon that runs the length of the muscle.

In a **pennate muscle** (*penna*, feather), the fascicles pull at an angle. Because of the arrangement of the fascicles, a contracting pennate muscle does not move its tendon as far as a parallel muscle would. However, a pennate muscle contains more muscle fibers than a parallel muscle of the same size. Therefore, a contracting pennate muscle produces more tension than a parallel muscle of the same size.

Not all pennate muscles have the same structure. When all the muscle cells are arranged on the same side of the tendon, the muscle is **unipennate (Figure 9.12e)**. A long muscle that extends the fingers, the extensor digitorum, is an example of a unipennate muscle. In a **bipennate muscle**, the muscle fibers are on both sides of the tendon **(Figure 9.12f)**. The rectus femoris, a prominent thigh muscle that helps extend the knee, is a bipennate muscle. In a **multipennate muscle**, the tendon branches within the muscle **(Figure 9.12g)**. The triangular deltoid that covers the superior surface of the shoulder joint is an example of a multipennate muscle.

Circular Muscles

▶ **KEY POINT** Circular muscles guard entrances and exits of long internal passageways, such as the digestive and urinary tracts.

In a **circular muscle**, or **sphincter** (SFINK-ter), the fibers are concentrically arranged around an opening **(Figure 9.12h)**. When the muscle contracts, the diameter of the opening decreases. An example is the orbicularis oris of the mouth.

9.6 CONCEPT CHECK

9 Which muscle arrangement more closely resembles the anatomical characteristics of a single muscle fiber: multipennate or parallel?

See the blue Answers tab at the back of the book.

9.7 | Muscle Terminology

▶ **KEY POINT** Terms describing skeletal muscle may tell you the attachments of the muscle or the muscle's location, action, or shape. It is important to *understand* the terminology used to describe a muscle rather than simply memorizing it.

Terms used to name muscles relative to directions, specific body regions, structural characteristics, and actions are presented in **Table 9.2**.

Origins and Insertions

▶ **KEY POINT** Knowing which end of a muscle is the origin and which is the insertion is less important than remembering *where* the two ends attach and understanding *what* the muscle does when it contracts.

Each muscle begins at an origin, ends at an insertion, and contracts to produce a specific action. The **origin** of a muscle usually remains stationary, and the **insertion** moves, or the origin is proximal to the insertion. For example, the triceps brachii inserts on the olecranon and originates closer to the shoulder. These determinations are based on normal movement in the anatomical position. *Part of the fun of studying the muscular system is that you can actually do the movements and think about the muscles involved. (Laboratory discussions of the muscular system often resemble a poorly organized aerobics class.)* When the origins and insertions cannot be determined easily on the basis of movement or position, we use other criteria:

- If a muscle extends between a broad aponeurosis and a narrow tendon, the aponeurosis is the origin, and the tendon is attached to the insertion.
- If there are several tendons at one end and just one at the other, there are multiple origins and a single insertion.

Actions

▶ **KEY POINT** When a skeletal muscle contracts to produce a movement, it plays one of four roles: agonist, antagonist, synergist, or fixator. These roles can and do change as the movement changes.

Almost all skeletal muscles either originate or insert on the skeleton. When a muscle moves a part of the skeleton, that movement involves any of the movements that occur at a joint. Before reading further, review the description of dynamic motion and the structural classification of synovial joints in **Spotlight Figure 8.2** and the discussion of planes of motion illustrated in **Figure 8.3** to **8.5**. ⮌ pp. 207–211

We can describe actions in two ways:

1. Refer to the bone region affected. Thus, the biceps brachii performs "flexion of the forearm."
2. Specify the joint involved. Thus, the biceps brachii performs "flexion of (or at) the elbow."

Both methods work, and each has its advantages, *but we use the second method when describing muscle actions*.

Muscles are grouped into four categories according to their **primary actions**:

1. An **agonist**, or **prime mover**, is a muscle whose contraction is mostly responsible for producing a particular movement, such as flexion at the elbow.
2. An **antagonist** is a muscle whose action opposes that of the agonist. For example, if the agonist produces flexion, the antagonist produces extension. When an agonist contracts to produce a particular movement, the antagonist stretches, but usually doesn't relax completely. Instead, its tension is adjusted to control the speed and smoothness of the movement. For example, the biceps brachii acts as an agonist when it contracts, flexing the elbow. The triceps brachii, on the opposite side of the humerus, is the

antagonist and acts to stabilize the flexion movement and to produce the opposing action, extension of the elbow.

3. When a **synergist** (*syn–*, together, + *ergon*, work) contracts, it assists the agonist in performing that action. Synergists provide additional pull near the insertion or stabilize the origin. Their importance in assisting a particular movement changes as the movement progresses. Synergists are often most useful at the start, when the agonist is stretched and relatively weak. For example, the latissimus dorsi and the teres major extend the arm. With the arm pointed at the ceiling, the muscle fibers of the massive latissimus dorsi are at maximum stretch and are parallel to the humerus. In this position, the latissimus dorsi cannot develop much tension. But because the teres major originates on the scapula, it can contract more efficiently and assist the latissimus dorsi in starting an extension movement. The importance of the teres major decreases as the extension proceeds. In this example, the latissimus dorsi is the agonist and the teres major is the synergist.

4. When agonists and antagonists contract simultaneously, they are acting as **fixators**, stabilizing a joint and creating an immovable base. For example, flexors and extensors of the wrist contract simultaneously to stabilize the wrist when muscles of the hand contract to firmly grasp an object in the fingers.

Names of Skeletal Muscles

▶ **KEY POINT** The name of a skeletal muscle provides important clues about that specific muscle. *Learn* the names of muscles and what they mean rather than simply memorizing them.

Skeletal muscles are named according to several criteria introduced in **Table 9.2**. Some names refer to the orientation of the muscle fibers. For example, rectus means "straight," and rectus muscles are parallel muscles whose fibers run along the longitudinal axis of the body. Because there are several rectus muscles, the name includes a second term that refers to a precise region of the body. The *rectus abdominis* is on the abdomen, and the *rectus femoris* is on the thigh. Other directional indicators include transversus and oblique for muscles whose fibers run across or at an oblique angle to the longitudinal axis of the body, respectively.

Other muscles were named after specific and unusual structural features. A *biceps* muscle has two tendons of origin (*bi–*, two, + *caput*, head), the *triceps* has three, and the *quadriceps* four. Shape is often an important clue in the muscle's name; *trapezius* (tra-PĒ-zē-us), *deltoid*, *rhomboideus* (rom-BOYD-ē-us), and *orbicularis* (or-bik-ū-LA-ris) refer to prominent muscles that look like a trapezoid, triangle, rhomboid, and circle, respectively. Long muscles are called *longus* (long)

Table 9.2 | Muscle Terminology

Terms Indicating Direction Relative to Axes of the Body	Terms Indicating Specific Regions of the Body*	Terms Indicating Structural Characteristics of the Muscle	Terms Indicating Actions
Anterior (front) Externus (superficial) Extrinsic (outside) Inferioris (inferior) Internus (deep, internal) Intrinsic (inside) Lateralis (lateral) Medialis/medius (medial, middle) Oblique (angular) Posterior (back) Profundus (deep) Rectus (straight, parallel) Superficialis (superficial) Superioris (superior) Transversus (transverse)	Abdominis (abdomen) Anconeus (elbow) Auricularis (auricle of ear) Brachialis (brachium) Capitis (head) Carpi (wrist) Cervicis (neck) Cleido–/–clavius (clavicle) Coccygeus (coccyx) Costalis (ribs) Cutaneous (skin) Femoris (femur) Genio– (chin) Glosso–/–glossal (tongue) Hallucis (great toe) Ilio– (ilium) Inguinal (groin) Lumborum (lumbar region) Nasalis (nose) Nuchal (back of neck) Oculo– (eye) Oris (mouth) Palpebrae (eyelid) Pollicis (thumb) Popliteus (behind knee) Psoas (loin) Radialis (radius) Scapularis (scapula) Temporalis (temples) Thoracis (thoracic region) Tibialis (tibia) Ulnaris (ulna) Uro– (urinary)	**ORIGIN** Biceps (two heads) Triceps (three heads) Quadriceps (four heads) **SHAPE** Deltoid (triangle) Orbicularis (circle) Pectinate (comblike) Piriformis (pear-shaped) Platys– (flat) Pyramidal (pyramid) Rhomboideus (rhomboid) Serratus (serrated) Splenius (bandage) Teres (long and round) Trapezius (trapezoid) **OTHER FEATURES** Alba (white) Brevis (short) Gracilis (slender) Lata (wide) Latissimus (widest) Longissimus (longest) Longus (long) Magnus (large) Major (larger) Maximus (largest) Minimus (smallest) Minor (smaller) Tendinosus (tendinous) Vastus (great)	**GENERAL** Abductor Adductor Depressor Extensor Flexor Levator Pronator Rotator Supinator Tensor **SPECIFIC** Buccinator (trumpeter) Risorius (laugher) Sartorius (like a tailor)

*For other regional terms, refer to **Figure 1.8**, p. 15, which identifies anatomical landmarks.

or *longissimus* (longest), and *teres* muscles are both round and long. Short muscles are called *brevis*; large ones *magnus* (big), *major* (bigger), or *maximus* (biggest); and small ones are called *minor* (smaller) or *minimus* (smallest).

Muscles visible at the body surface are external and often called *externus* or *superficialis* (superficial), whereas those lying beneath are internal, termed *internus* or *profundus*. Superficial muscles that position or stabilize an organ are called *extrinsic* muscles; those that operate within the organ are called *intrinsic* muscles.

Many muscle names identify their origins and insertions. In these cases, the first part of the name indicates the origin and the second part the insertion. For example, the *genioglossus* originates at the chin (*geneion*) and inserts in the tongue (*glossa*).

Action names, such as *flexor*, *extensor*, and *adductor*, indicate the primary function of the muscle. These are such common actions that the names also include other clues about the appearance or location of the muscle. For example, the *extensor carpi radialis longus* is a long muscle found along the radial (lateral) border of the forearm. When it contracts, its primary function is extension at the wrist.

A few muscles are named after the specific movements associated with special occupations or habits. For example, the *sartorius* (sar-TŌR-ē-us) is active when crossing the legs. Before sewing machines were invented, a tailor would sit on the floor cross-legged; the name of the muscle was derived from *sartor*, the Latin word for "tailor." On the face, the *buccinator* (BUK-si-nā-tor) compresses the cheeks, as when you purse your lips and blow forcefully. *Buccinator* translates as "trumpet player." Finally, another facial muscle, the *risorius* (ri-SOR-ē-us), was supposedly named after the mood expressed. However, the Latin term *risor* means "laughter," while a more appropriate description for the effect would be "grimace."

9.7 CONCEPT CHECK

10 What type of muscle is a synergist?

11 What does the name *flexor digitorum longus* tell you about this muscle?

See the blue Answers tab at the back of the book.

9.8 Levers and Pulleys: A System Designed for Movement

▶ KEY POINT In many instances, the tendon of insertion for a muscle passes around a bony projection, changing the direction of the tendon and the force generated by the muscle. When this happens, the anatomical projection is called an anatomical pulley. Anatomical pulleys make it easier to move parts of the body.

The force, speed, or direction of movement that a muscle contraction produces can be modified by attaching the muscle to a lever. A **lever** is a rigid structure—such as a board, a crowbar, or a bone—that moves on a fixed point called the **fulcrum**. In the body, each bone is a lever and each joint a fulcrum. In addition to levels and fulcrums, mechanical pulleys can change the direction of a force to accomplish a task more easily and efficiently. In the body, tendons act like lines that convey the forces produced by muscle contraction. The presence of bones or bony processes can change the path that a tendon takes. Bony structures that change the direction of applied forces are called **anatomical pulleys**. **Spotlight Figure 9.13** explains and illustrates how the human body utilizes levers and anatomical pulleys.

9.8 CONCEPT CHECK

12 Arrange the three types of levers found in the human body in order of decreasing frequency.

13 Would removing your patella affect the amount of force that your quadriceps muscles could produce?

See the blue Answers tab at the back of the book.

9.9 Aging and the Muscular System

▶ KEY POINT Starting at approximately age 60, skeletal muscle strength begins to decline for a number of reasons.

As the body ages, all muscle tissues undergo a general reduction in size and power. The effects of aging on the muscular system are summarized here.

- Skeletal muscle fibers become smaller in diameter. This reduction in size reflects primarily a decrease in the number of myofibrils. In addition, the muscle fibers contain less ATP, glycogen reserves, and myoglobin. The overall effect is a reduction in muscle strength and endurance and a tendency to fatigue quickly. Because cardiovascular performance also decreases with age, blood flow to active muscles does not increase with exercise as rapidly as it does in younger people.
- Skeletal muscles become smaller in diameter and less elastic. Aging skeletal muscles develop increasing amounts of fibrous connective tissue within the endomysium and perimysium, a process called **fibrosis**. Fibrosis makes the muscle less flexible, and collagen fibers begin to restrict movement and circulation.
- Tolerance for exercise decreases. A lower tolerance for exercise results in part from the tendency for rapid fatigue and in part from less ability to eliminate the heat generated during muscular contraction.
- Ability to recover from muscular injuries decreases. As we age, the number of myosatellite cells steadily decreases and the amount of fibrous tissue increases. As a result, when tissue is injured, repair capabilities are limited, so instead of tissue repair, scar tissue usually forms.

9.9 CONCEPT CHECK

14 Why does a physician recommend that her 60-year-old patient continue to do stretching exercises throughout the remainder of his life?

See the blue Answers tab at the back of the book.

SPOTLIGHT

FIGURE 9.13

Levers and Pulleys

Levers

Skeletal muscles do not work in isolation. For muscles attached to the skeleton, the muscle's insertion on a bone determines the force, speed, and range of movement produced. In this arrangement, the bone is a **lever**, a rigid structure that moves on a fixed point called the **fulcrum** (F). The seesaw shown at the right demonstrates an example of lever action. The fulcrum is the fixed center, the board acts as the lever, and each child is either the **applied force** (AF)—the effort produced by the muscle contraction—or the **load** (L), which is the weight that opposes the effort. In the body, each bone is a lever and each joint is a fulcrum. The three classes of levers in the body are illustrated below.

First-Class Lever

In a first-class lever, the applied force and the load are on opposite sides of the fulcrum. This lever changes the amount of force transmitted to the load and alters the direction and speed of movement. There are very few first-class levers in the human body.

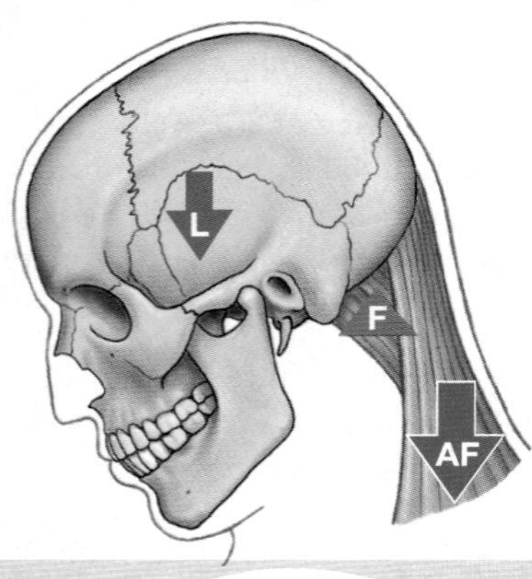

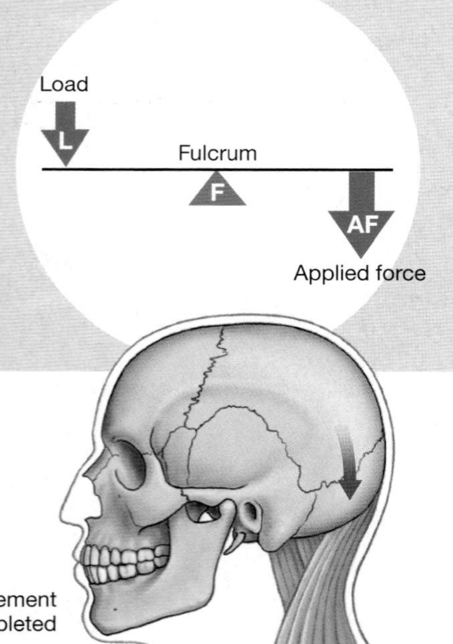

Second-Class Lever

In a second-class lever, the load is between the applied force and the fulcrum. This arrangement increases force at the expense of distance and speed; the direction of movement remains unchanged. There are very few second-class levers in the body.

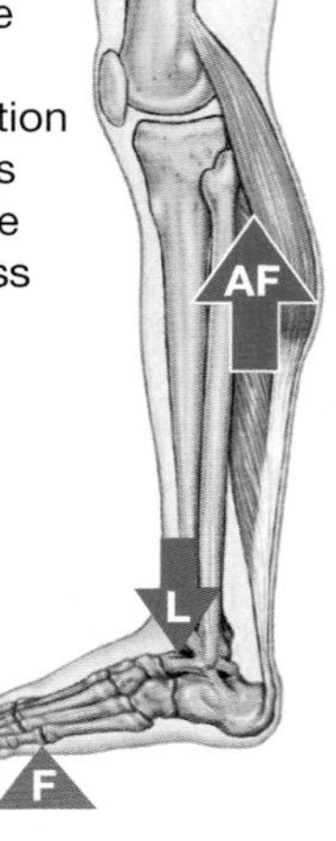

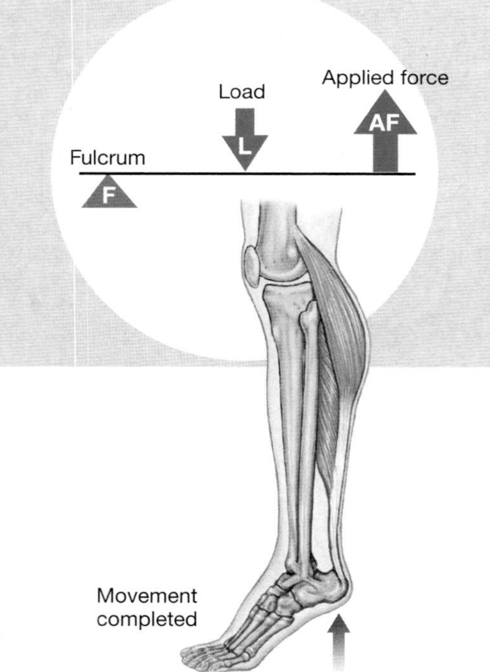

Third-Class Lever

In a third-class lever, which is the most common lever in the body, the applied force is between the load and the fulcrum. This arrangement increases speed and distance moved but requires a larger applied force.

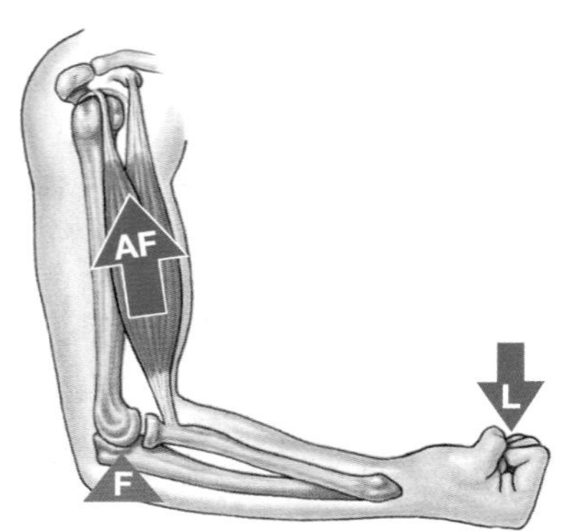

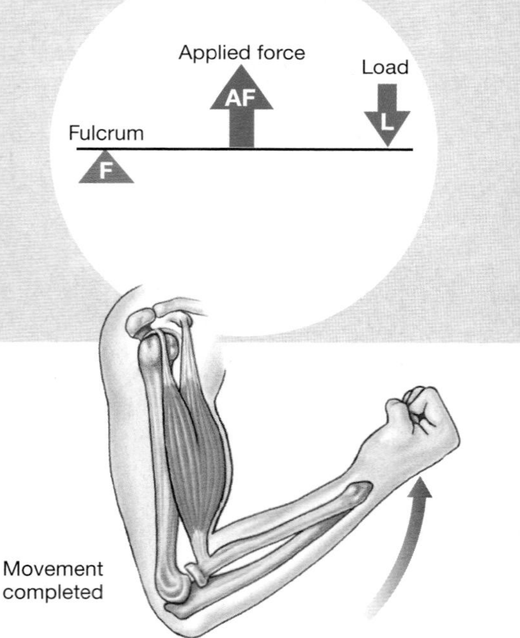

Pulleys

In the body, tendons act like lines that convey the forces produced by muscle contraction. The path taken by a tendon may be changed by the presence of bones or bony processes. These bony structures, which change the direction of applied forces, are called **anatomical pulleys**.

Mechanical pulleys are often used to change the direction of a force in order to accomplish a task more easily and efficiently. On a sailboat, a sailor pulls down on a rope to raise the sail. The sail goes up because a pulley at the top of the mast changes the direction of the force applied to the rope. Similarly, a flag goes up a flagpole when you pull the line down because the line passes through a pulley at the top of the pole.

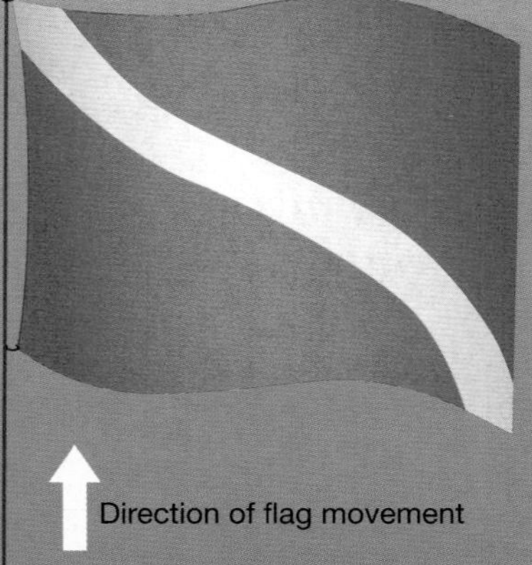

The Lateral Malleolus as an Anatomical Pulley

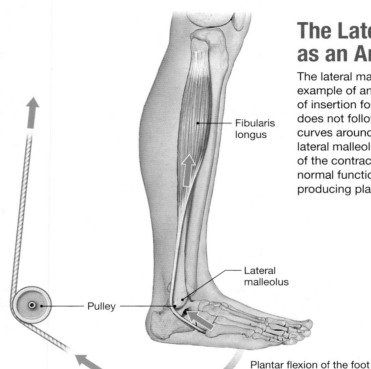

The lateral malleolus of the fibula is an example of an anatomical pulley. The tendon of insertion for the fibularis longus muscle does not follow a direct path. Instead, it curves around the posterior margin of the lateral malleolus of the fibula. This redirection of the contractile force is essential to the normal function of the fibularis longus in producing plantar flexion at the ankle.

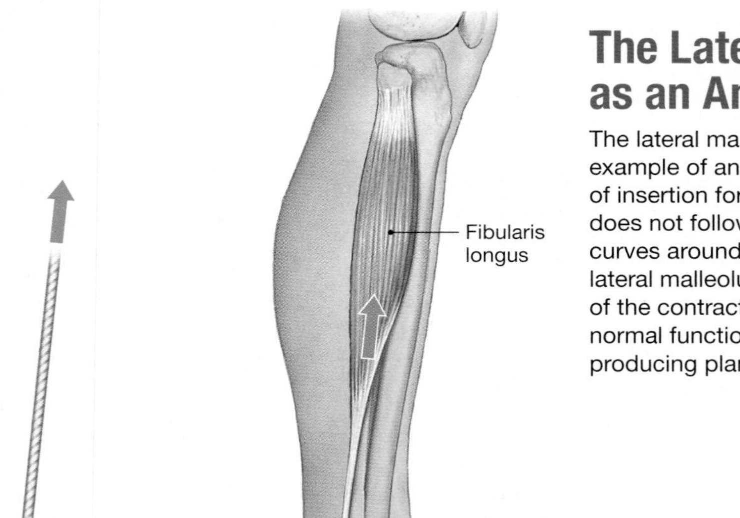

The Patella as an Anatomical Pulley

The patella is another example of an anatomical pulley. The quadriceps femoris is a group of anterior thigh muscles with four heads. They attach to the patella by the quadriceps tendon. The patella is, in turn, attached to the tibial tuberosity by the patellar ligament. The quadriceps femoris muscles produce extension at the knee by this two-link system. The quadriceps tendon pulls on the patella in one direction throughout the movement, but the direction of force applied to the tibia by the patellar ligament changes constantly as the movement proceeds.

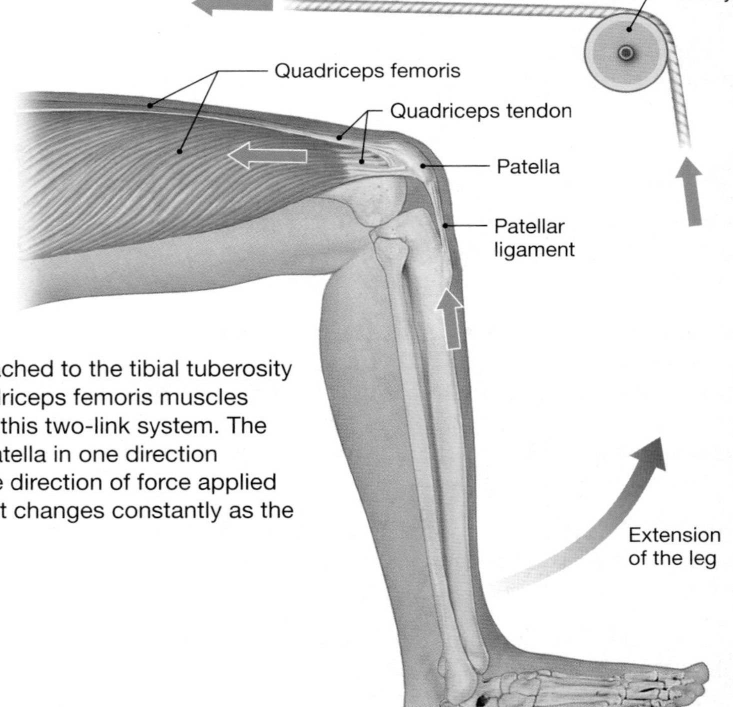

Study Outline

Introduction p. 236

- There are three types of muscle tissue: **skeletal muscle**, **cardiac muscle**, and **smooth muscle**. The **muscular system** includes all the skeletal muscle tissue that can be controlled voluntarily.

9.1 | Functions of Skeletal Muscle p. 236

- **Skeletal muscles** attach to bones directly or indirectly and perform these functions: (1) produce skeletal movement, (2) maintain posture and body position, (3) support soft tissues, (4) regulate entry and exit of materials, and (5) maintain body temperature.

9

9.2 | Anatomy of Skeletal Muscles p. 236

Gross Anatomy p. 236

- Each muscle fiber is wrapped by three concentric layers of connective tissue: an **epimysium**, a **perimysium**, and an **endomysium**. At the ends of the muscle are **tendons** or **aponeuroses** that attach the muscle to other structures. *(See Figure 9.1.)*

- Communication between a neuron and a muscle fiber occurs across the **neuromuscular junction**. *(See Figure 9.2.)*

Microanatomy of Skeletal Muscle Fibers p. 238

- A skeletal muscle cell has a plasma (cell) membrane, or **sarcolemma**; cytoplasm, or **sarcoplasm**; and an internal membrane system, or **sarcoplasmic reticulum (SR)**, similar to the endoplasmic reticulum of other cells. *(See Figure 9.3.)*

- A skeletal muscle cell is large and multinucleate. Deep indentations of the sarcolemma into the sarcoplasm of the skeletal muscle cell are called **transverse (T) tubules**. The T tubules carry the electrical impulse (action potential) that stimulates contraction into the sarcoplasm.

Myofibrils and Myofilaments p. 238

- The sarcoplasm contains numerous **myofibrils**. Protein filaments inside a myofibril are organized into repeating functional units called **sarcomeres**.

- Myofibrils are made up of **myofilaments**; there are **thin filaments** and **thick filaments**. *(See Figures 9.3–9.6.)*

9.3 | Muscle Contraction p. 243

The Sliding Filament Theory p. 243

- The **sliding filament theory** of muscle contraction explains how a muscle fiber exerts tension (a pull) and shortens. *(See Spotlight Figure 9.7.)*

- The contraction process involves **active sites** on thin filaments and **cross-bridges** of the thick filaments. Sliding involves a cycle of "attach, pivot, detach, and return" for the myosin bridges. At rest, the necessary interactions are prevented by the associated proteins, **tropomyosin** and **troponin**, on the thin filaments. *(See Figure 9.6 and Spotlight Figure 9.7.)*

- Contraction is an active process, but elongation of a muscle fiber is a passive process that can occur either through elastic forces or through the movement of other, opposing muscles.

- The amount of tension produced during a contraction is proportional to the degree of overlap between thick and thin filaments. *(See Spotlight Figure 9.7.)*

Neural Control of Muscle Fiber Contraction p. 243

- Neural control of muscle function involves a link between release of chemicals by the neurons and electrical activity in the sarcolemma that initiates a contraction.

- Each muscle fiber is controlled by a neuron at a neuromuscular junction the synapse includes the **axon terminal**, **synaptic vesicles**, and the **synaptic cleft**. **Acetylcholine (ACh)** release stimulates the motor end plate and generates action potentials that spread across the sarcolemma. **Acetylcholinesterase (AChE)** breaks down ACh and limits the duration of stimulation. *(See Figures 9.2 and 9.8.)*

Muscle Contraction: A Summary p. 243

- The steps involved in contraction are as follows: ACh release from synaptic vesicles → binding of ACh to the motor end plate → generation of an action potential in the sarcolemma → conduction of the action potential along T tubules → release of calcium ions by the SR → exposure of active sites on thin filaments → cross-bridge formation and contraction. *(See Figure 9.9.)*

9.4 | Motor Units and Muscle Control p. 247

- The number and size of a muscle's **motor units** indicate how precisely controlled its movements are. *(See Figure 9.10.)*

- A single momentary muscle contraction is called a **muscle twitch** and is the response to a single stimulus.

- Each muscle fiber either contracts completely or does not contract at all. This characteristic is the **all or none principle**.

Muscle Tone p. 247

- Even when a muscle is at rest, motor units are randomly stimulated so that a constant tension is maintained in the attached tendon. This resting tension in a skeletal muscle is called **muscle tone**. Resting muscle tone stabilizes bones and joints.

Muscle Hypertrophy p. 247

- Excessive repeated stimulation to produce near-maximal tension in skeletal muscle can lead to **hypertrophy** (enlargement) of the stimulated muscles.

Muscle Atrophy p. 248

- Inadequate stimulation to maintain resting muscle tone causes muscles to become flaccid and undergo **atrophy**.

9.5 | Types of Skeletal Muscle Fibers p. 248

- The three types of skeletal muscle fibers are *fast fibers*, *slow fibers*, and *intermediate fibers*. *(See Figure 9.11.)*
- **Fast fibers** are large in diameter; they contain densely packed myofibrils, large glycogen reserves, and relatively few mitochondria. They produce rapid and powerful contractions of relatively brief duration.

- **Slow fibers** are only about half the diameter of fast fibers and take three times as long to contract after stimulation. Slow fibers are specialized to continue contracting for extended periods.

- **Intermediate fibers** are very similar to fast fibers, although they have a greater resistance to fatigue.

Distribution of Fast, Slow, and Intermediate Fibers p. 249

- The percentage of fast, slow, and intermediate fibers varies from one skeletal muscle to another. Muscles contain a mixture of fiber types, but the fibers within one motor unit are of the same type. The percentage of fast versus slow fibers in each muscle is genetically determined and can change depending on the type and amount of physical exercise.

9.6 | Organization of Skeletal Muscle Fibers p. 249

- A muscle is classified according to the arrangement of fibers and fascicles as a *parallel muscle*, *convergent muscle*, *pennate muscle*, or *circular muscle* (*sphincter*).

Parallel Muscles p. 250

- In a **parallel muscle**, the fascicles are parallel to the long axis of the muscle. Most of the skeletal muscles in the body are parallel muscles. *(See Figure 9.12a–c.)*

Convergent Muscles p. 251

- In a **convergent muscle**, the muscle fibers are based over a broad area, but all the fibers come together at a common attachment site. *(See Figure 9.12d.)*

Pennate Muscles p. 251

- In a **pennate muscle**, one or more tendons run through the body of the muscle, and the fascicles form an oblique angle to the tendon. Contraction of pennate muscles generates more tension than that of parallel muscles of the same size. A pennate muscle may be **unipennate**, **bipennate**, or **multipennate**. *(See Figure 9.12e–g.)*

Circular Muscles p. 251

- In a **circular muscle** (**sphincter**), the fibers are concentrically arranged around an opening. *(See Figure 9.12h.)*

9.7 | Muscle Terminology p. 251

Origins and Insertions p. 251

- Each muscle may be identified by its **origin**, **insertion**, and **primary action**. Typically, the *origin* remains stationary and the *insertion* moves, or the origin is proximal to the insertion. Muscle contraction produces a specific action.

Actions p. 251

- A muscle may be classified as an **agonist** (**prime mover**), an **antagonist**, a **synergist**, or a **fixator**.

Names of Skeletal Muscles p. 252

- The names of muscles often provide clues to their location, orientation, or function. *(See Table 9.2.)*

9.8 | Levers and Pulleys: A Systems Design for Movement p. 253

- A **lever** is a rigid structure that moves on a fixed point called a **fulcrum**. Levers can change the direction, speed, or distance of muscle movements and can modify the force applied to the movement.
- Levers may be classified as **first-class**, **second-class**, or **third-class levers**; third-class levers are the most common type of lever in the body. *(See Spotlight Figure 9.13.)*
- Bony structures that change the direction of a muscle's contractile force are called **anatomical pulleys**. *(See Spotlight Figure 9.13.)*

9.9 | Aging and the Muscular System p. 253

- Aging reduces the size, elasticity, and power of all muscle tissues. Exercise tolerance and the ability to recover from muscular injuries decrease as the body ages.

Chapter Review

For answers, see the blue Answers tab at the back of the book.

Level 1 Reviewing Facts and Terms

1. Active sites on actin become available for binding when
   (a) calcium binds to troponin.
   (b) troponin binds to tropomyosin.
   (c) calcium binds to tropomyosin.
   (d) actin binds to troponin.
2. The function of a neuromuscular junction is to
   (a) generate new muscle fibers.
   (b) facilitate chemical communication between a neuron and a muscle fiber.
   (c) unite motor branches of nerves from different muscle fibers.
   (d) provide feedback about muscle activity to sensory nerves.
3. The direct energy supply for skeletal muscle production is
   (a) derived from fat, carbohydrate, and cholesterol.
   (b) independent of the supply of oxygen.
   (c) ATP.
   (d) infinite, as long as muscle activity is required.
4. Another name for a muscle that is a prime mover is
   (a) agonist. (b) antagonist.
   (c) synergist. (d) fixator.
5. Each of the following changes in skeletal muscles is a consequence of aging except
   (a) muscle fibers become smaller in diameter.
   (b) muscles become less elastic.
   (c) muscle fibers increase glycogen reserves..
   (d) the number of myosatellite cells decreases.
6. Interactions between actin and myosin filaments of the sarcomere are responsible for
   (a) muscle fatigue.
   (b) conducting neural information to the muscle fiber.
   (c) muscle contraction.
   (d) the striated appearance of skeletal muscle.
7. The theory that explains muscle contraction is formally known as the
   (a) muscle contraction theory.
   (b) striated voluntary muscle theory.
   (c) rotating myosin head theory.
   (d) sliding filament theory.
8. The bundle of collagen fibers at the end of a skeletal muscle that attaches the muscle to bone is called a(n)
   (a) fascicle. (b) tendon.
   (c) ligament. (d) epimysium.
9. Label the illustrations to the right as first-class, second-class, or third-class lever.

(a) ____________________

(b) ____________________

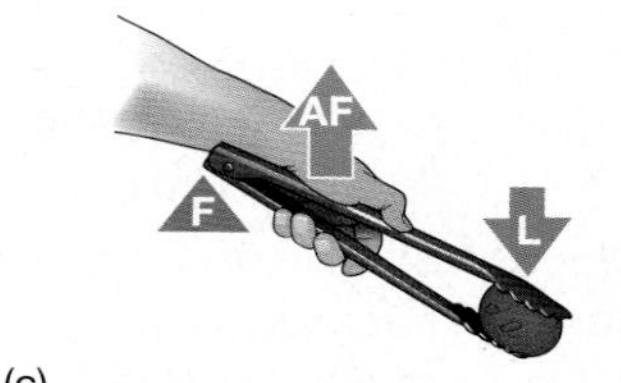

(c) ____________________

Level 2 Reviewing Concepts

1. To lessen the rate at which muscles fatigue during a contraction, motor units are activated
   (a) to less than their peak tension each time they contract.
   (b) in a stepwise fashion.
   (c) on a rotating basis.
   (d) quickly, to complete the contraction before they fatigue.
2. The ability to recover from injuries in older people decreases because
   (a) the number of myosatellite cells decreases with age.
   (b) myosatellite cells become smaller in size.
   (c) the amount of fibrous tissue in the muscle increases.
   (d) both a and c.
3. In which of the following is the ratio of motor neurons to muscle fibers the greatest?
   (a) large muscles of the arms
   (b) postural muscles of the back
   (c) muscles that control the eye
   (d) leg muscles
4. If a person is cold, a good way to warm up is to exercise. What is the mechanism of this warming?
   (a) Moving faster prevents the person from feeling the cold air because it moves past him or her more quickly.
   (b) Exercise moves blood faster, and the friction keeps tissues warm.
   (c) Muscle contraction uses ATP, and using this energy generates heat, which warms the body.
   (d) The movement of the actin and myosin filaments during the contraction generates heat, which warms the body.
5. Summarize the basic sequence of events that occurs at a neuromuscular junction.
6. A motor unit from a skeletal muscle contains 1500 muscle fibers. Would this muscle be involved in fine, delicate movements or powerful, gross movements? Explain.
7. What is the role of the zone of overlap in producing tension in a skeletal muscle?

Level 3 Critical Thinking

1. Several anatomy students take up weight lifting and bodybuilding. After several months, they notice physical changes such as increased muscle mass, lean body weight, and greater muscular strength. What anatomical mechanism is responsible for these changes?
2. Within the past 10–20 years, several countries have initiated the practice of performing leg muscle biopsies on track athletes to determine whether the athletes are better suited for sprints or long-distance events. What anatomical fact is the basis of this practice?

MasteringA&P™

Access more chapter study tools online in the Study Area:

- Chapter Quizzes, Chapter Practice Test, Clinical Cases, and more!
- Practice Anatomy Lab (PAL) PAL™
- A&P Flix for anatomy topics A&PFlix™

CLINICAL CASE WRAP-UP

A Case of Asymmetrical Development

Abdul is a victim of paralytic poliomyelitis, or polio. Polio is a highly infectious viral disease that is spread through human-to-human contact. The virus enters through the mouth, particularly in areas with poor sanitation, and multiplies in the intestines. From there, it invades the spinal cord, causing the neuronal death of motor neurons controlling skeletal muscles. With no motor neuron stimulation, the skeletal muscles become flaccid, smaller, and weaker; in other words, they atrophy. This is called a flaccid paralysis. Spinal reflexes also disappear.

There is no cure for polio, but there are safe and effective vaccines. Thanks to a national immunization campaign, the last case of naturally occurring polio in the United States was in 1979. However, polio, which only infects humans, is still affecting children in Afghanistan, Pakistan, and some African countries.

1. How would you describe the resting muscle tone in the muscles of Abdul's left hip girdle, thigh, and leg?
2. Do you think the majority of Abdul's muscle fibers that have been affected by polio are fast fibers or slow fibers?

See the blue Answers tab at the back of the book.

Related Clinical Terms

botulism: A severe, potentially fatal paralysis of skeletal muscles, resulting from the consumption of a bacterial toxin.

muscular dystrophies: A varied collection of inherited diseases that produce progressive muscle weakness and deterioration.

myasthenia gravis: A general muscular weakness resulting from a reduction in the number of ACh receptors on the motor end plate.

myopathy: Disease of muscle tissue.

RICE (rest, ice, compression, and elevation): Acronym for the standard treatment for muscle injuries, bruises, strains, and sprains.

10

The Muscular System

Axial Musculature

Learning Outcomes

These Learning Outcomes correspond by number to this chapter's sections and indicate what you should be able to do after completing the chapter.

10.1 Describe the location and function of the four groups of axial muscles. p. 260

10.2 Identify the six subgroups of the muscles of the head and neck and explain how they differ in their origins, insertions, actions, and innervations. p. 262

10.3 List the three layers of the muscles of the vertebral column and explain how the muscles differ in their origins, insertions, actions, and innervations. p. 270

10.4 Identify the muscles of the oblique and rectus groups and explain how they differ in their origins, insertions, actions, and innervations. p. 273

10.5 Name the muscles of the perineal region and pelvic diaphragm and explain how they differ in their origins, insertions, actions, and innervations. p. 277

CLINICAL CASE

Waking with a Crooked Smile

Sarah, a university student, went to an outdoor winter carnival last night. During the event, she noticed excessive tearing from her left eye and drooling from the left side of her mouth, which she attributed to the cold. The next morning, while gazing in the mirror, Sara realizes her smile is crooked. She is able to move the right side of her face normally, but the left side droops, and she is unable to wrinkle her forehead or close her left, still-watery eye. She can neither pucker her lips to whistle or blow, nor brush her teeth on the left side of her mouth without having to lift her lip with her hand. Her skin appears perfectly smooth, with no crow's feet or laugh lines (nasolabial folds) on the left side of her face.

Sarah begins to panic. Could she have had a stroke overnight? To find out, turn to the Clinical Case Wrap-Up on p. 281.

THE MUSCULAR SYSTEM, like the skeletal system, is divided into axial and appendicular divisions. The **axial musculature** originates on the axial skeleton. It positions the head and vertebral column and helps breathing by moving the rib cage. The **appendicular musculature** inserts onto and stabilizes or moves the appendicular skeleton.

Figures 10.1 and **10.2** illustrate the major axial and appendicular muscles of the human body. These are the superficial muscles, which are relatively large. Superficial muscles cover deeper, smaller muscles that cannot be seen unless the overlying muscles are reflected (that is, cut and pulled out of the way). For the sake of clarity, figures in this chapter that show deeper muscles will indicate whether superficial muscles have been reflected. To help you understand the relationships between skeletal muscles and skeletal bones, the figures include skeleton icons that show the origins and insertions of certain muscles in each group. Origins are shown in red, insertions in blue.

The tables in this chapter list the origin, insertion, and action of each muscle as well as the innervation of individual muscles. The term **innervation** refers to the supply of motor nerves that control each skeletal muscle or the nerve supply to a particular structure.

10.1 The Four Groups of Axial Muscles

▶ **KEY POINT** The four groups of axial muscles are the (1) muscles of the head and neck that are *not* associated with the vertebral column, (2) muscles that move the vertebral column (also referred to as the intrinsic muscles of the back), (3) muscles of the rib cage and the lateral walls of the abdominal and pelvic cavities, and (4) muscles that form the floor of the pelvic cavity.

To help you fully understand the origins, insertions, and actions of skeletal muscles, review (1) the appropriate skeletal figures in Chapters 6 and 7, (2) the four primary actions of skeletal muscles, and (3) **Spotlight Figure 9.13.** ⊃ pp. 254–255

Figure 10.1 Superficial Skeletal Muscles, Anterior View. A diagrammatic view of the major axial and appendicular muscles.

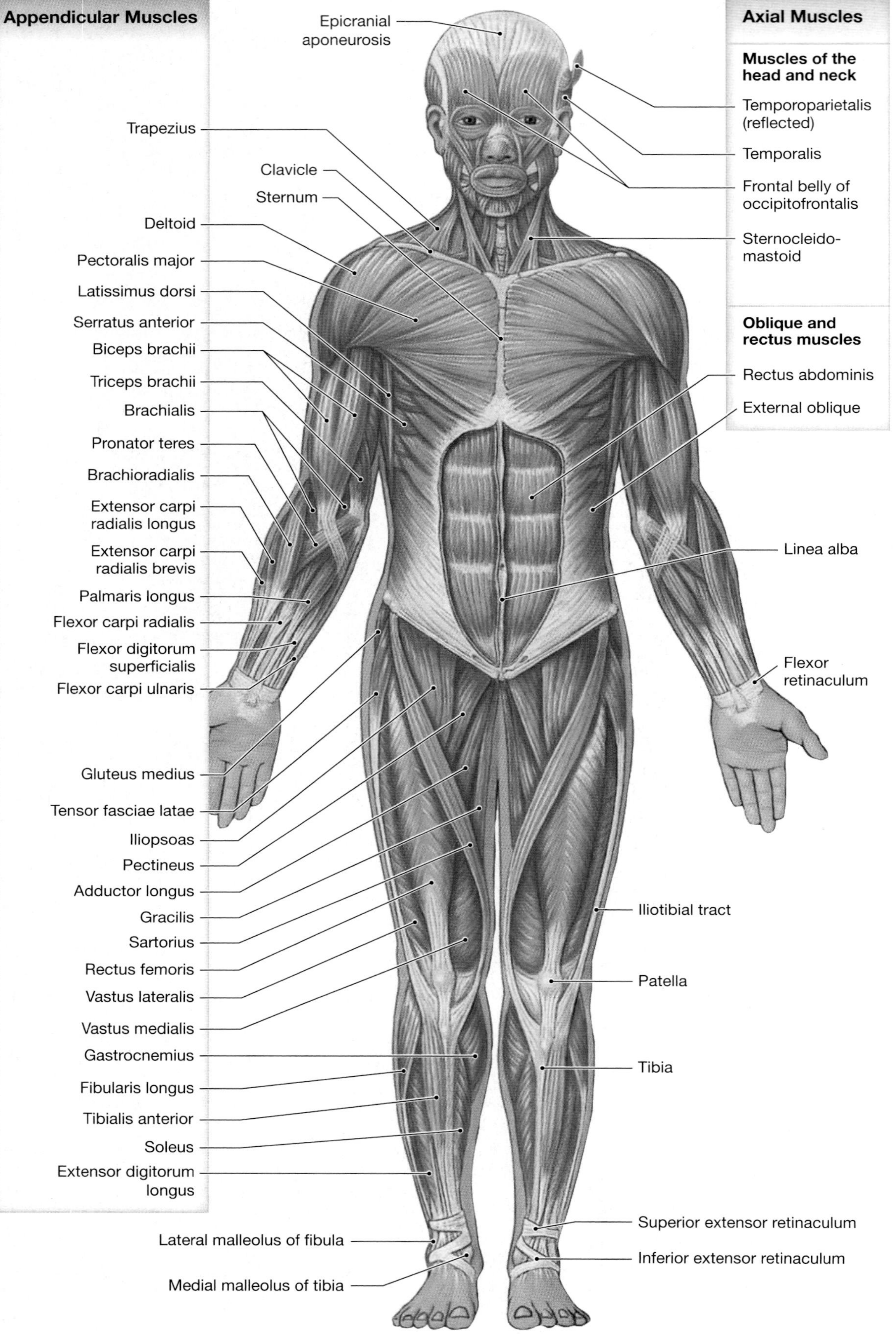

Figure 10.2 Superficial Skeletal Muscles, Posterior View. A diagrammatic view of the major axial and appendicular muscles.

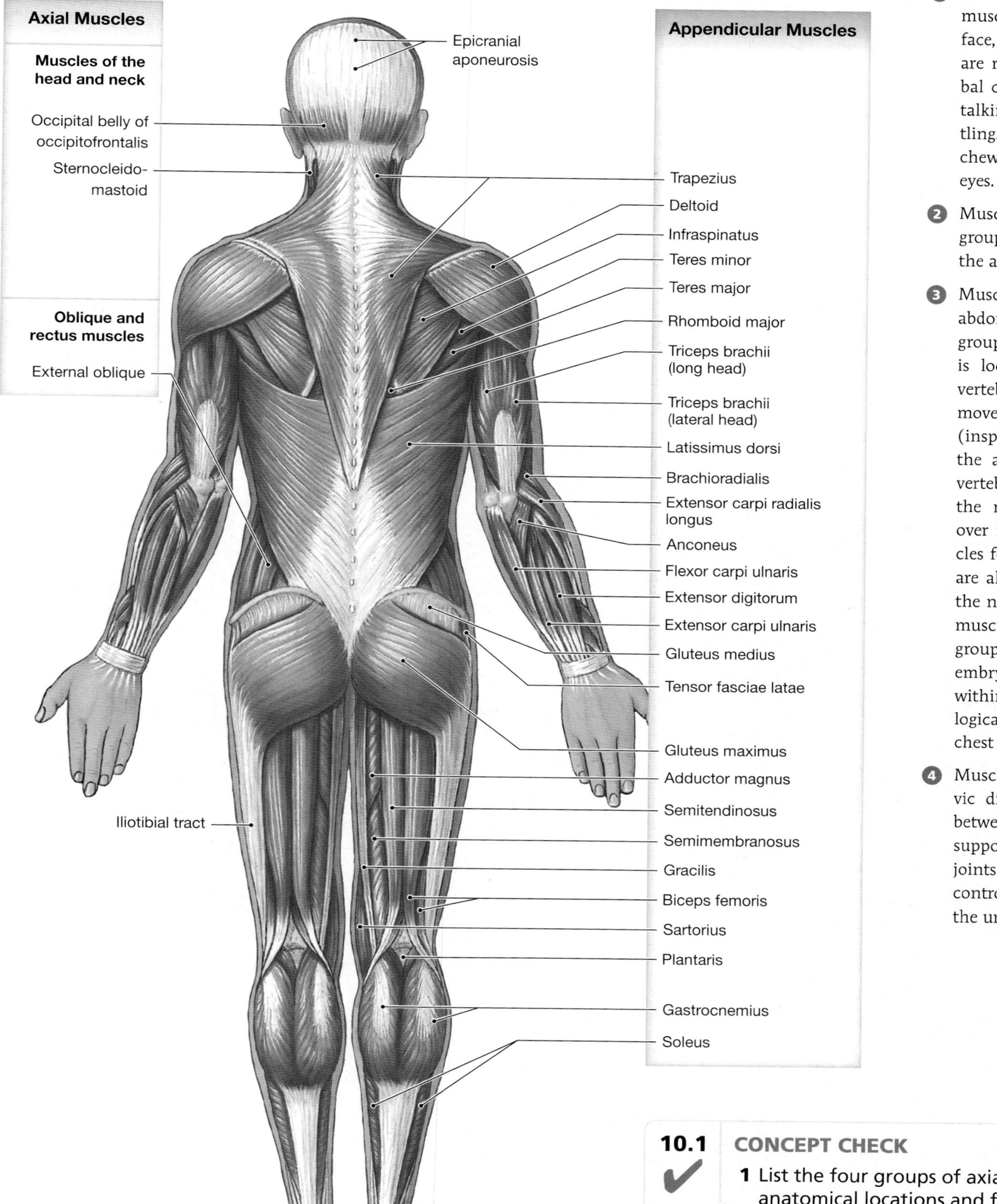

There are four groups of axial muscles:

1. Muscles of the head and neck. These muscles include those that move the face, tongue, larynx, and eyes. They are responsible for verbal and nonverbal communication, such as laughing, talking, frowning, smiling, and whistling. This group is also involved in chewing, swallowing, and moving the eyes.
2. Muscles of the vertebral column. This group includes flexors and extensors of the axial skeleton.
3. Muscles that form the walls of the abdominal and pelvic cavities. This group, the oblique and rectus muscles, is located between the first thoracic vertebra and the pelvis. These muscles move the chest wall during breathing (inspiration and expiration), compress the abdominal cavity, and rotate the vertebral column. In the thoracic area, the ribs separate these muscles, but over the abdominal surface, the muscles form broad muscular sheets. There are also oblique and rectus muscles in the neck. Although they do not form a muscular wall, they are included in this group because they share a common embryological origin. The diaphragm is within this group because it is embryologically linked to other muscles of the chest wall.
4. Muscles of the perineal region and pelvic diaphragm. These muscles extend between the sacrum and pelvic girdle to support organs of the pelvic cavity, flex joints of the sacrum and coccyx, and control movement of materials through the urethra and anus. **pp. 185, 190**

10

10.1 ✔ CONCEPT CHECK

1 List the four groups of axial muscles and give their anatomical locations and functions.

See the blue Answers tab at the back of the book.

10.2 | Muscles of the Head and Neck

▶ **KEY POINT** The muscles of the scalp and face are thin, sheetlike muscles that have one or more attachments to the connective tissue of the head or face. All the other muscles of the head and neck attach to bones of the skull and/or neck and move the eye, jaw, tongue, pharynx, or larynx.

The muscles of the head and neck are subdivided into several groups. The muscles of facial expression, eye movement (extra-ocular muscles), mastication (chewing), the tongue, and the pharynx originate on the skull or hyoid bone. The anterior muscles of the neck are concerned primarily with changing the position of the larynx, hyoid bone, and floor of the mouth. (Other muscles involved in sight and hearing originate on the skull and are discussed in Chapter 18.)

Muscles of Facial Expression

10

▶ **KEY POINT** The muscles of facial expression are divided into five groups (mouth, eyes, scalp, nose, and neck), and are all innervated by the facial nerve (VII).

The muscles of facial expression originate on the surface of the skull (**Figures 10.3** and **10.4**). At their insertions, the collagen fibers of the epimysium are continuous with the collagen fibers of the dermis of the skin and the superficial fascia. When these fibers contract, the skin moves, allowing us to convey a particular emotion or expression. The seventh cranial nerve, the facial nerve, innervates these muscles. **Table 10.1** summarizes the characteristics of the muscles of facial expression.

The largest group of facial muscles moves the mouth (**Figure 10.3**). The **orbicularis oris** (OR-is) compresses (or purses) the lips and thus is called the "kissing muscle." Other muscles move the lips and the corners of the mouth. The **buccinator** has two functions related to feeding (in addition to its importance in playing instruments like the trumpet). During chewing, the buccinator works with the muscles of mastication to move food back across the teeth from the space inside the cheeks. In infants, the buccinator produces the suction required for suckling at the breast.

Other muscles of facial expression control movements of the eyebrows and eyelids, the scalp, the nose, and the external ear. The scalp, or epicranium

Figure 10.3 Muscles of the Head and Neck, Part I

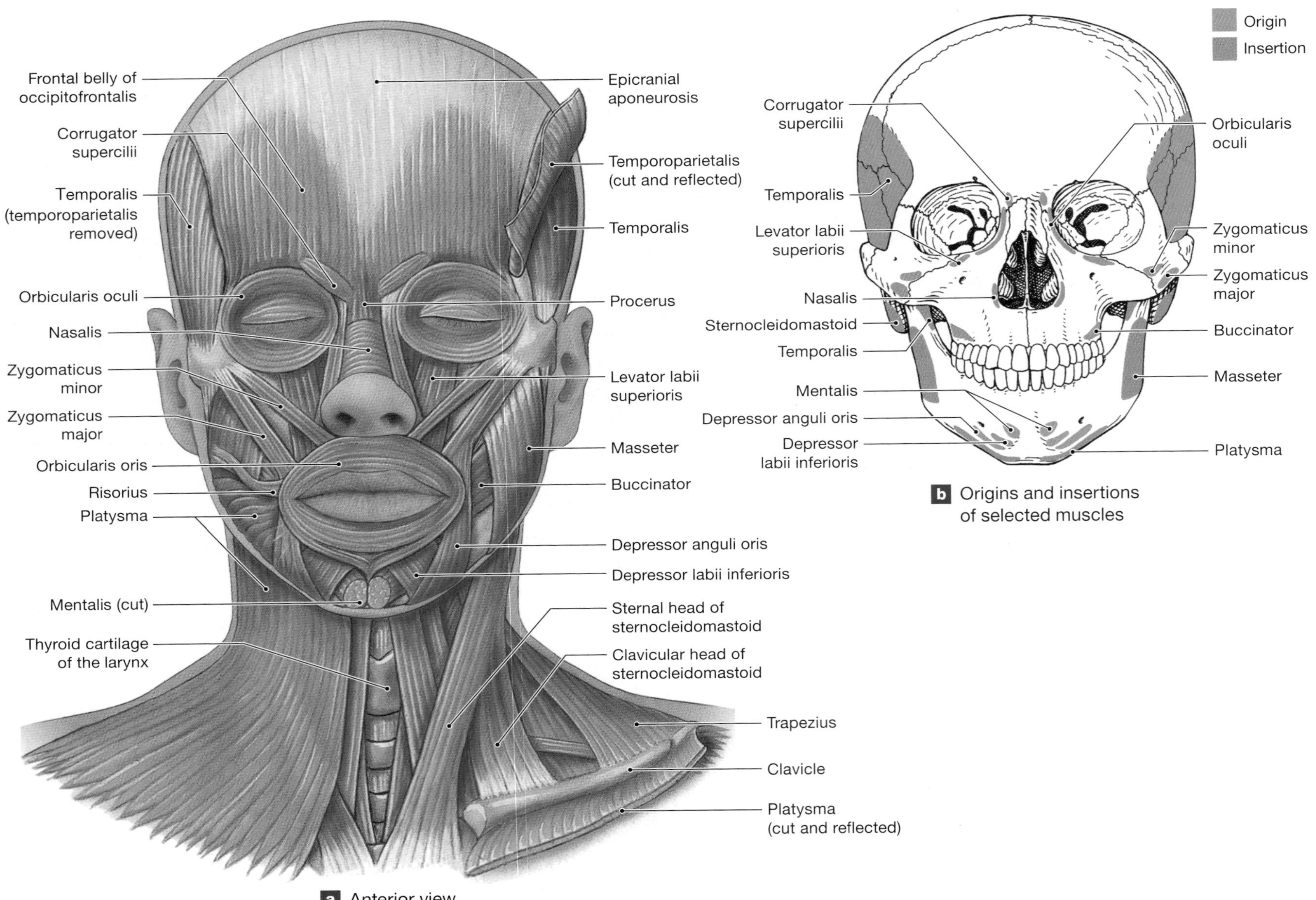

Figure 10.4 Muscles of the Head and Neck, Part II

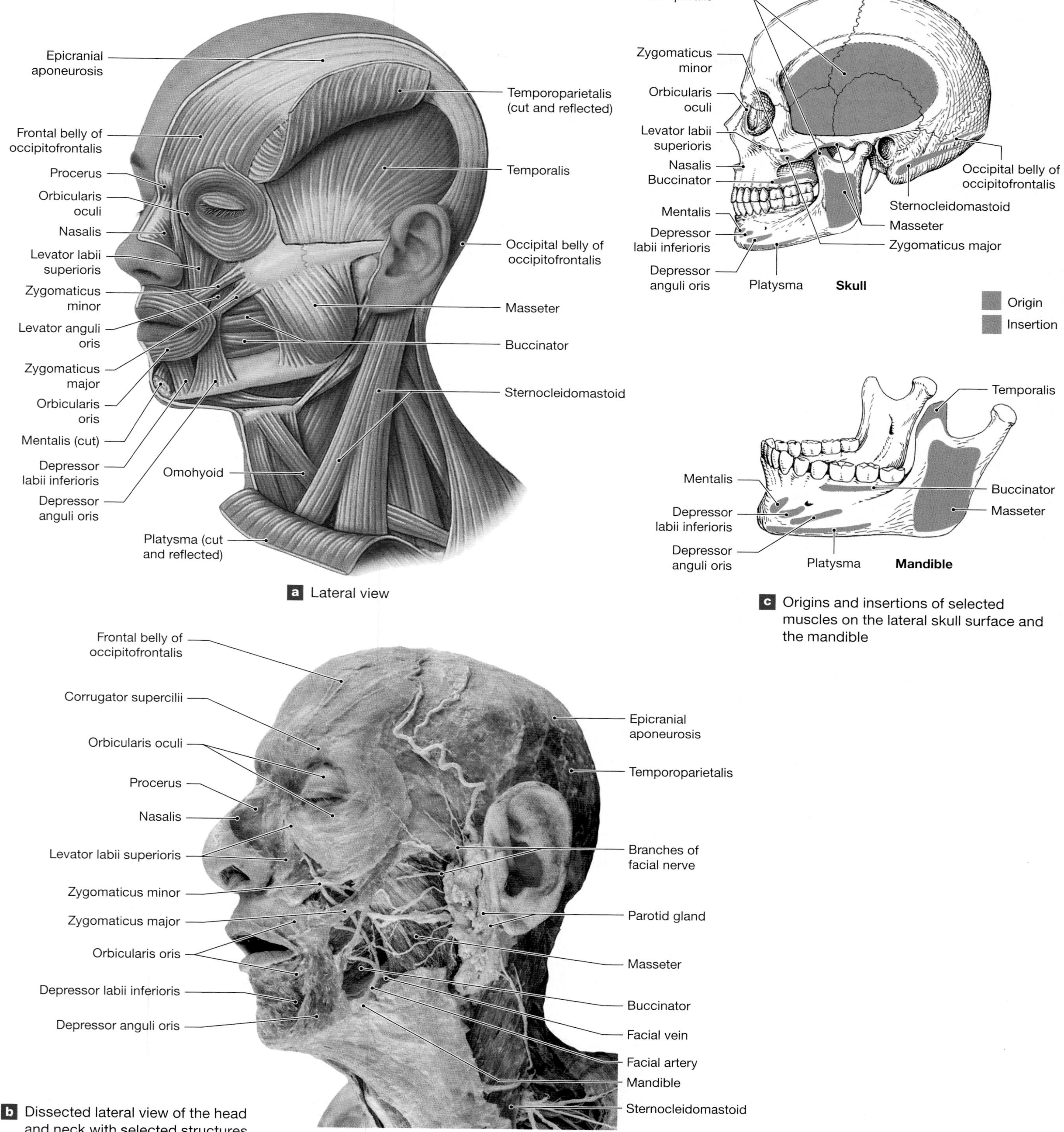

a Lateral view

c Origins and insertions of selected muscles on the lateral skull surface and the mandible

b Dissected lateral view of the head and neck with selected structures

Table 10.1 | Muscles of Facial Expression

Region/Muscle	Origin	Insertion	Action	Innervation
MOUTH				
Buccinator	Alveolar processes of maxilla and mandible opposite the molar teeth	Blends into fibers of orbicularis oris	Compresses cheeks	Facial nerve (VII)
Depressor labii inferioris	Mandible between the anterior midline and the mental foramen	Skin of lower lip	Depresses and helps evert lower lip	Facial nerve (VII)
Levator labii superioris	Maxilla and zygomatic bone, superior to the infra-orbital foramen	Orbicularis oris	Elevates and everts upper lip	Facial nerve (VII)
Mentalis	Incisive fossa of mandible	Skin of chin	Elevates, everts, and protrudes lower lip	Facial nerve (VII)
Orbicularis oris	Maxilla and mandible	Lips	Compresses, purses lips	Facial nerve (VII)
Risorius	Fascia surrounding parotid salivary gland	Angle of mouth	Draws corner of mouth laterally	Facial nerve (VII)
Levator anguli oris	Canine fossa of the maxilla inferior to the infra-orbital foramen	Skin at and below angle of mouth	Raises corner of mouth	Facial nerve (VII)
Depressor anguli oris	Anterolateral surface of mandibular body	Skin at angle of mouth	Depresses and draws the corner of mouth laterally	Facial nerve (VII)
Zygomaticus major	Zygomatic bone near the zygomaticotemporal suture	Fibers of levator anguli oris, orbicularis oris, and other muscles at angle of mouth	Elevates corner of mouth and draws it laterally	Facial nerve (VII)
Zygomaticus minor	Zygomatic bone posterior to zygomaticomaxillary suture	Upper lip	Elevates upper lip	Facial nerve (VII)
EYE				
Corrugator supercilii	Medial end of superciliary arch	Eyebrow	Pulls skin inferiorly and medially; wrinkles brow	Facial nerve (VII)
Levator palpebrae superioris	Inferior aspect of lesser wing of the sphenoid superior to and anterior to optic canal	Upper eyelid	Elevates upper eyelid	Oculomotor nerve (III)[a]
Orbicularis oculi	Medial margin of orbit	Skin around eyelids	Closes eye	Facial nerve (VII)
NOSE				
Procerus	Lateral nasal cartilages and the aponeuroses covering the inferior portion of the nasal bones	Aponeurosis at bridge of nose and skin of forehead	Moves nose, changes position, shape of nostrils; draws medial angle of eyebrows inferiorly	Facial nerve (VII)
Nasalis	Maxilla and alar cartilage of nose	Bridge of nose	Compresses bridge of nose; depresses tip; elevates corners of nostrils	Facial nerve (VII)
SCALP (EPICRANIUM)[b]				
Occipitofrontalis				
Frontal belly	Epicranial aponeurosis	Skin of eyebrow and bridge of nose	Raises eyebrows, wrinkles forehead	Facial nerve (VII)
Occipital belly	Superior nuchal line and adjacent region of mastoid portion of the temporal bone	Epicranial aponeurosis	Tenses and retracts scalp	Facial nerve (VII)
Temporoparietalis	Fascia around external ear	Epicranial aponeurosis	Tenses scalp, moves auricle of ear	Facial nerve (VII)
NECK				
Platysma	Fascia covering the superior parts of the pectoralis major and deltoid	Mandible and skin of cheek	Tenses skin of neck, depresses mandible	Facial nerve (VII)

[a] This muscle originates in association with the extra-ocular muscles, so its innervation is unusual, as discussed in Chapter 16.
[b] Includes the epicranial aponeurosis, temporoparietalis, and occipitofrontalis.

(ep-i-KRĀ-ne-um; *epi-*, on, + *kranion*, skull), contains the **temporoparietalis** and the **occipitofrontalis** (**Figures 10.3** and **10.4**). The temporoparietalis tenses the scalp and moves the ear, and the occipitofrontalis raises the eyebrows and retracts the scalp. The occipitofrontalis has two muscle bellies, the frontal belly and the occipital belly, separated by a collagenous sheet called the epicranial aponeurosis. The superficial **platysma** (pla-TIZ-ma; *platys*, flat) covers the anterior surface of the neck, extending from the base of the neck to the periosteum of the mandible and the fascia at the corners of the mouth (**Figures 10.3** and **10.4**). The platysma tenses the skin of the neck and depresses the mandible.

Extra-ocular Muscles

▶**KEY POINT** The extra-ocular muscles move the eyeballs. The oculomotor, trochlear, and abducens nerves (III, IV, and VI) innervate these muscles.

Six **extra-ocular muscles**, sometimes called the *oculomotor* (ok-ū-lō-MŌ-ter) or *extrinsic eye muscles*, originate on the orbit, insert onto the sclera (white of the eye) just posterior to the cornea, and change the position of each eye. The **inferior rectus**, **medial rectus**, **superior rectus**, **lateral rectus**, **inferior oblique**, and **superior oblique** move the eyes in the direction indicated by their names (**Figure 10.5** and **Table 10.2**).

Figure 10.5 Extra-ocular Muscles

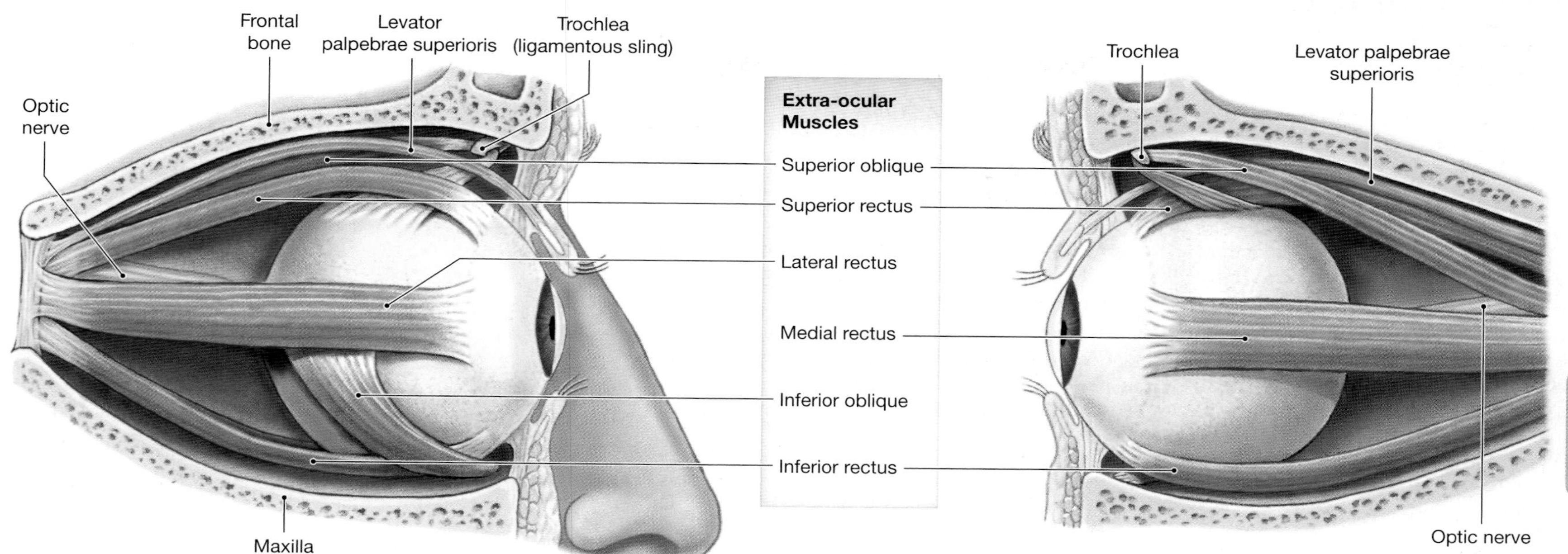

a Muscles on the lateral surface of the right eye.

b Muscles on the medial surface of the right eye.

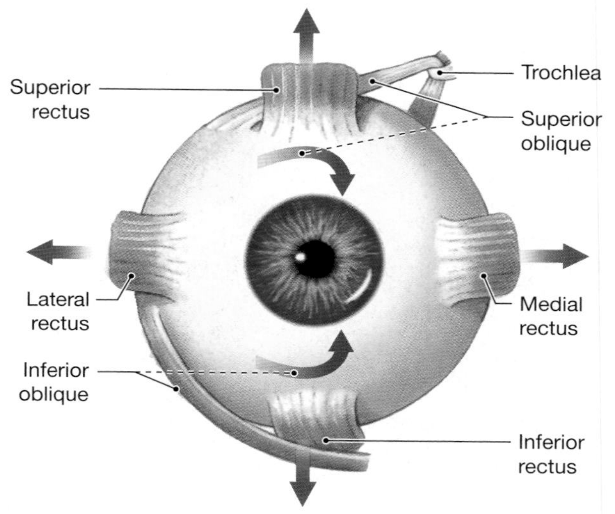

c Anterior view of the right eye showing the directions of eye movement produced by contractions of the individual muscles.

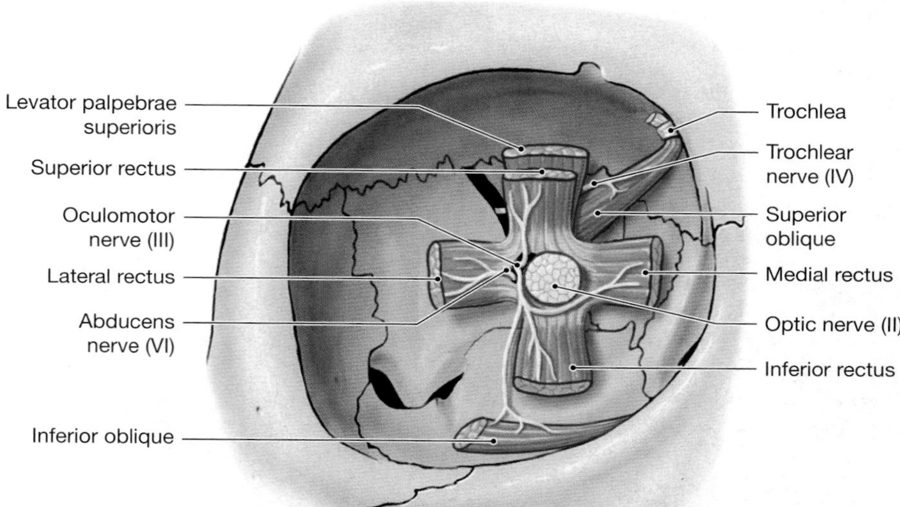

d Anterior view of the right orbit showing origins and innervations.

Table 10.2 | Extra-ocular Muscles

Muscle	Origin	Insertion	Action	Innervation
Inferior rectus	Sphenoid around optic canal	Inferior, medial surface of eyeball	Eye looks down	Oculomotor nerve (III)
Medial rectus	Sphenoid around optic canal	Medial surface of eyeball	Eye looks medially	Oculomotor nerve (III)
Superior rectus	Sphenoid around optic canal	Superior surface of eyeball	Eye looks up	Oculomotor nerve (III)
Lateral rectus	Sphenoid around optic canal	Lateral surface of eyeball	Eye looks laterally	Abducens nerve (VI)
Inferior oblique	Maxilla at anterior portion of orbit	Inferior, lateral surface of eyeball	Eye rolls, looks up and laterally	Oculomotor nerve (III)
Superior oblique	Sphenoid around optic canal	Superior, lateral surface of eyeball	Eye rolls, looks down and laterally	Trochlear nerve (IV)

The superior and inferior rectus also cause a slight medial movement of the eye, whereas the superior and inferior oblique cause a slight lateral movement. To roll the eye straight up, you contract the superior rectus and the inferior oblique; to roll the eye straight down, you contract the inferior rectus and the superior oblique. The third (oculomotor), fourth (trochlear), and sixth (abducens) cranial nerves innervate the extra-ocular muscles. The intrinsic eye muscles—smooth muscles inside the eyeball—control pupil diameter and lens shape. (These muscles are discussed in Chapter 18.)

CLINICAL NOTE

Botox

Botulinum toxin, produced by the bacterium *Clostridium botulinum*, is known by its brand name, Botox. Botox is injected directly into the muscles of facial expression to temporarily paralyze the muscles and eliminate unwanted lines or wrinkles. Facial wrinkles occur at right angles to the action lines of the facial muscles. For example, to smooth the horizontal wrinkle lines of the forehead, the frontal belly of the occipitofrontalis is injected. Injection of the corrugator supercilii and procerus smooths the vertical lines between the eyebrows. Injecting the lateral aspect of the orbicularis oculi eases crow's feet of the eyes, and injecting the levator labii superioris and zygomaticus minor and major softens the nasolabial fold lines. A side effect of Botox injection is loss of facial expression.

TIPS & TOOLS

An easy way to remember the innervations of the extra-ocular muscles is $[(\mathbf{LR_6})(\mathbf{SO_4})]_3$: Cranial nerve 6 innervates the lateral rectus (LR); cranial nerve 4 innervates the superior oblique (SO); and cranial nerve 3 innervates all the rest.

Muscles of Mastication

▶ **KEY POINT** All the muscles of mastication attach to the mandible and are innervated by the trigeminal nerve (V). The temporalis and masseter are superficial muscles, and the pterygoids are deep.

The muscles of mastication (chewing) move the mandible at the temporomandibular joint (**Figure 10.6** and **Table 10.3**). ↺ p. 212 The large **masseter** (ma-SĒ-ter), the most powerful muscle in this group, elevates the mandible. The **temporalis** (tem-po-RA-lis) assists in elevating the mandible, whereas the

Figure 10.6 Muscles of Mastication. The muscles of mastication move the mandible during chewing.

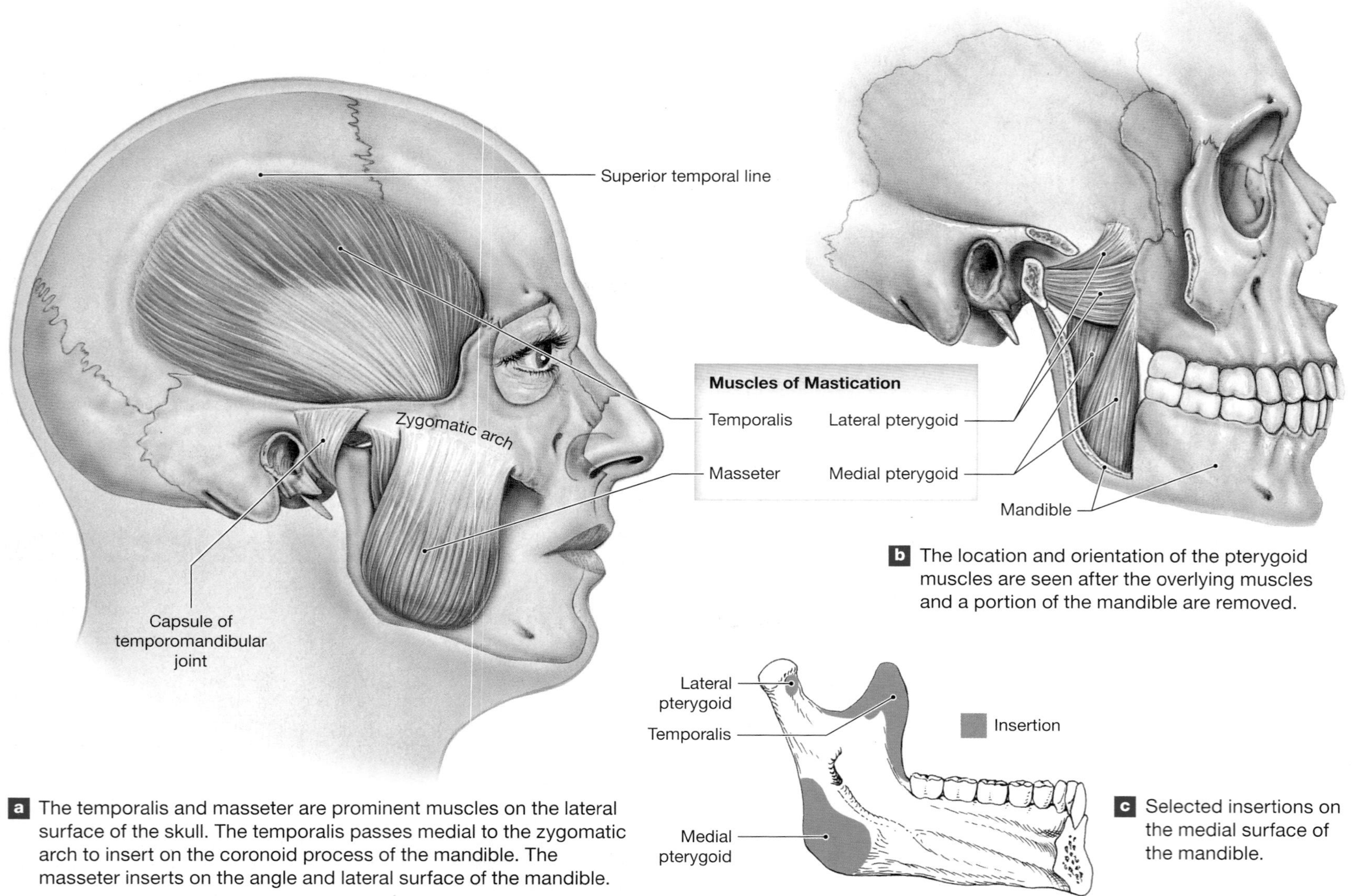

a The temporalis and masseter are prominent muscles on the lateral surface of the skull. The temporalis passes medial to the zygomatic arch to insert on the coronoid process of the mandible. The masseter inserts on the angle and lateral surface of the mandible.

b The location and orientation of the pterygoid muscles are seen after the overlying muscles and a portion of the mandible are removed.

c Selected insertions on the medial surface of the mandible.

Table 10.3 | **Muscles of Mastication**

Muscle	Origin	Insertion	Action	Innervation
Masseter	Aponeurosis and bones of the zygomatic arch	Lateral surface and angle of mandibular ramus	Elevates mandible and closes jaws, assists in protracting and retracting mandible and moving mandible side to side	Trigeminal nerve (V), mandibular branch
Temporalis	Along temporal lines of skull	Coronoid process of mandible and the anterior border of the mandibular ramus	Elevates mandible and closes jaws, assists in retracting and moving mandible from side to side	Trigeminal nerve (V), mandibular branch
Pterygoids				
Medial pterygoid	Medial surface of the lateral pterygoid plate and adjacent portions of palatine bone and maxilla	Medial surface of mandibular ramus	Elevates the mandible and closes the jaws, or moves mandible side to side	Trigeminal nerve (V), mandibular branch
Lateral pterygoid	Lateral surface of the lateral pterygoid plate and greater wing of sphenoid	Anterior part of the neck of the mandibular condyle	Opens jaws, protrudes mandible, or moves mandible side to side	Trigeminal nerve (V), mandibular branch

10

medial and lateral **pterygoids** (TER-i-goyd) elevate, protract, or slide the mandible from side to side. These movements maximize efficient use of the teeth while chewing or grinding foods. The fifth cranial nerve (V), the trigeminal nerve, innervates all the muscles of mastication.

Muscles of the Tongue

▶ **KEY POINT** Muscles of the tongue are divided into two groups: intrinsic and extrinsic tongue muscles. The names of the extrinsic muscles of the tongue all end with the suffix *–glossus* (tongue) and begin with one of the following prefixes: *genio–* (chin), *hyo–* (hyoid bone), *palato–* (palate), or *stylo–* (styloid process). Therefore, the names of these muscles tell you their origins and insertions.

Intrinsic tongue muscles originate and insert within the tongue; the extrinsic tongue muscles originate from structures outside the tongue and insert into the tongue. The **genioglossus** originates at the chin, the **hyoglossus** at the hyoid bone, the **palatoglossus** at the palate, and the **styloglossus** at the styloid process (**Figure 10.7**). These extrinsic tongue muscles move the tongue in the delicate and complex patterns necessary for speech. They also move food within the mouth as you prepare to swallow. The intrinsic tongue muscles, located entirely within the tongue, help with these activities. The 12th cranial nerve, the hypoglossal nerve (XII), innervates all the intrinsic and extrinsic muscles of the tongue except for the palatoglossus muscle. This muscle is innervated by the 10th cranial nerve, the vagus nerve (X). (**Table 10.4**).

Muscles of the Pharynx

▶ **KEY POINT** As with the muscles of the tongue, the names of the muscles of the pharynx tell you their origins and insertions: *palato–* means palate; *salingo–*, tube (in this instance, auditory tube); *stylo–*, styloid process; and *veli–*, membranous structure (in this instance, the soft palate).

The paired pharyngeal muscles start the swallowing process. The **pharyngeal constrictors** (**superior**, **middle**, and **inferior**) begin the process of moving a chewed mass of food (bolus) into the esophagus. The **palatopharyngeus**

Figure 10.7 Muscles of the Tongue. The left mandibular ramus has been removed to show the muscles on the left side of the tongue.

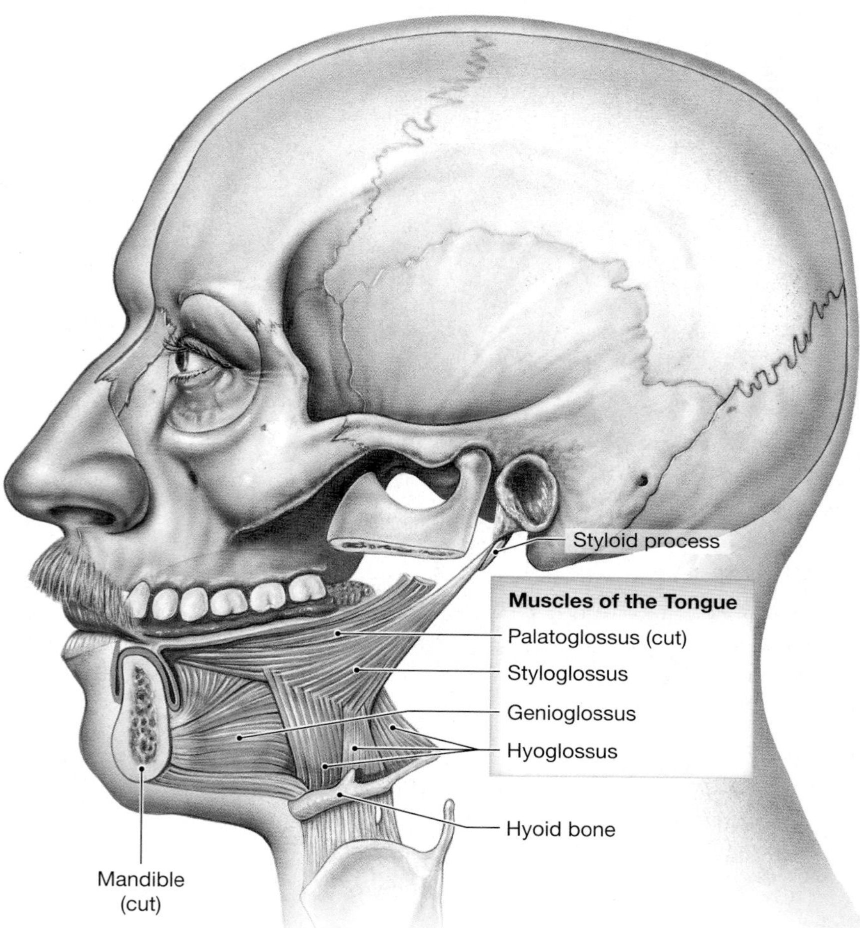

Table 10.4 | **Muscles of the Tongue**

Muscle	Origin	Insertion	Action	Innervation
Genioglossus	Medial surface of mandible around chin	Body of tongue, hyoid bone	Depresses and protracts tongue	Hypoglossal nerve (XII)
Hyoglossus	Body and greater horn of hyoid bone	Side of tongue	Depresses and retracts tongue	Hypoglossal nerve (XII)
Palatoglossus	Anterior surface of soft palate	Side of tongue	Elevates tongue, depresses soft palate	Branch of pharyngeal plexus (X)
Styloglossus	Styloid process of temporal bone	Along the side to tip and base of tongue	Retracts tongue, elevates sides	Hypoglossal nerve (XII)

(pal-āt-ō-far-IN-jē-us), **salpingopharyngeus** (sal-pin-gō-far-IN-jē-us), and **stylopharyngeus** (stī-lō-far-IN-jē-us) elevate the larynx. Together, they are known as the laryngeal elevators. The two **palatal muscles**, the **tensor veli palatini** and the **levator veli palatini**, raise the soft palate and adjacent portions of the pharyngeal wall and open the entrance to the auditory (Eustachian) tube. With the help of these muscles, a person can adjust to air pressure changes when flying or SCUBA diving by swallowing repeatedly, which opens and "pops" the ears. The glossopharyngeal (IX) and vagus (X) cranial nerves innervate the pharyngeal muscles. These muscles are illustrated in **Figure 10.8**; also see **Table 10.5**.

Figure 10.8 Muscles of the Pharynx. Pharyngeal muscles start the swallowing process.

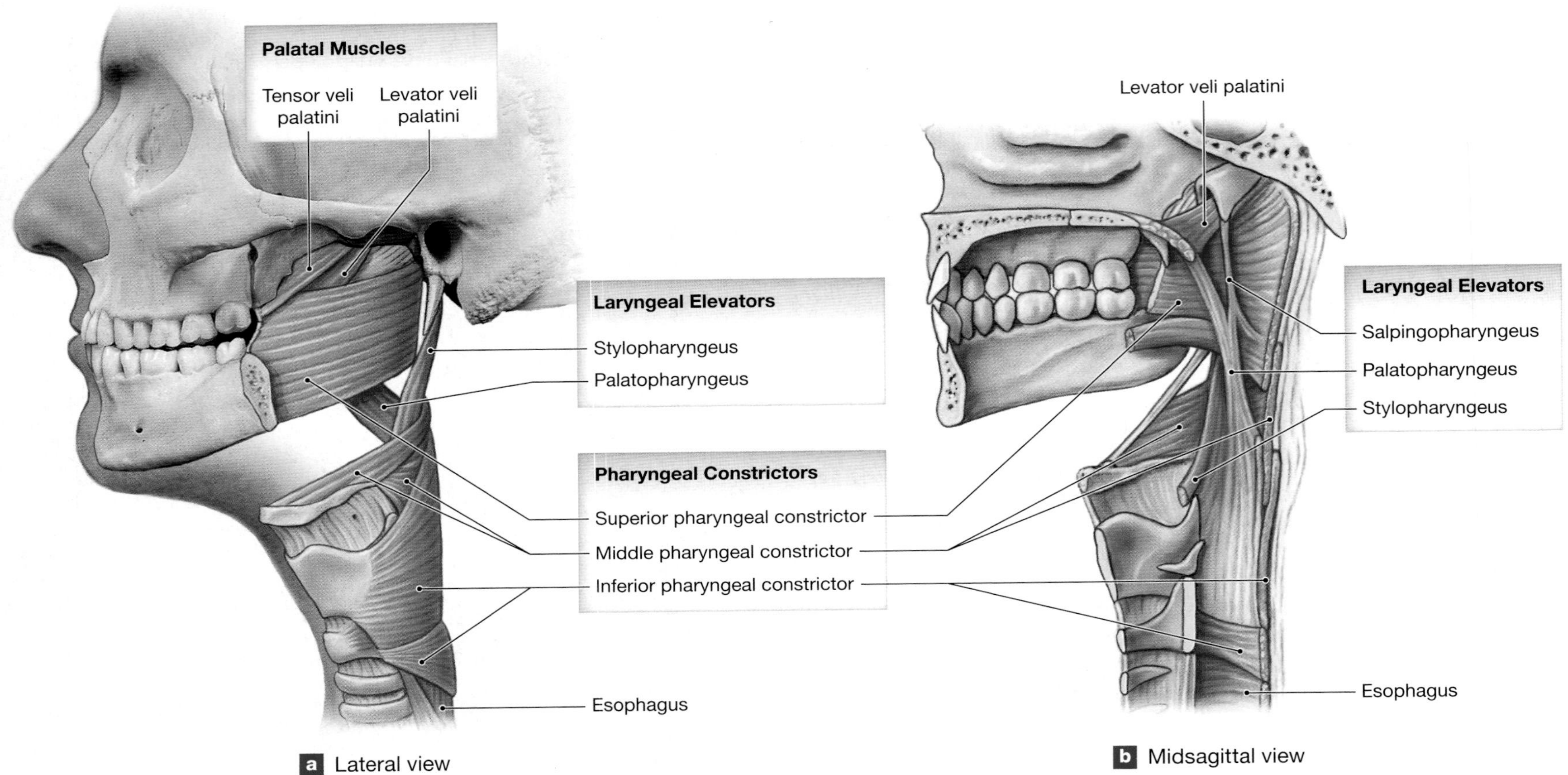

Table 10.5 | Muscles of the Pharynx

Muscle	Origin	Insertion	Action	Innervation
Pharyngeal Constrictors			Constrict pharynx to propel bolus into esophagus	Branches of pharyngeal plexus (N X)
Superior constrictor	Pterygoid process of sphenoid, medial surfaces of mandible, and the side of the tongue	Median raphe attached to occipital bone		Branches of pharyngeal plexus (N X)
Middle constrictor	Horns of hyoid bone	Median raphe		Branches of pharyngeal plexus (N X)
Inferior constrictor	Cricoid and thyroid cartilages of larynx	Median raphe		Branches of pharyngeal plexus (N X)
Laryngeal Elevators*			Elevate larynx	Branches of pharyngeal plexus (N IX and X)
Palatopharyngeus	Soft and hard palates	Thyroid cartilage		N X
Salpingopharyngeus	Cartilage around the inferior portion of the auditory tube	Thyroid cartilage		N X
Stylopharyngeus	Styloid process of temporal bone	Thyroid cartilage		N IX
Palatal Muscles				
Levator veli palatini	Petrous part of temporal bone, tissues around the auditory tube	Soft palate	Elevate soft palate	Branches of pharyngeal plexus (N X)
Tensor veli palatini	Sphenoidal spine, pterygoid process, and tissues around the auditory tube	Soft palate	Elevate soft palate	N V

*Assisted by the thyrohyoid, geniohyoid, stylohyoid, and hyoglossus muscles, discussed in Tables 10.4 and 10.6.

Anterior Muscles of the Neck

▶ **KEY POINT** The anterior muscles of the neck control the position of the larynx, depress the mandible, tense the floor of the mouth, and provide a stable foundation for muscles of the tongue and pharynx.

Figures 10.3, 10.4, and **10.9** illustrate the anterior muscles of the neck, and **Table 10.6** lists the origins, insertions, actions, and innervations of these muscles.

The anterior neck muscles that position the larynx are called *extrinsic* muscles, while those that affect the vocal cords are termed *intrinsic* muscles. (The vocal cords will be discussed in Chapter 24.) Additionally, the muscles of the neck are either *suprahyoid* or *infrahyoid* based on their location relative to the hyoid bone. The **digastric** (dī-GAS-trik) has two bellies (*di*–, two, + *gaster*, stomach). One belly originates on the mandible and inserts onto the hyoid bone; the other originates on the temporal bone and inserts onto the hyoid bone (**Figure 10.9**). This muscle opens the mouth by depressing the mandible. The anterior belly is superficial to the broad, flat **mylohyoid** (mī-lō-HĪ-oyd), which supports the floor of the mouth (**Figure 10.9**). The **geniohyoid**, superior to the mylohyoid muscle, provide additional support to the mouth floor.

The **stylohyoid** (stī-lō-HĪ-oyd) originates on the styloid process of the skull and inserts onto the hyoid bone (**Figure 10.9**). As its name indicates, the **sternocleidomastoid** (ster-nō-klī-dō-MAS-toid) originates from the sternum (*sterno*–) and the clavicle (*cleido*–) and inserts onto the mastoid process of the skull (**Figures 10.3, 10.4, 10.9**, and **Table 10.6**). *(Refer to Chapter 12,* **Figures 12.1** *and* **12.2a**, *to identify this structure from the body surface, and to* **Figures 12.9** *and* **12.10** *to visualize this structure in a cross section of the body at the levels of C_2 and T_2.)* The **omohyoid** (ō-mō-HĪ-oyd) attaches to the scapula, clavicle, first rib, and hyoid bone (**Figure 10.9**).

The remaining anterior muscles of the neck are straplike muscles connecting the sternum and the thyroid cartilage of the larynx (**sternothyroid**), the sternum and hyoid bone (**sternohyoid**), and the thyroid cartilage of the larynx and hyoid bone (**thyrohyoid**) (**Figure 10.9**).

10.2 CONCEPT CHECK

2 Where do muscles of facial expression originate?

3 What is the importance of the pharyngeal muscles?

See the blue Answers tab at the back of the book.

10

Figure 10.9 Anterior Muscles of the Neck

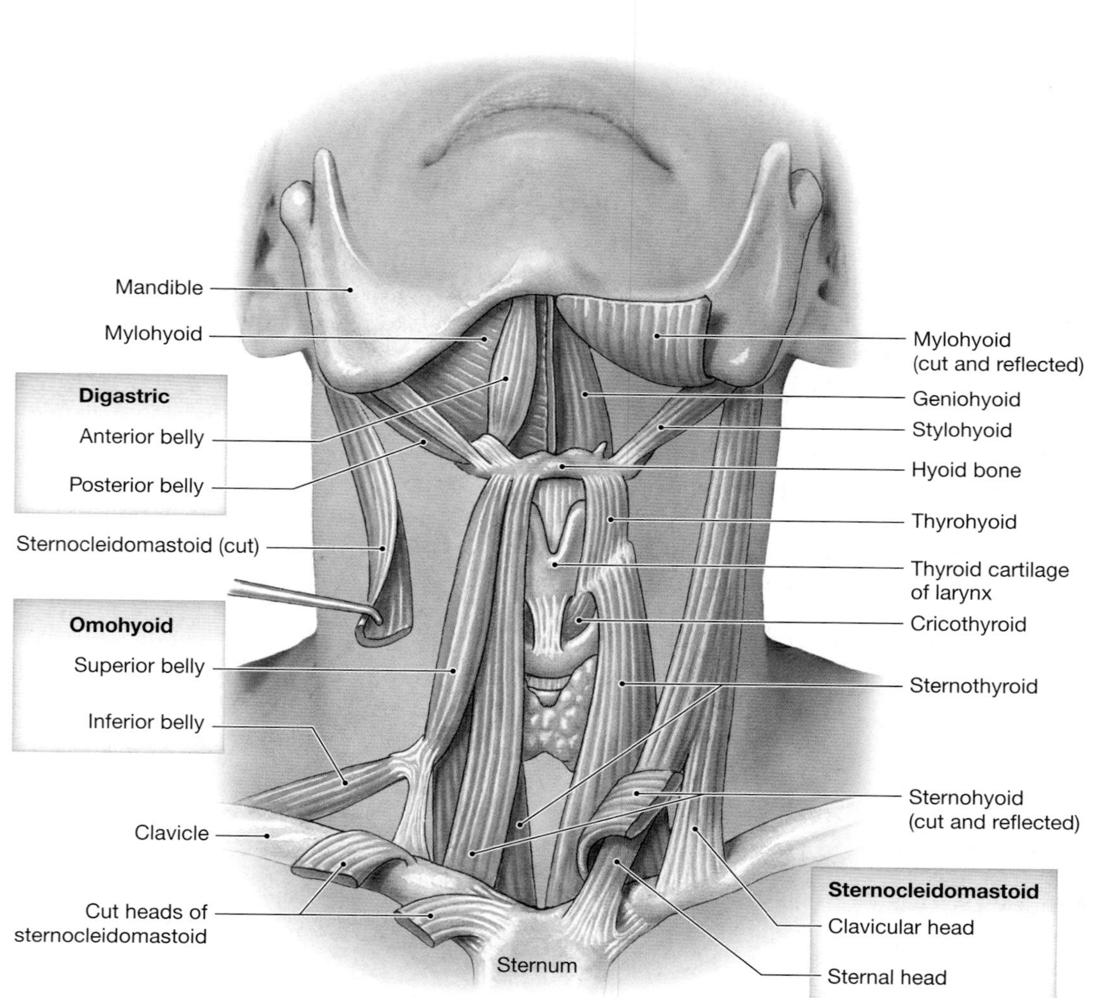

a Anterior view.

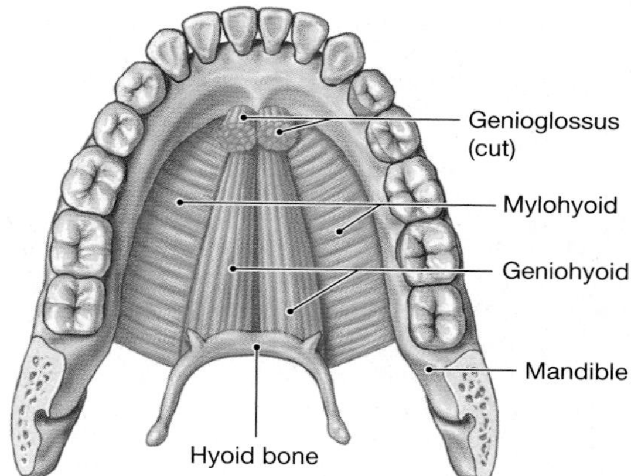

b Muscles that form the floor of the oral cavity, superior view.

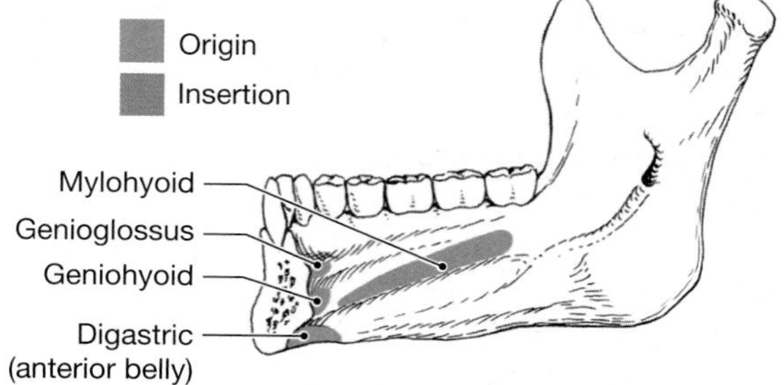

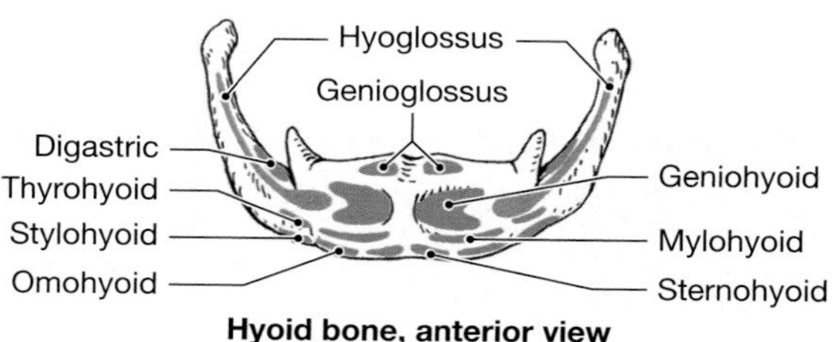

c Origins and insertions on the mandible and hyoid.

Table 10.6 | **Anterior Muscles of the Neck**

Muscle	Origin	Insertion	Action	Innervation
Digastric				
Anterior belly	From inferior surface of mandible at chin	Hyoid bone	Depresses mandible, opening mouth, and/or elevates larynx	Trigeminal nerve (V), mandibular branch
Posterior belly	From mastoid region of temporal bone	Hyoid bone	Depresses mandible, opening mouth, and/or elevates larynx	Facial nerve (VII)
Geniohyoid	Medial surface of mandible at chin	Hyoid bone	Depresses mandible, opening mouth, and/or elevates larynx, and retracts hyoid bone	Cervical nerve C_1 via hypoglossal nerve (XII)
Mylohyoid	Mylohyoid line of mandible	Median connective tissue band (raphe) that runs to hyoid bone	Elevates floor of mouth, elevates hyoid bone, and/or depresses mandible	Trigeminal nerve (V), mandibular branch
Omohyoid*	Superior border of the scapula near the suprascapular notch	Hyoid bone	Depresses hyoid bone and larynx	Cervical spinal nerves C_2–C_3
Sternohyoid	Clavicle and manubrium	Hyoid bone	Depresses hyoid bone and larynx	Cervical spinal nerves C_1–C_3
Sternothyroid	Posterior surface of manubrium and first costal cartilage	Thyroid cartilage of larynx	Depresses hyoid bone and larynx	Cervical spinal nerves C_1–C_3
Stylohyoid	Styloid process of temporal bone	Hyoid bone	Elevates larynx	Facial nerve (VII)
Thyrohyoid	Thyroid cartilage of larynx	Hyoid bone	Elevates larynx, depresses hyoid bone	Cervical spinal nerves C_1–C_2 via hypoglossal nerve (XII)
Sternocleidomastoid		Mastoid region of skull and lateral portion of superior nuchal line	Together, they flex the neck; alone, one side bends neck toward shoulder and turns face to opposite side	Accessory nerve (XI) and cervical spinal nerves (C_2–C_3) of cervical plexus
Clavicular head	Attaches to sternal end of clavicle			
Sternal head	Attaches to manubrium			

*Superior and inferior bellies, united at central tendon anchored to clavicle and first rib.

10.3 | Muscles of the Vertebral Column

▶**KEY POINT** The muscles of the back are arranged into three layers (superficial, intermediate, and deep). The muscles of the first two layers are the **extrinsic back muscles**. *These muscles are innervated by the anterior rami of the associated spinal nerves* and extend from the axial skeleton to the upper limb or the rib cage. The intermediate layer of the extrinsic back muscles consists of the serratus posterior. These muscles assist in moving the ribs during breathing. The deepest muscles of the back are the **intrinsic** (or *true*) **back muscles**. *These muscles are innervated by the posterior rami of the spinal nerves*, and they interconnect, move, and stabilize the vertebrae.

Muscles of the vertebral column are covered by more superficial back muscles and include many posterior extensors but few anterior flexors. We divide these muscles into extrinsic and intrinsic muscles of the back. *Extrinsic* back muscles are associated with upper extremity and shoulder movement. *Intrinsic* back muscles are fully contained (origin, belly, and insertion) within and act upon the back. The intrinsic (*true*) back muscles are arranged into superficial, intermediate, and deep layers (**Figure 10.10** and **Table 10.7**). These muscle layers are found lateral to the vertebral column, within the space between the spinous processes and the transverse processes of the vertebrae. Although these muscles extend from the sacrum to the skull overall, each muscle group is composed of numerous separate muscles of varying length.

The Superficial Layer of the Intrinsic Back Muscles

▶**KEY POINT** The superficial layer of the intrinsic back muscles is the splenius group. These posterior neck muscles extend, rotate, and laterally flex the cervical vertebrae.

The splenius muscles are the most superficial intrinsic back muscles. The **splenius capitis** originate on the ligamentum nuchae and the spines of C_7 and T_1 to T_4 and insert onto the skull. The **splenius cervicis** originate on the ligamentum nuchae and the spines of T_3 to T_6 vertebrae and insert onto C_1 to C_3. These two muscle groups extend and laterally flex the neck.

The Intermediate Layer of the Intrinsic Back Muscles

▶**KEY POINT** The intermediate layer of the intrinsic back muscles is the erector spinae group. The name tells us that they keep the spine erect.

The intermediate layer consists of the **erector spinae**. These muscles have a wide range of origins on the vertebral column, and the names of the individual muscles indicate where each muscle group inserts. For example, a *capitis* muscle inserts on the skull, whereas *cervicis* indicates an insertion on the upper cervical vertebrae. The erector spinae muscles are subdivided into the **spinalis**, **longissimus**, and **iliocostalis** muscle groups (**Figure 10.10a,b**). *(Refer to Chapter 12,* **Figures 12.9, 12.10, 12.12, 12.13,** *and* **12.14** *to visualize these structures in cross section of the body at the levels of C_2, T_2, T_{12}, and L_5.)* The spinalis group is the most medial of the three groups, and the iliocostalis is the most lateral. During bilateral contraction (when muscles on both sides contract), the erector spinae extend the vertebral column; unilateral contraction laterally flexes and rotates the vertebral column to the ipsilateral side.

The Deep Layer of the Intrinsic Back Muscles

▶**KEY POINT** The deepest layer of the intrinsic back muscles is composed of transversospinales muscles, which are the smallest and weakest true back muscles. These muscles extend and laterally flex the vertebral column to the same side and rotate the vertebral column to the contralateral side.

The deepest layer of true back muscles interconnects and stabilizes the vertebrae. This group of muscles, termed the **transversospinales muscles**, consists of the **semispinalis**, **multifidus**, **rotatores**, **interspinales**, and

Table 10.7 | **Muscles of the Vertebral Column**

Group/Muscle	Origin	Insertion	Action	Innervation
SUPERFICIAL LAYER				
Splenius (splenius capitis, splenius cervicis)	Spinous processes and ligaments connecting inferior cervical and superior thoracic vertebrae	Mastoid process, occipital bone of skull, superior cervical vertebrae	The two sides act together to extend neck; either alone rotates and laterally flexes neck to that side	Cervical spinal nerves
INTERMEDIATE LAYER (ERECTOR SPINAE)				
Spinalis Group				
Spinalis cervicis	Inferior portion of ligamentum nuchae and spinous process of C_7	Spinous process of axis and C_3–C_4	Extends neck	Cervical spinal nerves
Spinalis thoracis	Spinous processes of T_{11} and T_{12} and L_1 and L_2	Spinous processes of superior thoracic vertebrae	Extends vertebral column	Thoracic and lumbar spinal nerves
Longissimus Group				
Longissimus capitis	Transverse processes of inferior cervical and superior thoracic vertebrae	Mastoid process of temporal bone	The two sides act together to extend neck; either alone rotates and laterally flexes neck to that side	Cervical and thoracic spinal nerves
Longissimus cervicis	Transverse processes of superior thoracic vertebrae	Transverse processes of middle and superior cervical vertebrae	The two sides act together to extend neck; either alone rotates and laterally flexes neck to that side	Cervical and thoracic spinal nerves
Longissimus thoracis	Broad aponeurosis and at transverse processes of inferior thoracic and superior lumbar vertebrae; joins iliocostalis	Transverse processes of superior thoracic and lumbar vertebrae and inferior surfaces of lower 10 ribs	Extension of vertebral column; alone, each produces lateral flexion to that side	Thoracic and lumbar spinal nerves
Iliocostalis Group				
Iliocostalis cervicis	Superior borders of vertebrosternal ribs near the angles	Transverse processes of C_4–C_6	Extends or laterally flexes neck, elevates ribs	Cervical and superior thoracic spinal nerves
Iliocostalis thoracis	Superior borders of ribs 6–12 medial to the angles	Superior ribs and transverse process of C_7	Stabilizes thoracic vertebrae in extension	Thoracic spinal nerves
Iliocostalis lumborum	Iliac crest, sacral crests, and lumbar spinous processes	Inferior surfaces of ribs 6–12 near their angles	Extends vertebral column, depresses ribs	Inferior thoracic nerves and lumbar spinal nerves
DEEP MUSCLES OF THE SPINE (TRANSVERSOSPINALES)				
Semispinalis				
Semispinalis capitis	Processes of lower four cervical and superior six or seven thoracic vertebrae	Occipital bone, between nuchal lines	Together, the two sides extend neck; alone, each extends and laterally flexes neck and turns head to opposite side	Cervical spinal nerves
Semispinalis cervicis	Transverse processes of T_1–T_5 or T_6	Spinous processes of C_2–C_5	Extends vertebral column and rotates toward opposite side	Cervical spinal nerves
Semispinalis thoracis	Transverse processes of T_6–T_{10}	Spinous processes of C_6–T_4	Extends vertebral column and rotates toward opposite side	Thoracic spinal nerves
Multifidus	Sacrum and transverse process of each vertebra	Spinous processes of the third or fourth more superior vertebra	Extends vertebral column and rotates toward opposite side	Cervical, thoracic, and lumbar spinal nerves
Rotatores (cervicis, thoracis, and lumborum)	Transverse processes of the vertebrae in each region (cervical, thoracic, and lumbar)	Spinous process of adjacent, more superior vertebra	Extends vertebral column and rotates toward opposite side	Cervical, thoracic, and lumbar spinal nerves
Interspinales	Spinous process of each vertebra	Spinous processes of more superior vertebra	Extends vertebral column	Cervical, thoracic, and lumbar spinal nerves
Intertransversarii	Transverse processes of each vertebra	Transverse process of more superior vertebra	Lateral flexion of vertebral column	Cervical, thoracic, and lumbar spinal nerves
SPINAL FLEXORS				
Longus capitis	Transverse processes of C_4–C_6	Base of the occipital bone	Together, the two sides flex the neck; alone, each rotates head to that side	Cervical spinal nerves
Longus colli	Anterior surfaces of cervical and superior thoracic vertebrae	Transverse processes of superior cervical vertebrae	Flexes and/or rotates neck; limits hyperextension	Cervical spinal nerves
Quadratus lumborum	Iliac crest and iliolumbar ligament	Last rib and transverse processes of lumbar vertebrae	Together, they depress ribs; alone, each produces lateral flexion of vertebral column; fixes floating ribs (11 and 12) during forced exhalation; stabilizes diaphragm during inhalation	Thoracic and lumbar spinal nerves

intertransversarii (Figure 10.10a,b). *(Refer to Chapter 12, Figures 12.9 and 12.10 to visualize these structures in a cross section of the body at the levels of C_2 and T_2.)* These are all short muscles that work in various combinations to produce limited extension and lateral flexion of the vertebral column. They also rotate the vertebral column to the contralateral side. The transversospinales are also important in making delicate adjustments in the positions of individual vertebrae. If injured, these muscles start a cycle of pain → muscle stimulation → contraction → pain, which leads to pressure on adjacent spinal nerves, sensory losses, and limited mobility. Many of the warm-up and stretching exercises recommended before physical exercise prepare these small but very important muscles for their supporting roles.

Spinal Flexors

▶ **KEY POINT** There are only a few spinal flexors, in part because many of the large trunk muscles flex the vertebral column when they contract. Another reason is that most of the body weight lies anterior to the vertebral column, so gravity tends to flex the spine.

Figure 10.10 Muscles of the Vertebral Column. Selected origins and insertions are shown.

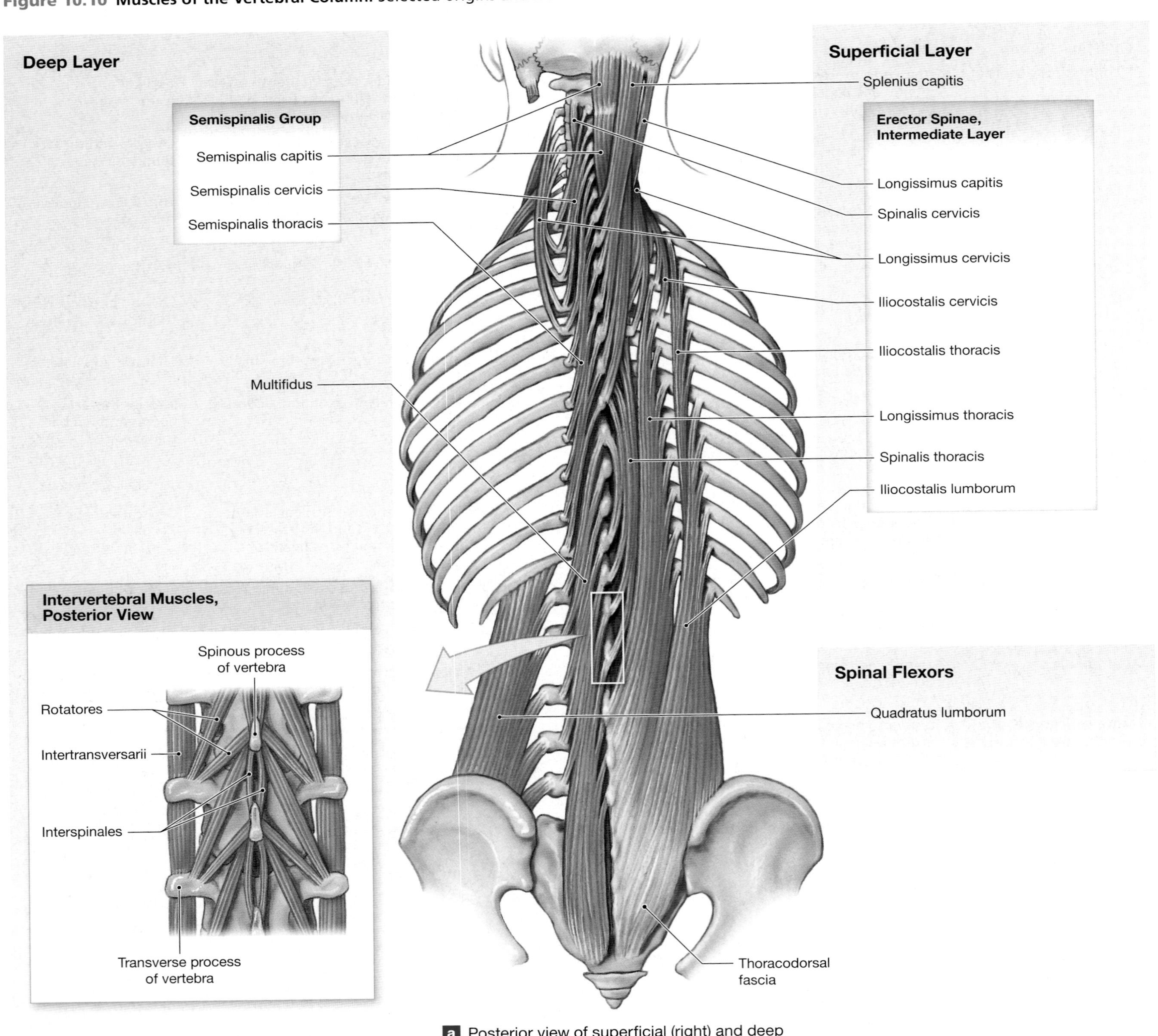

a Posterior view of superficial (right) and deep (left) muscles of the vertebral column

Spinal flexors are found on the anterior surface of the vertebral column. In the neck, the **longus capitis** and **longus colli** rotate and flex the neck, depending on whether the muscles of one or both sides are contracting **(Figure 10.10c)**. *(Refer to Chapter 12,* **Figure 12.9** *to visualize these structures in a cross section of the body at the level of* C_2*.)* In the lumbar region, the large **quadratus lumborum** laterally flex the vertebral column and depress the ribs **(Figure 10.10a)**.

Figure 10.10 (continued)

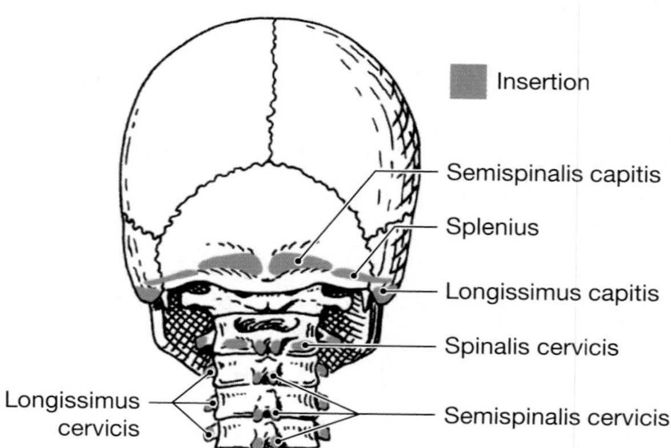

b Posterior view of the skull and cervical spine showing selected muscle insertions

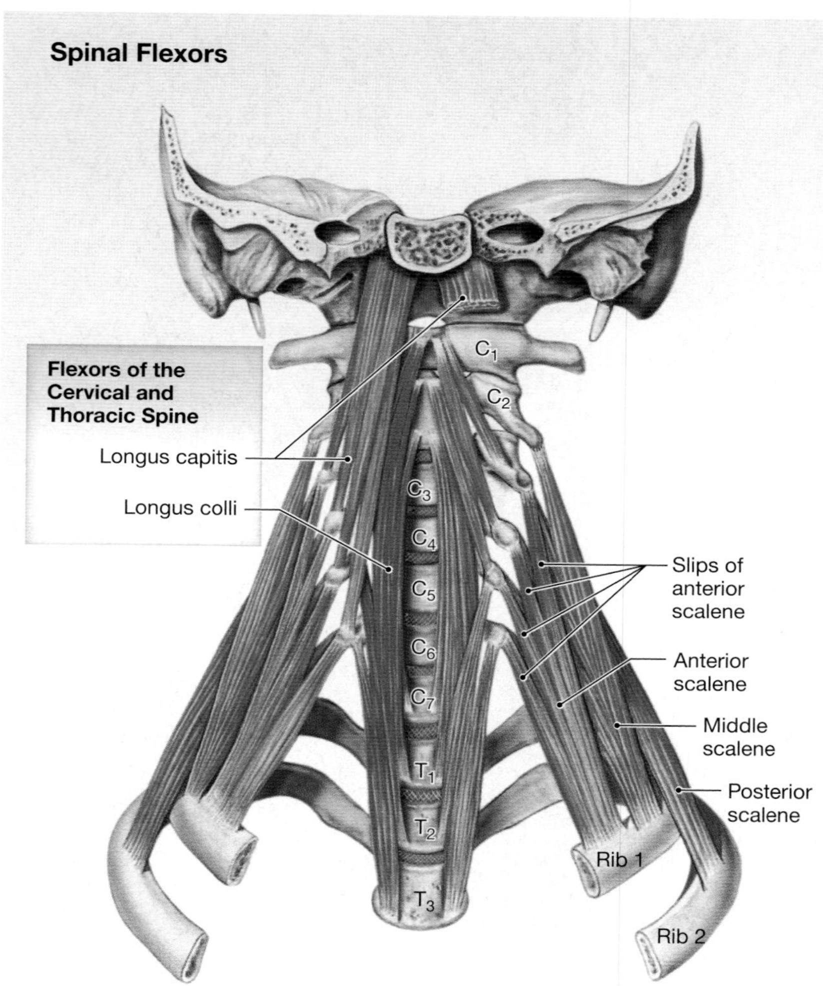

c Muscles on the anterior surfaces of the cervical and superior thoracic vertebrae

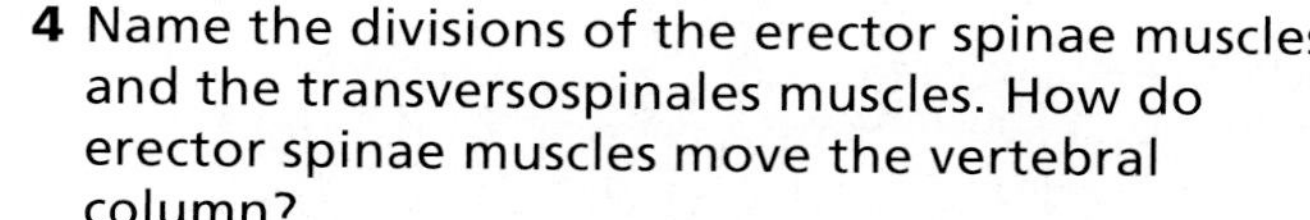

10.3 CONCEPT CHECK

4 Name the divisions of the erector spinae muscles and the transversospinales muscles. How do erector spinae muscles move the vertebral column?

5 List two reasons why there are fewer intrinsic back muscles that flex the vertebral column than intrinsic back muscles that extend the vertebral column.

See the blue Answers tab at the back of the book.

10.4 Oblique and Rectus Muscles

▶ **KEY POINT** The oblique muscles compress underlying structures and rotate the vertebral column. The rectus muscles are important flexors of the vertebral column and are antagonists to the erector spinae muscles.

The oblique and rectus muscles lie between the vertebral column and the anterior midline (**Figures 10.10c, 10.11**, and **10.12b** and **Table 10.8**). These muscles are divided into three groups: cervical, thoracic, and abdominal.*

The oblique muscles include the **scalenes** (SKĀ-lēn) of the neck and the **intercostal** (in-ter-KOS-tul) and **transversus muscles** of the thorax. The **anterior**, **middle**, and **posterior scalenes** elevate the first two ribs and flex the neck **(Figure 10.10c)**. In the thorax, the oblique muscles lie between the ribs and are called intercostal muscles. The **external intercostal muscles** cover the **internal intercostal muscles (Figure 10.11b)**. Both sets of intercostal muscles aid in breathing movements of the ribs. A small **transversus thoracis** crosses the inner surface of the rib cage and is separated from the pleural cavity by the parietal pleura, a serous membrane. ⮌ **pp. 19–21, 71**

The abdominal oblique muscles show the same muscular pattern as in the thorax: three layers of muscles, with each layer arranged in a manner similar to that of the thoracic cavity. These muscles are the **external** and **internal obliques** (also called the *abdominal obliques*) and the **transversus abdominis** (ab-DOM-i-nis) **(Figure 10.11b–d)**. The muscle arrangement in these three layers strengthens the abdominal wall.

The **rectus abdominis** originates at the xiphoid process and inserts onto the pubic bone. A band of fibrous connective tissue called the **linea alba** (*white line*) divides this muscle longitudinally. The transverse **tendinous inscriptions** are bands of fibrous connective tissue that divide this muscle into four repeated segments **(Figure 10.11)**.

The surface anatomy of the oblique and rectus muscles of the thorax and abdomen is shown in **Figures 10.11a** and **12.3**. *(Refer to Chapter 12,* **Figures 12.13** *and* **12.14**, *to visualize these structures in a cross section of the body at the levels of* T_{12} *and* L_5*.)*

*We group the oblique and rectus muscles of the trunk and the diaphragm together because of their common embryological origins.

Figure 10.11 The Oblique and Rectus Muscles

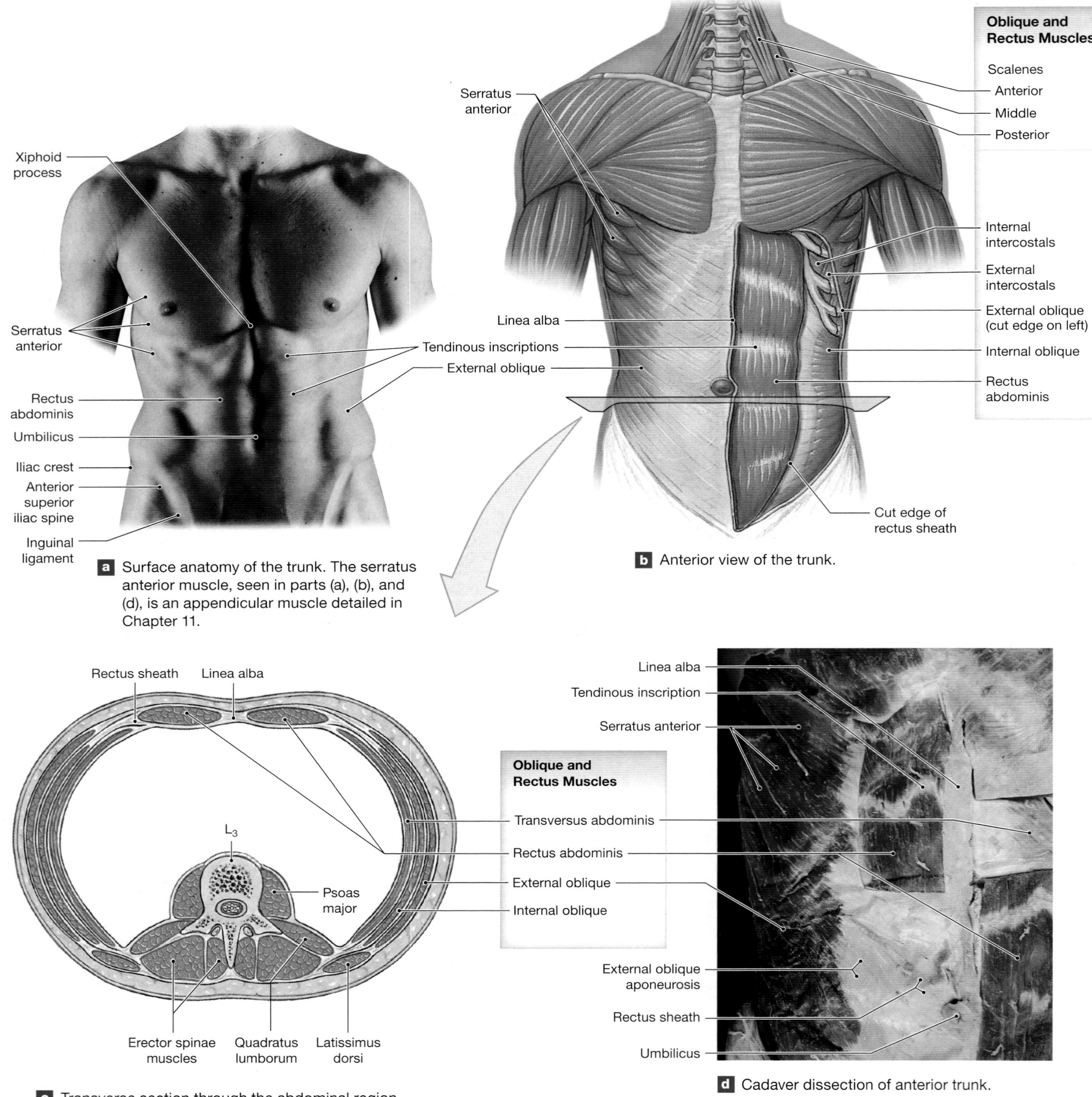

a Surface anatomy of the trunk. The serratus anterior muscle, seen in parts (a), (b), and (d), is an appendicular muscle detailed in Chapter 11.

b Anterior view of the trunk.

c Transverse section through the abdominal region.

d Cadaver dissection of anterior trunk.

Table 10.8 | **Oblique and Rectus Muscles**

Group/Muscle	Origin	Insertion	Action	Innervation
OBLIQUE GROUP				
Cervical Region				
Scalene anterior	Transverse and costal processes C_3–C_6	Superior surface of first rib	Elevate ribs and/or flex neck; one side bends neck and rotates to the opposite side	Cervical spinal nerves
Scalenus middle	Transverse and costal processes of atlas (C_1) and C_3–C_7	Superior surface of first rib	Elevate ribs and/or flex neck; one side bends neck and rotates to the same side	Cervical spinal nerves
Scalenus posterior	Transverse and costal processes C_4–C_6	Superior surface of second rib	Elevate ribs and/or flex neck; one side bends neck and rotates to the same side	Cervical spinal nerves
Thoracic Region				
External intercostals	Inferior border of each rib	Superior border of more inferior rib	Elevate ribs	Intercostal nerves (branches of thoracic spinal nerves)
Internal intercostals	Superior border of each rib	Inferior border of the more superior rib	Depress ribs	Intercostal nerves (branches of thoracic spinal nerves)
Transversus thoracis	Posterior surface of sternum	Cartilages of ribs	Depress ribs	Intercostal nerves (branches of thoracic spinal nerves)
Serratus Posterior				
Superior	Spinous processes of C_7–T_3 and ligamentum nuchae	Superior borders of ribs 2–5 near angles	Elevate ribs, enlarges thoracic cavity	Thoracic nerves (T_1–T_4)
Inferior	Aponeurosis from spinous processes of T_{11}–L_2 or L_3	Inferior borders of ribs 9–12	Pull ribs inferiorly; also pulls outward, opposing diaphragm	Thoracic nerves (T_9–T_{12})
Abdominal Region				
External oblique	External and inferior borders of ribs 5–12	External oblique aponeuroses extending to linea alba and iliac crest	Compress abdomen; depress ribs; flex, laterally flexes, or rotates vertebral column to the opposite side	Intercostal nerves 5–12, iliohypogastric, and ilioinguinal nerves
Internal oblique	Thoracolumbar fascia, inguinal ligament, and iliac crest	Inferior surfaces of ribs 9–12, costal cartilages 8–10, linea alba, and pubis	As above, but rotates vertebral column to same side	Intercostal nerves 5–12, iliohypogastric, and ilioinguinal nerves
Transversus abdominis	Cartilages of ribs 7–12, iliac crest, and thoracolumbar fascia	Linea alba and pubis	Compress abdomen	Intercostal nerves 5–12, iliohypogastric, and ilioinguinal nerves
RECTUS GROUP				
Cervical region	Includes the geniohyoid, omohyoid, sternohyoid, sternothyroid, and thyrohyoid in **Table 10.6**			
Thoracic region				
Diaphragm	Xiphoid process, ribs 7–12 and associated costal cartilages, and anterior surfaces of lumbar vertebrae	Central tendon sheet	Contraction expands thoracic cavity, compresses abdominopelvic cavity	Phrenic nerves (C_3–C_5)
Abdominal region				
Rectus abdominis	Superior surface of pubis around symphysis	Inferior surfaces of cartilages (ribs 5–7) and xiphoid process of sternum	Depress ribs, flex vertebral column and compress abdomen	Intercostal nerves (T_7–T_{12})

The Diaphragm

▶ **KEY POINT** The diaphragm is a dome-shaped sheet of skeletal muscle separating the thoracic and abdominal cavities. Contracting the diaphragm flattens the muscle, increasing the volume of the thoracic cavity and causing inhalation.

The **diaphragm** is a major muscle of breathing (**Figure 10.12**). Contracting the diaphragm increases the volume of the thoracic cavity, aiding inhalation. Relaxing the diaphragm decreases the volume of the thoracic cavity, aiding exhalation. (The muscles of breathing are examined in Chapter 24).

10.4 **CONCEPT CHECK**

6 Damage to the external intercostal muscles interferes with what important process?

See the blue Answers tab at the back of the book.

Figure 10.12 The Diaphragm. This muscular sheet separates the thoracic cavity from the abdominopelvic cavity.

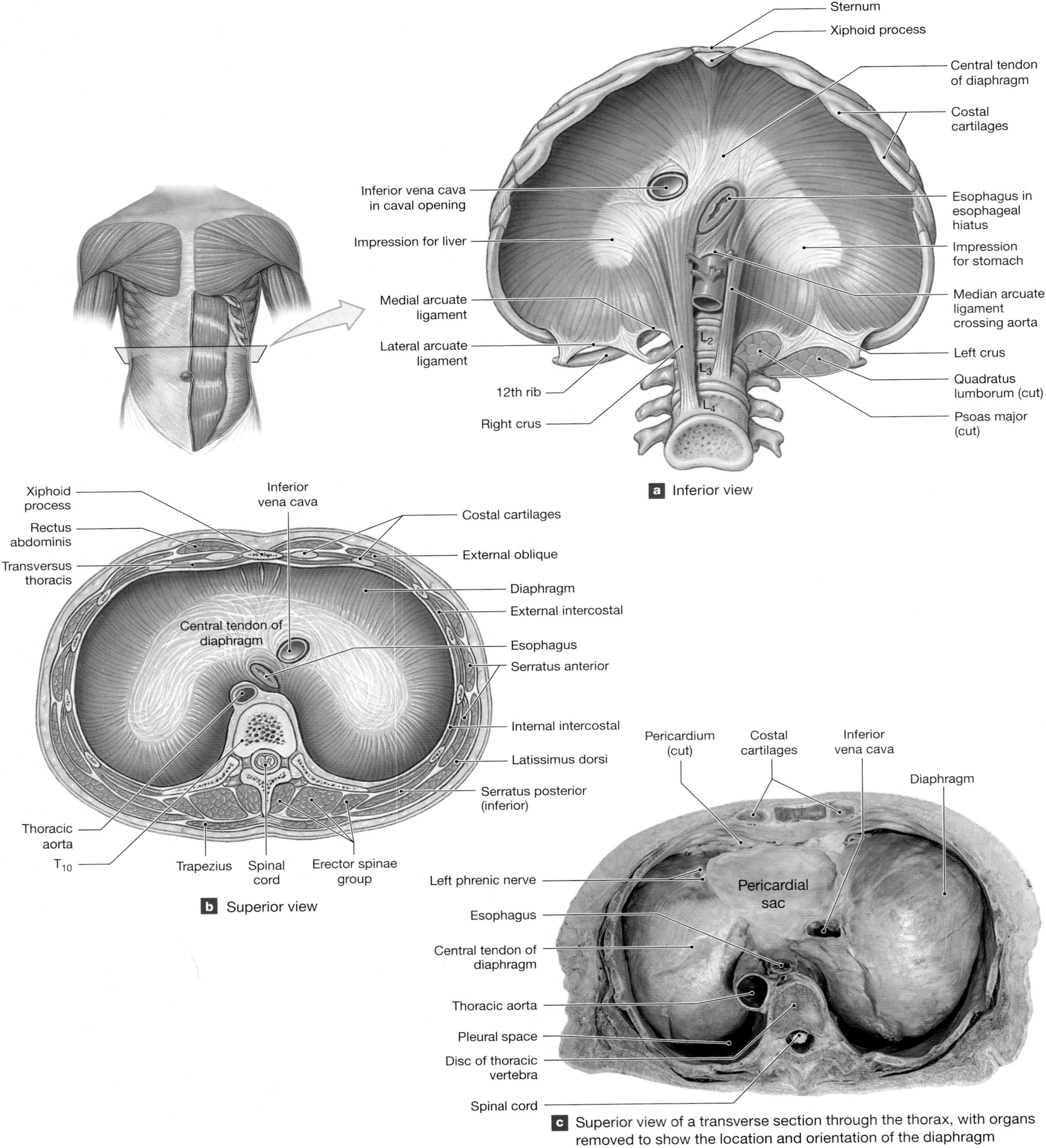
Sternum
Xiphoid process
Central tendon of diaphragm
Costal cartilages
Inferior vena cava in caval opening
Esophagus in esophageal hiatus
Impression for liver
Impression for stomach
Medial arcuate ligament
Median arcuate ligament crossing aorta
Lateral arcuate ligament
Left crus
Quadratus lumborum (cut)
12th rib
Right crus
Psoas major (cut)
L2
L3
L4
a Inferior view
Xiphoid process
Inferior vena cava
Costal cartilages
Rectus abdominis
External oblique
Transversus thoracis
Diaphragm
External intercostal
Central tendon of diaphragm
Esophagus
Serratus anterior
Internal intercostal
Latissimus dorsi
Serratus posterior (inferior)
Thoracic aorta
T10
Trapezius
Spinal cord
Erector spinae group
b Superior view
Pericardium (cut)
Costal cartilages
Inferior vena cava
Diaphragm
Left phrenic nerve
Pericardial sac
Esophagus
Central tendon of diaphragm
Thoracic aorta
Pleural space
Disc of thoracic vertebra
Spinal cord
c Superior view of a transverse section through the thorax, with organs removed to show the location and orientation of the diaphragm

10.5 Muscles of the Perineal Region and the Pelvic Diaphragm

▶ **KEY POINT** The muscles of the perineal region and pelvic diaphragm extend from the sacrum and coccyx to the ischium and pubis. These muscles support the organs of the pelvic cavity, flex the joints of the sacrum and coccyx, and control the movement of materials through the urethra and anus.

The inferior margins of the pelvis are the boundaries of the **perineal region** (the pelvic floor and associated structures). A line drawn between the ischial tuberosities divides the perineal region into two triangles: an anterior or **urogenital triangle**, and a posterior or **anal triangle**. The superficial muscles of the anterior triangle are the muscles of the external genitalia. They are superficial to deeper muscles that strengthen the pelvic floor and encircle the urethra. An even more extensive muscular sheet, the **pelvic diaphragm**, forms the muscular foundation of the anal triangle (**Figure 10.13a,b** and **Tables 10.9** and **10.10**).

These muscles do not completely close the pelvic outlet because the urethra, vagina, and anus pass through them to open to the external surface. Muscular sphincters surround their openings and control voluntary urination and defecation. Muscles, nerves, and blood vessels also pass through the pelvic outlet as they travel to or from the lower limbs. Selected origins and insertions are shown in **Figure 10.13c**.

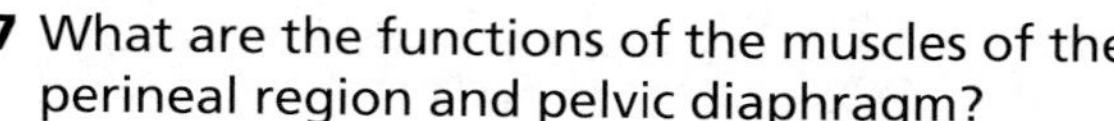

10.5 CONCEPT CHECK

7 What are the functions of the muscles of the perineal region and pelvic diaphragm?

See the blue Answers tab at the back of the book.

10

Table 10.9 | Muscles of the Perineal Region

Group/Muscle	Origin	Insertion	Action	Innervation
UROGENITAL TRIANGLE				
Superficial Muscles				
Bulbospongiosus				
Male	Perineal body (central tendon of perineal region) and median raphe	Corpus spongiosum, perineal membrane, and corpus cavernosum	Compress base, stiffen penis, eject urine or semen	Pudendal nerve, perineal branch (S_2–S_4)
Female	Perineal body (central tendon of perineal region)	Bulb of vestibule, perineal membrane, body of clitoris, and corpus cavernosum	Compress and stiffen clitoris, narrow vaginal opening	Pudendal nerve, perineal branch (S_2–S_4)
Ischiocavernosus	Ramus and tuberosity of ischium	Corpus cavernosum of penis or clitoris; also to ischiopubic ramus (in female only)	Compress and stiffen penis or clitoris, helping to maintain erection	Pudendal nerve, perineal branch (S_2–S_4)
Superficial transverse perineal muscle	Ischial ramus	Central tendon of perineal region	Stabilize central tendon of perineal region	Pudendal nerve, perineal branch (S_2–S_4)
Deep Muscles				
Deep transverse perineal muscle	Ischial ramus	Perineal body	Stabilize central tendon of perineal region	Pudendal nerve, perineal branch (S_2–S_4)
External urethral sphincter				
Male	Ischial and pubic rami	To median raphe at base of penis; inner fibers encircle urethra	Closes urethra, compresses prostate and bulbo-urethral glands	Pudendal nerve, perineal branch (S_2–S_4)
Female	Ischial and pubic rami	To median raphe; inner fibers encircle urethra	Closes urethra; compresses vagina and greater vestibular glands	Pudendal nerve, perineal branch (S_2–S_4)

Table 10.10 | Muscles of the Pelvic Diaphragm

Group/Muscle	Origin	Insertion	Action	Innervation
ANAL TRIANGLE				
Coccygeus	Ischial spine	Lateral, inferior borders of the sacrum and coccyx	Flex coccygeal joints, elevate and support pelvic floor	Inferior sacral nerves (S_4–S_5)
Levator ani				
Iliococcygeus	Ischial spine, pubis	Coccyx and median raphe	Tense floor of pelvis, support pelvic organs, flex coccygeal joints, elevate and retract anus	Pudendal nerve (S_2–S_4)
Pubococcygeus	Inner margins of pubis	Coccyx and median raphe	Tense floor of pelvis, support pelvic organs, flex coccygeal joints, elevate and retract anus	Pudendal nerve (S_2–S_4)
External anal sphincter	Via tendon from coccyx	Encircles anal opening	Closes anal opening	Pudendal nerve; hemorrhoidal branch (S_2–S_4)

Figure 10.13 **Muscles of the Perineal Region**

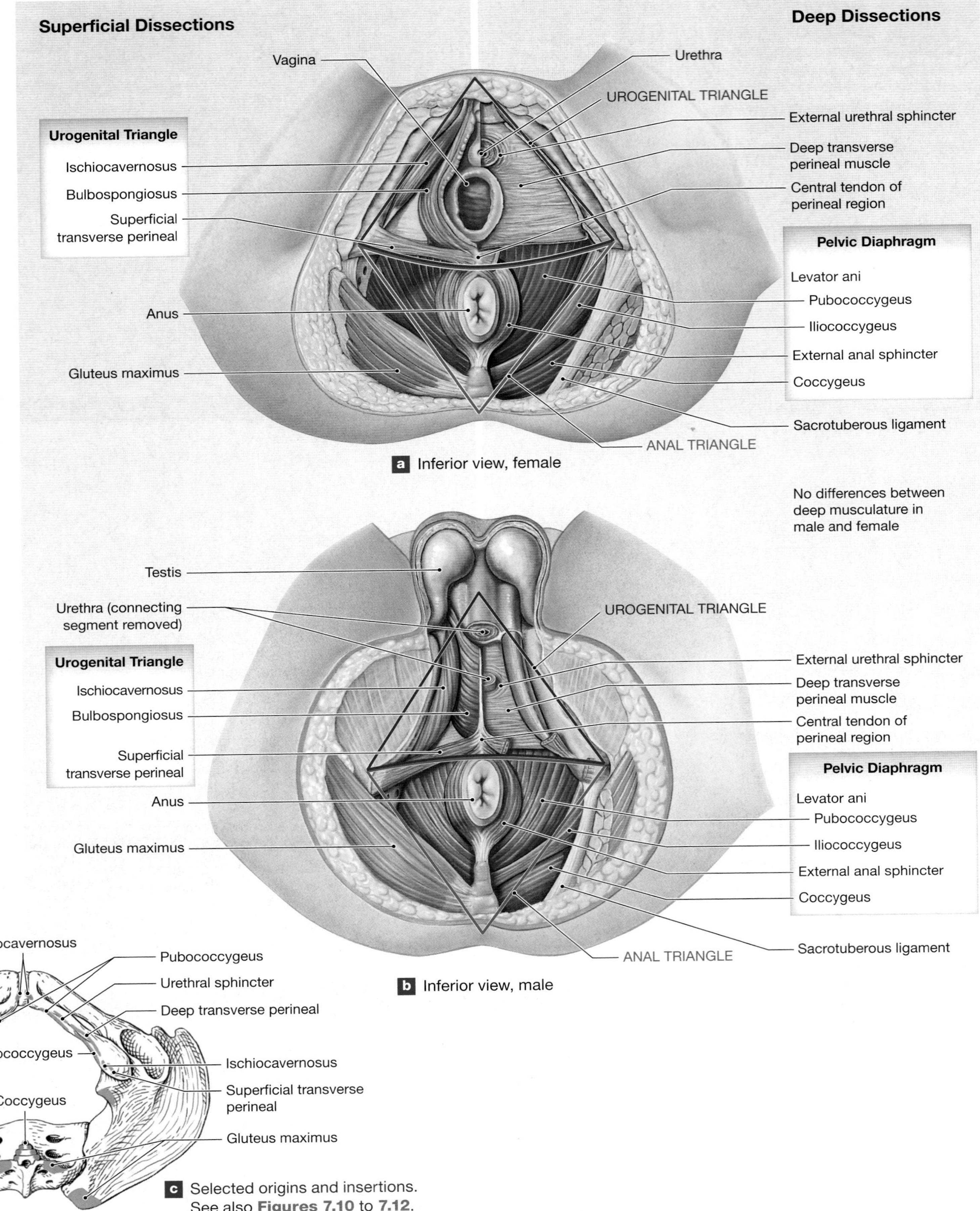

a Inferior view, female

b Inferior view, male

c Selected origins and insertions. See also **Figures 7.10** to **7.12**.

10

Study Outline

Introduction p. 260

- The muscular system, like the skeletal system, is divided into axial and appendicular divisions. The **axial musculature** arises from and inserts on the axial skeleton. It positions the head and spinal column and helps move the rib cage, which makes breathing possible.

10.1 The Four Groups of Axial Muscles p. 260

- The axial muscles are organized into four groups based on their location, function, or both: (1) muscles of the head and neck, (2) muscles of the vertebral column, (3) oblique and rectus muscles, including the diaphragm, and (4) muscles of the perineal region and pelvic diaphragm. *(See Figures 10.1 and 10.2.)*
- Organization of muscles into the four groups includes descriptions of innervation. **Innervation** refers to the identity of the nerve that controls a given muscle.

10.2 Muscles of the Head and Neck p. 262

- Muscles of the head and neck are divided into several groups: (1) muscles of facial expression, (2) extrinsic eye muscles, (3) muscles of mastication, (4) muscles of the tongue, (5) muscles of the pharynx, and (6) anterior muscles of the neck.

Muscles of Facial Expression p. 262

- The muscles of facial expression originate on the surface of the skull. The largest group is associated with the mouth; it includes the **orbicularis oris** and **buccinator**. The frontal and occipital bellies of the **occipitofrontalis** control movements of the eyebrows, forehead, and scalp. The **platysma** tenses skin of the neck and depresses the mandible. *(See Figures 10.3 and 10.4 and Table 10.1.)*

Extra-ocular Muscles p. 264

- The six **extra-ocular eye muscles** (*oculomotor muscles*) control eye position and movements. These muscles are the **inferior**, **lateral**, **medial**, and **superior rectus** and the **superior** and **inferior oblique**. *(See Figure 10.5 and Table 10.2.)*

Muscles of Mastication p. 266

- The muscles of mastication (chewing) act on the mandible. They are the **masseter**, **temporalis**, and **pterygoid**. *(See Figure 10.6 and Table 10.3.)*

Muscles of the Tongue p. 267

- The muscles of the tongue are necessary for speech and swallowing, and they assist in mastication. These muscles are the **genioglossus**, **hyoglossus**, **palatoglossus**, and **styloglossus**. *(See Figure 10.7 and Table 10.4.)*

Muscles of the Pharynx p. 267

- Muscles of the pharynx are important in beginning the swallowing process. These muscles include the **pharyngeal constrictors**, the laryngeal elevators (**palatopharyngeus**, **salpingopharyngeus**, and **stylopharyngeus**), and the **palatal muscles**, which raise the soft palate. *(See Figure 10.8 and Table 10.5.)*

Anterior Muscles of the Neck p. 269

- The anterior muscles of the neck control the position of the larynx, depress the mandible, and provide a foundation for the muscles of the tongue and pharynx. These include the **digastric**, **mylohyoid**, **stylohyoid**, and **sternocleidomastoid**. *(See Figures 10.3, 10.4, 10.9, 12.1, 12.2a, 12.9, and 12.10 and Table 10.6.)*

10.3 Muscles of the Vertebral Column p. 270

- The muscles of the back are arranged into three distinct layers: superficial, intermediate, and deep.
- Only the deepest of these layers is composed of the **intrinsic** (or true) **back muscles**. These intrinsic back muscles are innervated by the posterior rami of the spinal nerves, and they interconnect the vertebrae. *(See Figures 10.10, 12.9, 12.10, 12.12, 12.13, and 12.14 and Table 10.7.)*

The Superficial Layer of the Intrinsic Back Muscles p. 270

- The superficial layer contains the **splenius** of the neck and upper thorax.

The Intermediate Layer of the Intrinsic Back Muscles p. 270

- The intermediate group is composed of the **erector spinae** of the trunk.

The Deep Layer of the Intrinsic Back Muscles p. 270

- The deep layer is composed of the **transversospinales**, which consists of the **semispinalis group** and the **multifidus**, **rotatores**, **interspinales**, and **intertransversarii muscles**. These muscles interconnect and stabilize the vertebrae.

Spinal Flexors p. 272

- Other muscles of the vertebral column are the **longus capitis** and **longus colli**, which rotate and flex the neck, and the **quadratus lumborum** muscles in the lumbar region, which flex the spine and depress the ribs. *(See Figure 10.10c and Table 10.7.)*

10.4 Oblique and Rectus Muscles p. 273

- The oblique and rectus muscles lie between the vertebral column and the anterior midline. The abdominal oblique muscles **(external oblique** and **internal oblique)** compress underlying structures or rotate the vertebral column; the **rectus abdominis** is a flexor of the vertebral column.
- The oblique muscles of the neck and thorax include the **scalenes**, the **intercostals**, and the **transversus** muscles. The **external intercostals** and **internal intercostals** are important for breathing because they move the ribs. *(See Figures 10.11b–d, 10.12, 12.13, and 12.14 and Table 10.8.)*

The Diaphragm p. 275

- The **diaphragm** is important in breathing. It separates the abdominopelvic and thoracic cavities. *(See Figure 10.12.)*

10.5 Muscles of the Perineal Region and the Pelvic Diaphragm p. 277

- Muscles of the perineal region and pelvic diaphragm extend from the sacrum and coccyx to the ischium and pubis. These muscles (1) support the organs of the pelvic cavity, (2) flex the joints of the sacrum and coccyx, and (3) control the movement of materials through the urethra and anus.
- The **perineal region** (the pelvic floor and associated structures) is divided into an anterior **urogenital triangle** and a posterior **anal triangle**. The pelvic floor consists of the consists of the **pelvic diaphragm** and surrounding muscles. *(See Figure 10.13 and Tables 10.9 and 10.10.)*

Chapter Review

For answers, see the blue Answers tab at the back of the book.

Level 1 Reviewing Facts and Terms

Match each numbered item with the most closely related lettered item.

1. spinalis .... [ ]
2. perineal region .... [ ]
3. buccinator .... [ ]
4. extra-ocular .... [ ]
5. intercostals .... [ ]
6. stylohyoid .... [ ]
7. inferior rectus .... [ ]
8. temporalis .... [ ]
9. platysma .... [ ]
10. styloglossus .... [ ]

(a) compresses cheeks
(b) elevates larynx
(c) tenses skin of neck
(d) pelvic floor/associated structures
(e) elevates mandible
(f) move ribs
(g) retracts tongue
(h) extends neck
(i) eye muscles
(j) makes eye look down

11. Which of the following muscles does not compress the abdomen?
    (a) diaphragm (b) internal intercostal
    (c) external oblique (d) rectus abdominis

12. The muscle that arises from the pubis is the
    (a) internal oblique.
    (b) rectus abdominis.
    (c) transversus abdominis.
    (d) scalene.

13. Label the muscles of the head and neck on the figure below.

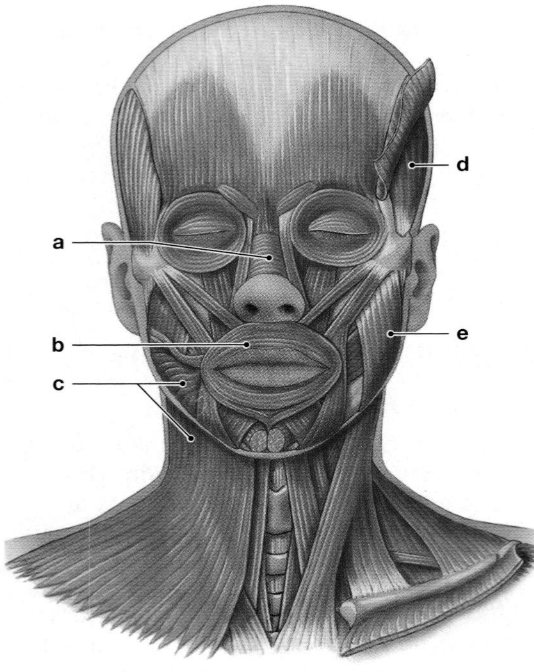

(a) __________ (b) __________
(c) __________ (d) __________
(e) __________

14. The iliac crest is the origin of the
    (a) quadratus lumborum.
    (b) iliocostalis cervicis.
    (c) longissimus cervicis.
    (d) splenius.

15. Which of the following describes the action of the digastric muscle?
    (a) elevates the larynx
    (b) elevates the larynx and depresses the mandible
    (c) depresses the larynx
    (d) elevates the mandible

16. Which of the following muscles inserts on the rib cartilages?
    (a) diaphragm
    (b) external intercostal
    (c) transversus thoracis
    (d) scalene

17. Label the muscles of the back on the figure below.

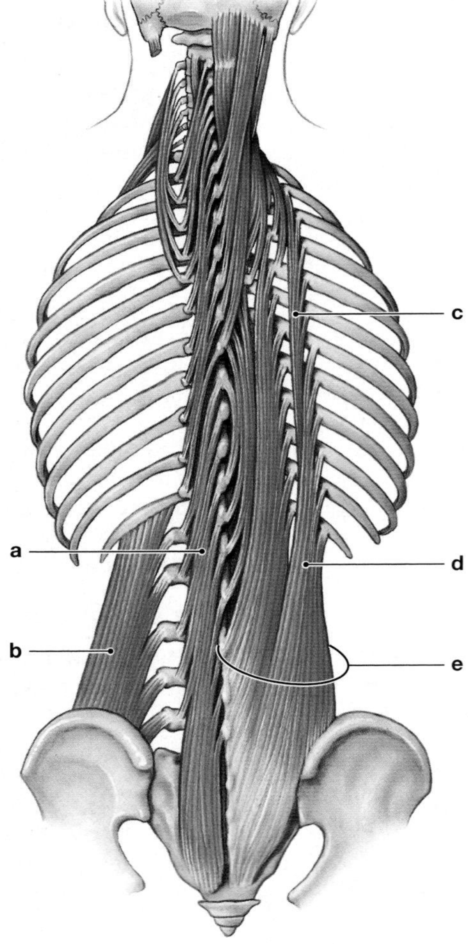

(a) __________
(b) __________
(c) __________
(d) __________
(e) __________

18. Some of the muscles of the tongue are innervated by
    (a) the hypoglossal nerve (N XII).
    (b) the trochlear nerve (N IV).
    (c) the abducens nerve (N VII).
    (d) both b and c.

19. Which of the following is *not* true of the muscles of the pelvic floor?
    (a) They extend between the sacrum and the pelvic girdle.
    (b) They form the perineal region.
    (c) They "fine-tune" the movements of the thigh relative to the pelvis.
    (d) They encircle the openings in the pelvic outlet.

20. The scalenes originate on the
    (a) transverse and costal processes of cervical vertebrae.
    (b) inferior border of the previous rib.
    (c) rib cartilages.
    (d) thoracolumbar fascia and iliac crest.

21. Which cranial nerve is most likely to have been damaged if a person cannot move the right eye to look laterally?
    (a) oculomotor nerve
    (b) trigeminal nerve
    (c) facial nerve
    (d) abducens nerve

Level 2 Reviewing Concepts

1. During abdominal surgery, the surgeon makes a cut through the muscle directly to the right of the linea alba. What is that muscle?
   (a) digastric
   (b) external oblique
   (c) rectus abdominis
   (d) scalene

2. Ryan hears a loud noise and quickly raises his eyes to look upward in the direction of the sound. To accomplish this action, he must use his _____ muscles.
   (a) superior rectus
   (b) inferior rectus
   (c) superior oblique
   (d) lateral rectus

3. Which of the following muscles plays no role in swallowing?
   (a) superior constrictor
   (b) pterygoids
   (c) palatopharyngeus
   (d) stylopharyngeus

4. Which of the following features are common to the muscles of mastication?
   (a) They share innervation through the oculomotor nerve.
   (b) They are also muscles of facial expression.
   (c) They move the mandible at the temporomandibular joint.
   (d) They enable a person to smile.

5. The muscles of the vertebral column include many posterior extensors but few anterior flexors. Why?
6. What role do the muscles of the tongue play in swallowing?
7. What is the effect of contracting the internal oblique?
8. What are the functions of the anterior muscles of the neck?
9. What is the function of the diaphragm? Why is it included in the axial musculature?
10. What muscles are involved in controlling the position of the head on the vertebral column?

Level 3 Critical Thinking

1. How do the muscles of the anal triangle control the functions of this area?
2. Mary sees Jill coming toward her and immediately contracts her frontalis and procerus muscles. Is Mary glad to see Jill? How can you tell?

MasteringA&P™

Access more chapter study tools online in the Study Area:

- Chapter Quizzes, Chapter Practice Test, Clinical Cases, and more!
- Practice Anatomy Lab (PAL) PAL™
- A&P Flix for anatomy topics A&PFlix™

10

CLINICAL CASE WRAP-UP

Waking with a Crooked Smile

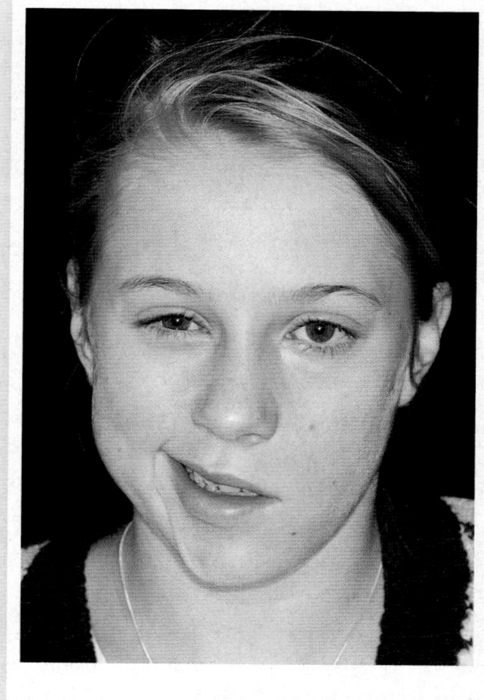

Sarah, along with 40,000 other Americans annually, is suffering from Bell's palsy, a sudden paralysis of the muscles of facial expression innervated by the facial nerve (N VII), usually on one side of the face. All of the facial muscles listed in **Table 10.1**, with the exception of the levator palpebrae superioris, can be affected by Bell's palsy. This peripheral palsy is typically caused by a viral infection causing inflammation of the facial nerve as it passes through the tight tunnel of the internal acoustic meatus and the stylomastoid foramen of the temporal bone. Pregnant women in their third trimester and patients with diabetes are more susceptible to Bell's palsy.

Interestingly, the muscles of facial expression do not atrophy nearly as fast as skeletal muscles that insert on bone. A majority of Bell's palsy patients recover within several weeks or months. Recovery is most often complete, with no visible residual weakness. Occasionally, a slightly crooked smile can remain for life.

1. Why is a patient suffering from Bell's palsy unable to *close* the affected eye willingly, yet is still able to *open* the eye?
2. Why is Sarah unable to whistle or blow?

See the blue Answers tab at the back of the book.

Related Clinical Terms

paraspinal neuromuscular syndrome: Neuromuscular disorder in which paraspinal muscle weakness may lead to dropped head syndrome or bent spine syndrome.

diaphragmatic hernia: The protrusion of abdominal contents into the thoracic cavity through a weakness in the diaphragm.

direct inguinal hernia: The protrusion of abdominal contents through the abdominal wall in a location slightly medial to the inguinal canal.

hernia: The protrusion of abdominal contents through a weak spot in the surrounding muscular wall.

indirect inguinal hernia: The entrance of abdominal contents into the inguinal canal.

stiff person syndrome: Rare, progressive neurological disorder characterized by painful muscle spasms and muscle stiffness of the spine and lower extremities.

11

The Muscular System

Appendicular Musculature

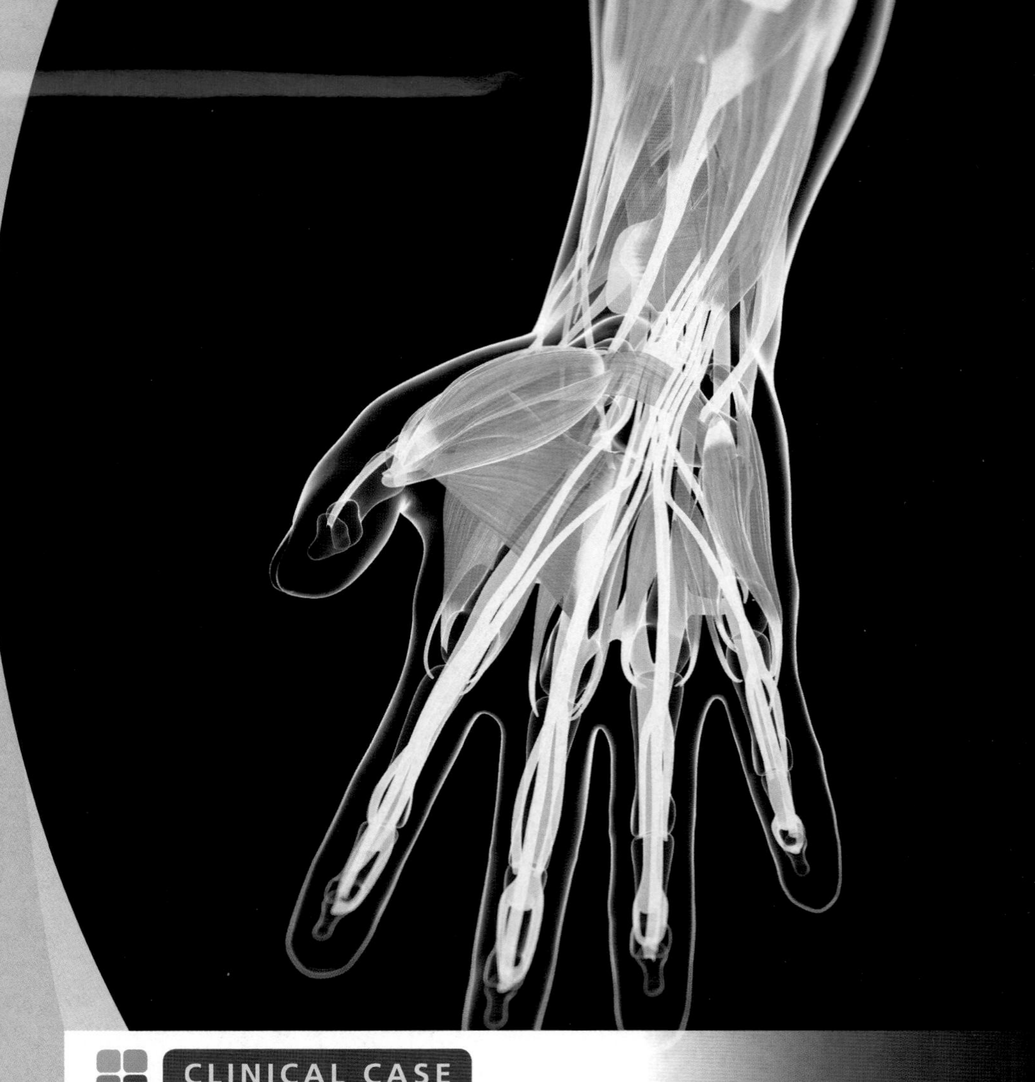

Learning Outcomes

These Learning Outcomes correspond by number to this chapter's sections and indicate what you should be able to do after completing the chapter.

11.1 Describe how the action produced by a muscle at a joint depends on the joint structure and the muscle location relative to the axis of movement at the joint. p. 283

11.2 Identify and locate the muscles of the pectoral girdle and upper limb, including their origins, insertions, actions, and innervations. p. 286

11.3 Identify and locate the compartments of the arm and forearm and the muscles within each of these compartments. p. 301

11.4 Identify and locate the muscles of the pelvic girdle and lower limb, including their origins, insertions, actions, and innervations. p. 303

11.5 Identify and locate the compartments of the thigh and leg and the muscles within each of these compartments. p. 319

CLINICAL CASE

Hamstrung

David is a 25-year-old avid water-skier. He is skiing behind a speedboat with both feet strapped into a wake board when he starts to fall to his right. He resists the fall and digs into the board with his right foot. Suddenly, he feels an excruciating pop below his right gluteal region and falls into the water.

David is unable to swim; his right lower extremity feels paralyzed. Fortunately, he is wearing a personal flotation device, and within minutes he is rescued. Back at the dock, David cannot flex his knee at all, but can get around by locking his quads, "hiking" his hip (knee in extension), and doing an "abduction swing" to place his right foot forward while leaning on a friend for support. He cannot sit on his right side. The next evening, massive bruising appears on his entire posterior thigh and knee. Three weeks later, the bruising and swelling begin to lessen, but David can feel a ball of muscle behind his right knee.

One month post-injury, David still cannot flex his knee. He visits an orthopedist, who immediately schedules him for surgery.

What kind of injury has David sustained? To find out, turn to the Clinical Case Wrap-Up on p. 324.

THE **APPENDICULAR MUSCULATURE** is the focus of this chapter. There are two major groups of appendicular muscles: (1) the muscles of the pectoral girdle and upper limb and (2) the muscles of the pelvic girdle and lower limb (**Figure 11.1**). The upper limb has a large range of motion (amount of movement that occurs at a joint) because of the muscular connections between the pectoral girdle and the axial skeleton. These muscular connections also act as shock absorbers. For example, when you jog, you can perform delicate hand movements at the same time because the appendicular muscles absorb the shocks and bounces in your stride. In contrast, the pelvic girdle transfers weight from the axial skeleton to the lower limb. The emphasis is on strength rather than mobility, and the anatomical features that strengthen the joints limit the range of movement of the lower limbs.

Before proceeding, you may want to review origins, insertions, and actions of appendicular muscles (see page 251) and levers and pulleys (see page 254).

11.1 Factors Affecting Appendicular Muscle Function

▶ **KEY POINT** As in Chapter 10, information about the origin, insertion, and action of each muscle is summarized in tables. These tables also contain information about the innervation of individual muscles.

Do not become frustrated by the details in this chapter; remember to relate the anatomical information to the muscle functions. *The goal of anatomy isn't rote memorization—it's understanding.* Use the origins and insertions to predict the action of a muscle. Then reverse the process, and use the action of the muscles to predict the origins and insertions. **Spotlight Figure 11.2** gives important information about how a muscle's action lines help you predict the actions of that muscle.

Figure 11.1 The Two Major Groups of Appendicular Muscles

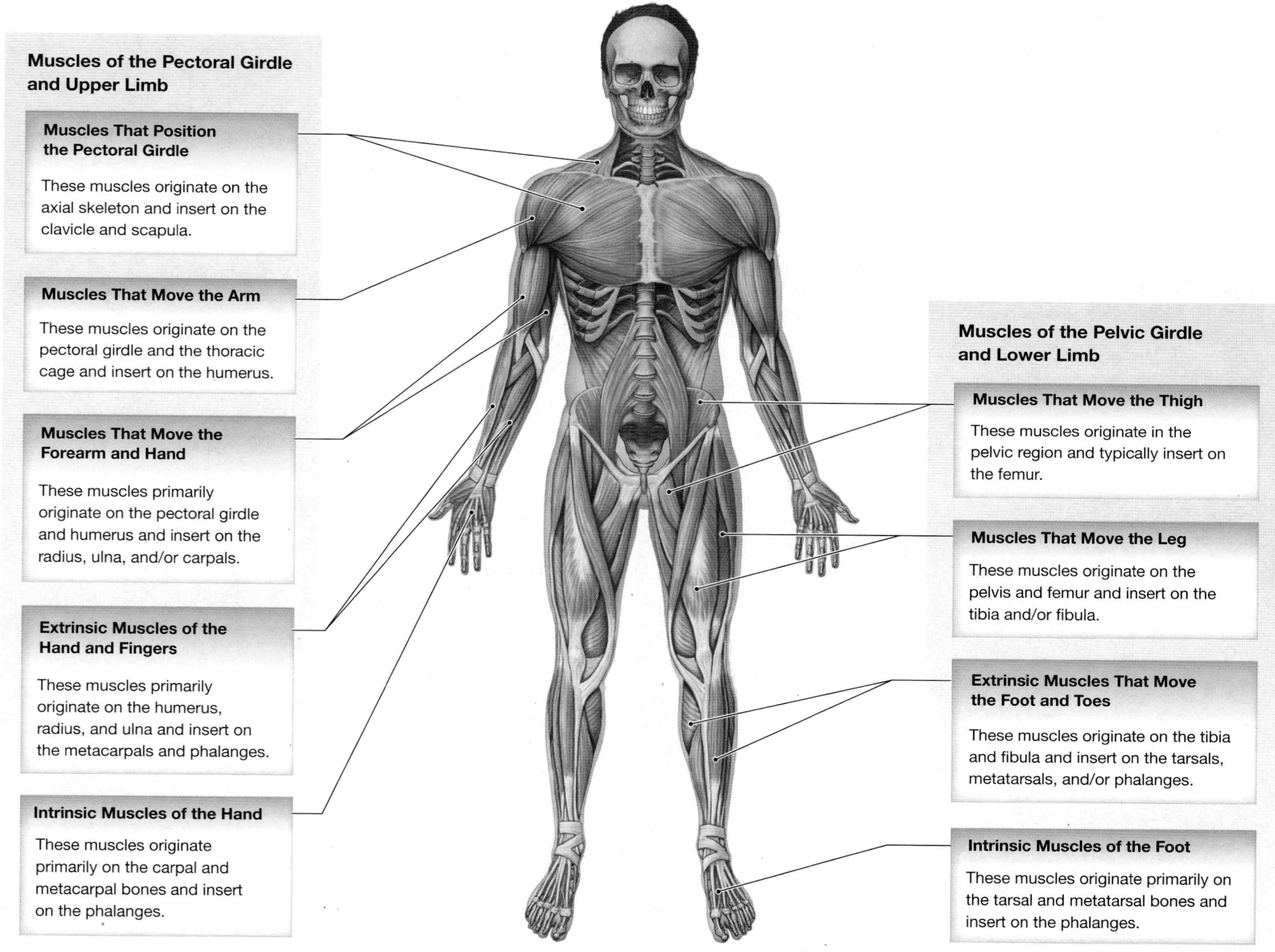

Muscles of the Pectoral Girdle and Upper Limb
Muscles That Position the Pectoral Girdle
These muscles originate on the axial skeleton and insert on the clavicle and scapula.
Muscles That Move the Arm
These muscles originate on the pectoral girdle and the thoracic cage and insert on the humerus.
Muscles That Move the Forearm and Hand
These muscles primarily originate on the pectoral girdle and humerus and insert on the radius, ulna, and/or carpals.
Extrinsic Muscles of the Hand and Fingers
These muscles primarily originate on the humerus, radius, and ulna and insert on the metacarpals and phalanges.
Intrinsic Muscles of the Hand
These muscles originate primarily on the carpal and metacarpal bones and insert on the phalanges.
Muscles of the Pelvic Girdle and Lower Limb
Muscles That Move the Thigh
These muscles originate in the pelvic region and typically insert on the femur.
Muscles That Move the Leg
These muscles originate on the pelvis and femur and insert on the tibia and/or fibula.
Extrinsic Muscles That Move the Foot and Toes
These muscles originate on the tibia and fibula and insert on the tarsals, metatarsals, and/or phalanges.
Intrinsic Muscles of the Foot
These muscles originate primarily on the tarsal and metatarsal bones and insert on the phalanges.

SPOTLIGHT

FIGURE 11.2

Factors Affecting Appendicular Muscle Function

The Anatomy of the Shoulder Joint

When a muscle or part of a large muscle contracts, it pulls the insertion in a specific direction. The direction of force (muscle pull) is called the **action line**. The movement that results depends on the anatomy of the joint and its axes of movement. Knowing the range of movement helps you understand or predict the muscle action at a specific joint.

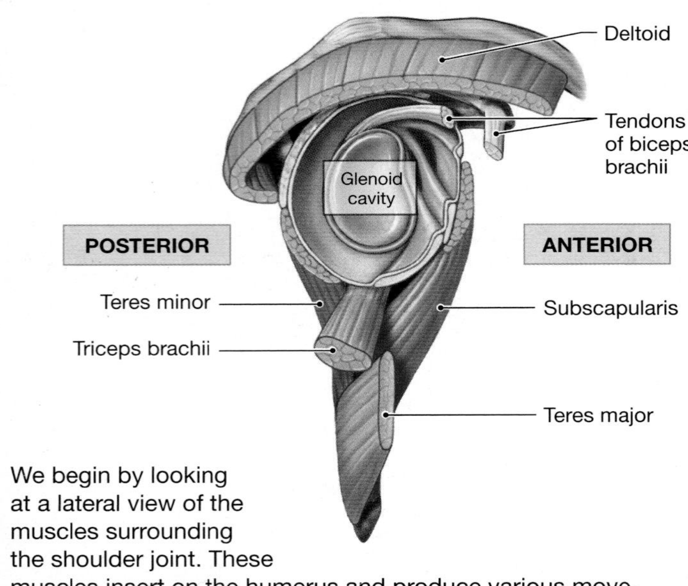

We begin by looking at a lateral view of the muscles surrounding the shoulder joint. These muscles insert on the humerus and produce various movements. The humerus has been removed in these figures.

Flexion and Extension

Shoulder joint muscles with action lines crossing the anterior aspect are flexors, while those crossing the posterior aspect are extensors.

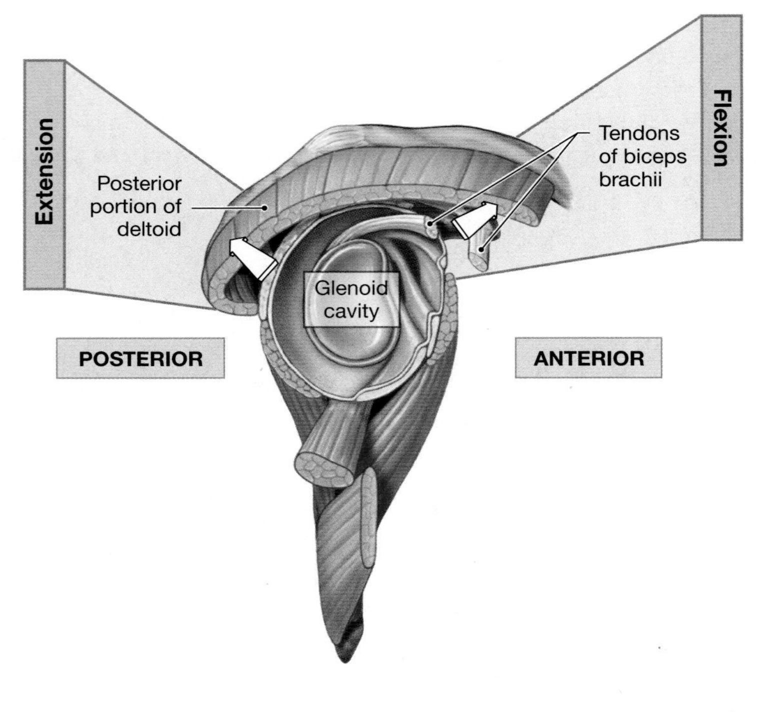

Adduction and Abduction

Muscles with action lines crossing the superior aspect of the shoulder joint are abductors, while those crossing the inferior aspect are adductors.

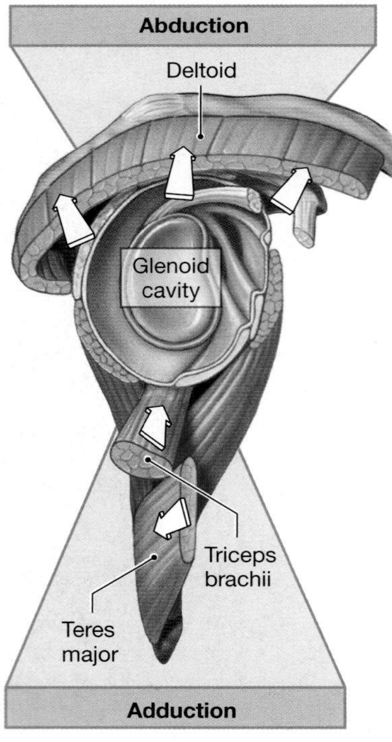

Medial and Lateral Rotation

Muscles with action lines crossing the anterior aspect of the shoulder joint medially rotate and flex, while those crossing the posterior aspect laterally rotate and extend.

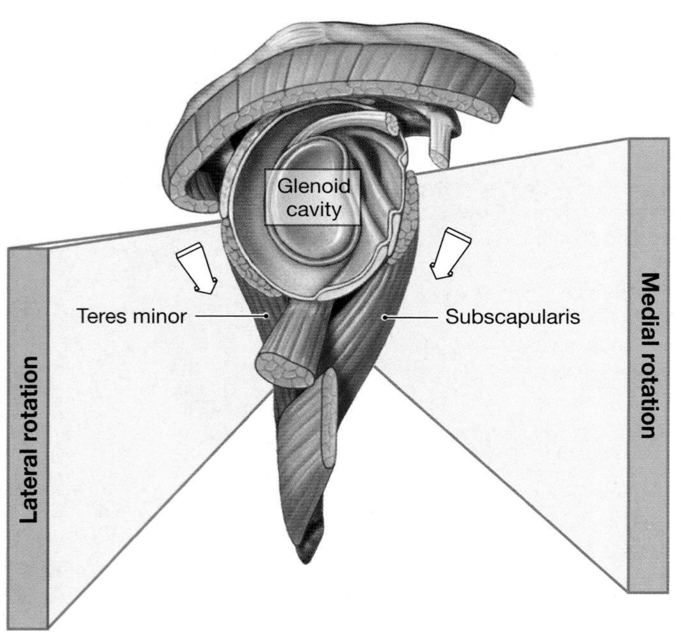

Action Lines at the Shoulder Joint

Here is a superficial lateral view showing the action lines of the deltoid, biceps brachii, and triceps brachii. Analyzing how these action lines cross the shoulder helps you determine the actions of these muscles on the humerus.

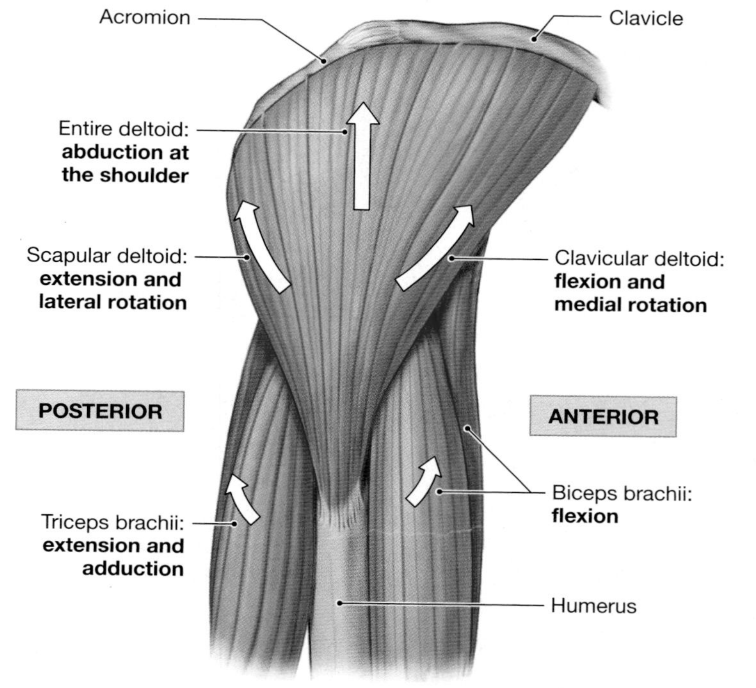

Spurt and Shunt Muscles

The location of a muscle's insertion, relative to the joint's axis of movement, gives further details about the muscle's function at that joint. The primary action of a muscle that inserts close to the joint is movement of that joint. Such a muscle is termed a **spurt muscle**. However, a muscle that inserts farther from the joint will stabilize that joint in addition to moving that joint. This type of muscle is a synergist and is termed a **shunt muscle**.

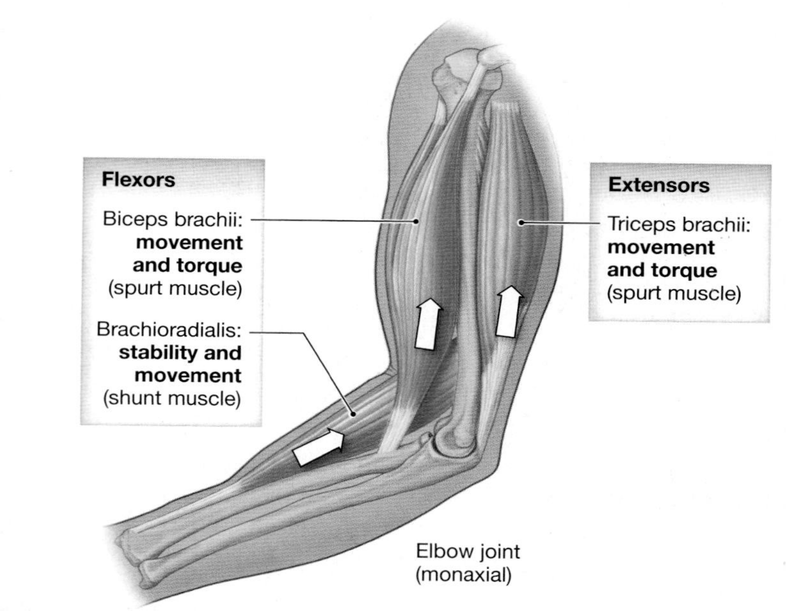

Action Lines at the Hip Joint

The hip joint, like the shoulder joint, is a multiaxial synovial joint that flexes, extends, adducts, abducts, medially rotates, and laterally rotates. The movement at the joint depends on the anatomy of the joint and its axes of movement.

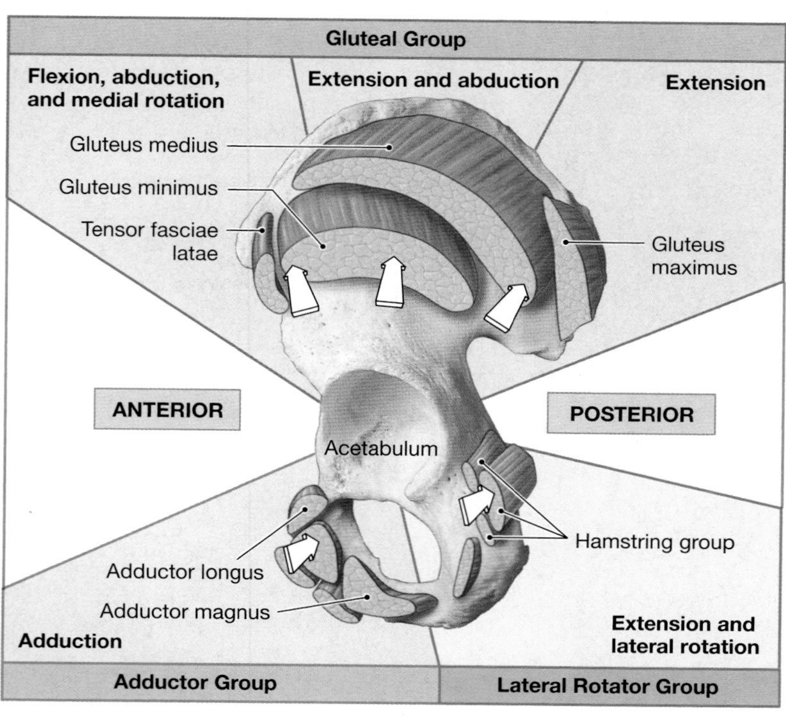

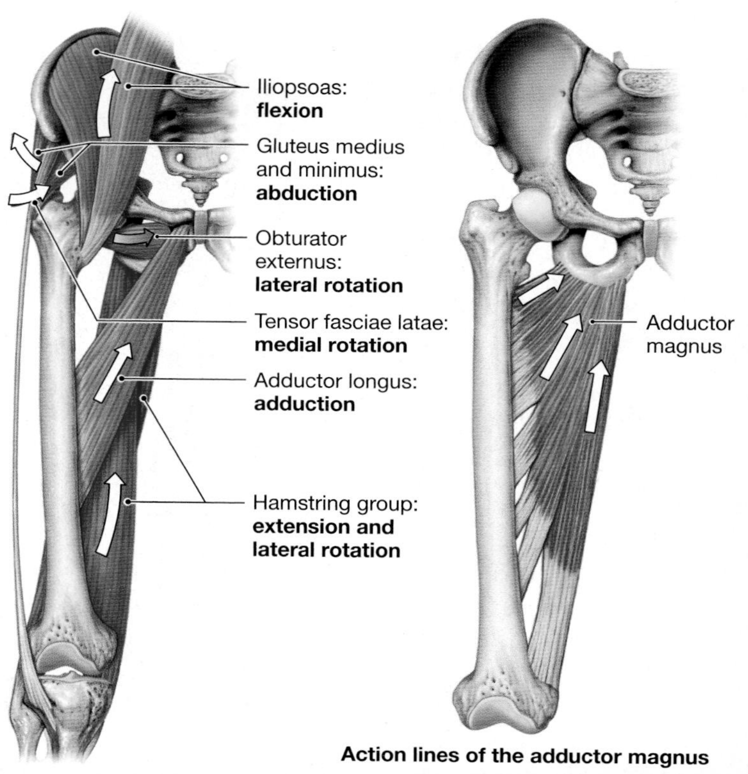

Action lines of the adductor magnus

11.1  **CONCEPT CHECK**

1 Define the action line of a muscle and explain how it is used to determine the action of a muscle at a particular joint.

See the blue Answers tab at the back of the book.

11.2 Muscles of the Pectoral Girdle and Upper Limb

▶ **KEY POINT** The most important function of the muscles of the pectoral girdle and upper limb is to position the hand so it can perform a desired task.

The muscles of the pectoral girdle and upper limb are divided into four groups:

1. muscles that position the pectoral girdle,
2. muscles that move the arm,
3. muscles that move the forearm and hand, and
4. muscles that move the hand and fingers.

Muscles That Position the Pectoral Girdle

▶ **KEY POINT** Muscles of the pectoral girdle, or extrinsic muscles of the shoulder, originate from the axial skeleton and insert onto bones of either the pectoral girdle or arm. Moving these muscles maximizes range of motion of the shoulder joint.

Figure 11.3 Superficial and Deep Muscles of the Neck, Shoulder, and Back. Posterior view of the axial muscles of the back and neck and the appendicular musculature of the pectoral girdle and proximal portion of the upper limb.

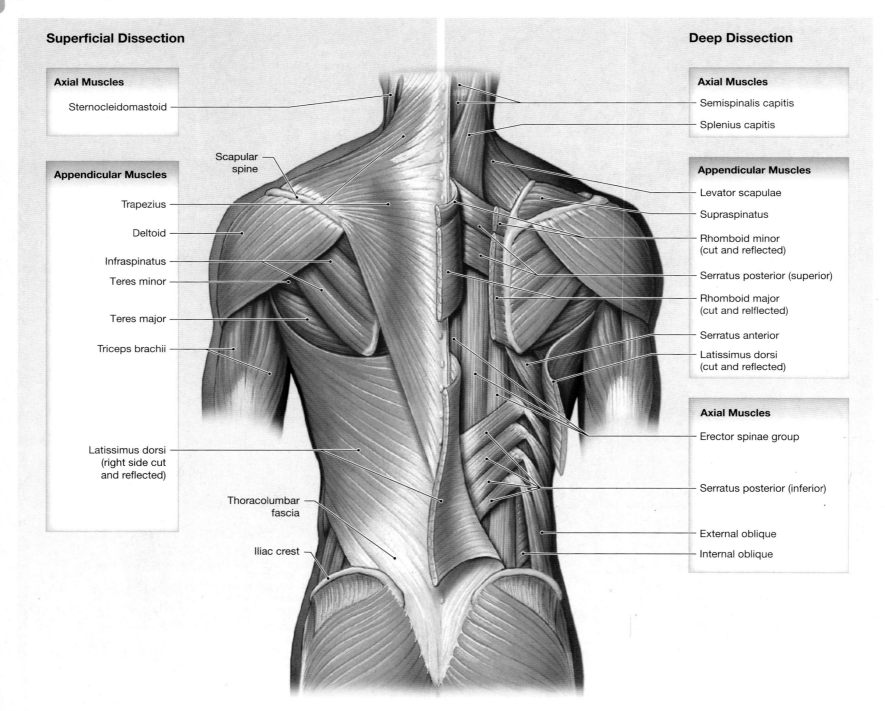

Muscles that position the pectoral girdle are shown in **Figures 11.3** to **11.6** and **Table 11.1**. *(Refer to Chapter 12,* **Figures 12.2** *and* **12.3**, *to identify these anatomical structures from the body surface.)*

The large **trapezius** (tra-PĒ-zē-us) covers the back and portions of the neck and base of the skull, forming a broad diamond shape **(Figures 11.3** and **11.5)**. These muscles originate on bones and connective tissue along the neck and back and insert onto the clavicles and the scapular spines.

More than one nerve innervates the trapezius. Because specific regions of the trapezius can contract independently, this muscle has a wide variety of actions **(Table 11.1)**. *(Refer to Chapter 12,* **Figures 12.10**, *to identify this structure in a cross section of the body at the level of* T_2.)

Removing the trapezius reveals the **rhomboids** (ROM-boyd) and the **levator scapulae** (SKAP-ū-lē) **(Figures 11.3** and **11.5)**. These muscles attach to the posterior surfaces of the cervical and thoracic vertebrae. They insert along the vertebral border of each scapula, between the superior and inferior angles. Contracting the rhomboids adducts (retracts) the scapula, pulling it toward the center of the back. Contracting the rhomboids also downwardly rotates the scapula, moving the glenoid cavity inferiorly and the inferior angle of the scapula medially and superiorly **(Figures 7.4** and **7.5)**. ⮌ pp. 175–176 *(Refer to Chapter 12,* **Figure 12.10**, *to identify this structure in a cross section of the body at the level of* T_2.) Contracting the levator scapula elevates the scapula, as in shrugging the shoulders.

The **serratus** (se-RĀ-tus) **anterior** originates along the anterior and superior surfaces of several ribs on the lateral wall of the chest **(Figures 11.5** and **11.6)**. This fan-shaped muscle inserts on the anterior surface of the vertebral border of the scapula. Contracting the serratus anterior abducts the scapula and moves the glenoid cavity of the scapula anteriorly.

The **subclavius** (sub-KLĀ-vē-us; *sub-*, below, + *clavius*, clavicle) and **pectoralis** (pek-tō-RA-lis) **minor** lie deep to the pectoralis major **(Figures 11.4** and **11.6)**. The subclavius originates from the first rib and inserts onto the inferior border of the clavicle. Contracting the subclavius

Figure 11.4 Superficial and Deep Muscles of the Trunk and Proximal Limb. Anterior view of the axial muscles of the trunk and the appendicular musculature associated with the pectoral girdle and the proximal portion of the upper limb.

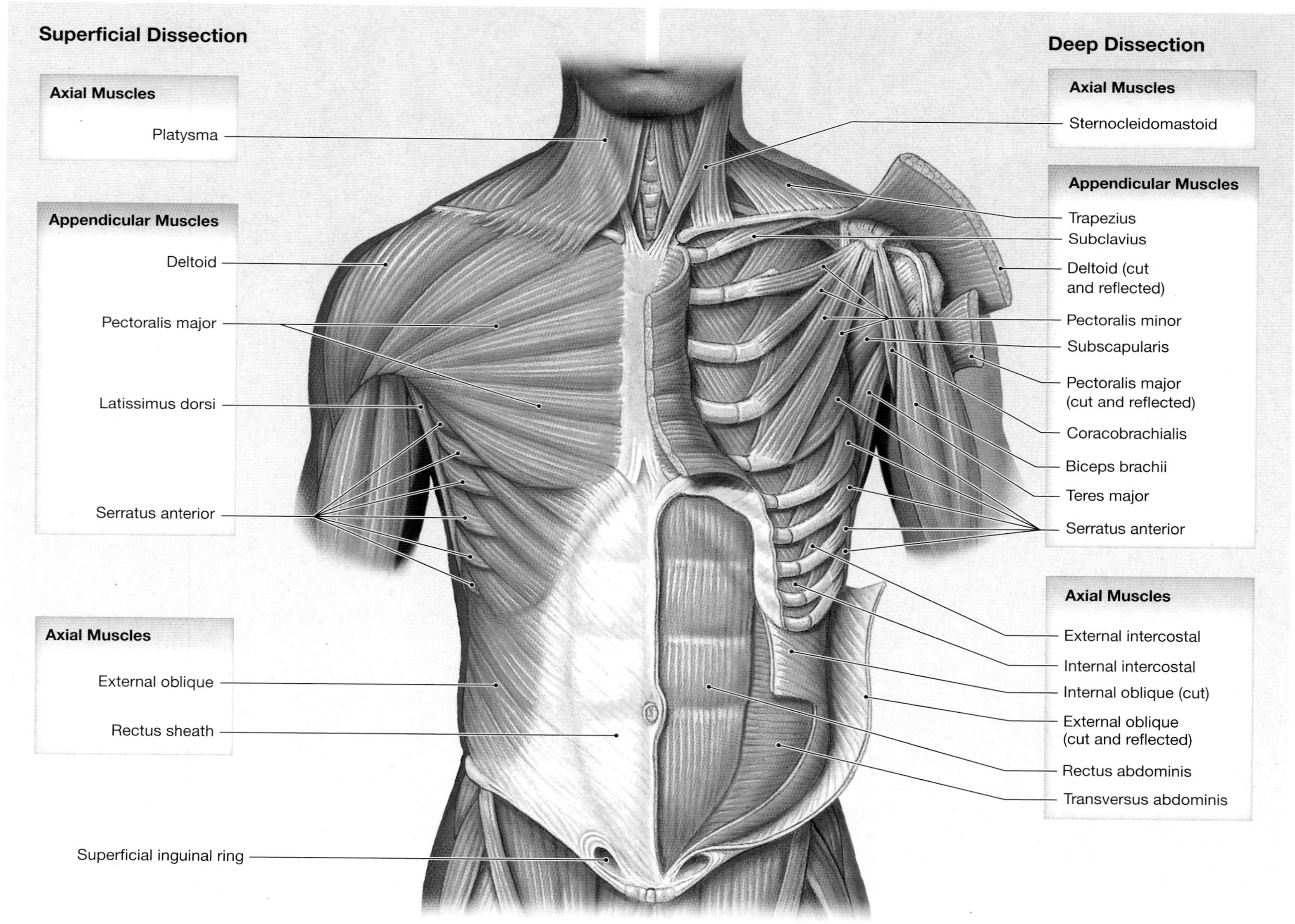

Figure 11.5 **Muscles That Position the Pectoral Girdle, Part I.** Posterior view showing superficial muscles and deep muscles of the pectoral girdle.

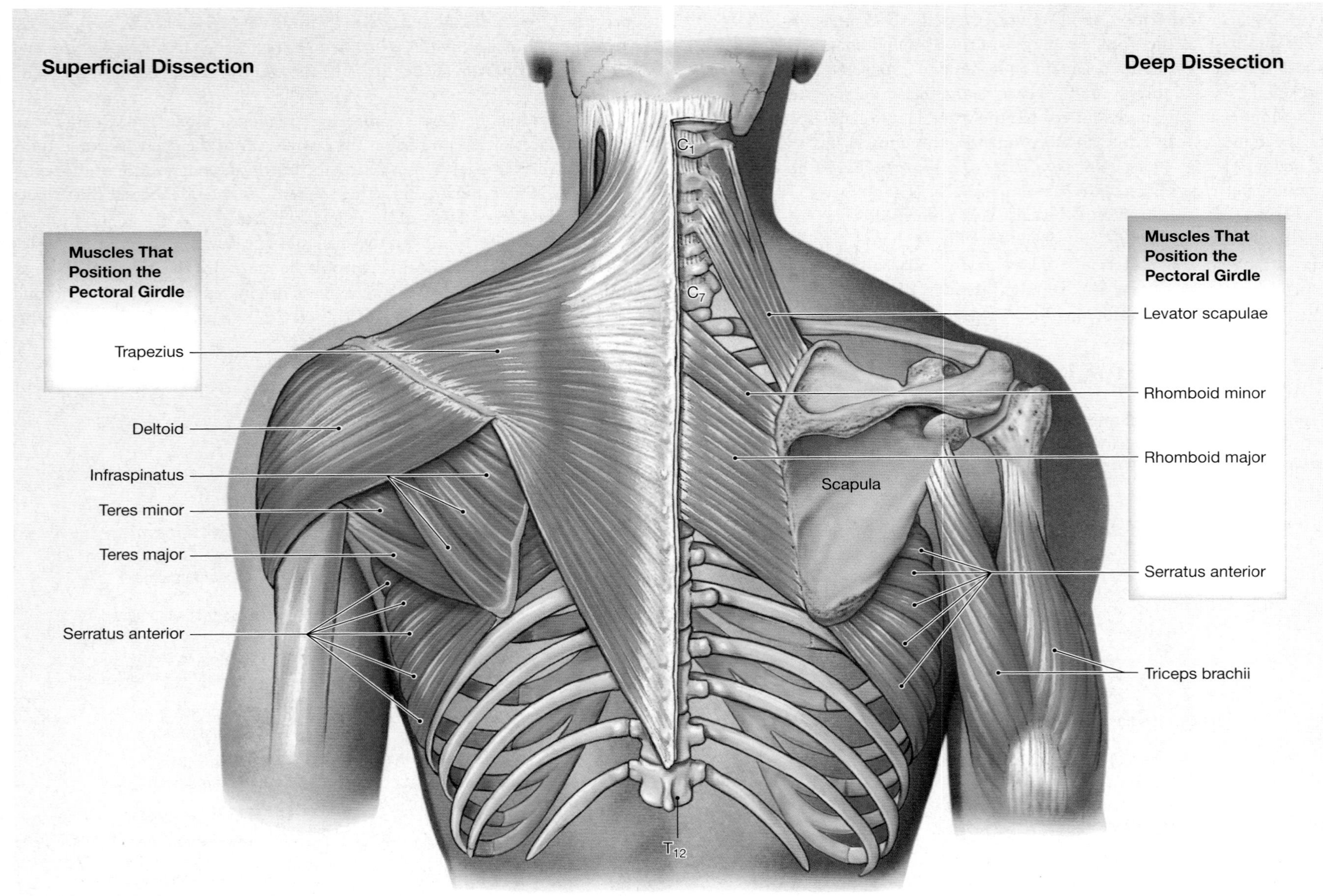

depresses and protracts the lateral (scapular) end of the clavicle. Because ligaments connect the lateral end of the clavicle to the scapula and shoulder joint, those structures also move when the subclavius contracts. The pectoralis minor originates from ribs 3 to 5 (or 2 to 4, depending on the individual) and inserts onto the coracoid process of the scapula (**Figure 11.6**). *(Refer to Chapter 12,* **Figure 12.10***, to identify this structure in a cross section of the body at the level of* T_2*.)* Contracting the pectoralis minor moves the scapula in the same manner as the subclavius. **Table 11.1** identifies the muscles that move the pectoral girdle and the nerves that innervate those muscles.

Muscles That Move the Arm

▶ **KEY POINT** Muscles that move the arm are termed intrinsic muscles of the shoulder. They originate from the pectoral girdle, insert onto the humerus, and move the arm at the shoulder joint. These muscles move the arm and position the hand for a desired task.

Muscles that move the arm are easiest to remember when they are grouped by their primary actions. **Figures 11.4** and **11.7a** show anterior muscles; posterior muscles are shown in **Figures 11.3** and **11.7b**. **Table 11.2** summarizes information on these muscles.

Figure 11.6 Muscles That Position the Pectoral Girdle, Part II. Anterior view showing superficial muscles and deep muscles of the pectoral girdle. Selected origins and insertions are detailed.

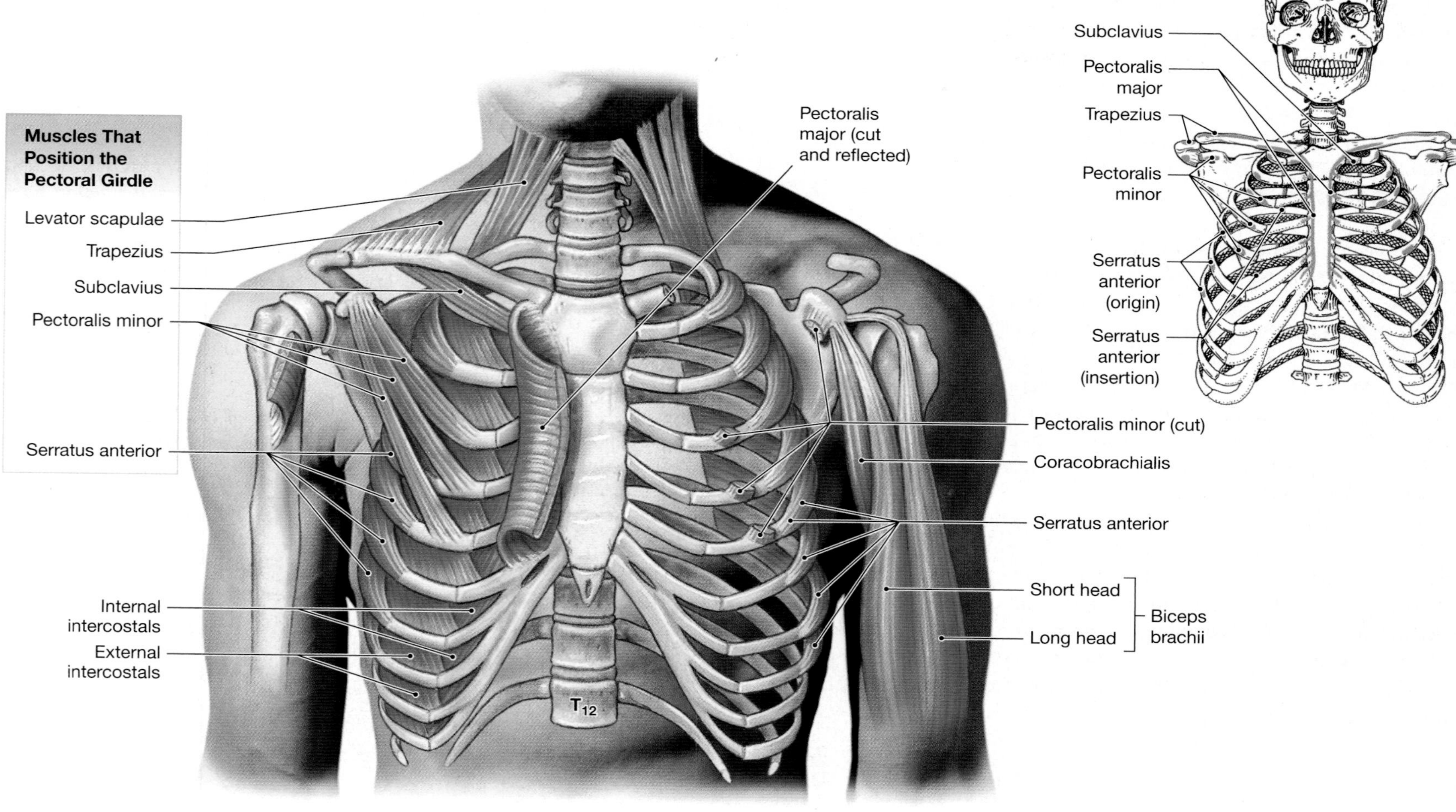

11

Table 11.1 | Muscles That Position the Pectoral Girdle

Muscle	Origin	Insertion	Action	Innervation
Levator scapulae	Transverse processes of first four cervical vertebrae	Vertebral border of scapula near superior angle and medial end of scapular spine	Elevates scapula	Cervical nerves C_3–C_4 and dorsal scapular nerve (C_5)
Pectoralis minor	Anterior surfaces and superior margins of ribs 3–5 or 2–4 and the fascia covering the associated external intercostal muscles	Coracoid process of scapula	Depresses and protracts shoulder; rotates scapula so glenoid cavity moves inferiorly (downward rotation); elevates ribs if scapula is stationary	Medial pectoral nerve (C_8, T_1)
Rhomboid major	Ligamentum nuchae and the spinous processes of vertebrae T_2 to T_5	Vertebral border of scapula from spine to inferior angle	Adducts and performs downward rotation of the scapula	Dorsal scapular nerve (C_5)
Rhomboid minor	Spinous processes of vertebrae C_7–T_1	Vertebral border of scapula	Adducts and performs downward rotation of the scapula	Dorsal scapular nerve (C_5)
Serratus anterior	Anterior and superior margins of ribs 1–8, 1–9, or 1–10	Anterior surface of vertebral border of scapula	Protracts shoulder; rotates scapula so glenoid cavity moves superiorly (upward rotation)	Long thoracic nerve (C_5–C_7)
Subclavius	First rib	Clavicle (inferior border of middle 1/3)	Depresses and protracts shoulder	Nerve to subclavius (C_5–C_6)
Trapezius	Occipital bone, ligamentum nuchae, and spinous processes of thoracic vertebrae	Clavicle and scapula (acromion and scapular spine)	Depends on active region and state of other muscles; may elevate, retract, depress, or rotate scapula upward and/or clavicle; can also extend neck when the position of the shoulder is fixed	Accessory nerve (XI)

The **deltoid** is the prime mover for abducting the arm, but the **supraspinatus** (sū-pra-spī-NĀ-tus) is a synergist at the start of this movement. The **subscapularis** and **teres** (TER-ēz) **major** rotate the arm medially, whereas the **infraspinatus** (in-fra-spī-NĀ-tus) and **teres minor** are antagonistic to that action, rotating the arm laterally. All of these muscles originate on the scapula. The **coracobrachialis** (kor-a-kō-brā-kē-A-lis) (**Figure 11.7a**) is the only muscle attached to the scapula that flexes and adducts the arm at the shoulder joint. *(Refer to Chapter 12,* **Figures 12.2, 12.4**, *and* **12.5**, *to identify these anatomical structures from the body surface, and* **Figure 12.10** *to identify these structures in a cross section of the body at the level of* T_2.)

The **pectoralis major** originates from the cartilages of ribs 2 to 6 and inserts onto the crest of the greater tubercle of the humerus. The pectoralis major flexes, adducts, and medially rotates the humerus at the shoulder joint. The **latissimus dorsi** (la-TIS-i-mus DOR-sē) has a wide variety of origins and inserts onto the intertubercular sulcus of the humerus (**Figures 11.3, 11.4**, and **11.7b**). The latissimus dorsi flexes, adducts, and medially rotates the humerus at the shoulder joint. *(Refer to Chapter 12,* **Figures 12.2a, 12.3b**, *and* **12.5**, *to identify these anatomical structures from the body surface, and* **Figure 12.10** *to identify these structures in a cross section of the body at the level of* T_2.)

The shoulder is a mobile but weak joint. The tendons of the supraspinatus, infraspinatus, subscapularis, and teres minor join with the connective tissue of the shoulder joint capsule and form the **rotator cuff**. The rotator cuff supports and strengthens the joint capsule of the shoulder. Powerful, repetitive arm movements common in many sports (such as pitching a fastball for many innings) place considerable strain on the muscles of the rotator cuff, often causing tendon damage, muscle strains, bursitis, and other painful injuries.

TIPS & TOOLS

To remember the muscles of the rotator cuff, think SITS: **S**upraspinatus, **I**nfraspinatus, **T**eres minor, and **S**ubcapularis.

Spotlight Figure 11.2 on pp. 284–285 discussed how a muscle's action lines help you predict the actions of that muscle. Specifically, the action lines of the biceps brachii, triceps brachii, and deltoid were shown in relation to the shoulder joint. What follows are examples of how to use action lines to predict the actions of these three muscles:

- Although the biceps brachii does not insert on the humerus, the biceps brachii is a flexor of the shoulder because its action lines pass anterior to the axis of the shoulder joint.
- The triceps brachii does not insert onto the humerus either, but the triceps brachii is an extensor of the shoulder because its action lines pass posterior to the axis of the shoulder joint.
- The action line of the clavicular, or anterior, portion of the deltoid also crosses anterior to the axis of the shoulder joint as it inserts onto the humerus. This portion of the deltoid flexes and medially rotates the shoulder. The action line of the scapular, or posterior, portion of the deltoid passes posterior to the axis of the shoulder joint. The scapular portion of the deltoid extends and laterally rotates the shoulder. Contracting the entire deltoid abducts the shoulder because the action line for the muscle as a whole passes superior and lateral to the axis of the joint.

Table 11.2 | Muscles That Move the Arm

Muscle	Origin	Insertion	Action	Innervation
Coracobrachialis	Coracoid process	Medial margin of shaft of humerus	Adduction and flexion at shoulder	Musculocutaneous nerve (C_5–C_7)
Deltoid	Clavicle and scapula (acromion and adjacent scapular spine)	Deltoid tuberosity of humerus	*Whole muscle*: abduction of shoulder; *anterior part*: flexion and medial rotation of humerus; *posterior part*: extension and lateral rotation of humerus	Axillary nerve (C_5–C_6)
Supraspinatus	Supraspinous fossa of scapula	Greater tubercle of humerus	Abduction at shoulder	Suprascapular nerve (C_5)
Infraspinatus	Infraspinous fossa of scapula	Greater tubercle of humerus	Lateral rotation at shoulder	Suprascapular nerve (C_5–C_6)
Subscapularis	Subscapular fossa of scapula	Lesser tubercle of humerus	Medial rotation at shoulder	Subscapular nerve (C_5–C_6)
Teres major	Inferior angle of scapula	Medial lip of intertubercular sulcus of humerus	Extension and medial rotation at shoulder	Lower subscapular nerve (C_5–C_6)
Teres minor	Lateral border of scapula	Greater tubercle of humerus	Lateral rotation and adduction at shoulder	Axillary nerve (C_5)
Triceps brachii (long head)	See Table 11.3			
Biceps brachii	See Table 11.3			
Latissimus dorsi	Spinous processes of inferior thoracic and all lumbar and sacral vertebrae, ribs 8–12, and thoracolumbar fascia	Floor of intertubercular sulcus of the humerus	Extension, adduction, and medial rotation at shoulder	Thoracodorsal nerve (C_6–C_8)
Pectoralis major	Cartilages of ribs 2–6, body of sternum, and inferior, medial portion of clavicle	Crest of greater tubercle and lateral lip of intertubercular sulcus of humerus	Flexion, adduction, and medial rotation at shoulder	Medial and lateral pectoral nerves (C_5–T_1)

Figure 11.7 Muscles That Move the Arm

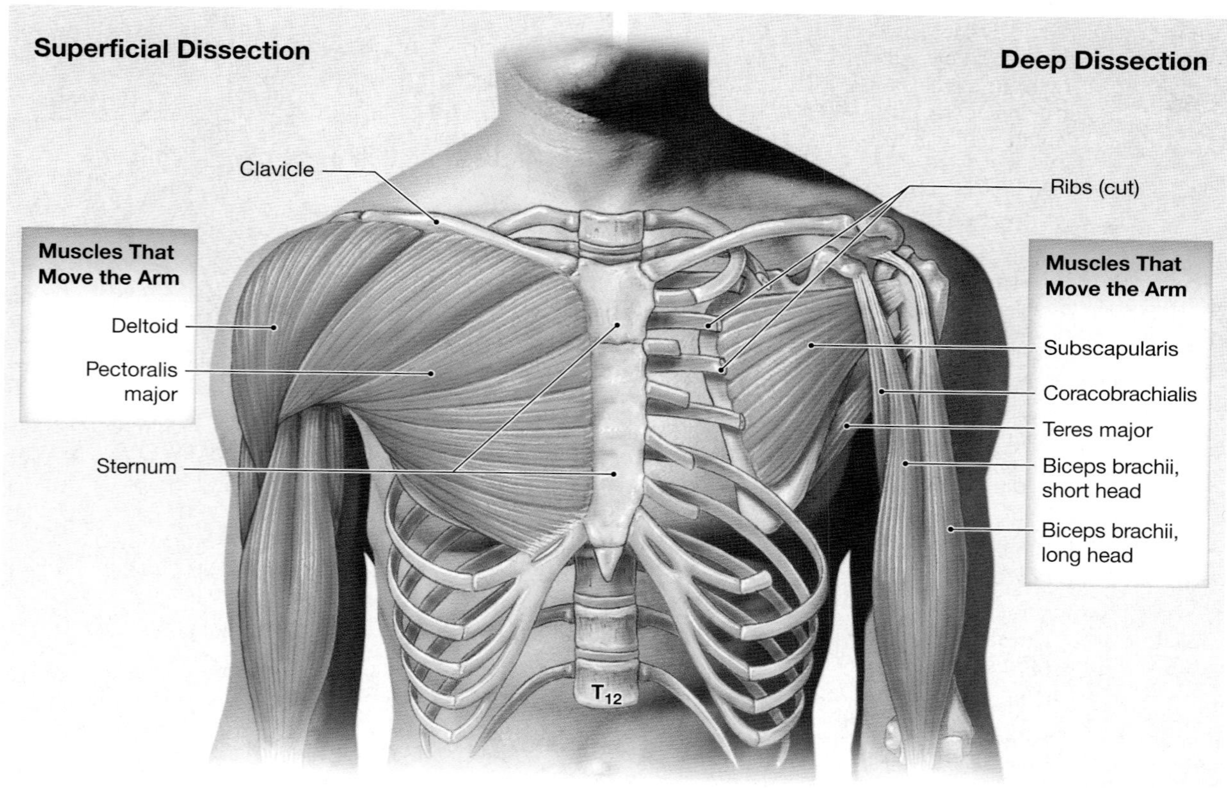

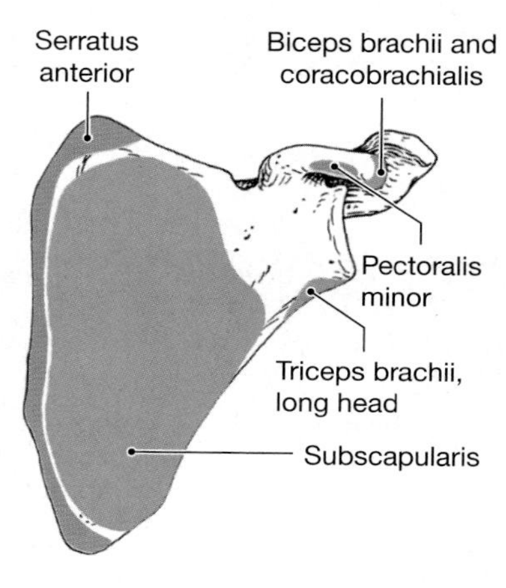

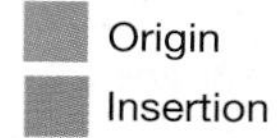

a Anterior view.

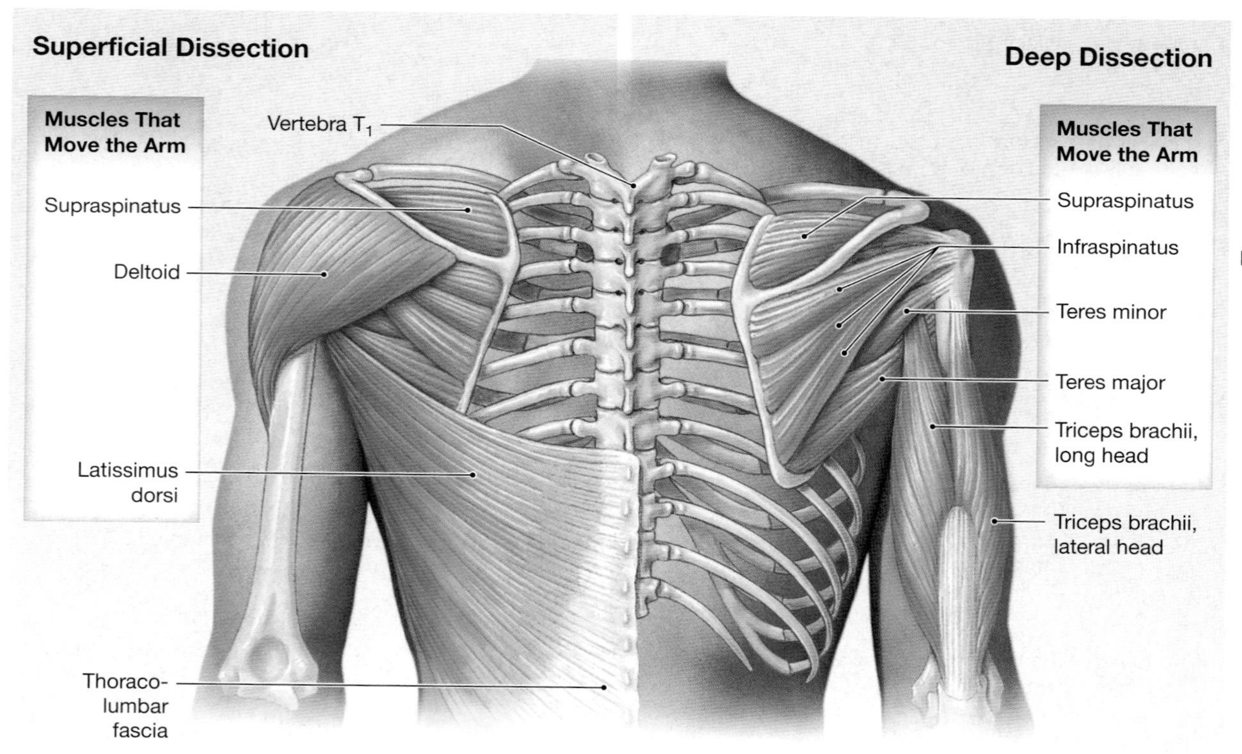

b Posterior view.

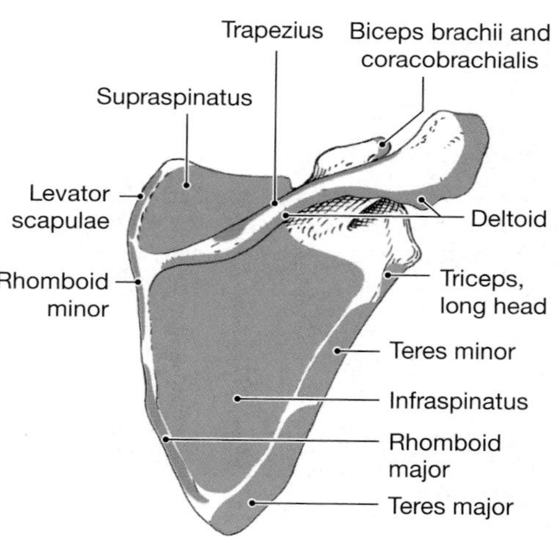

c Anterior and posterior views of the scapula showing selected origins and insertions.

11

Muscles That Move the Forearm and Hand

▶ KEY POINT All of the anterior muscles of the arm are innervated by the musculocutaneous nerve, and all of the posterior muscles of the arm are innervated by the radial nerve. In the forearm, all but two anterior forearm muscles (flexor carpi ulnaris and the ulnar portion of flexor digitorum profundus) are innervated by the median nerve. All of the posterior muscles of the forearm are innervated by the radial nerve.

Most of the muscles that move the forearm and hand originate on the humerus and insert on the forearm and wrist. *There are two noteworthy exceptions:*

1. The *long head* of the **triceps brachii** (TRĪ-seps BRĀ-kē-ī) originates on the scapula and inserts on the olecranon.
2. The *long head* of the **biceps brachii** originates on the scapula and inserts on the radial tuberosity of the radius **(Figures 11.4–11.8, 11.10)**.

Table 11.3 | Muscles That Move the Forearm and Hand

Muscle	Origin	Insertion	Action	Innervation
ACTION AT THE ELBOW				
FLEXORS				
Biceps brachii	*Short head* from the coracoid process; *long head* from the supraglenoid tubercle (both on the scapula)	Radial tuberosity	Flexion at elbow and shoulder; supination of forearm and hand by lateral rotation of radius at radioulnar joints	Musculocutaneous nerve (C_5–C_6)
Brachialis	Distal half of the anterior surface of the humerus	Ulnar tuberosity and coronoid process	Flexion at elbow	Musculocutaneous nerve (C_5–C_6) and radial nerve (C_7–C_8)
Brachioradialis	Ridge superior to the lateral epicondyle of humerus	Lateral aspect of styloid process of radius	Flexion at elbow	Radial nerve (C_6–C_8)
EXTENSORS				
Anconeus	Posterior surface of lateral epicondyle of humerus	Lateral margin of olecranon and ulnar shaft	Extension at elbow	Radial nerve (C_6–C_8)
Triceps brachii				
Lateral head	Superior, lateral margin of humerus	Olecranon of ulna	Extension at elbow	Radial nerve (C_6–C_8)
Long head	Infraglenoid tubercle of scapula	Olecranon of ulna	Extension at elbow, plus extension and adduction at shoulder	Radial nerve (C_6–C_8)
Medial head	Posterior surface of humerus, inferior to radial groove	Olecranon of ulna	Extension at elbow	Radial nerve (C_6–C_8)
PRONATORS/SUPINATORS				
Pronator quadratus	Anterior and medial surfaces of distal ulna	Anterolateral surface of distal portion of radius	Pronates forearm and hand by medial rotation of radius at radioulnar joints	Median nerve (C_8–T_1)
Pronator teres	Medial epicondyle of humerus and coronoid process of ulna	Middle of lateral surface of radius	Pronates forearm and hand by medial rotation of radius at radioulnar joints, plus flexion at elbow	Median nerve (C_6–C_7)
Supinator	Lateral epicondyle of humerus and ridge near radial notch of ulna	Anterolateral surface of radius distal to the radial tuberosity	Supinates forearm and hand by lateral rotation of radius at radioulnar joints	Deep radial nerve (C_6–C_8)
ACTION AT THE WRIST				
FLEXORS				
Flexor carpi radialis	Medial epicondyle of humerus	Bases of second and third metacarpal bones	Flexion and abduction at wrist	Median nerve (C_6–C_7)
Flexor carpi ulnaris	Medial epicondyle of humerus; adjacent medial surface of olecranon and anteromedial portion of ulna	Pisiform, hamate, and base of fifth metacarpal bone	Flexion and adduction at wrist	Ulnar nerve (C_8–T_1)
Palmaris longus	Medial epicondyle of humerus	Palmar aponeurosis and flexor retinaculum	Flexion at wrist	Median nerve (C_6–C_7)
EXTENSORS				
Extensor carpi radialis longus	Lateral supracondylar ridge of humerus	Base of second metacarpal bone	Extension and abduction at wrist	Radial nerve (C_6–C_7)
Extensor carpi radialis brevis	Lateral epicondyle of humerus	Base of third metacarpal bone	Extension and abduction at wrist	Radial nerve (C_6–C_7)
Extensor carpi ulnaris	Lateral epicondyle of humerus; adjacent dorsal surface of ulna	Base of fifth metacarpal bone	Extension and adduction at wrist	Deep radial nerve (C_6–C_8)

Figure 11.8 Muscles That Move the Forearm and Hand, Part I. Relationships among the muscles of the right upper limb are shown.

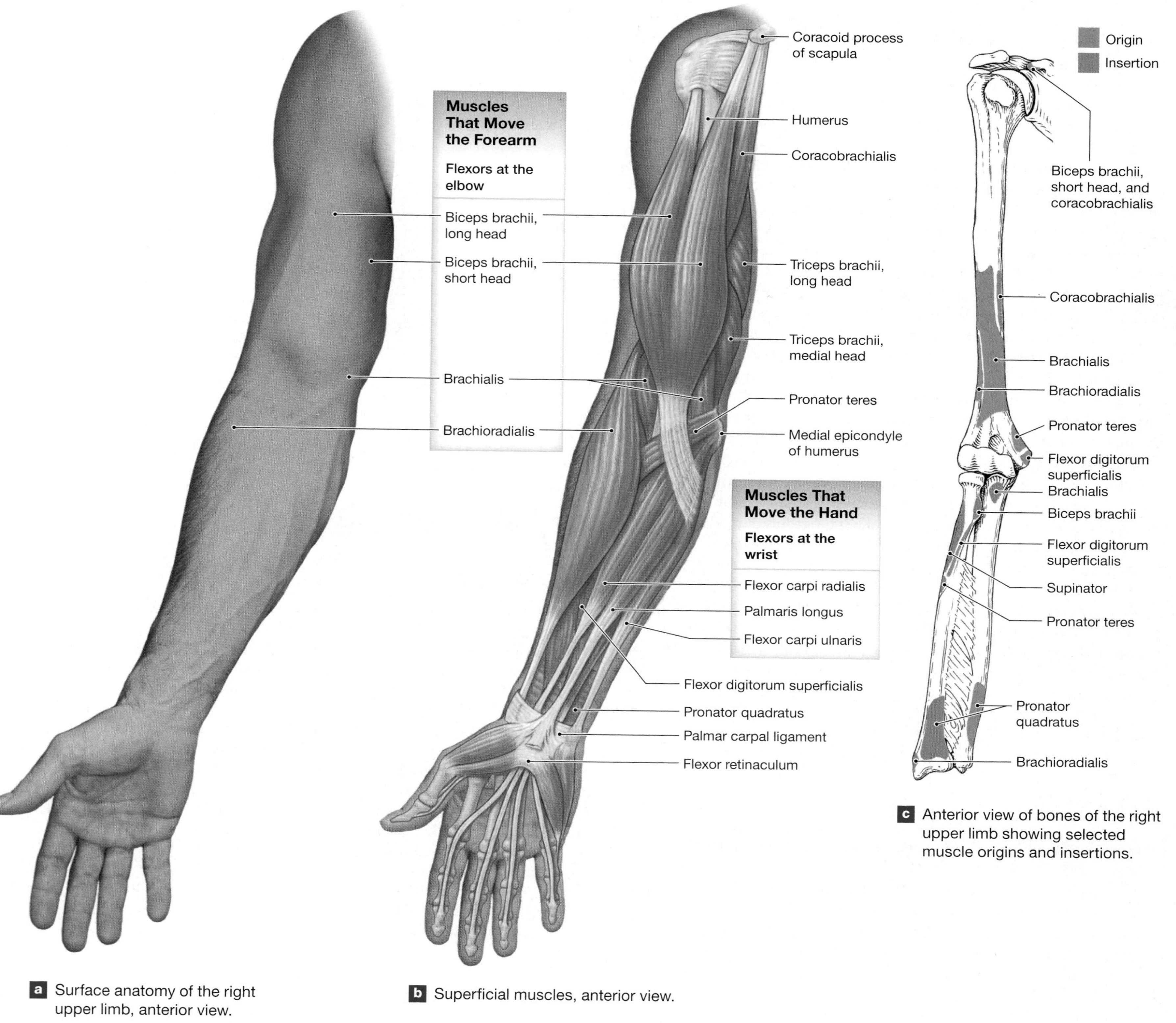

a Surface anatomy of the right upper limb, anterior view.

b Superficial muscles, anterior view.

c Anterior view of bones of the right upper limb showing selected muscle origins and insertions.

11

The triceps brachii and biceps brachii are examples of muscles of the arm that exert actions at more than one joint. Contracting the triceps brachii extends and adducts the shoulder and also extends the elbow. Contracting the biceps brachii flexes the shoulder and also flexes the elbow and supinates the forearm. Although these muscles exert an action at the shoulder, their primary (most important) actions are at the elbow.

The biceps brachii is also an example of how the position of the body affects the action of a muscle: When the forearm is pronated, the biceps brachii cannot contract as forcefully as when the forearm is supinated due to the position of the muscle's insertion.

The **brachialis** (BRĀ-kē-a-lis) and **brachioradialis** (BRĀ-kē-ō-rā-dē-a-lis) also flex the elbow. The **anconeus** (an-KŌ-nē-us) and the triceps brachii are antagonists to this action. The **flexor carpi ulnaris**, **flexor carpi radialis**, and **palmaris longus** are superficial muscles that work together to flex the wrist **(Figures 11.8, 11.10, and 11.11a)**. The flexor carpi radialis also abducts the wrist, while the flexor carpi ulnaris adducts the wrist. The **extensor carpi radialis** and the **extensor carpi ulnaris** also have an antagonistic action: The extensor carpi radialis extends and abducts the wrist, and the extensor carpi ulnaris extends and adducts the wrist.

The **pronator teres** and the **supinator muscle** are antagonistic muscles that originate on the humerus and the ulna. They insert on the radius and rotate the forearm without flexing or extending the elbow. The **pronator quadratus** originates on the ulna and assists the pronator teres in opposing the supination actions of the supinator muscle and the biceps brachii. **Figure 11.9** shows the muscles involved in pronation and supination (medial and lateral rotation). Note how the radius changes position as the pronator teres and pronator quadratus contract. A bursa prevents abrasion against the tendon as the tendon of the biceps brachii rolls under the radius during pronation.

The origins, insertions, and innervations of the muscles that move the forearm and hand are listed in **Table 11.3**. (*Refer to Chapter 12,* **Figures 12.4** *and* **12.5**, *to identify these anatomical structures from the body surface.*) As you study the muscles in **Table 11.3** note that extensor muscles typically lie along the posterior and lateral surfaces of the forearm, and flexors are on the anterior and medial surfaces. Many of the muscles that move the forearm and hand can be seen from the body surface **(Figures 11.8a, 11.10a, 12.4, and 12.5)**.

TIPS & TOOLS

Here is a simple trick to remember the four anterior superficial forearm muscles originating from the medial epicondyle of the humerus. Hold both arms out, palms touching. Then slide your right hand proximally until your palm reaches your elbow with your fingers pointing toward your wrist. With each finger representing one of the four muscles, think PFPF: **P**ronator teres (index finger), **F**lexor carpi radialis (middle finger), **P**almaris longus (ring finger), and **F**lexor carpi ulnaris (little finger).

Muscles That Move the Hand and Fingers

▶ **KEY POINT** The position of the wrist affects the functioning of the hand. Many muscles of the forearm, therefore, affect the actions of the wrist because (1) all of the muscles that flex or extend the wrist originate on the humerus, radius, and/or ulna and (2) many muscles that flex or extend the fingers originate on the radius and/or ulna.

Figure 11.9 Muscles Involved in Supination and Pronation. Deep muscles involved with supination and pronation.

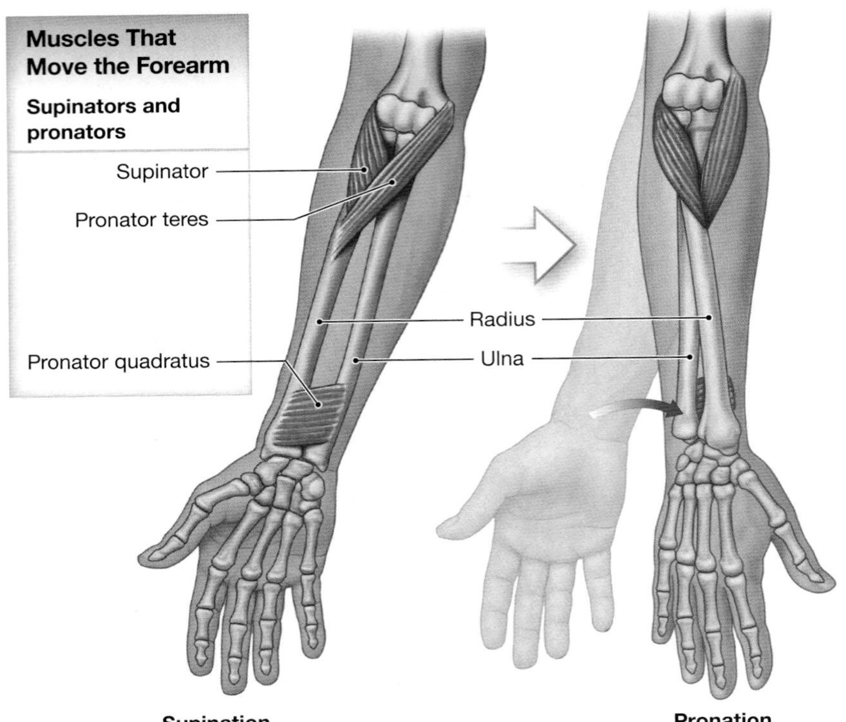

CLINICAL NOTE

Sports Injuries

Strains (stretching or tearing of muscle or tendons that attach muscle to bone) and **sprains** (stretching or tearing of ligaments that connect bone to bone) are by far the most common type of sports injuries. They can range from mild to severe with complete disruption of muscle, tendons, or ligaments.

The knee is the most commonly injured joint. The patellar ligament or quadriceps tendon can inflame, causing tendinitis, or it can rupture. Major ligaments of the knee joint, particularly the anterior cruciate ligament (ACL), can be sprained or torn. The menisci can be damaged or torn.

Tendinitis can affect both sides of the elbow. Lateral humeral epicondylitis, involving the origin of the wrist extensors, is commonly known as "tennis elbow." Medial humeral epicondylitis, involving the origin of the wrist flexors, is commonly known as "pitcher's elbow."

Bony injuries range from bruising, bleeding beneath the periosteum, to stress fractures or acute fractures. Joints, particularly the shoulder, can dislocate.

Figure 11.10 Muscles That Move the Forearm and Hand, Part II. Relationships among the muscles of the right upper limb are shown.

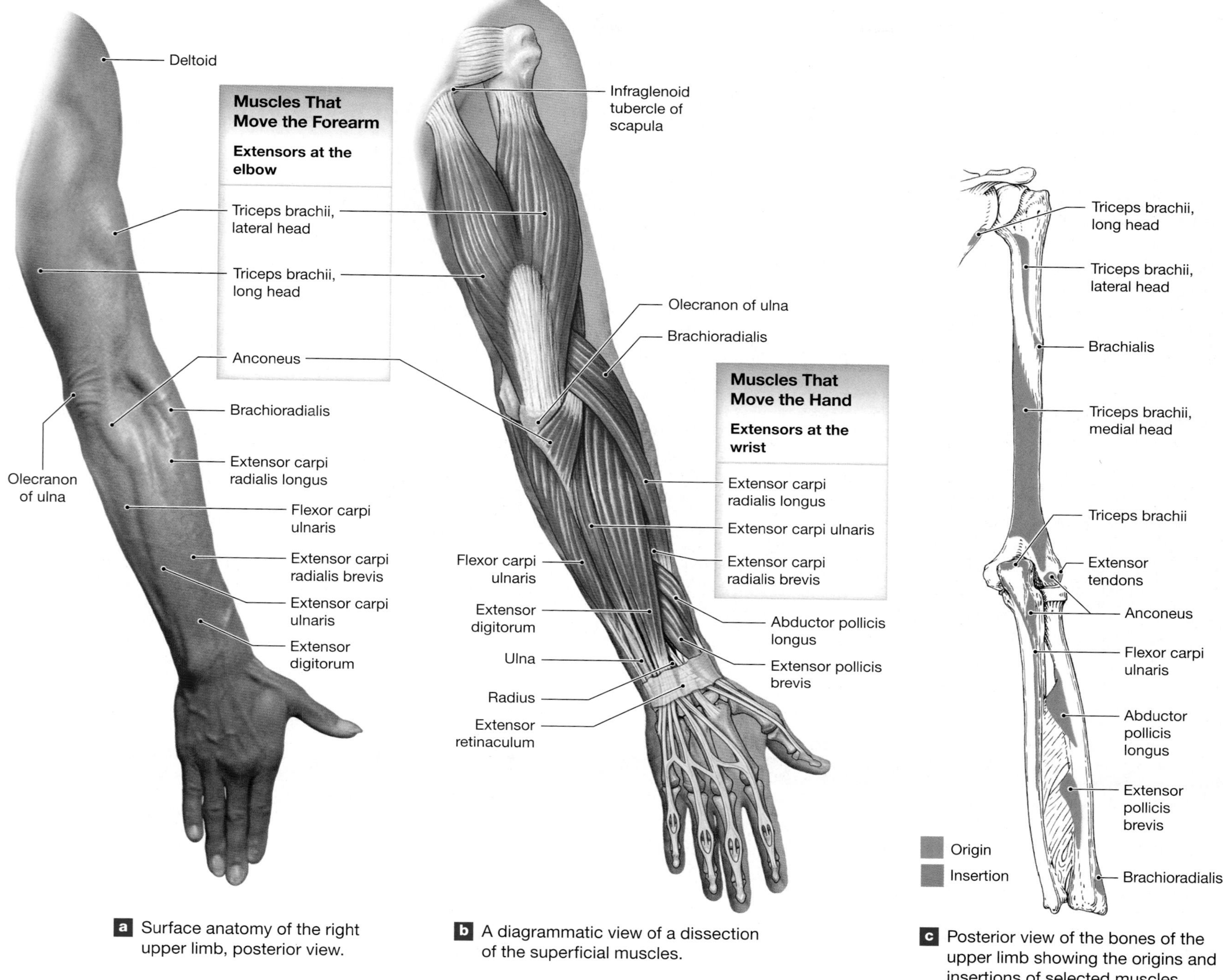

a Surface anatomy of the right upper limb, posterior view.

b A diagrammatic view of a dissection of the superficial muscles.

c Posterior view of the bones of the upper limb showing the origins and insertions of selected muscles.

11

Extrinsic Muscles of the Hand

Several superficial and deep muscles of the forearm flex and extend the joints of the fingers **(Table 11.4)**. These muscles provide strength and gross motor control of the hand and fingers and are called *extrinsic muscles of the hand*. *(Refer to Chapter 12, **Figures 12.4 and 12.5**, to identify these anatomical structures from the body surface.)*

Only the tendons of the extrinsic muscles of the hand cross the wrist joint. These are large muscles, so to ensure maximum mobility of the wrist and hand, the tendons of these muscles must be kept clear of the wrist joints **(Figures 11.8, 11.10, and 11.11)**. The tendons crossing the posterior and anterior surfaces of the wrist pass through **synovial tendon sheaths**, elongated bursae that reduce friction. **Figures 11.8b** and **11.11a–c** show these muscles in an anterior view, and **Figures 11.10b**, **11.11d–f**, and **11.12a** show them in a posterior view.

11

The fascia of the forearm thickens on the posterior surface of the wrist to form a wide band of connective tissue, the **extensor retinaculum** (ret-i-NAK-ū-lum) **(Figure 11.12a)**. The extensor retinaculum holds the tendons of the extensor muscles in place. The fascia also thickens on the anterior surface, forming another wide band of connective tissue, the **flexor retinaculum**, which holds the tendons of the flexor muscles in place **(Figure 11.12c)**.

TIPS & TOOLS

Look ahead to page 298. Here are some mnemonics to help you remember the muscles of the hand and their actions:

- PAD: **P**almar (hand) muscles **AD**uct
- DAB: **D**orsal (hand) muscles **AB**duct
- Muscles of the hand: "All For One And One For All." **A**bductor digiti minimi, **F**lexor digiti minimi, **O**pponens digit minimi, **A**dductor pollicis, **O**pponens pollicis, **F**lexor pollicis brevis, **A**bductor pollicis brevis

Inflammation of the retinacula and tendon sheaths restricts movement and irritates the median nerve, a sensory and motor nerve that innervates the hand. This condition, known as carpal tunnel syndrome, causes chronic pain.

CLINICAL NOTE

Carpal Tunnel Syndrome

Carpal tunnel syndrome is caused by inflammation and swelling of the flexor tendon sheaths within the carpal tunnel, deep to the flexor retinaculum within the palm. In addition to the long finger flexor tendons, the median nerve travels through this tunnel. Because it is the most vulnerable structure within this tight space, the median nerve can experience ischemia (is-KĒ-mē-a), an interruption of its blood supply, causing malfunction and pain. The pain manifests as tingling or numbness in digits 1,2 and 3, and the lateral half of digit 4. There is also abductor pollicis brevis weakness. Surgical treatment involves incising the flexor retinaculum to relieve pressure and create space for the nerve.

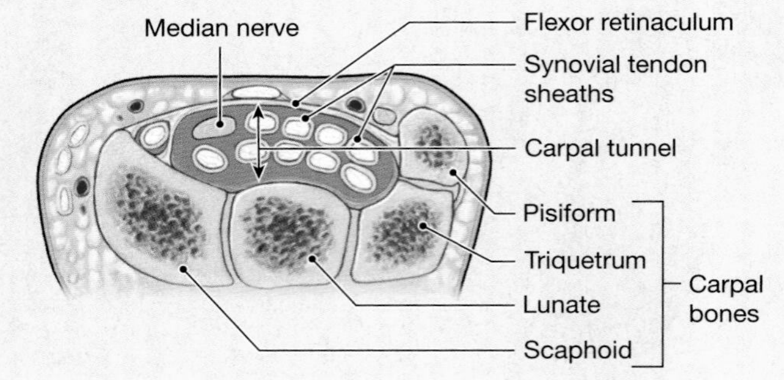

Table 11.4 | Muscles That Move the Hand and Fingers

Muscle	Origin	Insertion	Action	Innervation
Abductor pollicis longus	Proximal dorsal surfaces of ulna and radius	Lateral margin of first metacarpal bone and trapezium	Abduction at joints of thumb and wrist	Deep radial nerve (C_6–C_7)
Extensor digitorum	Lateral epicondyle of humerus	Posterior surfaces of the phalanges, digits 2–5	Extension at finger joints and wrist	Deep radial nerve (C_6–C_8)
Extensor pollicis brevis	Shaft of radius distal to origin of abductor pollicis longus; interosseous membrane	Base of proximal phalanx of thumb	Extension at joints of thumb; abduction at wrist	Deep radial nerve (C_6–C_7)
Extensor pollicis longus	Posterior and lateral surfaces of ulna; interosseous membrane	Base of distal phalanx of thumb	Extension at joints of thumb; abduction at wrist	Deep radial nerve (C_6–C_8)
Extensor indicis	Posterior surface of ulna; interosseous membrane	Posterior surface of proximal phalanx of index finger (2), with tendon of extensor digitorum	Extension and adduction at joints of index finger	Deep radial nerve (C_6–C_8)
Extensor digiti minimi	Via extensor tendon to lateral epicondyle of humerus and from intermuscular septa	Posterior surface of proximal phalanx of little finger	Extension at joints of little finger; extension at wrist	Deep radial nerve (C_6–C_8)
Flexor digitorum superficialis	Medial epicondyle of humerus; coronoid process of ulna and adjacent anterior surfaces of ulna and radius	To bases of middle phalanges of digits 2–5	Flexion at proximal interphalangeal, metacarpophalangeal, and wrist joints	Median nerve (C_7–T_1)
Flexor digitorum profundus	Medial and posterior surfaces of ulna, medial surfaces of coronoid process; interosseous membrane	Bases of distal phalanges of digits 2–5	Flexion at distal interphalangeal joints, and, to a lesser degree, proximal interphalangeal joints and wrist	Anterior interosseous branch of median nerve and ulnar nerve (C_8–T_1)
Flexor pollicis longus	Anterior shaft of radius; interosseous membrane	Base of distal phalanx of thumb	Flexion at joints of thumb	Median nerve (C_8–T_1)

Figure 11.11 Extrinsic Muscles That Move the Hands and Fingers

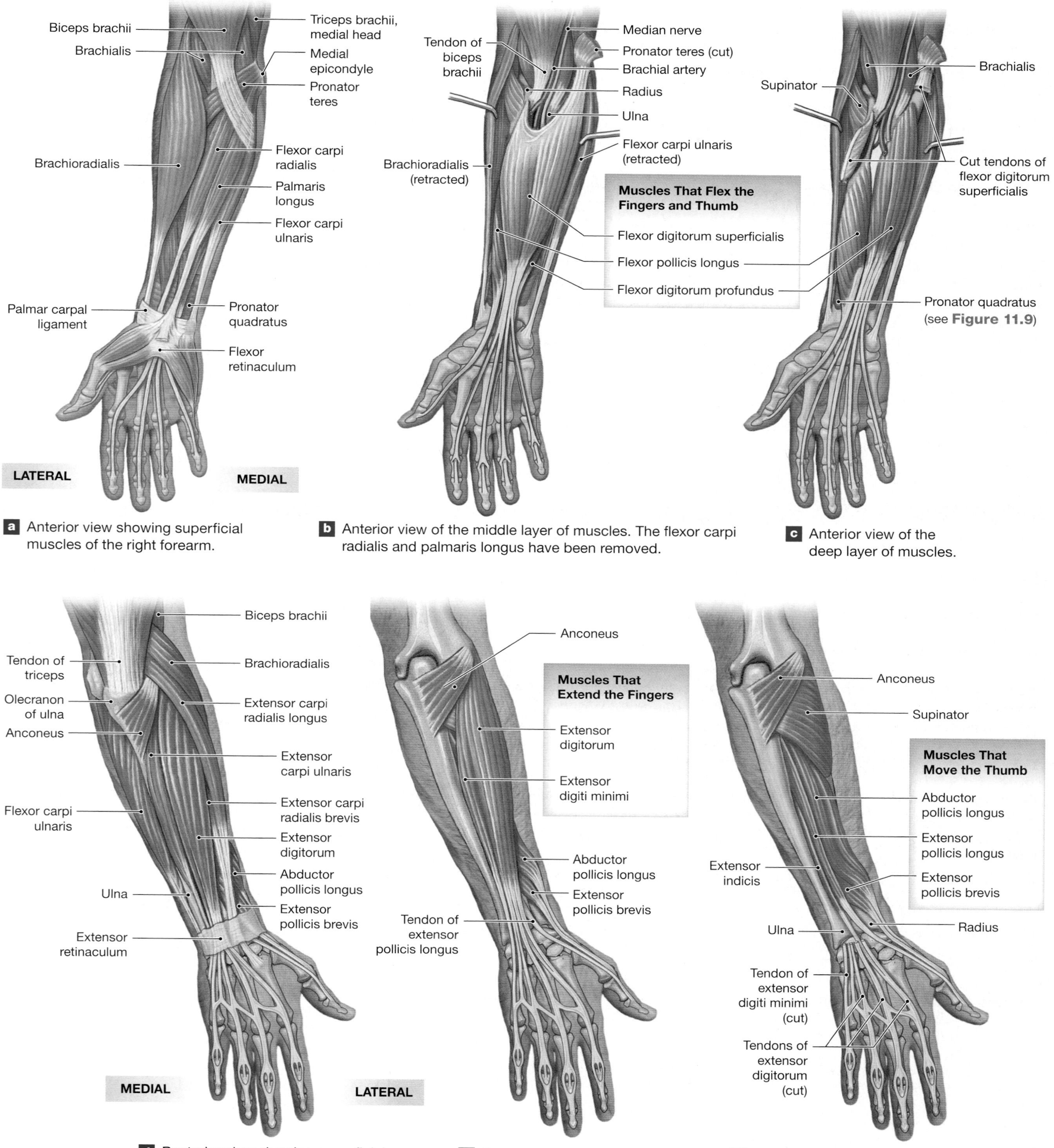

Biceps brachii
Brachialis
Triceps brachii, medial head
Medial epicondyle
Pronator teres
Brachioradialis
Flexor carpi radialis
Palmaris longus
Flexor carpi ulnaris
Palmar carpal ligament
Pronator quadratus
Flexor retinaculum
LATERAL
MEDIAL

a Anterior view showing superficial muscles of the right forearm.

Tendon of biceps brachii
Median nerve
Pronator teres (cut)
Brachial artery
Radius
Ulna
Flexor carpi ulnaris (retracted)
Brachioradialis (retracted)

Muscles That Flex the Fingers and Thumb

Flexor digitorum superficialis
Flexor pollicis longus
Flexor digitorum profundus

b Anterior view of the middle layer of muscles. The flexor carpi radialis and palmaris longus have been removed.

Supinator
Brachialis
Cut tendons of flexor digitorum superficialis
Pronator quadratus (see **Figure 11.9**)

c Anterior view of the deep layer of muscles.

Biceps brachii
Tendon of triceps
Brachioradialis
Olecranon of ulna
Extensor carpi radialis longus
Anconeus
Extensor carpi ulnaris
Flexor carpi ulnaris
Extensor carpi radialis brevis
Extensor digitorum
Abductor pollicis longus
Ulna
Extensor pollicis brevis
Extensor retinaculum
MEDIAL
LATERAL

d Posterior view showing superficial muscles of the right forearm.

Anconeus

Muscles That Extend the Fingers

Extensor digitorum
Extensor digiti minimi

Abductor pollicis longus
Extensor pollicis brevis
Tendon of extensor pollicis longus

e Posterior view of the middle layer of muscles.

Anconeus
Supinator

Muscles That Move the Thumb

Abductor pollicis longus
Extensor pollicis longus
Extensor pollicis brevis

Extensor indicis
Ulna
Radius
Tendon of extensor digiti minimi (cut)
Tendons of extensor digitorum (cut)

f Posterior view of the deep layer of muscles.

Intrinsic Muscles of the Hand

Fine motor control of the hand involves small *intrinsic muscles of the hand* that originate on the carpal and metacarpal bones (**Figures 11.12** and **11.13**). These intrinsic muscles are responsible for (1) flexion and extension of the fingers at the metacarpophalangeal joints, (2) abduction and adduction of the the fingers at the metacarpophalangeal joints, and (3) opposition and reposition of the thumb. No muscles originate on the phalanges, and only tendons extend across the distal joints of the fingers. **Table 11.5** lists the origins, insertions, and actions of the intrinsic muscles of the hand.

The four **lumbricals** originate on the tendons of the flexor digitorum profundus muscle in the palm of the hand. They insert onto the tendons of the extensor digitorum muscle. These muscles flex the metacarpophalangeal joints and extend the interphalangeal joints of the fingers.

11

The four **dorsal interossei** abduct the fingers. The **abductor digiti minimi** abducts the little finger, and the **abductor pollicis brevis** abducts the thumb. The **adductor pollicis** adducts the thumb, and the four **palmar interossei** adduct the fingers at the metacarpophalangeal joints.

Opposition of the thumb refers to flexing and medially rotating the thumb at the carpometacarpal joint and touching any other digit on the same hand. The **opponens pollicis** allows this action. Two extrinsic muscles of the hand, the extensor pollicis longus and the abductor pollicis longus reposition the thumb (see **Table 11.4**).

11.2 CONCEPT CHECK

2 Through which structures do the tendons that cross the posterior and anterior surfaces of the wrist pass before reaching their insertion points?

3 Name the thickened fascia on the posterior surface of the wrist that forms a wide band of connective tissue.

4 What is the primary muscle that abducts the arm at the shoulder joint?

5 Injury to the flexor carpi ulnaris impairs what two movements?

6 Identify the muscles that rotate the radius without flexing or extending the elbow.

See the blue Answers tab at the back of the book.

Table 11.5 | Intrinsic Muscles of the Hand

Muscle	Origin	Insertion	Action	Innervation
Adductor pollicis	Metacarpal and carpal bones	Proximal phalanx of thumb	Adduction of thumb	Ulnar nerve, deep branch (C_8–T_1)
Opponens pollicis	Trapezium and flexor retinaculum	First metacarpal bone	Opposition of thumb	Median nerve (C_6–C_7)
Palmaris brevis	Palmar aponeurosis	Skin of medial border of hand	Moves skin on medial border toward midline of palm	Ulnar nerve, superficial branch (C_8)
Abductor digiti minimi	Pisiform	Proximal phalanx of little finger	Abduction of little finger and flexion at its metacarpophalangeal joint	Ulnar nerve, deep branch (C_8–T_1)
Abductor pollicis brevis	Transverse carpal ligament, scaphoid and trapezium	Radial side of base of proximal phalanx of thumb	Abduction of thumb	Median nerve (C_6–C_7)
Flexor pollicis brevis*	Flexor retinaculum, trapezium, capitate, palmar ligaments of distal row of carpal bones, and ulnar side of first metacarpal	Radial and ulnar sides of proximal phalanx of thumb	Flexion and adduction of thumb	Branches of median and ulnar nerves
Flexor digiti minimi brevis	Hook of the hamate and flexor retinaculum	Proximal phalanx of little finger	Flexion at fifth metacarpophalangeal joint	Ulnar nerve, deep branch (C_8–T_1)
Opponens digiti minimi	Hook of the hamate and flexor retinaculum	Fifth metacarpal bone	Flexion at metacarpophalangeal joint; brings digit into opposition with thumb	Ulnar nerve, deep branch (C_8–T_1)
Lumbrical (4)	The four tendons of flexor digitorum profundus	Tendons of extensor digitorum to digits 2–5	Flexion at metacarpophalangeal joints; extension at proximal and distal interphalangeal joints	Median nerve (lumbricals 1, 2); ulnar nerve, deep branch (lumbricals 3, 4)
Dorsal interosseus (4)	Each originates from opposing faces of two metacarpal bones (I and II, II and III, III and IV, IV and V)	Bases of proximal phalanges of digits 2–4	Abduction at metacarpophalangeal joints of digits 2–4, flexion at metacarpophalangeal joints; extension at interphalangeal joints	Ulnar nerve, deep branch (C_8–T_1)
Palmar interosseus (4)	Sides of metacarpal bones II, IV, and V	Bases of proximal phalanges of digits 2, 4, and 5	Adduction at metacarpophalangeal joints of digits 2, 4, and 5; flexion at metacarpophalangeal joints; extension at interphalangeal joints	Ulnar nerve, deep branch (C_8–T_1)

*The portion of the flexor pollicis brevis originating on the first metacarpal bone is sometimes called the *first palmar interosseus*, which inserts on the ulnar side of the proximal phalanx and is innervated by the ulnar nerve.

Figure 11.12 Intrinsic Muscles, Tendons, and Ligaments of the Hand, Part I. Anatomy of the right wrist and hand.

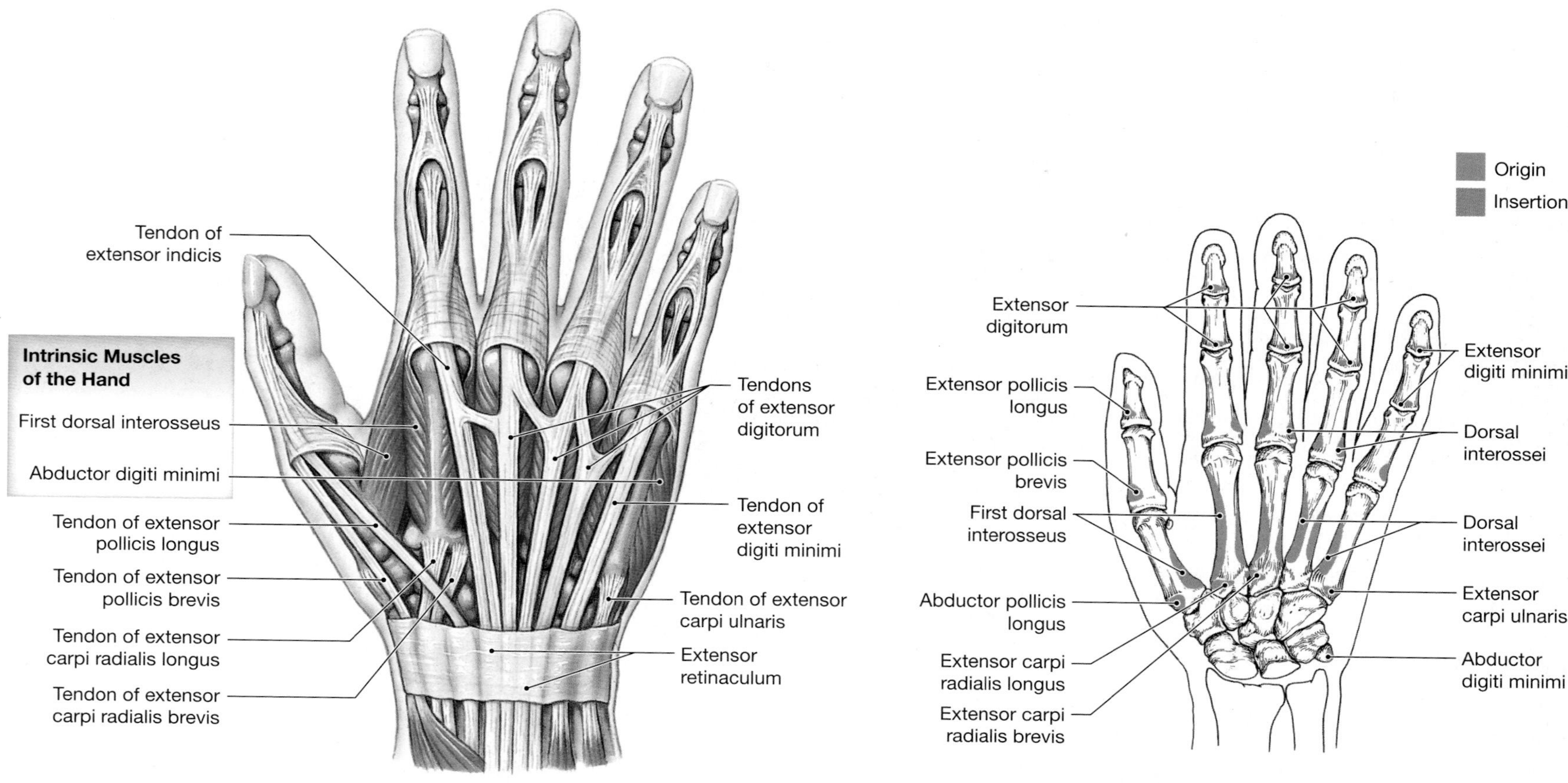

a Posterior (dorsal) view.

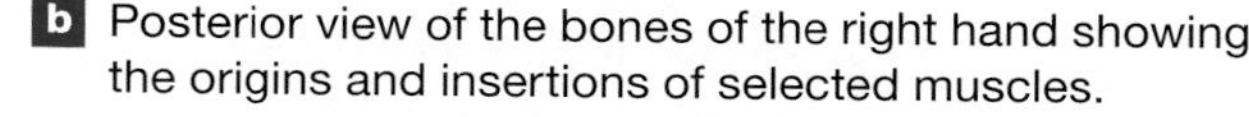

b Posterior view of the bones of the right hand showing the origins and insertions of selected muscles.

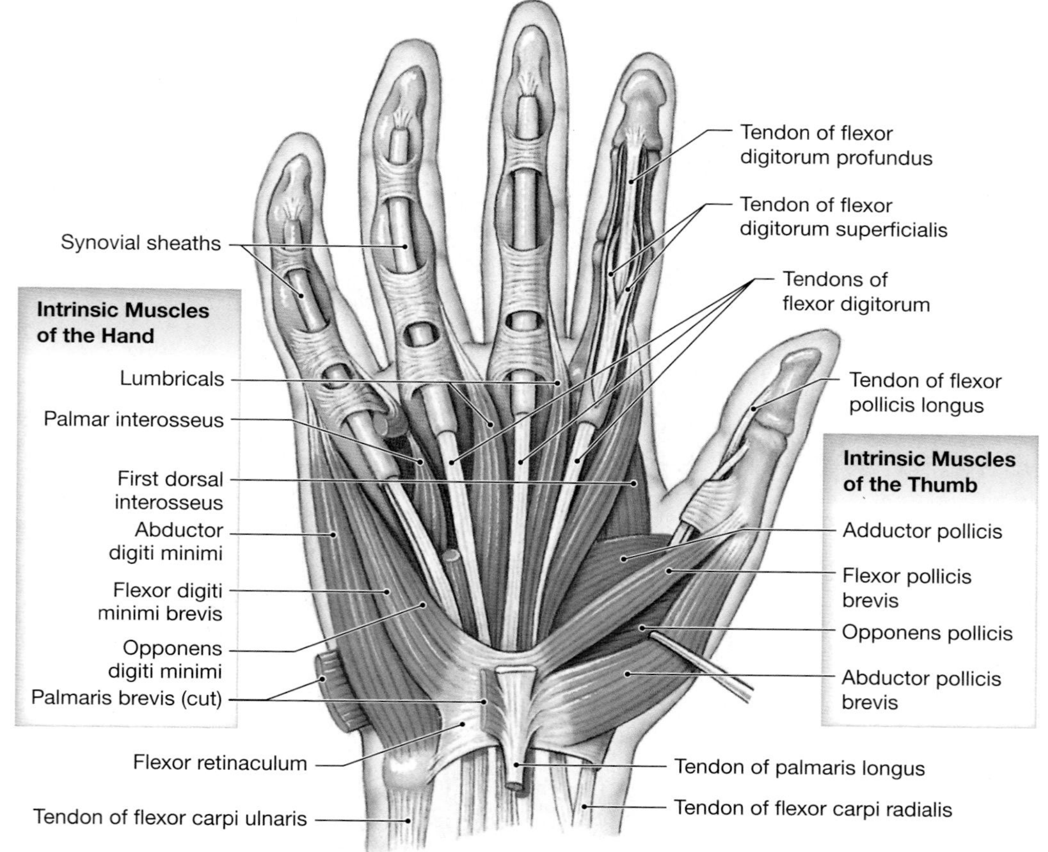

c Anterior (palmar) view.

Origin
Insertion
Flexor digitorum profundus
Flexor digitorum superficialis
Palmar interossei
Abductor digiti minimi
Palmar interossei
Opponens digiti minimi
Flexor carpi ulnaris
Abductor digiti minimi
Opponens digiti minimi
Adductor pollicis
Flexor pollicis longus
Adductor pollicis
Opponens pollicis
Abductor pollicis brevis
Flexor pollicis brevis

d Anterior view of the bones of the right hand, showing the origins and insertions of selected muscles.

Figure 11.13 Intrinsic Muscles, Tendons, and Ligaments of the Hand, Part II

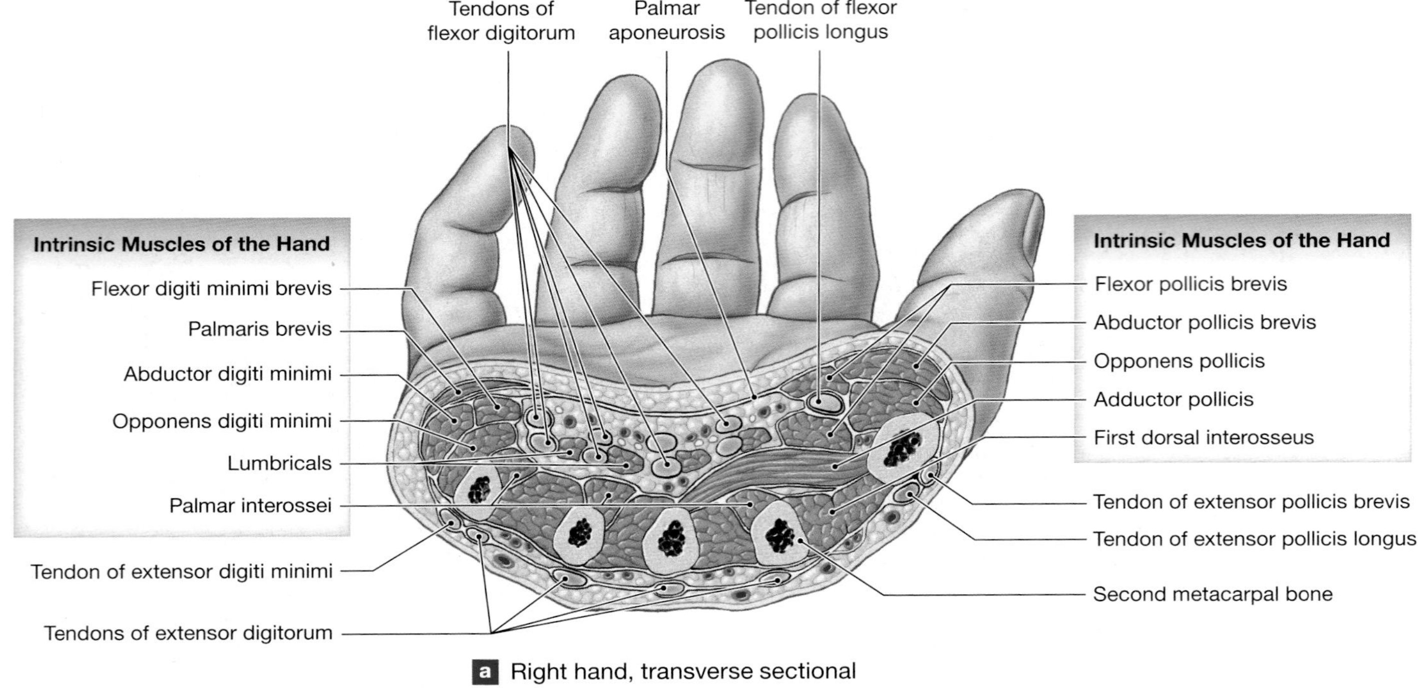

a Right hand, transverse sectional view through metacarpal bones

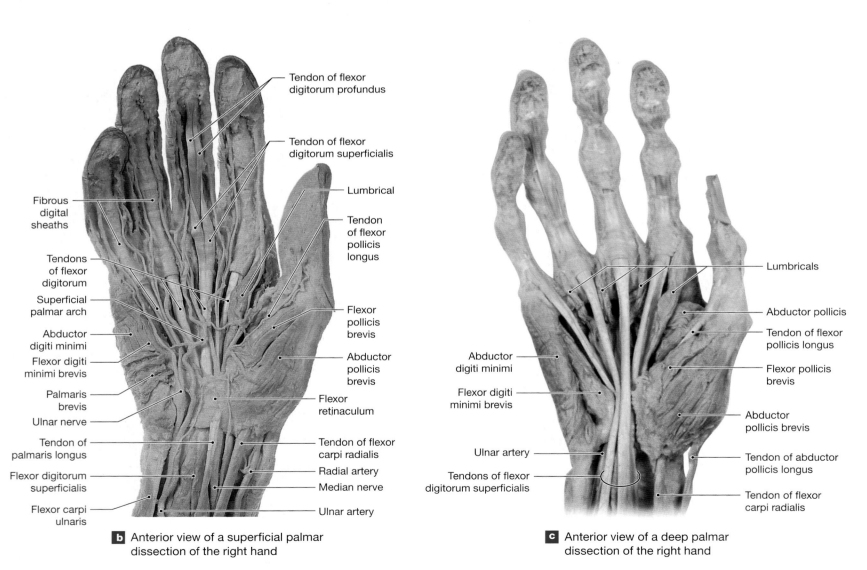

Tendon of flexor digitorum profundus
Tendon of flexor digitorum superficialis
Lumbrical
Tendon of flexor pollicis longus
Flexor pollicis brevis
Abductor pollicis brevis
Flexor retinaculum
Tendon of flexor carpi radialis
Radial artery
Median nerve
Ulnar artery
Fibrous digital sheaths
Tendons of flexor digitorum
Superficial palmar arch
Abductor digiti minimi
Flexor digiti minimi brevis
Palmaris brevis
Ulnar nerve
Tendon of palmaris longus
Flexor digitorum superficialis
Flexor carpi ulnaris

b Anterior view of a superficial palmar dissection of the right hand

Lumbricals
Abductor pollicis
Tendon of flexor pollicis longus
Flexor pollicis brevis
Abductor pollicis brevis
Tendon of abductor pollicis longus
Tendon of flexor carpi radialis
Abductor digiti minimi
Flexor digiti minimi brevis
Ulnar artery
Tendons of flexor digitorum superficialis

c Anterior view of a deep palmar dissection of the right hand

11.3 Compartments and Sectional Anatomy of the Arm and Forearm

▶ **KEY POINT** The deep fascia of the arm connects to the periosteum of the humerus, dividing the arm into (1) the anterior (flexor) compartment and (2) the posterior (extensor) compartment. The muscles within each of these compartments receive blood from a common artery and are innervated by a common nerve.

Chapter 3 introduced the types of fasciae in the body and explored how these dense connective tissue layers provide a structural framework for the soft tissues of the body. ⮌ pp. 73–74 There are three types of fasciae: (1) the superficial fascia, a layer of areolar tissue deep to the skin; (2) the deep fascia, a dense fibrous layer bound to the capsule, periosteum, epimysium, and other fibrous sheaths surrounding internal organs; and (3) the subserous fascia, a layer of areolar tissue separating a serous membrane from adjacent structures.

The connective tissue fibers of the deep fascia support and interconnect adjacent skeletal muscles while allowing independent muscle movement. If muscles are similar in orientation, action, and range of motion, they are tightly interconnected by the deep fascia. Such muscles are difficult to separate during dissection. However, if the orientations and actions of two muscles are different, they are less tightly interconnected and easier to separate on dissection.

In the limbs, the muscles are tightly packed together around the bones. The superficial fascia, deep fascia, and periosteum are closely interconnected. As a result, the muscles of the limb are separated into **compartments**.

The deep fascia of the arm forms the **lateral intermuscular septum** and the **medial intermuscular septum** (**Figure 11.14a,b**). The lateral intermuscular septum extends along the lateral surface of the humerus from the lateral epicondyle to the deltoid tuberosity. The medial intermuscular septum is shorter, extending along the medial surface of the humerus from the medial epicondyle to the insertion of the coracobrachialis. These two pieces of connective tissue divide the arm into an **anterior compartment**, or *flexor compartment*, and a **posterior compartment**, or *extensor compartment* (**Figures 11.14** and **11.15**). The biceps brachii, coracobrachialis, and brachialis are in the anterior compartment; the triceps brachii is the only muscle in the posterior compartment. The major blood vessels, lymphatics, and nerves of the arm are in the connective tissue between the anterior and posterior compartments.

The deep fascia and the antebrachial interosseous membrane divide the forearm into four compartments: (1) **superficial anterior compartment**, (2) **deep anterior compartment**, (3) **lateral compartment**, and (4) **posterior compartment** (**Figures 11.14c,d** and **11.15**). **Table 11.6** lists the structures within each compartment of the upper limb.

11.3 CONCEPT CHECK

7 There are five muscles in the superficial anterior compartment of the forearm. Name these muscles, moving medially to laterally.

8 Give the name of the only nerve that innervates all posterior compartment muscles of the arm and forearm.

See the blue Answers tab at the back of the book.

Table 11.6 | Compartments of the Upper Limb

Compartment	Muscles	Blood Vessels*,†	Nerves‡
ARM			
Anterior compartment	Biceps brachii Brachialis Coracobrachialis	Brachial artery Inferior ulnar collateral artery Superior ulnar collateral artery Brachial veins	Median nerve Musculocutaneous nerve Ulnar nerve
Posterior compartment	Triceps brachii	Deep brachial artery	Radial nerve
FOREARM			
Anterior compartment **Superficial**	Flexor carpi radialis Flexor carpi ulnaris Flexor digitorum superficialis Palmaris longus Pronator teres	Radial artery Ulnar artery Anterior interosseous artery Anterior ulnar recurrent artery Posterior ulnar recurrent artery	Median nerve Ulnar nerve Anterior interosseous nerve Ulnar nerve Median nerve
Deep	Flexor digitorum profundus Flexor pollicis longus Pronator quadratus		
Lateral compartment§	Brachioradialis Extensor carpi radialis brevis Extensor carpi radialis longus	Radial artery	Radial nerve
Posterior compartment	Abductor pollicis longus Anconeus Extensor carpi ulnaris Extensor digitorum Extensor digiti minimi Extensor indicis Extensor pollicis brevis Extensor pollicis longus Supinator	Posterior interosseous artery Posterior ulnar recurrent artery	Posterior interosseous nerve

*Cutaneous vessels are not listed.
†Only large, named vessels are listed.
‡Cutaneous nerves are not listed.
§Contains what is sometimes called the radial, or antero-external, group of muscles.

Figure 11.14 Musculoskeletal Compartments of the Upper Limb

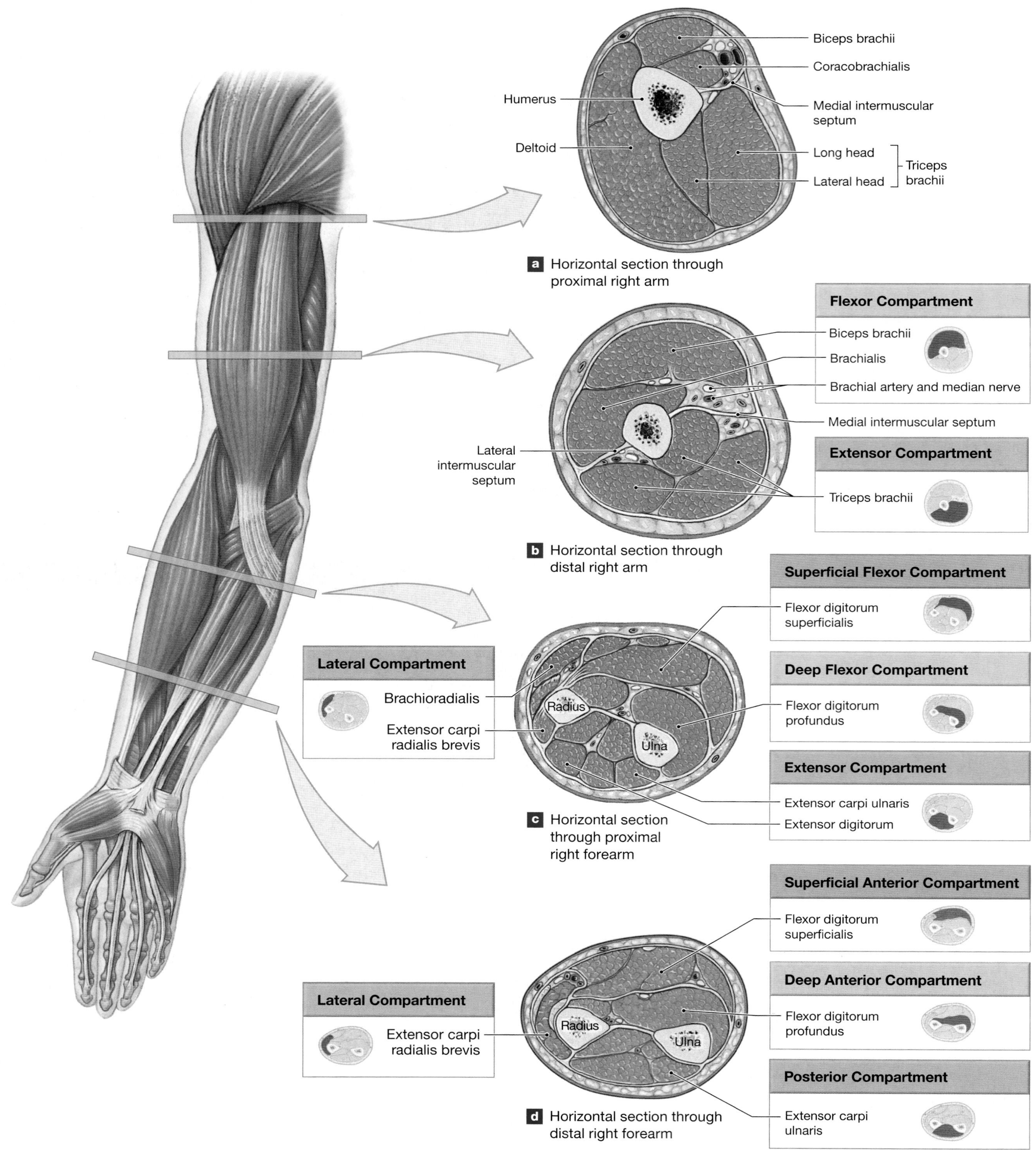
Humerus
Deltoid
Biceps brachii
Coracobrachialis
Medial intermuscular septum
Long head
Lateral head
Triceps brachii
a Horizontal section through proximal right arm
Flexor Compartment
Biceps brachii
Brachialis
Brachial artery and median nerve
Medial intermuscular septum
Lateral intermuscular septum
Extensor Compartment
Triceps brachii
b Horizontal section through distal right arm
Superficial Flexor Compartment
Flexor digitorum superficialis
Lateral Compartment
Brachioradialis
Extensor carpi radialis brevis
Radius
Ulna
Deep Flexor Compartment
Flexor digitorum profundus
Extensor Compartment
Extensor carpi ulnaris
Extensor digitorum
c Horizontal section through proximal right forearm
Superficial Anterior Compartment
Flexor digitorum superficialis
Lateral Compartment
Extensor carpi radialis brevis
Radius
Ulna
Deep Anterior Compartment
Flexor digitorum profundus
Posterior Compartment
Extensor carpi ulnaris
d Horizontal section through distal right forearm

Figure 11.15 **Dissection of the Right Upper Limb, with Sectional Views of the Arm and Forearm**

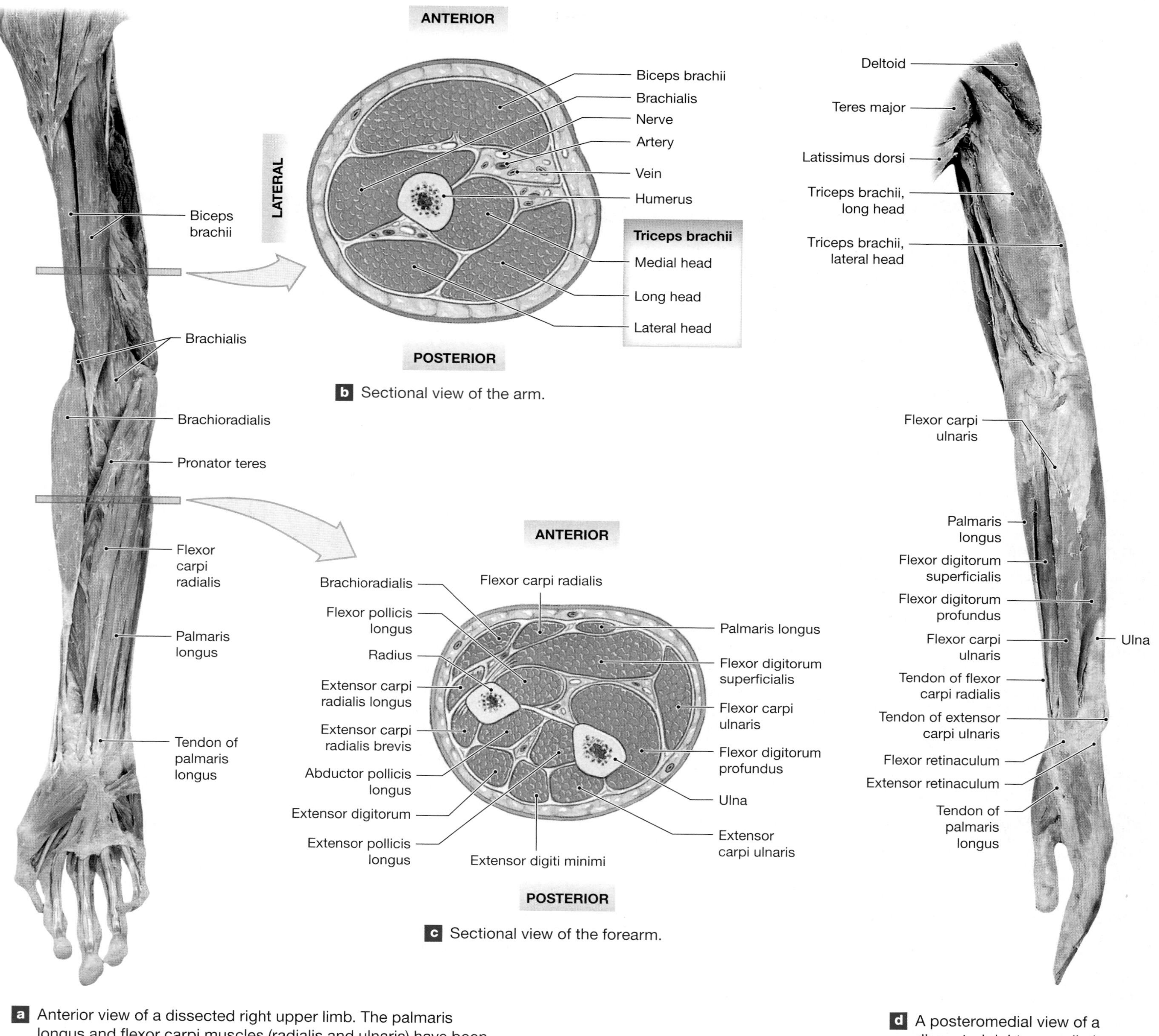

b Sectional view of the arm.

c Sectional view of the forearm.

a Anterior view of a dissected right upper limb. The palmaris longus and flexor carpi muscles (radialis and ulnaris) have been partly removed, and the flexor retinaculum has been cut.

d A posteromedial view of a dissected right upper limb.

11.4 Muscles of the Pelvic Girdle and Lower Limb

▶ **KEY POINT** The muscles of the pelvic girdle, lower limbs, and feet contract in a coordinated manner, allowing the body to move from place to place.

The attachments between the pelvic girdle and axial skeleton allow very little movement. The muscles that influence the position of the pelvis are discussed in Chapter 10. ↄ **pp. 277–278** The muscles of the lower limbs are larger and more powerful than those of the upper limbs. There are three groups of muscles in the lower limb: (1) muscles that move the thigh, (2) muscles that move the leg, and (3) muscles that move the foot and toes **(Figure 11.1)**.

Muscles That Move the Thigh

▶ **KEY POINT** Many of the muscles that move the thigh originate on the ilium and insert onto the femur or the iliotibial tract.

The large, powerful muscles that move the thigh originate on the pelvis. These muscles include the gluteal group, lateral rotator group, adductor group, and iliopsoas group (**Table 11.7**).

Three **gluteal muscles** cover the lateral surface of the ilium (**Spotlight Figure 11.2** and **Figures 11.16** and **11.17**). *(Refer to Chapter 12,* **Figure 12.6c**, *to identify these anatomical structures from the body surface.)* The **gluteus maximus** is the largest and most superficial of the gluteal muscles. It originates on the posterior gluteal line and parts of the iliac crest; the sacrum, coccyx, and associated ligaments; and the thoracolumbar fascia. This muscle extends and laterally rotates the thigh at the hip. The gluteus maximus shares an insertion with the **tensor fasciae latae** (TEN-sor FASH-ē-ē LĀ-tē),

Table 11.7 | **Muscles That Move the Thigh**

Muscle	Origin	Insertion	Action	Innervation
GLUTEAL GROUP				
Gluteus maximus	Iliac crest, posterior gluteal line, and lateral surface of ilium; sacrum, coccyx, and thoracolumbar fascia	Iliotibial tract and gluteal tuberosity of femur	Extension and lateral rotation at hip; helps stabilize the extended knee; abduction at the hip (superior fibers only)	Inferior gluteal nerve (L_5–S_2)
Gluteus medius	Anterior iliac crest, lateral surface of ilium between posterior and anterior gluteal lines	Greater trochanter of femur	Abduction and medial rotation at hip	Superior gluteal nerve (L_4–S_1)
Gluteus minimus	Lateral surface of ilium between inferior and anterior gluteal lines	Greater trochanter of femur	Abduction and medial rotation at hip	Superior gluteal nerve (L_4–S_1)
Tensor fasciae latae	Iliac crest and lateral surface of anterior superior iliac spine	Iliotibial tract	Extension of the knee and lateral rotation of the leg acting through the iliotibial tract; abduction* and medial rotation of the thigh	Superior gluteal nerve (L_4–S_1)
LATERAL ROTATOR GROUP				
Obturators (externus and internus)	Lateral and medial margins of obturator foramen	Trochanteric fossa of femur (externus); medial surface of greater trochanter (internus)	Lateral rotation and abduction of hip; help to maintain stability and integrity of the hip	Obturator nerve (externus: L_3–L_4) and special nerve from sacral plexus (internus: L_5–S_2)
Piriformis	Anterolateral surface of sacrum	Greater trochanter of femur	Lateral rotation and abduction of hip; help to maintain stability and integrity of the hip	Branches of sacral nerves (S_1–S_2)
Gemelli (superior and inferior)	Ischial spine (superior gemellus) and ischial tuberosity (inferior gemellus)	Medial surface of greater trochanter via tendon of obturator internus	Lateral rotation and abduction of hip; help to maintain stability and integrity of the hip	Nerves to obturator internus and quadratus femoris
Quadratus femoris	Lateral border of ischial tuberosity	Intertrochanteric crest of femur	Lateral rotation of hip	Special nerves from sacral plexus (L_4–S_1)
ADDUCTOR GROUP				
Adductor brevis	Inferior ramus of pubis	Linea aspera of femur	Adduction and flexion at hip	Obturator nerve (L_3–L_4)
Adductor longus	Inferior ramus of pubis, anterior to adductor brevis	Linea aspera of femur	Adduction, flexion, and medial rotation at hip	Obturator nerve (L_3–L_4)
Adductor magnus	Inferior ramus of pubis posterior to adductor brevis and ischial tuberosity	Linea aspera and adductor tubercle of femur	Whole muscle produces adduction at the hip; anterior part produces flexion and medial rotation; posterior part produces extension	Obturator and sciatic nerves
Pectineus	Superior ramus of pubis	Pectineal line inferior to lesser trochanter of femur	Flexion and adduction at hip	Femoral nerve (L_2–L_4)
Gracilis	Inferior ramus of pubis	Medial surface of tibia inferior to medial condyle	Flexion and medial rotation at knee; adduction and medial rotation at hip	Obturator nerve (L_3–L_4)
ILIOPSOAS GROUP				
Iliacus	Iliac fossa	Femur distal to lesser trochanter; tendon fused with that of psoas major	Flexion at the hip and, when working with the psoas major, flexes the intervertebral joints.	Femoral nerve (L_2–L_3)
Psoas major	Anterior surfaces and transverse processes of vertebrae (T_{12}–L_5)	Lesser trochanter in company with iliacus	Flexion at hip and/or lumbar intervertebral joints	Branches of the lumbar plexus (L_2–L_3)

*Research results have raised significant questions regarding the role of the tensor fasciae latae in abducting the thigh at the hip.

which originates on the iliac crest and lateral surface of the anterior superior iliac spine. Together, these muscles pull on the **iliotibial** (il-ē-ō-TIB-ē-al) **tract**, a band of collagen fibers that extends along the lateral surface of the thigh and inserts on the tibia. This tract braces the lateral surface of the knee and stabilizes the knee when a person balances on one foot.

The **gluteus medius** and **gluteus minimus** originate anterior to the gluteus maximus and insert on the greater trochanter of the femur **(Figures 11.16a–c** and **11.17b)**. Both abduct and medially rotate the thigh at the hip. The anterior gluteal line on the lateral surface of the ilium marks the boundary between the gluteus medius and gluteus minimus.

Figure 11.16 Muscles That Move the Thigh, Part I

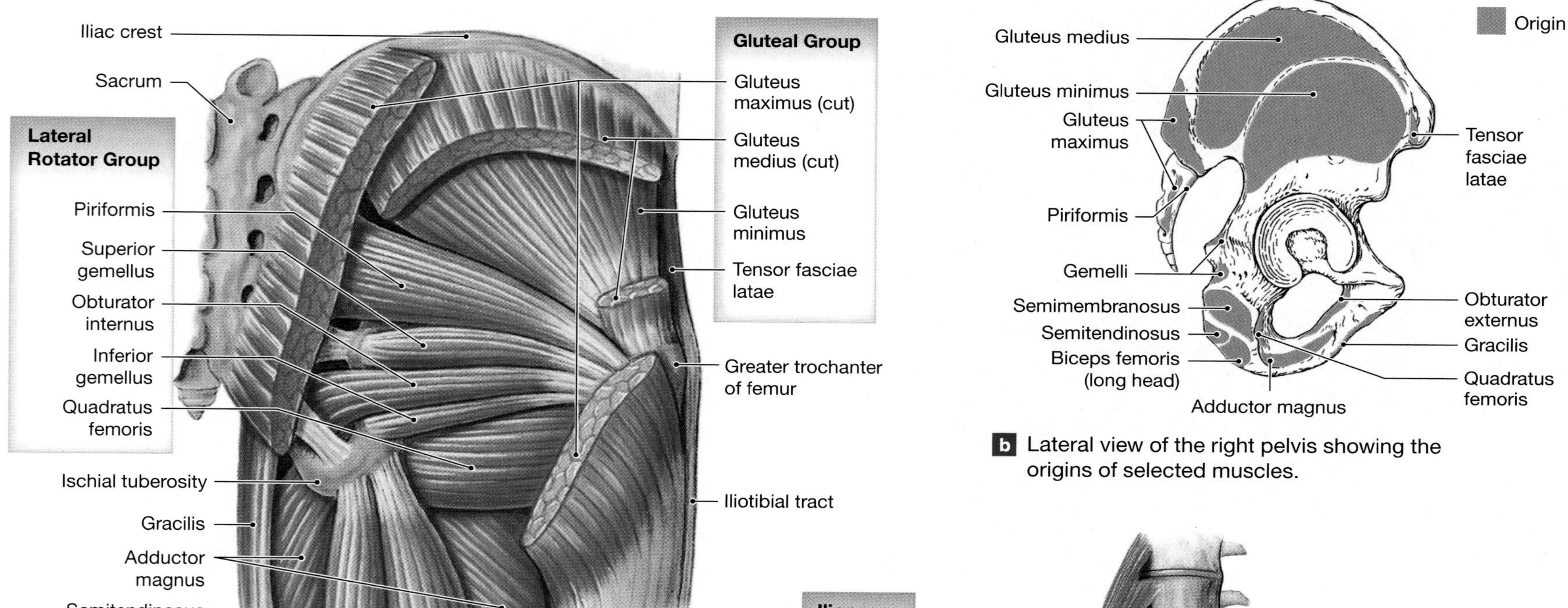

a Posterior view of pelvis showing deep dissections of the gluteal muscles and lateral rotators. For a superficial view of the gluteal muscles, see **Figures 11.2**, **11.18**, and **11.19**.

b Lateral view of the right pelvis showing the origins of selected muscles.

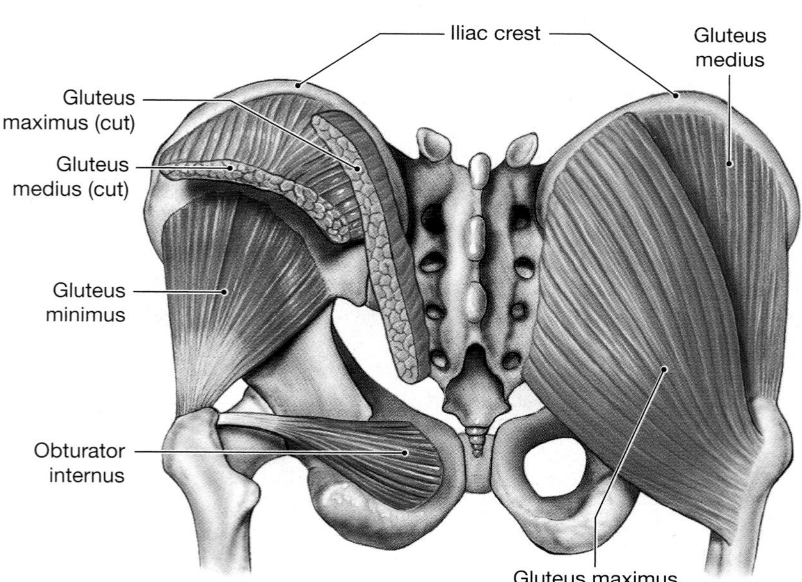

c Posterior view of the gluteal and lateral rotator muscles; the gluteus maximus has been removed to show the deeper muscles.

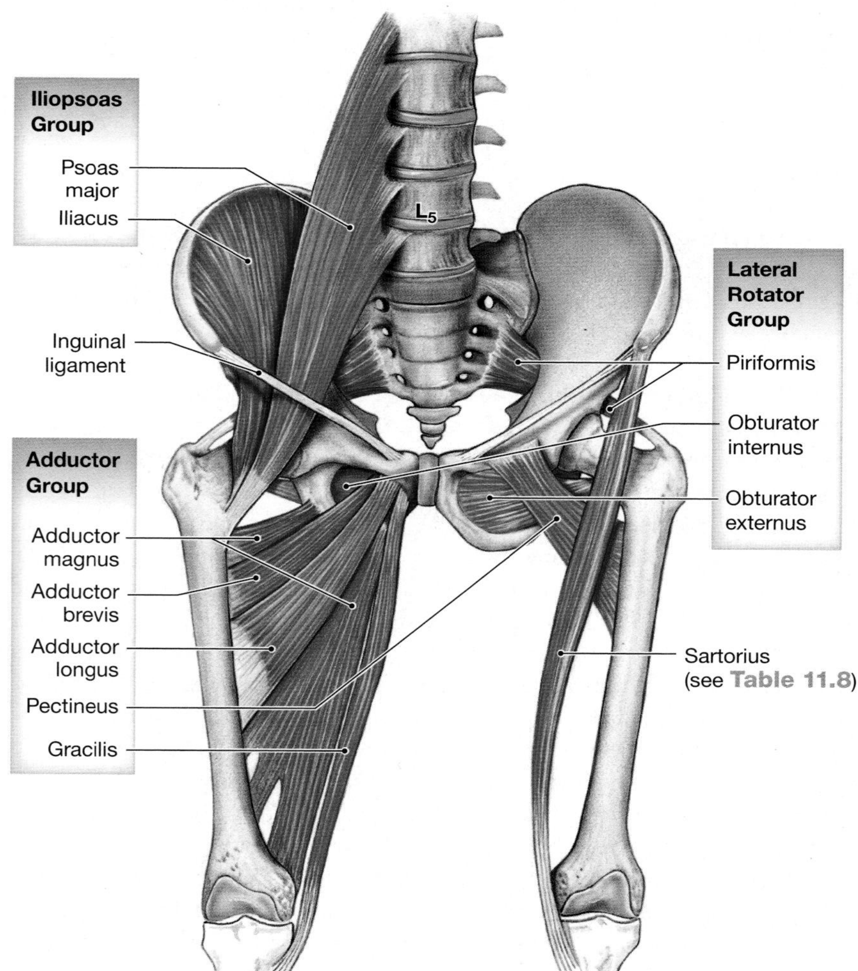

d Anterior view of the iliopsoas and the adductor group.

Figure 11.17 Muscles That Move the Thigh, Part II

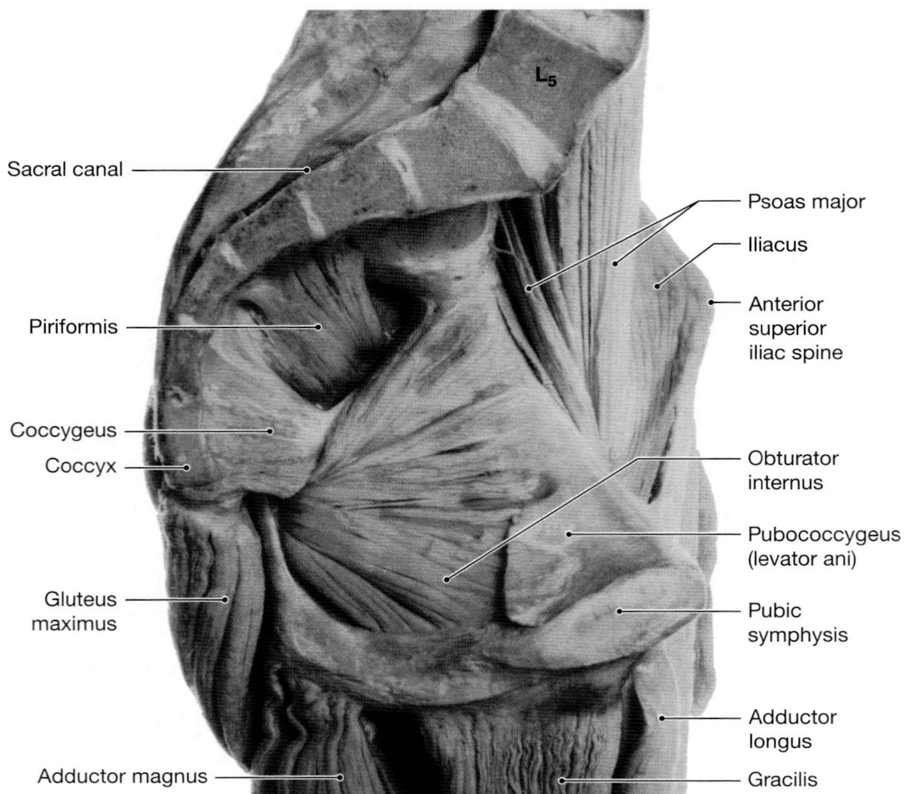

a Muscles and associated structures seen in a sagittal section through the pelvis

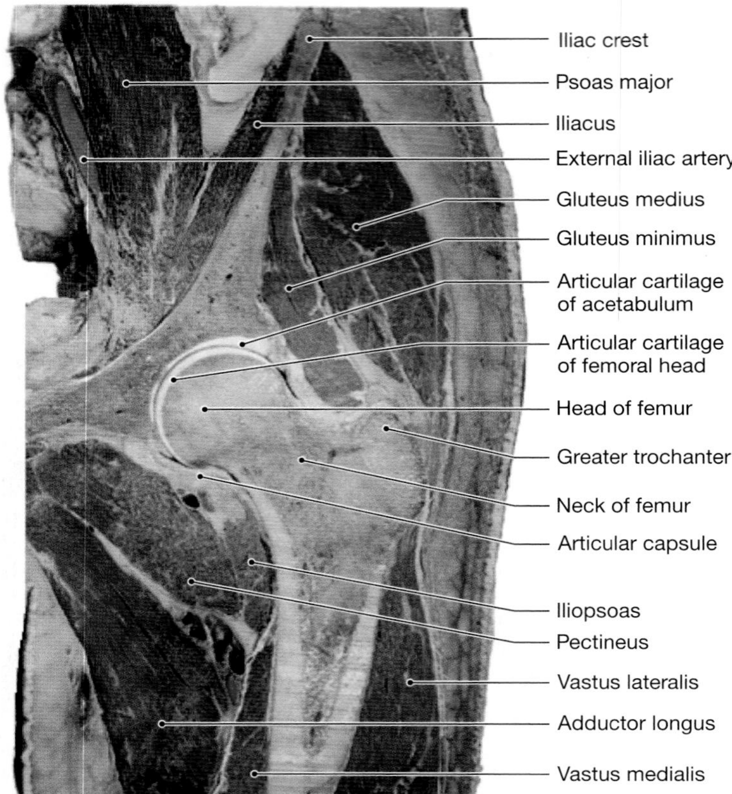

b Coronal section through the hip showing the hip joint and surrounding muscles

The **lateral rotators** laterally rotate the thigh at the hip **(Figures 11.16** and **11.17a)**. In addition, the **piriformis** (pir-i-FOR-mis) **obturator muscles**, and the **gemelli muscles** abduct the thigh at the hip. The dominant lateral rotators of this group are the piriformis, obturator externus, and obturator internus.

The **adductors** are found inferior to the acetabulum **(Figures 11.16** and **11.17)**. *(Refer to Chapter 12,* **Figures 12.6a** *and* **12.7a**, *to identify these anatomical structures from the body surface.)* The **adductor magnus**, **adductor brevis**, **adductor longus**, **pectineus** (pek-TI-nē-us), and **gracilis** (GRAS-i-lis) all originate on the pubis. Except for the gracilis, all of these muscles insert on the linea aspera, a ridge along the posterior surface of the femur. (The gracilis inserts on the tibia.) Their actions are varied. All of the adductors except the adductor magnus originate both anterior and inferior to the hip, so they are flexors, adductors, and medial rotators of the thigh at the hip. The adductor magnus adducts, flexes, and medially rotates, or extends and laterally rotates, the thigh at the hip, depending on which region of the muscle is stimulated. When an athlete pulls a groin muscle, he or she has torn one of these adductor muscles.

The medial surface of the pelvis is dominated by a single pair of muscles: the psoas major and iliacus. The **psoas** (SŌ-us) **major** originates on the inferior thoracic and lumbar vertebrae and inserts onto the lesser trochanter of the femur. The tendon of the psoas major muscle joins with the tendon of the **iliacus** (il-Ē-a-kus), which originates on the iliac fossa. These two muscles are powerful flexors of the hip, and they pass deep to the *inguinal ligament*. They are often referred to together as the **iliopsoas** (i-lē-ō-SŌ-us) **(Figures 11.16d** and **11.17)**.

One way to organize the diverse muscles is to group them by their orientation around the hip. Muscles that originate on the pelvis and insert on the femur produce characteristic movements determined by their position relative to the acetabulum. These action lines around the acetabulum were presented in **Spotlight Figure 11.2** on pp. 284–285.

Muscles That Move the Leg

▶ **KEY POINT** The femoral nerve innervates the muscles that extend the leg at the knee; the sciatic nerve innervates the muscles that flex the leg at the knee.

Muscles that move the leg are detailed in **Figures 11.18** to **11.21** and **Table 11.8**. You can use the relationships between the action lines and the axis of the knee joint to predict the actions of the muscles that move the leg at the knee. However, *the anterior/posterior orientation of the muscles that move the leg is reversed*. This is related to the rotation of the limb during embryological development (see Chapter 28, Embryology and Human Development). Therefore:

- Muscles that have action lines passing anteriorly to the axis of the knee joint, such as the quadriceps femoris, extend the knee.
- Muscles that have action lines passing posteriorly to the axis of the knee joint, such as the hamstrings, flex the knee.

Most of the extensor muscles originate on the femur and extend along the anterior and lateral surfaces of the thigh **(Figures 11.18, 11.20a,b,** and **11.21)**. Flexor muscles originate on the pelvis and extend along the posterior and medial surfaces of the thigh **(Figures 11.19, 11.20c,** and **11.21)**. *(Refer to Chapter 12,* **Figure 12.7a,b**, *to identify these anatomical structures from the body surface.)*

Collectively, the knee extensors are called the **quadriceps femoris**, or the *quadriceps muscles*. Three of the quadriceps muscles, the **vastus muscles** (**vastus lateralis**, **vastus medialis**, and **vastus intermedius**), originate on the femur, and the **rectus femoris** originates on the anterior inferior iliac spine. All of these muscles insert onto the tibial tuberosity by the quadriceps tendon, patella, and patellar ligament. The three vastus muscles surround the rectus femoris the same way a bun surrounds a hot dog. The vastus lateralis, vastus medialis, and vastus intermedius extend the knee. Because the rectus femoris originates on the anterior inferior iliac spine of the pelvis, it crosses the hip and the knee joints, so it flexes the hip and extends the knee.

The flexors of the knee are the **biceps femoris**, **semimembranosus** (sem-ē-mem-bra-NŌ-sus), **semitendinosus** (sem-ē-ten-di-NŌ-sus), and **sartorius** (sar-TOR-ē-us). These muscles originate on the pelvis and insert on the tibia and fibula **(Figures 11.18 to 11.21)**. Because the long head of the biceps femoris and the semimembranosus and semitendinosus originate on the pelvis inferior and posterior to the acetabulum, they also cross the hip joint and, therefore, extend the hip. These muscles are often called the "hamstrings."

The sartorius is the only knee flexor that originates superior to the acetabulum. It inserts on the medial aspect of the tibia. The sartorius flexes, abducts, and laterally rotates the hip and also flexes the knee.

Figure 11.18 Muscles That Move the Leg, Anterior Views

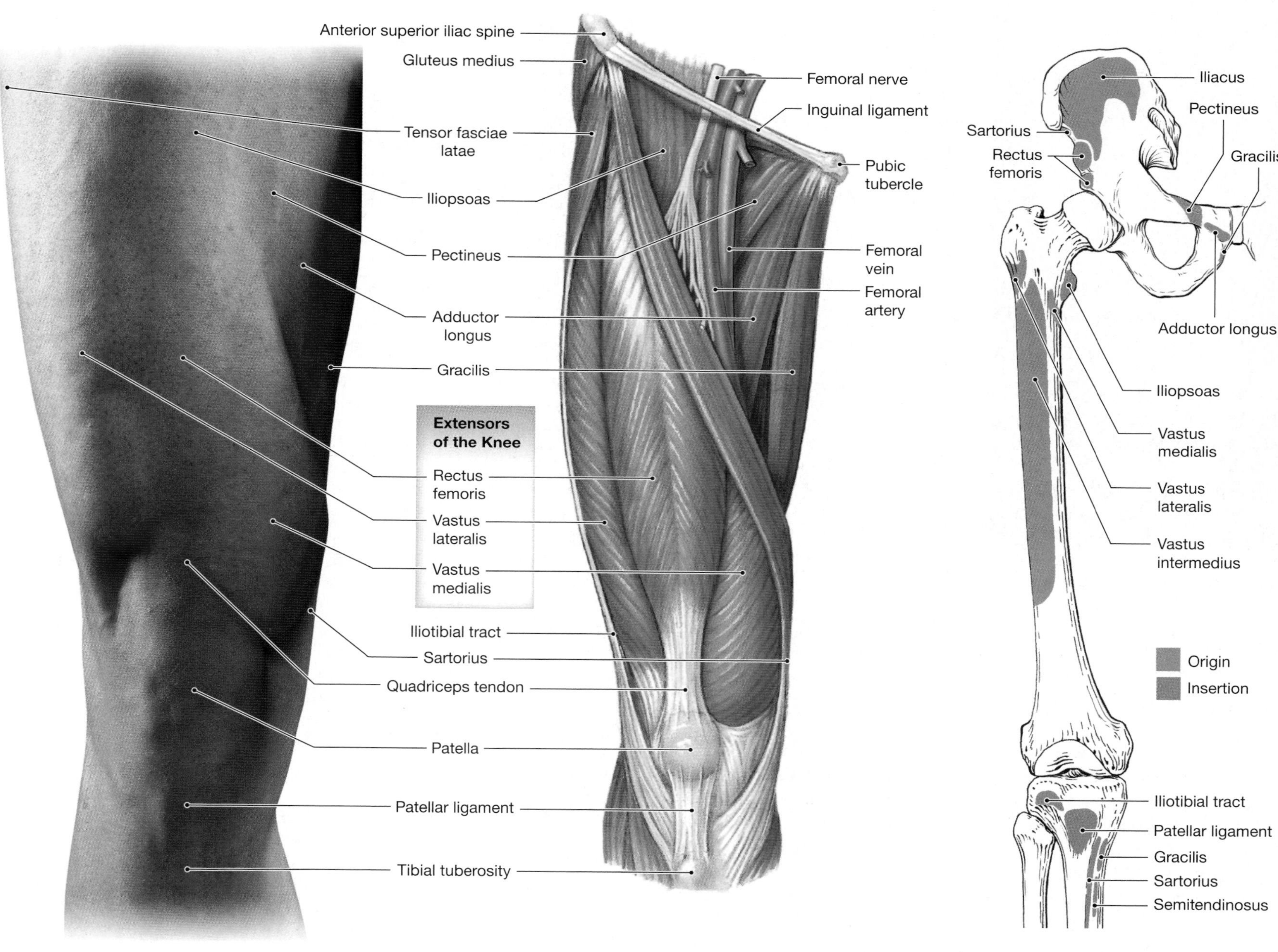

a Surface anatomy, anterior view, of the right thigh.

b Anterior view of the superficial muscles of the right thigh.

c Anterior view of the bones of the right lower limb showing the origins and insertions of selected muscles.

Figure 11.19 **Muscles That Move the Leg, Posterior Views**

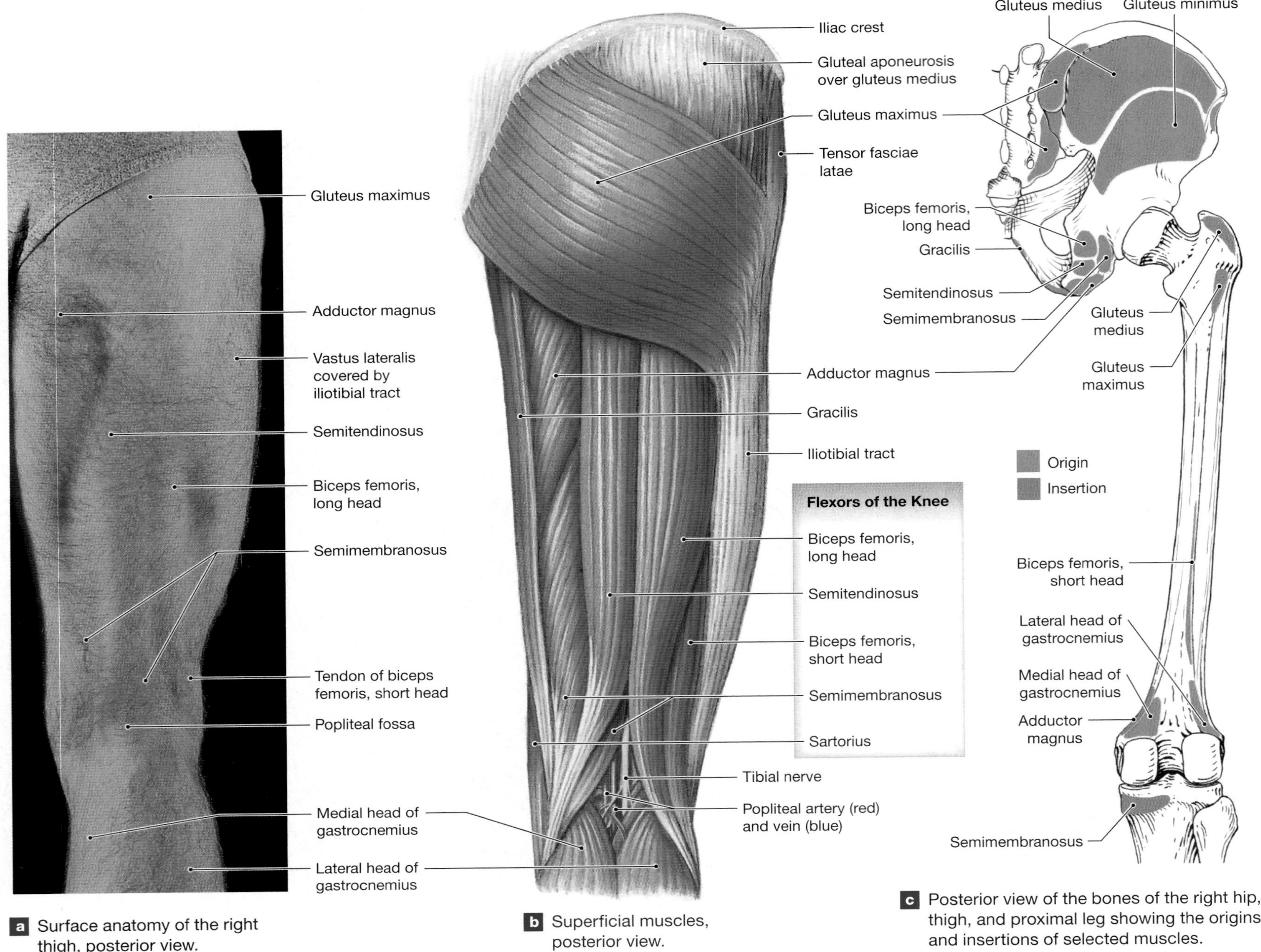

a Surface anatomy of the right thigh, posterior view.

b Superficial muscles, posterior view.

c Posterior view of the bones of the right hip, thigh, and proximal leg showing the origins and insertions of selected muscles.

As noted in Chapter 8, the knee joint can be locked at full extension by a slight lateral rotation of the tibia. The small **popliteus** (pop-LI-tē-us) originates on the femur near the lateral condyle and inserts on the posterior tibial shaft **(Figure 11.19d,e)**. When the knee starts to flex, this muscle contracts and medially rotates the tibia, unlocking the knee joint.

Figure 11.19a shows the surface anatomy of the posterior thigh and landmarks for some of the knee flexors. The sectional view of the thigh shows the four muscles of the quadriceps femoris **(Figure 11.20b)**. The vastus intermedius nestles against the femur and lies deep to the other three muscles of the quadriceps. The large mass of the vastus lateralis is visible in the lateral view of the thigh because part of it lies deep to the iliotibial tract **(Figure 11.21b)**.

TIPS & TOOLS

When you are studying the hamstring muscles, the following tips will help you remember the insertions and origins of the biceps femoris, semitendinosus, and semimembranosus:

- Biceps femoris: **Bi** means "two heads," but it should also remind you that this muscle has **two origins** (ischial tuberosity and linea aspera of the femur) and **two insertions** (head of the fibula and the lateral condyle of the tibia).
- There are **three** hamstrings—and each of these muscles has **three** actions.

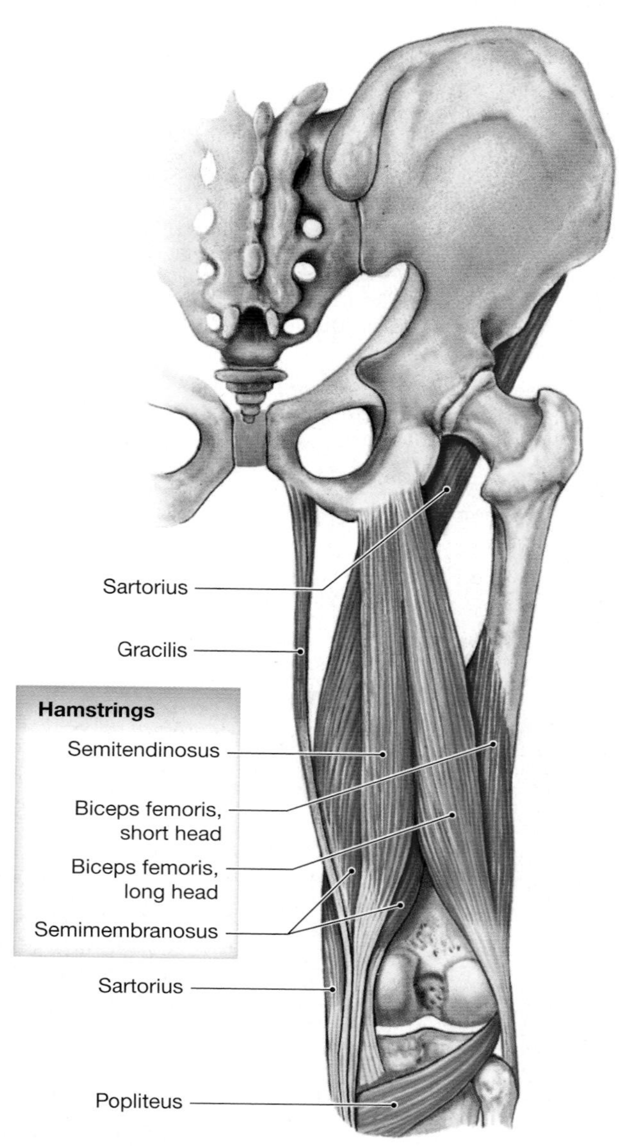

d Deep muscles, posterior view.

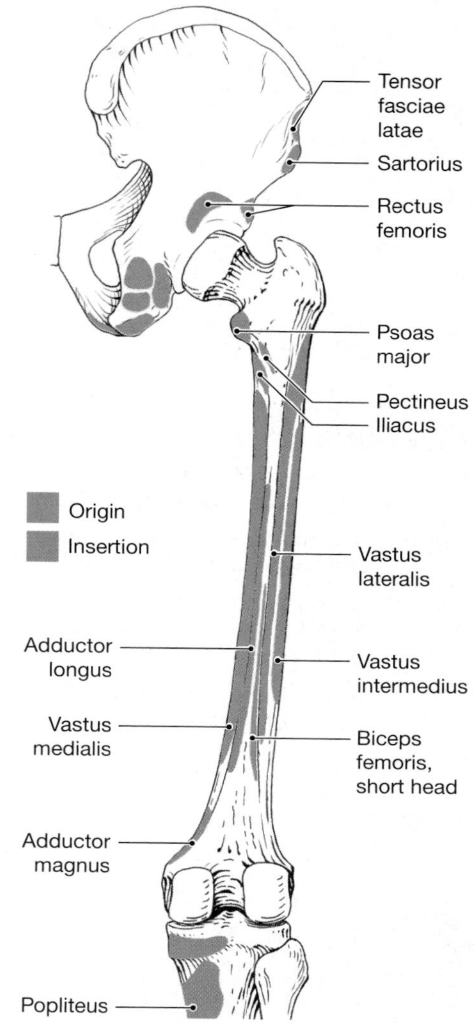

e Posterior view of the bones of the right hip, thigh, and proximal leg showing the origins and insertions of selected muscles.

Muscles That Move the Foot and Toes

▶ **KEY POINT** Muscles of the posterior compartment of the leg plantar flex the foot and flex the toes and are innervated by the tibial nerve. The two lateral compartment muscles plantar flex and evert the foot and are innervated by the superficial fibular nerve. The anterior compartment muscles dorsiflex the foot and extend the toes and are innervated by the deep fibular nerve.

Extrinsic Muscles of the Foot

Extrinsic muscles of the foot move the foot and toes. **Figures 11.22** to **11.24** show the extrinsic muscles of the foot, and **Table 11.9** presents additional information on these muscles. *(Refer to Chapter 12,* **Figure 12.7**, *to identify these anatomical structures from the body surface.)*

The large **gastrocnemius** (gas-trok-NĒ-mē-us; *gastroknemia,* calf of leg) and the underlying **soleus** (SŌ-lē-us) are plantar flexors of the foot **(Figures 11.22a,b,d** and **11.24)**. The soleus is a synergist to the gastrocnemius, increasing the speed and force of the plantar flexion. The gastrocnemius originates on the medial and lateral condyles of the femur. A sesamoid bone, called the fabella, is sometimes found in the tendon of the lateral head of the gastrocnemius. The gastrocnemius and soleus insert onto the **calcaneal tendon** (commonly called the "Achilles tendon").

The two **fibularis longus** and **fibularis brevis** (*peroneus longus* and *peroneus brevis*) lie partially deep to the gastrocnemius and soleus **(Figure 11.22b–d)**. These muscles plantar flex and evert the ankle. The **tibialis** (tib-ē-A-lis) **anterior** dorsiflexes and inverts the foot and is an antagonist to the gastrocnemius **(Figures 11.23** and **11.24)**. Muscles that flex or extend the toes originate on the tibia, the fibula, or both **(Figures 11.22** to **11.24)**. Large tendon sheaths surround the tendons of the tibialis anterior, extensor digitorum longus, and extensor hallucis longus where they cross the ankle joint. The **superior extensor retinaculum** and **inferior extensor retinaculum** stabilize these tendon sheaths **(Figures 11.23a** and **11.24)**.

Table 11.8 | Muscles That Move the Leg

Muscle	Origin	Insertion	Action	Innervation
FLEXORS OF THE KNEE				
Biceps femoris	Ischial tuberosity *(long head)* and linea aspera of femur *(short head)*	Head of fibula, lateral condyle of tibia	Flexion at knee; extension and lateral rotation at hip	Sciatic nerve; tibial portion (S_1–S_3 to long head) and common fibular branch (L_5–S_2 to short head)
Semimembranosus	Ischial tuberosity	Posterior surface of medial condyle of tibia	Flexion at knee; extension and medial rotation at hip	Sciatic nerve (tibial portion L_5–S_2)
Semitendinosus	Ischial tuberosity	Proximal, medial surface of tibia near insertion of gracilis	Flexion at knee; extension and medial rotation at hip	Sciatic nerve (tibial portion L_5–S_2)
Sartorius	Anterior superior iliac spine	Medial surface of tibia near tibial tuberosity	Flexion at knee; abduction, flexion, and lateral rotation at hip	Femoral nerve (L_2–L_3)
Popliteus	Lateral condyle of femur	Posterior surface of proximal tibial shaft	Medial rotation of tibia (or lateral rotation of femur) at knee; flexion at knee	Tibial nerve (L_4–S_1)
EXTENSORS OF THE KNEE				
Rectus femoris	Anterior inferior iliac spine and superior acetabular rim of ilium	Tibial tuberosity via quadriceps tendon, patella, and patellar ligament	Extension at knee; flexion at hip	Femoral nerve (L_2–L_4)
Vastus intermedius	Anterolateral surface of femur and linea aspera (distal half)	Tibial tuberosity via quadriceps tendon, patella, and patellar ligament	Extension at knee	Femoral nerve (L_2–L_4)
Vastus lateralis	Anterior and inferior to greater trochanter of femur and along linea aspera (proximal half)	Tibial tuberosity via quadriceps tendon, patella, and patellar ligament	Extension at knee	Femoral nerve (L_2–L_4)
Vastus medialis	Entire length of linea aspera of femur	Tibial tuberosity via quadriceps tendon, patella, and patellar ligament	Extension at knee	Femoral nerve (L_2–L_4)

Figure 11.20 Muscles That Move the Leg, Sectional Views

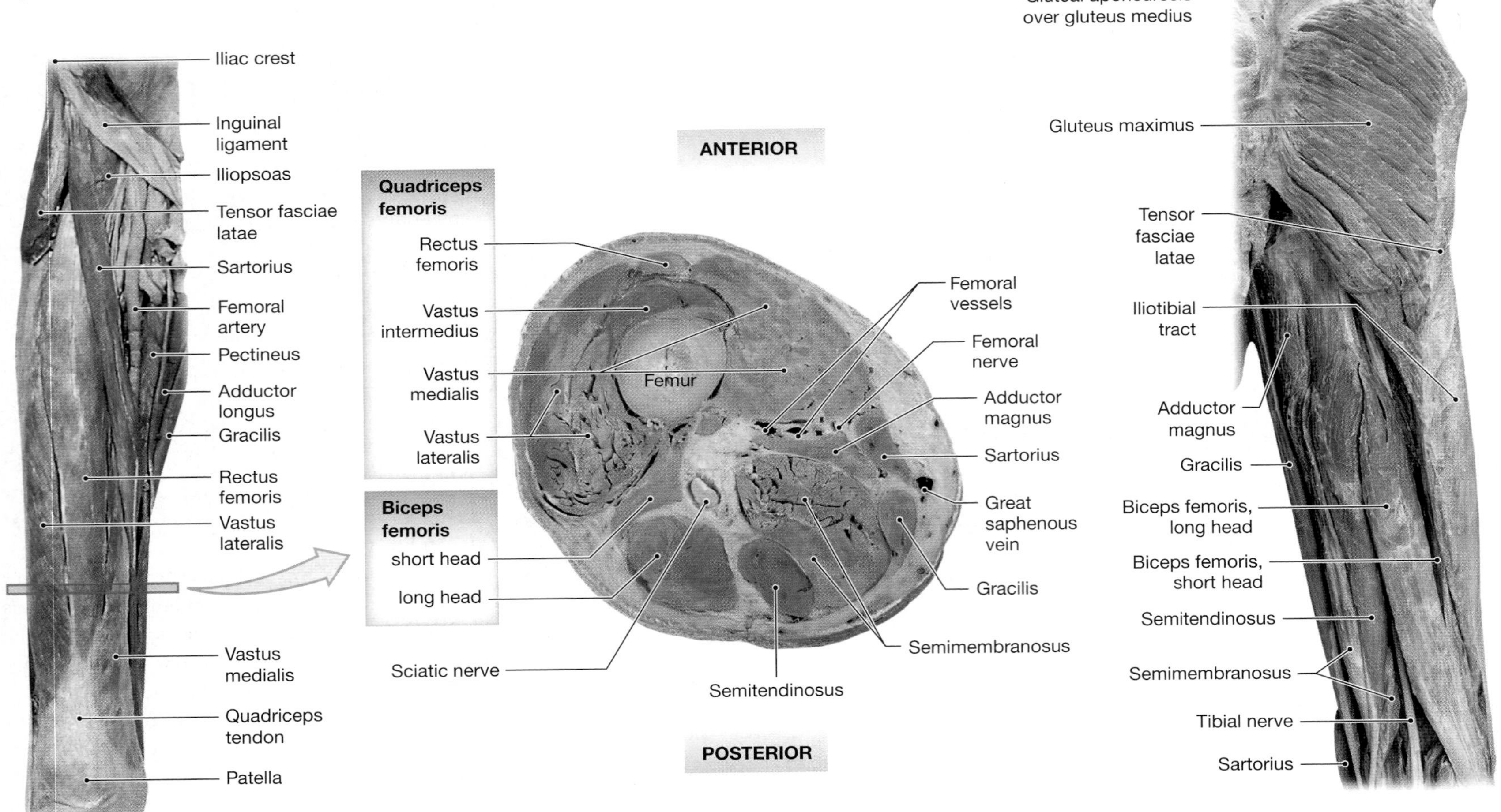

Iliac crest
Inguinal ligament
Iliopsoas
Tensor fasciae latae
Sartorius
Femoral artery
Pectineus
Adductor longus
Gracilis
Rectus femoris
Vastus lateralis
Vastus medialis
Quadriceps tendon
Patella

a Dissection of anterior right thigh

ANTERIOR
Quadriceps femoris
Rectus femoris
Vastus intermedius
Vastus medialis
Vastus lateralis
Biceps femoris
short head
long head
Sciatic nerve
Femur
Femoral vessels
Femoral nerve
Adductor magnus
Sartorius
Great saphenous vein
Gracilis
Semimembranosus
Semitendinosus
POSTERIOR

b Transverse section of the right thigh

Gluteal aponeurosis over gluteus medius
Gluteus maximus
Tensor fasciae latae
Iliotibial tract
Adductor magnus
Gracilis
Biceps femoris, long head
Biceps femoris, short head
Semitendinosus
Semimembranosus
Tibial nerve
Sartorius

c Dissection of posterior thigh and proximal leg

Figure 11.21 **Muscles that Move the Leg, Medial and Lateral Views**

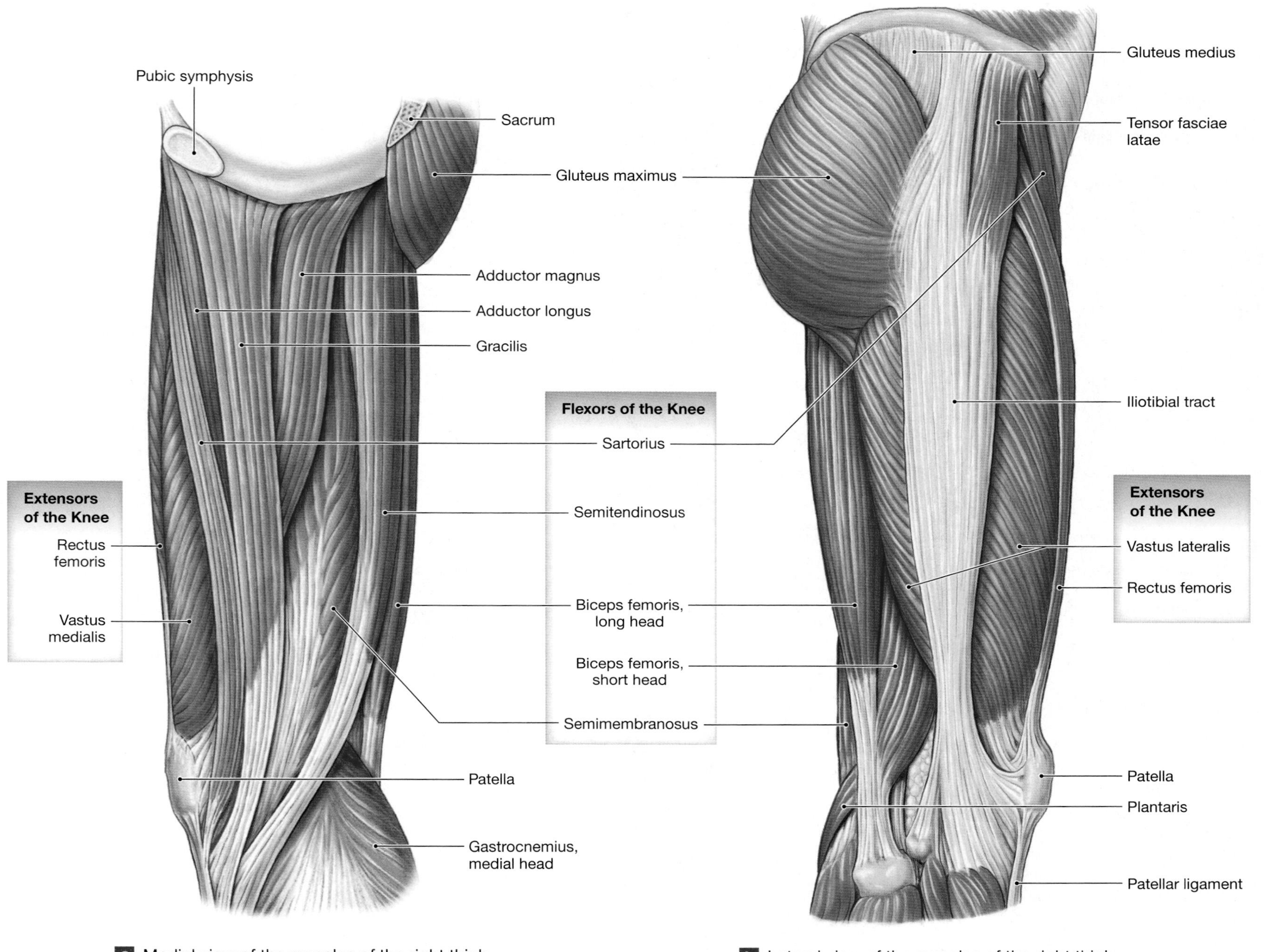

a Medial view of the muscles of the right thigh

b Lateral view of the muscles of the right thigh

11

Figure 11.22 Extrinsic Muscles That Move the Foot and Toes, Posterior Views

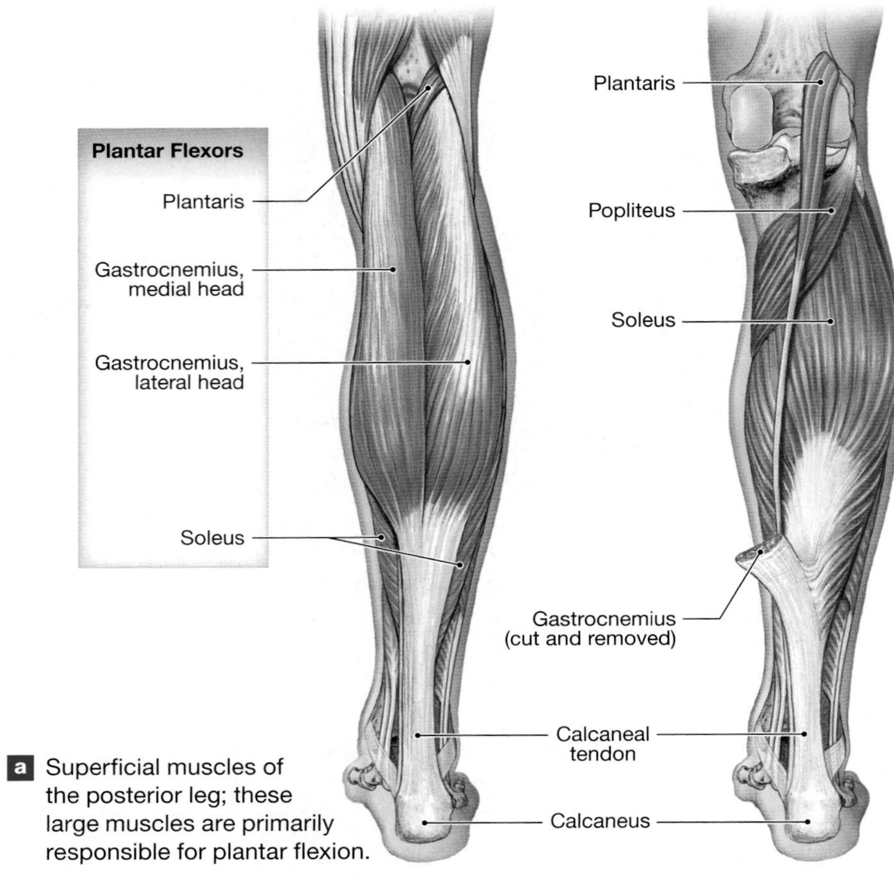

a Superficial muscles of the posterior leg; these large muscles are primarily responsible for plantar flexion.

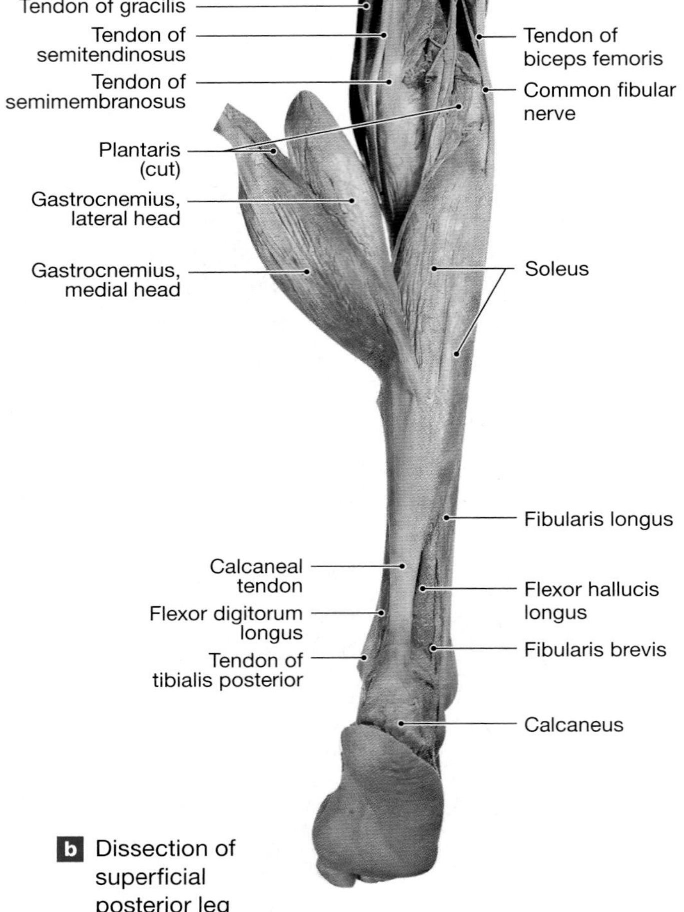

b Dissection of superficial posterior leg muscles.

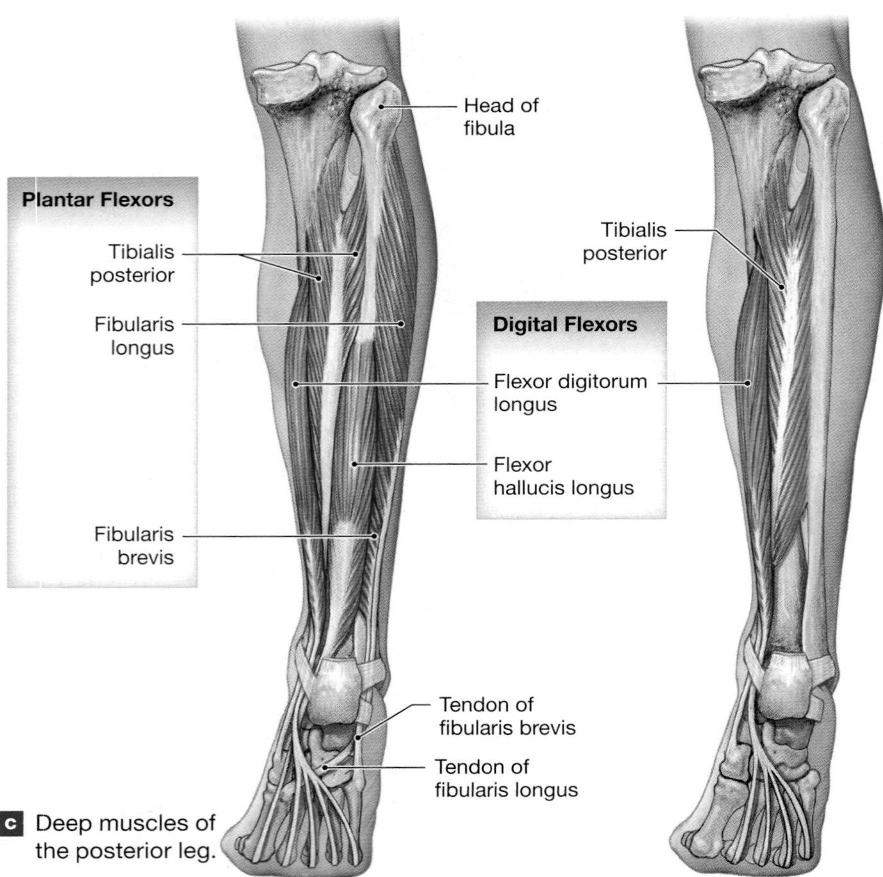

c Deep muscles of the posterior leg.

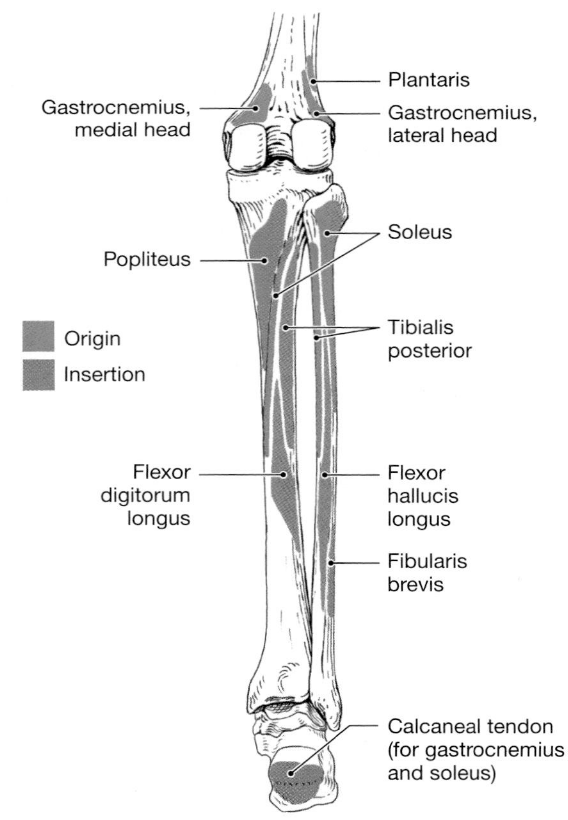

d A posterior view of the bones of the right leg and foot showing the origins and insertions of selected muscles. For sectional views of the leg, see **Figure 11.27**.

Figure 11.23 Extrinsic Muscles That Move the Foot and Toes, Anterior Views

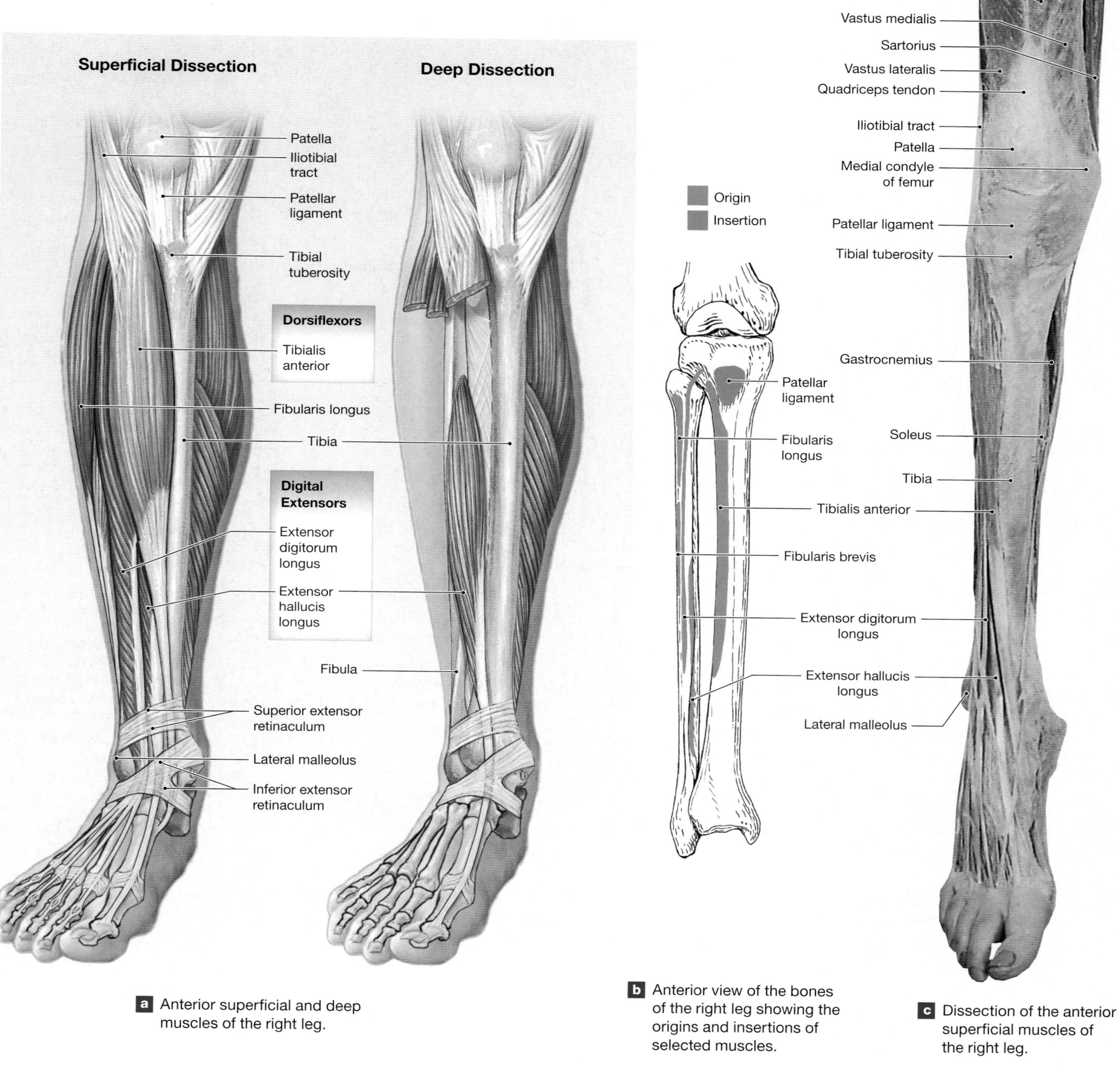

a Anterior superficial and deep muscles of the right leg.

b Anterior view of the bones of the right leg showing the origins and insertions of selected muscles.

c Dissection of the anterior superficial muscles of the right leg.

Figure 11.24 **Extrinsic Muscles That Move the Foot and Toes, Medial and Lateral Views**

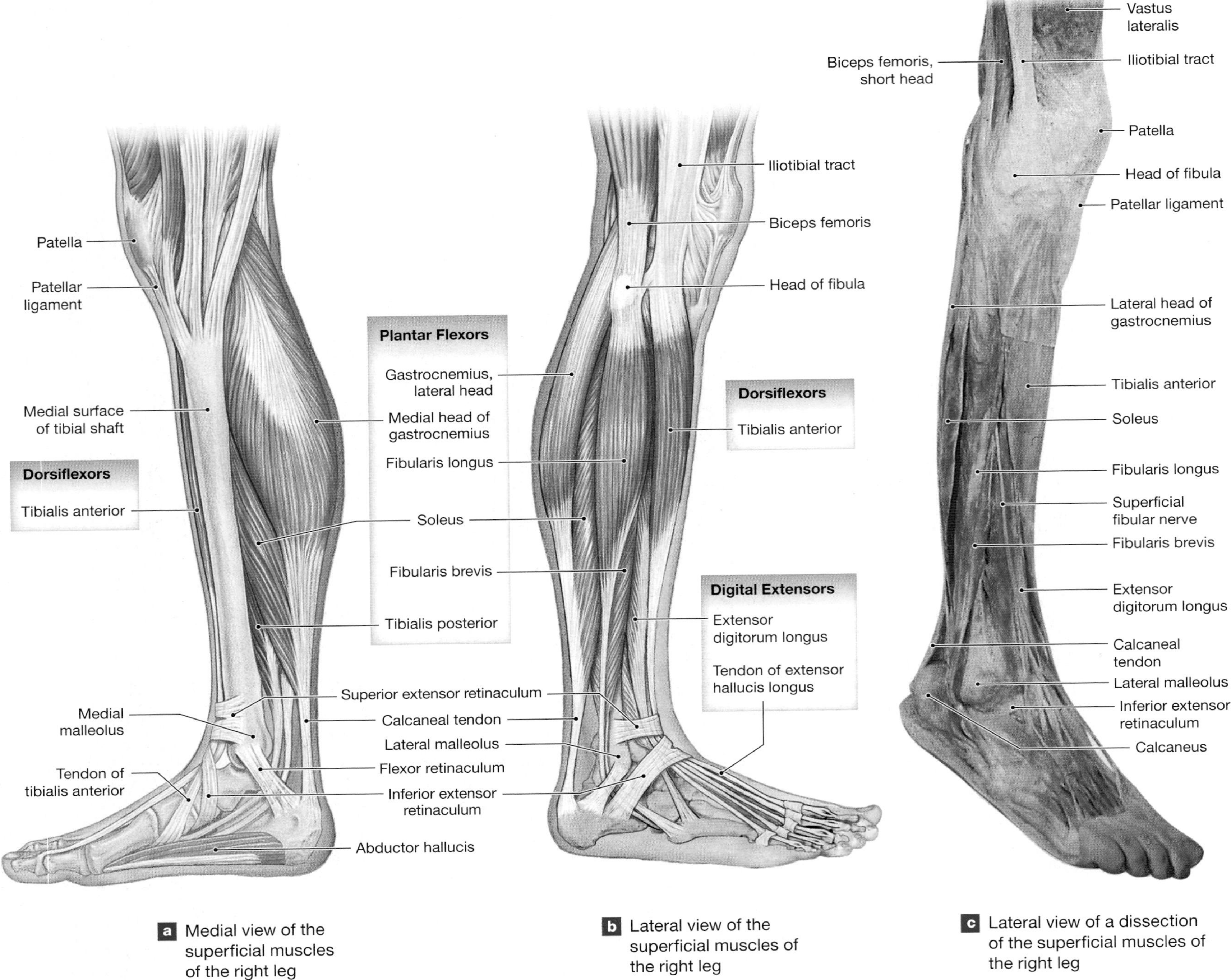

a Medial view of the superficial muscles of the right leg

b Lateral view of the superficial muscles of the right leg

c Lateral view of a dissection of the superficial muscles of the right leg

Table 11.9 | **Extrinsic Muscles That Move the Foot and Toes**

Muscle	Origin	Insertion	Action	Innervation
ACTION AT THE ANKLE				
DORSIFLEXORS				
Tibialis anterior	Lateral condyle and proximal shaft of tibia	Base of first metatarsal bone and medial cuneiform	Dorsiflexion at ankle; inversion of foot	Deep fibular nerve (L_4–S_1)
PLANTAR FLEXORS				
Gastrocnemius	Femoral condyles	Calcaneus via calcaneal tendon	Plantar flexion at ankle; flexion at knee	Tibial nerve (S_1–S_2)
Fibularis brevis	Midlateral margin of fibula	Base of fifth metatarsal bone	Eversion of foot and plantar flexion at ankle	Superficial fibular nerve (L_4–S_1)
Fibularis longus	Head and proximal shaft of fibula	Base of first metatarsal bone and medial cuneiform	Eversion of foot and plantar flexion at ankle; supports ankle; supports longitudinal and transverse arches	Superficial fibular nerve (L_4–S_1)
Plantaris	Lateral supracondylar ridge	Posterior portion of calcaneus	Plantar flexion at ankle; flexion at knee	Tibial nerve (L_4–S_1)
Soleus	Head and proximal shaft of fibula, and adjacent posteromedial shaft of tibia	Calcaneus via calcaneal tendon (with gastrocnemius)	Plantar flexion at ankle; postural muscle when standing	Sciatic nerve, tibial branch (S_1–S_2)
Tibialis posterior	Interosseous membrane and adjacent shafts of tibia and fibula	Navicular, all three cuneiforms, cuboid, second, third, and fourth metatarsal bones	Inversion of foot; plantar flexion at ankle	Sciatic nerve, tibial branch (S_1–S_2)
ACTION AT THE TOES				
DIGITAL FLEXORS				
Flexor digitorum longus	Posteromedial surface of tibia	Inferior surface of distal phalanges, toes 2–5	Flexion of joints of toes 2–5; plantar flexes ankle	Tibial branch (L_5–S_1)
Flexor hallucis longus	Posterior surface of fibula	Inferior surface, distal phalanx of great toe	Flexion at joints of great toe; plantar flexes ankle	Tibial branch (L_5–S_1)
DIGITAL EXTENSORS				
Extensor digitorum longus	Lateral condyle of tibia, anterior surface of fibula	Superior surfaces of phalanges, toes 2–5	Extension of toes 2–5; dorsiflexes ankle	Deep fibular nerve (L_5–S_1)
Extensor hallucis longus	Anterior surface of fibula	Superior surface, distal phalanx of great toe	Extension at joints of great toe; dorsiflexes ankle	Deep fibular nerve (L_5–S_1)

11

The **tibialis posterior** originates on the shaft of the tibia and fibula and inserts onto the navicular, all three cuneiform bones, the cuboid, and second, third, and fourth metatarsal bones (**Figures 11.22c,d** and **11.24a**). It inverts and plantar flexes the ankle. When stressed by repetitive use such as long-distance running, the muscle attachments along the tibial and fibular surfaces can get irritated and inflamed, causing a condition called "shin splints."

TIPS & TOOLS

The following mnemonic will help you remember what structures pass posterior to the medial malleolus of the tibia *from anterior to posterior*: "Tom, Dick, And Harry." **T**ibialis posterior, flexor **D**igitorum longus, posterior tibial **A**rtery, flexor **H**allucis longus

Intrinsic Muscles of the Foot

The small intrinsic muscles that flex and extend the toes originate on the tarsal and metatarsal bones of the foot (**Figures 11.25** and **11.26** and **Table 11.10**). Flexor muscles originating from the anterior border of the calcaneus maintain the longitudinal arch of the foot.

As in the hand, the small interossei (singular, *interosseus*) of the foot originate on the lateral and medial surfaces of the metatarsal bones. The four **dorsal interossei** abduct the metatarsophalangeal joints of toes 3 and 4, and the three **plantar interossei** adduct the metatarsophalangeal joints of toes 3–5.

Three intrinsic muscles of the foot move the great toe (hallux): The **flexor hallucis brevis** flexes the great toe. The **adductor hallucis** adducts it, and the **abductor hallucis** abducts it.

There are more intrinsic muscles of the foot that flex the joints of the toes than muscles that extend the toes. The **flexor digitorum brevis** the **quadratus plantae**, and the four **lumbricals** flex the joints of toes 2–5. The **flexor digiti minimi brevis** flexes toe 5. The **extensor digitorum brevis** extends the toes. This muscle assists the extensor hallucis longus in extending the great toe and assists the extensor digitorum longus in extending toes 2–4 (see **Table 11.9**). The extensor digitorum brevis is the only intrinsic muscle found on the dorsum of the foot.

The **flexor digitorum brevis**, **abductor digiti minimi**, and **quadratus plantae** all originate on the medial tubercle of the calcaneus (**Figures 11.25b–e** and **11.26c**). The **plantar aponeurosis**, or *plantar fascia*, is superficial to these muscles. This region of muscle attachment commonly becomes inflamed and tender from walking or running. This condition, called plantar fasciitis, results in tenderness on the sole of the foot. Without treatment, the aponeurosis and muscles will tear away from the calcaneus, leading to a bony thickening on the calcaneus called a heel spur.

CLINICAL NOTE

Calcaneal Tendon Rupture

The calcaneal tendon is the strongest, thickest tendon in the body. It connects the powerful gastrocnemius and soleus to the posterior calcaneus. The gastrocnemius crosses both the knee and the ankle and is reinforced by the soleus. Sudden, explosive contractions, as in jumping or pushing off, exert great pressure on the calcaneal tendon, which can result in spontaneous rupture. The patient, often a middle-aged, weekend athlete, describes a sensation of having been kicked in the back of the leg and experiences a sudden inability to run or jump. Surgical repair is often the treatment of choice.

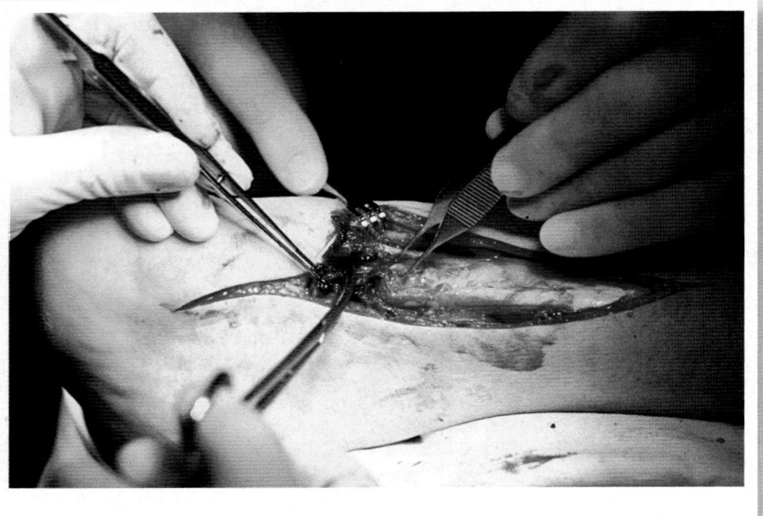

11

Figure 11.25 Intrinsic Muscles That Move the Foot and Toes, Part I

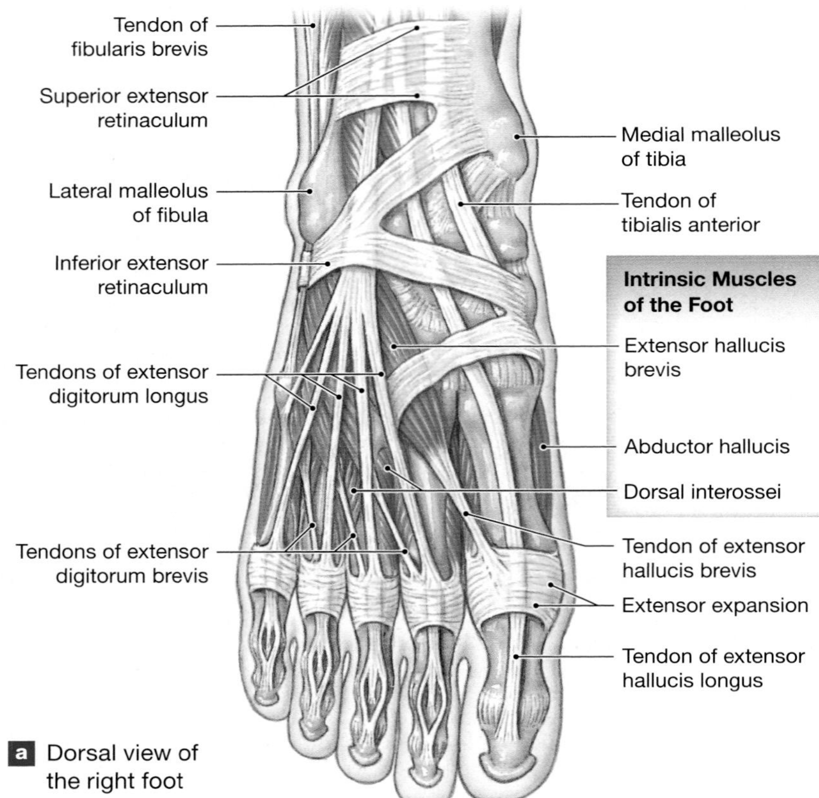

a Dorsal view of the right foot

Table 11.10 | Intrinsic Muscles of the Foot

Muscle	Origin	Insertion	Action	Innervation
Extensor digitorum brevis	Calcaneus (superior and lateral surfaces)	Dorsal surface of toes 1–4	Extension at metatarsophalangeal joints of toes 1–4	Deep fibular nerve (S_1–S_2)
Abductor hallucis	Calcaneus (tuberosity on inferior surface)	Medial side of proximal phalanx of great toe	Abduction at metatarsophalangeal joint of great toe	Medial plantar nerve (S_2–S_3)
Flexor digitorum brevis	Calcaneus (tuberosity on inferior surface)	Sides of middle phalanges, toes 2–5	Flexion of proximal interphalangeal joints of toes 2–5	Medial plantar nerve (S_2–S_3)
Abductor digiti minimi	Calcaneus (tuberosity on inferior surface)	Lateral side of proximal phalanx, toe 5	Abduction and flexion at metatarsophalangeal joint of toe 5	Lateral plantar nerve (S_2–S_3)
Quadratus plantae	Calcaneus (medial, inferior surfaces)	Tendon of flexor digitorum longus	Flexion at joints of toes 2–5	Lateral plantar nerve (S_2–S_3)
Lumbricals (4)	Tendons of flexor digitorum longus	Insertions of extensor digitorum longus	Flexion at metatarsophalangeal joints; extension at interphalangeal joints of toes 2–5	Medial plantar nerve (1), lateral plantar nerve (2–4)
Flexor hallucis brevis	Cuboid and lateral cuneiform	Proximal phalanx of great toe	Flexion at metatarsophalangeal joint of great toe	Medial plantar nerve (L_4–S_5)
Adductor hallucis	Bases of metatarsal bones II–IV and plantar ligaments	Proximal phalanx of great toe	Adduction and flexion at metatarsophalangeal joint of great toe	Lateral plantar nerve (S_1–S_2)
Flexor digiti minimi brevis	Base of metatarsal bone V	Lateral side of proximal phalanx of toe 5	Flexion at metatarsophalangeal joint of toe 5	Lateral plantar nerve (S_1–S_2)
Dorsal interossei (4)	Sides of metatarsal bones	Medial and lateral sides of toe 2; lateral sides of toes 3 and 4	Abduction at metatarsophalangeal joints of toes 3 and 4; flexion of metatarsophalangeal joints and extension at the interphalangeal joints of toes 2 through 4	Lateral plantar nerve (S_1–S_2)
Plantar interossei (3)	Bases and medial sides of metatarsal bones	Medial sides of toes 3–5	Adduction of metatarsophalangeal joints of toes 3–5; flexion of metatarsophalangeal joints and extension at interphalangeal joints	Lateral plantar nerve (S_1–S_2)

Figure 11.25 **(continued)**

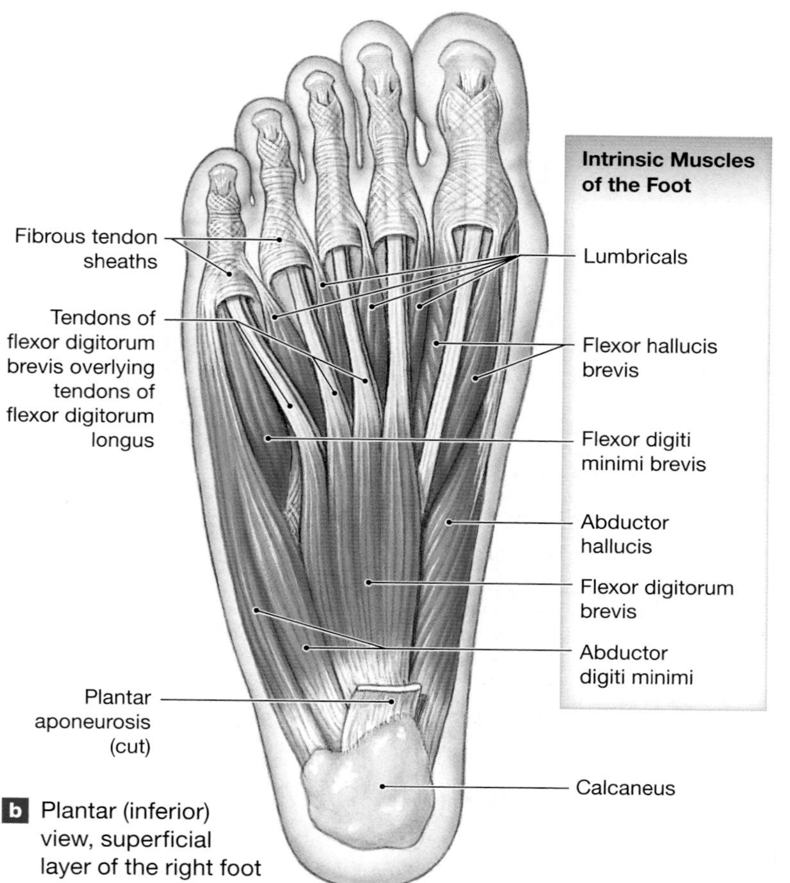

b Plantar (inferior) view, superficial layer of the right foot

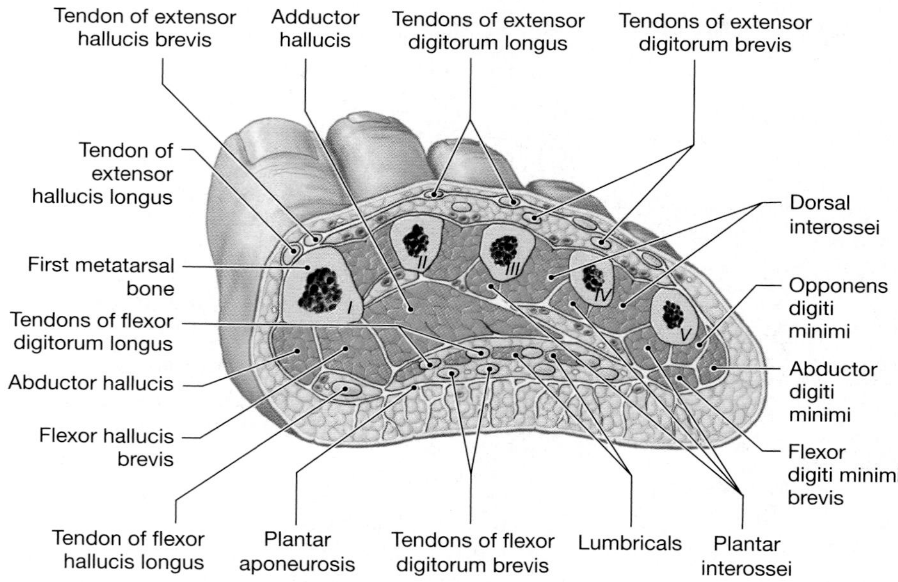

c Right foot, sectional view through the metatarsal bones

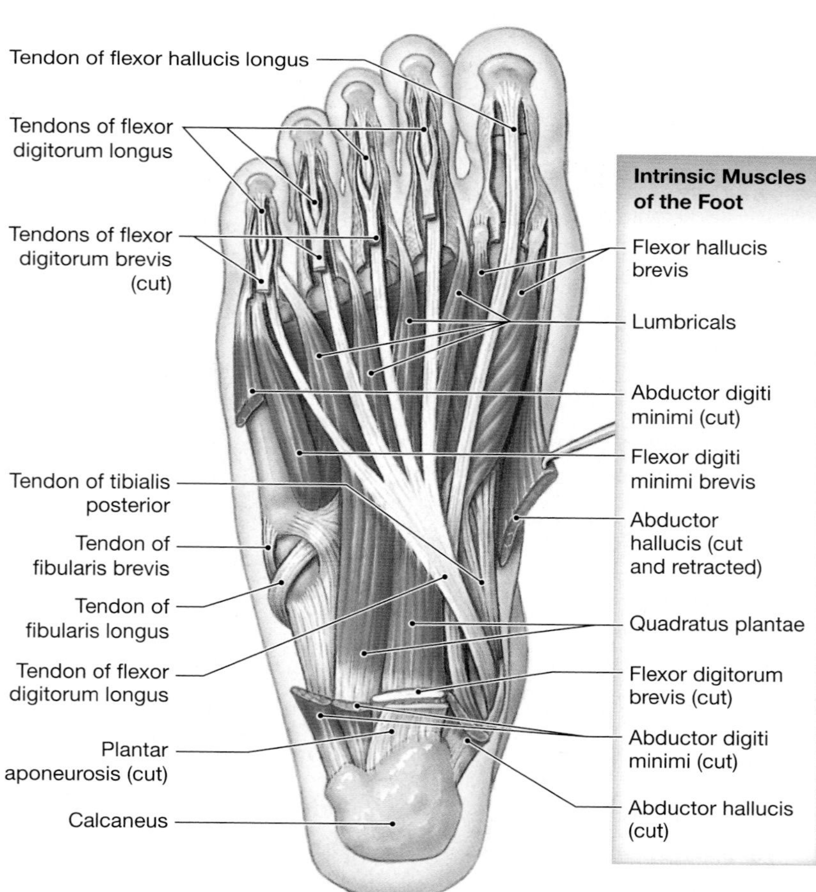

d Plantar (inferior) view, deep layer of the right foot

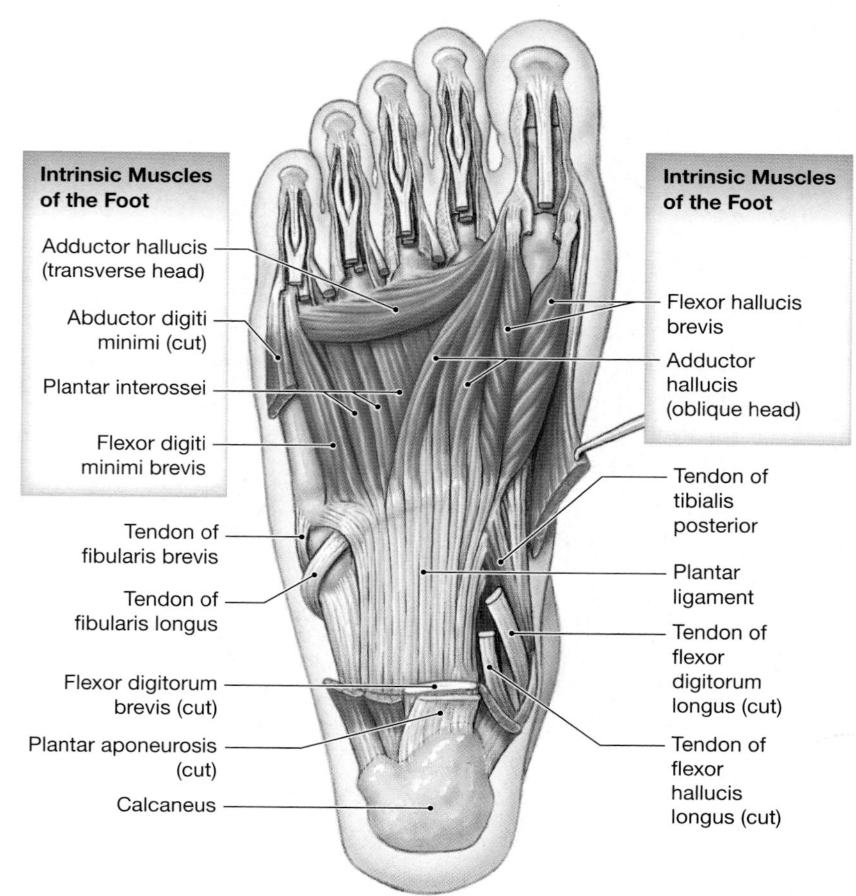

e Plantar (inferior) view, deepest layer of the right foot

11

Figure 11.26 Intrinsic Muscles That Move the Foot and Toes, Part II

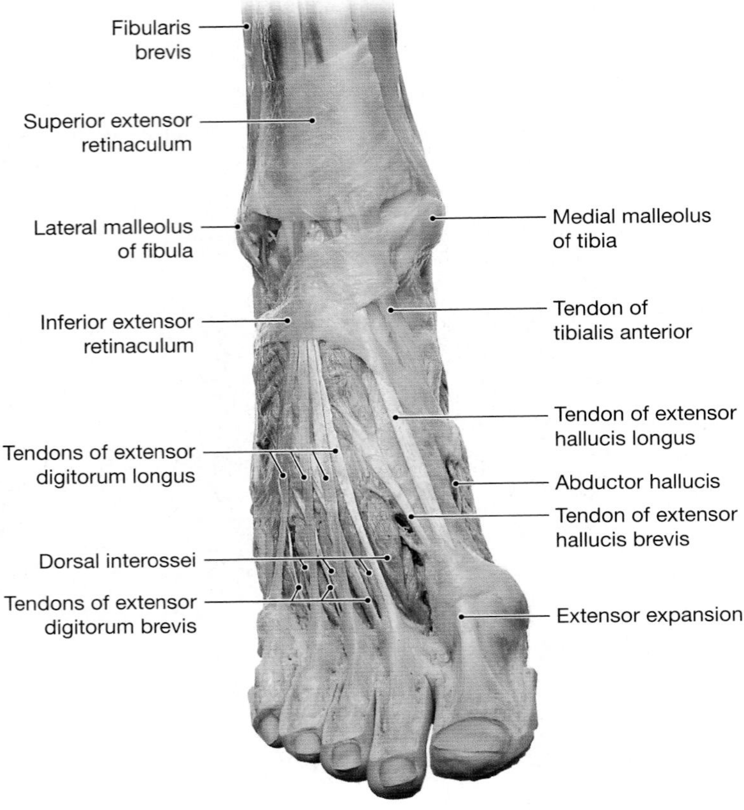

a Dorsal view of the right foot.

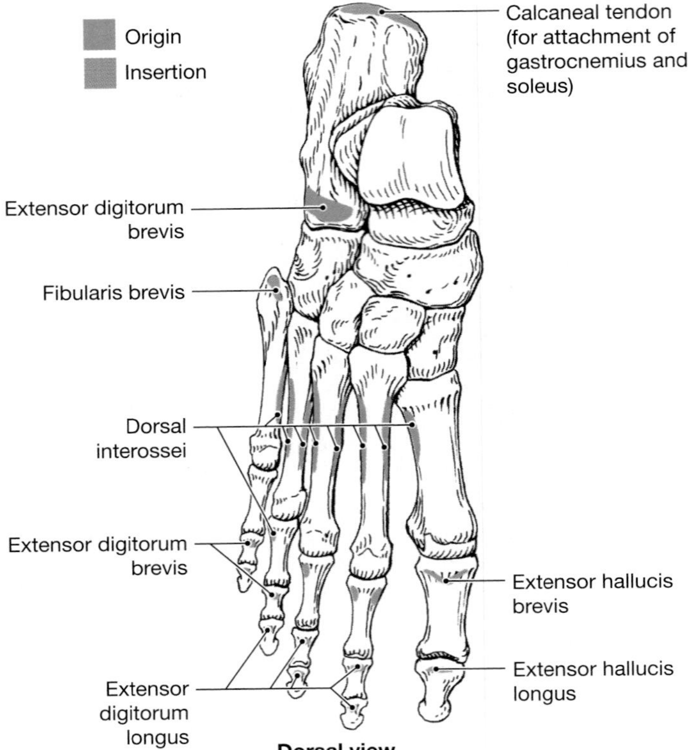

b Dorsal (superior) view of the bones of the right foot showing the origins and insertions of selected muscles.

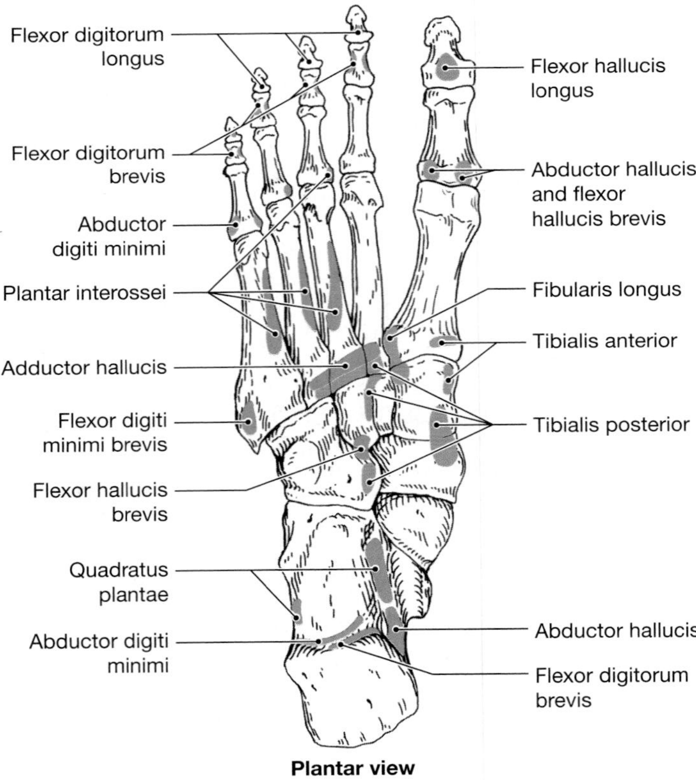

c Plantar (inferior) view of the bones of the right foot showing the origins and insertions of selected muscles.

11.4 CONCEPT CHECK

9 Which muscles that move the thigh at the hip are innervated by the femoral nerve?

10 What is the collective name for the knee extensors?

11 Which of the extrinsic muscles of the foot originate on the fibula?

12 List the muscles that evert the foot.

13 List the intrinsic muscles of the foot that originate on the calcaneus.

See the blue Answers tab at the back of the book.

11.5 Compartments and Sectional Anatomy of the Thigh and Leg

▶ **KEY POINT** As in the arm and forearm, the deep fascia in the thigh and leg extends between the bones and the superficial fascia and separates the soft tissues of the limb into separate compartments.

Figures 11.27 and **11.28c,d** show the compartments of the leg, and **Figure 11.28a,b** shows the compartments of the thigh. The **medial** and **lateral intermuscular septa** of the thigh extend outward from the femur, separating adjacent muscle groups. The thigh is divided into **anterior**, **posterior**, and **medial** (*adductor*) **compartments** (**Figure 11.28a,b**). The anterior compartment contains the tensor fasciae latae, sartorius, and the quadriceps group. The posterior compartment contains the hamstrings, and the medial compartment contains the gracilis, pectineus, obturator externus, adductor longus, adductor brevis, and adductor magnus (**Table 11.11**).

CLINICAL NOTE

Compartment Syndrome

Injuries (particularly fractures) to the extremities cause bleeding and swelling within anatomical compartments. If the pressure within a compartment exceeds the pressure in the blood vessels within that compartment, blood flow is reduced and the muscles become ischemic. Nerves within the compartments suffer irreversible damage within 2–4 hours. Ischemic muscles suffer irreversible damage after 6 hours. Taking anabolic steroids seems to worsen compartment syndrome. Emergency surgical treatment involves cutting the fascia that forms the compartments to alleviate pressure and reestablish blood flow.

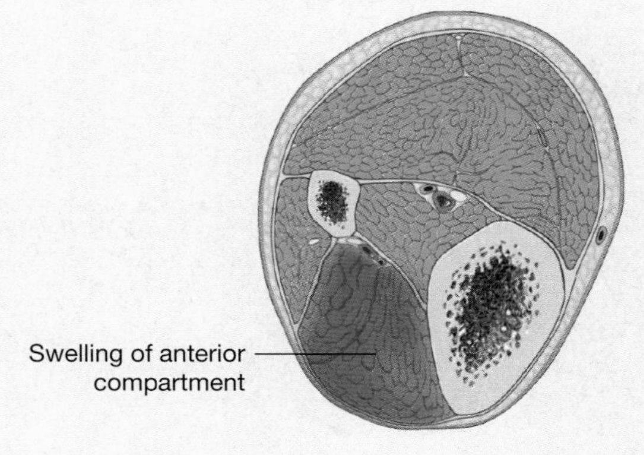

Figure 11.27 Musculoskeletal Compartments of the Leg, Anterior View

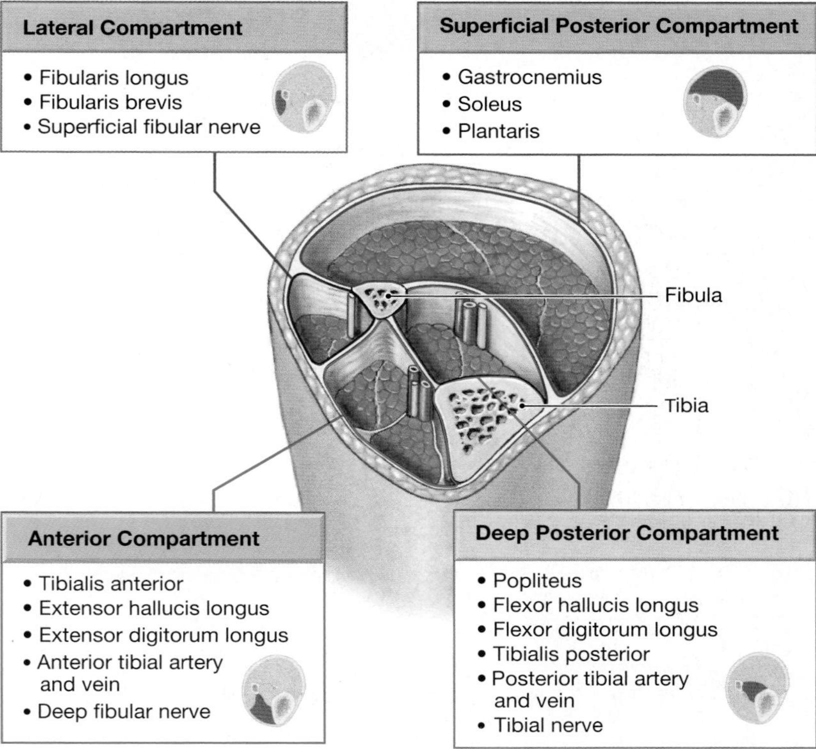

The tibia and fibula, crural interosseous membrane, and septa in the leg create four major compartments: an **anterior compartment**, a **lateral compartment**, and **superficial** and **deep posterior compartments** (**Figures 11.27** and **11.28c,d**). The anterior compartment contains muscles that dorsiflex the ankle, extend the toes, and invert and evert the ankle. The muscles of the lateral compartment evert and plantar flex the ankle. The superficial muscles of the posterior compartment plantar flex the ankle, and the deep muscles of the posterior compartment plantar flex the ankle and have other actions on the joints of the foot and the toes. **Table 11.11** lists the muscles and other structures within these compartments.

11.5 CONCEPT CHECK

14 List the functions of the muscles within the lateral compartment of the leg.

15 The posterior surface of the leg is composed of superficial and deep compartments. List the muscles of the deep posterior compartment of the leg and their functions.

16 Which compartment of the thigh contains the muscles that adduct the thigh at the hip?

17 List the muscles of the superficial posterior compartment of the leg and their functions.

See the blue Answers tab at the back of the book.

Figure 11.28 Musculoskeletal Compartments of the Lower Limb

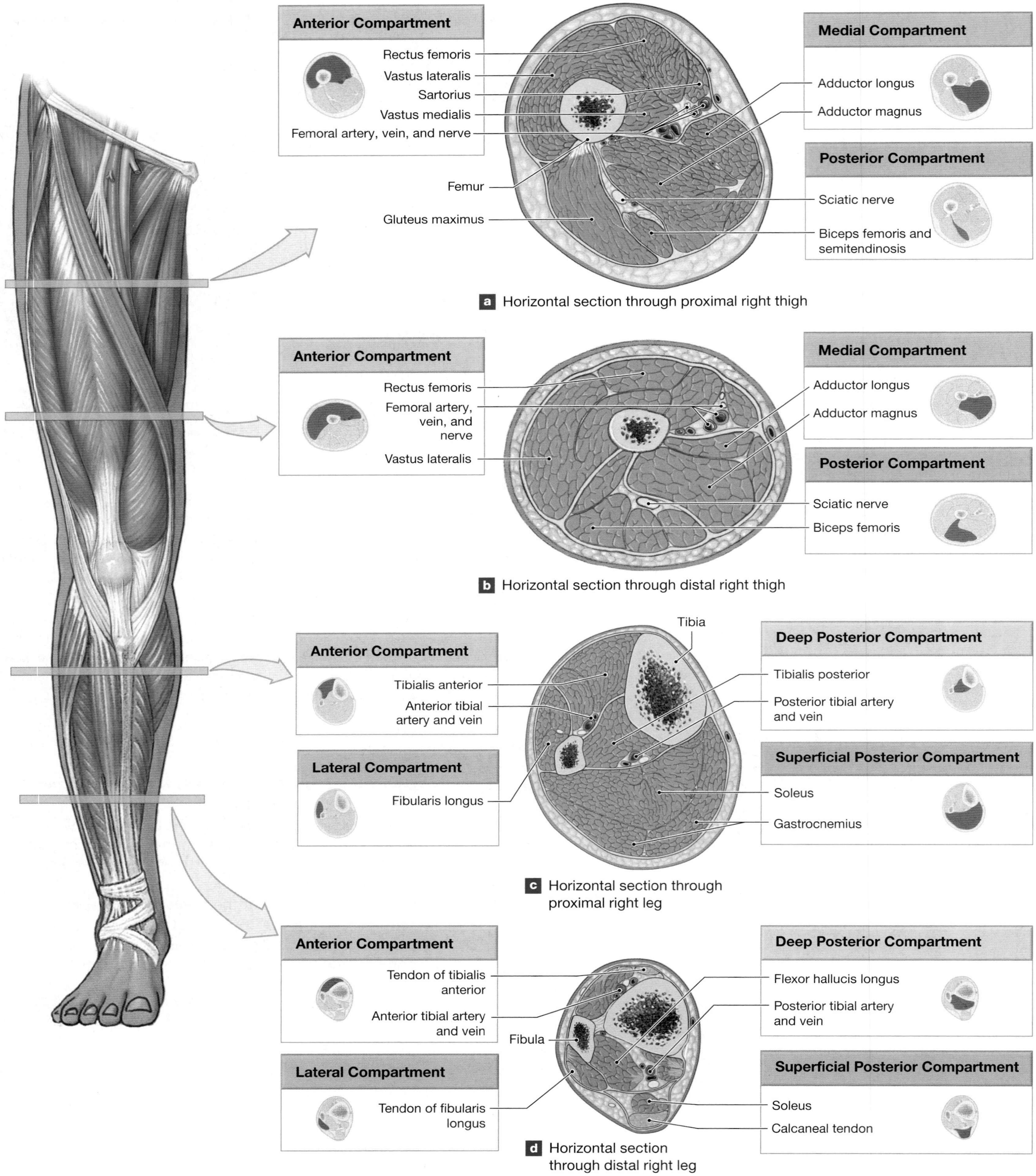

Anterior Compartment

Rectus femoris

Vastus lateralis

Sartorius

Vastus medialis

Femoral artery, vein, and nerve

Femur

Gluteus maximus

Medial Compartment

Adductor longus

Adductor magnus

Posterior Compartment

Sciatic nerve

Biceps femoris and semitendinosis

a Horizontal section through proximal right thigh

Anterior Compartment

Rectus femoris

Femoral artery, vein, and nerve

Vastus lateralis

Medial Compartment

Adductor longus

Adductor magnus

Posterior Compartment

Sciatic nerve

Biceps femoris

b Horizontal section through distal right thigh

Tibia

Anterior Compartment

Tibialis anterior

Anterior tibial artery and vein

Lateral Compartment

Fibularis longus

Deep Posterior Compartment

Tibialis posterior

Posterior tibial artery and vein

Superficial Posterior Compartment

Soleus

Gastrocnemius

c Horizontal section through proximal right leg

Anterior Compartment

Tendon of tibialis anterior

Anterior tibial artery and vein

Fibula

Lateral Compartment

Tendon of fibularis longus

Deep Posterior Compartment

Flexor hallucis longus

Posterior tibial artery and vein

Superficial Posterior Compartment

Soleus

Calcaneal tendon

d Horizontal section through distal right leg

Table 11.11 | **Compartments of the Lower Limb**

Compartment	Muscles	Blood Vessels	Nerves
THIGH			
Anterior compartment	Iliopsoas Iliacus Psoas major Psoas minor Quadriceps femoris Rectus femoris Vastus intermedius Vastus lateralis Vastus medialis Sartorius	Femoral artery Femoral vein Deep femoral artery Lateral circumflex femoral artery	Femoral nerve Saphenous nerve
Medial compartment	Pectineus Adductor brevis Adductor longus Adductor magnus Gracilis Obturator externus	Obturator artery Obturator vein Deep femoral artery Deep femoral vein	Obturator nerve
Posterior compartment	Biceps femoris Semimembranosus Semitendinosus	Deep femoral artery Deep femoral vein	Sciatic nerve
LEG			
Anterior compartment	Extensor digitorum longus Extensor hallucis longus Fibularis tertius Tibialis anterior	Anterior tibial artery Anterior tibial vein	Deep fibular nerve
Lateral compartment	Fibularis brevis Fibularis longus		Superficial fibular nerve
Posterior compartment			
Superficial	Gastrocnemius Plantaris Soleus		
Deep	Flexor digitorum longus Flexor hallucis longus Popliteus Tibialis posterior	Posterior tibial artery Fibular artery Fibular vein Posterior tibial vein	Tibial nerve

Study Outline

Introduction p. 283

- The **appendicular musculature** stabilizes the pectoral and pelvic girdles and moves the upper and lower limbs.

11.1 Factors Affecting Appendicular Muscle Function p. 283

- A muscle of the appendicular skeleton may cross one or more joints between its origin and insertion. The position of the muscle as it crosses a joint helps determine the action of that muscle. *(See Spotlight Figure 11.2.)*
- The primary action of a muscle whose insertion is close to a joint is to produce movement, whereas a muscle whose insertion is farther from a joint helps stabilize that joint.

11.2 Muscles of the Pectoral Girdle and Upper Limb p. 286

- Four groups of muscles are associated with the pectoral girdle and upper limb: (1) muscles that position the pectoral girdle, (2) muscles that move the arm, (3) muscles that move the forearm and hand, and (4) muscles that move the hand and fingers.

Muscles That Position the Pectoral Girdle p. 286

- The **trapezius** covers the back and parts of the neck to the base of the skull. The trapezius affects the position of the pectoral (shoulder) girdle, head, and neck. *(See Figures 11.3–11.6, 12.2, 12.3, and 12.10 and Table 11.1.)*
- Deep to the trapezius, the **rhomboid** adduct the scapula, and the **levator scapulae** elevates the scapula. Both insert on the scapula. *(See Figures 11.3, 11.5, and 12.10 and Table 11.1.)*
- The **serratus anterior**, which abducts the scapula and swings the shoulder anteriorly, originates along the anterior superior surfaces of several ribs. *(See Figures 11.5 and 11.6 and Table 11.1.)*
- Two deep chest muscles arise along the anterior surfaces of the ribs. Both the **subclavius** and the **pectoralis minor** depress and protract the shoulder. *(See Figures 11.4, 11.6, and 12.10 and Table 11.1.)*

Muscles That Move the Arm p. 288

- The **deltoid** and the **supraspinatus** produce abduction at the shoulder. The **subscapularis** and the **teres major** rotate the arm medially, whereas the **infraspinatus** and **teres minor** rotate the arm laterally. The supraspinatus, infraspinatus, subscapularis, and teres minor are known as the **rotator cuff**. The **coracobrachialis** flexes and adducts the shoulder. *(See Figures 11.3, 11.4, 11.7, 12.2, 12.4, 12.5, and 12.10 and Table 11.2.)*
- The **pectoralis major** flexes the shoulder, and the **latissimus dorsi** extends it. Both muscles adduct and medially rotate the arm. *(See Figures 11.3, 11.4, 11.7, 12.2a, 12.3b, 12.5, and 12.10 and Table 11.2.)*

Muscles That Move the Forearm and Hand p. 292

11

- The primary actions of the **biceps brachii** and the **triceps brachii** (*long head*) affect the elbow joint. The biceps brachii flexes the elbow and supinates the forearm, and the triceps brachii extends the elbow. Both have a secondary effect on the arm. *(See Figures 11.4–11.8, 11.10 12.4, and 12.5 and Table 11.3.)*
- The **brachialis** and **brachioradialis** flex the elbow. The **anconeus** and the triceps brachii oppose this action. The **flexor carpi ulnaris**, the **flexor carpi radialis**, and the **palmaris longus** are superficial muscles of the forearm that cooperate to flex the wrist. Additionally, the flexor carpi ulnaris adducts the wrist, and the flexor carpi radialis abducts it. The **extensor carpi radialis** and **extensor carpi ulnaris** extend and abduct the wrist. The **pronator teres** and **pronator quadratus** pronate the forearm without flexion or extension at the elbow; their action is opposed by the **supinator muscle**. *(See Figures 11.8–11.10a, 12.4, and 12.5 and Table 11.3.)*

Muscles That Move the Hand and Fingers p. 294

- Extrinsic muscles of the hand provide strength and gross motor control of the fingers. Intrinsic muscles provide fine motor control of the fingers and hand.
- The extrinsic muscles of the hand flex and extend the finger joints. *(See Figures 11.8–11.12, 12.4, and 12.5 and Table 11.4.)*
- Fine motor control of the hand involves small intrinsic muscles of the hand. *(See Figures 11.12 and 11.13 and Table 11.5.)*

11.3 Compartments and Sectional Anatomy of the Arm and Forearm p. 301

- The deep fascia of the upper limb separates the soft tissues into separate compartments. The arm consists of an anterior compartment and a posterior compartment. The forearm consists of four compartments: superficial anterior, deep anterior, lateral, and posterior.
- The relationships of the deeper muscles of the arm and forearm are best seen in sectional views. *(See Figures 11.14 and 11.15 and Table 11.6.)*

11.4 Muscles of the Pelvic Girdle and Lower Limb p. 303

- Three groups of muscles are associated with the pelvis and lower limb: (1) muscles that move the thigh, (2) muscles that move the leg, and (3) muscles that move the foot and toes.

Muscles That Move the Thigh p. 304

- **Gluteal muscles** cover the lateral surface of the ilium. The largest is the **gluteus maximus** which extends and laterally rotates the hip. It shares an insertion with the **tensor fasciae latae** which flexes, abducts, and medially rotates the hip. Together, these muscles pull on the **iliotibial tract** to laterally brace the knee. *(See Spotlight Figure 11.2, Figures 11.16, 11.17, and 12.6c and Table 11.7.)*
- The **piriformis** and the **obturator** are the most dominant lateral rotators.
- The **adductor group** (**adductor magnus**, **adductor brevis**, **adductor longus**, **pectineus**, and **gracilis**) adduct the hip. Individually, they can produce various other movements, such as medial or lateral rotation and flexion or extension at the hip. *(See Figures 11.16, 11.17, 12.6a, and 12.7a and Table 11.7.)*
- The **psoas major** and the **iliacus** merge to form the **iliopsoas**, a powerful hip flexor. *(See Figures 11.16d, 11.17 and Table 11.7.)*

Muscles That Move the Leg p. 306

- Extensor muscles of the knee lie along the anterior and lateral surfaces of the thigh; flexor muscles lie along the posterior and medial surfaces of the thigh. Flexors and adductors originate on the pelvic girdle, whereas most extensors originate on the femur.
- Collectively, the knee extensors are known as the **quadriceps femoris**. This group includes the **vastus intermedius**, **vastus lateralis**, **vastus medialis**, and **rectus femoris**. *(See Figures 11.18–11.21 and 12.7a,b and Table 11.8.)*
- The flexors of the knee include the **biceps femoris**, **semimembranosus**, and **semitendinosus** (these "hamstrings" also extend the hip), and the **sartorius**. The **popliteus** medially rotates the tibia (or laterally rotates the femur) to unlock the knee joint. *(See Figures 11.18–11.21 and 12.7a,b and Table 11.8.)*

Muscles That Move the Foot and Toes p. 309

- Extrinsic muscles move the foot and toes.
- The **gastrocnemius** and **soleus** produce plantar flexion. The large **tibialis anterior** opposes the gastrocnemius and dorsiflexes the ankle. The **fibularis** produces eversion as well as plantar flexion. *(See Figures 11.22 and 11.24 and Table 11.9.)*
- Smaller muscles of the leg position the foot and move the toes.
- Muscles originating on the tarsal and metatarsal bones provide precise control of the phalanges. *(See Figures 11.25 and 11.26 and Table 11.10.)*

11.5 Compartments and Sectional Anatomy of the Thigh and Leg p. 319

- In addition to the functional approach used in this chapter, many anatomists study the muscles of the lower limb in groups determined by their position within **compartments**.
- The thigh has **anterior**, **medial**, and **posterior compartments**; the leg has an **anterior**, a **lateral**, and **superficial** and **deep posterior compartments**. *(See Figures 11.27 and 11.28 and Table 11.11.)*

Chapter Review

For answers, see the blue Answers tab at the back of the book.

Level 1 Reviewing Facts and Terms

Match each numbered item with the most closely related lettered item.

1. rhomboid muscles .... [ ]
2. latissimus dorsi .... [ ]
3. infraspinatus .... [ ]
4. brachialis .... [ ]
5. supinator .... [ ]
6. flexor retinaculum .... [ ]
7. gluteal muscles .... [ ]
8. iliacus .... [ ]
9. gastrocnemius .... [ ]
10. tibialis anterior .... [ ]
11. interossei .... [ ]

(a) abducts the toes
(b) flexes hip and/or lumbar spine
(c) adduct (retract) scapula
(d) connective tissue bands
(e) plantar flexion at ankle
(f) originates on ilium
(g) flexes elbow
(h) dorsiflexes ankle and inverts foot
(i) lateral rotation of humerus at shoulder
(j) supinates forearm
(k) extends, adducts, medially rotates humerus at shoulder

12. The powerful extensors of the knee are the
(a) hamstrings.
(b) quadriceps.
(c) iliopsoas.
(d) tensor fasciae latae.

13. Which of the following is not a muscle of the rotator cuff?
(a) supraspinatus
(b) subclavius
(c) subscapularis
(d) teres minor

14. Which of the following does not originate on the humerus?
(a) anconeus
(b) biceps brachii
(c) brachialis
(d) triceps brachii, lateral head

15. Which of the following muscles is a flexor of the elbow?
(a) biceps brachii
(b) brachialis
(c) brachioradialis
(d) all of the above

16. The muscle that causes opposition of the thumb is the
(a) adductor pollicis.
(b) extensor digitorum.
(c) abductor pollicis.
(d) opponens pollicis.

Level 2 Reviewing Concepts

1. Damage to the pectoralis major would interfere with the ability to
(a) extend the elbow.
(b) abduct the humerus.
(c) adduct the humerus.
(d) elevate the scapula.

2. Which of the following muscles abducts the hip?
(a) pectineus
(b) psoas
(c) obturator internus
(d) piriformis

3. The tibialis anterior is a dorsiflexor of the foot. Which of the following muscles is an antagonist to that action?
(a) flexor digitorum longus
(b) gastrocnemius
(c) flexor hallucis longus
(d) all of the above

4. If you bruised your gluteus maximus, you would expect to experience discomfort when
(a) flexing the knee.
(b) extending the hip.
(c) abducting the hip.
(d) doing all of the above.

5. The biceps brachii exerts actions upon three joints. What are these joints and what are the actions?

6. What muscle supports the knee laterally and becomes greatly enlarged in ballet dancers because of the need for flexion and abduction at the hip?

7. When a dancer is stretching the muscles of a leg by placing the heel over a barre (horizontal bar at waist level), which groups of muscles are stretched?

8. What is the function of the intrinsic muscles of the hand?

9. How does the tensor fasciae latae act synergistically with the gluteus maximus?

10. What are the main functions of the flexor and extensor retinacula of the wrist and ankle?

Level 3 Critical Thinking

1. Describe how the hand muscles function to enable you to hold a pencil when you write.

2. While playing soccer, Jerry pulls his hamstrings. As a result of the injury, he has difficulty flexing and medially rotating his thigh. Which muscle(s) of the hamstring group did he probably injure?

3. While unloading the trunk of her car, Linda pulls a muscle and, as a result, has difficulty moving her arm. The doctor in the emergency room tells her that she pulled her pectoralis major. Linda tells you that she thought the pectoralis major was a chest muscle and doesn't understand what that has to do with her arm. What should you tell her?

 CLINICAL CASE WRAP-UP

Hamstrung

David sustained a complete avulsion (tearing away) of his hamstring tendons from their origin on the ischial tuberosity. When the tendons were forcibly detached or ripped from the ischial tuberosity, the muscle bellies contracted to about 70 percent of their resting length. With no proximal attachment, they remained contracted.

In surgery, David is placed prone on the operating table. The surgeon makes a "7"-shaped incision crossing the gluteal fold and extending down the posterior right thigh. Because it is difficult to get tendons to grow back to bone after they have torn away from their attachment, the surgeon must prepare the ischial tuberosity. This is done by scraping away the relatively avascular cortical bone down to the bleeding, cancellous bone. This roughening encourages attachment of the tendon to the healing bone. The surgeon pulls the tendon mass of the long head of the biceps femoris, semitendinosus, and semimembranosus proximally to the ischial tuberosity while David's knee is flexed. The surgeon attaches the tendons with bone-anchoring sutures.

Postoperatively, David is placed in a knee flexion sling (a device that keeps the knee flexed) that he will wear for 10 weeks.

1. What other muscles, in addition to the hamstrings, assist in flexing the knee?

2. Why does flexing the knee bring the avulsed hamstring tendons back up to the ischial tuberosity?

See the blue Answers tab at the back of the book.

11

Related Clinical Terms

contracture: Muscle shortening due to prolonged muscle contractions.

cubital tunnel syndrome: Ulnar nerve entrapment occurs when the ulnar nerve in the arm becomes compressed or irritated.

disuse atrophy: Muscle wasting caused by immobilization, such as casting or being bedridden.

myoglobinuria: Excretion of myoglobin in the urine, caused by muscle trauma or muscle ischemia ("blood starvation").

myositis: Muscle inflammation.

myotonia: Delayed muscle relaxation after a strong contraction, due to abnormality of the muscle membrane, specifically the ion channels.

12

Surface Anatomy and Cross-Sectional Anatomy

Learning Outcomes

These Learning Outcomes correspond by number to this chapter's sections and indicate what you should be able to do after completing the chapter.

12.1 Locate prominent skeletal landmarks and muscle contours for each major region of the body. p. 326

12.2 Visualize and understand the three-dimensional relationships of anatomical structures within the head, thorax, abdomen, and pelvis. p. 334

CLINICAL CASE

Breathing Through Your Neck

Katie, a first-year surgical intern, is walking into the cafeteria when she notices a commotion. A young woman is lying unconscious on the floor, looking cyanotic (bluish skin and mucous membranes from lack of oxygen). "She started choking and couldn't breathe or talk! I tried the Heimlich maneuver (firm abdominal thrusts to dislodge an airway obstruction), but it didn't work," says a man leaning over her.

Katie's initial assessment tells her that the woman is unresponsive and not breathing. Death is near—unless Katie can bypass the airway obstruction and get some air into her lungs. "Call 911!" she shouts as she grabs a clean steak knife and straw from the closest table.

"Thank goodness for my knowledge of surface anatomy," Katie thinks as she feels for the woman's **thyroid cartilage** or "Adam's apple." Sliding her finger down the midline, she feels a small space and then the **cricoid cartilage.** This space between the thyroid cartilage and cricoid cartilage is the **median cricothyroid ligament**. Spotting a mug of steaming tea on the table, Katie dunks the steak knife to clean it and prepares to make an incision.

Can Katie save this woman's life? To find out, turn to the Clinical Case Wrap-Up on p. 337.

THIS CHAPTER considers anatomy from two perspectives. Section 12.1 focuses on anatomical structures that we can identify from the body surface, and Section 12.2 views anatomical structures in cross section. Our detailed examination of anatomy in this chapter demonstrates the structural and functional relationships between the skeletal and muscular systems. The photographs give a visual tour of the entire body, highlighting skeletal landmarks and muscle contours.

12.1 Surface Anatomy: A Regional Approach

▶ **KEY POINT** We can locate many structures from their surface features and appearance—for example, superficial skeletal and muscular structures, tendons, ligaments, and veins. We can also use surface features to locate deeper anatomical structures.

12

Surface anatomy is the study of anatomical landmarks on the exterior of the human body. Chapter 1 introduced surface anatomy. ⮌ p. 2

We study surface anatomy using an approach based on anatomic regions of the body: head and neck; thorax; abdomen; shoulder and arm; arm, forearm, and wrist; pelvis and thigh; and leg and foot.

We present this information using photographs of living people. As you can see, we used living models with very little body fat, since subcutaneous fat hides many anatomical landmarks. Locating surface landmarks involves estimating their location and then palpating for specific structures.

Head and Neck

Figure 12.1 The Head and Neck

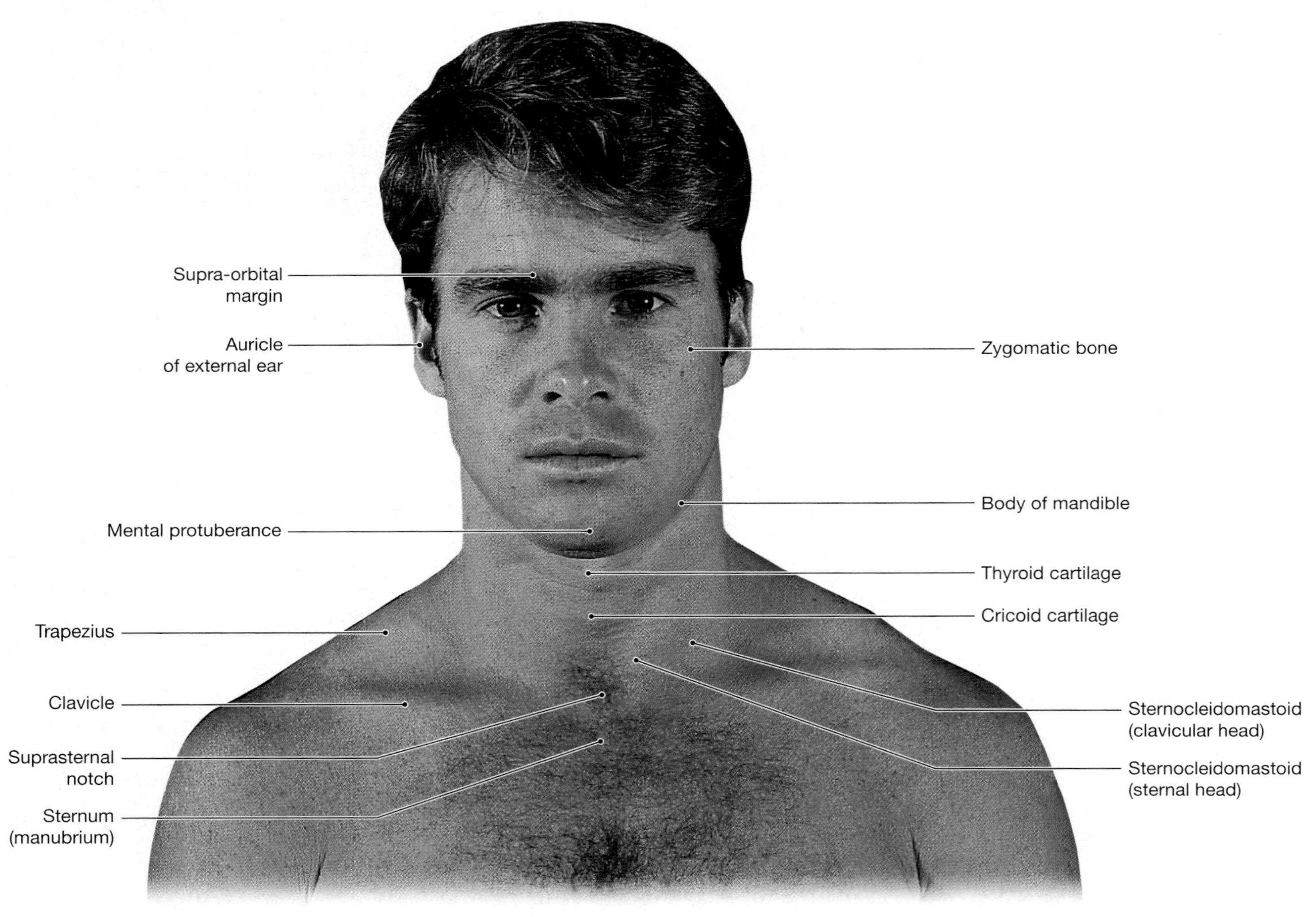

a Anterior view

Figure 12.1 *(continued)*

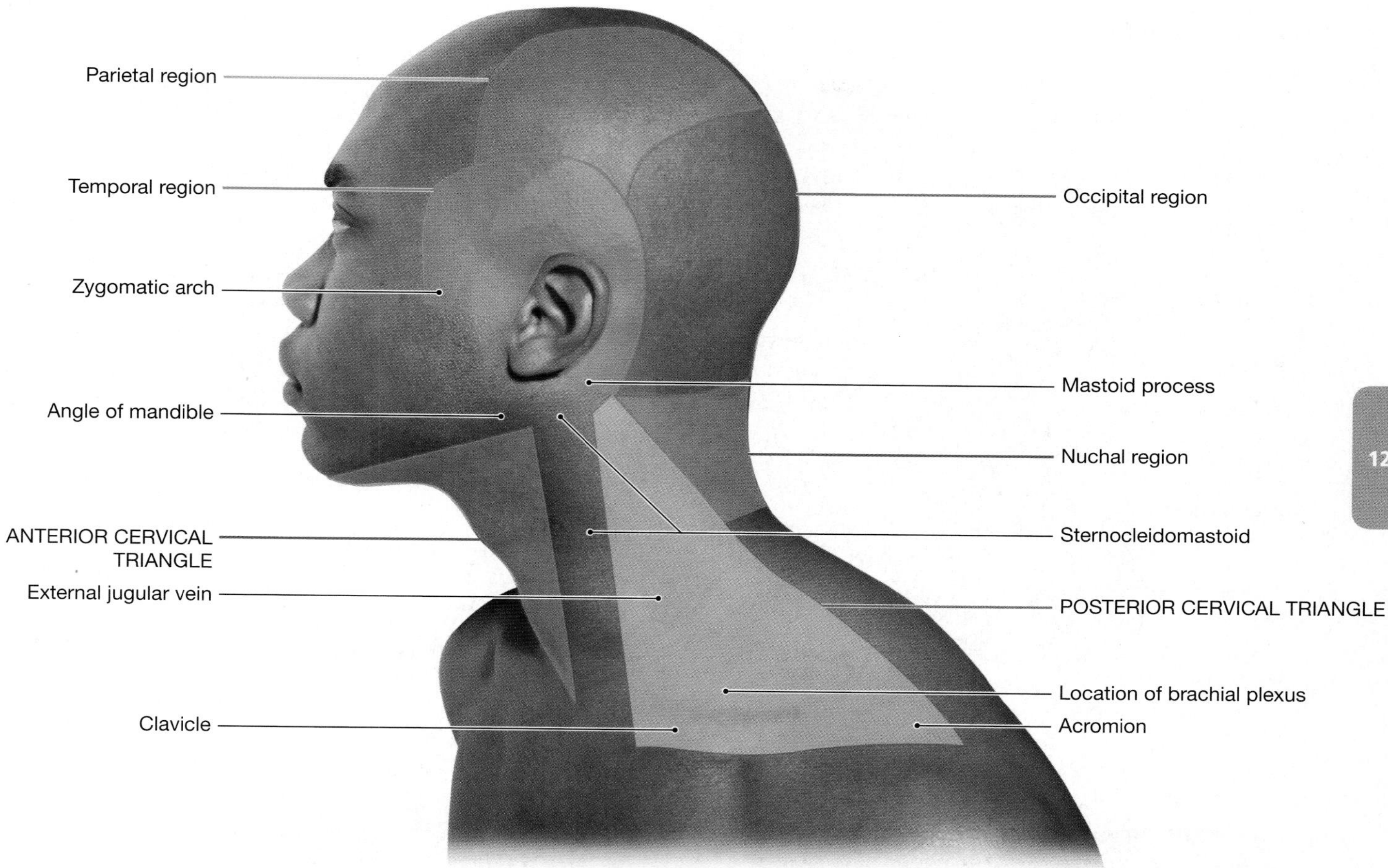

b The posterior cervical triangle and the larger regions of the head and neck

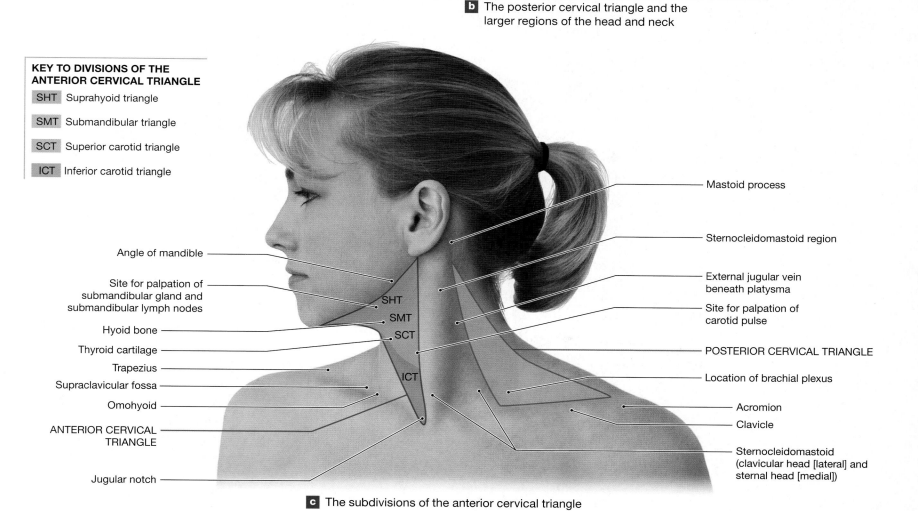

KEY TO DIVISIONS OF THE ANTERIOR CERVICAL TRIANGLE

SHT Suprahyoid triangle
SMT Submandibular triangle
SCT Superior carotid triangle
ICT Inferior carotid triangle

Angle of mandible
Site for palpation of submandibular gland and submandibular lymph nodes
Hyoid bone
Thyroid cartilage
Trapezius
Supraclavicular fossa
Omohyoid
ANTERIOR CERVICAL TRIANGLE
Jugular notch
SHT
SMT
SCT
ICT
Mastoid process
Sternocleidomastoid region
External jugular vein beneath platysma
Site for palpation of carotid pulse
POSTERIOR CERVICAL TRIANGLE
Location of brachial plexus
Acromion
Clavicle
Sternocleidomastoid (clavicular head [lateral] and sternal head [medial])

c The subdivisions of the anterior cervical triangle

Thorax

Figure 12.2 The Thorax

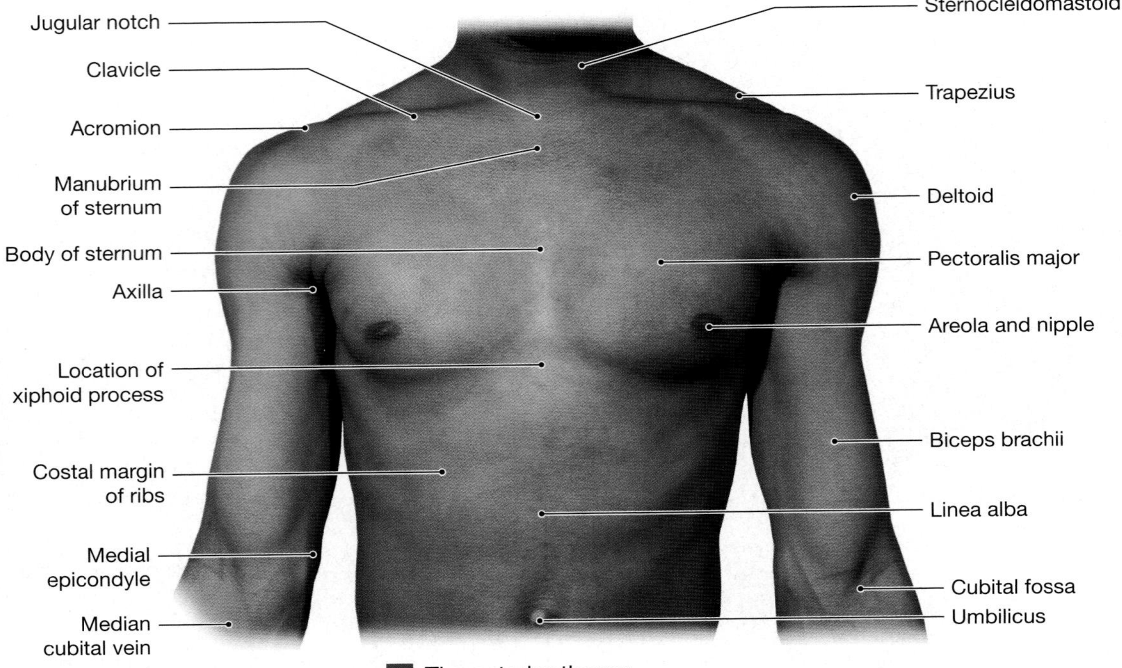

a The anterior thorax

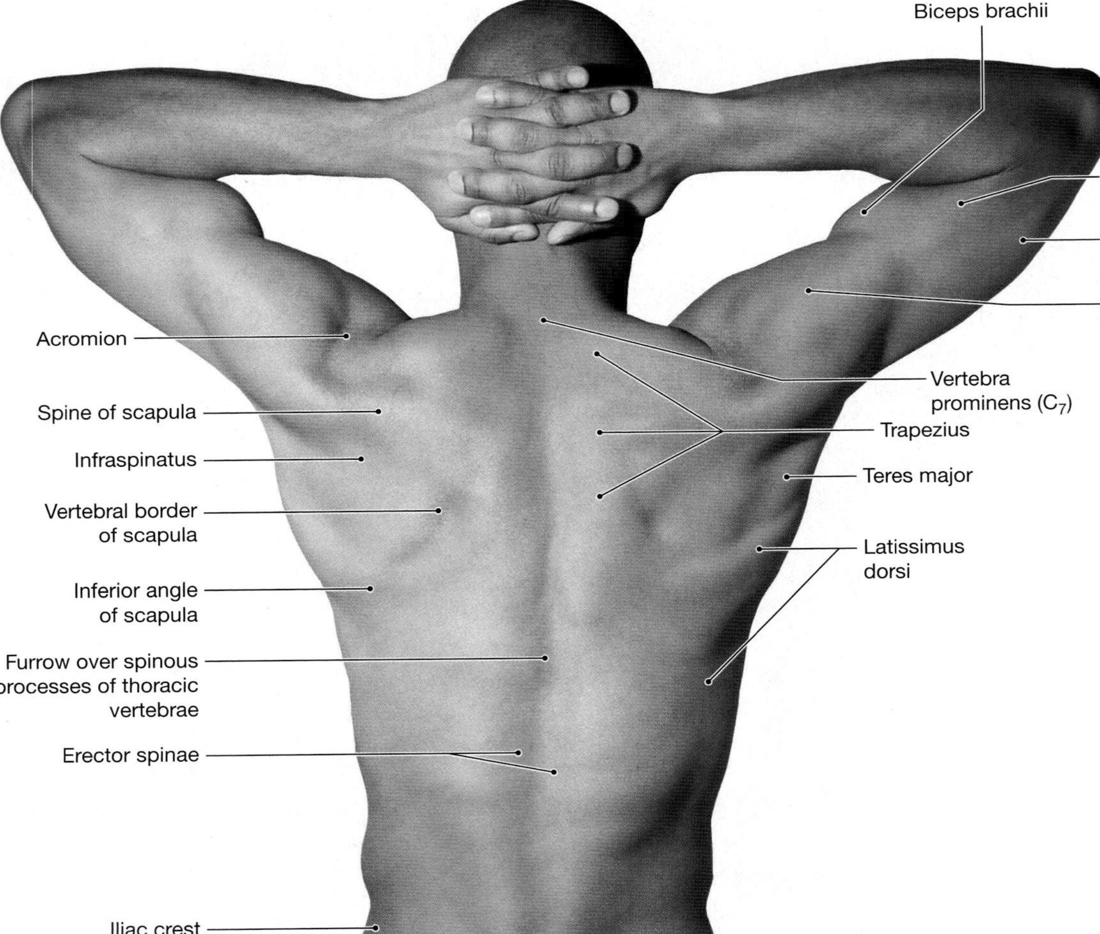

b The back and shoulder regions

CLINICAL NOTE

Heart Sounds

The red dots indicate the best locations for hearing the sounds made by the heart valves.

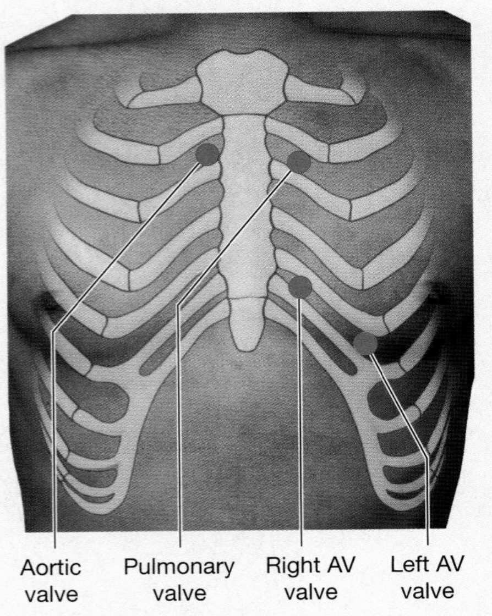

CLINICAL NOTE

Lumbar Puncture

To perform a lumbar puncture (also called a "spinal tap"), a health professional inserts a hollow needle into the intervertebral spaces between the L_3 and L_4 vertebrae or the L_4 and L_5 vertebrae to withdraw a sample of cerebrospinal fluid for testing. These areas are where the intervertebral spaces are largest.

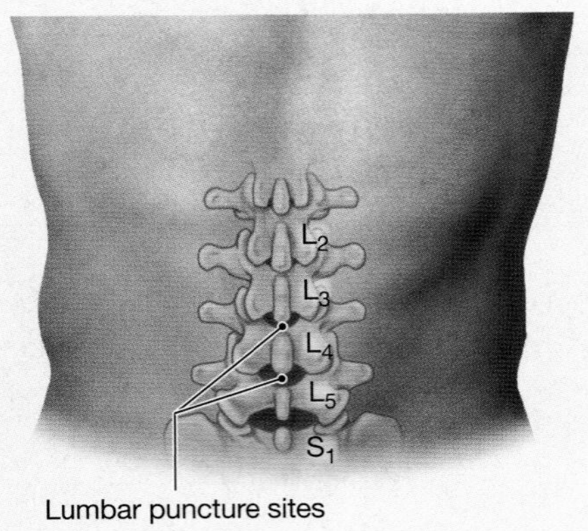

Abdomen

Figure 12.3 The Abdominal Wall

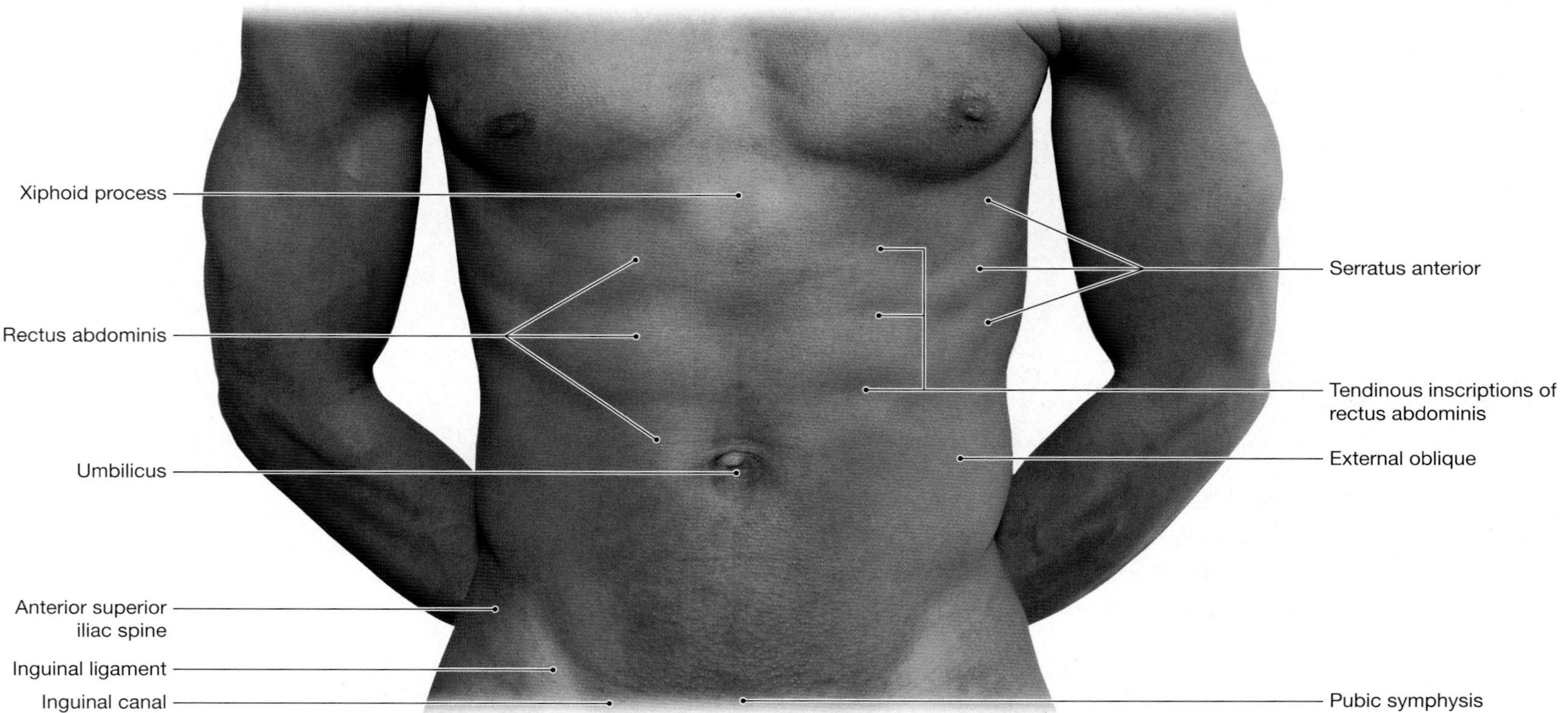

a The anterior abdominal wall

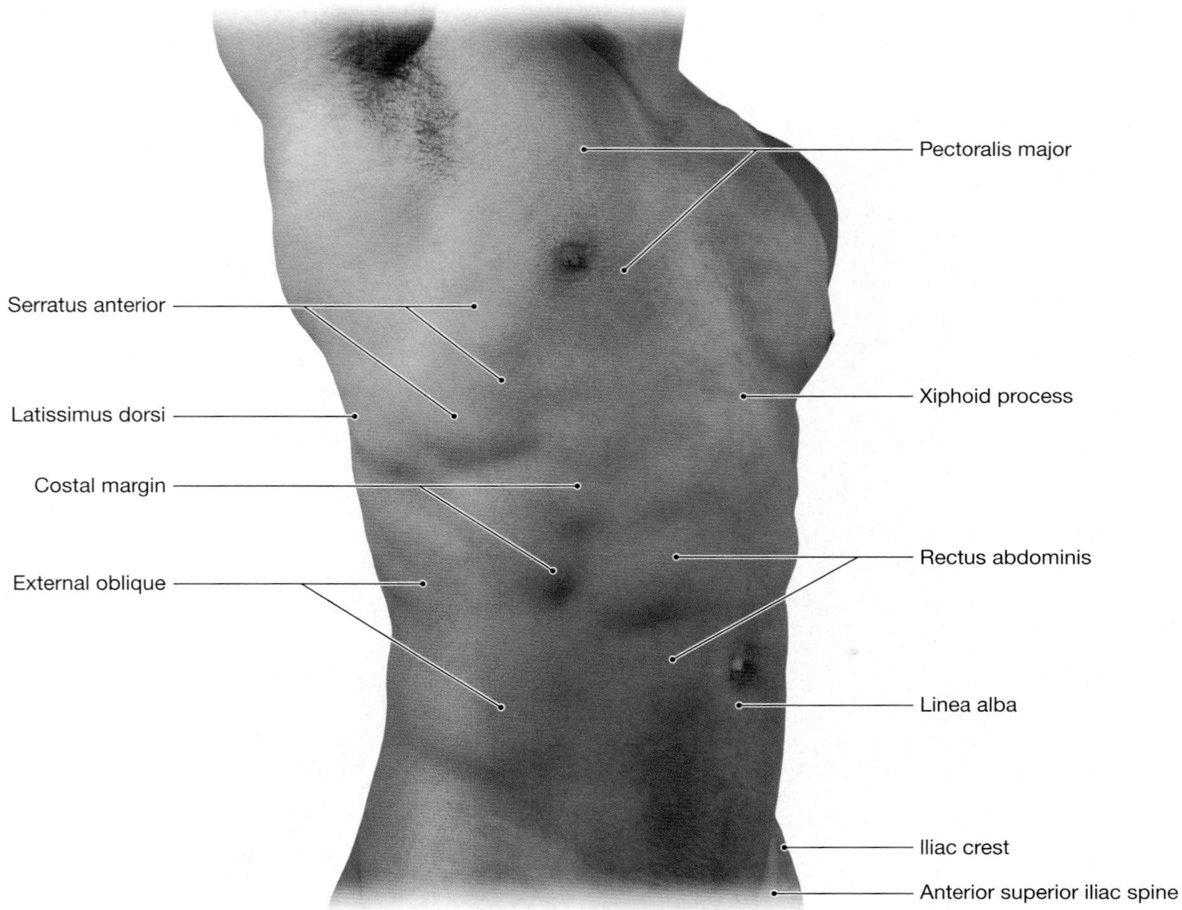

b Anterolateral view of the abdominal wall

Shoulder and Arm

Figure 12.4 The Shoulder and Arm

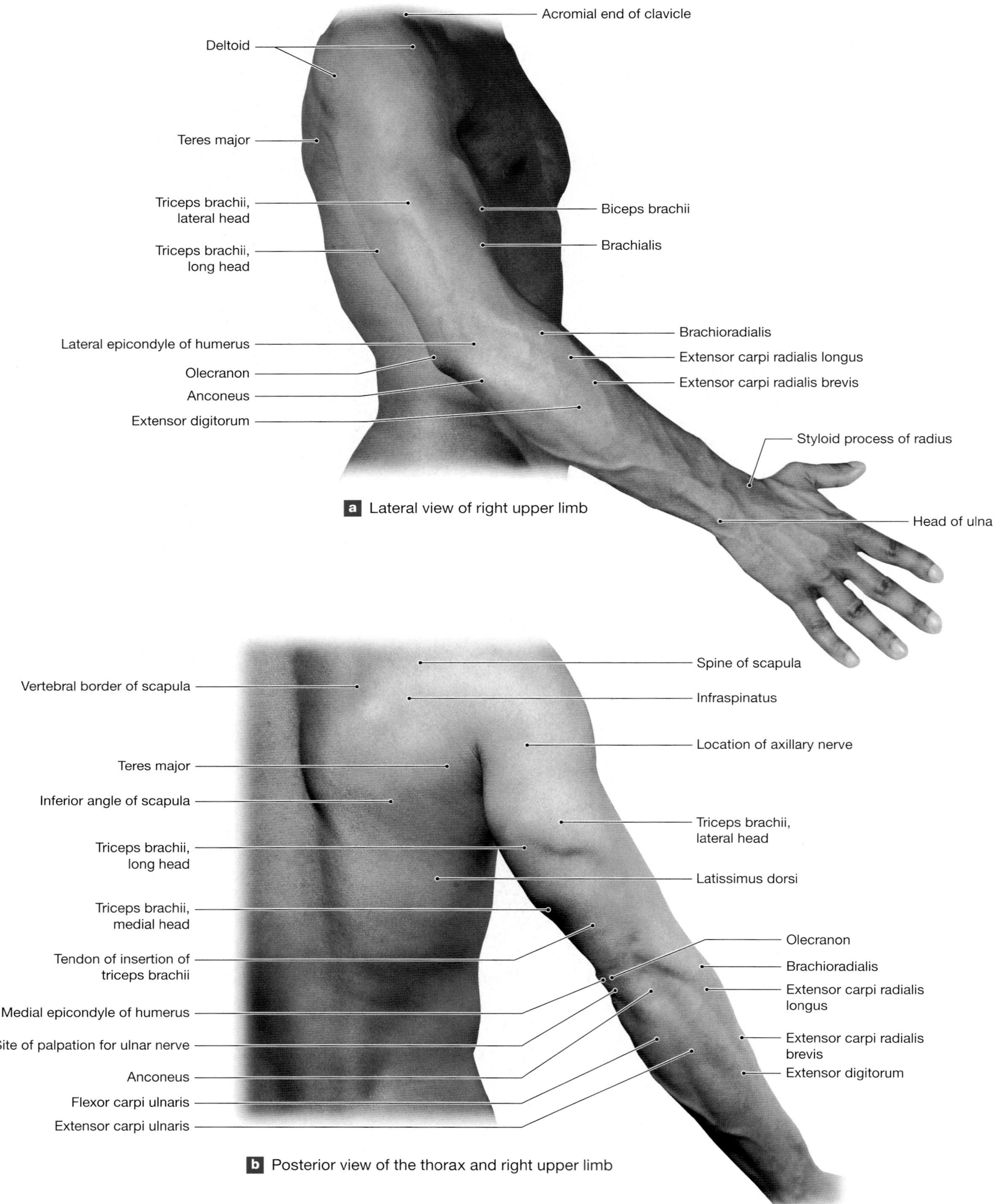

a Lateral view of right upper limb

b Posterior view of the thorax and right upper limb

Arm, Forearm, and Wrist

Figure 12.5 The Arm, Forearm, and Wrist

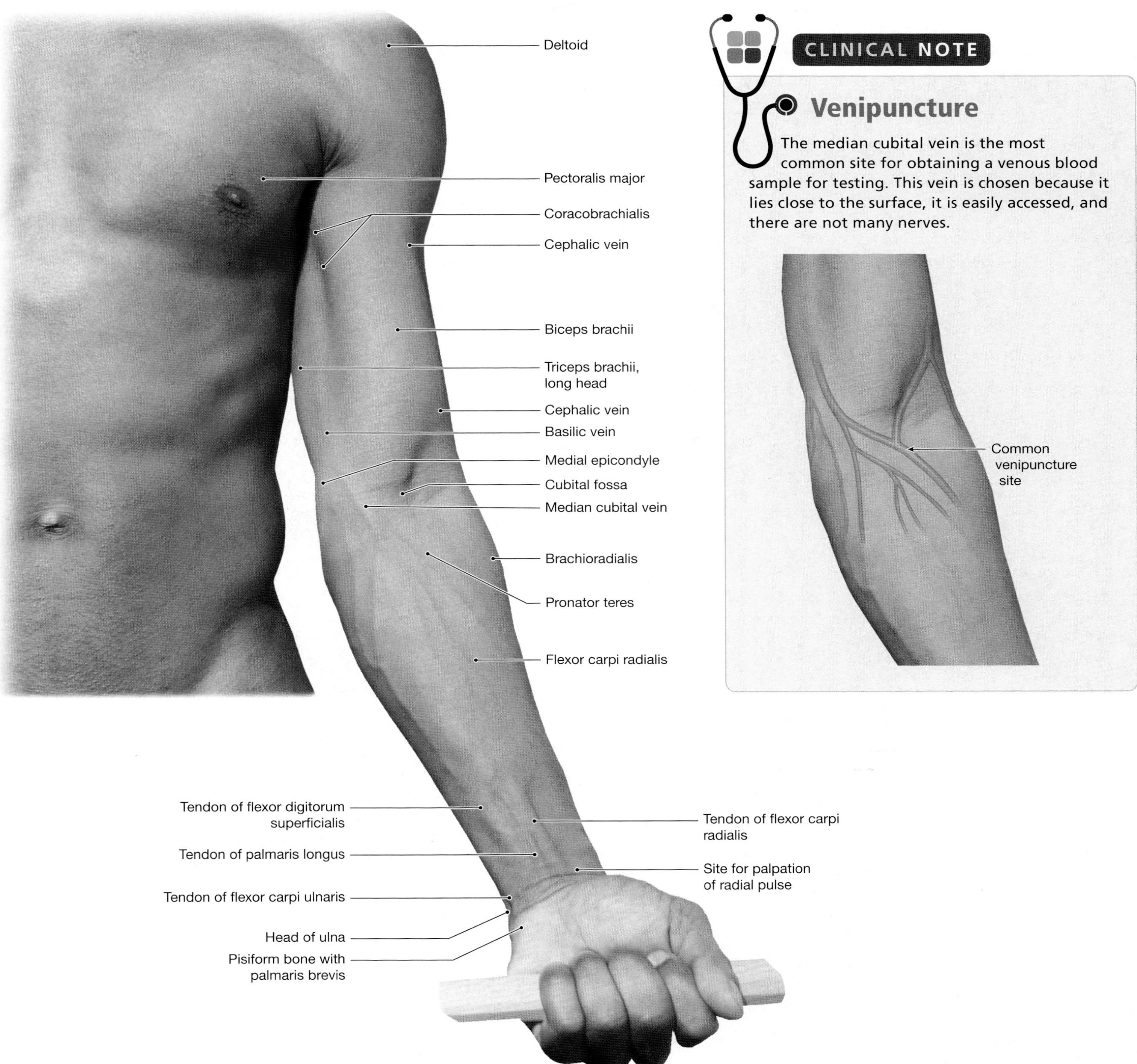

Deltoid
Pectoralis major
Coracobrachialis
Cephalic vein
Biceps brachii
Triceps brachii, long head
Cephalic vein
Basilic vein
Medial epicondyle
Cubital fossa
Median cubital vein
Brachioradialis
Pronator teres
Flexor carpi radialis
Tendon of flexor digitorum superficialis
Tendon of palmaris longus
Tendon of flexor carpi ulnaris
Head of ulna
Pisiform bone with palmaris brevis
Tendon of flexor carpi radialis
Site for palpation of radial pulse
CLINICAL NOTE
Venipuncture
The median cubital vein is the most common site for obtaining a venous blood sample for testing. This vein is chosen because it lies close to the surface, it is easily accessed, and there are not many nerves.
Common venipuncture site

12

Pelvis and Thigh

Figure 12.6 The Pelvis and Thigh. The inguinal ligament, medial border of the sartorius and border of the adductor longus form the boundaries of the femoral triangle.

CLINICAL NOTE

Femoral Artery

The femoral artery is an important site for locating a pulse or inserting a catheter for a variety of medical procedures.

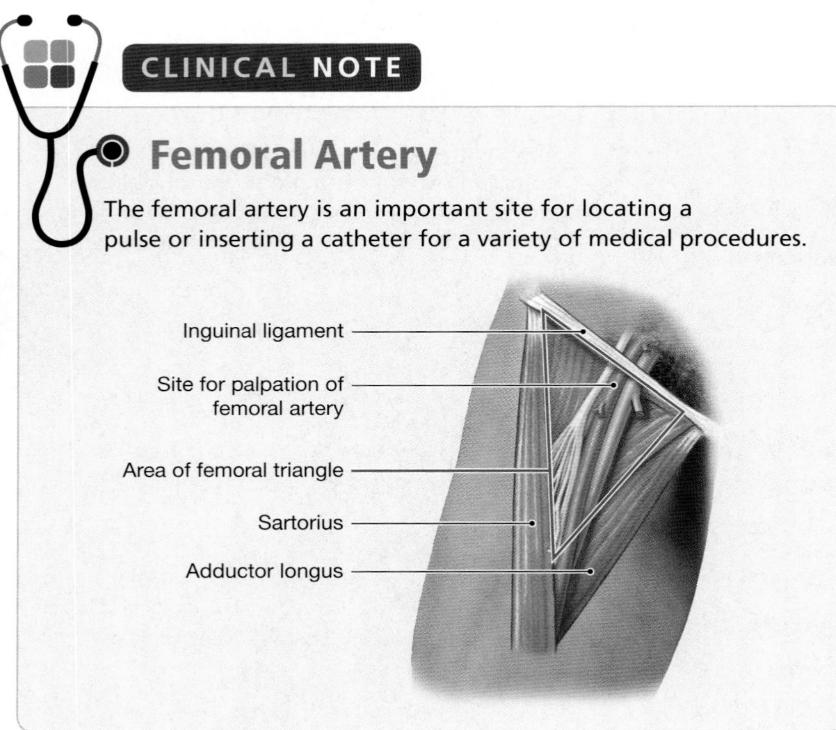

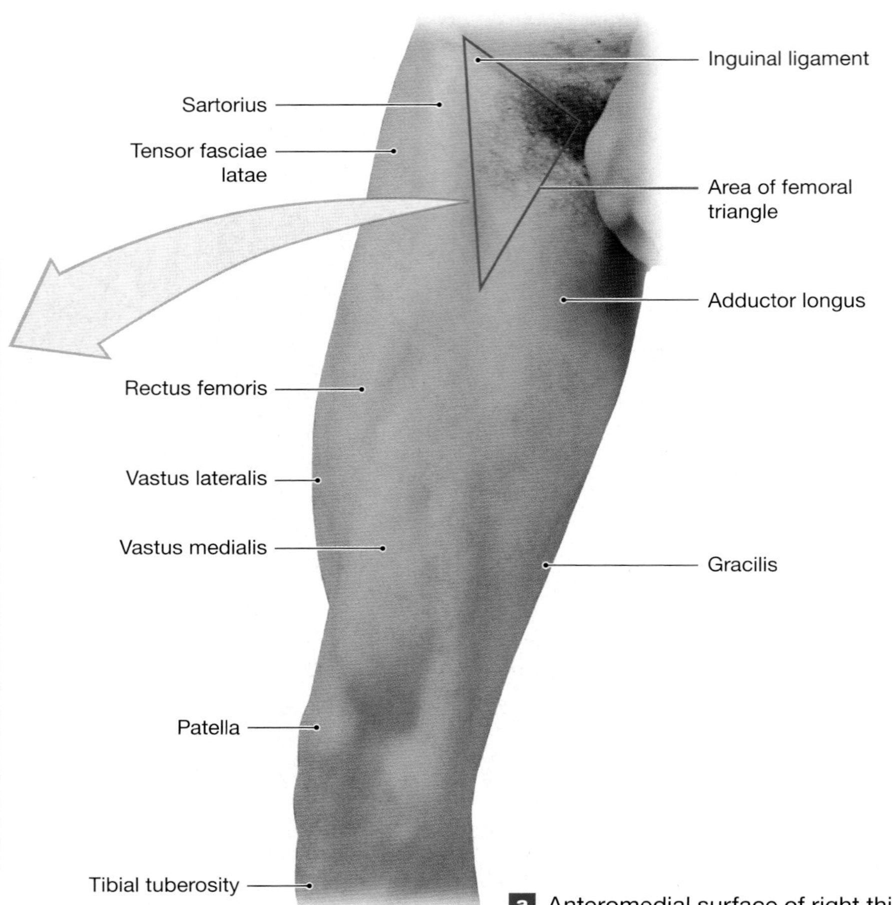

a Anteromedial surface of right thigh

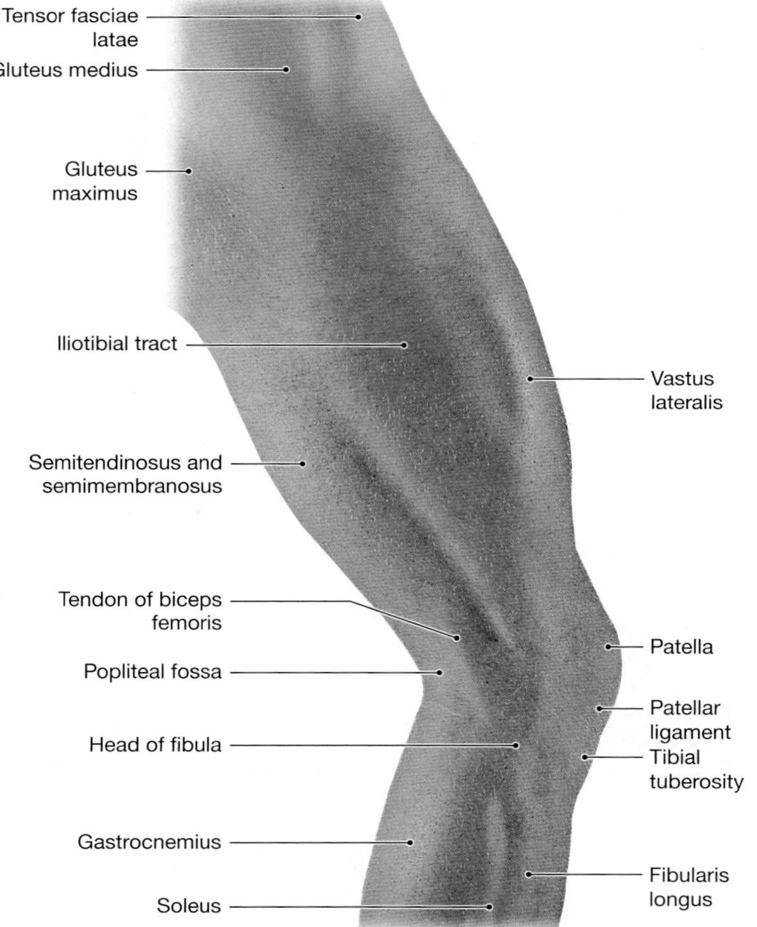

b Lateral surface of right thigh and gluteal region

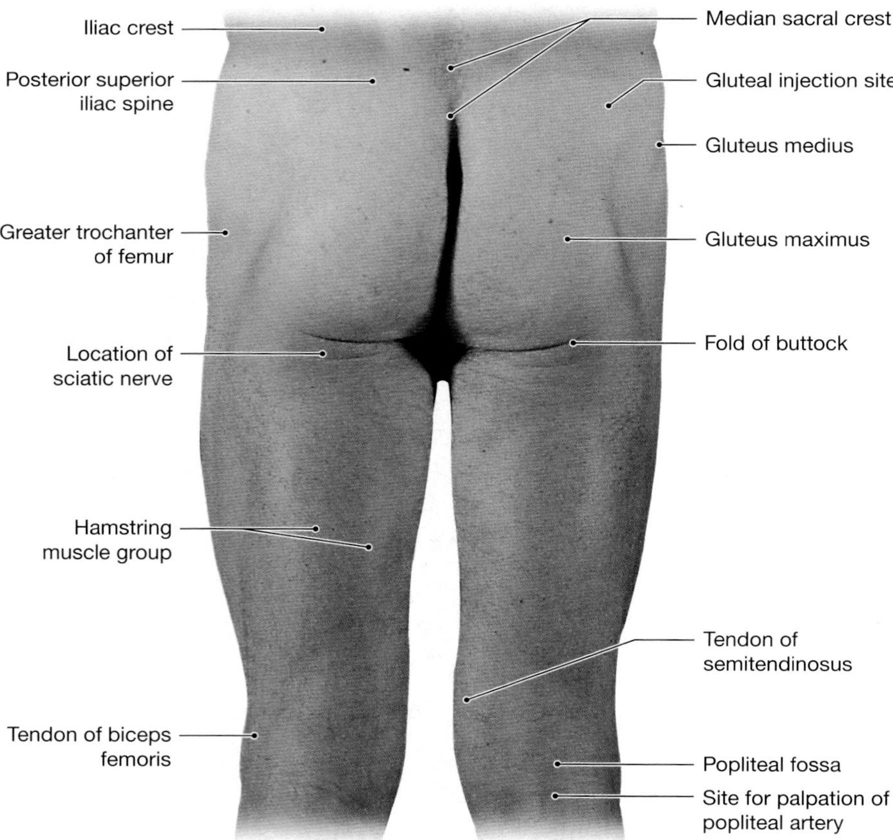

c Posterior surfaces of thigh and gluteal region

Leg and Foot

Figure 12.7 The Leg and Foot

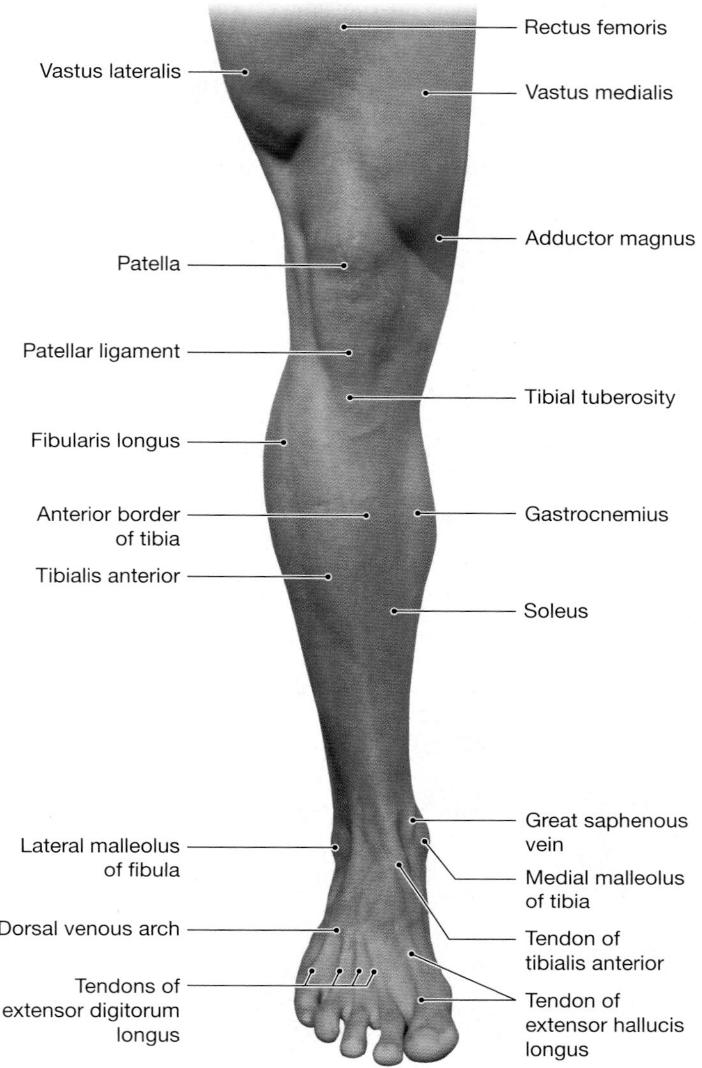

a Right thigh, knee, leg, and foot, anterior view

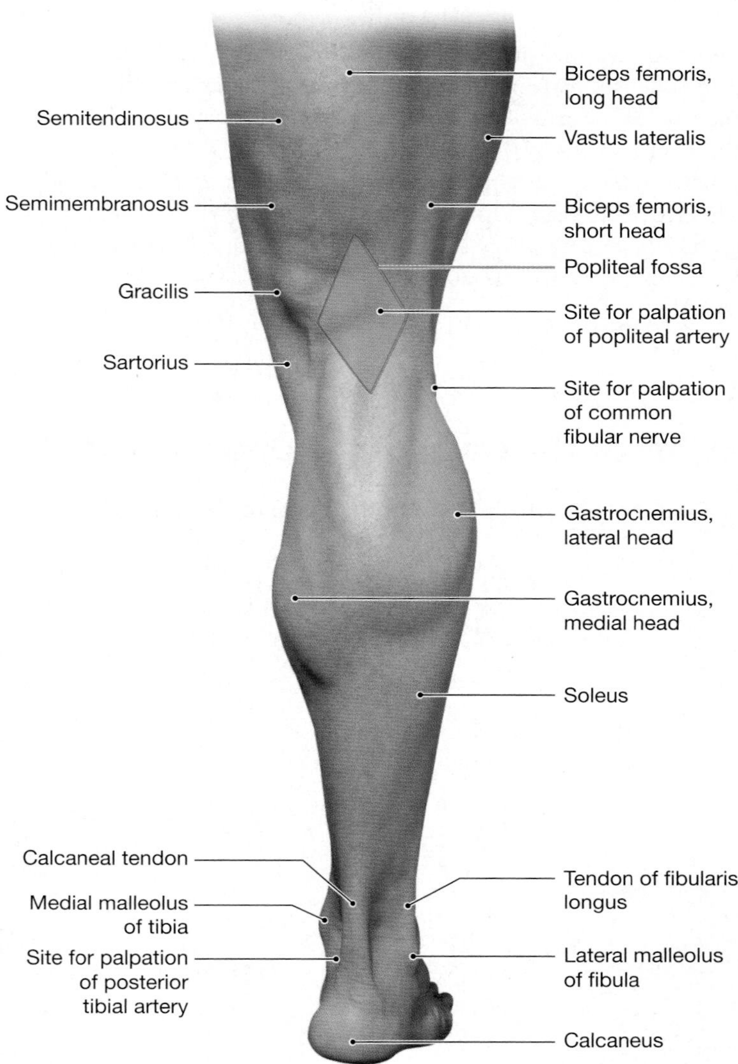

b Right thigh, knee, leg, and foot, posterior view

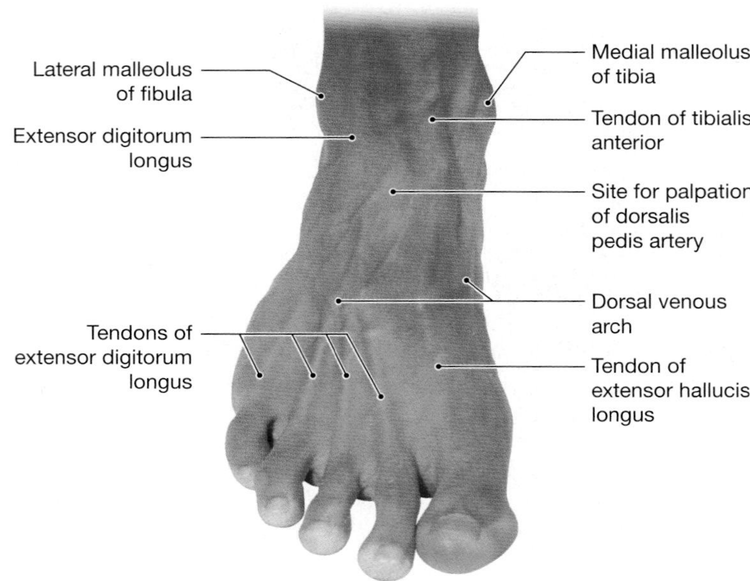

c Right ankle and foot, anterior view

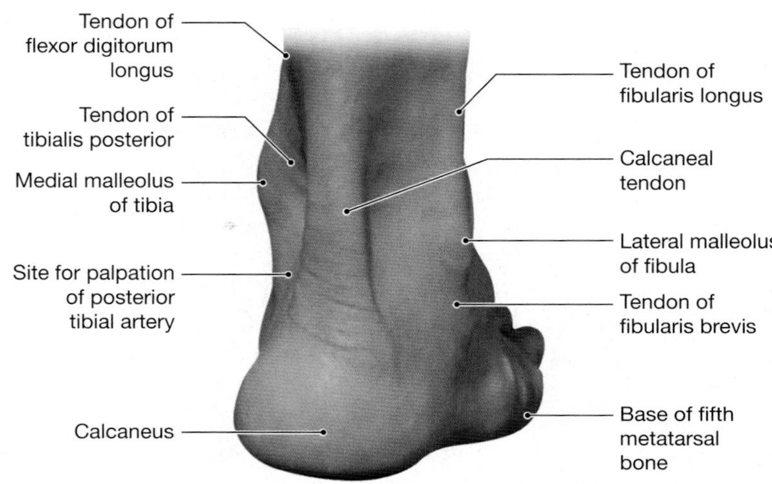

d Right ankle and foot, posterior view

12.2 | Cross-Sectional Anatomy

▶ **KEY POINT** Today's anatomy students must visualize and understand the three-dimensional relationships of anatomical structures in a wide variety of cross-sectional formats.

We define the cross-sectional plane (also termed the *transverse* or *horizontal plane*) as a plane oriented perpendicular to the longitudinal axis of the part of the body being studied. In **cross-sectional anatomy**, the body is divided into superior and inferior sections by the cross-sectional plane. ⮌ p. 18

The techniques used to view anatomical structures have changed dramatically within the last 10–20 years. Visualizing the human body in cross section is an intriguing and challenging way to study. We can use a variety of methods to view the body in cross section. ⮌ pp. 20–21

The cross-sectional images in Section 12.2 come from the National Library of Medicine's *The Visible Human Project.** As you view them, remember these points:

- The cross sections in this chapter are all inferior-view images, so they are viewed as if you are standing at the individual's feet and looking toward the head.
- The anterior surface is at the top of the image, and the posterior surface is at the bottom.
- In this method of presentation, structures on the right side of the body appear on the left side of the image.

Cross Section at the Level of the Optic Chiasm

Figure 12.8 Cross Section of the Head at the Level of the Optic Chiasm

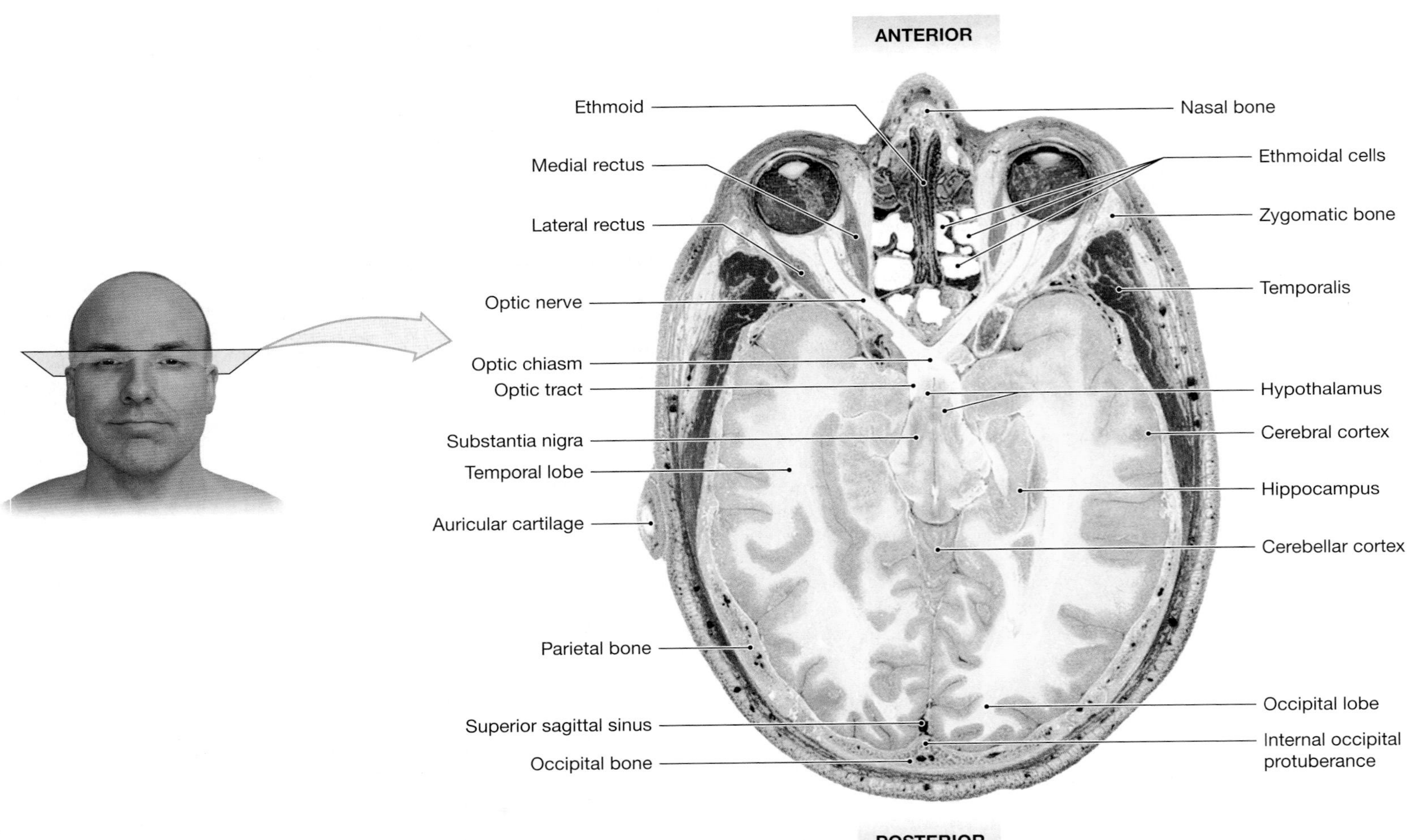

*To learn more about *The Visible Human Project*, go to http://www.nlm.nih.gov/research/visible/visible_human.html.

Cross Section at the Level of Vertebra C_2

Figure 12.9 Cross Section of the Head at the Level of Vertebra C_2

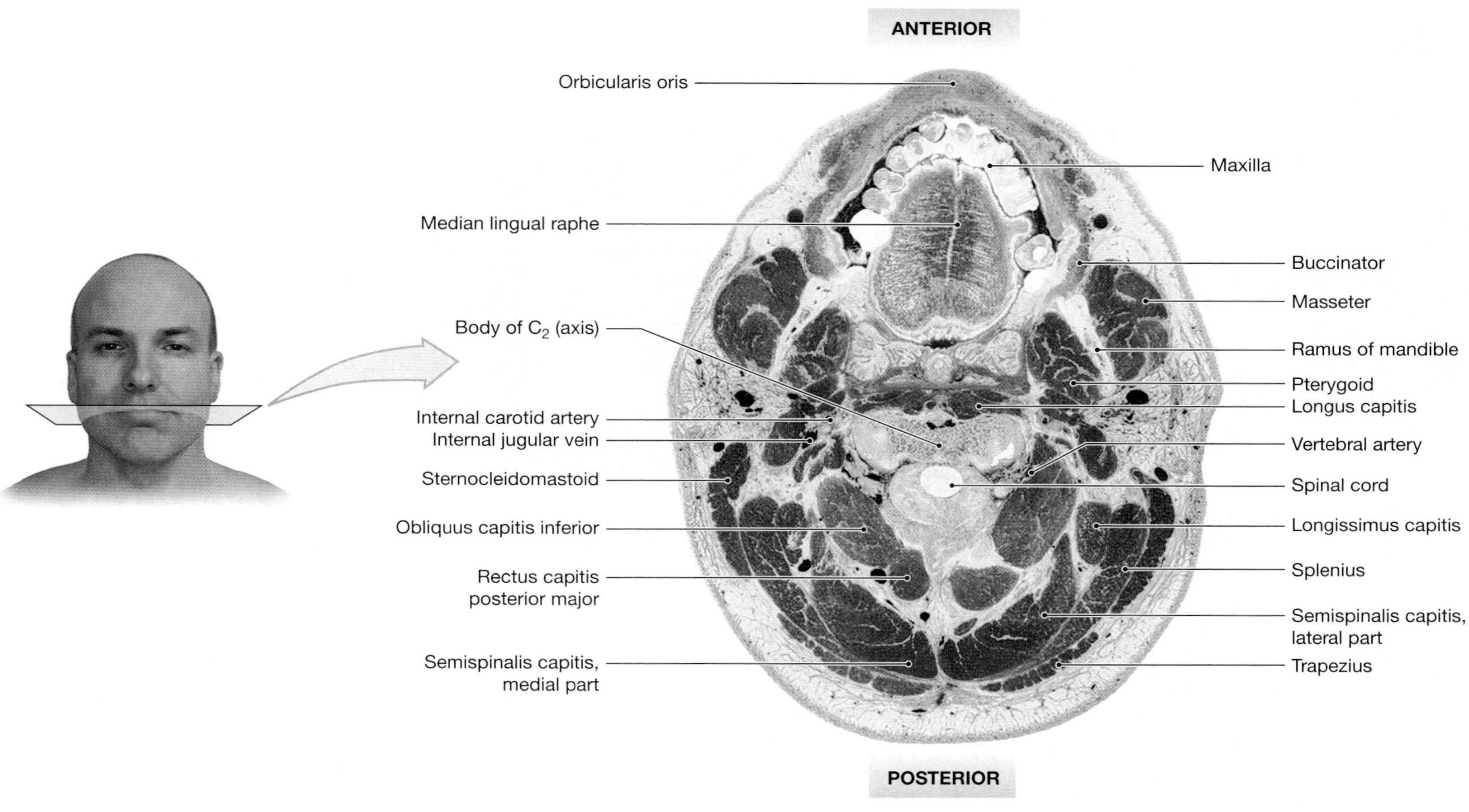

Cross Section at the Level of Vertebra T_2

Figure 12.10 Cross Section at the Level of Vertebra T_2

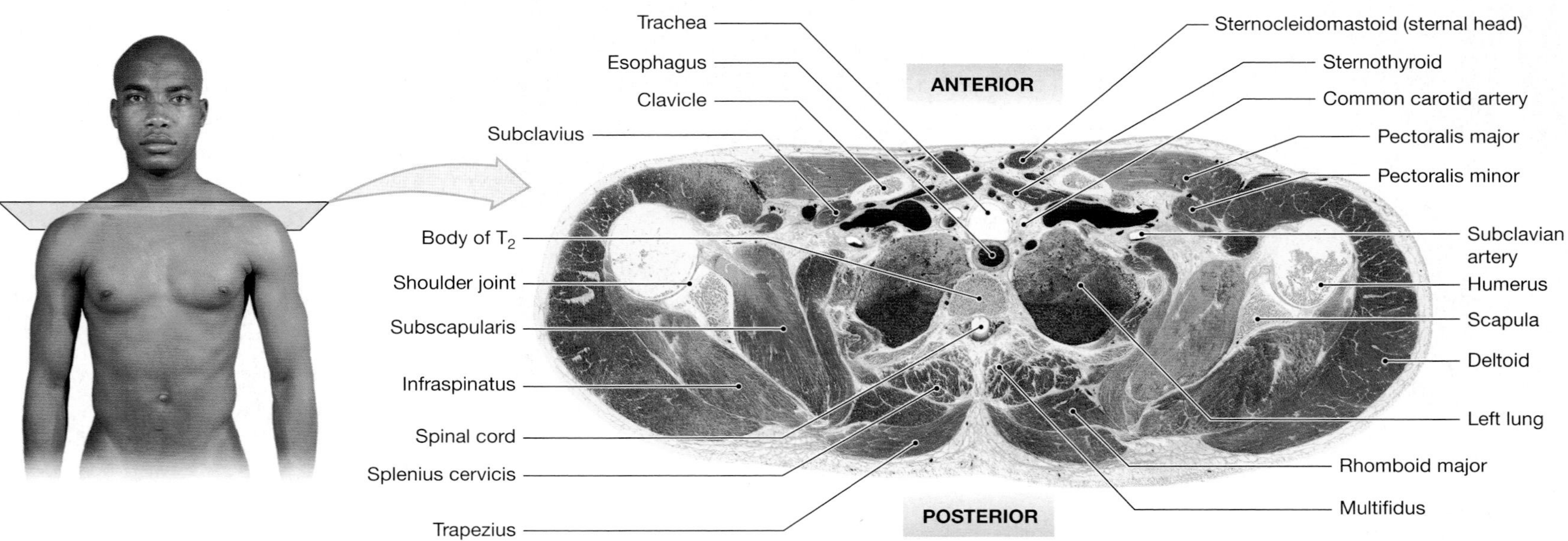

12

Cross Section at the Level of Vertebra T_8

Figure 12.11 Cross Section at the Level of Vertebra T_8

12

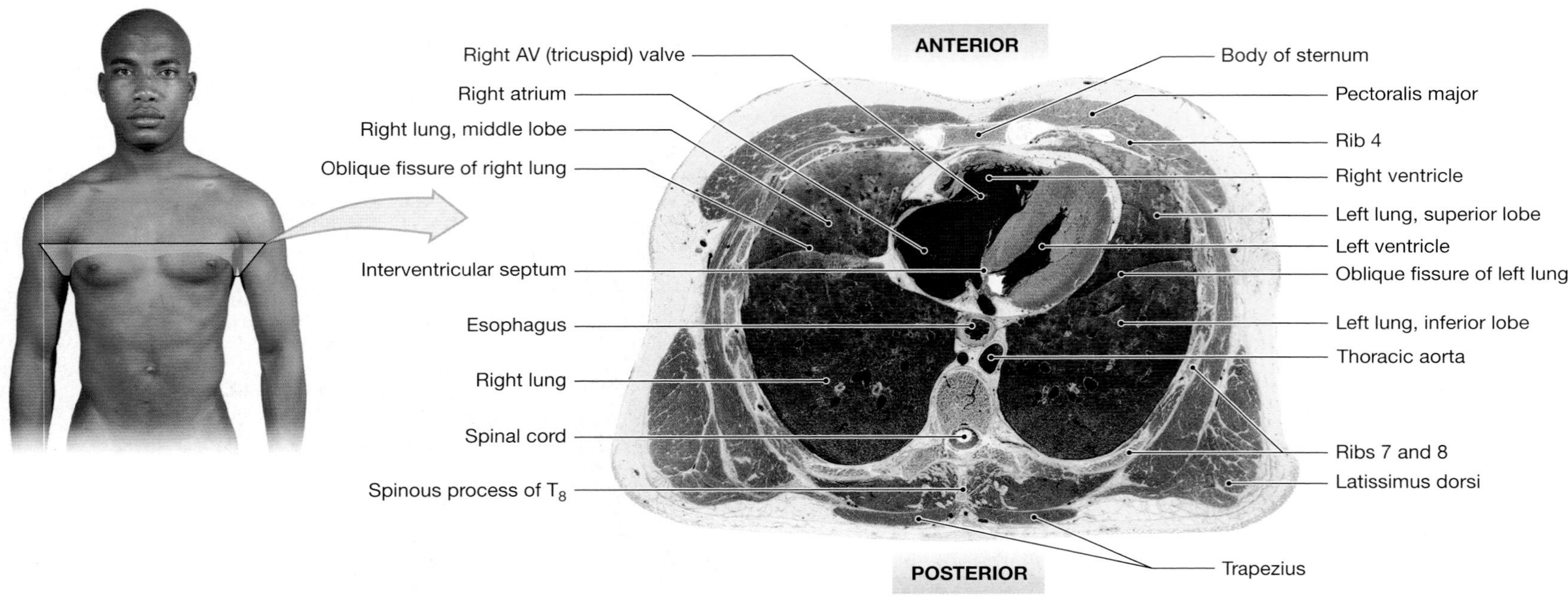

Cross Section at the Level of Vertebra T_{10}

Figure 12.12 Cross Section at the Level of Vertebra T_{10}

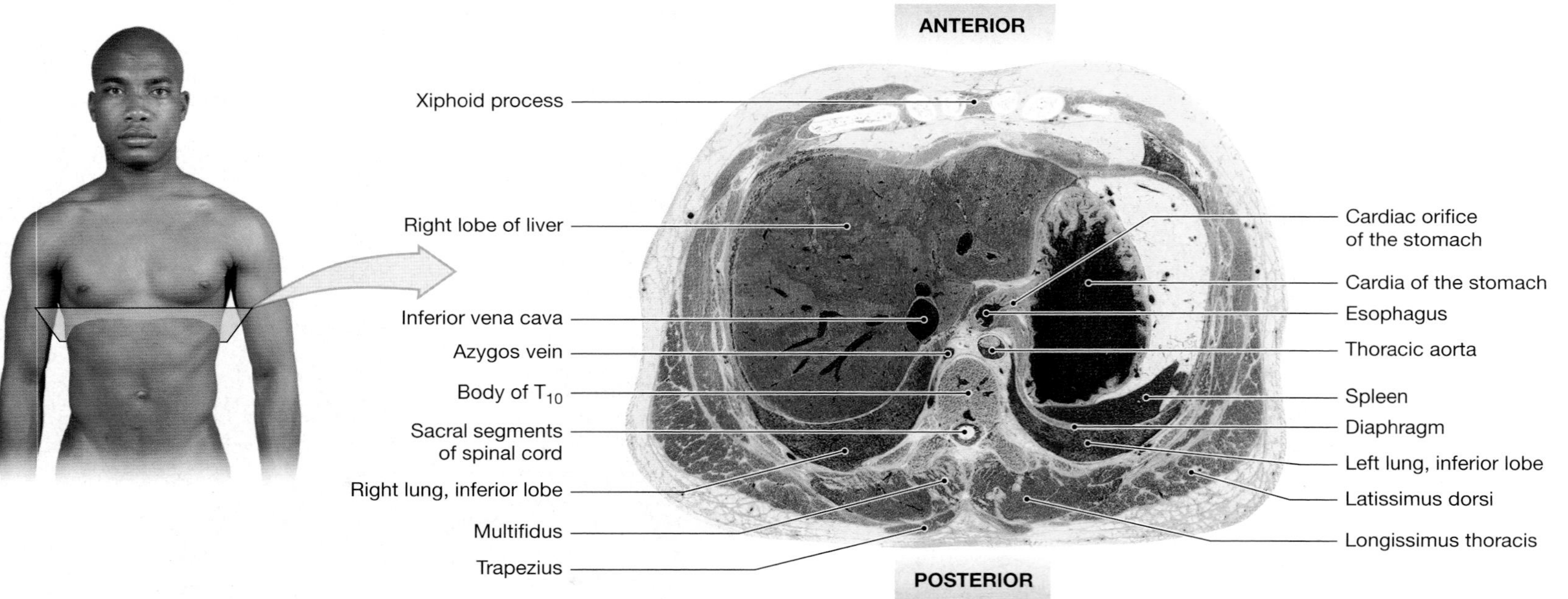

Cross Section at the Level of Vertebra T_{12}

Figure 12.13 Cross Section at the Level of Vertebra T_{12}

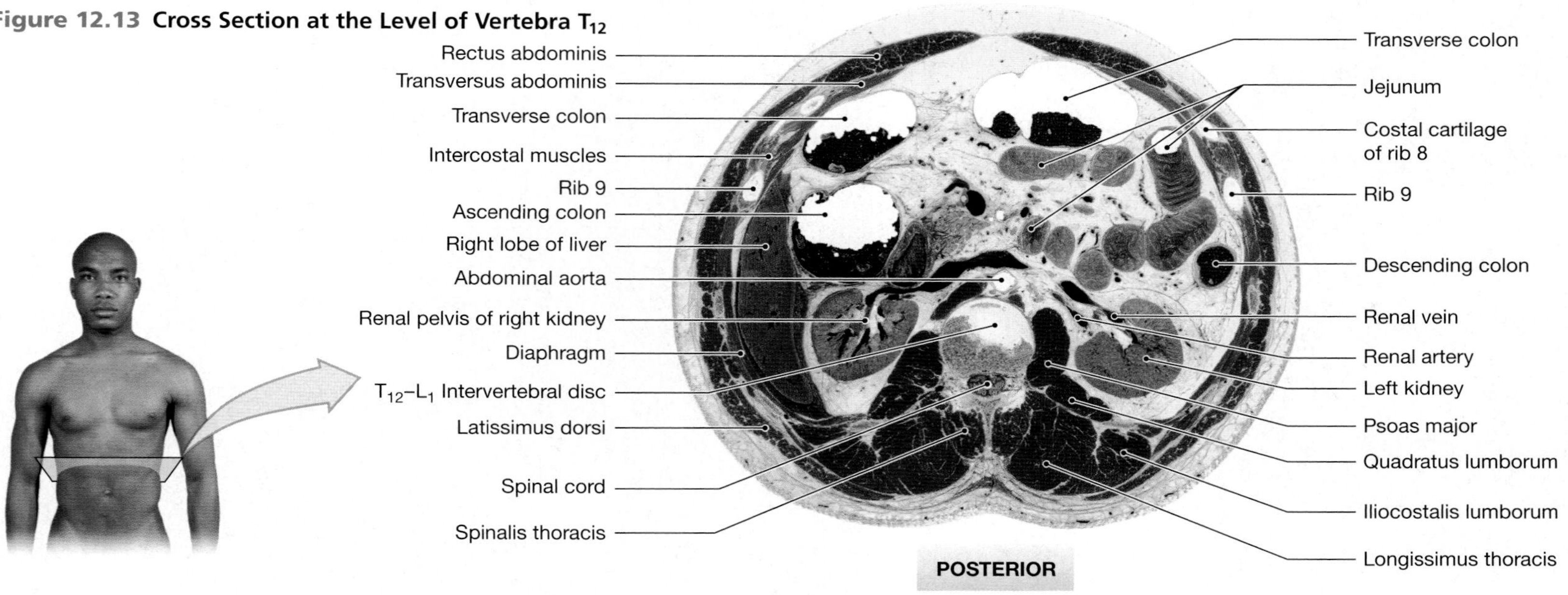

Cross Section at the Level of Vertebra L_5

Figure 12.14 Cross Section at the Level of Vertebra L_5

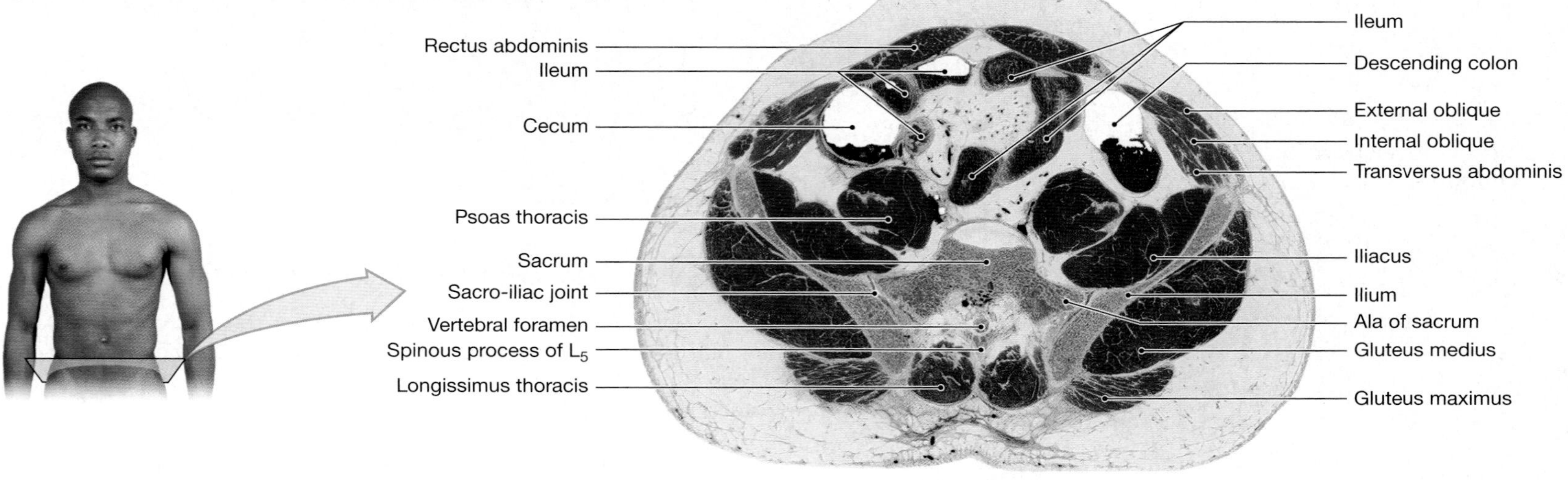

CLINICAL CASE WRAP-UP

Breathing Through Your Neck

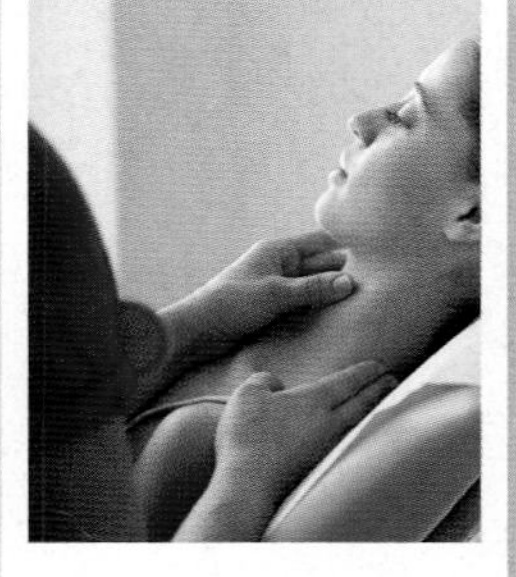

If this were an elective procedure with anesthesia and sterile conditions, Katie would make the incision in the woman's trachea in a slightly different place—lower and closer to the sternal notch, in a more convenient location for the patient. This procedure is a **tracheostomy**, or surgical opening into the trachea.

However, this is a life-threatening emergency. By palpating the patient's anterior neck, Katie finds the exact locations of the thyroid cartilage and cricoid cartilage, allowing her to locate the median cricothyroid ligament. In crisis situations, this soft ligamentous space, surrounded by a "cartilage cage," is the safest place to make a "blind" incision in the airway. This procedure is called a cricothyroidotomy, or cricothyrotomy. There are few blood vessels and nerves that cross this space, and it is safely superior to thyroid gland.

Using the steak knife, Katie makes a one-inch horizontal incision directly over the cricothyroid membrane. Katie inserts her index finger into the wound and feels the "soft spot" directly below her fingertip. She then makes a horizontal incision into the trachea and inserts the straw about two inches. Immediately, the unconscious woman takes a gasping breath, desperately sucking in air through the straw. Katie has saved a life today.

1. In which cervical triangle of the neck is a tracheostomy performed?
2. Where is the best place to *practice* feeling the median cricothyroid ligament?

See the blue Answers tab at the back of the book.

13 The Nervous System

Nervous Tissue

Learning Outcomes

These Learning Outcomes correspond by number to this chapter's sections and indicate what you should be able to do after completing the chapter.

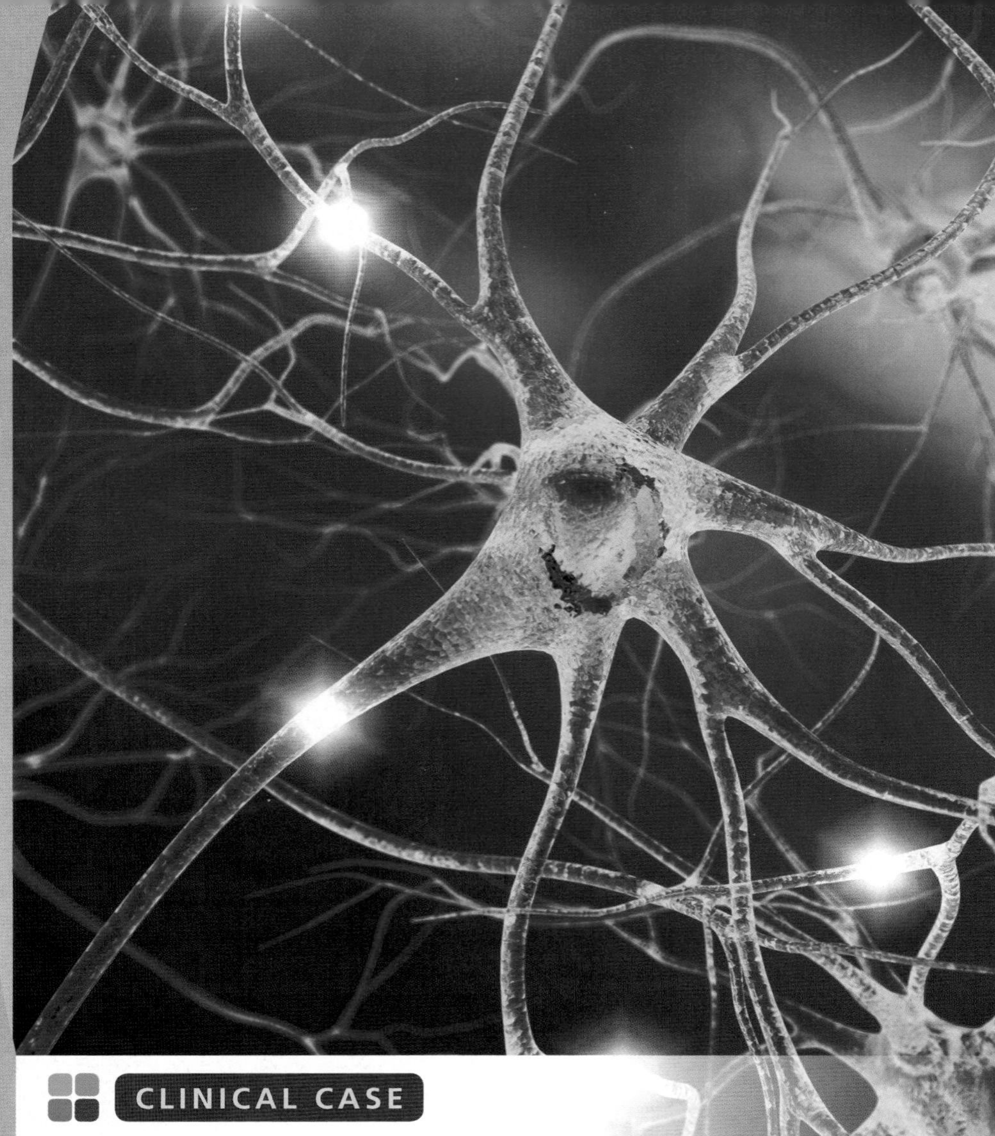

CLINICAL CASE

When Nerves Become Demyelinated

Nicole awakens early to get to the ski slopes before they get crowded. As she puts her feet on the floor, she feels an unusual tingling and numbness in her toes. But she hasn't been snowboarding in 2 weeks due to an upper respiratory infection (cold), so she decides a little numbness in her feet isn't going to stop her now.

By the time Nicole arrives at the ski lifts, the numbness in her feet has climbed to above her ankles. As she puts on her boots, her fingertips start to tingle and feel numb. As she reaches the top of the mountain, her ankles feel weak, causing her to fall when dismounting the ski lift chair. Still determined, she makes her way down the slope. At the bottom, she admits that her knees are too weak to continue snowboarding. She has never experienced anything like this, and she's terrified.

What is causing Nicole's ascending numbness and weakness? To find out, ***turn to the Clinical Case Wrap-Up on p. 359.***

ALTHOUGH OUR NERVOUS SYSTEM is often compared to a computer, it is more complicated and versatile. Both depend on electrical activity for the rapid flow of information and processing. However, unlike a computer, portions of the brain rework their electrical connections as new information arrives and learning occurs.

Along with the endocrine system (discussed in Chapter 19), the nervous system controls and adjusts the activities of other systems. Both the nervous system and endocrine system require chemical communication with target tissues and organs, and they often work together. The nervous system provides swift but brief responses to stimuli by temporarily changing the activities of other organ systems. The response appears in a few milliseconds—but the effects disappear almost as quickly. In contrast, endocrine system responses develop much more slowly than nervous system responses, but they last much longer—hours, days, or years.

This chapter considers the structure and function of nervous tissue. Subsequent chapters build on this foundation as we explore the organization of the brain, spinal cord, higher-order functions, and our senses of sight, smell, hearing, balance, touch, and pain.

13.1 An Overview of the Nervous System

▶ **KEY POINT** The nervous system is subdivided into the central nervous system (CNS) and peripheral nervous system (PNS). The CNS is composed of the brain and spinal cord. The PNS has afferent and efferent divisions. The afferent division consists of somatic and visceral sensory nerves. The efferent division consists of the somatic and autonomic nervous systems; the autonomic nervous system consists of the parasympathetic and sympathetic divisions.

The **nervous system** consists of all the **nervous tissue** in the body. The nervous system has two main anatomical subdivisions: central and peripheral (**Figure 13.1**).

The **central nervous system (CNS)** is composed of the brain and spinal cord. The CNS processes and coordinates sensory input and motor output. It is also the location of higher functions, including intelligence, memory, learning, and emotion. Early in development, the CNS begins as a mass of nervous tissue organized into a hollow tube. As development continues, the tube's central cavity decreases in size, and the thickness of the tube's walls and the diameter of the enclosed space varies from one region to another. The narrow central cavity of the developing spinal cord is called the central canal; the expanded chambers of the brain, called ventricles, are continuous with the central canal. A clear, watery fluid, called cerebrospinal fluid (CSF), fills the central canal and ventricles and surrounds the CNS.

The **peripheral nervous system (PNS)** consists of all the peripheral nerves and nervous tissue outside the CNS. The PNS provides sensory information to the CNS and carries motor commands from the CNS to peripheral tissues and systems. The PNS has two divisions, afferent and efferent (**Figure 13.2**). The **afferent division** of the PNS carries sensory information to the CNS. The afferent division begins at **receptors** that monitor specific characteristics of the environment. There are many forms of receptors. A receptor may be a sensory process, a specialized cell or cluster of cells, or a complex sense organ (such as the eye). The stimulation of a receptor carries information to the CNS. The afferent division also delivers information provided by special sense organs, such as the eye and ear. The **efferent division** of the PNS carries motor commands from the CNS to muscles and glands. The efferent division begins inside the CNS and ends at an **effector**: a muscle cell, gland cell, or another cell specialized to perform specific functions.

Both divisions have somatic and visceral components. The afferent division carries information from **somatic** sensory receptors that monitor skeletal muscles, joints, and the skin and from **visceral** sensory receptors that monitor other internal structures, such as smooth muscle, cardiac muscle, glands, and respiratory and digestive organs. The efferent division consists of the **somatic nervous system (SNS)**, which controls skeletal muscle contractions, and the **autonomic nervous system (ANS)**, or *visceral motor system*, which regulates smooth muscle, cardiac muscle, and glandular activity.

The activities of the somatic nervous system may be voluntary or involuntary. Voluntary contractions of our skeletal muscles are under conscious control; you exert voluntary control over your arm muscles as you raise a glass of

Figure 13.1 The Nervous System. The nervous system consists of all the nervous tissue in the body.

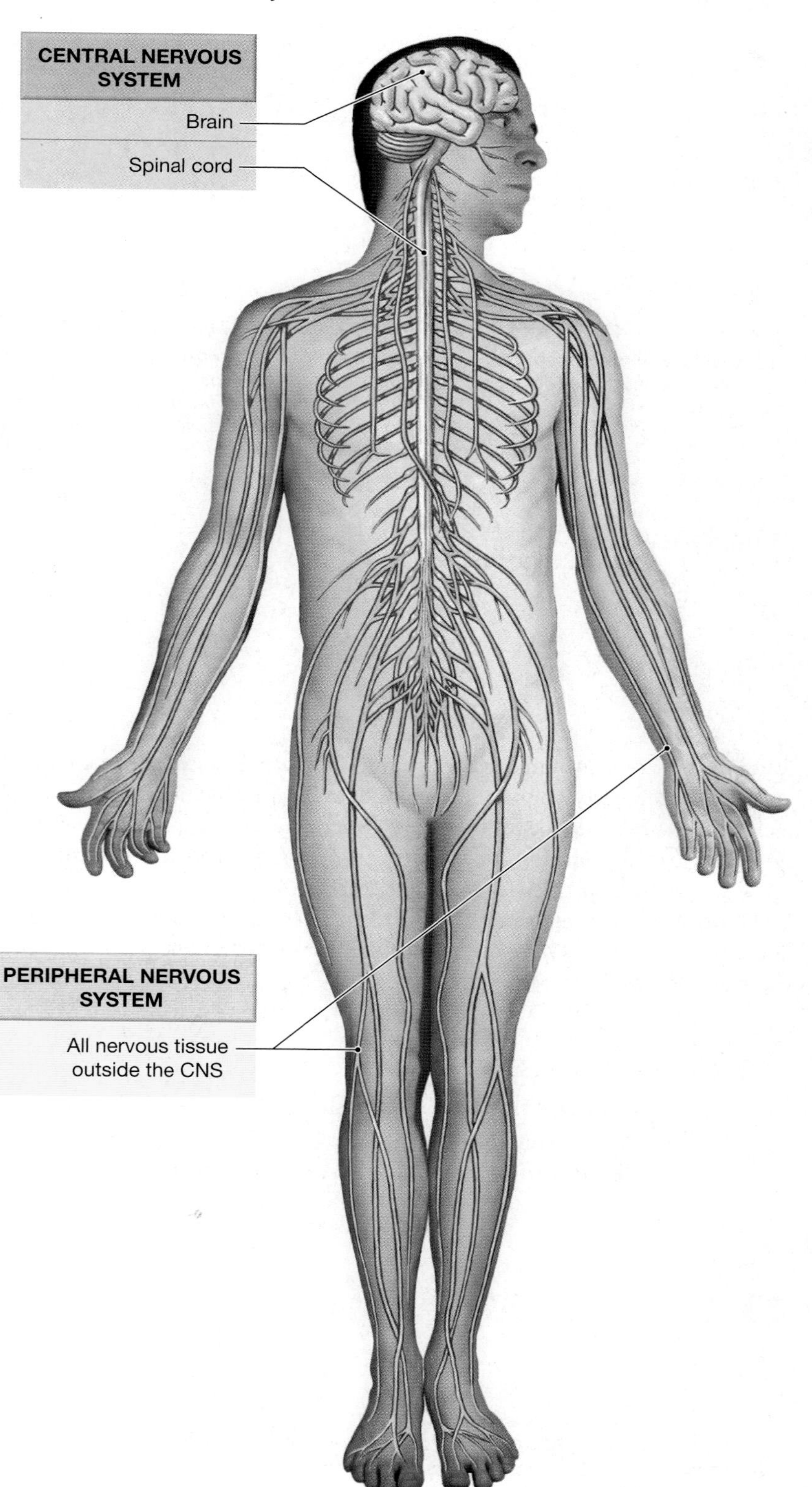

13

Figure 13.2 A Functional Overview of the Nervous System. This diagram shows the relationship between the CNS and PNS and the functions and components of the afferent and efferent divisions.

CENTRAL NERVOUS SYSTEM (CNS)
(brain and spinal cord)

Information processing

Sensory information within **afferent division**

Motor commands within **efferent division**

PERIPHERAL NERVOUS SYSTEM (PNS)
(nervous tissue outside CNS)

includes

Somatic nervous system (SNS)

Autonomic nervous system (ANS)

Parasympathetic division

Sympathetic division

Special sensory receptors monitor smell, taste, vision, balance, and hearing

Visceral sensory receptors monitor cardiovascular, respiratory, digestive, urinary, and reproductive systems

Somatic sensory receptors monitor skeletal muscles, joints, body position, skin surface; provide position sense and touch, pressure, pain, and temperature sensations

RECEPTORS

Skeletal muscle

- Smooth muscle
- Cardiac muscle
- Glands
- Adipose tissue

EFFECTORS

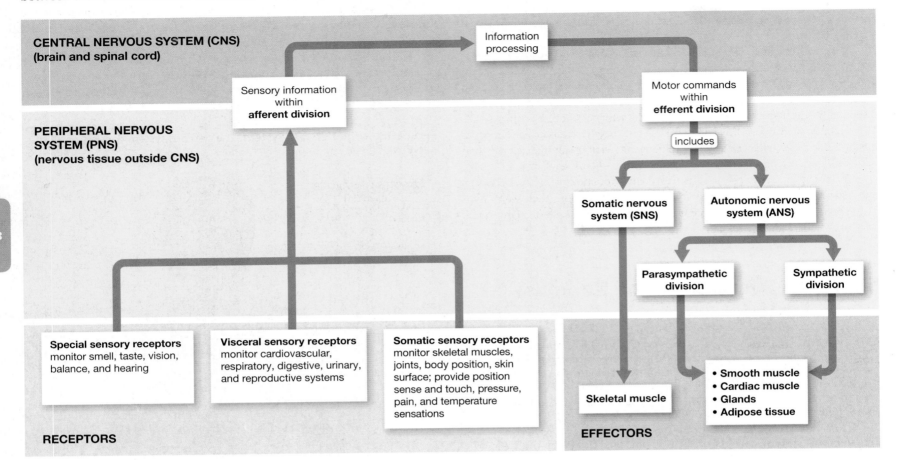

water to your lips. Involuntary contractions are not under conscious control; if you accidentally place your hand on a hot stove, you will withdraw it immediately, even before you notice the pain. The activities of the autonomic nervous system are usually outside our awareness or control, such as heartbeat, digestive processes, and our instinctive response to threatening situations.

Figure 13.3 summarizes the most important concepts and terms introduced in this chapter. Our discussion of the nervous system begins at the cellular level, with the histology of nervous tissue.

13.1 CONCEPT CHECK

1 What are the two subdivisions of the nervous system?

2 What are the two subdivisions of the efferent division of the peripheral nervous system?

See the blue Answers tab at the back of the book.

13.2 Cellular Organization in Nervous Tissue

▶ **KEY POINT** The nervous system contains only two types of cells: neurons and neuroglia. Neurons transfer and process information. Neuroglia isolate the neurons, provide a supporting framework, help maintain the intercellular environment, and act as phagocytes.

The structure of a **neuron** (*neuro*, nerve) was introduced in Chapter 3. ↺ p. 76 A "typical" neuron has a **cell body**, or *soma* (**Figure 13.4**). The region around the nucleus is the **perikaryon** (per-i-KAR-ē-on; *karyon*, nucleus). The cell body typically has several branching **dendrites** (sensory processes). In the CNS, dendrites are highly branched. Each branch has fine processes called **dendritic spines** that receive information from other neurons. Dendritic spines often comprise 80–90 percent of the neuron's total surface area.

The cell body is attached to an elongated **axon**, or *nerve fiber*, that ends at one or more **axon terminals**. The neuron communicates with another cell at these axon terminals.

Nervous tissue contains approximately 100 billion **neuroglia** (nū-RŌG-lē-a; *glia*, glue), also termed *glial cells*—approximately five times the number of neurons. Neuroglia are smaller than neurons, and, unlike neurons, they retain the ability to divide.

13.2 CONCEPT CHECK

3 What are the two distinct cell types found within nervous tissue?

4 Which of the two cell types found within nervous tissue has retained the ability to divide?

See the blue Answers tab at the back of the book.

Figure 13.3 An Introduction to Nervous System Terminology

MAJOR ANATOMICAL AND FUNCTIONAL DIVISIONS

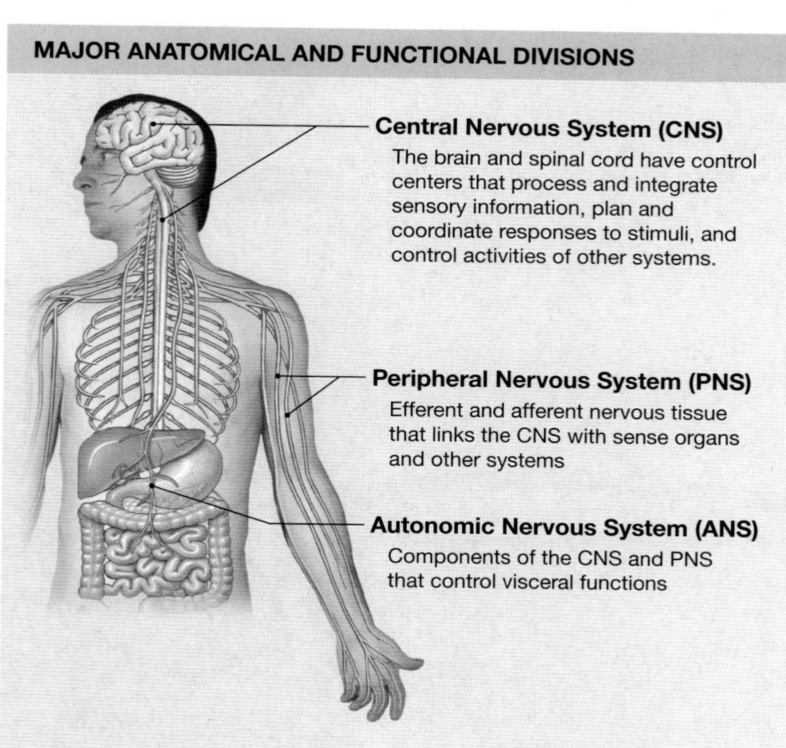

Central Nervous System (CNS)
The brain and spinal cord have control centers that process and integrate sensory information, plan and coordinate responses to stimuli, and control activities of other systems.

Peripheral Nervous System (PNS)
Efferent and afferent nervous tissue that links the CNS with sense organs and other systems

Autonomic Nervous System (ANS)
Components of the CNS and PNS that control visceral functions

GROSS ANATOMY

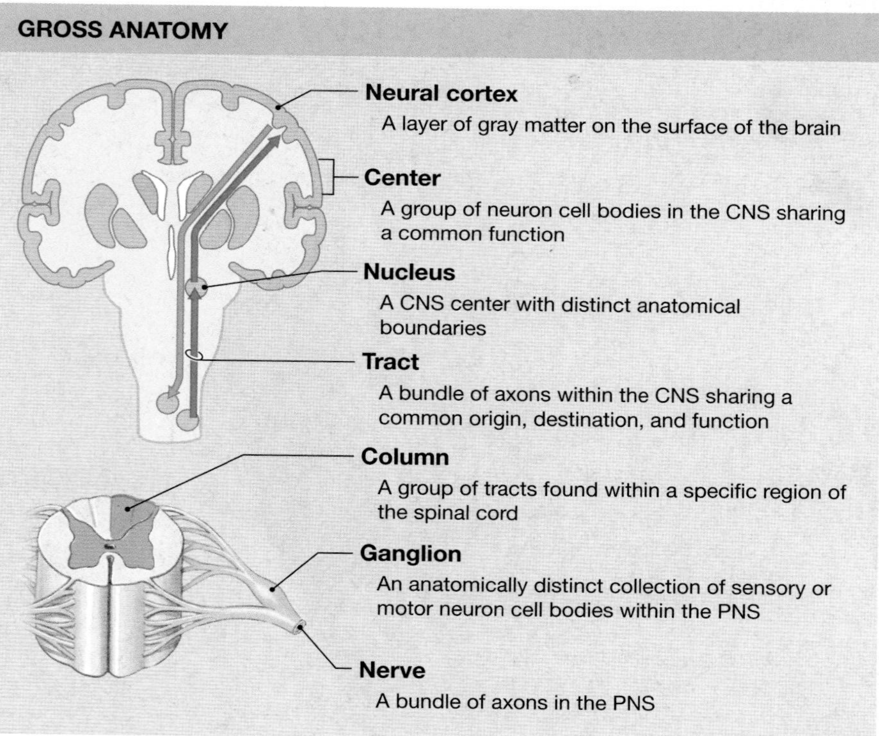

Neural cortex
A layer of gray matter on the surface of the brain

Center
A group of neuron cell bodies in the CNS sharing a common function

Nucleus
A CNS center with distinct anatomical boundaries

Tract
A bundle of axons within the CNS sharing a common origin, destination, and function

Column
A group of tracts found within a specific region of the spinal cord

Ganglion
An anatomically distinct collection of sensory or motor neuron cell bodies within the PNS

Nerve
A bundle of axons in the PNS

HISTOLOGY

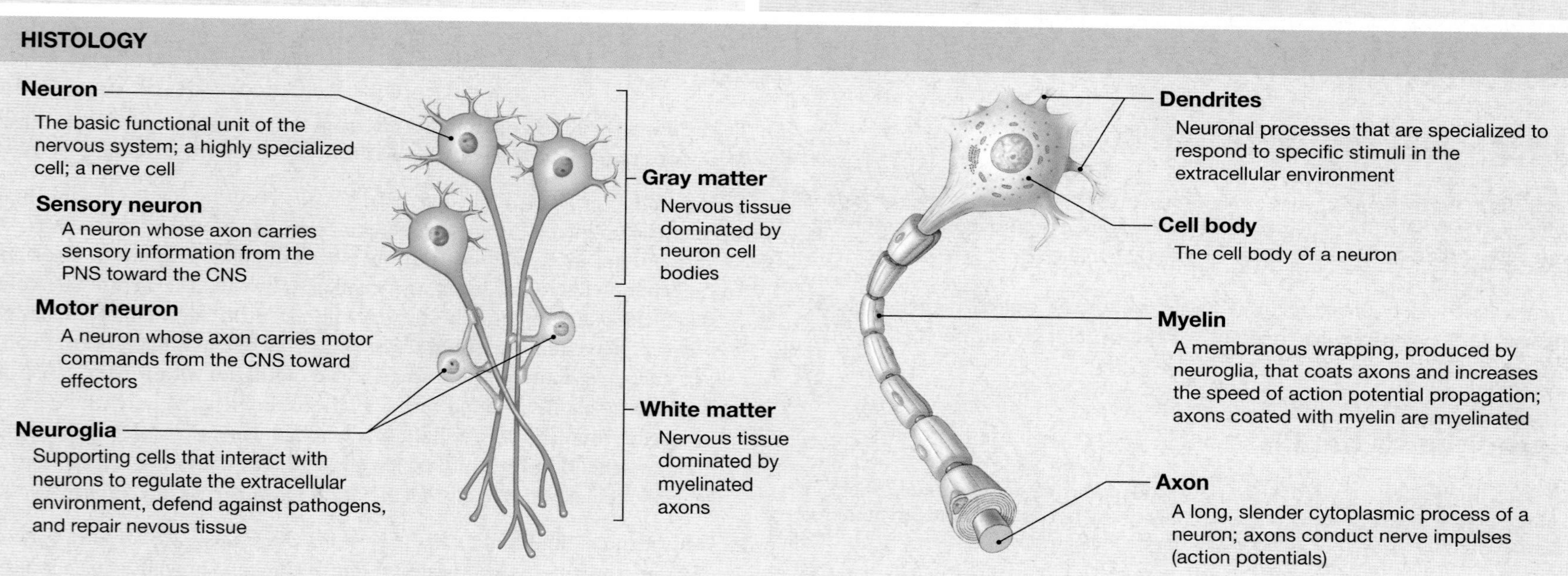

Neuron
The basic functional unit of the nervous system; a highly specialized cell; a nerve cell

Sensory neuron
A neuron whose axon carries sensory information from the PNS toward the CNS

Motor neuron
A neuron whose axon carries motor commands from the CNS toward effectors

Neuroglia
Supporting cells that interact with neurons to regulate the extracellular environment, defend against pathogens, and repair nevous tissue

Gray matter
Nervous tissue dominated by neuron cell bodies

White matter
Nervous tissue dominated by myelinated axons

Dendrites
Neuronal processes that are specialized to respond to specific stimuli in the extracellular environment

Cell body
The cell body of a neuron

Myelin
A membranous wrapping, produced by neuroglia, that coats axons and increases the speed of action potential propagation; axons coated with myelin are myelinated

Axon
A long, slender cytoplasmic process of a neuron; axons conduct nerve impulses (action potentials)

FUNCTIONAL CATEGORIES

Receptors
A specialized cell, dendrite, or organ that responds to specific stimuli in the extra-cellular environment and whose stimulation alters the level of activity in a sensory neuron

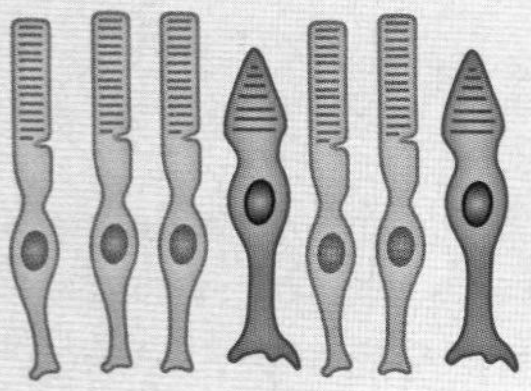

Effectors
A muscle, gland, or other specialized cell or organ that responds to neural stimulation by altering its activity and producing a specific effect

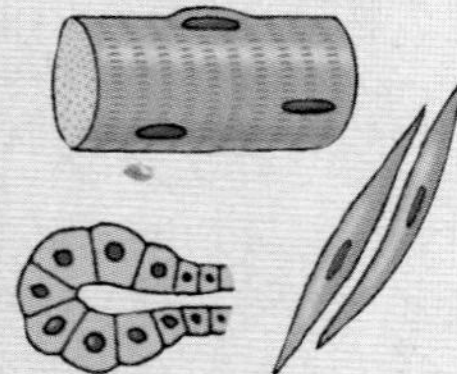

Reflexes
A rapid, stereotyped response (always the same) to a specific stimulus

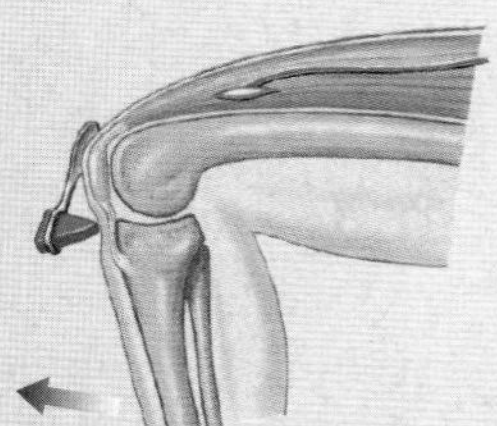

Other Functional Terminology

Somatic: Pertaining to the control of skeletal muscle activity (somatic motor) or sensory information from skeletal muscles, tendons, and joints (somatic sensory)

Visceral: Pertaining to the innervation of visceral organs, such as digestion, circulation, etc. (visceral motor) or sensory information from visceral organs (visceral sensory)

Voluntary: Under direct conscious control

Involuntary: Not under direct conscious control

Subconscious: Relating to centers in the brain that operate outside a person's conscious awareness

Action Potential: Sudden, transient changes in the membrane potential that are propagated along the surface of an axon or sarcolemma

Figure 13.4 **A Review of Neuron Structure.** The relationship of the four parts of a neuron (dendrites, cell body, axon, and axon terminals); the functional activities of each part and the normal direction of action potential conduction are shown.

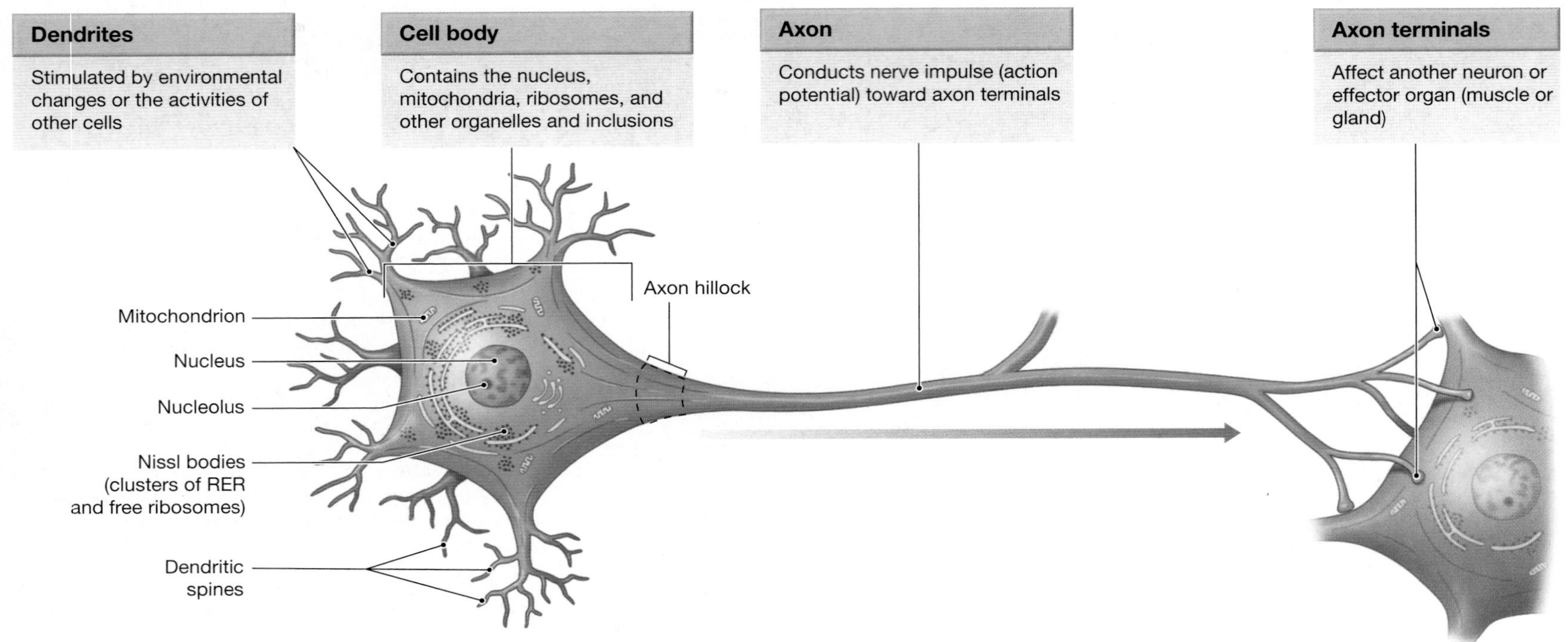

Dendrites
Stimulated by environmental changes or the activities of other cells
Cell body
Contains the nucleus, mitochondria, ribosomes, and other organelles and inclusions
Axon
Conducts nerve impulse (action potential) toward axon terminals
Axon terminals
Affect another neuron or effector organ (muscle or gland)
Axon hillock
Mitochondrion
Nucleus
Nucleolus
Nissl bodies (clusters of RER and free ribosomes)
Dendritic spines

13

13.3 | Neuroglia

▶ **KEY POINT** The four types of neuroglia in the CNS are astrocytes, oligodendrocytes, microglia, and ependymal cells. The two types of neuroglia in the PNS are satellite cells and Schwann cells.

The greatest variety of neuroglia is found within the central nervous system. **Figure 13.5** compares the functions of the major types of neuroglia in the CNS and PNS.

Neuroglia of the CNS

▶ **KEY POINT** Neuroglia of the CNS surround CNS neurons and hold them in place, isolate neurons from each other, supply oxygen and nutrients to neurons, destroy pathogens, and remove dead or damaged neurons.

There are four types of neuroglia within the central nervous system: astrocytes, oligodendrocytes, microglia, and ependymal cells. These cell types are distinguished by size, intracellular organization, the presence of specific cytoplasmic processes, and staining properties **(Figures 13.5 to 13.7)**.

Astrocytes

The largest and most numerous neuroglia are the **astrocytes** (AS-trō-sīts; *astro-*, star, + *-cyte*, cell) **(Figures 13.5 and 13.6)**. Astrocytes' functions, although still poorly understood, are varied:

- Control the interstitial environment. Astrocytes have a large number of cytoplasmic processes, termed pedicels (or *feet*). These processes increase their surface area, thus aiding the uptake of ions, neurotransmitters, or metabolic by-products accumulating around the neurons, which enables them to control the chemical content of the interstitial space. The cytoplasmic processes also contact the surfaces of adjacent neurons, often enclosing the entire neuron and isolating it from changes in the chemical composition of the interstitial space.
- Maintain the blood brain barrier. Nervous tissue is physically and biochemically isolated from the general circulation of the body because hormones or other chemicals in the blood could disrupt neuron function. The endothelial cells lining the capillaries within the central nervous system are quite impermeable and therefore control the chemical exchange between blood and interstitial fluid. Astrocytes' cytoplasmic processes contact and cover most of the surface of the capillaries within the CNS. This cytoplasmic blanket around the capillaries is interrupted only where other neuroglia contact the capillary walls. Chemicals secreted by astrocytes maintain the **blood brain barrier (BBB)** that isolates the CNS from the general circulation. (We discuss the blood brain barrier in Chapter 16.)
- Form a three-dimensional framework for the CNS. Astrocytes are packed with microfilaments that extend across the cell, providing mechanical strength and forming a structural framework that supports the neurons of the brain and spinal cord.
- Repair damaged nervous tissue. Astrocytes make structural repairs, stabilizing the tissue and preventing further injury by producing scar tissue at the injury site.
- Guide neuron development. In the embryonic brain, astrocytes appear to be involved in directing the growth and interconnection of developing neurons through the secretion of chemicals known as neurotropic factors.

Oligodendrocytes

A second type of neuroglia within the CNS is the **oligodendrocyte** (ōl-i-gō-DEN-drō-sīt; *oligo-*, few). Like astrocytes, these cells possess slender cytoplasmic extensions. However, oligodendrocytes have smaller cell bodies and fewer and shorter cytoplasmic processes **(Figures 13.5 and 13.6)**.

Figure 13.5 **The Classification of Neuroglia.** This flowchart summarizes the categories and functions of the various types of neuroglia.

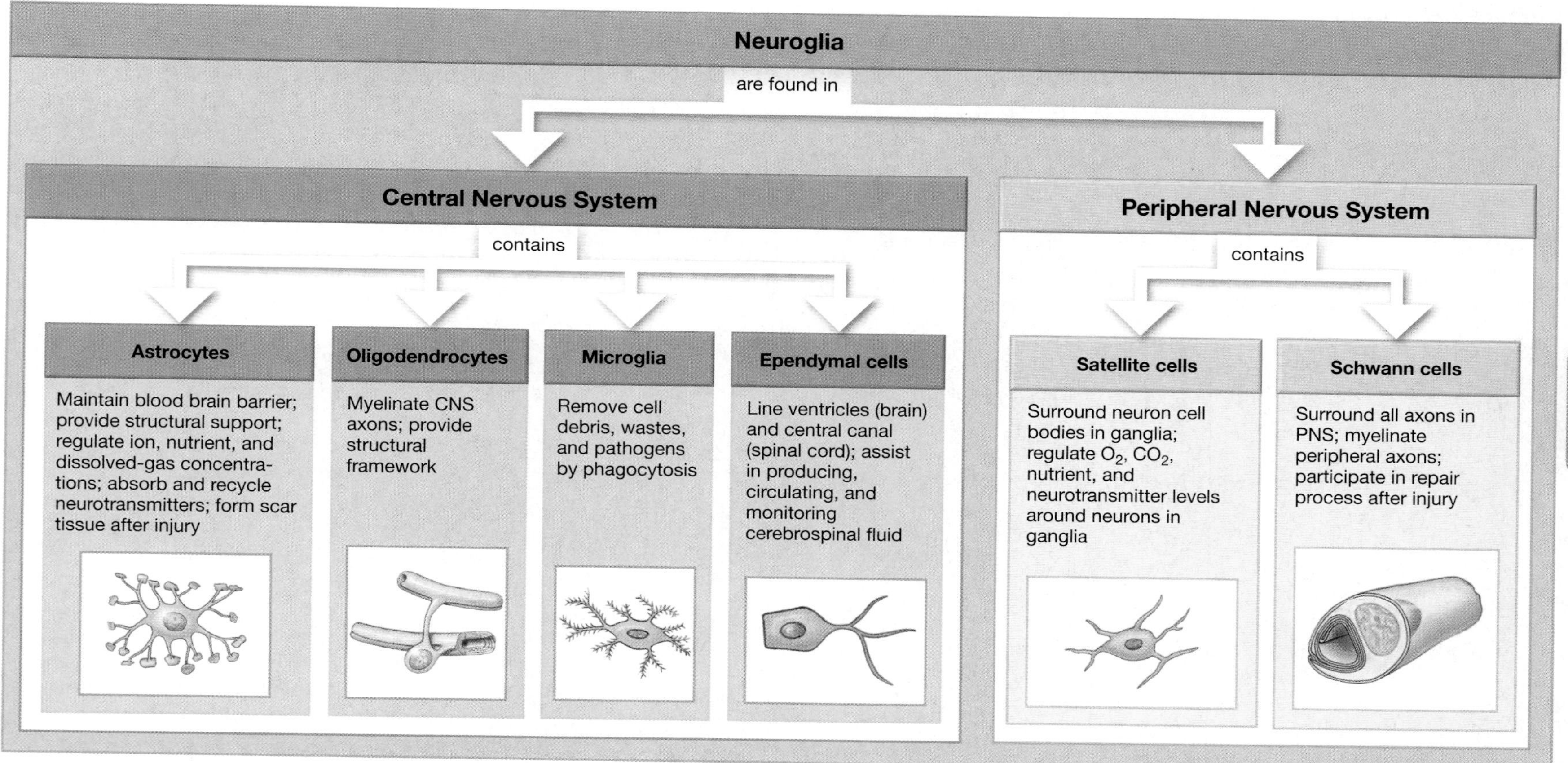

Oligodendrocyte processes contact the axons or cell bodies of neurons and tie clusters of axons together, improving the performance of neurons by wrapping axons in myelin, a material with insulating properties.

Oligodendrocytes cooperate in forming the myelin sheath along the entire length of a myelinated axon within the CNS (Figure 13.6 and **Spotlight Figure 13.9**). The large areas wrapped in myelin are called **internodes** (*inter-*, between). Small gaps between the myelin sheaths produced by adjacent oligodendrocytes are called **myelin sheath gaps**, or *nodes of Ranvier* (rahn-vē-Ā). Any region of the CNS dominated by myelinated axons is called **white matter**, and any region dominated by neuron cell bodies, dendrites, and unmyelinated axons is called **gray matter**.

Microglia

The smallest neuroglia possess slender cytoplasmic processes with many fine branches (Figures 13.5 and 13.6). These cells, called **microglia** (mī-KRŌ-glē-a), appear early in development through the division of mesodermal stem cells. The stem cells producing microglia originate in the bone marrow and are related to stem cells that produce tissue macrophages and monocytes of the blood. Microglia migrate into the CNS as it forms and remain within the nervous tissue, acting as a roving security force. They are the phagocytic cells of the CNS, engulfing cellular debris and wastes. Microglia also protect the CNS by phagocytosing viruses, microorganisms, and tumor cells. Only 5 percent of the CNS neuroglia are microglia, but when the CNS is infected or injured, this percentage increases dramatically.

Ependymal Cells

A cellular layer called the **ependyma** (e-PEN-di-mah) lines the ventricles of the brain and central canal of the spinal cord (Figures 13.5 to 13.7). These chambers and passageways are filled with **cerebrospinal fluid (CSF)**. This fluid also surrounds the brain and spinal cord, providing a protective cushion and transporting dissolved gases, nutrients, wastes, and other materials. (Chapter 16 discusses the composition, formation, and circulation of CSF.)

Ependymal cells are cuboidal to columnar in form. Unlike typical epithelial cells, ependymal cells have slender processes that branch extensively and make direct contact with neuroglia in the surrounding nervous tissue (Figure 13.7a). Ependymal cells may act as receptors monitoring the composition of the CSF. In the adult, cilia and microvilli are found on the apical surface of the ependymal cells lining the spinal cord and the lateral and fourth ventricles of the brain (Figure 13.7b). Ependymal cells lining the third ventricle lack cilia. The cilia help the CSF circulate, and the microvilli are involved in the absorption of CSF.

Neuroglia of the PNS

▶ **KEY POINT** Satellite cells and Schwann cells of the PNS have functions similar to those of the astrocytes and oligodendroglia of the CNS.

Neuron cell bodies in the PNS are clustered together in structures called **ganglia** (singular, *ganglion*). Axons are bundled together and wrapped in connective tissue, forming **peripheral nerves**. The processes of neuroglia insulate all neuron cell bodies and axons in the PNS from their surroundings. The two neuroglia types involved are satellite cells and Schwann cells.

Satellite Cells

Satellite cells surround neuron cell bodies in peripheral ganglia (Figure 13.8). Satellite cells regulate the exchange of nutrients and waste products between the neuronal cell body and extracellular fluid. They also isolate the neuron from stimuli not intended to pass from neuron to neuron.

Figure 13.6 Histology of Nervous Tissue in the CNS. A diagrammatic view of nervous tissue in the spinal cord, showing relationships between neurons and neuroglia.

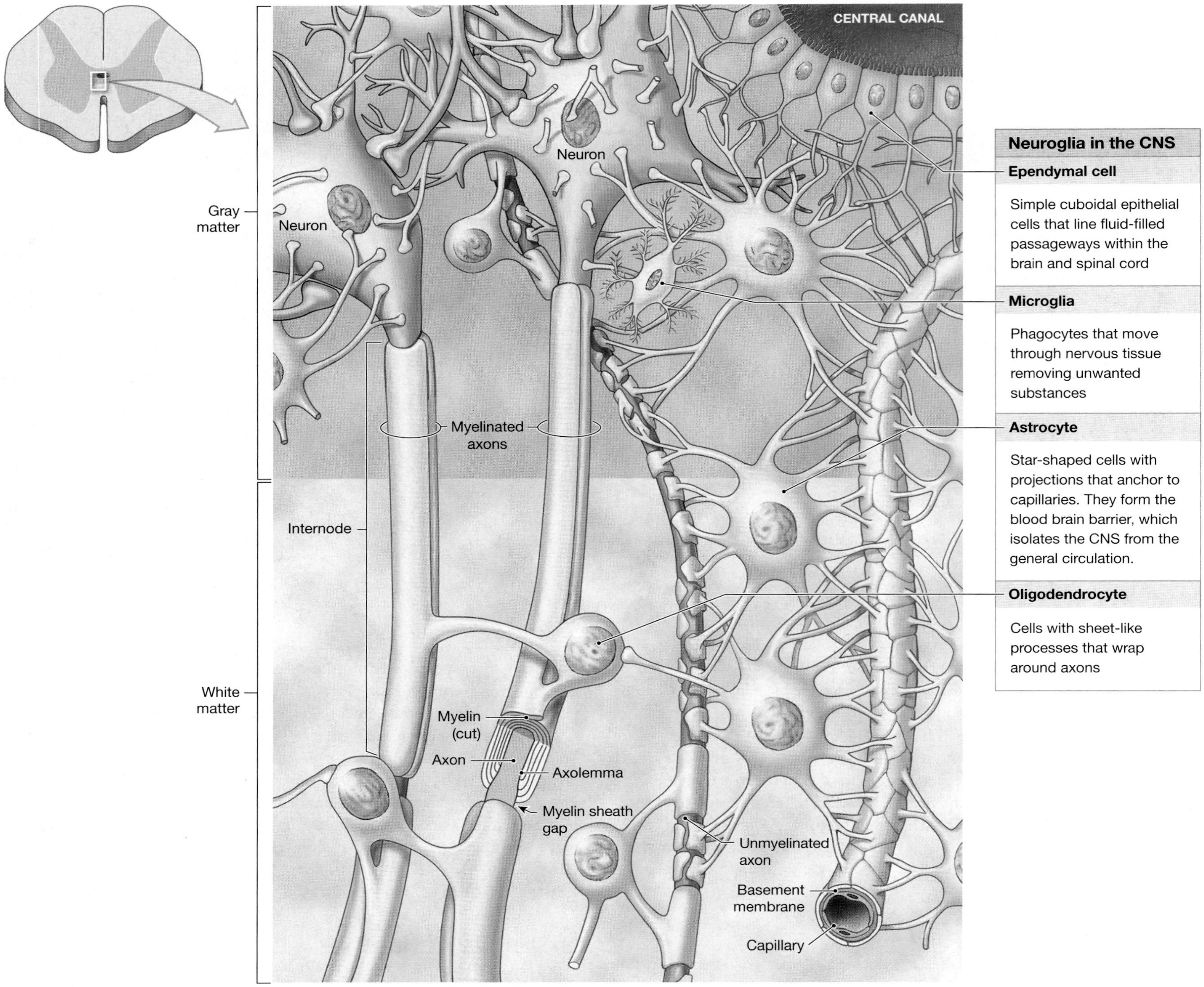

CENTRAL CANAL
Gray matter
White matter
Neuron
Neuron
Myelinated axons
Internode
Myelin (cut)
Axon
Axolemma
Myelin sheath gap
Unmyelinated axon
Basement membrane
Capillary
Neuroglia in the CNS
Ependymal cell
Simple cuboidal epithelial cells that line fluid-filled passageways within the brain and spinal cord
Microglia
Phagocytes that move through nervous tissue removing unwanted substances
Astrocyte
Star-shaped cells with projections that anchor to capillaries. They form the blood brain barrier, which isolates the CNS from the general circulation.
Oligodendrocyte
Cells with sheet-like processes that wrap around axons

Schwann Cells

Every peripheral axon, whether myelinated or unmyelinated, is covered by **Schwann cells**, or *neurolemmocytes*. The plasma membrane of an axon is the **axolemma** (*lemma*, husk). The cytoplasmic covering provided by the Schwann cells is the **neurolemma** (nū-RŌ-LEM-a). **Spotlight Figure 13.9** outlines the physical relationships between Schwann cells and myelinated and unmyelinated peripheral axons.

13.3 CONCEPT CHECK

✔

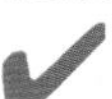

5 Specifically, what cells help maintain the blood brain barrier (BBB)?

6 What is the name of the membranous coating formed by oligodendrocytes around CNS axons ?

See the blue Answers tab at the back of the book.

Figure 13.7 The Ependyma. The ependyma is a cellular layer that lines the ventricles of the brain and the central canal of the spinal cord.

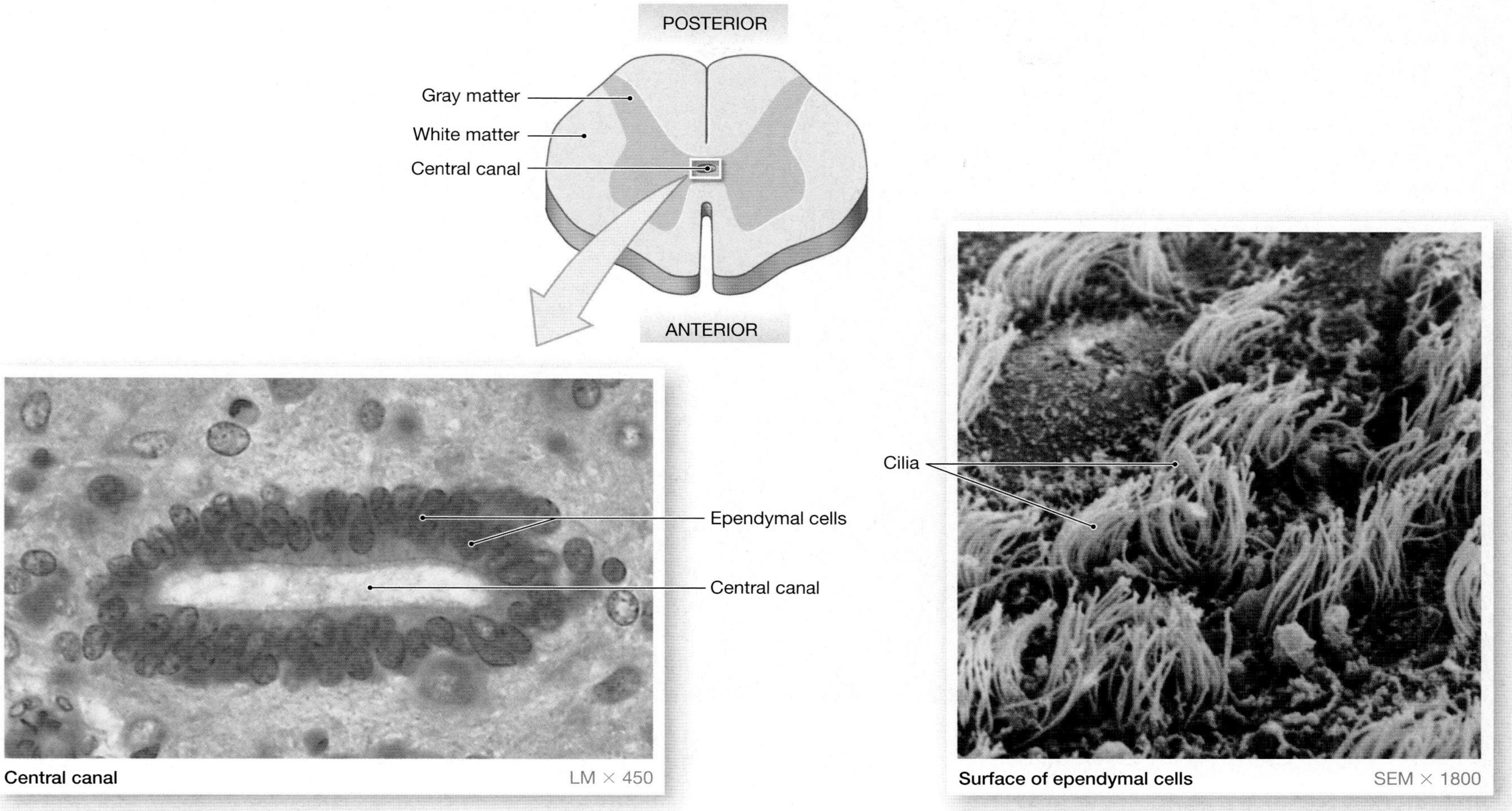

POSTERIOR

Gray matter

White matter

Central canal

ANTERIOR

Ependymal cells

Central canal

Central canal LM × 450

a Light micrograph of ependymal cells lining the central canal.

Cilia

Surface of ependymal cells SEM × 1800

b An SEM of the ciliated surface of the ependyma from the central canal.

Figure 13.8 Satellite Cells and Peripheral Neurons. Satellite cells surround neuron cell bodies in peripheral ganglia.

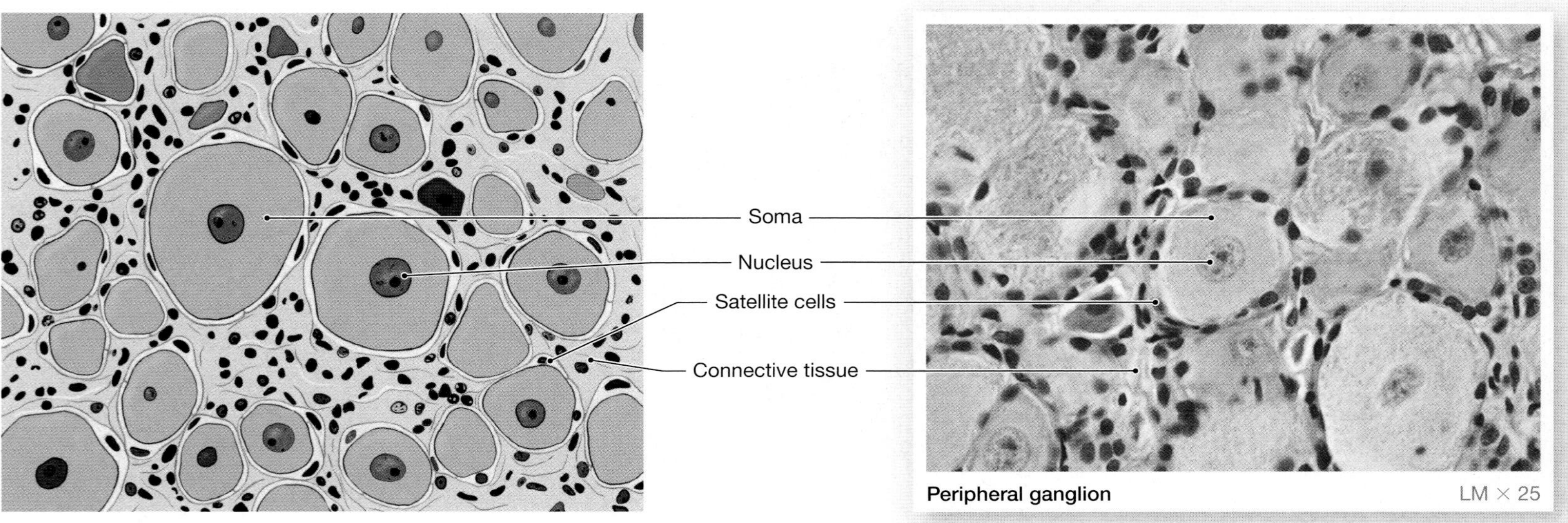

SPOTLIGHT

FIGURE 13.9

Myelination

All peripheral nervous system axons are shielded from contact with interstitial fluids by neuroglia called **Schwann cells.** Schwann cells ensheath many axons in a layered phospholipid covering known as **myelin**. An axon wrapped in myelin is said to be **myelinated**. Myelin improves the conduction speed of an **action potential**, or *nerve impulse*, along the axon. An individual Schwann cell produces myelin by wrapping itself around the axon as diagrammed below.

Axon Myelination in the PNS

1 In myelinating a peripheral axon, a Schwann cell first encloses a segment of the axon within a groove of its cytoplasm.

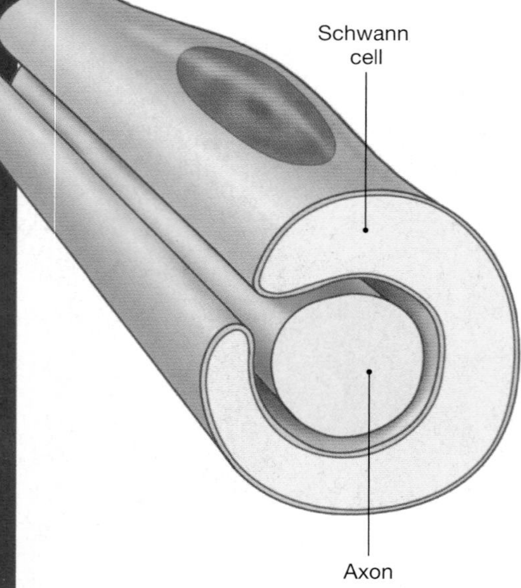

2 The Schwann cell then rotates around the axon.

3 As the Schwann cell rotates, the inner membranous layers are compressed and the cytoplasm is forced into more superficial layers. When completed, the myelin sheath consists only of the phospholipid bilayers of the plasma membrane, with the Schwann cell nucleus and cytoplasm at the surface.

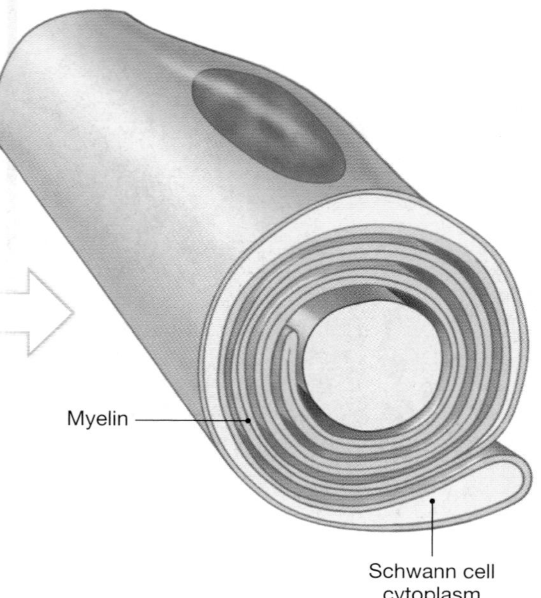

A Myelinated Axon in the PNS

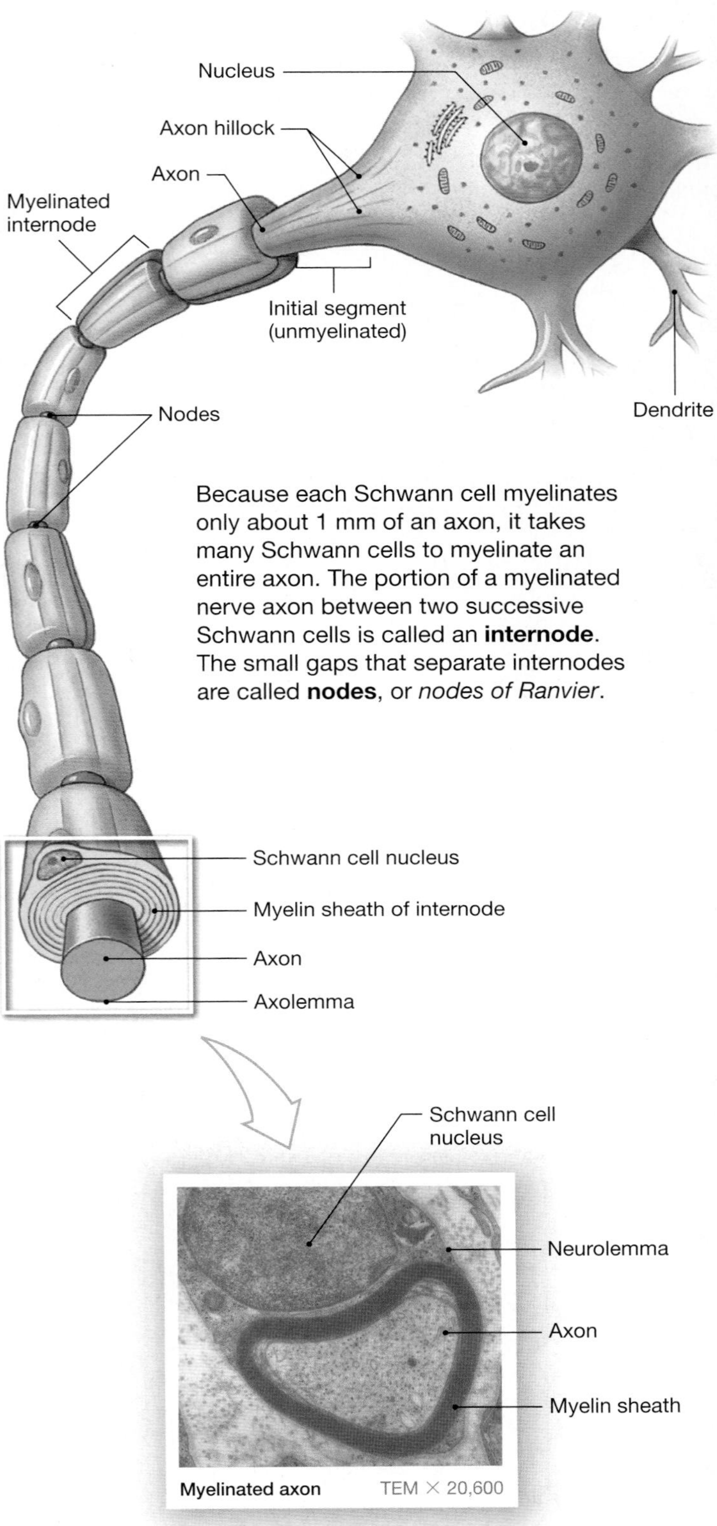

Because each Schwann cell myelinates only about 1 mm of an axon, it takes many Schwann cells to myelinate an entire axon. The portion of a myelinated nerve axon between two successive Schwann cells is called an **internode**. The small gaps that separate internodes are called **nodes**, or *nodes of Ranvier*.

In this cross section of a myelinated axon, the myelin sheath appears as concentric dense lines around the axon.

An Unmyelinated Axon in the PNS

A single Schwann cell shields multiple unmyelinated axons within superficial grooves. However, a chain of Schwann cells is still needed to enclose an entire axon.

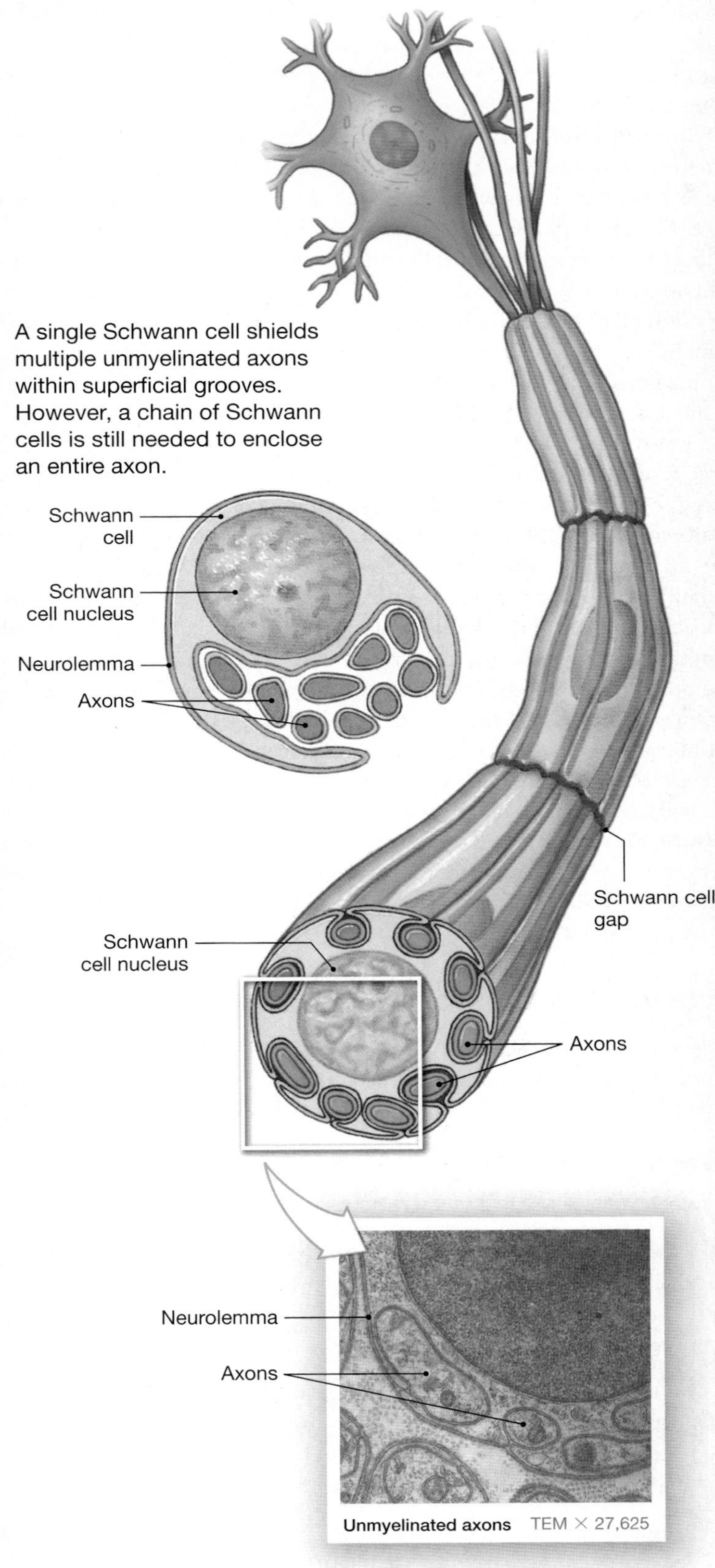

13.4 Neurons

▶ **KEY POINT** Neurons, the functional units of the nervous system, transmit information from one part of the nervous system to another by electrical impulses.

13

The cell body of a representative neuron contains a large, round nucleus with a prominent nucleolus. The cytoplasm of a neuron is the **perikaryon**. The cytoskeleton of the perikaryon contains **neurofilaments** and **neurotubules**, which are similar to the intermediate filaments and microtubules of other types of cells. Bundles of neurofilaments, called **neurofibrils**, extend into the dendrites and axons, providing internal support.

The perikaryon contains organelles that provide energy and synthesize organic materials. The mitochondria, free and fixed ribosomes, and membranes of the rough endoplasmic reticulum (RER) give the perikaryon a coarse, grainy appearance. Mitochondria generate ATP to meet the high energy demands of an active neuron; the ribosomes and RER synthesize proteins. Some areas of the perikaryon contain clusters of free ribosomes and RER. These regions, which stain a dark color, are called **Nissl bodies** (or *chromatophilic substance*), after the German neurologist Franz Nissl, who first described them. Nissl bodies give a gray color to areas containing neuronal cells bodies—the gray matter seen in gross dissection of the brain or spinal cord **(Figure 13.10)**.

Most neurons lack a centrosome. In other cells, the centrioles of the centrosome form the spindle fibers that move chromosomes during cell division. Neurons lose their centrosomes and centrioles during differentiation and therefore are unable to undergo cell division. Neurons lost to injury or disease cannot be replaced.

An axon is a long cytoplasmic process capable of propagating an action potential. In a multipolar neuron, a specialized region, the **axon hillock**, connects the **initial segment** (base) of the axon to the cell body. The **axoplasm** (AK-sō-plazm), or cytoplasm of the axon, contains neurofibrils, neurotubules, lysosomes, mitochondria, numerous small vesicles, and various enzymes. An axon may branch along its length, producing side branches called **collaterals (Figure 13.14b)**. The axon and collaterals end in fine terminal extensions called **telodendria** (tel-ō-DEN-drē-a; *telo-*, end, + *dendron*, tree). The telodendria end in **axon terminals**, where the neuron contacts another cell **(Figure 13.10)**. Organelles, nutrients, synthesized molecules, and wastes move between the cell body and the axon terminals by a process termed **axoplasmic transport**. This complex process consumes energy and relies on the neurofibrils of the axon and its branches.

Neuron Classification

▶ **KEY POINT** Neurons are classified into four structural groups based on the number of processes that extend from the cell body and into three functional groups based on their roles in the CNS and PNS.

Structural Classification of Neurons

Figure 13.11 shows the four structural groups of neurons.

- **Anaxonic** (an-ak-SON-ik) **neurons** are small. In these neurons it is very difficult to distinguish dendrites from axons **(Figure 13.11a)**. Anaxonic neurons are found only in the CNS and in special sense organs, and their functions are poorly understood.
- **Bipolar neurons** have a number of fine dendrites that fuse to form a single dendrite. The cell body lies between this single dendrite and the single axon **(Figure 13.11b)**. Bipolar neurons relay sensory information concerning sight, smell, and hearing. Their axons are not myelinated.
- **Pseudounipolar** (SŪ-dō-yū-ne-PŌ-lar) **neurons** have continuous dendrites and axons, and the cell body lies off to one side **(Figure 13.11c)**. In these neurons, the initial segment lies where the dendrites converge, and the rest of the process is usually considered an axon. Sensory neurons of the peripheral nervous system are pseudounipolar, and their axons may be myelinated.
- **Multipolar neurons** have several dendrites and a single axon with one or more branches **(Figure 13.11d)**. Multipolar neurons are the most common type of neuron in the CNS. An example of a multipolar neuron is a motor neuron that connects the CNS to skeletal muscles.

Functional Classification of Neurons

Neurons can be classified into three functional groups: (1) sensory neurons, (2) motor neurons, and (3) interneurons **(Figure 13.12)**.

Figure 13.10 Anatomy of a Representative Neuron. A neuron has a cell body (soma), some branching dendrites, and a single axon.

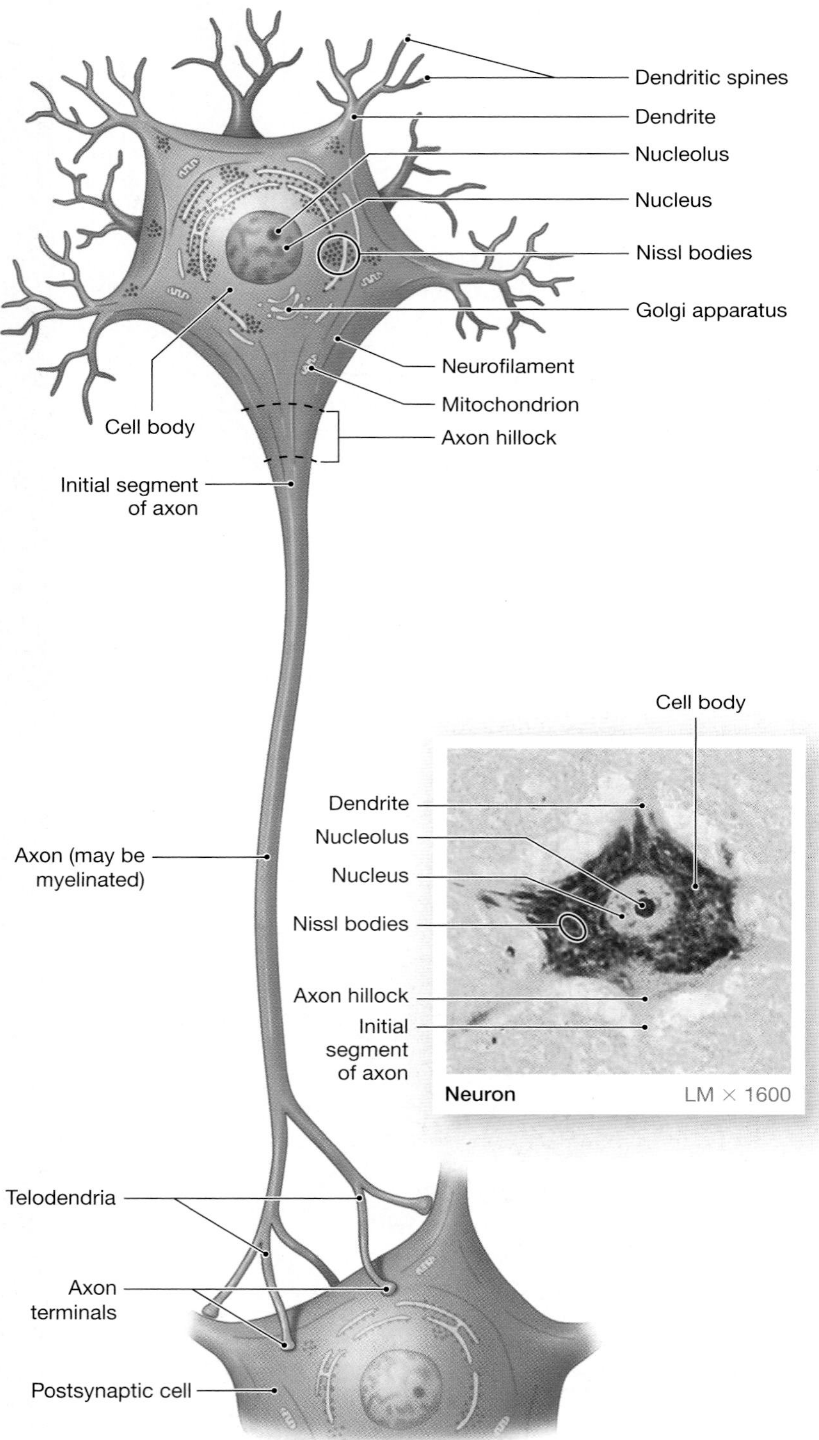

Figure 13.11 A Structural Classification of Neurons. This classification is based on the placement of the cell body and the number of associated processes.

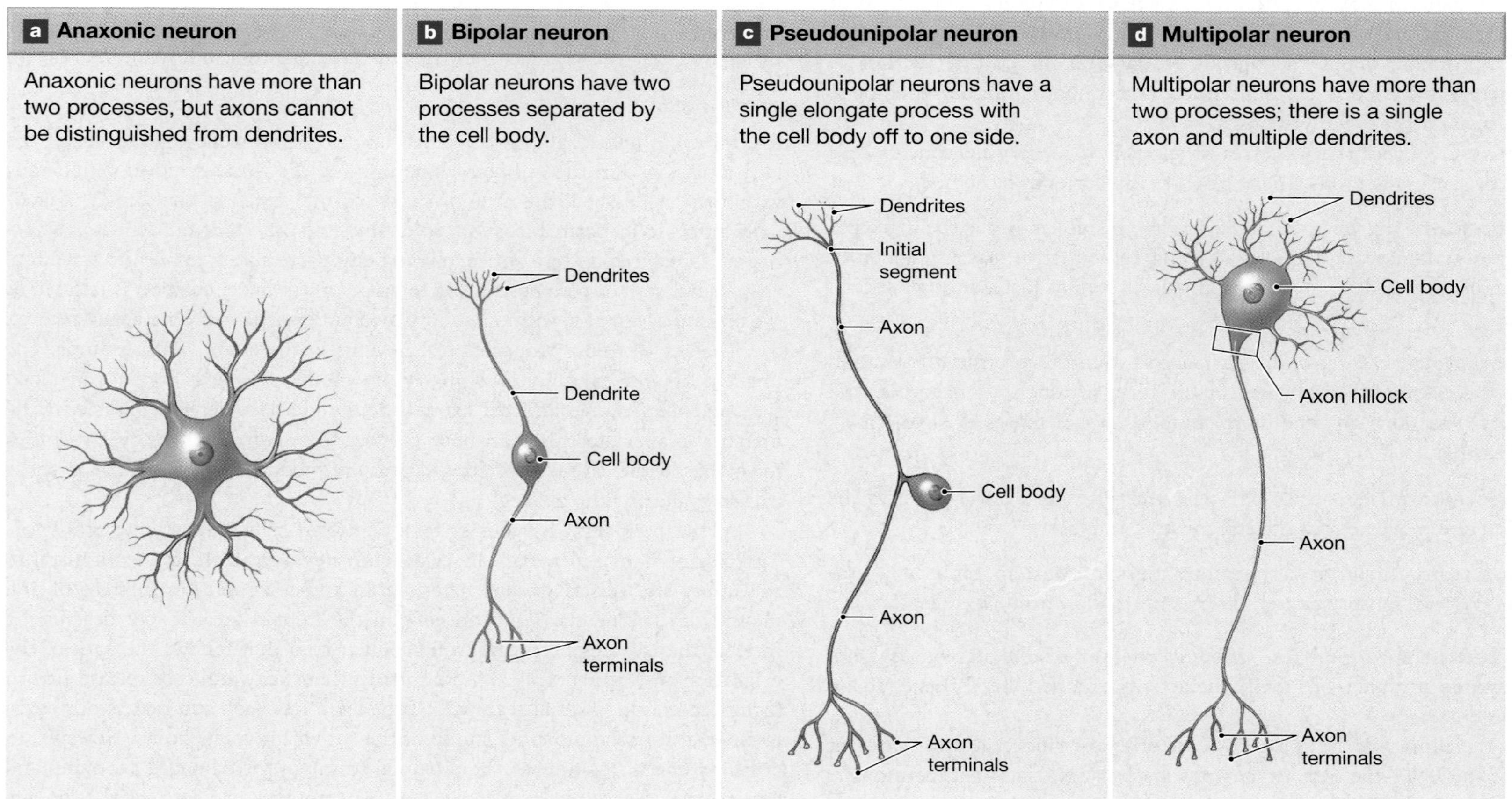

Figure 13.12 A Functional Classification of Neurons. Neurons are classified functionally into three categories: (1) sensory neurons that detect stimuli in the PNS and send information to the CNS, (2) motor neurons to carry instructions from the CNS to peripheral effectors, and (3) interneurons in the CNS that process sensory information and coordinate motor activity.

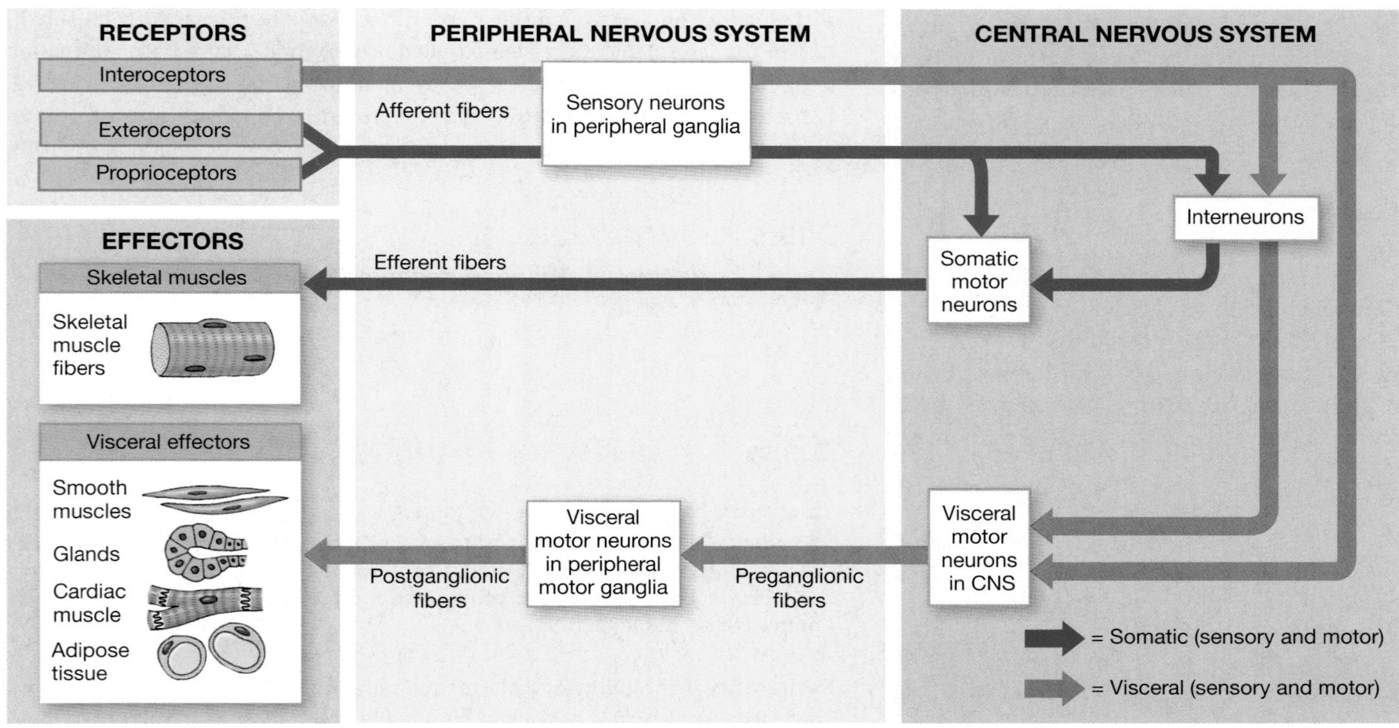

Sensory Neurons Forming the afferent division of the PNS, **sensory neurons** deliver information about the external or internal environment to the CNS. The axons of sensory neurons, called **afferent fibers**, extend between a sensory receptor and the spinal cord or brain. **Somatic sensory neurons** transmit information about the outside world and our position within it. **Visceral sensory neurons** transmit information about internal conditions and the status of other organ systems.

Receptors are either the processes of specialized sensory neurons or cells monitored by sensory neurons. There are three categories of receptors:

- **Interoceptors** (IN-ter-ō-SEP-ters; *intero-*, inside) monitor the digestive, respiratory, cardiovascular, urinary, and reproductive systems and provide sensations of deep pressure and pain as well as taste, another special sense.
- **Exteroceptors** (EKS-ter-ō-SEP-ters; *extero-*, outside) provide information about the external environment in the form of touch, temperature, and pressure sensations and the more complex special senses of sight, smell, and hearing.
- **Proprioceptors** (PRŌ-prē-ō-SEP-ters; *proprius*, one's own) monitor the position and movement of skeletal muscles and joints.

13

Somatic sensory neurons carry information from exteroceptors and proprioceptors. Visceral sensory neurons carry information from interoceptors.

Motor Neurons Forming the efferent division of the nervous system, **motor neurons** stimulate or modify the activity of a peripheral tissue, organ, or organ system.

Axons traveling away from the CNS are **efferent fibers**. The two efferent divisions of the PNS—the somatic nervous system (SNS) and autonomic nervous system (ANS)—differ in the way they innervate peripheral effectors. The SNS consists of all the **somatic motor neurons** innervating skeletal muscles. The cell bodies of these motor neurons lie inside the CNS, and their axons extend to the neuromuscular junctions that control skeletal muscles. Most of the activities of the SNS are consciously controlled.

The autonomic nervous system (discussed in Chapter 17) consists of all the **visceral motor neurons** innervating peripheral effectors other than skeletal muscles. There are two groups of visceral motor neurons: One group has cell bodies inside the CNS, and the other has cell bodies in peripheral ganglia. The neurons inside the CNS control the neurons in the peripheral ganglia, and these neurons in turn control the peripheral effectors. Axons extending from the CNS to a ganglion are called **preganglionic fibers**. Axons connecting the ganglion cells with peripheral effectors are **postganglionic fibers**. This arrangement clearly distinguishes the autonomic (visceral motor) system from the somatic motor system. We have little conscious control over the activities of the ANS.

Interneurons Located between the sensory and motor neurons within the brain and spinal cord, **interneurons** analyze sensory input and coordinate motor output. The more complex the response to a stimulus, the greater the number of interneurons involved. Interneurons are classified as **excitatory** (releasing excitatory neurotransmitters) or **inhibitory** (releasing inhibitory neurotransmitters) based on their effects on the postsynaptic membranes of other neurons.

13.4 CONCEPT CHECK

7 Examination of a tissue sample shows pseudounipolar neurons. Are these more likely to be sensory neurons or motor neurons?

See the blue Answers tab at the back of the book.

13.5 | Regeneration of Nervous Tissue

▶ **KEY POINT** Regeneration of nervous tissue refers to the repair of complete neurons, neuroglia, axons, myelin, or synapses. Regeneration in the CNS differs from that in the PNS in mechanism, speed of regeneration, and amount of regeneration that occurs.

A neuron has limited ability to recover after an injury. Following an injury, the Nissl bodies within the cell body disappear and the nucleus moves peripherally within the cell body. If the neuron regains normal function, the soma will gradually return to its normal location. Sometimes the oxygen or nutrient supply to a neuron is reduced, as in a stroke, or mechanical pressure is applied to a neuron, as in spinal cord or peripheral nerve injuries. Unless the circulation is restored or the pressure is removed within a short period of time, the neuron may not recover.

The key to recovery appears to be events in the axon. If, for example, the pressure applied during a crushing injury produces a local decrease in blood flow and oxygen, the affected axonal membrane becomes unexcitable. If the pressure is alleviated after an hour or two, the neuron will recover within a few weeks. More severe or prolonged pressure produces effects similar to those caused by cutting the axon.

In the peripheral nervous system, Schwann cells play an important role in repairing damaged nerves. In **Wallerian degeneration**, the axon distal to the injury site (distal stump) deteriorates, and macrophages migrate in and phagocytize the debris. Schwann cells in the injured area do not degenerate; instead, they divide and form a solid cellular cord that follows the path of the original axon (**Figure 13.13**). Additionally, these Schwann cells release growth factors promoting axonal regrowth. If the axon has been cut, new axons begin to emerge from the proximal stump of the cut within a few hours. However, in crushing or tearing injuries, the proximal stump of the damaged axon dies and regresses for 1 centimeter or more, and the sprouting of new axonal segments is usually delayed for 1 or more weeks. As the neuron continues to recover, the axon grows into the injury site, and the Schwann cells wrap around it.

If the axon continues to grow into the periphery within the appropriate cord of Schwann cells, it may reestablish its normal synaptic contacts. If it stops growing or wanders off in a new direction, normal function will not return. The growing axon will arrive at its appropriate destination if the damaged proximal and distal stumps remain in contact after the injury. When an entire peripheral nerve is damaged, only a small number of axons will successfully reestablish normal synaptic contacts. As a result, nerve function will be permanently impaired.

In the central nervous system, limited regeneration occurs, but the situation is more complicated because (1) many more axons are likely to be involved, (2) astrocytes produce scar tissue that can prevent axon growth across the damaged area, and (3) astrocytes release chemicals that block the regrowth of axons.

13.5 CONCEPT CHECK

8 What is Wallerian degeneration, and where does it occur?

See the blue Answers tab at the back of the book.

13.6 | The Nerve Impulse

▶ **KEY POINT** Excitability is the ability of a plasma membrane to respond to an adequate stimulus and generate an action potential. Plasma membranes of skeletal muscle fibers, cardiac muscle cells, some gland cells, and the axolemma of most neurons (including all multipolar and pseudounipolar neurons) are examples of excitable membranes.

Excitability is the ability of a plasma membrane to conduct electrical impulses. The plasma membranes of skeletal muscle fibers and most neurons are excitable.

Figure 13.13 Wallerian Degeneration and Nerve Regeneration

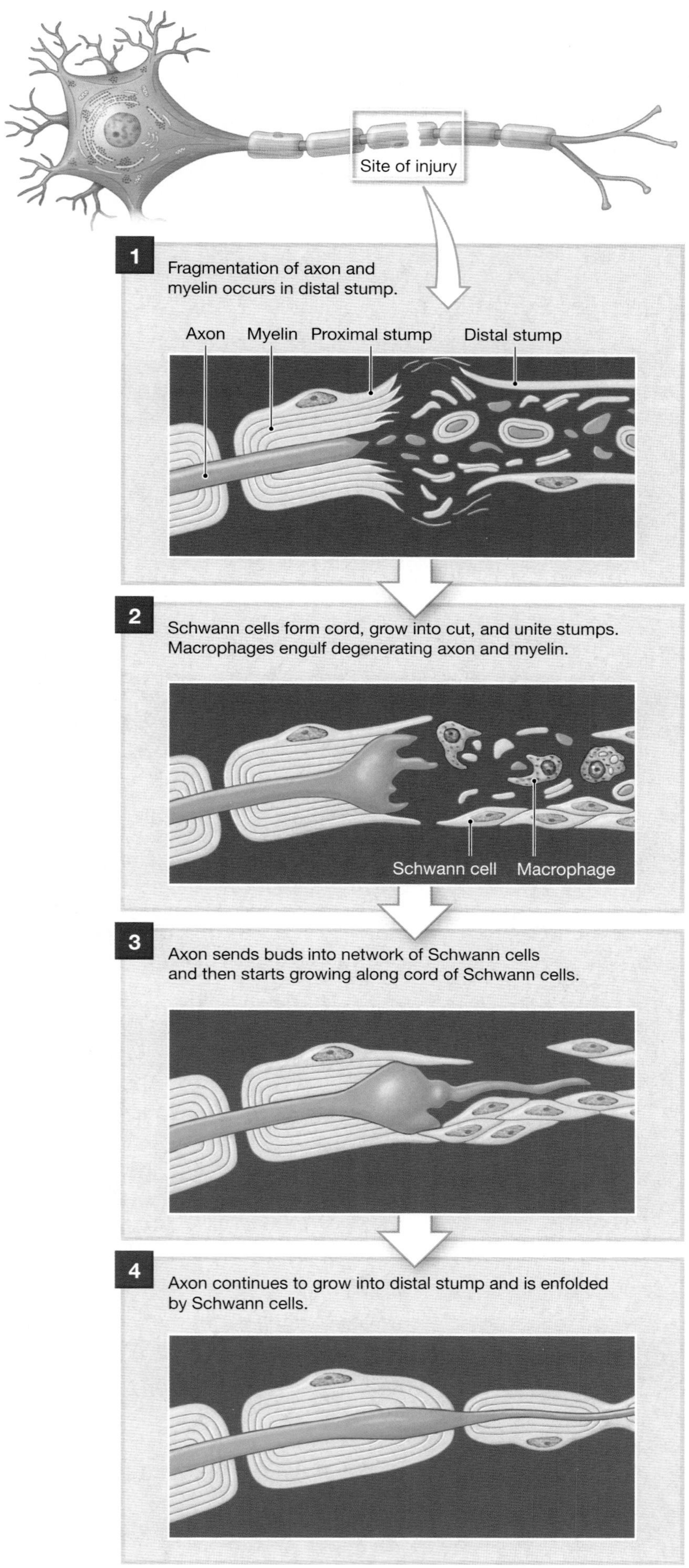

Site of injury
1
Fragmentation of axon and myelin occurs in distal stump.
Axon
Myelin
Proximal stump
Distal stump
2
Schwann cells form cord, grow into cut, and unite stumps. Macrophages engulf degenerating axon and myelin.
Schwann cell
Macrophage
3
Axon sends buds into network of Schwann cells and then starts growing along cord of Schwann cells.
4
Axon continues to grow into distal stump and is enfolded by Schwann cells.

An **action potential**, or *nerve impulse*, is the change in membrane potential that develops after the axolemma is stimulated to a level known as the threshold. The initiation and conduction of an action potential along the surface of an axon are summarized as follows:

- All cells possess a **membrane potential**, which is the resting electrical charge of the cell's plasma membrane. The membrane potential results from the uneven distribution of positive and negative ions across the plasma membrane.
- When a stimulus is applied to the axon, it produces a temporary change in the permeability of the axolemma in a localized area of the axon.
- This temporary change in permeability changes the membrane potential of the axon. If the stimulus is sufficient to start an action potential, the stimulus is termed a **threshold stimulus**.
- When an action potential develops in one location of an axon, it will **propagate** (spread) along the length of the axon toward the axon terminal.
- Once started, the rate of impulse conduction depends on the axon's properties, such as its diameter and whether it is myelinated. The larger the diameter of the axon, the more rapidly the impulse will be conducted. In addition, a myelinated axon conducts impulses five to seven times faster than an unmyelinated axon. The largest myelinated axons, with diameters ranging from 4 to 20 μm, conduct nerve impulses at speeds close to 140 m/s (300 mph), while small, unmyelinated fibers (less than 2 μm in diameter) conduct impulses at speeds below 1 m/s (2 mph).

13.6 CONCEPT CHECK

9 Two axons are tested for conduction speeds. One conducts action potentials at 50 m/s, the other at 1 m/s. Which axon is myelinated?

10 Define excitability.

See the blue Answers tab at the back of the book.

13.7 Synaptic Communication

▶ **KEY POINT** A synapse between neurons involves an axon terminal and (1) a dendrite (axodendritic), (2) cell body (axosomatic), or (3) axon (axoaxonic). At an axon terminal, a nerve impulse triggers events at a synapse that transfers the information to another neuron or cell.

A **synapse** is a site of intercellular communication between a neuron and another cell. A synapse may be chemical (*vesicular*) or electrical (*nonvesicular*).

Chemical Synapses

▶ **KEY POINT** Chemical synapses, or *vesicular synapses*, are the most abundant type of synapse. These synapses involve the passage of neurotransmitters between cells.

At a typical chemical synapse (also termed a *vesicular synapse*), a neurotransmitter released at the presynaptic membrane of an axon terminal binds to receptor proteins on the postsynaptic membrane. The neurotransmitter then triggers a temporary change in the membrane potential of the postsynaptic cell. Only the presynaptic membrane releases neurotransmitter, so communication occurs in one direction only: from the presynaptic neuron to the postsynaptic neuron **(Figure 13.14b)**.

Neuromuscular junctions (described in Chapter 9) are chemical synapses releasing the neurotransmitter acetylcholine (ACh). (More than 50 different neurotransmitters have been identified, but acetylcholine is the best known.) All somatic neuromuscular junctions utilize ACh; it is also released at many chemical synapses in the CNS and PNS. The general sequence of events is similar, regardless of the location of the synapse or the type of neurotransmitter:

- Arrival of the action potential at the axon terminal causes the release of neurotransmitter from secretory vesicles by exocytosis at the presynaptic membrane.
- The neurotransmitter diffuses across the synaptic cleft (the space between the axon and the postsynaptic surface) and binds to receptors on the postsynaptic membrane.
- Receptor binding changes the permeability of the postsynaptic membrane; the resulting effect may be excitatory or inhibitory. In general, excitatory effects promote the generation of action potentials, whereas inhibitory effects reduce the ability to generate action potentials.
- If the excitation is sufficient, receptor binding leads to the generation of an action potential in the axon (if the postsynaptic cell is a neuron) or sarcolemma (if the postsynaptic cell is a skeletal muscle fiber).
- The effects of one action potential on the postsynaptic membrane are short-lived because the neurotransmitter molecules are either enzymatically broken down or reabsorbed. To prolong or enhance the effects, additional action potentials must arrive at the axon terminal, and additional molecules of ACh must be released into the synaptic cleft.

There may be thousands of chemical synapses on the cell body of a single neuron **(Figure 13.14c)**. Many are active at any given moment, releasing a variety of different neurotransmitters. Some will have excitatory effects, others inhibitory effects. The activity of the receptive neuron depends on the sum of all the excitatory and inhibitory stimuli influencing the axon hillock at any given moment.

Electrical Synapses

▶ **KEY POINT** Electrical synapses, or *nonvesicular synapses*, are relatively rare and are found between neurons in both the CNS and PNS. At these synapses, the presynaptic and postsynaptic membranes form a gap junction.

Figure 13.14 The Structure of a Synapse. A synapse is the site of communication between a neuron and another cell.

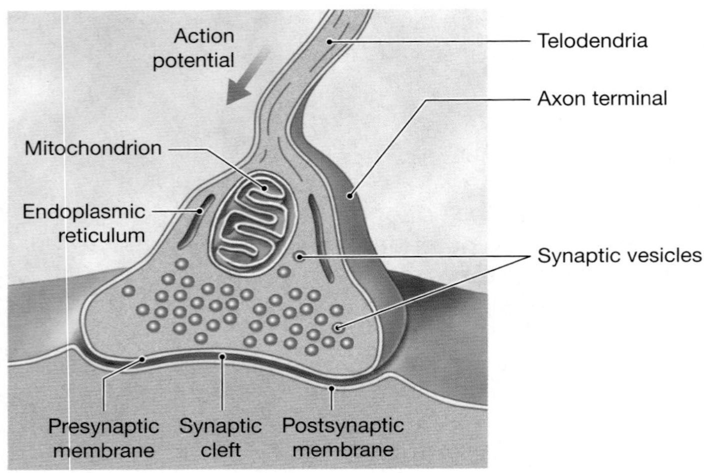

a Structure of a typical synapse.

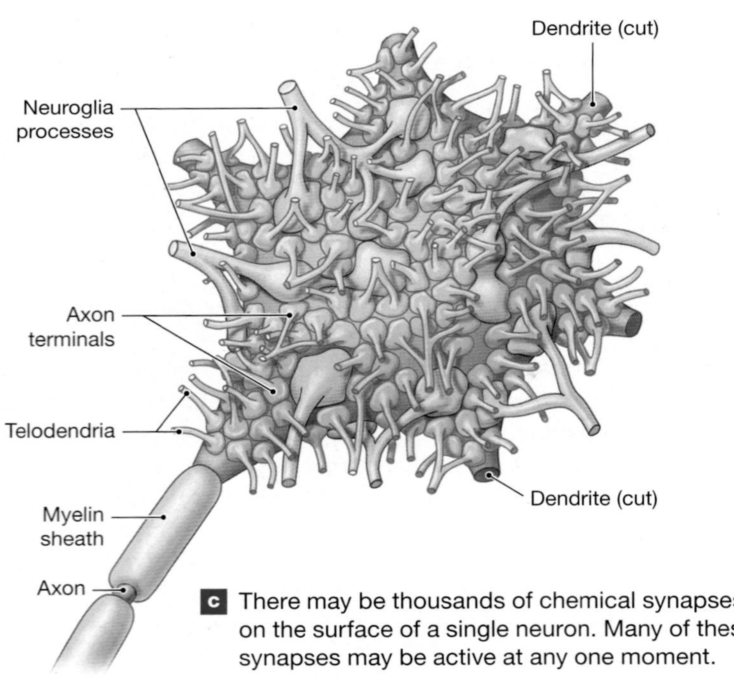

c There may be thousands of chemical synapses on the surface of a single neuron. Many of these synapses may be active at any one moment.

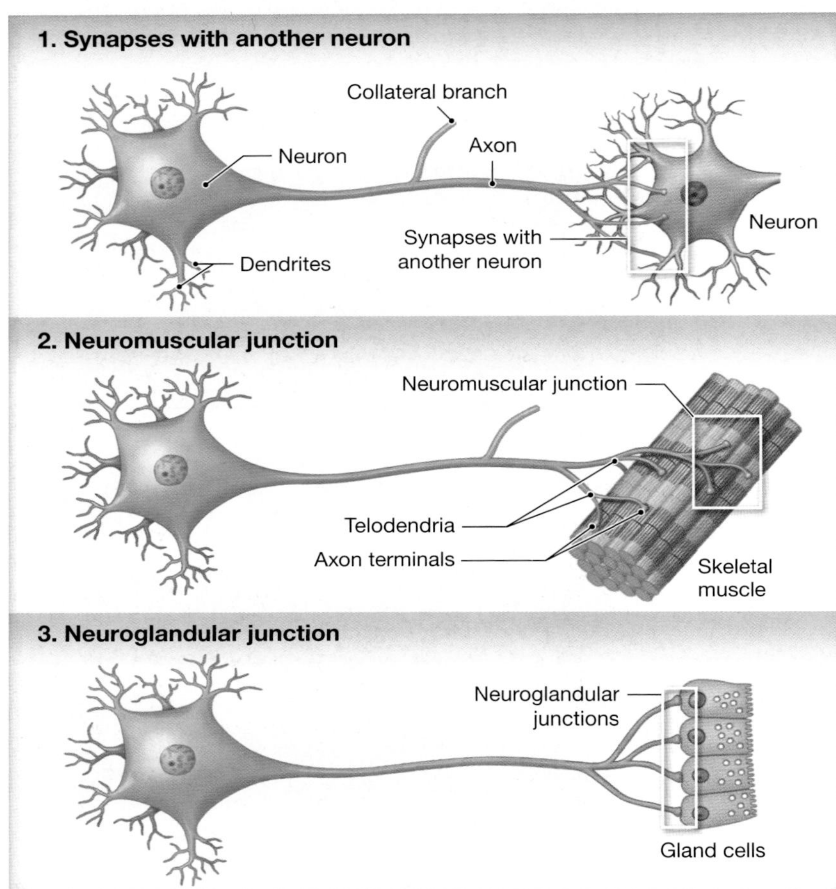

b A neuron may innervate (1) other neurons, (2) skeletal muscle fibers, or (3) gland cells. A single neuron would not innervate all three.

At an **electrical synapse** (also termed a *nonvesicular synapse*), the presynaptic and postsynaptic membranes are bound tightly together, and communicating junctions permit the passage of ions between the two cells. Because the cells are linked in this way, they function as if they share a common membrane, and the nerve impulse crosses from one neuron to the next without delay. In contrast to chemical synapses, electrical synapses can convey nerve impulses in either direction.

13.7  **CONCEPT CHECK**

11 Myasthenia gravis is a disease that decreases the amount of acetylcholine released from the axon terminal. What effect, if any, would this have on the generation of an action potential?

See the blue Answers tab at the back of the book.

13.8 | Neuron Organization and Processing

▶ **KEY POINT** Neurons are the basic building blocks of the nervous system. The billions of neurons within the CNS are organized into a much smaller number of neuronal pools.

A **neuronal pool** is a group of interconnected neurons with specific functions. *Neuronal pools are defined on the basis of function rather than anatomy.* They may be diffuse, involving neurons in several different regions of the brain, or localized, with all the neurons restricted to one specific location in the brain or spinal cord. Each neuronal pool has a limited number of input sources and output destinations, and the pool may contain both excitatory and inhibitory neurons.

The basic "wiring pattern" in a neuronal pool is called a **neural circuit**. A neural circuit has one of the following functions:

- **Divergence** is the spread of information from one neuron to several neurons or from one pool to multiple pools (**Figure 13.15a**). Divergence allows the broad distribution of a specific input, as when sensory neurons bring information into the CNS. The information is then distributed to neuronal pools throughout the spinal cord and brain. For example, visual information arriving from the eyes reaches your consciousness at the same time it is distributed to areas of the brain that subconsciously control posture and balance.
- In **convergence**, several neurons synapse on the same postsynaptic neuron (**Figure 13.15b**). Convergence permits the variable control of motor neurons by providing a mechanism for their voluntary and involuntary control. For example, the movements of your diaphragm and ribs are controlled by respiratory centers in the brain that operate outside of your awareness. However, the same motor neurons can also can be controlled voluntarily, as when you take a deep breath and hold it. Two different neuronal pools are involved, both synapsing on the same motor neurons.
- **Serial processing** relays information in a stepwise sequence from one neuron to another or from one neuronal pool to the next (**Figure 13.15c**). Serial processing occurs when sensory information is relayed from one processing center in the brain to another. For example, pain sensations on their way to your consciousness make stops at two neuronal pools along the pain pathway.
- **Parallel processing** occurs when several neurons or neuronal pools are processing the same information at one time (**Figure 13.15d**). Thanks to parallel processing, many different responses occur simultaneously. For example, stepping on a sharp object stimulates sensory neurons that distribute the information to a number of neuronal pools. As a result of parallel processing, you might withdraw your foot, shift your weight, move your arms, feel the pain, and shout, "Ouch!" at about the same time.
- **Reverberation** uses positive feedback. In this arrangement, collateral axons extend back toward the source of an impulse and further stimulate the presynaptic neurons (**Figure 13.15e**). Once a reverberating circuit is activated, it will continue to function until synaptic fatigue or an inhibitory stimulus breaks the cycle. Reverberation can occur within a single neuronal pool, or it may involve a series of interconnected pools. Highly complicated examples of reverberation among neuronal pools in the brain may help maintain consciousness, muscular coordination, and normal breathing.

We will discuss these and other "wiring patterns" as we consider the organization of the spinal cord and brain in subsequent chapters.

Figure 13.15 Organization of Neuronal Pools

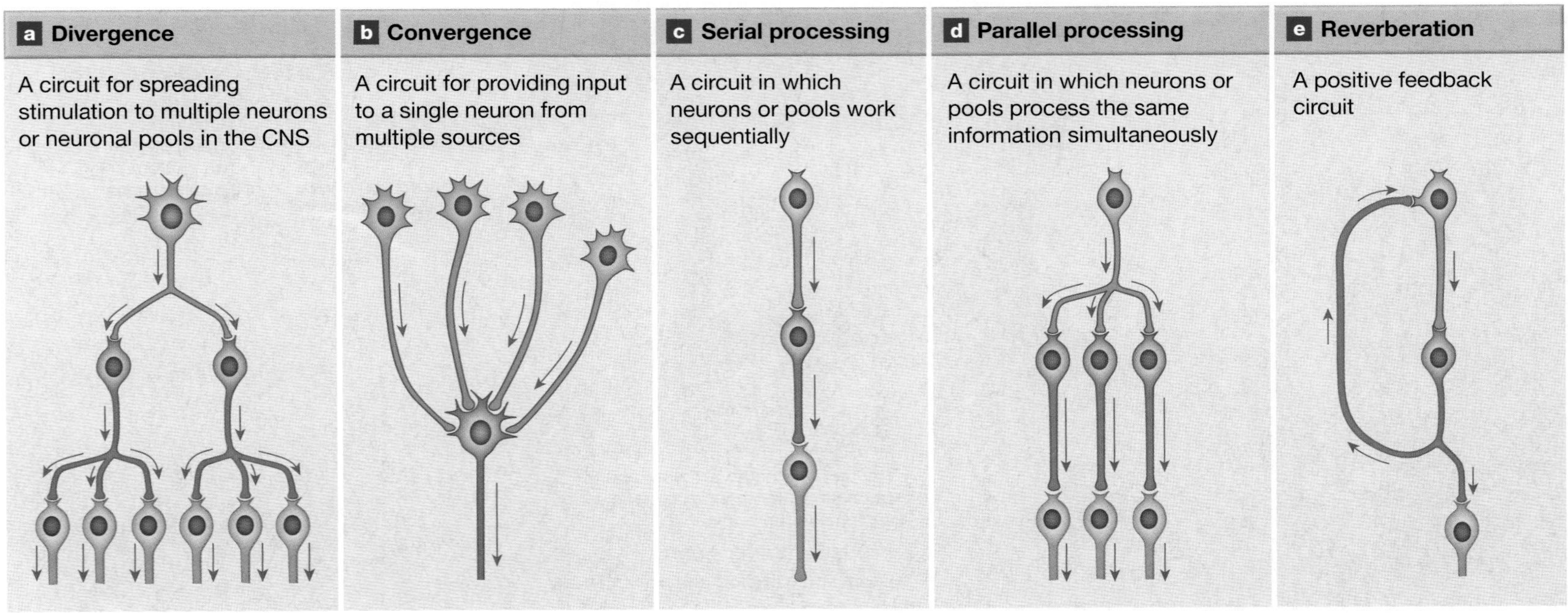

a Divergence

A circuit for spreading stimulation to multiple neurons or neuronal pools in the CNS

b Convergence

A circuit for providing input to a single neuron from multiple sources

c Serial processing

A circuit in which neurons or pools work sequentially

d Parallel processing

A circuit in which neurons or pools process the same information simultaneously

e Reverberation

A positive feedback circuit

13.8 CONCEPT CHECK

12 Distinguish between a neuronal pool whose function is divergence and a neuronal pool whose function is convergence.

See the blue Answers tab at the back of the book.

13.9 Anatomical Organization of the Nervous System

▶ **KEY POINT** The functions of the nervous system depend on interactions between neurons in neuronal pools. The most complex neural processing occurs in the CNS.

Arriving sensory information and outgoing motor commands are carried by the peripheral nervous system (PNS). Axons and cell bodies in the CNS and PNS are not randomly scattered. Instead, they form masses or bundles with distinct anatomical boundaries. **Figures 13.16** and **13.3** (p. 341) summarize the anatomical organization of the nervous system.

In the peripheral nervous system (PNS):

- The cell bodies of sensory neurons and visceral motor neurons are found in **ganglia**.
- Axons are bundled together in **nerves**, with spinal nerves connected to the spinal cord, and cranial nerves connected to the brain.

In the central nervous system (CNS):

- A collection of neuron cell bodies with a common function is called a **center**. A center with a distinct anatomical boundary is called a **nucleus**. A layer of gray matter called the neural cortex covers portions of the brain surface. The term higher centers refers to the most complex integration centers, nuclei, and cortical areas of the brain.
- The white matter of the CNS contains bundles of axons that share common origins, destinations, and functions. These bundles are called tracts. Tracts in the spinal cord form larger groups called columns.
- The centers and pathways that link the brain with the rest of the body are called tracts. For example, ascending (sensory) tracts distribute information from peripheral receptors to processing centers in the brain. Descending (motor) tracts begin at CNS centers concerned with motor control and end at the effectors they control.

Figure 13.16 Anatomical Organization of the Nervous System. An introduction to the terms commonly used when describing neuroanatomy.

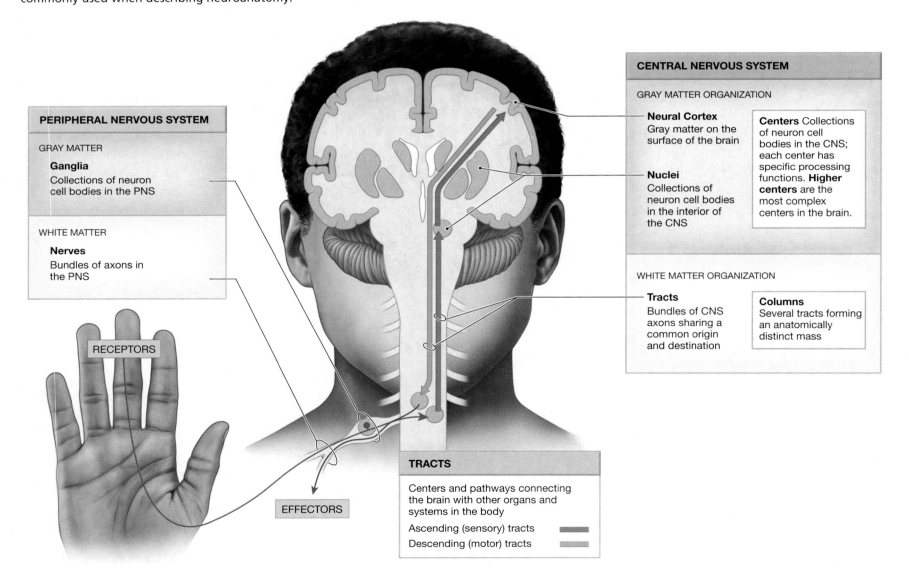

- A single neuron may have thousands of synapses on its cell body. The activity of the neuron depends on the summation of all of the excitatory and inhibitory stimuli arriving at any given moment at the axon hillock.

Electrical Synapses p. 352

- **Electrical synapses** (*nonvesicular synapses*) are found between neurons in the CNS and PNS, although they are rare. At these synapses, the plasma membranes of the presynaptic and postsynaptic cells are tightly bound together, and the cells function as if they shared a common plasma membrane. Electrical synapses transmit information more rapidly than chemical synapses and may be bidirectional.

13.8 Neuron Organization and Processing p. 353

- Neurons can be classified into **neuronal pools**. The **neural circuits** of these neuronal pools may show (1) divergence, (2) convergence, (3) serial processing, (4) parallel processing, or (5) reverberation. *(See Figure 13.15.)*
- **Divergence** is the spread of information from one neuron to several neurons or from one pool to several pools. This facilitates the widespread distribution of a specific input. *(See Figure 13.15a.)*
- **Convergence** is the presence of synapses from several neurons on one postsynaptic neuron. It permits the variable control of motor neurons. *(See Figure 13.15b.)*
- **Serial processing** is a pattern of stepwise information processing from one neuron to another or from one neuronal pool to the next. This is the way sensory information is relayed between processing centers in the brain. *(See Figure 13.15c.)*
- **Parallel processing** is a pattern that processes information by several neurons or neuronal pools at once. Many different responses occur at the same time. *(See Figure 13.15d.)*
- **Reverberation** occurs when neural circuits use positive feedback to continue the activity of the circuit. Collateral axons establish a circuit to continue to stimulate presynaptic neurons. *(See Figure 13.15e.)*

13.9 Anatomical Organization of the Nervous System p. 354

- Nervous system functions depend on interactions between neurons in neuronal pools. Almost all complex processing steps occur inside the brain and spinal cord. *(See Figure 13.16.)*
- Neuronal cell bodies and axons in both the PNS and CNS are organized into masses or bundles with distinct anatomical boundaries. *(See Figure 13.16.)*
- In the PNS, ganglia contain the cell bodies of sensory and visceral motor neurons. Axons in nerves occur within spinal nerves to the spinal cord and cranial nerves to the brain. *(See Figure 13.3.)*
- In the CNS, cell bodies are organized into **centers**; a center with discrete boundaries is called a **nucleus**. The **neural cortex** is the gray matter that covers portions of the brain. It is called a **higher center** to reflect its involvement in complex activities. White matter has bundles of axons called **tracts**. Tracts organize into larger units, called **columns**. The centers and tracts that link the brain and body are pathways. Sensory (ascending) pathways carry information from peripheral receptors to the brain; motor (descending) pathways extend from CNS centers concerned with motor control to the associated skeletal muscles. *(See Figures 13.3 and 13.16.)*

13

Chapter Review

For answers, see the blue Answers tab at the back of the book.

Level 1 Reviewing Facts and Terms

Match each numbered item with the most closely related lettered item.

1. afferent division ______
2. effector ______
3. astrocyte ______
4. oligodendrocyte ______
5. axon hillock ______
6. collaterals ______
7. bipolar neurons ______
8. proprioceptors ______
9. reverberation ______
10. ganglia ______

(a) positive feedback
(b) connects initial segment to cell body
(c) sensory information
(d) monitor position/movement of joints
(e) myelin
(f) one dendrite
(g) neuron cell bodies in PNS
(h) blood brain barrier (BBB)
(i) side branches of axons
(j) skeletal muscle cells

11. Which of the following is not a function of the neuroglia?
(a) support
(b) information processing
(c) secretion of cerebrospinal fluid
(d) phagocytosis

12. Neuroglia found surrounding the cell bodies of peripheral neurons are
(a) astrocytes.
(b) ependymal cells.
(c) microglia.
(d) satellite cells.

13. The most important function of the cell body of a neuron is to
(a) allow communication with another neuron.
(b) support the neuroglia.
(c) generate an electrical charge.
(d) house organelles that produce energy and synthesize organic molecules.

14. Fill in the blanks below with the proper structural classification for these neurons.

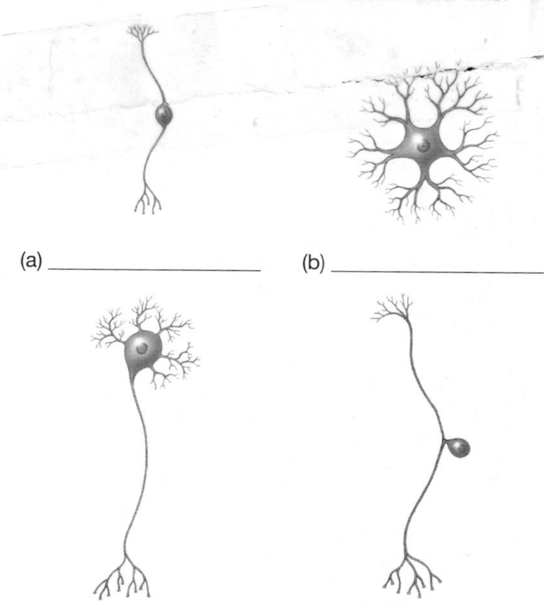

15. Axons terminate in a series of fine extensions known as
(a) telodendria.
(b) synapses.
(c) collaterals.
(d) hillocks.

16. Which of the following activities or sensations are not monitored by interoceptors?
(a) urinary activities
(b) digestive system activities
(c) visual activities
(d) cardiovascular activities

Level 2 Reviewing Concepts

13

1. Patterns of interactions between neurons include which of the following?
(a) divergence
(b) parallel processing
(c) reverberation
(d) all of the above

2. Which neuronal tissue cell type is likely to be malfunctioning if the blood brain barrier (BBB) is no longer adequately protecting the brain?
(a) ependymal cells
(b) astrocytes
(c) oligodendrocytes
(d) microglia

3. Developmental problems in the growth and interconnections of neurons in the brain reflect problems with the
(a) afferent neurons.
(b) microglia.
(c) astrocytes.
(d) efferent neurons.

4. What purpose do collaterals serve in the nervous system?

5. How does exteroceptor activity differ from interoceptor activity?

6. What is the purpose of the blood brain barrier (BBB)?

7. Differentiate between CNS and PNS functions.

8. Distinguish between the somatic nervous system and the autonomic nervous system.

9. Why is an electrical (nonvesicular) synapse more efficient than a chemical (vesicular) synapse? Why is it less versatile?

10. Differentiate between serial and parallel processing.

Level 3 Critical Thinking

1. In multiple sclerosis, there is progressive and intermittent damage to the myelin sheath of peripheral nerves. This results in poor motor control of the affected area. Why does destruction of the myelin sheath affect motor control?

2. An 8-year-old girl cut her elbow when she fell while skating. The injury caused only minor muscle damage but partially severed a nerve in her arm. What is likely to happen to the severed axons of this nerve, and will the little girl regain normal function of the nerve and the muscles it controls?

3. Eve is diagnosed with spinal meningitis. Her attending physician informs her father that high doses of antibiotics will be needed to treat Eve's condition. Her father assumes this is due to the severity of the disease. Is he correct? If not, why are such high doses required to treat Eve's condition?

CLINICAL CASE WRAP-UP

When Nerves Are Demyelinated

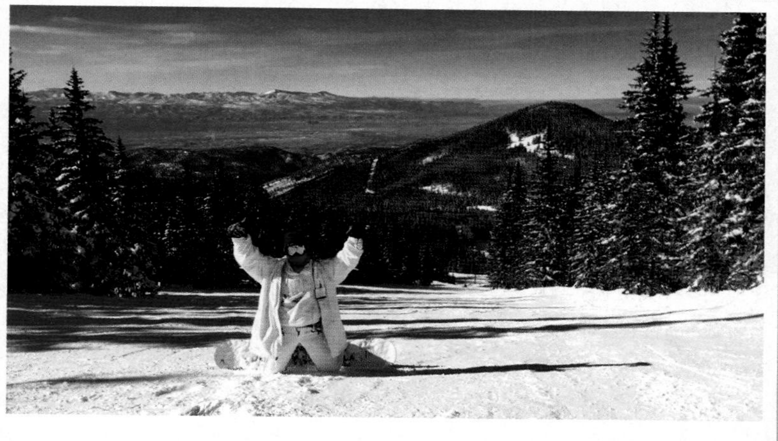

Nicole is suffering from Guillain-Barré syndrome, an autoimmune condition. Nicole's immune system has attacked the myelin sheaths of her peripheral nervous system. The bodies (somas) of PNS nerve cells are located in the central nervous system. The myelinated axons of the PNS connect the cell bodies to peripheral receptors or effectors (skeletal muscle fibers). Without myelin insulation, the nerve impulses travel slowly and inefficiently. The poor afferent conduction causes Nicole's tingling and numbness; the poor efferent conduction causes her progressive motor weakness.

Named after the two French physicians who first described it, Guillain-Barré syndrome is rare and not contagious, and we don't know what triggers it. It often follows a respiratory or gastrointestinal infection or influenza immunization.

Later that day, Nicole's numbness and weakness spread to her muscles of respiration. She is rushed to a hospital intensive care unit, where she is connected to a ventilator that can breathe for her. However, within 3 months, Nicole has recovered full sensation and 90 percent of her motor strength, thanks to new Schwann cells that have re-myelinated her peripheral neuronal axons. She looks forward to snowboarding again next season.

1. Would the nerve conduction velocity of Nicole's PNS be faster or slower during the acute phase of her Guillain-Barré syndrome?
2. Would the syndrome affect Nicole's autonomic nervous system, too? What might you notice clinically?

See the blue Answers tab at the back of the book.

13

Related Clinical Terms

anesthetic: An agent that produces a local or general loss of sensation or pain.

excitotoxicity: Continuous and exaggerated stimulation by a neurotransmitter, especially for the excitatory neurotransmitter, glutamate.

neurotoxin: A compound that disrupts normal nervous system function by interfering with the generation or propagation of action potentials.

14

The Nervous System

The Spinal Cord and Spinal Nerves

Learning Outcomes

These Learning Outcomes correspond by number to this chapter's sections and indicate what you should be able to do after completing the chapter.

- **14.1** Discuss the structure and functions of the spinal cord. p. 361
- **14.2** Locate the spinal meninges, and compare and contrast their structure and function. p. 361
- **14.3** Discuss the structure and location of the gray matter and white matter, and compare and contrast the roles of both in processing and relaying sensory and motor information. p. 364
- **14.4** Identify the regional groups of spinal nerves. p. 366
- **14.5** Define the term nerve plexus and compare and contrast the four main spinal nerve plexuses. p. 367
- **14.6** Describe the structures and steps involved in a neural reflex. p. 374

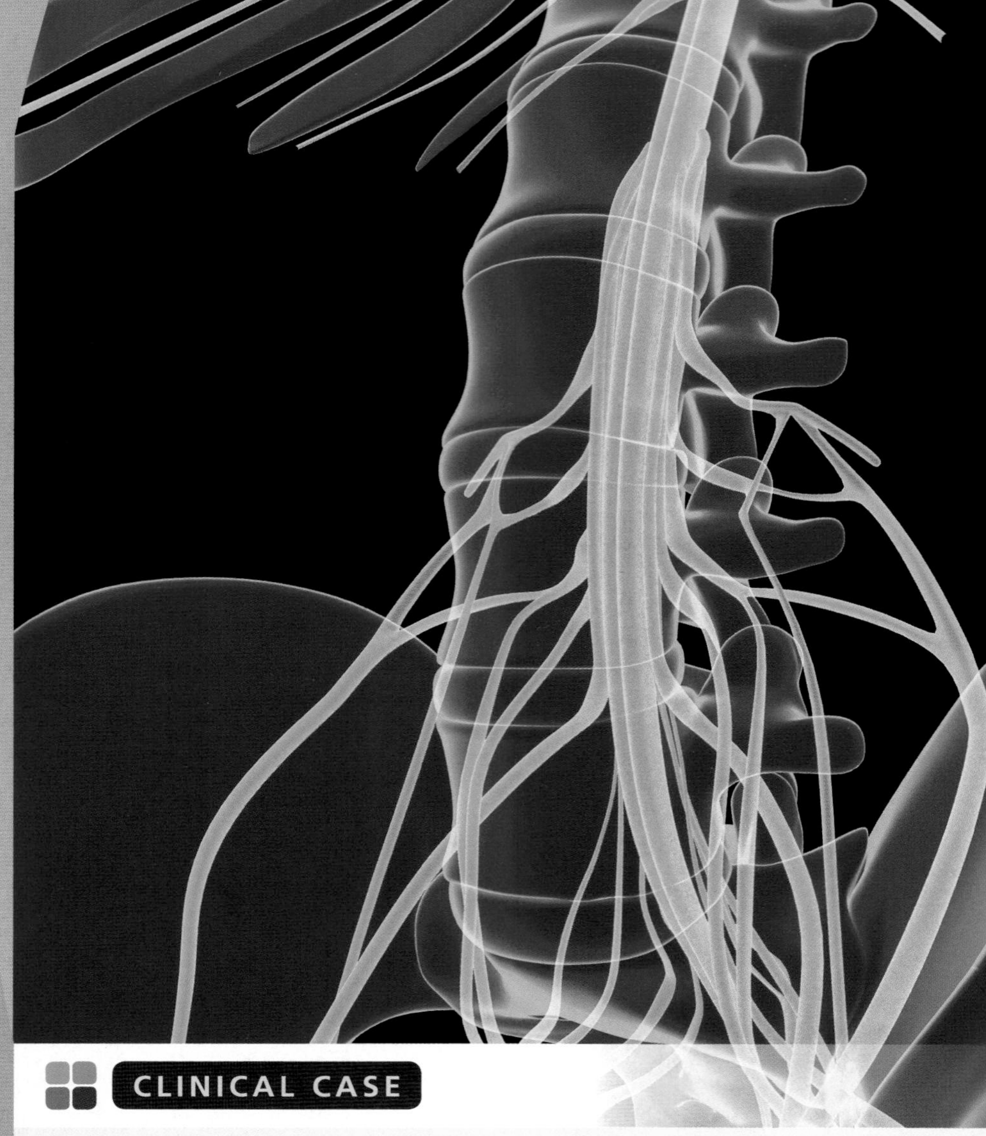

CLINICAL CASE

A Case of the Bends

Bob is an avid SCUBA (self-contained underwater breathing apparatus) diver. He is diving in Belize, where the fish and corals are distractingly beautiful. He is at the end of his third dive for the day, too deep and with not much air left. He knows he should ascend no faster than 10 meters per second with frequent decompression stops, but he has limited air pressure and is still 30 meters deep. To avoid drowning, he quickly swims to the surface.

At the surface, Bob takes a huge gulp of air and then feels severe back pain. He has trouble swimming to the boat. By the time he gets there, his feet are tingling and getting numb. He is too weak to climb into the boat and has to be lifted aboard. Lying on the bottom of the boat, speeding to the shore, he notices his numbness has climbed to the level of his umbilicus and he is unable to move his legs.

What has happened to Bob? Will he survive? To find out, turn to the Clinical Case Wrap-Up on p. 385.

THE CENTRAL NERVOUS SYSTEM (CNS) CONSISTS of the spinal cord and brain. The spinal cord and brain are anatomically connected but have a significant degree of functional independence. The spinal cord is more than just a pathway for information traveling to or from the brain. Although the spinal cord relays sensory information to the brain, it also integrates and processes information on its own. This chapter describes the anatomy of the spinal cord and examines its integrative and processing activities.

14.1 | Gross Anatomy of the Spinal Cord

▶ **KEY POINT** The spinal cord is continuous with the brain and ends at the conus medullaris. The diameter of the spinal cord is largest in the cervical region and smallest in the sacral and coccygeal regions.

The adult spinal cord extends from the foramen magnum of the skull to the inferior border of the first lumbar vertebra (L_1) (**Figure 14.1a,b**). The posterior surface of the spinal cord has a shallow longitudinal groove, the **posterior median sulcus**. The **anterior median fissure** is a deep crease on the anterior surface of the cord (**Figure 14.1c**). Each region of the spinal cord (cervical, thoracic, lumbar, and sacral) contains several tracts, which are bundles of axons sharing functional and structural characteristics. **Figure 14.1b,c** provides a series of sectional views demonstrating the variations in the gray matter and white matter along the length of the spinal cord. (Gray matter is composed of the cell bodies of neuroglia, neurons, and unmyelinated neuronal processes; white matter is composed of myelinated and unmyelinated neuronal processes.)

The amount of gray matter increases substantially in those segments of the spinal cord that are concerned with the sensory and motor innervation of the limbs (**Figure 14.1b,c**). These areas contain interneurons that are responsible for (1) relaying arriving sensory information and (2) coordinating the activities of the somatic motor neurons that control the complex muscles of the limbs. These expanded areas of the spinal cord form the **cervical enlargement**, which supplies nerves to the pectoral girdle and upper limbs, and the **lumbosacral enlargement**, which supplies nerves to the pelvis and lower limbs (**Figure 14.1a**). Caudal to the lumbosacral enlargement, the spinal cord tapers and forms a cone-shaped tip called the **conus medullaris**, which is located at or inferior to the level of the first lumbar vertebra (L_1). Extending within the vertebral canal from the inferior tip of the conus medullaris is the **filum terminale** ("terminal thread"). The filum terminale extends from L_1 to the dorsum of the coccyx, where it connects the spinal cord to the first coccygeal vertebra (**Figure 14.1a,b**).

The entire spinal cord is divided into 31 segments. A letter and number designation identifies each segment. For example, C_3 is the third cervical segment (**Figure 14.1a,b**).

Every spinal segment is associated with a pair of **dorsal root ganglia** that contain the cell bodies of sensory neurons. *(The only exceptions are at C_1 and the first coccygeal vertebra, Co_1, where some people lack dorsal roots and the associated dorsal root ganglia.)* These sensory ganglia lie between the pedicles of adjacent vertebrae. On both sides of the spinal cord the **dorsal roots** contain the afferent axons of the sensory neurons in the dorsal root ganglion (**Figures 14.1c** and **14.2a,b**). Anterior to the dorsal root, a **ventral root** leaves the spinal cord. The ventral root contains the efferent axons of somatic motor neurons and, at some levels, efferent visceral motor neurons that control peripheral effectors. The dorsal and ventral roots of each segment enter and leave the vertebral canal between adjacent vertebrae at the intervertebral foramina. The dorsal roots are thicker than the ventral roots.

Distal to each dorsal root ganglion, the sensory and motor fibers form a single **spinal nerve** that exits from the intervertebral foramina (**Figures 14.1c** and **14.2a**). Spinal nerves are classified as **mixed nerves** because they contain both afferent (sensory) and efferent (motor) fibers. **Figure 14.2a,b** shows the spinal nerves as they emerge from intervertebral foramina.

The spinal cord continues to grow until approximately age 4. Until then, the growth of the spinal cord keeps pace with the growth of the vertebral column, and the segments of the spinal cord are aligned with the corresponding vertebrae. The ventral and dorsal roots are short and leave the vertebral canal through the adjacent intervertebral foramina.

After age 4, the vertebral column continues to grow, but the spinal cord does not. This vertebral growth carries the dorsal root ganglia and spinal nerves farther and farther away from their original position. As a result, the dorsal and ventral roots gradually elongate. The adult spinal cord extends only to the level of the first or second lumbar vertebra; thus spinal cord segment S_2 lies at the level of vertebra L_1 (**Figure 14.1a,b**).

When seen in gross dissection, the filum terminale and the long ventral and dorsal roots are called the **cauda equina** (KAW-da ek-WĪ-na; *cauda*, tail, + *equus*, horse) because this structure reminded early anatomists of a horse's tail (**Figure 14.1a**).

14.1 CONCEPT CHECK

1 What structure contains the cell bodies of sensory neurons?

2 The filum terminale and the long dorsal and ventral roots that extend caudal to the conus medullaris form what anatomical structure?

See the blue Answers tab at the back of the book.

14.2 | Spinal Meninges

▶ **KEY POINT** The vertebral column isolates the spinal cord from the external environment. The delicate neural tissues also must be protected from the surrounding vertebral canal. Specialized membranes known as the spinal meninges provide protection, physical stability, and shock absorption for the spinal cord.

The **spinal meninges** (men-IN-jēz) cover and protect the spinal cord and spinal nerve roots (**Figure 14.2**). Blood vessels branching within the meninges deliver oxygen and nutrients to the spinal cord. At the foramen magnum of the skull, the spinal meninges are continuous with the **cranial meninges** surrounding the brain. There are three meningeal layers: the dura mater, arachnoid mater, and pia mater. (Chapter 16 describes the cranial meninges, which have the same three layers.)

The Dura Mater

▶ **KEY POINT** The dura mater is a tough, fibrous layer that forms the outermost covering of the spinal cord and brain.

The **dura mater** (DŪ-ra MĀ-ter; *dura*, hard, + *mater*, mother) of the spinal cord consists of a layer of dense irregular connective tissue (**Figure 14.2a–c**). A simple squamous epithelium covers the inner and outer surfaces of the dura mater. The outer epithelium is not attached to the bony walls of the vertebral canal, and the resulting space is called the **epidural space** (**Figure 14.2b**). The epidural space contains areolar tissue, blood vessels, and adipose tissue.

The dura mater attaches to the edge of the foramen magnum of the skull, the second and third cervical vertebrae, the sacrum, and the posterior longitudinal ligament. These attachments stabilize the spinal cord within the vertebral canal. Caudally, the spinal dura mater tapers and forms a dense cord of collagen fibers that blend with the filum terminale, forming the **coccygeal ligament**. The coccygeal ligament extends the length of the sacral canal and fuses with the periosteum of the sacrum and coccyx. Lateral support of the spinal cord is provided by the connective tissues within the epidural space and by the extensions of the dura mater accompanying the spinal nerve roots as they pass through the intervertebral foramina (**Figure 14.2b,c**).

Figure 14.1 Gross Anatomy of the Spinal Cord. The spinal cord extends inferiorly from the base of the brain along the vertebral canal.

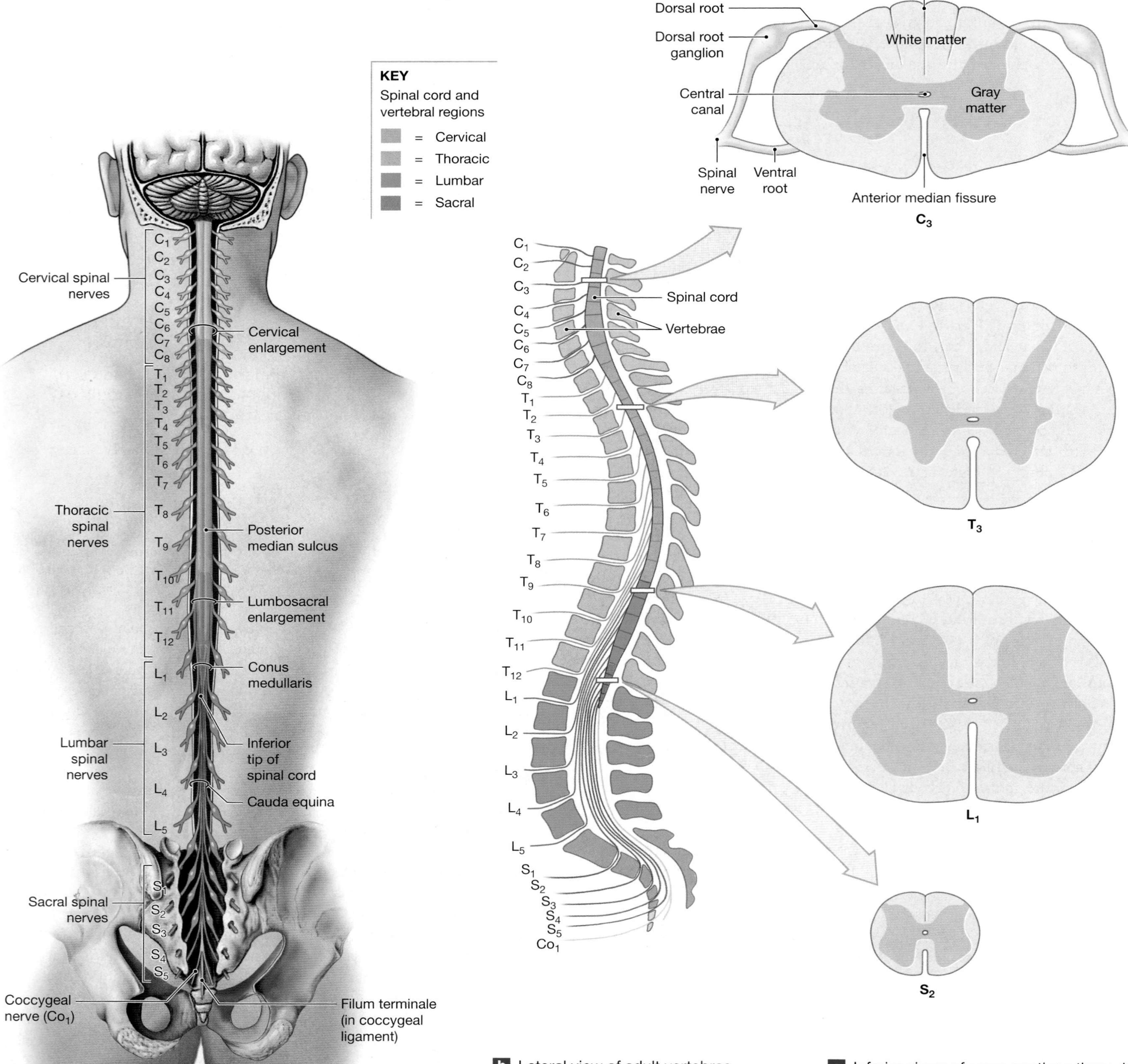

a Superficial anatomy and orientation of the adult spinal cord. The numbers to the left identify the spinal nerves and indicate where the nerve roots leave the vertebral canal.

b Lateral view of adult vertebrae and spinal cord. Note that the spinal cord segments for S_1–S_5 are level with the T_{12}–L_1 vertebrae.

c Inferior views of cross sections through representative segments of the spinal cord showing the arrangement of gray and white matter.

Figure 14.2 The Spinal Cord and Spinal Meninges

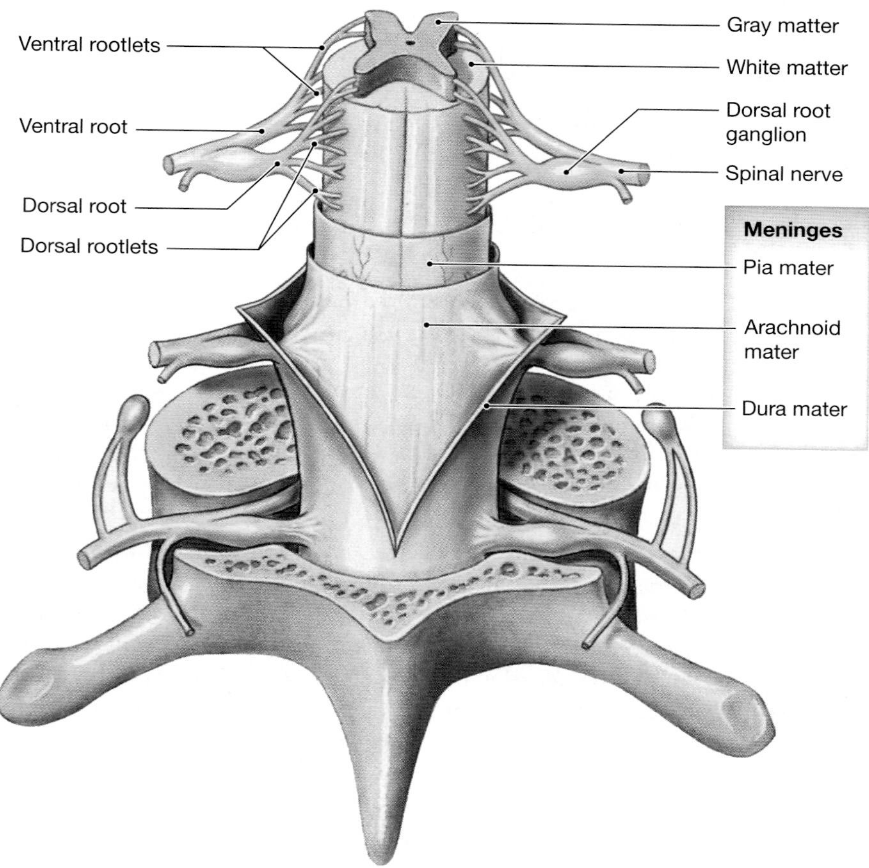

a Posterior view of the spinal cord shows the meningeal layers, superficial landmarks, and distribution of gray and white matter.

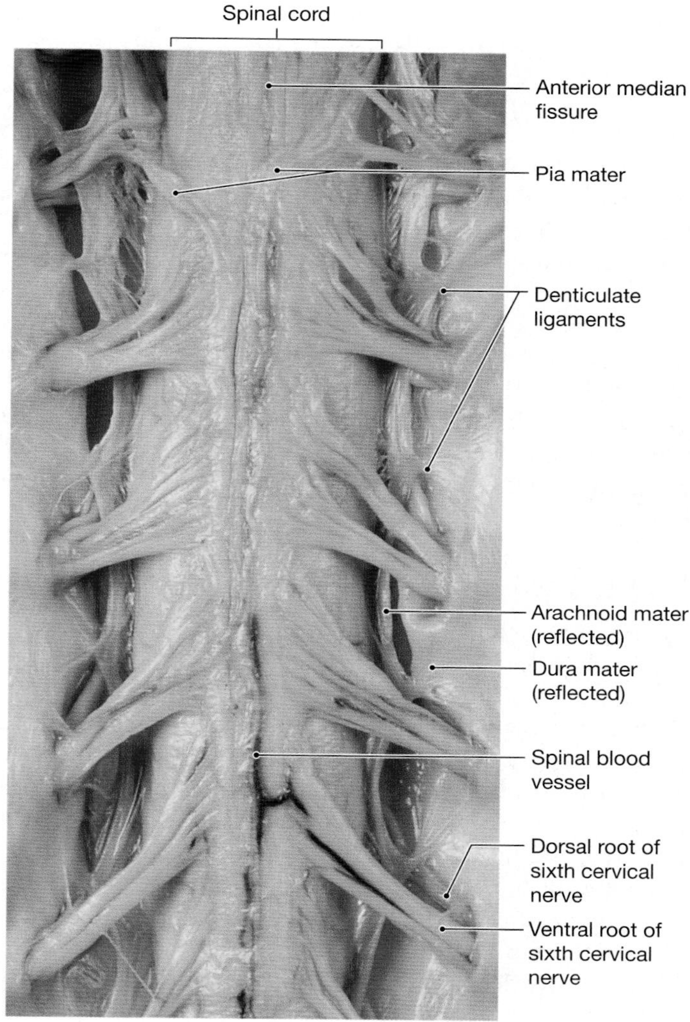

c Anterior view of spinal cord shows meninges and spinal nerves. For this view, the dura and arachnoid membranes have been cut longitudinally and retracted (pulled aside); notice the blood vessels that run in the subarachnoid space bound to the outer surface of the delicate pia mater.

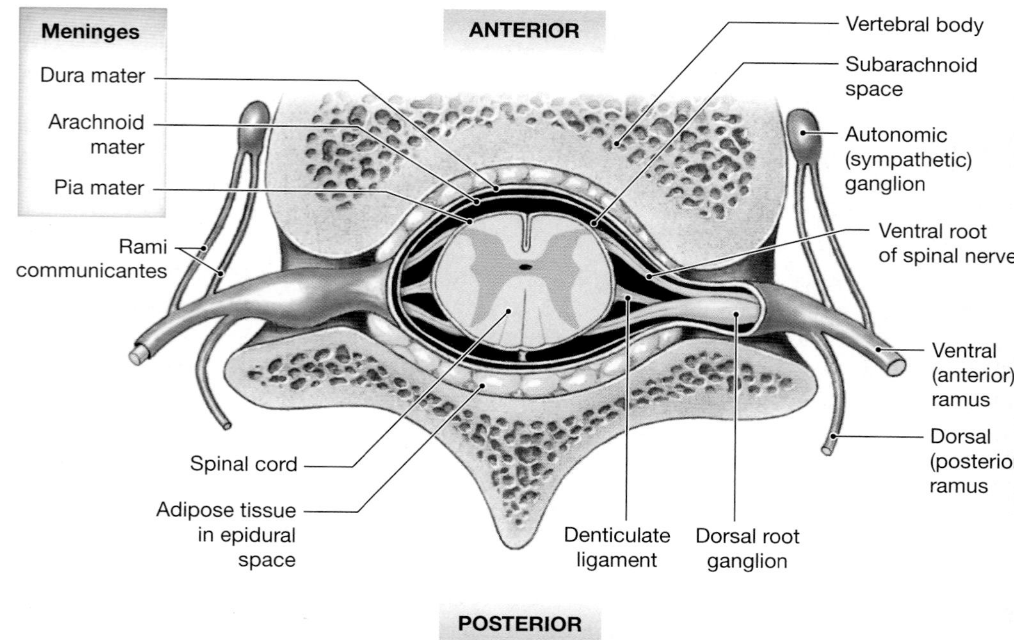

b Sectional view through the spinal cord and meninges shows the peripheral distribution of the spinal nerves.

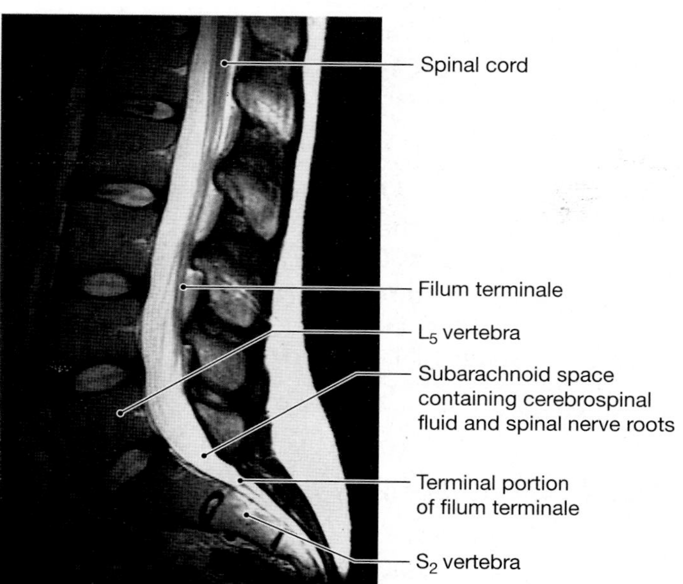

d An MRI scan of the inferior portion of the spinal cord shows its relationship to the vertebral column.

The Arachnoid Mater

▶ **KEY POINT** The arachnoid mater, the middle meningeal layer, is composed of a simple squamous epithelium. The arachnoid mater lines the inner surface of the dura mater.

In a cadaver, a narrow **subdural space** separates the dura mater from the deeper meninges of the spinal cord. However, in a living person this space does not exist, and the inner surface of the dura is lined by the outer surface of the **arachnoid** (a-RAK-noyd; *arachne*, spider) **mater** (**Figure 14.2a–c**). The **subarachnoid space** separates the arachnoid mater from the innermost layer, the pia mater. This space contains **cerebrospinal fluid (CSF)**, which is a shock absorber and a diffusion medium for dissolved gases, nutrients, chemical messengers, and waste products. Bundles of fibers known as arachnoid trabeculae extend from the inner surface of the arachnoid mater to the outer surface of the pia mater. The subarachnoid space of the spinal meninges is easily accessed between L_3 and L_4 (**Figure 14.2** and Clinical Note on p. 375) for the clinical examination of CSF or for administering anesthetics. (Chapter 16 discusses the subarachnoid space and the role of cerebrospinal fluid.)

The Pia Mater

▶ **KEY POINT** The surface of the CNS is covered with a connective tissue membrane, the pia mater. The pia mater closely follows the contours of the spinal cord.

Deep to the subarachnoid space is the **pia mater** (*pia*, delicate, + *mater*, mother), the innermost meningeal layer (**Figure 14.2a–c**). The elastic and collagen fibers of the pia mater are interwoven with those of the arachnoid trabeculae. The blood vessels supplying the spinal cord are found within the pia mater. The pia mater is firmly bound to the underlying neural tissue, conforming to its bulges and fissures. The surface of the spinal cord consists of a thin layer of astrocytes, and cytoplasmic extensions of these neuroglia lock the collagen fibers of the spinal pia mater in place.

Paired **denticulate ligaments** are located along the length of the spinal cord. These structures, which are found between the dorsal and ventral roots of the spinal nerves, are extensions of the spinal pia mater, and they connect the pia mater and arachnoid mater to the dura mater of the spinal cord. The denticulate ligaments begin at the foramen magnum of the skull, and they prevent side-to-side and downward movement of the spinal cord. At the inferior tip of the conus medullaris, the connective tissue fibers of the spinal pia mater form the filum terminale (**Figure 14.2b–d**).

The spinal meninges surround the dorsal and ventral roots of the spinal nerves. As seen in **Figure 14.2a,b**, the meningeal membranes are continuous with the connective tissues surrounding the spinal nerves and their peripheral branches.

14.2 CONCEPT CHECK

3 Identify the location of the cerebrospinal fluid that surrounds the spinal cord.

4 List the three meninges that cover the spinal cord and the spinal nerve roots in order from deep to superficial.

See the blue Answers tab at the back of the book.

14.3 Sectional Anatomy of the Spinal Cord

▶ **KEY POINT** The spinal cord contains a central mass of gray matter containing the cell bodies of neuroglia and the cell bodies of neurons and a peripheral region of white matter containing myelinated and unmyelinated axons.

The anterior median fissure and the posterior median sulcus divide the spinal cord into left and right halves (**Figure 14.3**). There is a central, H-shaped mass of **gray matter** containing the cell bodies of neuroglia and neurons (**Figures 14.1c** and **14.3a,b**). The gray matter surrounds the narrow **central canal**, which is located in the horizontal bar of the H. Gray matter called **horns** project toward the outer surface of the spinal cord (**Figure 14.3a,b**). The peripheral **white matter** contains myelinated and unmyelinated axons organized into tracts and columns. ↺ pp. 346–347, 354

Organization of Gray Matter

▶ **KEY POINT** Sensory nuclei in the gray matter receive and send sensory information from peripheral receptors. Motor nuclei in the gray matter send motor commands to peripheral effectors.

The cell bodies of neurons within the gray matter of the spinal cord are organized into groups called **nuclei** that have specific functions. **Sensory nuclei** receive and relay sensory information from peripheral receptors, such as touch receptors in the skin. **Motor nuclei** send motor commands to peripheral effectors, such as skeletal muscle.

Sensory nuclei and motor nuclei within the central gray matter of the spinal cord extend for a considerable distance along the length of the spinal cord (**Figure 14.3b**). A frontal section along the axis of the central canal separates the sensory (dorsal) nuclei from the motor (ventral) nuclei. The **posterior (dorsal) horns** contain somatic and visceral sensory nuclei, and the **anterior (ventral) horns** contain somatic motor neurons. **Lateral horns** (*intermediate horns*) are found *only* between segments T_1 and L_2 and contain visceral motor neurons. The **gray commissures** (*commissura*, a joining together) contain axons decussating (crossing) from one side of the cord to the other (**Figure 14.3a,b**). There are two gray commissures, one posterior to the central canal and one anterior to the central canal.

The motor nuclei within each horn are highly organized (**Figure 14.3b,c**). Nerves innervating skeletal muscles of more proximal structures (such as the trunk and shoulder) are located more medially within the gray matter than nuclei innervating the skeletal muscles of more distal structures (forearm and hand).

The size of the anterior horns varies depending on the number of skeletal muscles innervated by that segment. Therefore, the anterior horns are largest in cervical and lumbar regions of the spinal cord, regions that control the muscles of the upper and lower limbs.

Organization of White Matter

▶ **KEY POINT** White matter is organized into columns, and each column is organized into tracts. Ascending tracts carry sensory information toward the brain, and descending tracts carry motor commands into the spinal cord.

Although the general pattern of gray matter and white matter is the same throughout the spinal cord, the amount of white matter decreases as you move caudally within the spinal cord. White matter is divided into regions, or **columns** (also termed *funiculi*; singular, *funiculus*) (**Figure 14.3c**). The **posterior white columns** are located between the posterior horns and the posterior median sulcus. The **anterior white columns** are located between the anterior horns and the anterior median fissure, and they are interconnected by the **anterior white commissure**. **Lateral white columns**, between the anterior and posterior columns, are composed of white matter.

Each column contains **tracts** composed of axons sharing functional and structural characteristics. A specific tract carries either sensory information or motor commands, and the axons within a tract are uniform in diameter, myelination, and conduction speed. All the axons within a tract relay information in the same direction. Small commissural tracts carry sensory or motor signals between segments of the spinal cord; other, larger tracts connect the spinal cord with the brain.

Figure 14.3 Sectional Organization of the Spinal Cord

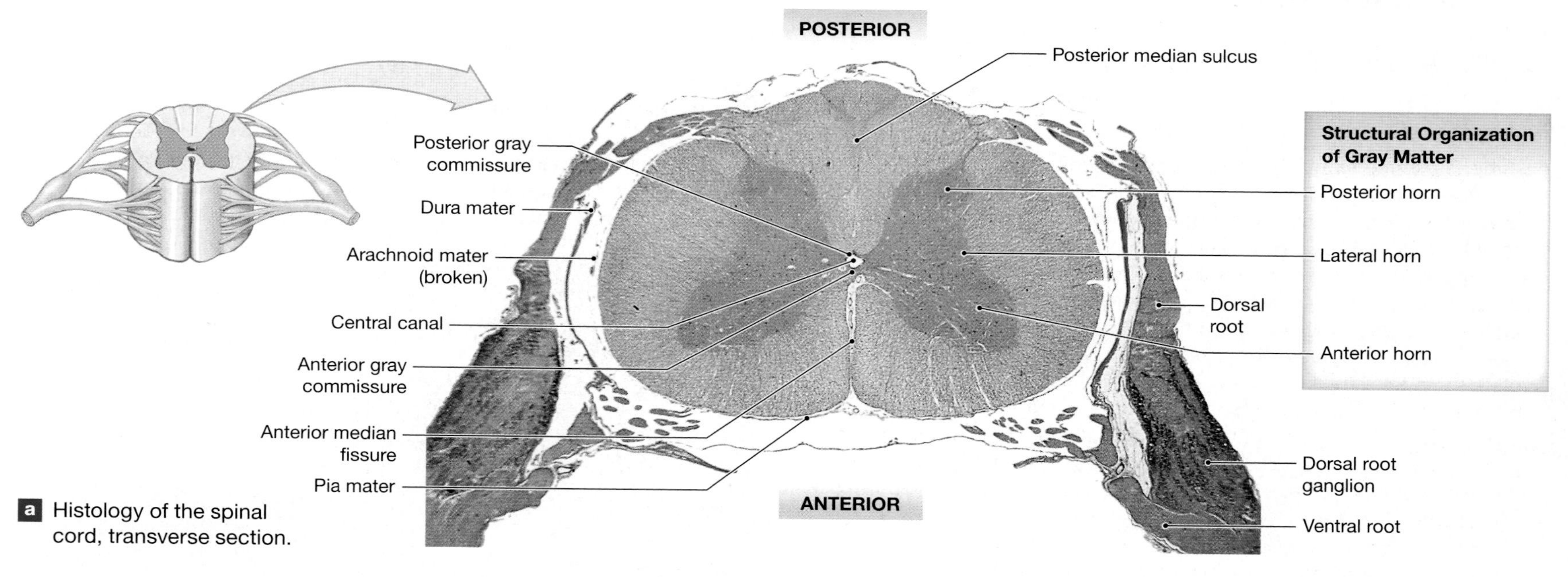

a Histology of the spinal cord, transverse section.

Posterior median sulcus
From dorsal root
Posterior horn
Posterior gray commissure
Lateral horn
Anterior horn
Anterior gray commissure
Anterior median fissure
To ventral root
Somatic
Visceral
Visceral
Somatic
Functional Organization of Gray Matter
Sensory nuclei
Motor nuclei

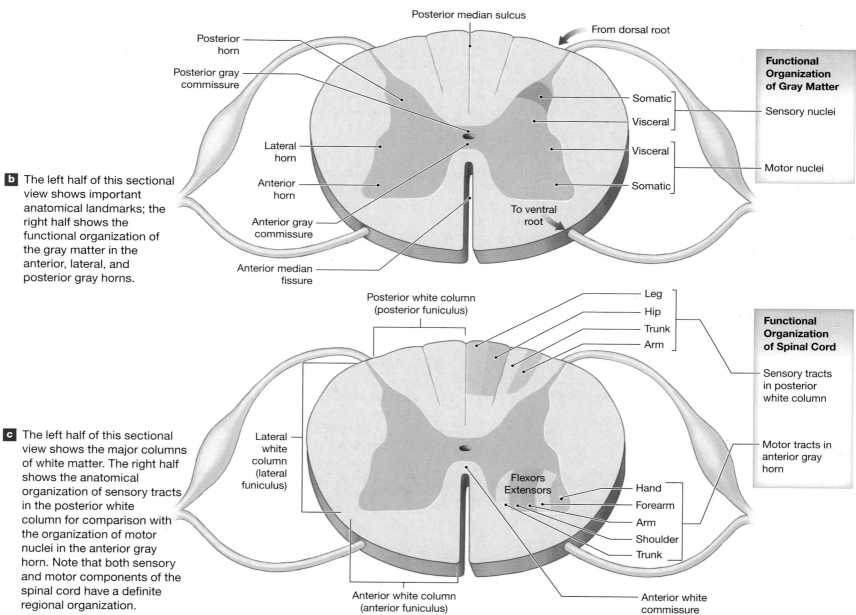

b The left half of this sectional view shows important anatomical landmarks; the right half shows the functional organization of the gray matter in the anterior, lateral, and posterior gray horns.

Posterior white column (posterior funiculus)
Leg
Hip
Trunk
Arm
Functional Organization of Spinal Cord
Sensory tracts in posterior white column
Lateral white column (lateral funiculus)
Flexors
Extensors
Hand
Forearm
Arm
Shoulder
Trunk
Motor tracts in anterior gray horn
Anterior white column (anterior funiculus)
Anterior white commissure

c The left half of this sectional view shows the major columns of white matter. The right half shows the anatomical organization of sensory tracts in the posterior white column for comparison with the organization of motor nuclei in the anterior gray horn. Note that both sensory and motor components of the spinal cord have a definite regional organization.

CLINICAL NOTE

Spinal Cord Injuries

Spinal cord injuries are often the result of blunt trauma (auto accidents, falls) or penetrating trauma (gunshot wounds, shrapnel). Immediate medical care after the injury is critical for recovery.

Physical examination can pinpoint the location of the injury. The level of sensory loss is indicated by the surface **dermatomes** that are affected (look ahead to **Figure 14.6**). Dermatomes also indicate the level of motor loss.

Spinal cord injuries are classified as incomplete or complete. Incomplete lesions include spinal cord concussion, contusion, or compression without anatomical disruption of the spinal cord itself. Some recovery from incomplete lesions can be expected. The best indicator of an incomplete spinal cord injury is **sacral sparing**, or intact sacral nerves. Muscle movement of the great toe (S_1), sensory preservation in the perianal area (S_2–S_4), and motor control of the anal sphincter (S_2–S_4) indicate an incomplete injury with potential return of function. No functional return within the first 24 hours indicates a complete spinal injury with a poor prognosis for functional return.

Spinal shock following a spinal cord injury is a period when all neurologic activity below the level of injury is lost, including motor, sensory, reflex, and autonomic function. This loss of nerve conduction is due to disrupted cellular potassium ion flow. Spinal shock may last for a few weeks and is clinically considered at an end when sacral reflexes return.

The level of spinal cord injury determines residual function. If the lesion is at C_3 or above, all of the muscles of respiration, including the diaphragm, are paralyzed and artificial ventilation is required. If the lesion is lower in the cervical cord, then shoulder, arm, forearm, and some hand function may be spared. Paralysis of all four limbs is called **quadriplegia**. Thoracic cord lesions spare the upper extremities but affect the trunk and abdomen, lower extremities, bowel, bladder, and pelvic function. This is known as **paraplegia**. Injuries in the lumbar spine cause some impairments in the hips and lower extremities, bowel, bladder, and sexual function.

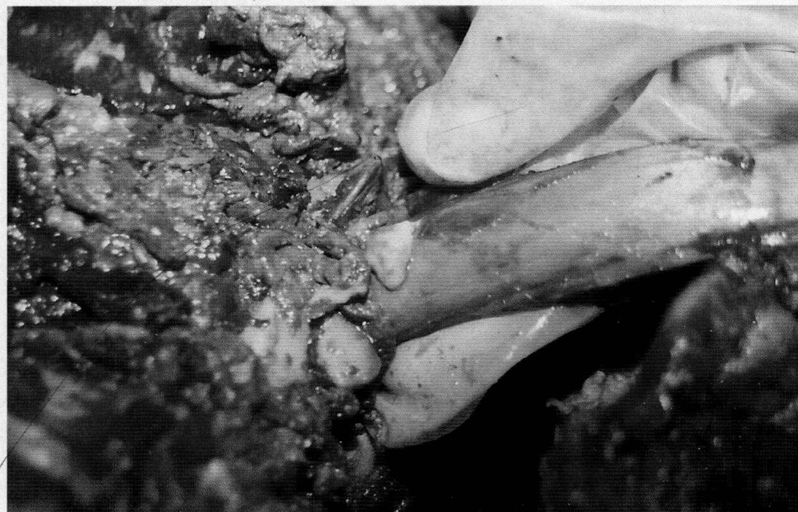

Autopsy dissection specimen of a traumatic spinal cord injury. Sharp fracture fragments have caused a small tear in the dura mater, arachnoid mater, and pia mater allowing nerve roots of the cauda equina to protrude.

Ascending tracts carry sensory information toward the brain, and **descending tracts** carry motor commands into the spinal cord. Within each column, the tracts are separated according to the destination of the motor information or the source of the sensory information. As a result, the tracts show a regional organization similar to that found in the nuclei of the gray matter (**Figure 14.3b,c**). (Chapter 15 discusses the major CNS tracts.)

14.3 CONCEPT CHECK

5 A patient with polio—a viral infection of motor neurons—has lost the use of his leg muscles. In what area of his spinal cord would you expect to find the infected motor neurons?

6 What is the difference between ascending tracts and descending tracts in the white matter?

See the blue Answers tab at the back of the book.

14.4 | Spinal Nerves

▶ **KEY POINT** We identify spinal nerves by their association with adjacent vertebrae. Three layers of connective tissue surround each peripheral nerve: an outer epineurium, a central perineurium, and an inner endoneurium.

There are 31 pairs of spinal nerves: 8 cervical, 12 thoracic, 5 lumbar, 5 sacral, and 1 coccygeal spinal nerve. We identify each pair by its association with an adjacent vertebra. Every spinal nerve has a regional number (refer to **Figure 14.1**, p. 362).

In the cervical region, the first pair of spinal nerves, C_1, exits between the skull and the first cervical vertebra. For this reason, *cervical nerves take their names from the vertebra immediately inferior to them*. In other words, cervical nerve C_2 exits from the vertebral column superior to vertebra C_2, and the same system is used for the rest of the cervical spinal nerves.

The transition from this identification method occurs between the last cervical and first thoracic vertebrae. The spinal nerve lying between these two vertebrae is designated C_8 (**Figure 14.1b**). Thus, there are *seven* cervical vertebrae but *eight* cervical nerves. *Spinal nerves caudal to the first thoracic vertebra take their names from the vertebra immediately superior to them*. Thus, the spinal nerve T_1 emerges immediately inferior to vertebra T_1, spinal nerve T_2 exits inferior to vertebra T_2, and so forth.

Three layers of connective tissue surround each peripheral nerve: an outer epineurium, a central perineurium, and an inner endoneurium (**Figure 14.4**). The arrangement of these layers is comparable to the connective tissue layers in skeletal muscles. The **epineurium** is a tough fibrous sheath forming the outermost layer of a peripheral nerve. It consists of dense irregular connective tissue mainly composed of collagen fibers and fibrocytes. At each intervertebral foramen, the epineurium of a spinal nerve is continuous with the dura mater of the spinal cord.

The **perineurium** is composed of collagenous fibers, elastic fibers, and fibrocytes. The perineurium divides the nerve into a series of compartments that contain bundles of axons. A single bundle of axons is known as a **fascicle**, or **fasciculus**. Peripheral nerves must be isolated and protected from the chemical components of the interstitial fluid and the general circulation. The connective tissue fibers and fibrocytes of the perineurium serve this function, forming the blood–nerve barrier.

The **endoneurium** surrounding each individual axon is composed of loose, irregularly arranged connective tissue containing delicate collagen and elastic connective tissue fibers and a few isolated fibrocytes. Capillaries pierce the epineurium and perineurium and branch in the endoneurium, providing oxygen and nutrients to the axons and Schwann cells of the nerve.

Figure 14.4 **Anatomy of a Peripheral Nerve.** A peripheral nerve consists of an outer epineurium enclosing a variable number of fascicles (bundles of nerve fibers). The fascicles are wrapped by the perineurium, and within each fascicle the individual axons, which are wrapped in Schwann cells, are surrounded by the endoneurium.

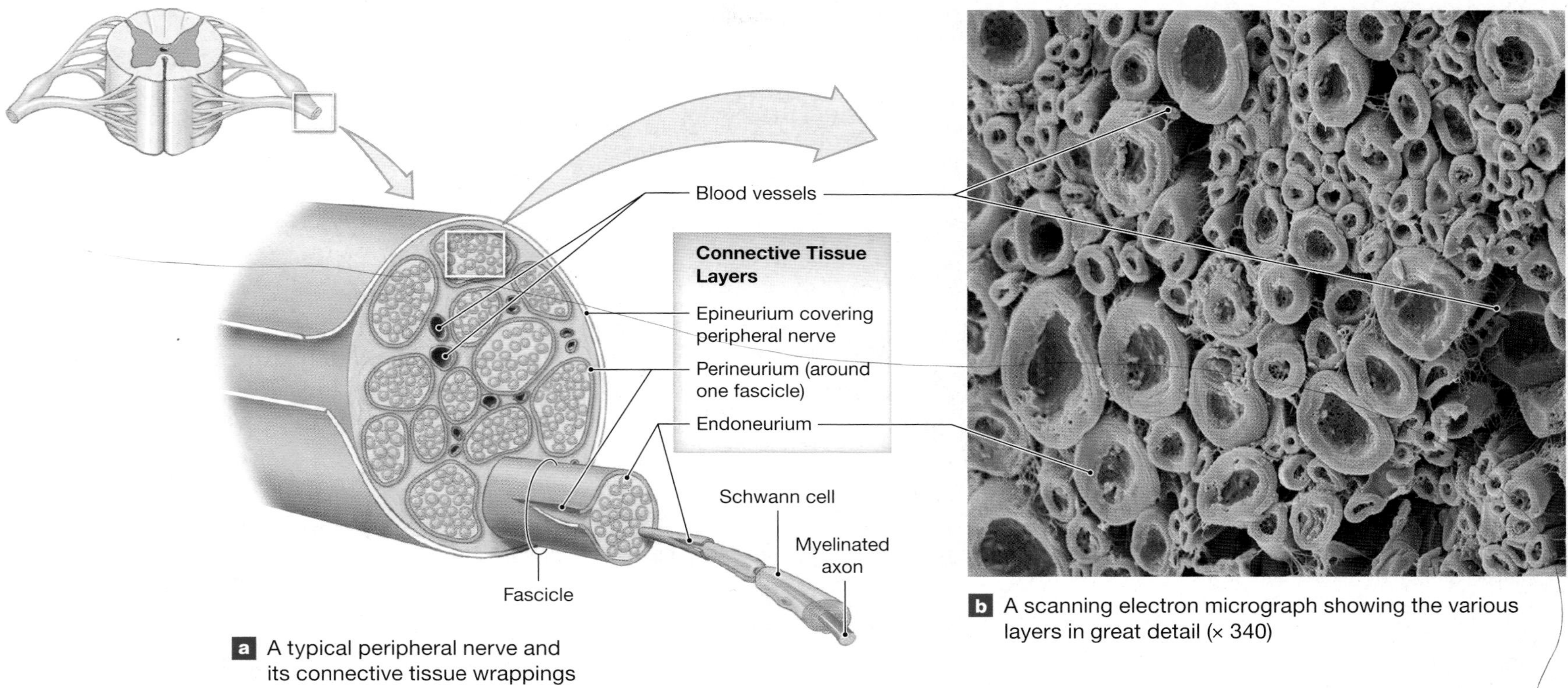

a A typical peripheral nerve and its connective tissue wrappings

b A scanning electron micrograph showing the various layers in great detail (× 340)

Peripheral Distribution of Spinal Nerves

▶ **KEY POINT** All spinal nerves have two branches (dorsal ramus and ventral ramus). Spinal nerves T_1 to L_2 have four branches: dorsal and ventral rami plus a white ramus communicans and a gray ramus communicans.

As the dorsal and ventral roots of a spinal nerve pass through an intervertebral foramen, they unite to form the spinal nerve. Distally, the spinal nerve divides into several branches. All spinal nerves form two branches: a dorsal ramus and a ventral ramus. *Spinal nerves* T_1 *to* L_2 *have four branches*: (1) a white ramus (*ramus*, branch) communicans, (2) a gray ramus communicans, (3) a dorsal ramus, and (4) a ventral ramus (**Figure 14.2**).

The rami communicantes carry visceral motor fibers to and from a nearby **autonomic ganglion** associated with the sympathetic division of the ANS. (We will discuss the sympathetic division in Chapter 17.) Because preganglionic axons are myelinated, the branch carrying those fibers to the ganglion has a light color, and it is known as the **white ramus communicans** (*white communicating ramus*). Two groups of unmyelinated postganglionic fibers leave the ganglion. Those innervating glands and smooth muscles in the body wall or limbs form a second branch, the **gray ramus communicantes** (*gray communicating ramus*), that rejoins the spinal nerve. The gray ramus is proximal to the white ramus. Preganglionic or postganglionic fibers innervating internal organs do not rejoin the spinal nerves. Instead, they form a series of separate autonomic nerves, such as the splanchnic nerves, involved with regulating the activities of organs in the abdominopelvic cavity.

The **dorsal (posterior) ramus** of each spinal nerve receives sensory innervation from, and sends motor innervation to, the skeletal muscles of the back. The relatively large **ventral (anterior) ramus** supplies the ventrolateral body surface, structures in the body wall, and the limbs.

The distribution of the sensory fibers within the dorsal and ventral rami illustrates the segmental division of labor along the length of the spinal cord (**Figure 14.5b**). Each pair of spinal nerves supplies a specific region of the skin, an area known as a **dermatome** (**Figure 14.6**). Dermatomes are clinically important because damage to either a spinal nerve or dorsal root ganglion will produce a characteristic loss of sensation in specific areas of the skin.

14.4 CONCEPT CHECK

7 Describe, in order from outermost to innermost, the three connective tissue layers surrounding each peripheral nerve.

8 Distinguish between a white ramus and a gray ramus.

See the blue Answers tab at the back of the book.

14.5 Nerve Plexuses

▶ **KEY POINT** There are four major nerve plexuses: cervical, brachial, lumbar, and sacral.

The distribution pattern illustrated in **Figure 14.5** applies to spinal nerves T_1–L_2. White rami communicantes are found only in these segments; however, gray rami communicantes, dorsal rami, and ventral rami are characteristic of all spinal nerves.

The dorsal rami provide roughly segmental sensory innervation, as evidenced by the pattern of dermatomes. The segmental alignment isn't exact, because the boundaries are imprecise, and there is some overlap between adjacent dermatomes. In adult spinal cord segments controlling the skeletal musculature of the neck and the upper and lower limbs, the ventral rami do not proceed directly to their peripheral targets. Instead, during embryonic development the ventral rami of adjacent spinal nerves blend their fibers and produce a series of compound nerve trunks. Such a complex, interwoven network of nerves is called a **nerve plexus** (PLEK-sus; "braid").

Nerve plexuses form during embryonic development as small skeletal muscles fuse with their neighbors to form larger muscles. Although the anatomical

Figure 14.5 **Peripheral Distribution of Spinal Nerves.** Diagrammatic view illustrating the distribution of fibers in the major branches of a representative thoracic spinal nerve.

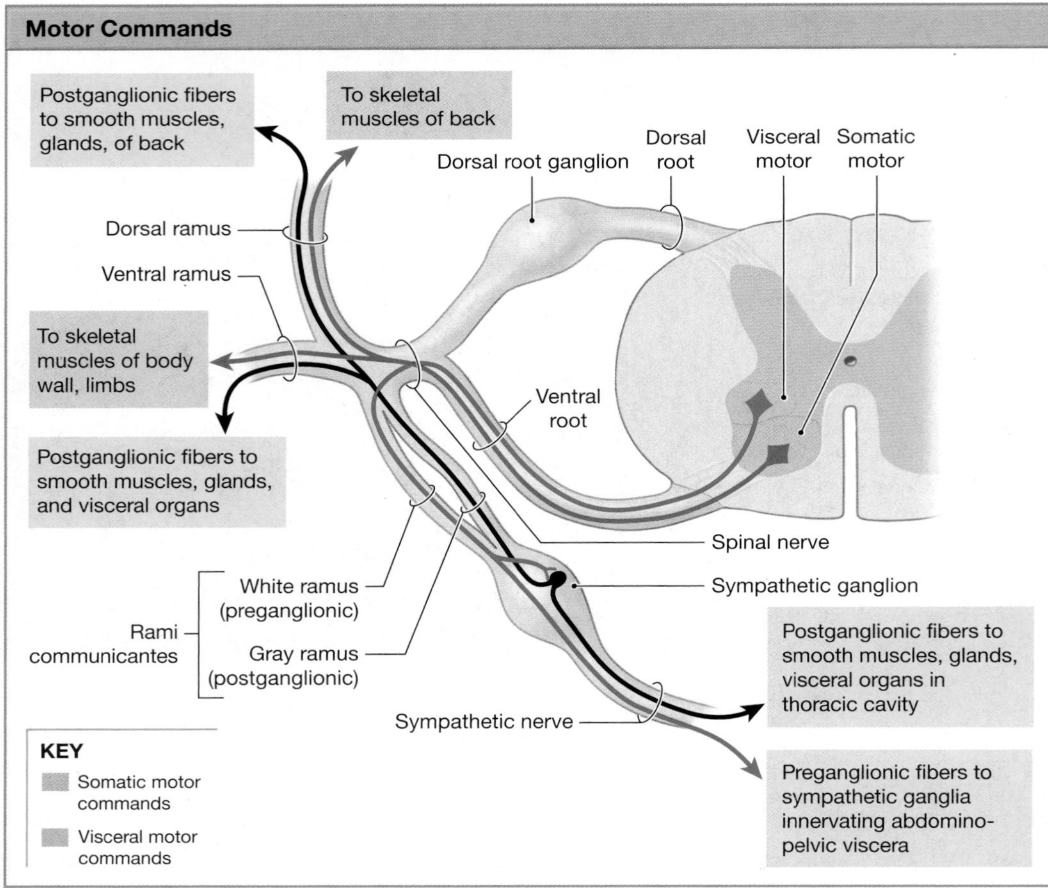

a The distribution of motor neurons in the spinal cord and motor fibers within the spinal nerve and its branches

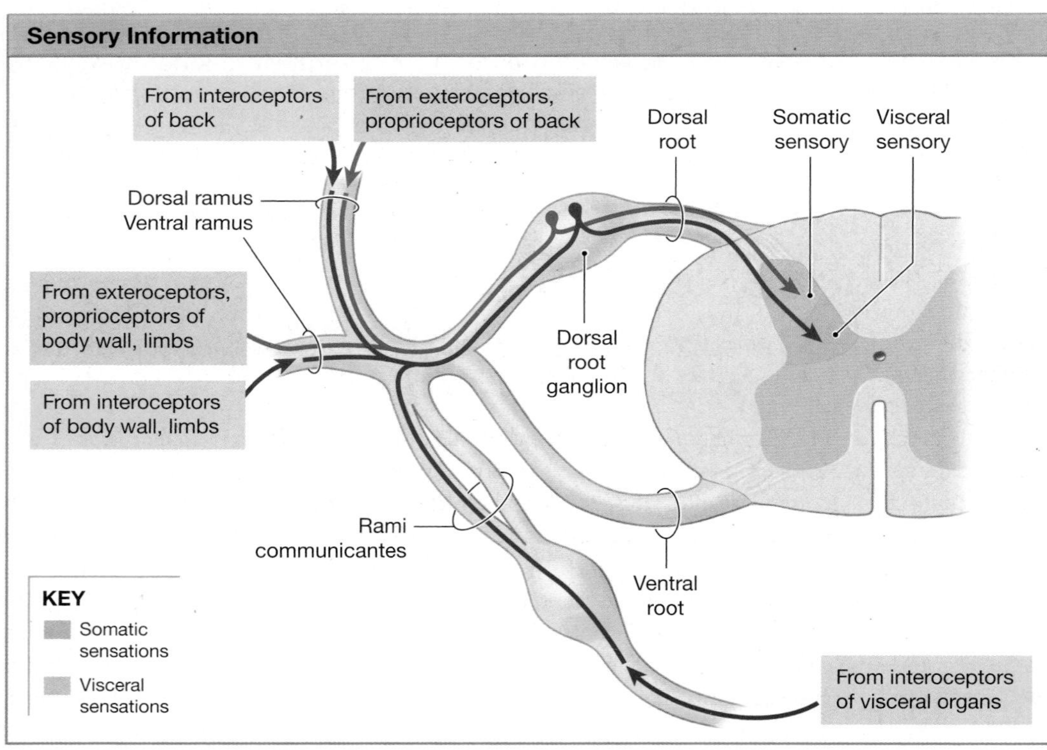

b A comparable view showing the distribution of sensory neurons and sensory fibers

boundaries between the embryonic muscles disappear, the original pattern of innervation remains intact. Therefore, "nerves" innervating these muscles in the adult contain sensory and motor fibers from the ventral rami that innervated the embryonic muscles, forming nerve plexuses. These plexuses are the cervical plexus, brachial plexus, lumbar plexus, and sacral plexus **(Figure 14.7)**.

The Cervical Plexus

▶ **KEY POINT** The cervical plexus consists of cutaneous branches that innervate areas of the head, neck, and chest. These are the largest branches of the cervical plexus. It also includes smaller muscular branches that innervate the muscles of the larynx.

Spotlight Figure 14.8 and **Figure 14.9c** illustrate the anatomy of the **cervical plexus**, which is composed of cutaneous and muscular branches of the ventral rami of spinal nerves C_1–C_4 and some nerve fibers from C_5.

Figure 14.6 **Dermatomes.** Anterior and posterior distribution of dermatomes; the related spinal nerves are indicated for each dermatome.

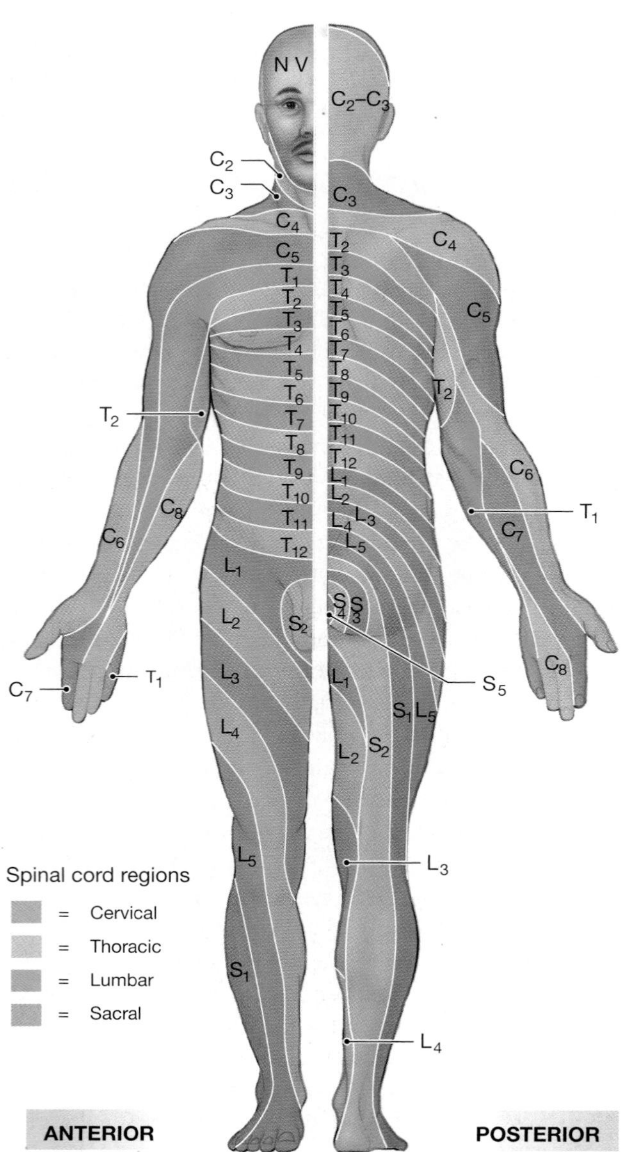

Figure 14.7 Peripheral Nerves and Nerve Plexuses

Cervical plexus (C₁–C₅)

The cervical plexus innervates the muscles of the neck and the diaphragm.

Brachial plexus (C₅–T₁)

The brachial plexus innervates the muscles of the pectoral girdles and upper limbs.

Lumbosacral Plexus

The lumbosacral plexus innervates the muscles of the pelvic girdle and lower limbs.

Lumbar plexus (T₁₂–L₄)

Sacral plexus (L₄–S₄)

C₁
C₂
C₃
C₄
C₅
C₆
C₇
C₈
T₁
T₂
T₃
T₄
T₅
T₆
T₇
T₈
T₉
T₁₀
T₁₁
T₁₂
L₁
L₂
L₃
L₄
L₅
S₁
S₂
S₃
S₄
S₅
Co₁

Major Nerves of the Cervical Plexus

Axillary nerve

Musculocutaneous nerve

Radial nerve

Ulnar nerve

Median nerve

Thoracic nerves

Major Nerves of the Lumbar Plexus

Iliohypogastric nerve

Ilio-inguinal nerve

Genitofemoral nerve

Femoral nerve

Obturator nerve

Lateral femoral cutaneous nerve

Major Nerves of the Sacral Plexus

Superior ⎤
 ⎬ Gluteal nerves
Inferior ⎦

Pudendal nerve

Sciatic nerve

Saphenous nerve

Common fibular nerve

Tibial nerve

Medial sural cutaneous nerve

14

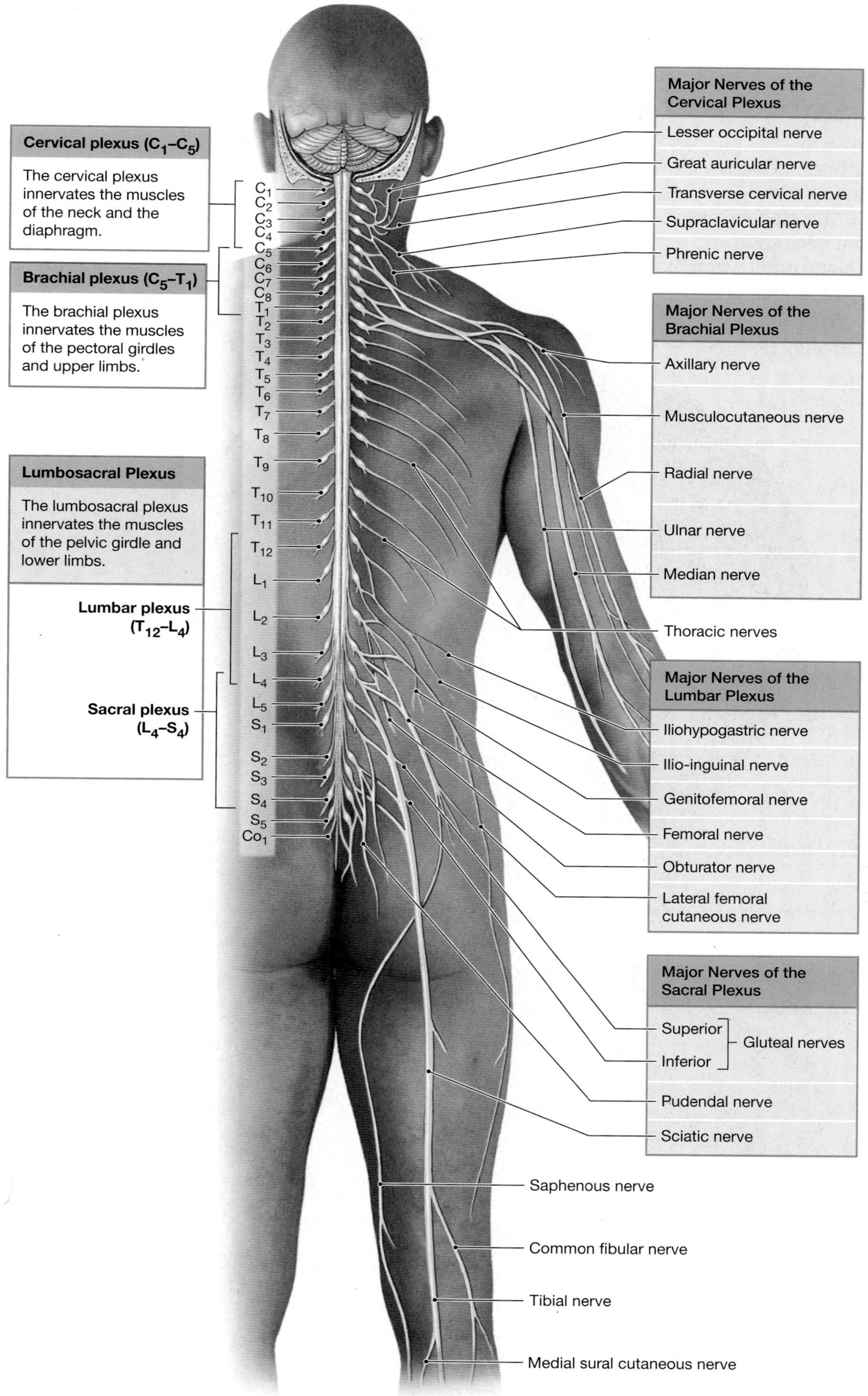

SPOTLIGHT

FIGURE 14.8
The Cervical and Brachial Plexuses

The Cervical Plexus

The cervical plexus consists of cutaneous and muscular branches from the ventral rami of spinal nerves C_1–C_4 and some fibers from C_5. The plexus lies deep to the sternocleidomastoid and anterior to the middle scalene and levator scapulae. The cutaneous branches of this plexus innervate areas on the head, neck, and chest.

Cranial Nerves

The cervical plexus supplies small branches to the hypoglossal nerve and the accessory nerve through C_1.

Hypoglossal nerve (XII)

Accessory nerve (XI)

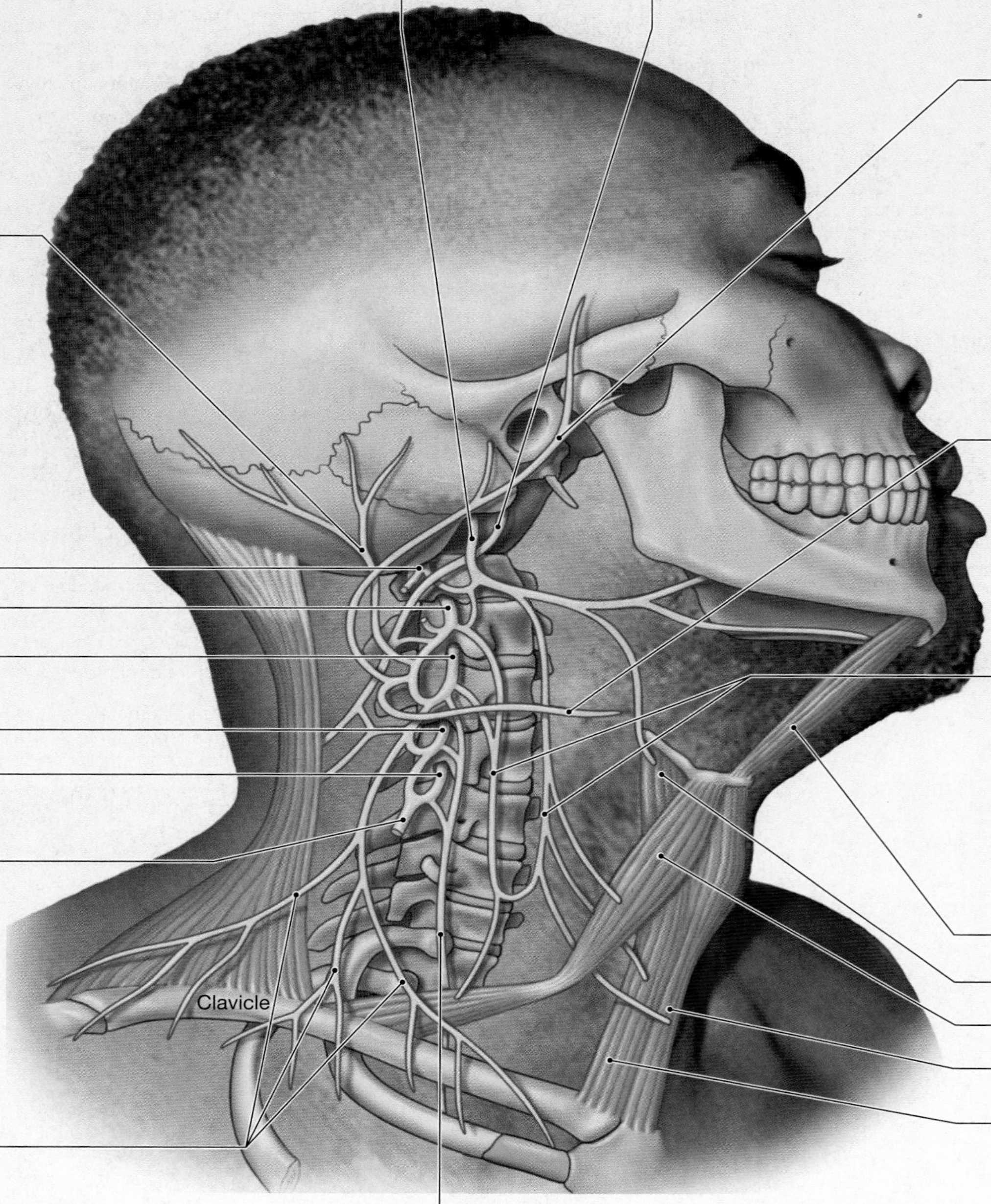

Great Auricular Nerve

The great auricular nerve arises from C_2 and C_3 and crosses anteriorly to the sternocleidomastoid and travels toward the parotid gland, where it divides. This nerve receives sensory information from the skin over the gland, the posterior aspect of the ear, and skin of the neck.

Lesser Occipital Nerve

The lesser occipital nerve originates from C_2 and receives sensory information from the skin of the neck and the scalp posterior and superior to the ear.

Nerve Roots of the Cervical Plexus

The cervical plexus originates from the ventral rami of the second, third, fourth, and fifth cervical nerves.

C_1

C_2

C_3

C_4

C_5

Transverse Cervical Nerve

The transverse cervical nerve originates from C_3 and C_4 and receives sensory input from the skin of the anterior triangle of the neck.

Ansa Cervicalis

The ansa cervicalis originates from branches of C_1–C_3 (and sometimes C_4) and travels inferiorly with fibers from cranial nerve XII. The ansa cervicalis innervates five of the extrinsic laryngeal muscles.

Geniohyoid

Thyrohyoid

Omohyoid

Sternohyoid

Sternothyroid

Nerves to Rhomboids and Serratus Anterior

Motor fibers originating at C_5 innervate the rhomboids (major and minor) and a portion of the serratus anterior.

Clavicle

Supraclavicular Nerves

The supraclavicular nerves originate from C_3 and C_4 as a common trunk. This trunk receives sensory input from the skin of the neck and shoulder.

Phrenic Nerve

The phrenic nerve, which provides sensory information from, and motor innervation to, the diaphragm, originates from C_4, with minor contributions from C_3 and C_5.

The Brachial Plexus

The brachial plexus is larger and more complex than the cervical plexus. It originates from the ventral rami of spinal nerves C_5–T_1 and innervates the pectoral girdle and upper limb. The ventral rami converge to form the superior, middle, and inferior trunks.

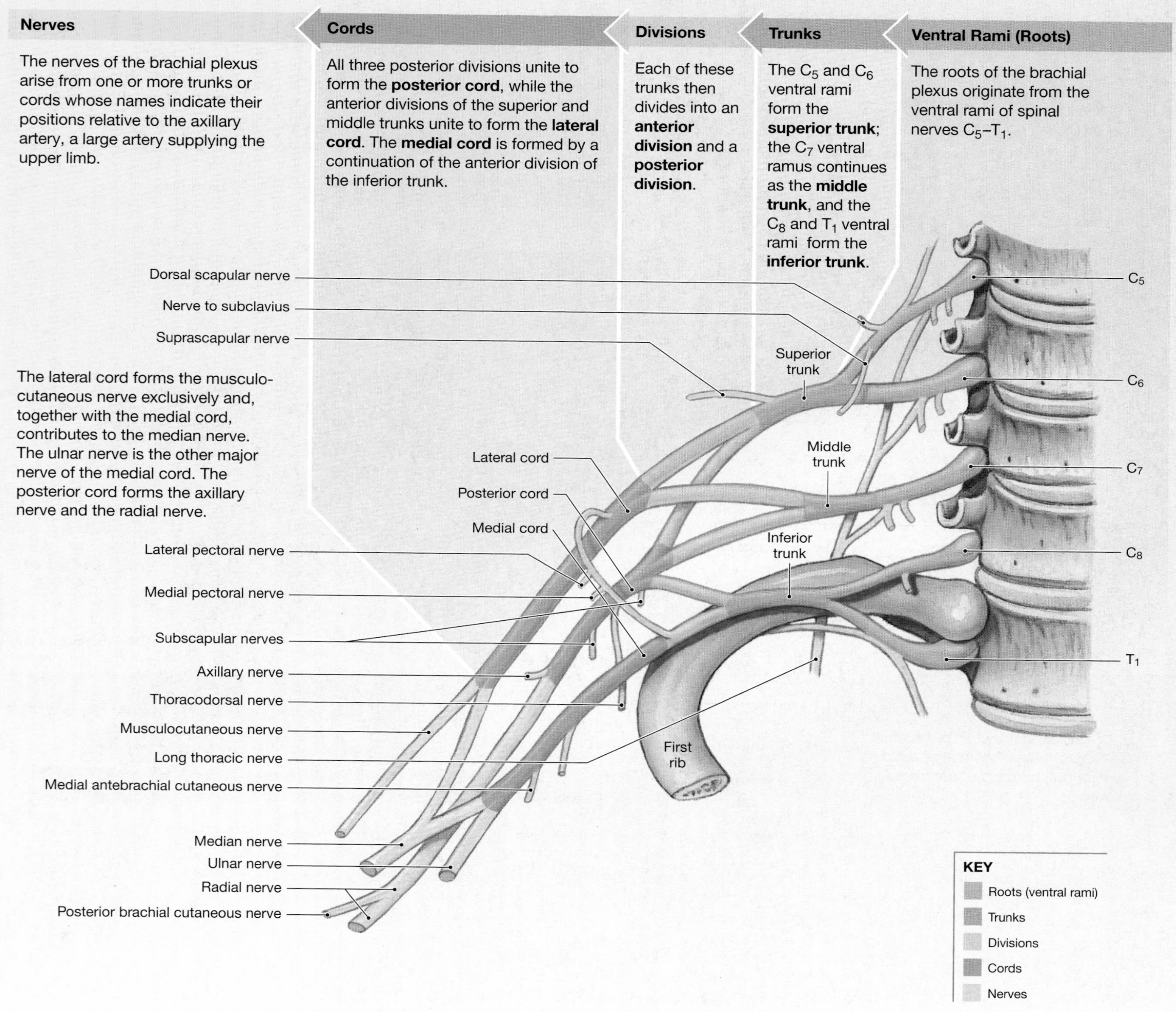

Nerves
The nerves of the brachial plexus arise from one or more trunks or cords whose names indicate their positions relative to the axillary artery, a large artery supplying the upper limb.
Cords
All three posterior divisions unite to form the **posterior cord**, while the anterior divisions of the superior and middle trunks unite to form the **lateral cord**. The **medial cord** is formed by a continuation of the anterior division of the inferior trunk.
Divisions
Each of these trunks then divides into an **anterior division** and a **posterior division**.
Trunks
The C_5 and C_6 ventral rami form the **superior trunk**; the C_7 ventral ramus continues as the **middle trunk**, and the C_8 and T_1 ventral rami form the **inferior trunk**.
Ventral Rami (Roots)
The roots of the brachial plexus originate from the ventral rami of spinal nerves C_5–T_1.
The lateral cord forms the musculocutaneous nerve exclusively and, together with the medial cord, contributes to the median nerve. The ulnar nerve is the other major nerve of the medial cord. The posterior cord forms the axillary nerve and the radial nerve.
Dorsal scapular nerve
Nerve to subclavius
Suprascapular nerve
Superior trunk
Middle trunk
Inferior trunk
Lateral cord
Posterior cord
Medial cord
Lateral pectoral nerve
Medial pectoral nerve
Subscapular nerves
Axillary nerve
Thoracodorsal nerve
Musculocutaneous nerve
Long thoracic nerve
Medial antebrachial cutaneous nerve
Median nerve
Ulnar nerve
Radial nerve
Posterior brachial cutaneous nerve
First rib
C_5
C_6
C_7
C_8
T_1
KEY
Roots (ventral rami)
Trunks
Divisions
Cords
Nerves

Figure 14.9 The Brachial Plexus

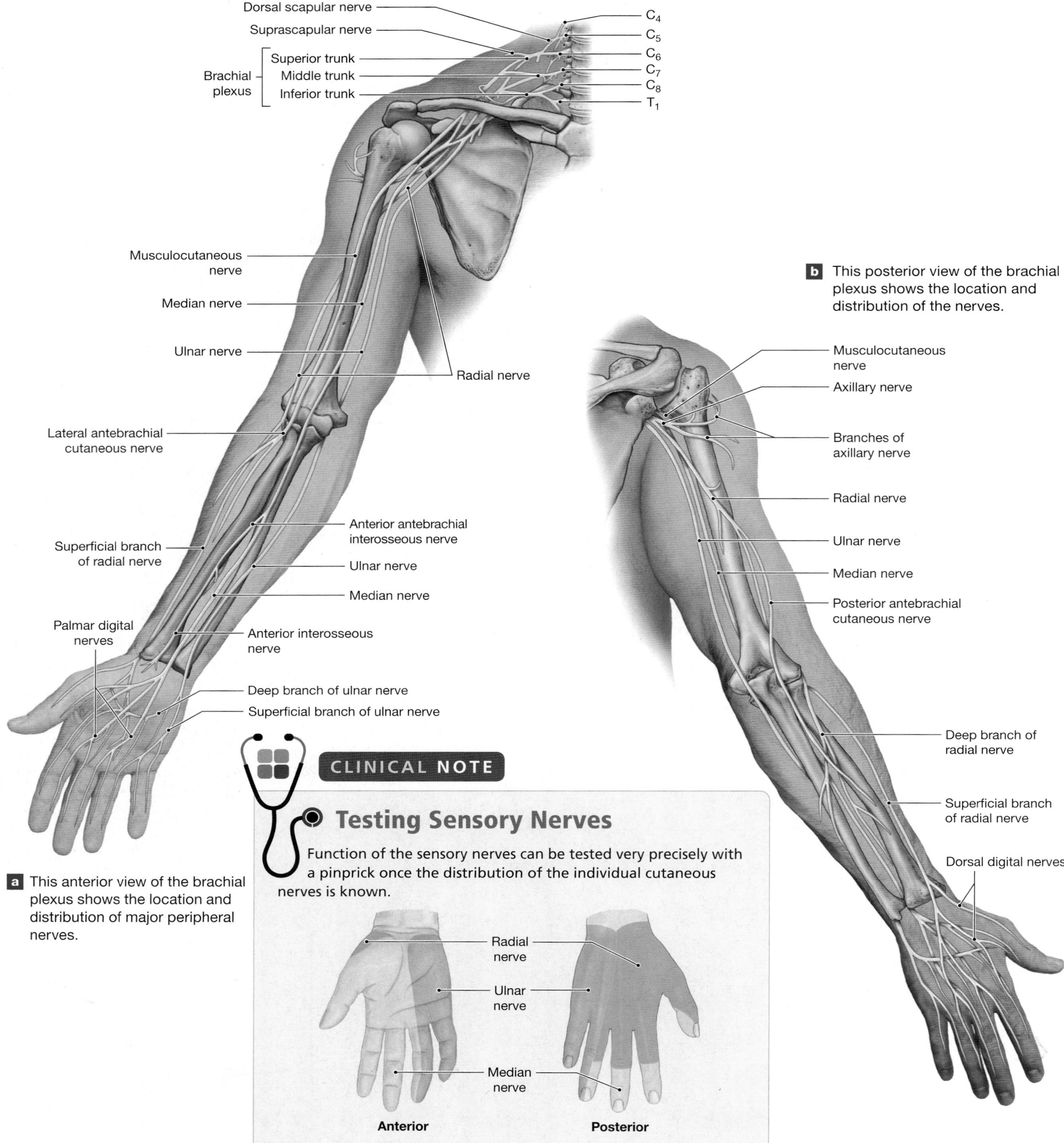

a This anterior view of the brachial plexus shows the location and distribution of major peripheral nerves.

b This posterior view of the brachial plexus shows the location and distribution of the nerves.

CLINICAL NOTE

Testing Sensory Nerves

Function of the sensory nerves can be tested very precisely with a pinprick once the distribution of the individual cutaneous nerves is known.

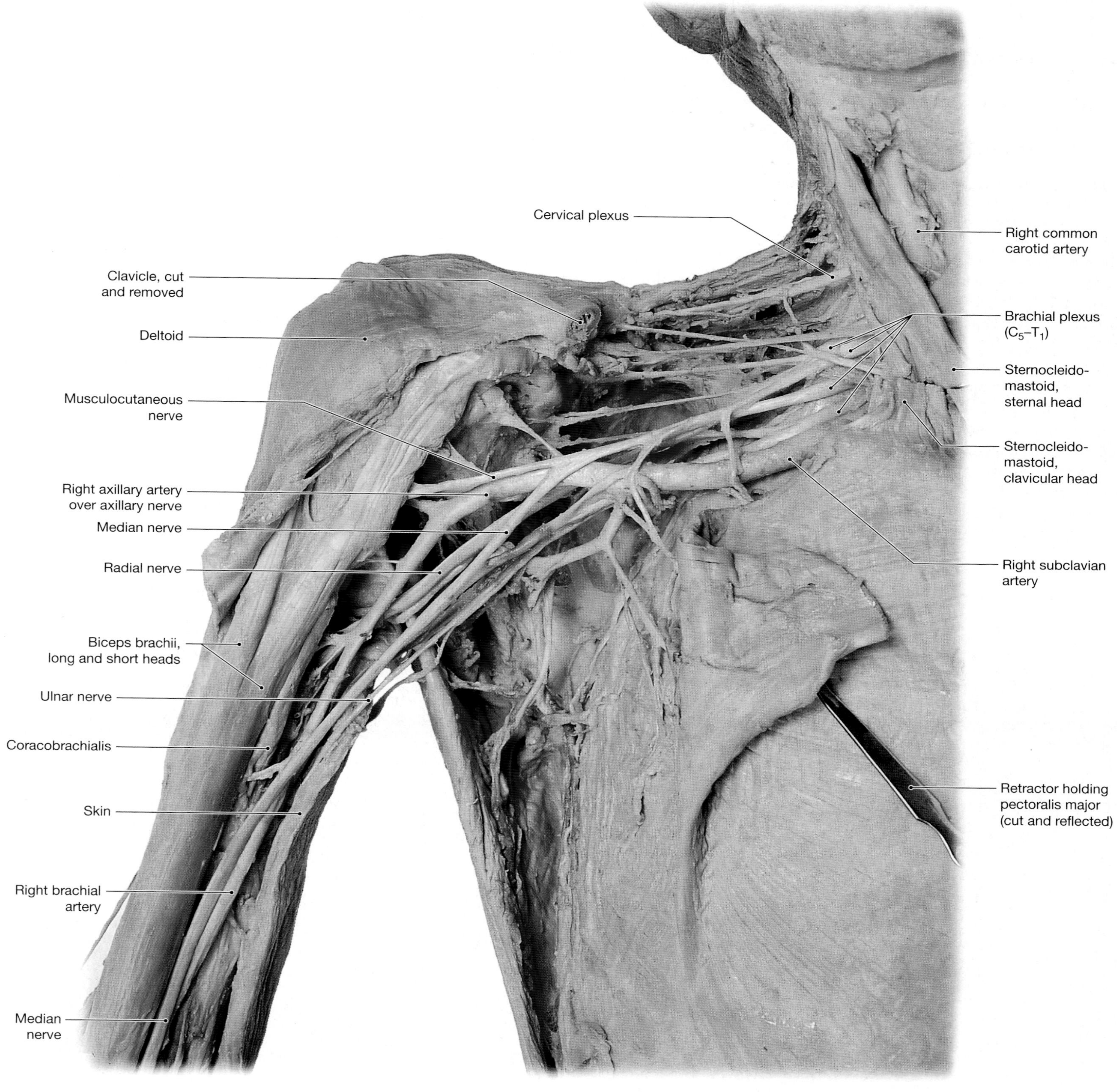

c This dissection shows the major nerves originating from the cervical and brachial plexuses.

The Brachial Plexus

▶ **KEY POINT** The brachial plexus arises from the rami of spinal nerves C_5–T_1 and innervates the pectoral girdle and upper limb.

Spotlight Figure 14.8 and **Figure 14.9** illustrate the anatomy of the **brachial plexus**, which is larger and more complex than the cervical plexus. **Table 14.1** provides further information about the nerves of the brachial plexus.

The Lumbar and Sacral Plexuses

▶ **KEY POINT** The muscles of the pelvis and lower limb, like the muscles of the pectoral girdle and upper limb, are innervated by nerves from two plexuses—the lumbar and sacral plexuses.

The **lumbar plexus** and the **sacral plexus** originate from the lumbar and sacral segments of the spinal cord. The ventral rami of these nerves innervate the pelvis and lower limb. Because the ventral rami of both plexuses innervate muscles of the lower limb, the lumbar and sacral plexuses are often referred to as the *lumbosacral plexus*. **Spotlight Figure 14.10** and **Figure 14.11** illustrate the anatomy of the lumbar plexus and sacral plexus.

Although dermatomes provide clues to the location of injuries along the spinal cord, loss of sensation at the skin does not pinpoint the site of injury because dermatome boundaries are not clearly defined lines (**Figure 14.6**, p. 368). We can draw more exact conclusions from the loss of motor control based on the origin and distribution of the peripheral nerves originating at nerve plexuses. When assessing motor performance, we must distinguish between the conscious ability to control motor activities and the performance of automatic, involuntary motor responses.

14.5 CONCEPT CHECK

9 Injury to which of the nerve plexuses would interfere with the ability to breathe?

10 Which nerve plexus may have been damaged if motor activity in the arm and forearm are affected by injury?

See the blue Answers tab at the back of the book.

14.6 Reflexes

▶ **KEY POINT** Reflexes help preserve homeostasis by enabling us to respond rapidly to changes in the internal or external environment.

Conditions inside or outside the body change unexpectedly. A **reflex** is an immediate involuntary motor response to a specific stimulus (**Figures 14.12** to **14.15**). Reflexes help preserve homeostasis by making rapid adjustments in the function of organs or organ systems. A reflexive response seldom varies—activation of a particular reflex *always* produces the same motor response. The neural "wiring" of a single reflex is called a **reflex arc**. A reflex arc begins at a receptor and ends at a peripheral effector, such as a muscle or gland cell. **Figure 14.12** illustrates the five steps involved in a reflex arc:

Step 1: Stimulation and activation of receptor. There are many types of sensory receptors (the general categories of sensory receptors were introduced in Chapter 13). Each receptor has a characteristic range of sensitivity; some receptors, such as pain receptors, respond to almost any stimulus.

Table 14.1 | The Brachial Plexus

Spinal Segments	Nerve(s)	Distribution
C_4–C_6	Nerve to subclavius	Subclavius
C_5	Dorsal scapular nerve	Rhomboid and levator scapulae
C_5–C_7	Long thoracic nerve	Serratus anterior
C_5, C_6	Suprascapular nerve	Supraspinatus and infraspinatus; sensory from shoulder joint and scapula
C_5–T_1	Pectoral nerves (medial and lateral)	Pectoralis muscles
C_5, C_6	Subscapular nerves	Subscapularis and teres major
C_6–C_8	Thoracodorsal nerve	Latissimus dorsi
C_5, C_6	Axillary nerve	Deltoid and teres minor; sensory from skin of shoulder
C_8, T_1	Medial antebrachial cutaneous nerve	Sensory from skin over anterior, medial surface of arm and forearm
C_5–T_1	Radial nerve	Many extensor muscles on the arm and forearm (triceps brachii, anconeus, extensor carpi radialis, extensor carpi ulnaris, and brachioradialis; supinator digital, extensor muscles, and abductor pollicis via the *deep branch;* sensory from skin over the posterolateral surface of the limb through the *posterior brachial cutaneous nerve* (arm), *posterior antebrachial cutaneous nerve* (forearm), and the *superficial branch* (radial portion of hand)
C_5–C_7	Musculocutaneous nerve	Flexor muscles on the arm (biceps brachii, brachialis, and coracobrachialis; sensory from skin over lateral surface of the forearm through the *lateral antebrachial cutaneous nerve*
C_6–T_1	Median nerve	Flexor muscles on the forearm (flexor carpi radialis and palmaris longus; pronator quadratus and pronator teres; radial half of flexor digitorum profundus, digital flexors (through the *anterior interosseous nerve*); sensory from skin over anterolateral surface of the hand
C_8, T_1	Ulnar nerve	Flexor carpi ulnaris, ulnar half of flexor digitorum profundus, adductor pollicis, and small digital muscles through the *deep branch*; sensory from skin over medial surface of the hand through the *superficial branch*

CLINICAL NOTE

Localized Peripheral Neuropathies

Generalized peripheral neuropathies affect the entire PNS. They include the very common diabetic neuropathy as well as the uncommon Guillain-Barré syndrome (see Clinical Case, p. 359). Localized peripheral neuropathies, or *peripheral nerve palsies*, are characterized by regional losses of sensory and motor function as a result of nerve trauma or compression.

Brachial plexus birth palsy, also known as *Erb's palsy*, is due to nerve trauma during difficult deliveries. Trying to quickly deliver a baby's shoulders after its head is delivered can stretch the head away from the resistant shoulder, tearing the upper ventral rami, or nerve roots, of the brachial plexus. This causes loss of shoulder sensation and motion but preserves function of the forearm and hand (innervated by lower nerve roots).

You are probably familiar with some pressure palsies. Sitting on your foot can cause it to "fall asleep" with resolving numbness and "pins-and-needles" paresthesias. Using a crutch that is too long and exerts pressure on the posterior cord of the brachial plexus can cause **radial nerve palsy** resulting in wrist drop (difficulty lifting the wrist). **Carpal tunnel syndrome** is an entrapment neuropathy that responds well to surgical release. If you bump the medial side of your elbow, you may experience temporary **ulnar nerve palsy** that causes transient dysfunction of the ulnar nerve posterior to the medial humeral epicondyle, commonly known as hitting your "funny bone." A protruding lumbar disc exerting pressure on lumbosacral nerve roots in the cauda equina often causes **sciatica**, a common peripheral neuropathy. The resulting sensory and motor loss can define the exact location of disc pressure.

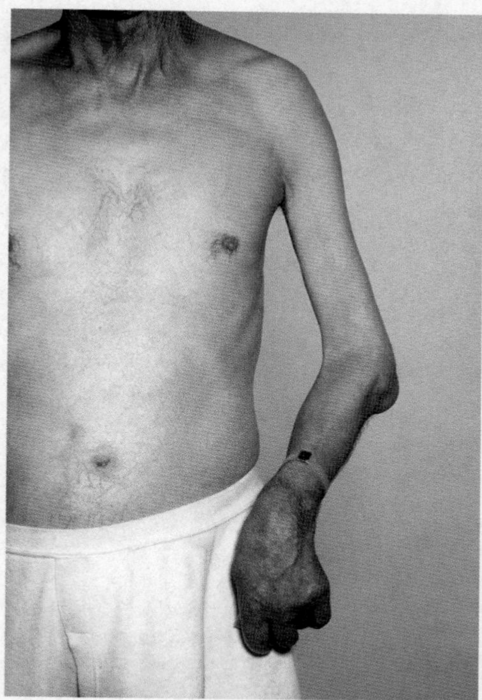

Patient with brachial plexus birth palsy (Erb's palsy). Note the paralysis of the brachial region and shoulder girdle.

CLINICAL NOTE

Lumbar Puncture and Spinal Anesthesia

A lumbar puncture, or spinal tap, is a medical procedure performed to collect cerebrospinal fluid (CSF) for diagnostic purposes. Under sterile conditions using local anesthesia, a clinician inserts a hollow needle into the subarachnoid space to extract a sample of CSF. The needle is inserted into the largest intervertebral spaces, usually between L_3 and L_4 or L_4 and L_5. There is no more spinal cord at this level, only lumbosacral peripheral nerve roots that can easily roll out of harm's way. The fluid collected can be studied for infection, blood, or metabolic wastes.

Epidural anesthesia can be a very effective obstetrical anesthesia, blocking pain but preserving some motor function, important when it comes time to "push." A small catheter is threaded into the posterior space between the bony lumbar vertebra and the dura mater. Continuous anesthesia can be dripped into this space, providing pain relief for the duration of labor and delivery.

Spinal anesthesia can be used when both sensory and motor function should be blocked, as in lower extremity surgery. A clinician injects an anesthetic agent directly into the CSF at the lower lumbar level, temporarily blocking sensory and motor activity.

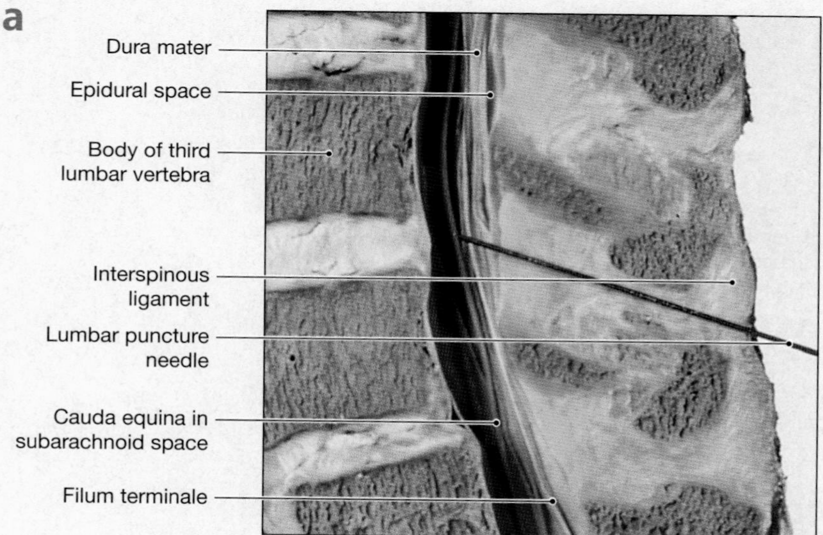

The lumbar puncture needle should be inserted in the midline between the third and fourth lumbar vertebral spines, pointing at a superior angle toward the umbilicus. The needle should puncture the dura mater and enter the subarachnoid space, near the nerves of the cauda equina, to obtain a sample of CSF.

SPOTLIGHT

FIGURE 14.10

The Lumbar and Sacral Plexuses

The lumbar plexus and sacral plexus originate from the lumbar and sacral segments of the spinal cord, respectively. The nerves originating from these plexuses innervate the pelvic girdle and lower limbs. Because the ventral rami of both plexuses are distributed to the lower limbs and spinal nerves L_4 and L_5 are involved, these two plexuses are collectively referred to as the lumbosacral plexus.

The Lumbar Plexus

The lumbar plexus originates from the ventral rami of T_{12}–L_4. The major nerves of the plexus are the lateral femoral cutaneous nerve, the genitofemoral nerve, and the femoral nerve.

Nerve Roots of Lumbar Plexus

The lumbar plexus is formed by the ventral rami of T_{12}–L_4.

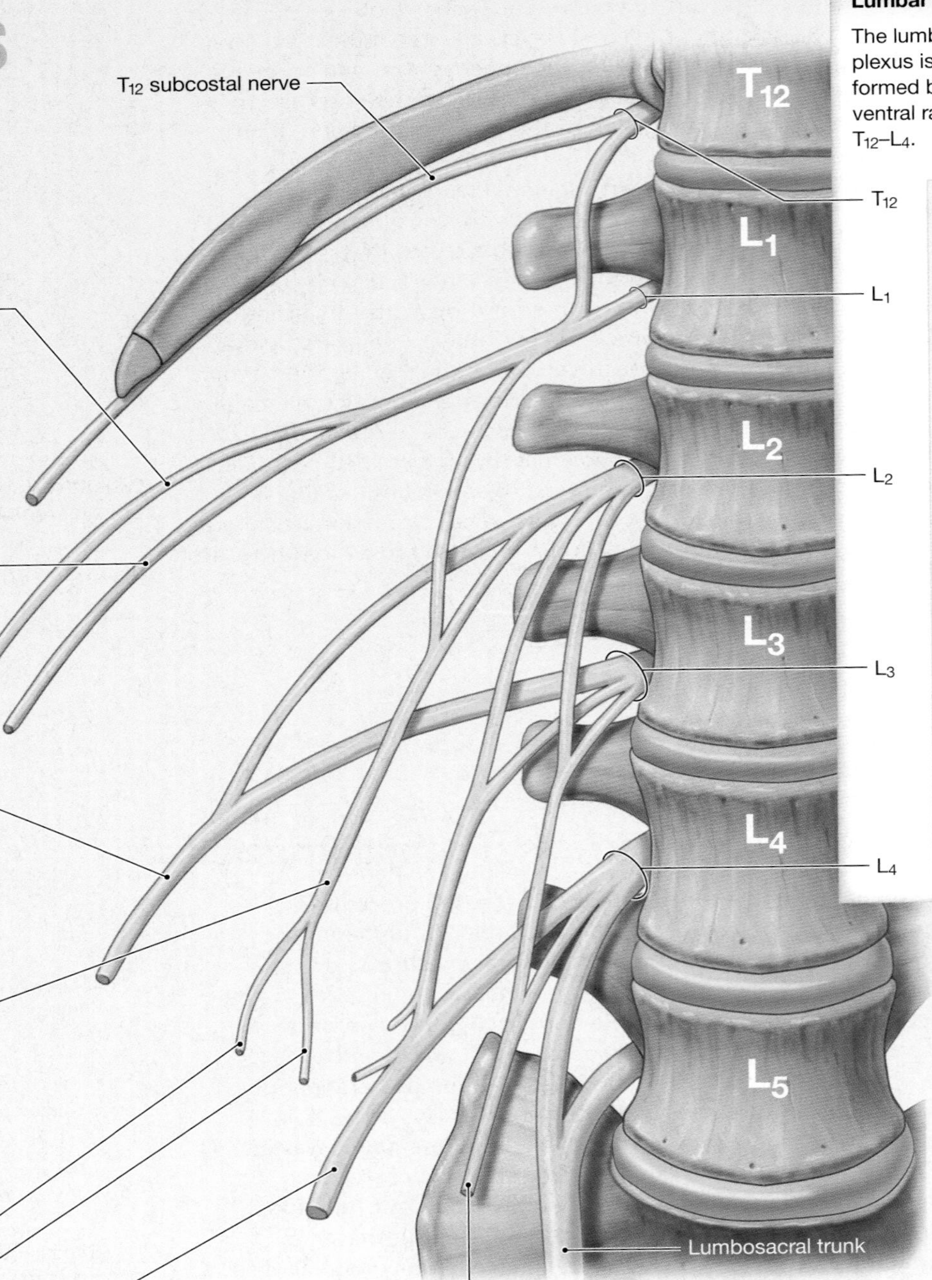

Iliohypogastric Nerve

The iliohypogastric nerve originates from the ventral rami of T_{12} and L_1. It innervates the external and internal oblique and transverse abdominis. It receives sensory information from the skin over the inferior abdomen and the buttocks.

Ilio-inguinal Nerve

The ilio-inguinal nerve originates from the ventral ramus of L_1. It innervates the external and internal oblique and transverse abdominis. It receives sensory information from the skin over the superior and medial thigh and portions of the external genitalia.

Lateral Femoral Cutaneous Nerve

The lateral femoral cutaneous nerve originates from the ventral rami of L_2 and L_3. It receives sensory information from the skin over the anterior, lateral, and posterior thigh.

Genitofemoral Nerve

The genitofemoral nerve originates from the ventral rami of L_1 and L_2. It receives sensory information from the skin over the anteromedial surface of the thigh and portions of the external genitalia.

Branches of genitofemoral nerve:

Femoral branch

Genital branch

Femoral Nerve

The femoral nerve originates from the ventral rami of L_2–L_4. It innervates the quadriceps femoris, sartorius, pectineus, and iliopsoas. It receives sensory information from the skin of the anteromedial surface of the thigh and the medial surface of the leg and foot.

Obturator Nerve

The obturator nerve originates from the ventral rami of L_2–L_4. It innervates the gracilis and obturator externus, and the adductor magnus, brevis, and longus. It receives sensory information from the medial surface of the thigh.

The Sacral Plexus

The sacral plexus is formed by the ventral rami of L_4–S_4. Part of the ventral ramus of L_4 and the ventral ramus of L_5 form the lumbosacral trunk, which joins the sacral plexus. The five major nerves of the sacral plexus are discussed below.

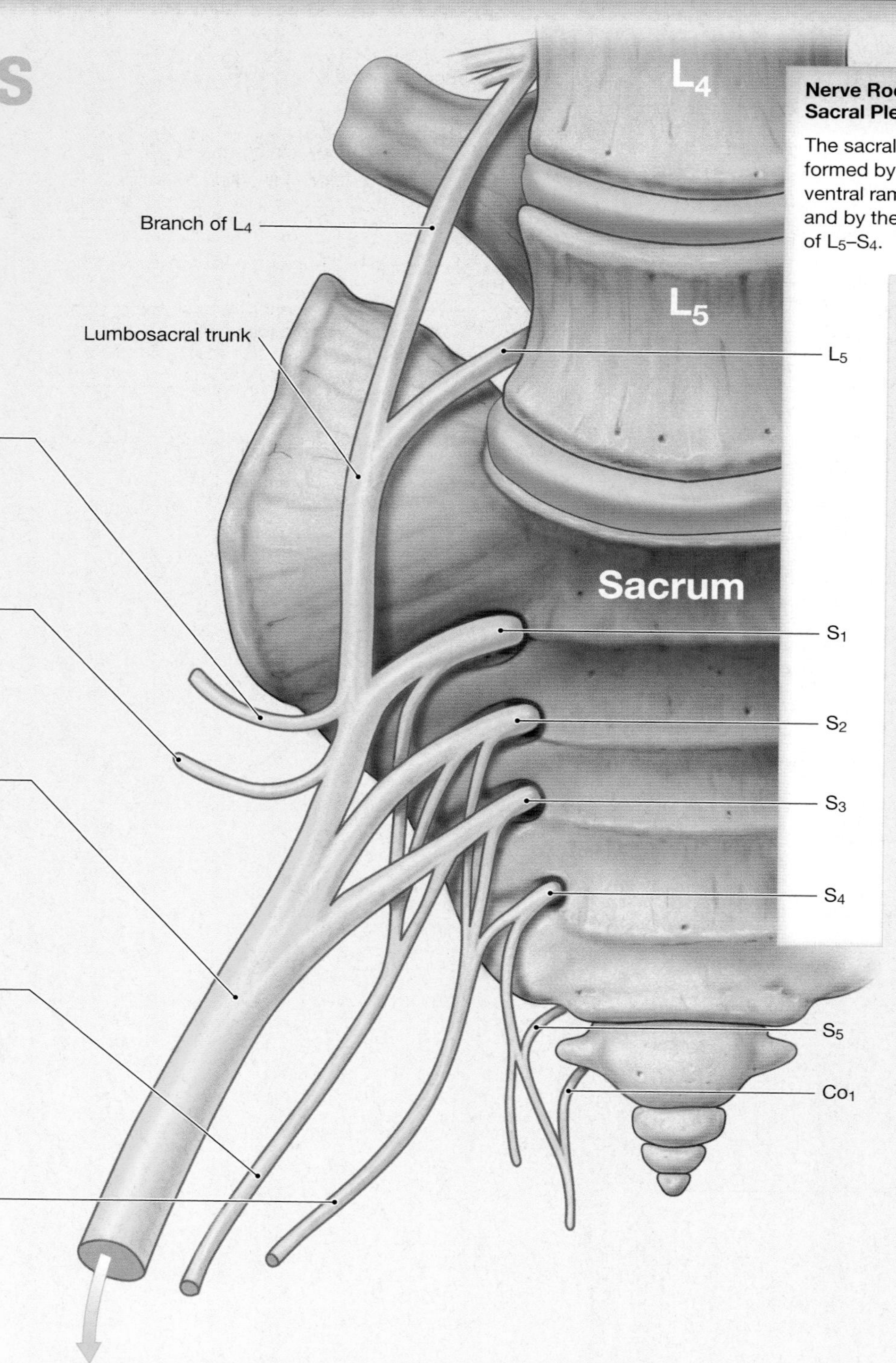

Nerve Roots of Sacral Plexus

The sacral plexus is formed by part of the ventral ramus of L_4, and by the ventral rami of L_5–S_4.

Superior Gluteal Nerve

The superior gluteal nerve originates from the ventral rami of L_4–S_1. It innervates the gluteus minimus, gluteus medius, and tensor fasciae latae.

Inferior Gluteal Nerve

The inferior gluteal nerve originates from the ventral rami of L_5–S_2. It innervates the gluteus maximus.

Sciatic Nerve

The sciatic nerve is the largest nerve in the body. It originates from the ventral rami of L_4–S_3 and innervates the semimembranosus, semitendinosus, and adductor magnus.

Posterior Femoral Cutaneous Nerve

The posterior femoral cutaneous nerve originates from the ventral rami of S_1–S_3. It receives sensory information from the perineum and the posterior surface of the thigh and leg.

Pudendal Nerve

The pudendal nerve originates from the ventral rami of S_2–S_4. It innervates muscles of the perineum, including the urogenital diaphragm and the external anal and urethral sphincters. It receives sensory information from the external genitalia and related skeletal muscles (the bulbospongiosus and ischiocavernosus).

Branches of the Sciatic Nerve

The sciatic nerve branches into the tibial and common fibular nerves near the popliteal fossa. The common fibular nerve then divides into the superficial and deep fibular nerves. (See **Figure 14.11**)

Tibial Nerve: Innervates the flexors of the knee and plantar flexors of the ankle; flexors of the toes; and skin over the posterior surface of the leg and the plantar surface of the foot.

Fibular Nerves: Innervate the short head of the biceps femoris (common fibular), tibialis anterior and extensors of the toes (deep fibular), and the fibularis brevis and fibularis longus (superficial fibular). The common fibular nerve receives information from the anterior surface of the leg and skin over the lateral portion of the foot (through the sural nerve).

Figure 14.11 Peripheral Nerves Originating from the Lumbar and Sacral Plexuses

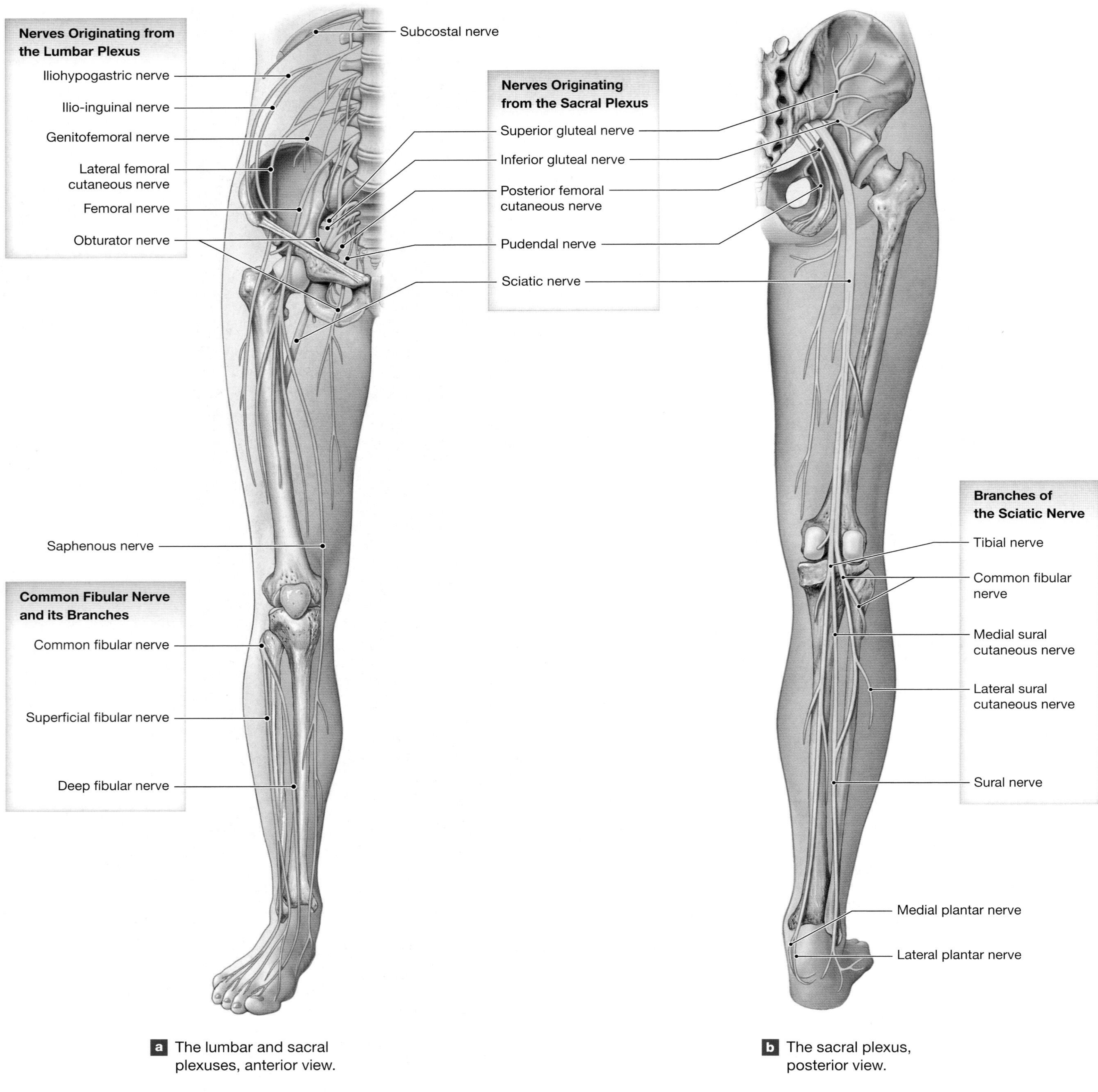

a The lumbar and sacral plexuses, anterior view.

b The sacral plexus, posterior view.

Figure 14.11 ***(continued)***

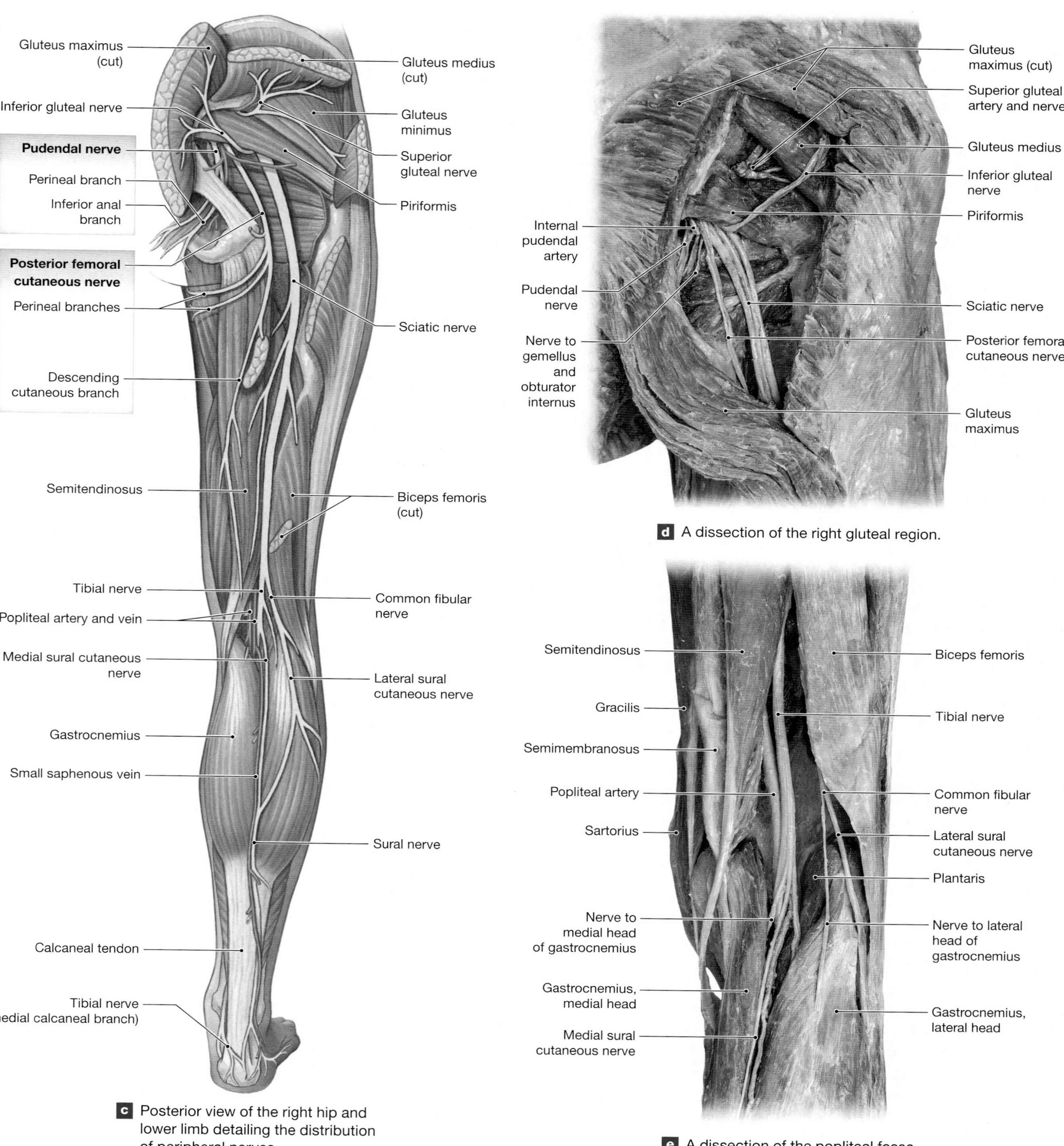

c Posterior view of the right hip and lower limb detailing the distribution of peripheral nerves.

d A dissection of the right gluteal region.

e A dissection of the popliteal fossa.

Figure 14.12 A Reflex Arc. This diagram illustrates the five steps involved in a neural reflex.

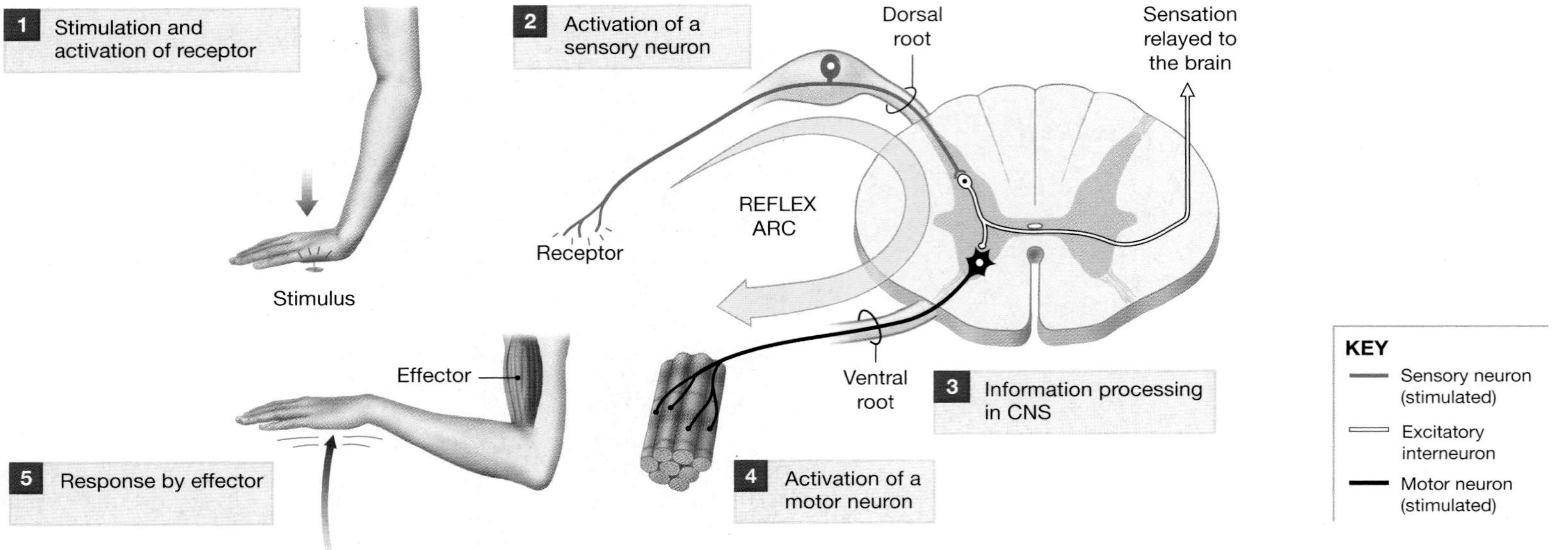

14

These receptors, which are the dendrites of sensory neurons, are stimulated by pressure, temperature extremes, physical damage, or exposure to abnormal chemicals. Other receptors, such as visual, auditory, or taste receptors, are specialized cells that respond to only a limited range of stimuli.

Step 2: Activation of a sensory neuron. Information is carried in the form of an action potential along an afferent fiber. In this case, the axon conducts the action potential into the spinal cord through one of the dorsal roots.

Step 3: Information processing in CNS. Information processing begins when the neurotransmitter released by the axon terminals of a sensory neuron reaches the postsynaptic membrane of a motor neuron or interneuron. In the simplest reflexes, this processing is performed by the motor neuron controlling the peripheral effectors. In more complex reflexes, one or more interneurons are located between the sensory and motor neurons, and both serial and parallel processing occur. This type of information processing selects the appropriate motor response through the activation of specific motor neurons.

Step 4: Activation of a motor neuron. When a motor neuron is stimulated to threshold, it conducts an action potential through the ventral root of a spinal nerve to the peripheral effector organ.

Step 5: Response by effector. Activation of the motor neuron causes a response by a peripheral effector, such as a skeletal muscle or gland. Reflexes play an important role in opposing potentially harmful changes in the internal or external environment.

Classification of Reflexes

▶ **KEY POINT** Reflexes are classified according to (1) their development (innate or acquired), (2) their response (somatic or visceral), (3) the complexity of the circuit (monosynaptic or polysynaptic), and (4) their processing site (spinal or cranial).

Figure 14.13 compares the four different criteria used to classify reflexes: development, response, complexity of circuit, and processing site. These categories are not mutually exclusive; they represent different ways of describing a single reflex.

In the simplest reflex arc, a **monosynaptic reflex**, a sensory neuron synapses directly on a motor neuron **(Figure 14.14a)**. Transmission of information across a chemical synapse always involves a synaptic delay, but with only one synapse, the delay between stimulus and response is minimized.

A **polysynaptic reflex** is more complex and has a longer delay between the stimulus and response **(Figure 14.15b)**. The length of the delay depends on the number of synapses involved. Polysynaptic reflexes produce far more complicated responses because the interneurons control several different muscle groups. The motor responses in a polysynaptic reflex are extremely complicated; for example, stepping on a sharp object not only causes withdrawal of the foot, but also initiates all the muscular adjustments needed to prevent a fall. Such complex responses result from the interactions between multiple interneuron pools.

Spinal Reflexes

The best-known spinal reflex is the **stretch reflex**. It is a simple monosynaptic reflex providing the automatic regulation of skeletal muscle length **(Figure 14.15a)**. The stimulus stretches a relaxed muscle, activating a sensory neuron and triggering the contraction of that muscle. The stretch reflex also adjusts autonomic muscle tone, increasing or decreasing it in response to information provided by the stretch receptors of muscle spindles **(Figure 14.15a)**. (We will discuss muscle spindles in Chapter 18.)

The most familiar stretch reflex is the **patellar reflex** (also known as the *knee jerk reflex*). In this reflex, a sharp tap on the patellar ligament stretches muscle spindles in the quadriceps femoris **(Figure 14.15b)**. Because the stimulus is so brief, the reflexive contraction is unopposed and produces a noticeable kick. Clinicians use this reflex to check the status of the lower segments of the spinal cord. A normal patellar reflex indicates that spinal nerves and spinal segments L_2–L_4 are undamaged.

The stretch reflex is an example of a **postural reflex**, a reflex that maintains normal upright posture. Postural muscles usually have a firm muscle tone and extremely sensitive stretch receptors. As a result, very fine adjustments are continually being made; you are not aware of the cycles of contraction and relaxation that occur.

Higher Centers and Integration of Reflexes

Reflexive motor activities occur automatically, without instructions from higher centers in the brain. However, higher centers can have a profound effect on reflex performance. For example, higher centers within the brain enhance or suppress spinal reflexes by modifying the information carried in descending tracts that synapse on interneurons and motor neurons throughout the spinal cord.

Motor control therefore involves a series of interacting levels. At the lowest level are monosynaptic reflexes that are rapid and seldom change. At the highest level are centers in the brain that can modify reflexive motor patterns.

Figure 14.13 The Classification of Reflexes. Four different methods are used to classify reflexes.

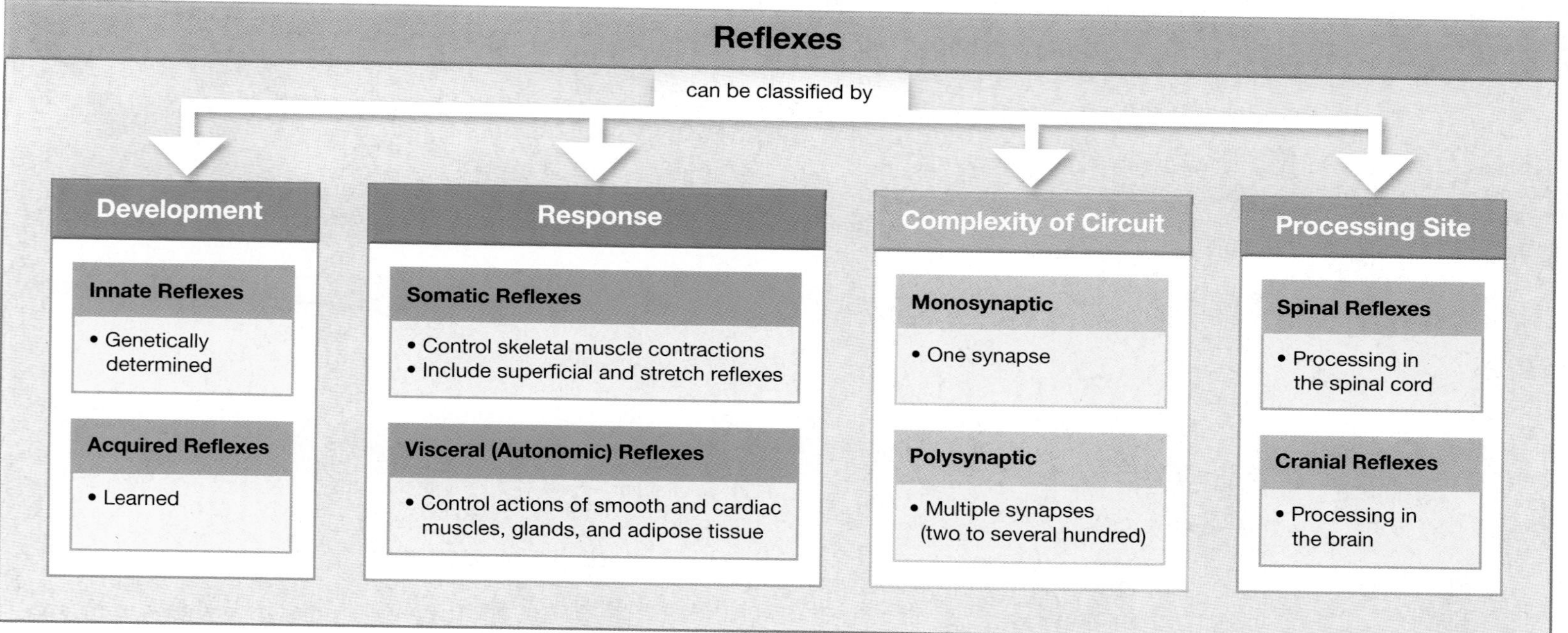

Reflexes
can be classified by
Development
Innate Reflexes
• Genetically determined
Acquired Reflexes
• Learned
Response
Somatic Reflexes
• Control skeletal muscle contractions
• Include superficial and stretch reflexes
Visceral (Autonomic) Reflexes
• Control actions of smooth and cardiac muscles, glands, and adipose tissue
Complexity of Circuit
Monosynaptic
• One synapse
Polysynaptic
• Multiple synapses (two to several hundred)
Processing Site
Spinal Reflexes
• Processing in the spinal cord
Cranial Reflexes
• Processing in the brain

14

Figure 14.14 Neural Organization and Simple Reflexes. A comparison of monosynaptic and polysynaptic reflexes.

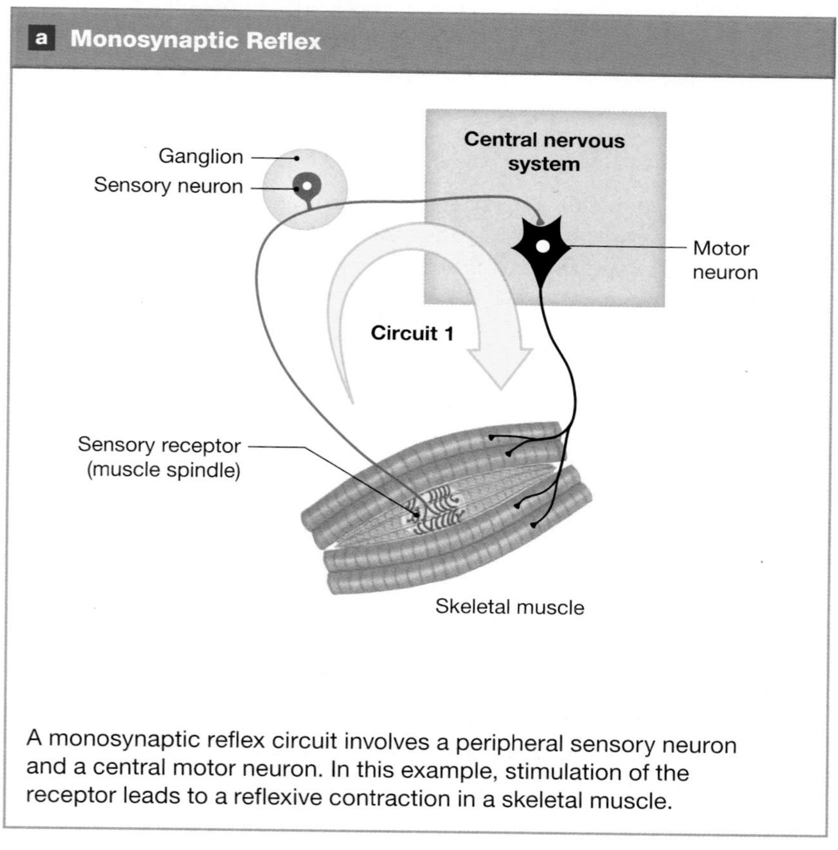

A monosynaptic reflex circuit involves a peripheral sensory neuron and a central motor neuron. In this example, stimulation of the receptor leads to a reflexive contraction in a skeletal muscle.

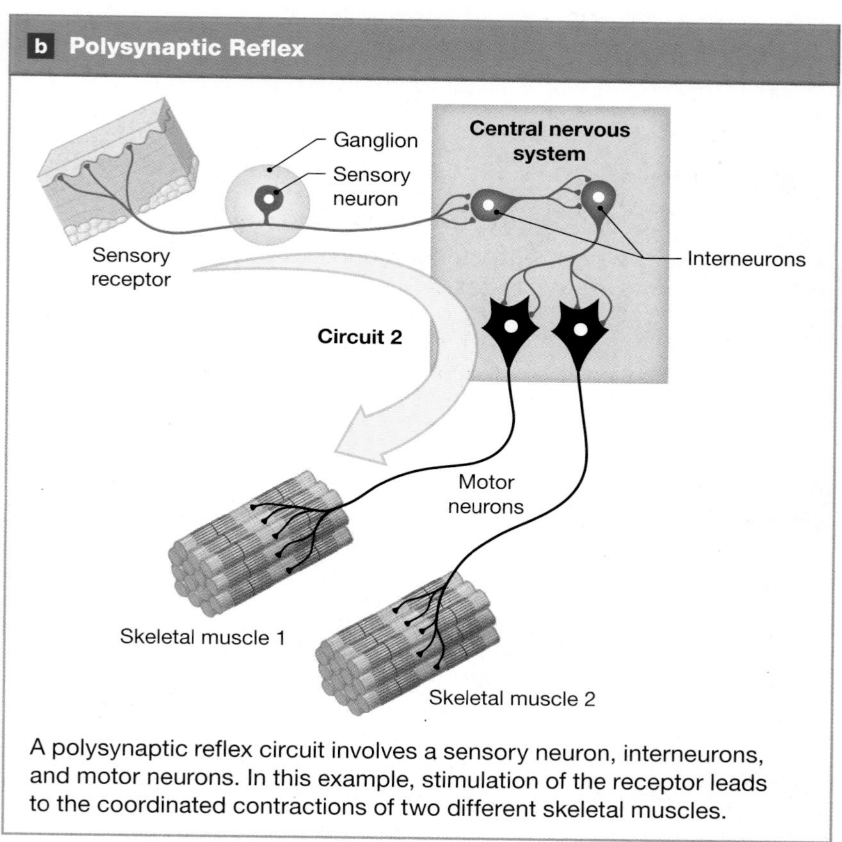

A polysynaptic reflex circuit involves a sensory neuron, interneurons, and motor neurons. In this example, stimulation of the receptor leads to the coordinated contractions of two different skeletal muscles.

Figure 14.15 Stretch Reflexes

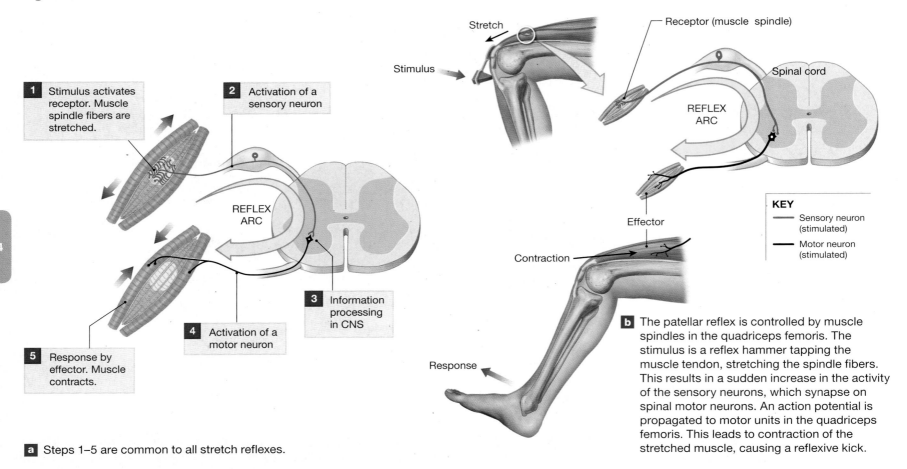

a Steps 1–5 are common to all stretch reflexes.

b The patellar reflex is controlled by muscle spindles in the quadriceps femoris. The stimulus is a reflex hammer tapping the muscle tendon, stretching the spindle fibers. This results in a sudden increase in the activity of the sensory neurons, which synapse on spinal motor neurons. An action potential is propagated to motor units in the quadriceps femoris. This leads to contraction of the stretched muscle, causing a reflexive kick.

14.6 CONCEPT CHECK

11 List, in order, the five steps in a reflex arc.

12 Distinguish between a monosynaptic and a polysynaptic reflex.

See the blue Answers tab at the back of the book.

EMBRYOLOGY SUMMARY

The adult spinal cord retains the segmental organization of the developing embryo, with 31 segments and 31 pairs of spinal nerves. For a summary of the development of the spinal cord and spinal nerves, see Chapter 28 (Embryology and Human Development).

Study Outline

Introduction p. 361

- The central nervous system (CNS) consists of the spinal cord and brain. Although they are connected, they have some functional independence. The spinal cord integrates and processes information on its own, in addition to relaying information to and from the brain.

14.1 Gross Anatomy of the Spinal Cord p. 361

- The adult spinal cord has a **posterior median sulcus** (shallow) and an **anterior median fissure** (wide). It includes localized **enlargements** (**cervical** and **lumbar**), which are expanded regions where there is increased gray matter to provide innervation of the limbs. *(See Figures 14.1 and 14.2.)*
- The adult spinal cord extends from the foramen magnum to L_1. The spinal cord tapers to a conical tip, the **conus medullaris**. The **filum terminale** (a strand of fibrous tissue) originates at this tip and extends through the vertebral canal to the second sacral vertebra, ultimately becoming part of the coccygeal ligament. *(See Figures 14.1 and 14.2.)*
- The spinal cord has 31 segments, each associated with a pair of **dorsal root ganglia** (containing sensory neuron cell bodies) and pairs of **dorsal roots** and **ventral roots**. The first cervical and first coccygeal nerves are exceptions, in that the dorsal roots are absent in many individuals. *(See Figures 14.1 and 14.2.)*
- Sensory and motor fibers unite as a single **spinal nerve** distal to each dorsal root ganglion. Spinal nerves emerge from the intervertebral foramina and are **mixed nerves** since they contain both sensory and motor fibers. *(See Figures 14.1 and 14.2.)*
- The **cauda equina** is the inferior extension of the ventral and dorsal roots and the filum terminale in the vertebral canal. *(See Figure 14.1.)*

14.2 Spinal Meninges p. 361

- The **spinal meninges** are a series of specialized membranes that provide physical stability and shock absorption for neural tissues of the spinal cord; the **cranial meninges** are membranes that surround the brain (Chapter 16). There are three meningeal layers: dura mater, arachnoid mater, and pia mater. *(See Figure 14.2.)*

The Dura Mater p. 361

- The spinal **dura mater** is the tough, fibrous outermost layer that covers the spinal cord; caudally, it forms the **coccygeal ligament** with the filum terminale. The **epidural space** separates the dura mater from the inner walls of the vertebral canal. *(See Figure 14.2.)*

The Arachnoid Mater p. 364

- Internal to the inner surface of the dura mater is the **subdural space**. When present, it separates the dura mater from the middle meningeal layer, the **arachnoid mater**. Internal to the arachnoid mater is the **subarachnoid space**, which has a network of collagen and elastic fibers, the arachnoid trabeculae. This space also contains **cerebrospinal fluid (CSF)**, which acts as a shock absorber and a diffusion medium for dissolved gases, nutrients, chemical messengers, and wastes. *(See Figure 14.2.)*

The Pia Mater p. 364

- The **pia mater** is the innermost meningeal layer. It is bound firmly to the underlying nervous tissue. Paired **denticulate ligaments** are supporting fibers extending laterally from the spinal cord surface, binding the spinal pia mater and arachnoid mater to the dura mater to prevent side-to-side or inferior movement of the spinal cord. *(See Figure 14.2.)*

14.3 Sectional Anatomy of the Spinal Cord p. 364

- The central **gray matter** surrounds the **central canal** and contains cell bodies of neurons and neuroglia. The gray matter projections toward the outer surface of the spinal cord are called **horns**. The peripheral **white matter** contains myelinated and unmyelinated axons in tracts and columns. *(See Figures 14.1–14.3.)*

Organization of Gray Matter p. 364

- Neuron cell bodies in the spinal cord gray matter are organized into groups, termed **nuclei**. The **posterior horns** contain somatic and visceral sensory nuclei, while nuclei in the **anterior horns** are involved with somatic motor control. The **lateral horns** contain visceral motor neurons. The **gray commissures**, posterior and anterior to the central canal, contain the axons of interneurons that cross from one side of the cord to the other. *(See Figure 14.3.)*

Organization of White Matter p. 364

- The white matter is divided into six **columns** (*funiculi*), each of which contains **tracts**. **Ascending tracts** relay information from the spinal cord to the brain, and **descending tracts** carry information from the brain to the spinal cord. *(See Figure 14.3.)*

14.4 Spinal Nerves p. 366

- There are 31 pairs of spinal nerves; each is identified through its association with an adjacent vertebra (cervical, thoracic, lumbar, sacral, and coccygeal). *(See Figure 14.1.)*

- Each spinal nerve is wrapped in a series of connective tissue layers. The outermost layer, the **epineurium**, is a dense network of collagen fibers; the middle layer, the **perineurium**, partitions the nerve into a series of bundles (**fascicles**) and forms the blood–nerve barrier; and the inner layer, the **endoneurium**, is composed of delicate connective tissue fibers that surround individual axons. *(See Figure 14.4.)*

Peripheral Distribution of Spinal Nerves p. 367

- The first branch of each spinal nerve in the thoracic and upper lumbar regions is the **white ramus communicans**, which contains myelinated axons going to an **autonomic ganglion**. Two groups of unmyelinated fibers exit this ganglion: a **gray ramus communicans**, carrying axons that innervate glands and smooth muscles in the body wall or limbs back to the spinal nerve, and an autonomic nerve carrying fibers to internal organs. Collectively, the white and gray rami are termed the rami communicantes. *(See Figures 14.2 and 14.5.)*

- Each spinal nerve has both a **dorsal (posterior) ramus** (provides sensory/motor innervation to the skin and muscles of the back) and a **ventral (anterior) ramus** (supplies ventrolateral body surface, body wall structures, and limbs). Each pair of spinal nerves monitors a region of the body surface, an area called a **dermatome**. *(See Figures 14.2, 14.5, and 14.6.)*

14

14.5 Nerve Plexuses p. 367

- A **nerve plexus** is a complex, interwoven network of nerves. The four major plexuses are the cervical, brachial, lumbar, and sacral plexuses. *(See Spotlight Figures 14.8 and 14.10, Figures 14.2, 14.7, 14.9, and 14.11, and Table 14.1.)*

The Cervical Plexus p. 368

- The **cervical plexus** consists of the ventral rami of C_1–C_4 and some fibers from C_5. Muscles of the neck are innervated; some branches extend into the thoracic cavity to the diaphragm. The phrenic nerve is the major nerve in this plexus. *(See Figures 14.7 and 14.9c and Spotlight Figure 14.8.)*

The Brachial Plexus p. 374

- The **brachial plexus** innervates the pectoral girdle and upper limbs by the ventral rami of C_5–T_1. The nerves in this plexus originate from cords or trunks: **superior**, **middle**, and **inferior trunks** give rise to the **lateral cord**, **medial cord**, and **posterior cord**. *(See Spotlight Figure 14.8, Figures 14.7 and 14.9, and Table 14.1.)*

The Lumbar and Sacral Plexuses p. 374

- Collectively, the **lumbar plexus** and **sacral plexus** originate from the posterior abdominal wall and ventral rami of nerves supplying the pelvic girdle and lower limb. The lumbar plexus contains fibers from spinal segments T_{12}–L_4, and the sacral plexus contains fibers from spinal segments L_4–S_4. *(See Spotlight Figure 14.10 and Figure 14.11.)*

14.6 Reflexes p. 374

- A **reflex** is a rapid, automatic, involuntary motor response to stimuli. Reflexes help preserve homeostasis by rapidly adjusting the functions of organs or organ systems. *(See Figures 14.12–14.15.)*

- A **reflex arc** is the neural "wiring" of a single reflex. *(See Figure 14.12.)*

- A receptor is a specialized cell that monitors conditions in the body or external environment. Each receptor has a characteristic range of sensitivity.

- There are five steps involved in a reflex arc: (1) stimulation and activation of a receptor; (2) activation of a sensory neuron; (3) information processing in the CNS; (4) activation of a motor neuron; and (5) response by an effector. *(See Figure 14.12.)*

Classification of Reflexes p. 380

- Reflexes are classified by (1) their development (innate, acquired); (2) their motor response (somatic, visceral); (3) the complexity of the neural circuit (monosynaptic, polysynaptic); and (4) their processing site (spinal, cranial). *(See Figures 14.13 and 14.4.)*
- Innate reflexes are genetically determined. Acquired reflexes are learned following repeated exposure to a stimulus. *(See Figure 14.13.)*
- Reflexes processed in the brain are cranial reflexes. In a spinal reflex, the important interconnections and processing occur inside the spinal cord. *(See Figure 14.13.)*
- Somatic reflexes control skeletal muscle contractions, and visceral (*autonomic*) reflexes control the activities of smooth and cardiac muscles and glands. *(See Figure 14.13.)*
- A **monosynaptic reflex** is the simplest reflex. A sensory neuron synapses directly on a motor neuron that acts as the processing center. A **polysynaptic reflex** has at least one interneuron located between the sensory afferent and the motor efferent. Thus, polysynaptic reflexes have a longer delay between stimulus and response. *(See Figures 14.13 and 14.14.)*
- Spinal reflexes range from simple monosynaptic reflexes (involving only one segment of the cord) to more complex polysynaptic reflexes (in which many segments of the cord interact to produce a coordinated motor response). *(See Figures 14.14 and 14.15.)*
- A **stretch reflex** is a monosynaptic reflex that automatically regulates skeletal muscle length and muscle tone. The sensory receptors involved are stretch receptors of muscle spindles. *(See Figure 14.15.)*
- A **patellar reflex** is the familiar *knee jerk*, wherein a tap on the patellar ligament stretches the muscle spindles in the quadriceps femoris. *(See Figure 14.15.)*
- A **postural reflex** is a stretch reflex that maintains normal upright posture.
- Higher centers in the brain can enhance or inhibit reflex motor patterns based in the spinal cord.

Chapter Review

For answers, see the blue Answers tab at the back of the book.

Level 1 Reviewing Facts and Terms

Match each numbered item with the most closely related lettered item.

1. ventral root ________
2. epidural space ________
3. white matter ________
4. fascicle ________
5. dermatome . ________
6. phrenic nerve ________
7. brachial plexus ________
8. obturator nerve ________
9. reflex ________
10. pudendal nerve ________

a. tracts and columns
b. specific region of body surface
c. cervical plexus
d. motor neuron axons
e. sacral plexus
f. lumbar plexus
g. single bundle of axons
h. involuntary motor response
i. loose connective tissue, adipose tissue
j. pectoral girdle/upper extremity

11. Label the following structures on the accompanying diagram of a cross section of the spinal cord.
(a) ventral root
(b) dorsal root ganglion
(c) anterior median fissure
(d) white matter
(e) posterior median sulcus

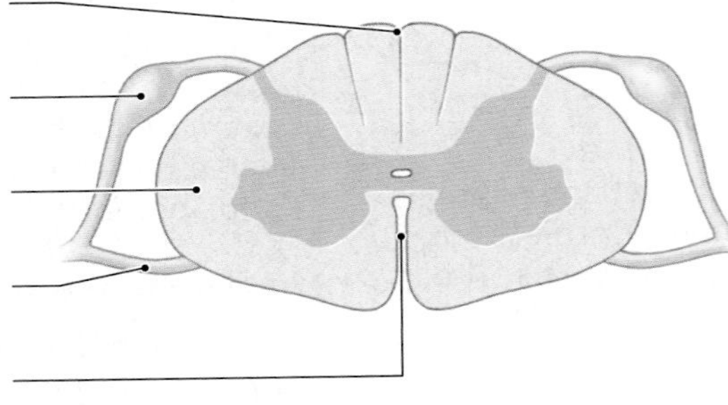

(a) ________
(b) ________
(c) ________
(d) ________
(e) ________

12. Sensory and motor innervations of the skin of the lateral and anterior surfaces of the body are provided by the
(a) white rami communicantes.
(b) gray rami communicantes.
(c) dorsal ramus.
(d) ventral ramus.

13. The brachial plexus
(a) innervates the shoulder girdle and the upper extremity.
(b) is formed from the ventral rami of spinal nerves C_5–T_1.
(c) is the source of the musculocutaneous, radial, median, and ulnar nerves.
(d) all of the above

14. The middle layer of connective tissue that surrounds each peripheral nerve is the
(a) epineurium. (b) perineurium.
(c) endoneurium. (d) endomysium.

15. The expanded area of the spinal cord that supplies nerves to the pectoral girdle and upper limbs is the
(a) conus medullaris.
(b) filum terminale.
(c) lumbosacral enlargement.
(d) cervical enlargement.

16. Spinal nerves are called mixed nerves because
(a) they contain sensory and motor fibers.
(b) they exit at intervertebral foramina.
(c) they are associated with a pair of dorsal root ganglia.
(d) they are associated with dorsal and ventral roots.

17. The gray matter of the spinal cord is dominated by
(a) myelinated axons only.
(b) cell bodies of neurons and neuroglia.
(c) unmyelinated axons only.
(d) Schwann cells and satellite cells.

18. The ________ is a strand of fibrous tissue that provides longitudinal support as a component of the coccygeal ligament.
(a) conus medullaris
(b) filum terminale
(c) cauda equina
(d) dorsal root

19. Axons crossing from one side of the spinal cord to the other within the gray matter are found in the
(a) anterior horns.
(b) white commissures.
(c) gray commissures.
(d) lateral horns.

20. The paired structures that contain cell bodies of sensory neurons and are associated with each segment of the spinal cord are the
    (a) dorsal rami.
    (b) ventral rami.
    (c) dorsal root ganglia.
    (d) ventral root ganglia.

21. The deep crease on the anterior surface of the spinal cord is the
    (a) posterior median sulcus.
    (b) posterior median fissure.
    (c) anterior median sulcus.
    (d) anterior median fissure.

Level 2 Reviewing Concepts

1. What nerve is likely to transmit pain when a person receives an intramuscular injection into the deltoid region of the arm?
   (a) ulnar nerve
   (b) radial nerve
   (c) intercostobrachial nerve
   (d) upper lateral cutaneous nerve of the arm

2. Which of the following actions would be compromised if a person suffered an injury to lumbar spinal segments L_3 and L_4?
   (a) performing a plié (shallow knee bend) in ballet
   (b) sitting cross-legged in the lotus position (lateral side of the foot on the medial side of opposite thigh)
   (c) riding a horse
   (d) all of the above

3. Tingling and numbness in the palmar region of the hand could be caused by
   (a) compression of the median nerve in the carpal tunnel.
   (b) compression of the ulnar nerve.
   (c) compression of the radial artery.
   (d) irritation of the structures that form the superficial arterial loop.

4. What is the role of the meninges in protecting the spinal cord?

5. How does a reflex differ from a voluntary muscle movement?

6. If the dorsal root of the spinal cord were damaged, what would be affected?

7. Why is response time in a monosynaptic reflex much faster than response time in a polysynaptic reflex?

8. Why are there eight cervical spinal nerves but only seven cervical vertebrae?

9. What prevents side-to-side movements of the spinal cord?

10. Why is it important that a spinal tap be done between the third and fourth lumbar vertebrae?

Level 3 Critical Thinking

1. The incision that allows access to the abdominal cavity involves cutting the sheath of the rectus abdominis. This muscle is always retracted laterally, never medially. Why?

2. Cindy is in an automobile accident and injures her spinal cord. She has lost feeling in her right hand, and her doctor tells her that it is the result of swelling compressing a portion of her spinal cord. Which part of her cord is likely to be compressed?

3. Karen falls down a flight of stairs and suffers spinal cord damage due to hyperextension of the cord during the fall. The injury results in edema of the spinal cord with resulting compression of the anterior horn cells of the spinal region. What signs would you expect to observe as a result of this injury?

MasteringA&P™

Access more chapter study tools online in the Study Area:

- Chapter Quizzes, Chapter Practice Test, Clinical Cases, and more!
- Practice Anatomy Lab (PAL) 
- A&P Flix for anatomy topics **A&PFlix™**

14

CLINICAL CASE WRAP-UP

A Case of the Bends

The compressed air in Bob's SCUBA tank is the same mixture found in sea-level air, including 78 percent nitrogen. During his last dive, Bob was inhaling three to four times more compressed air with each breath to overcome the surrounding water pressure. At sea level, we release all nitrogen through the lungs by exhaling, but this is not possible under water, so blood and tissues become loaded with dissolved nitrogen. Ascending too quickly causes this nitrogen to form gas bubbles that can clog arteries and kill cells distal to the blockage. This dangerous condition, decompression sickness, is nicknamed "the bends" because sufferers often double over with pain.

Nitrogen bubbles have occluded the tiny arteries of the pia mater feeding Bob's spinal cord at the T_{10} level (see **Figure 14.2c**). Bob's sensory neurologic loss (paresthesia) is at the dermatome level of T_{10}, matching his motor loss. He has no motor control over his hips or legs, and he has lost bowel and bladder control. We can describe Bob's condition as T_{10} paraplegia.

Bob is given 100 percent oxygen by mask as the boat speeds to the closest hyperbaric oxygen chamber, which, luckily, is nearby. Breathing 100 percent oxygen at pressure in the hyperbaric chamber gives Bob the best chance for a full recovery.

Within 3 months Bob's paraplegia has completely resolved. Thankful for this, he is happy to enjoy the sea from the surface from now on.

1. If Bob's numbness went from his nipples to his toes, what dermatome level would this describe?
2. If Bob's numbness and weakness included shoulder girdle, arm, forearm, hand, and everything distally, where should you expect to find the blockage in his spinal cord circulation?

See the blue Answers tab at the back of the book.

Related Clinical Terms

areflexia: Absence of reflexes.

Brown-Sequard syndrome: Loss of sensation and motor function that results from unilateral spinal cord lesions. Proprioception loss and weakness occur ipsilateral to the lesion, while pain and temperature loss occur contralateral.

cordotomy: Any operation of the spinal cord.

hemiparesis: Slight paralysis or weakness affecting one side of the body.

Kernig's sign: Symptom of meningitis where patient cannot extend the leg at the knee due to stiffness in the hamstring muscles.

nerve conduction study: Test that stimulates certain nerves and records their ability to send an impulse to the muscle; it can indicate where any blockage of the nerve pathway exists; often performed along with electromyography (EMG).

nerve growth factor: A peptide that promotes the growth and maintenance of neurons. Other factors that are important to neuron growth and repair include BDNF, NT-3, NT-4, and GAP-43.

neurosis: A functional nervous system disease or a nerve system disorder in which no lesion is evident.

paraplegia: Paralysis involving a loss of motor control of the lower, but not the upper, limbs.

tabes dorsalis: Slow, progressive degeneration of the myelin layer of the sensory neurons of the spinal cord that occurs in the tertiary (third) phase of syphilis. Common signs and symptoms are pain, weakness, diminished reflexes, unsteady gait, and loss of coordination.

15

The Nervous System

Sensory and Motor Tracts of the Spinal Cord

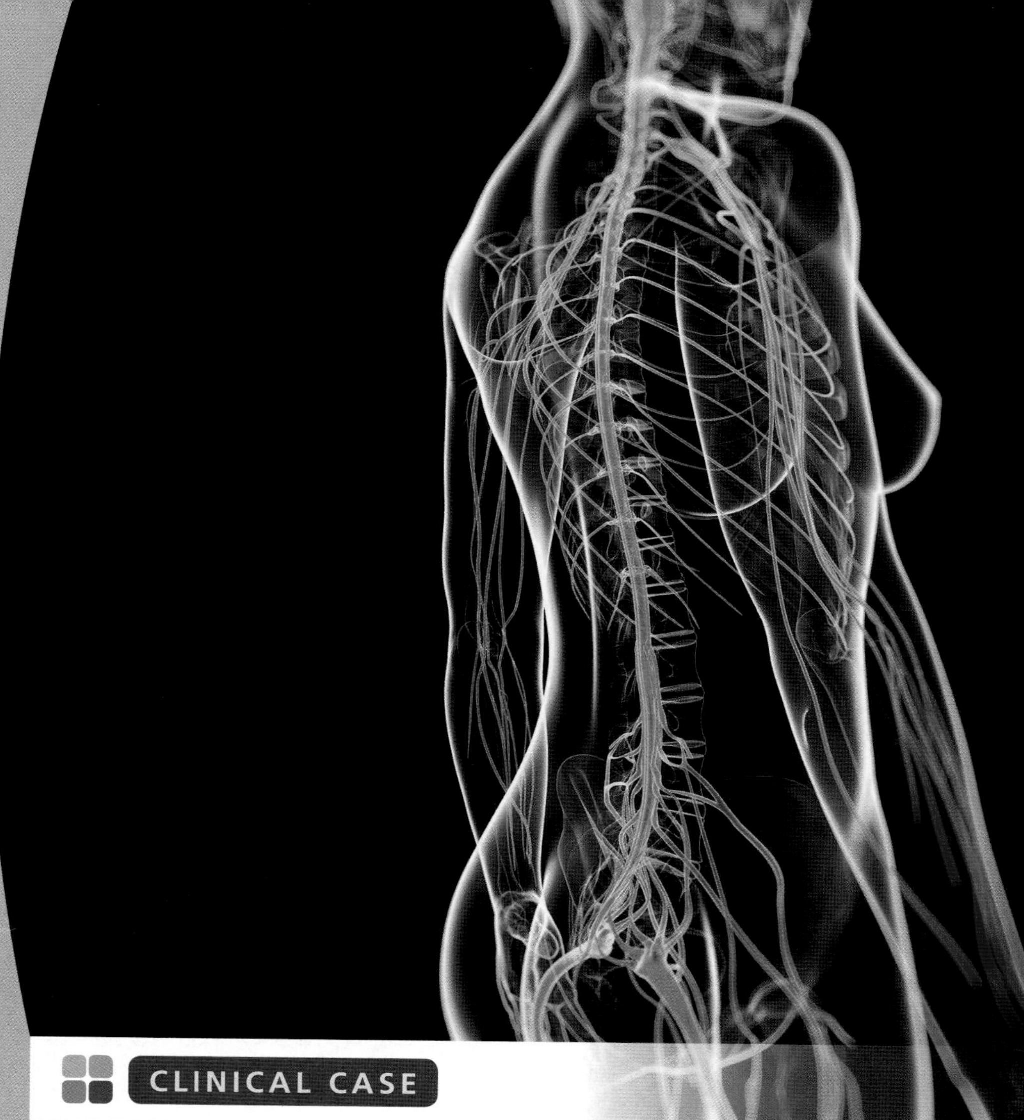

Learning Outcomes

These Learning Outcomes correspond by number to this chapter's sections and indicate what you should be able to do after completing the chapter.

15.1 Explain how the anatomical name of a spinal tract tells you where the tract begins and ends within the CNS. p. 388

15.2 List and then compare and contrast the sensory tracts of the spinal cord. p. 388

15.3 List and then compare and contrast the motor tracts of the spinal cord. p. 394

15.4 Identify the centers in the brain that interact to determine somatic motor output. p. 398

CLINICAL CASE

Amyotrophic Lateral Sclerosis

Stephen Hawking was a 21-year-old physics student at Oxford University in England when he first noticed he was growing increasingly clumsy. He felt muscle cramping and twitching (fasciculations) in his hands and feet. Soon he began slurring his speech and had trouble swallowing. In 1963 he was diagnosed with amyotrophic lateral sclerosis (ALS) and was given 2 years to live.

Most ALS patients die 3 to 5 years after diagnosis. In spite of his progressive neurologic disease, Hawking continued his remarkable career as a theoretical physicist and best-selling author. One of his many books, *A Brief History of Time,* was on the British best-seller list for 237 weeks.

By the late 1960s, Hawking was using a motorized wheelchair and communicating through computerized speech, controlled by his last functioning skeletal muscles, those in his right cheek. After nearly dying of pneumonia in 1985, he had a tracheostomy. A feeding tube (gastrostomy) provides his nutrition.

How does ALS produce complete paralysis without affecting touch, sight, smell, taste, or intellect? To find out, turn to the Clinical Case Wrap-Up on p. 402.

WHEN YOU PLAN A TRIP, you choose your route based on the location of your destination. You may vary your route depending on the time of day, traffic congestion, road construction, and so forth. When necessary, you map your route in advance using software or apps.

The routes of information flowing into and out of the central nervous system have also been mapped, but the diagram is much more complex than any road map. At any given moment, millions of sensory neurons are delivering information to different locations within the CNS, and millions of motor neurons are controlling or adjusting the activities of peripheral effectors. Afferent sensory and efferent motor information travels by several different routes depending on where the information is coming from, where it is going to, and the priority level of the information.

15.1 Organization and Patterns of Spinal Cord Tracts

▶ **KEY POINT** The tracts within the spinal cord are organized into two categories: (1) long, ascending sensory fibers originating at a sensory receptor and ending in the cerebral cortex, cerebellum, or brainstem and (2) long, descending motor fibers originating in the brain and ending in the spinal cord or brainstem.

Tracts within the spinal cord relay sensory and motor information between the peripheral nervous system and higher centers within the central nervous system. Each ascending (sensory) or descending (motor) tract consists of a chain of neurons and associated nuclei. Wherever synapses relay signals from one neuron to another, information is processed. The number of synapses varies from one tract to another. For example, a sensory tract ending in the cerebral cortex involves three neurons, while a sensory tract ending in the cerebellum involves just two neurons.

In this chapter we focus on the major sensory and motor tracts of the spinal cord. These tracts are paired in the spinal cord, and the axons within each tract are grouped according to the region of the body innervated.

All tracts involve both the brain and spinal cord. *The name of a tract often indicates its origin and destination.* If the name *begins* with *spino–*, the tract *starts* in the spinal cord and *ends* in the brain; it must therefore be an ascending tract that carries sensory information. The *last* part of the name indicates the tract's destination. For example, the *spinocerebellar tract* begins in the spinal cord and ends in the cerebellum. If the name ends in *–spinal*, the tract *starts* in the brain and *ends* in the spinal cord; it is a descending tract that carries motor commands. Then the first part of the name indicates the nucleus or cortical area of the brain where the tract originates. For example, the *vestibulospinal tract* starts in the vestibular nucleus and ends in the spinal cord. **Figure 15.1** and **Spotlight Figure 15.2** explain the organization and general anatomical pattern of sensory and motor spinal cord tracts.

15.1 CONCEPT CHECK

1 Compare and contrast the general organization and anatomical pattern of the (a) ascending sensory tracts and (b) descending motor tracts within the spinal cord.

See the blue Answers tab at the back of the book.

15.2 Sensory Tracts

▶ **KEY POINT** Ascending tracts carry sensory information. Sensory information carried by a tract that synapses in the thalamus is raised to a conscious level. Sensory information carried by a tract that does not synapse within the thalamus remains at a subconscious level.

The three major somatosensory tracts are the posterior columns, spinothalamic tracts, and spinocerebellar tracts **(Figure 15.3)**. Sensory receptors monitor conditions both inside the body and in the external environment. When stimulated, a receptor sends information to the central nervous system. This sensory information, called a **sensation**, arrives in the form of an action potential in an afferent (sensory) fiber.

The complexity of the response to a particular stimulus varies considerably depending on (1) where processing occurs and (2) where the motor response is initiated. Chapter 16 describes the brain and the various centers within the brain that receive sensory information or initiate motor impulses traveling down the spinal cord to effector organs. Chapter 17 describes the distribution of visceral sensory information and considers reflexive responses to visceral sensations, and Chapter 18 examines the origins of sensations and the pathways that relay special sensory information, such as olfaction (smell) or vision, to conscious and subconscious processing centers in the brain. **Table 15.1** summarizes the three major somatosensory tracts: (1) the posterior columns, (2) the spinothalamic tracts, and (3) the spinocerebellar tracts.

The Posterior Columns

▶ **KEY POINT** The posterior columns carry proprioceptive information about the type of stimulus, the exact site of stimulation, and when the stimulus stops. In other words, the posterior columns carry information that tells you "what," "where," and "when" for these sensations.

The **posterior columns**, also termed the *dorsal columns* or the *medial lemniscal pathway*, carry highly localized information from receptors in the skin and musculoskeletal system about proprioception (limb position), fine touch, pressure, and vibration **(Figures 15.1, Spotlight Figure 15.2,** and **15.3a)**.

The axons of the first-order neurons enter the CNS through the dorsal roots of spinal nerves and the sensory roots of cranial nerves. Axons from the dorsal roots of spinal nerves entering the spinal cord inferior to T_6 travel superiorly within the **gracile fasciculus** (also termed the *fasciculus gracilis*) of the spinal cord. Those entering the spinal cord at or superior to T_6 ascend within the **cuneate fasciculus** (also termed the *fasciculus cuneatus*) of the spinal cord.

Three neurons are involved in the transmission of information from the periphery to the cerebrum: first-order, second-order, and third-order neurons. The first-order neurons of the gracile fasciculus enter the spinal cord by the dorsal root of the spinal nerve and synapse with the second-order neurons within the gracile nucleus (also termed the *nucleus gracilis*) in the medulla oblongata. First-order neurons of the fasciculus cuneatus synapse with the second-order neurons within the cuneate nucleus (also termed the *nucleus cuneatus*) in the medulla oblongata. All of the second-order neurons in the posterior columns immediately **decussate** (*cross over*) to the contralateral (*opposite side*) of the medulla and ascend to the thalamus on the opposite side of the brain. These second-order neurons form the **medial lemniscus** (*lemniskos*, ribbon). As the information travels toward the thalamus, neurons carrying the same type of sensory information (fine touch, pressure, and vibration) collected by cranial nerves V, VII, IX, and X enter the medial lemniscus.

The ventral posterolateral nucleus (*VPL*) of the thalamus integrates sensory information carried within the posterior columns. The VPL sorts the sensory information according to the region of the body involved and sends it to specific regions of the primary somatosensory cortex. As a result of this sorting, we "know" the nature of the stimulus and its location. If it is relayed to another part of the somatosensory cortex, we will perceive the sensation as having originated in a different part of the body. For example, sensory information from the little finger on your right hand is sent to a specific part of the primary somatosensory cortex, and sensory information from your left knee is sent to another part of the somatosensory cortex. Our understanding of a particular stimulus as touch, rather than as temperature or pain, is due to the processing of the sensory information within the thalamus.

Figure 15.1 A Cross-sectional View Indicating the Locations of the Major Ascending (Sensory) Tracts in the Spinal Cord. For information about these tracts, see Table 15.1. Descending (motor) tracts are shown in detail in dashed outline; these tracts are identified in **Spotlight Figure 15.2**.

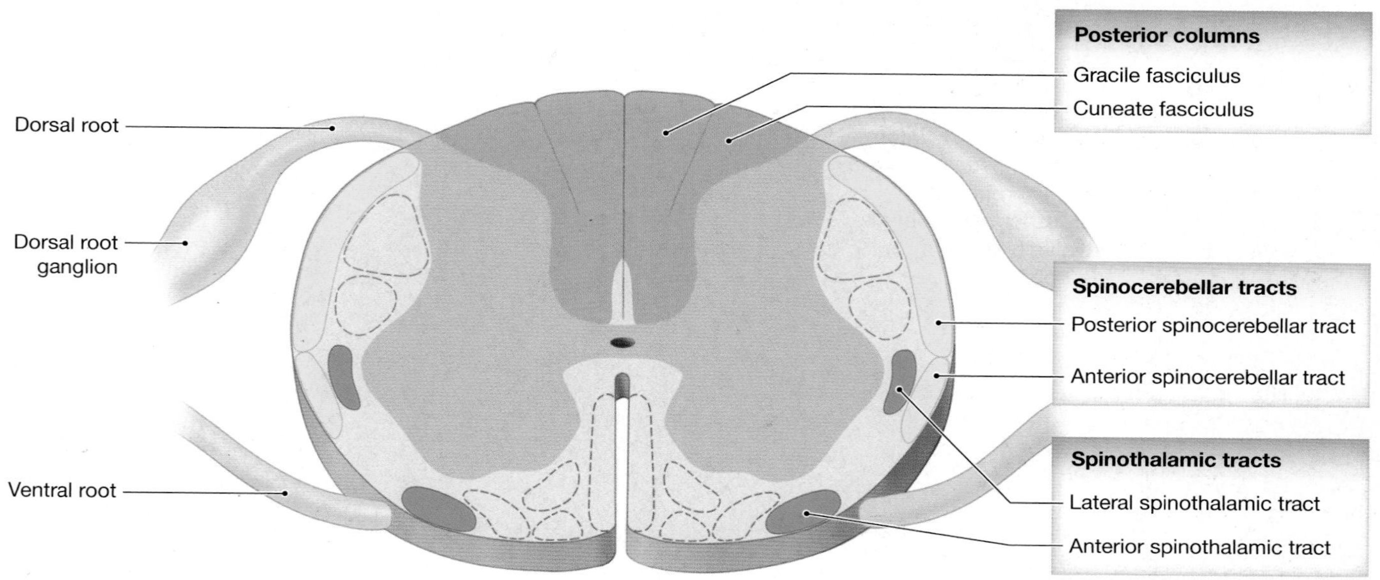

15

TABLE 15.1 | Major Ascending (Sensory) Tracts and the Sensory Information They Provide

		Location of Neuron Cell Bodies				
Tract	**Sensations**	**First-Order**	**Second-Order**	**Third-Order**	**Final Destination**	**Site of Decussation**
POSTERIOR COLUMNS						
Gracile fasciculus	Proprioception, fine touch, pressure, and vibration from levels inferior to T_6	Dorsal root ganglia of lower body; axons enter CNS in dorsal roots and ascend within gracile fasciculus	Gracile nucleus of medulla oblongata; axons decussate before entering medial lemniscus	Ventral posterolateral nucleus of thalamus	Primary somatosensory cortex on side opposite stimulus	Axons of second-order neurons, before joining medial lemniscus
Cuneate fasciculus	Proprioception, fine touch, pressure, and vibration from levels at or superior to T_6	Dorsal root ganglia of upper body; axons enter CNS in dorsal roots and ascend within cuneate fasciculus	Cuneate nucleus of medulla oblongata; axons decussate before entering medial lemniscus	Ventral posterolateral nucleus of thalamus	Primary somatosensory cortex on side opposite stimulus	Axons of second-order neurons, before joining medial lemniscus
SPINOTHALAMIC TRACTS						
Lateral spinothalamic tracts	Pain and temperature sensations	Dorsal root ganglia; axons enter CNS in dorsal roots and enter posterior horn	In posterior horn; axons enter lateral spinothalamic tract on contralateral side	Ventral posterolateral nucleus of thalamus	Primary somatosensory cortex on side opposite stimulus	Axons of second-order neurons, at level of entry
Anterior spinothalamic tracts	Crude touch and pressure sensations	Dorsal root ganglia; axons enter CNS in dorsal roots and enter posterior horn	In posterior horn; axons enter anterior spinothalamic tract on opposite side	Ventral posterolateral nucleus of thalamus	Primary somatosensory cortex on side opposite stimulus	Axons of second-order neurons, at level of entry
SPINOCEREBELLAR TRACTS						
Posterior spinocerebellar tracts	Proprioception	Dorsal root ganglia; axons enter CNS in dorsal roots	In posterior horn; axons enter posterior spinocerebellar tract on same side	Not present	Cerebellar cortex on side of stimulus	None
Anterior spinocerebellar tracts	Proprioception	Dorsal root ganglia; axons enter CNS in dorsal roots	In same spinal segment; axons enter anterior spinocerebellar tract on same or opposite side	Not present	Cerebellar cortex, primarily on side of stimulus	Axons of most second-order neurons cross before entering tract and then cross again within cerebellum

SPOTLIGHT

FIGURE 15.2

Organization of Spinal Cord Tracts

Sensory Tracts

Most of the processing of sensory information occurs in the spinal cord, brainstem or thalamus. About 1 percent of the information from afferent fibers reaches the cerebral cortex and our conscious awareness. However, the information arriving at the sensory cortex is organized so that we can determine the source and nature of the stimulus with great precision.

First-order, Second-order, and Third–order Neurons in Ascending Tracts

Sensory tracts delivering somatic sensory information to the sensory cortex of the cerebral or cerebellar hemispheres involve a chain of neurons.

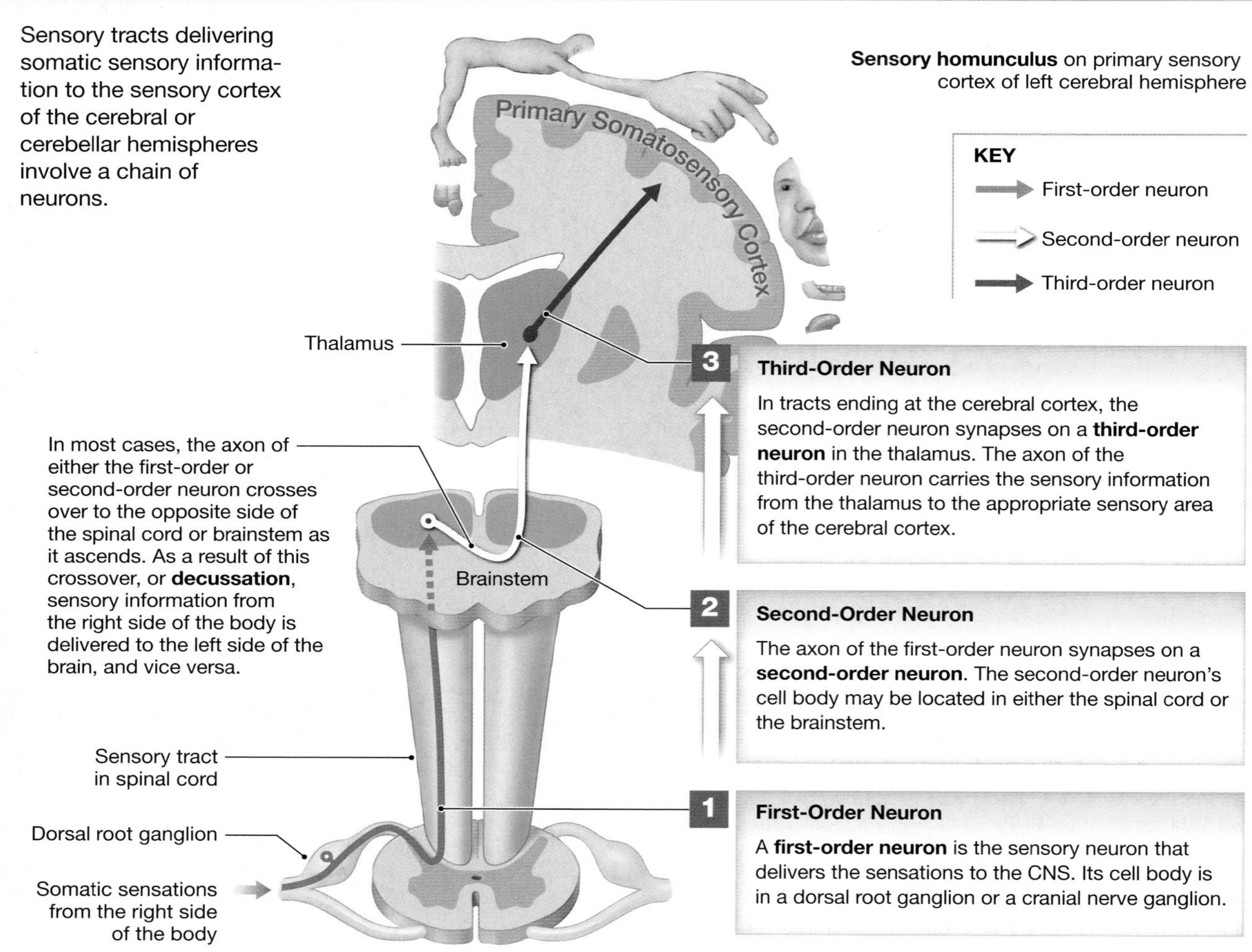

Sensory homunculus on primary sensory cortex of left cerebral hemisphere
Primary Somatosensory Cortex
KEY
First-order neuron
Second-order neuron
Third-order neuron
Thalamus
In most cases, the axon of either the first-order or second-order neuron crosses over to the opposite side of the spinal cord or brainstem as it ascends. As a result of this crossover, or decussation, sensory information from the right side of the body is delivered to the left side of the brain, and vice versa.
Brainstem
Sensory tract in spinal cord
Dorsal root ganglion
Somatic sensations from the right side of the body

3 **Third-Order Neuron**

In tracts ending at the cerebral cortex, the second-order neuron synapses on a **third-order neuron** in the thalamus. The axon of the third-order neuron carries the sensory information from the thalamus to the appropriate sensory area of the cerebral cortex.

2 **Second-Order Neuron**

The axon of the first-order neuron synapses on a **second-order neuron**. The second-order neuron's cell body may be located in either the spinal cord or the brainstem.

1 **First-Order Neuron**

A **first-order neuron** is the sensory neuron that delivers the sensations to the CNS. Its cell body is in a dorsal root ganglion or a cranial nerve ganglion.

Neuron Arrangement within Sensory Tracts

Neurons within the sensory tracts are arranged according to three anatomical principles.

1 **Sensory Modality Arrangement**

Sensory fibers are arranged within the spinal cord according to the type of sensory information carried by the individual neurons. In other words, information dealing with fine touch will be carried within one sensory tract, while information dealing with pain will be carried within another.

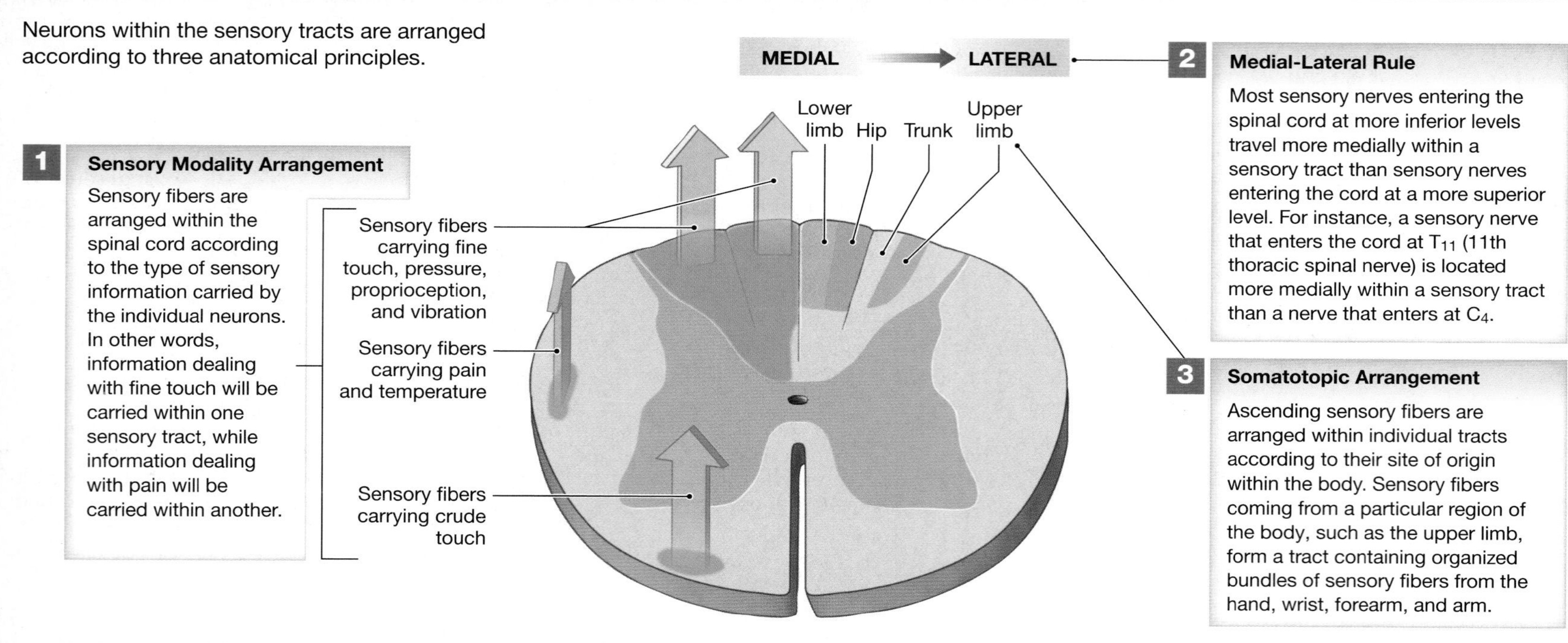

2 **Medial-Lateral Rule**

Most sensory nerves entering the spinal cord at more inferior levels travel more medially within a sensory tract than sensory nerves entering the cord at a more superior level. For instance, a sensory nerve that enters the cord at T_{11} (11th thoracic spinal nerve) is located more medially within a sensory tract than a nerve that enters at C_4.

3 **Somatotopic Arrangement**

Ascending sensory fibers are arranged within individual tracts according to their site of origin within the body. Sensory fibers coming from a particular region of the body, such as the upper limb, form a tract containing organized bundles of sensory fibers from the hand, wrist, forearm, and arm.

Motor Tracts

The central nervous system issues motor commands in response to information from sensory systems. These commands are distributed by the somatic nervous system and the autonomic nervous system. The somatic nervous system (SNS) issues somatic motor commands that direct skeletal muscle contractions. The autonomic nervous system (ANS), or visceral motor system, innervates visceral effectors, such as smooth muscles, cardiac muscle, adipose tissue and glands.

Organization of Motor Tracts

Somatic motor tracts always involve at least two motor neurons: an **upper motor neuron**, whose cell body lies in the CNS, and a **lower motor neuron** located in a motor nucleus of the brainstem or spinal cord. Activity in the upper motor neuron excites or inhibits the lower motor neuron. The axon of the lower motor neuron extends to skeletal muscle fibers, and it is only capable of exciting skeletal muscle fibers.

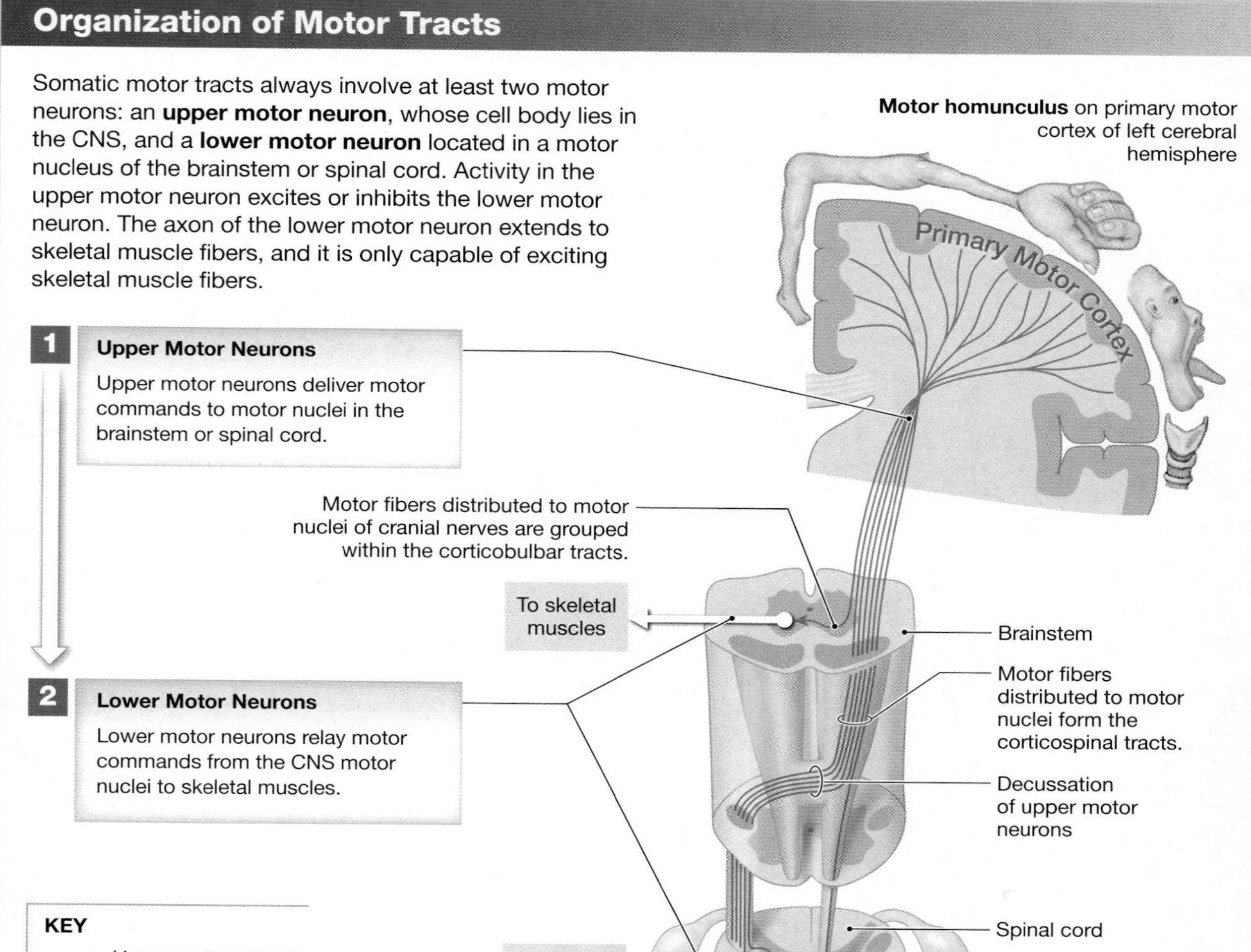

Descending Motor Tracts

Subconscious and conscious motor commands control skeletal muscles through descending motor tracts within the spinal cord.

Motor Tracts Carrying Subconscious Motor Commands

The **rubrospinal tract** automatically adjusts upper limb position and muscle tone during voluntary movements.

The **medial reticulospinal tract** originates in the reticular formation, a diffuse network of neurons in the brainstem. The functions of the tract vary depending on which area of the reticular formation is stimulated.

The **tectospinal tracts** control reflexive changes in position in response to auditory or visual stimuli.

The **vestibulospinal tract** carries motor commands that reflexively control posture and balance.

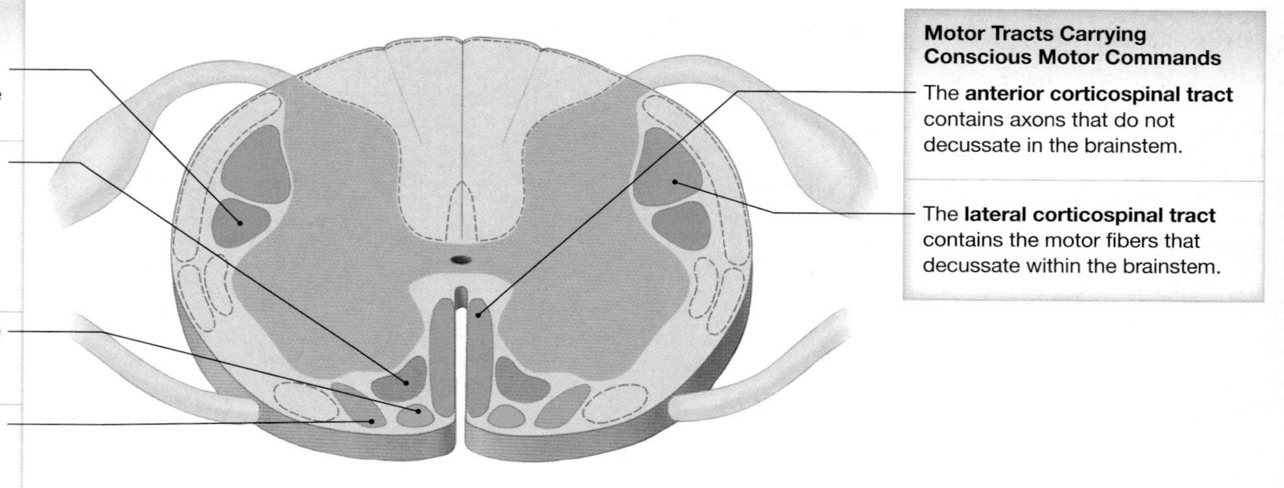

Motor Tracts Carrying Conscious Motor Commands
The **anterior corticospinal tract** contains axons that do not decussate in the brainstem.
The **lateral corticospinal tract** contains the motor fibers that decussate within the brainstem.

Figure 15.3 The Posterior Column, Spinothalamic, and Spinocerebellar Sensory Tracts. Diagrammatic comparison of first-, second-, and third-order neurons in ascending tracts. For clarity, this figure shows only the tract for sensations originating on the right side of the body.

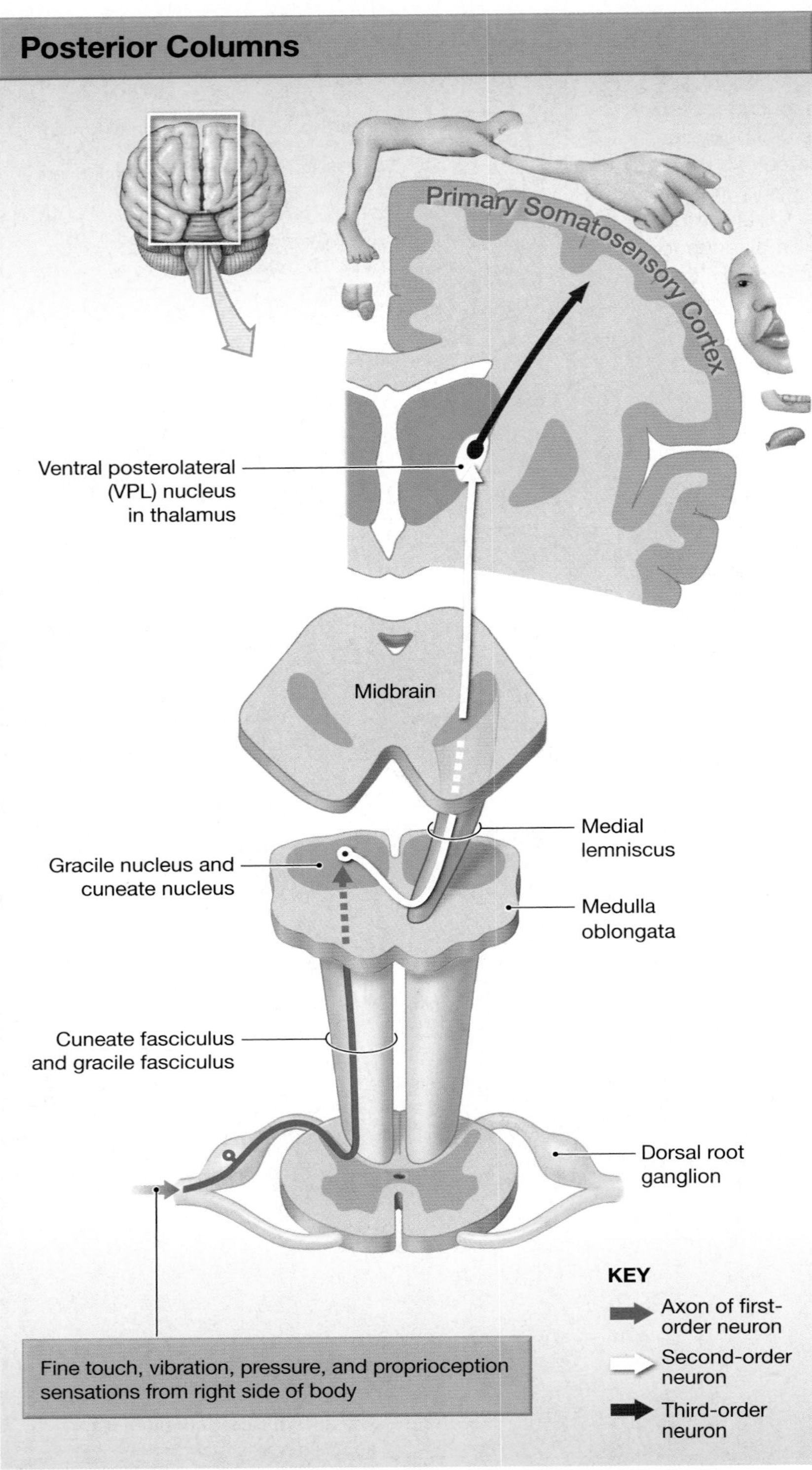

a The posterior columns carry sensory information to the primary somatosensory cortex on the opposite side of the body. The tracts decussate in the medulla.

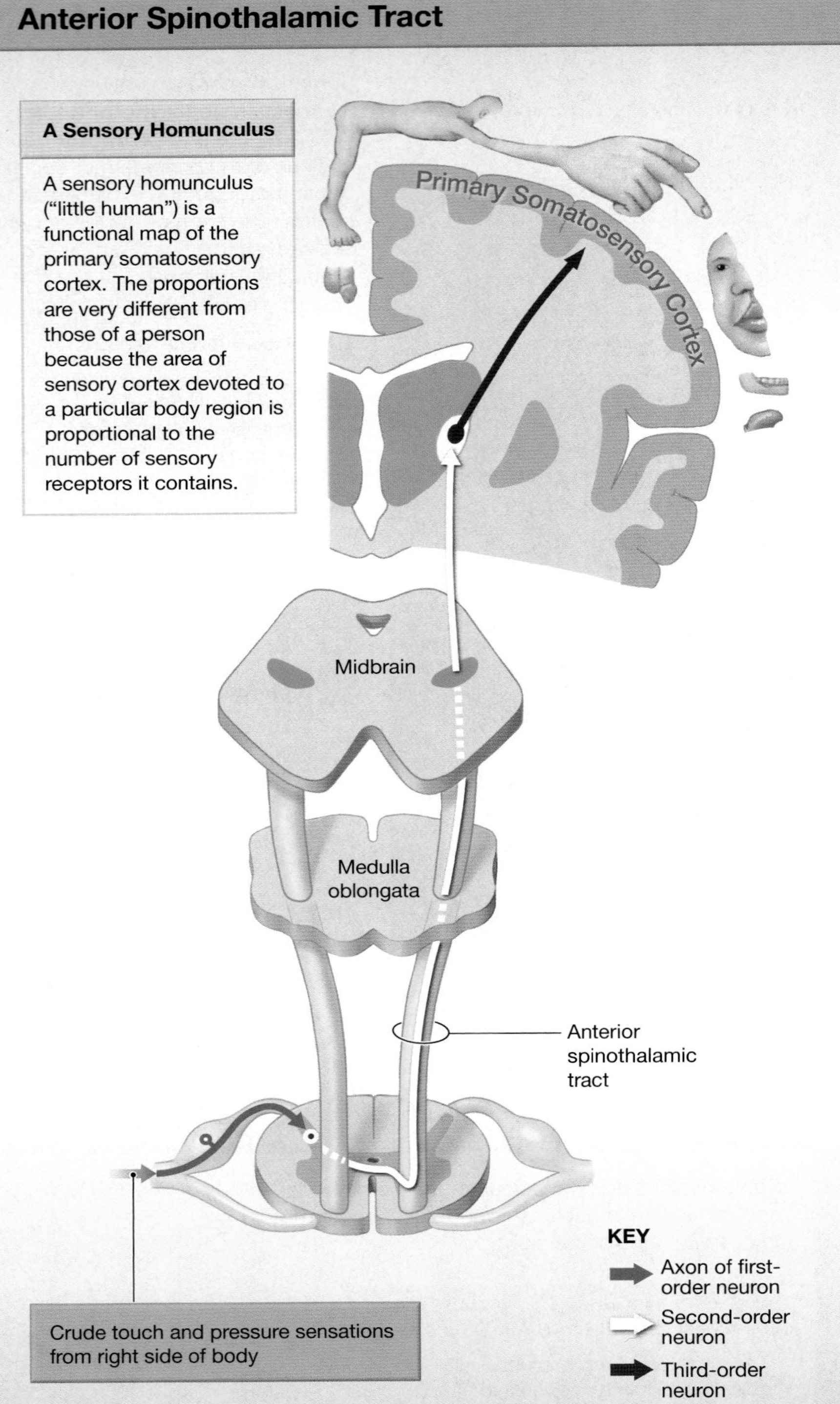

A Sensory Homunculus

A sensory homunculus ("little human") is a functional map of the primary somatosensory cortex. The proportions are very different from those of a person because the area of sensory cortex devoted to a particular body region is proportional to the number of sensory receptors it contains.

b The anterior spinothalamic tract carries sensory information to the primary somatosensory cortex on the opposite side of the body. The decussation occurs in the spinal cord at the level of entry.

15

Figure 15.3 (*continued*)

Lateral Spinothalamic Tract

Primary Somatosensory Cortex

Midbrain

Medulla oblongata

Lateral spinothalamic tract

Spinal cord

Pain and temperature sensations from right side of body

KEY

Axon of first-order neuron

Second-order neuron

Third-order neuron

c The lateral spinothalamic tract carries sensations of pain and temperature to the primary sensory cortex on the opposite side of the body. The decussation occurs in the spinal cord, at the level of entry.

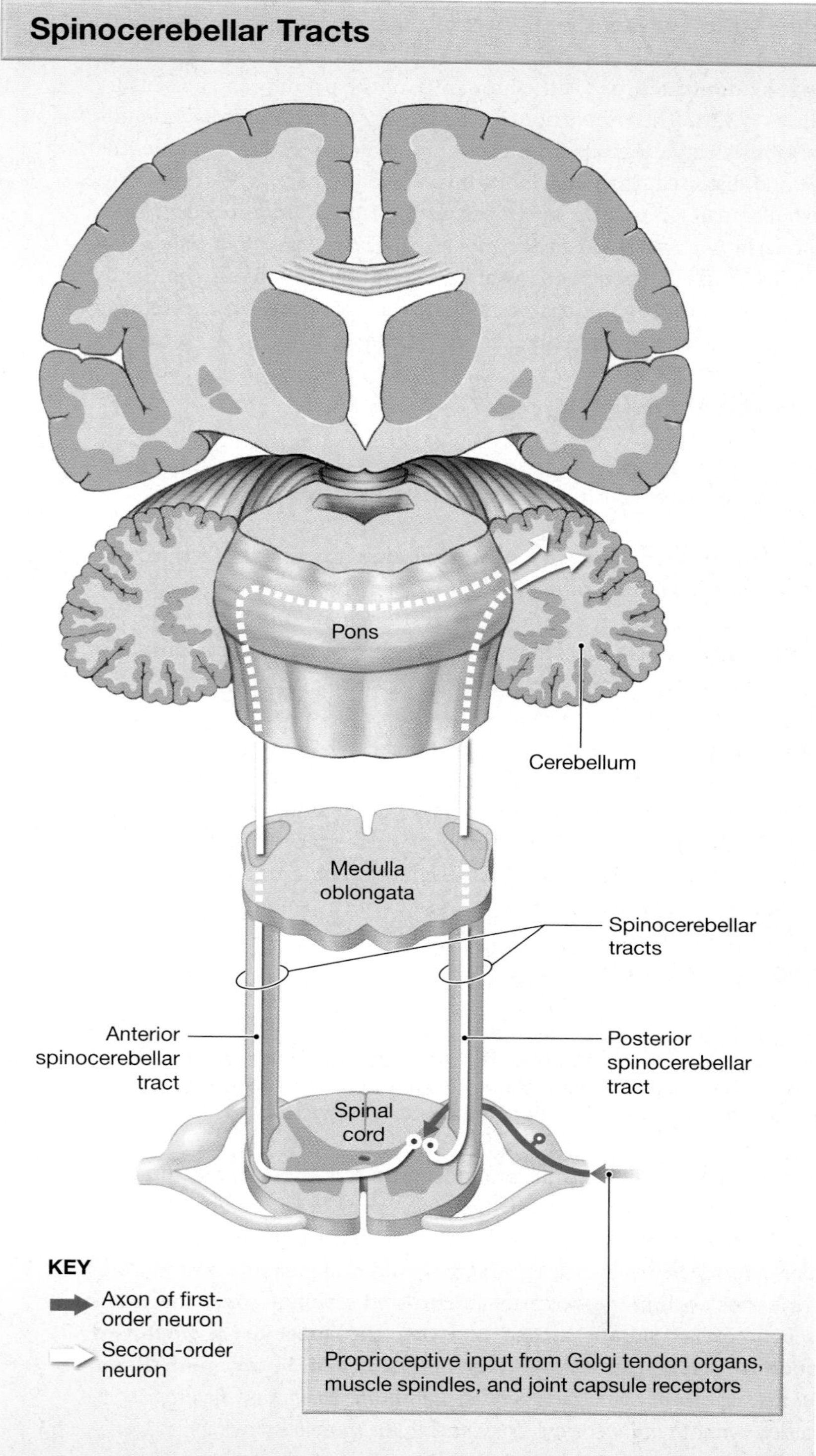

d The spinocerebellar tracts carry proprioceptive information to the cerebellum. (Only one tract is detailed on each side, although each side has both tracts.)

If the cerebral cortex were damaged, a person could still sense touch because the thalamic nuclei remain intact. However, the person would not be able to determine what was being touched because of the damage to the primary somatosensory cortex.

If a site on the primary sensory cortex is electrically stimulated, the person reports feeling sensations in a specific part of the body. By electrically stimulating the cortical surface, investigators have been able to create a sensory map, called a **sensory homunculus** ("little human"), of the primary somatosensory cortex (**Figure 15.3**). The proportions of body parts on the sensory homunculus are obviously very different from those of a real person. For example, the face is huge and distorted, with enormous lips, while the back is relatively tiny. *These distortions occur because the area of sensory cortex relating to a particular region of the body is proportional to the number of sensory receptors that region contains, not to its absolute size.* For instance, the lips have tens of thousands of touch receptors, whereas the back has far fewer; therefore, on the sensory homunculus the face and lips are larger than the back.

The Spinothalamic Tracts

▶ **KEY POINT** The spinothalamic tracts carry sensations of pain, temperature, and "crude" sensations of touch and pressure.

Figures 15.1 to **15.3b,c** show the **spinothalamic tracts** (also termed the *anterolateral system*). Like the posterior columns, the spinothalamic tracts have first-, second-, and third-order neurons. The first-order neurons enter the spinal cord and synapse with the second-order neurons within the posterior horns. The axons of the second-order neurons decussate to the contralateral side of the spinal cord and travel superiorly within the **anterior** and **lateral spinothalamic tracts**. These second-order neurons enter the ventral posterolateral nuclei of the thalamus, where they synapse with the third-order neurons. The third-order neurons carry the information to the primary somatosensory cortex.

Table 15.1 summarizes the destination of these tracts. **Figure 15.3b,c** shows the routes for the anterior and lateral spinothalamic tracts on the right side of the body, but they are present on both sides of the spinal cord.

The Spinocerebellar Tracts

▶ **KEY POINT** The spinocerebellar tracts are two-neuron tracts, having *only* first-order and second-order neurons. The first- and second-order neurons of these tracts do not synapse in the thalamus. As a result, a person is not aware of the sensory information carried in the spinocerebellar tracts.

The **spinocerebellar tracts** transmit proprioceptive sensations about the position of muscles, tendons, and joints of the lower limbs to the cerebellum. This information is essential for the fine coordination of body movements. The axons of first-order sensory neurons enter the spinal cord and then synapse on second-order neurons within the posterior horns of the spinal cord. The axons of these second-order neurons ascend in either the **anterior** or **posterior spinocerebellar tracts** (**Spotlight Figure 15.2** and **Figures 15.1** and **15.3d**). *Because neither of these tracts synapse within the thalamus, the proprioceptive information carried within them is not raised to the conscious level.*

Axons that cross over to the opposite side of the spinal cord enter the anterior spinocerebellar tract and ascend to the cerebellum by way of the superior cerebellar peduncle. These fibers then decussate a second time within the cerebellum to terminate in the cerebellum on the ipsilateral (same) side as the original stimulus. The functional significance of this "double cross" is not known.[1] The posterior spinocerebellar tract carries axons that do not decussate to the opposite side of the spinal cord. These axons enter the cerebellum by way of the inferior cerebellar peduncle.

[1]The anterior spinocerebellar tract also contains relatively small numbers of uncrossed axons in addition to axons that cross over and terminate in the contralateral cerebellum.

15.2 CONCEPT CHECK

2 As a result of pressure on her spinal cord, Jill cannot feel touch or pressure on her legs. What spinal tract is being compressed?

See the blue Answers tab at the back of the book.

15.3 | Motor Tracts

▶ **KEY POINT** All descending motor tracts are two-neuron systems. The upper motor neuron originates within the brain, travels inferiorly, and synapses with the lower motor neuron in the lateral or anterior horn of the spinal cord.

The central nervous system issues motor commands in response to information provided by sensory systems. These efferent motor commands are distributed to effector organs by either the somatic nervous system (SNS) or the autonomic nervous system (ANS). The SNS innervates skeletal muscles (**Figure 15.4a**). The ANS, or *visceral motor system*, innervates visceral effectors, such as smooth muscle, cardiac muscle, glands, and adipocytes (**Figure 15.4b**).

Conscious and subconscious motor commands control skeletal muscles by traveling over several descending motor tracts. **Spotlight Figure 15.2** shows the positions of the descending motor tracts within the spinal cord. The neural activity within these motor tracts is monitored and adjusted by the basal nuclei and cerebellum, higher motor centers discussed in Chapter 16.

The Corticospinal Tracts

▶ **KEY POINT** The corticospinal tracts include the corticobulbar, lateral corticospinal, and anterior corticospinal tracts. They are the most important descending motor tracts controlling the voluntary, fine motor movements of the upper and lower limbs.

The **corticospinal tracts**, sometimes called the *pyramidal tracts*, provide conscious, voluntary control over skeletal muscles (**Spotlight Figure 15.2** and **Figure 15.5**). This system begins at the pyramidal cells of the primary motor cortex. The axons of these upper motor neurons descend into the brainstem and spinal cord and synapse on lower motor neurons in the anterior horn that control skeletal muscles. In general, the corticospinal tract is a direct motor system: The upper motor neurons synapse directly on the lower motor neurons. However, the corticospinal tract also works indirectly, as it innervates other motor centers of the subconscious motor pathways.

There are three pairs of descending pyramidal tracts: (1) corticobulbar tracts, (2) lateral corticospinal tracts, and (3) anterior corticospinal tracts.

The Corticobulbar Tracts

Axons of the upper motor neurons of the **corticobulbar** (kor-ti-kō-BUL-bar; bulbar, brainstem) **tracts** synapse on the lower motor neurons in the motor nuclei of cranial nerves III, IV, V, VI, VII, IX, XI, and XII (**Figure 15.5** and **Table 15.2**). The corticobulbar tracts provide conscious control of the skeletal muscles that move the eye, jaw, and face and some muscles of the neck and pharynx. The corticobulbar tracts also innervate several motor centers involved in the subconscious control of skeletal muscle.

The Lateral and Anterior Corticospinal Tracts

Axons of the upper motor neurons of the **corticospinal tracts** synapse on the lower motor neurons in the anterior horns of the spinal cord (**Spotlight Figure 15.2** and **Figure 15.5**). As they descend, the corticospinal tracts are visible on the ventral surface of the medulla oblongata as a pair of thick bands, the **pyramids**. Within the medulla, about 85 percent of the axons cross the

Figure 15.4 Motor Tracts in the CNS and PNS. Organization of the somatic and autonomic nervous systems.

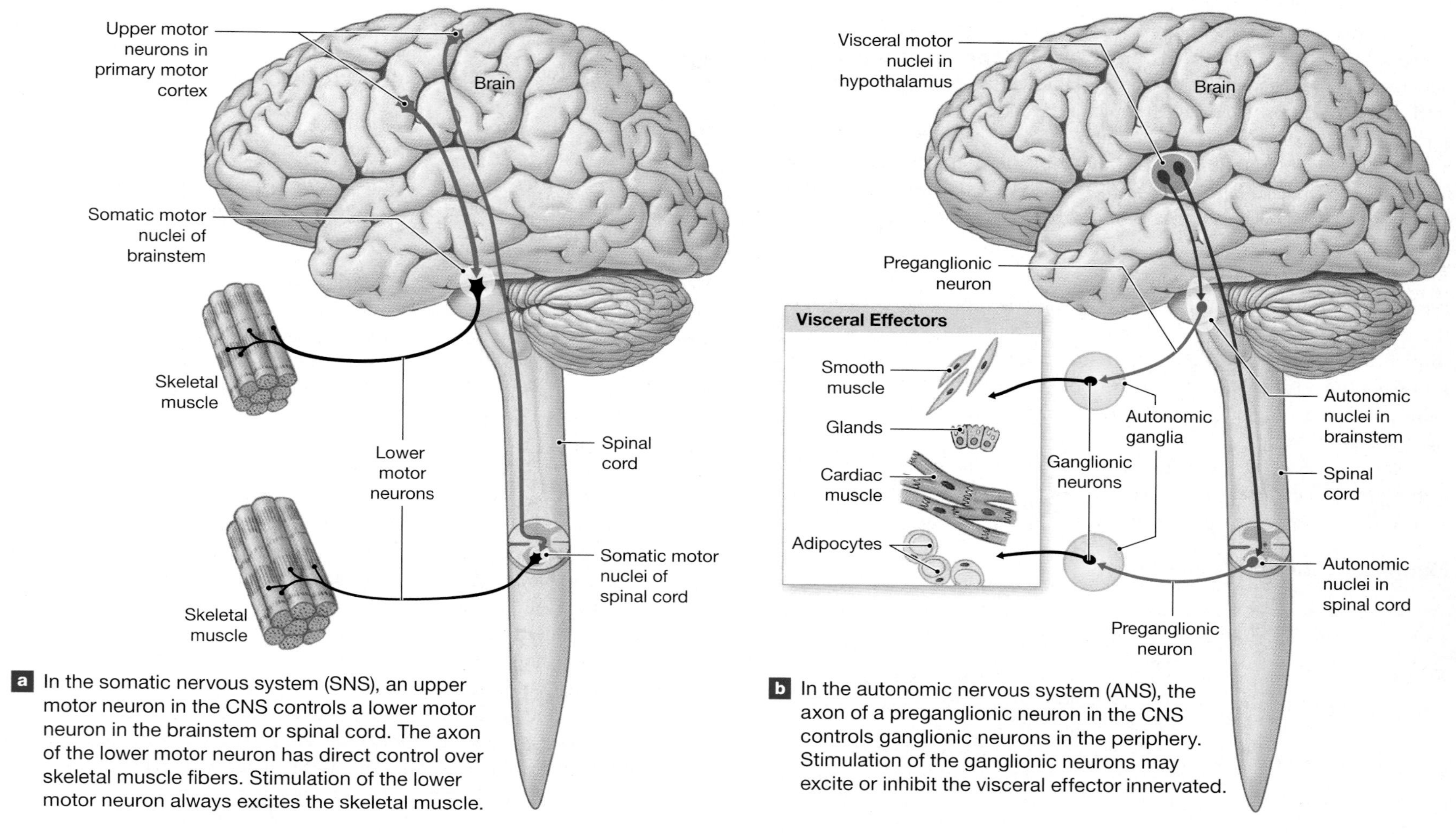

a In the somatic nervous system (SNS), an upper motor neuron in the CNS controls a lower motor neuron in the brainstem or spinal cord. The axon of the lower motor neuron has direct control over skeletal muscle fibers. Stimulation of the lower motor neuron always excites the skeletal muscle.

b In the autonomic nervous system (ANS), the axon of a preganglionic neuron in the CNS controls ganglionic neurons in the periphery. Stimulation of the ganglionic neurons may excite or inhibit the visceral effector innervated.

15

TABLE 15.2 | Principal Descending (Motor) Tracts and the General Functions of the Associated Nuclei in the Brain

Tract	Location of Upper Motor Neuron	Destination	Site of Decussation	Action
CORTICOSPINAL TRACTS				
Corticobulbar tracts	Primary motor cortex (cerebral hemisphere)	Lower motor neurons of cranial nerve nuclei in brain	Brainstem	Conscious motor control of skeletal muscles
Lateral corticospinal tracts	Primary motor cortex (cerebral hemisphere)	Lower motor neurons of anterior horns of spinal cord	Pyramids of medulla oblongata	Conscious motor control of skeletal muscles
Anterior corticospinal tracts	Primary motor cortex (cerebral hemisphere)	Lower motor neurons of anterior horns in cervical and upper thoracic segments	Level of lower motor neuron	Conscious motor control of skeletal muscles
SUBCONSCIOUS MOTOR PATHWAYS				
Vestibulospinal tracts	Vestibular nucleus (at border of pons and medulla oblongata)	Lower motor neurons of anterior horns of spinal cord	None (uncrossed)	Subconscious regulation of balance and muscle tone
Tectospinal tracts	Tectum (mesencephalon: superior and inferior colliculi)	Lower motor neurons of anterior horns (cervical spinal cord only)	Brainstem (midbrain)	Subconscious regulation of eye, head, neck, and upper limb position in response to visual and auditory stimuli
Medial reticulospinal tracts	Reticular formation (network of nuclei in brainstem)	Lower motor neurons of anterior horns of spinal cord	None (uncrossed)	Subconscious regulation of reflex activity
Rubrospinal tracts	Red nuclei of midbrain	Lower motor neurons of anterior horns of spinal cord	Brainstem (midbrain)	Subconscious regulation of upper limb muscle tone and movement

midline and form the descending **lateral corticospinal tracts** on the opposite side of the spinal cord. The lateral corticospinal tract synapses on lower motor neurons in the anterior horns at *all levels* of the spinal cord.

The other 15 percent of the corticospinal neurons continue to descend uncrossed within the spinal cord, forming the **anterior corticospinal tracts**. Before the upper motor neurons of the anterior corticospinal tract synapse with the lower motor neuron in the anterior horns, they decussate to the opposite side of the spinal cord in the anterior white commissure. The anterior corticospinal tract controls skeletal muscles of the neck, shoulder, and upper limb and *therefore does not descend farther than the cervical and superior thoracic regions of the spinal cord*. Table 15.2 summarizes these tracts and their actions.

The Motor Homunculus

Pyramidal cells in a specific portion of the primary motor cortex control the contraction of specific peripheral skeletal muscles. Like the somatosensory cortex, the primary motor cortex has been mapped, creating a **motor homunculus**. Figure 15.5 shows the motor homunculus of the left cerebral hemisphere and the corticospinal tracts controlling skeletal muscles on the right side of the body.

Like the sensory homunculus, the proportions of the motor homunculus are quite different from those of the actual body. This is because the size of the motor area of the cortex is proportional to the number of motor units controlling a specific region of the body. For example, the hands, face, and tongue, all of which are capable of complex movements, appear very large, while the trunk is relatively small. However, the sensory and motor homunculi differ in other respects because some highly sensitive regions, such as the soles of the feet, contain few motor units, while some areas with an abundance of motor units, such as the eye muscles, are not particularly sensitive.

Figure 15.5 The Corticospinal Tracts

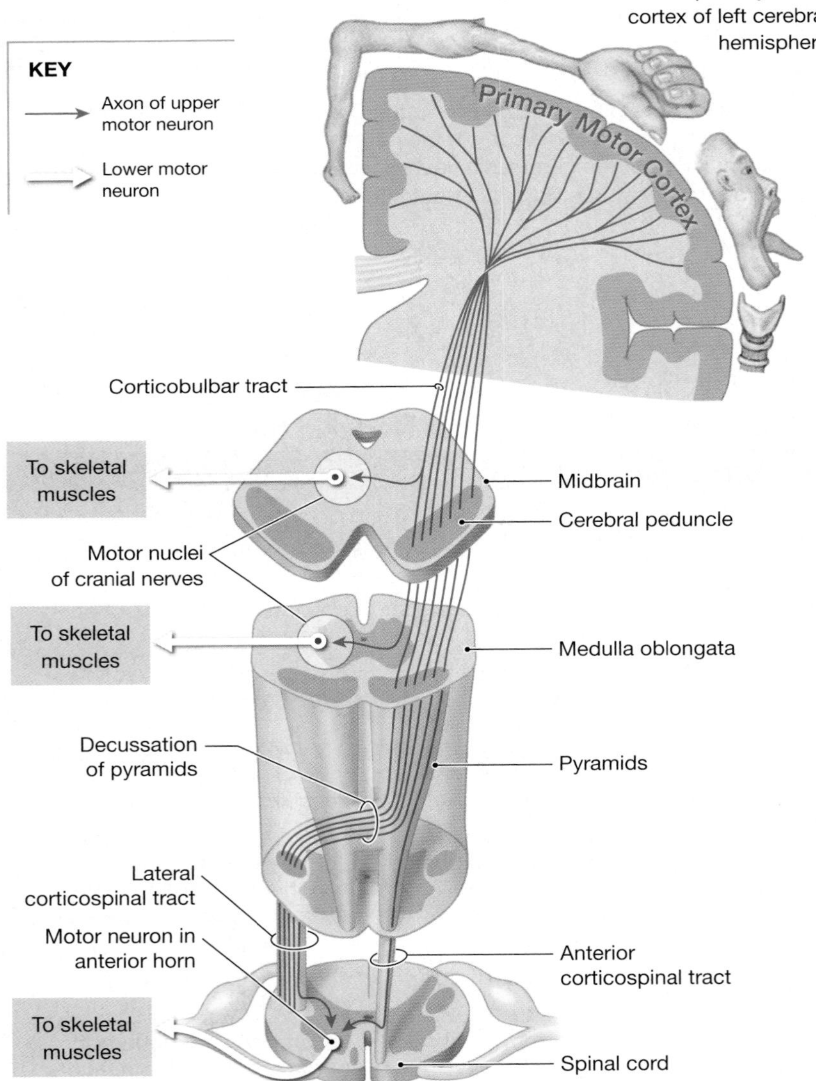

The Subconscious Motor Pathways

▶ **KEY POINT** Several centers in the cerebrum, diencephalon, and brainstem control somatic motor movements at a subconscious level. The motor pathways carrying this information are termed the subconscious motor pathways.

Several centers within the brain issue subconscious motor commands. Their associated motor tracts were known for a long time as the extrapyramidal system because anatomists thought that the extrapyramidal system operated independent of, and in parallel to, the pyramidal system. It is more appropriate, however, to group these nuclei and tracts in terms of their primary functions: The vestibulospinal, tectospinal, and medial reticulospinal tracts help control gross movements of the trunk and proximal limb muscles, and the rubrospinal tracts help control distal limb muscles.

These **subconscious motor pathways** modify or direct skeletal muscle contractions by stimulating or inhibiting lower motor neurons. It is important to remember that the axons of upper motor neurons in these pathways synapse on the same lower motor neurons innervated by the corticospinal tracts. This means that the different motor pathways interact (1) within the brain, through interconnections between the primary motor cortex and motor centers in the brainstem, and (2) through excitatory or inhibitory interactions at the level of the lower motor neurons.

The vestibulospinal, tectospinal, and medial reticulospinal tracts transmit action potentials that control muscle tone and gross movements of the neck, trunk, and proximal limb muscles. The upper motor neurons of these tracts are located in the vestibular nuclei, the superior and inferior colliculi, and the reticular formation (**Spotlight Figure 15.2** and Figure 15.6).

The Vestibulospinal Tracts

The vestibular nuclei receive information from cranial nerve VIII, the vestibulocochlear nerve. These sensory neurons receive information from receptors in the internal ear that monitor the position and movement of the head. The upper motor neurons within the vestibular nuclei respond to changes in the orientation of the head, sending motor commands to the lower motor neurons in the spinal cord that alter muscle tone, extension, and position of the neck, eyes, head, and limbs. The primary goal of this motor system is maintaining posture and balance. The descending, lower motor neurons in the spinal cord form the **vestibulospinal tracts (Spotlight Figure 15.2)**.

The Tectospinal Tracts

The superior and inferior colliculi are located in the tectum, or roof, of the midbrain. The colliculi receive visual (superior colliculi) and auditory (inferior colliculi) sensory information, and these nuclei coordinate and control reflexive responses to visual and auditory stimuli. The axons of upper motor neurons from the superior and inferior colliculi descend in the **tectospinal tracts (Spotlight Figure 15.2)**. These axons cross to the opposite side immediately, before descending to synapse on lower motor neurons in the brainstem or spinal cord. Axons in the tectospinal tracts cause reflexive changes in the position of the head, neck, and upper limbs in response to bright lights, sudden movements, or loud noises.

The Medial Reticulospinal Tracts

The reticular formation is a loosely organized network of neurons that extends throughout the brainstem. The reticular formation receives input from almost

Figure 15.6 Nuclei of Subconscious Motor Pathways. Cutaway view showing the location of major nuclei whose motor output is carried by subconscious pathways.

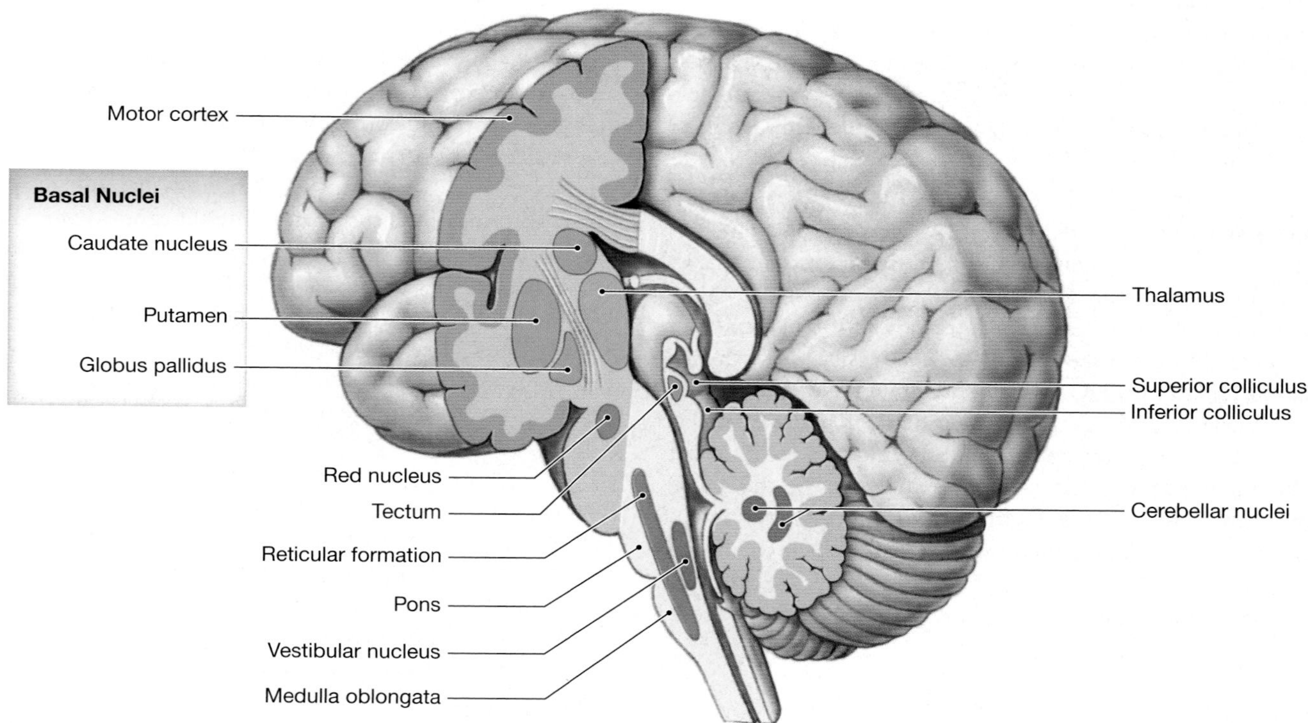

15

CLINICAL NOTE

Multiple Sclerosis

Multiple sclerosis (MS) is the most common demyelinating neurologic disease that affects the central nervous system. The body's own immune system attacks and destroys the myelin that insulates the axons of nerves within the CNS, including the brain, optic nerves, and spinal cord (see **Figure 13.6**). The demyelinated areas are known as *plaques* and when active show up as inflammatory areas within the CNS. Old plaques become scars, also inhibiting neurologic function.

Symptoms of MS depend on where the lesions (pathologic tissue changes) occur. Lesions in the cerebrum and cerebellum can cause tremors and problems with balance, speech, or coordination. Lesions in the optic nerves cause sight difficulties such as blurred or double vision. Lesions in the spinal motor nerve tracts cause muscle weakness, spasticity, and bowel or bladder problems. Plaques in spinal sensory nerve tracts cause numbness or burning sensations. There are at least four distinct clinical patterns of the disease.

Multiple sclerosis affects millions of people worldwide, most between the age of 15 and 60. It is most prevalent among Caucasians of European ancestry and affects women much more than men. Although the cause is unknown, it is believed to be a combination of genetic, infectious, and environmental factors. A cluster of genes on chromosome 6 may play a role. MS is more common in temperate climates and has been linked to several viruses, including the Epstein-Barr virus. Treatments focus on relieving symptoms and delaying the progression of the disease.

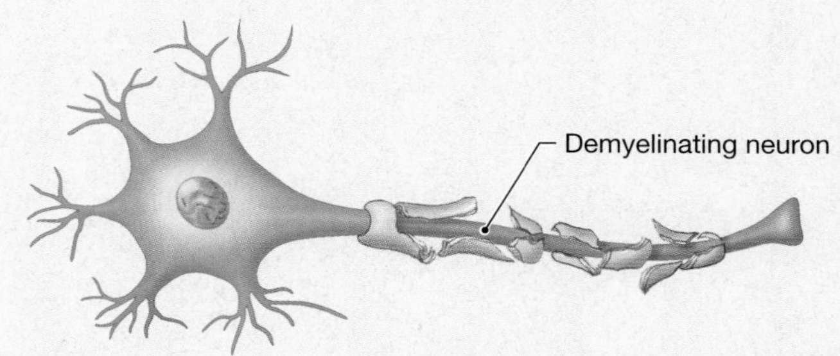

every ascending and descending tract. It also has extensive connections with every other part of the brain. Axons of the reticular formation's upper motor neurons descend in the **medial reticulospinal tracts** (also termed the *reticulospinal tracts*) without crossing to the opposite side **(Spotlight Figure 15.2)**. The effects of reticular formation stimulation are determined by what part is stimulated. For example, the stimulation of upper motor neurons in one portion of the reticular formation produces eye movements, while stimulation of another portion activates respiratory muscles.

The Rubrospinal Tracts

The **rubrospinal tracts** (*ruber*, red) control muscle tone and movements of the distal portions of the upper limbs **(Spotlight Figure 15.2)**. The information carried by these tracts stimulates flexor muscles and inhibits extensor muscles. The upper motor neurons of these tracts, located within the red nuclei of the mesencephalon, cross to the opposite side of the brain and descend into the spinal cord. The rubrospinal tracts are small and extend only to the cervical spinal cord. They control motor movement of the distal muscles of the upper limbs, and their effect is quite insignificant compared to that of the lateral corticospinal tracts. However, the rubrospinal tracts can be important in maintaining some motor control and muscle tone in the upper limbs if the lateral corticospinal tracts are damaged.

Table 15.2 reviews the major motor tracts.

15.3 CONCEPT CHECK

3 Through which of the motor tracts would the following commands travel: (a) reflexive change of head position due to bright lights, (b) automatic alterations in limb position to maintain balance?

See the blue Answers tab at the back of the book.

15.4 Levels of Somatic Motor Control

▶ **KEY POINT** Lower motor neuron activity is determined in two ways: (1) descending motor pathways regulate the activity of lower motor neurons, and (2) higher centers within the brain influence the activity of descending motor pathways. These higher centers within the brain are influenced by sensory input and other motor centers within the brain.

Ascending information is relayed from one nucleus or center to another in a series of steps. For example, somatic sensory information from the spinal cord goes from a nucleus in the medulla oblongata to a nucleus in the thalamus before it reaches the primary somatosensory cortex. Information processing occurs at each step along the way, blocking, reducing, or increasing our conscious awareness of the stimulus.

These processing steps are important, but they take time. Every synapse means another delay. Conduction time and synaptic delay means that it takes several milliseconds to relay information from a peripheral receptor to the primary somatosensory cortex. Additional time will pass before the primary motor cortex orders a voluntary motor response.

However, this delay is not dangerous because relay stations within the spinal cord and brainstem issue interim motor commands. While the conscious mind is still processing the information, reflexes provide an immediate response that will be "fine-tuned" a few milliseconds later. For example, if you touch a hot stove top, in the few milliseconds it takes for you to become consciously aware of the danger, you could be severely burned. However, thanks to a withdrawal reflex coordinated in the spinal cord, you jerk your hand from the hot stove before that happens. Voluntary motor responses, such as shaking your hand, stepping back, and crying out, occur somewhat later. In this case, the reflex of removing your hand was directed by neurons in the spinal cord; it was then supplemented by a voluntary response controlled by the cerebral cortex. The spinal reflex provided a rapid, automatic, stereotyped (preprogrammed) response that prevented serious injury. The cortical response was more complex and required more time to prepare and execute.

Nuclei in the brainstem also are involved in a variety of complex reflexes. Some of these nuclei receive sensory information and generate appropriate motor responses. These motor responses involve either the direct control of motor neurons or the regulation of reflex centers in other parts of the brain. **Figure 15.7** illustrates the various levels of somatic motor control from simple spinal reflexes to complex patterns of movement.

All levels of somatic motor control affect the activity of lower motor neurons. Reflexes coordinated in the spinal cord and brainstem are the simplest mechanisms of motor control. Higher levels perform more elaborate processing. As you move superiorly from the medulla oblongata to the cerebral cortex, the motor patterns become increasingly complex. For example, the respiratory rhythmicity center of the medulla oblongata sets a basic breathing rate. However, centers in the pons adjust that rate in response to commands received from the hypothalamus (subconscious) or cerebral cortex (conscious).

15.4 CONCEPT CHECK

4 Which neurons, upper motor or lower motor, are affected more by somatic motor control?

See the blue Answers tab at the back of the book.

Figure 15.7 **Somatic Motor Control**

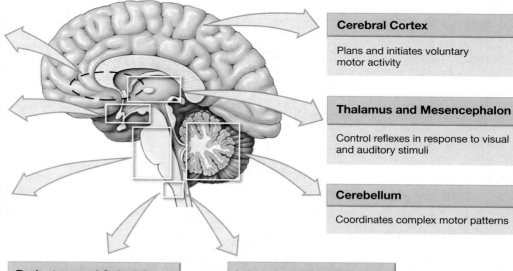

Basal Nuclei

Modify voluntary and reflexive motor patterns at the subconscious level

Hypothalamus

Controls reflex motor patterns related to eating, drinking, and sexual activity; modifies respiratory reflexes

Pons and Superior Medulla Oblongata

Control reflexes of balance and complex respiratory activity

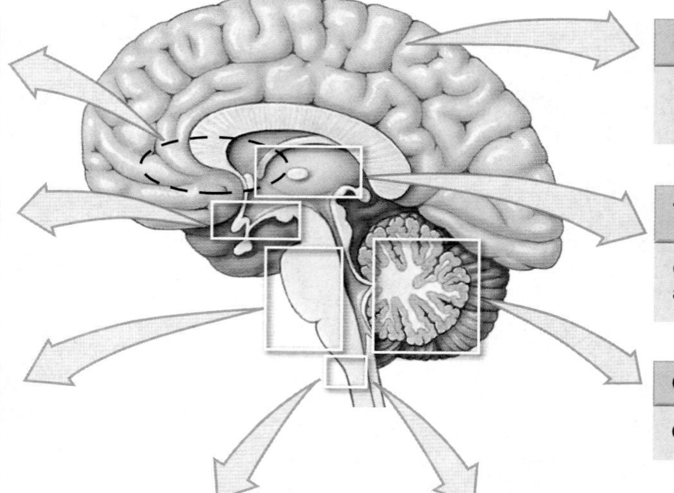

Cerebral Cortex

Plans and initiates voluntary motor activity

Thalamus and Mesencephalon

Control reflexes in response to visual and auditory stimuli

Cerebellum

Coordinates complex motor patterns

Brainstem and Spinal Cord

Control simple cranial and spinal reflexes

Inferior Medulla Oblongata

Controls basic respiratory reflexes

a Somatic motor control involves a series of levels, with simple spinal and cranial reflexes at the bottom and complex voluntary motor patterns at the top.

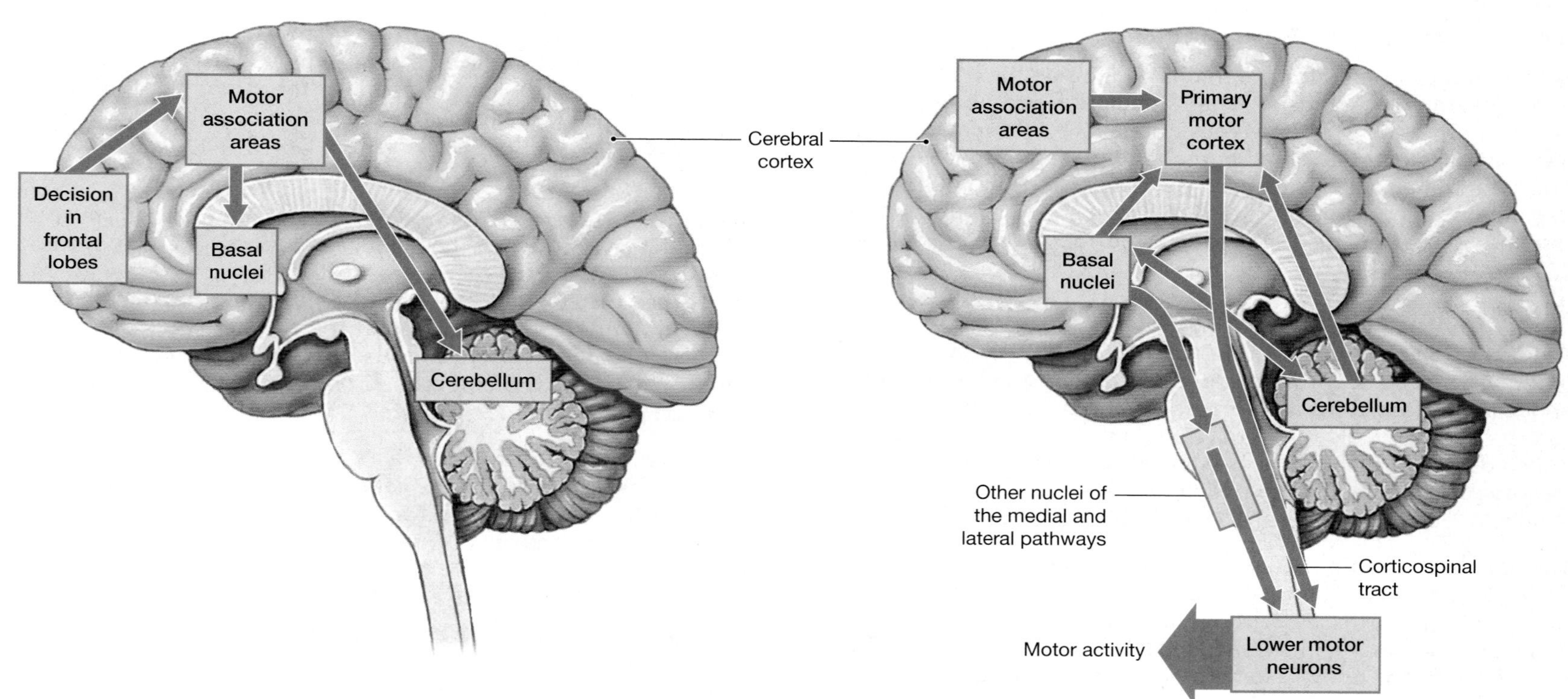

b The planning stage: When a conscious decision to move is made, information is relayed from the frontal lobes to motor association areas. These areas then relay the information to the cerebellum and basal nuclei.

c Movement: As the movement begins, the motor association areas send instructions to the primary motor cortex. Feedback from the basal nuclei and cerebellum modifies those commands, and output along the conscious and subconscious pathways directs involuntary adjustments in position and muscle tone.

Study Outline

Introduction p. 388

- Information passes continually between the brain, spinal cord, and peripheral nerves. Sensory information is delivered to CNS processing centers, and motor neurons control and adjust peripheral effectors.
- Tracts relay sensory and motor information between the CNS, the PNS, and peripheral organs and systems. Ascending (sensory) and descending (motor) tracts contain a chain of neurons and associated nuclei.

15.1 Organization and Patterns of Spinal Cord Tracts p. 388

- Sensory neurons that deliver the sensations to the CNS are termed **first-order neurons**. **Second-order neurons** are the CNS neurons on which the first-order neurons synapse. These neurons synapse on a **third-order neuron** in the thalamus. The axon of either the first-order or second-order neuron **decussates** (crosses) to the opposite side of the CNS. Thus, the right cerebral hemisphere receives sensory information from the left side of the body and the left cerebral hemisphere receives information from the right side of the body. *(See Figure 15.1 and Spotlight Figure 15.2.)*

15

15.2 Sensory Tracts p. 388

- Sensory receptors monitor conditions both inside the body and in the external environment. When stimulated, a receptor passes information to the central nervous system. This information, called a **sensation**, arrives as action potentials in an afferent (sensory) fiber. The complexity of the response to a particular stimulus depends in part on where processing occurs and where the motor response is initiated. *(See Figures 15.1 and 15.3, Spotlight Figure 15.2, and Table 15.1.)*

The Posterior Columns p. 388

- The **posterior columns** carry fine touch, pressure, and proprioceptive (position) sensations. The axons ascend within the **gracile fasciculus** and **cuneate fasciculus** and synapse in the gracile nucleus and cuneate nucleus within the medulla oblongata. This information is then relayed to the thalamus via the **medial lemniscus**. **Decussation** occurs as the second-order neurons enter the medial lemniscus. *(See Figures 15.1 and 15.3 and Spotlight Figure 15.2.)*
- The nature of any stimulus and its location is known because the information projects to a specific portion of the primary somatosensory cortex. Perceptions of sensations such as touch depend on processing in the thalamus. The precise localization is provided by the primary somatosensory cortex. A functional map of the primary somatosensory cortex is called a **sensory homunculus**. *(See Spotlight Figure 15.2 and Figure 15.3.)*

The Spinothalamic Tracts p. 394

- The **spinothalamic tracts** carry poorly localized sensations of touch, pressure, pain, and temperature. The axons of the second-order neurons decussate in the spinal cord and ascend in the **anterior** and **lateral spinothalamic tracts** to the ventral posterolateral nuclei of the thalamus. *(See Figure 15.3b,c and Table 15.1.)*

The Spinocerebellar Tracts p. 394

- The **posterior** and **anterior spinocerebellar tracts** carry subconscious sensations to the cerebellum concerning the position of muscles, tendons, and joints. *(See Figure 15.3d and Table 15.1.)*

15.3 Motor Tracts p. 394

- Motor commands from the CNS are issued in response to sensory system information. These commands are distributed by either the somatic nervous system (SNS) for skeletal muscles or the autonomic nervous system (ANS) for visceral effectors.
- Somatic motor tracts always involve an **upper motor neuron** (whose cell body lies in a CNS processing center) and a **lower motor neuron** (located in a motor nucleus of the brainstem or spinal cord). Autonomic motor control requires a **preganglionic neuron** (in the CNS) and a **ganglionic neuron** (in a peripheral ganglion). *(See Spotlight Figure 15.2 and Figure 15.4.)*

The Corticospinal Tracts p. 394

- The neurons of the primary motor cortex are pyramidal cells; the **corticospinal tracts** provide a rapid, direct mechanism for voluntary skeletal muscle control. The corticospinal tracts synapse on motor neurons in the anterior horns of the spinal cord and control movement in the neck and trunk and some coordinated movements in the axial skeleton. They consist of three pairs of descending motor tracts: (1) the corticobulbar tracts, (2) the lateral corticospinal tracts, and (3) the anterior corticospinal tracts. *(See Spotlight Figure 15.2, Figures 15.4 and 15.5, and Table 15.2.)*
- Axons in the **corticobulbar tracts** synapse on lower motor neurons in the motor nuclei of cranial nerves III, IV, V, VI, VII, IX, XI, and XII. The corticobulbar tracts end at the motor nuclei of cranial nerves controlling eye movements, facial muscles, tongue muscles, and neck and superficial back muscles. *(See Spotlight Figure 15.2, Figure 15.5, and Table 15.2.)*
- Axons in the **lateral** and **anterior corticospinal tracts** are visible along the ventral side of the medulla oblongata as a pair of thick elevations, the **pyramids**, where most of the axons decussate to enter the descending lateral corticospinal tracts. The remaining axons are uncrossed here and enter the anterior corticospinal tracts. These fibers will cross inside the anterior commissure before they synapse on motor neurons in the anterior horns. *(See Spotlight Figure 15.2, Figure 15.5, and Table 15.2.)*
- The activity of pyramidal cells in a specific portion of the primary motor cortex will result in the contraction of specific peripheral muscles. The cortical areas have been mapped out in diagrammatic form, creating a **motor homunculus**. *(See Figure 15.5.)*

The Subconscious Motor Pathways p. 396

- The **subconscious motor pathways** consist of several centers that may issue motor commands as a result of processing performed at an unconscious, involuntary level. These pathways can modify or direct somatic motor patterns. Their outputs may descend in (1) the vestibulospinal, (2) the tectospinal, (3) the reticulospinal, or (4) the rubrospinal tracts. *(See Spotlight Figure 15.2, Figure 15.6, and Table 15.2.)*
- The vestibular nuclei receive sensory information from inner ear receptors through N VIII. These nuclei issue motor commands to maintain posture and balance. The fibers descend through the **vestibulospinal tracts**. *(See Spotlight Figure 15.2, Figure 15.6, and Table 15.2.)*
- Commands carried by the **tectospinal tracts** change the position of the eyes, head, neck, and arms in response to bright lights, sudden movements, or loud noises. *(See Spotlight Figure 15.2, Figure 15.6, and Table 15.2.)*

- Motor commands carried by the **medial reticulospinal tracts** vary according to the region stimulated. The reticular formation receives inputs from almost all ascending and descending pathways and from numerous interconnections with the cerebrum, cerebellum, and brainstem nuclei. *(See Spotlight Figure 15.2, Figure 15.6, and Table 15.2.)*

15.4 | Levels of Somatic Motor Control p. 398

- Ascending sensory information is relayed from one nucleus or center to another in a series of steps. Information processing occurs at each step along the way. Processing steps are important but time-consuming. Nuclei in the spinal cord, brainstem, and cerebrum work together in various complex reflexes. *(See Figure 15.7.)*

Chapter Review

For answers, see the blue Answers tab at the back of the book.

Level 1 Reviewing Facts and Terms

Match each numbered item with the most closely related lettered item.

1. decussation .... ☐
2. sensory .... ☐
3. interneuron .... ☐
4. posterior column .... ☐
5. spinothalamic .... ☐
6. spinocerebellar .... ☐
7. corticospinal system .... ☐
8. tectospinal tracts .... ☐
9. rubrospinal tract .... ☐

a. second-order
b. pain, temperature, crude touch, pressure
c. voluntary control skeletal muscle
d. subconscious control of distal limb musculature
e. afferent
f. information about "what," "where," and "when"
g. unconscious proprioception
h. crossover
i. position change—noise-related

10. Axons ascend the posterior column to reach the
    (a) gracile nucleus and cuneate nucleus.
    (b) ventral nucleus of the thalamus.
    (c) posterior lobe of the cerebellum.
    (d) medial nucleus of the thalamus.
11. Which of the following is true of the spinothalamic tract?
    (a) Its neurons synapse in the anterior horn of the spinal cord.
    (b) It carries sensations of touch, pressure, and temperature from the brain to the periphery.
    (c) It transmits sensory information to the brain, where decussation occurs in the thalamus.
    (d) None of the above is correct.
12. Which of the following are spinal tracts within the subconscious motor pathways?
    (a) vestibulospinal tracts
    (b) tectospinal tracts
    (c) medial reticulospinal tracts
    (d) all of the above
13. Axons of the corticospinal tract synapse at
    (a) motor nuclei of cranial nerves.
    (b) motor neurons in the anterior horns of the spinal cord.
    (c) motor neurons in the posterior horns of the spinal cord.
    (d) motor neurons in ganglia near the spinal cord.
14. Give the anatomical names of the spinal tracts indicated on the following diagram.

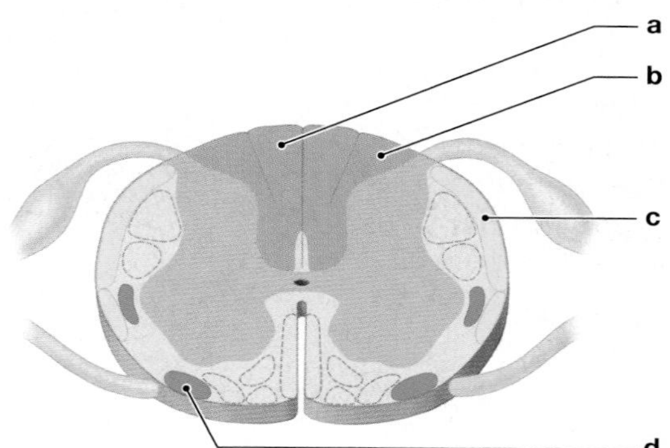

(a) ____________
(b) ____________
(c) ____________
(d) ____________

Level 2 Reviewing Concepts

1. What symptoms would you associate with damage to the gracile nucleus on the right side of the medulla oblongata?
   (a) inability to perceive fine touch from the left lower limb
   (b) inability to perceive fine touch from the right lower limb
   (c) inability to direct fine motor activities involving the left shoulder
   (d) inability to direct fine motor activities involving the right shoulder
2. Describe the function of first-order neurons in the CNS.
3. Why do the proportions of the sensory homunculus differ from those of the body?
4. What is the primary role of the cerebral nuclei in the function of the subconscious motor pathways?
5. Compare the actions directed by motor commands in the vestibulospinal tracts with those in the medial reticulospinal tracts.

Level 3 Critical Thinking

1. Cindy has a biking accident and injures her back. She is examined by a doctor who notices that Cindy cannot feel pain sensations (a pinprick) from her left hip and lower limb, but she has normal sensation elsewhere and has no problems with the motor control of her limbs. The doctor tells Cindy that he thinks a portion of the spinal cord may be compressed and that this is responsible for her symptoms. Where in the spinal cord might the problem be located, and why?
2. As a result of a snowboarding accident, John is unable to feel fine touch, pressure, vibration, or proprioception in his left upper limb distal to his wrist. He has no other sensory deficits. A neurologist suspects that a portion of his spinal cord has been damaged. Where in the spinal cord might the problem be located, and why?

15

CLINICAL CASE WRAP-UP

Amyotrophic Lateral Sclerosis

Amyotrophic (muscle atrophy, weakness) lateral sclerosis is a progressive, fatal neurological disease that attacks the neurons responsible for controlling voluntary muscles. It belongs to the category of **motor neuron diseases**, characterized by the gradual degeneration and death of motor neurons.

ALS involves the lateral columns of the spinal cord, the corticospinal motor tracts, corticobulbar tracts, anterior horn cells, and bulbar motor nuclei or a combination of these. The motor neurons in the brain, brainstem, and spinal cord deteriorate and die, causing the skeletal muscles they innervate to waste away. All skeletal muscles are affected, including the muscles of chewing, swallowing, speaking, and breathing. This disease affects both upper motor neurons and lower motor neurons and the lateral corticospinal tracts that connect them. At autopsy, the lateral portion of the spinal cord looks wasted and feels hard, or sclerosed.

ALS does not affect the ascending sensory tracts of the spinal cord. The posterior columns, anterior and lateral spinothalamic tracts, and spinocerebellar tracts remain intact and functioning. This preserves intellect and the senses of touch, pressure, vibration, proprioception, sight, smell, and taste.

As with most neurodegenerative diseases, nobody knows what causes ALS. This rare but devastating disorder usually appears between ages 40 and 60. Men are affected slightly more often than women, soldiers slightly more often than civilians. A small percentage of cases seem to have a genetic basis. Nobody can explain how Stephen Hawking has survived over 55 years with ALS, but the world of physics is thankful.

1. Because amyotrophic lateral sclerosis affects both upper motor neurons and lower motor neurons, where else would you expect to see cell death, besides the lateral corticospinal tracts?
2. Would you expect the dorsal root ganglia to be appear normal or abnormal in ALS?

See the blue Answers tab at the back of the book.

15

Related Clinical Terms

primary lateral sclerosis (PLS): A slowly progressive degenerative disorder of the motor neurons of the cerebral cortex, resulting in widespread weakness.

lower motor neuron dysarthria: A speech disturbance caused by dysfunction of the motor nuclei and the lower pons or medulla, or other neural connections, central and peripheral to the muscles involved with speech.

16

The Nervous System

The Brain and Cranial Nerves

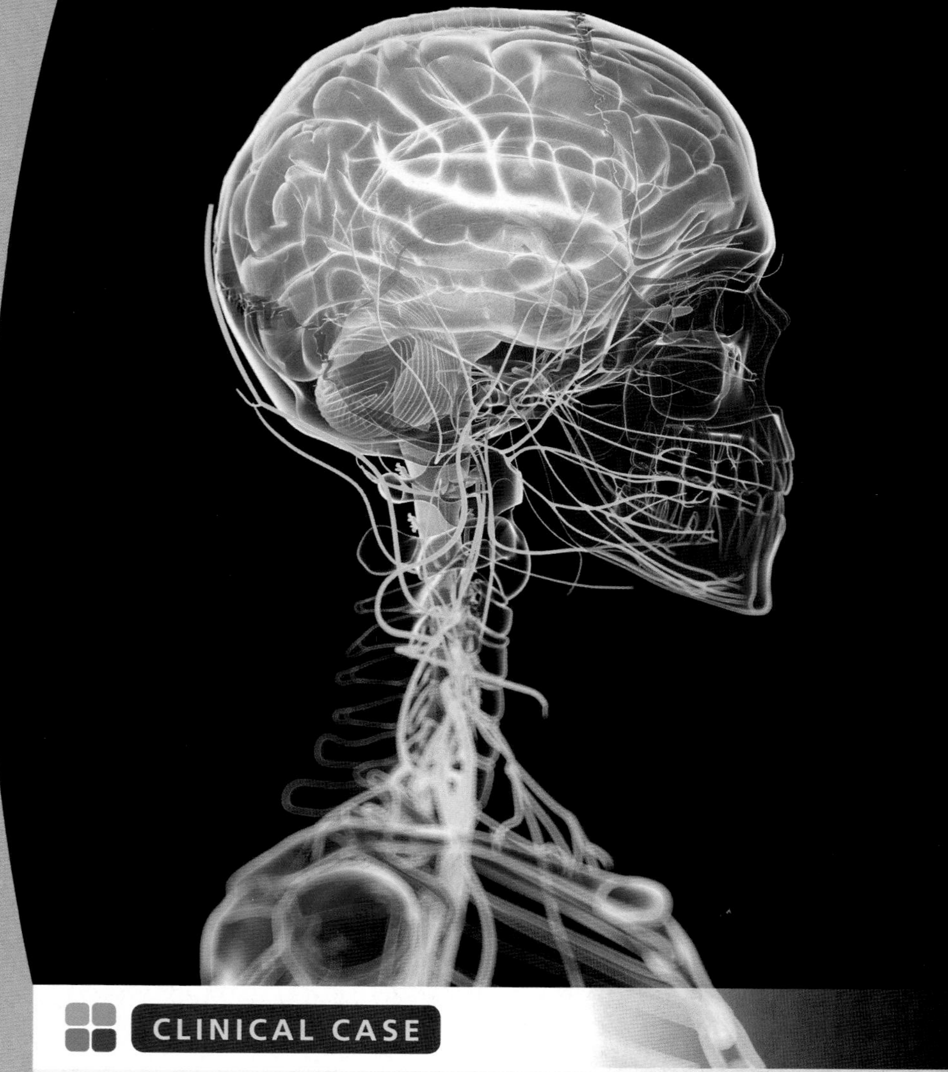

Learning Outcomes

These Learning Outcomes correspond by number to this chapter's sections and indicate what you should be able to do after completing the chapter.

16.1 Identify the major regions of the brain and explain their functions. p. 404

16.2 Compare and contrast the structures that protect and support the brain. p. 406

16.3 Identify the anatomical structures of the medulla oblongata and describe their functions. p. 413

16.4 List the pontine structures and functions. p. 414

16.5 Identify the features of the mesencephalon (midbrain) and describe their functions. p. 415

16.6 Identify the anatomical structures that form the thalamus and hypothalamus and list their functions. p. 417

16.7 Identify the components of the cerebellum and list their functions. p. 422

16.8 Compare and contrast motor, sensory, association, and limbic areas of the cerebrum. p. 422

16.9 Compare and contrast the 12 pairs of cranial nerves. p. 434

CLINICAL CASE

A Neuroanatomist's Stroke of Insight

Dr. Jill Taylor, a neuroanatomist, is 37 and at the top of her field. One morning she develops a throbbing headache behind her left eye. She then notices that her thoughts and movements are slowing down. Soon she realizes her right arm is paralyzed, and she is barely able to call for help. When she arrives at the hospital, she cannot walk, talk, read, write, or recall anything. She feels her spirit surrender and braces for death.

Dr. Taylor awakes later that day, shocked to be alive. She still cannot speak or understand speech, or recognize or use numbers. She can, however, appreciate the irony of her situation: a neuroscientist (scientist who studies the brain) witnessing her very own brain emergency, an evolving cerebrovascular accident (CVA) or stroke. Doctors perform open brain surgery to remove a large blood clot that was pressing on the left side of her brain near her language area.

Will Dr. Taylor recover? To find out, turn to the Clinical Case Wrap-Up on p. 448.

THE BRAIN HAS A COMPLEX, THREE-DIMENSIONAL STRUCTURE and performs a bewildering array of functions. People often compare the brain to a computer. Like the brain, a computer receives incoming information, files and processes this information, and directs appropriate responses. However, even the most sophisticated computer lacks the characteristics of a single neuron! One neuron can process information from up to 200,000 different sources at the same time, and there are billions of neurons in your nervous system. As a result of this incredible processing ability, the brain is the source of our dreams, passions, plans, memories, and behaviors. Everything we do and everything we are results from the brain's activity.

The brain is far more complex than the spinal cord, and it responds to stimuli with greater adaptability. That adaptability results from the tremendous number of neurons and neuronal pools in the brain and their complex interconnections. The brain contains approximately 20 billion neurons, and each neuron receives and processes information from thousands of synapses at one time. Excitatory and inhibitory interactions between the neuronal pools ensure that our responses can vary to meet changing circumstances. But adaptability has a price: A response cannot be immediate, precise, and adaptable all at the same time. Adaptability requires multiple synapses and processing steps, and every synapse adds to the delay between stimulus and response. Spinal reflexes, however, provide an *immediate* response that can be fine-tuned by more adaptable but slower processing centers in the brain. Let's begin with the brain and cranial nerves.

16

16.1 An Introduction to the Organization of the Brain

▶ **KEY POINT** The adult human brain is compact but contains almost 95 percent of all our nervous tissue.

The brain's external appearance gives few clues to its complexity and importance. Although an adult brain can be easily held in both hands, it contains almost 95 percent of the body's nervous tissue. A freshly removed brain is gray externally, and its internal tissues are tan to pink. Overall, the brain has the consistency of medium-firm tofu.

Embryology of the Brain

▶ **KEY POINT** The brain starts as a small, hollow tube that forms three primary vesicles: the prosencephalon, mesencephalon, and rhombencephalon. The prosencephalon and rhombencephalon subdivide to form secondary brain vesicles.

Chapter 28 presents a detailed explanation of the development of the brain. However, a brief explanation now will help you understand adult brain structure and organization.

The central nervous system begins as a hollow neural tube with a fluid-filled internal cavity called the neurocoel. As development proceeds, this simple passageway expands to form enlarged chambers called ventricles. In the fourth week of development, three areas in the cephalic portion of the neural tube expand rapidly. This expansion creates three **primary brain vesicles** named for their positions: the **prosencephalon** (prōs-en-SEF-a-lon; *proso*, forward, + *enkephalos*, brain), or "forebrain"; the **mesencephalon** (mez-en-SEF-a-lon; *mesos*, middle), or "midbrain"; and the **rhombencephalon** (rom-ben-SEF-a-lon), or "hindbrain." **Table 16.1** summarizes the developmental changes of the three primary divisions of the brain.

The prosencephalon and rhombencephalon subdivide further, forming **secondary brain vesicles**. The prosencephalon forms the **telencephalon** (tel-en-SEF-a-lon; *telos*, end) and the **diencephalon**. The telencephalon forms the cerebrum. The cerebrum possesses the paired cerebral hemispheres that dominate the superior and lateral surfaces of the adult brain. The hollow diencephalon has a roof (the epithalamus), walls (the left and right thalamus), and a floor (the hypothalamus). As the posterior end of the neural tube closes, secondary bulges, the optic vesicles, extend laterally from the sides of the diencephalon. The developing brain also bends, forming creases that mark the boundaries between the ventricles.

The mesencephalon does not subdivide, but its walls thicken and the neurocoel becomes a narrow passageway with a diameter similar to that of the central canal of the spinal cord.

The portion of the rhombencephalon closest to the mesencephalon forms the **metencephalon** (met-en-SEF-a-lon; *meta*, after). The anterior part of the metencephalon develops into the pons, and the posterior portion becomes the cerebellum. The part of the rhombencephalon closer to the spinal cord becomes the **myelencephalon** (mī-el-en-SEF-a-lon; *myelon*, spinal cord), which forms the medulla oblongata.

Major Regions and Landmarks

▶ **KEY POINT** The adult brain consists of six major regions: medulla oblongata, pons, mesencephalon (midbrain), diencephalon, cerebellum, and cerebrum.

There are six major regions in the adult brain, as shown in **Figure 16.1**. Together, the medulla oblongata, the pons, and the mesencephalon[1] are referred to as the **brainstem**. The brainstem contains important processing centers that relay information to and from the cerebrum or cerebellum.

The Medulla Oblongata

The **medulla oblongata** (also termed the "medulla") connects the brainstem to the spinal cord. The inferior portion of the medulla oblongata resembles the spinal cord, and the superior portion has a thin, membranous covering. The medulla oblongata relays sensory information to the thalamus and to other centers within the brainstem. In addition, it contains major centers regulating autonomic functions, such as heart rate, blood pressure, and digestive activities.

The Pons

Immediately superior to the medulla is the **pons**. Pons means "bridge," and the pons connects the cerebellum to the brainstem. The pons contains nuclei controlling somatic and visceral motor functions.

The Mesencephalon (Midbrain)

Nuclei in the **mesencephalon**, or *midbrain*, process visual and auditory information and coordinate reflexive somatic motor responses to visual and auditory stimuli. This region also contains centers that maintain levels of consciousness.

The Diencephalon

The **diencephalon** (dī-en-SEF-a-lon; *dia*, through) lies deep within the brain and attaches the mesencephalon to the cerebrum. The diencephalon has three subdivisions:

1. The **epithalamus** contains the hormone-secreting pineal gland, an endocrine structure.
2. The right **thalamus** (THAL-a-mus; plural, *thalami*) and left thalamus are sensory information relay and processing centers.
3. The **hypothalamus** (*hypo–*, below) is a visceral control center. A narrow stalk connects the hypothalamus to the **pituitary gland**, also called the *hypophysis*. The hypothalamus contains centers for emotions, autonomic nervous system function, and hormone production. It is an important link between the nervous and endocrine systems.

[1]Some sources consider the brainstem to include the diencephalon. We will use the more restrictive definition here.

Table 16.1 | **Development of the Human Brain** *(See also Chapter 28 for embryological summary)*

Primary Brain Vesicles (3-week embryo)	Secondary Brain Vesicles (6-week embryo)	Brain Regions at Birth	Ventricles
Prosencephalon	Telencephalon	Cerebrum	Lateral ventricle
	Diencephalon	Diencephalon	Third ventricle
Mesencephalon	Mesencephalon	Midbrain	Cerebral aqueduct
Rhombencephalon	Metencephalon	Cerebellum and pons	Fourth ventricle
	Myelencephalon	Medulla oblongata	Fourth ventricle

Spinal cord

16

Figure 16.1 Major Regions of the Brain. An introduction to brain regions and their major functions.

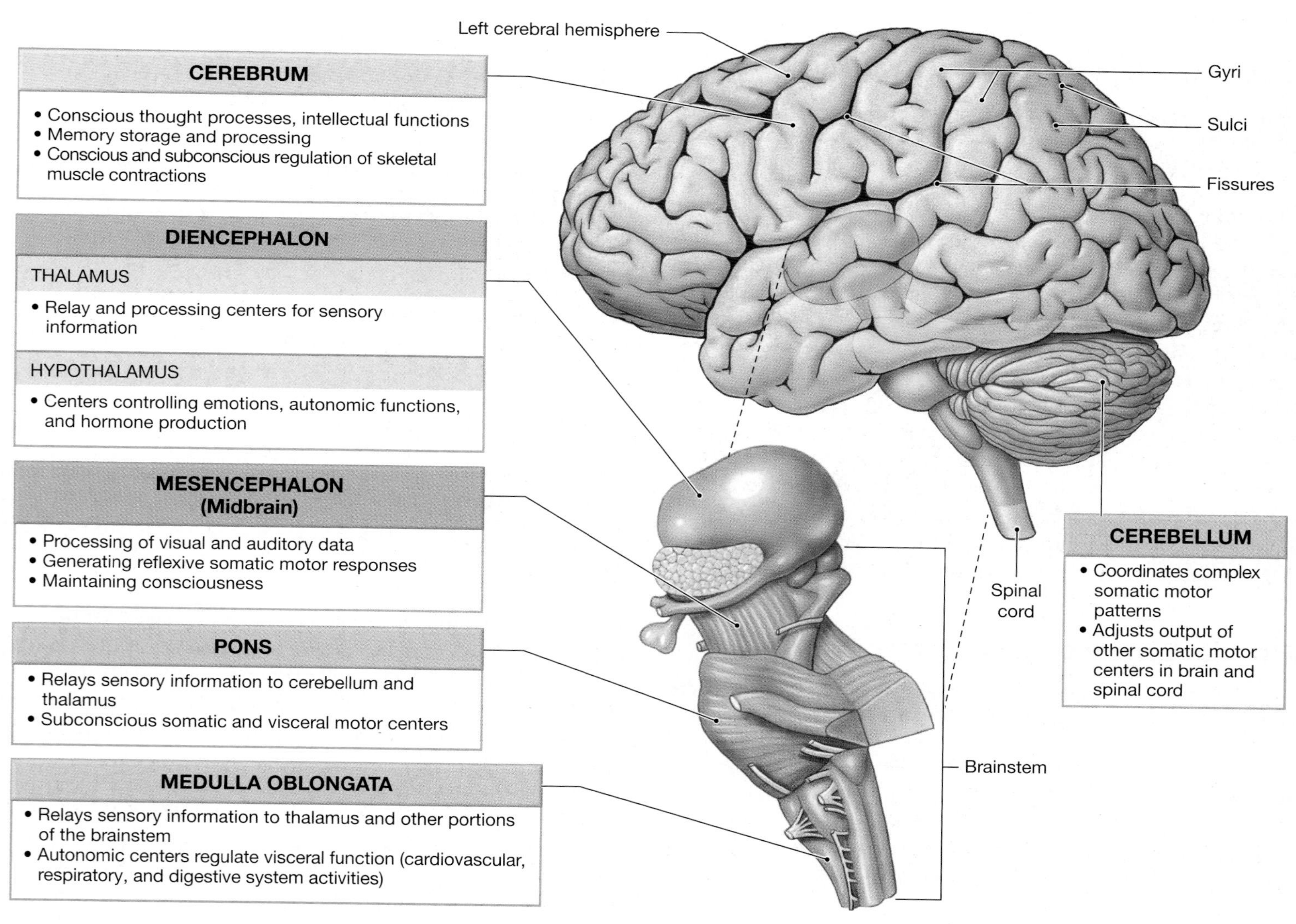

Left cerebral hemisphere
Gyri
Sulci
Fissures
CEREBRUM
• Conscious thought processes, intellectual functions
• Memory storage and processing
• Conscious and subconscious regulation of skeletal muscle contractions
DIENCEPHALON
THALAMUS
• Relay and processing centers for sensory information
HYPOTHALAMUS
• Centers controlling emotions, autonomic functions, and hormone production
MESENCEPHALON (Midbrain)
• Processing of visual and auditory data
• Generating reflexive somatic motor responses
• Maintaining consciousness
PONS
• Relays sensory information to cerebellum and thalamus
• Subconscious somatic and visceral motor centers
MEDULLA OBLONGATA
• Relays sensory information to thalamus and other portions of the brainstem
• Autonomic centers regulate visceral function (cardiovascular, respiratory, and digestive system activities)
Spinal cord
CEREBELLUM
• Coordinates complex somatic motor patterns
• Adjusts output of other somatic motor centers in brain and spinal cord
Brainstem

TIPS & TOOLS

To visualize the relationships among these structures, think of the diencephalon as a shoe box with the top of the box being the epithalamus, the left and right sides are the thalami, the bottom is the hypothalamus, and the enclosed space is the third ventricle.

The Cerebellum

The **cerebellum** (ser-e-BEL-um) is the second largest part of the brain, and its two small hemispheres lie posterior to the pons and inferior to the cerebral hemispheres. The cerebellum automatically adjusts motor activities based on sensory information and memories of learned movement.

The Cerebrum

The **cerebrum** (ser-Ē-brum or SER-e-brum) is the largest part of the brain. It is divided into large, paired **cerebral hemispheres** separated by the **longitudinal fissure**. The surface of the cerebrum is called the cerebral cortex, and it is composed of gray matter. Folds, termed **sulci**, convolute the surface of the cerebral cortex. These sulci separate the intervening ridges, termed **gyri**. The cerebrum is divided into lobes by a number of the larger sulci. The names of the lobes are derived from the bones of the cranium immediately superficial to them. For instance, the parietal bones of the skull are immediately superficial to the parietal lobes of the cerebrum. Conscious thought processes, intellectual functions, memory storage and retrieval, and complex motor patterns originate in the cerebrum.

16

Gray Matter and White Matter Organization

▶ **KEY POINT** In the brain, there is an inner region of gray matter surrounded by tracts of white matter. The fluid-filled ventricles and passageways of the brain correspond to the central canal of the spinal cord.

The general distribution of the gray matter in the brain resembles the distribution of the gray matter in the spinal cord: Tracts of white matter surround an inner region of gray matter. The gray matter of the brain forms nuclei—clusters of spherical, oval, or irregularly shaped neuronal cell bodies. However, the arrangement of gray matter and white matter in the brain is not as predictable as in the spinal cord. In the cerebrum and cerebellum, the white matter is covered by a superficial layer of gray matter called the **cortex** (*cortex*, rind or bark).

The term "higher centers" refers to cortical areas of the cerebrum, cerebellum, nuclei of the diencephalon, and mesencephalon. Output from these higher centers modifies the activities of nuclei in the lower (more caudal) brainstem and spinal cord. The nuclei and cortical areas of the brain receive sensory information and issue motor commands to peripheral effectors indirectly, through the spinal cord and spinal nerves, or directly, through the cranial nerves.

The Ventricular System of the Brain

▶ **KEY POINT** Ventricles are fluid-filled chambers within the brain. They are filled with cerebrospinal fluid and lined by ependymal cells.

There are four **ventricles** (VEN-tri-kels) in the adult brain: one within each cerebral hemisphere (making two), a third within the diencephalon, and a fourth between the pons and cerebellum and extending into the superior portion of the medulla oblongata. Each ventricle is filled with cerebrospinal fluid and lined by ependymal cells. **Figure 16.2** shows the position and orientation of the ventricles.

The **lateral ventricles** in the cerebral hemispheres have a complex shape. The two lateral ventricles are separated by a thin partition, the **septum pellucidum**. Each lateral ventricle has a body, anterior horn, posterior horn, and inferior horn. The body lies within the parietal lobe of the cerebrum, and the anterior horn extends into the frontal lobe. The posterior horn projects into the occipital lobe, and the inferior horn curves laterally within the temporal lobe. There is no direct connection between the two lateral ventricles. However, each of the two lateral ventricles communicates with the **third ventricle** of the diencephalon through an **interventricular foramen** (or *foramen of Monro*).

The mesencephalon has a slender canal called the **cerebral aqueduct**, or *aqueduct of the midbrain*. This passageway connects the third ventricle with the **fourth ventricle**. In the inferior portion of the medulla oblongata, the fourth ventricle narrows and is continuous with the central canal of the spinal cord. The cerebrospinal fluid (CSF) circulates from the ventricles into the central canal of the spinal cord. From there, the CSF moves through the foramina in the roof of the fourth ventricle into the subarachnoid space surrounding the brain and spinal cord. Before we talk about the circulation of the CSF, let's discuss the organization of the cranial meninges and how they differ from the spinal meninges (introduced in Chapter 14, ↩ pp. 361–364).

16.1 CONCEPT CHECK

1 List the six major regions in the adult brain.
2 Describe the ventricles and the cells that line them.

See the blue Answers tab at the back of the book.

16.2 Protection and Support of the Brain

▶ **KEY POINT** The human brain is an extremely delicate organ that must be protected from trauma. It also has a high demand for nutrients and oxygen and therefore has an extensive blood supply. However, the brain must be isolated from substances in the blood that could interfere with its complex operations.

Protection, support, and nourishment of the brain involve the (1) bones of the skull, (2) cranial meninges, (3) blood brain barrier, (4) cerebrospinal fluid, and (5) rich blood supply.

The Cranial Meninges

▶ **KEY POINT** The meninges and CSF stabilize the shape and position of the brain. The meninges are attached to the brain and the internal surfaces of the skull. In addition, the brain "floats" in the surrounding CSF, reducing the effects of forces that would severely damage the brain if the CSF were not present.

The brain lies cradled within the cranium of the skull **(Figure 16.3)**. The cranial bones protect the brain, but they also pose a threat. The brain is like a person driving a car, and the car is like the bones of the skull. If the car hits a tree, the car protects the driver from contact with the tree. However, serious injury will occur unless a seat belt or airbag protects the driver from contact with the interior of the car.

Within the cranial cavity, the **cranial meninges** protect the brain. The meninges surround the brain and act as shock absorbers, preventing contact with the skull bones **(Figure 16.3a)**. The cranial meninges are continuous with the spinal meninges, and they have the same three layers: dura mater (outermost), arachnoid mater (middle), and pia mater (innermost).

Figure 16.2 Ventricles of the Brain. The ventricles contain cerebrospinal fluid, which transports nutrients, chemicals, and wastes.

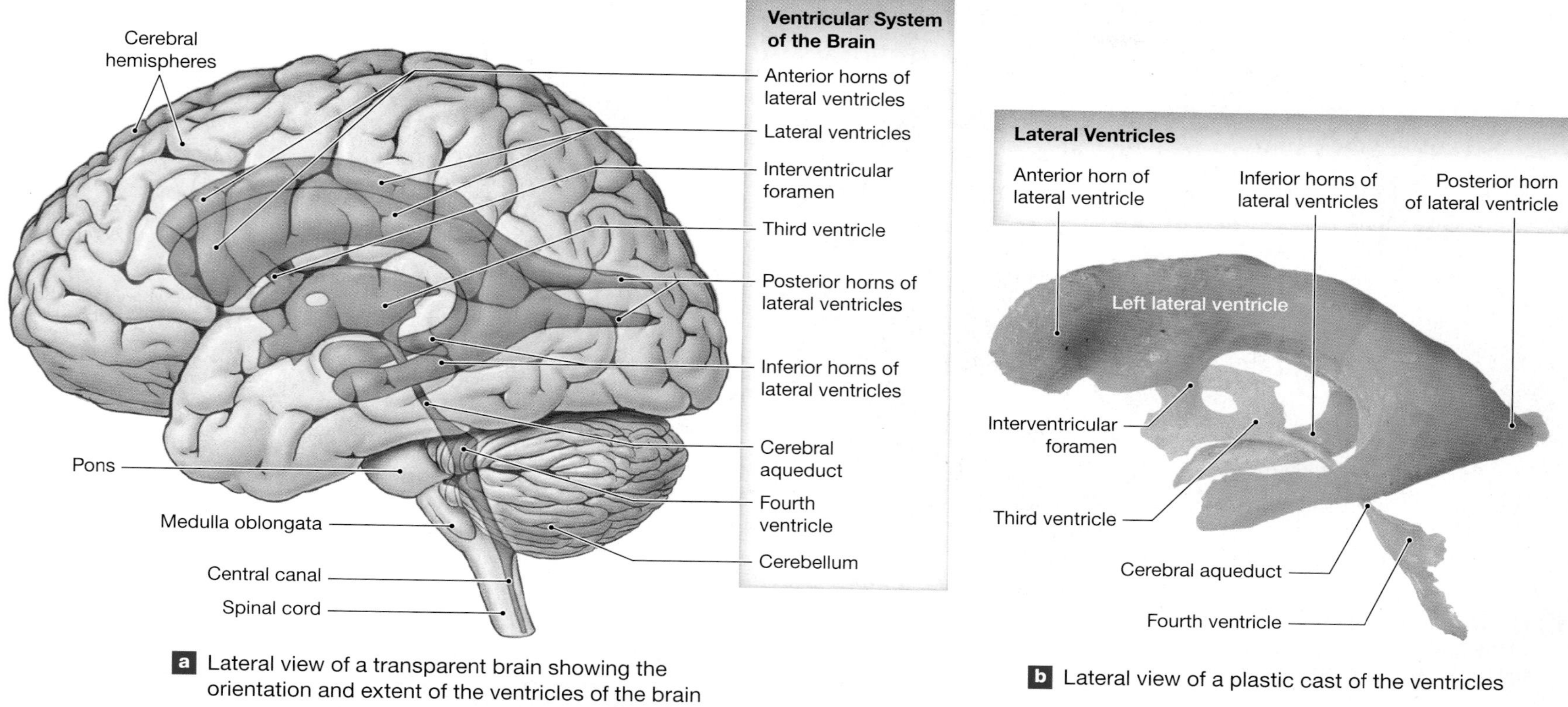

a Lateral view of a transparent brain showing the orientation and extent of the ventricles of the brain

b Lateral view of a plastic cast of the ventricles

16

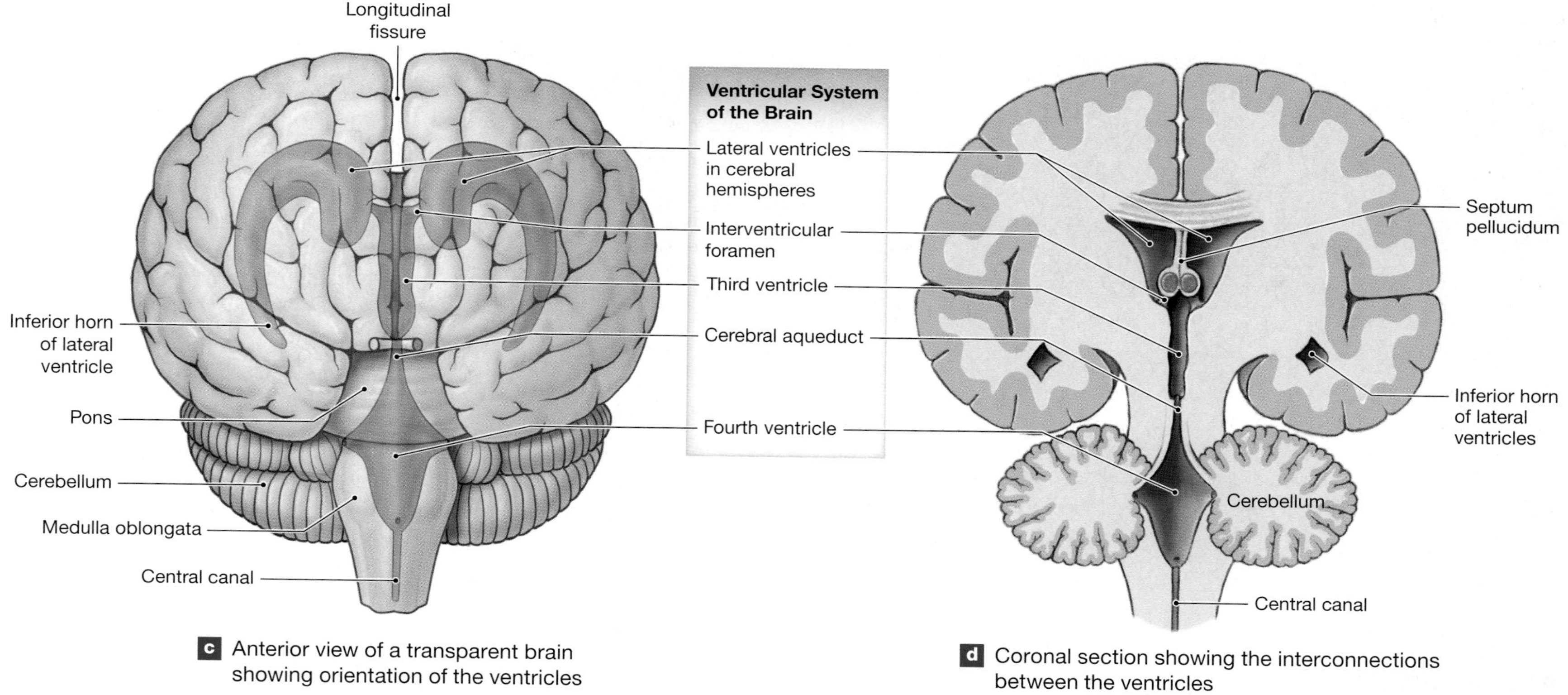

c Anterior view of a transparent brain showing orientation of the ventricles

d Coronal section showing the interconnections between the ventricles

Figure 16.3 **Relationships among the Brain, Cranium, and Meninges**

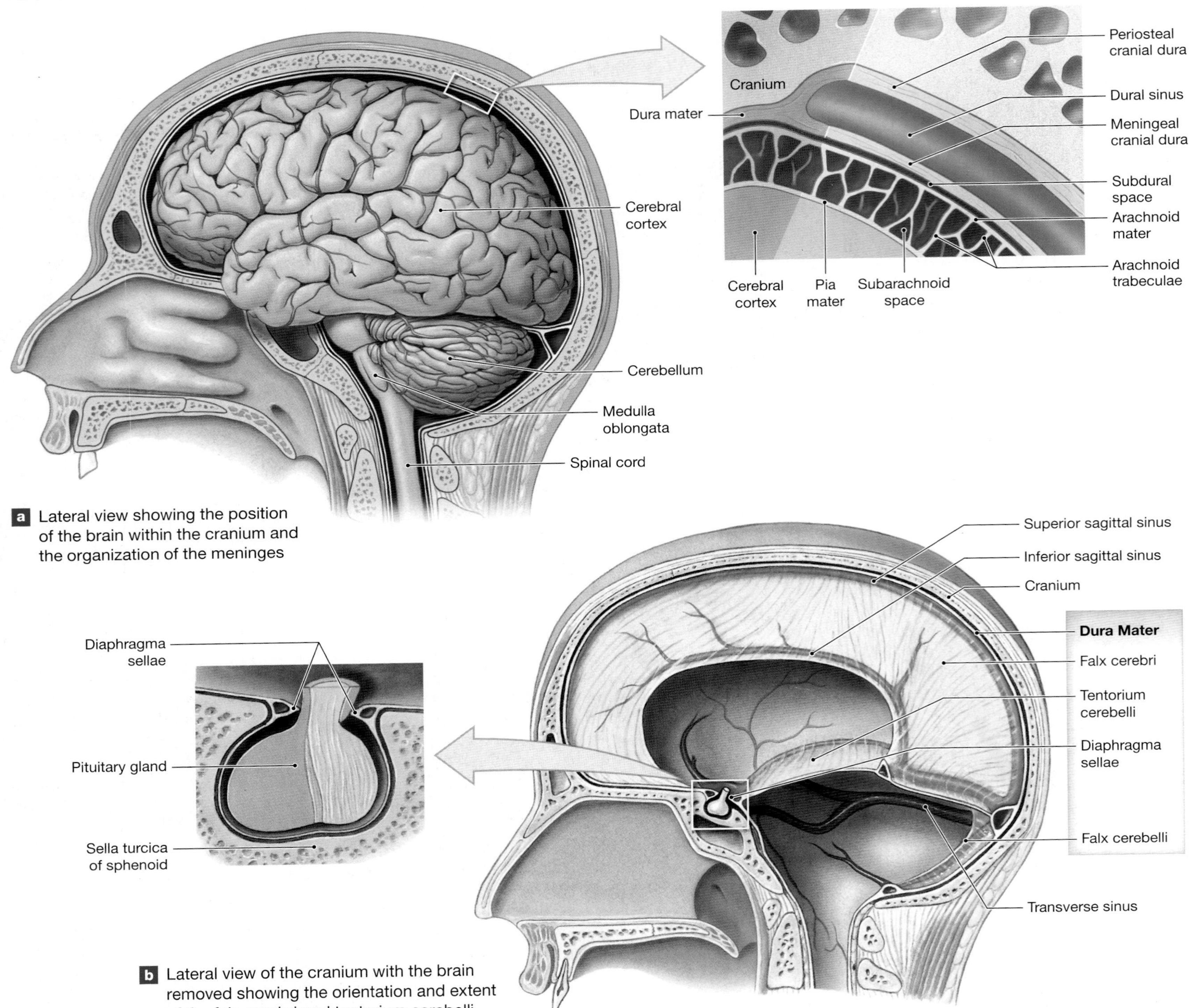

a Lateral view showing the position of the brain within the cranium and the organization of the meninges

b Lateral view of the cranium with the brain removed showing the orientation and extent of the falx cerebri and tentorium cerebelli

The Dura Mater

The cranial **dura mater** has two fibrous layers. The outermost layer, or **periosteal cranial dura**, is fused to the periosteum lining the cranial bones **(Figure 16.3a)**. The innermost layer is the **meningeal cranial dura**. In many areas the periosteal and meningeal dura are separated by a space containing interstitial fluid and blood vessels, including large collecting veins called **dural sinuses**. Veins of the brain empty into the sinuses, which deliver blood to the internal jugular veins in the neck.

At four locations the meningeal cranial dura folds and extends deep into the cranial cavity. These folds subdivide the cranial cavity, support the brain, and limit its movement **(Figures 16.3b, 16.4, and 16.5)**.

1. The **falx cerebri** (falks ser-Ē-brē; *falx*, curving or sickle-shaped) is a fold of dura mater found between the cerebral hemispheres in the longitudinal fissure. The falx cerebri attaches to the crista galli (anteriorly) and the internal occipital crest and tentorium cerebelli (posteriorly). Two large venous sinuses, the **superior sagittal sinus** and the **inferior sagittal sinus**, lie within this dural fold.

❷ The **tentorium cerebelli** (ten-TŌ-rē-um ser-e-BEL-ē; *tentorium*, covering) supports and protects the two occipital lobes of the cerebrum. It also separates the cerebellar hemispheres from the cerebrum. The tentorium cerebelli extends across the cranium at right angles to the falx cerebri. The **transverse sinus** lies within the tentorium cerebelli.

❸ The **falx cerebelli** divides the two cerebellar hemispheres. It extends midsagitally, inferior to the tentorium cerebelli.

❹ The **diaphragma sellae** is a small segment of the dura mater lining the sella turcica of the sphenoid **(Figure 16.3b)**. The diaphragma sellae anchors the dura mater to the sphenoid and surrounds the base of the pituitary gland.

The Arachnoid Mater

The cranial **arachnoid mater** is a delicate membrane covering the brain between the superficial dura mater and the deeper pia mater. In most anatomical preparations, a narrow **subdural space** separates the dura mater and the cranial arachnoid mater **(Figure 16.3a)**. And, as we saw in the spinal cord, it is highly probable that such a space does not exist in a living person. ⮌ p. 364

Deep to the arachnoid mater is the **subarachnoid space**, which contains a delicate, weblike meshwork of collagen and elastic fibers linking the arachnoid mater to the underlying pia mater. Superficially, fingerlike extensions of the cranial arachnoid mater penetrate the dura mater and project into the venous sinuses of the superior sagittal sinus **(Figures 16.3b, 16.4, and 16.5)**. These projections are called **arachnoid granulations**. Here, cerebrospinal fluid flows past the bundles of connective tissue fibers (termed the arachnoid trabeculae), crosses the arachnoid mater, and enters the venous circulation. The cranial arachnoid mater acts as a roof over the cranial blood vessels, and the underlying pia mater forms a floor. Cerebral arteries and veins are supported by the arachnoid trabeculae and surrounded by cerebrospinal fluid.

The Pia Mater

The cranial **pia mater** is tightly attached to the surface contours of the brain, sticking to its contours and lining the sulci. The pia is anchored to the surface of the brain by the processes of astrocytes. The cranial pia mater is a highly vascular membrane. It acts as a floor to support the large cerebral blood vessels as they branch and follow the convolutions of the brain, supplying superficial areas of cerebral cortex with blood **(Figures 16.4 and 16.5)**.

Figure 16.4 The Cranial Meninges, Part I. A superior view of a dissection of the cranial meninges.

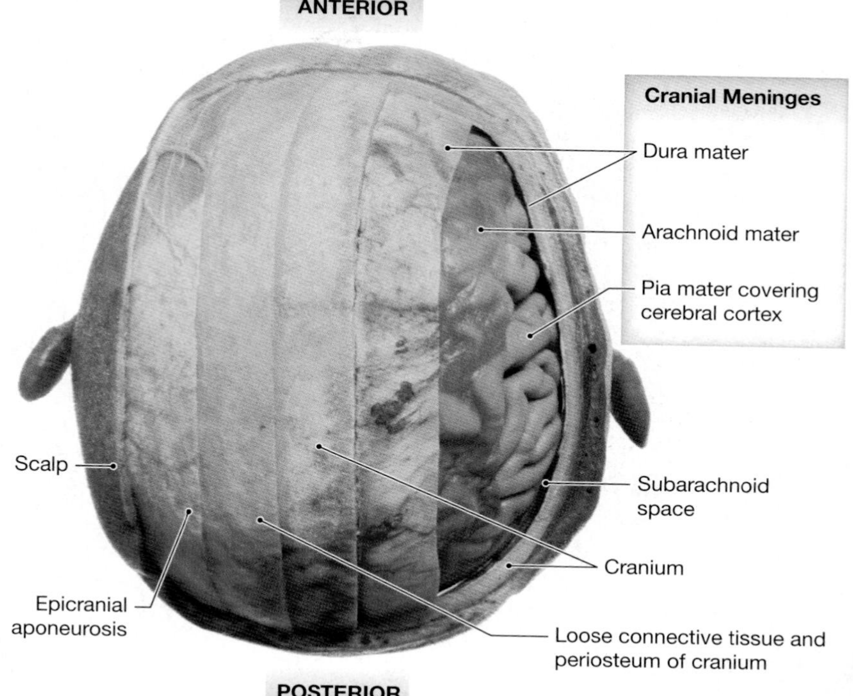

The Blood Brain Barrier

▶ **KEY POINT** Nervous tissue in the CNS has an extensive blood supply, but it is also isolated from the general circulation by the blood brain barrier. This barrier provides a mechanism to maintain a constant environment within the CNS, which is necessary for its proper control and functioning.

The **blood brain barrier (BBB)** is formed by capillary endothelial cells that are extensively interconnected by tight junctions. These junctions prevent materials from diffusing between the cells. As a result, only lipid-soluble compounds diffuse across the endothelial cell membranes and into the interstitial fluid of the brain and spinal cord. The endothelial cells of these capillaries have very few pinocytotic vesicles. This further limits the movement of large-molecular-weight compounds into the CNS. Passive or active transport mechanisms are required for the passage of water-soluble compounds across these capillary walls. Many different transport proteins are involved, and their activities are quite specific. In addition, astrocytes are in close contact with the CNS capillaries. These neuroglia secrete chemicals restricting the permeability of the endothelial cells of brain capillaries.

Endothelial transport across the BBB is selective and directional. Neurons have a constant need for glucose, and this need must be met regardless of the concentration in blood and interstitial fluid. Even when the circulating glucose level is low, endothelial cells continue to transport glucose from the blood to the interstitial fluid of the brain. In contrast, the amino acid glycine is a neurotransmitter, and its concentration in nervous tissue must be kept much lower than that in the circulating blood. Endothelial cells actively absorb this compound from the interstitial fluid of the brain and secrete it into the blood.

There are four regions within the brain where the BBB is notably different from that of the rest of the brain:

❶ Portions of the hypothalamus, where the capillary endothelium has an increased permeability. This increased permeability exposes hypothalamic nuclei to circulating hormones and permits hypothalamic hormones to diffuse into the circulation.

❷ Capillaries in the pineal gland, which are very permeable. The pineal gland, an endocrine structure, is located in the roof of the diencephalon. This increased capillary permeability allows pineal secretions into the general circulation.

❸ Capillaries at a choroid plexus. In the roof of the third and fourth ventricles, the pia mater supports extensive capillary networks that project into the ventricles of the brain, forming the choroid plexus. The choroid plexus is the site of CSF production. Transport activities of the specialized ependymal cells in this plexus maintain the blood–CSF barrier.

❹ Capillaries in the posterior lobe of the pituitary gland, which is continuous with the floor of the hypothalamus. At this site, the hormones antidiuretic hormone and oxytocin, produced by hypothalamic neurons, are released into the circulation.

Figure 16.5 **The Cranial Meninges, Part II**

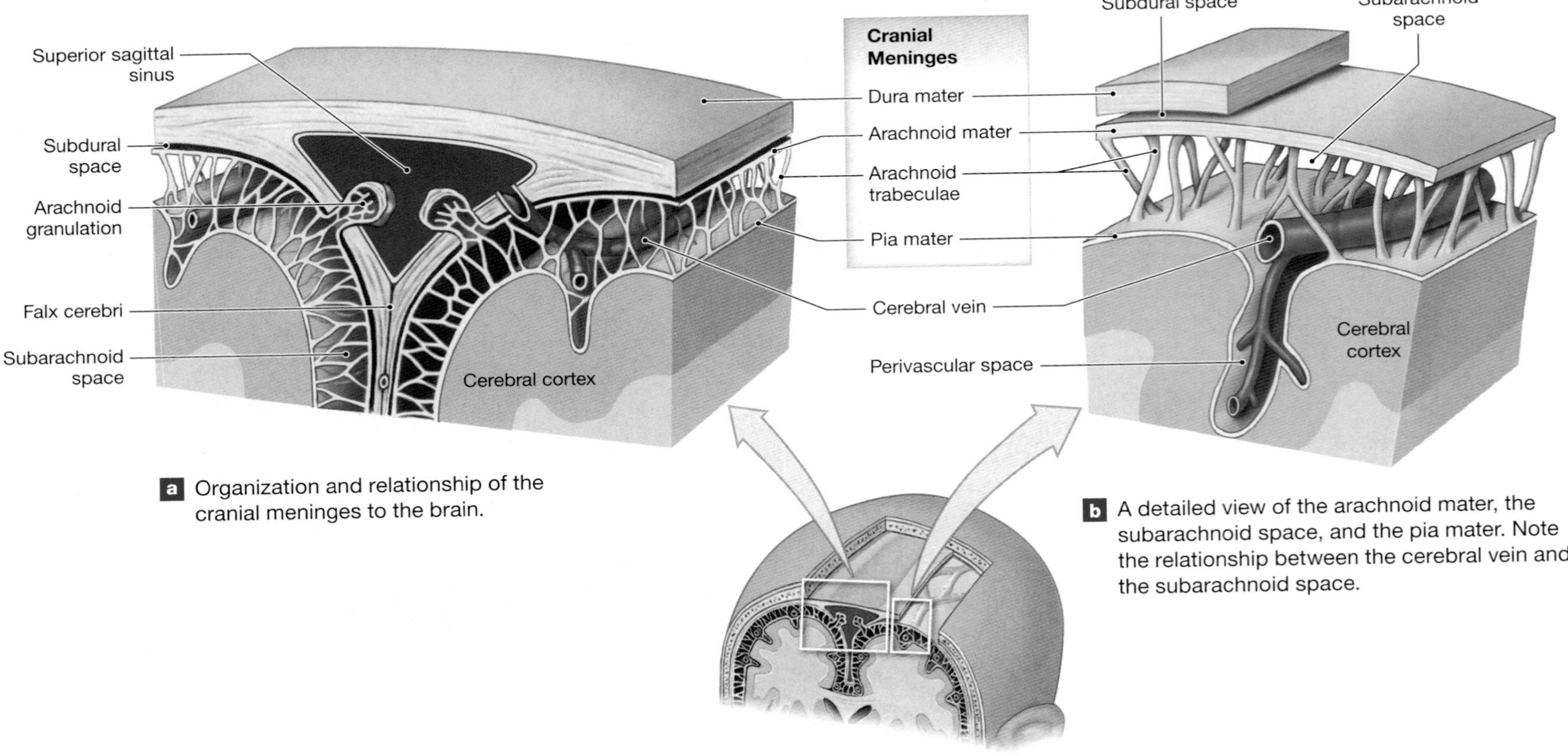

a Organization and relationship of the cranial meninges to the brain.

b A detailed view of the arachnoid mater, the subarachnoid space, and the pia mater. Note the relationship between the cerebral vein and the subarachnoid space.

16

Cerebrospinal Fluid

▶ **KEY POINT** Cerebrospinal fluid (CSF) surrounds, supports, and cushions the brain. It also transports nutrients, wastes, and chemicals.

Cerebrospinal fluid (CSF) completely surrounds and bathes the exposed surfaces of the central nervous system. It has several important functions, including the following:

- Preventing contact between delicate neural structures and the surrounding bones.
- Supporting the brain. In essence, the brain is suspended inside the cranium, floating in the cerebrospinal fluid. A human brain weighs about 1400 g in air but, when supported by the cerebrospinal fluid, weighs only about 50 g.
- Transporting nutrients, chemicals, and wastes. Except at the choroid plexus, the ependymal lining is freely permeable. As a result, CSF is in constant chemical communication with the interstitial fluid of the CNS. Because diffusion occurs freely between the interstitial fluid and CSF, changes in the pH or chemical composition of the interstitial fluid of the CNS alters the composition of the CSF. As noted in Chapter 14, a spinal tap provides useful clinical information concerning CNS injury, infection, or disease.

Formation of CSF

All the ventricles contain a **choroid plexus** (*choroid*, vascular coat; *plexus*, network), which is a combination of specialized ependymal cells and highly permeable capillaries **(Figure 16.6a)**. Two choroid plexuses are found in the roof of the third ventricle and extend through the interventricular foramina into the lateral ventricles. These plexuses cover the floors of the lateral ventricles. In the lower brainstem, a region of the choroid plexus in the roof of the fourth ventricle projects between the cerebellum and the pons.

The choroid plexus produces cerebrospinal fluid. The capillaries are fenestrated and highly permeable, but large, highly specialized ependymal cells cover the capillaries. The ependymal cells use both active and passive transport mechanisms to secrete CSF into the ventricles. The choroid plexus also removes wastes from the CSF and fine-tunes its composition over time. There are many differences in composition between CSF and blood plasma (blood with the cellular elements removed). For example, blood plasma contains high concentrations of suspended proteins, but CSF does not. There are also differences in the concentrations of individual ions and in the levels of amino acids, lipids, and wastes **(Figure 16.6b)**.

Circulation of CSF

The choroid plexus produces CSF at a rate of about 500 mL/day. The total volume of CSF at any given moment is approximately 140–270 mL. This means that the entire volume of CSF is replaced roughly every 8 to 12 hours, and the rate of removal normally keeps pace with the rate of production.

CSF produced in the lateral ventricles flows into the third ventricle through the interventricular foramina. From there, CSF flows into the cerebral aqueduct. Most of the CSF reaching the fourth ventricle enters the subarachnoid space by passing through the paired **lateral apertures** and a single **median aperture** in the membranous roof of the fourth ventricle. (A relatively small quantity of cerebrospinal fluid circulates between the fourth

CLINICAL NOTE

Traumatic Brain Injuries

Traumatic Brain Injury (TBI) can result from any hit to the head that is hard enough to affect brain function. Fifty percent of TBIs result from motor vehicle accidents, and they are the most common cause of death and disability in young people. In elderly people, TBIs may result from falls. Penetrating brain injuries are always accompanied by skull fracture. Traumatic brain injuries due to blasts have become the "specialty injury" of war. From military injuries, we have learned that immediate treatment must focus on restoring blood supply (perfusion, oxygenation) to the brain.

A **concussion** is a mild TBI that may be accompanied by a period of unconsciousness. Signs and symptoms of concussion include headache or neck pain, nausea, ringing in the ears, dizziness, and fatigue. Severe concussion can involve bruising, bleeding, or tearing of brain tissue. Symptoms of severe TBI include vomiting or nausea, convulsions or seizures, slurred speech, weakness or numbness in arms and legs, dilated pupils, and inability to wake up. Treatment and outcome depend on the severity of the injury, but the most important treatments for concussion are rest and avoiding further trauma.

Chronic traumatic encephalopathy (CTE) results from repeated sports-related head trauma. It has long been recognized in boxers and has its own name, dementia pugilistica (*pugil,* boxer), or boxer's dementia. The prevalence of CTE is still not known, and currently there is no cure.

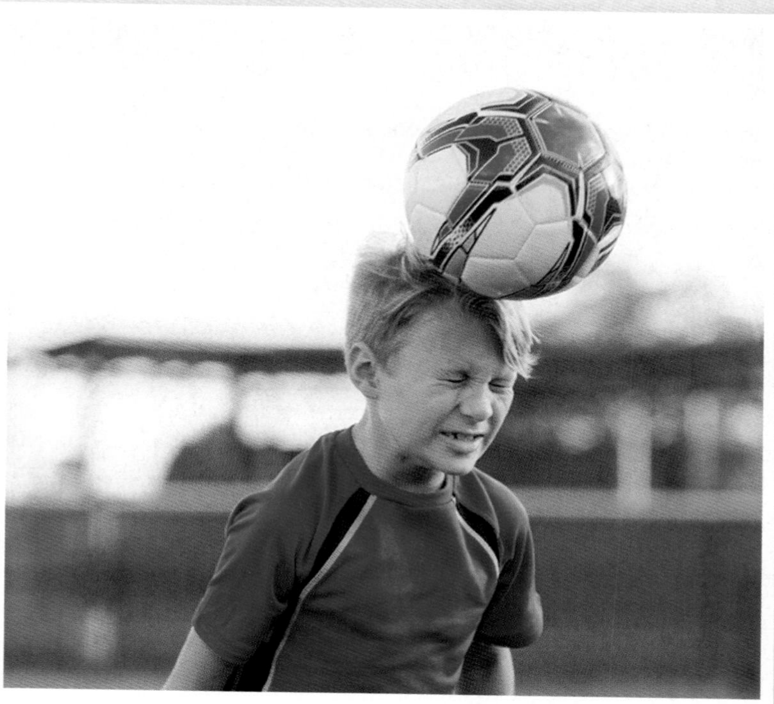

An **epidural hematoma** is the accumulation of blood between the inner table of the skull and the dura mater. Most epidural hematomas result from a skull fracture, and most are located in the temporoparietal region where a skull fracture crosses the path of the middle cerebral artery (see **Figure 22.12**). An arterial epidural hematoma often requires emergency surgery to relieve pressure on the brain that involves drilling a hole in the skull and allowing blood to drain.

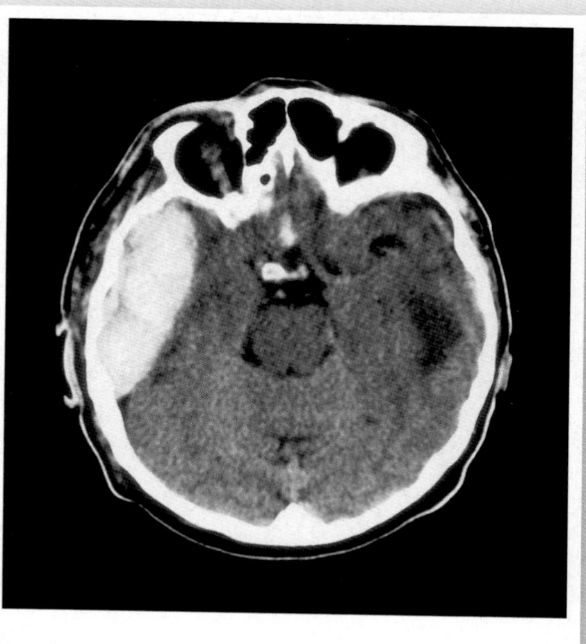

Epidural hemorrhage

A **subdural hematoma** is the accumulation of blood between the dura mater and the arachnoid mater. Subdural hematomas are often venous bleeds. People taking blood thinners are more susceptible to subdural hematomas.

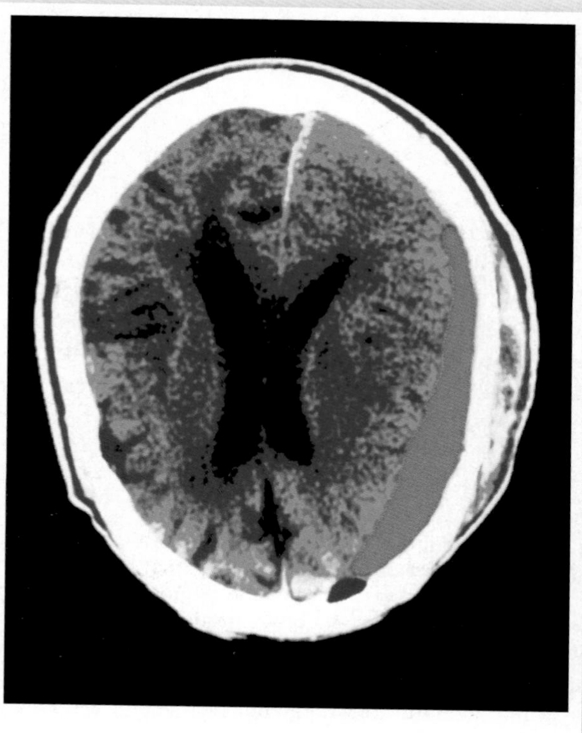

Subdural hemorrhage

16

Figure 16.6 The Choroid Plexus and Blood Brain Barrier

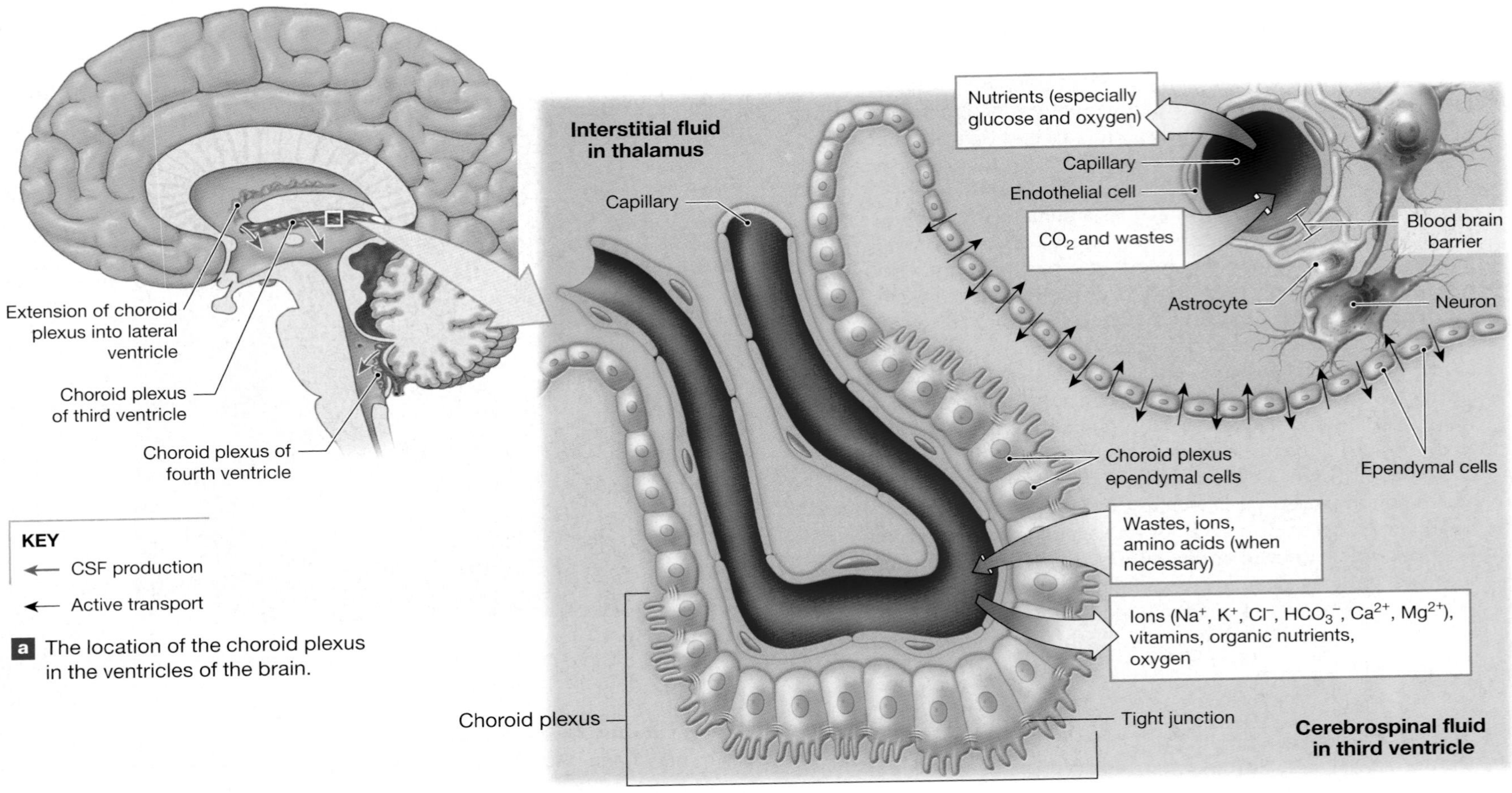

Extension of choroid plexus into lateral ventricle
Choroid plexus of third ventricle
Choroid plexus of fourth ventricle

KEY
← CSF production
← Active transport

a The location of the choroid plexus in the ventricles of the brain.

Interstitial fluid in thalamus
Capillary
Nutrients (especially glucose and oxygen)
Capillary
Endothelial cell
CO_2 and wastes
Blood brain barrier
Astrocyte
Neuron
Choroid plexus ependymal cells
Ependymal cells
Wastes, ions, amino acids (when necessary)
Ions (Na^+, K^+, Cl^-, HCO_3^-, Ca^{2+}, Mg^{2+}), vitamins, organic nutrients, oxygen
Choroid plexus
Tight junction
Cerebrospinal fluid in third ventricle

b The structure and function of the choroid plexus. The ependymal cells are a selective barrier, actively transporting nutrients, vitamins, and ions into the CSF. When necessary, these cells also actively remove ions or compounds from the CSF to stabilize its composition.

ventricle and the central canal of the spinal cord.) CSF continuously flows through the subarachnoid space surrounding the brain, and movements of the vertebral column move it around the spinal cord and cauda equina (**Figure 16.4**). Cerebrospinal fluid then reenters the circulation through the **arachnoid granulations** (**Figures 16.5a, 16.6,** and **16.7**). If the normal circulation of CSF is interrupted, a variety of clinical problems may appear.

The Blood Supply to the Brain

▶ **KEY POINT** An interruption of the blood supply to the brain for 10 seconds causes a person to lose consciousness. After 20 seconds, electrical activity ceases, and after a few minutes, irreversible damage begins. Because of this high demand for oxygen, blood vessels in the brain, particularly the gray matter, are arranged in a dense meshwork.

Neurons have a high demand for energy but lack energy reserves in the form of carbohydrates or lipids. In addition, neurons lack myoglobin and have no way to store oxygen. Therefore, their energy demands are met by an extensive vascular supply. Arterial blood reaches the brain through the internal carotid arteries (major arteries of the neck) and the vertebral arteries (arteries within the transverse foramina of the cervical vertebrae). Most of the venous blood from the brain leaves the cranium in the internal jugular veins (major veins of the neck), which drain the dural sinuses. (Chapter 22 discusses the arterial blood supply to the brain and the veins leaving the brain.)

Cerebrovascular diseases are circulatory disorders that interfere with the normal blood supply to the brain. The particular distribution of the vessel involved determines the symptoms, and the degree of oxygen or nutrient starvation determines the severity. A **cerebrovascular accident (CVA)**, or *stroke*, occurs when the blood supply to a portion of the brain is shut off. Affected neurons begin to die in a matter of minutes.

16.2 CONCEPT CHECK

3 Identify the four extensions of the innermost layer of the dura mater into the cranial cavity that stabilize and support the brain.

4 What is the function of the blood brain barrier?

See the blue Answers tab at the back of the book.

Figure 16.7 Circulation of Cerebrospinal Fluid. Sagittal section indicating the sites of formation and the routes of circulation of cerebrospinal fluid.

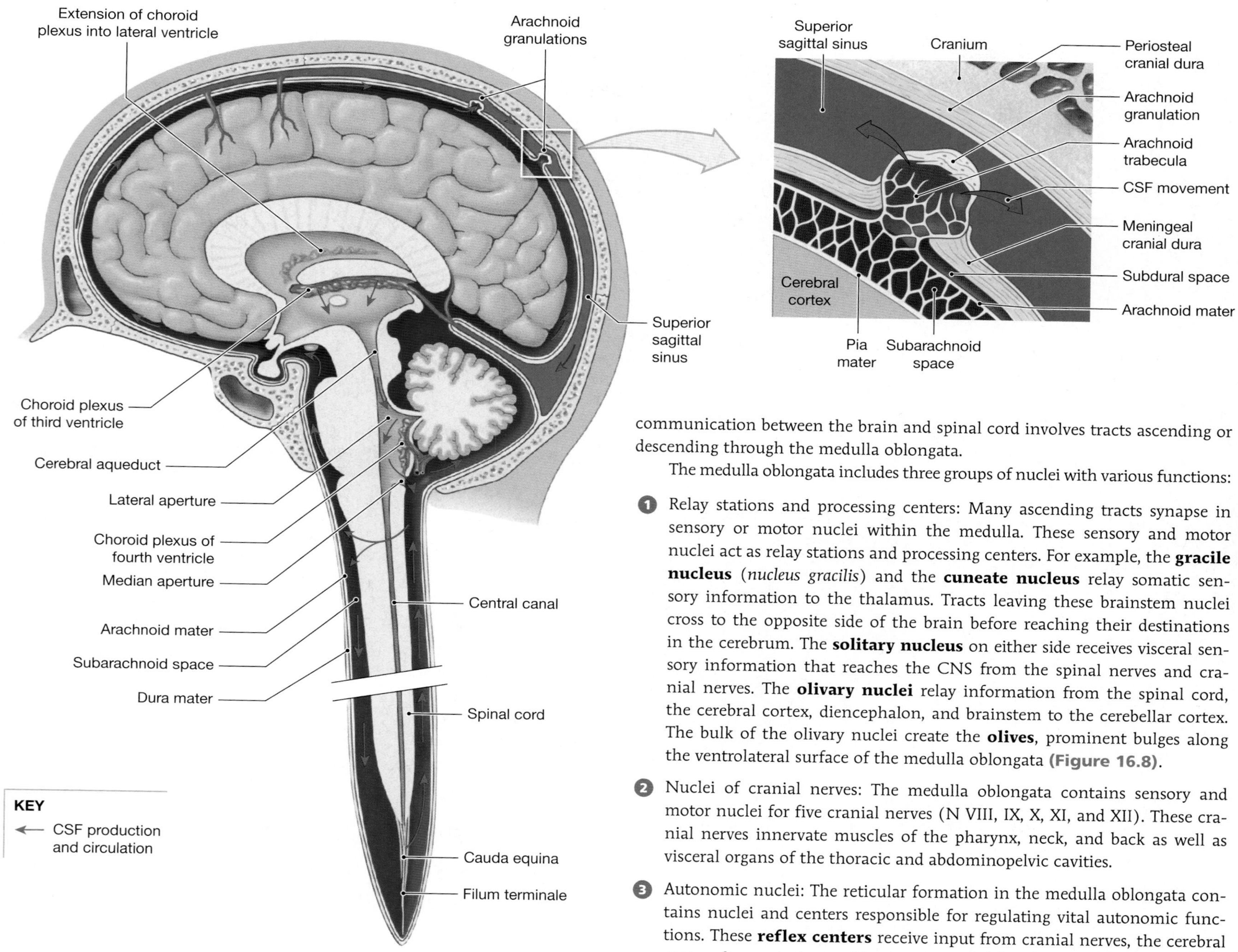

16.3 The Medulla Oblongata

▶ **KEY POINT** The spinal cord connects to and is continuous with the brainstem at the medulla oblongata. The medulla corresponds to the embryonic myelencephalon.

Figures 16.1, 16.13, 16.14, and **16.17** show the external appearance of the **medulla oblongata**, or *medulla*. The important nuclei and centers are diagrammed in **Figure 16.8** and detailed in **Table 16.2**.

Figure 16.13 shows the medulla oblongata in midsagittal section. The caudal portion resembles the spinal cord and has a rounded shape and a narrow central canal. Closer to the pons, the central canal becomes enlarged and continuous with the fourth ventricle.

The medulla oblongata connects the brain with the spinal cord, and many of its functions are directly related to this connection. For example, all communication between the brain and spinal cord involves tracts ascending or descending through the medulla oblongata.

The medulla oblongata includes three groups of nuclei with various functions:

1. Relay stations and processing centers: Many ascending tracts synapse in sensory or motor nuclei within the medulla. These sensory and motor nuclei act as relay stations and processing centers. For example, the **gracile nucleus** (*nucleus gracilis*) and the **cuneate nucleus** relay somatic sensory information to the thalamus. Tracts leaving these brainstem nuclei cross to the opposite side of the brain before reaching their destinations in the cerebrum. The **solitary nucleus** on either side receives visceral sensory information that reaches the CNS from the spinal nerves and cranial nerves. The **olivary nuclei** relay information from the spinal cord, the cerebral cortex, diencephalon, and brainstem to the cerebellar cortex. The bulk of the olivary nuclei create the **olives**, prominent bulges along the ventrolateral surface of the medulla oblongata (**Figure 16.8**).
2. Nuclei of cranial nerves: The medulla oblongata contains sensory and motor nuclei for five cranial nerves (N VIII, IX, X, XI, and XII). These cranial nerves innervate muscles of the pharynx, neck, and back as well as visceral organs of the thoracic and abdominopelvic cavities.
3. Autonomic nuclei: The reticular formation in the medulla oblongata contains nuclei and centers responsible for regulating vital autonomic functions. These **reflex centers** receive input from cranial nerves, the cerebral cortex, diencephalon, and brainstem, and their output adjusts the activities of one or more peripheral systems. Major centers include the following:
   - The **cardiovascular centers** adjust heart rate, strength of cardiac contractions, and the flow of blood through peripheral tissues. The cardiovascular centers are subdivided into cardiac (*kardia*, heart) and vasomotor (*vas*, canal) centers, but their anatomical boundaries are difficult to determine.
   - The **respiratory rhythmicity centers** set the basic pace for breathing. Inputs from the apneustic and pneumotaxic centers within the pons regulate their activity (**Figure 16.9**).

16.3 CONCEPT CHECK

5 What three types of nuclei are found within the medulla?

See the blue Answers tab at the back of the book.

Figure 16.8 **The Medulla Oblongata**

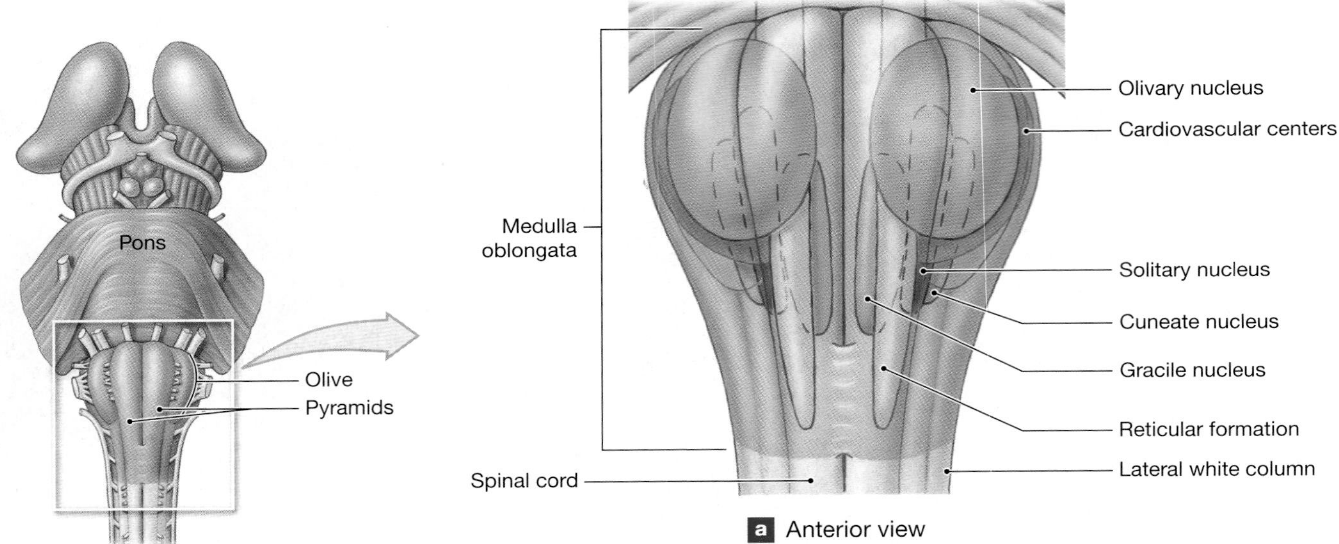

a Anterior view

Attachment to membranous roof of fourth ventricle

Lateral white column

Posterior white columns

Posterior median sulcus

Spinal cord

b Posterolateral view

Table 16.2 **The Medulla Oblongata**

Region/Nucleus	Function
GRAY MATTER	
Olivary nuclei	Relay information from the spinal cord, the red nuclei, other midbrain centers, and the cerebral cortex to the vermis of the cerebellum
Reflex centers	
Cardiovascular centers	Regulate heart rate and force of contraction, and distribution of blood flow
Respiratory rhythmicity centers	Set the rate of respiratory movements
Gracile nucleus **Cuneate nucleus**	Relay somatic information to the ventral posterior nuclei of the thalamus
Other nuclei/centers	Sensory and motor nuclei of five cranial nerves; relaying ascending information from the spinal cord to higher centers
Reticular formation	Contains nuclei and centers that regulate vital autonomic nervous system functions
WHITE MATTER	
Ascending and descending tracts	Link the brain with the spinal cord

16.4 The Pons

▶ **KEY POINT** The pons contains sensory and motor nuclei for four cranial nerves, nuclei involved with involuntary control of respiration, and nuclei that process and relay cerebellar signals. The pons also contains ascending, descending, and transverse tracts.

The pons extends from the medulla oblongata to the mesencephalon, forming a prominent bulge on the anterior surface of the brainstem. The pons and cerebellum are separated by the fourth ventricle. The pons is attached to the cerebellum by three cerebellar peduncles. Important features and regions of the pons are shown in **Figures 16.1, 16.9, 16.13**, and **16.14** and listed in **Table 16.3**. The pons contains the following:

- Sensory and motor nuclei for four cranial nerves. Four cranial nerves within the pons (N V, N VI, N VII, and N VIII) innervate jaw muscles, the anterior surface of the face, one of the extra-ocular muscles (the lateral rectus), and organs of hearing and equilibrium in the internal ear.
- Nuclei regulating the involuntary control of respiration. On each side of the brain, the reticular formation of the pons contains two respiratory centers, the apneustic center and the pneumotaxic center. These centers regulate the activity of the respiratory rhythmicity center in the medulla oblongata.
- Nuclei that process and relay cerebellar commands. Information entering the cerebellum by the middle cerebellar peduncles passes through the pons. The middle cerebellar peduncles are connected to the **transverse fibers** of the pons that cross its anterior surface.

Figure 16.9 **The Pons**

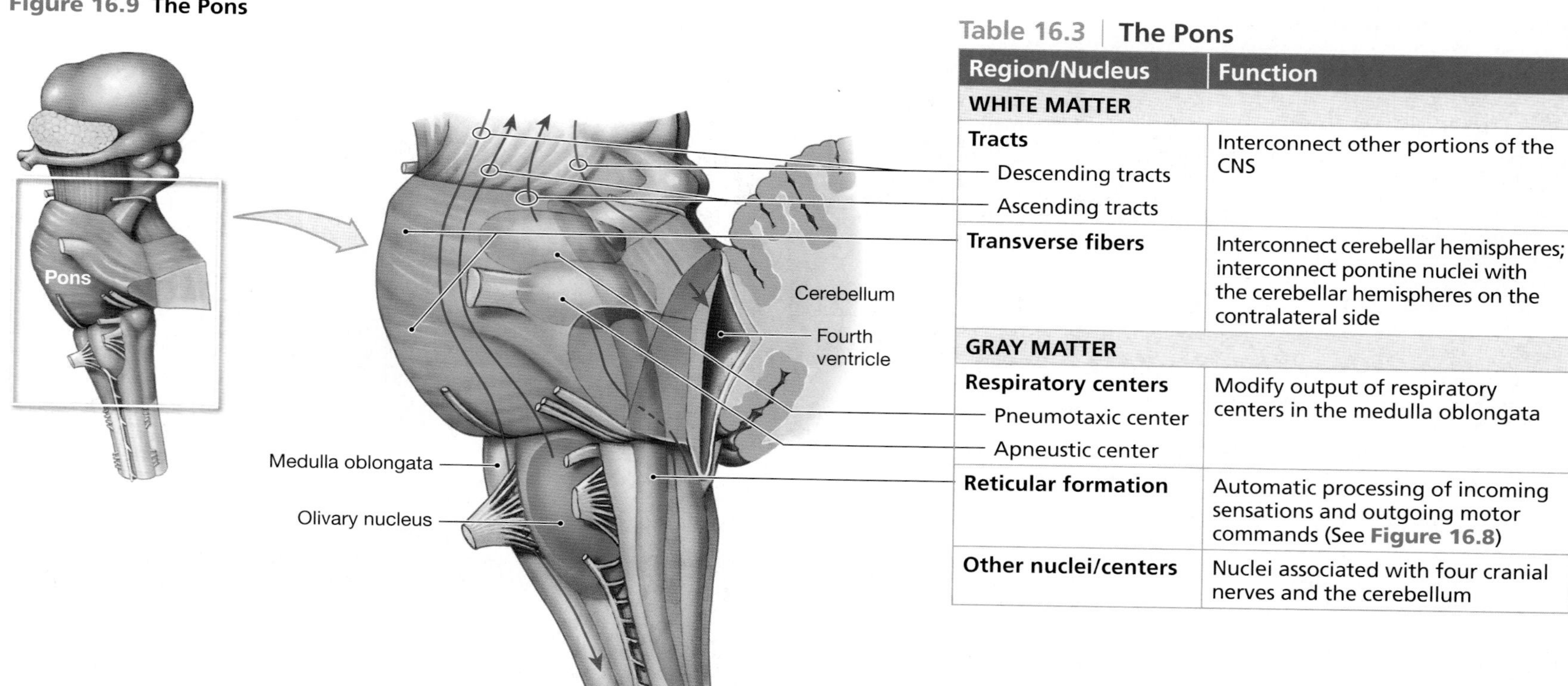

Table 16.3 | **The Pons**

Region/Nucleus	Function
WHITE MATTER	
Tracts Descending tracts Ascending tracts	Interconnect other portions of the CNS
Transverse fibers	Interconnect cerebellar hemispheres; interconnect pontine nuclei with the cerebellar hemispheres on the contralateral side
GRAY MATTER	
Respiratory centers Pneumotaxic center Apneustic center	Modify output of respiratory centers in the medulla oblongata
Reticular formation	Automatic processing of incoming sensations and outgoing motor commands (See **Figure 16.8**)
Other nuclei/centers	Nuclei associated with four cranial nerves and the cerebellum

- Ascending, descending, and transverse tracts. The longitudinal tracts of the pons connect the pons to other portions of the CNS. The anterior cerebellar peduncles contain efferent cerebellar tracts. These fibers permit communication between the cerebellar hemispheres of opposite sides. The inferior cerebellar peduncles contain both afferent and efferent tracts connecting the cerebellum with the medulla oblongata.

16.4 CONCEPT CHECK

6 Name the fibers that pass through the pons and cite their functions.

See the blue Answers tab at the back of the book.

16.5 | The Mesencephalon (Midbrain)

▶**KEY POINT** The mesencephalon (midbrain) contains nuclei that process visual and auditory stimuli. The mesencephalon also contains major nuclei of the reticular formation.

Figures 16.1, 16.13, and **16.14** show the external anatomy of the mesencephalon. **Figure 16.10** and **Table 16.4** detail its major nuclei. Its surface posterior to the cerebral aqueduct is called the roof, or **tectum**, of the mesencephalon. This region contains two pairs of sensory nuclei (superior colliculi and inferior colliculi) known as the **corpora quadrigemina** (KŌR-pō-ra qua-dri-JEM-i-na). These nuclei are relay stations concerned with processing visual and auditory sensations. The two **superior colliculi** (singular, *colliculus*, ko-LIK-Ū-lus; "small hill") receive visual input from the lateral geniculate of the thalamus on the ipsilateral (same) side. The two **inferior colliculi** receive auditory input from nuclei in the medulla oblongata; some of this information is forwarded to the medial geniculate on the ipsilateral side.

The mesencephalon also contains the major nuclei of the reticular formation. Stimulation of this region produces a variety of involuntary motor responses. Each side of the mesencephalon contains a pair of nuclei, the red nucleus and the substantia nigra **(Figure 16.10)**. The **red nucleus** has numerous blood vessels, giving it a rich red color. This nucleus processes information from the cerebrum and cerebellum and issues involuntary motor commands that maintain muscle tone and limb position. The **substantia nigra** (NĪ-grah; "black") lies lateral to the red nucleus. The gray matter in this region contains darkly pigmented cells, giving it a black color. The substantia nigra regulates the motor output of the basal nuclei.

The nerve fiber bundles on the ventrolateral surfaces of the mesencephalon are the **cerebral peduncles** (*peduncles*, little feet) **(Figures 16.10b** and **16.14)**. They contain ascending fibers that synapse in the thalamic nuclei and descending fibers of the corticospinal pathway carrying voluntary motor commands from the primary motor cortex of each cerebral hemisphere.

16.5 CONCEPT CHECK

7 The corpora quadrigemina are found within what portion of the mesencephalon, and what are the functions of these structures?

See the blue Answers tab at the back of the book.

Figure 16.10 **The Mesencephalon**

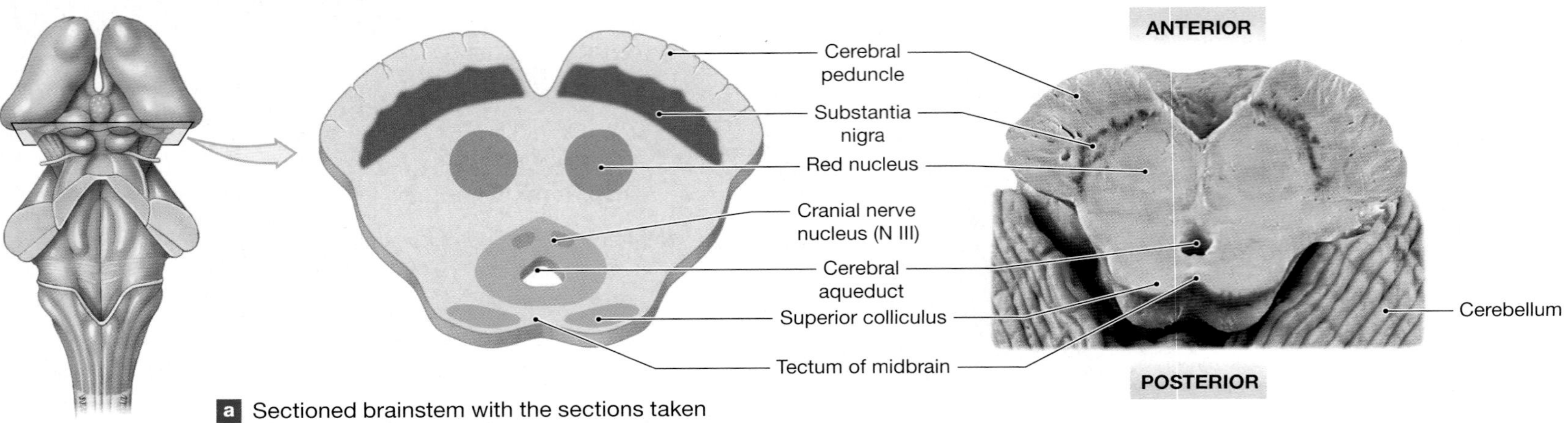

a Sectioned brainstem with the sections taken at the level indicated in the icon.

Pineal gland
Thalamus

b Posterior view of the diencephalon and brainstem. This view is drawn as if transparent, to show the positions of important nuclei.

Thalamus
Corpora Quadrigemina
Superior colliculi
Inferior colliculi
Cerebral peduncle
Trochlear nerve (IV)
Superior cerebellar peduncle
Fourth ventricle
Reticular formation in floor of fourth ventricle

c Posterior view of the diencephalon and brainstem.

Table 16.4 **The Mesencephalon**

Region/Nucleus	Function
GRAY MATTER	
Tectum (roof)	
Superior colliculi	Integrate visual information with other sensory input; initiate reflex responses to visual stimuli
Inferior colliculi	Integrate auditory information with medial geniculate nuclei; initiate reflex responses to auditory stimuli
Walls and floor	
Substantia nigra	Regulates activity in the basal nuclei
Red nuclei	Involuntary control of background muscle tone and limb position
Reticular formation	Automatic processing of incoming sensations and outgoing motor commands; can initiate motor responses to stimuli; helps maintain consciousness
Other nuclei/centers	Nuclei associated with two cranial nerves (N III, N IV)
WHITE MATTER	
Cerebral peduncles	Connect primary motor cortex with motor neurons in brain and spinal cord; carry ascending sensory information to thalamus

16.6 | The Diencephalon

▶ **KEY POINT** The diencephalon contains only about 2 percent of the gray matter of the CNS. However, it is extremely important and has widespread connections. Almost all of the motor and sensory systems of the CNS synapse within the diencephalon.

The diencephalon connects the brainstem to the cerebral hemispheres. It consists of the epithalamus, the left and right thalamus, and the hypothalamus. **Figures 16.1, 16.13, 16.14, 16.21c**, and **16.22** show the position of the diencephalon and its relationship to other landmarks in the brain.

The Epithalamus

▶ **KEY POINT** The epithalamus, the roof of the third ventricle, contains the pineal gland.

The epithalamus is shown in **Figures 16.10b, 16.12a,** and **16.13a**. The membranous, anterior portion of the epithalamus contains a choroid plexus that extends through the interventricular foramina into the lateral ventricles. The posterior portion of the epithalamus contains the **pineal gland**, an endocrine structure that secretes the hormone **melatonin**. Melatonin is thought to regulate day-night cycles, with other possible effects on reproductive function. (Chapter 19 describes the role of melatonin.)

The Thalamus

▶ **KEY POINT** The thalamus is the largest mass of nuclei within the entire CNS, and the connections within the thalamus are more diverse than any other part of the CNS. Nuclei in the thalamus integrate and relay sensory and motor information.

Most of the nervous tissue in the diencephalon is concentrated in the left thalamus and right thalamus (**Figure 16.11**). The two thalami form the walls of the diencephalon and surround the third ventricle (**Figures 16.13** and **16.21a,b**). The thalamic nuclei provide integration and relay centers for sensory and motor pathways. All ascending sensory information from the spinal cord (other than the spinocerebellar tracts) and cranial nerves (other than the olfactory nerve) synapses in the thalamic nuclei before the information is relayed to the cerebrum or brainstem. The thalamus is the final synapse for ascending sensory information that is projected to the primary somatosensory cortex. The thalami act as information filters, processing all of the incoming sensory information and then passing on only a small portion to the cerebrum or brainstem. The thalamus also acts as a relay station that coordinates motor activities at the conscious and subconscious levels.

The third ventricle separates the two thalami. When viewed in midsagittal section, the thalamus extends from the anterior commissure to the inferior base of the pineal gland (**Figure 16.13a**). A medial projection of gray matter, the **interthalamic adhesion**, or *massa intermedia*, extends into the ventricle from the thalamus on either side (see **Figure 16.22a**).

The thalamus on each side bulges laterally, away from the third ventricle and anteriorly toward the cerebrum (**Figures 16.11, 16.12, 16.13b, 16.21a,b,** and **16.22**). The lateral border of each thalamus is formed by the fibers of the internal capsule. Within each thalamus is a mass of several interconnected thalamic nuclei.

Functions of Thalamic Nuclei

The thalamic nuclei process sensory and motor information and then relay it to the basal nuclei and cerebral cortex. The five major groups of thalamic nuclei, shown in **Figure 16.11** and **Table 16.5**, are the following:

16

1. The **anterior nuclei** are part of the limbic system, and they play a role in emotions, memory, and learning. They relay information from the hypothalamus and hippocampus to the cingulate gyrus.
2. The **medial nuclei** provide an awareness of emotional states. They connect the basal nuclei and emotion centers in the hypothalamus with the prefrontal cortex of the cerebrum. These nuclei also integrate sensory information arriving at other portions of the thalamus before relaying it to the frontal lobes of the cerebrum.

Figure 16.11 The Thalamus

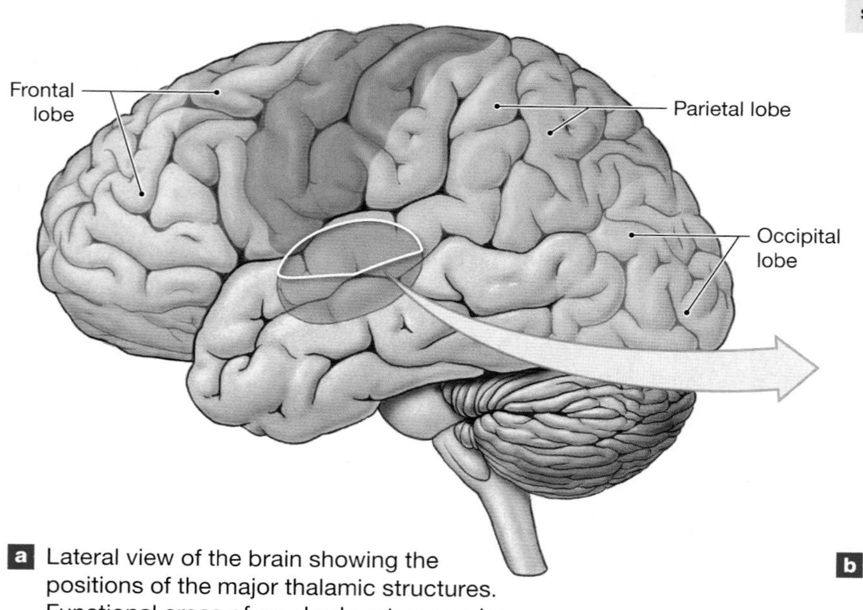

a Lateral view of the brain showing the positions of the major thalamic structures. Functional areas of cerebral cortex are also indicated, with colors corresponding to those of the associated thalamic nuclei.

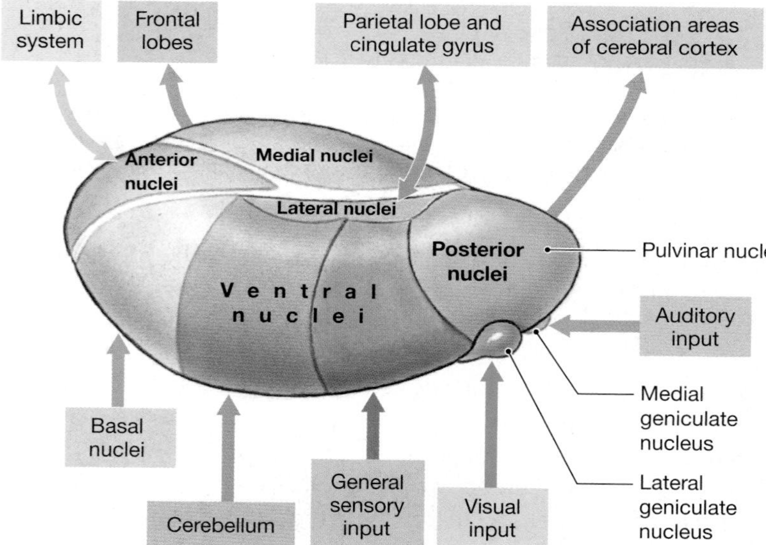

b Enlarged view of the thalamic nuclei of the left side. The color of each nucleus or group of nuclei matches the color of the associated cortical region. The boxes either provide examples of the types of sensory input relayed to the basal nuclei and cerebral cortex or indicate the existence of important feedback loops involved with emotional states, learning, and memory.

Table 16.5 | The Thalamus

Structure/Nuclei	Function
Anterior group	Part of the limbic system
Medial group	Integrates sensory information and other data arriving at the thalamus and hypothalamus for projection to the frontal lobes of the cerebral hemispheres
Ventral group	Projects sensory information to the primary somatosensory cortex of the parietal lobe; relays information from cerebellum and basal nuclei to motor areas of cerebral cortex
Posterior group	
Pulvinar	Integrates sensory information for projection to association areas of cerebral cortex
Lateral geniculate nuclei	Project visual information to the visual cortex of occipital lobe
Medial geniculate nuclei	Project auditory information to the auditory cortex of temporal lobe
Lateral group	Forms feedback loops involving the cingulate gyrus (emotional states) and the parietal lobe (integration of sensory information)

3 The **ventral nuclei** relay information from the basal nuclei of the cerebrum and the cerebellum to somatic motor areas of the cerebral cortex. Ventral group nuclei also relay sensory information about touch, pressure, pain, temperature, and proprioception (position) to the sensory areas of the cerebral cortex.

4 The **posterior nuclei** include the pulvinar and the geniculate nuclei. The **pulvinar nuclei** integrate sensory information and then project it to the association areas of the cerebral cortex. The **lateral geniculate** (je-NIK-ū-lāt; genicula, little knee) **nucleus** of each thalamus receives visual information from the eyes. Efferent fibers project to the visual cortex and to the mesencephalon. The **medial geniculate nuclei** relay auditory information to the auditory cortex from receptors of the inner ear.

5 The **lateral nuclei** are relay stations in feedback loops that adjust activity in the cingulate gyrus and parietal lobe. These nuclei have an impact on emotional states and the integration of sensory information.

The Hypothalamus

▶ **KEY POINT** The hypothalamus contains centers involved with emotions and visceral processes affecting the cerebrum and parts of the brainstem. It also controls a wide variety of autonomic functions and functionally links the nervous and endocrine systems.

The hypothalamus forms the floor of the third ventricle. The hypothalamus extends from the area superior to the **optic chiasm**, where the optic nerves from the eyes arrive at the brain, to the posterior margins of the mamillary bodies (**Figure 16.12**). Posterior to the optic chiasm, the **infundibulum** (in-fun-DIB-ū-lum; *infundibulum*, funnel) extends inferiorly, connecting the hypothalamus to the pituitary gland. In life, the diaphragma sellae surrounds the infundibulum as it enters the hypophyseal fossa of the sphenoid.

A midsagittal section of the hypothalamus (**Figures 16.12** and **16.13a**) shows the **tuberal area** (*tuber*, swelling) between the infundibulum and the mamillary bodies. The tuberal area contains nuclei that control the functioning of the pituitary gland.

Functions of the Hypothalamus

The hypothalamus contains a variety of important control and integrative centers (**Figure 16.12b** and **Table 16.6**). Hypothalamic centers continually receive sensory information from the cerebrum, brainstem, and spinal cord. Hypothalamic neurons also detect and respond to changes in the CSF and interstitial fluid composition. Because of the high permeability of the capillaries in this region, these centers also respond to stimuli in the circulating blood. Hypothalamic functions include the following:

- Subconscious control of skeletal muscle contractions: By stimulating appropriate centers in other portions of the brain, hypothalamic nuclei direct somatic motor patterns associated with the emotions of rage, pleasure, pain, and sexual arousal.
- Control of autonomic function: Hypothalamic centers adjust and coordinate the activities of autonomic centers in other parts of the brainstem that are concerned with regulating heart rate, blood pressure, respiration, and digestive functions.
- Coordination of activities of the nervous and endocrine systems: The hypothalamus controls the nervous and endocrine systems by inhibiting or stimulating endocrine cells within the pituitary gland.
- Secretion of hormones: The **supraoptic nucleus** secretes antidiuretic hormone, which regulates water loss by the kidneys. The **paraventricular nucleus** secretes oxytocin, which stimulates smooth muscle contractions in the uterus and prostate gland and myoepithelial cell contractions in the mammary glands. Both hormones are transported along axons down the infundibulum into the posterior portion of the pituitary gland and are released into the circulation.
- Production of emotions and behavioral drives: Specific hypothalamic centers produce sensations that lead to changes in voluntary or involuntary behavior patterns. For example, stimulation of the thirst center produces the desire to drink.
- Coordination between voluntary and autonomic functions: When you are facing a stressful situation, your heart rate and respiratory rate go up and your body prepares for an emergency. These autonomic adjustments occur because cerebral activities are monitored by the hypothalamus. The autonomic nervous system (ANS) is a division of the peripheral nervous system. The ANS consists of two divisions: sympathetic and parasympathetic. The sympathetic division stimulates tissue metabolism, increases alertness, and prepares the body to respond to emergencies; the parasympathetic division promotes sedentary activities and conserves body energy. (Chapter 17 discusses these divisions and their relationships.)
- Regulation of body temperature: The **pre-optic area** of the hypothalamus controls physiological responses to changes in body temperature.
- Control of circadian rhythms: The **suprachiasmatic nucleus** coordinates daily cycles of activity that are linked to the day-night cycle. This nucleus receives direct input from the retina of the eye, and its output adjusts the activities of other hypothalamic nuclei, the pineal gland, and the reticular formation.

16.6 CONCEPT CHECK

8 What area of the diencephalon is stimulated by changes in body temperature?

9 Which region of the diencephalon helps coordinate somatic motor activities?

See the blue Answers tab at the back of the book.

Figure 16.12 The Hypothalamus

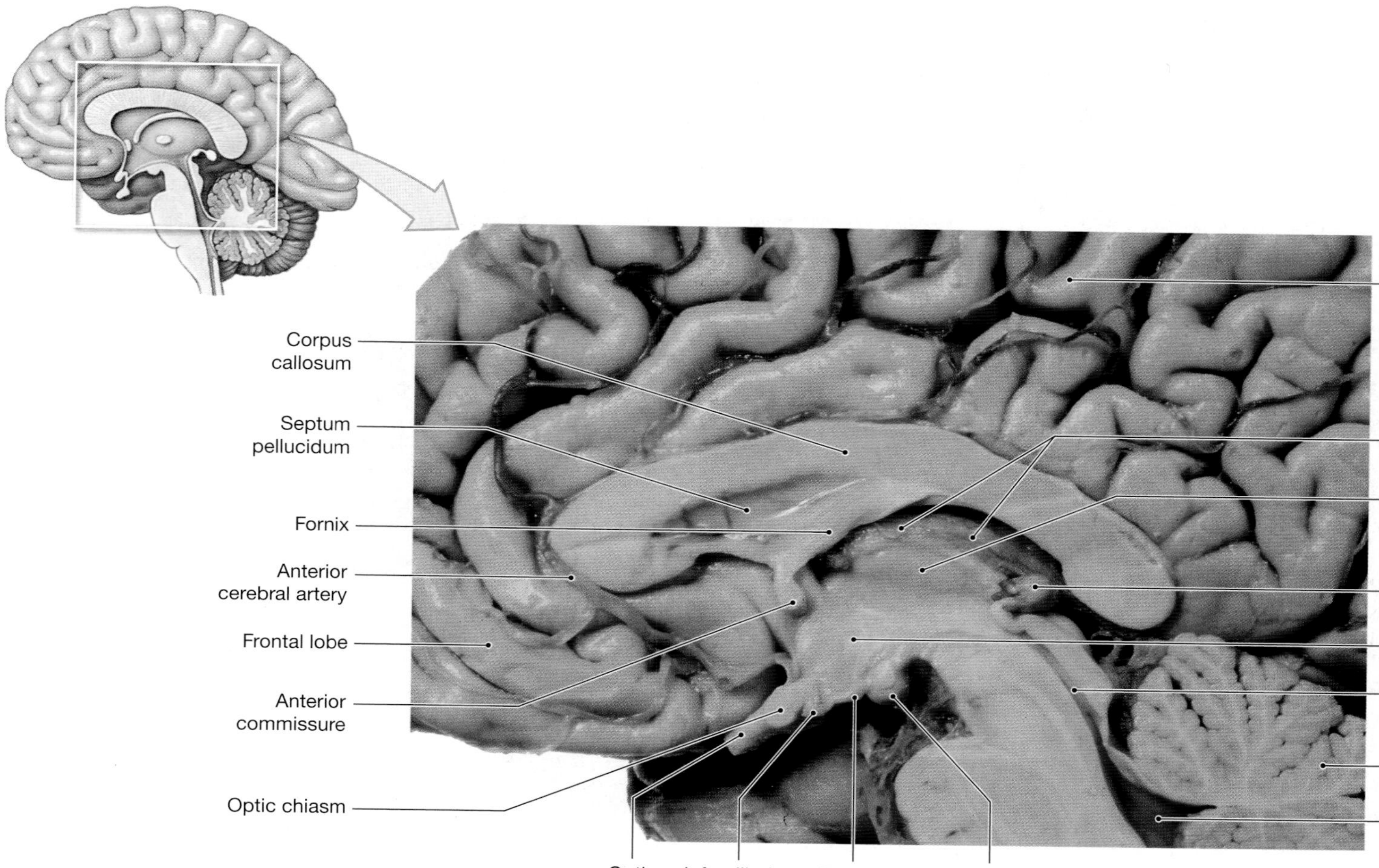

a Midsagittal section through the brain showing the major features of the diencephalon and adjacent portions of the brainstem

b Enlarged view of the hypothalamus showing the locations of major nuclei and centers

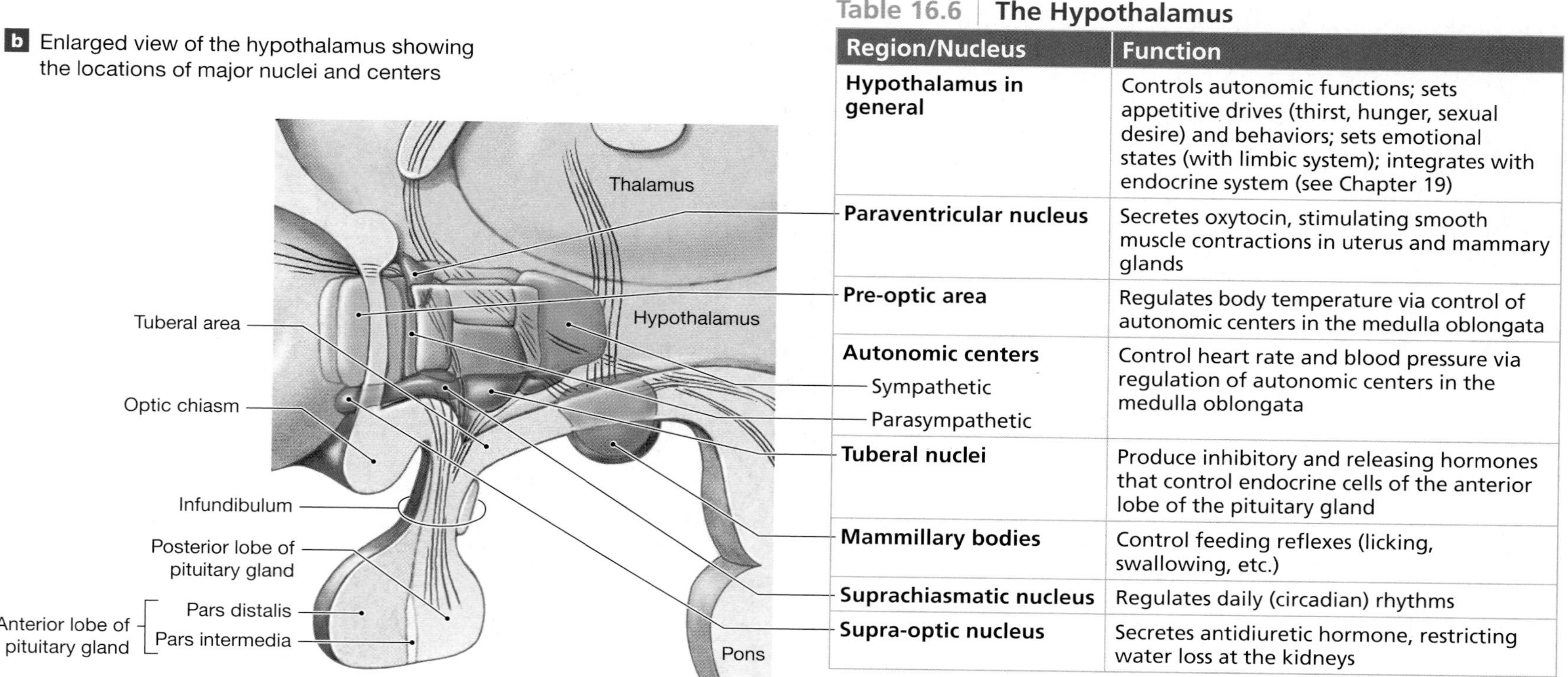

Table 16.6 | The Hypothalamus

Region/Nucleus	Function
Hypothalamus in general	Controls autonomic functions; sets appetitive drives (thirst, hunger, sexual desire) and behaviors; sets emotional states (with limbic system); integrates with endocrine system (see Chapter 19)
Paraventricular nucleus	Secretes oxytocin, stimulating smooth muscle contractions in uterus and mammary glands
Pre-optic area	Regulates body temperature via control of autonomic centers in the medulla oblongata
Autonomic centers Sympathetic Parasympathetic	Control heart rate and blood pressure via regulation of autonomic centers in the medulla oblongata
Tuberal nuclei	Produce inhibitory and releasing hormones that control endocrine cells of the anterior lobe of the pituitary gland
Mammillary bodies	Control feeding reflexes (licking, swallowing, etc.)
Suprachiasmatic nucleus	Regulates daily (circadian) rhythms
Supra-optic nucleus	Secretes antidiuretic hormone, restricting water loss at the kidneys

Figure 16.13 **Sectional Views of the Brain**

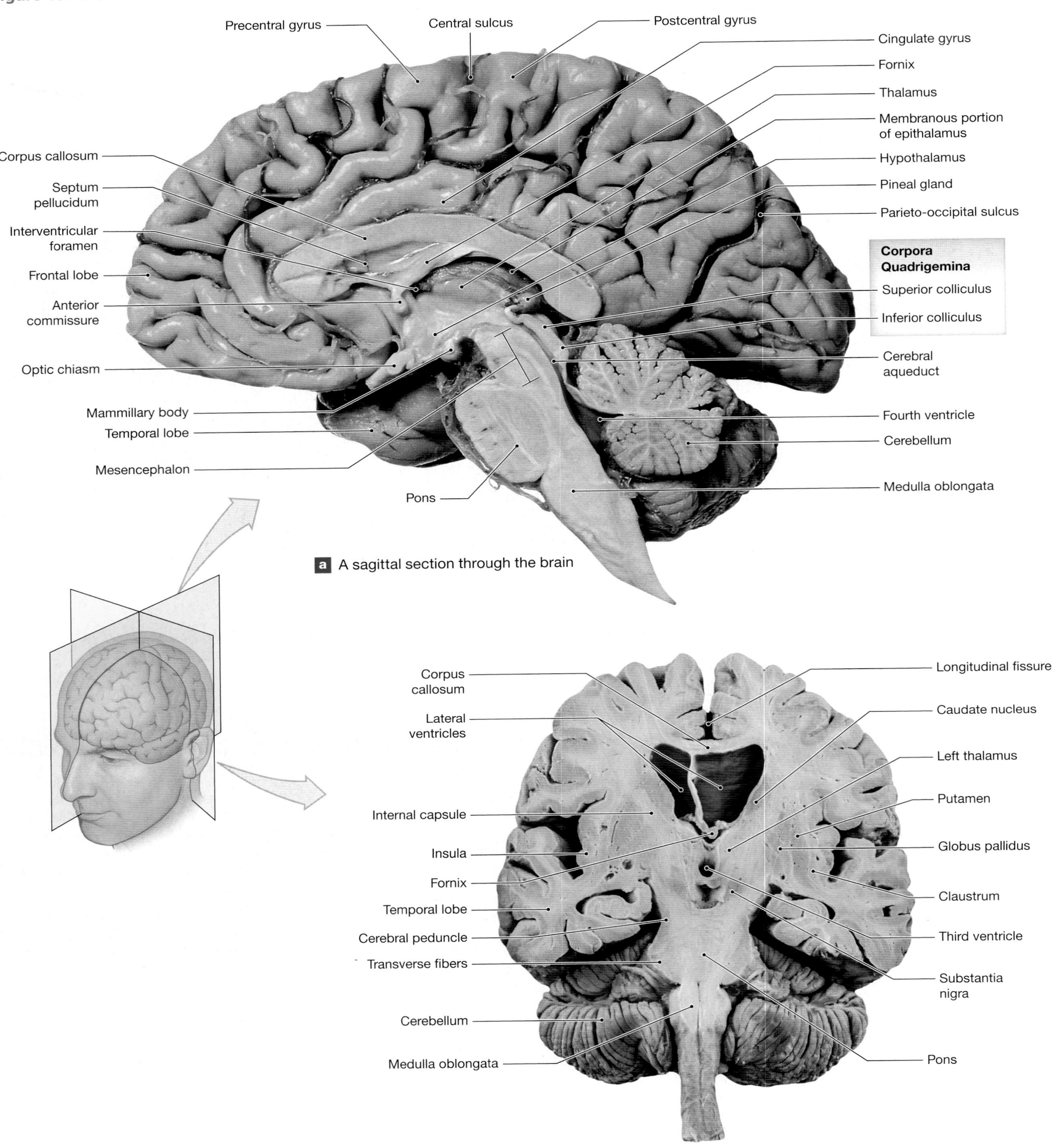

a A sagittal section through the brain

b A coronal section through the brain

Figure 16.14 **The Diencephalon and Brainstem**

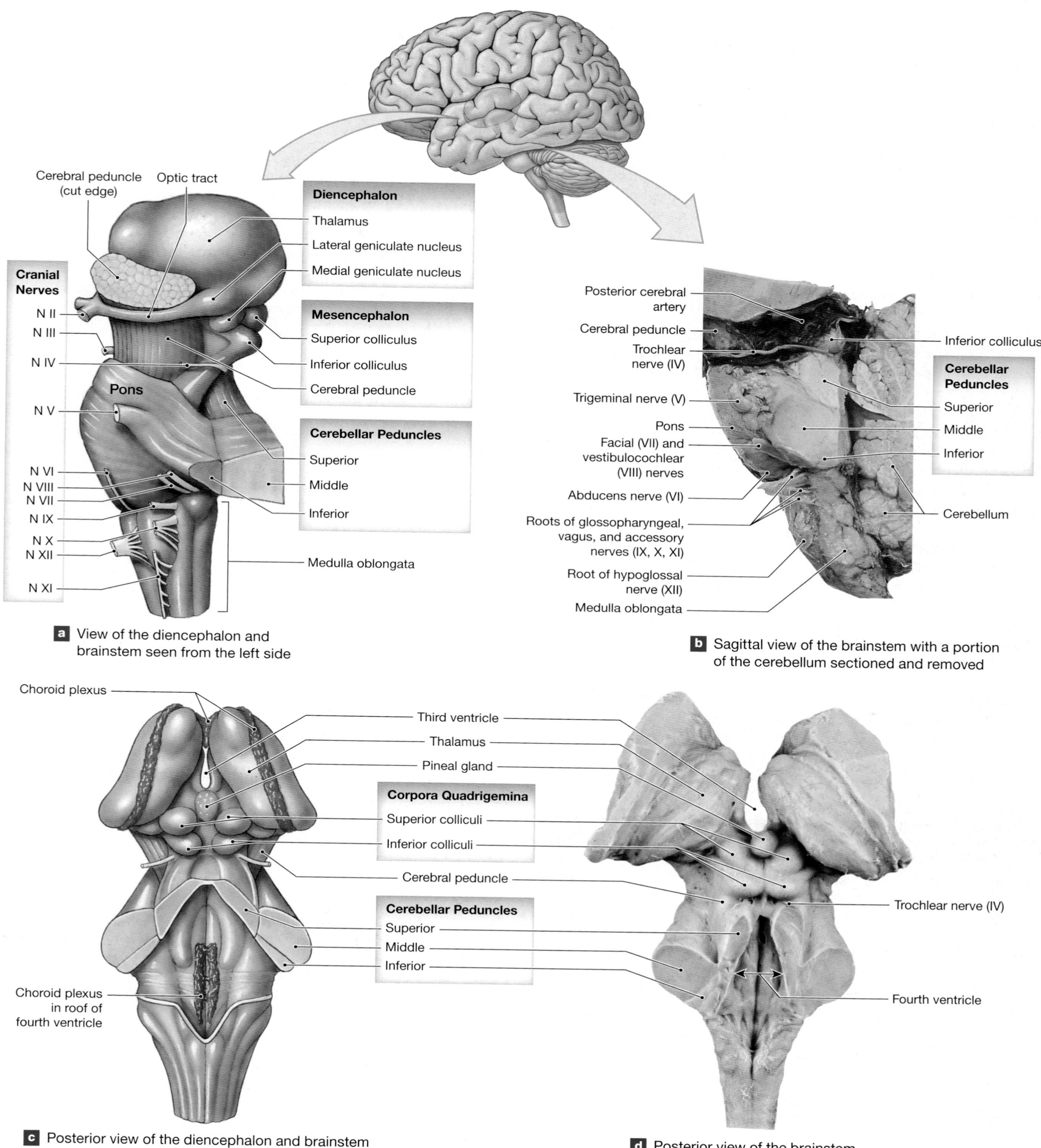

a View of the diencephalon and brainstem seen from the left side

b Sagittal view of the brainstem with a portion of the cerebellum sectioned and removed

c Posterior view of the diencephalon and brainstem

d Posterior view of the brainstem

16.7 | The Cerebellum

▶ **KEY POINT** The cerebellum has connections with the cerebral cortex, internal ear, and spinal cord. Each cerebellar hemisphere regulates and coordinates muscular activity *only* on its ipsilateral side.

The cerebellum has two **cerebellar hemispheres**, each with a highly folded surface composed of cerebellar cortex (**Figures 16.15, 16.16**, and **16.17**). These folds are termed the **folia** (FŌ-lē-a) **of the cerebellum**. Each hemisphere consists of two **lobes, anterior** and **posterior**, which are separated by the **primary fissure**. Along the midline, a narrow band of cortex, known as the **vermis** (VER-mis; "worm"), separates the cerebellar hemispheres. Slender **flocculonodular** (flok-ū-lō-NOD-ū-lar) **lobes** lie anterior and inferior to the cerebellar hemisphere. The anterior and posterior lobes assist in the planning, execution, and coordination of limb and trunk movements. The flocculonodular lobe is important for balance and eye movements. **Table 16.7** summarizes the structures of the cerebellum and their functions.

The cerebellar cortex contains huge, highly branched **Purkinje** (pur-KIN-jē) **cells** (**Figure 16.15b**). Purkinje cells have massive pear-shaped cell bodies with numerous large dendrites fanning out into the gray matter of the cerebellar cortex. Axons from the basal portion of these Purkinje cells project deep into the white matter to reach the **cerebellar nuclei**. Internally, the white matter of the cerebellum forms a branching array that, in sectional view, resembles a tree. Anatomists call it the **arbor vitae**, or "tree of life."

The cerebellum receives proprioceptive information (indicating body position) from the spinal cord and monitors all proprioceptive, visual, tactile, balance, and auditory sensations received by the brain. Most axons carrying sensory information do not synapse in the cerebellar nuclei but pass through the deeper layers of the cerebellar cortex to end near the cortical surface. There, they synapse with the dendritic processes of the Purkinje cells. Tracts containing the axons of Purkinje cells then relay motor commands to nuclei within the cerebrum and brainstem.

Motor commands issued by the cerebral cortex pass through the pons before reaching the cerebellum. A relatively small portion of these afferent fibers synapse within cerebellar nuclei before projecting to the cerebellar cortex.

Tracts linking the cerebellum with the brainstem, cerebrum, and spinal cord leave the cerebellar hemispheres as the superior, middle, and inferior cerebellar peduncles (**Figures 16.13a,b, 16.14c,d**, and **16.15b**). The **superior cerebellar peduncles** link the cerebellum with nuclei in the mesencephalon, diencephalon, and cerebrum. The **middle cerebellar peduncles** connect the cerebellar hemispheres with sensory and motor nuclei in the pons. The **inferior cerebellar peduncles** connect the cerebellum and nuclei in the medulla oblongata and carry ascending and descending cerebellar tracts from the spinal cord.

The cerebellum is an automatic processing center that has two primary functions:

1. Adjusting the postural muscles of the body: The cerebellum coordinates rapid, automatic adjustments that maintain balance and equilibrium. These alterations in muscle tone and position are made by modifying the activity of the red nucleus.
2. Programming and fine-tuning voluntary and involuntary movements: The cerebellum stores memories of learned movement patterns. These functions are performed indirectly by regulating activity along motor tracts involving the cerebral cortex, basal nuclei, and motor centers in the brainstem.

Table 16.7 | The Cerebellum

Region/Nucleus	Function
GRAY MATTER	
Cerebellar cortex	Subconscious coordination and control of ongoing movements of body parts
Cerebellar nuclei	Subconscious coordination and control of ongoing movements of body parts
WHITE MATTER	
Arbor vitae	Connects cerebellar cortex and nuclei with cerebellar peduncles
Cerebellar peduncles	
Superior	Link the cerebellum with mesencephalon, diencephalon, and cerebrum
Middle	Contain transverse fibers and carry communications between the cerebellum and pons
Inferior	Link the cerebellum with the medulla oblongata and spinal cord

CLINICAL NOTE

Cerebellar Dysfunction

The cerebellum assists in skeletal muscle coordination and fine movements, balance, and equilibrium. Abnormal cerebellar function causes unsteady gait, tremors, jerky movements of the arms or legs, slurred speech, and nystagmus (rapid, continuous eye movements). In other words, cerebellar dysfunction looks like failing a roadside sobriety test.

Causes of cerebellar dysfunction include cerebral palsy (brain injury before, during, or soon after birth), hereditary conditions, stroke, multiple sclerosis (causing cerebellar plaques), infections (meningitis), tumors, trauma, vitamin deficiencies, drugs, or toxins. **Ataxia** is the medical term for loss of muscle coordination in the arms or legs due to cerebellar dysfunction.

Acute alcohol intoxication (drunkenness) produces a loss of coordination that mimics cerebellar ataxia. Chronic alcoholism kills cerebellar neurons, causing permanent cerebellar ataxia.

16.7 CONCEPT CHECK

10 Name the three structures that link the cerebellum with the cerebrum, brainstem, and spinal cord.

11 What are the two primary functions of the cerebellum?

See the blue Answers tab at the back of the book.

16.8 | The Cerebrum

▶ **KEY POINT** The cerebrum is the largest region of the brain. It consists of the paired cerebral hemispheres, which rest on the diencephalon and brainstem. Conscious thought processes and all intellectual functions originate in the cerebral hemispheres.

Much of the cerebrum is involved in processing somatic sensory and motor information. Somatosensory information relayed to the cerebrum reaches our conscious awareness, and cerebral neurons exert direct (voluntary) or indirect (involuntary) control over somatic motor neurons. Most visceral sensory processing and visceral motor (autonomic) control occur at centers elsewhere in the brain, usually outside our conscious awareness. **Figures 16.16** and **16.17** provide additional perspective on the cerebrum and its relationships with other regions of the brain.

Figure 16.15 **The Cerebellum**

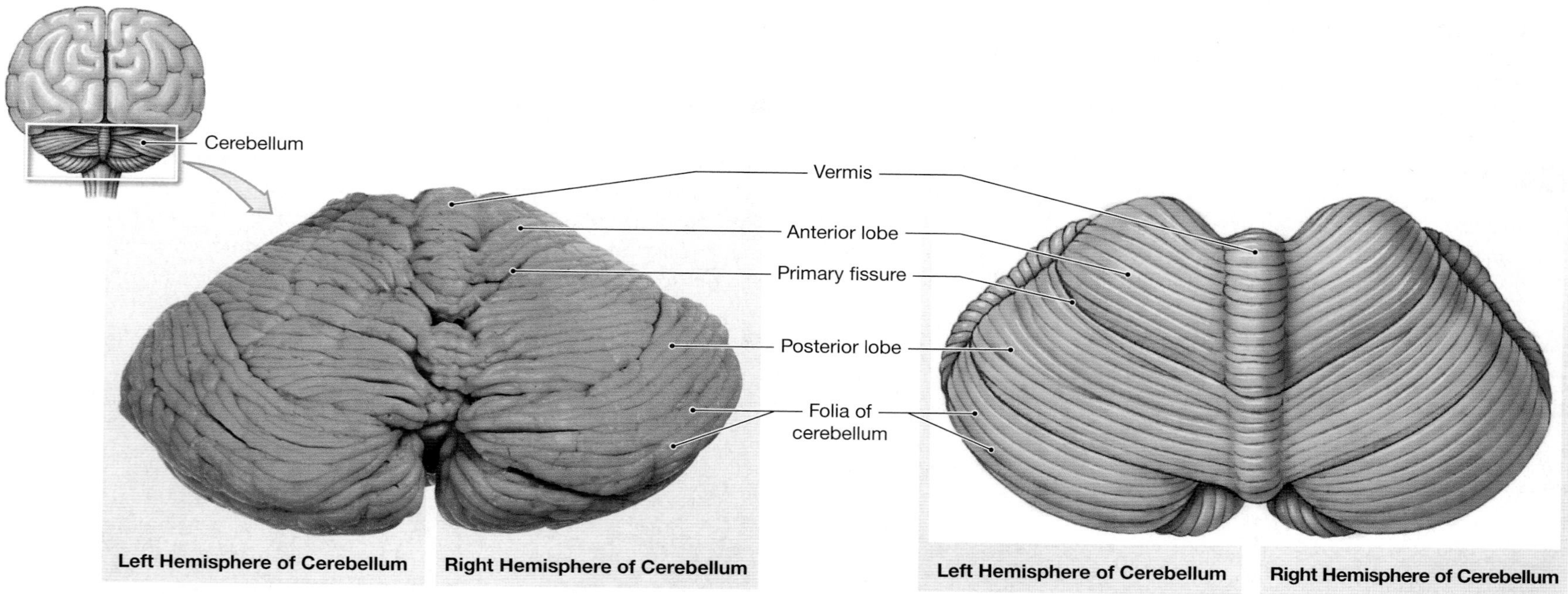

a Superior surface of the cerebellum. This view shows major anatomical landmarks and regions.

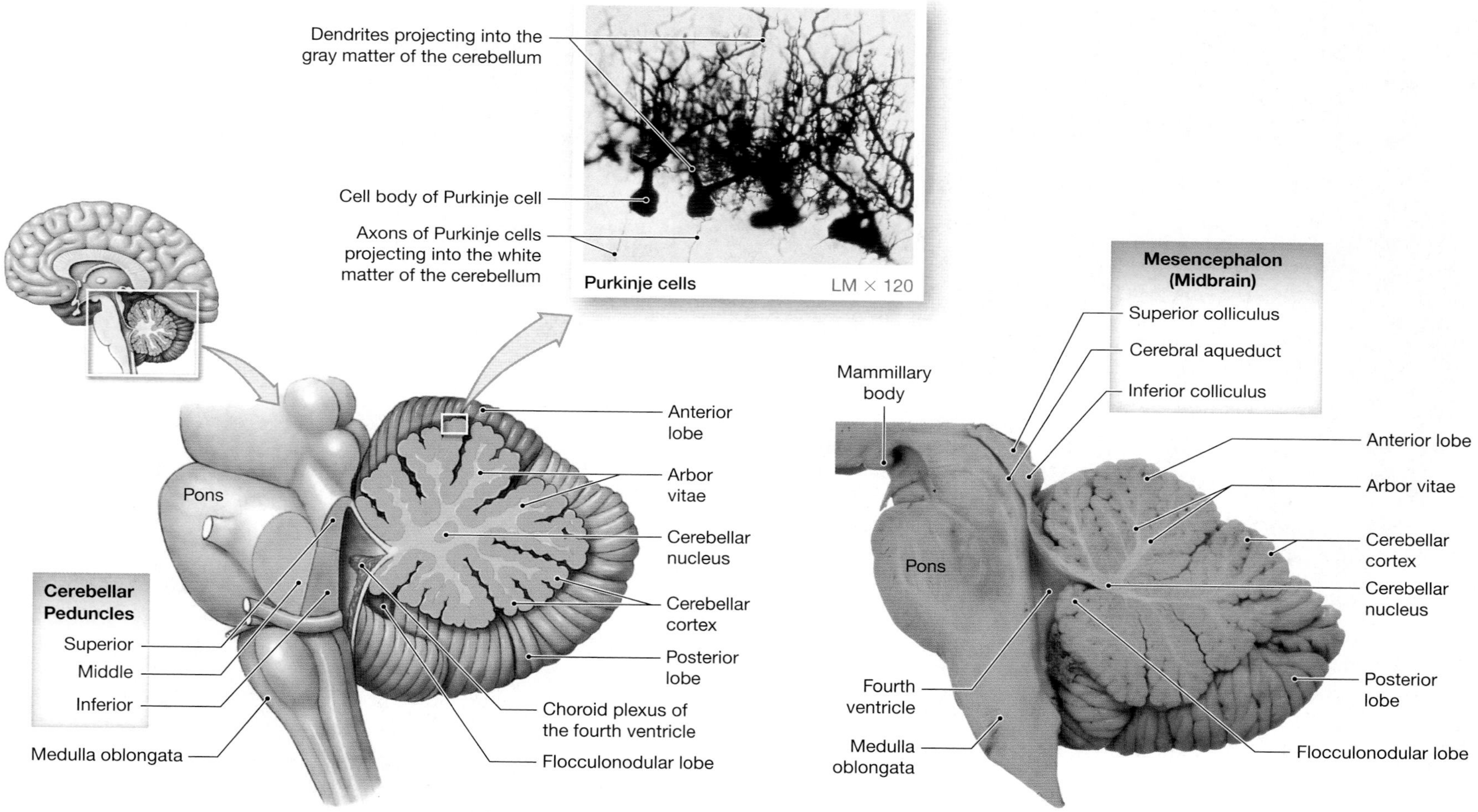

b Sagittal view of the cerebellum showing the arrangement of gray matter and white matter. Purkinje cells are seen in the photomicrograph; these large neurons are found in the cerebellar cortex.

The Cerebral Hemispheres

▶ **KEY POINT** While the two cerebral hemispheres appear identical anatomically, they have some functional differences. A region of the cerebral cortex may have more than one function.

A thick blanket of superficial gray matter (cerebral cortex) covers the paired cerebral hemispheres that form the superior and lateral surfaces of the cerebrum **(Figure 16.17)**. The cortical surface is marked by gyri and sulci. The gyri increase the surface area of the cerebral hemispheres, providing space for additional cortical neurons. The cerebral cortex performs the most complicated neural functions, and these analytical and integrative activities require large numbers of neurons. The brain and cranium have both enlarged in the course of human evolution, but the cerebral cortex has grown more than the rest of the brain.

The Cerebral Lobes

The deep **longitudinal fissure** separates the two cerebral hemispheres **(Figure 16.16)**. Each hemisphere is divided into **lobes** named after the overlying bones of the skull. A deep groove, the **central sulcus**, extends laterally from the longitudinal fissure. The **frontal lobe** is anterior to the central sulcus, and the **lateral sulcus** marks the frontal lobe's inferior border. The region inferior to the lateral sulcus is the **temporal lobe**. Reflecting (pulling back)

Figure 16.16 The Cerebral Hemispheres, Part I. The cerebral hemispheres are the largest part of the adult brain.

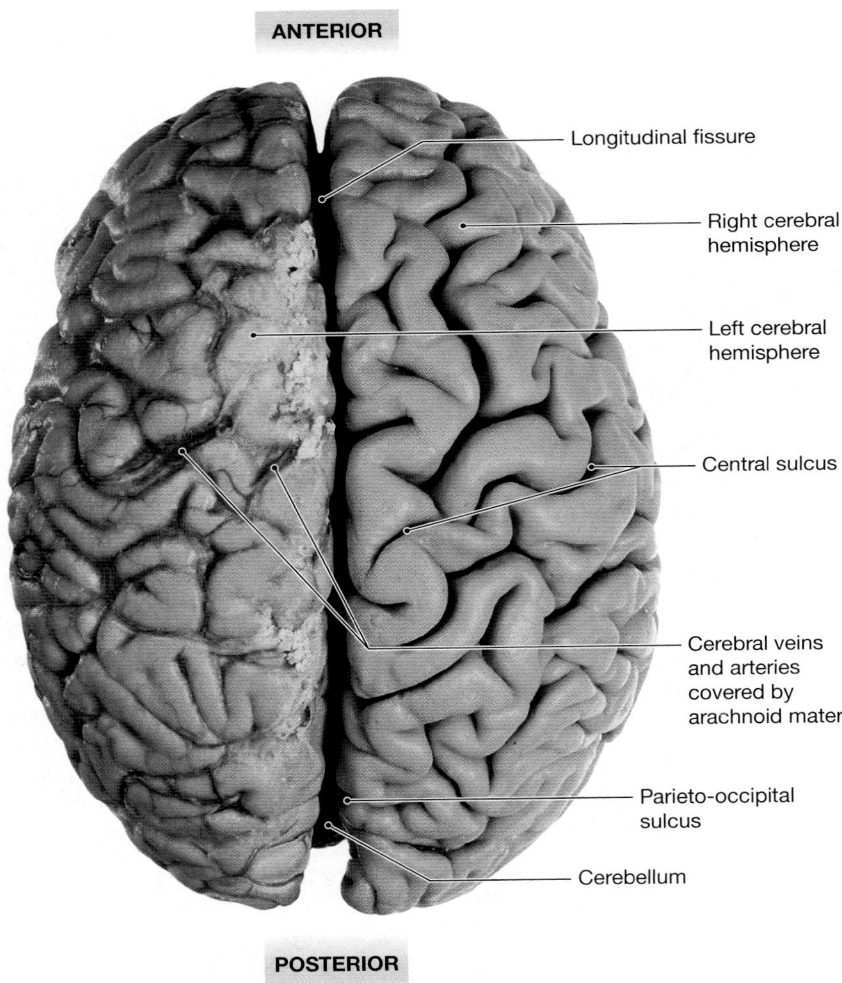

a Superior view.

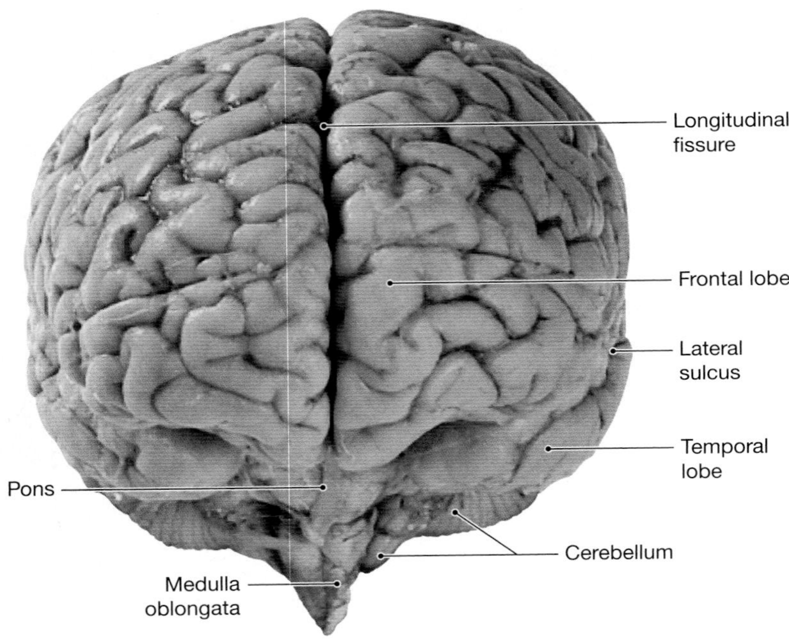

b Anterior view.

Parietal lobe
Longitudinal fissure
Left cerebral hemisphere
Occipital lobe
Cerebellar hemispheres
Medulla oblongata
Right cerebral hemisphere

c Posterior view. Note the relatively small size of the cerebellar hemispheres.

this lobe to the side exposes the **insula** (IN-sū-la), a hidden "island" of cortex (**Figure 16.17**). The **parietal lobe** extends posteriorly from the central sulcus to the **parieto-occipital sulcus**. The region posterior to the parieto-occipital sulcus is the **occipital lobe** (**Figure 16.16**).

Each lobe contains functional regions whose boundaries are not defined. Some of these functional regions process sensory information, while others are responsible for motor commands. Note the following three points about the cerebral lobes:

1. Each cerebral hemisphere receives sensory information from and generates motor commands to the contralateral (opposite) side of the body. Therefore, the left hemisphere controls the right side, and the right hemisphere controls the left side. This crossing over has no known functional significance.
2. The two hemispheres have some functional differences.
3. The assignment of a specific function to a specific region of the cerebral cortex is imprecise. Any one region may have several different functions. Some aspects of cortical function, such as consciousness, cannot easily be assigned to any single region.

Our understanding of brain function is still incomplete, and not every anatomical feature has a known function. However, it is clear from studies on metabolic activity and blood flow that a normal individual uses all portions of the brain.

Motor and Sensory Areas of the Cerebral Cortex

Conscious thought processes and all intellectual functions originate in the cerebral hemispheres. However, much of the cerebrum is involved with the processing of somatic sensory and motor information. **Figure 16.17a** and **Table 16.8** detail the major motor and sensory regions of the cerebral cortex. The central sulcus separates the motor and sensory portions of the cortex. The **precentral gyrus** of the frontal lobe forms the anterior margin of the central sulcus. The surface of this gyrus is the **primary motor cortex**. Neurons of the primary motor cortex direct voluntary movements by controlling somatic motor neurons in the brainstem and spinal cord. The neurons of the primary motor cortex are called **pyramidal cells**, and the pathway that provides voluntary motor control is known as the corticospinal pathway, or *pyramidal system*. ⮌ pp. 394–395

The **postcentral gyrus** of the parietal lobe forms the posterior margin of the central sulcus, and its surface contains the **primary somatosensory cortex**. The posterior columns and spinothalamic tracts provide the neurons in this region with sensory information from touch, pressure, pain, and temperature receptors. ⮌ pp. 388–394 We are consciously aware of these sensations because the sensory information has been relayed to the primary somatosensory cortex. At the same time, neurons deliver information to the basal nuclei and other centers. As a result, sensory information is monitored at both conscious and unconscious levels.

Sensory information concerning sensations of sight, sound, and smell arrives at other portions of the cerebral cortex. The **visual cortex** of the occipital lobe receives visual information, and the **auditory cortex** and **olfactory cortex** of the temporal lobe receive information concerned with hearing and smelling, respectively. The **gustatory cortex** lies in the anterior portion of the insula and adjacent portions of the frontal lobe. This region receives information from taste receptors of the tongue and pharynx. The regions of the cerebral cortex involved with special sensory information are shown in **Figure 16.17a**.

Table 16.8 | The Cerebral Cortex

Region (Lobe)	Function
FRONTAL LOBE	
Primary motor cortex	Conscious control of skeletal muscles
PARIETAL LOBE	
Primary somatosensory cortex	Conscious perception of touch, pressure, vibration, pain, temperature, and taste
OCCIPITAL LOBE	
Visual cortex	Conscious perception of visual stimuli
TEMPORAL LOBE	
Auditory cortex and olfactory cortex	Conscious perception of auditory and olfactory stimuli
ALL LOBES	
Association areas	Integration and processing of sensory data; processing and initiation of motor activities

Association Areas

Each of the sensory and motor regions of the cortex is connected to a nearby **association area** (**Figure 16.17a**). The term "association area" is used for regions of the cerebrum involved with *integrating and understanding* sensory or motor information. These areas do not directly receive sensory information, and they do not generate motor commands. Instead, they *interpret* sensory input arriving elsewhere in the cerebral cortex. The association areas then plan, prepare for, and help coordinate motor output. For example, the **somatosensory association area** allows you to understand the size, form, and texture of an object, and the **premotor cortex** uses memories of learned movement patterns to coordinate motor activities.

16

The functional distinctions between the sensory and motor association areas are most evident after a localized brain injury. For example, an individual with a damaged **visual association area** may see letters quite clearly but may not be able to recognize or interpret them. This person would scan the lines of a printed page and see rows of symbols that convey no meaning. People with damage to the area of the premotor cortex concerned with coordination of eye movements can understand written letters and words but cannot read them because their eyes cannot follow the lines on a printed page.

Higher-Order Functions

▶ **KEY POINT** The cerebral cortex performs higher-order functions that involve complex communication within the cerebral cortex and between the cerebral cortex and other areas of the brain.

Higher-order functions have the following characteristics:

- They are performed by the cerebral cortex.
- They involve complex interconnections and communication between areas within the cerebral cortex and between the cerebral cortex and other areas of the brain.
- They involve both conscious and unconscious information processing.
- They are not part of the programmed "wiring" of the brain; therefore, the functions are subject to modification and adjustment over time.

First, let's identify the cortical areas involved and discuss functional differences between the right and left hemispheres. We will then briefly consider the mechanisms of memory, learning, and consciousness.

Figure 16.17 The Cerebral Hemispheres, Part II. Lobes and functional regions.

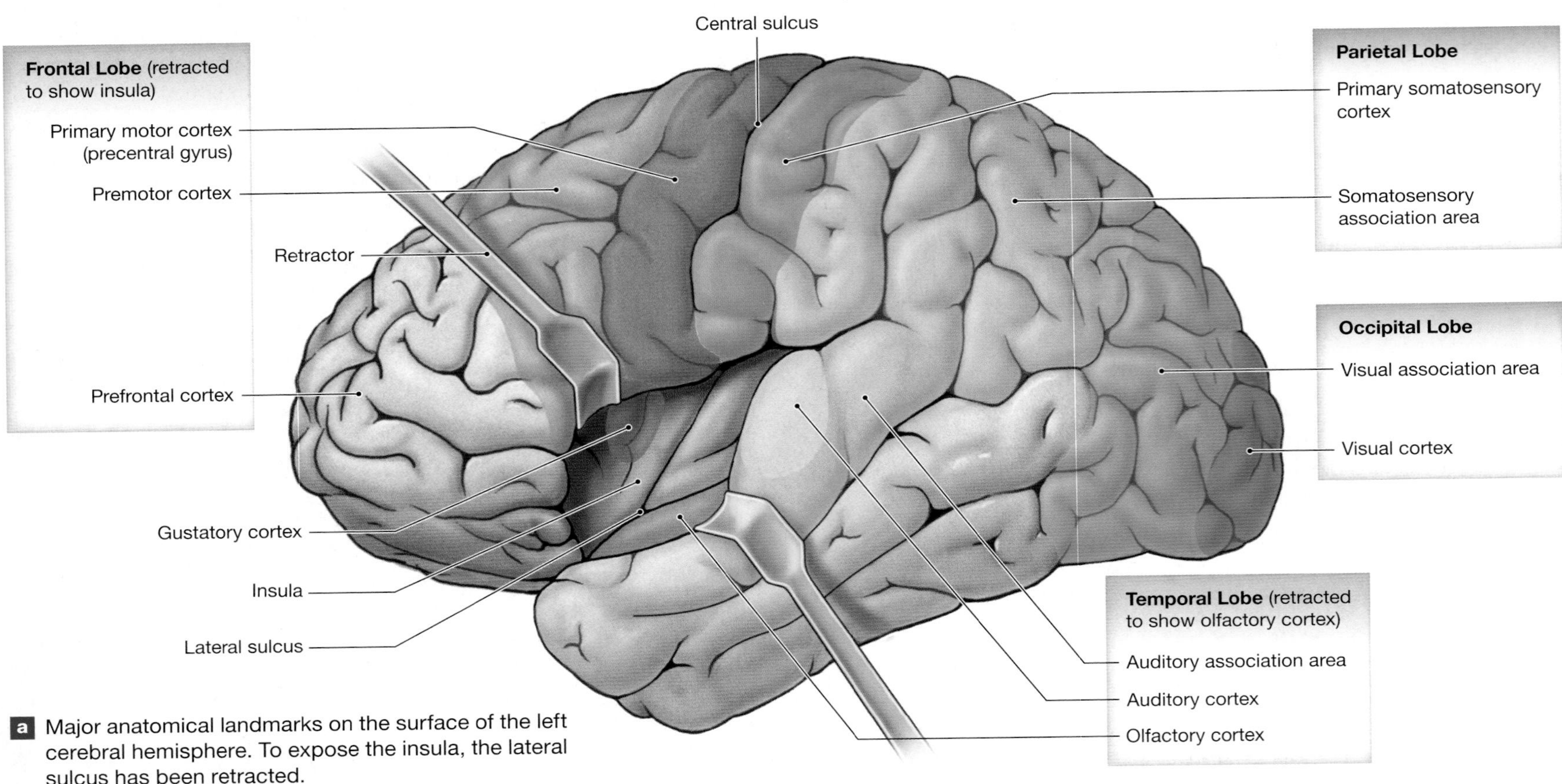

16

a Major anatomical landmarks on the surface of the left cerebral hemisphere. To expose the insula, the lateral sulcus has been retracted.

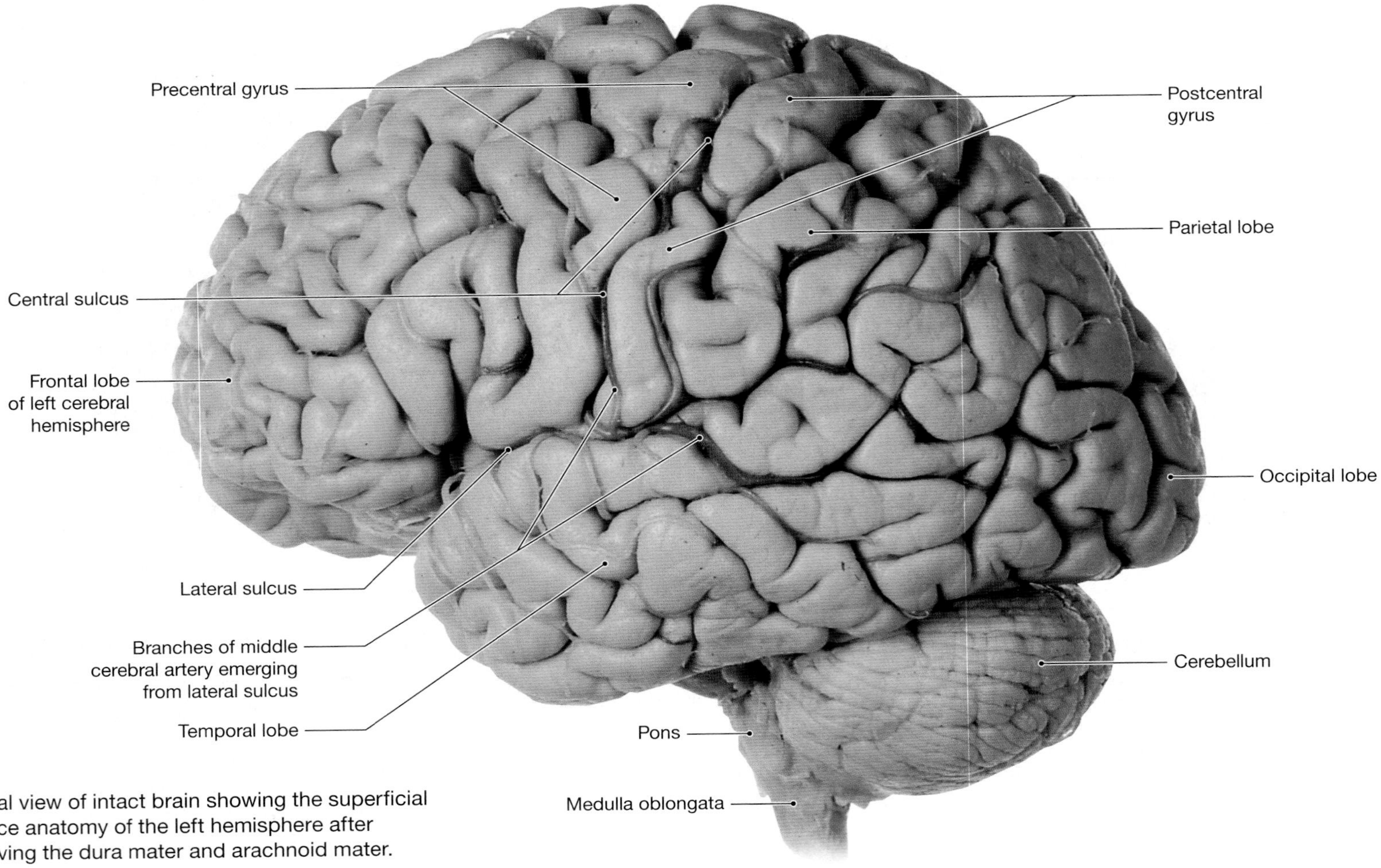

Precentral gyrus

Postcentral gyrus

Parietal lobe

Central sulcus

Frontal lobe of left cerebral hemisphere

Occipital lobe

Lateral sulcus

Branches of middle cerebral artery emerging from lateral sulcus

Cerebellum

Temporal lobe

Pons

Medulla oblongata

b Lateral view of intact brain showing the superficial surface anatomy of the left hemisphere after removing the dura mater and arachnoid mater.

CLINICAL NOTE

Microcephaly and Hydrocephalus

Microcephaly is a birth defect in which the head circumference is much smaller than expected for age and sex. As a fetus develops, the skull enlarges in response to growth and development of the brain. If the brain does not develop normally or stops developing, the entire head will be small. The lifelong disabilities that result from microcephaly can range from minimal to so severe that the baby will be blind, deaf, and unable to sit, swallow, or learn.

Often the cause of microcephaly is unknown. Some cases result from genetic abnormalities. Certain infections, including rubella, HIV, herpes, and syphilis, are known to cause microcephaly. Several drugs and toxic chemicals, particularly alcohol, are associated with it. Recently, clusters of microcephaly cases in areas affected by the Zika virus have led to concerns about a causal relationship. Zika virus is spread to people through the bite of an infected *Aedes aegypti* mosquito and other methods of transmission: sexual intercourse, blood transfusion, and lab exposure.

Zika virus was cultured from the placenta of this child, born in Brazil.

Hydrocephalus is a condition marked by an excessive accumulation of cerebrospinal fluid (CSF) within the brain ventricles. This puts pressure on brain tissue and, in infants with fontanelles and open skull sutures, enlarges the head. The CSF that circulates through the ventricular system and flows through the subarachnoid space should eventually re-enter the circulation through the arachnoid granulations. When this circulation is disrupted, the fluid accumulates, enlarging the head and sometimes causing brain damage.

Hydrocephalus can result from inherited genetic abnormalities or developmental problems. Other causes include encephalitis (a brain infection), meningitis (infection of the meninges covering the brain), trauma, and tumors. Acquired hydrocephalus, which can occur at any age after birth, can be due to head injuries, strokes, infections, tumors, and bleeding in the brain.

Untreated hydrocephalus is usually fatal. Treatment involves surgically inserting a shunt into the ventricle to divert the excess CSF to another part of the body for absorption.

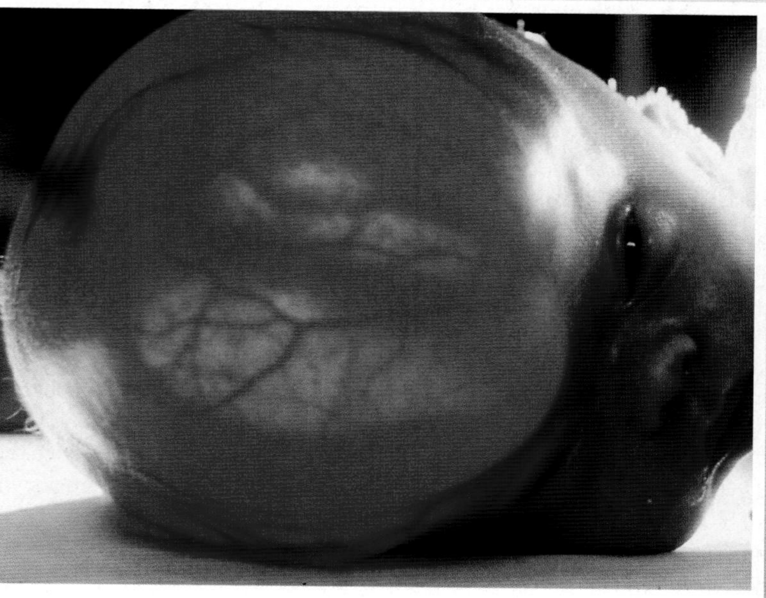

This infant has severe hydrocephalus, a condition usually caused by impaired circulation and removal of cerebrospinal fluid. CSF buildup distorts the brain and enlarges the cranium.

Integrative Centers of the Cerebral Cortex

Integrative centers receive information from many association areas and direct extremely complex motor activities, as well as performing analytical functions. For example, the **prefrontal cortex**, or *prefrontal association area*, of the frontal lobe integrates information from sensory association areas and performs intellectual functions, such as predicting the consequences of possible responses. These centers are located in the lobes and cortical areas of both cerebral hemispheres.

Centers such as the prefrontal cortex, Broca's area, and Wernicke's area are concerned with processes such as mathematical computation, speech, writing, and understanding spatial relationships. They are largely restricted to either the left or the right hemisphere. Although the two hemispheres look almost identical, they have different functions, a phenomenon called hemispheric lateralization. The corresponding regions on the opposite hemisphere are also active, but their functions are less well defined.

The Specialized Language Areas in the Brain

Language processing is much more complicated than previously thought, and brain imaging studies have shown that language processing occurs in both hemispheres and varies from one person to the next.

Two important cortical areas with varying functions related to human language are Wernicke's area and the speech center (**Figure 16.18**). Both are primarily associated with the left cerebral hemisphere. **Wernicke's area** is near the auditory cortex and is associated with language comprehension. This analytical center receives information from the sensory association areas and plays an important role in your personality by integrating sensory information and coordinating access to visual and auditory memories.

The **speech center** (also termed *Broca's area*) is near the motor cortex and is associated with speech production. The speech center regulates the patterns of breathing and vocalization needed for normal speech. This

Figure 16.18 Integrative Regions of the Cerebral Cortex

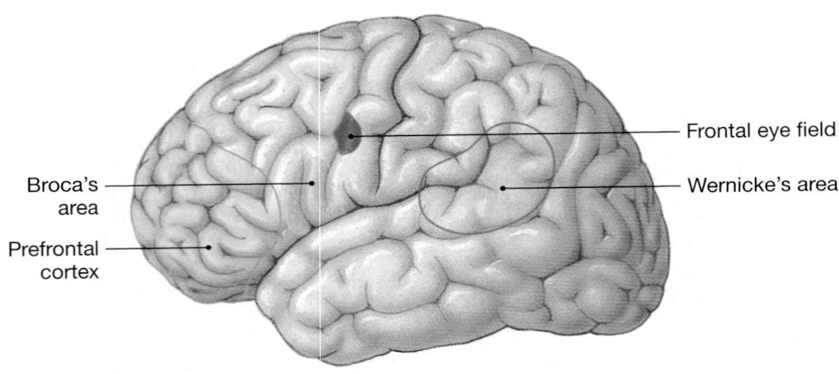

regulation involves coordinating the activities of the respiratory muscles, the laryngeal and pharyngeal muscles, and the muscles of the tongue, cheeks, lips, and jaws. A person with damage to the speech center area can make sounds but not words.

The motor commands issued by the motor speech center are adjusted by feedback from the auditory association area, also called the **receptive speech area**. Damage to the related sensory areas can cause a variety of speech-related problems. Some affected individuals have difficulty speaking, although they know exactly which words to use. Others talk constantly but use all the wrong words.

16

The Prefrontal Cortex

The **prefrontal cortex** of the frontal lobe coordinates information relayed from all of the cortical association areas. In doing so, it performs such abstract intellectual functions as predicting the consequences of events or actions. The prefrontal cortex does not fully develop until the early 20s. Damage to the prefrontal cortex leads to difficulties in estimating temporal relationships between events. Questions such as "How long ago did this happen?" or "What happened first?" become difficult to answer.

The prefrontal cortex has extensive connections with other cortical areas and with other portions of the brain. Feelings of frustration, tension, and anxiety are generated as the prefrontal cortex interprets ongoing events and makes predictions about future situations or consequences. If the connections between the prefrontal cortex and other brain regions are physically severed, the frustrations, tensions, and anxieties are removed. During the middle of the 20th century, a rather drastic procedure, called **prefrontal lobotomy**, was used to "cure" a variety of mental illnesses, especially those associated with violent or antisocial behavior. After a lobotomy, the patient would no longer be concerned about what had previously been a major problem, whether psychological (hallucinations) or physical (severe pain). However, the individual was often equally unconcerned about tact, decorum, and toilet training. Since then, drugs that target specific pathways and regions of the CNS have been developed, so lobotomies are no longer used to change behavior.

Hemispheric Lateralization

As mentioned, each of the two cerebral hemispheres is responsible for specific functions that are not ordinarily performed by the opposite hemisphere, a type of specialization known as **hemispheric lateralization**. In most people, the left hemisphere contains the specialized language areas of the brain and is responsible for language-based skills. For example, reading, writing, and speaking depend on processing done in the left cerebral hemisphere. In addition, the premotor cortex involved with the control of hand movements is larger on the left side for right-handed individuals than for left-handed ones. The left hemisphere is also important in performing analytical tasks, such as mathematical calculations and logical decision-making.

The right cerebral hemisphere analyzes sensory information and relates the body to the sensory environment. Interpretive centers in this hemisphere permit you to identify familiar objects by touch, smell, sight, or taste. For example, the right hemisphere plays a dominant role in recognizing faces and in understanding three-dimensional relationships. It is also important in analyzing the emotional context of a conversation—for instance, distinguishing between the threat "Get lost!" and the question "Get lost?" Individuals with a damaged right hemisphere may be unable to add emotional inflections to their own words.

Left-handed people represent about 9 percent of the human population. In most cases, the primary motor cortex of the right hemisphere controls motor function for the dominant (left) hand. However, the centers involved with speech and analytical function are in the left hemisphere, just as they are in right-handed people. Interestingly, an unusually high percentage of musicians and artists are left-handed. Additionally, the more a person favors one hand over the other, the stronger the connection with the other side of the brain. This suggests that as a species, we became left- or right-handed when we started developing language about 100,000 years ago.

The Central White Matter

▶ **KEY POINT** The central white matter of the cerebrum carries afferent information between areas of the cerebral cortex and between the cerebral cortex and other brain regions.

The **central white matter** is covered by the gray matter of the cerebral cortex (**Figure 16.19**). It contains myelinated fibers forming bundles that connect one cortical area to another or that connect areas of the cortex to other regions of the brain. These bundles include the following:

CLINICAL NOTE

Damage to the Specialized Language Areas

Aphasia (*a*–, without, + *phasia*, speech) is a neurological condition caused by damage to the portions of the brain that are responsible for language. It is caused by an acquired lesion of the brain.

People with **expressive aphasia**, also called *motor aphasia*, have trouble speaking words and sentences, while those with **receptive aphasia**, also called **sensory aphasia**, have trouble understanding spoken or written language. **Global aphasia** involves extensive damage to Broca's area in which all aspects of speech and communication are impaired. **Nominal aphasia**, the least severe form, involves trouble finding the correct word for particular objects, people, places, or events.

Stroke victims with lesions in the language area area can often recover slowly over years. A speech disorder known as stuttering may be associated with underactivity in the language area.

Dyslexia (*lexis,* diction) is the most common reading learning disability and often affects spelling as well. Dyslexia can affect intelligent people—many famous people, including Albert Einstein, are dyslexic. Dyslexia has a familial, inherited basis. It is neurobiological in origin, reflecting a deficit in the specialized language areas of the brain. Adult-onset dyslexia is usually the result of brain injury or dementia.

Figure 16.19 **The Central White Matter.** Shown are the major groups of axon fibers and tracts of the central white matter.

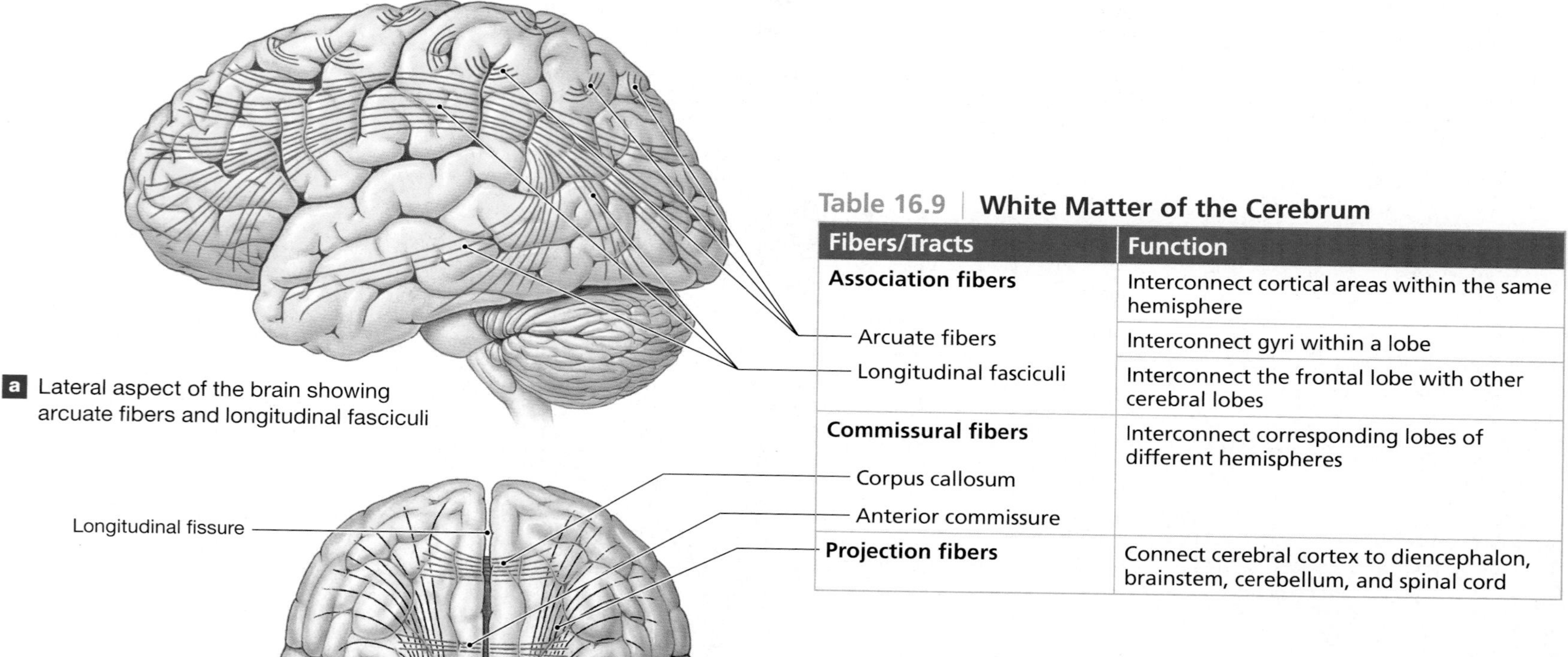

a Lateral aspect of the brain showing arcuate fibers and longitudinal fasciculi

b Anterior view of the brain showing commissural and projection fibers

Table 16.9 | **White Matter of the Cerebrum**

Fibers/Tracts	Function
Association fibers	Interconnect cortical areas within the same hemisphere
Arcuate fibers	Interconnect gyri within a lobe
Longitudinal fasciculi	Interconnect the frontal lobe with other cerebral lobes
Commissural fibers	Interconnect corresponding lobes of different hemispheres
Corpus callosum	
Anterior commissure	
Projection fibers	Connect cerebral cortex to diencephalon, brainstem, cerebellum, and spinal cord

- **Association fibers**: These fibers interconnect areas of cortex within a single cerebral hemisphere. The shortest association fibers are called **arcuate** (AR-kū-at) **fibers** because they curve in an arc to pass from one gyrus to another. The longer association fibers are organized into discrete bundles. The **longitudinal fasciculi** connect the frontal lobe to the other lobes of the same hemisphere.
- **Commissural fibers**: These fibers link the two cerebral hemispheres together. A dense band of commissural (kom-I-sūr-al; *commissura*, a crossing over) fibers permits communication between the two hemispheres. Prominent commissural bundles linking the cerebral hemispheres include the **corpus callosum** and the **anterior commissure**.
- **Projection fibers**: These fibers link the cerebrum with other regions of the brain and the spinal cord. All ascending and descending axons must pass through the diencephalon on their way to or from sensory, motor, or association areas of the cerebral cortex. In gross dissection, the afferent fibers and efferent fibers look alike, and the entire collection of fibers is known as the **internal capsule**.

Table 16.9 summarizes the names and functions of these groups.

The Basal Nuclei

▶ **KEY POINT** The basal nuclei are paired masses of gray matter within the cerebral hemispheres. They are embedded within the central white matter, and the projection and commissural fibers travel around or between these nuclei.

The **basal nuclei** are masses of gray matter that lie within each hemisphere deep to the floor of the lateral ventricle. They are embedded in the white matter of the cerebrum. The radiating projection fibers and commissural fibers travel around or between these nuclei (**Figure 16.20**).

Historically, the basal nuclei have been considered part of a larger functional group known as the basal ganglia. This group included the basal nuclei of the cerebrum and the associated motor nuclei in the diencephalon and mesencephalon. The basal nuclei include the caudate nucleus and lentiform nucleus.

The **caudate** (KOW-dāt) **nucleus** has a large head and a slender, curving tail that follows the curve of the lateral ventricle. The head of the caudate nucleus lies anterior to the **lentiform nucleus**. The lentiform nucleus consists of a lateral **putamen** (pū-TĀ-men) and a medial **globus pallidus** (GLŌ-bus PAL-ih-dus; "pale globe"). The term *corpus striatum* (striated body) has been used to refer to the caudate and lentiform nuclei, or to the caudate nucleus and putamen. The name refers to the striated (striped) appearance of the internal capsule as its fibers pass among these nuclei. The **claustrum** is a thin layer of gray matter lying close to the putamen. The amygdaloid body, part of the limbic system, lies anterior to the tail of the caudate nucleus and inferior to the lentiform nucleus (**Figure 16.20**).

Functions of the Basal Nuclei

The basal nuclei (1) subconsciously control and integrate skeletal muscle tone, (2) coordinate learned movement patterns, and (3) process, integrate, and relay information from the cerebral cortex to the thalamus. These nuclei do not initiate particular movements. However, once a movement is under way, the basal nuclei coordinate the movement by providing the general pattern and rhythm. This is especially true for movements of the trunk and proximal limb muscles. Next, let's look at some functions assigned to specific basal nuclei.

Caudate Nucleus and Putamen When you walk, the caudate nucleus and putamen control the cycles of arm and leg movements that occur between the time you decide to start walking and the time you decide to stop.

Figure 16.20 The Basal Nuclei

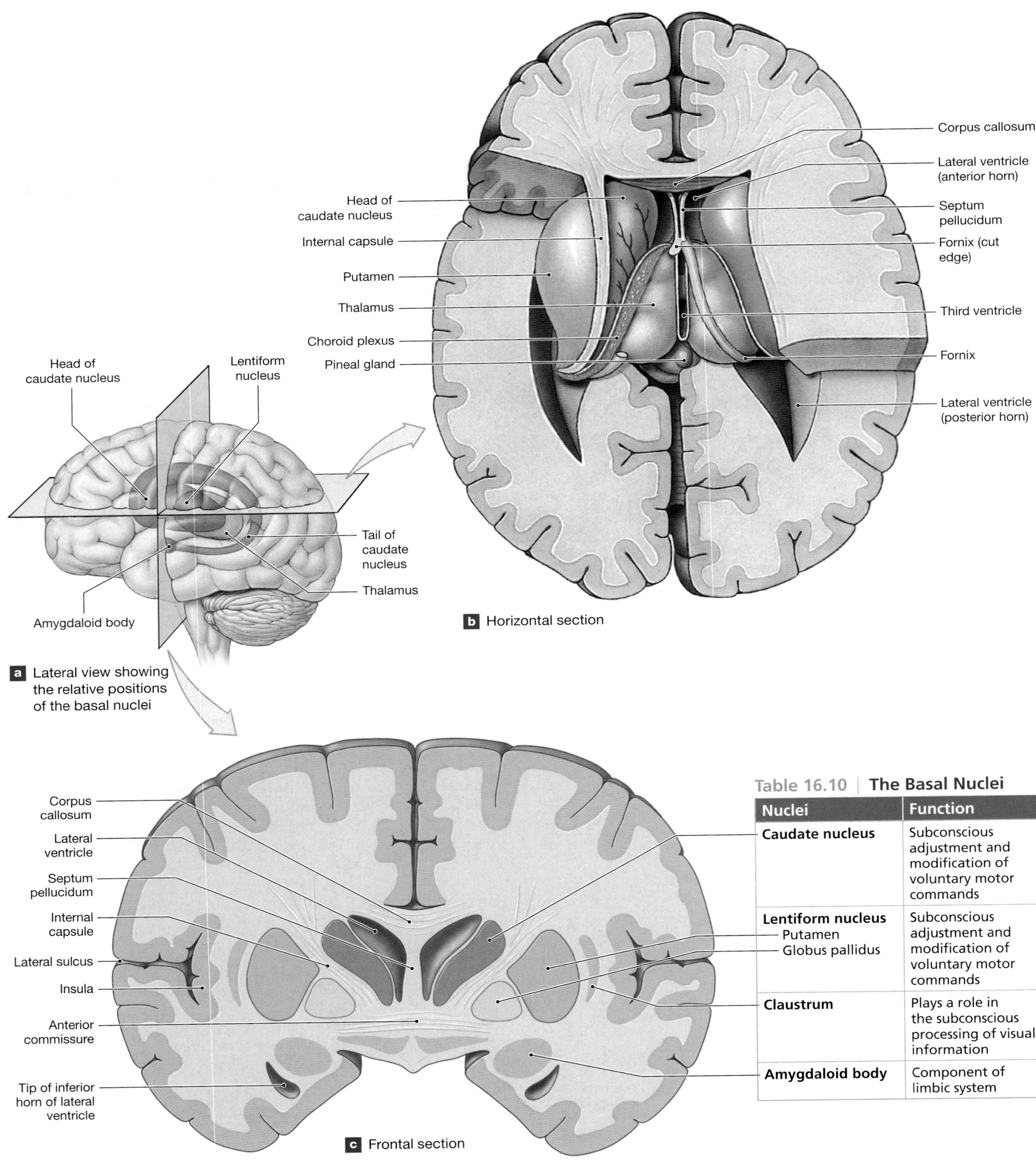

a Lateral view showing the relative positions of the basal nuclei

b Horizontal section

c Frontal section

Table 16.10 | **The Basal Nuclei**

Nuclei	Function
Caudate nucleus	Subconscious adjustment and modification of voluntary motor commands
Lentiform nucleus Putamen Globus pallidus	Subconscious adjustment and modification of voluntary motor commands
Claustrum	Plays a role in the subconscious processing of visual information
Amygdaloid body	Component of limbic system

Globus Pallidus The globus pallidus controls and adjusts muscle tone, particularly in the appendicular muscles, to set body position in preparation for a voluntary movement. For example, when you decide to pick up an object, the globus pallidus positions your shoulder and stabilizes your arm as you consciously reach and grasp with your forearm, wrist, and hand. The functions of other basal nuclei are poorly understood.

Other nuclei, while not anatomically part of the basal nuclei, are closely tied to the functions of the basal nuclei. These structures include the substantia nigra and the subthalamic nuclei. **Table 16.10** summarizes these relationships and the functions of the basal nuclei.

The Limbic System

▶ **KEY POINT** Emotions, thoughts, and behaviors are closely linked in the human brain. The term limbic system is given to those portions of the brain concerned with emotions and behaviors that are ultimately related to the preservation of the individual and the preservation of the species.

The **limbic** (LIM-bik; limbus, border) **system** includes nuclei and tracts along the border between the cerebrum and diencephalon. The functions of the limbic system include (1) establishing emotional states and related behavioral drives, (2) linking the conscious, intellectual functions of the cerebral cortex with the unconscious and autonomic functions of other portions of the brain, and (3) facilitating memory storage and retrieval. *This system is a functional grouping rather than an anatomical one*, and the limbic system includes components of the cerebrum, diencephalon, and mesencephalon (**Figure 16.21** and **Table 16.11**).

The amygdaloid body (**Figures 16.20a,c** and **16.21b**) is thought to act as an integration center between the limbic system, the cerebrum, and various sensory systems. The **limbic lobe** of the cerebral hemisphere consists of the gyri and deeper structures adjacent to the diencephalon. The **cingulate** (SIN-gū-lāt; *cingulum*, girdle or belt) **gyrus** is superior to the corpus callosum. The **dentate gyrus** and the adjacent **parahippocampal** (pa-ra-hip-ō-KAM-pal) **gyrus** conceal an underlying nucleus, the **hippocampus**, which

Table 16.11 | Components of the Limbic System

Overall functions of limbic system	Processing of memories, creation of emotional states, drives, and associated behaviors
Cerebral components	
Cortical areas	Limbic lobe (cingulate gyrus, dentate gyrus, and parahippocampal gyrus)
Nuclei	Hippocampus, amygdaloid body
Tracts	Fornix
Diencephalic components	
Thalamus	Anterior nuclei
Hypothalamus	Centers concerned with emotions, appetites (thirst, hunger), and related behaviors (**Table 16.6**)
Other components	
Reticular formation	Network of interconnected nuclei throughout brainstem

16

Figure 16.21 The Limbic System

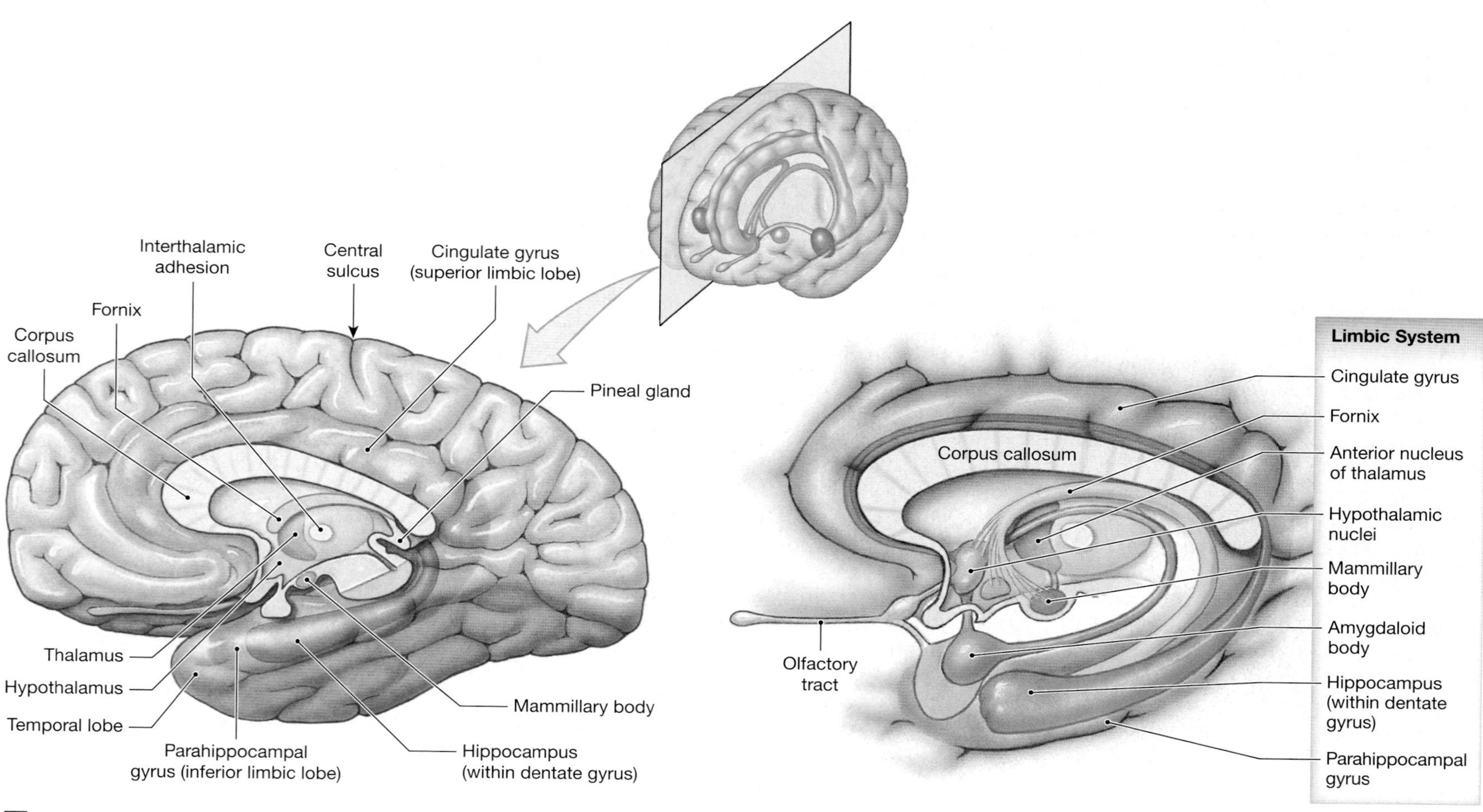

a Sagittal section through the cerebrum showing the cortical areas associated with the limbic system. The parahippocampal and dentate gyri are shown as if transparent so that deeper limbic components can be seen.

b Additional details concerning the three-dimensional structure of the limbic system.

lies deep within the temporal lobe (see **Figures 16.20** and **16.21**). Early anatomists thought this nucleus resembled a seahorse (*hippocampus*); it plays an essential role in learning and the storage of long-term memories.

The **fornix** (FŌR-niks; "arch") **(Figure 16.13)** is a tract of white matter connecting the hippocampus with the hypothalamus. Exiting the hippocampus, the fornix curves medially and superiorly, passing inferior to the corpus callosum. It then forms an arch and curves anteriorly, ending in the hypothalamus. Many of the fibers of the fornix end in the **mammillary** (MAM-i-lar-ē; *mammilla*, breast) **bodies**, which are prominent nuclei in the floor of the hypothalamus. The mammillary bodies contain motor nuclei that control reflex movements associated with eating, such as chewing, licking, and swallowing.

Several other nuclei in the thalamus and hypothalamus of the diencephalon are components of the limbic system. The **anterior nucleus** of the thalamus relays visceral sensations from the hypothalamus to the cingulate gyrus. Experimental stimulation of the hypothalamus has localized a number of important centers responsible for the emotions of rage, fear, pain, sexual arousal, and pleasure.

Stimulation of the hypothalamus also produces heightened alertness and generalized excitement. This response is caused by stimulation of the **reticular formation**, an interconnected network of brainstem nuclei. Stimulation of adjacent portions of the hypothalamus or thalamus depresses reticular activity, resulting in generalized weariness or actual sleep.

16.8  **CONCEPT CHECK**

12 Identify the lobes of each cerebral hemisphere and cite their functions.

13 List and describe the three major groups of axons in the central white matter.

See the blue Answers tab at the back of the book.

CLINICAL NOTE

Alzheimer's Disease

Alzheimer's disease is a chronic, progressive illness characterized by memory loss and impairment of higher-order cerebral functions including abstract thinking, judgment, and personality. It is the most common cause of senile dementia, or senility. Symptoms may appear at age 50–60 or later, although the disease occasionally affects younger individuals. Alzheimer's disease has widespread impact. An estimated 4 million people in the United States have Alzheimer's—including roughly 3 percent of those from age 65 to 70, with the number doubling for every five years of aging until nearly 50 percent of those over age 85 have some form of the condition. Over 230,000 victims require nursing home care, and Alzheimer's disease causes more than 53,000 deaths each year.

Most cases of Alzheimer's disease are associated with large concentrations of neurofibrillary tangles and plaques in the nucleus basalis, hippocampus, and parahippocampal gyrus. These brain regions are directly associated with memory processing. It remains to be determined whether these deposits cause Alzheimer's disease or are secondary signs of ongoing metabolic alterations with an environmental, hereditary, or infectious basis.

In Down syndrome and in some inherited forms of Alzheimer's disease, mutations affecting genes on either chromosome 21 or a small region of chromosome 14 lead to increased risk of the early onset of the disease. Other genetic factors certainly play a major role. The late-onset form of Alzheimer's disease has been traced to a gene on chromosome 19 that codes for proteins involved in cholesterol transport.

Diagnosis involves excluding metabolic and anatomical conditions that can mimic dementia, a detailed history and physical, and an evaluation of mental functioning. Initial symptoms are subtle: moodiness, irritability, depression, and a general lack of energy. These symptoms are often ignored, overlooked, or dismissed. Elderly relatives are viewed as eccentric or irascible and are humored whenever possible.

As the condition progresses, however, it becomes more difficult to ignore or accommodate. An individual with Alzheimer's disease has difficulty making decisions, even minor ones. Mistakes—sometimes dangerous ones—are made, through either bad judgment or forgetfulness. For example, the person might light the gas burner, place a pot on the stove, and go into the living room. Two hours later, the pot, still on the stove, melts and starts a fire.

As memory losses continue, the problems become more severe. The individual may forget relatives, his or her home address, or how to use the telephone. The memory loss commonly starts with an inability to store long-term memories, followed by the loss of recently stored memories. Eventually, basic long-term memories, such as the sound of the individual's own name, are forgotten. The loss of memory affects both intellectual and motor abilities, and a person with severe Alzheimer's disease has difficulty performing even the simplest motor tasks.

Individuals with Alzheimer's disease show a pronounced decrease in the number of cortical neurons, especially in the frontal and temporal lobes. This loss is correlated with inadequate ACh production in the nucleus basalis of the cerebrum. Axons leaving that region project throughout the cerebral cortex; when ACh production declines, cortical function deteriorates.

There is no cure for Alzheimer's disease, but a few medications and supplements slow its progress in many patients and reduce the need for nursing home care. The antioxidants vitamin E and ginkgo biloba and the B vitamins of folate, B_6, and B_{12} help some patients and may delay or prevent the disease. Drugs that increase glutamate levels (a neurotransmitter in the brain) also give some additional benefit. Various toxicities and side effects determine what combination of drugs is used. In mice, a vaccine has reduced tangles and plaques in the brain and improved maze-running ability. A preliminary trial of a human vaccine was stopped because cases of immune encephalitis developed in some treated patients. Modification of the vaccine may eliminate this problem, allowing further study of this new approach.

Figure 16.22 **Origins of the Cranial Nerves**

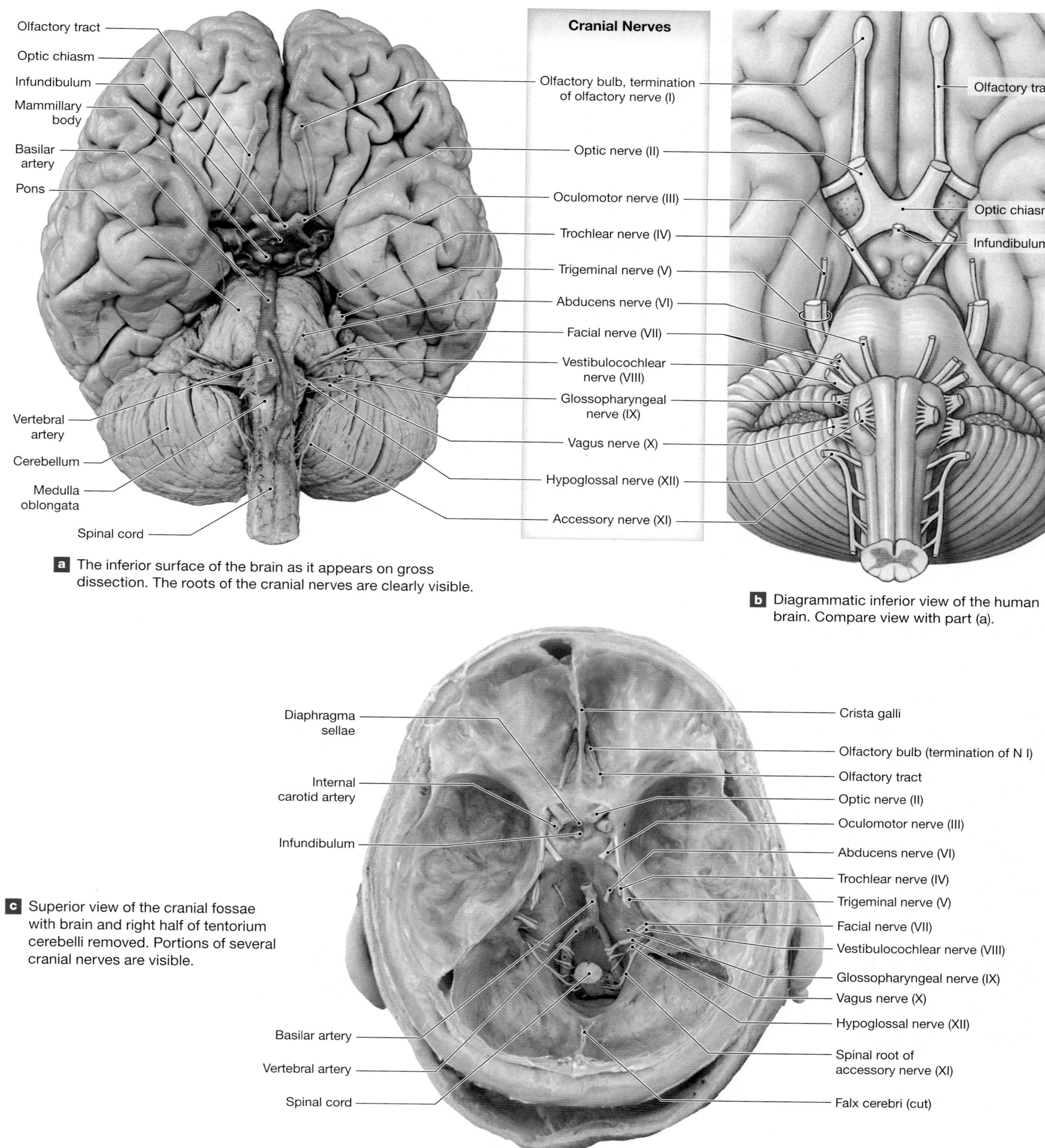

a The inferior surface of the brain as it appears on gross dissection. The roots of the cranial nerves are clearly visible.

b Diagrammatic inferior view of the human brain. Compare view with part (a).

c Superior view of the cranial fossae with brain and right half of tentorium cerebelli removed. Portions of several cranial nerves are visible.

16.9 | The Cranial Nerves

▶ **KEY POINT** Twelve pairs of cranial nerves, numbered N I through N XII, are found on the ventrolateral surface of the brain. Each cranial nerve is named according to its appearance or function.

The 12 pairs of cranial nerves are numbered according to their position along the longitudinal axis of the brain, beginning at the cerebrum **(Figure 16.22)**. In scientific writing, Roman numerals are usually used, with the prefix N or CN. We will use N, which is generally preferred by neuroanatomists and clinical neurologists. If the full name of the cranial nerve is given, then only the Roman numeral is necessary, for example, optic nerve (II).

Each cranial nerve attaches to the brain near the associated sensory or motor nuclei. The sensory nuclei act as processing and integration centers, with the neurons relaying sensory information either to other nuclei or to processing centers within the cerebral cortex or cerebellar cortex. The motor nuclei receive convergent inputs from higher centers or from other nuclei along the brainstem.

Cranial nerves are classified as sensory, special sensory, motor, or mixed (sensory and motor). This is a useful method of classification, but it is based on the primary function, and a cranial nerve can have important secondary functions. Two examples are worth noting:

1. As elsewhere in the PNS, a nerve containing tens of thousands of motor fibers to a skeletal muscle will also contain sensory fibers from proprioceptors in that muscle. These sensory fibers are assumed to be present in the cranial nerves innervating skeletal muscles, but are ignored in the primary classification of the nerve.
2. In addition to their other functions, several cranial nerves (N III, N VII, N IX, and N X) distribute autonomic fibers to peripheral ganglia, just as spinal nerves deliver them to ganglia along the spinal cord. The presence of small numbers of autonomic fibers will be noted (and discussed further in Chapter 17) but are ignored in the classification of the nerve.

16

The Olfactory Nerves (I)

Primary function: Special sensory (smell)

Origin: Receptors of olfactory epithelium

Pass through: Cribriform plate of ethmoid ⮌ p. 138

Destination: Olfactory bulbs

The first pair of cranial nerves **(Figure 16.23)** carries special sensory information from specialized neurons in the epithelium covering the roof of the nasal cavity, the superior nasal conchae of the ethmoid, and the superior parts of the nasal septum. Axons from these sensory neurons collect to form 20 or more bundles that penetrate the cribriform plate of the ethmoid. These bundles are components of the **olfactory nerves (I)**. Almost at once these bundles enter the **olfactory bulbs**, neural masses on either side of the crista galli. The olfactory afferents synapse within the olfactory bulbs. The axons of the postsynaptic neurons proceed to the cerebrum along the slender **olfactory tracts (Figures 16.22** and **16.23)**.

Because the olfactory tracts look like typical peripheral nerves, early anatomists misidentified them as the first cranial nerve. Later studies demonstrated that the olfactory tracts and bulbs are part of the cerebrum, but by then the numbering system was already established.

The olfactory nerves are the only cranial nerves attached directly to the cerebrum. The rest originate or terminate within nuclei of the diencephalon or brainstem, and the ascending sensory information synapses in the thalamus before reaching the cerebrum.

Figure 16.23 The Olfactory Nerve

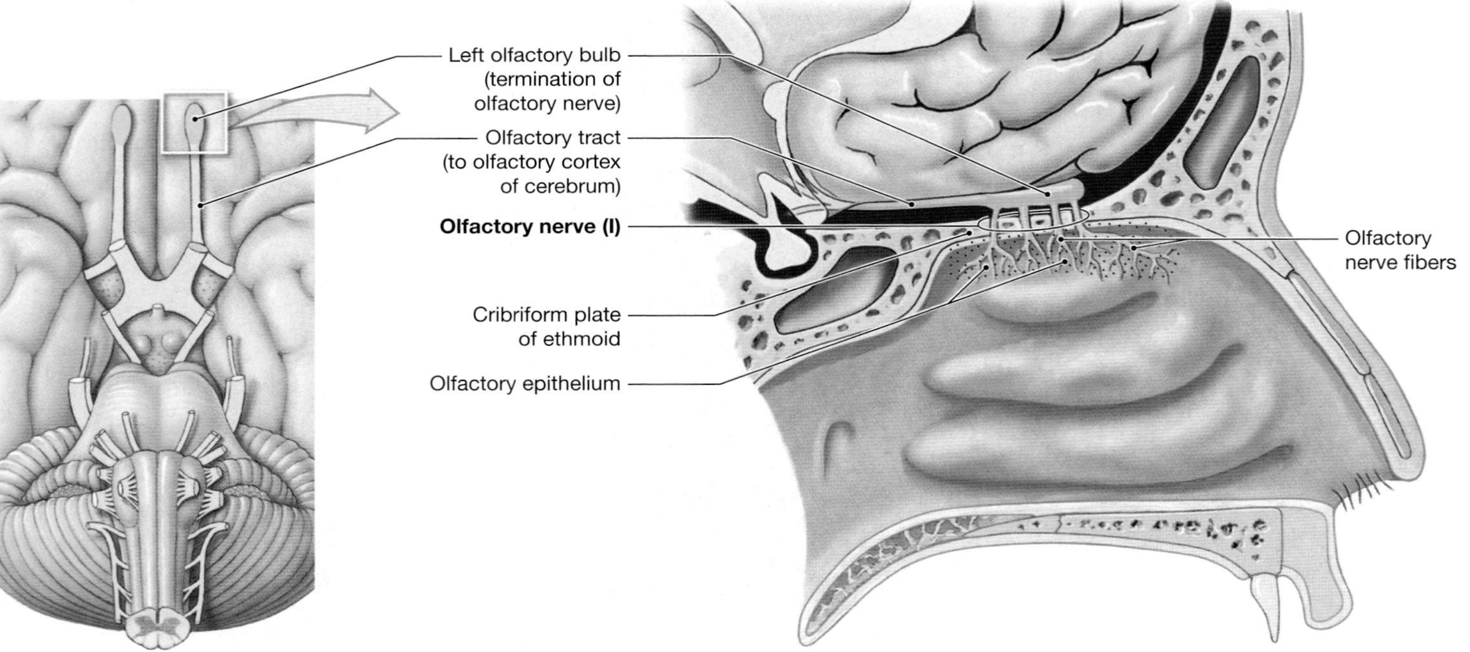

The Optic Nerves (II)

Primary function: Special sensory (vision)

Origin: Retina of eye

Pass through: Optic canal of sphenoid ⮌ p. 138

Destination: Diencephalon

The **optic nerves (II)** carry visual information from special sensory ganglia in the eyes. These nerves pass through the optic canals of the sphenoid before joining at the ventral and anterior margin of the diencephalon, forming the **optic chiasm** (*chiasma*, a crossing) **(Figure 16.24)**. At the optic chiasm, the medial fibers from each optic nerve cross over to the opposite side of the brain. However, the lateral fibers from each tract stay on the same side of the brain. The axons continue to the lateral geniculate nuclei of the thalamus as the **optic tracts** **(Figures 16.22 and 16.24)**. After synapsing in the lateral geniculate nuclei, projection fibers end in the occipital lobe of the brain. This pathway results in each cerebral hemisphere receiving visual information from the lateral half of the retina of the eye on that side and from the medial half of the retina of the eye on the opposite side. A few axons in the optic tracts bypass the lateral geniculate nuclei and synapse in the superior colliculi of the mesencephalon. (We will consider this pathway in Chapter 18.)

Figure 16.24 The Optic Nerve

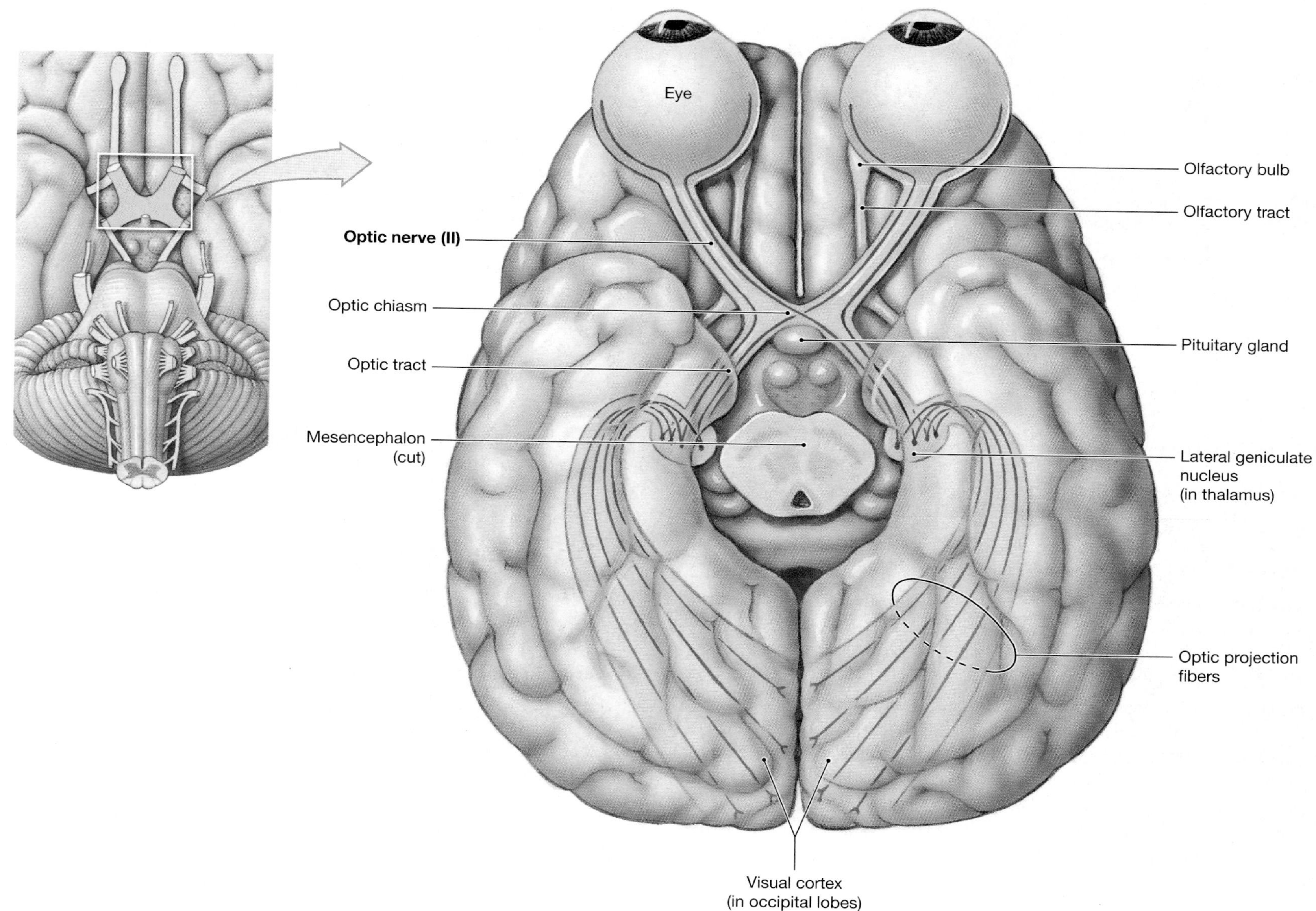

16

Figure 16.25 **Cranial Nerves Controlling the Extra-ocular Muscles**

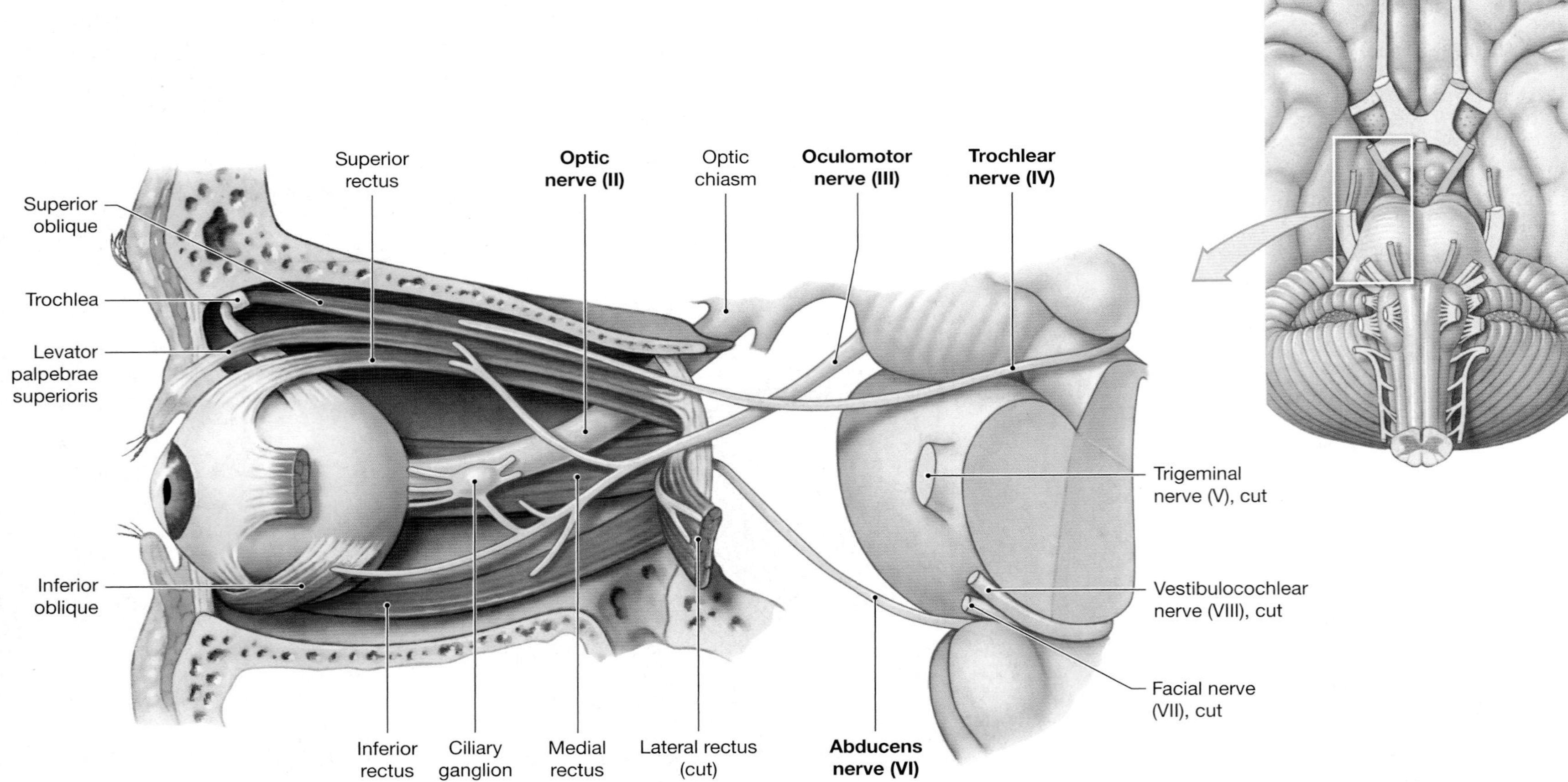

The Oculomotor Nerves (III)

Primary function: Motor, eye movements

Origin: Mesencephalon

Pass through: Superior orbital fissure of sphenoid ⮌ p. 148

Destination: Somatic motor: Superior, inferior, and medial rectus muscles; the inferior oblique; the levator palpebrae superioris ⮌ p. 265

Visceral motor: Intrinsic eye muscles

The mesencephalon contains motor nuclei controlling the third and fourth cranial nerves. The **oculomotor nerves (III)** emerge from the ventral surface of the mesencephalon **(Figure 16.22)** and enter the posterior orbital wall at the superior orbital fissure. The oculomotor nerves **(Figure 16.25)** innervate four of the six extra-ocular muscles and the levator palpebrae superioris, which raises the upper eyelid.

The oculomotor nerve also contains preganglionic autonomic fibers that synapse in the **ciliary ganglion**. The ganglionic neurons exiting the ciliary ganglion innervate intrinsic eye muscles. These muscles change the diameter of the pupil, adjusting the amount of light entering the eye, and change the shape of the lens to focus images on the retina.

The Trochlear Nerves (IV)

Primary function: Motor, eye movements

Origin: Mesencephalon

Pass through: Superior orbital fissure of sphenoid ⮌ p. 148

Destination: Superior oblique ⮌ p. 265

The **trochlear** (TRŌK-lē-ar; *trochlea*, pulley) **nerves (IV)** are the smallest cranial nerves. They innervate the superior obliques of the eyes **(Figure 16.25)**. The motor nuclei are found in the ventrolateral portions of the mesencephalon. The motor fibers emerge from the surface of the tectum and enter the orbit through the superior orbital fissure.

The Trigeminal Nerves (V)

Primary function: Mixed (sensory and motor); ophthalmic and maxillary divisions: sensory; mandibular division: mixed

Origin: Ophthalmic division (sensory): Orbital structures, nasal cavity, skin of forehead, superior eyelid, eyebrow, and part of the nose

Maxillary division (sensory): Inferior eyelid, upper lip, gums, and teeth; cheek; nose, palate, and part of the pharynx

Mandibular division (mixed): Sensory from lower gums, teeth, and lips; palate and tongue (part); motor from motor nuclei of pons.

Pass through: Ophthalmic division: superior orbital fissure; maxillary division: foramen rotundum; mandibular division: foramen ovale ⮌ p. 148

Destination: Ophthalmic, maxillary, and mandibular nerves: Sensory nuclei in the pons; mandibular nerve also innervates muscles of mastication (chewing)

The pons contains the nuclei of three cranial nerves (N V, N VI, and N VII) and contributes to the control of a fourth (N VIII). The **trigeminal** (trī-JEM-i-nal) (trigeminus, three fold) **nerves (V)** are the largest cranial nerves (**Figure 16.26**). These mixed nerves provide sensory information from the head and face and motor control to the muscles of mastication. Sensory and motor roots originate on the lateral surface of the pons. The sensory branch is larger, and the **semilunar ganglion** (*trigeminal ganglion*) contains the cell bodies of the sensory neurons. As its name implies, the trigeminal nerve has three major divisions; the small motor root contributes to only one of the three.

Branch 1. The **ophthalmic division** of the trigeminal nerve carries only afferent, sensory information. This nerve carries sensory information from orbital structures, the nasal cavity and sinuses, and the skin of the forehead, eyebrows, eyelids, and nose. It leaves the cranium through the superior orbital fissure, then branches within the orbit.

Branch 2. The **maxillary division** of the trigeminal nerve also carries only afferent, sensory information. It carries sensory information from the lower eyelid, upper lip, cheek, and nose and from deeper sensory structures of the upper gums and teeth, the palate, and portions of the pharynx. The maxillary division leaves the cranium at the foramen rotundum and enters the floor of the orbit through the inferior orbital fissure. A major branch of the maxillary, the infra-orbital nerve, passes through the infra-orbital foramen to supply adjacent portions of the face.

Branch 3. The **mandibular division** is the largest division of the trigeminal nerve, and it contains efferent motor fibers and afferent sensory fibers. This branch exits the cranium through the foramen ovale. The motor fibers of the mandibular division innervate the muscles of mastication. The sensory fibers carry proprioceptive information from those muscles and also carry sensory information from (1) the skin of the temples, (2) the lateral surfaces, gums, and teeth of the mandible, (3) the salivary glands, and (4) the anterior portions of the tongue.

Fibers of the trigeminal nerve travel to the ciliary, pterygopalatine, submandibular, and otic ganglia. These are autonomic ganglia whose neurons innervate structures of the face. *The trigeminal nerve does not contain visceral motor fibers, and all of its fibers pass through these ganglia without synapsing.* However, branches of other cranial nerves, such as the facial nerve, are intermingled with those of the trigeminal nerve. These fibers synapse within these ganglia. The postganglionic autonomic fibers then travel with the trigeminal nerve to peripheral structures.

16

Figure 16.26 The Trigeminal Nerve

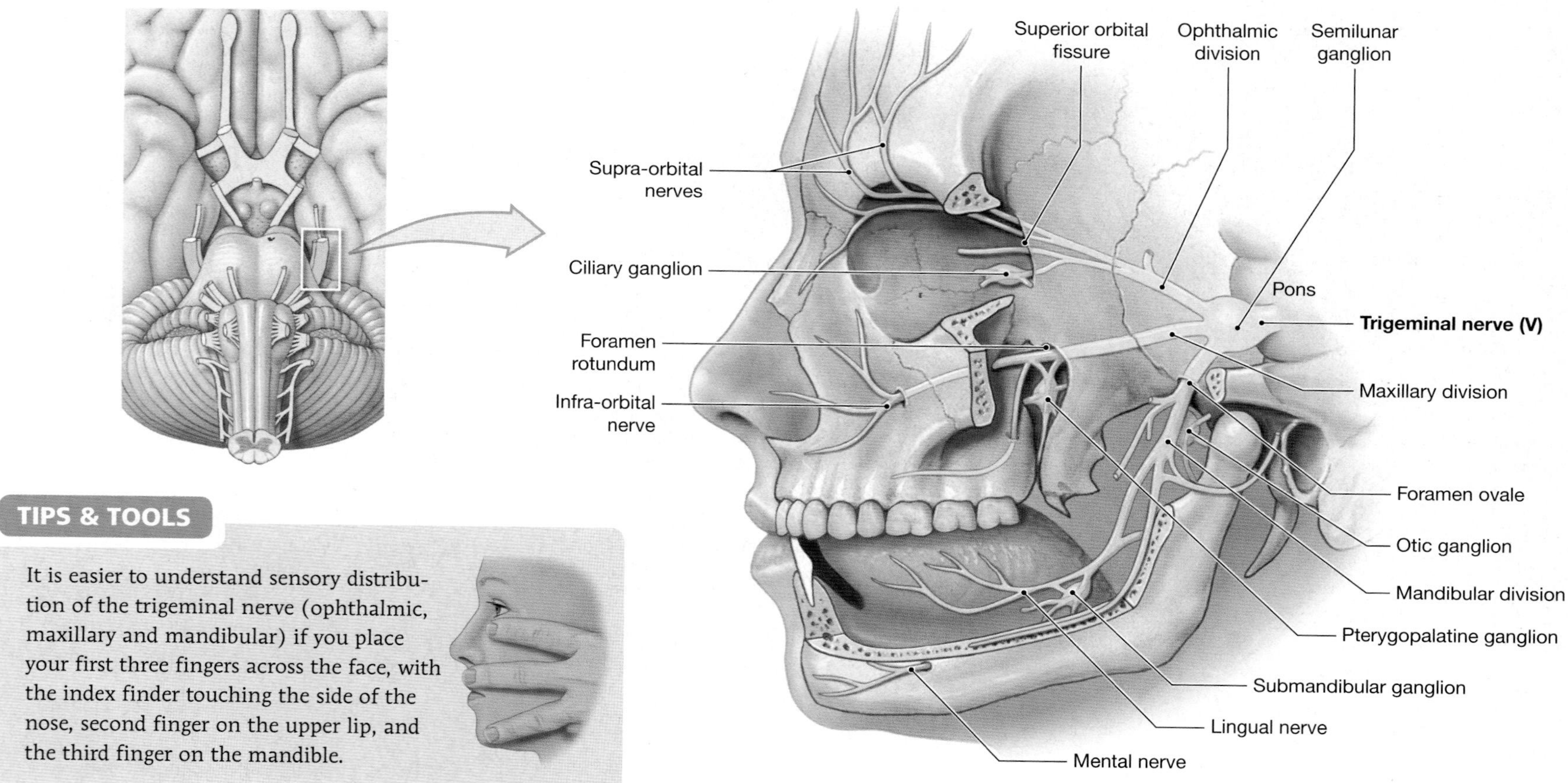

TIPS & TOOLS

It is easier to understand sensory distribution of the trigeminal nerve (ophthalmic, maxillary and mandibular) if you place your first three fingers across the face, with the index finder touching the side of the nose, second finger on the upper lip, and the third finger on the mandible.

CLINICAL NOTE

Tic Douloureux

Tic douloureux, or *trigeminal neuralgia*, is characterized by episodes of severe facial pain, often accompanied by a facial spasm or tic. The pain distribution is unilateral and follows the sensory distribution of the maxillary and/or mandibular divisions of the trigeminal nerve (V) (see **Figure 16.26**). The pain, lasting from a few seconds to a few minutes, is sudden and intense. Even a light touch to the face or mouth can trigger an episode. The pain is so severe that people become afraid to talk, eat, or move during attacks.

Tic douloureux generally appears in middle age or later and affects women more than men. Often it is caused by pressure on the trigeminal nerve (V) by a blood vessel or tumor, or injury to the nerve from surgery or trauma. People with multiple sclerosis are affected more frequently and at an earlier age. Treatment includes medications and surgical procedures to relieve the pressure or disruption of the affected branches of the nerve.

The Abducens Nerves (VI)

Primary function: Motor, eye movements

Origin: Pons

Pass through: Superior orbital fissure of sphenoid p. 148

Destination: Lateral rectus p. 265

The **abducens** (ab-DŪ-senz) **nerves (VI)** innervate the lateral rectus, the sixth of the extrinsic eye muscles. This muscle moves the eyeball laterally. The nerves emerge from the inferior surface of the brain at the junction of the pons and the medulla oblongata. They reach the orbit through the superior orbital fissure in company with the oculomotor and trochlear nerves **(Figure 16.25)**.

The Facial Nerves (VII)

Primary function: Mixed (sensory and motor)

Origin: Sensory: taste receptors on anterior two-thirds of tongue; motor: motor nuclei of pons

Pass through: Internal acoustic meatus of temporal bone, along facial canal to reach stylomastoid foramen p. 137

16

Figure 16.27 The Facial Nerve

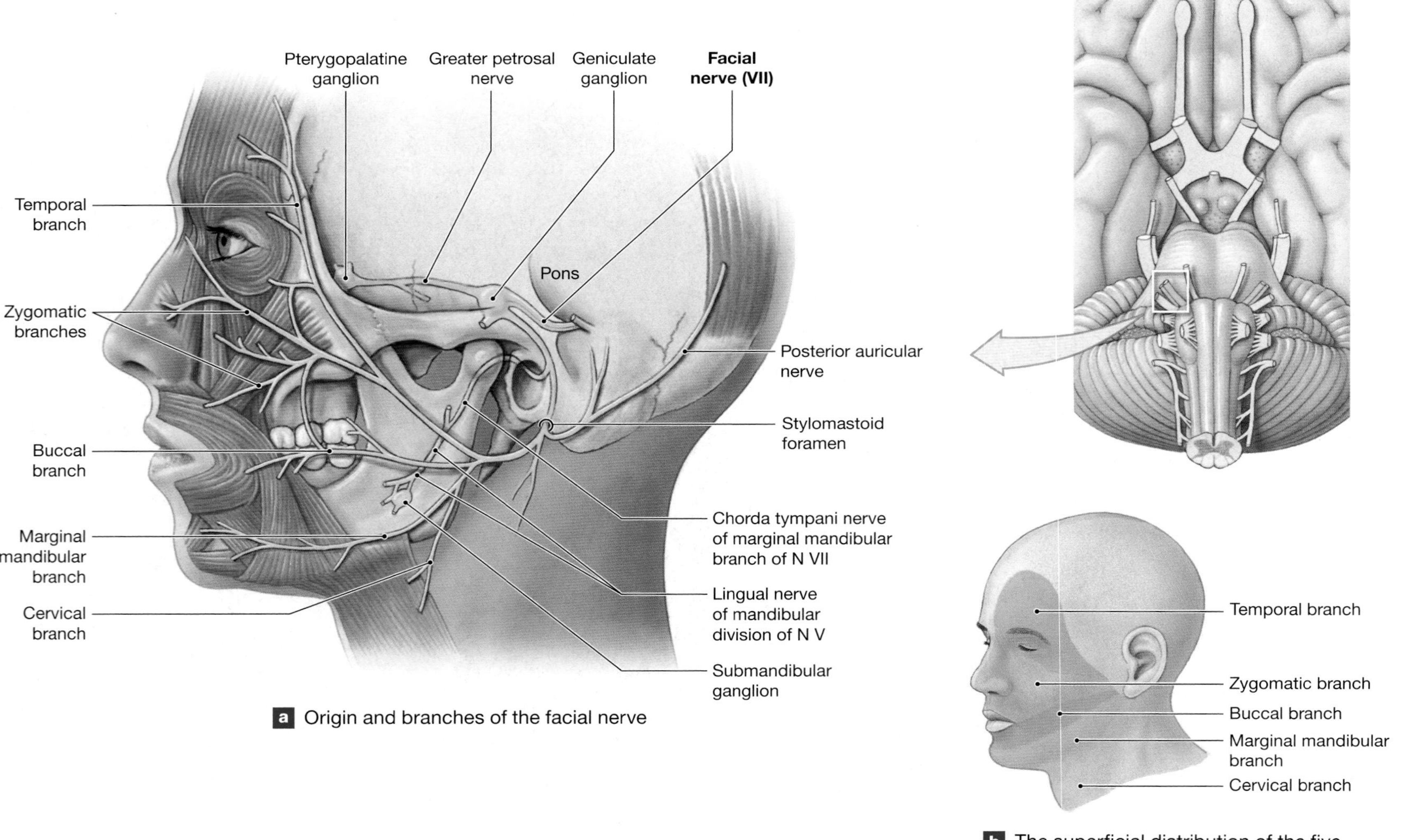

a Origin and branches of the facial nerve

b The superficial distribution of the five major branches of the facial nerve

Destination: Sensory nuclei of pons

Somatic motor: muscles of facial expression

Visceral motor: lacrimal (tear) gland and nasal mucous glands via pterygopalatine ganglion; submandibular and sublingual glands via submandibular ganglion

The **facial nerves (VII)** are mixed nerves. The cell bodies of the sensory neurons are located in the **geniculate ganglion**, and the motor nuclei are in the pons. The sensory and motor roots combine and form large nerves that pass through the internal acoustic meatus of each temporal bone **(Figure 16.27)**. The nerves then pass through the facial canal to reach the face through the stylomastoid foramen. ↩ p. 137 The sensory neurons monitor proprioceptors in the facial muscles, provide deep pressure sensations over the face, and receive taste information from receptors along the anterior two-thirds of the tongue. Somatic motor fibers innervate the superficial muscles of the scalp and face and deep muscles near the ear.

The facial nerves carry preganglionic autonomic fibers to the pterygopalatine and submandibular ganglia:

- Pterygopalatine ganglion: The preganglionic fibers within the greater petrosal nerve synapse in the pterygopalatine ganglion. Postganglionic fibers from this ganglion innervate the lacrimal gland and small glands of the nasal cavity and pharynx.
- Submandibular ganglion: To reach the submandibular ganglion, autonomic fibers leave the facial nerve and travel along the mandibular division of the trigeminal nerve. Postganglionic fibers from this ganglion innervate the submandibular and sublingual (*sub–*, under, + *lingua*, tongue) glands.

The Vestibulocochlear Nerves (VIII)

Primary function: Special sensory: balance and equilibrium (vestibular division) and hearing (cochlear division)

Origin: Receptors of the internal ear (vestibule and cochlea)

CLINICAL NOTE

Bell's Palsy

Bell's Palsy results from an inflammation of the facial nerve that is probably related to viral infection. Involvement of the facial nerve (VII) can be deduced from symptoms of paralysis of facial muscles on the affected side and loss of taste sensations from the anterior two-thirds of the tongue. The individual does not show prominent sensory deficits, and the condition is usually painless. In most cases, Bell's palsy "cures itself" after a few weeks or months, but this process can be accelerated by early treatment with corticosteroids and antiviral drugs.

Pass through: Internal acoustic meatus of the temporal bone ↩ pp. 138, 139

Destination: Vestibular and cochlear nuclei of pons and medulla oblongata

The **vestibulocochlear nerves (VIII)** exit the brain lateral to the origin of the facial nerves **(Figure 16.28)**. These nerves enter each internal acoustic meatus with the facial nerves. There are two separate bundles of sensory fibers within each vestibulocochlear nerve. (1) The **vestibular nerve** is the larger of the two bundles. It originates at the receptors of the vestibule, the portion of the inner ear concerned with balance. The cell bodies of the sensory neurons are located within an adjacent sensory ganglion, and their axons travel to the **vestibular nuclei** of the medulla oblongata. These afferents convey information concerning position, movement, and balance. (2) The **cochlear** (KŌK-lē-ar; cochlea, snail shell) **nerve** monitors hearing receptors in the cochlea of the internal ear. The cell bodies of these sensory neurons are located within a peripheral ganglion, and their axons synapse within the **cochlear nuclei** of the medulla oblongata. Axons leaving the vestibular and cochlear nuclei relay the sensory information to other centers or initiate reflexive motor responses. (Chapter 18 will discuss balance and the sense of hearing.)

Figure 16.28 The Vestibulocochlear Nerve

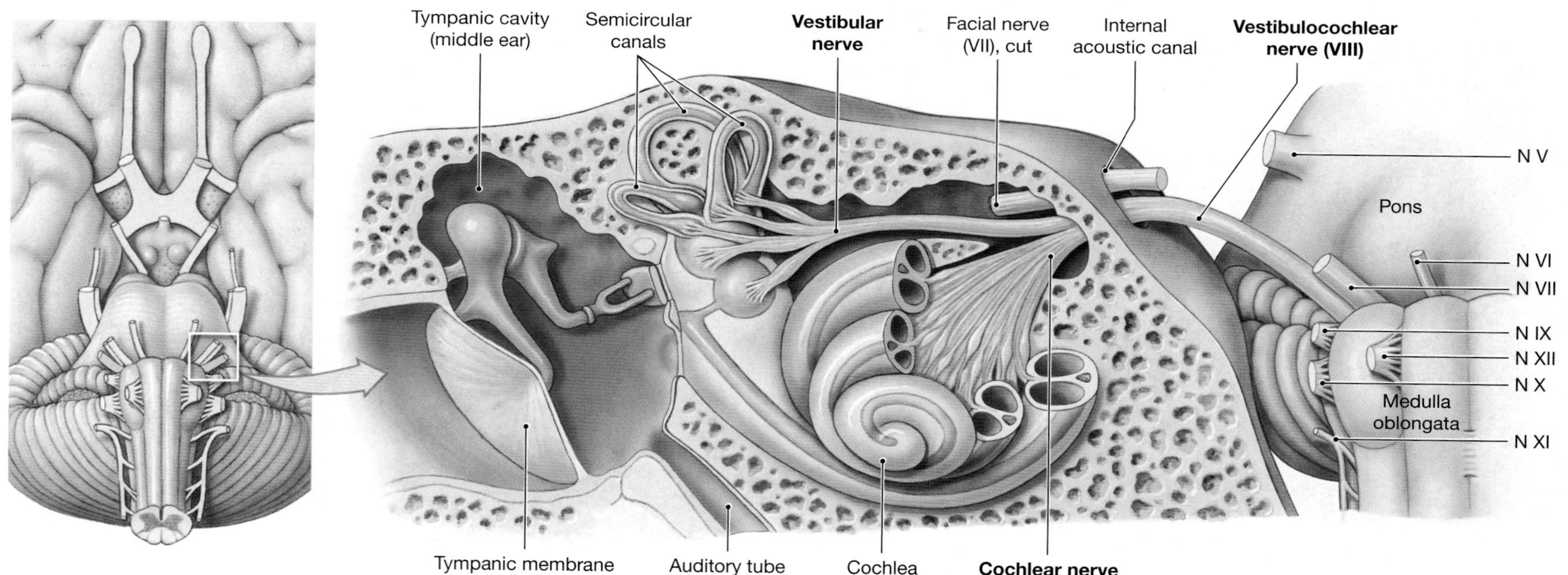

The Glossopharyngeal Nerves (IX)

Primary function: Mixed (sensory and motor)

Origin: Sensory: Posterior one-third of the tongue, part of the pharynx and palate, the carotid arteries of the neck; motor: motor nuclei of medulla oblongata

Pass through: Jugular foramen between occipital and temporal bones ⮌ pp. 137–138

Destination: Sensory fibers: Sensory nuclei of medulla oblongata

Somatic motor: Pharyngeal muscles involved in swallowing

Visceral motor: Parotid gland, after synapsing in the otic ganglion

The **glossopharyngeal** (glos-ō-fah-RIN-jē-al; *glossum*, tongue) **nerves (IX)** innervate the tongue and pharynx. Each glossopharyngeal nerve passes through the cranium by the jugular foramen along with N X and N XI **(Figure 16.29)**.

Each glossopharyngeal nerve is a mixed nerve, but sensory fibers are most abundant. The sensory neurons are in the **superior** (*jugular*) **ganglion** and the **inferior** (*petrosal*) **ganglion**. The afferent fibers carry general sensory information from the lining of the pharynx and the soft palate to a nucleus in the medulla oblongata. These nerves also provide taste sensations from the posterior third of the tongue. Additionally, they have special receptors monitoring the blood pressure and dissolved-gas concentrations within major blood vessels.

The somatic motor fibers innervate the pharyngeal muscles involved in swallowing. Visceral motor fibers synapse in the otic ganglion, and postganglionic fibers innervate the parotid gland of the cheek.

The Vagus Nerves (X)

Primary function: Mixed (sensory and motor)

Origin: Visceral sensory: Pharynx (part), auricle, external acoustic meatus, diaphragm, and visceral organs in thoracic and abdominopelvic cavities
Visceral motor: Motor nuclei in the medulla oblongata

Pass through: Jugular foramen between occipital and temporal bones ⮌ pp. 137–138

Destination: Sensory fibers: Sensory nuclei and autonomic centers of medulla oblongata

Somatic motor: Muscles of the palate and pharynx

Visceral motor: Respiratory, cardiovascular, and digestive organs in the thoracic and abdominal cavities

The **vagus** (VĀ-gus) **nerves (X)** arise immediately inferior to the glossopharyngeal nerves. As the name suggests (*vagus*, wanderer), the vagus nerves branch extensively. **Figure 16.30** shows only the general pattern of distribution.

Sensory neurons are located within the **superior** (*jugular*) **ganglion** and the **inferior** (*nodose*; NŌ-dōs) **ganglion**. The vagus nerve provides somatic sensory information from the ear and the diaphragm and special sensory information from pharyngeal taste receptors. The majority of the vagal afferents provide visceral sensory information from receptors of the esophagus, respiratory tract, and abdominal viscera. Vagal afferents are vital to the autonomic control of visceral function, but because the information often fails to reach the cerebral cortex, we are seldom aware of the sensations they provide.

Figure 16.29 The Glossopharyngeal Nerve

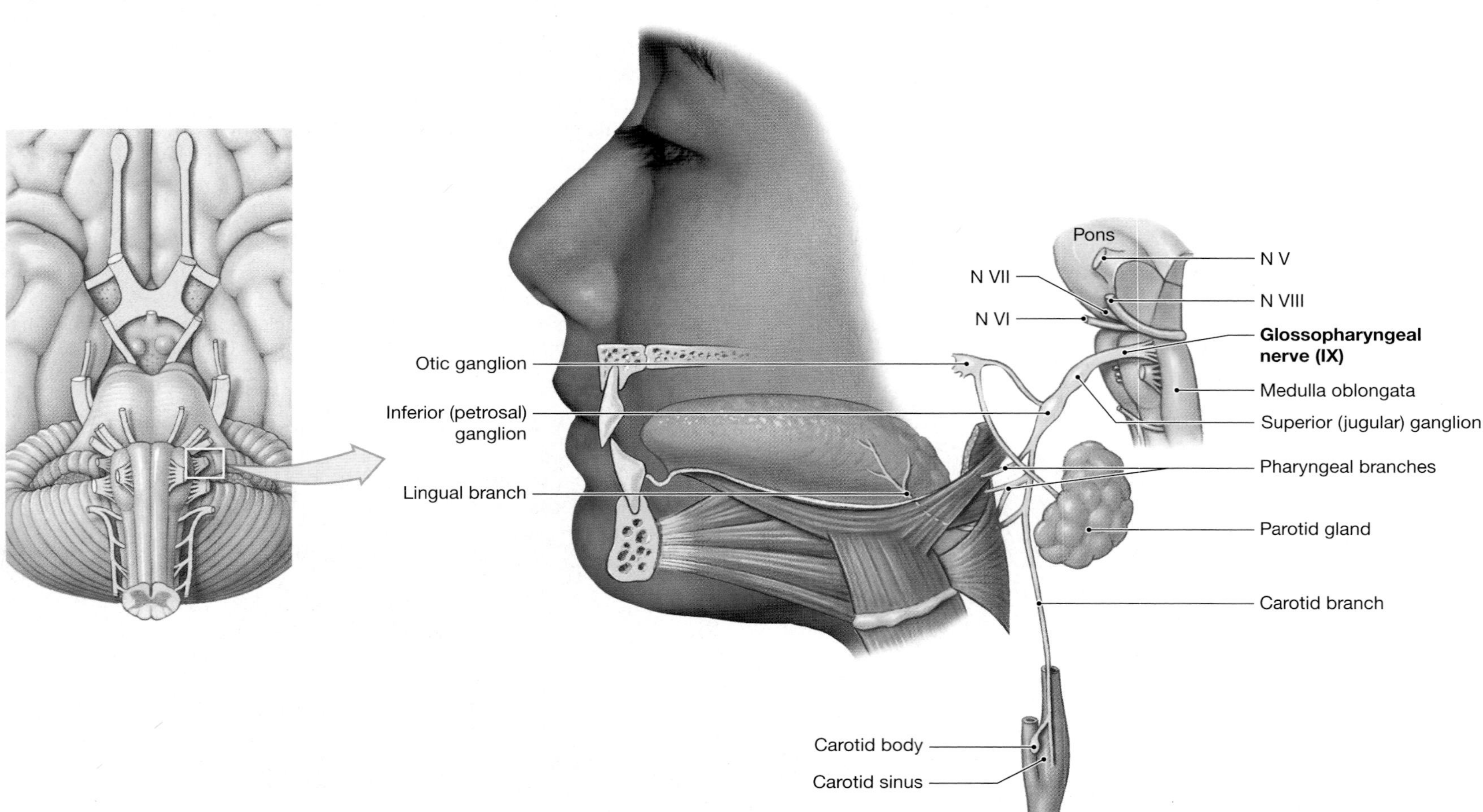

Figure 16.30 The Vagus Nerve

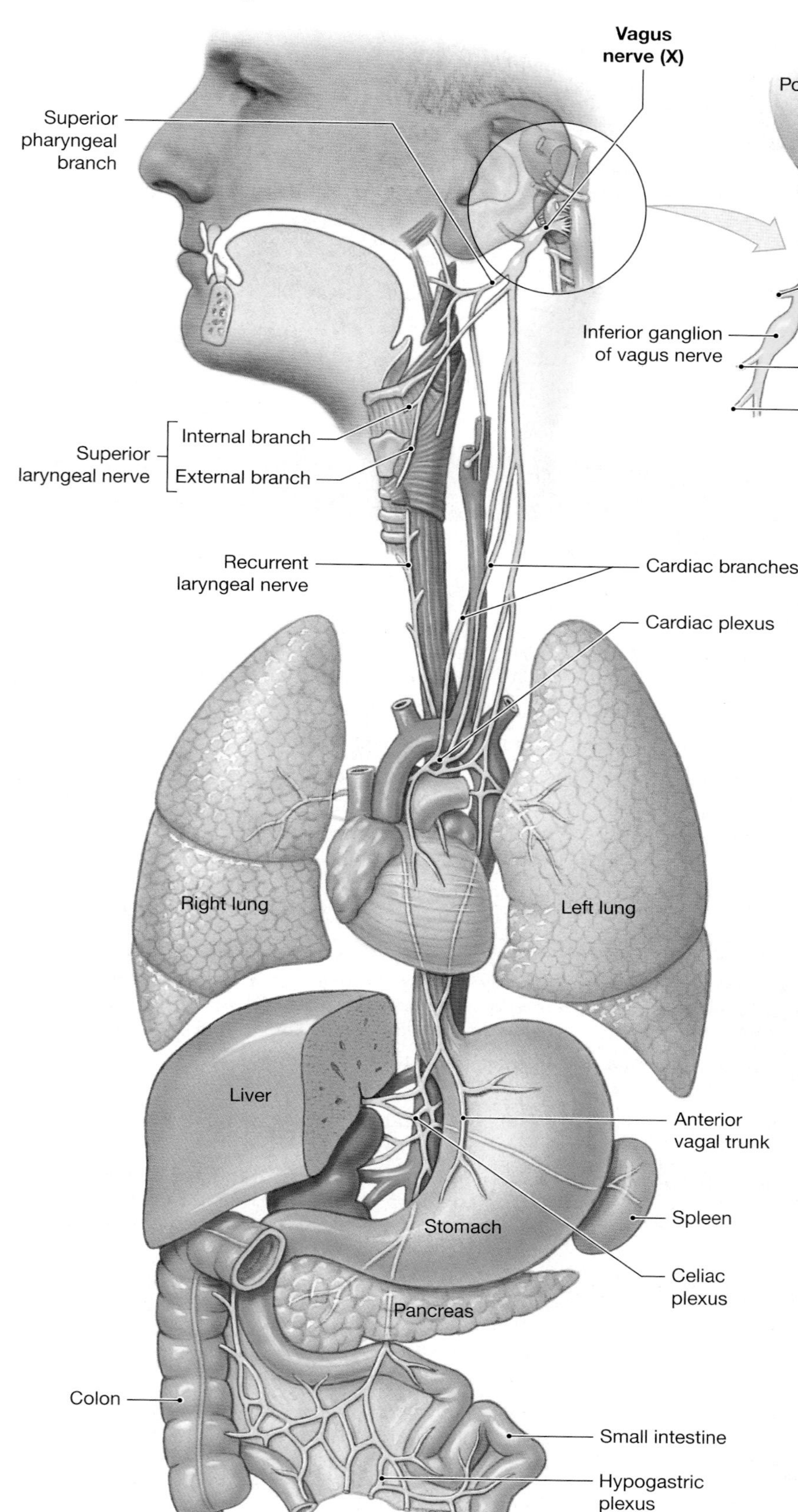

The motor components of the vagus nerve are equally diverse. The vagus nerve carries preganglionic autonomic (parasympathetic) fibers that affect the heart and control smooth muscles and glands within the areas monitored by its sensory fibers, including the respiratory tract, stomach, intestines, and gallbladder.

The Accessory Nerves (XI)

Primary function: Motor

Origin: Motor nuclei of spinal cord and medulla oblongata

Pass through: Jugular foramen between occipital and temporal bones pp. 137–138

Destination: Internal branch innervates voluntary muscles of palate, pharynx, and larynx; external branch controls sternocleidomastoid and trapezius muscles

The **accessory nerves (XI)** differ from other cranial nerves because some of the motor fibers originate in the lateral portions of the anterior horns of the first five cervical segments of the spinal cord (**Figure 16.31**). These fibers from each accessory nerve form the *spinal root of the accessory nerve*. The spinal root enters the cranium through the foramen magnum. Once within the skull, the spinal root unites with motor fibers of the *cranial root*. The cranial root originates at a nucleus in the medulla oblongata and leaves the cranium through the jugular foramen. Each accessory nerve consists of two branches:

1. The **internal branch** joins the vagus nerve and innervates the voluntary swallowing muscles of the soft palate and pharynx and the intrinsic muscles that control the vocal cords.
2. The **external branch** innervates the sternocleidomastoid and trapezius muscles of the neck and back. The motor fibers of this branch originate in the anterior horns of C_1 to C_5.

The Hypoglossal Nerves (XII)

Primary function: Motor, tongue movements

Origin: Motor nuclei of the medulla oblongata

Pass through: Hypoglossal canal of occipital bone ⊃ pp. 137–138

Destination: Muscles of the tongue

The **hypoglossal** (hī-pō-GLOS-al) **nerves (XII)** leave the cranium through the hypoglossal canals of the occipital bone. They then curve inferiorly, anteriorly, and superiorly to innervate the skeletal muscles of the tongue **(Figure 16.31)**. These nerves provides voluntary motor control over movements of the tongue.

Table 16.12 summarizes the basic distribution and function of each cranial nerve.

TIPS & TOOLS

The following mnemonic will help you remember the cranial nerves: "Oh, Once One Takes The Anatomy Final, Very Good Vacations Are Heavenly." **O**lfactory, **O**ptic, **O**culomotor, **T**rochlear, **T**rigeminal, **A**bducens, **F**acial, **V**estibulocochlear, **G**lossopharyngeal, **V**agus, **A**ccessory, **H**ypoglossal.

Table 16.12 | The Cranial Nerves

Cranial Nerve (#)	Sensory Ganglion	Branches and Divisions	Primary Function	Foramen	Innervation
Olfactory (I)			Special sensory	Cribriform plate	Olfactory epithelium
Optic (II)			Special sensory	Optic canal	Retina of eye
Oculomotor (III)			Motor	Superior orbital fissure	Inferior, medial, superior rectus, inferior oblique, and levator palpebrae; intrinsic muscles of eye
Trochlear (IV)			Motor	Superior orbital fissure	Superior oblique
Trigeminal (V)	Semilunar		Mixed		Areas associated with the jaws
		Ophthalmic	Sensory	Superior orbital fissure	Orbital structures, nasal cavity, skin of forehead, upper eyelid, eyebrows, nose (part)
		Maxillary	Sensory	Foramen rotundum	Lower eyelid; upper lip, gums, and teeth; cheek, nose (part), palate, and pharynx (part)
		Mandibular	Mixed	Foramen ovale	*Sensory* from lower gums, teeth, lips; palate (part) and tongue (part). *Motor* to muscles of mastication
Abducens (VI)			Motor	Superior orbital fissure	Lateral rectus
Facial (VII)	Geniculate		Mixed	Internal acoustic meatus to facial canal; exits at stylomastoid foramen	*Sensory* from taste receptors on anterior two-thirds of tongue; *motor* to muscles of facial expression, lacrimal gland, submandibular gland, sublingual glands
Vestibulocochlear (Acoustic) (VIII)		Cochlear Nerve	Special sensory	Internal acoustic meatus	Cochlea (receptors for hearing)
		Vestibular Nerve	Special sensory	As above	Vestibule (receptors for motion and balance)
Glossopharyngeal (IX)	Superior (jugular) and inferior (petrosal)		Mixed	Jugular foramen	*Sensory* from posterior third of tongue; pharynx and palate (part); carotid body (monitors blood pressure, pH, and levels of respiratory gases). *Motor* to pharyngeal muscles, parotid gland
Vagus (X)	Superior (jugular) and inferior (nodose)		Mixed	Jugular foramen	*Sensory* from pharynx; auricle and external acoustic meatus; diaphragm; visceral organs in thoracic and abdominopelvic cavities. *Motor* to palatal and pharyngeal muscles, and visceral organs in thoracic and abdominopelvic cavities
Accessory (XI)		Internal branch	Motor	Jugular foramen	Skeletal muscles of palate, pharynx, and larynx (with branches of the vagus nerve)
		External branch	Motor	Jugular foramen	Sternocleidomastoid and trapezius
Hypoglossal (XII)			Motor	Hypoglossal canal	Tongue musculature

CLINICAL NOTE

Dementia and Alzheimer's Disease

In dementia, damage to brain cells interferes with their ability to communicate with each other. This can cause difficulties with memory, judgment, behavior, thinking, emotions, and movement. There are many different types of dementia.

By far the most common dementia is Alzheimer's disease, which accounts for about 70 percent of cases. This is a chronic, progressive illness characterized by memory loss and impairment of higher-order cerebral functions. Symptoms may appear at age 50–60 or later, although there is an inherited form that becomes symptomatic earlier. Alzheimer's disease affects about one-third of people over age 85. Microscopic examination of patients' brain tissue reveals amyloid plaques (fragments of beta-amyloid peptide mixed with additional proteins and pieces of nerve cells). There is no cure for Alzheimer's disease, but some medications can slow its progression temporarily.

Vascular dementia, or "mini-stroke" dementia, accounts for about 10 percent of dementia cases. Smokers and people with high blood pressure or high cholesterol are particularly prone to this disease. Symptoms of impaired judgment or inability to make decisions, plan, and organize are more prominent than memory loss. Brain changes are caused by areas of cell death due to vascular impairment.

Lewy body dementia often presents with sleep disturbances, visual hallucinations, muscle rigidity, memory loss, and problems with thinking. The course of the disease is more rapid than Alzheimer's disease, and it can be difficult to diagnose because symptoms are similar to Alzheimer's disease. People with Lewy body dementia can have short periods of clarity even during end-stage disease. Microscopic examination shows abnormal clumps of the protein alpha-synuclein destroying neurons.

Frontotemporal dementia affects neurons in the frontal and temporal lobes. It typically involves changes in personality and behavior and difficulties with language.

It is important to medically investigate dementia. Some treatable conditions can cause dementia, including depression, medication side effects, thyroid problems, vitamin deficiencies, and excessive use of alcohol.

16

Figure 16.31 The Accessory and Hypoglossal Nerves

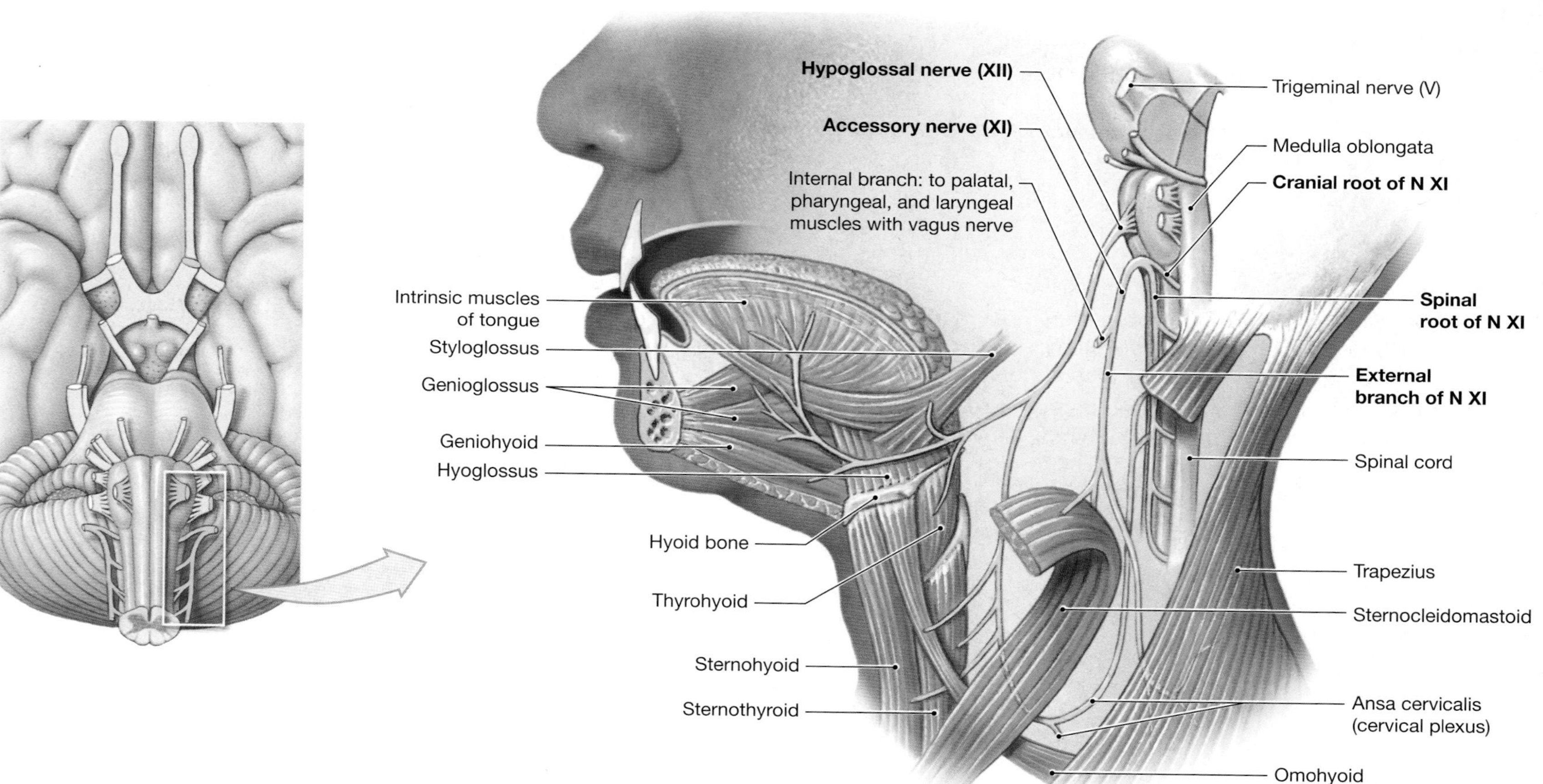

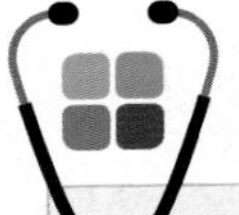

CLINICAL NOTE

Cranial Reflexes

Cranial reflexes are reflex arcs that involve the sensory and motor fibers of cranial nerves. Examples of cranial reflexes are discussed in later chapters, and this section will simply provide an overview and general introduction.

Table 16.13 lists representative examples of cranial reflexes and their functions. These reflexes are clinically important because they provide a quick and easy method for observing the condition of cranial nerves and specific nuclei and tracts in the brain.

Cranial somatic reflexes are seldom more complex than the somatic reflexes of the spinal cord. Table 16.13 includes four somatic reflexes: *the corneal reflex*, the *tympanic reflex*, the *auditory reflex*, and the *vestibulo-ocular reflex*. These reflexes are often used to check for damage to the cranial nerves or processing centers involved. The brain stem contains many reflex centers that control visceral motor activity. Many of these reflex centers are in the medulla oblongata, and they can direct very complex visceral motor responses to stimuli. These visceral reflexes are essential to the control of respiratory, digestive, and cardiovascular functions.

16

Table 16.13 | Cranial Reflexes

Reflex	Stimulus	Afferents	Central Synapse	Efferents	Response
SOMATIC REFLEXES					
Corneal reflex	Contact with corneal surface	N V	Motor nuclei for N VII	N VII	Blinking of eyelids
Tympanic reflex	Loud noise	N VIII	Inferior colliculi (midbrain)	N VII	Reduced movement of auditory ossicles
Auditory reflexes	Loud noise	N VIII	Motor nuclei of brainstem and spinal cord	N III, N IV, N VI, N VII, N X, cervical nerves	Eye and/or head movements triggered by sudden sounds
Vestibulo-ocular reflexes	Rotation of head	N VIII	Motor nuclei controlling extra-ocular muscles	N III, N IV, N VI	Opposite movement of eyes to stabilize field of vision
VISCERAL REFLEXES					
Pupillary reflex	Light striking photoreceptors in one eye	N II	Superior colliculi	N III	Constriction of ipsilateral pupil
Consensual light reflex	Light striking photoreceptors in one eye	N II	Superior colliculi	N III	Constriction of both pupils

16.9 CONCEPT CHECK

14 John is experiencing problems in moving his tongue. His doctor tells him the problems are due to pressure on a cranial nerve. Which cranial nerve is involved?

15 Bruce has lost the ability to detect tastes on the tip of his tongue. What cranial nerve is involved?

See the blue Answers tab at the back of the book.

EMBRYOLOGY SUMMARY

For a summary of the development of the brain and cranial nerves, see Chapter 28 (Embryology and Human Development).

Study Outline

Introduction p. 404

- The brain is far more complex than the spinal cord; its complexity makes it adaptable but slower in response than spinal reflexes.

16.1 An Introduction to the Organization of the Brain p. 404

Embryology of the Brain p. 404

- The brain forms from three swellings at the superior tip of the developing neural tube: the **prosencephalon**, **mesencephalon**, and **rhombencephalon**. *(See Table 16.1 and Embryology Summary in Chapter 28.)*

Major Regions and Landmarks p. 404

- There are six regions in the adult brain: cerebrum, diencephalon, mesencephalon, pons, cerebellum, and medulla oblongata. *(See Figure 16.1.)*
- Conscious thought, intellectual functions, memory, and complex motor patterns originate in the **cerebrum**. *(See Figure 16.1.)*
- The roof of the **diencephalon** is the **epithalamus**; the walls are the **thalami**, which contain relay and processing centers for sensory data. The floor is the **hypothalamus**, which contains centers involved with emotions, autonomic function, and hormone production. *(See Figure 16.1.)*
- The **mesencephalon** (*midbrain*) processes visual and auditory information and generates involuntary somatic motor responses. *(See Figure 16.1.)*
- The **pons** connects the cerebellum to the brainstem and is involved with somatic and visceral motor control. *(See Figure 16.1.)*
- The **cerebellum** adjusts voluntary and involuntary motor activities on the basis of sensory data and stored memories. *(See Figure 16.1.)*
- The spinal cord connects to the brain at the **medulla oblongata**, which relays sensory information and regulates autonomic functions. *(See Figure 16.1.)*
- The brain contains extensive areas of neural **cortex**, a layer of gray matter on the surfaces of the cerebrum and cerebellum that covers underlying white matter.
- The central passageway of the brain expands to form chambers called **ventricles**. Cerebrospinal fluid (CSF) continually circulates from the ventricles and central canal of the spinal cord into the subarachnoid space of the meninges that surround the CNS. *(See Figure 16.2.)*

16.2 Protection and Support of the Brain p. 406

The Cranial Meninges p. 406

- The **cranial meninges**—the **dura mater**, **arachnoid mater**, and **pia mater**—are continuous with the spinal meninges that surround the spinal cord. However, they have anatomical and functional differences. *(See Figures 14.2c,d and 16.3–16.5.)*
- Folds of dura mater stabilize the position of the brain within the cranium and include the **falx cerebri**, **tentorium cerebelli**, **falx cerebelli**, and **diaphragma sellae**. *(See Figures 16.3–16.5.)*

The Blood Brain Barrier p. 409

- The **blood brain barrier (BBB)** isolates nervous tissue from the general circulation.
- The blood–brain barrier remains intact throughout the CNS except in portions of the hypothalamus, in the pineal gland, and at the choroid plexus in the membranous roof of the diencephalon and medulla.

Cerebrospinal Fluid p. 410

- **Cerebrospinal fluid (CSF)** (1) cushions delicate neural structures, (2) supports the brain, and (3) transports nutrients, chemical messengers, and wastes.
- The **choroid plexus** is the site of cerebrospinal fluid production. *(See Figure 16.6.)*
- Cerebrospinal fluid reaches the subarachnoid space via the **lateral apertures** and a **median aperture**. Diffusion across the **arachnoid granulations** into the **superior sagittal sinus** returns CSF to the venous circulation. *(See Figures 14.2b–d and 16.4–16.7.)*

The Blood Supply to the Brain p. 412

- Arterial blood reaches the brain through the internal carotid arteries and the vertebral arteries. Venous blood leaves primarily in the internal jugular veins.

16

16.3 The Medulla Oblongata p. 413

- The medulla oblongata connects the brain to the spinal cord. It contains the **gracile nucleus** and the **cuneate nucleus**, which are processing centers, and the **olivary nuclei**, which relay information from the spinal cord, cerebral cortex, and brainstem to the cerebellar cortex. Its **reflex centers**, including the **cardiovascular centers** and the **respiratory rhythmicity centers**, control or adjust the activities of peripheral systems. *(See Figures 16.1, 16.8, 16.9, 16.13, 16.14, and 16.7 and Table 16.2.)*

16.4 The Pons p. 414

- The pons contains (1) sensory and motor nuclei for four cranial nerves, (2) nuclei concerned with involuntary control of respiration, (3) nuclei that process and relay cerebellar commands arriving over the middle cerebellar peduncles, and (4) ascending, descending, and transverse tracts. *(See Figures 16.1, 16.9, 16.13, and 16.14 and Table 16.3.)*

16.5 The Mesencephalon (Midbrain) p. 415

- The **tectum** (roof) of the mesencephalon contains two pairs of nuclei, the **corpora quadrigemina**. On each side, the **superior colliculus** receives visual inputs from the thalamus, and the **inferior colliculus** receives auditory data from the medulla oblongata. The **red nucleus** integrates information from the cerebrum and issues involuntary motor commands related to muscle tone and limb position. The **substantia nigra** regulates the motor output of the basal nuclei. The **cerebral peduncles** contain ascending fibers headed for thalamic nuclei and descending fibers of the corticospinal pathway that carry voluntary motor commands from the primary motor cortex of each cerebral hemisphere. *(See Figures 12.8, 16.1, 16.10, 16.13, and 16.14 and Table 16.4.)*

16.6 The Diencephalon p. 417

- The diencephalon provides the switching and relay centers necessary to integrate the sensory and motor pathways. *(See Figures 16.1, 16.13, 16.14, 16.21c, and 16.22.)*

The Epithalamus p. 417

- The epithalamus forms the roof of the diencephalon. It contains the hormone-secreting **pineal gland**. *(See Figures 16.12a and 16.13a.)*

The Thalamus p. 417

- The thalamus is the principal and final relay point for ascending sensory information and coordinates voluntary and involuntary somatic motor activities. *(See Figures 16.11–16.13, 16.21, and 16.22 and Table 16.5.)*

The Hypothalamus p. 418

- The hypothalamus contains important control and integrative centers. It can (1) control involuntary somatic motor activities, (2) control autonomic function, (3) coordinate activities of the nervous and endocrine systems, (4) secrete hormones, (5) produce emotions and behavioral drives, (6) coordinate voluntary and autonomic functions, (7) regulate body temperature, and (8) control circadian cycles of activity. *(See Figures 12.8, 16.12, and 16.13a and Table 16.6.)*

16.7 The Cerebellum p. 422

16

- The cerebellum oversees the body's postural muscles and programs and tunes voluntary and involuntary movements. The **cerebellar hemispheres** consist of neural cortex formed into folds, or **folia**. The surface can be divided into the **anterior** and **posterior lobes**, the **vermis**, and the **flocculonodular lobes**. *(See Figures 16.15–16.17 and Table 16.7.)*

16.8 The Cerebrum p. 422

The Cerebral Hemispheres p. 424

- The cortical surface contains **gyri** (elevated ridges) separated by **sulci** (shallow depressions) or deeper grooves (**fissures**). The **longitudinal fissure** separates the two cerebral hemispheres. The **central sulcus** marks the boundary between the **frontal lobe** and the **parietal lobe**. Other sulci form the boundaries of the **temporal lobe** and the **occipital lobe**. *(See Figures 16.1, 16.16, and 16.17.)*

- Each cerebral hemisphere receives sensory information from and generates motor commands to the opposite side of the body. There are significant functional differences between the two; thus, the assignment of a specific function to a specific region of the cerebral cortex is imprecise.

- The **primary motor cortex** of the **precentral gyrus** directs voluntary movements. The **primary somatosensory cortex** of the **postcentral gyrus** receives somatic sensory information from touch, pressure, pain, taste, and temperature receptors. *(See Figure 16.17a and Table 16.8.)*

- **Association areas**, such as the **visual association area** and **premotor cortex**, control our ability to understand sensory information. *(See Figure 16.17a and Table 16.8.)*

Higher-Order Functions p. 425

- Higher-order functions have four characteristics: (1) They are performed by the cerebral cortex; (2) they involve complex interconnections and communication between areas of the cerebral cortex as well as other areas of the brain; (3) they involve both conscious and unconscious information processing; and (4) they are not part of the programmed wiring of the brain. *(See Figures 16.17a, 16.18, and 16.19.)*

- "Higher-order" **integrative centers** receive information from many different association areas and direct complex motor activities and analytical functions.

- The portion of the cerebral cortex that receives all information from the sensory association areas is termed **Wernicke's area**. This is present in only one hemisphere, typically the left. *(See Figure 16.18.)*

- Efferents from the general interpretive area target the **speech center**. This is a motor area that regulates the patterns of breathing and vocalization needed for speech. *(See Figure 16.18.)*

- The part of the frontal lobe that is the most complex brain area is the **prefrontal cortex**. This area performs complicated learning and reasoning functions. *(See Figures 16.17 and 16.18.)*

- The left hemisphere contains the general interpretive and speech centers and is responsible for language-based skills. The right hemisphere is concerned with spatial relationships and analysis.

The Central White Matter p. 428

- The **central white matter** contains three major groups of axons: (1) **association fibers** (tracts that interconnect areas of neural cortex within a single cerebral hemisphere), (2) **commissural fibers** (tracts connecting the two cerebral hemispheres), and (3) **projection fibers** (tracts that link the cerebrum with other regions of the brain and spinal cord). *(See Figure 16.19 and Table 16.9.)*

The Basal Nuclei p. 429

- The **basal nuclei** within the central white matter include the **caudate nucleus**, **globus pallidus**, and **putamen**. They control muscle tone and coordinate learned movement patterns and other somatic motor activities. *(See Figure 16.20 and Table 16.10.)*

The Limbic System p. 431

- The **limbic system** includes the amygdaloid body, **cingulate gyrus**, **dentate gyrus**, **parahippocampal gyrus**, **hippocampus**, and **fornix**. The **mammillary bodies** control reflex movements associated with eating. The functions of the limbic system involve emotional states and related behavioral drives. *(See Figures 16.13, 16.20, and 16.21 and Table 16.11.)*

- The **anterior nucleus** relays visceral sensations, and stimulating the **reticular formation** produces heightened awareness and a generalized excitement.

16.9 The Cranial Nerves p. 434

- There are 12 pairs of cranial nerves. Each nerve attaches to the brain near the associated sensory or motor nuclei on the ventrolateral surface of the brain. *(See Figure 16.22.)*

The Olfactory Nerves (I) p. 434

- The **olfactory nerves** (I) carry sensory information responsible for the sense of smell. The olfactory afferents synapse within the **olfactory bulbs**. *(See Figure 16.23.)*

The Optic Nerves (II) p. 435

- The **optic nerves** (II) carry visual information from special sensory receptors in the eyes. *(See Figures 12.8 and 16.24.)*

The Oculomotor Nerves (III) p. 436

- The **oculomotor nerves** (III) are the primary source of innervation for the extra-ocular muscles that move the eyeball. *(See Figure 16.25.)*

The Trochlear Nerves (IV) p. 436

- The **trochlear nerves** (IV), the smallest cranial nerves, innervate the superior oblique of the eye. *(See Figure 16.25.)*

The Trigeminal Nerves (V) p. 437

- The **trigeminal nerves** (V), the largest cranial nerves, are mixed nerves with **ophthalmic**, **maxillary**, and **mandibular** branches. *(See Figure 16.26.)*

The Abducens Nerves (VI) p. 438

- The **abducens nerves** (VI) innervate the sixth extrinsic oculomotor muscles, the lateral rectus of each eye. *(See Figure 16.25.)*

The Facial Nerves (VII) p. 438

- The **facial nerves** (VII) are mixed nerves controlling muscles of the scalp and face. They provide pressure sensations over the face and receive taste information from the tongue. *(See Figure 16.27.)*

The Vestibulocochlear Nerves (VIII) p. 439

- The **vestibulocochlear nerves** (VIII) contain the **vestibular nerves**, which monitor sensations of balance, position, and movement, and the **cochlear nerves**, which monitor hearing receptors. *(See Figure 16.28.)*

The Glossopharyngeal Nerves (IX) p. 440

- The **glossopharyngeal nerves** (IX) are mixed nerves that innervate the tongue and pharynx and control the action of swallowing. *(See Figure 16.29.)*

The Vagus Nerves (X) p. 440

- The **vagus nerves** (X) are mixed nerves that are vital to the autonomic control of visceral function and have a variety of motor components. *(See Figure 16.30.)*

The Accessory Nerve (XI) p. 441

- Each **accessory nerve** (XI) has an **internal branch**, which innervates voluntary swallowing muscles of the soft palate and pharynx, and an **external branch**, which controls muscles associated with the pectoral girdle. *(See Figure 16.31.)*

The Hypoglossal Nerves (XII) p. 442

- Each **hypoglossal nerve** (XII) provides voluntary motor control over tongue movements. *(See Figure 16.31.)*

Chapter Review

For answers, see the blue Answers tab at the back of the book.

Level 1 Reviewing Facts and Terms

1. Label the following structures on the midsagittal section of the brain below.
   - pons
   - corpus callosum
   - cerebellum
   - thalamus
   - cerebral aqueduct

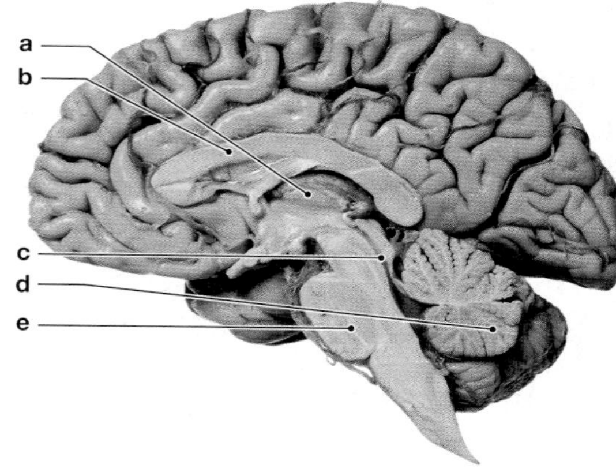

(a) __________
(b) __________
(c) __________
(d) __________
(e) __________

2. In contrast with those of the brain, responses of the spinal reflexes
   (a) are fine-tuned.
   (b) are immediate.
   (c) require many processing steps.
   (d) are stereotyped.

3. The primary link between the nervous and the endocrine systems is the
   (a) hypothalamus.
   (b) pons.
   (c) mesencephalon.
   (d) medulla oblongata.

4. Cranial blood vessels pass through the space directly deep to the
   (a) dura mater.
   (b) pia mater.
   (c) arachnoid granulations.
   (d) arachnoid mater.

5. The only cranial nerves that are attached to the cerebrum are the
   (a) optic.
   (b) oculomotor.
   (c) trochlear.
   (d) olfactory.

6. The anterior nuclei of the thalamus
   (a) are part of the limbic system.
   (b) are connected to the pituitary gland.
   (c) produce the hormone melatonin.
   (d) receive impulses from the optic nerve.

7. The cortex inferior to the lateral sulcus is the
   (a) parietal lobe.
   (b) temporal lobe.
   (c) frontal lobe.
   (d) occipital lobe.

8. Lying within each hemisphere inferior to the floor of the lateral ventricles is/are the
   (a) anterior commissures.
   (b) motor association areas.
   (c) auditory cortex.
   (d) basal nuclei.

9. Nerve fiber bundles on the ventrolateral surface of the mesencephalon are the
   (a) tegmenta.
   (b) corpora quadrigemina.
   (c) cerebral peduncles.
   (d) superior colliculi.

10. Efferent tracts from the hypothalamus
    (a) control involuntary motor activities.
    (b) control autonomic function.
    (c) coordinate activities of the nervous and endocrine systems.
    (d) do all of the above.

11. The diencephalic components of the limbic system include the
    (a) limbic lobe and hippocampus.
    (b) fornix.
    (c) amygdaloid body and parahippocampal gyrus.
    (d) thalamus and hypothalamus.

Level 2 Reviewing Concepts

1. Swelling of the jugular vein as it leaves the skull could compress which of the following cranial nerves?
   (a) N I, N IV, and N V
   (b) N IX, N X, and N XI
   (c) N II, N IV, and N VI
   (d) N VIII, N IX, and N XII

2. Why can the brain respond to stimuli with greater versatility than the spinal cord?

3. Which lobe and specific area of the brain would be affected if a person could no longer cut designs from construction paper?

4. Impulses from proprioceptors must pass through specific nuclei before arriving at their destination in the brain. What are the nuclei, and what is the destination of this information?
5. Which nuclei are involved in the coordinated movement of the head in the direction of a loud noise?
6. Which cranial nerves are responsible for all aspects of eye function?
7. If a person has poor emotional control and difficulty in remembering past events, what area of the brain might be damaged or have a lesion?
8. Why is the blood brain barrier less intact in the hypothalamus?

Level 3 Critical Thinking

1. Rose falls down a flight of stairs and bumps her head several times. Soon after, she develops a headache and blurred vision. Diagnostic tests at the hospital reveal an epidural hematoma in the temporoparietal area. The hematoma is pressing against the brainstem. What other signs and symptoms might she experience as a result of the injury?
2. If a person who has sustained a head injury passes out several days after the incident occurred, what would you suspect to be the cause of the problem, and how serious might it be?

MasteringA&P™

Access more chapter study tools online in the Study Area:

- Chapter Quizzes, Chapter Practice Test, Clinical Cases, and more!
- Practice Anatomy Lab (PAL) PAL™
- A&P Flix for anatomy topics A&PFlix™

CLINICAL CASE WRAP-UP

A Neuroanatomist's Stroke of Insight

While her stroke affected the left side of Dr. Taylor's brain, the right side continued functioning. Because language and thoughts are typically controlled in the left hemisphere (the dominant hemisphere of a right-handed person), Dr. Taylor "sat in an absolutely silent mind" for the first month. Since the center for mathematical calculation is situated in the left hemisphere, she had to learn to use numbers all over again. And because the primary motor cortex governing the right side of the body resides in the precentral gyrus of the left hemisphere, she had to learn to use her right arm again. Full recovery took 8 years.

The stroke destroyed some brain cells, but others were able to form new neuronal connections. Neuroplasticity, this ability of nerve cells to make new connections, allows the brain to reorganize itself after injury.

Dr. Taylor wants anatomy students to know two things. First, "if you study the brain, you will never be bored." Second, "if you treat stroke patients like they will recover, they are more likely to recover." She has written a best-selling memoir about her experience, *My Stroke of Insight: A Brain Scientist's Personal Journey.*

1. How would you know, based on signs and symptoms, which side of Dr. Taylor's brain was injured by the stroke?
2. What is neuroplasticity, and why was it important in Dr. Taylor's recovery?

See the blue Answers tab at the back of the book.

Related Clinical Terms

attention deficit hyperactivity disorder (ADHD): Disorder occurring mainly in children characterized by hyperactivity, inability to concentrate, and impulsive or inappropriate behavior.

Creutzfeldt-Jakob disease (CJD): A rare, degenerative, invariably fatal brain disorder that is marked by rapid mental deterioration. The disease, which is caused by a prion (an infectious protein particle), typically starts by causing mental and emotional problems, then progresses to affect motor skills, such as walking and talking.

delirium: An acutely disturbed state of mind that occurs in fever, intoxication, and other disorders and is characterized by restlessness, hallucinations, and incoherence of thought and speech.

Glasgow coma scale: The most widely used scoring system to quantify the level of consciousness of a victim of a traumatic brain injury. It rates three functions: eye opening, verbal response, and motor response.

hydrocephalus: A condition marked by an excessive accumulation of cerebrospinal fluid within the brain ventricles.

microcephaly: A birth defect in which the head circumference is much smaller than expected for the age and sex of the child.

migraine: A type of headache marked by severe debilitating head pain lasting several hours or longer.

myoclonus: A quick, involuntary muscle jerk or contraction; persistent myoclonus usually indicates a nervous system disorder.

pallidectomy: The destruction of all or part of the globus pallidus by chemicals or freezing; used in the treatment of Parkinson's disease.

prosopagnosia: The inability to recognize other humans by their faces.

psychosis: A severe mental disorder in which thought and emotions are so impaired that contact with reality is lost.

stupor: A state of near-unconsciousness or insensibility.

transient ischemic attack (TIA): An episode in which a person has stroke-like symptoms that last less than 24 hours and result in no permanent injury to the brain, but may be a warning sign of the potential for a major stroke.

17

The Nervous System

Autonomic Nervous System

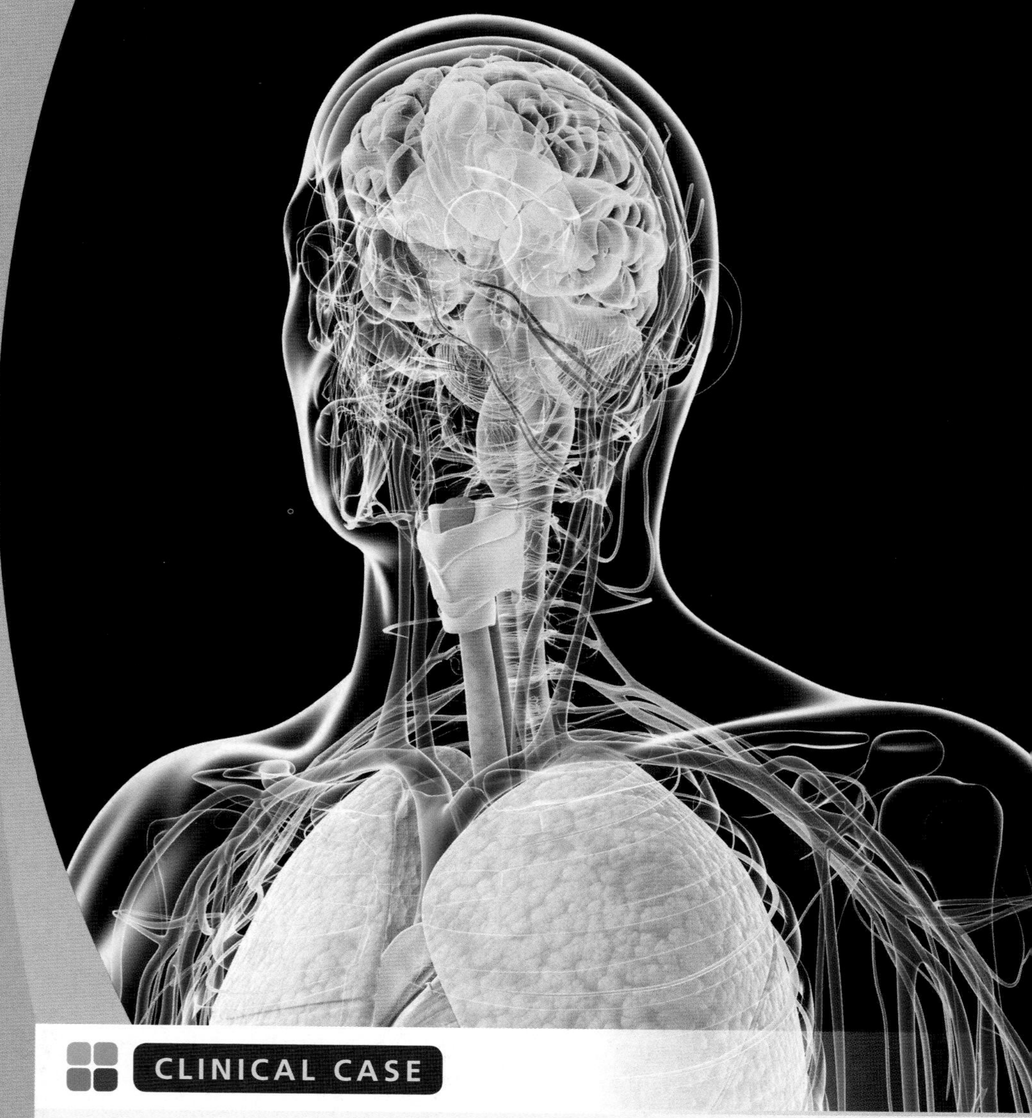

Learning Outcomes

These Learning Outcomes correspond by number to this chapter's sections and indicate what you should be able to do after completing the chapter.

17.1 Compare and contrast the somatic and autonomic nervous systems. p. 450

17.2 Summarize the anatomy and physiology of the sympathetic nervous system. p. 450

17.3 Summarize the anatomy and physiology of the parasympathetic nervous system. p. 459

17.4 Summarize the concept of dual innervation within the autonomic nervous system. p. 462

CLINICAL CASE

First Day of Anatomy Lab

It is Tim's first day of anatomy lab, and he is excited. He has a new white lab coat and a great anatomy textbook, and he's ready to go.

The first thing Tim notices upon entering the lab is the body on the table. A cadaver certainly resembles a human being, but there is no mistaking the two. There are bright lights and new odors.

Tim suddenly feels hot and sweaty. There is a vague discomfort in his chest and he feels nauseated and light-headed. His instructor says something, but he can't understand the words. He tries to respond, but is unable to speak. His vision blurs and he hears ringing in his ears. The last thing Tim remembers is a sensation of profound weakness.

Tim hits the floor with a thud, landing in a crumpled, prone position. His instructor rolls him over, loosens his belt, and props his legs up on a chair. Within 10 seconds Tim regains consciousness, but he feels exhausted. He also feels embarrassed.

What happened to Tim? To find out, turn to the Clinical Case Wrap-Up on p. 470.

OUR CONSCIOUS THOUGHTS, PLANS, AND ACTIONS are only a tiny fraction of the activities of the nervous system. When all consciousness is eliminated, such as when we sleep, vital homeostatic processes continue virtually unchanged. Longer, deeper states of unconsciousness are not more dangerous, as long as nourishment is provided. People who have suffered severe brain injuries have survived in a coma for decades. Survival is possible under such conditions because the **autonomic nervous system (ANS)** makes routine adjustments in physiological systems. The ANS regulates body temperature and coordinates cardiovascular, respiratory, digestive, excretory, and reproductive functions.

This chapter examines the anatomical structure and subdivisions of the autonomic nervous system. Each subdivision has a characteristic anatomical and functional organization. First, we describe the sympathetic and parasympathetic divisions. Then we examine the way these divisions maintain and adjust various organ systems to meet the body's ever-changing physiological needs.

17.1 A Comparison of the Somatic and Autonomic Nervous Systems

▶ **KEY POINT** The peripheral nervous system has two subdivisions: the somatic nervous system and the autonomic nervous system. The somatic nervous system is associated with voluntary movements of skeletal muscle and involves afferent sensory and efferent motor nerves. The autonomic nervous system is associated with involuntary activities of smooth muscles, glands, and visceral structures and involves afferent sensory and efferent motor nerves.

Let's begin by comparing the autonomic nervous system (ANS), which innervates visceral effectors, with the somatic nervous system (SNS), whose lower motor neurons innervate skeletal muscles. Like the SNS, the ANS has afferent and efferent neurons. Also like the SNS, the afferent sensory information of the ANS is processed in the central nervous system, and then efferent impulses are sent to effector organs. However, in the ANS, the afferent pathways originate in visceral receptors, and the efferent pathways connect to visceral effector organs, such as smooth muscle and glands.

In addition to the difference in receptor and effector organ location, the autonomic nervous system, composed of the sympathetic and parasympathetic divisions, differs from the somatic nervous system in the arrangement of the efferent neurons. In the ANS, the axon of a visceral motor neuron within the central nervous system (CNS) innervates a second neuron located in a peripheral ganglion. This second neuron innervates the peripheral effector. Visceral motor neurons in the CNS send short, myelinated axons, called **preganglionic fibers**, to synapse on a group of neurons located within a **ganglion** (pleural, *ganglia*) (also termed *autonomic ganglion*) located outside the CNS. Axons leaving the ganglia are relatively long and are unmyelinated. These axons are called **postganglionic fibers** because they carry impulses away from the ganglion. Postganglionic fibers innervate peripheral tissues and organs, such as cardiac and smooth muscle, adipose tissue, and glands. **Figure 17.1** and **Spotlight Figure 17.2** on pg. 425.

Sympathetic and Parasympathetic Subdivisions of the ANS

▶ **KEY POINT** The sympathetic division is most active during times of exertion, stress, and sexual climax, often termed the "fight-or-flight" reaction. The parasympathetic division is most active during sexual arousal and periods of "rest and digest."

The two divisions of the ANS often have opposing effects; if the sympathetic division causes excitation, the parasympathetic division causes inhibition. However, this is not always the case because (1) the two divisions may work independently, with some structures innervated by only one division, and (2) the two divisions may work together, each controlling one stage of a complex process. In general, the parasympathetic division is most active under resting conditions, while the sympathetic division is most active during times of exertion, stress, or emergency. **Spotlight Figure 17.2** introduces the anatomy of the **sympathetic division** (also termed the *thoracolumbar division*) and the **parasympathetic division** (also termed the *craniosacral* division).

The autonomic nervous system also includes a third division—the **enteric nervous system (ENS)**. The enteric nervous system is an extensive network of neurons located within the walls of the digestive tract. Although the ENS is affected by the sympathetic and parasympathetic divisions, many complex visceral reflexes are initiated and coordinated locally, without instructions from the central nervous system. In this chapter, we focus on the sympathetic and parasympathetic divisions that integrate and coordinate visceral functions throughout the body. We consider the activities of the ENS when we discuss visceral reflexes later in this chapter, and again when we discuss the digestive system in Chapter 25.

17.1 CONCEPT CHECK

1. Describe preganglionic fibers and neurons and postganglionic fibers and neurons.
2. Name the neurotransmitter released by most postganglionic sympathetic fibers.
3. Where do the preganglionic fibers of the parasympathetic division of the ANS originate?

See the blue Answers tab at the back of the book.

17.2 The Sympathetic Division

▶ **KEY POINT** The sympathetic division of the autonomic nervous system operates through a series of interconnected neurons. Efferent sympathetic neurons originate from thoracic and lumbar spinal nerves and synapse with neurons in the peripheral nervous system at a series of sympathetic ganglia.

Figure 17.1 and **Spotlight Figure 17.2** shows an overall organization of the sympathetic division. Preganglionic neurons are *only* located between segments T_1 and L_2 of the spinal cord. The cell bodies of these neurons occupy the lateral horns of the spinal cord between T_1 and L_2, and their axons enter the ventral roots of those segments. The ganglionic neurons are in three locations:

1. **Sympathetic chain ganglia** (also called *paravertebral ganglia*) are on both sides of the vertebral column. Postganglionic fibers exiting these ganglia innervate effector organs in the body wall, head and neck, limbs, and inside the thoracic cavity.
2. **Collateral ganglia** (also known as *prevertebral ganglia*) are anterior to the vertebral column. Postganglionic fibers exiting these ganglia innervate effector organs in the abdominopelvic cavity.
3. Specialized sympathetic neurons are located in the interior of the adrenal gland, known as the **adrenal medulla**. The adrenal medulla is a modified sympathetic ganglion. These ganglionic neurons have very short axons. When stimulated, they release neurotransmitters into the bloodstream for distribution throughout the body as hormones.

Sympathetic Chain Ganglia

▶ **KEY POINT** The preganglionic neurons of the sympathetic division form synapses with ganglionic neurons within the sympathetic trunk. The sympathetic trunk is found on each side of the vertebral column.

Figure 17.1 **Organization of the Sympathetic Division of the ANS.** This diagram highlights the relationships between preganglionic and ganglionic neurons and between ganglionic neurons and target organs.

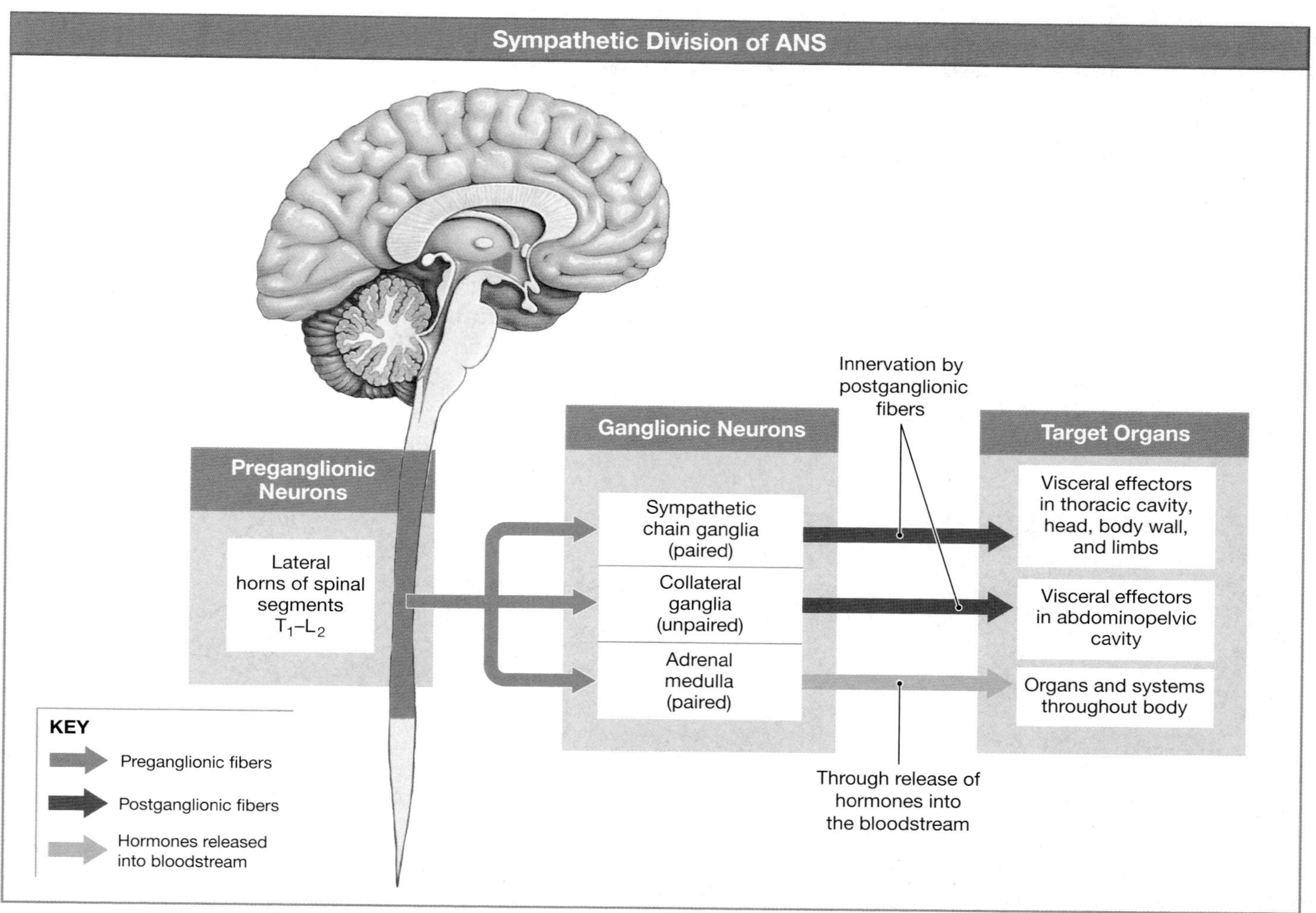

Spotlight Figure 17.3 outlines the anatomy of the sympathetic chain ganglia. Each sympathetic chain ganglion has 3 cervical, 11–12 thoracic, 2–5 lumbar, and 4–5 sacral sympathetic ganglia and 1 coccygeal sympathetic ganglion. Numbers may vary because adjacent ganglia may fuse. For example, the coccygeal ganglia from both sides usually fuse to form a single median ganglion, the ganglion impar, (look ahead to **Figure 17.4** on page 456), while the inferior cervical and first thoracic ganglia from both sides occasionally fuse to form a stellate ganglion. *Preganglionic sympathetic neurons are only found in segments T_1–L_2 of the spinal cord, and the spinal nerves of these segments have both white rami communicantes (preganglionic fibers) and gray rami communicantes (postganglionic fibers).* The neurons in the cervical, inferior lumbar, and sacral sympathetic chain ganglia are innervated by preganglionic fibers extending along the length of the chain. In turn, these chain ganglia provide postganglionic fibers, through the gray rami, to the cervical, lumbar, and sacral spinal nerves.

Every spinal nerve along the entire length of the spinal cord has a pair of gray rami communicantes carrying sympathetic postganglionic fibers. About 8 percent of the axons in each spinal nerve are sympathetic postganglionic fibers. The dorsal and ventral rami of the spinal nerves provide extensive sympathetic innervation to structures in the body wall and limbs. In the head, postganglionic fibers leaving the cervical sympathetic ganglia supply the regions and structures innervated by cranial nerves III, VII, IX, and X **(Figure 17.4)**.

Collateral Ganglia

▶ **KEY POINT** Splanchnic nerves carry visceral efferent motor fibers and visceral afferent sensory fibers. Postganglionic neurons within the collateral ganglia send postganglionic fibers to viscera within the abdominal and pelvic cavities.

The abdominopelvic viscera receive sympathetic innervation by sympathetic preganglionic fibers that synapse in separate collateral ganglia. These fibers pass through the sympathetic chain without synapsing. They form the paired **splanchnic** (SPLANK-nik) **nerves**, which lie in the posterior wall of the abdominal cavity. They originate as paired ganglia, but the two usually fuse **(Figure 17.4)**.

Anatomy of the Collateral Ganglia

The splanchnic nerves (greater, lesser, lumbar, and sacral) innervate three collateral ganglia. Preganglionic fibers from the seven inferior thoracic segments end at the **celiac** (SĒ-lē-ak) **ganglion** and the **superior mesenteric ganglion**. These ganglia are located within an extensive, weblike network of nerve fibers termed an autonomic plexus (plural, *plexuses*). Preganglionic fibers from the lumbar segments form splanchnic nerves that end at the **inferior mesenteric ganglion**. The sacral splanchnic nerves end in the hypogastric plexus, an autonomic network supplying pelvic organs and the external genitalia.

SPOTLIGHT

FIGURE 17.2

An Introduction to the Autonomic Nervous System

The autonomic nervous system (ANS) is composed of two divisions: the sympathetic division and the parasympathetic division.

Autonomic Nervous System

Sympathetic Division (Thoracolumbar Division)

Preganglionic fibers from both the thoracic and upper lumbar spinal segments synapse in ganglia near the spinal cord. These axons and ganglia are part of the **sympathetic division**, or *thoracolumbar division*, of the autonomic nervous system (ANS). This division is often called the “fight-or-flight” system because an increase in sympathetic activity generally stimulates tissue metabolism, increases alertness, and prepares the body to deal with emergencies.

Preganglionic Neurons

Preganglionic neurons are located in the lateral horns of spinal segments T_1–L_2.

Ganglia

Ganglia are located near the spinal cord. Preganglionic fibers release acetylcholine (ACh), which is excitatory and stimulates ganglionic neurons.

Target Organs

Most postganglionic fibers release norepinephrine (NE) at neuroeffector junctions. The effect is usually excitatory but may vary depending on the nature of the receptor on the target cell’s plasma membrane.

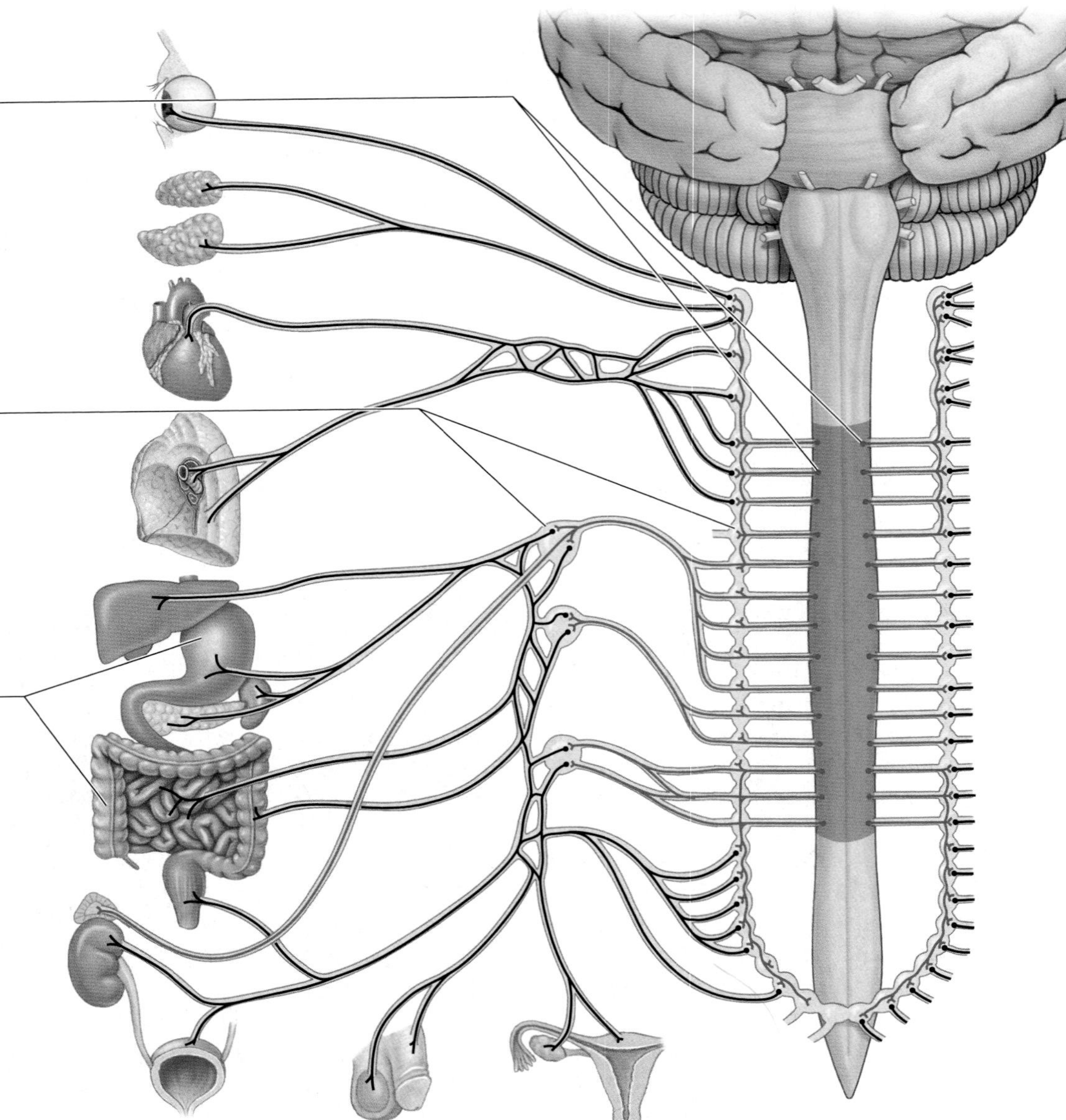

Response
“Fight or flight”

Parasympathetic Division (Craniosacral Division)

Preganglionic fibers originating in either the brainstem (cranial nerves III, VII, IX, and X) or the sacral spinal cord are part of the **parasympathetic division**, or *craniosacral division*, of the ANS. The preganglionic fibers synapse on neurons of terminal ganglia, located close to the target organs, or **intramural ganglia** (*murus*, wall), within the tissues of the target organs. This division is often called the "rest-and-digest" system because it conserves energy and promotes sedentary activities, such as digestion.

Preganglionic Neurons

Preganglionic neurons are located in the brainstem and in the lateral portion of the anterior horns of spinal segments S_2–S_4.

Ganglia

Ganglia are in or near the target organ. Preganglionic fibers release acetylcholine (ACh), which is excitatory and stimulates ganglionic neurons.

Target Organs

All postganglionic fibers release ACh at neuroeffector junctions. The effect is usually inhibitory, but may vary depending on the nature of the receptor on the target cell's plasma membrane.

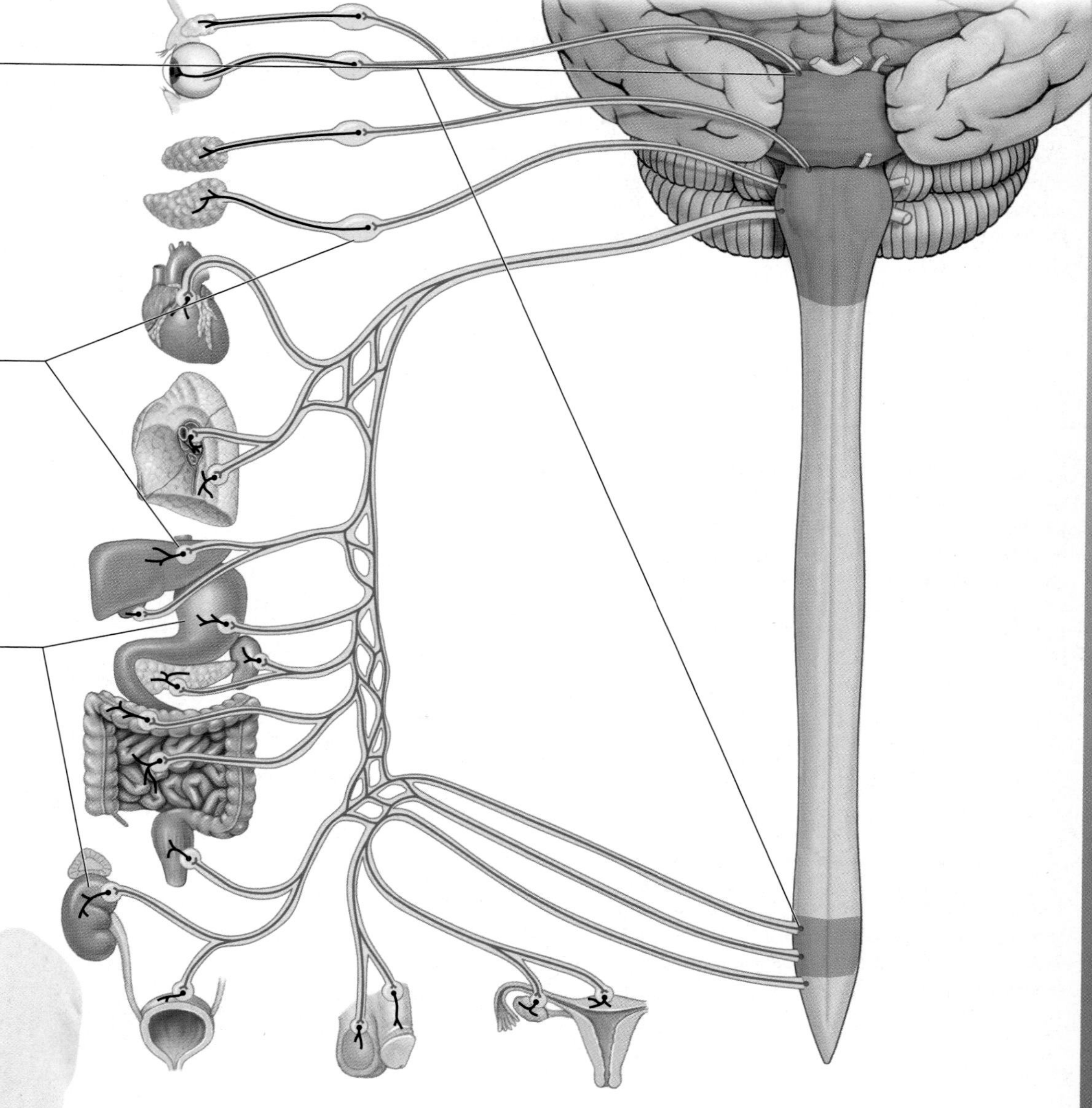

Response
"Rest and digest"

FIGURE 17.3

A Review of the Sympathetic Nervous System

The simplest way to understand the **sympathetic branch** of the autonomic nervous system (ANS) is to describe the distribution and the various pathways taken by the efferent visceral motor nerves. As you review the sympathetic branch, keep the following five anatomical details in mind: (1) The preganglionic fibers of the sympathetic branch are short and myelinated; the postganglionic fibers are long and unmyelinated. (2) The sympathetic branch of the ANS goes almost everywhere in the body: head, body walls, limbs, and viscera. (3) The postganglionic fibers reach the structures by (a) following arteries into the head, (b) accompanying somatic spinal nerves and their branches, or (c) following autonomic nerves to the viscera. (4) The presynaptic fibers of the sympathetic branch of the ANS exit from the CNS only by the ventral roots of spinal nerves T_1 to L_2. (5) Somatic motor output and somatic sensory input *are also contained* within spinal nerves T_1 to L_2.

Anatomy of the Preganglionic Neurons, Rami Communicantes, and Ganglionic Neurons

The cell bodies of the **preganglionic neurons** are located in the lateral horns of the spinal cord between T_1 and L_2. Their axons enter the ventral roots of the spinal nerves at these segments. Each ventral root joins the corresponding dorsal root, which carries afferent sensory fibers, to form a spinal nerve. As the spinal nerve exits the intervertebral foramen, a **white ramus communicans** branches from the spinal nerve. The white ramus communicans carries myelinated preganglionic fibers of the sympathetic branch of the ANS into a nearby sympathetic chain ganglion.

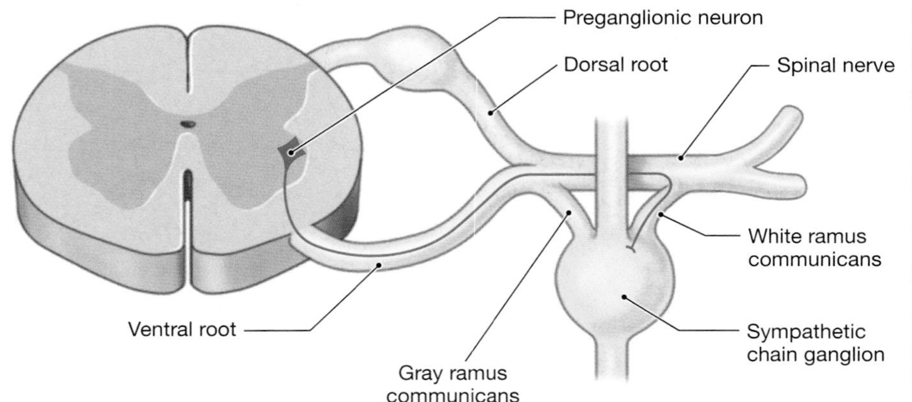

The preganglionic neuron synapses with the cell body of the **ganglionic neuron** in a sympathetic chain ganglion. Unmyelinated postganglionic fibers then leave the sympathetic chain by the **gray ramus communicans**, re-enter the spinal nerve, and proceed to their peripheral targets within spinal nerves. These postganglionic fibers will innervate structures in the body wall, such as the sweat glands of the skin or the smooth muscles in superficial blood vessels.

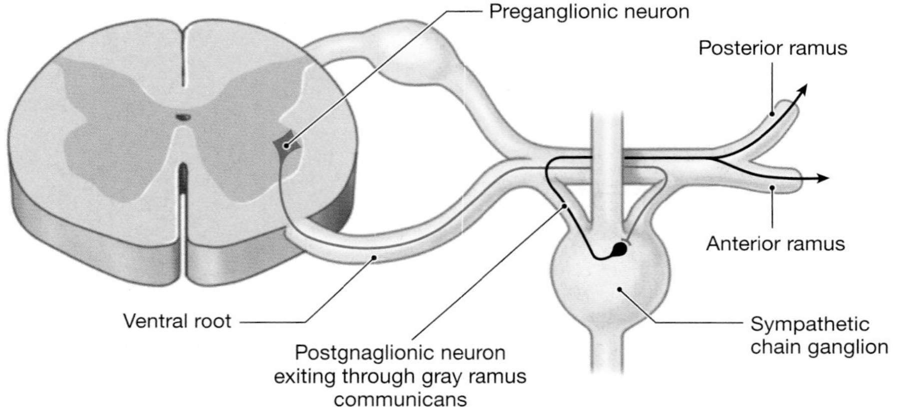

It is important to remember that efferent somatic motor neurons and afferent somatic sensory neurons are also found within *all* spinal nerves, including those spinal nerves exiting the spinal cord between T_1 and L_2.

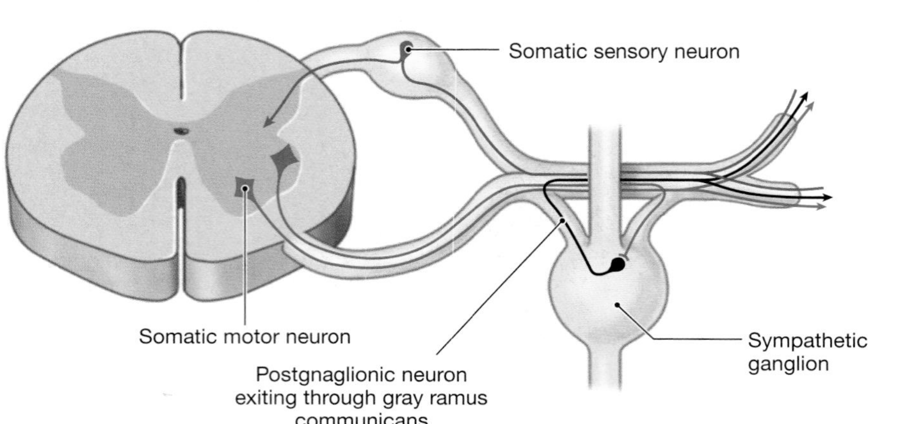

Autonomic Nervous System: What Happens in the Ganglionic Chain

Postganglionic fibers innervating visceral organs in the thoracic cavity, such as the heart and esophagus, will exit the spinal nerve by the white ramus communicans and enter the sympathetic chain ganglion. The preganglionic neuron will then exit the chain ganglion *without synapsing with a postganglionic neuron*. These preganglionic neurons will proceed to a collateral ganglion and synapse with the ganglionic neuron there. Fibers from the postganglionic neuron will then proceed directly to their peripheral targets as sympathetic nerves. These nerves are usually named after their primary targets, such as the cardiac nerves and esophageal nerves.

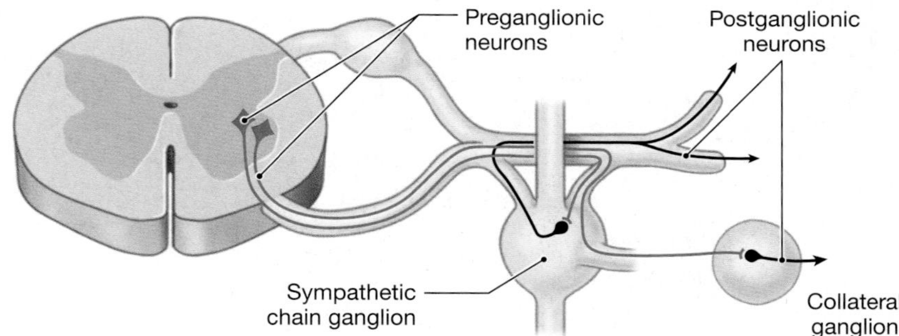

Some fibers entering the sympathetic chain ganglion will ascend within the sympathetic chain and synapse with a ganglionic neuron at a higher level. Fibers from the postganglionic neuron will then exit the ganglionic chain by the gray ramus communicans, re-enter the spinal nerve, and innervate its target organ.

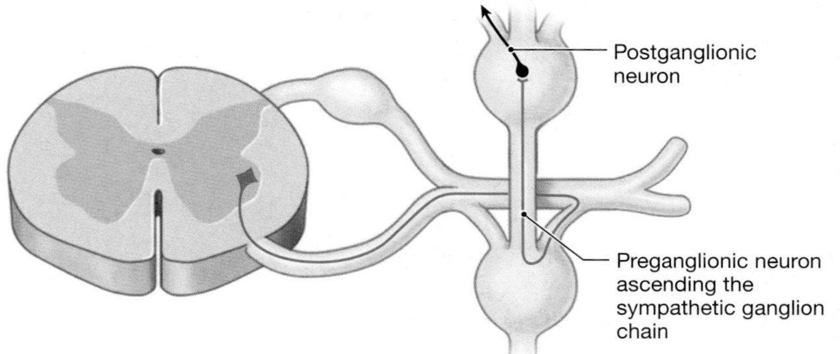

Some fibers entering the sympathetic chain ganglion will descend within the sympathetic chain and synapse with a ganglionic neuron at a lower level. Fibers from the postganglionic neuron will then exit the ganglionic chain by the gray ramus communicans, re-enter the spinal nerve, and innervate its target organ.

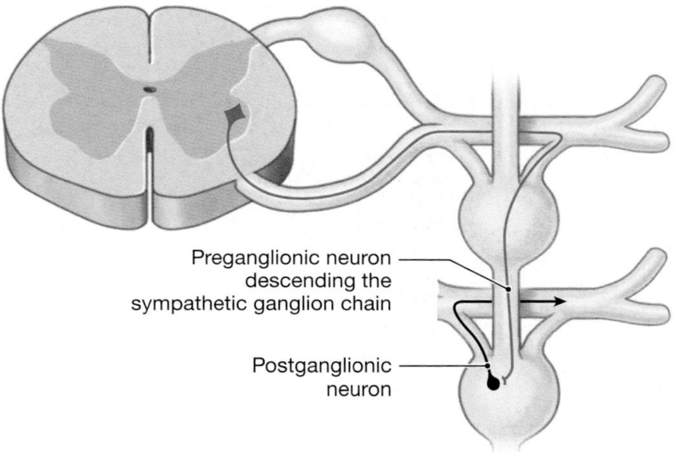

Preganglionic fibers regulating the activities of the abdominopelvic viscera originate at preganglionic neurons in the inferior thoracic and superior lumbar segments of the spinal cord. *These fibers pass through the sympathetic chain without synapsing* and enter a collateral ganglion. Preganglionic fibers then converge to form the **greater**, **lesser**, and **lumbar splanchnic** (SPLANK-nik) **nerves** in the posterior wall of the abdominal cavity. Splanchnic nerves from both sides of the body converge on the collateral ganglia. Collateral ganglia are located anterior and lateral to the descending aorta. These ganglia are most often single, rather than paired, structures.

Other preganglionic neurons will pass through the sympathetic chain without synapsing and proceed to the adrenal medulla.

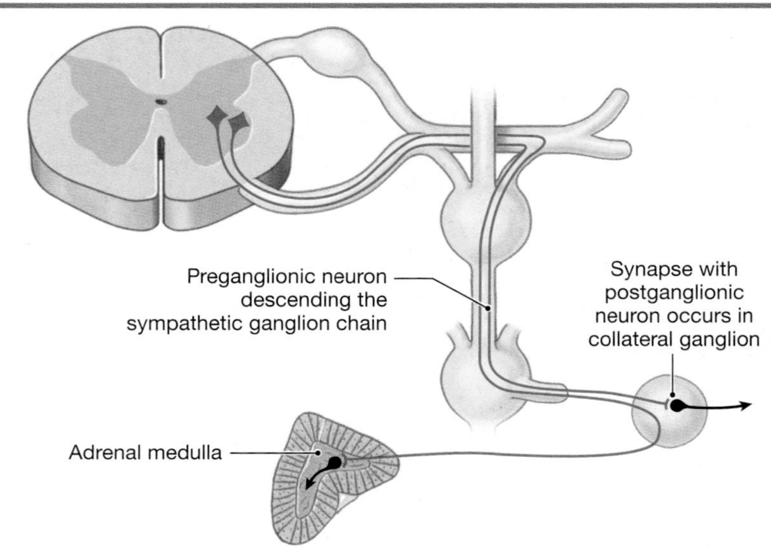

Figure 17.4 Anatomical Distribution of Sympathetic Postganglionic Fibers. The left side of this figure shows the distribution of sympathetic postganglionic fibers through the gray rami and spinal nerves. The right side shows the distribution of preganglionic and postganglionic fibers innervating visceral organs. However, *both* innervation patterns are found on *each* side of the body.

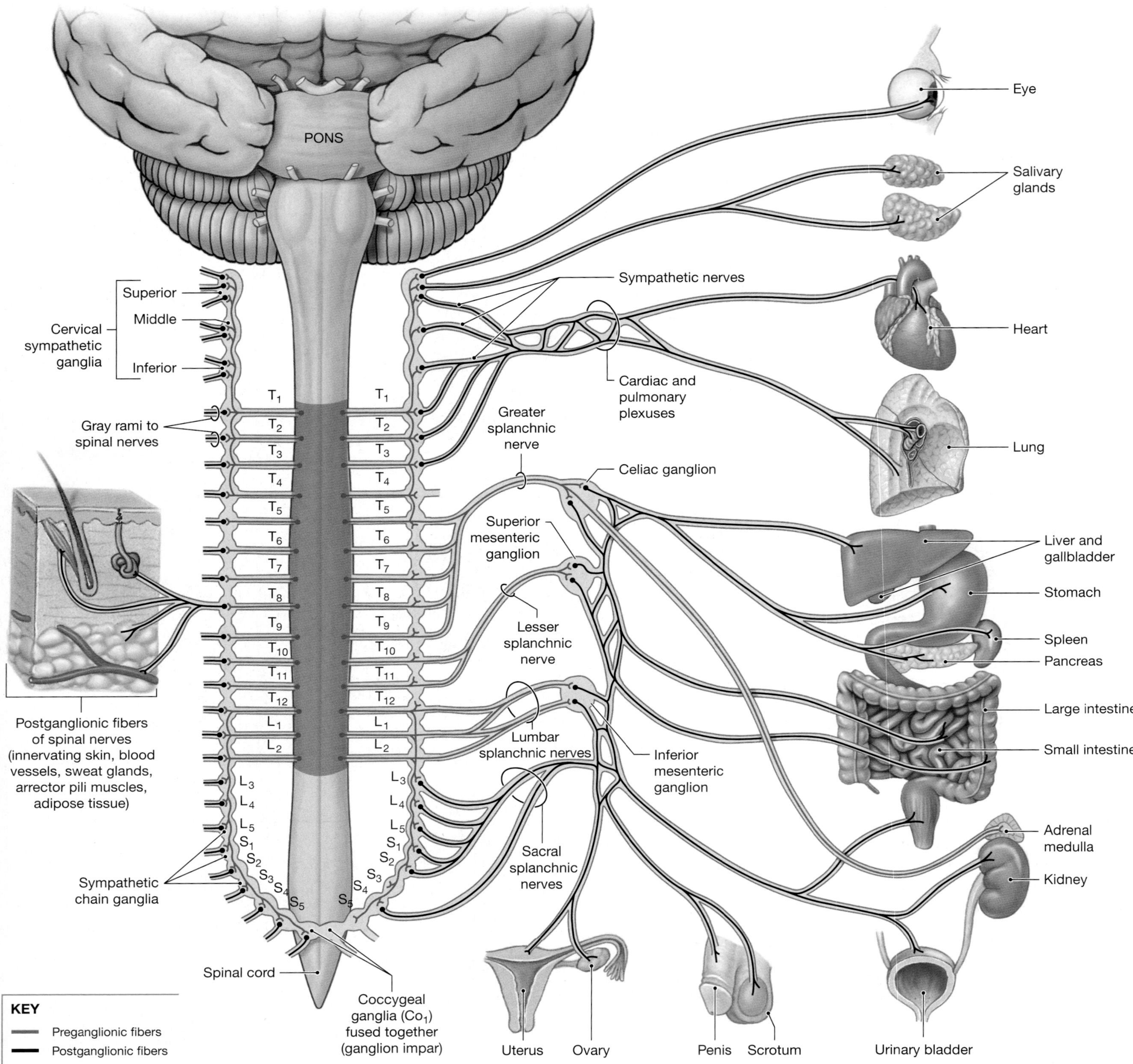

17

The Celiac Ganglion The celiac ganglion is located at the base of the celiac trunk. Postganglionic fibers from the celiac ganglion innervate the stomach, duodenum, liver, gallbladder, pancreas, spleen, and kidney. The celiac ganglion varies considerably in appearance and often consists of a pair of interconnected masses of gray matter.

The Superior Mesenteric Ganglion The superior mesenteric ganglion is located at the base of the superior mesenteric artery. Postganglionic fibers from the superior mesenteric ganglion innervate the small intestine and the initial segments of the large intestine.

The Inferior Mesenteric Ganglion The inferior mesenteric ganglion is located at the base of the inferior mesenteric artery. Postganglionic fibers from this ganglion innervate the terminal portions of the large intestine, the kidney and bladder, and the sex organs.

Adrenal Medulla

▶ **KEY POINT** The cells of the adrenal medulla secrete epinephrine and norepinephrine following stimulation by sympathetic preganglionic neurons.

Some preganglionic fibers originating between T_5 and T_8 pass through the sympathetic chain and the celiac ganglion without synapsing and proceed to the **adrenal medulla** (plural, *medullae*). There they synapse on modified neurons that perform an endocrine function (**Spotlight Figure 17.3** and **Figures 17.4** and **17.5**). When stimulated, these modified neurons release the neurotransmitters epinephrine (E) and norepinephrine (NE) into an extensive network of capillaries. These neurotransmitters function as hormones, exerting their effects in other regions of the body. Epinephrine, also called *adrenaline*, accounts for 75–80 percent of the secretory output; the rest is norepinephrine (*noradrenaline*).

The circulating blood distributes these hormones throughout the body, changing the metabolic activities of many different cells. In general, the effects resemble those produced by the stimulation of sympathetic postganglionic fibers. They differ, however, in two ways: (1) Cells not innervated by sympathetic postganglionic fibers are affected by circulating levels of epinephrine and norepinephrine only if they possess receptors for these molecules; and (2) the effects last much longer than those produced by direct sympathetic innervation, because the released hormones continue to diffuse out of the circulating blood for an extended period.

Effects of Sympathetic Stimulation

▶ **KEY POINT** The sympathetic division of the ANS changes tissue and organ activities by releasing norepinephrine at peripheral synapses and by releasing epinephrine and norepinephrine from the adrenal medulla.

Sympathetic motor fibers innervating specific effectors, such as smooth muscle fibers in blood vessels of the skin, are activated in reflexes not involving other peripheral effectors. In a crisis, however, the entire division responds. This event, called **sympathetic activation**, affects peripheral tissues and alters CNS activity. Sympathetic centers in the hypothalamus control sympathetic activation.

When sympathetic activation occurs, we experience the following:

- Increased alertness, through stimulation of the reticular activating system, causing us to feel "on edge"
- A feeling of energy and euphoria, often associated with a disregard for danger and temporary insensitivity to painful stimuli
- Increased activity in the cardiovascular and respiratory centers of the pons and medulla oblongata, leading to increased heart rate and contraction strength, elevations in blood pressure, breathing rate, and depth of respiration
- A general elevation in muscle tone through stimulation of the extrapyramidal system, so that we look tense and may even begin to shiver
- The mobilization of energy reserves through the accelerated breakdown of glycogen in muscle and liver cells and the release of lipids by adipose tissues

17

Figure 17.5 Adrenal Medulla

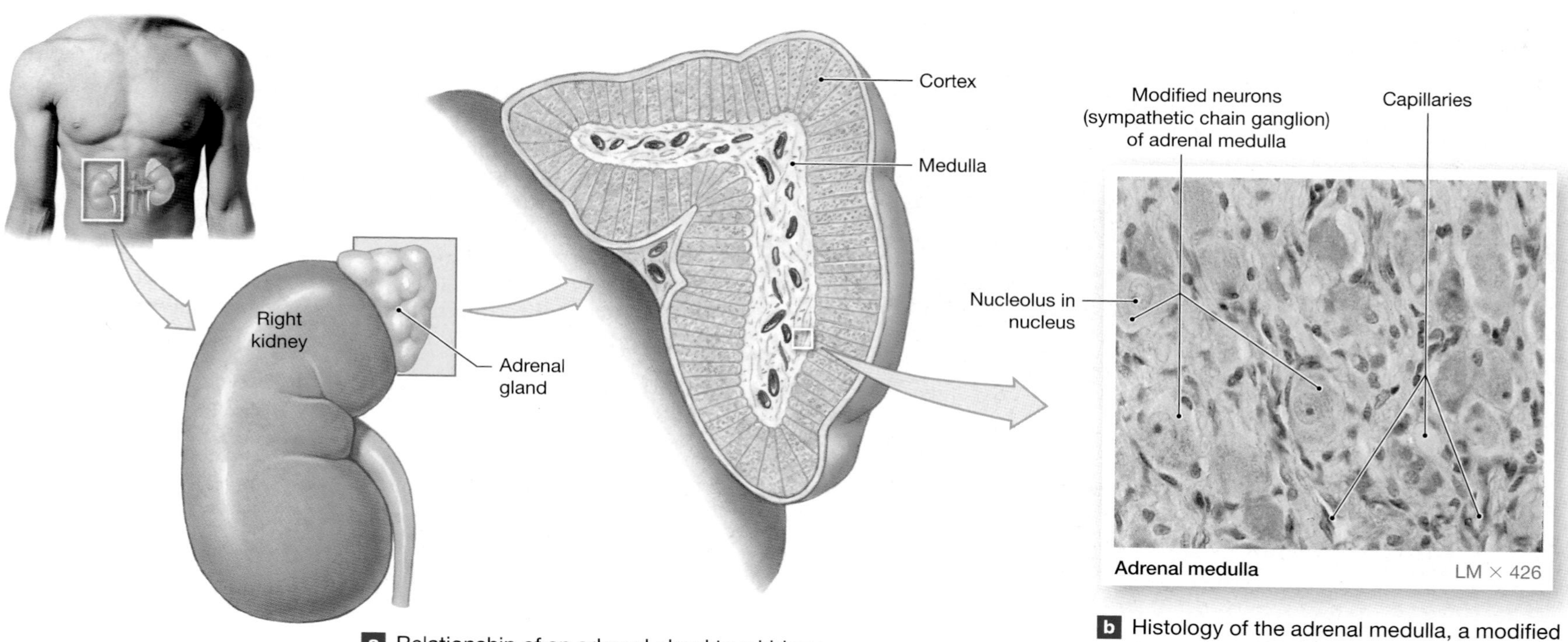

a Relationship of an adrenal gland to a kidney

b Histology of the adrenal medulla, a modified sympathetic chain ganglion

These changes, coupled with the peripheral changes already discussed, complete the preparations necessary for us to handle stressful and potentially dangerous situations. We will now consider the cellular basis for the general effects of sympathetic activation on peripheral organs.

Sympathetic Activation and Neurotransmitter Release

▶ **KEY POINT** Sympathetic preganglionic fibers release acetylcholine (ACh) at cholinergic synapses. Postganglionic fibers release norepinephrine (NE) at adrenergic, neuroeffector junctions.

The acetylcholine (ACh) released by cholinergic, preganglionic neurons during sympathetic activation always stimulates the ganglionic neurons. This leads to postganglionic fibers releasing norepinephrine (NE) at neuroeffector junctions. These neuroeffector junctions are adrenergic, sympathetic terminals. The sympathetic division also contains a small but significant number of ganglionic neurons that release ACh, rather than NE, at their neuroeffector junctions. For example, ACh is released at sympathetic neuroeffector junctions in the body wall, in the skin, and within skeletal muscles.

Figure 17.6 shows a typical sympathetic neuroeffector junction. This type of synaptic junction forms an extensive branching network rather than ending in a single axon terminal (as seen in a skeletal neuromuscular junction). Each branch resembles a string of beads, and each bead, or **varicosity**, is packed with mitochondria and neurotransmitter vesicles. These varicosities pass along or near the surfaces of many effector cells. A single axon may supply 20,000 varicosities, which can affect dozens of surrounding cells. Receptor proteins are scattered across most plasma membranes, and there are no specialized postsynaptic membranes.

17

Figure 17.6 Sympathetic Postganglionic Nerve Endings. A diagrammatic view of sympathetic neuroeffector junctions.

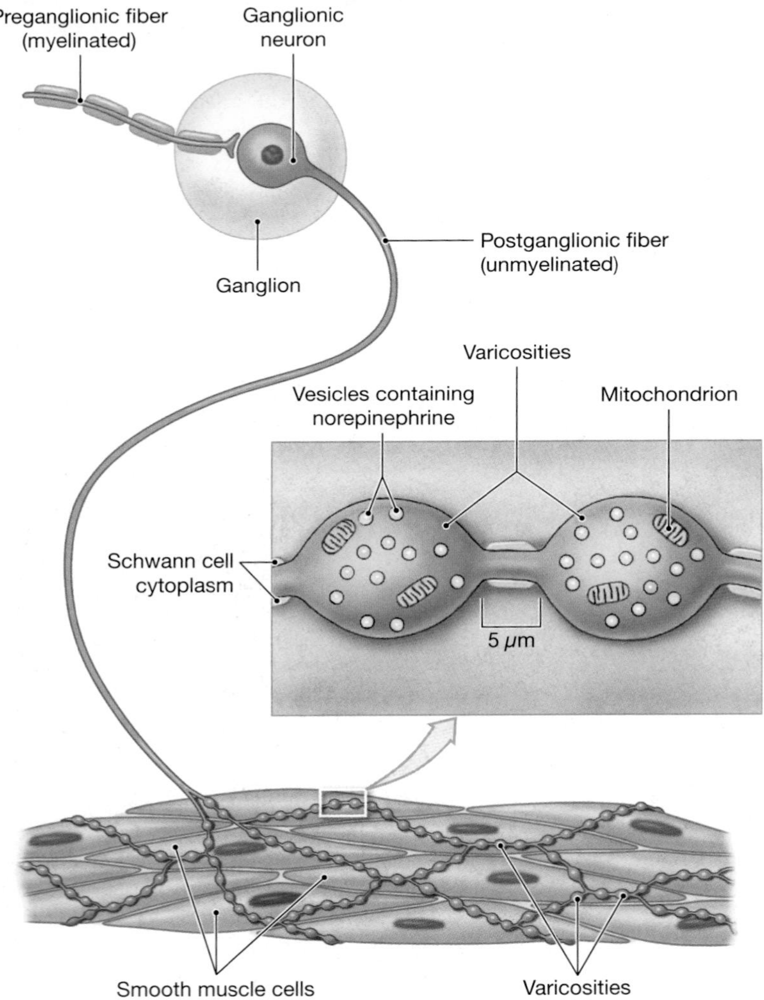

The effects caused by the neurotransmitter released at a varicosity last for only a few seconds before the neurotransmitter is reabsorbed, broken down by enzymes, or removed by diffusion into the bloodstream. In contrast, the effects of the epinephrine and norepinephrine secreted by the adrenal medullae last much longer because (1) the bloodstream does not contain the enzymes required to break down epinephrine or norepinephrine, and (2) most tissues contain relatively low concentrations of these enzymes. As a result, stimulation of the adrenal medulla causes widespread effects that continue for a relatively long time. For example, tissue concentrations of epinephrine may remain elevated for as long as 30 seconds, and the effects may last for several minutes.

Plasma Membrane Receptors and Sympathetic Function

▶ **KEY POINT** The effects of sympathetic stimulation result from the interaction between epinephrine or norepinephrine and plasma membrane receptors.

There are two classes of sympathetic receptors sensitive to epinephrine and norepinephrine: **alpha receptors** and **beta receptors**. Each of these classes of receptors has two or three subtypes. The diversity of receptors and the varying combinations found on plasma membranes account for the wide variety in target organ responses to sympathetic stimulation. In general, epinephrine stimulates both classes of receptors, while norepinephrine primarily stimulates alpha receptors.

A Summary of the Sympathetic Division

In summary:

- The sympathetic division of the ANS includes two sympathetic chains resembling a string of beads, one on each side of the vertebral column; three collateral ganglia anterior to the spinal column; and two adrenal medullae.
- Preganglionic fibers are short because the ganglia are close to the spinal cord. The postganglionic fibers are longer and extend a considerable distance before reaching their target organs. (In the case of the adrenal medullae, very short axons from modified ganglionic neurons end at capillaries that carry their secretions to the bloodstream.)
- The sympathetic division shows extensive divergence; a single preganglionic fiber may innervate as many as 32 ganglionic neurons in several different ganglia. As a result, a single sympathetic motor neuron inside the CNS controls a variety of peripheral effectors and produces a complex and coordinated response.
- All preganglionic neurons release ACh at their synapses with ganglionic neurons. Most of the postganglionic fibers release norepinephrine, but a few release ACh.
- The effector response depends on the function of the plasma membrane receptor activated when epinephrine or norepinephrine binds to either alpha or beta receptors.

Table 17.1 (look ahead to page 464) summarizes the characteristics of the sympathetic division of the ANS.

CLINICAL NOTE

Sympathetic Function: Too Little, Too Much

Too little: **Horner syndrome** is a condition where the sympathetic postganglionic innervation to one side of the face becomes interrupted. This may be due to a tumor, infection, injury, or trauma to the brachial plexus. The affected side of the face becomes flushed as vascular tone decreases, and there is no sweating. The pupil on that side becomes markedly constricted, the eyelid droops, and the eye appears to retreat into the orbit.

Too much: The sympathetic nervous system (SNS) prepares the body for "fight or flight" by constricting the superficial blood vessels, reserving blood for the muscles and brain, where it is needed in emergencies. **Raynaud's disease** (*Raynaud's phenomenon*) is a condition in which the SNS temporarily initiates peripheral vasoconstriction of the small arteries in the fingers and toes in response to cold temperatures or stress. The cause is unknown, and it affects more women than men. Smoking, caffeine, and drugs that cause vasoconstriction can make it worse.

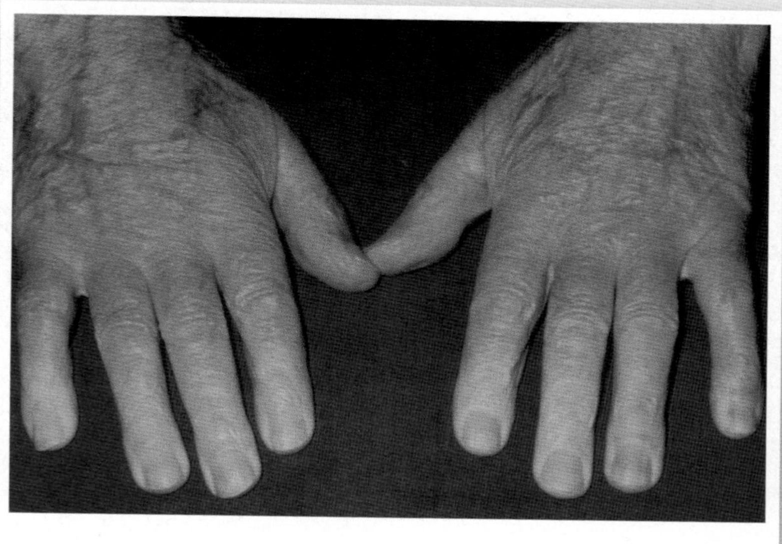

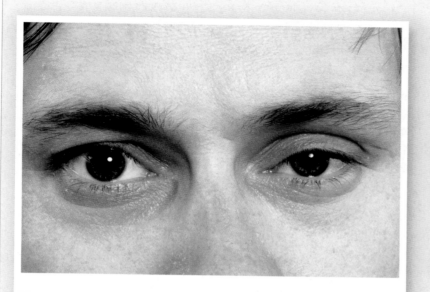

17.2 CONCEPT CHECK

4 Where do the nerve fibers that synapse in the collateral ganglia originate?

5 Individuals with high blood pressure may be given a medication that blocks beta receptors. How would this medication help their condition?

6 Describe sympathetic chain ganglia and collateral ganglia.

See the blue Answers tab at the back of the book.

17.3 | The Parasympathetic Division

▶ **KEY POINT** The parasympathetic division of the autonomic nervous system operates through a series of interconnected neurons. Efferent parasympathetic neurons originate from cranial nerves III, VII, IX, and X and sacral spinal nerves S_2–S_4 and synapse with neurons within or near the innervated organ.

The parasympathetic division of the ANS **(Figure 17.7)** consists of the following:

- Preganglionic neurons located in the brainstem and in sacral segments of the spinal cord. The mesencephalon (midbrain), pons, and medulla oblongata contain autonomic nuclei associated with cranial nerves III, VII, IX, and X. In the sacral segments of the spinal cord, the autonomic nuclei lie in spinal segments S_2–S_4.
- Ganglionic neurons located in peripheral ganglia within or adjacent to the target organs. Preganglionic fibers of the parasympathetic division do not diverge as extensively as do those of the sympathetic division. A typical preganglionic fiber synapses on six to eight ganglionic neurons. These neurons are all located in the same ganglion, and their postganglionic fibers influence the same target organ. The ganglion may be a terminal ganglion (near the target organs) or an intramural ganglion (within the tissues of target organs). As a result, the effects of parasympathetic stimulation are more specific and localized than those of the sympathetic division.

Organization and Anatomy of the Parasympathetic Division

▶ **KEY POINT** The parasympathetic division originates from cranial nerves and sacral spinal nerves.

Parasympathetic preganglionic fibers leave the brain in cranial nerves III (oculomotor), VII (facial), IX (glossopharyngeal), and X (vagus). The fibers in N III, N VII, and N IX control visceral structures in the head. These

Figure 17.7 Organization of the Parasympathetic Division of the ANS. This diagram summarizes the relationships between preganglionic and ganglionic neurons and between ganglionic neurons and target organs.

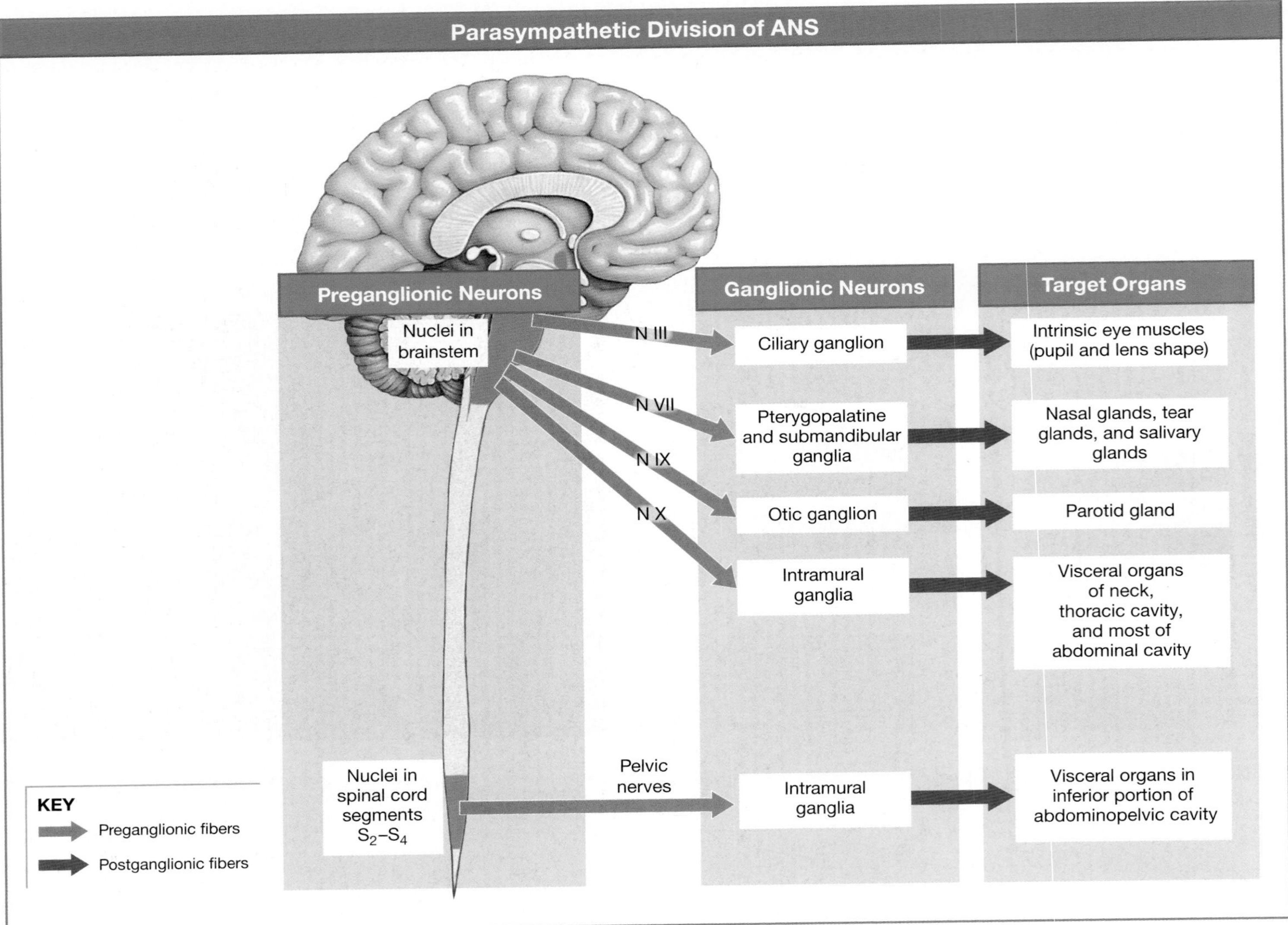

17

preganglionic fibers synapse in the **ciliary**, **pterygopalatine**, **submandibular**, and **otic ganglia**. Short postganglionic fibers then continue to their peripheral targets. The vagus nerve (X) provides preganglionic parasympathetic innervation to intramural ganglia within viscera in the thoracic and abdominopelvic cavities, traveling as far as the last segments of the large intestine. The vagus nerve alone provides roughly 75 percent of all parasympathetic outflow **(Figure 17.8)**.

The sacral parasympathetic outflow does not join the ventral rami of the spinal nerves. ⮌ pp. 366–368 Instead, the preganglionic fibers form distinct **pelvic nerves** that innervate intramural ganglia in the kidney and urinary bladder, the terminal portions of the large intestine, and the sex organs.

General Functions of the Parasympathetic Division

The following is a partial listing of the major effects produced by the parasympathetic division:

- Constriction of the pupils, which restricts the amount of light entering the eyes and aids focusing on nearby objects
- Secretion by digestive glands, including salivary glands, gastric glands, duodenal and other intestinal glands, the pancreas, and the liver
- Secretion of hormones promoting nutrient absorption by peripheral cells
- Increased smooth muscle activity along the digestive tract
- Stimulation and coordination of defecation
- Contraction of the urinary bladder during urination
- Constriction of the respiratory passageways
- Reduction in heart rate and force of contraction
- Sexual arousal and stimulation of sexual glands in both sexes

These functions center on relaxation, food processing, and energy absorption. Stimulation of the parasympathetic division leads to an increase in the nutrient content within the blood. Cells throughout the body respond to this increase by absorbing nutrients and using them to support growth and other anabolic activities.

Figure 17.8 Anatomical Distribution of the Parasympathetic Output. Preganglionic fibers exit the CNS through either cranial nerves or pelvic nerves. The pattern of target organ innervation is similar on each side of the body, although only nerves on the left side are illustrated.

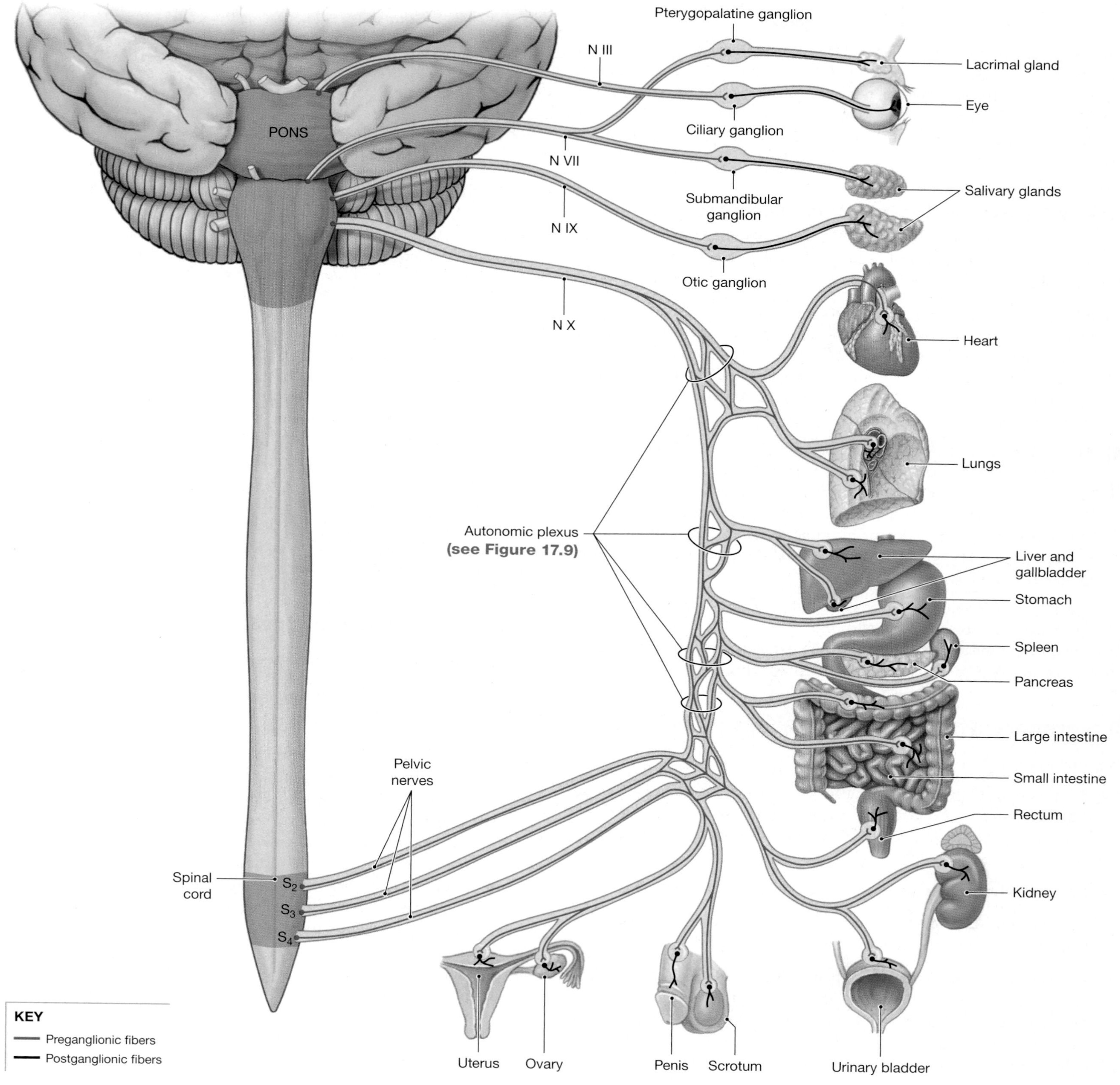

Parasympathetic Activation and Neurotransmitter Release

▶ **KEY POINT** All preganglionic and postganglionic fibers of the parasympathetic division release acetylcholine at their synapses and neuroeffector junctions.

Parasympathetic neuroeffector junctions are small, with narrow synaptic clefts. The effects of stimulation are short-lived because most of the acetylcholine released is inactivated by acetylcholinesterase (AChE) within the synapse. Any ACh diffusing into the surrounding tissues is deactivated by AChE. As a result, the effects of parasympathetic stimulation are quite localized and last a few seconds at most.

Plasma Membrane Receptors and Responses

The parasympathetic division uses the same neurotransmitter, ACh, at all of its synapses (neuron-to-neuron) and neuromuscular or neuroglandular junctions. Two types of ACh receptors are found on postsynaptic plasma membranes:

1. **Nicotinic** (nik-ō-TIN-ik) **receptors** are on the surfaces of all ganglionic neurons of both the parasympathetic and sympathetic divisions, as well as at neuromuscular synapses of the somatic nervous system. Exposure to ACh always causes excitation of the ganglionic neuron or muscle fiber through the opening of chemically gated Na^+ channels in the postsynaptic membrane.
2. **Muscarinic** (mus-ka-RIN-ik) **receptors** are found at all cholinergic neuromuscular or neuroglandular junctions in the parasympathetic division, as well as at the few cholinergic neuroeffector junctions in the sympathetic division. Stimulation of muscarinic receptors produces longer-lasting effects than does stimulation of nicotinic receptors. The response, which reflects the activation or inactivation of specific enzymes, may be either excitatory or inhibitory.

The names nicotinic and muscarinic indicate the chemical compounds that stimulate these receptor sites. Nicotinic receptors bind nicotine, a powerful component of tobacco smoke. Muscarinic receptors are stimulated by muscarine, a toxin produced by some poisonous mushrooms.

A Summary of the Parasympathetic Division

In summary:

- The parasympathetic division includes visceral motor nuclei in the brainstem associated with four cranial nerves (III, VII, IX, and X). Autonomic nuclei lie in the lateral portions of the anterior horns in sacral segments S_2–S_4.
- The ganglionic neurons are located in terminal or intramural ganglia.
- The parasympathetic division innervates structures in the head and organs in the thoracic and abdominopelvic cavities.
- All parasympathetic neurons are cholinergic. Release of acetylcholine by preganglionic neurons stimulates nicotinic receptors on ganglionic neurons, and the effect is always excitatory. The release of ACh at neuroeffector junctions stimulates muscarinic receptors, and the effects may be either excitatory or inhibitory, depending on the nature of the enzymes activated when ACh binds to the receptor.
- The effects of parasympathetic stimulation are brief and restricted to specific organs and sites.

Table 17.1 (p. 464) summarizes the characteristics of the parasympathetic division of the ANS.

17.3 CONCEPT CHECK

7 Where are intramural ganglia located?

8 Why does sympathetic stimulation have such widespread effects?

See the blue Answers tab at the back of the book.

17.4 Relationship between the Sympathetic and Parasympathetic Divisions

▶ **KEY POINT** Most organs innervated by the autonomic nervous system are innervated by both the sympathetic and parasympathetic branches. Typically, one division will increase activity of the organ, and the other will decrease the organ's activity.

The sympathetic division has a widespread impact, reaching visceral organs and tissues throughout the body. The parasympathetic division modifies the activity of structures innervated by specific cranial nerves and pelvic nerves. This includes the visceral organs within the thoracic and abdominopelvic cavities. Although some of these organs are innervated by only one autonomic division (sympathetic *or* parasympathetic), most vital organs receive **dual innervation**—that is, they are innervated by both the sympathetic and parasympathetic divisions.

Where dual innervation exists, the two divisions often have opposite, or antagonistic, effects. Dual innervation is most common in the digestive tract, the heart, and the lungs. For example, sympathetic stimulation *decreases* digestive tract motility, while parasympathetic stimulation *increases* its motility.

Anatomy of Dual Innervation

▶ **KEY POINT** Although the parasympathetic and sympathetic branches of the autonomic nervous system exit the CNS from different regions, parasympathetic and sympathetic fibers are often found within the same peripheral ganglia or plexus.

In the head, parasympathetic postganglionic fibers from the ciliary, pterygopalatine, submandibular, and otic ganglia accompany the cranial nerves to their peripheral destinations. Sympathetic innervation reaches the same structures by traveling directly from the superior cervical ganglia of the sympathetic chain.

In the thoracic and abdominopelvic cavities, the sympathetic postganglionic fibers intermix with parasympathetic preganglionic fibers at a series of plexuses (**Figure 17.9**). These are the cardiac plexus, pulmonary plexus, esophageal plexus, celiac plexus, inferior mesenteric plexus, and hypogastric plexus. Nerves leaving these plexuses travel with the blood vessels and lymphatics supplying visceral organs.

Autonomic fibers entering the thoracic cavity intersect at the **cardiac plexus** and the **pulmonary plexus**. These plexuses contain both sympathetic fibers innervating the heart and parasympathetic fibers innervating the heart and lungs. The **esophageal plexus** contains descending branches of the vagus nerve and splanchnic nerves leaving the sympathetic chain ganglia on each side.

Parasympathetic preganglionic fibers of the vagus nerve follow the esophagus as it enters the abdominopelvic cavity. There the parasympathetic fibers join the network of the **celiac plexus** (*solar plexus*). The celiac plexus and an associated smaller plexus, the **inferior mesenteric plexus**, innervate viscera within the abdominal cavity. The **hypogastric plexus** contains the parasympathetic

Figure 17.9 The Peripheral Autonomic Plexuses

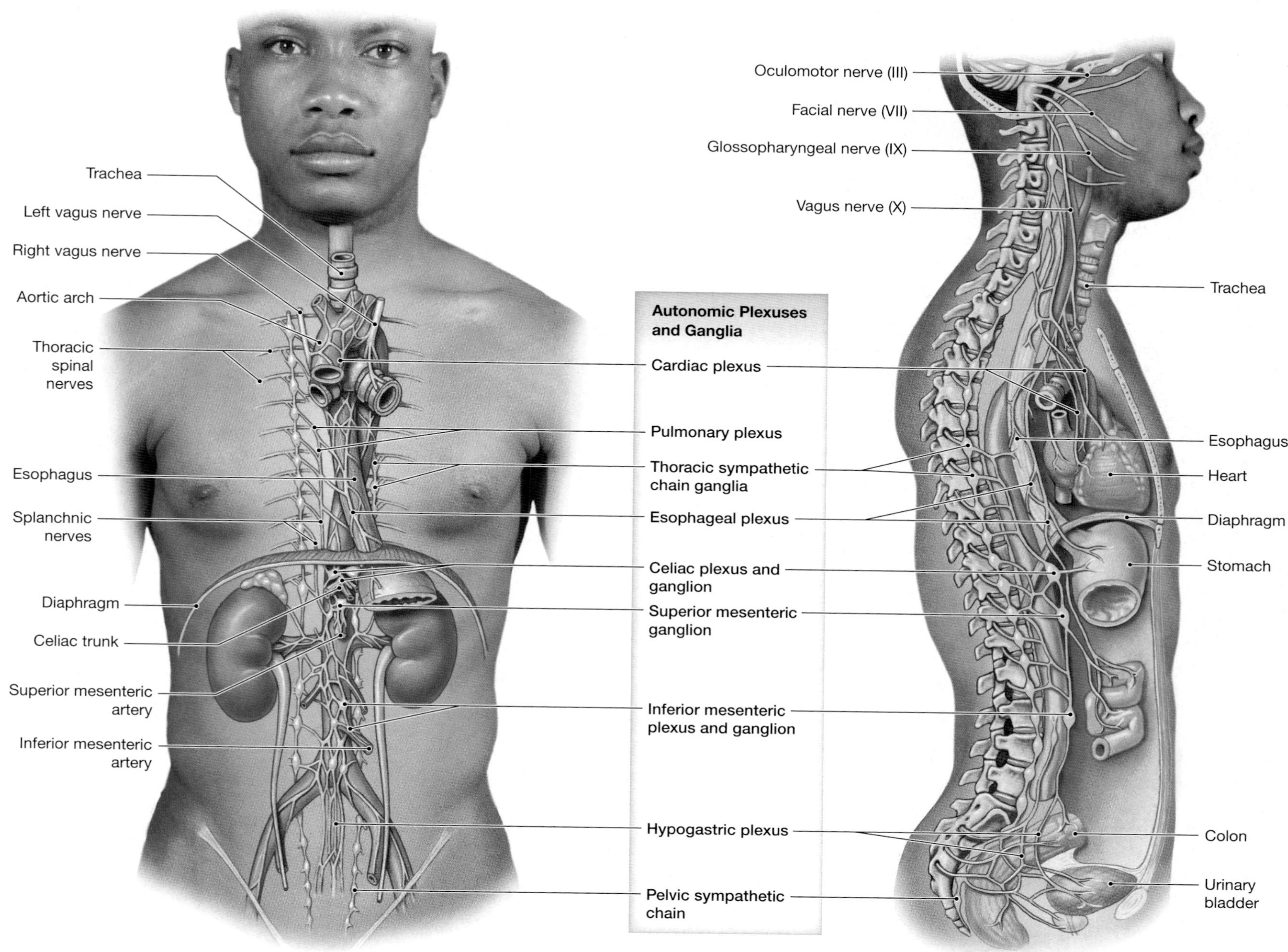

a A diagrammatic view of the distribution of ANS plexuses in the thoracic cavity (cardiac, esophageal, and pulmonary plexuses) and the abdominopelvic cavity (celiac, inferior mesenteric, and hypogastric plexuses)

b A sectional view of the autonomic plexuses

outflow of the pelvic nerves, sympathetic postganglionic fibers from the inferior mesenteric ganglion, and sacral splanchnic nerves from the sympathetic chain. The hypogastric plexus innervates the digestive, urinary, and reproductive organs of the pelvic cavity.

Figure 17.10 and **Table 17.1** compare key features of the sympathetic and parasympathetic divisions of the autonomic nervous system.

Visceral Reflexes

▶ **KEY POINT** Visceral reflexes play an important role in regulating and coordinating the activities of various organs in the digestive system. All autonomic, visceral reflexes are polysynaptic, with at least one synapse in the CNS and another in an autonomic nervous system ganglion.

Visceral reflexes are autonomic reflexes initiated in the viscera (**Figure 17.11**). They provide automatic motor responses that can be modified, facilitated, or inhibited by higher centers, especially those of the hypothalamus.

For example, when a light is shone in one of your eyes, a visceral reflex constricts the pupils of both eyes. In darkness, your pupils dilate. The motor nuclei directing pupillary constriction or dilation are also controlled by hypothalamic centers concerned with emotional states. For example, when you are queasy or nauseated, your pupils constrict; when you are sexually aroused, your pupils dilate.

All visceral reflexes are polysynaptic (having more than one synapse). Each **visceral reflex arc** is made up of a receptor, a sensory nerve, a processing center (one or more interneurons) in the CNS, and two

Figure 17.10 A Comparison of the Sympathetic and Parasympathetic Divisions. This diagram compares fiber length (preganglionic and postganglionic), the general location of ganglia, and the primary neurotransmitter released by each division of the autonomic nervous system.

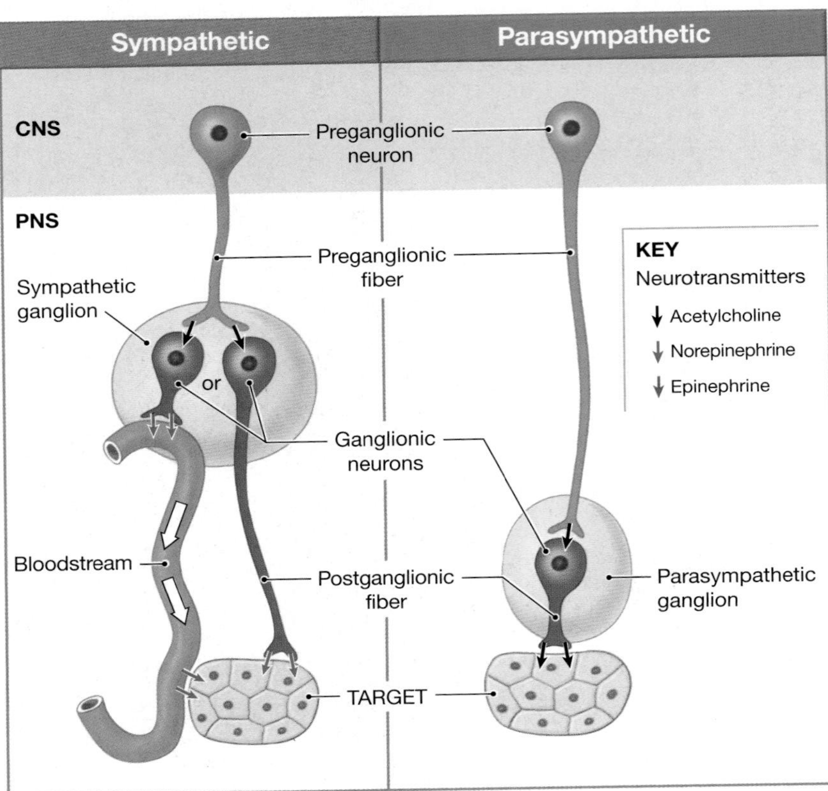

visceral motor neurons (preganglionic and ganglionic). Afferent (sensory) nerves deliver information to the CNS along spinal nerves, cranial nerves, and the autonomic nerves innervating peripheral effectors **(Figure 17.11)**.

Visceral reflexes are either long reflexes or short reflexes. **Long reflexes** of the autonomic nervous system resemble the polysynaptic reflexes introduced

Figure 17.11 Visceral Reflexes. Visceral reflexes have the same basic components as somatic reflexes, but all visceral reflexes are polysynaptic.

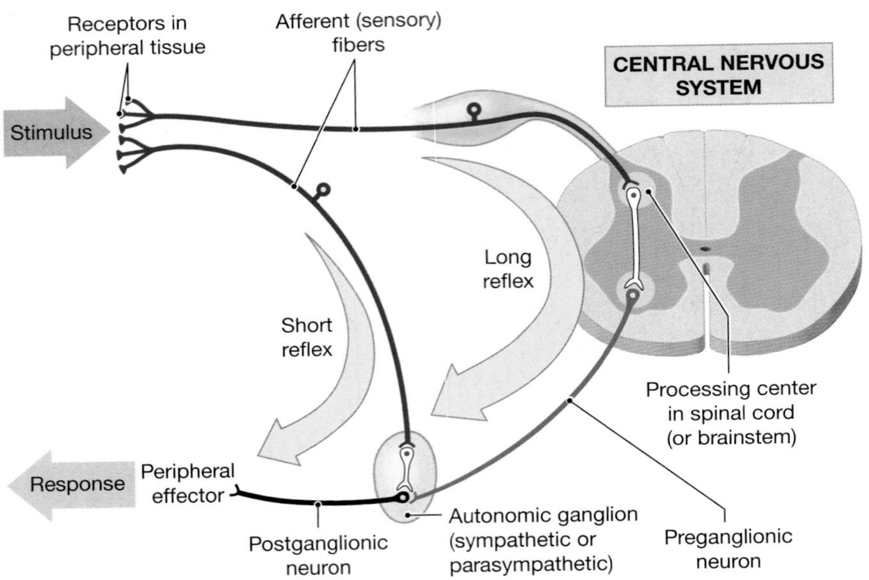

in Chapter 14. ↺ p. 380 Visceral sensory neurons deliver information to the CNS by the dorsal roots of spinal nerves, within the sensory branches of cranial nerves, and within the autonomic nerves innervating visceral effectors. The processing steps involve interneurons within the CNS. The motor neurons of these long reflexes are located within the brainstem or spinal cord. The ANS carries the motor commands to the appropriate visceral effectors after a synapse within a peripheral autonomic ganglion.

Short reflexes bypass the CNS entirely. They involve sensory neurons and interneurons with neuronal somas located within autonomic ganglia. The interneurons synapse on ganglionic neurons, and the motor commands are distributed by postganglionic fibers. Short reflexes control very simple motor responses with localized effects.

In general, short reflexes control patterns of activity in one part of a target organ, while long reflexes coordinate the activities of the entire organ. **Table 17.2** summarizes information concerning important long and short visceral reflexes.

Table 17.1 | A Comparison of the Sympathetic and Parasympathetic Divisions of the ANS

Characteristic	Sympathetic Division	Parasympathetic Division
Location of CNS Visceral Motor Neurons	Lateral horns of spinal segments T_1–L_2	Brainstem and spinal segments S_2–S_4
Location of PNS Ganglia	Sympathetic chain ganglia; collateral ganglia (celiac, superior mesenteric, and inferior mesenteric) located anterior and lateral to the descending aorta	Within the tissues of the target organs (intramural ganglion) or located close to the target organ (terminal ganglion)
Preganglionic Fibers		
Length	Relatively short, myelinated	Relatively long, myelinated
Neurotransmitter released	Acetylcholine	Acetylcholine
Postganglionic Fibers		
Length	Relatively long, unmyelinated	Relatively short, unmyelinated
Neurotransmitter released	Usually norepinephrine	Always acetylcholine
Neuroeffector Junction	Varicosities and enlarged terminal knobs that release transmitter near target cells	Neuroeffector junctions that release transmitter to special receptor surface
Degree of Divergence from CNS to Ganglion Cells	Approximately 1:32	Approximately 1:6
General Functions	Stimulate metabolism, increase alertness, prepare for emergency "fight-or-flight" response	Promote relaxation, nutrient uptake, energy storage ("rest and digest")

Table 17.2 | **Representative Visceral Reflexes**

Reflex	Stimulus	Response	Comments
PARASYMPATHETIC REFLEXES			
Gastric and intestinal reflexes *(See Chapter 25)*	Pressure and physical contact with food materials	Smooth muscle contractions that propel food materials and mix food with secretions	Mediated by the vagus nerve (X)
Defecation *(See Chapter 25)*	Distention of rectum	Relaxation of internal anal sphincter	Requires voluntary relaxation of external anal sphincter
Urination *(See Chapter 26)*	Distention of urinary bladder	Contraction of urinary bladder walls, relaxation of internal urethral sphincter	Requires voluntary relaxation of external urethral sphincter
Direct light and consensual light reflexes *(See Chapter 18)*	Bright light shining in eye(s)	Constriction of pupils of both eyes	Mediated by oculomotor nerve (III)
Swallowing reflex *(See Chapter 25)*	Movement of food and drink into superior pharynx	Smooth muscle and skeletal muscle contractions	Coordinated by swallowing center in medulla oblongata
Vomiting reflex *(See Chapter 25)*	Irritation of digestive tract lining	Reversal of normal smooth muscle action to eject contents	Coordinated by vomiting center in medulla oblongata
Coughing reflex *(See Chapter 24)*	Irritation of respiratory tract lining	Sudden explosive ejection of air	Coordinated by coughing center in medulla oblongata
Baroreceptor reflex *(See Chapter 21)*	Sudden rise in blood pressure in carotid artery	Reduction in heart rate and force of contraction	Coordinated in cardiac center in medulla oblongata
Sexual arousal *(See Chapter 27)*	Erotic stimuli (visual or tactile)	Increased glandular secretions, sensitivity	
SYMPATHETIC REFLEXES			
Cardioacceleratory reflex *(See Chapter 21)*	Sudden decline in blood pressure in carotid artery	Increase in heart rate and force of contraction	Coordinated in cardiac center in medulla oblongata
Vasomotor reflexes *(See Chapter 22)*	Changes in blood pressure in major arteries	Changes in diameter of peripheral blood vessels	Coordinated in vasomotor center in medulla oblongata
Pupillary reflex *(See Chapter 18)*	Low light level reaching visual receptors	Dilation of pupil	Mediated by oculomotor nerve (III)
Emission and ejaculation (in males) *(See Chapter 27)*	Erotic stimuli (tactile)	Contraction of seminal glands and prostate, and skeletal muscle contractions that eject semen	Ejaculation involves the contractions of the bulbospongiosus muscles

17

CLINICAL NOTE

Urinary Bladder Dysfunction Following Spinal Cord Injury

Urinary tract problems are of great concern for people with spinal cord injuries (SCIs). Normally, when the bladder is full, coordinated visceral (autonomic) reflexive contractions of the detrusor and relaxation of the involuntary internal urethral sphincter are accompanied by voluntary (somatic) relaxation of the external urethral sphincter. Afferent sensory nerves from stretch receptors in the wall of the urinary bladder enter the spinal cord between L_1 and S_4. Efferent parasympathetic nerves S_2–S_4 innervate the detrusor and the internal involuntary sphincter. Increased parasympathetic nervous activity causes the bladder to contract and the involuntary urethral sphincter to relax, producing the urge to urinate. Efferent sympathetic nerves L_1 and L_2 inhibit bladder contraction and cause the involuntary urethral sphincter to contract, preventing the passage of urine. The pudendal nerves (S_2–S_4) of the somatic nervous system regulate relaxation of the voluntary skeletal muscle of the external urethral sphincter, allowing urination.

After an SCI, the kidneys are unaffected, but messages can no longer move between the bladder and brain. During the immediate period of spinal shock, which lasts from days to several weeks, there is no somatic or visceral bladder activity and the bladder is completely flaccid, requiring continuous catheter drainage. After spinal shock subsides, depending on the level and completeness of the SCI, bladder and sphincter muscles may be weak, spastic, or poorly coordinated. An injury involving motor neurons T_{12} or above will result in a reflexive, spastic bladder that empties without control. Disruption of the reflex arc in spinal nerves S_2–S_4 results in a flaccid bladder that cannot contract and therefore cannot empty.

Involuntary urination is known as incontinence. To maintain urinary tract health following an SCI, an indwelling urinary catheter or intermittent catheterization to empty the bladder is necessary.

CLINICAL NOTE

Dysautonomia

Dysautonomia is the umbrella term used to describe conditions resulting from abnormal functioning of the autonomic nervous system. It can be caused by failure of the sympathetic or parasympathetic components of the ANS, but it can also involve excessive activities of the ANS. Most of these conditions involve heart rate irregularity, unstable blood pressure, and in extreme cases, death. The cause is unknown and currently there is no cure for any dysautonomia. Treatments are symptomatic and aimed at improving quality of life. Examples of common dysautonomia conditions that result from sympathetic failure include multiple system atrophy (MSA), neurocardiogenic syncope (NCS), and postural orthostatic tachycardia syndrome (POTS). Excessive sympathetic activity conditions include hypertension (high blood pressure) and tachycardia (abnormally rapid heart rate).

- Multiple system atrophy (MSA) is a neurodegenerative disease similar to Parkinson's disease. MSA is a fatal disorder that progresses rapidly and typically affects people aged 40 and older. It is characterized by nerve cell loss, gliosis (astrocyte overgrowth), accumulation of abnormal tubular structures in the cytoplasm and nucleus of oligodendrocytes and in the intermediolateral columns of the spinal cord. Signs and symptoms include ataxia (loss of full control of body movements) and autonomic failure. People with MSA are usually bedridden within 2 years of diagnosis and die within 5-10 years.
- Neurocardiogenic syncope (NCS) is the most common dysautonomia. People with NCS have fainting spells. In its mildest form, individuals will faint once or twice in a lifetime. In more severe forms, individuals will faint several times per day, often falling. People with severe forms of NCS often suffer broken bones and brain injury from falling after fainting.
- Postural orthostatic tachycardia syndrome (POTS) is characterized by lightheadedness, fainting, tachycardia, angina (chest pain), dyspnea (shortness of breath), tremors, exercise intolerance, and temperature sensitivity. It primarily affects women and signs and symptoms are similar to those of congestive heart failure and chronic obstructive pulmonary disease (COPD).

In most organs, long reflexes are most important in regulating visceral activities, but this is not the case with the digestive tract and its associated glands. Here, short reflexes provide most of the control and coordination for normal function, and the neurons involved form the **enteric nervous system**. Parasympathetic innervation by visceral motor neurons can stimulate and coordinate various digestive activities, but the enteric nervous system can control digestive activities independent of the central nervous system.

17.4 CONCEPT CHECK

9 What are visceral reflexes?

10 Name three plexuses in the abdominopelvic cavity.

See the blue Answers tab at the back of the book.

CLINICAL NOTE

Sexual Dysfunction After Spinal Cord Injury

Sexual dysfunction is another concern after SCI. Men normally have two types of erections. A **psychogenic erection** results from sexual thoughts or visual stimulation. The brain responds with arousal messages through spinal nerves T_{10}–L_2. A **reflex erection** is involuntary and results from physical stimulation of the genital region causing afferent signals to travel to the spinal cord via the pudendal nerves.

The parasympathetic nervous system controls erection by way of sacral segments S_2–S_4. Most men with an SCI above S_4 are able to have a reflex erection with physical stimulation. Medications can enhance a reflex erection. Ejaculation is controlled by the sympathetic nervous system and is less predictable. In general, erections are more likely with incomplete spinal cord injuries than with complete injuries.

Study Outline

Introduction p. 450

- The **autonomic nervous system (ANS)** regulates body temperature and coordinates cardiovascular, respiratory, digestive, excretory, and reproductive functions. Routine physiological adjustments to systems are made by the autonomic nervous system operating at the subconscious level.

17.1 A Comparison of the Somatic and Autonomic Nervous Systems p. 450

- The autonomic nervous system, like the somatic nervous system, has afferent and efferent neurons. However, in the ANS, the afferent pathways originate in visceral receptors, and the efferent pathways connect to visceral effector organs.

- In addition to the difference in receptor and effector organ location, the ANS differs from the SNS in the arrangement of the neurons connecting the central nervous system to the effector organs. Visceral motor neurons in the CNS send axons (**preganglionic fibers**) to synapse on ganglionic neurons, whose cell bodies are located in autonomic ganglia outside the CNS. The axon of the ganglionic neuron is a **postganglionic fiber** that innervates peripheral organs. *(See Figure 17.1 and Spotlight Figure 17.2.)*

Sympathetic and Parasympathetic Subdivisions of the ANS p. 450

- There are two major subdivisions in the ANS: the sympathetic division and the parasympathetic division. *(See Spotlight Figure 17.2.)*

- Visceral efferents from the thoracic and lumbar segments form the **sympathetic** (*thoracolumbar*) **division** ("fight-or-flight" system) of the ANS. Generally, it stimulates tissue metabolism, increases alertness, and prepares the body to deal with emergencies. Visceral efferents leaving the brainstem and sacral segments form the **parasympathetic** (*craniosacral*) **division** ("rest-and-digest" system). Generally, it conserves energy and promotes sedentary activities. *(See Spotlight Figure 17.2.)*

- Both divisions affect target organs via neurotransmitters. Plasma membrane receptors determine whether the response will be stimulatory or inhibitory. Generally, neurotransmitter effects are as follows: (1) All preganglionic terminals release acetylcholine (ACh) and are excitatory; (2) all postganglionic parasympathetic terminals release ACh and effects may be excitatory or inhibitory; and (3) most postganglionic sympathetic terminals release norepinephrine (NE) and effects are usually excitatory.

17.2 | The Sympathetic Division p. 450

- The **sympathetic division** consists of preganglionic neurons between spinal cord segments T_1 and L_2, ganglionic neurons in ganglia near the vertebral column, and specialized neurons within the adrenal gland. *(See Spotlight Figures 17.2 and 17.3, and Figures 17.1, 17.4, and 17.10.)*

- There are two types of sympathetic ganglia: **sympathetic chain ganglia** and **collateral ganglia** (prevertebral ganglia).

Sympathetic Chain Ganglia p. 450

- Between spinal segments T_1 and L_2, each ventral root gives off a white ramus communicans with preganglionic fibers to a sympathetic chain ganglion. These preganglionic fibers tend to undergo extensive divergence before they synapse with the ganglionic neuron. The synapse occurs within the sympathetic chain ganglia, within one of the collateral ganglia, or within the adrenal medulla. Preganglionic fibers run between the sympathetic chain ganglia and interconnect them. Postganglionic fibers targeting visceral effectors in the body wall enter the gray ramus communicans to return to the spinal nerve for distribution, whereas those that target thoracic cavity structures form autonomic nerves that go directly to their visceral destination. *(See Spotlight Figures 17.2 and 17.3 and Figure 17.4.)*

- There are 3 cervical, 11–12 thoracic, 2–5 lumbar, and 4–5 sacral ganglia and 1 coccygeal sympathetic ganglion in each sympathetic chain. Every spinal nerve has a gray ramus communicans that carries sympathetic postganglionic fibers. In summary: (1) Only thoracic and superior lumbar ganglia receive preganglionic fibers by way of white rami; (2) the cervical, inferior lumbar, and sacral chain ganglia receive preganglionic innervation from collateral fibers of sympathetic neurons; and (3) every spinal nerve receives a gray ramus communicans from a ganglion of the sympathetic chain. *(See Figure 17.4.)*

Collateral Ganglia p. 451

- The abdominopelvic viscera receive sympathetic innervation via preganglionic fibers that pass through the sympathetic chain to synapse within collateral ganglia. The preganglionic fibers that innervate the collateral ganglia form the **splanchnic nerves** (greater, lesser, lumbar, and sacral). *(See Spotlight Figure 17.3 and Figures 17.4 and 17.9.)*

- The splanchnic nerves innervate the hypogastric plexus and three collateral ganglia: (1) the celiac ganglion, (2) the superior mesenteric ganglion, and (3) the inferior mesenteric ganglion. *(See Figures 17.4 and 17.9.)*

- The **celiac ganglion** innervates the stomach, duodenum, liver, pancreas, spleen, and kidney; the **superior mesenteric ganglion** innervates the small intestine and initial segments of the large intestine; and the **inferior mesenteric ganglion** innervates the kidney, bladder, sex organs, and terminal portions of the large intestine. *(See Figures 17.4 and 17.9.)*

Adrenal Medulla p. 457

- Some preganglionic fibers do not synapse as they pass through both the sympathetic chain and collateral ganglia. Instead, they enter one of the adrenal glands and synapse on modified neurons within the **adrenal medulla**. These cells release norepinephrine (NE) and epinephrine (E) into the circulation, causing a prolonged sympathetic stimulation effect. *(See Spotlight Figure 17.3 and Figures 17.4 and 17.5.)*

Effects of Sympathetic Stimulation p. 457

- In a crisis, the entire division responds, an event called **sympathetic activation**. Its effects include increased alertness, a feeling of energy and euphoria, increased cardiovascular and respiratory activity, general increase in muscle tone, and mobilization of energy reserves.

Sympathetic Activation and Neurotransmitter Release p. 458

- Stimulation of the sympathetic division has two distinctive results: the release of norepinephrine (or in some cases acetylcholine) at neuroeffector junctions and the secretion of epinephrine and norepinephrine into the general circulation. *(See Figure 17.6.)*

Plasma Membrane Receptors and Sympathetic Function p. 458

- There are two classes of sympathetic receptors that are stimulated by both norepinephrine and epinephrine: **alpha receptors** and **beta receptors**.

- Most postganglionic fibers release norepinephrine, but a few release acetylcholine. Postganglionic fibers innervating sweat glands and blood vessels of skeletal muscles release acetylcholine (ACh).

A Summary of the Sympathetic Division p. 458

- The sympathetic division has the following characteristics: (1) two segmentally arranged sympathetic chains lateral to the vertebral column, three collateral ganglia anterior to the vertebral column, and two adrenal medullae; (2) preganglionic fibers are relatively short, except for those of the adrenal medullae, while postganglionic fibers are quite long; (3) extensive divergence typically occurs, with a single preganglionic fiber synapsing with many ganglionic neurons in different ganglia; (4) all preganglionic fibers release ACh, while most postganglionic fibers release NE; and (5) effector response depends on the nature and activity of the receptor. *(See Table 17.1.)*

17.3 The Parasympathetic Division p. 459

- The **parasympathetic division** consists of (1) preganglionic neurons in the brainstem and in sacral segments of the spinal cord and (2) ganglionic neurons in peripheral ganglia located within or immediately next to target organs. *(See Figures 17.7 and 17.8 and Table 17.1.)*

Organization and Anatomy of the Parasympathetic Division p. 459

- Preganglionic fibers leave the brain in cranial nerves III (oculomotor), VII (facial), IX (glossopharyngeal), and X (vagus). *(See Spotlight Figure 17.2 and Figures 17.7 and 17.8.)*
- Parasympathetic fibers in the oculomotor, facial, and glossopharyngeal nerves help control visceral structures in the head, and they synapse in the **ciliary**, **pterygopalatine**, **submandibular**, and **otic ganglia**. Fibers in the vagus nerve supply preganglionic parasympathetic innervation to intramural ganglia within structures in the thoracic and abdominopelvic cavity. *(See Figures 17.7 and 17.8.)*
- Preganglionic fibers leaving the sacral segments form **pelvic nerves** that innervate intramural ganglia in the kidney, bladder, latter parts of the large intestine, and sex organs. *(See Figure 17.8.)*

General Functions of the Parasympathetic Division p. 460

- The effects produced by the parasympathetic division include pupil constriction, digestive gland secretion, hormone secretion for nutrient absorption, increased digestive tract activity, defecation activities, urination activities, respiratory passageway constriction, reduced heart rate, and sexual arousal. These general functions center on relaxation, food processing, and energy absorption.

17

Parasympathetic Activation and Neurotransmitter Release p. 462

- All of the parasympathetic preganglionic and postganglionic fibers release ACh at synapses and neuroeffector junctions. The effects are short-lived because of the actions of enzymes at the postsynaptic plasma membrane and in the surrounding tissues.
- Two different types of ACh receptors are found in postsynaptic plasma membranes. **Nicotinic receptors** are located on ganglion cells of both divisions of the ANS and at neuromuscular synapses. Exposure to ACh causes excitation by opening plasma membrane channels. **Muscarinic receptors** are located at neuroeffector junctions in the parasympathetic division and those cholinergic neuroeffector junctions in the sympathetic division. Stimulation of muscarinic receptors produces a longer-lasting effect than does stimulation of nicotinic receptors.

A Summary of the Parasympathetic Division p. 462

- The parasympathetic division has the following characteristics: (1) It includes visceral motor nuclei associated with cranial nerves III, VII, IX, and X and sacral segments S_2–S_4; (2) ganglionic neurons are located in terminal or intramural ganglia near or within target organs, respectively; (3) it innervates areas serviced by cranial nerves and organs in the thoracic and abdominopelvic cavities; (4) all parasympathetic neurons are cholinergic—the postganglionic neurons are also cholinergic and are further subdivided as being either muscarinic or nicotinic receptors; and (5) effects are usually brief and restricted to specific sites. *(See Figure 17.10 and Table 17.1.)*

17.4 Relationship between the Sympathetic and Parasympathetic Divisions p. 462

- The sympathetic division has widespread influence, reaching visceral and somatic structures throughout the body. *(See Figure 17.4 and Table 17.1.)*
- The parasympathetic division innervates only visceral structures serviced by cranial nerves or lying within the thoracic and abdominopelvic cavities. Organs with **dual innervation** receive instructions from both divisions. *(See Figure 17.10 and Table 17.1.)*

Anatomy of Dual Innervation p. 462

- In body cavities, the parasympathetic and sympathetic nerves intermingle to form a series of characteristic nerve plexuses (nerve networks), which include the **cardiac**, **pulmonary**, **esophageal**, **celiac**, **inferior mesenteric**, and **hypogastric plexuses**. *(See Figure 17.9.)*
- Important anatomical and physiological differences exist between the sympathetic and parasympathetic divisions of the autonomic nervous system. *(See Figure 17.10 and Table 17.1.)*

Visceral Reflexes p. 463

- **Visceral reflexes** are the simplest functions of the ANS and are classified as either **long reflexes** or **short reflexes**. They provide automatic motor responses that can be modified, facilitated, or inhibited by higher centers, especially in the hypothalamus. *(See Figure 17.11 and Table 17.2.)*

Chapter Review

For answers, see the blue Answers tab at the back of the book.

Level 1 Reviewing Facts and Terms

Match each numbered item with the most closely related lettered item.

1. preganglionic ..... ☐
2. thoracolumbar ..... ☐
3. parasympathetic ..... ☐
4. prevertebral ..... ☐
5. paravertebral ..... ☐
6. acetylcholine ..... ☐
7. epinephrine ..... ☐
8. sympathetic ..... ☐
9. splanchnic ..... ☐
10. crisis ..... ☐

(a) all preganglionic fibers
(b) preganglionic fibers to collateral ganglia
(c) first neuron
(d) collateral ganglia
(e) adrenal medulla
(f) sympathetic activation
(g) sympathetic division
(h) terminal ganglia
(i) sympathetic chain ganglia
(j) long postganglionic fiber

11. Visceral motor neurons in the CNS
    (a) are ganglionic neurons.
    (b) are in the dorsal root ganglion.
    (c) have unmyelinated axons except in the lower thoracic region.
    (d) send axons to synapse on peripherally located ganglionic neurons.
12. Splanchnic nerves
    (a) are formed by parasympathetic postganglionic fibers.
    (b) include preganglionic fibers that go to collateral ganglia.
    (c) control sympathetic function of structures in the head.
    (d) connect one chain ganglion with another.
13. Which of the following ganglia belong to the sympathetic division of the ANS?
    (a) otic ganglion
    (b) sphenopalatine ganglion
    (c) sympathetic chain ganglia
    (d) all of the above

14. Preganglionic fibers of the ANS sympathetic division originate in the
    (a) cerebral cortex of the brain.
    (b) medulla oblongata.
    (c) brainstem and sacral spinal cord.
    (d) thoracic and lumbar spinal segments.
15. The neurotransmitter at all synapses and neuroeffector junctions in the parasympathetic division of the ANS is
    (a) epinephrine.
    (b) cyclic AMP.
    (c) norepinephrine.
    (d) acetylcholine.
16. The large cells in the adrenal medulla, which resemble neurons in sympathetic ganglia,
    (a) are located in the adrenal cortex.
    (b) release acetylcholine into blood capillaries.
    (c) release epinephrine and norepinephrine into blood capillaries.
    (d) have no endocrine functions.
17. Sympathetic preganglionic fibers are characterized as being
    (a) short in length and unmyelinated.
    (b) short in length and myelinated.
    (c) long in length and myelinated.
    (d) long in length and unmyelinated.
18. All preganglionic autonomic fibers release _____ at their axon terminals, and the effects are always _____.
    (a) norepinephrine; inhibitory
    (b) norepinephrine; excitatory
    (c) acetylcholine; excitatory
    (d) acetylcholine; inhibitory
19. Postganglionic fibers of autonomic neurons are usually
    (a) myelinated.
    (b) unmyelinated.
    (c) larger than preganglionic fibers.
    (d) located in the spinal cord.
20. The white ramus communicans
    (a) carries the postganglionic fibers to the effector organs.
    (b) arises from the dorsal root of the spinal nerves.
    (c) has fibers that do not diverge.
    (d) carries the preganglionic fibers into a nearby sympathetic chain ganglion.

Level 2 Reviewing Concepts

1. Cutting the ventral root of the spinal nerve at L_2 would interrupt the transmission of what type of information?
   (a) voluntary motor output
   (b) ANS motor output
   (c) sensory input
   (d) a and b
2. Damage to the ventral roots of the first five thoracic spinal nerves on the right side of the body would interfere with the ability to
   (a) dilate the right pupil.
   (b) dilate the left pupil.
   (c) contract the right biceps brachii.
   (d) contract the left biceps brachii.
3. What anatomical mechanism is involved in causing a person to blush?
   (a) Blood flow to the skin is increased by parasympathetic stimulation.
   (b) Sympathetic stimulation relaxes vessel walls, increasing blood flow to the skin.
   (c) Parasympathetic stimulation decreases skin muscle tone, allowing blood to pool at the surface.
   (d) Sympathetic stimulation increases respiratory oxygen uptake, making the blood brighter red.
4. If the visceral signal from the small intestine does not reach the spinal cord, which structures might be damaged?
   (a) preganglionic neurons
   (b) white rami communicantes
   (c) gray rami communicantes
   (d) none of the above
5. The effects of epinephrine and norepinephrine released by the adrenal glands last longer than those of either chemical when released at neuroeffector junctions. Why?
6. Why are the effects of parasympathetic stimulation more specific and localized than those of the sympathetic division?
7. How do sympathetic chain ganglia differ from both collateral ganglia and intramural ganglia?
8. Compare the general effects of the sympathetic and parasympathetic divisions of the ANS.
9. Describe the general organization of the pathway for visceral motor output.

Level 3 Critical Thinking

1. In some severe cases, a person suffering from stomach ulcers may need to have surgery to cut the branches of the vagus nerve that innervates the stomach. How would this help the problem?
2. Kassie is stung on the neck by a wasp. Because she is allergic to wasp venom, her throat begins to swell and her respiratory passages constrict. Which would be more helpful in relieving her symptoms: acetylcholine or epinephrine? Why?

 CLINICAL CASE WRAP-UP

First Day of Anatomy Lab

Tim has suffered a vasovagal (*vaso*, vascular, + *vagal*, vagus nerve) loss of consciousness, commonly known as fainting. Environmental triggers, including the sight and smell of the cadaver, caused a momentary malfunction of his autonomic nervous system (ANS). The sympathetic division of his ANS failed him, while his parasympathetic division went into overdrive.

Parasympathetic stimulation of the vagus nerve caused it to release ACh at the cardiac plexus. Meanwhile, there was no counteracting stimulation from the sympathetic nervous system. This slowed Tim's heart rate, made his heart's contractions less forceful, and lowered his blood pressure. Blood flow to his brain decreased, causing the fainting episode.

Tim's rapid recovery begins as soon as the sympathetic division of the ANS takes over. Norepinephrine is released at the cardiac plexus and throughout the body. His heart rate and blood pressure increase, along with his level of consciousness.

"Don't worry, Tim," says the instructor. "Everybody is allowed one fainting episode. And everybody can overcome this. Next time, just lie down before you fall down."

1. Why does the instructor tell Tim to lie down the next time he feels faint?
2. In addition to the cardiac plexus, where else in the body would the sympathetic nervous system act during Tim's recovery?

See the blue Answers tab at the back of the book.

17

Related Clinical Terms

parasympathetic blocking agents: Drugs that target the muscarinic receptors at neuromuscular or neuroglandular junctions.

sympathetic blocking agents: Drugs that bind to receptor sites, preventing a normal response to neurotransmitters or drugs that mimic the effects of sympathomimetic stimulation.

18 The Nervous System

General and Special Senses

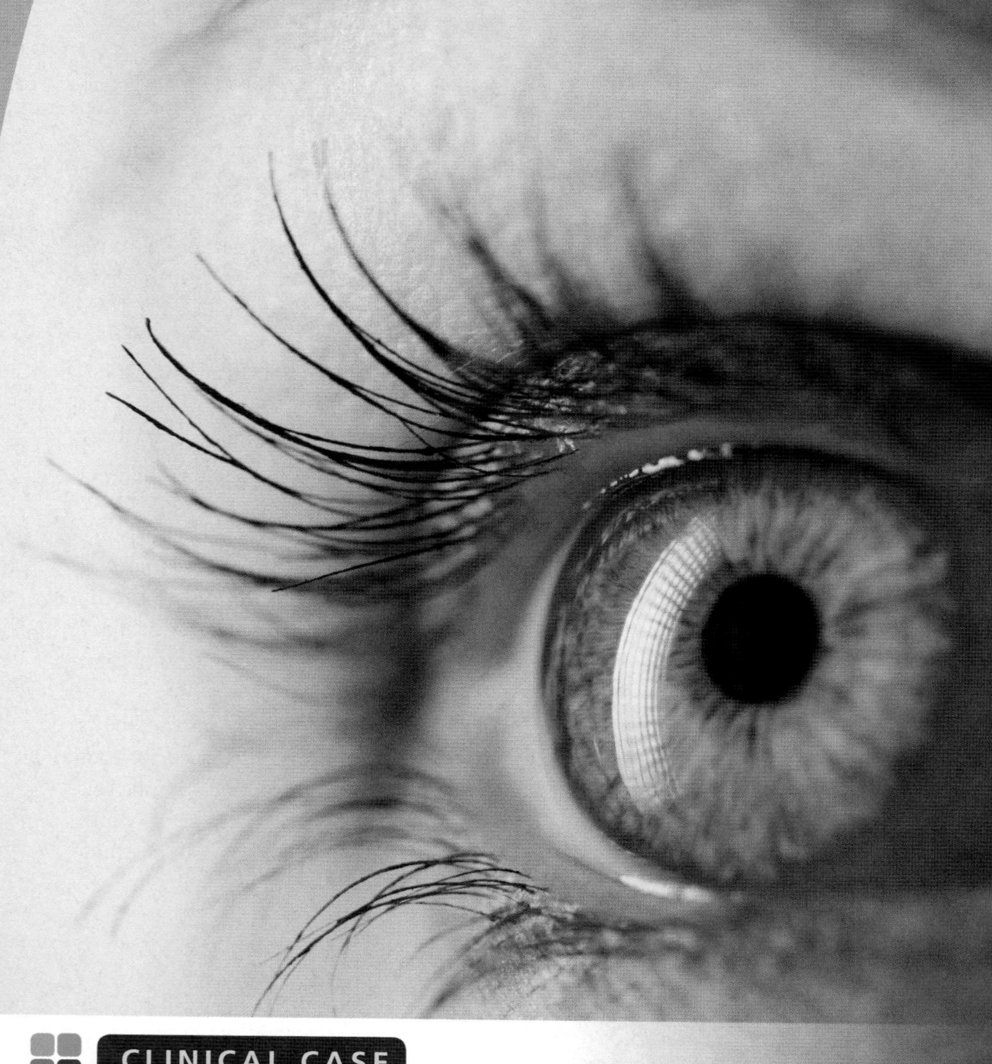

Learning Outcomes

These Learning Outcomes correspond by number to this chapter's sections and indicate what you should be able to do after completing the chapter.

CLINICAL CASE

Why Am I So Dizzy?

John, a young attorney, has always suffered from motion sickness. He never liked amusement park rides or even riding in a car. Recently, his allergies have been acting up and he has had a "stuffy head." This morning, as he is leaving for work, he has a sudden episode of vertigo (spinning sensation) so severe he cannot walk, and he falls to the floor. There is a constant roaring buzz (tinnitus) in his left ear, and it has a sensation of fullness. He feels nauseated and realizes he can't go to work today.

John has never experienced anything like this before. Frightened, he calls his doctor, who immediately refers him to an otolaryngologist (an ear, nose, and throat [ENT] doctor). While waiting for his appointment, he lies down and sleeps for over 3 hours. He awakens in time for his ENT appointment, but when he stands up, he realizes that his vertigo, tinnitus, and nausea have resolved. Now he feels almost silly to be going to the doctor.

What happened to John? Will it happen again? ***To find out, turn to the Clinical Case Wrap-Up on p. 504.***

EVERY PLASMA MEMBRANE FUNCTIONS as a receptor for the cell because it responds to changes in the extracellular environment. Plasma membranes differ in their sensitivities to specific electrical, chemical, and mechanical stimuli. For example, a hormone that stimulates a neuron may have no effect on an osteocyte because the plasma membranes of neurons and osteocytes contain different receptor proteins. A **sensory receptor** is a specialized cell or cell process monitoring conditions in the body or external environment. Stimulation of the receptor directly or indirectly alters the production of action potentials in a sensory neuron. pp. 348–351

The sensory information arriving at the central nervous system (CNS) is called a **sensation**; a **perception** is a conscious awareness of a sensation. **General senses** refer to sensations of temperature, pain, touch, pressure, vibration, and proprioception (body position). General sensory receptors are distributed throughout the body. These sensations arrive at the primary somatosensory cortex by pathways we described previously.

The **special senses** are smell, taste, equilibrium (balance), hearing, and vision. The specialized receptors providing these sensations are structurally more complex than those of the general senses. These receptors are localized within complex **sense organs**, such as the eye or ear. The information is then transmitted to centers throughout the brain.

Sensory receptors keep the nervous system updated on changes in the body's internal environment and the surrounding external environment. The nervous system controls and coordinates the body's swift responses to specific stimuli. This chapter begins by summarizing receptor function and basic concepts in sensory processing. We then apply this information to each of the general and special senses.

18.1 | Receptors

▶ **KEY POINT** Each receptor has a specific sensitivity. This specificity results from the anatomy of the receptor or from the presence of accessory cells or structures that minimize its exposure to other stimuli.

The simplest receptors in the human body are the dendrites of sensory neurons, called **free nerve endings**. They can be stimulated by many different stimuli. For example, free nerve endings providing the sensation of pain may respond to chemical stimulation, pressure, temperature changes, or physical damage. In contrast, the receptors of the eye are surrounded by accessory cells that normally prevent their stimulation by anything other than light. This characteristic receptor sensitivity is termed **receptor specificity**.

The area monitored by a single receptor cell is its **receptive field** (Figure 18.1). Whenever a sufficiently strong stimulus arrives in the receptive field, the CNS receives the information. Our ability to localize a stimulus depends on the size of the receptive field: *The larger the receptive field, the harder it is to localize the stimulus*. For example, a touch receptor on your back may have a receptive field 7 cm (2.5 in.) in diameter, while a receptor on your tongue may have a receptive field less than a millimeter in diameter. As a result, we can describe a light touch on the back only generally, but we can be very precise about the location of a stimulus on the tongue.

Figure 18.1 Receptors and Receptive Fields. Each receptor monitors a specific area known as the receptive field.

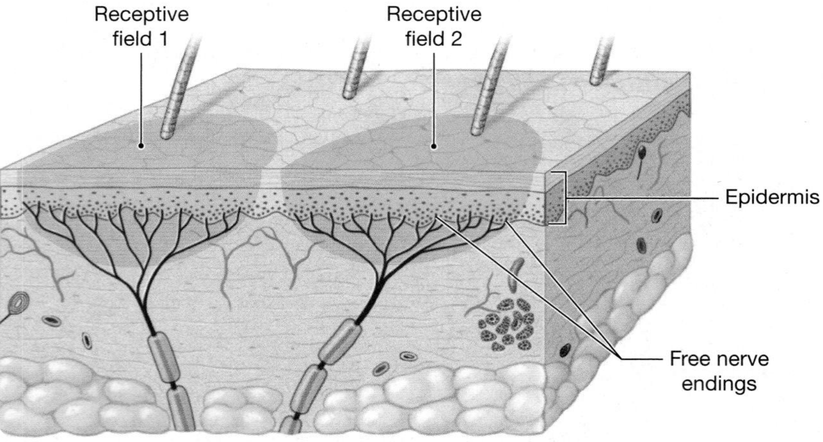

An arriving stimulus can take many different forms—it may be a physical force, such as pressure, a dissolved chemical, a sound, or a beam of light. Regardless of the nature of the stimulus, sensory information is sent to the CNS as an action potential, which is an electrical event. The CNS processes and interprets the incoming information at the conscious and subconscious levels.

Interpretation of Sensory Information

▶ **KEY POINT** Specific receptors detect sensory information and transmit it to the spinal cord or brainstem. There, specific tracts, organized to carry specific sensations, carry this information to specialized areas of the brain, where the information is interpreted.

Sensory information arriving at the CNS is transmitted up the spinal cord and into the brain according to (1) the location where the sensory information was detected and (2) the nature of the stimulus that was detected. Axons relay information from point A (the receptor) to point B in the CNS (a neuron at a specific site in the cerebral cortex) along the sensory pathways discussed in Chapter 15. Each pathway carries information for a specific sensation, such as touch, pressure, or vision, from receptors in a specific part of the body. All other characteristics of the stimulus are determined by the pattern of the action potentials in the afferent fibers. This **sensory coding** provides information about the strength, duration, variation, and movement of the stimulus.

Tonic receptors are sensory neurons that are always active. The photoreceptors of the eye and receptors monitoring body position are two examples of tonic receptors. Other receptors are normally inactive, but become active for a short time whenever there is a change in the conditions they are monitoring. These **phasic receptors** provide information on the intensity and rate of change of a stimulus. Touch and pressure receptors in the skin are examples of phasic receptors. Some receptors combine phasic and tonic coding. Such receptors convey extremely complicated sensory information; receptors monitoring the positions and movements of joints are in this category.

Central Processing and Adaptation

▶ **KEY POINT** As you slip your hand into a bucket of hot water, you are immediately aware of the water's temperature. However, if you keep your hand in the bucket, you become less and less aware of the water's temperature, even though the temperature hasn't changed much, if at all. This process is called adaptation.

Adaptation is a reduction in sensitivity in the presence of a constant stimulus. Adaptation occurs in the peripheral and central nervous systems, where it is called, respectively, peripheral and central adaptation.

- When a receptor or sensory neuron alters its level of activity, **peripheral** (*sensory*) **adaptation** occurs. The receptor responds strongly at first, but then the activity along the afferent fiber gradually declines because of synaptic fatigue. This response is characteristic of phasic receptors, which are also called **fast-adapting receptors**. Tonic receptors show little or no peripheral adaptation, so they are called **slow-adapting receptors**.
- A few seconds after exposure to a new smell, conscious awareness of the stimulus virtually disappears, although the sensory neurons within the nose are still quite active. This is **central adaptation**, a process involving

nuclei along the sensory pathways within the CNS. At the subconscious level, central adaptation restricts the amount of information arriving at the cerebral cortex. As we discussed in Chapter 15, most of the incoming sensory information is processed in centers along the spinal cord or brainstem. This processing triggers involuntary reflexes. Because of central adaptation, only about 1 percent of the information traveling over afferent fibers reaches the cerebral cortex and our conscious awareness.

Sensory Limitations

Sensory receptors provide a constant detailed picture of our bodies and our surroundings. This picture is, however, incomplete for several reasons:

- Humans do not have receptors for every possible stimulus.
- Our receptors have characteristic ranges of sensitivity.
- The CNS must interpret a stimulus. Our perception of a particular stimulus is interpretation and not always reality.

18.1 CONCEPT CHECK

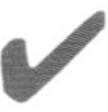

1. What different types of stimuli may activate free nerve endings?
2. Contrast tonic and phasic receptors.
3. What is a sensation?
4. Which sensations are grouped as "general senses"?

See the blue Answers tab at the back of the book.

18.2 | The General Senses

▶ **KEY POINT** Receptors for the general senses are scattered throughout the body and are simple in structure.

General sensory receptors are divided into three classes: exteroceptors, proprioceptors, and interoceptors. **Exteroceptors** provide information about the external environment; **proprioceptors** monitor body position; and **interoceptors** monitor conditions inside the body.

A more detailed classification system using the nature of the stimulus detected divides general sensory receptors into four classes:

1. **Nociceptors** (nō-sē-SEP-torz; *noceo*, hurt) respond to a variety of stimuli associated with tissue damage. Activation of these receptors causes the sensation of pain.
2. **Thermoreceptors** respond to changes in temperature.
3. **Mechanoreceptors** are stimulated or inhibited by physical distortion, contact, or pressure on their plasma membranes.
4. **Chemoreceptors** monitor the chemical composition of body fluids and respond to the presence of specific molecules.

Each receptor class has distinct structural and functional characteristics. Some mechanoreceptors are identified by the name of the person who first discovered or named them. Such terms are called eponyms (*commemorative names*). Anatomists have proposed more standardized names to replace the eponyms, but no standardization or consensus exists. More importantly, *none* of the alternative names have been widely accepted in the primary literature (professional, technical, or clinical journals or reports). To avoid confusion, we will use eponyms in this chapter whenever there is no generally accepted alternative.

Nociceptors

▶ **KEY POINT** Nociceptors (pain receptors) are free nerve endings with large receptive fields. They are common in the skin, joint capsules, and periostea of bones and around the walls of blood vessels. Very few nociceptors are found in other deep tissues or in most organs of the abdominopelvic cavities.

There are three types of **nociceptors**: (1) receptors sensitive to extreme temperature, (2) receptors sensitive to physical damage, and (3) receptors sensitive to dissolved chemicals, such as those released by injured cells. However, very strong temperature, pressure, and/or chemical stimuli will excite all three receptor types.

Sensations of **fast pain**, or *pricking pain*, are produced by deep cuts or similar injuries. Painful sensations cease only after tissue damage has ended. However, central adaptation may reduce pain *perception* while the pain receptors are still stimulated.

Slow pain (*burning pain*) sensations result from the same types of injuries as fast pain. However, sensations of slow pain begin later and persist longer. For example, a cut on the hand produces an immediate awareness of fast pain, followed somewhat later by the ache of slow pain. Slow pain sensations activate the reticular formation and thalamus. You become aware of the pain but only have a general idea of the area affected.

Pain sensations from visceral organs reach the spinal cord through visceral sensory nerves. These nerves enter the spinal cord through the dorsal roots of spinal nerves. Visceral pain sensations are perceived as originating in more superficial regions that are innervated by these same spinal nerves. The precise mechanism responsible for this **referred pain** remains to be determined, but several clinical examples are shown in **Figure 18.2**.

Figure 18.2 Referred Pain. Pain sensations originating in visceral organs are often perceived as involving specific regions of the body surface innervated by the same spinal nerves.

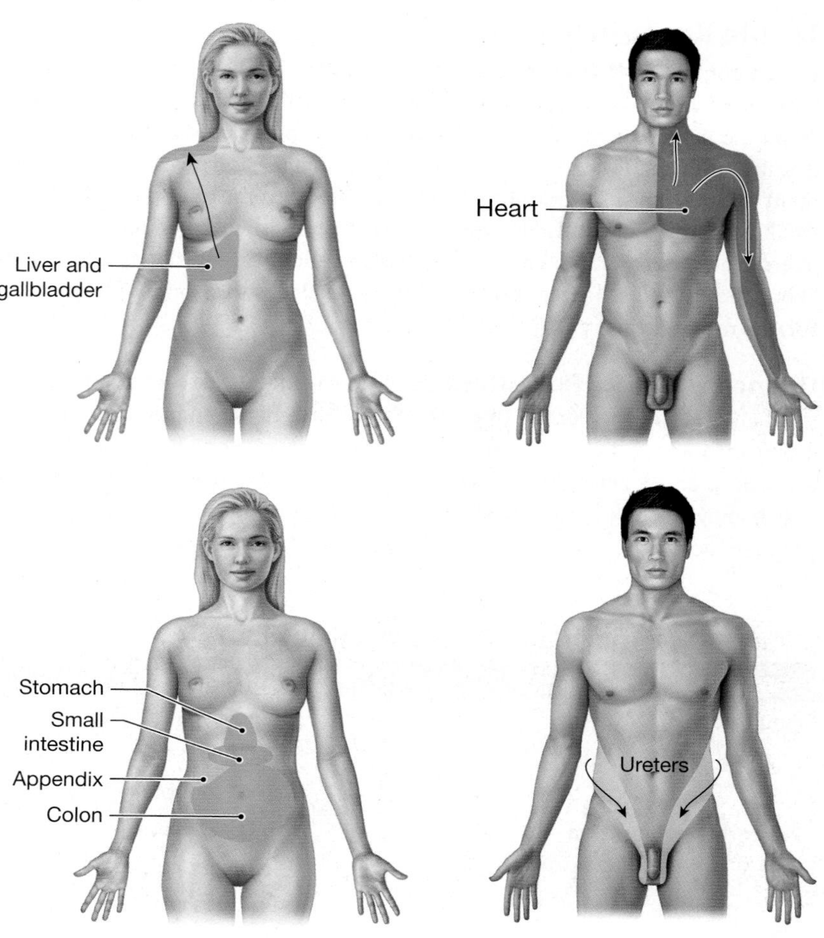

Thermoreceptors

▶ **KEY POINT** Thermoreceptors detect changes in temperature. There are two classes: cold and warm, and they are found in the dermis of the skin, skeletal muscles, the liver, and the hypothalamus.

There are two types of **thermoreceptors**: cold and warm. Cold receptors are three or four times more numerous than warm receptors. The receptors are free nerve endings that detect changes in temperature. There are no known structural differences between cold and warm thermoreceptors.

Temperature sensations are conducted in the spinothalamic tracts—the same pathways carrying pain sensations. Temperature sensations are sent to the reticular formation, the thalamus, and the primary somatosensory cortex.

Thermoreceptors are phasic receptors. They are very active when the temperature is changing, but quickly adapt to a stable temperature. When you enter an air-conditioned classroom on a hot summer day, the temperature change seems extreme at first, but you quickly become comfortable as adaptation occurs.

Mechanoreceptors

▶ **KEY POINT** Mechanoreceptors are sensitive to stimuli that stretch, compress, twist, or distort their cell membranes. There are three classes: tactile receptors, baroreceptors, and proprioceptors.

18

There are three classes of mechanoreceptors: (1) tactile receptors detect sensations of touch, pressure, and vibration; (2) baroreceptors (bar-ō-rē-SEP-torz; *baro-*, pressure) detect pressure changes in the walls of blood vessels and in portions of the digestive, reproductive, and urinary tracts; and (3) proprioceptors detect the positions of joints and skeletal muscles and are the most complex of the general sensory receptors.

Tactile Receptors

The anatomy of **tactile receptors** ranges from free nerve endings to specialized sensory complexes with accessory cells and supporting structures (**Figure 18.3**). Some tactile receptors, such as **fine touch** and **pressure receptors**, provide detailed information about exact location, shape, size, texture, and movement of the stimulation. These receptors are extremely sensitive and have small receptive fields. In contrast, other receptors, such as **crude touch** and **pressure receptors**, provide poor localization and little additional information about the stimulus. Tactile receptors are subdivided into two groups: unencapsulated receptors and encapsulated receptors.

Unencapsulated Receptors Free nerve endings are common in the papillary layer of the dermis (**Figure 18.3a**). Free nerve endings are also associated with hair follicles. The free nerve endings of the **root hair plexus** detect distortions and movements across the body surface (**Figure 18.3b**). When the hair is moved, the movement of the follicle distorts the sensory dendrites and produces action potentials in the afferent fiber. These receptors adapt rapidly, so they detect initial contact and subsequent movements. In sensitive areas, the dendritic branches penetrate the epidermis and contact **Merkel cells** in the stratum basale of the skin (**Figure 18.3c**). Each Merkel cell communicates with a sensory neuron across a vesicular synapse that has an expanded nerve terminal known as a **tactile disc**. Merkel cells are sensitive to fine touch and pressure. They are tonically active, are extremely sensitive, and have narrow receptive fields.

Encapsulated Receptors Large, oval **tactile corpuscles** (*Meissner's corpuscles*) are found where the sense of touch is well developed, such as at the eyelids, lips, fingertips, nipples, and external genitalia (**Figure 18.3d**). The dendrites are highly coiled and are surrounded by modified Schwann cells. A fibrous capsule surrounds the entire complex and anchors it to the dermis of the skin. Tactile corpuscles detect light touch, movement, and vibration. They are phasic receptors and adapt to stimulation within a second after contact.

Bulbous corpuscles (*Ruffini corpuscles*) are located in the dermis and are sensitive to pressure and distortion of the skin. They are tonically active and show little or no adaptation. The capsule surrounds a core of collagen fibers that are continuous with those of the surrounding dermis. Dendrites within the capsule are interwoven around the collagen fibers (**Figure 18.3e**). Any tension or distortion of the dermis tugs or twists the fibers within the capsule, and this change stretches or compresses the dendrites and alters the activity in the myelinated afferent fiber.

Lamellar corpuscles (*Pacinian corpuscles*) are large encapsulated receptors (**Figure 18.3f**). The dendritic process lies within a series of concentric cellular layers. These layers shield the dendrite from sources of stimulation other than direct pressure. Lamellar corpuscles are most sensitive to pulsing or vibrating stimuli, but they also respond to deep pressure. Although both lamellar corpuscles and bulbous corpuscles respond to pressure, the lamellar corpuscles adapt rapidly while the bulbous corpuscles do not undergo adaptation. These receptors are scattered throughout the dermis of the entire body, but they are most common in the dermis of the fingers, breasts, and external genitalia. They are also located in the superficial and deep fasciae, mesenteries and the periostea surrounding bones, joint capsules, the pancreas, and the walls of the urethra and urinary bladder.

Table 18.1 summarizes the functions and characteristics of these six tactile receptors. Tactile sensations are transmitted to the CNS in the posterior columns and spinothalamic tracts.

Baroreceptors

Baroreceptors are stretch receptors that monitor changes in the stretch of organ walls and, therefore, the pressure within that organ (**Figure 18.4**). Each receptor consists of free nerve endings that branch within the elastic tissues in the walls of distensible organs, such as a blood vessel, or portions of the respiratory, digestive, or urinary tracts. When the pressure changes, the elastic walls of these blood vessels or organs stretch or recoil. This movement distorts the dendritic branches and alters the rate of action potential generation. Baroreceptors respond immediately to a change in pressure.

Table 18.1 | Touch and Pressure Receptors

Sensation	Receptor	Responds to
Fine touch	Free nerve ending	Light contact with skin
	Tactile disc	Light contact with skin
	Root hair plexus	Initial contact with hair shaft
Pressure and vibration	Tactile corpuscle	Initial contact and low-frequency vibrations
	Lamellar corpuscle	Initial contact (deep) and high-frequency vibrations
Deep pressure	Bulbous corpuscle	Stretching and distortion of the dermis

Figure 18.3 Tactile Receptors in the Skin. The location and general histological appearance of six important tactile receptors.

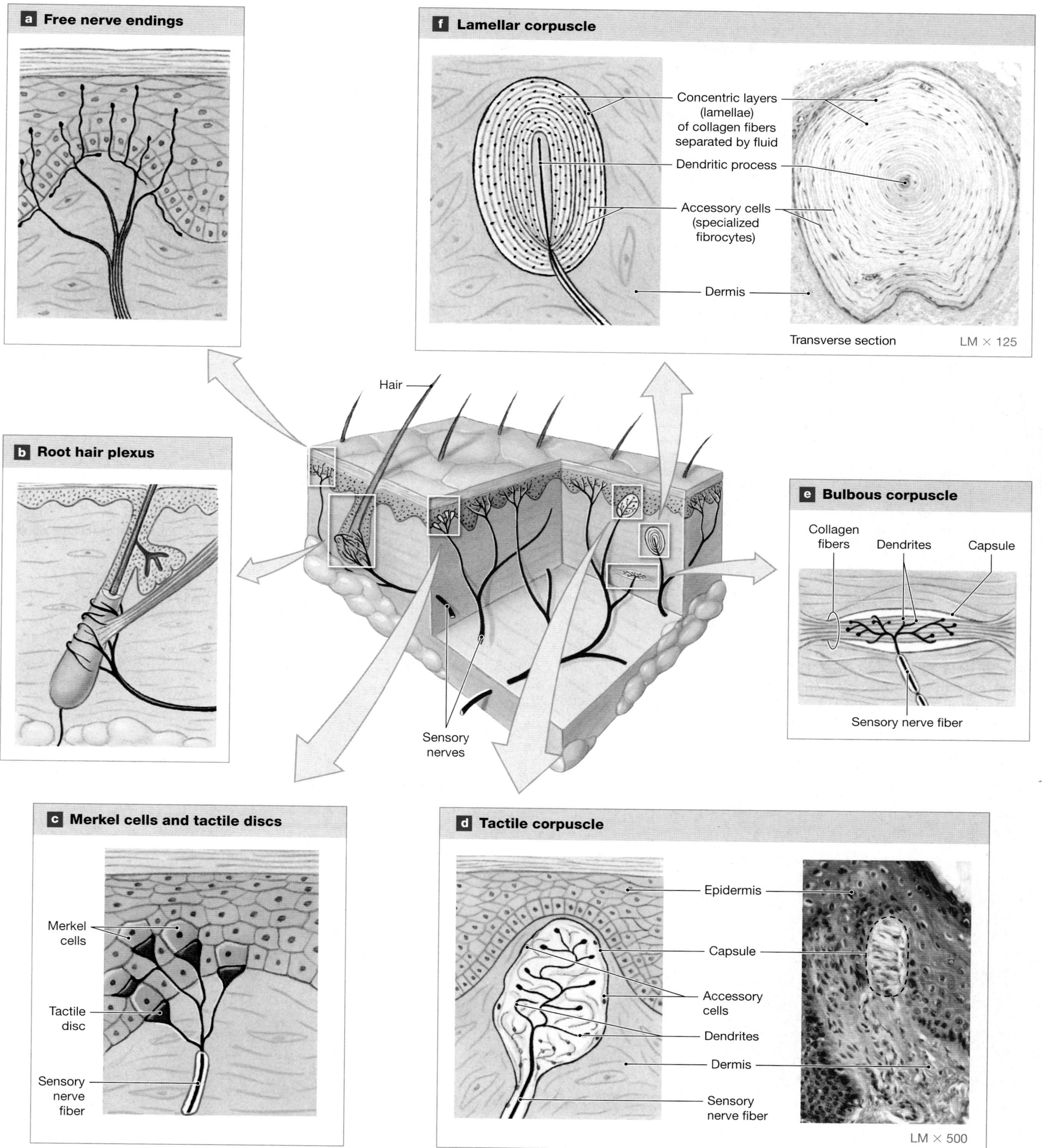

18

Proprioceptors

Proprioceptors monitor the position of joints, the tension in tendons and ligaments, and the extent of muscular contraction. Generally, proprioceptors do not adapt to constant stimulation. Muscle spindles are proprioceptors monitoring the length of skeletal muscles. **Golgi tendon organs** monitor the tension in tendons during muscle contraction. Joint capsules are richly supplied with free nerve endings that monitor tension, pressure, and movement at the joint. Your sense of body position results from the integration of information from these proprioceptors with sensory information from the internal ear.

Figure 18.4 Baroreceptors and the Regulation of Autonomic Functions. Baroreceptors provide information essential to the regulation of autonomic activities, including respiration, digestion, urination, and defecation.

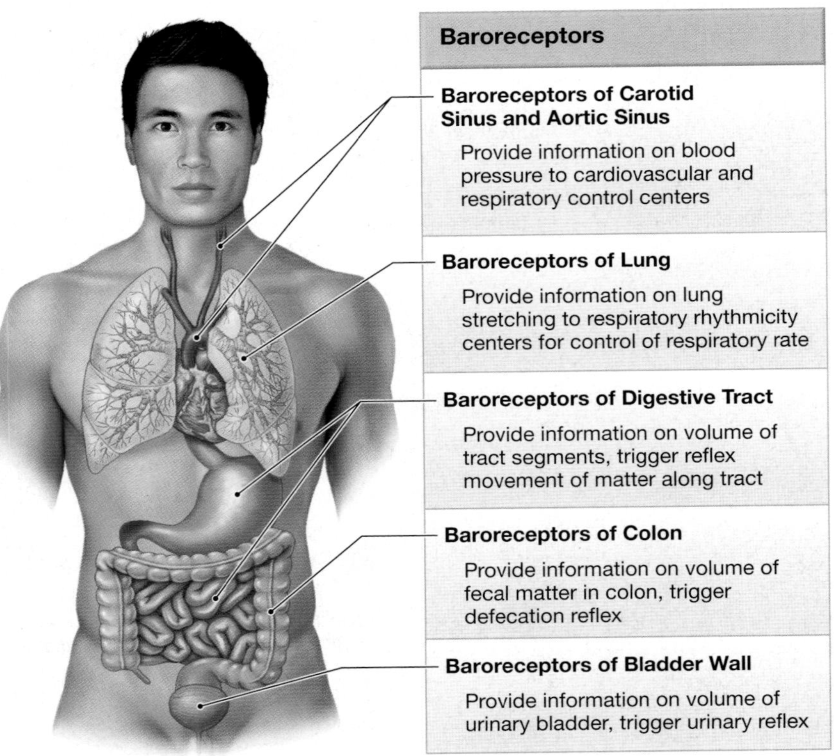

Baroreceptors
Baroreceptors of Carotid Sinus and Aortic Sinus
Provide information on blood pressure to cardiovascular and respiratory control centers
Baroreceptors of Lung
Provide information on lung stretching to respiratory rhythmicity centers for control of respiratory rate
Baroreceptors of Digestive Tract
Provide information on volume of tract segments, trigger reflex movement of matter along tract
Baroreceptors of Colon
Provide information on volume of fecal matter in colon, trigger defecation reflex
Baroreceptors of Bladder Wall
Provide information on volume of urinary bladder, trigger urinary reflex

Chemoreceptors

Chemoreceptors are specialized neurons that detect changes in the concentration of specific chemicals or compounds. Chemoreceptors respond only to water-soluble and lipid-soluble substances dissolved in body fluids. **Figure 18.5** shows the locations and functions of important chemoreceptors.

18.2 CONCEPT CHECK

5 When the nociceptors in your hand are stimulated, what sensation do you perceive?

6 What would happen to an individual if the information from proprioceptors in the lower limbs were blocked from reaching the CNS?

7 What are the three classes of mechanoreceptors?

See the blue Answers tab at the back of the book.

18.3 Olfaction (Smell)

▶ **KEY POINT** The two olfactory organs—located in the nasal cavity on either side of the nasal septum—contain olfactory receptors that give us our sense of olfaction (smell).

Our paired **olfactory organs** give us our special sense of **olfaction**, or *smell* (**Figure 18.6**). The olfactory organs consist of the following:

- A specialized **olfactory epithelium** containing bipolar **olfactory sensory neurons**, **supporting cells**, and **basal epithelial cells** (*stem cells*).

Figure 18.5 Chemoreceptors. Chemoreceptors are found both inside the CNS, on the ventrolateral surfaces of the medulla oblongata, and in the aortic and carotid bodies. These receptors are involved in the autonomic regulation of respiratory and cardiovascular function. The micrograph shows the histological appearance of the chemoreceptive neurons in the carotid body.

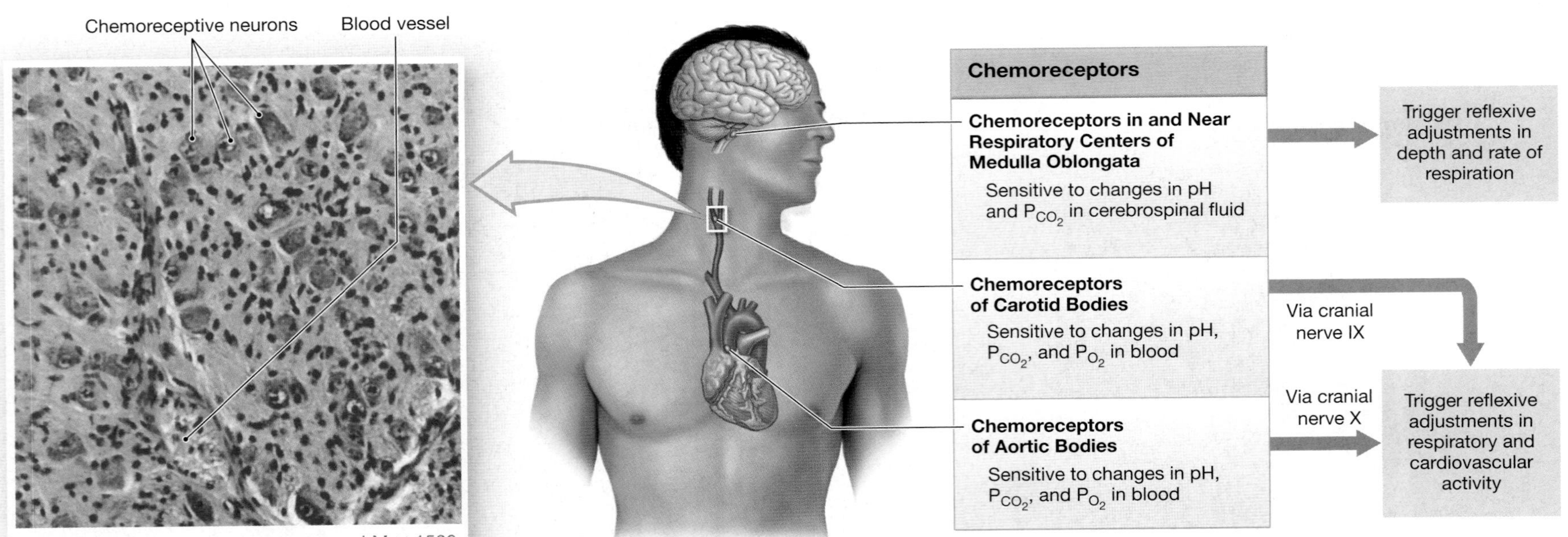

18

- A layer of loose connective tissue, the lamina propria, located deep to the olfactory epithelium. The lamina propria contains (1) **olfactory glands** (*Bowman's glands*), which produce a thick, pigmented mucus, (2) blood vessels, and (3) nerves.

The olfactory epithelium covers the inferior surface of the cribriform plate and the superior portions of the nasal septum and superior nasal conchae of the ethmoid. ↺ pp. 149–150 When air enters through the nose, the nasal conchae produce turbulent airflow. This brings airborne compounds into contact with the mucus produced by the olfactory glands. Once compounds have reached the olfactory mucus, water-soluble and lipid-soluble materials must diffuse into the mucus before they can stimulate the olfactory sensory neurons.

Olfactory Sensory Neurons

▶ **KEY POINT** Olfactory sensory neurons are specialized nerve cells of the olfactory nerve.

Olfactory sensory neurons are highly modified nerve cells. The dendritic portion of each olfactory sensory neuron forms a prominent bulb that projects beyond the epithelial surface and into the nasal cavity **(Figure 18.6b)**. That projection serves as a base for approximately 20 cilia that extend into the surrounding mucus, exposing them to chemical compounds that have dissolved into the mucus. Olfactory reception occurs on the surface of these olfactory cilia. When the odorous substance binds to its receptor on the cilium, the receptor membrane depolarizes. This depolarization triggers an action potential in the axon of the olfactory receptor. Approximately 10 to 20 million olfactory sensory neurons are packed into an area of roughly 5 cm^2.

Olfactory Pathways

▶ **KEY POINT** The olfactory pathway is a two-neuron pathway. Olfactory sensations are the only sensations that travel directly to the cerebral cortex without synapsing in the thalamus first.

The olfactory system is very sensitive. As few as four molecules of an odor-producing substance can activate an olfactory sensory neuron. However, the activation of an afferent fiber does not guarantee conscious awareness of the substance. Considerable convergence occurs along the olfactory pathway, and inhibition at one or more synapses can prevent an olfactory sensation from reaching the olfactory cortex.

The olfactory pathway is a two-neuron system. The first-order neuron is composed of axons leaving the olfactory epithelium, forming the first cranial nerve (I). These axons form 20 or more bundles that penetrate the cribriform plate of the ethmoid bone.

The first-order neuron synapses with the second-order neurons within the olfactory bulbs **(Figure 18.6b)**. The axons of the second-order neurons travel within the olfactory tract to reach the olfactory cortex, the hypothalamus, and portions of the limbic system.

Olfactory stimulation reaches the cerebral cortex directly. Certain smells trigger profound emotional and behavioral responses, such as memories, because olfactory information is also distributed to the limbic system and hypothalamus.

Olfactory Discrimination

▶ **KEY POINT** Sensory information is processed extensively within the olfactory bulb, and this is thought to be the first step in olfactory discrimination. Unlike other neurons, olfactory sensory neurons continuously regenerate throughout life.

The olfactory system can detect subtle differences between thousands of chemical stimuli. We know that there are upward of 50 different "primary smells." No apparent structural differences exist among the olfactory sensory neurons, but the epithelium as a whole contains neurons with distinctly different sensitivities. The CNS interprets the smell on the basis of the overall pattern of receptor activity. The processing of olfactory sensory input occurs at several locations within the CNS, with most of the processing occurring within the olfactory cortex.

18

Figure 18.6 The Olfactory Organs

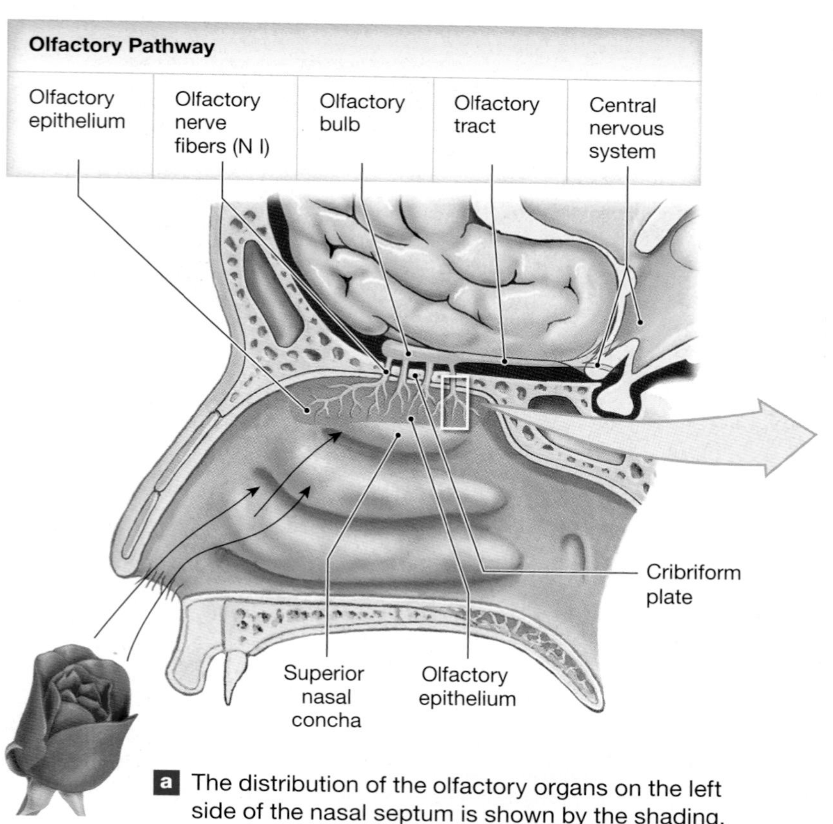

a The distribution of the olfactory organs on the left side of the nasal septum is shown by the shading.

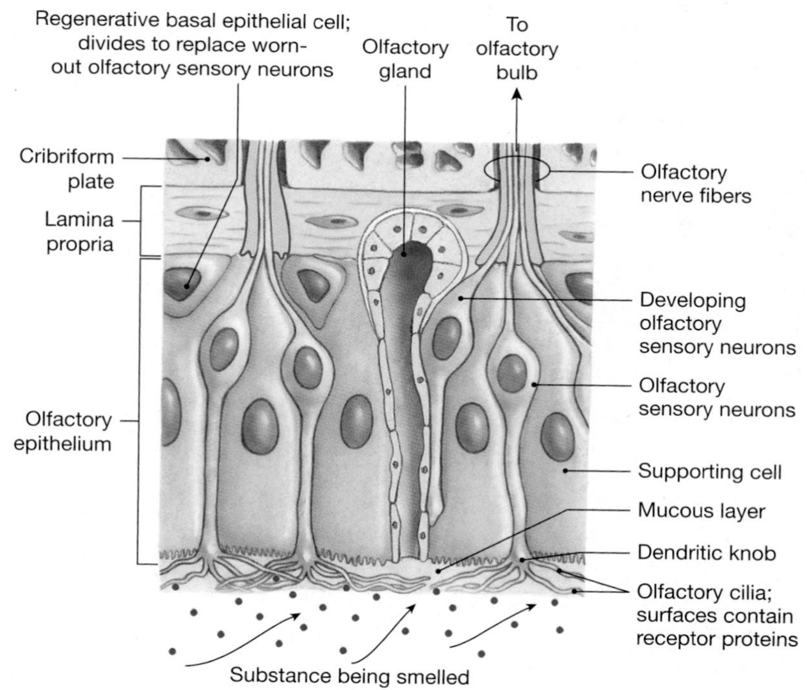

b A detailed view of the olfactory epithelium.

Olfactory sensory neurons have a life span of 4 to 6 weeks. Despite the continual replacement the total number of olfactory sensory neurons declines with age, and those that remain become less sensitive. As a result, elderly people have difficulty detecting odors in low concentrations. This decline in sensory neurons explains why Grandpa's aftershave smells so strong: He must apply more to be able to smell it.

18.3 CONCEPT CHECK

8 You and your roommate are studying for an anatomy exam. He insists that the olfactory pathway must pass through the thalamus because he is aware of the aroma of the hot coffee in his coffee cup. Is your roommate correct or incorrect? Explain.

See the blue Answers tab at the back of the book.

18.4 Gustation (Taste)

▶ **KEY POINT** Our special sense of gustation (taste) combines recognition and response to a diverse array of compounds. In adults, the taste buds contain the main gustatory epithelial cells.

Gustation, or *taste*, provides information about the foods and liquids we eat and drink. **Gustatory epithelial cells**, or *taste receptors* are found in taste buds that are distributed over the tongue surface and adjacent portions of the pharynx and larynx **(Figure 18.7a)**. These cells are stimulated by dissolved food molecules. This stimulation leads to action potentials that are sent to the gustatory cortex for interpretation.

The tongue surface has numerous epithelial projections called **lingual papillae** (pa-PIL-ē; *papilla*, nipple-shaped mound). The four types of lingual papillae are the **filiform** (*filum*, thread) **papillae**, **fungiform** (*fungus*, mushroom) **papillae**, **vallate** (VAL-āt) **papillae**, and **foliate** (FŌ-lē-āt) **papillae**. There are regional differences in the distribution of the papillae **(Figure 18.7a,b)**.

Figure 18.7 Gustatory Reception

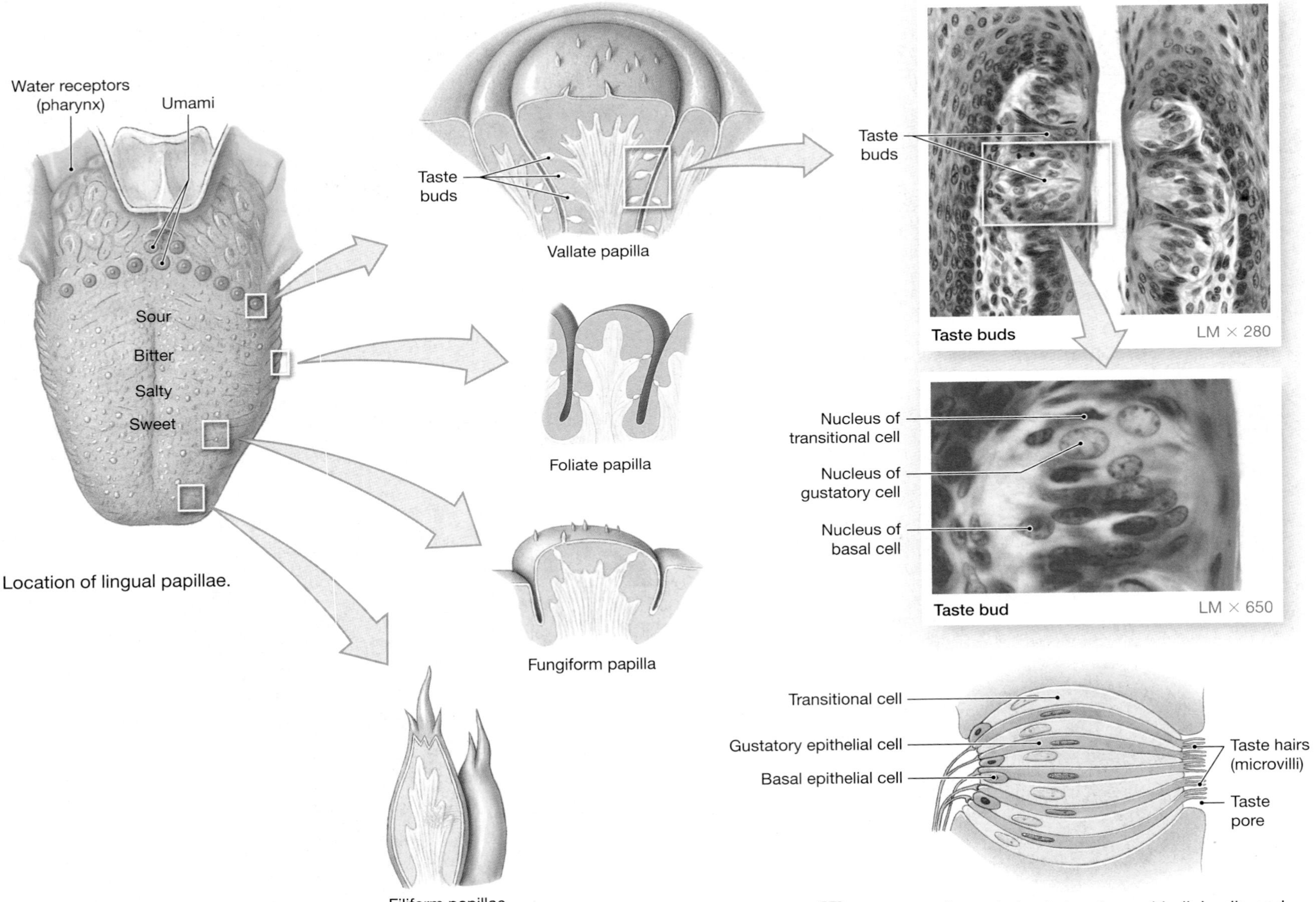

a Location of lingual papillae.

b Papillae on the surface of the tongue.

c Histology of a taste bud showing epithelial cells and supporting cells. The diagrammatic view shows details of the taste pore not visible in the light micrograph.

Gustatory Epithelial Cells (Taste Receptors)

▶ **KEY POINT** Taste buds contain sensory structures called gustatory epithelial cells. Dissolved chemicals stimulate these gustatory cells, triggering action potentials in the afferent sensory fiber.

Each **taste bud** contains about 40–100 gustatory epithelial cells, specialized epithelial cells, and many small stem cells called basal cells. The basal cells continually divide to produce daughter cells that mature in three stages: basal, transitional, and mature. The mature cells of the last stage are the **gustatory epithelial cells** (**Figure 18.7c**).

Each gustatory cell extends microvilli, sometimes called *taste hairs*, into the surrounding fluids through a narrow opening, the **taste pore**. A gustatory epithelial cell lives only 10–12 days before it is replaced.

Gustatory Pathways

▶ **KEY POINT** The gustatory pathway is a three-neuron pathway, with synapses in the medulla, thalamus, and cerebral cortex. The sensory input of taste is unique in that afferent information is carried over three different cranial nerves.

Gustatory information is carried by cranial nerves VII (facial), IX (glossopharyngeal), and X (vagus) (**Figure 18.8**). The facial nerve innervates all the taste buds located on the anterior two-thirds of the tongue, from the tip to the line of the vallate papillae. The glossopharyngeal nerve innervates the vallate papillae and the posterior one-third of the tongue. The vagus nerve innervates taste buds scattered on the surface of the epiglottis.

The sensory afferent fibers carried by these cranial nerves synapse in the solitary nucleus of the medulla oblongata. The axons of the postsynaptic neurons then enter the medial lemniscus. There, the neurons join axons that carry somatic sensory information on touch, pressure, and proprioception. After another synapse in the thalamus, the information is sent to the appropriate portions of the gustatory cortex of the insula.

You have a conscious perception of taste as the brain correlates information received from the taste buds with other sensory data. Information about the texture of food, along with taste-related sensations such as "peppery," comes from sensory afferent fibers in the trigeminal cranial nerve (V).

Gustatory Discrimination

▶ **KEY POINT** There is no universal classification system for taste sensations. Taste sensations may include sweet, salty, sour, bitter, umami, and water.

There are thought to be four **primary taste sensations**: sweet, salty, sour, and bitter. Although these do indeed represent distinct perceptions that are generally agreed on, they do not fully describe the range of tastes we experience. For example, in describing a particular taste, people may use very different terms, such as fatty, starchy, metallic, pungent, or astringent. In addition, other cultures consider different tastes to be "primary." Two additional tastes have been detected in humans:

- **Umami** (oo-MAH-mē) is a pleasant taste imparted by the amino acid glutamate. The distribution of umami receptors is not known in detail, but they are present in taste buds of the vallate papillae.
- Water is described as flavorless by most people. However, research on humans and other vertebrates has demonstrated the presence of **water receptors**, especially in the pharynx. Their sensory output is processed in the hypothalamus and affects several systems that affect water balance and blood pressure.

One of the limiting factors in studying gustatory reception is that it is very difficult to quantify tastes scientifically. Gustatory cells providing each of the primary sensations have been identified, and their plasma membrane characteristics and permeabilities differ. How a relatively small number of receptor types provides such a rich and diverse sensory experience remains to be determined.

The threshold for neuron stimulation varies for each of the primary taste sensations. In addition, taste receptors respond more readily to unpleasant than to pleasant stimuli. For example, we are almost a thousand times more sensitive to acids, which give a sour taste, than to either sweet or salty chemicals, and we are a hundred times more sensitive to bitter compounds than to acids. This sensitivity has survival value, because acids damage the mucous membranes of the mouth and pharynx, and many dangerous biological toxins taste bitter.

Our tasting abilities change with age. We begin life with more than 10,000 taste buds, but the number declines dramatically starting around age 50. The sensory loss becomes especially significant as aging individuals experience a decline in the number of olfactory sensory neurons. As a result, many elderly people complain that their food tastes bland and unappetizing, whereas children often find the same foods too spicy.

Figure 18.8 Gustatory Pathways. Three cranial nerves (VII, IX, and X) carry gustatory information to the gustatory cortex of the cerebrum.

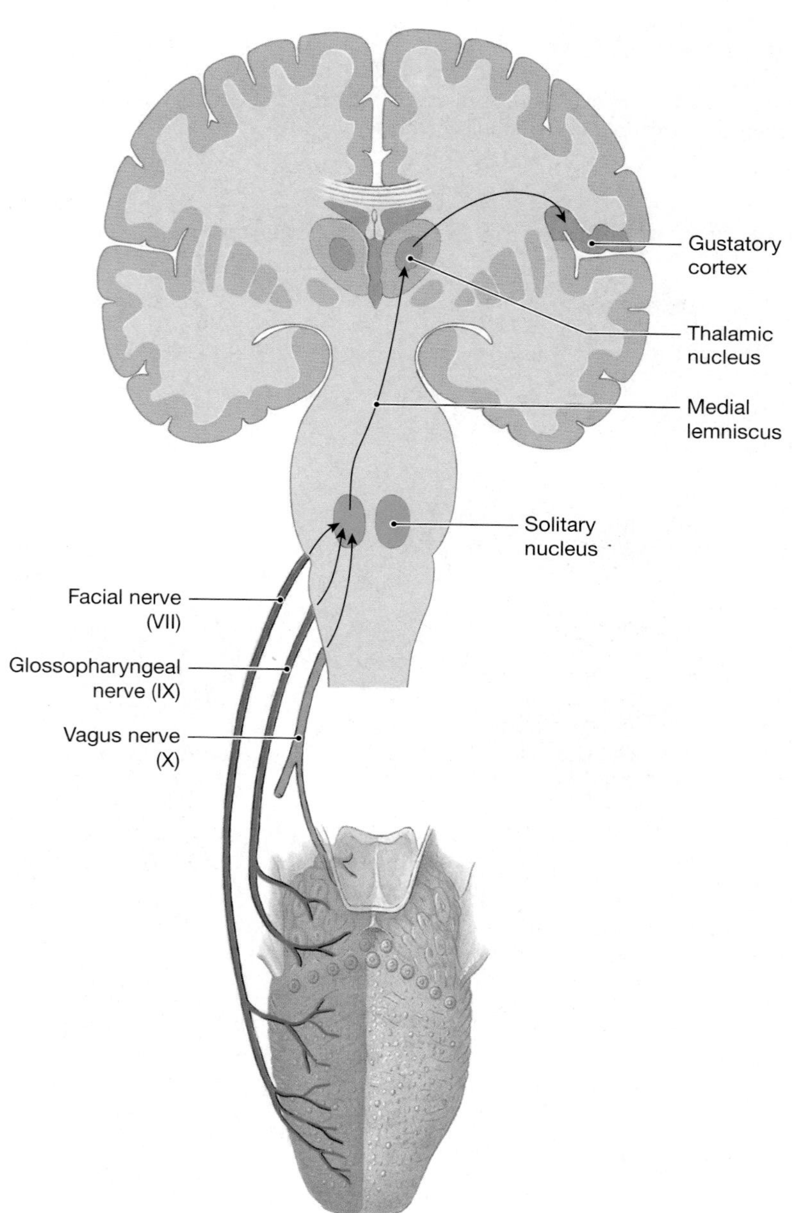

18.4 CONCEPT CHECK

9 Why does food taste bland when you have a cold?

10 Where are taste receptors located?

11 List the four types of lingual papillae.

See the blue Answers tab at the back of the book.

18.5 Equilibrium and Hearing

▶ **KEY POINT** The ear is associated with the detection of two special senses: equilibrium and hearing.

The ear is divided into three anatomical regions: the external ear, the middle ear, and the internal ear **(Figure 18.9)**. The external ear is the visible portion of the ear, and it collects and directs sound waves to the eardrum. The middle ear is a chamber located within the petrous portion of the temporal bone. Structures within the middle ear amplify sound waves and transmit them to an appropriate portion of the internal ear. The internal ear contains the sensory organs for equilibrium (balance) and hearing.

The External Ear

▶ **KEY POINT** The external ear directs sound into the external acoustic meatus and protects the delicate tympanic membrane.

The **external ear** includes the outer fleshy auricle, which surrounds a passageway called the **external acoustic meatus**. The auricle protects the opening passageway and provides directional sensitivity to the ear by directing sound inward to the **tympanic membrane** (*eardrum*). The tympanic membrane is a thin, semitransparent connective tissue sheet separating the external ear from the middle ear **(Figures 18.9** and **18.10)**.

The tympanic membrane is very delicate. The auricle and the narrow external acoustic meatus protect the tympanic membrane from injury. In addition, numerous small, outwardly projecting hairs and modified sweat glands called **ceruminous glands** line the external acoustic meatus. The hairs trap debris and provide tactile sensitivity through their root hair plexuses. The waxy secretion of the ceruminous glands, called **cerumen**, slows the growth of microorganisms and reduces the chances of infection.

The Middle Ear

▶ **KEY POINT** The middle ear consists of the tympanic cavity, which contains the auditory ossicles. The middle ear connects the tympanic membrane with the receptor complex of the internal ear.

The **middle ear** contains an air-filled space called the **tympanic cavity**, in which the auditory ossicles are located. **(Figure 18.10)**. The tympanic cavity also communicates with the nasopharynx through the auditory tube and with the mastoid air cells through a number of small and variable connections. The **auditory tube** (also called the *pharyngotympanic tube* or *Eustachian tube*) is approximately 4.0 cm long and penetrates the petrous part of the temporal bone. The connection to the tympanic cavity is narrow and supported by elastic cartilage. The opening into the nasopharynx is broad and funnel-shaped.

Figure 18.9 Anatomy of the Ear. A general orientation to the external, middle, and internal ear.

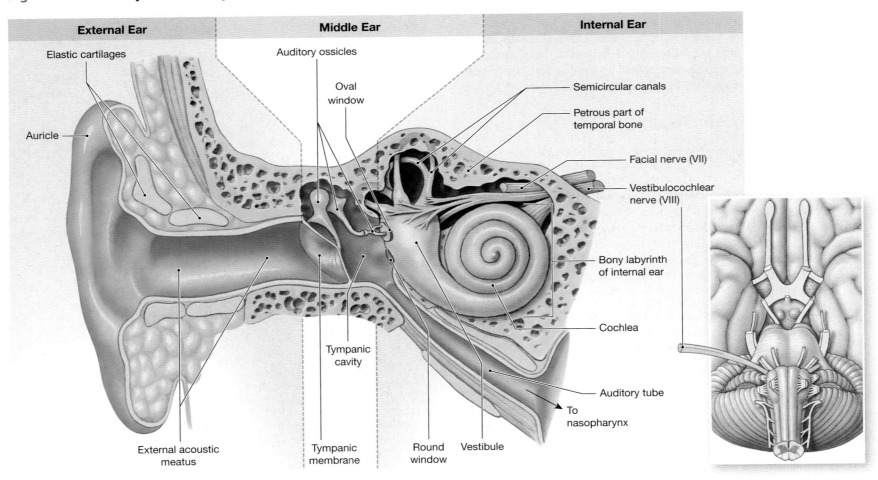

Figure 18.10 The Middle Ear

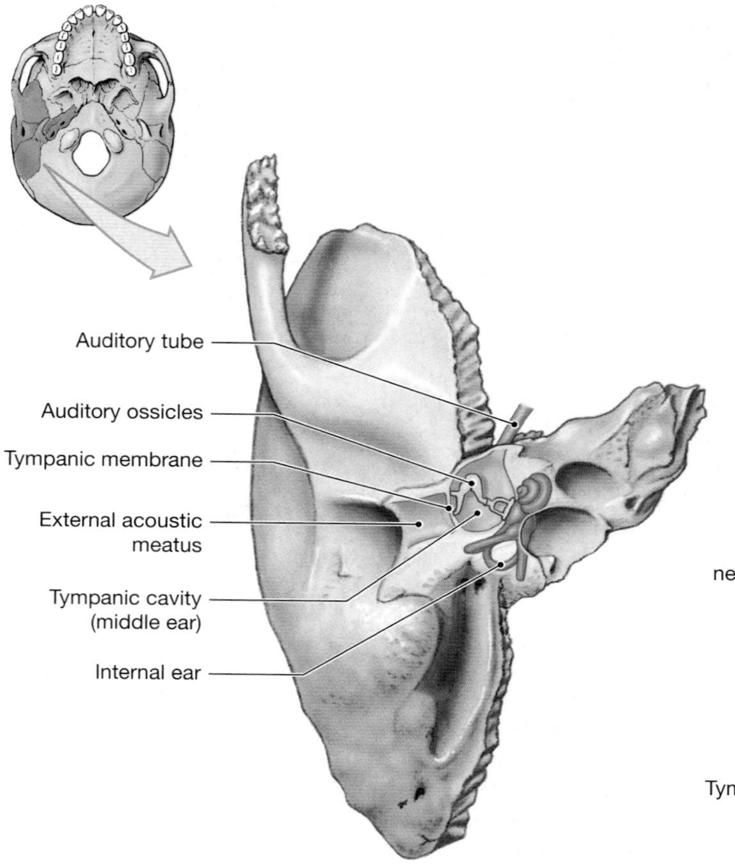

Auditory tube
Auditory ossicles
Tympanic membrane
External acoustic meatus
Tympanic cavity (middle ear)
Internal ear

a Inferior view of the right temporal bone drawn, as if transparent, to show the location of the middle and internal ear

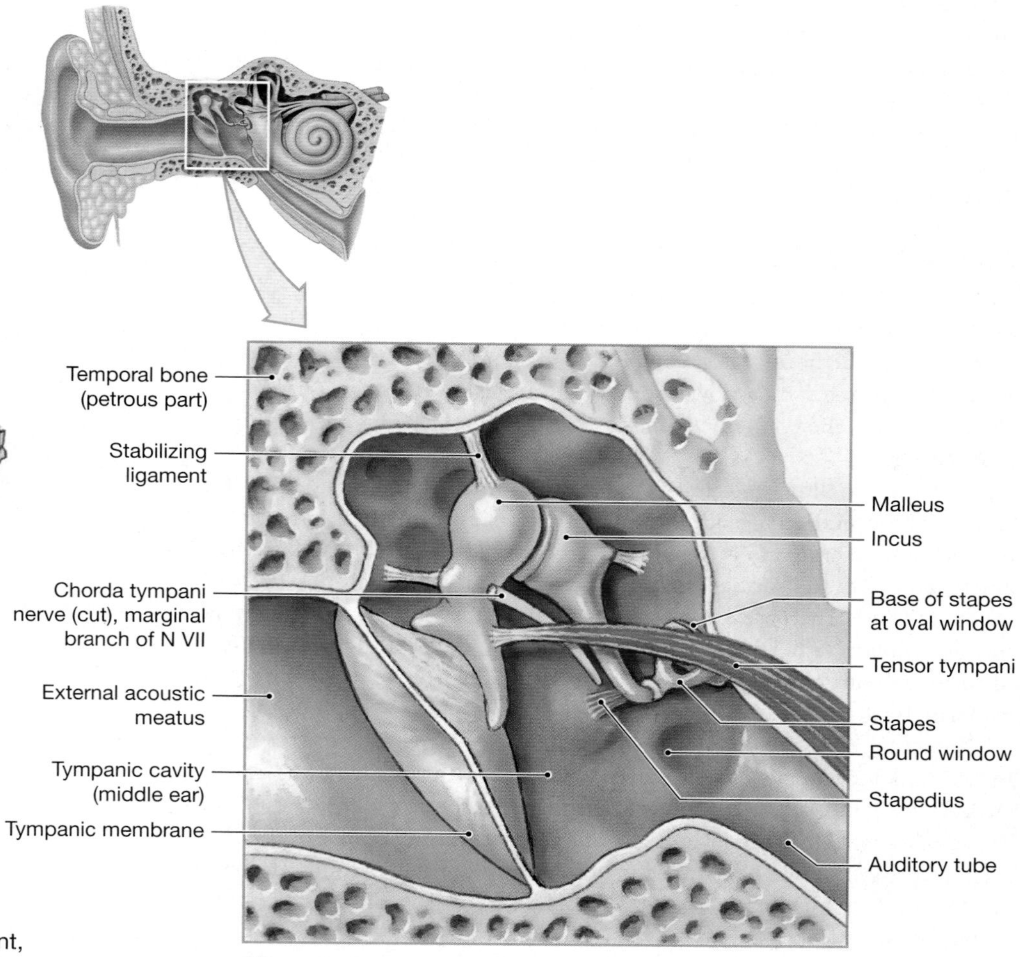

Temporal bone (petrous part)
Stabilizing ligament
Chorda tympani nerve (cut), marginal branch of N VII
External acoustic meatus
Tympanic cavity (middle ear)
Tympanic membrane
Malleus
Incus
Base of stapes at oval window
Tensor tympani
Stapes
Round window
Stapedius
Auditory tube

b Structures within the middle ear cavity

Incus
Malleus
Points of attachment to tympanic membrane
Stapes
Base of stapes

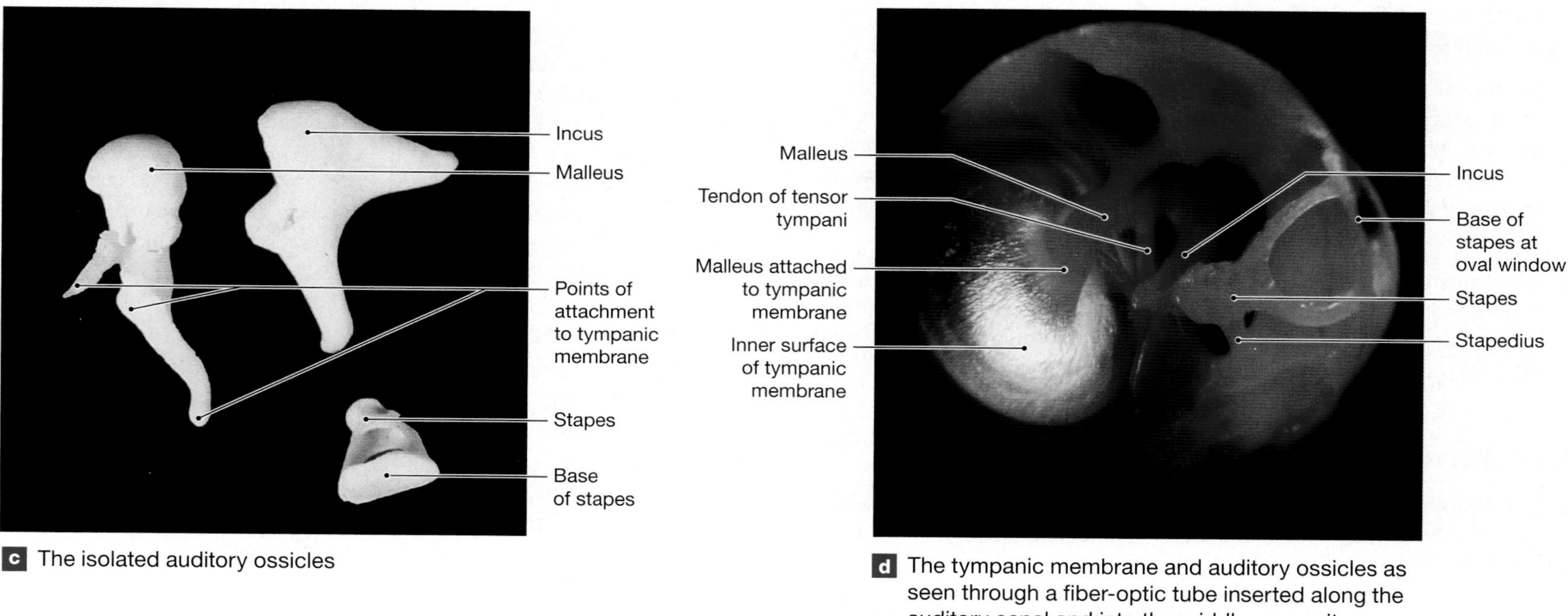

c The isolated auditory ossicles

Malleus
Tendon of tensor tympani
Malleus attached to tympanic membrane
Inner surface of tympanic membrane
Incus
Base of stapes at oval window
Stapes
Stapedius

d The tympanic membrane and auditory ossicles as seen through a fiber-optic tube inserted along the auditory canal and into the middle ear cavity

The auditory tube helps equalize the pressure in the middle ear with external, atmospheric pressure. Pressure must be equal on both sides of the tympanic membrane or there will be a painful distortion of the membrane. Unfortunately, the auditory tube also allows microorganisms to travel from the nasopharynx into the middle ear. Invasion of microorganisms can lead to a middle ear infection known as otitis media. Such infections are common in children because their auditory tubes are short and broad as compared to those of adults.

The Auditory Ossicles

Three tiny ear bones, the **auditory ossicles**, are located within the tympanic cavity. These bones, the smallest bones in the body, connect the tympanic membrane with the receptor complex of the internal ear (**Figure 18.10**). They are connected by synovial joints and act as levers, transferring sound vibrations from the tympanic membrane to a fluid-filled chamber within the internal ear.

The three auditory ossicles are the malleus, incus, and stapes. The lateral surface of the **malleus** (*malleus*, hammer) attaches to the interior surface of the tympanic membrane at three points. The middle bone, the **incus** (*incus*, anvil), connects the medial surface of the malleus to the **stapes** (STĀ-pēz; *stapes*, stirrup). The base, or footplate, of the stapes almost completely fills the oval window, a hole in the bony wall of the middle ear cavity. An annular ligament extends between the base of the stapes and the bony margins of the oval window.

Vibration of the tympanic membrane converts arriving sound waves into mechanical movements. These movements vibrate the malleus; the malleus vibrates the incus; the incus vibrates the stapes. The movement of the stapes causes vibrations in the fluid contents of the internal ear. Because of the alignment of the synovial joints connecting these three bones, an in-out movement of the tympanic membrane produces a forceful rocking motion at the stapes, amplifying the movement of the tympanic membrane.

We can hear very faint sounds because of this amplification. But this degree of magnification is also a problem when we are exposed to very loud noises. Two small muscles inside the tympanic cavity protect the eardrum and ossicles from violent movements under very noisy conditions (**Figure 18.10b,d**):

1. The **tensor tympani** (TEN-sor tim-PAN-ē) is a short muscle originating on the petrous part of the temporal bone. It inserts on the "handle" of the malleus. When the tensor tympani contracts, the malleus is pulled medially, stiffening the tympanic membrane. This increased stiffness reduces the amount of possible movement. The tensor tympani is innervated by motor fibers of the mandibular division of the trigeminal nerve (V).

2. The **stapedius** (sta-PĒ-dē-us) is innervated by the facial nerve (VII). The stapedius originates from the posterior wall of the tympanic cavity and inserts onto the stapes. Contraction of the stapedius pulls the stapes, reducing movement of the stapes at the oval window.

The Internal Ear

▶ **KEY POINT** The internal ear contains two labyrinths, one within the other. The membranous labyrinth lies within the osseous labyrinth and contains the sensory receptors for equilibrium and hearing.

The sensory receptors for equilibrium and hearing are located within the **internal ear** (**Figures 18.9** and **18.11**). These receptors are found within a collection of fluid-filled tubes and chambers known as the **membranous labyrinth** (*labyrinthos*, network of canals). The membranous labyrinth is filled with a fluid called **endolymph** (EN-dō-limf). The sensory receptors of the internal ear function only when exposed to the unique ionic composition of the endolymph. (Endolymph has a relatively high potassium ion concentration and a relatively low sodium ion concentration. Typical extracellular fluids have high sodium and low potassium ion concentrations.)

Figure 18.11 Structural Relationships of the Internal Ear. Flowchart showing internal ear structures and spaces, their contained fluids, and what stimulates these receptors.

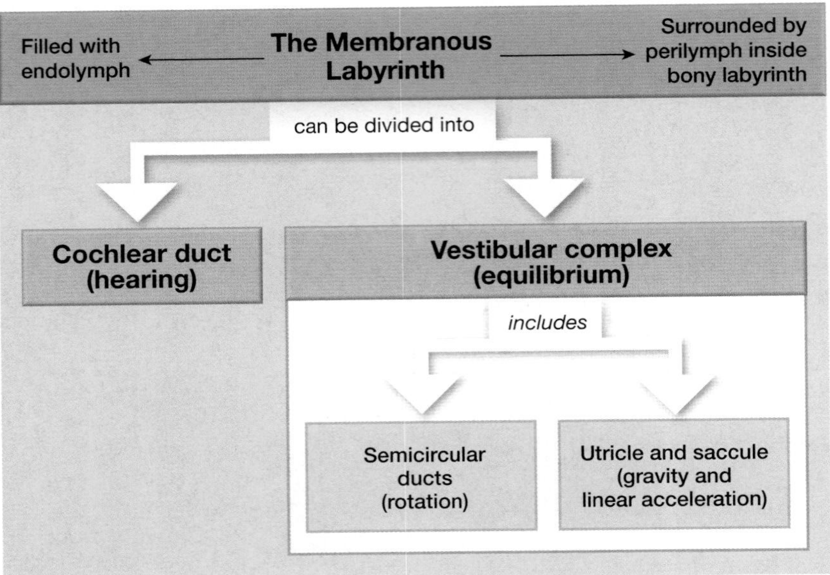

The **bony labyrinth** is located within the temporal bone. It surrounds and protects the membranous labyrinth (**Figure 18.12**). A fluid called **perilymph** (PER-i-limf) flows between the bony and membranous labyrinths. The chemical composition of perilymph closely resembles that of cerebrospinal fluid.

The bony labyrinth is subdivided into the **vestibule** (VES-ti-būl), the **semicircular canals**, and the **cochlea** (KOK-lē-a; *cochlea*, snail shell) (**Figure 18.12a**). The structures and air spaces of the external ear and middle ear work together to capture and transmit sound to the cochlea.

The vestibule and semicircular canals form the vestibular complex, because the fluid-filled chambers of the vestibule are continuous with those of the semicircular canals.

The cavity within the vestibule contains a pair of membranous sacs, the **utricle** (Ū-tre-kl) and the **saccule** (SAK-ūl). Hair cells in the utricle and saccule provide position and linear movement sensations. Those in the semicircular canals are stimulated by rotation of the head.

The Cochlear Duct and Hearing

The cochlea contains the **cochlear duct**, a slender, elongated portion of the membranous labyrinth (**Figure 18.12a**). The cochlear duct is located between a pair of perilymph-filled chambers, and the entire complex turns around a central bony hub.

The outer walls of the perilymph-filled chambers consist of dense bone except at two small areas near the base of the cochlear spiral. The **round window**, the more inferior of the two openings, is a thin, membranous partition that separates the perilymph of the cochlear chambers from the air-filled middle ear. The **oval window**, the more superior of the two openings, is in the cochlear wall (**Figure 18.10b–d**). The base of the stapes almost completely fills the oval window. The annular ligament extends between the edges of the base and the margins of the oval window, completing the seal. When a sound vibrates the tympanic membrane, the malleus and incus conduct these

Figure 18.12 **Semicircular Canals and Ducts.** The orientation of the bony labyrinth within the petrous part of each temporal bone.

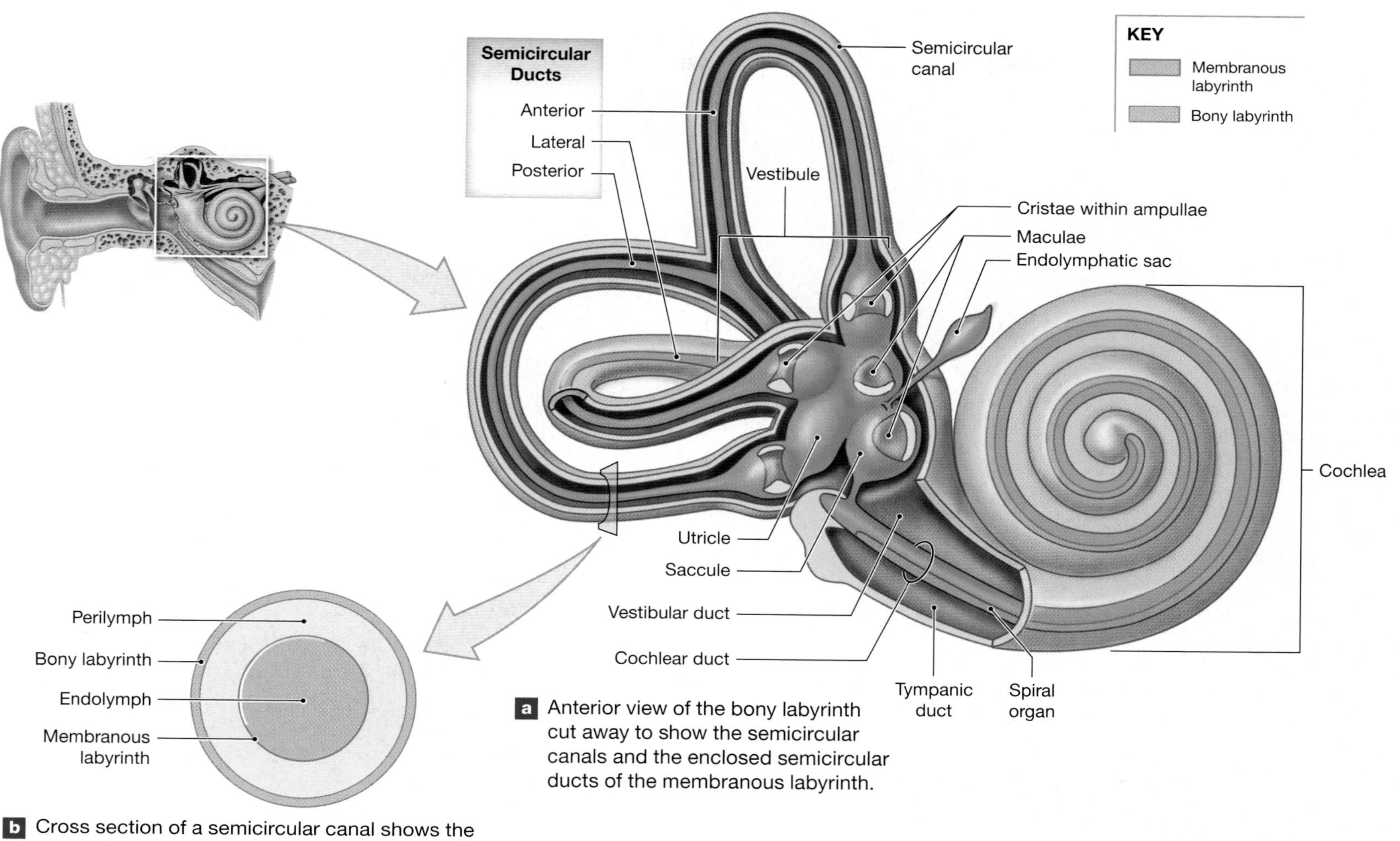

a Anterior view of the bony labyrinth cut away to show the semicircular canals and the enclosed semicircular ducts of the membranous labyrinth.

b Cross section of a semicircular canal shows the orientation of the bony labyrinth, perilymph, membranous labyrinth, and endolymph.

vibrations to the stapes. The movements of the stapes cause the oval window to vibrate. The vibrations of the oval window are conducted to the perilymph of the internal ear. This process ultimately leads to the stimulation of receptors within the cochlear duct, and we hear the sound.

The sensory receptors of the internal ear are called **hair cells**, which are monitored by sensory afferent fibers (Figure 18.13d). These receptor cells and sensory afferent fibers are surrounded by **supporting cells**. The free surface of each hair cell has 80–100 long stereocilia.

Hair cells are highly specialized mechanoreceptors. They are sensitive to the movement of their stereocilia. Their ability to provide hearing sensations in the cochlea depends on the presence of accessory structures that restrict the sources of stimulation. We will discuss the importance of these accessory structures as we discuss hair cell function in the next section.

The Vestibular Complex and Equilibrium

The **vestibular complex**, a part of the internal ear, provides equilibrium sensations by detecting rotation, gravity, and acceleration. It consists of the semicircular canals, the utricle, and the saccule.

The Semicircular Canals The **anterior**, **posterior**, and **lateral semicircular canals** are continuous with the vestibule (Figure 18.13a). Each semicircular canal surrounds a **semicircular duct**. The anterior, posterior, and lateral semicircular ducts are continuous with the utricle. Each duct contains an **ampulla** (plural, *ampullae*), an expanded region that contains the sensory receptors. The region in the wall of the ampulla that contains the hair cells is the **ampullary crest**.

Each hair cell in the vestibule has 80–100 long stereocilia that resemble microvilli. Each hair cell also has a single large cilium called a **kinocilium** (Figure 18.13d). These hair cells do not actively move their kinocilia and stereocilia. However, when an external force pushes against these processes, the distortion of the plasma membrane alters the rate at which a hair cell releases neurotransmitters.

The kinocilia and stereocilia of the hair cells are embedded in a gelatinous structure, the **ampullary cupula** (KŪ-pū-la). The ampullary cupula has a density very close to that of the surrounding endolymph, so it "floats" above the receptor surface, nearly filling the ampulla. When the head rotates in the plane of the duct, the endolymph moves along the axis of the duct. This movement of the endolymph pushes the ampullary cupula, distorting the receptor processes (Figure 18.13c). Fluid movement in one direction stimulates the hair cells, and movement in the opposite direction inhibits them. When the endolymph stops moving, the elastic nature of the cupula makes it "bounce back" to its normal position.

Figure 18.13 The Function of the Semicircular Ducts, Part I

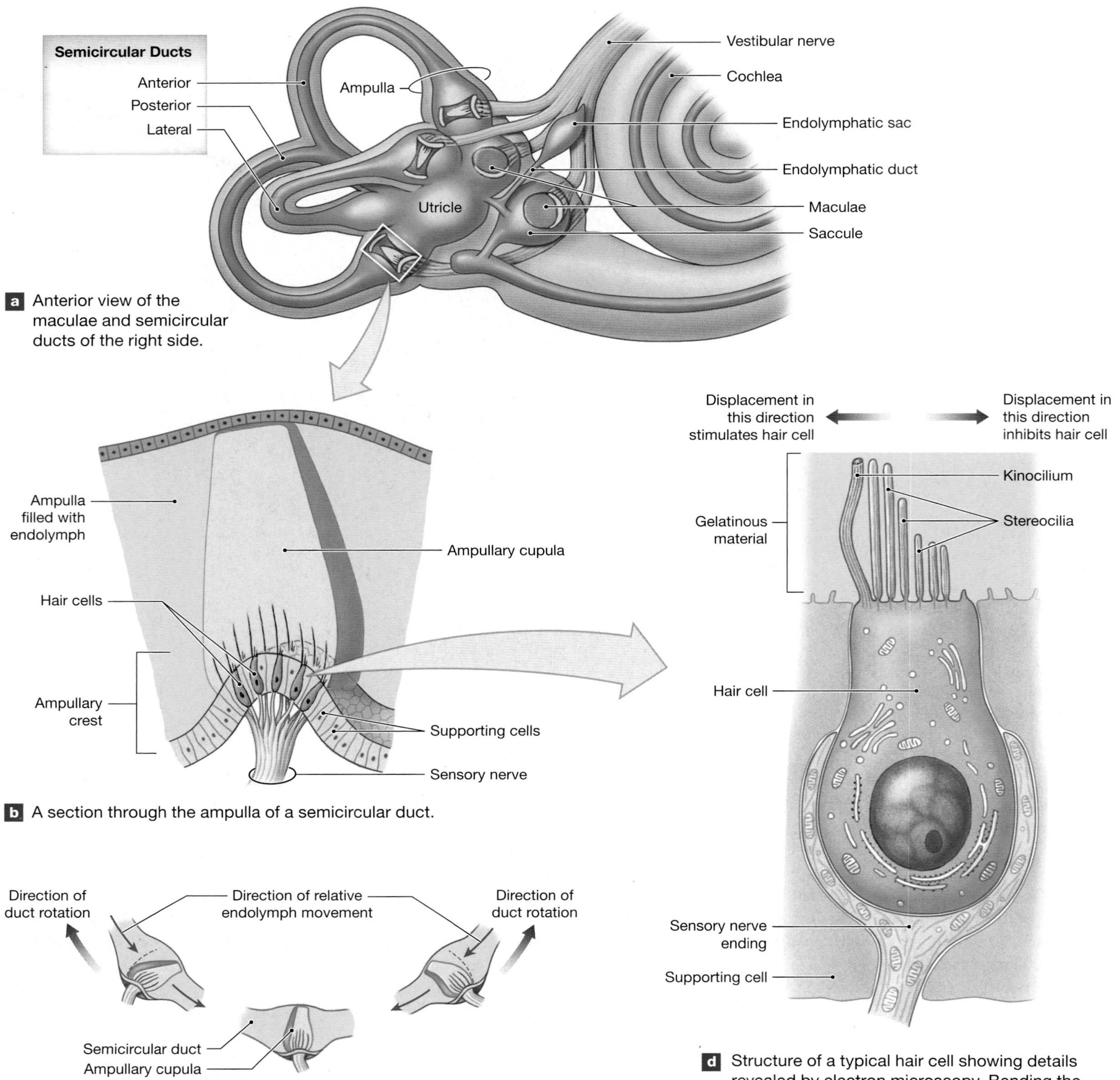

a Anterior view of the maculae and semicircular ducts of the right side.

b A section through the ampulla of a semicircular duct.

c Endolymph movement along the length of the duct moves the ampullary cupula and stimulates the hair cells.

d Structure of a typical hair cell showing details revealed by electron microscopy. Bending the stereocilia toward the kinocilium depolarizes the cell and stimulates the sensory neuron. Displacement in the opposite direction inhibits the sensory neuron.

We can analyze even the most complex movement in terms of motion in three rotational planes. The receptors within each semicircular duct respond to one of these rotational movements **(Figure 18.14)**:

- A rotation in the anterior-posterior plane, such as nodding "yes," stimulates the hair cells of the anterior duct.
- A rotation in the horizontal plane, as in shaking your head "no," stimulates the hair cells of the lateral semicircular duct.
- A rotation in the coronal plane, such as tilting your head from side to side, stimulates the hair cells of the posterior duct.

The Utricle and Saccule A slender passageway that is continuous with the narrow **endolymphatic duct** connects the utricle and saccule **(Figure 18.13a)**. The endolymphatic duct ends in a closed cavity, the **endolymphatic sac**. The endolymphatic sac projects through the dura mater lining the temporal bone and into the subdural space, where a capillary network surrounds it. Portions of the cochlear duct continually secrete endolymph, and excess fluid returns to the general circulation at the endolymphatic sac.

The hair cells of the utricle and saccule are clustered in the oval **maculae** (MAK-ū-lē; *macula*, spot) **(Figures 18.13a** and **18.15)**. The macula of the utricle is sensitive to changes in horizontal movement. The macula of the saccule is sensitive to changes in vertical movement. As in the ampullae, the processes of the hair cells are embedded in a gelatinous structure, the otolithic membrane. This membrane's surface contains densely packed calcium carbonate crystals termed **otoliths** (Ō-tō-lith; "ear stones"). The complex as a whole (gelatinous layer and otoliths) is called an **otolithic membrane (Figure 18.15a)**.

Figure 18.14 The Function of the Semicircular Ducts, Part II

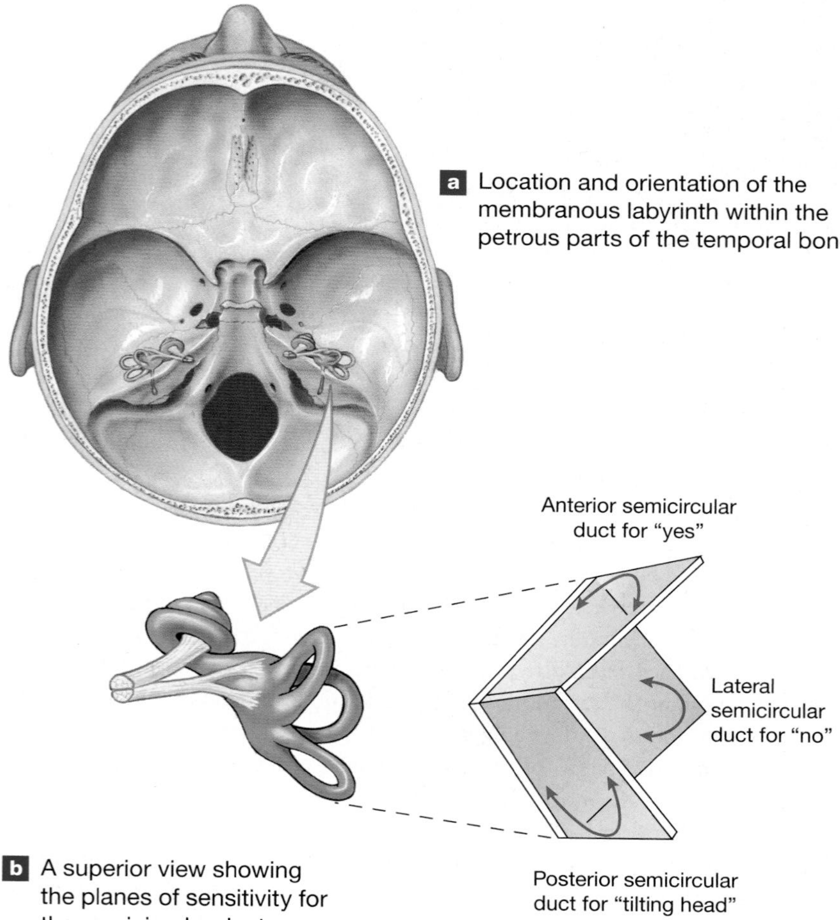

a Location and orientation of the membranous labyrinth within the petrous parts of the temporal bones

b A superior view showing the planes of sensitivity for the semicircular ducts

When your head is in the upright position, the otoliths sit atop the otolithic membrane of the macula of the utricle. The weight presses down on the macular surfaces, pushing the sensory hairs down rather than to one side or another. When your head is tilted, the pull of gravity on the otoliths pulls them to the side, distorting the hair cell processes and stimulating the macular receptors. This change in receptor activity tells the CNS that your head is no longer level **(Figure 18.15c)**.

A similar mechanism accounts for your perception of linear acceleration when you are in a car that speeds up suddenly. The otoliths lag behind due to their inertia, and the effect on the hair cells is comparable to tilting your head back. Under normal circumstances, your nervous system distinguishes between the sensations of tilting and linear acceleration by integrating vestibular sensations with visual information. Many amusement park rides confuse your sense of equilibrium by combining rapid rotation with changes in position and acceleration while providing restricted or misleading visual information.

Pathways for Vestibular Sensations Sensory neurons located within the adjacent **vestibular ganglia** monitor the hair cells of the vestibule and semicircular ducts. Sensory fibers from these ganglia form the **vestibular nerve**. These fibers synapse on neurons within the vestibular nuclei located at the boundary between the pons and medulla oblongata. The two vestibular nuclei

- integrate the sensory information concerning balance and equilibrium arriving from each side of the head;
- relay information from the vestibular complex to the cerebellum;
- relay information from the vestibular apparatus to the motor nuclei of the extra-ocular muscles of the eye;
- relay information from the vestibular complex to the cerebral cortex, providing a conscious sense of position and movement; and
- send commands to other motor nuclei in the brainstem and spinal cord.

The reflexive motor commands issued by the vestibular nuclei are distributed to the motor nuclei for cranial nerves III, IV, VI, and XI. These cranial nerves are involved with eye, head, and neck movements. Motor instructions adjusting peripheral muscle tone related to the reflexive movements of the head or neck descend within the **vestibulospinal tracts** of the spinal cord. **Figure 18.16** illustrates these pathways.

Hearing

▶ **KEY POINT** Vibrations of the oval window are conducted to the perilymph of the internal ear. Hearing is accomplished by the bending of hair cell receptors located within the spiral organ.

The Cochlea

The bony cochlea coils around a central hub, or **modiolus** (mō-DĪ-ō-lus) **(Figure 18.17)**. (There are usually 2.5 turns in the cochlear spiral.) The modiolus encloses the **spiral ganglion**. The spiral ganglion contains the cell bodies of the sensory neurons that monitor the receptors in the cochlear duct. In a sectional view, the **cochlear duct** lies between a pair of perilymphatic chambers, or *scalae*: the **scala vestibuli** (SKĀ-luh ves-TIB-yū-lē), or *vestibular duct*, and the **scala tympani** (TIM-pa-nē), or *tympanic duct*. The outer surfaces of all three ducts are encased by the bony labyrinth everywhere except at the oval window (the base of the scala vestibuli) and the round window (the base of the scala tympani). These scalae form one long and continuous perilymphatic chamber because they are interconnected at the tip of the spiral-shaped cochlea. This chamber begins at the oval window; extends through the scala

18

Figure 18.15 The Maculae of the Utricles

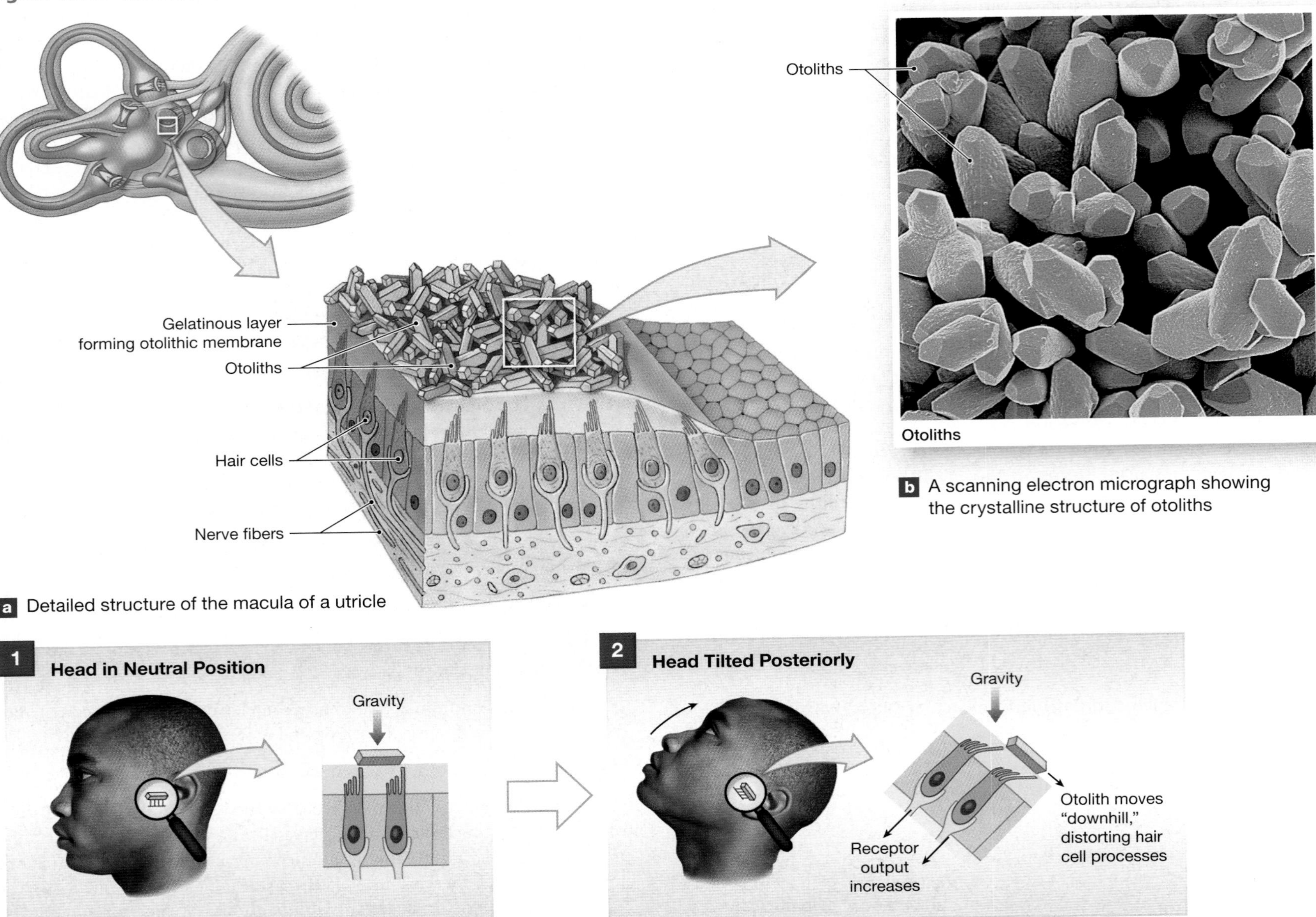

a Detailed structure of the macula of a utricle

b A scanning electron micrograph showing the crystalline structure of otoliths

c Diagrammatic view of changes in otolith position during tilting of the head

vestibuli, around the top of the cochlea, and along the scala tympani; and ends at the round window.

The Spiral Organ The hair cells of the cochlear duct are located in the **spiral organ**, or *organ of Corti* **(Figure 18.17b–e)**. This sensory structure rests on the **basilar membrane**. The basilar membrane separates the cochlear duct from the scala tympani. The hair cells of the spiral organ are arranged in two longitudinal rows: an inner and an outer row. These hair cells lack kinocilia, and their stereocilia contact the overlying **tectorial** (tek-TOR-ē-al; *tectum*, roof) **membrane**. This membrane is firmly attached to the inner wall of the cochlear duct. When a portion of the basilar membrane vibrates up and down, the stereocilia of the hair cells are bent.

Sound Detection

Hearing is the detection of sound, which consists of pressure waves conducted through air or water. Sound waves enter the external acoustic meatus and are transmitted to the tympanic membrane by the auditory ossicles. The tympanic membrane vibrates in response to sound waves with frequencies between approximately 20 and 20,000 Hz—the hearing range in a young child; this range decreases with age.

Movement of the stapes at the oval window applies pressure to the perilymph of the scala vestibuli. Liquids, such as the perilymph and endolymph of the ear, cannot be compressed. If you squeeze one part of a water-filled balloon, it bulges somewhere else. Because the cochlea is encased in bone, pressure applied at the oval window sets up pressure waves that travel through the entire cochlea until they reach the round window, causing it to bulge outward. So, when the base of the stapes moves inward at the oval window, the membrane spanning the round window bulges outward.

The frequency (pitch) of the sound we hear is determined by *what part* of the basilar membrane moves the most. Movement of the stapes sets up pressure waves in the perilymph. These waves distort the cochlear duct and the spiral organ, stimulating the hair cells. High-frequency (high-pitch) sounds affect the basilar membrane near the oval window; the lower the frequency of the sound, the farther away from the oval window the distortion will be.

Figure 18.16 **Neural Pathways for Equilibrium Sensations**

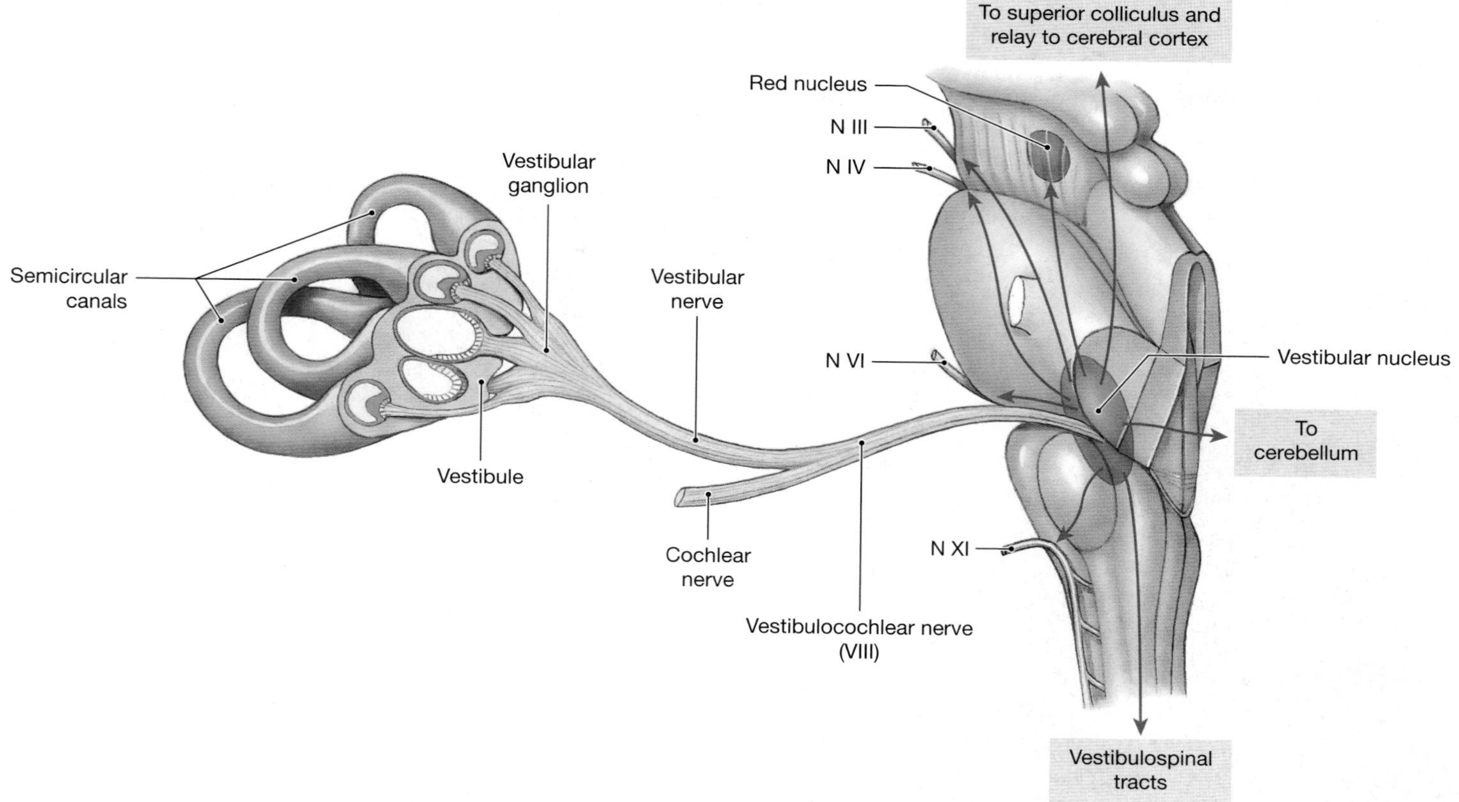

The intensity (volume) of a sound is determined by *the amount* of distortion in the basilar membrane. The actual amount of movement of the basilar membrane at a given location depends on the amount of force applied to the oval window. This relationship provides a mechanism for detecting the intensity of a sound. Very high-intensity (loud) sounds produce hearing loss by breaking the stereocilia off the surfaces of the hair cells. The tensor tympani and stapedius contract in response to dangerously loud sounds. This reflex contraction occurs in less than 0.1 second, but this may not be fast enough to prevent damage and related hearing loss.

Table 18.2 summarizes the steps involved in translating a sound wave into an auditory sensation.

Table 18.2 | **Steps in the Production of an Auditory Sensation**

1. Sound waves arrive at the tympanic membrane.
2. Movement of the tympanic membrane causes displacement of the auditory ossicles.
3. Movement of the stapes at the oval window establishes pressure waves in the perilymph of the scala vestibuli.
4. The pressure waves distort the basilar membrane on their way to the round window of the scala tympani.
5. Vibration of the basilar membrane causes hair cells to vibrate against the tectorial membrane, resulting in hair cell stimulation and neurotransmitter release.
6. Information concerning the frequency and intensity of stimulation is relayed to the CNS over the cochlear nerve.

Auditory Pathways

▶ KEY POINT The auditory pathway is a four-neuron pathway ending in the auditory cortex of the cerebrum. A large number of collateral fibers terminate in motor nuclei in the brainstem and initiate auditory reflexes of varying kinds.

Hair cell stimulation activates sensory neurons whose cell bodies are in the adjacent spiral ganglion. Their afferent fibers form the **cochlear nerve**. The anatomical organization of the auditory pathway has some unique features: This pathway (1) follows the smaller branch of the vestibulocochlear nerve, (2) involves four neurons, (3) involves several nuclei within various regions of the brainstem, and (4) has considerable branching and interconnections between brainstem nuclei.

CLINICAL NOTE

Hearing Loss

Conductive hearing loss results from conditions affecting the external or middle ear that block the normal transfer of vibrations from the external ear to the auditory ossicles. Individuals with conductive hearing loss have a reduced ability to hear faint sounds. Causes include impacted earwax, infection, or perforated tympanic membrane.

Sensorineural hearing loss is due to a problem within the cochlea or somewhere along the auditory pathway. Prolonged loud noise or certain antibiotics can damage the hair cells within the cochlear duct. If a pregnant woman develops certain infections, such as toxoplasmosis, rubella (German measles), or herpes, her baby may be born with sensorineural hearing loss.

Presbycusis, or age-related hearing loss, develops as a consequence of aging. High-pitched sounds are difficult to hear, but low-pitched sounds are still audible. Presbycusis affects both ears equally, and 40–50 percent of people aged 75 and older have this type of hearing loss.

Figure 18.17 The Cochlea and Spiral Organ

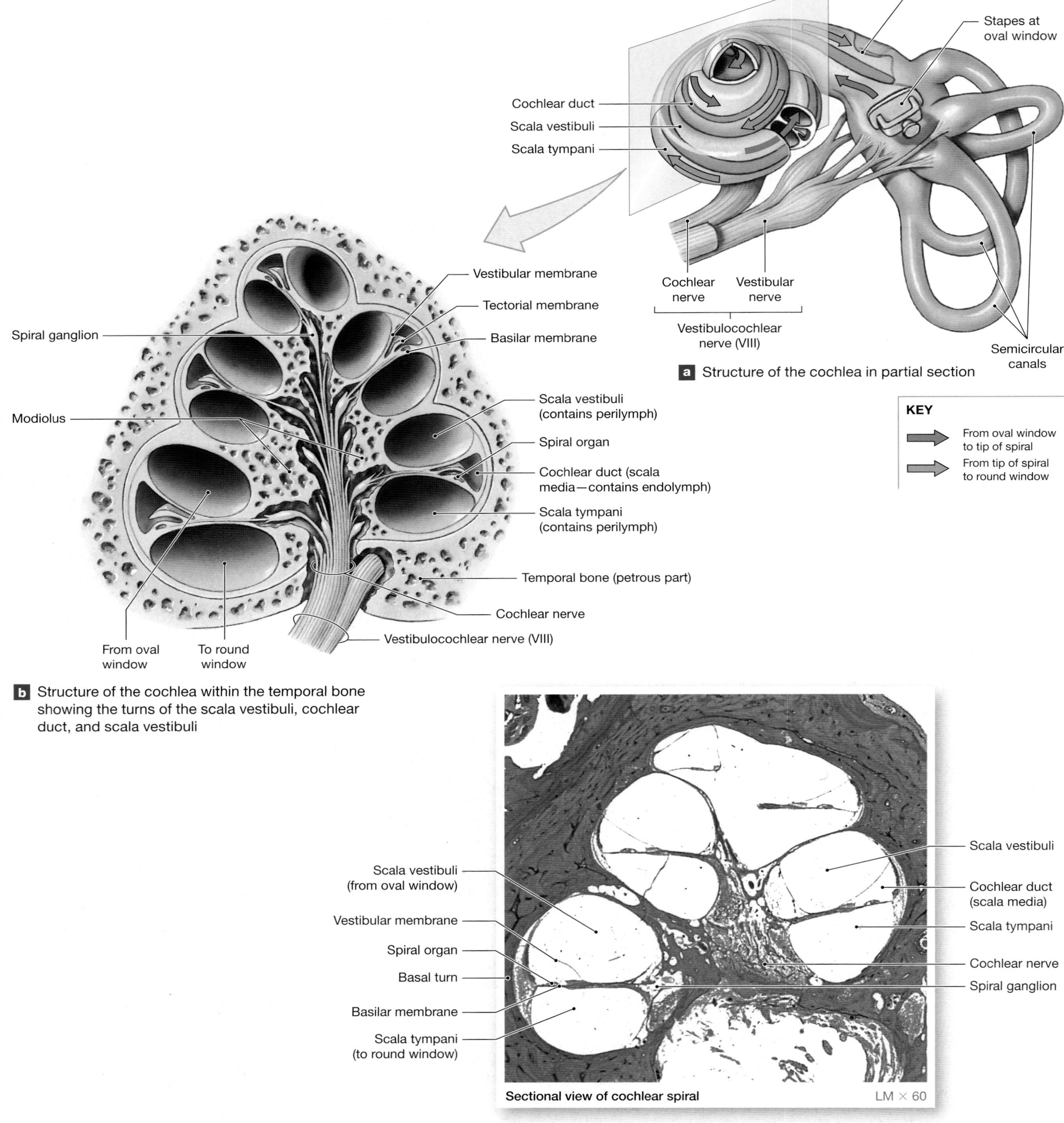

a Structure of the cochlea in partial section

b Structure of the cochlea within the temporal bone showing the turns of the scala vestibuli, cochlear duct, and scala vestibuli

c Histology of the cochlea showing many of the structures in part (b)

18

Figure 18.17 *(continued)*

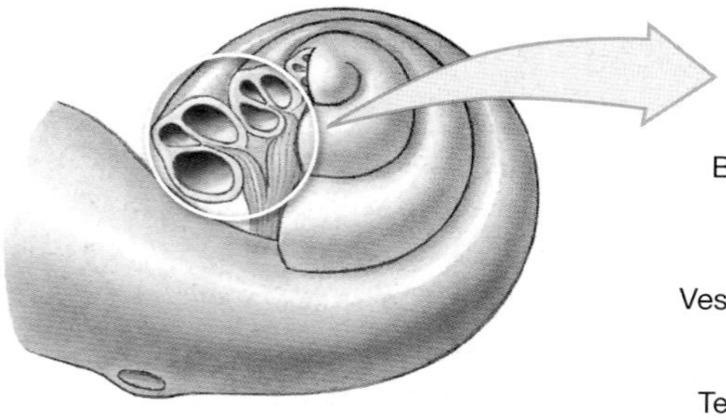

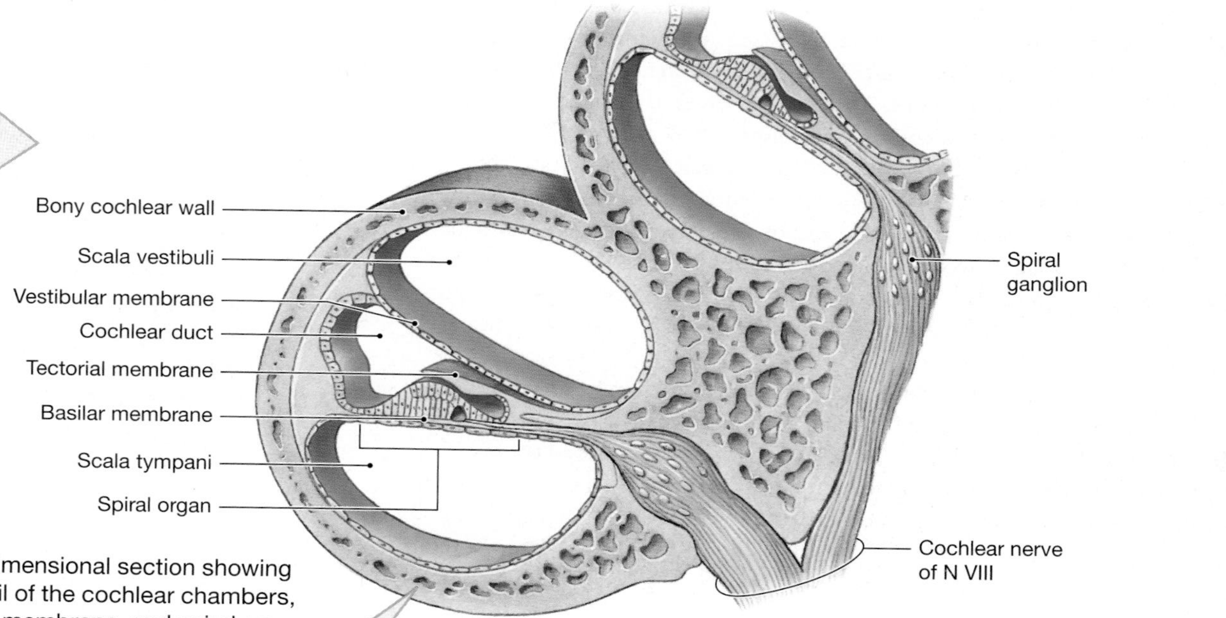

Bony cochlear wall
Scala vestibuli
Vestibular membrane
Cochlear duct
Tectorial membrane
Basilar membrane
Scala tympani
Spiral organ
Spiral ganglion
Cochlear nerve of N VIII

d Three-dimensional section showing the detail of the cochlear chambers, tectorial membrane, and spiral organ

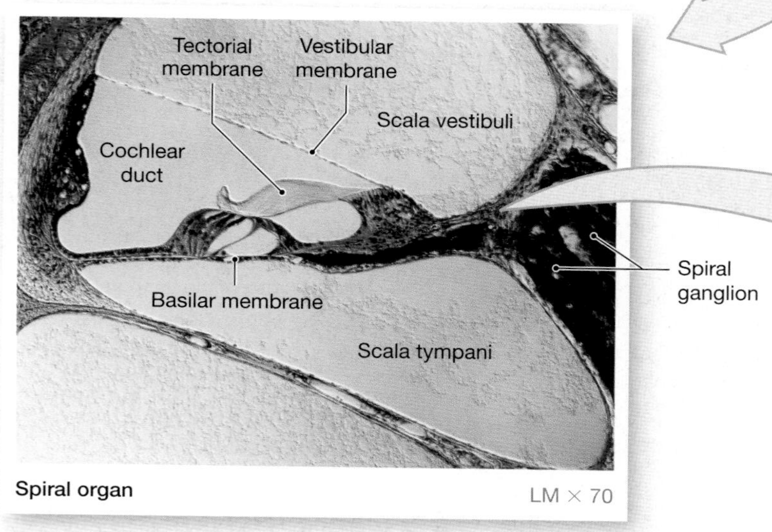

Tectorial membrane
Vestibular membrane
Scala vestibuli
Cochlear duct
Basilar membrane
Scala tympani
Spiral ganglion

Spiral organ LM × 70

e Histological section through the spiral organ

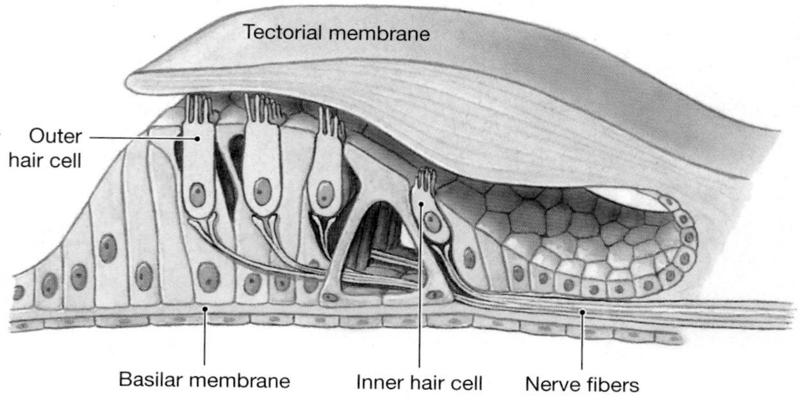

Tectorial membrane
Outer hair cell
Basilar membrane
Inner hair cell
Nerve fibers

f Enlarged view of the receptor hair cell complex of the spiral organ

Stereocilia of inner hair cells
Stereocilia of outer hair cells

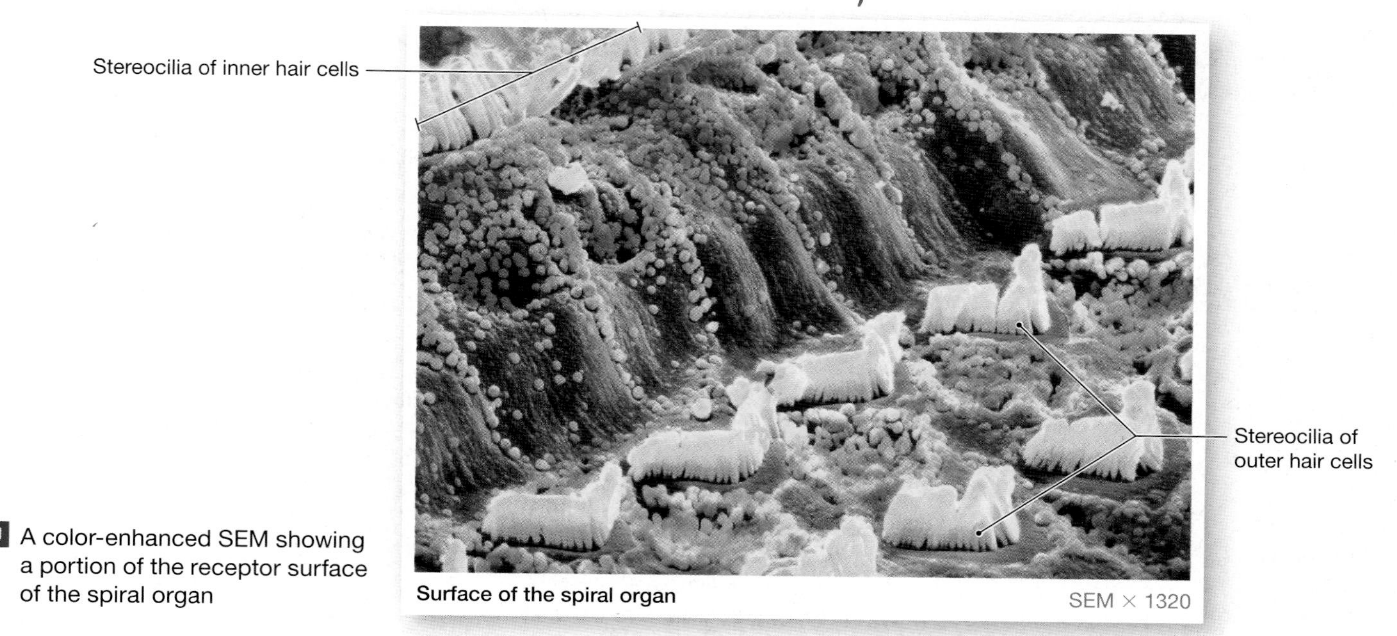

Surface of the spiral organ SEM × 1320

g A color-enhanced SEM showing a portion of the receptor surface of the spiral organ

Figure 18.18 summarizes the auditory pathway. Stimulation of hair cells along the basilar membrane activates sensory neurons whose cell bodies are in the adjacent spiral ganglion. ① The afferent fibers of those neurons form the **cochlear nerve**. These axons enter the medulla oblongata, where they synapse at the **cochlear nucleus** on that side. ② From there, information ascends to *both* **superior olivary nuclei** of the pons and *both* inferior colliculi of the midbrain. ③ This midbrain processing center coordinates a variety of unconscious motor responses to acoustic stimuli, including auditory reflexes that involve skeletal muscles of the head, face, and trunk. ④ These reflexes automatically change the position of your head in response to a sudden loud noise. You usually turn your head and your eyes toward the source of the sound.

Before reaching the cerebral cortex and your awareness, ascending auditory sensations synapse in the medial geniculate nucleus of the thalamus. ⑤ Projection fibers then deliver the information to the **auditory cortex** of the temporal lobe. ⑥ Information travels to the cortex over labeled lines: High-frequency sounds activate one portion of the cortex, low-frequency sounds another. In effect, the auditory cortex contains a map of the spiral organ. So, information about *frequency,* translated into information about *position* on the basilar membrane, is projected in that form onto the auditory cortex. There it is interpreted to produce your subjective sensation of pitch.

18.5 CONCEPT CHECK

12 You are exposed unexpectedly to very loud noises. What happens within the tympanic cavity to protect the tympanic membrane from damage?

13 As you shake your head "no," you are aware of this head movement. How are these sensations detected?

14 How does loss of stereocilia from the hair cells of the spiral organ affect hearing?

15 Distinguish between the cochlear duct and scala tympani.

See the blue Answers tab at the back of the book.

Figure 18.18 Pathways for Auditory Sensations.

18
6
Projection fibers then deliver the information to specific locations within the auditory cortex of the temporal lobe.
Low-frequency sounds
High-frequency sounds
Thalamus
To ipsilateral auditory cortex
1
Stimulation of hair cells at a specific location along the basilar membrane activates sensory neurons.
Cochlea
Low-frequency sounds
High-frequency sounds
5
Ascending sound information goes to the medial geniculate nucleus.
Vestibular nerve
4
The inferior colliculi direct a variety of unconscious motor responses to sounds.
To reticular formation and motor nuclei of cranial nerves
2
Sensory neurons carry the sound information in the cochlear nerve to the cochlear nuclei.
Superior olivary nucleus
Vestibulocochlear nerve (VIII)
3
Information ascends from each cochlear nucleus to the superior olivary nuclei of the pons and the inferior colliculi of the midbrain.
Cochlear nucleus
KEY
First-order neuron
Second-order neuron
Third-order neuron
Fourth-order neuron
Motor output to spinal cord through the tectospinal tracts

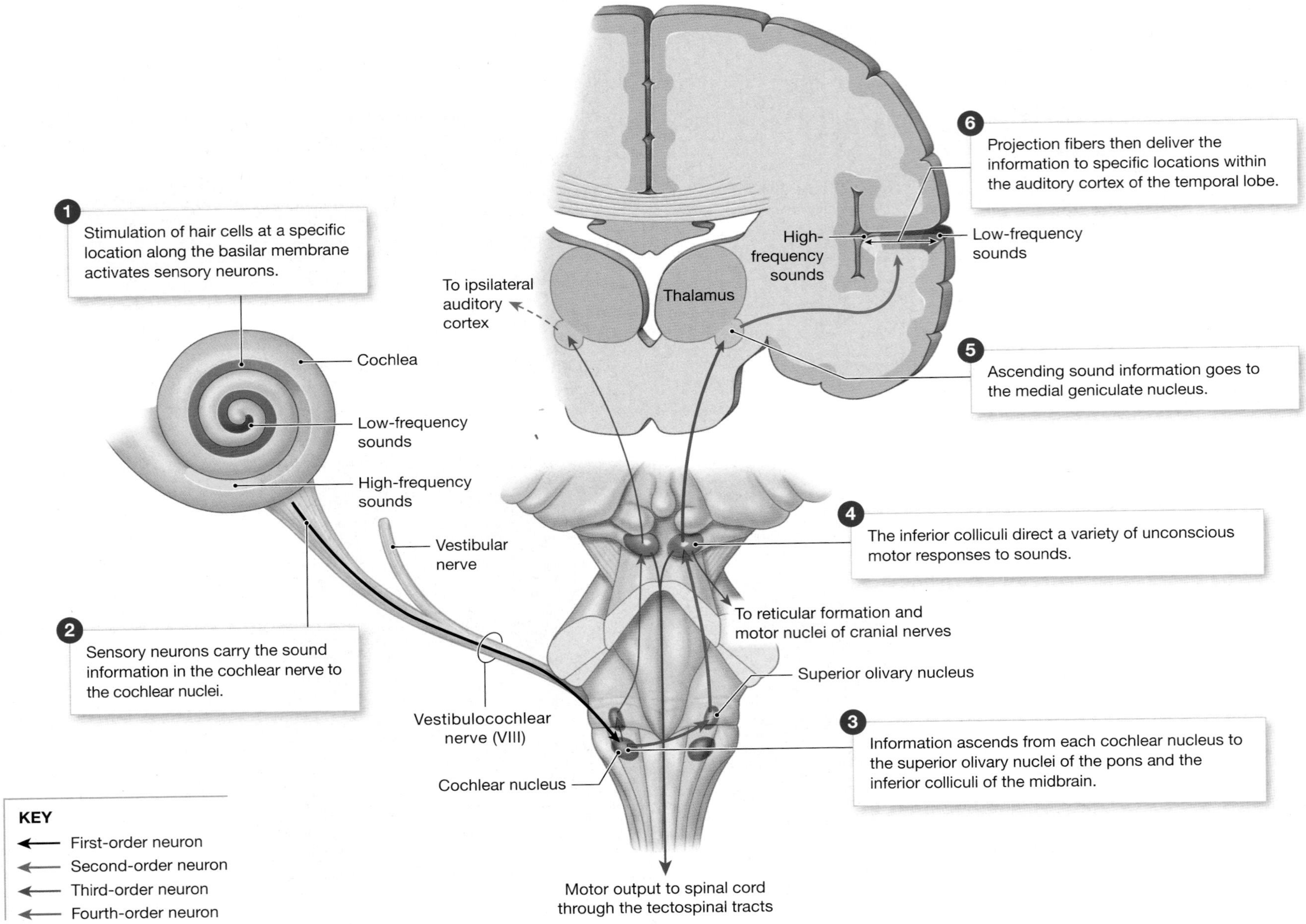

18.6 Vision

▶ **KEY POINT** Humans rely more on vision than on any other special sense. As a result, the visual cortex is several times larger than the cortical area for any other special sense.

Our visual receptors are contained in elaborate structures, the eyes, which enable us to detect light and create detailed visual images. We begin our discussion with the accessory structures of the eye.

Accessory Structures of the Eye

▶ **KEY POINT** The accessory structures of the eye include the eyelids, superficial epithelium, and lacrimal apparatus.

The **accessory structures** of the eye protect, lubricate, and support the eye (**Figure 18.19**). They include the eyelids, the superficial epithelium of the eye, and the structures associated with producing, secreting, and removing tears.

Eyelids

The **eyelids**, or *palpebrae* (pal-PĒ-brē), are a continuation of the skin. The eyelids act like windshield wipers; their continual blinking keeps the surface lubricated and free from dust and debris. They also protect the delicate surface of the eye by closing firmly. The **palpebral fissure** is the gap between the upper and lower eyelids. The upper and lower eyelids are connected at the **medial angle** (*medial canthus*) and the **lateral angle** (*lateral canthus*) (**Figure 18.19**). The **eyelashes** along the margins of the eyelids are very strong hairs. Sensory structures in the root hair plexus monitor each eyelash. Movement of the hair triggers a blinking reflex, which prevents foreign matter (including insects) from reaching the surface of the eye.

The eyelashes are associated with large sebaceous glands, called **tarsal glands**, on the inner surface of the eyelid. They secrete a lipid-rich product that keeps the eyelids from sticking together. At the medial angle, glands within the **lacrimal caruncle** (KAR-un-kul), a mass of soft tissue, produce the thick secretions that form the gritty deposits sometimes found at the edge of the eye after a good night's sleep (**Figure 18.19a**).

Occasionally, bacteria infect these various glands. A cyst, or *chalazion* (kah-LĀ-zē-on; "small lump"), results from infection of a tarsal gland. A painful, localized swelling known as a *sty* results from infection in a sebaceous gland of an eyelash.

A thin layer of stratified squamous epithelium covers the visible surface of the eyelid. Deep to the subcutaneous layer, the eyelids are supported and strengthened by broad sheets of connective tissue called the **tarsal plate** (**Figure 18.19b**). The muscle fibers of the orbicularis oculi and the levator palpebrae superioris lie between the tarsal plate and the skin (**Figures 18.19b** and **18.20**). These skeletal muscles close the eyelids (orbicularis oculi) and raise the upper eyelid (levator palpebrae superioris). ↺ pp. 262, 264–265

The **conjunctiva** (kon-junk-TĪ-va; "uniting" or "connecting") is a layer of epithelium covering the inner surface of the eyelids and the outer surface of the eye (**Figure 18.21b,e**). The conjunctiva is a mucous membrane covered by a specialized stratified squamous epithelium. The **palpebral conjunctiva** covers the inner surface of the eyelids, and the **bulbar conjunctiva** (*ocular conjunctiva*) covers the anterior surface of the eye. A continuous supply of fluid washes over the surface of the eyeball, lubricating the conjunctiva. A superficial lubricant produced by the mucous cells within the epithelium and the

Figure 18.19 Accessory Structures of the Eye, Part I

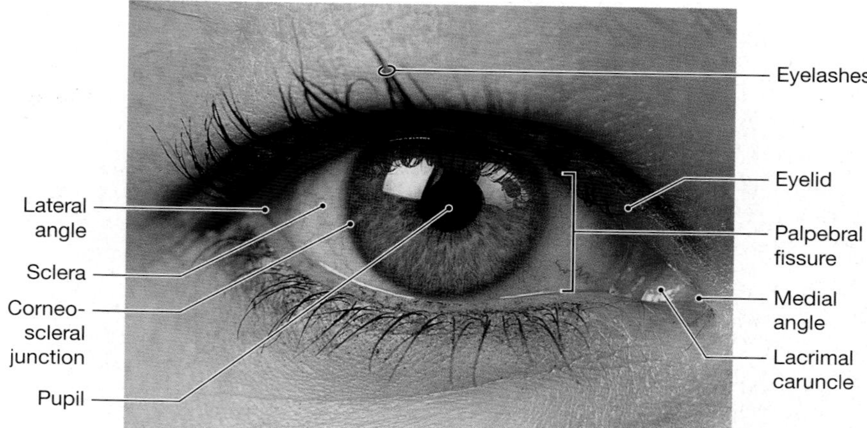

a Superficial anatomy of the right eye and its accessory structures

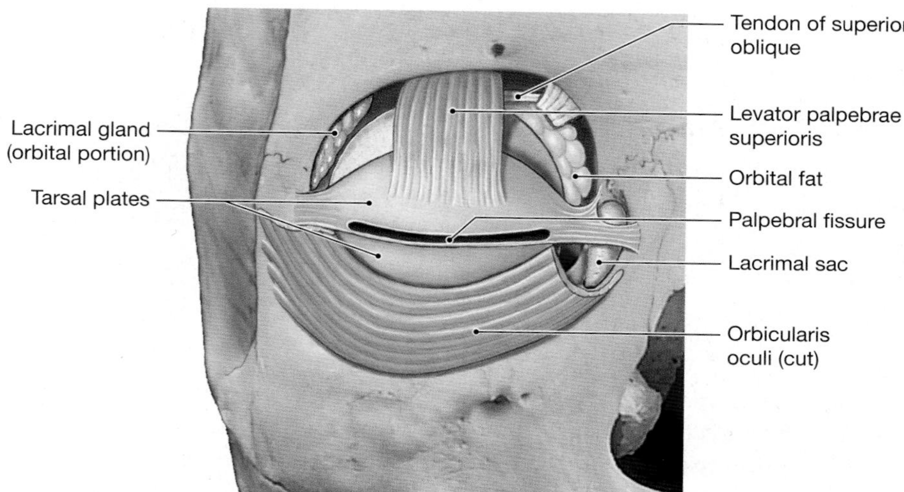

b Diagrammatic representation of a superficial dissection of the right orbit

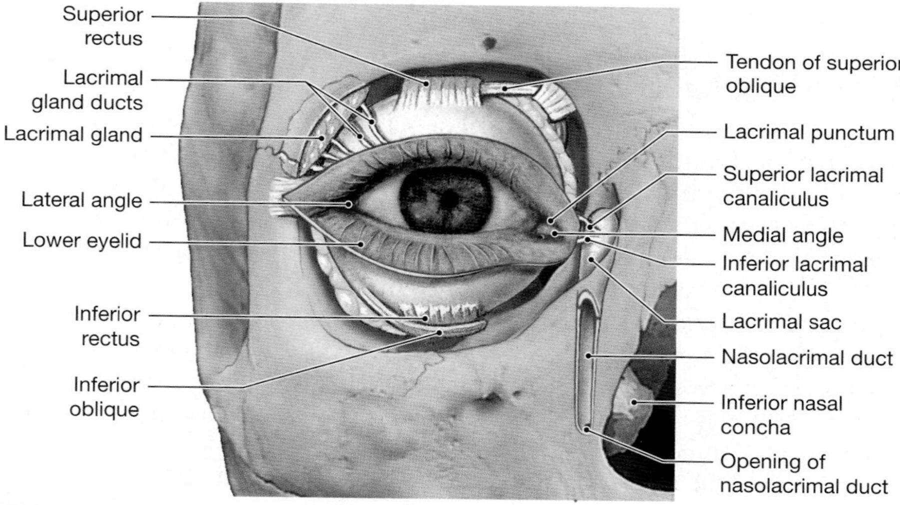

c Diagrammatic representation of a deeper dissection of the right eye showing its position within the orbit and its relationship to accessory structures, especially the lacrimal apparatus

Figure 18.20 Accessory Structures of the Eye, Part II. A superior view of structures within the right orbit.

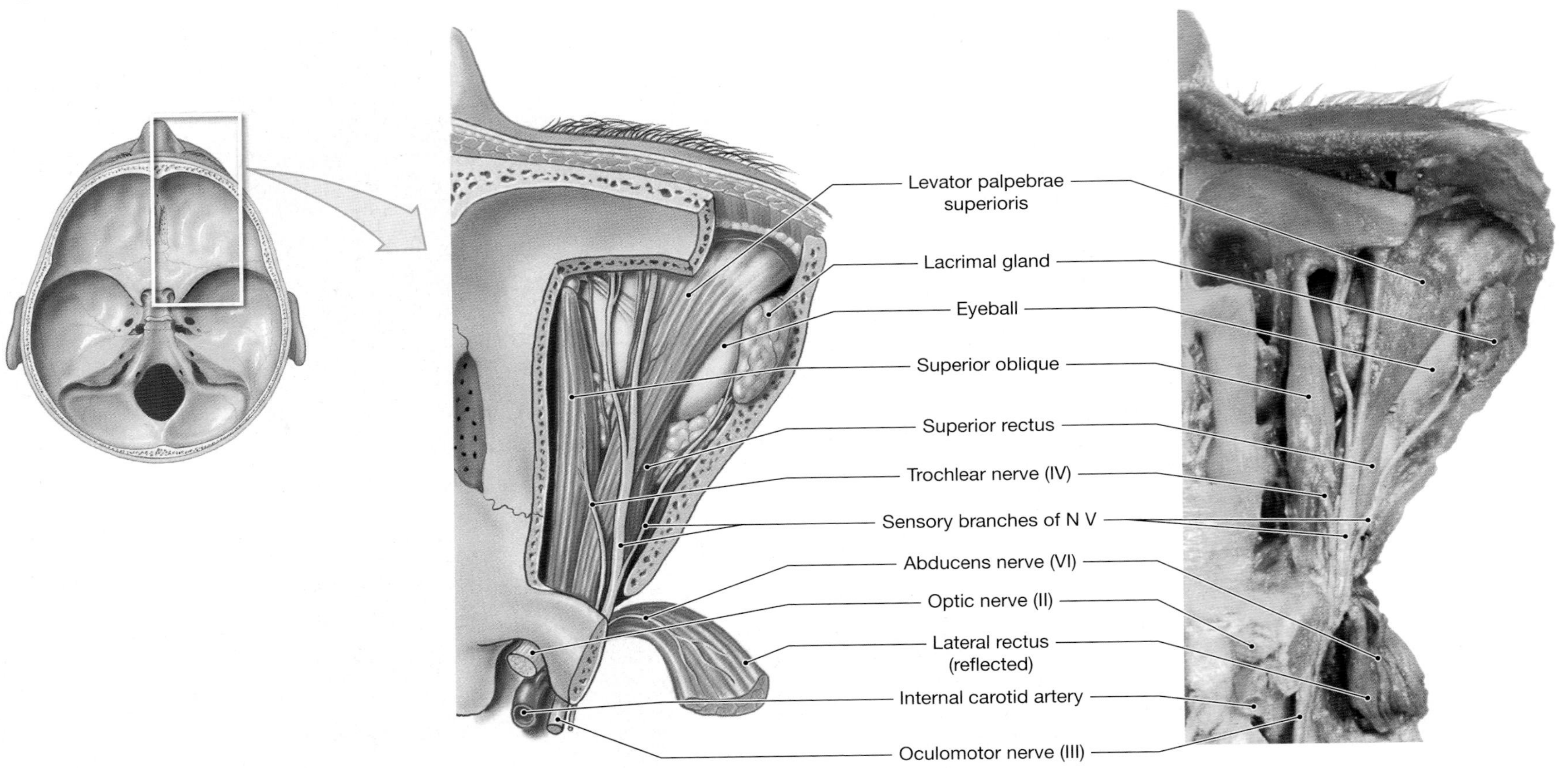

18

various accessory glands provides a superficial lubricant that prevents friction and drying of the conjunctival surfaces.

The stratified epithelium covering the conjunctiva changes to a very thin and delicate squamous epithelium that covers the transparent **cornea** (KOR-nē-a) of the eye. There are no specialized sensory receptors monitoring the surface of the eye. However, there are a large number of free nerve endings associated with the accessory structures of the eye. These free nerve endings have very broad sensitivities.

The Lacrimal Apparatus

A constant flow of tears keeps the conjunctival surfaces moist and clean. Tears reduce friction, remove debris, prevent bacterial infection, and provide nutrients and oxygen to the epithelium. The **lacrimal apparatus** produces, distributes, and removes tears. The lacrimal apparatus of each eye consists of (1) a lacrimal gland, (2) superior and inferior lacrimal canaliculi, (3) a lacrimal sac, and (4) a nasolacrimal duct **(Figures 18.19b,c** and **18.20)**.

The pocket created where the conjunctiva of the eyelid connects with the conjunctiva of the eye is called the **fornix** (FOR-niks). The lateral portion of the superior fornix receives 10–12 ducts from the **lacrimal gland**, or *tear gland*. The lacrimal gland is approximately the size and shape of an almond, measuring roughly 12–20 mm (0.5–0.75 in.). It sits within a depression in the frontal bone within the orbit, superior and lateral to the eyeball **(Figure 18.20)**.

The lacrimal gland provides the key ingredients and most of the volume of the tears that bathe the conjunctival surfaces. Lacrimal gland secretions are watery and slightly alkaline and contain the enzyme **lysozyme** and antibodies, which attack microorganisms.

The lacrimal gland produces 1 ml of tears a day. Once the lacrimal secretions reach the ocular surface, they mix with the products of accessory glands and the oily secretions of the tarsal glands and sebaceous glands. (The secretions of the sebaceous glands produce a superficial "oil slick" that lubricates the eye and slows the evaporation of the tears.)

Blinking sweeps the tears across the surface of the eye. Tears accumulate at the medial angle in an area called the lacrimal lake, or *lake of tears*. Two small pores, the **superior** and **inferior lacrimal puncta** (singular, *punctum*), drain the lacrimal lake. The lacrimal lake empties into the **lacrimal canaliculi**, which run along grooves in the surface of the lacrimal bone. These canaliculi lead to the **lacrimal sac**, which is located within the lacrimal groove of the lacrimal bone. From the lacrimal sac, the **nasolacrimal duct** extends along the nasolacrimal canal formed by the lacrimal bone and the maxilla. The nasolacrimal duct delivers the tears to the inferior meatus, a narrow passageway inferior and lateral to the inferior nasal concha.

The Eye

▶ **KEY POINT** The wall of the eye consists of three layers: the outer fibrous layer, the intermediate vascular layer, and the inner layer (retina).

The eyeball lies within the orbit of the skull along with (1) the extra-ocular muscles, (2) the lacrimal gland, and (3) the cranial nerves and blood vessels supplying the eye and adjacent portions of the orbit and face **(Figure 18.21e,f)**. Slightly smaller than a Ping-Pong ball, each eye weighs approximately 8 g (0.28 oz) and is approximately 24 mm (almost 1 in.) in diameter. A mass of **orbital fat** provides padding and insulation for the eye.

The wall of the eye is made up of three distinct layers **(Figure 18.21a)**: an outer fibrous layer, an intermediate vascular layer, and an inner layer. The eyeball is hollow, and the interior is divided into two cavities. The large **posterior cavity** (also called the *vitreous chamber*) contains the gelatinous vitreous body. The smaller **anterior cavity** has two chambers, the **anterior chamber**

CLINICAL NOTE

Disorders of the Eye

Refractive error occurs when the shape of the eye prevents light from focusing directly on the retina. This can be caused by an eyeball that is too long or too short, by changes in the cornea, or by stiffening of the lens. If incoming light focuses in front of the retina, near vision is normal but far vision is blurry. This is known as **myopia**, or nearsightedness. If the light focuses behind the retina, far vision is normal but near vision is blurry. This is **hyperopia**, or farsightedness. **Presbyopia** is farsightedness due to age-related stiffening of the lens, preventing it from focusing up close. Glasses or surgery can correct refractive errors.

Conjunctivitis, also known as "pink eye," is an inflammation of the conjunctiva. If conjunctivitis is caused by a bacterial or viral infection, it can be quite contagious. It can also result from allergies, environmental irritants, or contact lenses. If conjunctivitis is due to a bacterial infection, it usually responds quickly to antibiotic drops or ointment.

Corneal abrasions are the most common eye injury. Trauma, dust, foreign bodies, or contact lenses can scratch the cornea. If the cornea becomes severely damaged or diseased, it can turn opaque, causing blindness. A corneal transplant, using a fresh cadaver graft, can restore vision. Check your driver's license for your own donor status.

A child with conjunctivitis

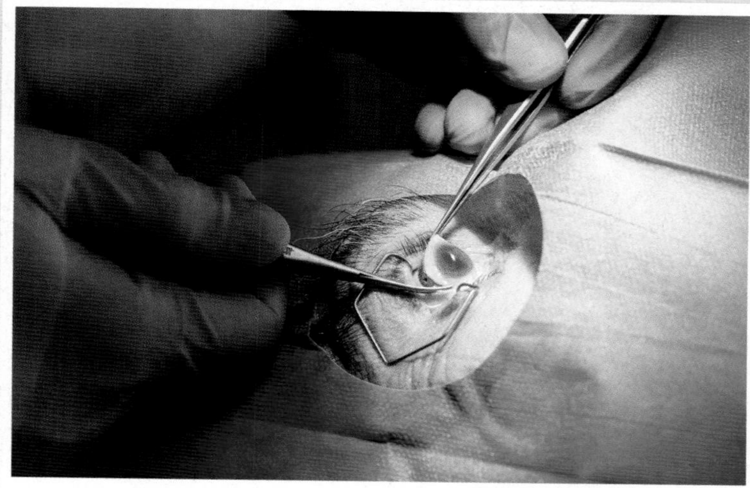

Cataract surgery to replace lens

Glaucoma is an eye disease marked by increased intra-ocular pressure that can cause blindness by damaging the optic nerve. If the flow of aqueous humor is blocked, fluid builds up, causing glaucoma. The cause of blockage is not known, although the disorder can be inherited. Eye exams check intra-ocular pressure by bouncing a tiny blast of air off the surface of the eye and measuring the deflection produced. Glaucoma can be treated medically and surgically to control vision loss.

Cataracts are a clouding of the lens until it can no longer transmit a clear image to the retina. Senile cataracts are related to aging. In rare cases, cataracts are congenital or induced by some medications. Cataract surgery takes about 15 minutes. Using a microscope, the surgeon removes the clouded lens and replaces it with an artificial lens. The new lens can be manufactured to correct for myopia and/or hyperopia. More than half of Americans over 80 have a cataract or have had cataract surgery.

Retinal detachment is a condition in which the retina pulls away from its nourishing choroid backing, losing its blood supply. If this is not treated immediately, permanent blindness will result.

Macular degeneration is a leading cause of blindness in Americans aged 60 and older. It is due to age-related deterioration of the central portion of the retina, the macula. When the macula degenerates, central vision is lost. Treatments can slow down vision loss, but currently there is no cure.

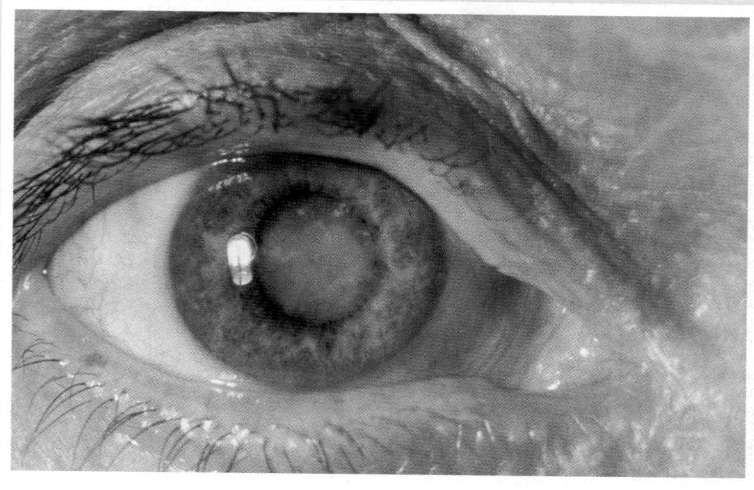

Cataract with characteristic lens clouding

Visual field of a person with macular degeneration

Figure 18.21 Sectional Anatomy of the Eye

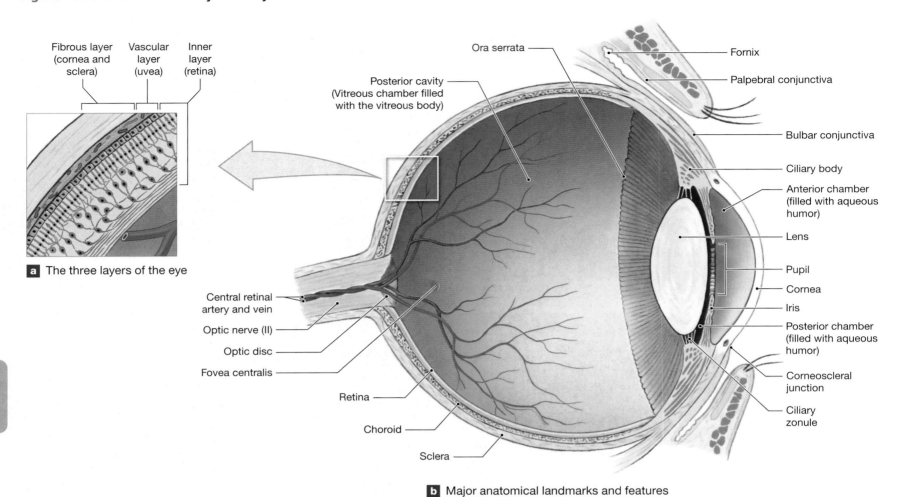

Fibrous layer (cornea and sclera)
Vascular layer (uvea)
Inner layer (retina)

a The three layers of the eye

Posterior cavity (Vitreous chamber filled with the vitreous body)
Ora serrata
Fornix
Palpebral conjunctiva
Bulbar conjunctiva
Ciliary body
Anterior chamber (filled with aqueous humor)
Lens
Pupil
Cornea
Iris
Posterior chamber (filled with aqueous humor)
Corneoscleral junction
Ciliary zonule
Central retinal artery and vein
Optic nerve (II)
Optic disc
Fovea centralis
Retina
Choroid
Sclera

b Major anatomical landmarks and features in a diagrammatic view of the left eyeball

Dilator pupillae (radial)
Pupil
Sphincter pupillae (sphincter)
Constrictors contract
Dilators contract

c The action of pupillary muscles and changes in pupillary diameter

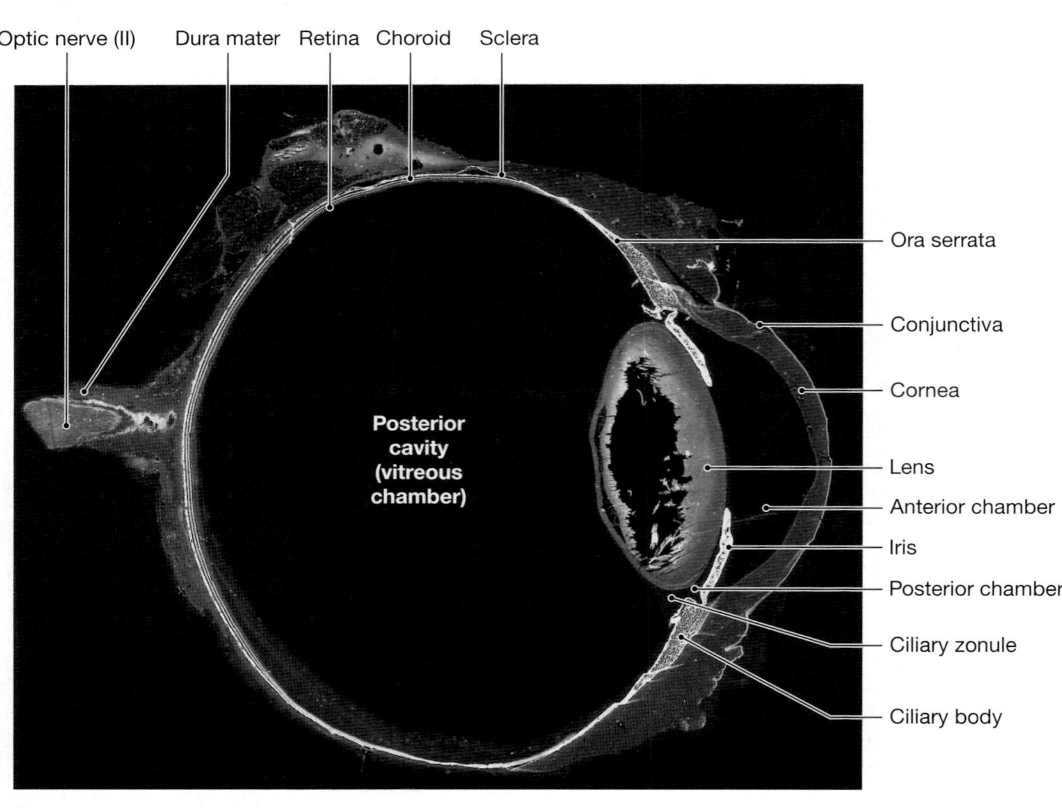

d Sagittal section through the eye

Figure 18.21 *(continued)*

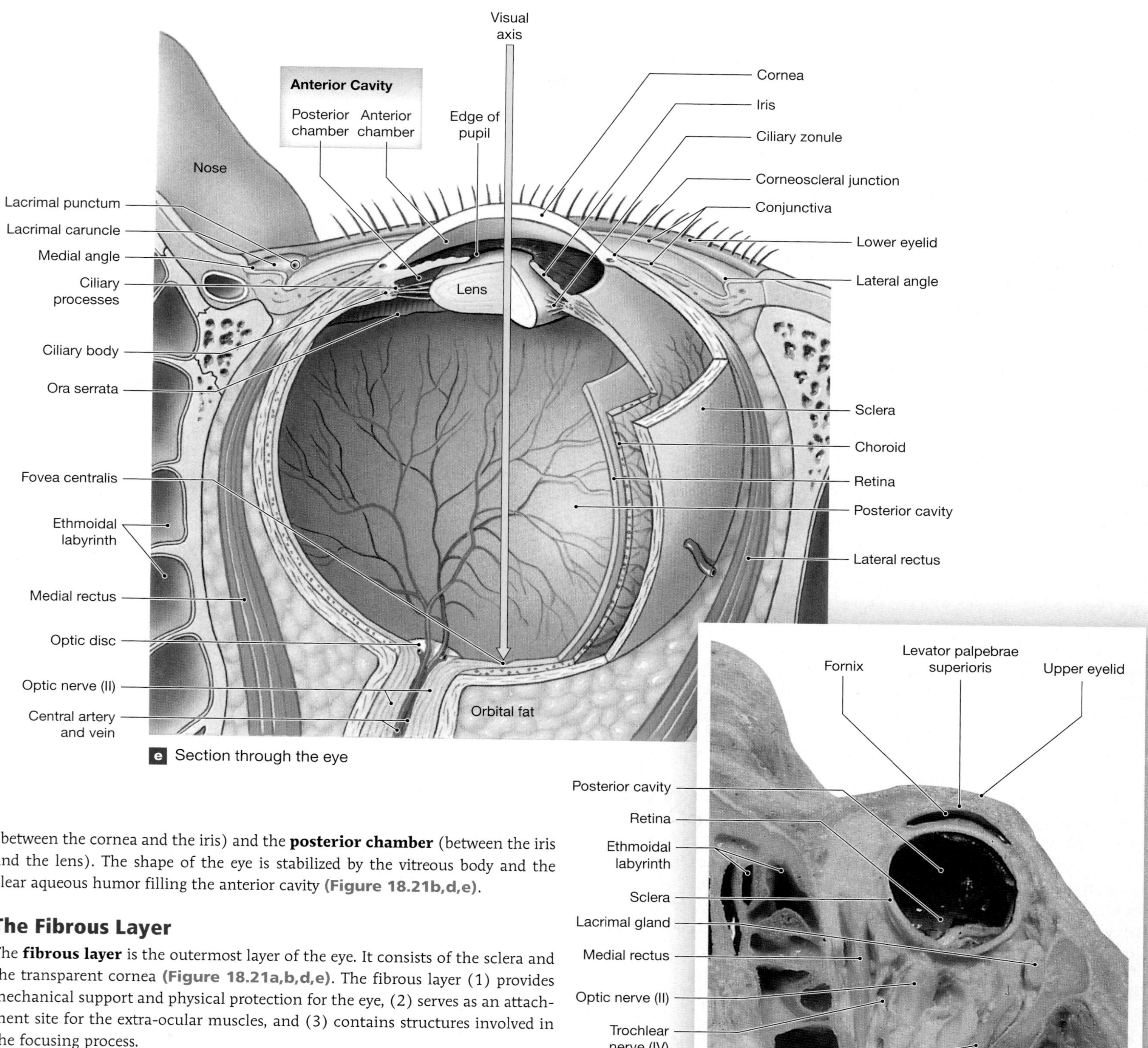

e Section through the eye

f Horizontal section, superior view

(between the cornea and the iris) and the **posterior chamber** (between the iris and the lens). The shape of the eye is stabilized by the vitreous body and the clear aqueous humor filling the anterior cavity **(Figure 18.21b,d,e)**.

The Fibrous Layer

The **fibrous layer** is the outermost layer of the eye. It consists of the sclera and the transparent cornea **(Figure 18.21a,b,d,e)**. The fibrous layer (1) provides mechanical support and physical protection for the eye, (2) serves as an attachment site for the extra-ocular muscles, and (3) contains structures involved in the focusing process.

The **sclera** (SKLER-a), or "white of the eye," covers most of the ocular surface. The sclera consists of dense, fibrous connective tissue containing collagen and elastic fibers. The sclera is thickest at the posterior portion of the eye, near the exit of the optic nerve. It is thinnest over the anterior surface. The six extra-ocular muscles insert onto the sclera. The collagen fibers of the tendons of insertion are interwoven with the collagen fibers of the fibrous layer **(Figure 18.20)**.

The anterior surface of the sclera contains small blood vessels and nerves that penetrate the sclera to reach internal structures. The network of small vessels interior to the bulbar conjunctiva generally does not carry enough blood to lend an obvious color to the sclera. On close inspection, however, the vessels are visible as red lines against the white background of collagen fibers.

The transparent cornea is part of the fibrous layer, and it is continuous with the sclera. A delicate stratified squamous epithelium covers the surface of the cornea. Deep to that epithelium, the cornea consists of multiple layers of collagen fibers. The precise alignment of the collagen fibers within these layers makes the cornea transparent. A simple squamous epithelium separates the innermost layer of the cornea from the anterior chamber of the eye.

The **corneoscleral junction**, also termed the *corneal limbus*, forms the junction between the cornea and the sclera. The cornea is avascular, and there are no blood vessels between the cornea and the overlying conjunctiva. As a result, the superficial epithelial cells obtain oxygen and nutrients from the tears flowing across their free surfaces. The innermost epithelial layer receives its nutrients from the aqueous humor within the anterior chamber. The cornea also has numerous free nerve endings, making it the most sensitive portion of the eye. This sensitivity is important because corneal damage can cause blindness even if the rest of the eye—photoreceptors included—is perfectly normal.

The Vascular Layer

The **vascular layer**, or *uvea*, contains numerous blood vessels, lymphatics, and the intrinsic eye muscles. The functions of this layer include (1) providing a route for blood vessels and lymphatics supplying tissues of the eye, (2) regulating the amount of light entering the eye, (3) secreting and reabsorbing the aqueous humor circulating within the eye, and (4) controlling the shape of the lens, which is an essential part of the focusing process. The vascular layer includes the iris, the ciliary body, and the choroid (**Figures 18.21a,b,d,e** and **18.22**).

18

The Iris The **iris** is seen through the transparent corneal surface. The iris contains blood vessels, pigment cells, and two layers of smooth muscle cells that are part of the intrinsic eye muscles. Contraction of these muscles changes the diameter of the **pupil**, the central opening of the iris. One group of smooth muscle fibers forms a series of concentric circles around the pupil (**Figure 18.21c**). The diameter of the pupil decreases when these **sphincter pupillae** contract. A second group of smooth muscles extends radially from the edge of the pupil. Contraction of these **dilator pupillae** enlarges the pupil. The autonomic nervous system controls these antagonistic muscles: parasympathetic activation constricts the pupil, and sympathetic activation dilates the pupil.

The body of the iris consists of connective tissue. The posterior surface of the iris is covered by an epithelium containing pigment cells. Pigment cells are also present within the connective tissue of the iris and in the epithelium covering its anterior surface. The density and distribution of these pigment cells determine the color of the eye. When there are no pigment cells in the body of the iris, light passes through it and bounces off the inner surface of the pigmented epithelium. The eye appears blue. Individuals with green, brown, or black eyes have increasing numbers of melanocytes in the body and on the surface of the iris.

The Ciliary Body At its periphery, the iris attaches to the anterior portion of the **ciliary body**. The ciliary body begins at the junction of the cornea and sclera and extends posteriorly to the **ora serrata** (Ō-ra ser-RĀ-ta; "serrated mouth") (**Figures 18.21b,d,e** and **18.22b**). The ciliary body consists of the **ciliary muscle**, a muscular ring projecting into the interior of the eye. The epithelium has numerous folds, called **ciliary processes**. The **ciliary zonule** (*suspensory ligament*) is the ring of fibers that attaches the lens to the ciliary processes. These connective tissue fibers hold the lens in place—posterior to the iris and in the center of the pupil. As a result, any light passing through the pupil also passes through the lens.

The Choroid The **choroid** (KOR-oyd) is a vascular layer that separates the fibrous layer and the inner layer posterior to the ora serrata. An extensive capillary network within the choroid delivers oxygen and nutrients to the retina.

Figure 18.22 The Lens and Chambers of the Eye

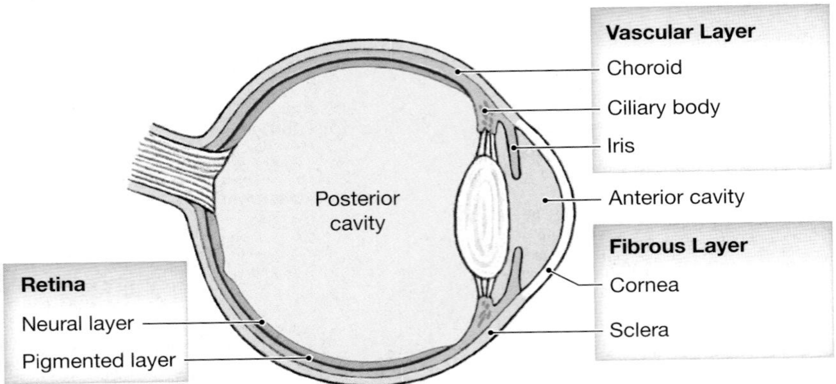

a The lens is suspended between the posterior cavity and the posterior chamber of the anterior cavity.

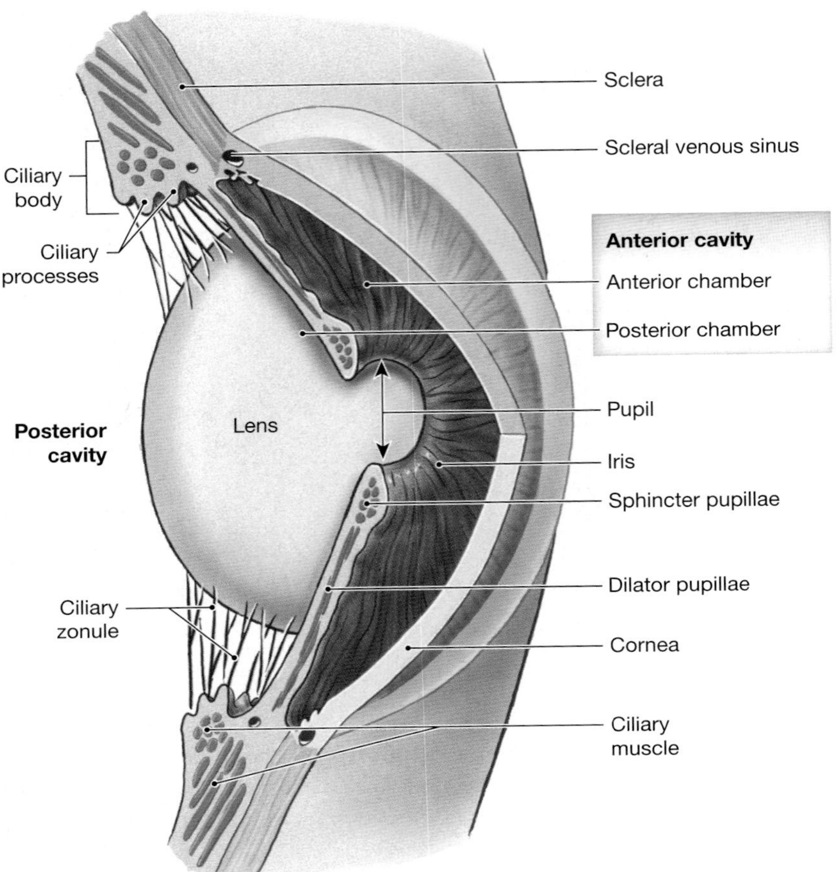

b Its position is maintained by the ciliary zonule that attaches the lens to the ciliary body.

The choroid also contains scattered melanocytes, which are especially dense near the sclera (**Figure 18.21a,b,d,e**). The innermost portion of the choroid attaches to the outer retinal layer.

The Inner Layer (Retina)

The **inner layer**, or **retina**, consists of two distinct layers: an outer thin lining called the **pigmented layer** and a thicker inner **neural layer**, which contains the visual receptors and associated neurons (**Figures 18.21** and **18.23**).

The pigmented layer absorbs light after it passes through the retina and has important biochemical interactions with retinal photoreceptors. The neural layer contains (1) the photoreceptors that respond to light, (2) supporting cells and

Figure 18.23 **Retinal Organization**

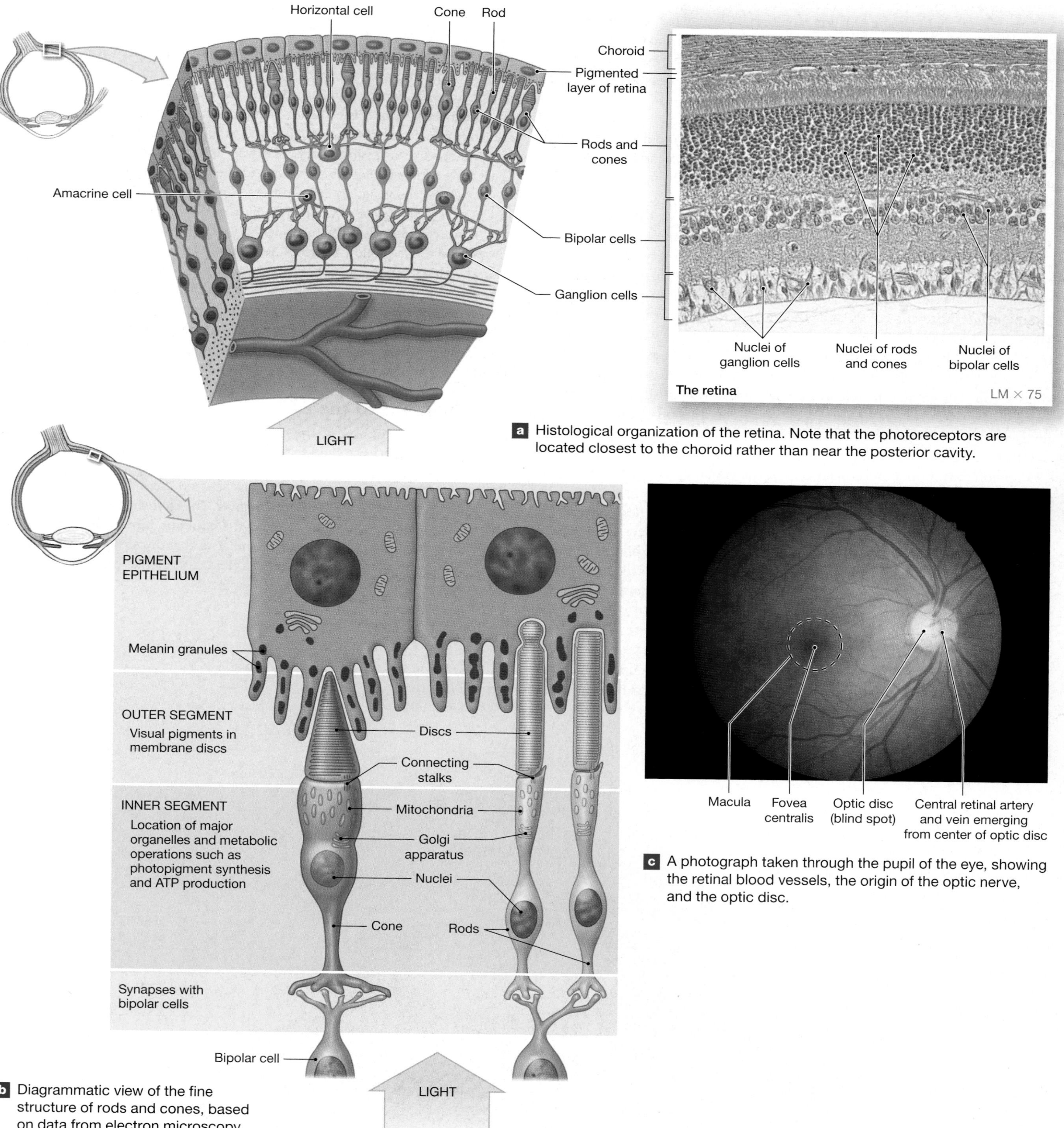

a Histological organization of the retina. Note that the photoreceptors are located closest to the choroid rather than near the posterior cavity.

b Diagrammatic view of the fine structure of rods and cones, based on data from electron microscopy.

c A photograph taken through the pupil of the eye, showing the retinal blood vessels, the origin of the optic nerve, and the optic disc.

neurons that perform preliminary processing and integration of visual information, and (3) blood vessels that supply the tissues lining the posterior cavity.

The neural and pigmented layers are normally very close together, but not tightly interconnected. The pigmented layer continues over the ciliary body and iris. The neural layer, however, only extends anteriorly as far as the ora serrata. The neural layer forms a cup that establishes the posterior and lateral boundaries of the posterior cavity (**Figure 18.21b,d,e,f**).

Retinal Organization There are approximately 130 million photoreceptors in the retina. Each photoreceptor monitors a specific location on the retinal surface. A visual image results from processing the information provided by all of the receptors.

The retina contains several layers of cells (**Figure 18.23a,b**). The outermost layer, closest to the pigmented layer, contains the visual receptors. There are two types of **photoreceptors**, the cells that detect light: rods and cones. **Rods** do not detect different colors of light. They are very light-sensitive and enable us to see in dimly lit rooms, at twilight, or in pale moonlight. **Cones** detect different colors and, therefore, provide us with color vision. There are three types of cones, and their stimulation in various combinations provides the detection of different colors. Cones give us sharper, clearer images, but they require more intense light than rods. If you sit outside at sunset, you will be able to tell when your visual system shifts from cone-based vision (clear images in full color) to rod-based vision (relatively grainy images in black and white).

18

Rods and cones are not evenly distributed across the retina. Approximately 125 million rods form a broad band around the periphery of the retina. As you move toward the center of the retina, the density of rods gradually decreases. In contrast, most of the roughly 6 million cones are concentrated in the area where a visual image arrives after it passes through the cornea and lens. This region, known as the **macula** (MAK-yū-luh; "spot"), has no rods. The highest concentration of cones occurs at the center of the macula, an area called the **fovea centralis** (FŌ-vē-uh; "shallow depression"), or simply the *fovea*. The fovea centralis is the site of sharpest color vision. When you look directly at an object, its image falls on this portion of the retina. (**Figures 18.21b,e** and **18.23c**).

The rods and cones synapse with neurons called **bipolar cells** (**Figure 18.23a,b**). Stimulation of rods and cones alters their rates of neurotransmitter release, and this alters the activity of the bipolar cells. **Horizontal cells** at this same level form a network that inhibits or promotes communication between the photoreceptors and bipolar cells. Bipolar cells synapse within the layer of **ganglion cells**. **Amacrine** (AM-a-krin) **cells**, also found at this level, modulate communication between bipolar and ganglion cells. The ganglion cells are the first-order neurons in the optic pathway and are the only cells in the retina generating action potentials that travel to the brain in the optic pathway.

Axons of the ganglion cells converge on the **optic disc**, a circular region just medial to the fovea centralis. The optic disc is the origin of the optic nerve (II). From this point, axons turn, penetrate the wall of the eye, and proceed toward the diencephalon (**Figure 18.21b,e**). The central retinal artery and central retinal vein pass through the center of the optic nerve and emerge on the surface of the optic disc (**Figures 18.21e** and **18.23c**). There are no photoreceptors or other retinal structures at the optic disc. Because light striking this area goes unnoticed, it is commonly called the **blind spot**. You do not notice a blank spot in your visual field because involuntary eye movements keep the visual image moving and allow the brain to fill in the missing information.

The Chambers of the Eye

Recall that the chambers of the eye are the anterior, posterior, and vitreous chambers. The anterior and posterior chambers are filled with aqueous humor.

Aqueous Humor Interstitial fluids pass between the epithelial cells of the ciliary processes and enter the posterior chamber. The epithelial cells alter the composition of the interstitial fluid, forming the **aqueous humor** (**Figure 18.24**). The composition of aqueous humor is similar to that of cerebrospinal fluid. The circulation of the aqueous humor forms a fluid cushion and provides an important route for transporting nutrients and wastes.

Figure 18.24 The Circulation of Aqueous Humor. Aqueous humor secreted at the ciliary body circulates through the posterior and anterior chambers as well as into the posterior cavity (arrows) before it is reabsorbed through the scleral venous sinus.

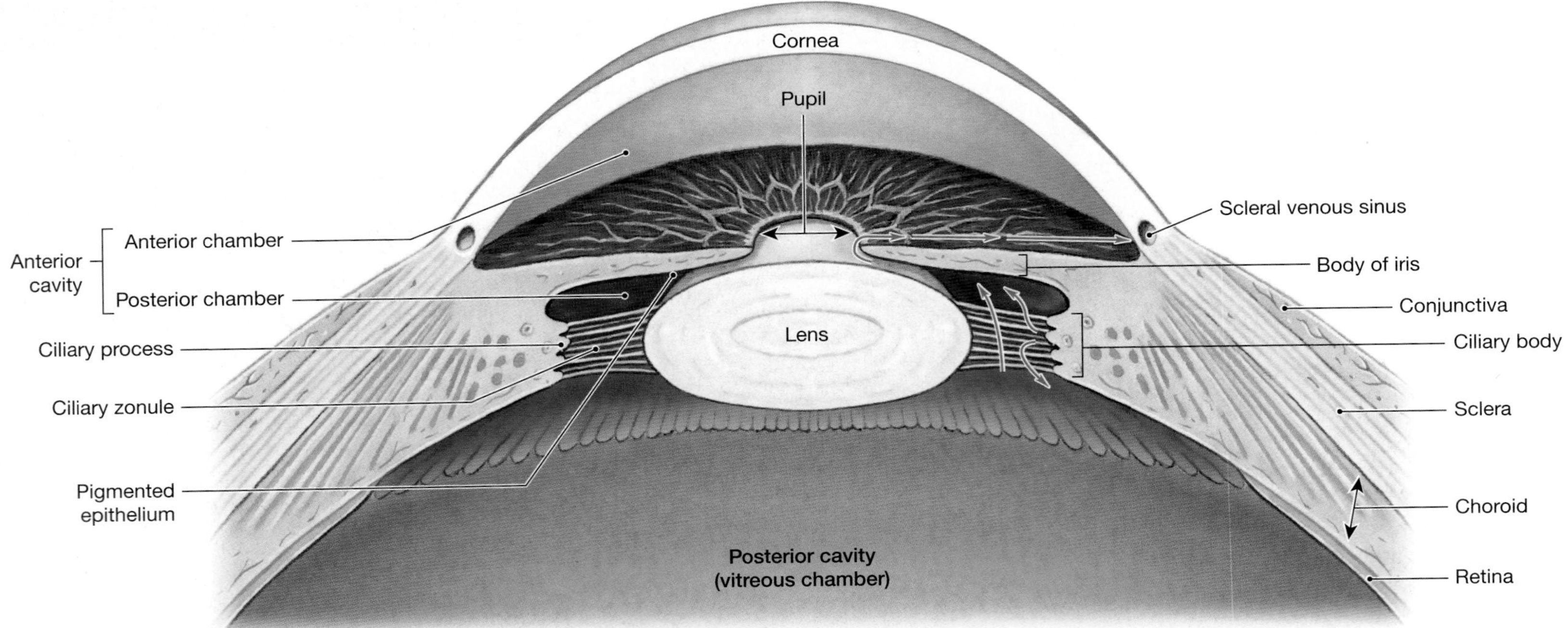

CLINICAL NOTE

Nystagmus

Nystagmus, also known as "dancing eyes," is an involuntary, rapid, repetitive movement of the eyes. The condition can be congenital and appear in early childhood, or it can be acquired, appearing later. It can also be caused by neurological problems, chronic alcohol use, or certain medications. Nystagmus can also be a sign of another condition, such as stroke or multiple sclerosis. Physicians often check for nystagmus by asking patients to follow a moving finger with their eyes.

Aqueous humor returns to the anterior chamber near the edge of the iris. After diffusing through the local epithelium, it passes into the **scleral venous sinus**, or the *canal of Schlemm*, which communicates with the veins of the eye.

The **lens** lies posterior to the cornea. It is held in place by the ciliary zonule that originates on the ciliary body of the choroid **(Figure 18.24)**. The lens and the ciliary zonule form the anterior boundary of the vitreous chamber. This chamber contains the **vitreous body**, a gelatinous mass sometimes called the *vitreous humor*. The vitreous body maintains the shape of the eye, supports the posterior surface of the lens, and supports the retina by pressing the neural layer against the pigmented layer. Aqueous humor produced in the posterior chamber freely diffuses through the vitreous body and across the retinal surface.

The Lens

The lens focuses the visual image on the retinal photoreceptors by changing its shape. The lens consists of precisely organized layers of cells **(Figures 18.21b,d,e, 18.22b,** and **18.24)**. A dense, fibrous capsule covers the entire lens. Many of the capsular fibers are elastic, and unless an outside force is applied, they contract and make the lens spherical. The capsular fibers intermingle with those of the ciliary zonule around the edges of the lens.

As tension in the ciliary zonule increases, the lens flattens **(Figures 18.21b,d,e, 18.22b,** and **18.24)**. With the lens in this position, the eye is able to focus on distant objects. When the ciliary muscles contract, the ciliary body moves toward the lens. This movement reduces the tension in the ciliary zonule, and the elastic lens assumes a more spherical shape. With the lens in this position, the eye is able to focus on nearby objects.

Visual Pathways

▶ **KEY POINT** The visual pathway is a three-neuron pathway, beginning in the ganglion cells of the retina and ending in the visual cortex of the cerebrum.

Each rod and cone cell monitors a specific receptive field. A visual image results from processing information provided by the entire receptor population. A significant amount of processing occurs in the retina before the information is sent to the brain because of interactions between the various cell types.

The right and left optic nerves reach the diencephalon after a partial decussation at the **optic chiasm (Figure 18.25)**. At the optic chiasm approximately half of the fibers proceed toward the lateral geniculate nucleus of the same

18

Figure 18.25 Anatomy of the Visual Pathways, Part I. A superior view of a horizontal section through the head at the level of the optic chiasm.

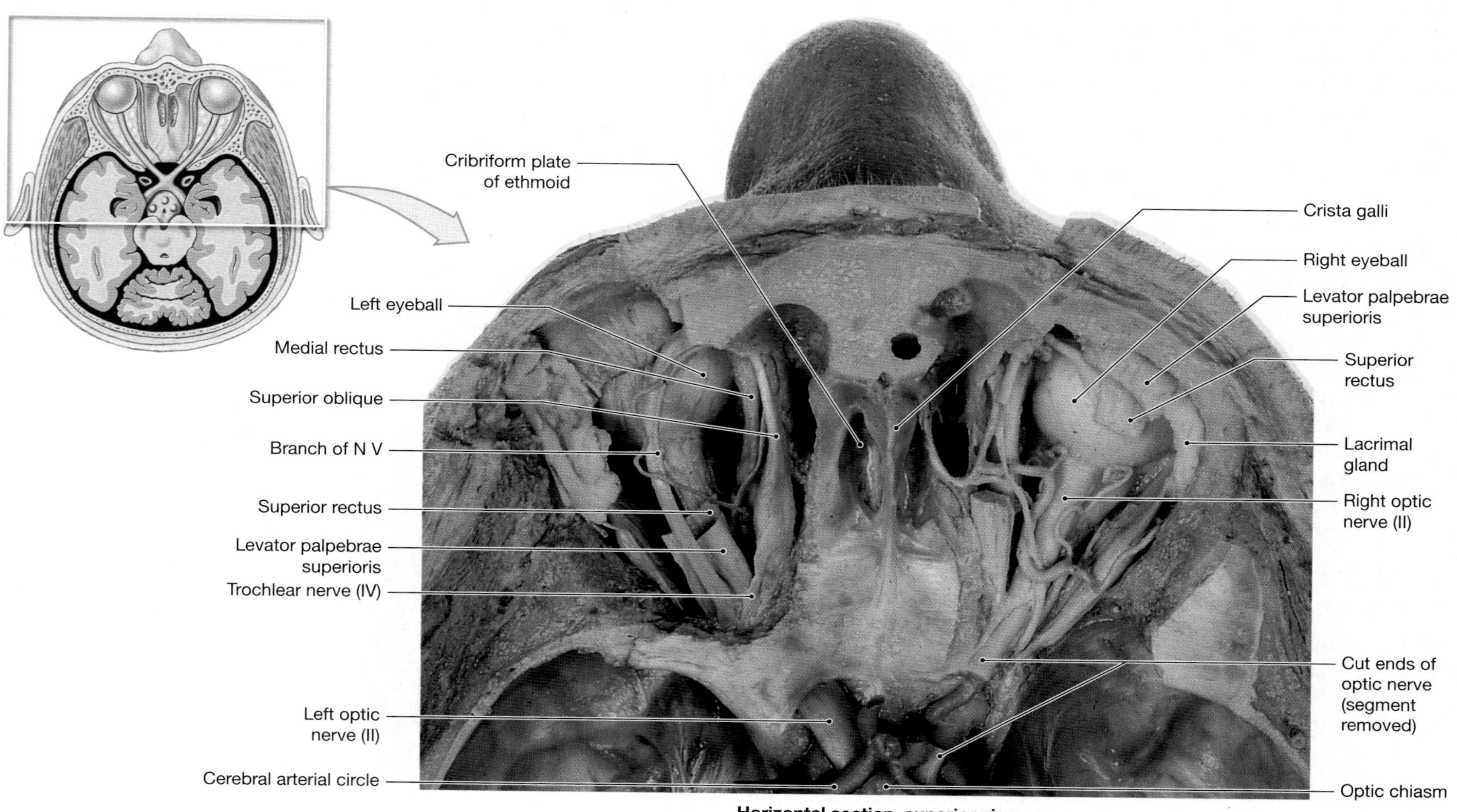

side of the brain, while the other half cross over to reach the lateral geniculate nucleus of the contralateral side **(Figure 18.26)**. Visual information from the left half of each retina synapses at the lateral geniculate nucleus of the left side; information from the right half of each retina goes to the right side. The lateral geniculate nuclei act as a switching center, relaying visual information to reflex centers in the brainstem and cerebral cortex. Some visual information, however, bypasses the lateral geniculate nuclei. For instance, reflexes controlling eye movement are triggered by information that bypasses the lateral geniculate nuclei to synapse in the superior colliculi.

Cortical Integration

The sensation of vision results from the integration of information arriving at the visual cortex of the occipital lobes of the cerebral hemispheres. The visual cortex contains a sensory map of the entire field of vision. As we saw with the primary sensory cortex, the map does not exactly duplicate the relative areas within the sensory field.

Each eye also receives a slightly different image because (1) the foveae are 2–3 inches apart, and (2) the nose and eye socket block the view of the opposite side. The association and integrative areas of the cortex compare the two images and use them to provide us with depth perception **(Figure 18.26)**. The partial crossover occurring at the optic chiasm ensures that the visual cortex receives a *composite* picture of the entire visual field.

The Brainstem and Visual Processing

Many centers in the brainstem receive visual information from the lateral geniculate nuclei or collaterals arising from the optic tracts. Collaterals bypassing the lateral geniculate nuclei synapse in either the superior colliculus or hypothalamus **(Figure 18.26)**. The superior colliculus of the mesencephalon issues motor commands controlling subconscious eye, head, or neck movements in response to visual stimuli. Visual inputs to the **suprachiasmatic** (soo-pra-kī-az-MA-tic) **nucleus** of the hypothalamus and the endocrine cells of the pineal gland affect the function of other brainstem nuclei. These nuclei establish a daily pattern of visceral activity that is tied to the day-night cycle. This **circadian rhythm** (*circa*, about, + *dies*, day) affects metabolic rate, endocrine function, blood pressure, digestive activities, the awake-asleep cycle, and other physiological processes.

18.6 CONCEPT CHECK

16 What layer of the eye is the first to be affected by inadequate tear production?

17 If the intra-ocular pressure becomes abnormally high, which structures of the eye are affected and how are they affected?

18 Would a person born without cones in her eyes be able to see? Explain.

19 Explain what ciliary processes are and what they do.

See the blue Answers tab at the back of the book.

EMBRYOLOGY SUMMARY

For a summary of the development of the special organs, see Chapter 28 (Embryology and Human Development).

Figure 18.26 Anatomy of the Visual Pathways, Part II. At the optic chiasm, a partial crossover of nerve fibers occurs. As a result, each hemisphere receives visual information from the lateral half of the retina of the eye on that side and from the medial half of the retina of the eye on the opposite side. Visual association areas integrate this information to develop a composite picture of the entire visual field.

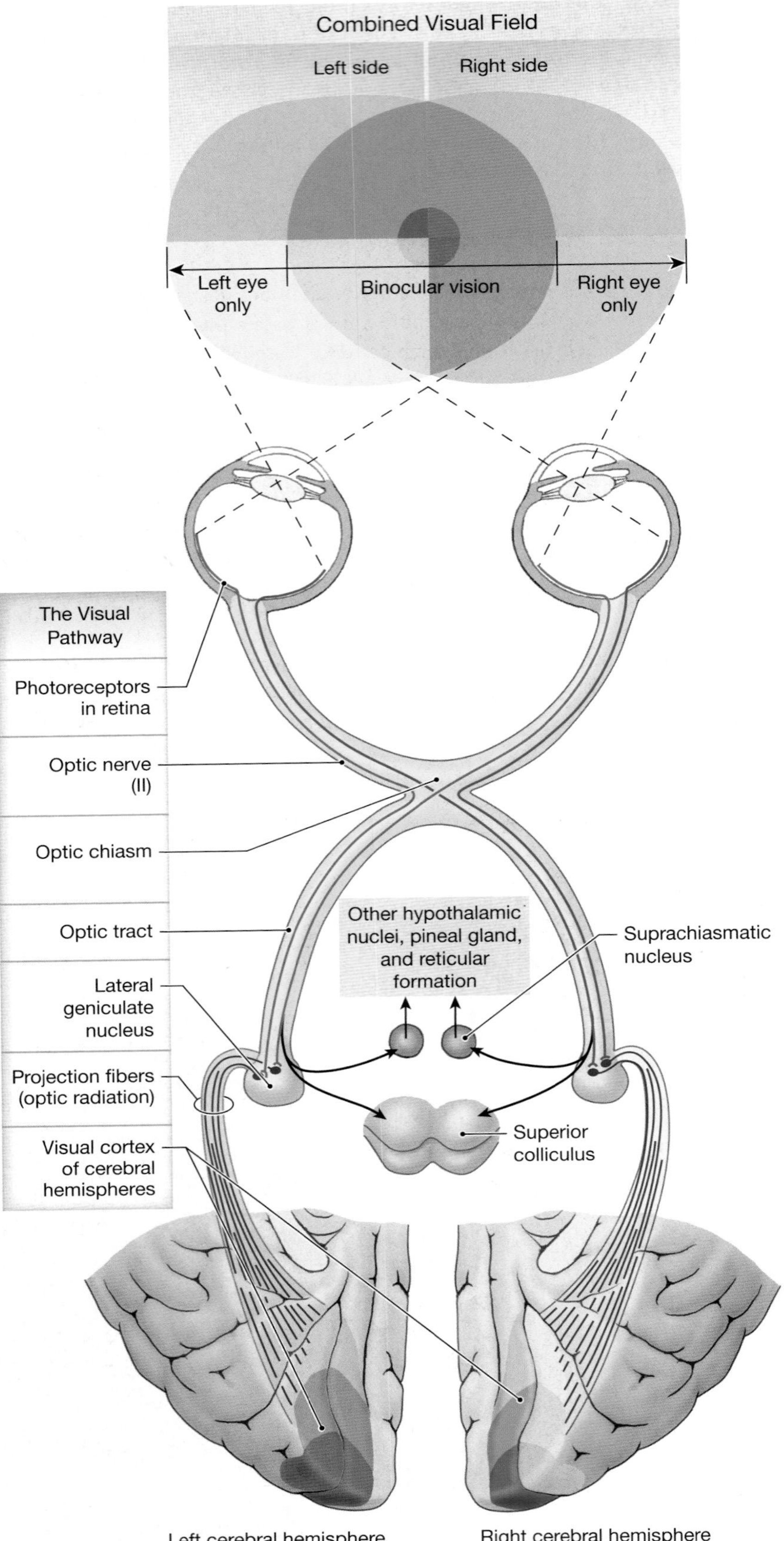

Study Outline

Introduction p. 472

- The **general senses** are temperature, pain, touch, pressure, vibration, and proprioception; receptors for these sensations are distributed throughout the body. Receptors for the **special senses** (olfaction, gustation, equilibrium, hearing, **and** vision) are located in specialized areas, or **sense organs**. A sensory receptor is a specialized cell that when stimulated sends a **sensation** to the CNS.

18.1 | Receptors p. 472

- **Receptor specificity** allows each receptor to respond to particular stimuli. The simplest receptors are **free nerve endings**; the area monitored by a single receptor cell is the **receptive field**. *(See Figure 18.1.)*

Interpretation of Sensory Information p. 472

- **Tonic receptors** are always sending signals to the CNS; **phasic receptors** become active only when the conditions that they monitor change.

Central Processing and Adaptation p. 472

- **Adaptation** (a reduction in sensitivity in the presence of a constant stimulus) may involve changes in receptor sensitivity (**peripheral**, or *sensory*, **adaptation**) or inhibition along the sensory pathways (**central adaptation**). **Fast-adapting receptors** are phasic; **slow-adapting receptors** are tonic.

- The information provided by our sensory receptors is incomplete because (1) we do not have for every stimulus; (2) our receptors have limited ranges of sensitivity; and (3) a stimulus produces a neural event that must be interpreted by the CNS.

18.2 | The General Senses p. 473

- Receptors are classified as **exteroceptors** if they provide information about the external environment, **proprioceptors** if they monitor body position, and **interoceptors** if they monitor conditions inside the body.

Nociceptors p. 473

- **Nociceptors** respond to a variety of stimuli usually associated with tissue damage. There are two types of these painful sensations: **fast** (*pricking*) pain and **slow** (*burning*) **pain**. *(See Figures 18.2 and 18.3a.)*

Thermoreceptors p. 474

- **Thermoreceptors** respond to changes in temperature. They conduct sensations along the same pathways that carry pain sensations.

Mechanoreceptors p. 474

- **Mechanoreceptors** respond to physical distortion, contact, or pressure on their cell membranes: **tactile receptors** to touch, pressure, and vibration; **baroreceptors** to pressure changes in the walls of blood vessels and the digestive, reproductive, and urinary tracts; and **proprioceptors** (muscle spindles) to positions of joints and muscles. *(See Figures 18.3 and 18.4.)*

- **Fine touch** and **pressure receptors** provide detailed information about a source of stimulation; **crude touch** and **pressure receptors** are poorly localized. Important tactile receptors include **free nerve endings**, the **root hair plexus**, **Merkel cells** and **tactile discs**, **tactile corpuscles** (*Meissner's corpuscles*), **bulbous corpuscles** (*Ruffini corpuscles*), and **lamellar corpuscles** (*Pacinian corpuscles*). *(See Figure 18.3 and Table 18.1.)*

- **Baroreceptors** (stretch receptors) monitor changes in pressure; they respond immediately but adapt rapidly. Baroreceptors in the walls of major arteries and veins respond to changes in blood pressure. Receptors along the digestive tract help coordinate reflex activities of digestion. *(See Figure 18.4.)*

- **Proprioceptors** monitor the position of joints, tension in tendons and ligaments, and the state of muscular contraction.

Chemoreceptors p. 476

- In general, **chemoreceptors** respond to water-soluble and lipid-soluble substances that are dissolved in the surrounding fluid. They monitor the chemical composition of body fluids. *(See Figure 18.5.)*

18.3 | Olfaction (Smell) p. 476

- The **olfactory organs** contain the **olfactory epithelium** with **olfactory sensory neurons** (neurons sensitive to chemicals dissolved in the overlying mucus), **supporting cells**, and **basal epithelial** (*stem*) **cells**. Their surfaces are coated with the secretions of the **olfactory glands**. *(See Figure 18.6.)*

Olfactory Receptors p. 477

- The olfactory neurons are modified nerve cells. *(See Figure 18.6b.)*

18

Olfactory Pathways p. 477

- The olfactory system has extensive limbic and hypothalamic connections that help explain the emotional and behavioral responses that can be produced by certain smells. *(See Figure 18.6.)*

Olfactory Discrimination p. 477

- The olfactory system can make subtle distinctions between thousands of chemical stimuli; the CNS interprets the smell.

- The olfactory receptor population shows considerable turnover and is the only known example of neuronal replacement in the adult human. The total number of receptors declines with age.

18.4 | Gustation (Taste) p. 478

- **Gustation**, or *taste*, provides information about the food and liquids that we consume.

Taste Receptors p. 479

- **Gustatory epithelial cells** are clustered in **taste buds**. These cells extend *taste hairs* through a narrow **taste pore**. *(See Figure 18.7.)*

- Taste buds are associated with epithelial projections (**lingual papillae**). *(See Figure 18.7a,b.)*

Gustatory Pathways p. 479

- The taste buds are monitored by cranial nerves VII, IX, and X. The afferent fibers synapse within the solitary nucleus before proceeding to the thalamus and cerebral cortex. *(See Figure 18.8.)*

Gustatory Discrimination p. 479

- The **primary taste sensations** are sweet, salty, sour, bitter, **umami**, and water.

- There are individual differences in the sensitivity to specific tastes. The number of taste buds and their sensitivity decline with age. *(See Figure 18.8.)*

18.5 | Equilibrium and Hearing p. 480

The External Ear p. 480

- The **external ear** includes the auricle, which surrounds the entrance to the **external acoustic meatus** that ends at the **tympanic membrane**, or *eardrum*. (*See Figures 18.9 and 18.10.*)

The Middle Ear p. 480

- In the **middle ear**, the **tympanic cavity** encloses and protects the **auditory ossicles**, which connect the tympanic membrane with the receptor complex of the internal ear. The tympanic cavity communicates with the nasopharynx via the **auditory tube**. *(See Figures 18.9 and 18.10.)*
- The **tensor tympani** and **stapedius** contract to reduce the amount of motion of the tympanic membrane when very loud sounds arrive. *(See Figures 18.9 and 18.10b,d.)*

The Internal Ear p. 482

- The senses of equilibrium and hearing are provided by the receptors of the **internal ear** (housed within fluid-filled tubes and chambers known as the **membranous labyrinth**). Its chambers and canals contain **endolymph**. The **bony labyrinth** surrounds and protects the membranous labyrinth. The bony labyrinth can be subdivided into the **vestibule** and **semicircular canals** (providing the sense of equilibrium) and the **cochlea** (providing the sense of hearing). *(See Figures 18.9–18.17.)*
- The vestibule includes a pair of membranous sacs, the **utricle** and **saccule**, whose receptors provide sensations of gravity and linear acceleration. The cochlea contains the **cochlear duct**, an elongated portion of the membranous labyrinth. *(See Figure 18.12.)*
- The basic receptors of the internal ear are **hair cells** whose surfaces support stereocilia. Hair cells provide information about the direction and strength of varied mechanical stimuli. *(See Figure 8.13d.)*
- The **anterior**, **posterior**, and **lateral semicircular ducts** are continuous with the utricle. Each contains an **ampulla** with sensory receptors. Here the cilia contact a gelatinous **ampullary cupula**. *(See Figures 18.13 and 18.14.)*
- The utricle and saccule are connected by a passageway continuous with the **endolymphatic duct**, which terminates in the **endolymphatic sac**. In the saccule and utricle, hair cells cluster within **maculae**, where their cilia contact **otoliths** consisting of densely packed mineral crystals in a gelatinous matrix. When the head tilts, the mass of each otolith shifts, and the resulting distortion in the sensory hairs signals the CNS. *(See Figure 18.15.)*
- The vestibular receptors activate sensory neurons of the **vestibular ganglia**. The axons form the **vestibular nerve** synapsing within the vestibular nuclei. (*See Figure 18.16.*)

Hearing p. 485

- Sound waves travel toward the tympanic membrane, which vibrates; the auditory ossicles conduct the vibrations to the base of the stapes at the oval window. Movement at the oval window applies pressure first to the perilymph of the **scala vestibuli** (*vestibular duct*). This pressure is passed on to the perilymph in the **scala tympani** (*tympanic duct*). *(See Figure 18.17.)*
- Pressure waves distort the **basilar membrane** and push the hair cells of the **spiral organ**, or *organ of Corti*, against the **tectorial membrane**. *(See Figure 18.17 and Table 18.2.)*

Auditory Pathways p. 487

- The sensory neurons for hearing are located in the **spiral ganglion** of the cochlea. Their afferent fibers form the **cochlear nerve** that synapses at the **cochlear nucleus**. *(See Figure 18.18.)*

18.6 | Vision p. 491

Accessory Structures of the Eye p. 491

- The **accessory structures** of the eye include the **eyelids**, which are separated by the **palpebral fissure**. The **eyelashes** line the palpebral margins. **Tarsal glands**, which secrete a lipid-rich product, line the inner margins of the eyelids. Glands at the **lacrimal caruncle** produce other secretions. *(See Figure 18.19.)*
- An epithelium called the **conjunctiva** covers most of the exposed surface of the eye; the **bulbar conjunctiva** (*ocular conjunctiva*) covers the anterior surface of the eye, and the **palpebral conjunctiva** lines the inner surface of the eyelids. The **cornea** is transparent. *(See Figure 18.21.)*
- The secretions of the **lacrimal gland** bathe the conjunctiva; these secretions are slightly alkaline and contain **lysozymes** (enzymes and antibodies that attack bacteria). Tears collect in the lacrimal lake. The tears reach the inferior meatus of the nasal cavity after passing through the **lacrimal puncta**, the **lacrimal canaliculi**, the **lacrimal sac**, and the **nasolacrimal duct**. Collectively, these structures constitute the **lacrimal apparatus**. *(See Figures 18.19–18.21.)*

The Eye p. 492

- The eye has three layers: an outer fibrous layer, a vascular layer (uvea), and an inner layer (retina).
- The **fibrous layer** includes most of the ocular surface, which is covered by the **sclera** (a dense, fibrous connective tissue of the fibrous layer); the **corneoscleral junction**, or *corneal limbus*, is the border between the sclera and the cornea. *(See Figure 18.21.)*
- The **vascular layer** includes the **iris**, the **ciliary body**, and the **choroid**. The iris forms the boundary between the anterior and posterior chambers. The ciliary body contains the **ciliary muscle** and the **ciliary processes**, which attach to the **ciliary zonule** of the lens. *(See Figures 18.21 and 18.22.)*
- The **inner layer** (**retina**) consists of an outer **pigmented layer** and an inner **neural layer**; the latter contains visual receptors and associated neurons. *(See Figures 18.21 and 18.23.)*
- There are two types of **photoreceptors** (visual receptors of the retina). **Rods** provide black-and-white vision in dim light; **cones** provide color vision in bright light. Cones are concentrated in the **macula**; the **fovea centralis** is the area of sharpest vision. *(See Figures 18.21 and 18.23.)*
- The direct line to the CNS proceeds from the photoreceptors to **bipolar cells**, then to **ganglion cells**, and to the brain via the optic nerve. **Horizontal cells** and **amacrine cells** modify the signals passed between other retinal components. *(See Figure 18.23a,b.)*
- The **aqueous humor** continuously circulates within the eye and re-enters the circulation after diffusing through the walls of the anterior chamber and into the **scleral venous sinus** (*canal of Schlemm*). *(See Figure 18.24.)*

- The **lens**, held in place by the ciliary zonule, lies posterior to the cornea and forms the anterior boundary of the vitreous chamber. This chamber contains the **vitreous body**, a gelatinous mass that helps stabilize the shape of the eye and support the retina. The lens focuses a visual image on the retinal receptors. *(See Figures 18.21 and 18.24.)*

Visual Pathways p. 499

- Each photoreceptor monitors a specific receptive field. The axons of ganglion cells converge on the **optic disc** and proceed along the optic tract to the **optic chiasm**. *(See Figures 18.21b,e, 18.23, 18.25, and 18.26.)*
- From the optic chiasm, after a partial decussation, visual information is relayed to the lateral geniculate nuclei. From there the information is sent to the visual cortex of the occipital lobes. *(See Figure 18.26.)*
- Visual inputs to the **suprachiasmatic nucleus** and the pineal gland affect the function of other brainstem nuclei. These nuclei establish a visceral **circadian rhythm** that is tied to the day-night cycle and affects other metabolic processes. *(See Figure 18.26.)*

Chapter Review

For answers, see the blue Answers tab at the back of the book.

Level 1 Reviewing Facts and Terms

1. Using the diagram, fill in the blanks below with the proper anatomical terms.

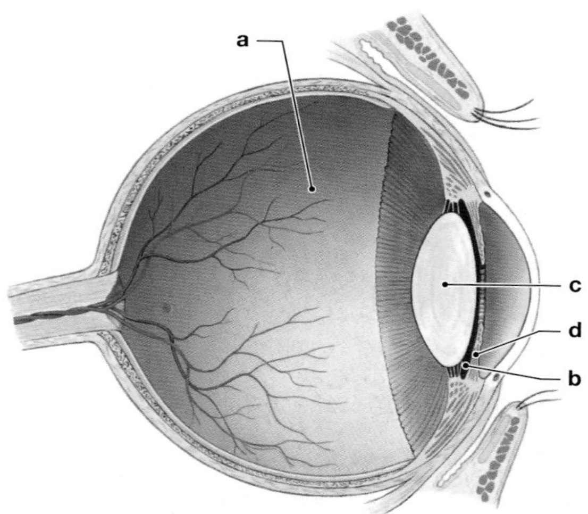

(a) ___________
(b) ___________
(c) ___________
(d) ___________

2. A receptor that is especially common in the superficial layers of the skin and that responds to pain is a
(a) proprioceptor.
(b) baroreceptor.
(c) nociceptor.
(d) mechanoreceptor.

3. Fine touch and pressure receptors provide detailed information about
(a) the source of the stimulus.
(b) the shape of the stimulus.
(c) the texture of the stimulus.
(d) all of the above.

4. Receptors in the saccule and utricle provide sensations of
(a) balance and equilibrium.
(b) hearing.
(c) vibration.
(d) gravity and linear acceleration.

5. Deep to the subcutaneous layer, the eyelids are supported by broad sheets of connective tissues, collectively termed the
(a) eyelids.
(b) tarsal plate.
(c) chalazion.
(d) medial angle.

6. The inner layer
(a) consists of three distinct layers.
(b) contains the photoreceptors.
(c) forms the iris.
(d) all of the above.

7. The semicircular canals include which of the following?
(a) dorsal and ventral
(b) lateral, middle, and medial
(c) anterior, posterior, and lateral
(d) spiral, upright, and reverse

8. Mechanoreceptors that detect pressure changes in the walls of blood vessels as well as in portions of the digestive, reproductive, and urinary tracts are
(a) tactile receptors.
(b) baroreceptors.
(c) proprioceptors.
(d) free nerve receptors.

9. Pupillary muscle groups are controlled by the ANS. Parasympathetic activation causes pupillary _____, and sympathetic activation causes _____.
(a) dilation; constriction
(b) dilation; dilation
(c) constriction; dilation
(d) constriction; constriction

10. Auditory information about the frequency and intensity of stimulation is relayed to the CNS over the cochlear nerve, a division of cranial nerve
(a) IV.
(b) VI.
(c) VIII.
(d) X.

11. Fill in the blanks below with the proper anatomical terms.

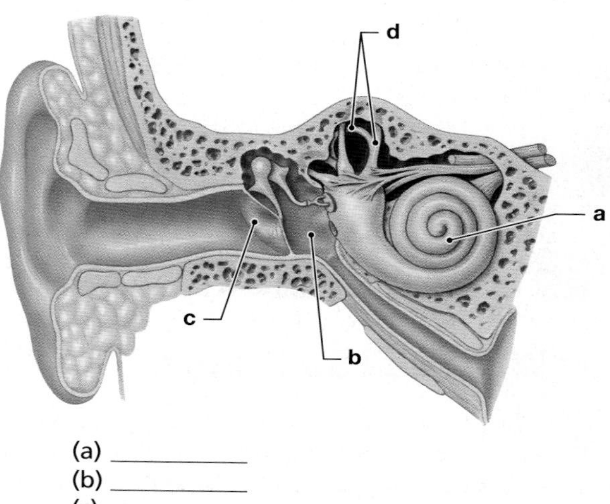

(a) ___________
(b) ___________
(c) ___________
(d) ___________

Level 2 Reviewing Concepts

1. Why is a more severe burn less painful initially than a less serious burn of the skin?
(a) The skin's nociceptors are burned away and cannot transmit pain sensations to the CNS.
(b) A severe burn overwhelms the nociceptors, and they adapt rapidly so no more pain is felt.
(c) A mild skin burn registers pain from pain receptors and many other types simultaneously.
(d) A severe burn is out of the range of sensitivity of most pain receptors.

2. How do the tensor tympani and stapedius affect the functions of the ear?
   (a) They do not affect hearing, but play an important role in equilibrium.
   (b) They increase the cochlea's sensitivity to vibration produced by incoming sound waves.
   (c) They regulate the opening and closing of the auditory tube.
   (d) They dampen excessively loud sounds that could harm sensitive auditory hair cells.
3. A person salivates when anticipating eating a tasty confection. Would this physical response enhance taste or olfaction? If so, why?
   (a) No, it would not enhance either taste or olfaction.
   (b) Salivation permits foods to slide through the oral cavity more easily; it has no effect on taste or smell.
   (c) Additional moisture would enhance the ability of molecules to be dissolved and to enter the taste pores more readily and thus enhance taste; similar changes would enhance olfaction.
   (d) Only the sense of taste would be enhanced.

4. What is receptor specificity?
5. What could stimulate the release of an increased quantity of neurotransmitter by a hair cell into the synapse with a sensory neuron?
6. What are the functions of hair cells in the internal ear?
7. What is sensory adaptation?
8. What type of information about a stimulus does sensory coding provide?
9. What would be the consequence of damage to the lamellar corpuscles of the arm?
10. What is the structural relationship between the bony labyrinth and the membranous labyrinth?

Level 3 Critical Thinking

1. Beth has surgery to remove some polyps (growths) from her sinuses. After she heals from the surgery, she notices that her sense of smell is not as keen as it was before the surgery. Provide an accurate explanation.
2. Jared is 10 months old, and his pediatrician diagnoses him with otitis media. What does the physician tell Jared's mother?
3. What happens to reduce the effectiveness of your sense of taste when you have a cold?

CLINICAL CASE WRAP-UP

Why Am I So Dizzy?

John's ENT doctor listens to his story, does a quick hearing test, and looks into his ears with an otoscope. "Your physical examination seems normal now," she tells John. "But we both know what happened to you is not normal. I think you have suffered an episode of Ménière's disease."

Prosper Ménière, a French physician in the mid-19th century, first described an episodic condition involving vertigo, tinnitus, and hearing loss. Ménière's disease is a disorder of the internal ear. The cause is unknown, but it may be due to a buildup of endolymph in the internal ear. Because endolymph circulates throughout the semicircular canals, utricle, saccule, and cochlea, abnormal buildup can affect both balance and hearing. This excessive pressure pushes on the hair cells attached to the walls of the ampullae, sending abnormal messages to the sensory receptors. The otoliths within the maculae of the vestibule move abnormally as well, contributing to the sense of vertigo. The hair cells of the cochlear duct within the spiral organ send abnormal sensory signals, causing tinnitus and temporary hearing loss.

John's doctor recommends a diet low in salt and prescribes a diuretic ("water pill") to decrease fluid volume in case of another attack. She also prescribes an anti-vertigo medication to keep on hand. She knows that John's Ménière's disease is likely to recur episodically.

1. When John's doctor looked into his external auditory canal, could she actually see into his internal ear?
2. Does Ménière's disease affect the conduction of sound through the middle ear? Explain.

See the blue Answers tab at the back of the book.

Related Clinical Terms

ageusia: A rare inability to taste. More common is hypogeusia, a disorder in which the person affected has trouble distinguishing between tastes.

mydriasis: Dilation of the pupils of the eye induced by medical eye drops or caused by disease.

otalgia: Pain in the ear; an earache.

photophobia: An oversensitivity to light possibly leading to tearing, discomfort, or pain. Causes include abrasions to the corneal area, inflammation, disease, and some medications.

tinnitus: A buzzing, whistling, or ringing sound heard in the absence of an external stimulus. Causes include injury, disease, inflammation, or some drugs.

vertigo: A feeling that you are dizzily spinning or that things are dizzily turning about you. Vertigo is usually caused by a problem with the internal ear, but can also be due to vision problems.

19

The Endocrine System

Learning Outcomes

These Learning Outcomes correspond by number to this chapter's sections and indicate what you should be able to do after completing the chapter.

19.1 Define hormone and target tissue; name the major classes of hormones based on structure, and describe an endocrine positive feedback response. p. 507

19.2 Describe the anatomy of the anterior lobe and posterior lobe of the pituitary gland and discuss the hormones they release. p. 508

19.3 Describe the anatomy of the thyroid gland and discuss the actions of the hormones released from this gland. p. 509

19.4 Describe the anatomy of the parathyroid gland and discuss the actions of the hormone released from this gland. p. 515

19.5 Compare and contrast the anatomy of the thymus in a prepubescent individual with that in an adult individual. p. 516

19.6 List the cell types of the adrenal cortex and adrenal medulla and describe the hormones produced by each cell type. p. 516

19.7 Discuss the functions of the hormones produced by the kidneys and the heart. p. 518

19.8 List the four cell types of the endocrine pancreas and the hormones produced by each cell type. p. 518

19.9 Compare and contrast the hormones produced by the testes and ovaries. p. 520

19.10 List the functions of the hormone produced by pinealocytes. p. 521

19.11 List two major functional changes that occur in the endocrine system as a person ages. p. 524

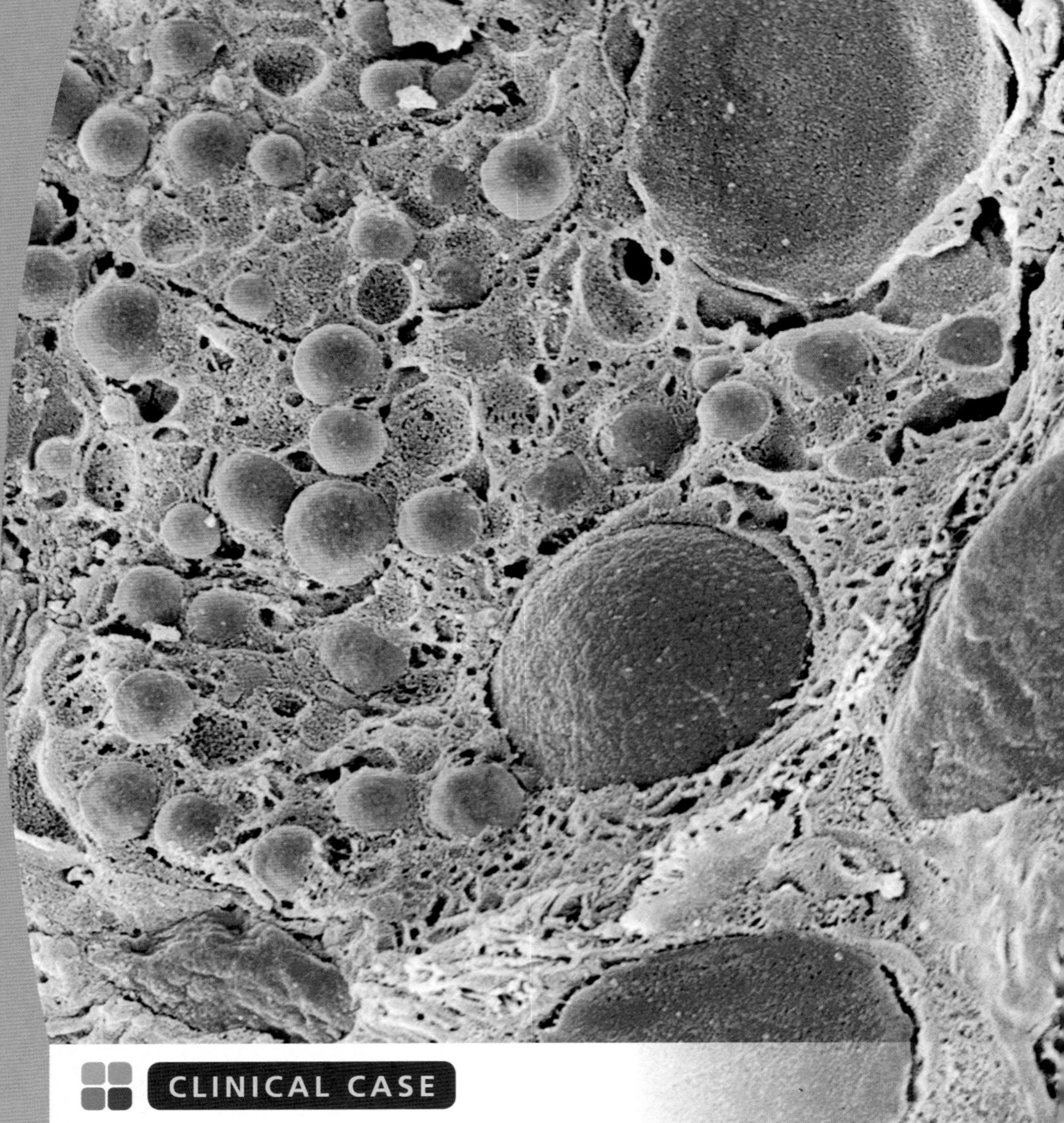

CLINICAL CASE

Why Am I So Cold and Tired?

Kathy, a 50-year-old laboratory technician, has always been an active person. After raising her three children and now working at a job she really enjoys, she should be on top of the world. Instead, she feels cold and tired all the time. She has gained 20 pounds over the past year. Her libido (sex drive) has been in the basement. In addition to depression, memory loss, and chronic constipation, she is not happy about changes in her physical appearance. Her hair has become coarse and dry and falls out, no matter what products she uses. Her skin is rough and dry. Her face is puffy and her voice has become hoarse. Instead of her menstrual periods lessening with menopause, they have become heavier than normal and are unpredictable.

Kathy's doctor took some blood for testing during her last visit. Now, at her follow-up appointment, Kathy learns that her heart rate is slow and her cholesterol is high. As she wonders how her metabolism has gotten so screwed up, her doctor says, "Kathy, the rest of these blood tests tell me exactly what is going on with you, and, I am happy to say, we can fix every problem you have."

What is causing Kathy's metabolic problems? ***To find out, turn to the Clinical Case Wrap-Up on p. 527.***

HOMEOSTASIS INVOLVES coordinating the activities of the various organs and systems throughout the body. At any given moment, cells of the nervous and endocrine systems are working together, monitoring and adjusting the body's physiological activities. The activities of these two systems are coordinated closely, and their effects are usually complementary. The nervous system produces short-term (lasting a few seconds) responses to environmental stimuli. In contrast, the endocrine system produces long-term responses by releasing chemicals directly into the bloodstream to be distributed throughout the body. These chemicals, called hormones (meaning "to excite"), alter the metabolic activities of many different tissues and organs simultaneously. The hormonal effects may not be apparent immediately, but they can last for days. This makes the endocrine system effective in regulating ongoing processes such as growth and development.

At the gross anatomy level, the nervous and endocrine systems are easily distinguished. Yet when their functions are analyzed, these two systems are difficult to separate. For example, the adrenal medulla is a modified sympathetic ganglion whose neurons secrete epinephrine and norepinephrine into the bloodstream. Therefore, the adrenal medulla is an endocrine structure that is functionally part of the nervous system. The hypothalamus, however, which is anatomically part of the brain, secretes various hormones and is functionally part of the endocrine system.

This chapter describes the anatomy and physiology of the endocrine system. We will also consider the interactions between the endocrine and nervous systems.

19.1 An Overview of the Endocrine System

▶ **KEY POINT** There are three classes of hormones: amino acid derivatives, peptide hormones, and lipid derivatives. All three types influence cellular operations by changing the types, activities, or quantities of key cytoplasmic enzymes.

The **endocrine system** includes all the endocrine cells and tissues of the body that produce hormones. **Endocrine cells** are glandular secretory cells that release hormones directly into the interstitial fluids, lymphatic system, or blood. These **hormones** are chemical messengers that stimulate specific cells or tissues into action. The major endocrine glands and the hormones they release are shown in **Figure 19.1**.

Figure 19.1 The Endocrine System. Location of endocrine glands and endocrine cells and the major hormones produced by each gland.

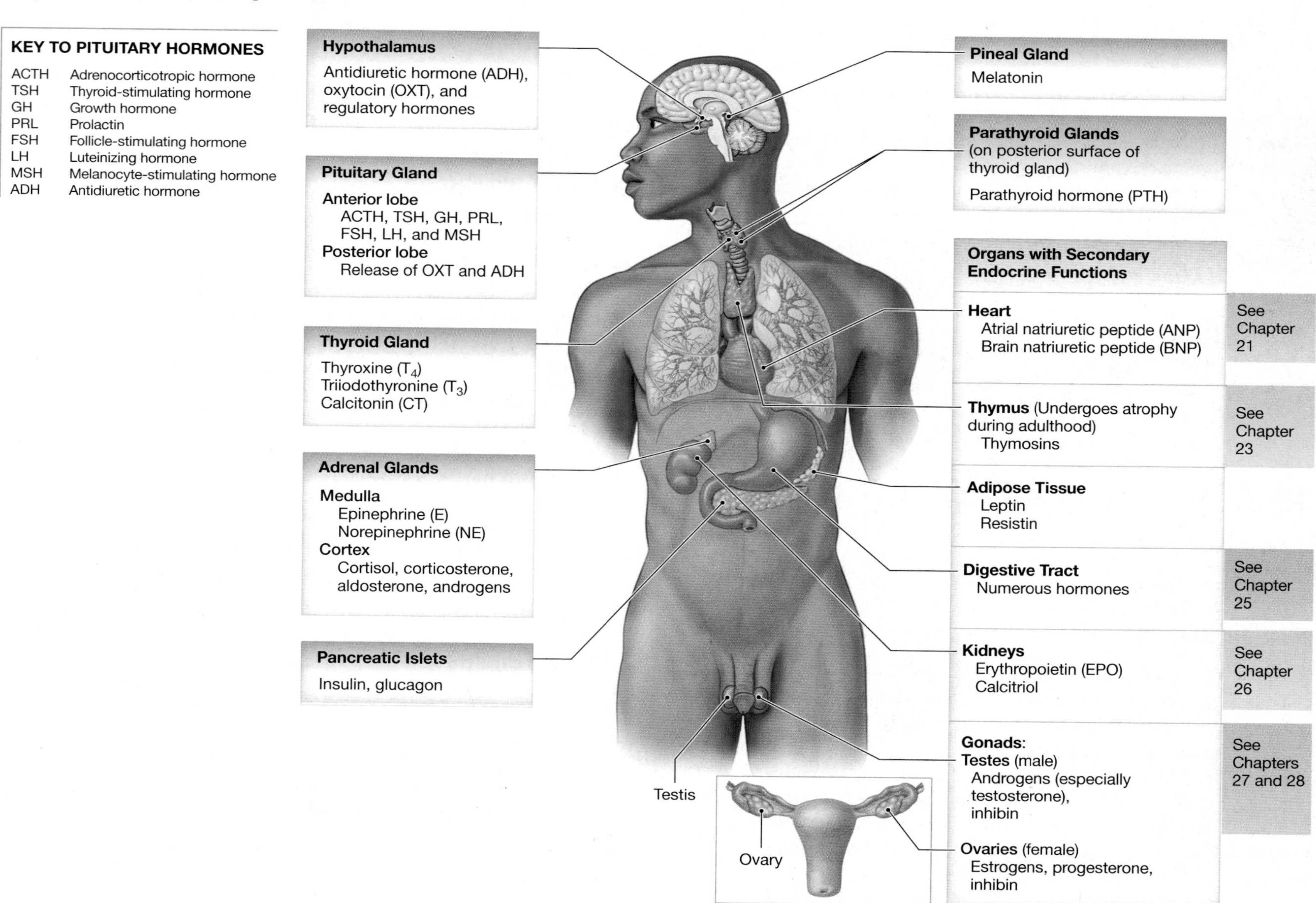

19

Hormones regulate the metabolic operations of target cells—peripheral cells that respond to the presence of that hormone. Hormones are organized into three main classes based on their chemical structure:

1. **Amino acid derivatives** are small molecules that are structurally similar to amino acids. Examples include (1) derivatives of tyrosine, such as the **thyroid hormones** released by the thyroid gland and the **catecholamines** (epinephrine, norepinephrine, and dopamine) released by the adrenal medulla; and (2) derivatives of tryptophan, such as **melatonin** synthesized by the pineal gland.
2. **Peptide hormones**, the largest group of hormones, are chains of amino acids. The hormones released by the pituitary gland are examples of peptide hormones.
3. **Lipid derivatives** can be divided into two groups: eicosanoids and steroid hormones. Eicosanoids (Ī-kō-sa-noydz) are small molecules with a five-carbon ring at one end. Most body cells release eicosanoids, which coordinate cellular activities and affect enzymatic processes (such as blood clotting). Steroid hormones are lipids that are structurally similar to cholesterol. Reproductive organs and the adrenal cortex release steroid hormones.

Enzymes control all cellular activities and metabolic reactions. Hormones influence cellular operations by changing the types, activities, or quantities of key cytoplasmic enzymes. Each hormone has **target cells**, specific cells that have receptors needed to bind hormones and respond to their presence. Endocrine activity is triggered by (1) humoral stimuli (changes in the composition of the extracellular fluid), (2) hormonal stimuli (the arrival or removal of a specific hormone), or (3) neural stimuli (the arrival of neurotransmitters at neuroglandular junctions). In most cases, negative feedback regulates endocrine responses.

Positive feedback responses regulate some endocrine processes that must be rushed to completion. In these instances, the secretion of a hormone produces an effect that further stimulates hormone release. An example of positive feedback occurs during labor and delivery: Oxytocin release causes uterine smooth muscle contractions, and the uterine contractions further stimulate oxytocin release.

19.1 CONCEPT CHECK

1 What is a target cell? What is the relationship between a hormone and its target cell?

See the blue Answers tab at the back of the book.

19.2 Hypothalamus and the Pituitary Gland

▶ KEY POINT The hypothalamus coordinates the activity of the pituitary gland. The pituitary gland is composed of glandular epithelial cells and neural secretory tissue. The pituitary is subdivided into the posterior lobe and the anterior lobe.

The **hypothalamus** provides the highest level of endocrine control. It integrates the activities of the nervous and endocrine systems (neural stimuli) and has a close relationship with both lobes of the pituitary gland. The **pituitary gland**, or **hypophysis** (hī-POF-i-sis), weighs one-fifth of an ounce (~6 g) and is the most compact chemical factory in the body. This small, oval gland, about the size and weight of a small grape, lies inferior to the hypothalamus within the sella turcica, a depression in the sphenoid. The **infundibulum** (in-fun-DIB-ū-lum) extends from the hypothalamus inferiorly to the posterior and superior surfaces of the pituitary gland. The diaphragma sellae is a dural sheet that encircles the infundibulum and holds the pituitary gland in position within the sella turcica. The hypothalamus regulates the functions of both the anterior and posterior lobes of the pituitary gland. It also integrates the activities of the nervous and endocrine systems in three ways (**Spotlight Figure 19.2**).

Hypophyseal Portal System

The hypothalamus controls the production of hormones in the anterior lobe of the pituitary gland by secreting specific regulatory hormones. At the median eminence, a swelling near the attachment of the infundibulum, hypothalamic neurons release regulatory hormones. Before leaving the hypothalamus, capillary networks unite to form a series of larger vessels that spiral around the infundibulum to reach the anterior lobe. In the anterior lobe, these vessels form a second capillary network that branches among the endocrine cells. This particular network is the **hypophyseal** (hī-pō-FIZ-ē-al) **portal system** (**Spotlight Figure 19.2**).

The Anterior Lobe of the Pituitary Gland

▶ KEY POINT The anterior lobe of the pituitary gland secretes seven hormones. These hormones regulate other endocrine glands and stimulate some nonendocrine tissues.

The **anterior lobe** of the pituitary gland, also called the **adenohypophysis** (ad-e-nō-hī-POF-i-sis), contains a variety of endocrine cells. The anterior lobe has three regions: the **pars distalis**, **pars tuberalis**, and **pars intermedia**. An extensive capillary network radiates through these regions, giving every endocrine cell immediate access to the bloodstream. Hormones of the anterior lobe include thyroid-stimulating hormone (TSH), adrenocorticotropic hormone (ACTH), growth hormone (GH), prolactin (PRL), follicle-stimulating hormone (FSH), and luteinizing hormone (LH). Melanocyte-stimulating hormone (MSH) is produced by the pars intermedia (**Spotlight Figure 19.2** and Figure 19.3).

1. **Thyroid-stimulating hormone (TSH)** is secreted by cells of the pars distalis called thyrotropes. TSH targets the thyroid gland and triggers the release of thyroid hormones.
2. **Adrenocorticotropic hormone (ACTH)** stimulates the release of steroid hormones from the adrenal cortex. ACTH specifically targets cells producing hormones called **glucocorticoids** (glū-kō-KOR-ti-koyds) **(GC)**. Glucocorticoids affect glucose metabolism. The cells secreting ACTH are called corticotropes.
3. **Follicle-stimulating hormone (FSH)** is secreted by cells called gonadotropes. FSH promotes the development of oocytes (female gametes) within the ovaries of mature women. The development of oocytes occurs within structures called follicles. FSH also stimulates the secretion of **estrogens** (ES-trō-jens) by follicular cells within the ovary. Estrogens, which are steroid hormones, are female sex hormones. Estradiol is the most important estrogen. In men, FSH secretion supports sperm production in the testes.
4. **Luteinizing** (LOO-tē-in-ī-zing) **hormone (LH)** induces ovulation in women. LH promotes the secretion of **progestins** (prō-JES-tinz) by the ovary. Progestins are steroid hormones that prepare the body for pregnancy. Progesterone is the most important progestin. In men, LH stimulates the production of male sex hormones called **androgens** (AN-drō-jenz; *andros*, man) by the interstitial cells of the testes. Testosterone is the most important androgen. Because FSH and LH regulate the activities of the male and female sex organs (gonads), they are called **gonadotropins** (gō-nad-ō-TRŌ-pinz).
5. **Prolactin** (prō-LAK-tin; *pro-*, before, + *lac*, milk) **(PRL)** is secreted by cells called lactotropes. PRL stimulates the development of the mammary glands and the production of milk. The mammary glands are regulated by

the interaction of a number of other hormones, including estrogen, progesterone, growth hormone, glucocorticoids, and hormones produced by the placenta. However, PRL exerts the greatest effect on the glandular cells. The functions of prolactin in males are poorly understood.

6 **Growth hormone (GH)** is also called *human growth hormone* (*HGH*) or *somatotropin* (*soma*, body). GH stimulates cell growth and replication by accelerating the rate of protein synthesis. Cells called somatotrophs secrete GH. Every tissue responds to GH to some degree. However, GH promotes protein synthesis and cellular growth in bone and muscle cells. Liver cells also respond to GH by synthesizing and releasing somatomedins. Somatomedins are peptide hormones that stimulate protein synthesis and cell growth in skeletal muscle fibers, cartilage cells, and many other target cells. Children unable to produce adequate concentrations of growth hormone have pituitary growth failure, sometimes called pituitary dwarfism. These individuals do not experience the steady growth and maturation that precede and accompany puberty.

7 **Melanocyte-stimulating hormone (MSH)** is the only hormone released by the pars intermedia. MSH increases the rate of melanin production and distribution in the melanocytes of the skin. MSH is secreted by corticotropes (also termed *ACTH cells*) only during fetal development, in young children, in pregnant women, and in some disease states.

The Posterior Lobe of the Pituitary Gland

▶ **KEY POINT** The posterior lobe of the pituitary gland is an extension of the central nervous system. The posterior lobe stores and releases two hormones produced by the hypothalamus: antidiuretic hormone and oxytocin.

The **posterior lobe** of the pituitary gland, also called the **neurohypophysis** (nū-rō-hī-POF-i-sis), contains the axons of hypothalamic neurons. The posterior lobe does not have a portal system. The inferior hypophyseal artery delivers blood to it, and the hypophyseal veins carry blood and hormones away. The posterior lobe releases two hormones: antidiuretic hormone (ADH) and oxytocin (OXT) (**Spotlight Figure 19.2** and **Figure 19.3**).

1 **Antidiuretic hormone (ADH)**, or *vasopressin*, is released in response to a variety of stimuli, including (1) a rise in the concentration of electrolytes in the blood or (2) a fall in blood volume or blood pressure. ADH decreases the amount of urine produced by the kidneys. ADH also causes constriction of peripheral blood vessels, which elevates blood pressure.

2 **Oxytocin** (ok-sē-TŌ-sin; *oxy–*, quick, + *tokos*, childbirth) stimulates the contractions of (1) smooth muscle cells in the uterus and (2) contractile (myoepithelial) cells surrounding the secretory cells of the mammary glands. Stimulation of uterine muscles by oxytocin is required for normal labor and childbirth. After birth, the suckling of an infant at the breast stimulates the release of oxytocin into the blood. Oxytocin then stimulates contraction of the myoepithelial cells in the mammary glands, causing the discharge of milk from the nipple. In males, oxytocin causes smooth muscle contractions in the ductus deferens and prostate.

19.2 CONCEPT CHECK

2 Which brain region controls production of hormones in the pituitary gland?

3 Identify the two regions of the pituitary gland and describe how hormone release is controlled for each.

See the blue Answers tab at the back of the book.

CLINICAL NOTE

Diabetes Insipidus

There are several different forms of diabetes, all characterized by excessive urine production (polyuria). Although diabetes can be caused by physical damage to the kidneys, most forms are the result of endocrine abnormalities. The two most important forms are diabetes insipidus, considered here, and diabetes mellitus, considered later.

Diabetes insipidus develops when the neurohypophysis, or posterior lobe of the pituitary gland, no longer releases adequate amounts of antidiuretic hormone (ADH). Water conservation by the kidneys is impaired, and excessive amounts of water are lost in the urine. As a result, an individual with diabetes insipidus is constantly thirsty, but fluids consumed are not retained by the body. Mild cases may not require treatment, as long as fluid and electrolyte intake keep pace with urinary losses. In severe diabetes insipidus, the fluid losses can reach 10 liters per day, and fatal dehydration will occur unless treatment is provided. Administering a synthetic form of ADH, desmopressin acetate (DDAVP), in a nasal spray concentrates the urine and reduces urine volume. The drug enters the bloodstream after diffusing through the nasal epithelium. It is also an effective treatment for bed-wetting if used at bedtime.

19.3 | The Thyroid Gland

▶ **KEY POINT** Shaped like a butterfly, the thyroid gland has two main lobes and an extensive blood supply.

The **thyroid gland** is located on the anterior surface of the trachea (*windpipe*) inferior to the **thyroid** ("shield-shaped") and **cricoid** (KRĪ-koyd; "ring-shaped") **cartilages** of the larynx (**Figure 19.4a**). Because of its location, you can feel the thyroid gland with your fingers. The size of the thyroid gland varies considerably, depending on heredity, environment, and nutrition. When something goes wrong with the gland it often enlarges and becomes prominent.

The thyroid gland has a deep red color because of the large number of blood vessels supplying it. The thyroid is supplied with blood from two sources: a pair of superior thyroid arteries, which are branches of the external carotid arteries (major arteries in the neck), and a pair of inferior thyroid arteries, branches of the thyrocervical trunks. Venous drainage of the gland is through (1) the superior and middle thyroid veins, which drain into the internal jugular veins (major veins of the neck), and (2) the inferior thyroid veins, which drain into the brachiocephalic veins (major veins that form the superior vena cava).

The thyroid gland has two main **lobes**, giving it a butterfly-like appearance. The two lobes are joined by a slender connection, the **isthmus** (IS-mus). The superior portions of the gland extend over the lateral surface of the trachea toward the inferior border of the thyroid cartilage. Inferiorly, the thyroid gland extends to the second or third cartilage ring of the trachea. A thin connective tissue capsule attaches the thyroid gland to the trachea. The capsule of the thyroid extends inward and is continuous with the connective tissue partitions that divide the glandular tissue and surround the thyroid follicles.

SPOTLIGHT

FIGURE 19.2

Neuroendocrine Integration: The Hypothalamus and Pituitary Gland

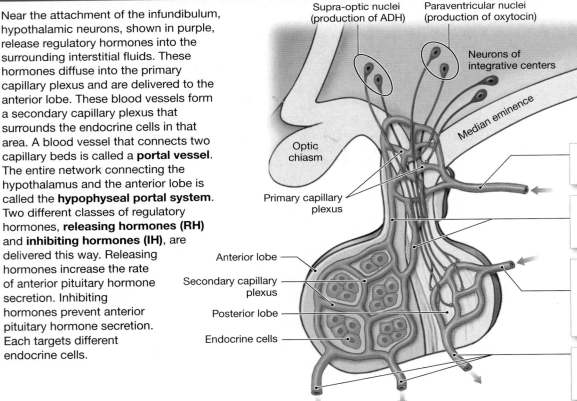

The Hypothalamus and Endocrine Regulation

Coordinating centers in the hypothalamus integrate the activities of the nervous and endocrine systems in three different ways.

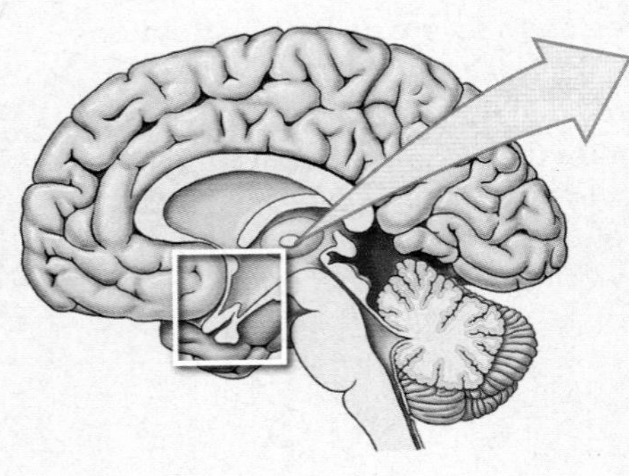

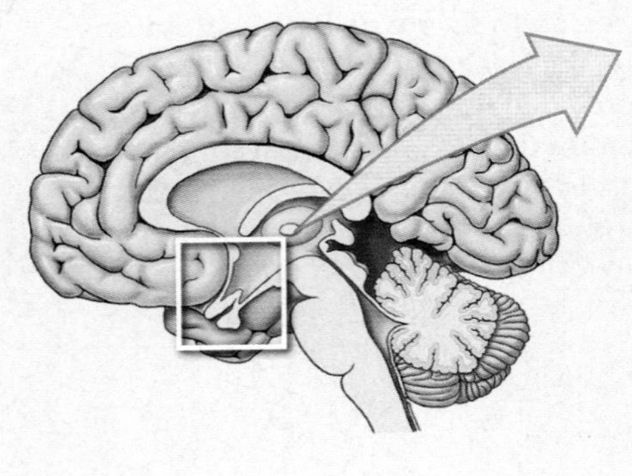

1 Hypothalamic neurons produce the hormones **antidiuretic hormone (ADH)** and **oxytocin (OXT)**. After transport along their axons, these hormones are released from the posterior lobe of the pituitary gland.

2 Integrative centers in the hypothalamus release **regulatory hormones**. These regulatory hormones control the activity of the anterior lobe of the pituitary gland. Regulatory hormones reach their targets by the hypophyseal portal system, detailed below.

3 Autonomic centers in the hypothalamus control hormone secretion from the adrenal gland by sympathetic preganglionic motor neurons.

HYPOTHALAMUS

Preganglionic motor fibers

Infundibulum (connection between hypothalamus and the pituitary gland)

Anterior lobe of pituitary gland (adenohypophysis)

Posterior lobe of pituitary gland (neurohypophysis)

Adrenal cortex

Adrenal medulla

Adrenal gland

Secretion of multiple hormones that control other endocrine organs

Release of ADH and OXT. These hormones are called **neurosecretions** because they are produced and released by neurons.

Secretion of epinephrine (E) and norepinephrine (NE)

The Hypophyseal Portal System

Near the attachment of the infundibulum, hypothalamic neurons, shown in purple, release regulatory hormones into the surrounding interstitial fluids. These hormones diffuse into the primary capillary plexus and are delivered to the anterior lobe. These blood vessels form a secondary capillary plexus that surrounds the endocrine cells in that area. A blood vessel that connects two capillary beds is called a **portal vessel**. The entire network connecting the hypothalamus and the anterior lobe is called the **hypophyseal portal system**. Two different classes of regulatory hormones, **releasing hormones (RH)** and **inhibiting hormones (IH)**, are delivered this way. Releasing hormones increase the rate of anterior pituitary hormone secretion. Inhibiting hormones prevent anterior pituitary hormone secretion. Each targets different endocrine cells.

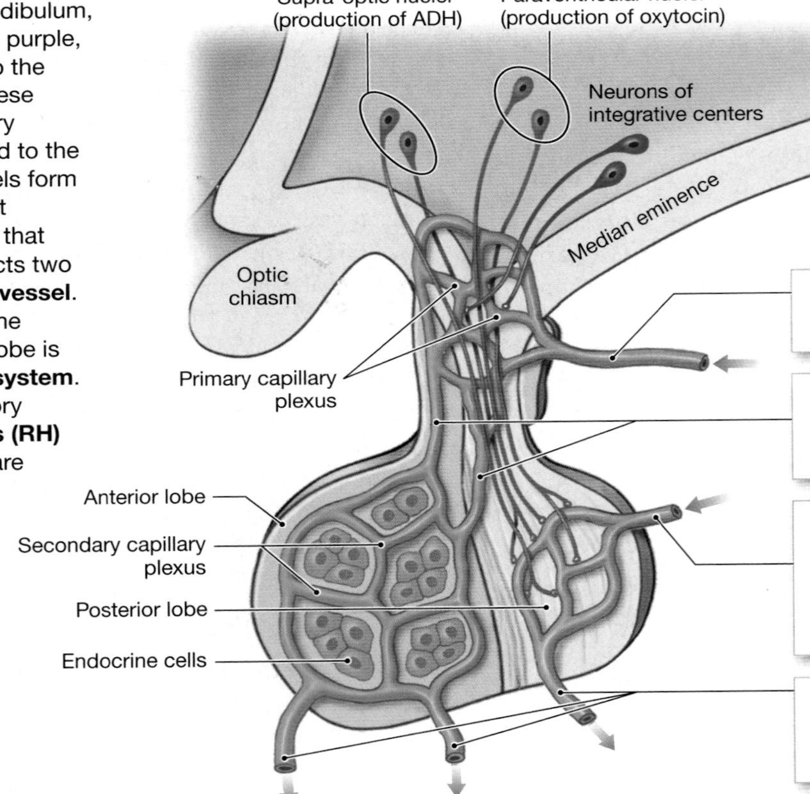

The superior hypophyseal artery delivers blood to the primary capillary plexus.

The portal vessels deliver blood containing regulatory hormones to the secondary capillary plexus within the anterior lobe.

The inferior hypophyseal artery delivers blood to a capillary plexus in the posterior lobe. This plexus picks up the ADH and OXT released by the axons of the hypothalamic neurons shown in green.

Hypophyseal veins carry blood containing the pituitary hormones for delivery to the rest of the body.

The Pituitary Gland

The pituitary gland has two lobes: the **anterior lobe**, or **adenohypophysis**, and the **posterior lobe**, or **neurohypophysis**. The anterior lobe is subdivided into three regions: (1) a large **pars distalis**, which represents the major portion of the pituitary gland; (2) a slender **pars intermedia**, which forms a narrow band adjacent to the posteror lobe; and (3) an extension called the **pars tuberalis**, which wraps around the adjacent portion of the infundibulum.

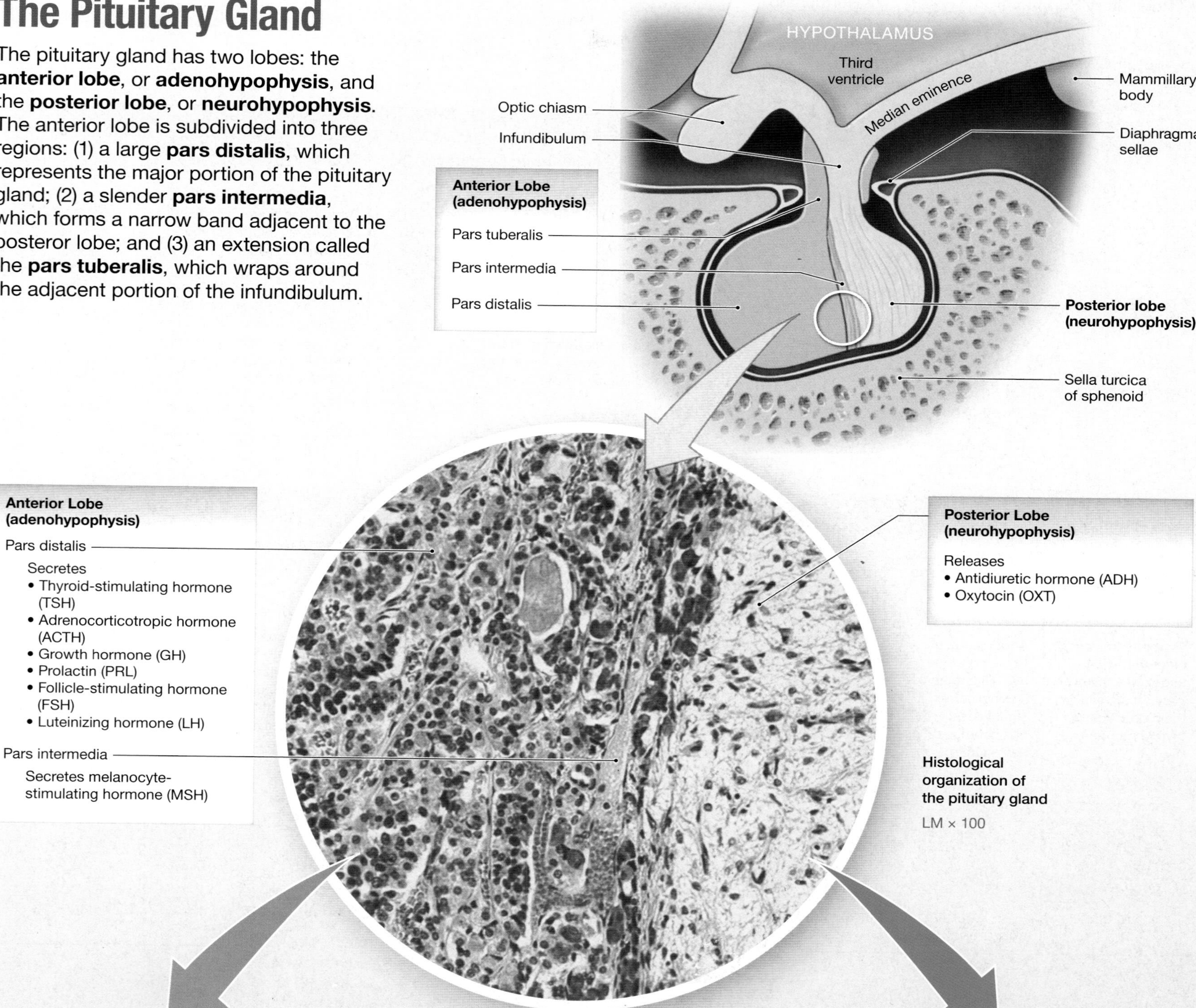

Histological organization of the pituitary gland

LM × 100

The Anterior Lobe

The richly vascularized anterior lobe of the pituitary contains five different cell types and produces seven different hormones. The pars distalis produces growth hormone, which has widespread effects on metabolism, plus five hormones that regulate hormone production by other endocrine glands. The small pars intermedia produces a single hormone, MSH, which stimulates melanocyte production of melanin. Hormones regulating the secretion of other glands are termed tropic hormones, and their names indicate their activities. For example, thyroid-stimulating hormone (TSH) targets the thyroid gland and triggers the release of thyroid hormones.

The Posterior Lobe

The posterior lobe of the pituitary contains the axons and axon terminals of roughly 50,000 hypothalamic neurons. The neuronal cell bodies are either in the supra-optic or paraventricular nuclei of the hypothalamus. The supra-optic nuclei produce ADH, and the paraventricular nuclei produce OXT.

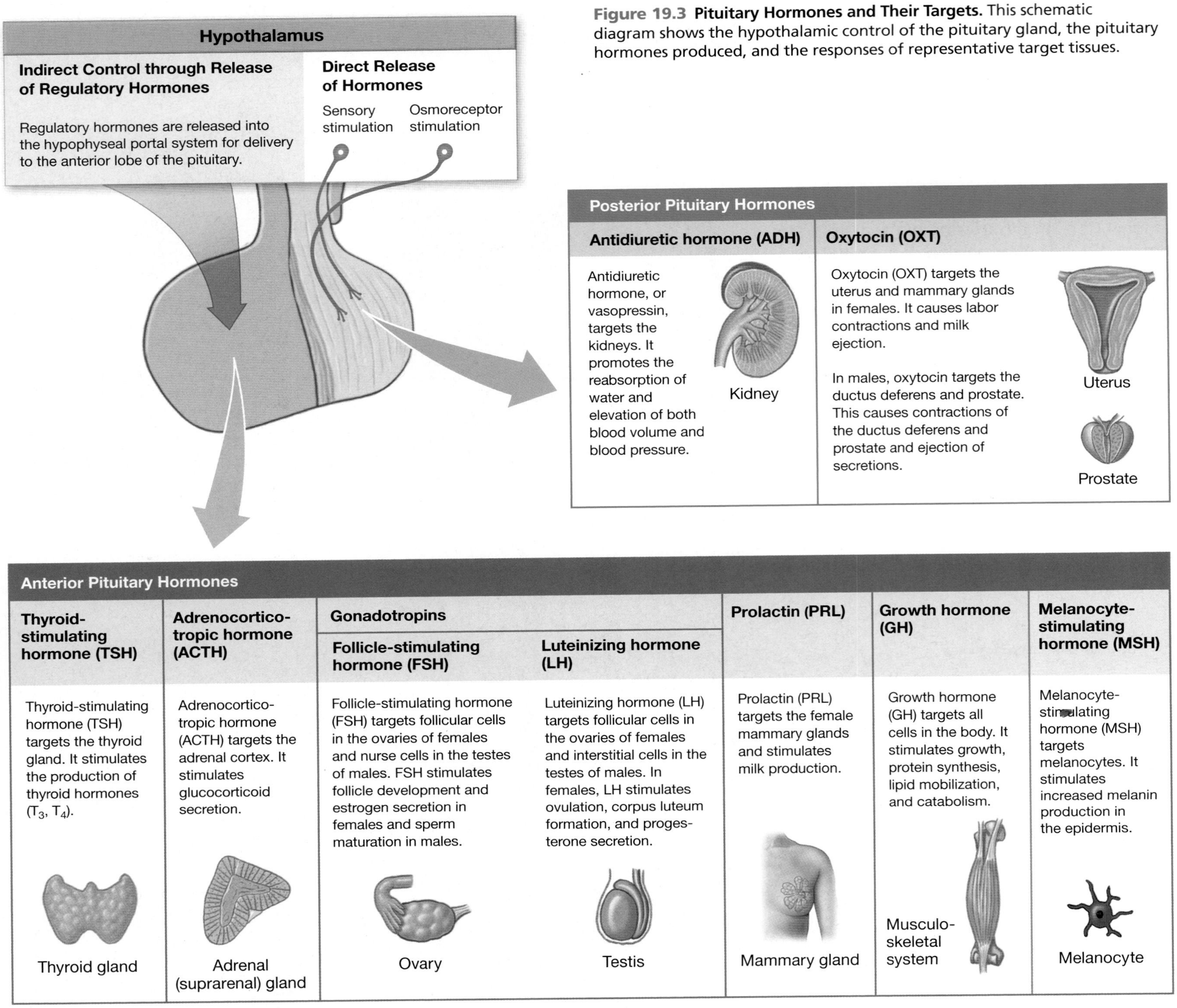

Figure 19.3 Pituitary Hormones and Their Targets. This schematic diagram shows the hypothalamic control of the pituitary gland, the pituitary hormones produced, and the responses of representative target tissues.

Hypothalamus

Indirect Control through Release of Regulatory Hormones	Direct Release of Hormones
Regulatory hormones are released into the hypophyseal portal system for delivery to the anterior lobe of the pituitary.	Sensory stimulation; Osmoreceptor stimulation

Posterior Pituitary Hormones

Antidiuretic hormone (ADH)	Oxytocin (OXT)
Antidiuretic hormone, or vasopressin, targets the kidneys. It promotes the reabsorption of water and elevation of both blood volume and blood pressure. Kidney	Oxytocin (OXT) targets the uterus and mammary glands in females. It causes labor contractions and milk ejection. In males, oxytocin targets the ductus deferens and prostate. This causes contractions of the ductus deferens and prostate and ejection of secretions. Uterus; Prostate

Anterior Pituitary Hormones

Thyroid-stimulating hormone (TSH)	Adrenocorticotropic hormone (ACTH)	Gonadotropins: Follicle-stimulating hormone (FSH)	Gonadotropins: Luteinizing hormone (LH)	Prolactin (PRL)	Growth hormone (GH)	Melanocyte-stimulating hormone (MSH)
Thyroid-stimulating hormone (TSH) targets the thyroid gland. It stimulates the production of thyroid hormones (T_3, T_4).	Adrenocorticotropic hormone (ACTH) targets the adrenal cortex. It stimulates glucocorticoid secretion.	Follicle-stimulating hormone (FSH) targets follicular cells in the ovaries of females and nurse cells in the testes of males. FSH stimulates follicle development and estrogen secretion in females and sperm maturation in males.	Luteinizing hormone (LH) targets follicular cells in the ovaries of females and interstitial cells in the testes of males. In females, LH stimulates ovulation, corpus luteum formation, and progesterone secretion.	Prolactin (PRL) targets the female mammary glands and stimulates milk production.	Growth hormone (GH) targets all cells in the body. It stimulates growth, protein synthesis, lipid mobilization, and catabolism.	Melanocyte-stimulating hormone (MSH) targets melanocytes. It stimulates increased melanin production in the epidermis.
Thyroid gland	Adrenal (suprarenal) gland	Ovary	Testis	Mammary gland	Musculoskeletal system	Melanocyte

Thyroid Follicles and Thyroid Hormones

▶ **KEY POINT** The functional unit of the thyroid gland is the thyroid follicle. The follicular epithelium has two cell types: T thyrocytes and C thyrocytes.

Thyroid follicles manufacture, store, and secrete thyroid hormones. A simple cuboidal epithelium composed of **T thyrocytes** (also termed *follicular cells*) lines the follicle (**Figure 19.4b,c**). The activity of the gland determines the shape and size of the follicular epithelium. An inactive gland will have a very low, simple cuboidal epithelium, while a highly active gland will have a simple columnar epithelium. The T thyrocytes surround a **follicle cavity**, which contains **colloid**, a viscous fluid containing large quantities of suspended proteins. The structure of a thyroid follicle is spherical, similar to a tennis ball. The "fuzzy" part of the tennis ball represents the basement membrane of the follicular epithelium. The rubber wall of the tennis ball represents the follicular epithelium, and the hollow central portion of the tennis ball represents the follicle cavity.

A network of capillaries surrounds each follicle. These capillaries deliver nutrients and regulatory hormones to the follicular cells and remove their secretory products and metabolic wastes.

Figure 19.4 Anatomy and Histological Organization of the Thyroid Gland

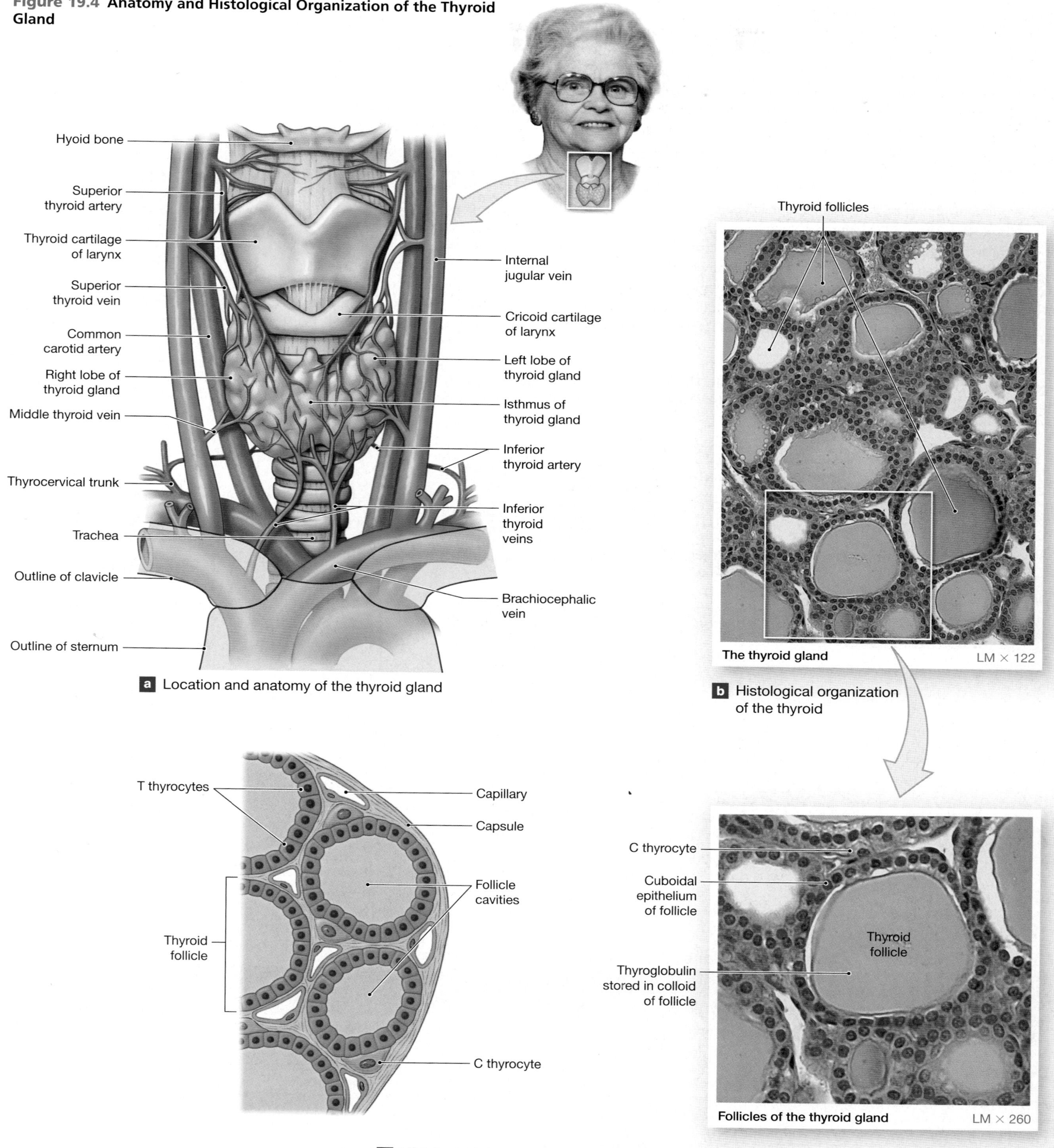

a Location and anatomy of the thyroid gland

b Histological organization of the thyroid

c Histological details of the thyroid gland showing thyroid follicles and both cell types in the follicular epithelium

19

The T thyrocytes are protein secretors. Therefore, they have large numbers of mitochondria and an extensive rough endoplasmic reticulum. Follicle cells synthesize a globular protein called **thyroglobulin** (thī-rō-GLOB-ū-lin). The thyroglobulin is then secreted into the colloid of the thyroid follicle. Thyroglobulin contains molecules of the amino acid tyrosine, the building block of thyroid hormones. The T thyrocytes actively transport iodide ions (I^-) into the cell from the interstitial fluid. The iodide is converted to an atom of iodine (I^o). This reaction, which occurs at the apical membrane surface, also attaches two iodine atoms to the tyrosine molecules of the thyroglobulin molecule within the follicle cavity. The thyroid hormone **thyroxine** (thī-ROK-sēn), also called *tetraiodothyronine*, or T_4, contains four iodine atoms. A related molecule called **triiodothyronine (T_3)**, contains three iodine atoms. *The thyroid gland is the only endocrine gland that stores its hormone product extracellularly.*

The concentration of thyroid-stimulating hormone (TSH) circulating in the blood regulates both the synthesis and release of thyroid hormones (Figure 19.5). TSH stimulates the active transport of iodide into the follicle cells and is released in response to **thyrotropin-releasing hormone (TRH)** from the hypothalamus. TRH is released into the hypophyseal portal system. Under the influence of TSH, follicle cells remove thyroglobulin from the follicles by endocytosis. Next, lysosomal enzymes break down the thyroglobulin, releasing molecules of T_3 and T_4. These hormones diffuse out of the cell and enter the bloodstream. About 90 percent of all thyroid secretion is T_4; T_3 is secreted in comparatively small amounts. These two thyroid hormones increase the rate of cellular metabolism and oxygen consumption in almost every body cell (Table 19.1).

The C Thyrocytes of the Thyroid Gland

▶ **KEY POINT** C thyrocytes lie within the basement membrane of the thyroid follicles. These cells secrete calcitonin, which helps regulate calcium concentrations in blood.

The thyroid also contains a second type of endocrine cell. These cells are **C (clear) thyrocytes**, or *parafollicular cells.* They lie sandwiched between the cuboidal follicle cells and their basement membrane. They are larger than

Figure 19.5 The Regulation of Thyroid Secretion. This negative feedback loop is responsible for the homeostatic control of thyroid hormone release. TRH = thyrotropin-releasing hormone; TSH = thyroid-stimulating hormone.

19

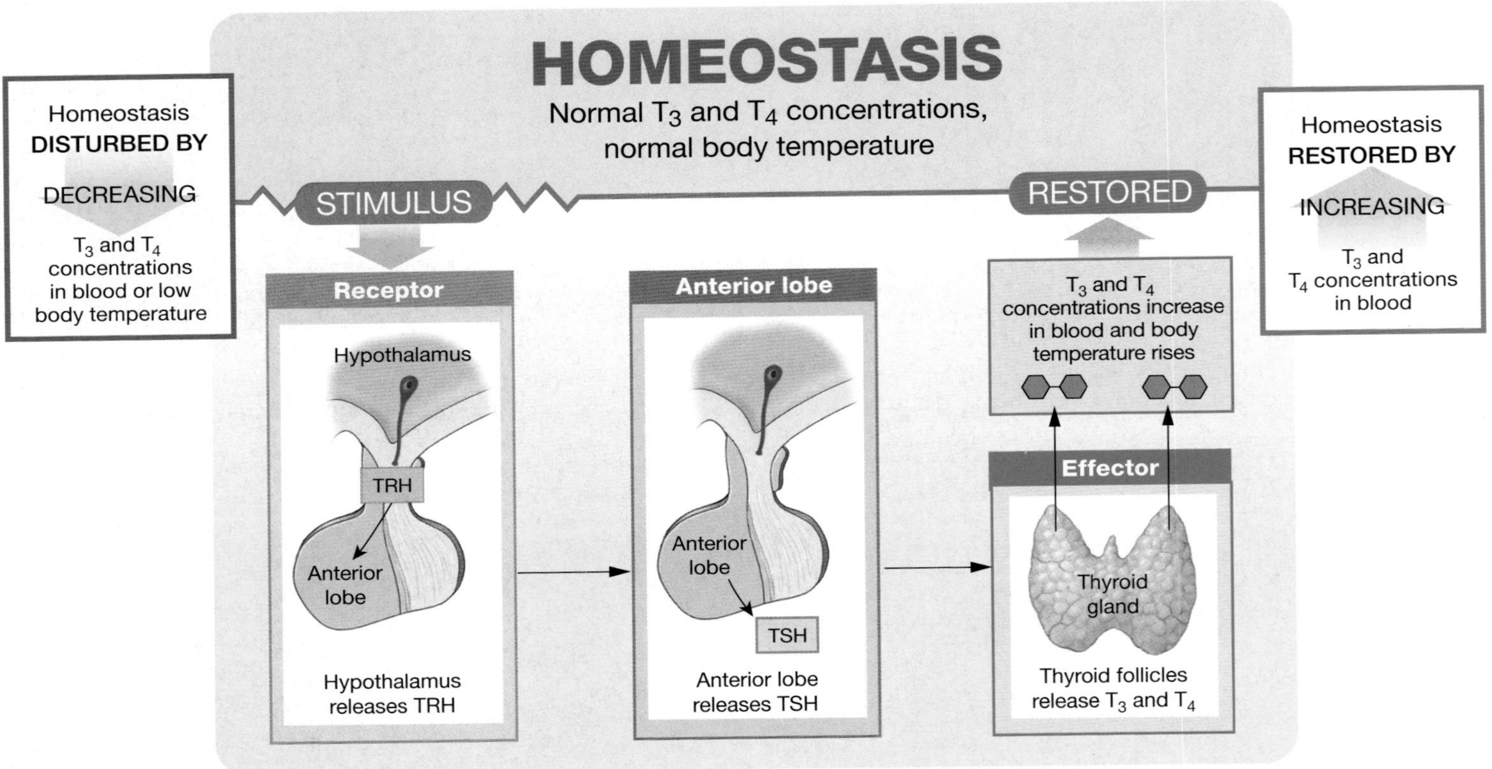

Table 19.1 | Hormones of the Thyroid Gland, Parathyroid Glands, and Thymus

Gland/Cells	Hormones	Targets	Effects
THYROID GLAND			
T thyrocytes	Thyroxine (T_4), triiodothyronine (T_3)	Most cells	Increase energy utilization; increase oxygen consumption, growth, and development
C thyrocytes	Calcitonin (CT)	Bone and kidneys	Decreases calcium ion concentrations in body fluids; uncertain significance in healthy nonpregnant adults
PARATHYROID GLAND			
Parathyroid cells	Parathyroid hormone (PTH)	Bone and kidneys	Increases calcium ion concentrations in body fluids; increases bone mass
THYMUS			
Epithelial reticular cells	Thymosins *(see Chapter 23)*	Lymphocytes	Maturation and functional competence of immune system

Figure 19.6 Anatomy and Histological Organization of the Parathyroid Glands. There are usually four separate parathyroid glands bound to the posterior surface of the thyroid gland.

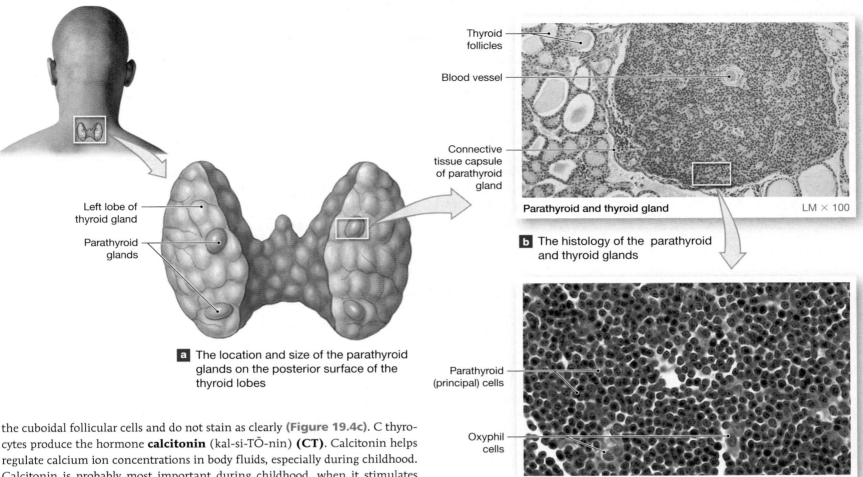

Left lobe of thyroid gland

Parathyroid glands

a The location and size of the parathyroid glands on the posterior surface of the thyroid lobes

Thyroid follicles

Blood vessel

Connective tissue capsule of parathyroid gland

Parathyroid and thyroid gland LM × 100

b The histology of the parathyroid and thyroid glands

Parathyroid (principal) cells

Oxyphil cells

Parathyroid gland LM × 600

c A histological section showing parathyroid cells and oxyphil cells of the parathyroid gland

the cuboidal follicular cells and do not stain as clearly **(Figure 19.4c)**. C thyrocytes produce the hormone **calcitonin** (kal-si-TŌ-nin) **(CT)**. Calcitonin helps regulate calcium ion concentrations in body fluids, especially during childhood. Calcitonin is probably most important during childhood, when it stimulates skeletal growth. It also appears to be important in reducing the loss of bone mass (1) during prolonged starvation and (2) in the late stages of pregnancy. The actions of calcitonin are outlined in **Table 19.1**.

19.3 CONCEPT CHECK

4 Identify the hormones of the thyroid gland.

See the blue Answers tab at the back of the book.

19.4 The Parathyroid Glands

▶ **KEY POINT** There are typically four parathyroid glands located on the posterior surface of the thyroid: two superior and two inferior parathyroid glands. They produce parathyroid hormone, which regulates calcium and phosphate levels in the bloodstream.

The four pea-sized, reddish brown **parathyroid glands** are typically located on the posterior surfaces of the thyroid gland, embedded within the connective tissue of the thyroid gland **(Figure 19.6a)**. The parathyroid glands are surrounded by a connective tissue capsule. The connective tissue from the capsule extends into the interior of the gland, forming small, irregular lobules.

Blood is supplied to the superior parathyroid glands by the superior thyroid arteries and to the inferior pair by the inferior thyroid arteries. The venous drainage for the parathyroid glands is the same as for the thyroid.

There are two types of cells in the parathyroid gland: parathyroid cells and oxyphil cells. **Parathyroid cells** (also termed *principal cells*) produce **parathyroid hormone (PTH)**. The functions of the **oxyphil cells** are unknown. They do not appear until after puberty, and then their numbers increase with age **(Figure 19.6b,c)**.

Like the C thyrocytes of the thyroid, the parathyroid cells monitor the concentration of calcium ions in the bloodstream. When the calcium concentration falls below normal, the parathyroid cells secrete PTH. PTH increases blood levels of Ca^{2+} by stimulating osteoblasts to secrete a growth factor that binds to osteoclasts. This growth factor results in an increase is osteoclast activity, causing blood calcium concentration to increase. PTH also reduces urinary excretion of calcium ions, and it stimulates the production of calcitriol, a kidney hormone that promotes intestinal absorption of calcium. PTH levels remain elevated until blood Ca^{2+} concentrations return to normal **(Table 19.1)**. PTH has been shown to be effective in reducing the progress of osteoporosis in elderly people.

19.4 CONCEPT CHECK

5 Removal of the parathyroid glands would result in a decrease in the blood of what important mineral?

See the blue Answers tab at the back of the book.

19.5 | The Thymus

▶ **KEY POINT** Located within the superior mediastinum of the thoracic cavity, the thymus is surrounded by connective tissue that divides it into lobules. The thymus produces several hormones that are important for the body's immune defenses.

The **thymus** lies just posterior to the sternum within the mediastinum of the thoracic cavity **(Figure 19.1)**. ↺ p. 507 In newborn infants and young children, the thymus is relatively large, extending from the base of the neck to the superior border of the heart. Although its relative size decreases as a child grows, the thymus continues to slowly enlarge. The thymus reaches its maximum size just before puberty, weighing approximately 40 g. After puberty it gradually diminishes in size; by age 50 the thymus weighs less than 12 g.

The thymus produces several hormones important in developing and maintaining immune defenses **(Table 19.1)**. **Thymosin** (thī-MŌ-sin) was the name originally given to a thymic extract that promoted the development and maturation of lymphocytes, the white blood cells responsible for immunity. Researchers have determined that "thymosin" is actually a blend of several different hormones (thymosin-1, thymopoietin, thymopentin, thymulin, thymic humoral factor, and IGF-1).

Although researchers do not totally agree, it has been suggested that the gradual decrease in the size and secretory abilities of the thymus may make the elderly more susceptible to disease. (We discuss the histological organization of the thymus and the functions of the various "thymosins" in Chapter 23.)

19.5 CONCEPT CHECK

6 Describe the anatomical location of the thymus.

See the blue Answers tab at the back of the book.

19.6 | The Adrenal Glands

▶ **KEY POINT** The two adrenal glands are located at the superior poles of the kidneys. They are subdivided into an adrenal cortex and an adrenal medulla. The cells of the adrenal cortex secrete steroid hormones, and the cells of the adrenal medulla secrete catecholamines.

A yellow, pyramid-shaped **adrenal gland**, or **suprarenal gland** (sū-pra-RĒ-nal; *supra-*, above, + *-renal*, kidney), is attached firmly to the superior border of each kidney by a dense, fibrous capsule **(Figure 19.7a)**. Each adrenal gland nestles among the kidney, the diaphragm, and the major arteries and veins running along the posterior wall of the abdominopelvic cavity. The adrenal glands project into the peritoneal cavity, and their anterior surfaces are covered by a layer of parietal peritoneum. Like the other endocrine glands, the adrenal glands are highly vascularized. Branches of the renal artery, the inferior phrenic artery, and a direct branch from the aorta (the middle adrenal artery) supply blood to each adrenal gland. The adrenal veins carry blood away from the adrenal glands.

The adrenal gland weighs approximately 7.5 g. It is usually heavier in men than in women, but the size varies greatly as secretory demands change. Each adrenal gland is divided structurally and functionally into two regions: a superficial **cortex** and an inner **medulla (Figure 19.7b,c)**.

The Adrenal Cortex

▶ **KEY POINT** The adrenal cortex is subdivided into three zones: (1) the outer zona glomerulosa, which produces mineralocorticoids; (2) the middle zona fasciculata, which produces glucocorticoids; and (3) the inner zona reticularis, which secretes androgens.

The yellowish color of the adrenal cortex is due to stored lipids, especially cholesterol and various fatty acids. The **adrenal cortex** produces more than 24 different steroid hormones. These **corticosteroids** (also called *adrenocortical steroids*) are vital. If the adrenal glands are destroyed or removed, corticosteroids must be administered or the person will not survive. The hormones affect metabolic operations by determining which genes are transcribed in their target cells and at what rates.

Deep to the capsule, the cortex is divided into three zones: (1) an outer zona glomerulosa, (2) a middle zona fasciculata, and (3) an inner zona reticularis **(Figure 19.7c)**. Each zone synthesizes different steroid hormones **(Table 19.2)**. All of the cortical cells have an extensive smooth endoplasmic reticulum (SER) for the manufacture of lipid-based steroids. The large amount of SER in these cells is in contrast to the large amount of rough endoplasmic reticulum (RER) found in protein-secreting cells, such as those of the anterior pituitary lobe or thyroid gland.

The Zona Glomerulosa

The **zona glomerulosa** (glō-mer-ū-LŌ-sa) is the outermost cortical region **(Figure 19.7c)**. A glomerulus is a little ball or knot, and the endocrine cells form densely packed clusters in this zone. This zone, which occupies approximately 15 percent of the adrenal cortex, extends from the capsule to the deeper zona fasciculata.

The zona glomerulosa produces **mineralocorticoids (MCs)**. These steroid hormones affect the electrolyte composition of body fluids. **Aldosterone** (al-DOS-ter-ōn) is the main mineralocorticoid. It stimulates the conservation of sodium ions (Na^+) and the elimination of potassium ions (K^+). It causes the retention of sodium ions by the kidneys, sweat glands, salivary glands, and pancreas and prevents sodium loss in the urine, sweat, saliva, and digestive secretions. A loss of K^+ accompanies this retention of Na^+. Aldosterone secretion occurs when the zona glomerulosa is stimulated by (1) a decrease in blood Na^+ levels, (2) an increase in blood K^+ levels, or (3) the arrival of angiotensin II, a hormone produced by the kidneys.

The Zona Fasciculata

Deep to the zona glomerulosa is the **zona fasciculata** (fa-sik-ū-LA-ta; *fasciculus*, little bundle), which occupies approximately 80 percent of the cortex **(Figure 19.7c)**. The cells in this zone are larger and contain more lipids than those of the zona glomerulosa. The lipid droplets give the cytoplasm a pale, foamy appearance. The cells of the zona fasciculata form cords that radiate outward like a sunburst from the innermost zona reticularis. Adjacent cords are separated by flattened blood vessels (sinusoids) with fenestrated walls.

ACTH from the anterior lobe of the pituitary gland stimulates steroid production in the zona fasciculata. This zone produces steroid hormones known as **glucocorticoids (GCs)** because of their effects on glucose metabolism. **Cortisol** (KOR-ti-sol; also called *hydrocortisone*) and **corticosterone** (kor-ti-KOS-ter-ōn) are the most important glucocorticoids secreted by the adrenal cortex. The liver converts some of the circulating cortisol to **cortisone**, another active glucocorticoid. These hormones speed up the rates of glucose synthesis and glycogen formation, especially within the liver.

The Zona Reticularis

The deepest zone of the adrenal cortex is the **zona reticularis** (re-tik-ū-LAR-is; *reticulum*, network) **(Figure 19.7c)**. The cells of the zona reticularis are much smaller than the other cells of the adrenal medulla. This characteristic makes the boundary between the zona reticularis and the renal medulla easy to distinguish. The zona reticularis, the smallest of the three zones, occupies approximately 5 percent of the adrenal cortex. The cells of the zona reticularis form a folded, branching network with an extensive capillary supply.

Figure 19.7 Anatomy and Histological Organization of the Adrenal Gland

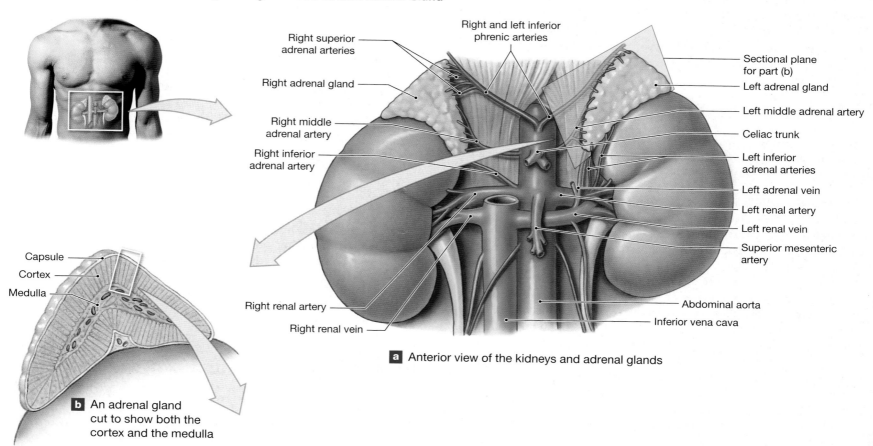

Right superior adrenal arteries
Right adrenal gland
Right middle adrenal artery
Right inferior adrenal artery
Right and left inferior phrenic arteries
Sectional plane for part (b)
Left adrenal gland
Left middle adrenal artery
Celiac trunk
Left inferior adrenal arteries
Left adrenal vein
Left renal artery
Left renal vein
Superior mesenteric artery
Right renal artery
Right renal vein
Abdominal aorta
Inferior vena cava

a Anterior view of the kidneys and adrenal glands

Capsule
Cortex
Medulla

b An adrenal gland cut to show both the cortex and the medulla

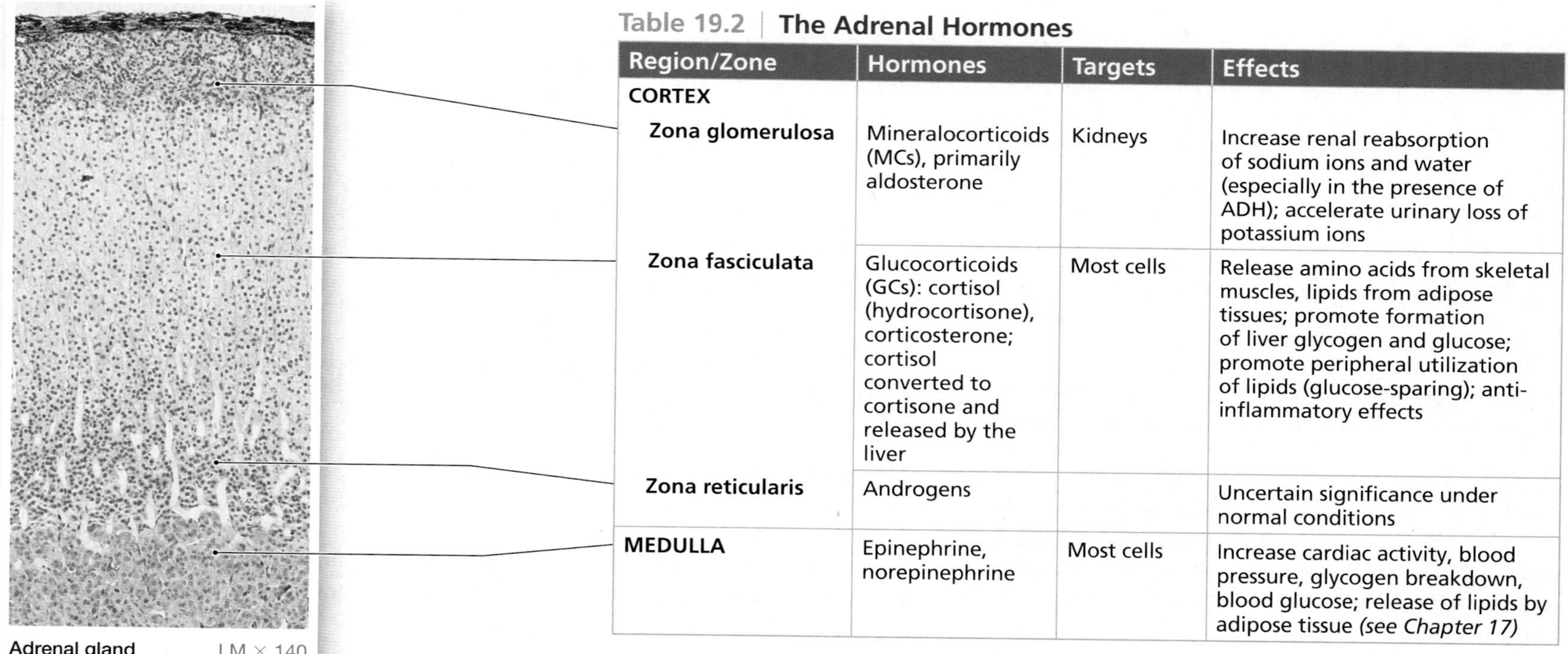

Table 19.2 | The Adrenal Hormones

Region/Zone	Hormones	Targets	Effects
CORTEX			
Zona glomerulosa	Mineralocorticoids (MCs), primarily aldosterone	Kidneys	Increase renal reabsorption of sodium ions and water (especially in the presence of ADH); accelerate urinary loss of potassium ions
Zona fasciculata	Glucocorticoids (GCs): cortisol (hydrocortisone), corticosterone; cortisol converted to cortisone and released by the liver	Most cells	Release amino acids from skeletal muscles, lipids from adipose tissues; promote formation of liver glycogen and glucose; promote peripheral utilization of lipids (glucose-sparing); anti-inflammatory effects
Zona reticularis	Androgens		Uncertain significance under normal conditions
MEDULLA	Epinephrine, norepinephrine	Most cells	Increase cardiac activity, blood pressure, glycogen breakdown, blood glucose; release of lipids by adipose tissue *(see Chapter 17)*

c Histology of the adrenal gland showing identification of the major regions and the hormones produced

The zona reticularis secretes small amounts of sex hormones called androgens. Adrenal androgens stimulate the development of pubic hair in boys and girls before puberty. In women, the adrenal androgens promote muscle mass, stimulate blood cell formation, and support the sex drive. Adrenal androgens are less important in men because the testes produce androgens in relatively large amounts.

The Adrenal Medulla

▶ **KEY POINT** The chromaffin cells of the adrenal medulla are innervated by preganglionic neurons of the sympathetic nervous system. When stimulated, chromaffin cells release catecholamines.

The **adrenal medulla** is pale gray or pink, due in part to the many blood vessels in the area (**Figure 19.7b,c**). The cells of the adrenal medulla, **chromaffin cells**, are large, rounded cells that resemble the neurons in a sympathetic ganglion. Chromaffin cells are innervated by preganglionic sympathetic fibers. Sympathetic activation of the adrenal medulla by the splanchnic nerves triggers the secretory activity of these modified ganglionic neurons. ⮌ p. 455

The adrenal medulla contains two populations of secretory cells, one secreting epinephrine (*adrenaline*), and the other norepinephrine (*noradrenaline*); both are catecholamines. ⮌ pp. 455–459 The medulla secretes three times more epinephrine than norepinephrine. Secretion of these catecholamines speeds up the use of cellular energy and the mobilization of energy reserves. This combination increases muscular strength and endurance (**Table 19.2**). The metabolic changes following catecholamine release peak 30 seconds after stimulation of the adrenal medulla and last for several minutes. As a result, the effects produced by stimulating the adrenal medulla outlast the other effects of sympathetic activation.

19

19.6 CONCEPT CHECK

7 A disorder of the adrenal gland prevents the retention of sodium ions in body fluids. Which region of the gland is affected, and what hormone is deficient?

See the blue Answers tab at the back of the book.

19.7 Endocrine Functions of the Kidneys and Heart

▶ **KEY POINT** Some cells within the kidneys and heart function as endocrine glands. Their secretions regulate blood pressure and blood volume.

The kidneys produce the enzyme renin and two hormones: the peptide erythropoietin and the steroid calcitriol. When **renin** enters the circulation it converts circulating **angiotensinogen**, an inactive protein produced by the liver, to **angiotensin I**. This compound is then converted to **angiotensin II** within the capillaries of the lungs. Angiotensin II stimulates the secretion of aldosterone by the adrenal cortex.

The kidneys release the peptide hormone **erythropoietin** (e-rith-rō-POY-e-tin) **(EPO)** in response to low oxygen levels in the kidneys. EPO stimulates red blood cell production in the red bone marrow. This increases blood volume and improves oxygen delivery to tissues.

Calcitriol is a steroid hormone secreted by the kidneys in response to the presence of parathyroid hormone (PTH) levels in the bloodstream. Calcitriol synthesis depends on the availability of cholecalciferol (vitamin D_3), which may be synthesized in the skin or absorbed from the diet. The liver converts cholecalciferol to an intermediary product that is released into the circulation and absorbed by the kidneys. The kidneys then convert this product into calcitriol. The best-known function of calcitriol is stimulating the absorption of calcium and phosphate ions along the digestive tract. Since PTH stimulates the release of calcitriol, PTH has an indirect effect on intestinal calcium absorption.

Increased blood pressure or blood volume stimulates cardiac muscle cells to produce **atrial natriuretic peptide (ANP)**. In general, the effects of ANP oppose those of angiotensin II—they promote the loss of Na^+ and water by the kidneys and inhibit renin release and ADH and aldosterone secretion. The net result is a reduction in both blood volume and blood pressure.

19.7 CONCEPT CHECK

8 EPO is a hormone that is often used for "blood doping" by endurance athletes in order to obtain a competitive advantage. How might increased levels of EPO help an endurance athlete?

See the blue Answers tab at the back of the book.

19.8 The Pancreas and Other Endocrine Tissues of the Digestive System

▶ **KEY POINT** The pancreas and other digestive system organs produce important hormones that aid digestion and other body functions.

The pancreas, the lining of the digestive tract, and the liver produce exocrine secretions that are essential for the normal digestion of food. Although the autonomic nervous system influences the pace of digestive activities, most digestive processes are controlled locally by individual organs.

The various digestive organs communicate with one another using hormones, which we will discuss in Chapter 25. In this section, we focus on one "accessory" digestive organ, the pancreas. The pancreas produces hormones that affect metabolic operations throughout the body.

The Pancreas

▶ **KEY POINT** The pancreas is a mixed gland with both exocrine and endocrine activities. The exocrine portion of the pancreas is a serous gland, while the endocrine portion is composed of pancreatic islets scattered throughout the whole organ.

The **pancreas** is both an exocrine and an endocrine gland. It lies within the abdominopelvic cavity within the first fold of the small intestine (duodenum), close to the stomach (**Figure 19.8a**). It is a slender, pale organ with a lumpy appearance. The adult pancreas is approximately 20 to 25 cm (8 to 10 in.) in length and weighs 80 g (2.8 oz). (We consider the detailed anatomy of the pancreas in Chapter 25.)

The **exocrine pancreas** makes up approximately 99 percent of the pancreatic volume. This segment of the pancreas produces large quantities of a digestive enzyme-rich fluid that enters the digestive tract through a network of secretory ducts.

The **endocrine pancreas** consists of small groups of endocrine cells scattered throughout the gland. These groups of cells are known as **pancreatic islets**. Even though the pancreatic islets account for only 1 percent of the pancreatic cell population, there are approximately 2 million pancreatic islets in a normal pancreas (**Figure 19.8b**).

Figure 19.8 Anatomy and Histological Organization of the Pancreas. This organ, which is dominated by exocrine cells, contains clusters of endocrine cells known as the pancreatic islets.

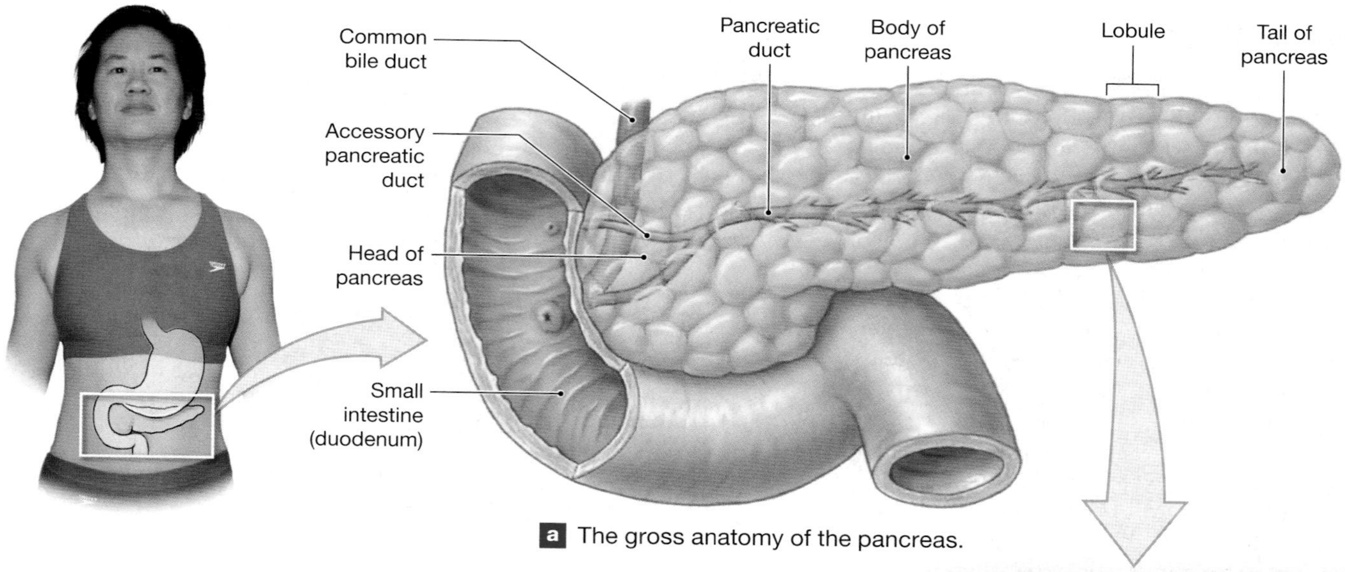

Common bile duct
Accessory pancreatic duct
Head of pancreas
Small intestine (duodenum)
Pancreatic duct
Body of pancreas
Lobule
Tail of pancreas

a The gross anatomy of the pancreas.

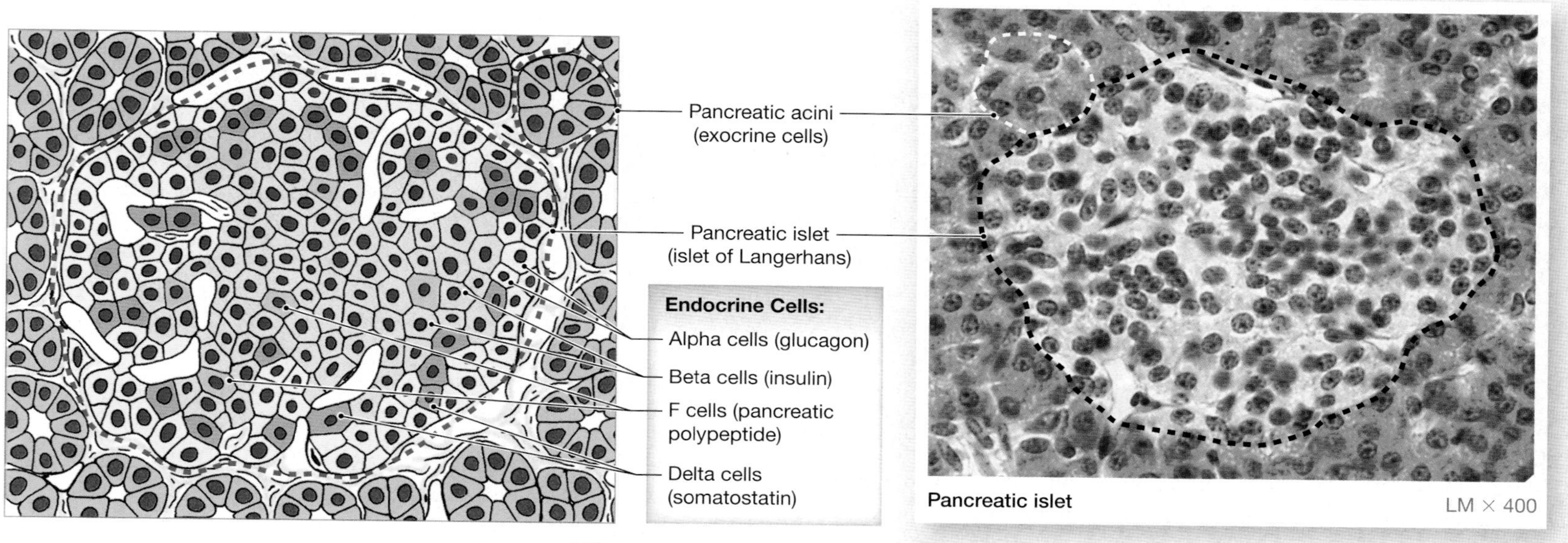

Pancreatic acini (exocrine cells)
Pancreatic islet (islet of Langerhans)
Endocrine Cells:
Alpha cells (glucagon)
Beta cells (insulin)
F cells (pancreatic polypeptide)
Delta cells (somatostatin)
Pancreatic islet LM × 400

b Histology of the pancreatic islets.

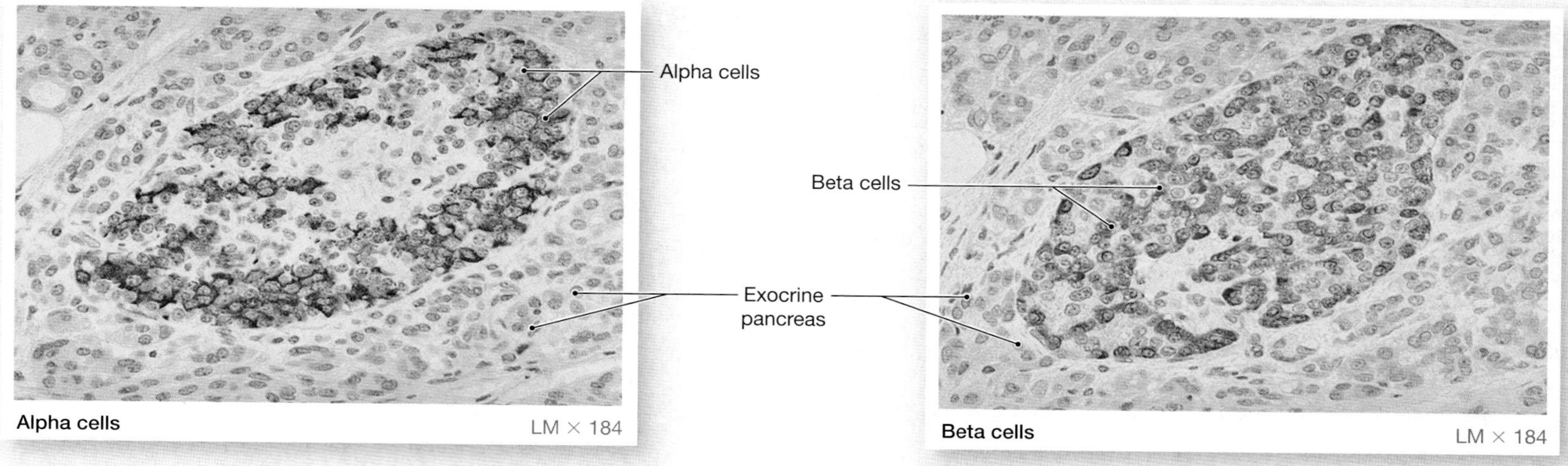

c Special histological staining techniques are used to differentiate between alpha cells and beta cells in pancreatic islets.

19

CLINICAL NOTE

Diabetes Mellitus

Diabetes mellitus (MEL-i-tus; *mellitum*, honey) is characterized by glucose concentrations that are high enough to overwhelm the reabsorption capabilities of the kidneys. (The presence of abnormally high glucose levels in the blood in general is called hyperglycemia.) Glucose appears in the urine (glycosuria), and urine production generally becomes excessive (polyuria).

Diabetes mellitus can be caused by genetic abnormalities, and some of the genes responsible have been identified. Mutations that result in inadequate insulin production, the synthesis of abnormal insulin molecules, or the production of defective receptor proteins produce comparable symptoms. Under these conditions, obesity accelerates the onset and severity of the disease. Diabetes mellitus can also result from other pathological conditions, injuries, immune disorders, or hormonal imbalances. There are two major types of diabetes mellitus: **insulin-dependent (type 1) diabetes** and **non-insulin-dependent (type 2) diabetes**. Type 1 diabetes can be controlled with varying success through the administration of insulin by injection or infusion by an insulin pump. Dietary restrictions are most effective in treating type 2 diabetes.

Probably because glucose levels cannot be stabilized adequately, even with treatment, individuals with diabetes mellitus commonly develop chronic medical problems. These problems arise because the tissues involved are experiencing an energy crisis—in essence, most of the tissues are responding as they would during chronic starvation, breaking down lipids and even proteins because they are unable to absorb glucose from their surroundings. Among the most common examples of diabetes-related medical disorders are the following:

- The proliferation of capillaries and hemorrhaging at the retina may cause partial or complete blindness. This condition is called diabetic retinopathy.
- Changes in the clarity of the lens of the eye occur, producing cataracts.
- Small hemorrhages and inflammation at the kidneys cause degenerative changes that can lead to kidney failure. This condition, called diabetic nephropathy, is the primary cause of kidney failure. Treatment with drugs that improve blood flow to the kidneys can slow the progression to kidney failure.
- A variety of neural problems appear, including peripheral neuropathies and abnormal autonomic function. These disorders, collectively termed diabetic neuropathy, are probably related to disturbances in the blood supply to neural tissues.
- Degenerative changes in cardiac circulation can lead to early heart attacks. For a given age group, heart attacks are three to five times more likely in diabetic individuals than in nondiabetic people.
- Other changes in the vascular system can disrupt normal blood flow to the distal portions of the limbs. For example, a reduction in blood flow to the feet can lead to tissue death, ulceration, infection, and loss of toes or a major portion of one or both feet.

Like other endocrine tissues, an extensive, fenestrated capillary network surrounds the islets. Two major arteries supply blood to the pancreas: the pancreaticoduodenal and pancreatic arteries. Venous blood returns to the hepatic portal vein. (Chapter 22 discusses circulation to and from major organs.) The islets are innervated by the autonomic nervous system through branches from the celiac plexus. ⮌ p. 464

Each pancreatic islet contains four cell types:

1. **Alpha cells** produce the hormone **glucagon** (GLŪ-ka-gon). Glucagon increases the blood glucose level by increasing the rates of glycogen breakdown and glucose release by the liver **(Figure 19.8b)**.
2. **Beta cells** produce the hormone **insulin** (IN-su-lin), which lowers the blood glucose level. Insulin increases the rate of glucose uptake and utilization by most body cells **(Figure 19.8b)**.
3. **Delta cells** produce the hormone **somatostatin** (*growth hormone-inhibiting hormone*). This hormone inhibits the production and secretion of glucagon and insulin and slows the rates of food absorption and enzyme secretion by the digestive tract.
4. **F cells** produce the hormone **pancreatic polypeptide (PP)**. PP inhibits gallbladder contractions and regulates the production of some pancreatic enzymes. It may control the rate of nutrient absorption by the digestive tract.

Pancreatic alpha and beta cells are sensitive to blood glucose concentrations, and they are not under the direct control of the nervous system or other endocrine glands. When the blood glucose level increases, beta cells secrete insulin. When the blood glucose level decreases, alpha cells secrete glucagon. **Table 19.3** summarizes the major hormones of the pancreas.

19.8 CONCEPT CHECK

9 Hormones are released from what portion of the pancreas (often termed the endocrine pancreas?)

See the blue Answers tab at the back of the book.

19.9 Endocrine Tissues of the Reproductive System

The endocrine tissues of the reproductive system are restricted primarily to the male and female gonads—the testes and ovaries, respectively. (Chapter 27 describes the anatomy of the reproductive organs.)

Testes

▶ **KEY POINT** In the testes, interstitial cells produce androgens, especially testosterone. Nurse cells release inhibin.

In males, the **interstitial cells** of the testes produce male hormones known as androgens. The most important androgen is **testosterone** (tes-TOS-ter-ōn). This hormone (1) promotes the production of functional sperm, (2) maintains the secretory glands of the male reproductive tract, (3) influences secondary sexual characteristics, and (4) stimulates muscle growth **(Table 19.4)**. During embryonic development, the production of testosterone affects the anatomical development of the hypothalamic nuclei of the CNS, which will later influence sexual behaviors.

Table 19.3 | **Hormones of the Pancreas**

Structure/Cells	Hormone	Primary Targets	Effects
PANCREATIC ISLETS			
Alpha cells	Glucagon	Liver, adipose tissues	Mobilization of lipid reserves; glucose synthesis and glycogen breakdown in liver; elevation of blood glucose concentrations
Beta cells	Insulin	All cells except those of brain, kidneys, and digestive tract epithelium and RBCs	Facilitation of uptake of glucose by cells; stimulation of lipid and glycogen formation and storage; decrease in blood glucose concentrations
Delta cells	Somatostatin	Alpha and beta cells, digestive tract	Inhibition of secretion of insulin and glucagon
F cells	Pancreatic polypeptide (PP)	Gallbladder and pancreas, possibly gastrointestinal tract	Inhibition of gallbladder contractions; regulation of production of some pancreatic enzymes; may control nutrient absorption

Nurse cells in the testes support the differentiation and physical maturation of sperm. Under FSH stimulation, these cells secrete the hormone **inhibin**. It inhibits the secretion of FSH at the anterior lobe of the pituitary. Throughout adult life, inhibin and FSH interact to maintain sperm production at normal levels.

Ovaries

▶ **KEY POINT** In the ovaries, follicular cells produce estrogens (especially estradiol) and inhibin.

In the ovaries, oocytes begin their maturation into female gametes (sex cells) within specialized structures called **follicles**. Stimulation by FSH triggers the maturation process. Follicular cells surrounding the oocytes produce estrogens, especially **estradiol**. These steroid hormones support the maturation of the oocytes and stimulate the growth of the uterine lining (**Table 19.4**). FSH also stimulates active follicles to secrete inhibin. Inhibin decreases FSH release from the anterior lobe of the pituitary gland.

After ovulation has occurred, the remaining follicular cells reorganize into a **corpus luteum** (LOO-tē-um) that releases a mixture of estrogens and progestins, especially **progesterone** (prō-JES-ter-ōn). Progesterone accelerates the movement of the oocyte along the uterine tube and prepares the uterus for the arrival of the developing embryo.

Table 19.4 summarizes the reproductive hormones.

19.9 CONCEPT CHECK ✔

10 What hormones are produced by the testes, and what are their functions?

See the blue Answers tab at the back of the book.

19.10 | The Pineal Gland

▶ **KEY POINT** The pineal gland, which is part of the epithalamus of the CNS, contains two types of endocrine cells: pinealocytes and interstitial cells. Pinealocytes secrete melatonin, and interstitial cells resemble neuroglia.

The small, red, pinecone-shaped **pineal gland**, or *pineal body*, is part of the epithalamus (**Figure 19.1** and look ahead to **Figure 19.9** on page 524). ⊃ p. 417 The pineal gland contains neurons and interstitial cells that resemble the neuroglia of the CNS. The pineal gland also contains special secretory cells called **pinealocytes** (PIN-ē-al-ō-sīts).

Pinealocytes synthesize the hormone **melatonin** (mel-a-TŌN-in). Melatonin is derived from molecules of the neurotransmitter serotonin. Melatonin slows the maturation of sperm, oocytes, and reproductive organs. It does this by inhibiting the production of a hypothalamic releasing factor that stimulates FSH and LH secretion. Collaterals from the visual pathways enter the pineal gland and affect the rate of melatonin production.

Table 19.4 | **Hormones of the Reproductive System**

Structure/Cells	Hormone	Primary Targets	Effects
TESTES			
Interstitial cells	Androgens	Most cells	Support functional maturation of sperm; protein synthesis in skeletal muscles; male secondary sex characteristics and associated behaviors
Nurse cells	Inhibin	Anterior lobe of pituitary gland	Inhibits secretion of FSH
OVARIES			
Follicular cells	Estrogens (especially estradiol)	Most cells	Support follicle maturation; female secondary sex characteristics and associated behaviors
	Inhibin	Anterior lobe of pituitary gland	Inhibits secretion of FSH
Corpus luteum	Progestins (especially progesterone)	Uterus, mammary glands	Prepare uterus for implantation; prepare mammary glands for secretory functions
	Relaxin	Pubic symphysis, uterus, mammary glands	Loosens pubic symphysis; relaxes uterine (cervical) muscles; stimulates mammary gland development

CLINICAL NOTE

Endocrine Disorders

Endocrine disorders may develop for a variety of reasons, including abnormalities in the endocrine gland, the endocrine or neural regulatory mechanisms, or the target tissues. For example, a hormone level may rise because its target organs are becoming less responsive, because a tumor has formed among the gland cells, or because something has interfered with the normal feedback control mechanism. When naming endocrine disorders, clinicians use the prefix *hyper-* when referring to excessive hormone production and *hypo-* when referring to inadequate hormone production.

Table 19.5 | Clinical Implications of Endocrine Malfunctions

Hormone	Underproduction Syndrome	Principal Symptoms	Overproduction Syndrome	Principal Symptoms
Growth hormone (GH)	Pituitary growth failure (children)	Retarded growth, abnormal fat distribution, low blood glucose hours after a meal	Gigantism (children), acromegaly (adults)	Excessive growth in stature of a child or in face and hands in an adult
Antidiuretic hormone (ADH)	Diabetes insipidus	Polyuria	SIADH (syndrome of inappropriate ADH secretion)	Increased body water content and hyponatremia
Thyroxine (T_4), triiodothyronine (T_3)	Myxedema (in adults); infantile hypothyroidism	Low metabolic rate, body temperature; impaired physical and mental development	Graves disease	High metabolic rate, body temperature; tachycardia; weight loss
Parathyroid hormone (PTH)	Hypoparathyroidism	Muscular weakness, neurological problems, tetany due to low blood calcium concentrations	Hyperparathyroidism	Neurological, mental, muscular problems due to high blood calcium concentrations; weak and brittle bones
Insulin	Diabetes mellitus	High blood glucose; impaired glucose utilization; dependence on lipids for energy; glucosuria; ketosis	Excess insulin production or administration	Low blood glucose levels, possibly causing coma
Mineralocorticoids (MCs)	Hypoaldosteronism	Polyuria; low blood volume; high blood potassium concentrations	Aldosteronism	Increased body weight due to water retention; low blood potassium concentrations
Glucocorticoids (GCs)	Addison disease	Inability to tolerate stress, mobilize energy reserves, maintain normal blood glucose concentrations	Cushing disease	Excessive breakdown of tissue proteins and lipid reserves; impaired glucose metabolism
Epinephrine (E), norepinephrine (NE)	None identified		Pheochromocytoma	High metabolic rate, body temperature, and heart rate; elevated blood glucose levels; other symptoms comparable to those of excessive autonomic stimulation
Estrogens (female)	Hypogonadism	Sterility; lack of secondary sexual characteristics	Androgenital syndrome	Overproduction of androgens by zona reticularis of adrenal cortex leading to masculinization
			Precocious puberty	Early production of developing follicles and estrogen secretion
	Menopause	Cessation of ovulation		
Androgens (male)	Hypogonadism, eunuchoidism	Sterility; lack of secondary sexual characteristics	Gynecomastia	Abnormal production of estrogens, sometimes due to adrenal or intestinal cell tumors, leading to breast enlargement
			Precocious puberty	Early production of androgens, leading to premature physical development and behavioral changes

Melatonin production increases at night and decreases during the day. This cycle is apparently important in regulating our circadian rhythms, our natural awake-asleep cycles. p. 418 This hormone is also a powerful antioxidant that might help protect CNS tissues from free radicals generated by active neurons and neuroglia.

19.10 CONCEPT CHECK

11 You are scheduled to fly from Chicago, Illinois, to Beijing, China, across multiple time zones, for a business trip. A colleague suggests taking melatonin tablets to counteract "jet lag." What might be a possible action of these tablets in preventing jet lag?

See the blue Answers tab at the back of the book.

CLINICAL NOTE (*continued*)

Most endocrine disorders are the result of problems within the endocrine gland itself. The typical result is hyposecretion, the production of inadequate levels of a particular hormone. Hyposecretion may be caused by the following:

- Metabolic factors: Hyposecretion may result from a deficiency in some key substrate needed to synthesize the hormone in question. For example, hypothyroidism can be caused by inadequate dietary iodine levels or by exposure to drugs that inhibit iodine transport or utilization at the thyroid gland.
- Physical damage: Any condition that interrupts the normal circulatory supply or that physically damages the endocrine cells may cause them to become inactive immediately or after an initial surge of hormone release. If the damage is severe, the gland can become permanently inactive. For instance, temporary or permanent hypothyroidism can result from infection or inflammation of the gland (thyroiditis), from the interruption of normal blood flow, or from exposure to radiation as part of treatment for cancer of the thyroid gland or adjacent tissues. The thyroid gland can also be damaged in an autoimmune disorder that results in the production of antibodies that attack and destroy normal follicle cells.

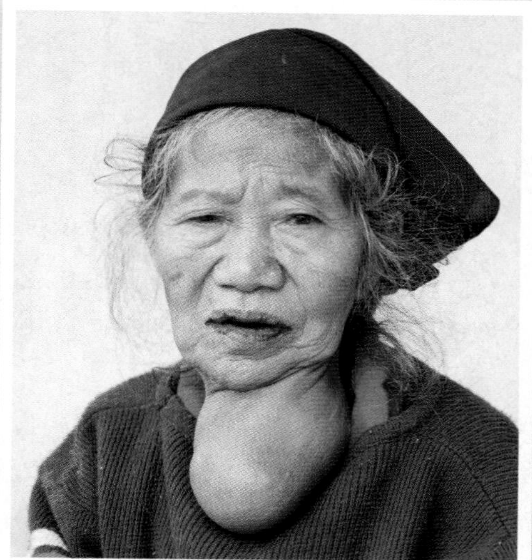
Enlarged thyroid gland

- Congenital disorders: An individual may be unable to produce normal amounts of a particular hormone because (1) the gland itself is too small, (2) the required enzymes are abnormal, (3) the receptors that trigger secretion are relatively insensitive, or (4) the gland cells lack the receptors normally involved in stimulating secretory activity.

Endocrine abnormalities can also be caused by the presence of abnormal hormonal receptors in target tissues. In such a case, the gland involved and the regulatory mechanisms are normal, but the peripheral cells are unable to respond to the circulating hormone. The best example of this type of abnormality is type 2 diabetes, in which peripheral cells do not respond normally to insulin.

Many of these disorders produce distinctive anatomical features or abnormalities that are evident on a physical examination (Table 19.5).

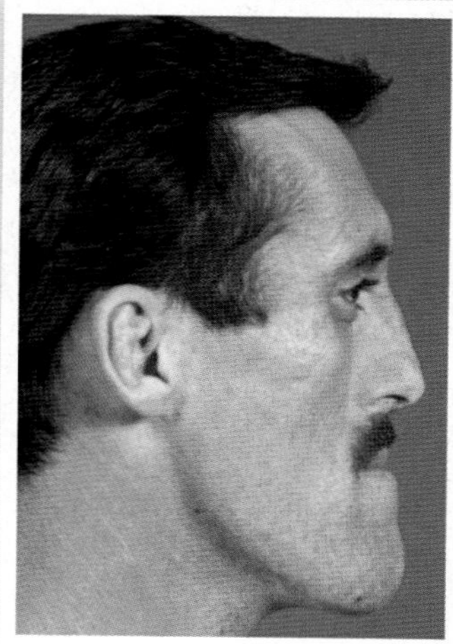
Acromegaly

Acromegaly

Acromegaly results from the overproduction of growth hormone after the epiphyseal plates have fused. Bone shapes change, and cartilaginous areas of the skeleton enlarge. Note the broad facial features and the enlarged lower jaw.

Infantile hypothyroidism

Infantile hypothyroidism results from thyroid hormone insufficiency in infancy.

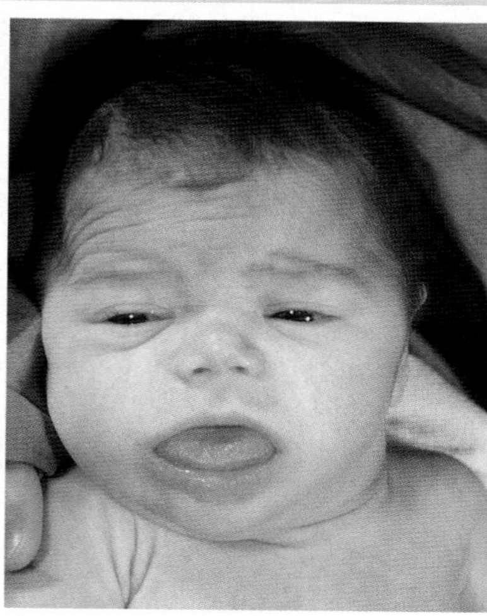
Infantile hypothyroidism

Enlarged Thyroid Gland

An enlarged thyroid gland, or goiter, is usually associated with thyroid hyposecretion due to nutritional iodine insufficiency.

Addison Disease

Addison disease is caused by hyposecretion of corticosteroids, especially glucocorticoids. Pigment changes result from stimulation of melanocytes by ACTH, which is structurally similar to MSH.

Cushing Disease

Cushing disease is caused by hypersecretion of glucocorticoids. Lipid reserves are mobilized, and adipose tissue accumulates in the cheeks and at the base of the neck.

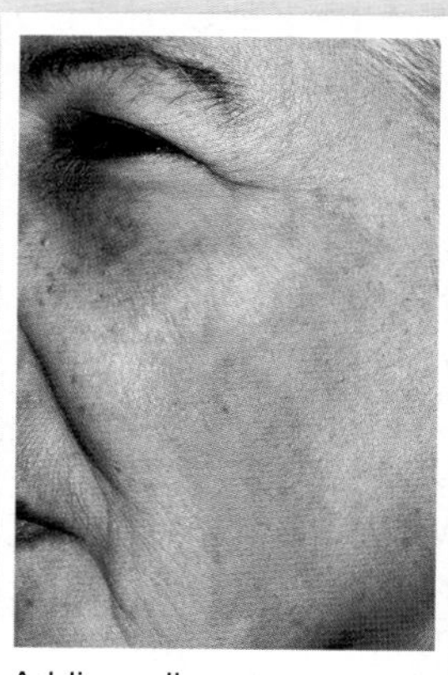
Addison disease

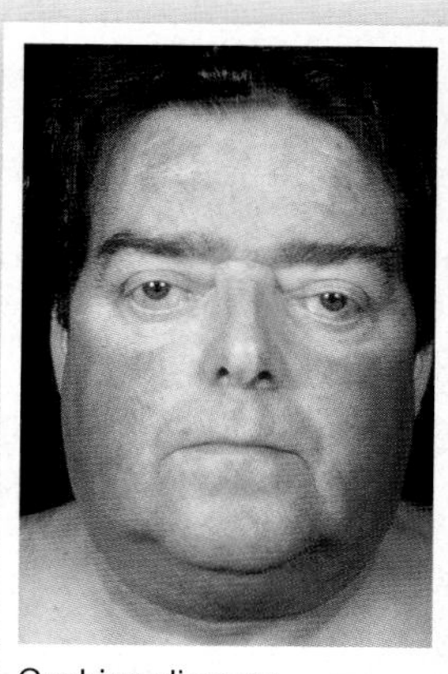
Cushing disease

19

Figure 19.9 Anatomy and Histological Organization of the Pineal Gland

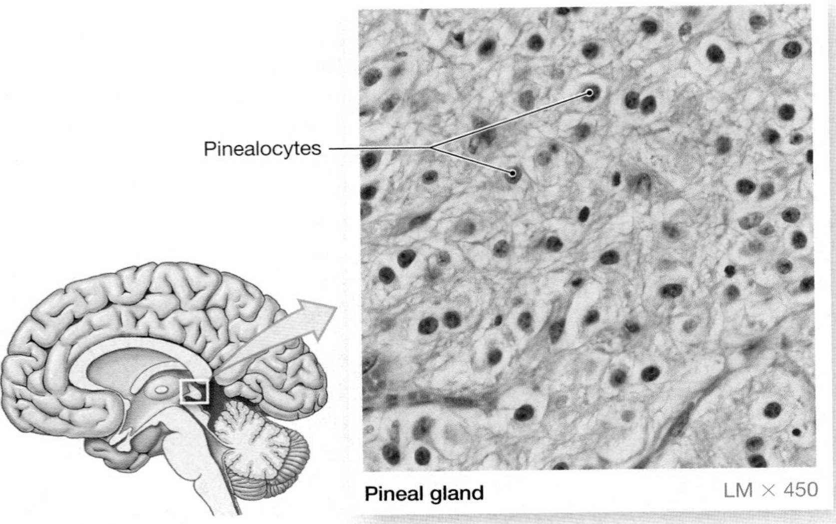

EMBRYOLOGY SUMMARY

For a summary of the development of the endocrine system, see Chapter 28 (Embryology and Human Development).

19.11 Hormones and Aging

The endocrine system shows relatively few functional changes with advancing age. The most dramatic exceptions are (1) the changes in reproductive hormone levels at puberty and (2) the decline in the concentration of reproductive hormones at menopause in women. It is interesting to note that age-related changes in other tissues affect their abilities to respond to hormonal stimulation. As a result, most tissues may become less responsive to circulating hormones, even though hormone concentrations remain normal.

19.11 CONCEPT CHECK

12 Which hormone(s) of the endocrine system show the most dramatic decline in concentration as a result of aging?

See the blue Answers tab at the back of the book.

Study Outline

Introduction p. 507

- The nervous and endocrine systems work together in a complementary way to monitor and adjust physiological activities for the regulation of homeostasis.
- In general, the nervous system performs short-term "crisis management," while the endocrine system regulates longer-term, ongoing metabolic processes. Endocrine cells release chemicals called hormones that alter the metabolic activities of many different tissues and organs simultaneously.

19.1 An Overview of the Endocrine System p. 507

- The **endocrine system** consists of all endocrine cells and tissues. They release their secretory products into the lymphatic system or blood. *(See Figure 19.1.)*
- **Hormones** can be divided into three groups based on chemical structure: **amino acid derivatives**, **peptide hormones**, and **lipid derivatives.** There are two groups of lipid derivatives: eicosanoid and steroid hormones.
- Cellular activities and metabolic reactions are controlled by enzymes. Hormones exert their effects by modifying the activities of **target cells** (cells that are sensitive to that particular hormone).
- Endocrine activity can be controlled by (1) neural activity, (2) positive feedback (rare), or (3) complex negative feedback mechanisms.

19.2 Hypothalamus and the Pituitary Gland p. 508

- The **hypothalamus** regulates endocrine and neural activities. It (1) controls the output of the adrenal (suprarenal) medulla, an endocrine component of the sympathetic division of the ANS; (2) produces two hormones of its own (ADH and oxytocin), which are released from the posterior lobe of the pituitary gland; and (3) controls the activity of the anterior lobe of the pituitary through the production of **regulatory hormones** and **inhibiting hormones**. *(See Spotlight Figure 19.2.)*
- The **pituitary gland** (**hypophysis**) releases nine important peptide hormones. Two are synthesized in the hypothalamus and released at the posterior lobe of the pituitary and seven are synthesized in the anterior lobe of the pituitary. *(See Spotlight Figure 19.2 and Figure 19.3.)*

The Anterior Lobe of the Pituitary Gland p. 508

- The **anterior lobe** (**adenohypophysis**) is subdivided into the large **pars distalis**, the slender **pars intermedia**, and the **pars tuberalis**. The entire anterior lobe is highly vascularized.
- In the floor of the hypothalamus in the tuberal area, neurons release regulatory factors into the surrounding interstitial fluids. Endocrine cells in the anterior lobe are controlled by releasing factors, inhibiting factors (hormones), or some combination of the two. These secretions enter the circulation through **fenestrated** capillaries that contain open spaces between their epithelial cells. Blood vessels, called **portal vessels**, form an unusual vascular arrangement that connects the hypothalamus and anterior lobe of the pituitary gland. This complex is the **hypophyseal portal system**. It ensures that all of the blood entering the portal vessels will reach the intended target cells before returning to the general circulation. *(See Spotlight Figure 19.2.)*
- Important hormones released by the pars distalis are (1) **thyroid-stimulating hormone** (**TSH**), which triggers the release of thyroid hormones; (2) **adrenocorticotropic hormone** (**ACTH**), which stimulates the release of **glucocorticoids** by the adrenal gland; (3) **follicle-stimulating hormone** (**FSH**), which stimulates **estrogen** secretion (*estradiol*) and egg development in women and sperm production in men; (4) **luteinizing hormone** (**LH**), which causes ovulation and production of **progestins** (*progesterone*) in women and **androgens** (*testosterone*) in men (together, FSH and LH are called **gonadotropins**); (5) **prolactin** (**PRL**), which stimulates the development of the mammary glands and the production of milk; and (6) **growth hormone** (**GH**, or **somatotropin**), which stimulates cells' growth and replication. *(See Spotlight Figure 19.2 and Figure 19.3.)*
- **Melanocyte-stimulating hormone** (**MSH**), released by the pars intermedia, stimulates melanocytes to produce melanin.

The Posterior Lobe of the Pituitary Gland p. 509

- The **posterior lobe** (**neurohypophysis**) contains the axons of some hypothalamic neurons. Neurons within the supra-optic and paraventricular nuclei manufacture **antidiuretic hormone** (**ADH**) and **oxytocin**, respectively. ADH decreases the amount of water lost at the kidneys. It is released in response to a rise in the concentration of electrolytes in the blood or a fall in blood volume. In women, oxytocin stimulates smooth muscle cells in the uterus and contractile cells in the mammary glands. It is released in response to stretched uterine muscles and/or suckling of an infant. In men, it stimulates ductus deferens and prostatic smooth muscle contractions. *(See Spotlight Figure 19.2 and Figure 19.3.)*

19.3 | The Thyroid Gland p. 509

- The **thyroid gland** lies inferior to the **thyroid cartilage** of the larynx. It consists of two **lobes** connected by a narrow **isthmus**. *(See Figure 19.4a.)*

Thyroid Follicles and Thyroid Hormones p. 512

- The thyroid gland contains numerous **thyroid follicles**. Cells of the follicles manufacture **thyroglobulin** and store it within the **colloid** (a viscous fluid containing suspended proteins) in the **follicle cavity**. The cells also transport iodine from the extracellular fluids into the cavity, where it complexes with tyrosine residues of the thyroglobulin molecules to form thyroid hormones. *(See Figure 19.4b,c and Table 19.1.)*
- When stimulated by TSH, the follicular cells reabsorb the thyroglobulin, break down the protein, and release the thyroid hormones, **thyroxine** (T_4) and **triiodothyronine** (T_3), into the circulation. *(See Figure 19.5.)*

The C Thyrocytes of the Thyroid Gland p. 514

- The **C thyrocytes** of the thyroid follicles produce **calcitonin** (**CT**), which helps lower calcium ion concentrations in body fluids by inhibiting osteoclast activities and stimulating calcium ion excretion at the kidneys. *(See Figure 19.4c.)*
- Actions of calcitonin are opposed by those of the parathyroid hormone produced by the parathyroid glands. *(See Table 19.1.)*

19.4 | The Parathyroid Glands p. 515

- Four **parathyroid glands** are embedded in the posterior surface of the thyroid gland. The **parathyroid** (principal) **cells** of the parathyroid produce **parathyroid hormone** (**PTH**) in response to lower-than-normal concentrations of calcium ions. Oxyphil cells of the parathyroid have no known function. *(See Figures 19.4a and 19.6 and Table 19.1.)*
- PTH (1) stimulates osteoclast activity, (2) stimulates osteoblast activity to a lesser degree, (3) reduces calcium loss in the urine, and (4) promotes calcium absorption in the intestine by stimulating calcitriol production. *(See Table 19.1.)*
- The parathyroid glands and the C thyrocytes of the thyroid gland maintain calcium ion levels within relatively narrow limits. *(See Figure 19.6c and Table 19.1.)*

19.5 | The Thymus p. 516

- The **thymus**, embedded in a connective tissue mass in the thoracic cavity, produces several hormones that stimulate the development and maintenance of normal immunological defenses. *(See Figure 19.1 and Table 19.1.)*
- **Thymosins** produced by the thymus promote the development and maturation of lymphocytes.

19.6 | The Adrenal Glands p. 516

- A single **adrenal** (**suprarenal**) **gland** rests on the superior border of each kidney. Each adrenal gland is surrounded by a fibrous capsule and is subdivided into a superficial **cortex** and an inner **medulla**. *(See Figure 19.7.)*

The Adrenal Cortex p. 516

- The cortex of the adrenal gland manufactures steroid hormones called **corticosteroids** (*adrenocortical steroids*). The cortex can be subdivided into three separate areas. (1) The outer **zona glomerulosa** releases **mineralocorticoids** (**MCs**), principally **aldosterone**, which restrict sodium and water losses at the kidneys, sweat glands, digestive tract, and salivary glands. The zona glomerulosa responds to the presence of the hormone angiotensin II, which appears after the enzyme renin has been secreted by kidney cells exposed to a decrease in blood pressure. (2) The middle **zona fasciculata** produces **glucocorticoids** (**GC**), notably **cortisone** and **cortisol**. All of these hormones accelerate the rates of both glucose synthesis and glycogen formation, especially in liver cells. (3) The inner **zona reticularis** produces small amounts of sex hormones called androgens. The significance of the small amounts of androgens produced by the adrenal glands remains uncertain. *(See Figure 19.7c and Table 19.2.)*

The Adrenal Medulla p. 518

- Each medulla of the adrenal gland contains clusters of **chromaffin cells**, which resemble sympathetic ganglia neurons. They secrete either epinephrine (75–80 percent) or norepinephrine (20–25 percent). These catecholamines trigger cellular energy utilization and the mobilization of energy reserves (see Chapter 17). *(See Figure 19.7b,c and Table 19.2.)*

19

19.7 | Endocrine Functions of the Kidneys and Heart p. 518

- Endocrine cells in both the kidneys and heart produce hormones that are important for the regulation of blood pressure and blood volume, blood oxygen levels, and calcium and phosphate ion absorption.
- The kidney produces the enzyme renin and the peptide hormone erythropoietin when blood pressure or blood oxygen levels in the kidneys decline, and it secretes the steroid hormone calcitriol when parathyroid hormone is present. **Renin** catalyzes the conversion of circulating **angiotensinogen** to **angiotensin I**. In lung capillaries, it is converted to **angiotensin II**, the hormone that stimulates the production of aldosterone in the adrenal cortex. **Erythropoietin** (**EPO**) stimulates red blood cell production by the bone marrow. **Calcitriol** stimulates the absorption of both calcium and phosphate in the digestive tract.
- Specialized muscle cells of the heart produce **atrial natriuretic peptide** (**ANP**) when blood pressure or blood volume becomes excessive. These hormones stimulate water and sodium ion loss at the kidneys, eventually reducing blood volume.

19.8 | The Pancreas and Other Endocrine Tissues of the Digestive System p. 518

- The lining of the digestive tract, the liver, and the pancreas produce exocrine secretions that are essential to the normal breakdown and absorption of food.

The Pancreas p. 518

- The **pancreas** is a nodular organ occupying a space between the stomach and small intestine. It contains both exocrine and endocrine cells. The **exocrine pancreas** secretes an enzyme-rich fluid into the lumen of the digestive tract. Cells of the **endocrine pancreas** form clusters called **pancreatic islets**. Each islet contains four cell types: **Alpha cells** produce **glucagon** to raise blood glucose levels; **beta cells** secrete **insulin** to lower blood glucose levels; **delta cells** secrete **somatostatin** (**growth hormone–inhibiting hormone**) to inhibit the production and secretion of glucagon and insulin; and **F cells** secrete **pancreatic polypeptide** (**PP**) to inhibit gallbladder contractions and regulate the production of some pancreatic enzymes. PP may also help control the rate of nutrient absorption by the GI tract. *(See Figure 19.8 and Table 19.3.)*

- Insulin lowers blood glucose by increasing the rate of glucose uptake and utilization by most body cells; glucagon raises blood glucose levels by increasing the rates of glycogen breakdown and glucose synthesis in the liver. Somatostatin reduces the rates of hormone secretion by alpha and beta cells and slows food absorption and enzyme secretion in the digestive tract. *(See Table 19.3.)*

19.9 Endocrine Tissues of the Reproductive System p. 520

Testes p. 520

- The **interstitial cells** of the male testes produce androgens. **Testosterone** is the most important androgen. It promotes the production of functional sperm, maintains reproductive-tract secretory glands, influences secondary sexual characteristics, and stimulates muscle growth. *(See Table 19.4.)*

- The hormone **inhibin**, produced by **nurse cells** in the testes, interacts with FSH from the anterior lobe of the pituitary gland to maintain sperm production at normal levels.

Ovaries p. 521

- Oocytes develop in **follicles** in the female ovary; follicle cells surrounding the oocytes produce estrogens, especially **estradiol**. Estrogens support the maturation of the oocytes and stimulate the growth of the uterine lining. Active follicles secrete inhibin, which suppresses FSH release by negative feedback. *(See Table 19.4.)*

- After ovulation, the follicle cells remaining within the ovary reorganize into a **corpus luteum**, which produces a mixture of estrogens and progestins, especially **progesterone**. Progesterone facilitates the movement of a fertilized egg through the uterine tube to the uterus and stimulates the preparation of the uterus for implantation. *(See Table 19.4.)*

19.10 The Pineal Gland p. 521

- The **pineal gland** (*pineal body*) contains secretory cells called **pinealocytes**, which synthesize **melatonin**. Melatonin slows the maturation of sperm, eggs, and reproductive organs by inhibiting the production of FSH- and LH-releasing factors from the hypothalamus. Additionally, melatonin may establish circadian rhythms. *(See Figures 19.1 and 19.9.)*

19.11 Hormones and Aging p. 524

- The endocrine system shows relatively few functional changes with advancing age. The most dramatic endocrine changes are the rise in reproductive hormone levels at puberty and the decline in reproductive hormone levels at menopause.

19

Chapter Review

For answers, see the blue Answers tab at the back of the book.

Level 1 Reviewing Facts and Terms

Match each numbered item with the most closely related lettered item.

1. target cellls ______
2. hypothalamus ______
3. ADH ______
4. prolactin ______
5. FSH ______
6. colloid ______
7. oxyphil ______
8. thymosin ______
9. chromaffin cells ______
10. melatonin ______

(a) unknown function
(b) stimulates milk production
(c) regulated by hormones
(d) pineal gland
(e) norepinephrine release
(f) decreases water loss
(g) lymphocyte maturation
(h) stimulates estrogen secretion
(i) produces releasing hormone
(j) viscous fluid with stored hormones

11. The hormone that targets the thyroid gland and triggers the release of thyroid hormone is
(a) follicle-stimulating hormone (FSH).
(b) thyroid-stimulating hormone (TSH).
(c) adrenocorticotropic hormone (ACTH).
(d) luteinizing hormone (LH).

12. Blood vessels that supply or drain the thyroid gland include which of the following?
(a) superior thyroid artery
(b) inferior thyroid artery
(c) superior, inferior, and middle thyroid veins
(d) all of the above

13. How does aging affect the function of the endocrine system?
(a) It is relatively much less affected than most other systems.
(b) Hormone production increases to offset diminished response by receptors.
(c) Endocrine function of the reproductive system is the most affected by increasing age.
(d) Hormone production by the thyroid gland suffers the greatest decline with age.

14. Endocrine organs can be controlled by
(a) hormones from other endocrine glands.
(b) direct neural stimulation.
(c) changes in the composition of extracellular fluid.
(d) all of the above.

15. Reduced fluid losses in the urine due to retention of sodium ions and water are a result of the action of
(a) antidiuretic hormone.
(b) calcitonin.
(c) aldosterone.
(d) cortisone.

16. When blood glucose levels decrease,
(a) insulin is released.
(b) glucagon is released.
(c) peripheral cells stop taking up glucose.
(d) aldosterone is released to stimulate these cells.

17. Hormones released by the kidneys include
(a) calcitriol and erythropoietin.
(b) ADH and aldosterone.
(c) epinephrine and norepinephrine.
(d) cortisol and cortisone.

18. The mineral required for normal thyroid function is
(a) magnesium. (b) potassium.
(c) iodine. (d) calcium.

19. A structure known as the corpus luteum secretes
(a) testosterone. (b) progesterone.
(c) aldosterone. (d) cortisone.

Level 2 Reviewing Concepts

1. If a person has too few or defective lymphocytes, which gland might be at fault?
   (a) thyroid (b) thymus
   (c) pituitary (d) pineal
2. Reductions in cardiac activity, blood pressure, ability to process glycogen, and blood glucose level and release of lipids by adipose tissues are collectively symptoms of a defective
   (a) pituitary gland. (b) adrenal cortex.
   (c) pancreas. (d) adrenal medulla.
3. Discuss the functional differences between the endocrine and nervous systems.
4. Hormones can be divided into three groups on the basis of chemical structure. What are these groups?
5. Describe the primary targets and effects of testosterone.
6. What effects do thyroid hormones have on body tissues?
7. Why is normal parathyroid function essential in maintaining normal calcium ion levels?
8. Describe the role of melatonin in regulating reproductive function.
9. What is the significance of the capillary network within the hypophysis?

Level 3 Critical Thinking

1. How could a pituitary tumor result in the production of excess amounts of growth hormone?
2. Endocrine abnormalities rarely, if ever, result in only a single change to a person's metabolism. What two endocrine abnormalities would result in excessive thirst *and* excessive urination?
3. Hypothyroidism (insufficient thyroid hormone production by the thyroid gland) can be caused by a problem at the level of the hypothalamus and pituitary gland or at the level of the thyroid. Explain how this is medically possible.
4. How do kidney and heart hormones regulate blood pressure and volume?

MasteringA&P™

Access more chapter study tools online in the Study Area:

- Chapter Quizzes, Chapter Practice Test, Clinical Cases, and more!
- Practice Anatomy Lab (PAL) PAL™
- A&P Flix for anatomy topics A&PFlix™

CLINICAL CASE WRAP-UP

Why Am I So Cold and Tired?

Kathy is suffering from **hypothyroidism** (deficient thyroid hormone production). Her thyroid function has decreased to the point where her metabolism and oxygen consumption have slowed down.

Because of her constellation of symptoms, Kathy's doctor drew blood to check her thyroid function. Recall that the hypothalamus secretes thyrotropin-releasing hormone (TRH), which stimulates the adenohypophysis (anterior lobe of the pituitary gland) to produce and release thyroid-stimulating hormone (TSH). Normally, the thyroid responds to TSH by releasing thyroxine (T_4) and triiodothyronine (T_3) into the bloodstream. These thyroid hormones increase the rate of cellular metabolism and O_2 consumption in nearly every cell in the body. Without them, metabolism slows everywhere.

At some time in the past, Kathy must have had a silent inflammation of the thyroid gland that caused damage or death to enough hormone-producing cells to cause a thyroid gland failure. Hashimoto's thyroiditis, a common autoimmune thyroid disease, is a likely cause.

Kathy's normal hypothalamus continuously reads her low T_4 and produces more TRH. Her pituitary, responding to the increased TRH in her blood, produces more TSH. Kathy's blood level of TSH is very high, indicating that her pituitary is responding normally to a low blood level of T_4.

The good news for Kathy is that her condition is easily treatable with a dose of levothyroxine every morning. This medication is a pure synthetic form of T_4. A few months later, Kathy is feeling better and many of the changes in her physical appearance have reversed.

1. If Kathy's thyroxine (T_4) level is low, would her triiodothyronine (T_3) level to be low as well?
2. What would be the problem if Kathy's T_3 and thyroid-stimulating hormone (TSH) levels were both low?

See the blue Answers tab at the back of the book.

Related Clinical Terms

adrenalectomy: Surgical removal of an adrenal gland.

Hashimoto's disease: Also known as chronic lymphocytic thyroiditis, a disorder that affects the thyroid gland, causing the immune system to attack it. It is the most common cause of hypothyroidism in the United States.

thyroidectomy: Surgical removal of all or part of the thyroid gland.

thyroid function tests: Blood and radionuclide tests to determine thyroid gland activity.

virilism: A disorder in females in which there is development of secondary male sexual characteristics such as hirsutism and lowered voice caused by a number of conditions that affect hormone regulation.

20

The Cardiovascular System

Blood

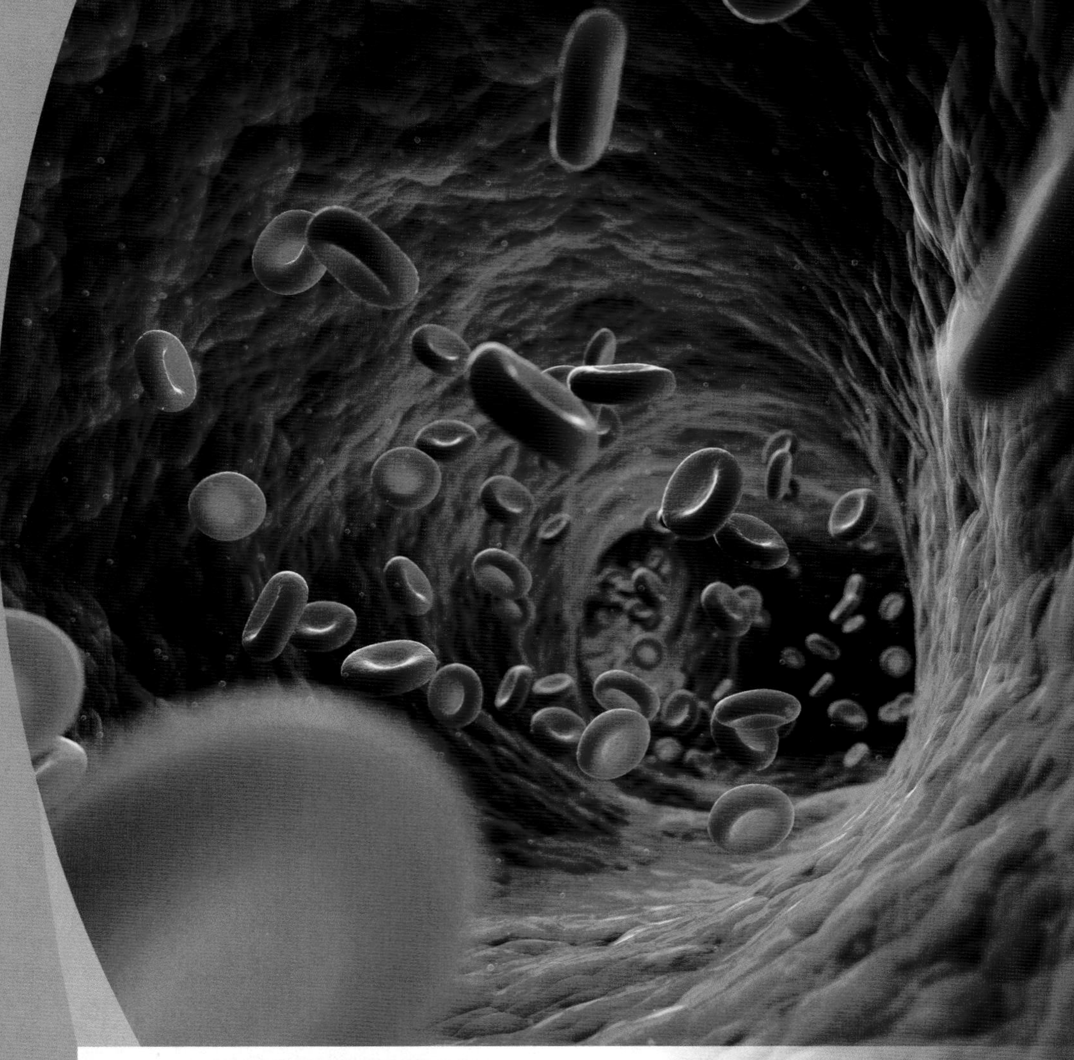

Learning Outcomes

These Learning Outcomes correspond by number to this chapter's sections and indicate what you should be able to do after completing the chapter.

20.1 Compare and contrast the components of blood and plasma. p. 529

20.2 Compare and contrast the formed elements of blood. p. 531

20.3 List the cells involved in erythropoiesis and leukopoiesis. p. 539

A Surplus of WBCs

Danny, the youngest of five children, is a very active three-year-old. That is, until recently. He has an upper respiratory infection with bilateral otitis media (ear ache) that just won't respond to antibiotics. He has lost weight, has a low-grade fever, looks pale, gets frequent nosebleeds, and tires easily. He complains that his "bones hurt."

At his last appointment, Danny's pediatrician ordered blood work. Last week Danny's mother was told he needed more tests, including a bone marrow biopsy (extracting spongy bone from the posterior ileum with a large needle under sedation) and a lumbar puncture. Now Danny and his mom are back at the doctor's office.

"I am sorry to tell you that Danny is quite sick," explains the doctor. "His blood and bone marrow have way too many white blood cells."

"But I thought white blood cells fight infections! Why won't Danny's ears clear up?" asks his mom. The doctor explains, "Sadly, these white cells are not good for anything."

What is going on with Danny? To find out, turn to the Clinical Case Wrap-Up on p. 544.

THE LIVING BODY IS IN CONSTANT CHEMICAL COMMUNICATION with the external environment. The lining of the digestive tract absorbs nutrients; gases diffuse across the delicate epithelium of the lungs; wastes are excreted in the feces, urine, saliva, bile, and sweat. These chemical exchanges occur at specialized sites or organs because all parts of the body are linked together by the cardiovascular system.

We can compare the cardiovascular system to the cooling system of a car. The components include a circulating fluid (blood), a pump (the heart), and an assortment of conducting pipes (a network of blood vessels). The three chapters on the cardiovascular system focus on each of these components. This chapter discusses the nature of the circulating blood. Chapter 21 discusses the structure and function of the heart, and Chapter 22 discusses the network of blood vessels and the integrated functioning of the cardiovascular system. You will then be ready for Chapter 23, which discusses the lymphatic (*lymphoid*) system, whose vessels and organs are structurally and functionally linked to the cardiovascular system.

20.1 Functions and Composition of the Blood

▶ **KEY POINT** Blood is a fluid connective tissue that circulates through the cardiovascular system. Blood consists of plasma and formed elements (red blood cells, white blood cells, and platelets).

Blood is a fluid connective tissue distributing nutrients, oxygen, and hormones to each of the roughly 75 trillion cells in the human body. Blood also carries metabolic wastes to the kidneys for excretion and transports specialized cells that defend peripheral tissues from infection and disease. Table 20.1 details the functions of the blood. The services performed by the blood are absolutely essential—so much so that any cell deprived of circulation may die within minutes.

Blood is normally confined to the circulatory system. Figure 20.1 and Table 20.2 outline the composition of blood. Blood consists of the following two components:

1. **Plasma** (PLAZ-mah) is the liquid component of blood. It has a density only slightly greater than water. Plasma contains dissolved proteins and other solutes (nutrients, electrolytes, and wastes).
2. **Formed elements** are blood cells (red blood cells and white blood cells) and cell fragments (platelets) suspended in the plasma. **Red blood cells (RBCs)** transport oxygen and carbon dioxide. **White blood cells (WBCs)** are components of the immune system and are less numerous than RBCs. **Platelets** (PLĀT-lets) are small, membrane-enclosed packets of cytoplasm containing enzymes and clotting factors, proteins that play a role in blood clotting.

Whole blood is a mixture of plasma and formed elements. Its components can be separated, or **fractionated**, for clinical purposes. Whole blood is sticky, cohesive, and resistant to flow. These characteristics determine the **viscosity** of a solution. Solutions are compared with pure water, which has a viscosity of 1.0. Plasma has a viscosity of 1.5, but whole blood is 5 times as viscous as water. Its high viscosity results from interactions among dissolved proteins, formed elements, and water molecules in plasma.

Adult males typically have more blood than do adult females. We can estimate blood volume in liters for a person of either sex by calculating 7 percent of the body weight in kilograms. For example, a 75-kg (165-lb) person would have a blood volume of approximately 5.25 liters (5.4 quarts).

Blood is slightly alkaline, with a pH range between 7.35 and 7.45 and a temperature of 38°C (100.4°F), slightly higher than normal body temperature of 37°C. Clinicians use the terms **hypovolemic** (hī-pō-vō-LĒ-mik), **normovolemic** (nor-mō-vō-LĒ-mik), and **hypervolemic** (hī-per-vō-LĒ-mik) to refer to low, normal, and excessive blood volumes, respectively. Low or high blood volumes are potentially dangerous. For example, a hypervolemic condition, as seen in kidney failure, causes fluid retention and places severe stress on the heart, which must push the extra fluid around the circulatory system.

Plasma

▶ **KEY POINT** Ninety-two percent of the volume of plasma is water. Water is the solvent for a variety of materials, including dissolved gases, electrolytes, nutrients, wastes, regulatory substances, and proteins.

Plasma is approximately 55 percent of the volume of whole blood, and water accounts for 92 percent of the plasma volume. These are average values; actual concentrations vary depending on (1) the region of the cardiovascular system or area of the body sampled and (2) the ongoing activity within that particular region. Figure 20.1 and Table 20.2 summarize the composition of plasma.

Differences between Plasma and Interstitial Fluid

▶ **KEY POINT** Plasma and interstitial fluid have similar concentrations of ions but different concentrations of dissolved gases and proteins. Plasma proteins include albumins, globulins, and fibrinogen.

Table 20.1 | Functions of the Blood

Functions
1. Transport dissolved gases, bringing oxygen from the lungs to the tissues and carrying carbon dioxide from the tissues to the lungs.
2. Distribute nutrients absorbed from the digestive tract or released from storage in adipose tissue or the liver.
3. Transport metabolic wastes from peripheral tissues to sites of excretion, especially the kidneys.
4. Deliver enzymes and hormones to specific target tissues.
5. Stabilize the pH and electrolyte composition of interstitial fluids throughout the body. By absorbing, transporting, and releasing ions as it circulates, blood helps prevent regional variations in the ion concentrations of body tissues. An extensive array of buffers enables the bloodstream to deal with the acids generated by tissues, such as the lactic acid produced by skeletal muscles.
6. Prevent fluid losses through damaged vessels or at other injury sites. The **clotting reaction** seals the breaks in the vessel walls, preventing changes in blood volume that could seriously affect blood pressure and cardiovascular function.
7. Defend against toxins and pathogens. Blood transports white blood cells, specialized cells that migrate into peripheral tissues to fight infections or remove debris, and delivers antibodies, special proteins that attack invading organisms or foreign compounds. The blood also collects toxins, such as those produced by infection, and delivers them to the liver and kidneys, where they can be inactivated or excreted.
8. Stabilize body temperature by absorbing and redistributing heat. Active skeletal muscles and other tissues generate heat, and the bloodstream carries it away. When body temperature is too high, blood flow to the skin increases, as does the rate of heat loss across the skin surface. When body temperature is too low, warm blood is directed to the most temperature-sensitive organs. These changes in circulatory flow are controlled and coordinated by the cardiovascular centers in the medulla oblongata.

Figure 20.1 The Composition of Whole Blood. The percentage ranges for white blood cells indicate the normal variation seen in a count of 100 white blood cells in a healthy individual.

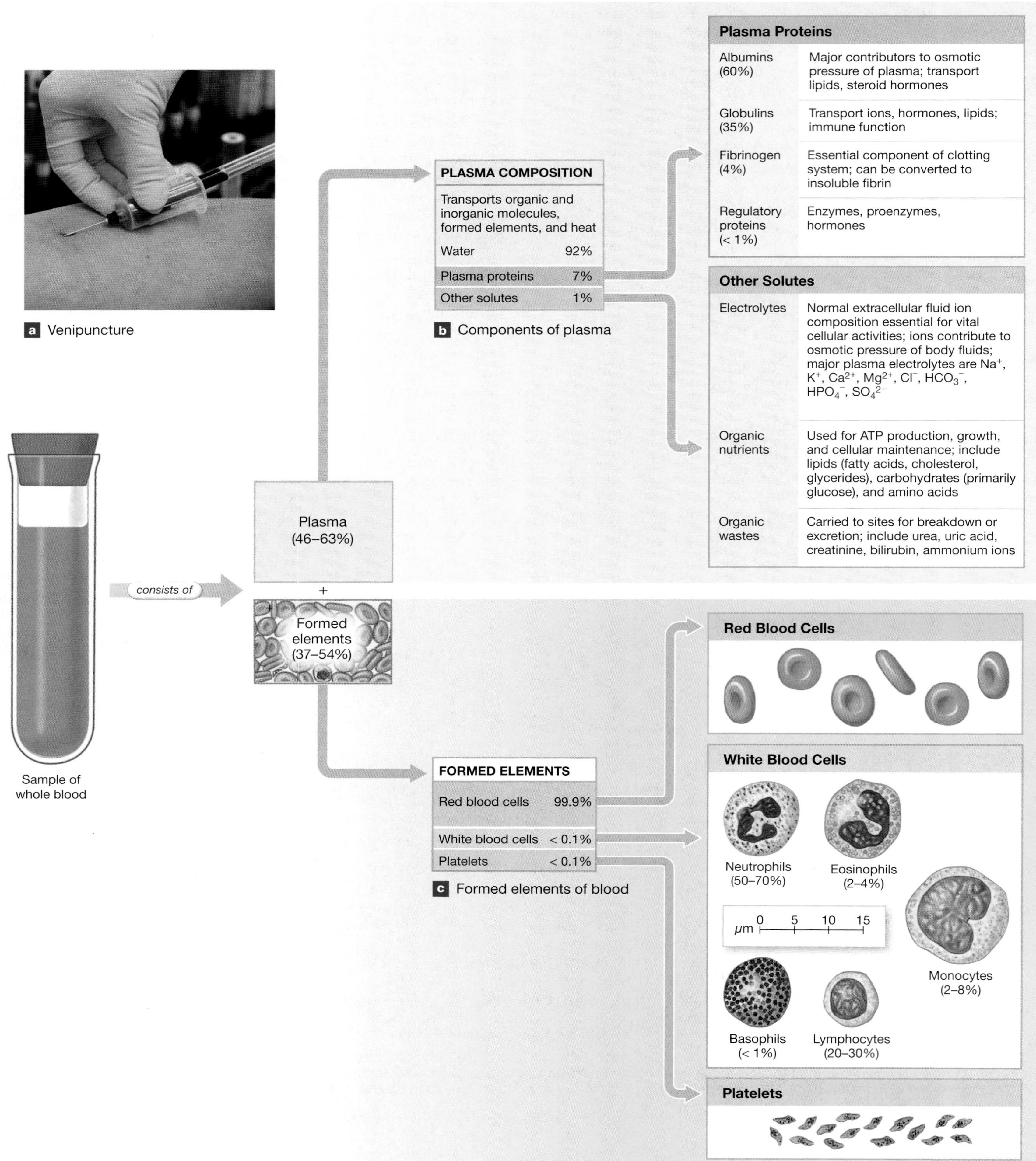

a Venipuncture
Sample of whole blood
consists of
Plasma (46–63%)
+
Formed elements (37–54%)
PLASMA COMPOSITION
Transports organic and inorganic molecules, formed elements, and heat
Water 92%
Plasma proteins 7%
Other solutes 1%
b Components of plasma
Plasma Proteins
Albumins (60%)
Major contributors to osmotic pressure of plasma; transport lipids, steroid hormones
Globulins (35%)
Transport ions, hormones, lipids; immune function
Fibrinogen (4%)
Essential component of clotting system; can be converted to insoluble fibrin
Regulatory proteins (< 1%)
Enzymes, proenzymes, hormones
Other Solutes
Electrolytes
Normal extracellular fluid ion composition essential for vital cellular activities; ions contribute to osmotic pressure of body fluids; major plasma electrolytes are Na^+, K^+, Ca^{2+}, Mg^{2+}, Cl^-, HCO_3^-, HPO_4^-, SO_4^{2-}
Organic nutrients
Used for ATP production, growth, and cellular maintenance; include lipids (fatty acids, cholesterol, glycerides), carbohydrates (primarily glucose), and amino acids
Organic wastes
Carried to sites for breakdown or excretion; include urea, uric acid, creatinine, bilirubin, ammonium ions
FORMED ELEMENTS
Red blood cells 99.9%
White blood cells < 0.1%
Platelets < 0.1%
c Formed elements of blood
Red Blood Cells
White Blood Cells
Neutrophils (50–70%)
Eosinophils (2–4%)
μm 0 5 10 15
Monocytes (2–8%)
Basophils (< 1%)
Lymphocytes (20–30%)
Platelets

20

Table 20.2 | **Composition of Whole Blood**

Component	Significance
PLASMA	
Water	Dissolves and transports organic and inorganic molecules, distributes blood cells, and transfers heat
Electrolytes	Normal extracellular fluid ion composition essential for vital cellular activities
Nutrients	Used for energy production, growth, and maintenance of cells
Organic wastes	Carried to sites of breakdown or excretion
Proteins	
Albumins	Major contributor to osmotic concentration of plasma; transport some lipids
Globulins	Transport ions, hormones, lipids
Fibrinogen	Essential component of clotting system; can be converted to insoluble fibrin
FORMED ELEMENTS	
Red blood cells (erythrocytes)	Transport gases (oxygen and carbon dioxide)
White blood cells (leukocytes)	Defend body against pathogens; remove toxins, wastes, and damaged cells
Platelets	Participate in clotting response

Interstitial fluid is formed from the blood plasma. Therefore, the two fluids are very similar. For example, the ion concentrations in plasma are similar to those of interstitial fluid but very different from the ion concentrations in cytoplasm. Two main differences between plasma and interstitial fluid involve the concentrations of dissolved gases and proteins:

1. Concentrations of dissolved oxygen and carbon dioxide: The concentration of dissolved oxygen in plasma is higher than that of interstitial fluid. As a result, oxygen diffuses out of the bloodstream and into peripheral tissues. The carbon dioxide concentration in interstitial fluid is higher than that of plasma, so carbon dioxide diffuses out of the tissues and into the bloodstream.
2. Concentration of dissolved proteins: Plasma contains significant quantities of dissolved proteins, while interstitial fluid does not. The large size and globular shapes of these plasma proteins prevent them from crossing capillary walls, and they remain trapped within the cardiovascular system.

The Plasma Proteins

Plasma proteins make up about 7 percent of plasma (**Figure 20.1**). One hundred milliliters of human plasma normally contain 6–7.8 g of soluble proteins. There are three major classes of plasma proteins: albumins (al-BŪ-minz), globulins (GLOB-ū-linz), and fibrinogen (fī-BRIN-ō-jen):

1. **Albumins** make up about 60 percent of the plasma proteins. Thus, they are the major contributors to osmotic pressure. They are also important in transporting fatty acids, thyroid hormones, some steroid hormones, and other substances.
2. **Globulins** are the second most abundant proteins and make up about 35 percent of plasma proteins. Globulins include both immunoglobulins and transport globulins. **Immunoglobulins** (im-ū-nō-GLOB-ū-linz), also called **antibodies**, aid in the body's defense. **Transport globulins** bind small ions, hormones, or compounds that either are insoluble or will be filtered out of the blood by the kidneys.
3. **Fibrinogen** makes up about 4 percent of plasma proteins. This protein is the largest of the plasma proteins and is essential for normal blood clotting. Under certain conditions fibrinogen molecules interact, forming large, insoluble strands of fibrin (FĪ-brin). Fibrin is the basic framework of a blood clot (look ahead to **Figure 20.7**, p. 537). If steps are not taken to prevent clotting in a blood sample, fibrinogen (a soluble protein) will be converted to fibrin (an insoluble protein). To prevent clotting, clotting proteins are removed, leaving a fluid known as **serum**.

Both albumins and globulins attach to lipids that are not water-soluble, such as triglycerides, fatty acids, or cholesterol. These protein-lipid combinations, called **lipoproteins** (līp-ō-PRŌ-tēnz), dissolve in plasma and can be transported as insoluble lipids to peripheral tissues.

The liver synthesizes and releases more than 90 percent of the plasma proteins. Because the liver is the primary source of plasma proteins, liver disorders alter the composition and functions of the blood. For example, some liver diseases lead to uncontrolled bleeding due to the inadequate synthesis of fibrinogen and other plasma proteins involved in blood clotting.

20.1  **CONCEPT CHECK**

1. How would slow blood flow affect the stability of your body's temperature?
2. If a person is diagnosed as being hypovolemic, how would blood pressure be affected?
3. Why does whole blood have such a high viscosity?

See the blue Answers tab at the back of the book.

20.2 | Formed Elements

▶ **KEY POINT** The formed elements are red blood cells (erythrocytes), white blood cells (leukocytes), and platelets. There are two major classes of leukocytes: granular and agranular.

The major cellular components of blood are erythrocytes (red blood cells) and leukocytes (white blood cells). There are two major classes of leukocytes: granular (with cytoplasmic granules) and agranular (without cytoplasmic granules). In addition, blood contains platelets, noncellular formed elements that function in clotting. **Table 20.3** summarizes the formed elements of blood.

Red Blood Cells (RBCs)

▶ **KEY POINT** Red blood cells (RBCs, or erythrocytes) are anucleate, biconcave discs. RBCs contain hemoglobin, a protein specialized for oxygen transport.

Red blood cells (RBCs), or **erythrocytes** (e-RITH-rō-sīts; *erythros*, red), account for a little less than half of the total blood volume (**Figure 20.1**). The percentage of whole blood containing formed elements is called the **hematocrit** (hē-MA-tō-krit). The hematocrit in an average adult man is 45 (range: 40–54); the hematocrit in an average adult woman is 42 (range: 37–47). Whole blood contains approximately 1000 times more red blood cells than white blood cells. Therefore, the hematocrit is a good approximation of the volume of red blood cells. As a result, hematocrit values are reported as the **volume of packed red cells (VPRC)**, or, simply, the **packed cell volume (PCV)**.

The number of red blood cells in an average person is staggering. One microliter (μl), or cubic millimeter (mm^3), of whole blood from a man contains approximately 5.4 million RBCs. A microliter of blood from a woman contains approximately 4.8 million RBCs. There are approximately 260 million RBCs in a single drop of whole blood, and 25 trillion (2.5×10^{13}) RBCs in the blood of an average adult!

Table 20.3 | A Review of the Formed Elements of the Blood

Formed Elements	Abundance (average per μl)	Characteristics	Functions	Remarks
Red blood cells (Erythrocytes)	5.2 million (range: 4.2–6.3 million)	Biconcave disc without a nucleus, mitochondria, or ribosomes; red color due to presence of hemoglobin molecules	Transport oxygen from lungs to tissues, and carbon dioxide from tissues to lungs	120-day life expectancy; amino acids and iron recycled; produced in red bone marrow
White blood cells (Leukocytes)	7000 (range: 5000–10,000)			
Granular leukocytes (Granulocytes)				
Neutrophils	4150 (range: 1800–7300) differential count: 50–70%	Round cell; nucleus resembles a series of beads; cytoplasm contains large, pale inclusions	Phagocytic; engulf pathogens or debris in tissues	Survive minutes to days, depending on activity; produced in red bone marrow
Eosinophils	165 (range: 0–700) differential count: 2–4%	Round cell; nucleus usually in two lobes; cytoplasm contains large granules that stain bright orange-red with acid dyes	Attack anything that is labeled with antibodies; important in fighting parasitic infections; suppress inflammation	Produced in red bone marrow
Basophils	44 (range: 0–150) differential count: <1%	Round cell; nucleus usually cannot be seen because of dense, purple-blue granules in cytoplasm	Enter damaged tissues and release histamine and other chemicals	Assist mast cells of tissues in producing inflammation; produced in red bone marrow
Agranular leukocytes (Agranulocytes)				
Monocytes	456 (range: 200–950) differential count: 2–8%	Very large, kidney bean–shaped nucleus; abundant pale cytoplasm	Enter tissues to become free macrophages; engulf pathogens or debris	Primarily produced in red bone marrow
Lymphocytes	2185 (range: 1500–4000) differential count: 20–400%	Slightly larger than RBC; round nucleus; very little cytoplasm	Cells of lymphatic system, providing defense against specific pathogens or toxins	T cells attack directly; B cells form plasma cells that secrete antibodies; produced in red bone marrow and lymphatic tissues
Platelets	350,000 (range: 150,000–500,000)	Cytoplasmic fragments; contain enzymes and proenzymes; no nucleus	Hemostasis: clump together and stick to vessel wall (platelet phase); activate intrinsic pathway of coagulation phase	Produced by megakaryocytes in red bone marrow

Structure of RBCs

Red blood cells transport oxygen and carbon dioxide within the bloodstream. They are among the most specialized cells of the body, and their anatomical specialization is apparent when you compare them with "typical" body cells **(Figure 20.2a,b)**. Each red blood cell is a biconcave disc with a thin central region and a thick outer margin **(Figure 20.2c)**. The diameter of a typical RBC is 7.7 μm. It has a maximum thickness of about 2.85 μm, and the center narrows to about 0.8 μm.

A red blood cell's unusual biconcave shape gives each RBC a very large surface area. The large surface area permits rapid diffusion between the RBC cytoplasm and the surrounding plasma. As blood circulates from the capillaries of the lungs to the capillaries in the peripheral tissues and then back to the lungs, RBCs absorb and release respiratory gases. If you estimate the total surface area of all of the RBCs in the blood of a typical adult, you would get approximately 3800 m^2—which is *2000 times the total surface area of the entire body*!

Their biconcave shape also allows RBCs to form stacks, like dinner plates. These stacks, called **rouleaux** (rū-LŌ; singular, *rouleau*; "little rolls"), repeatedly form and come apart without damaging the cells involved. One rouleau can pass through a blood vessel a little larger than the diameter of a single red blood cell **(Figure 20.2d)**. In contrast, individual RBCs would bump into the walls, clump together, and form logjams that could block the vessel.

In addition, the slender shape of red blood cells gives them considerable strength and flexibility. They can bend and flex with ease. Changing their shape allows individual cells to squeeze through small-diameter, distorted, or compressed capillaries.

RBC Life Span and Circulation

During their differentiation and maturation, red blood cells lose most of their organelles, retaining only an extensive cytoskeleton. As a result, circulating RBCs lack mitochondria, endoplasmic reticula, ribosomes, and nuclei. (We describe the process of RBC formation in a later section.) Without mitochondria, these cells obtain energy only by anaerobic metabolism, and they rely on glucose obtained from the surrounding plasma. This internal anatomy ensures that absorbed oxygen will be carried to peripheral tissues and not "stolen" by mitochondria in the RBCs. In addition, because RBCs lack a nucleus, ER, and ribosomes, they cannot use oxygen and nutrients to replace damaged enzymes or structural proteins.

Red blood cells are exposed to severe stresses. A single trip through the entire cardiovascular system takes less than 30 seconds. In this time a single RBC must (1) stack into a rouleau, (2) contort and squeeze through capillaries, and then (3) join other RBCs in a headlong rush back to the heart for another trip. With all this wear and tear and no way to replace damaged enzymes or structural proteins, a typical RBC has a short life span—approximately 120 days. After traveling about 700 miles in 120 days, either the plasma membrane ruptures or phagocytic cells detect and destroy the aging cell. About 1 percent of circulating RBCs are replaced each day, and in the process approximately 3 million new ones enter the bloodstream *each second*!

RBCs and Hemoglobin

A developing red blood cell loses all intracellular organelles not involved with its primary functions: oxygen and carbon dioxide transport. A mature red blood

Figure 20.2 **Histology of Red Blood Cells**

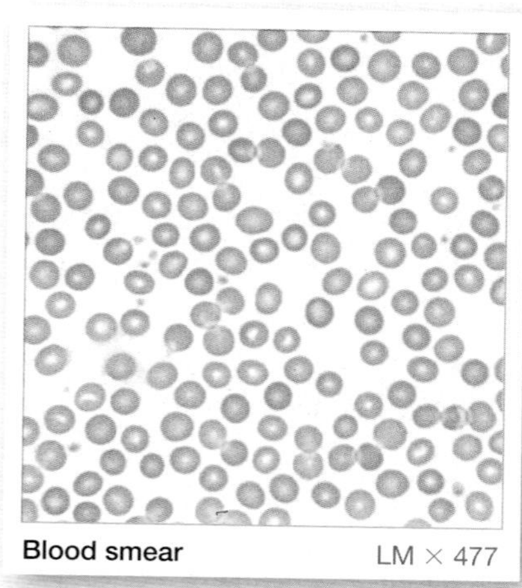

a In a histological blood smear, red blood cells appear as two-dimensional objects because they are flattened against the surface of the slide.

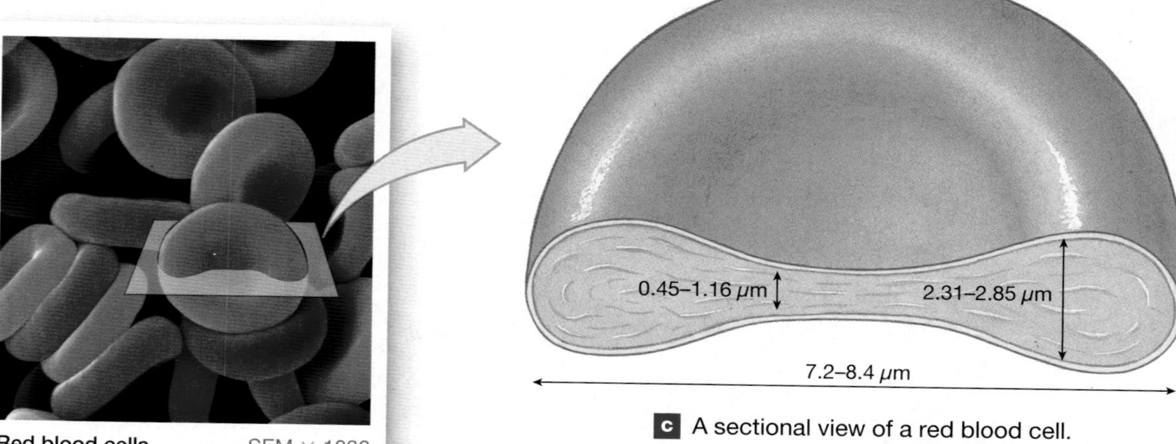

b A scanning electron micrograph of red blood cells reveals their three-dimensional structure quite clearly.

c A sectional view of a red blood cell.

Sectioned capillaries LM × 1430

Red blood cell

Rouleau (stacked RBCs)

Nucleus of endothelial cell

Blood vessels in longitudinal section

d When traveling through relatively narrow capillaries, red blood cells may stack like dinner plates, forming a rouleau.

cell consists of a plasma membrane surrounding cytoplasm that contains water (66 percent) and proteins (about 33 percent).

Hemoglobin (HĒ-mō-glō-bin) **(Hb)** molecules make up more than 95 percent of the RBC's proteins. Hemoglobin is responsible for the cell's ability to transport oxygen and carbon dioxide. It is a globular protein that contains a nonprotein red pigment. Oxygenated hemoglobin has a bright red color, while deoxygenated hemoglobin has a deep red color.

Each hemoglobin molecule has a complex shape, and this shape allows hemoglobin to carry oxygen. Each Hb molecule has two alpha (α) chains and two beta (β) chains of polypeptides. Each Hb chain contains a single molecule of **heme**, a nonprotein pigment that forms a ring **(Figure 20.3)**. Each heme unit holds an iron in such a way that the iron can interact with an oxygen molecule, forming oxyhemoglobin. Blood that contains RBCs filled with oxyhemoglobin is bright red.

The binding of an oxygen molecule to the iron in a heme unit is reversible. The iron–oxygen interaction is very weak. The iron and oxygen can easily dissociate without damaging the heme unit or the oxygen molecule. A hemoglobin molecule whose iron is not bound to oxygen is called deoxyhemoglobin. Blood containing RBCs filled with deoxyhemoglobin is dark red.

Each RBC contains about 280 million Hb molecules. Because one Hb molecule contains four heme units, each RBC can carry *more than a billion molecules of oxygen*!

Hemoglobin also carries 23 percent of the carbon dioxide transported in the blood. Rather than competing with oxygen by binding with the iron in hemoglobin, carbon dioxide binds to amino acids of the globin subunits. The binding of carbon dioxide to a globin subunit is just as reversible as the binding of oxygen to heme.

The amount of oxygen bound to hemoglobin depends mostly on the oxygen content of the plasma. When the plasma oxygen level is low, hemoglobin releases oxygen. Under these conditions, typical of peripheral capillaries, the plasma carbon dioxide level is elevated. The alpha and beta chains of hemoglobin then bind carbon dioxide, forming carbaminohemoglobin. In the lung capillaries, the plasma oxygen level is high and the carbon dioxide level is low. Upon reaching these capillaries, RBCs absorb oxygen (which is then bound to hemoglobin) and release carbon dioxide.

Blood Types

The presence or absence of specific surface antigens in the RBC plasma membrane determines an individual's **blood type**. These surface antigens are glycoproteins whose characteristics are genetically determined. The membrane of a red blood cell contains a number of **surface antigens**, or *agglutinogens* (a-glū-TIN-ō-jenz), exposed to the plasma. Red blood cells have at least 50 kinds of surface antigens, but three surface antigens are of particular importance: **A**, **B**, and **Rh** (or **D**).

The red blood cells of each person have a characteristic combination of surface antigens **(Figure 20.4)**. For example, **type A** blood has antigen A, **type B** has antigen B, **type AB** has both, and **type O** has neither A nor B. Individuals with these blood types are not evenly distributed throughout the world. The average percentages for various populations in the United States are given in **Table 20.4**.

The Rh blood group is based on the presence or absence of the Rh surface antigen. The term **Rh positive (Rh^+)** indicates the presence of this antigen, commonly called the Rh factor. The absence of this antigen is indicated as **Rh negative (Rh^-)**. In recording the complete blood type, the term Rh is usually omitted, and the types are reported as O negative (O^-), A positive (A^+), and so on.

Figure 20.3 The Structure of Hemoglobin. Hemoglobin consists of four globular protein subunits. Each subunit contains a single molecule of heme, a porphyrin ring surrounding a single ion of iron. It is the iron ion that reversibly binds to an oxygen molecule.

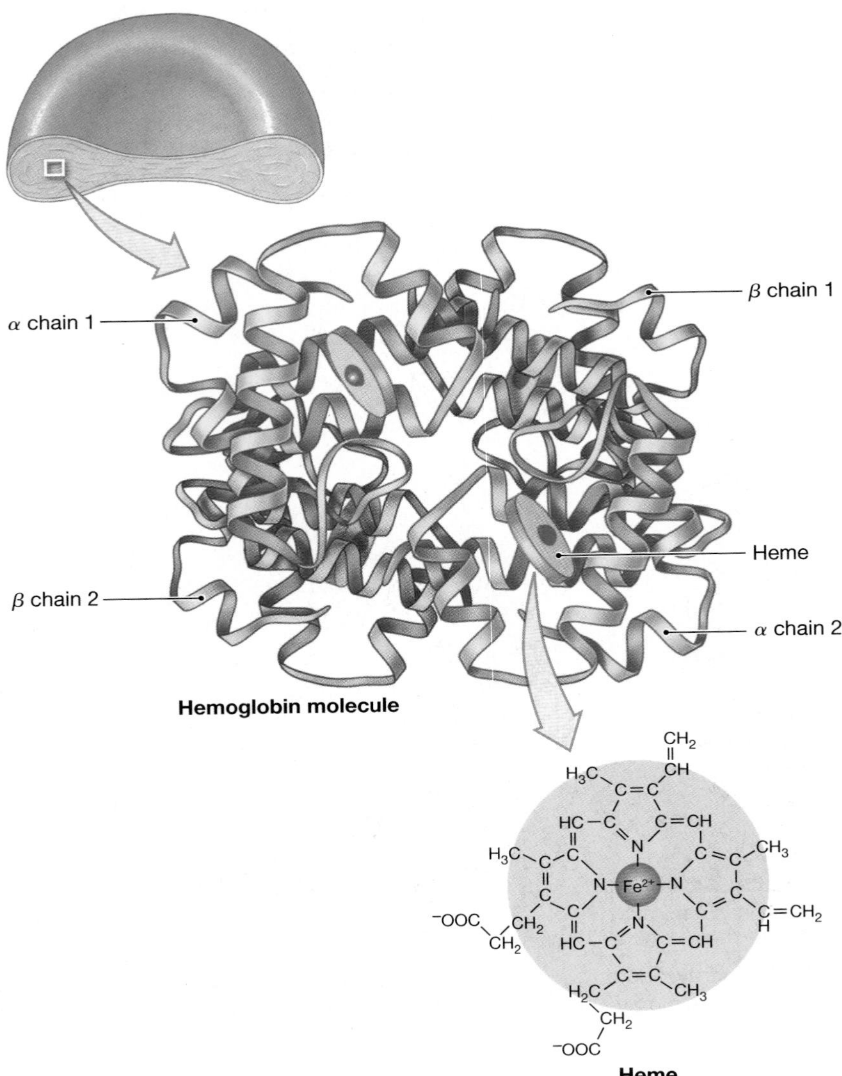

Antibodies and Cross-Reactions Your blood type must be checked before you give or receive blood. This is because your immune system ignores the surface antigens on your own red blood cells. (Chapter 23 will discuss this ability to recognize your own body cells.) However, your plasma contains antibodies that will attack "foreign" surface antigens. These antibodies are known as agglutinins (ah-GLŪ-ti-ninz). The blood of a type A, type B, or type O individual contains antibodies that attack foreign surface antigens (**Figure 20.4a**). For example, if you have type A blood, your plasma contains anti-B antibodies that will attack type B surface antigens (**Figure 20.4b**). If you are type B, your plasma contains anti-A antibodies. The RBCs of an individual with type O blood have neither A nor B surface antigens, and that person's plasma contains both anti-A and anti-B antibodies. A type AB individual has RBCs with both A and B surface antigens, and the plasma does not contain anti-A or anti-B antibodies. The presence of anti-A and/or anti-B antibodies is genetically determined. These antibodies are present throughout life, regardless of whether the person has ever been exposed to foreign RBCs.

In contrast, the plasma of an Rh-negative person does not contain anti-Rh antibodies. These antibodies are present only if the person has been **sensitized** by previous exposure to Rh^+ RBCs. Such exposure can occur accidentally during a transfusion, but it can also accompany a seemingly normal pregnancy involving an Rh^- mother and an Rh^+ fetus.

Table 20.4 | **Differences in Blood Group Distribution**

	Percentage with Each Blood Type				
Population	**O**	**A**	**B**	**AB**	**Rh^+**
United States					
Black American	49	27	20	4	95
Caucasian	45	40	11	4	85
Chinese American	42	27	25	6	100
Filipino American	44	22	29	6	100
Hawaiian	46	46	5	3	100
Hispanic American	57	31	10	2	92
Japanese American	31	39	21	10	100
Korean American	32	28	30	10	100
Native North American	79	16	4	<1	100
Native South American	100	0	0	0	100
Australian Aborigine	44	56	0	0	100

A **cross-reaction** occurs when an antibody meets its specific surface antigen (**Figure 20.4b**). First, the red blood cells clump together, a process called **agglutination** (a-glū-ti-NĀ-shun). The RBCs may also **hemolyze** (rupture). The clumps and fragments of RBCs under attack form drifting masses that can plug small blood vessels in the kidneys, lungs, heart, or brain, damaging or destroying affected tissues. To avoid cross-reactions, the blood types of the donor and recipient must be **compatible**. This involves choosing a donor whose blood cells will not undergo a cross-reaction with the plasma of the recipient.

White Blood Cells (WBCs)

▶ **KEY POINT** White blood cells (WBCs, or leukocytes) are classified into two groups: granular leukocytes (which possess cytoplasmic granules) and agranular leukocytes (which lack cytoplasmic granules).

Leukocytes (LŪ-kō-sīts; *leukos*, white), or **white blood cells (WBCs)**, are found circulating within the blood vessels of the cardiovascular system and scattered throughout peripheral tissues. Most of the leukocytes are found in peripheral tissues. White blood cells defend the body against invasion by pathogens and remove toxins, wastes, and abnormal or damaged cells. All WBCs are as large as or larger than RBCs (**Figure 20.5**). Leukocytes are divided into two groups based on the presence or absence of cytoplasmic granules: (1) **granular leukocytes**, or **granulocytes** (GRAN-ū-lō-sīts) and (2) **agranular leukocytes**, or **agranulocytes** (**Figure 20.5**).

A microliter of blood contains 5000–10,000 WBCs. A stained blood smear provides a **differential count** of the white blood cell population. The values obtained indicate the number of each type of cell in a sample of 100 white blood cells. **Table 20.3** shows the normal range for each cell type. The endings *-penia* and *-osis* indicate low and high numbers, respectively, of specific types of white blood cells. For example, *lymphopenia* means too few lymphocytes, and *lymphocytosis* means an unusually high number.

When body cells are infected or damaged, they release chemicals into the interstitial fluids that attract leukocytes. This attraction to chemical stimuli is called **chemotaxis**. WBCs are transported in the bloodstream to the invading pathogens, damaged tissues, and white blood cells already in the damaged tissues. They then cross the endothelial lining of a capillary by squeezing between adjacent endothelial cells in a process called **diapedesis**, or *emigration*.

Figure 20.4 Blood Typing. The relative frequencies of each blood type in the U.S. population are indicated in **Table 20.4**.

Type A	Type B	Type AB	Type O
Type A blood has RBCs with surface antigen A only.	**Type B** blood has RBCs with surface antigen B only.	**Type AB** blood has RBCs with both A and B surface antigens.	**Type O** blood has RBCs lacking both A and B surface antigens.
Surface antigen A	Surface antigen B		
If you have type A blood, your plasma contains anti-B antibodies, which will attack type B surface antigens.	If you have type B blood, your plasma contains anti-A antibodies, which will attack type A surface antigens.	If you have type AB blood, your plasma has neither anti-A nor anti-B antibodies.	If you have type O blood, your plasma contains both anti-A and anti-B antibodies.

a Blood type depends on the presence of surface antigens (agglutinogens) on RBC surfaces. The plasma contains antibodies (agglutinins) that will react with foreign surface antigens.

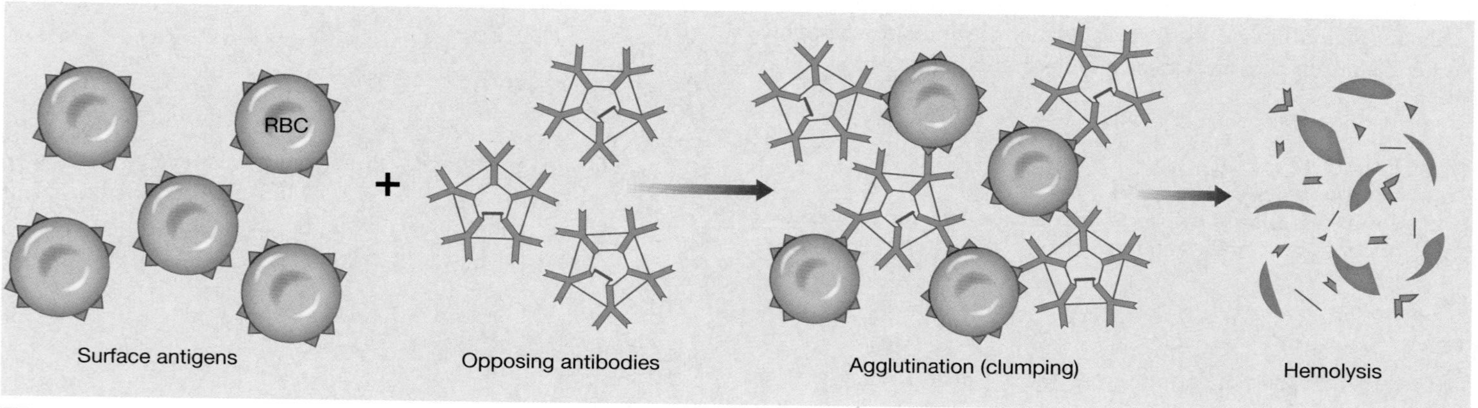

b In a cross-reaction, antibodies react with their target antigens, causing agglutination and hemolysis of the affected RBCs. In this example, anti-B antibodies encounter B surface antigens, which causes the RBCs bearing the B surface antigens to clump together and break up.

Granular Leukocytes

▶ **KEY POINT** On the basis of their staining characteristics, we classify granular leukocytes as neutrophils, eosinophils, or basophils. Neutrophils and eosinophils are phagocytic cells that are important in the immune response. Basophils release histamine and heparin.

Neutrophils

Fifty to seventy percent of circulating WBCs are **neutrophils** (NŪ-trō-filz). The cytoplasm of a neutrophil is packed with pale, neutral-staining granules containing lysosomal enzymes and bactericidal (bacteria-killing) compounds (**Figure 20.5**). Each mature neutrophil has a diameter nearly twice that of a red blood cell. The nucleus of a neutrophil is very dense and arranged into a series of lobes like beads on a string.

Neutrophils are highly mobile and are usually the first WBCs to arrive at an injury site. They are active phagocytes, specializing in attacking and digesting bacteria. Neutrophils have a short life span, surviving in the bloodstream for only 10 hours. A neutrophil dies after actively engulfing one to two dozen bacteria, but its death and breakdown release chemicals that attract other neutrophils to the site.

Eosinophils

Two to four percent of circulating WBCs are **eosinophils** (ē-ō-SIN-ō-filz; eosin + *phil*, loving). Their cytoplasmic granules stain with eosin, an acidic red dye. Eosinophils are similar in size to neutrophils but have deep red cytoplasmic granules and a bilobed (two-lobed) nucleus (**Figure 20.5**).

Eosinophils attack objects that are coated with antibodies. The number of eosinophils increases dramatically during an allergic reaction or parasitic infection. These WBCs release enzymes that reduce inflammation produced by mast cells and neutrophils, which controls the spread of inflammation to adjacent tissues.

Basophils

Basophils (BĀ-sō-filz) have numerous cytoplasmic granules that stain deep purple or blue with basic dyes (**Figure 20.5**). Basophils are rare, accounting

Figure 20.5 Histology of White Blood Cells. Histological comparison of leukocytes as seen in blood smears.

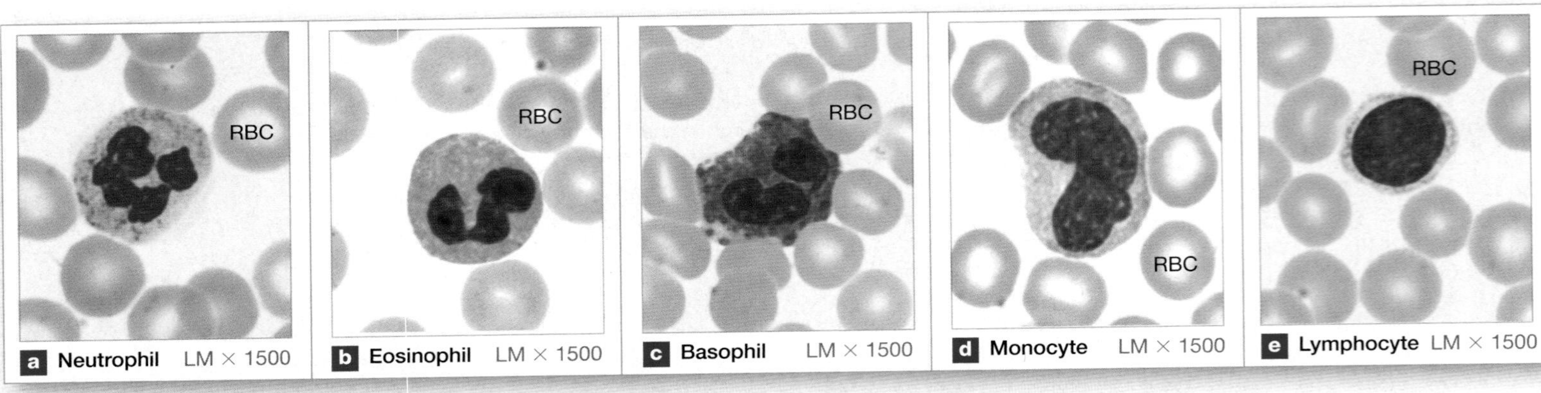

for less than 1 percent of the circulating WBC population. They migrate to sites of injury and cross the capillary endothelium into the damaged or infected tissues. There they release their granules into the interstitial fluids. The granules contain histamine and heparin. Histamine dilates blood vessels, and heparin prevents blood clotting. Stimulated basophils release these chemicals into the interstitial fluids to enhance the local inflammation initiated by mast cells. Although mast cells release the same compounds in damaged connective tissues, mast cells and basophils are distinct populations. Stimulated basophils also release other chemicals that attract eosinophils and other basophils to the area.

Agranular Leukocytes

▶ **KEY POINT** The two classes of agranular leukocytes—monocytes and lymphocytes—differ both structurally and functionally.

20

Monocytes

Two to eight percent of WBCs are **monocytes** (MON-ō-sīts). They are the largest white blood cells, with a diameter two to three times larger than an RBC. Each monocyte has a large oval or kidney bean-shaped nucleus (**Figure 20.5**).

Monocytes circulate for just a few days before entering peripheral tissues. Outside the bloodstream, monocytes are called *free macrophages* and are very different from the immobile *fixed macrophages* found in connective tissues. Free macrophages are highly mobile, phagocytic cells. They usually arrive at an injury site shortly after the first neutrophils. While phagocytizing, free and fixed macrophages release chemicals that attract and stimulate other monocytes and other phagocytic cells.

Macrophages also secrete substances that attract fibroblasts into the region. The fibroblasts then begin producing scar tissue to wall off the injured area.

Lymphocytes

Lymphocytes (LIM-fō-sīts) have very little cytoplasm, forming a thin halo around a large, round, purple-staining nucleus (**Figure 20.5**). Lymphocytes are slightly larger than RBCs and make up 20–30 percent of WBCs. Those within the bloodstream are a very small percentage of the entire lymphocyte population. Lymphocytes are the primary cells of the **lymphatic system**, a network of special vessels and organs distinct from, but connected to, those of the cardiovascular system. (We will discuss the lymphatic system in Chapter 23.)

Lymphocytes are responsible for specific immunity. Specific immunity is the ability of the body to attack invading pathogens or foreign proteins *on an individual basis*. The circulating blood contains three functional classes of lymphocytes. These classes cannot be distinguished using just a light microscope. They are as follows: (1) **T cells (T lymphocytes)** enter peripheral tissues and attack foreign cells directly. (2) **B cells (B lymphocytes)** differentiate into plasma cells (*plasmocytes*). Plasma cells secrete antibodies that attack foreign cells or proteins. (3) **Natural killer (NK) cells** are responsible for immune surveillance and the destruction of abnormal cells. These cells, sometimes known as *large granular lymphocytes*, are important in preventing cancer.

TIPS & TOOLS

The following mnemonic will help you remember the various white blood cell populations: "Never Let Monkeys Eat Bananas." **N**eutrophils, **L**ymphocytes, **M**onocytes, **E**osinophils, **B**asophils

Platelets

▶ **KEY POINT** Platelets are small, membrane-bound, anucleate cytoplasmic fragments that play an important role in blood clotting.

Platelets are flattened, membrane-enclosed cytoplasmic packets circulating in the bloodstream (look back to **Figure 20.1c**, p. 530). They are formed from cells called **megakaryocytes** (meg-a-KAR-ē-ō-sīts; *mega-*, big, + *karyon*, nucleus, + *cyte*, cell). Normal red bone marrow contains a number of megakaryocytes, which are enormous cells (up to 160 μm in diameter) with large nuclei (**Figure 20.6**). The dense nucleus is lobed or ring-shaped. The surrounding cytoplasm contains a Golgi apparatus, large numbers of ribosomes, and mitochondria. The plasma membrane communicates inwardly with an extensive membrane network that radiates throughout the peripheral cytoplasm.

During their development and growth, megakaryocytes manufacture structural proteins, enzymes, and membranes. They then begin to form platelets by shedding cytoplasm in the form of small membrane-enclosed packets that enter the circulation. A mature megakaryocyte gradually loses all of its cytoplasm, producing around 4000 platelets. Then phagocytes engulf its nucleus and break it down for recycling.

Platelets are continually replaced; an individual platelet circulates for 10–12 days before being phagocytized. A microliter of circulating blood contains approximately 350,000 platelets. At any moment one-third of the platelets in the body are held in the spleen and other vascular organs rather than in the circulating blood. These reserves are mobilized when a circulatory crisis occurs, such as severe bleeding.

Figure 20.6 Histology of Megakaryocytes and Platelet Formation. Histologically, megakaryocytes stand out in red bone marrow sections because of their enormous size and the unusual shape of their nuclei. These cells are continually shedding chunks of cytoplasm that enter the circulation as platelets.

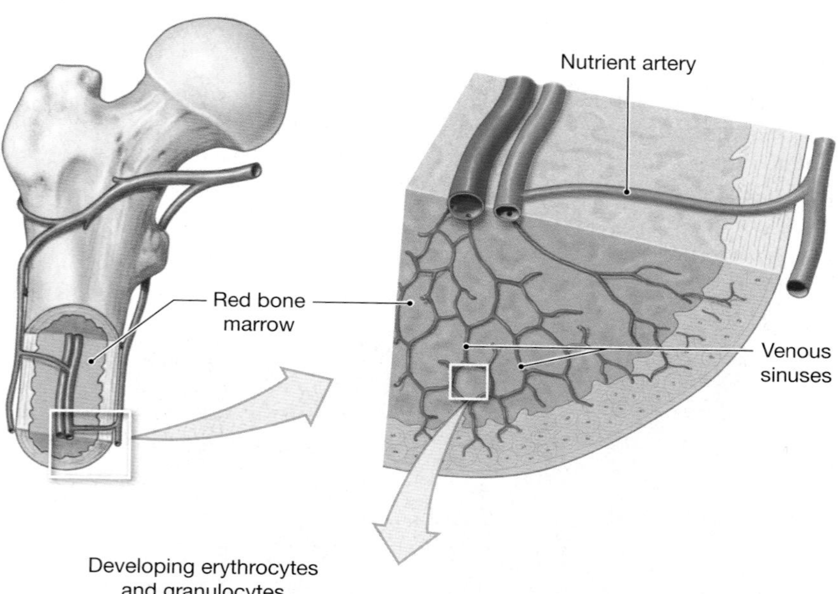

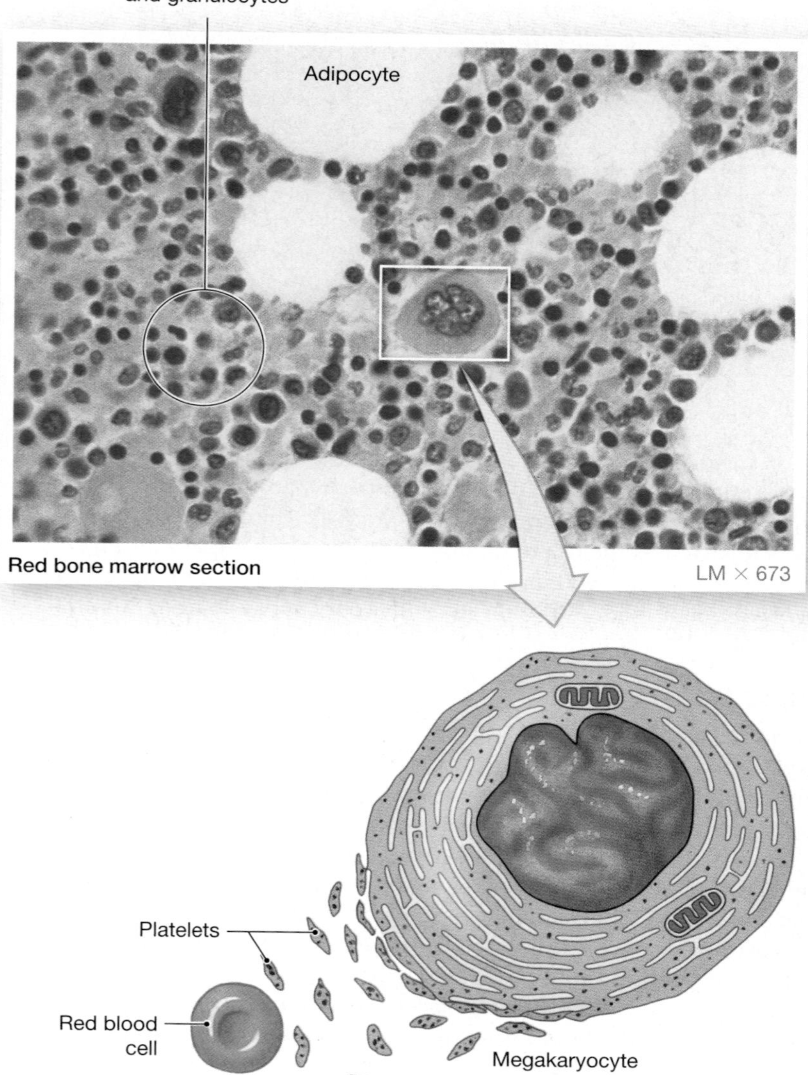

An abnormally low platelet count (80,000 per μl or less) is known as **thrombocytopenia** (throm-bō-sī-tō-PĒ-nē-a). This condition is caused by either inadequate platelet production or excessive platelet destruction. Symptoms include bleeding in the digestive tract and skin and occasional bleeding inside the CNS. In **thrombocytosis** (throm-bō-sī-TŌ-sis), an abnormally high platelet count, platelet counts often exceed 1,000,000 per μl. This indicates accelerated platelet formation in response to infection, inflammation, or cancer.

Platelets are only one participant in a vascular clotting system that also includes plasma proteins and cells and tissues of the circulatory system. **Hemostasis** (*haima*, blood, + *stasis*, halt) prevents the loss of blood through the walls of damaged vessels. This restricts blood loss and establishes a framework for tissue repairs. **Figure 20.7** shows a portion of a blood clot.

Hemostasis involves a complex chain of events, and a disorder that affects any one step can disrupt the entire process. In addition, there are general requirements; for example, a deficiency of calcium ions or vitamin K will interfere with virtually all aspects of hemostasis.

The functions of platelets include the following:

- Initiating and controlling the clotting process by releasing enzymes and other factors at the appropriate times.
- Clumping together at an injury site to form a platelet plug that slows the rate of blood loss while clotting occurs.
- Containing actin and myosin filaments that interact and shorten during the clotting process. After a blood clot has formed, the contraction of platelets shrinks the clot and pulls together the cut edges of the vessel wall.

Figure 20.7 Structure of a Blood Clot. A colorized scanning electron micrograph showing the network of fibers that forms the framework of the clot. Red blood cells trapped in the clot add to its mass and give it a red color.

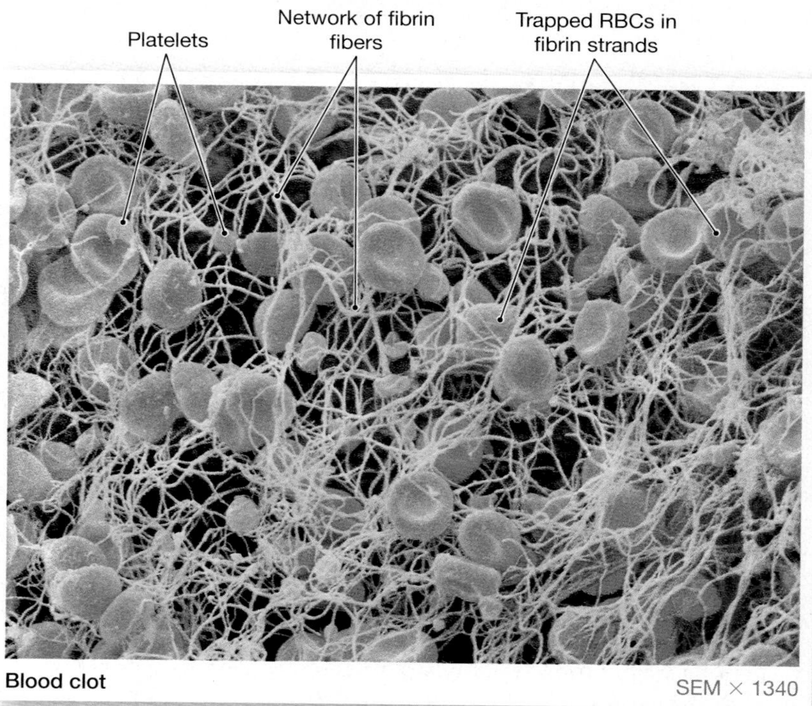

CLINICAL NOTE

Disorders of the Blood, Blood Doping, and Treatments for Blood Disorders

Sickle Cell Disease

Sickle cell disease, an inherited blood disorder, results from a mutation affecting the amino acid sequence in the beta chains of the hemoglobin (Hb) molecule. When the defective hemoglobin releases its bound oxygen, the Hb molecules cluster into rods, and the red blood cells become stiff and curved. These "sickled" RBCs become trapped in capillaries, reducing oxygen in peripheral tissues and possibly damaging organs. This is called a sickle cell crisis.

Today, sickle cell disease affects approximately 0.2 percent of Black Americans and 0.07 to 0.1 percent of Hispanic Americans. Transfusions of healthy blood can temporarily prevent complications, and treatment of affected infants with antibiotics reduces deaths due to infections.

Sickle cell disease may have developed in areas where malaria is endemic. Sickled cells infected with malaria parasites collapse and prevent the parasites from multiplying, giving people with sickle cell disease or people who are carriers for sickle cell disease immunity and thus a protective advantage against malaria.

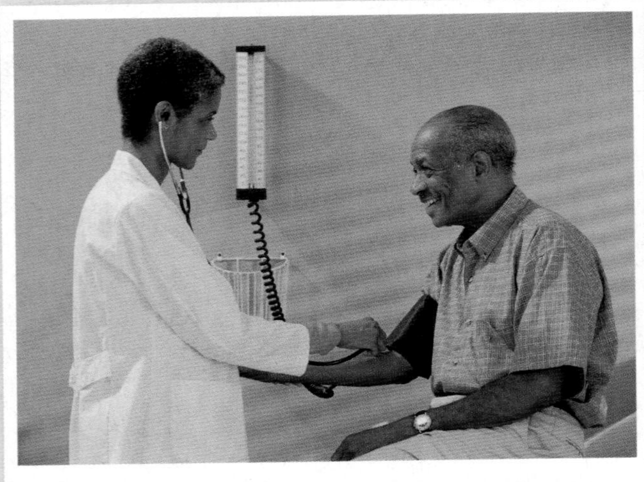

Anemia and Polycythemia

Anemia (a-NĒ-mē-a) is any condition in which the oxygen-carrying capacity of the blood is reduced, diminishing the delivery of oxygen to peripheral tissues. This reduction may be asymptomatic or cause a variety of symptoms, including lethargy, weakness, or premature muscle fatigue. Anemia may exist because the hematocrit is abnormally low or the amount of hemoglobin in the RBCs is reduced. Standard laboratory tests can differentiate between the various forms of anemia on the basis of the number, size, shape, and hemoglobin content of red blood cells. Treatment depends on the specific type.

Polycythemia is an abnormally high number of RBCs in circulating blood. Polycythemia vera is a chronic form of polycythemia of unknown cause characterized by bone marrow hyperplasia, increased blood volume, and splenomegaly. The hematocrit may reach 80–90, at which point the tissues become oxygen-starved because red blood cells block the smaller vessels.

Most cases of polycythemia involve people aged 60–80. There are several treatment options, but no cure as yet. The simplest treatment is periodic phlebotomy, or drawing off a unit of blood.

Hemophilia

Hemophilia (hē-mō-FĒL-ē-a) is an inherited blood disorder characterized by inadequate production of clotting factors. The condition affects about 1 person in 10,000, and most are males. The severity of hemophilia varies. In severe cases, even the slightest physical stress causes excessive bleeding, and internal bleeding occurs spontaneously in joints and around muscles.

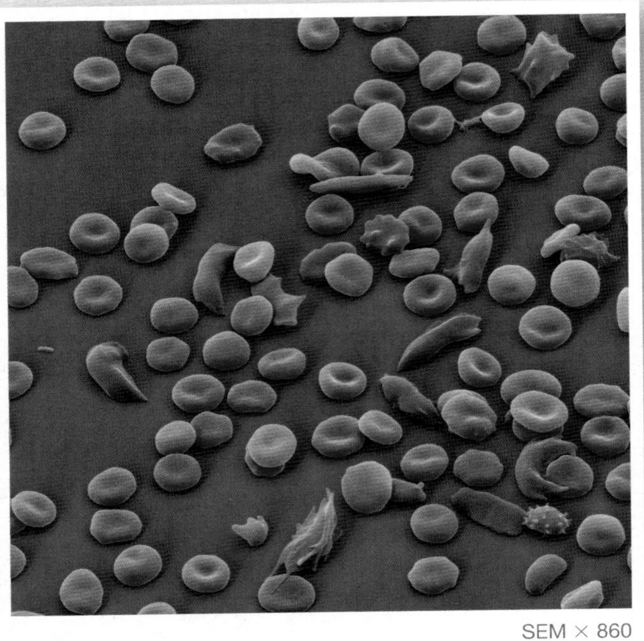
SEM × 860

Transfusions of clotting factors often reduce or control the symptoms of hemophilia, but plasma samples from many individuals must be combined to obtain enough clotting factors. This process is expensive and increases the risk of blood-borne infections. Gene-splicing techniques can be used to manufacture the clotting factor most often involved (factor VIII). Although currently supplies are limited, this procedure should eventually provide a safer treatment.

Blood Doping

Blood doping is the use of RBC transfusions or RBC-enhancing drugs (erythropoietin) to improve athletic performance. Blood doping in various forms has become widespread among competitive athletes.

One procedure entails removing whole blood from the athlete in the weeks before an event. The packed red cells are separated from the plasma and stored. In the meantime, the athlete's red bone marrow replaces the lost cells. Immediately before the event, the packed red cells are reinfused, increasing the hematocrit. The objective is to raise the oxygen-carrying capacity of the blood, thereby increasing performance. However, this also places the athlete's heart under tremendous strain. The long-term effects are unknown, but this practice carries a significant risk of stroke or heart attack. Training at high altitudes is a safer and currently acceptable alternative.

CLINICAL NOTE (*continued*)

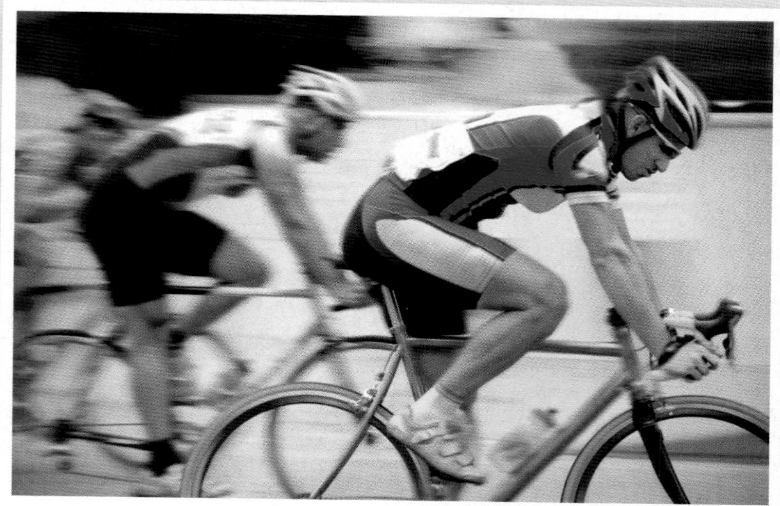

Treatment of Blood Disorders

Transfusions

A **transfusion** is the transfer of blood or blood components from one person to another. Transfusions of whole blood are most often used to restore blood volume after massive hemorrhaging has occurred. An **exchange transfusion** is the removal of most of a person's blood and simultaneous replacement with an equal amount from a donor. This may be necessary to treat acute drug poisoning or hemolytic disease of the newborn (maternal-child Rh incompatibility).

The blood is obtained under sterile conditions from carefully screened donors. It is tested for infectious bacteria and viruses and discarded if pathogens are detected. Whole blood is treated to prevent clotting and stabilize the red blood cells and then is refrigerated. Chilled whole blood remains usable for around 3.5 weeks. For longer storage, the blood must be fractionated. The red blood cells are separated from the plasma, and if necessary they may be frozen after treatment with a special antifreeze solution. The plasma can then be stored chilled, frozen, or freeze-dried. This procedure permits long-term storage of rare blood types that might otherwise be unavailable for emergencies.

Fractionated blood has many uses. **Packed red blood cells (PRBCs)**, whole blood without the plasma, are preferred for cases of anemia, in which blood volume may be close to normal but its oxygen-carrying capabilities are low. Plasma may be administered to patients who are losing massive amounts of fluid, such as after severe burns.

In the United States, the demand for blood or blood components often exceeds the supply. Moreover, there is increasing concern about the risk of contaminated blood infecting recipients with hepatitis viruses or HIV (the virus that causes AIDS). As a result, transfusion practices have changed. In general, fewer units of blood are now administered. There has also been an increase in **autologous transfusion**, in which blood is removed from a person, stored, and later transfused back into the same person when needed, such as during surgery. Moreover, technology permits the reuse of blood "lost" during surgery. The blood is collected and filtered, the platelets are removed, and the remainder of the blood is reinfused into the patient.

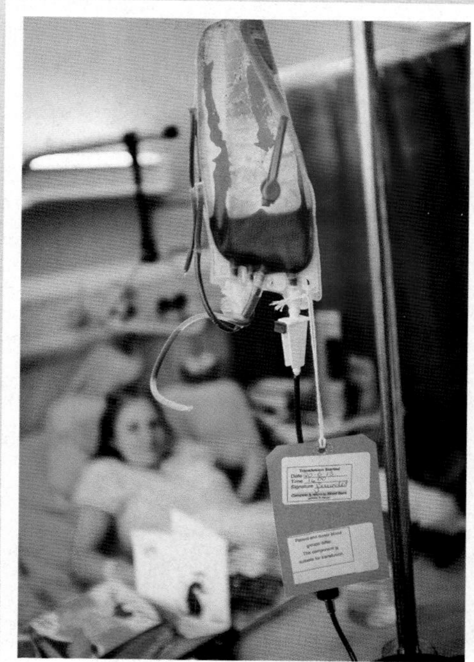

Plasma Expanders

Plasma expanders are solutions used for transfusion in hemorrhage or shock as a substitute for plasma. They increase blood volume temporarily, over a period of hours, while preparing for a transfusion of whole blood. Plasma expanders contain large carbohydrate molecules, rather than dissolved proteins, to maintain proper osmolarity. Although these carbohydrates are not metabolized, phagocytes gradually remove them and the blood volume steadily decreases. Plasma expanders are easily stored and transported, and their sterile preparation ensures that there are no problems with viral or bacterial contamination. Although they provide a temporary solution to hypovolemia (low blood volume), plasma expanders do not increase the amount of oxygen delivered to peripheral tissues.

20.2 CONCEPT CHECK

4 If the hematocrit value of a woman's blood is 42, what percentage of red blood cells is present in her blood?

5 How does the shape of red blood cells aid in blood flow?

6 What type of white blood cells appear in the greatest number in an infected cut?

7 What is the function of the granules in basophils?

See the blue Answers tab at the back of the book.

20.3 | Hemopoiesis

Hemopoiesis (hēm-ō-poy-Ē-sis) is the process of blood cell formation. Stem cells, called **hematopoietic stem cells**, ultimately give rise to all blood cells by a process outlined in **Figure 20.8**.

Erythropoiesis

▶ **KEY POINT** In adults, erythropoiesis occurs primarily within the red marrow of bones. Erythropoietin (EPO) regulates this process.

Figure 20.8 The Origins and Differentiation of Formed Elements. Hematopoietic stem cells give rise to both myeloid and lymphatic stem cells. Myeloid stem cells produce progenitor cells, which divide to produce the various classes of blood cells.

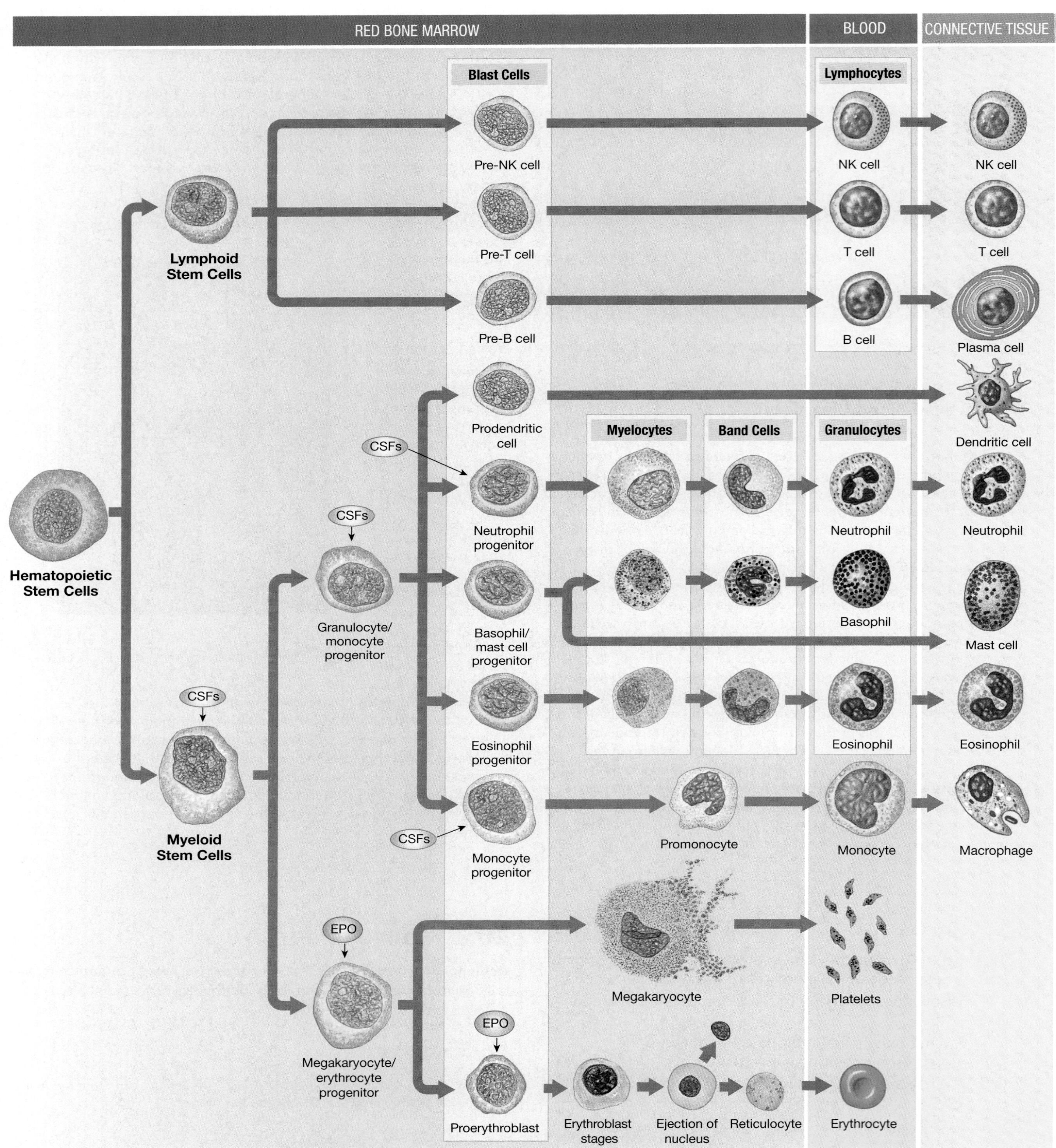

20

Erythropoiesis (e-rith-rō-poy-Ē-sis) refers specifically to the formation of erythrocytes (**Figure 20.8**). Erythropoiesis begins early in embryonic development. In adults, the **red bone marrow** is the primary site of blood cell formation. Red bone marrow is found in portions of the vertebrae, sternum, ribs, skull, scapulae, pelvis, and proximal limb bones. Under extreme conditions the fatty **yellow bone marrow** in other bones can be converted to red bone marrow. For example, this conversion occurs after a severe blood loss, increasing the rate of red blood cell formation. For normal erythropoiesis, the RBC-forming tissues must receive adequate supplies of amino acids, iron, and **vitamin B_{12}**, a vitamin obtained from dairy products and meat.

Erythropoiesis is regulated by the hormone **erythropoietin (EPO)**. ⮌ p. 518 Erythropoietin is a glycoprotein formed by the kidneys and liver under hypoxic (low-oxygen) conditions. Erythropoietin has two major effects:

1. It stimulates cell division rates in erythroblasts and in the stem cells that produce erythroblasts.
2. It speeds up the maturation of RBCs, mainly by accelerating Hb synthesis. Under maximum EPO stimulation, red bone marrow can increase RBC formation tenfold, to about 30 million cells per second.

A maturing red blood cell passes through a series of developmental stages as seen in **Figure 20.8**.

Leukopoiesis

▶ **KEY POINT** Leukopoiesis is the formation of white blood cells. In adults, leukopoiesis occurs in the red bone marrow.

The process of WBC production is called **leukopoiesis** (lū-kō-POY-eh-sis). It begins early in embryonic development. Stem cells responsible for leukopoiesis originate in the red bone marrow (**Figure 20.8**). Stem cells responsible for producing lymphocytes, a process called **lymphopoiesis**, also originate in the red bone marrow. Some lymphocytes are derived from lymphoid stem cells that remain in red bone marrow. These lymphocytes differentiate into either B cells or NK cells.

Many of the lymphoid stem cells that produce lymphocytes migrate from the red bone marrow to peripheral **lymphatic tissues**, including the thymus, spleen, and lymph nodes. As a result, lymphocytes are produced in these organs as well as in the red bone marrow. Lymphoid stem cells migrating to the thymus mature into T cells. (We will consider formation of lymphocytes in more detail in Chapter 23.)

Factors regulating lymphocyte maturation are not completely understood. However, prior to maturity, hormones of the thymus gland promote the differentiation and maintenance of T cell populations. Several hormones, called **colony-stimulating factors (CSFs)**, are involved in regulating other WBC populations. Commercially available CSFs are now used to stimulate the production of WBCs in individuals undergoing cancer chemotherapy.

20.3 CONCEPT CHECK

8 What are the two main effects of erythropoietin?
9 What is the function of hematopoietic stem cells?
10 What is the next stage of red blood cell maturation after ejection of the nucleus?
11 What cell sheds cytoplasmic packets to produce platelets?

See the blue Answers tab at the back of the book.

Study Outline

20

Introduction p. 529

- The cardiovascular system provides a mechanism for the rapid transport of nutrients, waste products, and cells within the body.

20.1 | Functions and Composition of the Blood p. 529

- **Blood** is a specialized connective tissue. Its functions include (1) transporting dissolved gases, (2) transporting and distributing nutrients, (3) transporting metabolic wastes, (4) transporting and delivering enzymes and hormones, (5) stabilizing the pH and electrolyte composition of interstitial fluids, (6) restricting fluid losses through damaged vessels or injuries via the **clotting reaction**, (7) defending the body against toxins and pathogens, and (8) stabilizing body temperature by absorbing and redistributing heat. *(See Figure 20.1 and Tables 20.1 and 20.2.)*
- Blood consists of two components: **plasma**, the liquid matrix of blood, and **formed elements**, which are **red blood cells** (**RBCs** or **erythrocytes**), **white blood cells** (**WBCs** or **leukocytes**), and **platelets**. The plasma and formed elements constitute **whole blood**, which can be **fractionated** for analytical or clinical purposes. *(See Figure 20.1 and Table 20.2.)*
- There are 4–6 liters of whole blood in an average adult. The terms **hypovolemic**, **normovolemic**, and **hypervolemic** refer to low, normal, and excessive blood volume, respectively.

Plasma p. 529

- Plasma accounts for about 55 percent of the volume of blood; roughly 92 percent of plasma is water. *(See Figure 20.1 and Table 20.2.)*

Differences between Plasma and Interstitial Fluid p. 529

- Plasma differs from interstitial fluid because it has a higher dissolved oxygen concentration and large numbers of dissolved proteins. There are three classes of plasma proteins: albumins, globulins, and fibrinogen. *(See Figures 20.1 and 20.7 and Table 20.2.)*
- **Albumins** constitute about 60 percent of plasma proteins. **Globulins** constitute roughly 35 percent of plasma proteins: They include **immunoglobulins** (**antibodies**), which attack foreign proteins and pathogens, and **transport globulins**, which bind ions, hormones, and other compounds. **Fibrinogen** molecules function in the clotting reaction by interacting to form fibrin; removing fibrinogen from plasma leaves a fluid called **serum**. When albumins or globulins become attached to lipids, they form lipoproteins, which are carried in the circulatory system until the lipids are delivered to the tissues. *(See Table 20.2.)*

20.2 | Formed Elements p. 531

Red Blood Cells (RBCs) p. 531

- **Red blood cells** (**RBCs**), or **erythrocytes**, account for slightly less than half the blood volume. The **hematocrit** value indicates the percentage of whole blood occupied by cellular elements. Since blood contains about 1000 RBCs for each WBC, this value closely approximates the volume of RBCs. *(See Figures 20.2–20.4.)*

- RBCs transport oxygen and carbon dioxide within the bloodstream. They are highly specialized cells with large surface-to-volume ratios. Each RBC is a biconcave disc. This shape gives RBCs a large surface area, allowing for rapid diffusion of gases and the ability to form stacks (called **rouleaux**) that can pass easily through small vessels.
- Because RBCs lack mitochondria, ribosomes, and nuclei, they are unable to perform normal maintenance operations, so they usually degenerate after about 120 days in the circulation. Damaged or dead RBCs are recycled by phagocytes. *(See Figure 20.2 and Table 20.3.)*
- Molecules of **hemoglobin** (**Hb**) account for more than 95 percent of the RBCs' proteins. Hemoglobin gives RBCs the ability to transport oxygen and carbon dioxide. Hemoglobin is a globular protein formed from four subunits. Each subunit contains a single molecule of **heme**, which holds an iron ion that can reversibly bind an oxygen molecule. At the lungs, carbon dioxide diffuses out of the blood and oxygen diffuses into the blood. In the peripheral tissues, the opposite occurs: Oxygen diffuses out of the blood and carbon dioxide diffuses into the blood. *(See Figure 20.3.)*
- One's **blood type** is determined by the presence or absence of specific **surface antigens** (*agglutinogens*) in the RBC plasma membrane: **A**, **B**, and **Rh** (**D**). **Type A** blood has surface antigen A, **type B** blood has surface antigen B, **type AB** has both, and **type O** has neither. **Rh-positive** blood has the Rh surface antigen, and **Rh-negative** does not. Antibodies specific to these surface antigens are called agglutinins. Antibodies within a person's plasma will react with RBCs bearing foreign surface antigens, causing a **cross-reaction**. *(See Figure 20.4 and Table 20.4.)*

20

White Blood Cells (WBCs) p. 534

- **White blood cells** (**WBCs**), or **leukocytes**, defend the body against pathogens and remove toxins, wastes, and abnormal or damaged cells. The two classes of WBCs are granular leukocytes (granulocytes) and agranular leukocytes (agranulocytes). *(See Figure 20.5.)*
- A stained blood smear provides a **differential count** of the white blood cell population. The word endings *–penia* and *–osis* are used to indicate low or high numbers, respectively, of specific types of white blood cells.
- Leukocytes show **chemotaxis** (the attraction to specific chemicals) and **diapedesis**, or *emigration* (the ability to move through vessel walls).

Granular Leukocytes p. 535

- **Granular leukocytes** (**granulocytes**) are subdivided into **neutrophils**, **eosinophils**, and **basophils**. Fifty to seventy percent of circulating WBCs are neutrophils, which are highly mobile phagocytes. The much less common eosinophils are phagocytic cells, which are attracted to foreign compounds that have reacted with circulating antibodies. The relatively rare basophils migrate to damaged tissues and release histamines, aiding the inflammation response. *(See Figure 20.5 and Table 20.3.)*

Agranular Leukocytes p. 536

- **Agranular leukocytes** (**agranulocytes**) are subdivided into **monocytes** and **lymphocytes**. Monocytes migrating into peripheral tissues become free macrophages, which are highly mobile, phagocytic cells. Lymphocytes, the primary cells of the **lymphatic system**, consist of **T cells** (which enter peripheral tissues and attack foreign cells directly), **B cells** (which produce antibodies), and **NK cells** (which destroy abnormal tissue cells). *(See Figure 20.5 and Table 20.3.)*

Platelets p. 536

- **Platelets** are not cells but are membrane-enclosed packets of cytoplasm.
- **Megakaryocytes** are enormous cells in the bone marrow that release packets of cytoplasm (platelets) into the circulating blood. The functions of platelets include (1) transporting chemicals important to the clotting process, (2) forming a temporary patch in the walls of damaged blood vessels, and (3) causing contraction after a clot has formed in order to reduce the size of the break in the vessel wall. *(See Figures 20.6 and 20.7 and Table 20.3.)*

20.3 | Hemopoiesis p. 539

- **Hemopoiesis** is the process of blood cell formation. Stem cells called **hematopoietic** divide to form all of the blood cells. *(See Figure 20.8.)*

Erythropoiesis p. 539

- **Erythropoiesis**, the formation of erythrocytes, occurs mainly within the **red bone marrow** in adults. RBC formation increases under the influence of **erythropoietin** (**EPO**). Stages in RBC development include erythroblasts and reticulocytes. *(See Figure 20.8.)*

Leukopoiesis p. 541

- **Leukopoiesis**, the formation of white blood cells, begins in red bone marrow. Stem cells in red bone marrow produce granular leukocytes and monocytes. Stem cells responsible for **lymphopoiesis** (production of lymphocytes) also originate in red bone marrow, but many migrate to peripheral lymphatic tissues. *(See Figure 20.8.)*
- Factors that regulate lymphocyte maturation are not completely understood. Several **colony-stimulating factors** (**CSFs**) are involved in regulating other WBC populations.

Chapter Review

For answers, see the blue Answers tab at the back of the book.

Level 1 Reviewing Facts and Terms

1. The five figures to the right are labeled a through e. Fill in the blanks with the proper identification.
   (a) __________
   (b) __________
   (c) __________
   (d) __________
   (e) __________

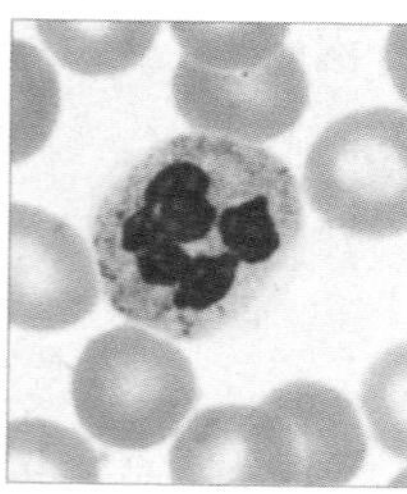
a

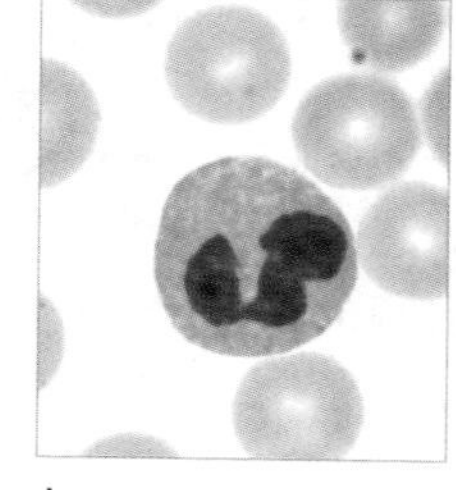
b

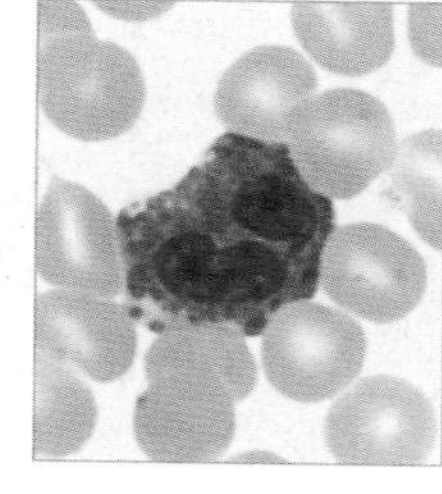
c

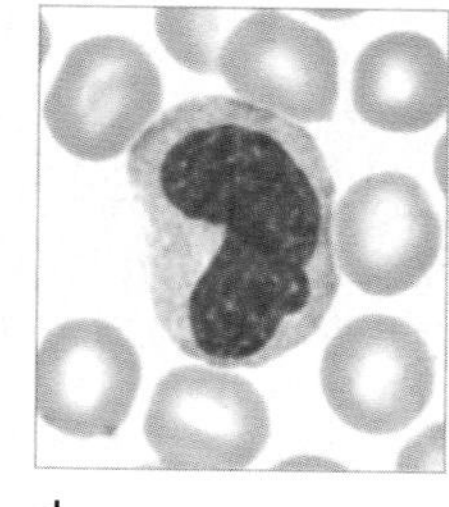
d

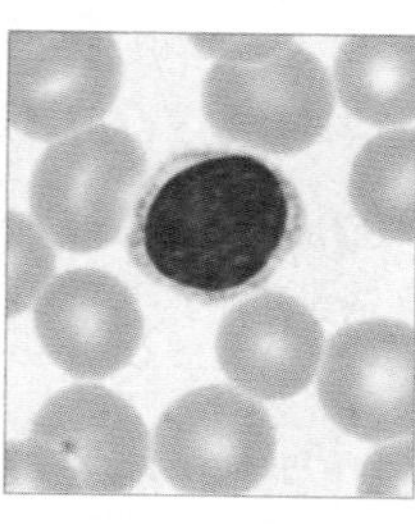
e

2. Functions of the blood include
   (a) transport of nutrients and wastes.
   (b) regulation of pH and electrolyte concentrations.
   (c) restricting fluid loss.
   (d) all of the above.
3. The most common formed elements in the blood are the
   (a) platelets.
   (b) white blood cells.
   (c) proteins.
   (d) red blood cells.
4. The most abundant proteins in blood are
   (a) globulins.
   (b) albumins.
   (c) fibrinogens.
   (d) lipoproteins.
5. The major classes of white blood cells include
   (a) erythrocytes and platelets.
   (b) granular and agranular cells.
   (c) fibrinogens and collagen fibers.
   (d) macromolecules and colloids.
6. Stem cells responsible for the production of white blood cells originate in the
   (a) liver.
   (b) thymus.
   (c) spleen.
   (d) red bone marrow.
7. Which of the following statements concerning red blood cells (RBCs) is *not* true?
   (a) RBCs are biconcave discs.
   (b) RBCs lack mitochondria.
   (c) RBCs have a large nucleus.
   (d) RBCs can form stacks called rouleaux.
8. The primary function of hemoglobin is to
   (a) store iron.
   (b) transport glucose.
   (c) give RBCs their color.
   (d) carry oxygen to peripheral tissues.
9. People with type A blood have
   (a) A surface antigens on their red blood cells.
   (b) B surface antigens in their plasma.
   (c) anti-A antibodies in their plasma.
   (d) anti-O antibodies in their plasma.
10. The white blood cells that increase in number during an allergic reaction or in response to parasitic infections are the
    (a) neutrophils.
    (b) eosinophils.
    (c) basophils.
    (d) monocytes.
11. Platelets are
    (a) large cells that lack a nucleus.
    (b) small cells that lack a nucleus.
    (c) fragments of cells.
    (d) small cells with an irregular-shaped nucleus.

Level 2 Reviewing Concepts

1. How does the reaction of an Rh-positive or Rh-negative blood type differ from that of types A, B, and O?
   (a) There are no significant differences; these blood types react all the same way.
   (b) The blood of an Rh-positive individual contains Rh-positive surface antigens, and the blood of an Rh-negative individual contains Rh-negative surface antigens.
   (c) The blood of an Rh-negative individual contains anti-Rh surface antigens only if he or she has been sensitized by previous exposure to Rh-positive erythrocytes.
   (d) The response is greater in a manner inverse to the amount of different Rh blood administered to the individual.
2. Why does the lack of mitochondria make an erythrocyte more efficient at transporting oxygen?
   (a) Since an erythrocyte transports gases passively, mitochondria would be useless, occupying valuable space within the cell.
   (b) Mitochondria require a large amount of energy to function, and so an erythrocyte lacking them has more energy to transport oxygen.
   (c) Since an erythrocyte transports gases passively, ATP is not needed for active transport processes, and therefore mitochondria are not needed and the energy mitochondria require is used to transport oxygen.
   (d) Without mitochondria, the erythrocyte will not use the oxygen it absorbs and can therefore carry all of it to peripheral tissues.
3. Iron deficiency results in which of the following?
   (a) decreased leukocyte count
   (b) decreased monocyte count
   (c) anemia
   (d) polycythemia
4. What is the volume of packed red cells, and why is it sometimes called "packed cell volume"?
5. What is the function of the clotting reaction?
6. What is the fate of megakaryocytes?
7. What are lipoproteins, and what is their function in the blood?
8. Can a person with type O blood receive type AB blood? Why or why not?

Level 3 Critical Thinking

1. Why do athletes, several months before a competition, often move to elevations higher than those at which they will compete?
2. Mononucleosis is a disease that can cause an enlarged spleen because of increased numbers of phagocytic and other cells. Common symptoms include pale complexion, a tired feeling, and a lack of energy, sometimes to the point of not being able to get out of bed. What causes these symptoms?

CLINICAL CASE WRAP-UP

A Surplus of WBCs

Danny is suffering from acute lymphocytic (*lymphoblastic*) leukemia (ALL), a cancer of lymphocytes. It is characterized by increased numbers of abnormal WBCs in the red bone marrow and peripheral blood. In ALL, cancerous lymphatic stem cells take over the red bone marrow. It is the most common type of childhood cancer.

Normally, lymphoblasts mature into lymphocytes before being released into the blood. But in Danny's case, many abnormal, immature lymphoblasts are released. These cells do not function as normal lymphocytes, and they build up and crowd out healthy cells. Danny has too few red blood cells and very few platelets. This is why he is so pale and bleeds easily. The abnormal lymphoblasts have also invaded his spinal cord, brain, thymus, liver, and spleen.

The good news is that with aggressive treatment, Danny has a 95 percent chance of going into remission. Chemotherapy will be given orally, intravenously, and intrathecally (directly into his cerebrospinal fluid). And, if all else fails, Danny has four siblings who could donate bone marrow for a transplant. Even though Danny's chemotherapy will make him bald for a while, bald is beautiful.

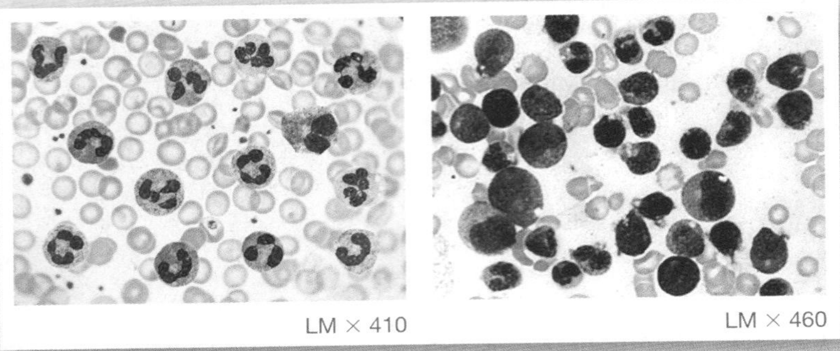

Danny's peripheral blood smear shows very few RBCs and many immature lymphoblasts.

1. If Danny were suffering from acute myeloid leukemia, what type of white cell would you predominantly see in his blood smear?
2. How many white blood cells and what percentage of lymphocytes are seen in the blood of healthy children?

See the blue Answers tab at the back of the book.

Related Clinical Terms

blood bank: Place where blood is collected, typed, separated into components, stored, and prepared for transfusion to recipients.

dyscrasia: An abnormal condition, especially of the blood.

hematology: The science concerned with the medical study of blood and blood-producing organs.

myeloproliferative disorders: A group of slow-growing blood cancers, including chronic myelogenous leukemia, characterized by large numbers of abnormal RBCs, WBCs, or platelets growing and spreading in the bone marrow and the peripheral blood.

phlebotomist: Medical technician who extracts blood by venipuncture for treatment or for laboratory analysis.

septicemia: Systemic toxic illness due to bacterial invasion of the bloodstream from a local infection. Signs and symptoms include chills, fever, and exhaustion. The disorder is treated with massive doses of antibiotics. Also known as blood poisoning.

thrombolytic: An agent that causes the breakup of a thrombus (clot).

21 The Cardiovascular System

The Heart

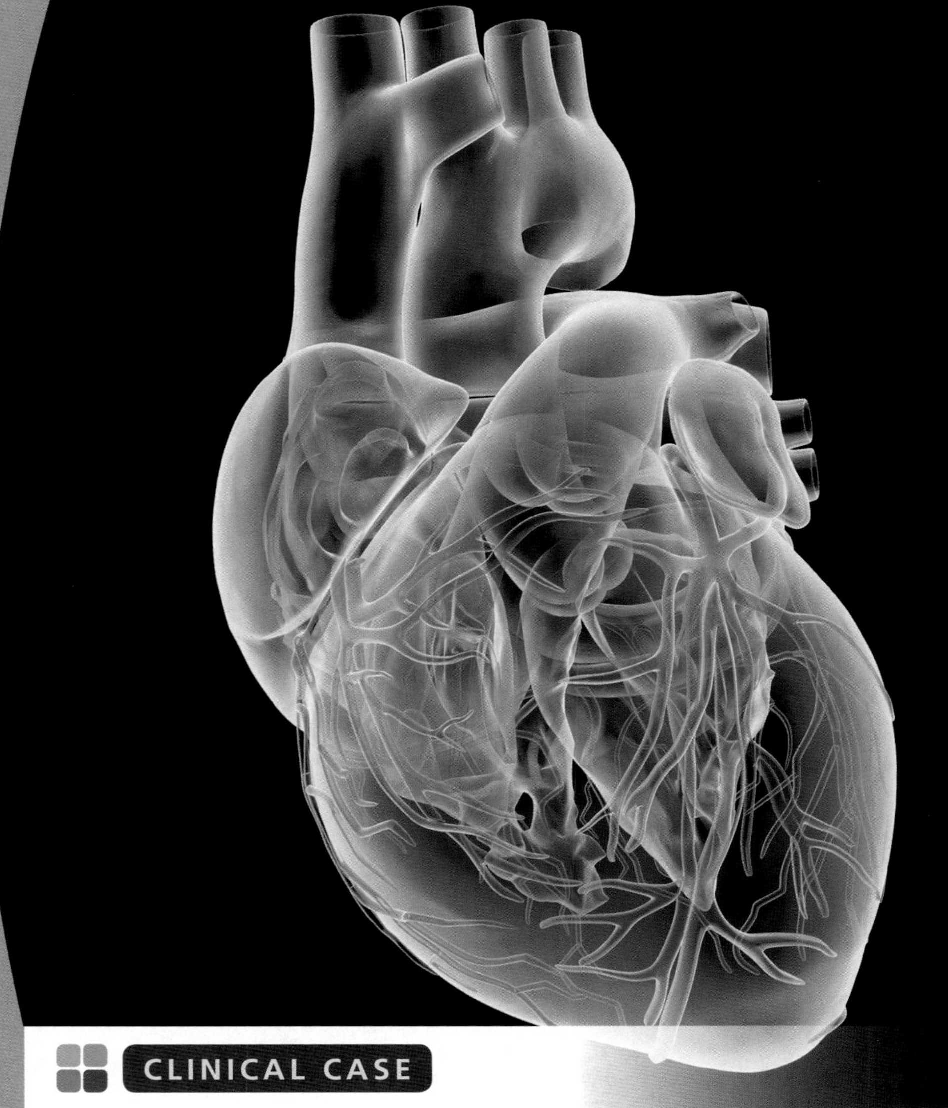

Learning Outcomes

These Learning Outcomes correspond by number to this chapter's sections and indicate what you should be able to do after completing the chapter.

21.1 Compare and contrast the pulmonary and systemic circuits. p. 546

21.2 Outline the anatomy of the pericardium. p. 546

21.3 Describe the macroscopic and microscopic anatomy of the heart wall. p. 548

21.4 Outline how the heart is orientated within the thoracic cavity and explain the superficial anatomy of the heart. p. 550

21.5 Compare and contrast the anatomy of the four chambers of the heart. p. 553

21.6 Compare and contrast the anatomy of the right and left coronary arteries. p. 555

21.7 Explain what pacemaker cells are and the role they play in coordinating the cardiac cycle. p. 559

21.8 Outline the events of the cardiac cycle. p. 562

21.9 Outline how the heart rate is modified by the autonomic nervous system. p. 562

CLINICAL CASE

A Broken Heart

A busy attorney and primary caregiver for her terminally ill mother, 57-year-old Ellen can't remember the last time she could relax. Today her mother has been rushed to the hospital again. As her mother is wheeled away for yet another x-ray, Ellen is nearly paralyzed with grief. Suddenly, she becomes weak, dizzy, and nauseated. She feels a heavy, crushing pain in her chest and begins sweating profusely. Her heart is thumping as if it's skipping beats. She collapses into a hospital chair and vomits onto the floor.

A nurse comes to the door, takes one look at Ellen, and calls a code blue—a warning that someone is experiencing a cardiopulmonary arrest. As doctors rush in and apply cardiac monitors to her chest, Ellen loses consciousness.

What is happening to Ellen? To find out, turn to the Clinical Case Wrap-Up on p. 566.

EVERY LIVING CELL RELIES on the surrounding interstitial fluid as a source of oxygen and nutrients and as a place to dispose of wastes. Continuous exchange between the interstitial fluid and the circulating blood stabilizes the levels of gases, nutrients, and wastes in the interstitial fluid. This constant blood supply is essential for homeostasis. If blood flow stops, oxygen and nutrients in the blood are quickly depleted, wastes cannot be discarded, and hormones and white blood cells cannot reach their targets. Therefore, all the functions of the cardiovascular system depend on the **heart**, because it is the heart that keeps blood moving. This muscular organ beats approximately 100,000 times each day. Each year, your heart pumps more than 1.5 million gallons of blood—enough to fill 200 train tank cars!

To appreciate your heart's pumping abilities, turn on the faucet in your kitchen and open it all the way. You would have to leave that faucet on for *45 years* to deliver an amount of water equal to the volume of blood pumped by the heart in an average lifetime.

The nervous system closely monitors and regulates the performance of the heart to ensure that gas, nutrient, and waste levels in peripheral tissues remain within normal limits, whether you are sleeping peacefully, reading a book, or running a marathon. As a result of the nervous system's monitoring and regulation, the volume of blood pumped by the heart varies widely, ranging from 5 to 30 liters per minute.

In this chapter we examine the structural features that enable your heart to perform so reliably. We will then examine the mechanisms that regulate cardiac activity to meet your body's ever-changing needs.

21.1 An Overview of the Cardiovascular System

▶ **KEY POINT** The cardiovascular system is composed of the heart and a network of blood vessels. The blood vessel network is divided into a pulmonary circuit and a systemic circuit. Each of these circuits begins and ends at the heart.

The heart is a small organ—roughly the size of your fist. It has four muscular chambers, the right and left **atria** (Ā-trē-a; singular, *atrium*; "chamber") and right and left **ventricles** (VEN-tri-kls; "little belly"). These four chambers work together, pumping blood through a network of blood vessels that connect the heart to peripheral tissues.

21

The network of vessels is divided into two circuits: the pulmonary circuit and the systemic circuit. The **pulmonary circuit** carries carbon dioxide–rich blood from the heart to the gas exchange surfaces of the lungs and returns oxygen-rich blood to the heart. The **systemic circuit** transports oxygen-rich blood from the heart to the rest of the body's cells and returns carbon dioxide-rich blood back to the heart. The right atrium receives blood from the systemic circuit, and the right ventricle pumps blood into the pulmonary circuit. The left atrium receives blood from the pulmonary circuit, and the left ventricle pumps blood into the systemic circuit. With each heartbeat, the atria contract first, followed by the ventricles. The two ventricles contract at the same time and eject equal volumes of blood into the pulmonary and systemic circuits.

Each circuit begins and ends at the heart, and blood flows through these circuits in sequence. Thus, blood returning to the heart from the systemic circuit must complete the pulmonary circuit before re-entering the systemic circuit. The blood vessels of both circuits are arteries, veins, and capillaries. **Arteries** transport blood away from the heart; **veins** return blood to the heart. **Capillaries** are small, thin-walled vessels connecting the smallest arteries and veins (**Figure 21.1**). Capillaries are called exchange vessels because their thin walls permit exchange of nutrients, dissolved gases, and wastes between the blood and surrounding tissues.

21.1 CONCEPT CHECK ✔

1 What are the two circuits of the cardiovascular system, and what are their functions?

See the blue Answers tab at the back of the book.

Figure 21.1 A Generalized View of the Pulmonary and Systemic Circuits. Blood flows through separate pulmonary and systemic circuits, driven by the pumping of the heart. Each circuit begins and ends at the heart and contains arteries, capillaries, and veins. Arrows indicate the direction of blood flow within each circuit.

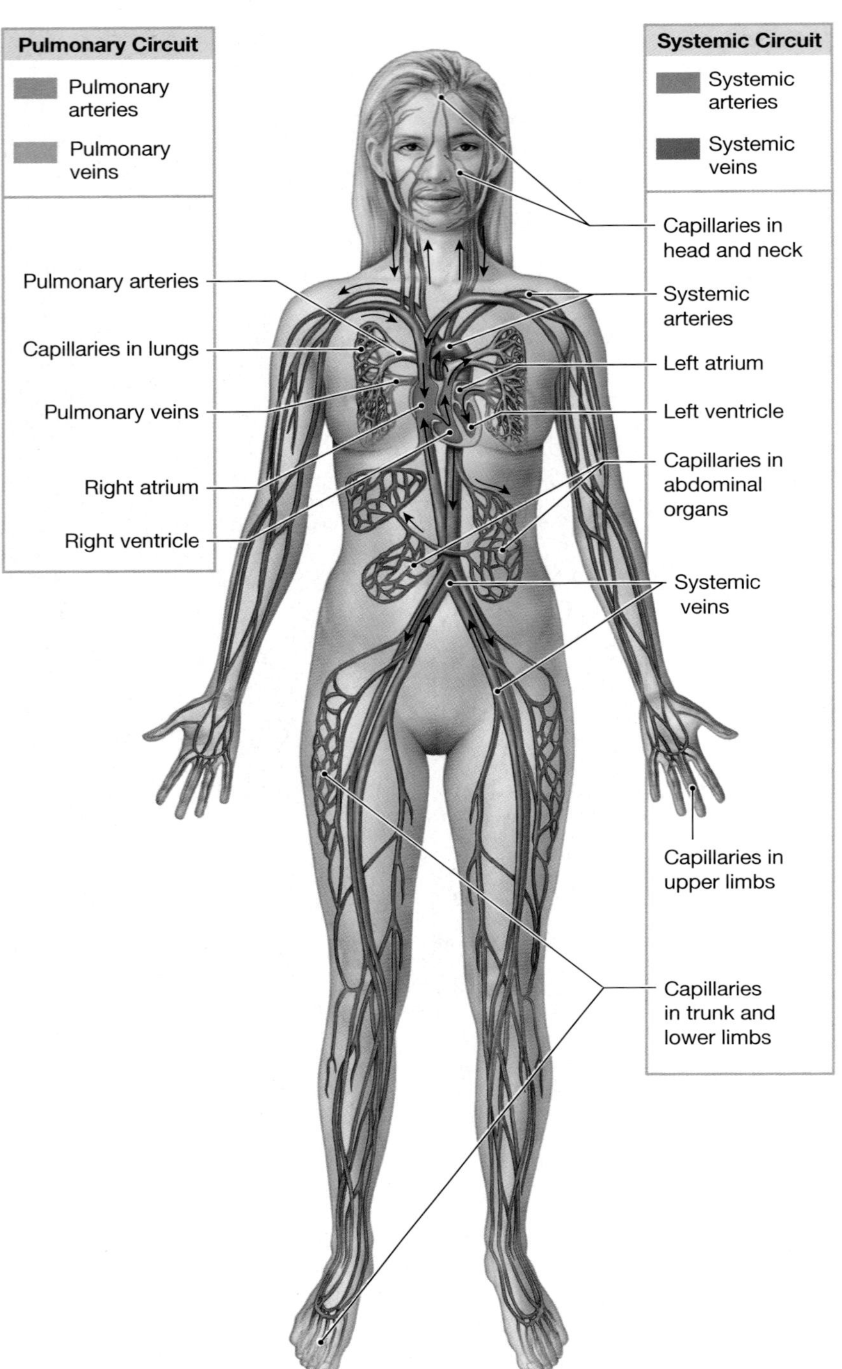

21.2 The Pericardium

▶ **KEY POINT** The heart sits within the mediastinum, located between the two pleural cavities. The pericardium, which surrounds the heart, is composed of a fibrous pericardium and a serous pericardium.

The heart is located near the anterior chest wall, directly posterior to the sternum. The mediastinum contains the great vessels, which are attached at the base of the heart, as well as the thymus, esophagus, and trachea (**Figure 21.2a,c**).

The **pericardium** surrounds the heart and is composed of two parts: an outer fibrous pericardium and an inner serous pericardium (**Figure 21.2b**).

Figure 21.2 Location of the Heart in the Thoracic Cavity. The heart is located within the middle portion of the mediastinum, immediately posterior to the sternum.

Trachea
Thyroid gland
Right lung
First rib (cut)
Left lung
Base of heart
Diaphragm
Parietal pericardium (cut)
Apex of heart

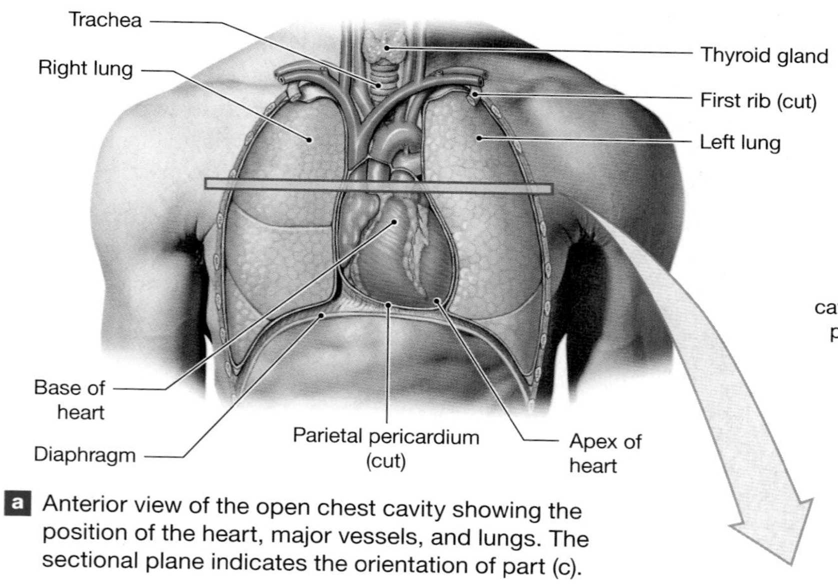

a Anterior view of the open chest cavity showing the position of the heart, major vessels, and lungs. The sectional plane indicates the orientation of part (c).

Cut edge of parietal layer
Pericardial cavity containing pericardial fluid
Cut edge of epicardium (visceral layer)
Fibrous attachment to diaphragm
Air space (corresponds to pericardial cavity)
Balloon

b Relationships between the heart and the pericardial cavity. The pericardial cavity surrounds the heart like the balloon surrounds the fist (right).

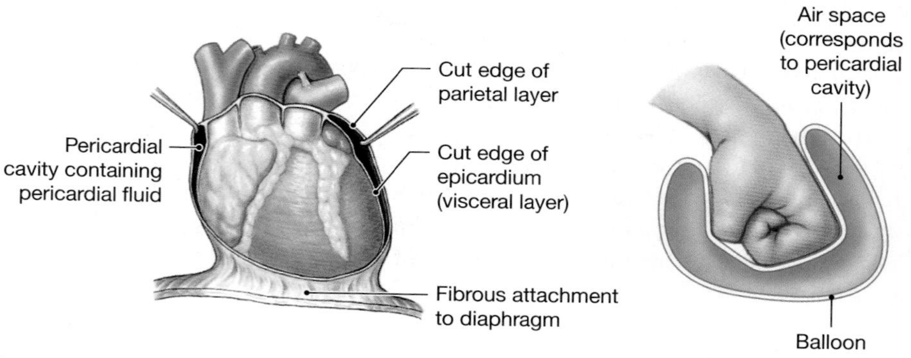

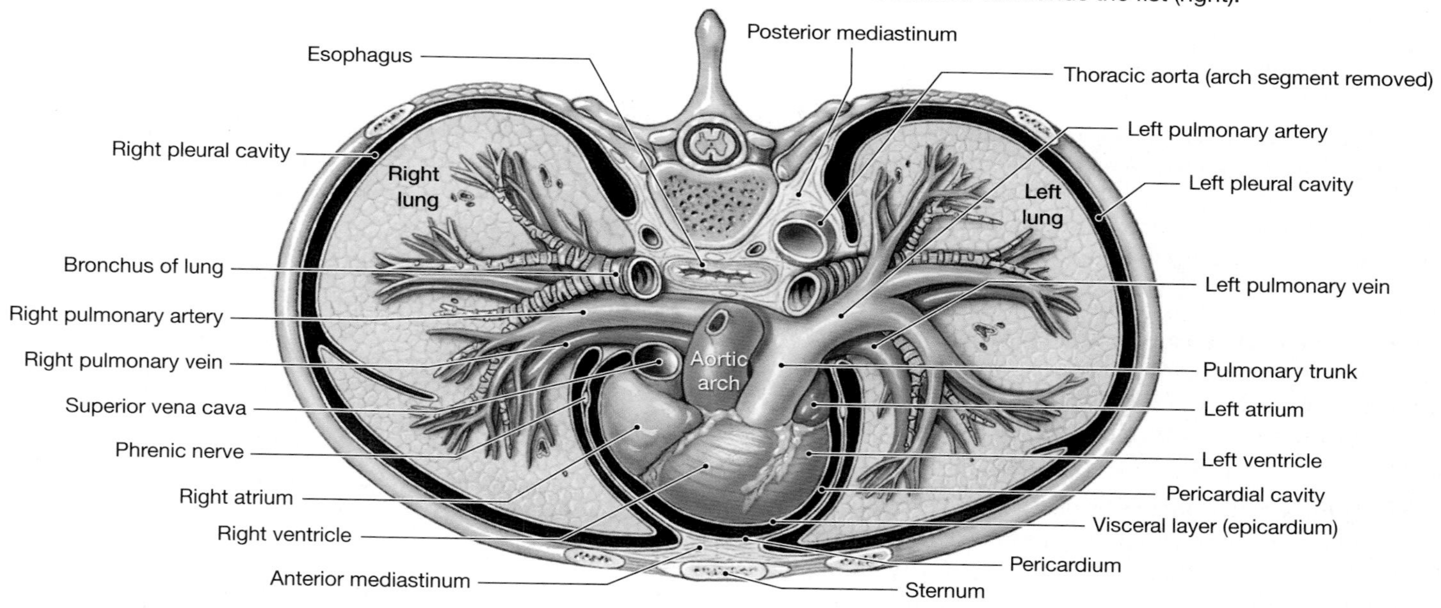

Posterior mediastinum
Esophagus
Thoracic aorta (arch segment removed)
Left pulmonary artery
Right pleural cavity
Right lung
Left lung
Left pleural cavity
Bronchus of lung
Left pulmonary vein
Right pulmonary artery
Right pulmonary vein
Aortic arch
Pulmonary trunk
Superior vena cava
Left atrium
Phrenic nerve
Left ventricle
Right atrium
Pericardial cavity
Right ventricle
Visceral layer (epicardium)
Pericardium
Anterior mediastinum
Sternum

c A superior view of the organs in the mediastinum; portions of the lungs have been removed to reveal blood vessels and airways. The heart is located in the anterior part of the mediastinum, immediately posterior to the sternum.

Pericardial cavity
Body of sternum
Right ventricle
Parietal layer of pericardium
Interventricular septum
Papillary muscle of left ventricle
Right atrium
Parietal pleura
Right pleural cavity
Left pleural cavity
Superior vena cava
Rib (cut)
Left AV valve
Bronchi
Left atrium
Left lung
Esophagus
Right lung
Descending aorta
Body of vertebra
Spinal cord

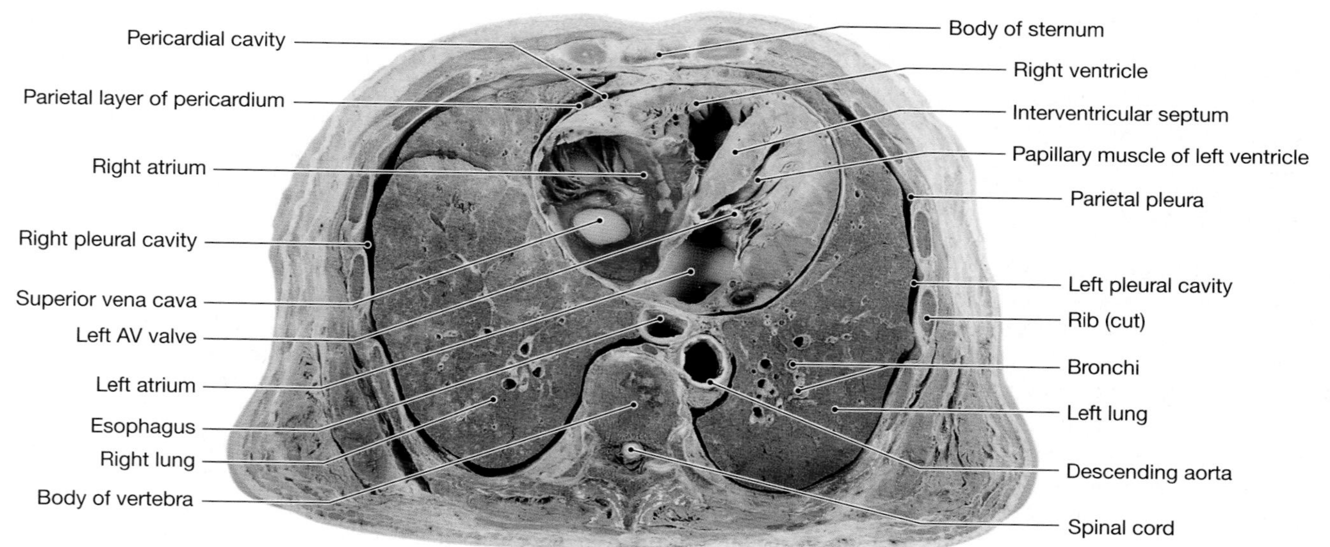

d View of a horizontal section through the trunk at the level of vertebra T_8.

The **fibrous pericardium** is composed of a dense network of collagen fibers that stabilize the position of the heart and associated vessels within the mediastinum.

The lining of the pericardium is the **serous pericardium**. This two-layered membrane is composed of an outer **parietal layer** and an inner **visceral layer**. The visceral layer is also known as the *epicardium*. The potential, fluid-filled space between these two serous layers is the **pericardial cavity** (Figure 21.2b–d). The pericardial cavity normally contains up to 50 mL of **pericardial fluid**, secreted by the pericardial membranes. This fluid acts as a lubricant, reducing friction between the opposing visceral and parietal surfaces as the heart beats. To visualize the relationship between the heart and the pericardial cavity, imagine pushing your fist toward the center of a large, partially inflated balloon. The balloon represents the pericardium, and your fist represents the heart.

21.2 CONCEPT CHECK

2 What is the pericardial cavity?

See the blue Answers tab at the back of the book.

21.3 | Structure of the Heart Wall

▶ KEY POINT The wall of the heart is composed of three layers: from superficial to deep: the epicardium, myocardium, and endocardium.

A section through the wall of the heart reveals three distinct layers: (1) an outer epicardium, (2) a middle myocardium, and (3) an inner endocardium (Figure 21.3).

1. The **visceral layer of the serous pericardium** (**epicardium**) covers the surface of the heart. The epicardium has two layers: a mesothelium and an underlying, supporting layer of areolar tissue. The **parietal layer of the serous pericardium** consists of an outer dense fibrous layer and an inner mesothelium.
2. The **myocardium** is cardiac muscle tissue that forms the atria and ventricles. Associated with the myocardium are the cardiac muscle cells, connective tissues, blood vessels, and nerves. The atrial myocardium is quite thin and is organized into layers forming figure eights as they pass from one atrium to the other. The ventricular myocardium is much thicker, and its muscle orientation changes from layer to layer. The most superficial ventricular muscles wrap around both ventricles. Deeper layers spiral around and between the individual ventricles.
3. The **endocardium** covers the inner surfaces of the heart, including those of the heart valves. This simple squamous epithelium is continuous with the endothelium of the attached great vessels.

Cardiac Muscle Tissue

▶ KEY POINT Cardiac muscle is striated. Cardiac muscle cells have a single central nucleus, numerous mitochondria, and large amounts of stored glycogen; intercalated discs form junctions between adjacent cardiac muscle cells.

The histological characteristics of cardiac muscle tissue give the heart its unique functional properties. We introduced cardiac muscle tissue in Chapter 3, where we compared it with other types of muscle. Cardiac muscle cells are relatively small, averaging 10–20 μm in diameter and 50–100 mm in length. A typical cardiac muscle cell has a single, centrally placed nucleus (Figure 21.3b–d).

Although they are much smaller, cardiac muscle cells resemble skeletal muscle fibers. Each cardiac muscle cell contains organized myofibrils, and the arrangement of their sarcomeres produces striations. However, cardiac muscle cells differ from skeletal muscle fibers in several important respects:

- Cardiac muscle cells are almost totally dependent on aerobic respiration to obtain the energy needed to contract. Therefore, the sarcoplasm of a cardiac muscle cell contains hundreds of mitochondria and large reserves of myoglobin to store oxygen. The sarcoplasm of cardiac muscle cells also contains large amounts of glycogen and lipid inclusions as energy reserves.
- The T tubules of cardiac muscle cells are shorter than those of skeletal muscle cells. In addition, T tubules do not form triads with the sarcoplasmic reticulum.
- Cardiac muscle has a larger number of blood vessels, even more than in red skeletal muscle tissue.
- Cardiac muscle cells contract without nervous system stimulation. (We will discuss the mechanism of cardiac muscle contraction later in the chapter.)
- Cardiac muscle cells are interconnected by specialized cell junctions called intercalated discs (Figure 21.3c–e).

The Intercalated Discs

Cardiac muscle cells are connected to neighboring cells at specialized cell junctions called **intercalated** (in-TER-ka-lā-ted) **discs**. Intercalated discs are unique to cardiac muscle tissue (Figure 21.3c–e). The arrangement of these specialized cell-to-cell junctions and the extensive interlocking of the adjacent cardiac plasma membranes give intercalated discs a jagged appearance. Features of intercalated discs include the following:

- The plasma membranes of two cardiac muscle cells are bound together by desmosomes. This locks the cells together and prevents them from separating during contractions.
- Intercalated discs possess a specialized junction termed a fascia adherens. Actin filaments in cardiac muscle cells anchor firmly to the plasma membrane at the fascia adherens within the intercalated disc. As a result, the intercalated disc ties together the actin filaments of the adjacent cells, and the two muscle cells "pull together" with maximum efficiency.
- Cardiac muscle cells are also connected by gap junctions (communicating junctions). Ions and small molecules move between cells at gap junctions, creating a direct electrical connection between the two muscle cells. As a result, the stimulus for contraction—an action potential—moves from one cardiac muscle cell to another as if the sarcolemmae were continuous.

Because cardiac muscle cells are mechanically, chemically, and electrically connected to one another, cardiac muscle tissue functions like a single, enormous muscle cell. The contraction of any one cell will trigger the contraction of several others, and the contraction will spread throughout the myocardium. For this reason, cardiac muscle has been called a **functional syncytium** (sin-SI-shē-um; *syn*, together + *kytos*, cell).

The Cardiac Skeleton

▶ KEY POINT The cardiac skeleton of the heart is the connective tissue upon which the heart is built. It supports and reinforces the heart, distributes heart muscle contractions, isolates atrial and ventricular muscle cells, and gives the heart elasticity.

The connective tissues of the heart include large numbers of reticular, collagen, and elastic fibers. Each cardiac muscle cell is wrapped in a strong, elastic sheath, and adjacent cells are tied together by fibrous cross-links, or "struts." In turn, each muscle layer has a fibrous wrapping, and fibrous sheets separate the superficial and deep muscle layers. These connective tissue layers are continuous with bands of dense connective tissue that (1) encircle the bases of the pulmonary trunk and

Figure 21.3 Histological Organization of Muscle Tissue in the Heart Wall

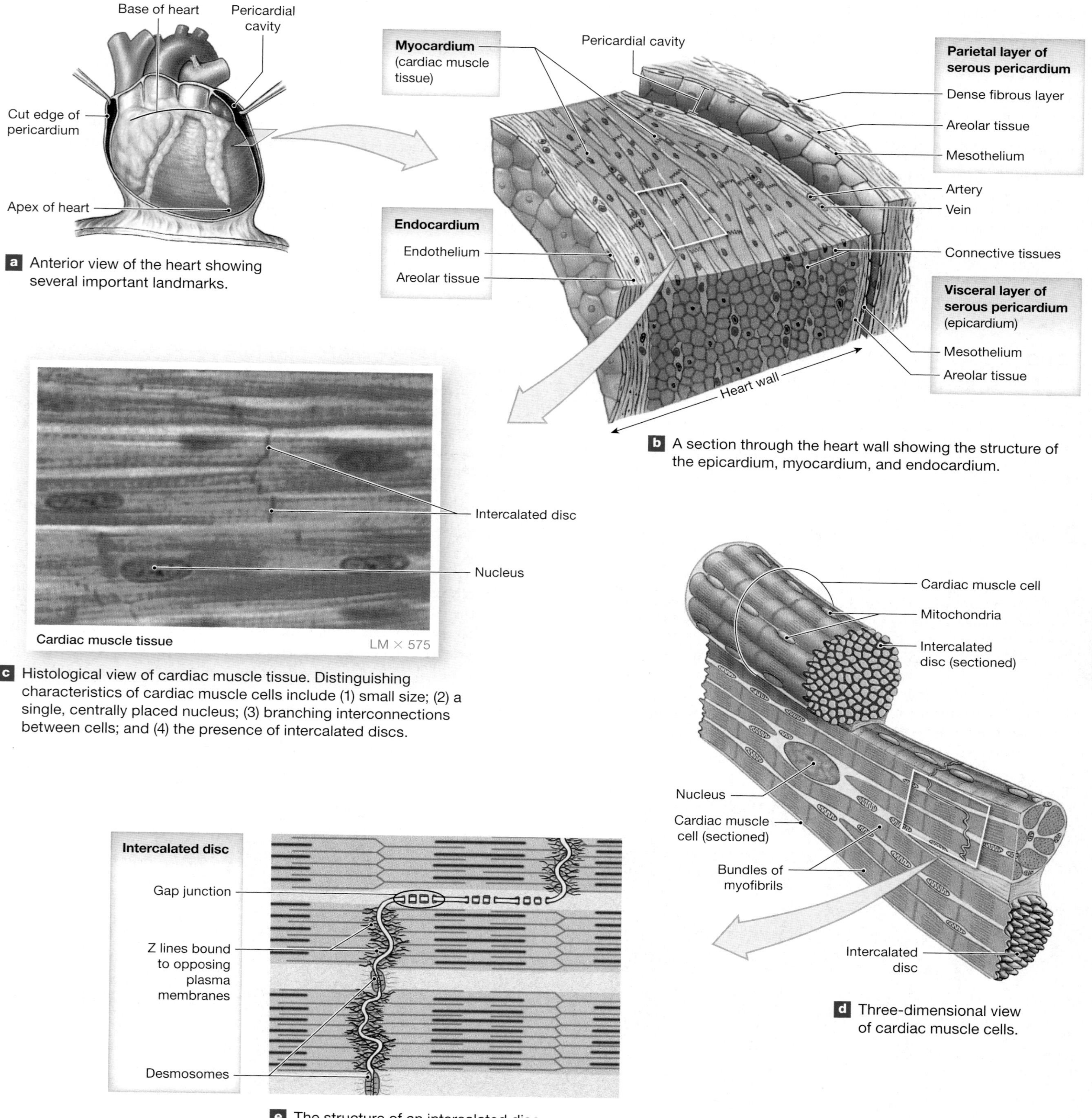

a Anterior view of the heart showing several important landmarks.

b A section through the heart wall showing the structure of the epicardium, myocardium, and endocardium.

c Histological view of cardiac muscle tissue. Distinguishing characteristics of cardiac muscle cells include (1) small size; (2) a single, centrally placed nucleus; (3) branching interconnections between cells; and (4) the presence of intercalated discs.

d Three-dimensional view of cardiac muscle cells.

e The structure of an intercalated disc.

aorta, (2) encircle the valves of the heart, (3) connect the fibrous rings surrounding the openings for the heart valves, and (4) extend into the cardiac muscle that separates the atria and ventricles. This extensive connective tissue network is called the **cardiac skeleton** of the heart (look ahead to **Figure 21.9** on p. 556).

Functions of the cardiac skeleton include the following:

- Stabilizing the positions of the muscle cells and valves in the heart.
- Providing physical support for the cardiac muscle cells and the blood vessels and nerves within the myocardium.
- Distributing the forces of contraction.
- Reinforcing the valves and helping prevent overexpansion of the heart.
- Providing the elasticity that returns the heart to its original shape after each contraction.
- Physically isolating the atrial muscle cells from the ventricular muscle cells. (As you will see in a later section, this isolation is vital for the coordination of cardiac contractions.)

21.3 CONCEPT CHECK

3 How could you distinguish a histological slide of cardiac muscle tissue from one of skeletal muscle tissue?

4 How are cardiac muscle cells connected to their neighbors?

See the blue Answers tab at the back of the book.

21.4 Orientation and Superficial Anatomy of the Heart

▶ **KEY POINT** The heart sits slightly to the left of the midline within the mediastinum. The heart is rotated to the left, causing the right atrium and right ventricle to be located more anteriorly within the thoracic cavity than the left atrium and left ventricle.

Advertisements and cartoons often show the heart at the center of the chest. However, a midsagittal section does not cut the heart in half. This is because the heart (1) lies slightly to the left of the midline, (2) sits at an angle to the longitudinal axis of the body, and (3) is rotated to the left.

The heart is located within the mediastinum, between the two lungs. Because the heart lies slightly to the left of midline, the cardiac notch within the medial surface of the left lung is deeper than the cardiac notch in the medial surface of the right lung.

The **base** of the heart is the broad, superior portion of the heart, where it is attached to the major arteries and veins of the systemic and pulmonary circuits. The base of the heart begins at the origins of the major vessels and the superior surfaces of the two atria. Thinking back to our balloon analogy, the base of the heart corresponds to your wrist (**Figure 21.2b**). The base sits posterior to the sternum, approximately at the third costal cartilage (**Figure 21.4**). The **apex** (Ā-peks) of the heart is the inferior, pointed tip of the heart and is formed mainly by the left ventricle. It points laterally. The apex reaches the fifth intercostal space and extends to the left of the midline.

The heart is rotated slightly to the left. Therefore, the base forms the **superior border** of the heart. The **right border** of the heart is formed by the right atrium. The left ventricle and a small portion of the left atrium form the **left border**. The left border extends to the apex, where it meets the **inferior border**. The apex is formed mainly by the left ventricle, and the inferior border is formed mainly by the inferior wall of the right ventricle.

The **anterior surface**, or *sternocostal* (ster-nō-KOS-tal) *surface*, of the heart faces the anterior thoracic wall and consists mostly of the wall of the right ventricle and some of the left ventricle (**Figures 21.5a** and **21.6a**). The **posterior surface**, at the base, is formed by the left atrium and a small portion of the right atrium. The **diaphragmatic surface** of the heart is composed mainly of the posterior, inferior surfaces of the right and left ventricles (**Figures 21.5b** and **21.6b**).

External grooves, or sulci, of the heart show the approximate borders of the four internal chambers of the heart (**Figures 21.5** and **21.6**). A shallow **interatrial groove** separates the two atria. The deeper **coronary sulcus** marks the border between the atria and the ventricles. On the anterior surface the **anterior interventricular sulcus** separates the left and right ventricles. The **posterior interventricular sulcus** separates the left and right ventricles

Figure 21.4 Position and Orientation of the Heart. The location of the heart within the thoracic cavity and the borders of the heart.

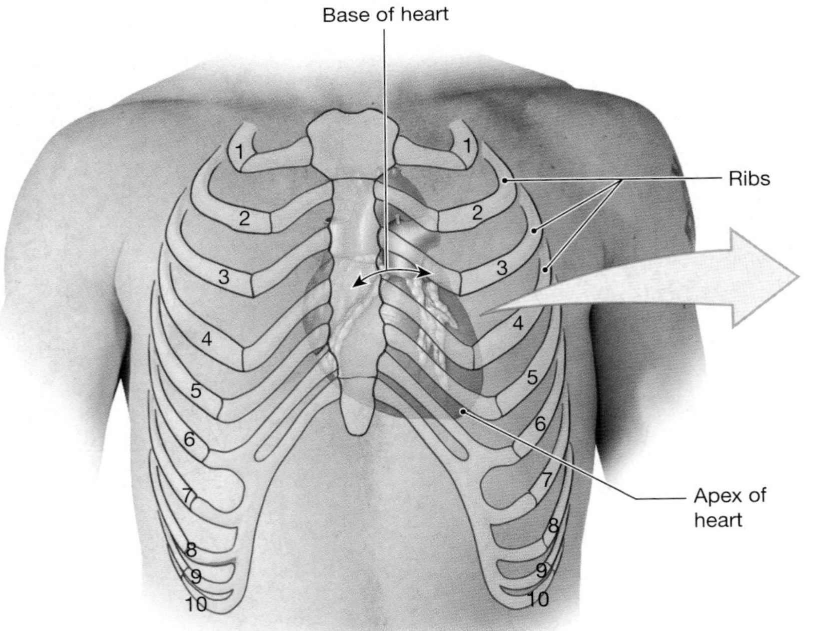

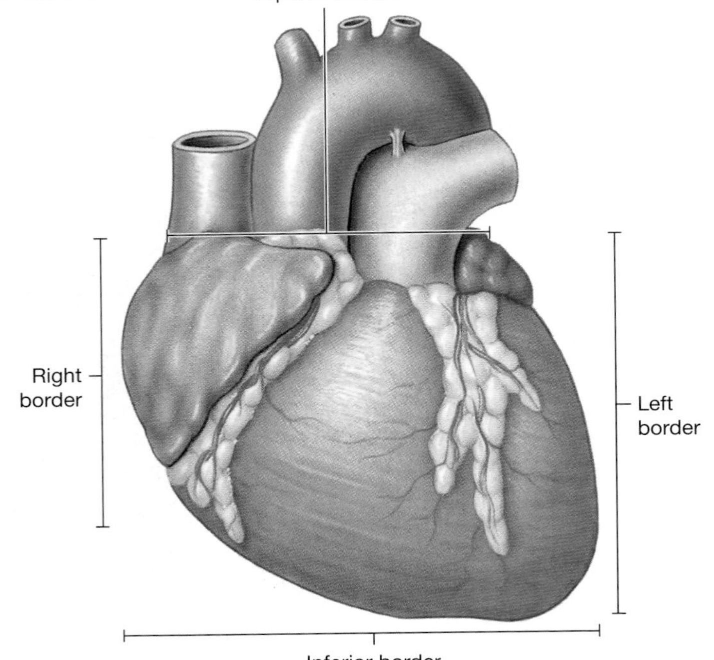

Figure 21.5 Superficial Anatomy of the Heart, Part I

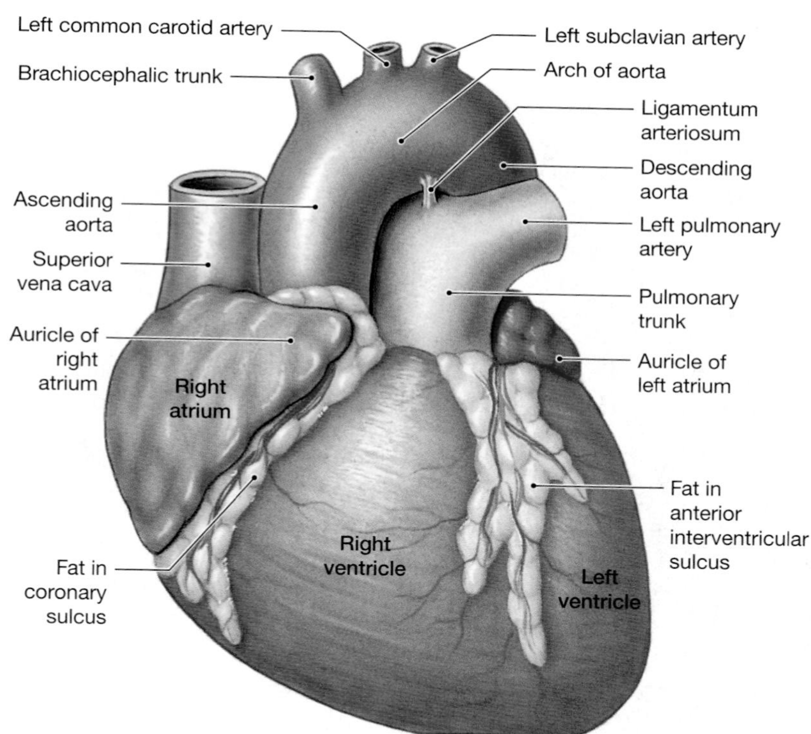

a Anterior view of the heart and great vessels

b Posterior view of the heart and great vessels

on the posterior surface. A considerable amount of adipose tissue is usually found in the connective tissue of the epicardium at the coronary sulcus and the interventricular sulci. In fresh or preserved hearts, this fat must be removed to expose the underlying grooves. These sulci also contain coronary arteries and veins—the arteries and veins supplying blood to the cardiac muscle.

The atria and the ventricles have very different functions. The atria receive venous blood that continues flowing into the ventricles. The ventricles propel blood to the peripheral tissues and the lungs. These functional differences are linked to external and internal structural differences between the right and left sides of the heart **(Figures 21.5, 21.6, and 21.7)**.

Because the heart sits at an angle, the right atrium is anterior, inferior, and to the right of the left atrium. The left atrium curves posteriorly and forms most of the posterior surface of the heart superior to the coronary sulcus. Both atria have relatively thin, muscular walls. When the atria are not filled with

21

Figure 21.6 Superficial Anatomy of the Heart, Part II

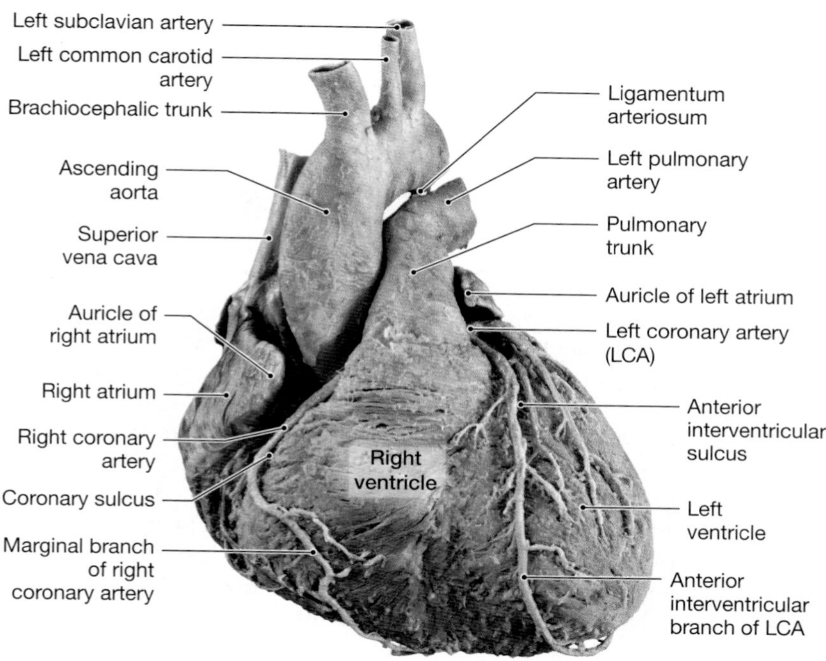

a Anterior view of a dissected heart and great vessels

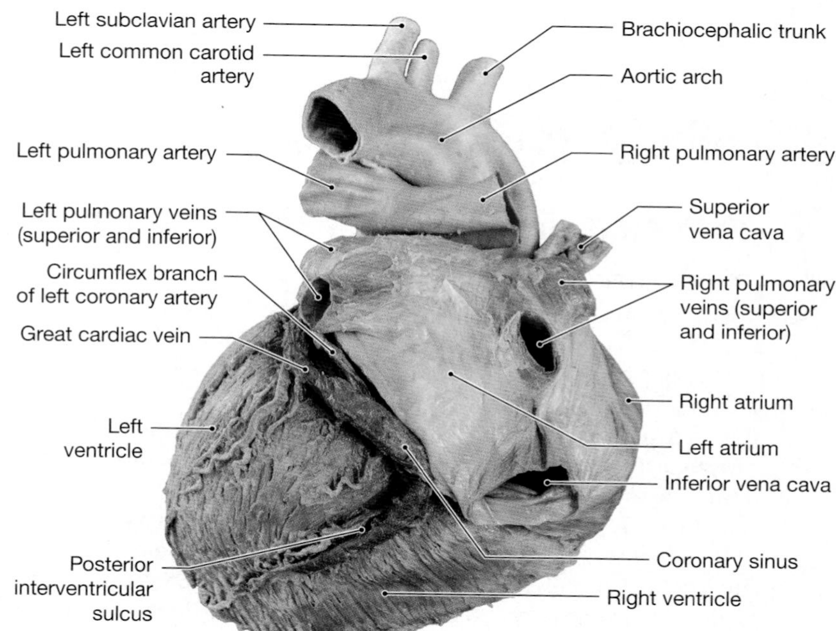

b Posterior view of the heart and great vessels

Figure 21.7 Sectional Anatomy of the Heart, Part I

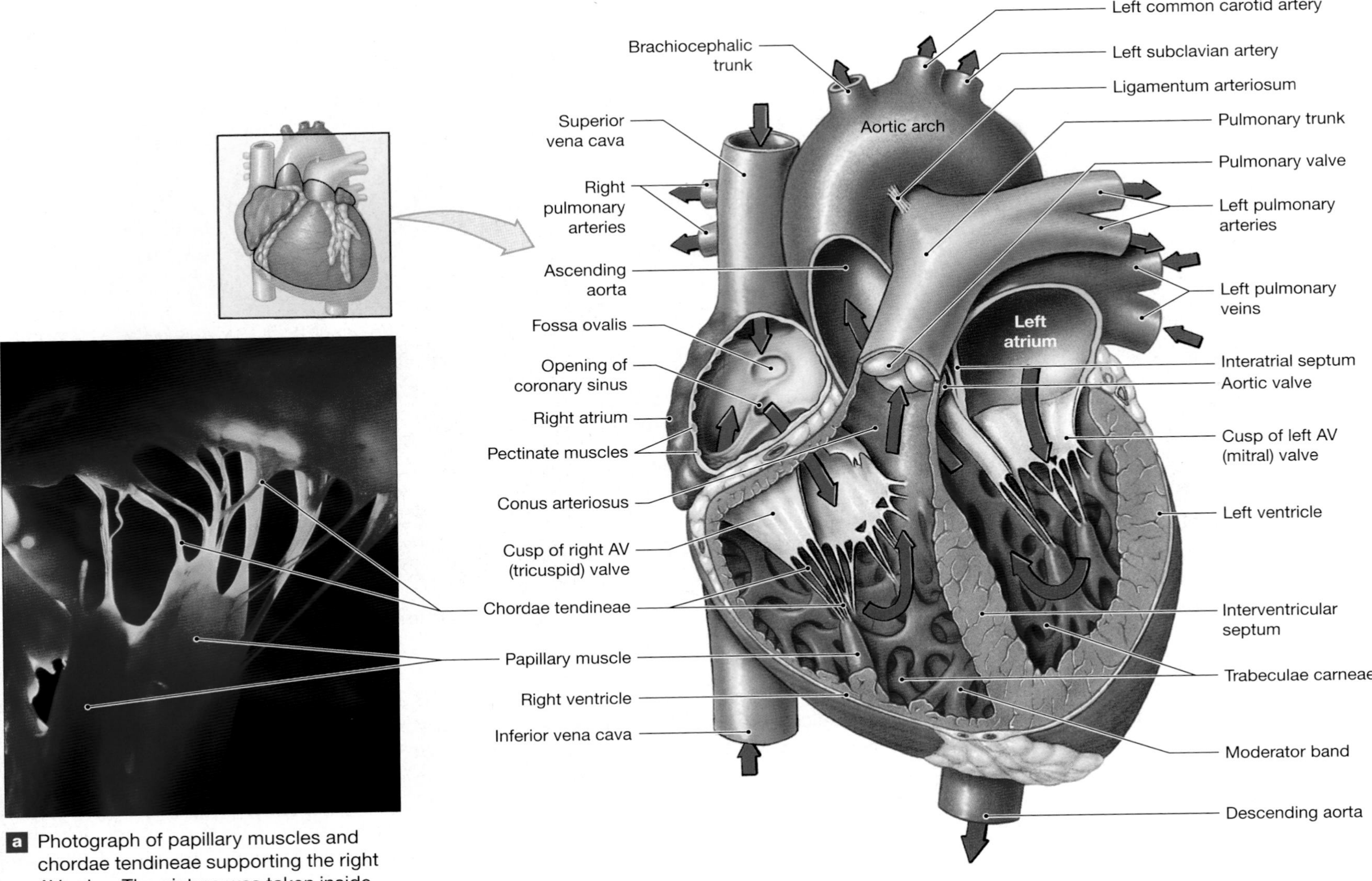

a Photograph of papillary muscles and chordae tendineae supporting the right AV valve. The picture was taken inside the right ventricle, looking toward a light shining from the right atrium.

b Frontal section of the relaxed heart showing the major landmarks and the path of blood flow (arrows) through the atria and ventricles.

21

CLINICAL NOTE

Pericarditis, Myocarditis, and Epicarditis

Pericarditis refers to infection or inflammation of the pericardium surrounding the heart. The most frequent cause is a viral infection. Bacterial pericarditis can be a complication of tuberculosis. Other possible causes include cancer, which can invade the pericardium, and kidney failure, which causes uremic pericarditis. Pericarditis can also be due to severe chest trauma, with bleeding into the pericardial sac.

Myocarditis is an inflammation of the myocardium. Myocarditis is commonly caused by a viral infection and can affect young, healthy people. Myocarditis can also result from autoimmune disease, environmental toxins, alcohol, certain medications, and chemotherapy agents and radiation frequently used in breast cancer therapy. If enough cardiac muscle cells are damaged, chronic cardiomyopathy with impaired pumping power can result. Blood clots may form in the heart, possibly leading to heart attack or stroke. Chronic cardiomyopathy may require a heart transplant.

Endocarditis indicates inflammation of the endocardium. Endocarditis is almost always the result of a bacterial infection, but may also be the result of a fungal infection. Hearts with cardiac birth defects or damaged or abnormal valves are particularly susceptible to endocarditis. Endocarditis begins when bacteria enter the bloodstream and settle on the endocardium. This can happen following dental surgery or in the hospital following placement of a central venous access line. Another common cause is unsterile self-injection of drugs. Endocarditis can destroy heart valves, requiring their surgical replacement.

blood, the anterior portion of each atrium deflates and becomes a rather lumpy and wrinkled flap. This is called an **auricle** (AW-ri-kel; *auris*, ear) because it reminded early anatomists of the external ear.

21.4 CONCEPT CHECK

5 What is the name of the groove separating the atria from the ventricles?

6 What is the base of the heart, and what structures does it include?

See the blue Answers tab at the back of the book.

21.5 Internal Anatomy and Organization of the Heart

▶ **KEY POINT** The interatrial septum separates the two atria of the heart, and the interventricular septum separates the two ventricles. The two atria receive blood, and the two ventricles pump it away from the heart. Four sets of valves prevent the backflow of blood, ensuring that blood flows from the atria into the ventricles and from the ventricles into the great arteries exiting the heart.

Figures 21.7 and **21.8** show the internal anatomy of the atria and ventricles. The **interatrial septum** (*septum*, wall) separates the atria, and the **interventricular septum** separates the ventricles. Blood flows from each atrium into the ventricle of the same side. The **valves** are folds of endocardium extending into the openings between the atria and ventricles. These valves open and close to prevent the backflow of blood, maintaining a one-way flow of blood from the atria into the ventricles.

The atria collect blood returning to the heart and then deliver it to the attached ventricle. The functional demands placed on the right and left atria are very similar, and the two chambers look almost identical. However, the demands placed on the right and left ventricles are very different. As a result, there are significant anatomical differences between the two ventricles.

The Right Atrium

▶ **KEY POINT** The right atrium receives oxygen-poor blood from the systemic and coronary circuits by the superior vena cava, inferior vena cava, and coronary sinus. This blood flows from the right atrium into the right ventricle.

The right atrium receives oxygen-poor (deoxygenated) venous blood from the systemic circuit by the **superior vena cava** (VĒ-na CĀ-va) and the **inferior vena cava** (**Figures 21.5, 21.6, 21.7b,** and **21.8**). The superior vena cava opens into the posterior, superior portion of the right atrium. It receives venous blood from the head, neck, upper limbs, and chest. The inferior vena cava opens into the posterior, inferior portion of the right atrium. It receives venous blood from the tissues and organs of the abdominal and pelvic cavities and the lower limbs.

The veins of the heart itself, called coronary veins, collect blood from the heart wall and deliver it to the coronary sinus (**Figures 21.5b** and **21.6b**). The coronary sinus opens into the posterior wall of the right atrium, inferior to the opening of the inferior vena cava.

Figure 21.8 Sectional Anatomy of the Heart, Part II

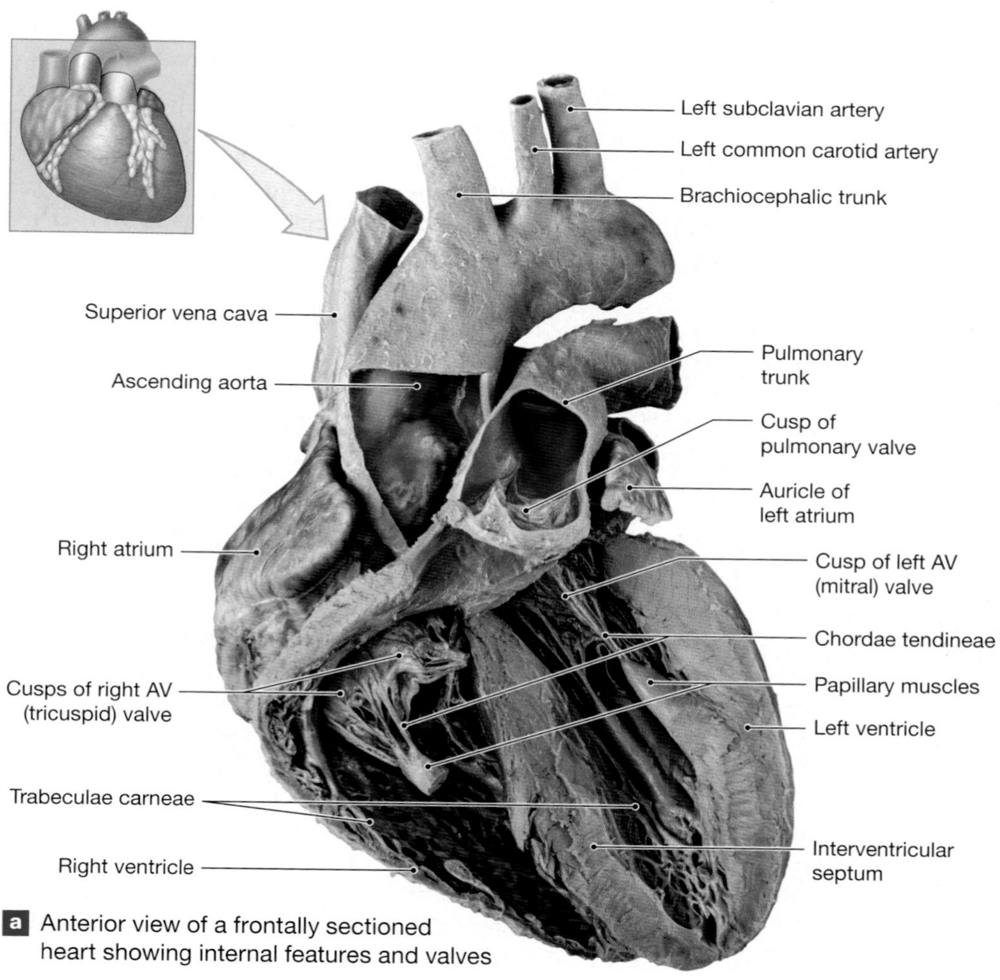

a Anterior view of a frontally sectioned heart showing internal features and valves

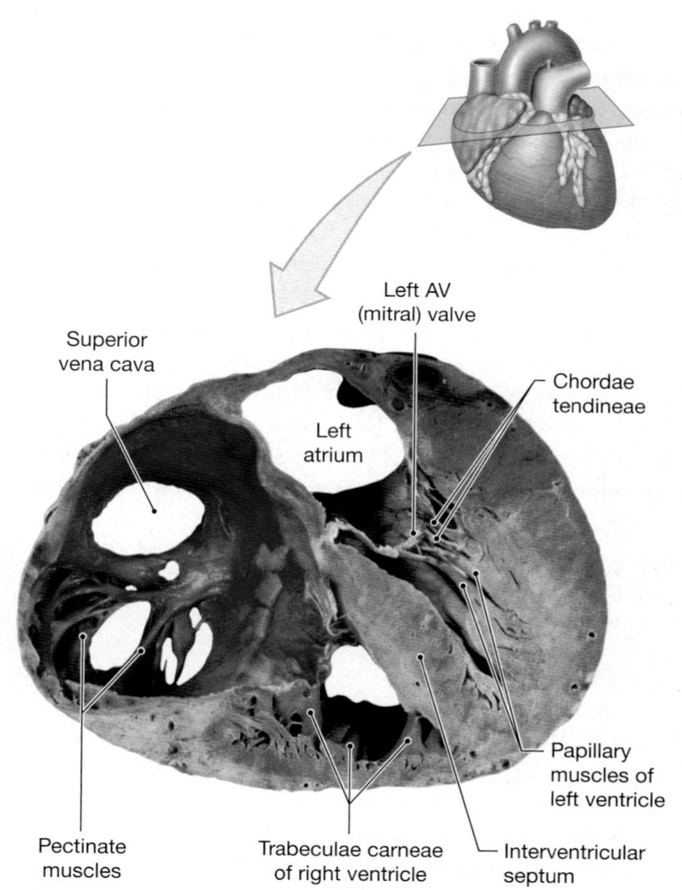

b Inferior view of a horizontal section through the heart at the level of vertebra T_8

The **pectinate muscles** (*pectin*, comb) extend along the inner surface of the right auricle and across the anterior wall of the right atrium. The **interatrial septum** separates the right and left atria. From the fifth week of embryonic development until birth, there is an oval opening in this septum. This opening is called the **foramen ovale**. (See the Embryology Summaries in Chapter 28.) The foramen ovale allows blood to flow directly from the right atrium into the left atrium while the lungs are developing. At birth, the lungs expand and begin functioning, and the foramen ovale closes. Within 3 months, it is permanently sealed. A small depression called the **fossa ovalis** remains at this site in the adult heart.

Occasionally, the foramen ovale remains patent (open). As a result, blood passes from the left atrium into the right atrium and recirculates into the pulmonary circuit. This reduces the efficiency of systemic circulation and elevates blood pressure in the pulmonary vessels. If not corrected, this leads to cardiac enlargement, fluid buildup in the lungs, and eventual heart failure.

The Right Ventricle

▶ **KEY POINT** Oxygen-poor blood flows from the right atrium through the right AV valve and into the right ventricle. Blood flows from the right ventricle into the pulmonary trunk to enter the pulmonary circuit.

Oxygen-poor venous blood travels from the right atrium into the right ventricle. In doing so, the blood passes through an opening guarded by three fibrous flaps. These flaps, or **cusps**, form the **right atrioventricular (AV) valve**, or *tricuspid valve* (trī-KUS-pid; *tri*, three) (**Figures 21.7** and **21.8**). On one side, the cusps are attached to the cardiac skeleton of the heart. Their free edges are attached to connective tissue fibers called **chordae tendineae** (KOR-dē TEN-di-nē-ē; "tendinous cords"). These fibers arise from the **papillary** (PAP-i-ler-ē) **muscles**—cone-shaped muscular projections of the inner surface of the right ventricle. The chordae tendineae limit the movement of the cusps when the valve closes. This prevents backflow of blood from the right ventricle into the right atrium.

The internal surface of the right ventricle contains a series of irregular muscular ridges called the **trabeculae carneae** (tra-BEK-ū-lē CAR-nē-ē; *carneus*, fleshy). The **moderator band** is a muscular ridge that extends horizontally from the inferior portion of the interventricular septum and connects to the anterior papillary muscle.

21

The superior end of the right ventricle tapers to the **conus arteriosus**, a smooth-walled, cone-shaped pouch. The conus arteriosus ends at the **pulmonary valve** (*pulmonary semilunar valve*). This valve consists of three thick semilunar (half moon–shaped) cusps. As blood is pumped out of the right ventricle, it passes through this valve and enters the **pulmonary trunk**. The pulmonary trunk is the first vessel of the pulmonary circuit. The pulmonary valve prevents the backflow of blood into the right ventricle when that chamber relaxes. From the pulmonary trunk, blood flows into both the **left** and **right pulmonary arteries** (**Figures 21.5** to **21.8**). These vessels branch repeatedly within the lungs before supplying the pulmonary capillaries, where gas exchange occurs.

The Left Atrium

▶ **KEY POINT** The left and right pulmonary veins carry oxygen-rich blood from the lungs to the left atrium. Blood flows from the left atrium through the left AV valve into the left ventricle.

Oxygen enters the bloodstream at the pulmonary capillaries. The oxygen-rich (oxygenated) blood flows from the pulmonary capillaries into small veins. These ultimately unite to form four pulmonary veins, usually two for each lung. These **left** and **right pulmonary veins** empty into the posterior portion of the left atrium (**Figures 21.5, 21.6,** and **21.7b**). The left atrium differs from the right atrium in that (1) the left atrium is more cuboidal in shape; (2) the left auricle is longer, narrower, and more hook-shaped; and (3) all of the pectinate muscles of the left atrium are contained within the left auricle.

As blood flows from the left atrium into the left ventricle, it passes through the **left atrioventricular (AV) valve**, also known as the *mitral* (MĪ-tral; *mitre*, bishop's hat) or *bicuspid valve*. This valve has two cusps compared to the three seen in the right AV valve. The left AV valve permits the flow of oxygen-rich blood from the left atrium into the left ventricle, but prevents blood flow in the reverse direction.

The Left Ventricle

▶ **KEY POINT** The left ventricle has the thickest wall of any heart chamber. Its extra-thick myocardium enables the left ventricle to develop enough pressure to force blood around the entire systemic circuit.

The wall of the left ventricle is approximately three times thicker than the wall of the right ventricle. Contractions of the left ventricle must produce enough pressure to push the blood through the entire systemic circuit. The right ventricle, in contrast, has a relatively thin wall. It only has to develop enough pressure to push blood to the lungs and then back to the heart, a total distance of only about 30 cm (1 ft). (*Refer to Chapter 12,* **Figure 12.11**, *to identify these structures in a cross section of the body at the level of* T_8.)

The internal organization of the left ventricle closely resembles that of the right ventricle (**Figures 21.7b** and **21.8a**). However, (1) its trabeculae carneae are more prominent than they are in the right ventricle; (2) there is no moderator band; and (3) since the left AV valve has only two cusps, there are two large papillary muscles rather than three.

Blood leaving the left ventricle passes through the **aortic valve** (*aortic semilunar valve*) into the **ascending aorta**. The arrangement of the cusps in the aortic valve is similar to that in the pulmonary valve. Small, saclike dilations of the base of the ascending aorta occur next to each cusp of the aortic valve. These sacs, called **aortic sinuses**, prevent the individual cusps from sticking to the wall of the aorta when the valve opens. The right and left coronary arteries, which deliver blood to the myocardium, originate at the aortic sinuses. The aortic valve prevents the backflow of blood into the left ventricle once it has been pumped out of the heart and into the systemic circuit.

From the ascending aorta, blood flows into the **aortic arch** and then into the **descending aorta** (**Figures 21.5, 21.6,** and **21.7b**). The pulmonary trunk is attached to the aortic arch by the ligamentum arteriosum, a fibrous band of connective tissue that is left over from an important fetal blood vessel that once linked the pulmonary and systemic circuits. (Chapter 22 will describe cardiovascular changes that occur at birth.)

Structural Differences between the Right and Left Ventricles

▶ **KEY POINT** The anatomical differences between the right and left ventricles reflect their functional differences.

The best way to view the anatomical differences between the right and left ventricles is in three-dimensional or sectional views (**Figures 12.11, 21.7b,** and **21.8**). The lungs are close to the heart, and the pulmonary blood vessels are short and wide. As a result, the right ventricle does not need to work very hard to push blood through the pulmonary circuit. Accordingly, the muscular wall of the right ventricle is relatively thin. In sectional view it resembles a pouch attached to the massive wall of the left ventricle (**Figures 12.11** and **21.8b**).

Contraction of the right ventricle moves it toward the wall of the left ventricle, which compresses the blood within the right ventricle. The rising pressure forces the blood through the pulmonary valve and into the pulmonary trunk. This contraction moves blood very efficiently with minimal effort, but it develops relatively low pressure. Low pressure is all that is needed to move blood around the pulmonary circuit. Higher pressures would actually be dangerous because the pulmonary capillaries are very delicate. Pressures as high as those in systemic capillaries would damage the pulmonary vessels and force fluid into the alveoli of the lungs.

An identical pumping arrangement would not work for the left ventricle. Four to six times more force must be generated to push blood through the systemic circuit. The left ventricle, which has an extremely thick muscular wall, is round in cross section (**Figures 12.11** and **21.8b**). When the left ventricle contracts, it shortens and narrows, and (1) the distance between the base and apex decreases, and (2) the diameter of the left ventricle chamber decreases. Imagine the effects of simultaneously squeezing and rolling up the end of a toothpaste tube and you have the idea. The pressure generated is more than enough to force open the aortic valve and eject blood into the ascending aorta. As the powerful left ventricle contracts, it also bulges into the right ventricular cavity. This action improves the pumping efficiency of the right ventricle. Individuals with severe damage to the right ventricle may survive because the contraction of the left ventricle helps push blood into the pulmonary circuit.

The Structure and Function of Heart Valves

▶ **KEY POINT** The atrioventricular valves and the pulmonary and aortic valves prevent regurgitation of blood into the atria and ventricles. The valves open and close due to pressure changes as the heart pumps blood.

Figures 21.7 to **21.9** and **Spotlight Figure 21.11** illustrate the structure and function of the four heart valves.

The atrioventricular (AV) valves are located between the atria and the ventricles. Each AV valve has four components: (1) a ring of connective tissue attached to the cardiac skeleton of the heart, (2) connective tissue cusps, which close the opening between the heart chambers, (3) chordae tendineae that attach the margins of the cusps to papillary muscles, and (4) the papillary muscles that tense the chordae tendineae. Tension in the papillary muscles and chordae tendineae keeps the cusps from swinging farther and opening into the atria. Thus, the chordae tendineae and papillary muscles are essential to prevent the **regurgitation**, or *backflow*, of blood into the atria each time the ventricles contract.

The **pulmonary valve** (*pulmonary semilunar valve*) is located at the junction between the right ventricle and the pulmonary artery, and the **aortic valve** (*aortic semilunar valve*) is located at the junction between the left ventricle and the ascending aorta. These valves lack chordae tendineae because the three symmetrical cusps support one another like the legs of a tripod.

CLINICAL NOTE

Heart Murmurs

A **heart murmur** is a soft sound—such as a whooshing or swishing—made by turbulent blood flow in or near the heart during a heartbeat. Most murmurs are "innocent," meaning they have no clinical significance. Pregnant women often develop innocent murmurs because their blood volume and cardiac output increase while their cardiac valves remain the same size. Congenital cardiac defects, including atrial septal and ventricular septal defects (holes in the interatrial septum and interventricular septum), cause blood to flow in the wrong direction through the septum during systole, generating a murmur.

Stenotic valves (valves that do not open all the way) or incompetent valvular heart disease, such as valvular stenosis and valvular insufficiency, occur when the valves do not function properly. **Valvular stenosis** is a narrowing of the valve opening that reduces the amount of blood flow. **Valvular insufficiency** is a regurgitation of blood that results from incomplete valve closure.

Serious valvular abnormalities interfere with cardiac function, and the timing and intensity of the related heart sounds provide useful diagnostic information. Health professionals use an instrument called a **stethoscope** (STETH-ō-scōp) to listen to normal and abnormal heart sounds. Heart sounds may be muffled because they must pass through the pericardium, surrounding tissues, and chest wall. As a result, stethoscope placement does not always correspond to the position of the valve being listened to.

21.5 CONCEPT CHECK

7 What would happen if there were no valves between the atria and ventricles?

8 What prevents the AV valves from opening back into the atria?

See the blue Answers tab at the back of the book.

21.6 | Coronary Blood Vessels

▶ **KEY POINT** The coronary circulation supplies blood to the muscle tissue of the heart. The right and left coronary arteries are the first vessels to branch from the ascending aorta.

The heart works continuously, and cardiac muscle cells require reliable supplies of oxygen and nutrients. The **coronary circulation** supplies blood to the muscle tissue of the heart. During maximum exertion, the oxygen demand rises considerably, and blood flow to the heart may increase to nine times resting levels.

The coronary circulation includes an extensive network of coronary blood vessels (**Figure 21.10**). Considerable variations often occur between individuals. The descriptions given here are considered the typical pattern.

The left and right **coronary arteries** originate at the base of the ascending aorta, within the aortic sinus. They are the first branches from this vessel. Blood pressure here is the highest found anywhere in the systemic circuit, and this pressure guarantees a continuous flow of blood to meet the demands of active cardiac muscle tissue.

The Right Coronary Artery

▶ **KEY POINT** The right coronary artery circles the heart to the right (when viewed from above) within the coronary sulcus. In addition to the atrial branches, the right coronary artery gives off two major branches: the right marginal branch and the right posterior interventricular branch.

The **right coronary artery (RCA)** branches off the ascending aorta and turns to the right. This vessel lies within the coronary sulcus and passes between the right auricle and the pulmonary trunk. Sixty percent of the time, the right coronary artery is the dominant coronary artery—the coronary artery that gives off a posterior interventricular branch. Although variations occur, the branches of the right coronary artery typically supply blood to (1) the right atrium, (2) a portion of the left atrium, (3) the interatrial septum, (4) the entire right ventricle, (5) a variable portion of the left ventricle, (6) the postero-inferior one-third of the interventricular septum, and (7) portions of the conducting system (sinoatrial node) of the heart. **Figure 21.10** shows the major branches.

The right coronary artery gives off **atrial branches** as it curves across the anterior surface of the heart. These branches supply the right atrium and a portion of the left atrium with blood.

Near the right border of the heart, the **right marginal branch** is formed. This vessel extends toward the apex of the heart along the anterior surface of the right ventricle. The right marginal branch supplies the right atrium, interatrial septum, and right ventricle with blood. As the right coronary artery

Figure 21.9 **Valves of the Heart.** Red (oxygenated) and blue (deoxygenated) arrows indicate blood flow into or out of a ventricle. Black arrows indicate blood flow into an atrium, and green arrows indicate ventricular contraction.

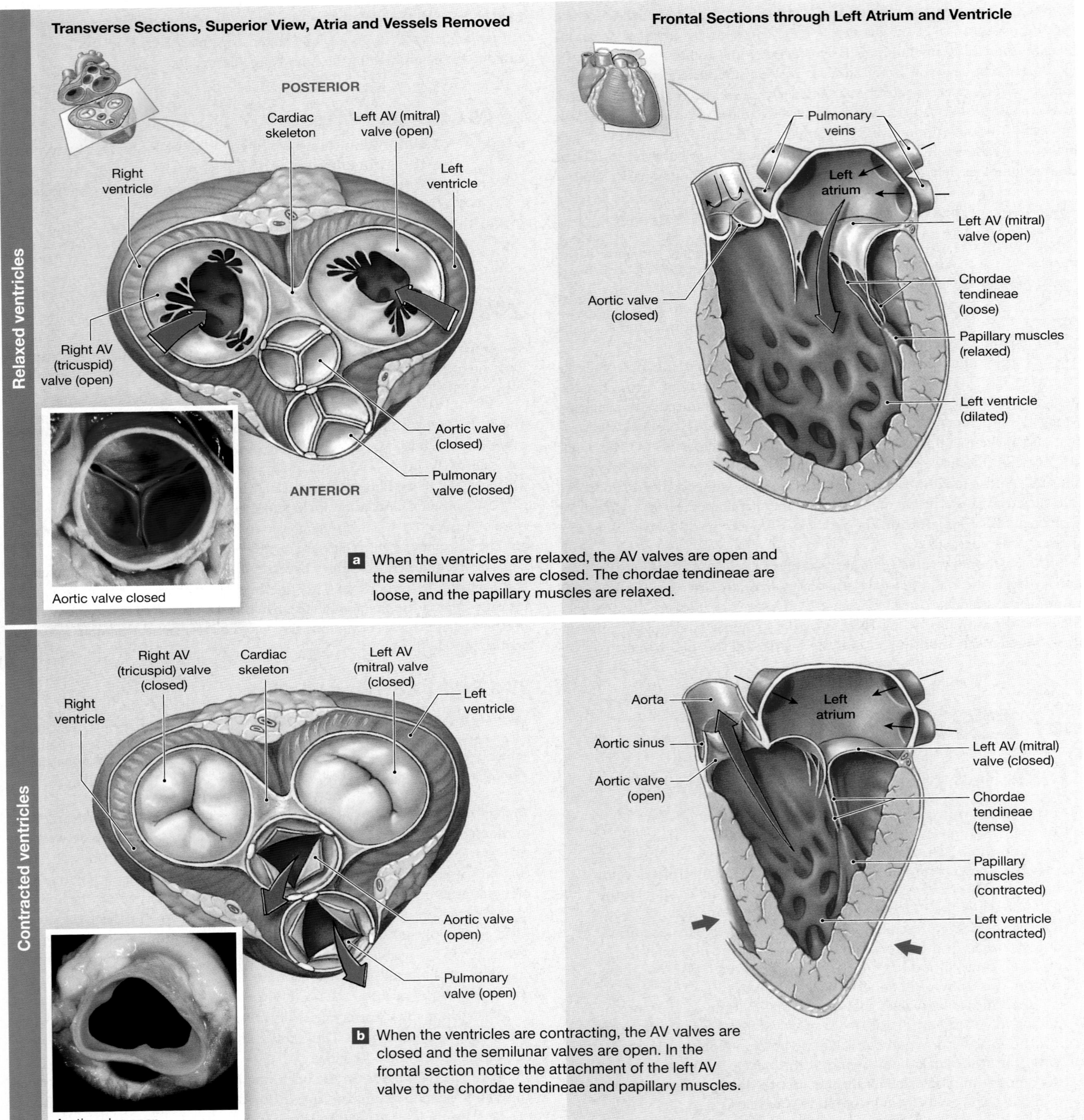

a When the ventricles are relaxed, the AV valves are open and the semilunar valves are closed. The chordae tendineae are loose, and the papillary muscles are relaxed.

b When the ventricles are contracting, the AV valves are closed and the semilunar valves are open. In the frontal section notice the attachment of the left AV valve to the chordae tendineae and papillary muscles.

Figure 21.10 Coronary Circulation

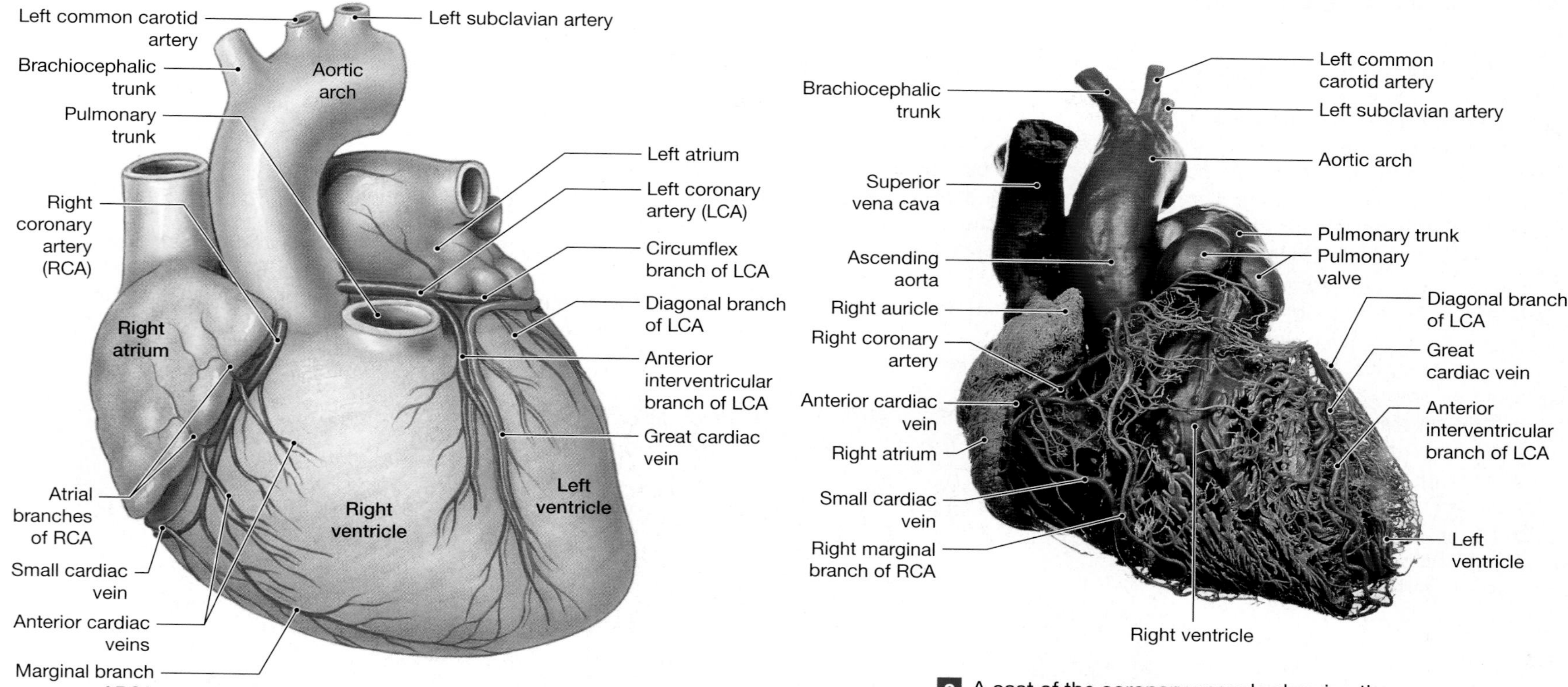

a Coronary vessels supplying the anterior surface of the heart.

c A cast of the coronary vessels showing the complexity and extent of the coronary circulation. Coronary vessels are also seen in **Figure 21.6**.

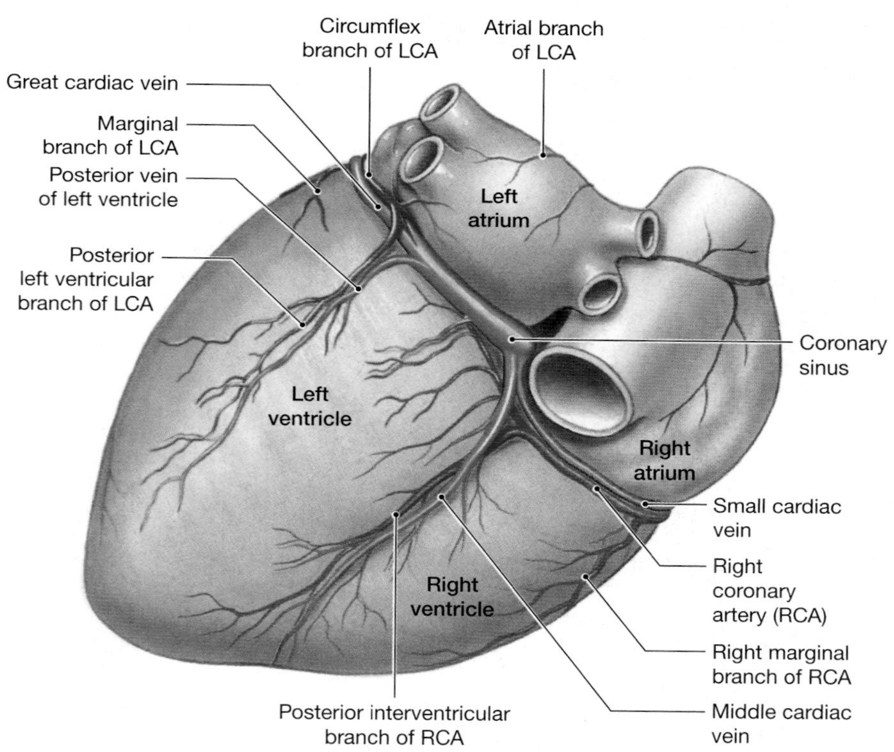

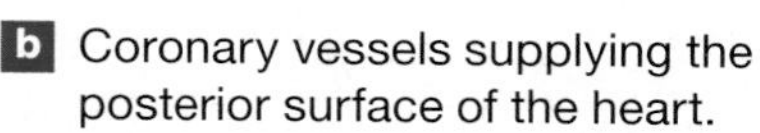

b Coronary vessels supplying the posterior surface of the heart.

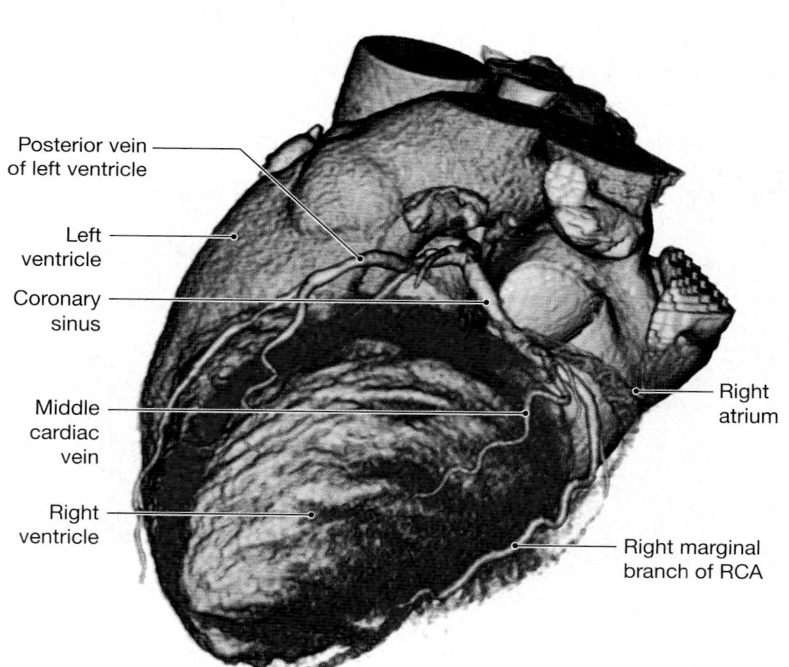

d Spiral scan of the heart showing the coronary veins and coronary sinus. [Courtesy of TeraRecon, Inc.]

21

CLINICAL NOTE

Coronary Artery Disease and Myocardial Infarction

Coronary Artery Disease (CAD) is the leading cause of death for men and women in the United States. It occurs when cardiac arteries become hardened and narrowed due to the buildup of cholesterol, calcium, and plaque. This narrowing of coronary arteries, termed **atherosclerosis**, decreases blood supply to the heart muscle and leads to **cardiac ischemia** (is-KĒ-mē-a). Cardiac ischemia often causes heart pain known as **angina** (an-JĪ-na), which is usually manifested as a crushing, smothering chest discomfort that may radiate to the back, neck, jaw, or left arm.

If a blood clot forms at the site of a plaque and suddenly blocks the artery completely, a **myocardial infarction (MI)**, or *heart attack*, can occur. The insufficient blood supply causes necrotic damage, called an **infarct**. The affected heart muscle then scars, which affects the heart's ability to contract properly. This impairs the cardiac conduction system and can lead to arrhythmias (heartbeat irregularities) and heart failure. Treatment involves restoring blood flow using a thrombolytic (clot dissolver) or coronary angioplasty (surgical repair of blood vessel). Sometimes atherosclerosis develops so slowly that collateral circulation (alternative blood supply growing around a blockage) will develop and can save some cardiac muscle in the event of a sudden coronary thrombosis (stationary blood clot).

Many causes of CAD are preventable, including smoking, obesity, hypertension, hypercholesterolemia (high blood cholesterol), diabetes, and sedentary lifestyle.

Percutaneous (through a puncture) **balloon angioplasty** involves threading a catheter past an area of atherosclerosis, inflating a balloon to compress the plaque against the vessel wall, then inserting a metal stent, or sleeve, to keep the area open. A surgical procedure in which damaged sections of coronary arteries are replaced with new venous graftings, known as a **coronary artery bypass graft (CABG)**, can also be done. Such procedures typically use the great saphenous vein of the leg or the internal thoracic artery of the chest wall for the graft.

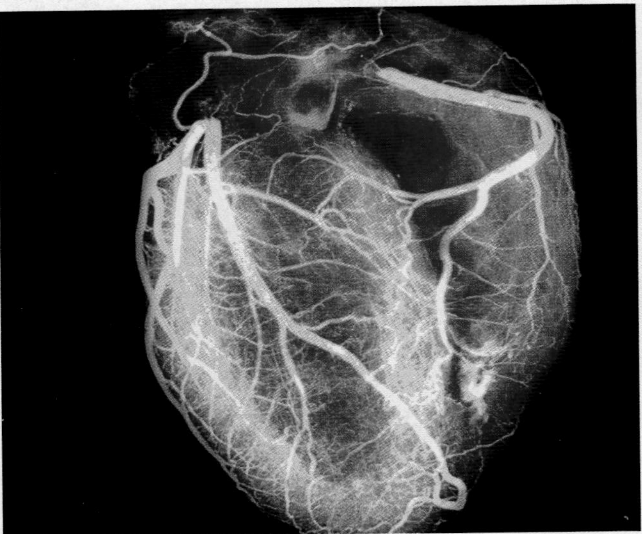

Normal circulation

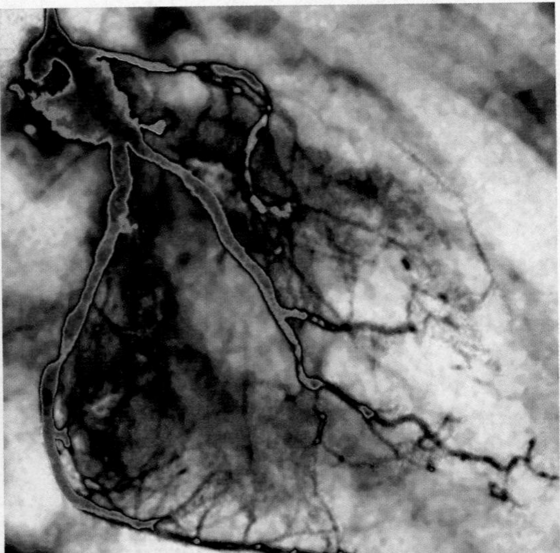

Restricted circulation

This is a digital subtraction angiography (DSA) scan showing normal ventricular circulation.

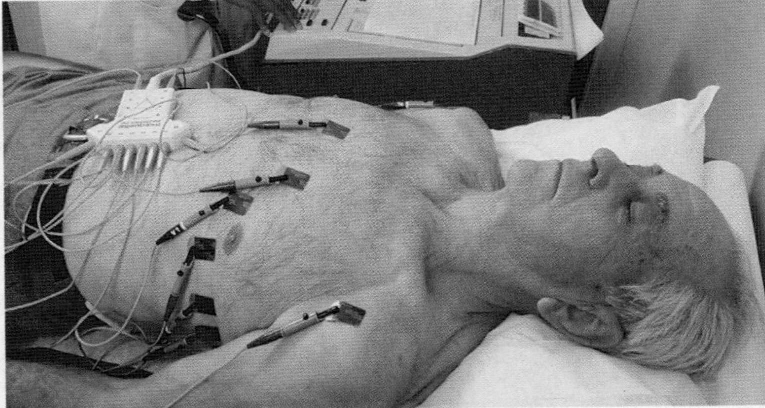

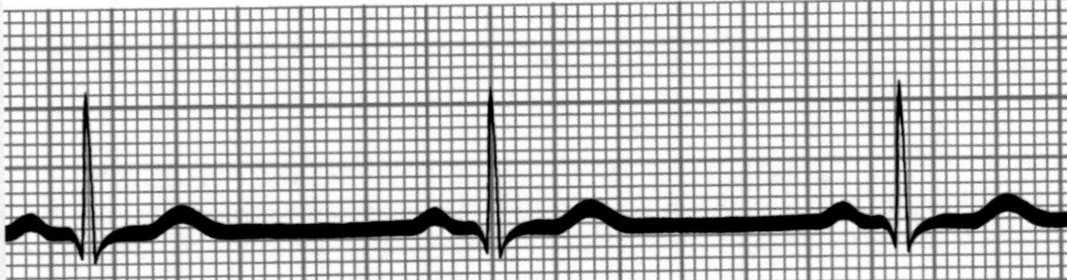

ECG rhythm strip

An electrocardiogram (ECG) can demonstrate cardiac ischemia and myocardial infarctions.

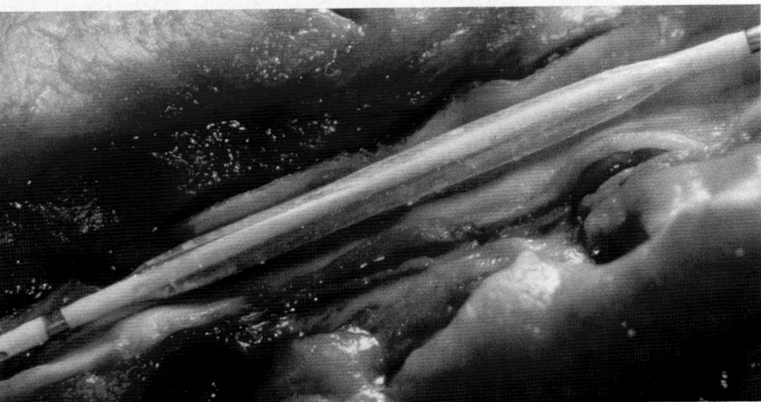

Balloon angioplasty

continues across the posterior surface of the heart it gives off the **posterior interventricular branch**. This branch, or *posterior descending artery*, continues toward the apex of the heart within the posterior interventricular sulcus. This branch supplies blood to the interventricular septum and adjacent portions of the ventricles.

A small branch near the base of the right coronary artery penetrates the atrial wall to reach the sinoatrial (SA) node, also known as the *cardiac pacemaker*. A small branch to the atrioventricular (AV) node, another part of the conducting system of the heart, originates from the right coronary artery near the posterior interventricular branch.

The Left Coronary Artery

▶ **KEY POINT** The left coronary artery circles the heart to the left (when viewed from above) within the coronary sulcus. The left coronary artery has a larger diameter and supplies more of the heart with blood than the right coronary artery. The left coronary artery gives off four major branches: the anterior interventricular, circumflex, left marginal, and posterior interventricular branches.

In most individuals, the lumen of the **left coronary artery (LCA)** is larger in diameter than the lumen of the right coronary artery. It supplies blood to (1) most of the left ventricle, (2) a small segment of the right ventricle, (3) most of the left atrium, and (4) the anterior two-thirds of the interventricular septum.

As the left coronary artery reaches the anterior surface of the heart, it forms the anterior interventricular branch and the circumflex branch **(Figure 21.10)**. The **anterior interventricular branch**, or *left anterior descending branch*, is a large artery running along the anterior surface within the anterior interventricular sulcus. This artery supplies the anterior ventricular myocardium and the anterior two-thirds of the interventricular septum. Often, small branches of the anterior interventricular branch of the left coronary artery are continuous with those of the posterior interventricular branch of the right coronary artery.

The **circumflex branch** curves to the left within the coronary sulcus. As it circles toward the posterior surface of the heart, it gives rise to one or more diagonal branches that supply portions of the left ventricle with blood. In most individuals the left coronary artery forms the **left marginal branch**. Typically, this vessel reaches the apex of the heart and supplies much of the left ventricle. Upon reaching the posterior surface of the heart, the right coronary artery forms the **posterior left ventricular branch**. This branch is small and quite variable; in some individuals it is totally absent. If this branch is very small or absent, it is often replaced by the **posterior interventricular branch** of the circumflex artery. In addition, distal portions of the circumflex branch often meet and fuse with small branches of the right coronary artery.

As mentioned previously, interconnections between coronary arteries are visible in various locations on the ventricles of the heart. Such interconnections are called **anastomoses** (a-nas-tō-MŌ-sez; *anastomosis*, outlet). Because the arteries are interconnected in this way, blood supply to the ventricular muscle remains relatively constant, regardless of pressure fluctuations within the left and right coronary arteries.

The Coronary Veins

▶ **KEY POINT** The heart is drained by the coronary sinus and the anterior cardiac veins. The coronary sinus and the anterior cardiac veins empty into the right atrium.

The **great cardiac vein** and **middle cardiac vein** collect blood from smaller veins draining the myocardial capillaries. They deliver venous blood to the **coronary sinus**, a large, thin-walled vein lying in the posterior portion of the coronary sulcus **(Figures 21.5b, 21.6b, and 21.10a,b,d)**. The coronary sinus drains into the right atrium inferior to the opening of the inferior vena cava.

Cardiac veins that empty into the great cardiac vein or the coronary sinus include (1) the **posterior vein of the left ventricle**, draining the area served by the circumflex branch of the left coronary artery, (2) the **middle cardiac vein**, draining the area supplied by the posterior interventricular branch of the left coronary artery, and (3) the **small cardiac vein**, which receives blood from the posterior surfaces of the right atrium and ventricle. The **anterior cardiac veins** drain the anterior surface of the right ventricle. These vessels empty directly into the right atrium, bypassing the coronary sinus.

21.6 CONCEPT CHECK

9 Describe the right coronary artery (RCA).

10 The left coronary artery supplies what parts of the heart with blood?

See the blue Answers tab at the back of the book.

21.7 The Coordination of Cardiac Contractions

▶ **KEY POINT** Each contraction cycle of the heart follows a precise sequence: The atria contract first and then the ventricles. Nodal cells and conducting fibers coordinate the contractions that make up each cycle.

The function of any pump is to (1) develop pressure and (2) move a volume of fluid in a specific direction at an acceptable speed. The heart works in cycles of contraction (systole) and relaxation (diastole), and the pressure within each chamber rises and falls within each cycle **(Spotlight Figure 21.11)**. The AV, aortic, and pulmonary valves help ensure a one-way flow of blood. Blood will flow out of an atrium only when the AV valve is open and atrial pressure is greater than ventricular pressure. Likewise, blood will flow from a ventricle into the aorta or pulmonary trunk only when the aortic or pulmonary valve is open and ventricular pressure is greater than arterial pressure. The proper functioning of the heart depends on the proper timing of atrial and ventricular contractions. The pacemaking and conduction systems of the heart provide this required timing.

Unlike skeletal muscle, cardiac muscle tissue contracts on its own, without neural or hormonal stimulation; even a heart removed for a heart transplant will continue to beat unless it is kept chilled in a preservation solution. This ability to generate and conduct impulses is called autorhythmicity. Neural or hormonal stimuli alter the basic rhythm of the contractions.

Cardiac contractions are coordinated by specialized conducting cells. These specialized cardiac muscle cells do not contract like the other cells within the heart. There are two distinct populations of these cells: (1) **Nodal cells** establish the rate of cardiac contraction, and (2) **conducting cells** distribute the contractile stimulus to the myocardium **(Spotlight Figure 21.11)**.

21.7 CONCEPT CHECK

11 Briefly describe the two types of specialized cardiac muscle cells.

See the blue Answers tab at the back of the book.

SPOTLIGHT

FIGURE 21.11

The Conducting System and the Cardiac Cycle

The Conducting System of the Heart

Cardiac muscle tissue contracts on its own, in the absence of neural or hormonal stimulation. This inherent ability to generate and conduct impulses is called **automaticity** or **autorhythmicity**. The conducting system of the heart is a network of specialized cardiac muscle cells responsible for initiating and distributing the stimulus to contract.

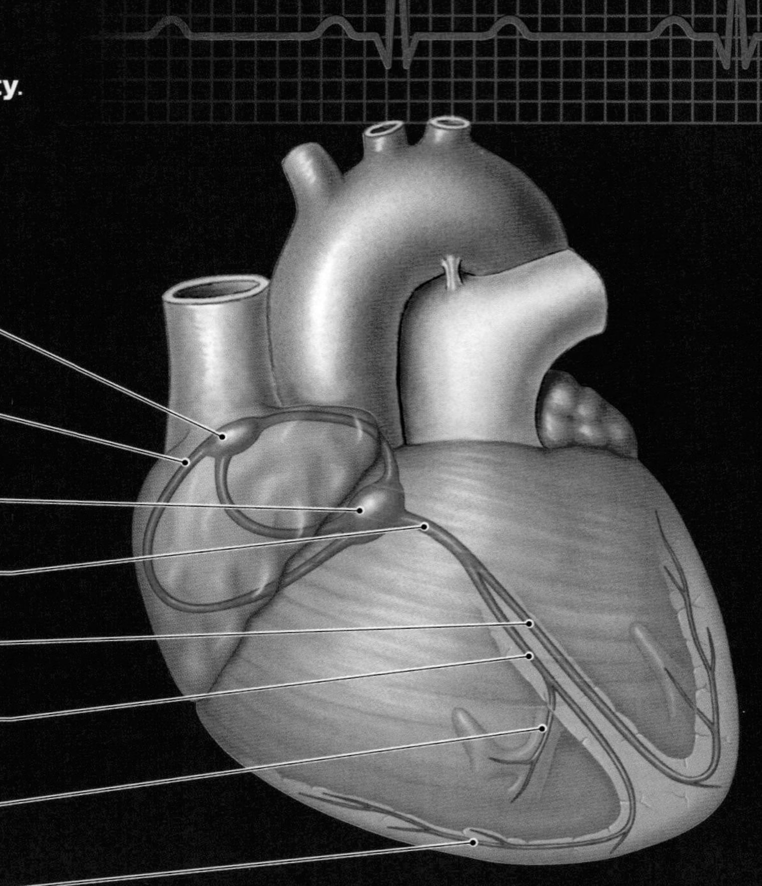

Components of the Conducting System

Sinoatrial (SA) node	contains pacemaker cells that initiate the electrical impulse that results in a heartbeat
Internodal pathways	are conducting fibers in the atrial wall that conduct the impulse to the AV node while simultaneously stimulating cardiac muscle cells of both atria
Atrioventricular (AV) node	slows the electrical impulse when it arrives from the internodal pathways
AV bundle	conducts the impulse from the AV node to the bundle branches
Left bundle branch	extends toward the apex of the heart and then radiates across the inner surface of the left ventricle
Right bundle branch	extends toward the apex of the heart and then radiates across the inner surface of the right ventricle
Moderator band	relays the stimulus through the ventricle to the papillary muscles, which tense the chordae tendineae before the ventricles contract
Purkinje fibers	convey the impulses very rapidly to the contractile cells of the ventricular myocardium

Movement of Electrical Impulses through the Conducting System

Each cardiac contraction follows a precise sequence: The atria contract first and then the ventricles. The following sequence illustrates the spread of electrical activity through the heart and shows how the conducting system coordinates the contractions of the cardiac cycle.

1

Time = 0

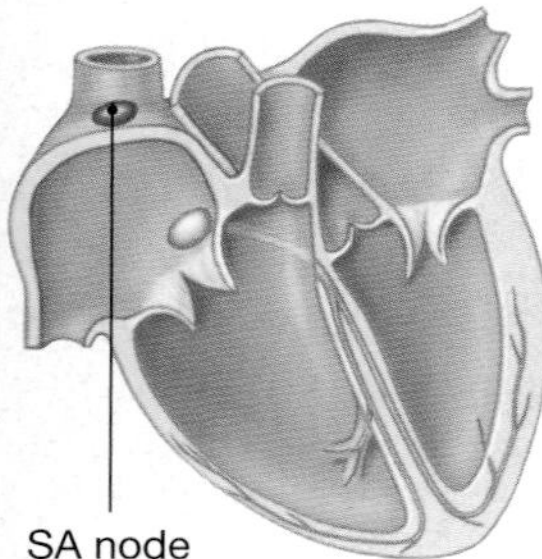

The SA node depolarizes and atrial activation begins.

2

Elapsed time = 50 msec

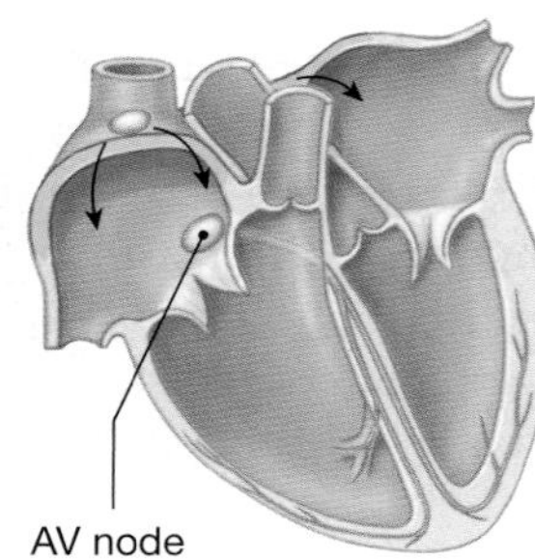

Depolarization spreads across the atrial surfaces and reaches the AV node.

3

Elapsed time = 150 msec

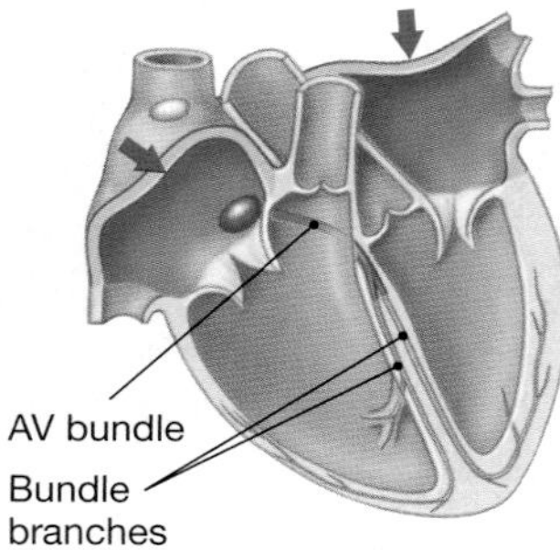

The AV node delays the spread of electrical activity to the AV bundle by 100 msecs. Atrial contraction begins.

4

Elapsed time = 175 msec

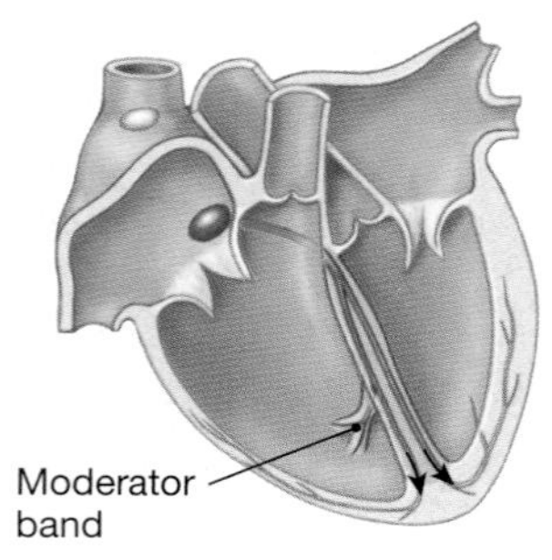

Impulses travel along the AV bundle within the interventricular septum to the apex of the heart. Impulses also spread to the papillary muscles of the right ventricle by the moderator band.

5

Elapsed time = 225 msec

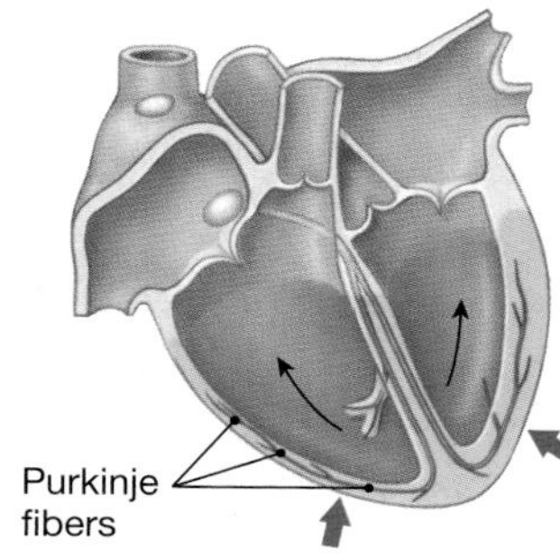

The impulse is distributed by Purkinje fibers (*subendocardial branches*) and relayed throughout the ventricular myocardium. Atrial contraction is completed and ventricular contraction begins.

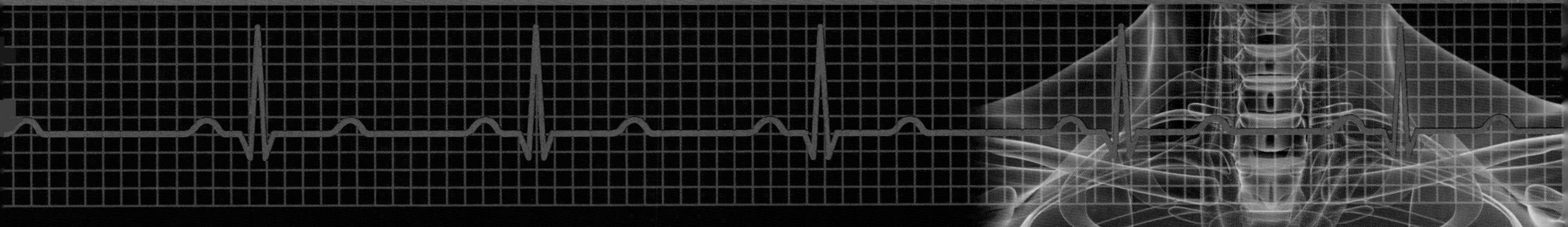

The Phases of the Cardiac Cycle

The period between the start of one heartbeat and the beginning of the next is a single **cardiac cycle**. The cardiac cycle therefore includes alternate periods of contraction and relaxation. For any one chamber in the heart, the cardiac cycle can be divided into two phases. During contraction, or **systole**, a chamber ejects blood either into another heart chamber or into an arterial trunk. Systole is followed by the second phase: relaxation, or **diastole**.

In this illustration, black arrows indicate the flow of blood or movement of valves; green arrows indicate contractions; and the red-hued areas indicate which heart chambers are in systole.

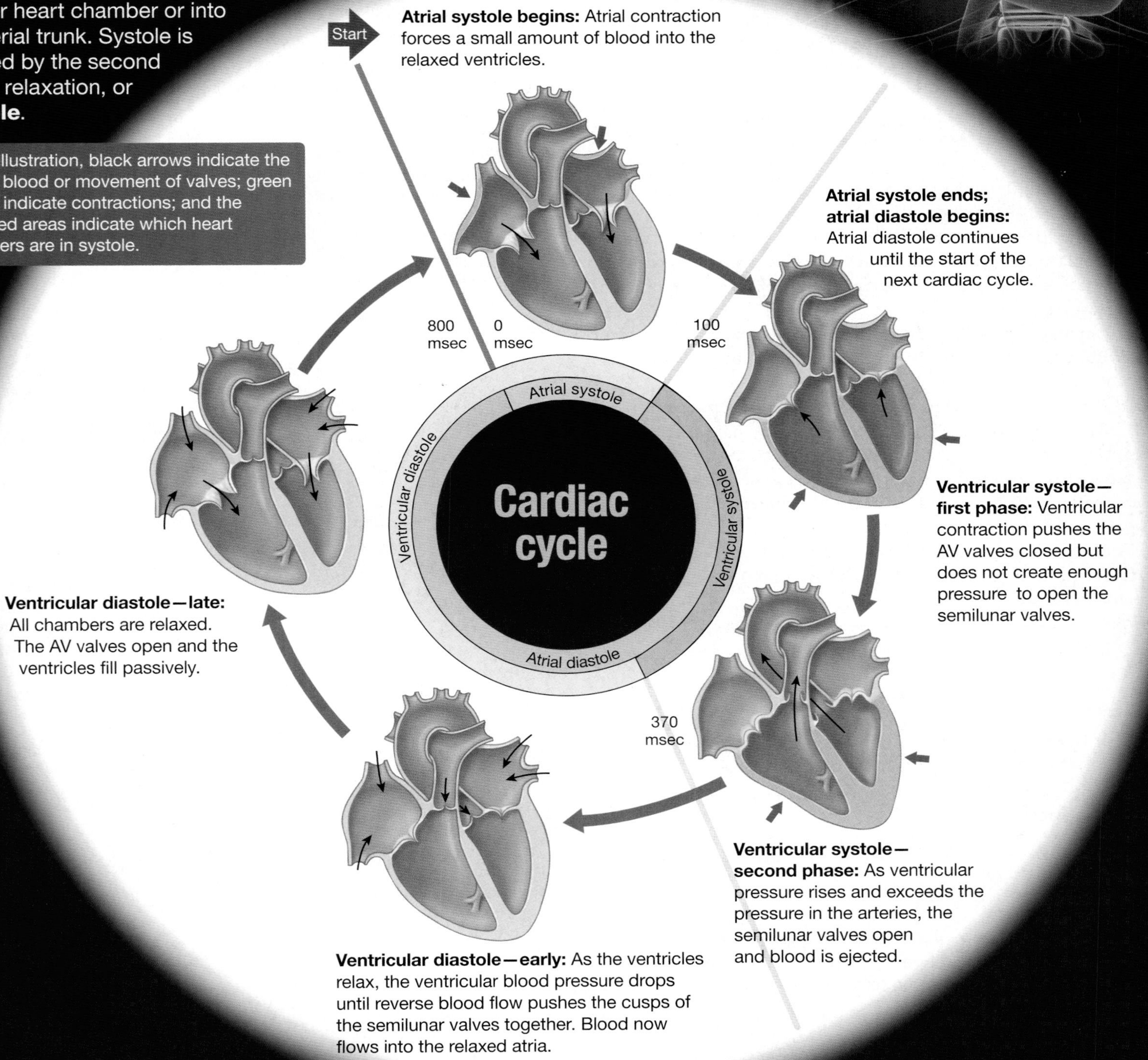

Start
Atrial systole begins: Atrial contraction forces a small amount of blood into the relaxed ventricles.
Atrial systole ends; atrial diastole begins: Atrial diastole continues until the start of the next cardiac cycle.
Ventricular systole—first phase: Ventricular contraction pushes the AV valves closed but does not create enough pressure to open the semilunar valves.
Ventricular systole—second phase: As ventricular pressure rises and exceeds the pressure in the arteries, the semilunar valves open and blood is ejected.
Ventricular diastole—early: As the ventricles relax, the ventricular blood pressure drops until reverse blood flow pushes the cusps of the semilunar valves together. Blood now flows into the relaxed atria.
Ventricular diastole—late: All chambers are relaxed. The AV valves open and the ventricles fill passively.
800 msec
0 msec
100 msec
370 msec
Atrial systole
Ventricular systole
Atrial diastole
Ventricular diastole
Cardiac cycle

21.8 | The Cardiac Cycle

▶ KEY POINT Nodal cells within the heart spontaneously depolarize to threshold, generating an action potential. These action potentials spread throughout the conducting system of the heart to the cardiac muscle cells, initiating a contraction and a cardiac cycle.

The plasma membranes of nodal cells possess unique qualities that allow these cells to spontaneously depolarize to threshold. In addition, nodal cells possess intercellular junctions that electrically couple these cells to one another, to the conducting fibers, and to cardiac muscle cells. When a nodal cell depolarizes, it generates an action potential. The action potential travels through the conducting system of the heart and reaches all the cardiac muscle tissue, causing a contraction and a **cardiac cycle**, or a complete heartbeat. In this way, nodal cells determine the heart rate.

Not all nodal cells depolarize at the same rate. The normal rate of contraction is determined by which nodal cells reach threshold first. The impulse they produce brings all other nodal cells to threshold. These rapidly depolarizing cells are called **pacemaker cells**. They are found in the **sinoatrial*** (sī-nō-Ā-trē-al) **node (SA node)**, or **cardiac pacemaker**. The SA node is located in the posterior wall of the right atrium, near the entrance of the superior vena cava **(Spotlight Figure 21.11)**. These pacemaker cells depolarize spontaneously and rapidly, generating 80–100 action potentials per minute.

Each time the SA node generates an impulse, it produces a heartbeat. Therefore, the resting heart rate is 80–100 beats per minute (bpm). However, any factor that changes either the SA node's resting potential or the rate of spontaneous depolarization alters the heart rate. For example, nodal cell activity is affected by the activity of the autonomic nervous system. When parasympathetic neurons release acetylcholine (ACh), the rate of spontaneous depolarization slows, and the heart rate decreases. In contrast, when sympathetic neurons release norepinephrine (NE), the rate of depolarization increases, and the heart rate increases. Under normal resting conditions, parasympathetic activity reduces the heart rate from the inherent nodal rate of 80–100 impulses per minute to a more leisurely 70–80 beats per minute.

A number of clinical problems result from abnormal pacemaker cell function. **Bradycardia** (brăd-ē-KAR-dē-a; *bradys*, slow) indicates a slower-than-normal heart rate, whereas **tachycardia** (tak-ē-KAR-dē-a; *tachys*, swift) refers to a faster-than-normal heart rate. In clinical practice the definition varies depending on the normal resting heart rate and conditioning of the individual.

The cells of the SA node are electrically connected to those of the larger **atrioventricular** (ā-trē-ō-ven-TRIK-ū-lar) **node (AV node)** through conducting fibers in the atrial walls.

Spotlight Figure 21.11 summarizes the events of the cardiac cycle.

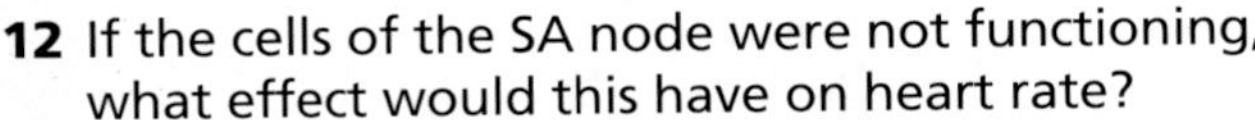

21.8 CONCEPT CHECK

12 If the cells of the SA node were not functioning, what effect would this have on heart rate?

13 How do pacemaker cells coordinate cardiac muscle contractions?

See the blue Answers tab at the back of the book.

*The second edition of the *Terminologia Anatomica* uses the term *sinuatrial node*. The authors of this edition of *Human Anatomy* have chosen to continue to use the commonly accepted term *sinoatrial node*.

21.9 | Autonomic Control of Heart Rate

▶ KEY POINT The pacemaker cells of the SA node determine the intrinsic heart rate, which is modified by the sympathetic and parasympathetic divisions of the autonomic nervous system.

The sympathetic and parasympathetic divisions of the autonomic nervous system (ANS) innervate the heart through the cardiac plexus. ⮌ pp. 462–463 Both the sympathetic and parasympathetic divisions of the ANS innervate the SA and AV nodes. These divisions also innervate the atrial and ventricular cardiac muscle cells and smooth muscle in the walls of cardiac blood vessels **(Figure 21.12)**.

In section 21.8 we discussed the effects of NE and ACh on nodal tissues, and we summarize these effects here:

- NE release increases heart rate and contraction force by stimulating the beta receptors on nodal cells and contractile cells.
- ACh release decreases heart rate and contraction force by stimulating the muscarinic receptors on nodal cells and contractile cells.

The cardiac centers of the medulla oblongata contain the ANS centers for cardiac control. Stimulation of the **cardioacceleratory center** activates sympathetic neurons; the nearby **cardioinhibitory center** activates parasympathetic neurons. The cardiac centers receive inputs from higher centers, especially from the parasympathetic and sympathetic headquarters in the hypothalamus.

Figure 21.12 The Autonomic Innervation of the Heart. Cardiac centers in the medulla oblongata modify heart rate and cardiac output through the vagus nerves (parasympathetic) and through the cardiac nerves (sympathetic).

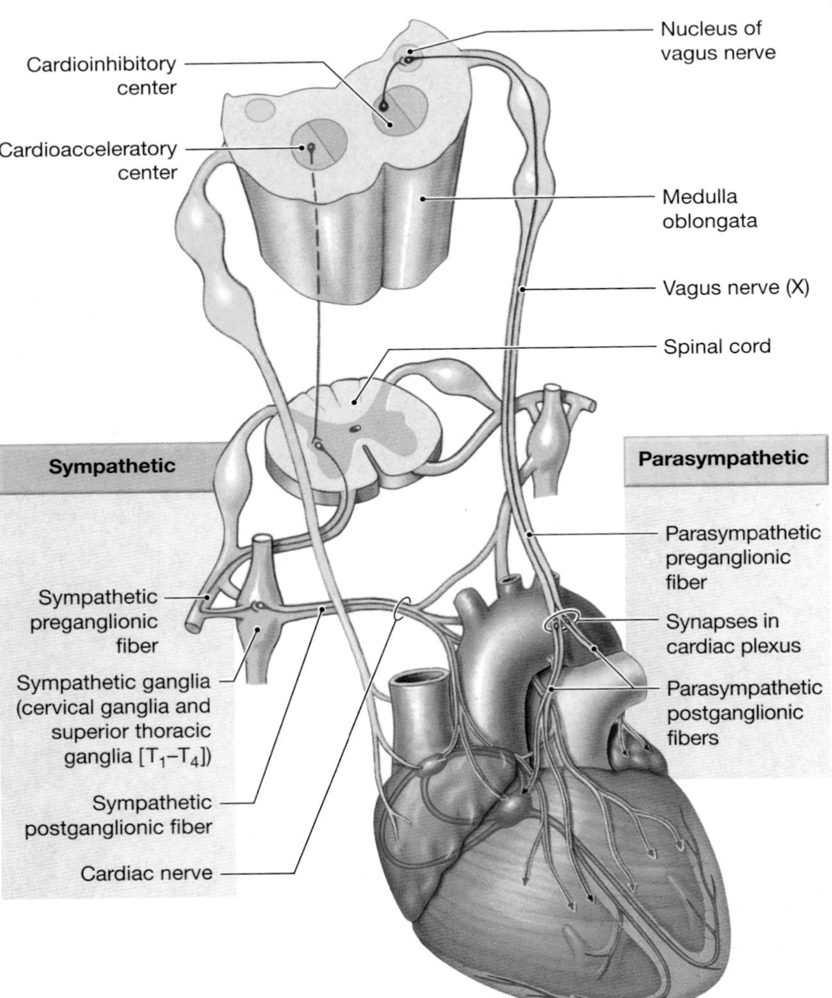

Sensory information about the status of the cardiovascular system arrives at the cardiac centers via visceral sensory fibers from the glossopharyngeal (IX) and vagus (X) nerves. These afferent fibers originate in baroreceptors sensitive to blood pressure and chemoreceptors sensitive to dissolved gas concentrations. The cardiac centers respond very quickly to this information in order to adjust cardiac performance and maintain adequate circulation to vital organs, such as the brain. For example, a drop in blood pressure or an increase in carbon dioxide concentration indicates that the heart must work harder to meet the demands of peripheral tissues. The cardiac centers then respond by increasing the heart rate and force of contraction by activating the sympathetic nervous system.

TIPS & TOOLS

To remember the effect of the sympathetic nervous system on cardiac performance, think SSS: **S**ympathetic input **S**peeds and **S**trengthens the heartbeat.

21.9 CONCEPT CHECK

14 The sympathetic and parasympathetic branches of the autonomic nervous system have different effects on nodal tissues within the heart. What are these effects?

See the blue Answers tab at the back of the book.

EMBRYOLOGY SUMMARY

For a summary of the development of the cardiovascular system, see Chapter 28 (Embryology and Human Development).

Study Outline

Introduction p. 546

- All the tissues and fluids in the body rely on the cardiovascular system to maintain homeostasis. The proper functioning of the cardiovascular system depends on the activity of the **heart**, which can vary its pumping capacity depending on the needs of the peripheral tissues.

21.1 An Overview of the Cardiovascular System p. 546

- The cardiovascular system is subdivided into two closed circuits that occur in series. Each circuit functions individually in series, while the two circuits together function in parallel. The **pulmonary circuit** carries oxygen-poor blood from the heart to the lungs and back, and the **systemic circuit** transports oxygen-rich blood from the heart to the rest of the body and back. **Arteries** carry blood away from the heart; **veins** return blood to the heart. **Capillaries** are tiny vessels between the smallest arteries and veins. *(See Figure 21.1.)*
- The heart contains four chambers: the right atrium and ventricle, and the left atrium and ventricle. The **atria** collect blood returning to the heart, and the **ventricles** discharge blood into vessels to leave the heart.

21.2 The Pericardium p. 546

- The heart is surrounded by the **pericardium** and lies within the anterior portion of the mediastinum, which separates the two pleural cavities. The pericardial cavity is lined by the **pericardium** and contains a small amount of lubricating fluid, called the **pericardial fluid**. The **visceral layer of serous pericardium** (*epicardium*) covers the heart's outer surface, and the **parietal layer of serous pericardium** lines the inner surface. *(See Figures 12.11 and 21.2.)*

21.3 Structure of the Heart Wall p. 548

- The heart wall contains three layers: the **visceral layer of serous pericardium (epicardium)**, the **myocardium** (the muscular wall of the heart), and the **endocardium** (the epithelium lining the inner surfaces of the heart). *(See Figure 21.3.)*

Cardiac Muscle Tissue p. 548

- The bulk of the heart consists of the muscular myocardium. Cardiac muscle cells, which are smaller than skeletal muscle cells, are almost totally dependent on aerobic respiration. *(See Figure 21.3.)*
- Cardiac muscle cells are interconnected by **intercalated discs**, which both convey the force of contraction from cell to cell and conduct action potentials. Intercalated discs join cardiac muscle cells through desmosomes, myofibrils, and gap junctions. Because cardiac muscle cells are connected in this way, they function like a single, enormous cell. *(See Figure 21.3.)*

The Cardiac Skeleton p. 548

- The internal connective tissue of the heart is called the **cardiac skeleton**. *(See Figures 21.3b and 21.9.)*
- The cardiac skeleton of the heart functions to stabilize the heart's contractile cells and valves; support the muscle cells, blood vessels, and nerves; distribute the forces of contraction; add strength and elasticity; and physically isolate the atria from the ventricles.

21.4 Orientation and Superficial Anatomy of the Heart p. 550

- The great vessels are connected to the superior end of the heart at the **base**. The inferior, pointed tip of the heart is the **apex**. *(See Figures 21.2b and 21.4.)*
- The heart sits at an angle to the longitudinal axis of the body and presents the following **borders**: **superior**, **inferior**, **left**, and **right**. *(See Figure 21.4.)*
- The heart has the following surfaces: The **anterior surface**, or *sternocostal surface*, of the heart faces the anterior thoracic wall and consists mostly of the wall of the right ventricle (two-thirds) and some of the left ventricle (one-third). The **posterior surface**, at the base, is formed by the left atrium and a small portion of the right atrium. The **diaphragmatic surface** of the heart is composed mainly of the posterior, inferior surfaces of the right and left ventricles. *(See Figures 21.5 and 21.6.)*
- The division of the heart into four chambers produces external landmarks that are visible as grooves or sulci on the surface of the heart. The **interatrial groove** separates the two atria, while the **coronary sulcus** separates the atria from the ventricles. Other shallower depressions include the **anterior interventricular sulcus** and the **posterior interventricular sulcus**. *(See Figures 21.5 and 21.6.)*
- The **auricle** (atrial appendage) is an expandable extension of the atrium.

21.5 Internal Anatomy and Organization of the Heart p. 553

- The atria are separated by the **interatrial septum**, and the ventricles are divided by the **interventricular septum**. The openings between the atria and ventricles contain folds of connective tissue covered by endocardium; these **valves** maintain a one-way flow of blood. *(See Figures 21.7 and 21.8.)*

The Right Atrium p. 553

- The right atrium receives blood from the systemic circuit through two great veins, the **superior vena cava** and **inferior vena cava**. The atrial walls contain prominent muscular ridges, the **pectinate muscles**. The **coronary veins** return blood to the coronary sinus, which opens into the right atrium. During embryonic development an opening called the **foramen ovale** penetrates the interatrial septum. This opening closes after birth, leaving a depression termed the **fossa ovalis**. *(See Figures 21.5–21.8.)*

The Right Ventricle p. 554

- Blood flows from the right atrium into the right ventricle through the **right atrioventricular (AV) valve**, or *tricuspid valve*. This valve consists of three **cusps** of fibrous tissue braced by the tendinous **chordae tendineae** that are connected to **papillary muscles**. *(See Figures 21.7 and 21.8.)*
- Blood leaving the right ventricle enters the **pulmonary trunk** after passing through the **pulmonary valve**. The pulmonary trunk divides to form the **left** and **right pulmonary arteries**. *(See Figures 21.5–21.8.)*

The Left Atrium p. 554

- The left atrium receives oxygenated blood from the **left** and **right pulmonary veins**; it has thicker walls than those of the right atrium. *(See Figures 21.5, 21.6, and 21.7b.)*

21

- Blood leaving the left atrium flows into the left ventricle through the **left atrioventricular (AV) valve** (*mitral valve*).

The Left Ventricle p. 554

- The left ventricle is the largest and thickest of the four chambers because it must pump blood to the entire body. Blood leaving the left ventricle passes through the **aortic valve** and into the systemic circuit via the **ascending aorta**. Blood passes from the ascending aorta through the **aortic arch** and into the **descending aorta**. *(See Figures 12.11, 21.5, 21.6, 21.7b, and 21.8.)*

Structural Differences between the Right and Left Ventricles p. 554

- The right ventricle has thin walls and develops low pressure when pumping into the pulmonary circuit to and from the adjacent lungs. Functionally, low pressure is necessary because the pulmonary capillaries at the gas exchange surfaces of the lungs are very delicate. The left ventricle has a thick wall because it pumps blood throughout the systemic circuit. *(See Figures 12.11, 21.7b, and 21.8).*

The Structure and Function of Heart Valves p. 555

- The AV valves have four components: (1) a ring of connective tissue attached to the cardiac skeleton of the heart, (2) cusps, (3) chordae tendineae, and (4) papillary muscles.
- There are two semilunar valves, the **aortic valve** and the **pulmonary valve**, guarding the exits of the left and right ventricles. *(See Figures 21.7–21.9.)*
- Valves normally permit blood flow in only one direction, preventing **regurgitation** (backflow) of blood.

21.6 Coronary Blood Vessels p. 555

- The **coronary circulation** supplies blood to the muscles of the heart to meet the high oxygen and nutrient demands of cardiac muscle cells.
- The **coronary arteries** originate at the base of the ascending aorta, and each gives rise to two branches.

The Right Coronary Artery p. 555

- The **right coronary artery (RCA)** gives rise to both a **right marginal branch** and a **posterior interventricular branch**. *(See Figure 21.10.)*

The Left Coronary Artery p. 559

- The **left coronary artery (LCA)** gives rise to both a **circumflex branch** and an **anterior interventricular branch**. Interconnections between arteries called **anastomoses** ensure a constant blood supply. *(See Figure 21.10.)*

The Coronary Veins p. 559

- The **great** and **middle cardiac veins** carry blood from the coronary capillaries to the **coronary sinus**. *(See Figure 21.10.)*
- Other cardiac veins that empty into the great cardiac vein or the coronary sinus are the **posterior vein of the left ventricle**, draining the areas served by the circumflex branch of the LCA; the middle cardiac vein, draining the areas supplied by the posterior interventricular branch of the LCA; and the **small cardiac vein**, draining blood from the posterior surfaces of the right atrium and ventricle. *(See Figures 21.5b, 21.6b, and 21.10b,d.)*
- The **anterior cardiac veins** drain the anterior surface of the right ventricle and empty directly into the right atrium.

21.7 The Coordination of Cardiac Contractions p. 559

- Cardiac muscle tissue contracts on its own, without neural or hormonal stimulation. This is called **autorhythmicity**.
- **Nodal cells** establish the rate of cardiac contraction, and **conducting cells** distribute the contractile stimulus to the general myocardium. *(See Spotlight Figure 21.11.)*

21.8 The Cardiac Cycle p. 562

- Nodal cells depolarize spontaneously and determine the heart rate.
- **Pacemaker cells** found in the **sinoatrial (SA) node** (**cardiac pacemaker**) normally establish the rate of contraction. *(See Spotlight Figure 21.11.)*
- From the SA node, the stimulus travels over the **internodal pathways** to the **atrioventricular (AV) node**, then to the **AV bundle**, which divides into a **right** and **left bundle branch**. From there **Purkinje fibers** convey the impulses to the ventricular myocardium. *(See Spotlight Figure 21.11.)*
- The **cardiac cycle** consists of periods of **atrial** and **ventricular systole** (contraction) and atrial and ventricular **diastole** (relaxation/filling). *(See Spotlight Figure 21.11.)*

21.9 Autonomic Control of Heart Rate p. 562

- The basic heart rate is established by the pacemaker cells, but it can be modified by the ANS. Norepinephrine produces an increase in heart rate and force of contraction, while acetylcholine produces a decrease in heart rate and contraction strength.
- The **cardioacceleratory center** in the medulla oblongata activates sympathetic neurons; the **cardioinhibitory center** governs the activities of the parasympathetic neurons. The cardiac centers receive inputs from higher centers and from receptors monitoring blood pressure and the concentrations of dissolved gases in the blood. *(See Figure 21.12.)*

Chapter Review

For answers, see the blue Answers tab at the back of the book.

Level 1 Reviewing Facts and Terms

Match each numbered item with the most closely related lettered item.

1. cardiac muscle cells ☐
2. bradycardia ☐
3. diastole ☐
4. coronary circulation ☐
5. visceral layer of serous pericardium ☐
6. systole ☐
7. myocardium ☐
8. right pulmonary vein ☐
9. superior vena cava ☐
10. parietal pericardium ☐

(a) vein to the left atrium
(b) covers outer surface of the heart
(c) supplies blood to heart muscle
(d) lines inner surface of fibrous pericardium
(e) slow heart rate
(f) functional syncytium
(g) muscular wall of the heart
(h) relaxation phase of the cardiac cycle
(i) vein to the right atrium
(j) contraction phase of the cardiac cycle

11. The potential space between the parietal and visceral layers of serous pericardium is the
(a) pleural cavity.
(b) peritoneal cavity.
(c) abdominopelvic cavity.
(d) pericardial cavity.

12. Give the anatomical names of the structures labeled on the diagram below.

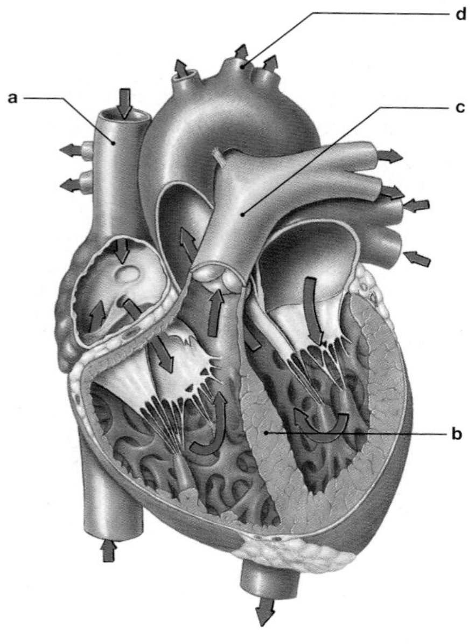

(a) __________ (b) __________
(c) __________ (d) __________

13. Which of the following is *not* true of intercalated discs?
(a) They provide additional strength from cells bound together by tight junctions.
(b) They have a smooth junction between the plasma membrnaes of apposed muscle cells.
(c) Myofibrils of the interlocking muscle fibers are anchored at the membrane.
(d) The cardiac muscle fibers at the intercalated discs are connected by gap junctions.

14. The heart is innervated by
(a) only parasympathetic nerves.
(b) only sympathetic nerves.
(c) both sympathetic and parasympathetic nerves.
(d) only splanchnic nerves.

15. The pacemaker cells of the heart are located in
(a) the SA node.
(b) the wall of the left ventricle.
(c) the Purkinje fibers.
(d) both the left and right ventricles.

16. The two main branches of the right coronary artery (RCA) are the
(a) circumflex branch and the left marginal branch.
(b) anterior interventricular branch and the left anterior descending branch.
(c) right marginal branch and the posterior interventricular branch.
(d) great and middle cardiac veins.

Level 2 Reviewing Concepts

1. The cardiac skeleton of the heart has which *two* of the following functions?
(a) It physically isolates the muscle fibers of the atria from those of the ventricles.
(b) It maintains the normal shape of the heart.
(c) It helps distribute the forces of cardiac contraction.
(d) It allows more rapid contraction of the ventricles.
(e) It strengthens and helps prevent overexpansion of the heart.

2. If the papillary muscles fail to contract,
(a) blood will not enter the atria.
(b) the ventricles will not pump blood.
(c) the AV valves will not close properly.
(d) the aortic and pulmonary valves will not open.

3. If there were damage to the sympathetic innervation to the heart, what would happen to the heart rate under the influence of the remaining autonomic nervous system stimulation?
(a) It would increase.
(b) It would not change.
(c) It would decrease.
(d) It would first increase and then decrease.

4. How is cardiac muscle similar to skeletal muscle?
5. Why do the aortic and pulmonary valves lack muscular braces like those found in AV valves?
6. Define a pacemaker cell, and list the group of cells that normally serve as the heart's pacemaker, as well as those other cells that have the potential to serve as a pacemaker.
7. What is the function of the pericardial fluid?
8. Which chamber of the heart has the thickest walls? Why are its walls so thick?
9. Why are nodal cells unique? What is their function?
10. Describe the function of the SA node in the cardiac cycle. How does this function differ from that of the AV node?

Level 3 Critical Thinking

1. Harvey has a heart murmur in his left ventricle that produces a loud "gurgling" sound at the beginning of systole. What do you suspect to be the cause of this sound?
2. Lee is brought to the emergency room suffering from a cardiac arrhythmia. In the ER he begins to exhibit tachycardia and as a result loses consciousness. His wife asks you why he lost consciousness. What would you tell her?
3. If the cardiac centers detect an abundance of oxygen in the blood, what chemical is likely to be released?

CLINICAL CASE WRAP-UP

A Broken Heart

Ellen is unconscious but appears to be in regular cardiac sinus rhythm. However, an ECG (electrocardiogram) shows changes that resemble a myocardial infarction (heart attack).

She is rushed to the cardiac catheterization laboratory, where a catheter is inserted into her right brachial artery and threaded "upstream" into her aorta and left ventricle. The cardiologist injects dye to outline the inner aspect of her left ventricle, which appears abnormally large and balloon-shaped. Her left ventricular apical wall does not move with ventricular systole, and this hypokinesia (decreased motion) is causing decreased cardiac output. The catheter is withdrawn back through Ellen's aortic valve and more dye is injected into her aortic sinus. Her coronary vessels appear normal.

The cardiologist identifies Ellen's case as takotsubo cardiomyopathy—a name based on the characteristic shape of her ventricle at end systole, which resembles a Japanese fisherman's octopus pot, or *tako-tsubo*. Another name for this condition is broken heart syndrome, because many patients, like Ellen, are suffering severe emotional stress. Ellen's sympathetic nervous system is releasing norepinephrine (NE) in "overdrive," affecting her heart.

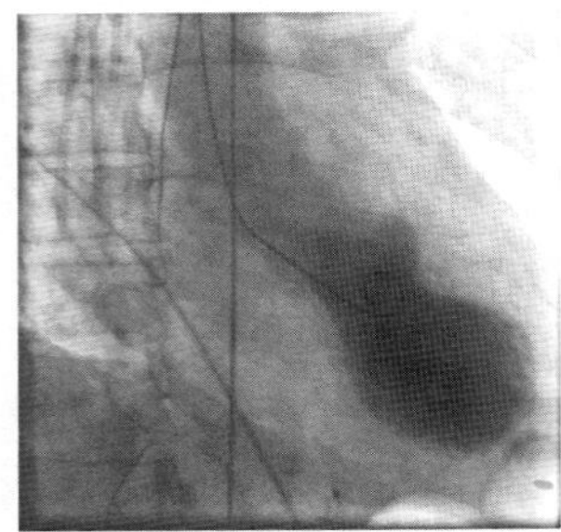

The best news for Ellen is that she is expected to make a complete recovery, simply with supportive care.

1. Why did Ellen feel like she was having a heart attack?
2. If Ellen's cardiologist wanted to inject dye into her right atrium and ventricle, would an artery or vein in her antecubital fossa be best to use? Explain.

See the blue Answers tab at the back of the book.

Related Clinical Terms

21

artificial pacemaker: A small battery-operated device that keeps a person's heart beating in a regular rhythm. It may be permanently implanted or for temporary usage, it may be an external device.

asystole: The absence of cardiac activity with no contraction and no output.

automated external defibrillator (AED): A device that, when applied, automatically checks the function of the heart. Upon detecting a condition that may respond to an electric shock, it delivers a shock to restore normal heartbeat rhythm.

cardiomegaly: An enlarged heart, which is a sign of some other condition such as stress, weakening of the heart muscle, coronary artery disease, heart valve problems, or abnormal heart rhythms.

cardiomyoplasty: A surgical procedure that uses stimulated latissimus dorsi to assist with cardiac function. The latissimus dorsi is relocated and wrapped around the left and right ventricles and stimulated to contract during cardiac systole by means of an implanted burst-stimulator.

congestive heart failure: The heart condition of weakness, edema, and shortness of breath caused by the inability of the heart to maintain adequate blood circulation in the peripheral tissues and lungs.

fibrillation: Fast twitching of the heart muscle fibers with little or no movement of the muscle as a whole. Atrial fibrillation occurs in the atria of the heart and is characterized by chaotic quivers and irregular ventricular beating, with both atria and ventricles being out of sync.

heart block: Delay in the normal electrical pulses that cause the heart to beat.

palpitation: Irregular and rapid beating of the heart.

percutaneous transluminal coronary angioplasty (PTCA): The surgical use of a balloon-tipped catheter to enlarge a narrowed artery.

sick sinus syndrome: A group of heart rhythm disorders or problems in which the sinoatrial node does not work properly to regulate the heart rhythms.

22

The Cardiovascular System

Vessels and Circulation

Learning Outcomes

These learning outcomes correspond by number to this chapter's sections and indicate what you should be able to do after completing the chapter.

22.1 Compare and contrast the histology of an elastic artery, muscular artery, arteriole, capillary, venule, and vein. p. 568

22.2 Compare the percentage of blood that is normally contained within the heart, arteries, and capillaries to the percentage that is normally contained within the veins. p. 574

22.3 Outline three important functional patterns that describe the distribution of blood vessels within the body. p. 575

22.4 Outline the path blood would take as it passes from the right atrium of the heart to the lungs and back to the right atrium. p. 576

22.5 Identify the major arteries of the systemic circuit and the areas and organs supplied by each vessel. p. 576

22.6 Identify the major veins of the systemic circuit and the areas and organs drained by each vessel. p. 590

22.7 Describe the major cardiovascular changes that occur at birth and explain their functional significance. p. 596

22.8 Outline the age-related changes that occur in the cardiovascular system. p. 598

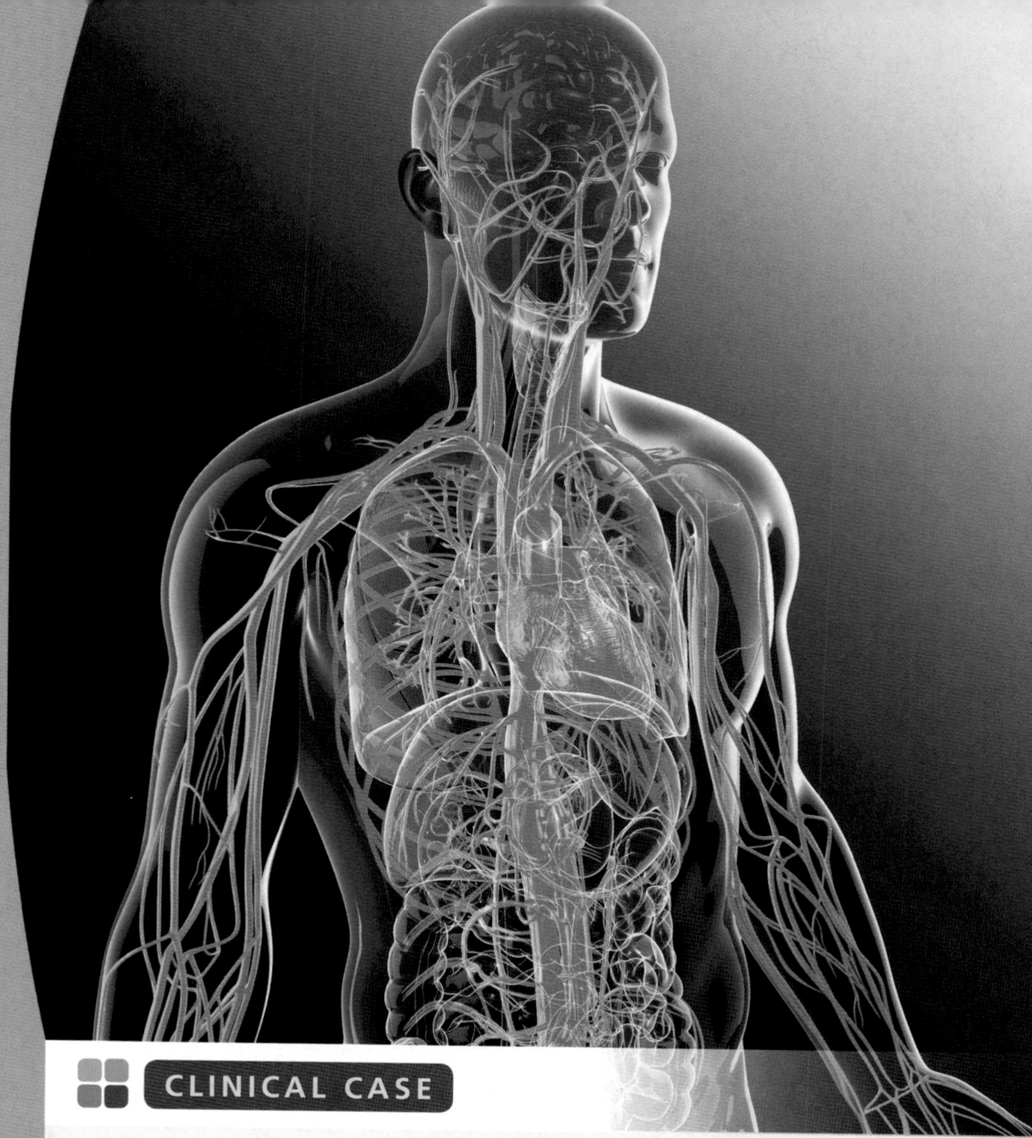

CLINICAL CASE

In the Absence of Capillaries

Rodrigo, a college freshman, is facing second semester finals. For the past few weeks he has noticed intermittent numbness and weakness in his right hand, sometimes extending up to his shoulder. Now, during finals week, he has pounding headaches in the left side of his head.

Rodrigo is in the library with his study group when he suddenly falls to the floor, unconscious. His body stiffens and jerks violently, and he loses bladder control. Another student recognizes this as a seizure. She calls 911 and stays with him, padding his head with her coat and preventing him from hurting himself.

After a few minutes, Rodrigo slowly opens his eyes but is confused and unable to speak. The rescue squad loads him on a stretcher and transports him to the emergency room. After interviewing Rodrigo's study buddy, the emergency room physician sends Rodrigo for a magnetic resonance imaging (MRI) scan to inspect his brain.

What is happening to Rodrigo? To find out, turn to the Clinical Case Wrap-Up on p. 601.

THE CARDIOVASCULAR SYSTEM is a closed system that circulates blood throughout the body. There are two groups of blood vessels: The pulmonary circuit supplies the lungs, and the systemic circuit supplies the rest of the body.

The heart pumps blood into the pulmonary and systemic circuits simultaneously. The pulmonary circuit begins at the pulmonary valve and ends at the entrance to the left atrium. Pulmonary arteries branch from the pulmonary trunk and carry blood to the lungs for gas exchange. The systemic circuit begins at the aortic valve and ends at the entrance to the right atrium. Systemic arteries branch from the aorta and distribute blood to all other organs for nutrient, gas, and waste exchange.

After blood vessels enter an organ, further branching occurs, forming smaller and smaller blood vessels. This creates several *hundred million* tiny arteries providing blood to more than *10 billion* capillaries, each barely the diameter of a single red blood cell. These capillaries form extensive branching networks. Estimates of the combined length of all capillaries in the body, if they were placed end to end, range from 5000 to 25,000 miles. In other words, the capillaries in your body could at least cross the continental United States and perhaps circle the globe!

All chemical and gaseous exchange between the blood and interstitial fluid takes place across capillary walls. Tissue cells rely on capillary diffusion to obtain nutrients and oxygen and remove metabolic wastes. Blood leaving the capillary networks enters a network of small veins that gradually merge to form larger vessels. These larger vessels ultimately drain into either the pulmonary veins (pulmonary circuit) or the inferior or superior vena cava (systemic circuit).

Our initial discussion in this chapter focuses on the histological and anatomical organization of arteries, capillaries, and veins. We then identify the major blood vessels and circulatory routes of the cardiovascular system.

22.1 Histological Organization of Blood Vessels

22

▶ **KEY POINT** All blood vessels within the cardiovascular system, with the exception of capillaries, follow the same histological organization. The walls of arteries, arterioles, veins, and venules contain three layers: intima, media, and adventitia.

The walls of arteries and veins contain three distinct layers, from superficial to deep: (1) an outer adventitia, (2) a middle media, and (3) an inner intima **(Spotlight Figure 22.1)**.

1. The outer **adventitia** (ad-ven-TISH-a) (*tunica externa*) forms a connective tissue sheath around the vessel. This layer is very thick and is composed chiefly of collagen fibers, with scattered bands of elastic fibers. In veins, this layer is usually thicker than the media. The fibers of the adventitia blend into those of adjacent tissues, stabilizing and anchoring the blood vessel.

2. The **media** (*tunica media*), the middle layer, contains concentric layers of smooth muscle tissue supported by a framework of loose connective tissue. The smooth muscle cells of the media encircle the lumen of the blood vessel. When stimulated by the sympathetic branch of the autonomic nervous system, these smooth muscles contract, reducing the luminal diameter of the blood vessel. This process is called **vasoconstriction**. Relaxation of the smooth muscles increases the diameter of the lumen, a process called **vasodilation** (vaz-ō-dī-LĀ-shun). In addition to responding to the autonomic nervous system, these smooth muscles contract or relax in response to local stimuli, such as changes in pH, pCO_2, or pO_2. Any change in vessel diameter affects both blood pressure and blood flow. Collagen fibers bind the media to both the intima and adventitia. Arteries have a thin band of elastic fibers, called the **external elastic membrane**, located between the media and adventitia.

3. The **intima** (*tunica intima*) is the innermost layer of a blood vessel. It is composed of the endothelial lining of the vessel and an underlying layer of connective tissue containing variable amounts of elastic fibers. In arteries, the outer margin of the intima contains a thick layer of elastic fibers called the **internal elastic membrane**. In the largest arteries, the connective tissue within this layer is extensive, and the intima is thicker than in smaller arteries.

The combination of muscular and elastic components gives arteries and veins considerable strength. It also permits alterations in vessel luminal diameter as blood pressure or blood volume change. However, the vessel walls are too thick to allow diffusion between the bloodstream and surrounding tissues, or even between the blood and the tissues of the vessel itself. Instead, the walls of large vessels contain small arteries and veins that supply the smooth muscle cells, fibroblasts, and fibrocytes of the media and adventitia. These blood vessels are the **vasa vasorum** ("vessels of vessels").

Distinguishing Arteries from Veins

▶ **KEY POINT** In a histological section, the walls of arteries are thicker than those of veins, arteries retain their circular shape, and arterial endothelial linings look pleated.

Arteries and veins in the same region of the body typically lie side by side within a sheath of connective tissue **(Spotlight Figure 22.1)**. We can distinguish arteries from veins by the following characteristics:

- Vessel walls: The walls of arteries are thicker than those of veins. The media of an artery contains more smooth muscle and elastic fibers than a vein. These contractile and elastic components resist the pressure generated by the heart as it forces blood into the circuit.
- Vessel lumen: The walls of an artery will contract, constricting the lumen. Therefore, when viewed during dissection or in sectional view, the lumen of an artery appears smaller than an accompanying vein. Because arterial walls are thicker and stronger, they retain their circular shape in histological sections. In contrast, cut veins collapse, and in a histological section they often look flattened or distorted.
- Vessel lining: The endothelial lining of an artery cannot contract, so when an artery constricts, its endothelium folds. Therefore, the sectioned arteries have a pleated appearance. The lining of a vein lacks these folds.
- Valves: Veins typically contain valves—internal structures that prevent the backflow of blood toward the capillaries. Arteries do not have valves.

Arteries

▶ **KEY POINT** There are three types of arteries: elastic arteries, muscular arteries, and arterioles.

In traveling from the heart to peripheral capillaries, blood passes through a series of arteries of smaller and smaller diameter: elastic arteries, muscular arteries, and arterioles **(Spotlight Figure 22.1)**.

Elastic Arteries

Elastic arteries are large vessels with luminal diameters of up to about 2.5 cm. They transport large volumes of blood away from the heart. Examples include the pulmonary artery and the aorta and their major branches, such as the brachiocephalic trunk and carotid, subclavian, and common iliac arteries. The intima in an elastic artery is relatively thick and is composed of an endothelial lining and an underlying basal lamina of collagen and elastic connective tissue fibers. A layer of elastic tissue, the internal elastic membrane, separates the intima from the media.

The walls of elastic arteries are extremely resilient because the tunica media contains a high density of elastic fibers and relatively few smooth muscle cells. As a result, elastic arteries can tolerate the pressure changes of the cardiac cycle. Smooth muscle cells present within the media of elastic arteries do not contract in response to sympathetic or local stimulation.

During ventricular systole, pressure within the arterial system rises rapidly, and the elastic arteries stretch. During ventricular diastole, blood pressure falls, and the elastic fibers recoil to their original dimensions. Their stretching cushions the sudden rise in pressure during ventricular systole, and their recoiling slows the drop in pressure during ventricular diastole and forces blood onward toward the capillaries.

Muscular Arteries

Muscular arteries (*medium-sized arteries*) transport blood to the body's skeletal muscles and internal organs. A typical muscular artery has a luminal diameter of approximately 4 mm. Most of the visible, named arteries are muscular arteries, but many muscular arteries are too small to identify with the naked eye.

In a muscular artery, the media is thicker and contains significantly more smooth muscle cells than in an elastic artery **(Spotlight Figure 22.1)**. In addition, a prominent internal elastic membrane separates the intima from the media. The sympathetic division of the ANS and local stimulation control the luminal diameter of muscular arteries. By constricting or relaxing the smooth muscle in the media, the autonomic nervous system regulates blood flow to each organ independently.

Arterioles

Arterioles (ar-TĒR-ē-ōlz) are considerably smaller than muscular arteries. They have an average luminal diameter of about 30 μm and can only be seen with a microscope. They have a thin adventitia. The media is composed of one or two layers of smooth muscle cells that may not form a complete layer **(Spotlight Figure 22.1)**. Smaller muscular arteries and all arterioles change their luminal diameter in response to local conditions or sympathetic or endocrine stimulation. Arterioles control blood flow between arteries and capillaries.

Elastic and muscular arteries are seamlessly interconnected, and vessel characteristics change gradually as the vessels get farther away from the heart. For example, the largest muscular arteries contain a considerable amount of elastic tissue, while the smallest resemble heavily muscled arterioles.

Capillaries

▶ **KEY POINT** The wall of a typical capillary is composed of one to three circularly arranged endothelial cells and the underlying basil lamina. These thin walls allow the exchange of nutrients and waste products.

Capillaries are the smallest and most delicate blood vessels **(Spotlight Figure 22.1)**. They are the only blood vessels whose walls allow the exchange of nutrients and wastes between the blood and the surrounding interstitial fluids. Because the walls are thin, the diffusion distances are short. As a result, the exchange between the blood and interstitial fluids occurs quickly. In addition,

CLINICAL NOTE

Arteriosclerosis

Arteriosclerosis is hardening of the arteries. Healthy arteries are flexible and elastic, but with aging and wear, arterial walls harden and stiffen. There are several types of arteriosclerosis.

Atherosclerosis is a disease of the arteries characterized by the deposition of fatty plaques on the inner arterial walls. It is the most common type of arteriosclerosis and begins with damage to the endothelial lining. The plaques damage the arterial walls and restrict blood flow. Atherosclerosis develops gradually, likely beginning early in life.

Atherosclerosis can affect blood vessels anywhere in the body. When the coronary vessels are involved, it is termed **coronary artery disease (CAD)**. It can lead to **angina** (heart pain) and a **myocardial infarction** (heart attack). If atherosclerosis develops in the carotid arteries and smaller arteries leading to the brain, it can produce a **transient ischemic attack (TIA)**, or mini stroke. The blockage in a TIA is temporary, but if the condition progresses, it can lead to a stroke that causes permanent brain damage. Atherosclerosis in the arteries serving the extremities, particularly the lower legs, can lead to **peripheral artery disease (PAD)**. Intermittent claudication (limping), ischemia, and pain in the muscles (chiefly the calf muscles) brought on by walking is a common symptom. If PAD progresses and blood supply is lost completely, gangrene can result. If atherosclerosis affects the arteries that lead to the kidneys, it can lead to high blood pressure followed by kidney failure.

Some contributors to atherosclerosis cannot be prevented—for instance, aging, family history, and certain inflammatory diseases. However, many factors are preventable. The worst and most preventable contributor is smoking. Other controllable risk factors include high blood pressure, which causes early and progressive arterial damage, and diabetes, which produces widespread and rapid damage. Obesity and a sedentary lifestyle play an important role. High cholesterol and high triglycerides encourage the formation of atherosclerotic plaques.

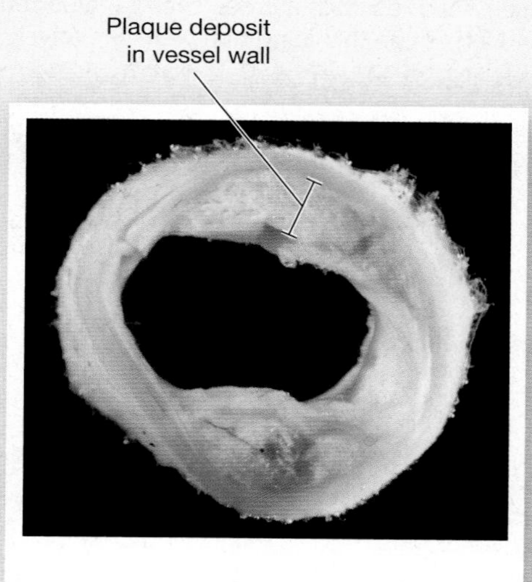

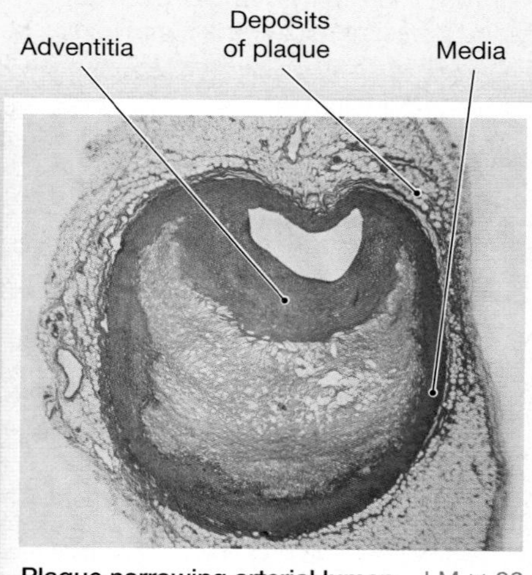

Plaque narrowing arterial lumen LM × 28

SPOTLIGHT

FIGURE 22.1

The Structure of Blood Vessels

Histological Organization of Blood Vessels

The walls of arteries and veins contain three distinct layers: (1) an outer adventitia, (2) a middle media, and (3) an inner intima.

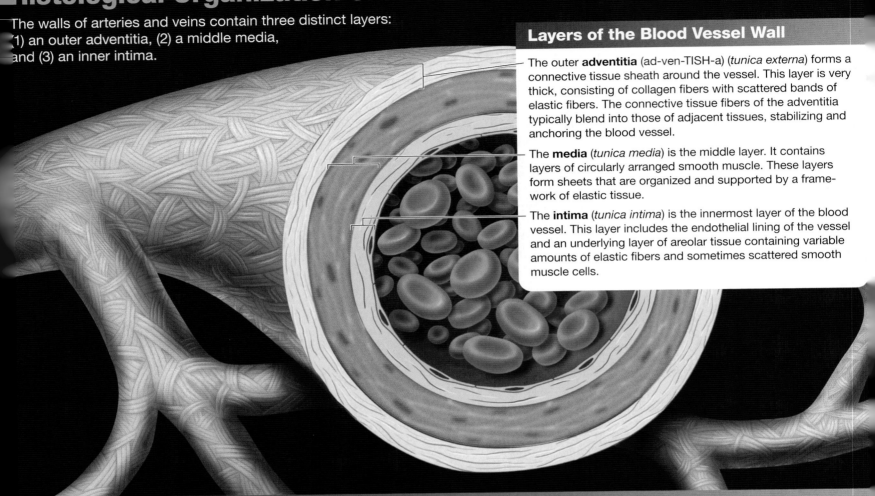

Layers of the Blood Vessel Wall

The outer **adventitia** (ad-ven-TISH-a) (*tunica externa*) forms a connective tissue sheath around the vessel. This layer is very thick, consisting of collagen fibers with scattered bands of elastic fibers. The connective tissue fibers of the adventitia typically blend into those of adjacent tissues, stabilizing and anchoring the blood vessel.

The **media** (*tunica media*) is the middle layer. It contains layers of circularly arranged smooth muscle. These layers form sheets that are organized and supported by a framework of elastic tissue.

The **intima** (*tunica intima*) is the innermost layer of the blood vessel. This layer includes the endothelial lining of the vessel and an underlying layer of areolar tissue containing variable amounts of elastic fibers and sometimes scattered smooth muscle cells.

Histological Comparison of Arteries and Veins

In general, when comparing two adjacent vessels, the walls of arteries are thicker than those of the corresponding veins, and the lumen is relatively smaller. In arteries the adventitia is usually thinner than the media; in veins the adventitia is typically the thickest layer of the vessel wall. The media of an artery contains more smooth muscle and elastic fibers than that of a vein. When the wall of an artery is no longer stretched by blood pressure it constricts and the endothelium wrinkles as the luminal diameter decreases. As a result, in a cross section the arterial lining has a pleated appearance. In contrast, when venous blood pressure falls, veins simply collapse, and in section the lining of a vein is relatively smooth.

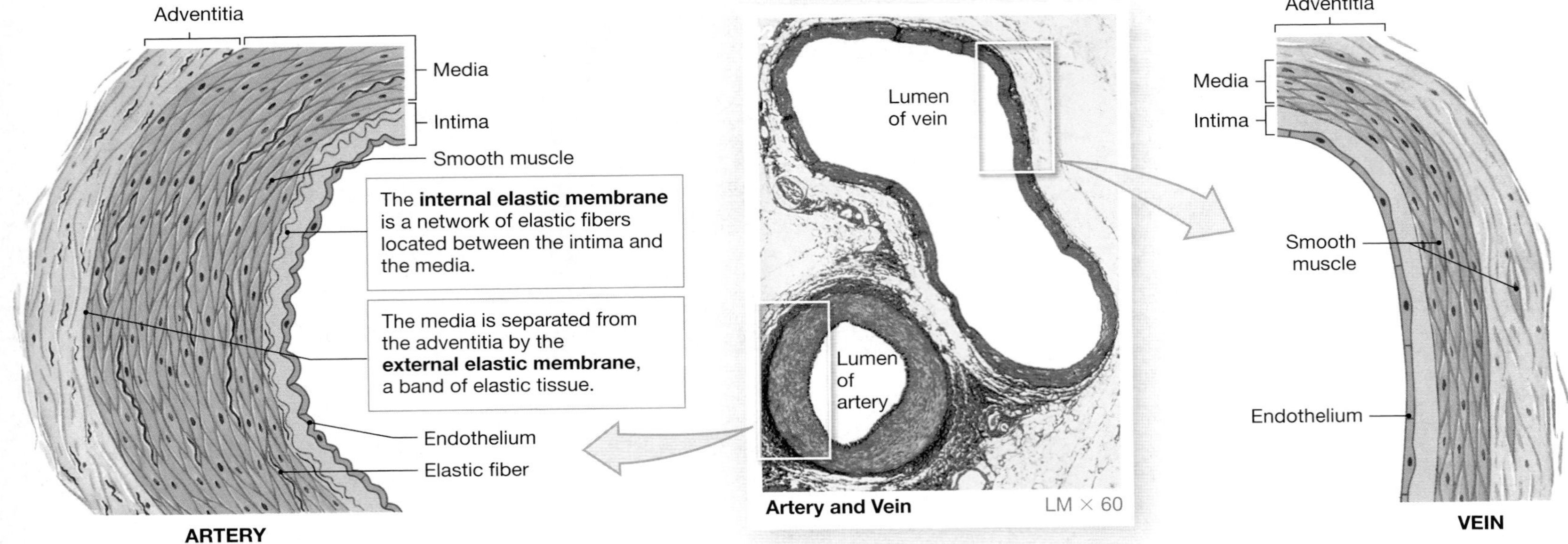

Structural Differences between Arteries and Veins

VEINS

ARTERIES

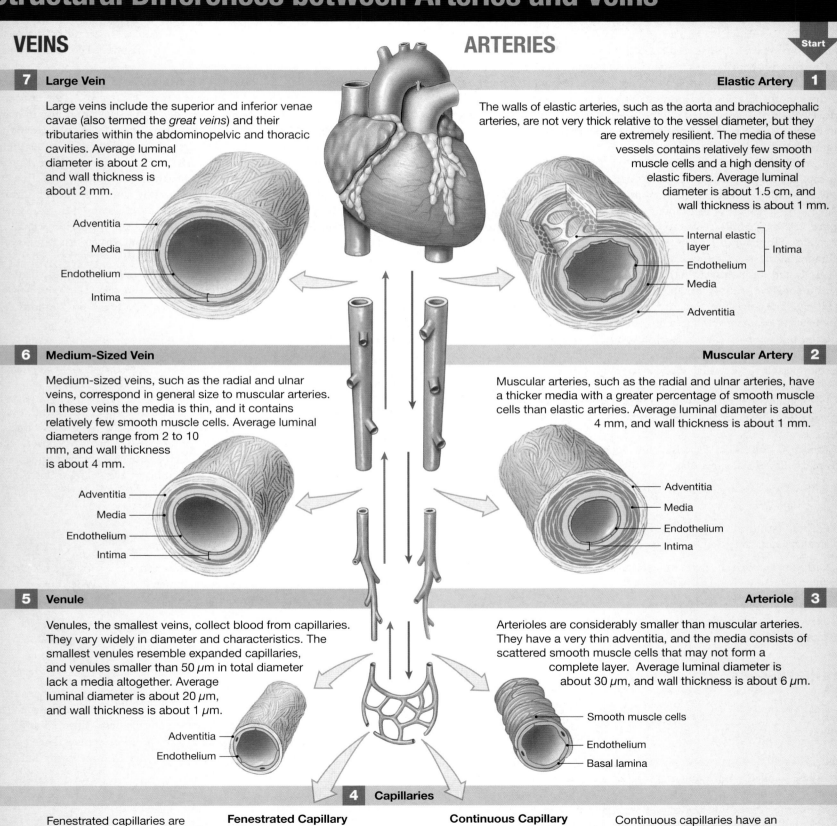

Start

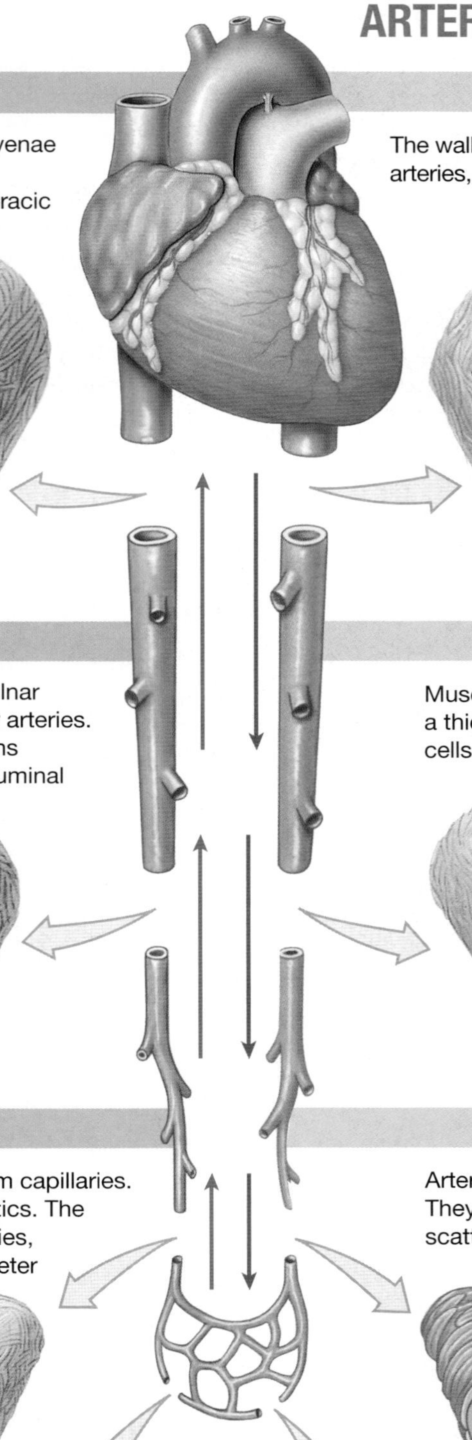

1 Elastic Artery

The walls of elastic arteries, such as the aorta and brachiocephalic arteries, are not very thick relative to the vessel diameter, but they are extremely resilient. The media of these vessels contains relatively few smooth muscle cells and a high density of elastic fibers. Average luminal diameter is about 1.5 cm, and wall thickness is about 1 mm.

Internal elastic layer
Endothelium
Intima
Media
Adventitia

2 Muscular Artery

Muscular arteries, such as the radial and ulnar arteries, have a thicker media with a greater percentage of smooth muscle cells than elastic arteries. Average luminal diameter is about 4 mm, and wall thickness is about 1 mm.

Adventitia
Media
Endothelium
Intima

3 Arteriole

Arterioles are considerably smaller than muscular arteries. They have a very thin adventitia, and the media consists of scattered smooth muscle cells that may not form a complete layer. Average luminal diameter is about 30 μm, and wall thickness is about 6 μm.

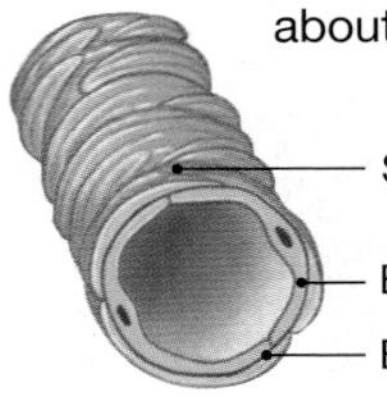

Smooth muscle cells
Endothelium
Basal lamina

4 Capillaries

Fenestrated capillaries are capillaries that contain "windows," or pores in their walls, due to an incomplete or perforated endothelial lining.

Fenestrated Capillary

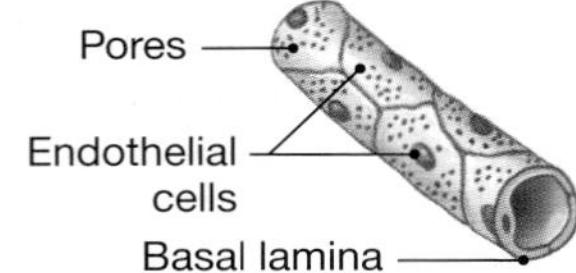

Continuous Capillary

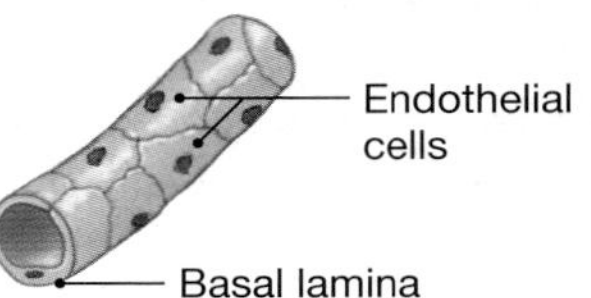

Continuous capillaries have an endothelium that completely surrounds the lumen. Tight junctions and desmosomes connect the endothelial cells. They have an average luminal diameter of 8 μm.

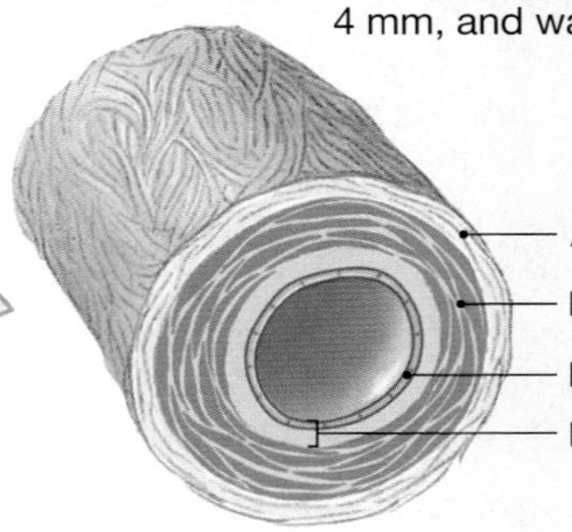

5 Venule

Venules, the smallest veins, collect blood from capillaries. They vary widely in diameter and characteristics. The smallest venules resemble expanded capillaries, and venules smaller than 50 μm in total diameter lack a media altogether. Average luminal diameter is about 20 μm, and wall thickness is about 1 μm.

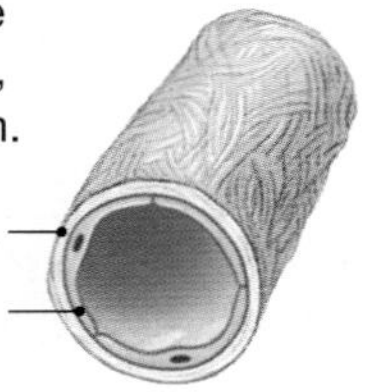

Adventitia
Endothelium

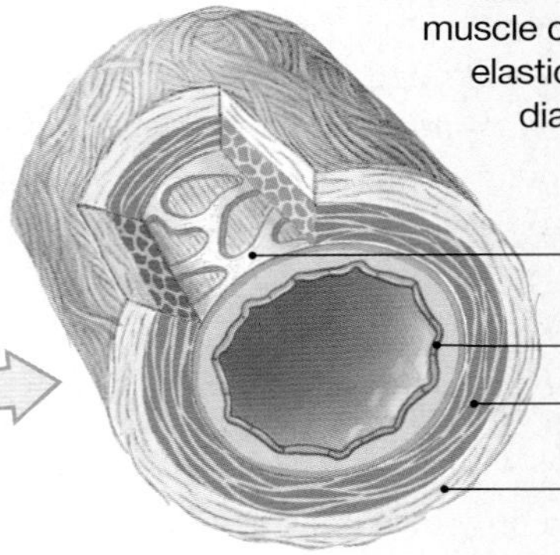

6 Medium-Sized Vein

Medium-sized veins, such as the radial and ulnar veins, correspond in general size to muscular arteries. In these veins the media is thin, and it contains relatively few smooth muscle cells. Average luminal diameters range from 2 to 10 mm, and wall thickness is about 4 mm.

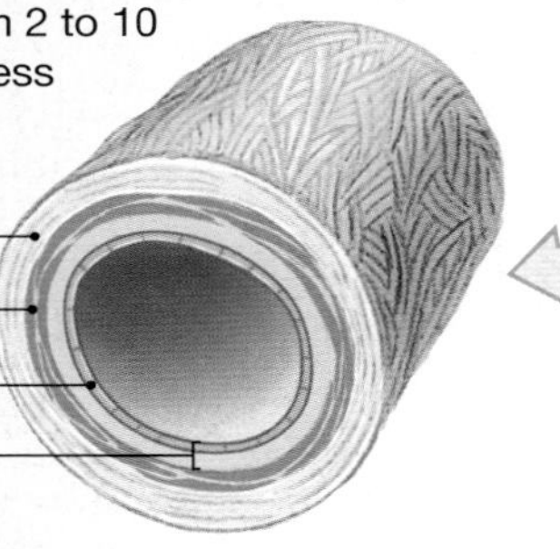

Adventitia
Media
Endothelium
Intima

7 Large Vein

Large veins include the superior and inferior venae cavae (also termed the *great veins*) and their tributaries within the abdominopelvic and thoracic cavities. Average luminal diameter is about 2 cm, and wall thickness is about 2 mm.

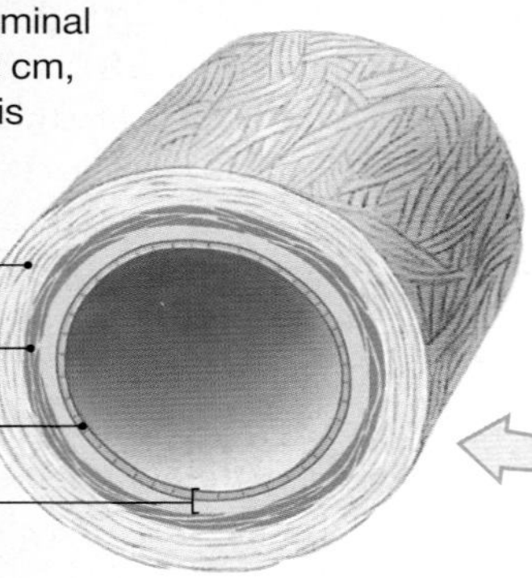

Adventitia
Media
Endothelium
Intima

blood flows slowly through capillaries, allowing sufficient time for diffusion or active transport of materials across the capillary walls. Some substances cross the capillary walls by diffusing across the endothelial cell lining. Other substances pass through gaps between adjacent endothelial cells. The fine structure of each capillary determines its ability to regulate the two-way exchange of substances between blood and interstitial fluid.

A typical capillary wall consists of one to three endothelial cells sitting on a delicate basil lamina. The average luminal diameter of a capillary is only 8 μm, close to that of a single red blood cell.

Continuous capillaries are found in all tissues except epithelia and cartilage. In continuous capillaries, the endothelium forms a complete lining, and the endothelial cells are connected by tight junctions and desmosomes **(Figure 22.2a,d)**. **Fenestrated** (FEN-es-trā-ted; *fenestra*, window) **capillaries** contain "windows," or pores in their walls, due to an incomplete or perforated endothelial lining **(Figure 22.2b,e)**.

A single endothelial cell wraps all the way around the lumen of a continuous capillary **(Figure 22.2d)**. The walls of a fenestrated capillary are far more permeable, with a "Swiss cheese" appearance, and the pores allow molecules as large as peptides and small proteins to pass into or out of the circulation **(Figure 22.2b,e)**. This type of capillary permits very rapid exchange of fluids and solutes. Examples of fenestrated capillaries include the choroid plexus of the brain and the capillaries in the endocrine glands, gallbladder, kidney, pancreas, and some locations within the intestines.

Sinusoids (SĪ-nu-soydz), or *discontinuous capillaries*, resemble fenestrated capillaries that are flattened and irregularly shaped. In addition to being fenestrated, sinusoids commonly have gaps between adjacent endothelial cells, and the basil lamina is either thinner or absent **(Figure 22.2c)**. (In some organs, such as the liver, there is no basil lamina.) As a result, sinusoids permit the free exchange of water and solutes, such as plasma proteins, between blood and interstitial fluid. Blood moves through sinusoids slowly, maximizing the time available for absorption and secretion across the sinusoidal walls. Sinusoids are found in the liver, bone marrow, spleen, and adrenal glands.

Four mechanisms are responsible for the exchange of materials across the walls of capillaries and sinusoids:

1. Diffusion across the capillary endothelial cells (lipid-soluble materials, gases, and water by osmosis)
2. Diffusion through gaps between adjacent endothelial cells (water and small solutes; larger solutes in the case of sinusoids)
3. Diffusion through the pores in fenestrated capillaries and sinusoids (water and solutes)
4. Vesicular transport by endothelial cells (endocytosis at luminal side, exocytosis at basal side), water, and specific bound and unbound solutes ⮌ pp. 32–33

Capillary Beds

▶ **KEY POINT** A capillary bed is a network of capillaries supplying blood to a specific organ or area of the body. Blood flow through a capillary bed can vary considerably.

Capillaries are the site of nutrient, waste, O_2, and CO_2 exchange with interstitial tissues. A collection of capillaries is called a **capillary bed**. A single arteriole gives rise to dozens of capillaries, which empty into several venules. Blood flows from the arterioles to the venules at a constant rate, but the blood flow within a capillary bed varies considerably. Smooth muscle cells within an arteriole cycle between contracting and relaxing, perhaps a dozen times each minute. As a result, blood flow occurs in a series of pulses rather than a steady stream. The effect is that blood reaches the venules by one route at one time and by a different route one moment later. This process is controlled at the tissue level and is called **autoregulation**.

There are also mechanisms modifying the circulatory supply to the entire capillary bed. More than one muscular artery supplies the capillary beds within an area. The arteries, called **collaterals**, enter the area and fuse together rather than forming a series of arterioles. The interconnection is termed an **arterial anastomosis**. Arterial anastomoses are found in the brain, heart, stomach, and other organs or regions with high circulatory demands. This arrangement guarantees a reliable blood supply to the tissues. If one artery becomes blocked, another will supply blood to the capillary bed.

Arteriovenous (ar-tēr-ē-ō-VĒ-nus) **anastomoses** are direct connections between arterioles and venules (look ahead to **Figure 22.3**, p. 574). Arteriovenous anastomoses are common in visceral organs and joints where a change in body position hinders blood flow through one vessel or another. Smooth muscles in the walls of these vessels contract or relax to regulate the amount of blood reaching the capillary bed. For example, when the arteriovenous anastomoses are dilated, blood will bypass the capillary bed and flow directly into the venous circulation.

In some forms of connective tissue, passageways known as **thoroughfare channels** connect arterioles with venules. Lined with smooth muscle, their contractions and relaxations constrict or dilate, thus narrowing or widening the diameter of the vessel and regulating blood flow. Other structures, such as **metarterioles** (also called *precapillary sphincters*), guard the entrances to capillaries in mesenteries **(Figure 22.3)**. Within mesenteric circulation, metarterioles represent thoroughfare channels. However, the existence of metarterioles outside mesenteries has not been supported in the literature.

Veins

▶ **KEY POINT** Veins return blood to the heart. Veins are classified as venules, medium-sized veins, or large veins.

Veins collect blood from all tissues and organs and return it to the heart. The walls of veins are thinner than those of corresponding arteries because the blood pressure in veins is lower than in arteries. We classify veins on the basis of their size. Even though their walls are thinner, in general veins have larger luminal diameters than their corresponding arteries. (Review **Spotlight Figure 22.1** to compare arteries and veins.)

TIPS & TOOLS

One of the best ways to histologically distinguish arteries from veins is to compare the vessels' wall thickness and lumen diameter. When an artery and vein have approximately the same luminal diameter, the artery will have a thicker wall than the vein.

Venules

Venules, the smallest veins, collect blood from capillaries. The smallest venules, termed postcapillary venules, resemble expanded capillaries. Venules smaller than 50 μm in total diameter lack a complete media, possessing only isolated smooth muscle cells deep to the intima. The media of venules larger than 50 μm in diameter contains one or two layers of smooth muscle cells, but it is thin and dominated by connective tissue.

Medium-Sized Veins

The luminal diameter of a **medium-sized vein** ranges from 2 to 10 mm. Most deep veins (radial, tibial, and popliteal veins) are in this category. A medium-sized vein is typically found in the same connective tissue sheath and side by side with a muscular artery.

Figure 22.2 Structure of Capillaries and Sinusoids

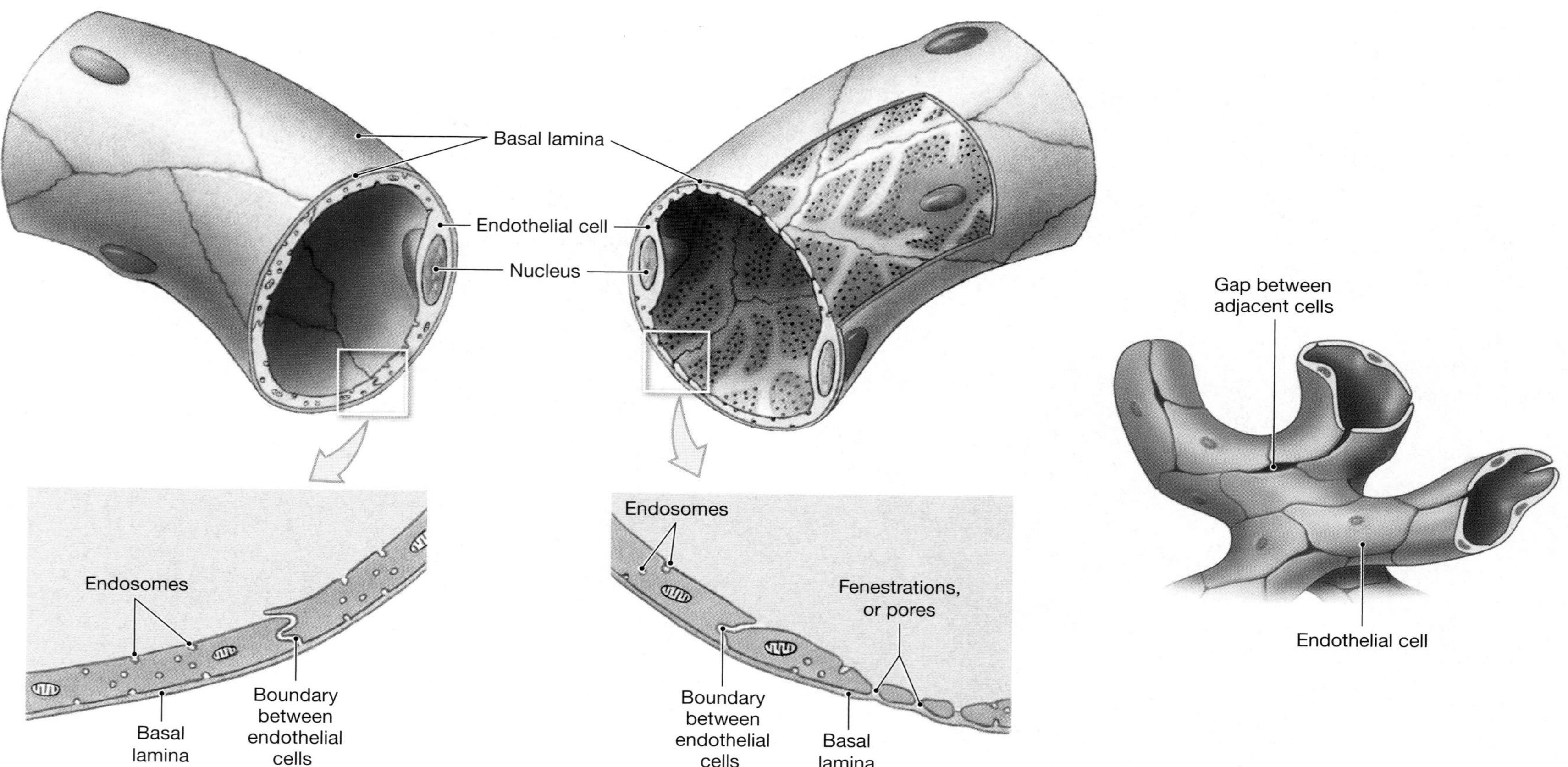

a This diagrammatic view of a continuous capillary shows the structure of its wall.

b This diagrammatic view of a fenestrated capillary shows the structure of its wall.

c Diagrammatic view of a sinusoid showing gaps between endothelial cells.

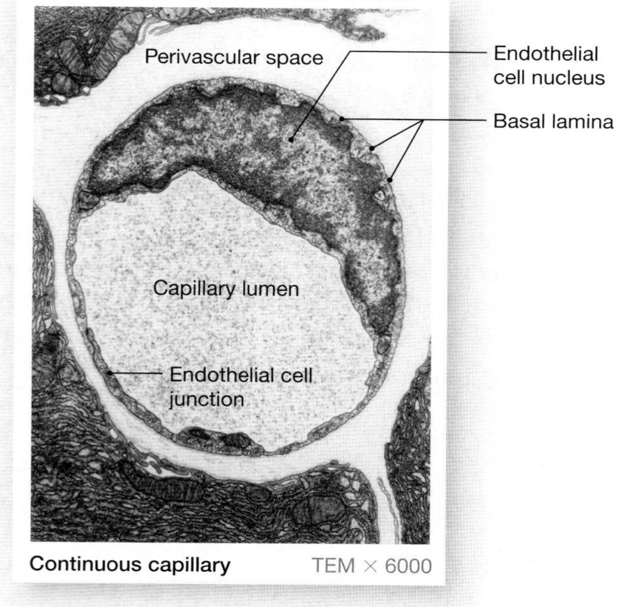

Continuous capillary TEM × 6000

d The TEM shows a cross section through a continuous capillary. A single endothelial cell forms a complete wall around this portion of the capillary.

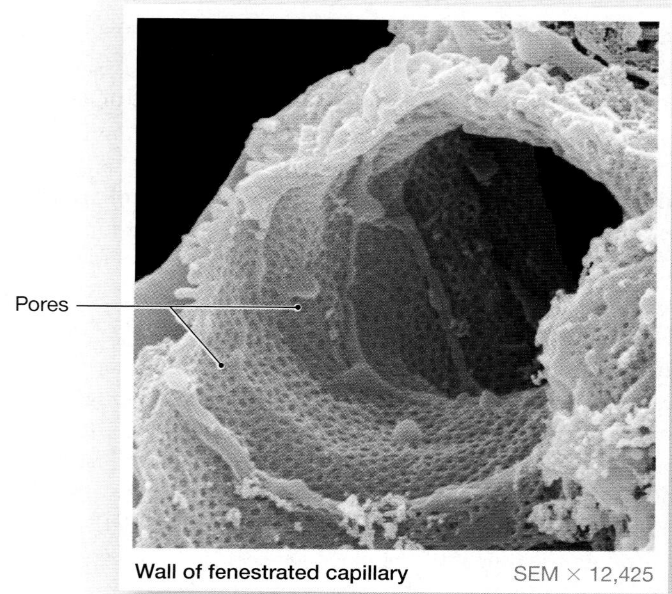

Wall of fenestrated capillary SEM × 12,425

e A SEM shows the wall of a fenestrated capillary. The pores are gaps in the endothelial wall permitting the passage of large volumes of fluid and solutes.

22

Figure 22.3 **Organization of a Capillary Bed**

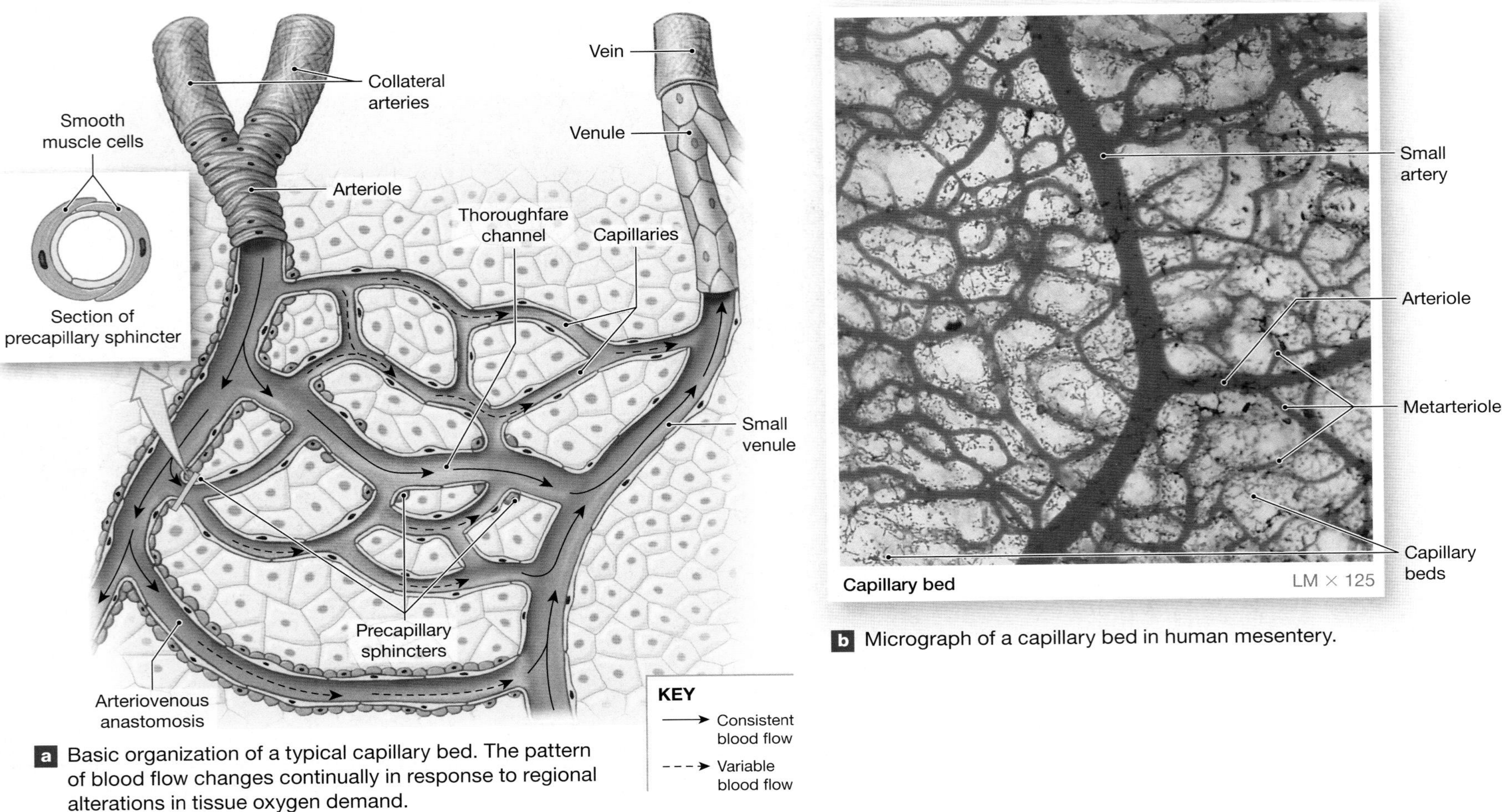

a Basic organization of a typical capillary bed. The pattern of blood flow changes continually in response to regional alterations in tissue oxygen demand.

b Micrograph of a capillary bed in human mesentery.

22

The three-layer organization of vein walls is most visible in a medium-sized vein. The intima is composed of an endothelium resting on a basil lamina. Several layers of smooth muscle cells are found within the media, along with a considerable amount of collagen and elastic connective tissue. The adventitia is typically thicker than the media and contains elastic and collagenous connective tissue fibers. Valves are a characteristic feature of medium-sized veins.

Venous Valves

The blood pressure in venules and medium-sized veins is so low that it cannot overcome the force of gravity. For example, when you are standing, blood returning from your feet must overcome the pull of gravity to ascend to your heart. In the limbs, medium-sized veins contain one-way **valves** that form from infoldings of the intima (**Figure 22.4**). These valves act like the valves in the heart, preventing the backflow of blood. Valves compartmentalize the blood within the veins, dividing the weight of the blood between the compartments. As long as the valves function normally, any movement in the surrounding skeletal muscles squeezes the blood toward the heart. This mechanism is called a skeletal muscle pump.

Large Veins

Large veins include the superior and inferior venae cavae and tributaries such as the subclavian, renal, mesenteric, and portal veins within the abdominopelvic and thoracic cavities. The intima and media are small and difficult to distinguish from one another in large veins. The thickest layer is the adventitia. It contains collagen and elastic fibers and isolated, longitudinally arranged smooth muscle cells. Large veins do not have valves, but changes in pressure within body cavities help move blood toward the heart.

22.1 CONCEPT CHECK

1. Examination of a section of tissue shows several small, thin-walled vessels with very little smooth muscle tissue in the media. What type of vessels are these?
2. Why are valves found in veins but not in arteries?
3. The femoral artery is an example of which type of artery?

See the blue Answers tab at the back of the book.

22.2 | The Distribution of Blood

▶ **KEY POINT** The veins normally contain most of our blood volume (65 to 70 percent), with the rest distributed among the heart, arteries, and capillaries.

The total blood volume is unevenly distributed among arteries, veins, and capillaries (**Figure 22.5**). The heart, arteries, and capillaries contain approximately 30–35 percent of the blood volume (about 1.5 L of whole blood), and the venous system contains the rest (65–70 percent, or about 3.5 L).

Because the walls of veins are thinner and contain more elastic tissue and less smooth muscle than those of arteries, veins are much more

Figure 22.4 **Function of Valves in the Venous System.** Valves in the walls of medium-sized veins prevent the backflow of blood. Venous compression caused by the contraction of adjacent skeletal muscles creates pressure (shown by arrows) that assists in maintaining venous blood flow. Changes in body position and the thoraco-abdominal pump may provide additional assistance.

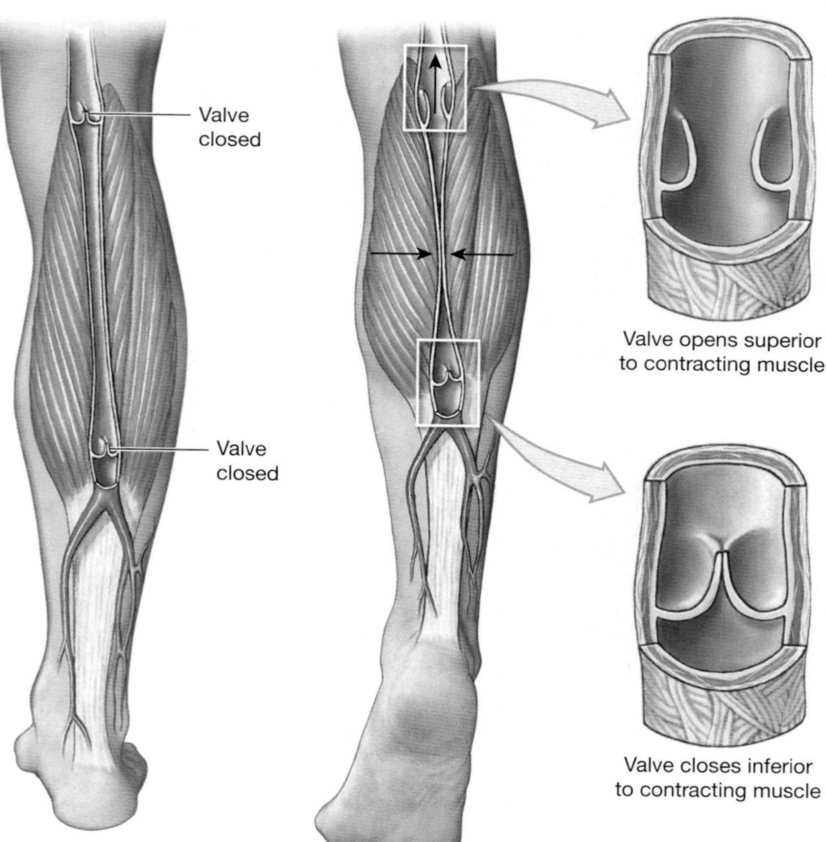

Figure 22.5 **The Distribution of Blood in the Cardiovascular System**

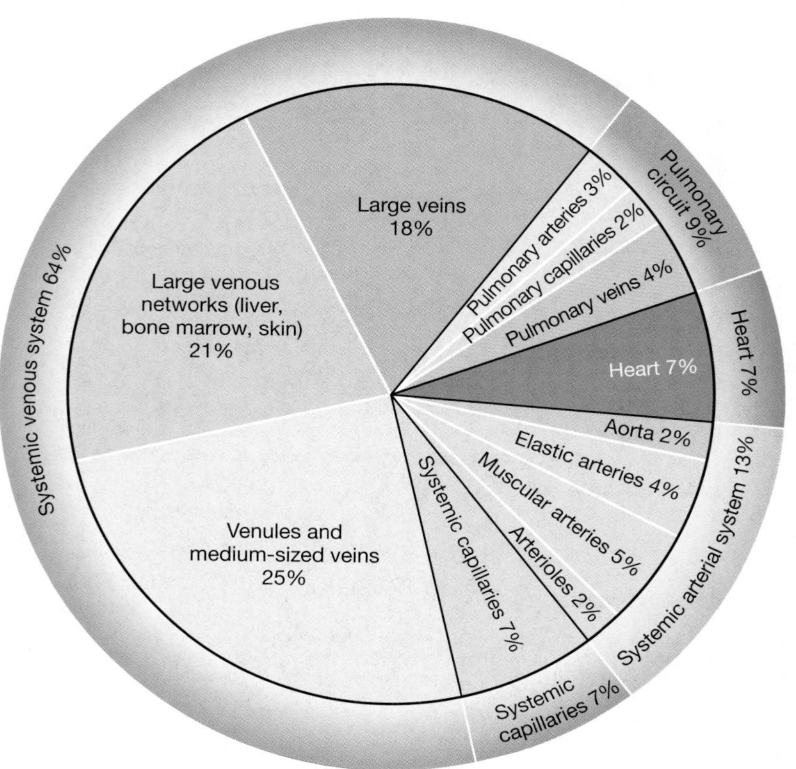

distensible (stretchable) than arteries. For a given rise in blood pressure, a typical vein stretches about eight times as much as a corresponding artery. The capacitance of a blood vessel is the relationship between the volume of blood it contains and the blood pressure, and veins are called **capacitance vessels**. Because veins have high capacitance, they act as **blood reservoirs**, which can accommodate large changes in blood volume. If the blood volume rises or falls, the elastic walls stretch or recoil, respectively, changing the volume of blood in the venous system.

If serious hemorrhaging (blood loss) occurs, sympathetic nerves stimulate smooth muscle cells in the walls of medium-sized veins in the systemic system to constrict. This process, called **venoconstriction** (vē-nō-kon-STRIK-shun) reduces the volume of blood within the venous system, increasing the volume within the arterial system and capillaries. Reducing the amount of blood in the venous system maintains the blood volume within the arterial system at near-normal levels despite a significant blood loss. As a result, blood flow to active skeletal muscles and delicate organs, such as the brain, can be increased or maintained.

22.2 CONCEPT CHECK

4 What is venoconstriction, and how does it contribute to maintaining an adequate amount of blood volume in the arterial circuit following blood loss?

See the blue Answers tab at the back of the book.

22.3 Blood Vessel Distribution

▶ **KEY POINT** The pulmonary circuit transports blood between the heart and the lungs. The systemic circuit transports blood between the heart and all other tissues.

The blood vessels of the body are divided into the pulmonary circuit and the systemic circuit (**Figure 22.6**). The pulmonary circuit transports blood between the heart and the lungs, a relatively short distance. The arteries and veins of the systemic circuit transport blood between the heart and all other tissues, a round trip that involves much longer distances. There are functional and structural differences between the vessels in these circuits. For example, blood pressure within the pulmonary circuit is relatively low, and the walls of pulmonary arteries are thinner than those of systemic arteries.

As you read the rest of this chapter, note three important functional patterns:

1. The peripheral distribution of arteries and veins on the left and right sides of the body is usually identical, except near the heart, where the largest vessels connect to the atria or ventricles.
2. A single vessel may have several different names as it crosses specific anatomical boundaries, making accurate anatomical descriptions possible when the vessel extends far into the periphery.
3. Anastomoses are often found between arteries and veins. These connections reduce the impact of temporary or even permanent occlusion (blockage) of a single blood vessel.

Figure 22.6 **An Overview of the General Pattern of Circulation**

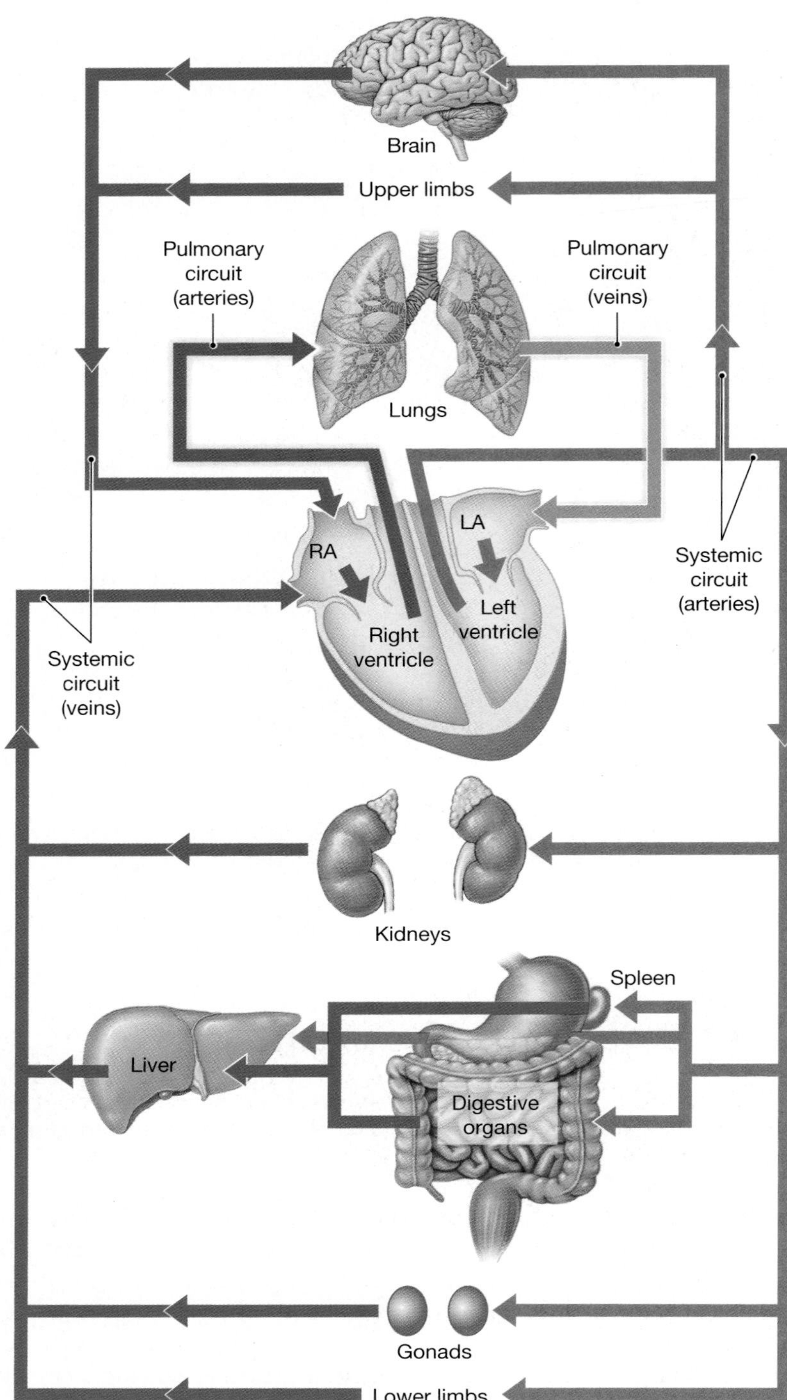

22.3 CONCEPT CHECK

5 Explain why one vessel may have several different anatomical names as it progresses from the abdominal aorta to the great toe on the left foot.

See the blue Answers tab at the back of the book.

22.4 The Pulmonary Circuit

▶ **KEY POINT** The pulmonary circuit carries deoxygenated blood from the heart to the lungs and returns oxygenated blood from the lungs to the heart.

Blood entering the right atrium is returning from capillary beds in the peripheral tissues and myocardium where oxygen was released and carbon dioxide absorbed. After traveling through the right atrium and right ventricle, blood passes through the pulmonary valve and enters the pulmonary trunk, the start of the pulmonary circuit. This circuit, which contains approximately 9 percent of the total blood volume at any given moment, begins at the pulmonary valve and ends at the entrance to the left atrium. In the pulmonary circuit, oxygen is replenished and carbon dioxide is released. Compared with the systemic circuit, the pulmonary circuit is short: The base of the pulmonary trunk and the lungs are only about 15 cm (6 in.) apart **(Figure 22.7a)**.

The arteries of the pulmonary circuit carry deoxygenated (oxygen-poor) blood, which is different from the arteries of the systemic circuit. (For this reason, color-coded diagrams usually show the pulmonary arteries in blue, the same color as systemic veins.) The pulmonary trunk curves over the superior border of the heart, where it divides into the **left** and **right pulmonary arteries**. These large arteries enter the lungs before branching repeatedly, giving rise to smaller and smaller arteries. The smallest branches, the pulmonary arterioles, provide blood to capillary networks surrounding small air pockets, or **alveoli** (al-VĒ-ō-lī; *alveolus,* sac), in the lungs. The walls of the alveoli are thin enough to allow gas exchange between the capillary blood and inspired air. As oxygenated (oxygen-rich) blood leaves the alveolar capillaries, it enters venules that merge to form larger and larger vessels carrying blood toward the **pulmonary veins**. These four veins, typically two from each lung, empty into the left atrium, to complete the pulmonary circuit **(Figure 22.7a)**. **Figure 22.7b** is a spiral scan of the heart showing the vessels of the pulmonary circuit and their relationship to the heart and lungs in a living person.

22.4 CONCEPT CHECK

6 Name the blood vessels that enter and exit the lungs, and indicate whether they contain primarily oxygenated or deoxygenated blood.

See the blue Answers tab at the back of the book.

22.5 Systemic Arteries

▶ **KEY POINT** Systemic arteries carry oxygenated blood to all parts of the body. In the periphery, oxygen diffuses from capillaries into the interstitial space, and carbon dioxide diffuses from the interstitial space into the capillaries.

The systemic circuit begins at the aortic valve and ends at the entrance to the right atrium. It supplies the capillary beds in all parts of the body not supplied by the pulmonary circuit. At any given moment, the systemic circuit contains about 84 percent of total blood volume.

Figure 22.8, an overview of the arterial system, shows the locations of major systemic arteries. **Figures 22.9** to **22.17** detail the distribution of these vessels and their branches. Note that because we have separate figures for arteries and veins, we have not included the terms artery(ies) and vein(s) in the labels. As a result, you will see the singular form of the name, such as intercostal, whether the leader points to one intercostal artery or several intercostal arteries. Notice that several large arteries are called *trunks*. To avoid confusion, these names are shown in their entirety. Because the descriptions that follow focus on major branches found on both sides of the body, the terms right and left appear only when both vessels are labeled.

Figure 22.7 The Pulmonary Circuit

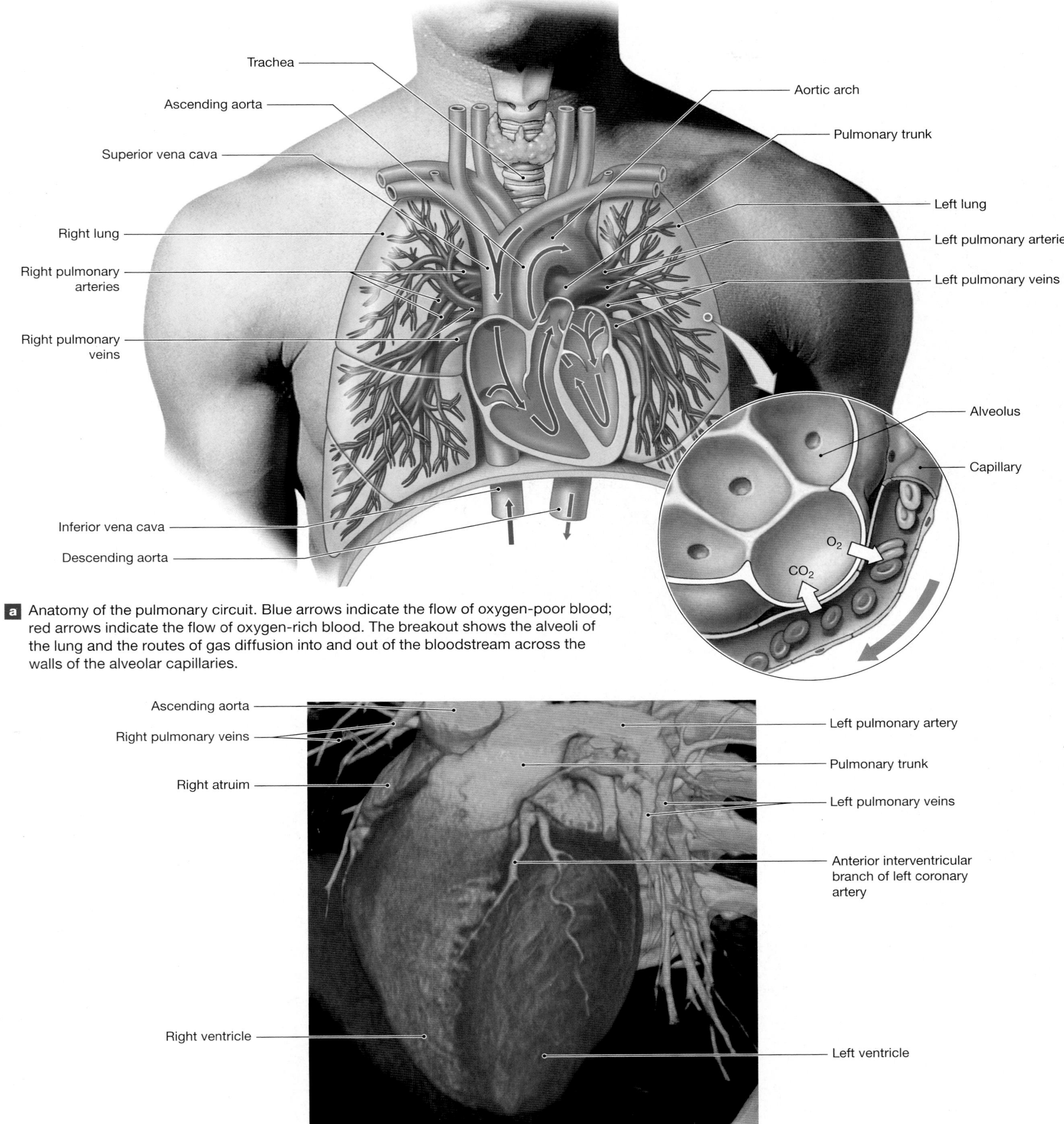

a Anatomy of the pulmonary circuit. Blue arrows indicate the flow of oxygen-poor blood; red arrows indicate the flow of oxygen-rich blood. The breakout shows the alveoli of the lung and the routes of gas diffusion into and out of the bloodstream across the walls of the alveolar capillaries.

b Spiral 3-D volume rendered scan of the heart and major vessels.

22

Figure 22.8 **An Overview of the Systemic Arterial System**

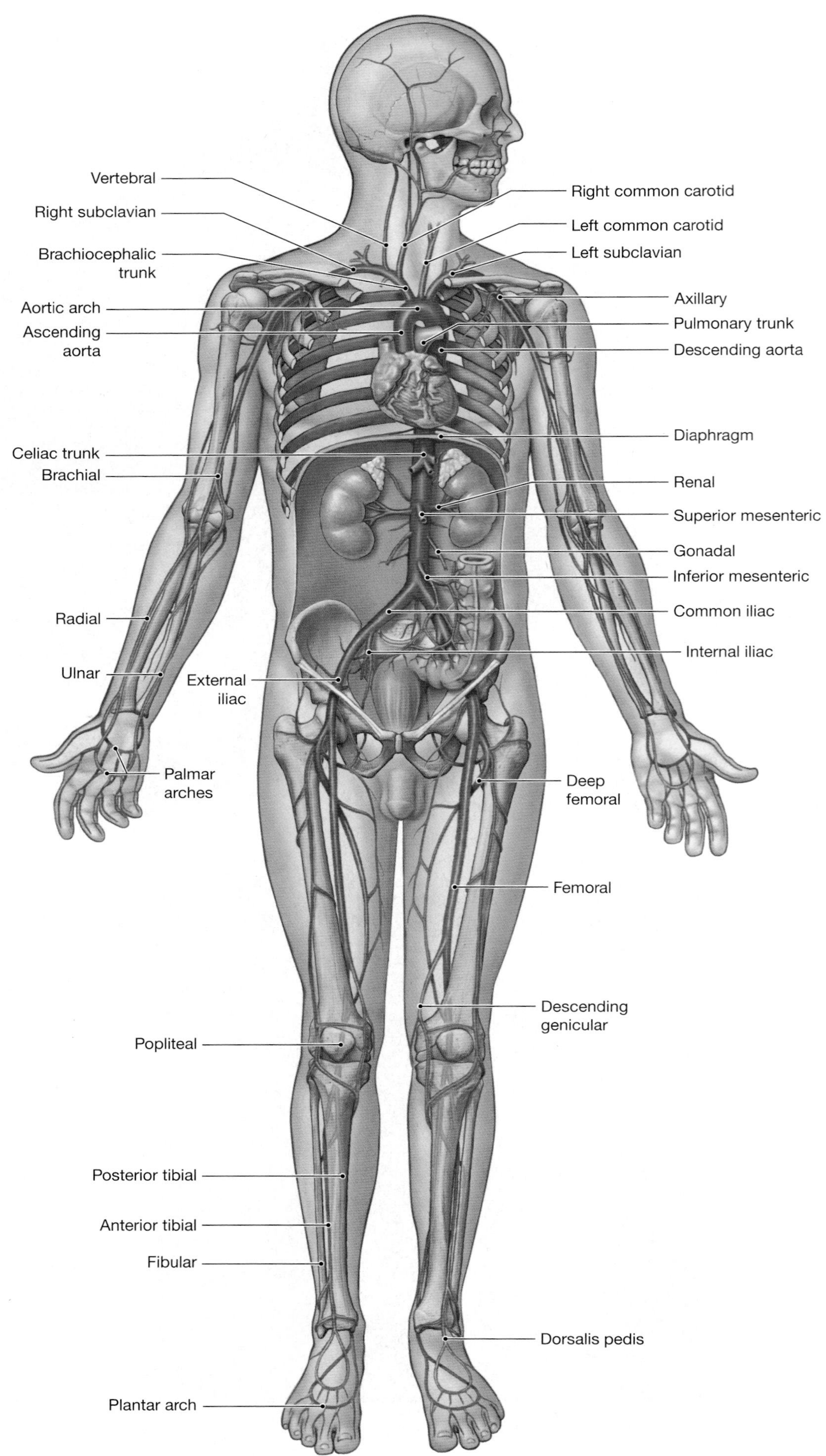

22

The Aorta

▶ **KEY POINT** The aorta, the first artery of the systemic circuit, exits from the left ventricle of the heart. It is subdivided into the ascending aorta, aortic arch, and descending aorta.

The Ascending Aorta and the Aortic Arch

The **aorta** begins at the aortic valve of the left ventricle (**Figure 21.7b**, p. 552, and **Figure 22.8**). The first segment of the aorta is the **ascending aorta**. The left and right coronary arteries originate at the base of the ascending aorta, just superior to the aortic valve. (The distribution of coronary vessels was described in Chapter 21 and illustrated in **Figure 21.10**.) ⮌ p. 557

The **aortic arch** (also termed the *arch of the aorta*) curves like a cane handle, passing superiorly and posteriorly across the superior surface of the heart. It connects the ascending aorta with the descending aorta. Three elastic arteries originate along the aortic arch (**Figures 22.8** to **22.10**). These arteries, the **brachiocephalic** (brā-kē-ō-se-FAL-ik) **trunk**, the **left common carotid artery**, and the **left subclavian** (sub-CLĀ-vē-an) **artery**, deliver blood to the head, neck, shoulders, and upper limbs. The brachiocephalic trunk ascends for a short distance where, posterior to the right sternoclavicular joint, it divides and forms the **right subclavian artery** and the **right common carotid artery**. There is only one brachiocephalic trunk; the left common carotid and left subclavian arteries arise separately from the aortic arch. However, in terms of their peripheral distribution of blood, the vessels on the left side are mirror images of those on the right side. **Figure 22.10** illustrates the major branches of these arteries.

TIPS & TOOLS

To remember the branches of the aorta, think of your ABCs: **A**ortic arch, **B**rachiocephalic trunk, **C**ommon carotid artery, **S**ubclavian artery.

The Subclavian Arteries

The subclavian arteries arch over the apex (superior tip) of the right and left lungs. These vessels supply the upper limbs, chest wall, shoulders, back, brain, and spinal cord with blood (**Figures 22.8** to **22.10**). Three major branches arise from the subclavian arteries before they leave the thoracic cavity: (1) a **thyrocervical trunk**, which supplies blood to muscles and other tissues of the neck, shoulder, and upper back; (2) an **internal thoracic artery**, supplying the pericardium and anterior wall of the chest; and (3) a **vertebral artery**, which supplies blood to the brain and spinal cord.

Figure 22.9 Aortic Angiogram. This angiogram shows the ascending aorta, the aortic arch, the descending aorta, the brachiocephalic trunk (branching into the right subclavian and right common carotid arteries), and the left subclavian and left common carotid arteries.

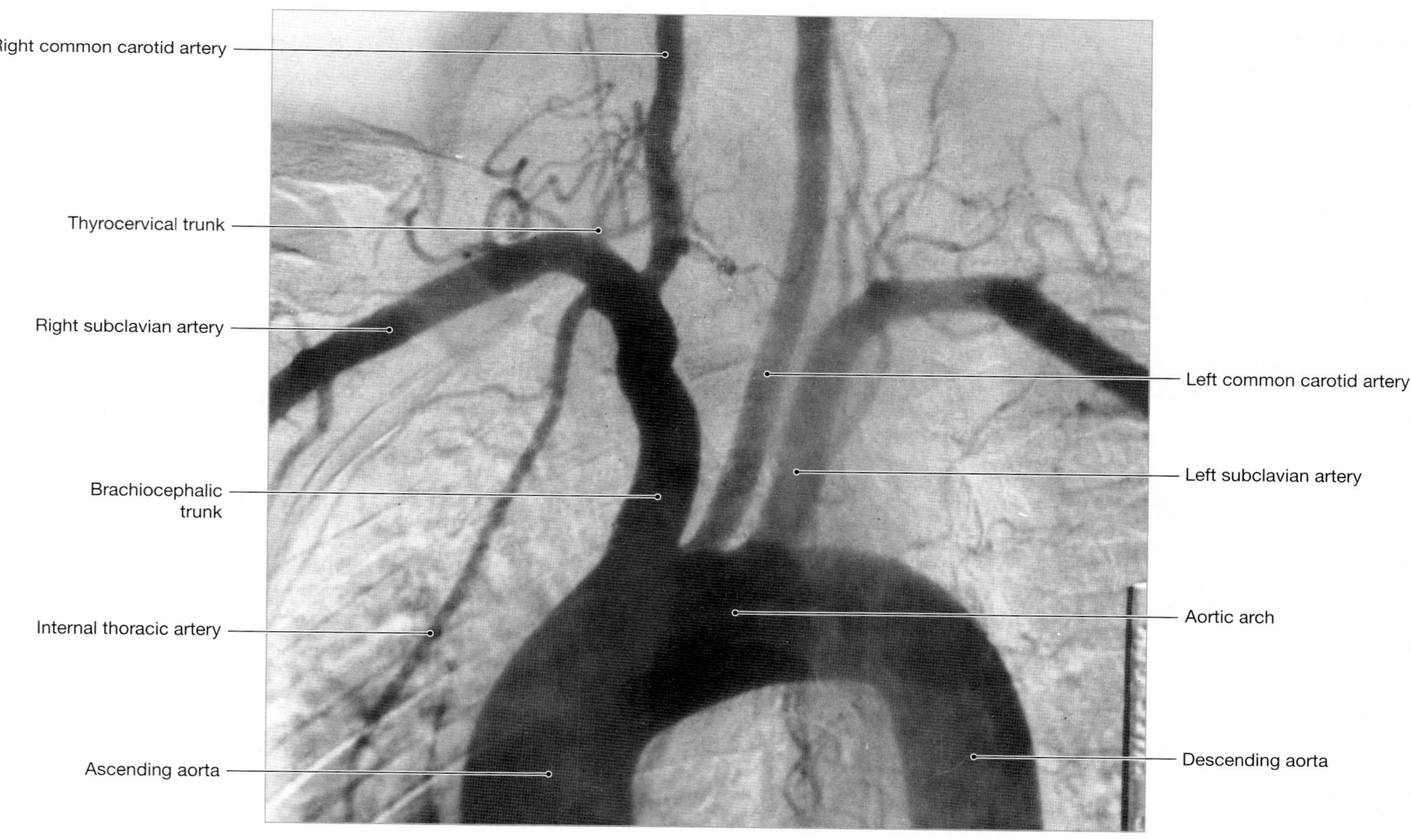

22

Figure 22.10 Arteries of the Chest and Upper Limb

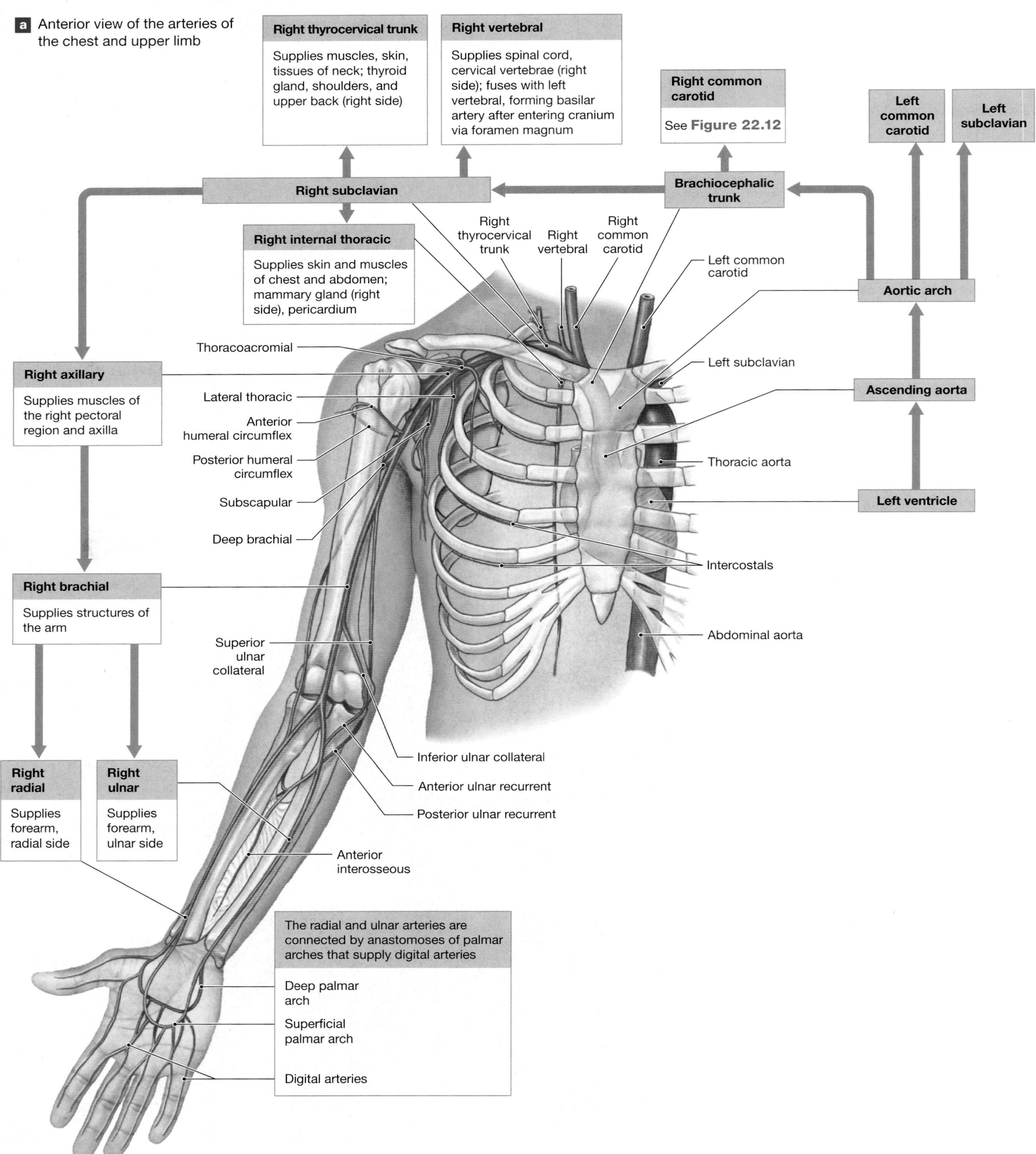

a Anterior view of the arteries of the chest and upper limb
Right thyrocervical trunk
Supplies muscles, skin, tissues of neck; thyroid gland, shoulders, and upper back (right side)
Right vertebral
Supplies spinal cord, cervical vertebrae (right side); fuses with left vertebral, forming basilar artery after entering cranium via foramen magnum
Right common carotid
See Figure 22.12
Left common carotid
Left subclavian
Right subclavian
Brachiocephalic trunk
Right internal thoracic
Supplies skin and muscles of chest and abdomen; mammary gland (right side), pericardium
Right thyrocervical trunk
Right vertebral
Right common carotid
Left common carotid
Aortic arch
Thoracoacromial
Right axillary
Supplies muscles of the right pectoral region and axilla
Lateral thoracic
Anterior humeral circumflex
Posterior humeral circumflex
Subscapular
Deep brachial
Left subclavian
Ascending aorta
Thoracic aorta
Left ventricle
Intercostals
Right brachial
Supplies structures of the arm
Abdominal aorta
Superior ulnar collateral
Inferior ulnar collateral
Anterior ulnar recurrent
Posterior ulnar recurrent
Right radial
Supplies forearm, radial side
Right ulnar
Supplies forearm, ulnar side
Anterior interosseous
The radial and ulnar arteries are connected by anastomoses of palmar arches that supply digital arteries
Deep palmar arch
Superficial palmar arch
Digital arteries

22

Figure 22.10 **(continued)**

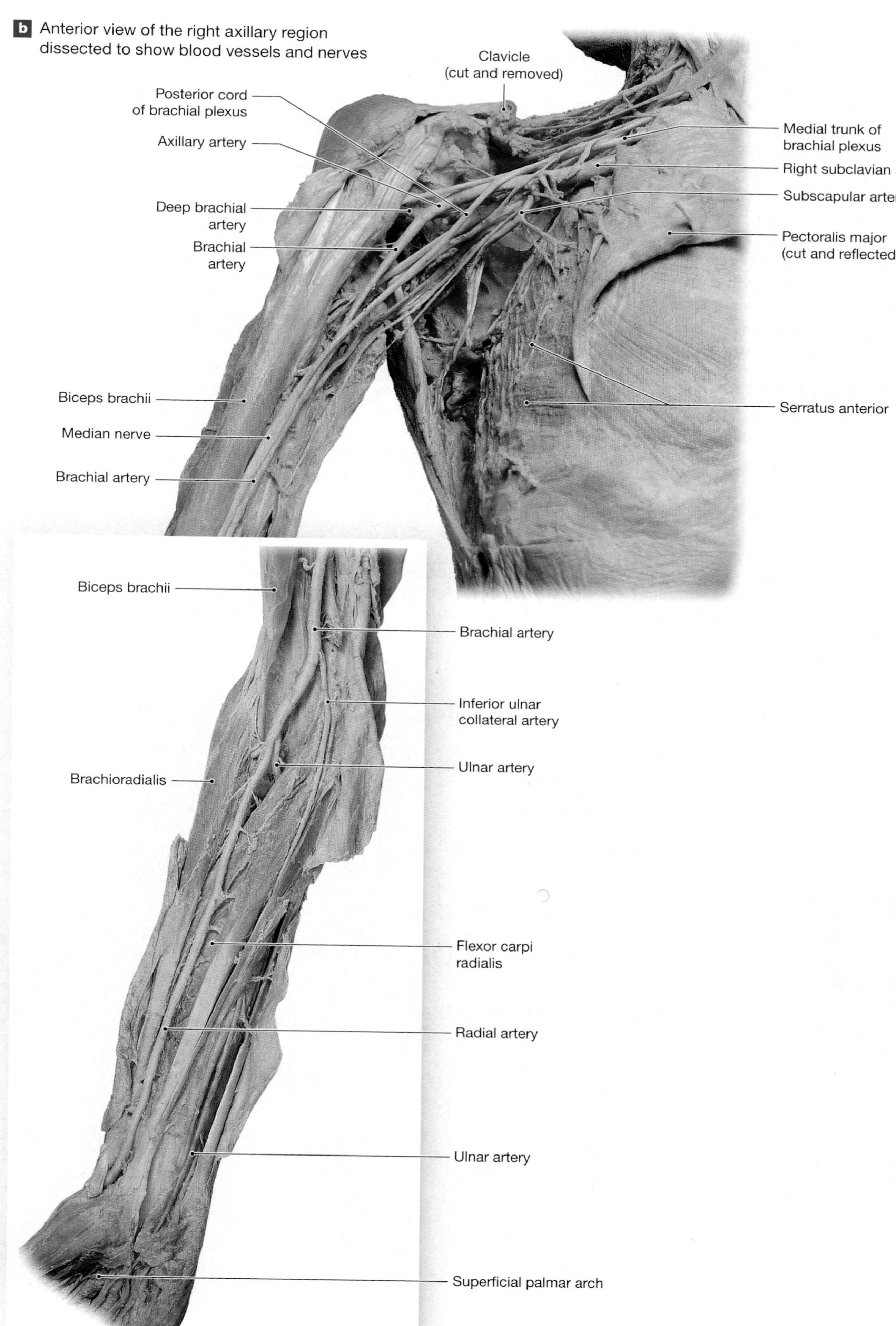

b Anterior view of the right axillary region dissected to show blood vessels and nerves

c Anterior view of the right forearm dissected to show the main arteries

22

After leaving the thoracic cavity and passing over the first rib, each subclavian artery becomes an **axillary artery**. The axillary arteries supply blood to the muscles of the pectoral region and axilla. The axillary artery crosses the axilla and enters the arm, where it gives rise to the **humeral circumflex arteries**. These vessels supply structures near the head of the humerus. Distally, the axillary artery becomes the **brachial artery**, which supplies blood to the upper limb.

The first branch from the brachial artery is the **deep brachial artery**. This vessel supplies deep structures along the posterior surface of the arm. The **ulnar collateral arteries** and the **ulnar recurrent arteries** branch from the brachial artery as it passes distally in the arm. They supply the area around the elbow. As it approaches the coronoid fossa of the humerus, the brachial artery divides into the **radial** and **ulnar arteries**. These arteries follow the radius and ulna distally to the wrist, supplying blood to the forearm. At the wrist, these arteries anastomose to form a **superficial palmar arch** and a **deep palmar arch**. These arches supply blood to the palm of the hand and to the **digital arteries** of the thumb and fingers.

The Carotid Arteries and the Blood Supply to the Brain

The common carotid arteries ascend deep in the tissues of the neck, just lateral to the trachea (windpipe). At the level of the larynx, each common carotid artery divides into an **external carotid artery** and an **internal carotid artery**. The **carotid sinus** is located at the base of the internal carotid artery. Often, it extends inferiorly along a portion of the common carotid artery **(Figures 22.11** and **22.12)**. The carotid sinus contains baroreceptors and chemoreceptors involved in cardiovascular regulation. The external carotid arteries supply blood to the structures of the neck, pharynx, esophagus, larynx, lower jaw, and face. The internal carotid arteries enter the skull through the carotid canals of the temporal bones, delivering blood to the brain. Each internal carotid artery ascends to the level of the optic nerves, where it divides into three branches: (1) an **ophthalmic artery** supplying the eye, (2) an **anterior cerebral artery** supplying the frontal and parietal lobes of the brain, and (3) a **middle cerebral artery** supplying the midbrain and lateral surfaces of the cerebral hemispheres **(Figures 22.12** and **22.13)**.

The brain is extremely sensitive to changes in its circulatory supply. An interruption of circulation for several seconds produces unconsciousness, and after 4 minutes there may be permanent neural damage. Such interruptions in circulation are rare because blood reaches the brain through two sources: the vertebral arteries and the internal carotid arteries. The **vertebral arteries** branch from the subclavian arteries and ascend within the transverse foramina of the cervical vertebrae. The vertebral arteries enter the cranium at the foramen magnum. There, the two vertebral arteries fuse along the ventral surface of the medulla oblongata and form the **basilar artery**. The basilar artery continues on the ventral surface of the brain along the pons. The basilar artery branches many times before dividing into the **posterior cerebral arteries**. The posterior cerebral arteries branch into the **posterior communicating arteries (Figures 22.11, 22.12,** and **22.13a,b)**.

Figure 22.11 Major Arteries of the Neck. This dissection of the anterior neck shows the position and appearance of the major arteries in this region. In this dissection, part of the right clavicle, first rib, and manubrium of the sternum have been removed, along with the inferior portion of the right internal jugular vein.

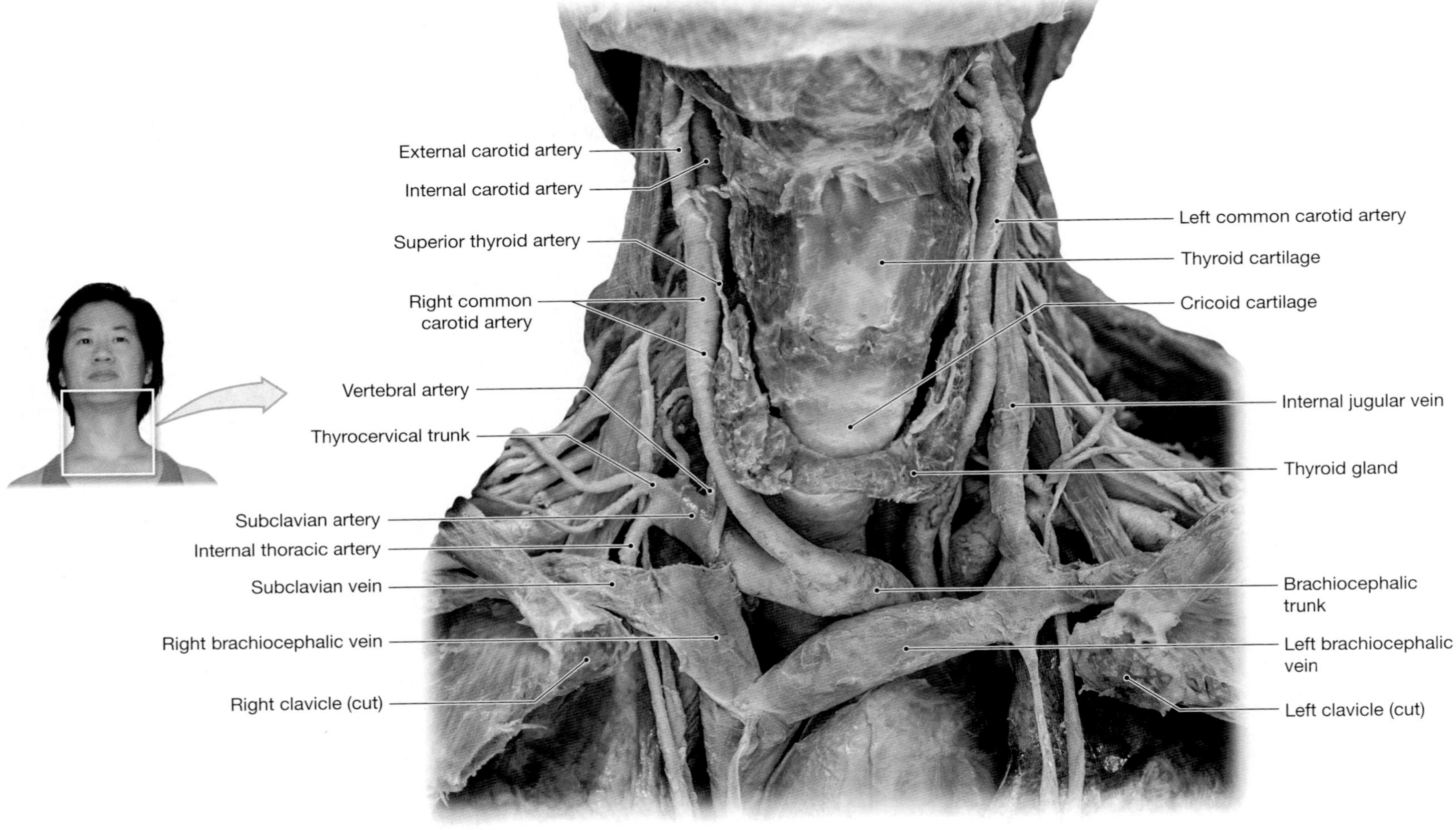

22

Figure 22.12 Arteries of the Neck and Head

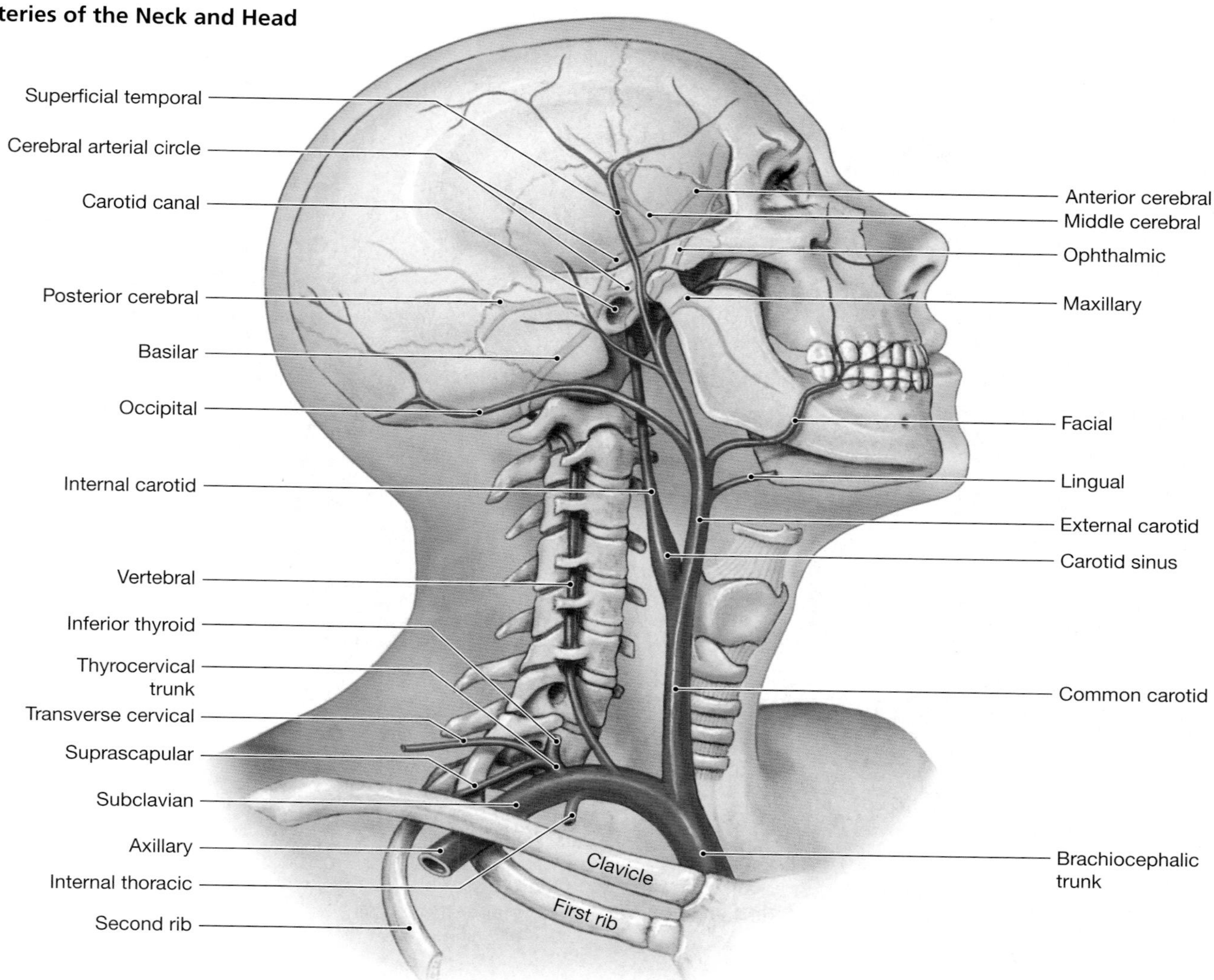

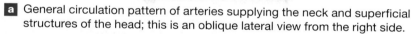

a General circulation pattern of arteries supplying the neck and superficial structures of the head; this is an oblique lateral view from the right side.

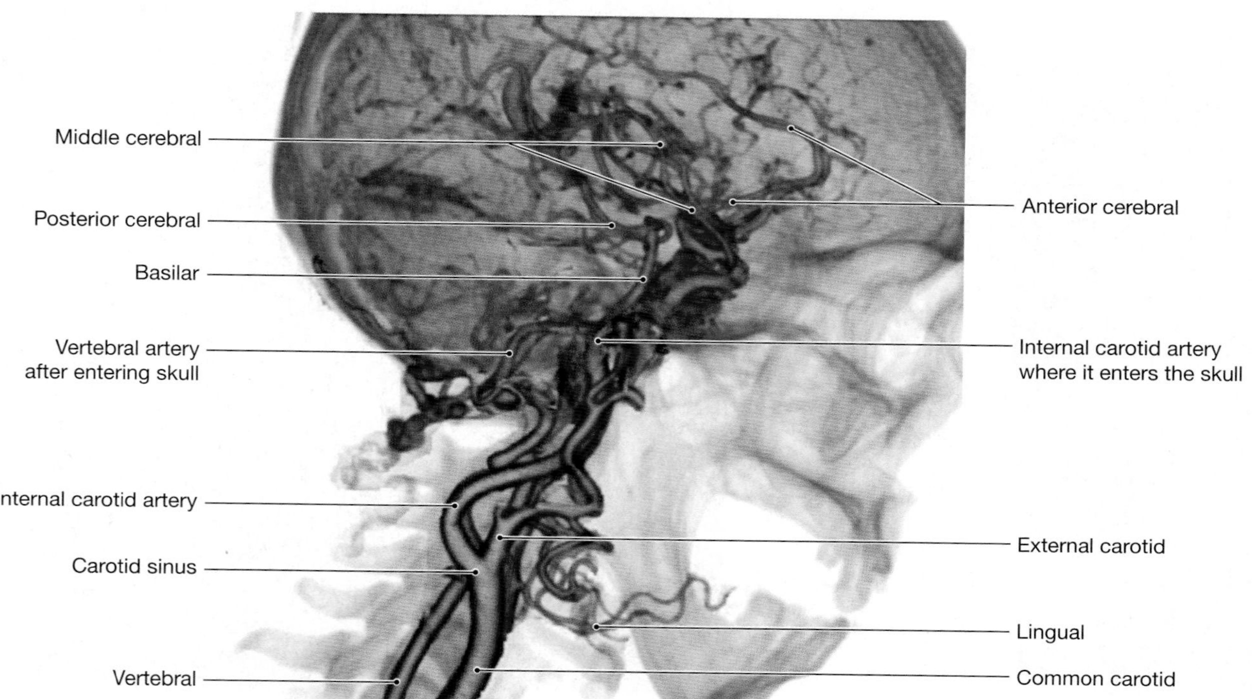

b Spiral 3-D volume rendered scan of the arteries supplying the neck and head. [Courtesy of TeraRecon, Inc.]

Figure 22.13 The Arterial Supply to the Brain

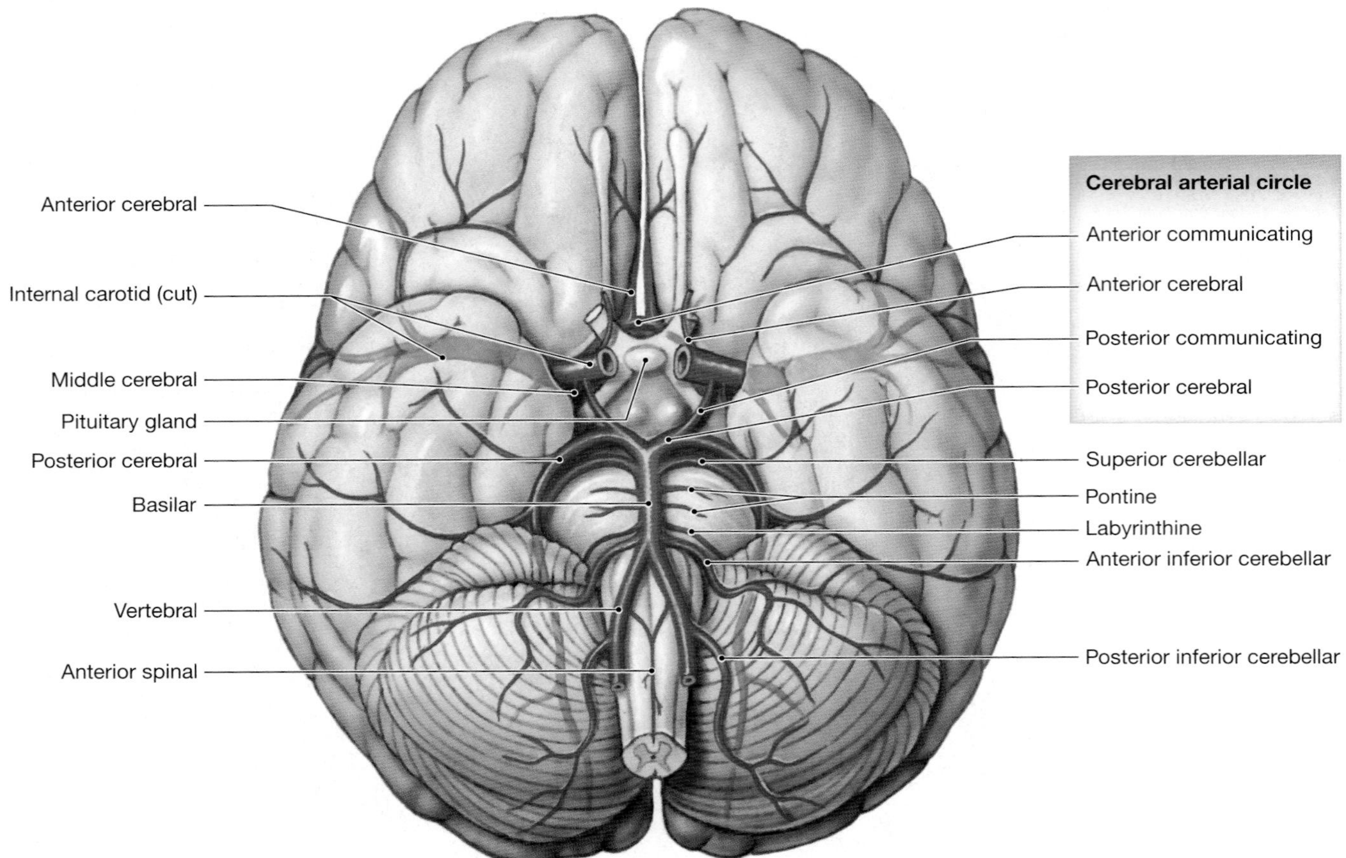

a An inferior view of the brain showing the distribution of arteries. See **Figure 22.19b** for a comparable view of the veins on the inferior surface of the brain.

22

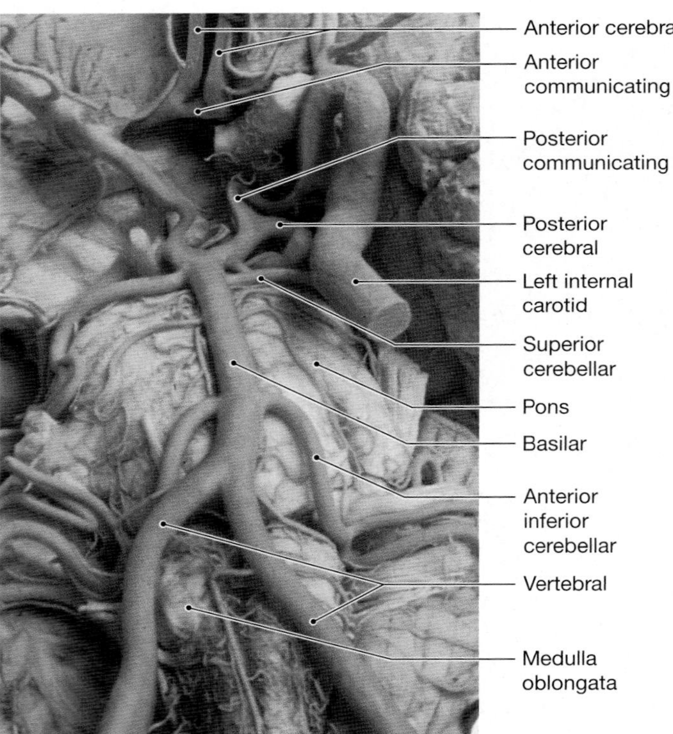

b The arteries on the inferior surface of the brain; the vessels have been injected with red latex, making them easier to see.

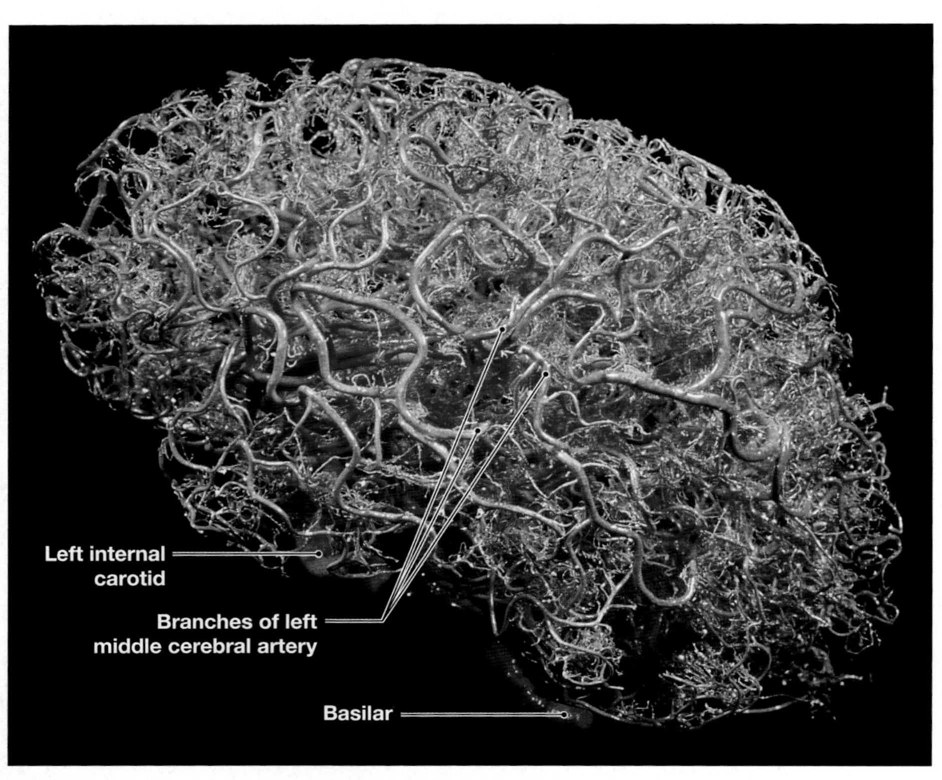

c A lateral view of the arteries supplying the brain. This is a corrosion cast: The vessels have been injected with latex, and then the brain tissue was dissolved and removed in an acid bath.

CLINICAL NOTE

Repair of an Aortic Aneurysm

An **aneurysm** is a bulge, or weakness, that develops in the wall of an artery. An aneurysm can occur in arteries anywhere in the body, including the brain and heart. Here we focus on aneurysms in the aorta.

The wall of the aorta is very elastic, able to stretch and recoil as it adapts to blood flow. An aortic aneurysm is defined by its anatomic location as a **thoracic aortic aneurysm**, either ascending or descending, or an **abdominal aortic aneurysm**.

Although some abdominal aortic aneurysms can be palpated or auscultated (heard with a stethoscope) in thin individuals, most do not cause symptoms early on and are discovered as incidental findings with x-ray or ultrasound studies. Many primary care physicians believe people over 65 with a family history of aortic aneurysm, and men over 65 who have ever smoked, should have a one-time ultrasound screening for abdominal aortic aneurysms.

If an aortic aneurysm suddenly ruptures, the prognosis is very poor, with a mortality rate of over 90 percent. Some doctors recommend preventive surgical treatment of aneurysms greater than 5 cm in diameter that are enlarging with time.

A traditional **open surgical repair** is performed through a large midline incision with the monitored, anesthetized patient in the supine position. The aorta is cross-clamped above and below the aneurysm to stop blood flow long enough for repair. The aneurysm sac is opened and a long, synthetic tube graft is inserted across the defect and sutured in place. The clamps are released and the aorta is wrapped around the graft and sutured closed.

A less-invasive **endovascular repair** is sometimes possible, depending on the location of the aneurysm and whether major arterial branches are involved. This process involves making a small incision into the femoral artery, passing a guide wire to the aneurysm site, and sliding a sheath over the wire. A stent graft can then be inserted through the femoral artery and advanced up the aorta to the aneurysm. The stent graft is then expanded and attached to the internal wall of the aorta, preventing it from expanding further.

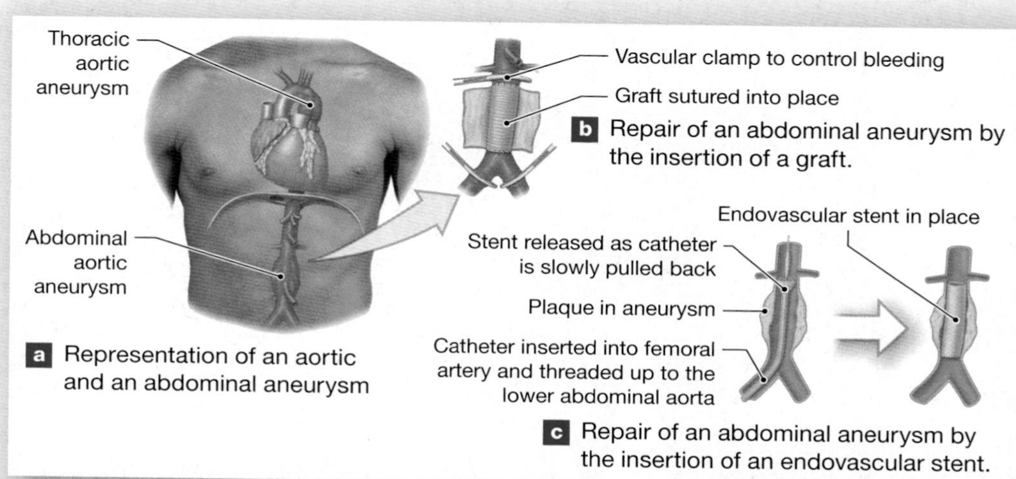

a Representation of an aortic and an abdominal aneurysm

b Repair of an abdominal aneurysm by the insertion of a graft.

c Repair of an abdominal aneurysm by the insertion of an endovascular stent.

22

The internal carotid arteries branch to form the arteries supplying the anterior half of the cerebrum. The rest of the brain receives blood from the vertebral arteries. This circulatory pattern often changes because the internal carotid arteries and the basilar artery are interconnected. They form a ring-shaped anastomosis called the **cerebral arterial circle**, or *circle of Willis*, which encircles the infundibulum of the pituitary gland (**Figure 22.13a,b**). With this arrangement the brain can receive blood from the carotids *and/or* the vertebrals, reducing the likelihood of interrupted circulation.

The Descending Aorta

▶ **KEY POINT** The descending aorta is divided into the thoracic aorta and abdominal aorta.

The **descending aorta** is continuous with the aortic arch. The diaphragm divides the descending aorta into a superior **thoracic aorta** and an inferior **abdominal aorta** (**Figure 22.8**, p. 578). **Figure 22.14** shows the branches of the thoracic aorta and abdominal aorta.

The Thoracic Aorta

The thoracic aorta begins at the level of vertebra T_1 and lies slightly to the left of midline. Gradually the thoracic aorta shifts toward the midline, exits from the posterior mediastinum, and passes through the diaphragm at the level of vertebra T_{12} (**Figure 22.14**). The thoracic aorta supplies blood to organs of the thorax (except the heart), the muscles of the chest and the diaphragm, and the thoracic portion of the spinal cord.

The branches of the thoracic aorta are grouped anatomically as either visceral or parietal. **Visceral branches** supply the organs of the chest. The **bronchial arteries** supply the conducting passageways of the lungs, **pericardial arteries** supply the pericardium, **mediastinal arteries** supply general mediastinal structures, and **esophageal arteries** supply the esophagus. **Parietal branches** of the thoracic aorta supply the chest wall. The **intercostal arteries** supply the chest muscles and the vertebral column area, and the **superior phrenic** (FREN-ik) **arteries** deliver blood to the superior surface of the muscular diaphragm that separates the thoracic and abdominopelvic cavities.

The Abdominal Aorta

The abdominal aorta begins immediately inferior to the diaphragm and descends slightly to the left of the vertebral column, posterior to the peritoneal cavity (**Figures 22.14** and **22.15**). It is often surrounded by a cushion of adipose tissue. At the level of vertebra L_4, the abdominal aorta splits into the right and left common iliac arteries. These vessels supply blood to deep pelvic structures and the lower limbs. The region where the aorta splits is called the **terminal segment of the aorta**.

Figure 22.14 **Major Arteries of the Trunk**

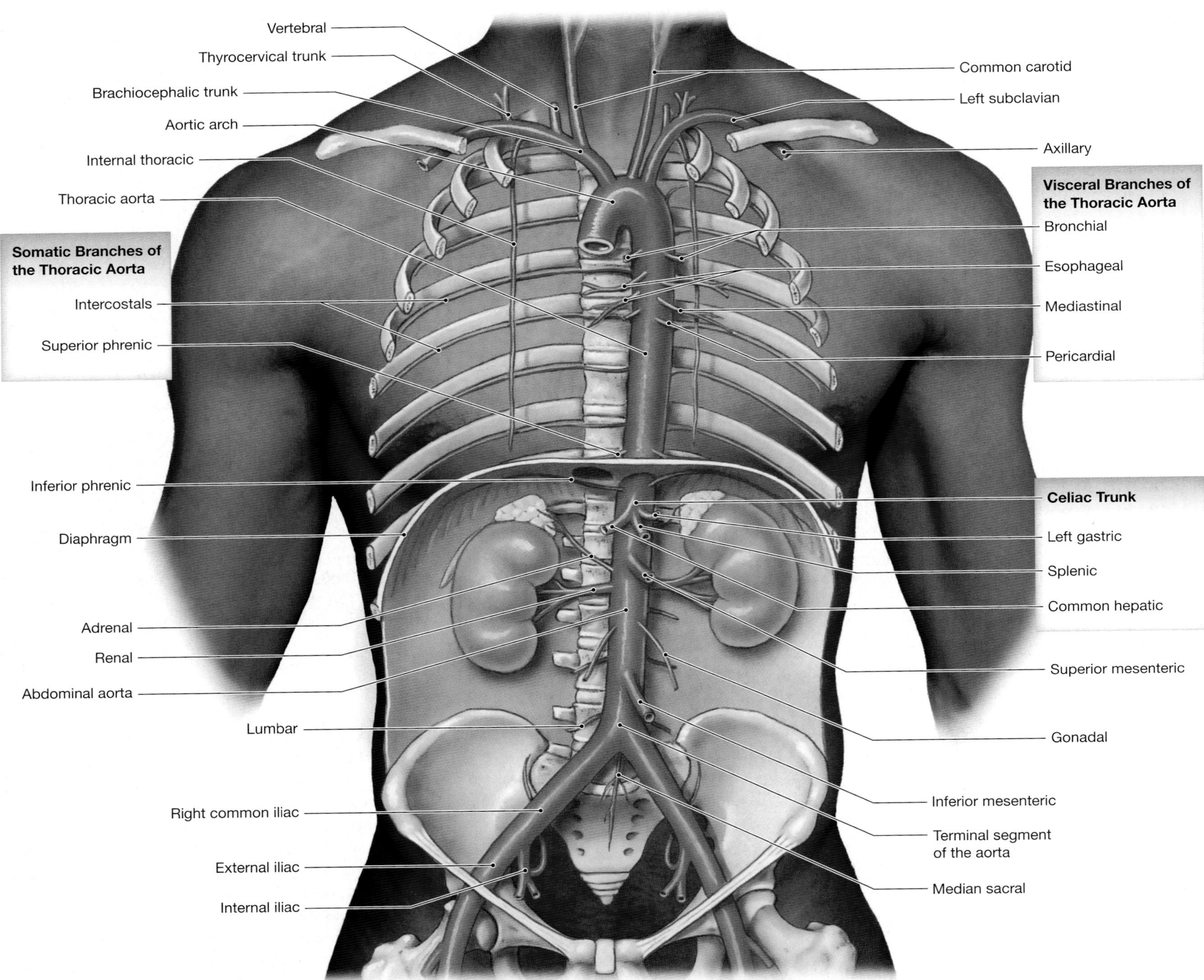

Vertebral
Thyrocervical trunk
Brachiocephalic trunk
Aortic arch
Internal thoracic
Thoracic aorta
Somatic Branches of the Thoracic Aorta
Intercostals
Superior phrenic
Inferior phrenic
Diaphragm
Adrenal
Renal
Abdominal aorta
Lumbar
Right common iliac
External iliac
Internal iliac
Common carotid
Left subclavian
Axillary
Visceral Branches of the Thoracic Aorta
Bronchial
Esophageal
Mediastinal
Pericardial
Celiac Trunk
Left gastric
Splenic
Common hepatic
Superior mesenteric
Gonadal
Inferior mesenteric
Terminal segment of the aorta
Median sacral

22

The abdominal aorta delivers blood to all abdominopelvic organs and structures. The major branches to visceral organs are unpaired. These vessels originate on the anterior surface of the abdominal aorta and extend into the mesenteries to reach the visceral organs. Branches to the body wall, the kidneys, and other structures outside the peritoneal cavity are paired, and they originate along the lateral surfaces of the abdominal aorta. **Figure 22.14** shows the major arteries of the trunk, with the thoracic and abdominal organs removed.

The abdominal aorta gives rise to three unpaired arteries: (1) the celiac trunk, (2) the superior mesenteric artery, and (3) the inferior mesenteric artery **(Figures 22.14** and **22.15)**.

1. The **celiac** (SĒ-lē-ak) **trunk** supplies blood to the liver, stomach, esophagus, gallbladder, duodenum, pancreas, and spleen. The celiac artery divides into three branches:
   - the **left gastric artery**, which supplies the stomach and inferior portion of the esophagus;
   - the **splenic artery**, which supplies the spleen and arteries to the stomach (left gastro-epiploic artery) and pancreas (pancreatic arteries); and
   - the **common hepatic artery**, which supplies blood to the liver (hepatic artery proper), stomach (right gastric artery), gallbladder (cystic artery), and duodenal area (gastroduodenal, right gastro-epiploic, and superior pancreaticoduodenal arteries).
2. The **superior mesenteric** (mez-en-TER-ik) **artery** arises about 2.5 cm inferior to the celiac trunk. It divides and supplies blood to the pancreas and duodenum (inferior pancreaticoduodenal artery), small intestine (intestinal arteries), and most of the large intestine (right colic, middle colic, and ileocolic arteries).
3. The **inferior mesenteric artery** arises about 5 cm superior to the terminal segment of the aorta. It supplies the terminal portions of the colon (left colic and sigmoid arteries) and the rectum (rectal arteries).

The abdominal aorta also gives rise to five paired arteries: (1) the inferior phrenics, (2) the adrenals, (3) the renals, (4) the gonadals, and (5) the lumbars.

1. The **inferior phrenic arteries** supply the inferior surface of the diaphragm and the inferior portion of the esophagus.
2. The **adrenal arteries** originate on either side of the aorta near the base of the superior mesenteric artery. Each adrenal artery supplies an adrenal gland, which caps the superior part of a kidney.
3. The short (about 7.5 cm) **renal arteries** originate along the posterolateral surface of the abdominal aorta, about 2.5 cm inferior to the superior mesenteric artery. They travel posterior to the peritoneal lining to reach the adrenal glands and kidneys. (We discuss branches of the renal arteries in Chapter 26.)
4. The **gonadal** (gō-NAD-al) **arteries** originate between the superior and inferior mesenteric arteries. In males, they are called **testicular arteries** and are long, thin arteries supplying blood to the testes and scrotum. In females, they are termed **ovarian arteries** and supply blood to the ovaries, uterine tubes, and uterus. The distribution of gonadal vessels (both arteries and veins) differs in males and females.
5. Small **lumbar arteries** arise on the posterior surface of the aorta and supply the vertebrae, spinal cord, and abdominal wall.

Arteries of the Pelvis and Lower Limbs

At vertebra L_4, the terminal segment of the abdominal aorta divides to form the **right** and **left common iliac** (IL-ē-ak) **arteries** and the small **median sacral artery**. These arteries carry blood to the pelvis and lower limbs **(Figures 22.14, 22.16,** and **22.17)**. The common iliac arteries travel along the inner surface of the ilium, descending posterior to the cecum and sigmoid colon. At the level of the lumbosacral joint, each common iliac divides to form an **internal iliac artery** and an **external iliac artery**. The internal iliac arteries enter the pelvic cavity and supply the urinary bladder, internal and external walls of the pelvis, external genitalia, and medial side of the thigh. The major branches of the internal iliac artery are the superior gluteal, internal pudendal, obturator, and lateral sacral arteries. In females, these vessels also supply the uterus and vagina. The external iliac arteries supply blood to the lower limbs, and they are much larger in diameter than the internal iliac arteries.

Arteries of the Thigh and Leg The external iliac artery crosses the surface of the iliopsoas and penetrates the abdominal wall, exiting the abdomen between the anterior superior iliac spine and the pubic symphysis. It emerges on the anteromedial surface of the thigh as the **femoral artery**. Approximately 5 cm distal to its exit from the abdominal cavity, the **deep femoral artery** branches off its lateral surface **(Figure 22.16)**. The **medial** and **lateral circumflex arteries** branch off the deep femoral artery. The femoral, deep femoral, medial, and lateral circumflex arteries supply the muscles of the thigh and the ventral and lateral skin of the thigh.

The femoral artery continues inferiorly and posterior to the femur. At the popliteal fossa, the femoral artery gives off a branch, the **descending genicular artery**. This vessel supplies blood to the medial aspect of the knee. The femoral artery continues distally and becomes the **popliteal** (pop-LIT-ē-al) **artery** as it passes through the adductor magnus **(Figure 22.17)**. The popliteal artery crosses the popliteal fossa and then branches, forming the **posterior tibial artery** and the **anterior tibial artery**. The posterior tibial artery travels distally, deep to the soleus, supplying the skin and posterior muscles of the leg. The anterior tibial artery passes between the tibia and fibula, emerging on the anterior surface of the tibia. As it descends toward the foot, it supplies blood to the skin and muscles of the anterior portion of the leg. The posterior tibial artery gives rise to the **fibular artery** or *peroneal artery*, approximately 4 cm distal to its formation. The fibular artery travels distally, lying posterior to the fibula. It supplies blood to the skin and muscles on the lateral side of the leg.

Arteries of the Foot When the anterior tibial artery reaches the ankle, it becomes the **dorsalis pedis artery**. The dorsalis pedis branches repeatedly, supplying the ankle and dorsal portion of the foot **(Figure 22.16)**.

At the ankle, the posterior tibial artery divides to form the **medial** and **lateral plantar arteries**. They supply blood to the plantar surface of the foot. These arteries connect to the dorsalis pedis artery by a pair of anastomoses. This connection links the **dorsal arch** (*arcuate arch*) to the **plantar arch**. Small arteries branching off these arches supply the distal portions of the foot and the toes.

22.5 CONCEPT CHECK

7 What regions of the body receive their blood from the carotid arteries?

8 Which artery is found at the biceps region of the right arm?

9 What artery does the external iliac artery become after leaving the abdominal cavity?

10 Does damage to the internal carotid arteries always result in brain damage? Why or why not?

See the blue Answers tab at the back of the book.

Figure 22.15 Arteries of the Abdomen

Abdominal aorta

Branches of the Common Hepatic Artery

Hepatic artery proper

Cystic

Gastroduodenal

Right gastric

Right gastro-epiploic

Superior pancreaticoduodenal

Duodenal

Ascending colon

Superior Mesenteric Artery

Inferior pancreaticoduodenal

Middle colic (cut)

Right colic

Ileocolic

Intestinal

Right external iliac

Right internal iliac

Liver

Stomach

Spleen

Pancreas

Small intestine

Sigmoid colon

Rectum

Inferior vena cava

Thoracic aorta

Celiac Trunk

Left gastric

Common hepatic

Splenic

Branches of the Splenic Artery

Left gastro-epiploic

Pancreatic

Inferior Mesenteric Artery

Left colic

Sigmoid

Rectal

Left external iliac

a Major arteries supplying the abdominal viscera

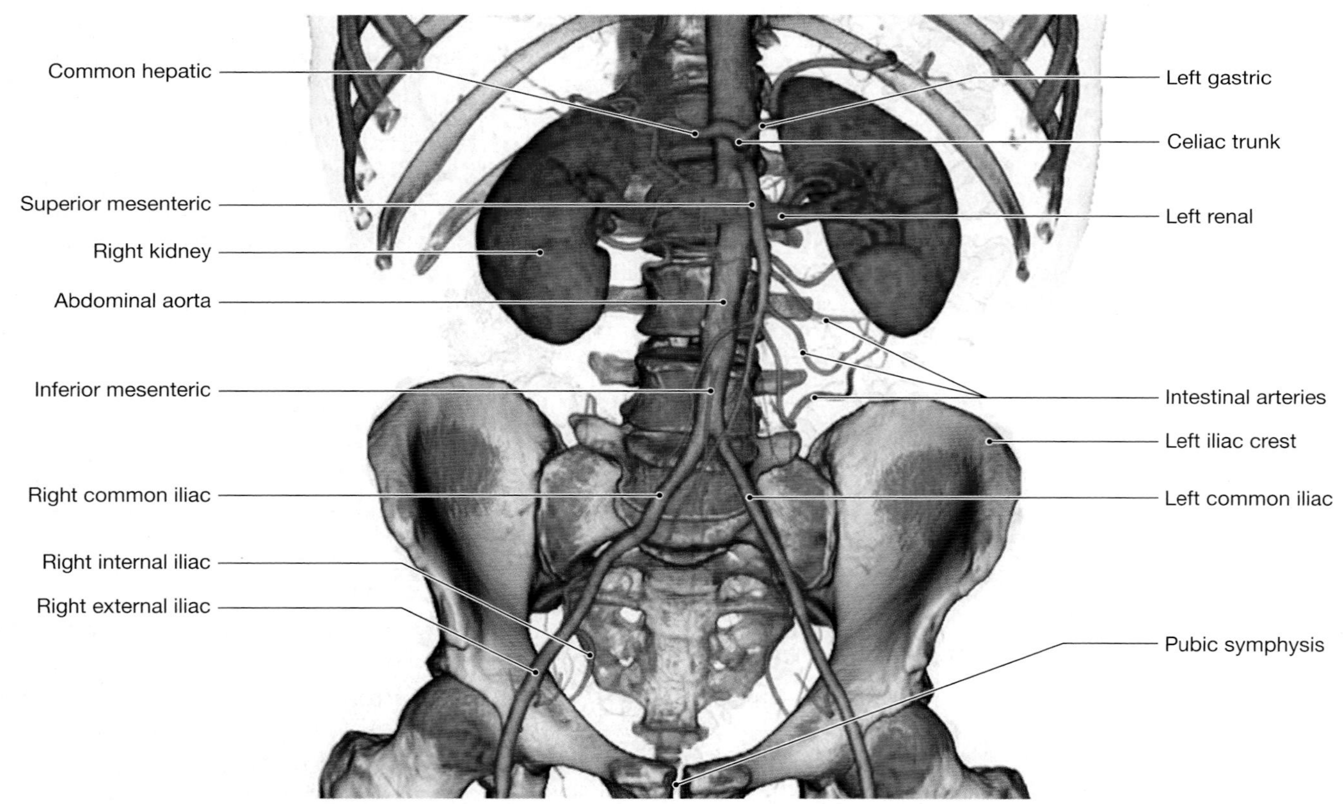

b Spiral 3-D volume rendered scan of the abdominal aorta and its branches [Courtesy of TeraRecon, Inc.]

Figure 22.16 Major Arteries of the Lower Limb, Part I

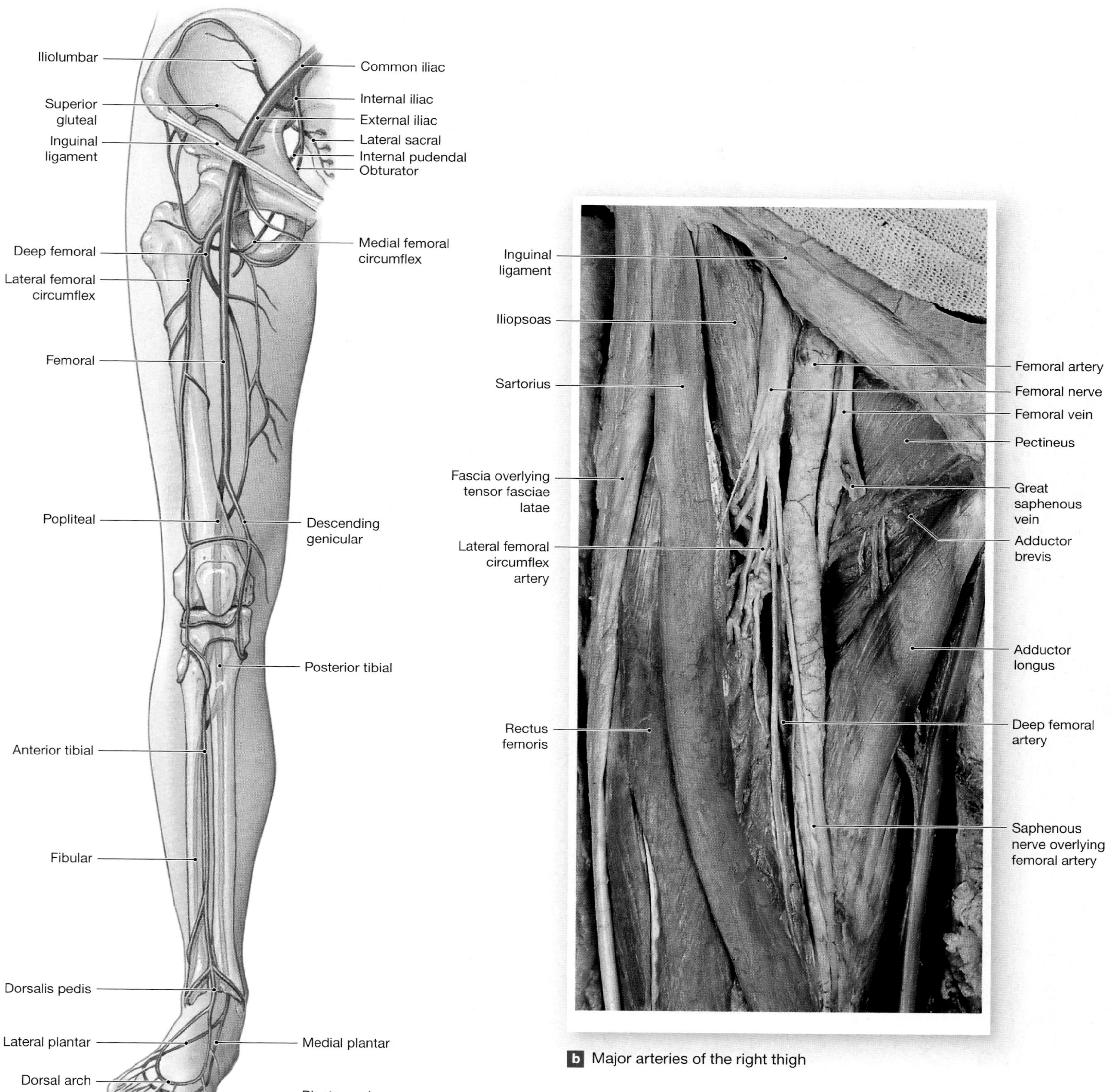

a Anterior view of the arteries supplying the right lower limb

b Major arteries of the right thigh

22

Figure 22.17 Major Arteries of the Lower Limb, Part II

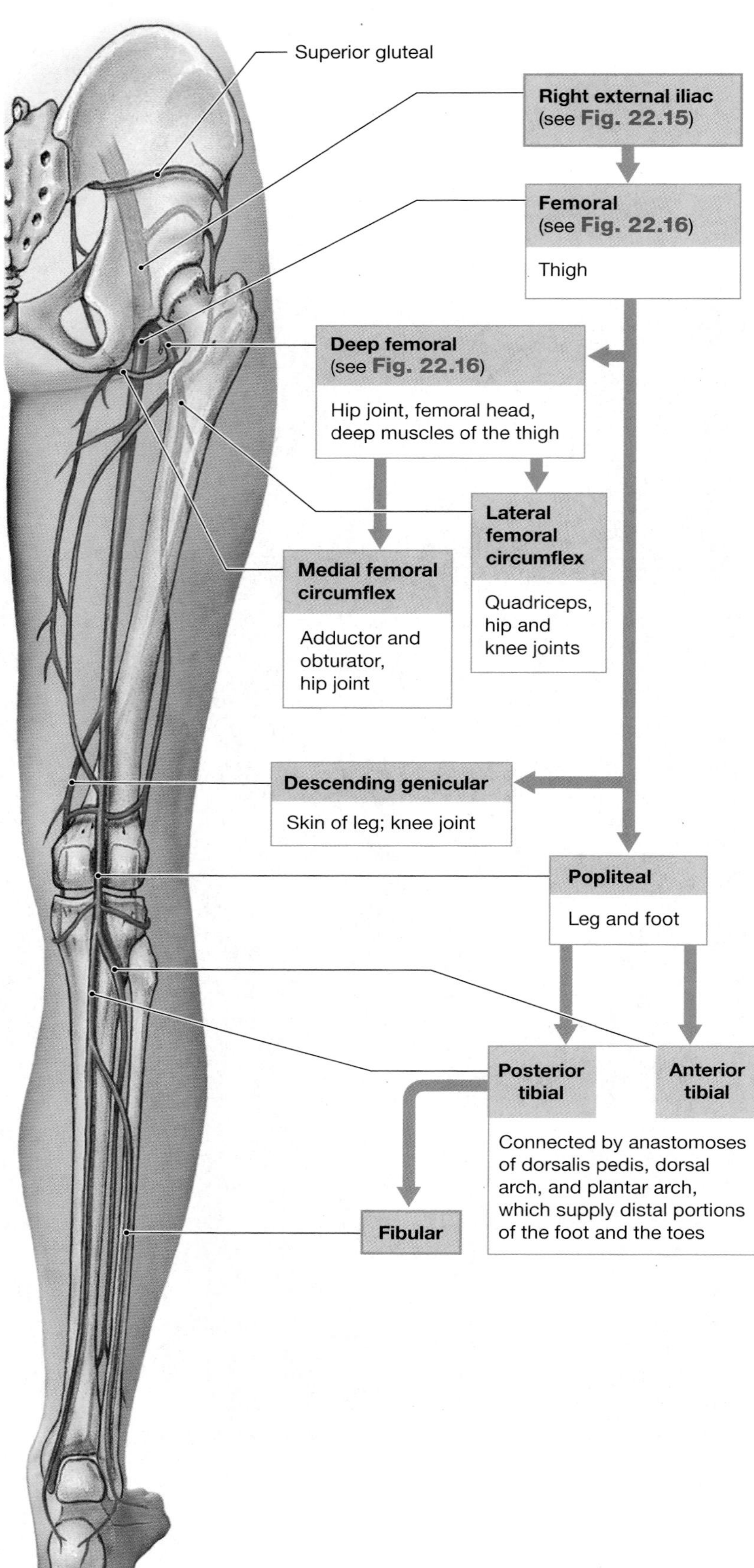

Posterior view of the arteries supplying the right lower limb

22.6 | Systemic Veins

▶ **KEY POINT** Systemic veins drain all of the areas supplied by the arterial branches of the systemic circuit. The systemic veins ultimately unite to form either the superior vena cava or the inferior vena cava, both of which empty into the right atrium of the heart.

Veins collect blood from the body's tissues and organs in an elaborate venous network. This network drains into the right atrium of the heart by the superior and inferior venae cavae. Arteries and veins typically run side by side, and in many cases they have comparable names (**Figure 22.18**). For example, the axillary artery runs alongside the axillary vein. In addition, arteries and veins often travel in a connective tissue sheath containing peripheral nerves that have the same names and innervate the same structures.

There are two significant differences between the arterial and venous systems in the neck and limbs:

1. Venous drainage is variable. Significant variations commonly occur in the number *and* location of veins draining the neck and limbs. *Figures in this chapter represent the most common location of veins in these regions.*
2. Arteries in the neck and limbs are not found at the body surface. Instead, they are deep, protected by bones and surrounding soft tissues. In contrast, the neck and limbs usually have *two sets* of peripheral veins, one superficial and the other deep. Superficial veins are so close to the surface that they are clearly visible. It is easy to obtain blood samples from these veins, and most blood tests are performed on venous blood collected from the superficial veins of the upper limb (usually the antecubital surface).

This dual venous drainage plays an important role in controlling body temperature. When body temperature becomes abnormally low, the arterial blood supply to the skin is reduced, the superficial veins are bypassed, and blood flows back to the trunk in the deep veins. When overheating occurs, the blood supply to the skin increases, and the superficial veins dilate. This is one reason why superficial veins in the arms and legs become prominent during heavy exercise.

The branching pattern of peripheral veins is more variable than that of arteries. Arterial pathways are usually direct, because developing arteries grow toward active tissues. By the time blood reaches the venous system, pressures are low, and routing variations make little functional difference. The following discussion is based on the most common arrangement of veins.

The Superior Vena Cava

▶ **KEY POINT** The superior vena cava receives blood from the head, neck, chest, shoulders, and upper limbs.

All systemic veins (except the cardiac veins, which drain into the coronary sinus) drain into either the superior vena cava or the inferior vena cava. The **superior vena cava** receives blood from the tissues and organs of the head, neck, chest, shoulders, and upper limbs (**Figure 22.18**).

Venous Return from the Cranium

Numerous veins drain the cerebral hemispheres. The **superficial cerebral veins** empty into a network of dural sinuses. These include the superior and inferior sagittal sinuses, the petrosal sinuses, the occipital sinus, the left and right transverse sinuses, and the straight sinus (**Figure 22.19**). The **superior sagittal sinus** lies within the falx cerebri. ↺ p. 408 The majority of the

Figure 22.18 An Overview of the Systemic Venous System

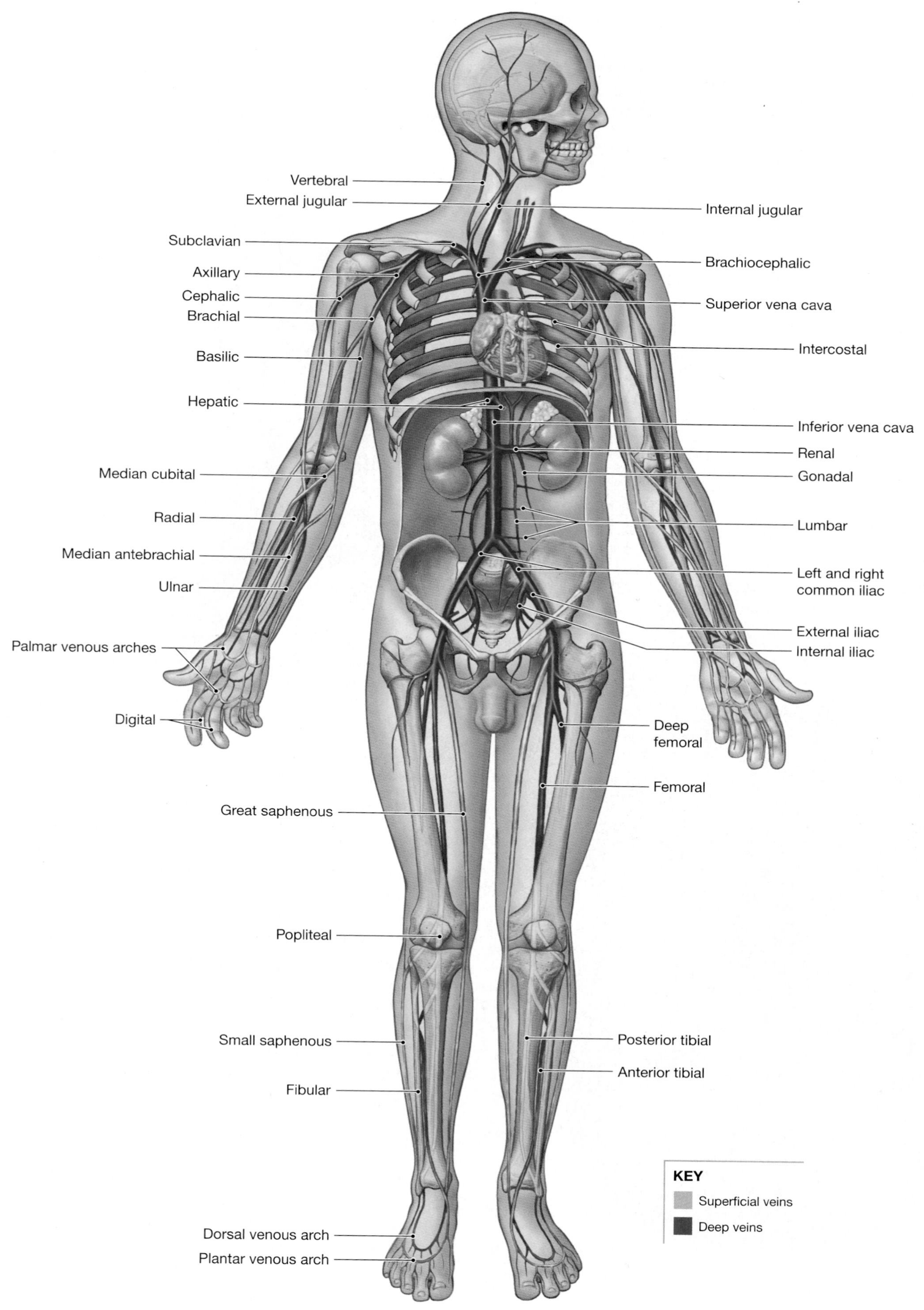
Vertebral
External jugular
Internal jugular
Subclavian
Axillary
Cephalic
Brachial
Brachiocephalic
Superior vena cava
Intercostal
Basilic
Hepatic
Inferior vena cava
Renal
Gonadal
Median cubital
Radial
Median antebrachial
Ulnar
Lumbar
Left and right common iliac
Palmar venous arches
External iliac
Internal iliac
Digital
Deep femoral
Femoral
Great saphenous
Popliteal
Small saphenous
Posterior tibial
Anterior tibial
Fibular
KEY
Superficial veins
Deep veins
Dorsal venous arch
Plantar venous arch

22

Figure 22.19 **Major Veins of the Head and Neck**

Superior sagittal sinus
Superficial cerebral veins
Inferior sagittal sinus
Great cerebral
Straight sinus
Petrosal sinuses
Right transverse sinus
Occipital sinus
Sigmoid sinus
Occipital
Vertebral
External jugular
Right subclavian
Axillary
Clavicle
First rib
Temporal
Deep cerebral
Cavernous sinus
Maxillary
Facial
Internal jugular
Right brachiocephalic
Left brachiocephalic
Superior vena cava
Internal thoracic

a An oblique lateral view of the head and neck showing the major superficial and deep veins.

22

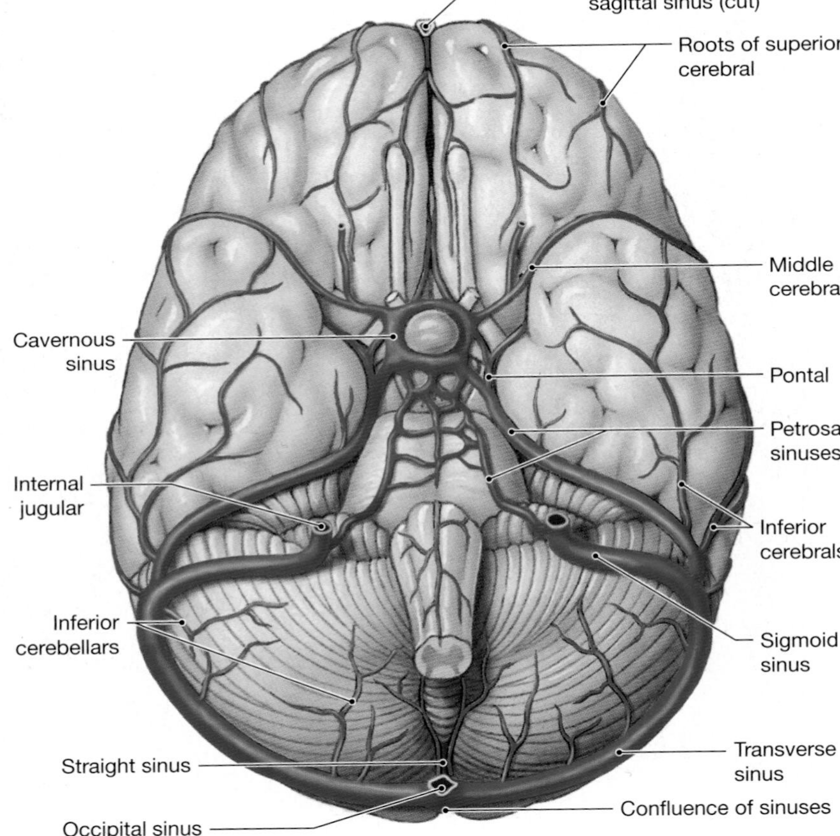

b An inferior view of the brain showing the major veins. Compare with the arterial supply to the brain shown in **Figure 22.13a**.

internal cerebral veins collect inside the brain and form the **great cerebral vein**. This vessel collects blood from the interior of the cerebral hemispheres and the choroid plexus and delivers it to the **straight sinus**. Other cerebral veins and numerous small veins from the orbit of the eye drain into the **cavernous sinus**. Blood from the cavernous sinus reaches the internal jugular vein through the **petrosal sinuses**.

The venous sinuses converge within the dura mater in the region of the lambdoid suture. The **left** and **right transverse sinuses** form at the **confluence of sinuses** near the base of the petrous part of the temporal bone. Each drains into a **sigmoid sinus**. Each sigmoid sinus continues as the **internal jugular vein**, which exits the skull at the jugular foramen. The internal jugular vein descends parallel to the common carotid artery in the neck **(Figure 22.19)**.

Vertebral veins drain the cervical spinal cord and the posterior surface of the skull. These vessels descend within the transverse foramina of the cervical vertebrae, adjacent to the vertebral arteries. The vertebral veins empty into the brachiocephalic veins of the chest.

Superficial Veins of the Head and Neck The superficial veins of the head converge to form the **temporal**, **facial**, and **maxillary veins** **(Figure 22.19)**. The temporal and maxillary veins drain into the **external jugular vein**. The facial vein drains into the internal jugular vein. A broad anastomosis between the external and internal jugular veins, at the angle of the mandible, provides dual venous drainage of the face, scalp, and cranium. The

Figure 22.20 **The Venous Drainage of the Trunk and Upper Limb**

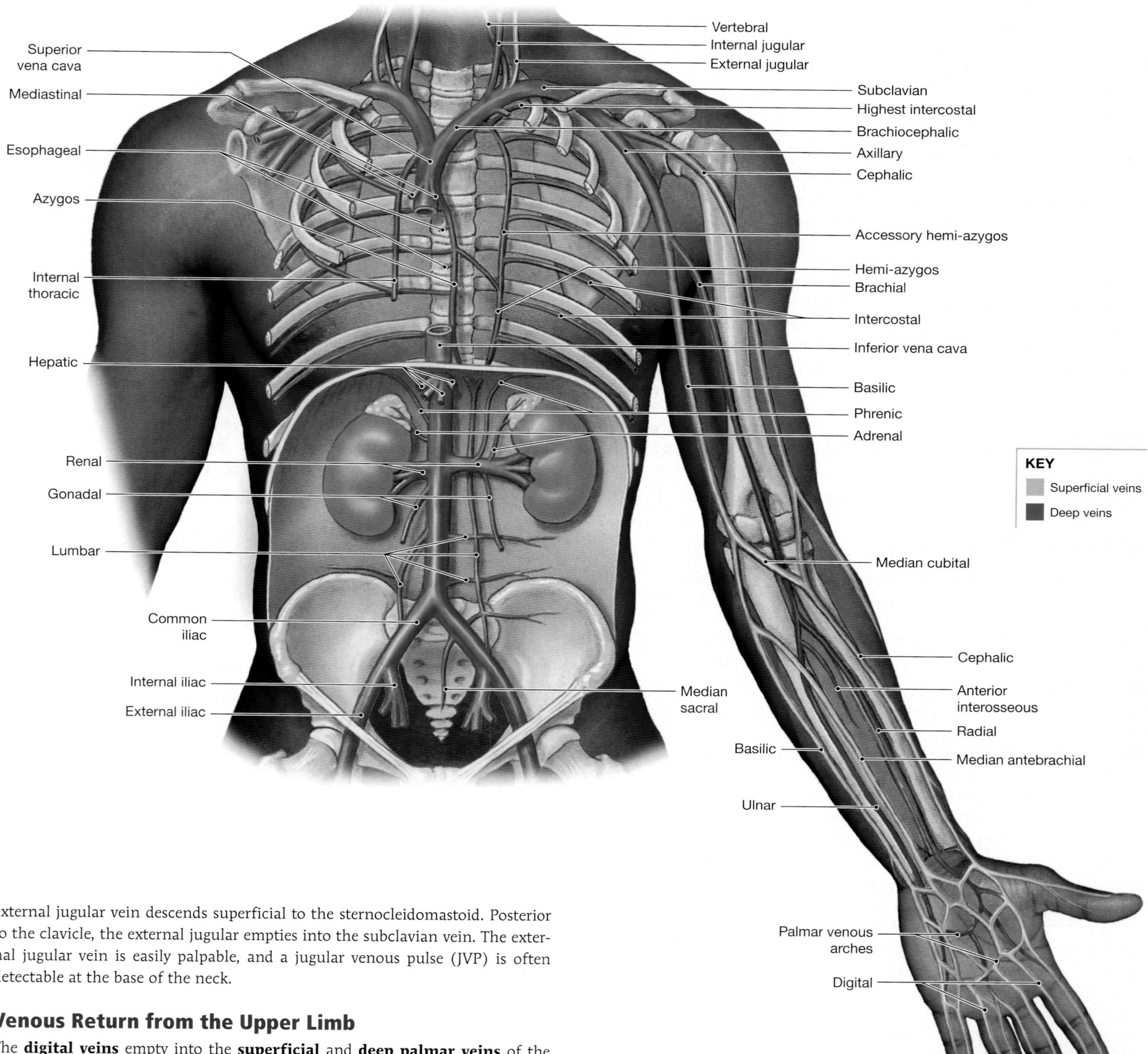

external jugular vein descends superficial to the sternocleidomastoid. Posterior to the clavicle, the external jugular empties into the subclavian vein. The external jugular vein is easily palpable, and a jugular venous pulse (JVP) is often detectable at the base of the neck.

Venous Return from the Upper Limb

The **digital veins** empty into the **superficial** and **deep palmar veins** of the hand. These interconnect to form the **palmar venous arches** (Figure 22.20). The superficial arch empties into the **cephalic vein**, the **median antebrachial vein**, and the **basilic vein**. The cephalic vein travels proximally on the radial side of the forearm. The median antebrachial vein and the basilic vein both travel proximally on the ulnar side of the forearm. Anterior to the elbow is the superficial **median cubital vein**, which interconnects the cephalic and basilic veins. Venous blood samples are typically collected from the median cubital vein.

The deep palmar veins drain into the **radial vein** and the **ulnar vein**. After crossing the elbow, these veins fuse with the anterior interosseous vein to form the **brachial vein**. The brachial vein lies parallel to the brachial artery. As the brachial vein continues proximally toward the trunk, it receives blood from the basilic vein, which lies on the medial surface of the biceps brachii. The brachial vein enters the axilla as the **axillary vein** (Figure 22.20).

The Formation of the Superior Vena Cava

The cephalic vein joins the axillary vein, forming the **subclavian vein**. The subclavian vein passes over the superior surface of the first rib, deep to the clavicle, and into the thoracic cavity. The subclavian merges with the external and internal jugular veins, creating the **brachiocephalic vein** (Figure 22.20). The brachiocephalic vein receives blood from the vertebral vein, which drains the posterior portion of the skull and the spinal cord. At the level of the first and second ribs, the left and right brachiocephalic veins merge and form the superior vena cava. Close to the point of fusion, the **internal thoracic vein** empties into the brachiocephalic vein.

The **azygos** (AZ-ī-gos) **vein** merges with the superior vena cava. This vein ascends from the lumbar region over the right side of the vertebral column to enter the thoracic cavity through the diaphragm. The azygos vein joins the superior vena cava at the level of vertebra T_2. On the left side, the azygos receives blood from the smaller **hemi-azygos vein**, which may also drain into the left brachiocephalic vein through the highest **intercostal vein**.

The azygos and hemi-azygos veins are the chief collecting vessels of the thorax. They receive blood from (1) **intercostal veins**, which in turn receive blood from the chest muscles, (2) **esophageal veins**, which drain blood from the inferior portion of the esophagus, and (3) smaller veins draining other mediastinal structures.

The hemi-azygos vein may also drain into the **left superior intercostal vein**, a tributary of the left brachiocephalic vein, by way of a small **accessory hemi-azygos vein**. The accessory hemi-azygos vein starts at the fourth or fifth intercostal space and travels inferiorly on the left side of the vertebral column. It drains the left superior portion of the thoracic cavity. At T_8, it crosses to the right side of the thoracic cavity and empties into the azygos vein.

The Inferior Vena Cava

▶ **KEY POINT** The inferior vena cava ascends to the right of the aorta, draining structures supplied with blood by the arteries of the systemic circulation.

The **inferior vena cava** collects most of the venous blood from organs inferior to the diaphragm. (A small amount reaches the superior vena cava by the azygos and hemi-azygos veins).

Veins Draining the Lower Limb

Blood leaving capillaries in the sole of each foot collects into a network of **plantar veins**. Blood in the **plantar venous arch** flows into the deep veins of the leg: the **anterior tibial vein**, the **posterior tibial vein**, and the **fibular** (or *peroneal*) **vein** (Figure 22.21). The **dorsal venous arch** collects blood from capillaries on the dorsal surface of the foot and the digital veins of the toes. There are extensive anastomoses between the plantar arch and dorsal arch, and the blood in these arches can easily flow into either the superficial or deep veins.

The dorsal venous arch is drained by two superficial veins, the **great saphenous** (sa-FĒ-nus; *saphenes,* prominent) **vein** and the **small saphenous vein**. (During coronary artery bypass graft surgery, surgeons often remove the great saphenous vein to replace blocked coronary vessels.) The longest vein in the body, the great saphenous travels proximally along the medial aspect of the leg and thigh, draining into the femoral vein near the hip joint. The small saphenous vein arises from the dorsal venous arch and travels proximally along the posterior and lateral aspect of the calf. It then enters the popliteal fossa, where it drains into the popliteal vein. The **popliteal vein** is formed by the union of the anterior tibial, posterior tibial, and fibular veins (Figure 22.21). The popliteal vein is easily palpated in the popliteal fossa adjacent to the adductor magnus. Once it leaves the popliteal fossa, the popliteal vein becomes the **femoral vein**, which travels proximally next to the femoral artery.

Immediately before entering the abdominal wall, the femoral vein receives blood from three vessels: (1) the great saphenous vein, (2) the **deep femoral vein**, and (3) the **femoral circumflex vein**. The deep femoral vein collects blood from deeper structures of the thigh, and the femoral circumflex vein drains the area around the neck and head of the femur. The femoral vein penetrates the body wall and enters the pelvic cavity as the **external iliac vein**.

Veins Draining the Pelvis

The external iliac veins receive blood from the lower limbs, pelvis, and lower abdomen. As each external iliac vein travels across the inner surface of the ilium, it merges with the **internal iliac vein**. The internal iliac veins drain the organs of the pelvis. The **gluteal**, **internal pudendal**, **obturator**, and **lateral sacral veins** unite and form the internal iliac veins (Figure 22.21). The union of the external and internal iliac veins forms the **common iliac vein**. The **median sacral vein**, which drains the region supplied by the median sacral artery, empties into the left common iliac. The left and right common iliac veins ascend at an oblique angle and, anterior to vertebra L_5, they unite to form the inferior vena cava (Figure 22.21).

Veins Draining the Abdomen

Vessels emptying into the inferior vena cava drain the abdominal wall, gonads, liver, kidneys, adrenal glands, and diaphragm. The hepatic portal vein drains the visceral organs within the abdominal cavity. The inferior vena cava, which is retroperitoneal, ascends parallel and to the right of the aorta. Blood from the inferior vena cava flows into the right atrium, where it mixes with venous blood from the superior vena cava. This blood enters the right ventricle and is pumped into the pulmonary circuit for oxygenation at the lungs.

The abdominal portion of the inferior vena cava collects blood from six major veins (Figure 22.20):

1. **Lumbar veins** drain the lumbar portion of the abdomen. These veins are connected to the azygos vein (right side) and hemi-azygos vein (left side), which empty into the superior vena cava.
2. **Gonadal veins** (*ovarian* or *testicular* veins) drain the ovaries or testes. The right gonadal vein empties into the inferior vena cava; the left gonadal drains into the left renal vein.
3. **Hepatic veins** leave the liver and empty into the inferior vena cava at the level of vertebra T_{10}.
4. **Renal veins** collect blood from the kidneys. These are the largest vessels draining into the inferior vena cava.
5. **Adrenal veins** drain the adrenal glands. Usually, only the right adrenal vein drains into the inferior vena cava, and the left drains into the left renal vein.
6. **Phrenic veins** drain the diaphragm. Only the right phrenic vein drains into the inferior vena cava; the left drains into the left renal vein.

The Hepatic Portal System

The **hepatic portal system** begins in the capillaries of the digestive organs and ends in the liver sinusoids. The liver is the *only* digestive organ draining directly into the inferior vena cava. Blood leaving the capillary beds supplied by the celiac, superior mesenteric, and inferior mesenteric arteries flows into the veins of the *hepatic portal system* instead of traveling directly to the inferior vena cava. As we mentioned in Chapter 19, a blood vessel connecting two capillary beds is called a portal vessel, and the network is a portal system. ↺ pp. 508–510 Blood flowing in the hepatic portal system is quite different from blood in other systemic veins because it contains substances absorbed from the stomach and intestines. For example, levels of blood glucose and amino acids in the hepatic portal vein often exceed those found anywhere else in the cardiovascular system. The hepatic portal system delivers these and other absorbed compounds directly to the liver for storage, metabolic conversion, or excretion (Figure 22.22).

The largest vessel of the hepatic portal system is the **hepatic portal vein**. It delivers venous blood to the liver. It receives blood from three large veins draining organs within the peritoneal cavity:

1. The **inferior mesenteric vein** collects blood from capillaries in the inferior portion of the large intestine. The **left colic vein** and the **superior rectal veins**, which drain the descending colon, sigmoid colon, and rectum, empty into the inferior mesenteric vein.
2. The **splenic vein** collects blood from the inferior mesenteric vein and veins from the spleen, the lateral border of the stomach (left vein), and the pancreas (pancreatic veins).
3. The **superior mesenteric vein** collects blood from veins draining the stomach (right gastro-epiploic vein), the small intestine (intestinal and pancreaticoduodenal veins), and two-thirds of the large intestine (ileocolic, right colic, and middle colic veins).

The hepatic portal vein forms through the fusion of the superior mesenteric and splenic veins. Of the two, the superior mesenteric vein normally contributes the greater volume of blood and most of the nutrients. As the hepatic

Figure 22.21 The Venous Drainage of the Lower Limb

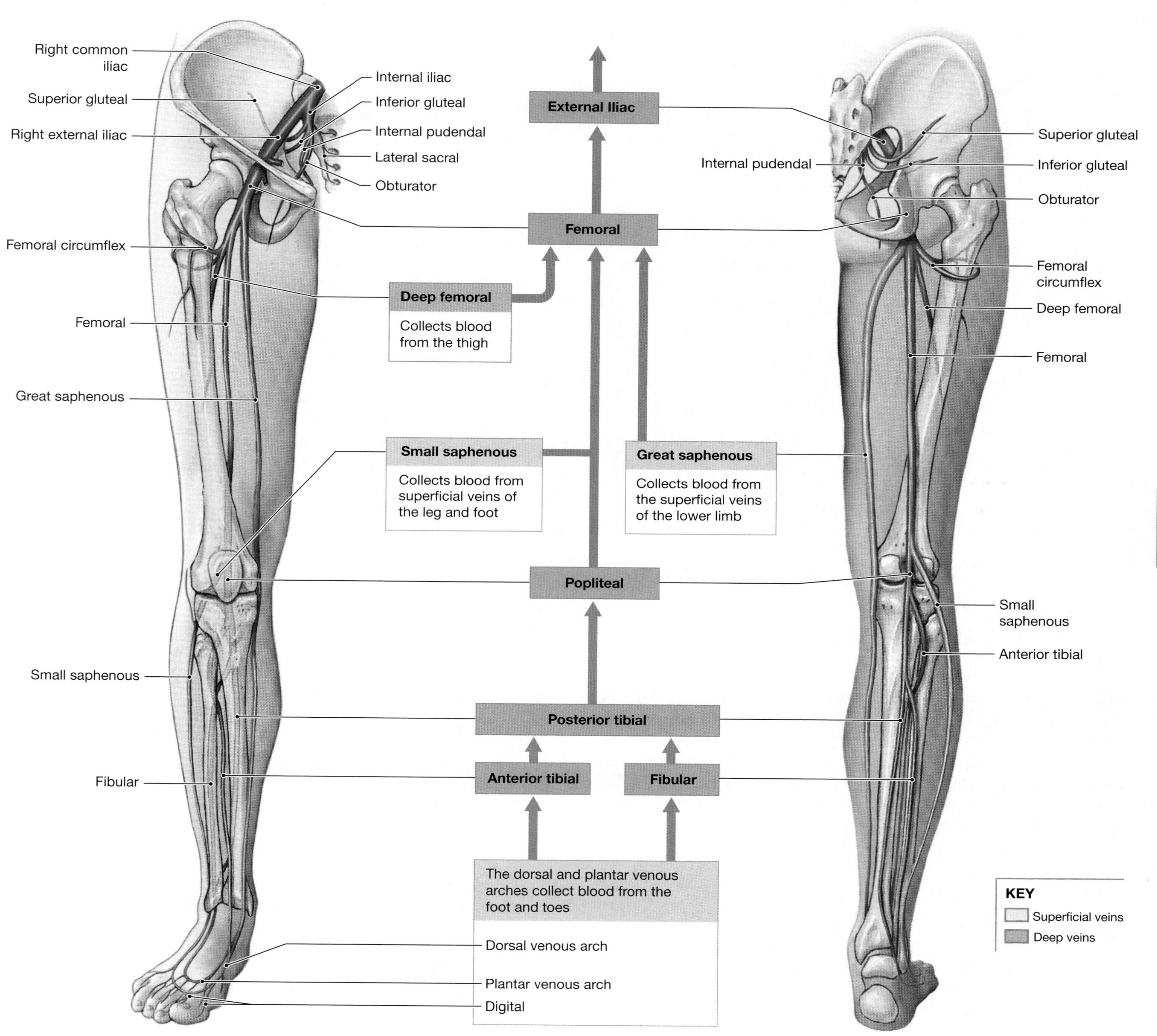

a Anterior view showing the veins of the right lower limb

b Posterior view showing the veins of the right lower limb

22

Figure 22.22 The Hepatic Portal System

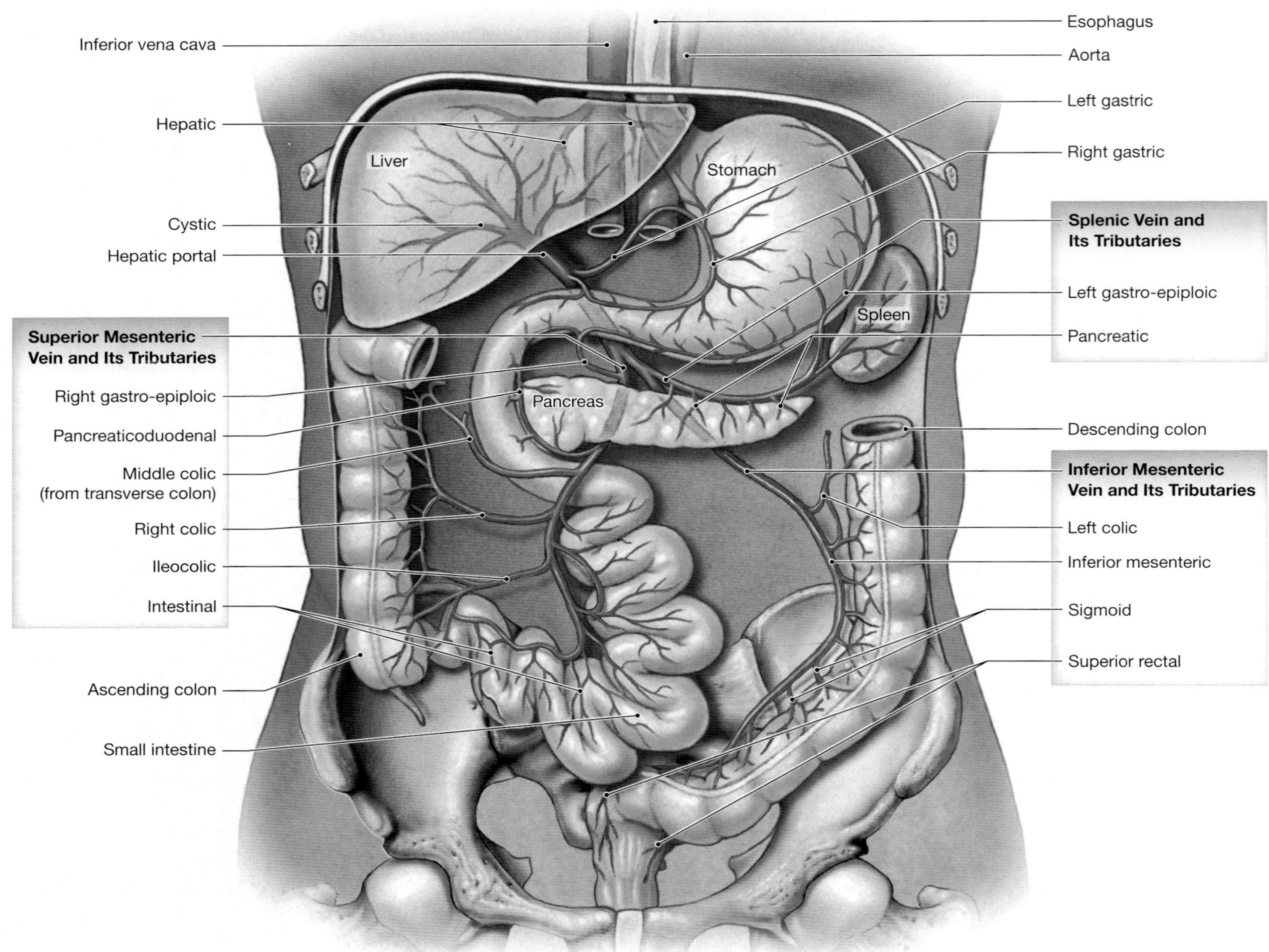

portal vein proceeds to the liver, it also receives blood from the **gastric veins** and **cystic veins**. The gastric veins drain the medial border of the stomach, and the cystic veins drain the gallbladder.

After passing through the liver sinusoids, blood flows into the hepatic veins, which empty into the inferior vena cava **(Figure 22.22)**. Because blood goes to the liver first, the composition of blood in the systemic circuit remains relatively stable, regardless of the digestive activities under way.

22.6 CONCEPT CHECK

11 It is 110°F outside, and you are very hot. What changes have occurred in your veins and why?

12 Which major vein receives blood from the head, neck, chest, shoulders, and upper limbs?

13 Why does blood leaving the intestines first go to the liver?

See the blue Answers tab at the back of the book.

22.7 Cardiovascular Changes at Birth

▶ **KEY POINT** During development, the fetus receives its nutrients and oxygen, and excretes its wastes and carbon dioxide, by diffusion through the placenta. With a newborn's first breath, lung function, gastrointestinal activity, and circulatory patterns change.

There are significant differences between the fetal and adult cardiovascular systems. These differences reflect the different sources of respiratory and nutritional support. The fetal lungs are collapsed and nonfunctional, and the digestive tract has nothing to digest. The placenta, a complex organ regulating the exchange between the fetal and maternal bloodstreams, provides all fetal respiratory and nutritional needs.

Two **umbilical arteries** branch off the internal iliac arteries of the fetus **(Figure 22.23)**. These vessels enter the umbilical cord and deliver blood to the placenta. Blood returns to the fetus from the placenta in the single **umbilical vein**. The umbilical vein brings oxygen and nutrients to the developing fetus. The umbilical vein drains into the **ductus venosus**, which is connected to a network

Figure 22.23 Changes in Fetal Circulation at Birth

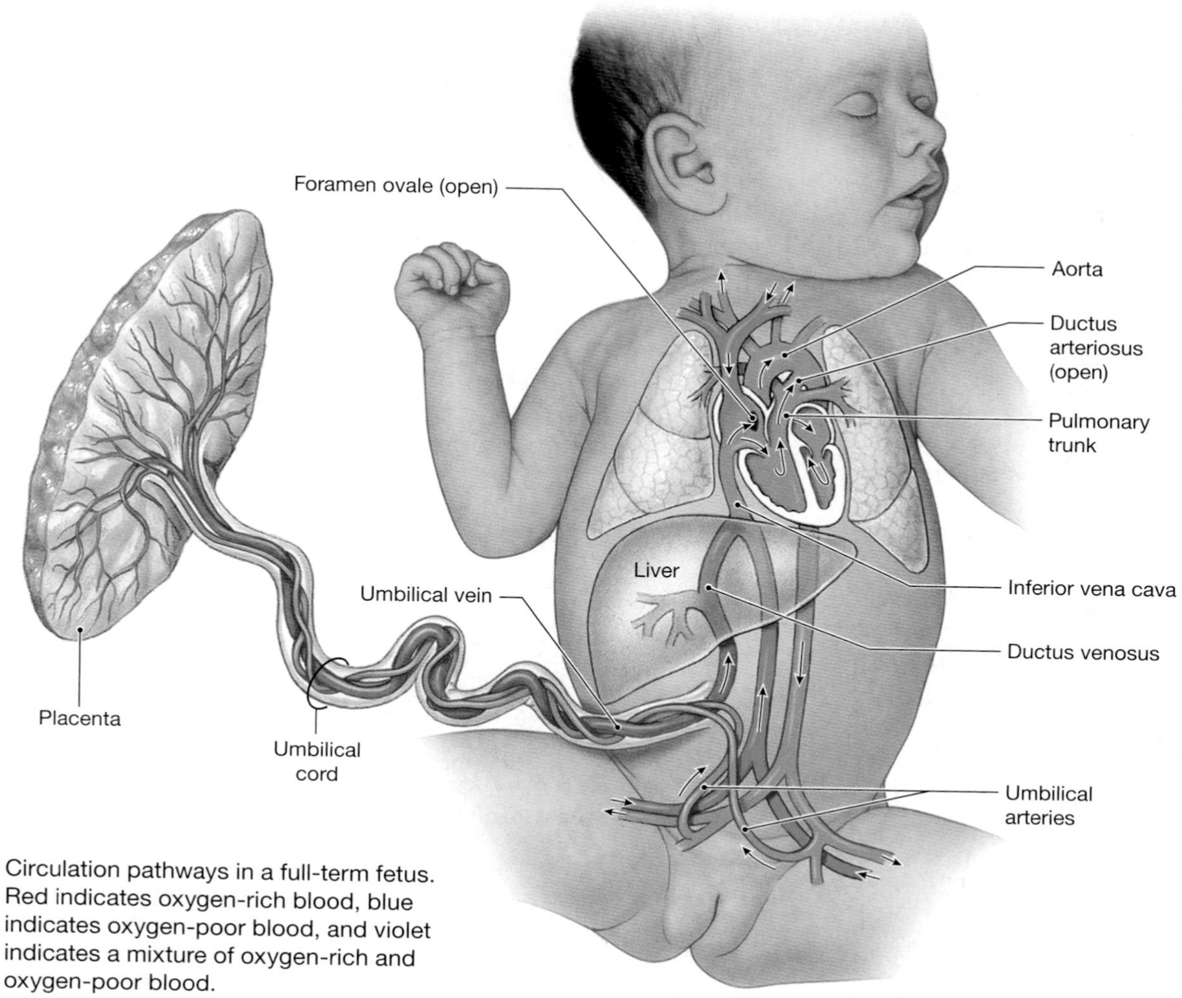

Foramen ovale (open)
Aorta
Ductus arteriosus (open)
Pulmonary trunk
Liver
Umbilical vein
Inferior vena cava
Ductus venosus
Placenta
Umbilical cord
Umbilical arteries

a Circulation pathways in a full-term fetus. Red indicates oxygen-rich blood, blue indicates oxygen-poor blood, and violet indicates a mixture of oxygen-rich and oxygen-poor blood.

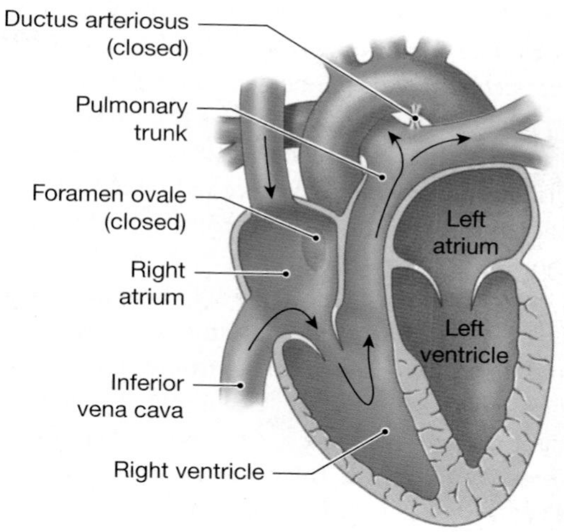

Ductus arteriosus (closed)
Pulmonary trunk
Foramen ovale (closed)
Right atrium
Inferior vena cava
Right ventricle
Left atrium
Left ventricle

b Blood flow through the heart of the newborn.

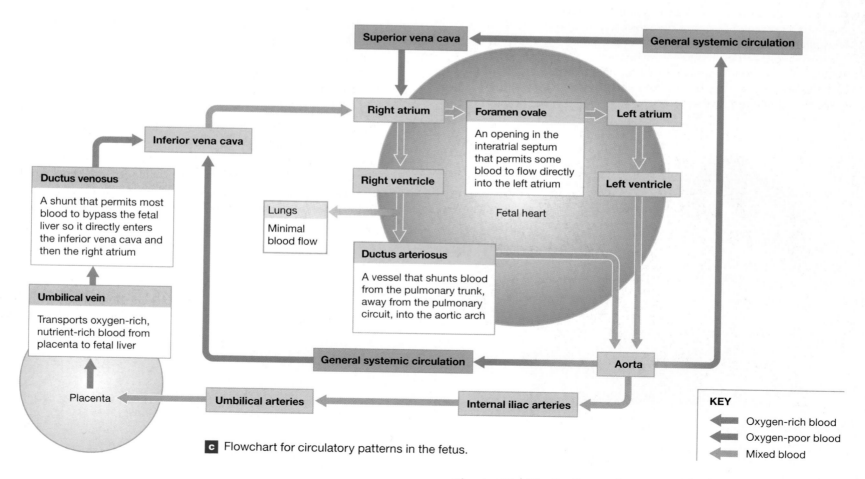

Superior vena cava
General systemic circulation
Right atrium
Foramen ovale
An opening in the interatrial septum that permits some blood to flow directly into the left atrium
Left atrium
Inferior vena cava
Right ventricle
Left ventricle
Ductus venosus
A shunt that permits most blood to bypass the fetal liver so it directly enters the inferior vena cava and then the right atrium
Lungs
Minimal blood flow
Fetal heart
Ductus arteriosus
A vessel that shunts blood from the pulmonary trunk, away from the pulmonary circuit, into the aortic arch
Umbilical vein
Transports oxygen-rich, nutrient-rich blood from placenta to fetal liver
General systemic circulation
Aorta
Placenta
Umbilical arteries
Internal iliac arteries

KEY
Oxygen-rich blood
Oxygen-poor blood
Mixed blood

c Flowchart for circulatory patterns in the fetus.

of veins. When the placental connection is broken at birth, blood flow ceases in the umbilical vessels, and they soon degenerate.

Although the interatrial and interventricular septa develop early in fetal life, the interatrial partition is functionally incomplete up to the time of birth. The **foramen ovale** *(interatrial opening)* is associated with a long flap that acts as a valve. Blood can flow freely from the right atrium into the left atrium, but any backflow closes the valve and isolates the two chambers from one another. Thus, blood entering the heart at the right atrium can bypass the pulmonary circuit. A second, shorter circuit exists between the pulmonary and aortic trunks. This connection, the **ductus arteriosus**, consists of a short, muscular vessel.

With the lungs collapsed, the capillaries are compressed and little blood flows through the lungs. During diastole, blood enters the right atrium and flows into the right ventricle, but it also passes into the left atrium through the foramen ovale. About 25 percent of the blood arriving at the right atrium bypasses the pulmonary circuit in this way. In addition, more than 90 percent of the blood leaving the right ventricle passes through the ductus arteriosus and enters the systemic circuit, rather than continuing to the lungs.

At birth, dramatic changes occur. When the infant takes its first breath, the lungs and pulmonary vessels expand. The smooth muscle in the ductus arteriosus contracts. This isolates the pulmonary and aortic trunks, and blood begins flowing through the pulmonary circuit. As pressures rise in the left atrium, the valvular flap closes the foramen ovale and completes the cardiovascular remodeling **(Figure 22.23a,b)**. In adults, the interatrial septum bears a shallow depression, the **fossa ovalis**, marking the site of the original foramen ovale. The remnants of the ductus arteriosus remain as a fibrous cord, the **ligamentum arteriosum**.

If the proper circulatory changes do not occur at birth or shortly thereafter, problems will eventually develop. The severity of the problem varies depending on which connection remains open and the size of the opening. Treatment may involve surgical closure of the foramen ovale, the ductus arteriosus, or both. Other forms of congenital heart defects result from abnormal cardiac development or inappropriate connections between the heart and major arteries and veins.

CLINICAL NOTE

Congenital Cardiovascular Problems

Congenital cardiovascular problems serious enough to represent a threat to homeostasis are relatively rare. They usually reflect abnormal formation of the heart or problems with the interconnections between the heart and the great vessels. Most of these conditions can be surgically corrected, although multiple surgeries may be required and life expectancy may be shortened in more severe defects.

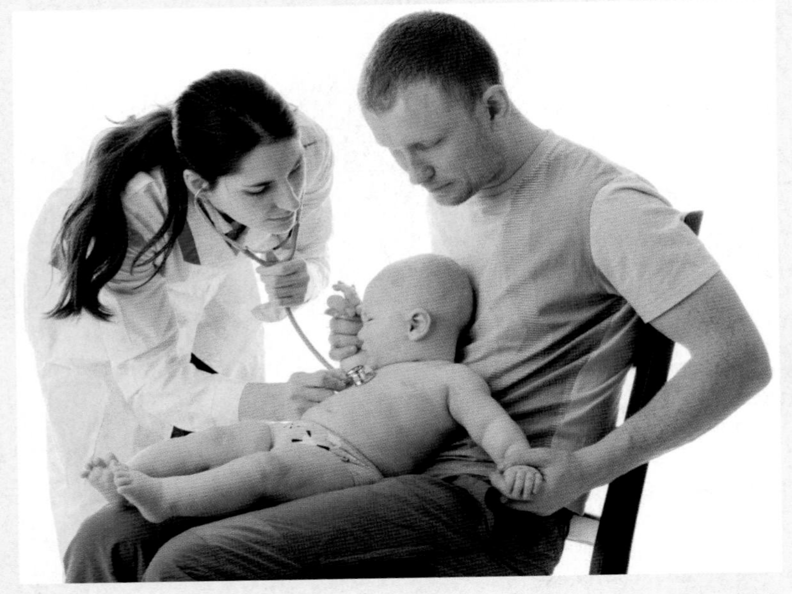

22.7  **CONCEPT CHECK**

14 What changes occur in the heart and major vessels of a newborn at birth?

See the blue Answers tab at the back of the book.

22.8 Aging and the Cardiovascular System

▶ **KEY POINT** Age-related changes occur in the composition of the blood and in the anatomy and functioning of the heart and blood vessels.

The capabilities of the cardiovascular system gradually decline with age. As you age, your cardiovascular system undergoes major changes, as outlined below:

- Age-related changes in the blood often include (1) decreased hematocrit, (2) constriction or blockage of peripheral veins by a **thrombus** (stationary blood clot), which can become detached, pass through the heart, and become wedged in a small artery (commonly the lungs), causing a **pulmonary embolism**, and (3) pooling of blood in the veins of the legs because valves work less efficiently.
- Age-related anatomical changes in the heart include (1) a reduction in the maximum cardiac output, (2) decreased elasticity of the cardiac skeleton, (3) progressive atherosclerosis that can restrict coronary circulation, (4) replacement of damaged cardiac muscle cells by scar tissue, and (5) changes in the activities of pacemaker and conducting cells.
- Age-related changes in blood vessels, some of which are related to arteriosclerosis, include (1) decreased elasticity of arteries and veins; (2) walls of arteries becoming less tolerant of sudden increases in pressure, which can lead to an **aneurysm** (an enlarged blood vessel), which may cause a stroke, heart attack, or massive blood loss, depending on the vessel involved; (3) calcium salts depositing onto weakened vascular walls, increasing the risk of a stroke or heart attack; and (4) thrombi forming at atherosclerotic plaques.

22.8  **CONCEPT CHECK**

15 Why is a reduction in elasticity of the arteries with age dangerous?

See the blue Answers tab at the back of the book.

EMBRYOLOGY SUMMARY

For a summary of the development of the cardiovascular system, see Chapter 28 (Embryology and Human Development).

Study Outline

Introduction p. 568

- The cardiovascular system is a closed system with two circulatory patterns: a pulmonary circuit and a systemic circuit.
- Blood flows through a network of arteries, veins, and capillaries.

22.1 Histological Organization of Blood Vessels p. 568

- The walls of arteries and veins contain three layers: the **intima** (*tunica intima*) (the innermost layer), the **media** (*tunica media*) (the middle layer), and the **adventitia** (*tunic externa*) (the connective tissue sheath around the vessel). *(See Spotlight Figure 22.1.)*

Distinguishing Arteries from Veins p. 568

- In general, the walls of arteries are thicker than those of veins. The endothelial lining of an artery cannot contract, so it appears as folds. Arteries constrict when blood pressure does not distend them; veins constrict very little. *(See Spotlight Figure 22.1.)*

Arteries p. 568

- The arterial system includes the large **elastic arteries**, **muscular** (*medium-sized*) **arteries**, and smaller **arterioles**. *(See Spotlight Figure 22.1.)*

Capillaries p. 569

- Capillaries are the smallest blood vessels and the only blood vessels whose walls permit exchange between blood and interstitial fluid. Capillaries may be **continuous** (the endothelium is a complete lining) or **fenestrated** (the endothelium contains "windows"). **Sinusoids** are specialized fenestrated capillaries found in selected tissues (such as the liver) that allow very slow blood flow. *(See Spotlight Figure 22.1 and Figure 22.2.)*

Capillary Beds p. 572

- Capillaries form interconnected networks called **capillary beds**. A **metarteriole** adjusts the blood flow into each capillary. **Thoroughfare channels** provide the means of arteriole-venule communication. *(See Figure 22.3.)*

Veins p. 572

- **Veins** collect oxygen-poor blood from the tissues and organs and return it to the heart. The venous system consists of **venules**, **medium-sized veins**, and **large veins**. The arterial system is a high-pressure system; blood pressure in veins is much lower. **Valves** in veins prevent the backflow of blood. *(See Spotlight Figure 22.1 and Figure 22.4.)*

22.2 The Distribution of Blood p. 574

- While the heart, arteries, and capillaries usually contain about 30–35 percent of the blood volume, most of the blood volume is in the venous system (65–70 percent). *(See Figure 22.5.)*

22.3 Blood Vessel Distribution p. 575

- The blood vessels of the body can be divided into those of the **pulmonary circuit** (between the heart and lungs) and the **systemic circuit** (from the heart to all organs and tissues). *(See Figure 22.6.)*

22.4 The Pulmonary Circuit p. 576

- The arteries of the pulmonary circuit carry oxygen-poor blood. The pulmonary circuit includes the **pulmonary trunk**, the **left** and **right pulmonary arteries**, and the **pulmonary veins**, which empty into the left atrium. *(See Figure 22.7.)*

22.5 Systemic Arteries p. 576

- The systemic circuit begins at the aortic valve and ends at the entrance to the right atrium. It supplies the capillary beds in all parts of the body not supplied by the pulmonary circuit. *(See Figure 22.8.)*

The Aorta p. 579

- The **ascending aorta** gives rise to the coronary circulation.
- The **aortic arch** continues as the **descending aorta**. Three large arteries arise from the aortic arch to collectively supply the head, neck, shoulder, and upper limbs: the **brachiocephalic trunk**, the **left common carotid artery**, and the **left subclavian artery**. *(See Figures 22.8–22.17.)*
- The brachiocephalic trunk gives rise to the **right subclavian artery** and the **right common carotid artery**. These arteries supply the right upper limb and portions of the right shoulder, neck, and head. *(See Figures 22.10–22.12.)*
- Each subclavian artery exits the thoracic cavity to become the **axillary artery**, which enters the arm to become the **brachial artery**. The brachial arteries and their branches supply blood to the upper limbs. *(See Figure 22.10.)*
- Each common carotid artery divides into an **external carotid artery** and an **internal carotid artery**. The external carotids and their branches supply blood to structures in the neck and face. The internal carotids and their branches enter the skull to supply blood to the brain and eyes. The brain also receives blood from the vertebral arteries. The **vertebral arteries** and the internal carotids form the **cerebral arterial circle** (or *circle of Willis*), which ensures the blood supply to the brain. *(See Figures 22.11–22.13.)*

The Descending Aorta p. 585

- The **descending aorta** superior to the diaphragm is termed the **thoracic aorta** and that inferior to it the **abdominal aorta**. The thoracic aorta and its branches supply blood to the thorax and thoracic viscera. The abdominal aorta and its branches supply blood to the abdominal wall, abdominal viscera, pelvic structures, and lower limbs. The three unpaired arteries are the **celiac trunk**, the **superior mesenteric artery**, and the **inferior mesenteric artery**. The celiac trunk divides into the **left gastric artery**, the **common hepatic artery**, and the **splenic artery**. Paired arteries include the **adrenal arteries**, the **renal arteries**, the **lumbar arteries**, and the **gonadal arteries**. *(See Figures 22.8–22.16.)*
- Arteries in the neck and limbs are deep beneath the skin; in contrast, there are usually two sets of peripheral veins—one superficial and one deep. This dual venous drainage is important for controlling body temperature. *(See Figures 22.8 and 22.18.)*
- Arteries of the pelvis and lower limbs include the **right** and **left common iliac arteries**, which branch to form the **external** and **internal iliac arteries**. The **femoral** and **deep femoral arteries** supply the lower limb. *(See Figure 22.16.)*

22.6 Systemic Veins p. 590

- Veins collect blood from the body's tissues and organs in an elaborate venous network that drains into the right atrium of the heart through the superior and inferior venae cavae. *(See Figure 22.18.)*

The Superior Vena Cava p. 590

- The **superior vena cava** receives blood from the head, neck, chest, shoulders, and upper limbs. *(See Figures 22.18–22.20.)*

The Inferior Vena Cava p. 594

- The **inferior vena cava** collects most of the venous blood from organs and structures inferior to the diaphragm that are not drained by the hepatic portal vein. *(See Figures 22.20–22.22.)*
- Any blood vessel connecting two capillary beds is called a portal vessel, and the network of blood vessels comprises a portal system. Blood leaving the capillaries supplied by the celiac trunk, superior, and inferior mesenteric arteries flows into the **hepatic portal system**. Blood in the hepatic portal system is unique compared to that of the other systemic veins because portal blood contains high concentrations of nutrients. These substances are collected from the digestive organs through the vessels of the portal system and are transported directly to the liver for processing. *(See Figure 22.22.)*

22.7 Cardiovascular Changes at Birth p. 596

- During fetal development, the **umbilical arteries** carry blood to the placenta. It returns via the **umbilical vein** and enters a network of vascular sinuses in the liver. The **ductus venosus** collects this blood and returns it to the inferior vena cava. *(See Figure 22.23 and Development of the Cardiovascular System in Chapter 28.)*
- At this time, the interatrial septum is incomplete and the **foramen ovale** allows the passage of blood from the right atrium to the left atrium. The **ductus arteriosus** also permits the flow of blood between the pulmonary trunk and the aortic arch. At birth or shortly thereafter, as the pulmonary circuit becomes functional, these connections normally close, forming a depression called the **fossa ovalis**, where the foramen ovale was, and the **ligamentum arteriosum**, where the ductus arteriosus used to be. *(See Figure 22.23 and Development of the Cardiovascular System in Chapter 28.)*

22.8 Aging and the Cardiovascular System p. 598

- Age-related changes occur in the blood, heart, and blood vessels. Blood changes include increased risk of **thrombus** (stationary blood clot).
- Age-related changes in blood vessels are often related to arteriosclerosis. Inelastic arterial walls may lead to an **aneurysm**.

Chapter Review

For answers, see the blue Answers tab at the back of the book.

Level 1 Reviewing Facts and Terms

Match each numbered item with the most closely related lettered item.

1. elastic arteries ____
2. thrombus ____
3. collaterals ____
4. renal veins ____
5. external iliac arteries ____
6. alveoli ____
7. carotid arteries ____
8. subclavian arteries ____
9. capillary bed ____
10. muscular arteries ____

(a) deliver blood to the head
(b) transport blood to the body's skeletal muscles and internal organs
(c) arteries whose smooth muscles do not contract
(d) stationary blood clot
(e) network of capillaries
(f) arteries that supply a capillary network
(g) supply blood to the upper limbs
(h) small air sacs
(i) supply blood to the lower limbs
(j) collect blood from the kidneys

11. Identify the major arteries in the diagram below.

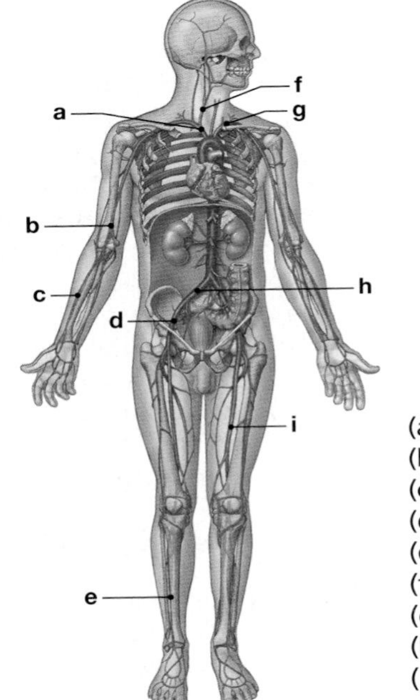

(a) ____
(b) ____
(c) ____
(d) ____
(e) ____
(f) ____
(g) ____
(h) ____
(i) ____

12. Identify the major veins in the diagram below.

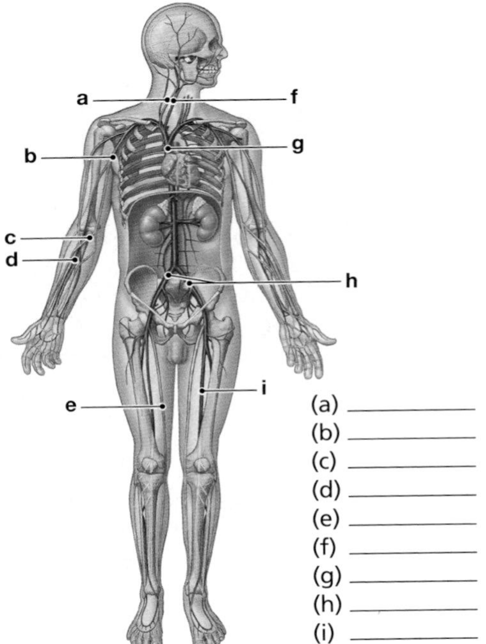

(a) ____
(b) ____
(c) ____
(d) ____
(e) ____
(f) ____
(g) ____
(h) ____
(i) ____

13. Compared with arteries, veins
    (a) are more elastic.
    (b) have thinner walls.
    (c) have more smooth muscle in their media.
    (d) have a pleated endothelium.
14. Capillaries that have a complete lining are called
    (a) continuous capillaries.
    (b) fenestrated capillaries.
    (c) sinusoidal capillaries.
    (d) sinusoids.
15. The only blood vessels whose walls permit exchange between the blood and the surrounding interstitial fluids are the
    (a) arteries.
    (b) arterioles.
    (c) veins.
    (d) capillaries.
16. Blood flow through the capillaries is regulated by the
    (a) arterial anastomosis.
    (b) thoroughfare channel.
    (c) vasa vasorum.
    (d) metarteriole.

Level 2 Reviewing Concepts

1. A major difference between the arterial and venous systems is that
    (a) arteries are usually more superficial than veins.
    (b) in the limbs there is dual venous drainage.
    (c) veins are usually less branched than the arteries.
    (d) veins exhibit a more orderly pattern of branching in the limbs.
2. You would expect to find fenestrated capillaries in
    (a) the choroid plexus.
    (b) skeletal muscles.
    (c) cardiac muscle.
    (d) the spleen.
3. Why does the endothelial lining of a constricted artery appear to have pleats?
    (a) Spaces between the endothelial cells allow the lining to sag when the artery is constricted.
    (b) The endothelial lining cannot contract, so it is folded in pleats when the artery contracts.
    (c) The expansion regions of the artery are folded when the artery constricts.
    (d) The vasa vasorum contract irregularly.
    (e) None of the above is correct.
4. What are some examples of elastic arteries?
5. Where are sinusoids found?
6. What are arteriovenous anastomoses?
7. What are the functions of venous valves in the limbs?
8. What three elastic arteries originate along the aortic arch?
9. From which regions of the body does the superior vena cava receive blood?
10. What is the function of the foramen ovale in a fetal heart?

Level 3 Critical Thinking

1. Why can it be dangerous for a person to squeeze pimples in the upper nasal and eyebrow region?
2. John loves to soak in hot tubs and whirlpools. One day he decides to raise the temperature in his hot tub as high as it will go. After a few minutes in the water, he feels faint, passes out, and nearly drowns. Luckily, he is saved by a bystander. Explain what happened.
3. Millie's grandfather suffers from congestive heart failure. When she visits him, she notices that his ankles and feet appear swollen. She asks you why this occurs. What would you tell her?

CLINICAL CASE WRAP-UP

In the Absence of Capillaries

Rodrigo has an arteriovenous malformation (AVM) of his left middle cerebral artery. An AVM is a tangle of abnormal blood vessels connecting arteries and veins with no capillary bed between them. Because a branch of his left middle cerebral artery connects directly to a small, thin-walled vein, blood flows quickly and directly from the artery to the vein, never slowing enough to supply the surrounding tissues with oxygen. This deprives his left primary motor cortex (precentral gyrus) and primary somatosensory cortex (postcentral gyrus) of blood, causing numbness and weakness in his right hand. The higher arterial pressure forces blood directly into the lower pressure, thin-walled veins, causing them to rupture and leading to a hemorrhagic stroke (brain bleed). This caused his sudden seizure.

Rodrigo was likely born with this blood vessel "mistake" that began during fetal development. His AVM has grown into a large tangle of abnormal blood vessels that presses on brain tissue.

The good news? Rodrigo's arteriovenous malformation can be corrected with open surgery (an incision to repair the aneurysm), endovascular embolization (a procedure that cuts off blood supply to a particular region), or radiation therapy to eliminate this tangle of vessels before it causes more trouble.

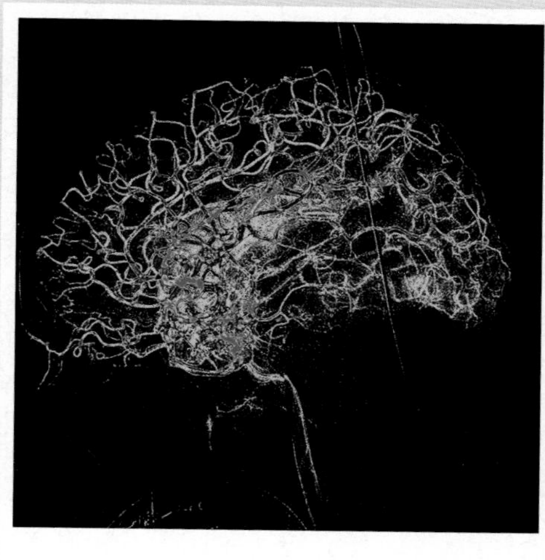

1. Can water, gases, or wastes be exchanged between the blood and interstitial fluids anywhere within the arteriovenous malformation? Explain.
2. Can arteriovenous malformations form in other parts of the body besides the brain? Explain.

See the blue Answers tab at the back of the book.

Related Clinical Terms

angiogram: An x-ray of a blood vessel that becomes visible due to a prior injection of dye into the subject's bloodstream.

carotid sinus massage: A procedure that involves rubbing the large part of the arterial wall at the point where the common carotid artery divides into its two main branches.

deep vein thrombosis (DVT): A blood clot in a major vein, usually in the legs. They often occur after extended periods of inactivity, such as long airplane flights. The clot can break free and travel as an embolus to the lungs, where it can cause respiratory distress or failure.

intermittent claudication: A limp that results from cramping leg pain that is typically caused by obstruction of the arteries.

normotensive: Having normal blood pressure.

orthostatic hypotension: A form of low blood pressure that occurs when you stand up from sitting or lying down. It can cause dizziness or a light-headed feeling.

phlebitis: Inflammation of a vein.

Raynaud's phenomenon: A condition resulting in the discoloration of the fingers and/or the toes when a person is subjected to changes in temperature or to emotional stress.

sclerotherapy: The treatment of varicose veins in which an irritant is injected to cause inflammation, coagulation of blood, and a narrowing of the blood vessel wall.

sounds of Korotkoff: Distinctive sounds, caused by turbulent arterial blood flow, heard through the stethoscope while measuring blood pressure.

sphygmomanometer: A device that measures blood pressure using an inflatable cuff placed around a limb.

syncope: A temporary loss of consciousness due to a sudden drop in blood pressure.

thrill: A vibration felt in a blood vessel that usually occurs due to abnormal blood flow. It is also often noticed at the fistula of a hemodialysis patient.

thrombophlebitis: An inflammation in a vein associated with the formation of a thrombus (clot).

vascular murmur: Periodic abnormal sounds heard upon auscultation that are produced as a result of turbulent blood flow.

white coat hypertension: A short-term increase in blood pressure triggered by the sight of medical personnel in white coats or other medical attire.

23 The Lymphatic System

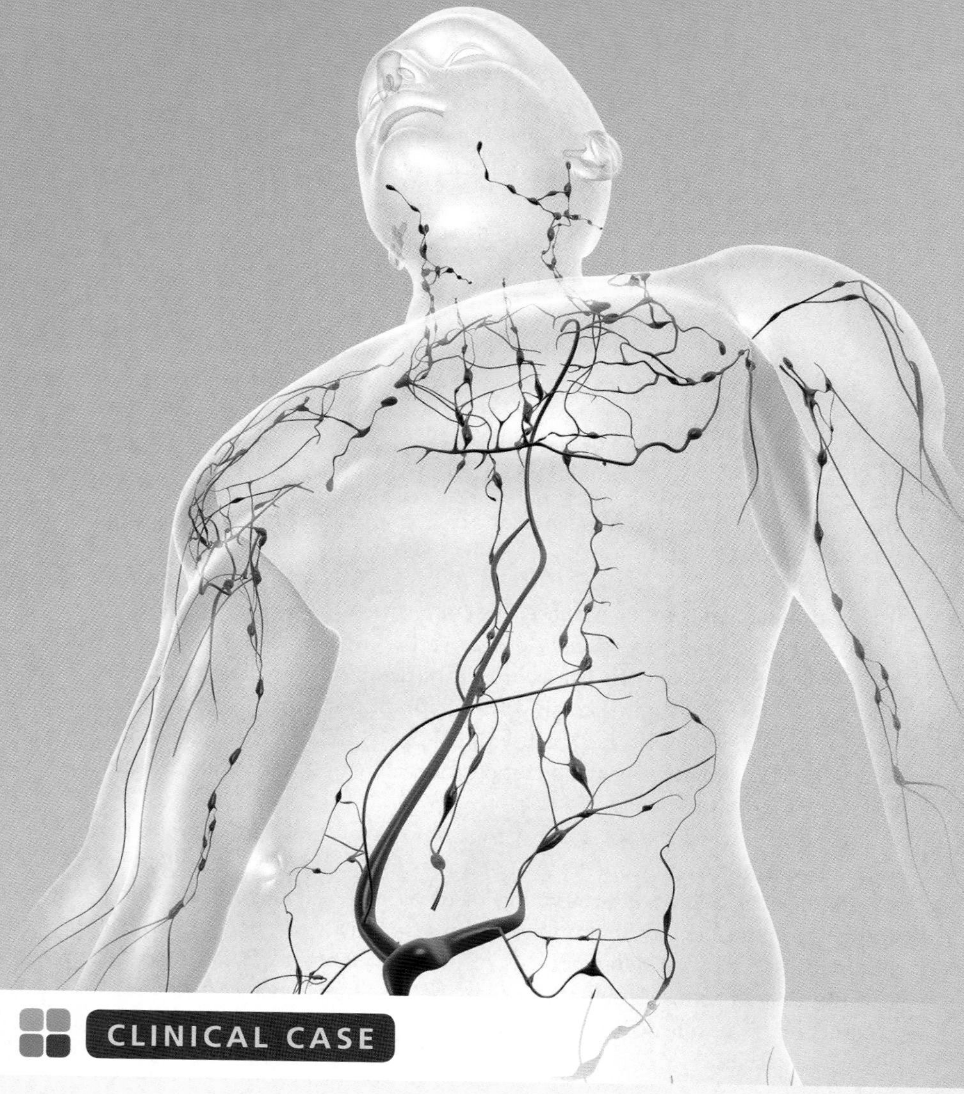

Learning Outcomes

These learning outcomes correspond by number to this chapter's sections and indicate what you should be able to do after completing the chapter.

- **23.1** List the major functions of the lymphatic system. p. 604
- **23.2** Compare and contrast a lymphatic capillary and a vascular capillary. p. 605
- **23.3** Compare and contrast the thoracic duct and the right lymphatic duct. p. 606
- **23.4** Compare and contrast the different classes of lymphocytes. p. 607
- **23.5** Define the term "lymphatic nodule" and give two examples of where lymphatic nodules are found within the body. p. 610
- **23.6** Compare and contrast the anatomical structure of a lymph node to that of the thymus and spleen. p. 612
- **23.7** Describe the effects of aging on the lymphatic system and immune surveillance. p. 619

CLINICAL CASE

Fighting a Zoonotic Disease

Grace, a young medical student from central Africa, watched her brother die of Ebola virus disease after he had caught, prepared, and eaten some large fruit bats. Now, 10 days after her brother died, Grace is burning with fever and has a severe headache, weakness, fatigue, and muscle pain to the point that she can barely move. She is vomiting and has bloody diarrhea; her nose and gums are also bleeding.

Ebola virus disease is transmitted to people from wild animals. Diseases such as this, in which an infection is spread between humans and animals, are called zoonotic diseases. Grace is very ill, and as of yet, there are no Ebola vaccines. However, there are several immunological and drug therapies under development.

Will Grace survive the Ebola virus disease when only supportive care is available? To find out, turn to the Clinical Case Wrap-Up on p. 622.

THE WORLD IS NOT ALWAYS KIND to the human body. Injuries and pathogens (disease-causing agents) have the potential to cause us great harm. Remaining alive and healthy involves the massive, combined effort of many different organs and systems. The lymphatic system plays the primary role in this ongoing struggle.

In this chapter we describe the anatomical organization of the lymphatic system and how it interacts with other systems and tissues to defend the body.

23.1 An Overview of the Lymphatic System

▶ **KEY POINT** The lymphatic system, functionally part of the circulatory and immune systems, is composed of lymphatic tissue, lymphatic organs, and lymphatic vessels. Lymphatic vessels transport lymph.

The **lymphatic system**, or *lymphoid* system,* which is functionally part of the circulatory and immune systems, has several components **(Figure 23.1)**. It monitors and transports the fluid connective tissue called lymph. pp. 61, 65 The vessels carrying lymph are called lymphatic vessels, and the cells suspended within the lymph (and blood) are lymphocytes. Specialized lymphatic tissues and lymphatic organs adjust the composition of lymph and produce lymphocytes of various kinds.

Lymphatic vessels originate in peripheral tissues and deliver lymph to the venous system. Lymph consists of (1) interstitial fluid that resembles blood plasma but has a lower concentration of proteins, (2) lymphocytes, which are the cells responsible for the immune response, and (3) macrophages, which are phagocytic cells of various types. pp. 61, 62 Lymphatic vessels often begin within or pass through lymphatic tissues and lymphatic organs, structures that contain a large number of lymphocytes, macrophages, and (in many cases) lymphatic stem cells.

Functions of the Lymphatic System

▶ **KEY POINT** The lymphatic system produces, maintains, and distributes lymphocytes; maintains blood volume and composition; and provides an alternate route for hormones, nutrients, and wastes.

The lymphatic system has the following functions:

- Produces, maintains, and distributes lymphocytes: Lymphocytes are essential to the normal defense mechanisms of the body. They are produced and stored within lymphatic organs, such as the spleen, thymus, and red bone marrow. Lymphatic tissues and organs are classified as either primary or secondary. **Primary lymphatic structures** are sites where lymphocytes are produced and mature. They include the red bone marrow and the adult thymus. **Secondary lymphatic structures** are sites where lymphocytes are activated to produce additional lymphocytes of the same type. For example, activated B cells divide within a secondary lymphatic structure to produce the additional B cells needed to fight off an infection. Secondary lymphatic structures are located at "the front lines," where invading pathogens are first encountered. Examples include tonsils and lymph nodes.
- Maintains normal blood volume and eliminates local variations in the chemical composition of the interstitial fluid: The blood pressure at the proximal end of a systemic capillary is approximately 35 mm Hg. The blood pressure within every systemic capillary forces water and solutes out of the plasma and into the interstitial fluid **(Figure 23.2)**. Although the net movement of fluid from plasma into interstitial fluid at any one capillary is small, the total volume is substantial. Approximately 3.6 L, or 72 percent of our total blood volume, enters the interstitial fluid each day. Under normal circumstances this movement goes unnoticed because the lymphatic vessels return an equal volume of interstitial fluid back into the bloodstream. As a result, there is a constant movement of fluid from the bloodstream into the tissues and then back to the bloodstream through the lymphatic vessels. This circulation of fluid eliminates regional differences in the composition of interstitial fluid. Because so much fluid moves through the lymphatic system each day, a break in a major lymphatic vessel can cause a rapid and potentially fatal decrease in blood volume.

Figure 23.1 Lymphatic System. An overview of the arrangement of lymphatic vessels, lymph nodes, and lymphatic organs.

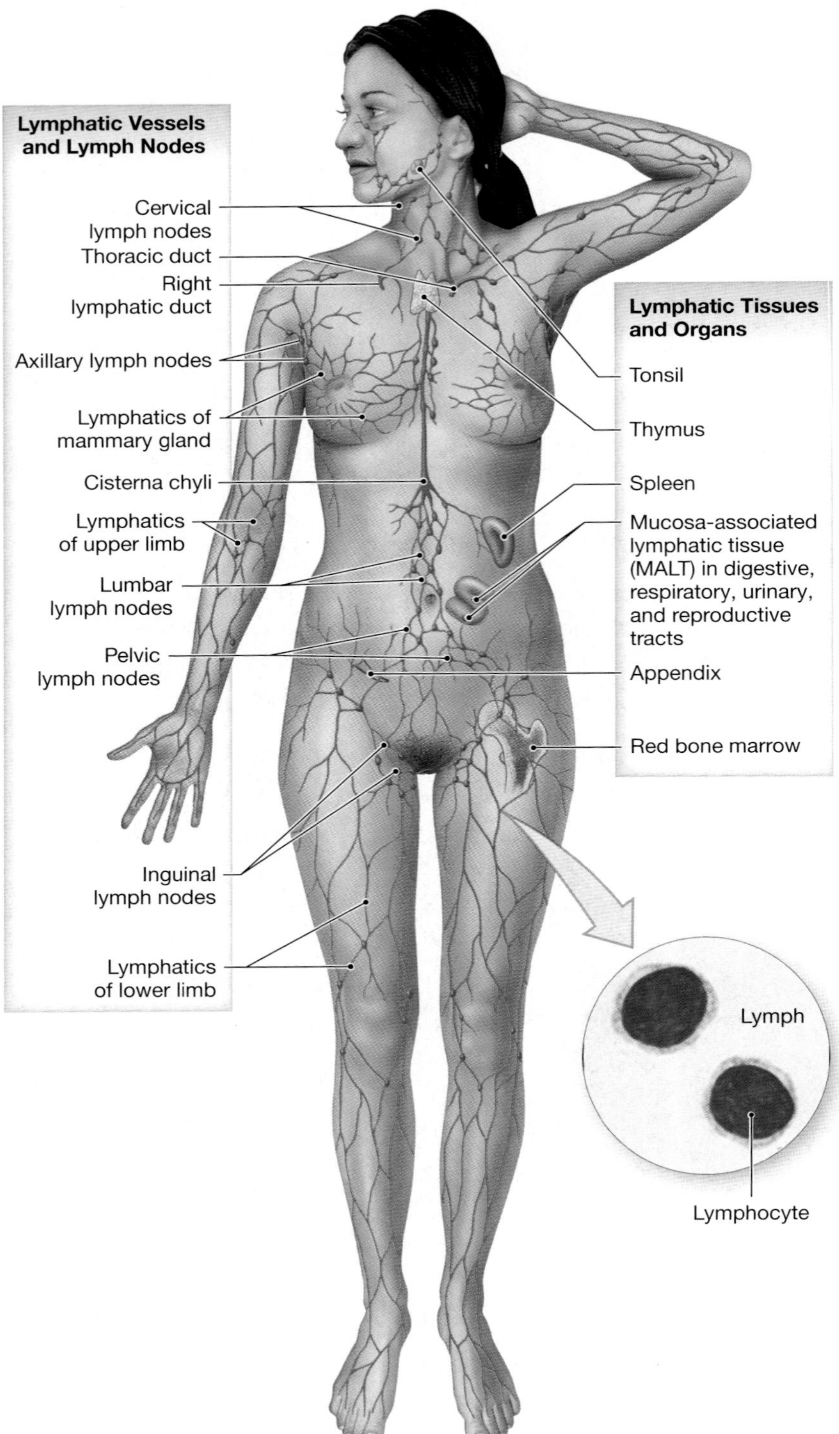

*The terms "lymphatic" and "lymphoid" have long been used interchangeably, and this synonymy is reflected in the terms endorsed by the *Terminologia Anatomica* and *Terminologia Histologica*.

- Provides an alternate route for transporting hormones, nutrients, and wastes: Hormones, nutrients, and wastes are distributed from their tissues of origin to the general circulation. For example, lipids absorbed by the digestive tract are carried to the bloodstream by lymphatic vessels rather than by absorption across capillary walls.

23.1 CONCEPT CHECK

1 What are the main functions of the lymphatic system?

2 Would the rupture of a major lymphatic vessel be fatal? Why or why not?

See the blue Answers tab at the back of the book.

23.2 Structure of Lymphatic Vessels

▶ **KEY POINT** Lymphatic capillaries are the smallest lymphatic vessels. As lymph moves from peripheral tissues toward the heart, the diameter of lymphatic vessels increases.

The blood vessels of the cardiovascular system form a complete circuit, beginning and ending at the heart. However, **lymphatic vessels** (or *lymphatics*) carry lymph *only from peripheral tissues to the venous system*. As with blood vessels, the lymphatic vessels range in size from small-diameter lymphatic capillaries to large-diameter collecting vessels.

Lymphatic Capillaries

▶ **KEY POINT** The anatomy of lymphatic capillaries is significantly different from the anatomy of vascular, or *blood*, capillaries, and lymphatic capillaries are more permeable.

The lymphatic network, which branches through peripheral tissues, begins with closed tubes called **lymphatic capillaries**. Lymphatic capillaries differ from vascular capillaries in several ways:

1. Lymphatic capillaries have larger luminal diameters.
2. They have thinner walls, and their endothelial cells lack a continuous basal lamina.
3. They have a flat or irregular outline.
4. They have collagenous anchoring filaments that extend from the incomplete basal lamina to the surrounding connective tissue. These filaments help keep the passageways open when interstitial pressure increases.
5. They have greater permeability because their endothelial cells overlap instead of being tightly bound to one another **(Figure 23.2)**.

Figure 23.2 Lymphatic Capillaries. Lymphatic capillaries are close-ended vessels that begin in areas of loose connective tissue.

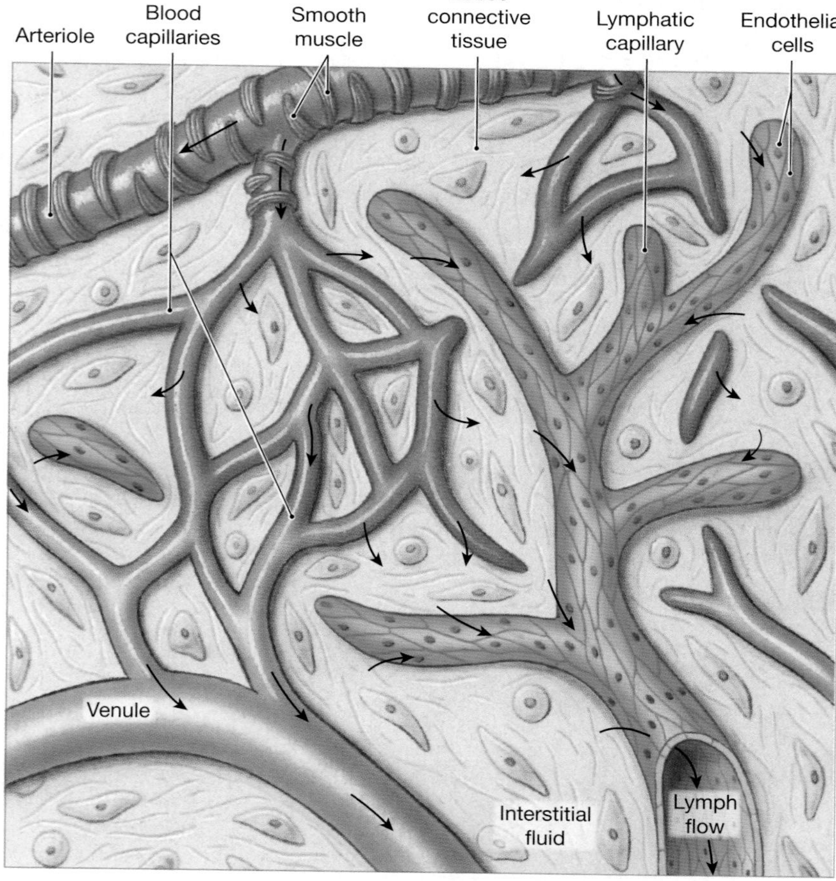

a A three-dimensional view of the association of blood capillaries and lymphatic capillaries. Arrows show the net flow of blood, interstitial fluid, and lymph.

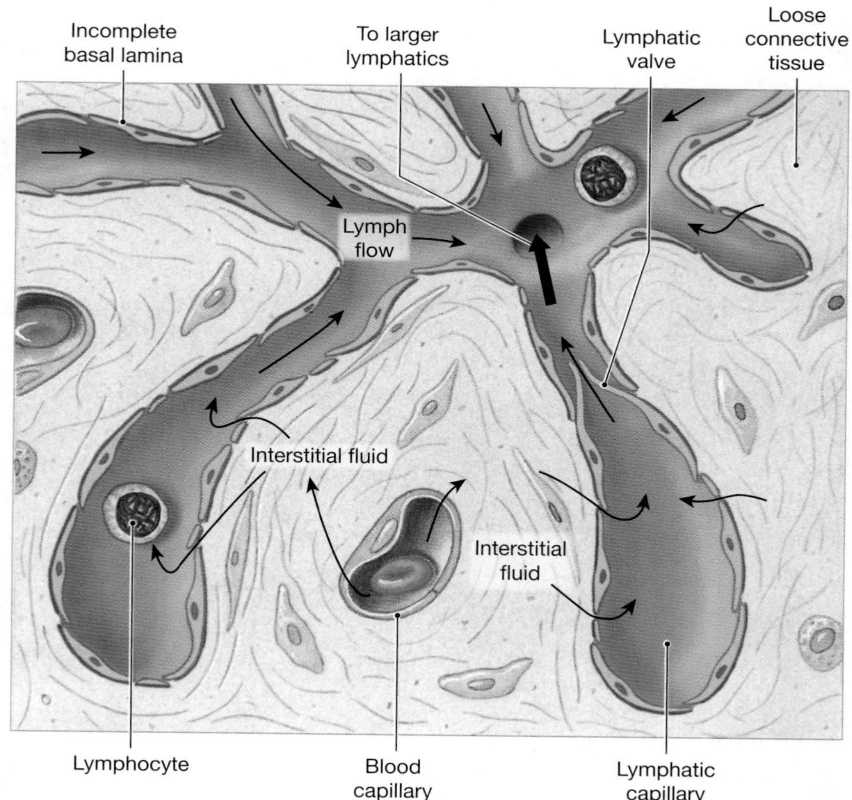

b Sectional view through a cluster of lymphatic capillaries.

The endothelial cells of a lymphatic capillary are not bound tightly together, but they do overlap. The region of overlap acts as a one-way valve. It permits fluids and solutes (including proteins) to enter, along with viruses, bacteria, and cell debris, but it prevents them from returning to the intercellular spaces. Lymphatic capillaries are present in most tissues. However, they are especially numerous in the connective tissue deep to skin and mucous membranes and in the mucosa and submucosa of the digestive tract. Lymphatic capillaries in the small intestine, called lacteals, transport lipids absorbed by the digestive tract. Lymphatic capillaries are absent in areas without a blood supply, such as the cornea of the eye.

Larger Lymphatic Vessels

▶ **KEY POINT** Like lymphatic capillaries, lymphatic vessels differ histologically from the veins of the systemic and pulmonary circuits.

Lymph flows from the lymphatic capillaries into larger lymphatic vessels that lead toward the trunk. Lymphatic vessels differ histologically from veins in that (1) their walls are thinner; (2) they have wider lumens; and (3) there are no clear boundaries between their three layers.

Small to medium-sized lymphatic vessels, like veins, have internal valves. The valves are close together, and at each valve the lymphatic vessel bulges noticeably. This gives these vessels a beaded appearance **(Figure 23.3)**. As in the venous system, the valves within lymphatic vessels prevent the backflow of lymph, especially in the limbs. Pressure within the lymphatic system is very low, and the valves are essential to maintaining normal lymph flow toward the thoracic cavity. Skeletal muscle contractions help move lymph through the lymphatic vessels.

When a lymphatic vessel is compressed or blocked or its valves are damaged, lymph drainage slows or stops in the affected area. Fluid continues to leave the vascular, or *blood*, capillaries in that region, but the lymphatic system is no longer able to remove it. As a result, the interstitial fluid volume and pressure gradually increase. The affected tissues become swollen, causing a condition called lymphedema.

Lymphatic vessels are found in close association with blood vessels **(Figure 23.3a,c)**. In living tissues, characteristic color differences are apparent. Arteries are usually bright red, veins dark red, and lymphatics a pale golden color.

23.2 CONCEPT CHECK

3 You are looking through a microscope at a cross section of two capillaries. The first capillary has a thinner wall and a larger lumen than the second. Which capillary is probably a lymphatic capillary?

See the blue Answers tab at the back of the book.

23.3 | Major Lymph-Collecting Vessels

▶ **KEY POINT** The superficial and deep lymphatics converge to form lymphatic trunks. These trunks unite to form two large lymphatic ducts, which drain into the venous circulation.

The superficial lymphatics and deep lymphatics collect lymph from lymphatic capillaries. **Superficial lymphatics** anastomose freely and are more numerous than superficial veins. They travel adjacent to superficial veins and are found in the following locations:

Figure 23.3 Lymphatic Vessels and Valves. Valves in lymphatic vessels prevent backflow of lymph.

Lymphatic valve
Lymphatic vessel
LM × 63

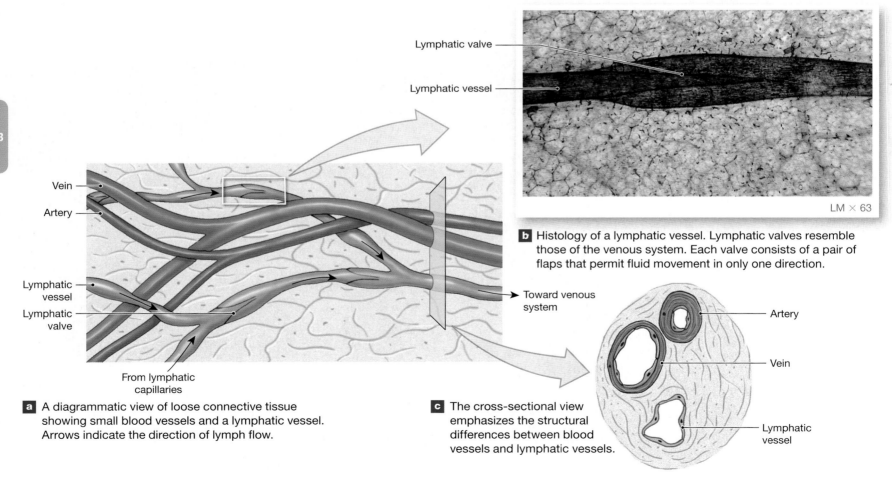

b Histology of a lymphatic vessel. Lymphatic valves resemble those of the venous system. Each valve consists of a pair of flaps that permit fluid movement in only one direction.

Vein
Artery
Lymphatic vessel
Lymphatic valve
From lymphatic capillaries
Toward venous system

a A diagrammatic view of loose connective tissue showing small blood vessels and a lymphatic vessel. Arrows indicate the direction of lymph flow.

Artery
Vein
Lymphatic vessel

c The cross-sectional view emphasizes the structural differences between blood vessels and lymphatic vessels.

- The subcutaneous layer deep to the skin
- The loose connective tissues of the mucous membranes lining the digestive, respiratory, urinary, and reproductive tracts
- The loose connective tissues of the serous membranes lining the pleural, pericardial, and peritoneal cavities

Deep lymphatics are large lymphatic vessels accompanying the deep arteries and veins. Deep lymphatics collect lymph from skeletal muscles and other organs of the neck, limbs, and trunk, as well as visceral organs in the thoracic and abdominopelvic cavities.

The superficial and deep lymphatics converge to form larger vessels called **lymphatic trunks**. The lymphatic trunks are the (1) lumbar trunks, (2) intestinal trunks, (3) bronchomediastinal trunks, (4) subclavian trunks, and (5) jugular trunks. The lymphatic trunks empty into two large collecting vessels, the thoracic duct and the right lymphatic duct. These ducts drain into the venous circulation **(Figures 23.4** and **23.5)**.

The Thoracic Duct

▶ **KEY POINT** The thoracic duct and its tributaries collect lymph from the abdomen, pelvis, lower limbs, left side of the head, neck, and thoracic cavity. The thoracic duct drains into the left subclavian vein.

The **thoracic duct** collects lymph from both sides of the body inferior to the diaphragm and from the left side of the body superior to the diaphragm. The thoracic duct begins as the **cisterna chyli** (KĪ-lē), an expanded saclike chamber **(Figures 23.4** and **23.5a)**. The cisterna chyli begins inferior to the diaphragm at the level of vertebra L_2. Here the deep lumbar and intestinal trunks come together to form a singular lymphatic vessel that lies anterior to the vertebral column, slightly to the right of midline. The cisterna chyli receives lymph from the abdomen, pelvis, and lower limbs by way of the right and left lumbar trunks and intestinal trunk.

As the cisterna chyli passes through the diaphragm with the aorta, at an opening known as the aortic hiatus, it becomes the thoracic duct. At the level of vertebra T_5, the thoracic duct shifts left, coming to lie along the left side of the esophagus. The thoracic duct ascends to the level of the left clavicle. It collects lymph from the left bronchomediastinal trunk, the left subclavian trunk, and the left jugular trunk. The thoracic duct empties into the left subclavian vein near the left internal jugular vein **(Figures 23.4** and **23.5)**. Lymph collected from the left side of the head, neck, and thorax, as well as lymph from the entire body inferior to the diaphragm, re-enters the venous circulation in this way.

Figure 23.4 Major Lymphatic Vessels of the Trunk. Anterior view of a dissection of the thoracic duct and adjacent blood vessels. The thoracic and abdominopelvic organs have been removed.

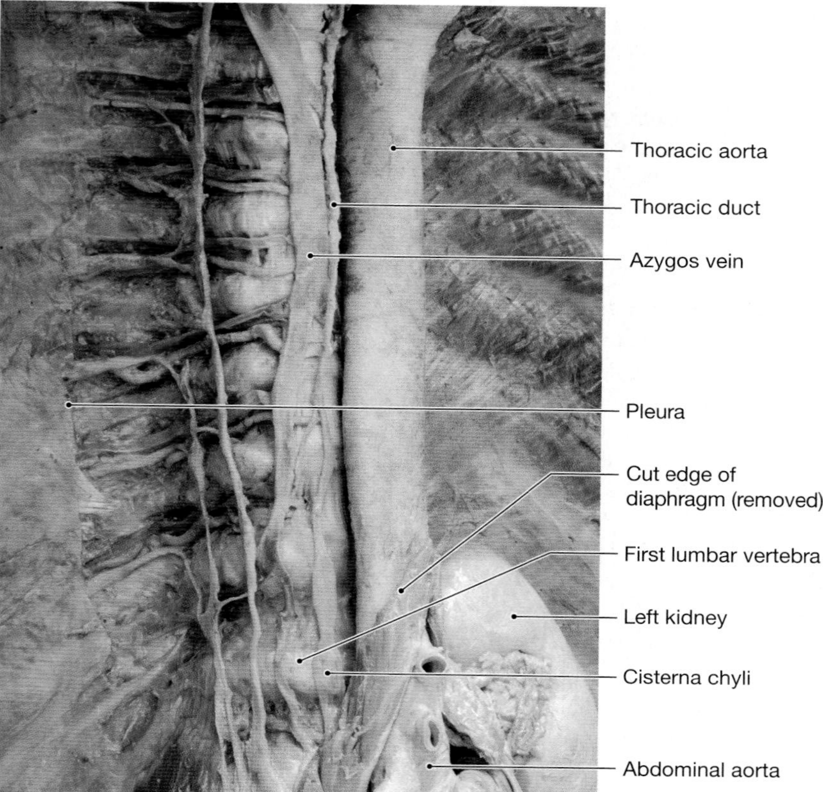

The Right Lymphatic Duct

▶ **KEY POINT** The right lymphatic duct collects lymph from the right side of the thoracic cavity. It drains into the venous system near the junction of the right internal jugular and right subclavian veins.

The **right lymphatic duct** is quite small, and its anatomy varies from person to person. It collects lymph from the right side of the body superior to the diaphragm. The right lymphatic duct receives lymph from smaller lymphatic vessels that converge in the region of the right clavicle. This duct empties into the venous system at or near the junction of the right internal jugular and right subclavian veins **(Figure 23.5)**.

23.3 **CONCEPT CHECK**

4 List the five lymphatic trunks.

5 List the areas drained by the thoracic duct and the right lymphatic duct.

See the blue Answers tab at the back of the book.

23.4 | Lymphocytes

▶ **KEY POINT** Lymphocytes, the primary cells of the lymphatic system, are responsible for specific immunity.

Lymphocytes are responsible for specific immunity. ⮌ p. 536 They respond by a combination of physical and chemical attack to (1) invading organisms, such as bacteria and viruses. (2) abnormal body cells, such as virus-infected cells or cancer cells, and (3) foreign proteins, such as the toxins released by some bacteria.

Lymphocytes travel throughout the body, circulating in the bloodstream and lymph. They exit the bloodstream and move through the interstitial space and peripheral tissues, eventually returning to the bloodstream through the lymphatic system. The time spent within the lymphatic system varies; a lymphocyte may remain within a lymphatic organ for hours, days, or even years. However, when a lymphocyte encounters an invading pathogen or foreign protein, it initiates an immune response.

Types of Lymphocytes

▶ **KEY POINT** The three classes of lymphocytes (T cells, B cells, and NK cells) have distinctive biochemical and functional characteristics.

There are three classes of lymphocytes in the blood: **T cells** (**t**hymus-dependent), **B cells** (**b**one marrow–derived), and **NK cells** (**n**atural **k**iller) **(Spotlight Figure 23.6)**. These cells have few or no structural differences. They are classified based on their biochemical and functional characteristics. Adaptive (specific) defenses result from the coordinated activities of T cells and B cells, also known as T lymphocytes and B lymphocytes, respectively.

Figure 23.5 Lymphatic Ducts and Lymphatic Drainage

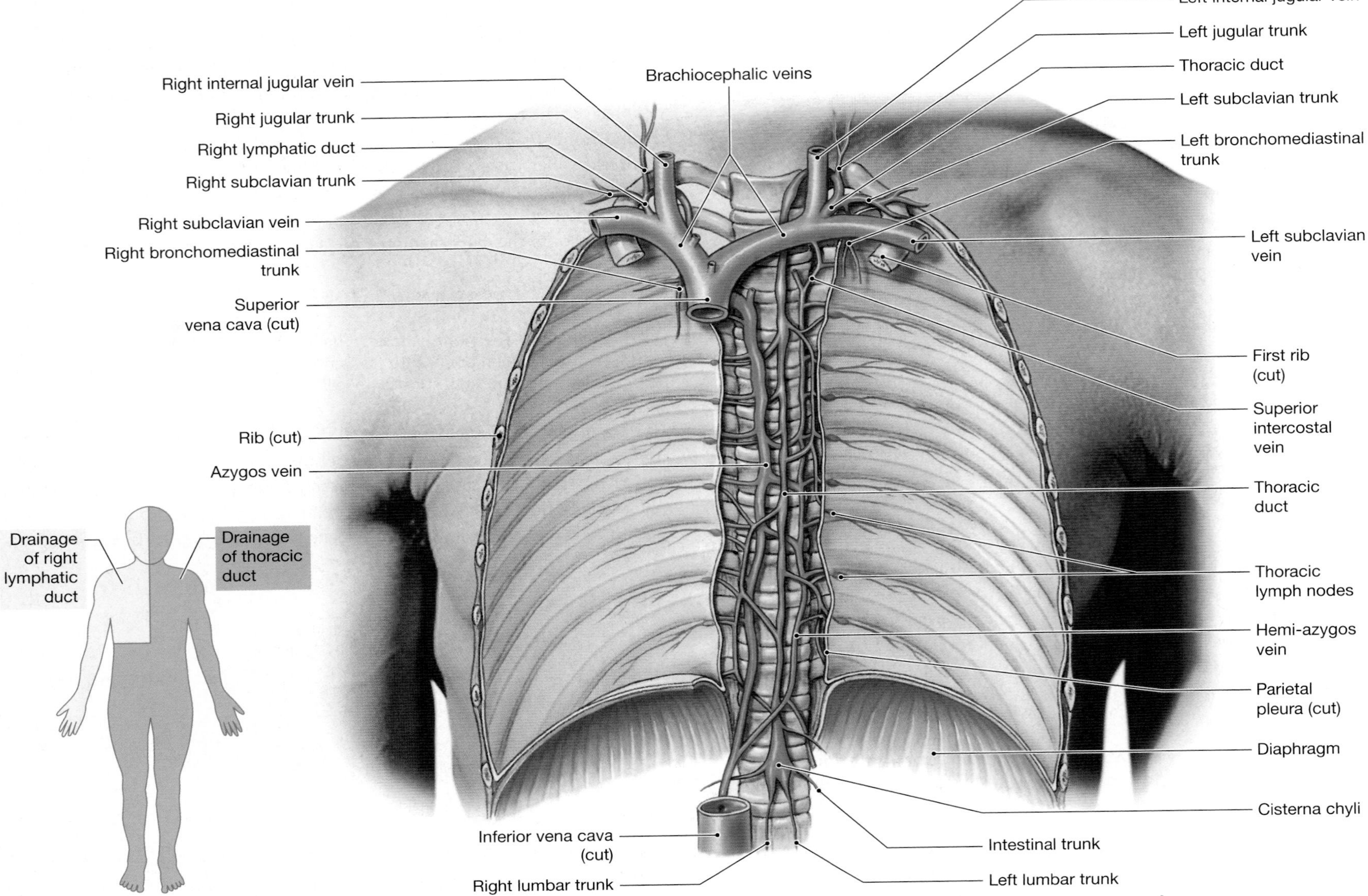

b The thoracic duct collects lymph from tissues inferior to the diaphragm and from the left side of the upper body. The right lymphatic duct drains the right half of the body superior to the diaphragm.

a The collecting system of lymph vessels, lymph nodes, and major lymphatic collecting ducts and their relationship to the brachiocephalic veins.

23

T Cells

Approximately 80 percent of circulating lymphocytes are T cells. T cells originate within the red bone marrow. However, they must migrate to the thymus to become activated, or immunocompetent. Under proper stimulation, T cells differentiate into several types of cells, which attack antigens (foreign substances) and help to increase the immune response. T cell types include cytotoxic T cells, helper T cells, regulatory T cells, and memory T cells.

Cytotoxic T cells are involved in direct cellular attack and provide **cell-mediated immunity**. These cells enter peripheral tissues and attack antigens physically and chemically. **Helper T cells** stimulate the responses of both T cells and B cells. Helper T cells are important to the immune response because they must activate B cells before the B cells can produce antibodies. **Regulatory T cells** are a subset of T cells that moderate the immune response. **Memory T cells** respond to antigens they have already encountered by cloning (producing identical cellular copies) more lymphocytes to ward off the invader. If the same antigens eliminated in a primary response reappear, a rapid secondary response is initiated by memory cells. The interplay between helper T cells and regulatory T cells helps establish and control the sensitivity of the immune response. Other types of regulatory T cells also take part in the immune response. For example, suppressor/inducer T cells suppress B cell activity but stimulate other T cells.

B Cells

B cells originate and become immunocompetent within the red bone marrow. They account for 10–15 percent of circulating lymphocytes. Activated B cells divide, producing daughter cells that differentiate into **plasma cells** (*plasmocytes*) and **memory B cells**. Plasma cells are responsible for producing and secreting **antibodies**. ⮌ p. 536

When an antibody binds an antigen, this triggers a chain of events leading to the destruction, neutralization, or elimination of the antigen. Antibodies are also known as **immunoglobulins** (im-ū-nō-GLOB-ū-lins). B cells provide **antibody-mediated immunity**, a defense within body fluids.

SPOTLIGHT

FIGURE 23.6
Lymphocyte Formation and Immunity

Lymphatic tissue formation, called lymphocytopoiesis, involves the red bone marrow, thymus, and peripheral lymphatic tissues. Hematopoietic stem cells in the red bone marrow produce lymphatic stem cells with two distinct fates.

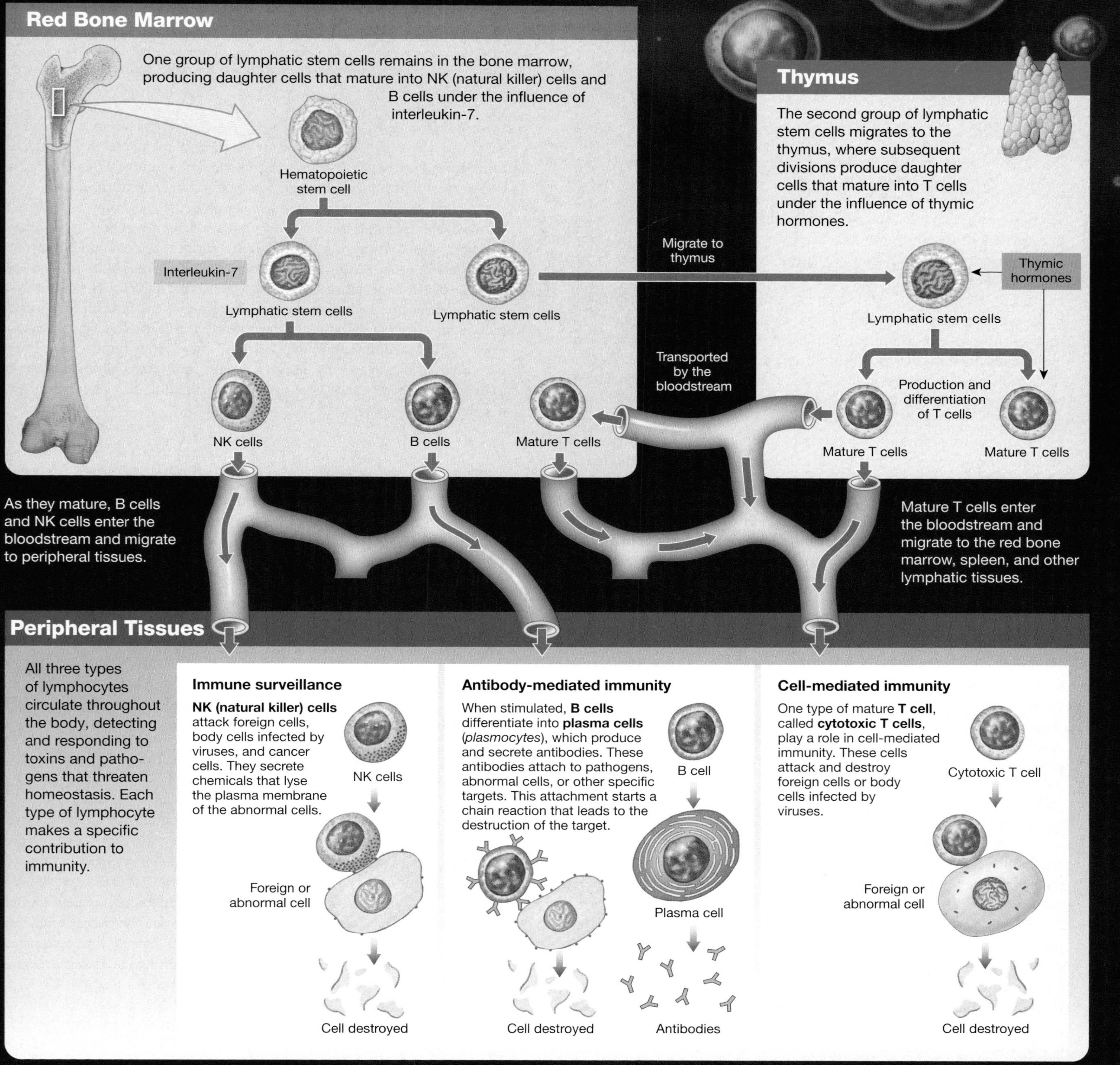

Red Bone Marrow
One group of lymphatic stem cells remains in the bone marrow, producing daughter cells that mature into NK (natural killer) cells and B cells under the influence of interleukin-7.
Hematopoietic stem cell
Interleukin-7
Lymphatic stem cells
Lymphatic stem cells
NK cells
B cells
Mature T cells
Migrate to thymus
Transported by the bloodstream
Thymus
The second group of lymphatic stem cells migrates to the thymus, where subsequent divisions produce daughter cells that mature into T cells under the influence of thymic hormones.
Thymic hormones
Lymphatic stem cells
Production and differentiation of T cells
Mature T cells
Mature T cells
As they mature, B cells and NK cells enter the bloodstream and migrate to peripheral tissues.
Mature T cells enter the bloodstream and migrate to the red bone marrow, spleen, and other lymphatic tissues.
Peripheral Tissues
All three types of lymphocytes circulate throughout the body, detecting and responding to toxins and pathogens that threaten homeostasis. Each type of lymphocyte makes a specific contribution to immunity.
Immune surveillance
NK (natural killer) cells attack foreign cells, body cells infected by viruses, and cancer cells. They secrete chemicals that lyse the plasma membrane of the abnormal cells.
NK cells
Foreign or abnormal cell
Cell destroyed
Antibody-mediated immunity
When stimulated, B cells differentiate into plasma cells (plasmocytes), which produce and secrete antibodies. These antibodies attach to pathogens, abnormal cells, or other specific targets. This attachment starts a chain reaction that leads to the destruction of the target.
B cell
Plasma cell
Cell destroyed
Antibodies
Cell-mediated immunity
One type of mature T cell, called cytotoxic T cells, play a role in cell-mediated immunity. These cells attack and destroy foreign cells or body cells infected by viruses.
Cytotoxic T cell
Foreign or abnormal cell
Cell destroyed

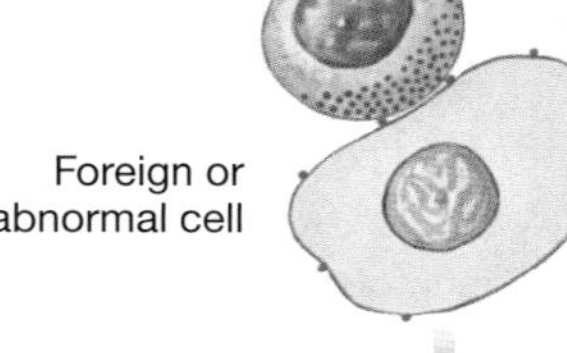

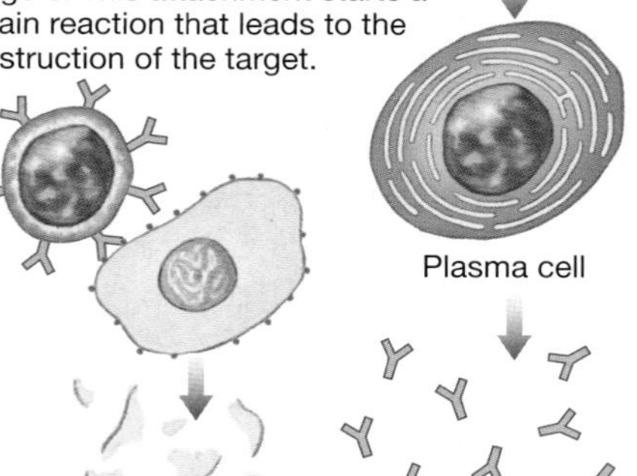

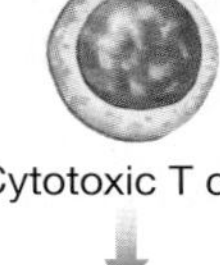

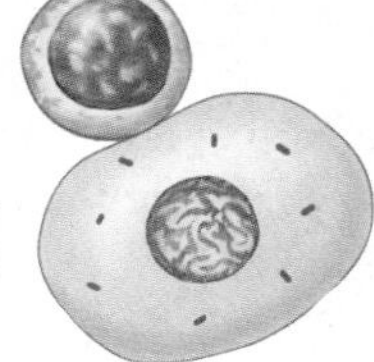

This type of immunity is also called *humoral* ("liquid") *immunity* because antibodies are found in body fluids. Memory B cells perform the same role in antibody-mediated immunity that memory T cells perform in cell-mediated immunity. Memory B cells become activated only if the antigen appears again in the body at a later date.

NK Cells

The remaining 5–10 percent of circulating lymphocytes are **natural killer cells**, or NK cells. NK cells attack foreign cells, normal cells infected with viruses, and cancer cells that appear in normal tissues. The constant monitoring of peripheral tissues by NK cells and cytotoxic T cells is called **immune surveillance**.

Lymphocytes and the Immune Response

▶ **KEY POINT** The immune response destroys or inactivates pathogens, abnormal cells, and foreign molecules such as toxins. The body has two different ways to do this: direct attack by activated T cells and antibodies released by plasma cells derived from activated B cells.

After an antigen appears, the first step in the **immune response** is often the phagocytosis of the antigen by a macrophage. The macrophage then incorporates pieces of the antigen into its plasma membrane. This allows the macrophage to "present" the antigen to T cells. The T cells respond to that particular antigen by the process of cell-mediated immunity. These lymphocytes respond because their plasma membranes contain receptors capable of binding only to that specific antigen. When binding occurs, the T cells become activated and begin to divide. Some daughter cells differentiate into cytotoxic T cells, and others into helper T cells. The helper T cells activate B cells. Some helper T cells become memory T cells, which will differentiate further if they encounter this same antigen at a later date.

Your immune system has no way of anticipating which antigens it will actually encounter. The strategy is to prepare for *any* antigen that might appear. During development, the differentiation of cells within the lymphatic system produces an enormous number of lymphocytes, each with its own antigen sensitivities. Among the trillion or more lymphocytes in the human body, there are millions of different lymphocyte populations. Each population consists of several thousand cells prepared to recognize a specific antigen.

A lymphocyte becomes activated when it binds an antigen. It then divides and produces more lymphocytes sensitive to that particular antigen. Some of the lymphocytes function immediately to eliminate the antigen, and others (the cells) will be ready if the antigen reappears at a later date. This immunocompetence provides an immediate defense and ensures an even more massive and rapid response if the antigen reappears in the body.

Distribution and Life Span of Lymphocytes

▶ **KEY POINT** The distribution of lymphocytes within the body varies. Lymphocytes typically live longer than any other cellular formed element in blood.

The ratio of B cells to T cells varies, depending on the tissue or organ. For example, B cells are seldom found in the thymus, and T cells outnumber B cells in the blood by a ratio of 8:1. In the spleen this ratio changes to one B cell for every T cell. The ratio changes yet again to one B cell for every three T cells in the red bone marrow.

The lymphocytes within these organs are visitors, not residents. Lymphocytes continually move throughout the body. They wander through a tissue and then enter a blood vessel or lymphatic vessel for transport to another site. For example, a wandering T cell spends about 30 minutes circulating in blood, then spends 15–20 hours within a lymph node. B cells move more slowly. A typical B cell spends around 30 hours in a lymph node before moving to another location.

In general, lymphocytes have relatively long life spans, far longer than other formed elements in the blood. Roughly 80 percent survive for 4 years, and some last 20 years or more. The process of lymphocytopoiesis maintains normal lymphocyte populations.

Lymphocytopoiesis: Lymphocyte Production

▶ **KEY POINT** Lymphocytopoiesis occurs within the red bone marrow and thymus. One type of lymphatic stem cell remains within red bone marrow and produces NK cells and B cells, while the other type migrates to the thymus to ultimately produce T cells.

Lymphocytopoiesis (lim-fō-SĪ-tō-poy-Ē-sis), the formation of lymphocytes, occurs in the red bone marrow, thymus, and peripheral lymphatic tissue. **Spotlight Figure 23.6** shows the relationships of these lymphatic tissues with respect to lymphocyte production, maturation, and distribution.

As lymphocytes migrate through peripheral tissues, they retain the ability to divide. Lymphocytic division produces daughter cells of the same type and with sensitivity to the same specific antigen. For example, a dividing B cell produces other B cells, not T cells or NK cells. This ability to increase the number of lymphocytes of a specific type is important for the success of the immune response. If that ability is compromised, the individual is unable to mount an effective defense against infection and disease. For example, the disease AIDS (acquired immune deficiency syndrome) results from infection with a virus that selectively destroys T cells. Individuals with AIDS are likely to be killed by bacterial or viral infections that would be overcome easily by a normal immune system.

23.4 CONCEPT CHECK

6 Which type of lymphocyte is the most common?

7 John becomes infected with a pathogen. A few months later, while relatively healthy, he is exposed to the same pathogen. Will he definitely get sick again? Why or why not?

8 Circulating lymphocytes retain the ability to divide. Why is this important?

See the blue Answers tab at the back of the book.

23.5 | Lymphatic Tissues

▶ **KEY POINT** Lymphatic tissues include diffuse lymphatic tissue and lymphatic nodules.

Lymphatic tissues are connective tissues dominated by lymphocytes. In **diffuse lymphatic tissue**, lymphocytes are loosely aggregated within connective tissue, such as the mucous membrane of the respiratory and urinary tracts. **Lymphatic nodules** are aggregations of lymphocytes contained within a supporting framework of reticular cells and fibers. Lymphatic nodules are oval in shape and are often found within the wall of various segments of the digestive tract **(Figure 23.7a,b)**. Because they lack a capsule, the boundaries of the nodules are indistinct. They have a pale, central zone, called a **germinal center**, which contains activated, dividing lymphocytes **(Figure 23.7)**. Lymphatic nodules average 1 mm in diameter.

Figure 23.7 **Histology of Lymphatic Tissues**

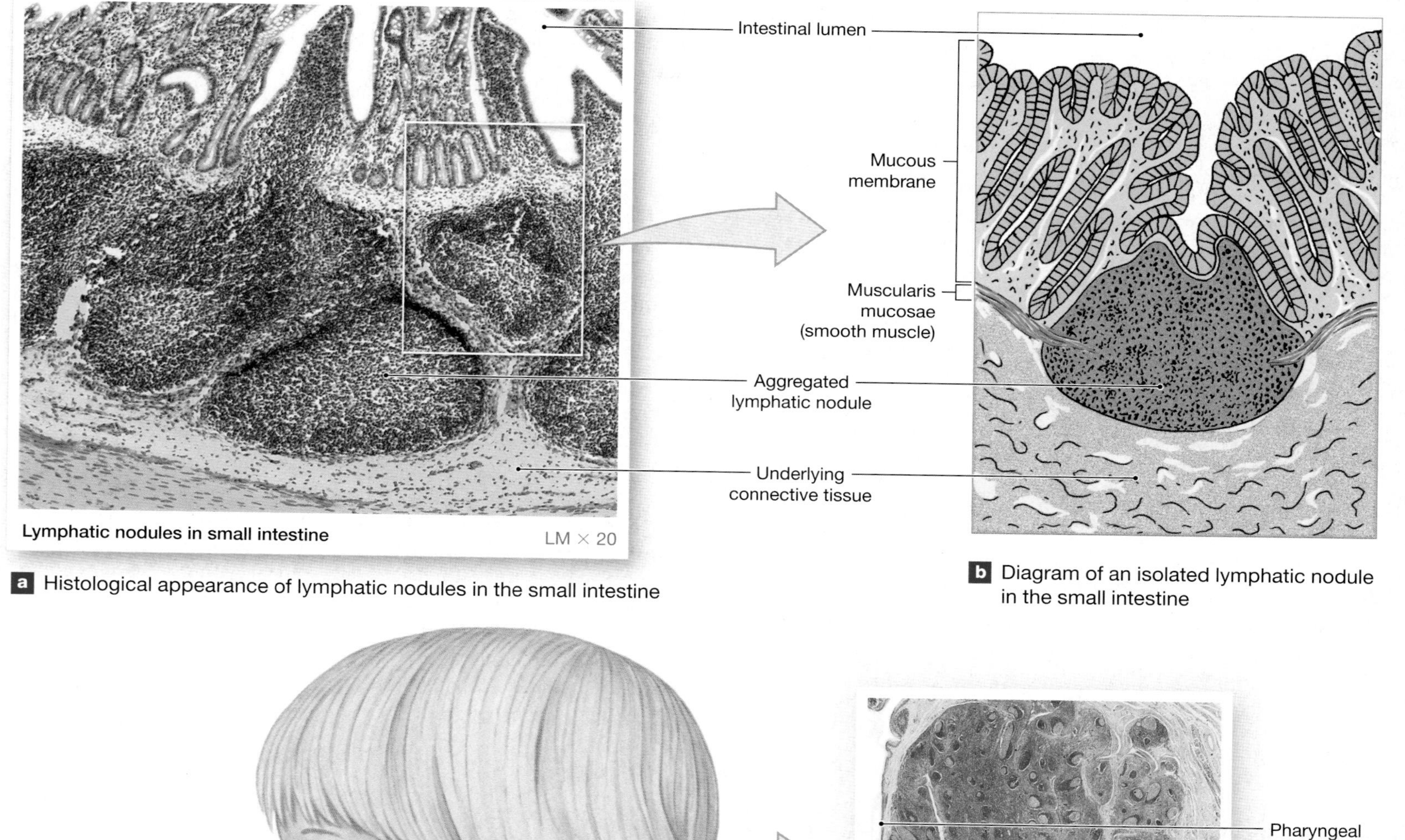

a Histological appearance of lymphatic nodules in the small intestine

b Diagram of an isolated lymphatic nodule in the small intestine

c The location of the tonsils and the histological organization of a single tonsil

CLINICAL NOTE

Infected Lymphatic Nodules

Tonsils are the first lymphatic nodules to encounter invading bacteria or viruses that enter through the mouth. Especially in children, tonsils can be overwhelmed by infectious invasion, causing **tonsillitis**. Pharyngeal tonsils that become chronically infected can form an abscess and become a source of infection themselves. This can contribute to bacteria entering the bloodstream by passing through the lymphatic capillaries and vessels to the venous system, causing bacteremia (bacteria in the bloodstream). Chronically infected tonsils can swell, causing difficulty swallowing or difficulty with breathing, particularly during sleep.

In the early stages of tonsillitis, antibiotics can usually control the infection. However, if a peritonsillar (around the tonsils) abscess forms, if difficulty breathing or swallowing occurs, or if a child experiences more than five to seven episodes of tonsillitis per year, **tonsillectomy** (surgical removal of the tonsils) is considered.

Tonsils are most active immunologically before puberty. After puberty, healthy tonsils atrophy and may disappear by late adulthood.

The collection of lymphatic tissues within the epithelia of the digestive, respiratory, urinary, and reproductive systems is called the **mucosa-associated lymphoid tissue (MALT)** (**Figure 23.7a,b**). Large nodules in the wall of the pharynx are called **tonsils** (**Figure 23.7c**). The lymphocytes within the tonsils gather and remove pathogens that are ingested in food or inspired in air. There are usually five tonsils:

- A single **pharyngeal tonsil**, often called the *adenoid*, located in the posterior superior wall of the nasopharynx
- A pair of **palatine tonsils**, located at the posterior, inferior margin of the oral cavity along the boundary of the pharynx to the soft palate
- A pair of **lingual tonsils**, which are not visible because they are located deep to the mucosa at the base of the tongue

Clusters of lymphatic nodules in the mucosal lining of the small intestine are known as **aggregated lymphoid nodules**, or *Peyer's patches*. In addition, the walls of the appendix, a tube-shaped pouch originating near the junction between the small and large intestines, contain a mass of fused lymphoid nodules.

The lymphocytes within these lymphoid nodules are not always able to destroy bacterial or viral invaders that have crossed the epithelium of the digestive tract. An infection may then develop; familiar examples include tonsillitis and appendicitis.

23.5 **CONCEPT CHECK**

9 What is the name given to the clusters of lymphoid nodules found in the mucosal lining of the small intestine?

See the blue Answers tab at the back of the book.

23.6 | Lymphatic Organs

Lymphatic organs are separated from surrounding tissues by a fibrous connective tissue capsule. Lymphatic organs include the lymph nodes, thymus, and spleen.

Lymph Nodes

▶ **KEY POINT** Lymph nodes are small, oval lymphatic organs surrounded by a fibrous connective tissue capsule. They are widely distributed throughout the body.

A **lymph node** is approximately 1–25 mm in diameter. Lymph nodes are surrounded by a dense, fibrous connective tissue capsule and are widely distributed throughout the body. Fibrous extensions from the capsule called trabeculae extend into the interior of the node (**Figure 23.8**).

Blood vessels and nerves enter and exit the lymph node at the **hilum** (**Figure 23.8**). Each lymph node has both afferent lymphatic vessels and efferent lymphatic vessels. The afferent lymphatic vessels bring lymph to the node from peripheral tissues. The vessels penetrate the capsule on the side opposite the hilum. Lymph from the afferent lymphatics flows through the lymph node within a network of sinuses, open passageways with incomplete walls. The lymph node interior is divided into an outer cortex, an inner medulla, and a region between the two called the paracortex. Lymph first enters a subcapsular space that contains a meshwork of branching reticular fibers, macrophages, and dendritic cells. **Dendritic cells** are involved in starting an immune response. Lymph passes through the subcapsular space and then flows through the **cortex** of the node. The periphery of the cortex contains B cells within germinal centers similar to those of lymphoid nodules.

Figure 23.8 Structure of a Lymph Node. Lymph nodes are covered by a dense, fibrous, connective tissue capsule. Lymphatic vessels and blood vessels penetrate the capsule to reach the lymphatic tissue within. Note that there are several afferent lymphatic vessels and only one efferent vessel.

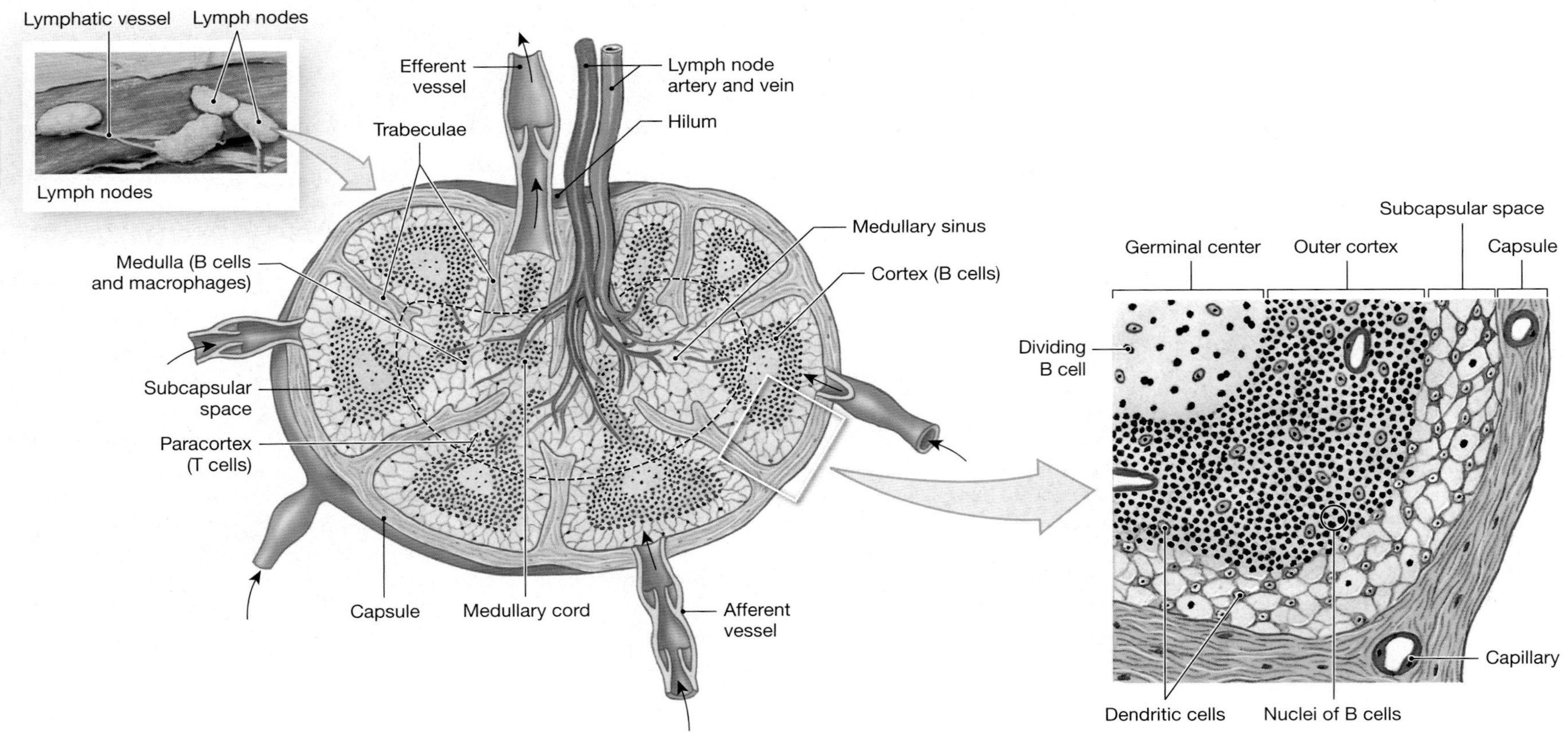

Lymph then continues through lymph sinuses toward an ill-defined area known as the **paracortex**, which is dominated by T cells. Here, lymphocytes leave the bloodstream and enter the lymph node by crossing the walls of blood vessels.

After flowing through the sinuses of the paracortex, lymph continues into the core, or **medulla**. The medulla contains B cells and macrophages. Elongated masses of dense lymphoid tissue between the sinuses in the medulla are known as **medullary cords**. Lymph passes through a network of sinuses in the medulla and then enters the efferent lymphatics at the hilum.

Lymph nodes function like water filters, filtering and purifying lymph before it reaches the venous system. As lymph flows through a lymph node, 99 percent or more of the antigens present in the arriving lymph are removed. Fixed macrophages in the lymphatic sinuses engulf debris or pathogens in the lymph as it flows past. Antigens removed in this way are then processed and "presented" to nearby T cells. Other antigens stick to reticular fibers or to the surfaces of dendritic cells, where they stimulate T cell activity.

The largest lymph nodes are found where peripheral lymphatics connect with a lymphatic trunk, such as the base of the neck, the axillae, and the groin (**Figures 23.1** and **23.5**, p. 604 and 608, and **Figures 23.9** to **23.11**). These nodes are often called *lymph glands*. "Swollen glands" indicate inflammation or infection of peripheral structures.

Distribution of Lymphatic Tissues and Lymph Nodes

Lymphatic tissues and lymph nodes are distributed in areas susceptible to injury or invasion. If you wanted to protect your house against intrusion, you might guard all doors and windows and perhaps keep a big dog indoors. The distribution of lymphatic tissues and lymph nodes is based on a similar strategy:

- The **cervical lymph nodes** filter lymph originating in the head and neck, including lymphatic vessels within the meninges of the brain (**Figure 23.9**).
- The **axillary lymph nodes** filter lymph arriving at the trunk from the upper limbs (**Figure 23.10a**). In women, the axillary nodes also drain lymph from the mammary glands (**Figure 23.10b**).
- The **popliteal lymph nodes** filter lymph arriving at the thigh from the leg, and the **inguinal lymph nodes** filter lymph arriving at the trunk from the lower limbs (**Figure 23.11**).
- The **thoracic lymph nodes** filter lymph from the lungs, respiratory passageways, and mediastinal structures (**Figure 23.5**, p. 608).

Figure 23.9 Lymphatic Drainage of the Head and Neck. Position of the lymphatic vessels and nodes that drain the head and neck regions.

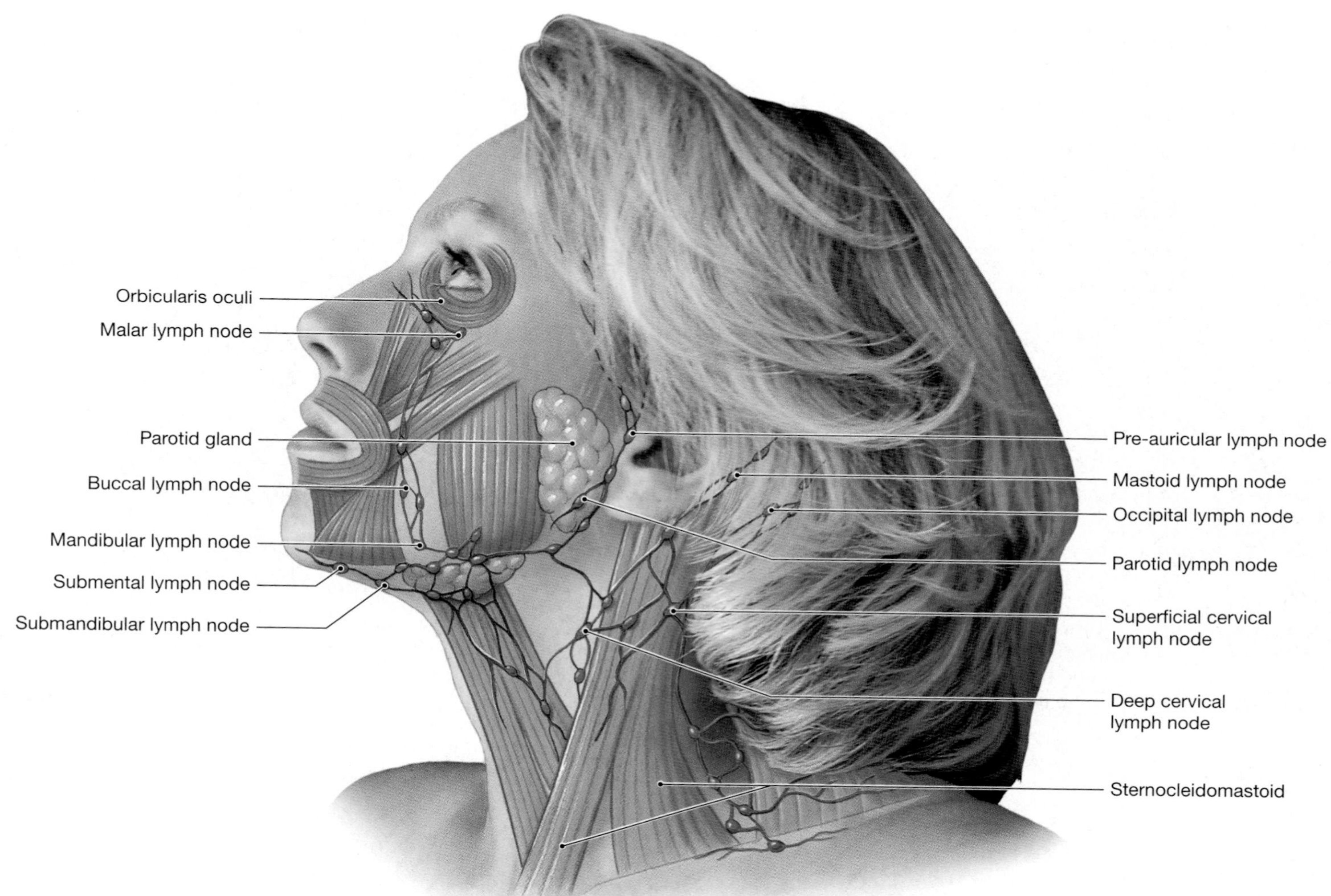

23

Figure 23.10 Lymphatic Drainage of the Upper Limb

Deltoid
Deltopectoral lymph node
Pectoralis major
Axillary lymph nodes
Cephalic vein
Basilic vein
Supratrochlear lymph node

a Superficial lymphatic vessels and nodes that drain the upper limb and chest of a male

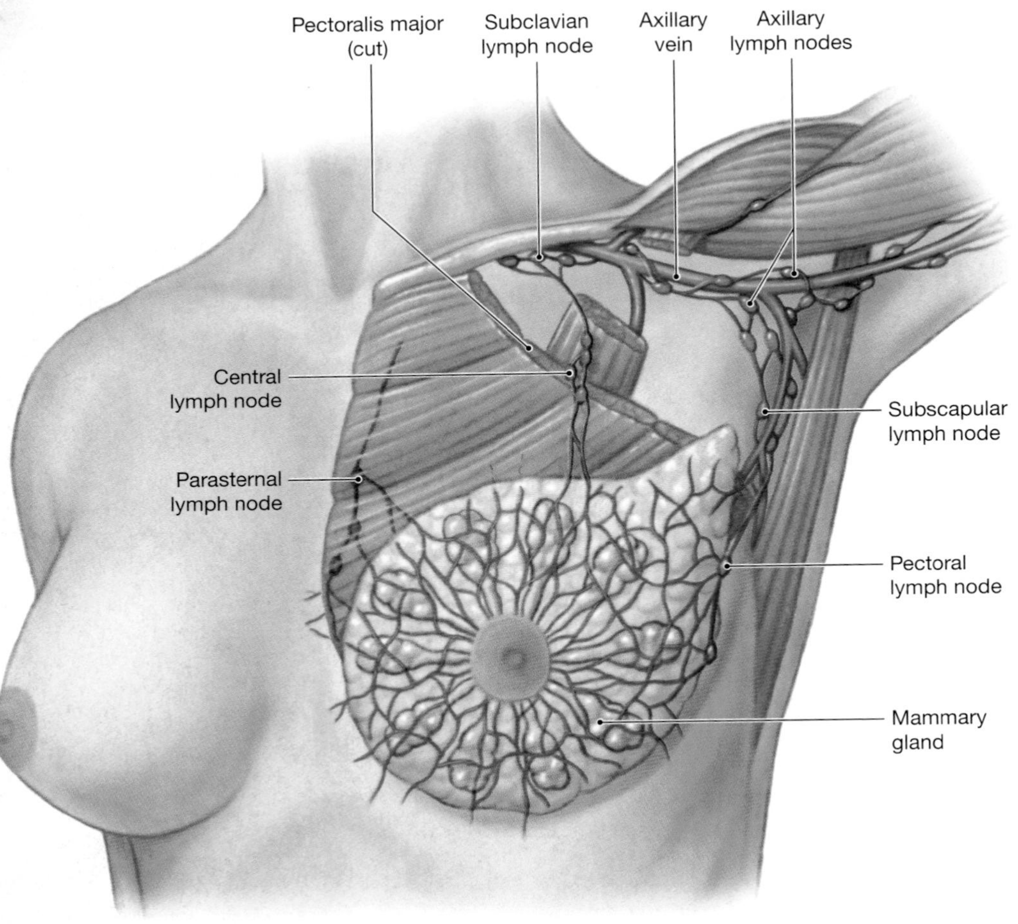

b Superficial and deeper lymphatic vessels and nodes of the upper limb and chest of a female

HIV/AIDS

Human Immunodeficiency Virus (HIV) is a retrovirus, or an RNA virus. It targets human helper T cells by binding to their surface, invading the T cells, and taking over the host's replication machinery to make more copies of itself. HIV can destroy entire "families" of helper T cells known as "cluster of differentiation 4," or **CD4**, cells. With no CD4 cells, the diseases that these families of helper T cells normally fight can easily take over, causing opportunistic infections (diseases that occur because resistance is lowered). A normal CD4 cell count is 500–1,600 cells per cubic millimeter of blood. If the CD4 cell count falls below 200, the disease has progressed to **acquired immunodeficiency syndrome (AIDS)**.

Unlike most viruses, the human body can never get rid of HIV completely, so once infected, the person is infected for life. Several classes of antiretroviral agents, each acting on a different stage of the HIV life cycle, are used to treat HIV. With good medical care and early treatment, HIV can be controlled and progression to AIDS prevented.

Figure 23.11 Lymphatic Drainage of the Lower Limb

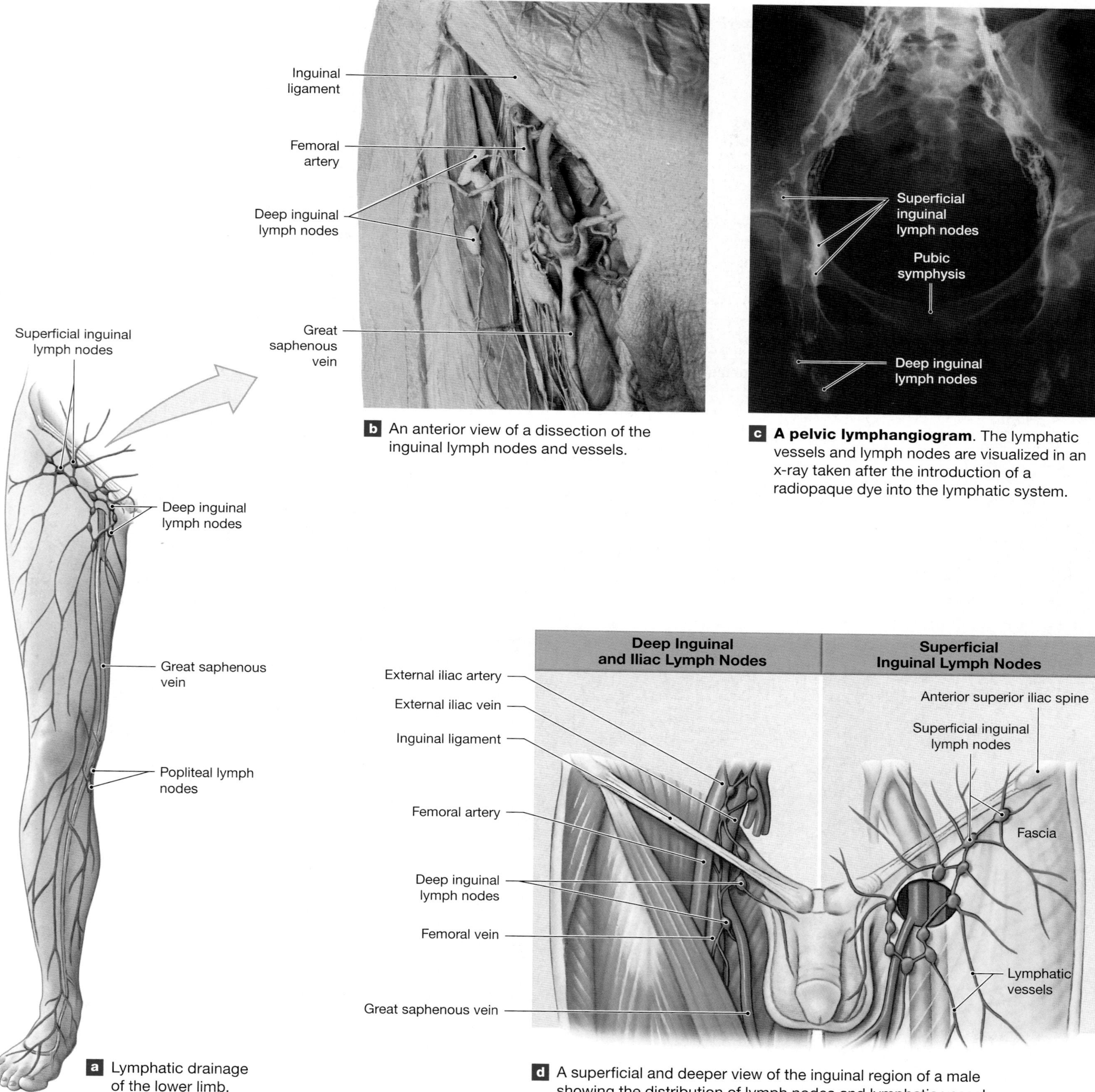

a Lymphatic drainage of the lower limb.

b An anterior view of a dissection of the inguinal lymph nodes and vessels.

c **A pelvic lymphangiogram**. The lymphatic vessels and lymph nodes are visualized in an x-ray taken after the introduction of a radiopaque dye into the lymphatic system.

d A superficial and deeper view of the inguinal region of a male showing the distribution of lymph nodes and lymphatic vessels.

- The **abdominal lymph nodes** filter lymph arriving from the urinary and reproductive systems.
- The **intestinal lymph nodes** and the **mesenteric lymph nodes** filter lymph originating from the digestive tract **(Figure 23.12)**.

The Thymus

▶ **KEY POINT** The thymus contains developing T cells, reticular cells, and thymic corpuscles.

The **thymus** lies posterior to the manubrium of the sternum in the superior portion of the mediastinum. The thymus reaches its greatest size relative to body size in the first year or two after birth. However, it reaches its maximum size during puberty, when it weighs between 30 and 40 g. After puberty the thymus gradually shrinks, and its functional cells are replaced by fibrous connective tissue fibers and fat. This degenerative process is called involution of the thymus.

The capsule covering the thymus divides it into two **thymic lobes** **(Figure 23.13a,b)**. Fibrous **septa** extend inward from the capsule and divide the lobes into **lobules** **(Figure 23.13b,c)**. Each lobule averages 2 mm in width and consists of a dense outer **cortex** and a diffuse and paler-staining central **medulla**.

The cortex contains rapidly dividing lymphatic stem cells. These stem cells produce daughter cells that mature into T cells and migrate into the medulla. During the maturation process, any T cells that are sensitive to normal tissue antigens are destroyed. While they are within the thymus, the surviving T cells do not participate in the immune response; they remain inactive until they enter the blood. The capillaries of the thymus resemble those of the CNS in that they form a **blood thymus barrier**. The blood thymus barrier does not permit free exchange between the circulation and the interstitial fluid of the thymus; this prevents premature stimulation of the developing T cells by circulating antigens.

Epithelial reticular cells are scattered among the lymphocytes of the thymus. These cells produce thymic hormones that promote the differentiation of functional T cells. In the medulla, reticular cells cluster together in concentric layers, forming distinctive structures known as **thymic corpuscles** **(Figure 23.13d)**. The function of thymic corpuscles remains unknown.

Figure 23.12 Lymph Nodes of the Large Intestine and Associated Mesenteries

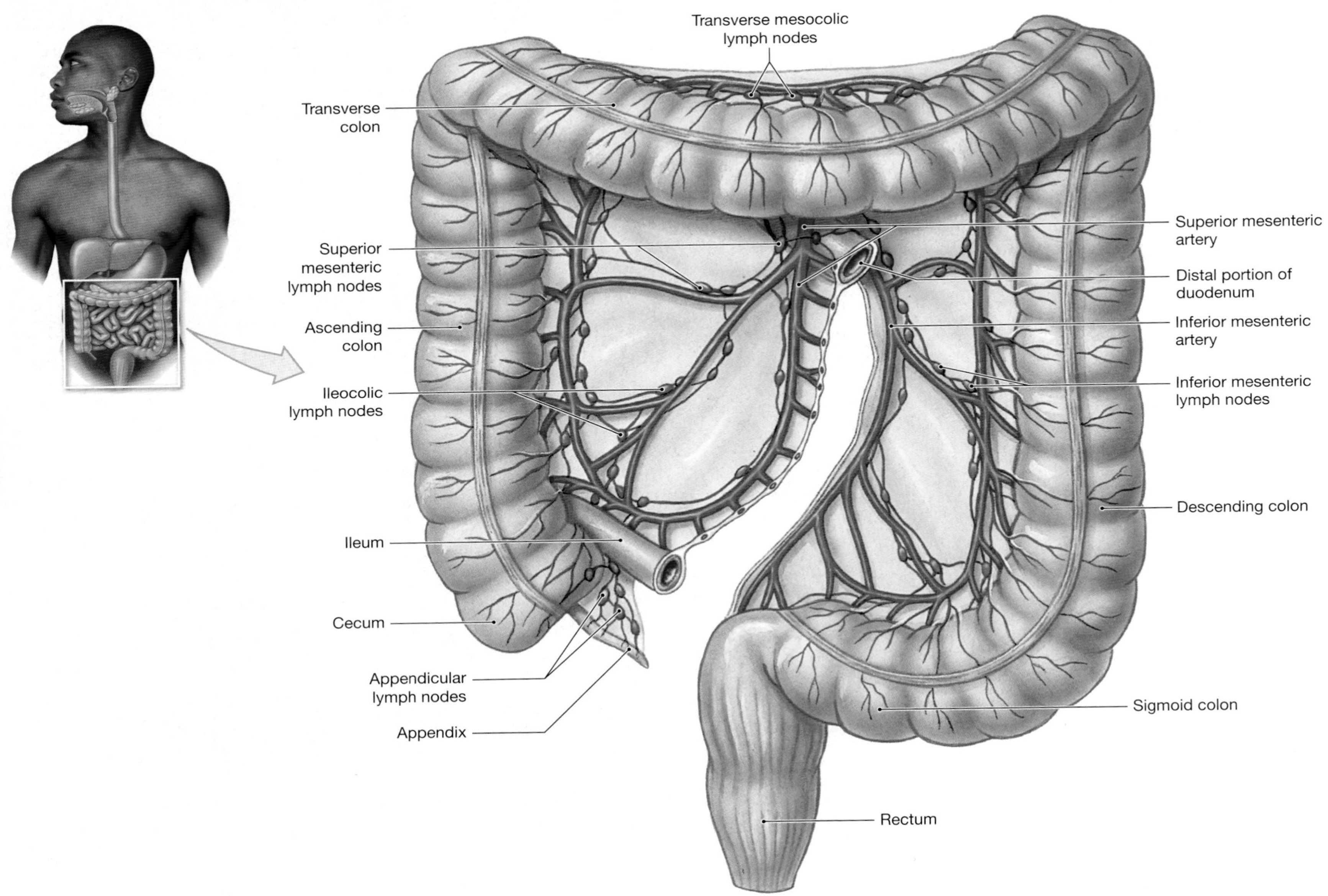

Figure 23.13 **Anatomy and Histological Organization of the Thymus**

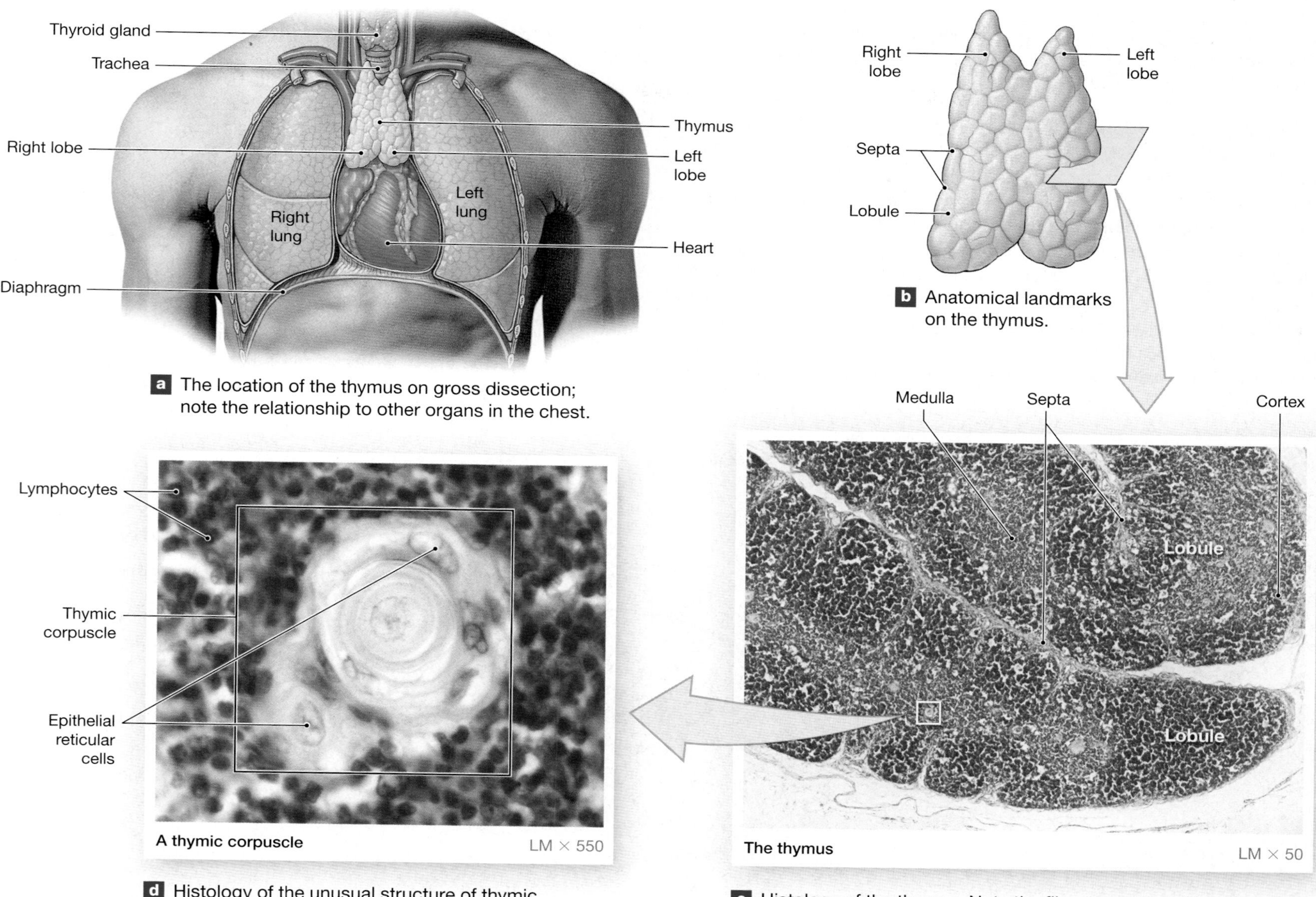

a The location of the thymus on gross dissection; note the relationship to other organs in the chest.

b Anatomical landmarks on the thymus.

c Histology of the thymus. Note the fibrous septa dividing the thymic tissue into lobules resembling interconnected lymphatic nodules.

d Histology of the unusual structure of thymic corpuscles. The small cells in view are lymphocytes in various stages of development.

23

CLINICAL NOTE

Lymphatic Vessels and Metastatic Cancer

Cancer can appear in lymph nodes in two ways: within the lymph nodes or by the lymphatic vessels and bloodstream. Cancer that starts in the lymph nodes it is called lymphoma. More often, cancer starts somewhere else (primary tumor) and then spreads to lymph nodes. When cancer cells break away from the primary tumor, they travel through the bloodstream to the lymphatic system. Most of the escaped cancer cells die or are killed by NK cells before they can start growing. However, one or two might settle in new areas and begin to grow, forming **metastatic tumors**.

Cancers that often spread hematogenously (through the bloodstream) include hepatomas (primary liver tumors) and sarcomas (connective tissue tumor). Cancers that tend to metastasize through the lymphatic system include carcinomas (cancers derived from epithelial cells, chiefly glandular or squamous), such as breast and ovarian cancer, and melanomas (malignant tumors derived from melanin-forming cells). Eventually, all lymph ends up in the venous system, so cancers that start in the lymphatic system eventually travel through the blood.

The Spleen

▶ **KEY POINT** The spleen removes damaged red blood cells, stores iron, and initiates immune responses by B cells and T cells in response to antigens. The spleen contains red pulp and white pulp.

The largest lymphatic organ in the body, the **spleen**, is around 12 cm (5 in.) long and weighs up to 160 g (5.6 oz). The spleen is located on the left side, along the lateral border of the stomach. It lies between the ninth and eleventh ribs and is attached to the lateral border of the stomach by a piece of mesentery, the **gastrosplenic ligament** (**Figure 23.14a**).

On gross dissection the spleen has a deep red color because it contains a large amount of blood. The spleen performs functions for the blood similar to those performed by the lymph nodes for lymph, including (1) removing abnormal blood cells and other blood components through phagocytosis, (2) storing iron recycled from the breakdown of red blood cells, and (3) initiating immune responses by B cells and T cells in response to antigens in the circulating blood.

Surfaces of the Spleen

The spleen is soft, and its shape reflects the structures around it. It lies wedged between the stomach, the left kidney, and the muscular diaphragm. The **diaphragmatic surface** (**Figure 23.14a**) is smooth and convex and conforms to the shape of the diaphragm and body wall. The inferior **visceral surface** contains indentations that follow the shapes of the stomach (the **gastric area**) and left kidney (the **renal area**) (**Figure 23.14b**). The **splenic artery**, **splenic vein**, and lymphatic vessels draining the spleen enter and leave at the hilum (**Figure 23.14b**).

Histology of the Spleen

The spleen is surrounded by a capsule containing collagen and elastic fibers. The cellular components deep to the capsule make up the **pulp** of the spleen (**Figure 23.14c**). The **red pulp** forms splenic cords, which contain large quantities of red blood cells. The **white pulp** forms lymphatic nodules. The splenic artery enters at the hilum and branches to produce

Figure 23.14 Anatomy and Histological Organization of the Spleen

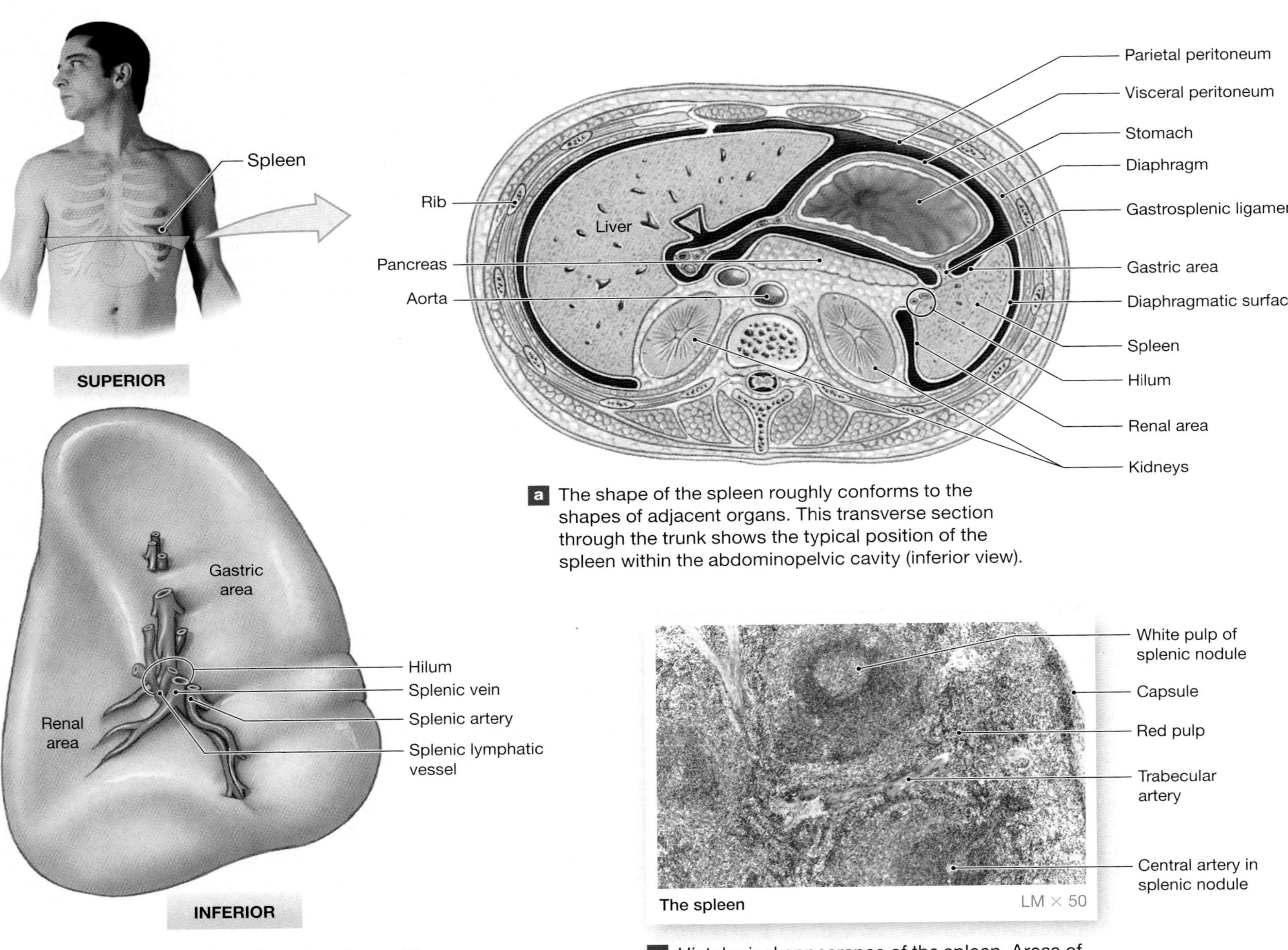

a The shape of the spleen roughly conforms to the shapes of adjacent organs. This transverse section through the trunk shows the typical position of the spleen within the abdominopelvic cavity (inferior view).

b External appearance of the visceral surface of the intact spleen showing major anatomical landmarks. This view should be compared with that of part (a).

c Histological appearance of the spleen. Areas of white pulp are dominated by lymphocytes. Areas of red pulp contain a large number of red blood cells.

23

CLINICAL NOTE

Lymphoma

Lymphoma is a group of blood cancers that develop in the lymphatic system. Lymphoma begins when a mutation within a lymphocyte in a lymph node or lymphoid tissue causes it to grow out of control and produce abnormal daughter cells. These abnormal lymphocytes accumulate in lymphatic tissues, crowding out normal lymphocytes and impeding their function, and the function of the lymphatic system breaks down.

There are two main types of lymphoma, **Hodgkin lymphoma** and **non-Hodgkin lymphoma**. Hodgkin lymphoma, named for the physician who identified the disease, is a cancer of the blood and bone marrow. About 10 percent of lymphomas are this type. It is characterized by an abnormal type of B cell called Reed-Sternberg cells and is associated with Epstein-Barr virus infection. Hodgkin lymphoma affects people in their 20s and people over 55, most commonly in the United States, Canada, and northern Europe. It is rare in Asia. If one identical twin has Hodgkin lymphoma, the other twin has a very high risk of developing it as well. Hodgkin lymphoma is very treatable, and most cases can be cured.

Non-Hodgkin lymphoma (NHL) is a diverse group that affects the lymphatic system and in some cases involves the bone marrow and blood. About 90 percent of lymphomas are non-Hodgkin lymphomas. About 85 percent of NHL cases affect B cells. The remaining NHLs involve T cells and NK cells. It is described as either indolent (slow-growing) or aggressive (fast-growing) and can occur at any age, but it most often affects people over age 60. Although the cause of NHL is unknown, certain chemicals, including benzene, and nuclear radiation exposure have been implicated. This disease also develops in immunocompromised individuals.

Symptoms of lymphoma include enlarged lymph nodes that do not resolve, unexplained weight loss, fever, night sweats, and unusual infections.

Treatment of lymphoma depends on the type and stage (how far it has spread) of the disease. Chemotherapy, radiation therapy, and immunotherapy, using the body's own immune system to attack cancer cells, are used to treat lymphoma. If these fail, a stem cell transplant can be considered.

a number of **trabecular arteries** radiating outward toward the capsule. These arteries branch extensively, and their arteriolar branches are surrounded by areas of white pulp. Capillaries discharge blood into the venous sinuses of the red pulp.

The cell population of the red pulp includes all the normal components of circulating blood, plus fixed and free macrophages. A network of reticular fibers forms the structural framework of the red pulp. Blood passes through this network and enters large sinusoids, which are lined by fixed macrophages. The sinusoids empty into small veins that merge to form trabecular veins, which continue toward the hilum.

This circulatory arrangement gives the phagocytes of the spleen the opportunity to identify and engulf damaged or infected cells in circulating blood. Macrophages are scattered throughout the red pulp, and the region surrounding the white pulp has a high concentration of lymphocytes and dendritic cells. Thus, any microorganisms or antigens in the blood quickly trigger an immune response.

23.6 CONCEPT CHECK

10 What is important about the placement of lymph nodes?

11 Why is there a blood thymus barrier in the capillaries of the thymus?

12 Why do lymph nodes often enlarge during an infection?

See the blue Answers tab at the back of the book.

23.7 | Aging and the Lymphatic System

▶ KEY POINT With advancing age, the lymphatic system becomes less effective at combating disease.

With advancing age, T cells become less responsive to antigens. As a result, fewer cytotoxic T cells respond to an infection. The number of helper T cells also declines, so B cells are less responsive and antibody levels do not rise as quickly after antigen exposure. Aging therefore increases susceptibility to viral and bacterial infections. For this reason, vaccinations for acute viral diseases, such as flu (influenza), are strongly recommended for older people. The increased incidence of cancer in elderly people reflects the fact that immune surveillance declines, so tumor cells are not eliminated as effectively.

23.7 CONCEPT CHECK

13 You have recently retired, and at your annual physical, your physician urges you to get the flu vaccine. Why?

See the blue Answers tab at the back of the book.

EMBRYOLOGY SUMMARY

For a summary of the development of the lymphatic system, see Chapter 28 (Embryology and Human Development).

23

Study Outline

Introduction p. 604

- The cells, tissues, and organs of the lymphatic system play a central role in the body's defenses against viruses, bacteria, and other microorganisms.

23.1 | An Overview of the Lymphatic System p. 604

- The **lymphatic system** includes a network of lymphatic vessels that carry **lymph** (a fluid similar to plasma but with a lower concentration of proteins). A series of lymphatic organs and lymphatic tissues are interconnected by the lymphatic vessels. *(See Figures 23.1 and 23.2.)*

Functions of the Lymphatic System p. 604

- The lymphatic system produces, maintains, and distributes lymphocytes (cells that attack invading organisms, abnormal cells, and foreign proteins). It helps maintain blood volume and eliminate local variations in the composition of the interstitial fluid. Lymphatic structures can be classified as **primary** (containing stem cells) or **secondary** (containing immature or activated lymphocytes).

23.2 | Structure of Lymphatic Vessels p. 605

Lymphatic Capillaries p. 605

- **Lymphatic vessels**, or *lymphatics*, carry lymph from peripheral tissues to the venous system. Lymph flows along a network of lymphatics that originate in the **lymphatic capillaries**. The endothelial cells of a lymphatic capillary overlap to act as a one-way valve, preventing fluid from returning to the intercellular spaces. *(See Figure 23.2b.)*

Larger Lymphatic Vessels p. 606

- Lymphatic vessels contain numerous internal valves to prevent backflow of lymph. *(See Figure 23.3.)*

23.3 | Major Lymph-Collecting Vessels p. 606

23

- Two sets of lymphatic vessels collect blood from the lymphatic capillaries: **superficial lymphatics** and **deep lymphatics**.

The Thoracic Duct p. 607

- The **thoracic duct** collects lymph from both sides of the body inferior to the diaphragm and from the left side of the body superior to the diaphragm. *(See Figures 23.4 and 23.5.)*

The Right Lymphatic Duct p. 607

- The **right lymphatic duct** collects lymph from the right side of the body superior to the diaphragm. *(See Figures 23.4 and 23.5.)*

23.4 | Lymphocytes p. 607

Types of Lymphocytes p. 607

- There are three classes of lymphocytes: **T cells** (thymus-dependent), **B cells** (bone marrow–derived), and **NK cells** (natural killer). *(See Spotlight Figure 23.6.)*

- **Cytotoxic T cells** attack foreign cells or body cells infected by viruses; they provide **cell-mediated immunity**. **Regulatory T cells** are a subset of T cells that moderate the immune response. **Memory T cells** remain "on reserve" to respond to antigens they have previously encountered. **Helper T cells** stimulate the responses of both T cells and B cells. *(See Spotlight Figure 23.6.)*

- B cells can differentiate into **plasma cells**, which produce and secrete **antibodies (immunoglobulins)** that react with specific chemical targets, or antigens. B cells are responsible for **antibody-mediated immunity**. **Memory B cells** are activated if the antigen appears again at a later date. *(See Spotlight Figure 23.6.)*

- **Natural killer cells**, or NK cells, attack foreign cells, normal cells infected with viruses, and cancer cells. They provide **immune surveillance**. *(See Spotlight Figure 23.6.)*

Lymphocytes and the Immune Response p. 610

- The goal of the **immune response** is to destroy or inactivate pathogens, abnormal cells, and foreign molecules such as toxins. Antigens are engulfed by macrophages, which then present the antigen to T cells so they can begin differentiating. The millions of different lymphocytes, which retain the ability to divide, allow the body to be prepared for any antigen. The ability to recognize antigens is called immunocompetence.

Distribution and Life Span of Lymphocytes p. 610

- The ratio of B cells to T cells in organs and tissues varies continuously. Lymphocytes have a relatively long life span.

Lymphocytopoiesis: Lymphocyte Production p. 610

- Lymphocytes continually migrate in and out of the blood through lymphatic tissues and organs. **Lymphocytopoiesis** (lymphocyte production) involves the red bone marrow, thymus, and peripheral lymphatic tissues. *(See Spotlight Figure 23.6.)*

23.5 | Lymphatic Tissues p. 610

- **Lymphatic tissues** are connective tissues dominated by lymphocytes. In a **lymphatic nodule**, the lymphocytes are densely packed in an area of loose connective tissue. Important lymphoid nodules are **aggregated lymphoid nodules** deep to the lining of the intestine, in the appendix, and the **tonsils** within the pharynx. *(See Figure 23.7.)*

23.6 | Lymphatic Organs p. 612

- Important **lymphatic organs** include the lymph nodes, thymus, and spleen. *(See Figures 23.1, 23.5, and 23.8–23.12.)*

Lymph Nodes p. 612

- **Lymph nodes** are encapsulated masses of lymphatic tissue. The **paracortex** is dominated by T cells; the **cortex** and **medulla** contain B cells arranged into **medullary cords**. Lymph glands are the largest lymph nodes, found where peripheral lymphatics connect with the trunk. *(See Figures 23.1, 23.5, and 23.8–23.12.)*

- Lymphatic tissues and nodes are located in areas particularly susceptible to injury or invasion by microorganisms.

- The **cervical lymph nodes**, **axillary lymph nodes**, **popliteal lymph nodes**, **inguinal lymph nodes**, **thoracic lymph nodes**, **abdominal lymph nodes**, **intestinal lymph nodes**, and **mesenteric lymph nodes** protect the vulnerable areas of the body. *(See Figures 23.4, 23.5, and 23.9–23.12.)*

The Thymus p. 616

- The **thymus** lies posterior to the manubrium, in the superior mediastinum. **Epithelial reticular cells** scattered among the lymphocytes produce thymic hormones. These hormones promote the differentiation of T cells. The **blood thymus barrier** does not allow free exchange between the interstitial fluid and the circulation, protecting the T cells from being prematurely activated. After puberty the thymus gradually decreases in size, a process called involution. *(See Figure 23.13.)*

The Spleen p. 618

- The adult **spleen** contains the largest mass of lymphatic tissue in the body. The spleen performs the same functions for the blood that lymph nodes perform for the lymph. The **diaphragmatic surface** of the spleen lies against the diaphragm; the **visceral surface** is against the stomach and kidney and contains a groove called the **hilum**. The cellular components form the **pulp** of the spleen. **Red pulp** contains large numbers of red blood cells, and areas of **white pulp** resemble lymphatic nodules. Macrophages are scattered throughout the red pulp, and the region surrounding the white pulp has a high concentration of lymphocytes and dendritic cells. *(See Figure 23.14.)*

23.7 | Aging and the Lymphatic System p. 619

- With aging, the immune system becomes less effective at combating disease.

Chapter Review

For answers, see the blue Answers tab at the back of the book.

Level 1 Reviewing Facts and Terms

Match each numbered item with the most closely related lettered item.

1. plasma cells .... ☐
2. spleen .... ☐
3. thymus .... ☐
4. cytotoxic T cells .... ☐
5. antibodies .... ☐
6. NK cells .... ☐
7. memory cells .... ☐
8. cisterna chyli .... ☐
9. lymphocytopoiesis .... ☐
10. B cells .... ☐

a. rapid secondary response
b. responsible for cell-mediated immunity
c. produce antibodies
d. aid in immune surveillance
e. contains developing T cells
f. immunoglobulins
g. responsible for antibody-mediated immunity
h. production of lymphocytes
i. saclike chamber of the thoracic duct
j. largest lymphatic organ in the body

11. Identify the structures of the lymphatic system in the following diagram.

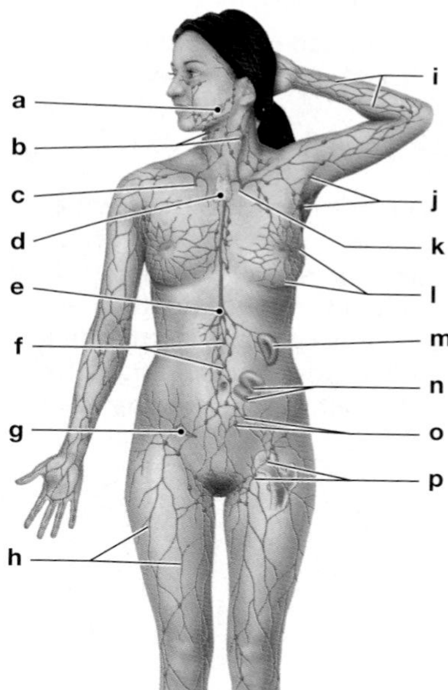

(a) ________
(b) ________
(c) ________
(d) ________
(e) ________
(f) ________
(g) ________
(h) ________
(i) ________
(j) ________
(k) ________
(l) ________
(m) ________
(n) ________
(o) ________
(p) ________

12. The lymphatic system is composed of
    (a) lymphatic vessels.
    (b) the spleen.
    (c) lymph nodes.
    (d) all of the above.
13. Compared to blood capillaries, lymphatic capillaries
    (a) have a basal lamina.
    (b) are smaller in diameter.
    (c) have walls of a smooth endothelial lining.
    (d) are larger in diameter.
14. Most of the lymph returns to the venous circulation by way of the
    (a) right lymphatic duct.
    (b) cisterna chyli.
    (c) hepatic portal vein.
    (d) thoracic duct.
15. Some cells known as plasma cells
    (a) are actively phagocytic.
    (b) destroy red blood cells.
    (c) produce proteins called antibodies.
    (d) are primarily found in red bone marrow.
16. _____ are large lymphatic nodules located in the walls of the pharynx.
    (a) Tonsils
    (b) Lymph nodes
    (c) Thymus glands
    (d) Thymic corpuscles

23

17. Areas of the spleen that contain large numbers of lymphocytes are known as
(a) white pulp.
(b) red pulp.
(c) adenoids.
(d) lymph nodes.

18. The red pulp of the spleen contains large numbers of
(a) macrophages.
(b) antibodies.
(c) neutrophils.
(d) lymphocytes.

19. The cells responsible for the production of circulating antibodies are
(a) NK cells.
(b) plasma cells.
(c) helper T cells.
(d) cytotoxic T cells.

20. The medullary cords of a lymph node contain
(a) cytotoxic T cells.
(b) regulatory T cells.
(c) NK cells.
(d) B cells.

21. Lymphocytes that attack foreign cells or body cells infected with viruses are
(a) B cells.
(b) helper T cells.
(c) cytotoxic T cells.
(d) suppressor T cells.

Level 2 Reviewing Concepts

1. If the thymus failed to produce thymic hormones, we would expect to see a decrease in the number of
(a) B lymphocytes.
(b) NK cells.
(c) cytotoxic T cells.
(d) neutrophils.

2. Lymph from the right arm, the right half of the head, and the right chest is received by the
(a) cisterna chyli.
(b) right lymphatic duct.
(c) right thoracic duct.
(d) aorta.

3. Blocking the antigen receptors on the surface of lymphocytes would interfere with
(a) phagocytosis of the antigen.
(b) those lymphocytes' ability to produce antibodies.
(c) antibody recognition.
(d) the ability of the lymphocytes to present antigen.

4. What is the function of the blood thymus barrier?
5. What major artery and vein pass through the hilum of the spleen?
6. From what areas of the body does the thoracic duct collect lymph?
7. Which type of lymphocyte is most common?
8. What is lymphedema?
9. What occurs in secondary lymphatic structures?
10. Where are aggregated lymphoid nodules, also termed *Peyer's patches*, found?

Level 3 Critical Thinking

1. Tom has just been exposed to the measles virus, and since he can't remember whether he has had measles before, he wonders if he is going to come down with the disease. He asks you how he can tell whether he has been previously exposed or is going to get sick before it actually happens. What should you tell him?
2. While walking along the street, you and your friend see an elderly woman whose left arm appears to be swollen to several times its normal size. Your friend wonders what might be its cause. You say that the woman may have had a radical mastectomy (the removal of a breast because of cancer). Explain the rationale behind your answer.
3. Paula's grandfather is diagnosed as having lung cancer. His physician orders biopsies of several lymph nodes from neighboring regions of the body, and Paula wonders why, since the cancer is in his lungs. What should you tell her?

CLINICAL CASE WRAP-UP

Fighting a Zoonotic Disease

Animal species that harbor zoonotic diseases, such as bats and bugs, are called reservoir hosts. According to the Centers for Disease Control, zoonotic diseases cause three out of five new human sicknesses each year in the United States.

As Grace makes her way to the closest hospital for rehydration and supportive care, her immune system is also taking action: T cells and B cells are activated and begin dividing. Direct attack by her activated cytotoxic T cells produces cell-mediated immunity. Meanwhile, under the regulation and coordination of helper T cells, B cells differentiate into plasma cells to provide antibody-mediated immunity. Grace's survival depends on her immune response outpacing her advancing Ebola virus infection.

With proper medical treatment, Grace's prognosis is good. Once she recovers, she plans to stay in the hospital and care for other Ebola victims.

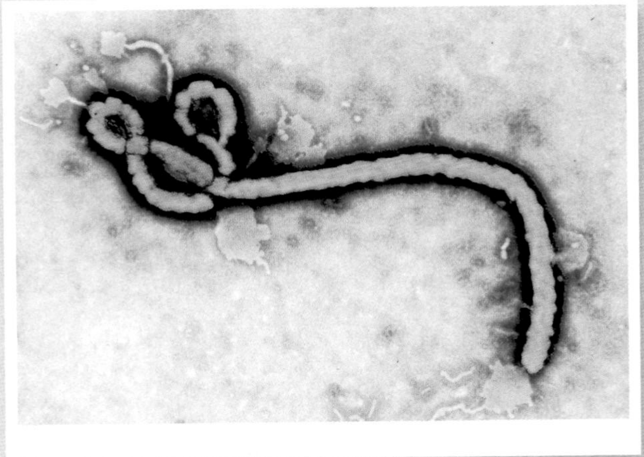

1. Will Grace's lymphatic system produce more or fewer lymphatic stem cells during her Ebola infection?
2. Will Grace be immune to Ebola once she recovers? Explain your answer.

See the blue Answers tab at the back of the book.

Related Clinical Terms

adenitis: Inflammation of the adenoid (pharyngeal tonsil).

allograft: Transplant between compatible recipient and donor of the same species.

anamnestic response: An immune response that is initiated by memory cells.

autograft: A transplant of tissue that is taken from the same person.

Burkitt's lymphoma: A malignant cancer of B lymphocytes.

congenital thymic aplasia: Congenital (present at birth) absence of the thymus and parathyroid glands and a deficiency of immunity.

Coombs test: A medical test to detect antibodies or complement in the blood.

dermatomyositis: An autoimmune disease characterized by inflammation of the skin and muscles.

eczema: A genetic inflammatory skin disorder, often with crusts, papules, and leaky eruptions.

host versus graft disease: A pathological condition in which cells from the transplanted tissue of a donor initiate an immune response, attacking the cells and tissue of the recipient.

hybridoma: A tissue culture composed of cancer cells fused to lymphocytes to mass produce a specific antibody.

immunology: Branch of biomedicine concerned with the structure and function of the immune system.

infectious mononucleosis: An acute disease caused by the Epstein–Barr virus, producing fever, swelling of the lymph nodes, sore throat, and increased lymphocytes in the bloodstream.

latex allergy: Hypersensitivity to products made of the sap of the rubber plant.

polymyositis: An autoimmune disease characterized by inflammation and atrophy of muscles.

sentinel node: The first lymph node to receive drainage from a tumor. It is used to determine if there is lymphatic metastasis in some types of cancer.

splenomegaly: Enlargement of the spleen.

systemic lupus erythematosus (SLE): An autoimmune disease in which a person's immune system attacks and injures its own organs and tissues in virtually every system of the body.

xenograft: A transplant that is made between two different species.

24

The Respiratory System

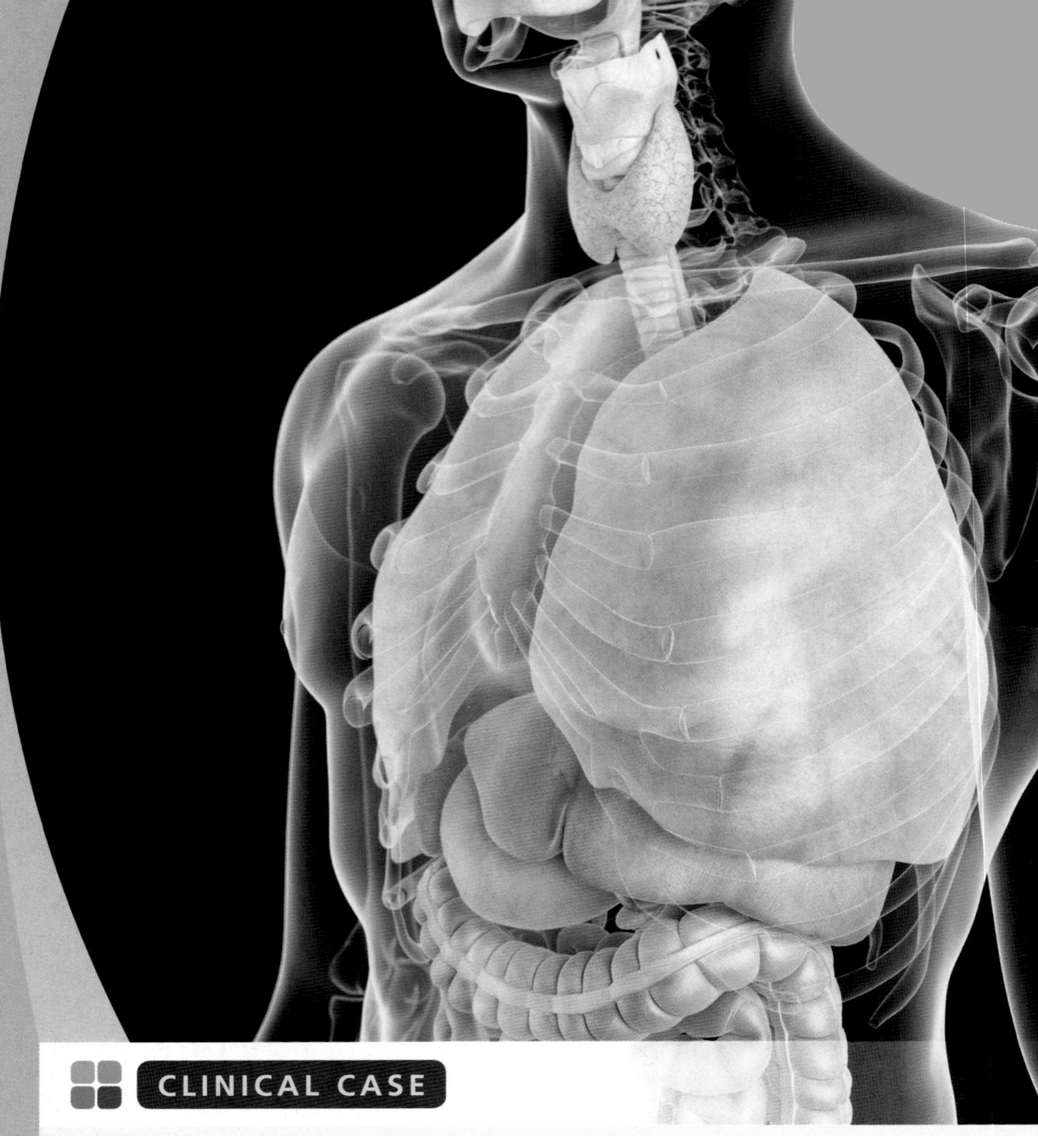

Learning Outcomes

These learning outcomes correspond by number to this chapter's sections and indicate what you should be able to do after completing the chapter.

CLINICAL CASE

How Long Should This Cough Last?

Andrea has been dissecting a cadaver with her lab partners for a month. Today, one of her partners, Bohdan, has an upper respiratory infection. Bohdan coughs, spreading aerosolized droplets across the dissecting table. Andrea inhales the droplets, which contain infectious rhinoviruses that cause common colds. The virus spreads across Andrea's respiratory mucosa, overcoming her natural immunity. Some viruses enter the epithelial cells of her pharynx and trachea, where they take over the cell's nuclear "machinery" and begin to multiply.

Three days later, the infected respiratory epithelial cells burst, releasing rhinoviruses to infect other cells throughout Andrea's body. She feels terrible, with a fever of 101°F, a runny nose (rhinitis), sore throat (pharyngitis), generalized fatigue, and an unrelenting cough, producing thick yellow mucus.

After a week, Andrea is still coughing. She sees her family doctor, insisting that she needs a prescription for antibiotics. The last time she had a weeklong cough, an urgent care doctor prescribed antibiotics, and her cough subsided after another 10 days.

Does Andrea need to take antibiotics again? To find out, turn to the Clinical Case Wrap-Up on p. 649.

MOST CELLS OBTAIN MOST OF THEIR ENERGY through aerobic metabolism, which requires oxygen and produces carbon dioxide. The respiratory system facilitates the exchange of oxygen and carbon dioxide between the air and blood. As blood circulates, it carries oxygen from the lungs to peripheral tissues, picks up carbon dioxide (CO_2) produced by these tissues, and transports CO_2 to the lungs for elimination.

We begin our study of the respiratory system by describing the structures that conduct air from the external environment to the gas exchange surfaces in the lungs. We then discuss the mechanics of breathing and neural control of respiration.

24.1 An Overview of the Respiratory System and Respiratory Tract

▶ **KEY POINT** The respiratory system includes all structures from the nose to the alveoli. The respiratory system is divided into upper and lower respiratory systems.

The **respiratory system** consists of the nose, nasal cavity and paranasal sinuses, pharynx, larynx (voice box), trachea (windpipe), and smaller conducting passageways to the alveoli. Alveoli are the gas exchange surfaces within the lungs (**Figure 24.1**).

Anatomically, the respiratory system is divided into the upper and lower respiratory systems. The **upper respiratory system** consists of the paranasal sinuses, nasal conchae, nose, nasal cavity, and nasopharynx. These passageways "condition the air" by filtering, warming, and humidifying it, thereby protecting the delicate conduction and exchange surfaces of the lower respiratory system from debris, pathogens, and environmental extremes. The **lower respiratory system** includes the larynx, trachea, bronchi, lungs, bronchioles, and alveoli of the lungs.

The **respiratory tract** consists of a conducting portion, which extends from the entrance to the nasal cavity to the bronchioles, and a respiratory portion, which includes the respiratory bronchioles and the **alveoli** (**Figure 24.1**). The process of filtering, warming, and humidifying inhaled air begins at the entrance to the upper respiratory system and continues through the rest of the conducting portion. By the time the air reaches the alveoli of the lungs, most

Figure 24.1 Structures of the Respiratory System

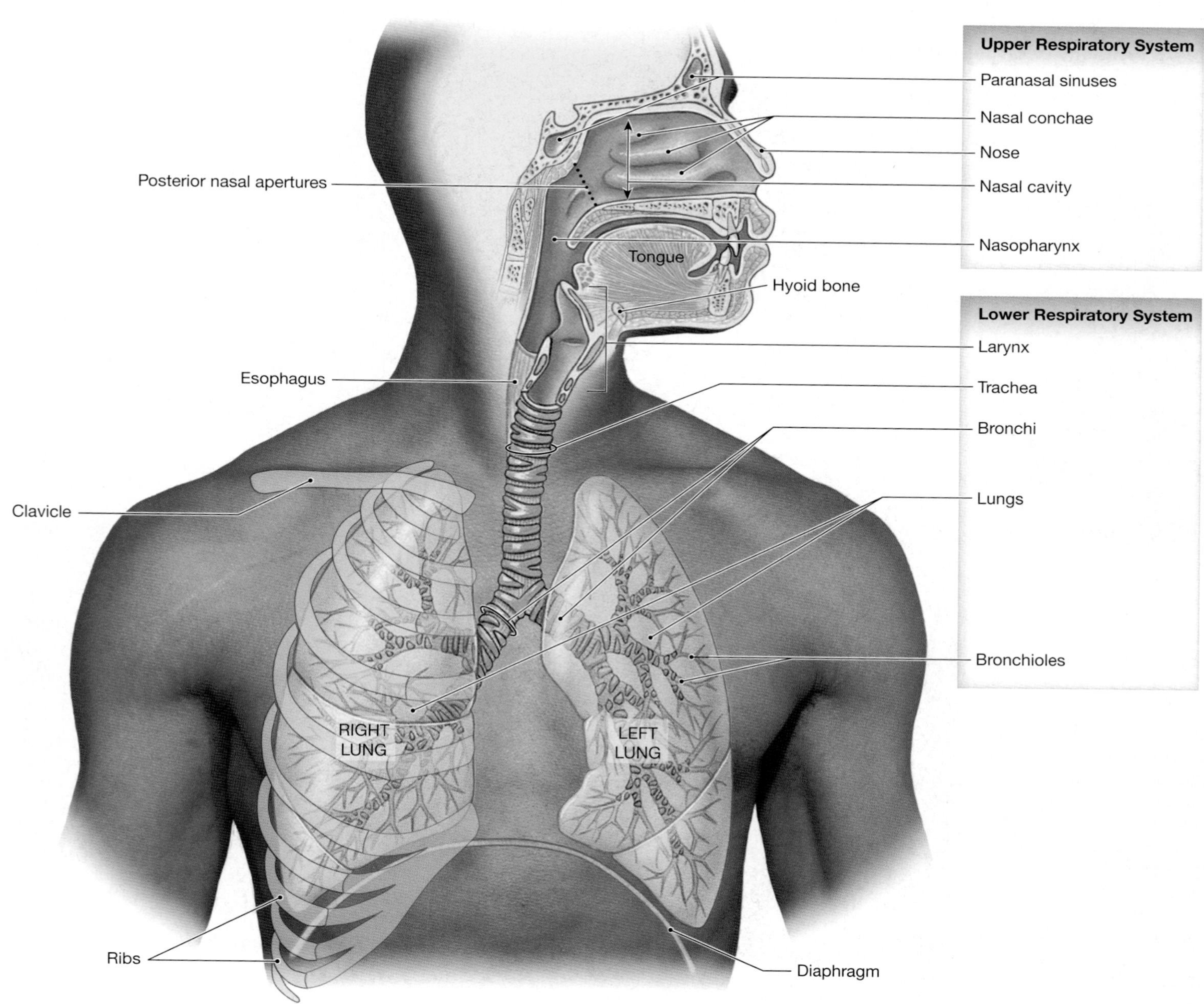

24

foreign particles and pathogens have been removed, and its humidity and temperature are within acceptable limits. The success of this conditioning process is due to the properties of the respiratory epithelium.

Functions of the Respiratory System

▶ **KEY POINT** Functions of the respiratory system include conducting air, providing an area for gas exchange, protecting respiratory surfaces, producing sounds, and helping regulate blood pressure and body fluid pH.

The respiratory system has the following basic functions:

- providing an extensive area for gas exchange between air and circulating blood;
- moving air to and from the exchange surfaces of the lungs;
- protecting respiratory surfaces from dehydration, temperature changes, and other environmental variations;
- defending the respiratory system and other tissues from pathogens;
- producing sounds involved in speaking, singing, or nonverbal communication;
- helping regulate blood volume, blood pressure, and body fluid pH.

The respiratory system performs these functions in conjunction with the cardiovascular and lymphatic systems, selected skeletal muscles, and the nervous system.

The Respiratory Epithelium

▶ **KEY POINT** The "typical" respiratory epithelium is a pseudostratified, ciliated columnar epithelium with mucous cells. As you move distally within the respiratory tract, the epithelium changes, resulting in a simple squamous epithelium specialized for gas exchange.

The **respiratory epithelium** is a pseudostratified ciliated columnar epithelium with numerous mucous cells (**Figure 24.2**). The respiratory epithelium lines the entire respiratory tract except for the inferior portions of the pharynx, the smallest conducting passages, and the alveoli. A stratified squamous epithelium lines the inferior portion of the pharynx, protecting it from abrasion and chemicals. This portion of the pharynx conducts air to the larynx and also conveys food to the esophagus.

Mucous cells in the epithelium and mucous glands in the lamina propria produce a sticky mucus that bathes the exposed surfaces. In the nasal cavity, cilia sweep that mucus and any trapped debris or microorganisms toward the pharynx, where it is swallowed and exposed to the acids and enzymes in the

Figure 24.2 Histology of the Respiratory Epithelium

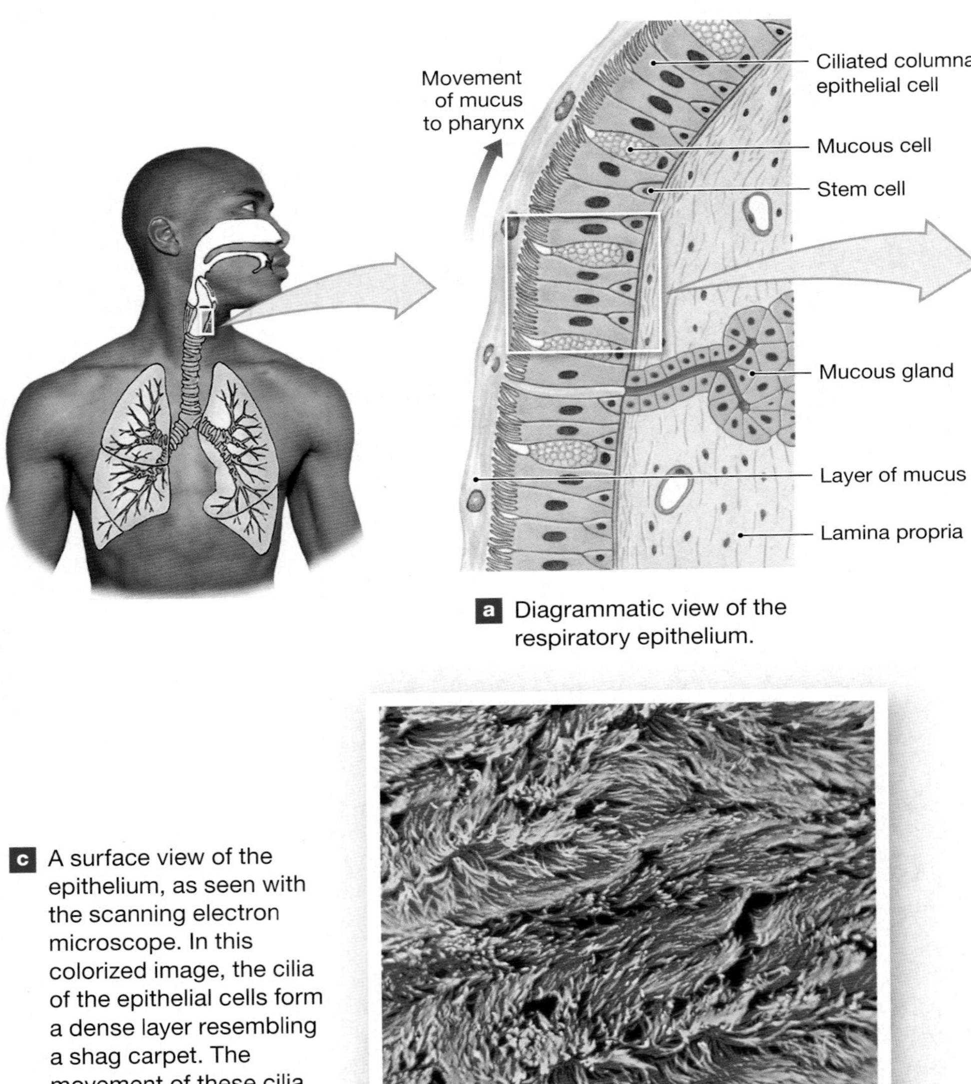

a Diagrammatic view of the respiratory epithelium.

c A surface view of the epithelium, as seen with the scanning electron microscope. In this colorized image, the cilia of the epithelial cells form a dense layer resembling a shag carpet. The movement of these cilia propels mucus across the epithelial surface.

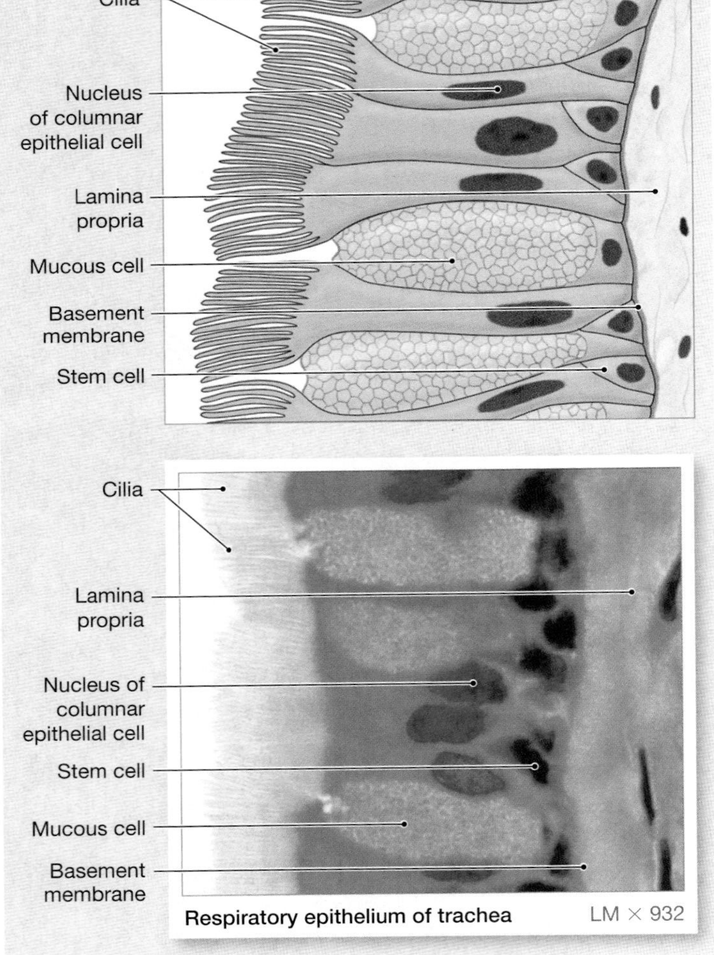

b Histology of respiratory epithelium.

CLINICAL NOTE

Cystic Fibrosis

Cystic Fibrosis (CF) is a hereditary disorder of the exocrine glands. A mutation in a chloride regulator gene on chromosome 7 causes the production of abnormally thick mucus. This thick mucus obstructs passageways of the pancreatic and bile ducts, intestines, and bronchial tree. Within the respiratory tract, breathing is impeded and the mucociliary escalator is impaired, leading to frequent infections. The average lifespan for people with CF who live into adulthood is 37 years.

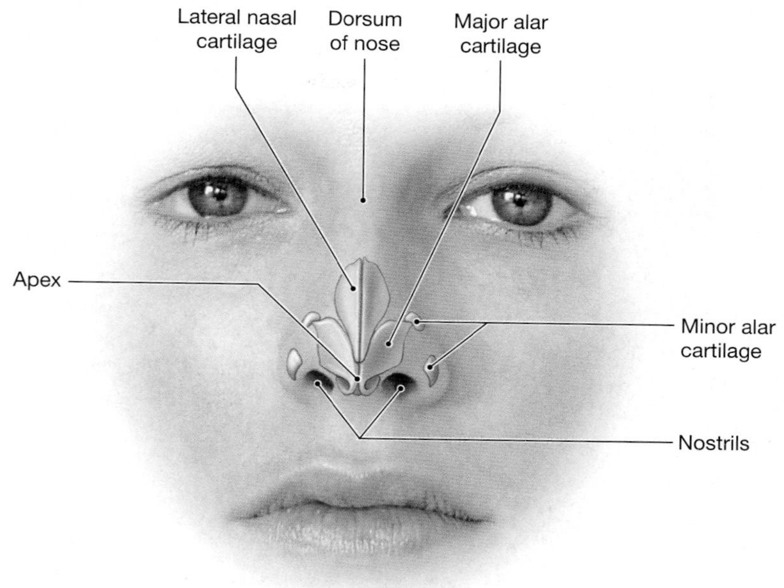

Figure 24.3 The Upper Respiratory System, Part I. The nasal cartilages and external landmarks on the nose.

stomach. In the lower respiratory system, the cilia beat toward the pharynx, moving a carpet of mucus in that direction and cleaning the respiratory surfaces. This mechanism is the mucociliary escalator.

Contamination of the inspired air by pathogens or debris will damage the delicate surfaces of the respiratory system. However, the air entering the respiratory system is filtered to remove contaminants. The respiratory filtration mechanisms form the **respiratory defense system**. In the nasal cavity, almost all particles larger than $10\ \mu m$ are removed from the inspired air. Stiff hairs within the nasal cavity, termed vibrissae, remove larger particles. The mucus of the nasopharynx or secretions of the pharynx trap smaller particles before they enter smaller passageways of the conducting portion. Unpleasant stimuli—such as noxious vapors, dust and debris, allergens, and pathogens—can rapidly increase the rate of mucus production. The familiar symptoms of the "common cold" result when any of more than 200 types of viruses invade the respiratory epithelium.

Filtration, warming, and humidification of inhaled air occur throughout the conducting portion of the respiratory system (that portion of the respiratory system that does not participate in gas exchange). However, the greatest changes occur within the nasal cavity. Breathing through the mouth eliminates much of the initial filtration, heating, and humidifying of the inspired air. Patients breathing on a respirator (mechanical ventilator), which uses a tube to conduct air directly into the trachea, must receive air that has been externally filtered and humidified to prevent alveolar damage.

24.1 CONCEPT CHECK

1 List the main functions of the respiratory system.
2 What is the mucociliary escalator?

See the blue Answers tab at the back of the book.

24.2 | The Upper Respiratory System

The Nose and Nasal Cavity

▶ **KEY POINT** The nose is the primary pathway for incoming air. The structures of the nose and nasal cavity are specialized for conditioning the air.

The **nose** is the primary passageway for air entering the respiratory system. (Chapter 6 introduced the bones, cartilages, and sinuses associated with the nose.) Air enters the nose through the paired **nostrils**, or **nares** (NA-rēz). These open into the **nasal vestibule** (VES-ti-būl) and the **nasal cavity** (**Figures 24.3** and **24.4a**). The nasal vestibule is the portion of the nasal cavity enclosed by the flexible tissues of the nose. It is supported by thin, paired lateral and alar cartilages (**Figure 24.3**). The epithelium of the nasal vestibule contains coarse hairs that extend across the external nares. Large airborne particles such as sand, sawdust, and even insects are trapped in these hairs and are prevented from entering the nasal cavity.

The **nasal septum** separates the right and left portions of the nasal cavity. The fusion of the perpendicular plate of the ethmoid and the vomer forms the bony portion of the nasal septum (**Figure 24.4b**). The anterior portion of the nasal septum is formed of hyaline cartilage. This cartilaginous plate supports the **dorsum** (*bridge*) **of the nose** and the **apex** (*tip*) **of the nose**.

The maxillae, nasal and frontal bones, ethmoid, and sphenoid form the lateral and superior walls of the nasal cavity. Mucus produced in the paranasal sinuses, aided by tears draining through the nasolacrimal ducts, keep the surface of the nasal cavity moist and clean. The superior portion, or olfactory region, of the nasal cavity consists of the areas lined by olfactory epithelium: (1) the inferior surface of the cribriform plate, (2) the superior nasal conchae of the ethmoid, and (3) the superior portion of the nasal septum.

The superior, middle, and inferior nasal conchae, or *turbinated bones*, project toward the nasal septum from the lateral walls of the nasal cavity. Air flows between adjacent conchae and through the **superior**, **middle**, or **inferior meatuses** (mē-Ā-tus-es; *meatus*, passage) as it passes from the nasal vestibule to the narrow **posterior nasal apertures** (or *choanae*; kō-AN-ē) (**Figure 24.4**). The incoming air bounces off the conchae and churns around like water flowing over rapids. This air turbulence brings small airborne particles into contact with the mucus lining the nasal cavity. The turbulence allows extra time for warming and humidifying incoming air and creating air currents that bring olfactory stimuli to the olfactory receptors.

A bony **hard palate**, formed by the maxillae and palatine bones, forms the floor of the nasal cavity and separates the oral and nasal cavities. A fleshy **soft palate** extends posterior to the hard palate. This marks the boundary between the superior nasopharynx and the rest of the pharynx (**Figures 24.4a** and **24.5**). The nasal cavity opens into the nasopharynx at the posterior nasal apertures.

Figure 24.4 The Upper Respiratory System, Part II

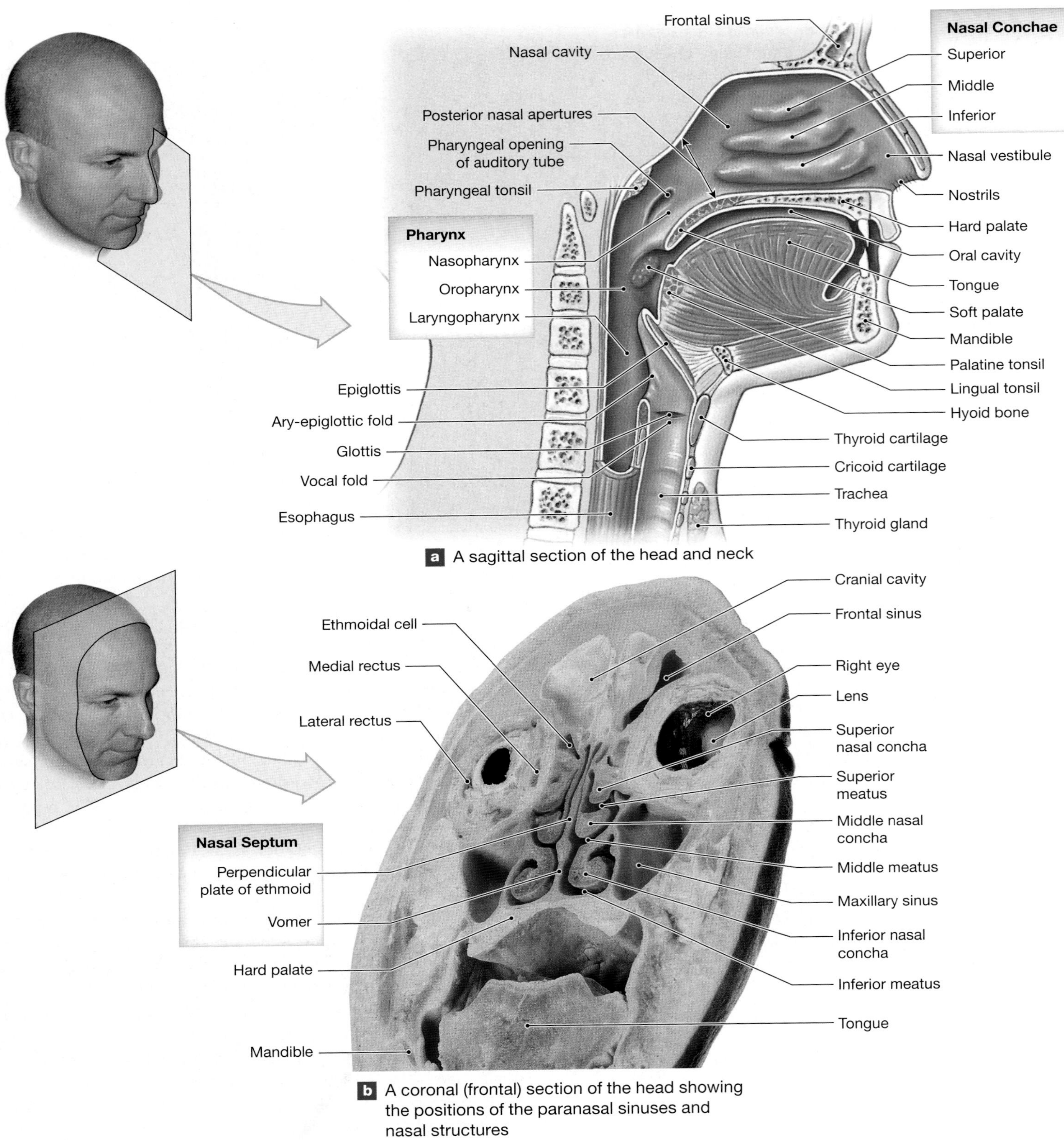

a A sagittal section of the head and neck

b A coronal (frontal) section of the head showing the positions of the paranasal sinuses and nasal structures

24

The Pharynx

▶ **KEY POINT** The pharynx, a common passageway shared by the digestive and respiratory systems, is subdivided into the nasopharynx, oropharynx, and laryngopharynx.

The **pharynx** (FAR-inks) connects the nose, mouth, and throat. The digestive and respiratory systems share the pharynx. It extends between the posterior nasal apertures and the entrances to the trachea and esophagus. The curving superior and posterior walls are attached to the axial skeleton, but the lateral walls are flexible and muscular. The pharynx is divided into three regions: nasopharynx, oropharynx, and laryngopharynx (**Figures 24.4a** and **24.5**).

The Nasopharynx

The **nasopharynx** (nā-zō-FAR-inks) is the superior portion of the pharynx. It is connected to the posterior portion of the nasal cavity by the posterior nasal apertures. The soft palate separates it from the oral cavity (**Figures 24.4a** and **24.5**).

A typical respiratory epithelium lines the nasopharynx. The pharyngeal tonsil is located on the posterior wall of the nasopharynx; the lateral walls contain the **pharyngeal opening** of the auditory tube (**Figure 24.4a**).

The Oropharynx

The **oropharynx** (ōr-ō-FAR-inks; *oris*, mouth) extends between the soft palate and the base of the tongue at the level of the hyoid bone. Like the posterior and inferior portions of the nasopharynx, the posterior portion of the oral cavity communicates directly with the oropharynx (**Figures 24.4a** and **24.5**). The epithelium changes from a pseudostratified ciliated columnar epithelium to a nonkeratinized (*mucosal type*) stratified squamous epithelium at the boundary between the nasopharynx and oropharynx.

The posterior margin of the soft palate supports the dangling **uvula** (Ū-vū-la) and two pairs of muscular pharyngeal arches, the **posterior arch** and the **anterior arch**.

The Laryngopharynx

The narrow **laryngopharynx** (la-RING-gō-far-inks) includes the region of the pharynx lying between the hyoid bone and the entrance to the esophagus (**Figures 24.4a** and **24.5**). Like the oropharynx, the laryngopharynx is lined with a stratified squamous epithelium that resists abrasion, chemicals, and pathogens.

24.2 CONCEPT CHECK

3 What are the three subdivisions of the pharynx?

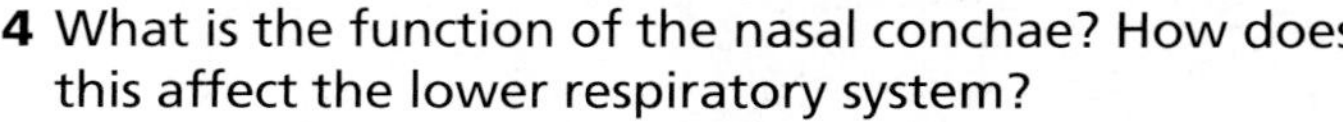

4 What is the function of the nasal conchae? How does this affect the lower respiratory system?

See the blue Answers tab at the back of the book.

Figure 24.5 The Upper Respiratory System, Part III. The nasal cavity and pharynx as seen in a sagittal section of the head and neck.

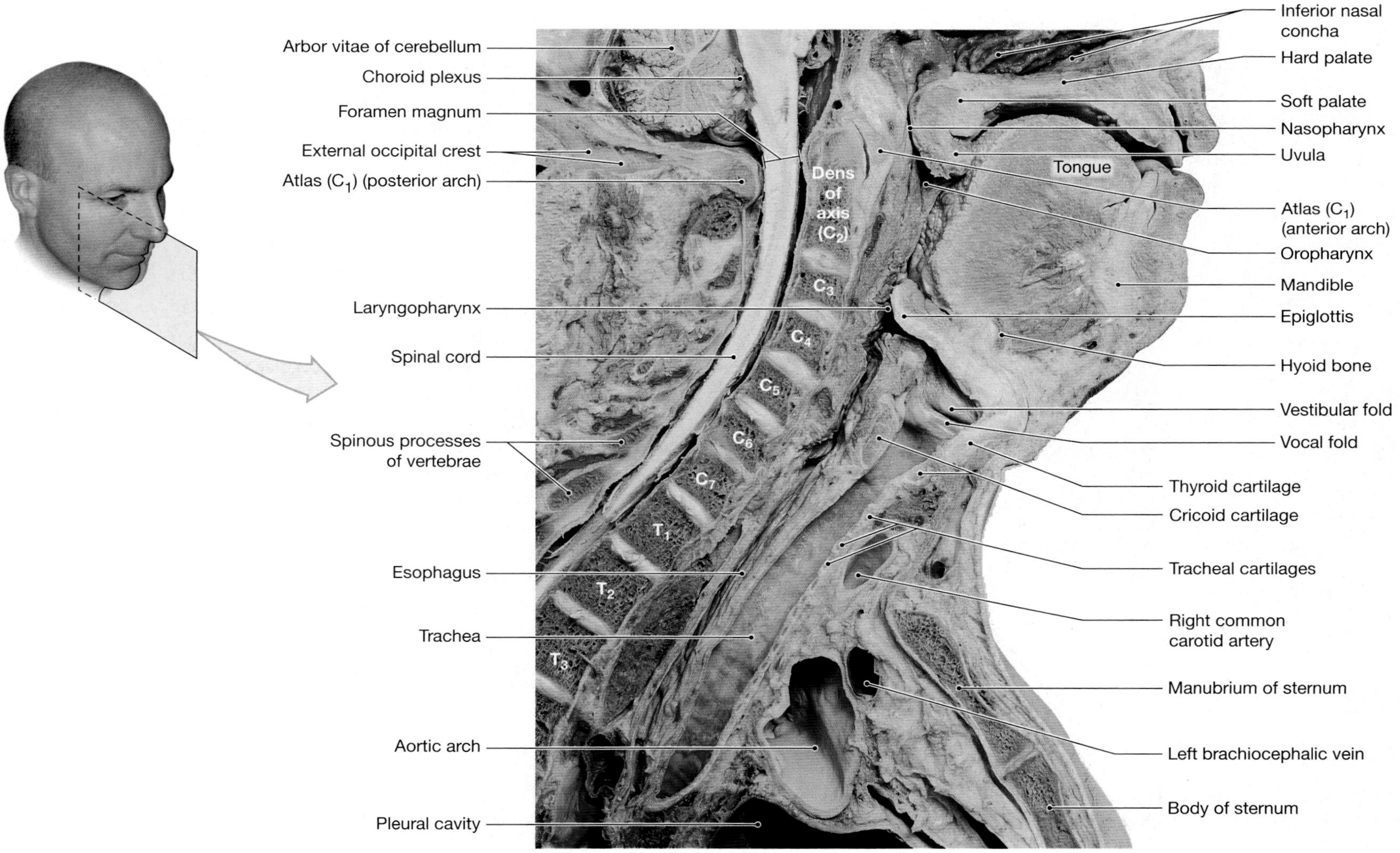

Figure 24.6 **Anatomy of the Larynx**

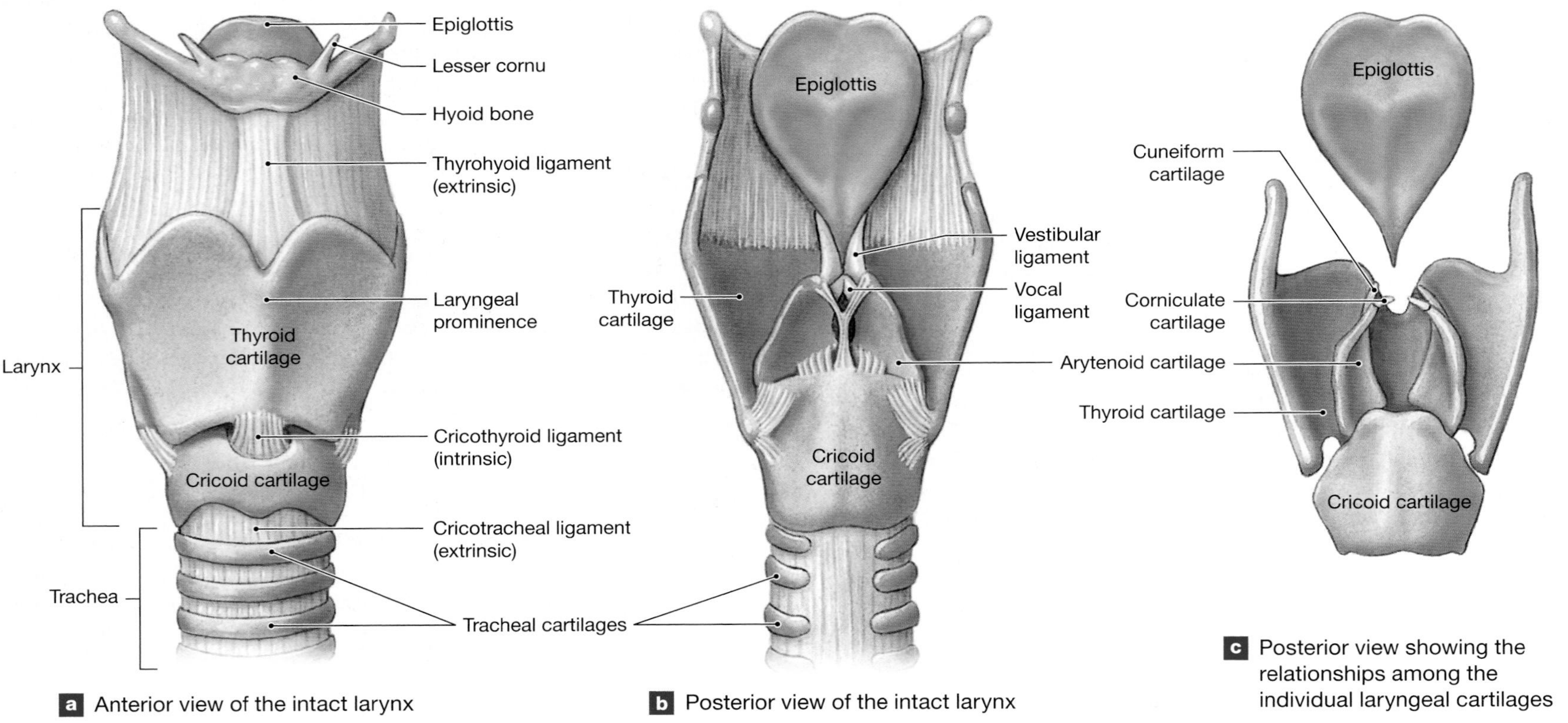

a Anterior view of the intact larynx

b Posterior view of the intact larynx

c Posterior view showing the relationships among the individual laryngeal cartilages

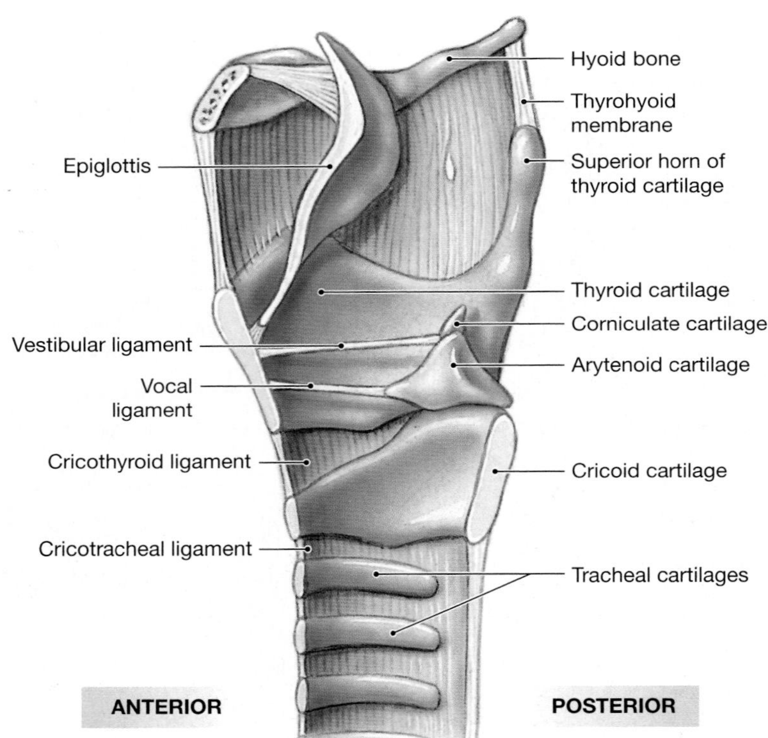

d Sagittal section of the intact larynx

24.3 | The Lower Respiratory System

The Larynx

▶ **KEY POINT** The larynx is supported by three unpaired pieces of cartilage and three paired pieces of cartilage. The larynx contains the true vocal folds and false vocal folds.

Inhaled air leaves the pharynx by passing through a slit-like opening, the **glottis** (GLOT-is) **(Figure 24.4a)**. The **larynx** (LAR-inks) begins between the level of vertebra C_3 to C_5 and ends at the level of vertebra C_6 or C_7. Ligaments and skeletal muscles stabilize the cartilaginous walls of the cylindrical larynx.

Cartilages of the Larynx

Three large unpaired cartilages form the body of the larynx: the thyroid cartilage, the cricoid cartilage, and the epiglottis **(Figure 24.6)**. The thyroid and cricoid cartilages are hyaline cartilages; the epiglottic cartilage is elastic cartilage.

The Thyroid Cartilage The largest laryngeal cartilage is the **thyroid** ("shield-shaped") **cartilage**. It forms most of the anterior and lateral walls of the larynx **(Figure 24.6a,b)**. The thyroid cartilage, when viewed in sagittal section, is incomplete posteriorly. The **laryngeal prominence**, or *Adam's apple*, is located on the anterior surface of this cartilage. This structure is easily seen and felt.

The inferior surface of the thyroid cartilage articulates with the cricoid cartilage. The superior surface of the thyroid cartilage has ligamentous attachments to the epiglottis and smaller laryngeal cartilages.

The Cricoid Cartilage The **cricoid cartilage** (KRĪ-koyd; "ring-shaped") forms a complete circle. Its posterior portion is greatly expanded and supports the posterior surface of the larynx. The cricoid and thyroid cartilages protect the glottis and the entrance to the trachea. The broad surfaces are sites for the attachment of important laryngeal muscles and ligaments. Ligaments attach the inferior surface of the cricoid cartilage to the first cartilage of the trachea **(Figure 24.6a,c)**. The superior surface of the cricoid cartilage articulates with the small paired arytenoid cartilages.

The Epiglottis The shoehorn-shaped **epiglottis** (ep-i-GLOT-is) projects superior to the glottis **(Figures 24.4a, 24.5,** and **24.6b,c,d)**. The epiglottic cartilage supporting it has ligamentous attachments to the inner and superior borders of the thyroid cartilage and the hyoid bone, respectively. During swallowing, the larynx is elevated, and the epiglottis folds back over the glottis. This prevents both liquids and solid food from entering the respiratory passageways.

Paired Laryngeal Cartilages The larynx also contains three pairs of smaller cartilages: the arytenoid, corniculate, and cuneiform cartilages. The arytenoids and corniculates are hyaline cartilages; the cuneiforms are elastic cartilages (**Figure 24.6b–d** and **24.7**).

- The paired **arytenoid** (ar-i-TĒ-noyd; "ladle-shaped") **cartilages** articulate with the posterior, superior border of the cricoid cartilage (**Figure 24.6b–d**).
- The small **corniculate** (kor-NIK-ū-lāt; "horn-shaped") **cartilages** articulate with the arytenoid cartilages (**Figures 24.6c,d** and **24.7**). The corniculate and arytenoid cartilages are important in opening and closing the glottis and producing sound.
- Elongated, curving **cuneiform** (kū-NĒ-i-form; "wedge-shaped") **cartilages** lie within the ary-epiglottic fold, tissue that extends between the lateral aspect of each arytenoid cartilage and the epiglottis (**Figures 24.6c** and **24.7**).

Laryngeal Ligaments

A series of **intrinsic ligaments** binds all nine cartilages together, forming the larynx (**Figure 24.6a,b,d**). **Extrinsic ligaments** attach the thyroid cartilage to the hyoid bone and the cricoid cartilage to the trachea. The **vestibular ligaments** and the **vocal ligaments** extend between the thyroid cartilage and the arytenoids.

Folds of laryngeal epithelium project into the glottis, covering the vestibular and vocal ligaments. The vestibular ligaments lie within the superior pair of folds, known as the **vestibular folds** (**Figures 24.5, 24.6b,** and **24.7**). The vestibular folds are relatively inelastic. They help prevent foreign objects from entering the glottis and provide protection for the more delicate **vocal folds**.

Because the vocal ligament is a band of elastic tissue, the vocal folds are also highly elastic. The vocal folds are involved with producing sounds, so they are known as the **true vocal cords**. Because the vestibular folds play no part in sound production, they are called the **false vocal cords**.

Sound Production Air passing through the glottis vibrates the vocal folds and produces sound waves. The pitch of the sound produced depends on the diameter, length, and tension of the vocal folds. The diameter and length are related to the size of the larynx. The tension is controlled by the contraction of skeletal muscles. This changes the positions of the thyroid and arytenoid cartilages. When the distance between the thyroid and arytenoid cartilages increases, the vocal folds tense and the pitch rises. When the distance decreases, the vocal folds relax and the pitch falls.

Children have slender, short vocal folds, so their voices are high-pitched. At puberty the larynx of a male enlarges more than that of a female. The true vocal cords of an adult male are thicker and longer, and they produce lower tones than those of an adult female.

The entire larynx is involved in sound production because its walls vibrate, creating a composite sound. Amplification and echoing of the sound occur within the pharynx, oral cavity, nasal cavity, and paranasal sinuses. The final production of distinct sounds depends on voluntary movements of the tongue, lips, and cheeks.

The Laryngeal Musculature

The larynx is associated with two different groups of muscles: the intrinsic laryngeal muscles and the extrinsic laryngeal muscles. The **intrinsic laryngeal muscles** are totally contained within the larynx and have two functions. One group regulates tension in the vocal folds. The second set opens and closes the glottis. Those regulating vocal fold tension insert upon the thyroid, arytenoid, and corniculate cartilages. Those opening or closing the glottis insert onto the arytenoid cartilages.

The **extrinsic laryngeal muscles** connect the larynx to neighboring structures and move it up and down during swallowing and speaking. (Chapter 10 considered these muscles.)

The extrinsic and intrinsic muscles cooperate to prevent food and liquids from entering the glottis during swallowing. Before you swallow, food is crushed and chewed, forming a pasty mass known as a bolus. Extrinsic muscles elevate the larynx, bending the epiglottis over the entrance to the glottis. This causes the bolus to glide across the epiglottis, rather than falling into the larynx (**Figure 24.8**). While this movement is under way, intrinsic muscles close the glottis. If particles or liquids touch the surfaces of the vestibular or vocal folds, the coughing reflex is triggered. Coughing prevents the material from entering the glottis.

Figure 24.7 The Vocal Cords

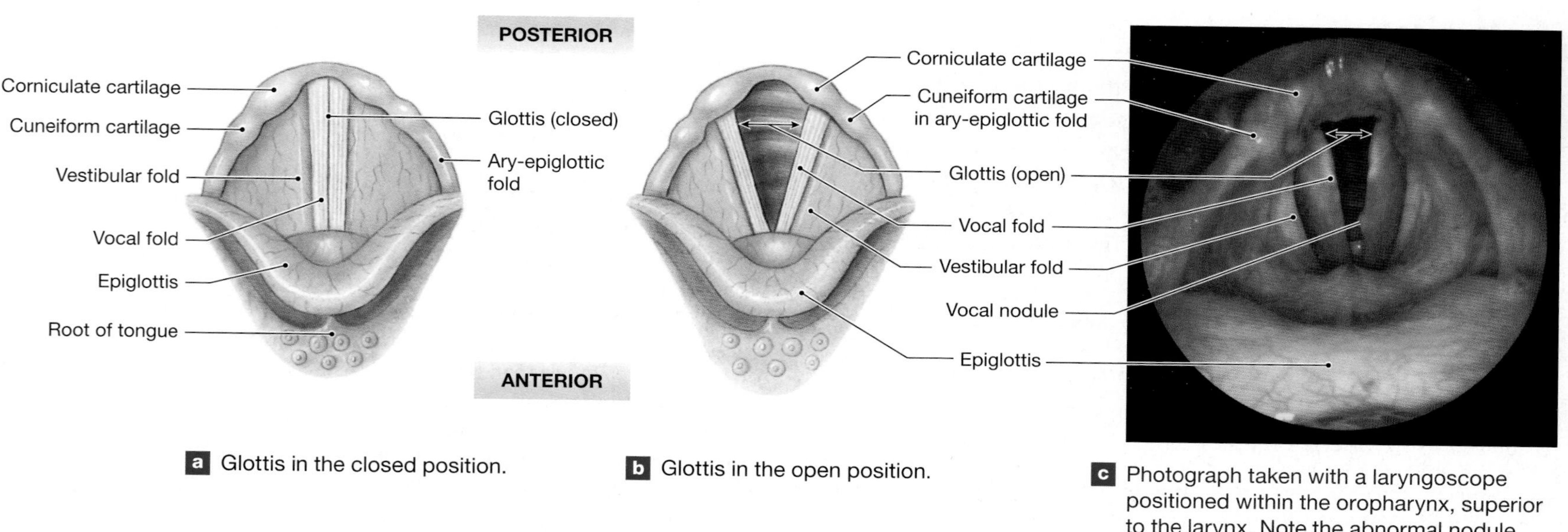

a Glottis in the closed position.

b Glottis in the open position.

c Photograph taken with a laryngoscope positioned within the oropharynx, superior to the larynx. Note the abnormal nodule.

Figure 24.8 The Swallowing Process. During swallowing, the elevation of the larynx folds the epiglottis over the glottis, steering the bolus into the esophagus.

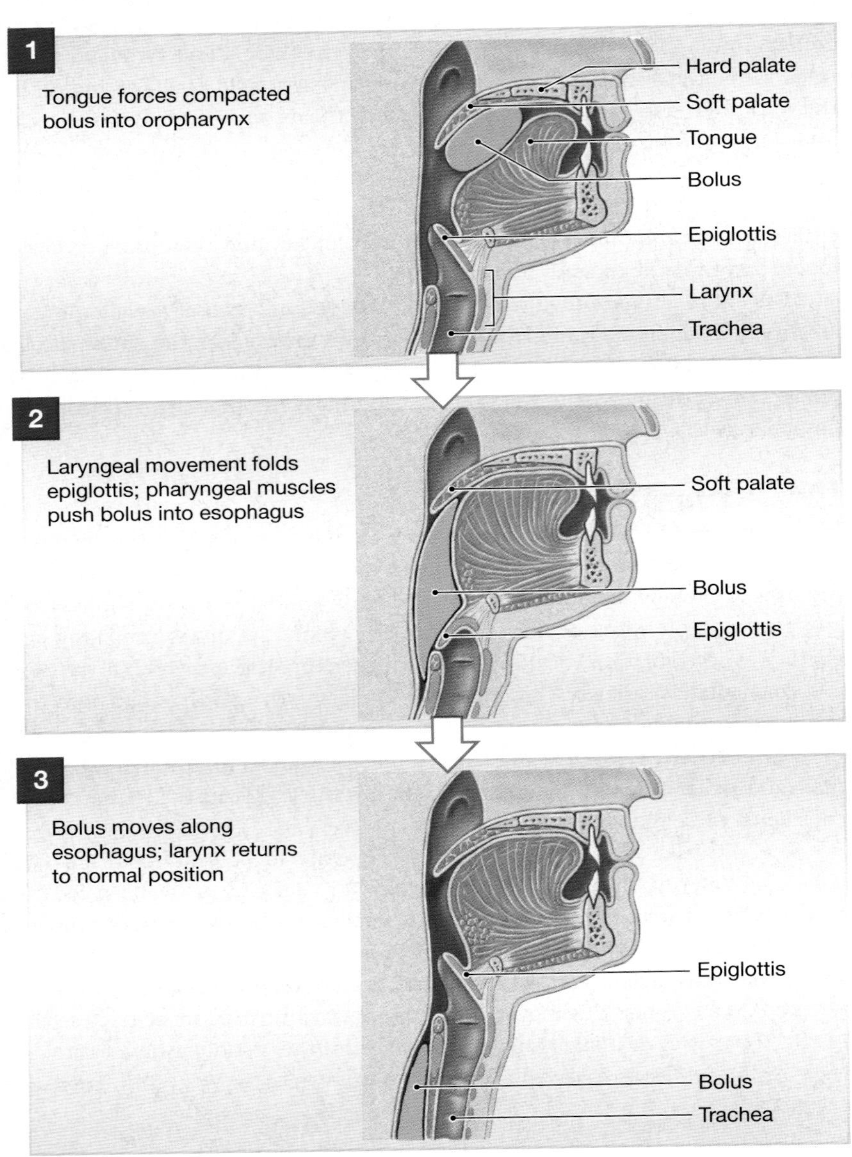

1
Tongue forces compacted bolus into oropharynx
Hard palate
Soft palate
Tongue
Bolus
Epiglottis
Larynx
Trachea
2
Laryngeal movement folds epiglottis; pharyngeal muscles push bolus into esophagus
Soft palate
Bolus
Epiglottis
3
Bolus moves along esophagus; larynx returns to normal position
Epiglottis
Bolus
Trachea

24

24.3 CONCEPT CHECK

5 What are the functions of the thyroid cartilage?

6 What is the function of the epiglottis?

7 Laurel uses voluntary muscle contraction to shorten the distance between her thyroid and arytenoid cartilages. What happens to the pitch of her voice?

8 How would the absence of intrinsic laryngeal muscles affect swallowing?

See the blue Answers tab at the back of the book.

24.4 The Trachea

▶ **KEY POINT** The trachea starts at the cricoid cartilage in the neck, enters the mediastinum, and ends at its division into the right and left bronchi. It sits anterior to the esophagus and is lined with a typical respiratory epithelium.

The **trachea** (TRĀ-kē-a), or "windpipe," is a tough, flexible tube with a diameter of about 2.5 cm (1 in.) and a length of approximately 11 cm (4.25 in.) **(Figure 24.9)**. The trachea begins anterior to vertebra C_6, attached to the cricoid cartilage by connective tissue. It ends within the mediastinum at T_5, where it branches to form the right and left main bronchi.

The lining of the trachea is a pseudostratified, ciliated columnar epithelium with mucous cells. This respiratory epithelium sits on a layer of loose connective tissue called the **lamina propria** (LA-mi-na PRŌ-prē-a) **(Figure 24.2a, p. 626)**. The lamina propria separates the respiratory epithelium from the deeper hyaline cartilages. The combination of epithelium and lamina propria is an example of a **mucosa** (mū-KŌ-sa), or **mucous membrane.** ⮌ p. 71

A layer of connective tissue, the **submucosa** (sub-mū-KŌ-sa), is deep to the lamina propria. The submucosa contains tracheal glands that secrete mucus onto the epithelial surface through secretory ducts.

Deep to the submucosa, the trachea contains 15–20 **tracheal cartilages** **(Figure 24.9)**. Elastic **annular ligaments** attach each tracheal cartilage to its adjacent cartilages. The tracheal cartilages stiffen the tracheal walls and protect the airway. They also prevent its collapse or overexpansion as pressures change in the respiratory system.

Each tracheal cartilage is a C-shaped piece of hyaline cartilage. The closed portion of the C protects the anterior and lateral surfaces of the trachea. The open portions face posteriorly, toward the esophagus **(Figure 24.9b)**. Because the cartilages are not complete circles, the posterior tracheal wall changes shape easily during swallowing. This permits the passage of large masses of food along the esophagus.

An elastic ligament and a band of smooth muscle, the **trachealis**, connect the open ends of each tracheal cartilage **(Figure 24.9b)**. Contraction of the trachealis decreases the diameter of the tracheal lumen. This increases the resistance to airflow. Sympathetic activation relaxes the trachealis, increasing the diameter of the trachea, lowering resistance to airflow and making it easier to move large volumes of air through the trachea.

24.4 CONCEPT CHECK

9 The cartilages reinforcing the trachea are C-shaped rather than complete rings. How does this shape help with swallowing while still protecting the trachea?

10 What type of epithelium can be observed in the trachea?

11 How are tracheal cartilages involved in airflow?

See the blue Answers tab at the back of the book.

24.5 The Main Bronchi

▶ **KEY POINT** The trachea divides into the right and left main bronchi.

At T_5, the trachea branches within the mediastinum, giving rise to the **right** and **left main bronchi** (BRONG-kī; singular, *bronchus*). An internal ridge, the **carina** (ka-RĪ-na) **of trachea**, lies between the entrances to the two main bronchi **(Figure 24.9a)**. The histology of the main bronchi is the same as the trachea. The right main bronchus supplies the right lung, and the left supplies the left lung. The right main bronchus is shorter, has a larger diameter, and descends toward the lung at a steeper angle than the left. For these reasons, most foreign objects that enter the trachea find their way into the right bronchus rather than the left.

Each main bronchus travels to and through the **hilum** (HĪ-lum) of the lung before branching further **(Figure 24.10)**. The hilum is a groove on the mediastinal surface of the lung and, in addition to the entrance of the main bronchi, is the site for the entry and exit of the pulmonary vessels and nerves. A meshwork of

Figure 24.9 Anatomy of the Trachea and Main Bronchi

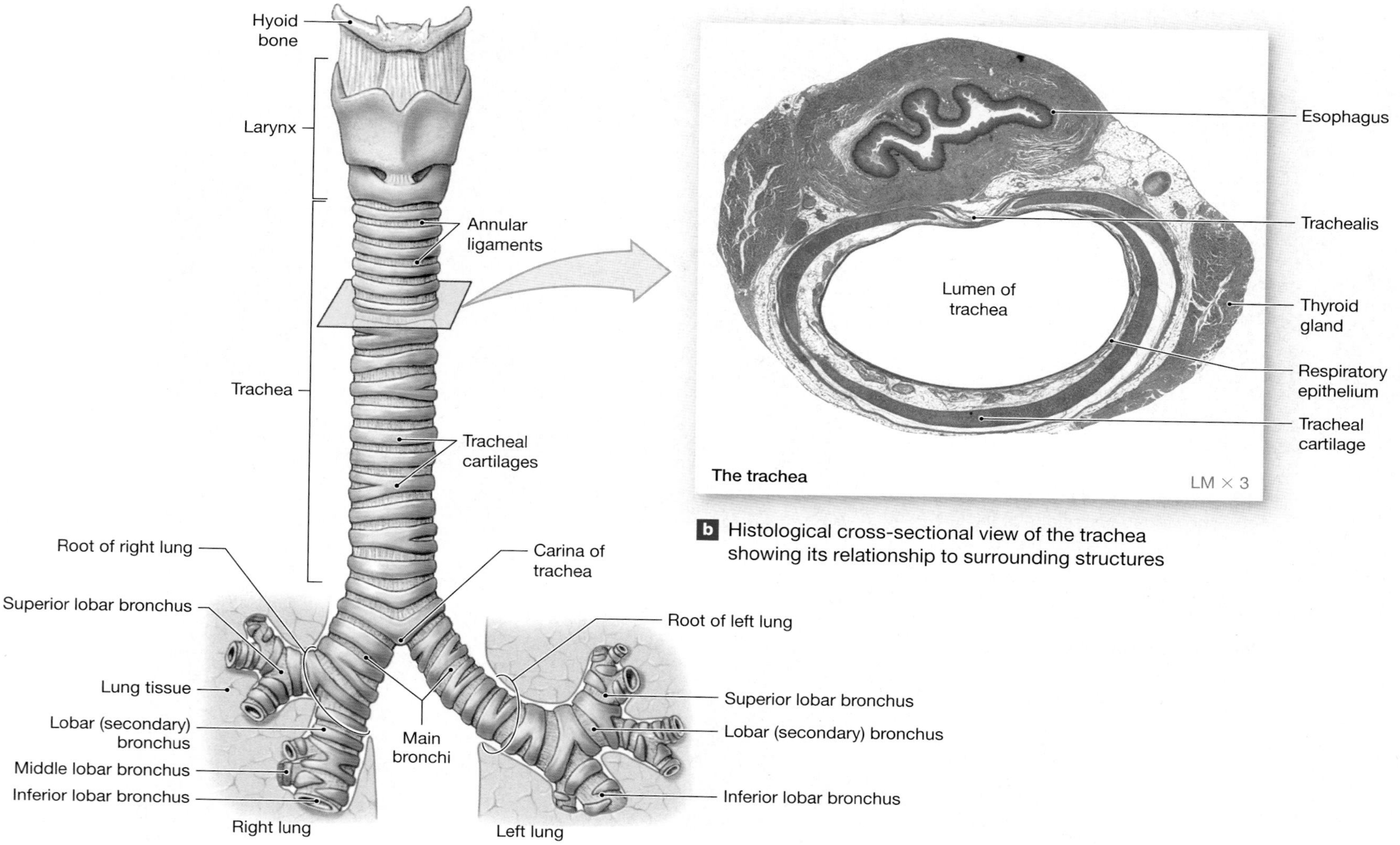

a Anterior view showing the plane of section for part (b)

b Histological cross-sectional view of the trachea showing its relationship to surrounding structures

dense connective tissue anchors all of these structures to the lung. This meshwork is called the **root** of the lung **(Figure 24.9a)**. The root of the lung is attached to the mediastinum and holds the major nerves, vessels, and lymphatics in place. The roots of the lungs are located anterior to vertebrae T_5 (right) and T_6 (left).

24.5  **CONCEPT CHECK**

12 How can you distinguish the right main bronchus from the left main bronchus?

See the blue Answers tab at the back of the book.

24.6 | The Lungs

▶ KEY POINT Each lung is cone-shaped. Its apex extends superior to the attachment of the first rib to the sternum, and its base sits on the diaphragm.

The left and right lungs are situated in the left and right pleural cavities **(Figure 24.10)**. Each lung is a blunt cone with the tip, or **apex**, pointing superiorly. The apex on each side extends into the base of the neck, superior to the first rib. The broad concave inferior portion, or **base**, of each lung rests on the superior surface of the diaphragm.

Lobes and Fissures of the Lungs

▶ KEY POINT The right lung is divided into three lobes, and the left is divided into two. Each lobe is separated by a fissure.

The **lobes** of the lung are separated by deep fissures. The **right lung** has three lobes: **superior**, **middle**, and **inferior**. The **horizontal fissure** separates the superior and middle lobes. The **oblique fissure** separates the superior and inferior lobes. The **left lung** has only two lobes, **superior** and **inferior**, separated by the **oblique fissure** **(Figure 24.10)**. The right lung is shorter, broader, and heavier than the left. The right lung is shorter because the right half of the diaphragm rises higher than the left half to accommodate the liver within the abdominal cavity. The difference in width and weight is due to the heart and great vessels projecting into the left pleural cavity.

Lung Surfaces

▶ KEY POINT The surfaces of the lungs are named for the structures that come into contact with the lungs. Connective tissue from the root of the lung travels inward, forming trabeculae and septa that partition the lobes into smaller compartments.

The curving anterior **costal surface** of the lungs contacts the inner contours of the rib cage **(Figure 24.10a,b)**. The **mediastinal surface** contains the hilum of the lung **(Figure 24.10c)**. The mediastinal surface of both lungs have

Figure 24.10 Gross Anatomy of the Lungs

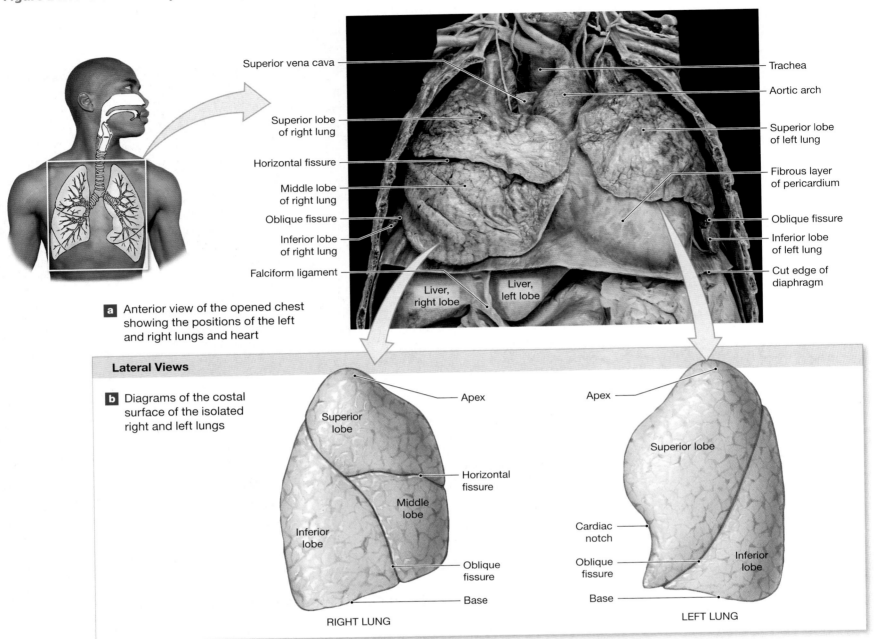

Superior vena cava
Trachea
Aortic arch
Superior lobe of right lung
Superior lobe of left lung
Horizontal fissure
Fibrous layer of pericardium
Middle lobe of right lung
Oblique fissure
Oblique fissure
Inferior lobe of right lung
Inferior lobe of left lung
Falciform ligament
Cut edge of diaphragm
Liver, right lobe
Liver, left lobe

a Anterior view of the opened chest showing the positions of the left and right lungs and heart

Lateral Views

b Diagrams of the costal surface of the isolated right and left lungs

Apex
Superior lobe
Horizontal fissure
Middle lobe
Inferior lobe
Oblique fissure
Base
RIGHT LUNG

Apex
Superior lobe
Cardiac notch
Oblique fissure
Inferior lobe
Base
LEFT LUNG

Medial Views

c Diagrams of the mediastinal surface of the isolated right and left lungs

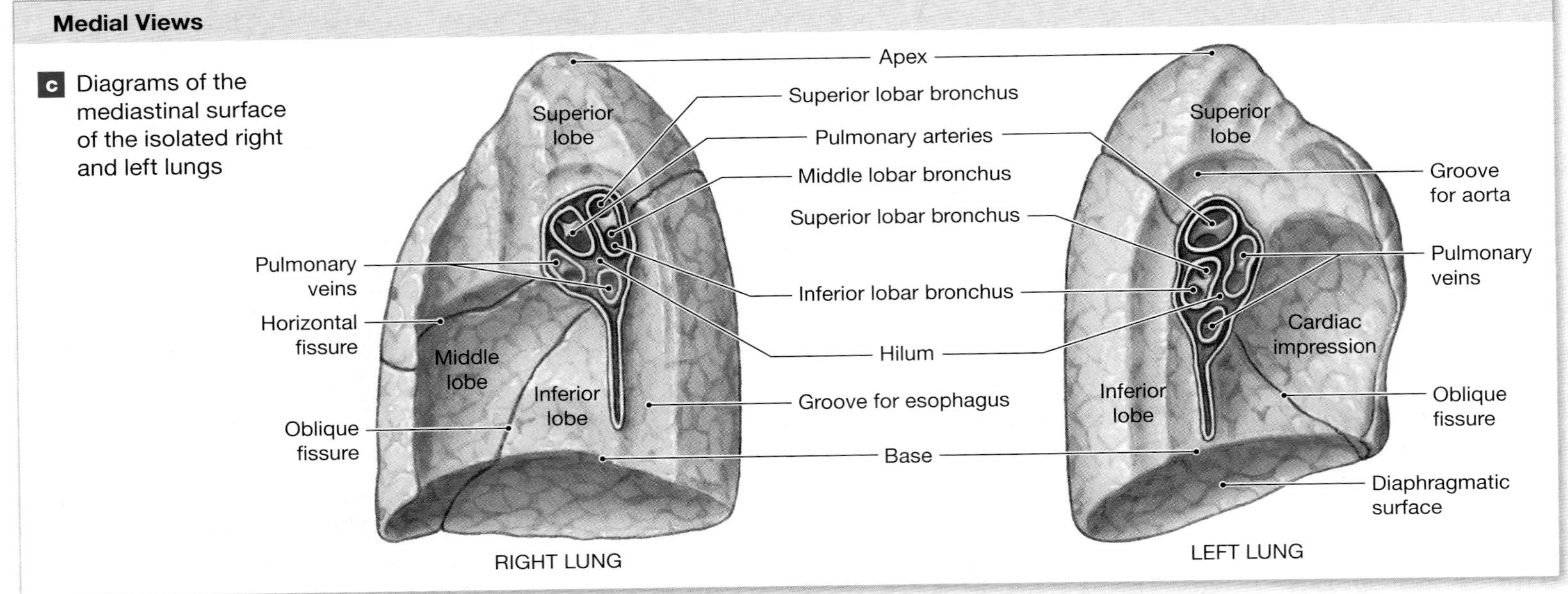

grooves marking the positions of the great vessels and the heart. The left lung has a larger cardiac impression because most of the heart is located to the left of midline. The medial margin of the right lung forms a vertical line when viewed anteriorly. The left, however, has an indentation, the **cardiac notch**, when viewed anteriorly.

The connective tissues of the root of each lung extend inward, forming fibrous partitions, or **trabeculae**. These trabeculae, which branch repeatedly, contain elastic fibers, smooth muscle, and lymphatic vessels. The branching of the trabeculae divides the lobes into smaller and smaller compartments. The branches of the conducting passageways, pulmonary vessels, and nerves of the lungs follow these trabeculae to their peripheral destinations.

The terminal connective tissue partitions are called **interlobular septa**. These septa divide the lung into **pulmonary lobules** (LOB-ūlz). Each lobule is supplied by branches of the pulmonary arteries, pulmonary veins, and respiratory passageways. The connective tissues of the septa are continuous with those of the visceral pleura.

The Main Bronchi

▶ **KEY POINT** After each main bronchus enters the lung, it divides into smaller and smaller lobar and segmental bronchi.

The main bronchi and their branches form the bronchial tree. As the main bronchi enter the lungs, they divide to form smaller passageways (**Figures 24.9, 24.11,** and **24.12**).

After entering the lung, each main bronchus divides, forming **lobar bronchi**, also known as *secondary bronchi*. Lobar bronchi then divide, forming **segmental,** or *tertiary,* **bronchi**. The branching pattern differs in the right and left lungs. Each segmental bronchus supplies air to a single **bronchopulmonary segment** (**Figure 24.12a,b**). There are 10 segmental bronchi and 10 bronchopulmonary segments in the right lung. In the developing fetus, the left lung also starts out with 10 bronchopulmonary segments. However, one or more of these segments fuse, resulting in eight or nine bronchopulmonary segments in the adult.

The walls of the main, lobar, and segmental bronchi have a typical respiratory epithelium but contain less and less cartilage with each division (**Figures 24.11** and **24.13**). The walls of lobar and segmental bronchi contain cartilage plates arranged around the lumen. These cartilages serve the same purpose as the rings of cartilage in the trachea and main bronchi.

Branches of the Right Main Bronchus

The right lung has three lobes. The right main bronchus divides into three lobar bronchi: a **superior lobar bronchus**, a **middle lobar bronchus**, and an **inferior lobar bronchus** (**Figure 24.9a**). Each lobar bronchus delivers air to one of the lobes of the right lung (**Figure 24.12**).

Branches of the Left Main Bronchus

The left lung has two lobes. The left main bronchus divides into two branches: a **superior lobar bronchus** and an **inferior lobar bronchus** (**Figures 24.9a, 24.11,** and **24.12**).

Branches of the Lobar Bronchi

The lobar bronchi in each lung divide, forming segmental bronchi. In the right lung there are 10 segmental bronchi. Three supply the superior lobe, two supply the middle lobe, and five supply the inferior lobe. The left lung usually has nine segmental bronchi. Four supply the superior lobe and five supply the inferior lobe (**Figure 24.12a,d**). The segmental bronchi deliver air to the bronchopulmonary segments of the lungs.

Figure 24.11 Bronchi and Bronchioles. For clarity, the degree of branching has been reduced; an airway branches approximately 23 times before reaching the level of a lobule.

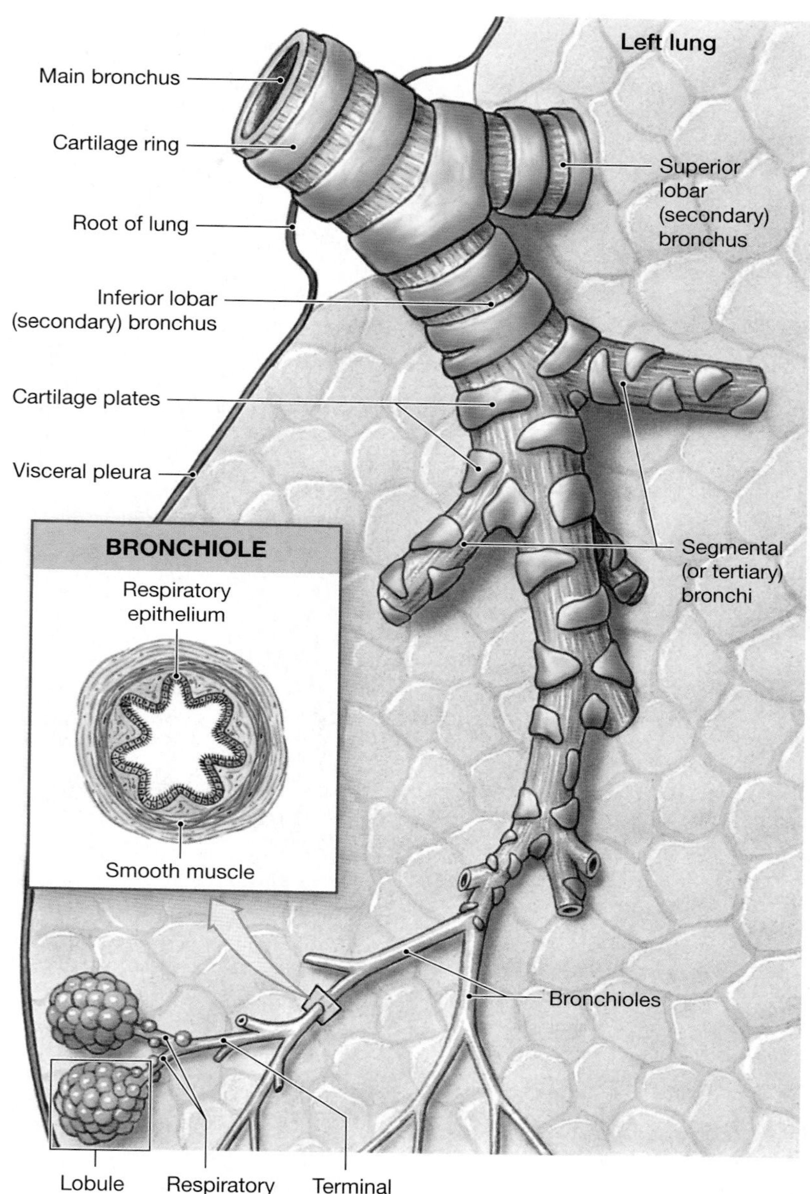

The Bronchopulmonary Segments

Each lobe of the lung can be divided into smaller units called **bronchopulmonary segments**. Each segment consists of the lung tissue associated with a single segmental bronchus. The bronchopulmonary segments have names that correspond to those of the associated segmental bronchi (**Figure 24.12a,b,d**).

The Bronchioles

▶ **KEY POINT** Each segmental bronchus branches several times, ultimately ending in terminal bronchioles. The walls of terminal bronchioles lack cartilage, and their epithelium is not the typical respiratory epithelium.

Each segmental bronchus branches several times within the bronchopulmonary segment, forming many **bronchioles**. These bronchioles then branch into the smallest-diameter conducting branches, called **terminal bronchioles**. Nearly 6500 terminal bronchioles arise from each segmental bronchus. The walls of the

Figure 24.12 The Bronchial Tree and Divisions of the Lungs

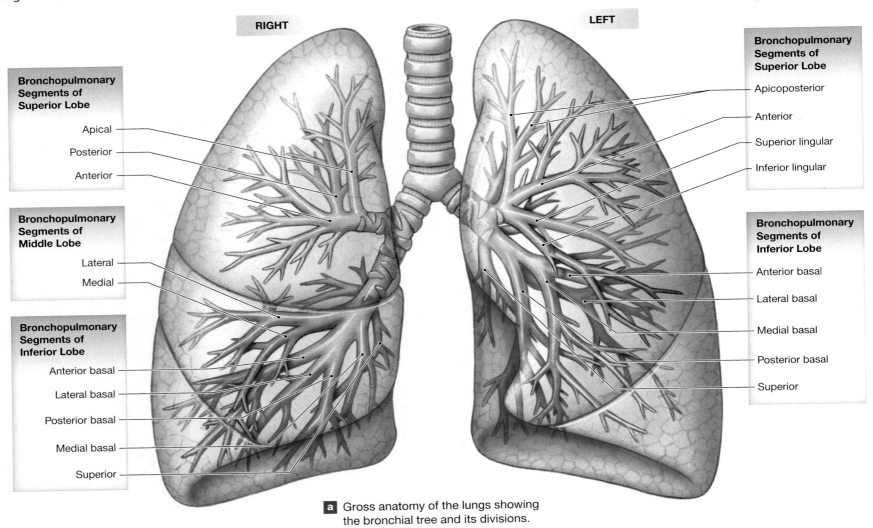

RIGHT

LEFT

Bronchopulmonary Segments of Superior Lobe

Apical
Posterior
Anterior

Bronchopulmonary Segments of Middle Lobe

Lateral
Medial

Bronchopulmonary Segments of Inferior Lobe

Anterior basal
Lateral basal
Posterior basal
Medial basal
Superior

Bronchopulmonary Segments of Superior Lobe

Apicoposterior
Anterior
Superior lingular
Inferior lingular

Bronchopulmonary Segments of Inferior Lobe

Anterior basal
Lateral basal
Medial basal
Posterior basal
Superior

a Gross anatomy of the lungs showing the bronchial tree and its divisions.

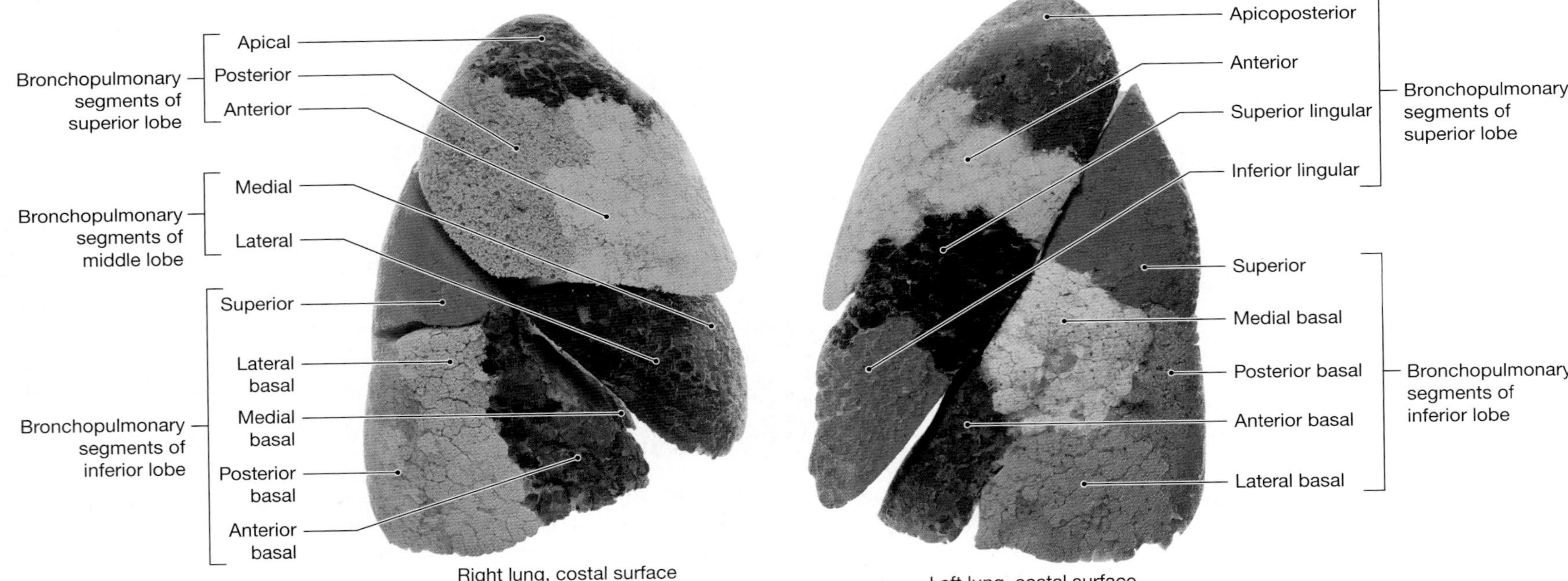

b Isolated left and right lungs have been colored to show the distribution of the bronchopulmonary segments.

Figure 24.12 **(*continued*)**

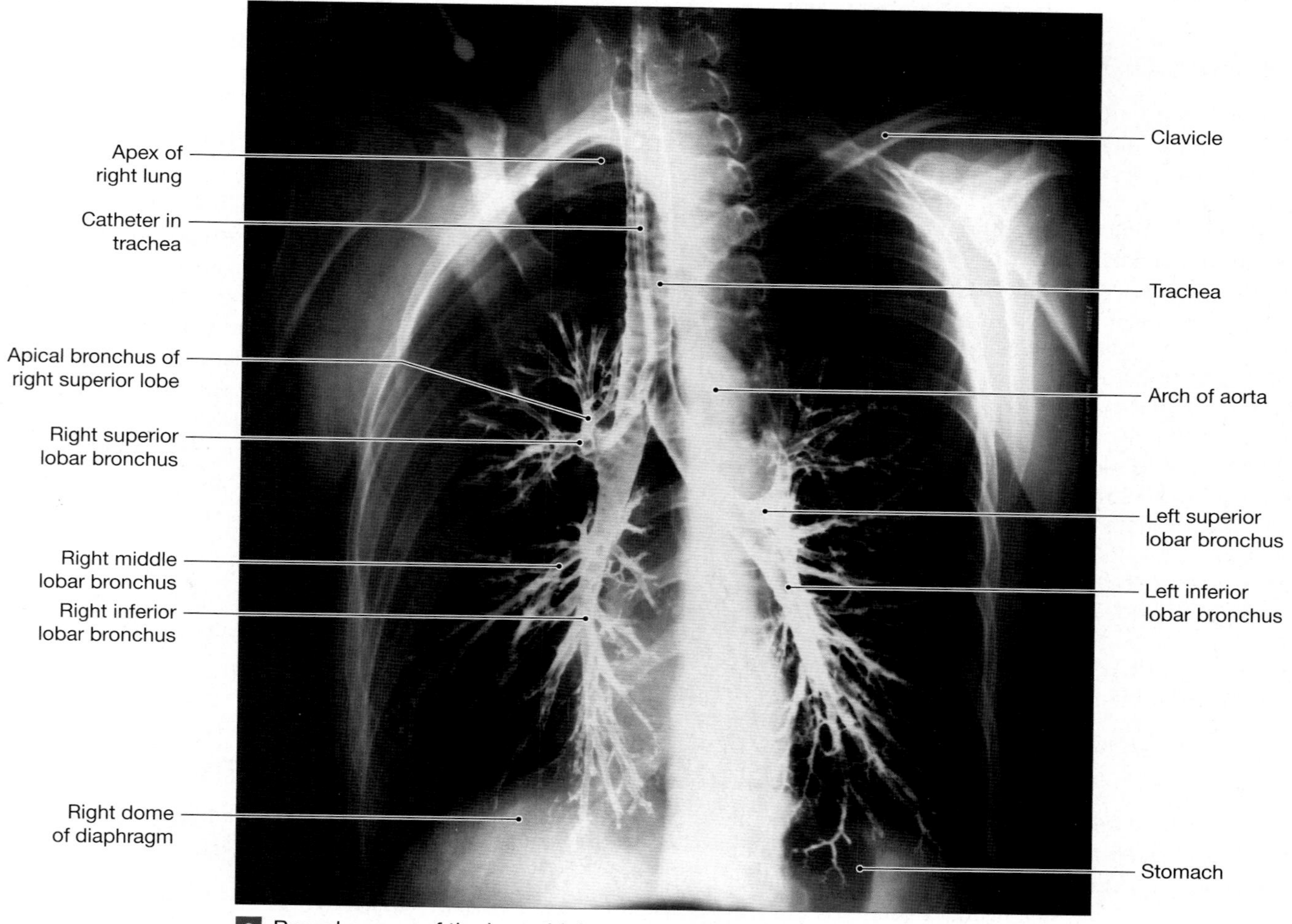

c Bronchogram of the bronchial tree, anteroposterior view.

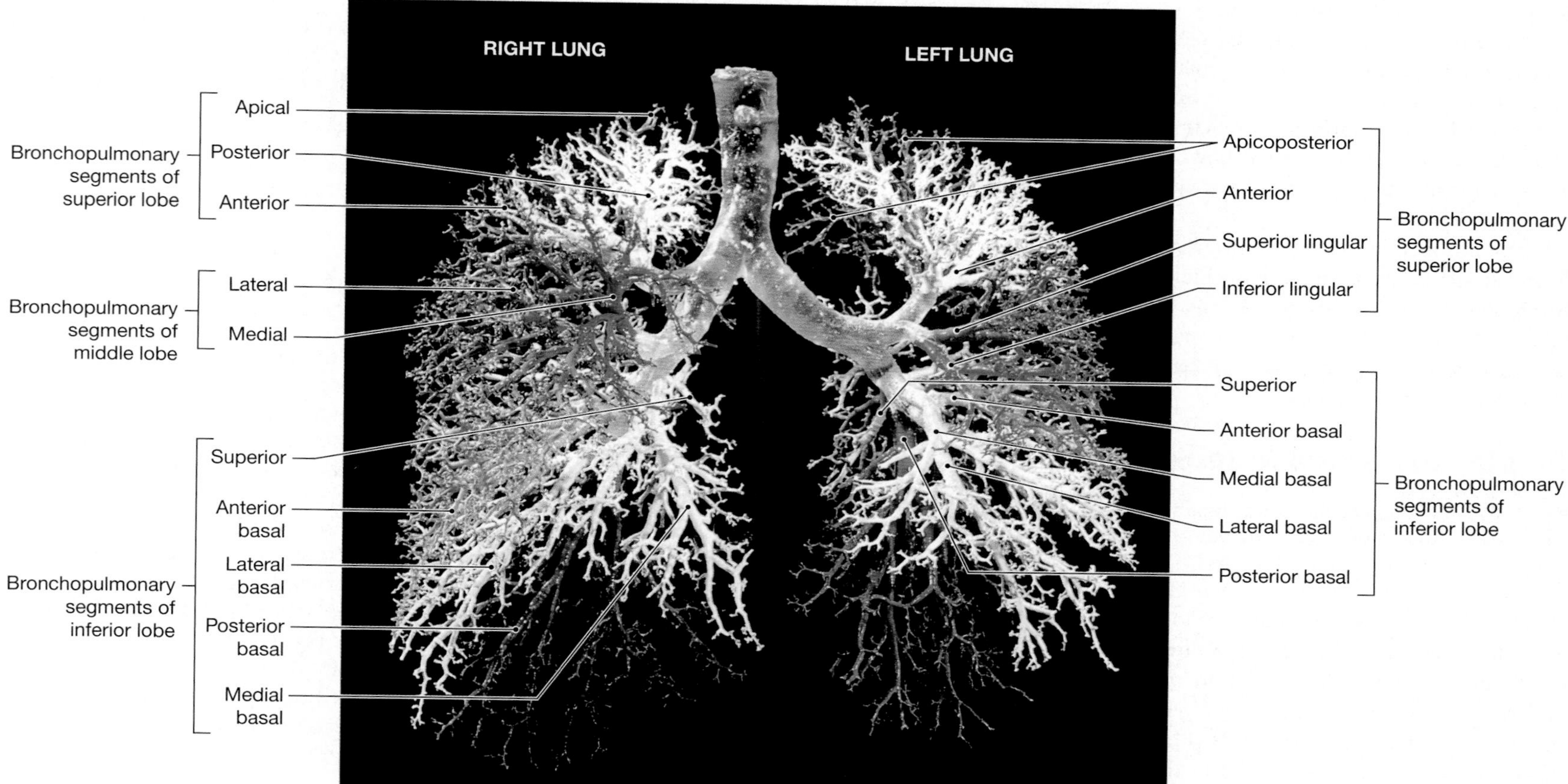

d Plastic cast of the adult bronchial tree. All of the branches in a given bronchopulmonary segment have been painted the same color.

24

CLINICAL NOTE

Tracheal Blockage

Foreign objects, usually food, that become lodged in the trachea or larynx are usually expelled by coughing. If the individual can speak, the airway is still open, and no emergency measures are needed beyond careful observation. If the person cannot breathe or speak but is still conscious, this is a life-threatening emergency requiring immediate action.

The **heimlich** (HĪM-lik) **maneuver** is an action designed to expel an obstruction from the trachea by applying compression to the abdomen just inferior to the diaphragm. This is done by grasping the fist from behind with the other hand, and forcefully thrusting inward and upward to force the diaphragm upward, forcing air into the trachea to dislodge the object.

If the blockage remains, professionally qualified rescuers may perform a **tracheostomy** (trā-kē-OS-tō-mē). This procedure involves making an incision through the anterior tracheal wall and inserting a tube inferior to the larynx.

A tracheostomy may also be required if (1) the larynx becomes blocked by inflammation, sustained laryngeal spasms, severe allergic reactions, burns, or smoke inhalation; (2) a portion of the trachea has been crushed; (3) the epiglottis is severely inflamed; or (4) part of the trachea has been removed during treatment of laryngeal cancer.

terminal bronchioles are continuous and lack cartilage and mucous cells. The proximal segments are lined by a pseudostratified, ciliated columnar epithelium. The more distal segments are lined with a simple columnar ciliated epithelium.

Terminal bronchiole walls contain a significant amount of smooth muscle (**Figures 24.11a** and **24.13a,b**). The autonomic nervous system regulates the activity of the smooth muscle layer in the terminal bronchioles. Sympathetic activation and the release of epinephrine by the adrenal medullae cause **bronchodilation**, which increases the luminal diameter of the airways, reducing resistance to airflow. Parasympathetic stimulation causes **bronchoconstriction**, a reduction in the luminal diameter of the airways, increasing resistance to airflow. Excessive parasympathetic stimulation, as in asthma, can almost completely prevent airflow along the terminal bronchioles.

Each terminal bronchiole delivers air to a single pulmonary lobule. Within the lobule, the terminal bronchiole branches into several **respiratory bronchioles**. These are the thinnest and most delicate branches of the bronchial tree. Their walls are incomplete and lined by a simple cuboidal epithelium that lacks cilia (**Figure 24.13b**). Terminal bronchioles deliver air to the exchange surfaces of the lungs.

24

Alveolar Ducts and Alveoli

▶ **KEY POINT** Alveolar ducts are respiratory pathways with incomplete walls and alveoli at their most distal segment. Gas exchange occurs across the respiratory membrane.

Respiratory bronchioles are connected to individual alveoli and to multiple alveoli along regions called **alveolar ducts**. Alveolar ducts end at **alveolar sacs**, common chambers connected to multiple **alveoli** (singular, *alveolus*). Each lung contains about 150 million alveoli, giving the lungs an open, spongy appearance. Alveolar ducts are lined with simple squamous epithelium. Isolated smooth muscle cells are found in the interalveolar septa of these alveolar ducts, and an extensive network of capillaries surrounds each alveolus (**Figures 24.13** and **24.14**). The capillaries are surrounded by a network of elastic fibers. This elastic tissue maintains the positions of the alveoli, alveolar ducts, and respiratory bronchioles. During inspiration these elastic fibers stretch. During expiration they recoil, reducing the size of the alveoli and assisting the process of expiration.

The Alveolus and the Blood Air Barrier

The alveolar cell layer consists mainly of simple squamous epithelium (**Figure 24.14a,b**). The squamous epithelial cells, called **type I alveolar cells,** are unusually thin and are the sites of gas diffusion. Roaming **alveolar macrophages** patrol the epithelial surface. They engulf any particles that have eluded other defenses. Large **type II alveolar cells** are scattered among the squamous cells. Type II alveolar cells produce **surfactant** (sur-FAK-tant), an oily secretion containing phospholipids and proteins. Surfactant plays a key role in keeping the alveoli open. Surfactant interacts with water molecules, reducing surface tension and preventing the collapse of the alveoli.

Gas exchange occurs across the three-layered **blood air barrier** of the alveoli. This barrier is made up of (1) the alveolar cell layer, (2) the capillary endothelial layer, and (3) the fused basement membrane between them (**Figure 24.14c**).

Gas exchange can take place quickly and efficiently at the blood air barrier, as only a very short distance separates alveolar air from blood. Diffusion proceeds very rapidly because the distance is short and both oxygen and carbon dioxide are small, lipid-soluble molecules. In addition, the surface area of the blood air barrier is very large.

The Blood Supply to the Lungs

▶ **KEY POINT** The lungs receive blood from two sources: the pulmonary circuit and the bronchial arteries of the systemic circuit.

The respiratory exchange surfaces receive blood from arteries of the pulmonary circuit. The pulmonary arteries enter the lungs at the hilum and branch with the bronchi as they approach the lobules. Each lobule receives an arteriole, and a venule and a network of capillaries surround each alveolus as part of the blood air barrier. Blood from the alveolar capillaries passes through the pulmonary venules and then enters the pulmonary veins, which deliver the blood to the left atrium.

CLINICAL NOTE

Lung Cancer

Lung cancer is an aggressive class of malignancies; more people die of it than any other type of cancer. Lung cancer affects the epithelial cells lining conducting passageways, mucous glands, or alveoli. Signs and symptoms generally do not appear until tumors restrict airflow or compress adjacent structures. Chest pain, shortness of breath, a cough or a wheeze, and weight loss are common. Evidence has shown that 85–90 percent of all lung cancers are the direct result of cigarette smoking. Treatment varies with the cellular organization of the tumor and whether **metastasis** (cancer cell migration) has occurred. If detected when the disease is still localized (within the lungs), the five-year survival rate is 54 percent. Sadly, only 15 percent of lung cancers are diagnosed at this early stage. If the disease has already metastasized, the five-year survival rate is less than 5 percent. Surgery, radiation, or chemotherapy may be involved.

Figure 24.13 Bronchi and Bronchioles

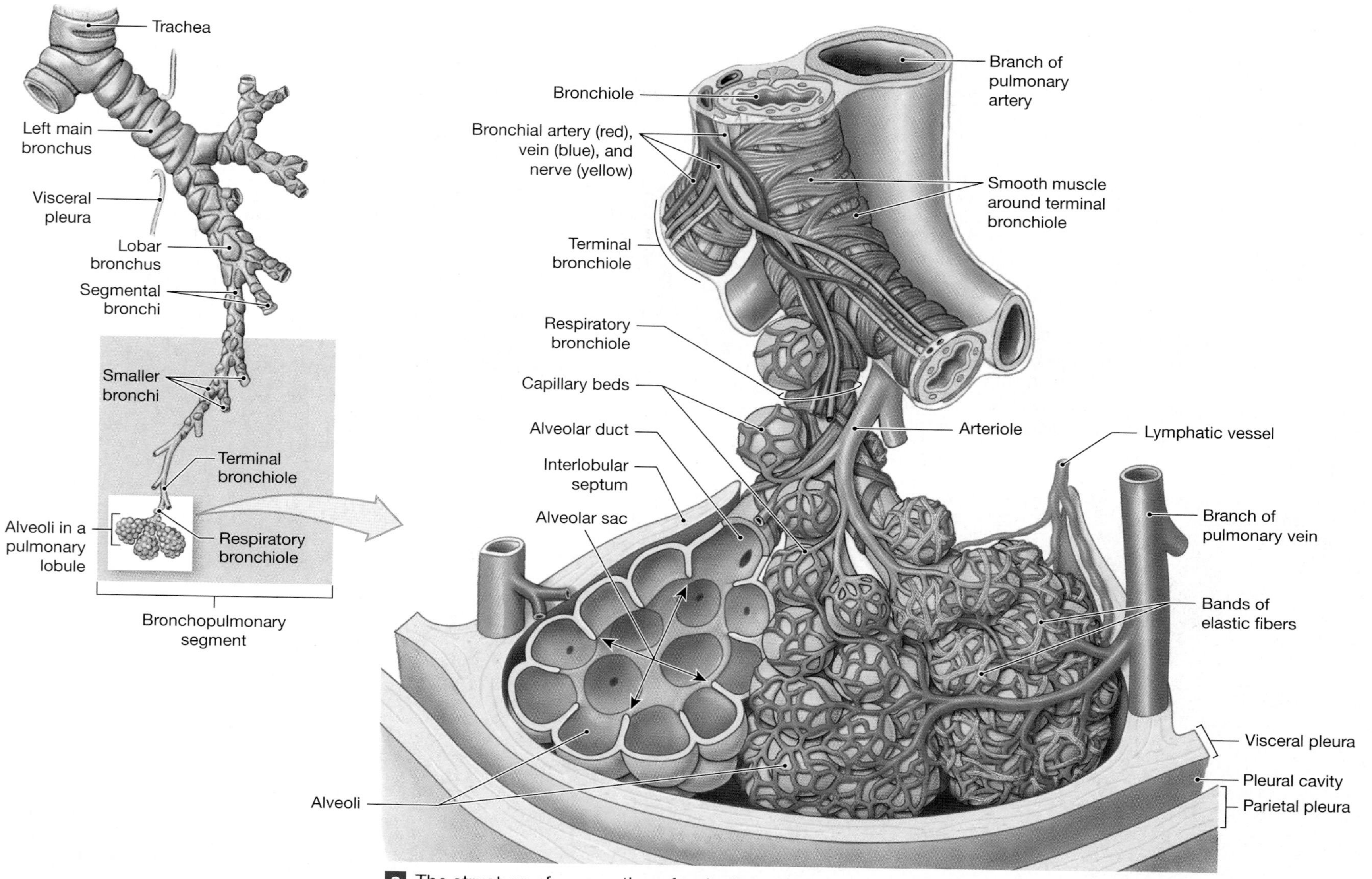

Trachea
Left main bronchus
Visceral pleura
Lobar bronchus
Segmental bronchi
Smaller bronchi
Terminal bronchiole
Alveoli in a pulmonary lobule
Respiratory bronchiole
Bronchopulmonary segment
Bronchiole
Bronchial artery (red), vein (blue), and nerve (yellow)
Terminal bronchiole
Respiratory bronchiole
Capillary beds
Alveolar duct
Interlobular septum
Alveolar sac
Alveoli
Branch of pulmonary artery
Smooth muscle around terminal bronchiole
Arteriole
Lymphatic vessel
Branch of pulmonary vein
Bands of elastic fibers
Visceral pleura
Pleural cavity
Parietal pleura

a The structure of one portion of a single pulmonary lobule

Alveoli
Respiratory bronchiole
Alveolar sac
Alveolar duct
Arteriole
Histology of the lung
LM × 14

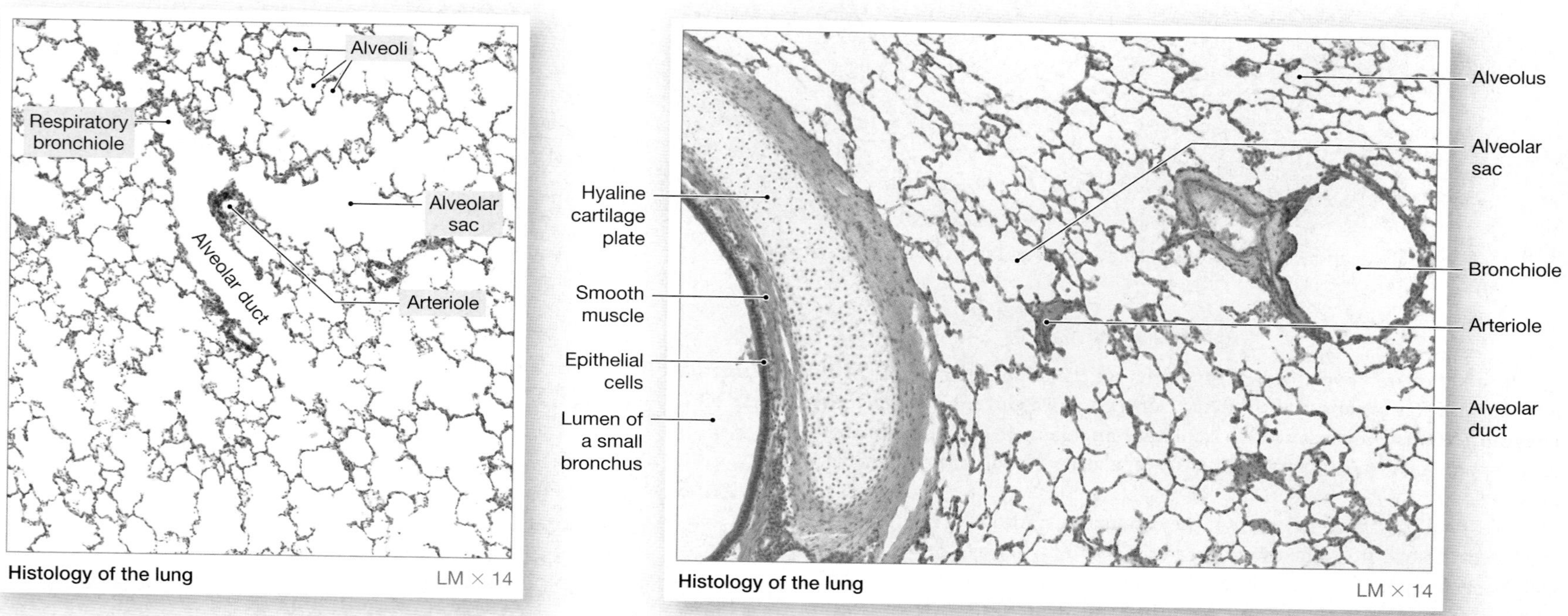

b Low-power micrograph of lung tissue

Hyaline cartilage plate
Smooth muscle
Epithelial cells
Lumen of a small bronchus
Alveolus
Alveolar sac
Bronchiole
Arteriole
Alveolar duct
Histology of the lung
LM × 14

c Histological section of the lung showing a small bronchus and bronchiole

Figure 24.14 Alveolar Organization

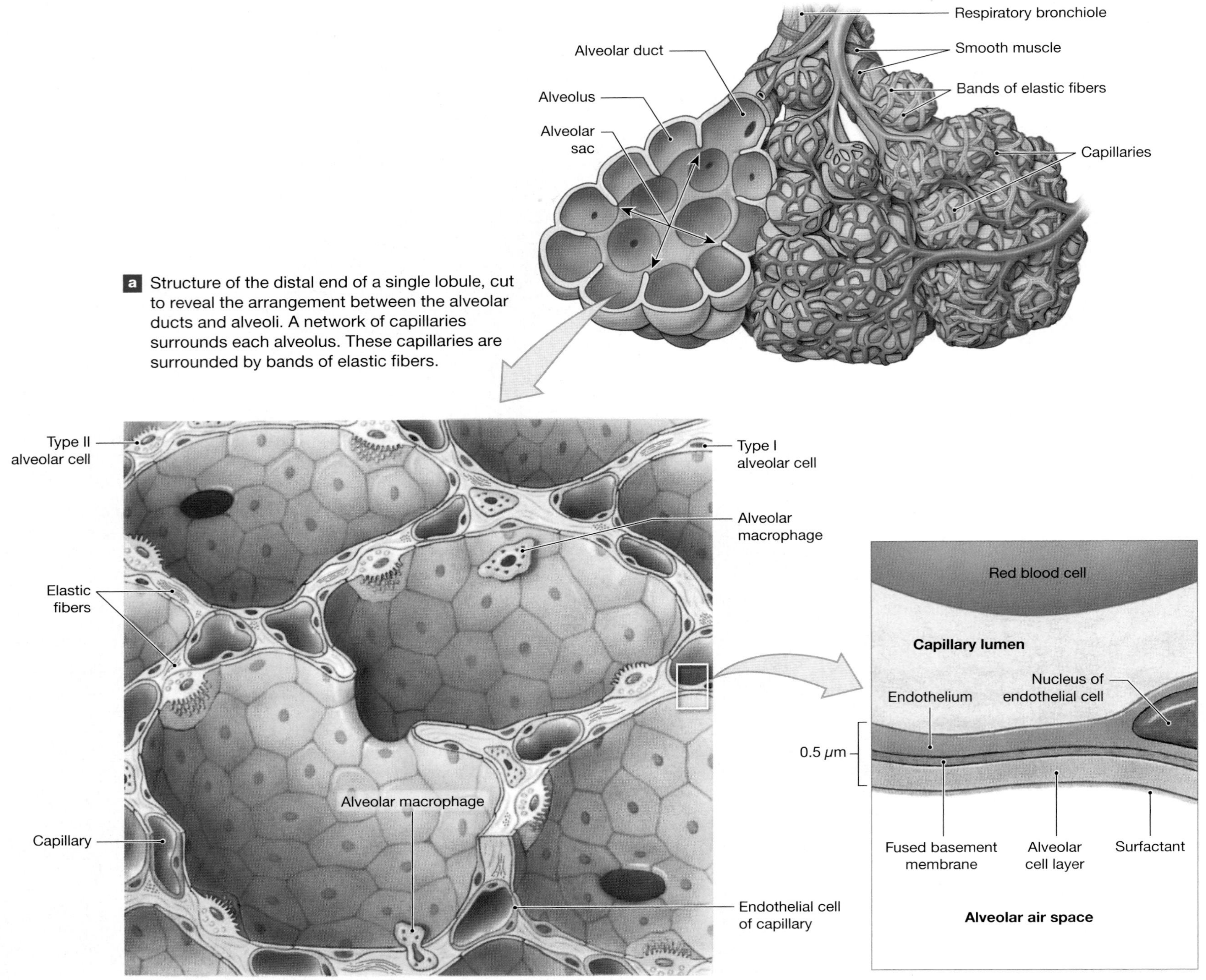

Respiratory bronchiole
Smooth muscle
Alveolar duct
Bands of elastic fibers
Alveolus
Alveolar sac
Capillaries

a Structure of the distal end of a single lobule, cut to reveal the arrangement between the alveolar ducts and alveoli. A network of capillaries surrounds each alveolus. These capillaries are surrounded by bands of elastic fibers.

Type II alveolar cell
Type I alveolar cell
Alveolar macrophage
Elastic fibers
Alveolar macrophage
Capillary
Endothelial cell of capillary

b Diagrammatic sectional view of alveolar structure and the respiratory membrane.

Red blood cell
Capillary lumen
Nucleus of endothelial cell
Endothelium
0.5 μm
Fused basement membrane
Alveolar cell layer
Surfactant
Alveolar air space

c The blood air barrier.

In addition to providing for gas exchange, the endothelial cells of the alveolar capillaries are the primary source of **angiotensin-converting enzyme (ACE)**, which converts circulating angiotensin I to angiotensin II. This enzyme plays an important role in regulating blood volume and blood pressure.

The nasal passages and larynx of the conducting portions of the respiratory tract receive blood from the external carotid arteries. The thyrocervical trunks, which are branches of the subclavian arteries, supply blood to the inferior portions of the larynx and the trachea. The bronchial arteries supply blood to the bronchial tree and alveoli **(Figure 24.13a)**. The venous blood from these structures flows into the pulmonary veins, bypassing the rest of the systemic circuit and diluting the oxygen-rich blood leaving the alveoli.

24.6 CONCEPT CHECK

13 Why are there almost no cilia or mucous glands in the respiratory bronchiole?

14 What is the function of the surfactant produced by the type II alveolar cells?

See the blue Answers tab at the back of the book.

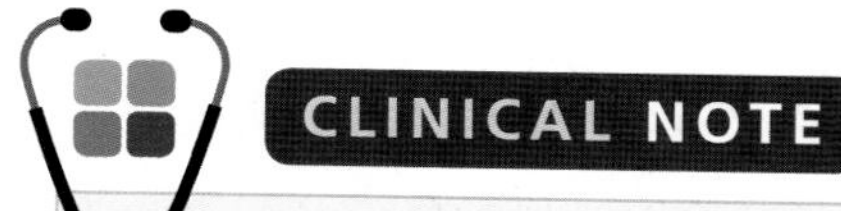

Chronic Obstructive Pulmonary Disease (COPD)

Chronic Obstructive Pulmonary Disease (COPD) is a group of disorders that restricts airflow and reduces alveolar ventilation. Types include asthma, chronic bronchitis, and emphysema.

Asthma

Asthma is an inflammatory disease of the lungs characterized by bronchial inflammation, spasms, and difficulty breathing. It can result from a hypersensitivity reaction (allergy), cold temperature, or exercise. During an asthmatic attack, smooth muscle spasms within the respiratory tract constrict and excessive mucus production make breathing difficult. Signs and symptoms include coughing, wheezing, shortness of breath, and chest tightness. The sensation of asthma has been described as breathing through a small straw.

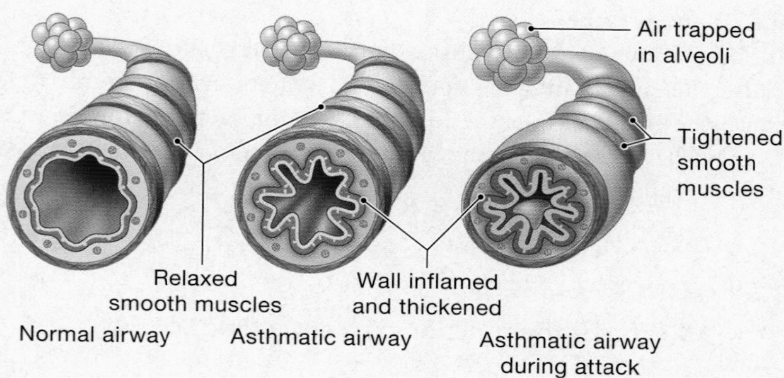

Allergic asthma is a reaction to environmental allergies such as animal dander, dust mites, pollen, or mold. Non-allergic triggers include smoke, pollution, and cold air. Exercise-induced asthma is a response to physical activity.

The majority of children who develop asthma do so before the age of 5. Although there is no cure for asthma, early diagnosis and aggressive treatment can manage the symptoms.

Chronic Bronchitis

Chronic bronchitis results from recurrent, long-term inflammation and swelling of the bronchi. This can lead to recurrent infections. There is excessive mucus production and a harsh cough with expectoration (coughing up mucus). To be classified as chronic bronchitis, the symptoms must occur on most days, for more than 3 months of the year, for 2 or more years in a row. The most common cause of chronic bronchitis is cigarette smoking. Chronic bronchitis causes low blood oxygenation, cyanosis (bluish skin coloration), and widespread inflammation, leading to the descriptive term "blue bloaters" for people with this condition.

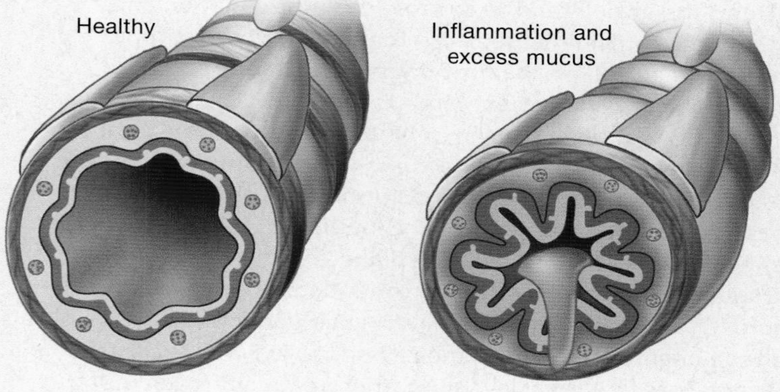

Emphysema

Emphysema is a disorder characterized by damaged and permanently enlarged alveoli, causing breathlessness. An emphysematous lung looks like an old, used sponge. The underlying issue is the destruction of alveolar surfaces and inadequate surface area for gas exchange. The alveoli expand and adjacent alveoli merge to form large sacks. Because the lung tissue has lost its elasticity, air is trapped during expiration. People with emphysema use a great deal of energy to breathe and can typically maintain near normal blood oxygenation, and their skin is usually pink, leading to the descriptive germ "pink puffers." The main cause of emphysema is smoking.

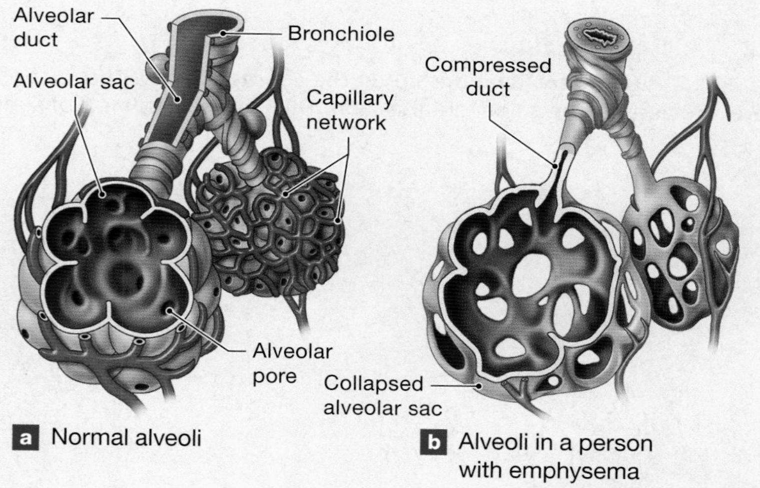

a Normal alveoli

b Alveoli in a person with emphysema

24

CLINICAL NOTE

Respiratory Distress Syndrome (RDS)

Neonatal Respiratory Distress Syndrome (RDS) is a breathing disorder of premature newborns in which the alveoli do not remain open because surfactant is absent or insufficient. Surfactant production begins at the end of the sixth month of fetal development; by the eighth month, surfactant has reached a level adequate to sustain respiratory function. Without enough surfactant, the alveoli tend to collapse during exhalation and stick together, forcing the newborn to inhale with extra force to reopen the alveoli on the next breath. The infant becomes exhausted quickly, and gas exchange is ineffective.

The risk is reduced if delivery can be delayed until the fetal lungs are developed. Most cases occur in babies born before 37–39 weeks of development. Other risk factors include problems with the baby's blood flow during delivery, multiple pregnancy (twins or more), rapid labor, or a mother with diabetes.

One treatment is administering continuous positive airway pressure (CPAP), a technique of delivering air under pressure to hold the alveoli open. This can keep the newborn alive until surfactant production rises to normal levels. Another treatment is administering synthetic surfactant through an endotracheal tube immediately after delivery and several times during the first few days of life until the condition resolves.

Adult respiratory distress syndrome (ARDS) is caused by inadequate surfactant production that leads to alveolar collapse on exhalation. It can result from severe respiratory infections or pulmonary injuries. The condition is treated by positive end-expiratory pressure (PEEP), a technique in which airway pressure greater than atmospheric pressure is achieved at the end of exhalation by a device. The PEEP procedure maintains life until the underlying problem can heal, but ARDS has a high mortality rate.

24.7 The Pleural Cavities and Pleural Membranes

▶ **KEY POINT** The pleural membrane consists of two layers: the parietal pleura and the visceral pleura.

The thoracic cavity has the shape of a broad cone. Its walls are the rib cage, and the muscular diaphragm forms the floor. The two pleural cavities are separated by the mediastinum **(Figures 24.10a** and **24.15)**. Each lung occupies a single pleural cavity, lined by a serous membrane, or **pleura** (PLOO-ruh; plural, *pleurae*). The pleural membrane consists of two continuous layers. The **parietal pleura** covers the inner surface of the thoracic wall, the superior surface of the diaphragm, and the surfaces of the mediastinum. The **visceral pleura** covers the outer surfaces of the lungs and extends into the fissures between the lobes. The space between the parietal and visceral pleurae is called the **pleural cavity**. ⮌ pp. 19–20 Each pleural cavity is actually a "potential space" rather than an open chamber because the parietal and visceral layers are in close contact. Both pleural membranes secrete a small amount of **pleural fluid**, a moist, slippery coating that lubricates. It reduces friction between the parietal and visceral surfaces as you breathe.

Inflammation of the pleurae, a condition called **pleurisy**, may (1) cause the membranes to produce and secrete excess pleural fluid or (2) cause the inflamed pleurae to adhere to one another, limiting movement. In either case, breathing becomes difficult, and prompt medical attention is required.

24.7 CONCEPT CHECK

15 What is the function of pleural fluid?

See the blue Answers tab at the back of the book.

Figure 24.15 Anatomical Relationships in the Thoracic Cavity. The relationships among thoracic structures are best seen in a horizontal section. This is an inferior view of a section taken at the level of T_8.

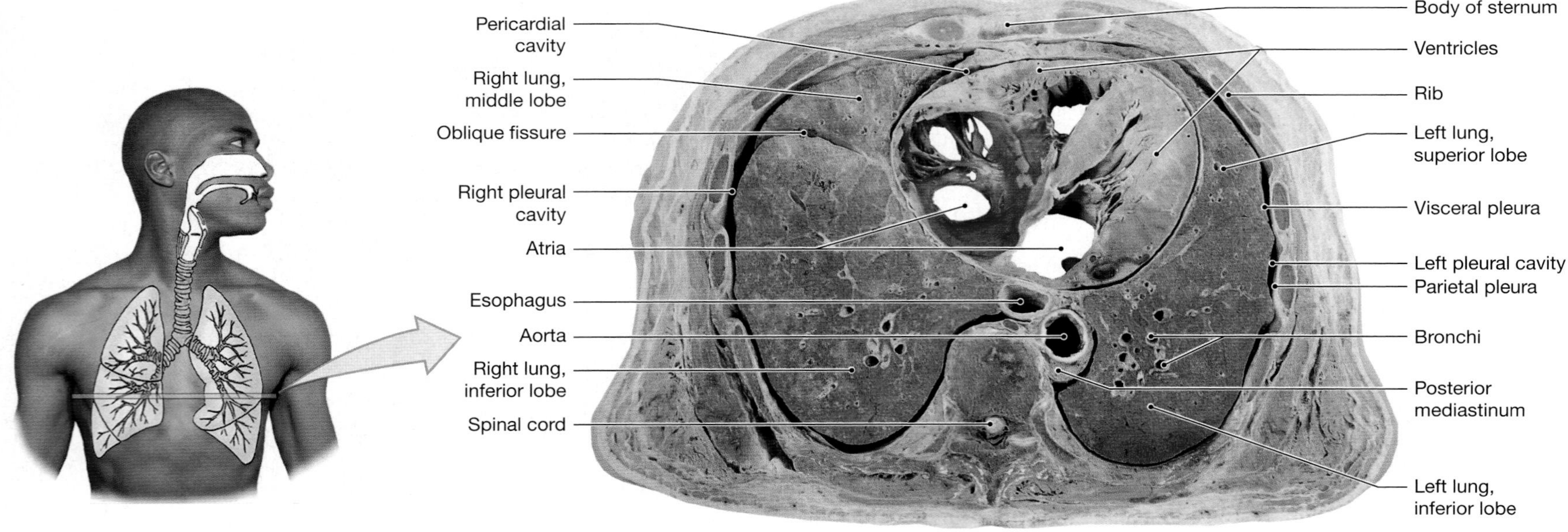

24.8 Respiratory Muscles and Pulmonary Ventilation

▶ **KEY POINT** Pulmonary ventilation (breathing) moves air into and out of the alveoli.

Pulmonary ventilation, or breathing, physically moves air into and out of the lungs. The function of pulmonary ventilation is to maintain adequate alveolar ventilation, the movement of air into and out of the alveoli. Alveolar ventilation prevents the buildup of carbon dioxide in the alveoli and ensures a continual supply of oxygen that keeps pace with absorption by the bloodstream.

Respiratory Muscles

▶ **KEY POINT** The diaphragm and external intercostal muscles are the primary muscles of respiration. Respiratory movements are classified as eupnea or hyperpnea.

Chapters 10 and 11 introduced the skeletal muscles involved in respiratory movements. The most important muscles involved in breathing are the diaphragm and the external intercostals. Respiratory muscles are classified as **primary respiratory muscles** and the **accessory respiratory muscles (Spotlight Figure 24.16)**.

Respiratory Movements

We use our respiratory muscles in various combinations, depending on the volume of air that must be moved into or out of the lungs. Respiratory movements are classified as eupnea or hyperpnea, based on whether expiration is passive or active **(Spotlight Figure 24.16)**.

During **eupnea**, expansion of the lungs stretches their elastic fibers. In addition, elevation of the rib cage stretches skeletal muscles and elastic fibers in the connective tissues of the body wall. When the inspiratory muscles relax, these elastic structures recoil, returning the diaphragm and rib cage to their original positions.

Eupnea may involve diaphragmatic breathing or costal breathing.

- During **diaphragmatic breathing**, or *deep breathing*, contraction of the diaphragm provides the necessary change in thoracic volume. Air is drawn into the lungs as the diaphragm contracts, and exhalation occurs when the diaphragm relaxes.
- In **costal breathing**, or *shallow breathing*, the thoracic volume changes because the rib cage changes shape. Inhalation occurs when contraction of the external intercostal muscles elevates the ribs and enlarges the thoracic cavity. Exhalation occurs when these muscles relax. During pregnancy women increasingly rely on costal breathing as the uterus enlarges and pushes the abdominal viscera against the diaphragm.

During **hyperpnea**, accessory muscles assist with inspiration, and expiration involves contraction of the transversus thoracis and internal intercostal muscles. When we breathe at maximum levels, such as during vigorous exercise, we also use our abdominal muscles in exhalation. Their contraction compresses the abdominal contents, pushing them up against the diaphragm, reducing the volume of the thoracic cavity.

24.8 CONCEPT CHECK

16 Athletes commonly try to catch their breath after strenuous exercise by standing with their hands on their thighs while breathing deeply. Is this the best way to increase the flow of air into their lungs? Explain.

See the blue Answers tab at the back of the book.

24.9 Respiratory Changes at Birth

▶ **KEY POINT** A developing fetus does not use its lungs for gas exchange. At birth, several significant changes occur in the anatomy of the respiratory system.

There are several important differences between the respiratory system of a fetus and that of a newborn infant. Before delivery, pulmonary arterial resistance is high because the pulmonary vessels are collapsed. The rib cage is compressed, and the lungs and conducting passageways contain only small amounts of fluid and no air.

At birth, the newborn infant takes a truly heroic first breath thanks to powerful contractions of the diaphragm and external intercostal muscles. The inspired air enters the passageways with enough force to push the fetal fluids out of the way and inflate the entire bronchial tree and most of the alveoli. The same drop in pressure that pulls air into the lungs pulls blood into the pulmonary circulation. The changes in blood flow cause the foramen ovale and the ductus arteriosus to close. ⮌ pp. 596–598

The newborn's first exhalation fails to completely empty the lungs because the rib cage does not return to its former, fully compressed state. The supporting cartilages and other connective tissues keep the conducting passageways open, and the surfactant covering the alveolar surfaces prevents their collapse. Subsequent breaths complete the inflation of the alveoli.

24.9 CONCEPT CHECK

17 Summarize the changes that occur in a newborn infant's cardiovascular and respiratory systems as a newborn infant starts to breathe.

See the blue Answers tab at the back of the book.

24.10 Respiratory Centers of the Brain

▶ **KEY POINT** Respiratory centers within the brain regulate the activities of the respiratory muscles by adjusting the frequency and depth of pulmonary ventilation.

Under normal conditions, oxygen absorption and carbon dioxide generation are matched because of the rates of oxygen delivery and carbon dioxide removal by the capillary beds. When these rates are not equal, the cardiovascular and respiratory systems make coordinated adjustments to meet the body's changing demands. The regulatory centers coordinating the responses by these systems are located in the pons and medulla oblongata.

The **respiratory centers** are three pairs of loosely organized nuclei in the reticular formation of the pons and medulla oblongata **(Figure 24.17)**. These nuclei regulate the activities of the respiratory muscles by adjusting the frequency and depth of pulmonary ventilation.

The **respiratory rhythmicity centers** set the basic pace and depth of respiration. They are subdivided into a **dorsal respiratory group (DRG)** and a **ventral respiratory group (VRG) (Figure 24.17)**. The dorsal respiratory group, or *inspiratory center*, controls motor neurons innervating the external intercostal muscles and the diaphragm. This group functions in every respiratory cycle, whether quiet or forced. The ventral respiratory group functions only during forced respiration. It innervates motor neurons controlling accessory muscles involved in active exhalation and maximal inhalation. The neurons involved with active exhalation are sometimes called an expiratory center.

The **apneustic** (ap-NŪ-stik) and **pneumotaxic** (nū-mō-TAKS-ik) **centers** of the pons are paired nuclei that adjust the output of the rhythmicity center **(Figure 24.17)**. This modifies the pace of respiration. These centers adjust the rate and depth of respiration in response to sensory stimuli or instructions from higher centers.

SPOTLIGHT

FIGURE 24.16

Respiratory Muscles and Pulmonary Ventilation

1 The Respiratory Muscles

The most important skeletal muscles involved in respiration (breathing) are the diaphragm and the external intercostal muscles. These muscles are the **primary respiratory muscles** and are active during normal breathing at rest. The **accessory respiratory muscles** become active when the depth and rate of breathing must be increased markedly, such as during exercise.

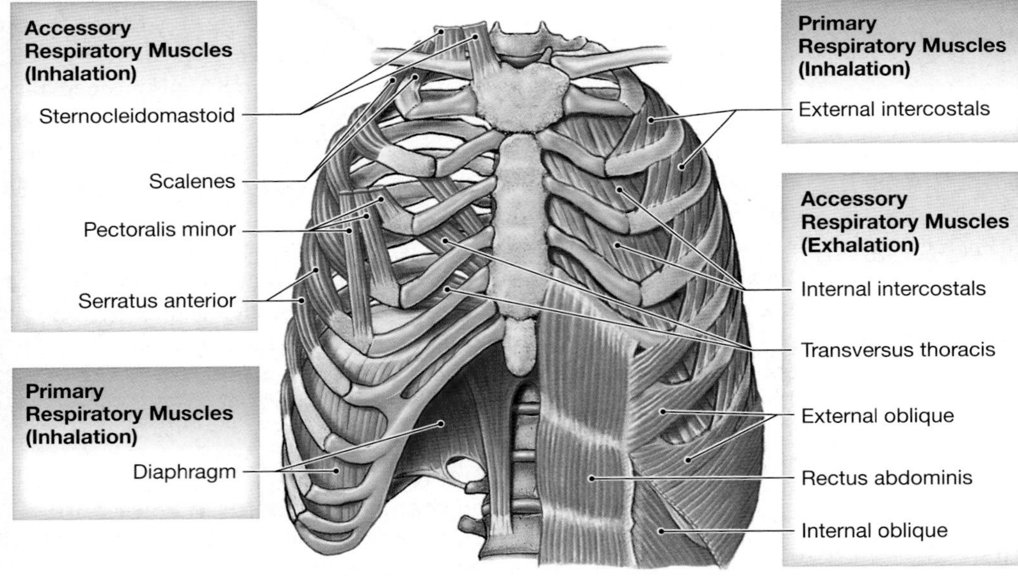

2 The Mechanics of Breathing

Air movement into and out of the respiratory system occurs by changing the volume of the lungs. This process is called **pulmonary ventilation**. The changes of volume in the lungs take place through the contraction of skeletal muscles—specifically, the diaphragm and those that insert on the rib cage. As the ribs are elevated or the diaphragm contracts and flattens, the volume of the thoracic cavity increases and air moves into the lungs. The outward movement of the ribs as they are elevated resembles the outward swing of a raised bucket handle.

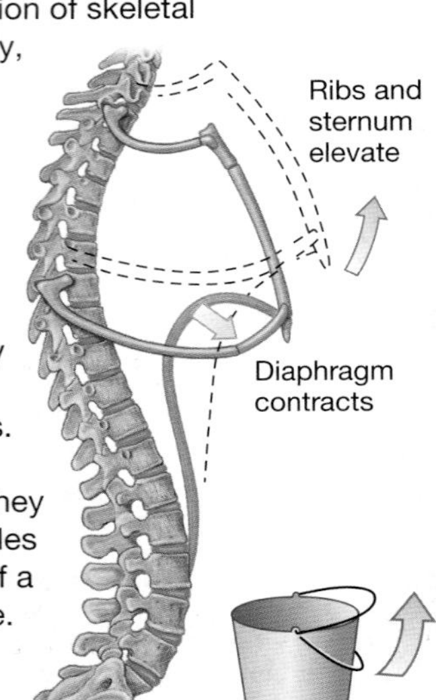

3 Respiratory Movements

Respiratory muscles may be used in various combinations, depending on the volume of air that must be moved in or out of the lungs. In **eupnea** (*quiet breathing*), inhalation involves muscular contractions, but exhalation is a passive process that relies on the elastic recoil of the ribs, muscles, and lungs. **Hyperpnea** (*forced breathing*) requires the assistance of the accessory respiratory muscles for achieving inhalation and exhalation.

KEY

- = Movement of rib cage
- = Movement of diaphragm
- = Muscle contraction

Inhalation

Inhalation is an active process. It primarily involves the diaphragm and the external intercostal muscles, with assistance from the accessory respiratory muscles as needed.

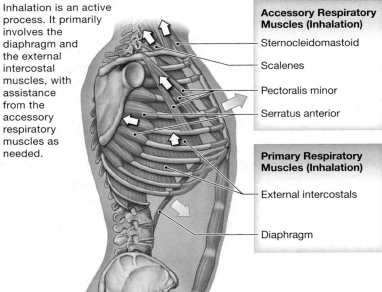

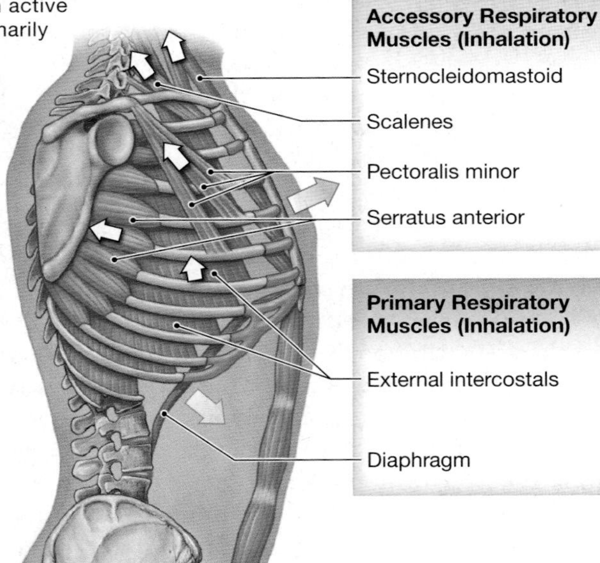

Exhalation

During forced exhalation, the tranversus thoracis and internal intercostal muscles actively depress the ribs, and the abdominal muscles compress the abdomen and push the diaphragm up.

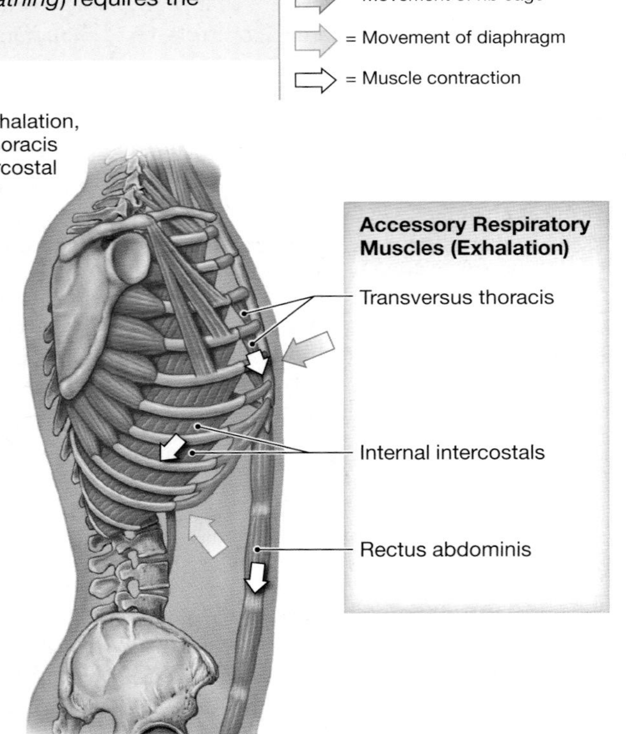

Normal breathing occurs automatically, without conscious control. Three different reflexes are involved in regulating respiration: (1) mechanoreceptor reflexes respond to changes in the volume of the lungs or to changes in arterial blood pressure; (2) chemoreceptor reflexes respond to changes in the P_{CO_2}, pH, and P_{O_2} of the blood and cerebrospinal fluid; and (3) protective reflexes respond to physical injury or irritation of the respiratory tract.

Figure 24.17 Respiratory Centers and Reflex Controls. This figure shows the relationships between the respiratory centers and other factors associated with respiration. Pathways for conscious control over respiratory muscles are not shown.

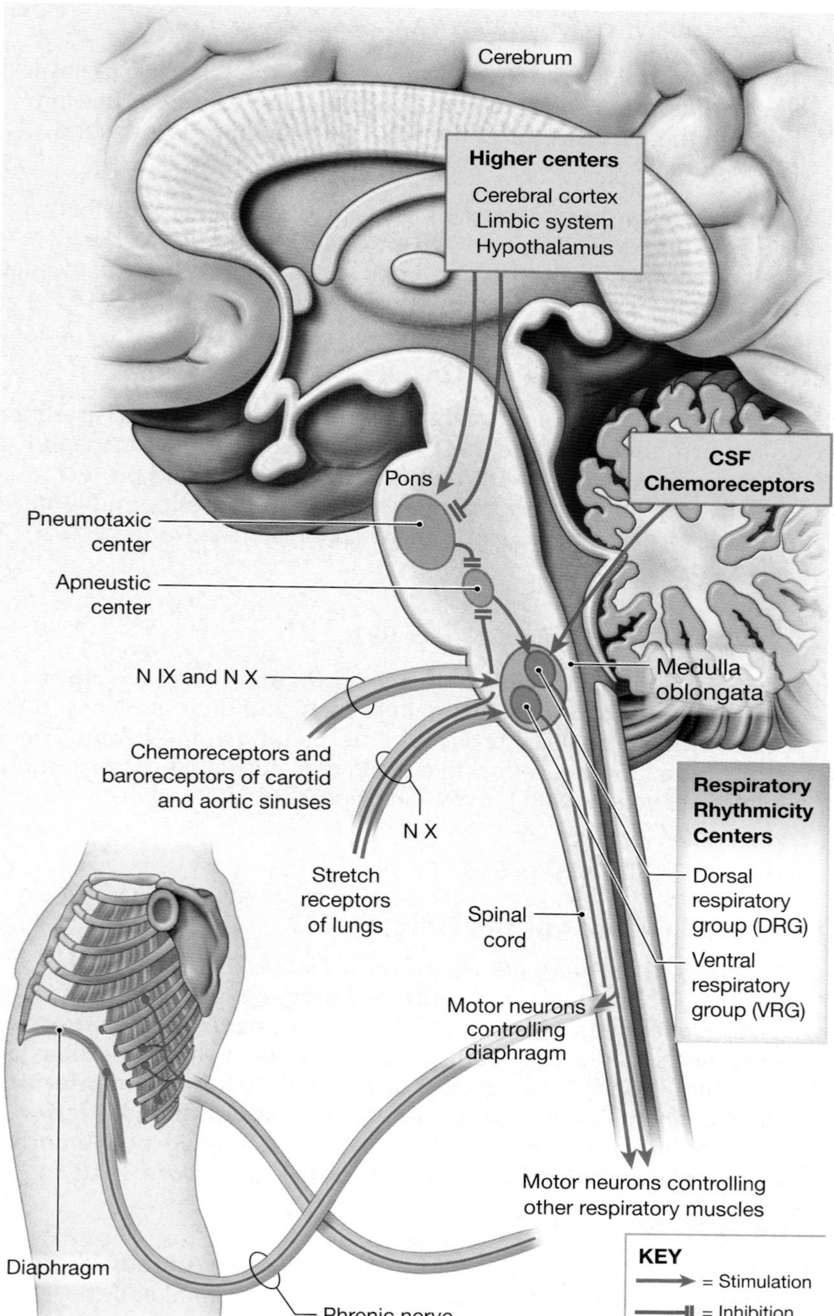

Higher centers influence respiration by their connections to the pneumotaxic center and by their direct influence on respiratory muscles. These higher centers are found in the cerebrum, especially the cerebral cortex, and in the hypothalamus. Pyramidal output provides conscious control over the respiratory muscles. However, these muscles receive most of their instructions by the extrapyramidal pathways. In addition, the respiratory centers are embedded in the reticular formation. Almost every sensory and motor nucleus has some connection with this complex. As a result, emotional and autonomic activities also affect the pace and depth of respiration.

24.10 CONCEPT CHECK

18 Stimulation of which respiratory center(s) would increase the rate and depth of breathing?

See the blue Answers tab at the back of the book.

24.11 | Aging and the Respiratory System

▶ **KEY POINT** Aging produces significant changes in the respiratory system that reduce its efficiency.

Many factors interact to reduce the efficiency of the respiratory system in elderly individuals. Four examples are particularly noteworthy:

1. With increasing age, elastic tissue throughout the body deteriorates. This reduces the lungs' ability to inflate and deflate.
2. Arthritic changes in the joints of the ribs and decreased flexibility of the costal cartilages restrict movements of the rib cage.
3. A stiffening and reduction in chest movement limits the respiratory volume. This restriction contributes to reducing exercise performance and capabilities.
4. Some degree of emphysema appears in many individuals after age 50. On average, roughly 1 square foot of the blood air barrier is lost each year after age 30. However, this varies widely, depending on lifetime exposure to tobacco smoke and other respiratory irritants.

24.11 CONCEPT CHECK

19 Your 80-year-old grandfather asks you why he has more difficulty catching his breath after his daily bicycle ride now as compared to just 10 years ago. How should you explain this to him?

See the blue Answers tab at the back of the book.

EMBRYOLOGY SUMMARY

For an introduction to the development of the respiratory system, see Chapter 28 (Embryology and Human Development).

Study Outline

Introduction p. 625

24.1 An Overview of the Respiratory System and Respiratory Tract p. 625

- The **respiratory system** includes the nose, nasal cavity and sinuses, pharynx, larynx, trachea, and conducting passageways leading to the exchange surfaces of the lungs. *(See Figure 24.1.)*
- The **upper respiratory system** consists of the paranasal sinuses, nasal conchae, nose, nasal cavity, and nasopharynx. These structures begin the process of filtration and humidification of the incoming air. The **lower respiratory system** includes the larynx, trachea, lungs, bronchi, bronchioles, and alveoli of the lungs.
- The **respiratory tract** consists of a conducting portion, which extends from the entrance to the nasal cavity to the bronchioles, and a respiratory portion, which includes the respiratory bronchioles and the **alveoli**. *(See Figure 24.1.)*

Functions of the Respiratory System p. 626

- The respiratory system (1) provides an area for gas exchange between air and circulating blood, (2) moves air to and from exchange surfaces, (3) protects respiratory surfaces, (4) defends the respiratory system and other tissues from pathogens, (5) permits vocal communication, and (6) helps regulate blood volume and pressure and body fluid pH.

The Respiratory Epithelium p. 626

- The **respiratory epithelium** lines the conducting portions of the respiratory system down to the level of the terminal bronchioles.
- The respiratory epithelium consists of a pseudostratified ciliated columnar epithelium with mucous cells. *(See Figure 24.2.)*
- The respiratory epithelium produces mucus that traps incoming particles. Underneath is the **lamina propria** (a layer of connective tissue); the combined respiratory epithelium and lamina propria form a **mucosa** (mucous membrane). *(See Figure 24.2.)*
- The **respiratory defense system** includes mucous glands and the mucus produced (which washes particles toward the stomach), alveolar macrophages, hairs, and cilia.

24

24.2 The Upper Respiratory System p. 627

The Nose and Nasal Cavity p. 627

- Air normally enters the respiratory system via the **nostrils**, or **nares**, which open into the **nasal cavity**. The **nasal vestibule** (entryway) of the nose is guarded by hairs that screen out large particles. *(See Figures 24.3, 24.4, and 24.5.)*
- Incoming air flows through the **superior**, **middle**, or **inferior meatuses** (narrow grooves) and bounces off the conchal surfaces. *(See Figure 24.4.)*
- The **hard palate** separates the oral and nasal cavities. The **soft palate** separates the superior nasopharynx from the oral cavity. The connections between the nasal cavity and nasopharynx represent the **posterior nasal apertures**. *(See Figures 24.3, 24.4, and 24.5.)*

The Pharynx p. 629

- The **pharynx** is a chamber shared by the digestive and respiratory systems. The **nasopharynx** is the superior part of the pharynx. The **oropharynx** is continuous with the oral cavity; the **laryngopharynx** includes the narrow zone between the hyoid and the entrance to the esophagus. *(See Figures 24.4a and 24.5.)*

24.3 The Lower Respiratory System p. 630

The Larynx p. 630

- Inhaled air passes through the **glottis** en route to the lungs; the **larynx** surrounds and protects the glottis. The **epiglottis** projects into the pharynx. *(See Figures 24.4, 24.5, 24.6, and 24.7.)*
- Two pairs of folds span the glottal opening: the relatively inelastic **vestibular folds** and the more delicate **vocal folds**. Air passing through the glottis vibrates the vocal folds and produces sound. *(See Figure 24.7.)*
- The **intrinsic laryngeal muscles** regulate tension in the vocal folds and open and close the glottis. The **extrinsic laryngeal muscles** position and stabilize the larynx. During swallowing, both sets of muscles help to prevent particles from entering the glottis. *(See Figure 24.8.)*

24.4 The Trachea p. 632

- The **trachea** ("windpipe") extends from the sixth cervical vertebra to the fifth thoracic vertebra. The **submucosa** contains C-shaped **tracheal cartilages** that stiffen the tracheal walls and protect the airway. The posterior tracheal wall can distort to permit large masses of food to move along the esophagus. *(See Figures 24.2a, 24.4a, 24.5, and 24.9.)*

24.5 The Main Bronchi p. 632

- The trachea branches within the mediastinum to form the **right** and **left main bronchi**. The main bronchi and their branches form the bronchial tree. Each bronchus enters a lung at the **hilum**. The **root** of the lung is a connective tissue mass including the bronchus, pulmonary vessels, and nerves. *(See Figure 24.9a.)*

24.6 The Lungs p. 633

Lobes and Fissures of the Lungs p. 633

- The **lobes** of the lungs are separated by fissures. The **right lung** has three lobes: **superior**, **middle**, and **inferior** with an **oblique fissure** separating superior and inferior and a **horizontal fissure** separating superior and middle. The right lung also has three lobar, or *secondary*, bronchi: the **superior lobar**, **middle lobar**, and **inferior lobar bronchi**. The **left lung** has two lobes: **superior** and **inferior** with an **oblique fissure** separating the lobes; and two lobar bronchi: **superior lobar** and **inferior lobar bronchi**. *(See Figure 24.10.)*

Lung Surfaces p. 633

- The **costal surface** of the lung follows the inner contours of the rib cage. The **mediastinal surface** contains a hilum, and the left lung bears the **cardiac notch**. *(See Figure 24.10.)*
- The connective tissues of the root of each lung extend inward as a series of **trabeculae** (partitions). These branches form **septa** that divide the lung into **lobules**.

The Main Bronchi p. 635

- The left and right main bronchi are outside the lung tissue. Branches within the lung are surrounded by bands of smooth muscle. *(See Figures 24.9, 24.11, 24.12, and 24.13.)*

- Each lung is further divided into smaller units called **bronchopulmonary segments**. These segments are named according to the associated segmental bronchi. The right lung contains 10 and the left lung usually contains 8–9 bronchopulmonary segments. *(See Figures 24.9, 24.11, and 24.12.)*

The Bronchioles p. 635

- Within the bronchopulmonary segments, each segmental bronchus ultimately gives rise to **terminal bronchioles** that supply individual lobules. *(See Figures 24.11 and 24.13.)*

Alveolar Ducts and Alveoli p. 638

- The **respiratory bronchioles** open into **alveolar ducts**; many **alveoli** are interconnected at each duct. *(See Figures 24.13 and 24.14.)*

- The **blood air barrier** consists of a simple squamous epithelium of **type I alveolar cells**; **type II alveolar cells** scattered in it produce an oily secretion (**surfactant**) that keeps the alveoli from collapsing. **Alveolar macrophages** patrol the epithelium and engulf foreign particles. *(See Figure 24.14.)*

The Blood Supply to the Lungs p. 638

- The respiratory exchange surfaces are extensively connected to the circulatory system via the vessels of the pulmonary circuit. *(See Figure 24.13a.)*

24.7 The Pleural Cavities and Pleural Membranes p. 642

- Each lung occupies a single **pleural cavity** lined by a **pleura** (serous membrane). The two types of pleurae are the **parietal pleura**, covering the inner surface of the thoracic wall, and the **visceral pleura**, covering the lungs. **Pleural fluid** is secreted by both pleural membranes. *(See Figures 24.10a and 24.15.)*

24.8 Respiratory Muscles and Pulmonary Ventilation p. 643

Respiratory Muscles p. 643

- **Pulmonary ventilation** is the movement of air into and out of the lungs. The most important respiratory muscles are the diaphragm and the external and internal intercostal muscles. Contraction of the diaphragm increases the volume of the thoracic cavity; the external intercostals may assist in inspiration by elevating the ribs; the internal intercostals depress the ribs and reduce the width of the thoracic cavity, thereby contributing to expiration. The **accessory respiratory muscles** become active when the depth and frequency of respiration must be increased markedly. Accessory muscles include the sternocleidomastoid, serratus anterior, transversus thoracis, scalene, pectoralis minor, oblique, and rectus abdominis muscles. *(See Spotlight Figure 24.16.)*

24.9 Respiratory Changes at Birth p. 643

- Before delivery, the fetal lungs are fluid-filled and collapsed. At the first breath, the lungs inflate and never collapse completely thereafter.

24.10 Respiratory Centers of the Brain p. 643

- The **respiratory centers** are three pairs of nuclei in the reticular formation of the pons and medulla. The **respiratory rhythmicity centers** set the pace for respiration. The **apneustic center** causes strong, sustained inspiratory movements, and the **pneumotaxic center** inhibits the apneustic center and the inspiratory center in the medulla oblongata. *(See Figure 24.17.)*

- Three different reflexes are involved in the regulation of respiration: (1) Mechanoreceptor reflexes respond to changes in the volume of the lungs or to changes in arterial blood pressure; (2) chemoreceptor reflexes respond to changes in the P_{CO2}, pH, and P_{O2} of the blood and cerebrospinal fluid; and (3) protective reflexes respond to physical injury or irritation of the respiratory tract. *(See Figure 24.17.)*

- Conscious and unconscious thought processes can also control respiratory activity by affecting the respiratory centers or controlling the respiratory muscles.

24.11 Aging and the Respiratory System p. 645

- The respiratory system is generally less efficient in elderly people because (1) elastic tissue deteriorates, reducing the lungs' ability to inflate and deflate; (2) movements of the chest cage are restricted by arthritic changes and decreased flexibility of costal cartilages; (3) a stiffing and reduction in chest movements limits respiratory volumne, reducing exercise performance and capabilities; and (4) some degree of emphysema is normal.

24

Chapter Review

For answers, see the blue Answers tab at the back of the book.

Level 1 Reviewing Facts and Terms

Match each numbered item with the most closely related lettered item.

1. extrinsic laryngeal musculature .... ☐
2. segmental bronchi .... ☐
3. nasopharynx .... ☐
4. vestibular folds .... ☐
5. lower respiratory system .... ☐
6. trachea .... ☐
7. dorsum of nose .... ☐
8. lobar bronchi .... ☐
9. upper respiratory system .... ☐
10. larynx .... ☐

(a) superior portion of the pharynx
(b) bridge of the nose
(c) nose, nasal cavity, paranasal sinuses, nasal conchae, and nasopharynx
(d) windpipe
(e) lobar bronchi
(f) contains and protects the glottis
(g) larynx, trachea, bronchi, bronchioles and alveoli of the lungs
(h) provide protection for the vocal folds
(i) positions and stabilizes the larynx
(j) segmental bronchi

11. Air entering the body is filtered, warmed, and humidified by the
    (a) upper respiratory tract.
    (b) lower respiratory tract.
    (c) lungs.
    (d) alveoli.

12. Surfactant
    (a) protects the surfaces of the lungs.
    (b) phagocytizes small particulates.
    (c) replaces mucus in the alveoli.
    (d) helps prevent the alveoli from collapsing.

13. Identify the structures of the respiratory system on the following diagram.

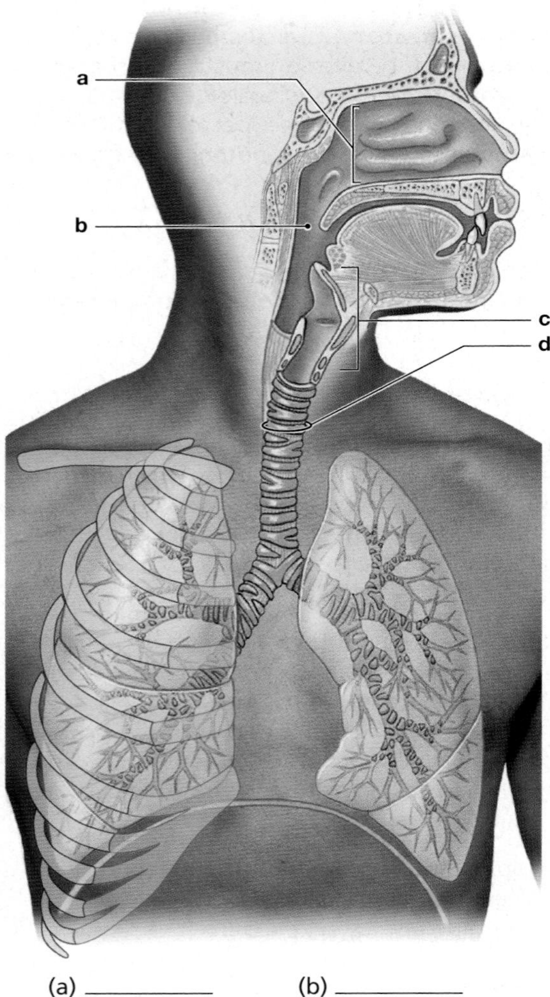

(a) __________ (b) __________
(c) __________ (d) __________

14. The portion of the nasal cavity contained within the flexible tissues of the external nose is the
    (a) nasopharynx.
    (b) nasal vestibule.
    (c) internal chamber.
    (d) glottis.

15. What aspect of laryngeal function would be impaired if the arytenoid and corniculate cartilages were damaged?
    (a) The air would not be able to enter.
    (b) The larynx would be unable to move up or downward during swallowing to ease the passing of a bolus.
    (c) Sound production would be impaired.
    (d) The person would be likely to choke.

16. The cartilage that serves as a base for the larynx is the
    (a) thyroid cartilage.
    (b) cuneiform cartilage.
    (c) corniculate cartilage.
    (d) cricoid cartilage.

17. The vocal folds are located in the
    (a) nasopharynx.
    (b) oropharynx.
    (c) larynx.
    (d) trachea.

18. The trachea
    (a) is lined by simple squamous epithelium.
    (b) is reinforced with C-shaped cartilages.
    (c) contains no mucous glands.
    (d) does not alter its diameter.

19. The cartilage in the walls of the lobar and segmental bronchi
    (a) have a completely different function than do those of the tracheal rings.
    (b) support the bronchi and assist in keeping the lumens open.
    (c) are unusual among cartilaginous tissues in that they are highly vascular.
    (d) assist directly in gas exchange by acting as baffles to direct the airflow.

Level 2 Reviewing Concepts

1. The epithelium of the nasal vestibule
    (a) does not assist with filtration.
    (b) contains many small short cilia.
    (c) contains hairs that prevent only the smallest particles from entering the nasal passages.
    (d) contains coarse hairs that guard the nasal opening from the entry of large particles and insects.

2. Why does a person with a cold often have a stuffed-up nose?
    (a) The response of the nasal epithelium to irritants, such as those that cause the common cold, is to secrete greater amounts of mucus than normal to trap the virus particles.
    (b) During the time that a person has a cold, air is easier to breathe if it is of much higher humidity than normal, and the presence of the cold viruses stimulates mucus production to increase that humidity.
    (c) Excess saliva is inhaled through the posterior nasal apertures when a person sneezes and fills the nasal cavity.
    (d) No known reason exists.

3. How does deep breathing differ from costal breathing?
    (a) In deep breathing, the thoracic volume changes because the rib cage changes shape.
    (b) It moves air into and out of the bronchial tree, which does not occur in costal breathing.
    (c) The diaphragm contracts in deep breathing, while it does not contract in costal breathing.
    (d) In deep breathing, muscles contract to assist in both inspiration and expiration.

4. How many lobes does the right lung have?

5. What is bronchodilation?

6. What is the function of the alveolar macrophages?

7. What is the function of the septa?

8. What do the intrinsic laryngeal muscles do?

9. Which paired laryngeal cartilages are involved with the opening and closing of the glottis?

10. What portion of the pharynx does the laryngopharynx include?

Level 3 Critical Thinking

1. What is the condition called asthma, and how can the symptoms be explained in anatomical terms?

2. A newborn infant was found dead, abandoned by the road. Among the many questions that the police would like to have answered is whether the infant was born dead or alive. After an autopsy, the medical examiner tells them that the infant was dead at birth. How could the medical examiner determine this?

24

CLINICAL CASE WRAP-UP

How Long Should This Cough Last?

Andrea thinks a typical cough should last only about a week. In reality, however, coughs caused by colds and flu tend to last about 18 days. Andrea believes that antibiotics cured her previous cough, but antibiotics only work against bacteria. If a virus caused her previous cough, her body's immune system likely resolved the cough on its own during the normal course of the illness. It is worth noting that viruses cause 90 percent of acute upper respiratory infections.

Fever is the body's innate immunity at work. Excessive mucus production, leading to a productive cough and runny nose, indicates that the immune system is working. The excessive mucus produced by the respiratory mucosa builds up and irritates the trachea and bronchi. This causes parasympathetic reflexive coughing. The explosive ejection of air and mucus against a suddenly open glottis helps clear away cold viruses. However, it can also spread these viruses to new hosts.

Andrea's doctor does not prescribe antibiotics because they are ineffective against viruses. However, she does advise Andrea to rest, drink plenty of fluids, and use a vaporizer and saline nasal spray to humidify the nasal passages. She coaches Andrea on "good coughing technique" to minimize the risk of spreading her cold to other people. For the duration of her cough, Andrea also wears a mask to avoid spreading this virus to her lab partners.

1. Should Andrea take medicine to suppress her acute, productive cough? Explain.
2. Which structures of Andrea's respiratory system seem to be infected by this rhinovirus?

See the blue Answers tab at the back of the book.

Related Clinical Terms

aspirate: To draw fluid from the body by suction; foreign material accidentally sucked into the lungs.

bronchography: A procedure in which radiopaque materials are introduced into the airways to improve x-ray imaging of the bronchial tree.

Cheyne-Stokes breathing: Hyperpnea (deep, fast breathing) alternating with apnea (absence of breathing).

dyspnea: The condition of labored breathing.

endotracheal tube: Tube that is passed through the mouth or nose to the trachea.

hemothorax: The condition of having blood in the pleural cavity.

hyperbaric oxygenation: Therapy to force more oxygen into the blood by use of a pressure chamber.

orthopnea: Condition in which one has breathing difficulty except when in an upright position.

rhinoplasty: Plastic surgery of the nose.

sputum: A mixture of saliva and mucus coughed up from the respiratory tract, often as the result of an infection.

wheeze: An audible whistling sound when breathing.

25 The Digestive System

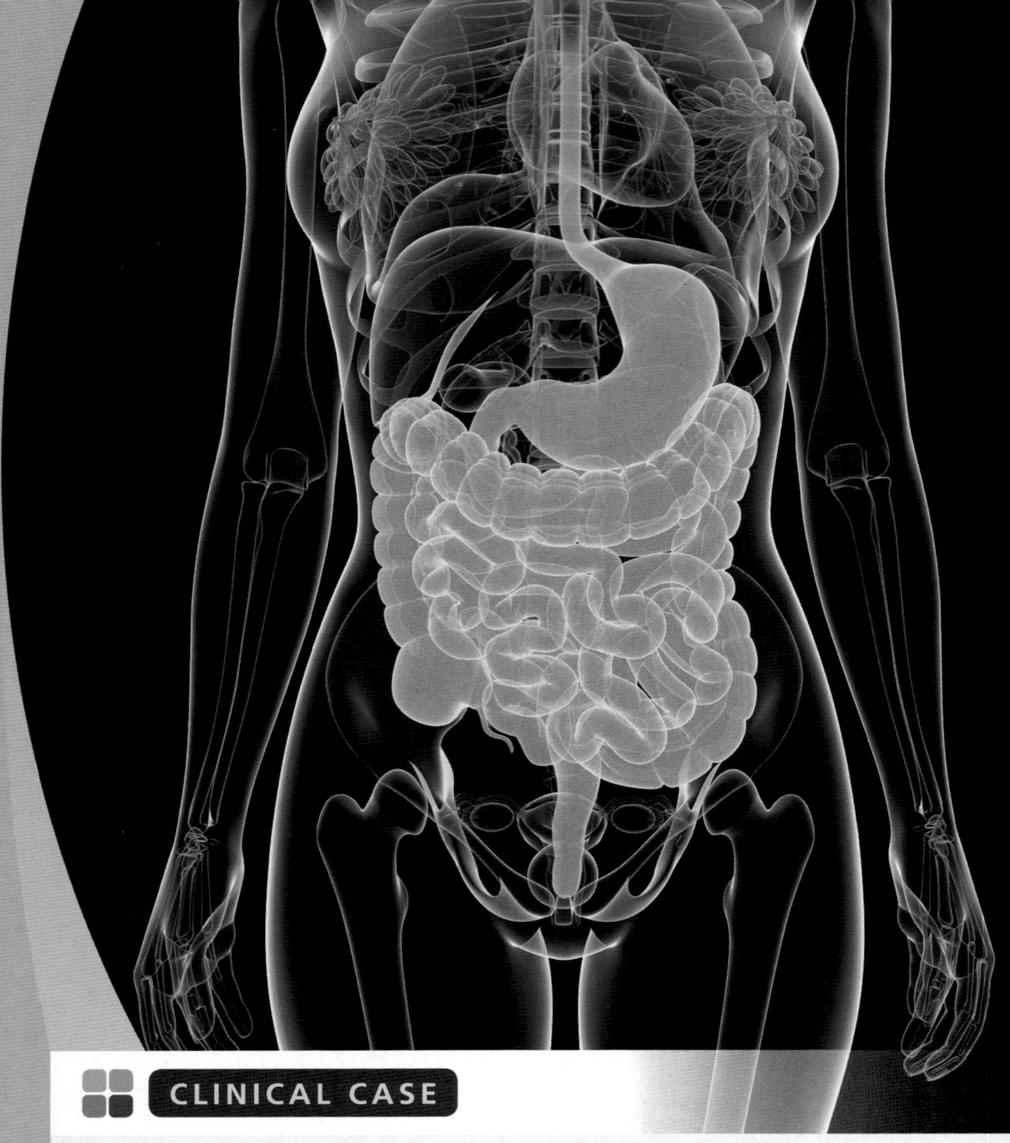

Learning Outcomes

These Learning Outcomes correspond by number to this chapter's sections and indicate what you should be able to do after completing the chapter.

25.1 Describe the histological organization of the four layers of the digestive tract. p. 651

25.2 Outline the gross and microscopic anatomy of the tongue, teeth, and salivary glands. p. 655

25.3 Describe the structure and function of the pharynx. p. 661

25.4 Outline the gross and microscopic anatomy of the esophagus. p. 661

25.5 Describe the gross and microscopic anatomy of the stomach. p. 663

25.6 Describe the gross and microscopic anatomy of the small intestine. p. 668

25.7 Describe the gross and microscopic anatomy of the large intestine and compare and contrast it to that of the small intestine. p. 672

25.8 Describe the anatomy of the liver, gallbladder, and pancreas. p. 675

25.9 Describe the changes in the digestive tract that occur with aging. p. 682

CLINICAL CASE

An Unusual Transplant

Tayvian has been hospitalized for 2 weeks and is taking antibiotics for a stubborn case of pneumonia. Now, although his cough has resolved, he has watery diarrhea (#7 on the Bristol stool chart) 10 times a day. The last few times it was bloody. He has no appetite and has crampy abdominal pain. He has a fever and is losing weight.

His doctor enters the room with results of tests performed on his feces. "I am afraid the normal, helpful bacteria that live in your colon have been killed off by the antibiotics that cured your pneumonia. Now harmful bacteria—*Clostridium difficile*—have taken over," reports the doctor. "They are producing toxins that are causing colitis (inflammation of the colon). The usual treatment is more antibiotics."

"Doctor," admits Tayvian, "I don't think I can take any more antibiotics. I feel worse than when I came in here."

The doctor replies, "Well, I have an idea: a transplant from one of your healthy, adult grandchildren. It is a little unusual, so you need to keep an open mind."

What kind of transplant can possibly help Tayvian? To find out, turn to the Clinical Case Wrap-Up on p. 686.

FEW OF US THINK ABOUT OUR DIGESTIVE SYSTEM unless it malfunctions, yet each day we spend conscious effort filling and emptying it. References to this system are part of our everyday language. We may have a "gut feeling" or find an opinion "hard to swallow."

A muscular tube called the **digestive tract** (or *alimentary canal*) and various **accessory organs** make up the digestive system. The digestive tract includes the oral cavity (*mouth*), pharynx, esophagus, stomach, small intestine, and large intestine. The accessory organs include the teeth, tongue, and various glandular organs, such as the salivary glands, liver, gallbladder, and pancreas. The accessory glandular organs secrete water, enzymes, buffers, and other components into ducts emptying into the digestive tract. As food enters and passes along the digestive tract, the secretions prepare nutrients for absorption across the digestive tract epithelium. The digestive tract and accessory organs work together to perform the following functions:

- Ingestion: Ingestion occurs when foods and liquids enter the digestive tract via the mouth.
- Mechanical processing: Most ingested solids undergo mechanical processing by the tongue and teeth before they are swallowed. Swirling, mixing, churning, and propulsive motions of the digestive tract provide additional mechanical processing after swallowing.
- Digestion: Digestion is the chemical and enzymatic breakdown of complex sugars, lipids, and proteins into small organic molecules that are absorbed by the digestive epithelium.
- Secretion: Digestion involves the action of acids, enzymes, and buffers produced by active secretion. Some of these secretions are produced by the lining of the digestive tract, but most are produced and secreted by accessory organs, such as the pancreas.
- Absorption: Absorption is the movement of organic molecules, electrolytes, vitamins, and water across the digestive epithelium and into the interstitial fluid of the digestive tract.
- Excretion: Excretion is the elimination from the body of the undigested residue of food and the waste products of metabolism. Waste products are secreted into the digestive tract, primarily by the accessory glands, especially the liver.
- Compaction: Compaction is the progressive dehydration of undigested materials and organic wastes prior to excretion from the body. The compacted material is called **feces**. **Defecation** (def-e-KĀ-shun) is the elimination of feces from the body through the anus.

The lining of the digestive tract plays an important defensive role by protecting surrounding tissues against (1) the corrosive effects of digestive acids and enzymes, (2) mechanical stresses, such as abrasion, and (3) pathogens that are swallowed with food or reside within the digestive tract.

In summary, the organs of the digestive system mechanically and chemically process food that is eaten and passed along the digestive tract. These activities reduce the solid, complex chemical structures of food into small molecules. The epithelium lining the digestive tract absorbs these small molecules for transfer to the circulating blood.

25.1 An Overview of the Digestive System

Figure 25.1 shows the major components of the digestive system. Although several of the organs have overlapping functions, each has certain areas of specialization and shows distinctive histological characteristics. Before discussing these specializations and distinctions, let's consider the basic structural characteristics common to all portions of the digestive tract.

Histological Organization of the Digestive Tract

▶ **KEY POINT** Most organs of the digestive tract have four histological layers: the (1) mucosa, (2) submucosa, (3) muscular layer, and (4) serosa. Just as the functions of the different organs of the digestive tract vary, these layers differ histologically from one organ to another.

The major layers of the digestive tract are the (1) mucosa, (2) submucosa, (3) muscular layer, and (4) serosa. Variations in the structure of these four layers from organ to organ are related to the specific functions of each organ (**Figure 25.2**).

The Mucosa

The **mucosa**, the inner lining of the digestive tract, is a **mucous membrane**, which is a layer of loose connective tissue covered by an epithelium moistened by glandular secretions. ⮌ p. 71 The **mucosal epithelium** is either a stratified or a simple epithelium, depending on the location and the stresses involved. For example, the oral cavity and esophagus are lined by a nonkeratinized stratified squamous epithelium that resists stress and abrasion. In contrast, the stomach, small intestine, and almost the entire large intestine encounter considerably less stress and abrasion than the oral cavity and esophagus. As a result, these structures have a simple columnar epithelium specialized for secretion and absorption.

Many segments of the digestive tract have **circular folds** or *plicae* (PLĪ-se; singular, *plica* [PLĪ-kā]) *circulares*. (**Figure 25.2**). These transverse or longitudinal folds of the mucosa increase the surface area available for absorption and secretion. In some regions of the digestive tract, circular folds are permanent features involving both the mucosa and submucosa. In other regions, circular folds are temporary features that disappear as the lumen fills, enabling the lumen to expand after a large meal. Ducts opening onto the epithelial surfaces carry the secretions of gland cells located either in the mucosa and submucosa or within accessory organs.

A layer of areolar connective tissue is found deep to the epithelium of the mucosa. This layer, the **lamina propria**, contains blood vessels, sensory nerve endings, lymphatic vessels, smooth muscle fibers, and scattered areas of lymphatic tissue. In most regions of the digestive tract the border of the mucosa is a narrow band of smooth muscle and elastic fibers. This band of smooth muscle is called the **muscularis** (mus-kū-LA-ris) **mucosae**. The smooth muscle fibers in the muscularis mucosae are arranged in two thin concentric layers—an internal circular layer and an external longitudinal layer (**Figure 25.2**). Contraction of these layers alters the shape of the lumen and moves the epithelial pleats and folds.

The Submucosa

The **submucosa** (sub-mū-KŌ-sa) is a layer of areolar connective tissue deep to the muscularis mucosae. Large blood and lymphatic vessels are found in this layer. In some regions of the digestive tract the submucosa also contains exocrine glands that secrete buffers and enzymes into the lumen. The submucosa contains a network of nerve fibers and scattered neuron cell bodies. These **submucous neural plexuses** innervate the mucosa. They contain sensory neurons, parasympathetic ganglia, and sympathetic postganglionic fibers (**Figure 25.2**).

The Muscular Layer

The **muscular layer** is a double layer of smooth muscle fibers deep to the submucosa. These smooth muscle fibers are arranged in internal circular and external longitudinal layers (**Figure 25.2**). These layers of smooth muscle mechanically process and propel materials along the digestive tract. These movements are coordinated by neurons of the **myenteric** (mī-en-TER-ik; *mys*, muscle, + *enteron*, intestine) **neural plexus**. This plexus, located between the two muscular layers, is composed of parasympathetic ganglia and sympathetic postganglionic fibers. Parasympathetic stimulation increases muscular tone and stimulates contractions, while sympathetic stimulation decreases muscular tone and promotes relaxation.

Figure 25.1 Components of the Digestive System. This figure introduces the major regions and accessory organs of the digestive tract, together with their primary functions.

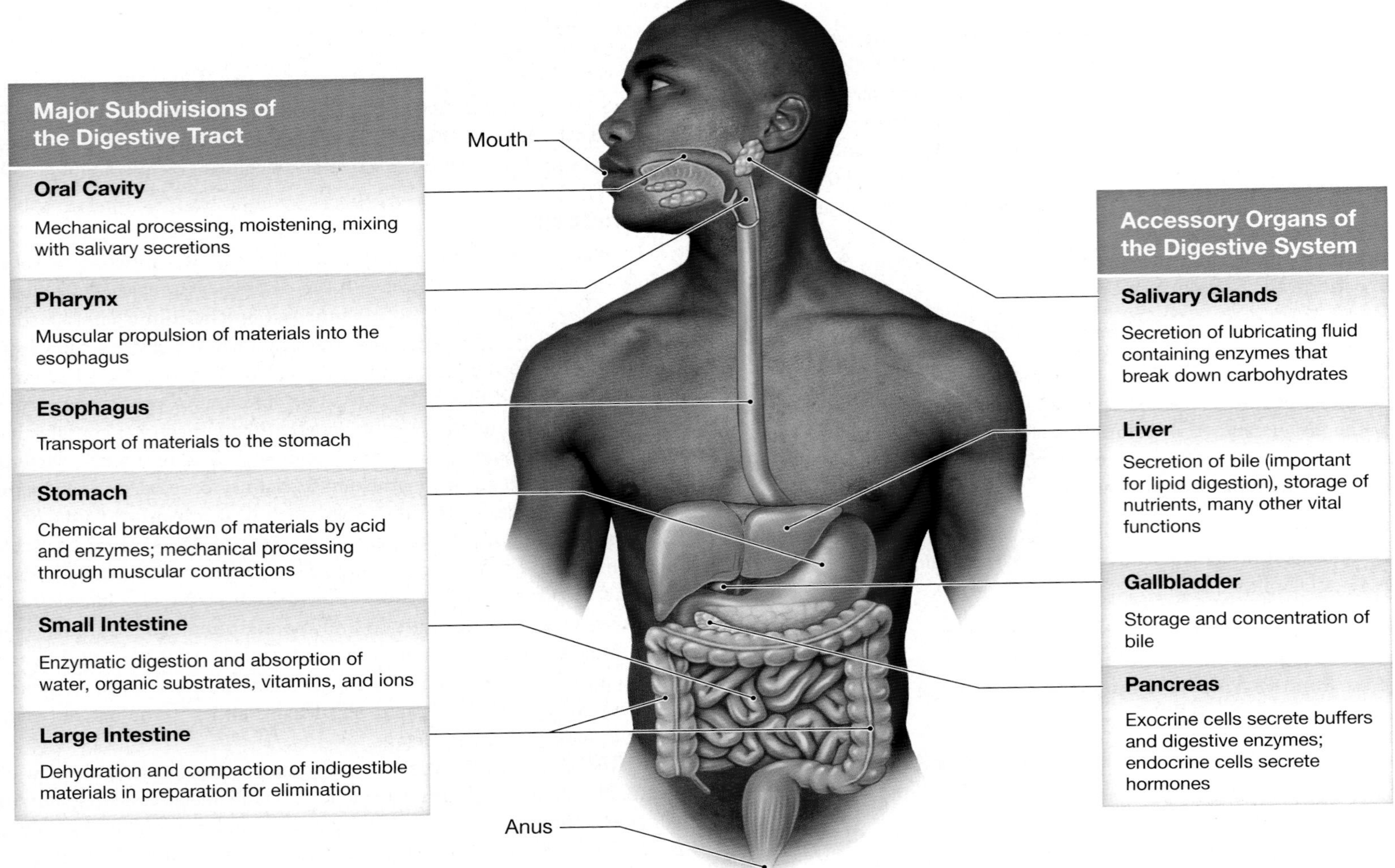

Major Subdivisions of the Digestive Tract
Oral Cavity
Mechanical processing, moistening, mixing with salivary secretions
Pharynx
Muscular propulsion of materials into the esophagus
Esophagus
Transport of materials to the stomach
Stomach
Chemical breakdown of materials by acid and enzymes; mechanical processing through muscular contractions
Small Intestine
Enzymatic digestion and absorption of water, organic substrates, vitamins, and ions
Large Intestine
Dehydration and compaction of indigestible materials in preparation for elimination
Mouth
Anus
Accessory Organs of the Digestive System
Salivary Glands
Secretion of lubricating fluid containing enzymes that break down carbohydrates
Liver
Secretion of bile (important for lipid digestion), storage of nutrients, many other vital functions
Gallbladder
Storage and concentration of bile
Pancreas
Exocrine cells secrete buffers and digestive enzymes; endocrine cells secrete hormones

At specific locations along the digestive tract there are thickened areas of the muscular circular layer. These localized thickenings form **sphincters**, or *valves*. The sphincters constrict to restrict movement or to ensure one-way passage of materials along the lumen.

The Serosa

Along regions of the digestive tract within the peritoneal cavity the muscular layer is covered by a serous membrane known as the **serosa** (**Figure 25.2**). There is no serosa, however, surrounding the muscular layer of the pharynx, esophagus, and rectum. Instead, the muscular layer is wrapped by a dense network of collagen fibers that firmly attaches the digestive tract to adjacent structures. This fibrous sheath is the **adventitia** (ad-ven-TISH-a).

Muscular Layers and the Movement of Digestive Materials

▶ **KEY POINT** The smooth muscle of the digestive tract contracts spontaneously, producing the rhythmic contractions of peristalsis and segmentation.

The muscularis mucosae and muscular layer of the digestive tract contain **smooth muscle**. Smooth muscle cells range from 5 to 10 μm in diameter and from 30 to 200 μm in length. They are surrounded by connective tissue, but unlike skeletal muscle, the collagen fibers do not form tendons or aponeuroses. In addition, unlike skeletal muscle cells, the contractile proteins of these smooth muscle cells are not organized into sarcomeres. Therefore, the smooth muscle cells lack striations. Although nonstriated, their contractions are as strong as those of skeletal or cardiac muscle cells.

Because the contractile filaments of smooth muscle cells are not rigidly organized, a stretched smooth muscle cell soon adapts to its new length and still retains the ability to contract on demand. This ability to tolerate extreme stretching is called plasticity. Plasticity is especially important for digestive organs that undergo great changes in volume, such as the stomach.

The smooth muscle cells of the digestive tract are involuntary muscle cells, and many have no autonomic nervous system motor innervation. **Pacesetter cells** are located in the muscularis mucosae and muscular layer that surround the lumen of the digestive tract. Pacesetter cells undergo spontaneous depolarization, triggering contraction of the smooth muscle. Gap junctions electrically connect the adjacent muscle cells. When one smooth muscle cell contracts, the contraction spreads in a wave that travels throughout the tissue. The initial stimulus may be a pacesetter cell or a motor neuron stimulating one of the smooth muscle cells in that region. It may also be a local response to chemicals, hormones, concentrations of oxygen or carbon dioxide, or physical factors such as extreme stretching or irritation.

Figure 25.2 **Histological Structure of the Digestive Tract**

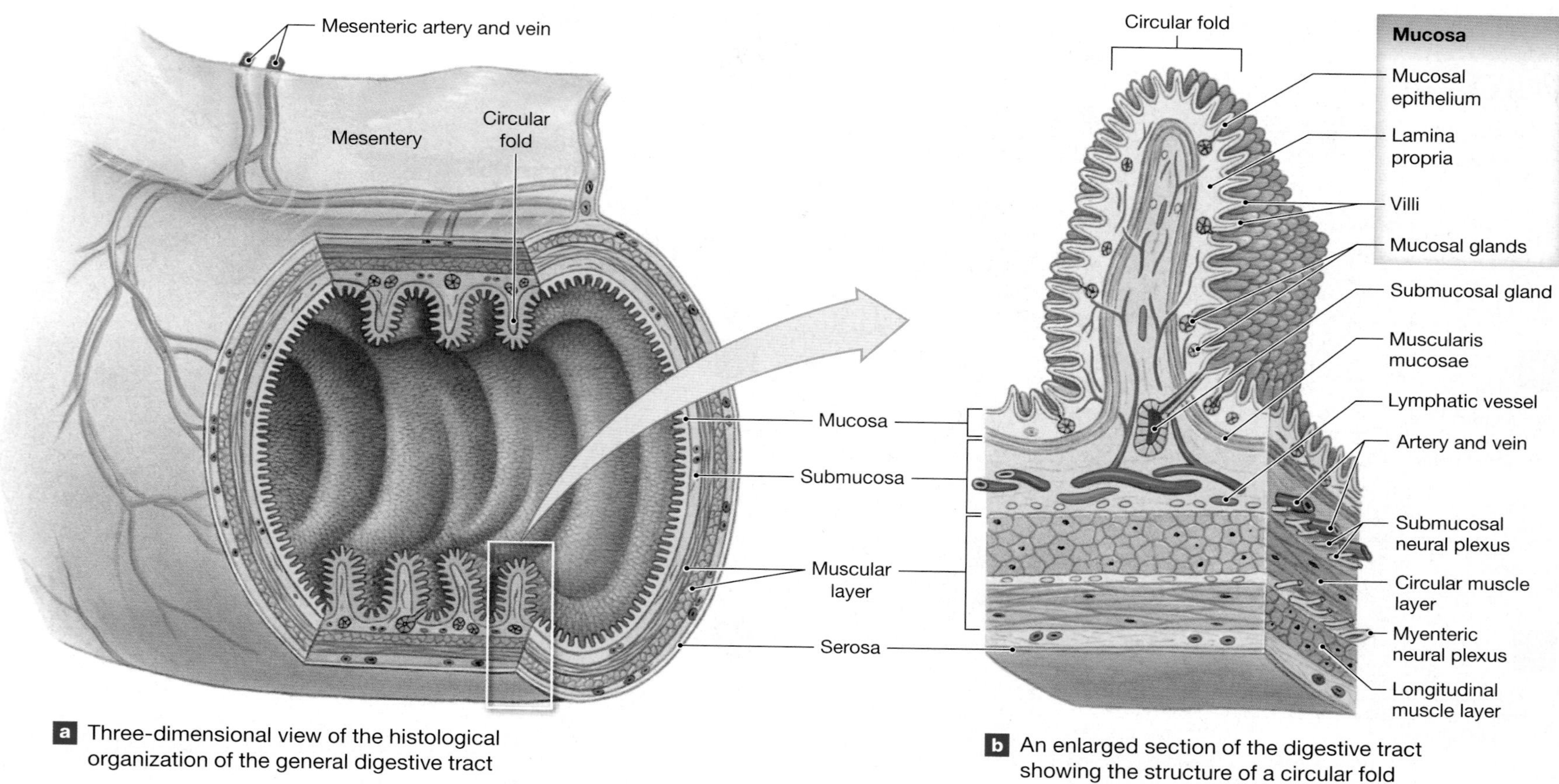

a Three-dimensional view of the histological organization of the general digestive tract

b An enlarged section of the digestive tract showing the structure of a circular fold

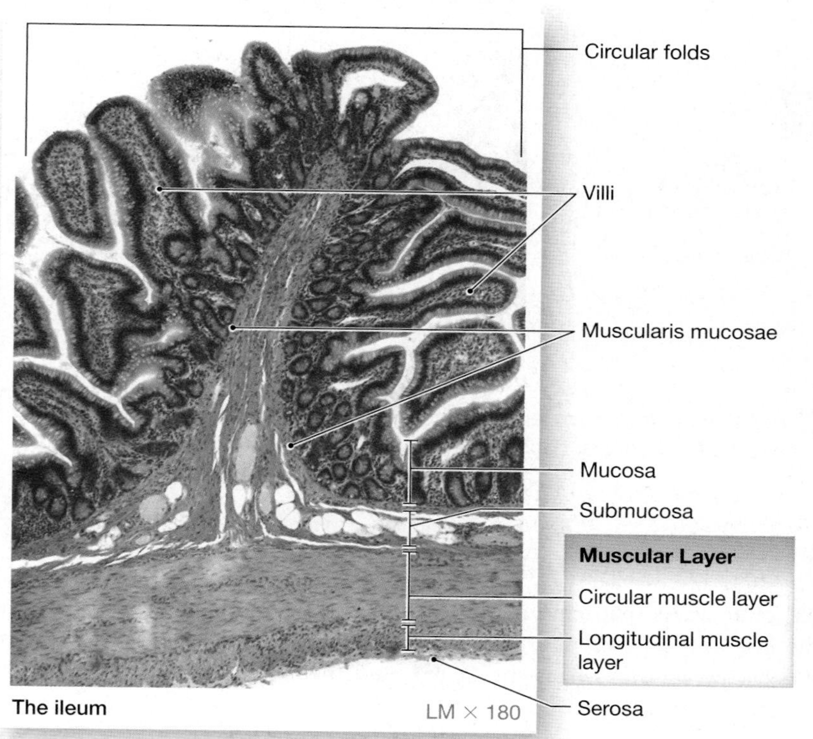

c Photomicrograph of the ileum showing aspects of the histological organization of the small intestine

Peristalsis and Segmentation

Contractions within the muscularis mucosae that move a **bolus** (a small oval mass of food) along the digestive tract are called **peristalsis** (per-i-STAL-sis). During a **peristaltic wave**, the circular muscles contract behind the digestive contents. Longitudinal muscles contract next, shortening adjacent segments. A wave of contraction in the circular muscles then forces the materials in the desired direction **(Figure 25.3a)**.

Most areas of the small intestine and some regions of the large intestine undergo **segmentation (Figure 25.3b)**. These movements churn and fragment the digestive materials, mixing the contents with intestinal secretions. They do not move the bolus in any particular direction.

Segmentation and peristaltic contractions are also triggered by hormones, chemicals, and physical stimulation. Afferent and efferent fibers of the glossopharyngeal, vagus, or pelvic nerves initiate peristaltic waves. Sensory receptors in the wall of the digestive tract trigger *local* peristaltic movements that are limited to a few centimeters of the digestive tract. These afferent fibers synapse within the myenteric neural plexus and produce *localized, short* **myenteric reflexes** that do not involve the CNS. The term enteric nervous system refers to the neural network coordinating these reflexes.

In general, short reflexes control activities in *one* region of the digestive tract. This control involves coordinating local peristalsis and triggering the secretion of digestive glands. Many neurons are involved—the enteric nervous system has about as many neurons and neurotransmitters as the spinal cord has.

Sensory information from receptors in the digestive tract is also distributed to the CNS, where it triggers *long* reflexes. Long reflexes involve interneurons and motor neurons in the CNS and provide a higher level of control over digestive and glandular activities. These reflexes generally control large-scale

Figure 25.3 **Peristalsis and Segmentation**

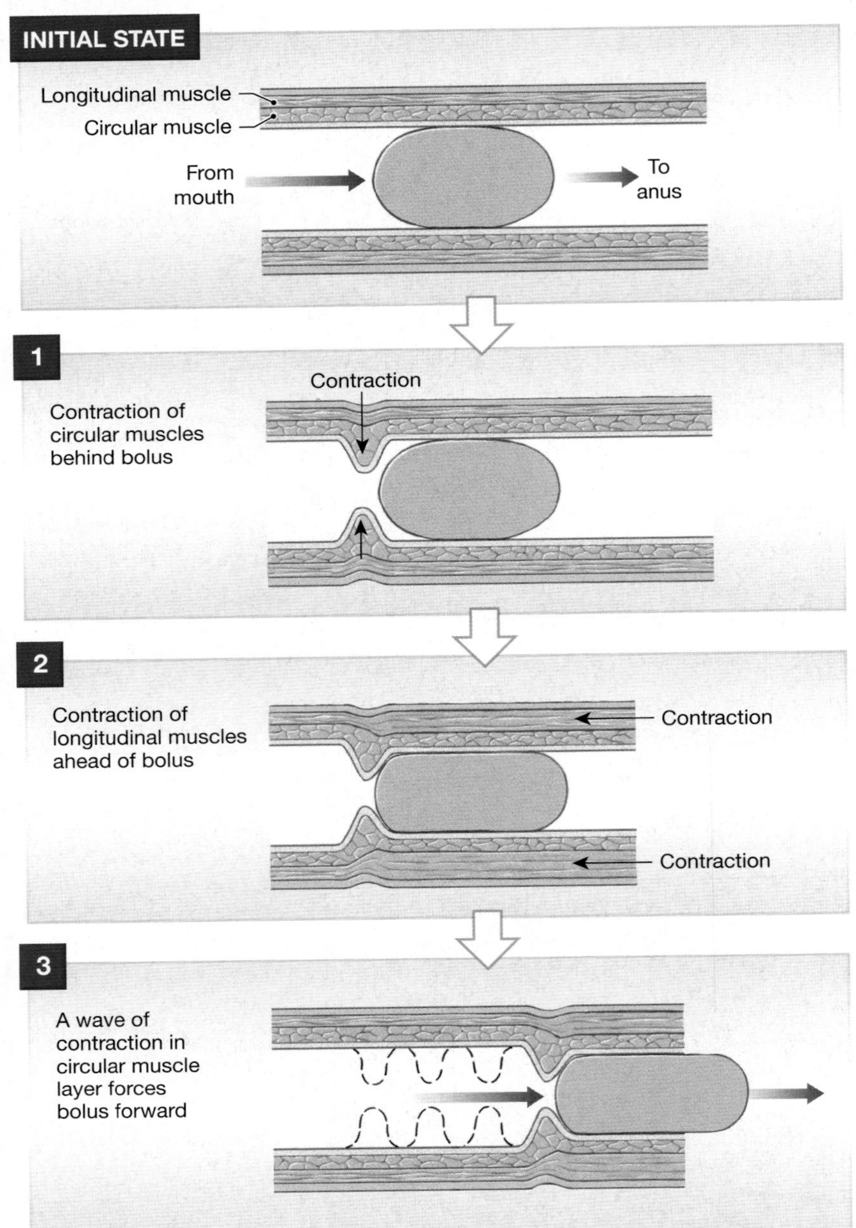

a Peristalsis propels materials along the length of the digestive tract by coordinated contractions of the circular and longitudinal layers.

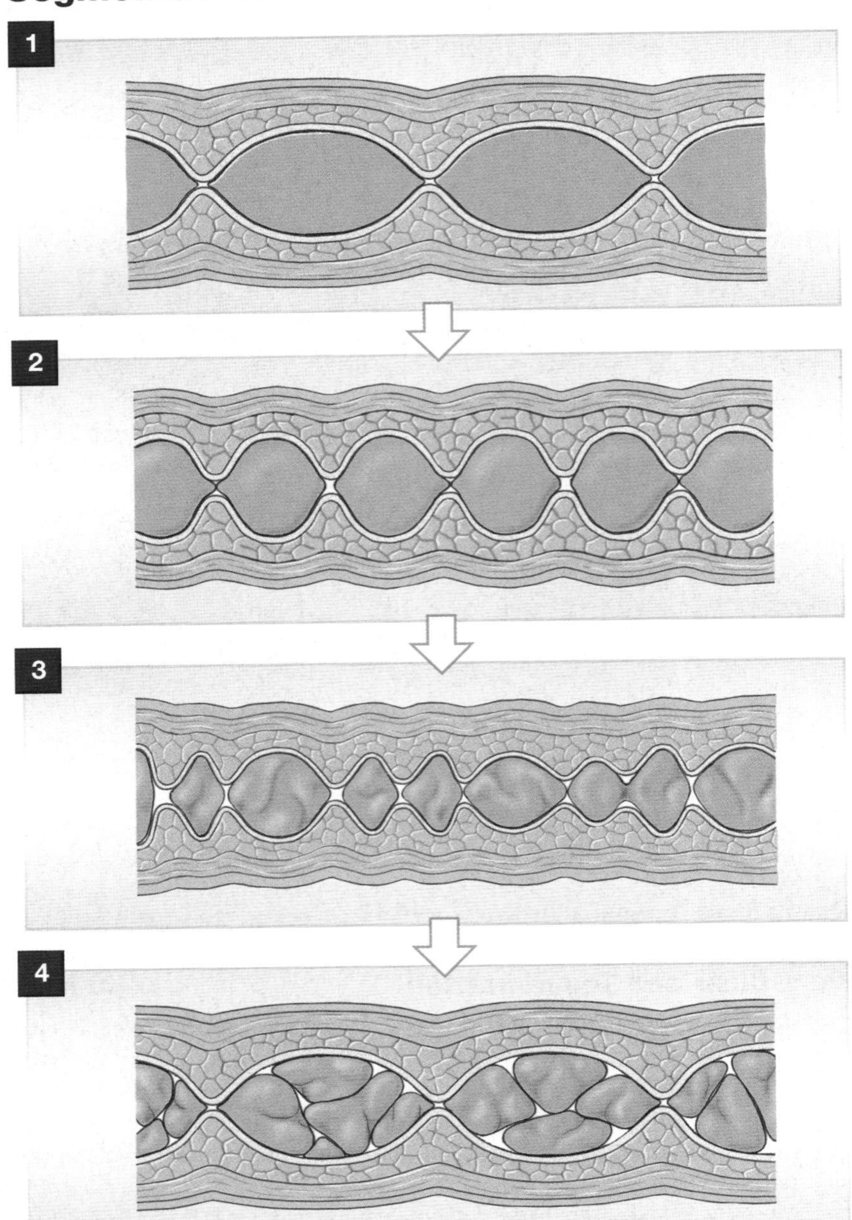

b Segmentation movements primarily involve the circular muscle layers. These activities churn and mix the contents of the digestive tract, but do not produce net movement in a particular direction.

25

peristaltic waves that move materials from one region of the digestive tract to another. Long reflexes may involve motor fibers in the glossopharyngeal, vagus, or pelvic nerves that synapse in the myenteric neural plexus.

TIPS & TOOLS

Squeezing toothpaste out of a tube is similar to peristalsis: Your squeezing hand (contracting circular muscles) forces toothpaste (the bolus) along and out of the tube (the digestive tract).

The Peritoneum

▶ **KEY POINT** The peritoneum is a serous membrane with two parts: the visceral peritoneum and parietal peritoneum. Mesenteries connect the visceral and parietal peritoneal membranes.

The peritoneum is a serous membrane with two parts. The **visceral peritoneum**, or serosa, is continuous with the **parietal peritoneum** lining the inner surfaces of the body wall. The organs of the abdominal cavity are often described as lying *within* the abdominal and peritoneal cavities. Abdominal organs demonstrate one or more of the following relationships with the peritoneal membranes:

- **Intraperitoneal** organs lie within the peritoneal cavity, in that they are covered on all sides by the visceral peritoneum. The stomach, liver and ileum are intraperitoneal organs.
- **Retroperitoneal** organs are covered by the parietal peritoneum on their anterior surface only, so they lie outside the peritoneal cavity. These organs typically do not develop from the embryonic gut. The kidneys, ureters, and abdominal aorta are retroperitoneal.
- **Secondarily retroperitoneal** organs are organs of the digestive tract that form as intraperitoneal organs but become retroperitoneal. The shift occurs during embryonic development as a portion of the visceral peritoneum fuses with the opposing parietal peritoneum. The pancreas and the distal two-thirds of the duodenum are secondarily retroperitoneal organs.

The peritoneum is a serosal membrane. It continually produces a watery fluid that lubricates the peritoneal surfaces. About 7 liters of fluid are secreted and reabsorbed each day. However, the volume within the peritoneal cavity at any one time is very small. Under unusual conditions, such as liver disease, heart failure, or electrolyte imbalance, the volume of peritoneal fluid increases markedly, resulting in a dangerous reduction in blood volume and distortion of visceral organs.

Mesenteries

Within the peritoneal cavity, most regions of the digestive tract are suspended by sheets of serous membrane connecting the parietal peritoneum with the visceral peritoneum. These **mesenteries** (MES-en-ter-ēz) are fused double sheets of peritoneal membrane **(Figure 25.2a)**. The areolar connective tissue between the two surfaces provides a route for the passage of blood vessels, nerves, and lymphatic vessels to and from the digestive tract. Mesenteries also stabilize the positions of the attached organs and prevent their entanglement during digestive movements or sudden changes in body position.

During development, dorsal and ventral mesenteries suspend the digestive tract and accessory organs within the peritoneal cavity **(Figure 25.4a)**. The ventral mesentery later disappears along most of the digestive tract, remaining only in two locations. One is on the ventral surface of the stomach, between the stomach and liver, forming the **lesser omentum** (Ō-MEN-tum; *omentum*, fat skin). The second is between the liver and the anterior abdominal wall and diaphragm, forming the falciform ligament **(Figure 25.4b–d)**. (*Although this peritoneal sheet is called a "ligament," it is not comparable to the ligaments interconnecting bones.*) (For additional information concerning the development of the digestive system, see the Embryology Summary in Chapter 28.)

As the digestive tract elongates, it twists and turns within the crowded peritoneal cavity. The dorsal mesentery of the stomach enlarges, forming a pouch extending inferiorly between the body wall and the anterior surface of the small intestine. This pouch is the **greater omentum** (**Figure 25.4b** and **Spotlight Figure 25.10**). The loose connective tissue within the mesentery of the greater omentum contains a thick layer of adipose tissue. The lipids in the adipose tissue are thought to have two possible functions: (1) serving as an important energy reserve and (2) providing insulation that minimizes heat loss across the anterior abdominal wall. The greater omentum also contains numerous lymph nodes. These protect the body from foreign proteins, toxins, or pathogens that evade the defenses of the digestive tract.

A thick mesentery, the **mesentery proper**, suspends all but the first 25 cm of the small intestine. This provides stability while permitting a certain amount of independent movement. The **mesocolon** is a mesentery attached to the large intestine. The middle portion of the large intestine (the transverse colon) is suspended by the **transverse mesocolon**. The sigmoid colon, which leads to the rectum and anus, is suspended by the **sigmoid mesocolon**. During embryonic development, the dorsal mesentery of the ascending colon, descending colon, and rectum fuses to the posterior body wall. This fused mesentery fixes them in position. These organs are now secondarily retroperitoneal, and the visceral peritoneum covers only their anterior surfaces and portions of their lateral surfaces **(Figure 25.4b–d)**.

25.1 CONCEPT CHECK

1. What are the components and functions of the mucosa of the digestive tract?
2. What are the functions of mesenteries?
3. What is the functional difference between peristalsis and segmentation?
4. What is important about the lack of organization in the contractile filaments of smooth muscle cells?

See the blue Answers tab at the back of the book.

CLINICAL NOTE

Ascites

ASCITES (a-SĪ-tēz) is the buildup of serous fluid within the peritoneal cavity, which creates a characteristic abdominal swelling. This excess fluid overwhelms the peritoneal cavity's reabsorption capacity. As more than the normal 100–400 mL of peritoneal fluid accumulates, the person develops symptoms including abdominal discomfort, heartburn, indigestion, shortness of breath, and lower back pain.

The most common cause of ascites is advanced liver disease, often called **cirrhosis** of the liver. This can result from alcohol abuse or infection with hepatitis C or B. Other causes include kidney failure, heart failure, and advanced cancer.

Treatment of ascites depends on its underlying cause. Sodium intake is restricted and diuretics are prescribed to increase water loss and to limit water retention.

25.2 The Oral Cavity

▶ **KEY POINT** The oral cavity functions in sensory analysis of food, mechanically processes and lubricates it, and performs limited digestion of carbohydrates and lipids.

Our exploration of the digestive tract follows the path of food from the mouth to the anus. The mouth opens into the oral cavity. The functions of the oral cavity include (1) sensory analysis of food before swallowing, (2) mechanical digestion through the actions of the teeth, tongue, and palatal surfaces, (3) lubrication by mixing ingested material with mucus and saliva, and (4) limited chemical digestion of carbohydrates by salivary amylase and lipids by lingual lipase.

Anatomy of the Oral Cavity

▶ **KEY POINT** Structures of the oral cavity include the oral mucosa, tongue, salivary glands, and teeth.

The **oral mucosa** lines the **oral cavity (Figure 25.5)**. The oral mucosa is a nonkeratinized stratified squamous epithelium that protects the mouth from abrasion. Buccal fat pads and the buccinator support the mucosa of the **cheeks**, which are the lateral walls of the oral cavity. Anteriorly, the mucosa of the cheeks is continuous with the **lips**. The space between the cheeks, lips, and

Figure 25.4 Mesenteries

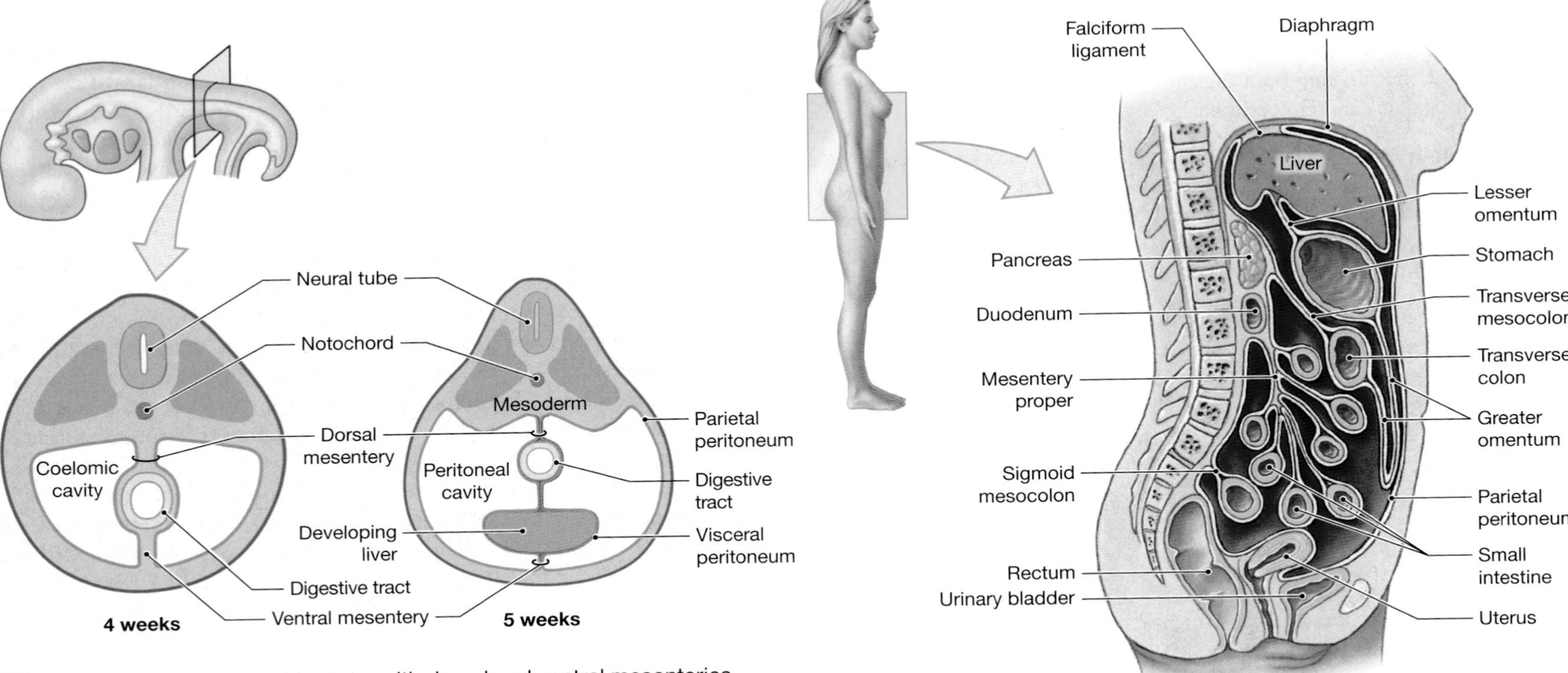

a Diagram of early embryonic state, with dorsal and ventral mesenteries supporting the digestive tract (left). A slightly later stage showing the development of the liver within the ventral mesentery (right). (For further information, see the **Embryology Summary** in **Chapter 28**.)

b Mesenteries of the abdominopelvic cavity, as seen in a diagrammatic sagittal section.

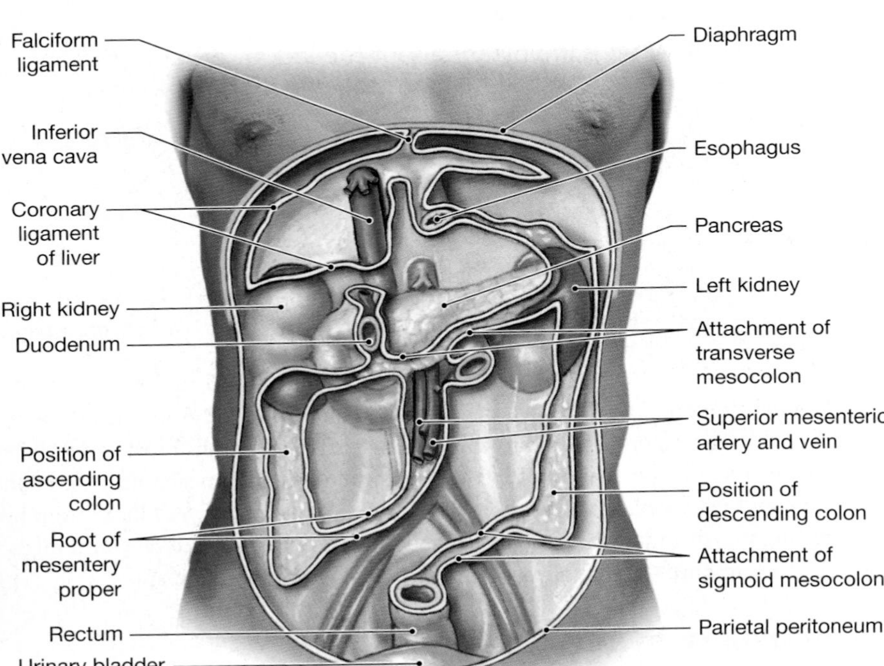

c Anterior view of the empty peritoneal cavity showing the attachment of mesenteries and visceral organs to the posterior wall of the abdominal cavity.

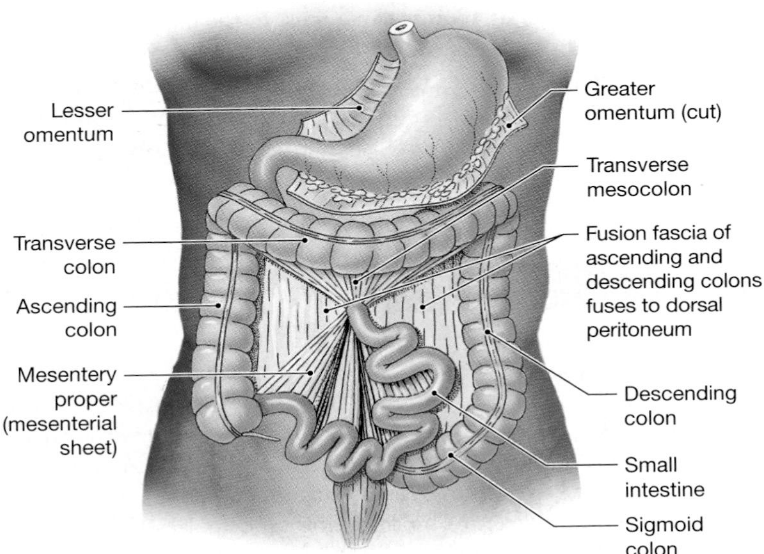

d The organization of mesenteries in the adult. This is a simplified view; the length of the small intestine has been greatly reduced.

25

Figure 25.5 **The Oral Cavity**

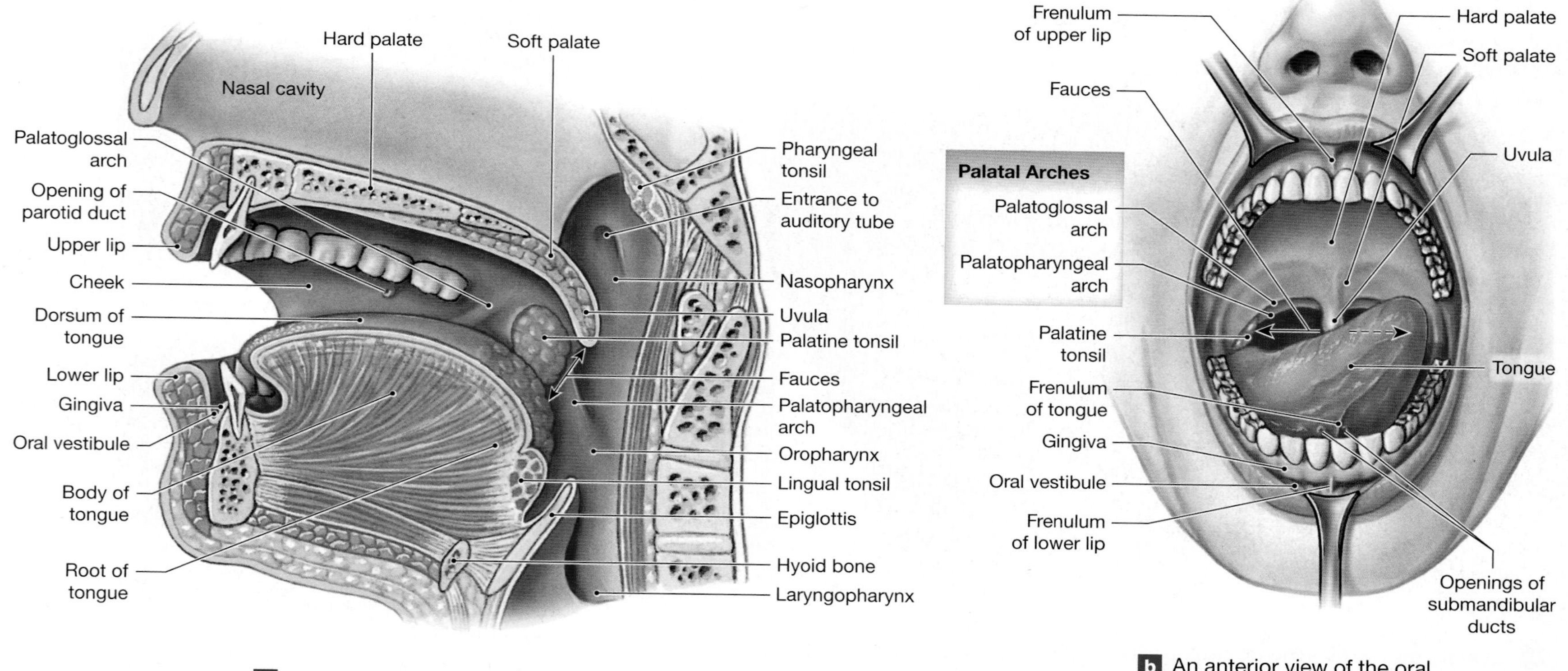

a The oral cavity in sagittal section

b An anterior view of the oral cavity through the open mouth

teeth is the **oral vestibule**. The ridges of oral mucosa, the **gingivae** (JIN-ji-vē), or *gums*, surround the base of each tooth on the alveolar processes of the maxillae and the alveolar part of the mandible.

The tongue forms the floor of the oral cavity. The mylohyoid, which is located inferior to the tongue, provides additional support to the floor of the oral cavity. ⮌ p. 269

The **hard palate** separates the oral cavity from the nasal cavity **(Figure 25.5a)**. The hard palate is formed by the palatine processes of the maxilla and the horizontal plates of the palatine bones.

The **soft palate** lies posterior to the hard palate. It separates the oral cavity from the nasopharynx and closes off the nasopharynx during swallowing. The finger-shaped **uvula** (Ū-vū-la) dangles from the center of the posterior margin of the soft palate. The uvula prevents food from entering the pharynx too soon. The two pairs of muscular **palatal arches** are found on each side of the uvula **(Figure 25.5)**:

1. The palatoglossal arches extend between the soft palate and the base of the tongue. Each arch consists of a mucous membrane and the underlying palatoglossus muscle and associated tissues.
2. The palatopharyngeal arches extend from the soft palate to the side of the pharynx. Each arch consists of a mucous membrane and the underlying palatopharyngeus muscle and associated tissues.

The palatine tonsils lie between the palatoglossal and palatopharyngeal arches. The space between the oral cavity and the pharynx, bounded by the soft palate and the base of the tongue, is called the **fauces** (FAW-sēz). **(Figure 25.5b)**.

The Tongue

The **tongue** has four primary functions: (1) mechanical digestion by compression, abrasion, and distortion, (2) manipulation to assist in chewing and to prepare food for swallowing, (3) sensory analysis by touch, temperature, and taste receptors. and (4) secretion of mucin and the enzyme *lingual lipase*. The tongue is divided into an anterior **body** and a posterior **root** **(Figure 25.5)**. The **dorsum of the tongue** contains numerous fine projections called lingual papillae. Each papilla has a thick layer of stratified squamous epithelium, which produces additional friction that helps the tongue move materials around. Additionally, taste buds are found along the edges of many papillae. **Figure 18.7** shows surface features and histological details of the tongue. ⮌ p. 478 A V-shaped line of circumvallate papillae roughly indicates the boundary between the body and the root of the tongue, which is situated in the pharynx.

The tongue's epithelium is flushed by water, mucin, and lingual lipase, which are secretions from small glands extending into the lamina propria. Lingual lipase begins the enzymatic breakdown of lipids, specifically triglycerides.

The epithelium covering the inferior surface of the tongue is thinner and more delicate than that of the dorsum. Along the inferior midline is the **frenulum of the tongue** (FREN-ū-lum; *frenulum*, small bridle). This thin fold of mucous membrane connects the body of the tongue to the mucosa of the oral floor. Ducts from two pairs of salivary glands open on each side of the frenulum, which prevents extreme movements of the tongue **(Figures 25.5b** and **25.6a)**. However, an overly restrictive frenulum hinders normal eating or speech. This condition, called ankyloglossia (ang-ki-lō-GLOS-ē-uh), can be corrected surgically.

The tongue contains two different groups of muscles, **extrinsic tongue muscles** and **intrinsic tongue muscles**. The hypoglossal nerve (XII) controls these muscles. The extrinsic muscles include the hyoglossus, styloglossus, genioglossus, and palatoglossus muscles. These muscles perform all gross movements of the tongue. The smaller intrinsic muscles change the shape of the tongue and assist the extrinsic muscles during precise movements, such as speech.

Figure 25.6 The Salivary Glands

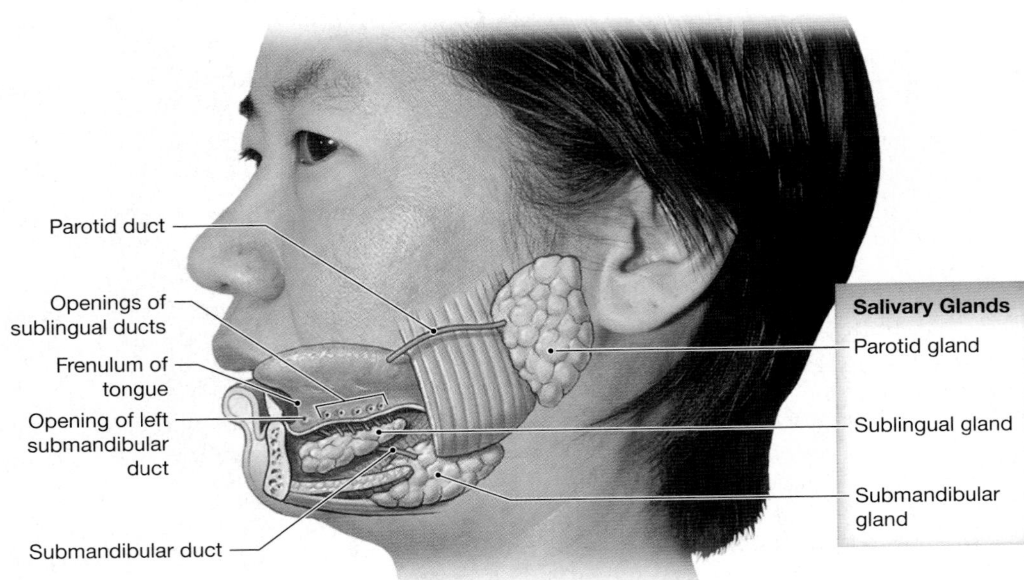

a Lateral view showing the relative positions of the salivary glands and ducts on the left side of the head. Much of the left half of the body and ramus of the mandible have been removed. For the positions of the ducts inside the oral cavity, see **Figure 25.5**.

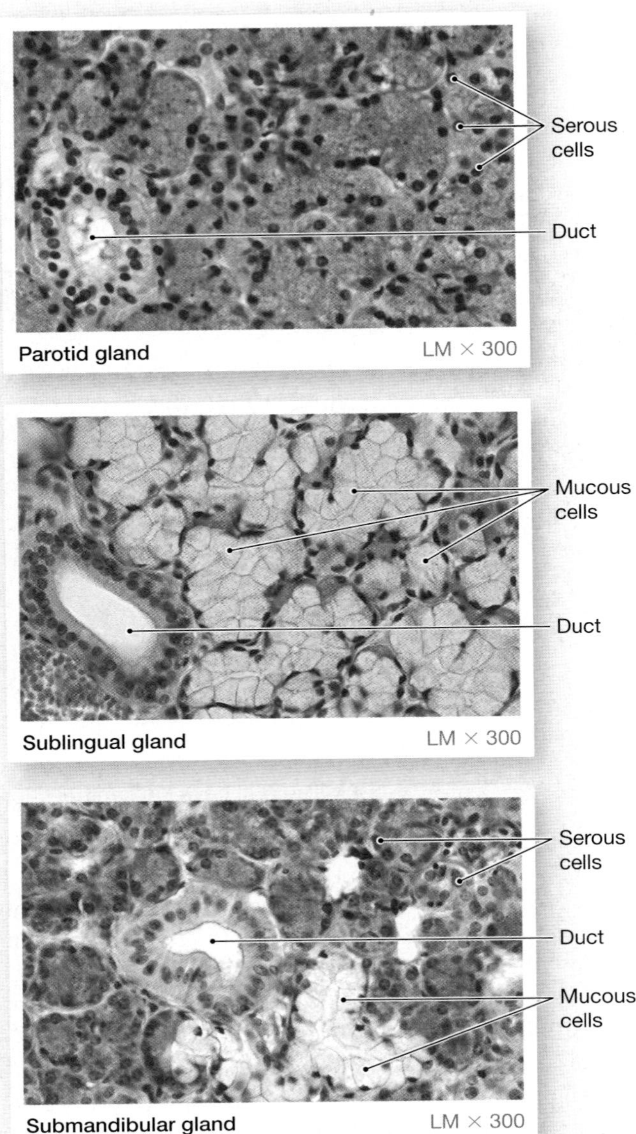

b Histological detail of the parotid, sublingual, and submandibular glands. The parotid gland produces saliva rich in enzymes. The gland is dominated by serous secretory cells. The sublingual gland produces saliva rich in mucin. This gland is dominated by mucous secretory cells. The submandibular gland produces saliva containing enzymes and mucin, and it contains both serous and mucous secretory cells.

Salivary Glands

Three pairs of salivary glands secrete into the oral cavity **(Figure 25.6)**:

1. The **parotid** (pa-ROT-id) **glands** are the largest salivary glands, weighing approximately 20 g. The superior, anterior portion of each parotid gland extends between the inferior surface of the zygomatic arch and the anterior margin of the sternocleidomastoid. The posterior portion extends from the mastoid process of the temporal bone anteriorly, crossing the superficial surface of the masseter. The secretions of each gland are drained by a **parotid duct**, which empties into the oral vestibule at the second upper molar **(Figures 25.5a** and **25.6a)**.
2. The mucous membrane of the floor of the mouth covers the **sublingual** (sub-LING-gwal) **glands**. Numerous **sublingual ducts** open along either side of the frenulum of the tongue **(Figure 25.6a)**.
3. The **submandibular glands** are found in the floor of the mouth along the medial surfaces of the mandible inferior of the mylohyoid line. The **submandibular ducts** open into the mouth on either side of the frenulum of the tongue, immediately posterior to the teeth **(Figure 25.5b)**. **Figure 25.6b** shows the histological appearance of a submandibular gland.

The saliva in the mouth is a mixture of all of the salivary glands' secretions. About 70 percent of the saliva originates in the submandibular salivary glands, 25 percent in the parotid salivary glands, and 5 percent in the sublingual salivary glands. Approximately 1.0–1.5 L of saliva is produced each day. It has a composition of 99.4 percent water, with the remainder being an assortment of ions, buffers, metabolites, and enzymes. Glycoproteins called mucin give saliva its lubricating action.

At mealtimes the production of large quantities of saliva lubricates the mouth, moistens the food, and dissolves chemicals that stimulate the taste buds. A continual background level of secretion flushes the oral surfaces and controls the populations of oral bacteria. Reduction or elimination of salivary secretions triggers a bacterial population explosion in the oral cavity. This proliferation rapidly leads to recurring infections and the progressive erosion of teeth and gums.

Regulation of Salivary Secretion

The autonomic nervous system controls the salivary gland secretions. Each salivary gland receives parasympathetic and sympathetic innervation. Food in the mouth triggers a salivary reflex by stimulating receptors monitored by the trigeminal nerve or by stimulating taste buds innervated by N VII, N IX, or N X. Parasympathetic stimulation speeds up secretion by all the salivary glands. As a result, you produce large amounts of saliva. The role of sympathetic innervation is unclear. Evidence suggests that it provokes the secretion of small amounts of very thick saliva.

The Teeth

The movements of the tongue pass food across the surfaces of the teeth. Teeth perform chewing, or **mastication** (mas-ti-KĀ-shun), of food. Mastication breaks down tough connective tissues in meat and plant fibers in vegetables. It also saturates food with salivary secretions and enzymes.

Figure 25.7a is a sectional view through an adult tooth. Most of each tooth consists of **dentine** (DEN-teen), or *dentin* (DEN-tin), a mineralized matrix similar to bone. However, unlike bone, dentine does not contain living cells. Instead, cells in the central **pulp cavity** extend cytoplasmic processes into the dentine. The pulp cavity is spongy and highly vascular. Blood vessels and nerves enter the pulp cavity by a narrow tunnel, the **root canal**, located at the base, or **root**, of the tooth. The **dental artery**, **dental vein**, and **dental nerve** enter the root canal through the **apical foramen** at the tip of the root.

The root of each tooth sits in a bony cavity or socket called the **tooth socket**, or **tooth alveolus**. Collagen fibers of the **periodontal** (per-ē-ō-DON-tal) **ligament** extend from the dentine of the root to the bone, creating a strong, fibrous articulation known as a gomphosis. A layer of **cement** covers the dentine of the root. This protects and firmly anchors the periodontal ligament. The cement is histologically similar to bone and less resistant to erosion than dentine.

The **neck** of the tooth marks the boundary between the root and the **crown**, the visible portion of the tooth projecting above the soft tissue of the gingiva. A shallow groove called the **gingival** (JIN-ji-val) **sulcus** surrounds the neck of each tooth. This prevents bacteria from accessing the lamina propria of the gingiva or the relatively soft cement of the root. If this attachment breaks down, bacteria can infect the gingiva, causing inflammation termed gingivitis.

A layer of **enamel** covers the dentine of the crown. Enamel forms the **occlusal surface** of each tooth, which is the biting surface that grinds food against the opposing tooth surface. Elevations or projections of the occlusal surface are called **cusps**. Enamel contains densely packed crystals of calcium phosphate and is the hardest biologically manufactured substance. Adequate amounts of calcium, phosphates, and vitamin D are essential during childhood if the enamel coating is to be complete and resistant to decay.

CLINICAL NOTE

Mumps

Mumps is caused by an infection with the mumps virus. Infection most often occurs between ages 5 and 9. Humans are the sole reservoir for the mumps virus, which is transmitted person to person via respiratory droplets and saliva.

The mumps virus usually targets the salivary glands, most often the parotid glands, causing them to become swollen and painful. In post-adolescent males, the virus may also infect the testes and cause sterility, or the pancreas, producing temporary or permanent diabetes. Severe cases may affect other organ systems, including the central nervous system. The first exposure stimulates antibody production and usually confers permanent immunity. An effective preventive mumps vaccine is available and has almost eliminated this disease in the United States, but it still appears in developing countries.

Types of Teeth The alveolar processes of the maxillae and the alveolar part of the mandible form the **maxillary dental arcade** and **mandibular dental arcade**, or *upper* and *lower dental arcades*, respectively. These arcades contain four types of teeth, each with specific functions **(Figure 25.7b)**:

1. **Incisors** (in-SĪ-zerz) are blade-shaped teeth found at the front of the mouth. Incisors clip and cut food—think of biting off the tip of a carrot stick.
2. The **canines**, or *cuspids* (KUS-pidz), are conical teeth with a sharp ridgeline and a pointed tip. They tear and slash food. We might bite off a tough piece of celery with our incisors, but we will then move it to one side to take advantage of the shearing action of the canines. Incisors and canines each have a single root.
3. **Premolars**, or *bicuspids* (bī-KUS-pidz), have flattened crowns with two prominent rounded cusps. They crush, mash, and grind. Premolars have one or two roots.
4. **Molars** have very large, flattened crowns with four to five prominent rounded cusps adapted for crushing and grinding. Molars in the upper jaw typically have three roots, whereas those in the lower jaw usually have two roots.

Sets of Teeth During development, two sets of teeth begin to form: deciduous and permanent. The **deciduous** (dē-SID-ū-us; *deciduus*, falling off) **teeth** appear first. Most children have 20 deciduous teeth. On each side of the upper or lower jaw, the deciduous teeth consist of two incisors, one canine, and a pair of deciduous molars for a total of 20 teeth **(Figure 25.7c,d)**. These teeth are later replaced by the **permanent teeth** of the larger adult jaws. As replacement proceeds, the periodontal ligaments and roots of the deciduous teeth erode. The deciduous teeth either fall out or are pushed aside by the **eruption**, or *emergence*, of the permanent teeth. The adult premolars take the place of the deciduous molars, while the adult molars extend the tooth rows as the jaw enlarges. Three additional molars appear on each side of the upper and lower jaws as the person ages. These molars extend the rows of teeth posteriorly and bring the permanent tooth count to 32.

The third molars, or *wisdom teeth*, may not erupt before age 21. Wisdom teeth may not erupt because they develop in inappropriate positions or because there is not enough space. **(Figure 25.7d,e)**.

Mastication The muscles of mastication close the jaws and slide or rock the lower jaw from side to side. During mastication, food is forced back and forth between the vestibule and the rest of the oral cavity, crossing and recrossing the occlusal surfaces of the teeth. This movement of food results in part from the action of the masticatory muscles. However, control would be impossible without the aid of the buccal, labial, and lingual muscles. Once the food has been shredded or torn to a satisfactory consistency and moistened with salivary secretions, the tongue begins compacting the debris into a bolus that can be swallowed.

25.2 CONCEPT CHECK

5 What type of epithelium lines the oral cavity?
6 What are the functions of saliva?
7 What nutrient begins its chemical breakdown in the mouth?
8 Pretend you are eating an apple. Summarize the actions of your teeth.

See the blue Answers tab at the back of the book.

Figure 25.7 Teeth

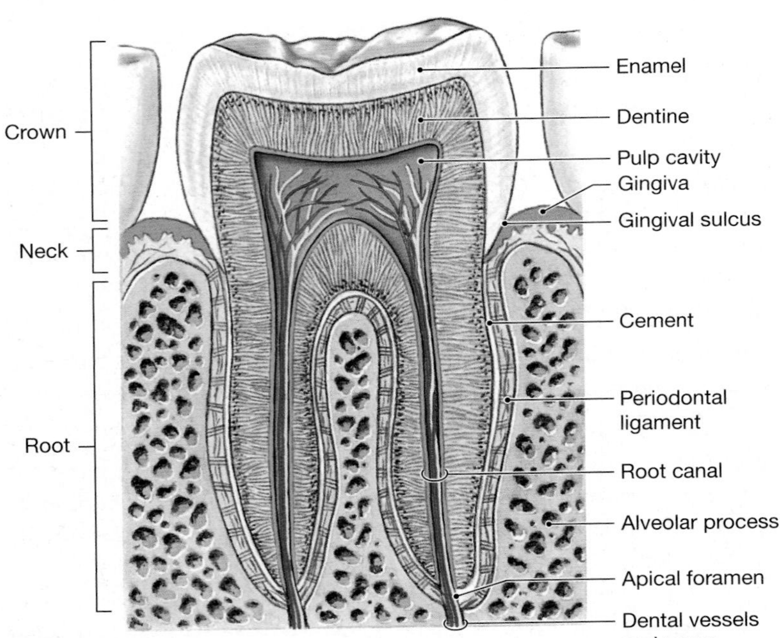

a Diagrammatic section through a typical adult tooth.

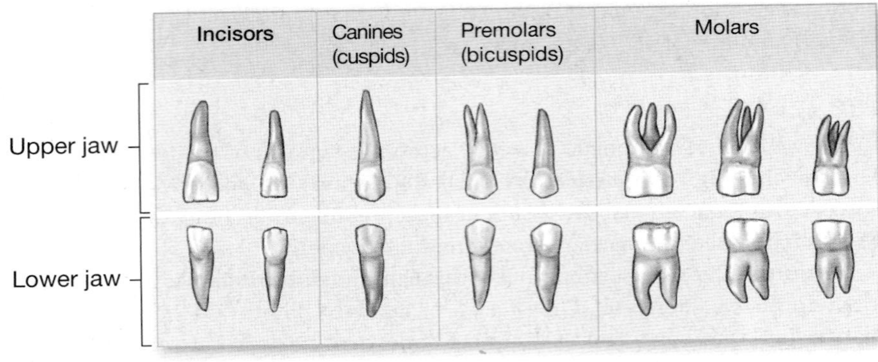

b The adult teeth.

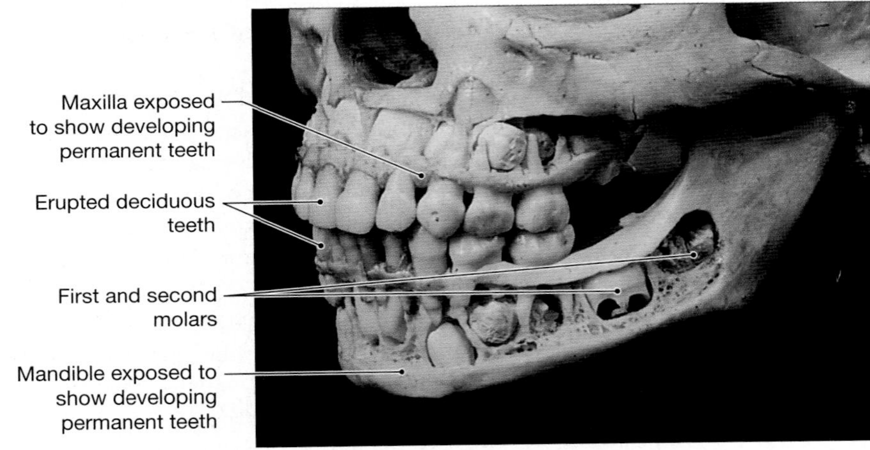

c The skull of a 4-year-old child, with the maxillae and mandible cut away to expose the unerupted permanent teeth.

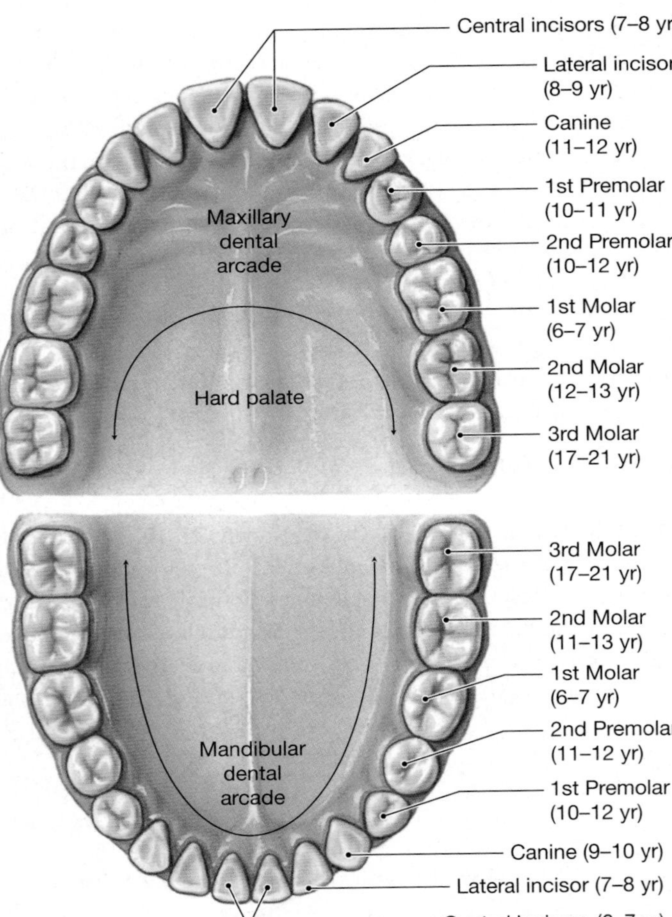

d The permanent teeth, with the age at eruption given in years.

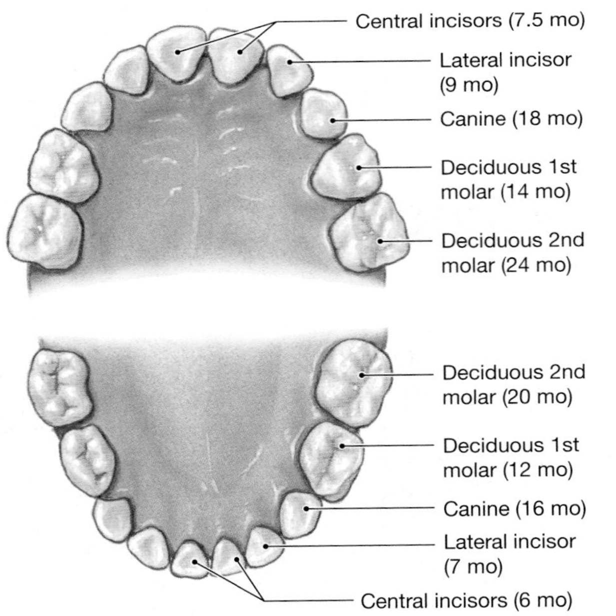

e The deciduous teeth with the age at eruption given in months.

25

25.3 | The Pharynx

▶ **KEY POINT** The pharynx is a common entrance to the respiratory system and the digestive tract. The process of swallowing begins in the oral cavity but continues in the pharynx.

The pharynx serves as a common passageway for food, liquids, and air. Chapter 24 described the epithelial lining and divisions of the pharynx: the nasopharynx, oropharynx, and laryngopharynx. ⮌ pp. 626–629 Deep to the lamina propria of the mucosa is a dense layer of elastic fibers, bound to the underlying skeletal muscles. Chapter 10 discussed the specific pharyngeal muscles involved in swallowing, which we summarize here. ⮌ pp. 267–268

- The superior, middle, and inferior pharyngeal constrictors push the bolus toward the esophagus.
- The palatopharyngeus and stylopharyngeus muscles elevate the larynx.
- The palatal muscles raise the soft palate and adjacent portions of the pharyngeal wall.

The pharyngeal muscles cooperate with muscles of the oral cavity and esophagus to initiate the swallowing process, or **deglutition** (dē-glū-TISH-un).

The Swallowing Process

▶ **KEY POINT** Swallowing is a complex process that is initiated voluntarily but continues involuntarily. There are three phases of swallowing: buccal, pharyngeal, and esophageal.

There are three phases of swallowing (**Figure 25.8**):

1. The **buccal phase** of swallowing starts when the bolus of food is compressed against the hard palate. The tongue retracts, which (a) forces the bolus into the pharynx, (b) helps the palatal muscles elevate the soft palate, and (c) isolates the nasopharynx. The buccal phase is voluntary. However, once the bolus enters the oropharynx, involuntary reflexes take over and move the bolus toward the stomach.
2. The **pharyngeal phase** begins when the bolus comes in contact with the palatal arches, the posterior pharyngeal wall, or both. The palatopharyngeus and stylopharyngeus muscles elevate the larynx. The folding of the epiglottis directs the bolus past the closed glottis. In less than a second, the pharyngeal constrictor muscles propel the bolus into the esophagus. As the bolus travels through the pharynx and into the esophagus, the respiratory centers are inhibited and breathing ceases.
3. The **esophageal phase** of swallowing starts when the **upper esophageal sphincter** opens. After passing through the open sphincter, peristaltic waves push the bolus through the esophagus. As the bolus approaches the stomach, the **lower esophageal sphincter** opens, and the bolus enters the stomach.

Figure 25.8 The Swallowing Process. This sequence, based on a series of x-rays, shows the stages of swallowing and the movement of materials from the mouth to the stomach.

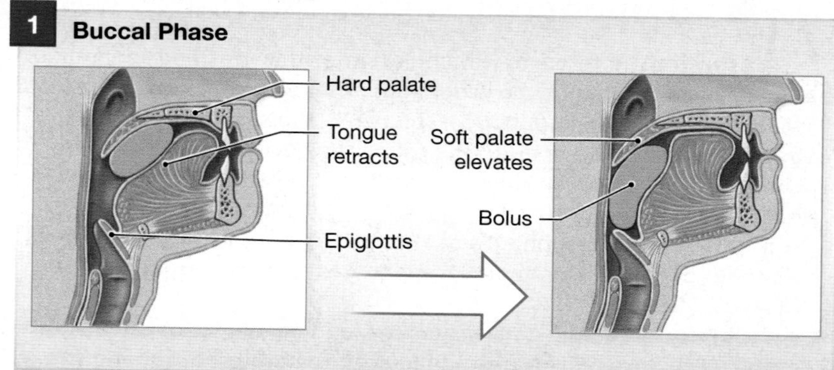

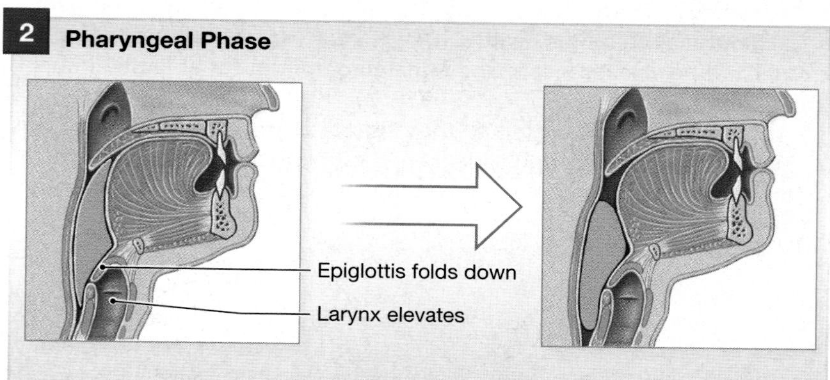

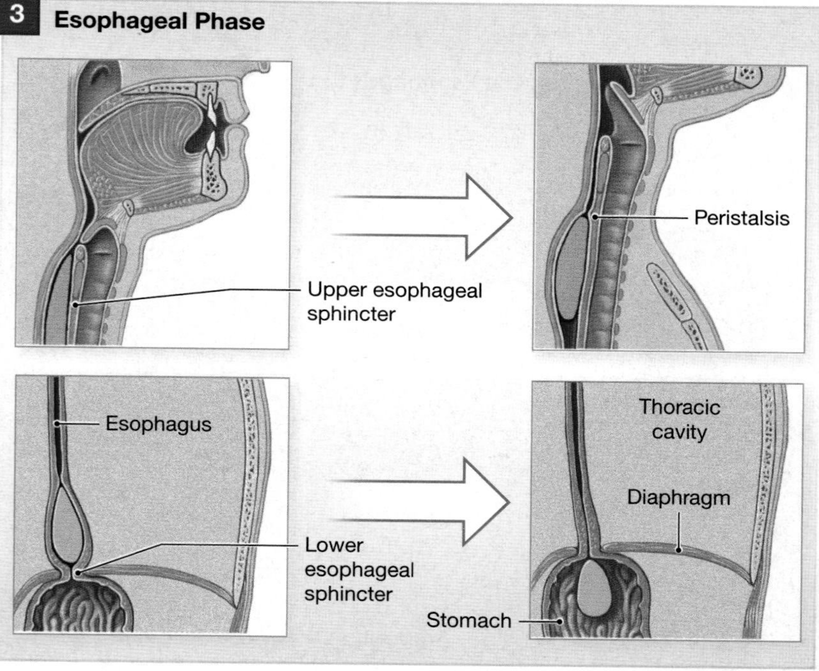

25.3 CONCEPT CHECK

9 List the six groups of muscles involved in the swallowing process.

10 What is occurring when the soft palate and larynx are elevated and the glottis closes?

See the blue Answers tab at the back of the book.

25.4 | The Esophagus

▶ **KEY POINT** The esophagus, a musculoelastic tube lying posterior to the trachea, empties into the stomach. Seven arterial branches supply the esophagus with blood.

The **esophagus** is a hollow muscular tube that transports foods and liquids to the stomach (**Figure 25.1**, p. 652). It lies posterior to the trachea slightly left of midline (**Figure 25.9**). It passes along the dorsal wall of the mediastinum in the thoracic cavity and enters the peritoneal cavity through an opening in

CLINICAL NOTE

Esophagitis, GERD, and Hiatal Hernia

Esophagitis refers to inflammation of the stratified squamous epithelial lining of the esophagus. The most common cause is failure of the lower esophageal sphincter, allowing reflux of the acidic stomach contents into the esophagus. This causes a burning chest discomfort, often called "heartburn," which has been compared to the pain of a heart attack. Inhaling the acidic stomach contents from the esophagus can lead to asthma. Chronic, recurrent episodes of reflux esophagitis leads to **gastroesophageal reflux disease (GERD)**.

Obesity, pregnancy, advancing age, and certain foods can all make esophagitis worse. Citrus fruits and juices, tomato products, and strong seasonings can irritate a damaged esophageal epithelium. Treatment includes maintaining a healthy weight, quitting smoking, changing the diet, eating smaller meals, and remaining upright for at least 2 hours after eating.

A **hiatal hernia** refers to the stomach or other abdominal organs pushing upward through a weakened esophageal hiatus of the diaphragm. Hiatal hernias are common in elderly people but are not always symptomatic. They can, however, contribute to GERD.

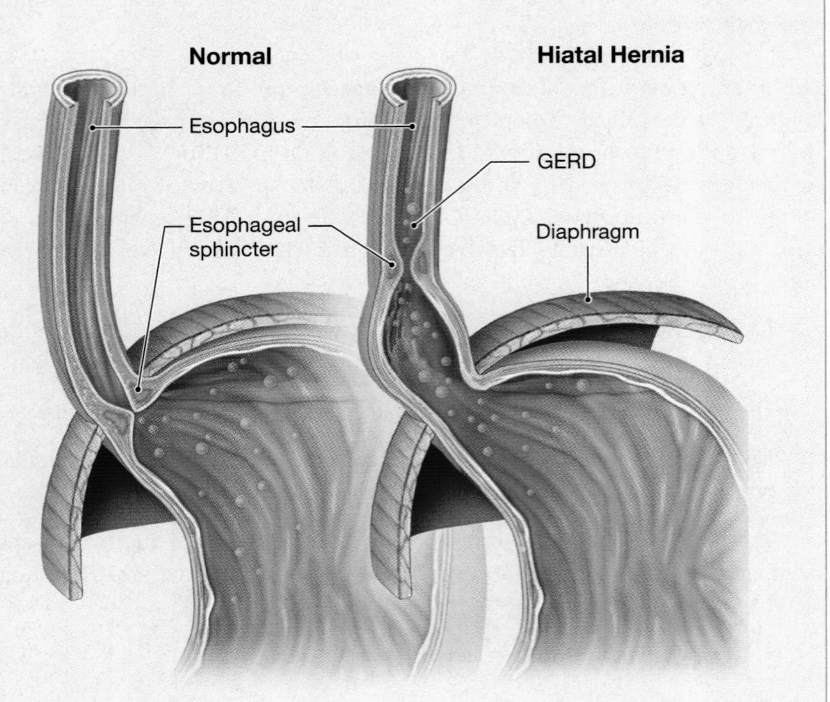

Figure 25.9 Histology of the Esophagus

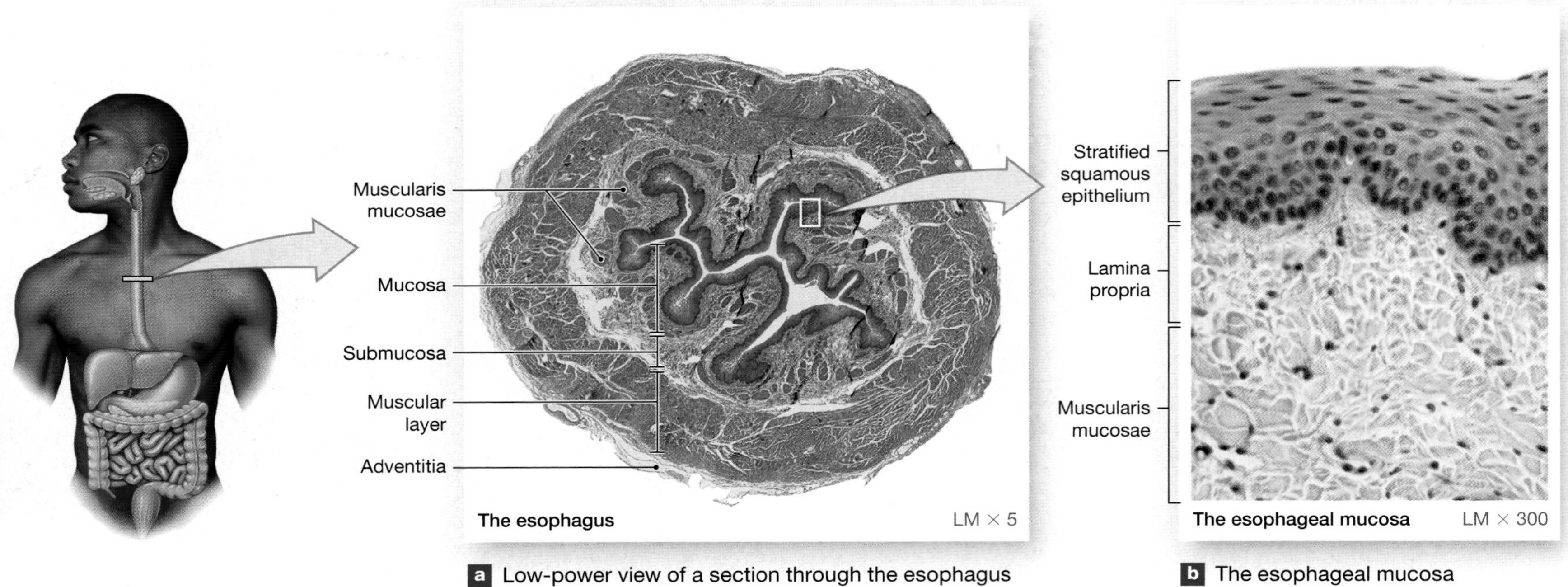

a Low-power view of a section through the esophagus

b The esophageal mucosa

the diaphragm, the **esophageal hiatus** (hī-Ā-tus). The esophagus empties into the stomach. The esophagus begins at the level of the cricoid cartilage anterior to vertebra C_6 and ends anterior to vertebra T_7.

The esophagus receives blood from (1) the esophageal arteries, branches of (2) the thyrocervical trunk and (3) external carotid arteries of the neck, (4) the bronchial arteries and (5) esophageal arteries of the mediastinum, (6) the inferior phrenic artery, and (7) the left gastric artery of the abdomen. Venous blood from the esophagus drains into the esophageal, inferior thyroid, azygos, and gastric veins. The vagus and sympathetic trunks innervate the esophagus via the esophageal plexus.

Neither the upper nor the lower portion of the esophagus has a well-defined sphincter muscle comparable to those located elsewhere along the digestive tract. Nevertheless, the terms upper esophageal sphincter and lower esophageal sphincter (or *cardiac sphincter*) are used to describe these regions because they are similar in function to other sphincters.

Histology of the Esophageal Wall

▶ **KEY POINT** The wall of the esophagus has mucosal, submucosal, and muscularis components but lacks a serosa. Six histological features distinguish it from other digestive structures.

The wall of the esophagus has mucosa, submucosa, and muscular layers comparable to those described earlier (**Figure 25.2**, p. 653). The esophageal wall has several distinctive features (**Figure 25.9**):

- The epithelium of the mucosa is an abrasion-resistant nonkeratinized (mucosal), stratified squamous epithelium.
- The mucosa and submucosa form large folds that run the length of the esophagus. These folds allow it to expand when a large bolus passes through. Muscle tone in the walls of the esophagus keeps the esophagus closed except during swallowing.
- The smooth muscle layer of the muscularis mucosae is very thin or absent near the pharynx. As you move inferiorly within the esophagus, it gradually thickens to 200–400 μm near the junction with the stomach. The muscularis mucosae of the esophagus is composed of only a single layer of longitudinal smooth muscle, which is different from the rest of the digestive tract.
- The submucosa contains scattered esophageal glands. (*The esophagus is only one of two regions of the digestive tract that contains submucosal glands; the other is the duodenum.*) These simple, branched, tubular glands produce a mucous secretion that lubricates the bolus and protects the epithelial surface.
- The muscular layer has inner circular and outer longitudinal muscle layers. In the superior third of the esophagus, both layers contain mostly skeletal muscle fibers and some isolated smooth muscle cells. In the middle third, there is an even mixture of skeletal and smooth muscle tissue. The inferior third contains only smooth muscle cells. Visceral reflexes control the skeletal muscle and smooth muscle in the esophagus; you do not have voluntary control over these contractions.
- An adventitia of connective tissue outside the muscular layer anchors the esophagus to the posterior body wall. Between the diaphragm and stomach, the esophagus is retroperitoneal. Peritoneum covers the anterior and left lateral surfaces.

25.4 CONCEPT CHECK

11 List the blood vessels that supply the esophagus with blood.

12 Why are the mucosa and submucosa of the esophagus folded?

See the blue Answers tab at the back of the book.

25.5 | The Stomach

The functions of the **stomach** are to (1) temporarily store ingested food, (2) mechanically digest food, and (3) chemically digest food through the action of acids and enzymes. This mixing of ingested substances with the acids and enzymes secreted by the glands of the stomach produces a viscous, strongly acidic, soupy mixture called **chyme** (kīm).

Anatomy of the Stomach

▶ **KEY POINT** The stomach is intraperitoneal and subdivided into four regions: the cardia, fundus, body, and pyloric part.

The stomach is intraperitoneal and shaped like an expanded letter J. The **anterior** and **posterior surfaces** are smooth and rounded. The stomach typically extends between the levels of vertebrae T_7 and L_3 (**Spotlight Figure 25.10**). It occupies the left hypochondriac, epigastric, and portions of the umbilical and left lumbar regions (**Figure 25.11**). The shape and size of the stomach are extremely variable from individual to individual and from one meal to the next.

The stomach is divided into four regions (**Spotlight Figure 25.10**):

1. The **cardia** (KAR-dē-a)—so named because of its proximity to the heart—is the superior, medial portion of the stomach immediately surrounding its junction with the esophagus. The esophageal lumen opens into the cardia at the **cardiac orifice**.
2. The **fundus** (FUN-dus) is the region of the stomach that projects superior to the junction between the esophagus and the stomach. The fundus contacts the inferior and posterior surfaces of the diaphragm.
3. The **body** of the stomach is the area between the fundus and the curve of the J. The largest region of the stomach, the body, functions as a mixing tank for ingested food and gastric secretions.
4. The **pyloric part** (pī-LOR-ik; *pyloros*, gatekeeper) forms the portion of the stomach between the body of the stomach and the duodenum (first segment of the small intestine). It is divided into a **pyloric antrum**, which is connected to the body, a **pyloric canal**, which empties into the duodenum, and the **pylorus** (pī-LOR-us), which is the muscular tissue surrounding the **pyloric orifice** *(stomach outlet)*. During digestion, the shape of the pyloric part changes often. A thickening of the circular layer of muscle within the pylorus, called the **pyloric sphincter**, regulates the release of chyme into the duodenum.

The volume of the stomach increases at mealtimes and decreases as chyme leaves the stomach and enters the small intestine. When the stomach is empty, **gastric folds**, or *gastric rugae,* (RŪ-gē; "wrinkles") appear. As the stomach fills, the gastric folds gradually flatten out until, at maximum distension, they almost disappear (**Spotlight Figure 25.10**).

Blood Supply to the Stomach

The three branches of the celiac trunk supply blood to the stomach (**Spotlight Figure 25.10**):

1. The left gastric artery supplies blood to the lesser curvature and cardia.
2. The splenic artery supplies the fundus directly and the greater curvature through the left gastro-epiploic artery.
3. The common hepatic artery branches into the right gastric, the right gastro-epiploic, and the gastroduodenal arteries. These vessels supply blood to the greater and lesser curvatures at the pylorus of the stomach. The gastric and gastro-epiploic veins drain blood from the stomach into the hepatic portal vein (**Figure 22.22**). ⮌ p. 596

FIGURE 25.10

Anatomy of the Stomach

The Stomach and Omenta

The J-shaped stomach is divided into four regions: the **fundus**, **cardia**, **body**, and **pyloric part**. The stomach has a short **lesser curvature**, forming the medial surface of the organ, and a long **greater curvature**, which forms the lateral surface. The mesenteries of the stomach, called **omenta**, connect the stomach to adjacent organs. The shape and size of the stomach are extremely variable from person to person and from one meal to the next.

Mesenteries of the Stomach

Lesser Omentum

The **lesser omentum** extends from the liver to the stomach and first segment of the proximal duodenum.

Hepatogastric Ligament

The **hepatogastric ligament** connects the liver to the lesser curvature of the stomach.

Hepatoduodenal Ligament

The **hepatoduodenal ligament** connects the liver to the proximal segment of the duodenum.

Greater Omentum

The **greater omentum** forms a large pouch that hangs like an apron from the greater curvature of the stomach.

Retractor

Lesser curvature (medial surface)

Liver

Fundus

Cardia

STOMACH

Body

Gall-bladder

Pylorus

Right kidney

Transverse colon

Esophagus

Diaphragm

Spleen

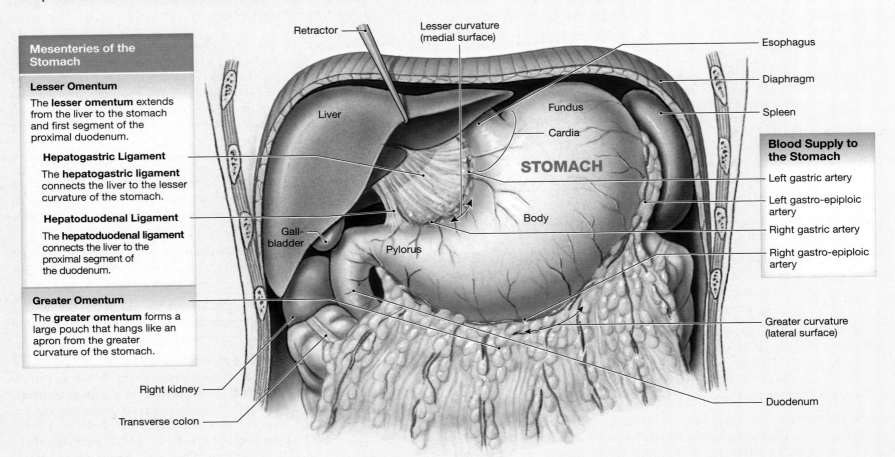

Blood Supply to the Stomach

Left gastric artery

Left gastro-epiploic artery

Right gastric artery

Right gastro-epiploic artery

Greater curvature (lateral surface)

Duodenum

This anterior view of a cadaver dissection shows the superior portion of the abdominal cavity after removal of the lesser omentum. Note the position and orientation of the stomach.

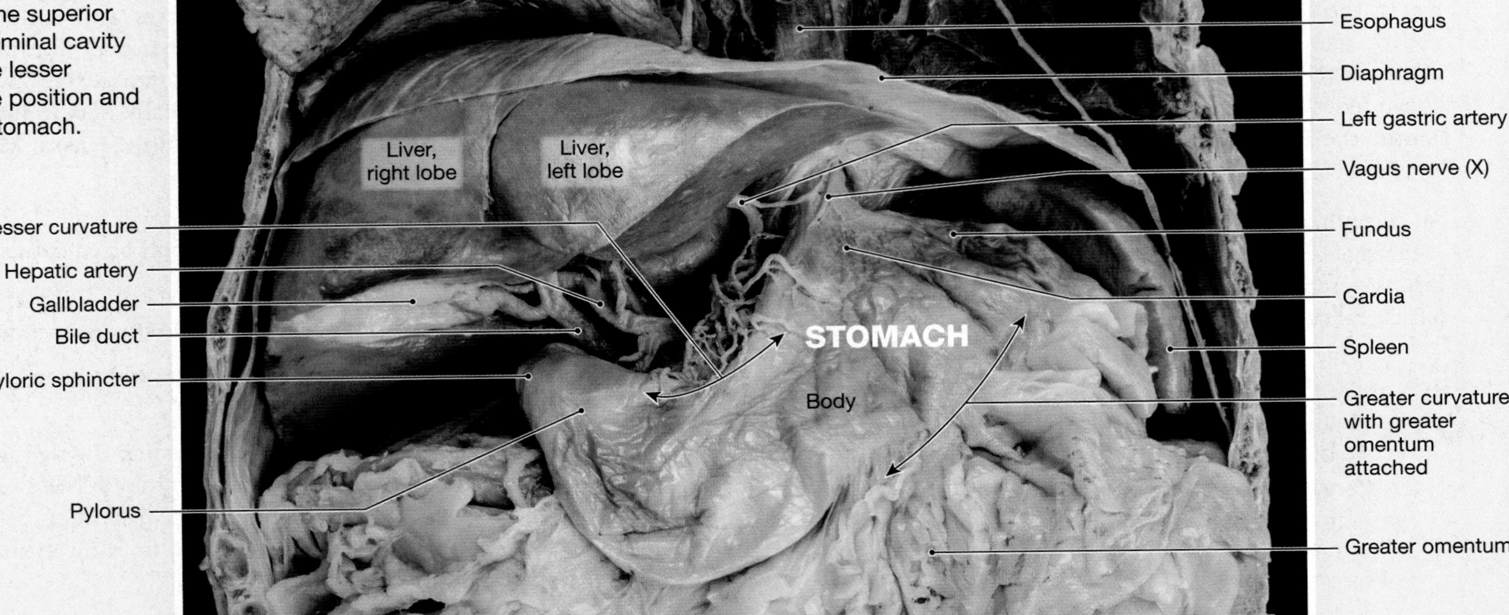

External and Internal Anatomy of the Stomach

The muscular layer of the stomach consists of a superficial longitudinal smooth muscle layer surrounding a deeper layer of circular muscle. This pattern is seen elsewhere along the digestive tract. However, the stomach also contains a third layer of smooth muscle—an inner oblique layer that strengthens the wall of the stomach and assists in the mixing and churning of the gastric contents. In a relaxed stomach, the mucosa has prominent longitudinal folds called **gastric folds**, or *gastric rugae*, which expand the gastric lumen during meals.

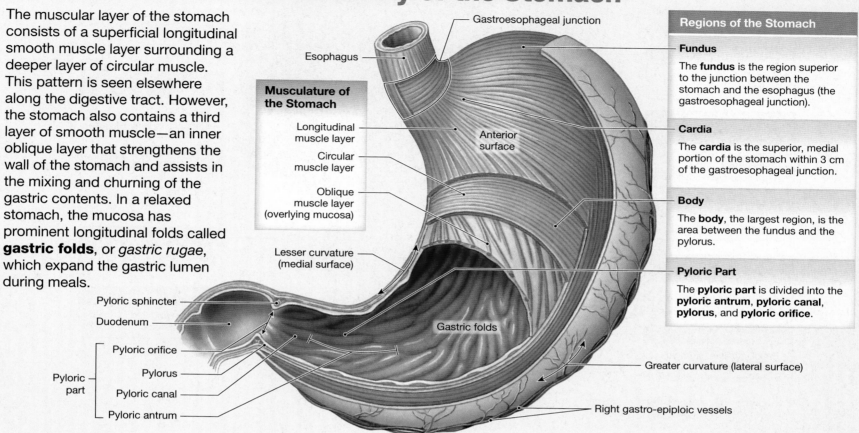

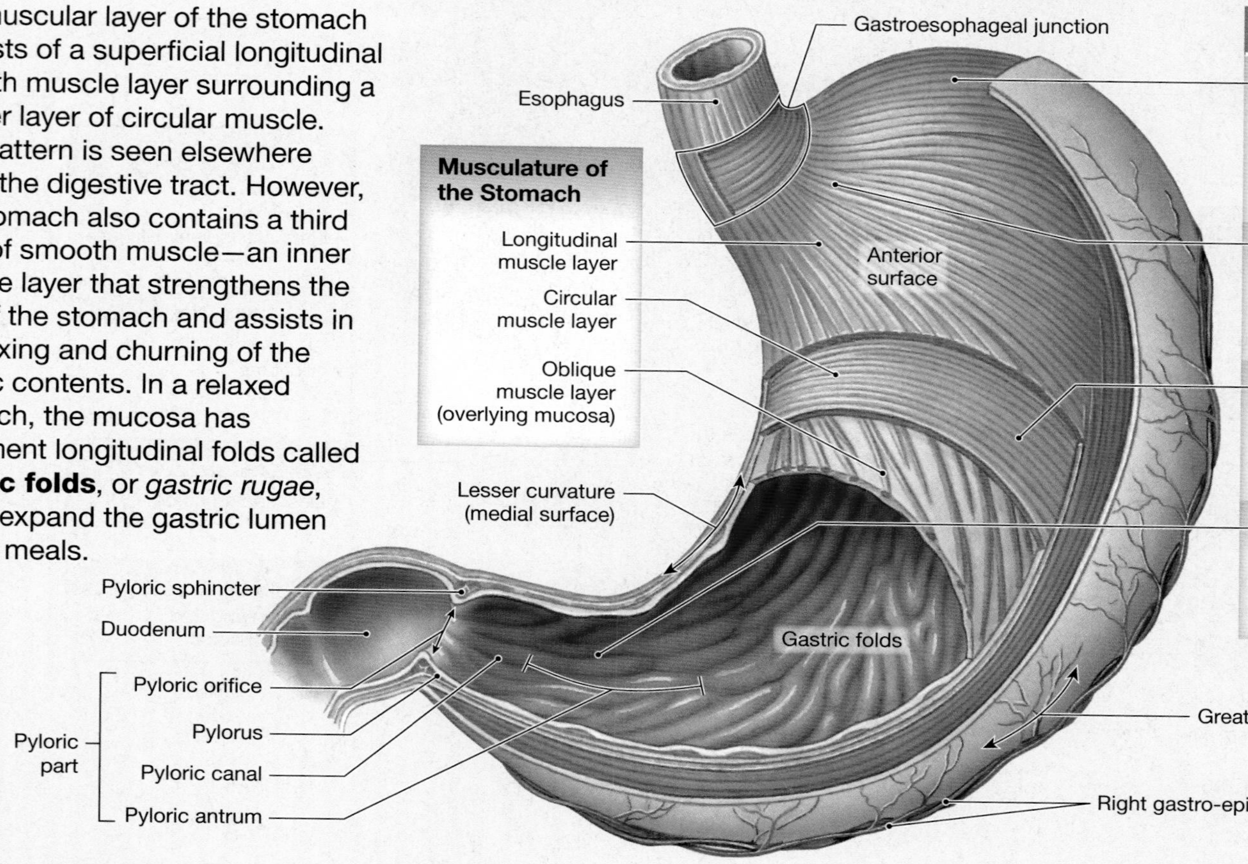

Gastroesophageal junction
Esophagus

Musculature of the Stomach

Longitudinal muscle layer
Circular muscle layer
Oblique muscle layer (overlying mucosa)

Anterior surface
Lesser curvature (medial surface)
Pyloric sphincter
Duodenum
Gastric folds
Pyloric part: Pyloric orifice, Pylorus, Pyloric canal, Pyloric antrum
Greater curvature (lateral surface)
Right gastro-epiploic vessels

Regions of the Stomach

Fundus

The **fundus** is the region superior to the junction between the stomach and the esophagus (the gastroesophageal junction).

Cardia

The **cardia** is the superior, medial portion of the stomach within 3 cm of the gastroesophageal junction.

Body

The **body**, the largest region, is the area between the fundus and the pylorus.

Pyloric Part

The **pyloric part** is divided into the **pyloric antrum**, **pyloric canal**, **pylorus**, and **pyloric orifice**.

CLINICAL NOTE

Endoscopy

An endoscope is a flexible fiberoptic tube with a light at one end and a camera at the other. It can be inserted into the digestive system for direct visualization. This procedure can be done with the patient lightly anesthetized or fully conscious. **Esophagogastroduodenoscopy (EGD)** refers to endoscopy of the upper gastrointestinal tract in which the endoscope is inserted through the mouth. In a **colonoscopy**, the endoscope is inserted through the anus to examine the colon.

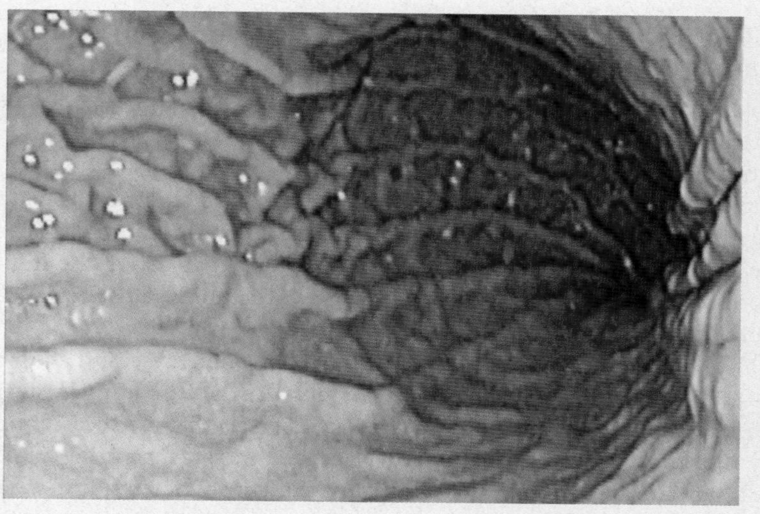

Figure 25.11 **Abdominal Regions and Planes**

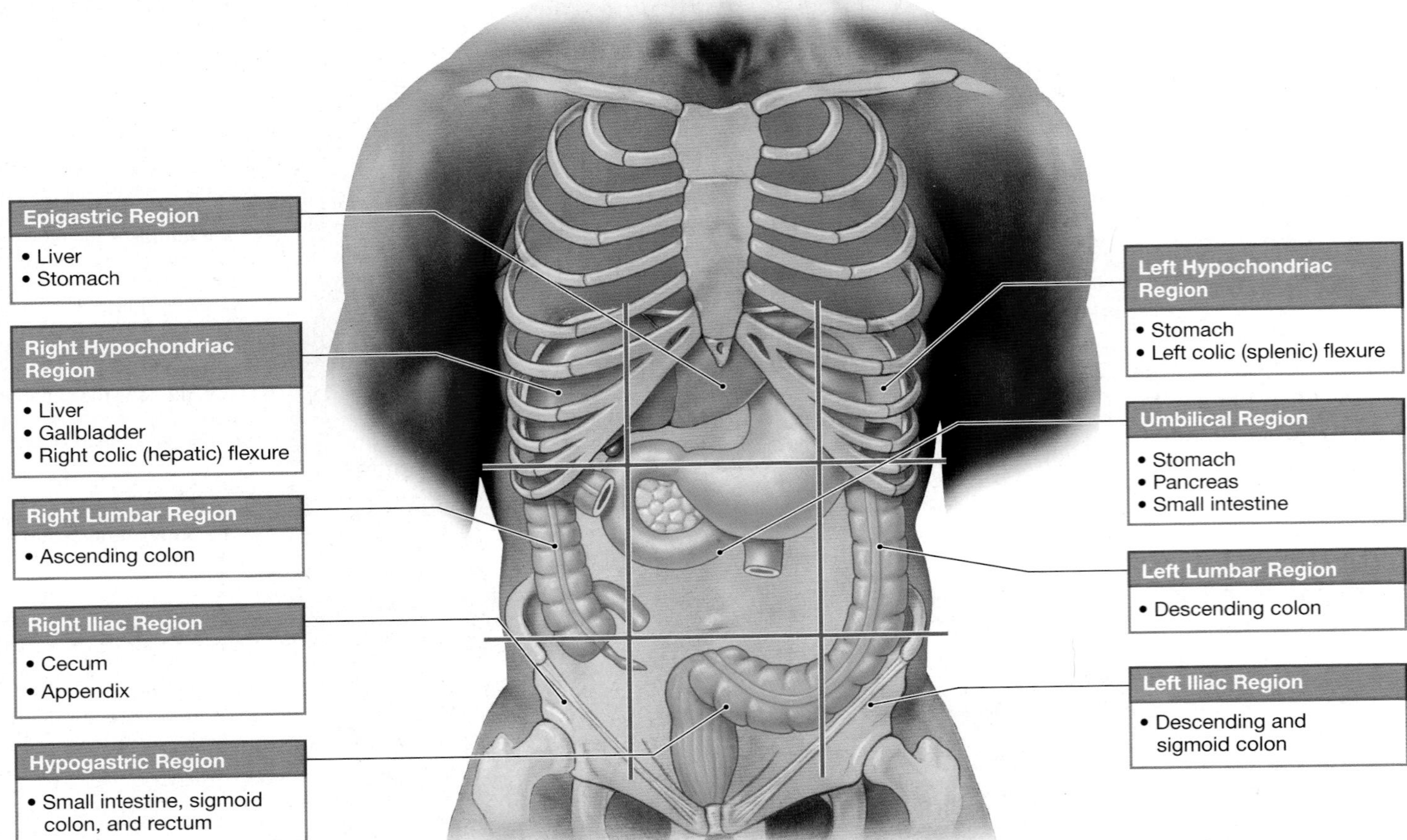

Musculature of the Stomach

The muscularis mucosae and muscular layer of the stomach contain extra layers of smooth muscle cells in addition to the usual circular and longitudinal layers. The muscularis mucosae generally contain an outer, circular layer of muscle cells. The muscular layer has an inner **oblique layer** of smooth muscle. The extra layers of smooth muscle strengthen the stomach wall and assist in the mixing and churning essential to the formation of chyme **(Spotlight Figure 25.10)**.

Histology of the Stomach

▶ **KEY POINT** The surface epithelium of the gastric mucosa is a simple columnar epithelium. Four different secretory cells are found within the glands of the stomach.

25

A simple columnar epithelium lines all regions of the stomach. The surface epithelium is composed of mucous surface cells. This secretory sheet of mucous surface cells produces a layer of mucus that covers the luminal surfaces of the stomach and protects the epithelium against acids and enzymes.

Shallow depressions, called **gastric pits**, open onto the gastric surface **(Figure 25.12)**. Regenerative stem cells are found at the base of each gastric pit. These cells actively divide to replace superficial cells that are shed continuously into the chyme. The continual replacement of the superficial epithelial cells provides an additional defense against the gastric contents. If stomach acid and digestive enzymes penetrate the mucous layers, any damaged epithelial cells are quickly replaced.

Gastric Secretory Cells

Each gastric pit communicates with several **gastric glands** that extend deep into the underlying lamina propria. Gastric glands are simple branched tubular glands **(Figure 25.12c,d)**. Gastric pits and gastric glands contain four types of secretory cells: mucous neck cells, parietal cells, G cells, and chief cells **(Figure 25.12c–f)**. The parietal and chief cells work together to secrete about 1500 ml of **gastric juice** each day.

Mucous Neck Cells **Mucous neck cells** are found in all regions of the stomach. These cells are columnar in shape, similar to the mucous surface cells. The apical cytoplasm is filled with a secretory product that is water-soluble and lubricates the stomach contents.

Parietal Cells **Parietal cells** are large round or pyramid-shaped cells. They are very numerous in the proximal portions of the gastric glands. Parietal cells secrete intrinsic factor, a glycoprotein that helps absorb vitamin B_{12} across the intestinal lining. Parietal cells also secrete hydrochloric acid (HCl). Hydrochloric acid lowers the pH of the gastric juice, kills microorganisms, helps break down plant cell walls and connective tissues in meat, and denatures proteins.

Enteroendocrine Cells and G Cells **Enteroendocrine** (en-ter-ō-EN-dō-krin) **cells** are quite numerous in the fundus of the stomach, but are only occasionally found within the cardia and pyloric part. In the fundus they are scattered among the parietal and chief cells. **G cells** are enteroendocrine cells found in the gastric pits of the pyloric part. They secrete the hormone **gastrin** (GAS-trin). Gastrin, which is released when food enters the stomach,

Figure 25.12 Histology of the Stomach Wall

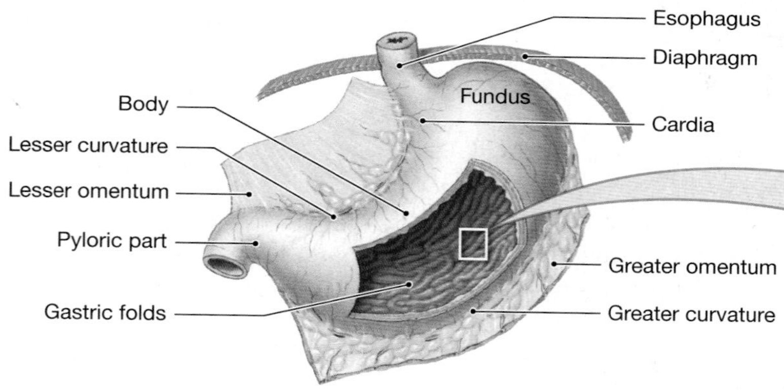

Esophagus
Diaphragm
Fundus
Cardia
Body
Lesser curvature
Lesser omentum
Pyloric part
Gastric folds
Greater omentum
Greater curvature

a Diagram of the stomach and mucosa.

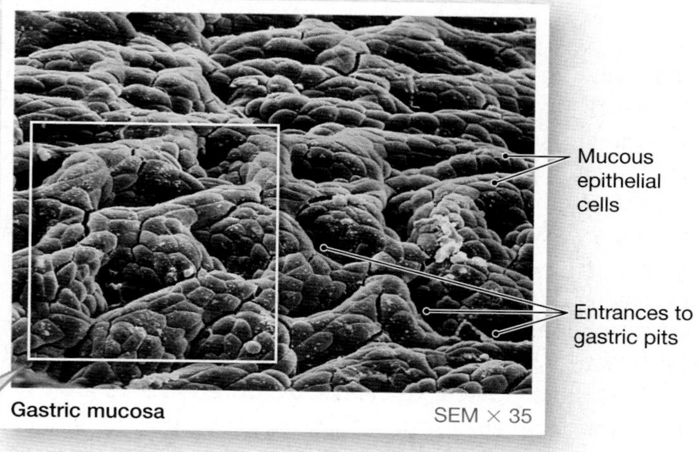

Mucous epithelial cells
Entrances to gastric pits
Gastric mucosa
SEM × 35

b Colorized SEM of the gastric mucosa.

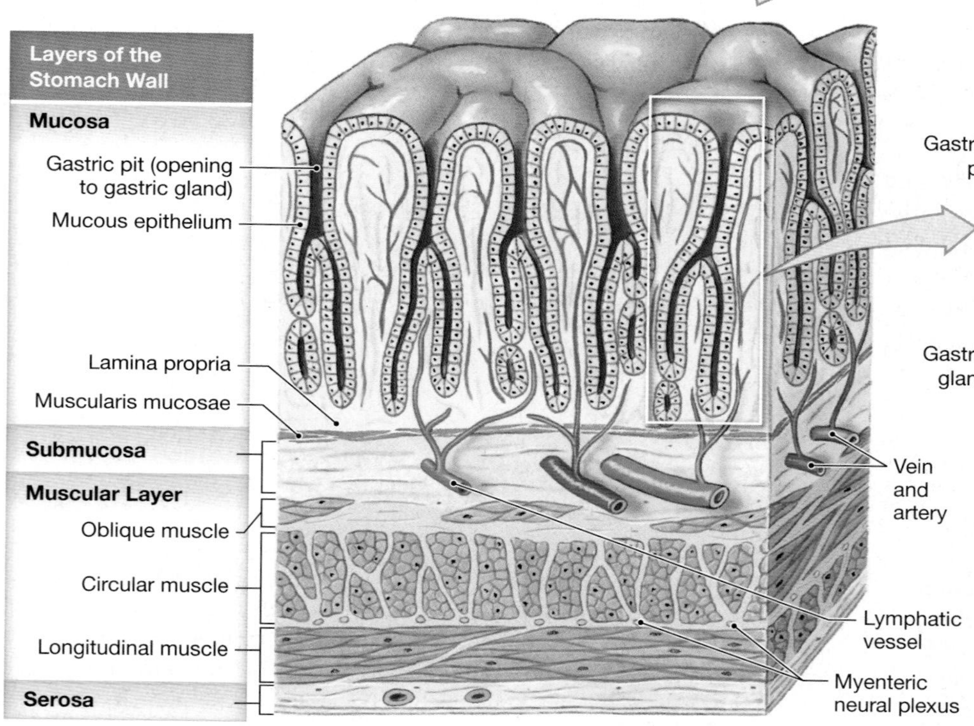

Layers of the Stomach Wall
Mucosa
Gastric pit (opening to gastric gland)
Mucous epithelium
Lamina propria
Muscularis mucosae
Submucosa
Muscular Layer
Oblique muscle
Circular muscle
Longitudinal muscle
Serosa
Vein and artery
Lymphatic vessel
Myenteric neural plexus

c Diagram of the stomach wall. This corresponds to a sectional view through the area indicated by the box in part (b).

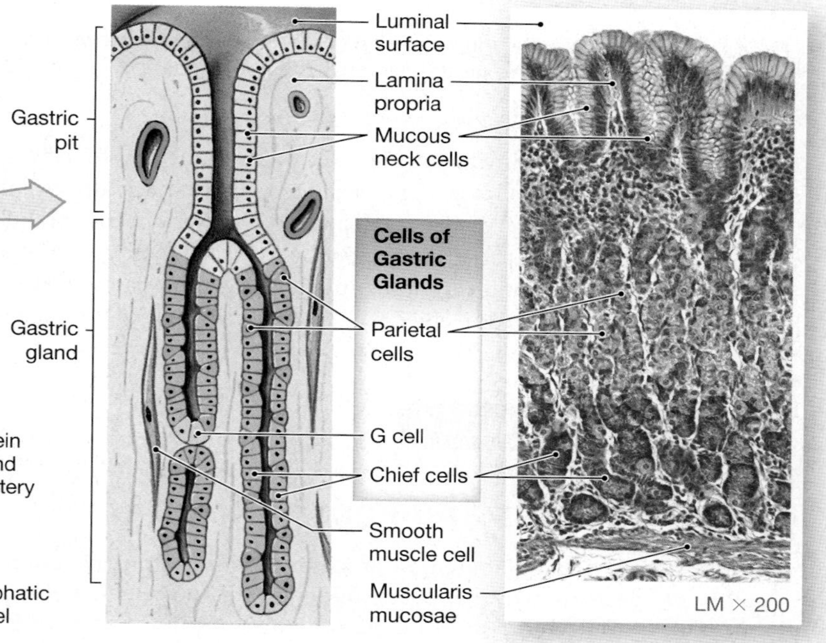

Gastric pit
Gastric gland
Luminal surface
Lamina propria
Mucous neck cells
Cells of Gastric Glands
Parietal cells
G cell
Chief cells
Smooth muscle cell
Muscularis mucosae
LM × 200

d Diagram of a gastric gland and micrograph of the gastric mucosa.

Mucous neck cells
Parietal cells
LM × 500

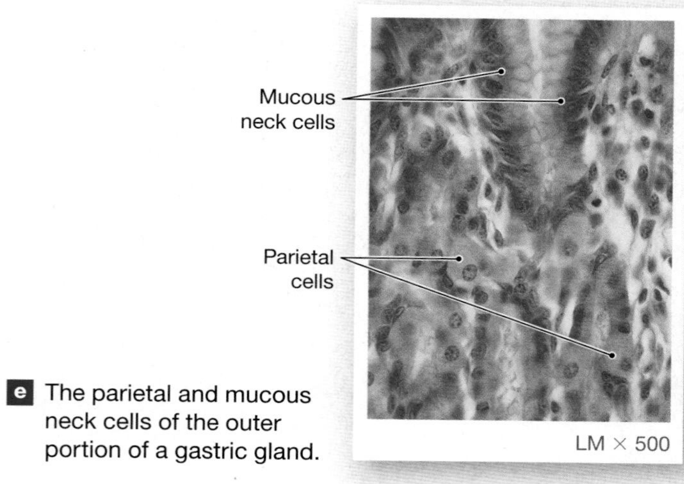

e The parietal and mucous neck cells of the outer portion of a gastric gland.

Parietal cells
Chief cells
LM × 500

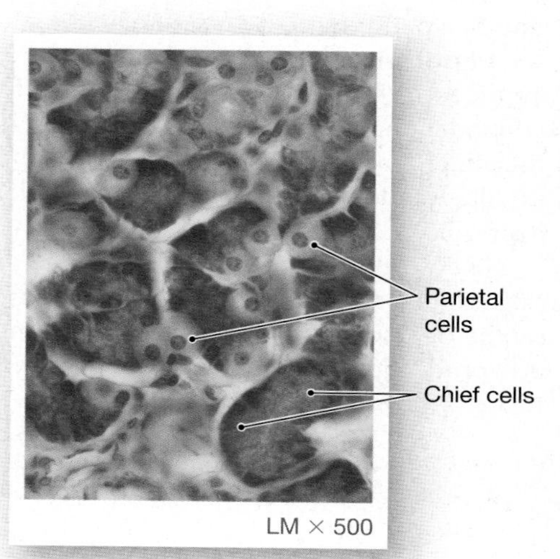

f Chief and parietal cells of the deepest portions of a gastric gland.

stimulates the secretory activity of both parietal and chief cells. It also promotes smooth muscle activity in the stomach wall that enhances mixing and churning activity.

Chief Cells Found only in the fundus, **chief cells** are most abundant in the base of a gastric gland. These cells are columnar in shape and secrete **pepsinogen** (pep-SIN-ō-jen), an inactive proenzyme. Acid in the gastric lumen converts pepsinogen to **pepsin**, an active proteolytic, or protein-digesting, enzyme. Pepsin functions most effectively at an acidic pH of 1.5–2.0. The stomachs of newborn infants (but not adults) produce **rennin** and **gastric lipase**, enzymes important for digesting milk. Rennin coagulates milk proteins, and gastric lipase initiates the digestion of milk fats.

The lining of the stomach does not directly absorb any nutrients. However, some salts, water, and lipid-soluble substances are indirectly absorbed in the stomach. For instance, alcohol and some drugs, such as aspirin and nonsteroidal anti-inflammatory drugs (NSAIDs), enter the lamina propria of the mucosa after damaging the surface epithelium. These substances then enter the bloodstream through the capillaries within the lamina propria.

Regulation of Gastric Activity

▶ **KEY POINT** The production of acid and enzymes by the gastric mucosa can be directly controlled by the central nervous system and indirectly regulated by local hormones.

Two mechanisms exert direct control over the production of acid and enzymes by the gastric mucosa. One regulatory mechanism involves the CNS and uses the vagus nerve (parasympathetic innervation) and branches of the celiac plexus (sympathetic innervation). For example, the sight or thought of food triggers motor output in the ganglionic parasympathetic fibers innervating parietal cells, chief cells, and mucous cells of the stomach. Stimulation increases the production of acids, enzymes, and mucus. The arrival of food in the stomach stimulates stretch receptors in the stomach wall and chemoreceptors in the mucosa. Reflexive contractions occur in the muscular layers of the stomach wall, and G cells release gastrin. Both parietal and chief cells respond to the presence of gastrin by accelerating their secretory activities. Parietal cells are especially sensitive to gastrin, so the rate of acid production increases more dramatically than the rate of enzyme secretion.

Sympathetic activation inhibits gastric activity. In addition, the small intestine releases two hormones that inhibit gastric secretion. **Secretin** (se-KRĒ-tin) and **cholecystokinin** (kō-lē-sis-tō-KĪ-nin) stimulate secretion by both the pancreas and liver. The depression of gastric activity is a secondary, but complementary, effect.

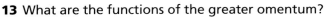

25.5 ✔ **CONCEPT CHECK**

13 What are the functions of the greater omentum?

14 How do the cells that line the stomach keep from becoming damaged in such an acidic environment?

15 What do chief cells secrete?

16 What hormone stimulates the secretion of parietal and chief cells?

See the blue Answers tab at the back of the book.

25.6 | The Small Intestine

▶ **KEY POINT** The gross and histological anatomy of the small intestine is specialized for absorption. The circular folds, villi, and microvilli increase its surface area.

CLINICAL NOTE

Gastritis and Peptic Ulcers

Gastritis is an inflammation or irritation of the stomach lining. Acute gastritis, which appears suddenly, is often due to a viral infection ("stomach flu") or food poisoning (ingesting contaminated food or beverages containing bacteria, parasites, or toxins). Fever, cramping, nausea, vomiting, and diarrhea often develop within 2 to 4 hours after eating the contaminated food.

Chronic gastritis is often caused by a bacterium, *Helicobacter pylori* (*H. pylori*), that infects the stomach. Untreated chronic gastritis can lead to a peptic ulcer.

Peptic ulcers are sores of the alimentary mucosa exposed to acidic gastric secretions. They are usually found in the stomach (gastric ulcers) or duodenum (duodenal ulcers). Excessive alcohol intake, smoking or chewing tobacco, aspirin and nonsteroidal anti-inflammatory drugs (NSAIDs), caffeine, severe illness, or radiation treatment can also cause peptic ulcers.

Endoscopy is often used to diagnose peptic ulcers. Some treatments can also be administered through an endoscope. *H. pylori* infections can be treated with antibiotics. Medications to neutralize acid or block acid production can help, and patients are advised to avoid alcohol, tobacco, aspirin, and NSAIDs.

Left untreated, an ulcer can erode through the epithelial layer and expose the lamina propria. Pain and bleeding will develop. Further invasion of gastric juices can actually perforate the stomach, causing its contents to leak into the peritoneum, causing a surgical emergency.

Peptic Ulcer Disease

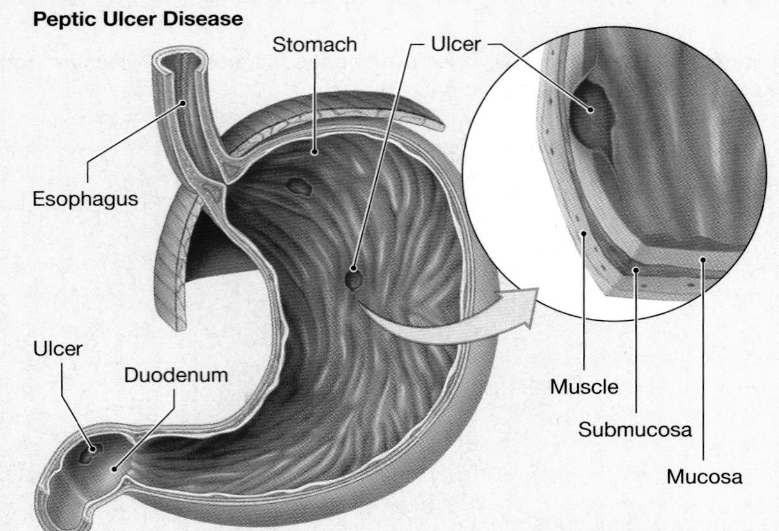

Ninety percent of nutrient absorption occurs in the **small intestine**, and most of the rest occurs in the proximal portion of the large intestine. The small intestine averages 6 m (20 ft) in length (range 5.0–8.3 m [5–25 ft]) and has an average diameter of 4 cm (1.6 in.) at its junction with the stomach and about 2.5 cm (1 in.) at its junction with the large intestine. The small intestine is found in all abdominal regions except the left hypochondriac and epigastric regions (**Figures 25.11**, p. 666, and **25.13**).

The small intestine fills much of the peritoneal cavity. It is held in place by mesenteries attached to the dorsal body wall. **Spotlight Figure 25.10** and **Figures 25.11** and **25.13** show the position of the small intestine in comparison to the other segments of the digestive tract.

The anatomy of the small intestine is specialized to increase its surface area for absorption and secretion. The intestinal lining has a series of ring-shaped projections called **circular folds** or *plicae circulares* (sir-kū-LAR-ēs) (**Figures 25.2a,b**, p. 653, **25.14a**, and **25.15**). Unlike the gastric folds in the stomach, each circular fold is a permanent feature of the intestinal lining. These folds do not disappear as the small intestine fills. Roughly 800 circular folds (about 2 per centimeter) are found along the length of the duodenum, jejunum, and proximal half of the ileum. The mucosa possesses intestinal villi, and each cell of the surface epithelium has small microvilli on its apical surface.

Regions of the Small Intestine

▶ **KEY POINT** The small intestine is subdivided into the duodenum, jejunum, and ileum.

The small intestine has three anatomical subdivisions: the duodenum, jejunum, and ileum (**Figure 25.13**).

The Duodenum

The shortest and widest segment of the small intestine is the **duodenum** (dū-ō-DĒ-num) (**Figure 25.13**). It is approximately 25 cm (10 in.) long. The duodenum is connected to the pylorus of the stomach. A ring of smooth muscle, the pyloric sphincter, regulates movement of chyme from the stomach into the duodenum. From its start at the pyloric sphincter, the duodenum curves in a C that surrounds the pancreas. The proximal 2.5 cm (1 in.) portion is intraperitoneal, while the rest is secondarily retroperitoneal and located between vertebrae L_1 and L_4 (**Figure 25.4b**, p. 656).

The duodenum is a "mixing bowl" that receives chyme from the stomach and digestive secretions from the pancreas and liver. Almost all essential digestive enzymes enter the small intestine from the pancreas.

The Jejunum

A sharp bend, the duodenojejunal flexure, marks the junction of the duodenum and the **jejunum** (jeh-JŪ-num). At this junction the small intestine re-enters the peritoneal cavity, becoming intraperitoneal and supported by a sheet of mesentery. The jejunum is about 2.5 m (8 ft) long. The bulk of chemical digestion and nutrient absorption occurs in the jejunum.

The Ileum

The third and last segment of the small intestine is the **ileum** (IL-ē-um). It is intraperitoneal and is the longest segment of the small intestine, averaging 3.5 m (12 ft) in length. The ileum ends at the ileocecal valve. This sphincter controls the flow of material from the ileum into the cecum of the large intestine. The ileocecal valve protrudes into the cecum.

There is no sharp anatomical distinction between the jejunum and ileum. However, the following subtle anatomical differences enable a surgeon to distinguish the jejunum from the ileum:

Figure 25.13 Regions of the Small Intestine. The color coding indicates the relative sizes and positions of the duodenum, jejunum, and ileum.

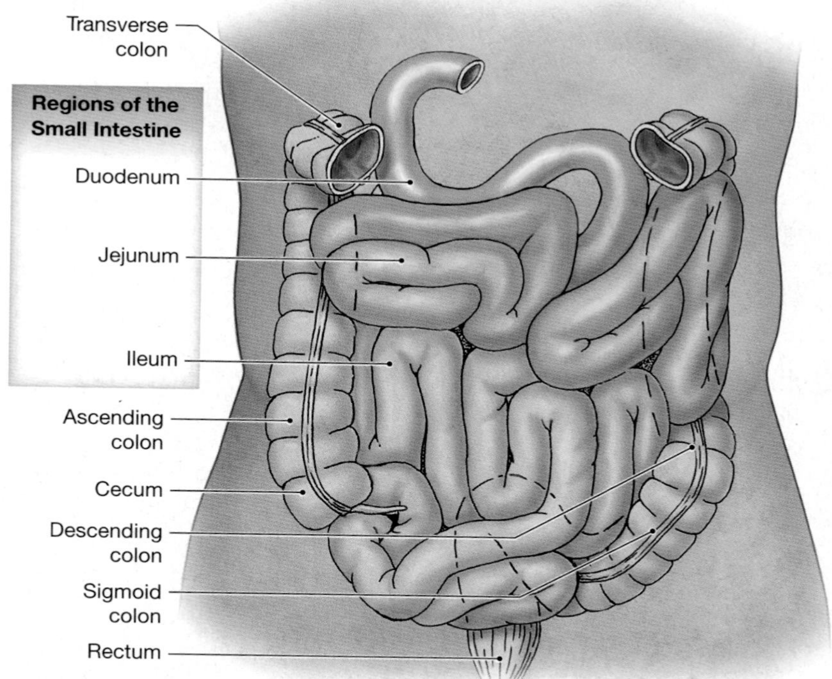

- The jejunum has a thicker wall and a larger diameter than the ileum.
- The jejunum is typically found in the umbilical region, while the ileum tends to lie in the hypogastric region of the abdominal cavity.
- The mesentery covering the ileum is typically thicker and contains more fat than does the mesentery covering the jejunum.
- The vasculature extending from the mesentery to the jejunum tends to be more straightforward and has fewer branches than the vasculature extending from the mesentery to the ileum.

Support of the Small Intestine

▶ **KEY POINT** Blood vessels, nerves, and lymphatic vessels pass through the mesentery proper to reach the small intestine.

The duodenum has no supporting mesentery. The proximal 2.5 cm is intraperitoneal and movable, but the rest is secondarily retroperitoneal and fixed in position. The jejunum and ileum are intraperitoneal and supported by an extensive, fan-shaped mesentery—the mesentery proper (**Figure 25.4**, p. 656). Blood vessels, lymphatic vessels, and nerves pass through the connective tissue of the mesentery to reach these segments of the small intestine. These blood vessels are branches of the superior mesenteric artery and tributaries of the superior mesenteric vein (**Figure 22.15**). ⮌ p. 588 The vagus nerve provides parasympathetic innervation; the superior mesenteric ganglion provides sympathetic innervation.

Histology of the Small Intestine

▶ **KEY POINT** The histology of the small intestine follows the "typical" structure of a hollow organ of the digestive tract: mucosa, muscularis mucosae, submucosa, muscular layer, and serosa.

Figure 25.14 Histology of the Intestinal Wall

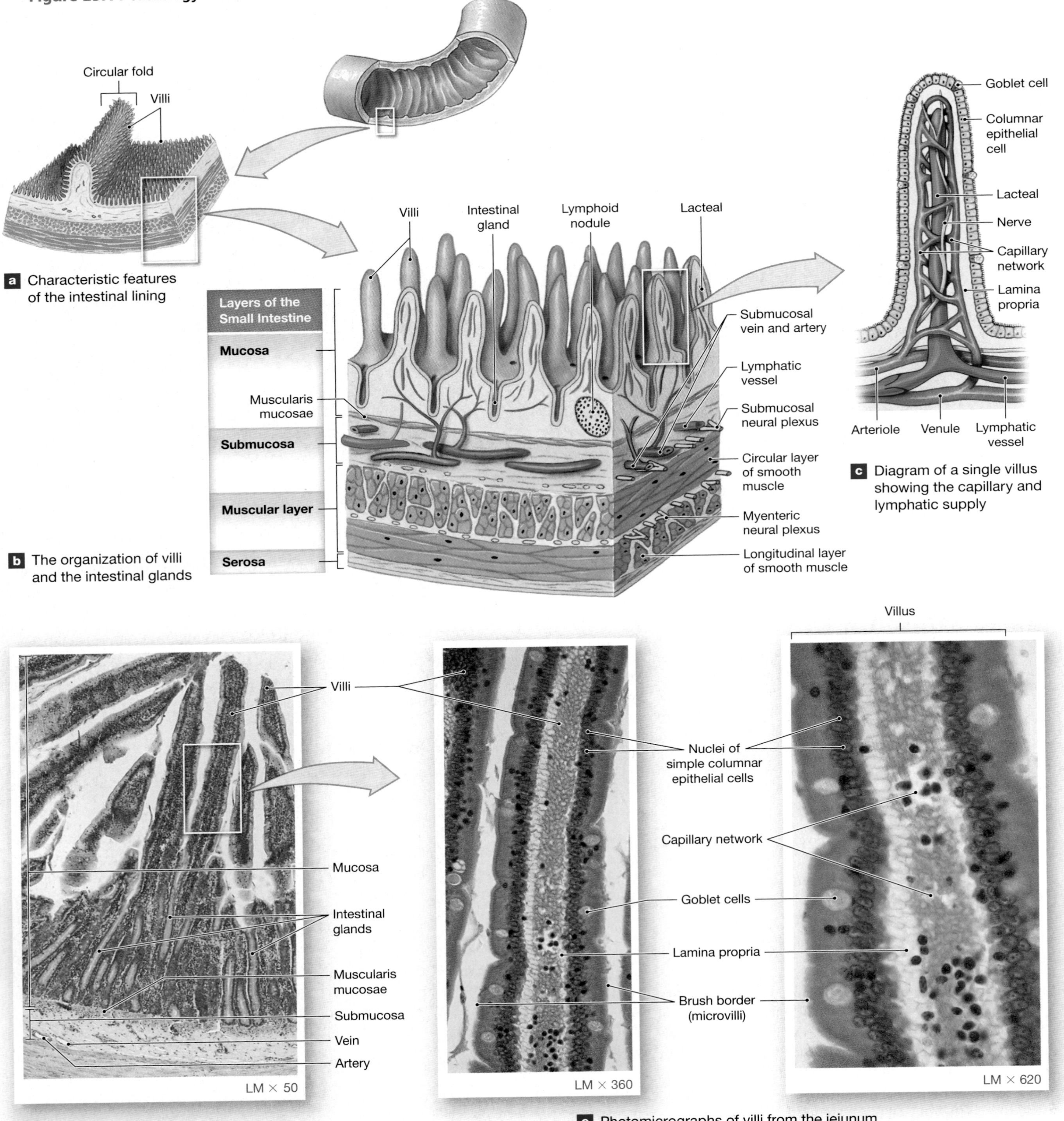

Circular fold
Villi

a Characteristic features of the intestinal lining

Layers of the Small Intestine
Mucosa
Muscularis mucosae
Submucosa
Muscular layer
Serosa
Villi
Intestinal gland
Lymphoid nodule
Lacteal
Submucosal vein and artery
Lymphatic vessel
Submucosal neural plexus
Circular layer of smooth muscle
Myenteric neural plexus
Longitudinal layer of smooth muscle

b The organization of villi and the intestinal glands

Goblet cell
Columnar epithelial cell
Lacteal
Nerve
Capillary network
Lamina propria
Arteriole
Venule
Lymphatic vessel

c Diagram of a single villus showing the capillary and lymphatic supply

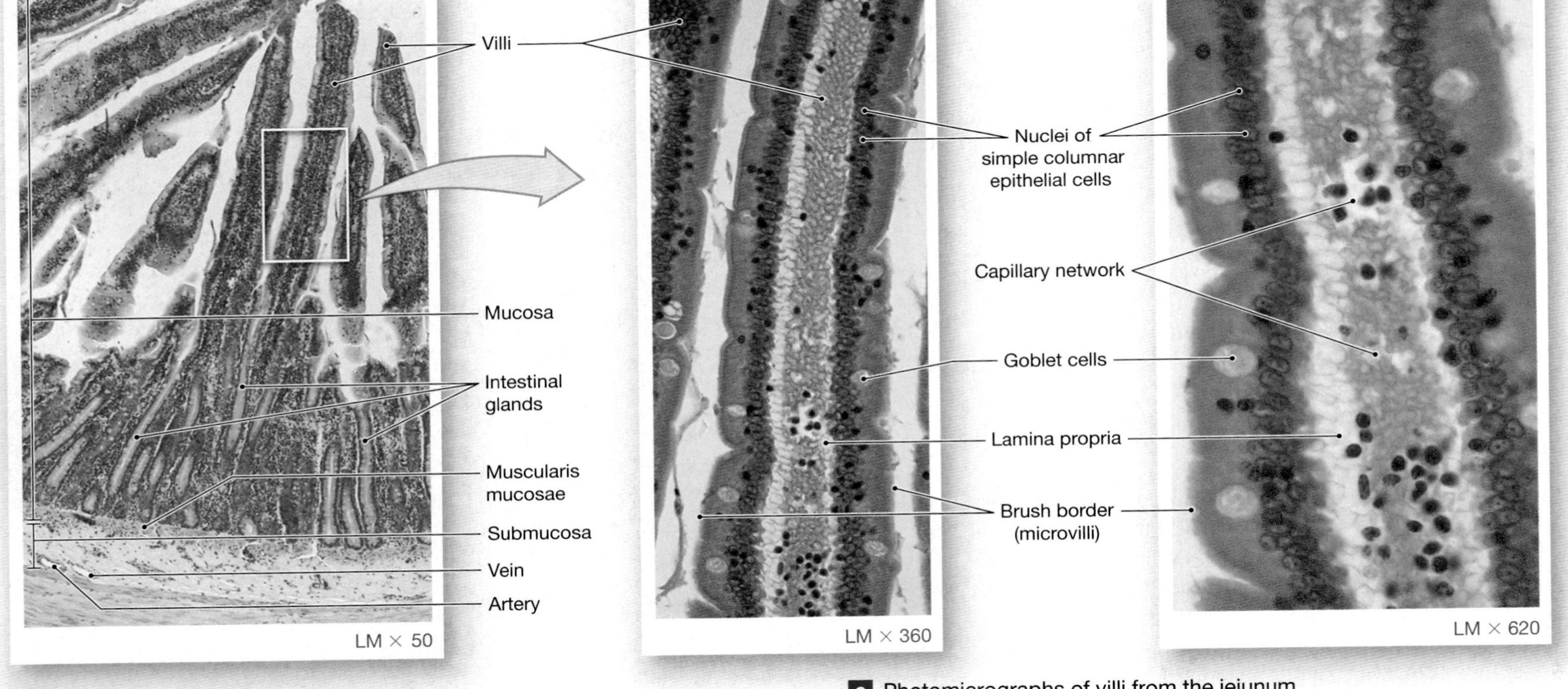

d Panoramic view of the wall of the small intestine showing mucosa with characteristic villi, submucosa, and muscularis layers

e Photomicrographs of villi from the jejunum

The Intestinal Epithelium

The mucosa of the small intestine forms fingerlike projections, **intestinal villi**, that project into the lumen (**Figures 25.14** and **25.15**). Each villus is covered by a simple columnar epithelium. The apical surfaces of the epithelial cells are covered with even smaller projections called **microvilli**. The circular folds, villi, and microvilli of the mucosa increase the surface area of the small intestine by a factor of more than 600, to approximately 2 million cm^2.

Intestinal Glands

Between the columnar epithelial cells, goblet cells secrete mucin onto the intestinal surfaces. At the bases of the villi are the entrances to the **intestinal glands** (also called *intestinal crypts*). These glandular pockets extend deep into the underlying lamina propria (**Figure 25.14b,d**). Near the base of each gland, stem cells continually divide, producing new generations of epithelial cells. These new cells move superficially toward the intestinal surface. They reach the top of the gland within a few days, renewing the epithelial surface and adding intracellular enzymes to the chyme.

Paneth cells at the base of the intestinal glands have a role in innate (nonspecific) immunity and release defensins and lysozyme. These secretions kill some bacteria and allow others to live, thereby establishing the flora of the intestinal lumen. Intestinal glands also contain enteroendocrine cells that produce several intestinal hormones.

The Lamina Propria

The lamina propria is the loose areolar connective tissue located within the core of each intestinal villus. The lamina propria contains numerous lymphatic cells, occasional lymphoid nodules, and an extensive network of capillaries that absorbs and carries nutrients to the hepatic portal circulation. Each villus also contains a central lymphatic vessel called a **lacteal** (LAK-tē-al; *lacteus*, milky) (**Figure 25.14b,c,e**). Lacteals transport materials that cannot enter blood capillaries. For example, absorbed fatty acids are assembled into protein–lipid packages that are too large to diffuse into the bloodstream. These packets, called chylomicrons, reach the venous circulation through the thoracic duct.

Figure 25.15 Regions of the Small Intestine. Diagrammatic view highlighting the distinguishing features of each region of the small intestine. The photographs show the gross anatomy of the intestinal lining in each region of the small intestine.

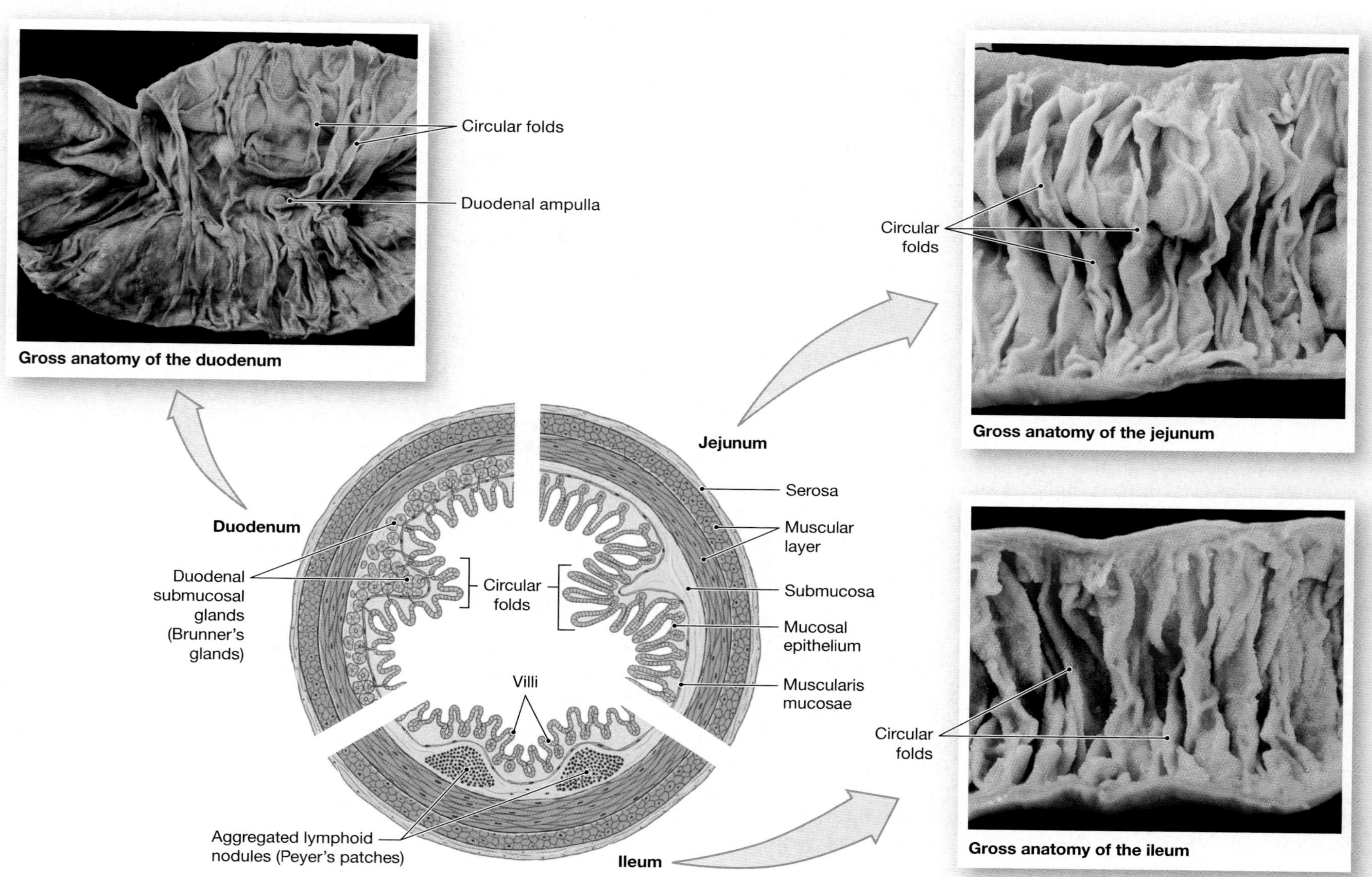

25

Regional Specializations

The duodenum, jejunum, and ileum of the small intestine each have histological specializations related to their primary functions. **Figure 25.15** shows representative sections from each region.

The Duodenum In addition to the intestinal glands, the submucosa contains **duodenal submucosal glands**, also known as *Brunner's glands* **(Figure 25.15)**. These glands produce large quantities of mucus when chyme arrives from the stomach. The mucus protects the epithelium from the acidity of chyme and also contains bicarbonate ions that help raise the pH of the chyme. Submucosal glands are most abundant in the proximal portion of the duodenum, and they decrease in numbers as you approach the jejunum. The pH of the intestinal contents changes from 1–2 to 7–8 as chyme moves from the duodenum to the jejunum. By the beginning of the jejunum, the extra mucus production is no longer needed.

Buffers and enzymes from the pancreas and bile from the liver enter the duodenum roughly halfway along its length. The bile duct from the liver and gallbladder and the pancreatic duct from the pancreas come together within the duodenal wall at a muscular chamber called the **duodenal ampulla** (am-PŪL-la). The duodenal ampulla opens into a small mound within the duodenum termed the **duodenal papilla.** The muscular **hepatopancreatic sphincter** encircles the lumen of the bile duct and, generally, the pancreatic duct and duodenal ampulla as well **(Figures 25.15** and **25.21a,d)**.

The Jejunum and Ileum Circular folds and villi are prominent in the proximal half of the jejunum **(Figures 25.14d,e** and **25.15)**. Most nutrient absorption occurs here. As you approach the end of the ileum, the folds and villi diminish in size and number. This reduction parallels the reduction in absorptive activity; most nutrient absorption has occurred before materials reach the ileum. The distal portions of the ileum lack circular folds. The lamina propria there contains 20–30 masses of lymphoid tissue called submucous **aggregated lymphoid nodules**, or *Peyer's patches*. These lymphoid tissues are most abundant in the terminal portion of the ileum, near the entrance to the large intestine. The lymphocytes in the aggregated lymphoid nodules protect the small intestine from bacteria that normally inhabit the large intestine.

Regulation of the Small Intestine

▶ **KEY POINT** The autonomic nervous system and local reflexes control the peristaltic contractions and secretory activity of the small intestine.

Weak peristaltic contractions slowly move materials along the length of the small intestine as absorption occurs. The movements of the small intestine are controlled by neural reflexes involving the submucosal and myenteric neural plexuses. Stimulation of the parasympathetic system increases the sensitivity of these reflexes and accelerates peristaltic contractions and segmentation movements. These contractions and movements promote the mixing of the intestinal contents and usually occur within a few centimeters of the original stimulus. When food enters the stomach, coordinated intestinal movements occur. These contractions move chyme away from the duodenum and toward the large intestine. At the same time, the ileocecal valve opens, allowing the passage of material into the large intestine.

Hormonal and CNS controls regulate the secretory output of the small intestine and accessory glands. The secretions of the small intestine are called **intestinal juice**. Secretory activities are triggered by local, short reflexes and parasympathetic (vagal) stimulation. Sympathetic stimulation inhibits secretion. Duodenal enteroendocrine cells produce secretin and cholecystokinin, hormones that coordinate the secretory activities of the stomach, duodenum, liver, and pancreas.

25.6  **CONCEPT CHECK**

17 Which histological features of the small intestine facilitate the digestion and absorption of nutrients?

18 What is the function of circular folds?

19 What are the functions of intestinal glands?

20 Which section of the small intestine serves as a "mixing bowl"?

See the blue Answers tab at the back of the book.

25.7 | The Large Intestine

▶ **KEY POINT** The large intestine begins where the ileum joins with an expanded pouch called the cecum. It ends at the anus. The large intestine absorbs water, electrolytes, and vitamins and compacts and stores the fecal waste material.

The horseshoe-shaped **large intestine** begins at the junction with the ileum and ends at the anus. The large intestine lies inferior to the stomach and liver and forms an almost complete frame around the small intestine (**Figures 25.1**, p. 652, **25.11**, p. 666, and **25.13**).

The large intestine is often called the **large bowel**. It has an average length of about 1.5 m (5 ft) and a width of 7.5 cm (3 in.). It is divided into three parts: (1) the cecum, (2) the colon, and (3) the rectum **(Figure 25.16)**.

The large intestine (1) absorbs water and electrolytes and compacts the intestinal contents into feces, (2) absorbs the important vitamins produced by bacteria, and (3) stores feces before defecation.

The large intestine receives blood from branches of the superior mesenteric and inferior mesenteric arteries. Venous blood is collected from the large intestine by the superior mesenteric and inferior mesenteric veins. ⮌ **pp. 594–596**

The Cecum and Appendix

▶ **KEY POINT** The cecum, which is intraperitoneal, is the first segment of the large intestine. Materials pass through the ileocecal valve as they move from the ileum into the cecum. The appendix is attached to the posteromedial surface of the cecum.

Material arriving from the ileum first enters an expanded pouch called the **cecum**. The ileum attaches to the medial surface of the cecum and opens into the cecum at the **ileocecal valve** **(Figure 25.16)**. The cecum collects and stores materials from the ileum and begins the process of compaction.

The slender, hollow **appendix**, or *vermiform appendix* (*vermis*, worm), is attached to the posteromedial surface of the cecum **(Figure 25.16)**. The appendix is normally about 9 cm (3.6 in.) long, but its size and shape are quite variable. A small mesentery called the **meso-appendix** connects the appendix to the ileum and cecum. The primary function of the appendix is as an organ of the lymphatic system. Inflammation of the appendix is known as *appendicitis*. Both the cecum and appendix are intraperitoneal.

The Colon

▶ **KEY POINT** The colon has a larger diameter and a thinner wall than the small intestine. The colon is subdivided into the ascending colon, transverse colon, descending colon, and sigmoid colon.

Refer to **Figure 25.16** as we describe several distinctive features of the **colon**:

- The wall of the colon forms a series of pouches, or **haustra** (HAWS-truh; singular, *haustrum*). The haustra permit considerable distension and elongation. Cutting into the intestinal lumen reveals that the creases between the haustra extend into the mucosal lining, producing a series of internal folds.

Figure 25.16 The Large Intestine

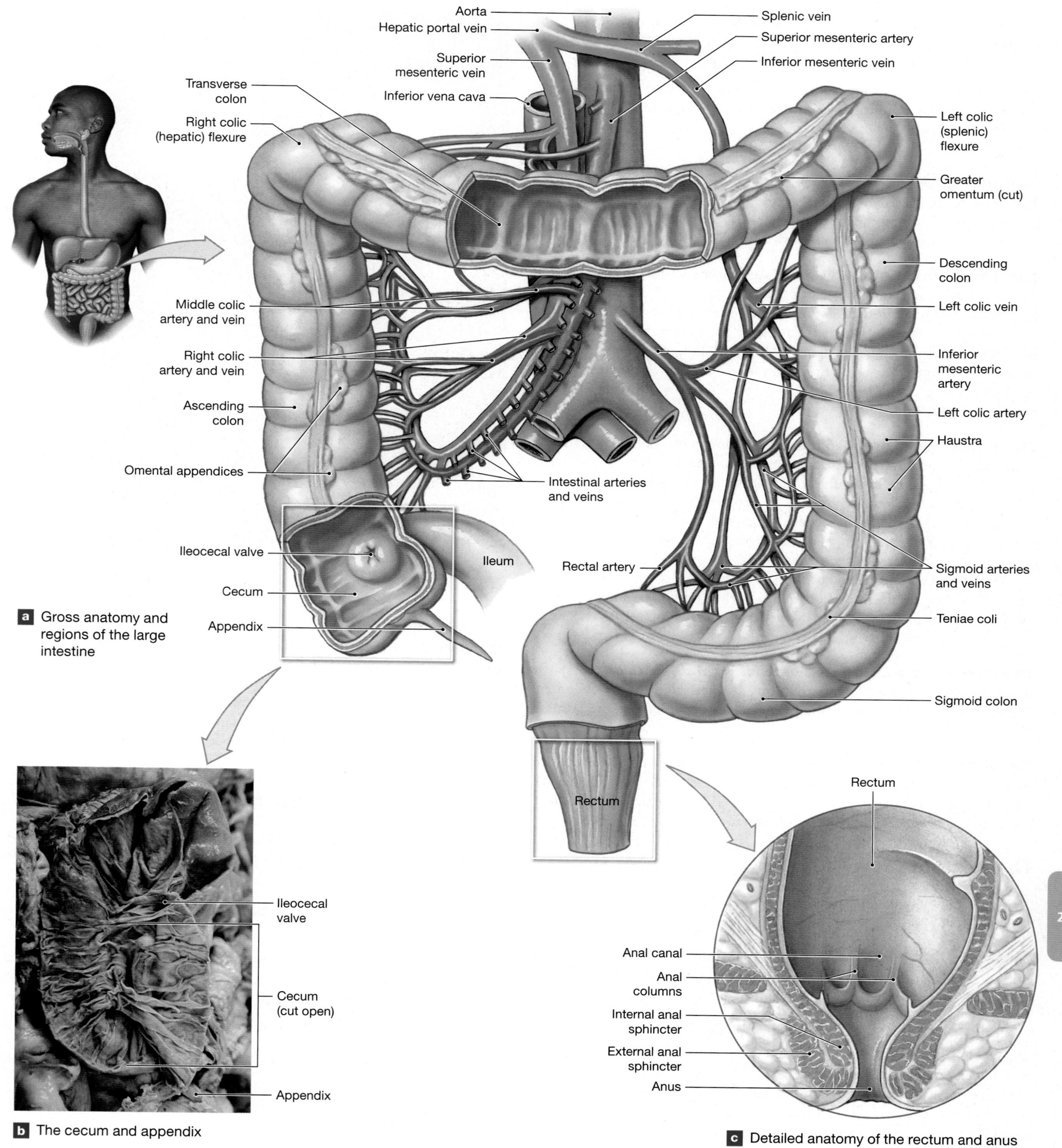

a Gross anatomy and regions of the large intestine

b The cecum and appendix

c Detailed anatomy of the rectum and anus

- Three separate longitudinal bands of smooth muscle—called the **teniae coli** (TĒ-nē-ē KŌ-lē)—run along the outer surfaces of the colon just deep to the serosa. These bands correspond to the outer, longitudinal layer of the muscular layer in other portions of the digestive tract. Muscle tone within the teniae coli is what creates the haustra.
- The serosa of the colon contains numerous teardrop-shaped sacs of fat called the **omental appendices**, or *fatty appendices of the colon* (**Figure 25.16a**).

Regions of the Colon

The colon is subdivided into four regions: the ascending colon, transverse colon, descending colon, and sigmoid colon (**Figure 25.16a**). You can see these regions in the radiograph in **Figure 25.17**.

The Ascending Colon The **ascending colon** begins at the superior border of the cecum. It ascends along the right lateral and posterior abdominal wall of the peritoneal cavity to the inferior surface of the liver. At this point the colon turns to the left at the **right colic flexure**, or *hepatic flexure*. This bend marks the end of the ascending colon and the beginning of the transverse colon. The ascending colon is secondarily retroperitoneal; visceral peritoneum covers its lateral and anterior surfaces (**Figures 25.4c,d**, p. 656, and **25.16a**).

The Transverse Colon The **transverse colon** curves anteriorly from the right colic flexure and crosses the abdomen from right to left. As the transverse colon crosses the abdominal cavity, its peritoneal relationship changes. The initial segment is intraperitoneal. It is supported by the transverse mesocolon and separated from the anterior abdominal wall by the layers of the greater omentum. As the transverse colon passes inferior to the greater curvature of the stomach it becomes secondarily retroperitoneal.

The gastrocolic ligament attaches the transverse colon to the greater curvature of the stomach. Near the spleen, the colon makes a right turn, termed the **left colic flexure**, or *splenic flexure*. It then proceeds caudally as the descending colon.

The Descending Colon The **descending colon**, which is secondarily retroperitoneal, proceeds inferiorly along the left side of the abdomen. At the iliac fossa, the descending colon curves and becomes the sigmoid colon.

The Sigmoid Colon The **sigmoid** (SIG-moyd; *sigmoides*, the Greek letter S) **colon** is an S-shaped segment that is only about 15 cm (6 in.) long. The sigmoid colon lies posterior to the urinary bladder, suspended from the sigmoid mesocolon. The sigmoid colon empties into the rectum (**Figures 25.4**, p. 656, and **25.16**).

The Rectum

▶ **KEY POINT** Feces enter the rectum from the sigmoid colon and exit by the anal canal and anus.

The **rectum** (REK-tum) forms the last 15 cm (6 in.) of the digestive tract. It is an expandable organ for the temporary storage of **feces** (**Figures 25.11**, p. 666, **25.16a,c**, and **25.17**). Movement of fecal material into the rectum triggers the urge to defecate.

The last portion of the rectum is the **anal canal**. It contains small longitudinal folds, the **anal columns**. The distal margins of these columns are joined by transverse folds that mark the boundary between the columnar epithelium of the proximal rectum and a stratified squamous epithelium like that in the oral cavity. The **anus** is the exit of the anal canal. There, the epidermis becomes keratinized and identical to the surface of the skin.

Figure 25.17 Anterior/Posterior Radiograph of the Colon

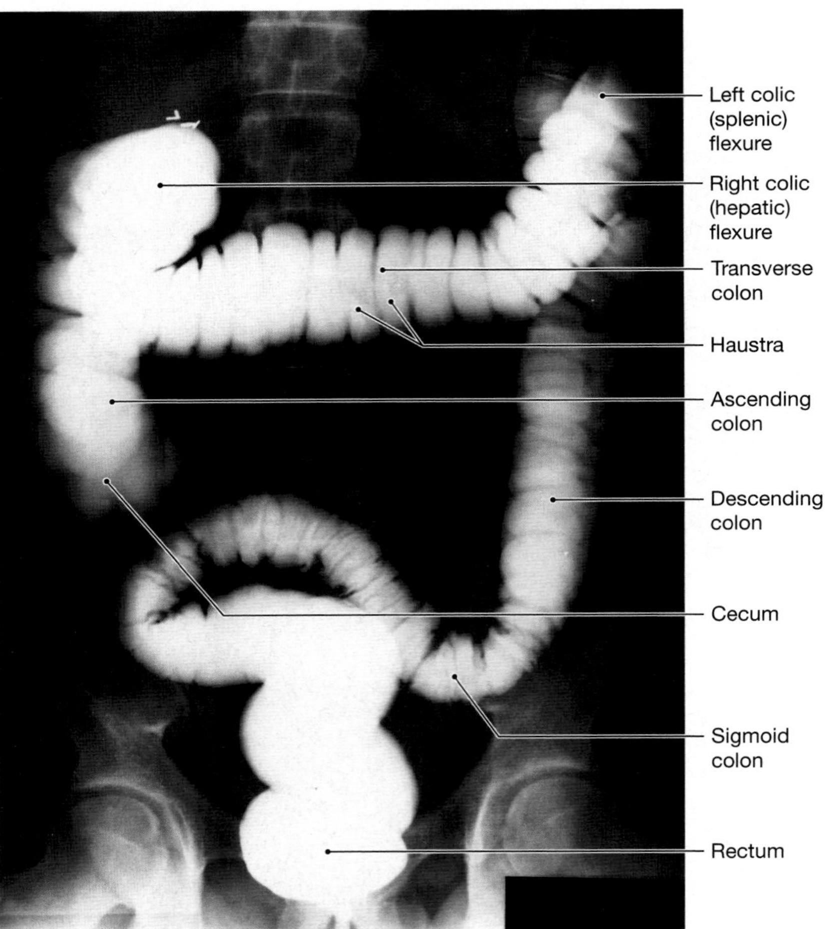

The circular layer of muscle in this region forms the **internal anal sphincter**. The smooth muscle fibers of this sphincter are not under voluntary control. The **external anal sphincter**, which guards the anus, encircles the distal portion of the anal canal. This sphincter, which consists of a ring of skeletal muscle fibers, is under voluntary control. The lamina propria and submucosa of the anal canal contain a network of veins. If venous pressures there rise too high due to straining during defecation or pregnancy, the veins can become distended, producing hemorrhoids.

Histology of the Large Intestine

▶ **KEY POINT** Unlike the small intestine, the large intestine lacks villi, has an increased number of goblet cells and deeper intestinal glands, and possesses teniae coli within the muscular layer.

Several histological characteristics distinguish the large intestine from the small intestine:

- The wall of the large intestine is much thinner than that of the small intestine, even though the diameter of the colon is roughly three times that of the small intestine.
- The large intestine lacks villi, which are characteristic of the small intestine.
- Goblet cells are more abundant in the mucosal epithelium of the large intestine.
- The intestinal glands of the large intestine are deeper than those of the small intestine, and they have more goblet cells (**Figure 25.18**). Secretion

occurs as a result of local reflexes involving the local neural plexuses. This produces large amounts of mucus, which lubricates the mucosa as undigested waste is compacted and passed along the large intestine.

- Large lymphoid nodules are scattered throughout the lamina propria and extend into the submucosa **(Figure 23.7a)**. ⮌ p. 611
- The muscular layer differs from that of other intestinal regions. The longitudinal layer has been reduced to the three muscular bands of the teniae coli. However, the mixing and propulsive contractions of the colon resemble those of the small intestine.

Regulation of the Large Intestine

▶ **KEY POINT** Ingested materials move slowly from the cecum to the transverse colon. Movement involves peristaltic activity and haustral churning.

Movement from the cecum to the transverse colon is very slow, allowing for water absorption to convert the already thick material into a sludgy paste. Peristaltic waves move material along the length of the colon. Segmentation movements, called **haustral churning**, mix the contents of adjacent haustra. When the stomach and duodenum are distended, signals are relayed over the intestinal neural plexuses. These signals cause powerful peristaltic contractions called **mass movements** that move material from the transverse colon through the rest of the large intestine. The contractions force feces into the rectum.

The rectal chamber is typically empty, except when a mass movement forces feces out of the sigmoid colon into the rectum. When feces move into the rectum, the distension on the rectal wall stimulates stretch receptors, initiating the defecation reflex. The defecation reflex relaxes the internal anal sphincter, and feces move into the anal canal. When the external anal sphincter is voluntarily relaxed, defecation can occur.

25.7 ✔ CONCEPT CHECK

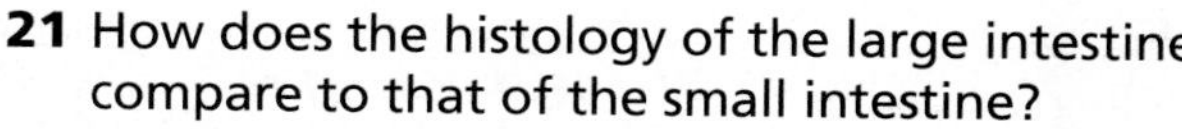

21 How does the histology of the large intestine compare to that of the small intestine?

22 List the four regions of the colon.

See the blue Answers tab at the back of the book.

25.8 | Accessory Digestive Organs

▶ **KEY POINT** The accessory organs of the digestive tract are the liver, gallbladder, and pancreas.

The accessory digestive organs are the liver, gallbladder, and pancreas. These organs produce and store enzymes and buffers essential for normal digestive function. In addition to their roles in digestion, the liver and pancreas have exocrine functions.

The Liver

▶ **KEY POINT** The liver (1) regulates metabolism, (2) regulates the composition of blood by removing old and damaged red blood cells and synthesizing plasma proteins, and (3) produces and secretes bile.

The **liver** is the largest visceral organ, a firm, reddish-brown organ weighing about 1.5 kg (3.3 lb) **(Figure 25.19)**. Most of its mass lies in the right hypochondriac

Figure 25.18 The Wall of the Large Intestine

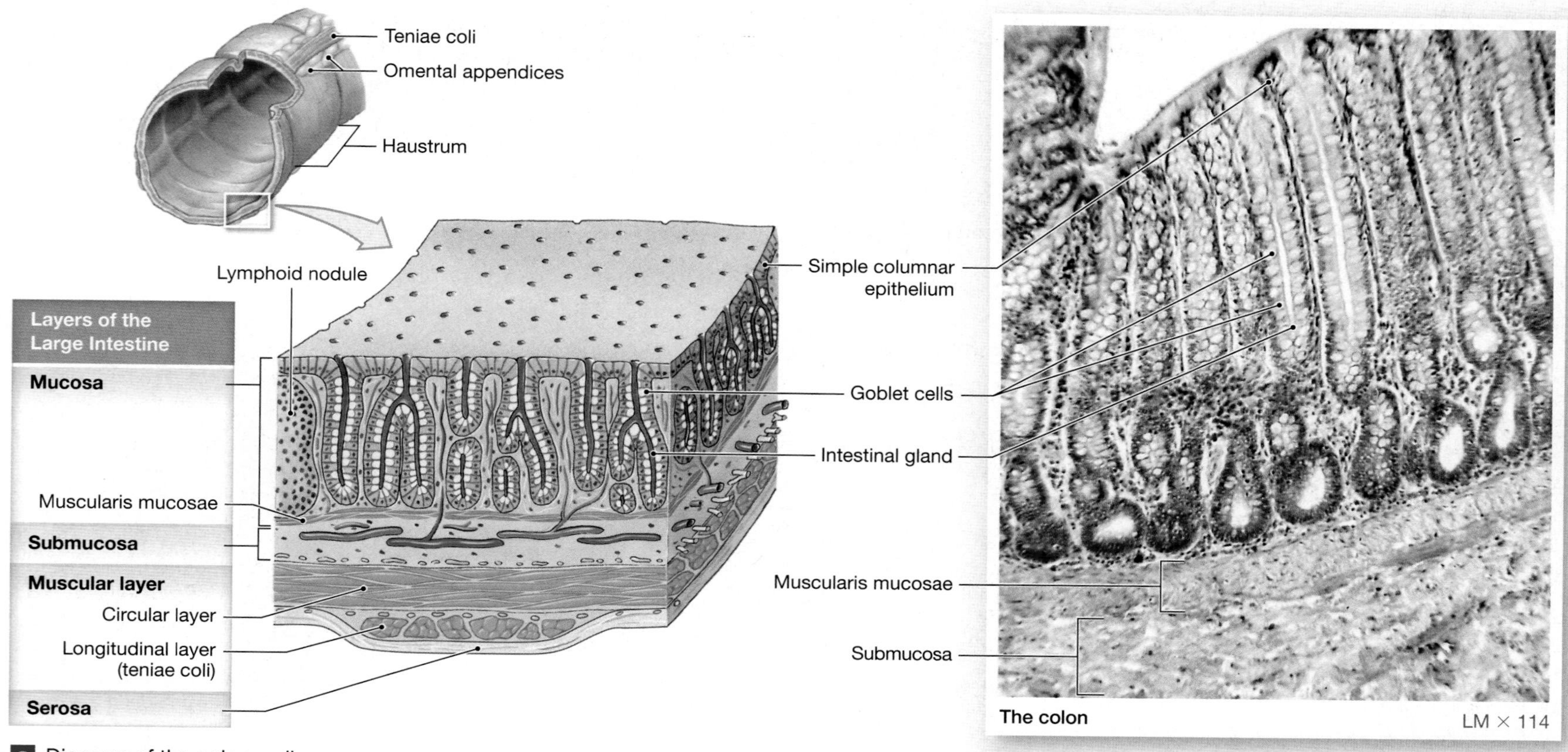

a Diagram of the colon wall

b Colon histology showing detail of mucosal and submucosal layers

Figure 25.19 Anatomy of the Liver

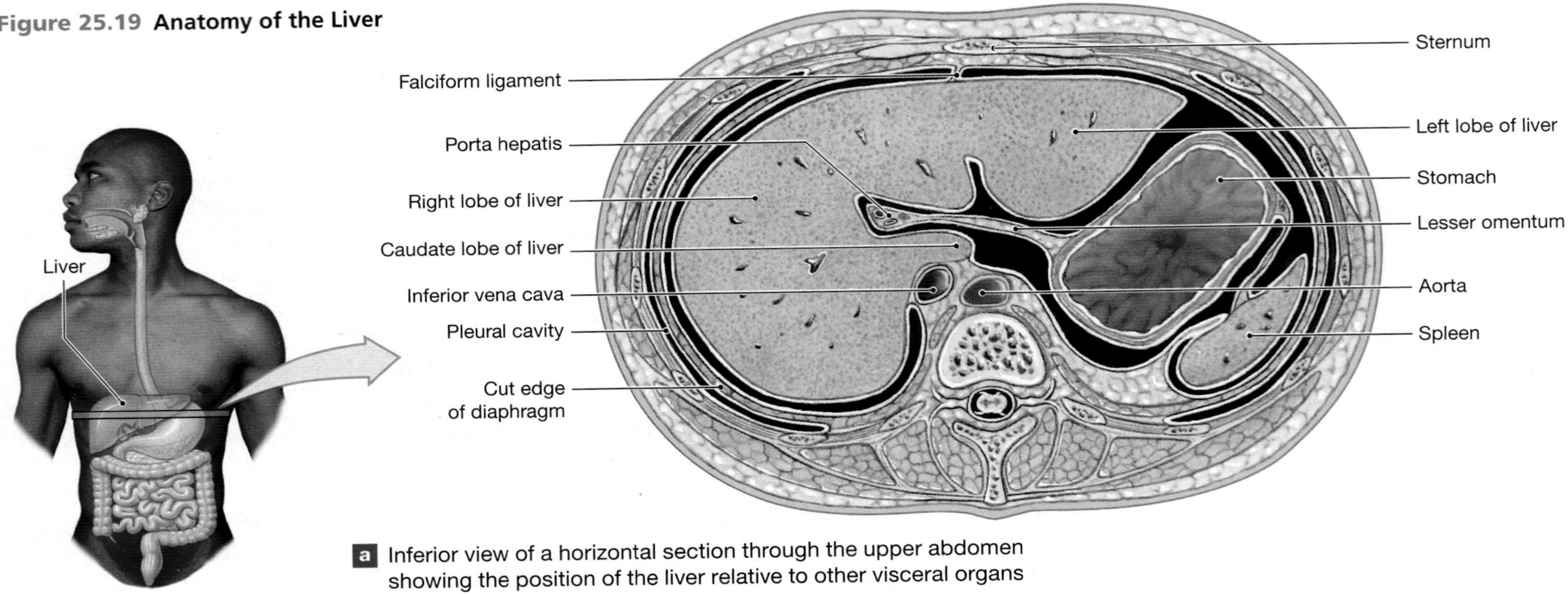

a Inferior view of a horizontal section through the upper abdomen showing the position of the liver relative to other visceral organs

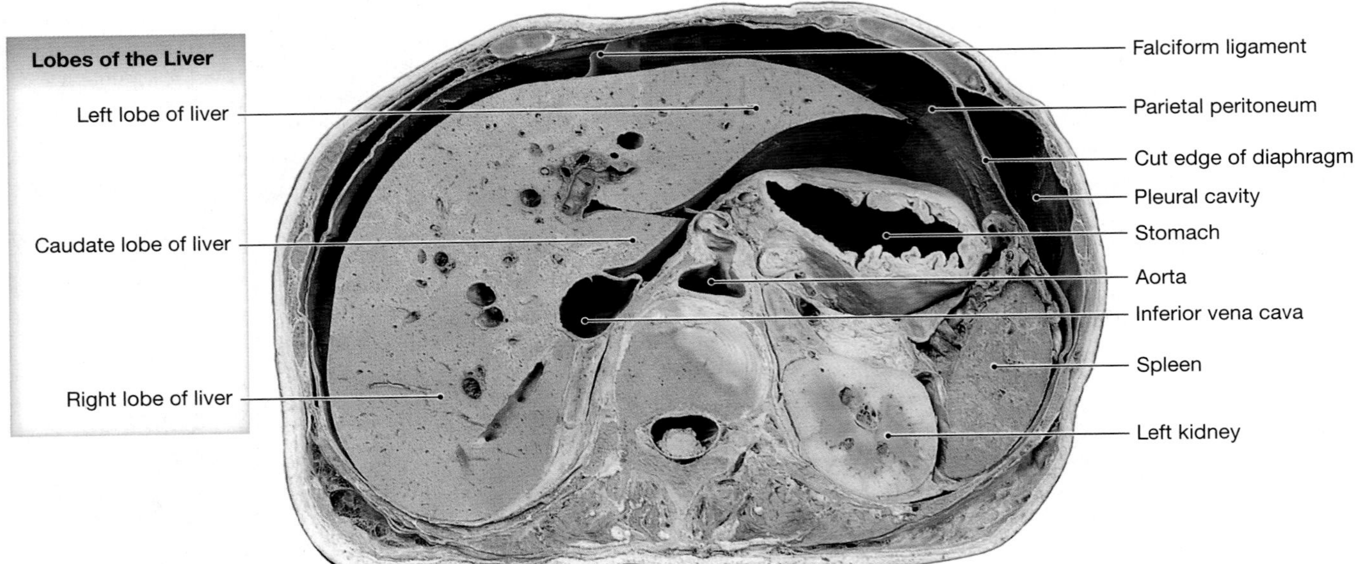

b Horizontal section through the upper abdomen showing structures illustrated in part (a)

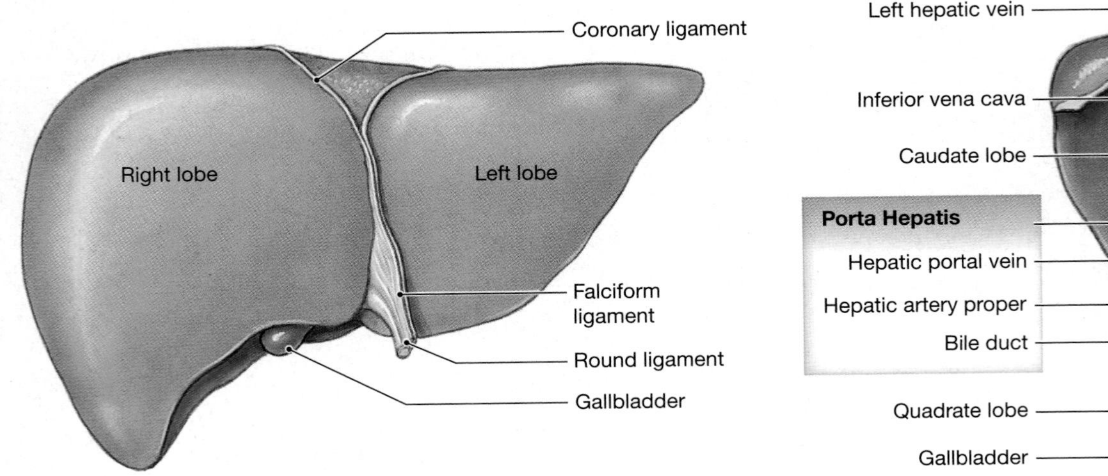

c Anatomical landmarks on the anterior surface of the liver

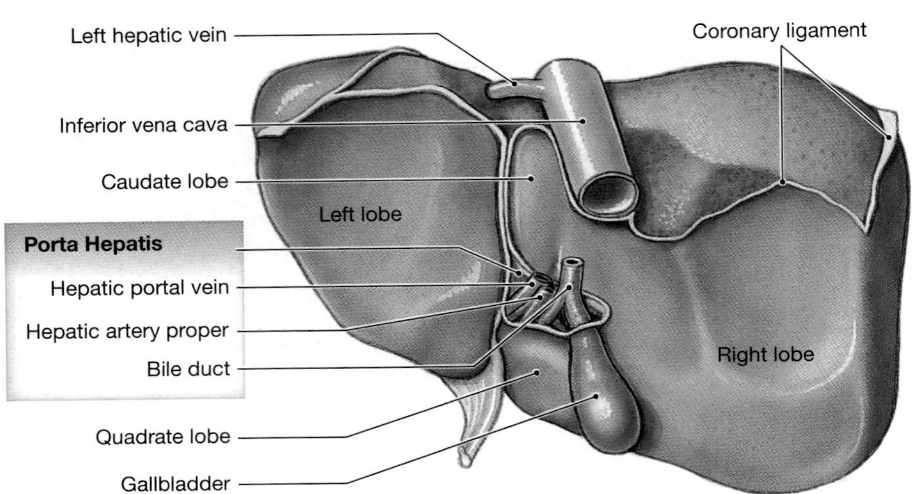

d The inferior and posterior surfaces of the liver

e Cast of the liver showing gallbladder, biliary ducts, and associated blood vessels as seen from the inferior and posterior surfaces

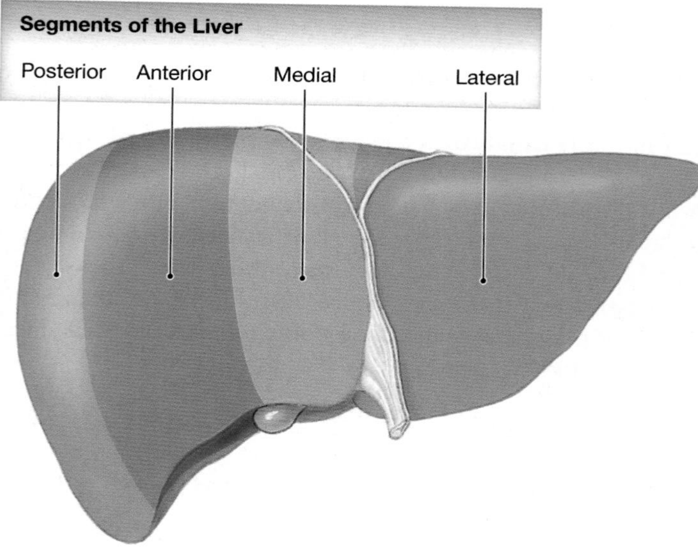

f Anterior surface of the liver showing approximate boundaries between the major segments

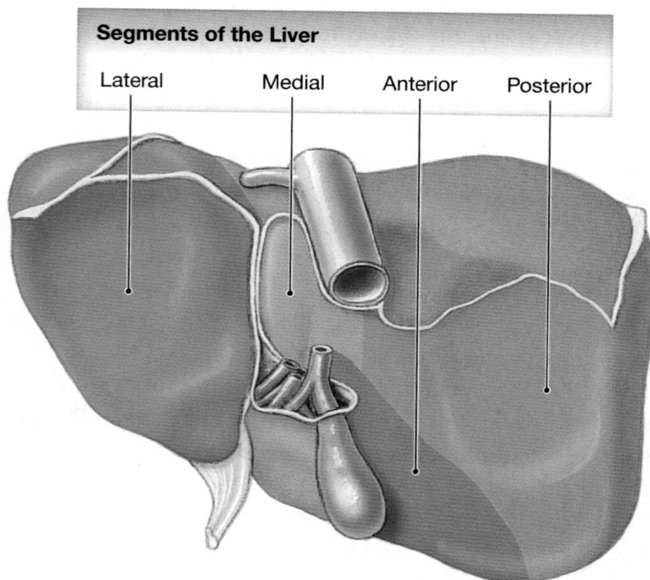

g The inferior and posterior surfaces of the liver showing approximate boundaries between the major segments

and epigastric regions, but it may extend into the left hypochondriac and umbilical regions as well. The versatile liver performs essential metabolic and synthetic functions:

- Metabolic regulation: The liver is an essential organ for regulating body metabolism. The liver regulates the circulating levels of carbohydrates, lipids, and amino acids. Blood draining all of the absorptive surfaces of the stomach and small and large intestines enters the hepatic portal system and flows into the liver. As a result, the liver cells extract absorbed nutrients and toxins from the blood before they reach the systemic circulation through the hepatic veins. Liver cells (called hepatocytes) monitor circulating levels of metabolites and make any necessary adjustments. Excess nutrients are removed and stored. Deficiencies are corrected by mobilizing stored reserves or performing appropriate synthetic activities. Circulating toxins and metabolic wastes are also removed. They are then inactivated, stored, or excreted. Finally, fat-soluble vitamins (A, D, K, and E) are absorbed and stored in the liver.
- Hematological regulation: The liver is the largest blood reservoir in the body, receiving about 25 percent of the cardiac output. As blood passes through the liver phagocytic cells remove old or damaged RBCs, cellular debris, and pathogens. Liver cells synthesize and secrete plasma proteins into the blood. Plasma proteins contribute to the osmotic concentration of the blood, transport nutrients, and establish the clotting and complement systems.
- Production of bile: **Bile** is synthesized by liver cells, stored in the gallbladder, and excreted into the lumen of the duodenum. Bile is mostly water, with small amounts of ions, bilirubin (a pigment derived from hemoglobin), and an assortment of lipids known as bile salts. The water and ions dilute and buffer acids in the chyme as it enters the small intestine. Bile salts make it possible for enzymes to break down lipids in the chyme into fatty acids suitable for absorption.

The liver is thought to have more than 200 functions. **Table 25.1** shows just a partial list. Any condition that severely damages the liver presents a serious threat to life. Liver cells do have a limited ability to regenerate after an injury. However, liver function will not fully recover unless the normal vascular patterns also regenerate.

Anatomy of the Liver

The liver is intraperitoneal. Deep to its visceral peritoneal layer is a tough fibrous capsule. On the anterior surface a ventral mesentery, the **falciform** (FAL-si-form) **ligament**, marks the division between the **left lobe** and **right lobe** of the liver (**Figure 25.19a–c**). The **round ligament**, or *ligamentum teres,* is a thickening of the inferior margin of the falciform ligament. This fibrous band marks the path of the degenerated fetal umbilical vein. The liver is suspended from the inferior surface of the diaphragm by the **coronary ligament**.

The shape of the liver conforms to its surroundings. Its **anterior surface** follows the smooth curve of the body wall (**Figure 25.19c**). The **posterior surface** has impressions from the stomach, small intestine, right kidney, and large intestine (**Figure 25.19d**). The superior, anterior, and posterior surfaces of the liver are called the diaphragmatic surfaces because of their anatomical relationships to the diaphragm. The inferior surface is called the visceral surface.

The liver has classically been described as having four lobes (**Figure 25.19d**). The impression left by the inferior vena cava marks the division between the right lobe and the small **caudate** (KAW-dāt) **lobe**. Inferior to the caudate lobe lies the **quadrate lobe**, sandwiched between the left lobe and the gallbladder.

Table 25.1 | **Major Functions of the Liver**

DIGESTIVE AND METABOLIC FUNCTIONS
Synthesize somatomedins
Synthesize and secrete bile
Store glycogen and lipid reserves
Maintain normal blood glucose, amino acid, and fatty acid concentrations
Synthesize and convert nutrient types (e.g., transaminate amino acids or convert carbohydrates to lipids)
Synthesize and release cholesterol bound to transport proteins
Inactivate toxins
Store iron reserves
Store fat-soluble vitamins
OTHER MAJOR FUNCTIONS
Synthesize plasma proteins
Synthesize clotting factors
Synthesize the inactive hormone angiotensinogen
Phagocytose damaged red blood cells (by stellate macrophages)
Store blood (major contributor to venous reserve)
Absorb and break down circulating hormones (including insulin and epinephrine) and immunoglobulins
Absorb and inactivate lipid-soluble drugs

The classical description of four lobes was based on the superficial topography of the liver. However, this description of the liver did not meet the needs of modern medical science, especially surgery. As a result, a more comprehensive system for describing the structure of the liver was developed. The new terminology subdivides the lobes of the liver into segments based on the major subdivisions of the hepatic artery, portal vein, and hepatic ducts. **Figure 25.19f** indicates the approximate boundaries between the segments; the actual boundaries cannot be determined without dissecting the liver.

The Blood Supply to the Liver Chapter 22 discussed the circulation to the liver (**Figures 22.15** and **22.22**). ⊃ pp. 588, 596

Two blood vessels deliver blood to the liver, the **hepatic artery proper** and the **hepatic portal vein** (**Figures 25.16a** and **25.19d**). Roughly one-third of the normal hepatic blood flow arrives through the hepatic artery; the rest arrives through the hepatic portal vein. Blood returns to the systemic circuit through the **hepatic veins**, which empty into the inferior vena cava. The arterial supply provides oxygen-rich blood to the liver, and the hepatic portal vein supplies nutrients and other chemicals absorbed from the intestine.

Histology of the Liver

Each lobe of the liver is divided by connective tissue into approximately 100,000 **liver lobules**, the basic functional units of the liver. **Figure 25.20** shows the histological organization and structure of a typical liver lobule.

The Liver Lobule In a liver lobule, the liver cells, or **hepatocytes** (HEP-a-tō-sīts), form a series of irregular plates arranged like the spokes of a wheel (**Figure 25.20**). Up to the age of 7, the plates are no more than two cells thick. After age 7 the plates are only one cell thick. The exposed apical and basal surfaces of the hepatocytes are covered with short microvilli. Blood entering the liver by the hepatic artery proper *and* the hepatic portal system drain into highly fenestrated capillaries termed hepatic sinusoids that surround the plates of hepatocytes. These hepatic sinusoids empty into the **central vein** (**Figure 25.20a–c**). The fenestrated walls of the sinusoids contain large openings that allow substances to pass out of the circulation and into the spaces surrounding the hepatocytes. The sinusoidal lining includes a large number of **stellate macrophages**, or *Kupffer* (KOOP-fer) cells. These phagocytic cells are part of the monocyte-macrophage system. They engulf pathogens, cell debris, and damaged blood cells. Stellate macrophages also store iron, some lipids, and heavy metals, such as tin or mercury, that are absorbed by the digestive tract.

Blood enters the liver sinusoids from small branches of the portal vein and hepatic artery proper. A typical lobule is hexagonal (six-sided) in cross section. There are six **portal triads**, one at each corner of the lobule. A portal triad contains three structures: (1) an interlobular vein, (2) an interlobular artery, and (3) an interlobular bile duct (**Figure 25.20**).

Branches from the arteries and veins deliver blood to the sinusoids of adjacent liver lobules (**Figure 25.20a,b**). As blood flows through the sinusoids, hepatocytes absorb solutes from the plasma and secrete materials such as plasma proteins. Blood then leaves the sinusoids and enters the central vein of the lobule. The central veins ultimately merge to form the hepatic veins, which then empty into the inferior vena cava.

Bile Secretion and Transport The hepatocytes form bile. It is secreted into a network of narrow channels between adjacent liver cells. These small channels, called **bile canaliculi** (singular, *canaliculus*), extend outward, away from the central vein (**Figure 25.20b**). The canaliculi connect with fine **bile ductules** (DUK-tūlz) that carry bile to an **interlobular bile duct** in the nearest portal triad (**Figure 25.20a,b**). The **right** and **left hepatic ducts** collect bile from all the bile ducts of the liver lobes. These ducts unite to form the **common hepatic duct** that leaves the liver. The bile in the common hepatic duct either flows into the bile duct that empties into the duodenal ampulla or enters the cystic duct that leads to the gallbladder (**Figure 25.21**).

The Gallbladder

▶ **KEY POINT** The gallbladder stores and concentrates bile before it is excreted into the small intestine.

The **gallbladder** is a hollow, pear-shaped, muscular sac. The gallbladder stores and concentrates bile before it is released into the small intestine. The gallbladder is located in a recess, or fossa, in the posterior surface of the liver's right lobe. Like the liver, the gallbladder is intraperitoneal.

The gallbladder is divided into three regions: the **fundus**, **body**, and **neck** (**Figure 25.21a,c**). The **cystic duct** exits the gallbladder and unites with the common hepatic duct to create the **bile duct** (**Figure 25.21a,c**). At the duodenum, a muscular **sphincter of ampulla** surrounds the lumen of the bile duct and the duodenal ampulla (**Figures 25.15**, p. 671, and **25.21b**). The duodenal ampulla opens into the duodenum at the duodenal papilla, a small raised projection. Contraction of this sphincter seals off the passageway and prevents bile from entering the small intestine.

The gallbladder has two major functions: storing and modifying bile. When the sphincter of the ampulla is closed, bile enters the cystic duct and flows into the gallbladder for storage. When filled to capacity, the gallbladder contains 40–70 ml of bile. As bile remains in the gallbladder, its composition gradually changes. Water is absorbed from the bile, and the bile salts and other components of bile become increasingly concentrated.

Bile ejection occurs under stimulation of the hormone cholecystokinin (CCK). Cholecystokinin is released into the bloodstream at the duodenum when chyme arrives containing large amounts of lipids and partially digested proteins. CCK relaxes the sphincter of the ampulla and contracts the gallbladder.

Figure 25.20 **Liver Histology**

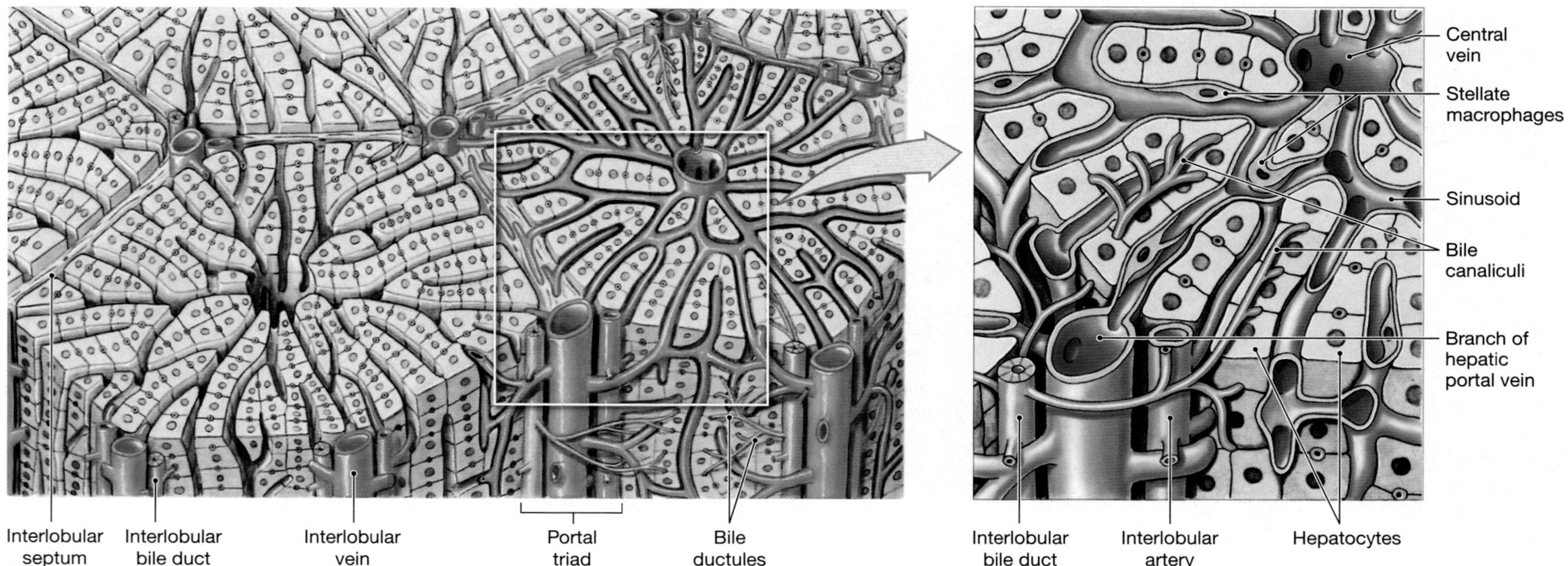

a Diagram of lobular organization.

b Magnified view showing the portal triad and central vein.

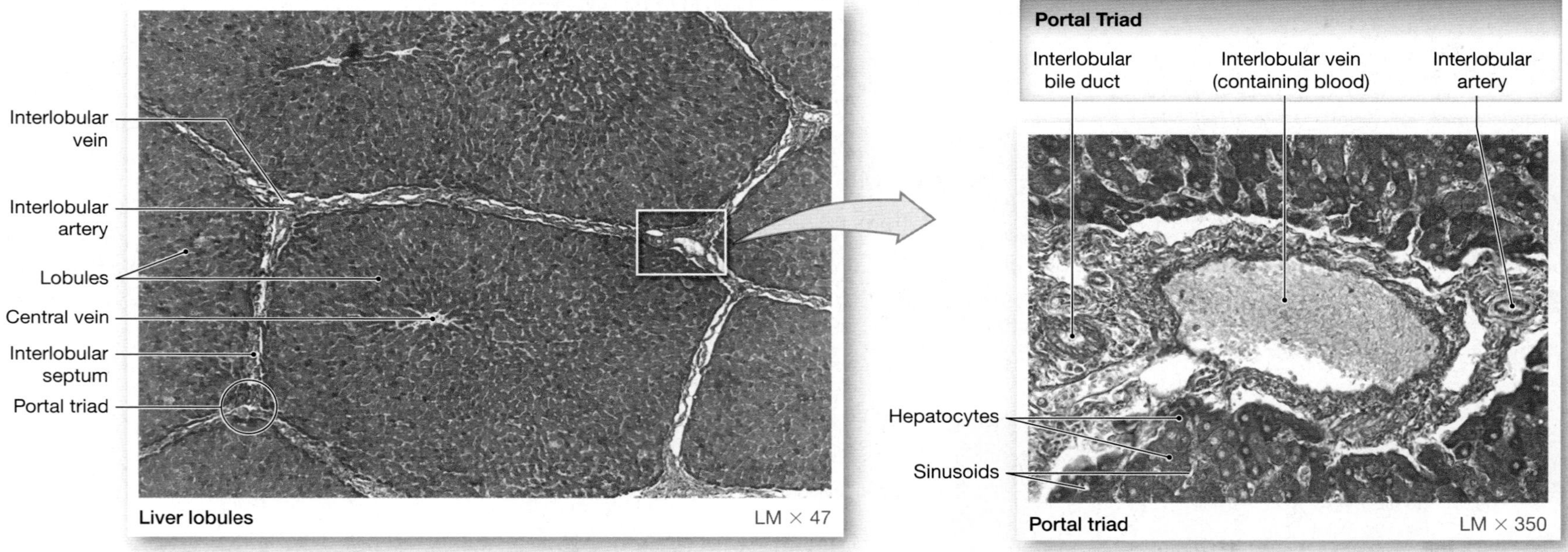

c Light micrograph showing representative mammalian liver lobules. Human liver lobules lack a distinct connective tissue boundary, making them difficult to distinguish in histological section.

d Light micrograph showing histological detail of portal triad.

Histological Organization of the Gallbladder

The wall of the gallbladder is composed of only the mucosa, lamina propria, muscular layer and serosa; a muscularis mucosae and submucosa are lacking. The mucosa has numerous folds that divide the surface into irregular **mucosal glands**. The lamina propria is composed of areolar connective tissue, and the muscular layer is composed of two interlacing layers of smooth muscle. The inner layer is composed mostly of longitudinally arranged smooth muscle, while the outer layer is composed mostly of circularly arranged smooth muscle. **(Figure 25.22)**.

The Pancreas

▶ **KEY POINT** The pancreas is a mixed gland, having both exocrine and endocrine functions. The exocrine pancreas produces digestive enzymes and buffers that are emptied into the duodenum.

The **pancreas** lies posterior to the stomach, extending laterally from the duodenum toward the spleen **(Figures 25.21a** and **25.23)**. The pancreas is an elongated, pinkish-gray organ, approximately 15 cm (6 in.) long and weighing around 80 g (3 oz). It has three subdivisions: head, body, and tail.

Figure 25.21 The Gallbladder and Associated Bile Ducts

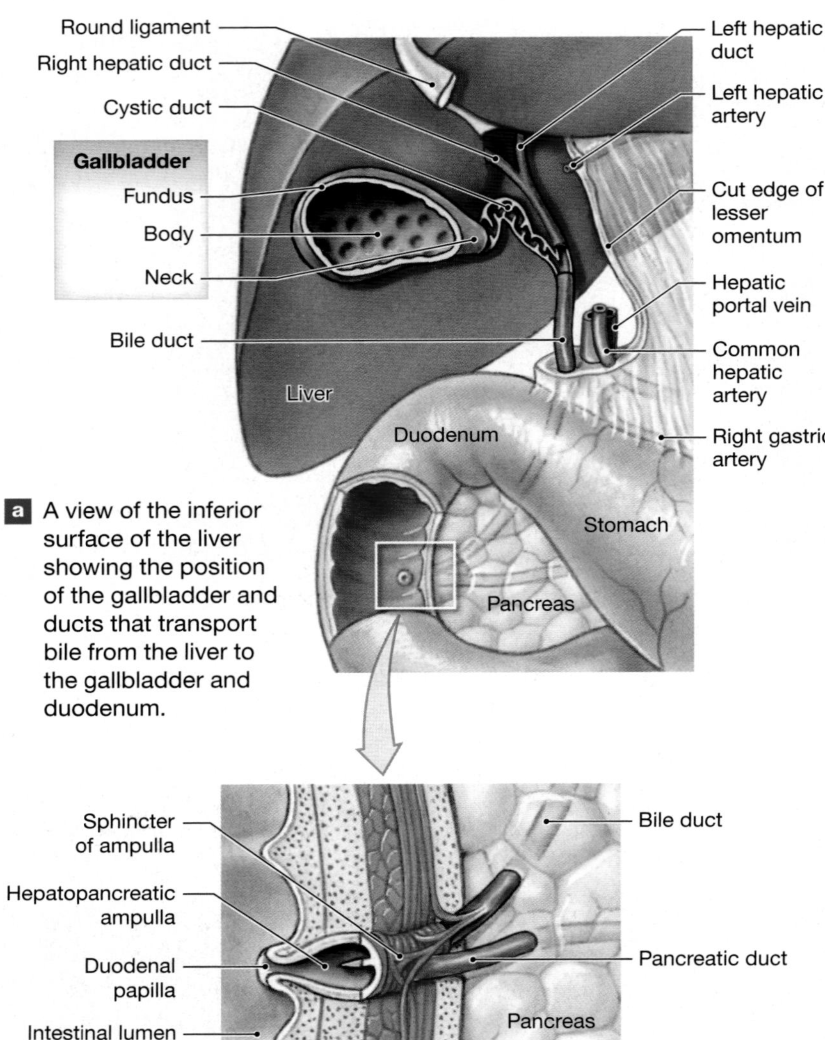

a A view of the inferior surface of the liver showing the position of the gallbladder and ducts that transport bile from the liver to the gallbladder and duodenum.

b A portion of the lesser omentum has been cut away to make it easier to see the relationships among the bile duct, the hepatic duct, and the cystic duct.

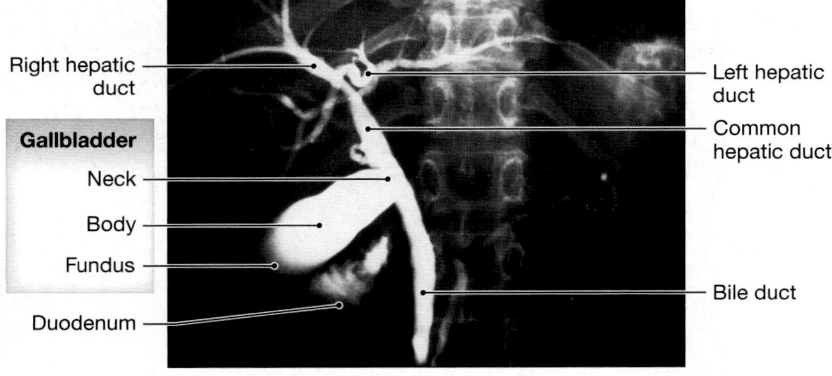

c A radiograph (cholangiogram, anterior-posterior view) of the gallbladder, biliary ducts, and pancreatic ducts.

The broad **head** of the pancreas lies within the loop formed by the initial segment of the duodenum. The slender **body** extends transversely to the left toward the spleen, and the **tail** is short and rounded. The pancreas is secondarily retroperitoneal, and it is firmly bound to the posterior wall of the abdominal cavity.

The surface of the pancreas has a lumpy, nodular texture. A thin, transparent connective tissue capsule wraps the pancreas. The pancreatic lobules, associated blood vessels, and excretory ducts are easily seen through the anterior capsule and the overlying layer of peritoneum.

The pancreas is a mixed gland, with both exocrine and endocrine functions. The exocrine portion of the pancreas produces digestive enzymes and buffers. (Chapter 19 discussed the endocrine function of the pancreas ⮌ pp. 518–520).

The large **pancreatic duct** delivers its exocrine secretions to the duodenal ampulla. Sometimes a small **accessory pancreatic duct**, or *duct of Santorini*, branches from the pancreatic duct before it leaves the pancreas **(Figures 25.21** and **25.23a)**. The pancreatic duct extends within the attached mesentery to reach the duodenum, where it meets the bile duct from the liver and gallbladder. The two ducts then empty into the duodenal ampulla, a chamber located roughly halfway along the length of the duodenum. When present, the accessory pancreatic duct usually empties into the duodenum at a separate duodenal papilla, outside the duodenal ampulla.

The splenic, superior mesenteric, and common hepatic arteries supply the pancreas with blood **(Figures 22.15a** and **25.23a)**. ⮌ p. 588 The **pancreatic arteries** and **pancreaticoduodenal arteries** (**superior** and **inferior**) are the major branches from these vessels. The splenic vein and its branches drain the pancreas **(Figure 22.22)**. ⮌ p. 596

Histological Organization of the Pancreas

Connective tissue septa divide the pancreatic tissue into distinct lobules **(Figure 25.23b,c)**. The blood vessels and tributaries of the pancreatic ducts are found within these connective tissue septa. The pancreas is an example of a compound tubulo-acinar gland. ⮌ p. 58 The secretory units of the exocrine pancreas are the **pancreatic acini** (AS-i-nī). Each pancreatic acinus is composed of a simple cuboidal epithelium. The pancreatic acini are the beginning of the duct system of the pancreas. These ducts drain into progressively larger ducts, which drain into the main pancreatic duct, which runs the length of the pancreas.

Figure 25.22 Histology of the gallbladder mucosa

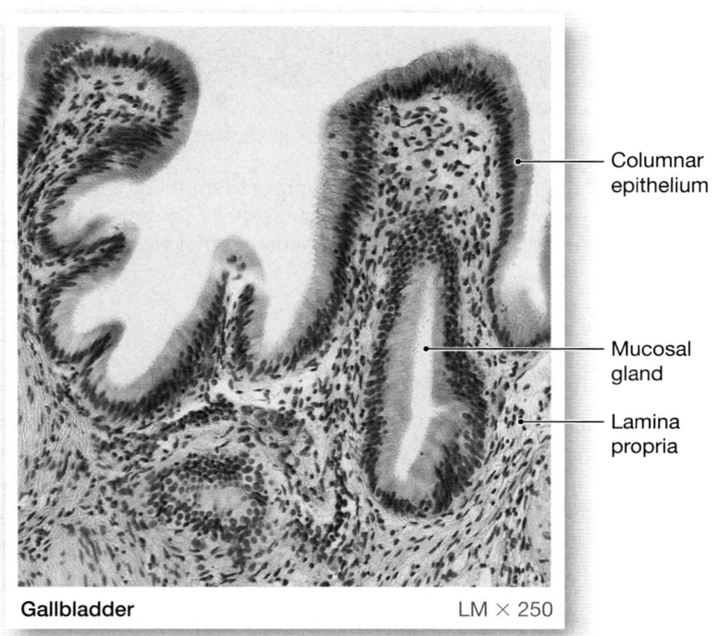

25

Figure 25.23 **The Pancreas**

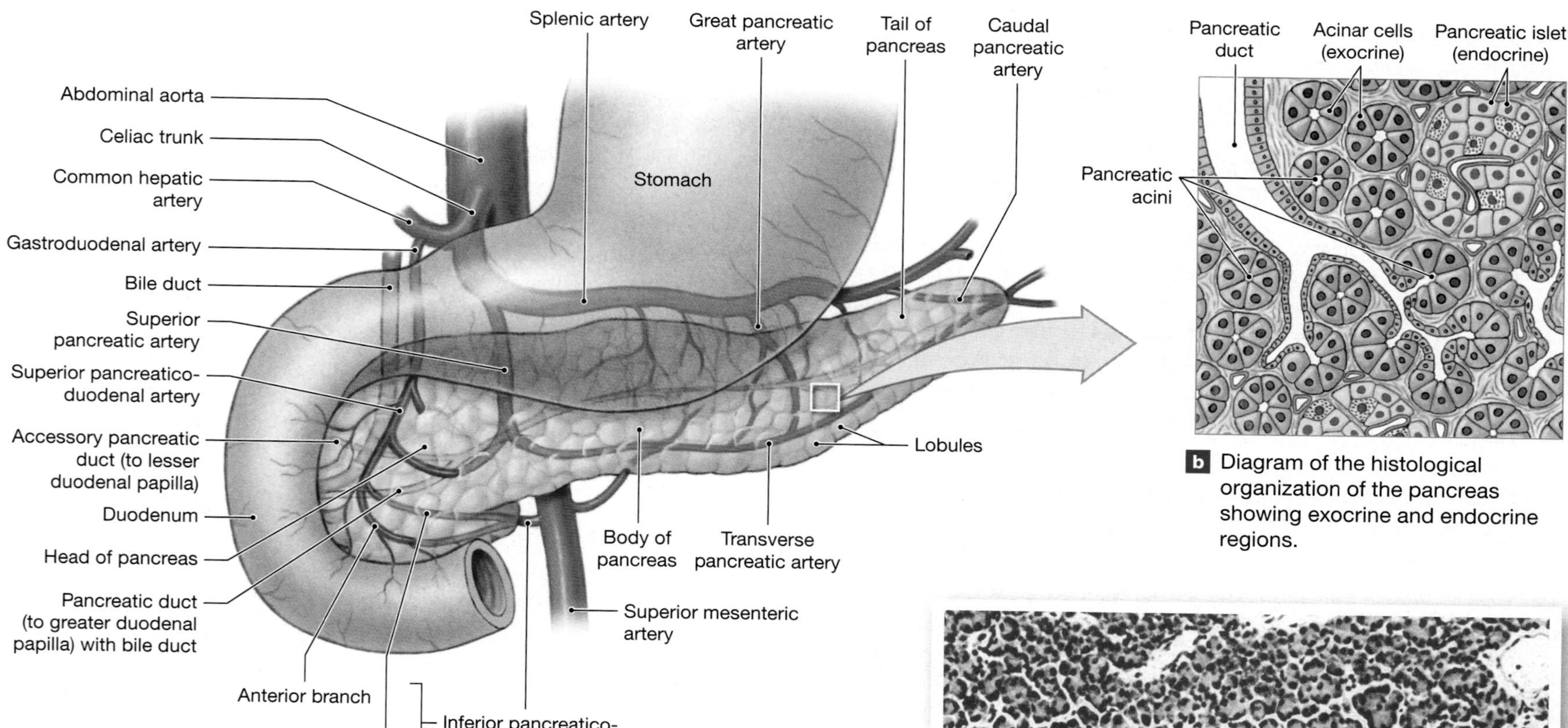

a Gross anatomy of the pancreas. The head of the pancreas is tucked into a curve of the duodenum that begins at the pylorus of the stomach.

b Diagram of the histological organization of the pancreas showing exocrine and endocrine regions.

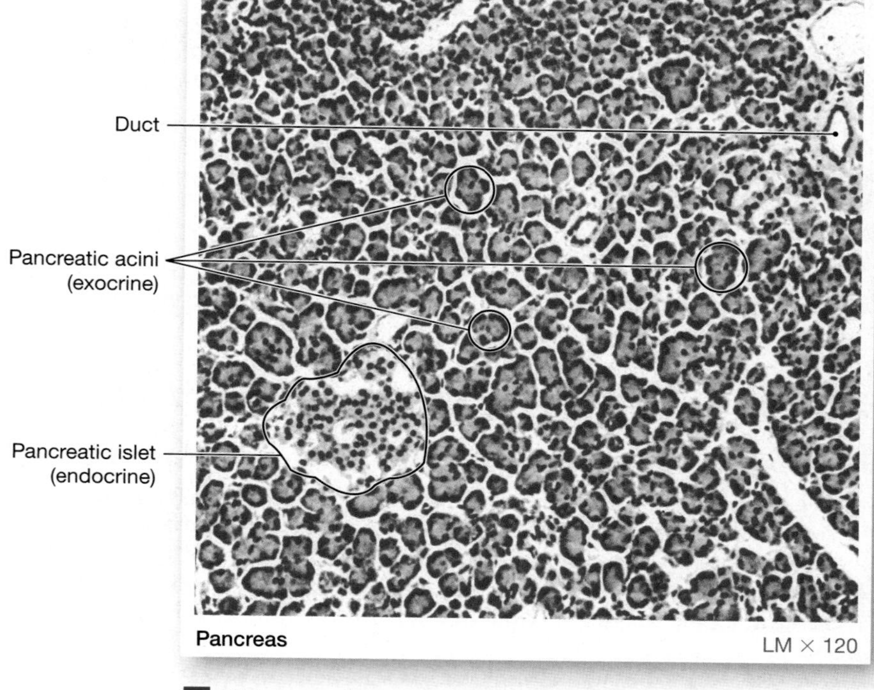

c Histology of the pancreas showing exocrine and endocrine cells.

The pancreatic acini produce **pancreatic juice**, a mixture of water, ions, and pancreatic digestive enzymes. These enzymes are released into the duodenum, where they break down ingested materials into small molecules suitable for absorption. The pancreatic ducts also secrete buffers (primarily sodium bicarbonate). These buffers neutralize the acid in chyme and stabilize the pH of the intestinal contents.

As we saw in Chapter 19, pancreatic islets are scattered between the acini. They account for approximately 1 percent of the cellular population of the pancreas and are responsible for the organ's endocrine function.

Pancreatic Enzymes

Pancreatic enzymes are classified according to their intended targets. **Lipases** (LĪ-pā-zez) digest lipids; **carbohydrases** (kar-bō-HĪ-drā-zez), such as pancreatic amylase, digest sugars and starches; **nucleases** break down nucleic acids; and **proteolytic** (prō-tē-ō-LIT-ik) **enzymes** break proteins apart. The proteolytic enzymes include **proteinases** and **peptidases**. Proteinases break apart large protein complexes, and peptidases break small peptide chains into individual amino acids.

The Regulation of Pancreatic Secretion

Hormones released by the duodenum trigger the secretion of pancreatic juice. When acid chyme arrives in the small intestine, secretin is released. This hormone triggers the production of the watery pancreatic juice containing buffers. Another duodenal hormone, cholecystokinin, stimulates the production and secretion of pancreatic enzymes.

25.8 ✔ CONCEPT CHECK

23 What are the three major functions of the liver?

24 What does contraction of the sphincter of ampulla accomplish?

25 What are the functions of the pancreas?

See the blue Answers tab at the back of the book.

25.9 | Aging and the Digestive System

▶ **KEY POINT** Age-related changes in the digestive system include accumulating damage, declining rate of stem cell division, decreased smooth muscle tone and activity, and increased rates of cancer.

Digestion and absorption are essentially normal in elderly individuals. However, many changes in the digestive system parallel age-related changes in other systems:

- The effects of cumulative damage become apparent. A familiar example is the gradual loss of teeth because of tooth decay or gingivitis. Cumulative damage can involve internal organs as well. Toxins such as alcohol, heavy metals, and other chemicals that are absorbed by the digestive tract are transported to the liver for processing or storage. The liver cells are not immune to these compounds, and chronic exposure can lead to *cirrhosis* or other types of liver disease.
- The rate of epithelial stem cell division declines. Stem cells in the epithelium divide less frequently with age, so tissue repair is less efficient. As a result, the digestive epithelium becomes more susceptible to damage by abrasion, acids, or enzymes. Peptic ulcers become more likely. In the mouth, esophagus, and anus, the stratified epithelium becomes thinner and more fragile.
- Smooth muscle tone decreases, general motility decreases, and peristaltic contractions are weaker. These changes slow down the movement of chyme and promote constipation. Sagging in the walls of haustra in the colon produces symptoms of diverticulitis. Straining to eliminate compacted fecal materials stresses the less resilient walls of blood vessels, producing hemorrhoids. A weakened cardiac sphincter can lead to esophageal reflux and frequent "heartburn."
- Cancer rates increase. Cancers are most common in organs where stem cells divide to maintain epithelial cell populations. Rates of colon cancer and stomach cancer rise in the elderly; oral and pharyngeal cancers are particularly common in elderly smokers.
- Changes in other systems affect the digestive system directly or indirectly. For example, reduced bone mass and calcium content in the skeleton result in the erosion of the tooth sockets and eventual tooth loss. The decline in olfactory and gustatory sensitivity with age leads to dietary changes that affect the entire body.

25.9 CONCEPT CHECK

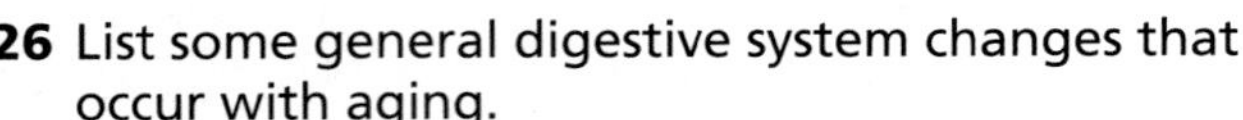

26 List some general digestive system changes that occur with aging.

See the blue Answers tab at the back of the book.

EMBRYOLOGY SUMMARY

For an introduction to the development of the digestive system, see Chapter 28 (Embryology and Human Development).

Study Outline

Introduction p. 651

- The **digestive system** consists of the muscular **digestive tract** and various **accessory organs**.

25.1 | An Overview of the Digestive System p. 651

Histological Organization of the Digestive Tract p. 651

- The major layers of the digestive tract are the **mucosa** (formed by the **mucosal epithelium** and **lamina propria**), **submucosa** (areolar tissue), the **muscular layer** (a region of smooth muscle fibers), and (in the peritoneal cavity) a serous membrane called the **serosa**. *(See Figures 25.1 and 25.2.)*

Muscular Layers and the Movement of Digestive Materials p. 652

- The **smooth muscle** cells of the digestive tract are capable of plasticity, which is the ability to tolerate extreme stretching. The digestive system contains visceral smooth muscle tissue, in which the muscle cells are arranged in sheets and contain no motor innervation. The presence of **pacesetter cells** allows for rhythmic waves of contraction that spread through the entire muscular sheet.
- The muscular layer propels materials through the digestive tract through the contractions of **peristalsis**. **Segmentation** movements in areas of the small intestine churn digestive materials. *(See Figure 25.3.)*

The Peritoneum p. 654

- The serosa, also known as the **visceral peritoneum**, is continuous with the **parietal peritoneum** that lines the inner surfaces of the body wall.
- Fused double sheets of peritoneal membrane called **mesenteries** suspend portions of the digestive tract. *(See Figures 25.2a and 25.4 and Spotlight Figure 25.10.)*
- Organs of the abdominal cavity may have a variety of relationships with the peritoneum, including **intraperitoneal**, **retroperitoneal**, and **secondarily retroperitoneal**.

25.2 | The Oral Cavity p. 655

- The functions of the oral cavity include (1) sensory analysis of potential foods, (2) mechanical processing using the teeth, tongue, and palatal surfaces, (3) lubrication by mixing with mucus and salivary secretions, and (4) digestion by salivary enzymes. Structures of the oral cavity include the tongue, salivary glands, and teeth. *(See Figure 25.5.)*

Anatomy of the Oral Cavity p. 655

- The **oral cavity** is lined by a stratified squamous epithelium. The **hard** and **soft palates** form the roof of the oral cavity. Other important features can be seen in *Figure 25.5*.
- The **tongue** aids in mechanical processing and manipulation of food as well as sensory analysis. The superior surface (**dorsum**) of the **body** of the tongue is covered with *papillae*. The inferior portion of the tongue contains a thin fold of mucous membrane called the **frenulum of the tongue**. **Intrinsic** and **extrinsic tongue muscles** are controlled by the hypoglossal nerve. *(See Figure 25.5.)*
- The **parotid**, **sublingual**, and **submandibular glands** discharge their secretions into the oral cavity. The parotid glands release salivary amylase, which begins the breakdown of carbohydrates. *(See Figures 25.5 and 25.6.)*

- **Saliva** lubricates the mouth, solubilizes food, dissolves chemicals, flushes the oral surfaces, and helps control bacteria. Salivation is usually controlled by the autonomic nervous system.
- **Dentine** forms the basic structure of a tooth. The **crown** is coated with **enamel**, and the **root** with **cement**. The **neck** marks the boundary between the root and the crown. The **periodontal ligament** anchors the tooth in an alveolar socket. **Mastication** (chewing) occurs through the contact of the opposing **occlusal surfaces** of the teeth. *(See Figure 25.7.)*
- There are four types of teeth, each with specific functions: **incisors**, for cutting; **canines** (*cuspids*), for tearing; **premolars** (*bicuspids*), for crushing; and **molars**, for grinding. *(See Figure 25.7b,c.)*
- The first set of teeth to appear are called **deciduous teeth**. These are replaced by the adult **permanent teeth**. The sequence of tooth **eruption** is presented in *Figure 25.7d*.
- Mastication forces the food across the surfaces of the teeth until it forms a **bolus** that can be swallowed easily.

25.3 | The Pharynx p. 661

- Skeletal muscles involved with swallowing include the pharyngeal constrictor muscles and the palatopharyngeus, stylopharyngeus, and palatal muscles.

The Swallowing Process p. 661

- **Deglutition** (swallowing) has three phases. The **buccal phase** begins with the compaction of a bolus and its movement into the pharynx. The **pharyngeal phase** involves the elevation of the larynx, reflection of the epiglottis, and closure of the glottis. Finally, the **esophageal phase** involves the opening of the **upper esophageal sphincter** and peristalsis moving the bolus down the esophagus to the **lower esophageal sphincter**. *(See Figure 25.8.)*

25.4 | The Esophagus p. 661

- The **esophagus** is a hollow muscular tube that transports food and liquid to the stomach through the **esophageal hiatus**, an opening in the diaphragm. *(See Figures 25.1 and 25.8.)*

Histology of the Esophageal Wall p. 663

- The wall of the esophagus is formed by mucosa, submucosa, muscular layer, and adventitia layers. *(See Figures 25.1, 25.2, and 25.9.)*

25.5 | The Stomach p. 663

- The **stomach** has three major functions: (1) bulk storage of ingested matter, (2) mechanical breakdown of resistant materials, and (3) chemical digestion through the disruption of chemical bonds using acids and enzymes.

Anatomy of the Stomach p. 663

- The stomach is divided into four regions: the **cardia**, **fundus**, **body**, and **pyloric part**. The **pyloric sphincter** guards the exit from the stomach. The mucosa and submucosa have longitudinal folds, called **gastric folds**. The muscular layer is formed of three bands of smooth muscle: a longitudinal layer, a circular layer, and an inner oblique layer. *(See Spotlight Figure 25.10 and Figures 25.4, 25.11, and 25.12.)*
- The mesenteries of the stomach are the **greater omentum** and the **lesser omentum**.
- Three branches of the celiac trunk supply blood to the stomach: the left gastric artery, the splenic artery, and the *common hepatic artery*. *(See Figures 22.15 and 22.22 and Spotlight Figure 25.10).*

Histology of the Stomach p. 666

- Simple columnar epithelia line all portions of the stomach. Shallow depressions, called **gastric pits**, contain the **gastric glands** of the fundus and body. **Parietal cells** secrete **intrinsic factor** and hydrochloric acid. **Chief cells** secrete **pepsinogen**, which acids in the gastric lumen convert to the enzyme **pepsin**. **G cells** of the stomach secrete the hormone **gastrin**. *(See Figure 25.12.)*

Regulation of Gastric Activity p. 668

- The production and secretion of **gastric juices** are directly controlled by the CNS and the celiac plexus. The release of the local hormones **secretin** and **cholecystokinin** inhibits gastric secretion but stimulates secretion by the pancreas and liver.

25.6 | The Small Intestine p. 668

Regions of the Small Intestine p. 669

- The **small intestine** includes the **duodenum**, **jejunum**, and **ileum**. *(See Figures 25.2a,b, 25.4b, 25.11, 25.13, 25.14a, and 25.15.)*

Support of the Small Intestine p. 669

- The superior mesenteric artery and superior mesenteric vein supply numerous branches to the segments of the small intestine. *(See Figures 22.15 and 22.22.)*
- The **mesentery proper** supports the branches of the superior mesenteric artery and vein, lymphatics, and nerves that supply the jejunum and ileum. *(See Figures 22.15, 22.22, and 25.4.)*

Histology of the Small Intestine p. 669

- The intestinal mucosa bears transverse folds called **circular folds**. The mucosa of the small intestine forms small projections, called **intestinal villi**, that increase the surface area for absorption. Each villus contains a terminal lymphatic called a **lacteal**. Pockets called **intestinal glands** house enteroendocrine, goblet, and stem cells. *(See Figures 25.4, 25.11, and 25.13–25.15.)*
- The regions of the small intestine have histological specializations that determine their primary functions. The duodenum (1) contains **duodenal submucosal glands** that aid the glands in producing mucus and (2) receives the secretions of the bile duct and pancreatic duct. The ileum contains large groups of **aggregated lymphoid nodules** (*Peyer's patches*) within the lamina propria. *(See Figures 25.1, 25.4, 25.11, 25.13, 25.15, and 25.21a,d.)*

Regulation of the Small Intestine p. 672

- **Intestinal juice** moistens the chyme, helps buffer acids, and dissolves digestive enzymes and the products of digestion.

25.7 | The Large Intestine p. 672

- The **large intestine** (**large bowel**) begins as a pouch inferior to the terminal portion of the ileum and ends at the anus. The main functions of the large intestine are to (1) reabsorb water and compact feces, (2) absorb vitamins produced by bacteria, and (3) store fecal material prior to defecation. *(See Figures 25.1, 25.4, 25.11, 25.13, and 25.16–25.18.)*
- The large intestine is divided into three parts: the **cecum**, the **colon**, and the **rectum**.

The Cecum and Appendix p. 672

- The **cecum** collects and stores materials arriving from the ileum. The ileum opens into the cecum with muscles encircling the opening forming the **ileocecal valve**. The **appendix** is attached to the cecum, and it functions as part of the lymphatic system. *(See Figure 25.16.)*

The Colon p. 672

- The **colon** has a larger diameter and a thinner wall than the small intestine. It bears **haustra** (pouches), the **teniae coli** (longitudinal bands of muscle), and **omental appendices**, or *fatty appendices of the colon* (fat aggregations within the serosa). *(See Figures 25.16 and 25.17.)*
- The colon is subdivided into four regions: **ascending**, **transverse**, **descending**, and **sigmoid**. *(See Figures 25.4, 25.16, and 25.17.)*

The Rectum p. 674

- The **rectum** terminates in the **anal canal** leading to the **anus**. **Internal** and **external anal sphincters** control the passage of fecal material to the anus. Distension of the rectal wall triggers the *defecation reflex*. *(See Figures 25.11, 25.16, and 25.17.)*

Histology of the Large Intestine p. 674

- The major histological features of the colon are lack of villi, abundance of goblet cells, and distinctive mucus-secreting intestinal glands. *(See Figures 23.7* and *25.18.)*

Regulation of the Large Intestine p. 675

- Movement from the cecum to the transverse colon occurs slowly via peristalsis and **haustral churning**. Movement from the transverse to the sigmoid colon occurs several times each day via **mass movements**.

25.8 | Accessory Digestive Organs p. 675

The Liver p. 675

- The **liver** performs metabolic and hematological regulation and produces **bile**. Its metabolic role is to regulate the concentrations of wastes and nutrients in the blood, and its hematological role is as a blood reservoir. *(See Figures 25.11, 25.19, and 25.20 and Table 25.1.)*
- In the classical topographical description of the liver, the organ is divided into four lobes: **left**, **right**, **quadrate**, and **caudate**. The gallbladder is located in a fossa within the **posterior surface** of the right lobe. New terminology has recently been adopted, based on subdivisions of the hepatic artery, portal vein, and hepatic ducts. *(See Figure 25.19a,b,c,d,f.)*
- The **hepatic artery proper** and **hepatic portal vein** supply blood to the liver. **Hepatic veins** drain blood from the liver and return it to the systemic circuit via the inferior vena cava. *(See Figures 22.15, 22.22, and 25.19d.)*
- Liver cells are specialized epithelial cells, termed **hepatocytes**. **Stellate macrophages** *(Kupffer cells)* are phagocytic cells that reside in the sinusoidal lining. The **liver lobule** is the basic functional unit of the liver. Each lobule is hexagonal in cross section and contains six **portal triads.** A portal triad consists of a branch of the hepatic portal vein, a branch of the hepatic artery proper, and a branch of the hepatic (bile) duct. **Bile canaliculi** carry bile to **bile ductules**, which carry bile to the **interlobular bile ducts**. The **left** and **right hepatic ducts**, which merge to form the **common hepatic duct**. *(See Figures 25.19–25.22.)*

The Gallbladder p. 678

- The **gallbladder** is a hollow muscular organ that stores and concentrates bile before excretion in the small intestine. Bile salts break apart large drops of lipids and make them accessible to digestive enzymes. Bile ejection occurs under stimulation of cholecystokinin (CCK).
- The gallbladder is divided into **fundus**, **body**, and **neck** regions. The **cystic duct** leads from the gallbladder to merge with the common hepatic duct to form the **bile duct**. *(See Figures 25.15, 25.19d, and 25.21.)*
- The wall of the gallbladder is composed of only the mucosa, lamina propria, muscular layer, and serosa. *(See Figure 25.21.)*

The Pancreas p. 679

- The **pancreas** is divided into **head**, **body**, and **tail** regions. The **pancreatic duct** penetrates the wall of the duodenum. Within each lobule, ducts branch repeatedly before ending in the **pancreatic acini** (blind pockets). An **accessory pancreatic duct** may be present. **Pancreatic juice** is discharged into the duodenum. *(See Figures 22.15, 25.16a, 25.21a, and 25.23.)*
- Pancreatic tissue consists of exocrine and endocrine portions. The bulk of the organ is exocrine in function, as the pancreatic acini secrete water, ions, and digestive enzymes into the small intestine. Pancreatic enzymes include **lipases**, **carbohydrases**, **nucleases**, and **proteolytic enzymes**. The major hormones produced by the endocrine portion are insulin and glucagon.

25.9 | Aging and the Digestive System p. 682

- Normal digestion and absorption occur in elderly individuals; however, changes in the digestive system reflect age-related changes in other body systems. These include a slowed rate of epithelial stem cell division, a decrease in smooth muscle tone, the appearance of cumulative damage, an increase in cancer rates, and numerous changes in other systems.

Chapter Review

For answers, see the blue Answers tab at the back of the book.

Level 1 Reviewing Facts and Terms

Match each numbered item with the most closely related lettered item.

1. segmentation .... [ ]
2. mesentery proper .... [ ]
3. cuspids .... [ ]
4. serosa .... [ ]
5. buccal fat pads .... [ ]
6. mastication .... [ ]
7. bicuspids .... [ ]
8. digestive tract .... [ ]
9. mesocolon .... [ ]
10. peristalsis .... [ ]

a. mesentery sheet suspending small intestine
b. propels materials through the digestive tract
c. churn and fragment digestive materials
d. canines
e. mechanical/chemical digestion of food
f. serous membrane covering muscularis externa
g. chewing
h. premolars
i. mesentery associated with the large intestine
j. form the cheeks

11. Label the digestive system structures in the following figure.

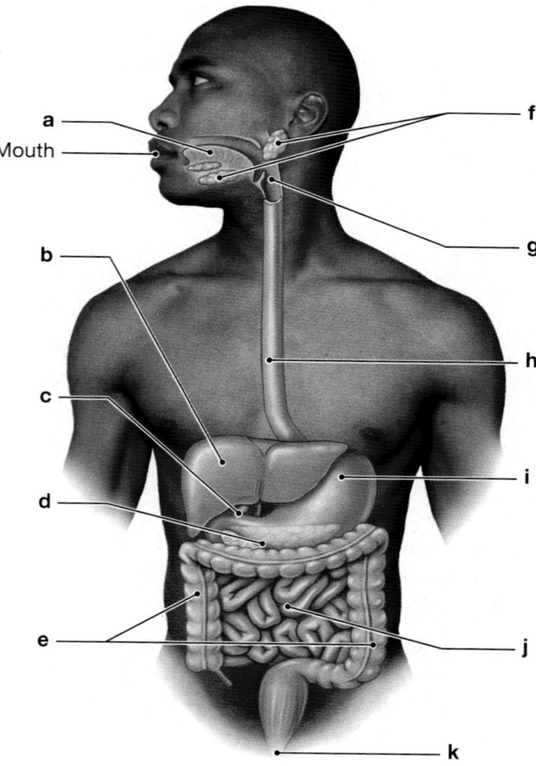

(a) ________ (b) ________
(c) ________ (d) ________
(e) ________ (f) ________
(g) ________ (h) ________
(i) ________ (j) ________
(k) ________

12. The actions involved in the mechanical processing of food include all but which of the following?
(a) absorption
(b) squashing foods with the tongue
(c) tearing or crushing foods with the teeth
(d) churning or swirling of the foods by the stomach

13. Digestion refers to
(a) the progressive dehydration of indigestible residue.
(b) the input of food into the digestive tract.
(c) the chemical breakdown of food.
(d) the absorption of nutrients into the gut.

14. Most of the digestive tract is lined by
(a) pseudostratified ciliated columnar epithelium.
(b) cuboidal epithelium.
(c) stratified squamous epithelium.
(d) simple columnar epithelium.

15. The _____ are double sheets of peritoneal membrane that hold some of the visceral organs in their proper position.
(a) serosa
(b) adventitia
(c) mesenteries
(d) fibrosa

16. The activities of the digestive system are regulated by
(a) hormones.
(b) parasympathetic neurons.
(c) sympathetic neurons.
(d) all of the above.

17. Label the four layers of the digestive tract in the following diagram.

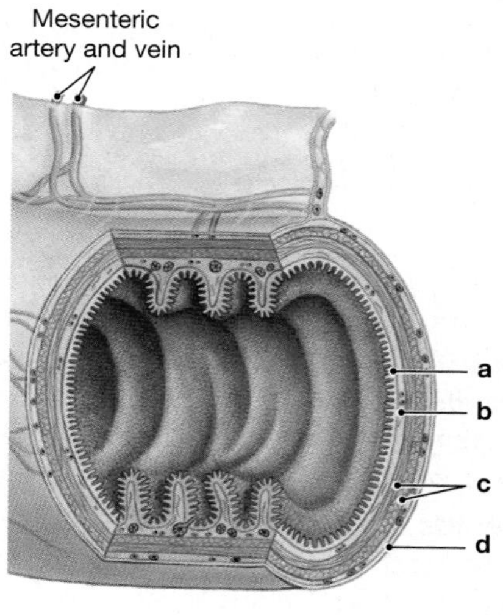

Mesenteric artery and vein

a
b
c
d

(a) ________
(b) ________
(c) ________
(d) ________

18. Sandwiched between the layer of circular and longitudinal muscle in the muscular layer is the
(a) mucosa.
(b) submucosa.
(c) muscularis mucosa.
(d) myenteric neural plexus.

19. The mesentery that suspends most of the small intestine is the
(a) mesentery proper.
(b) transverse mesentery.
(c) lesser omentum.
(d) greater omentum.

20. The passageway between the oral cavity and the pharynx is the
(a) uvula.
(b) fauces.
(c) palatoglossal arch.
(d) palatopharyngeal arch.

21. The ventral mesentery
(a) becomes the main attachment point for digestive organs in the peritoneal cavity in the adult.
(b) is highly glandular, but not vascular.
(c) contains and forms a pathway for the hepatic portal vein and its tributaries.
(d) none of the above

Level 2 Reviewing Concepts

1. Which of the following does not apply to the entire small intestine?
(a) It is the primary site of digestion and the absorption of approximately 80 percent of nutrients.
(b) It averages 6 meters in length.
(c) It is retroperitoneal.
(d) It has a series of transverse folds called circular folds.

2. In elderly individuals, the function of the digestive tract
(a) changes in ways that parallel age-related changes of most other systems of the body.
(b) is independent of the changes that occur in other systems.
(c) can be determined by a study of liver function.
(d) improves in efficiency, but not in the rate of digestion.

3. How would damage or removal of parts of the mesentery interfere with normal function of the small intestine?
(a) It would cause an increase in peristalsis.
(b) Hormone secretion would increase.
(c) The blood and nerve supply would suffer interference.
(d) The intestines would lose some of their motility.

4. What is the function of the lipase from the pancreas?
5. What is the function of the hepatopancreatic sphincter?
6. What does the gallbladder do with bile?
7. What is the function of stellate macrophages?
8. What is the last region of the colon before it reaches the rectum?
9. What is the function of the lacteals in the small intestine?
10. What triggers the release of gastrin?

Level 3 Critical Thinking

1. A murder suspect claims to have had dinner with the victim 4 hours before the latter was killed. The forensic scientist who performed the autopsy states that the suspect is lying, as it is clear that the victim had not eaten for more than 10 hours. How does the forensic specialist know this?
2. From the oral cavity to the anus, what six sphincters control movement of materials through the digestive tract? Over which do you have conscious, voluntary control, and why?
3. From the lumen outward, what six histological layers form the wall of the digestive tract?

CLINICAL CASE WRAP-UP

An Unusual Transplant

Normally, the lining of the colon is lubricated with mucus and populated with millions of beneficial bacteria. These "good" bacteria make up our normal colonic microbiome. They help us digest food and generate essential vitamins.

When antibiotics kill these normal intestinal bacteria, other strains of bacteria can replace them. Often, the replacing bacteria are harmful, producing toxins that irritate the intestinal lining and cause colitis. This is the case with *Clostridium difficile*, or "C. diff."

One treatment for stubborn *C. difficile* infections is a fecal microbiota transplant from a suitable donor—often a relative. Feces from a healthy donor contain populations of good bacteria that usually colonize the large intestine. With a minimal amount of processing, the donated feces are infused into the patient, often through a nasogastric (NG) tube inserted into the nose and passed through the stomach into the duodenum. These normal bacteria take up residence in the patient's large intestine, crowd out the harmful *C. difficile* bacteria, and resume their usual work of digestion and vitamin production.

Tayvian agreed to proceed with the fecal transplant. After the procedure, he recovered completely and left the hospital.

BRISTOL STOOL CHART		
Type 1	Separate hard lumps	Very constipated
Type 2	Lumpy and sausage like	Slightly constipated
Type 3	A sausage shape with cracks in the surface	Normal
Type 4	Like a smooth, soft sausage or snake	Normal
Type 5	Soft blobs with clear-cut edges	Lacking fibre
Type 6	Mushy consistency with ragged edges	Inflammation
Type 7	Liquid consistency with no solid pieces	Inflammation

1. If Tayvian's *Clostridium difficile* colitis was not treated with a fecal transplant, what kind of health problems might he continue to experience?
2. If you could look inside Tayvian's large intestine with a colonoscope, what would you see?

See the blue Answers tab at the back of the book.

Related Clinical Terms

Crohn disease: An incurable chronic inflammatory bowel disease that can affect any part of the digestive tract, from the mouth to the anus. The presence of strictures, fistulas, and fissures is common.

esophageal varices: Swollen and fragile esophageal veins that result from portal hypertension.

gastrectomy: The surgical removal of the stomach, generally to treat advanced stomach cancer.

irritable bowel syndrome (IBS): A common disorder affecting the large intestine, accompanied by cramping, abdominal pain, bloating, gas, diarrhea, and constipation.

pancreatic cancer: Malignancy of the pancreas that does not cause symptoms in its early stages, leading to late detection and a survival rate of only 4 percent.

periodontal disease: A loosening of the teeth within the alveolar sockets caused by erosion of the periodontal ligaments by acids produced through bacterial action.

polyps: Small growths with a stalk protruding from a mucous membrane that are usually benign.

pyloric stenosis: Uncommon condition in which the muscle of the lower end of the stomach enlarges and prevents food from entering the small intestine.

26

The Urinary System

Learning Outcomes

These Learning Outcomes correspond by number to this chapter's sections and indicate what you should be able to do after completing the chapter.

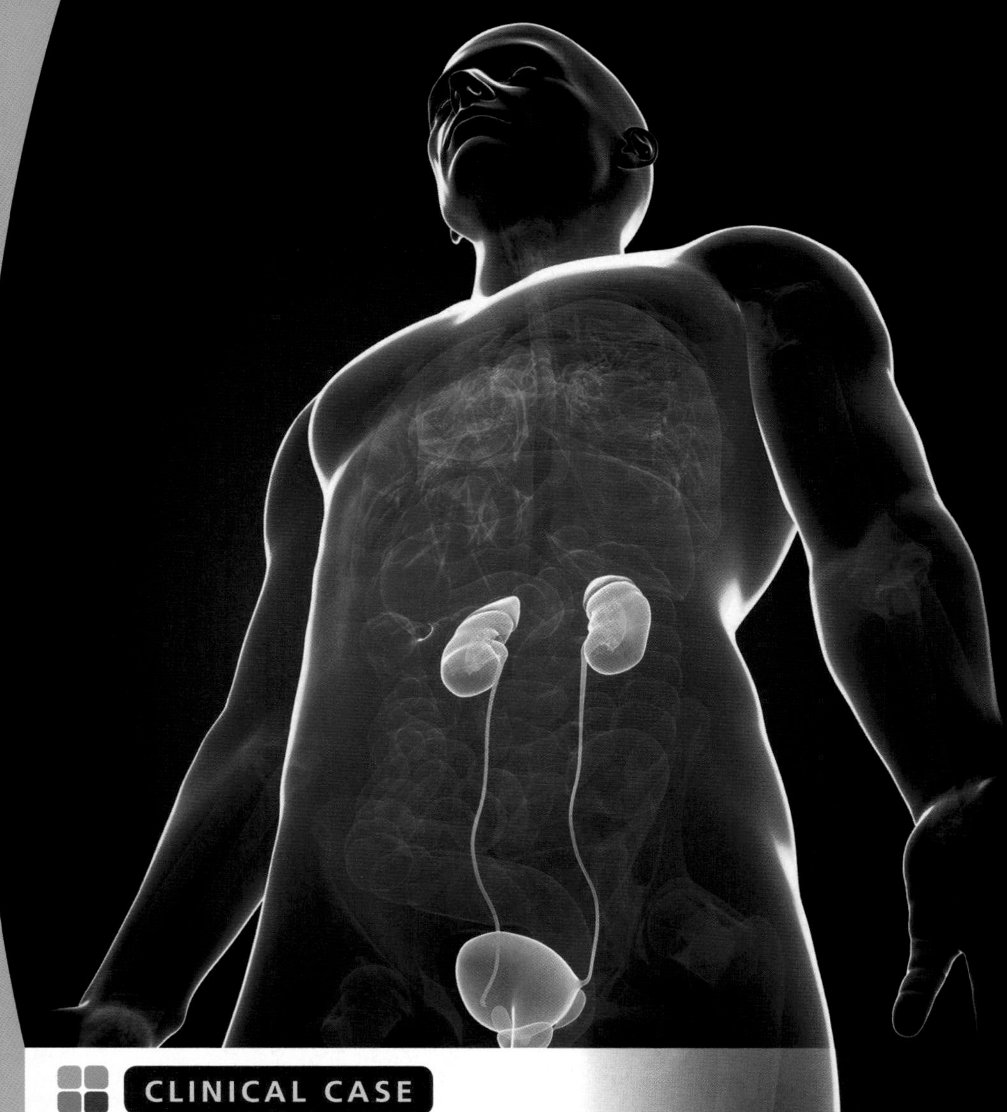

CLINICAL CASE

This Too Shall Pass

Jack, a senior at the University of Florida, is on the golf team. It has been a long, hot season. While taking a practice swing one day, he experiences pain in his left lumbar area. The pain is so sharp that he excuses himself and heads to the bathroom. Feeling an urge to urinate, Jack experiences another sharp pain that radiates to his hypogastric region. Jack notices his urine is pinkish. The next wave of pain brings him to his knees and makes him vomit.

Jack's pain is so severe he can't get up. He calls for help and is taken to the closest emergency room. He is doubled over with pain that radiates in waves from his left lumbar area just under his ribs to his groin. He urinates frequently, and now his urine looks bloody.

As Jack's doctor inserts an IV to infuse fluids, he explains that the only time he sees patients in such pain in this location is from "renal colic." "And if I am right, a CT scan will show what is going on."

What is happening to Jack? To find out, turn to the Clinical Case Wrap-Up on p. 706.

THE COORDINATED ACTIVITIES of the digestive, cardiovascular, respiratory, and urinary systems prevent "pollution" problems inside the body. The digestive tract absorbs nutrients from food, and the liver adjusts the nutrient concentration of the circulating blood. The cardiovascular system delivers these nutrients, along with oxygen from the respiratory system, to peripheral tissues. The cardiovascular system then carries the carbon dioxide and other wastes generated by active cells to sites of excretion. The carbon dioxide is eliminated at the lungs.

The **urinary system** removes most metabolic wastes generated by body cells. This chapter describes the organs of the urinary system and explains how the kidneys remove these wastes from the bloodstream to produce urine. The digestive and integumentary systems assist; however, their contributions are minor compared to that of the urinary system.

The urinary system has other essential functions, including the following:

- regulating blood plasma concentrations of sodium, potassium, chloride, calcium, and other ions by controlling the quantities excreted in the urine;
- regulating blood volume and blood pressure by (a) adjusting the volume of water lost in the urine, (b) releasing erythropoietin, and (c) releasing renin;
- working with the respiratory system to regulate blood pH;
- conserving valuable nutrients by preventing their excretion in the urine;
- eliminating metabolic wastes, especially urea, uric acid, toxic substances, and drugs;
- synthesizing calcitriol, a hormone derivative of vitamin D_3 that stimulates calcium ion absorption by the intestinal epithelium;
- helping the liver detoxify poisons; and
- during starvation, deaminating amino acids to be used by other tissues.

Urinary system activities are carefully regulated to keep the composition and concentration of solutes in the blood within acceptable limits. A disruption of any one of these functions may have immediate and potentially fatal consequences.

The urinary system includes the kidneys, ureters, urinary bladder, and urethra (**Figure 26.1**). The **kidneys** perform the excretory functions of this system. These organs produce **urine**, a fluid waste product containing water, ions, and small soluble compounds. Urine leaving the kidneys travels along the **urinary tract**. The urinary tract consists of the paired **ureters** (ū-RĒ-terz), the **urinary bladder**, and the **urethra** (ū-RĒ-thra). Urine is temporarily stored in the urinary bladder. When **urination**, or **micturition** (mik-tū-RISH-un), occurs, the muscular urinary bladder contracts to force urine through the urethra and out of the body.

26.1 | The Kidneys

▶ **KEY POINT** The retroperitoneal kidneys lie on the posterior abdominal wall. The adrenal glands lie on their superior surface, and three layers of connective tissue stabilize each kidney.

The kidneys lie lateral to the vertebral column between the last thoracic and third lumbar vertebrae on each side. Due to the presence of the liver on the right side of the abdominal cavity, the right kidney sits slightly lower than the left kidney (**Figures 26.1, 26.2**, and **26.3a**).

The anterior surface of the right kidney is covered by the liver, the right colic (hepatic) flexure, and the duodenum. The anterior surface of the left kidney is covered by the spleen, stomach, pancreas, jejunum, and left colic (splenic) flexure. The adrenal, or *suprarenal*, gland sits on the superior surface of each kidney. The kidneys, adrenal glands, and ureters, which lie against the muscles of the posterior body wall, are all retroperitoneal (**Figures 26.2b,c**, and **26.3**).

Figure 26.1 An Introduction to the Urinary System

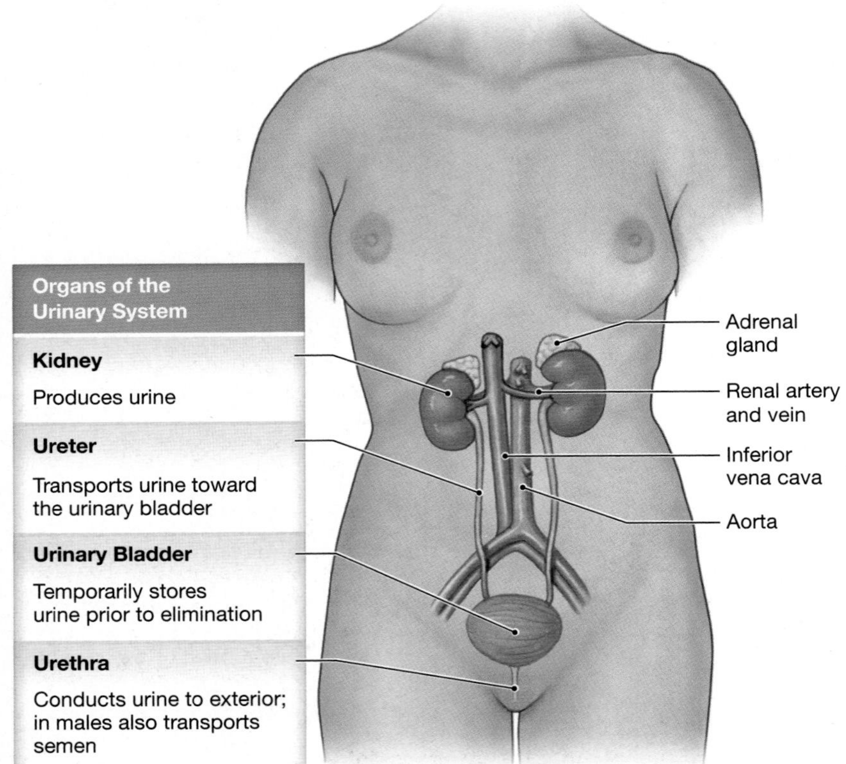

The parietal peritoneum contacts with adjacent visceral organs, and supporting connective tissues maintain the position of the kidneys. Three concentric layers of connective tissue (**Figure 26.2b,c**) protect and stabilize each kidney:

1. The **fibrous capsule** covers the outer surface of the entire organ. This layer of collagen fibers maintains the shape of the kidney and provides physical protection.
2. A layer of adipose tissue, the **perinephric fat** or **perirenal fat capsule** (*peri-*, around, + *renes*, kidneys) surrounds the fibrous capsule.
3. Collagen fibers extend outward from the inner fibrous capsule through the perinephric fat to a dense outer layer of connective tissue, the **renal fascia**. The renal fascia anchors the kidney to surrounding structures and attaches to the deep fascia surrounding the muscles of the posterior body wall. Another layer of adipose tissue, the pararenal (*para*, near) fat body, separates the posterior and lateral portions of the renal fascia from the body wall. Anteriorly, the renal fascia is attached to the peritoneum and to the anterior renal fascia of the opposite side.

This arrangement cushions the kidneys, preventing the day-to-day jolts of regular activity from disturbing normal kidney function. If the suspensory fibers stretch, or the amount of adipose tissue padding is reduced, the kidneys may become more vulnerable to traumatic injury.

Superficial Anatomy of the Kidney

▶ **KEY POINT** The hilum is the entry point for the renal artery and renal nerves and the exit point for the renal vein and ureter.

Each brownish-red kidney has the shape of a kidney bean. A typical adult kidney is approximately 10 cm (4 in.) in length, 5.5 cm (2.2 in.) in width, and 3 cm (1.2 in.) in thickness (**Figures 26.3** and **26.4**). A single kidney averages around 150 g (5.30 oz). Each kidney has an indentation on its medial surface called the **hilum**.

The fibrous capsule has inner and outer layers. The inner layer folds inward at the hilum and lines the **renal sinus**, an internal cavity within the kidney

Figure 26.2 The Urinary System

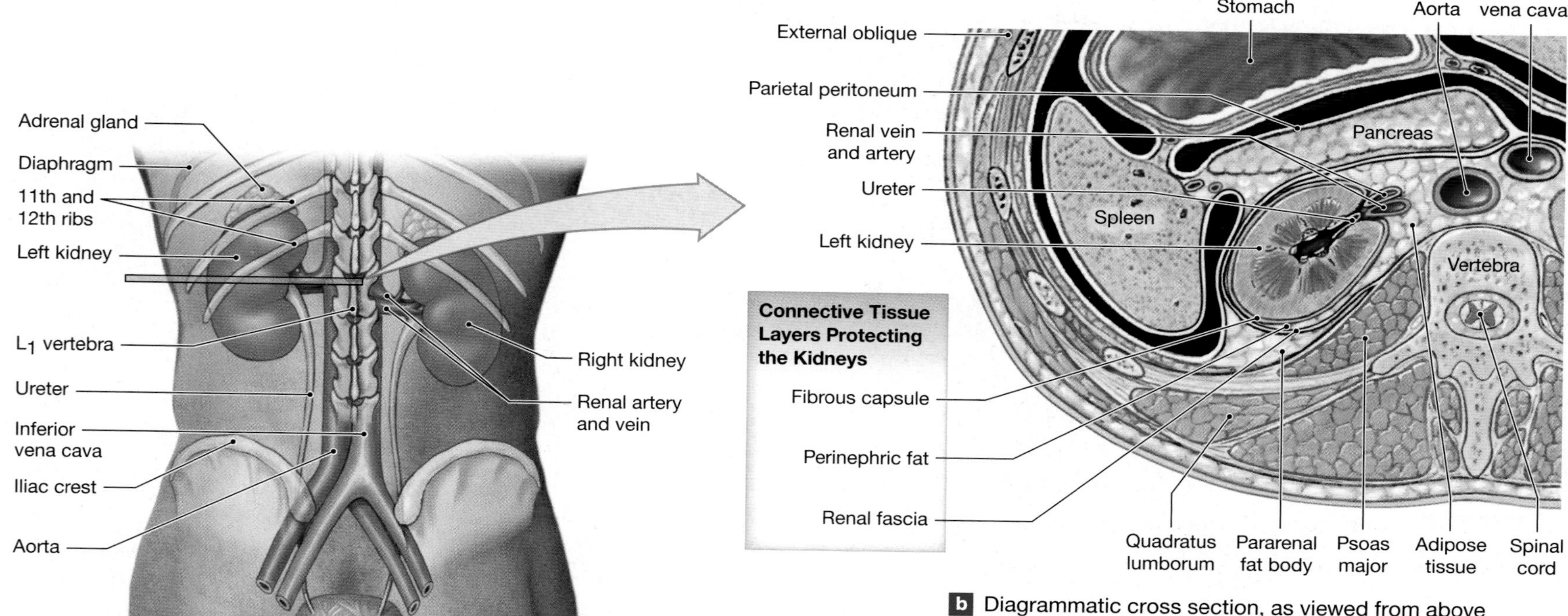

a A posterior view of the trunk showing the positions of the kidneys and other structures of the urinary system

b Diagrammatic cross section, as viewed from above

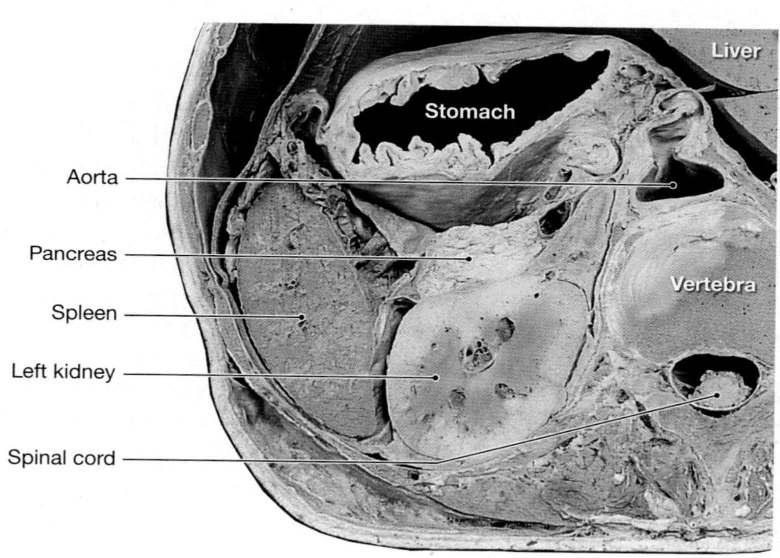

c Cross section, as viewed from above, at a level slightly superior to the plane of part (b)

(**Figure 26.4a**). Renal blood vessels, lymphatic vessels, nerves, and the ureter pass through the hilum and branch within the renal sinus. The thick, outer layer of the capsule extends across the hilum and stabilizes the position of these structures.

Sectional Anatomy of the Kidney

▶ **KEY POINT** The interior of each kidney contains a renal cortex, renal medulla, and renal sinus.

Deep to the renal capsule lies the **renal cortex** (**Figure 26.4a**). The renal cortex is granular and reddish-brown in color. The **renal medulla** is deep to the cortex. It consists of 6–18 distinct triangular structures, called **renal pyramids**. The base of each pyramid faces the cortex, and the tip, the **renal papilla**, projects into the renal sinus. Each pyramid has a series of fine grooves that come together at the papilla. Bands of cortical tissue, termed **renal columns**, extend into the medulla of the kidney and separate the adjacent renal pyramids. A **kidney lobe** contains a renal pyramid, the overlying area of renal cortex, and adjacent tissues of the renal columns.

Urine is produced in the kidney lobes. Ducts within each renal papilla empty urine into a cup-shaped drain, called a **minor calyx** (KĀ-liks). Four or five minor calyces (KĀL-i-sēz) merge to form a **major calyx**. The major calyces combine to form a large, funnel-shaped chamber, the **renal pelvis**. The renal pelvis is connected to the ureter, which drains the kidney (**Figure 26.4**).

Urine production begins in the **nephrons** (NEF-ronz), microscopic tubular structures. Two types of nephrons are found in the kidney: cortical nephrons (in the renal cortex) and juxtamedullary nephrons (closer to the renal medulla). Each kidney has roughly 1.25 million nephrons, with a combined length of about 145 km (90 miles).

The Blood Supply to the Kidneys

▶ **KEY POINT** The kidneys receive 20 to 25 percent of the total cardiac output. Afferent arterioles supply individual nephrons and efferent arterioles drain the nephrons.

Twenty to twenty-five percent of the total cardiac output, or about 1200 mL of blood, flows through the kidneys each minute. Each kidney receives blood from a **renal artery**. The renal artery branches from the lateral surface of the abdominal aorta near the superior mesenteric artery. It enters the renal sinus where it branches into the **segmental arteries** (**Figure 26.5**).

Figure 26.3 **The Urinary System in Gross Dissection**

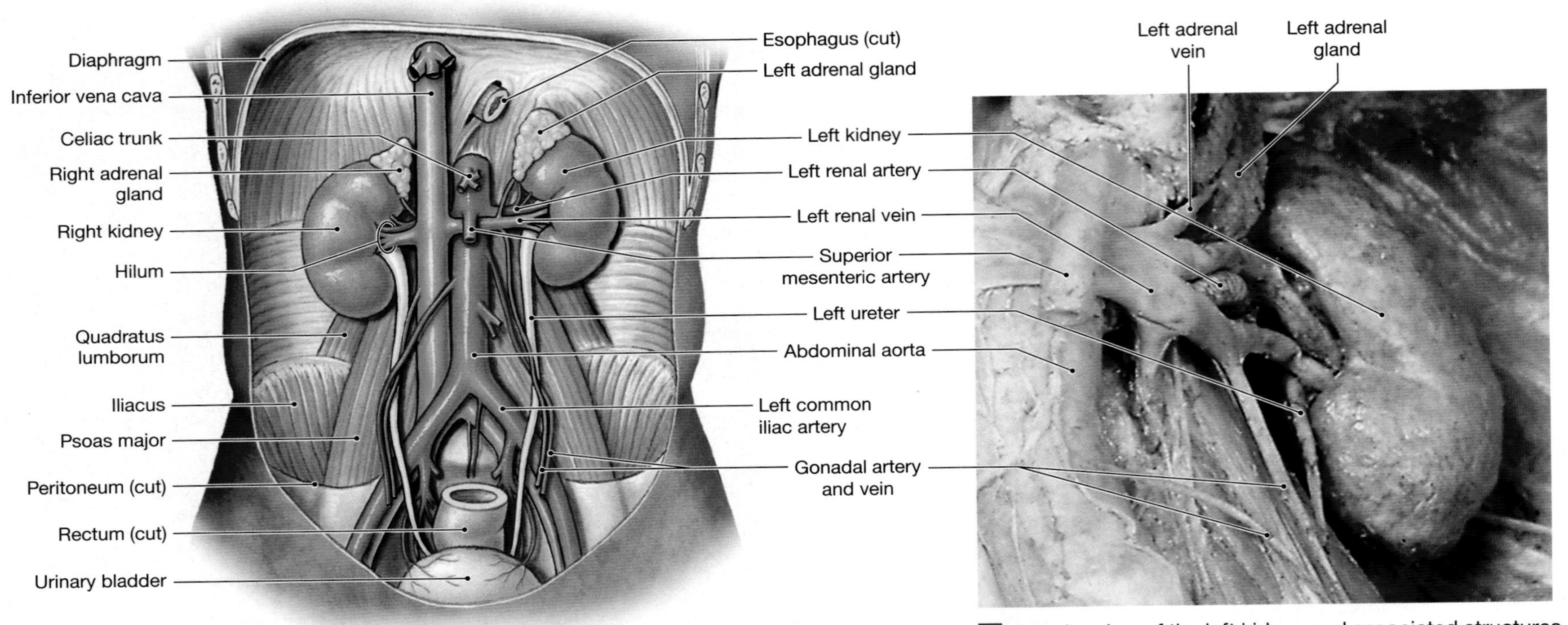

a Diagrammatic anterior view of the abdominopelvic cavity showing the kidneys, adrenal glands, ureters, urinary bladder, and blood supply to the kidneys

b Anterior view of the left kidney and associated structures

Segmental arteries divide into a series of **interlobar arteries** that radiate outward, extending through the renal columns between the renal pyramids into the cortex. The interlobar arteries supply blood to the **arcuate** (AR-kū-āt) **arteries,** which arch along the boundary between the renal cortex and the renal medulla. Each arcuate artery gives off a number of **cortical radiate arteries**, or *interlobular arteries.* These vessels supply portions of the adjacent kidney lobe. Numerous **afferent arterioles** branch from each cortical radiate artery to supply individual nephrons **(Figure 26.5b,c)**.

From the nephrons, blood enters a mirror image of the arterial distribution. A network of venules and small veins converges on the **cortical radiate** (*interlobular*) **veins**. The cortical radiate veins deliver blood to **arcuate veins** that empty into **interlobar veins**. The interlobar veins merge to form the **renal vein**; there are no segmental veins **(Figure 26.5)**. Many of these blood vessels are visible in corrosion casts of the kidneys and in renal angiograms **(Figures 26.6b** and **26.10b)**.

Innervation of the Kidneys

▶ **KEY POINT** Both branches of the ANS innervate the kidneys. Most of the fibers are sympathetic postganglionic fibers.

Urine production is regulated by altering nephron filtration rates. This is accomplished through a process termed autoregulation. Autoregulation (local blood flow regulation) involves reflexive changes in the diameters of the arterioles supplying the nephrons, thereby altering blood flow and filtration rates. **Renal nerves** innervate the kidneys and ureters.

Both branches of the autonomic nervous system innervate the kidneys. However, most of the nerve fibers are sympathetic postganglionic fibers from the celiac and inferior mesenteric ganglia. A renal nerve enters each kidney at the hilum. It then follows the branches of the renal artery to reach individual nephrons. The sympathetic innervation (1) adjusts rates of urine formation by changing blood flow at the nephron and (2) influences the composition of urine by stimulating the release of renin.

Histology of the Kidney

▶ **KEY POINT** The nephron is the structural and functional unit of the kidney. The tubular segments of the nephron are named according to their location (proximal or distal), thickness (thick or thin), and shape (convoluted or straight).

The nephron is the basic structural and functional unit of the kidney. We can view it only with a microscope. **Figure 26.7** shows a nephron that has been shortened and straightened out.

The Nephron and Collecting System

Each nephron consists of a renal corpuscle and a renal tubule. The **renal corpuscle** (KOR-pus-ul), a spherical structure consisting of the glomerular capsule, a cup-shaped chamber, and a capillary network called the **glomerulus** (glō-MER-ū-lus; plural, *glomeruli*). The renal tubule, a long tubular passageway, begins at the renal corpuscle. Each renal tubule empties into the collecting system, a series of tubes that carry tubular fluid away from the nephron.

Blood arrives at the glomerulus through the **afferent arteriole** and leaves by the **efferent arteriole (Figure 26.7)**. Filtration across the walls of the glomerulus produces a protein-free solution known as the **glomerular filtrate**, or simply *filtrate*. From the renal corpuscle, the filtrate enters a long tubular passageway that has three major subdivisions: (1) the **proximal convoluted tubule**, (2) the **nephron loop**, or *loop of Henle* (HEN-lē), and (3) the **distal convoluted tubule**.

Each nephron empties into the **collecting system**. A **connecting tubule** carries the filtrate from the distal convoluted tubule toward a nearby **collecting duct**. The collecting duct leaves the cortex and descends into the medulla. It carries filtrate toward a **papillary duct** that drains into the renal pelvis.

Nephrons from different locations within the kidney differ slightly in structure. Roughly 85 percent of all nephrons are **cortical nephrons**. These nephrons are located almost entirely within the superficial cortex of the kidney. In a cortical nephron, the nephron loop is relatively short, and the efferent arteriole delivers blood to a network of **peritubular capillaries**, which

Figure 26.4 Structure of the Kidney

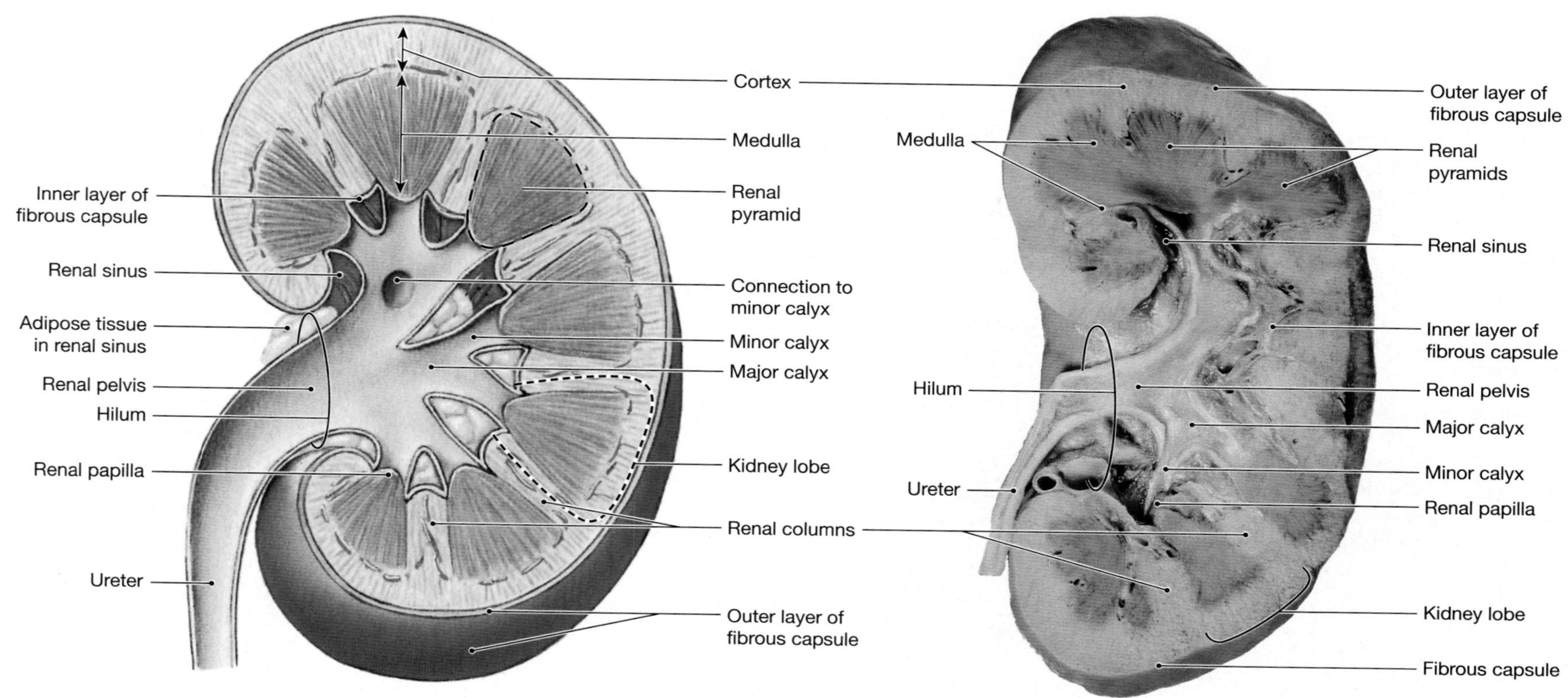

a Frontal section through the left kidney showing major structures. The outlines of a kidney lobe and a renal pyramid are indicated by dotted lines.

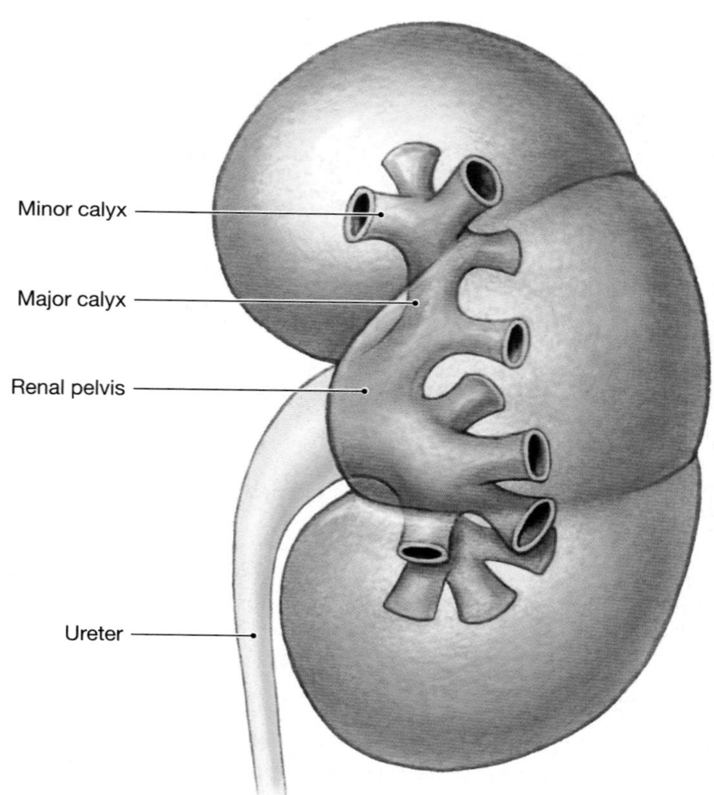

b Shadow drawing showing the arrangement of the calyces and renal pelvis within the kidney.

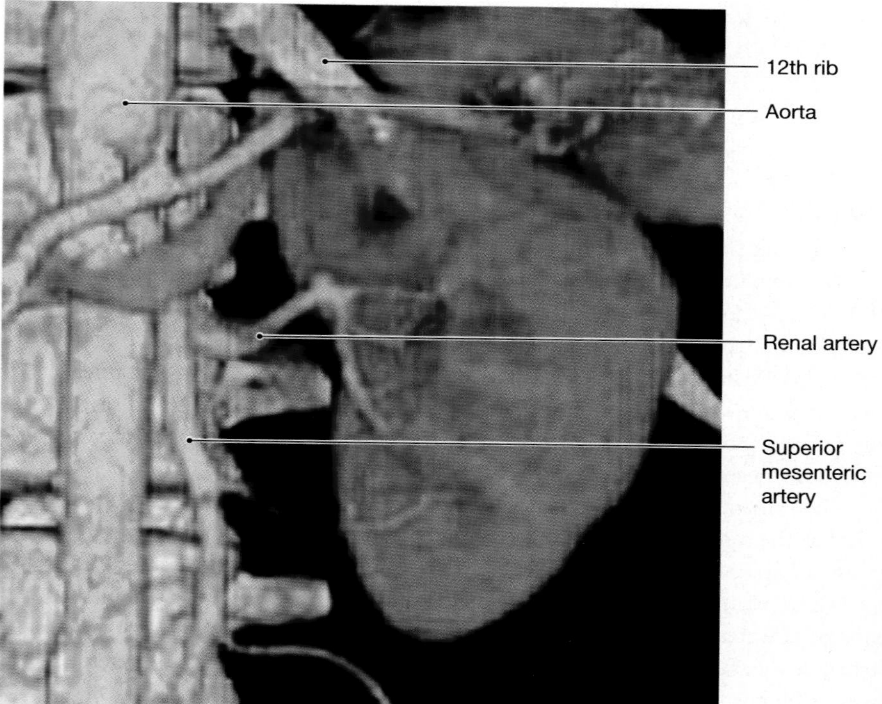

c Spiral scan of the left kidney.

26

Figure 26.5 Blood Supply to the Kidneys

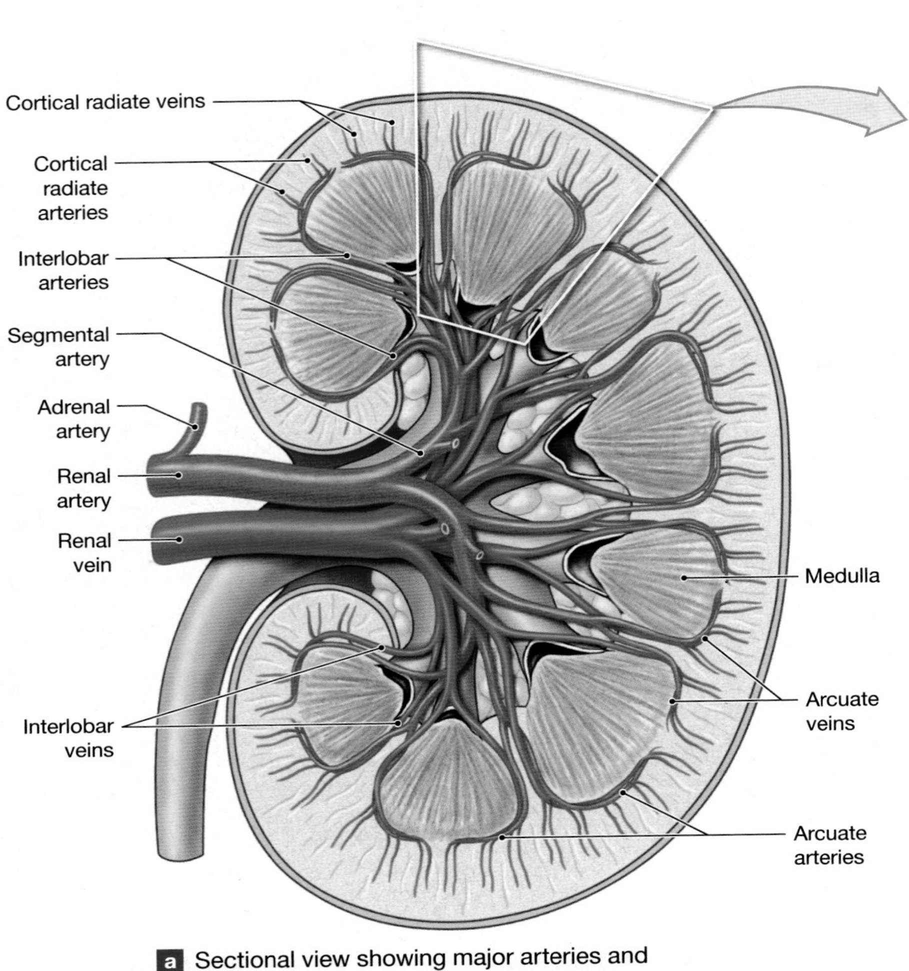

a Sectional view showing major arteries and veins. Compare with **Figures 26.3** and **26.8**.

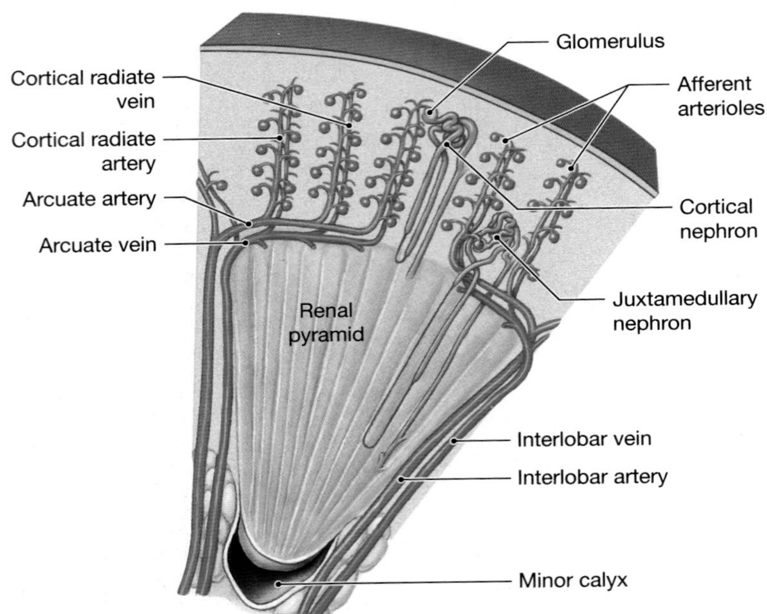

b Circulation in the renal cortex.

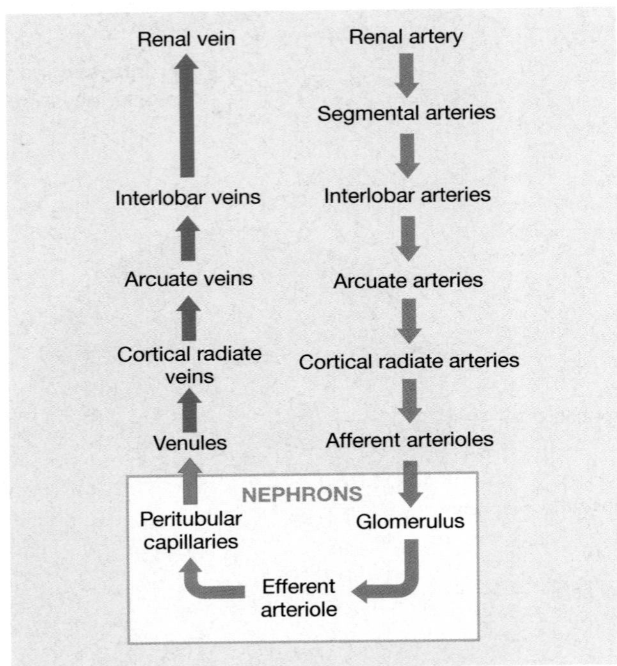

c The flowchart summarizes the pattern of renal circulation.

surround the entire renal tubule **(Figure 26.8a,e)**. These capillaries drain into small venules that carry blood to the interlobular veins **(Figure 26.5c)**.

The remaining 15 percent of nephrons are **juxtamedullary** (juks-ta-MED-ū-lar-ē; *juxta*, near) **nephrons**. These nephrons are located closer to the medulla, and they have longer nephron loops that extend deep into the renal pyramids **(Figure 26.8a,f)**. Because they are more numerous than juxtamedullary nephrons, cortical nephrons perform most of the reabsorptive and secretory functions of the kidneys. However, the juxtamedullary nephrons create the conditions necessary for producing concentrated urine.

The urine arriving at the renal pelvis is very different from the filtrate produced at the renal corpuscle. The passive process of filtration allows movement across a barrier based solely on the size of the solute. A filter with pores large enough to permit the passage of large metabolic wastes is unable to prevent the passage of water, ions, and other smaller metabolic molecules, such as glucose, fatty acids, or amino acids. The distal segments of the nephron are therefore responsible for:

- reabsorbing all the useful metabolic substrates from the filtrate,
- reabsorbing more than 80 percent of the water in the filtrate, and
- secreting into the filtrate waste products that were missed by the filtration process.

We will now examine each of the segments of a juxtamedullary nephron in greater detail.

The Renal Corpuscle

The renal corpuscle averages 150–250 μm in diameter. It includes the capillaries of the glomerulus and the **glomerular capsule** *(Bowman's capsule)*. The glomerulus projects into the glomerular capsule in the same way that the heart projects into the pericardial cavity. The outer wall of the capsule is made up of a simple squamous epithelium and is called the **capsular outer layer** (parietal layer). This layer ends at the **visceral layer**, which covers the glomerular capillaries. The visceral layer is composed of large cells with complex processes

Figure 26.6 **Renal Vessels and Blood Flow**

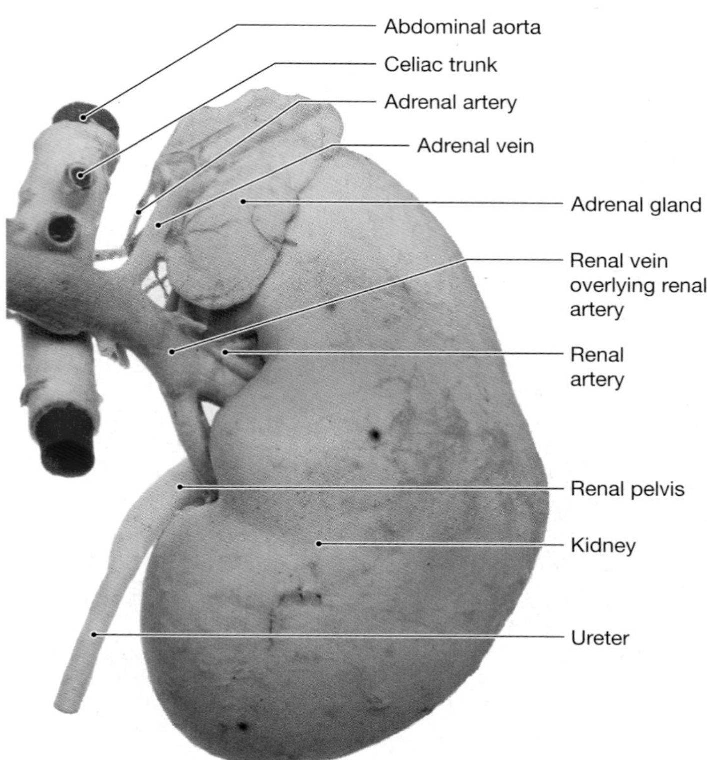

a The left kidney, ureter, and associated vessels. The vessels have been injected with latex to make them easier to see.

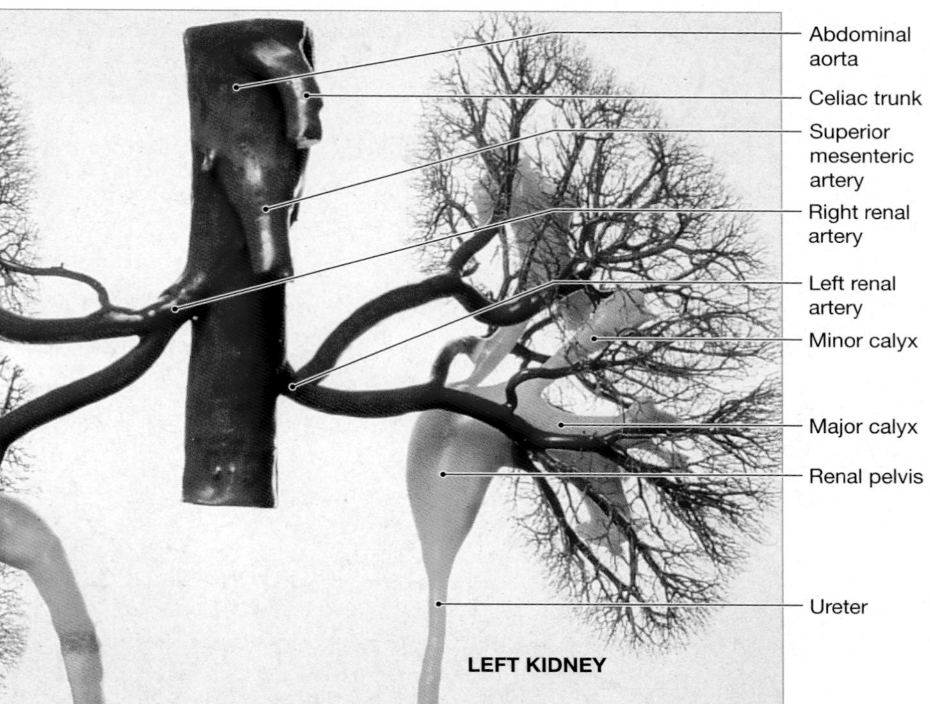

b Corrosion cast of the circulation and conducting passageways of the kidneys.

that wrap around the glomerular capillaries. These specialized cells are called **podocytes** (PŌD-ō-sīts; *podos*, foot, + *cyte*, cell). The **capsular space** separates the parietal and visceral epithelia **(Spotlight Figure 26.9)**.

The parietal and visceral epithelia are connected at the **vascular pole** of the renal corpuscle. At the vascular pole, the glomerular capillaries are connected to the afferent and efferent arterioles. Blood arrives at these capillaries through the afferent arteriole and exits in the smaller-diameter efferent arteriole **(Spotlight Figure 26.9)**.

Filtration occurs as blood pressure forces fluid and dissolved solutes out of the glomerular capillaries and into the capsular space. The resulting filtrate is very similar to plasma with the blood proteins removed. The filtration process at the renal corpuscle involves five filtration barriers **(Spotlight Figure 26.9)**:

1. The endothelial surface layer: The luminal surface (surface facing the lumen) of the endothelial cells of the glomerular capillaries has a thick, carbohydrate-rich meshwork. This glycocalyx limits filtration of large plasma proteins out of the glomerular capillaries.
2. The capillary endothelium: The glomerular capillaries are fenestrated capillaries with pores 60–100 nm (0.06–0.1 μm) in diameter. ⮌ pp. 571–572 These openings are small enough to prevent the passage of blood cells, but they are too large to prevent the diffusion of solutes, even those the size of smaller plasma proteins.
3. The basement membrane: The basement membrane surrounding the glomerular capillary endothelium is several times thicker than a typical capillary basement membrane. This prevents the passage of larger plasma proteins, but permits the movement of small plasma proteins, amino acids, glucose, and ions. Unlike basement membranes elsewhere in the cardiovascular system, the basement membrane of the glomerular capillaries encircles two or more capillaries. **Mesangial cells** are located between the endothelial cells of adjacent capillaries. Mesangial cells provide physical support for the capillaries and play a role in regulating glomerular blood flow and filtration. They do this by engulfing organic materials and regulating the diameters of the glomerular capillaries.
4. The glomerular epithelium: The podocytes have long cellular processes that wrap around the outer surfaces of the basement membrane of the glomerular capillaries. These delicate **secondary processes**, or "feet," are separated by narrow gaps called filtration slits. Because the filtration slits are very narrow, the filtrate entering the capsular space consists of water with dissolved ions, small molecules, and few if any plasma proteins.
5. The subpodocyte space: The subpodocyte space occupies approximately 60 percent of the filtration space of the glomerulus. This narrow space lies between the secondary processes of the podocytes and the soma of the podocytes. It assists the filtration slits of the glomerular epithelium in the filtration process.

In addition to metabolic wastes, the filtrate contains other organic compounds such as glucose, free fatty acids, amino acids, and vitamins. The proximal convoluted tubule reabsorbs these potentially useful materials.

The Proximal Convoluted Tubule

The **proximal convoluted tubule** (PCT) is the first part of the renal tubule. The PCT begins at the **tubular pole** of the renal corpuscle, opposite the vascular pole **(Spotlight Figure 26.9)**. The tubule is lined by a simple cuboidal epithelium with microvilli on the apical surface. These microvilli increase the surface area for reabsorption **(Figures 26.7** and **26.8a,b)**.

Figure 26.7 A Typical Nephron. A diagrammatic view showing the histological structure and the major functions of each segment of the nephron (violet) and collecting system (tan).

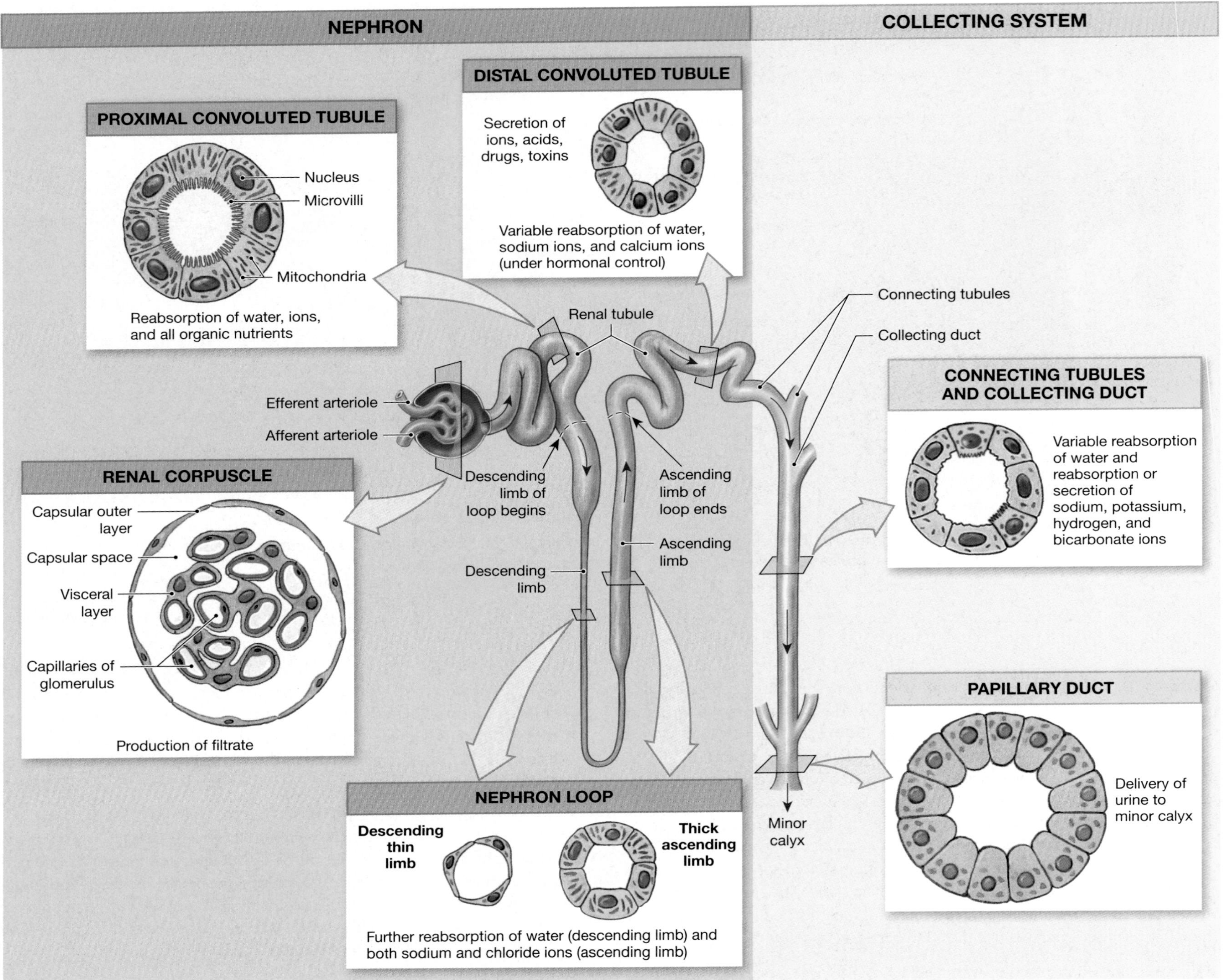

26

The primary function of the proximal convoluted tubule is reabsorption. Its cells actively reabsorb organic nutrients, ions, and plasma proteins (if any) from the filtrate. In addition, as the tubular fluid passes through the tubule, the epithelial cells reabsorb 60 percent of the sodium ions, chloride ions, and water. The tubule also actively reabsorbs potassium, calcium, magnesium, bicarbonate, phosphate, and sulfate ions. As these solutes are reabsorbed, osmotic forces pull water across the wall of the PCT and into the surrounding peritubular (interstitial) fluid.

The Nephron Loop

The proximal straight tubule ends as the renal tubule enters the medulla. This bend marks the start of the nephron loop **(Figure 26.8a,d)**. The **nephron loop** is divided into a **descending limb** and an **ascending limb**. Fluid in the ascending limb flows toward the renal cortex. The descending and ascending limbs of the nephron loop are lined with a simple squamous epithelium and are found within the deeper medulla. The ascending limb contains active transport mechanisms that pump sodium and chloride ions out of the tubular fluid. As a result, the medullary interstitial fluid contains an unusually high concentration of solutes. The limbs contain thin and thick segments. The terms *thick* and *thin* refer to the height of the epithelium, not the diameter of the lumen **(Figures 26.7 and 26.8a,d)**.

Solute concentration is usually expressed in terms of milliosmoles (mOsm). Near the base of the loop, in the deepest part of the medulla, the solute concentration of the interstitial fluid is four times that of plasma

Figure 26.8 Histology of the Nephron

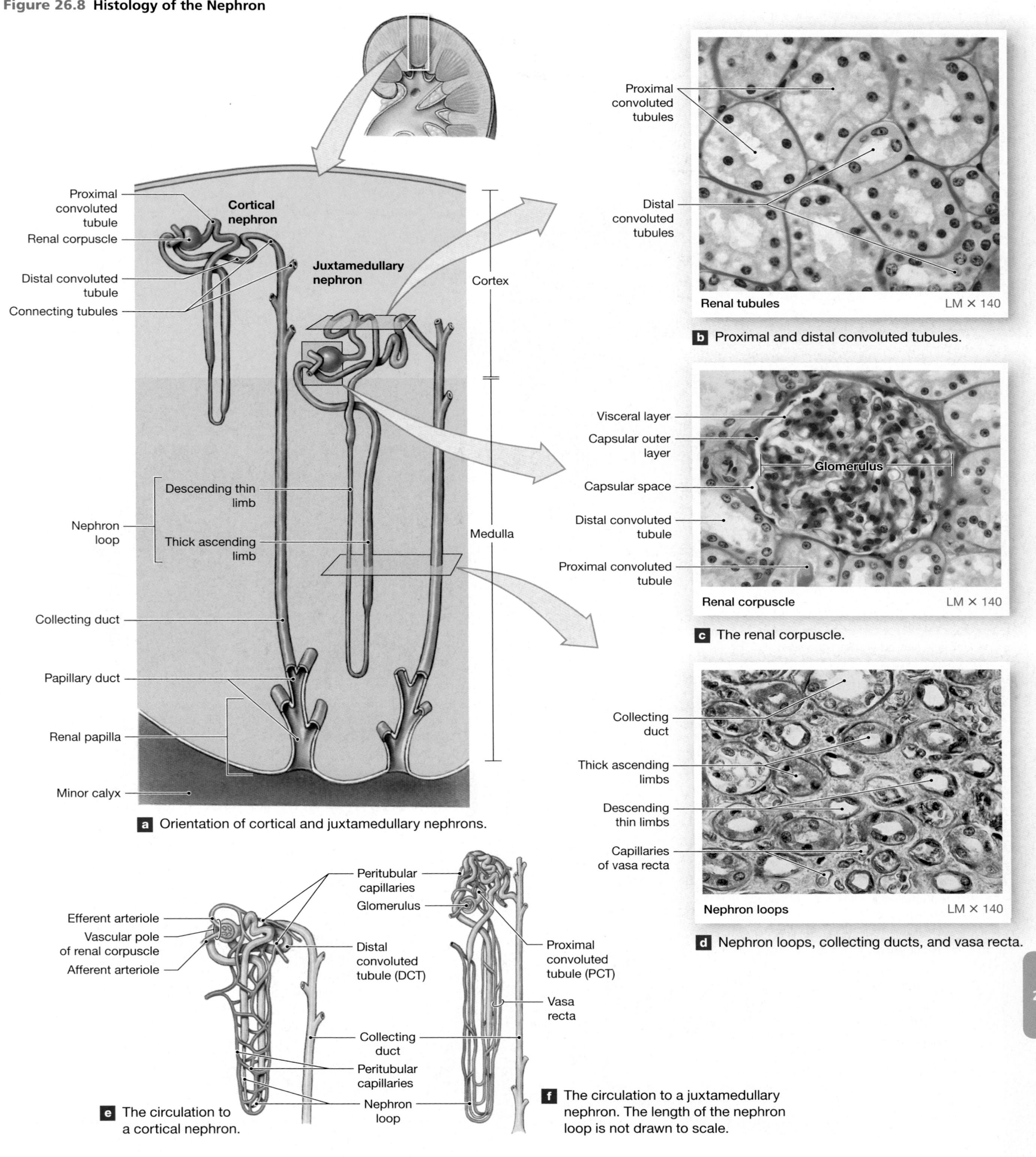

a Orientation of cortical and juxtamedullary nephrons.

b Proximal and distal convoluted tubules.

c The renal corpuscle.

d Nephron loops, collecting ducts, and vasa recta.

e The circulation to a cortical nephron.

f The circulation to a juxtamedullary nephron. The length of the nephron loop is not drawn to scale.

26

SPOTLIGHT

FIGURE 26.9
The Renal Corpuscle

The **renal corpuscle** consists of the glomerulus and the glomerular capsule (*Bowman's capsule*). Red arrows indicate the pathway of blood flow. Adjacent to the renal corpuscle is the juxtaglomerular complex, which is involved in the regulation of blood pressure and blood flow.

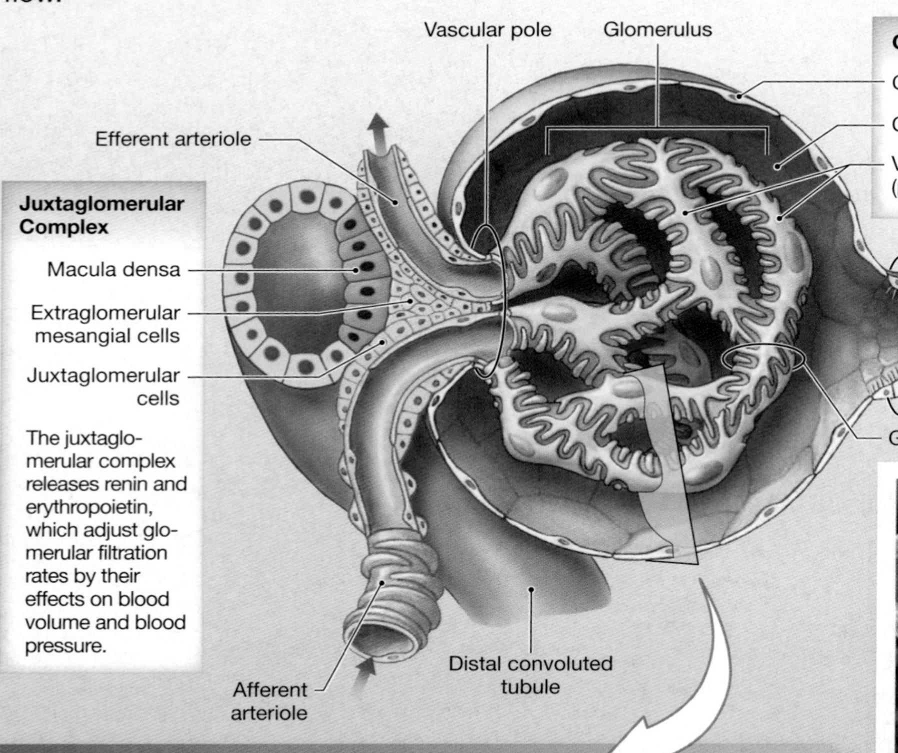

Filtration Apparatus

Plasma is filtered across the walls of the glomerulus and into the capsular space. The solution produced by this filtration process is called **glomerular filtrate**. This filtration process involves five barriers, which collectively form the filtration membrane.

Filtration Membrane

- Endothelial surface layer
- Capillary endothelium
- Basement membrane
- Glomerular epithelium
- Subpodocyte space

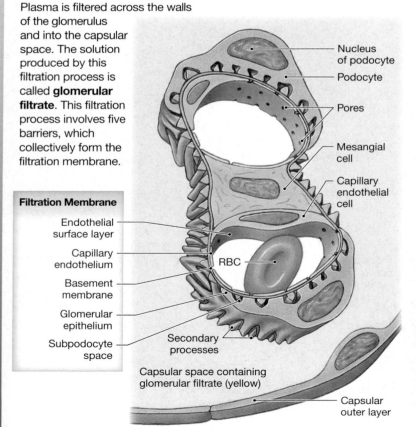

Nucleus of podocyte

Podocyte

Pores

Mesangial cell

Capillary endothelial cell

RBC

Secondary processes

Capsular space containing glomerular filtrate (yellow)

Capsular outer layer

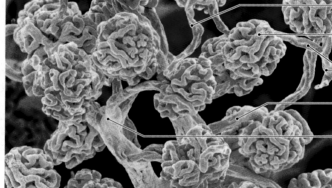

Efferent arteriole

Glomerulus

Peritubular capillaries

Afferent arteriole

Cortical radiate artery

Glomeruli and associated blood vessels SEM × 94

This colorized SEM shows the three-dimensional structure of several glomeruli and their associated blood vessels.

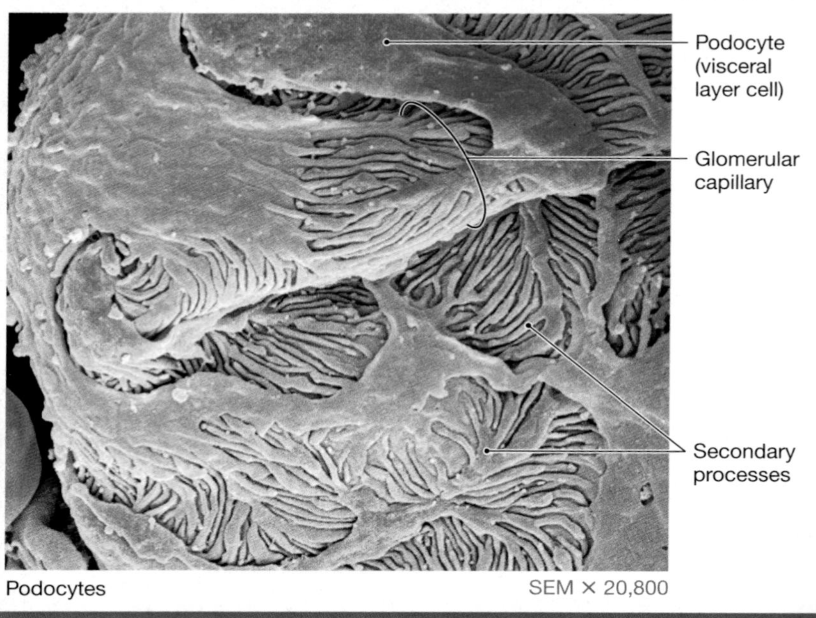

Podocytes SEM × 20,800

This colorized scanning electron micrograph of the glomerular surface shows individual podocytes and their secondary (foot) processes.

(1200 mOsm versus 300 mOsm). The descending thin and ascending thick limbs are freely permeable to water, but are impermeable to ions and other solutes. The high osmotic concentration surrounding the nephron loop results in an osmotic flow of water out of the nephron. The **vasa recta**, a group of slender capillaries, absorb the water and return it to the general circulation **(Figure 26.8f)**.

The net effect is that the nephron loop reabsorbs an additional 25 percent of the water from the tubular fluid and an even higher percentage of the sodium and chloride ions. Reabsorption in the PCT and nephron loop reclaims all of the organic nutrients, 85 percent of the water, and more than 90 percent of the sodium and chloride ions. The remaining water, ions, and all the organic wastes filtered at the glomerulus remain in the nephron loop and enter the distal convoluted tubule.

The Distal Convoluted Tubule

The **distal convoluted tubule (DCT)** ascends out of the medulla of the kidney and enters the cortex. The distal convoluted tubule passes between the afferent and efferent arterioles at the vascular pole **(Spotlight Figure 26.9)**.

In sectional view **(Figures 26.7** and **26.8b,c)**, the DCT differs from the PCT in three ways: (1) the DCT has a smaller diameter; (2) its epithelial cells lack microvilli; and (3) the boundaries between the epithelial cells in the DCT are easily seen. These anatomical characteristics reflect the major functional differences: The PCT is involved in reabsorption, while the DCT is involved in secretion.

The DCT is an important site for (1) active secretion of ions, acids, and other materials, (2) reabsorption of sodium and calcium ions from the tubular fluid, and (3) reabsorption of water, which helps concentrate the tubular fluid. The sodium transport activities of the distal tubule are controlled by circulating levels of aldosterone secreted by the adrenal cortex.

The Juxtaglomerular Complex The **juxtaglomerular complex** is a structure that helps regulate blood pressure and filtrate formation. Renin and erythropoietin elevate blood volume, hemoglobin levels, and blood pressure and restore normal rates of filtrate production. It is composed of three specialized cells **(Spotlight Figure 26.9)**:

1. Macula densa: The **macula densa** is a group of tall, closely packed epithelial cells in the distal convoluted tubule epithelium, immediately adjacent to the afferent arteriole, at the vascular pole of the glomerulus. They are thought to monitor the Na^+ concentration in the tubular fluid and regulate both the glomerular filtration rate and the release of renin from the juxtaglomerular cells.
2. Juxtaglomerular cells: The **juxtaglomerular cells** are modified smooth muscle cells in the wall of the afferent arteriole that secrete renin.
3. Extraglomerular mesangial cells: **Extraglomerular mesangial cells** are located in the triangular space between the afferent and efferent glomerular arterioles. These cells provide feedback control between the macula densa and the juxtaglomerular cells.

The Collecting System

The distal convoluted tubule is the last segment of the nephron. It opens into the collecting system, which consists of connecting tubules, collecting ducts, and papillary ducts **(Figure 26.8a,d)**. Every nephron is connected to a connecting tubule, and several connecting tubules are connected to a collecting duct **(Figure 26.8a,e,f)**. Several collecting ducts converge and empty into the larger papillary duct that empties into a minor calyx in the renal pelvis. The epithelium lining the collecting system begins as simple cuboidal cells in the connecting tubules and changes to a columnar epithelium in the collecting and papillary ducts **(Figure 26.7)**.

The collecting system transports tubular fluid from the nephron to the renal pelvis and makes final adjustments to its osmotic concentration and volume. Antidiuretic hormone (ADH) controls the permeability of the collecting system. This permeability change is significant because the collecting ducts pass through the medulla, where the nephron loop has established very high solute concentrations in the interstitial fluids. If collecting duct permeability is low, most of the tubular fluid reaching the collecting duct will flow into the renal pelvis and the urine will be dilute. However, if collecting duct permeability is high, this promotes the osmotic flow of water out of the duct into the medulla. This results in a small amount of highly concentrated urine. The higher the levels of circulating ADH, the greater the amount of water reabsorbed, and the more concentrated the urine.

26.1  **CONCEPT CHECK**

1 If your blood pressure is low, what changes will you see in the kidneys?

2 Trace the path of a drop of blood from the renal artery to a glomerulus and back to a renal vein.

3 Trace the path taken by filtrate in traveling from a glomerulus to a minor calyx.

4 Explain why filtration alone is not sufficient for urine production.

5 What is the function of the nephron loop?

See the blue Answers tab at the back of the book.

26.2 Structures for Urine Transport, Storage, and Elimination

▶ **KEY POINT** The ureters, urinary bladder, and urethra are responsible for the transport, storage, and elimination of urine.

Filtrate modification and urine production end when the fluid enters the minor calyx. The remaining parts of the urinary system (the ureters, urinary bladder, and urethra) transport, store, and eliminate the urine **(Figure 26.10c)**.

A transitional epithelium lines the minor and major calyces, renal pelvis, ureters, urinary bladder, and proximal portion of the urethra. This type of epithelium tolerates cycles of distension and relaxation without damage.

The Ureters

▶ **KEY POINT** The ureters—retroperitoneal muscular tubes connecting the kidneys to the urinary bladder—transport urine from the kidneys to the urinary bladder.

The ureters are a pair of retroperitoneal muscular tubes extending inferiorly from the kidneys to the urinary bladder **(Figures 26.1** and **26.2a**, pp. 688–689). Each ureter is approximately 30 cm (12 in.) long and begins as a funnel-shaped continuation of the renal pelvis **(Figures 26.3, 26.4**, pp. 690, 691, and **26.10c)**. The ureters pass inferiorly and medially over the psoas major. The anatomical locations of the ureters differ in men and women because of variations in the nature, size, and position of the reproductive organs **(Figure 26.11a,b)**.

Figure 26.10 Images of the Urinary System

Large intestine
Stomach
Liver
Ribs
Left kidney
Aorta
Right kidney

a A CT scan showing the position of the kidneys in a transverse section through the trunk as viewed from below. Such scans provide useful information concerning localized abnormalities or masses.

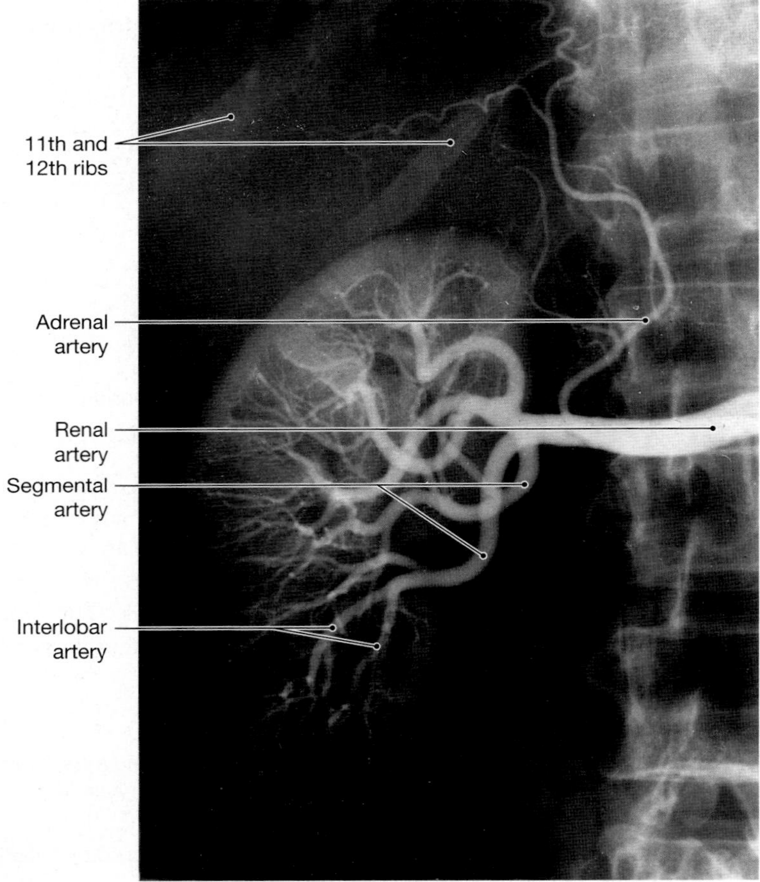

b An arteriogram of the right kidney. Arteriography enables visualization of arteries by x-ray imaging after injection of a radiopaque dye.

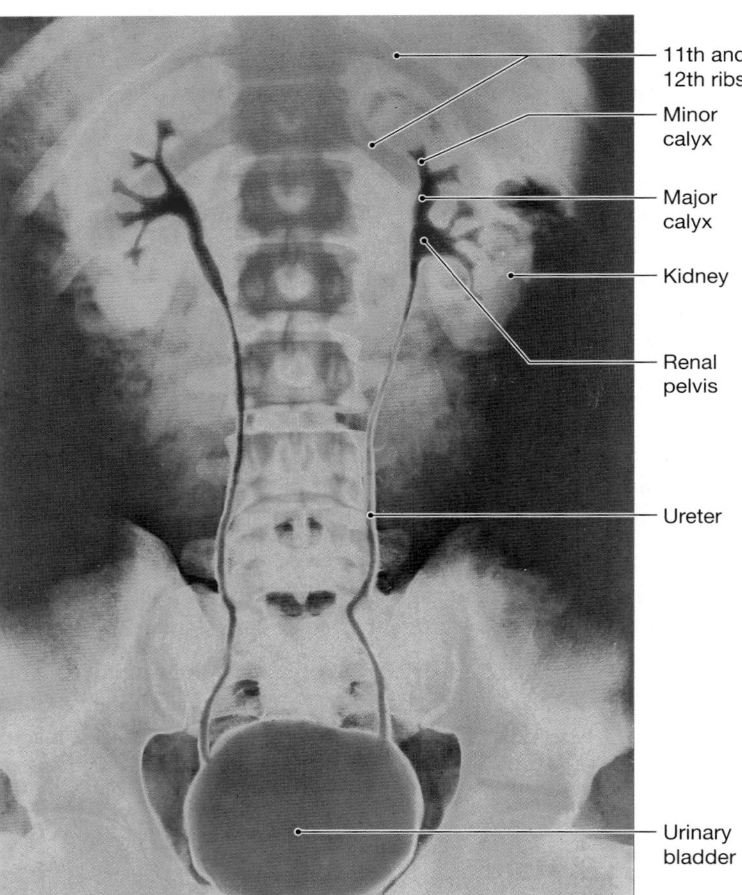

c This colorized x-ray was taken after intravenous injection of a radiopaque dye that was filtered into the urine. The image is known as a pyelogram, and the procedure is often called an IVP (intravenous pyelography).

26

CLINICAL NOTE

Treatment of Renal Failure

Every 30 minutes, your kidneys filter all the blood in your body, removing wastes and excess fluid and balancing electrolytes, minerals, and blood pH. The kidneys also produce renin to control blood pressure, erythropoietin to stimulate red blood cell production, and calcitriol to control blood calcium.

One healthy kidney is usually sufficient for living a healthy life. End-stage kidney disease occurs when 90 percent of kidney function has been lost. Nearly half a million Americans are on dialysis, the medical purification of blood needed when the kidneys are no longer functioning, and 200,000 have a functioning transplanted kidney. Over 100,000 are waiting for a transplant, but fewer than 20,000 transplants will be performed this year.

Acute renal failure is the sudden loss of kidney function, frequently due to multiple trauma, myocardial infarction, or other problems that reduce blood supply to the kidneys. Acute renal failure is often reversible, although temporary dialysis is required during recovery.

Chronic renal failure, the gradual loss of kidney function, may not become apparent until kidney function is significantly impaired. Diabetes and hypertension are the leading causes of chronic renal failure. Other causes include genetic diseases, infections, kidney stones, autoimmune diseases, and certain drugs, legal and illegal, including some antibiotics.

The first step in controlling kidney failure is a diet that strictly limits intake of fluids, protein, and minerals, including sodium (found not only in table salt, but also in processed meats and canned soups), potassium (found in many fruits, vegetables, meats, and milk), and phosphorus (found in dairy products, many vegetables, and carbonated beverages).

Dialysis is a treatment that performs some functions normally provided by a healthy kidney. It is required for survival once 90 percent of normal kidney function has been lost. Dialysis can remove wastes, salt, and extra water and can balance electrolytes, minerals, and blood pH. It cannot, however, produce renin, erythropoietin, or calcitriol.

Dialysis is performed in a dialysis center, usually three times per week, and takes a few hours. Venous access is required, usually through a surgically created shunt in the forearm. The shunt is connected to the dialysis machine, which cleanses the blood.

Peritoneal dialysis uses the peritoneal cavity as a dialysis membrane. A permanent intraperitoneal catheter is surgically placed to allow inflow and outflow of the dialysate (the fluid that will pick up extra fluid and wastes and balance electrolytes and minerals). Continuous ambulatory peritoneal dialysis is performed without machines by the patient, four or five times a day. A bag of dialysate, containing about 2 quarts, is infused into the peritoneal cavity through the catheter. Four or five hours later the fluid is drained back into a bag and discarded, and a new bag of fluid is infused. Automated peritoneal dialysis uses an exchange machine that cycles dialysate overnight during sleep.

The best way to treat renal failure is a kidney transplant, from either a living donor or a deceased donor within 48 hours after death. The recipient's nonfunctioning kidney(s) may be removed, especially if an infection is present. The transplanted kidney and ureter are usually placed extraperitoneally in the pelvic cavity (within the iliac fossa). The ureter is connected to the recipient's urinary bladder. The patient survival rate at 2 years after the transplant is greater than 90 percent, which is much better than the survival rate for those who are treated with dialysis.

The ureters penetrate the posterior wall of the urinary bladder. They pass through the bladder wall at an oblique angle, and the **ureteral orifices** are slit-like rather than rounded (**Figure 26.11c**). This shape prevents backflow of urine into the ureter and toward the kidneys when the urinary bladder contracts.

Histology of the Ureters

The wall of each ureter consists of three layers: (1) an inner mucosa lined by a transitional epithelium, (2) a middle muscular layer made up of inner longitudinal and outer circular layers of smooth muscle, and (3) an outer adventitia that is continuous with the fibrous capsule of the kidney and the parietal peritoneum of the abdominal wall (**Figure 26.12a**). Approximately every 30 seconds, peristaltic contractions of the muscular wall are triggered by stimulation of stretch receptors in the ureteral wall. These peristaltic contractions travel from the renal pelvis to the bladder and "milk" urine through the ureter.

The Urinary Bladder

▶ **KEY POINT** The urinary bladder stores urine temporarily. As the bladder fills, it moves superiorly into the peritoneal cavity and becomes intraperitoneal.

The urinary bladder is a hollow, muscular organ that serves as temporary storage for urine. In males, the base of the urinary bladder lies between the rectum and the pubic symphysis; in females, the base of the urinary bladder sits inferior to the uterus and anterior to the vagina. The dimensions of the urinary bladder vary, depending on the state of distension, but a full urinary bladder can contain about a liter of urine.

The peritoneum covers the superior surfaces of an empty urinary bladder. As the bladder fills, it displaces the parietal peritoneum from the anterior wall of the abdomen and becomes intraperitoneal. Several peritoneal folds stabilize the bladder's position. The **median umbilical ligament** extends from the anterior and superior border of the bladder toward the umbilicus. The **lateral umbilical ligaments** pass along the sides of the bladder and also reach the umbilicus. These fibrous cords are what is left of the two umbilical arteries that supplied blood to the placenta during embryonic and fetal development (**Figure 26.11b,c**). A series of tough connective tissue bands anchor the posterior, inferior, and anterior surfaces of the urinary bladder to the pelvic bones.

In sectional view, the mucosa lining the urinary bladder usually has folds, called rugae, that disappear as the bladder stretches and fills (**Figure 26.11c**). The triangular area extending from the openings of the ureters and the entrance to the urethra constitutes the **trigone** (TRĪ-gōn) of the urinary bladder.

Figure 26.11 Organs Responsible for the Conduction and Storage of Urine

Peritoneum
Left ureter
Rectum
Urinary bladder
Pubic symphysis
Prostate
External urethral sphincter
Spongy urethra
External urethral orifice
Urethra [see part c]

a Position of the ureter, urinary bladder, and urethra in the male

Rectum
Right ureter
Uterus
Peritoneum
Urinary bladder
Pubic symphysis
Internal urethral sphincter
Urethra
External urethral sphincter
Vagina
Vestibule

b Position of the ureter, urinary bladder, and urethra in the female

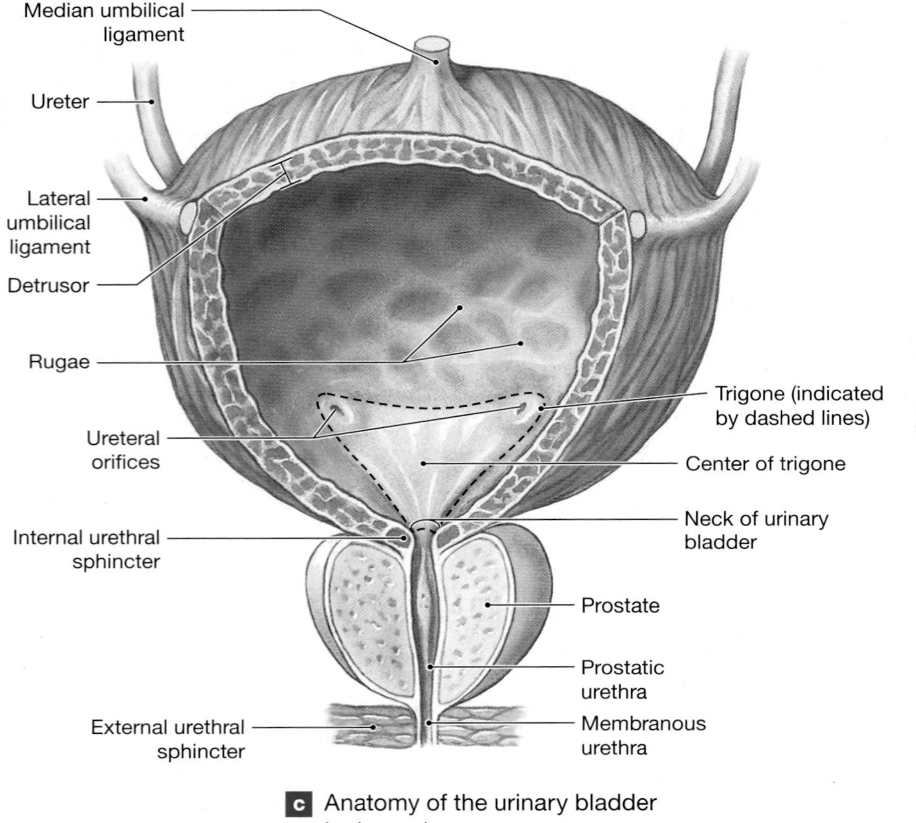

c Anatomy of the urinary bladder in the male

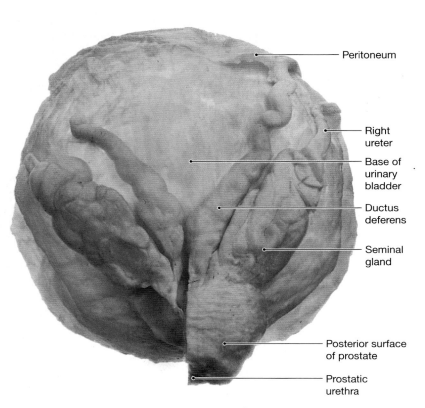

Peritoneum
Right ureter
Base of urinary bladder
Ductus deferens
Seminal gland
Posterior surface of prostate
Prostatic urethra

d Posterior view of the male urinary bladder and accessory reproductive structures

Figure 26.12 Histology of the Collecting and Transport Organs

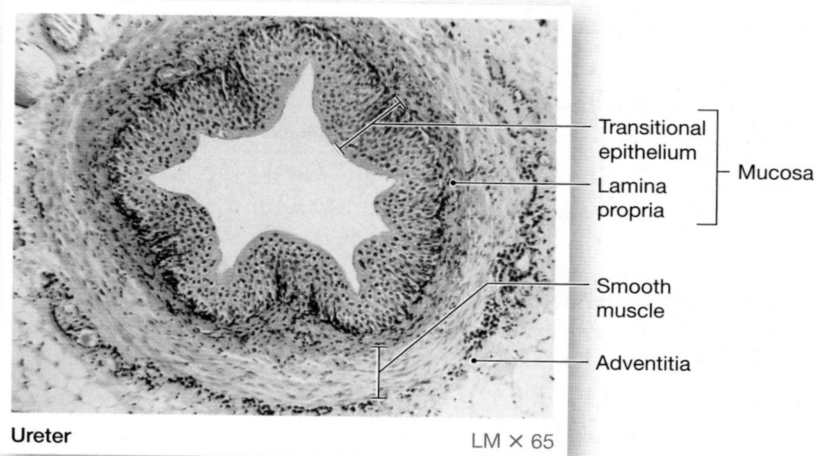

a Transverse section of the ureter. Note the thick layer of smooth muscle surrounding the lumen.

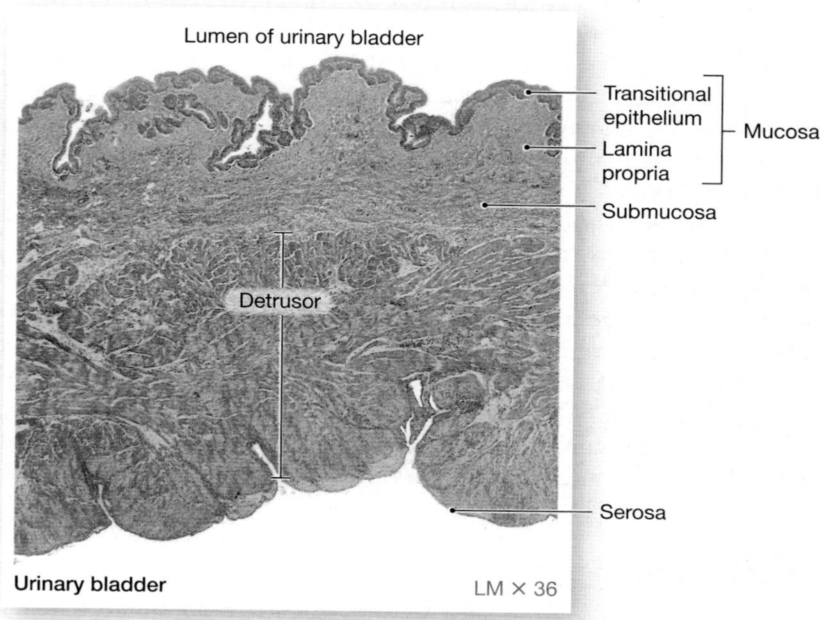

b The wall of the urinary bladder.

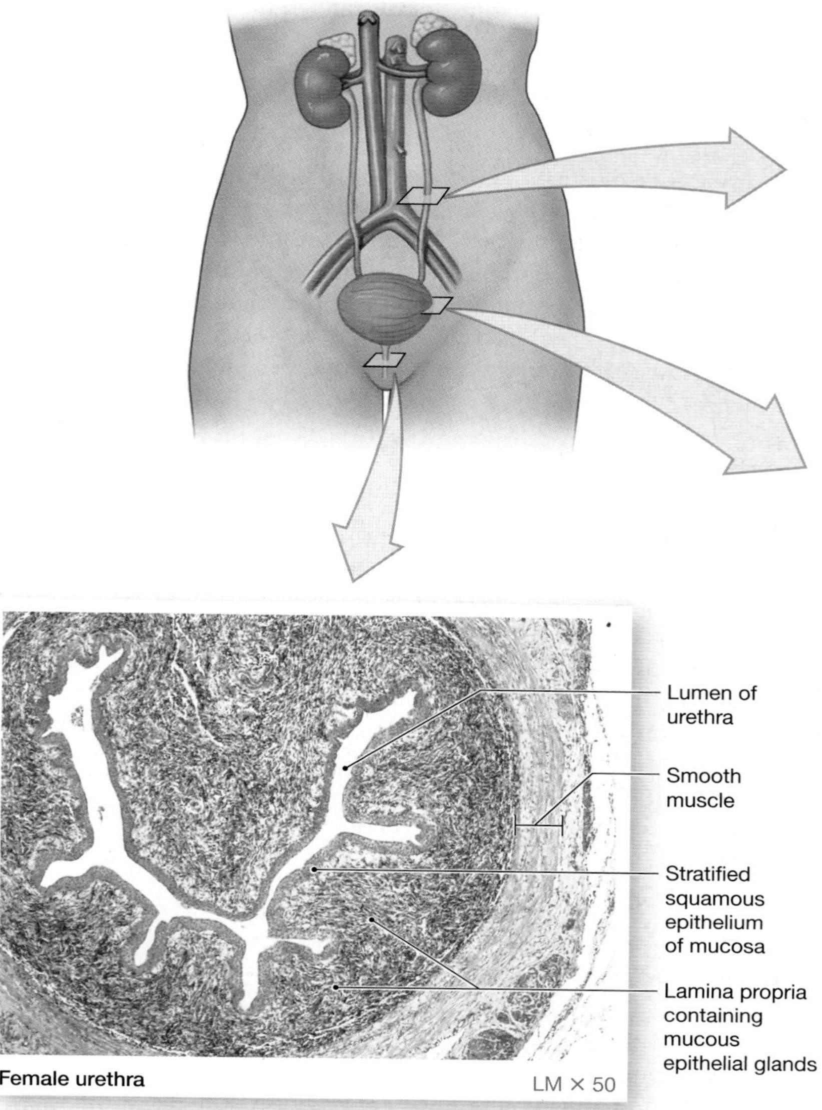

c A transverse section through the female urethra.

The mucosa here is smooth and very thick. The trigone acts as a funnel channeling urine into the urethra when the urinary bladder contracts.

The urethral entrance lies at the most inferior point in the bladder. The region surrounding the urethral opening, known as the **neck of the urinary bladder**, contains a muscular **internal urethral sphincter** (**Figure 26.11b,c**). The smooth muscle of the internal urethral sphincter provides involuntary control over the discharge of urine from the bladder. The urinary bladder is innervated by postganglionic fibers from ganglia in the hypogastric plexus and by parasympathetic fibers from intramural ganglia that are controlled by branches of the pelvic nerves.

Histology of the Urinary Bladder

The mucosa of the bladder, which contains prominent **rugae** (folds), is a transitional epithelium. Deep to the mucosa lies the connective tissue of the submucosa. The outer muscular layer consists of three layers: inner and outer longitudinal layers of smooth muscle, with a layer of circular smooth muscle sandwiched between (**Figure 26.12b**). Together, these layers form the powerful **detrusor** (dē-TRŪ-sor) of the urinary bladder. When the detrusor contracts, it compresses the urinary bladder and expels urine into the urethra. An adventitia forms the outer most layer of the urinary bladder.

The Urethra

▶ **KEY POINT** The urethra extends from the urinary bladder to the exterior. The male urethra is subdivided into three segments, while the female urethra is shorter and has only one segment.

The urethra extends from the neck of the urinary bladder to the exterior (**Figure 26.11c**). The female and male urethrae differ in length and in function. In the female, the urethra is very short, extending 3–5 cm (1–2 in.) from the bladder to the vestibule (**Figure 26.11b**). The external urethral opening, or **external urethral orifice**, is situated near the anterior wall of the vagina.

CLINICAL NOTE

Transitional Cell Carcinoma

Transitional cell carcinoma, the most common type of bladder cancer, begins in the innermost layer of transitional epithelium. Smoking is the most important risk factor for bladder cancer. Although toxins in cigarette smoke are filtered by the lungs and kidneys, the urine, which is stored in the bladder, may still contain some toxins. The hallmark of bladder cancer is hematuria, or bloody urine. Cystoscopy, inserting a thin, lighted endoscope made especially for bladders through the urethra into the bladder, is both a diagnostic and therapeutic tool.

In the male, the urethra extends from the neck of the urinary bladder to the tip of the penis, a distance that may be 18–20 cm (7–8 in.). The male urethra is subdivided into three portions: (1) the prostatic urethra, (2) the membranous urethra, and (3) the spongy urethra **(Figure 26.11a,c,d)**.

The **prostatic urethra** passes through the center of the prostate **(Figure 26.11c)**. The **membranous urethra** (also termed the *intermediate part of urethra*) is the short segment that penetrates the muscular floor of the pelvic cavity. The **spongy urethra**, or *penile* (PĒ-nīl) *urethra*, extends from the membranous urethra to the external urethral orifice at the tip of the penis **(Figure 26.11a)**.

Both sexes have a skeletal muscular band, called the **external urethral sphincter**, that acts as a valve **(Figure 26.11a–c)**. Branches of the hypogastric plexus control the contractions of both the external and internal urethral sphincters. Only the external urethral sphincter is under voluntary control, through the perineal branch of the pudendal nerve. The sphincter has a resting muscle tone and must be voluntarily relaxed to permit urination. The autonomic innervation of the external sphincter becomes important only if voluntary control is lacking, as in infants or in adults after spinal cord injuries. (See the section below on urinary reflexes.)

Histology of the Urethra

In females, the urethral lining is a transitional epithelium near the neck of the urinary bladder. A stratified squamous epithelium lines the rest of the urethra **(Figure 26.12c)**. The lamina propria contains an extensive network of veins, and concentric layers of smooth muscle surround the entire complex.

The histological organization of the male urethra varies along its length. As you proceed distally from the neck of the urinary bladder to the external urethral orifice, the epithelium changes. It starts as a transitional epithelium near the bladder, and then changes to pseudostratified columnar or stratified columnar, and then to stratified squamous near the external urethral orifice. The lamina propria is thick and elastic, and the mucosa is folded into longitudinal creases. Mucus-secreting cells are found in the epithelium, and in males the epithelial mucous glands form tubules extending into the lamina propria. The connective tissue of the lamina propria anchors the urethra to surrounding structures.

Urinary Reflexes: Urine Storage and Urine Voiding

▶ **KEY POINT** The urine voiding reflex coordinates the process of urination.

Urine reaches the urinary bladder by peristaltic contractions of the ureters. The urge to urinate generally appears when your urinary bladder contains about 200 mL of urine. Whether or not we urinate depends on an interplay between spinal reflexes and higher centers in the brain that provide conscious control over urination. Two spinal reflexes control urination (micturition): the urine storage reflex and the urine voiding reflex. These must be considered together, because urine storage and release involve simultaneous, coordinated activities.

Urine Storage Reflex

Urine storage occurs by spinal reflexes and the pontine storage center. When urine is being stored, low-frequency afferent impulses from stretch receptors in the urinary bladder act to (1) increase sympathetic activity (inhibit detrusor contraction and stimulate internal urethral sphincter contraction) and (2) stimulate contraction of the external urethral sphincter. In addition, the pontine storage center stimulates somatic motor neurons innervating the external urethral sphincter, causing contraction. These spinal reflexes and pontine responses promote urine storage (continence).

Urine Voiding Reflex

Urine voiding occurs by spinal reflexes and the pontine micturition center. The voiding reflex begins when high-frequency afferent impulses from stretch receptors in the urinary bladder stimulate interneurons that relay sensations to the pontine micturition center. This center initiates sacral spinal reflexes that (1) stimulate increased parasympathetic activity (detrusor contracts and internal urethral sphincter relaxes), (2) decrease sympathetic activity (internal urethral sphincter relaxes), and (3) decrease somatic motor nerve activity (external urethral sphincter relaxes). The pontine micturition center promotes urination through three spinal spinal reflexes to cause voiding (urination).

At the end of a typical urination, less than 10 mL of urine remains in the urinary bladder.

26.2 CONCEPT CHECK

6 An obstruction of a ureter by a kidney stone would interfere with the flow of urine between what two points?

7 Explain how the lining of the urinary bladder allows the bladder to become distended.

8 How is the urinary bladder held in place?

See the blue Answers tab at the back of the book.

26.3 Aging and the Urinary System

▶ **KEY POINT** Aging often results in one or more of four typical changes in the urinary system.

Aging is usually associated with an increased number of kidney problems. Age-related changes in the urinary system include the following:

1. A decline in the number of functional nephrons. Between ages 25 and 85, the total number of kidney nephrons drops by 30–40 percent.
2. A reduction in glomerular filtration. This reduction results from decreased numbers of glomeruli, cumulative damage to the filtration apparatus in the remaining glomeruli, and reductions in renal blood flow.
3. Reduced sensitivity to ADH. With age, the distal portions of the nephron and the entire collecting system become less responsive to ADH. With less

CLINICAL NOTE

Urinary Tract Infections

Urinary tract infections, or UTIs, result from bacteria, fungi, or viruses colonizing the urinary tract; the intestinal bacterium *Escherichia coli* is most often involved. Women are particularly susceptible to urinary tract infections because of the close proximity of the external urethral orifice to the anus. Sexual intercourse may also push bacteria into the urethra (since the female urethra is relatively short) toward the urinary bladder.

Symptomatic UTIs cause painful urination, known as dysuria. Other signs and symptoms include fever, foul- or strong-smelling urine, cloudy or bloody urine, frequent urination, or a persistent urge to urinate. In elderly people, the first symptom of a UTI is often confusion.

Urinary tract infections can be detected by the presence of bacteria and blood cells in the urine. If the urethral wall becomes inflamed, the condition is called **urethritis**; inflammation of the lining of the urinary bladder is **cystitis**. Many infections affect both regions to some degree. Urinary tract infections usually respond to antibiotic therapies. A mainstay of therapy is to keep the urine dilute (drink plenty of fluids) and the bladder empty (urinate often).

In untreated cases, bacteria may migrate along the ureters to the renal pelvis. The resulting inflammation of the renal pelvis produces **pyelitis** (pī-e-LĪ-tis). If the bacteria invade the renal cortex and medulla as well, **pyelonephritis** (pī-e-lō-ne-FRĪ-tis) results. Signs and symptoms of pyelonephritis include high fever, intense pain on the affected side, vomiting, diarrhea, and blood cells and pus in urine.

reabsorption of water and sodium ions, urination becomes more frequent, and daily fluid requirements increase.

4 Problems with urinary reflexes. Three factors are involved in age-related urination problems: (a) The external sphincter loses muscle tone and becomes less effective at voluntarily retaining urine. This leads to incontinence, often involving a slow leakage of urine. (b) The ability to control urination can be lost due to a stroke, Alzheimer's disease, or other CNS problems affecting the cerebral cortex or hypothalamus. (c) In males, urinary retention may develop if the prostate enlarges and compresses the urethra, restricting the flow of urine.

26.3 CONCEPT CHECK

9 Your 92-year-old grandfather is frustrated because he has to go to the bathroom more frequently now than he did when he was younger. How would you explain this to him?

See the blue Answers tab at the back of the book.

EMBRYOLOGY SUMMARY

For an introduction to the development of the urinary system, see Chapter 28 (Embryology and Human Development).

Study Outline

Introduction p. 688

- The functions of the urinary system include (1) regulating plasma concentrations of ions, (2) regulating blood volume and pressure by adjusting the volume of water lost and releasing erythropoietin and renin, (3) helping stabilize blood pH, (4) conserving nutrients, (5) eliminating organic wastes, and (6) synthesizing calcitriol.
- The **urinary system** includes the **kidneys**, **ureters**, **urinary bladder**, and **urethra**. The kidneys produce **urine** (a fluid containing water, ions, and soluble compounds); during **urination** (**micturition**) urine is forced out of the body. *(See Figure 26.1.)*

26.1 The Kidneys p. 688

- The kidneys are located on either side of the vertebral column between the last thoracic and third lumbar vertebrae. *(See Figures 26.1–26.3.)*
- The position of the kidneys in the abdominal cavity is maintained by (1) the overlying peritoneum, (2) contact with adjacent visceral organs, and (3) supporting connective tissues. The three concentric layers of connective tissue are the **fibrous capsule**, which covers the outer surface of the organ; the **perinephric fat**, which surrounds the fibrous capsule; and the **renal fascia**, which anchors the kidney to surrounding structures. *(See Figures 12.13 and 26.4.)*

Superficial Anatomy of the Kidney p. 688

- The ureter and renal blood vessels are attached to the **hilum** of the kidney. The inner layer of the fibrous capsule lines the **renal sinus**. *(See Figures 26.3 and 26.4.)*

Sectional Anatomy of the Kidney p. 689

- The kidney is divided into an outer **renal cortex**, a central **renal medulla**, and an inner **renal sinus**. The medulla contains 6–18 **renal pyramids**, whose tips, or **renal papillae**, project into the renal sinus. **Renal columns** composed of cortex separate adjacent pyramids. A **kidney lobe** contains a renal pyramid, the overlying area of renal cortex, and adjacent tissues of the renal columns. *(See Figure 26.4a.)*
- The **minor calyces** are continuous with the **major calyces**. These spaces lead into the **renal pelvis**, which is continuous with the **ureter**. *(See Figure 26.4a–c.)*

The Blood Supply to the Kidneys p. 689

- The vasculature of the kidneys includes the **renal**, **segmental**, **interlobar**, **arcuate**, and **cortical radiate arteries** to the **afferent arterioles** that supply the nephrons. From the nephrons, blood flows into the **cortical radiate**, **arcuate**, **interlobar**, and **renal veins**. *(See Figures 26.5, 26.6b, and 26.10b.)*

Innervation of the Kidneys p. 690

- The kidneys and ureters are innervated by **renal nerves**. Sympathetic activation regulates glomerular blood flow and pressure, stimulates renin release, and accelerates sodium ion and water reabsorption.

Histology of the Kidney p. 690

- The **nephron** (the basic functional unit in the kidney) consists of a renal corpuscle and a **renal tubule** that empties into the **collecting system**. From the **renal corpuscle**, the **tubular fluid** travels through the **proximal convoluted tubule**, the **nephron loop** (**loop of Henle**), and the **distal convoluted tubule**. It then flows through the **connecting tubule**, **collecting duct**, and **papillary duct** to reach the renal minor calyx. *(See Figure 26.7.)*

- Roughly 85 percent of the nephrons are **cortical nephrons** found within the cortex. The nephron loops are short, and the **efferent arteriole** provides blood to the **peritubular capillaries** that surround the renal tubules. The **juxtamedullary nephrons** are closer to the medulla, with their nephron loops extending deep into the renal pyramids. *(See Figures 26.5b,c and 26.8a,e,f.)*

- Nephrons are responsible for (1) producing filtrate, (2) reabsorbing organic nutrients, and (3) reabsorbing water and ions. The **capsular outer layer** lines the outer wall of the renal corpuscle. Blood arrives via the relatively large **afferent arteriole** and departs in the relatively small **efferent arteriole**. *(See Spotlight Figure 26.9.)*

- The renal corpuscle contains the capillary knot of the **glomerulus** and **glomerular capsule**. At the glomerulus, **podocytes** of the visceral epithelium wrap their "feet" around the capillaries. The **secondary processes** of the podocytes are separated by narrow **filtration slits**. The **capsular space** separates the **capsular outer layer** and **visceral layer**. The glomerular capillaries are *fenestrated capillaries*. The dense layer of the basement membrane is unusually thick. Blood arrives at the **vascular pole** of the renal corpuscle via the afferent arteriole and departs in the **efferent arteriole**. From the efferent arteriole, blood enters the peritubular capillaries and the **vasa recta** that follow the nephron loops in the medulla. *(See Spotlight Figure 26.9.)*

- The **proximal convoluted tubule (PCT)** actively reabsorbs nutrients, ions, plasma proteins, and electrolytes from the tubular fluid. The **nephron loop** includes a **descending limb** and an **ascending limb.** The ascending limb delivers fluid to the **distal convoluted tubule (DCT)**, which actively secretes ions, toxins, and drugs and reabsorbs sodium ions from the tubular fluid. *(See Figures 26.7 and 26.8 and Spotlight Figure 26.9.)*

- The **juxtaglomerular complex** is composed of the **macula densa**, **juxtaglomerular cells**, and the **extraglomerular mesangial cells**. The juxtaglomerular complex secretes the hormones renin and erythropoietin. *(See Spotlight Figure 26.9.)*

- The distal convoluted tubule opens into the **collecting system**. The collecting system consists of connecting tubules, collecting ducts, and papillary ducts. In addition to transporting fluid from the nephron to the renal pelvis, the collecting system adjusts the osmotic concentrations and volume. *(See Figures 26.7 and 26.8a,d–f.)*

26.2 Structures for Urine Transport, Storage, and Elimination p. 697

- Tubular fluid modification and urine production end when the fluid enters the minor calyx in the renal sinus. The rest of the urinary system (the ureters, urinary bladder, and urethra) is responsible for transporting, storing, and eliminating the urine. *(See Figure 26.10c.)*

The Ureters p. 697

- The ureters extend from the renal pelvis to the urinary bladder and are responsible for transporting urine to the bladder. The wall of each ureter consists of an inner mucosal layer, a middle muscular layer, and an outer connective tissue layer. *(See Figures 26.1–26.4, 26.10c, and 26.11.)*

The Urinary Bladder p. 699

- The **urinary bladder** is a hollow, muscular organ that stores urine. The bladder is stabilized by the **median umbilical ligament** and the **lateral umbilical ligaments**. Internal features include the **trigone**, the **neck**, and the **internal urethral sphincter**. The mucosal lining contains prominent **rugae** (folds). Contraction of the **detrusor** compresses the urinary bladder and expels the urine into the urethra. *(See Figures 26.11 and 26.12.)*

The Urethra p. 701

- The urethra extends from the neck of the urinary bladder to the exterior. In the female, the urethra is short and ends in the **external urethral orifice** (external urethral opening), and in the male, the urethra has **prostatic**, **membranous**, and **spongy** *(penile)* sections; the spongy urethra ends at the external urethral orifice. In both sexes, a circular band of skeletal muscles forms the **external urethral sphincter**, which is under voluntary control.

- The female urethral lining is usually a transitional epithelium near the urinary bladder; the rest is usually a stratified squamous epithelium. The urethral lining of males varies from a transitional epithelium at the urinary bladder, to a stratified columnar or a pseudostratified epithelium, and then to stratified squamous epithelium near the external urethral orifice. *(See Figures 26.11 and 26.12.)*

Urinary Reflexes: Urine Storage and Urine Voiding p. 702

- **Urination** is coordinated by the **urine storage reflex** and the **urine voiding reflex**, which are initiated by stretch receptors in the wall of the urinary bladder. Voluntary urination involves coupling these reflexes with the voluntary relaxation of the external urethral sphincter, which allows the relaxation of the internal urethral sphincter.

26.3 Aging and the Urinary System p. 702

- Kidney problems usually increase with age. Age-related changes in the urinary system include (1) declining number of functional nephrons, (2) reduced glomerular filtration, (3) reduced sensitivity to ADH, and (4) problems with urinary reflexes (urinary retention may develop in men whose prostates are enlarged).

Chapter Review

For answers, see the blue Answers tab at the back of the book.

Level 1 Reviewing Facts and Terms

Match each numbered item with the most closely related lettered item.

1. urination .... ☐
2. detrusor .... ☐
3. macula densa .... ☐
4. medulla .... ☐
5. hilum .... ☐
6. nephron .... ☐
7. cortex .... ☐
8. vasa recta .... ☐
9. fibrous capsule .... ☐
10. renal fascia .... ☐

(a) muscle of the urinary bladder
(b) basic functional unit of a kidney
(c) micturition
(d) fibrous tunic of the kidney
(e) outer region of kidney
(f) consists of 6–18 renal pyramids
(g) dense outer layer of the kidney
(h) series of capillaries
(i) region of the juxtaglomerular complex
(j) site of entry/exit for the renal artery and vein

11. Label the following structures on the diagram below.

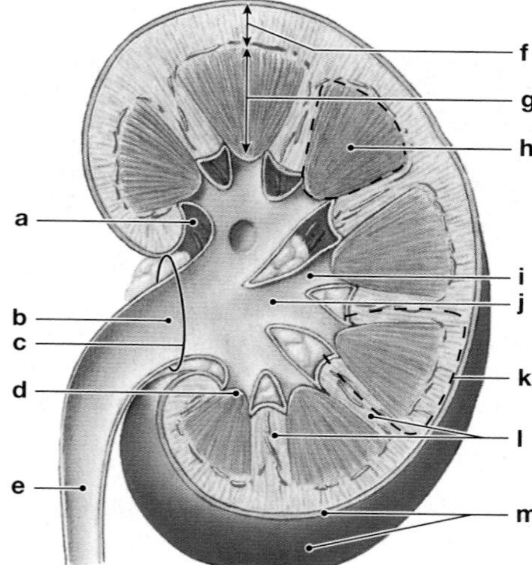

(a) ______
(b) ______
(c) ______
(d) ______
(e) ______
(f) ______
(g) ______
(h) ______
(i) ______
(j) ______
(k) ______
(l) ______
(m) ______

12. Each kidney is protected/stabilized by the
(a) fibrous capsule only.
(b) perinephric fat and fibrous capsule only.
(c) perinephric fat only.
(d) fibrous capsule, perinephric fat, and renal fascia.

13. The urinary system does all of the following *except*
(a) secrete excess glucose molecules.
(b) regulate blood volume.
(c) contribute to stabilizing blood pH.
(d) eliminate organic wastes.

14. The renal sinus is
(a) the innermost layer of kidney tissue.
(b) a conical structure located in the renal medulla.
(c) an internal cavity lined by the fibrous capsule and located inside the hilum.
(d) a large branch of the renal pelvis.

15. Which vessels form the plexus that supplies the proximal and distal convoluted tubules?
(a) segmental arteries
(b) peritubular capillaries
(c) cortical radiate arteries
(d) arcuate arteries

16. The process of filtration occurs at the
(a) proximal convoluted tubule.
(b) renal corpuscle.
(c) collecting duct.
(d) nephron loop.

17. The ability to form concentrated urine depends on the functions of the
(a) proximal convoluted tubule.
(b) distal convoluted tubule.
(c) glomerular capillaries.
(d) nephron loop.

18. The ureters and urinary bladder are lined by ________ epithelium.
(a) stratified squamous
(b) pseudostratified columnar
(c) simple cuboidal
(d) transitional

19. Each of the following is a normal component of urine *except*
(a) hydrogen ions.
(b) urea.
(c) large proteins.
(d) salts.

20. A ligament that extends from the anterior and superior border of the bladder to the umbilicus is the
(a) round ligament.
(b) square ligament.
(c) median umbilical ligament.
(d) lateral umbilical ligament.

21. The portion of the nephron that attaches to the collecting duct is the
(a) nephron loop.
(b) proximal convoluted tubule.
(c) distal convoluted tubule.
(d) glomerular capillaries.

Level 2 Reviewing Concepts

1. What is the significance of the slit-like, rather than rounded, openings of the entrance of the ureters into the bladder?
(a) They can distort more easily to permit urine to move in or out.
(b) The shape prevents urine backflow toward the ureters when the urinary bladder contracts.
(c) The opening is compressed between the middle and lateral umbilical ligaments because these structures support this part of the ureter.
(d) There is no significance; the shape occurs because of the position only.

2. Problems with the urination reflex in older individuals include
(a) increased tone in the sphincter muscles, preventing easy emptying of the urinary bladder.
(b) urinary retention in males because of prostate enlargement.
(c) increased sensitivity to ADH.
(d) increased retention of sodium ions.

3. How will kidney function be altered in a person who has more than one renal artery and vein per side, with the same overall volume of possible lumen space entering the kidney at the hilum?
(a) These kidneys will be able to handle blood at higher pressures in this individual than in others.
(b) As long as the arteries and veins in the individual are normal, kidney function will be normal.
(c) Greater blood flow will occur through these kidneys.
(d) These kidneys will be more flexible in the amounts of urine they can produce at any given time.

4. Where is the glomerulus located in a nephron?

5. What is unique about the glomerular epithelium?

6. What does the juxtaglomerular complex secrete?

7. What is the trigone of the urinary bladder?

26

8. Which urethral sphincter is under voluntary control?
9. What is the primary function of the proximal convoluted tubule?
10. Identify the purpose of rugae in the urinary bladder.

Level 3 Critical Thinking

1. Although neural control of the external urethral sphincter does not completely develop until 2 to 3 years of age, the internal urethral sphincter is functional at birth. Why is premature toilet training often a fruitless and frustrating experience for both parents and child?
2. Jennifer was a champion marathon runner until renal failure forced her to retire from running. During her last two marathons, lab tests confirmed she experienced acute kidney dysfunction. How does long-distance running cause this condition?
3. Why does a pregnant woman need to urinate more frequently than she does when she is not pregnant?

CLINICAL CASE WRAP-UP

This Too Shall Pass

Jack is suffering from a kidney stone. A kidney stone is a small, hard, sharp-edged deposit made of minerals, usually calcium and phosphate, that are normally dissolved in urine. Because it has been hot, Jack has lost fluid through sweating. Since he is hesitant to drink lots of water during a tournament, his urine has become so concentrated that minerals have crystallized and stuck together, forming a stone in his renal pelvis. As the stone moves into the ureter, it obstructs the flow of urine and stretches the collecting system behind it, causing flank pain. The abraded epithelial lining of the ureter is bleeding, causing hematuria. The muscular ureter continues to contract in peristaltic waves, trying to push the stone along, and causing more pain. As the stone moves, little by little, through the ureter where it passes over the psoas major muscles, Jack's pain radiates.

As Jack is waiting in the radiology department, he feels yet another wave of pain that radiates all the way down to his left testicle. Again feeling an urge to urinate, he uses a urinal and feels the pain radiate down to the tip of his urethra. He passes the tiny stone into the urinal.

"Well, congratulations, Jack," says his doctor. "You have saved yourself an x-ray study and possibly surgery. Now, the best thing you can do for yourself from now on is to drink, drink, drink!"

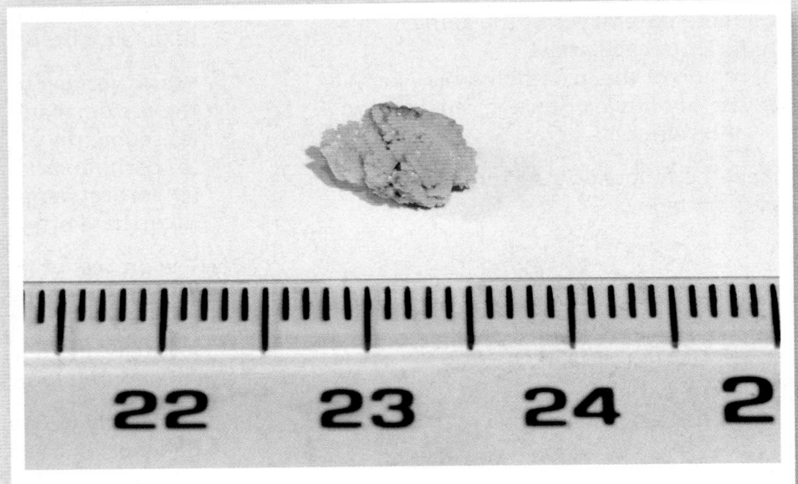

1. What would happen if Jack's kidney stone permanently blocked his ureter completely?
2. Why did Jack's doctor tell him to "drink, drink, drink"?

See the blue Answers tab at the back of the book.

Related Clinical Terms

azotemia: The condition characterized by excessive urea or other nitrogen-containing compounds in the blood.

bacteriuria: Abnormal presence of bacteria in the urine.

cystocele: Condition that occurs when the supportive tissue between a woman's bladder and vaginal wall weakens, stretches, and allows the bladder to bulge into the vagina.

enuresis: Involuntary urination, especially that of a child while asleep.

hydroureter: Dilation of a ureter caused by obstruction of urine flow.

nephroptosis: Condition in which the kidney is displaced downward from its usual and normal position; also called a floating kidney.

nephrotic syndrome: A kidney disorder that causes excessive protein to be excreted in the urine.

nephrotoxin: A toxin that has a specific harmful effect on the kidney.

polycystic kidney disease: An inherited abnormality that affects the development and structure of kidney tubules.

shock-wave lithotripsy: A noninvasive technique used to pulverize kidney stones by passing high-pressure shock waves through a water-filled tub in which the patient sits.

urologist: Physician who specializes in functions and disorders of the urinary system.

27

The Reproductive System

Learning Outcomes

These Learning Outcomes correspond by number to this chapter's sections and indicate what you should be able to do after completing the chapter.

27.1 Compare and contrast the general organization of the male and female reproductive systems. p. 708

27.2 Identify and describe the location, gross anatomy, and histology of the major structures of the male reproductive system. p. 708

27.3 Identify and describe the location, gross anatomy, and histology of the major structures of the female reproductive system. p. 719

27.4 Compare and contrast the age-related changes in the male and female reproductive systems. p. 733

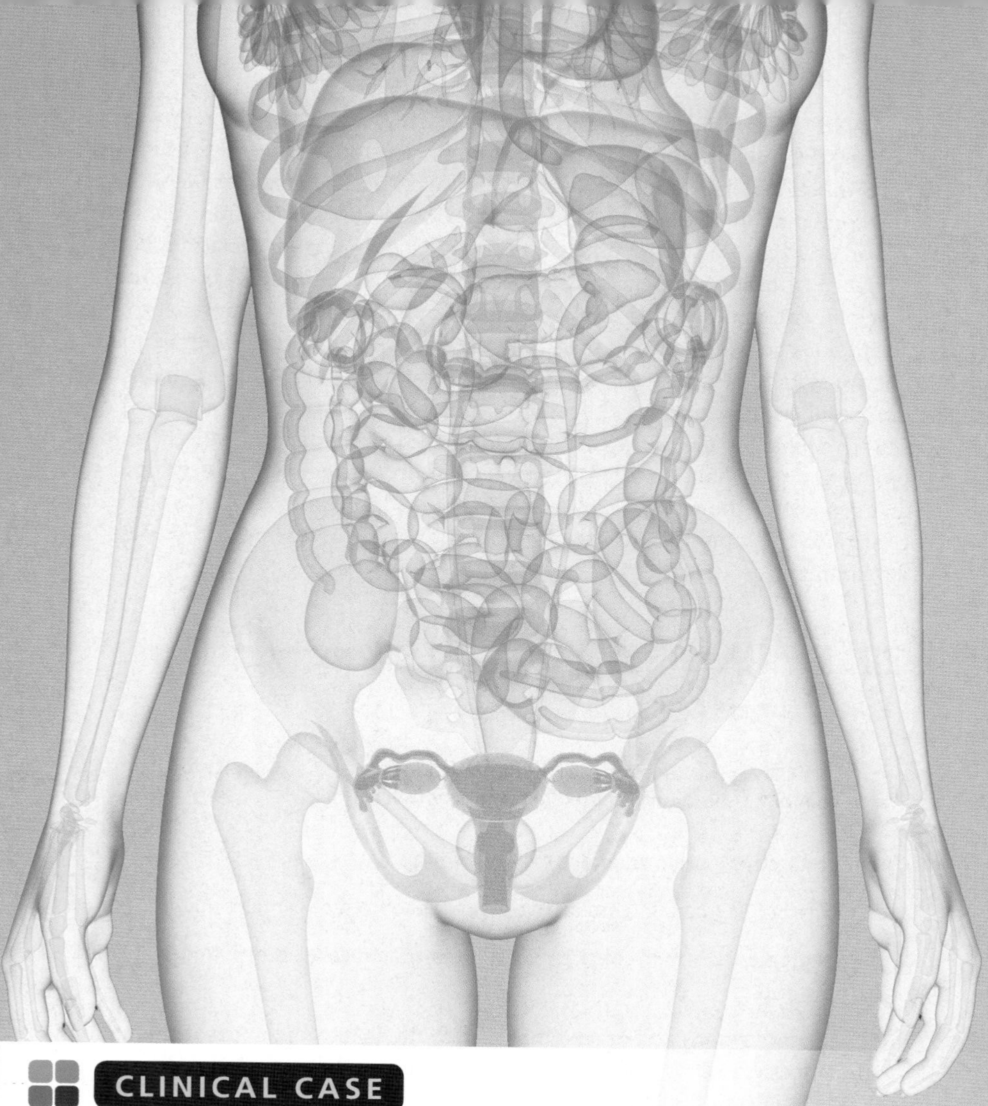

CLINICAL CASE

A Serious Game of Twister

Liam, a high school senior, is finishing his last football game as team quarterback. In the final quarter, as he gets up from a tackle, he feels a twinge of pain in the right side of his scrotum. On his way back to the locker room, he can't stand up straight. In the shower, he examines his scrotum but doesn't see any bruising or swelling. However, his right testis is higher than his left, and it's so tender he can barely get dressed. He is walking hunched over and feels nauseated.

The coach notices Liam's protective, slow gait and sees him stop to vomit in the trashcan. "What's going on, Liam? You played a great game tonight." "Oh, sorry, Coach. I am just a little sick to my stomach," replies Liam. "And this is embarrassing, but my right testicle is killing me. I don't remember getting hit there, but it is so painful I can't touch it."

His coach says, "Liam, you are going straight to the emergency room. Call your parents."

Why does his coach send Liam to the emergency room? To find out, turn to the Clinical Case Wrap-Up on p. 737.

SIMPLY PUT, THE HUMAN SPECIES has survived because of the reproductive system. The human reproductive system produces, stores, nourishes, and transports functional male and female reproductive cells called **gametes** (GAM-ēts). The combination of the genetic material provided by a **sperm** from the father and an **oocyte** (Ō-ō-sīt) from the mother occurs shortly after **fertilization**, or *conception*. Fertilization produces a **zygote** (ZĪ-gōt), a single cell whose growth, development, and repeated divisions will, in approximately 9 months, produce an infant who will grow and mature as part of the next generation. The reproductive system also produces sex hormones that affect the structure and function of all other systems.

This chapter describes the structures and mechanisms involved in the production and maintenance of gametes and, in the female, the development and support of the embryo and fetus. Chapter 28, the last chapter in the text, continues the story, beginning at fertilization and discussing the process of development.

27.1 Organization of the Reproductive System

▶ **KEY POINT** The reproductive system includes the gonads, reproductive tract, accessory glands and organs, and external genitalia.

The reproductive system includes the following:

- **Gonads** (GŌ-nadz), reproductive organs that produce gametes and hormones
- A reproductive tract consisting of ducts that receive, store, and transport the gametes
- Accessory glands and organs that secrete fluids into the ducts of the reproductive system or into other excretory ducts
- **External genitalia** (jen-i-TĀ-lē-a), perineal structures associated with the reproductive system

In adult males, the gonads are the **testes** (TES-tēz; singular, *testis*). The testes secrete sex hormones called androgens (principally testosterone) and produce half a billion sperm per day. After storage, mature sperm travel along a system of ducts where they are mixed with the secretions of several accessory glands. This forms a mixture known as semen (SĒ-men). During ejaculation (ē-jak-ū-LĀ-shun), semen is expelled from the body.

In adult females, the gonads are the **ovaries**. The ovaries typically produce only one immature gamete, or oocyte, per month. The oocyte travels within the uterine tubes that open into the muscular uterus (Ū-ter-us). A short passageway, the vagina (va-JĪ-na), connects the uterus with the exterior. During sexual intercourse, male ejaculation introduces semen into the vagina, and the sperm then ascend the female reproductive tract, where they may encounter an oocyte and begin the process of fertilization.

27.1  **CONCEPT CHECK**

1 What do the gonads produce?

See the blue Answers tab at the back of the book.

27.2 Anatomy of the Male Reproductive System

▶ **KEY POINT** The male reproductive system is composed of structures that produce semen and expel it from the body.

The major structures of the male reproductive system are shown in **Figure 27.1**. Sperm leave the testes and travel along a duct system that includes the epididymis (ep-i-DID-i-mus), ductus deferens (DUK-tus DEF-e-renz) (*vas deferens*), ejaculatory (ē-JAK-ū-la-tōr-ē) duct, and urethra before leaving the body. Accessory organs, notably the seminal (SEM-i-nal) glands, prostate (PROS-tāt), and bulbo-urethral (bul-bō-ū-RĒ-thral) glands, secrete into the ejaculatory ducts and urethra. The external genitalia include the scrotum (SKRŌ-tum), which encloses the testes, and the penis (PĒ-nis), an erectile organ through which the distal portion of the urethra passes.

The Testes

▶ **KEY POINT** The testes produce the male gametes. The testes develop within the abdomen of the fetus and descend, during development, to be housed within the scrotum outside the abdominopelvic cavity.

Each testis is roughly 5 cm (2 in.) long, 3 cm (1.2 in.) wide, and 2.5 cm (1 in.) thick. Each weighs 10–15 g (0.35–0.53 oz). The testes hang within the **scrotum**, a pouch of skin suspended inferior to the perineum and anterior to the anus (**Figures 27.1** and **27.3**).

Descent of the Testes

During development, the testes form inside the body cavity adjacent to the kidneys. As the fetus enlarges, the positions of these organs change. The testes gradually move inferiorly and anteriorly toward the anterior abdominal wall.

The **gubernaculum testis** is a cord of connective tissue and muscle fibers extending from the inferior part of each testis to the posterior wall of a small, inferior pocket of the peritoneum (**Figure 27.2**). As growth proceeds, each gubernaculum does not elongate, and the testes are held in position. During the seventh developmental month, growth continues rapidly, and circulating hormones stimulate contraction of the muscles of the gubernaculum testis. As a result, the position of the testes changes considerably as they move through the inguinal canal and come to lie within the scrotum, outside the body. As the testes change position, small pockets of the peritoneal cavity also develop. This process is known as the **descent of the testes**.

As each testis moves through the body wall, the ductus deferens and the testicular blood vessels, nerves, and lymphatic vessels accompany it. Together, these structures form the body of the spermatic cord.

The Spermatic Cords

The **spermatic cords** are paired structures extending between the abdominopelvic cavity and the testes. Each spermatic cord begins at the deep inguinal ring, extends through the inguinal canal (a passageway through the abdominal musculature), exits at the superficial inguinal ring, and descends to the testes within the scrotum (**Figure 27.3**). The spermatic cords form during the descent of the testes. Each spermatic cord consists of layers of fascia and muscle enclosing the ductus deferens and the blood vessels, nerves, and lymphatic vessels that supply the testes. The blood vessels include the **deferential artery**, a **testicular artery**, and the **pampiniform** (pam-PIN-i-form; *pampinus*, tendril, + *forma*, form) **plexus of the testicular vein**. The **ilio-inguinal** and **genitofemoral nerves** from the lumbar plexus provide innervation.

The **inguinal canals** are the narrow canals linking the scrotal chambers with the peritoneal cavity. These passageways usually close, but the presence of the spermatic cords within the canal creates weak points in the abdominal wall that remain throughout life. As a result, inguinal hernias (discussed in Chapter 10) are relatively common in males. The inguinal canals in females are very small, containing only the ilio-inguinal nerves and the round ligaments of the uterus. The abdominal wall is nearly intact, and inguinal hernias in women are rare.

Figure 27.1 The Male Reproductive System, Part I. The male reproductive system as seen in sagittal section. The diagrammatic view shows several intact organs on the left side; this should help you to interpret the cadaver section.

Prostatic urethra
Pubic symphysis
Urinary bladder
Ureter
Rectum

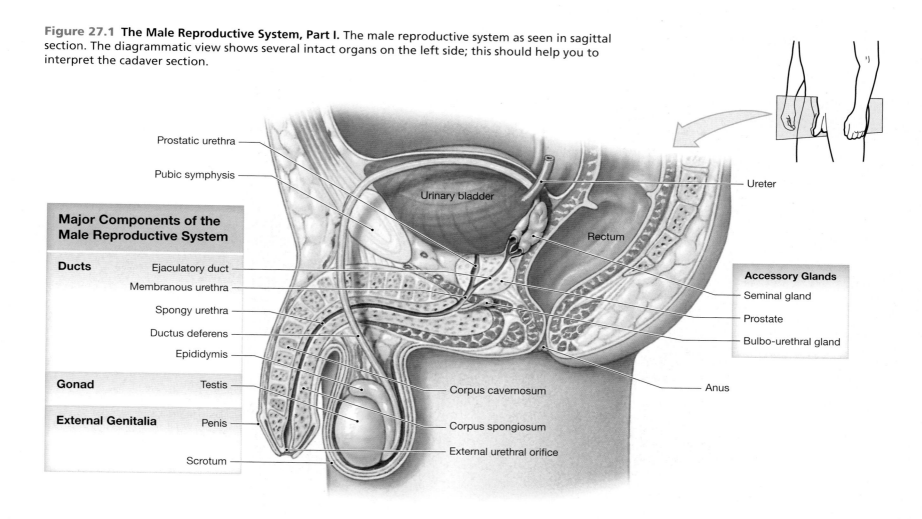

Major Components of the Male Reproductive System	
Ducts	Ejaculatory duct
	Membranous urethra
	Spongy urethra
	Ductus deferens
	Epididymis
Gonad	Testis
External Genitalia	Penis
	Scrotum

Accessory Glands
Seminal gland
Prostate
Bulbo-urethral gland

Anus
Corpus cavernosum
Corpus spongiosum
External urethral orifice

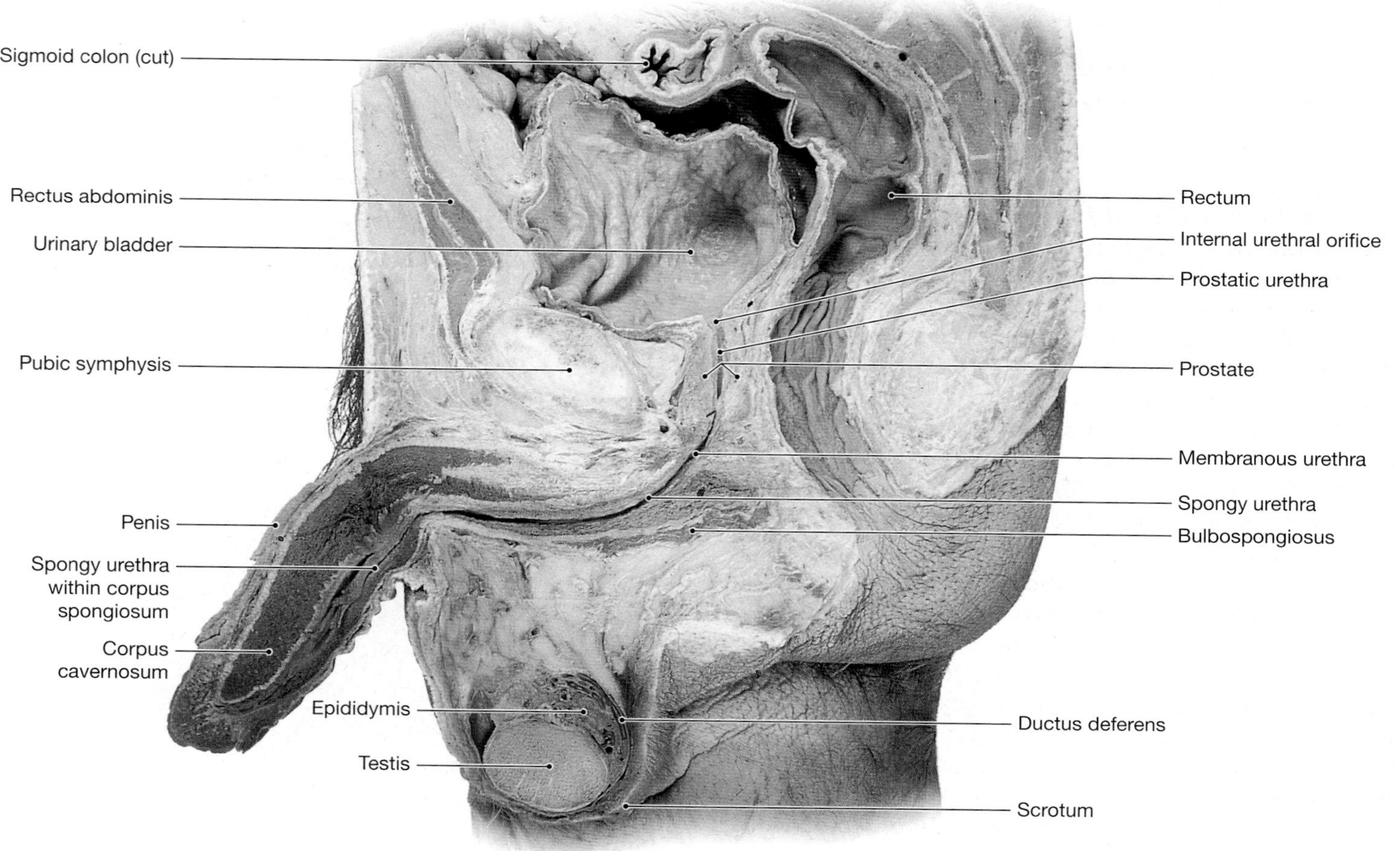

27

Figure 27.2 The Descent of the Testes

Peritoneum
Testis
Gubernaculum testis
Colon
5 mm
2 months
Umbilical cord
Penis
5 mm
3 months
Pubis
Testis
Scrotum
5 mm
Birth

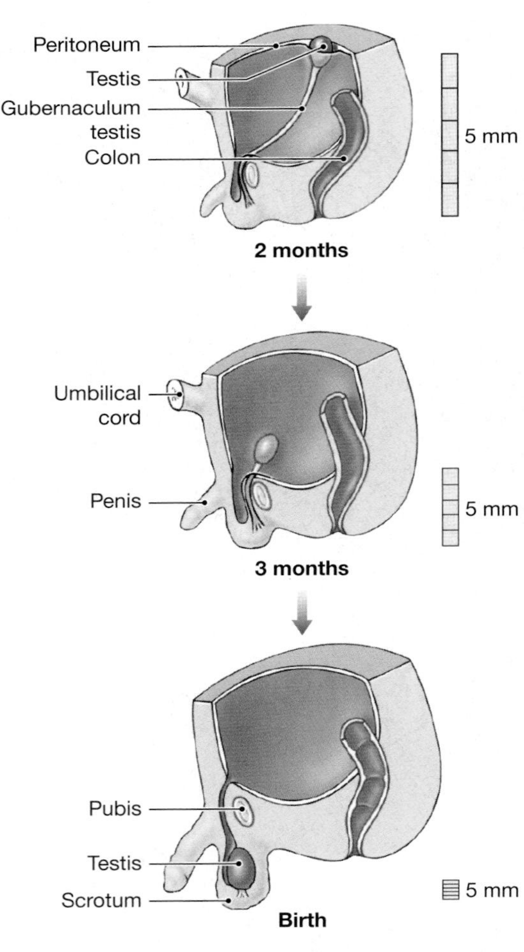

a Diagrammatic sectional views. Note the scale bar to the right of each image.

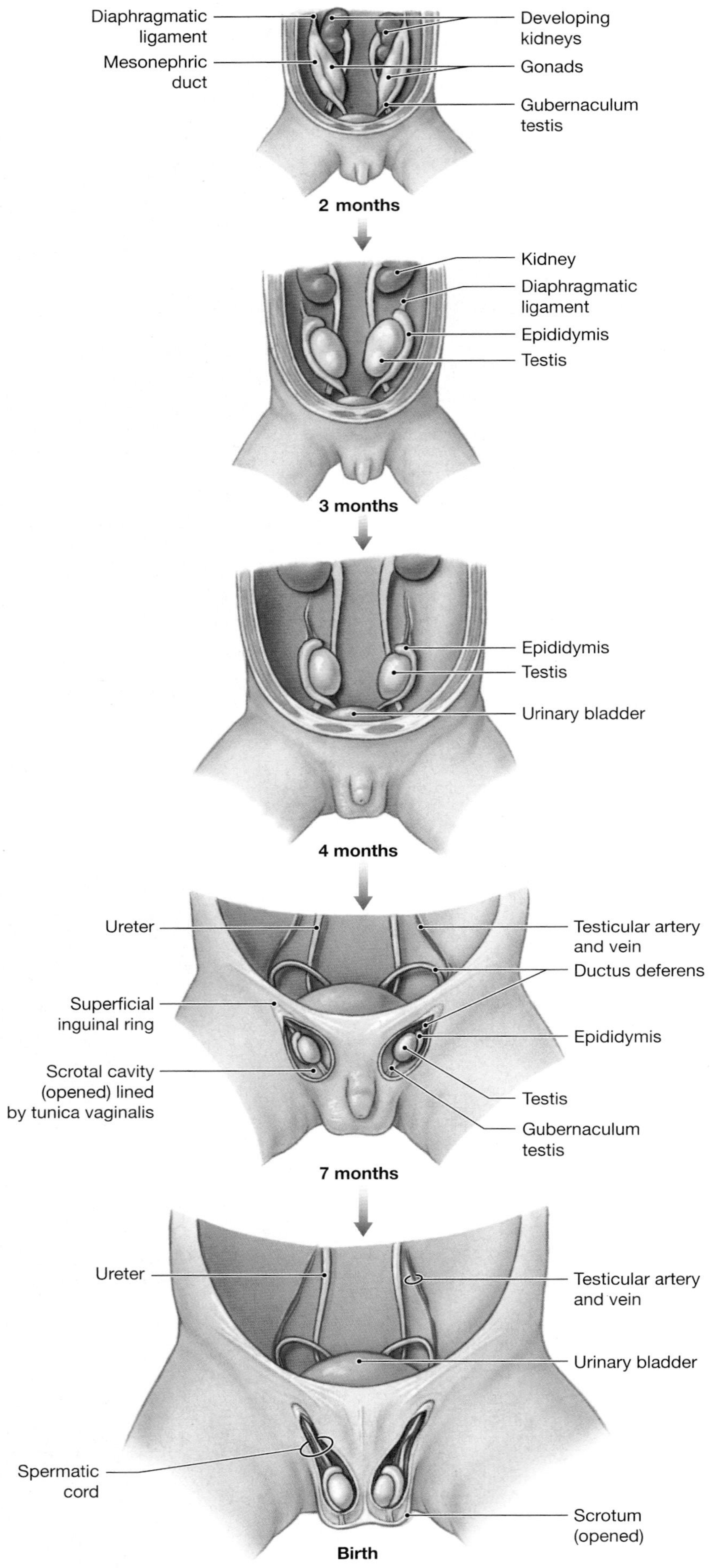

Diaphragmatic ligament
Mesonephric duct
Developing kidneys
Gonads
Gubernaculum testis
2 months
Kidney
Diaphragmatic ligament
Epididymis
Testis
3 months
Epididymis
Testis
Urinary bladder
4 months
Ureter
Superficial inguinal ring
Scrotal cavity (opened) lined by tunica vaginalis
Testicular artery and vein
Ductus deferens
Epididymis
Testis
Gubernaculum testis
7 months
Ureter
Testicular artery and vein
Urinary bladder
Spermatic cord
Scrotum (opened)
Birth

b Anterior views of the opened abdomen at representative stages in the descent of the testes.

The Scrotum and the Position of the Testes The scrotum is divided internally into two separate chambers. A raised thickening in the scrotal surface known as the **raphe** (RĀ-fē) **of the scrotum** divides it in two (**Figures 27.3** and **27.4a**). Each testis occupies a separate **scrotal cavity**. The **tunica vaginalis** (TŪ-ni-ka vaj-i-NAL-is), a serous membrane, lines the scrotal cavity and reduces friction between the opposing parietal (outer) layer and visceral (inner) layer.

The scrotum consists of a thin layer of skin and the underlying superficial fascia. Deep to the dermis of the scrotum are two layers of muscle. The most superficial is a layer of smooth muscle, the **dartos** (DAR-tōs) **muscle**. Resting muscle tone in the dartos muscle elevates the testes and causes wrinkling of the scrotal surface. Deep to the dartos muscle is a layer of skeletal muscle, the **cremaster** (kre-MAS-ter). Contraction of the cremaster, controlled by the cremasteric reflex, tenses the scrotum and pulls the testes closer to the body. Contraction occurs during sexual arousal and in response to temperature changes. Normal sperm development in the testes requires temperatures approximately 1.1°C (2°F) lower than those found elsewhere in the body. The cremaster moves the testes away from or toward the body as needed to maintain acceptable testicular temperatures. When air or body temperatures rise, the cremaster relaxes and the testes move away from the body. When the scrotum cools suddenly, as happens when jumping into a cold swimming pool, these muscles contract to pull the testes closer to the body and keep testicular temperatures from decreasing.

Figure 27.3 **The Male Reproductive System, Part II.** Diagrammatic anterior view of the gonads, external genitalia, and associated structures in a male.

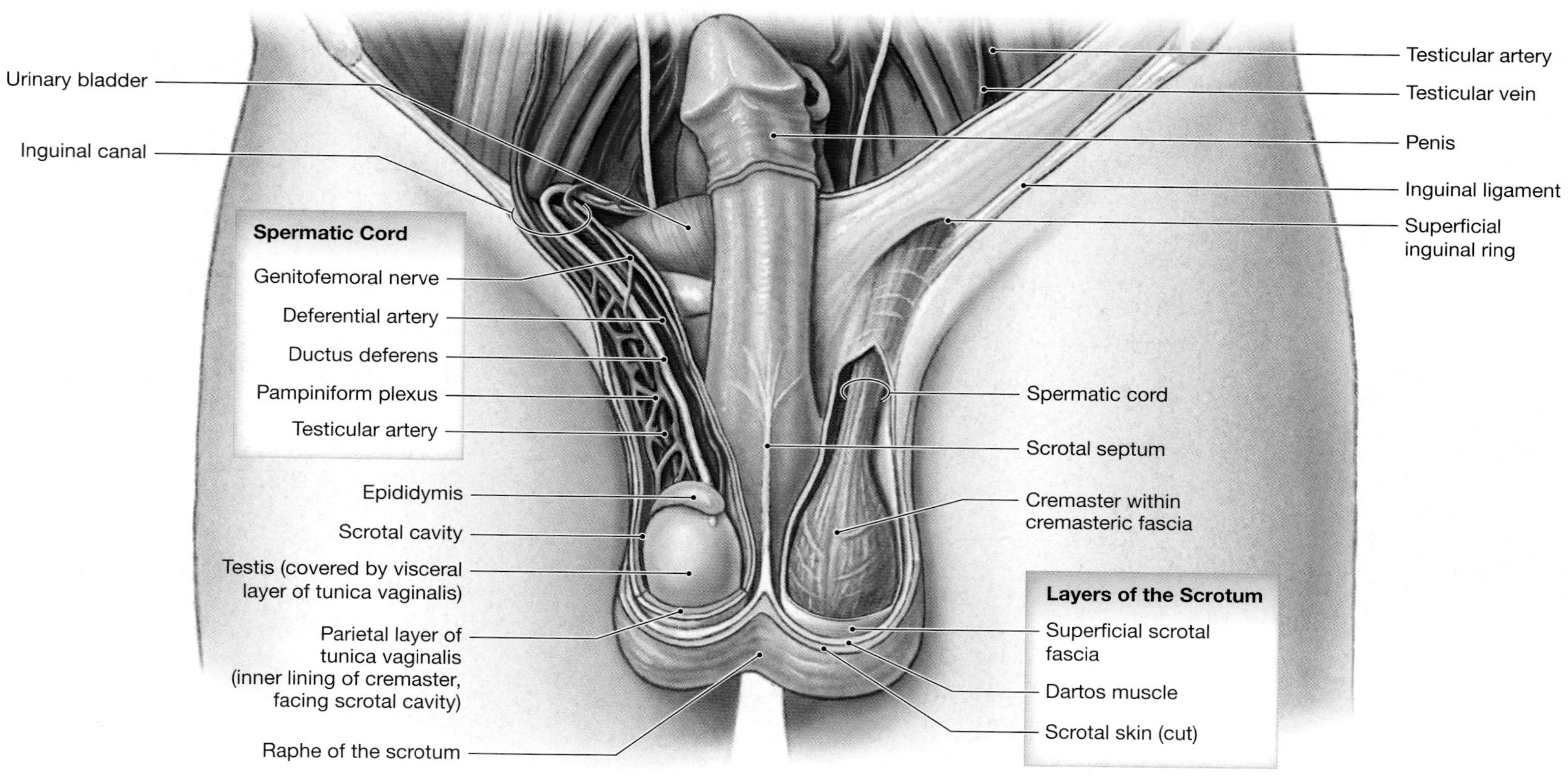

The scrotum is richly supplied with sensory and motor nerves from the hypogastric plexus and branches of the ilio-inguinal nerves, genitofemoral nerves, and pudendal nerves. The vascular supply to the scrotum includes the **internal pudendal arteries** (branches of the internal iliac arteries), the **external pudendal arteries** (branches of the femoral arteries), and the cremasteric branch of the **inferior epigastric arteries** (branches of the external iliac arteries). The names and distributions of the veins follow those of the arteries.

Structure of the Testes

The **tunica albuginea** (TŪ-ni-ka al-bū-JIN-ē-a) is a dense layer of fibrous connective tissue that surrounds the testis and is covered by the tunica vaginalis. This fibrous capsule is rich in collagen fibers that are continuous with those surrounding the adjacent epididymis. The collagen fibers of the tunica albuginea also extend into the interior of the testis, forming fibrous partitions termed the septa testis. These septa testis converge toward the **mediastinum of the testis**. The mediastinum supports the blood vessels and lymphatics supplying the testis and the ducts that collect and transport sperm into the epididymis (Figure 27.4).

Histology of the Testes

The septa testis partition the testis into compartments called **lobules**. Roughly 800 slender, tightly coiled **seminiferous** (sem-i-NIF-er-us) **tubules** are distributed among the lobules (Figure 27.4). Each tubule averages around 80 cm (31 in.) in length, and a typical testis contains approximately one-half mile of seminiferous tubules. Sperm production occurs within these tubules.

Each seminiferous tubule is U-shaped and connected to a single **straight tubule** that enters the mediastinum of the testis (Figures 27.4a and 27.7a). Within the mediastinum, straight tubules are extensively interconnected, forming a maze of passageways known as the **rete** (RĒ-tē; *rete*, a net) **testis**. Fifteen to twenty large **efferent ductules** connect the rete testis to the epididymis.

Because the seminiferous tubules are tightly coiled, histological preparations most often show them in transverse section (Figure 27.5c,d). A delicate capsule surrounds each tubule, and loose connective tissue fills the external spaces between the tubules. Within those spaces are numerous blood vessels and large **interstitial endocrine cells** (*Leydig cells*). Interstitial endocrine cells produce androgens, the dominant male sex hormones. Testosterone is the most important androgen, and functions to:

- stimulate spermatogenesis,
- promote the physical and functional maturation of sperm,
- maintain the accessory organs of the male reproductive tract,
- cause development of secondary sexual characteristics such as facial hair,
- stimulate growth and metabolism throughout the body, and
- influence brain development by stimulating sexual behaviors and sexual drive.

Spermatogenesis and Meiosis

Sperm are produced through the process of **spermatogenesis** (sper-ma-tō-JEN-e-sis) (Figure 27.5). Spermatogenesis begins at the outermost layer of cells in the seminiferous tubules. Stem cells called **spermatogonia** (sper-ma-tō-GŌ-nē-a) form during embryonic development, but remain dormant until puberty.

27

Figure 27.4 Structure of the Testes

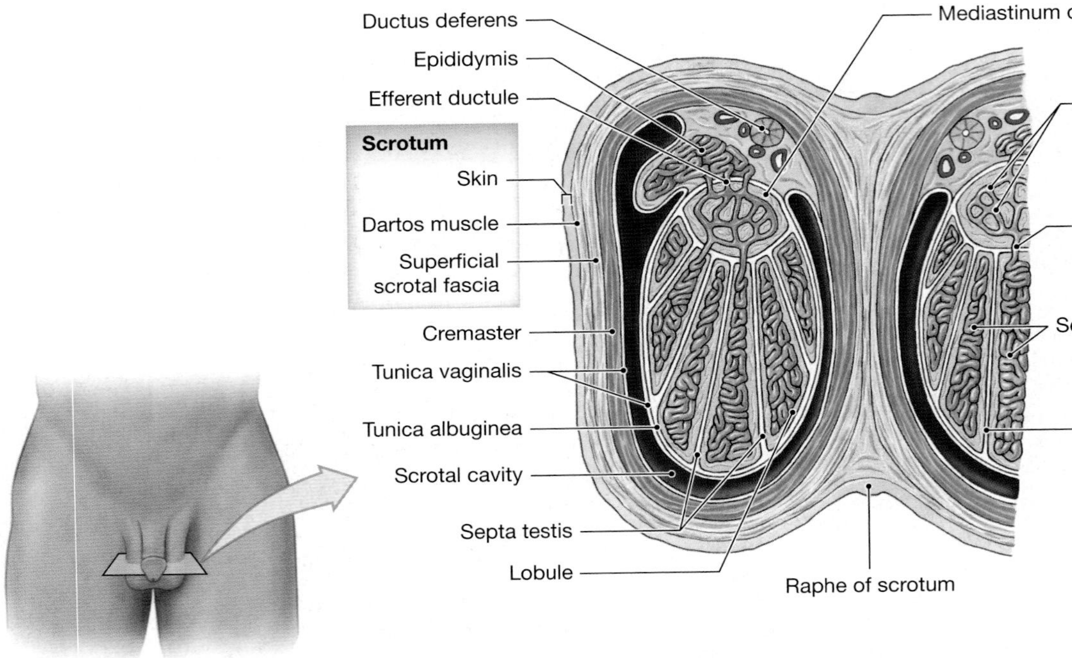

a Diagrammatic horizontal section showing the anatomical relationships of the testes within the scrotal cavities

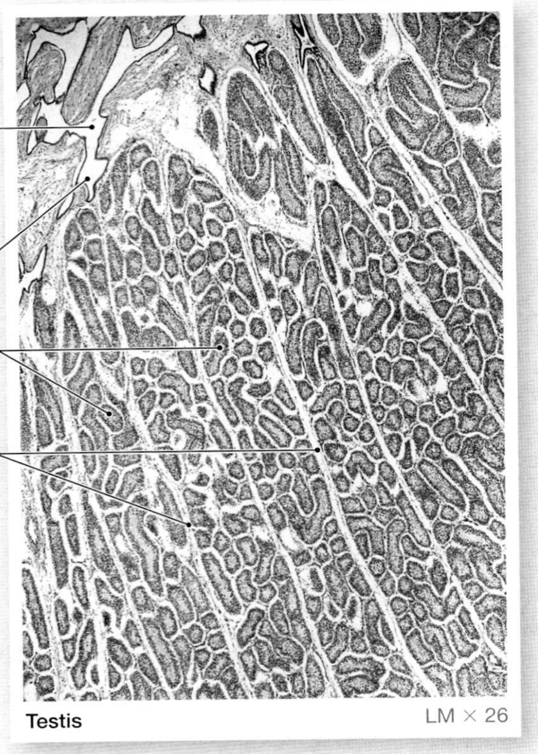

Testis LM × 26

b A section through a testis

Beginning at sexual maturation, spermatogonia continuously divide throughout the reproductive years. As each division occurs, one of the daughter cells remains in the outer layer of the seminiferous tubule as an undifferentiated stem cell. The other cell is pushed toward the lumen and differentiates into a **primary spermatocyte** (sper-MA-tō-sīt) that prepares to begin **meiosis** (mī-Ō-sis). Meiosis is a form of cell division that produces gametes containing half the normal number of chromosomes. Because they contain only one member of each pair of chromosomes, gametes are called **haploid** (HAP-loyd; *haplo*, single).

Mitosis and meiosis differ significantly in terms of the nuclear events. In mitosis, a single division produces two identical daughter cells, each containing 23 *pairs* of chromosomes (**Spotlight Figure 2.17,** p. 45). In meiosis, a pair of divisions produces four different, haploid gametes, each containing 23 *individual* chromosomes.

In the testes, in the first step in meiosis a primary spermatocyte divides to produce a pair of **secondary spermatocytes**. Each secondary spermatocyte divides to produce a pair of **spermatids** (SPER-ma-tidz). As a result, for every primary spermatocyte that enters meiosis, four spermatids are produced (**Figure 27.5**).

Spermatogenesis is directly stimulated by testosterone and indirectly stimulated by FSH (follicle-stimulating hormone). Interstitial endocrine cells of the testis produce testosterone in response to LH (luteinizing hormone). ⊃ pp. 508–512, 520–521

Spermiogenesis

In **spermiogenesis**, spermatids differentiate into physically mature **sperm** (**Figure 27.5**). During spermiogenesis, spermatids are embedded within the cytoplasm of large **nurse cells** (*Sertoli cells*). Nurse cells are attached to the basal lamina at the tubular capsule and extend toward the lumen between the spermatocytes undergoing meiosis. As spermiogenesis proceeds, the spermatids gradually develop into mature sperm. The detachment of the sperm from the nurse cell releases it into the lumen of the seminiferous tubule. This marks the end of spermiogenesis. The entire process, from start to finish, takes approximately 9 weeks.

Nurse Cells

Nurse (*Sertoli*) cells have five important functions:

1. Maintain the blood testis barrier: A **blood testis barrier**, comparable to the blood brain barrier, isolates the seminiferous tubules from the general circulation (**Figure 27.5c,d**). Several factors contribute to the blood testis barrier. Tight junctions between nurse cells isolate the luminal portion of the seminiferous tubule from the surrounding interstitial fluid. Transport of materials across the nurse cells is tightly regulated, which maintains a stable environment surrounding the spermatocytes and spermatids. The lumen of a seminiferous tubule contains a fluid high in androgens, estrogens, potassium, and amino acids, which is very different from interstitial fluid; the blood testis barrier is essential to preserving these differences. In addition, developing sperm contain sperm-specific antigens in their cell membranes. These antigens, not found in somatic plasma membranes, would be attacked by the immune system if the blood testis barrier did not prevent their detection.
2. Support mitosis and meiosis: Circulating follicle-stimulating hormone (FSH) and testosterone stimulate nurse cells. These stimulated nurse cells then promote the division of spermatogonia and the meiotic divisions of spermatocytes.
3. Support spermiogenesis: Spermiogenesis requires the presence of nurse cells. These cells surround and enfold the spermatids, providing nutrients and chemical stimuli that promote their development.

Figure 27.5 Histology of the Seminiferous Tubules

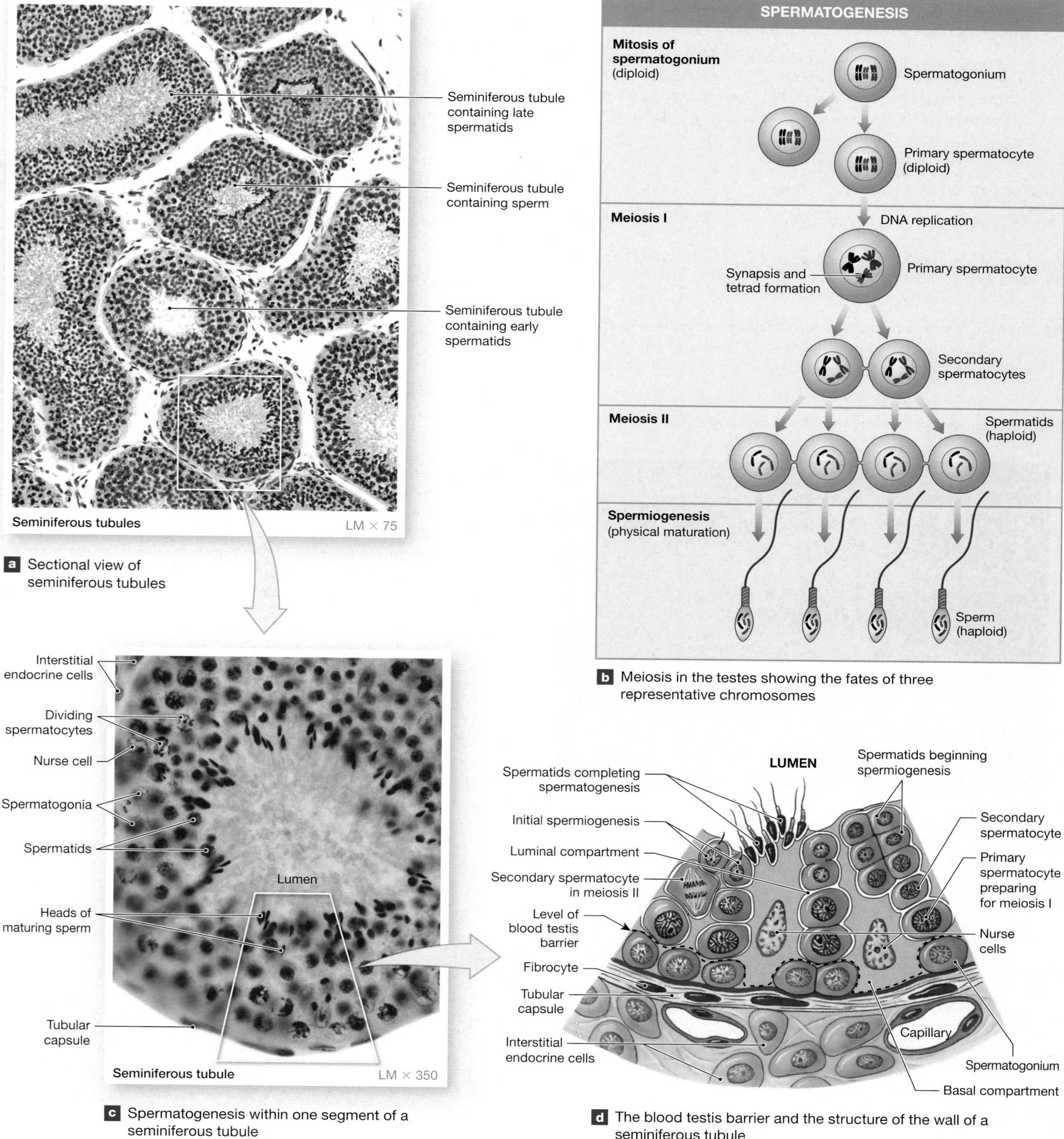

a Sectional view of seminiferous tubules

b Meiosis in the testes showing the fates of three representative chromosomes

c Spermatogenesis within one segment of a seminiferous tubule

d The blood testis barrier and the structure of the wall of a seminiferous tubule

4 Secrete inhibin: Nurse cells secrete a hormone called **inhibin** (in-HIB-in). Inhibin depresses the pituitary production of FSH and gonadotropin-releasing hormone (GnRH). By regulating FSH and GnRH secretion, nurse cells provide negative feedback control of spermatogenesis.

5 Secrete androgen-binding protein: **Androgen binding protein (ABP)** binds androgens in the fluid contents of the seminiferous tubules. This protein is thought to be important in elevating the concentration of androgens within the tubules and stimulating spermiogenesis. FSH stimulates the production of ABP.

Anatomy of a Sperm

▶ **KEY POINT** A sperm has three distinct regions: head, middle piece, and tail.

There are three distinct regions to each sperm: the head, middle piece, and tail (**Figure 27.6**).

1 The **head** of a sperm contains densely packed chromosomes. The tip of the head contains the **acrosome** (AK-rō-sōm). This is a membrane-bound, vesicular compartment containing enzymes involved in fertilization.

Figure 27.6 Spermiogenesis and Sperm Histology

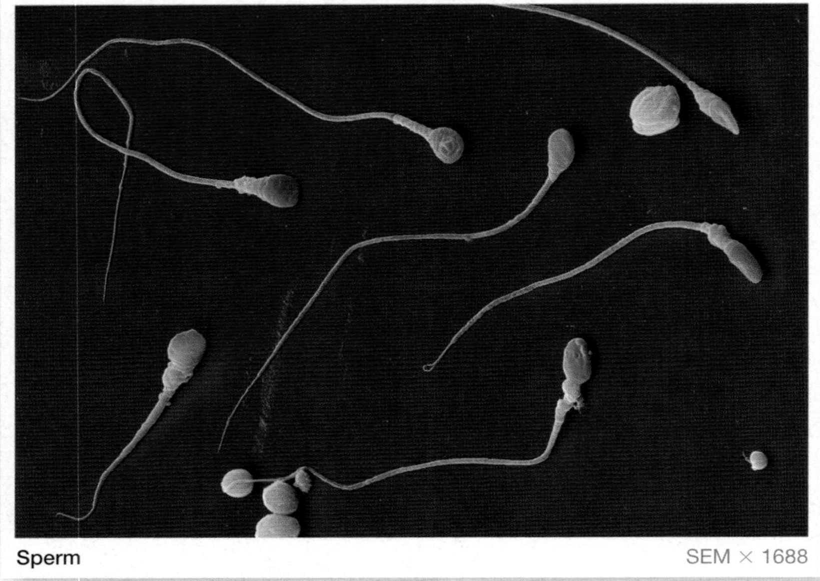

a Histology of human sperm

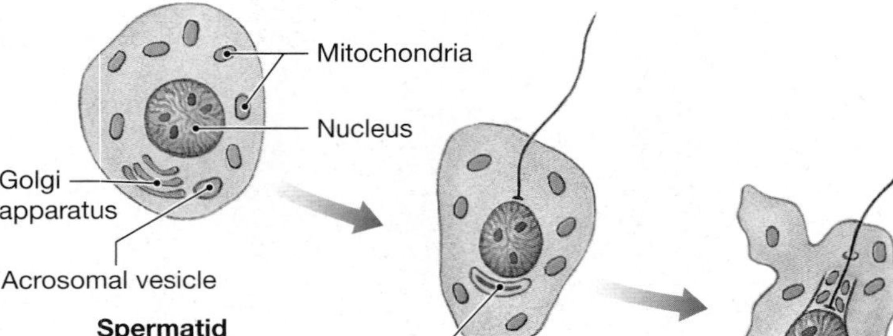

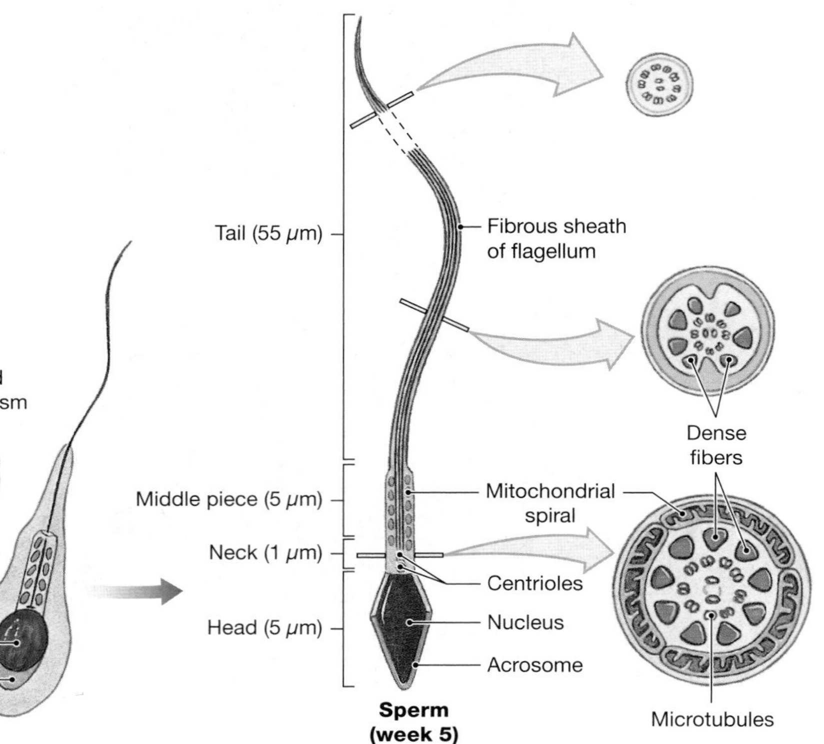

b Differentiation of a spermatid into a sperm

2 A short **neck** attaches the head to the **middle piece**. The neck contains both centrioles of the original spermatid. The distal centriole is connected to microtubules that are continuous throughout the middle piece and tail. Mitochondria that provide the ATP needed to move the tail are arranged in a spiral around the microtubules.

3 The **tail** is the only **flagellum** in the human body. The tail moves the sperm from one place to another. Unlike cilia, which beat in a predictable, waving fashion, the tail of a sperm has a whip-like, corkscrew motion.

Unlike other, less specialized cells, a mature sperm lacks an endoplasmic reticulum, a Golgi apparatus, lysosomes, peroxisomes, inclusions, and many other organelles. Because a sperm does not contain glycogen or other energy reserves, it must absorb nutrients (primarily fructose) from the surrounding fluid.

The Male Reproductive Tract

▶ **KEY POINT** The components of the male reproductive tract are concerned with the functional maturation, nourishment, storage, and transport of sperm.

The Epididymis

Late in their development, the spermatozoa detach from the nurse cells and are free within the lumen of the seminiferous tubule. Although they have most of the physical characteristics of mature sperm, they are still functionally immature and incapable of locomotion or fertilization. Fluid currents within the straight tubule and rete testis transport the sperm into the epididymis (**Figure 27.7a,b**). The lumen of the epididymis is lined by a distinctive pseudostratified columnar epithelium with long stereocilia (**Figure 27.7c,d**).

The **epididymis** is the start of the reproductive tract and lies along the posterior border of each testis (**Figures 27.1**, p. 709, **27.3**, p. 711, **27.4**, p. 712, and **27.7a**). It has a firm texture and can be felt through the scrotum.

Each epididymis is a tubule almost 7 m (23 ft) long, but it is coiled and twisted so it takes up very little space. The epididymis is subdivided into a head, body, and tail:

- The superior **head** receives sperm via the efferent ducts of the mediastinum of the testis.
- The **body** begins distal to the last efferent duct and extends inferiorly along the posterior margin of the testis.
- Near the inferior border of the testis, the number of convolutions decreases, marking the start of the **tail**. The tail reverses direction and, as it ascends, the tubular epithelium changes. The stereocilia disappear, and the epithelium becomes almost identical to that of the ductus deferens—the next segment of the male reproductive tract. The tail of the epididymis is the main storage site for sperm.

The epididymis has three major functions:

1. It monitors and adjusts the composition of the fluid produced by the seminiferous tubules. The pseudostratified columnar epithelial lining of the epididymis has distinctive stereocilia (**Figure 27.7c,d**). These stereocilia increase the surface area available for absorption and secretion into the tubular fluid.
2. It recycles damaged sperm. Cellular debris and damaged sperm are absorbed in the epididymis. The products of enzymatic breakdown are released into the surrounding interstitial fluids for pickup by the epididymal blood vessels.

Figure 27.7 The Epididymis

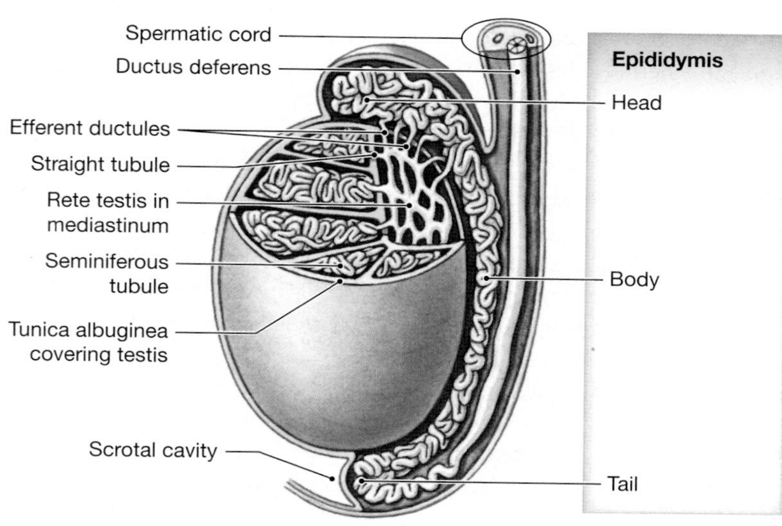

a Diagrammatic view of the testis and epididymis

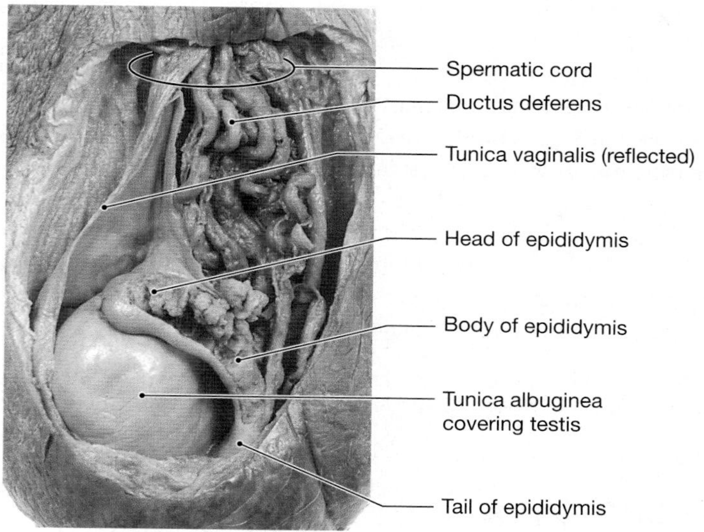

b Gross dissection of testis and epididymis

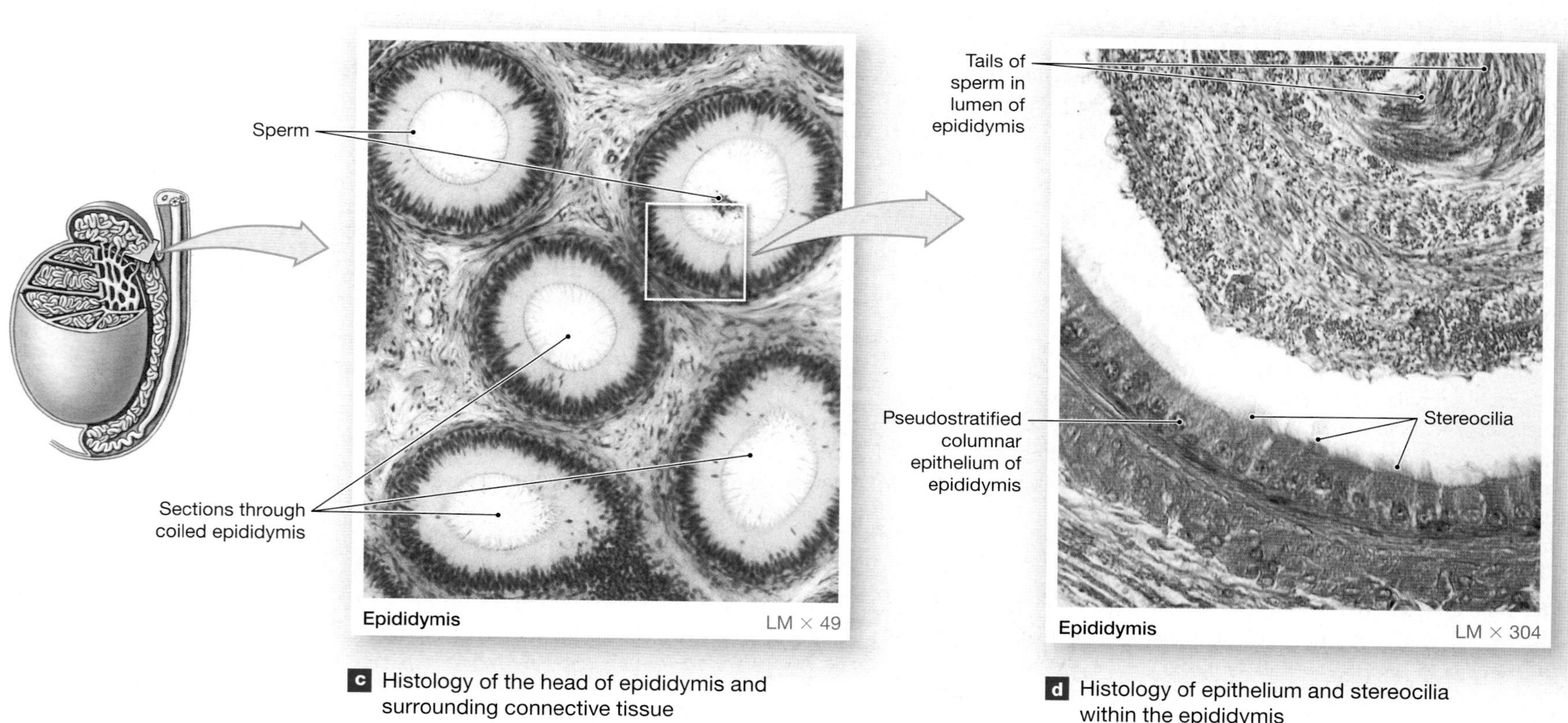

c Histology of the head of epididymis and surrounding connective tissue

d Histology of epithelium and stereocilia within the epididymis

CLINICAL NOTE

Testicular Cancer

Testicular cancer is usually a disease of young and middle-aged men. It almost always develops in the rapidly dividing stem cells, the spermatogonia. Often a painless lump is the first symptom.

The best defense against this disease is regular testicular self-examination. If found and treated early, testicular cancer is often curable. Treatment usually includes an orchiectomy (surgical removal of the testis) and chemotherapy.

3 It stores spermatozoa and facilitates their functional maturation. It takes about 2 weeks for a sperm to pass through the epididymis and complete its functional maturation in a protected environment. Although sperm leaving the epididymis are mature, they remain immobile. To become motile (actively swimming) and fully functional, they must undergo **capacitation**. This process normally occurs in two steps: (a) sperm become motile when mixed with secretions of the seminal glands, and (b) they become capable of successful fertilization when exposed to conditions inside the female reproductive tract, which changes the plasma membrane of the sperm. While the sperm remain in the male reproductive tract, a secretion of the epididymis prevents premature capacitation.

Rhythmic contractions of the stereocilia, and peristaltic contractions of the smooth muscle of the epididymis, push the sperm toward the ductus deferens.

The Ductus Deferens

The **ductus deferens**, or **vas deferens**, is 40–45 cm (16–18 in.) long. It begins at the end of the tail of the epididymis and ascends into the abdominopelvic cavity through the inguinal canal as part of the spermatic cord (**Figure 27.7a,b**). Inside the abdominal cavity, the ductus deferens passes posteriorly and inferiorly along the lateral surface of the bladder, passing medially and anteriorly to the ureters. The ductus deferens then descends along the posterior surface of the urinary bladder toward the superior and posterior margin of the prostate. Just before it reaches the prostate, the ductus deferens enlarges, forming an expanded portion termed the **ampulla** (am-PŪL-la) **of the ductus deferens** (**Figure 27.8a**).

The wall of the ductus deferens contains a thick layer of smooth muscle (**Figure 27.8b**). Peristaltic contractions in this layer propel spermatozoa and fluid along the duct, which is lined by a pseudostratified columnar epithelium with stereocilia. In addition to transporting sperm, the ductus deferens stores sperm for several months. During this time, the sperm remain in a temporary state of inactivity with low metabolic rates.

Each ampulla joins with an **excretory duct of the seminal gland** on that side. This marks the start of the **ejaculatory duct**. These relatively short passageways (2 cm, or less than 1 in.) penetrate the muscular wall of the prostate and empty into the prostatic urethra near the ejaculatory duct from the other side.

The Urethra

The male urethra extends from the urinary bladder to the tip of the penis, a distance of 15–20 cm (6–8 in.). It is divided into **prostatic**, **membranous** (*intermediate part of urethra*), and **spongy** regions (**Figures 27.1**, p. 709, and **27.9**, p. 718). In males, the urethra is a passageway used by both the urinary and reproductive systems.

The Accessory Glands

▶ **KEY POINT** Seminal fluid is a mixture of secretions from many different glands, each with distinctive biochemical characteristics.

In addition to the seminiferous tubules and the epididymis, the seminal glands, prostate, and bulbo-urethral glands contribute distinctive fluids to the semen (**Figure 27.8a**). Major functions of these glands include (1) activating the sperm, (2) providing the nutrients sperm need for motility, and (3) producing buffers that counteract the acidity of the urethral and vaginal environments.

The Seminal Glands

Each ductus deferens ends at the junction between the ampulla and the excretory duct draining the seminal gland on that side. The **seminal glands**, or **seminal vesicles**, are embedded in connective tissue on either side of the midline, sandwiched between the posterior wall of the urinary bladder and the anterior wall of the rectum. Each seminal gland is a tubular gland, around 15 cm (6 in.) in length. Each tubule has many short side branches, and the entire gland is coiled and folded into a compact, tapered mass roughly 5 cm by 2.5 cm (2 in. by 1 in.) (**Figure 27.9a**).

The seminal glands are lined with a pseudostratified columnar epithelium (**Figure 27.8c**). These are extremely active secretory glands, contributing about 60 percent of the volume of semen. Although the fluid has the same osmotic concentration as blood plasma, the composition is quite different. The secretion of the seminal glands contains prostaglandins, clotting proteins, and relatively high concentrations of fructose, which is easily metabolized by sperm to produce ATP.

Seminal fluid is discharged into the ductus deferens during ejaculation, when peristaltic contractions are under way in the ductus deferens, seminal glands, and prostate. These contractions are under the control of the sympathetic nervous system. When mixed with the secretions of the seminal glands, previously inactive but functional spermatozoa begin beating their flagella, becoming highly mobile.

The Prostate

The **prostate** is a small, muscular, rounded organ about 4 cm (1.6 in.) in diameter. The prostate encircles the prostatic urethra as it leaves the urinary bladder (**Figures 27.1, 27.8a**, and **27.9a**). The prostate contains 30–50 compound tubulo-alveolar glands (**Figure 27.8d**). The epithelial lining of these glands is generally a simple columnar epithelium. However, it may vary from a simple squamous to simple cuboidal to pseudostratified columnar epithelium. These glands are surrounded and wrapped in a thick blanket of smooth muscle fibers.

The prostate produces **prostatic fluid**, a weakly acidic secretion that contributes 20–30 percent of the volume of semen. This solution is rich in enzymes that nourish and prevent sperm coagulation in the vagina and contains **seminalplasmin** (sem-i-nal-PLAZ-min), an antibiotic that may help prevent urinary tract infections in males. Peristaltic contractions of the muscular wall eject these secretions into the prostatic urethra.

The Bulbo-urethral Glands

The paired **bulbo-urethral glands** (*Cowper's glands*) are located at the base of the penis (**Figures 27.8a** and **27.9a**). The bulbo-urethral glands are small, round glands, with diameters approaching 10 mm (less than 0.5 in.). The duct of each gland parallels the spongy urethra for 3–4 cm (1.2–1.6 in.) before emptying into the urethral lumen. The glands and ducts are lined by an epithelium that varies between simple cuboidal and simple columnar epithelium. These are compound tubulo-alveolar mucous glands that secrete a thick, sticky, alkaline mucus (**Figure 27.8e**). This secretion neutralizes any urinary acids that may remain in the urethra and lubricates the tip of the penis.

Figure 27.8 The Ductus Deferens and Accessory Glands

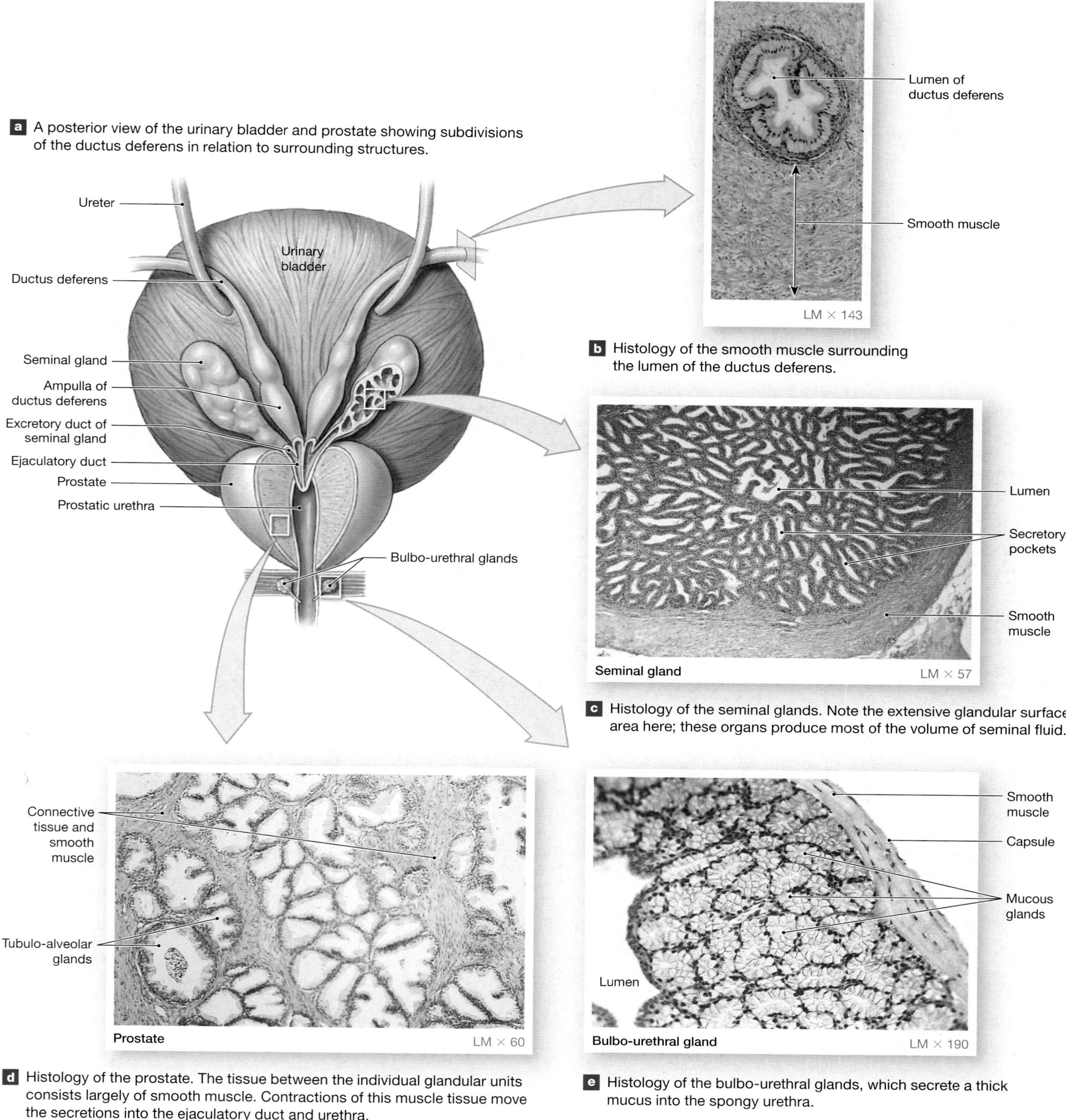

a A posterior view of the urinary bladder and prostate showing subdivisions of the ductus deferens in relation to surrounding structures.

b Histology of the smooth muscle surrounding the lumen of the ductus deferens.

c Histology of the seminal glands. Note the extensive glandular surface area here; these organs produce most of the volume of seminal fluid.

d Histology of the prostate. The tissue between the individual glandular units consists largely of smooth muscle. Contractions of this muscle tissue move the secretions into the ejaculatory duct and urethra.

e Histology of the bulbo-urethral glands, which secrete a thick mucus into the spongy urethra.

Figure 27.9 The Penis

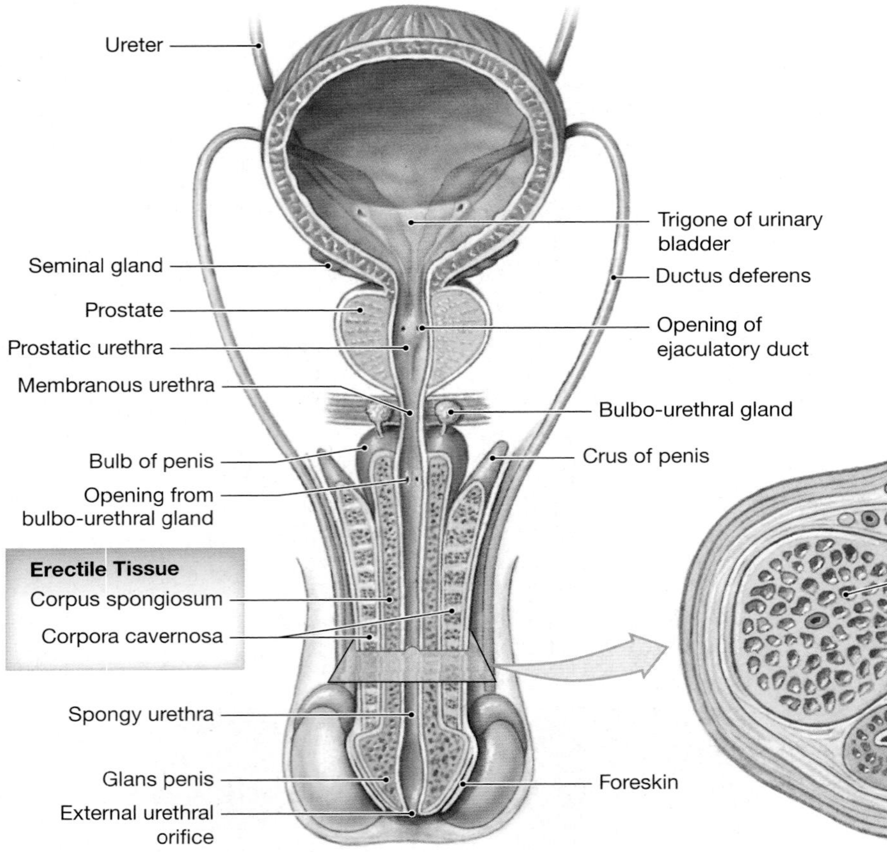

Ureter
Trigone of urinary bladder
Seminal gland
Ductus deferens
Prostate
Opening of ejaculatory duct
Prostatic urethra
Membranous urethra
Bulbo-urethral gland
Bulb of penis
Crus of penis
Opening from bulbo-urethral gland

Erectile Tissue
Corpus spongiosum
Corpora cavernosa

Spongy urethra
Glans penis
Foreskin
External urethral orifice

a Frontal section of the penis

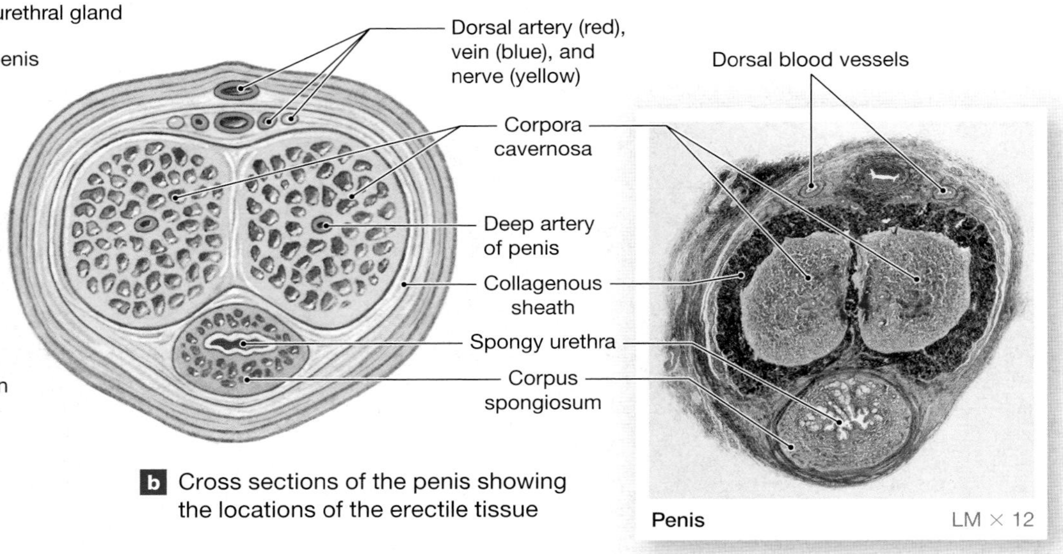

Dorsal artery (red), vein (blue), and nerve (yellow)
Corpora cavernosa
Deep artery of penis
Collagenous sheath
Spongy urethra
Corpus spongiosum

Dorsal blood vessels
Penis
LM × 12

b Cross sections of the penis showing the locations of the erectile tissue

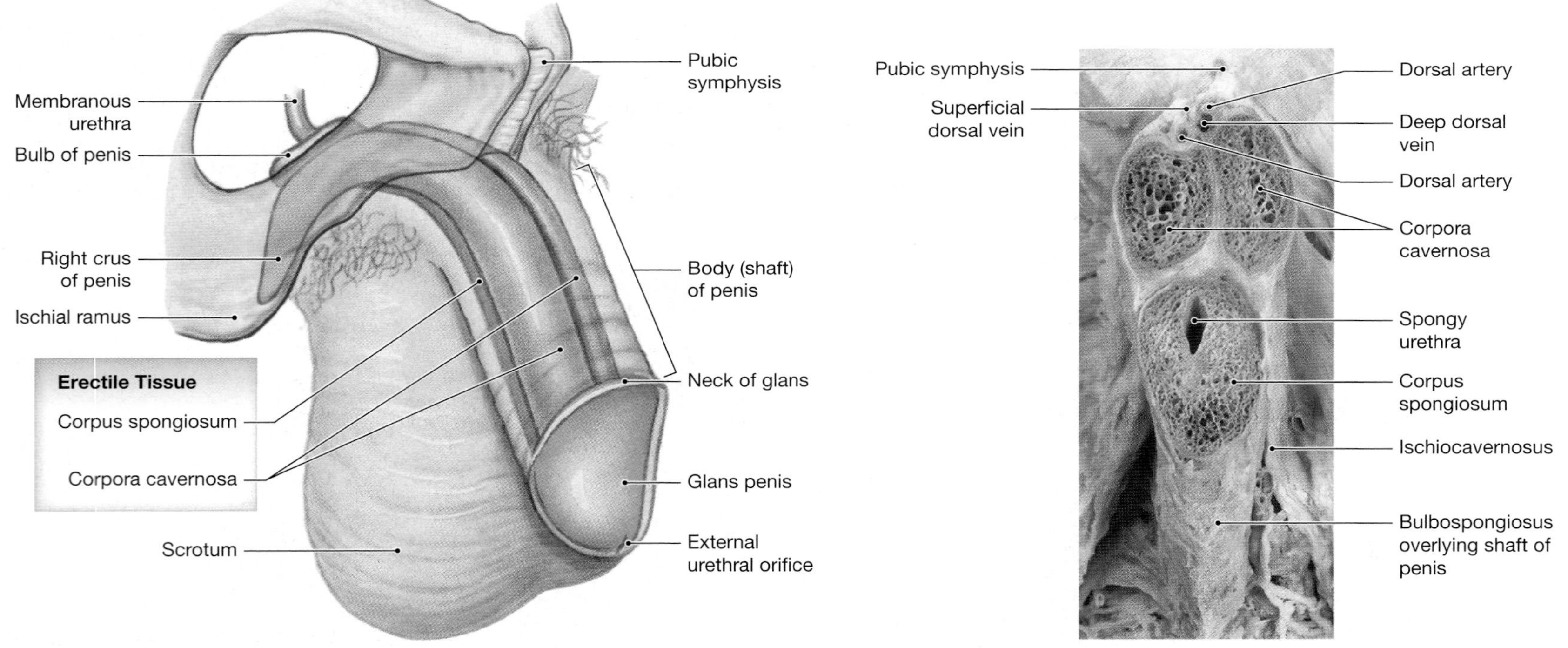

c Lateral and oblique view of the penis showing the orientation of the erectile tissues

d A cross section through the shaft of the penis

27

Semen

▶ **KEY POINT** Semen contains sperm, seminal fluid, and various enzymes.

A typical ejaculation releases 2–5 mL of **semen** (*ejaculate*). Semen contains the following:

1. Sperm: A normal **sperm count** ranges from 20 million to 100 million sperm per milliliter of semen.
2. Seminal fluid: The fluid component of semen is called **seminal fluid**. It is a mixture of glandular secretions with a distinctive ionic and nutrient composition. Seminal fluid contains the combined secretions of the seminal glands (60 percent), the prostate (30 percent), the nurse cells and epididymis (5 percent), and the bulbo-urethral glands (5 percent).
3. Enzymes: Several important enzymes are present in the seminal fluid, including (a) a protease that may help dissolve mucus in the vagina, (b) seminalplasmin, an antibiotic enzyme that kills a variety of bacteria, including *Escherichia coli*, (c) a prostatic enzyme that converts fibrinogen to fibrin, coagulating the semen within a few minutes after ejaculation, and (d) fibrinolysin, which liquefies the clotted semen after 15–30 minutes.

The Penis

▶ **KEY POINT** The penis is divided into three regions and contains the distal portion of the male urethra.

The **penis** is a tubular organ for sexual intercourse and conducting urine to the exterior. Main structures include the root, body, and glans penis (**Figure 27.9a,c**):

- The **root** of the penis attaches the penis to the rami of the ischia. This connection occurs within the urogenital triangle immediately inferior to the pubic symphysis.
- The **body** (**shaft**) of the penis is the tubular, movable portion. Masses of erectile tissue are found within the body.
- The **glans penis** is the expanded distal end surrounding the external urethral orifice.

The skin overlying the penis is generally hairless and contains more pigmentation than skin elsewhere on the body. The dermis contains a layer of smooth muscle, and the underlying loose connective tissue allows the skin to move without distorting underlying structures. The subcutaneous layer also contains superficial arteries, veins, and lymphatics, but relatively few fat cells.

A fold of skin, the **foreskin** or **prepuce** (PRĒ-pūs), surrounds the tip of the penis. The foreskin attaches to the narrow **neck of the glans** and continues distally over the glans penis. There are no hair follicles on the opposing surfaces. **Preputial** (prē-PŪ-shē-al) **glands** in the skin of the neck of the glans and the inner surface of the foreskin secrete a waxy material known as **smegma** (SMEG-ma). Unfortunately, smegma is an excellent nutrient source for bacteria. Mild inflammation and infections in this region are common, especially if the area is not washed thoroughly and frequently. One way to avoid such problems is **circumcision** (ser-kum-SIZH-un), the surgical removal of the foreskin.

Deep to the loose connective tissue underlying the dermis, a dense network of elastic fibers encircles the internal structures of the penis. Most of the body of the penis consists of three parallel cylindrical columns of **erectile tissue** (**Figure 27.9**). This erectile tissue is composed of a maze of vascular channels incompletely separated by partitions of elastic connective tissue and smooth muscle fibers. In the resting state, the arterial branches are constricted and the muscular partitions are tense. This combination restricts blood flow into the erectile tissue. During **erection**, the penis stiffens and elevates to an upright position, but the flaccid (nonerect) penis hangs inferior to the pubic symphysis and anterior to the scrotum.

On the posterior surface of the penis, the two cylindrical **corpora cavernosa** (KŌR-pōr-a ka-ver-NŌ-sa) are encircled by a dense collagenous sheath and separated by a thin septum. The corpora cavernosa diverge at their bases, forming the **crus of the penis**. Each crus is bound to the ischial ramus and pubis by tough connective tissue ligaments. The corpora cavernosa extend along the length of the penis as far as its neck. The erectile tissue within each corpus cavernosum surrounds a central artery, the **deep artery of the penis** (**Figure 27.9a–c**).

The **corpus spongiosum** (spon-jē-Ō-sum) surrounds the spongy urethra. This erectile body extends from the superficial fascia to the tip of the penis, where it expands to form the glans penis. The sheath surrounding the corpus spongiosum contains more elastic fibers than does that of the corpora cavernosa, and the erectile tissue contains a pair of small arteries. The **bulb of the penis** is the thickened, proximal end of the corpus spongiosum.

After erection has occurred, semen release involves a two-step process. During **emission** the sympathetic nervous system coordinates peristaltic contractions that sweep from the ductus deferens to the seminal glands to the prostate to the bulbo-urethral glands. These contractions mix the fluid components of the semen within the male reproductive tract. **Ejaculation** then occurs, the result of powerful, rhythmic contractions beginning in the ischiocavernosus and bulbospongiosus, two muscles of the pelvic floor. The ischiocavernosus inserts along the sides of the penis, and their contractions stiffen the organ. The bulbospongiosus wraps around the base of the penis, and its contraction pushes semen toward the external urethral orifice. Reflexes involving the inferior lumbar and superior sacral segments of the spinal cord control these contractions.

27.2 CONCEPT CHECK

2 What structures make up the body of the spermatic cord?
3 Why are inguinal hernias relatively common in males?
4 How is the location of the testes (outside the peritoneal cavity) important to the production of viable sperm?
5 What is the function of the blood testis barrier?
6 Trace the path of a developing sperm from the time it becomes detached from the nurse cells until it is released from the body.

See the blue Answers tab at the back of the book.

EMBRYOLOGY SUMMARY

For a summary of the development of the male reproductive system, see Chapter 28 (Embryology and Human Development).

27.3 Anatomy of the Female Reproductive System

▶ **KEY POINT** The female reproductive system produces oocytes, protects and supports a developing embryo and fetus, and nourishes the newborn infant. The ovaries are the female gonads.

The main structures of the female reproductive system are shown in **Figure 27.10**. Female oocytes form in the **ovaries**. They then leave the ovaries and travel along the **uterine tubes** (*Fallopian tubes*), where fertilization may occur.

Figure 27.10 The Female Reproductive System. The female pelvis and perineum in sagittal section.

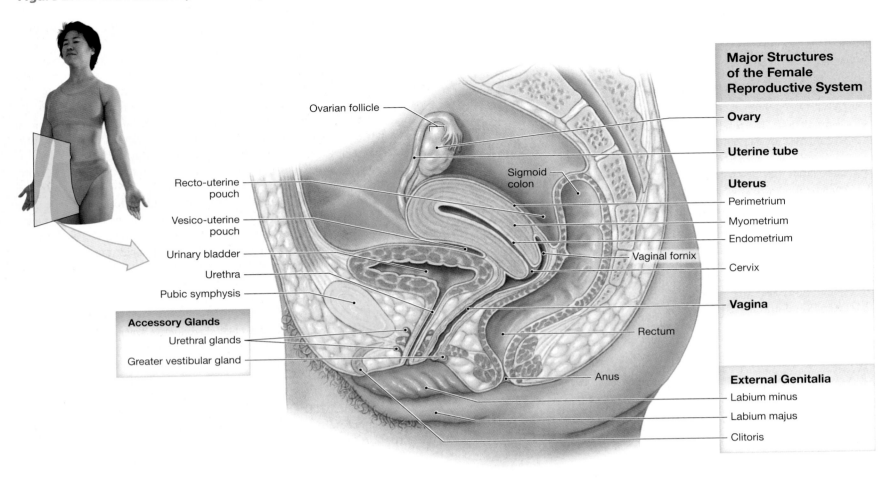

Ovarian follicle
Recto-uterine pouch
Vesico-uterine pouch
Urinary bladder
Urethra
Pubic symphysis
Sigmoid colon
Vaginal fornix
Rectum
Anus

Accessory Glands
Urethral glands
Greater vestibular gland

Major Structures of the Female Reproductive System

Ovary

Uterine tube

Uterus
Perimetrium
Myometrium
Endometrium
Cervix

Vagina

External Genitalia
Labium minus
Labium majus
Clitoris

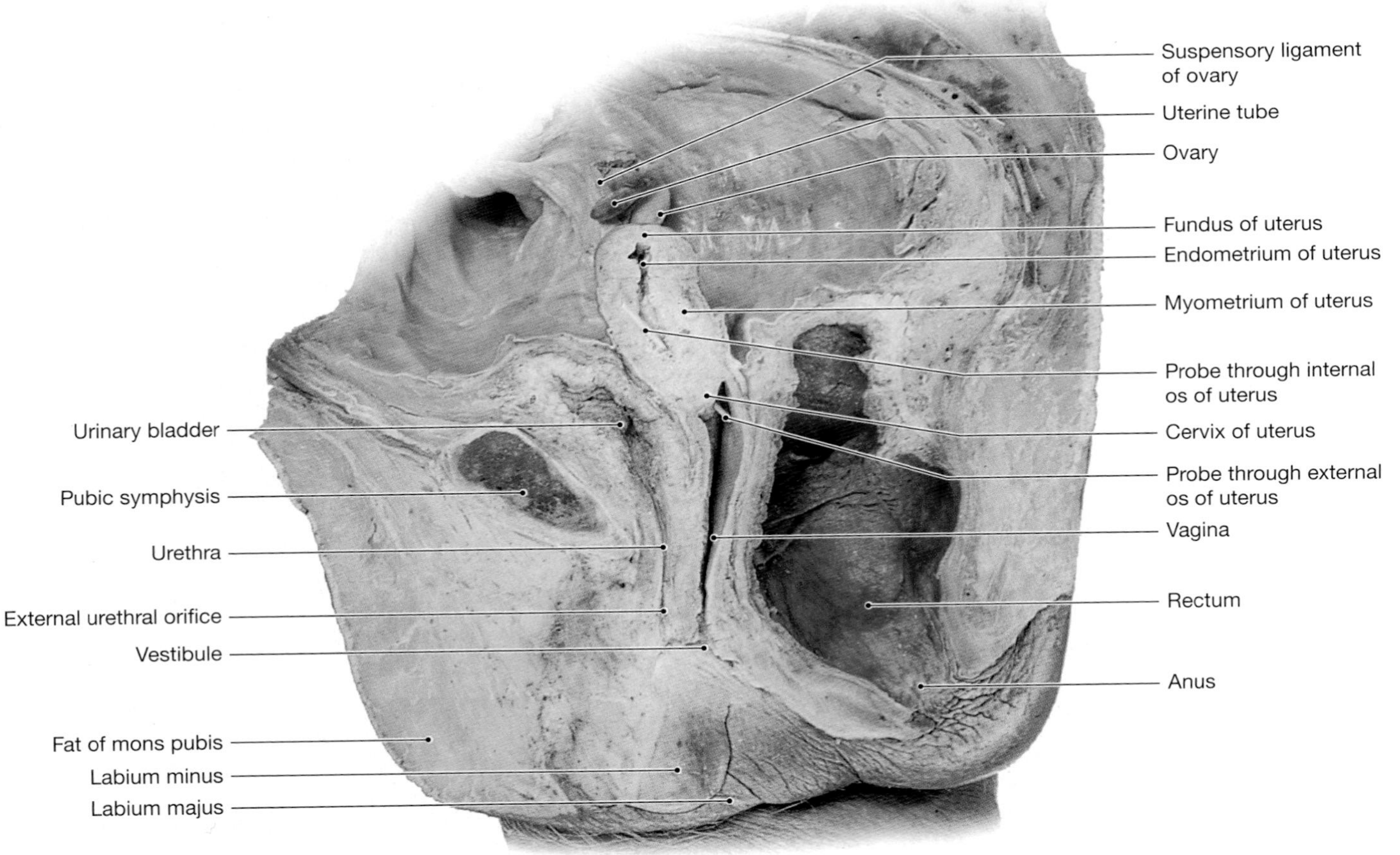

27

The oocytes eventually reach the **uterus**. The uterus opens into the **vagina**; the external opening of the vagina is surrounded by the female external genitalia. As in males, accessory glands secrete into the reproductive tract.

The **broad ligament** is a large mesentery enclosing the ovaries, uterine tubes, and uterus **(Figures 27.11** and **27.15)**. The uterine tubes are found along the superior border of the broad ligament. They open into the pelvic cavity lateral to the ovaries. The **mesosalpinx** (mez-ō-SAL-pinks) is the free edge of the broad ligament attached to each uterine tube. A thickened fold of the broad ligament, the **mesovarium** (mez-ō-VA-rē-um), supports and stabilizes the position of each ovary **(Figure 27.11a,b)**.

The broad ligament attaches to the sides and floor of the pelvic cavity, where it is continuous with the parietal peritoneum. The broad ligament subdivides the pelvic cavity. The pocket formed between the posterior wall of the uterus and the anterior surface of the colon is the **recto-uterine** (rek-tō-Ū-ter-in) **pouch**, while the pocket between the anterior wall of the uterus and the posterior wall of the urinary bladder is the **vesico-uterine** (ves-i-kō-Ū-ter-in) **pouch (Figures 27.10** and **27.11c)**. (These subdivisions are most apparent in sagittal section.)

Several other ligaments help support and stabilize the uterus and associated reproductive organs. These ligaments lie within the mesentery sheet of the broad ligament and are connected to the ovaries or uterus. The broad ligament limits side-to-side movement and rotation, and the other ligaments, discussed in the next section, prevent superior-inferior movement.

The Ovaries

▶**KEY POINT** The ovaries are small, paired organs located near the lateral walls of the pelvic cavity. The ovaries produce oocytes and secrete hormones.

A typical ovary is a flattened oval measuring approximately 5 cm long, 2.5 cm wide, and 8 mm thick (2 in. × 1 in. × 0.33 in.). Each ovary weighs 6–8 g (roughly 0.25 oz). These organs are responsible for producing oocytes and secreting hormones.

The ovaries have no peritoneal covering, but each ovary is stabilized by the mesovarium and by the ovarian ligament and suspensory ligament **(Figures 27.10, 27.11**, and **27.15)**. The **ovarian ligaments** extend from the lateral walls of the uterus, near the attachment of the uterine tube, to the medial surface of the ovary. The **suspensory ligament** extends from the lateral surface of the ovary past the open end of the uterine tube to the pelvic wall. The suspensory ligament contains the major blood vessels of the ovary: the **ovarian artery** and **ovarian vein**. These vessels are connected to the ovary at the **ovarian hilum**, where the ovary attaches to the mesovarium **(Figure 27.11)**.

Each ovary has a pink or yellowish coloration and a nodular consistency. Covering the surface of each ovary is a single layer of cuboidal epithelium called the **germinal epithelium**. It overlies a layer of dense connective tissue called the **tunica albuginea**. We divide the interior tissues, or **stroma**, of the ovary into a superficial cortex and a deeper medulla. Gametes are produced in the cortex **(Figure 27.11b)**.

The Ovarian Cycle and Oogenesis

Three interrelated processes help to ensure the continuation of our species from the female perspective: oogenesis, the ovarian cycle, and the uterine cycle. **Oogenesis** (ō-ō-JEN-e-sis) is the production of female gametes called oocytes. The **ovarian cycle** is the monthly series of events associated with the maturation of an oocyte. The **uterine cycle** involves a series of events that prepares the uterus for implantation of a fertilized oocyte. Oogenesis begins before birth, accelerates at puberty, and ends at menopause (end of menstruation). During fetal development, female reproductive stem cells called **oogonia** (ō-ō-GŌ-nē-uh; singular, *oogonium*) eventually produce diploid **primary oocytes** ($2n$). Specialized structures in the cortex of the ovaries called **ovarian follicles** (ō-VAR-ē-an FOL-i-klz) are the sites of both oocyte growth and oogenesis.

CLINICAL NOTE

Ovarian Cancer

Cancer of the ovary is not common, but it is deadly. Most often, it develops in the germinal epithelium. Currently, there is no screening examination for ovarian cancer. There are no symptoms in the early stage, and thus it often goes undetected until it has spread within the peritoneum. Advanced ovarian cancer is difficult to treat with a combination of surgery and chemotherapy. Signs and symptoms of advanced cancer include the following:

- Pain or a heavy feeling in the lower abdomen
- Abnormal bleeding from the vagina
- Abnormal periods
- Weight gain or loss
- Unexplained back pain that gets worse
- Gas, nausea, or loss of appetite

Women who have a full-term pregnancy before age 26 have a lower risk of ovarian cancer, and that risk goes down with each full-term pregnancy. Other factors that reduce a woman's risk include using oral contraceptives (birth control pills), having a tubal ligation (removing a piece of the uterine tubes), or having a hysterectomy (removing the uterus without removing the ovaries).

The risk of ovarian cancer increases with age. Most tumors develop after menopause. Like breast cancer, ovarian cancer can run in families, and the risk rises with an increased number of relatives with the disease. Inherited mutations in the genes BRCA1 and BRCA2, as well as other genes that are being studied, are linked to ovarian and breast cancer. BRCA1 is considered a "tumor suppressor" gene that makes proteins, which protect cells from growing abnormally. A BRCA1 gene mutation significantly increases a woman's lifetime risk of ovarian cancer. In comparison, the ovarian cancer lifetime risk for women with normal BRCA1 genes is less than 2 percent.

At puberty, rising levels of FSH trigger the start of the ovarian cycle. Each month thereafter, FSH stimulates some of the primordial follicles to undergo further development—a process called the ovarian cycle. Follicle development and the ovarian cycle are shown in **Figure 27.12**.

1. Primordial ovarian follicles: Primary oocytes are located in the outer portion of the ovarian cortex, near the tunica albuginea, in clusters called egg nests. A single squamous layer of follicle cells surrounds each primary oocyte within an egg nest. The primary oocyte and its follicle cells form a **primordial ovarian follicle**.
2. Formation of primary ovarian follicles: The ovarian cycle begins as activated primordial ovarian follicles develop into **primary ovarian follicles**, a process involving changes in the oocyte, follicle cells, and surrounding stroma. The follicular cells become cuboidal in shape and undergo repeated cell divisions. As a result, several layers of follicular cells develop around the oocyte. These follicle cells are called **granulosa cells**. The enlarging oocyte secretes proteins that form an extracellular coat called the **zona pellucida** (ZŌ-na pe-LŪ-si-da) around the oocyte. Granulosa cells continually provide the developing oocyte with nutrients. These cells possess microvilli that, along with microvilli from the oocyte, penetrate the zona pellucida.

Figure 27.11 The Ovaries, Uterine Tubes, and Uterus

Structures Stabilizing the Ovary
Ovarian ligament
Mesovarium
Suspensory ligament
Retractor
Ovarian artery and vein
Fimbriae
Uterine tube
Uterus
Infundibulum
Ovary
Broad ligament
Ureter
Uterosacral ligament
External os
Cervix
Vaginal rugae
Vaginal wall
Uterine tube
Ovarian hilum
Medulla
Corpus luteum
Cortex
Tunica albuginea
Mesenteries of the Ovary and Uterine Tube
Mesosalpinx
Mesovarium
Broad ligament
Egg nest
Germinal epithelium
Mature follicle

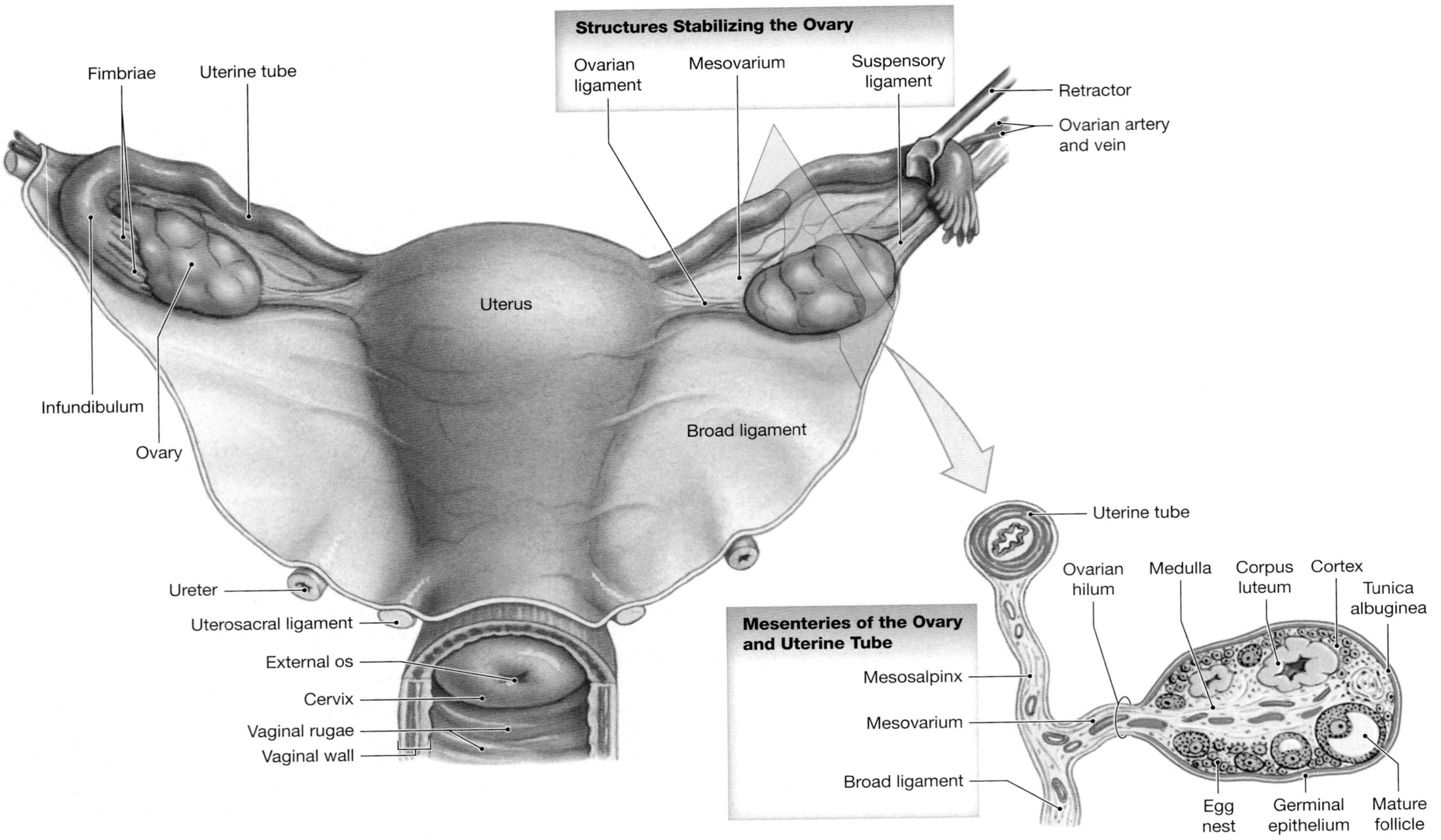

a Posterior view of the ovaries, uterine tubes, and uterus along with their supporting ligaments.

b Sectional view of the ovary, uterine tube, and associated mesenteries.

POSTERIOR
Supporting Ligaments of Uterus and Ovaries
Uterosacral ligament
Suspensory ligament of ovary
Cardinal ligaments (under broad ligament)
Ovarian ligament
Round ligament of uterus
Sigmoid colon
Ovary
Uterus
Urinary bladder
ANTERIOR
Uterine Tube
Infundibulum
Ampulla
Isthmus
Vesico-uterine pouch
Mesovarium
Recto-uterine pouch
Ovary
Ovarian ligament
POSTERIOR
Fundus of uterus
ANTERIOR

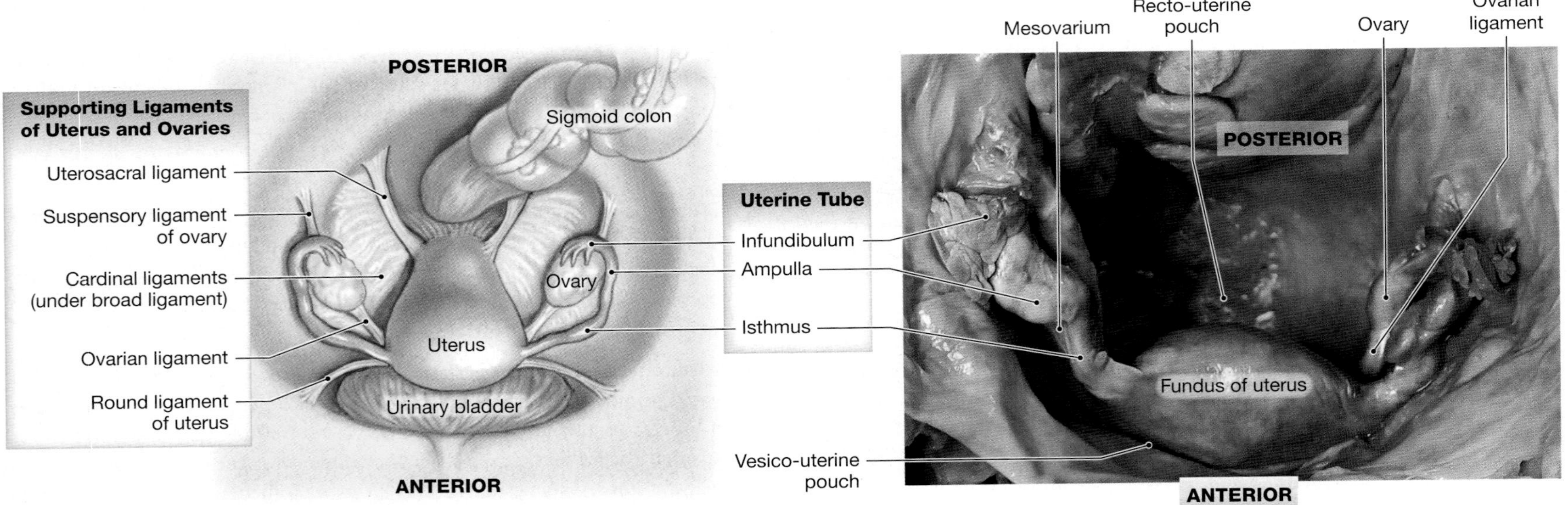

c Superior view of the female pelvic cavity showing supporting ligaments of uterus and ovaries. In the photo, the urinary bladder cannot be seen because it is covered by peritoneum.

27

Figure 27.12 Histological Summary of the Ovarian Cycle. Follicular development during the ovarian cycle. The nuclear events that occur in ovum production (oogenesis) are illustrated in **Figure 27.13**.

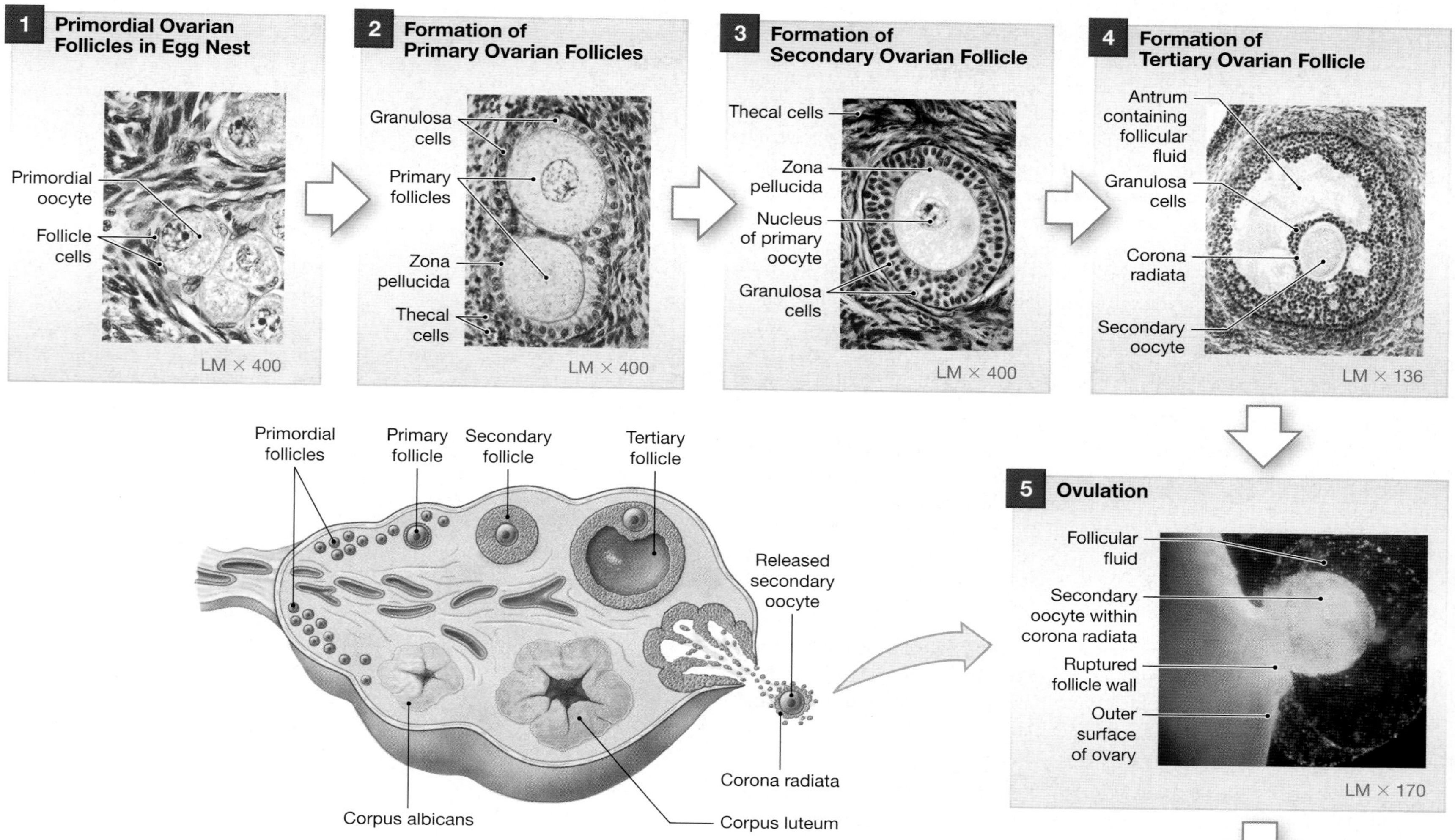

1 Primordial Ovarian Follicles in Egg Nest

Primordial oocyte
Follicle cells
LM × 400

2 Formation of Primary Ovarian Follicles

Granulosa cells
Primary follicles
Zona pellucida
Thecal cells
LM × 400

3 Formation of Secondary Ovarian Follicle

Thecal cells
Zona pellucida
Nucleus of primary oocyte
Granulosa cells
LM × 400

4 Formation of Tertiary Ovarian Follicle

Antrum containing follicular fluid
Granulosa cells
Corona radiata
Secondary oocyte
LM × 136

Primordial follicles
Primary follicle
Secondary follicle
Tertiary follicle
Released secondary oocyte
Corona radiata
Corpus albicans
Corpus luteum

5 Ovulation

Follicular fluid
Secondary oocyte within corona radiata
Ruptured follicle wall
Outer surface of ovary
LM × 170

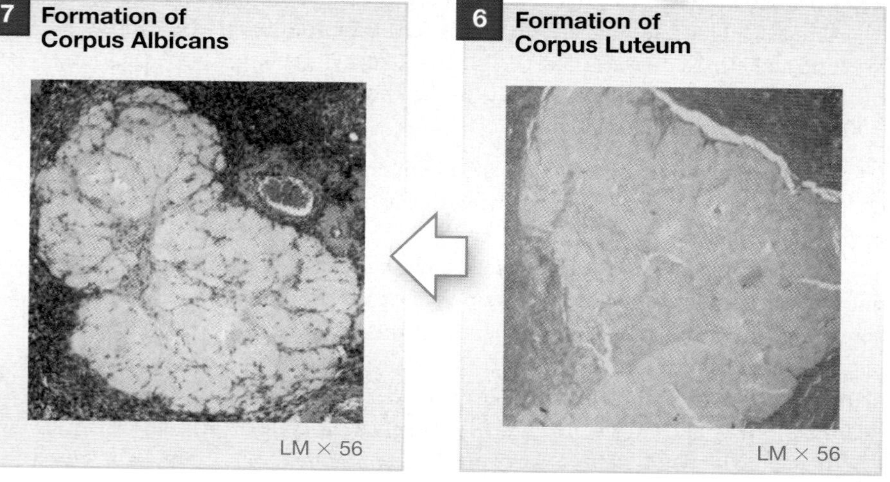

As the follicular cells enlarge and multiply, adjacent cells in the ovarian stroma form a layer of **thecal endocrine cells** around the follicle. Thecal endocrine cells and granulosa cells work together to produce the sex hormones called estrogens. The hormone **estradiol** (es-tra-DĪ-ol) is the most important estrogen, and it is the dominant hormone prior to ovulation. Estrogens have several important functions, including (a) stimulating bone and muscle growth, (b) maintaining female secondary sex characteristics, (c) affecting CNS activity, including sex-related behaviors and drives, (d) maintaining the function of reproductive glands and organs, and (e) initiating repair and growth of the uterine lining.

3 Formation of secondary ovarian follicles: Many primordial follicles develop into primary follicles, but only a few of the primary follicles mature further. This process is apparently under the control of a growth factor produced by the oocyte. The transformation begins as the wall of the follicle thickens. At this stage, the complex is known as a **secondary ovarian follicle**. The primary oocyte continues to grow slowly.

4 Formation of a tertiary ovarian follicle: During the next 2–3 months, the follicle wall continues to grow and the deeper follicular cells begin secreting follicular fluid. This fluid accumulates in small pockets that gradually expand between the inner and outer cellular layers of the follicle. The secondary follicle as a whole has doubled in size, and is now called a **tertiary follicle**. The tertiary follicle, also called a *vesicular follicle*, continues to grow and accumulate follicular fluid. Its large size creates a prominent bulge in the surface of the ovary. The oocyte and its covering of follicular cells project into the expanded central chamber of the follicle, the **antrum** (AN-trum). The antrum is surrounded by a mass of **granulosa cells**. The granulosa cells still associated with the secondary oocyte form a protective

27

layer known as the **corona radiata** (kō-RŌ-nuh rā-dē-AH-tuh). As the granulosa cells enlarge and multiply, adjacent cells in the ovarian stroma form a layer of **thecal endocrine cells** (*theca*, a box) around the follicle. Thecal endocrine cells and granulosa cells work together to produce the sex hormones called **estrogens**. Until this time, the primary oocyte has been suspended in prophase of the first meiotic division. That division is now completed. Although the nuclear events during oogenesis are identical to those in spermatogenesis, the cytoplasm of the primary oocyte is not evenly distributed **(Figure 27.13)**. Instead of producing two secondary oocytes, the first meiotic division yields a **secondary oocyte** and a small, nonfunctional **polar body**. The secondary oocyte proceeds to the metaphase stage of the second meiotic division. However, this stage will not be completed unless fertilization occurs. If fertilization does occur, the second meiotic division will be completed, producing an ovum and a nonfunctional polar body. Thus, instead of producing four equal-sized gametes, oogenesis produces a single ovum containing most of the cytoplasm of the primary oocyte and polar bodies that are simply containers for the "extra" chromosomes.

5 Ovulation: At **ovulation** (ōv-ū-LĀ-shun), the tertiary follicle releases the secondary oocyte. The distended follicular wall then ruptures, releasing the follicular contents, including the secondary oocyte, into the pelvic cavity. The sticky follicular fluid keeps the corona radiata attached to the surface of the ovary near the ruptured wall of the follicle. The oocyte is then moved into the uterine tube by contact with the fimbriae that extend from the tube's funnel-like opening, or by fluid currents produced by the cilia that line the tube.

Usually, only a single oocyte is released into the pelvic cavity at ovulation, although many primordial follicles may have developed into primary follicles, and several primary follicles may have been converted to secondary follicles. These follicles undergo follicular atresia.

6 Formation of corpus luteum: The empty tertiary follicle initially collapses, and ruptured vessels bleed. Under the stimulation of LH, the remaining granulosa cells then invade the area and proliferate to create the **corpus luteum**.

7 Formation of corpus albicans: The corpus luteum begins to degenerate about 12 days after ovulation (unless fertilization takes place). Progesterone and estrogen levels then decline markedly. Fibroblasts invade the nonfunctional corpus luteum, producing a knot of pale scar tissue called a **corpus albicans** (AL-bi-kanz).

The disintegration of the corpus luteum marks the end of the ovarian cycle. A new ovarian cycle then begins with the activation of another group of primordial follicles.

Another ovarian cycle begins immediately, because the decline in progesterone and estrogen levels that occurs as one cycle ends stimulates **gonadotropin-releasing hormone (GnRH)** production at the hypothalamus. This hormone triggers a rise in FSH and LH production by the anterior lobe of the pituitary gland, and this rise stimulates another period of follicle development.

The hormonal changes involved with the ovarian cycle affect the activities of other reproductive tissues and organs. In the uterus, the hormonal changes are responsible for the maintenance of the uterine cycle, which we discuss in a later section.

Figure 27.13 Meiosis and Ovum Production

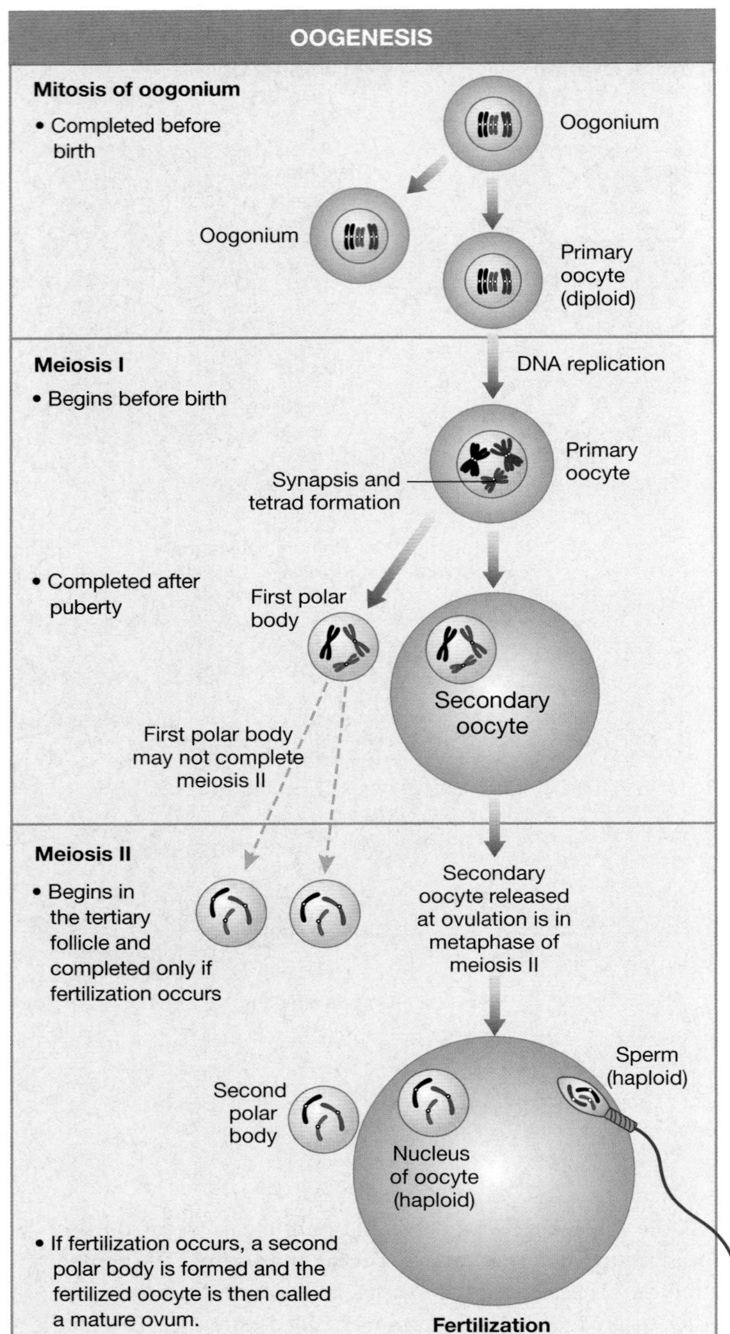

The Uterine Tubes

▶ **KEY POINT** The uterine tubes are hollow, muscular tubes that nourish the oocyte and transport it to the uterus. Each uterine tube is divided into four regions.

Each uterine tube is a hollow, muscular tube measuring roughly 13 cm (5 in.) in length **(Figures 27.10**, p. 720, **27.11**, p. 722, **27.14**, and **27.15)**. Each tube is divided into five regions:

1 **Fimbriae** are numerous fingerlike projections that drape over the ovary surface.

2 The **infundibulum** (in-fun-DIB-ū-lum) is the end closest to the ovary that forms an expanded funnel. The cells lining the inner surface of the infundibulum have cilia that beat toward the middle region of the uterine tube.

3 The **ampulla** is the middle region of the uterine tube. The smooth muscle layers in the wall of the ampulla thicken as we approach the uterus.

4 The **isthmus** (IS-mus) is a short region connecting the ampulla to the adjacent uterine wall.

5 The short **uterine part** is a continuation of the isthmus. It opens into the uterine cavity.

Histology of the Uterine Tube

The epithelium lining the uterine tube has both ciliated and nonciliated simple columnar cells. Concentric layers of smooth muscle surround the mucosa **(Figures 27.14b,c)**. Transport of materials along the uterine tube involves a combination of ciliary movement and peristaltic contractions in the walls of the uterine tube. A few hours before ovulation, sympathetic and parasympathetic nerves from the hypogastric plexus "turn on" this beating pattern. The uterine tubes transport the oocyte for final maturation and fertilization.

It normally takes 3–4 days for an oocyte to travel from the infundibulum to the uterine cavity. *If fertilization is to occur, the oocyte must encounter sperm during the first 12–24 hours of its passage.* Fertilization typically occurs in the ampulla of the uterine tube.

The uterine tube also provides a rich, nutritive environment containing lipids and glycogen. This mixture nourishes both sperm and a developing pre-embryo. Unfertilized oocytes degenerate in the terminal portions of the uterine tubes or within the uterus.

The Uterus

▶ **KEY POINT** The uterus provides physical protection, nutritional support, and waste removal for the developing embryo (weeks 1–8) and fetus (week 9 to delivery). Contractions in the muscular uterine wall help expel the fetus at the time of birth.

The empty uterus is a small, pear-shaped organ about 7.5 cm (3 in.) long with a maximum diameter of 5 cm (2 in.) **(Figures 27.10**, p. 720, **27.11**, p. 722, and **27.15)**. It weighs 30–40 g (1–1.4 oz). In its normal position, the uterus bends anteriorly near its base. This is called **anteflexion** (an-tē-FLEK-shun). In this position, the body of the uterus lies across the posterior and superior surfaces of the urinary bladder **(Figure 27.10)**. If the uterus bends backward toward the sacrum, it is termed **retroflexion** (re-trō-FLEK-shun). Retroflexion, which occurs in about 20 percent of adult women, has no clinical significance.

Suspensory Ligaments of the Uterus

In addition to the broad ligament, three pairs of suspensory ligaments stabilize the uterus: the uterosacral, round, and cardinal ligaments **(Figures 27.11** and **27.15a)**. The **uterosacral** (ū-ter-ō-SĀ-kral) **ligaments** extend from the lateral surfaces of the uterus to the anterior face of the sacrum, keeping the body of the uterus from moving inferiorly and anteriorly. The **round ligaments** begin at the lateral margins of the uterus just inferior to the bases of the uterine tubes. They extend anteriorly and pass through the inguinal canal before ending in the connective tissues of the external genitalia. These ligaments primarily restrict posterior movement of the uterus. The **cardinal ligaments** (also called the *transverse cervical ligaments*) extend from the base of the uterus and vagina to the lateral walls of the pelvis. These ligaments prevent inferior movement of the uterus. The skeletal muscles and fascia of the pelvic floor provide additional support.

Gross Anatomy of the Uterus

The uterine **body** is the largest region of the uterus **(Figure 27.15a)**. The **fundus** is the rounded portion of the body superior to the attachment of the uterine tubes. The body ends at a constriction known as the **isthmus** of uterus. The **cervix** (SER-viks) of the uterus is the inferior portion extending from the isthmus to the vagina.

Figure 27.14 The Uterine Tubes

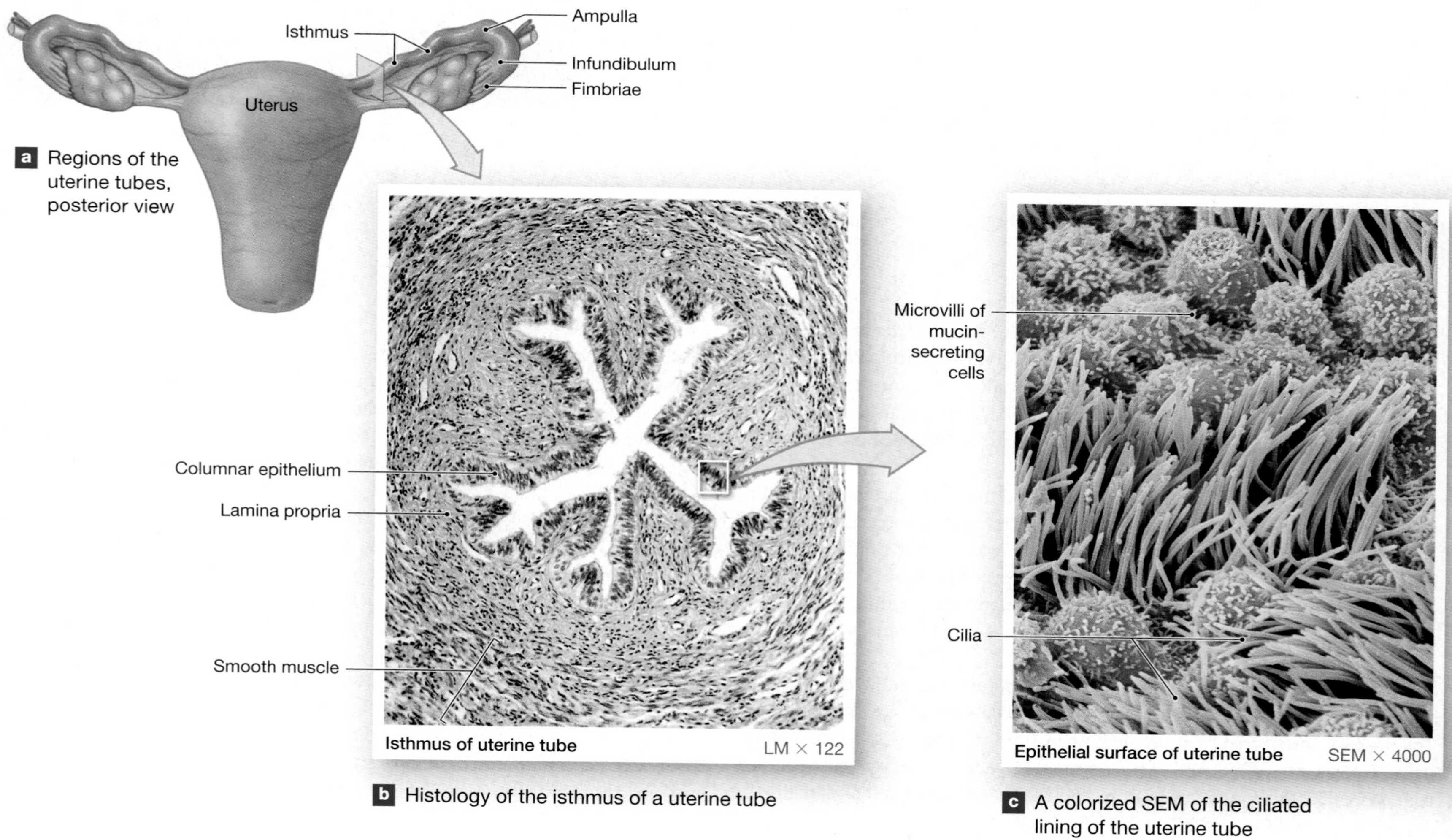

a Regions of the uterine tubes, posterior view

b Histology of the isthmus of a uterine tube

c A colorized SEM of the ciliated lining of the uterine tube

Figure 27.15 The Uterus

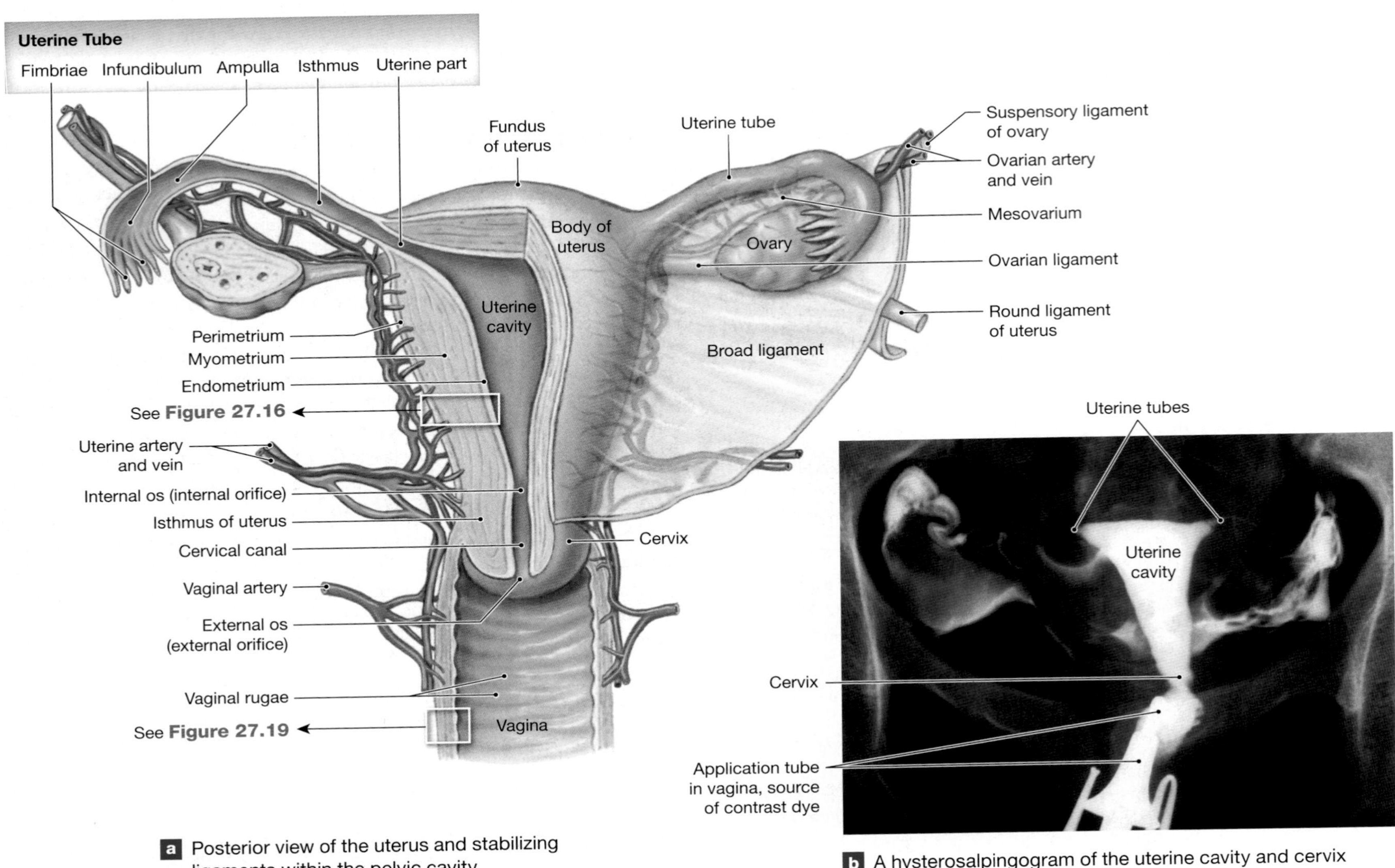

a Posterior view of the uterus and stabilizing ligaments within the pelvic cavity

b A hysterosalpingogram of the uterine cavity and cervix

The tubular cervix projects about 1.25 cm (0.5 in.) into the vagina. Within the vagina, the distal end of the cervix forms a curving surface surrounding the **external os** (*os*, opening or mouth), the external opening of the uterus. The external os leads into the **cervical canal**. The cervical canal is a constricted passageway that opens into the **uterine cavity** at the **internal os**, or internal opening (**Figure 27.15**).

The viscous mucus filling the cervical canal and covering the external os prevents the passage of bacteria from the vagina into the cervical canal. As ovulation approaches, the mucus becomes more watery. If the mucus does not change in consistency and remains viscous, sperm will have a difficult time entering the uterus, reducing the odds for successful fertilization. For this reason, treatment of female infertility may include drugs that make the cervical mucus more watery.

The uterus is supplied with branches of the uterine and ovarian arteries and veins. Numerous lymphatic vessels also supply each portion of the uterus. The uterus is innervated by autonomic fibers from the hypogastric plexus (sympathetic nervous system) and sacral segments S_3 and S_4 (parasympathetic nervous system). Sensory afferents from the uterus enter the spinal cord in the dorsal roots of spinal nerves $T_{11}$and T_{12}.

The Uterine Wall

The uterine wall has an outer, muscular **myometrium** (mī-ō-MĒ-trē-um; *myo-*, muscle, + *metra*, uterus) and an inner glandular **endometrium** (en-dō-MĒ-trē-um). The outer aspects of the fundus and the anterior and posterior surfaces of the uterine body are covered by a serous membrane continuous with the peritoneal lining. This incomplete serosal layer is called the **perimetrium** (**Figures 27.15a** and **27.16**).

Uterine glands open onto the endometrial surface. These glands extend deep into the lamina propria, almost all the way to the myometrium. Under the influence of estrogen, the uterine glands, blood vessels, and endothelium change with the various phases of the monthly uterine cycle. The glandular and vascular tissues of the endometrium support the physiological demands of a growing fetus.

The myometrium, the thickest portion of the uterine wall, is composed of three poorly defined layers of smooth muscle. The middle layer is the thickest and is composed of circularly or spirally arranged smooth muscle. The inner and outer layers are longitudinally arranged. The smooth muscle tissue of the myometrium provides much of the force needed to push a fetus out of the uterus and into the vagina.

Figure 27.16 The Uterine Wall

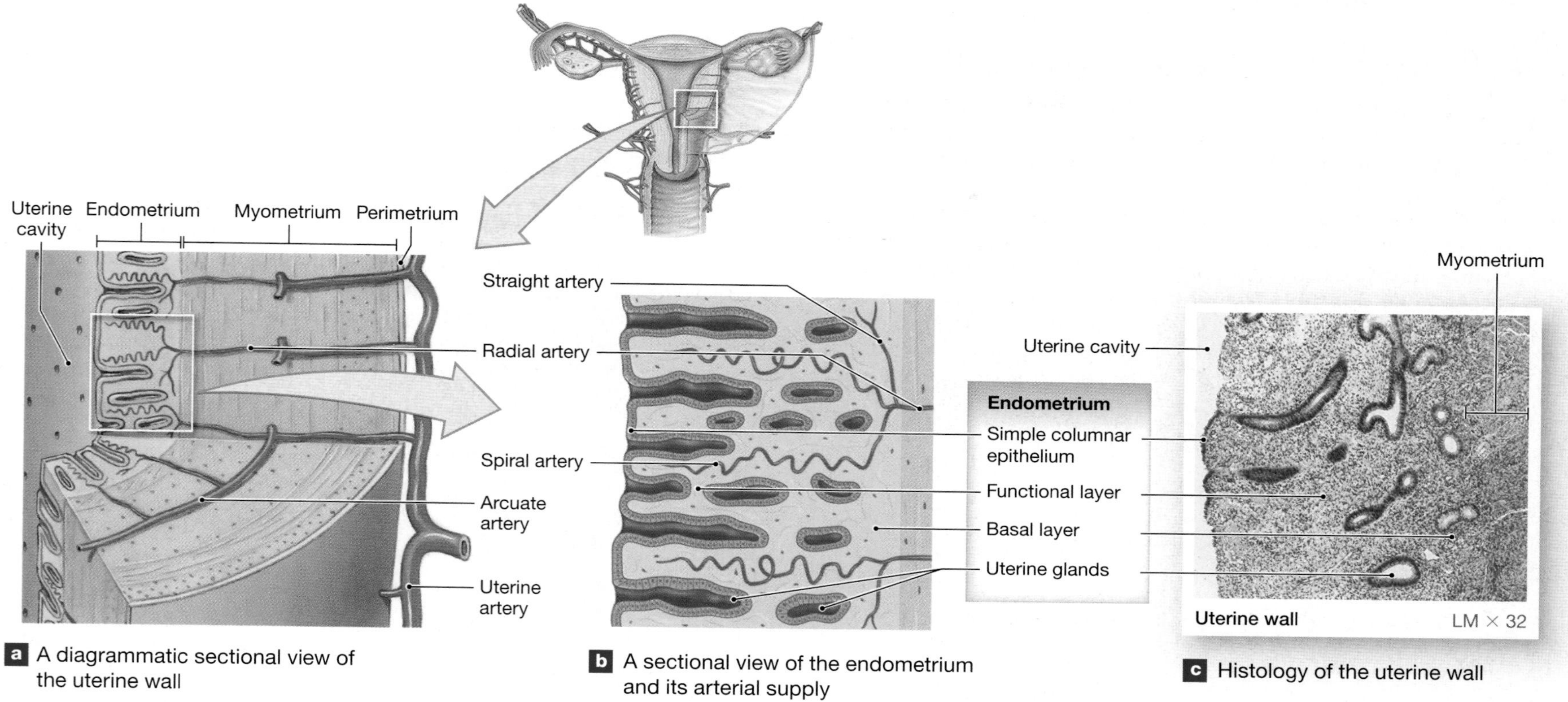

a A diagrammatic sectional view of the uterine wall

b A sectional view of the endometrium and its arterial supply

c Histology of the uterine wall

Blood Supply to the Uterus

The uterus receives blood from branches of the **uterine arteries**, which arise from branches of the internal iliac arteries, and the ovarian arteries, which arise from the abdominal aorta inferior to the renal arteries (**Figure 27.15a**). There are extensive interconnections among the arteries to the uterus. This arrangement helps guarantee a reliable flow of blood, despite changes in position and shape that accompany pregnancy.

Histology of the Uterus

The endometrium can be subdivided into an inner **functional layer**, the layer closest to the uterine cavity, and an outer **basil layer** adjacent to the myometrium. The functional layer contains most of the uterine glands and contributes most of the endometrial thickness. The basil layer attaches the endometrium to the myometrium and contains the terminal branches of the tubular glands (**Figure 27.16**).

Within the myometrium, branches of the uterine arteries form **arcuate arteries** that encircle the endometrium. **Radial arteries** branch from the arcuate arteries and supply both **straight arteries** and **spiral arteries** (**Figure 27.16a,b**). Straight arteries deliver blood to the basil layer of the endometrium, and spiral arteries supply the functional layer.

The structure of the basil layer remains relatively constant over time, but the structure of the functional layer undergoes cyclical changes in response to sexual hormone levels. These alterations produce the characteristic histological features of the uterine cycle.

The Uterine Cycle

The **uterine cycle**, or *menstrual* (MEN-strū-al) *cycle*, averages 28 days in length, but it can range from 21 to 35 days in normal individuals. The cycle is divided into three phases: (1) menstrual, (2) proliferative, and (3) secretory. The phases occur in response to hormones that regulate the ovarian cycle (**Figures 27.17** and **27.18**).

Menstrual phase The uterine cycle begins with the onset of the **menstrual phase**, an interval marked by the degeneration of the functional layer of the endometrium. The arteries nourishing the functional layer of the endometrium begin constricting, which weakens the arterial walls and reduces blood flow. As a result, the secretory glands and tissues of the functional layer begin to die. The weakened arterial walls rupture, and blood pours into the connective tissues of the functional layer. Blood cells and degenerating tissues break away from the endometrium and enter the uterine cavity to be shed as they pass through the external os and into the vagina. The process of endometrial sloughing, called **menstruation** (men-strū-Ā-shun) or **menses**, generally lasts from 1 to 7 days. During this time, about 35 to 50 ml (1.2–1.7 oz) of blood are lost (**Figure 27.17a**). Painful menstruation, or **dysmenorrhea**, may result either from uterine inflammation and contraction or from conditions involving adjacent pelvic structures.

Menstruation occurs when progestin and estrogen concentrations decrease at the end of the ovarian cycle. It continues until the next group of follicles has developed to the point where estrogen levels rise again (**Figure 27.18**).

The Proliferative Phase During menstruation, the circulatory supply to the basal layer of the endometrium, including the basal portions of the uterine glands, remains constant. As a result, the basil layer of the endometrium survives menstruation. In the days following the completion of menstruation, circulating levels of estrogens rise (**Figure 27.18**). Under the influence of these circulating estrogens, the epithelial cells of the glands multiply and spread across the endometrial surface. This restores the integrity of the uterine epithelium (**Figure 27.17b**). Continued growth and vascularization restore the functional layer of the endometrium. During the restoration, the endometrium is in the **proliferative phase**. This restoration occurs while the primary and secondary follicles are enlarging in the ovary. Estrogens secreted by the developing ovarian follicles stimulate and sustain the proliferative phase.

CLINICAL NOTE

Uterine Fibroids and Cancers

Uterine fibroids, the most common benign uterine tumors, are not cancerous. Fibroids form in the muscular myometrium. They may interfere with fertility and can cause heavy, painful periods or bleeding between periods.

Cervical cancer is one of the few cancers that is known to result from a viral infection. Specifically, certain types of human papillomavirus (HPV), transmitted very commonly through sexual contact, cause cervical cancer. If everyone could be vaccinated against HPV *before* becoming sexually active many cervical cancers could theoretically be eliminated. Early cervical cancer generally does not cause symptoms, but can be detected with a pelvic examination and a screening Pap smear (named after Dr. George Papanicolaou, an anatomist and cytologist). Pre-cancerous lesions and early cervical cancers have a good prognosis.

Endometrial cancer, the most common type of uterine cancer, arises within the endometrium. Most cases occur in women ages 50–70. Estrogen replacement therapy without the use of progesterone increases the risk for endometrial cancer. Post-menopausal bleeding is the most common sign and should always be investigated. A Pap smear may not detect endometrial cancer. If the cancer has not spread beyond the uterus and is treated with surgical hysterectomy, possibly with additional radiation therapy, the prognosis is good.

Figure 27.17 Histological Changes in the Uterine Cycle

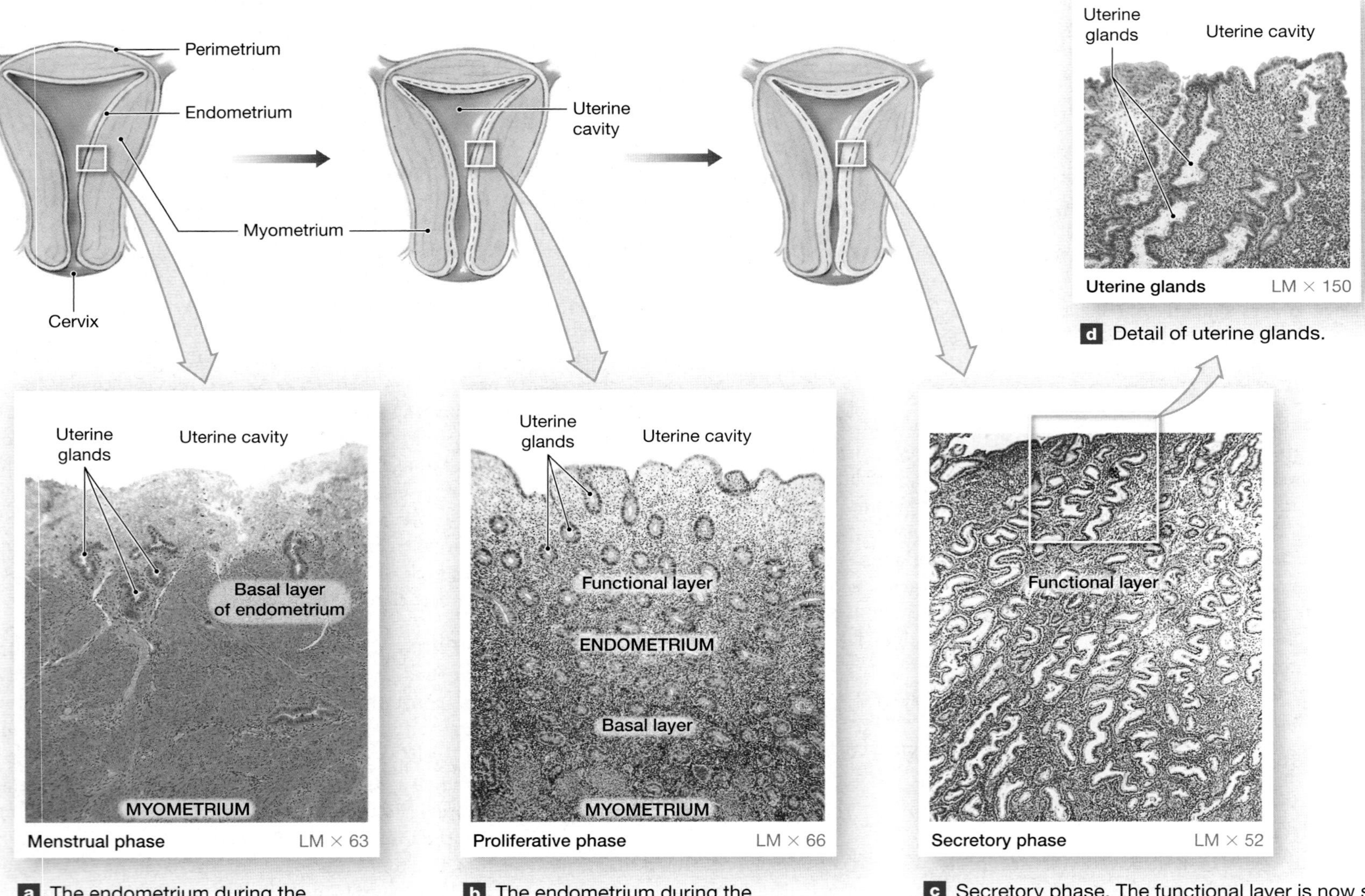

d Detail of uterine glands.

a The endometrium during the menstrual phase.

b The endometrium during the proliferative phase.

c Secretory phase. The functional layer is now so thick that at a magnification comparable to that of part (a) or part (b) you cannot capture the entire width of the endometrium in one image.

Figure 27.18 **The Hormonal Regulation of Female Reproductive Function**

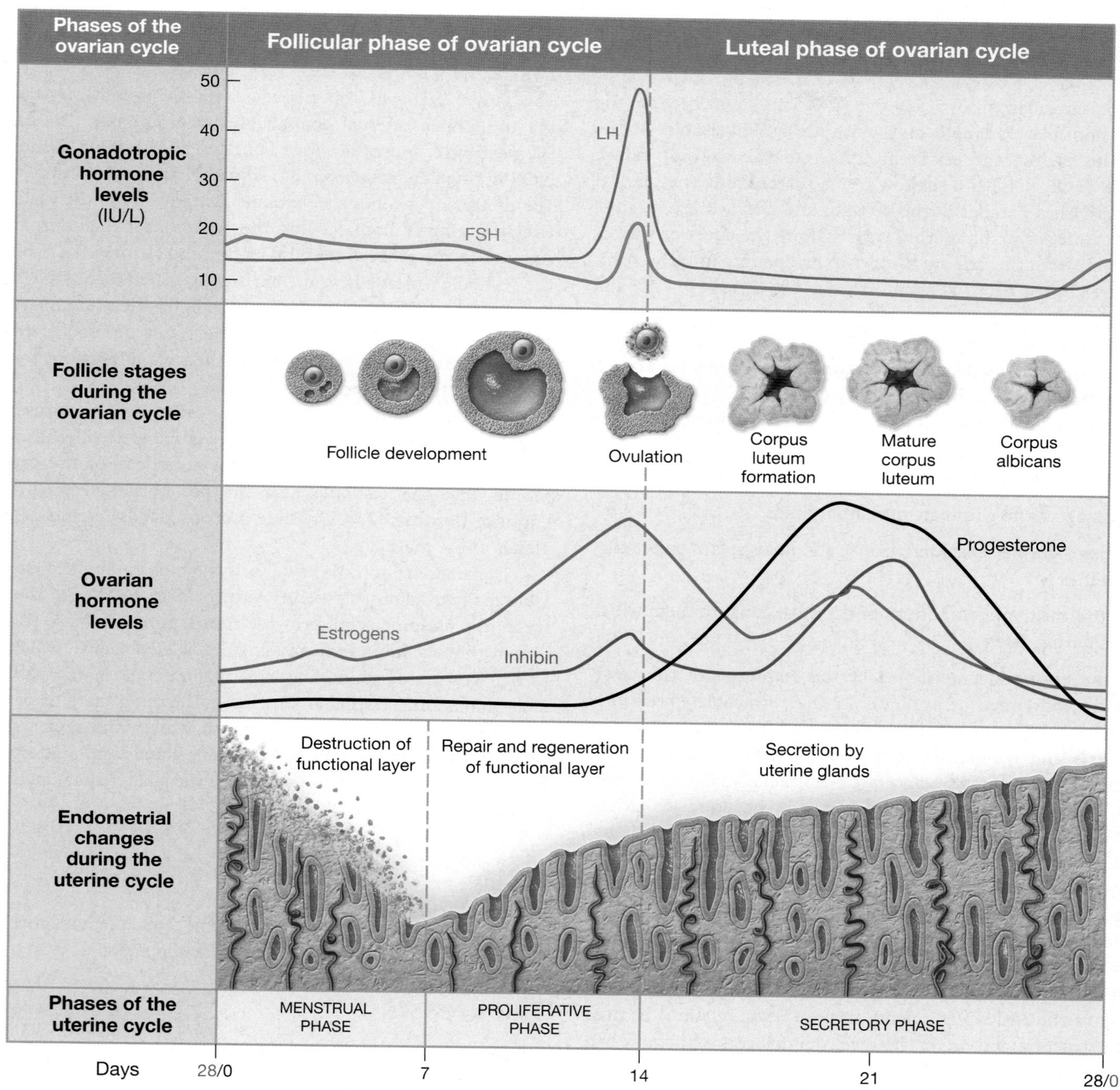

By the time ovulation occurs, the functional layer is several millimeters thick and highly vascularized. The mucous glands extend throughout the functional layer of the endometrium and are manufacturing a mucus rich in glycogen. The mucus is essential for the survival of the fertilized ovum through its earliest developmental stages.

The Secretory Phase During the **secretory phase** of the uterine cycle, the endometrial glands enlarge and accelerate their rate of secretion. Endometrial arteries elongate and spiral through the tissues of the functional layer. The combined stimulatory effects of progestins and estrogens from the corpus luteum cause this increased secretory activity **(Figures 27.17c** and **27.18)**. This phase begins at the time of ovulation and persists as long as the corpus luteum remains intact.

Endometrial secretory activities peak about 12 days after ovulation. Over the next day or two, glands become less active, and the uterine cycle ends as the corpus luteum stops producing stimulatory hormones. A new cycle then begins with the onset of menstruation and the disintegration of the functional layer. The secretory phase usually lasts 14 days. As a result, the date of ovulation can be determined after the fact by counting backward 14 days from the first day of menstruation.

Menarche and Menopause The first uterine cycle is termed **menarche** (me-NAR-kē), and usually occurs at age 11–12. The cycles continue until age 45–55, when **menopause** (MEN-ō-paws), the last uterine cycle, occurs. Between these two events, the uterine cycles are interrupted only by unusual circumstances such as illness, stress, starvation, or pregnancy. Irregular uterine cycles are characteristic of the first 2 years after menarche and the last 2 years before menopause.

The Vagina

▶ **KEY POINT** The vagina is an elastic, muscular tube extending from the cervix to the vestibule, a space bounded by the external genitalia.

The vagina has an average length of 7.5–9 cm (3–3.5 in.), but because the vagina is highly distensible, its length and width are quite variable. At the proximal end of the vagina, the cervix projects into the **vaginal canal**. The **vaginal fornix** (FOR-niks) is a shallow recess surrounding the cervical protrusion. The vagina lies parallel to the rectum, and the two are in close contact posteriorly. Anteriorly, the urethra travels along the superior wall of the vagina as it travels from the urinary bladder to its opening into the vestibule, a space bordered by the external genitalia (**Figures 27.10**, p. 720, and **27.11a**, p. 722).

The **vaginal branches** of the internal iliac arteries are the primary blood supply of the vagina. However, in some women, these vessels branch from the uterine arteries. Venous drainage is by the vaginal vein. Innervation is from the hypogastric plexus, sacral nerves S_2–S_4, and branches of the pudendal nerve.

The vagina has three major functions:

1. It serves as a passageway to eliminate menstrual fluids.
2. It receives the penis during sexual intercourse and holds sperm before they pass into the uterus.
3. In childbirth, it forms the lower portion of the birth canal through which the fetus passes during delivery.

The walls of the vagina are moistened by the secretions of the cervical glands and by the movement of water across the permeable epithelium. The muscular layer and the outer adventitial layer contain an extensive network of blood vessels and lymphatics (**Figures 27.15a** and **27.19**). The vagina and vestibule are separated by an elastic epithelial fold, the **hymen** (HĪ-men), which may partially or completely block the entrance to the vagina. The two bulbospongiosus muscles pass on either side of the vaginal orifice, and their contractions constrict the entrance. These muscles cover the bulb of the vestibule, a mass of erectile tissue on each side of the vaginal entrance (**Figure 27.20b**).

During development, the bulbs of the vestibule derive from the same embryonic tissues as the corpus spongiosum of the male (see the Embryology Summary in Chapter 28). The bulbs of the vestibule and corpus spongiosum are called homologous (ho-MOL-ō-gus; *homo*, same, + *logos*, relation) because they are similar in structure and origin; however, homologous structures can have very different functions.

The vagina contains a normal population of resident bacteria, and their growth is supported by the nutrients found in the cervical mucus. The metabolic activity of these bacteria creates an acid environment, which restricts the growth of many pathogenic organisms. An acid environment also inhibits sperm motility, and for this reason the buffers contained in seminal fluid are important for successful fertilization.

Histology of the Vagina

The vaginal lumen is lined by a mucosal stratified squamous epithelium (**Figure 27.19**) that, in the relaxed state, forms folds called **rugae** (**Figures 27.15a** and **27.19**). The underlying lamina propria is thick and elastic, and it contains small blood vessels, nerves, and lymph nodes. The vaginal mucosa is surrounded by an inner circular layer and an outer longitudinal layer of smooth muscle. Adjacent to the uterus, the vagina has a serosal covering continuous with the pelvic peritoneum. The rest of the vagina is covered by a highly elastic adventitia that blends in with the connective tissue of the surrounding structures.

The External Genitalia

▶ **KEY POINT** The vulva is composed of the vestibule, labia minora, clitoris, mons pubis, and labia majora.

The **vulva** (VUL-va), or *pudendum* (pū-DEN-dum), is the region enclosing the female external genitalia (**Figure 27.20a**). The vagina opens into the **vestibule**, a central space bounded by the **labia minora** (LĀ-bē-a mī-NOR-a; singular, *labium minus*). Smooth, hairless skin covers the labia minora. The urethra opens into the vestibule just anterior to the vaginal entrance. The **urethral glands** discharge into the urethra near the external urethral orifice. Anterior to the external urethral orifice, the **clitoris** (KLIT-ō-ris) projects into the vestibule. Internally, the clitoris contains erectile tissue homologous with the corpora cavernosa in males (see the Embryology Summary in Chapter 28). The clitoris engorges with blood during sexual arousal. A small erectile glans sits atop the organ, and extensions of the labia minora encircle the body of the clitoris, forming its **prepuce**.

A variable number of small **lesser vestibular glands** discharge their secretions onto the exposed surface of the vestibule, keeping it moistened. During arousal, a pair of ducts discharge the secretions of the **greater vestibular glands** into the vestibule near the posterolateral margins of the vaginal entrance (**Figure 27.10a**). These mucous glands resemble the bulbo-urethral glands of the male.

The anterior limit of the vulva is the mons pubis; the lateral limits are the labia majora. Adipose tissue beneath the skin anterior to the pubic symphysis forms the prominent bulge of the **mons pubis**. Adipose tissue also accumulates within the **labia majora** (singular, *labium majus*), which are homologous to the scrotum of males. The labia majora encircle and partially conceal the labia minora and vestibular structures. The outer margins of the labia majora are covered with the same coarse hair that covers the mons pubis, but the inner surfaces are relatively hairless. Sebaceous glands and scattered apocrine sweat glands secrete onto the inner surface of the labia majora, moistening and lubricating them.

EMBRYOLOGY SUMMARY

For a summary of the development of the female reproductive system, see Chapter 28 (Embryology and Human Development).

The Breasts

▶ **KEY POINT** Mammary glands—specialized accessory glands—are apocrine glands that produce milk to nourish a newborn infant.

At birth, the newborn infant cannot fend for itself, and several key systems have yet to complete development. Over the initial period of adjustment following birth, the infant gains nourishment from the milk secreted by the maternal **mammary gland** in each breast. Milk production, or **lactation** (lak-TĀ-shun), occurs in the mammary glands of the breasts, specialized accessory organs of the female reproductive system.

The breasts lie on either side of the chest in the subcutaneous tissue of the **pectoral fat pad** deep to the skin (**Figure 27.21a,b**). Each breast bears a small conical projection, the **nipple**. Here the ducts of the underlying mammary glands open onto the body surface. The region of reddish-brown skin surrounding each nipple is the **areola** (a-RĒ-ō-la). The presence of large sebaceous glands in the underlying dermis gives the areolar surface a granular texture.

The apocrine glandular tissue of the breast consists of a number of separate lobes, each containing several secretory lobules (**Figure 27.21a**). Ducts leaving the lobules converge, giving rise to a single **lactiferous** (lak-TIF-e-rus) **duct** in each lobe (**Figure 27.21a,c,d**). Near the nipple, that lactiferous duct

Figure 27.19 Histology of the Vaginal Wall

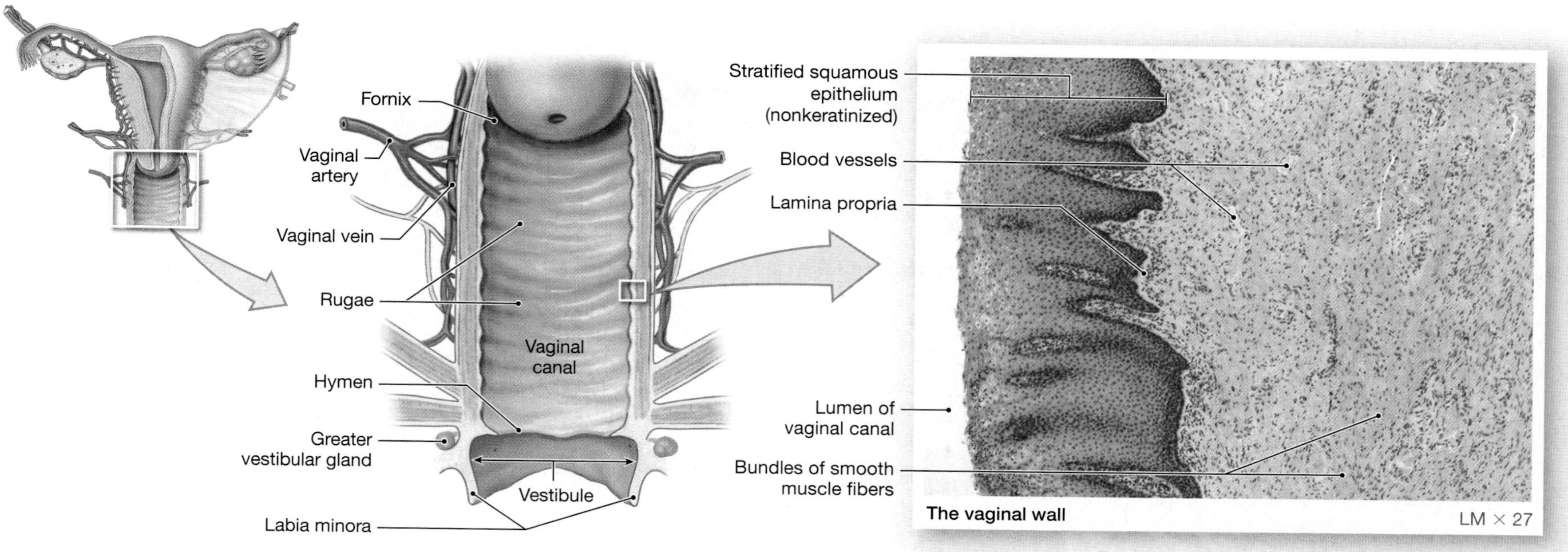

Fornix
Vaginal artery
Vaginal vein
Rugae
Vaginal canal
Hymen
Greater vestibular gland
Vestibule
Labia minora
Stratified squamous epithelium (nonkeratinized)
Blood vessels
Lamina propria
Lumen of vaginal canal
Bundles of smooth muscle fibers
LM × 27

The vaginal wall

Figure 27.20 The Female External Genitalia

External Genitalia
Mons pubis
Clitoris
Prepuce
Glans
Vestibule
Labia minora
Hymen (torn)
Labia majora
Urethral opening
Bulb of the vestibule
Vaginal entrance
Greater vestibular gland
Anus

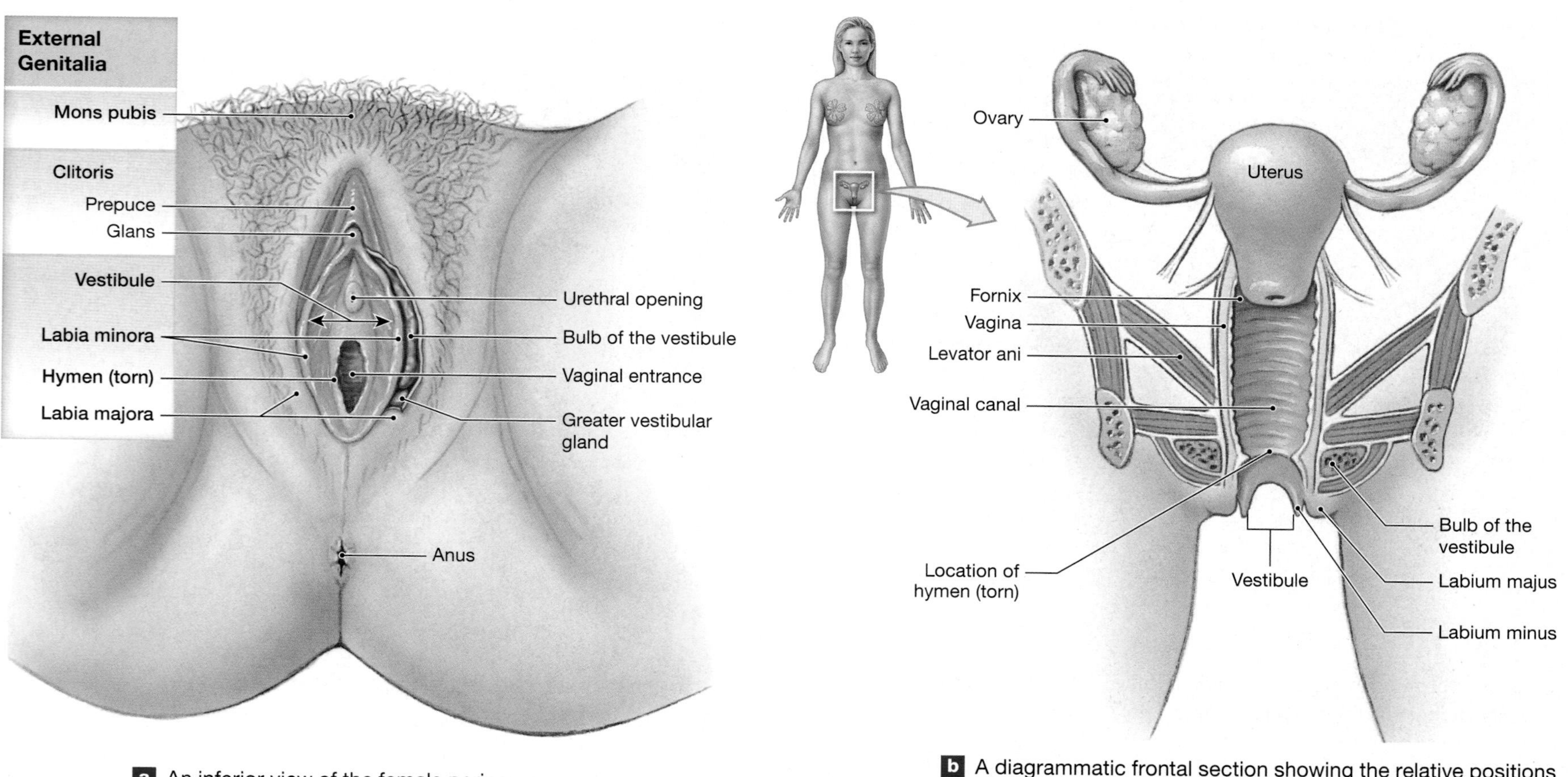

a An inferior view of the female perineum

Ovary
Uterus
Fornix
Vagina
Levator ani
Vaginal canal
Location of hymen (torn)
Vestibule
Bulb of the vestibule
Labium majus
Labium minus

b A diagrammatic frontal section showing the relative positions of the internal and external reproductive structures

Figure 27.21 The Mammary Glands

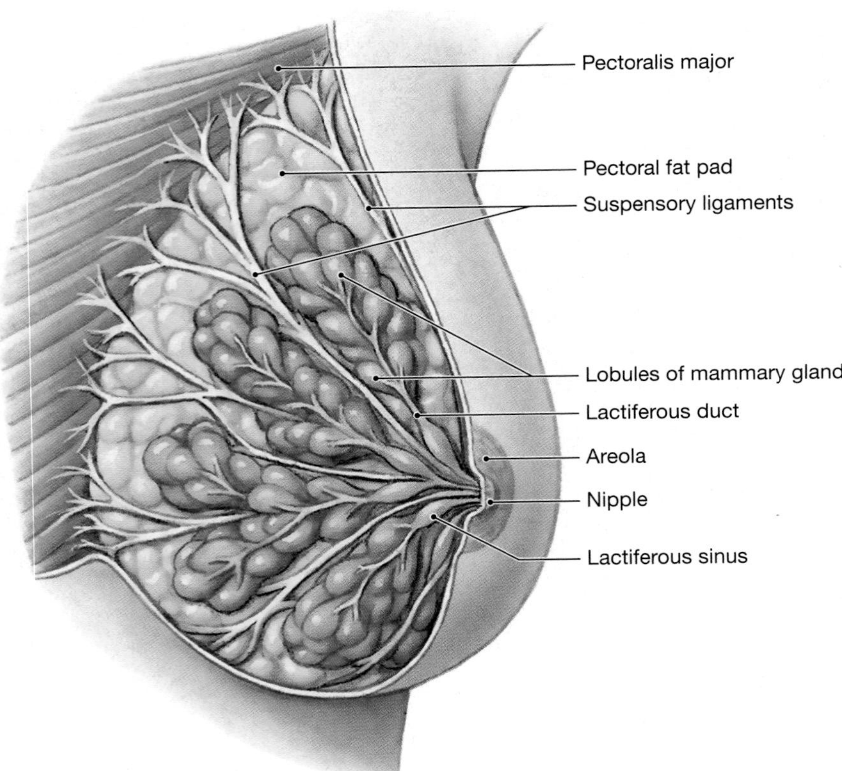

a Gross anatomy of the breast

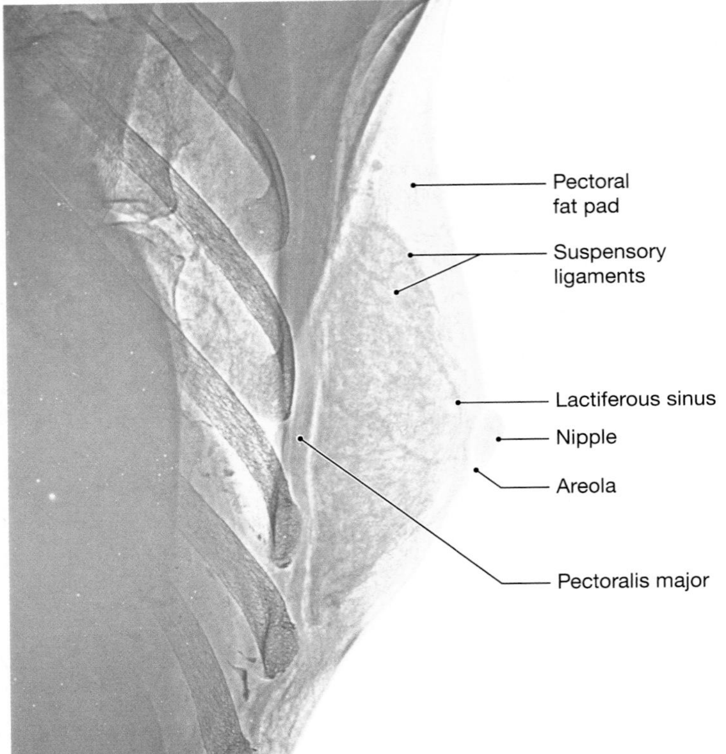

b Xeromammogram, a radiographic technique designed to show the tissue detail of the breast (mediolateral projection)

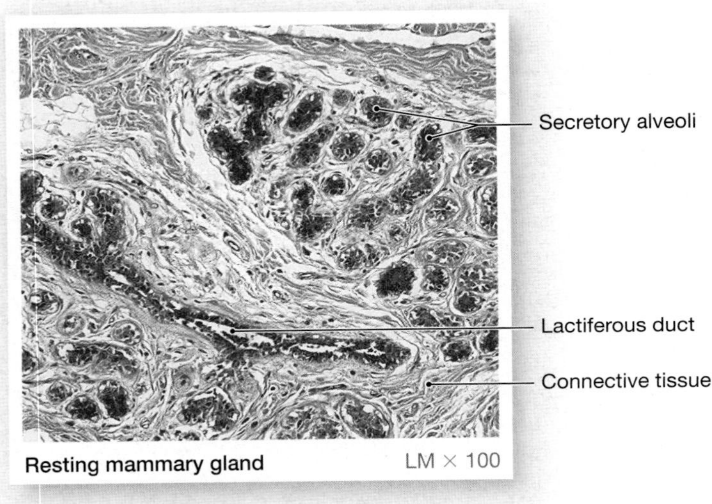

c Histology of a resting mammary gland

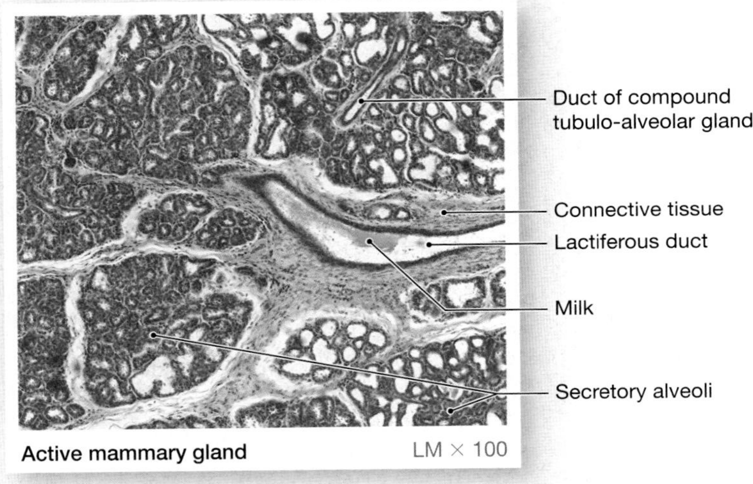

d Histology of an active mammary gland

expands, forming an expanded chamber called a **lactiferous sinus**. Fifteen to twenty lactiferous sinuses typically open onto the surface of each nipple. Dense connective tissue surrounds the duct system and forms partitions that extend between the lobes and lobules. These bands of connective tissue, known as the **suspensory ligaments of the breast**, originate in the dermis of the overlying skin. A layer of loose connective tissue separates the mammary complex from the underlying pectoralis muscles. Branches of the internal thoracic artery supply blood to each mammary gland. (Chapter 23 detailed the lymphatic drainage of the mammary gland.) ⮌ pp. 613–614

27

Development of the Mammary Glands During Pregnancy

Figure 27.21c,d compares the histological organization of resting and active mammary glands. The resting mammary gland is dominated by a duct system, rather than by active glandular cells. The size of the breast in a nonpregnant woman reflects the amount of adipose tissue in the breast rather than the amount of glandular tissue. The secretory apparatus does not develop until pregnancy occurs.

CLINICAL NOTE

Breast Cancer

One in eight American women will develop breast cancer at some time in her lifetime. Only 10 percent of breast cancers are due to inherited gene mutations, but if a woman has a BRCA1 or BRCA2 gene mutation, her lifetime risk of breast cancer rises significantly. Although rare, breast cancer can also afflict men.

Nobody knows what causes breast cancer, but maintaining a healthy weight, remaining physically active, and limiting alcohol intake can help guard against it. Regular screening exams, such as self-examination and mammography, are also important.

Signs of breast cancer include the following:

- A breast lump or thickening that feels different from surrounding tissue
- Bloody discharge from the nipple
- A change in the size, shape, or appearance of a breast
- Changes to the skin over the breast, such as dimpling, redness, flaking, or pitting (like the skin of an orange)
- A newly inverted nipple

Annual mammograms are recommended for women beginning in their mid-40s and continuing as long as a woman is in good health.

Most breast cancers begin in the lactiferous ducts (ductal carcinoma). Some start in the lobules (lobular carcinoma). Even though breasts come in different sizes, they have approximately the same number of ducts and lobules, with adipose tissue causing the differences in size.

Breast cancers are staged according to the size of the tumor, evidence of tumor spread to local lymph nodes, and evidence of distant metastases. Breast cancer spreads through the lymphatic system, so lymph nodes that drain the breast tissue are sampled to look for tumor cells. Breast tumors are graded based on how quickly they are likely to grow and spread. Additional tests look for proteins, known as receptors, in or on tumor cells that can attach to hormones. If the cancer cells are estrogen or progesterone receptor positive, they may respond to hormone therapy drugs. Researchers are now looking at gene patterns of breast tumors. All of these studies help direct treatment and determine prognosis.

Further mammary gland development requires a combination of hormones from the anterior lobe of the pituitary gland, including prolactin (PRL) and growth hormone (GH), and **human placental lactogen** (LAK-tō-jen) **(HPL)** from the placenta. ⮌ pp. 508–512 Under stimulation by these hormones, ducts of the mammary gland become mitotically active, and gland cells appear. By the end of the sixth month of pregnancy, the mammary glands are fully developed and begin producing secretions that are stored in the duct system. Milk is released when the infant begins to suck on the nipple. This stimulation causes the posterior lobe of the pituitary gland to release oxytocin. Oxytocin triggers the contraction of smooth muscles in the walls of the lactiferous ducts and sinuses, ejecting milk.

27.3 CONCEPT CHECK

7 What are the functions of the follicular cells?

8 How could scarring of the uterine tubes cause infertility?

9 During the proliferative phase of the uterine cycle, what activities occur in the ovarian cycle?

10 Would blockage of a single lactiferous sinus interfere with delivery of milk to the nipple? Explain.

11 What hormones stimulate lactation?

See the blue Answers tab at the back of the book.

27.4 | Aging and the Reproductive System

▶ **KEY POINT** The most striking age-related changes in the female reproductive system occur at menopause, while changes in the male reproductive system occur more gradually and over a longer period of time.

Menopause

Menopause is usually defined as the time when ovulation and menstruation cease. Menopause typically occurs at age 45–55, but in the years preceding it, the ovarian and menstrual cycles gradually become irregular. **Premature menopause** occurs before age 40. A shortage of primordial follicles is the underlying cause of premature menopause.

Menopause is accompanied by a rapid and sustained rise in the production of GnRH, FSH, and LH, while circulating concentrations of estrogen and progesterone decline. The decline in estrogen levels (1) reduces the size of the uterus, (2) reduces the size of the breasts, (3) thins the urethral and vaginal walls, and (4) weakens the connective tissue supporting the ovaries, uterus, and vagina. Reduced estrogen concentrations have also been linked to osteoporosis (brittle bones) and a variety of cardiovascular and neural effects, including "hot flashes," anxiety, and depression.

The Male Climacteric

Changes in the male reproductive system occur more gradually, over a period known as the **male climacteric** (klī-MAK-ter-ik). Circulating testosterone levels begin to decline between ages 50 and 60, coupled with increases in circulating levels of FSH and LH. Although sperm production continues (men can father children well into their 80s), there is a gradual reduction in sexual activity in older men, which may be linked to declining testosterone levels. Some clinicians suggest testosterone replacement therapy to enhance libido (sexual drive) in older men and women.

27.4 CONCEPT CHECK

12 A prolapsed uterus is a condition in which the position of the uterus changes, moving to a more inferior position within the pelvic cavity. Such a condition is more common in women who have undergone menopause. Why?

13 Why would some physicians recommend testosterone replacement therapy for males age 60 or older?

See the blue Answers tab at the back of the book.

Study Outline

Introduction p. 708

- The human reproductive system produces, stores, nourishes, and transports functional **gametes** (reproductive cells). **Fertilization** is the fusion of a **sperm** from the male and an immature **oocyte** from the female to form a **zygote** (fertilized egg).

27.1 Organization of the Reproductive System p. 708

- The reproductive system includes **gonads**, ducts, accessory glands and organs, and the **external genitalia**.
- In the male, the **testes** (gonads) produce sperm, which are expelled from the body in semen during **ejaculation**. The **ovaries** (gonads) of a sexually mature female produce an egg that travels along **uterine tubes** to reach the **uterus**. The **vagina** connects the uterus with the exterior.

27.2 Anatomy of the Male Reproductive System p. 708

- The **sperm** travel along the **epididymis**, the **ductus deferens**, the **ejaculatory duct**, and the **urethra** before leaving the body. Accessory organs (notably the **seminal glands**, **prostate**, and **bulbo-urethral glands**) secrete into the ejaculatory ducts and urethra. The **scrotum** encloses the testes, and the **penis** is an erectile organ. *(See Figures 27.1–27.9.)*

The Testes p. 708

- The testes hang within the scrotum, and each measures about 2 in. long and 1 in. in diameter.
- The **descent of the testes** through the inguinal canals occurs during development. Before this time, the testes are held in place by the **gubernaculum testis**. During the seventh developmental month, differential growth and contraction of the gubernaculum testis cause the testes to descend. *(See Figure 27.2.)*
- Layers of fascia, connective tissue, and muscle collectively form a sheath, the spermatic cord, which encloses the ductus deferens, the testicular artery and vein, the **pampiniform plexus**, and the **ilio-inguinal** and **genitofemoral nerves**. *(See Figure 27.3.)*
- The testes remain connected to the abdominal cavity through the **spermatic cords**. The **raphe of scrotum** marks the boundary between the two chambers in the scrotum. Each testis lies in its own **scrotal cavity**. *(See Figures 27.1–27.4.)*
- Contraction of the **dartos muscle** gives the scrotum a wrinkled appearance; the cremaster muscle pulls the testes closer to the body. The **tunica vaginalis** is a serous membrane that covers the **tunica albuginea**, the fibrous capsule that surrounds each testis. Septa extend from the tunica albuginea to the **mediastinum**, creating a series of lobules. Seminiferous tubules within each lobule are the sites of sperm production. From there, sperm pass through a **straight tubule** to the **rete testis**. **Efferent ductules** connect the rete testis to the epididymis. Between the seminiferous tubules, **interstitial endocrine cells** secrete male sex hormones, androgens. Seminiferous tubules contain **spermatogonia**, stem cells involved in **spermatogenesis** (the production of sperm). *(See Figures 27.1–27.5 and 27.7.)*
- The spermatogonia produce **primary spermatocytes**, **diploid** cells ready to undergo **meiosis**. Four spermatids are produced for every primary spermatocyte. The spermatids remain embedded within **nurse cells** while they mature into a **sperm**, a process called **spermiogenesis**. The nurse cells function to maintain the **blood testis barrier**, support spermatogenesis, support spermiogenesis, secrete **inhibin**, and secrete **androgen-binding protein**.

Anatomy of a Sperm p. 714

- Each sperm has a **head**, **neck**, **middle piece**, and **tail**. The tip of the head contains the **acrosome**. The **tail** consists of a single **flagellum**. Lacking most intracellular structures, the sperm must absorb nutrients from the environment. *(See Figure 27.6.)*

The Male Reproductive Tract p. 714

- After detaching from the nurse cells, the sperm are carried along fluid currents into the **epididymis**, which lies along the posterior border of the testis. It is subdivided into a head, body, and tail. The epididymis monitors and adjusts the composition of the fluid produced by the seminiferous tubules, acts as a recycling center for damaged sperm, and stores sperm and facilitates their functional maturation (a process called **capacitation**). *(See Figures 27.1, 27.3, 27.4, and 27.7.)*
- The **ductus deferens** (**vas deferens**) begins at the epididymis and passes through the inguinal canal as one component of the spermatic cord. Near the prostate, it enlarges to form the **ampulla**. The junction of the base of the seminal gland and the ampulla creates the **ejaculatory duct**, which empties into the urethra. The ductus deferens functions to transport and store spermatozoa. *(See Figures 27.1, 27.7, and 27.8.)*
- The urethra extends from the urinary bladder to the tip of the penis. It can be divided into three regions: the **prostatic** urethra, **membranous** urethra (also termed the *intermediate part of urethra*), and **spongy** urethra. *(See Figures 27.1 and 27.9.)*

The Accessory Glands p. 716

- The accessory glands function to activate and provide nutrients to the spermatozoa and to produce buffers to neutralize the acidity of the urethra and vagina.
- Each **seminal gland** (**seminal vesicle**) is an active secretory gland that contributes about 60 percent of the volume of semen; its secretions are high in fructose, which is easily used to produce ATP by sperm. The sperm become highly active after mixing with the secretions of the seminal glands. *(See Figures 27.1, 27.8, and 27.9.)*
- The **prostate** secretes a weakly acidic fluid (**prostatic fluid**) that accounts for 20–30 percent of the volume of semen. These secretions contain an antibiotic, **seminalplasmin**, which may help prevent urinary tract infections in men. *(See Figures 27.1, 27.8, and 27.9.)*
- Alkaline mucus secreted by the **bulbo-urethral glands** (*Cowper's glands*) has lubricating properties. *(See Figures 27.1, 27.8, and 27.9.)*

Semen p. 719

- A typical ejaculation releases 2–5 ml of **semen** *(ejaculate)*, which contains a **sperm count** of 20 to 100 million sperm per milliliter. The **seminal fluid** is a specific mixture of secretions of the accessory glands and contains important enzymes.

The Penis p. 719

- The **penis** can be divided into a **root**, **body** (**shaft**), and **glans**. The skin overlying the penis resembles that of the scrotum. The **foreskin** (**prepuce**) surrounds the tip of the penis. **Preputial glands** on the inner surface of the prepuce secrete **smegma**. Most of the body of the penis consists of three masses of **erectile tissue**. Beneath the superficial fascia, there are two **corpora cavernosa** and a single **corpus spongiosum** that surrounds the urethra. When the smooth muscles in the arterial walls relax, the erectile tissue becomes engorged with blood, producing an **erection**. *(See Figures 27.1 and 27.9.)*

27.3 Anatomy of the Female Reproductive System p. 719

- Principal structures of the female reproductive system include the **ovaries**, **uterine tubes**, **uterus**, **vagina**, and external genitalia. *(See Figures 27.10–27.21.)*
- The ovaries, uterine tubes, and uterus are enclosed within the **broad ligament** (an extensive mesentery). The **mesovarium** supports and stabilizes each ovary. *(See Figures 27.10, 27.11, and 27.15.)*

The Ovaries p. 721

- The ovaries are held in position by the **ovarian ligaments** and the **suspensory ligaments**. The **ovarian artery** and **vein** enter the ovary at the **ovarian hilum**. Each ovary is covered by a **tunica albuginea**. *(See Figures 27.10, 27.11, and 27.15.)*
- **Oogenesis** (gamete production) occurs monthly in **ovarian follicles** as part of the **ovarian cycle**. As **primordial ovarian follicles** develop into **primary ovarian follicles**, **thecal endocrine** and **granulosa cells** surrounding the oocyte release estrogens, most importantly **estradiol**. Follicular fluid encourages rapid growth during the formation of only a few **secondary ovarian follicles**. Finally, one **tertiary** (*vesicular*) **follicle** develops. The primary oocyte undergoes meiotic division, producing a secondary oocyte. At **ovulation**, a **secondary oocyte** surrounded by follicular cells (the **corona radiata**) is released through the ruptured ovarian wall. *(See Figures 27.12 and 27.13.)*
- The follicular cells remaining within the ovary form the **corpus luteum**, which produces progestins, mainly progesterone. If pregnancy does not occur, it degenerates into a **corpus albicans** of scar tissue. *(See Figure 27.12.)*
- The decline in progesterone and estrogen triggers the secretion of **GnRH**, which in turn triggers a rise in FSH and LH production, and the entire cycle begins again. *(See Figure 27.18.)*

The Uterine Tubes p. 724

- Each uterine tube has an expanded funnel, the **infundibulum**, with **fimbriae** (projections); an **ampulla**; an **isthmus**; and a **uterine part** that opens into the uterine cavity. *(See Figures 27.10, 27.11, 27.14, and 27.15.)*
- The uterine tube is lined with ciliated and nonciliated simple columnar epithelial cells, which aid in the transport of materials. For fertilization to occur, the oocyte must encounter sperm during the first 12–24 hours of its passage from the infundibulum to the uterus.

The Uterus p. 725

- The **uterus** provides mechanical protection and nutritional support to the developing embryo. Normally, the uterus bends anteriorly near its base (**anteflexion**). It is stabilized by the **broad ligament**, **uterosacral ligaments**, **round ligaments**, and the **cardinal ligaments**. *(See Figures 27.10, 27.11, and 27.15.)*
- The divisions of the uterus include the **body** (the largest portion), **fundus**, isthmus, **cervix**, **external os**, **uterine cavity**, **cervical canal**, and **internal os**. The uterine wall can be divided into an inner **endometrium**, a muscular **myometrium**, and a superficial **perimetrium**. *(See Figures 27.15 and 27.16.)*
- The uterus receives blood from the **uterine arteries**, which then branch and form extensive interconnections.
- A typical 28-day **uterine cycle** (*menstrual cycle*) begins with the **menstrual phase** and the destruction of the functional layer of the endometrium. This process of **menstruation**, or **menses**, continues from 1 to 7 days. *(See Figures 27.17 and 27.18.)*
- After the menstrual phase, the **proliferative phase** begins and the functional layer undergoes repair and thickens. During the **secretory phase** of the uterine cycle, the endometrial glands enlarge and accelerate their rate of secretion. Menstrual activity begins at **menarche** (first uterine cycle) and continues until **menopause**. *(See Figures 27.17 and 27.18.)*

The Vagina p. 730

- The **vagina** is an elastic, muscular tube extending between the uterus and external genitalia. The vagina serves as a passageway for menstrual fluids, receives the penis during sexual intercourse, and forms the lower portion of the birth canal. A thin epithelial fold, the **hymen**, partially blocks the entrance to the vagina. The vagina is lined by a stratified squamous epithelium which, when relaxed, forms **rugae** (folds). *(See Figures 27.10, 27.11, 27.15, and 27.19.)*

The External Genitalia p. 730

- The structures of the **vulva** (*pudendum*) include the **vestibule**, **labia minora**, **clitoris**, **prepuce**, and **labia majora**. The **lesser** and **greater vestibular glands** keep the area moistened in and around the vestibule. The fatty **mons pubis** creates the outer limit of the vulva. *(See Figures 27.10 and 27.20.)*

The Breasts p. 730

- The **mammary glands** lie in the subcutaneous layer beneath the skin of the chest and are the site of milk production, or **lactation**. The glandular tissue of the breast consists of secretory lobules. Ducts leaving the lobules converge into a single **lactiferous duct** and expand near the nipple, forming a **lactiferous sinus**. The ducts of underlying mammary glands open onto the body surface at the **nipple**. *(See Figure 27.21.)*
- Branches of the internal thoracic artery supply blood to each breast.
- Mammary glands develop during pregnancy under the influence of prolactin (PRL) and growth hormone (GH) from the anterior pituitary, as well as **human placental lactogen (HPL)** from the placenta.

27.4 | Aging and the Reproductive System p. 733

Menopause p. 733

- **Menopause** (the time when ovulation and menstruation cease) typically occurs around age 50. **Premature menopause** occurs before age 40. Production of GnRH, FSH, and LH rises, while circulating concentrations of estrogen and progestins decline.

The Male Climacteric p. 733

- The **male climacteric**, which occurs between ages 50 and 60, involves a decline in circulating testosterone levels and a rise in FSH and LH levels.

Chapter Review

For answers, see the blue Answers tab at the back of the book.

Level 1 Reviewing Facts and Terms

Match each numbered item with the most closely related lettered item.

1. interstitial endocrine cells .... ☐
2. inguinal canals .... ☐
3. mammary glands .... ☐
4. endometrium .... ☐
5. dartos .... ☐
6. scrotal cavity .... ☐
7. spermatogonia .... ☐
8. estradiol .... ☐
9. ampulla .... ☐
10. sperm tail .... ☐

(a) link from scrotal chamber to peritoneal cavity
(b) stem cells that produce sperm
(c) compartment of the scrotum containing a testis
(d) site of lactation
(e) flagellum
(f) inner layer of the uterine wall
(g) enlarged portion of the uterine tube
(h) important estrogen hormone
(i) responsible for the production of androgens
(j) layer of smooth muscle deep to the dermis of the scrotum

11. Identify the principal structures of the male reproductive system in the diagram below.

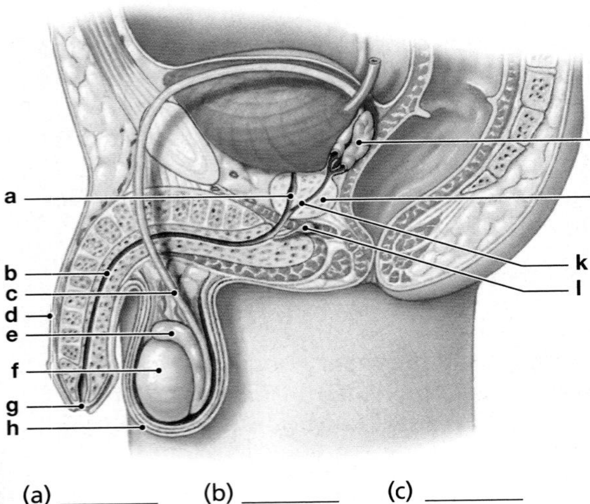

(a) ________ (b) ________ (c) ________
(d) ________ (e) ________ (f) ________
(g) ________ (h) ________ (i) ________
(j) ________ (k) ________ (l) ________

12. The reproductive system includes
    (a) gonads and external genitalia.
    (b) ducts that receive and transport the gametes.
    (c) accessory glands and organs that secrete fluids.
    (d) all of the above.

13. Accessory organs of the male reproductive system include all of the following *except*
    (a) seminal glands.
    (b) scrotum.
    (c) bulbo-urethral glands.
    (d) prostate.

14. Sperm production occurs in the
    (a) ductus deferens.
    (b) seminiferous tubules.
    (c) epididymis.
    (d) seminal glands.

15. The structure that enters the urethra is the
    (a) ductus deferens.
    (b) epididymis.
    (c) seminal gland.
    (d) ejaculatory duct.

16. The mediastinum of the testis
    (a) is ventrally located.
    (b) supports the blood and lymphatic vessels supplying the testis.
    (c) separates the left and right testis.
    (d) forms part of the external testicular capsule.

17. The structure that transports the ovum to the uterus is the
    (a) uterosacral ligament.
    (b) vagina.
    (c) uterine tube.
    (d) infundibulum.

18. Which of the following is *not* a supporting ligament of the uterus?
    (a) suspensory
    (b) cardinal
    (c) round
    (d) broad

19. During the proliferative phase of the menstrual cycle,
    (a) ovulation occurs.
    (b) a new functional layer is formed in the uterus.
    (c) secretory glands and blood vessels develop in the endometrium.
    (d) the old functional layer is shed.

20. The vagina is
    (a) a central space surrounded by the labia minora.
    (b) the inner lining of the uterus.
    (c) the inferior portion of the uterus.
    (d) a muscular tube extending between the uterus and the external genitalia.

21. Identify the main structures of the female reproductive system in the diagram below.

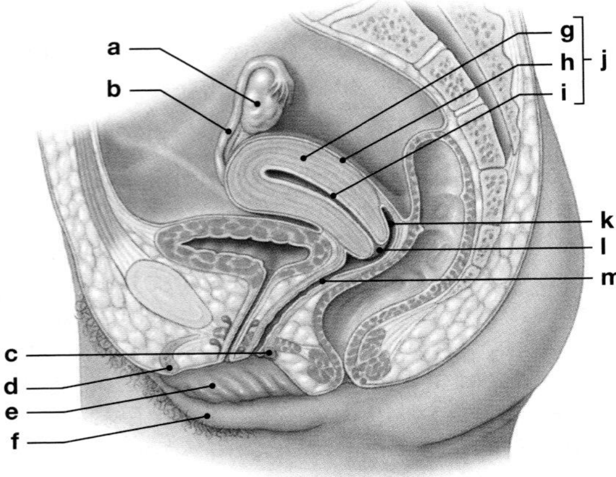

(a) ________ (b) ________ (c) ________
(d) ________ (e) ________ (f) ________
(g) ________ (h) ________ (i) ________
(j) ________ (k) ________ (l) ________
(m) ________

Level 2 Reviewing Concepts

1. How will penile function be altered if the blood supply to the corpora cavernosa is impaired?
   (a) The erect penis will be unable to become flaccid.
   (b) Erection will not occur.
   (c) Erection will occur, but ejaculation will not.
   (d) The urethra will be occluded.
2. What will happen to the function of the male reproductive tract if the testes fail to descend?
   (a) The testicular blood supply will not be adequate.
   (b) The spermatic cord will herniate through the inguinal canal.
   (c) Excess amounts of male sex hormones will be secreted in the abdominal cavity.
   (d) No viable sperm will be produced.
3. In some cases, a tumor of the breast can be detected by noticing a dimple in the skin over the area. Which breast structures are responsible for this phenomenon?
   (a) the tumor, in that it stimulates contraction of the smooth muscles of the skin
   (b) no breast structures; only the pectoralis major and pectoralis minor
   (c) the breast lymph nodes, which the tumor causes to contract
   (d) the suspensory ligaments of the breast, which are connected to the tumor
4. Which region of the sperm cell contains chromosomes?
5. What is the function of the acrosome?
6. Where is seminalplasmin produced, and what is its function?
7. How does follicular fluid benefit the development of a follicle?
8. Menstruation results in the loss of which layer of the endometrium?
9. When does descent of the testes normally begin?

Level 3 Critical Thinking

1. Jerry is in an automobile accident that severs his spinal cord at the L_3 level. After his recovery, he wonders if he will still be able to have an erection. What should you tell him?
2. In a condition known as endometriosis, endometrial cells proliferate within a uterine tube or within the peritoneal cavity. A major symptom of endometriosis is intermittent pain. Why does this occur?
3. Women are much more susceptible than men to both peritonitis and urinary tract infections as a result of sexual activity. Explain why, in terms of the differences between male and female anatomy.

CLINICAL CASE WRAP-UP

A Serious Game of Twister

Liam is suffering from acute testicular torsion. His right testis has twisted on its spermatic cord. This is squeezing off the blood supply, including the inflowing testicular artery and the outgoing pampiniform plexus of the testicular vein. The torsion has also twisted the ilio-inguinal and genitofemoral nerves.

Most males who experience testicular torsion have an inherited trait that allows the testicle to rotate freely inside the scrotum. Often torsion occurs during sleep, after testicular injury, or after vigorous activity, as in Liam's case.

The emergency room physician examines Liam and notes how tender his right testis is. It is visibly higher than his left testis and now is swollen. The physician also notes that the epididymis is located anteromedially, near the scrotal raphe, instead of its normal posterior location. Liam's cremasteric reflex is intact on the left, but not on the right—when his right inner thigh is stroked, the right testis does not retract toward his inguinal canal. The cremaster muscle is no longer working.

The physician uses bedside ultrasonography, an imaging technique using high-frequency sound waves, and determines that there is no blood supply to Liam's right testicle.

The consulting urologist quickly arrives. "Sorry about ruining your post-game night, Liam. But our best chance of saving your testis is to head straight to the operating room and untwist it."

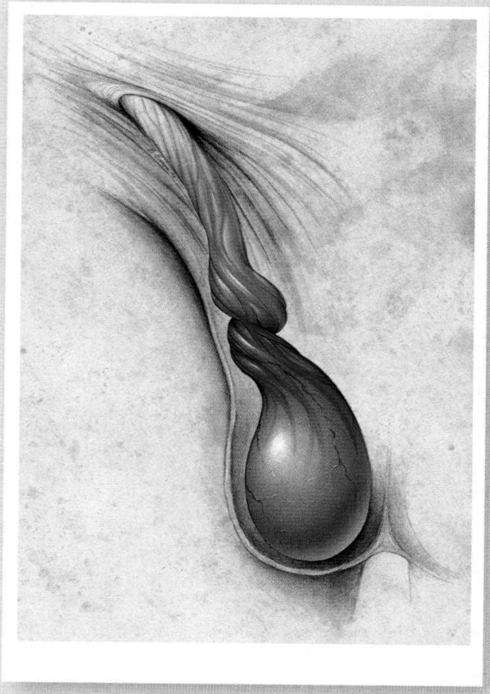

1. What other structure in Liam's right spermatic cord is being twisted with this testicular torsion?
2. Epididymitis is a painful inflammation of the epididymis that can mimic the signs and symptoms of testicular torsion. Would Liam's cremasteric reflex be intact if he were suffering from epididymitis? Explain.

See the blue Answers tab at the back of the book.

Related Clinical Terms

amenorrhea:: Menstruation not starting before the age of 16 or the stopping of menstruation for 6 months or more during the female reproductive age.

cervical dysplasia: Abnormal growth of epithelial cells in the uterine cervix; may progress to cancer.

endometriosis: The growth of endometrial tissue outside the uterus.

genital herpes: A sexually transmitted disease caused by a herpes virus and characterized by painful blisters in the genital area.

gonorrhea: A sexually transmitted bacterial disease caused by *Neisseria gonorrhoeae*. Commonly called "the clap."

gynecology: The branch of medicine that deals with the functions and diseases specific to women and girls affecting the reproductive system.

gynecomastia: Development of breast tissue in the male; a consequence of adrenal cortex hypersecretion of estrogens, certain drugs (cimetidine, spironolactone, and some chemotherapeutic agents), and marijuana use.

hydrocele: The accumulation of serous fluid in any body sac, but especially in the tunica vaginalis of the testis or along the spermatic cord.

menorrhagia: The condition of experiencing extremely heavy bleeding at menstruation.

oophorectomy: The surgical removal of one or both ovaries.

oophoritis: Inflammation of the ovaries.

orchitis: Inflammation of the testis.

polycystic ovary syndrome (PCOS): Most common endocrinopathy in women and most common cause of anovulatory infertility.

premature ejaculation: A common complaint of ejaculating semen sooner than the man desires while achieving orgasm during intercourse. An estimated 30 percent of men regularly experience the problem.

premenstrual dysphoric disorder (PMDD): A collection of physical and emotional symptoms that occurs 5 to 11 days before a woman's period begins, and goes away once menstruation starts. Over 150 signs and symptoms have been associated with the condition.

premenstrual syndrome: A condition occurring in the last half of the menstrual cycle after ovulation that is a combination of physical and mood disturbances that normally end with the onset of the menstrual flow.

pseudohermaphrodites: Individuals with external genitalia that do not "match" their gonads.

salpingitis: Inflammation of a uterine tube.

uterine prolapse: Condition that occurs when a woman's pelvic floor muscles and ligaments stretch and weaken and provide inadequate support for the uterus, which then descends into the vaginal canal.

vasectomy: The surgical removal of a segment of each ductus (vas) deferens, making it impossible for sperm to reach the distal portions of the male reproductive tract.

vulvovaginal candidiasis: A common female vaginal infection caused by the yeast *Candida*, usually *Candida albicans*.

28

The Reproductive System

Embryology and Human Development

Learning Outcomes

These Learning Outcomes correspond by number to this chapter's sections and indicate what you should be able to do after completing the chapter.

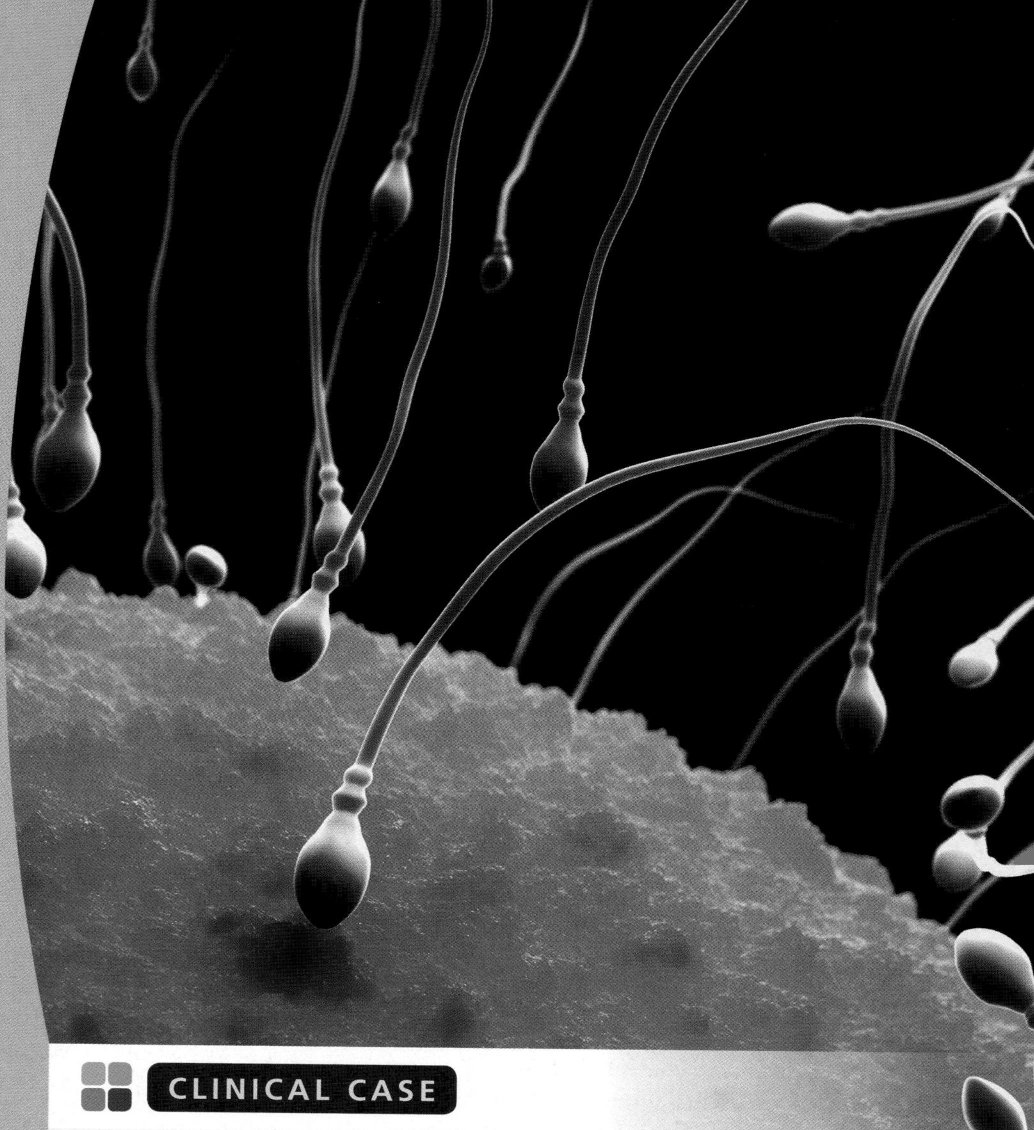

CLINICAL CASE

The Least-Alike Twins

Mayu and her twin, Ruchira, are in Twinsburg, Ohio for the Twins Days Festival, the largest annual celebration of twins in the world. There are thousands of identical twins in attendance, and many fraternal twins as well. Mayu and Ruchira are amazed to see so many twins who not only look alike, but also move, sound, and even smile alike.

As they walk through the festival grounds, Mayu and Ruchira notice tents set up to recruit twins for everything from FBI face-recognition software development to fingerprint and handwriting analysis. The festival is a popular event for "nurture-versus-nature" researchers who are looking for twins to enlist in their studies.

Another highlight of the festival is a ceremony that awards medals to twins in various categories, including the oldest, the youngest, the most alike, and the least alike. This year, the least-alike medals are awarded to two girls, one who is Caucasian, and the other who is of African descent.

What do you think is the story with those twins? To find out, turn to the Clinical Case Wrap-Up on p. 790.

DEVELOPMENT IS THE GRADUAL MODIFICATION of anatomical structures from conception to maturity. The changes observed during development are truly remarkable. What begins as a single cell slightly larger than the period at the end of this sentence becomes a human body containing trillions of cells organized into tissues, organs, and organ systems.

Differentiation is the formation of specialized cell types during development. Differentiation occurs through selective changes in genetic activity. An understanding of human development enhances our understanding of anatomical structures. Our discussion focuses on highlights of the developmental process. At the end of this chapter are Embryology Summaries that will help you visualize the development of specific body systems.

28.1 | An Overview of Development

▶ **KEY POINT** The process of development begins at fertilization and can be separated into two periods: prenatal and postnatal development.

In the process of **development**, a human being forms in the womb for about 9 months and grows to maturity in 15 to 20 years. The process of development is a continuum and can be separated into two periods characterized by specific anatomical changes. **Prenatal development** begins at **fertilization**, or *conception*, when the male gamete (sperm) and the female gamete (oocyte) fuse, and continues to delivery. **Embryology** (em-brē-OL-ō-jē) is the study of the developmental events occurring during prenatal development. **Postnatal development** begins at birth and continues to **maturity**, the state of full development or completed growth.

The period of prenatal development is often subdivided further. **Pre-embryonic development** involves the processes that occur in the first 2 weeks after fertilization, producing a pre-embryo that implants within the uterine lining.

Pre-embryonic development is followed by **embryonic development**, which includes the events during the third through eighth weeks. The development of the fetus, or **fetal development**, begins at the start of the ninth developmental week and continues until birth.

28.1 CONCEPT CHECK

1 You and your wife have just found out that she is pregnant. How should you explain the difference between prenatal development and postnatal development to the child's future grandparents?

See the blue Answers tab at the back of the book.

28.2 | Fertilization

▶ **KEY POINT** Fertilization involves the union of two haploid gametes, which produces one diploid zygote. Normally, fertilization occurs at the junction between the ampulla and isthmus of the uterine tube.

Fertilization involves the fusion of two haploid gametes, each containing 23 chromosomes. This process produces a zygote that contains 46 chromosomes, the normal diploid number in a human somatic cell. The functional roles and contributions of the male and female gametes are very different. The sperm delivers the paternal (father's) chromosomes to the site of fertilization. However, the female oocyte provides all the nourishment and genetic programming needed to support embryonic development for nearly a week after conception. For this reason, the volume of this gamete is much greater than that of the sperm. At fertilization, the diameter of the secondary oocyte is more than twice the entire length of the sperm (**Figure 28.1a**).

Fertilization typically takes place near the junction between the ampulla and isthmus of the uterine tube, usually within a day of ovulation. During this 24-hour period, the oocyte travels only a few centimeters, while the sperm travel from the vagina to the ampulla of the uterine tube. The sperm arriving in the vagina are already motile, but they must undergo **capacitation** within the female reproductive tract before they can fertilize an oocyte. ⮌ p. 716

It typically takes 30 minutes to 2 hours for sperm to travel from the vagina to the fertilization site. Contractions of the uterine musculature and currents produced by the cilia within the uterine tubes may help the sperm travel from the vagina to the fertilization site. It is still not an easy passage. Of the 200 million sperm introduced into the vagina from a typical ejaculate, only around 10,000 enter the uterine tube, and fewer than 100 reach the isthmus. A male with a sperm count below 20 million per milliliter is functionally sterile because too few sperm will survive to reach and fertilize the oocyte. While it is true that only one sperm fertilizes an oocyte, dozens of sperm are required for successful fertilization.

The Oocyte at Ovulation

▶ **KEY POINT** Three factors—at least—make the fertilization of an ovulated oocyte more difficult.

Fertilization of an oocyte is complicated by at least three factors (**Figure 28.1b**):

1. The oocyte is surrounded by a layer of follicle cells, the corona radiata.
2. Metabolic operations of the oocyte have been discontinued.
3. The oocyte leaving the follicle is in metaphase of the second meiotic division.

The corona radiata protects the oocyte as it passes through the ruptured follicular wall and into the infundibulum of the uterine tube. Before the physical process of fertilization can occur, a sperm must penetrate the corona radiata. This is aided by the acrosome reaction, which frees the enzymes contained in the acrosome of each sperm, including **hyaluronidase** (hī-uh-lū-RON-i-dās). Hyaluronidase breaks down the bonds between adjacent follicle cells. Dozens of sperm must release hyaluronidase before the connections between the follicle cells of the corona radiata break down enough to allow an intact sperm to reach the oocyte. When that sperm slips through the gap in the corona radiata, it must have an intact acrosome. The sperm then binds to sperm receptors in the zona pellucida. This binding triggers the rupture of the acrosome. When the sperm contacts the oocyte surface, their cell membranes fuse, and the sperm enters the **ooplasm**, the cytoplasm of the oocyte. This membrane fusion initiates **oocyte activation**, a series of changes in the metabolic activity of the oocyte. The metabolic rate of the oocyte rises suddenly, and immediate changes in the plasma membrane prevent fertilization by additional sperm. (If more than one sperm does penetrate the oocyte membrane, an event called polyspermy, normal development cannot occur.) Perhaps the most dramatic change in the oocyte is that it completes meiosis.

Pronucleus Formation and Amphimixis

▶ **KEY POINT** After the oocyte is activated and completes meiosis, the nuclear material of the oocyte and sperm reorganize, forming pronuclei that fuse to become a zygote.

Following oocyte activation and the completion of meiosis, the nuclear material within the ovum reorganizes as the **female pronucleus** (**Figure 28.1b**). At the same time, the nucleus of the sperm swells, becoming the **male pronucleus**. The male pronucleus migrates toward the center of the cell, and the two pronuclei fuse in a process called **amphimixis** (am-fi-MIK-sis). The cell is now a **zygote** that contains the normal diploid complement of 46 chromosomes, and fertilization is complete. This is the "moment of conception." The first cleavage yields two daughter cells. The cell divisions of cleavage subdivide the cytoplasm of the zygote.

Figure 28.1 Fertilization and Preparation for Cleavage

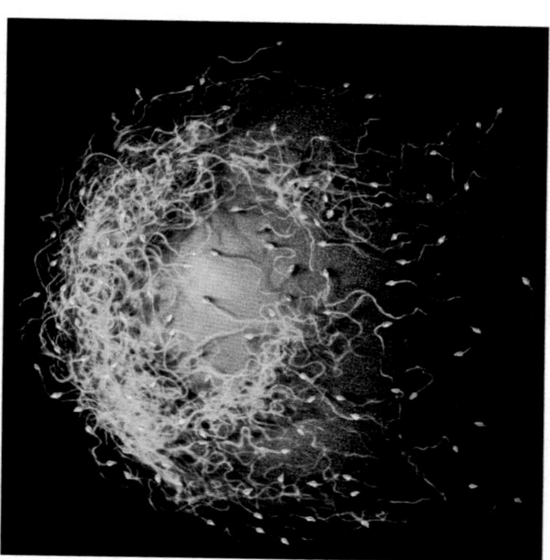

a A secondary oocyte surrounded by sperm

28.2 

CONCEPT CHECK

2 Can sperm arriving in the vagina perform fertilization immediately? Explain.

3 Why must large numbers of sperm reach the oocyte to accomplish fertilization?

4 As soon as the sperm enters the ooplasm, what happens to the oocyte?

See the blue Answers tab at the back of the book.

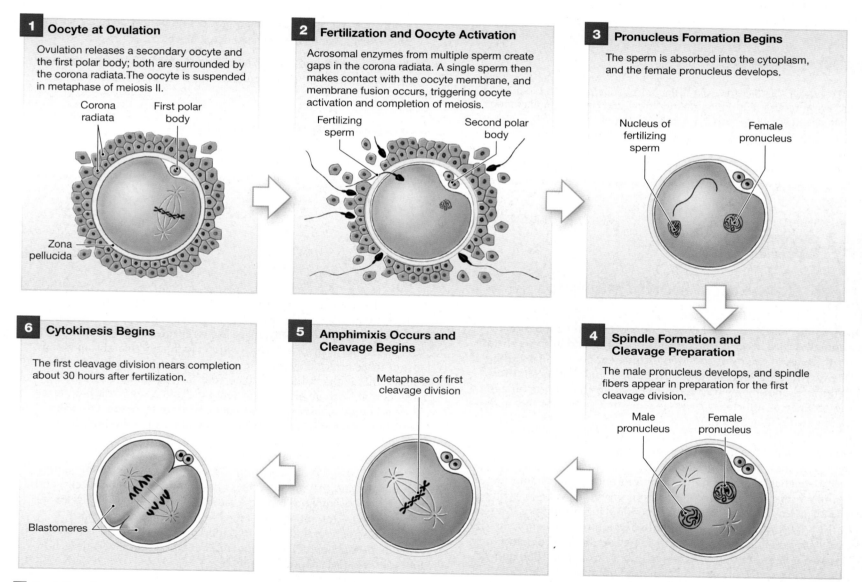

1 Oocyte at Ovulation

Ovulation releases a secondary oocyte and the first polar body; both are surrounded by the corona radiata. The oocyte is suspended in metaphase of meiosis II.

Corona radiata
First polar body
Zona pellucida

2 Fertilization and Oocyte Activation

Acrosomal enzymes from multiple sperm create gaps in the corona radiata. A single sperm then makes contact with the oocyte membrane, and membrane fusion occurs, triggering oocyte activation and completion of meiosis.

Fertilizing sperm
Second polar body

3 Pronucleus Formation Begins

The sperm is absorbed into the cytoplasm, and the female pronucleus develops.

Nucleus of fertilizing sperm
Female pronucleus

4 Spindle Formation and Cleavage Preparation

The male pronucleus develops, and spindle fibers appear in preparation for the first cleavage division.

Male pronucleus
Female pronucleus

5 Amphimixis Occurs and Cleavage Begins

Metaphase of first cleavage division

6 Cytokinesis Begins

The first cleavage division nears completion about 30 hours after fertilization.

Blastomeres

b Ovulation, fertilization, and cleavage preparation

28.3 | Prenatal Development

▶ **KEY POINT** Human prenatal development is divided into three trimesters, each lasting 3 months.

The time spent in prenatal development is known as **gestation** (jes-TĀ-shun). For convenience, the gestation period is considered as three integrated **trimesters**, each lasting 3 months:

- The **first trimester** is the period of pre-embryonic, embryonic, and early fetal development. During this time, the beginnings of all the major organ systems appear.
- The **second trimester** is dominated by the development of fetal organs and organ systems. Body proportions change, and by the end of the second trimester the fetus looks distinctively human.
- The **third trimester** is characterized by rapid fetal growth. Early in the third trimester most of the major organ systems become fully functional. An infant born 1 month or even 2 months prematurely has a reasonable chance of survival.

The First Trimester

▶ **KEY POINT** The first trimester is the most dangerous period in prenatal life. Four key processes occur during the first trimester: cleavage, implantation, placentation, and embryogenesis.

By the end of the first trimester (12th developmental week), the fetus is almost 75 mm (3 in.) long and weighs about 14 g (0.5 oz). Because of the complexity of the processes that occur during this time, only 40 percent of conceptions produce embryos that survive the first trimester. For this reason, pregnant women are warned to avoid drugs, alcohol, and other disruptive stresses during their first trimester in the hopes of preventing an error in the developmental processes under way.

Four general processes occur during the first trimester: cleavage, implantation, placentation, and embryogenesis.

1. **Cleavage** (KLĒV-ej) is a sequence of cell divisions beginning immediately after fertilization and ending when the zygote makes its first contact with the uterine wall. During this time the zygote becomes a **pre-embryo** that develops into a multicellular complex known as a **blastocyst** (BLAS-tō-sist). ⤺ pp. 78–81
2. **Implantation** begins when the blastocyst attaches to the endometrium of the uterus and continues as the blastocyst burrows into the uterine wall. A number of other important events take place as implantation proceeds, and these events set the stage for the formation of vital embryonic structures.
3. **Placentation** (plas-en-TĀ-shun) is marked by the formation of blood vessels around the edges of the blastocyst. This is the first step in the formation of the **placenta**, the anatomical structure linking the maternal and embryonic systems. The placenta provides respiratory and nutritional support essential for further prenatal development.
4. **Embryogenesis** (em-brē-ō-JEN-e-sis) is the formation of a viable embryo. This process establishes the foundations for all major organ systems.

Cleavage and Blastocyst Formation

Cleavage is a series of cell divisions that subdivides the cytoplasm of the zygote, producing an ever-increasing number of smaller and smaller daughter cells. The first cleavage produces a pre-embryo consisting of two identical cells called

CLINICAL NOTE

Congenital Malformations

Congenital malformations, also known as congenital abnormalities or birth defects, are structural or functional anomalies that occur during intra-uterine life and are present at birth. Some of the most common include congenital heart disease, Down syndrome, cleft lip and cleft palate, spina bifida, clubfoot deformity, and phenylketonuria, or PKU (which is characterized by the absence or deficiency of an enzyme that processes the essential amino acid phenylalanine).

Of the 2.75 million neonatal deaths that occur worldwide every year, 10 percent are due to congenital malformations. Consanguinity (denoting that the biological parents are related by blood) increases the prevalence of rare genetic conditions and nearly doubles the risk for neonatal death. Maternal infections, including rubella, syphilis, and Zika virus disease, can cause congenital malformations. Maternal nutritional status—particularly folate and iodine deficiency, obesity, and diabetes mellitus—is linked to some congenital problems; for instance, folate deficiency increases the risk of neural tube defects (spina bifida and anencephaly). Environmental factors, including maternal exposure to alcohol, tobacco, certain prescription medications, street drugs, and radiation during pregnancy, are also risk factors. Advanced maternal and paternal age increase the risk of chromosomal conditions such as Down syndrome (an extra chromosome 21, or trisomy 21).

Worldwide, 94 percent of severe congenital malformations occur in low- and middle-income countries. As many as half of all conceptions produce zygotes that do not survive the cleavage stage. At least 10 percent of fertilizations produce zygotes with abnormal chromosomes. Most of these affected zygotes never complete development. Of those embryos that implant, roughly 20 percent fail to complete 5 months of development, with an average survival time of 8 weeks. This pregnancy failure is known as a spontaneous abortion, or miscarriage. Less than 1 percent of newborns show chromosomal abnormalities that result from spontaneous mutations.

Preconception screening can identify people at risk for disorders that may be passed on to offspring. Other procedures, including amniocentesis (testing the amniotic fluid for abnormalities in fetal cells) and chorionic villus sampling (testing a sample of chorionic villi for fetal cells), can be performed in the first or second trimester. Neonatal (newborn) screening that looks for blood disorders, metabolism and hormone production defects, congenital deafness, and heart defects can facilitate life-saving treatments.

Because prenatal development involves so many complex, interdependent steps, malformations of some kind are statistically inevitable. The most reasonable public health approach to congenital malformations must include three essential elements: surveillance to track birth defects, research to identify causes, and prevention research and programs.

blastomeres (BLAS-tō-mērz). The first cleavage division is completed about 30 hours after fertilization **(Figure 28.2)**. Subsequent divisions occur at intervals of 10–12 hours. During the initial cleavage divisions, all the blastomeres undergo mitosis simultaneously. As the number of blastomeres increases, the time interval between divisions becomes less predictable.

After 3 days of cleavage, the pre-embryo is a solid ball of cells called the **morula** (MOR-ū-la; "mulberry"). After 5 days of cleavage, the blastomeres form a hollow ball, the **blastocyst**, with an inner cavity known as the **blastocoele** (BLAS-tō-sēl). The blastomeres are now no longer identical in size and shape. The outer layer of cells, separating the external environment from the blastocoele, is the **trophoblast** (TRŌ-fō-blast; *trophos*, food, + *blast*, precursor). These cells are responsible for providing food to the developing embryo. Trophoblast cells are the only cells of the pre-embryo that contact the uterine wall. In the blastocyst, a second group of cells, the **inner cell mass**, lies clustered at one end. These cells are exposed to the blastocoele on one side, but are insulated from contact with the outside environment by the trophoblast on the opposite side. The cells of the inner cell mass are the stem cells responsible for producing all the cells and cell types of the body.

Implantation

Immediately after fertilization, the zygote is 4 days away from the uterus. It arrives in the uterine cavity as a morula, and blastocyst formation occurs during the next 2 to 3 days. During this period the cells are absorbing nutrients from the fluid within the uterine cavity. The endometrial glands secrete this fluid, which is rich in glycogen. When fully formed, the blastocyst contacts the endometrium and implantation occurs, usually in the fundus or body of the uterus **(Figure 28.3)**.

Implantation begins as the surface of the blastocyst closest to the inner cell mass touches and adheres to the uterine lining (see Day 7, **Figure 28.3**). The trophoblast cells at the point of contact divide rapidly, making the trophoblast several layers thick. Near the endometrial wall, however, the plasma membranes separating the trophoblast cells disappear, creating a layer of cytoplasm with multiple nuclei (Day 8). This outer layer is called the **syncytiotrophoblast (Figure 28.3)**.

The syncytiotrophoblast erodes a path through the uterine epithelium by secreting hyaluronidase. This enzyme dissolves the proteoglycans between adjacent epithelial cells, just as hyaluronidase released by sperm dissolved the connections between cells of the corona radiata. At first, the erosion creates a gap in the uterine lining, but maternal epithelial cells migrate and divide and soon repair the surface. Once the blastocyst has lost contact with the uterine cavity, its further development occurs entirely within the functional layer of the endometrium.

As implantation proceeds, the syncytiotrophoblast enlarges and spreads into the surrounding endometrium (Days 8 and 9, **Figure 28.3**). This process results in the disruption and enzymatic digestion of uterine glands. The nutrients released are absorbed by the syncytiotrophoblast and distributed by diffusion across the underlying **cytotrophoblast** to the inner cell mass. These nutrients provide the energy needed for the early stages of embryo formation. Cytoplasmic extensions of the trophoblast grow around endometrial capillaries, and the capillary walls are destroyed. Maternal blood now flows slowly through trophoblastic channels called **lacunae** (singular, *lacuna*). Fingerlike **primary villi** extend away from the trophoblast into the surrounding endometrium. Each primary villus consists of an extension of syncytiotrophoblast with a core of cytotrophoblast. Over the next few days, the trophoblast begins breaking down larger endometrial veins and arteries, and blood flow through the lacunae increases.

Figure 28.2 Cleavage and Blastocyst Formation

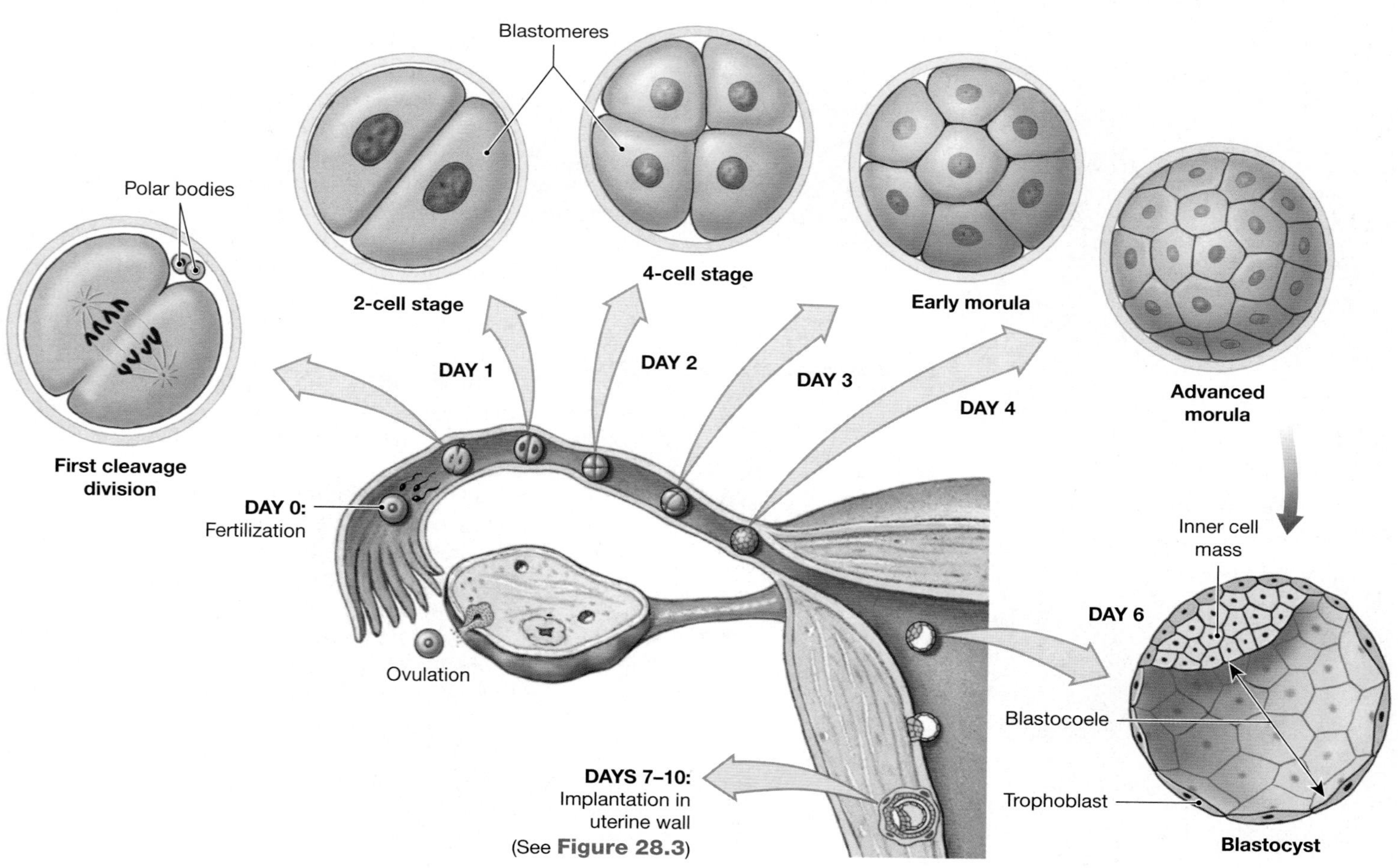

28

Formation of the Blastodisc In the early blastocyst stage, the inner cell mass has little visible organization. At the time of implantation, however, the inner cell mass has already started separating from the trophoblast. The separation gradually increases, creating a fluid-filled chamber called the **amniotic** (am-nē-OT-ik) **cavity** (Day 9, **Figure 28.3**, and Days 10–12, **Figure 28.4**). At this stage, the cells of the inner cell mass are organized into an oval sheet called a **blastodisc** (BLAS-tō-disk) that is two layers thick. At days 10 to 11, the blastodisc consists of two epithelial layers: the **epiblast** (EP-i-blast), which faces the amniotic cavity, and the **hypoblast** (HĪ-pō-blast), which is exposed to the contents of the blastocoele.

Figure 28.3 Stages in the Implantation Process

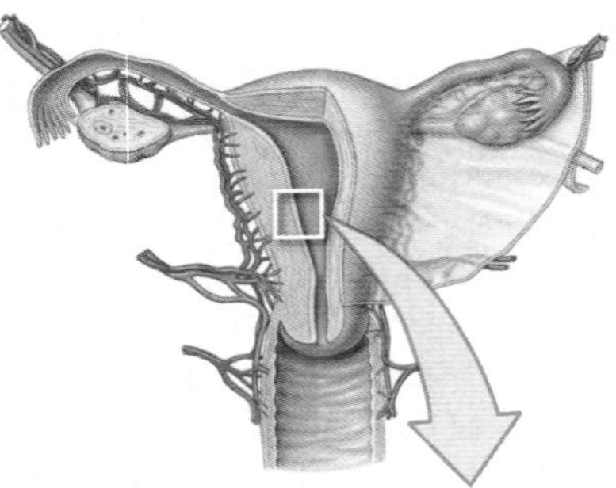

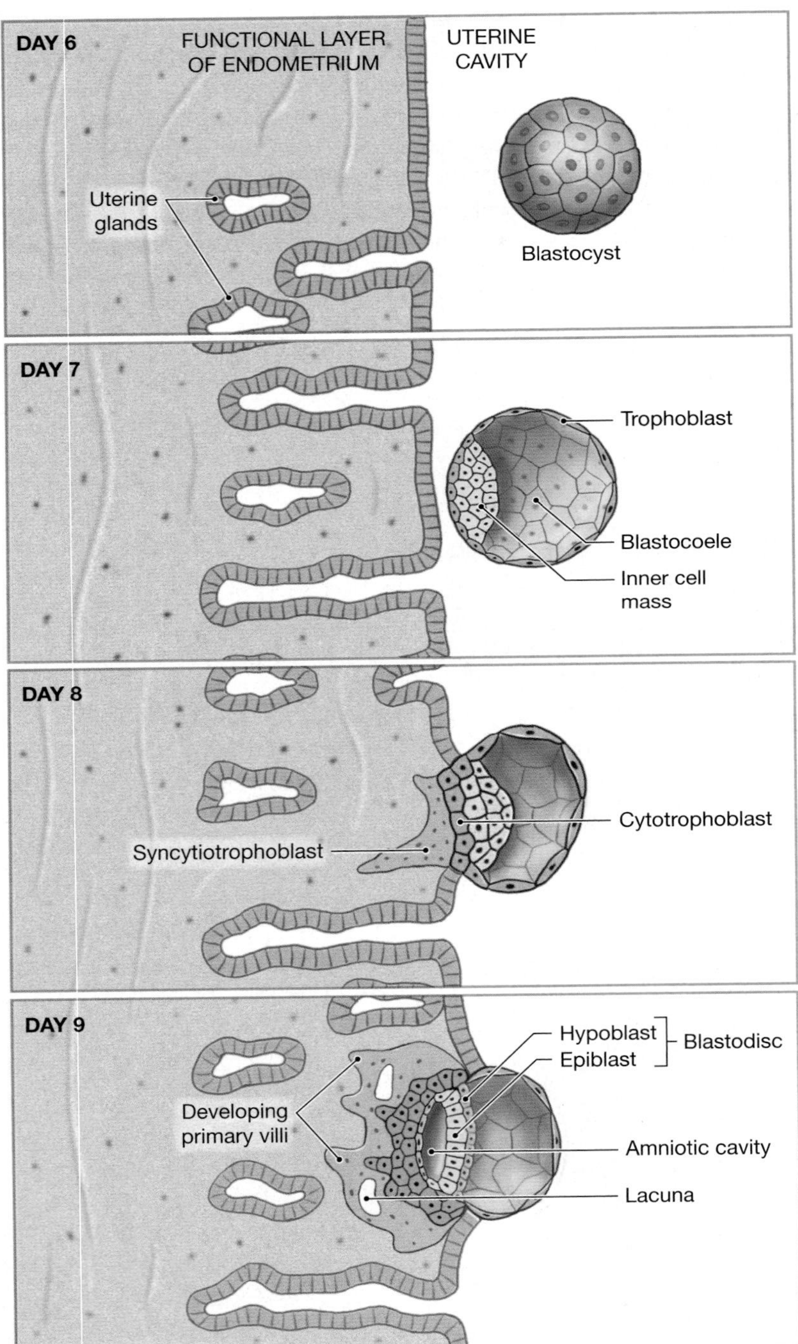

Gastrulation and Germ Layer Formation A few days later, a third layer begins forming through a cell migration process called **gastrulation** (gas-trū -LĀ-shun) (see Day 12, **Figure 28.4**). During gastrulation, cells in specific areas of the epiblast move toward the center of the blastodisc, toward a line known as the **primitive streak**. Once they arrive at the primitive streak, the migrating cells leave the surface and move between the epiblast and hypoblast. This process creates three distinct embryonic layers with very different fates called **germ layers**. Once gastrulation begins, the layer remaining in contact with the amniotic cavity is now called the **ectoderm**; the hypoblast is now the

CLINICAL NOTE

Teratogens and Abnormal Development

A teratogen (ter-A-tō-jen) is any agent that causes an abnormality of an embryo or fetus during prenatal development. The developing fetus is most vulnerable during the first trimester of pregnancy. Teratogenic agents include physical agents (high fever), maternal health factors (diabetes, obesity, hypertension), environmental chemicals (herbicides, industrial solvents), and infectious agents (rubella, Zika viral infections).

In the United States, the most frequent cause of preventable, abnormal fetal development is maternal use of tobacco, alcohol, and drugs (prescription, over-the-counter, and illegal). Smoking nearly doubles a mother's risk of delivering a low-birth-weight baby as a result of poor growth before birth, premature delivery, or both. Fetal alcohol syndrome (FAS) is the most common known nongenetic cause of growth deficiency, functional deficits, and intellectual impairment. Alcohol passes from the mother's blood through the placenta to the fetus. Because alcohol is broken down more slowly in a fetus than an adult, fetal alcohol levels remain high for a longer period of time. Developing embryos are very vulnerable to alcohol. Victims of FAS may have small heads, characteristic facial abnormalities, short stature, low weight, poor fine and gross motor coordination, speech and language delays, attention and memory deficits, vision or hearing problems, and heart, kidney, or bone abnormalities. The effects of FAS vary from child to child, but they are irreversible.

Maternal use of methamphetamine, heroin, codeine, oxycodone, and methadone produces some of the most difficult newborns to care for. These substances pass through the placenta, and the developing fetus becomes addicted. After birth, the newborn develops a constellation of withdrawal symptoms called neonatal abstinence syndrome. Neonatal intensive care nurses describe the high-pitched, unrelenting cry of a neonate in withdrawal as the worst sound on earth. Signs and symptoms include hyperreflexia (exaggerated reflexes), increased muscle tone, irritability, poor feeding, vomiting, diarrhea, sleep difficulties, and seizures. The central nervous system abnormalities, including learning and behavioral difficulties, last a lifetime.

endoderm; and the new third layer, located between the ectoderm and endoderm, is the **mesoderm**. (An Embryology Summary in Chapter 3 introduced the formation of the mesoderm and the developmental fates of these three germ layers.) ⊃ pp. 78–81 See **Table 28.1** for the contributions made by each germ layer to body systems. Gastrulation produces an oval, two-layered sheet known as the **embryonic disc**. This disc will form the body of the embryo, whereas all other cells of the blastocyst will be part of the extra-embryonic membranes.

Formation of Extra-embryonic Membranes In addition to forming body structures and organs, the germ layers also form four structures that extend outside the embryonic body. These structures, known as **extra-embryonic membranes**, are the (1) yolk sac (endoderm and mesoderm), (2) amnion (ectoderm and mesoderm), (3) allantois (endoderm and mesoderm), and (4) chorion (mesoderm and trophoblast). The extra-embryonic membranes support embryonic and fetal development by maintaining a consistent, stable environment and by providing access to the oxygen and nutrients carried by the maternal bloodstream. Despite their importance during prenatal development, they leave few traces of their existence in adult systems.

The Yolk Sac The **yolk sac** is the first extra-embryonic membrane to develop **(Figures 28.4** and **28.5)**. Migrating hypoblast cells spread out around the outer edges of the blastocoele to form a complete pouch suspended below the blastodisc **(Figure 28.4** and Week 2, **Figure 28.5)**. This pouch is already visible 10 days after fertilization. As gastrulation proceeds, mesodermal cells migrate around this pouch and complete the formation of the yolk sac. Blood vessels soon appear within the mesoderm, and the yolk sac becomes an important site of early blood cell formation.

The Amnion The ectodermal layer also expands, and ectodermal cells spread over the inner surface of the amniotic cavity. Mesodermal cells soon follow, forming a second, outer layer. This combination of ectoderm and mesoderm forms the **amnion** (AM-nē-on) **(Figure 28.5)**. As the embryo, and later the fetus, enlarges, this membrane continues to expand, increasing the size of the amniotic cavity. The amniotic cavity contains amniotic fluid, which surrounds and cushions the developing embryo or fetus **(Figure 28.5)**.

The Allantois The third extra-embryonic membrane begins as an outpocketing of the endoderm near the base of the yolk sac (Week 3, **Figure 28.5**). The free endodermal tip grows toward the wall of the blastocyst, surrounded by a mass of mesodermal cells. This sac of endoderm and mesoderm is the **allantois** (a-LAN-tō-is). Its base later gives rise to the urinary bladder. See the Embryology Summary: The Development of the Urinary System (pp. 783–784) for the formation of the allantois and its relationship to the urinary bladder.

The Chorion The mesoderm associated with the allantois spreads until it extends completely around the inside of the trophoblast. This forms a mesodermal layer underneath the trophoblast. This combination of mesoderm and trophoblast forms the **chorion** (KO-rē-on) **(Figure 28.5)**.

When implantation first occurs, the nutrients absorbed by the trophoblast easily reach the blastodisc by simple diffusion. But as the embryo and the trophoblastic complex enlarge, the distance between the two increases, and diffusion can no longer keep pace with the demands of the developing embryo. The chorion solves this problem, as blood vessels developing within the mesoderm link the embryo with the trophoblast. Circulation through those chorionic vessels begins early in the third week of development, when the heart starts beating.

Placentation

Placentation is the formation of the placenta from embryonic and maternal tissues. The appearance of blood vessels in the chorion is the first step in the formation of a functional placenta. By the third week of development, mesoderm extends along the core of each of the trophoblastic villi, forming **chorionic villi** in contact with maternal tissues **(Figure 28.5)**. These villi continue to enlarge and branch, forming an intricate network within the endometrium. Maternal blood vessels continue to be eroded, and maternal blood flows slowly through lacunae lined by the syncytiotrophoblast. Diffusion occurs between the maternal blood flowing through the lacunae and fetal blood flowing through vessels within the chorionic villi.

At first, the entire blastocyst is surrounded by chorionic villi. The chorion continues to enlarge, expanding like a balloon within the endometrium. By

Figure 28.4 Blastodisc Organization and Gastrulation

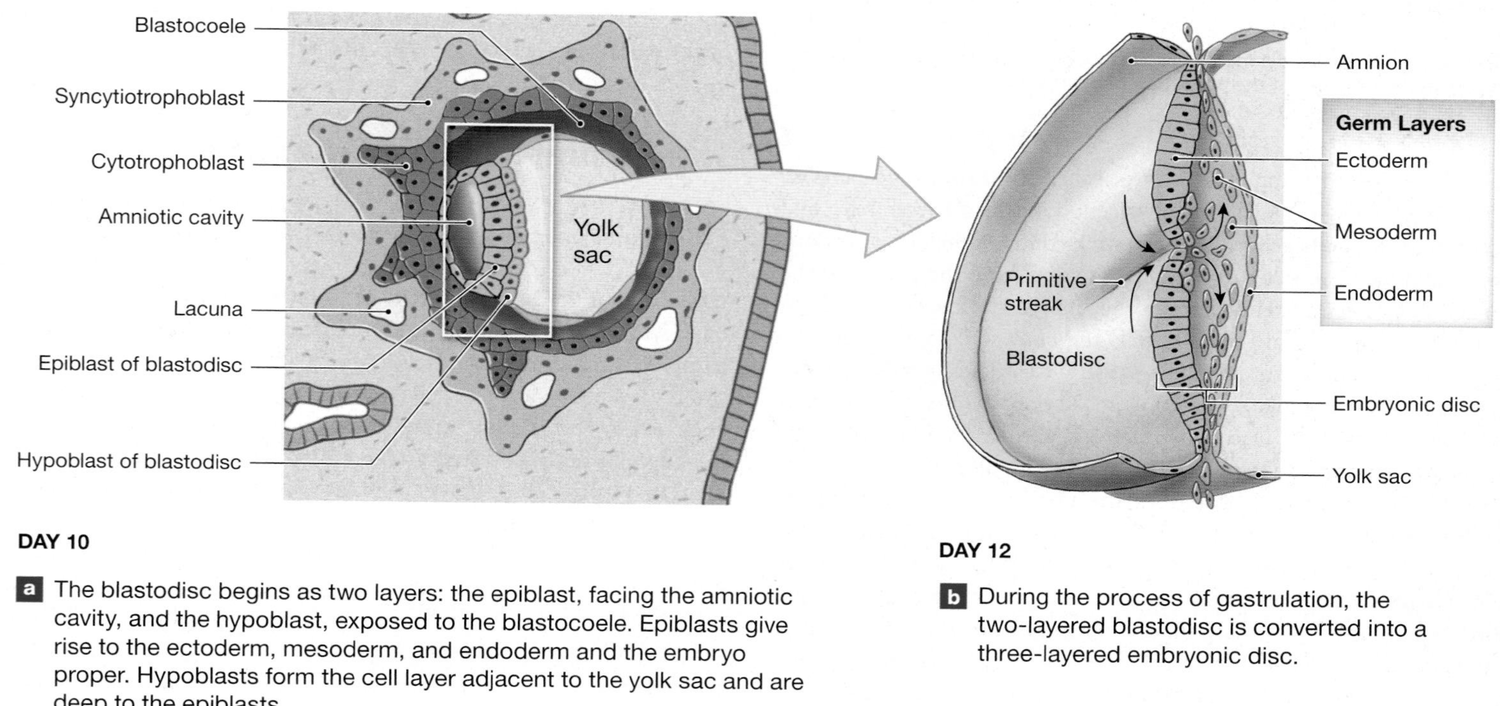

a The blastodisc begins as two layers: the epiblast, facing the amniotic cavity, and the hypoblast, exposed to the blastocoele. Epiblasts give rise to the ectoderm, mesoderm, and endoderm and the embryo proper. Hypoblasts form the cell layer adjacent to the yolk sac and are deep to the epiblasts.

b During the process of gastrulation, the two-layered blastodisc is converted into a three-layered embryonic disc.

Table 28.1 | The Fates of the Primary Germ Layers

System	Ectodermal Contributions
Integumentary system	Epidermis, hair follicles and hairs, nails, and glands communicating with the skin (apocrine and eccrine sweat glands, mammary glands, and sebaceous glands)
Skeletal system	Pharyngeal cartilages and their derivatives in the adult (portion of sphenoid, the auditory ossicles, the styloid processes of the temporal bones, the horns and superior rim of the hyoid bone)*
Nervous system	All nervous tissue, including brain and spinal cord
Endocrine system	Pituitary gland and adrenal medullae
Respiratory system	Mucous epithelium of nasal passageways
Digestive system	Mucous epithelium of oral cavity and anus, salivary glands
	Mesodermal Contributions
Integumentary system	Dermis, except for epidermal derivatives
Skeletal system	All components except some pharyngeal derivatives
Muscular system	All components
Endocrine system	Adrenal cortex and endocrine tissues of heart, kidneys, and gonads
Cardiovascular system	All components, including bone marrow
Lymphatic system	All components
Urinary system	The kidneys, including the nephrons and the initial portions of the collecting system
Reproductive system	The gonads and the adjacent portions of the duct systems
Miscellaneous	The linings of the body cavities (thoracic, pleural, pericardial, pleural peritoneal) and the connective tissues supporting all organ systems
	Endodermal Contributions
Endocrine system	Thymus, thyroid, and pancreas
Respiratory system	Respiratory epithelium (except nasal passageways) and associated mucous glands
Digestive system	Mucous epithelium (except mouth and anus), exocrine glands (except salivary glands), liver, pancreas
Urinary system	Urinary bladder and distal portions of the duct system
Reproductive system	Distal portions of the duct system; stem cells that produce gametes

*The neural crest is derived from ectoderm and contributes to the formation of the skull and the skeletal derivatives of the embryonic pharyngeal arches.

week 4, the embryo, amnion, and yolk sac are suspended within an expansive, fluid-filled chamber (**Figure 28.5**). The connection between the embryo and chorion is known as the **body stalk**. The body stalk contains the distal portions of the allantois and blood vessels carrying blood to and from the placenta. The narrow connection between the endoderm of the embryo and the yolk sac is the **yolk stalk**. For the formation of the yolk stalk and body stalk, see the Embryology Summary: The Development of the Digestive System later in this chapter.

The placenta does not enlarge indefinitely. Regional differences in placental organization develop as placental expansion forms a prominent bulge in the endometrial surface. The relatively thin portion of the endometrium covering the embryo and separating it from the uterine cavity is the **capsular decidua** (KAP-sū-lar dē-SID-yū-uh; decidus, a falling off), or *decidua capsularis*. This layer no longer exchanges nutrients, and the chorionic villi disappear in this region (**Figure 28.5**). Placental functions are now concentrated in the **basal decidua** or *decidua basalis*, a disc-shaped region located within the deepest portion of the endometrium. The rest of the endometrium, which has no contact with the chorion, is called the **parietal decidua** (pa-RĪ-e-tal dē-SID-yū-uh) or *decidua parietalis*. As the end of the first trimester approaches, the fetus moves farther away from the placenta (**Figure 28.5**). It remains connected by the **umbilical cord**, which contains the allantois, placental blood vessels, and yolk stalk.

The developing fetus is totally dependent on maternal organ systems for nourishment, respiration, and waste removal. The maternal systems perform these functions in addition to their normal operations. For example, the mother must absorb enough oxygen, nutrients, and vitamins for herself and her fetus, and she must eliminate all the generated wastes. Although this is not a burden over the initial weeks of gestation, as the fetus grows in subsequent trimesters, the demands placed on the mother become significant. In practical terms the mother must breathe, eat, and excrete for two.

Placental Circulation **Figure 28.6a** shows the circulation at the placenta near the end of the first trimester. Blood flows from the fetus to the placenta through the paired **umbilical arteries** and returns to the fetus in a single **umbilical vein**. Active and passive exchange between the fetal and maternal bloodstreams occurs at the chorionic villi (**Figure 28.6b**).

As noted in Chapter 27, the placenta also synthesizes important hormones that affect maternal and embryonic tissues. Human chorionic gonadotropin (HCG) production begins a few days after implantation. This hormone stimulates the corpus luteum so that it continues to produce progesterone throughout the early stages of the pregnancy. During the second and third trimesters, the placenta also releases progesterone, estrogens, human placental lactogen (HPL), and relaxin. These hormones are synthesized and released into the maternal circulation by the trophoblast.

Embryogenesis

Shortly after gastrulation begins, the body of the embryo begins to separate itself from the rest of the embryonic disc. This process, **embryogenesis**, begins as folding and differential growth of the embryonic disc produce a bulge that projects into the amniotic cavity. This projection is known as the **head fold**. Similar movements form a **tail fold** (Weeks 3 and 4, **Figure 28.5**). The **embryo** is now physically and developmentally separated from the rest of the blastodisc and the extra-embryonic membranes. The posterior and anterior surfaces and left and right sides of the embryo are now visible. **Figures 28.5** and **28.7** illustrate changes in proportions and appearance between the fourth week and the end of the first trimester.

The first trimester is a critical period for development because events during this time establish the basis for organ formation, a process called **organogenesis**. See **Table 28.2** for important milestones of organogenesis in each organ system.

The Second and Third Trimesters

▶**KEY POINT** During the second and third trimesters, the fetus grows rapidly and all of the major organ systems complete their development and become functional.

By the end of the first trimester, the foundations of all the major organ systems have formed (**Figure 28.7c**). During the second trimester, these systems will complete their functional development. By the end of the second trimester, the fetus weighs around 0.64 kg (1.4 lb). During these 3 months, the fetus, encircled by the amnion, grows faster than the surrounding placenta, and the mesodermal outer covering of the amnion fuses with the inner lining of the chorion. **Figure 28.8** shows a 4-month fetus as viewed with a fiber-optic endoscope and a 6-month fetus as seen in ultrasound.

During the third trimester, all the fetal organ systems become functional. Although the rate of growth begins to slow, this trimester sees the greatest weight gain. In the last 3 months of gestation, the fetus gains around 2.6 kg (5.7 lb), reaching a full-term weight of somewhere near 3.2 kg (7.05 lb). See **Table 28.2** and the Embryology Summaries at the end of this chapter for important events in organ system development during the second and third trimesters.

Figure 28.5 The Embryonic Membranes and Placenta Formation

Week 2

Migration of mesoderm around the inner surface of the trophoblast forms the chorion. Mesodermal migration around the outside of the amniotic cavity, between the ectodermal cells and the trophoblast, forms the amnion. Mesodermal migration around the endodermal pouch forms the yolk sac.

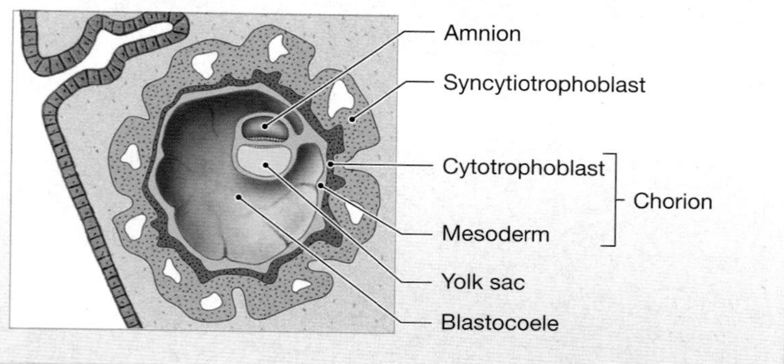

Week 3

The embryonic disc bulges into the amniotic cavity at the head fold. The allantois, an endodermal extension surrounded by mesoderm, extends toward the trophoblast.

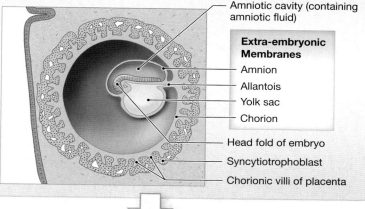

Week 4

The embryo now has a head fold and a tail fold. Constriction of the connections between the embryo and the surrounding trophoblast narrows the yolk stalk and body stalk.

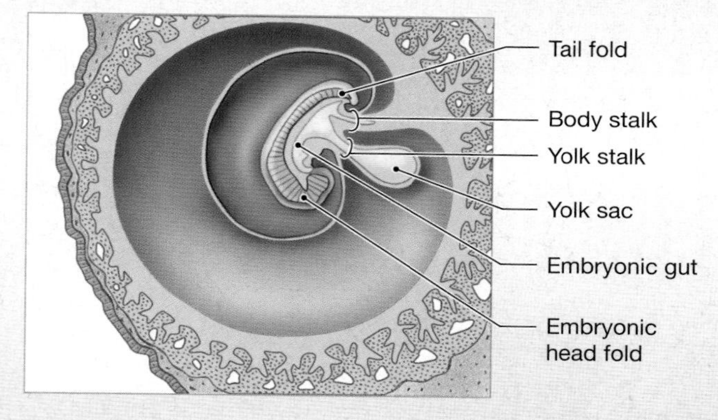

Week 5

The developing embryo and extra-embryonic membranes bulge into the uterine cavity. The trophoblast pushing out into the uterine cavity remains covered by endometrium but no longer participates in nutrient absorption and embryo support. The embryo moves away from the placenta, and the body stalk and yolk stalk fuse to form an umbilical stalk.

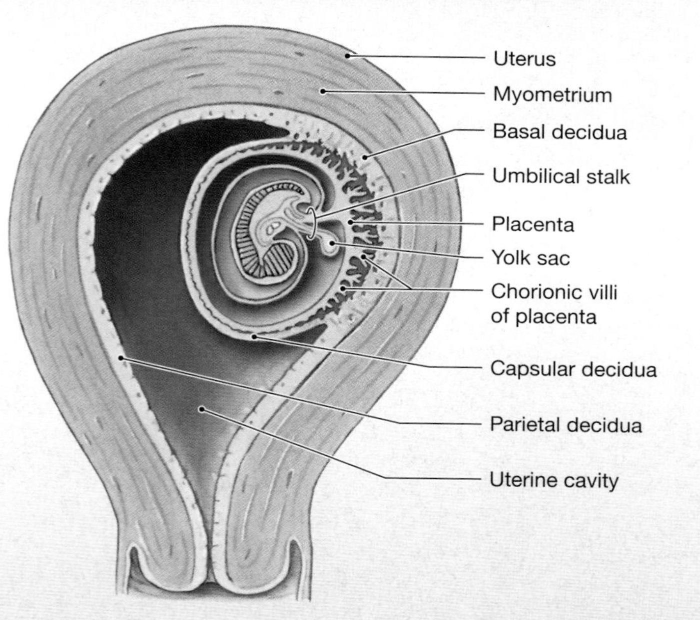

Week 10

The amnion has expanded greatly, filling the uterine cavity. The fetus is connected to the placenta by an elongated umbilical cord containing a portion of the allantois, blood vessels, and the remnants of the yolk stalk. A mucus plug forms, preventing bacteria from entering the uterus.

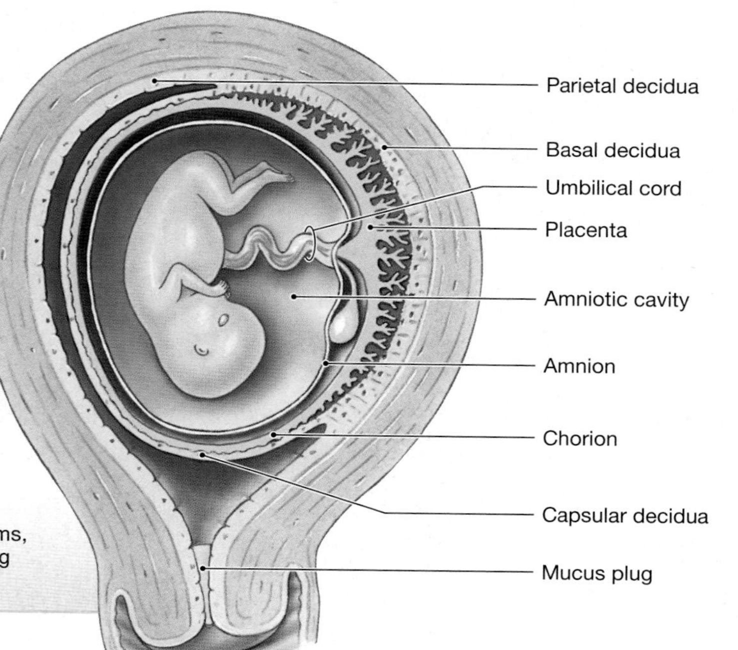

28

Figure 28.6 A Three-Dimensional View of Placental Structure

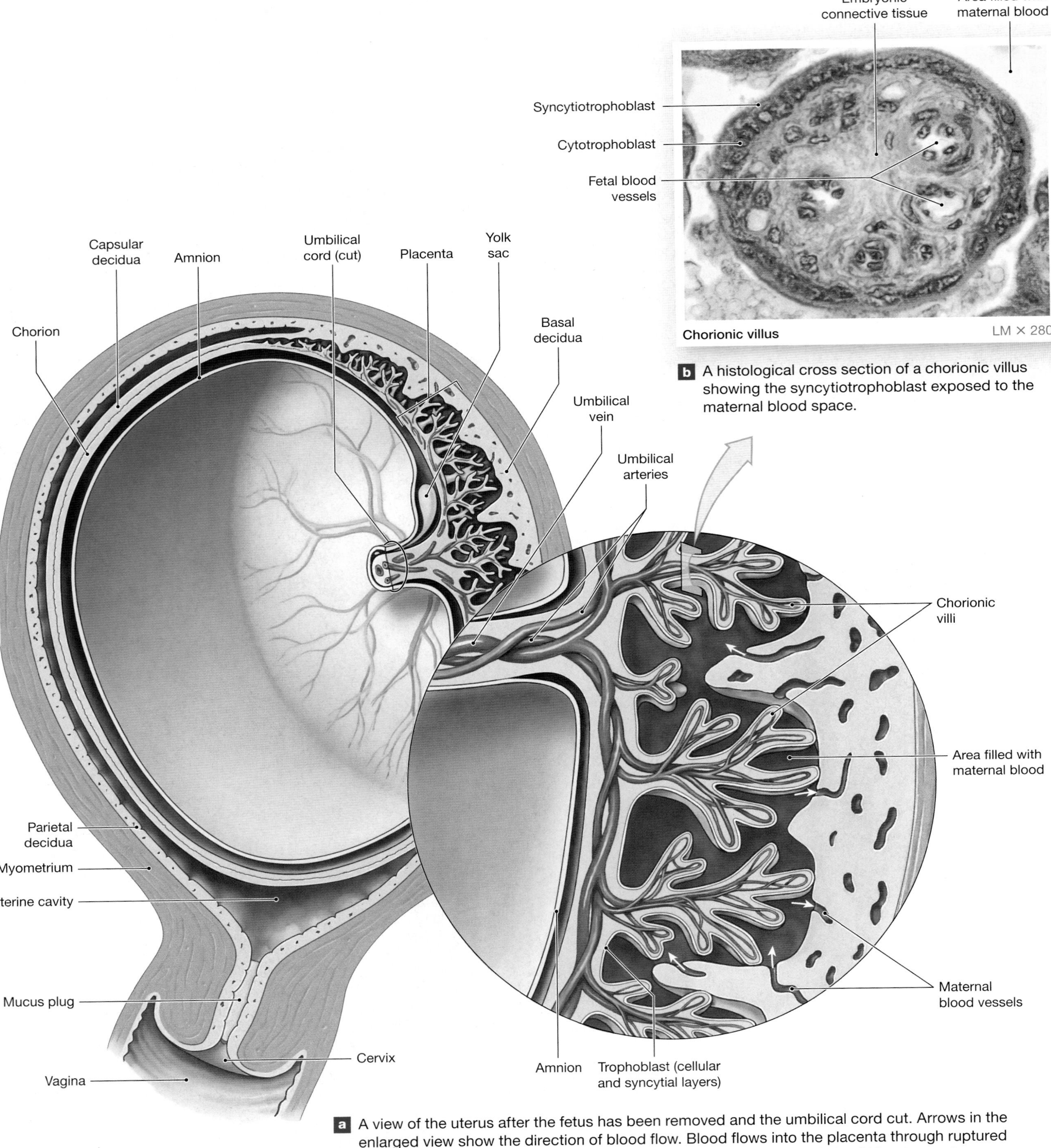

b A histological cross section of a chorionic villus showing the syncytiotrophoblast exposed to the maternal blood space.

a A view of the uterus after the fetus has been removed and the umbilical cord cut. Arrows in the enlarged view show the direction of blood flow. Blood flows into the placenta through ruptured maternal arteries and then flows around chorionic villi, which contain fetal blood vessels.

28

Figure 28.7 The First Trimester

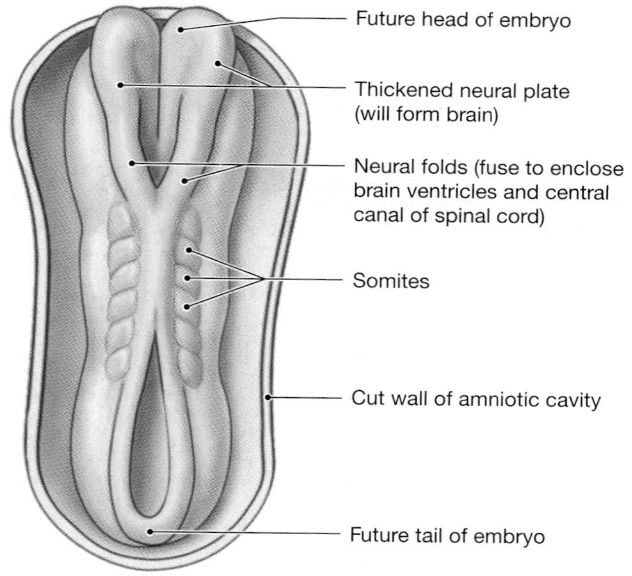

a Illustration of superior surface of human development at week 3 undergoing neurulation (neural tube formation)

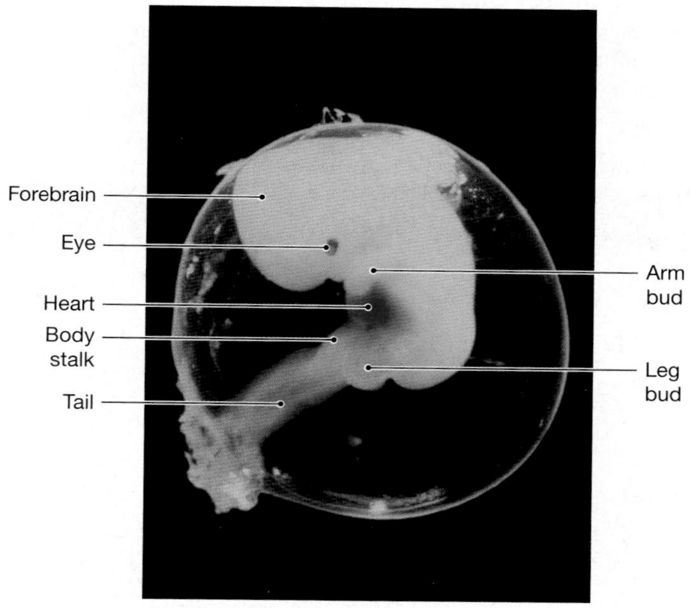

b Fiber-optic view of human development at week 4 (about 5 mm in size)

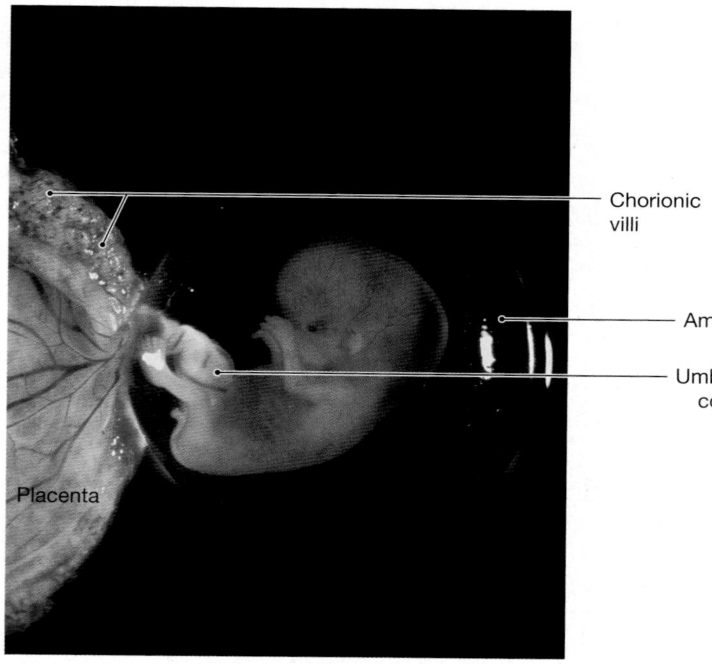

c Fiber-optic view of human development at week 8 (about 1.6 cm in size)

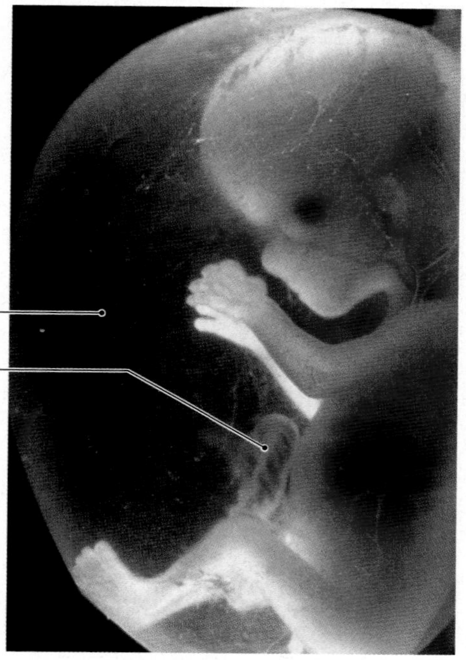

d Fiber-optic view of human development at week 12 (about 5.4 cm in size)

At the end of gestation, a typical uterus has undergone a tremendous increase in size. It grows from 7.5 cm (3 in.) to 30 cm (12 in.) long and contains almost 5 L of fluid. The uterus and its contents weigh roughly 10 kg (22 lb). This remarkable expansion in the size of the uterus occurs through the enlargement and elongation of existing smooth muscle fibers. **Figure 28.9** shows the position of the uterus, fetus, and placenta from 16 weeks to full term (9 months). When the pregnancy is at term, the uterus and fetus push many of the abdominal organs out of their normal positions **(Figure 28.9c)**.

28.3 CONCEPT CHECK

5 What is the fate of the inner cell mass of the blastocyst?
6 What is the function of the syncytiotrophoblast?
7 What systems does the mesodermal layer give rise to?
8 What are the functions of the placenta?

See the blue Answers tab at the back of the book.

Table 28.2 | **An Overview of Prenatal Development**

Background Material in Chapter 3:
Formation of Tissues (p. 78)
Development of Epithelia (p. 79)
Origins of Connective Tissues (p. 80)
Development of Organ Systems (p. 81)

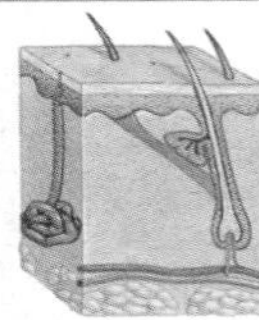
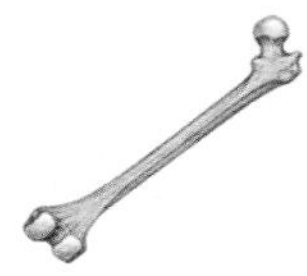
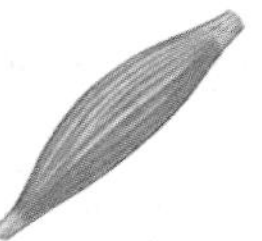
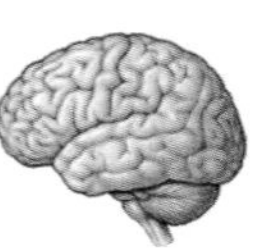
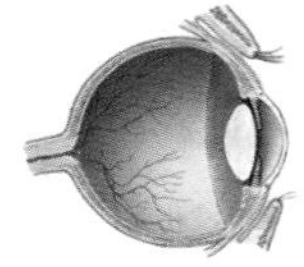

Note: (b) = begin(s) formation; (c) = complete(s) formation.

Gestational Age (months)	Size and Weight	Integumentary System	Skeletal System	Muscular System	Nervous System	Special Sense Organs
1	5 mm, 0.02 g		(b) Somite formation	(b) Somite formation	(b) Neural tube formation	(b) Eye and ear formation
2	1.6 cm, 1.0 g	(b) Formation of nail beds, hair follicles, sweat glands	(b) Formation of axial and appendicular cartilages	(c) Rudiments of axial musculature	(b) CNS, PNS organization, growth of cerebrum	(b) Formation of taste buds, olfactory epithelium
3	5.4 cm, 14 g	(b) Epidermal layers appear	(b) Ossification centers spreading	(c) Rudiments of appendicular musculature	(c) Basic spinal cord and brain structure	
4	11.6 cm, 100 g	(b) Formation of hair, sebaceous glands (c) Sweat glands	(b) Articulations (c) Facial and palatal organization	Fetus starts moving	(b) Rapid expansion of cerebrum	(c) Basic eye and ear structure (b) Peripheral recepto formation
5	16.4 cm, 300 g	(b) Keratin production, nail production			(b) Myelination of spinal cord	
6	30 cm, 600 g			(c) Perineal muscles	(b) CNS tract formation (c) Layering of cortex	
7	37.6 cm, 1.005 kg	(b) Keratinization, nail formation, hair formation				(c) Eyelids open, retina sensitive to light
8	42.4 cm, 1.702 kg		(b) Epiphyseal cartilage formation			(c) Taste receptors functional
9	47.4 cm, 3.2 kg					
Postnatal development		Hair changes in consistency and distribution	Formation and growth of epiphyseal cartilages continue	Muscle mass and control increase	Myelination, layering, CNS tract formation continue	
Embryological Summaries by System		Development of the Integumentary System (pp. 756–757)	Development of the Cranium (pp. 758–759) Development of the Vertebral Column (pp. 760–761) Development of the Appendicular Skeleton (pp. 762–763)	Development of the Muscular System (pp. 764–765)	Introduction to the Development of the Nervous System (p. 766) Development of the Spinal Cord and Spinal Nerves (pp. 767–768) Development of the Brain and Cranial Nerves (pp. 769–770)	Development of Special Sense Organs (pp. 771–772)

28

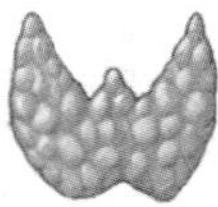 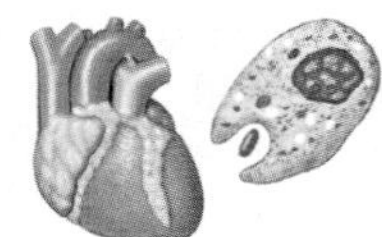 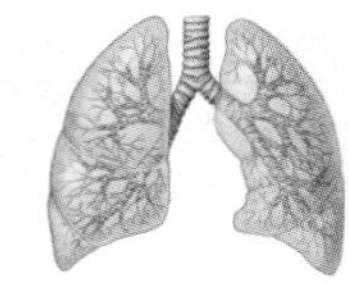 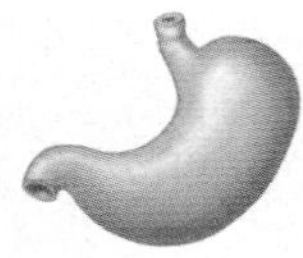 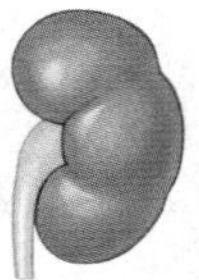 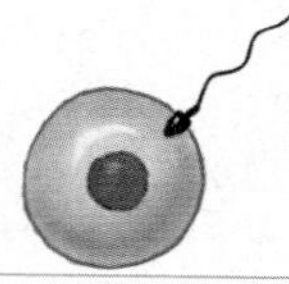

te: (b) = begin(s) formation; (c) = complete(s) formation.

Endocrine System	Cardiovascular and Lymphatic Systems	Respiratory System	Digestive System	Urinary System	Reproductive System
	(b) Heartbeat	(b) Trachea and lung formation	(b) Formation of intestinal tract, liver, pancreas (c) Yolk sac	(c) Allantois	
(b) Formation of thymus, thyroid, pituitary, adrenal glands	(c) Basic heart structure, major blood vessels, lymph nodes and ducts (b) Blood formation in liver	(b) Extensive bronchial branching into mediastinum (c) Diaphragm	(b) Formation of intestinal subdivisions, villi, salivary glands	(b) Kidney formation (adult form)	(b) Formation of mammary glands
(c) Thymus, thyroid gland	(b) Tonsils, blood formation in red bone marrow		(c) Gallbladder, pancreas		(b) Formation of definitive gonads, ducts, genitalia
	(b) Migration of lymphocytes to lymphoid organs: blood formation in spleen			(b) Degeneration of embryonic kidneys	
	(c) Tonsils	(c) Nostrils open	(c) Intestinal subdivisions		
(c) adrenal glands	(c) Spleen, liver, bone marrow	(b) Alveolar formation	(c) Epithelial organization, glands		
(c) Pituitary gland			(c) Intestinal plicae		(b) Descent of testes
		Complete pulmonary branching and alveolar formation		Complete nephron formation at birth	Descent of testes complete at or near time of birth
	Cardiovascular changes at birth; lymphatic system gradually becomes fully operational				
Development of the Endocrine System (**pp. 773–774**)	Development of the Heart (**p. 775**) Development of the Cardiovascular System (**pp. 776–777**) Development of the Lymphatic System (**p. 778**)	Development of the Respiratory System (**pp. 779–780**)	Development of the Digestive System (**pp. 781–782**)	Development of the Urinary System (**pp. 783–784**)	Development of the Reproductive System (**pp. 785–787**)

28

Figure 28.8 **The Second and Third Trimesters**

a A 4-month fetus seen through a fiber-optic endoscope (about 13.3 cm in size)

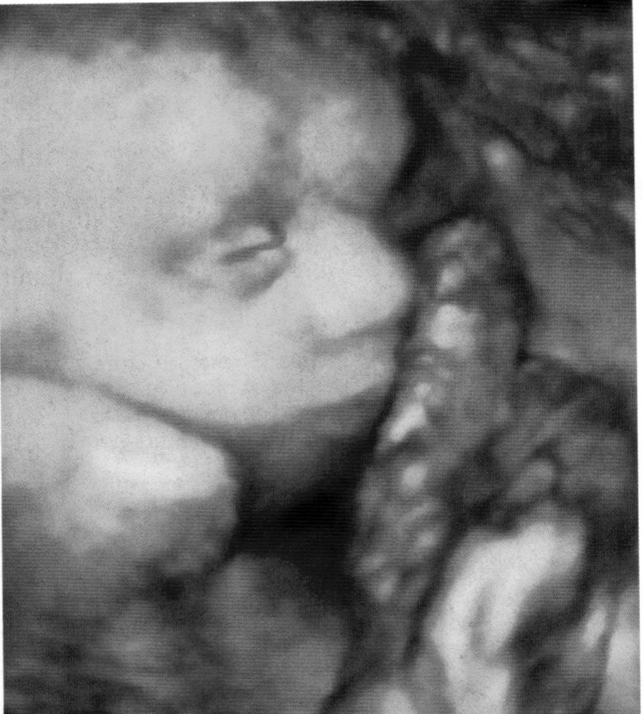

b Head of a 6-month-old fetus, revealed through ultrasound (about 30 cm in size)

28.4 | Labor and Delivery

▶ **KEY POINT** During labor, rhythmic uterine contractions move the fetus toward the cervical canal.

Parturition (par-tū-RISH-un), or *childbirth*, is the forcible expulsion of the fetus from the uterus through the cervix and vagina. This occurs by a series of strong, rhythmic uterine contractions called **labor**. Labor is triggered by the combination of increased oxytocin levels and increased uterine sensitivity to oxytocin. This combination stimulates labor contractions in the myometrium. During true labor, as opposed to the occasional uterine spasms of false labor, each labor contraction begins near the fundus of the uterus and sweeps in a wave toward the cervix. These contractions are strong and occur at regular intervals.

Stages of Labor

▶ **KEY POINT** Labor is divided into three stages: dilation, expulsion, and placental.

The Dilation Stage

The **dilation stage** begins with the onset of true labor, as the cervix dilates and the fetus begins to move down the cervical canal (**Figure 28.10**). This stage is highly variable in length but typically lasts 8 or more hours. At the start of the dilation stage, labor contractions last up to half a minute and occur once every 10–30 minutes. Late in the process, the amnion usually ruptures, an event sometimes referred to as the woman having her "water break."

The Expulsion Stage

The **expulsion stage** begins as the cervix, pushed open by the approaching fetus, completes its dilation to about 10 cm (4 in.) (**Figure 28.10**). Expulsion continues until the fetus has emerged completely from the vagina, a period

CLINICAL NOTE

Multiple Births

Multiple births (twins, triplets, quadruplets, and so forth) can take place for several reasons, including the growing use of fertility drugs and assisted reproductive technology. Another reason is the rise in the average age of women giving birth (older women are more likely to release more than one oocyte in a cycle). According to the CDC, the rate of twin births in the United States is 33.7 twin births out of every 1,000 deliveries. "Fraternal twins," or **dizygotic** (dī-zī-GOT-ik) **twins**, develop when two separate oocytes are ovulated and fertilized. Because chromosomes are shuffled during meiosis, the odds against any two zygotes from the same parents having identical genes exceed 1 in 8.4 million.

"Identical twins" or **monozygotic twins**, result either from the separation of blastomeres early in cleavage or from the splitting of the inner cell mass before gastrulation. In either event, the genetic makeup of the twins is identical because both formed from the same pair of gametes.

Triplets, quadruplets, and larger multiples can result from multiple ovulations, blastomere splitting, or some combination of the two. For unknown reasons, the rates of naturally occurring multiple births fall into a pattern: Twins occur in 1 of every 89 births, triplets in 1 of every 7921 births, quadruplets in 1 of every 704,969 births, and so forth. Taking fertility drugs that stimulate the maturation of abnormally high numbers of follicles can increase the incidence of multiple births.

Figure 28.9 The Growth of the Uterus and Fetus

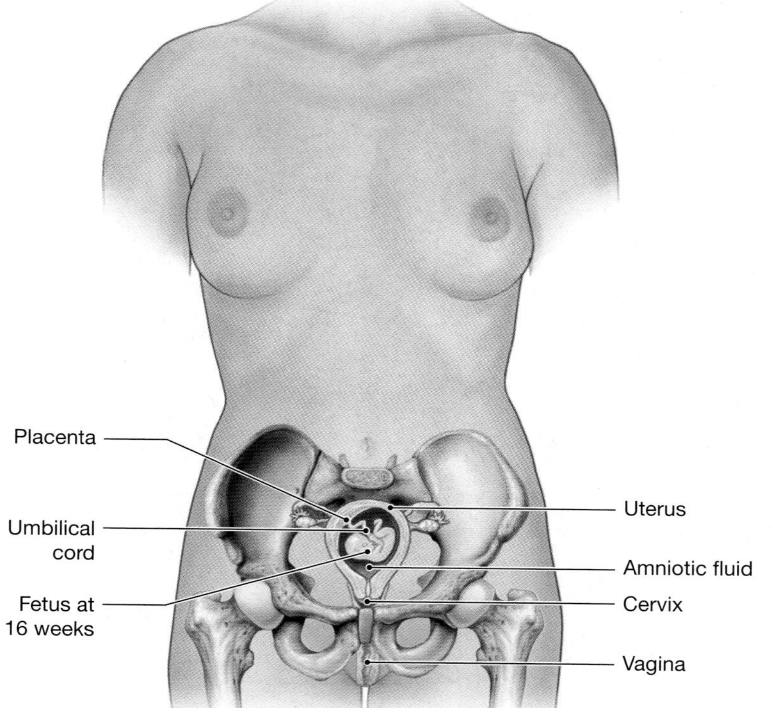

a Pregnancy at 16 weeks showing the positions of the uterus, fetus, and placenta.

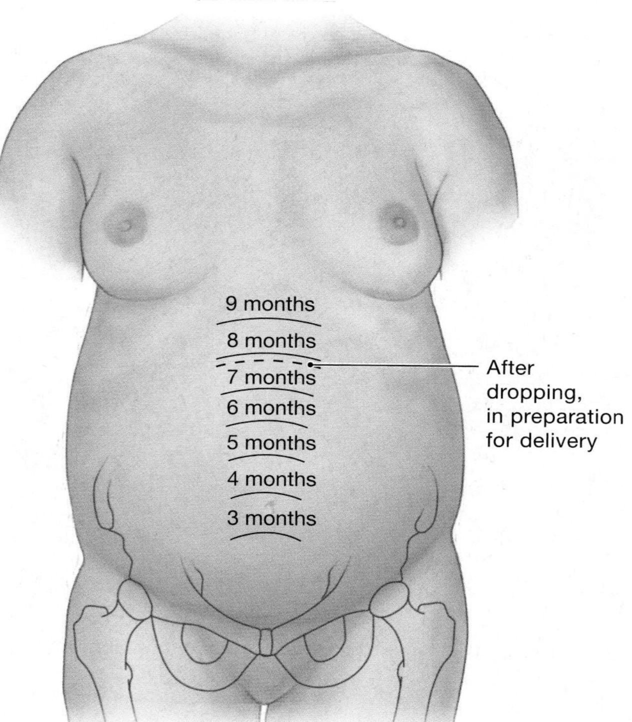

b Pregnancy at 3 months to 9 months, showing the superior-most position of the uterus within the abdomen.

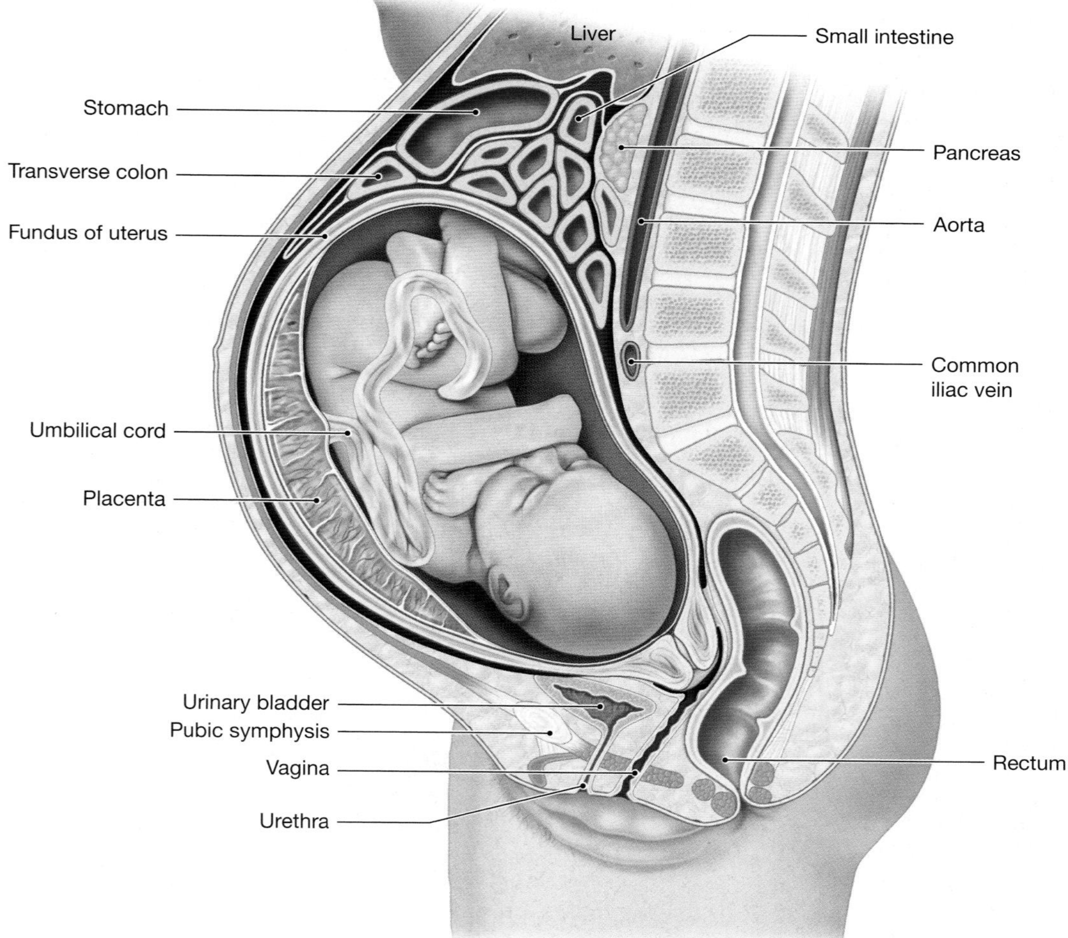

c Pregnancy at full term. Note the position of the uterus and fetus and the displacement of abdominal organs relative to part (d).

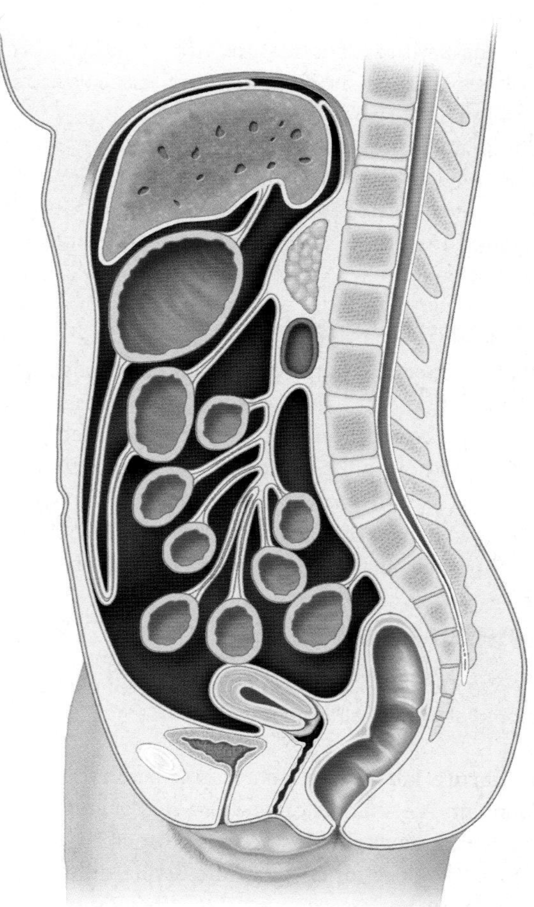

d Organ position and orientation in a nonpregnant female.

28

Figure 28.10 **The Stages of Labor**

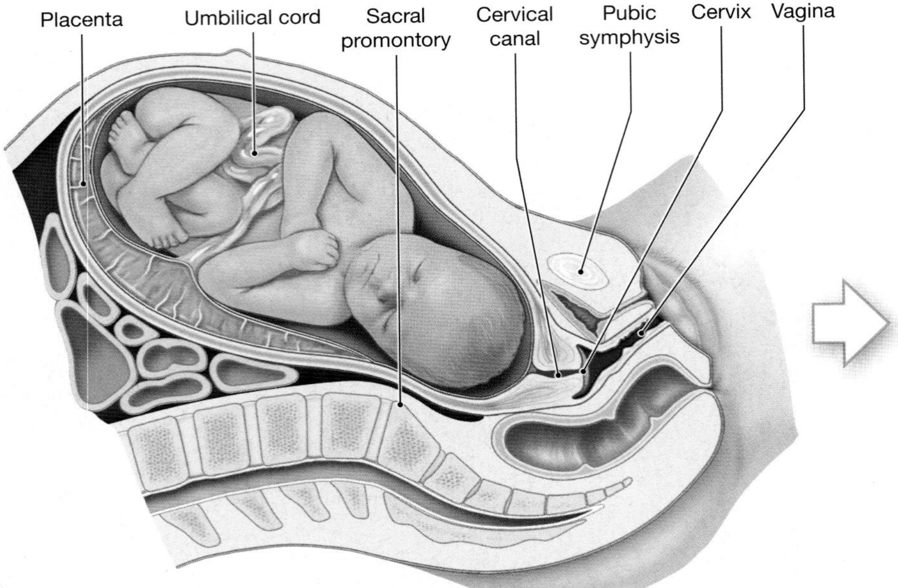

Fully developed fetus before labor begins

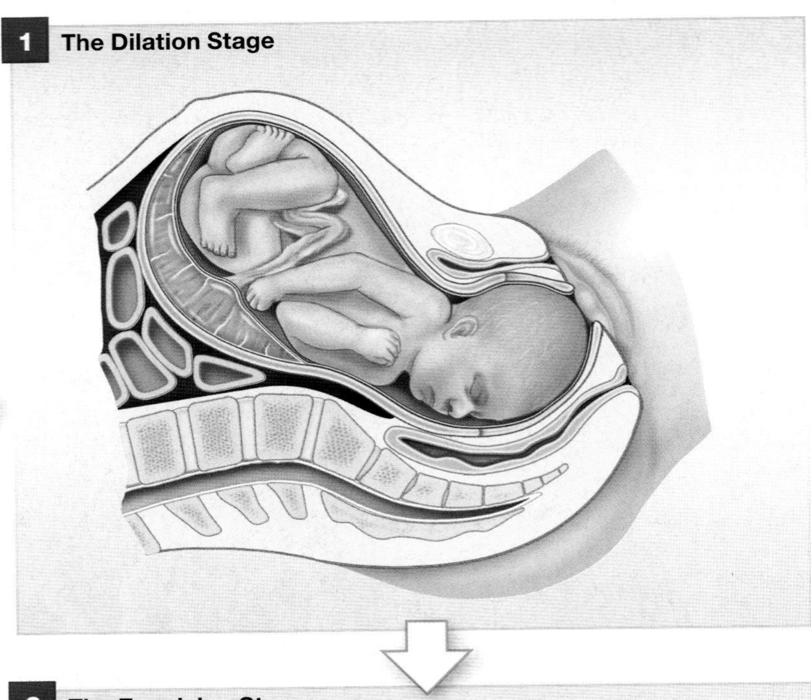

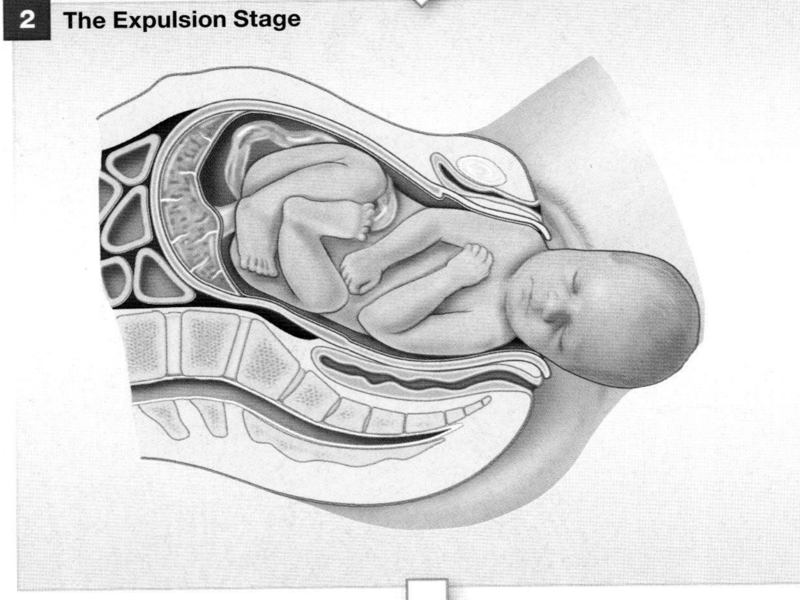

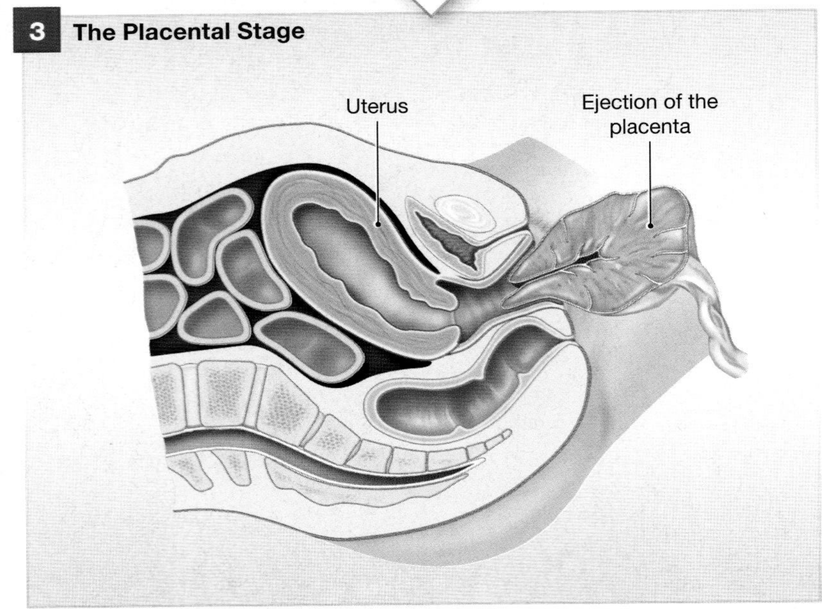

usually lasting less than 2 hours. The arrival of the newborn infant into the outside world is **delivery**, or *birth*.

If the vaginal canal is too small to permit the passage of the fetus, posing acute danger of perineal tearing, a clinician may temporarily enlarge the passageway by performing an **episiotomy** (eh-piz-ē-OT-ō-mē), an incision through the perineal musculature. After delivery, this surgical cut is repaired with sutures, a much simpler procedure than suturing the jagged edges associated with an extensive perineal tear.

The relative sizes of the fetal skull and the maternal pelvic outlet affect the ease and success of delivery. If progress is slow or complications arise, the infant may be delivered by **cesarean section**, or *C-section*. This involves making an incision through the abdominal wall and opening the uterus just enough to remove the infant.

The Placental Stage

During the **placental stage** of labor, muscle tension builds in the walls of the partially empty uterus, which gradually becomes smaller. This uterine contraction tears the connections between the endometrium and the placenta. Usually within an hour after delivery, the placental stage ends with the ejection of the placenta, or *afterbirth* (**Figure 28.10**). The disruption of the placenta is accompanied by loss of blood (as much as 500–600 ml). Because the maternal blood volume has increased during pregnancy, the loss is easily tolerated.

Premature Labor

▶ **KEY POINT** Premature labor occurs when true labor begins before the fetus has completed normal development.

In **premature labor**, true labor begins before the fetus has completed normal development. The chances of newborn survival are directly related to the infant's body weight at delivery. Most fetuses born at 25–27 weeks of gestation (a birth weight under 600 g, or 21.1 oz) die despite intensive neonatal care. Survivors have a high risk of developmental abnormalities. **Premature delivery** usually refers to birth at 28–36 weeks (a birth weight over 1 kg or 2.2 lb). With care, these newborns have a good chance of surviving and developing normally.

28.4 CONCEPT CHECK

9 Which stage of labor usually takes the longest?

10 What triggers the expulsion of the placenta?

See the blue Answers tab at the back of the book.

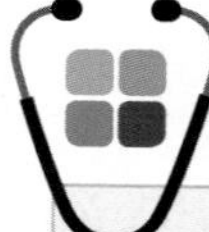

CLINICAL NOTE

Common Complications of Parturition

Parturition is a dangerous time for mothers and newborns. These are some of the more common complications that can occur.

In **fetal distress**, the fetus does not obtain enough oxygen, which typically manifests as a decrease in fetal heart rate. Depending on the stage of labor, this may require either an emergency cesarean section (C-section) or assisted delivery using forceps or a vacuum extractor.

Breech position, also known as "malpresentation," refers to the baby being positioned in the uterus buttocks or feet first. Because the baby's head is its largest part, the infant can get "stuck" in the birth canal if it comes feet first. This is an indication for a C-section.

Placenta previa means the placenta is covering the cervix. If labor progresses naturally, the placenta can separate too early, causing potentially deadly blood loss for mother and baby. A C-section is often scheduled before natural labor can begin.

Meconium aspiration is the intra-uterine aspiration by the fetus of amniotic fluid contaminated with meconium (first feces). If meconium is present in amniotic fluid, the fetus may inhale the feces, resulting in breathing difficulty. Risk factors for meconium aspiration include "aging" of the placenta if past delivery date, decreased oxygen to the fetus in the uterus, diabetes or high blood pressure in the mother, and long labor or a difficult delivery.

Cephalopelvic disproportion (CPD) is a condition in which the fetal head is too large to pass through the maternal pelvis, preventing a vaginal delivery. Cephalopelvic disproportion is the most common reason for a C-section.

28.5 The Neonatal Period

▶ **KEY POINT** During the neonatal period, the neonate continues to undergo physiological and anatomical changes.

Developmental processes do not cease at delivery, as the newborn infant has few of the anatomical, functional, or physiological characteristics of the mature adult. The **neonatal period** extends from the moment of birth to 1 month thereafter. During this period, a variety of physiological and anatomical changes occur as the fetus completes its transition to the status of a newborn infant, or **neonate**.

Before delivery, the transfer of dissolved gases, nutrients, waste products, hormones, and immunoglobulins occurred across the placental interface. At birth, the newborn infant must become relatively self-sufficient, with its own organs and organ systems performing respiration, digestion, and excretion. We can summarize the transition from fetus to neonate as follows:

1. At birth, the lungs are collapsed and filled with fluid. Filling them with air involves a massive and powerful inhalation.
2. When the lungs expand, changes in blood pressure and flow rates within the cardiovascular system alter the blood flow pattern. The ductus arteriosus closes, isolating the pulmonary and systemic trunks. The foramen ovale closes, separating the atria of the heart, which completes the separation of the pulmonary and systemic circuits. (Chapters 21 and 22 discussed these cardiovascular changes.)
3. Heart rates of 120–140 beats per minute and respiratory rates of 30 breaths per minute, rates considerably higher than those of adults, are normal in neonates.
4. Before birth, the digestive system is relatively inactive, although it does accumulate a mixture of bile secretions, mucus, and epithelial cells. This collection of debris, called meconium, is excreted in the first few days of life. During that period, the newborn begins to nurse.
5. As wastes build up in the arterial blood, the kidneys excrete them. Glomerular filtration is normal, but the neonate's kidneys cannot concentrate urine to any significant degree. As a result, urinary water losses are high, and neonatal fluid requirements are proportionally much greater than those of adults.
6. In the first few days after delivery, the neonate has little ability to control body temperature. As the infant grows larger and increases the thickness of its insulating subcutaneous adipose "blanket," its metabolic rate also rises. Daily and even hourly alterations in body temperature continue throughout childhood.

The next section of this chapter summarizes important aspects of the embryological development of each organ system. Table 28.2 on pages 750–751 summarizes major developmental landmarks in each trimester.

28.5 CONCEPT CHECK

11 Why does a newborn need a higher relative fluid intake than an adult?

12 While you are holding your newborn child for the first time, you notice that her heart rate is considerably higher than yours. Should you be worried? Explain.

See the blue Answers tab at the back of the book.

28.6 | Embryology of Organ Systems

The Development of the Integumentary System

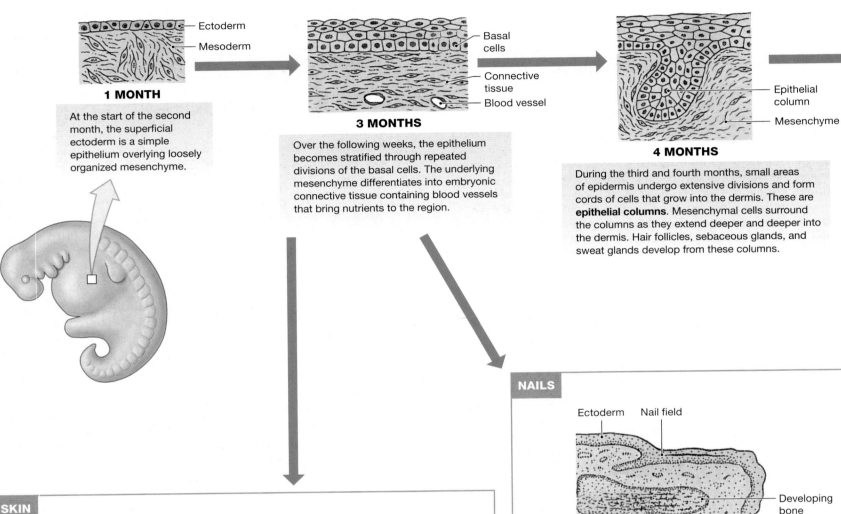

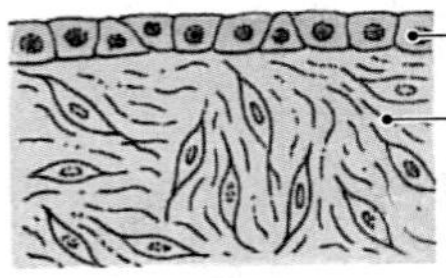

Ectoderm
Mesoderm

1 MONTH

At the start of the second month, the superficial ectoderm is a simple epithelium overlying loosely organized mesenchyme.

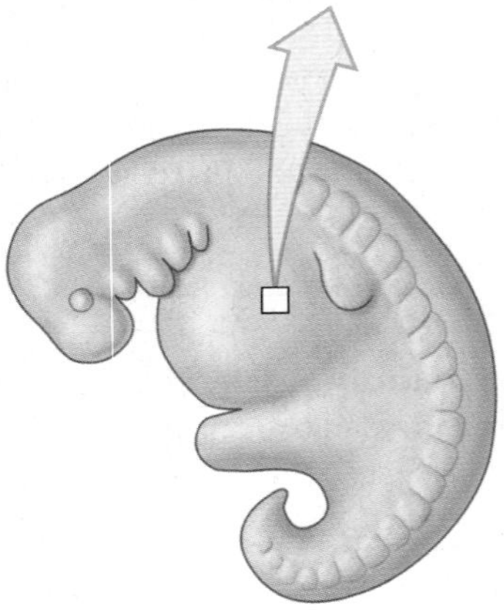

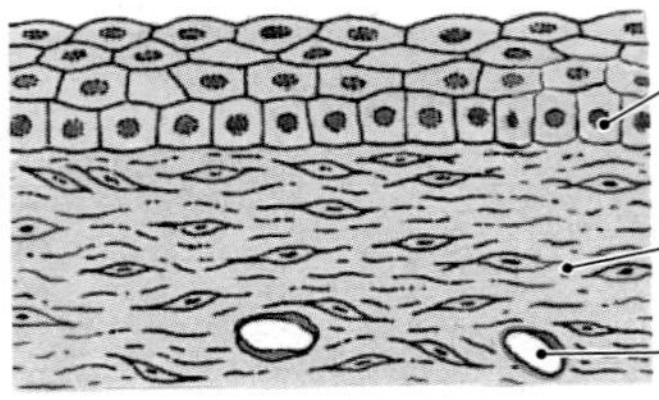

Basal cells
Connective tissue
Blood vessel

3 MONTHS

Over the following weeks, the epithelium becomes stratified through repeated divisions of the basal cells. The underlying mesenchyme differentiates into embryonic connective tissue containing blood vessels that bring nutrients to the region.

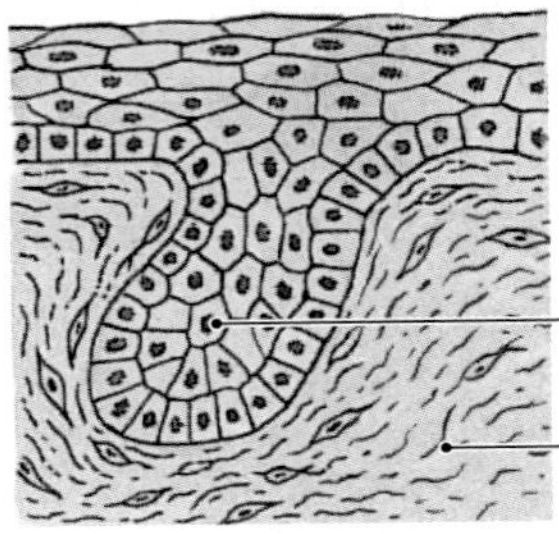

Epithelial column
Mesenchyme

4 MONTHS

During the third and fourth months, small areas of epidermis undergo extensive divisions and form cords of cells that grow into the dermis. These are **epithelial columns**. Mesenchymal cells surround the columns as they extend deeper and deeper into the dermis. Hair follicles, sebaceous glands, and sweat glands develop from these columns.

SKIN

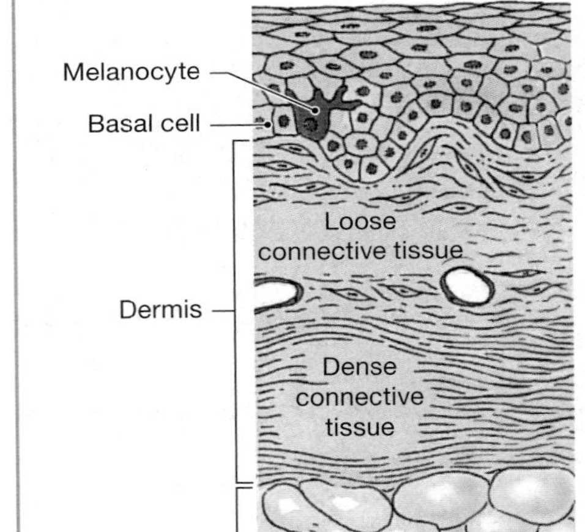

Melanocyte
Basal cell
Loose connective tissue
Dermis
Dense connective tissue
Subcutaneous layer

4 MONTHS

As basal cell divisions continue, the epithelial layer thickens and the basal layer forms irregular folds. Pigment cells called melanocytes migrate into the area and squeeze between the basal cells. The epithelium now resembles the epidermis of an adult.

The embryonic connective tissue differentiates into the dermis. Fibroblasts and other connective tissue cells form from mesenchymal cells or migrate into the area. The density of fibers increases. Loose connective tissue extends into the ridges, but a deeper, less vascular region is dominated by a dense, irregular collagen fiber network. Below the dermis the embryonic connective tissue develops into the subcutaneous layer, a layer of loose connective tissue.

NAILS

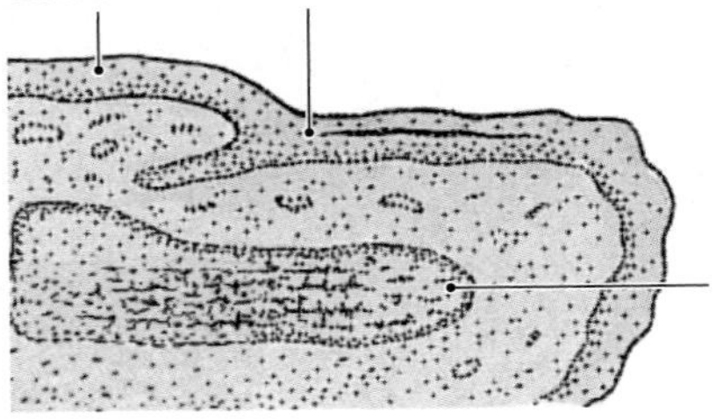

Ectoderm
Nail field
Developing bone

4 MONTHS

Nails begin as thickenings of the epidermis near the tips of the fingers and toes. These thickenings settle into the dermis, and the border with the general epidermis becomes distinct. Initially, nail production involves all of the basal cells of the nail field.

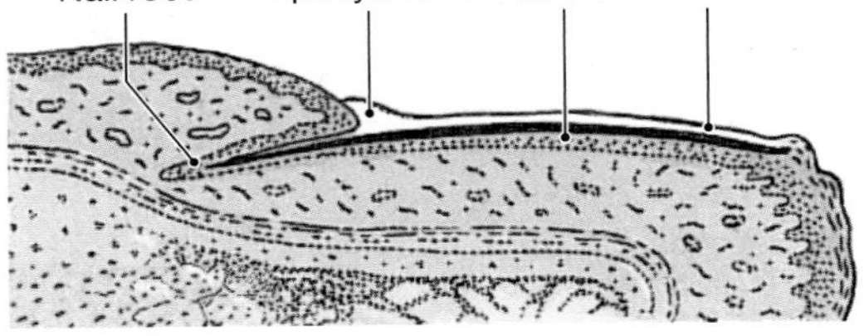

Nail root
Eponychium
Nail bed
Nail

BIRTH

By the time of birth, nail production is restricted to the nail root.

28

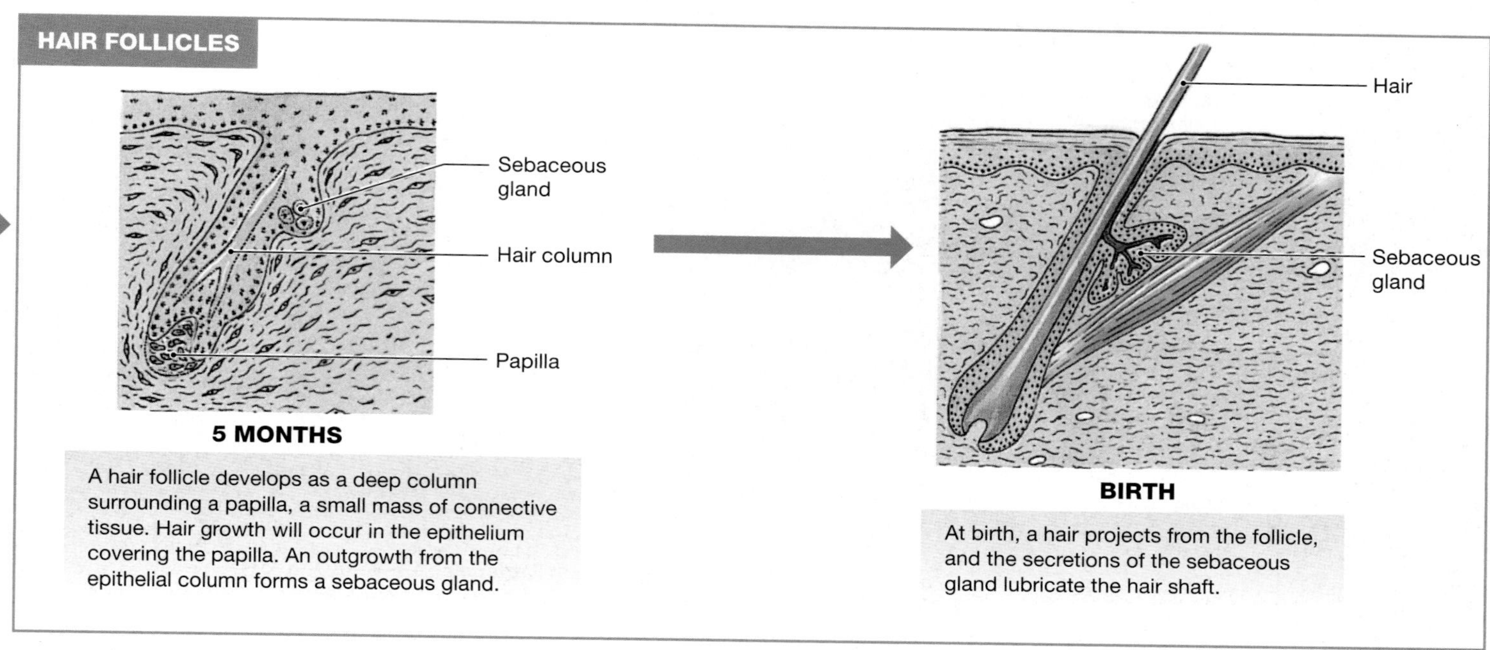

A hair follicle develops as a deep column surrounding a papilla, a small mass of connective tissue. Hair growth will occur in the epithelium covering the papilla. An outgrowth from the epithelial column forms a sebaceous gland.

At birth, a hair projects from the follicle, and the secretions of the sebaceous gland lubricate the hair shaft.

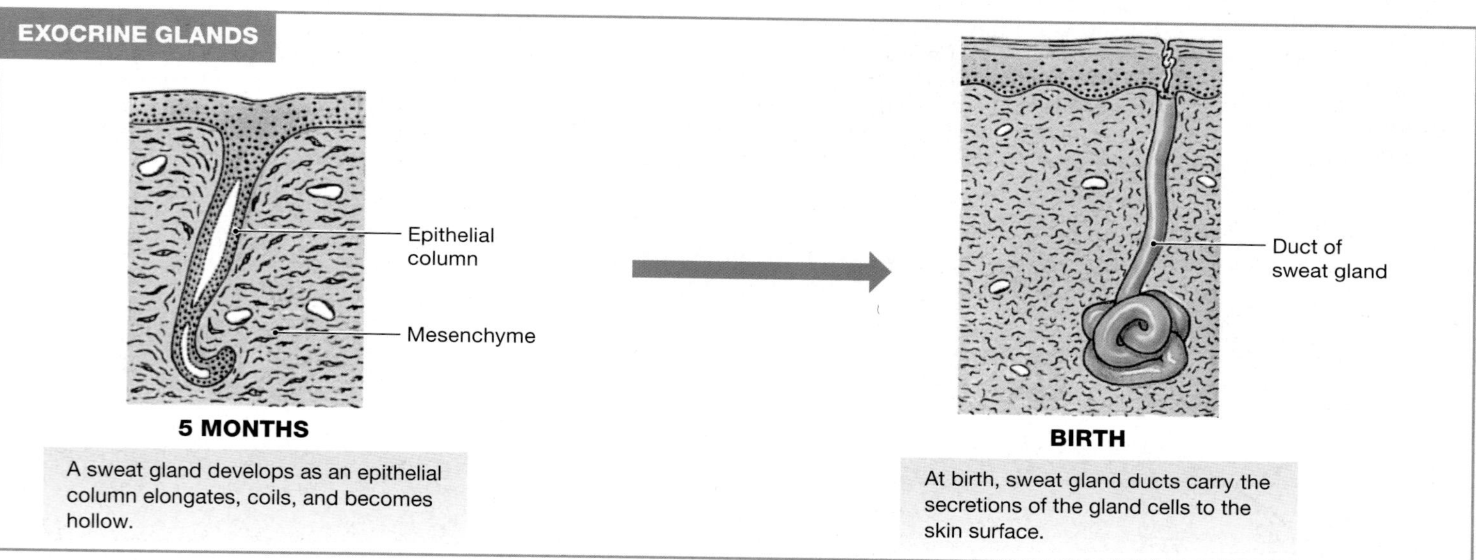

A sweat gland develops as an epithelial column elongates, coils, and becomes hollow.

At birth, sweat gland ducts carry the secretions of the gland cells to the skin surface.

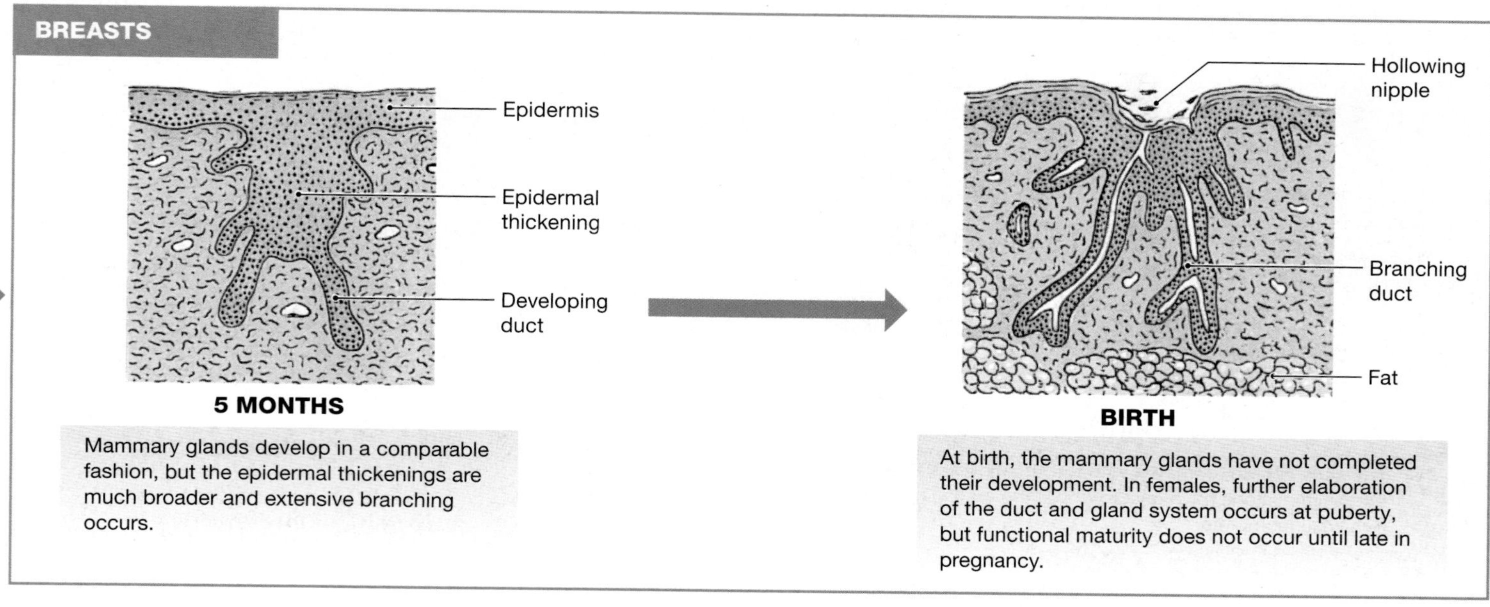

Mammary glands develop in a comparable fashion, but the epidermal thickenings are much broader and extensive branching occurs.

At birth, the mammary glands have not completed their development. In females, further elaboration of the duct and gland system occurs at puberty, but functional maturity does not occur until late in pregnancy.

The Development of the Cranium

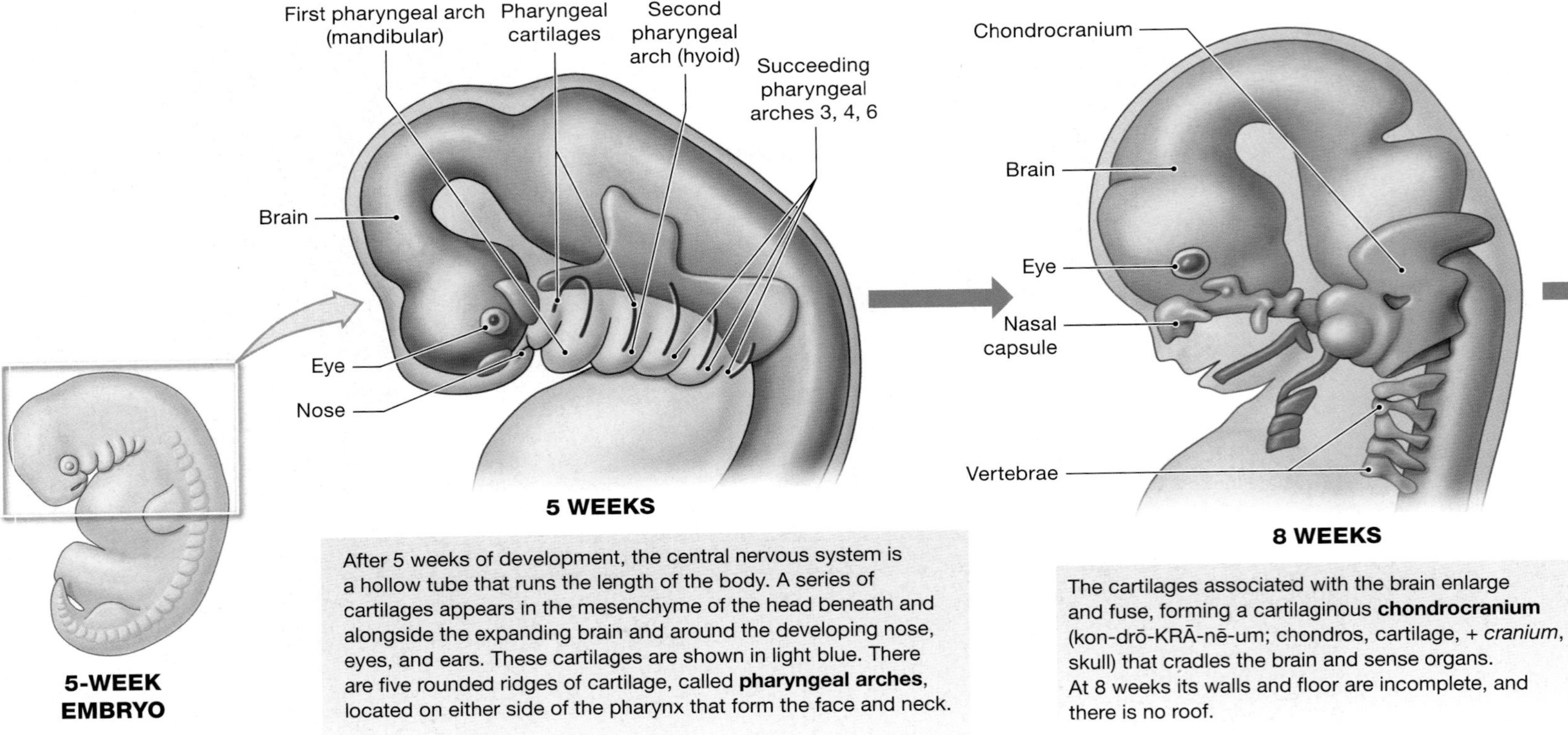

After 5 weeks of development, the central nervous system is a hollow tube that runs the length of the body. A series of cartilages appears in the mesenchyme of the head beneath and alongside the expanding brain and around the developing nose, eyes, and ears. These cartilages are shown in light blue. There are five rounded ridges of cartilage, called **pharyngeal arches**, located on either side of the pharynx that form the face and neck.

The cartilages associated with the brain enlarge and fuse, forming a cartilaginous **chondrocranium** (kon-drō-KRĀ-nē-um; chondros, cartilage, + *cranium*, skull) that cradles the brain and sense organs. At 8 weeks its walls and floor are incomplete, and there is no roof.

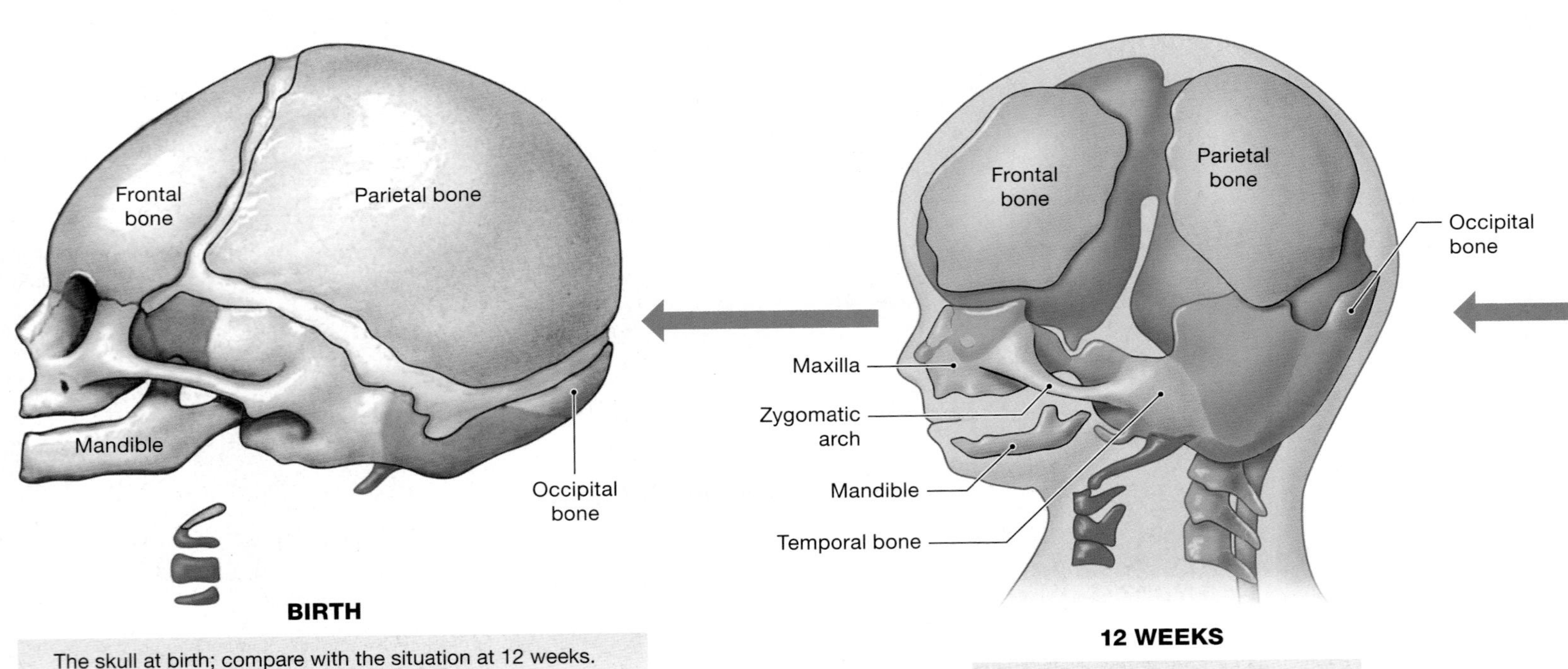

After 12 weeks ossification is well under way in the cranium and face.

The skull at birth; compare with the situation at 12 weeks. Extensive fusions have occurred, but the cranial roof remains incomplete.

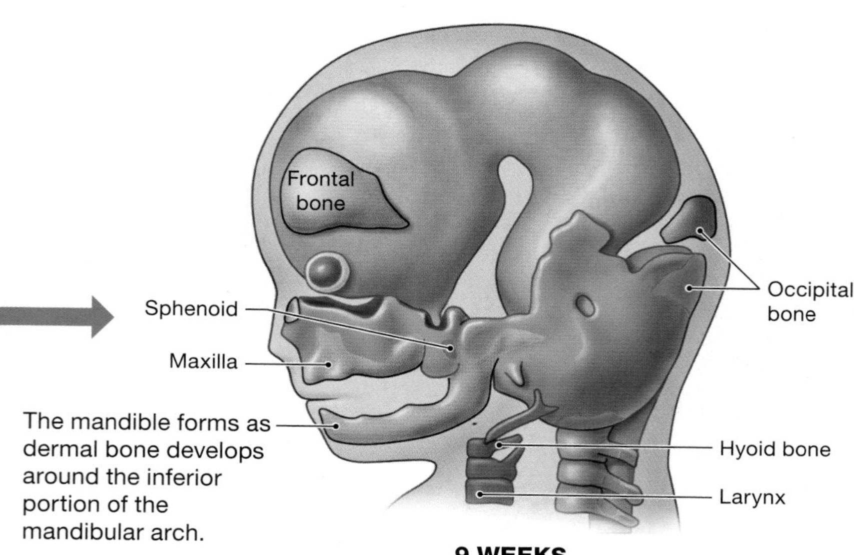

During the ninth week, numerous centers of endochondral ossification appear within the chondrocranium. These centers are shown in red. Gradually, the frontal and parietal bones of the cranial roof appear as intramembranous ossification begins in the overlying dermis. As these centers (beige) enlarge and expand, extensive fusions occur.

The dorsal portion of the mandibular arch fuses with the chondrocranium. The fused cartilages do not ossify; instead, osteoblasts begin covering them in dermal bone. On each side this covering fuses with a bone developing at the entrance to the nasal cavity, producing the two maxillae. Ossification centers in the roof of the mouth spread to form the palatine processes and later fuse with the maxillae.

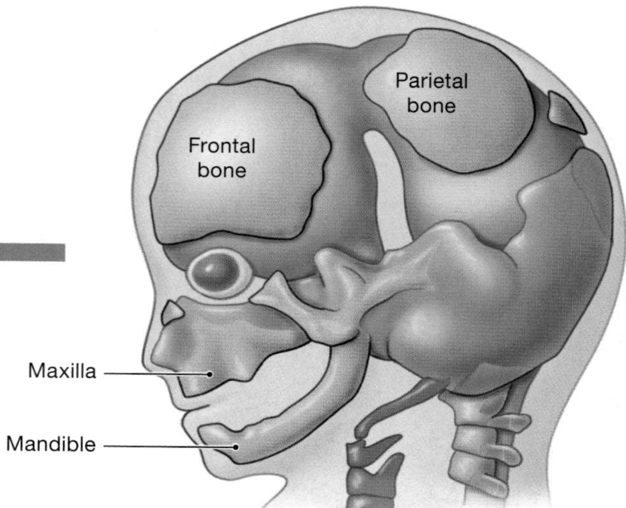

The second pharyngeal arch, or **hyoid arch**, forms near the temporal bones. Fusion of the superior tips of the hyoid with the temporals forms the styloid processes. The ventral portion of the hyoid arch ossifies as the hyoid bone. The third arch fuses with the hyoid, and the fourth and sixth arches form laryngeal cartilages.

CLINICAL NOTE

Development of Cleft Lip and Cleft Palate

If the overlying skin does not fuse normally, the result is a **cleft lip**. Cleft lips affect roughly one birth in a thousand. A split extending into the orbit and palate is called a **cleft palate**. Cleft palates are half as common as cleft lips. Both conditions can be corrected surgically.

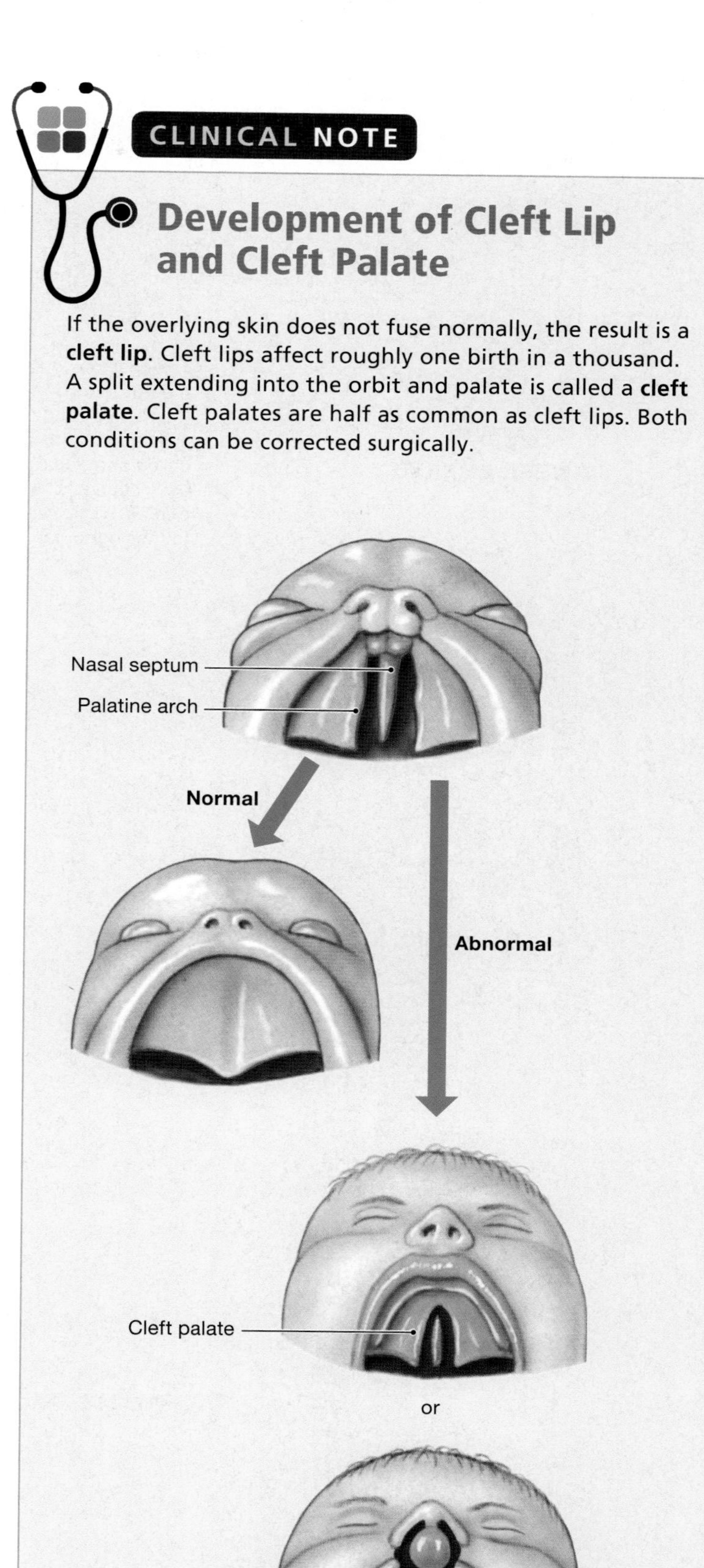

28

The Development of the Vertebral Column

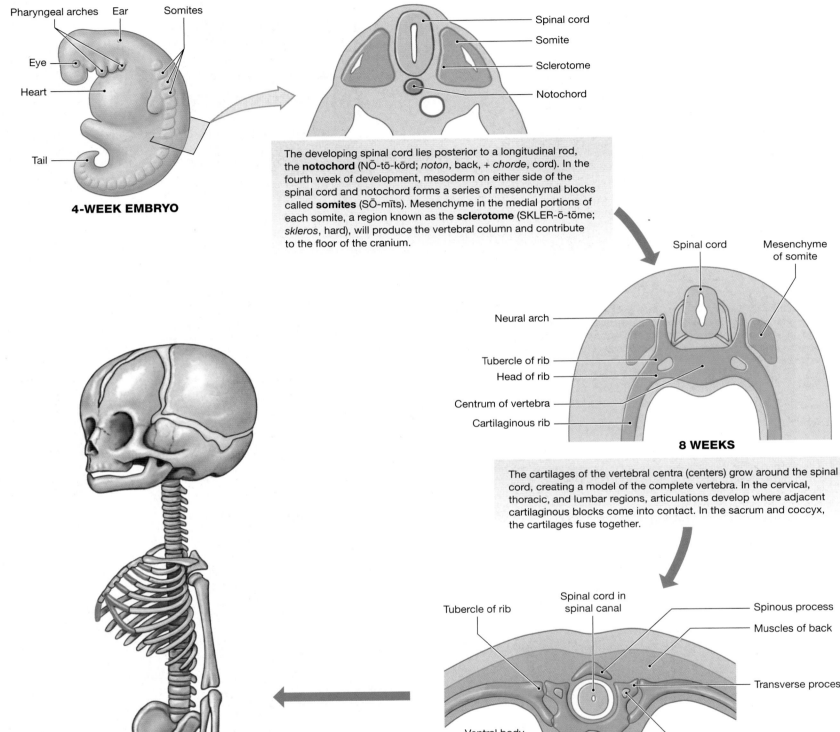

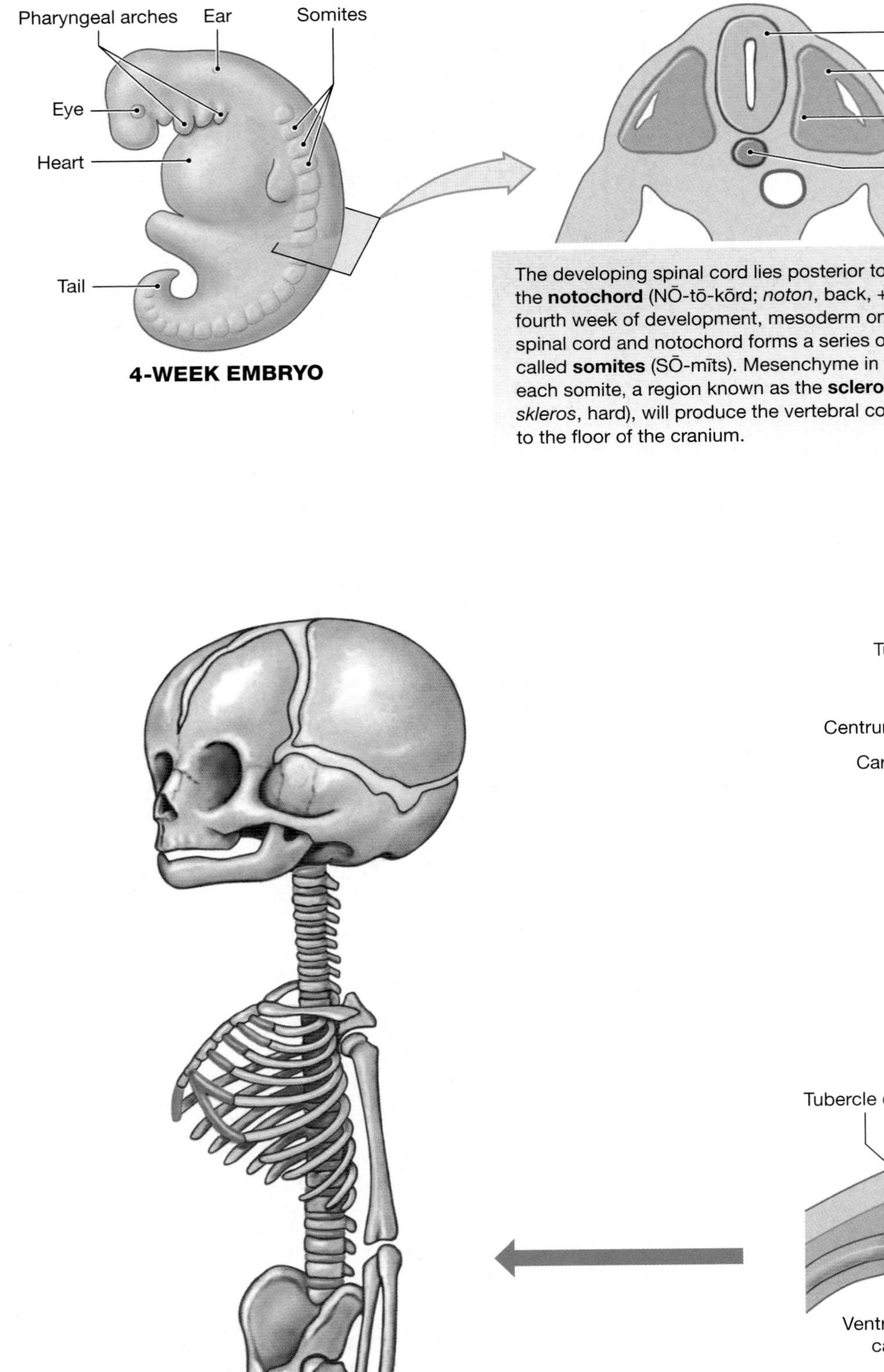

The developing spinal cord lies posterior to a longitudinal rod, the **notochord** (NŌ-tō-kōrd; *noton*, back, + *chorde*, cord). In the fourth week of development, mesoderm on either side of the spinal cord and notochord forms a series of mesenchymal blocks called **somites** (SŌ-mīts). Mesenchyme in the medial portions of each somite, a region known as the **sclerotome** (SKLER-ō-tōme; *skleros*, hard), will produce the vertebral column and contribute to the floor of the cranium.

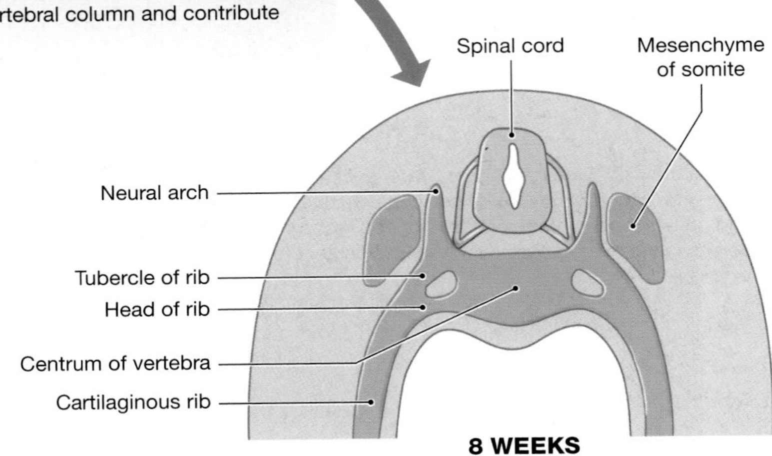

The cartilages of the vertebral centra (centers) grow around the spinal cord, creating a model of the complete vertebra. In the cervical, thoracic, and lumbar regions, articulations develop where adjacent cartilaginous blocks come into contact. In the sacrum and coccyx, the cartilages fuse together.

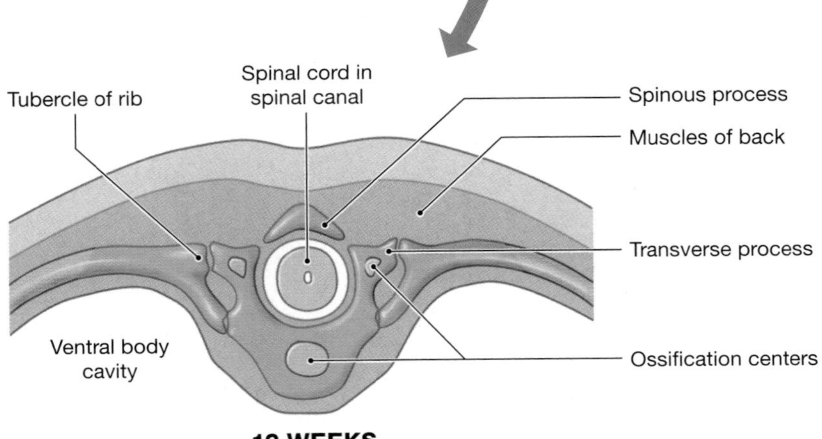

About the time the ribs separate from the vertebrae, ossification begins. Only the shortest ribs undergo complete ossification. In the rest, the distal portions remain cartilaginous, forming the costal cartilages. Several ossification centers appear in the sternum, but fusion gradually reduces the number.

At birth, the vertebrae and ribs are ossified, but many cartilaginous areas remain. For example, the anterior portions of the ribs remain cartilaginous. Additional growth will occur for many years; in vertebrae, the bases of the neural arches enlarge until ages 3–6, and the vertebral processes and vertebral bodies grow until ages 18–25.

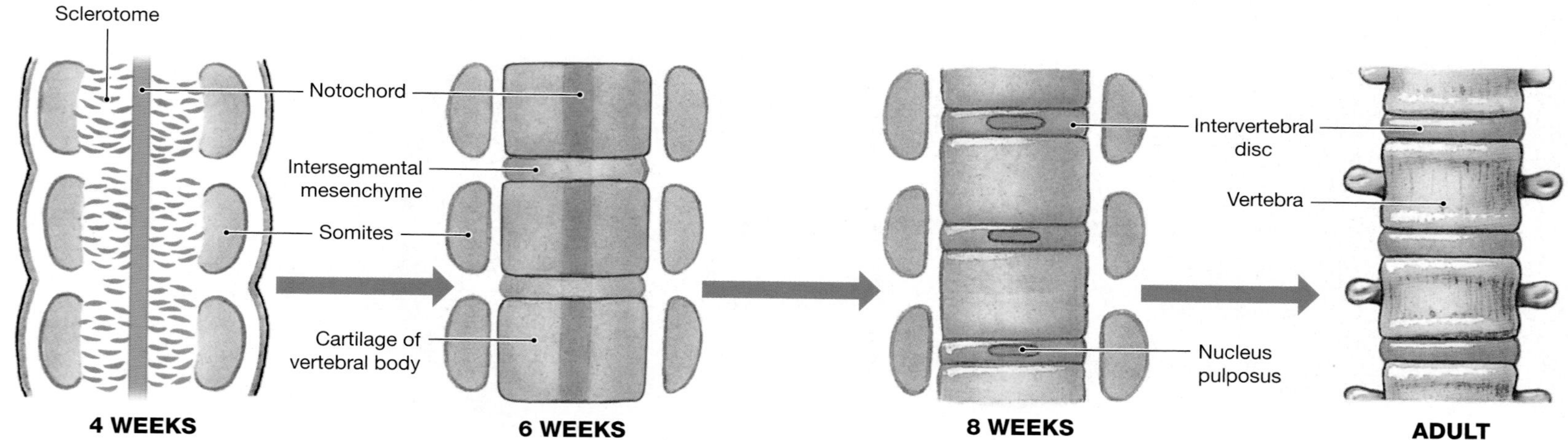

Cells of the sclerotomal segments migrate away from the somites and cluster around the notochord.

The migrating cells differentiate into chondroblasts and produce a series of cartilaginous blocks that surround the notochord. These cartilages, which will develop into the vertebral centra, are separated by patches of mesenchyme.

Expansion of the vertebral centra eventually eliminates the notochord, but it remains intact between adjacent vertebrae, forming the nucleus pulposus of the intervertebral discs. Later, surrounding mesenchymal cells differentiate into chondroblasts and produce the fibrous cartilage of the anulus fibrosus.

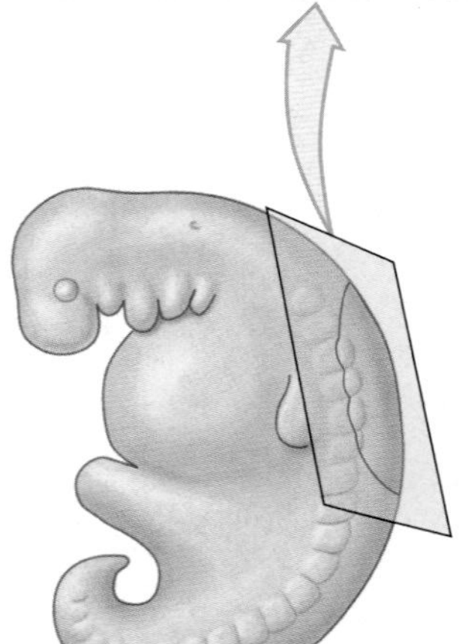

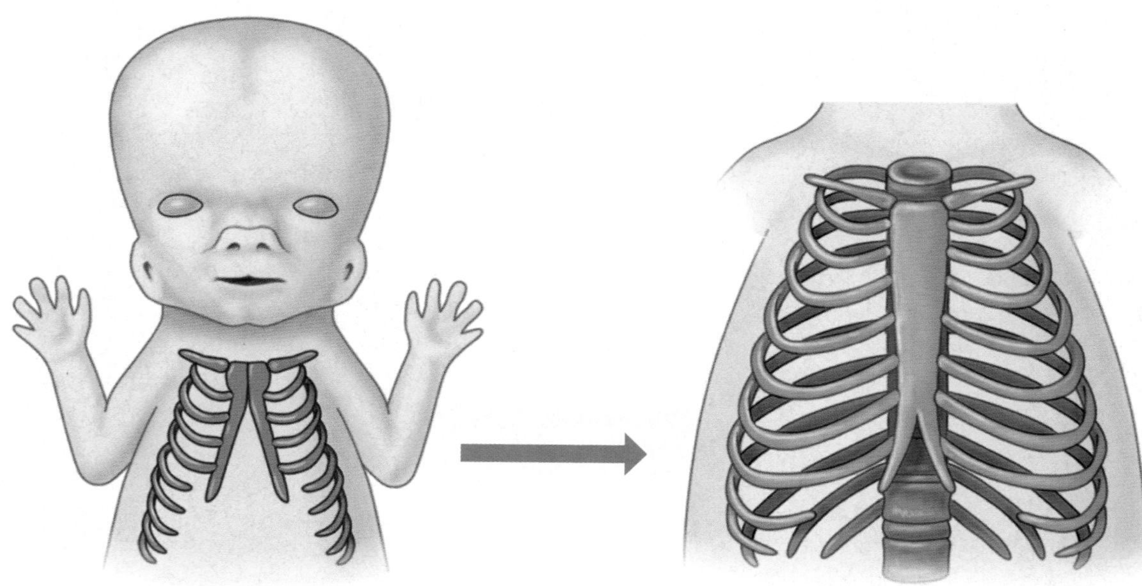

Rib cartilages expand away from the developing transverse processes of the vertebrae. At first they are continuous, but by week 8 the ribs have separated from the vertebrae. Ribs form at every vertebra, but in the cervical, lumbar, sacral, and coccygeal regions they remain small and later fuse with the growing vertebrae. The ribs of the thoracic vertebrae continue to enlarge, following the curvature of the body wall. When they reach the ventral midline, they fuse with the cartilages of the sternum.

The Development of the Appendicular Skeleton

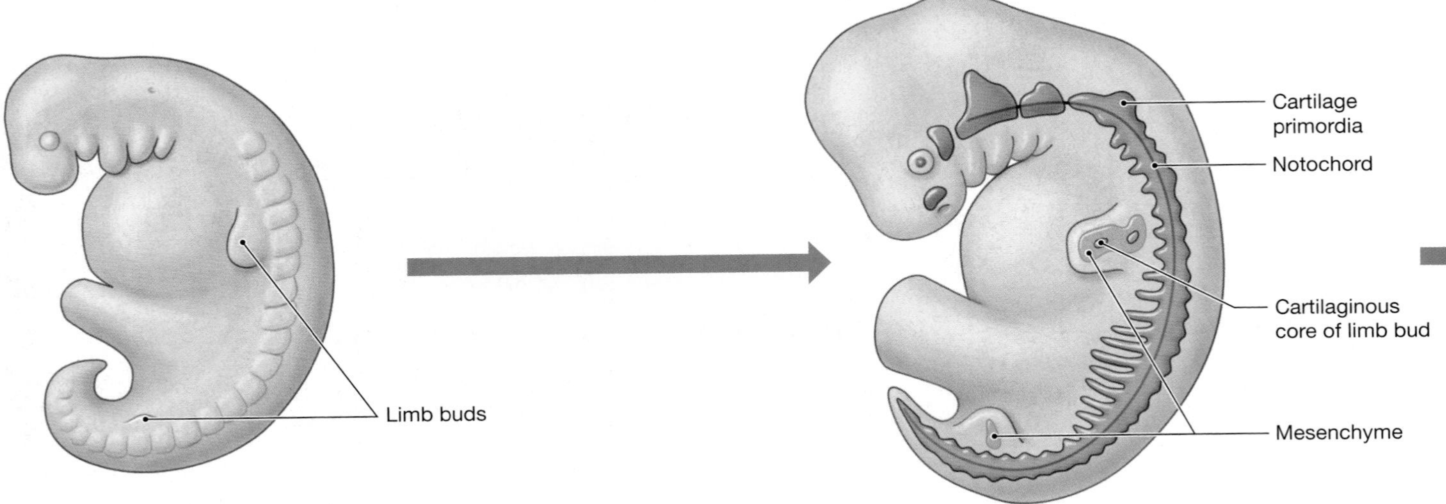

4 WEEKS

In the fourth week of development, ridges appear along the abdomen of the embryo, extending from just behind the throat to just before the anus. These ridges form as mesodermal cells and congregate beneath the ectoderm of the abdomen. Mesoderm gradually accumulates at the end of each ridge, forming two pairs of limb buds.

5 WEEKS

After 5 weeks of development, the pectoral limb buds have a cartilaginous core and scapular cartilages are developing in the mesenchyme of the trunk.

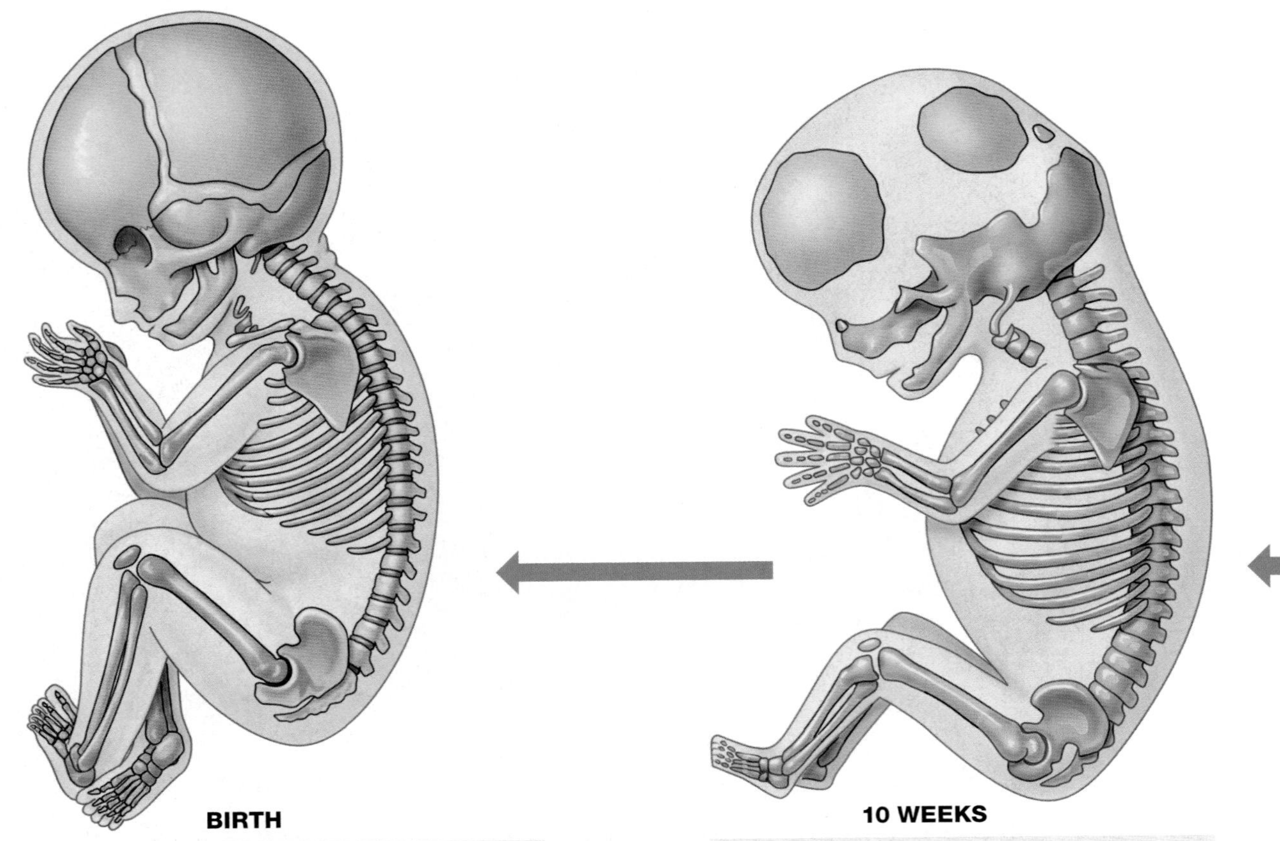

10 WEEKS

Ossification in the embryonic skeleton after approximately 10 weeks of development. The shafts of the limb bones are undergoing rapid ossification, but the distal bones of the carpus and tarsus remain cartilaginous.

BIRTH

The skeleton of a newborn infant. Note the extensive areas of cartilage (blue) in the humeral head, wrist, between the bones of the palm and fingers, and in the hips. Notice the appearance of the axial skeleton, with reference to the two previous Embryology Summaries.

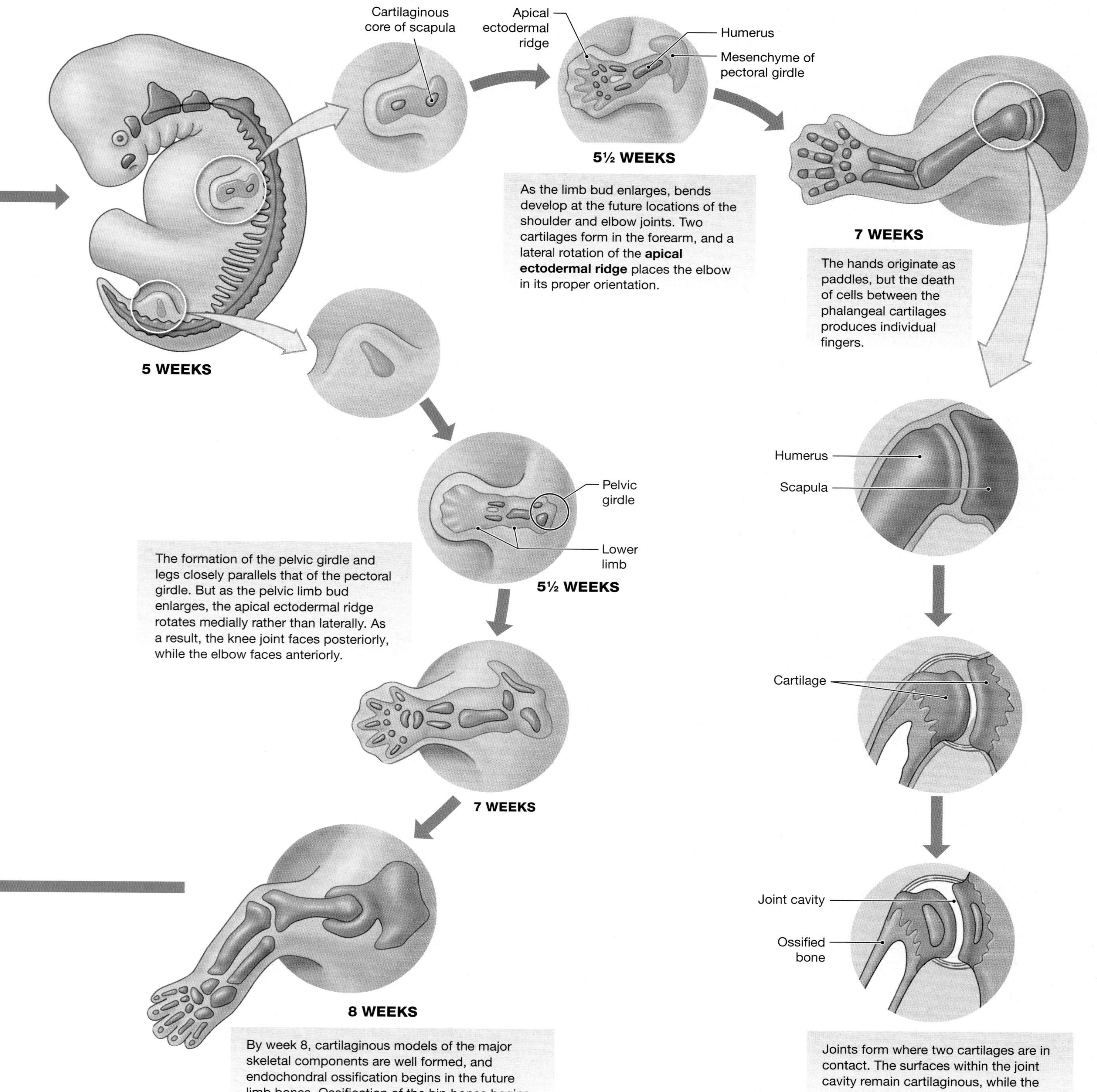
Cartilaginous core of scapula
Apical ectodermal ridge
Humerus
Mesenchyme of pectoral girdle
5½ WEEKS
As the limb bud enlarges, bends develop at the future locations of the shoulder and elbow joints. Two cartilages form in the forearm, and a lateral rotation of the **apical ectodermal ridge** places the elbow in its proper orientation.
7 WEEKS
The hands originate as paddles, but the death of cells between the phalangeal cartilages produces individual fingers.
5 WEEKS
Pelvic girdle
Lower limb
5½ WEEKS
The formation of the pelvic girdle and legs closely parallels that of the pectoral girdle. But as the pelvic limb bud enlarges, the apical ectodermal ridge rotates medially rather than laterally. As a result, the knee joint faces posteriorly, while the elbow faces anteriorly.
Humerus
Scapula
Cartilage
7 WEEKS
Joint cavity
Ossified bone
8 WEEKS
By week 8, cartilaginous models of the major skeletal components are well formed, and endochondral ossification begins in the future limb bones. Ossification of the hip bones begins at three separate centers that gradually enlarge.
Joints form where two cartilages are in contact. The surfaces within the joint cavity remain cartilaginous, while the rest of the bones undergo ossification.

The Development of the Muscles

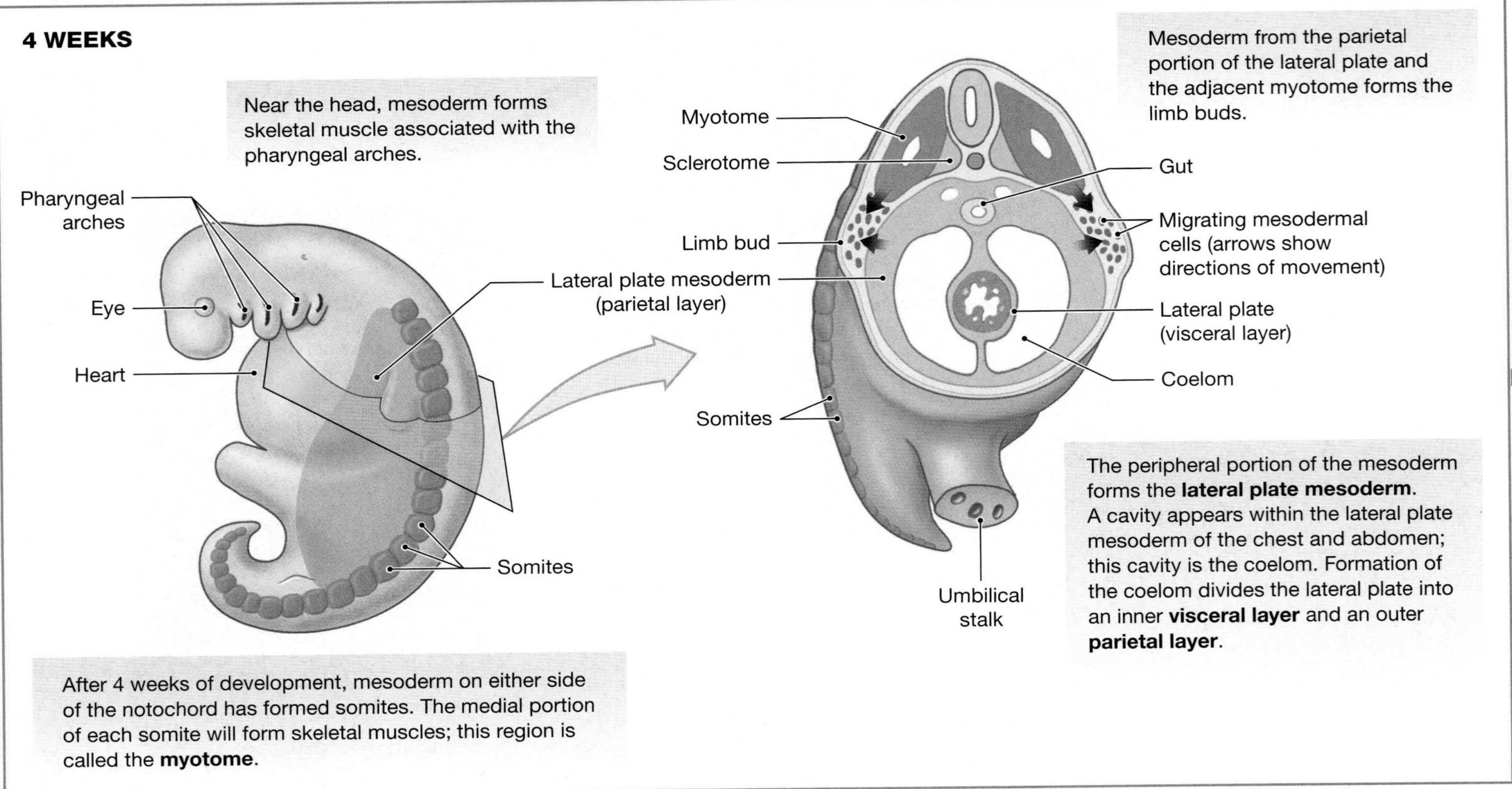

4 WEEKS
Near the head, mesoderm forms skeletal muscle associated with the pharyngeal arches.
Pharyngeal arches
Eye
Heart
Lateral plate mesoderm (parietal layer)
Somites
Myotome
Sclerotome
Limb bud
Somites
Umbilical stalk
Mesoderm from the parietal portion of the lateral plate and the adjacent myotome forms the limb buds.
Gut
Migrating mesodermal cells (arrows show directions of movement)
Lateral plate (visceral layer)
Coelom
The peripheral portion of the mesoderm forms the **lateral plate mesoderm.** A cavity appears within the lateral plate mesoderm of the chest and abdomen; this cavity is the coelom. Formation of the coelom divides the lateral plate into an inner **visceral layer** and an outer **parietal layer**.
After 4 weeks of development, mesoderm on either side of the notochord has formed somites. The medial portion of each somite will form skeletal muscles; this region is called the **myotome**.

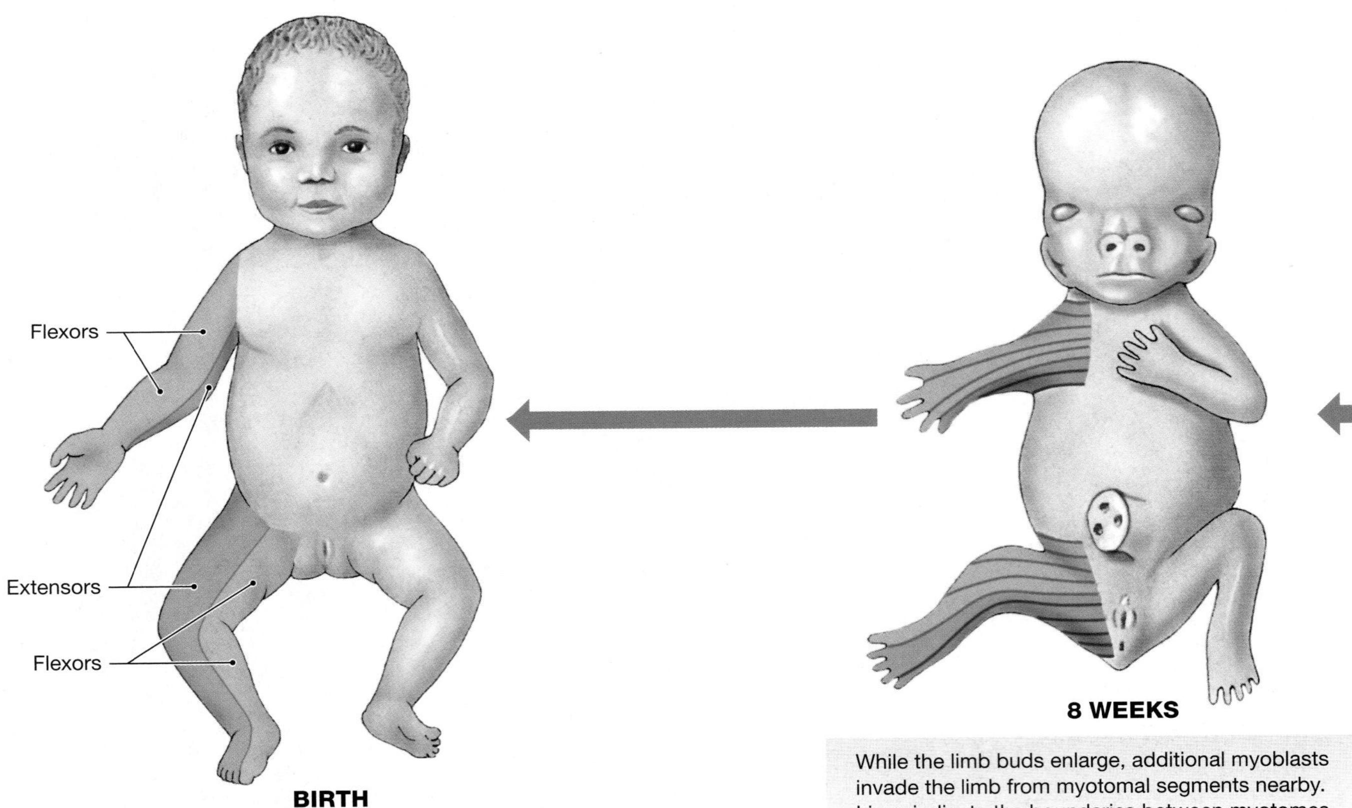

While the limb buds enlarge, additional myoblasts invade the limb from myotomal segments nearby. Lines indicate the boundaries between myotomes providing myoblasts to the limb.

Rotation of the upper limb bud and lower limb bud produces a change in the position of these masses relative to the body axis.

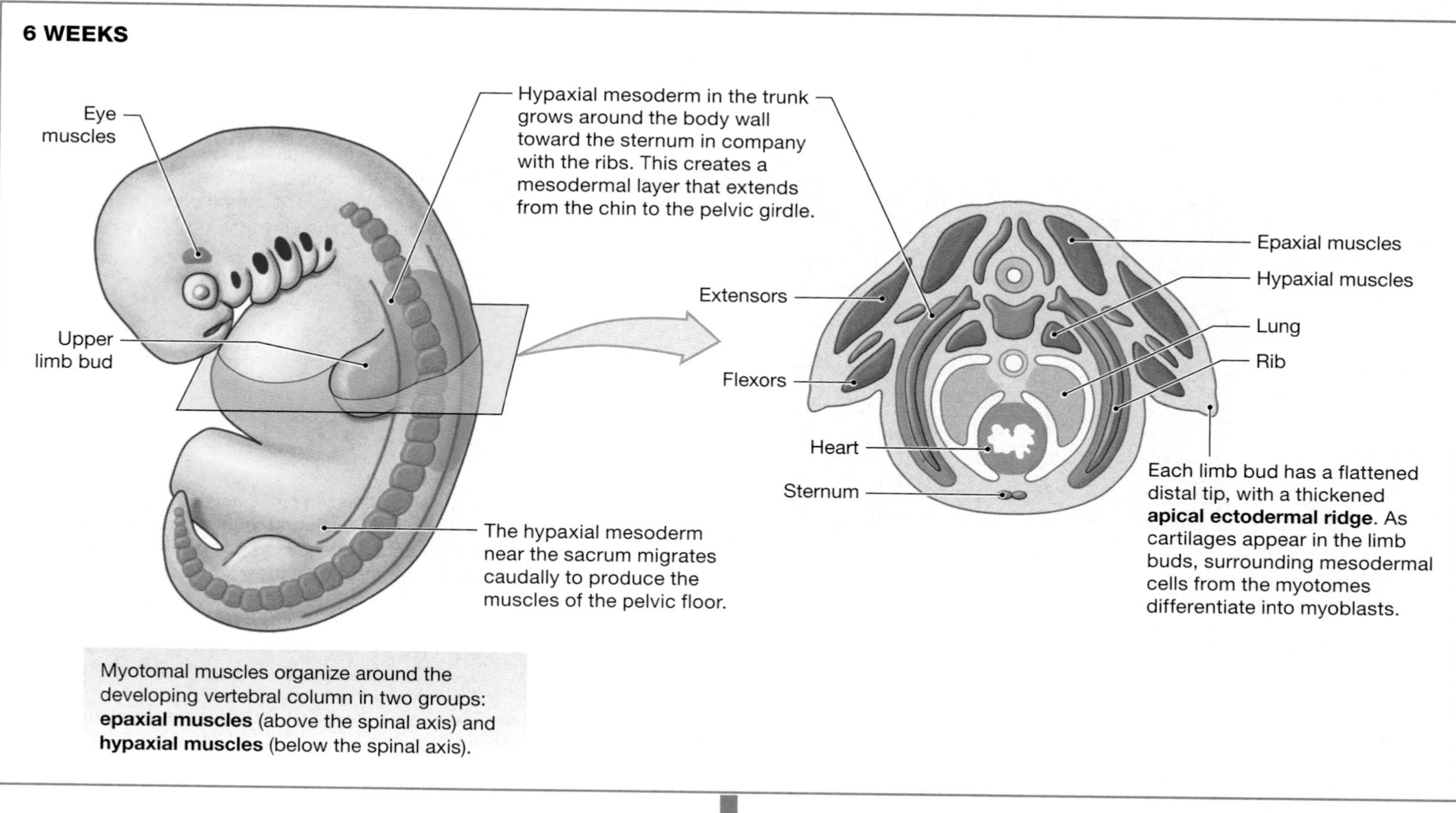

7 WEEKS

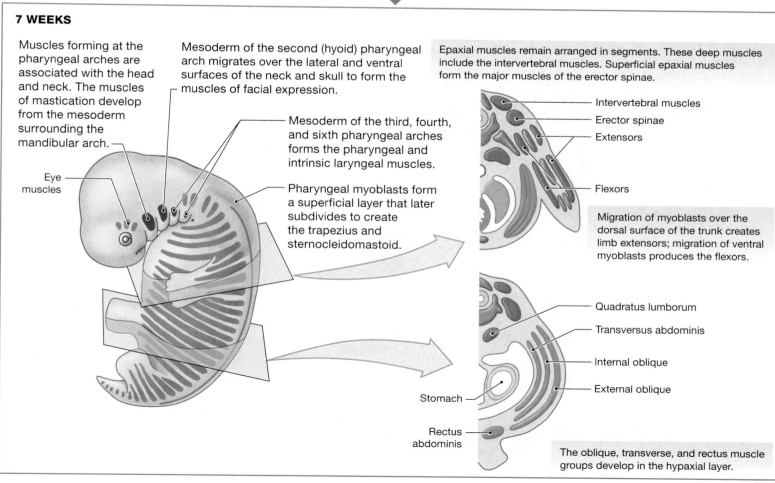

Muscles forming at the pharyngeal arches are associated with the head and neck. The muscles of mastication develop from the mesoderm surrounding the mandibular arch.

Mesoderm of the second (hyoid) pharyngeal arch migrates over the lateral and ventral surfaces of the neck and skull to form the muscles of facial expression.

Epaxial muscles remain arranged in segments. These deep muscles include the intervertebral muscles. Superficial epaxial muscles form the major muscles of the erector spinae.

Eye muscles

Mesoderm of the third, fourth, and sixth pharyngeal arches forms the pharyngeal and intrinsic laryngeal muscles.

Pharyngeal myoblasts form a superficial layer that later subdivides to create the trapezius and sternocleidomastoid.

Intervertebral muscles

Erector spinae

Extensors

Flexors

Migration of myoblasts over the dorsal surface of the trunk creates limb extensors; migration of ventral myoblasts produces the flexors.

Quadratus lumborum

Transversus abdominis

Internal oblique

External oblique

Stomach

Rectus abdominis

The oblique, transverse, and rectus muscle groups develop in the hypaxial layer.

28

The Early Development of the Nervous System

20 DAYS

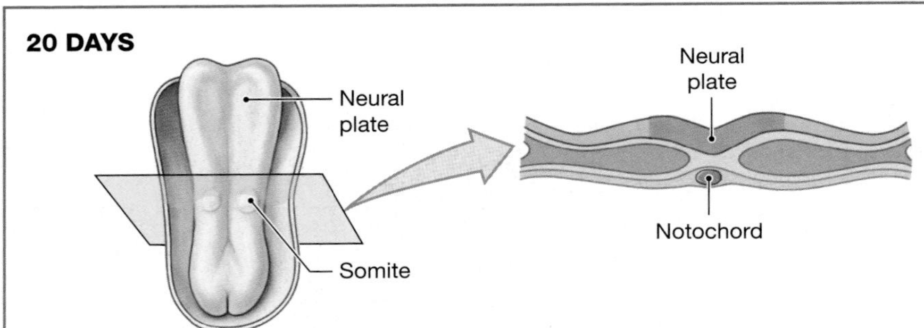

After 2 weeks of development, somites are appearing on either side of the notochord. The ectoderm near the midline thickens, forming an elevated neural plate. The **neural plate** is largest near the future head of the developing embryo.

21 DAYS

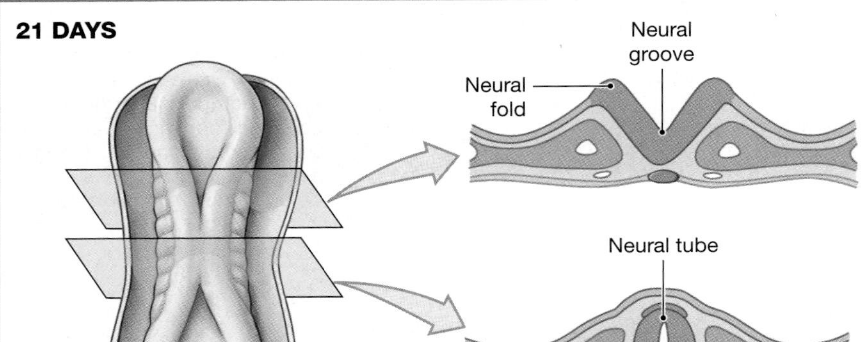

A crease develops along the axis of the neural plate, creating the **neural groove**. The edges, or **neural folds**, gradually move together. They first contact one another midway along the axis of the neural plate, near the end of the third week.

Where the neural folds meet, they fuse to form a cylindrical **neural tube** that loses its connection with the superficial ectoderm. The process of neural tube formation is called **neurulation**; it is completed in less than a week.

23 DAYS

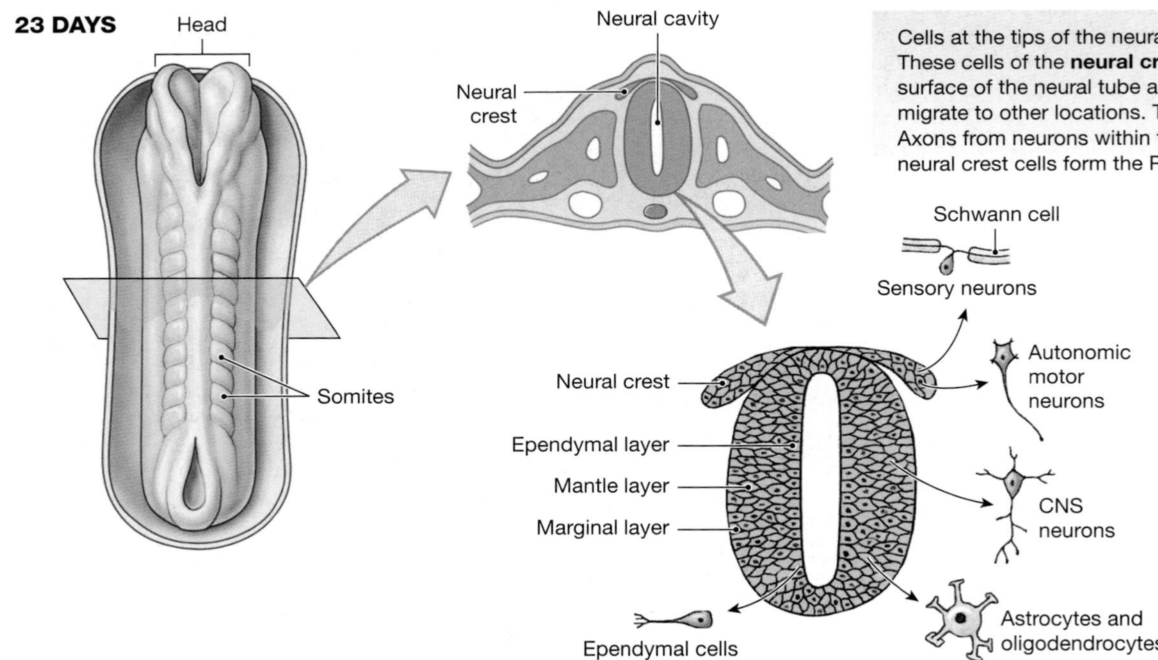

Cells at the tips of the neural folds do not form the neural tube. These cells of the **neural crest** at first remain between the dorsal surface of the neural tube and the ectoderm, but they later migrate to other locations. The neural tube becomes the CNS. Axons from neurons within the neural tube and the axons of neural crest cells form the PNS.

The first cells to appear in the mantle differentiate into neurons, while the last cells to arrive become astrocytes and oligodendrocytes. Further development of the CNS and PNS will be found in the Embryology Summaries later in this chapter.

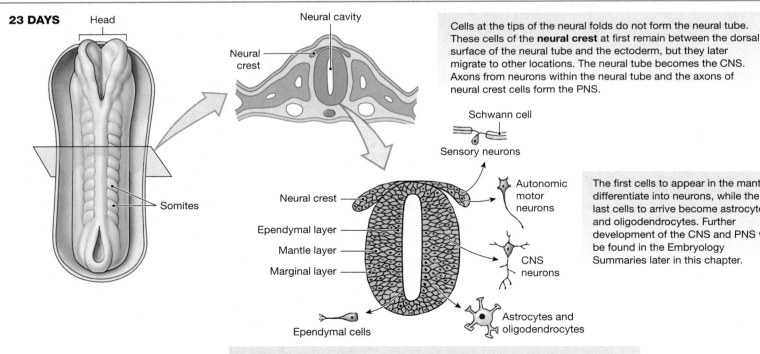

The neural tube increases in thickness as its epithelial lining undergoes repeated mitoses. By the middle of the fifth developmental week, there are three distinct layers. The **ependymal layer** lines the enclosed **neural cavity**. The ependymal cells continue their mitotic activities, and daughter cells create the surrounding **mantle layer**. Axons from developing neurons form a superficial **marginal layer**.

The Development of the Spinal Cord, Part I

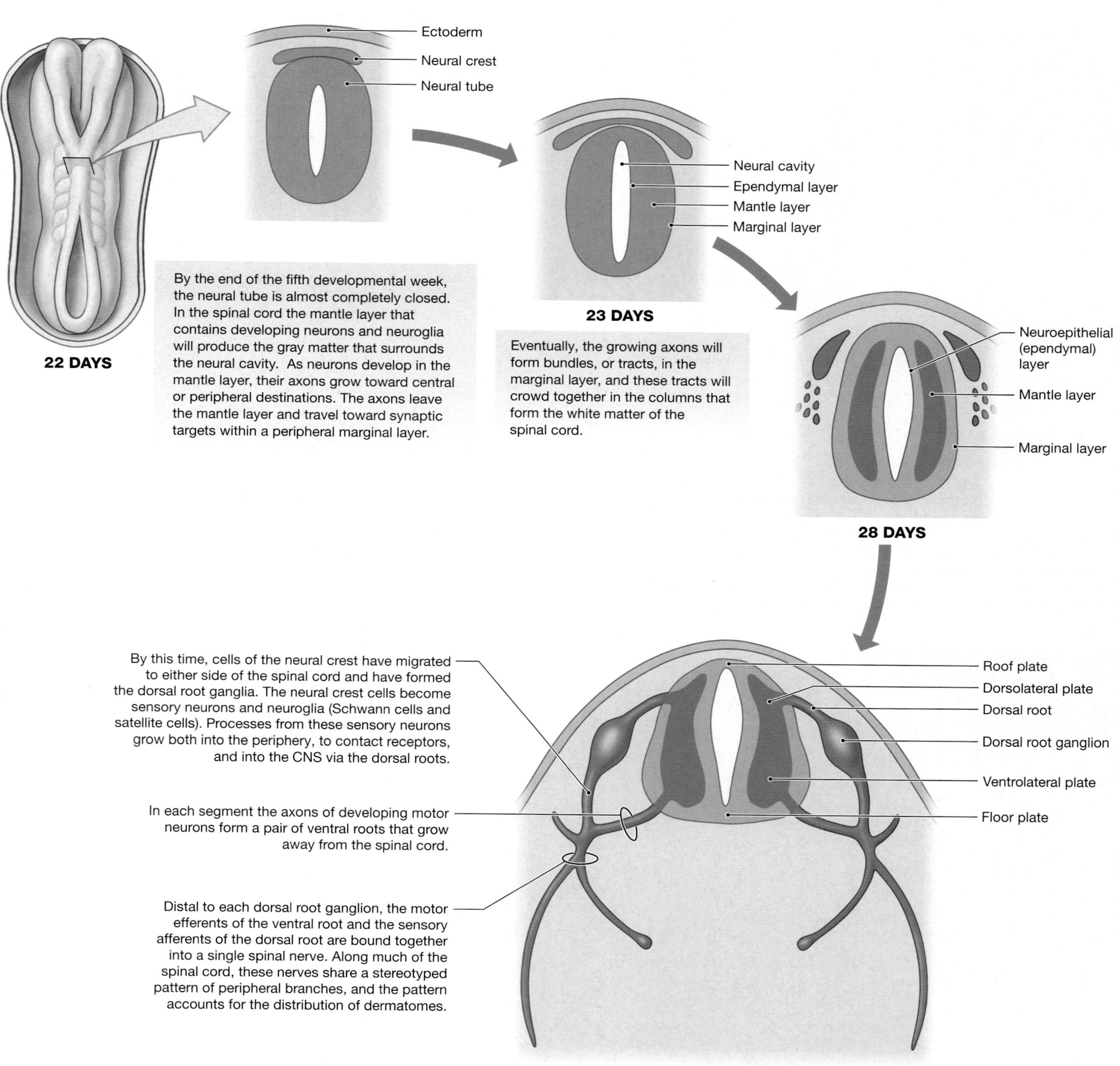

By the end of the fifth developmental week, the neural tube is almost completely closed. In the spinal cord the mantle layer that contains developing neurons and neuroglia will produce the gray matter that surrounds the neural cavity. As neurons develop in the mantle layer, their axons grow toward central or peripheral destinations. The axons leave the mantle layer and travel toward synaptic targets within a peripheral marginal layer.

Eventually, the growing axons will form bundles, or tracts, in the marginal layer, and these tracts will crowd together in the columns that form the white matter of the spinal cord.

By this time, cells of the neural crest have migrated to either side of the spinal cord and have formed the dorsal root ganglia. The neural crest cells become sensory neurons and neuroglia (Schwann cells and satellite cells). Processes from these sensory neurons grow both into the periphery, to contact receptors, and into the CNS via the dorsal roots.

In each segment the axons of developing motor neurons form a pair of ventral roots that grow away from the spinal cord.

Distal to each dorsal root ganglion, the motor efferents of the ventral root and the sensory afferents of the dorsal root are bound together into a single spinal nerve. Along much of the spinal cord, these nerves share a stereotyped pattern of peripheral branches, and the pattern accounts for the distribution of dermatomes.

As the mantle enlarges, the neural cavity becomes laterally compressed and relatively narrow. The relatively thin roof plate and floor plate will not thicken substantially, but the dorsolateral and ventrolateral plates enlarge rapidly. Neurons developing within the dorsolateral plate will receive and relay sensory information, while those in the ventrolateral region will develop into motor neurons.

The Development of the Spinal Cord, Part II

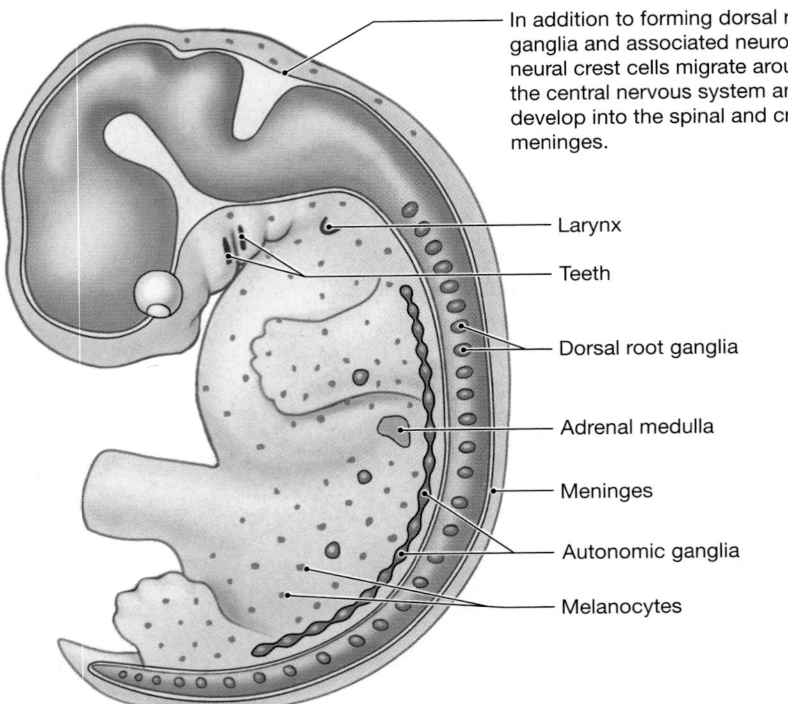
In addition to forming dorsal root ganglia and associated neuroglia, neural crest cells migrate around the central nervous system and develop into the spinal and cranial meninges.
Larynx
Teeth
Dorsal root ganglia
Adrenal medulla
Meninges
Autonomic ganglia
Melanocytes

7 WEEKS
(Distribution of neural crest cells)

Neural crest cells aggregate to form autonomic ganglia near the vertebral column and in peripheral organs. Migrating neural crest cells contribute to the formation of teeth and form the laryngeal cartilages, melanocytes of the skin, the skull, connective tissues around the eye, the intrinsic muscles of the eye, Schwann cells, satellite cells, and the adrenal medullae.

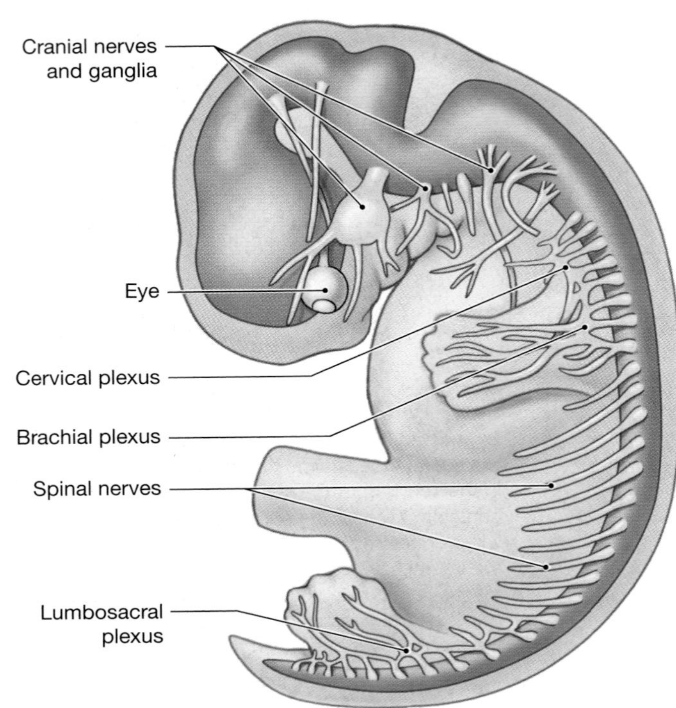
Cranial nerves and ganglia
Eye
Cervical plexus
Brachial plexus
Spinal nerves
Lumbosacral plexus

7 WEEKS
(Peripheral nerve distribution)

Several spinal nerves innervate each developing limb. When embryonic muscle cells migrate away from the myotome, the nerves grow along with them. If a large muscle in the adult is derived from several myotomal blocks, connective tissue partitions will often mark the original boundaries, and the innervation will always involve more than one spinal nerve.

CLINICAL NOTE

Developmental Abnormalities

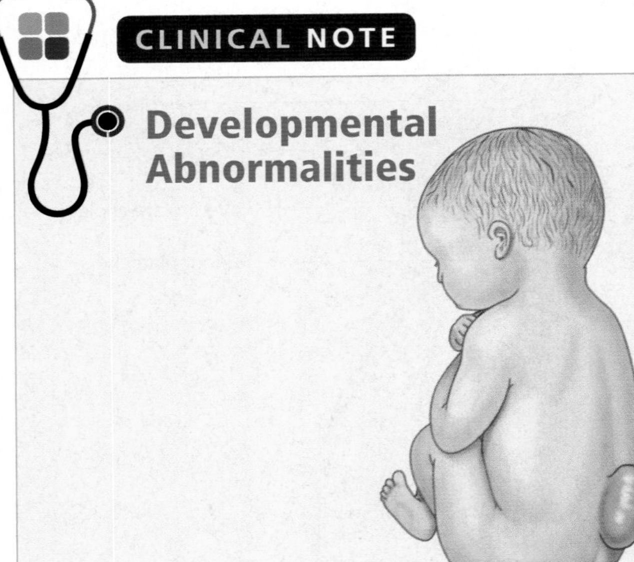

Spina bifida

Spina bifida (BI-fi-da) results when the developing vertebral laminae fail to unite due to abnormal neural tube formation at that site. The neural arch is incomplete, and the meninges bulge outward beneath the skin of the back. The extent of the abnormality determines the severity of the defects. In mild cases, the condition may pass unnoticed; extreme cases involve much of the length of the vertebral column.

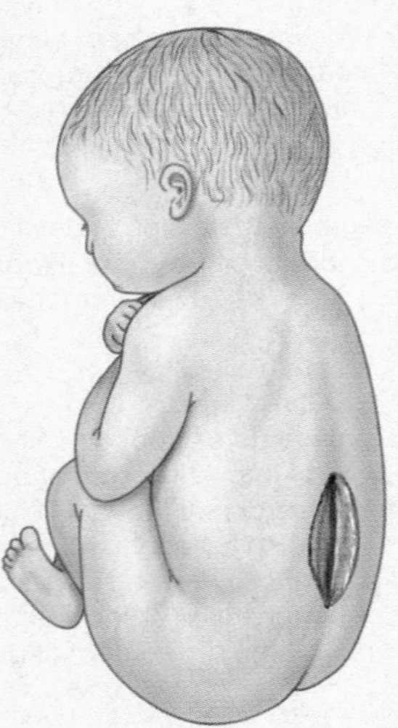

Neural tube defect

A **neural tube defect (NTD)** is a condition that is secondary to a developmental error in the formation of the spinal cord. Instead of forming a hollow tube, a portion of the spinal cord develops as a broad plate. This is often associated with spina bifida. Neural tube defects affect roughly one individual in 1000; prenatal testing can detect the existence of these defects with an 80–85 percent success rate.

The Development of the Brain, Part I

Before proceeding, briefly review the summaries of skull formation, vertebral column development, and development of the spinal cord in the previous Embryology Summaries.

23 DAYS

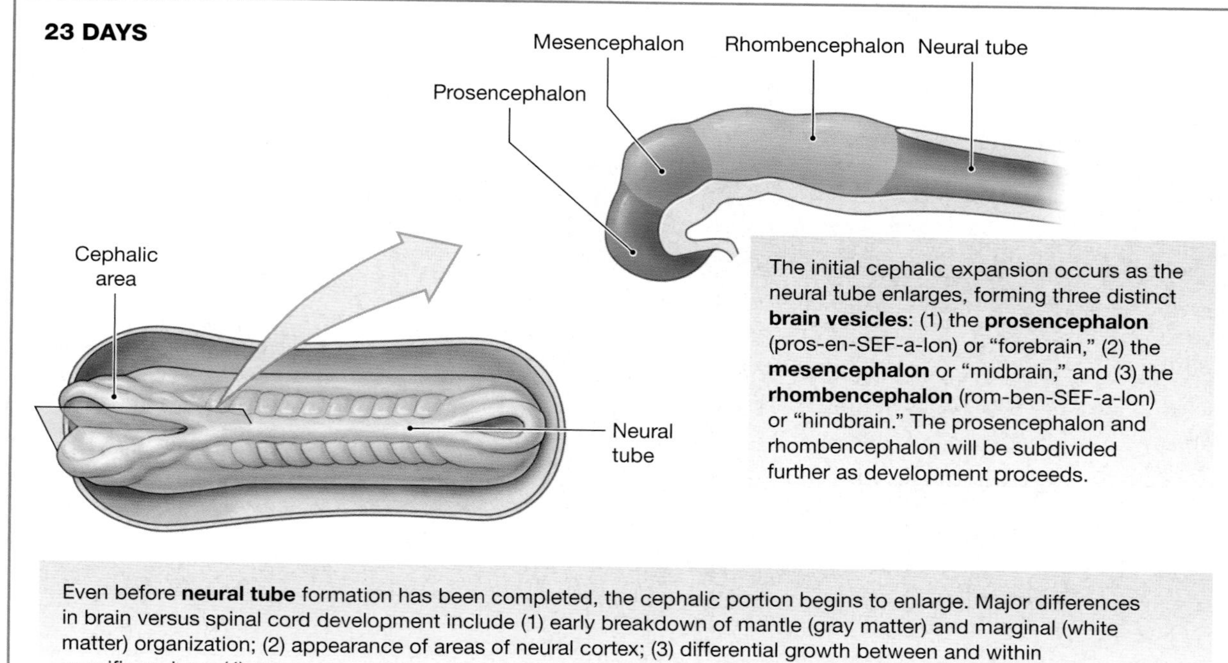

The initial cephalic expansion occurs as the neural tube enlarges, forming three distinct **brain vesicles**: (1) the **prosencephalon** (pros-en-SEF-a-lon) or "forebrain," (2) the **mesencephalon** or "midbrain," and (3) the **rhombencephalon** (rom-ben-SEF-a-lon) or "hindbrain." The prosencephalon and rhombencephalon will be subdivided further as development proceeds.

Even before **neural tube** formation has been completed, the cephalic portion begins to enlarge. Major differences in brain versus spinal cord development include (1) early breakdown of mantle (gray matter) and marginal (white matter) organization; (2) appearance of areas of neural cortex; (3) differential growth between and within specific regions; (4) appearance of characteristic bends and folds; and (5) loss of obvious segmental organization.

4 WEEKS

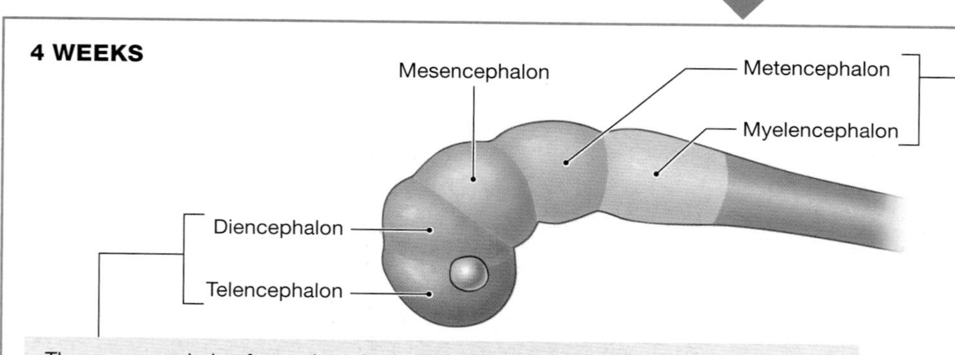

The rhombencephalon first subdivides into the **metencephalon** (met-en-SEF-a-lon; *meta*, after) and the **myelencephalon** (mī-el-en-SEF-a-lon; *myelon*, spinal cord).

The prosencephalon forms the **telecephalon** (tel-en-SEF-a-lon; *telos*, end, + *enkephalos*, brain) and the **diencephalon**. The telencephalon begins as a pair of swellings near the cephalic, posterolateral border of the prosencephalon.

5 WEEKS

Cranial nerves develop as sensory ganglia and link peripheral receptors with the brain, and motor fibers grow out of developing cranial nuclei. Special sensory neurons of cranial nerves I, II, and VIII develop in association with the developing receptors. The somatic motor nerves III, IV, and VI grow to the eye muscles; the mixed nerves (V, VII, IX, and X) innervate the **pharyngeal arches**.

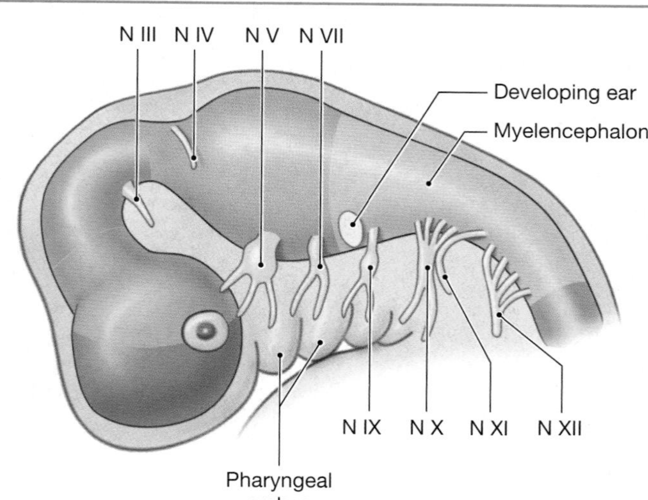

Development of the **mesencephalon** produces a small mass of nervous tissue with a constricted neural tube, the aqueduct of the midbrain.

As differential growth proceeds and the position and orientation of the embryo change, a series of bends, or **flexures** (FLEK-sherz), appears along the axis of the developing brain.

28

The Development of the Brain, Part II

8 WEEKS

The roofs of the diencephalon and myelencephalon fail to develop, leaving a thin ependymal layer in contact with the developing meninges. Blood vessels invading these regions form areas of the choroid plexus.

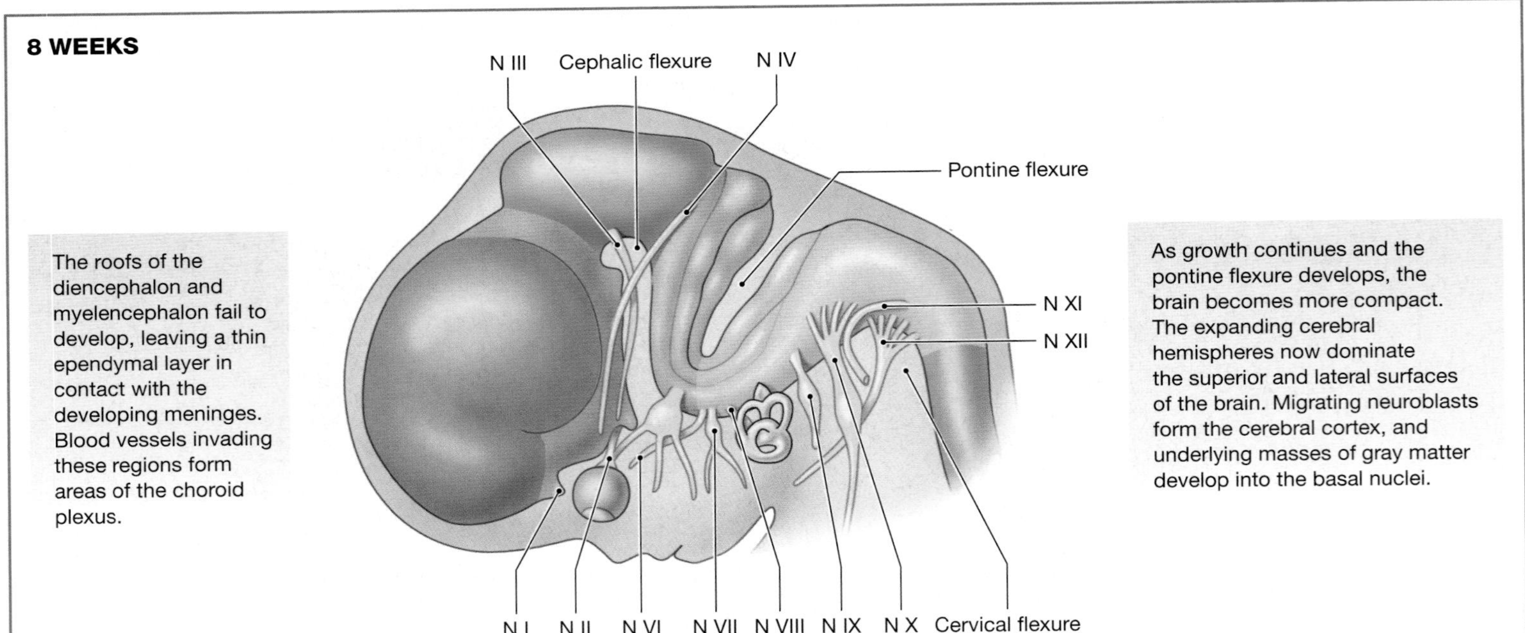

As growth continues and the pontine flexure develops, the brain becomes more compact. The expanding cerebral hemispheres now dominate the superior and lateral surfaces of the brain. Migrating neuroblasts form the cerebral cortex, and underlying masses of gray matter develop into the basal nuclei.

11 WEEKS

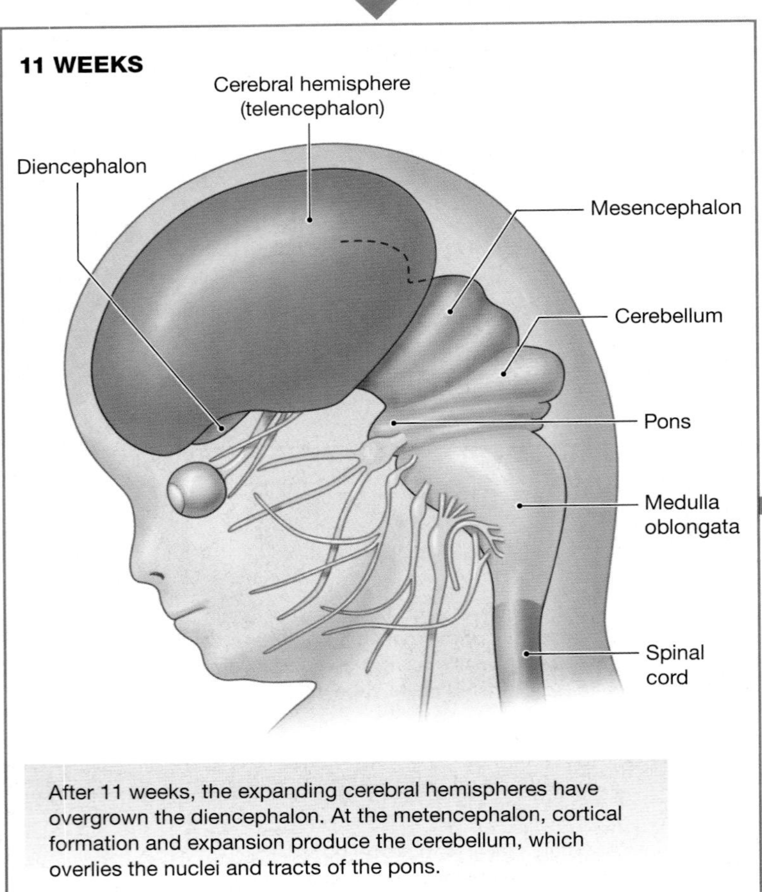

After 11 weeks, the expanding cerebral hemispheres have overgrown the diencephalon. At the metencephalon, cortical formation and expansion produce the cerebellum, which overlies the nuclei and tracts of the pons.

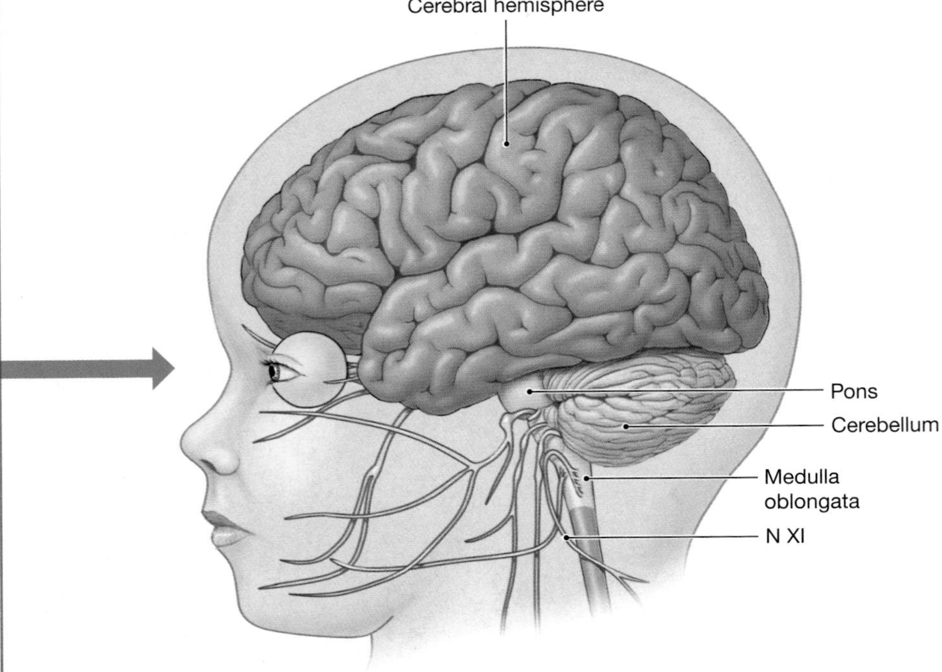

CHILD

The Development of Special Sense Organs, Part I

All special sense organs develop from the interaction between the epithelia and the developing nervous system of the embryo.

VISION

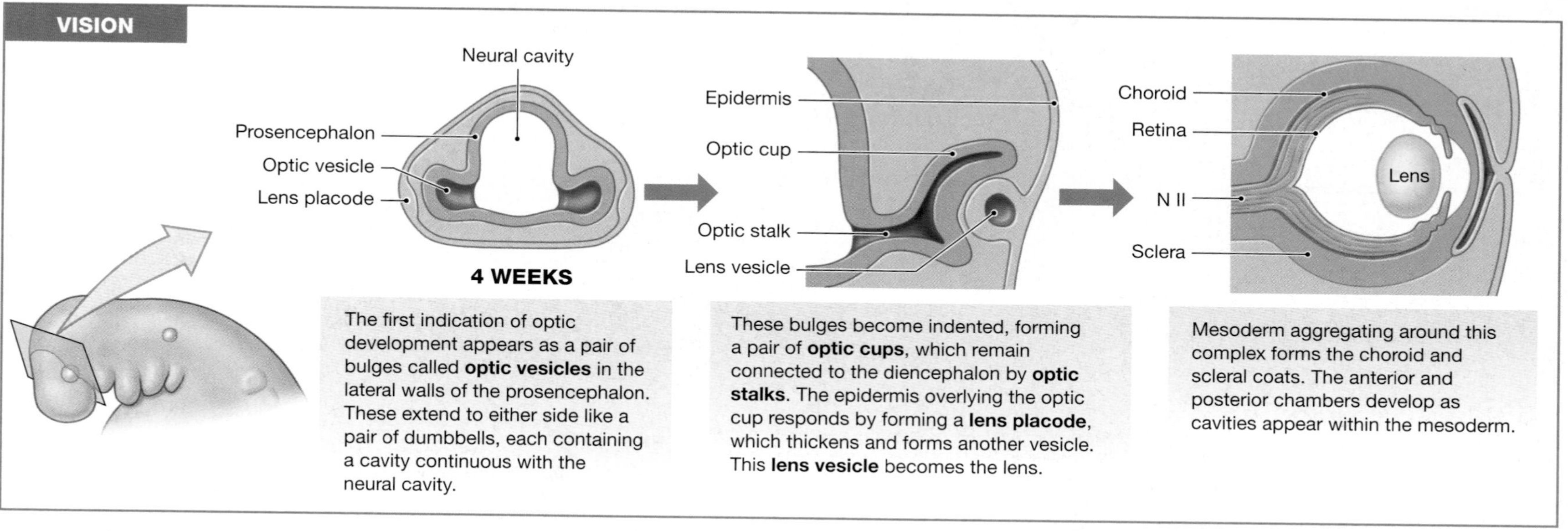

The first indication of optic development appears as a pair of bulges called **optic vesicles** in the lateral walls of the prosencephalon. These extend to either side like a pair of dumbbells, each containing a cavity continuous with the neural cavity.

These bulges become indented, forming a pair of **optic cups**, which remain connected to the diencephalon by **optic stalks**. The epidermis overlying the optic cup responds by forming a **lens placode**, which thickens and forms another vesicle. This **lens vesicle** becomes the lens.

Mesoderm aggregating around this complex forms the choroid and scleral coats. The anterior and posterior chambers develop as cavities appear within the mesoderm.

OLFACTION

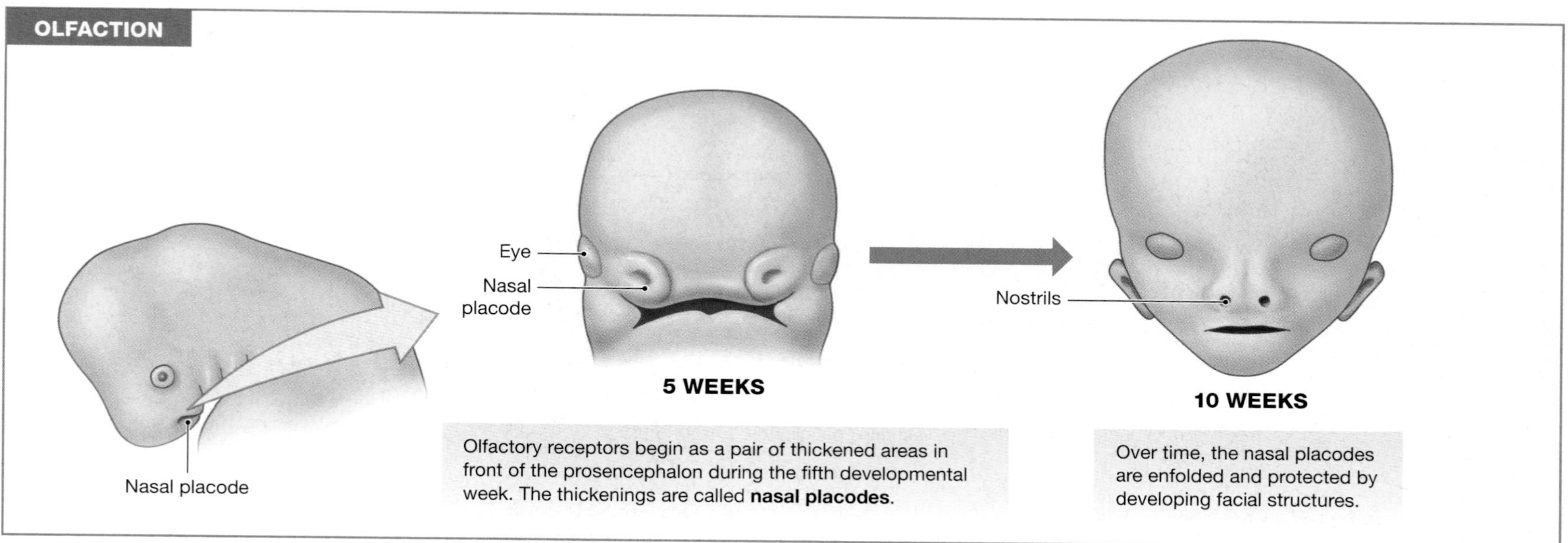

Olfactory receptors begin as a pair of thickened areas in front of the prosencephalon during the fifth developmental week. The thickenings are called **nasal placodes**.

Over time, the nasal placodes are enfolded and protected by developing facial structures.

GUSTATION

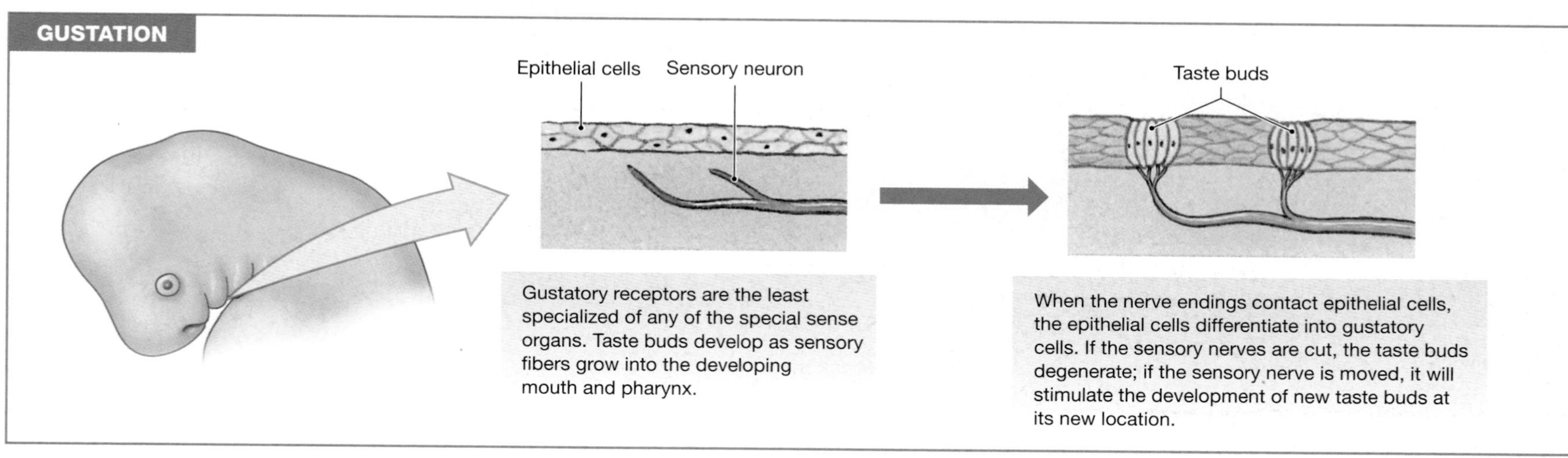

Gustatory receptors are the least specialized of any of the special sense organs. Taste buds develop as sensory fibers grow into the developing mouth and pharynx.

When the nerve endings contact epithelial cells, the epithelial cells differentiate into gustatory cells. If the sensory nerves are cut, the taste buds degenerate; if the sensory nerve is moved, it will stimulate the development of new taste buds at its new location.

The Development of Special Sense Organs, Part II

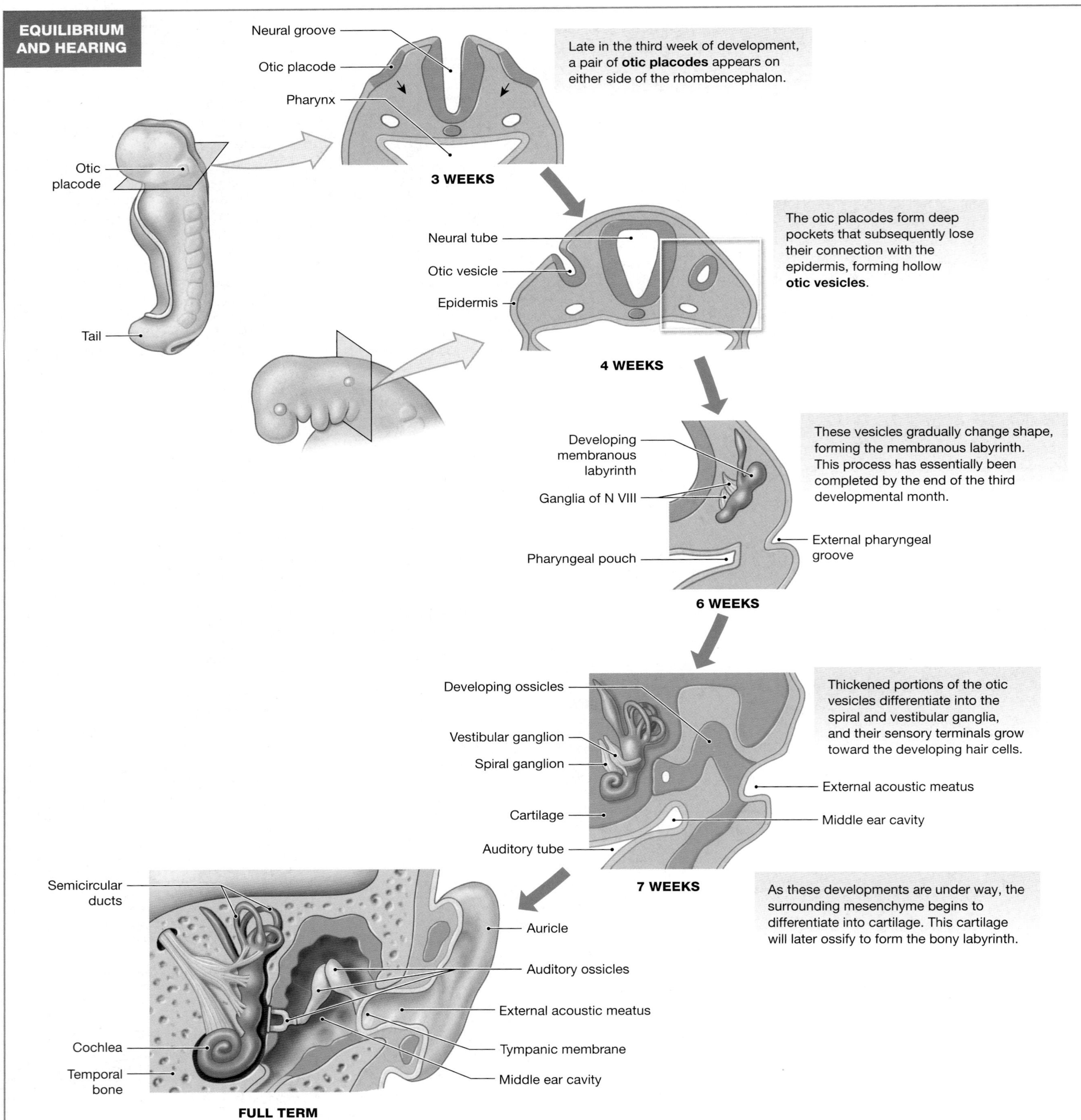

EQUILIBRIUM AND HEARING
Neural groove
Otic placode
Pharynx
Late in the third week of development, a pair of **otic placodes** appears on either side of the rhombencephalon.
Otic placode
Tail
3 WEEKS
Neural tube
Otic vesicle
Epidermis
The otic placodes form deep pockets that subsequently lose their connection with the epidermis, forming hollow **otic vesicles**.
4 WEEKS
Developing membranous labyrinth
Ganglia of N VIII
Pharyngeal pouch
External pharyngeal groove
These vesicles gradually change shape, forming the membranous labyrinth. This process has essentially been completed by the end of the third developmental month.
6 WEEKS
Developing ossicles
Vestibular ganglion
Spiral ganglion
Cartilage
Auditory tube
External acoustic meatus
Middle ear cavity
Thickened portions of the otic vesicles differentiate into the spiral and vestibular ganglia, and their sensory terminals grow toward the developing hair cells.
7 WEEKS
As these developments are under way, the surrounding mesenchyme begins to differentiate into cartilage. This cartilage will later ossify to form the bony labyrinth.
Semicircular ducts
Cochlea
Temporal bone
Auricle
Auditory ossicles
External acoustic meatus
Tympanic membrane
Middle ear cavity
FULL TERM

The Development of the Endocrine System, Part I

As noted in Chapter 3, all secretory glands, whether exocrine or endocrine, are derived from epithelia. Endocrine organs develop from epithelia (1) covering the outside of the embryo, (2) lining the digestive tract, and (3) lining the neural cavity.

PARATHYROID GLANDS AND THYMUS

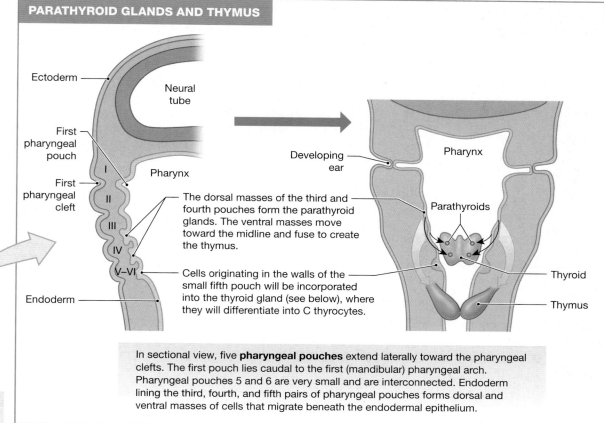

Ectoderm

Neural tube

First pharyngeal pouch

First pharyngeal cleft

I

II

III

IV

V–VI

Pharynx

Endoderm

Developing ear

Pharynx

Parathyroids

Thyroid

Thymus

The dorsal masses of the third and fourth pouches form the parathyroid glands. The ventral masses move toward the midline and fuse to create the thymus.

Cells originating in the walls of the small fifth pouch will be incorporated into the thyroid gland (see below), where they will differentiate into C thyrocytes.

In sectional view, five **pharyngeal pouches** extend laterally toward the pharyngeal clefts. The first pouch lies caudal to the first (mandibular) pharyngeal arch. Pharyngeal pouches 5 and 6 are very small and are interconnected. Endoderm lining the third, fourth, and fifth pairs of pharyngeal pouches forms dorsal and ventral masses of cells that migrate beneath the endodermal epithelium.

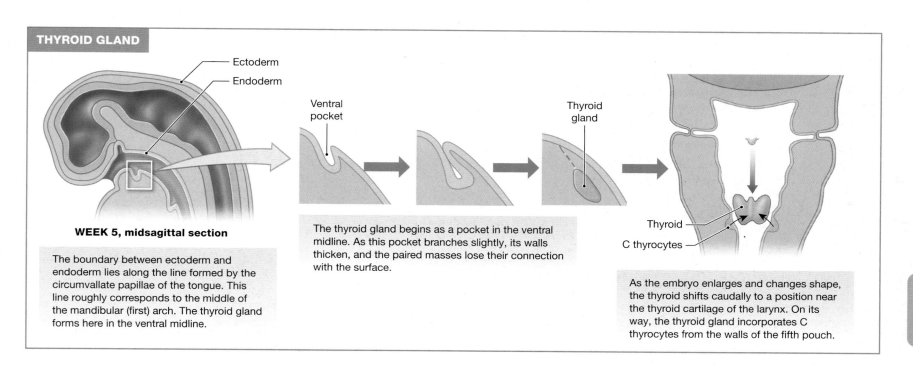

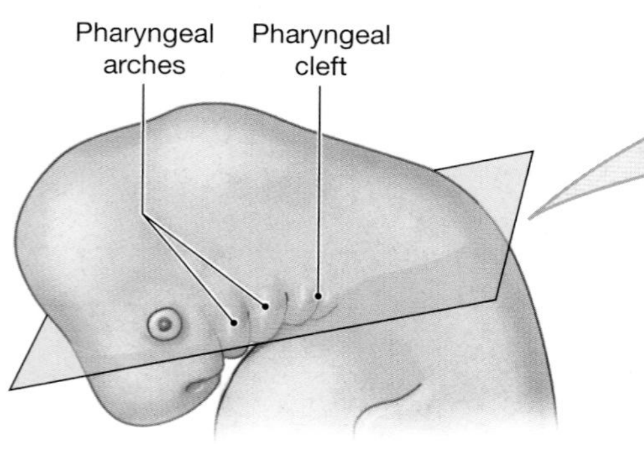

WEEK 5

The pharyngeal region of the embryo plays a particularly important role in endocrine development. After 4–5 weeks of development, the pharyngeal arches are well formed. Human embryos develop five or six pharyngeal arches, not all visible from the exterior. (Arch 5 may not appear or may form and degenerate almost immediately.) The five major arches (1–4, 6) are separated by **pharyngeal clefts**, deep ectodermal grooves.

THYROID GLAND

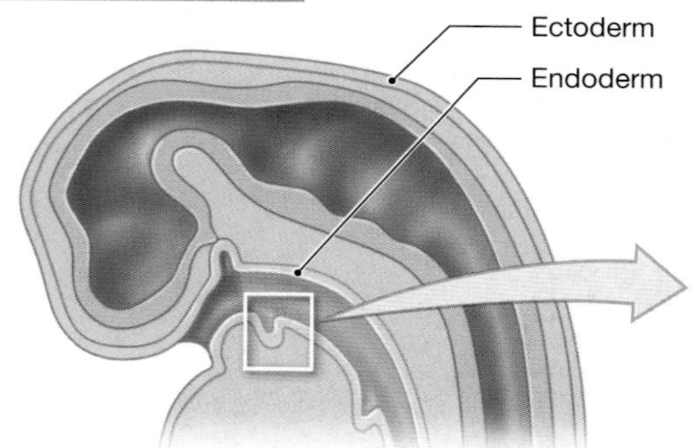

WEEK 5, midsagittal section

The boundary between ectoderm and endoderm lies along the line formed by the circumvallate papillae of the tongue. This line roughly corresponds to the middle of the mandibular (first) arch. The thyroid gland forms here in the ventral midline.

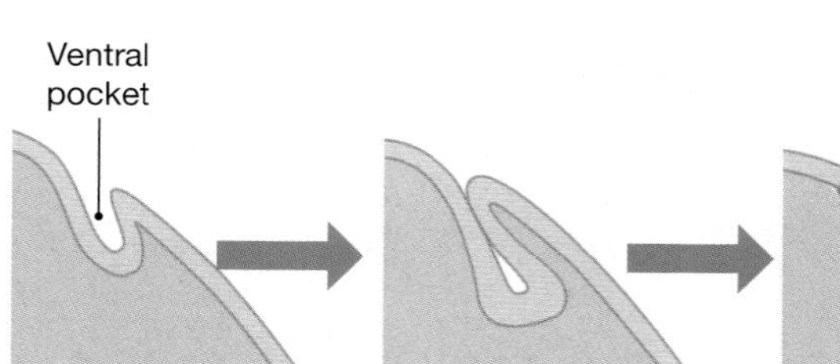

The thyroid gland begins as a pocket in the ventral midline. As this pocket branches slightly, its walls thicken, and the paired masses lose their connection with the surface.

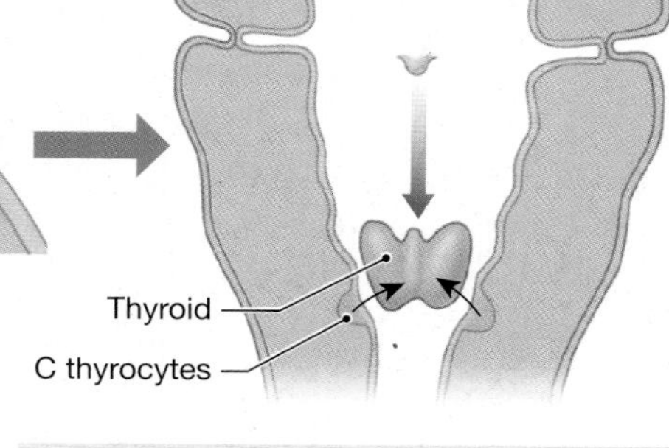

As the embryo enlarges and changes shape, the thyroid shifts caudally to a position near the thyroid cartilage of the larynx. On its way, the thyroid gland incorporates C thyrocytes from the walls of the fifth pouch.

The Development of the Endocrine System, Part II

PITUITARY GLAND

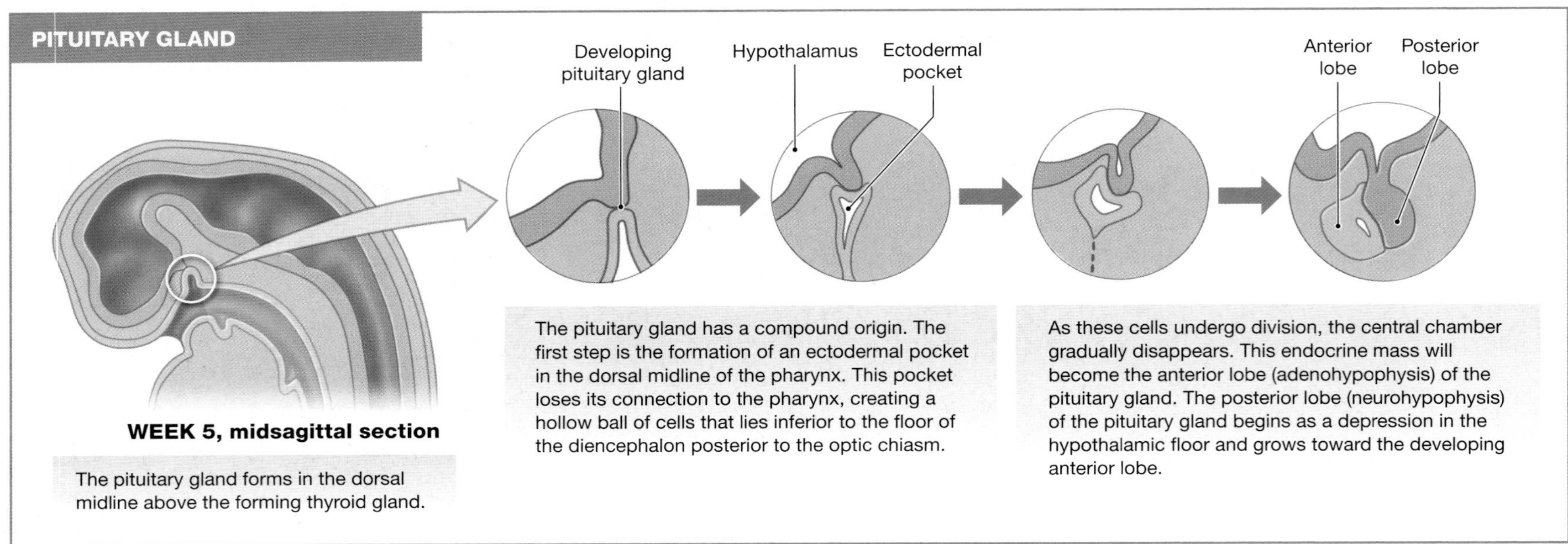

The pituitary gland forms in the dorsal midline above the forming thyroid gland.

The pituitary gland has a compound origin. The first step is the formation of an ectodermal pocket in the dorsal midline of the pharynx. This pocket loses its connection to the pharynx, creating a hollow ball of cells that lies inferior to the floor of the diencephalon posterior to the optic chiasm.

As these cells undergo division, the central chamber gradually disappears. This endocrine mass will become the anterior lobe (adenohypophysis) of the pituitary gland. The posterior lobe (neurohypophysis) of the pituitary gland begins as a depression in the hypothalamic floor and grows toward the developing anterior lobe.

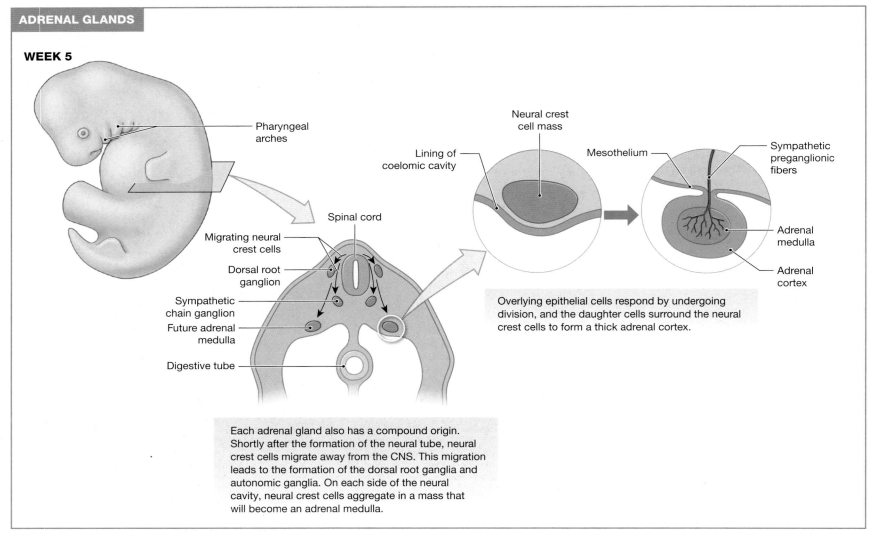

ADRENAL GLANDS

WEEK 5

Pharyngeal arches

Spinal cord

Migrating neural crest cells

Dorsal root ganglion

Sympathetic chain ganglion

Future adrenal medulla

Digestive tube

Neural crest cell mass

Lining of coelomic cavity

Mesothelium

Sympathetic preganglionic fibers

Adrenal medulla

Adrenal cortex

Overlying epithelial cells respond by undergoing division, and the daughter cells surround the neural crest cells to form a thick adrenal cortex.

Each adrenal gland also has a compound origin. Shortly after the formation of the neural tube, neural crest cells migrate away from the CNS. This migration leads to the formation of the dorsal root ganglia and autonomic ganglia. On each side of the neural cavity, neural crest cells aggregate in a mass that will become an adrenal medulla.

28

For additional details concerning the development of other endocrine organs, refer to the subsequent Embryology Summaries on the Lymphatic, Digestive, Urinary, and Reproductive systems.

The Development of the Heart

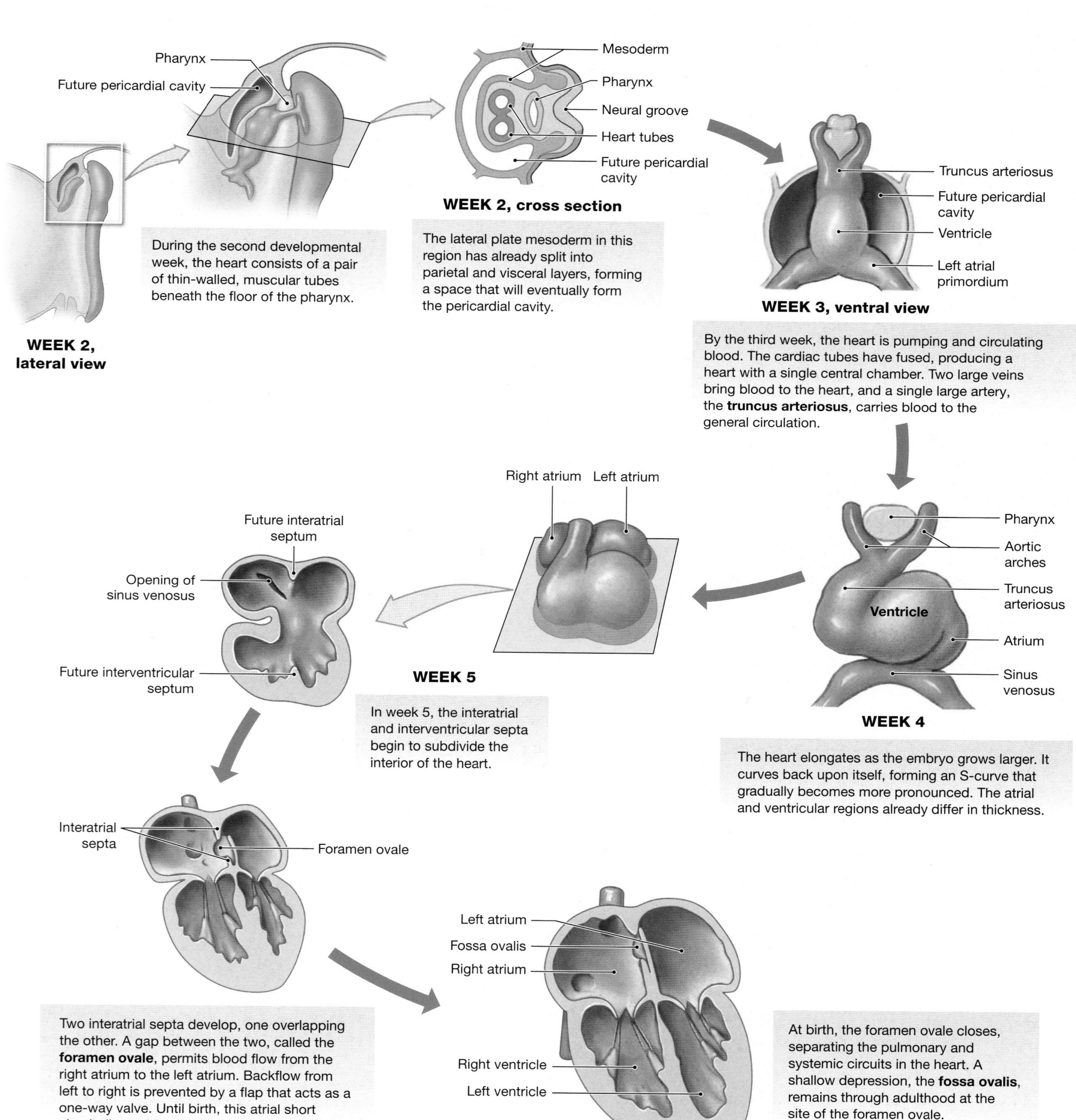

Pharynx
Future pericardial cavity
WEEK 2, lateral view
During the second developmental week, the heart consists of a pair of thin-walled, muscular tubes beneath the floor of the pharynx.
Mesoderm
Pharynx
Neural groove
Heart tubes
Future pericardial cavity
WEEK 2, cross section
The lateral plate mesoderm in this region has already split into parietal and visceral layers, forming a space that will eventually form the pericardial cavity.
Truncus arteriosus
Future pericardial cavity
Ventricle
Left atrial primordium
WEEK 3, ventral view
By the third week, the heart is pumping and circulating blood. The cardiac tubes have fused, producing a heart with a single central chamber. Two large veins bring blood to the heart, and a single large artery, the **truncus arteriosus**, carries blood to the general circulation.
Pharynx
Aortic arches
Truncus arteriosus
Ventricle
Atrium
Sinus venosus
WEEK 4
The heart elongates as the embryo grows larger. It curves back upon itself, forming an S-curve that gradually becomes more pronounced. The atrial and ventricular regions already differ in thickness.
Right atrium
Left atrium
Future interatrial septum
Opening of sinus venosus
Future interventricular septum
WEEK 5
In week 5, the interatrial and interventricular septa begin to subdivide the interior of the heart.
Interatrial septa
Foramen ovale
Two interatrial septa develop, one overlapping the other. A gap between the two, called the **foramen ovale**, permits blood flow from the right atrium to the left atrium. Backflow from left to right is prevented by a flap that acts as a one-way valve. Until birth, this atrial short circuit diverts blood from the pulmonary circuit.
Left atrium
Fossa ovalis
Right atrium
Right ventricle
Left ventricle
AGE 1 YEAR
At birth, the foramen ovale closes, separating the pulmonary and systemic circuits in the heart. A shallow depression, the **fossa ovalis**, remains through adulthood at the site of the foramen ovale.

The Development of the Cardiovascular System

We will follow the development of three major vessel complexes: (1) the aortic arches, (2) the venae cavae, and (3) the hepatic portal and umbilical vessels. (*In diagrams of prenatal circulation, arteries are shown in red and veins in blue regardless of the oxygenation of the blood they carry.*)

4 WEEKS

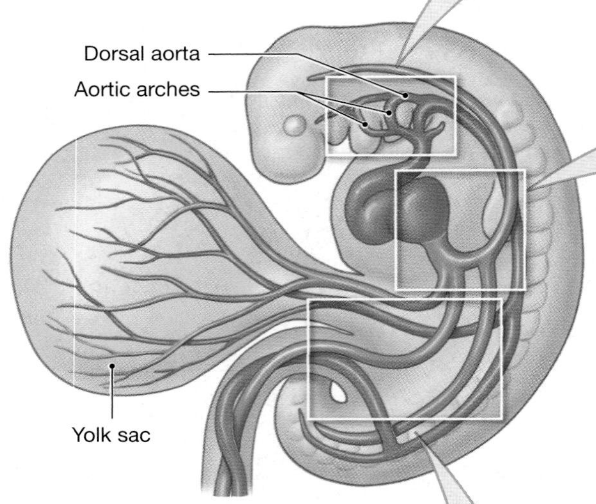

Dorsal aorta
Aortic arches
Yolk sac

THE AORTIC ARCHES

An **aortic arch** carries arterial blood through each of the pharyngeal arches. In the dorsal pharyngeal wall, these vessels fuse to create the **dorsal aortae**, which distributes blood throughout the body. The arches are usually numbered from 1 to 6, corresponding to the pharyngeal arches.

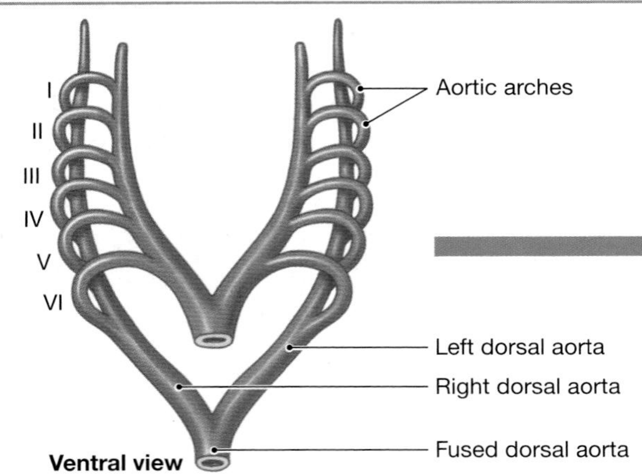

THE VENAE CAVAE

The early venous circulation draining the tissues of the body wall, limbs, and head centers around the paired **anterior cardinal veins, posterior cardinal veins,** and **subcardinal veins**.

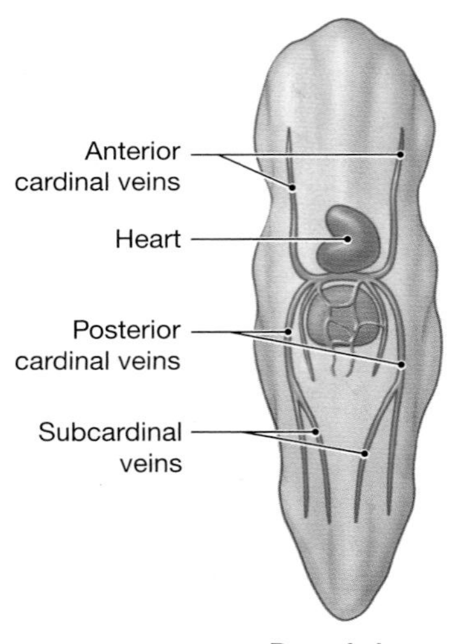

THE HEPATIC PORTAL AND UMBILICAL VESSELS

Paired **umbilical arteries** deliver blood to the placenta. At 4 weeks, paired **umbilical veins** return blood to capillary networks in the liver. Veins running along the length of the digestive tract have extensive interconnections.

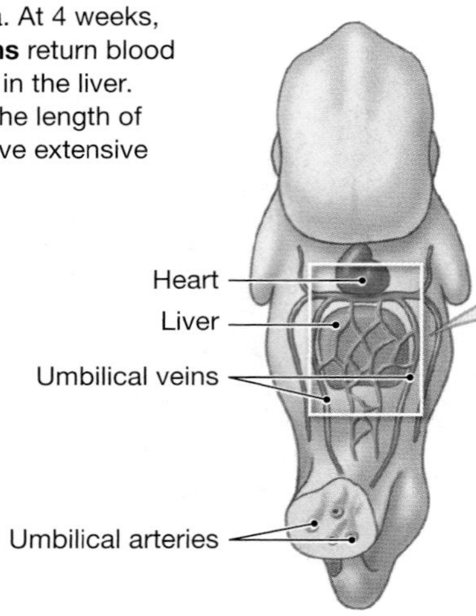

Heart
Liver
Umbilical veins
Umbilical arteries

4 WEEKS

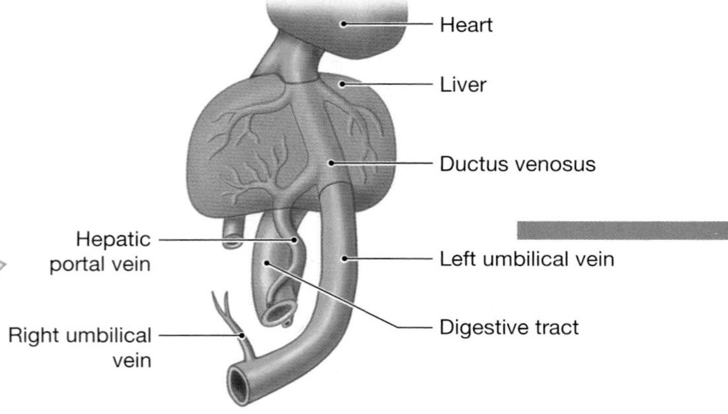

Heart
Liver
Ductus venosus
Hepatic portal vein
Left umbilical vein
Right umbilical vein
Digestive tract

12 WEEKS

By week 12, the right umbilical vein disintegrates, and the blood from the placenta travels along a single umbilical vein. The **ductus venosus** allows some venous blood to bypass the liver. The veins draining the digestive tract have fused, forming the hepatic portal vein.

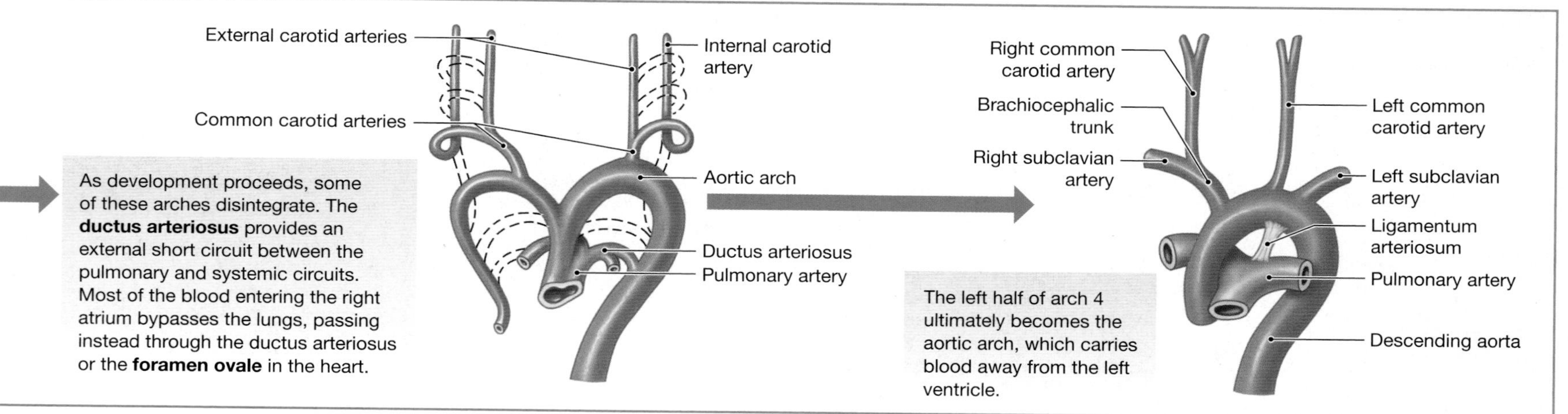

As development proceeds, some of these arches disintegrate. The **ductus arteriosus** provides an external short circuit between the pulmonary and systemic circuits. Most of the blood entering the right atrium bypasses the lungs, passing instead through the ductus arteriosus or the **foramen ovale** in the heart.

The left half of arch 4 ultimately becomes the aortic arch, which carries blood away from the left ventricle.

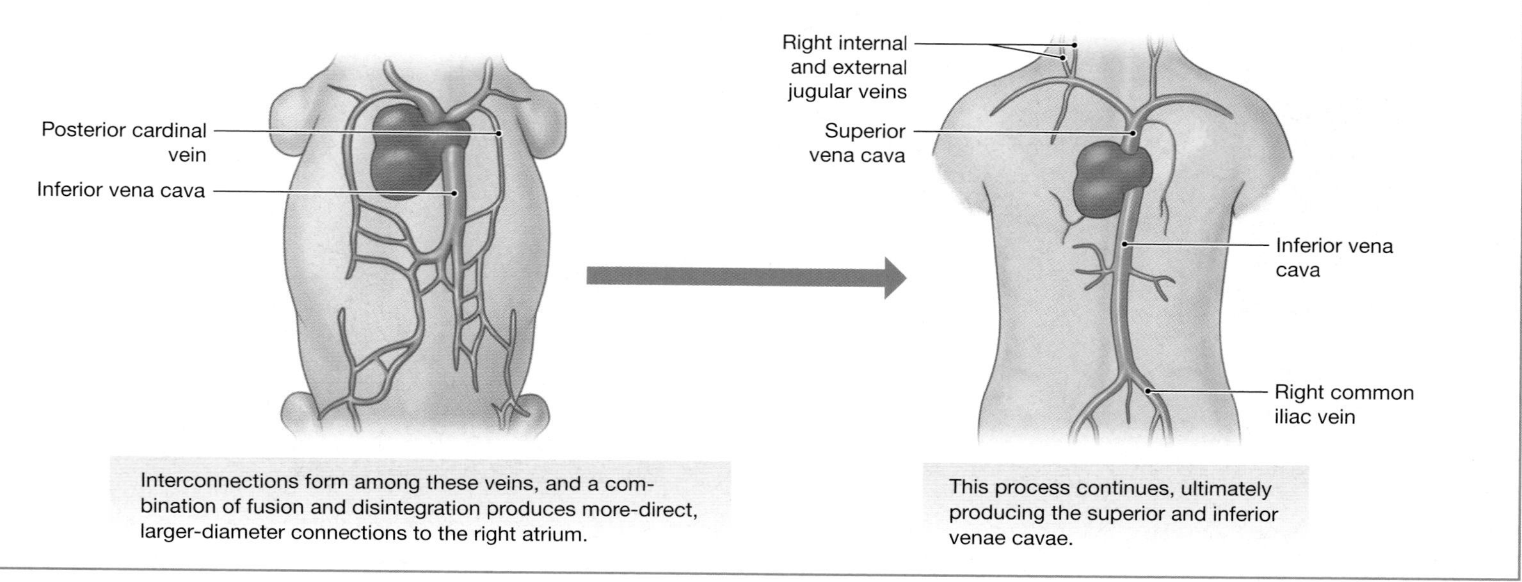

Interconnections form among these veins, and a combination of fusion and disintegration produces more-direct, larger-diameter connections to the right atrium.

This process continues, ultimately producing the superior and inferior venae cavae.

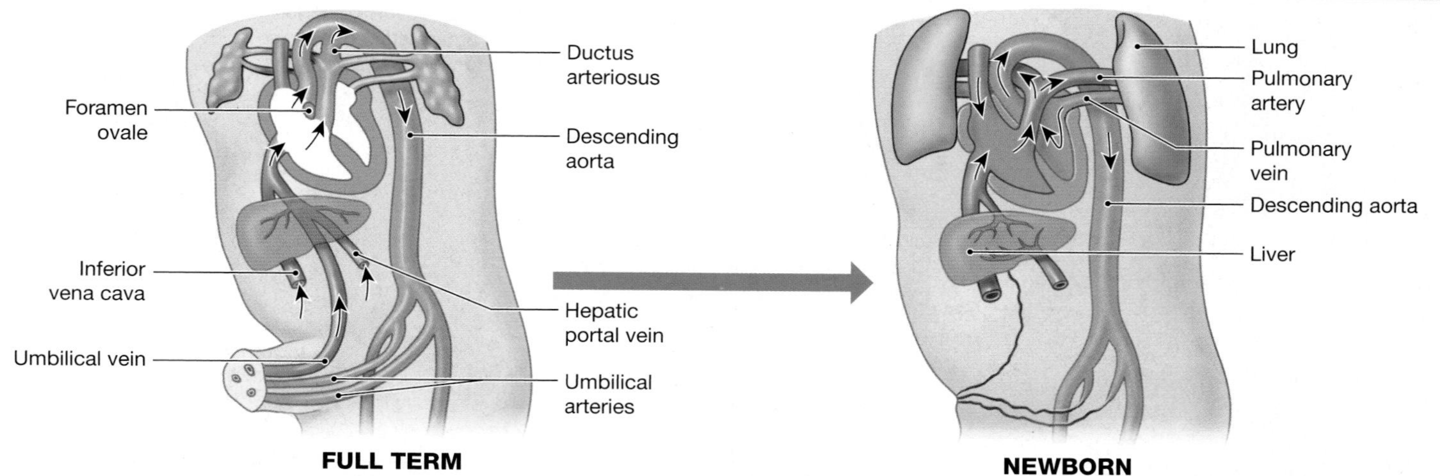

FULL TERM

NEWBORN

Shortly before birth, blood returning from the placenta travels through the liver in the ductus venosus to reach the inferior vena cava. Much of the blood delivered by the venae cavae bypasses the lungs by traveling through the foramen ovale and the ductus arteriosus.

At birth, pressures drop in the pleural cavities as the chest expands and the infant takes its first breath. The pulmonary vessels dilate, and blood flow to the lungs increases. Pressure falls in the right atrium, and the higher left atrial pressure closes the valve that guards the foramen ovale. Smooth muscles contract the ductus arteriosus, which ultimately converts to the **ligamentum arteriosum**, a fibrous strand.

The Development of the Lymphatic System

The development of the lymphatic vessels is closely tied to the formation of blood vessels.

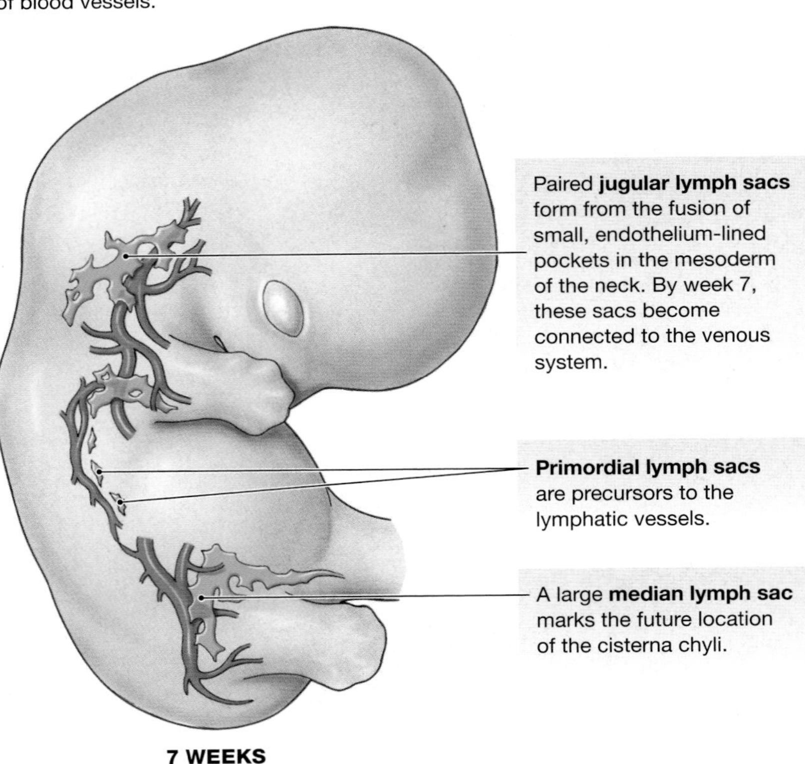

Paired **jugular lymph sacs** form from the fusion of small, endothelium-lined pockets in the mesoderm of the neck. By week 7, these sacs become connected to the venous system.
Primordial lymph sacs are precursors to the lymphatic vessels.
A large **median lymph sac** marks the future location of the cisterna chyli.
7 WEEKS

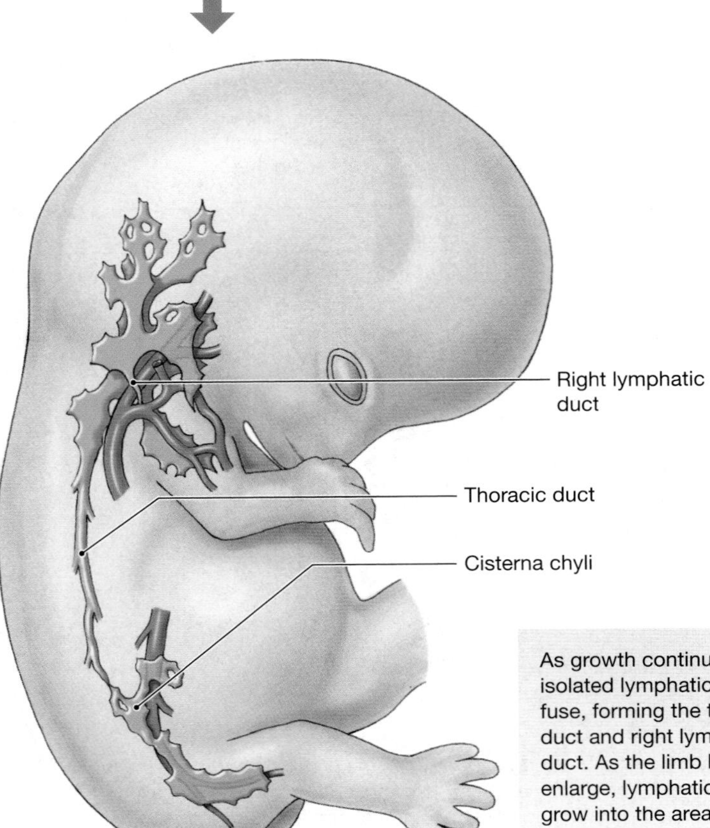

As growth continues, the isolated lymphatic sacs fuse, forming the thoracic duct and right lymphatic duct. As the limb buds enlarge, lymphatic vessels grow into the area along with developing arteries and veins.

THE THYMUS GLAND

The thymus forms from cells of the third pharyngeal pouch. These cells lose their connection with the epithelium and divide repeatedly. As the embryo changes shape, the thymic lobes are brought together near the midline of the chest. At birth, the thymus is relatively large, filling much of the anterior mediastinum.

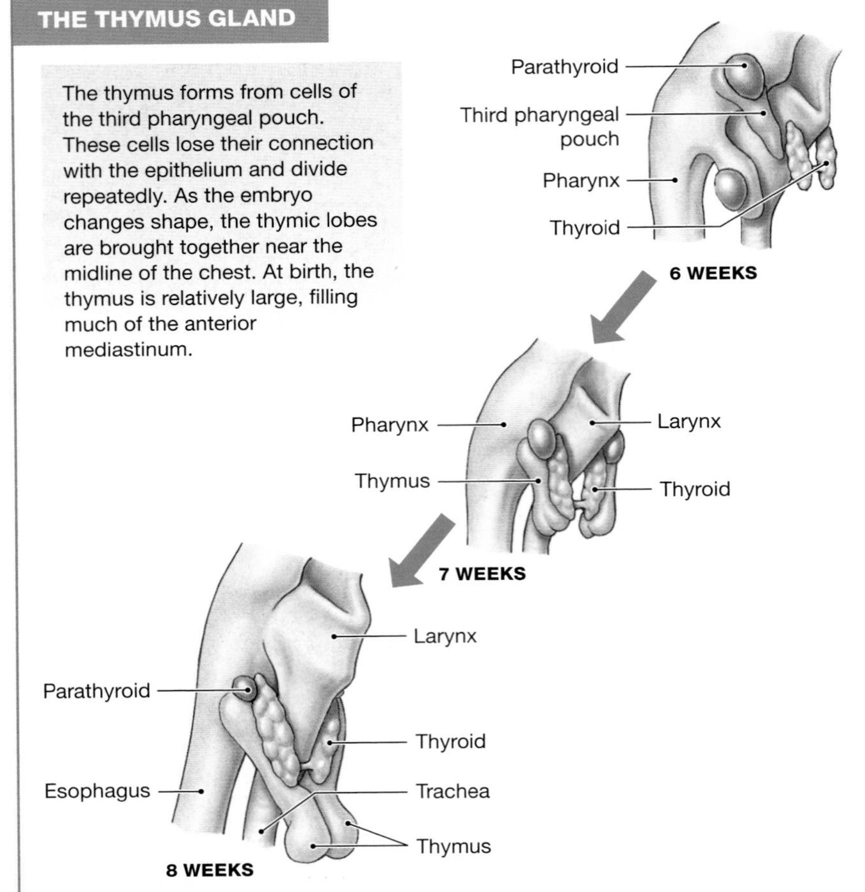

LYMPH NODE FORMATION

Small blood vessels grow into areas where lymphocytes cluster within developing lymphatic sacs. Connective tissue capsules form, and the internal organization of a lymph node gradually appears.

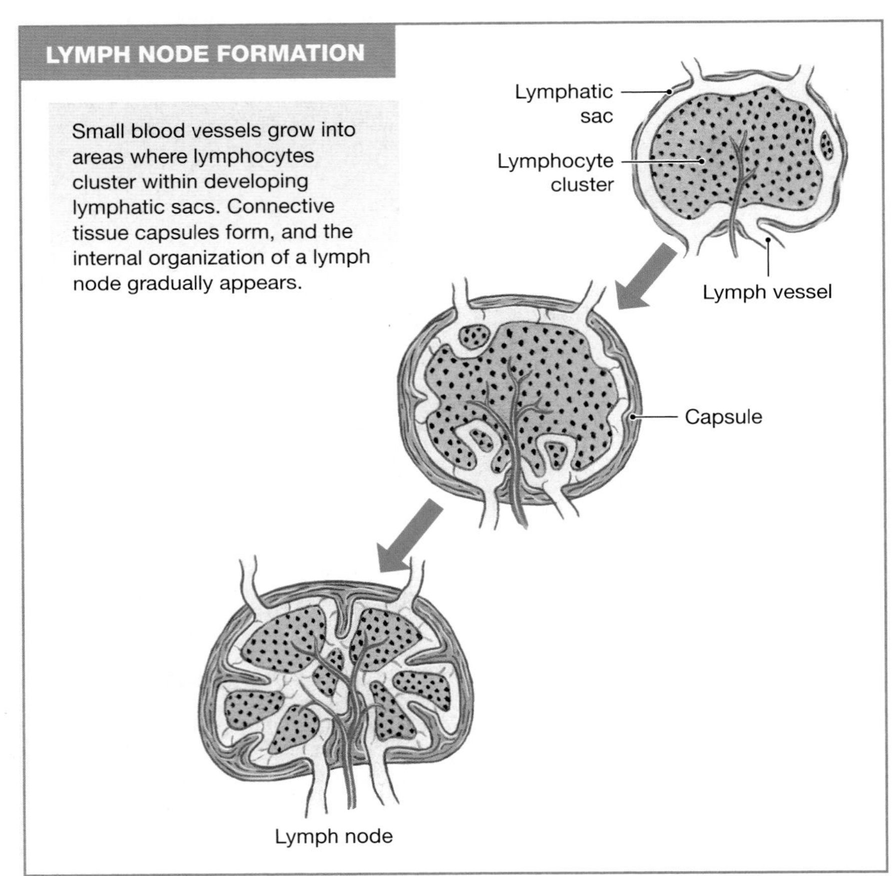

28

The Development of the Respiratory System, Part I

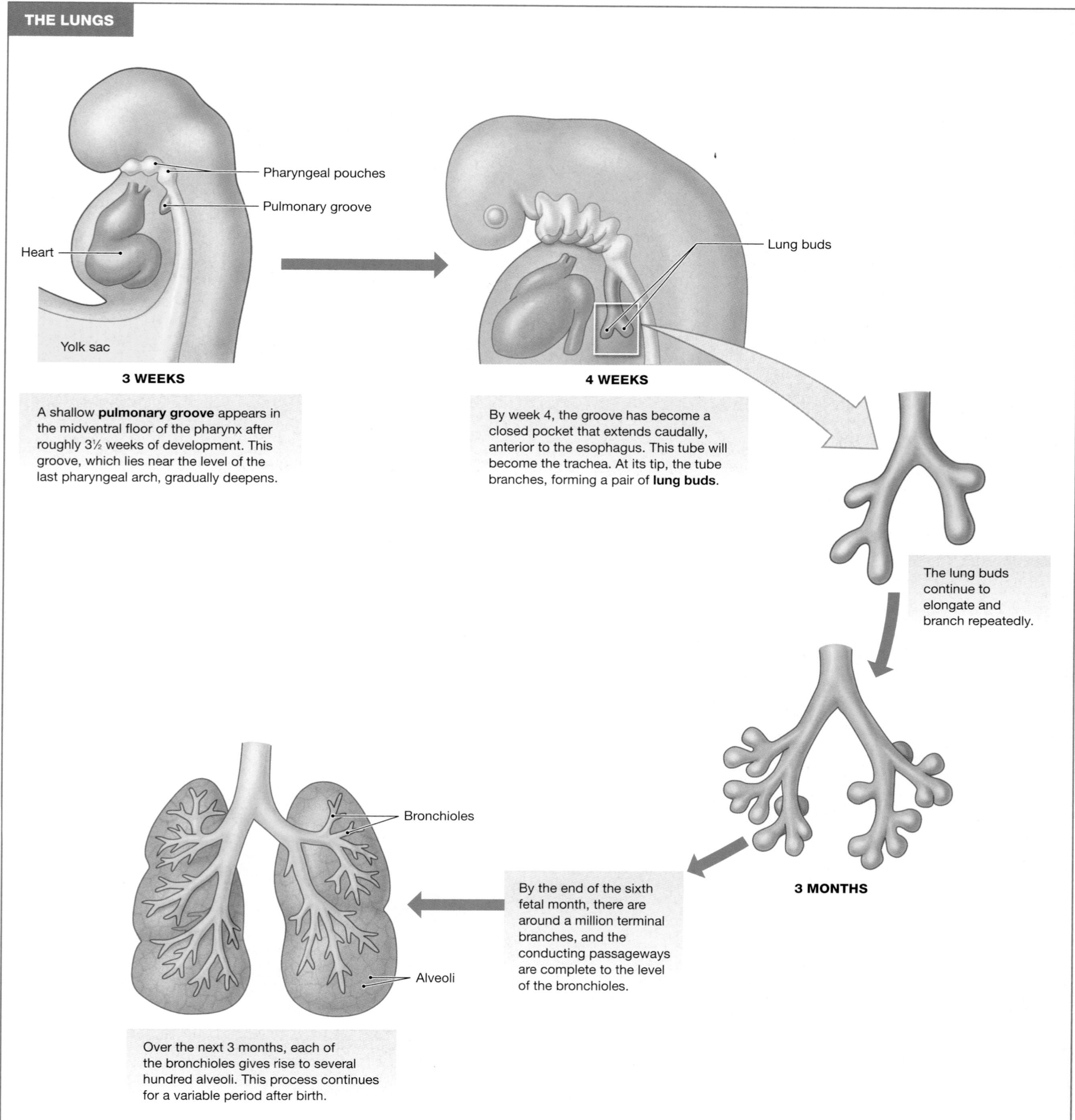

THE LUNGS
Pharyngeal pouches
Pulmonary groove
Heart
Yolk sac
3 WEEKS
A shallow **pulmonary groove** appears in the midventral floor of the pharynx after roughly 3½ weeks of development. This groove, which lies near the level of the last pharyngeal arch, gradually deepens.
Lung buds
4 WEEKS
By week 4, the groove has become a closed pocket that extends caudally, anterior to the esophagus. This tube will become the trachea. At its tip, the tube branches, forming a pair of **lung buds**.
The lung buds continue to elongate and branch repeatedly.
3 MONTHS
By the end of the sixth fetal month, there are around a million terminal branches, and the conducting passageways are complete to the level of the bronchioles.
Bronchioles
Alveoli
Over the next 3 months, each of the bronchioles gives rise to several hundred alveoli. This process continues for a variable period after birth.

The Development of the Respiratory System, Part II

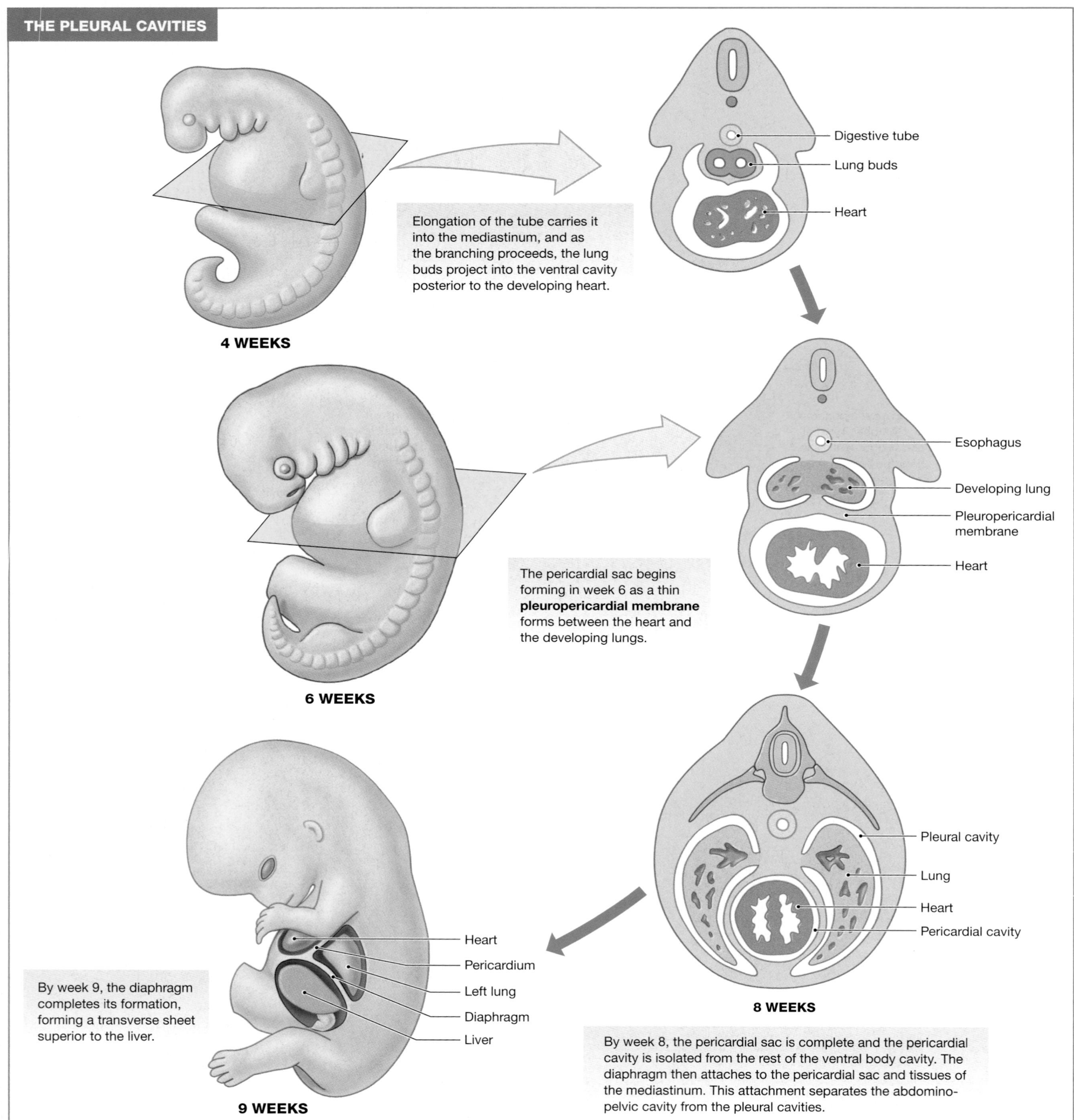
THE PLEURAL CAVITIES
4 WEEKS
Digestive tube
Lung buds
Heart
Elongation of the tube carries it into the mediastinum, and as the branching proceeds, the lung buds project into the ventral cavity posterior to the developing heart.
6 WEEKS
Esophagus
Developing lung
Pleuropericardial membrane
Heart
The pericardial sac begins forming in week 6 as a thin **pleuropericardial membrane** forms between the heart and the developing lungs.
8 WEEKS
Pleural cavity
Lung
Heart
Pericardial cavity
By week 8, the pericardial sac is complete and the pericardial cavity is isolated from the rest of the ventral body cavity. The diaphragm then attaches to the pericardial sac and tissues of the mediastinum. This attachment separates the abdominopelvic cavity from the pleural cavities.
9 WEEKS
Heart
Pericardium
Left lung
Diaphragm
Liver
By week 9, the diaphragm completes its formation, forming a transverse sheet superior to the liver.

The Development of the Digestive System, Part I

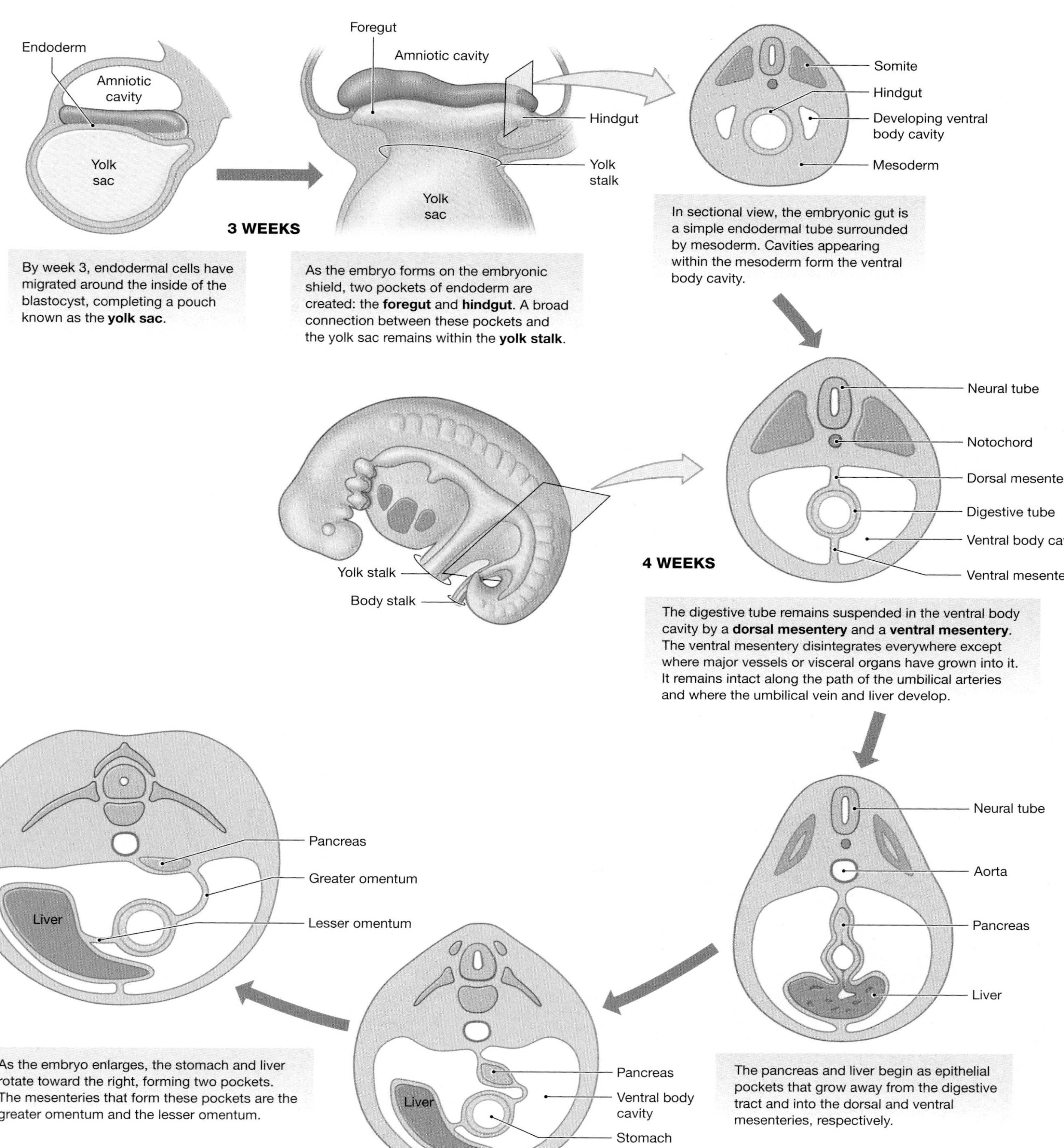

By week 3, endodermal cells have migrated around the inside of the blastocyst, completing a pouch known as the **yolk sac**.

As the embryo forms on the embryonic shield, two pockets of endoderm are created: the **foregut** and **hindgut**. A broad connection between these pockets and the yolk sac remains within the **yolk stalk**.

In sectional view, the embryonic gut is a simple endodermal tube surrounded by mesoderm. Cavities appearing within the mesoderm form the ventral body cavity.

The digestive tube remains suspended in the ventral body cavity by a **dorsal mesentery** and a **ventral mesentery**. The ventral mesentery disintegrates everywhere except where major vessels or visceral organs have grown into it. It remains intact along the path of the umbilical arteries and where the umbilical vein and liver develop.

The pancreas and liver begin as epithelial pockets that grow away from the digestive tract and into the dorsal and ventral mesenteries, respectively.

As the embryo enlarges, the stomach and liver rotate toward the right, forming two pockets. The mesenteries that form these pockets are the greater omentum and the lesser omentum.

The Development of the Digestive System, Part II

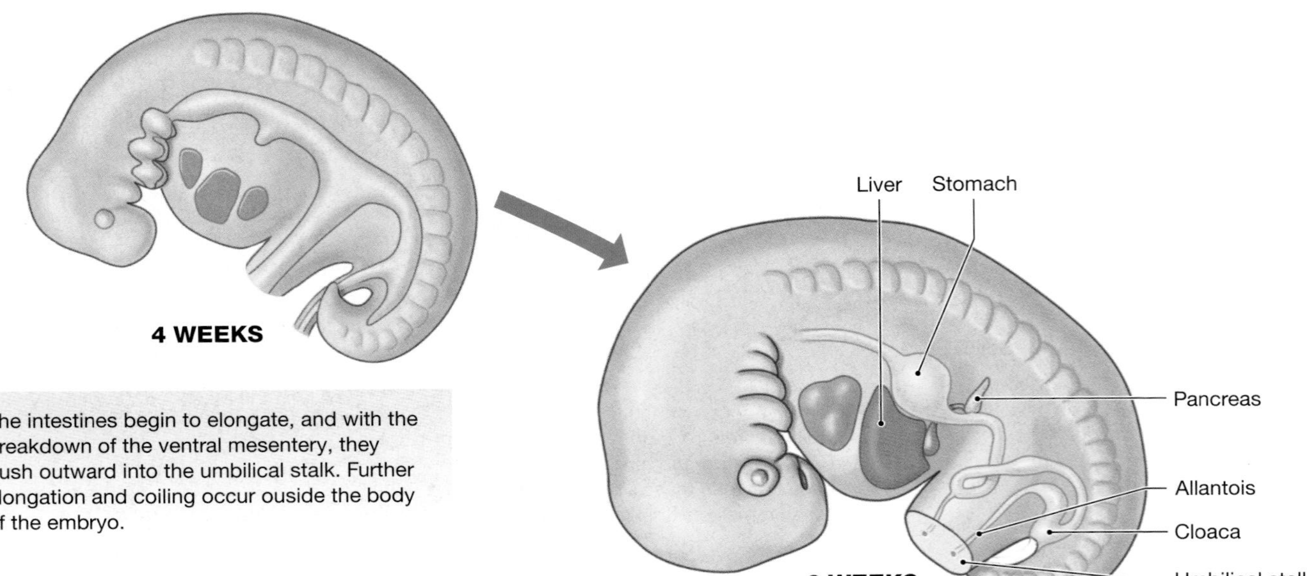

The intestines begin to elongate, and with the breakdown of the ventral mesentery, they push outward into the umbilical stalk. Further elongation and coiling occur ouside the body of the embryo.

The hindgut extends into the tail, where it forms a large chamber, the **cloaca**. A tubular extension of the cloaca, the **allantois** (a-LAN-tō-is; *allantos*, sausage), projects away from the body and into the **body stalk**. Fusion of the yolk stalk and body stalk will create the **umbilical stalk**, also known as the umbilical cord.

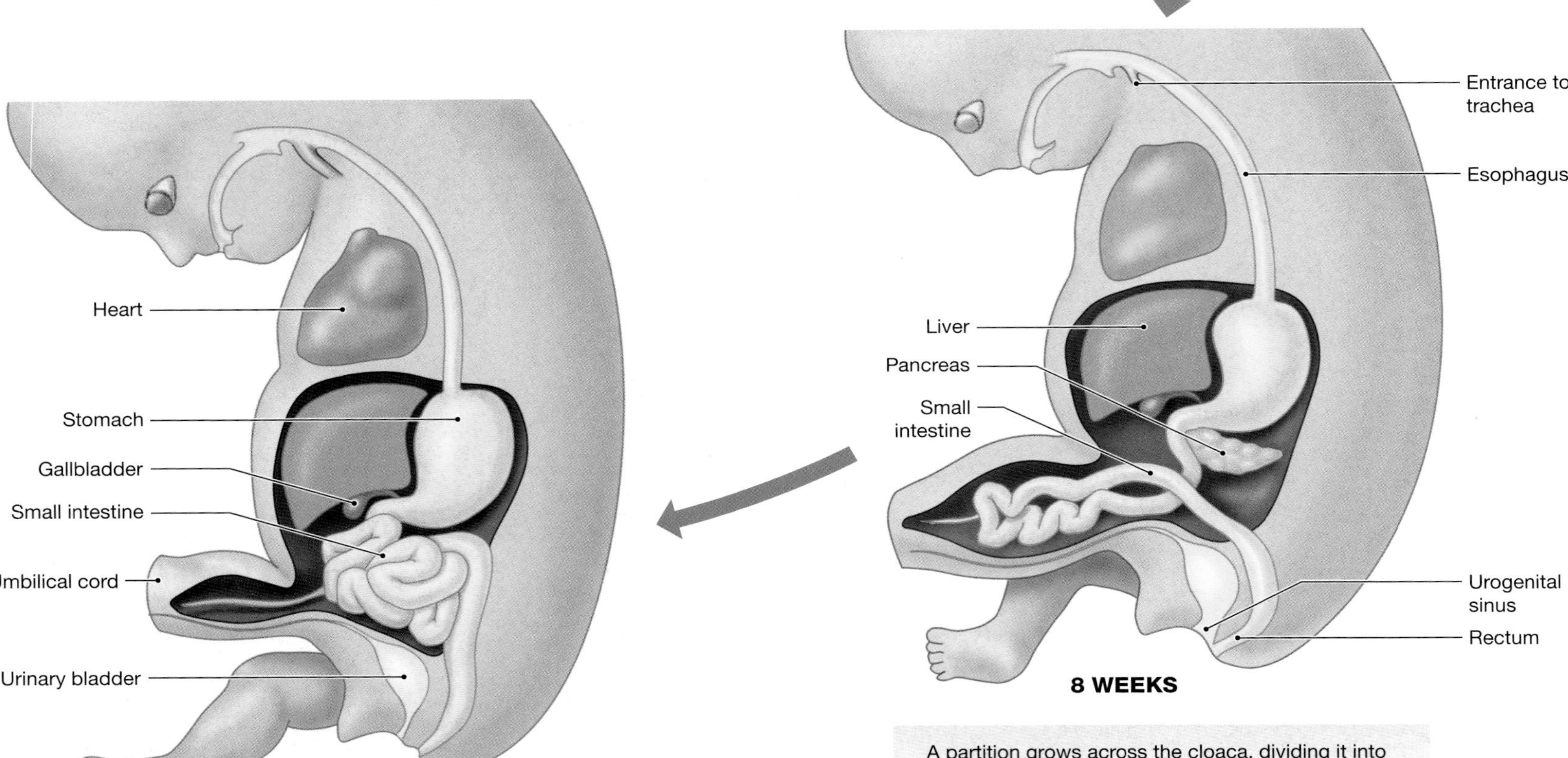

A partition grows across the cloaca, dividing it into a posterior rectum and an anterior **urogenital sinus** that retains a connection to the allantois.

By week 10, the intestines have begun moving back into the visceral cavity, although they continue to grow longer.

The Development of the Urinary System, Part I

The kidneys develop in stages along the axis of the **urogenital ridge**, a thickened area beneath the dorsolateral wall of the coelomic cavity.

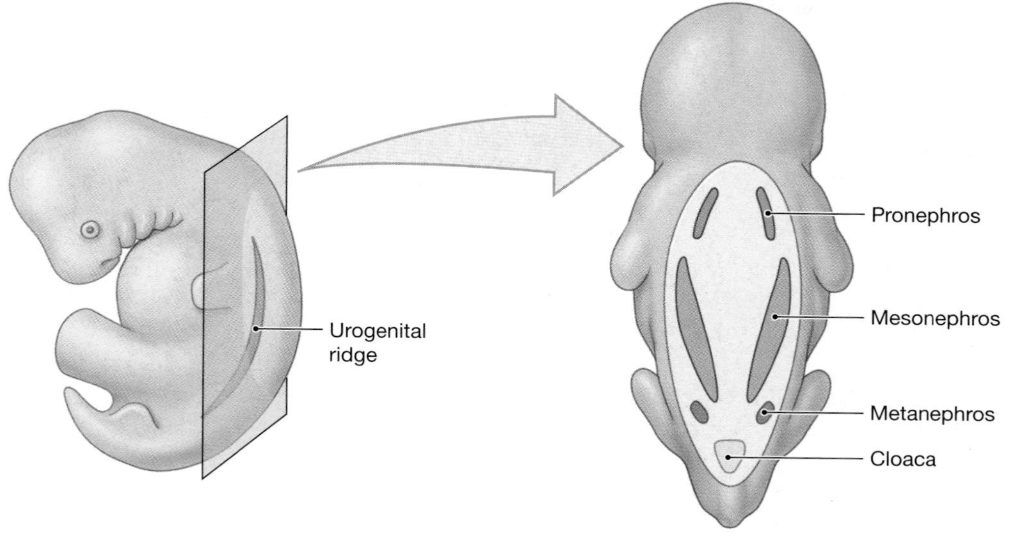

Kidney development proceeds along the cranial/caudal axis of this ridge, beginning with the formation of the **pronephros**, continuing along the **mesonephros**, and ending with the development of the **metanephros**.

3½ WEEKS

The pronephros consists of a series of tubules (generally 7 pairs) that appears within the **nephrotome**, the narrow band of mesoderm between the somites and the lateral ectodermal plate.

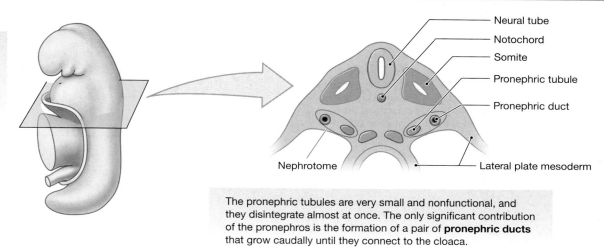

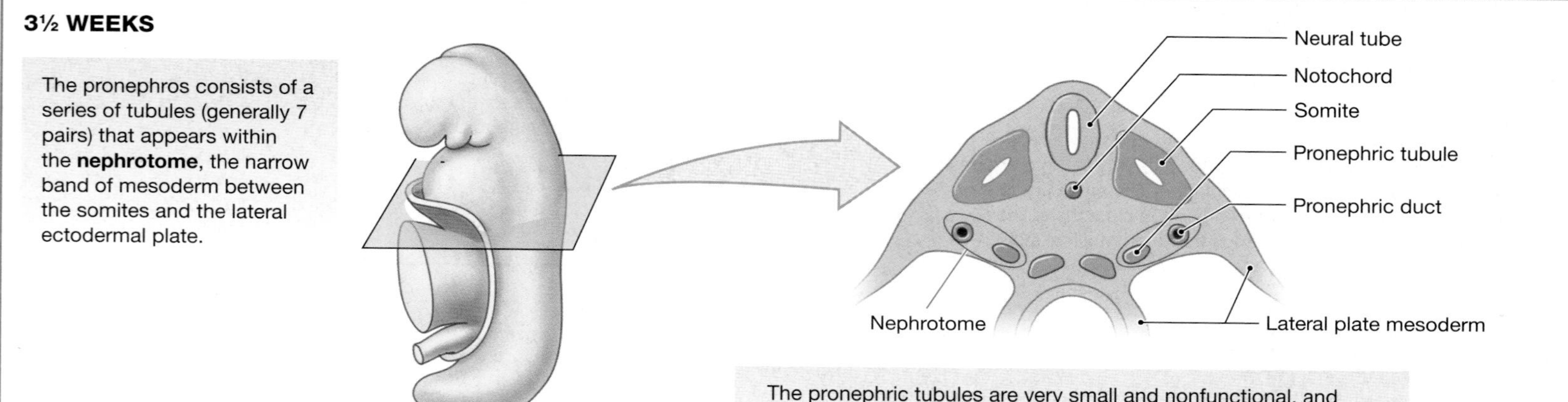

The pronephric tubules are very small and nonfunctional, and they disintegrate almost at once. The only significant contribution of the pronephros is the formation of a pair of **pronephric ducts** that grow caudally until they connect to the cloaca.

4 WEEKS

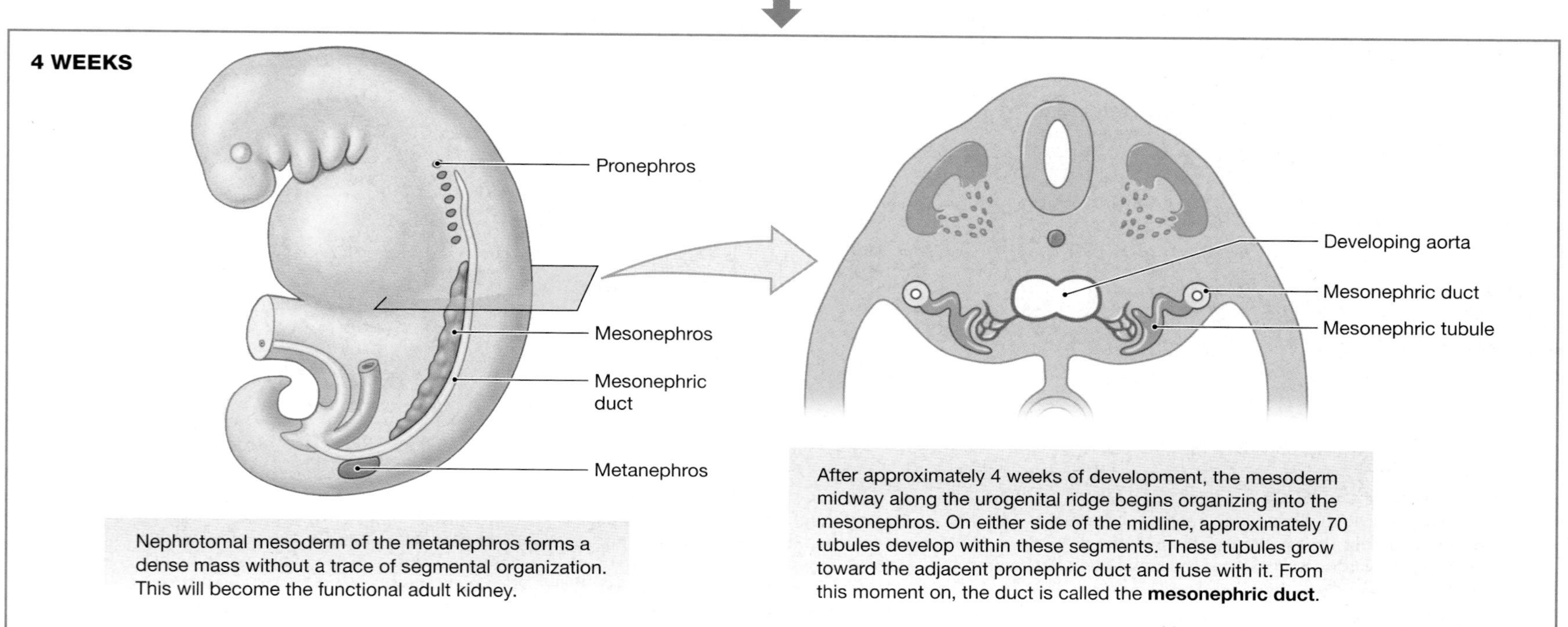

Nephrotomal mesoderm of the metanephros forms a dense mass without a trace of segmental organization. This will become the functional adult kidney.

After approximately 4 weeks of development, the mesoderm midway along the urogenital ridge begins organizing into the mesonephros. On either side of the midline, approximately 70 tubules develop within these segments. These tubules grow toward the adjacent pronephric duct and fuse with it. From this moment on, the duct is called the **mesonephric duct**.

The Development of the Urinary System, Part II

6 WEEKS

A **ureteric bud**, or *metanephric diverticulum*, forms in the wall of each mesonephric duct, and this closed tube elongates and branches within the adjacent metanephros. Tubules developing within the metanephros then connect to the terminal branches of the ureteric bud.

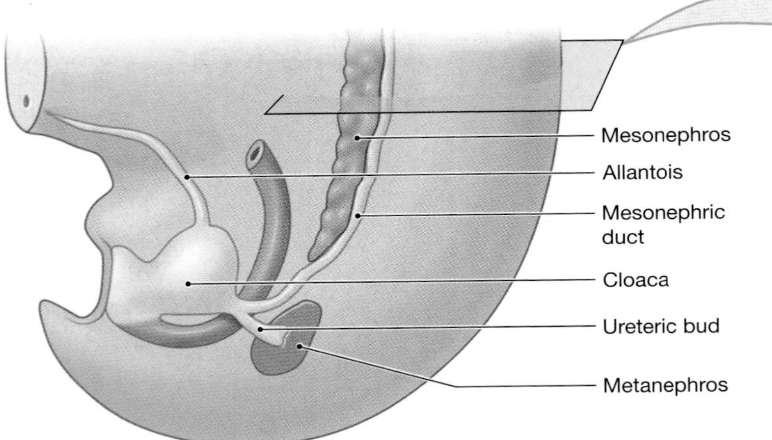

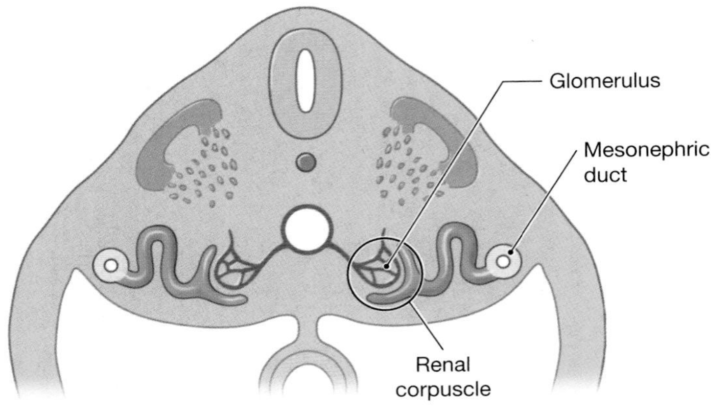

In each segment, a branch of the aorta grows toward the nephrotome, and the tubules form large nephrons with enormous glomeruli. Like the pronephros, the mesonephros does not persist, and when the last segments of the mesonephros are forming, the first are already beginning to degenerate.

Most of the metabolic wastes produced by the developing embryo are passed across the placenta to enter the maternal circulation. The small amount of urine produced by the kidneys accumulates within the cloaca and the **allantois**, an endoderm-lined sac that extends into the umbilical stalk.

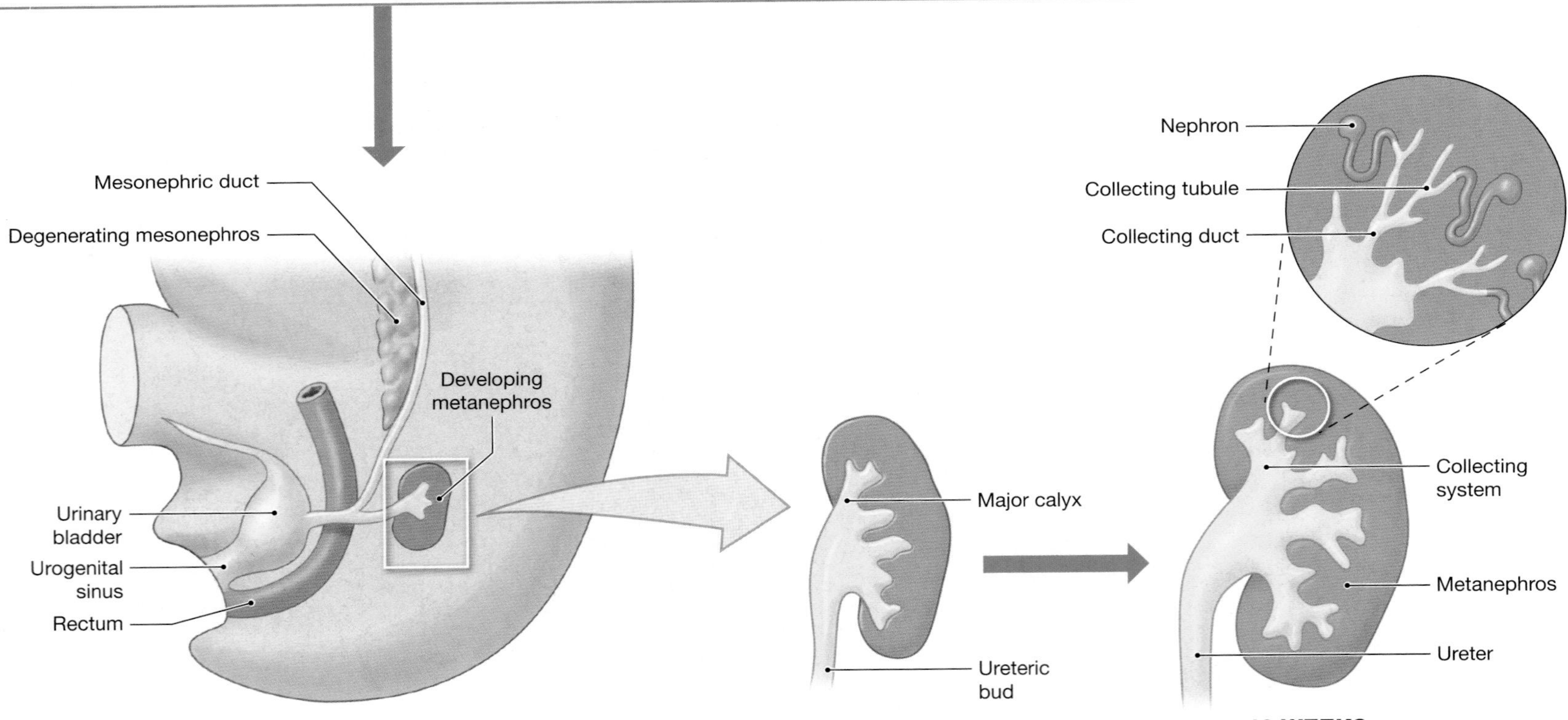

8 WEEKS

Near the end of the second developmental month, the cloaca is subdivided into a dorsal rectum and a ventral **urogenital sinus**. The proximal portions of the allantois persist as the urinary bladder, and the connection between the bladder and an opening on the body surface will form the urethra.

The ureteric bud branches within the metanephros, forming the calyces and the collecting system. The nephrons, which form within the mesoderm of the metanephros, tap into the collecting tubules.

12 WEEKS

The kidneys begin producing filtrate by the third developmental month. The filtrate does not contain waste products, as they are excreted by the placenta for removal and elimination by the maternal kidneys. The sterile filtrate mixes with the amniotic fluid and is swallowed by the fetus and reabsorbed across the lining of the digestive tract.

The Development of the Reproductive System

Sexually Indifferent Stages (Weeks 3–6)

DEVELOPMENT OF THE GONADS

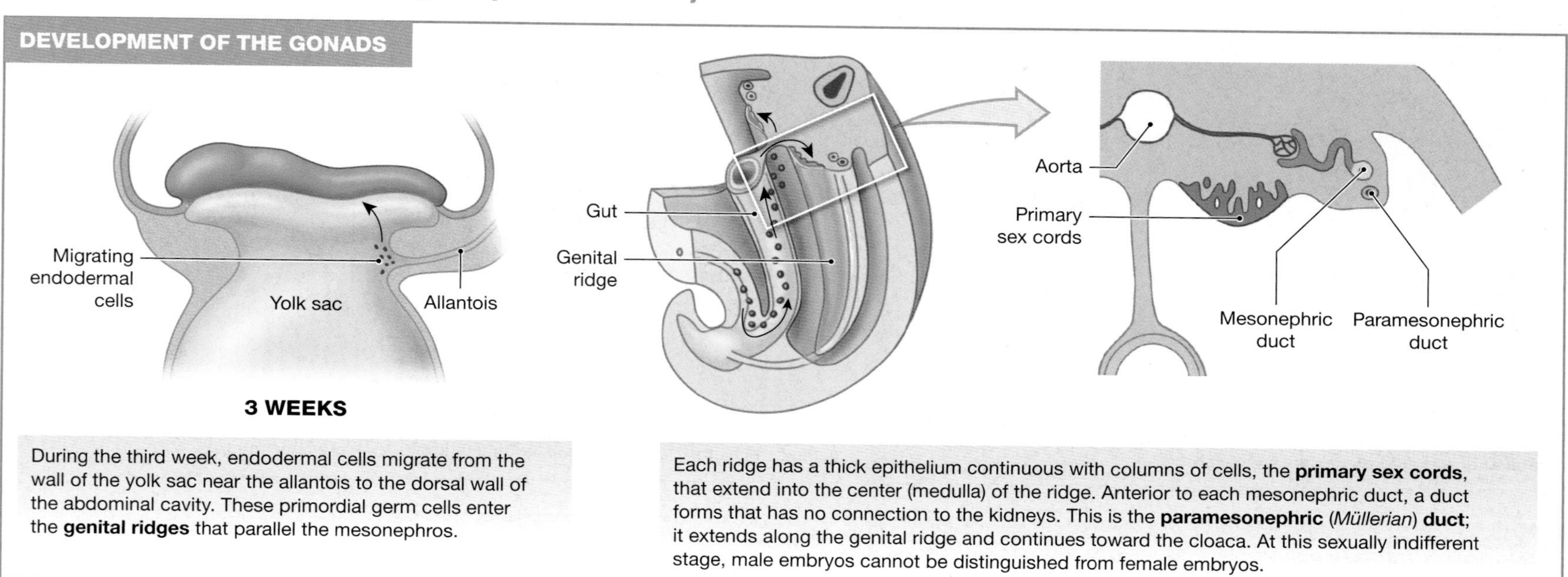

During the third week, endodermal cells migrate from the wall of the yolk sac near the allantois to the dorsal wall of the abdominal cavity. These primordial germ cells enter the **genital ridges** that parallel the mesonephros.

Each ridge has a thick epithelium continuous with columns of cells, the **primary sex cords**, that extend into the center (medulla) of the ridge. Anterior to each mesonephric duct, a duct forms that has no connection to the kidneys. This is the **paramesonephric** (*Müllerian*) **duct**; it extends along the genital ridge and continues toward the cloaca. At this sexually indifferent stage, male embryos cannot be distinguished from female embryos.

DEVELOPMENT OF DUCTS AND ACCESSORY ORGANS

Both sexes have mesonephric and paramesonephric ducts at this stage. Unless exposed to androgens, the embryo—regardless of its genetic sex—will develop into a female. In a normal male embryo, cells in the core (medulla) of the genital ridge begin producing testosterone sometime after week 6. Testosterone triggers the changes in the duct system and external genitalia.

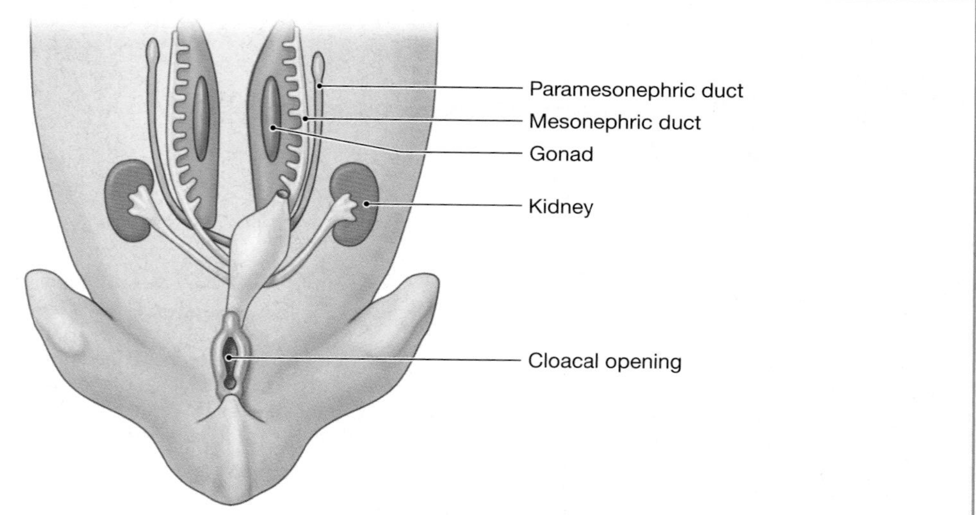

DEVELOPMENT OF EXTERNAL GENITALIA

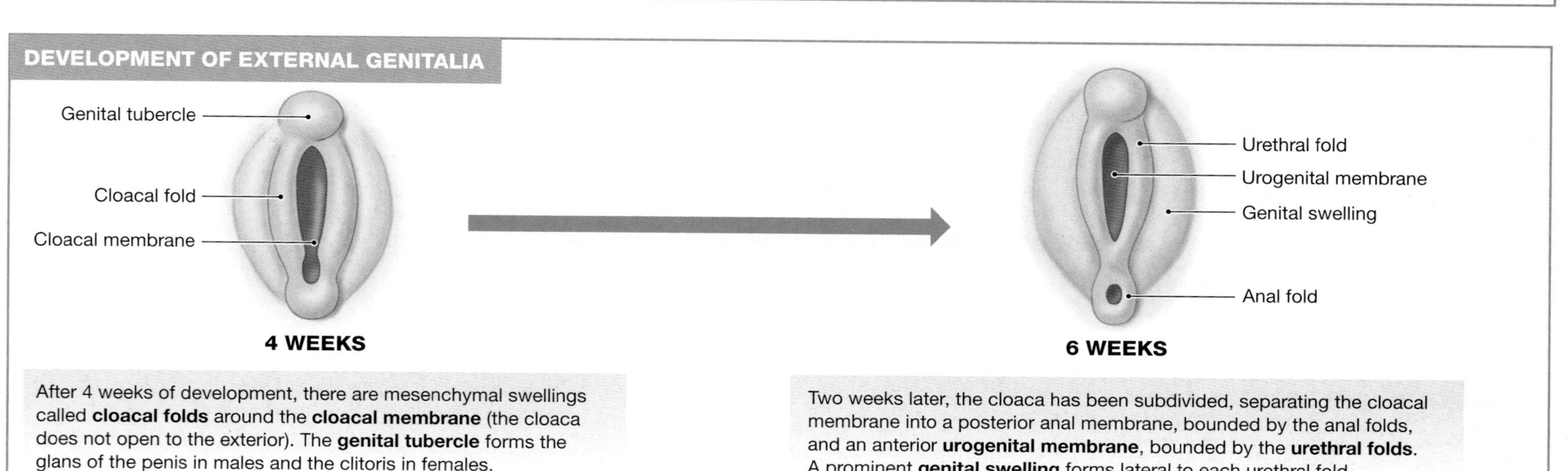

After 4 weeks of development, there are mesenchymal swellings called **cloacal folds** around the **cloacal membrane** (the cloaca does not open to the exterior). The **genital tubercle** forms the glans of the penis in males and the clitoris in females.

Two weeks later, the cloaca has been subdivided, separating the cloacal membrane into a posterior anal membrane, bounded by the anal folds, and an anterior **urogenital membrane**, bounded by the **urethral folds**. A prominent **genital swelling** forms lateral to each urethral fold.

The Development of the Reproductive System

Development of the Male Reproductive System

DEVELOPMENT OF THE TESTES

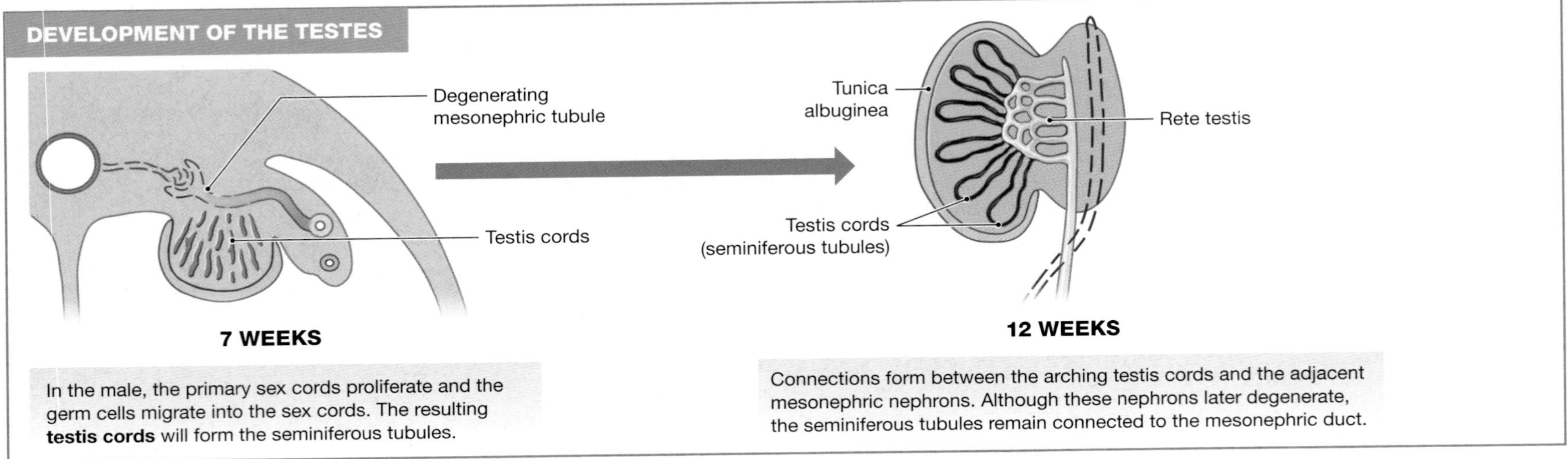

In the male, the primary sex cords proliferate and the germ cells migrate into the sex cords. The resulting **testis cords** will form the seminiferous tubules.

Connections form between the arching testis cords and the adjacent mesonephric nephrons. Although these nephrons later degenerate, the seminiferous tubules remain connected to the mesonephric duct.

DEVELOPMENT OF MALE DUCTS AND ACCESSORY ORGANS

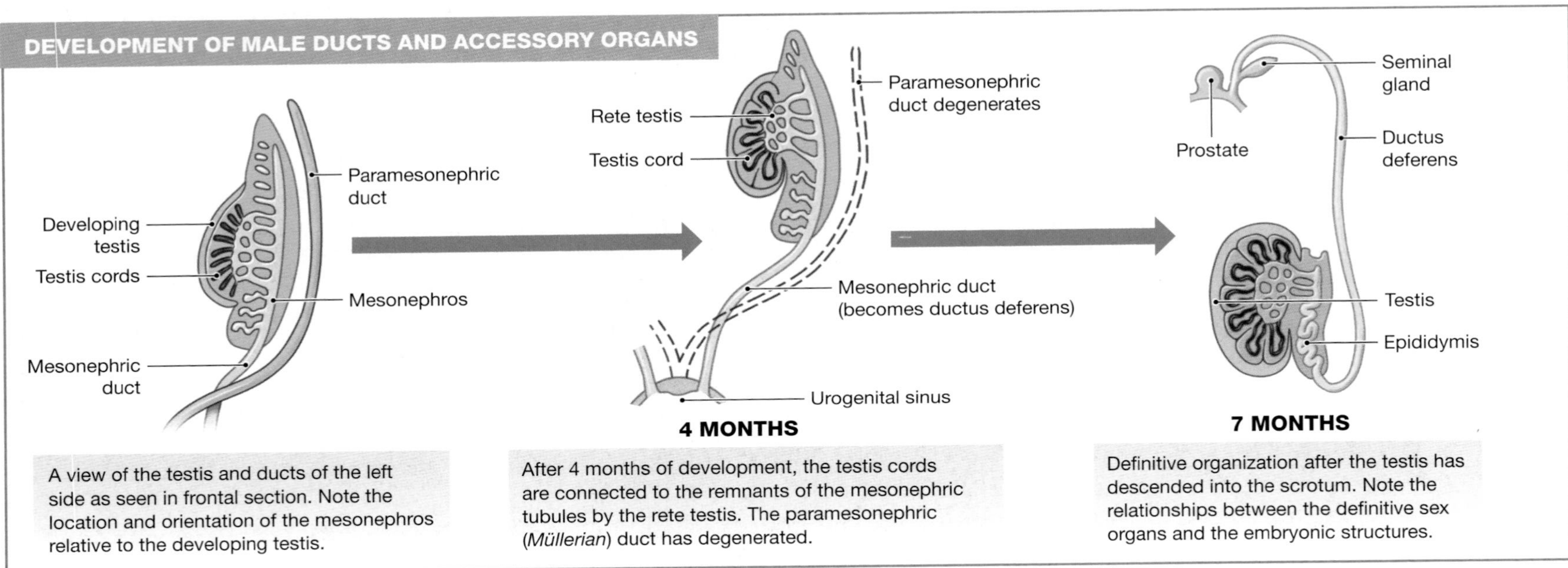

A view of the testis and ducts of the left side as seen in frontal section. Note the location and orientation of the mesonephros relative to the developing testis.

After 4 months of development, the testis cords are connected to the remnants of the mesonephric tubules by the rete testis. The paramesonephric (*Müllerian*) duct has degenerated.

Definitive organization after the testis has descended into the scrotum. Note the relationships between the definitive sex organs and the embryonic structures.

DEVELOPMENT OF MALE EXTERNAL GENITALIA

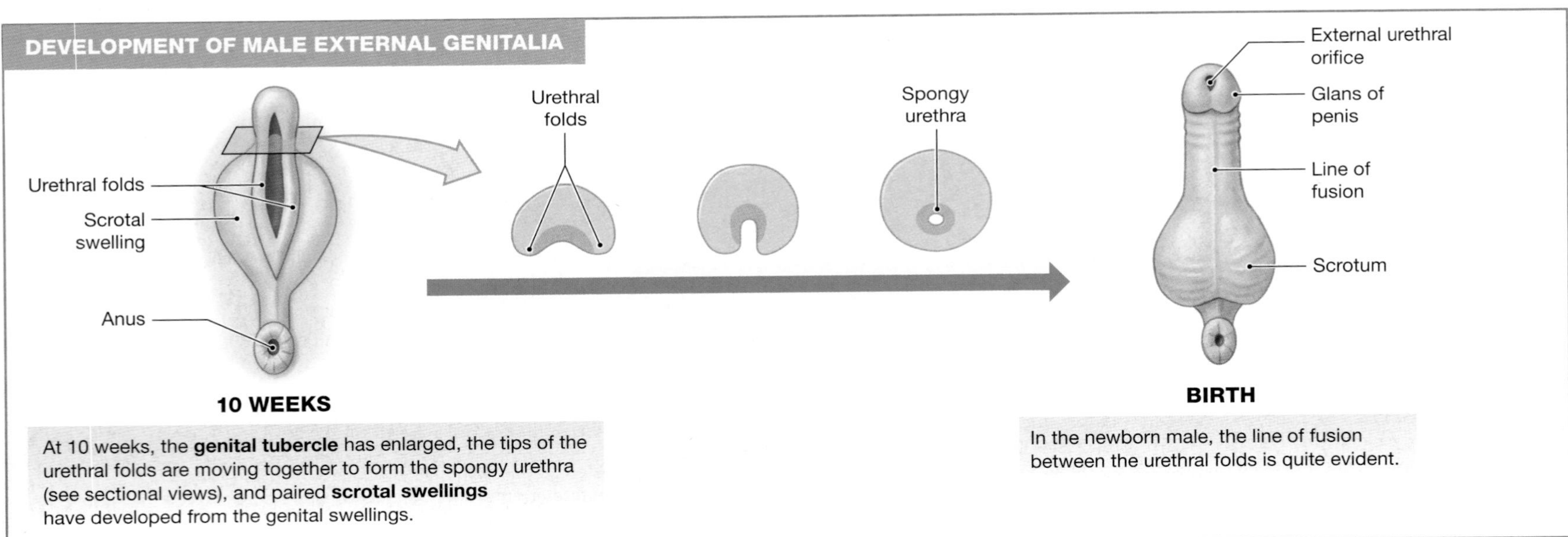

At 10 weeks, the **genital tubercle** has enlarged, the tips of the urethral folds are moving together to form the spongy urethra (see sectional views), and paired **scrotal swellings** have developed from the genital swellings.

In the newborn male, the line of fusion between the urethral folds is quite evident.

28

The Development of the Reproductive System

Development of the Female Reproductive System

DEVELOPMENT OF THE OVARIES

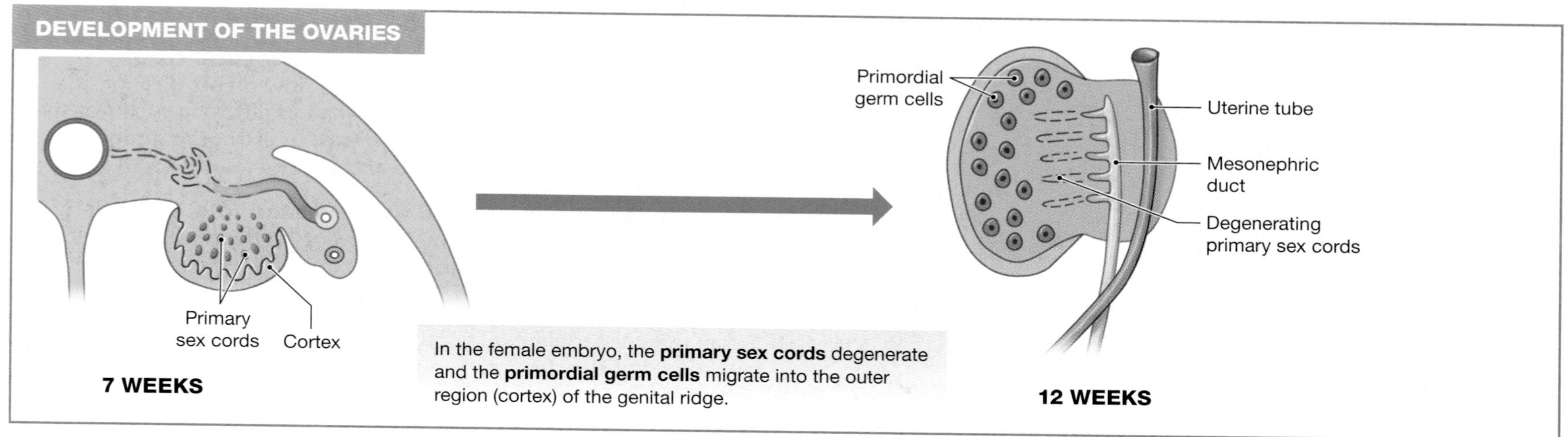

In the female embryo, the **primary sex cords** degenerate and the **primordial germ cells** migrate into the outer region (cortex) of the genital ridge.

DEVELOPMENT OF FEMALE DUCTS AND ACCESSORY ORGANS

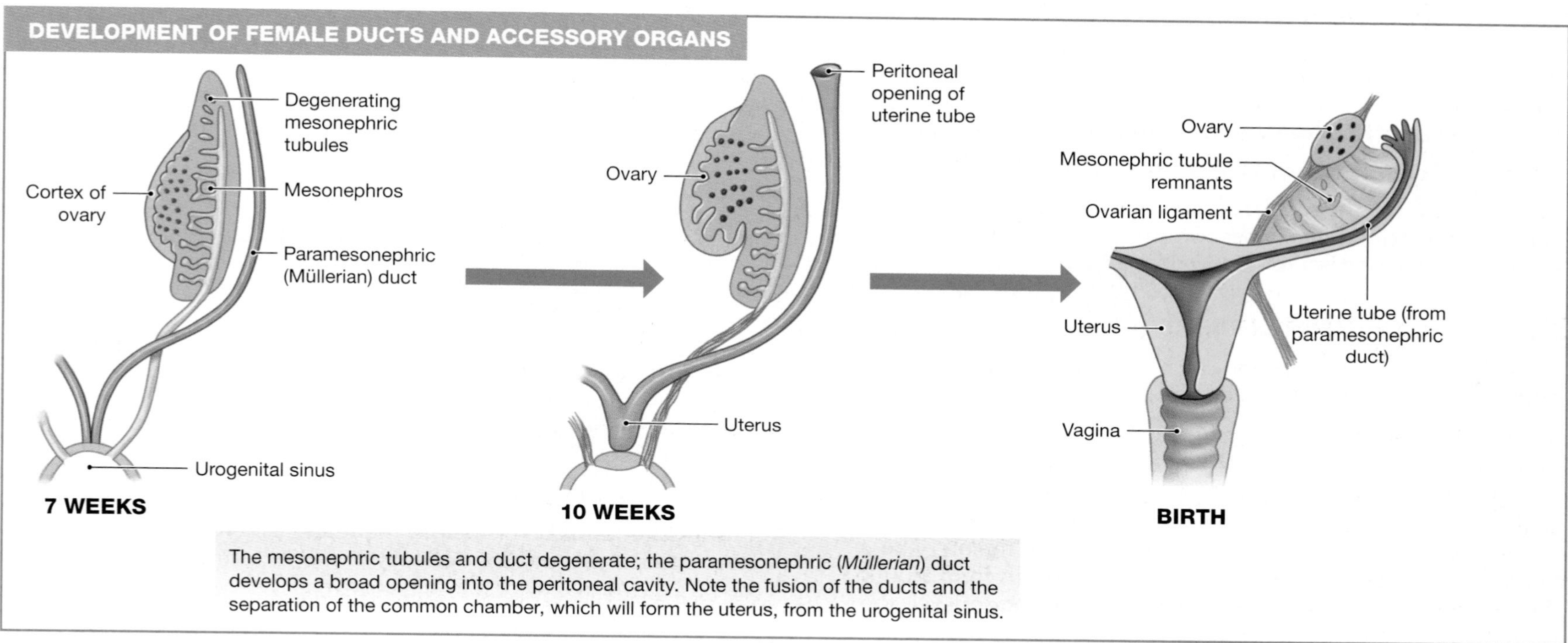

The mesonephric tubules and duct degenerate; the paramesonephric (*Müllerian*) duct develops a broad opening into the peritoneal cavity. Note the fusion of the ducts and the separation of the common chamber, which will form the uterus, from the urogenital sinus.

DEVELOPMENT OF FEMALE GENITALIA

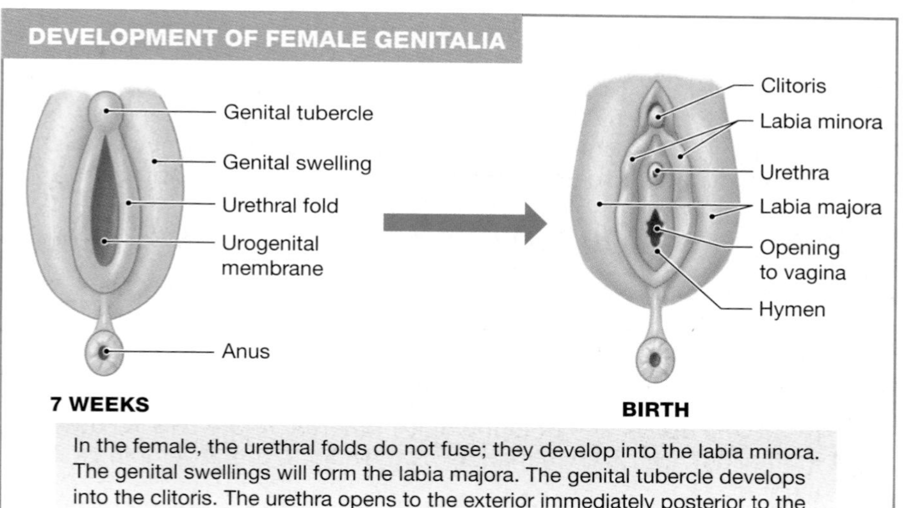

In the female, the urethral folds do not fuse; they develop into the labia minora. The genital swellings will form the labia majora. The genital tubercle develops into the clitoris. The urethra opens to the exterior immediately posterior to the clitoris. The hymen remains as an elaboration of the urogenital membrane.

COMPARISON OF MALE AND FEMALE GENITALIA

MALE	FEMALE
Penis	Clitoris
Corpora cavernosa	Erectile tissue
Corpus spongiosum	Bulbs of vestibule
Proximal shaft of penis	Labia minora
Spongy urethra	Vestibule
Bulbo-urethral glands	Greater vestibular glands
Scrotum	Labia majora

28

Study Outline

Introduction p. 740

- **Development** is the gradual modification of physical and physiological characteristics from conception to maturity. The formation of different cell types during development is called **differentiation**.

28.1 An Overview of Development p. 740

- Development begins at **fertilization** (*conception*) and can be divided into **prenatal** (before birth) and **postnatal** (birth to maturity) **development**.

28.2 Fertilization p. 740

- Fertilization normally occurs in or near the junction between the ampulla and isthmus of the uterine tube within a day after ovulation. Sperm cannot fertilize an oocyte until they have undergone capacitation. Around 200 million sperm are ejaculated into the vagina, and only around 100 will reach the ampulla. A man with fewer than 20 million sperm per milliliter is functionally sterile.

The Oocyte at Ovulation p. 740

- The acrosomes of the sperm release **hyaluronidase**, an enzyme that separates cells of the corona radiata and exposes the oocyte membrane. When a single sperm contacts that membrane, fertilization occurs and **oocyte activation** follows. *(See Figure 28.1.)*

Pronucleus Formation and Amphimixis p. 740

- During activation, the secondary oocyte completes meiosis. The **female pronucleus** then fuses with the **male pronucleus**, a process called **amphimixis**. *(See Figure 28.1.)*

28.3 Prenatal Development p. 742

- The 9-month **gestation** period can be divided into three **trimesters**, each 3 months in duration.

The First Trimester p. 742

- The **first trimester** is the most critical period in prenatal life. **Cleavage** subdivides the cytoplasm of the zygote in a series of mitotic divisions; the zygote becomes a **blastocyst**. During **implantation** the blastocyst burrows into the uterine endometrium. **Placentation** occurs as blood vessels form around the blastocyst and the **placenta** appears. **Embryogenesis** is the formation of a viable embryo.

- The blastocyst consists of an outer **trophoblast** and an **inner cell mass**, which is clustered at one end next to a hollow cavity (the **blastocoele**). *(See Figure 28.2 and Table 28.1.)*

- When the blastocyst adheres to the uterine lining, the trophoblast next to the endometrium undergoes changes and becomes a **syncytiotrophoblast**, which then erodes a path through the uterine epithelium. As the trophoblast enlarges and spreads, maternal blood flows through trophoblastic channels, or **lacunae**. The blastocyst organizes into layers and becomes a **blastodisc**. After **gastrulation**, the blastodisc contains an embryo composed of **endoderm**, **ectoderm**, and an intervening **mesoderm**. These **germ layers** help form four **extra-embryonic membranes**: the yolk sac, amnion, allantois, and chorion. *(See Figures 28.3–28.5 and Table 28.1.)*

- The **yolk sac** is an important site of blood cell formation. The **amnion** encloses fluid that surrounds and cushions the developing embryo. The base of the **allantois** later gives rise to the urinary bladder. Circulation within the vessels of the **chorion** provides a rapid-transit system linking the embryo with the trophoblast. *(See Figures 28.4 and 28.5.)*

- **Chorionic villi** extend outward into the maternal tissues, forming a branching network around which maternal blood flows. The **body stalk** and the **yolk stalk** connect the embryo with the chorion and the yolk sac, respectively. The **umbilical cord** connects the fetus to the placenta. Blood flow occurs through the umbilical arteries and the umbilical vein, and diffusion occurs at the chorionic villi. The placenta synthesizes HCG, estrogens, progestins, HPL, and relaxin. *(See Figures 28.5 and 28.6.)*

- The **embryo**, which has a **head fold** and a **tail fold**, will undergo critical changes in the first trimester. Events in the first 12 weeks establish the basis for **organogenesis** (organ formation). *(See Figures 28.5b,c and 28.7 and Table 28.2.)*

- Development of fully functional fetal organ systems is completed by the end of the third trimester. *(See Table 28.2 and Embryology Summaries.)*

The Second and Third Trimesters p. 746

- In the **second trimester**, the organ systems near functional completion and the fetus grows rapidly. During the **third trimester**, the organ systems become functional and the fetus undergoes its largest weight gain. *(See Figures 28.7–28.9, Table 28.2, and Embryology Summaries.)*

28.4 Labor and Delivery p. 752

- The goal of true labor is **parturition**, the forcible expulsion of the fetus.

Stages of Labor p. 752

- Labor can be divided into three stages: **dilation stage**, **expulsion stage**, and **placental stage**. The dilation stage, in which the cervix dilates, usually lasts 8 or more hours. The expulsion stage involves the *birth* (**delivery**) of the fetus. The placenta is ejected during the placental stage. *(See Figure 28.10.)*

Premature Labor p. 754

- **Premature labor** occurs before the fetus has completely developed. A **premature delivery** produces a newborn weighing over 1 kg, which may or may not survive.

28.5 The Neonatal Period p. 755

- The **neonatal period** extends from birth to 1 month of age.

- In the transition from fetus to **neonate**, the respiratory, circulatory, digestive, and urinary systems undergo tremendous changes as they begin functioning independently. The newborn's metabolic rate increases, and body temperature shifts hourly.

28.6 Embryology of Organ Systems p. 756

- Embryonic primordia and their derivatives have definite relationships to each other and to other developing structures. Because of these relationships, early anatomical patterns develop.

- These anatomical patterns persist into the newborn and adult. As a result, the derivatives of embryonic primordia relate to each other in the adult in the same manner as they did in the developing embryo.

Chapter Review

For answers, see the blue Answers tab at the back of the book.

Level 1 Reviewing Facts and Terms

Match each numbered item with the most closely related lettered item.

1. corona radiata ....
2. hyaluronidase ....
3. second trimester ....
4. expulsion stage ....
5. ooplasm ....
6. third trimester ....
7. first trimester ....
8. parturition ....
9. dilation stage ....
10. episiotomy ....

(a) begins with the onset of true labor
(b) begins as the cervix dilates completely
(c) birth
(d) characterized by rapid fetal growth
(e) incision through the perineal musculature
(f) layer of follicular cells around the oocyte
(g) enzyme in the acrosome
(h) cytoplasm of the oocyte
(i) time of embryonic/early fetal development
(j) characterized by the continued development of organ systems

11. Identify the two extra-embryonic membranes and the three different regions of the endometrium at week 10 of development in the following diagram.
(a) ______ (b) ______
(c) ______ (d) ______
(e) ______

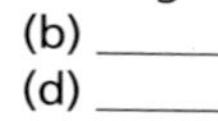

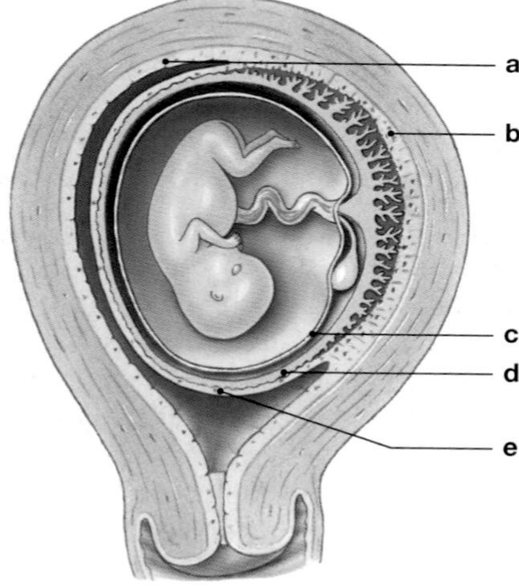

12. During amphimixis,
(a) sperm become capacitated.
(b) the ovum finishes meiosis II.
(c) the male and female pronuclei fuse.
(d) meiosis occurs.

13. During the first trimester of development,
(a) the fetal lungs begin to process air.
(b) the beginnings of all the major organ systems begin to appear.
(c) the organs and organ systems complete their development.
(d) the fastest period of fetal growth occurs.

14. The process of cell division that occurs after fertilization is called
(a) cleavage. (b) implantation.
(c) placentation. (d) embryogenesis.

15. A blastocyst is a(n)
(a) extra-embryonic membrane that forms blood vessels.
(b) solid ball of cells.
(c) hollow ball of cells.
(d) portion of the placenta.

16. During the second trimester, the most significant event that permits the mother to realize that she is carrying a fetus within her body is that
(a) the fetal brain begins to grow greatly.
(b) the fetus begins to move inside the uterus.
(c) eye and ear basic structures form.
(d) articulations between joints appear.

17. The hormone that is the basis for pregnancy tests is
(a) LH.
(b) progesterone.
(c) human chorionic gonadotropin (HCG).
(d) human placental lactogen (HPL).

18. The first stage of labor is the _____ stage.
(a) dilation (b) expulsion
(c) placental (d) decidual

19. The extra-embryonic membrane that forms the urinary bladder is the
(a) yolk sac.
(b) allantois.
(c) amnion.
(d) chorion.

20. During implantation, the
(a) syncytiotrophoblast erodes a path through the endometrium.
(b) inner cell mass begins to form the placenta.
(c) maternal blood vessels in the endometrium are walled off from the blastocyst.
(d) entire trophoblast becomes syncytial.

Level 2 Reviewing Concepts

1. What would be the fate of an embryo if the chorion failed to form properly?
(a) There would be an insufficient number of blood cells formed, and the infant would need a transfusion at birth.
(b) There would be no effect; the allantois would assume the function.
(c) There would be no protective cushion of fluid surrounding the embryo, which would be subjected to injury within the uterus.
(d) The embryo would be unable to receive nutrients and oxygen in sufficient amounts, and the pregnancy might terminate spontaneously.

2. What would be the effect of a direct connection between the blood supply of the chorionic villi and the basal decidua of the placenta?
(a) There would be normal development, as this is usually what occurs.
(b) An incompatibility reaction that could damage the fetus is possible.
(c) Fetal growth would be slowed.
(d) The fetus would grow faster than usual.

3. The stage of development that follows cleavage is the
(a) blastocyst.
(b) morula.
(c) trophoblast.
(d) blastocoele.

4. What is cleavage, and in which trimester does it occur?

5. What is the primitive streak?

6. Describe the effects of human chorionic gonadotropin. Where and when is it produced?

7. What occurs during amphimixis?

8. Which extra-embryonic membrane is important for blood cell formation?

9. Describe the process of organogenesis.

10. Why is the first trimester considered a critical period?

Level 3 Critical Thinking

1. Joe and Jane desperately want to have children, and although they have tried for 2 years, they have not been successful. Finally, each of them consults a physician, and it turns out that Joe has oligospermia (low sperm count). He confides to you that he doesn't understand why this would interfere with his ability to have children since he remembers from biology class that it only takes one sperm to fertilize an oocyte. What should you tell him?

2. The virus that causes rubella, or "German measles," has been recognized for decades as a powerful teratogen (agent that causes malformation of an embryo). During which trimester of pregnancy would it be most dangerous for pregnant women to be exposed to this pathogen? Why?

CLINICAL CASE WRAP-UP

The Least-Alike Twins

The winners of the prize for the "least-alike" twins are Mayu and Ruchira. The girls traveled from Moline, Illinois to Twinsburg to attend the Twins Days Festival.

Mayu's and Ruchira's parents used the services of a prestigious fertility clinic in the United States. Their mother's ovaries were hormonally stimulated, and the released secondary oocytes were harvested via laparoscopic surgery, a minimally invasive technique using a fiber-optic instrument inserted through the abdominal wall. These oocytes were then fertilized with sperm in the laboratory. On the sixth post-fertilization day—coinciding with the time a fertilized blastocyst would be finishing its journey through the uterine tube into the uterus—two blastocysts were transferred to the mother's uterus. Both had a successful implantation, placentation, embryogenesis, fetal development, and birth.

Why, then, do Mayu and Ruchira look so different? As it turned out, the technician performing the in vitro fertilization accidentally used a previously used pipette to transfer the sperm. The couple who visited the clinic just before Mayu's and Ruchira's parents were Afro-Caribbean, and the man's sperm was still present in the pipette. As a result, one oocyte was fertilized by sperm from a Caucasian man, and the other was fertilized by sperm from an African-Caribbean man.

1. Why were the fertilized blastocysts implanted in Mayu's and Ruchira's mother on the sixth day after ovulation?
2. Are Mayu and Ruchira monozygotic twins or dizygotic twins?

See the blue Answers tab at the back of the book.

Related Clinical Terms

eclampsia: A condition in which one or more convulsions occur in a pregnant woman suffering from high blood pressure, often followed by coma and posing a threat to the health of mother and baby.

infertility: The inability to achieve pregnancy after engaging in 1 year of appropriately timed intercourse.

in vitro fertilization: Fertilization outside the body, generally in a Petri dish.

gamete intrafallopian transfer (GIFT): An assisted reproductive procedure in which a woman's eggs are removed, mixed with sperm, and replaced into the woman's uterine tube where the fertilization takes place, rather than in the laboratory.

oligodactyly: The presence of fewer than five digits on one or more limbs at birth.

placenta abruptio: Condition in which there is separation of the placenta from the uterine site of implantation before delivery of the baby.

placenta previa: Condition during pregnancy in which the placenta is abnormally placed so as to totally or partially cover the cervix.

polydactyly: A physical abnormality in which the person has more than five fingers or toes on one or more limb.

pre-eclampsia: A condition in pregnancy characterized by sudden hypertension, albuminuria, and edema of the hands, feet, and face. It is the most common complication of pregnancy, affecting about 5 percent of pregnancies.

therapeutic cloning: A procedure that usually takes skin cells from a patient, and inserts a skin cell nucleus into a fertilized egg whose nucleus has been removed to create a new cell. That new cell divides repeatedly to form a blastocyst from which stem cells can be extracted to grow new tissue that is genetically matched to the patient.

Answers to Concept Checks, Chapter Reviews, and Clinical Case Wrap-Ups

1 FOUNDATIONS An Introduction to Anatomy

Concept Check

Page 2

1. Histologists study the structure and properties of tissue.
2. Tissues in combination form organs. Organs are anatomical units that have multiple functions.

Page 2

3. A gross anatomist investigates organ systems and their relationships to the body as a whole, while a histologist investigates the structure and properties of tissue.
4. An organ system is a group of organs that function together to produce coordinated effects. These coordinated effects are system-wide; therefore, the study of anatomy organized by organ systems is known as systemic anatomy.

Page 5

5. Surgical anatomy studies anatomical landmarks important for surgical procedures, while clinical anatomy focuses on anatomical features that may undergo recognizable pathological changes during illness.
6. Cross-sectional anatomy is a subspecialty of gross anatomy.

Page 7

7. Because there are interactions among organ systems, a patient exhibiting cyanosis may have a problem with his or her lungs or heart. A lung problem could cause insufficient oxygen intake, and a heart problem would result in insufficient blood delivery to the tissues.

Page 14

8. Differentiation is the gradual appearance of characteristic cellular specializations during development as the result of gene activation or repression.
9. These components are found within the integumentary system.

Page 19

10. If you broke your antebrachium, you broke your forearm.
11. Groin = inguen or inguinal region, buttock = gluteus or gluteal region, and hand = manus or manual region.
12. A midsagittal section would separate the two eyes.
13. Mesenteries are double sheets of serous membranes in the peritoneal cavity that provide blood supply, lubrication, support, and stability for organs (stomach, small intestine, and parts of the large intestine) while permitting limited movement.
14. The body cavity inferior to the diaphragm is the abdominopelvic cavity.

Chapter Review

Level 1 Reviewing Facts and Terms

1. e 2. d 3. f 4. g 5. b 6. c 7. a 8. a. frontal or coronal; b. sagittal; c. transverse, horizontal, or cross-sectional 9. a. pleural; b. pericardial; c. abdominal; d. pelvic 10. b 11. c 12. b 13. b

Level 2 Reviewing Concepts

1. a
2. All living organisms have the same basic functions: responsiveness, growth and differentiation, reproduction, movement, metabolism, and excretion.
3. See Figure 1.10.
4. b
5. b
6. In large organisms with specialized organ systems that perform absorption, respiration, and excretion in different regions of the body, there must be a means of internal transport for these products. The vascular system provides that transport. Passive processes such as diffusion and osmosis would occur too slowly to permit the organism to function and remain alive.

Level 3 Critical Thinking

1. A disruption of cellular division within the red bone marrow will result in the production of too few red blood cells (anemia) or too many red blood cells (polycythemia) within the vessels of the cardiovascular system. Anemia will affect the amount of oxygen carried to peripheral tissues, thereby affecting overall metabolism. Polycythemia will produce altered peripheral metabolism due to the "clogging" of blood vessels by the increased number of cells attempting to get through peripheral capillaries.
2. The body systems affected are the digestive, respiratory, and skeletal systems. The anatomical specialties involved include surface anatomy, regional anatomy studies, systemic anatomy, comparative anatomy, and developmental anatomy.

Answers to Clinical Case Wrap-Up

Page 26

1. The hip fracture is proximal to the more distal ankle fracture.
2. The spleen is located in the left hypochondriac region, or the left upper quadrant (LUQ). A clinical clue to this life-threatening injury is the multiple rib fractures on the left side.

2 FOUNDATIONS The Cell

Concept Check

Page 42

1. Selectively permeable is the term used to describe the permeability of the plasma membrane.
2. The three types of endocytosis are pinocytosis, phagocytosis, and receptor-mediated endocytosis. Pinocytosis is the formation of vesicles filled with extracellular fluid. Phagocytosis produces vesicles containing solid objects. Receptor-mediated endocytosis resembles pinocytosis but is very selective: The coated vesicles contain a specific target molecule in high concentrations bound to receptors on the membrane surface.
3. The two subdivisions of the cytoplasm are (1) the cytosol, which is the intracellular fluid containing dissolved nutrients, ions, proteins, and wastes, and (2) organelles, which are structures that perform specific functions within the cell.
4. Mitochondria produce ATP. The number of mitochondria in a particular cell varies depending on the cell's energy demand. If a cell contains many mitochondria, the energy demands for the cell are very high.
5. Cells in the ovaries and testes contain large amounts of smooth endoplasmic reticulum (SER) because it is used to synthesize steroid hormones.

Page 42

6. Communicating junctions are channeled proteins, which allow the passage of ions, small metabolites, and regulatory molecules from cell to cell. The passage of these substances from cell to cell helps to facilitate the coordination of cellular functions.
7. At tight junctions the lipid portions of two plasma membranes are tightly bound together.

Page 44

8. Cell division is a form of cellular reproduction that results in an increase in cell number.
9. Mitosis is a process that occurs during the division of somatic cells. It is the accurate duplication of the cell's genetic material and the distribution of one copy to each of the two new daughter cells. It is divided into prophase, metaphase, anaphase, and telophase.
10. Interphase can be divided into the G_1, S, and G_2 phases. In the G_1 phase, the cell manufactures organelles and cytosol to make two functional cells. During the S phase, the cell replicates its chromosomes. The G_2 phase is a time for last-minute protein synthesis before mitosis. An interphase cell in the G_0 phase is not preparing for mitosis, but is performing all other normal cell functions. Mitosis consists of four stages: prophase, metaphase, anaphase, and telophase. In prophase, the spindle fibers form and the nuclear membrane disappears; in metaphase, the chromatids line up along the metaphase plate; in anaphase, the chromatids separate and move to opposite poles of the spindle apparatus; and in telophase, the nuclear membrane reappears and chromosomes uncoil as the daughter cells separate through cytokinesis.

Chapter Review

Level 1 Reviewing Facts and Terms

1. e 2. g 3. h 4. b 5. d 6. i 7. c 8. f 9. a 10. b 11. c 12. d 13. See Figure 2.1 14. d 15. b 16. c 17. c 18. b

Level 2 Reviewing Concepts

1. The nuclear membrane partially separates the nucleoplasm from the cytosol, which is necessary to permit and promote specific nuclear functions.
2. The three basic concepts of the cell theory are the following: (1) Cells are the structural "building blocks" of all plants and animals; (b) cells are produced by the division of pre-existing cells, and (c) cells are the smallest structural units that perform all vital functions.
3. The three passive processes by which substances get into and out of cells are diffusion, osmosis, and facilitated diffusion.
4. Similarities: Both processes utilize carrier proteins. Differences: In facilitated diffusion, substances move down their concentration gradients (from high to low concentration), and this process does not utilize energy obtained by the splitting of ATP. In active transport, substances move against their concentration gradients (low to high concentration), a process requiring the energy provided by the splitting of ATP.
5. The three major factors that determine whether a substance can diffuse across a plasma membrane are the size of the molecule, the concentration gradient across the membrane, and the solubility of the molecule.

6. Organelles are structures that perform specific functions within the cell. Organelles can be divided into two broad categories: (1) nonmembranous organelles, which are always in contact with the cytoplasm, and (2) membranous organelles, which are surrounded by membranes that isolate their contents from the cytoplasm.
7. a. interphase; b. early prophase; c. late prophase; d. metaphase; e. anaphase; f. telophase; g. cytokinesis
8. The four general functions of the plasma membrane are the following: (1) physical isolation, (2) regulation of exchange with the environment, (3) sensitivity, and (4) structural support.

Level 3 Critical Thinking

1. Skin cells are tightly held together by abundant desmosomes, very strong cell junctions that can resist twisting and stretching or breaking apart.
2. Since the transport of the molecule is against the concentration gradient (that is, from lower to higher concentration) and requires energy to move across the membrane, this type of transport is called active transport.
3. Solution A must have initially had more solutes than solution B. As a result, water moved by osmosis across the selectively permeable membrane from side B to side A, increasing the fluid level on side A.

Answers to Clinical Case Wrap-Up

Page 48

1. No, Jessica's red blood cells have no mitochondria and therefore are not affected by her mitochondrial disease. Every other cell in her body, however, is affected.
2. Because the nucleus contains the genetic material, it had to be removed (enucleated) so that the surrogate's genetic material was not passed on to Jessica's offspring.

3 | FOUNDATIONS Tissues and Early Embryology

Concept Check

Page 58

1. No. A simple squamous epithelium does not provide enough protection against infection, abrasion, and dehydration and is not found in the skin surface.
2. In holocrine secretion, the entire gland cell is destroyed during the secretory process. Further secretion by the gland requires regeneration of cells to replace those lost during secretion.
3. During apocrine secretion, the secretory product and the apical portion of the cell cytoplasm are shed.
4. A simple columnar epithelium provides protection and may be involved in both absorption and secretion.

Page 70

5. The three basic components of connective tissue are specialized cells, extracellular protein fibers, and a fluid called ground substance.
6. Connective tissue proper refers to connective tissues with many types of cells and extracellular fibers in a syrupy ground substance. Supporting connective tissues (cartilage and bone) have a less diverse cell population and a matrix that contains closely packed fibers. The matrix is either gel-like (cartilage) or calcified (bone).
7. The two types of cells in connective tissue proper are fixed cells and wandering cells. Fixed cells include mesenchymal cells, fibroblasts, fibrocytes, fixed macrophages, adipocytes, and sometimes melanocytes. Wandering cells include free macrophages, mast cells, lymphocytes, plasmocytes, neutrophils, and eosinophils.

Page 72

8. Mucous membranes line passageways that communicate with the exterior and form a barrier that resists the entry of pathogens. Serous membranes line the subdivisions of the ventral body cavity and are composed of two parts: parietal and visceral portions. Minimizing friction between these two opposing surfaces is the primary function of a serous membrane. The cutaneous membrane (skin) covers the body. Synovial membranes line synovial joints. Synovial membranes secrete a fluid (synovial fluid) that nourishes and lubricates the surfaces of a synovial joint.

Page 74

9. Other names for the superficial fascia are *subcutaneous layer* and *hypodermis*. It separates the skin from underlying tissues and organs, provides insulation and padding, and lets the skin or underlying structures move independently.
10. The subserous fascia lies between the deep fascia and the serous membranes lining the body cavities. Because this layer separates the serous membranes from the deep fascia, movements of muscle or muscular organs do not seriously distort the delicate lining.

Page 74

11. Muscle tissue is specialized for contraction.
12. Cardiac and skeletal muscle are both forms of striated muscle. Cardiac muscle has a single, centrally located nucleus. Skeletal muscle cells have numerous, peripherally located nuclei. Smooth muscle is not striated and has a single, centrally located nucleus.

Page 76

13. Nervous tissue is specialized to conduct electrical impulses from one region of the body to another.
14. The two types of cells found within nervous tissue are neurons and neuroglia. Neurons conduct electrical impulses from one region of the body to another. Neuroglia provide support for neurons, regulate the composition of the intercellular tissue, and provide nutrients to neurons.

Page 76

15. As tissues age, repair and maintenance become less efficient, and the structure and chemical composition of many tissues change.

Chapter Review

Level 1 Reviewing Facts and Terms

1. j 2. h 3. a 4. c 5. f 6. g 7. i 8. b 9. e 10. d 11. a. transitional epithelium; b. tendon; c. hyaline cartilage; d. skeletal muscle; e. smooth muscle 12. c 13. c 14. d 15. d 16. c 17. a 18. d

Level 2 Reviewing Concepts

1. A single cell is the unit of structure and function, whereas groups of cells work together as units called tissues (collections of special cells and cell products that perform limited functions).
2. A tendon is made of cords of dense regular connective tissue that attach a skeletal muscle to a bone. Ligaments resemble tendons, but they usually connect one bone to another bone. They also contain significantly more elastic fibers than tendons.
3. Exocrine secretions occur in ducts that lead onto surfaces; endocrine secretions occur in the interstitial fluid and then enter into blood vessels.
4. The presence of the cilia at the surface of the epithelium means that the mucous layer, which rests on the epithelium surface, will be kept moving and the respiratory surface will be kept clean (or clear).
5. While a tendon is made of cords of dense regular connective tissue, an aponeurosis is made of a sheet or ribbon of connective tissue that resembles a flat tendon. Aponeuroses may cover the surface of muscles and assist superficial muscles in attaching to one another or to other structures.
6. b
7. d
8. The skin is separated from the muscles by a layer of loose connective tissue. The loose connective tissue provides padding and elastic properties that permit a considerable amount of independent movement of the two layers. Thus, pinching the skin does not affect the underlying muscle.
9. Stem cells are usually found in the deepest layer of the epithelium. They divide to replace cells lost or destroyed at the epithelial surface.

Level 3 Critical Thinking

1. The presence of DNA, RNA, and membrane components suggests that the cells were destroyed during the process of secretion. This is consistent with a holocrine type of secretion.
2. Mucus is produced by the respiratory epithelium in response to irritation from smoking. The cilia found on many cells of the respiratory system beat upward, thereby moving mucus produced by the respiratory epithelium up to the level of the esophagus, where it can be swallowed and eliminated. Destruction of the cilia prevents the elimination of this mucus. Therefore, coughing is the only mechanism by which the mucus can be removed.
3. Ischemia occurs when the blood flow to an organ has been reduced, preventing it from receiving oxygen. While ischemia can damage organs, skeletal muscle ischemia is not as life-threatening as cardiac muscle ischemia. Heart (myocardial) ischemia can lead to a heart attack and possible death.

Answers to Clinical Case Wrap-Up

Page 85

1. An abnormal protein affects the elastin fibers within the matrix of connective tissue proper. This results in abnormally long and thin digits, vision difficulty, and cardiovascular defects.
2. The abnormal elastin of Marfan's syndrome would also affect loose areolar connective tissue beneath the skin and beneath the epithelial lining of the digestive, respiratory, and urinary tracts, and blood vessels throughout the body.

4 | The Integumentary System

Concept Check

Page 88

1. Epithelium covers the surface of the integumentary system; connective tissue provides strength and flexibility; smooth muscle is found within the blood vessels of the dermis; and nervous tissue controls the activity of the smooth muscle and is also associated with the numerous sensory structures found within the integument.

Page 92

2. The third layer of the epidermis is the stratum granulosum.
3. Keratinization is the production of keratin by epidermal cells. It occurs in the stratum granulosum of the epidermis. Keratin fibers develop within cells of the stratum granulosum. As keratin fibers are produced, these cells become thinner and flatter, and their plasma membranes become thicker and less permeable. As these cells die, they form the densely packed layers of the stratum lucidum and stratum corneum.
4. The color of the epidermis is due to a combination of the dermal blood supply and variable quantities of two pigments: carotene and melanin.
5. Epidermal ridges are formed by the deeper layers of the epidermis that extend into the dermis, increasing the contact area between the two regions. Dermal papillae are projections of the dermis that extend between adjacent ridges.

Page 94

6. Tension lines are important because a cut that runs parallel to a tension line closes more easily, whereas a cut made at right angles to a tension line will sever elastic fibers in the skin and thus be pulled open as they recoil. A parallel cut will also heal faster and with less scarring than a cut at right angles to cleavage lines.
7. Arteries and veins supplying the skin form an interconnected network in the subcutaneous layer along the border with the reticular layer. This network is called the cutaneous plexus.

Page 95

8. The subcutaneous layer is important in stabilizing the position of the skin in relation to underlying tissues. The fat contained within the subcutaneous layer also serves as an energy reserve and a cushion.

Page 102

9. Hair serves numerous functions. The hair on top of the head protects the scalp from ultraviolet light, cushions a hit to the head, and provides insulation to the skull. Hairs at the entrance to the nostrils help prevent the entrance of foreign objects. Eyelashes protect foreign objects from entering the eyes. Because of the associated nervous tissue, the movement of even a single hair can be felt, and thus we can perceive when something brushes past our skin.
10. The two types of hairs are vellus hairs and terminal hairs.
11. Apocrine sweat glands produce a viscous, cloudy, and potentially odorous secretion that contains several kinds of organic compounds. Some of these compounds have an odor, and others produce an odor when metabolized by skin bacteria. Merocrine sweat glands produce a watery secretion that mainly contains sodium chloride, metabolites, and wastes. Merocrine secretions are generally known as sweat.
12. Prominent structures of the nail include the nail, free edge, lateral nail fold, lunula, eponychium, proximal nail fold, nail root, hyponychium, and nail body.

Page 102

13. Mechanical stress will cause the cells within the stratum basale to divide more rapidly, forming a callus.

Page 103

14. The epidermis thins; the number of Langerhans cells decreases; vitamin D_3 production decreases; melanocyte activity decreases; glandular activity decreases; blood supply to the dermis is decreased; hair follicles stop functioning; dermis becomes thinner; skin repairs proceed more slowly; secondary sexual characteristics in hair and body fat distribution change.

Chapter Review

Level 1 Reviewing Facts and Terms

1. e 2. f 3. d 4. g 5. c 6. h 7. b 8. i 9. a 10. a. epidermis; b. dermis; c. sebaceous gland; d. arrector pili muscle; e. lamellar corpuscle; f. sweat gland 11. a 12. d 13. a. nail root; b. eponychium; c. phalanx; d. hyponychium 14. b 15. b 16. b 17. c 18. a 19. a

Level 2 Reviewing Concepts

1. Fair-skinned individuals produce less melanin in their melanocytes, and are therefore less able to prevent the absorption of potentially damaging ultraviolet radiation. Since more UV radiation reaches the deeper skin layers, it may cause a greater amount of damage.
2. Mechanical stresses cause cells in the stratum basale to divide more rapidly, increasing the thickness of the epithelium.
3. If the skin stretches and then does not contract to its original size, it wrinkles and creases, creating a network of stretch marks.
4. Keratin is produced in large amounts in the stratum granulosum. Fibers of keratin interlock in the cells in this layer during the process of keratinization. The cells become thinner and flatter; subsequently, they dehydrate. Keratin is important in maintaining the structure of the outer part of the epidermis and making it water resistant. Additionally, keratin forms the basic structural component of hair and nails.
5. The subcutaneous layer, as compared to other body layers, (a) is quite elastic, (b) has a high density of blood vessels to facilitate the transport of the drug, and (c) has no vital organs.
6. These activities either reduce the amount of potentially odorous secretions on the surface of the skin or inhibit the activity of bacteria on these secretions.
7. A person who is cyanotic has a sustained reduced blood flow to the skin. As a result, the skin takes on a bluish coloration, called cyanosis, which is especially apparent in areas of thin skin, such as the lips or beneath the nails.
8. In elderly people, the blood supply to the dermis and the activity of sweat glands both decrease. Due to this combination, the elderly are less able to lose body heat and less able to cool themselves when exposed to high temperatures.

Level 3 Critical Thinking

1. Tension lines are important because a cut that runs parallel to a tension line closes more easily, whereas a cut made at right angles to a tension line will sever elastic fibers in the skin and thus be pulled open as the fibers recoil. A parallel cut will also heal faster and with less scarring than a cut at right angles to tension lines.
2. The palms of the hands and the soles of the feet have a thicker epidermis and an extra layer in the epidermis known as the stratum lucidum. This thicker epidermis slows down the rate of diffusion of the medication and significantly decreases its effectiveness.

Answers to Clinical Case Wrap-Up

Page 106

1. Because Martin did not adequately cleanse his skin with alcohol prior to his insulin injection, the bacteria living on Martin's skin were introduced into the subcutaneous layer by the needle.
2. The skin that is supplied by the clotted cutaneous and subpapillary plexuses is not likely to survive. It will most likely become necrotic and require further debridement and skin grafting.

5 THE SKELETAL SYSTEM
Osseous Tissue and Skeletal Structure

Concept Check

Page 113

1. If the ratio of collagen to hydroxyapatite in a bone increased, the bone would be more flexible and less strong.
2. Concentric layers of bone around a central canal are indicative of an osteon, or Haversian system. Osteons are found in compact bone. Since the ends (epiphyses) of long bones are primarily spongy (cancellous) bone, this sample most likely came from the shaft (diaphysis) of a long bone.
3. Since osteoclasts function in breaking down or demineralizing bone, the bone would have less mineral content and as a result would be weaker.
4. Fracture repair would be impeded.

Page 118

5. Long bones of the body, like the femur, have an epiphyseal cartilage separating the epiphysis from the diaphysis, as long as the bone is still growing in length. An x-ray would indicate whether the epiphyseal cartilage is still present. If it is, then growth is still occurring; if not, the bone has reached its adult length.
6. (1) In a fibrous connective tissue, osteoblasts secrete matrix components in an ossification center. (2) Growth occurs outward from the ossification center in small struts called spicules. (3) Over time, the bone assumes the shape of spongy bone; subsequent remodeling can produce compact bone.
7. The diameter of a bone enlarges through appositional growth at the outer surface. In this process, periosteal cells differentiate into osteoblasts and contribute to the growth of the bone matrix.
8. The epiphyseal cartilage is a relatively narrow band of cartilage separating the epiphysis from the diaphysis. It is located at the metaphysis. The continued growth of chondrocytes at the epiphyseal side and their subsequent replacement by bone at the diaphyseal side allow for an increase in the length of a developing bone.
9. Vitamins A, C, and D_3 regulate bone growth. Vitamin A stimulates osteoblast activity, and vitamin C is required for enzymatic reactions in collagen synthesis and osteoblast differentiation. Vitamin D_3 plays an important role in normal calcium and phosphate absorption into the blood. Calcitonin in children and pregnant females inhibits osteoclasts and increases the rate of calcium loss in the urine. Parathyroid hormone stimulates osteoclasts and osteoblasts, increases calcium absorption at the intestines (this action requires another hormone, calcitriol), and decreases the rate of urinary calcium loss. Growth hormone and thyroxine stimulate bone growth. Sex hormones stimulate osteoblasts to deposit bone.

Page 120

10. Bones increase in thickness in response to physical stress. One common type of stress that is applied to a bone is that produced by muscles. We would expect the bones of an athlete to be thicker after the addition of the extra muscle mass because of the greater stress that the muscle would apply to the bone.
11. In a 15-year-old, we would expect the rate of bone formation to exceed the rate of bone reabsorption, whereas in the 30-year-old, these rates would be approximately the same.

Page 125

12. Bone markings are often palpable at the surface; they provide reference points for orientation on associated soft tissues. For pathologists, bone markings can be used to estimate the size, weight, sex, and general appearance of an individual on the basis of incomplete remains.
13. A sesamoid bone is usually round, small, and flat. Irregular bones may have complex shapes with short, flat, notched, or ridged surfaces.
14. Sutural bones, also called *Wormian bones*, are small, flat, oddly shaped bones found between the flat bones of the skull in the sutures.

Page 125

15. An individual who has a diet low in calcium would see, over time, a reduction in bone mass due to a decreased deposition of bone by osteoblasts and an increase in reabsorption of bone by osteoclasts.

Chapter Review

Level 1 Reviewing Facts and Terms

1. a. articular surfaces of the proximal end of the bone; b. diaphysis; c. epiphysis; d. articular surfaces of the distal end of the bone 2. b 3. a 4. b 5. b 6. a 7. a 8. a 9. a 10. b

Level 2 Reviewing Concepts

1. c
2. a
3. In intramembranous ossification, bone develops from mesenchyme or fibrous connective tissue. In endochondral ossification, bone replaces an existing cartilage model.
4. At maturity, the rate of epiphyseal cartilage growth slows down, while the rate of osteoblast production increases, narrowing the region of the epiphyseal plate. When the plate disappears, the epiphysis and diaphysis of the bone grow together, and elongation stops.
5. Spongy bone is found where bones receive stresses from many directions. In the expanded ends of the long bones, the epiphyses, the trabeculae in the spongy bone are extensively cross-braced to withstand stress applied from different directions. In contrast, compact bone is heavier and is best suited for resisting stresses that are parallel to the shaft of a long bone.
6. A bone grows in diameter by appositional growth, wherein the outer surface of the bone grows as cells of the periosteum differentiate into osteoblasts that contribute to the bone matrix. Eventually, these cells become surrounded by matrix and differentiate into osteocytes. Additionally, the bone matrix is removed from the inner surface by osteoclasts to expand the medullary cavity.

7. The final repair is slightly thicker and stronger than the original bone.
8. A fractured bone requires a good supply of calcium and phosphate to manufacture new matrix. Therefore, a diet that is low in these minerals will hinder the rate at which a broken bone heals.
9. A sesamoid bone is usually small, round, and flat. It develops inside tendons and is most often encountered near joints at the knee, the hands, and the feet. Sutural (Wormian) bones are small, flat, oddly shaped bones found between the flat bones of the skull in the suture line.
10. Ossification is the process of replacing other tissues with bone, while calcification refers to the deposition of calcium salts within a tissue.

Level 3 Critical Thinking

1. At the fracture point, bleeding into the area creates a fracture hematoma. An internal callus forms as a network of spongy bone that unites the inner surfaces, and an external callus of cartilage and bone stabilizes the outer edges. Together, these are visible as the enlargement. A swelling initially marks the location of the fracture. Over time this region will be remodeled, and little evidence of the fracture will remain.
2. In children, the long bones are relatively supple and easily deformed because the relative proportion of collagenous fibers to ossified calcium and phosphate salts is higher. This makes a greenstick fracture more common in children than in adults. With increased age, the proportion of collagenous fibers to ossified calcium and phosphate salts decreases, making the bones more brittle, thereby decreasing the frequency of greenstick fractures.

Answers to Clinical Case Wrap-Up

Page 130

1. If Emily continued to run on a tibial cortical stress fracture, the fracture could spread across the entire tibial shaft, producing a complete transverse fracture.
2. If the stress fracture continued across the entire anterior cortex involving the cellular endosteum lining the medullary cavity, osteoprogenitor cells there would be recruited in the healing process.

6 THE SKELETAL SYSTEM Axial Division

Concept Check

Page 140

1. There are seven associated bones of the skull: six auditory ossicles and one hyoid bone.
2. The function of the cranial cavity is to protect the brain. The occipital, parietal, frontal, temporal, sphenoid, and ethmoid bones make up the cranial cavity.

Page 140

3. The coronal suture crosses the superior aspect of the skull, separating the frontal and parietal bones.
4. The lambdoid suture separates the occipital bone from the parietal bones.

Page 150

5. Each internal jugular vein passes through the jugular foramen, an opening between the occipital bone and the temporal bone.
6. The sella turcica is located in the sphenoid, and it contains the pituitary gland.
7. Nerve fibers to the olfactory bulb, which is involved with the sense of smell, pass through the cribriform plate from the nasal cavity. If the cribriform plate failed to form, these sensory nerves could not reach the olfactory bulbs, so the sense of smell (olfaction) would be lost.

Page 154

8. The 14 facial bones are the maxillae (2), zygomatic bones (2), nasal bones (2), lacrimal bones (2), inferior nasal conchae (2), palatine bones (2), vomer, and mandible. These bones protect and support the entrances to the respiratory and digestive tracts and provide extensive areas for skeletal muscle attachment.

Page 156

9. The paranasal sinuses make the skull lighter, produce mucus, and resonate when a person produces sound.
10. The orbital complex consists of portions of the frontal, maxilla, lacrimal, ethmoid, sphenoid, palatine, and zygomatic bones.

Page 156

11. Fontanelles are areas of fibrous connective tissue that connect the cranial bones of the fetal skull.
12. (1) The anterior fontanelle, the largest, lies at the intersection of the frontal, sagittal, and coronal sutures. (2) The posterior fontanelle is found at the junction between the lambdoid and sagittal sutures. (3) The sphenoidal fontanelles are at the junctions between the squamous sutures and the coronal suture. (4) The mastoid fontanelles are at the junctions between the squamous sutures and the lambdoid suture.

Page 165

13. The dens is found on the second cervical vertebra, or axis, which is located in the neck.
14. The vertebral column is divided into the cervical, thoracic, lumbar, sacral, and coccygeal regions. Distinguishing features are as follows: cervical: triangular foramen, bifid spinous process, transverse foramina; thoracic: round foramen, heart-shaped body, transverse facets, costal facets; lumbar: triangular foramen, oval-shaped, large robust body; sacral: five fused vertebrae; coccygeal: three to five small fused vertebrae.
15. The spinal curves from superior to inferior are the (1) cervical curve, (2) thoracic curve, (3) lumbar curve, and (4) sacral curve.

Page 168

16. Improper compression of the chest during CPR could result in a fracture of the ribs or sternum, especially at the xiphoid process.
17. The internal rib surface is concave, and the prominent costal groove along its inferior border marks the path of nerves and blood vessels.

Chapter Review

Level 1 Reviewing Facts and Terms

1. b 2. e 3. h 4. j 5. i 6. a 7. d 8. g 9. f 10. c 11. d 12. a 13. a. frontal bone; b. sphenoid; c. lambdoid suture; d. temporal process of the zygomatic bone; e. styloid process 14. a 15. a 16. a. spinous process; b. lamina; c. transverse process; d. pedicle 17. a 18. d 19. a 20. a 21. d

Level 2 Reviewing Concepts

1. The structure of the atlas is such that it provides more free space for the spinal cord than any other vertebra. This extra space helps ensure that the spinal cord is not impinged on during the extensive amount of motion that occurs here.
2. A prominent central depression between the wings of the sphenoid bone cradles the pituitary gland just inferior to the brain. This depression is called the hypophyseal fossa, and the bony enclosure is called the sella turcica.
3. The shifting of weight over the legs helps to move the center of gravity anteriorly, which will provide a more stable base for standing and walking.
4. The ligamentum nuchae is a large elastic ligament that begins at the vertebra prominens and extends cranially to an insertion along the external occipital crest. Along the way it attaches to the spinous processes of the other cervical vertebrae. When the head is held upright, this ligament is like the string on a bow, maintaining the cervical curve without muscular effort.
5. The mucous membranes of the paranasal sinuses respond to environmental stress by increasing the production of mucus. The mucus flushes irritants off the walls of the nasal cavities. A variety of stimuli produce this result, including sudden changes in temperature or humidity, irritating vapors, and bacterial or viral infections.
6. The body of the lumbar vertebrae is largest because these vertebrae bear the most weight.
7. The thick petrous portion of the temporal bone houses the internal ear structures that provide information about hearing and balance.
8. The foramina in the cribriform plate permit passage of the olfactory nerves, providing the sense of smell.

Level 3 Critical Thinking

1. An inflammation of the paranasal sinuses' mucous membranes increases the production of mucus. The maxillary sinuses are often involved because gravity does little to assist mucus drainage from these sinuses. Congestion increases, and the patient experiences headaches, a feeling of pressure in facial bones, and an ache in the teeth rooted in the maxillae.
2. The facial features referred to as cheekbones are made by the zygomatic processes of the temporal bones, which articulate with the temporal processes of the zygomatic bones inferior to the orbit. These features are prominent in the face as they act as protection for the orbits superior to them. The model complimented for the large appearance of her eyes would have to have larger than normal regions housing larger than normal eyeballs to truly have larger eyes. In all probability, the appearance of her eyes is due to the arrangement of her soft tissue structures and her makeup, not an unusual arrangement of the bones of the face.

Answers to Clinical Case Wrap-Up

Page 171

1. Basilar skull fractures can involve the occipital and temporal bones, as in Earnhardt's ring fracture, and can also involve the sphenoid and ethmoid bones.
2. The carotid canal and jugular vein are within the temporal bone. A fracture of the temporal bone would disrupt both the carotid artery and the jugular vein.

7 THE SKELETAL SYSTEM Appendicular Division

Concept Check

Page 184

1. The mobility of the scapula is affected because the clavicle attaches the scapula to the sternum and thus restricts the scapula's range of movement. If the clavicle is broken, the scapula will have a greater range of movement and will be less stable.
2. The radius is in a lateral position when the forearm is in the anatomical position.
3. The olecranon is the point of the elbow. During extension of the elbow, the olecranon swings into the olecranon fossa on the posterior surface of the humerus to prevent overextension.
4. The clavicle articulates with the manubrium of the sternum; this provides the only direct connection between the pectoral girdle and the axial skeleton.

Page 197

5. The three bones that make up the hip bone are the ilium, ischium, and pubis.
6. Although the fibula is not part of the knee joint and does not bear weight, it is an important point of attachment for many leg muscles. When the fibula is fractured, these muscles cannot function properly to move the leg, and walking is difficult and painful. The fibula also helps stabilize the ankle joint.
7. Mark has most likely fractured his calcaneus (heel bone).

8. The female pelvis is adapted for supporting the weight of the developing fetus and allowing the newborn to pass through the pelvic outlet during delivery. Compared to males, such adaptations of the female pelvis include the following: an enlarged pelvic outlet; less curvature on the sacrum and coccyx, which in the male arc anteriorly into the pelvic outlet; a wider, more circular pelvic inlet; a relatively broad, low pelvis; ilia that project farther laterally, but do not extend as far superior to the sacrum; and a broader pubic arch, with the inferior angle between the pubic bones greater than 100°.
9. During dorsiflexion, as when "digging in the heels," all of the body weight rests on the calcaneus. During plantar flexion, as when "standing on tiptoe," the talus and calcaneus transfer the weight to the metatarsal bones and phalanges through more anterior tarsal bones.

Page 198

10. See Table 7.1 for a complete listing of regions and features that are utilized to determine the sexual differences in the adult human skeleton.

Chapter Review

Level 1 Reviewing Facts and Terms

1. b 2. e 3. g 4. j 5. c 6. i 7. f 8. a 9. h 10. d 11. a. acromion; b. coracoid process; c. glenoid cavity; d. spine of scapula 12. c 13. a. olecranon; b. coronoid process; c. radial notch; d. ulnar tuberosity 14. a 15. d 16. c 17. a. anterior gluteal line; b. anterior superior iliac spine; c. posterior inferior iliac spine; d. superior pubic ramus 18. c

Level 2 Reviewing Concepts

1. To determine the age of a skeleton, one would consider some or all of the following: the fusion of the epiphyseal plates, the amount of mineral content, the size and roughness of bone markings, teeth, bone mass of the mandible, and intervertebral disc size.
2. Weight transfer occurs along the longitudinal arch of the foot. Ligaments and tendons maintain this arch by tying the calcaneus to the distal portions of the metatarsal bones. The lateral, calcaneal side of the foot carries most of the weight of the body while standing. This part of the arch has less curvature than the medial, talar portion.
3. Fractures of the medial portion of the clavicle are common because a fall on the palm of an outstretched hand produces compressive forces that are conducted to the clavicle and its articulation with the manubrium.
4. The tibia is part of the knee joint and is involved in the transfer of weight to the ankle and foot. The fibula is excluded from the knee joint and does not transfer weight to the ankle and foot.
5. The olecranon of the ulna is the point of the elbow. During full extension, this process swings into the olecranon fossa on the posterior surface of the humerus to prevent overextension of the forearm relative to the arm.
6. Body weight is passed to the metatarsal bones through the cuboid and the cuneiform bones.

Level 3 Critical Thinking

1. In osteoporosis, a decrease in the calcium content of the body leads to bones that are weak and brittle. Since the hip joint and leg bones must support the weight of the body, any weakening of these bones may result in insufficient strength to support the body mass, and as a result the bone will break under the great weight. The shoulder joint is not a load-bearing joint and is not subject to the same great stresses or strong muscle contractions as the hip joint. As a result, breaks in the bones of this joint occur less frequently.
2. Many skull characteristics would reveal the sex of the individual, but there are also characteristics in other skeletal structures, such as the robustness of the bones, the angles at which the pubic bones meet, the width of the pelvis, and the angles of the femurs.

Answers to Clinical Case Wrap-Up

Page 202

1. The acetabular fracture is within the lesser (true) pelvis.
2. The force incurred from the accident pushed the femoral head and greater trochanter of the femur into the acetabulum, fracturing the anterior and posterior segments. The greater trochanter is a strong process at the proximal and lateral part of the shaft of the femur. The femoral head is the articular surface at the proximal portion of the thigh bone. These structures jammed into the cup-shaped depression on the external surface of the hip bone called the acetabulum, fracturing it.

8 THE SKELETAL SYSTEM Articulations

Concept Check

Page 207

1. No relative movement occurs at a synarthrosis, whereas slight movement occurs at an amphiarthrosis.
2. Synovial fluid lubricates the joint, reduces friction, and distributes dissolved gases, nutrients, and wastes.

Page 212

3. Monaxial: pivot and hinge joints; biaxial: saddle and condylar joints; triaxial: ball-and-socket joint
4. (a) abduction, (b) supination, (c) flexion

Page 212

5. The temporomandibular joint is primarily a hinge joint, but the loose capsule and relatively flat articular surfaces also permit small gliding movements. These movements are important for chewing.
6. The articular disc of the temporomandibular joint extends horizontally and therefore divides the joint cavity into two separate chambers. As a result, the temporomandibular joint is really two synovial joints: one between the temporal bone and the articular disc, and the other between the articular disc and the mandible.

Page 216

7. The ligaments that stabilize the vertebral column are the anterior longitudinal ligament, posterior longitudinal ligament, ligament flava, interspinous ligament, and supraspinous ligament.
8. Symphysis joints are found between the vertebral bodies of C_2 and S_1. Zygapophysial joints are synovial joints found between the superior and inferior articulating processes of adjacent vertebrae.

Page 219

9. The glenoid labrum is a piece of fibrous cartilage that serves to enlarge and deepen the glenoid cavity of the scapula and therefore contribute to the stability of the shoulder joint.
10. Because the subscapular bursa is located in the shoulder joint, an inflammation of this structure (bursitis) would be more likely to be found in the tennis player. The condition is associated with repetitive motion that occurs at the shoulder, such as swinging a tennis racket. The jogger would be more at risk for injuries to the knee joint.

Page 221

11. The elbow joint is composed of the articulations between the trochlea of the humerus and the trochlear notch of the ulna and the capitulum of the humerus and the head of the radius.
12. No, the pain isn't originating in his elbow. The elbow joint, which does not allow for supination and pronation of the forearm, is the joint between (1) the humerus and the ulna and (2) the humerus and the radius. The pain originates in his forearm. The proximal and distal radio-ulnar joints allow for supination and pronation of the forearm.

Page 222

13. The six structures are the following: (1) palmar radiocarpal ligament, (2) dorsal radiocarpal ligament, (3) ulnar collateral ligament, (4) radial collateral ligament, (5) intercarpal ligaments, and (6) digitocarpal ligaments.
14. The types of movement are flexion/extension and adduction/abduction.

Page 224

15. The almost complete bony socket enclosing the head of the femur, the strong articular capsule, the stout supporting ligaments, and the dense muscular padding make the hip an extremely stable joint. Because of this stability, fractures of the femoral neck or between the trochanters are more common than hip dislocations.

Page 228

16. The cruciate ligaments are reinforcing ligaments in the knee. The anterior cruciate ligament (ACL) and posterior cruciate ligament (PCL) are intracapsular ligaments of the knee joint. They attach the intercondylar area of the tibia to the condyles of the femur. These ligaments limit the anterior and posterior movement of the femur and maintain the alignment of the femoral and tibial condyles.

Page 231

17. The tibiotalar joint is the primary weight-bearing joint in the ankle.
18. Three joints provide medial and lateral stability in the ankle joint: (1) the proximal tibiofibular joint, (2) the distal tibiofibular joint, and (3) the fibulotalar joint.

Page 231

19. Arthritis is joint inflammation and is a type of rheumatic disease that affects synovial joints. Arthritis always involves damage to the articular cartilages, but the specific cause can vary.

Chapter Review

Level 1 Reviewing Facts and Terms

1. i 2. d 3. h 4. c 5. g 6. b 7. f 8. a 9. e 10. d 11. b 12. a. plane; b. hinge; c. condylar; d. saddle; e. pivot; f. ball-and-socket 13. d 14. a. acromioclavicular ligament; b. subdeltoid bursa; c. coracoacromial ligament; d. coracoid process 15. c 16. a

Level 2 Reviewing Concepts

1. A joint cannot be both highly mobile and very strong. The greater the range of motion at a joint, the weaker it is. The stronger the joint, the lesser its range of motion. For example, a synarthrosis, which is the strongest type of joint, does not permit any movement.
2. Prior to fusion, the two parts of a single bone are united by a line of cartilage and are called a synchondrosis. Once the cartilaginous plate is obliterated, there is no more joint, and it becomes an immovable synostosis.
3. The tibiotalar joint involves the distal articular surface of the tibia, including the medial malleolus, the lateral malleolus of the fibula, and the trochlea and lateral articular facets of the talus. The malleoli, supported by ligaments of the ankle joint (the medial deltoid ligament and the three lateral ligaments) and associated fat pads, prevent the ankle bones from sliding from side to side.
4. Articular cartilages cover articulating surfaces of bones. They resemble hyaline cartilages elsewhere in the body, but they have no perichondrium, and the matrix contains more water than other cartilages have.
5. Factors that limit the range of movement of a joint include accessory ligaments and collagen fibers of the joint capsule, the shapes of the articulating surfaces that allow movement in some directions while preventing it in others, the tension in the tendons attached to the articulating bones, and the bulk of the muscles surrounding the joint.
6. The joint capsule that surrounds the entire synovial joint is continuous with the periostea of the articulating bones. Accessory ligaments are localized thickenings of the capsule. Extracapsular ligaments are on the outside of the capsule; intracapsular ligaments are found inside the capsule. In the humero-ulnar joint, the capsule is reinforced by strong ligaments. The radial collateral ligament stabilizes the lateral surface of the joint. The annular ligament binds the proximal radial head to the ulna. The medial surface of the joint is stabilized by the ulnar collateral ligament.

7. The edges of the bones are interlocked and bound together at the suture by dense connective tissue. A different type of synarthrosis binds each tooth to the surrounding bony socket. This fibrous connection is the periodontal ligament.
8. The movement of the wrist and hand from palm-facing-front (supination) to palm-facing-back is called pronation. Circumduction is a special type of angular motion that encompasses all types of angular motion: flexion, extension, adduction, and abduction.
9. As one ages, the water content of the nucleus pulposus within each disc decreases. Loss of water within the discs causes shortening of the vertebral column.

Level 3 Critical Thinking

1. The term "whiplash" is used to describe an injury wherein the body suddenly changes position, as in a fall or during rapid acceleration or deceleration. The balancing muscles are not strong enough to stabilize the head. A dangerous partial or complete dislocation of the cervical vertebrae can result, with injury to muscles and ligaments and potential injury to the spinal cord. It is called whiplash because the movement of the head resembles the cracking of a whip.
2. When the knee is flexed, it is able to move in response to a hit from the inside or outside (medial or lateral surfaces). However, when the foot is "planted," the knee is in the locked position. In this position, the medial and lateral collateral ligaments and the anterior cruciate ligaments are taut, thereby increasing their chance of injury.

Answers to Clinical Case Wrap-Up

Page 234

1. A synovial tendon sheath is one form of a bursa. The synovial tendon sheath surrounds and protects a tendon that passes over a bony surface.
2. The tendon of the semimembranosus joins with and reinforces the tibial collateral ligament of the knee joint.

9 | THE MUSCULAR SYSTEM Skeletal Muscle Tissue and Muscle Organization

Concept Check

Page 236

1. Shivering is an involuntary contraction of skeletal muscles that occurs when you are cold. These muscle contractions utilize ATP, which produces heat, which in turn helps to maintain body temperature.

Page 243

2. Skeletal muscle appears striated when viewed under the microscope because it is composed of the myofilaments actin and myosin, which are arranged in such a way as to produce a banded appearance in the muscle.
3. The proteins tropomyosin and troponin help regulate the interaction between actin and myosin.

Page 247

4. During contraction, the width of the A band remains the same, and the I band gets smaller.
5. Terminal cisternae are expanded chambers of the sarcoplasmic reticulum—they store calcium ions, which are required to initiate contraction within skeletal muscle cells. T tubules are sandwiched between terminal cisternae. A T tubule is a deep indentation of the sarcolemma, which conducts the stimulation into the interior of the cell. When an action potential arrives, it causes the release of calcium ions from the terminal cisternae, and this results in a muscle contraction.

Page 248

6. The size of a motor unit is an indication of how fine the control of movement is. In the eye muscles, where precise control is important, a motor neuron may control just a few fibers. In large leg muscles, where there is less precise control, a single motor neuron may control a thousand muscle fibers.
7. Muscle tone is involved in maintaining body position while standing. The administration of a chemical that blocks the action of acetylcholine would prevent action potentials crossing from nerves to skeletal muscles. ACh is needed to generate an action potential and thus is needed for a muscle contraction. If ACh is not available, the muscles could not contract, and there would be no muscle tone. Without muscle tone, the person falls.

Page 249

8. The sprinter requires large amounts of energy for a relatively short burst of activity. To supply this demand for energy, the muscles switch to anaerobic metabolism. Anaerobic metabolism is not as efficient in producing energy as is aerobic metabolism, and the process also produces acidic wastes. The combination of less energy and wastes contributes to fatigue. Marathon runners, on the other hand, derive most of their energy from aerobic metabolism, which is more efficient and does not produce the level of wastes that anaerobic metabolism does.

Page 251

9. The parallel arrangement more closely resembles the anatomical characteristics of a single muscle fiber.

Page 253

10. A synergist contracts to assist the agonist (prime mover) in performing a specific action.
11. This is a long muscle that flexes the joints of the finger.

Page 253

12. The order of decreasing frequency is third-class lever (most common), followed by second-class lever, and then first-class lever (least common).
13. Removing the patella would significantly decrease the amount of force that could be produced by the quadriceps muscles because the angle of attachment of the patellar ligament would have been altered.

Page 253

14. Aging skeletal muscles develop increasing amounts of fibrous connective tissue, a process called fibrosis. Fibrosis makes the muscles less flexible, and the collagen fibers can restrict movement and circulation. Continued stretching throughout life would lessen this effect.

Chapter Review

Level 1 Reviewing Facts and Terms

1. a 2. b 3. c 4. a 5. c 6. c 7. d 8. b 9. a. second-class lever; b. first-class lever; c. third-class lever

Level 2 Reviewing Concepts

1. c
2. d
3. c
4. c
5. An action potential (nerve impulse) arrives at the axon terminal of the neuromuscular synapse and causes acetylcholine to be released into the synaptic cleft. The acetylcholine released then binds to receptors on the sarcolemma surface, initiating a change in the local membrane potential. This change results in the generation of action potentials (electrical signals) that sweep over the surface of the sarcolemma.
6. A motor unit with 1500 fibers would be involved in powerful, gross movements. The greater the number of fibers in a motor unit, the more powerful the contraction and the less fine control exhibited by the motor unit.
7. In the zone of overlap, the thin filaments pass between the thick filaments. It is in this region that interaction between thick and thin filaments occurs to form cross-bridges so that contraction may occur and tension may be generated.

Level 3 Critical Thinking

1. Weight lifting requires anaerobic endurance. The students would want to develop fast fibers for short-term maximum strength. This could be achieved by engaging in activities that involve frequent, brief, but intensive workouts, such as with progressive-resistance machines. Repeated exhaustive stimulation will help the fast fibers develop more mitochondria and a higher concentration of glycolytic enzymes, as well as increase the size (hypertrophy) and strength of the muscle.
2. Muscle biopsies are used to determine the relative percentage of slow and fast muscle fibers within the leg muscles. An athlete with a higher percentage of white muscle has a greater chance of being a good sprinter than an athlete with a higher percentage of red muscle. Conversely, an athlete with a high percentage of red muscle has a greater chance of being a good distance runner than an athlete with a higher percentage of white muscle.

Answers to Clinical Case Wrap-Up

Page 258

1. The muscle tone in Abdul's left hip, thigh, and leg is virtually absent; the muscles can be described as flaccid.
2. The majority of Abdul's muscle fibers that have been affected by polio are fast fibers, or white fibers. Most muscles of the hip, thigh, and leg are dominated by fast fibers.

10 | THE MUSCULAR SYSTEM Axial Musculature

Concept Check

Page 261

1. The axial muscles are divided into four groups, based on location, function, or both. These groups are (1) the muscles of the head and neck, (2) the muscles of the vertebral column, (3) the oblique and rectus muscles, and (4) the muscles of the perineum and pelvic diaphragm. The muscles of the head and neck move the face, tongue, larynx, and eye. The muscles of the vertebral column flex, extend, and rotate the vertebral column. The oblique and rectus muscles move the chest wall during breathing (inspiration and expiration), compress the abdominal cavity, and rotate the vertebral column. The muscles of the perineum and pelvic diaphragm support the organs of the pelvic cavity, flex the joints of the sacrum and coccyx, and control the movement of materials through the urethra and anus.

Page 269

2. Muscles of facial expression originate on the surface of the skull.
3. The pharyngeal muscles are important for starting the swallowing process.

Page 273

4. The erector spinae muscles are the iliocostalis, longissimus, and spinalis muscles. The transversospinales muscles are the multifidus, semispinalis, rotatores, interspinales, and intertransversarii muscles. When the erector spinae muscles on both sides of the vertebral column contract together, they extend the vertebral column; when the erector spinae muscles on only one side contract, they laterally flex the vertebral column.
5. There are fewer intrinsic back muscles that flex the vertebral column because (1) many of the large trunk muscles flex the vertebral column when they contract and (2) most of the body weight is anterior to the vertebral column and so gravity tends to flex the vertebral column.

Page 275

6. Damage to the external intercostal muscles interferes with the process of inhalation during breathing.

Page 277

7. Muscles of the perineum and pelvic diaphragm support the organs of the pelvic cavity, flex the joints of the sacrum and coccyx, and control the movement of materials through the urethra and anus.

Chapter Review

Level 1 Reviewing Facts and Terms

1. h 2. d 3. a 4. i 5. f 6. b 7. j 8. e 9. c 10. g 11. b 12. b 13. a. nasalis; b. orbicularis oris; c. platysma; d. temporalis; e. masseter 14. a 15. b 16. c 17. a. multifidus; b. quadratus lumborum; c. iliocostalis thoracis; d. iliocostalis lumborum; e. erector spinae 18. a 19. c 20. a 21. d

Level 2 Reviewing Concepts

1. c
2. a
3. b
4. c
5. Gravity can help the trunk flex forward. The bulk of the mass of the body is anterior to the vertebral column, making the movement toward the anterior easier to make by gravity alone, so not as many anterior flexors are necessary.
6. All the extrinsic and intrinsic muscles of the tongue are involved in different aspects of the swallowing process. These muscles control the position of the chewed mass of food (bolus) on the tongue and position it correctly to begin swallowing.
7. Contracting the internal oblique compresses the abdomen, depresses the ribs, and flexes, bends to the side, or rotates the spine.
8. The anterior muscles of the neck control the position of the larynx, depress the mandible, tense the floor of the mouth, and provide a stable foundation for muscles of the tongue and pharynx.
9. The diaphragm is a major muscle of breathing. It is included in the axial musculature because it is embryologically linked to the other muscles of the chest wall.
10. The muscles involved in controlling the head on the vertebral column include the muscles of the erector spinae and multifidus groups and the sternocleidomastoid. These muscles control the position of the head, maintaining posture involuntarily and moving the head voluntarily.

Level 3 Critical Thinking

1. The muscles of the anal triangle form the posterior aspect of the perineum, a structure that has boundaries established by the inferior margins of the pelvis. A muscular sheet, the pelvic diaphragm, forms the muscular foundation of the anal triangle and extends as far as the pubic symphysis. A sphincter muscle in this region surrounds the opening to the anus.
2. The contraction of the frontalis and procerus muscles causes Mary to raise her eyebrows and wrinkle her forehead (frontalis) and to move her nose and change the position and shape of the nostrils. Raising her eyebrows and flaring her nostrils indicates that she has some concerns about meeting Jill.

Answers to Clinical Case Wrap-Up

Page 281

1. In Bell's palsy, all muscles innervated by the facial nerve (N VII), including the orbicularis oculi that closes the eye, are paralyzed. The only muscle of facial expression that is *not* innervated by the facial nerve is the levator palpebrae superioris, which opens the eye. Eye opening is unaffected.
2. All the muscles that move the mouth are affected in Bell's palsy. This includes the buccinators, depressor labii inferioris, mentalis orbicularis oris, risorius, levator anguli oris, and zygomaticus major and minor muscles. Sara is unable to purse her lips to whistle or blow when the left side of her mouth is paralyzed.

11 | THE MUSCULAR SYSTEM Appendicular Musculature

Concept Check

Page 286

1. The action line of a muscle is the line of applied tension of a muscle (the line between the origin and insertion of a muscle as it crosses a joint). Knowing the action line of a muscle helps you determine the action of that muscle.

Page 298

2. These tendons pass through synovial tendon sheaths before reaching their insertion points on the hand.
3. This thickened fascia is the extensor retinaculum.
4. The deltoid muscle is the primary abductor of the arm at the shoulder.
5. Injury to the flexor carpi ulnaris impairs wrist flexion and wrist adduction.
6. The pronator quadratus and supinator both rotate the radius without flexing or extending the elbow.

Page 301

7. Moving medially to laterally, the five muscles in the superficial anterior compartment of the forearm are the flexor carpi radialis, palmaris longus, flexor carpi ulnaris, pronator teres, and flexor digitorum superficialis.
8. The radial nerve innervates all of the posterior compartment muscles of the arm and forearm.

Page 318

9. The pectineus, gracilis, sartorius, rectus femoris, and iliacus are innervated by the obturator nerve.
10. The knee extensors are collectively known as the quadriceps femoris.
11. The fibularis longus and the fibularis brevis are extrinsic muscles of the foot that originate on the fibula.
12. The fibularis brevis and fibularis longus evert the foot.
13. The intrinsic muscles that originate on the calcaneus are the extensor digitorum brevis, abductor hallucis, flexor digitorum brevis, abductor digiti minimi, and quadratus plantae.

Page 319

14. The functions of the muscles within the lateral compartment of the leg are to evert and plantar flex the ankle.
15. The muscles of the deep compartment of the leg are the flexor digitorum longus (flexion of toes 2–5 and plantar flexion of the ankle), flexor hallucis longus (flexion of the big toe and plantar flexion of the ankle), popliteus (medial rotation of the tibia at the knee), and tibialis posterior (plantar flexion of the ankle and inversion of the foot).
16. The medial compartment of the thigh contains the muscles (adductor longus, adductor magnus, and other adductors) that adduct the thigh at the hip.
17. The muscles of the superficial posterior compartment of the leg are the gastrocnemius (plantar flexion of the ankle and flexion of the leg at the knee), soleus (plantar flexion of the ankle), and plantaris (plantar flexion of the ankle and flexion of the leg at the knee).

Chapter Review

Level 1 Reviewing Facts and Terms

1. c 2. k 3. i 4. g 5. j 6. d 7. f 8. b 9. e 10. h 11. a 12. b 13. b 14. b 15. d 16. d

Level 2 Reviewing Concepts

1. c
2. d
3. d
4. b
5. The biceps brachii exerts actions on the shoulder joint (flexion), elbow joint (flexion), and radio-ulnar joint (supination, also termed lateral or external rotation).
6. The muscle that becomes greatly enlarged in ballet dancers because of the need to flex and abduct the hip and that supports the knee laterally is the tensor fasciae latae.
7. The muscles that are most involved are the adductors of the thigh, which are stretched as the leg is supported at the hip and calcaneal tendon. Other muscles that may also be stretched are the hamstring group at the rear of the thigh, especially if the dancer leans forward over the barre.
8. The intrinsic muscles of the hand are involved in fine motor control of the hand and fingers.
9. Both the tensor fasciae latae and the gluteus maximus pull on the iliotibial tract, a band of collagen fibers that provides a lateral brace for the knee. The tensor fasciae latae is especially important when a person balances on one foot.
10. These retinacula (flexor retinaculum and extensor retinaculum) are sheets of connective tissue that encircle the wrist and ankle joints and act as a bracelet and anklet, respectively. The tendons of the hand and foot pass internal to them, allowing the hand and foot to change position as they move.

Level 3 Critical Thinking

1. The intrinsic muscles of the hand enable you to grip a pencil with precision, thereby allowing you to make the fine motor movements necessary for writing. Additionally, the large muscles of the arm maintain the proper position of the arm for writing.
2. Jerry probably injured the semimembranosus and/or the semitendinosus, since both of these muscles are primarily involved in flexing and medially rotating the thigh.
3. Although the pectoralis muscle is located across the chest, it inserts on the greater tubercle of the humerus, the large bone of the arm. When the muscle contracts, it contributes to flexion, adduction, and medial rotation of the humerus at the shoulder joint. All of these arm movements would be partly impaired if the muscle were damaged.

Answers to Clinical Case Wrap-Up

Page 324

1. In addition to the hamstrings (biceps femoris laterally, semitendinosus and semimembranosus medially), the sartorius muscle assists in knee flexion. The sartorius muscle is too weak to do the job by itself.
2. With the knee extended and the avulsed hamstrings shortened to 70 percent of their resting length, the proximal tendons would not reach the ischial tuberosity. Flexing the knee moves the insertion closer to the origin, helping the tendons reach their original attachment site.

12 | Surface Anatomy and Cross-Sectional Anatomy

Answers to Clinical Case Wrap-Up

Page 337

1. A tracheostomy is performed in the anterior cervical triangle.
2. The best place to practice feeling the median cricothyroid ligament is on yourself. Put your finger on your thyroid cartilage, or "Adam's apple." When you swallow, you should feel this thyroid cartilage move superiorly and the cricoid cartilage rise to meet your finger. The soft spot between these two prominent cartilages is the median cricothyroid ligament. This is the safest place to gain emergency access to the airway below an obstruction.

13 | THE NERVOUS SYSTEM Nervous Tissue

Concept Check

Page 340

1. The two anatomical subdivisions of the nervous system are the central nervous system and the peripheral nervous system.
2. The efferent division of the peripheral nervous system is divided into the somatic nervous system (SNS) and the autonomic nervous system (ANS).

Page 340

3. The two distinct cell types found within nervous tissue are neurons and neuroglia.
4. Neuroglia have retained the ability to divide.

Page 344

5. Astrocytes help maintain the blood brain barrier.
6. Myelin is the membranous coating formed by oligodendrocytes around CNS axons.

Page 350

7. Sensory neurons of the peripheral nervous system are usually pseudounipolar; thus, this tissue is most likely associated with a sensory organ.

Page 350

8. Wallerian degeneration is the process by which a peripheral nerve distal to an injury undergoes deterioration and regeneration.

Page 351

9. Myelination increases the conduction speed of a neuron, so the neuron that conducts action potentials at 50 m/s would be myelinated.
10. Excitability is the ability of a plasma membrane to respond to an adequate stimulus and generate an action potential.

Page 353

11. If the axon terminal cannot release the appropriate amount of acetylcholine at a chemical synapse, an action potential will not be generated at the postsynaptic membrane.

Page 354

12. Divergence is the spread of information from one neuron to several neurons or from one pool to multiple pools. Convergence occurs when several neurons synapse on the same postsynaptic neuron, or several neuronal pools synapse on one neuronal pool.

Page 355

13. A center is a collection of neuron cell bodies with a common function. Bundles of axons in the CNS that share common origins, destinations, and functions are called tracts. The centers and tracts that link the brain with the rest of the body are called pathways.

Chapter Review

Level 1 Reviewing Facts and Terms

1. c 2. j 3. h 4. e 5. b 6. i 7. f 8. d 9. a 10. g 11. b 12. d 13. d 14. a bipolar neuron; b. anaxonic neuron; c. multipolar neuron; d. pseudounipolar neuron 15.a 16. c.

Level 2 Reviewing Concepts

1. d
2. b
3. c
4. Collaterals enable a single neuron to communicate with several other cells at the same time.
5. Exteroceptors provide information about the external environment in the form of touch, temperature, pressure, sight, smell, hearing, and taste. Interoceptors monitor the digestive, respiratory, cardiovascular, urinary, and reproductive systems and provide sensations of taste, deep pressure, and pain.
6. The blood brain barrier (BBB) is needed to isolate nervous tissue from the general circulation because hormones or other chemicals normally present in the blood could have disruptive effects on neuron function.
7. The CNS is responsible for integrating, processing, and coordinating sensory data and motor commands. It is also the seat of higher functions, such as intelligence, memory, learning, and emotion. The PNS provides sensory information to the CNS and carries motor commands to peripheral tissues and systems.
8. The somatic nervous system controls skeletal muscle contractions, which may be voluntary or involuntary. The autonomic nervous system regulates smooth muscle, cardiac muscle, and glandular activity, usually outside our conscious awareness or control.
9. An electrical (nonvesicular) synapse is a more efficient carrier of impulses than is a chemical (vesicular) synapse because the two cells are linked by gap junctions and function as if they shared a common plasma membrane. However, chemical synapses are more versatile because the neuron membrane can be influenced by excitatory and inhibitory stimuli simultaneously. The plasma membrane at an electrical synapse simply passes the signal from one cell to another, whereas a chemical synapse integrates information arriving across multiple synapses.
10. In serial processing, information may be relayed in a stepwise sequence from one neuron to another or from one neuronal pool to the next. Parallel processing occurs when several neurons or neuronal pools are processing the same information at one time.

Level 3 Critical Thinking

1. Action potentials travel faster along myelinated fibers than along unmyelinated fibers. Destruction of the myelin sheath slows the time it takes for motor neurons to communicate with their effector muscles. This delay in response results in varying degrees of uncoordinated muscle activity. The situation is very similar to that of a newborn; the infant cannot control its arms and legs very well because the myelin sheaths are still being laid down for the first year. Since not all motor neurons to the same muscle may be demyelinated to the same degree, some fibers are slow to respond while others respond normally, producing contractions that are erratic and poorly controlled.
2. In the process known as Wallerian degeneration, the axons distal to the injury site deteriorate, and macrophages migrate in to phagocytize the debris. The Schwann cells in the area divide and form a solid cellular cord that follows the path of the original axon. As the neuron recovers, its axon grows into the injury site, and the Schwann cells wrap around it. If the axon continues to grow into the periphery alongside the appropriate cord of Schwann cells, it may eventually reestablish normal synaptic contacts. If it stops growing or wanders off in some new direction, normal function will not return.
3. No, Eve's father is not totally correct. In order to treat Eve's disease, any medication used will have to pass through the blood brain barrier (BBB). High doses of antibiotics will be needed in order to facilitate their passage through the BBB.

Answers to Clinical Case Wrap-Up

Page 359

1. The conduction velocity of Nicole's PNS would be slower during her acute phase of Guillain-Barré syndrome. As the Schwan cells divide and re-myelinate her peripheral nerve axons, the conduction velocity would return to normal.
2. The demyelination could affect Nicole's autonomic nervous system. The ANS regulates smooth muscle, cardiac muscle, and glandular activity. Nicole may experience constipation or cardiac arrhythmias as a result of this syndrome.

14 | THE NERVOUS SYSTEM The Spinal Cord and Spinal Nerves

Concept Check

Page 361

1. The cell bodies of sensory neurons are found within the dorsal root ganglion.
2. The cauda equina is formed from the filum terminale and the long dorsal and ventral roots that extend caudal to the conus medullaris.

Page 364

3. The cerebrospinal fluid that surrounds the spinal cord is found in the subarachnoid space, which lies between the arachnoid mater and the pia mater.
4. There are three meningeal layers surrounding the spinal cord and nerve roots. From deep (closest to the spinal cord) to superficial, these three layers are the pia mater, arachnoid mater, and dura mater.

Page 366

5. Since the poliovirus would be located in the somatic motor neurons, we would find it in the anterior horns of the spinal cord where the cell bodies of these neurons are located.
6. Ascending tracts carry sensory information toward the brain. Descending tracts carry motor commands into the spinal cord.

Page 367

7. The outermost layer is called the epineurium. It surrounds the entire nerve. The middle layer, or perineurium, divides the nerve into a series of compartments that contain bundles of axons. A single bundle is called a fascicle. The endoneurium is the innermost layer, and it surrounds individual axons.
8. The white and gray rami connect the spinal nerve to a nearby autonomic ganglion. The white ramus carries preganglionic axons that are myelinated from the nerve to the ganglion. The gray ramus carries postganglionic unmyelinated axons from the ganglion back to the spinal nerve.

Page 374

9. The phrenic nerves that innervate the diaphragm originate in the cervical plexus. Damage to this plexus or, more specifically, to the phrenic nerves would greatly interfere with the ability to breathe and possibly result in death by suffocation.
10. The brachial plexus may have been damaged.

Page 382

11. (1) Stimulation and activation of a receptor. (2) Activation of a sensory neuron. (3) Information processing in CNS. (4) Activation of a motor neuron. (5) Response by effector.
12. A monosynaptic reflex has a sensory neuron synapsing directly on a motor neuron. A polysynaptic reflex has more than one synapse between the stimulus and response.

Chapter Review

Level 1 Reviewing Facts and Terms

1. d 2. i 3. a 4. g 5. b 6. c 7. j 8. f 9. h 10. e 11. a. posterior median sulcus; b. dorsal root ganglion; c. white matter; d. ventral root; e. anterior median fissure 12. d 13. d; 14. b 15. d 16. a 17. b 18. b 19. c 20. c 21. d

Level 2 Reviewing Concepts

1. d
2. d
3. a
4. The meninges provide a tough protective covering, longitudinal physical stability, and a space for shock-absorbing fluid.
5. A reflex is an immediate involuntary response, whereas voluntary motor movement is under conscious control and is voluntary.
6. Incoming sensory information would be disrupted.
7. Transmission of information between neurons at synapses takes a finite amount of time. Thus, the more synapses in a reflex, the longer the delay between the stimulus and responses. In a monosynaptic reflex, there is only one synapse. It has the most rapid response time. In a polysynaptic reflex, there are multiple synapses, each contributing to the overall delay.
8. In the cervical region, the first pair of spinal nerves, C_1, exits between the skull and the first cervical vertebra. Thereafter, a numbered cervical spinal nerve exits after each cervical vertebra. For instance, nerve C_2 exits after vertebra C_1, nerve C_3 exits after vertebra C_2, and so on until nerve C_8, which exits after vertebra C_7.
9. The denticulate ligaments prevent side-to-side movements of the spinal cord.
10. The adult spinal cord extends only as far as vertebra L_1 or L_2. Inferior to this point in the vertebral foramen, the meningeal layers enclose the relatively sturdy components of the cauda equina and a significant quantity of CSF.

Level 3 Critical Thinking

1. The rectus abdominis muscle is always retracted laterally, as it is innervated by T_7–T_{12}, entering through the posterior aspect along the lateral margin. If the muscle were to be retracted medially, the nerves would be torn, thereby paralyzing the muscle.
2. The part of the cord that is most likely compressed is an ascending tract.
3. The neurons for the anterior horn of the spinal cord are somatic motor neurons that direct the activity of skeletal muscles. The lumbar segments of the spinal cord control the skeletal muscles that are involved with the control of the muscles of the hip and lower limb. As a result of the injury, Karen would have poor control of most lower limb muscles, a problem with walking (if she could walk at all), and (if she could stand) problems maintaining balance.

Answers to Clinical Case Wrap-Up

Page 385

1. If Bob's numbness began at his nipples, this would represent a T_4 dermatome.
2. If Bob's numbness and weakness include shoulder girdle, arm, forearm, and hand, you would expect to find the lesion in his cervical spinal cord.

15 THE NERVOUS SYSTEM Sensory and Motor Tracts of the Spinal Cord

Concept Check

Page 388

1. (a) Ascending sensory tracts have the following general organization and anatomical pattern: (1) If the sensory information is raised to a conscious level, the ascending tract has three neurons. The posterior columns are examples of this type of tract. The first-order neuron travels from the receptor to the spinal cord. Its cell body is found within the dorsal root ganglion of the spinal nerve. The first-order neuron will synapse with the second-order neuron either within the spinal cord or within the brainstem. The second-order neuron will travel to the thalamus, where it will synapse with the third-order neuron, which will travel to the cerebral cortex. Typically, either the first-order neuron or the second-order neuron will decussate to the opposite side. (2) If the sensory information is not raised to a conscious level, the ascending tract has two neurons; the two spinocerebellar tracts are examples of this type of sensory tract. The first-order neuron will synapse with the second-order neuron within the spinal cord. The second-order neuron will travel to the cerebellum on the same side. (b) Descending motor tracts are two-neuron tracts. (1) The upper motor neuron will begin in either the cerebral cortex or the brainstem. It will travel to either the brainstem or the spinal cord, where it will synapse with the lower motor neuron. (2) The lower motor neuron will then exit the brainstem or spinal cord and synapse with skeletal muscle in the periphery. (3) Most descending motor tracts will decussate, either in the medulla or spinal cord, to the opposite side.

Page 394

2. The gracile fasciculus in the posterior column of the spinal cord is being compressed. It is responsible for carrying information about touch and pressure from the lower part of the body to the brain.

Page 398

3. (a) tectospinal tracts, (b) vestibulospinal tracts

Page 398

4. Lower motor neurons are affected more by somatic motor control.

Chapter Review

Level 1 Reviewing Facts and Terms

1. h 2. e 3. a 4. f 5. b 6. g 7. c 8. i 9. d 10. a 11. a 12. d 13. b 14. a. gracile fasciculus; b. cuneate fasciculus; c. posterior spinocerebellar tracts; d. anterior spinothalamic tract

Level 2 Reviewing Concepts

1. a
2. The first-order neuron is the sensory neuron that delivers the sensations to the CNS.
3. The sensory homunculus is distorted because the area of sensory cortex devoted to a particular region is proportional to the number of sensory receptors the region contains, not to its absolute size.
4. The cerebral nuclei are processing centers that provide background patterns of movement involved in the performance of voluntary motor activities.
5. The vestibulospinal tracts direct the involuntary regulation of balance in response to sensations from the internal ear. The medial reticulospinal tracts direct the involuntary regulation of reflex activity and autonomic functions.

Level 3 Critical Thinking

1. The problem with Cindy's spinal cord is probably located in the lateral spinothalamic tract on the right side, somewhere around the level of spinal segment L_2. To figure this out, you would need to determine (1) what tract carries sensory information from the lower limb, (2) where it decussates, and (3) what spinal segments innervate the hip and lower limb, as detailed in the dermatome illustration in Chapter 14 (Figure 14.7).
2. The problem with John's spinal cord is probably located within the posterior columns, specifically, the most lateral portions of the cuneate fasciculus. To figure this out you would need to determine (1) what tract carries sensory information dealing with fine touch, vibration, pressure, and proprioception from the upper limb and (2) according to the medial-lateral rule, which portion of the involved tract would be injured that matches John's symptoms.

Answers to Clinical Case Wrap-Up

Page 402

1. Because ALS affects upper motor neurons you would expect to see cell death in the primary motor cortex of the brain.
2. Because ALS affects only motor neurons, you would expect the dorsal root ganglia—the location of sensory, ascending neurons—to be normal in appearance.

16 THE NERVOUS SYSTEM The Brain and Cranial Nerves

Concept Check

Page 406

1. The six major regions in the adult brain are the cerebrum, diencephalon, mesencephalon (midbrain), pons, cerebellum, and medulla oblongata.
2. The ventricles are four fluid-filled chambers that form from the neurocoel. These four chambers are the two lateral ventricles, the third ventricle, and the fourth ventricle. The lateral ventricles are located within each cerebral hemisphere. The third ventricle is located within the diencephalon, and the fourth begins between the pons and cerebellum. In the inferior portion of the medulla oblongata, the fourth ventricle narrows and becomes continuous with the central canal of the spinal cord. Ependymal cells line the ventricles.

Page 412

3. The four extensions of the dura mater are the falx cerebri, tentorium cerebelli, falx cerebelli, and diaphragma sellae.
4. The blood brain barrier functions to isolate the CNS from the general circulation.

Page 413

5. The three types of nuclei found within the medulla are the relay nuclei, cranial nerve nuclei, and autonomic nuclei.

Page 415

6. Ascending, descending, and transverse fibers pass through the pons. Ascending and descending fibers interconnect other portions of the CNS, and the transverse fibers interconnect the two hemispheres of the cerebellum and interconnect pontine nuclei with the cerebellar hemisphere on the opposite side.

Page 415

7. The corpora quadrigemina are found within the tectum of the mesencephalon (midbrain). The two superior colliculi are visual reflex centers, while the two inferior colliculi are auditory reflex centers.

Page 418

8. Changes in body temperature stimulate the pre-optic area of the hypothalamus, a division of the diencephalon.
9. The thalamus coordinates somatic motor activities at the conscious and subconscious levels.

Page 422

10. The superior, middle, and inferior cerebellar peduncles link the cerebellum to the cerebrum, brainstem, and spinal cord.
11. The two main functions of the cerebellum are adjusting the postural muscles of the body and programming and fine-tuning voluntary and involuntary movements of the body.

Page 432

12. The frontal lobe contains the primary motor cortex for voluntary muscular activities. Additionally, it contains the prefrontal cortex, which integrates information and performs abstract intellectual functions. The parietal lobe contains the primary somatosensory cortex, which receives somatic sensory information. The occipital lobe contains the visual cortex for conscious perception of visual stimuli. The temporal lobe contains the auditory cortex and olfactory cortex for the conscious perception of auditory and olfactory stimuli.
13. The three major groups of axons in the central white matter are (1) association fibers, tracts that interconnect areas of neural cortex within a single cerebral hemisphere; (2) commissural fibers, tracts that connect the two cerebral hemispheres; and (3) projection fibers, tracts that link the cerebrum with other regions of the brain and spinal cord.

Page 444

14. The cranial nerve responsible for tongue movements is the hypoglossal nerve (XII).
15. The facial nerve (VII) is involved with taste sensation.

Chapter Review

Level 1 Reviewing Facts and Terms

1. (a) thalamus; (b) corpus callosum; (c) cerebral aqueduct; (d) cerebellum; (e) pons 2. b 3. a 4. d 5. d 6. a 7. b 8. d 9. c 10. d 11. d

Level 2 Reviewing Concepts

1. b
2. The brain can respond with greater versatility because it includes many more interneurons, pathways, and connections than the tracts of the spinal cord.
3. Damage would have occurred in the premotor cortex of the frontal lobe.
4. Impulses from proprioceptors must pass through the olivary nuclei on their way to the cerebellum.
5. The nuclei involved in the coordinated movement of the head in the direction of a loud noise are the inferior colliculi.
6. The cranial nerves that collectively participate in eye function are N II, N III, N IV, N V, and N VI.
7. The person might have a lesion in the limbic system.
8. A less intact blood brain barrier suggests that the endothelium is extremely permeable. This permeability exposes hypothalamic nuclei to circulating hormones and permits the diffusion of hypothalamic hormones into the circulation.

Level 3 Critical Thinking

1. Increasing pressure in the cranium could compress important blood vessels, leading to further brain damage in areas not directly affected by the hematoma. Pressure on the brainstem could disrupt vital respiratory, cardiovascular, and vasomotor functions and possibly cause death. Pressure on the motor nuclei of the cranial nerves would lead to drooping eyelids and dilated pupils. Pressure on descending motor tracts would impair muscle function and decrease muscle tone in the affected areas of the body.
2. The person might have an epidural hemorrhage, an extremely serious injury that occurs when an artery—one of the meningeal vessels—breaks and allows blood to leak into the epidural space. Depending on the size of the break, more or less blood may enter the space, and at different rates of speed. Such injuries cause death in 100 percent of untreated cases and in approximately 50 percent of treated cases because the brain can be damaged by compression and starvation of the normal blood supply before the problem is discovered.

Answers to Clinical Case Wrap-Up

Page 448

1. Broca's area is usually located in the left hemisphere. Dr. Taylor was unable to speak or understand speech. The primary motor cortex that directs voluntary movement of the right side of the body is located in the left hemisphere, in the precentral gyrus of the cerebral cortex. She was unable to use her right arm. These facts indicate that the stroke affected the left side of her brain.
2. Neuroplasticity is the ability of nerve cells in the brain to make new connections to other nerve cells. Following trauma to the brain, as in a stroke, brain cells that are still alive but dormant can make new connections with other nerve cells. These changes are known as healing after a stroke.

17 THE NERVOUS SYSTEM Autonomic Nervous System

Concept Check

Page 450

1. Preganglionic fibers are short with myelinated axons; preganglionic neurons have their cell bodies in the CNS, and their axons project to ganglia in the PNS. Postganglionic fibers are long and unmyelinated; postganglionic neurons have their cell bodies in ganglia in the PNS, and their axons innervate effector cells.
2. Norepinephrine is the neurotransmitter released by most postganglionic sympathetic fibers.
3. Preganglionic fibers of the parasympathetic division of the ANS originate in either the brainstem (cranial nerves III, IX, and X) or the sacral spinal cord.

Page 459

4. The neurons that synapse in the collateral ganglia originate in the inferior thoracic and superior lumbar regions of the spinal cord and pass through the sympathetic chain ganglia without synapsing before reaching the collateral ganglia.
5. Blocking the beta receptors on cells would decrease or prevent sympathetic stimulation of those tissues. This would result in decreased heart rate and force of contraction and relaxation of the smooth muscle in the walls; the combination would lower blood pressure.
6. Sympathetic chain ganglia (paravertebral ganglia) are on both sides of the vertebral column. Neurons in these ganglia control effectors in the body wall, inside the thoracic cavity, and in the head and limbs. Collateral ganglia (prevertebral ganglia) are anterior to the vertebral column. Neurons on these ganglia innervate tissues and organs in the abdominopelvic cavity.

Page 462

7. Intramural ganglia are located in the tissues of their target organs.
8. Sympathetic stimulation has such widespread effects because (1) the extensive divergence of preganglionic fibers in the sympathetic division distributes sympathetic output to many different visceral organs and tissues simultaneously, and (2) the release of E and NE by the adrenal medullae affects tissues and organs throughout the body.

Page 465

9. Visceral reflexes are the simplest functional units in the autonomic nervous system. They provide automatic motor responses that can be modified, facilitated, or inhibited by higher centers, especially those of the hypothalamus.
10. The celiac plexus, the inferior mesenteric plexus, and the hypogastric plexus are three plexuses in the abdominopelvic cavity.

Chapter Review

Level 1 Reviewing Facts and Terms

1. c 2. g 3. h 4. d 5. i 6. a 7. e 8. j 9. b 10. f 11. d 12. b 13. c 14. d 15. d 16. c 17. b 18. c 19. b 20. d

Level 2 Reviewing Concepts

1. d
2. a
3. b
4. d
5. There are no enzymes to break down epinephrine and norepinephrine in the blood and very little in peripheral tissues.
6. The parasympathetic division innervates only visceral structures served by some cranial nerves or lying within the thoracic and/or abdominopelvic cavities. The sympathetic division has widespread impact due to extensive collateral branching of preganglionic fibers, which reach visceral organs and tissues throughout the body.
7. Sympathetic chain ganglia are innervated by preganglionic fibers from the thoracolumbar regions of the spinal cord, and they are interconnected by preganglionic fibers and axons from each ganglion in the chain innervating a particular body segment. The collateral ganglia are part of the abdominal autonomic plexus anterior to the vertebral column. Preganglionic sympathetic fibers innervate the collateral ganglia as splanchnic nerves. Intramural ganglia are part of the parasympathetic division. They are located near or within the tissues of the visceral organs.
8. The sympathetic division of the ANS stimulates metabolism, increases alertness, and prepares for emergency in "fight-or-flight." The parasympathetic division promotes relaxation, nutrient uptake, energy storage, and "rest-and-digest."
9. Visceral motor neurons, called preganglionic neurons, send their axons, called preganglionic fibers, from the CNS to synapse on ganglionic neurons, whose cell bodies are located in ganglia outside the CNS.

Level 3 Critical Thinking

1. Cutting off autonomic nervous system stimulation to the stomach through the vagus nerve decreases stimulation of digestive glands, thus reducing their secretion. This may diminish ulcers in the wall of the stomach.
2. Kassie should be treated with epinephrine. This would mimic sympathetic activation, which dilates air passageways in the lungs. The constriction of her respiratory passages would be alleviated.

Answers to Clinical Case Wrap-Up

Page 469

1. If Tim is already lying down, he can't hurt himself by falling. Also, lying down places his head level with or lower than his heart, so his brain will get more blood. Under parasympathetic stimulation, heart rate, force of contraction, and blood pressure decrease. The heart cannot push blood against gravity under the influence of the parasympathetic stimulation. If the head is level with or below the level of the heart, gravity need not be overcome.
2. In addition to the reflexive sympathetic activity at the heart and vascular system, the hypothalamus would initiate sympathetic activation. This would stimulate the reticular activating system, increasing alertness. Stimulation of the respiratory centers of the pons and medulla oblongata would increase breathing rate and depth.

18 | THE NERVOUS SYSTEM General and Special Senses

Concept Check

Page 473

1. Free nerve endings may be activated by chemical stimulation, pressure, temperature changes, or physical damage.
2. Tonic receptors are always active, whereas phasic receptors are normally inactive, but become active for a short time.
3. A sensation is the sensory information arriving at the CNS.
4. The general senses refer to sensations of temperature, pain, touch, pressure, vibration, and proprioception (body position).

Page 476

5. Since nociceptors are pain receptors, if they are stimulated, you perceive a sensation of pain in your affected hand.
6. Proprioceptors relay information about limb position and movement to the central nervous system, especially the cerebellum. Lack of this information would result in uncoordinated movements, and the individual would probably be unable to walk.
7. The three classes of mechanoreceptors are tactile receptors, baroreceptors, and proprioceptors.

Page 478

8. Your roommate is incorrect. The olfactory tract, contrary to other sensations that are raised to a conscious level, does not pass through the thalamus.

Page 480

9. When you have a cold, airborne molecules cannot reach the olfactory sensory neurons and meals taste dull and unappealing.
10. Gustatory epithelial cells are clustered in individual taste buds. Lingual papillae contain taste buds.
11. The tongue has four types of lingual papillae: filiform, fungiform, vallate, and foliate.

Page 490

12. Two small muscles contract to protect the tympanic membrane from loud noises. These are the tensor tympani and stapedius.
13. Shaking the head "no" stimulates the hair cells of the lateral semicircular duct. This stimulation is interpreted by the brain as a movement of the head.
14. Loss of stereocilia (as a result of constant exposure to loud noises, for instance) reduces hearing sensitivity and could eventually result in deafness.
15. The cochlear duct (scala media) is sandwiched between the scala vestibuli (vestibular duct) and scala tympani (tympanic duct). The hair cells for hearing are located in the cochlear duct.

Page 500

16. Inadequate tear production affects the cornea first. Since the cornea is avascular, the cells of the cornea must obtain oxygen and nutrients from the tear fluid that passes over its surface.
17. The two structures most affected by an abnormally high intra-ocular pressure are (1) the scleral venous sinus (the aqueous humor no longer has free access to this structure) and (2) the optic nerve (the nerve fibers of this structure are distorted, which affects visual perception).
18. An individual born without cones would be able to see only in black and white (monochromatic) and would have very poor visual acuity.
19. Ciliary processes are folds in the epithelium of the ciliary body. The ciliary body includes the ciliary muscle, which helps control the shape of the lens for near and far vision.

Chapter Review

Level 1 Reviewing Facts and Terms

1. (a) posterior cavity; (b) posterior chamber; (c) lens; (d) iris 2. c 3. d 4. a 5. b 6. b 7. c 8. b 9. c 10. c 11. (a) cochlea; (b) tympanic cavity; (c) tympanic membrane; (d) semicircular canals

Level 2 Reviewing Concepts

1. a
2. d
3. c
4. Receptor specificity means that each receptor has a characteristic sensitivity. For example, a touch receptor is very sensitive to pressure but relatively insensitive to chemical stimuli.
5. An increased quantity of neurotransmitter is released when the stereocilia of the hair cell are displaced toward the kinocilium.
6. The hair cells in the internal ear act as sensory receptors.
7. Sensory adaptation is a reduction in sensitivity in the presence of a constant stimulus. The receptor responds strongly at first, but thereafter the activity along the afferent fiber gradually declines, in part because of synaptic fatigue.
8. Sensory coding provides information about the strength, duration, variation, and movement of the stimulus.
9. An individual with damage to the lamellar corpuscles would have trouble feeling direct pressure, such as a pinch.
10. The bony labyrinth is a shell of dense bone. It surrounds and protects fluid-filled tubes and chambers known as the membranous labyrinth.

Level 3 Critical Thinking

1. In removing the polyps, some of the olfactory epithelium was probably damaged or destroyed. This would decrease the surface area available for the detection of odor molecules and thus the intensity of the stimulus. As a result, after the surgery it would take a larger stimulus to provide the same level of smell that could be detected before the surgery.
2. Jared has an infection of the middle ear, which is most often of bacterial origin and most commonly found in children and infants. During an upper respiratory infection pathogens usually gain access to the middle ear cavity through the auditory tube, which is shorter and more horizontally oriented in infants and children than in adults. As the infection progresses, the middle ear cavity can fill with pus. The increase in pressure in the middle ear cavity can eventually rupture the tympanic membrane. This condition can be treated with antibiotics.
3. When you have a cold, the mucous membranes of your nasal and oral cavities swell and may become plugged with mucus, preventing the molecules that form the odors of foods from reaching the olfactory epithelium at the superior aspect of the olfactory chamber. As much of what we perceive as taste is really olfaction, a reduced sense of smell decreases the appeal of foods.

Answers to Clinical Case Wrap-Up

Page 504

1. When looking into the external auditory canal, John's doctor can see only up to the tympanic membrane. The tympanic membrane is opaque, so she cannot see through it into the middle or internal ear.
2. No, it does not affect sound conduction. Because Ménière's disease is a phenomenon of the internal ear, specifically of endolymph, it does not affect the conduction of sound through the middle ear.

19 | The Endocrine System

Concept Check

Page 508

1. A target cell is a specific cell that that has receptors needed to bind hormones and respond to their presence. Hormones change cellular metabolic activities.

Page 509

2. The hypothalamus is the region of the brain responsible for regulating hormone secretion by the pituitary gland.
3. The two regions of the pituitary gland are the anterior lobe and the posterior lobe. The posterior lobe of the pituitary gland contains axon terminals of neurons whose cell bodies are in the hypothalamus. When these neurons are stimulated, their axon terminals release oxytocin or ADH. Most endocrine cells of the anterior lobe of the pituitary gland are controlled by the secretion of regulatory factors from the hypothalamus.

Page 515

4. Thyroxine (T_4), triiodothyronine (T_3), and calcitonin (CT) are hormones associated with the thyroid gland.

Page 515

5. Removal of the parathyroid glands would result in a decrease in the blood levels of calcium ions. This can be counteracted by increasing the amount of vitamin D and calcium in the diet.

Page 516

6. The thymus lies just posterior to the sternum within the mediastinum of the thoracic cavity.

Page 518

7. The region of the gland affected is the zona glomerulosa. The deficient hormone is aldosterone.

Page 518

8. EPO is used by some endurance athletes in "blood doping" in an attempt to increase red blood cell production, thus increasing the blood volume and its oxygen-carrying capacity.

Page 520

9. The pancreatic islets make up the endocrine pancreas. The primary hormones released are glucagon, insulin, somatostatin, and pancreatic polypeptide.

Page 521

10. The testes produce testosterone and inhibin. Testosterone promotes the production of functional sperm, maintains the secretory glands of the male reproductive tract, influences secondary sexual characteristics, and stimulates muscle growth. Inhibin depresses the secretion of FSH by the anterior lobe of the pituitary gland.

Page 522

11. Melatonin is thought to help regulate our circadian rhythms, which play a role in our sleep patterns. Taking melatonin might help counteract "jet lag" by allowing an international traveler to sleep more regularly during flight and immediately after crossing multiple time zones.

Page 524

12. The reproductive hormones show the most dramatic decline as we age.

Chapter Review

Level 1 Reviewing Facts and Terms

1. c 2. i 3. f 4. b 5. h 6. j 7. a 8. g 9. e 10. d 11. b 12. d 13. c 14. d 15. c 16. b 17. a 18. c 19. b

Level 2 Reviewing Concepts

1. b
2. b
3. The nervous system has localized, immediate, short-term effects on neurons, gland cells, muscle cells, and fat cells. The endocrine system has widespread, gradual, long-term effects on all tissues.
4. On the basis of their chemical structure, hormones can be divided into amino acid derivatives, peptide hormones, and lipid derivatives. There are two groups of lipid derivatives: eicosanoids and steroid hormones.
5. The primary targets of testosterone are most cells in the body. Its effects include supporting functional maturation of sperm, protein synthesis in skeletal muscles, male secondary sex characteristics, and associated behaviors.
6. Thyroid hormones increase energy utilization, oxygen consumption, and growth and development of cells.
7. Parathyroid glands produce parathyroid hormone (PTH) in response to low calcium concentrations. PTH increases calcium ion concentrations in body fluids by stimulating osteoclasts, inhibiting osteoblasts, reducing urinary excretion of calcium ions, and promoting intestinal absorption of calcium (through stimulation of calcitriol production by the kidneys) until blood calcium ion concentrations return to normal.
8. Melatonin slows the maturation of sperm, oocytes, and reproductive organs by inhibiting the production of a hypothalamic-releasing hormone that stimulates FSH and LH secretion.
9. The capillary network is part of the hypophyseal portal system. Capillaries in the hypothalamus absorb regulatory secretions from hypothalamic nuclei, then unite as portal vessels, which proceed to the anterior pituitary gland. There, the portal vessels branch into a second capillary network, where regulatory factors leave the vessels and stimulate endocrine cells in the anterior pituitary gland.

Level 3 Critical Thinking

1. Secretion of growth hormone (GH) by the pituitary gland is stimulated by growth hormone–releasing hormone and inhibited by GH–inhibiting hormone, or somatostatin. Most pituitary tumors resulting in exaggerated growth patterns, such as acromegaly, are tumors that result in excess growth of the somatotropes within the pituitary. This excessive growth of somatotropes within the pituitary will result in elevated blood levels of GH.
2. The two abnormalities are diabetes insipidus and diabetes mellitus. In diabetes insipidus, the posterior pituitary no longer produces ADH; dehydration consequently occurs, and increased urination is an outcome. In diabetes mellitus, there is an inadequate production of insulin, which results in an elevation of blood glucose levels and an increase in urine production.
3. Hypothyroidism can be caused by one of the following: (1) the hypothalamus isn't secreting enough releasing hormone to stimulate adequate production of TSH by the adenohypophysis gland, (2) the adenohypophysis gland cannot produce normal levels of TSH under normal stimulation, or (3) the thyroid gland is unable to respond normally to TSH stimulation.
4. Two kidney hormones act indirectly to increase blood pressure and volume, opposed by a hormone from the heart. First, renin, released by kidney cells, converts angiotensinogen from the liver into angiotensin I, which is converted by the capillaries of the lungs into angiotensin II. Angiotensin II stimulates suprarenal production of aldosterone, which in turn causes the kidney to retain sodium ions and water, thereby reducing fluid loss in the urine and increasing blood volume and pressure. Erythropoietin (EPO), the second kidney hormone, stimulates red blood cell production by bone marrow. Released when blood pressure or oxygen levels are low, EPO ultimately increases blood volume and its oxygen-carrying capacity. When blood pressure or volume becomes excessive, cardiac muscle in the right atrium produces atrial natriuretic peptide (ANP). ANP suppresses the release of ADH and aldosterone, stimulating water and sodium loss at the kidneys and gradually reducing blood volume and pressure.

Answers to Clinical Case Wrap-Up

Page 527

1. Yes, if Kathy's thyroid were not functioning well, both her T_3 and T_4 hormones would be low.
2. If Kathy's T_3 and TSH levels are both low, this would indicate pathology in her pituitary gland or her hypothalamus.

20 | THE CARDIOVASCULAR SYSTEM Blood

Concept Check

Page 531

1. Blood carries heat away from areas that are warm and distributes it to the skin when the body is too warm and to vital organs when the body is cold; marked slowing of flow would disrupt the body's ability to cool or warm itself properly.
2. A hypovolemic individual has a low blood volume and would therefore have abnormally low blood pressure.
3. Whole blood contains significant numbers of formed elements, such as red blood cells, white blood cells, and platelets. These components of the blood make it thicker and more resistant to flow.

Page 539

4. The hematocrit value closely approximates the percentage of red blood cells; thus, red blood cells account for 42 percent of her blood volume.
5. Red blood cells have the ability to stack and create a rouleau, which can pass through a tiny blood vessel more easily than could many separate red blood cells. In addition, red blood cells are flexible, which allows them to squeeze through small capillaries.
6. Neutrophils appear in the greatest number in an infected cut.
7. The granules contain histamine; its release promotes inflammation at the injury site.

Page 541

8. Erythropoietin increases the rate of both erythroblast cell division and stem cell division, and it speeds up the maturation of RBCs.
9. Hematopoietic stem cells produce all formed elements.
10. The ejection of the nucleus transforms an erythroblast into a reticulocyte.
11. Megakaryocytes, which are derived from hemopoietic stem cells, produce platelets.

Chapter Review

Level 1 Reviewing Facts and Terms

1. (a) basophil; (b) lymphocyte; (c) neutrophil; (d) eosinophil; (e) monocyte 2. d 3. d 4. b 5. b 6. d 7. c 8. d 9. a 10. b 11. c

Level 2 Reviewing Concepts

1. c
2. d
3. c
4. The volume of packed cells is a hematocrit. It is expressed as a percentage and closely approximates the volume of erythrocytes in the blood sample. As a result, the hematocrit value is often called the volume of packed red cells or simply the packed cell volume.
5. The clotting reaction seals the breaks in blood vessel walls, preventing changes in blood volume that could seriously affect blood pressure and cardiovascular function.
6. A mature megakaryocyte begins to shed its cytoplasm in small membrane-enclosed packets called platelets.
7. Lipoproteins are protein–lipid molecules that readily dissolve in plasma. Some lipoproteins transport insoluble lipids to peripheral tissues.
8. People with type O blood have anti-A and anti-B antibodies in their plasma. Thus, they could not receive blood from an AB donor because the RBCs in this blood type contain surface antigens A and B on their surface. A cross-reaction would occur.

Level 3 Critical Thinking

1. At higher elevations, each erythrocyte carries less oxygen because less oxygen is present in the atmosphere. This situation triggers erythrocytosis, the production of an increased number of erythrocytes in response to a large release of erythropoietin from oxygen-deprived tissues. The larger number of RBCs increases the total oxygen-carrying capacity of the blood to offset the lower saturation of each RBC. Thus, when the athlete moves back to a lower elevation, his or her blood can carry even more oxygen, increasing endurance and perhaps allowing better performance in competition.
2. A major function of the spleen is to destroy old, defective, and worn-out red blood cells. As the spleen increases in size, so does its capacity to eliminate red blood cells, and this produces anemia. The decreased number of red blood cells decreases the blood's ability to deliver oxygen to the tissues and thus the metabolism is slowed down. This accounts for the tired feeling and lack of energy. Because there are fewer red blood cells than normal, the blood circulating through the skin is not as red, and so the person has pale or white skin.

Answers to Clinical Case Wrap-Up

Page 544

1. If Danny had acute myeloid leukemia, his blood smear would contain a high number of immature myeloblast cells.
2. Normally, there are 5000 to 10,000 white blood cells per cubic microliter, and 30 percent of these are lymphocytes.

21 | THE CARDIOVASCULAR SYSTEM The Heart

Concept Check

Page 546

1. The two circuits of the cardiovascular system are the pulmonary circuit, which pumps blood to the lungs, and the systemic circuit, which pumps blood to the rest of the body.

Page 548

2. The pericardial cavity is a potential, fluid-filled space between the outer parietal layer and the inner visceral layer of the serous pericardium. It contains pericardial fluid, which acts as a lubricant, reducing friction between opposing surfaces.

Page 550

3. Cardiac muscle tissue contains a single, centrally placed nucleus, abundant mitochondria, relatively short T tubules, an extensive blood supply, and intercalated discs. Skeletal muscle tissue has multiple nuclei, fewer mitochondria, relatively long T tubules, a less extensive blood supply, and no intercalated discs.

4. Cardiac muscle cells are connected to their neighbors at specialized junctional sites termed intercalated discs. At an intercalated disc, the membranes are bound together by desmosomes, myofibrils of the cells are anchored to the membrane, and gap junctions connect the cells.

Page 553

5. The groove between the atria and ventricles is the coronary sulcus.
6. The base of the heart is the broad superior portion of the heart, and it includes the origins of the major arteries and veins and the superior surfaces of the two atria.

Page 555

7. The valves prevent backflow from the ventricles into the atria when the ventricles contract, so without them, the blood would rush back into the atria.
8. As the ventricles begin to contract, they force the AV valves to close, which in turn pull on the chordae tendineae. The chordae tendineae pull on the papillary muscles. The papillary muscles respond by contracting, opposing the force that is pushing the valves toward the atria.

Page 561

9. The right coronary artery branches off the ascending aorta. It passes around the right side of the heart in the coronary sulcus, giving branches to the right atrium and right ventricle, including the atrioventricular branches, the sinoatrial nodal branch, the right marginal artery, the atrioventricular nodal branch, and the posterior interventricular branch.
10. The left coronary artery supplies blood to (1) most the left ventricle, (2) some of the right ventricle, (3) most of the left atrium, and (4) the anterior two-thirds of the interventricular septum.

Page 561

11. The two types of specialized cardiac muscle cells are nodal cells and conducting cells. Nodal cells establish the rate of cardiac contraction, and conducting cells distribute the contractile stimulus to the myocardium.

Page 562

12. If the cells of the SA node were not functioning, the heart would still continue to beat, but at a slower rate, following the pace set by the AV node.
13. Because the plasma membranes of pacemaker cells depolarize spontaneously and are electrically connected to one another, to conducting fibers, and to cardiac muscle cells, contractions are able to travel through the conducting system and trigger the contractions of cardiac muscle cells. The timing of contraction is controlled by the rate of propagation and the distribution of the contractile stimulus to the cardiac muscle tissue in the atria versus the ventricles.

Page 563

14. Sympathetic neurons release norepinephrine (NE), which increases both the heart rate and force of contraction through stimulation of beta receptors on nodal cells and contractile cells. Parasympathetic neurons release acetylcholine (ACh), which decreases both the heart rate and force of contraction by stimulating muscarinic receptors on nodal cells and contractile cells.

Chapter Review

Level 1 Reviewing Facts and Terms

1. f 2. e 3. h 4. c 5. b 6. j 7. g 8. a 9. i 10. d 11. d 12. (a) superior vena cava; (b) interventricular septum; (c) pulmonary trunk; (d) left common carotid artery 13. b 14. c 15. a 16. c

Level 2 Reviewing Concepts

1. a and b
2. c
3. c
4. Cardiac muscle cells are like skeletal muscle fibers in that each cardiac muscle cell contains organized myofibrils, and the alignment of their sarcomeres gives the cardiac tissue a striated appearance.
5. The aortic and pulmonary valves do not require muscular braces because the arterial walls do not contract, and the relative positions of the cusps are stable.
6. A pacemaker cell is a cell that depolarizes spontaneously. The normal pacemaker cells (SA node) depolarize at the fastest rate. Other groups of cells that have the potential to serve as pacemaker cells are the atrioventricular node, AV bundle, right bundle branch, left bundle branch, and Purkinje fibers.
7. Pericardial fluid is secreted by pericardial membranes and acts as a lubricant, reducing friction between the opposing visceral and parietal pericardial surfaces.
8. The left ventricle has the thickest walls because it needs to exert so much force to push blood around the systemic circuit.
9. Nodal cells are unique because their plasma membranes depolarize spontaneously. Specific types of nodal cells called pacemaker cells are responsible for establishing the rate of cardiac contraction.
10. The SA node, which is composed of cells that exhibit rapid pacemaker potential, is the pacemaker of the heart. The AV node slows the impulse that signals contraction because its cells are smaller than those of the conduction pathway.

Level 3 Critical Thinking

1. It would appear that Harvey has a regurgitating mitral valve. When an AV valve fails to close properly, the blood flowing back into the atrium produces the abnormal heart sound, or murmur. If the sound is heard at the beginning of the systole, this would indicate the AV valve because this is the period when the valve is just closed and the blood in the ventricle is under increasing pressure; thus, the likelihood of backflow is the greatest. If the sound is heard at the end of systole or the beginning of diastole, it would indicate a regurgitating semilunar valve—in this case, the aortic semilunar valve.
2. During tachycardia, the heart beats at an abnormally fast rate. The faster the heart beats, the less time there is between contractions for it to fill with blood again. As a result, over a period of time, the heart fills with less and less blood and thus pumps less blood out. The stroke volume decreases, as does the cardiac output. When the cardiac output decreases to the point where not enough blood reaches the central nervous system, loss of consciousness occurs.
3. Acetylcholine is likely to be released, resulting in a decreased heart rate.

Answers to Clinical Case Wrap-Up

Page 566

1. Ellen felt like she was having a myocardial infarction because her heart was ischemic, or lacking a good blood supply. In left ventricular apical hypokinesis, there is not enough blood pressure in the proximal aorta to fill the cardiac arteries. This cardiac ischemia mimics a blocked cardiac artery, or cardiovascular disease.
2. If Ellen's cardiologist wanted to look at the right side of the heart, she could use a vein in the antecubital fossa, threading the catheter "with the stream" to the superior vena cava, into the right atrium, through the tricuspid valve, into the right ventricle.

22 THE CARDIOVASCULAR SYSTEM Vessels and Circulation

Concept Check

Page 574

1. The blood vessels are veins. Arteries and arterioles have a relatively large amount of smooth muscle tissue in a thick, well-developed tunica media.
2. Blood pressure in the arterial system pushes blood into the capillaries. Blood pressure on the venous side is very low, and other forces help keep the blood moving. Valves in the walls of venules and medium-sized veins permit blood flow in only one direction, toward the heart, preventing the backflow of blood toward the capillaries.
3. The femoral artery is a muscular artery.

Page 575

4. Venoconstriction is the contraction of the smooth muscle within the walls of veins. This helps reduce the volume of the venous system, which forces blood into the arterial system in an attempt to maintain normal arterial blood volume.

Page 576

5. This allows for accurate anatomical descriptions as the blood vessel extends from the aorta to the periphery (great toe on the left foot).

Page 576

6. The pulmonary arteries enter the lungs carrying deoxygenated blood, and the pulmonary veins leave the lungs carrying oxygenated blood.

Page 587

7. The carotid arteries supply blood to structures of the head and the neck, including the brain.
8. The right brachial artery is the artery at the biceps region.
9. The external iliac artery gives rise to the femoral artery in the thigh.
10. Damage to the internal carotid arteries does not always result in brain damage because the vertebral arteries also supply blood to the brain.

Page 596

11. The superficial veins are dilated to promote heat loss through the skin.
12. The superior vena cava receives blood from the head, neck, chest, shoulders, and upper limbs.
13. Blood from the intestines contains high amounts of glucose, amino acids, and other nutrients and toxins absorbed from the digestive tract. These are processed by the liver before the blood goes to the general systemic circuit in order to keep the composition of the blood in the body relatively stable.

Page 598

14. The changes in the heart and the major vessels that occur at birth are as follows: (1) The pulmonary vessels expand; (2) the ductus arteriosus contracts, forcing blood to flow through the pulmonary circuit; and (3) the valvular flap closes the foramen ovale.

Page 598

15. Elasticity of the arteries allows these vessels to expand with a sudden increase in pressure. A reduced ability to do so could result in a bulge or tear in the wall of the artery.

Chapter Review

Level 1 Reviewing Facts and Terms

1. c 2. d 3. f 4. j 5. i 6. h 7. a 8. g 9. e 10. b 11. (a) brachiocephalic trunk; (b) brachial artery; (c) radial artery; (d) external iliac artery; (e) anterior tibial artery; (f) right common carotid artery; (g) left subclavian artery; (h) common iliac artery; (i) femoral artery 12. (a) external jugular vein; (b) brachial vein; (c) median cubital vein; (d) radial vein; (e) great saphenous vein; (f) internal jugular vein; (g) superior vena cava; (h) common iliac veins; (i) femoral vein 13. b 14. a 15. d 16. d

Level 2 Reviewing Concepts

1. b
2. a
3. b
4. The aorta, brachiocephalic, pulmonary, common carotid, subclavian, and common iliac arteries are examples of elastic arteries.
5. Sinusoids are found in spleen, liver, bone marrow, and adrenal glands.
6. Arteriovenous anastomoses are direct connections between arterioles and venules.
7. Venous valves in the limbs prevent backflow and aid in the flow of blood back to the heart by compartmentalizing it.
8. The brachiocephalic, left common carotid, and left subclavian are elastic arteries that originate on the aortic arch.

9. The superior vena cava receives blood from the chest, head, neck, shoulders, and upper limbs.
10. Blood can flow directly from the right atrium to the left atrium, bypassing the pulmonary circuit.

Level 3 Critical Thinking

1. Blood channels connect the superficial circulation to the deep circulation draining into the cranial venous sinuses. Infections from the superficial region in the area of the eyes can be transported inward to anterior cranial sinuses, which could lead to meningitis. Therefore, it is important not to spread infection from the superficial to the deep region.
2. In response to the high temperature of the water, John's body shunted more blood to the superficial veins to decrease body temperature. The dilation of the superficial veins caused a shift in blood to the arms and legs and resulted in a decreased venous return. Because of the decreased venous return, the cardiac output decreased and less blood with oxygen was delivered to the brain. This caused John to feel light-headed and faint and nearly caused his demise.
3. In heart failure, the heart is not able to produce enough force to circulate the blood properly. The blood tends to pool in the extremities, and as more and more fluid accumulates in the capillaries, the blood hydrostatic pressure increases and the blood osmotic pressure decreases. The fluid accumulation exceeds the ability of the lymphatics to drain it, and, as a result, edema occurs and produces obvious swelling.

Answers to Clinical Case Wrap-Up

Page 601

1. No. With no capillary bed, there is no opportunity for exchange between blood and interstitial fluids within the area of the AVM. Capillaries are the only blood vessels whose walls permit the exchange of water, gases, or wastes between blood and surrounding interstitial fluids.
2. Yes. AVMs can form anywhere in the body. They are, however, most often found within the brain and spinal cord.

23 | The Lymphatic System

Concept Check

Page 605

1. The lymphatic system produces, maintains, and distributes lymphocytes, which are essential to the defense of the body.
2. Because lymph delivers so much of the body's fluid back to the bloodstream, a break in a major lymphatic vessel could mean a potentially fatal decrease in blood volume.

Page 606

3. The first capillary is probably a lymphatic capillary.

Page 607

4. The five lymphatic trunks are the lumbar trunks, intestinal trunks, bronchomediastinal trunks, subclavian trunks, and jugular trunks.
5. The thoracic duct drains both sides of the body inferior to the diaphragm and the left side of the body superior to the diaphragm. The right lymphatic duct drains the right side of the body superior to the diaphragm.

Page 610

6. T cells are the most common type of lymphocyte and account for about 80 percent of the body's circulating lymphocytes.
7. Some B cells differentiate into memory B cells, which will become activated when the antigen appears again. Therefore, John's body will be able to mount an immune response faster and more effectively, possibly warding off any symptoms of illness.
8. The body must be able to mount a rapid and powerful response to antigens, requiring the production of certain types of lymphocytes at certain times. Without this ability, the body's immune response would be slower and weaker and could fall victim to infection or disease.

Page 612

9. Aggregated lymphoid nodules (*Peyer's patches*) are found in the mucosa of the small intestine.

Page 619

10. Lymph nodes are strategically placed throughout the body in areas susceptible to injury or invasion.
11. The capillaries in the thymus do not allow free exchange between the interstitial fluid and the circulation. If they did, circulating antigens would prematurely stimulate the developing T cells.
12. Lymph nodes enlarge during an infection due to increased numbers of lymphocytes and phagocytes within the active nodes.

Page 619

13. It is important to get the flu vaccine because with advancing age, the lymphatic system becomes less effective at combating disease. T cells become less responsive to antigens; as a result, fewer cytotoxic T cells respond to an infection. Because the number of helper T cells is also reduced, B cells are less responsive, and antibody levels do not rise as quickly after antigen exposure. The net result is an increased susceptibility to viral and bacterial infection.

Chapter Review

Level 1 Reviewing Facts and Terms

1. c 2. j 3. e 4. b 5. f 6. d 7. a 8. i 9. h 10. g 11. (a) tonsil; (b) cervical lymph node; (c) right lymphatic duct; (d) thymus; (e) cisterna chyli; (f) lumbar lymph nodes; (g) appendix; (h) lymphatics of lower limb; (i) lymphatics of upper limb; (j) axillary lymph nodes; (k) thoracic duct; (l) lymphatics of mammary gland; (m) spleen; (n) mucosa-associated lymph tissue (MALT); (o) pelvic lymph nodes; (p) inguinal lymph nodes 12. d 13. d 14. d 15. c 16. a 17. a 18. a 19. b 20. d 21. c

Level 2 Reviewing Concepts

1. c
2. b
3. b
4. The blood thymus barrier prevents premature stimulation of developing T cells by circulating antigens.
5. The splenic artery and the splenic vein pass through the hilum of the spleen.
6. The thoracic duct collects lymph from areas of the body inferior to the diaphragm and from the left side of the body superior to the diaphragm.
7. T cells are the most common type of lymphocyte.
8. Lymphedema is swelling in the tissues as a result of damaged valves in lymphatic vessels or blocked lymphatic vessels.
9. Immature or activated lymphocytes divide to produce additional lymphocytes of the same type.
10. Aggregated lymphoid nodules (*Peyer's patches*) are found in the mucosal lining of the small intestine.

Level 3 Critical Thinking

1. If Tom has previously had the measles, there should be a significant amount of immunoglobulins in his blood shortly after the exposure, the result of an antibody-mediated immune response. If he has not previously had the disease and is in the early stages of a primary response, his blood might show an elevated level of antibodies.
2. In a radical mastectomy, lymph nodes in the nearby axilla and surrounding region are removed along with the cancerous breast to try to prevent the spread of cancer cells by the lymphatic system. Lymphatic vessels from the limb on the affected side are tied off, and because there is no place for the lymph to drain, over time lymphedema causes swelling of the limb.
3. Lymphatic capillaries are found in most regions of the body, and lymphatic capillaries offer little resistance to the passage of cancer cells, which often spread along them, using them as way stations.

Answers to Clinical Case Wrap-Up

Page 623

1. Grace's lymphatic system will be producing more lymphatic stem cells that mature into both T and B cells to fight her Ebola infection.
2. Grace will be immune to further Ebola infection because her lymphatic system has produced memory cells that will immediately recognize and kill any further Ebola viruses that appear.

24 | The Respiratory System

Concept Check

Page 627

1. The main functions of the respiratory system include providing an extensive area for gas exchange between air and circulating blood; moving air to and from the exchange surfaces of the lungs; protecting respiratory surfaces from dehydration, temperature changes, and other environmental variations; defending the respiratory system from pathogens; producing sounds involved in speaking, singing, or nonverbal communication; and helping regulate blood volume, blood pressure, and the control of body fluid pH.
2. The mucociliary escalator is present in the lower regions of the respiratory tract, where the cilia of the respiratory epithelium beat toward the pharynx, cleaning the respiratory passageways.

Page 629

3. The three subdivisions are the nasopharynx, oropharynx, and laryngopharynx.
4. The conchae cause turbulence in the inspired air. This slows air movement and brings the air into contact with the moist, warm walls of the nasal cavity. Turbulent airflow is essential for filtering, humidifying, and warming air and creating currents that bring olfactory stimuli to the olfactory sensory neurons. If the nasal cavity were a tubular passageway with straight walls, turbulence would be minimal.

Page 632

5. The thyroid cartilage protects the glottis and the opening to the trachea.
6. During swallowing, the epiglottis folds over the glottis, preventing food or liquids from entering the respiratory passageways.
7. The pitch of her voice gets lower.
8. The glottis could not close without the intrinsic laryngeal muscles, so food or liquids could enter the respiratory passageways.

Page 632

9. The tracheal cartilages are C-shaped to allow room for esophageal expansion when large masses of food or liquid are swallowed.
10. The trachea has a typical respiratory epithelium, which is pseudostratified, ciliated, columnar epithelial cells.
11. Tracheal cartilages prevent the overexpansion or collapse of the airways during respiration, thereby keeping the airway open and functional.

Page 633

12. The right main bronchus is shorter, has a larger diameter, and extends toward the lung at a steeper angle than the left main bronchus.

Page 640

13. Filtration and humidification are complete by the time air reaches this point, so the need for cilia and mucous glands is eliminated.
14. The surfactant coats the inner surface of each alveolus and helps to reduce surface tension and avoid the collapse of the alveoli.

Page 642

15. Pleural fluid reduces friction and provides lubrication between the parietal and visceral surfaces during breathing.

Page 643

16. No. In order to increase the flow of air into their lungs, athletes should stand upright and use their diaphragm, external intercostals, and accessory respiratory muscles to increase the size of their thoracic cavity as much as possible.

Page 643

17. During a baby's first breath, air is forced into the lungs due to the change in pressure. Fluids are pushed out of the way of the conducting passageways, and the alveoli immediately inflate with air. Pulmonary circulation becomes activated, and this closes the foramen ovale and the ductus arteriosus.

Page 645

18. Stimulation of the apneustic and pneumotaxic centers would result in an increased rate and depth of breathing.

Page 645

19. As your grandfather ages, his lungs do not inflate and deflate as easily due to the deterioration in the elastic tissue of his lungs. In addition, the changes in the costal cartilages and the joints between the ribs and vertebral column make expansion of the thoracic cavity more difficult with increased age. These changes alone make it more difficult for him to catch his breath.

Chapter Review

Level 1 Reviewing Facts and Terms

1. i 2. j 3. a 4. h 5. g 6. d 7. b 8. e 9. c 10. f 11. a 12. d 13. (a) nasal cavity; (b) nasopharynx; (c) larynx; (d) trachea 14. b 15. b 16. d 17. c 18. b 19. b

Level 2 Reviewing Concepts

1. d
2. a
3. c
4. The right lung has three lobes.
5. Bronchodilation is the enlargement of the airway lumen.
6. The alveolar macrophages phagocytize particulate matter.
7. The septa divide the lung into lobules.
8. One group regulates tension in the vocal folds, and the second group opens and closes the glottis.
9. The paired laryngeal cartilages involved with the opening and closing of the glottis are the corniculate and arytenoid cartilages.
10. The laryngopharynx includes the portion lying between the hyoid bone and the entrance to the esophagus.

Level 3 Critical Thinking

1. Asthma occurs when the conducting respiratory passageways are unusually sensitive and irritable, usually as a result of exposure to an antigen in the inspired air. The most important symptoms are edema and swelling of the walls of the passageways; the constriction of the smooth muscles in the walls of the bronchial tree, reducing lumen size; and accelerated production of mucus. Together, these factors reduce the ability of the lungs to function normally in air exchange.
2. Unless the infant was suffocated immediately when it was born, the first breath that it took would start to inflate the lungs, and some of the air would be trapped in the lungs. By placing the lungs in water to see if they float, the medical examiner can determine whether any air is in the lungs. Other measurements and tests could also be used to determine whether the infant had breathed at all (air in the lungs) or was dead at birth (lungs collapsed and a small amount of fluid).

Answers to Clinical Case Wrap-Up

Page 649

1. No. Andrea should not take a cough suppressant because her cough is serving a good purpose: As she expels the irritating mucus, she is also expelling the virus. This is a good thing and should not be suppressed. She should, however, be very careful to use tissues to cover her cough and dispose of the infective mucus.
2. Andrea's nose (runny), pharynx (sore throat), and trachea and bronchi (productive cough) are infected by this virus. Her bronchioles are likely infected as well.

25 | The Digestive System

Concept Check

Page 655

1. The components of the mucosa are (1) the mucosal epithelium (depending on location, it may be simple columnar or stratified squamous); (2) the lamina propria, areolar tissue underlying the epithelium; and (3) the muscularis mucosae, bands of smooth muscle fibers arranged in concentric layers. The mucosa of the digestive tract is an example of a mucous membrane, serving both absorptive and secretory functions.
2. Mesenteries provide an access route for the passage of blood vessels, nerves, and lymphatic vessels to and from the digestive tract. They also stabilize the relative positions of the attached organs.
3. Peristalsis is waves of muscular contractions that move substances the length of the digestive tube. Segmentation activities churn and mix the contents of the small and large intestines but do not produce net movement in a particular direction.
4. The lack of organization in the contractile filaments of smooth muscle cells allows the stretched smooth muscle to adapt to its new shape and still have the ability to contract when necessary.

Page 659

5. The oral cavity is lined by the oral mucosa, which is composed of nonkeratinized stratified squamous epithelium.
6. At mealtime, saliva lubricates the mouth and dissolves chemicals that stimulate the taste buds. Saliva contains the digestive enzyme salivary amylase, which begins the chemical breakdown of carbohydrates.
7. Carbohydrates are broken down by salivary amylase in the mouth.
8. The incisors cut away a section of the apple, which then enters the mouth. The canines tear at the rough skin and pulp of the apple. The apple then moves to the premolars and molars for thorough mashing and grinding before finally being swallowed.

Page 661

9. The six groups of muscles involved in the swallowing process are the superior, middle, and inferior pharyngeal constructors and the palatopharyngeus, stylopharyngeus, and palatal muscles.
10. The process that is being described is deglutition, or swallowing.

Page 663

11. The following blood vessels supply the esophagus with blood: (1) the esophageal arteries, branches of (2) the thyrocervical trunk and (3) external carotid arteries of the neck, (4) the bronchial arteries and (5) esophageal arteries of the mediastinum, (6) the inferior phrenic artery, and (7) the left gastric artery of the abdomen.
12. The mucosa and submucosa are folded because the longitudinal folds of the esophagus allow expansion of the esophagus when a large bolus passes through.

Page 668

13. The greater omentum provides support to the surrounding organs, pads the organs from the surfaces of the abdomen, provides an important energy reserve, and provides insulation.
14. The epithelium produces a carpet of mucus that covers the interior surfaces of the stomach, providing protection against the powerful acids and enzymes. Any cells that do become damaged are quickly replaced.
15. Chief cells secrete pepsinogen. In infants, they also secrete rennin and gastric lipase.
16. Gastrin, produced by G cells, stimulates the secretion of parietal and chief cells.

Page 672

17. The characteristic lining of the small intestine contains plicae circulares, which support intestinal villi. The villi are covered by a simple columnar epithelium whose apical surface is covered by microvilli. This arrangement increases the total area for digestion and absorption.
18. Circular folds line the intestine and greatly increase the surface area available for absorption.
19. Intestinal glands house the stem cells that produce new epithelial cells, which renew the epithelial surface and add intracellular enzymes to the chyme. In addition, intestinal glands contain cells that produce several intestinal hormones.
20. The duodenum acts as a mixing bowl for the chyme entering from the stomach.

Page 675

21. The large intestine is larger in diameter than the small intestine. The thin walls of the large intestine lack villi and have an abundance of goblet cells and intestinal glands.
22. The four regions of the colon are the ascending colon, transverse colon, descending colon, and sigmoid colon.

Page 681

23. The liver acts as a metabolic regulator by extracting nutrients and toxins from the blood before it enters the bloodstream. It also regulates the blood, serving as a blood reservoir, phagocytizing old or damaged RBCs, and synthesizing plasma proteins. Finally, the liver synthesizes and secretes bile.
24. Contraction of the sphincter of ampulla seals off the passageway between the gallbladder and the small intestine and keeps bile from entering the small intestine.
25. The pancreas produces digestive enzymes and buffers (exocrine functions) and hormones (endocrine functions).

Page 682

26. Some general digestive system changes that occur with aging include the following: the effects of cumulative damage becomes more apparent, the division rate of epithelial stem cells declines, smooth muscle tone decreases, and cancer rates increase.

Chapter Review

Level 1 Reviewing Facts and Terms

1. c 2. a 3. d 4. f 5. j 6. g 7. h 8. e 9. i 10. b 11. (a) oral cavity, teeth, tongue; (b) liver; (c) gallbladder; (d) pancreas; (e) large intestine; (f) salivary glands; (g) pharynx; (h) esophagus; (i) stomach; (j) small intestine; (k) anus 12. a 13. c 14. d 15. c 16. d 17. (a) mucosa; (b) submucosa; (c) muscular layer; (d) serosa 18. d 19. a 20. b 21. d

Level 2 Reviewing Concepts

1. c
2. a
3. c
4. Lipase attacks lipids.
5. The hepatopancreatic sphincter seals off the passageway between the gallbladder and the small intestine and prevents bile from entering the small intestine.
6. The gallbladder stores and concentrates bile.
7. Stellate macrophages engulf pathogens, cell debris, and damaged blood cells in the liver.
8. The last region of the colon before the rectum is the sigmoid colon.
9. Lacteals transport materials that could not enter local capillaries. These materials eventually reach the circulation via the thoracic duct.
10. Gastrin release is triggered by food entering the stomach.

Level 3 Critical Thinking

1. During the autopsy, the forensic scientist examined the stomach contents of the murder victim and found the stomach visually empty. As a full meal usually takes more than 4 hours to leave the stomach completely, it was obvious that the victim had not eaten dinner at the time indicated by the murder suspect.
2. The (1) upper esophageal, (2) lower esophageal, (3) pyloric, (4) ileocecal valve, (5) internal anal, and (6) external anal sphincters constrict the lumen of the digestive tract to control movement of materials through it. Only the last, the external anal sphincter, is composed of skeletal muscle with somatic motor innervation and therefore is under conscious, voluntary control.
3. The following six histological layers form the wall of the digestive tract: (1) mucosal epithelium, (2) lamina propria, (3) muscularis mucosae, (4) submucosa, (5) muscular layer, and (6) serosa or adventitia.

Answers to Clinical Case Wrap-Up

Page 686

1. If Tayvian's *Clostridium difficile* colitis was not successfully treated, he would continue to experience bloody diarrhea. This would result in progressive dehydration, abdominal pain, and blood loss that could eventually be fatal.
2. If you could look at the lining of Tayvian's large intestine with a colonoscope, the surface would look inflamed, swollen, and reddened. You would likely see areas of raw bleeding mucosa. You might even see areas of white cells, accumulated in pockets of pus, as Tayvian's immune system tries to fight this *C. difficile* infection.

26 | The Urinary System

Concept Check

Page 697

1. The kidneys will reabsorb more water, release renin, and release erythropoietin.
2. The route blood must take from the renal artery to the glomerulus is as follows: renal artery → segmental arteries → interlobar arteries → arcuate arteries → cortical radiate arteries → afferent arterioles → glomerulus. The route blood must take from the glomerulus to the renal vein is as follows: glomerulus → efferent arterioles → peritubular capillaries/vasa recta → venules → cortical radiate veins → arcuate veins → interlobar veins → renal vein.
3. Filtrate flows from the glomerulus to a minor calyx via the following route: glomerulus → capsular space → proximal convoluted tubule → descending limb of loop → nephron loop → ascending limb of loop → distal convoluted tubule → connecting tubule → collecting duct → papillary duct → minor calyx.
4. Filtration, which is a passive process, allows ions and molecules to pass based on their size. This means that if the pores are big enough to allow the passage of organic wastes, they are also big enough to allow the passage of water, ions, and other important organic molecules. This is why some elements of the filtrate must be actively reabsorbed before the production of urine is complete.
5. The nephron loop absorbs additional water from the tubular fluid and an even higher percentage of the sodium and chloride ions. This results from the high osmotic concentration of its surroundings in the medulla.

Page 702

6. An obstruction of the ureters would interfere with the passage of urine from the renal pelvis to the urinary bladder.
7. The mucosa lining of the urinary bladder has folds, called rugae, that allow the bladder to stretch when it is full.
8. The urinary bladder is held in place by the median umbilical ligament extending from the anterior and superior border and the lateral umbilical ligaments passing along the sides of the bladder.

Page 703

9. As your grandfather has aged, his urinary system has decreased its sensitivity to ADH, thereby requiring more frequent trips to the bathroom. In addition, the sphincter muscles have lost muscle tone and therefore have become less effective at voluntarily retaining urine.

Chapter Review

Level 1 Reviewing Facts and Terms

1. c 2. a 3. i 4. f 5. j 6. b 7. e 8. h 9. d 10. g 11. (a) renal sinus; (b) renal pelvis; (c) hilum; (d) renal papilla; (e) ureter; (f) cortex; (g) medulla; (h) renal pyramid; (i) minor calyx; (j) major calyx; (k) renal lobe; (l) renal column; (m) renal capsule 12. d 13. a 14. c 15. b 16. b 17. d 18. d 19. c 20. c 21. c

Level 2 Reviewing Concepts

1. b
2. b
3. b
4. The glomerulus is contained within the expanded chamber of the nephron (glomerular capsule).
5. The glomerular epithelium consists of large cells (podocytes) with "feet" that wrap around the glomerular capillaries.
6. The juxtaglomerular complex secretes two hormones: renin and erythropoietin.
7. The trigone is the triangular area of the urinary bladder bounded by the ureteral orifices and the entrance to the urethra.
8. The external urethral sphincter is under voluntary control.
9. The primary function of the proximal convoluted tubule is absorption.
10. The rugae in the urinary bladder allow it to expand as it fills with urine.

Level 3 Critical Thinking

1. Since the external urethral sphincter provides the only conscious, voluntary control of micturition, it is physically impossible for a child to restrict urine release by choice until the neural control of the muscle matures. The internal urethral sphincter is not under voluntary, somatic nervous system control. Before age 2, successful toilet training usually amounts to little more than the parents' learning to anticipate the timing of the child's micturition reflex.
2. Strenuous exercise, such as long-distance running, causes sympathetic activation, which stimulates powerful vasoconstriction of the afferent arterioles in the kidneys. Because these vessels deliver blood to the renal corpuscles, their constriction decreases the rate of glomerular filtration. Blood flow to the kidneys declines further as dilated peripheral blood vessels shunt blood away from the kidneys and to skeletal muscle during running. As the passage of blood through the kidney continues to fall due to water loss during exercise, the potential for renal dysfunction increases as a result of the prolonged periods of exercise associated with running marathons.
3. The increased mass of the uterus, which is located superior and posterior to the urinary bladder in females, presses downward on the bladder, increasing the feeling of pressure and triggering the micturition reflex, even when the bladder is only partially full. Also, as a woman comes closer to delivery of her infant, the baby might kick her in the bladder, which would stimulate the micturition reflex.

Answers to Clinical Case Wrap-Up

Page 706

1. If Jack's kidney stone permanently blocked his ureter completely, urine would back up in the kidney and eventually cause renal failure. If Jack's right kidney is healthy, this should be enough to sustain him.
2. Keeping his urine dilute is Jack's best defense against forming more kidney stones. When urine is concentrated, calcium can precipitate out of solution and form more stones.

27 | The Reproductive System

Concept Check

Page 708

1. Gonads produce gametes: sperm in males and oocytes in females.

Page 719

2. The ductus deferens, testicular blood vessels, nerves, and lymphatic vessels make up the body of the spermatic cord.
3. The inguinal canals, which are narrow canals linking the scrotal chambers with the peritoneal cavity, usually close, but the presence of the spermatic cords leaves weak points in the abdominal wall.
4. The temperature inside the peritoneal cavity is too high for the production of sperm, so the testes are located within the scrotal cavity, outside the peritoneal cavity, where temperatures are cooler.
5. The blood testis barrier isolates the inner portions of the seminiferous tubule from the surrounding interstitial fluid. Transport across the nurse cells is tightly regulated to maintain a very stable environment inside the tubule.
6. After becoming detached from the nurse cells, the sperm lies within the lumen of the seminiferous tubule. It will then travel in fluid currents along the straight tubule, through the rete testis, and into the epididymis, where it will remain for about 2 weeks, completing its functional maturation. Upon leaving the epididymis, the sperm will enter the ductus deferens, where it can be stored for several months. The sperm then enters the ejaculatory duct, which enters the urethra for passage out of the body.

Page 733

7. The follicular cells provide the developing oocyte with nutrients and release estrogens. They stimulate the growth of the follicle by secreting follicular fluid. Once ovulation occurs, the remaining follicular cells in the empty follicle create the corpus luteum.
8. Scarring in the uterine tubes can cause infertility by preventing the passage of an oocyte or a sperm.
9. Enlargement of the primary and secondary follicles in the ovary happens at the same time as the proliferative phase of the uterine cycle.
10. Blockage of a single lactiferous sinus would not interfere with milk moving to the nipple because each breast usually has between 15 and 20 lactiferous sinuses.
11. Prolactin, growth hormone, and human placental lactogen stimulate the mammary glands to become mitotically active, producing the gland cells, which are necessary for lactation.

Page 733

12. Menopause is accompanied by a decrease in blood levels of estrogen and progesterone, which results in a weakening of the connective tissue supporting the ovaries, uterus, and vagina, which would contribute to the development of a prolapsed uterus in postmenopausal women.
13. Testosterone replacement therapy is thought to enhance libido (sexual drive) in older men.

Chapter Review

Level 1 Reviewing Facts and Terms

1. i 2. a 3. d 4. f 5. j 6. c 7. b 8. h 9. g 10. e 11. (a) prostatic urethra; (b) spongy urethra; (c) ductus deferens; (d) penis; (e) epididymis; (f) testes; (g) external urethral orifice; (h) scrotum; (i) seminal gland; (j) prostate; (k) ejaculatory duct; (l) bulbo-urethral gland 12. d 13. b 14. b 15. d 16. b 17. c 18. a 19. b 20. d 21. (a) ovary; (b) uterine tube; (c) greater vestibular gland; (d) clitoris; (e) labium minus; (f) labium majus; (g) myometrium; (h) perimetrium; (i) endometrium; (j) uterus; (k) vaginal fornix; (l) cervix; (m) vagina

Level 2 Reviewing Concepts

1. b
2. d
3. d
4. The head of the sperm cell contains a nucleus with the chromosomes.
5. The acrosome contains enzymes involved in the primary steps of fertilization.
6. Seminalplasmin is produced by the prostate; it is an antibiotic that may help prevent urinary tract infections in males.
7. Follicular fluid causes the follicle to enlarge rapidly.
8. During menstruation, the functional layer of the endometrium is shed.
9. The testes normally begin to descend at the seventh developmental month.

Level 3 Critical Thinking

1. Yes, Jerry will still be able to have an erection. Forming an erection is a parasympathetic reflex that is controlled by the sacral region of the spinal cord (inferior to the injury). Tactile stimulation of the penis would initiate the parasympathetic reflex that controls erection. He will also be able to experience an erection by the sympathetic route, since this is controlled in the T_{12}–L_2 area of the cord (superior to the injury). Stimulation by higher centers could produce a decreased sympathetic tone in the vessels to the penis, resulting in an erection.
2. The endometrial cells have receptors for the hormones estrogen and progesterone and respond to them the same as they would if the cells were in the body of the uterus. Under the influence of estrogen, the cells proliferate at the beginning of the menstrual cycle and begin to develop glands and blood vessels, which then further develop under the control of progesterone. This dramatic change in tissue size and characteristics interferes with neighboring tissues by pressing on them or interrupting their functions in other ways. This interference causes periodic painful sensations.
3. Sexually transmitted pathogens can pass through the vagina, uterus, and uterine tubes and directly into the peritoneal cavity of a woman to cause peritonitis; males have no such access route. Because the female urethra is relatively short and opens near the vagina and anus, microbial infections of the urethra (urethritis) and urinary bladder (cystitis) are often caused by normal microbial inhabitants of the perineum that are moved upward through the urethra as an indirect consequence of sexual intercourse. Because the male urethra is longer, urinary tract infection is less likely.

Answers to Clinical Case Wrap-Up

Page 737

1. In addition to the blood vessels and the ilio-inguinal and genitofemoral nerves, testicular torsion also twists the ductus deferens with its layers of fascia and muscle.
2. Yes, the cremasteric reflex would still be intact if Liam were suffering from epididymitis. The genitofemoral nerve would still be working and innervating the cremaster muscle.

28 | THE REPRODUCTIVE SYSTEM Embryology and Human Development

Concept Check

Page 740

1. Prenatal development is the period from fertilization (conception) to delivery. Postnatal development begins at birth and continues to maturity, the state of full development or completed growth.

Page 741

2. The sperm is not ready for fertilization until it undergoes capacitation, which occurs in the female reproductive tract.
3. The oocyte is surrounded by a layer of follicular cells. The acrosome of the sperm contains hyaluronidase to break down the connections between adjacent follicular cells. One or two sperm is not enough, however, because at least dozens of sperm must release hyaluronidase in order for the connections to break down enough to allow fertilization.
4. The oocyte undergoes oocyte activation, which includes a sudden rise in metabolism, a change in the plasma membrane to prevent fertilization by other sperm, and the completion of meiosis.

Page 749

5. The inner cell mass of the blastocyst eventually develops into the embryo.
6. The syncytiotrophoblast erodes a path through the uterine epithelium. There is some digestion of uterine glands, which release nutrients that are absorbed by the syncytiotrophoblast and distributed by diffusion to the inner cell mass.
7. The mesodermal layer gives rise to the skeletal, muscular, endocrine, cardiovascular, lymphatic, urinary, and reproductive systems and the lining of body cavities; it also forms connective tissues.
8. The placenta contains the maternal arteries and single umbilical vein, permits nutrient exchange at the chorionic villi, and synthesizes hormones that are important to the mother and to the embryo.

Page 755

9. The dilation stage, in which the cervix dilates and the fetus slides down the cervical canal, is the longest stage of labor, usually taking 8 or more hours.
10. During the placental stage, the empty uterus contracts and decreases in size, which breaks the connection between the endometrium and the placenta, causing the expulsion of the placenta.

Page 755

11. As wastes build up in the arterial blood, the kidneys excrete them. Glomerular filtration is normal, but the urine cannot be concentrated to any significant degree. As a result, urinary water losses are high, and neonatal fluid requirements are much greater than those of adults.
12. There is no need to worry. Heart rates of 120–140 beats per minute and respiratory rates of 30 breaths per minute in neonates are normal even though they are considerably higher than those of adults.

Chapter Review

Level 1 Reviewing Facts and Terms

1. f 2. g 3. j 4. b 5. h 6. d 7. i 8. c 9. a 10. e 11. (a) parietal decidua; (b) basal decidua; (c) amnion; (d) chorion; (e) capsular decidua; 12. c 13. b 14. a 15. c 16. b 17. c 18. a 19. b 20. a

Level 2 Reviewing Concepts

1. d
2. b
3. b
4. Cleavage is a sequence of cell divisions that occur immediately after fertilization in the first trimester.
5. The primitive streak is the center line of the blastodisc where cells migrate and begin separating into germ layers.
6. Human chorionic gonadotropin is produced in the trophoblast cells shortly after implantation. It signals the corpus luteum to produce more progesterone.
7. Amphimixis is the fusion of the male and female pronuclei.
8. The yolk sac functions in the production of blood cells.
9. Organogenesis is the process of organ formation.
10. The first trimester is critical because events in the first 12 weeks establish the basis for organ formation.

Level 3 Critical Thinking

1. Although what Joe says is technically true—it takes only one sperm to fertilize an oocyte—the probability of this occurring if not enough sperm are deposited is very slim. Of the millions of sperm that enter the female reproductive tract, most are killed or disabled before they reach the uterus. The acid environment, temperature, and presence of immunoglobulins in the vaginal secretions are just a few of the factors responsible for the demise of so many sperm. Many sperm are not capable of making the complete trip. Once they arrive at the oocyte, the sperm must penetrate the corona radiata, and this requires the combined enzyme contributions of dozens or more sperm. If the ejaculate contains few sperm, it is likely that none will reach the oocyte, and fertilization will be impossible.
2. Because processes such as cleavage, gastrulation, and organogenesis occur during the first 3 months of pregnancy, embryonic development can be disrupted during the first trimester of pregnancy. Since each cell of the early embryo may give rise to multiple tissues, the earlier embryonic development is disrupted, the more structures are affected. Contraction of rubella during the first 10 weeks of pregnancy often causes congenital heart defects, eye cataracts, deafness, and mental disabilities in the developing infant. By the end of the first trimester, the basis for the formation of all the major organ systems has been established, so disruptions after this time will more likely affect only specific organs or body systems.

Answers to Clinical Case Wrap-Up

Page 790

1. Implanting the blastocysts on the sixth day after fertilization (ovulation) simulates the schedule in nature. This ensures the blastocysts are available for implantation in the uterine wall by days 7 through 10 after fertilization.
2. Mayu and Ruchira are neither monozygotic nor dizygotic twins. Monozygotic twins are genetically identical and develop from a single fertilized ovum and therefore have the same genetic makeup (DNA). Dizygotic twins develop from two different oocytes from the same biological mother and two different sperm from the same biological father. Mayu and Ruchira are genetic half-sisters. They have the same mother (different oocytes) and different biological fathers. Socially, they are fraternal twins.

Appendices

- Foreign Word Roots, Prefixes, Suffixes, and Combining Forms
- Eponyms in Common Use

Foreign Word Roots, Prefixes, Suffixes, and Combining Forms

Many of the words we use in everyday English have their roots in other languages, particularly Greek and Latin. This is especially true for anatomical terms, many of which were introduced into the anatomical literature by Greek and Roman anatomists. This list includes some of the foreign word roots, prefixes, suffixes, and combining forms that are part of many of the biological and anatomical terms in this text.

a-, *a-*, without: avascular
ab-, *ab*, from: abduct
-ac, *-akos*, pertaining to: cardiac
ad-, *ad*, to, toward: adduct
aden-, adeno-, *adenos*, gland: adenoid
af-, *ad*, toward: afferent
-al, *-alis*, pertaining to: brachial
-algia, *algos*, pain: neuralgia
ana-, *ana*, up, back: anaphase
andro-, *andros*, male: androgen
angio-, *angeion*, vessel: angiogram
anti-, ant-, *anti*, against: antibiotic
apo-, *apo*, from: apocrine
arachn-, *arachne*, spider: arachnoid
arthro-, *arthros*, joint: arthroscopy
-asis, -asia, state, condition: homeostasis
astro-, *aster*, star: astrocyte
atel-, *ateles*, imperfect: atelectasis
baro-, *baros*, pressure: baroreceptor
bi-, *bi-*, two: bifurcate
blast-, -blast, *blastos*, precursor: blastocyst
brachi-, *brachium*, arm: brachiocephalic
brady-, *bradys*, slow: bradycardia
bronch-, *bronchus*, windpipe, airway: bronchial
cardi-, cardio-, -cardia, *kardia*, heart: cardiac
-centesis, *kentesis*, puncture: thoracocentesis
cerebro-, *cerebrum*, brain: cerebrospinal
chole-, *chole*, bile: cholecystitis
chondro-, *chondros*, cartilage: chondrocyte
chrom-, chromo-, *chroma*, color: chromatin
circum-, *circum*, around: circumduction
-clast, *klastos*, broken: osteoclast
coel-, -coel, *koila*, cavity: coelom
contra-, *contra*, against: contralateral
cranio-, *cranium*, skull: craniosacral
cribr-, *cribrum*, sieve: cribriform
-crine, *krinein*, to separate: endocrine
cyst-, -cyst, *kystis*, sac: blastocyst
desmo-, *desmos*, band: desmosome
di-, *dis*, twice: disaccharide
dia-, *dia*, through: diameter
diure-, *diourein*, to urinate: diuresis
dys-, *dys-*, painful: dysmenorrhea
-ectasis, *ektasis*, expansion: atelectasis
ecto-, *ektos*, outside: ectoderm
ef-, *ex*, away from: efferent
emmetro-, *emmetros*, in proper measure: emmetropia
encephalo-, *enkephalos*, brain: encephalitis
end-, endo-, *endos*, inside: endometrium
entero-, *enteron*, intestine: enteric
epi-, *epi*, on: epimysium
erythema-, *erythema*, flushed (skin): erythematosis
erythro-, *erythros*, red: erythrocyte
ex-, *ex*, out, away from: exocytosis
ferr-, *ferrum*, iron: transferrin
-gen, -genic, *gennan*, to produce: mutagen
genicula-, *geniculum*, kneelike structure: geniculate
genio-, *geneion*, chin: geniohyoid
glosso-, -glossus, *glossus*, tongue: hypoglossal
glyco-, *glykys*, sugar: glycogen
-gram, *gramma*, record: myogram
-graph, -graphia, *graphein*, to write, record: electroencephalograph
gyne-, gyno-, *gynaikos*, woman: gynecologist
hem-, hemato-, *haima*, blood: hemopoiesis
hemi-, *hemi-*, half: hemisphere
hepato-, *hepaticus*, liver: hepatocyte
hetero-, *heteros*, other: heterosexual
histo-, *histos*, tissue: histology
holo-, *holos*, entire: holocrine
homeo-, homo-, *homos*, same: homeostasis
hyal-, hyalo-, *hyalos*, glass: hyaline
hydro-, *hydros*, water: hydrolysis
hyo-, *hyoeides*, U-shaped: hyoid
hyper-, *hyper*, above: hyperpolarization
ili-, ilio-, *ilium:* iliac
infra-, *infra*, beneath: infra-orbital
inter-, *inter*, between: interventricular
intra-, *intra*, within: intracapsular
ipsi-, *ipse*, itself: ipsilateral
iso-, *isos*, equal: isotonic
-itis, *-itis*, inflammation: dermatitis
karyo-, *karyon*, body: megakaryocyte
kerato-, *keros*, horn: keratin
kino-, -kinin, *kinein*, to move: bradykinin
lact-, lacto-, -lactin, *lac*, milk: prolactin
-lemma, *lemma*, husk: sarcolemma
leuko-, *leukos*, white: leukocyte
liga-, *ligare*, to bind together: ligase
lip-, lipo-, *lipos*, fat: lipoid
lyso-, -lysis, -lyze, *lysis*, dissolution: hydrolysis
mal-, *mal*, abnormal: malabsorption
mammilla-, *mammilla*, nipple: mammillary
mast-, masto-, *mastos*, breast: mastoid
mega-, *megas*, big: megakaryocyte
mero-, *meros*, part: merocrine
meso-, *mesos*, middle: mesoderm
meta-, *meta*, after, beyond: metaphase
mono-, *monos*, single: monocyte
morpho-, *morphe*, form: morphology
-mural, *murus*, wall: intramural
myelo-, *myelos*, marrow: myeloblast
myo-, *mys*, muscle: myofilament
natri-, *natrium*, sodium: natriuretic
neur-, neuro-, *neuron*, nerve: neuromuscular
oculo-, *oculus*, eye: oculomotor
oligo-, *oligos*, little, few: oligopeptide
-ology, *logos*, the study of: physiology
-oma, *-oma*, swelling: carcinoma
onco-, *onkos*, mass, tumor: oncology
-opia, *ops*, eye: myopia
-osis, *-osis*, state, condition: neurosis
osteon, osteo-, *os*, bone: osteocyte
oto-, *otikos*, ear: otoconia
para-, *para*, a pair: paraplegia
patho-, -path, -pathy, *pathos*, disease: pathology
pedia-, *paidos*, child: pediatrician
peri-, *peri*, around: perineurium
-phasia, *phasis*, speech: aphasia
-phil, -philia, *philus*, love: neutrophil
-phobe, -phobia, *phobos*, fear: claustrophobia
-phylaxis, *phylax*, a guard: prophylaxis
physio-, *physis*, nature: physiology
-plasia, *plasis*, formation: dysplasia
platy-, *platys*, flat: platysma
-plegia, *plege*, a blow, paralysis: hemiplegia
-plexy, *plessein*, to strike: apoplexy
podo-, *podon*, foot: podocyte
-poiesis, *poiesis*, making: hemopoiesis
poly-, *polys*, many: polysaccharide
presby-, *presbys*, old: presbyopia
pro-, *pro*, before: prophase
pterygo-, *pteryx*, wing: pterygoid
pulp-, *pulpa*, pulp: pulpitis
retro-, *retro*, backward: retroperitoneal
-rrhea, *rhein*, flow, discharge: diarrhea
sarco-, *sarkos*, flesh: sarcomere
scler-, sclero-, *skleros*, hard: sclera
semi-, *semis*, half: semitendinosus
-septic, *septikos*, putrid: antiseptic
-sis, state, condition: metastasis
som-, -some, *soma*, body: somatic
spino-, *spina*, spine, vertebral column: spinodeltoid
-stomy, *stoma*, mouth, opening: colostomy
stylo-, *stylus*, stake, pole: stylohyoid
sub-, *sub*, below: subcutaneous
syn-, *syn*, together: synthesis
tachy-, *tachys*, swift: tachycardia
telo-, *telos*, end: telophase
therm-, thermo-, *therme*, heat: thermoregulation
-tomy, *temnein*, to cut: appendectomy
trans-, *trans*, through: transudate
-trophic, -trophin, -trophy, *trophikos*, nourishing: hypertrophy
tropho-, *trophe*, nutrition: trophoblast
tropo-, *tropikos*, turning: troponin
uro-, -uria, *ouron*, urine: glycosuria

Eponyms in Common Use

Table A.1 | Eponyms

Eponym	Equivalent Terms	Individual Referenced
The Cellular Level of Organization (Chapter 2)		
Golgi apparatus	Dictyosome	Camillo Golgi (1844–1926), Italian histologist; shared Nobel Prize in 1906
Krebs cycle	Tricarboxylic or citric acid cycle	Hans Adolph Krebs (1900–1981), British biochemist; shared Nobel Prize in 1953
The Skeletal System (Chapters 5–8)		
Colles fracture		Abraham Colles (1773–1843), Irish surgeon
Haversian canals	Central canals, Osteonic canal	Clopton Havers (1650–1702), English anatomist and microscopist
Haversian systems	Osteons	Clopton Havers (1650–1702), English anatomist and microscopist
Pott's fracture	Bimalleolar fracture	Percivall Pott (1714–1788), English surgeon
Sharpey's fibers	Perforating fibers	William Sharpey (1802–1880), Scottish histologist and physiologist
Volkmann's canals	Perforating canals	Alfred Wilhelm Volkmann (1800–1877), German surgeon
Wormian bones	Sutural bones	Olas Worm (1588–1654), Danish anatomist
The Muscular System (Chapters 9–11)		
Achilles tendon	Calcaneal tendon	Achilles, hero of Greek mythology
Cori cycle	Lactic acid cycle	Carl Ferdinand Cori (1896–1984) and Gerty Theresa Cori (1896–1957), American biochemists; shared Nobel Prize in 1947
The Nervous System (Chapters 13–17)		
Broca's center	Speech center	Pierre Paul Broca (1824–1880), French surgeon
Foramina of Luschka	Lateral foramina	Hubert von Luschka (1820–1875), German anatomist
Foramen of Magendie	Median foramen	François Magendie (1783–1855), French physiologist
Foramen of Munro	Interventricular foramen	John Cummings Munro (1858–1910), American surgeon
Nissl bodies	Chromatophilic substance	Franz Nissl (1860–1919), German neurologist
Purkinje cells	Ganglionic layer of cerebellar cortex	Johannes E. Purkinje (1787–1869), Czechoslovakian physiologist
Nodes of Ranvier	Myelin sheath gap	Louis Antoine Ranvier (1835–1922), French physiologist
Island of Reil	Insula	Johann Christian Reil (1759–1813), German anatomist
Fissure of Rolando	Central sulcus	Luigi Rolando (1773–1831), Italian anatomist
Schwann cells	Neurolemma cells	Theodor Schwann (1810–1882), German anatomist
Aqueduct of Sylvius	Aqueduct of midbrain	Jacobus Sylvius (Jacques Dubois, 1478–1555), French anatomist
Sylvian fissure	Lateral sulcus	Franciscus Sylvius (Franz de la Boë, 1614–1672), Dutch anatomist
Pons Varolii	Pons	Costanzo Varolio (1543–1575), Italian anatomist
Sensory Function (Chapter 18)		
Organ of Corti	Spiral organ	Alfonso Corti (1822–1888), Italian anatomist
Eustachian tube	Auditory tube	Bartolomeo Eustachio (1520–1574), Italian anatomist
Golgi tendon organs	Tendon organs	*See* Golgi apparatus *under* The Cellular Level (Chapter 2)
Hertz (Hz)		Heinrich Hertz (1857–1894), German physicist
Meibomian glands	Tarsal glands	Heinrich Meibom (1638–1700), German anatomist
Meissner's corpuscles	Tactile corpuscles	Georg Meissner (1829–1905), German physiologist
Merkel's discs	Tactile discs	Friedrich Siegismund Merkel (1845–1919), German anatomist
Pacinian corpuscles	Lamellated corpuscles	Filippo Pacini (1812–1883), Italian anatomist
Ruffini corpuscles		Angelo Ruffini (1864–1929), Italian anatomist
Canal of Schlemm	Scleral venous sinus	Friedrich S. Schlemm (1795–1858), German anatomist
Zeis glands		Edward Zeis (1807–1868), German ophthalmologist
The Endocrine System (Chapter 19)		
Islets of Langerhans	Pancreatic islets	Paul Langerhans (1847–1888), German pathologist
Interstitial cells of Leydig	Interstitial cells	Franz von Leydig (1821–1908), German anatomist

Table A.1 | Eponyms (continued)

Eponym	Equivalent Terms	Individual Referenced
The Cardiovascular System (Chapters 20–22)		
Bundle of His	Atrioventricular bundle	Wilhelm His (1863–1934), German physician
Purkinje cells	*See under* The Nervous System (Chapters 13–17)	
Starling's law	Law of the heart	Ernest Henry Starling (1866–1927), English physiologist
Circle of Willis	Cerebral arterial circle	Thomas Willis (1621–1675), English physician
The Lymphatic System (Chapter 23)		
Hassall's corpuscles	Thymic corpuscles	Arthur Hill Hassall (1817–1894), English physician
Kupffer cells	Stellate reticuloendothelial cells	Karl Wilhelm Kupffer (1829–1902), German anatomist
Langerhans cells	Dendritic cells	*See* Islets of Langerhans *under* The Endocrine System (Chapter 19)
Peyer's patches	Aggregated lymphoid nodules	Johann Conrad Peyer (1653–1712), Swiss anatomist
The Respiratory System (Chapter 24)		
Adam's apple	Laryngeal prominence of thyroid cartilage	Biblical reference
Bohr effect		Cristian Bohr (1855–1911), Danish physiologist
Boyle's law		Robert Boyle (1621–1691), English physicist
Charles' law		Jacques Alexandre César Charles (1746–1823), French physicist
Dalton's law		John Dalton (1766–1844), English physicist
Henry's Law		William Henry (1775–1837), English chemist
The Digestive System (Chapter 25)		
Plexus of Auerbach	Myenteric plexus	Leopold Auerbach (1827–1897), German anatomist
Brunner's glands	Duodenal submucosal glands	Johann Conrad Brunner (1653–1727), Swiss anatomist
Crypts of Lieberkuhn	Intestinal crypts (Intestinal glands)	Johann Nathaniel Lieberkuhn (1711–1756), German anatomist
Plexus of Meissner	Submucosal plexus	*See* Meissner's Corpuscles *under* Sensory Function (Chapter 18)
Sphincter of Oddi	Hepatopancreatic sphincter	Ruggero Oddi (1864–1913), Italian physician
Peyer's patches	Aggregated lymphoid nodules	*See under* The Lymphatic System (Chapter 23)
Duct of Santorini	Accessory pancreatic duct	Giovanni Domenico Santorini (1681–1737), Italian anatomist
Stensen's duct	Parotid duct	Niels Stensen (1638–1686), Danish physician/priest
Ampulla of Vater	Duodenal ampulla	Abraham Vater (1684–1751), German anatomist
Wharton's duct	Submandibular duct	Thomas Wharton (1614–1673), English physician
Foramen of Winslow	Epiploic foramen	Jacob Benignus Winslow (1669–1760), French anatomist
Duct of Wirsung	Pancreatic duct	Johann Georg Wirsung (1600–1643), German physician
The Urinary System (Chapter 26)		
Bowman's capsule	Glomerular capsule	Sir William Bowman (1816–1892), English physician
Loop of Henle	Nephron loop	Friedrich Gustav Jakob Henle (1809–1885), German histologist
Littré glands	Urethral glands of male	Alexis Littre (1658–1726), French surgeon
The Reproductive System (Chapters 27–28)		
Bartholin's glands	Greater vestibular glands	Casper Bartholin, Jr. (1655–1738), Danish anatomist
Cowper's glands	Bulbo-urethral glands	William Cowper (1666–1709), English surgeon
Fallopian tube	Uterine tube/oviduct	Gabriele Falloppio (1523–1562), Italian anatomist
Graafian follicle	Tertiary follicle	Reijnier de Graaf (1641–1673), Dutch physician
Sertoli cells	Nurse cells (Sustentacular cells)	Enrico Sertoli (1842–1910), Italian histologist

Glossary

A

A bands: The dark-staining striations in the myofibrils of muscle fibers that make up the regions of overlapping thick and thin filaments.

abdominal cavity: The space bounded by the abdominal walls, diaphragm, and pelvis.

abdominopelvic cavity: The term used to refer to the general region bounded by the abdominal wall and the pelvis; it contains the peritoneal cavity and visceral organs.

abducens nerves (N VI): Cranial nerves, which innervate the lateral rectus muscle of the eye.

abduction: Movement away from the midline.

absorption: The active or passive uptake of gases, fluids, or solutes.

acetabulum: Fossa on lateral aspect of pelvis that accommodates the head of the femur.

acetylcholine (ACh): Chemical neurotransmitter in the brain and peripheral nervous system (PNS); dominant neurotransmitter in the PNS, released at neuromuscular junctions and synapses of the parasympathetic division.

acetylcholinesterase (AChE): Enzyme found in the synaptic cleft, bound to the postsynaptic membrane, and in tissue fluids; breaks down and inactivates acetylcholine (ACh) molecules; also called *cholinesterase*.

acinar: One of the minute grape-shaped secretory portions of an acinar gland.

acquired immunodeficiency syndrome (AIDS): A disease caused by the human immunodeficiency virus (HIV); characterized by the destruction of helper T cells and a resulting severe impairment of the immune response.

acromion: Continuation of the scapular spine that projects superior to the capsule of the shoulder joint.

acrosome: A membranous sac at the tip of a spermatozoon that contains hyaluronidase; also called *acrosomal cap*.

actin: Protein component of microfilaments; forms thin filaments in skeletal muscles and produces contractions of all muscles through interaction with thick (myosin) filaments; *see* **sliding filament theory.**

action potential: A propagated change in the membrane potential of excitable cells, initiated by a change in the membrane permeability to sodium ions: see nerve impulse.

active transport: The ATP-dependent absorption or excretion of solutes across a plasma membrane.

adaptation: A decrease in receptor sensitivity or perception after chronic stimulation; a change in pupillary size in response to changes in light intensity; physiological responses that produce acclimatization.

adduction: Movement toward the axis or midline of the body as viewed in the anatomical position.

adenohypophysis: The anterior lobe of the pituitary gland, also called the *anterior pituitary*.

adipocytes: Fat cells.

adipose tissue: Loose connective tissue dominated by adipocytes.

adrenal cortex: The superficial region of the adrenal gland that produces steroid hormones; also called the suprarenal cortex.

adrenal gland: A small endocrine gland that secretes steroids and catecholamines and is located superior to each kidney; also called *suprarenal gland*.

adrenal medulla: A modified sympathetic ganglion that secretes catecholamines into the blood during sympathetic activation; the core of the adrenal gland; also called *suprarenal medulla*.

adrenocorticotropic hormone (ACTH): Hormone that stimulates the production and secretion of glucocorticoids by the zona fasciculata of the adrenal cortex; released by the anterior pituitary in response to CRF.

adventitia: Superficial layer of connective tissue surrounding an internal organ; fibers are continuous with those of surrounding tissues, providing support and stabilization.

aerobic metabolism: The complete breakdown of organic substrates into carbon dioxide and water, by pyruvate; a process that yields large amounts of ATP but requires mitochondria and oxygen.

afferent arterioles: An arteriole bringing blood to the glomerulus of the kidney.

afferent fibers: Axons that carry sensory information to the central nervous system.

agglutination: Aggregation of red blood cells due to interactions between surface antigens and plasma agglutinins.

aggregated lymphoid nodules: Lymphoid nodules beneath the epithelium of the small intestine. Also called *Peyer's Patches.*

agonist: A muscle responsible for a specific movement; also called a *prime mover*.

albinism: Absence of pigment in hair and skin caused by inability of body to produce melanin.

albumins: The smallest of the plasma proteins; function as transport proteins and important in contributing to plasma oncotic pressure.

aldosterone: A mineralocorticoid produced by the zona glomerulosa of the adrenal cortex; stimulates the kidneys to conserve sodium and water; secreted in response to the presence of angiotensin II.

allantois: One of the extra-embryonic membranes; it provides vascularity to the chorion and is therefore essential to placenta formation; the proximal portion becomes the urinary bladder.

alpha (α) cells: Cells in the pancreatic islets that secrete glucagon.

alpha (α) receptors: Membrane receptors sensitive to norepinephrine or epinephrine; stimulation usually results in excitation of the target cell.

alveolar sacs: Air-filled chambers at the terminal ends of the alveolar ducts that give rise to the alveoli in the lung.

alveolus/alveoli: Terminal pockets at the end of the respiratory tree, lined by a simple squamous epithelium and surrounded by a capillary network; sites of gas exchange with the blood; a bony socket that holds the root of a tooth.

amacrine cells: Modified neurons in the retina that facilitate or inhibit communication between bipolar and ganglion cells.

amnion: One of the extra-embryonic membranes; surrounds the developing embryo/fetus.

amphiarthrosis: Articulation joint that permits a small degree of independent movement.

amphimixis: The fusion of male and female pronuclei following fertilization.

ampulla: A localized dilation in the lumen of a canal or passageway.

anabolism: The synthesis of complex organic compounds from simpler precursors.

anal canal: The distal portion of the rectum that contains the anal columns and ends at the anus.

anal triangle: The posterior subdivision of the perineum.

anaphase: Mitotic stage in which the paired chromatids separate and move toward opposite ends of the spindle apparatus.

anastomosis/anastomoses: The joining of two tubes, usually referring to a connection between two peripheral vessels without an intervening capillary bed.

anatomical position: An anatomical reference position, the body viewed from the anterior surface with the palms facing forward.

anatomical pulleys: Bones or bony structures which change the applied forces of skeletal muscles.

anatomy: The study of internal and external body structures and their physical relationship among other body parts.

androgens: Steroid sex hormones mainly produced by the interstitial endocrine cells of the testis and manufactured in small quantities by the adrenal cortex in both sexes.

anemia: The condition marked by a decrease in the hematocrit, the hemoglobin content of the blood, or both.

aneurysm: A weakening in the arterial wall causing an outpouching or enlargement of an artery.

angina: Chest pain.

angiotensin I: The hormone produced by the activation of angiotensinogen by renin; angiotensin-converting enzyme (ACE) converts angiotensin I into angiotensin II in lung capillaries.

angiotensin II: A hormone that causes an increase in systemic blood pressure, stimulates the secretion of aldosterone, promotes thirst, and causes the release of antidiuretic hormone; angiotensin-converting enzyme (ACE) in lung capillaries converts angiotensin I into angiotensin II.

angiotensin-converting enzyme (ACE): Enzyme necessary for the conversion of angiotensin I to angiotensin II.

angiotensinogen: Blood protein produced by the liver that is converted to angiotensin I by the enzyme renin.

antagonist: A muscle that opposes the movement of an agonist.
anteflexion: Normal position of the uterus, with the superior surface bent forward.
anterior: On or near the front, or ventral, surface of the body.
antibodies: Globular proteins produced by plasma cells that will bind to specific antigens and promote their destruction or removal from the body.
antidiuretic hormone (ADH): Hormone synthesized in the hypothalamus and secreted at the posterior pituitary; causes water retention at the kidneys and an increase in blood pressure.
antrum: A chamber or pocket.
anus: External opening of the anal canal.
aorta: The large, elastic artery that carries blood away from the left ventricle and into the systemic circuit.
aortic sinus: Space between the superior portion of each of the three aortic valve cusps and the dilated portion of the wall of the ascending aorta.
apex: A pointed tip, usually referring to a triangular object and positioned opposite a broad base.
aphasia: Inability to speak.
apocrine secretion: Mode of secretion where the glandular cell sheds portions of its cytoplasm.
aponeurosis/aponeuroses: Broad tendinous sheet(s) that may serve as the origin(s) or insertion(s) of a skeletal muscle.
appendicular skeleton: Bones of the upper limbs and lower limbs including the shoulder and pelvic girdles.
appendix: A closed sac connected to the cecum of the large intestine; also called *vermiform appendix*.
appositional growth: Enlargement by the addition of cartilage or bony matrix to the outer surface.
aqueous humor: Fluid similar to perilymph or cerebrospinal fluid that fills the anterior chamber of the eye.
arachnoid mater: The middle meninx (layer) that encloses cerebrospinal fluid and protects the central nervous system.
arachnoid granulations: Processes of the arachnoid mater that project into the superior sagittal sinus; sites where cerebrospinal fluid enters the venous circulation.
arachnoid mater: The middle meninx (layer) that encloses cerebrospinal fluid and protects the central nervous system.
areola: Pigmented area that surrounds the nipple of a breast.
areolar tissue: Loose connective tissue with an open framework.
arrector pili muscle: Smooth muscle whose contractions force hairs to stand erect.
arteries: Blood vessels that carry blood away from the heart and toward peripheral capillaries.
arterioles: Small arterial branches that deliver blood to a capillary network.
arteriosclerosis: Hardening of the arteries.
arteriovenous anastomoses: A natural communication, direct or indirect, between an artery and a vein.
arthritis: Inflammation of a joint.
articular capsule: Dense collagen fiber sleeve that surrounds a joint and provides protection and stabilization.
articulations: The places of union between two or more skeletal components; also called *joints*.
arytenoid cartilages: A pair of small cartilages in the larynx.
ascending tracts: Tracts carrying information from the spinal cord to the brain.
ascites: Overproduction and accumulation of peritoneal fluid.
association area: Cortical area of the cerebrum that is responsible for the integration of sensory inputs and/or motor commands.
asthma: Reversible constriction of smooth muscles around respiratory passageways, frequently caused by an allergic response.
astrocytes: One of the four types of neuroglia in the central nervous system; responsible for maintaining the blood brain barrier by the stimulation of endothelial cells.
ataxia: Inability to coordinate muscular movement.
atherosclerosis: Formation of fatty plaques in the walls of arteries, leading to circulatory impairment.
atlas: The first cervical vertebra; it articulates with the occipital bone.
atria: Thin-walled chambers of the heart that receive venous blood from the pulmonary or systemic circuits.
atrioventricularnode (AV node): Specialized cardiocytes that relay the contractile stimulus to the bundle of His, the bundle branches, the Purkinje fibers, and the ventricular myocardium; located at the boundary between the atria and ventricles.
atrophy: The wasting away of tissues from a lack of use, ischemia, or nutritional abnormalities.
auditory ossicles: The bones of the middle ear: malleus, incus, and stapes.
auditory tube: A passageway that connects the nasopharynx with the middle ear cavity; also called the *Eustachian tube or pharyngotympanic tube.*
auricle: A broad, flattened process that resembles the external ear; in the ear, the expanded, projecting portion that surrounds the external auditory meatus; also called *pinna*; in the heart, the externally visible flap formed by the collapse of the outer wall of a relaxed atrium.
autologous transfusion: Blood that has been removed, stored, and then infused back into the same individual.
autolysis: Destruction of a cell due to the rupture of lysosomal membranes in its cytoplasm.
autonomic ganglion: A collection of visceral motor neurons outside the central nervous system.
autonomic nervous system (ANS): Centers, nuclei, tracts, ganglia, and nerves involved in the unconscious regulation of visceral functions; includes components of the central nervous system and peripheral nervous system.
autoregulation: Changes in activity that maintain homeostasis in direct response to changes in the local environment; does not require neural or endocrine control.
avascular: Without blood vessels.
axial skeleton: Bones of the head, trunk, vertebral column, and thorax.
axis: The second cervical vertebra.
axolemma: The plasma membrane of an axon, continuous with the plasma membrane of the cell body and dendrites and distinct from any glial cell coverings.
axon: Elongate extension of a neuron that conducts an action potential away from the cell body and toward the synaptic terminals.
axoplasm: Cytoplasm within an axon.
axoplasmic transport: Transport by way of axoplasm flow toward the cell body (retrograde) or toward the axon terminal (anterograde).

B

B cells: Lymphocytes capable of differentiating into plasma cells, which produce antibodies.
baroreceptors: Receptors that have the ability to detect changes in pressure.
basal nuclei: Nuclei of the cerebrum that are involved in the regulation of somatic motor activity at the subconcious level.
basement membrane: A layer of filaments and fibers that attach an epithelium to the underlying connective tissue.
benign: Not malignant.
beta (β) receptors: Membrane receptors sensitive to epinephrine; stimulation may result in excitation or inhibition of the target cell.
beta cells: Cells of the pancreatic islets that secrete insulin in response to increased blood sugar concentrations.
bile: Exocrine secretion of the liver that is stored in the gallbladder and ejected into the duodenum.
bipennate muscle: A muscle whose fibers are arranged on either side of a common tendon.
blastocoele: Fluid-filled cavity within a blastocyst.
blastocyst: Early stage in the developing embryo, consisting of an outer trophoblast and an inner cell mass.
blastodisc: Later stage in the development of the inner cell mass; it includes the cells that will form the embryo.
blood brain barrier (BBB): Isolation of the central nervous system, from the general circulation; primarily the result of astrocyte regulation of capillary permeabilities.
blood testis barrier: Isolation of the seminiferous tubules from the general circulation, due to the activities of the nurse (*Sertoli*) cells.
bolus: A compact mass; usually refers to compacted ingested material (food) on its way to the stomach.
bone: A strong connective tissue containing specialized cells and a mineralized matrix of crystalline calcium phosphate and calcium carbonate; also called *bone tissue* and *osseous tissue*.
brachial plexus: Network formed by branches of spinal nerves C_5–T_1 en route to innervate the upper limb.
brachium: The arm.
bradycardia: An abnormally slow heart rate.
brainstem: The brain minus the cerebrum and cerebellum; midbrain, pons, and medulla oblongata.
bronchoconstriction: The reduction in size of the bronchial passages.
bronchodilation: The dilation of the bronchial passages; can be caused by sympathetic stimulation.
bulbo-urethral glands: Mucous glands at the base of the penis that secrete into the penile urethra; the equivalent of the greater vestibular glands of females; also called *Cowper's glands*.
bursae: Small sacs filled with synovial fluid that cushion adjacent structures and reduce friction.
bursitis: Painful inflammation of bursa(e).

C

calcaneal tendon: The large tendon that inserts on the calcaneus; tension on this tendon produces extension (plantar flexion) of the foot; also called *Achilles tendon*.

calcaneus: The heelbone, the largest of the tarsal bones.

calcification: The deposition of calcium salts within a tissue.

calcitonin (CT): The hormone secreted by C cells of the thyroid when calcium ion concentrations are abnormally high; restores homeostasis by increasing the rate of bone deposition and the rate of calcium loss by the kidneys.

callus: A localized thickening of the epidermis due to chronic physical stresses; a thickened area that forms at the site of a bone break as part of the repair process.

calvaria: The skullcap, formed of the frontal, parietal, and occipital bones.

canaliculi: Microscopic passageways between cells; bile canaliculi carry bile to bile ducts in the liver; in bone, canaliculi permit the diffusion of nutrients and wastes to and from osteocytes.

cancer: A malignant tumor that tends to undergo metastasis.

capacitation: Activation process that must occur before a sperm can successfully fertilize an oocyte; occurs in the vagina following ejaculation.

capillaries: Small blood vessels, located between arterioles and venules, whose thin walls permit the diffusion of gases, nutrients, and wastes between the plasma and interstitial fluids.

capitulum: General term for a small, elevated articular process; used to refer to the rounded distal surface of the humerus that articulates with the radial head.

cardia: The region of the stomach surrounding its connection with the esophagus.

cardiac cycle: One complete heartbeat, including atrial and ventricular systole and diastole.

cardiac muscle: Involuntary, striated muscle making up the myocardium of the heart.

cardiovascular centers: Poorly localized centers in the reticular formation of the medulla of the brain; includes cardioacceleratory, cardioinhibitory, and vasomotor centers.

cardiovascular system: Body system pertaining to the heart, blood, and blood vessels.

carotene: A yellow-orange pigment found in carrots and in green and orange leafy vegetables; a compound that the body can convert to vitamin A.

carotid sinus: A dilated segment of the internal carotid artery whose walls contain baroreceptors sensitive to changes in blood pressure.

cartilage: A connective tissue with a gelatinous matrix and an abundance of fibers.

catabolism: The breakdown of complex organic molecules into simpler components, accompanied by the release of energy.

catecholamines: Epinephrine, norepinephrine, dopamine, and related compounds.

cauda equina: Bundle of spinal nerve roots arising from the lumbosacral enlargement and medullary cone of the adult spinal cord; they extend caudally inside the vertebral canal en route to lumbar and sacral segments.

caudate nucleus: One of the basal nuclei involved with the subconscious control of skeletal muscular activity.

cell body: The body of a neuron; also called *soma*.

cells: The smallest living structural units in the human body.

central canal: Longitudinal canal in the center of an osteon that contains blood vessels and nerves, also called the *Haversian canal*; a passageway along the longitudinal axis of the spinal cord that contains cerebrospinal fluid.

central nervous system (CNS): The brain and spinal cord.

central sulcus: Groove in the surface of a cerebral hemisphere, between the primary sensory and primary motor areas of the cortex.

centriole: A cylindrical intracellular organelle composed of nine groups of microtubules, three in each group; functions in mitosis or meiosis by forming the basis of the spindle apparatus.

centrosome: Region of cytoplasm containing a pair of centrioles oriented at right angles to one another.

cerebellum: Posterior portion of the metencephalon, containing the cerebellar hemispheres; includes the arbor vitae, cerebellar nuclei, and cerebellar cortex.

cerebral hemispheres: Expanded portions of the cerebrum covered in neural cortex.

cerebrospinal fluid (CSF): Fluid bathing the internal and external surfaces of the central nervous system; secreted by the choroid plexus.

cerebrovascular accident (CVA): The occlusion of a blood vessel that supplies a portion of the brain, resulting in damage to the dependent neurons; also called *stroke*.

cerebrum: The largest portion of the brain, composed of the cerebral hemispheres; includes the cerebral cortex, the basal nuclei, and the internal capsule.

cerumen: Waxy secretion of ceruminous glands along the external acoustic meatus.

ceruminous glands: Integumentary glands that secrete cerumen.

cervical enlargement: Relative enlargement of the cervical portion of the spinal cord due to the abundance of central nervous system neurons involved with motor control of the arms.

cervical plexus: Network of nerves arising from the anterior rami of the first four cervical spinal nerves.

cervix: The inferior part of the uterus.

cesarean section: Surgical delivery of an infant via an incision through the lower abdominal wall and uterus.

chemoreceptors: Receptors that have the ability to detect changes in the concentrations of dissolved compounds or gases, mainly CO_2 and O_2.

chemotaxis: The attraction of phagocytic cells to the source of abnormal chemicals in tissue fluids.

chief cells: Gastric cells near the base of gastric glands that secrete pepsinogen.

cholecystokinin: Duodenal hormone that stimulates the contraction of the gallbladder and the secretion of enzymes by the exocrine pancreas; also called *pancreozymin*.

chondrocytes: Cartilage cells.

chondroitin sulfates: The predominant proteoglycans in cartilage, responsible for the gelatinous consistency of the matrix.

chordae tendineae: Fibrous cords that stabilize the position of the AV valves in the heart, preventing backflow during ventricular systole (contraction).

chorion: An extra-embryonic membrane, consisting of the trophoblast and underlying mesoderm, that forms the placenta.

choroid plexus: The vascular complex in the roof of the third and fourth ventricles of the brain, responsible for the production of cerebrospinal fluid.

choroid: Middle, vascular layer in the wall of the eye.

chromatin: Histological term referring to the grainy material visible in cell nuclei during interphase; the appearance of the DNA content of the nucleus when the chromosomes are uncoiled.

chromosome: Dense structures, composed of tightly coiled DNA strands and associated histones, that become visible in the nucleus when a cell prepares to undergo mitosis or meiosis; normal human somatic cells contain 46 chromosomes apiece.

chyme: A semifluid mixture of ingested food and digestive secretions that is found in the stomach as digestion proceeds.

ciliary body: A thickened region of the choroid that encircles the lens of the eye; it includes the ciliary muscle and the ciliary processes that support the ciliary zonule of the lens.

ciliary zonule: Series of fibers connecting the ciliary body with the lens of the eye.

cilium/cilia: Slender organelle/organelles that extends above the free surface of an epithelial cell. Multiple *motile cilia* generally undergo cycles of movement and are composed of a basal body and microtubules in a 9 + 2 array; a nonmotile, solitary *primary cilium* acts as an environmental sensor, and has microtubules arranged in a 9 + 0 array.

circumduction: A movement at a synovial joint in which the distal end of the bone moves in a circular direction, but the shaft does not rotate.

clavicle: The collarbone.

clitoris: A small erectile organ of the female that is the developmental equivalent of the male penis.

coccygeal ligament: Fibrous extension of the dura mater and filum terminale; provides longitudinal stabilization to the spinal cord.

coccyx: Terminal portion of the vertebral column, consisting of relatively tiny, fused vertebrae.

cochlea: Spiral portion of the bony labyrinth of the internal ear that surrounds the organ of hearing.

cochlear duct: The central membranous tube within the cochlea that is filled with endolymph and contains the spiral organ (*organ of Corti*); also called *scala media*.

collagen fibers: Strong, insoluble protein fibers common in connective tissues.

collateral ganglia: Sympathetic ganglia located anterior to the spinal column and separate from the sympathetic chain.

colon: The large intestine from the cecum to the rectum.

compact bone: Dense bone containing parallel osteons.

compound: A molecule containing two or more elements in combination.

concussion: A violent hit to the head with resulting loss of consciousness.

condyle: A rounded articular projection on the surface of a bone.

cones: Photoreceptosr responsible for sharp vision and color vision.

conjunctiva: A layer of stratified squamous epithelium that covers the inner surfaces of the eyelids and the anterior surface of the eye to the edges of the cornea.

conus medullaris: Conical tip of the spinal cord that gives rise to the filum terminale.
convergence: In the nervous system, the innervation of a single neuron by the axons from several neurons; this is most common along motor pathways.
coracoid process: A hook-shaped process of the scapula that projects above the anterior surface of the capsule of the shoulder joint.
cornea: Transparent portion of the fibrous layer of the anterior surface of the eye.
corneoscleral junction: Margin of the cornea that is overlapped by the sclera.
corniculate cartilages: A pair of small laryngeal cartilages.
corona radiata: A layer of follicle cells surrounding an oocyte at ovulation.
corpora quadrigemina: The superior and inferior colliculi of the mesencephalic tectum (roof) in the brain.
corpus albicans: The scar tissue that remains after degeneration of the corpus luteum at the end of a uterine cycle.
corpus callosum: Bundle of axons linking centers in the left and right cerebral hemispheres.
corpus luteum: The progesterone-secreting mass of follicle cells that develops in the ovary after ovulation.
corpus spongiosum: Mass of erectile tissue that surrounds the urethra in the penis and expands distally to form the glans.
cortex: Outer layer or region of an organ or bone.
corticobulbar tracts: Descending tracts that carry information/commands from the cerebral cortex to nuclei and centers in the brainstem.
corticospinal tracts: Descending tracts that carry motor commands from the cerebral cortex to the anterior gray horns of the spinal cord.
corticosteroids: Steroid hormones produced by the adrenal (suprarenal) cortex.
corticosterone: A corticosteroid secreted by the zona fasciculata of the adrenal (suprarenal) cortex; a glucocorticoid.
cortisol: One of the corticosteroids secreted by the zona fasciculata of the adrenal cortex; a glucocorticoid.
cranium: The braincase; the skull bones that surround the brain.
cribriform plate: Portion of the ethmoid of the skull that contains the foramina used by the axons of olfactory receptors en route to the olfactory bulbs of the cerebrum.
cricoid cartilage: Ring-shaped cartilage that forms the inferior margin of the larynx.
cricoid cartilages: Ring-shaped cartilages that form the inferior margin of the larynx.
crista/cristae: Ridge-shaped collection(s) of hair cells in the ampulla of a semicircular duct; the crista and cupula form a receptor complex sensitive to movement along the plane of the semicircular canal.
cross-bridges: The binding of myosin heads that project from the surface of thick filaments at the active sites of thin filaments in the presence of calcium ions.
cross-sectional anatomy: Referring to planar sections of an anatomic structure.
cuneiform cartilages: A pair of small cartilages in the larynx.
cutaneous membrane: The epidermis and papillary layer of the dermis.
cuticle: Layer of dead, keratinized cells surrounding the shaft of a hair; for nails, *see* **eponychium.**
cyanosis: Abnormal bluish coloration of the skin due to the presence of deoxygenated blood in vessels near the body surface.
cystic duct: A duct that carries bile between the gallbladder and the common bile duct.
cystitis: Inflammation of the urinary bladder.
cytology: The study of cells.
cytoplasm: The material between the plasma membrane and the nuclear membrane.
cytoskeleton: A network of microtubules and microfilaments in the cytoplasm.
cytosol: The fluid portion of the cytoplasm.
cytotoxic T cells: Lymphocytes involved in cell-mediated immunity that kill target cells by direct contact or by the secretion of lymphotoxins; also called *killer T cells* and *T_C cells*.

D

decussate: To cross over to the opposite side, usually referring to the crossover of the descending tracts of the corticospinal pathway on the ventral surface of the medulla oblongata.
defecation: The elimination of fecal wastes.
deglutition: Swallowing.
dendrites: Sensory processes of a neuron.
dendritic spines: Distinct outgrowths of nerve cell dendrites that are the sites of synaptic axodendritic contact; they are absent in motor neurons and numerous in pyramidal cells of the cerebral cortex and Purkinje cells of the cerebellar cortex.
denticulate ligaments: Supporting fibers that extend laterally from the surface of the spinal cord, tying the pia mater to the dura mater and providing lateral support for the spinal cord.
depression: Inferior (downward) movement of a body part.
dermal ridges: Surface ridges on the palms of the hands and the soles of the feet where the sweat pores open.
dermatome: A sensory region monitored by the posterior rami of a single spinal segment.
dermis: The connective tissue layer beneath the epidermis of the skin.
detrusor: Collectively, the three layers of smooth muscle in the wall of the urinary bladder.
development: Growth and the acquisition of increasing structural and functional complexity; includes the period from conception to maturity.
diabetes insipidus: Disorder of the pituitary gland characterized by polyuria (excessive urination) and polydipsia (excessive thirst) that results from inadequate production of antidiuretic hormone (ADH).
diabetes mellitus: Disorder characterized by polyuria (excessive urination) and glycosuria (glucose in the urine), most commonly due to the inadequate production or diminished sensitivity to insulin with a resulting increase of blood glucose levels.
diapedesis: Movement of white blood cells through the walls of blood vessels by migration between adjacent endothelial cells.
diaphragm: Any muscular partition; often used to refer to the respiratory muscle that separates the thoracic cavity from the abdominopelvic cavity.
diaphragma sellae: The fold of dura mater extending across the sella turcica and hypophyseal fossa.
diaphragmatic breathing: Inhalation or exhalation using only the diaphragm.
diaphysis: The shaft of a long bone.
diarthrosis: A synovial joint.
diencephalon: A division of the brain that includes the epithalamus, thalamus, and hypothalamus.
differential count: An estimate of the number of each type of white blood cell on the basis of a random sampling of 100 white blood cells.
differentiation: The gradual appearance of characteristic cellular specializations during development as the result of gene activation or repression.
diffusion: Passive molecular movement from an area of relatively high concentration to an area of relatively low concentration.
digestion: The chemical breakdown of ingested food into simple molecules that can be absorbed by the cells of the digestive tract.
digestive tract: An internal passageway that begins at the mouth and ends at the anus.
disease: A disorder of structure or function within a body system or organ.
distal convoluted tubule (DCT): Segment of the nephron closest to the collecting tubule and duct; an important site of active secretion.
divergence: In nervous tissue, the spread of excitation from one neuron to many neurons; an organizational pattern common along sensory pathways of the central nervous system.
dorsiflexion: Upward movement of the foot through flexion at the ankle.
ductus arteriosus: Vascular connection between the pulmonary trunk and the aorta that functions throughout fetal life; normally closes at birth or shortly thereafter, and persists as the ligamentum arteriosum.
ductus deferens: A passageway that carries sperm from the epididymis to the ejaculatory duct; also called the *vas deferens.*
duodenal ampulla: Chamber that receives bile from the common bile duct and pancreatic secretions from the pancreatic duct.
duodenal papilla: Conical projection from the inner surface of the duodenum that contains the opening of the duodenal ampulla.
duodenum: The proximal 25 cm (9.8 in.) of the small intestine that contains short villi and submucosal glands.
dura mater: The outermost membrane (meninx) of the cranial and spinal meninges.
dyslexia: Impaired ability to comprehend written words.
dysmenorrhea: Painful menstruation.

E

eccrine sweat glands: Sweat glands of the skin that produce a watery secretion.
ectoderm: One of the three primary germ layers; covers the surface of the embryo and gives rise to the nervous system, the epidermis and associated glands, and a variety of other structures.
effector: A peripheral gland or muscle cell innervated by a motor neuron.
efferent arteriole: An arteriole carrying blood away from the glomerulus of the kidney.
efferent fiber: An axon that carries impulses away from the central nervous system.

ejaculation: The ejection of semen from the penis as the result of muscular contractions of the bulbospongiosus and ischiocavernosus muscles.

ejaculatory duct: Short duct that passes within the walls of the prostate and connect the ductus deferens with the prostatic urethra.

elastin: Connective tissue fibers that stretch and rebound, providing elasticity to connective tissues.

elevation: Movement in a superior, or upward, direction.

embryo: The developmental stage beginning at fertilization and ending at the start of the third developmental month.

embryogenesis: The process of embryo formation.

embryology: The study of embryonic development, focusing on the first two months after fertilization.

enamel: Crystalline material similar in mineral composition to bone, but harder and without osteocytes, that covers the exposed surfaces of the teeth.

endocardium: The simple squamous epithelium that lines the heart and is continuous with the endothelium of the great vessels.

endochondral ossification: The conversion of a cartilaginous model to bone; the characteristic mode of formation for skeletal elements other than the bones of the cranium, the clavicles, and sesamoid bones.

endocrine glands: Gland that secretes hormones into the blood.

endocrine system: The endocrine (ductless) glands and organs of the body.

endocytosis: The movement of relatively large volumes of extracellular material into the cytoplasm via the formation of a membranous vesicle at the cell surface; includes pinocytosis and phagocytosis.

endoderm: One of the three primary germ layers; the layer on the undersurface of the embryonic disc that gives rise to the epithelia and glands of the digestive system, the respiratory system, and portions of the urinary system.

endolymph: Fluid contents of the membranous labyrinth (the saccule, utricle, semicircular ducts, and cochlear duct) of the internal ear.

endometrium: The mucous membrane lining the uterus.

endomysium: A delicate network of connective tissue fibers that surrounds individual muscle cells.

endoneurium: A delicate network of connective tissue fibers that surrounds individual nerve fibers.

endoplasmic reticulum: A network of membranous channels in the cytoplasm of a cell that functions in intracellular transport, synthesis, storage, packaging, and secretion.

endosteum: An incomplete cellular lining found on the inner (medullary) surfaces of bones.

endothelium: The simple squamous epithelium that lines blood and lymphatic vessels.

enteric nervous system (ENS): One of the three main divisions of the nervous system; it is an extensive nerve network in the walls of the digestive tract that initiates and coordinates digestive motility and secretions.

enteroendocrine cells: Endocrine cells scattered among the epithelial cells lining the digestive tract.

eosinophils: A granulocyte (WBC) with a lobed nucleus and red-staining granules; participates in the immune response and is especially important during allergic reactions.

ependyma: Layer of cells lining the ventricles and central canal of the central nervous system.

epiblast: The layer of the inner cell mass facing the amniotic cavity prior to gastrulation.

epicardium: Serous membrane covering the outer surface of the heart; also called the *visceral pericardium*.

epidermis: The epithelium covering the surface of the skin.

epididymis: Coiled duct that connects the rete testis to the ductus deferens; site of functional maturation of sperm.

epidural space: Space between the spinal dura mater and the walls of the vertebral foramen; contains blood vessels and adipose tissue; a frequent site of injection for regional anesthesia.

epiglottis: Blade-shaped flap of tissue, reinforced by cartilage, that is attached to the posterior and superior surface of the thyroid cartilage; it folds over the entrance to the larynx during swallowing.

epimysium: A dense layer of collagen fibers that surrounds a skeletal muscle, and is continuous with the tendons/aponeuroses of the muscle and with the perimysium.

epineurium: A dense investment of collagen fibers that surrounds a peripheral nerve.

epiphyses: The heads of long bones.

episiotomy: Surgical incision of the vulva to prevent laceration during delivery.

epithalamus: Small, posterior, medial area of the thalamus.

epithelium/epithelia: One of the four primary tissue types; a layer of cells that forms a superficial covering or an internal lining of a body cavity or vessel.

eponychium: A narrow zone of stratum corneum that extends across the surface of a nail at its exposed base; also called the **cuticle.**

erection: Stiffening of the penis prior to copulation due to the engorgement of the erectile tissues of the corpora cavernosa and the corpus spongiosum.

erythrocytes: Red blood cells (RBCs); have no nucleus and contain large quantities of hemoglobin.

erythropoiesis: Red blood cell formation.

erythropoietin (EPO): A hormone released by most tissues, and especially by the kidneys, when oxygen levels decrease; stimulates erythropoiesis (red blood cell formation) in red bone marrow.

esophagitis: Inflammation of the esophagus.

esophagus: A muscular tube that connects the pharynx to the stomach.

estradiol: The primary estrogen secreted by ovarian follicles.

estrogens: A class of steroid sex hormones that includes estradiol.

eversion: A turning outward.

exchange transfusion: Removal of most of a patient's blood followed by replacement of an equal amount of blood from donors.

excretion: The removal of wastes from the blood, tissues, or organs.

exocrine glands: Glands that secrete onto the body surface or into a passageway connected to the exterior.

exocytosis: The ejection of cytoplasmic materials by fusion of a membranous vesicle with the plasmamembrane.

extension: An increase in the angle between two articulating bones; the opposite of flexion.

extensor retinaculum: A thickening of the fascia of the forearm at the wrist or the leg at the ankle, forming a band of dense connective tissue that holds extensor muscle tendons in place.

external acoustic meatus: Passageway in the temporal bone that leads to the tympanic membrane of the middle ear.

external ear: The auricle, external acoustic meatus, and tympanic membrane.

exteroceptors: Sensory receptors in the skin, mucous membranes, and special sense organs that provide information about the external environment and our position within it.

extracellular fluid: All body fluid other than that contained within cells; includes plasma and interstitial fluid.

extra-embryonic membranes: The yolk sac, amnion, chorion, and allantois.

eyelids: Movable folds that cover the front of the eyeballs; also called *palpebrae*.

F

facilitated diffusion: Passive movement of a substance across a plasma membrane via a protein carrier.

falciform ligament: A sheet of mesentery that contains the ligamentum teres, the fibrous remains of the umbilical vein of the fetus.

falx cerebelli: Curving sheet of dura mater that projects forward from the internal occipital crest inferior to the tentorium.

falx cerebri: Curving sheet of dura mater that extends between the two cerebral hemispheres; encloses the superior sagittal sinus.

fascia: Connective tissue fibers, primarily collagenous, that form sheets or bands beneath the skin to attach, stabilize, enclose, and separate muscles and other internal organs.

fasciculus: A small bundle; usually refers to a collection of nerve axons or muscle fibers.

fauces: The passage from the mouth to the pharynx, bounded by the palatal arches, the soft palate, and the uvula.

feces: Waste products eliminated by the digestive tract at the anus; contain indigestible residue, bacteria, mucus, and epithelial cells.

femur: The long bone of the thigh.

fertilization: Fusion of a secondary oocyte and sperm to form a zygote.

fibrinogen: A plasma protein that is the soluble precursor of the insoluble protein fibrin.

fibroblasts: Cells of connective tissue proper that are responsible for the production of extracellular fibers and the secretion of the organic compounds of the extracellular matrix.

fibrocytes: Cells of connective tissue proper that are responsible for the maintenance of the extracellular fibers and ground substance of the extracellular matrix.

fibrous cartilage: Cartilage containing an abundance of collagen fibers; found around the edges of joints, in the intervertebral discs, the menisci of the knee, etc. Also termed *fibrocartilage*.

fibrous layer: The outermost layer of the eye, composed of the sclera and cornea; also called *fibrous tunic*.
fibula: The lateral, relatively small bone of the lower leg.
filiform papillae: Slender conical projections from the dorsal surface of the anterior two-thirds of the tongue.
filum terminale: A fibrous extension of the spinal cord that extends from the conus medullaris to the coccygeal ligament.
fimbriae: A fringes; used to describe the fingerlike processes that surround the entrance to the uterine tube.
flagellum/flagella: An organelle that is structurally similar to a motile cilium but is used to propel a cell through a fluid; found on sperm.
flexion: A movement at a joint that reduces the angle between two articulating bones; the opposite of extension.
follicles: Small secretory sacs or glands.
follicle-stimulating hormone (FSH): A hormone secreted by the anterior lobe of the pituitary; stimulates oogenesis (female) and spermatogenesis (male).
folliculitis: Inflammation of a follicle, such as a hair follicle of the skin.
fontanelles: Relatively soft, flexible, fibrous regions between two flat bones in the developing skull; also spelled *fontanels*.
fornix: An arch or the space bounded by an arch; in the brain, an arching tract that connects the hippocampus with the mammillary bodies; in the eye, a slender pocket located where the epithelium of the ocular conjunctiva folds back on itself as the palpebral conjunctiva; in the vagina, the shallow recess surrounding the protrusion of the cervix.
fourth ventricle: An elongated ventricle of the metencephalon (pons and cerebellum) and the myelencephalon (medulla oblongata) of the brain; the roof contains a region of choroid plexus.
fovea centralis: Portion of the retina providing the sharpest vision, with the highest concentration of cones; also called the *fovea*.
fracture: A break or crack in a bone.
frontal (coronal) plane: A vertical plane that divides the body into anterior and posterior portions.
fundus: The base of an organ such as the stomach, uterus, or gallbladder.
furuncle: A boil, resulting from the invasion and inflammation of a hair follicle or sebaceous gland.

G

gallbladder: The pear-shaped reservoir for bile after it is secreted by the liver.
gametes: Reproductive cells (sperm or oocytes) that contain half the normal chromosome complement.
ganglion/ganglia: Collection of neuron cell bodies in the peripheral nervous system.
gap junction: Connection between cells that permits electrical coupling.
gastric glands: Tubular glands of the stomach whose cells produce acid, enzymes, intrinsic factor, and hormones.
gastrin: Hormone produced by enteroendocrine cells of the stomach, when exposed to mechanical stimuli or vagal stimulation, and the duodenum, when exposed to chyme containing undigested proteins.
gastritis: Inflammation of the stomach.
gastrulation: The movement of cells of the inner cell mass that creates the three primary germ layers of the embryo.
germinal center: Pale region in the interior of lymphoid tissues or lymphoid nodules, where cell divisions occur that produce additional lymphocytes.
gestation: The period of intra-uterine development.
gingivae: The gums.
glenohumeral joint: Ball-and-socket join between the head of the humerus and the glenoid cavity if the scapula; also called the *shoulder joint*.
glenoid cavity: A rounded depression that forms the articular surface of the scapula at the shoulder joint.
globulins: Globular plasma proteins with a variety of important functions.
glomerular capsule: The expanded initial portion of the nephron that surrounds the glomerulus; also called *Bowman's capsule*.
glossopharyngeal nerves (N IX): Cranial nerves that provide sensations to the pharynx and posterior third of the tongue; they also carry motor fibers to the stylopharyngeus muscle.
glottis: Structure within the larynx that consists of the vocal folds and the rima glottidis.
glucagon: Hormone secreted by the alpha cells of the pancreatic islets; increases blood glucose concentrations.
glucocorticoids (GCs): Hormones secreted by the zona fasciculata of the adrenal (suprarenal) cortex to modify glucose metabolism; cortisol and corticosterone are important examples.
glycolysis: The anaerobic cytosolic breakdown of glucose into two 3-carbon molecules of pyruvate, with a net gain of 2 ATP molecules.
goblet cells: Goblet-shaped, mucus-producing, unicellular glands in certain epithelia of the digestive and respiratory tracts; they are called goblet cells in the mucosa of the small intestine, large intestine, terminal bronchioles, and conjunctiva; they are called mucous cells when they are found in the stomach mucosa, respiratory mucosa, and salivary glands.
Golgi apparatus: Cellular organelle consisting of a series of membranous plates that give rise to lysosomes and secretory vesicles.
gomphosis: A fibrous synarthrosis that binds a tooth to the bone of the jaw; *see* **periodontal ligament.**
gonadotropin-releasing hormone (GnRH): A hypothalamic releasing hormone that causes the secretion of both follicle-stimulating hormone and luteinizing hormone by the anterior lobe of the pituitary gland (adenohypophysis).
gonadotropins: Hormones that stimulate the gonads (testes or ovaries).
gonads: Reproductive organs that produce gametes and sex hormones
gracile fasciculus: The smaller, medial subdivision of the posterior funiculus.
gracile nucleus: The medial of the three nuclei of the dorsal column.
granular leukocytes: White blood cells containing granules that are visible with the light microscope; includes eosinophils, basophils, and neutrophils; also called *granulocytes*.
gray matter: Areas in the central nervous system dominated by nerve cell bodies, glial cells, and unmyelinated axons.
gray ramus communicans: A bundle of postganglionic sympathetic nerve fibers distributed to effectors in the body wall, skin, and limbs by way of a spinal nerve.
greater omentum: A large fold of the dorsal mesentery of the stomach that hangs in front of the intestines.
greater vestibular glands: Mucous glands in the vaginal walls that secrete into the vestibule; the equivalent of the bulbo-urethral glands of the male.
gross anatomy: The study of the structural features of the human body without the aid of a microscope.
gustation: The sense of taste.
gyri: Prominent folds or ridges of neural cortex on the surfaces of the cerebral hemispheres.

H

H band: The paler area in the center of the A band in striated muscle.
hair cells: Sensory cells of the internal ear.
hair follicles: Accessory structures of the integument; tubes lined by a stratified squamous epithelium that begin at the surface of the skin and end at the hair papilla.
hair root: A thickened, conical structure consisting of a connective tissue papilla and the overlying matrix; a layer of epithelial cells that produces the hair shaft.
hair: A keratinous strand produced by epithelial cells of the hair follicle.
hallux: The great toe.
haploid: Possessing half the normal number of chromosomes; a characteristic of gametes.
hard palate: The bony roof of the oral cavity, formed by the maxillae and palatine bones.
haustra: Saclike pouches along the length of the large intestine that result from tension in the teniae coli.
Heimlich maneuver: A technique for removing an airway blockage by external compression of the abdomen and forceful elevation of the diaphragm.
helper T cells: Lymphocytes whose secretions and other activities coordinate cell-mediated and antibody-mediated immunities; also called T_H *cells*.
hematocrit: The percentage of formed elements in a sample of blood; also called *volume of packed red cells (VPRC)* or *packed cell volume (PCV)*.
hematopoietic stem cells: Stem cells whose divisions produce each of the various populations of blood cells; also called *hemocytoblasts*.
heme: A porphyrin ring containing a central iron atom that can reversibly bind oxygen molecules; a component of the hemoglobin molecule.
hemoglobin (Hb): Protein composed of four globular subunits, each bound to a single molecule of heme; the protein found in red blood cells that gives them the ability to transport oxygen in the blood.
hemolyze: To break down red blood cells.
hemophilia: A congenital condition resulting from the inadequate synthesis of one of the clotting factors.
hemopoiesis: Blood cell formation and differentiation.
hemostasis: The stoppage of bleeding.
hemothorax: The entry of blood into one of the pleural cavities.
heparin: An anticoagulant released by activated basophils and mast cells.

hepatic portal vein: The vessel that carries blood from the intestinal capillaries to the sinusoids of the liver.
hepatocytes: Liver cells.
herniated disc: Rupture of the connective tissue sheath of the nucleus pulposus of an intervertebral disc.
hilum: A localized region where blood vessels, lymphatic vessels, nerves, and/or other anatomical structures are attached to an organ.
hippocampus: A portion of the limbic system that is concerned with the organization and storage of memories.
histamine: Chemical released by stimulated mast cells or basophils to initiate or enhance an inflammatory response.
histology: The study of tissues.
histones: Proteins associated with the DNA of the nucleus, and around which the DNA strands are wound.
holocrine secretion: Form of exocrine secretion where the secretory cell becomes swollen with vesicles and then ruptures.
homeostasis: The maintenance of a relatively constant internal environment.
hormones: Chemicals that are secreted by one cell and travel through the bloodstream to affect the activities of cells in another part of the body.
human immunodeficiency virus (HIV): The infectious agent that causes acquired immunodeficiency syndrome (AIDS).
human placental lactogen (HPL): Placental hormone that stimulates the functional development of the mammary glands.
hyaline cartilage: The most common type of cartilage; the matrix contains collagen fibers. Examples include the connections between the ribs and sternum, the tracheal and bronchial cartilages, and synovial cartilages.
hyaluronan: A carbohydrate component of proteoglycans in the matrix of many connective tissues; also called *hyaluronic acid*.
hyaluronidase: An enzyme that breaks down hyaluronic acid; produced by some bacteria and found in the acrosomeof sperm.
hydrocephalus: Condition resulting from excessive production or inadequate drainage of cerebrospinal fluid.
hymen: A membrane that forms during development and coversthe entrance to the vagina.
hyperextension: Extension of a body part past the anatomical position.
hypertrophy: Increase in the size of tissue without cell division.
hypervolemic: An abnormally high blood volume.
hypoblast: The undersurface of the inner cell mass that faces the blastocoele of the early embryo.
hyponychium: A thickening in the epidermis beneath the free edge of a nail.
hypophyseal portal system: Network of vessels that carry blood from capillaries in the hypothalamus to capillaries in the anterior lobe of the pituitary gland (adenohypophysis).
hypophysis: See **pituitary gland**.
hypothalamus: The floor of the diencephalon; region of the brain containing centers involved with the unconscious regulation of visceral functions, emotions, drives, and the coordination of neural and endocrine functions.
hypovolemic: An abnormally low blood volume.

I

I bands: The light bands in each side of Z lines in striated muscle.
ileocecal valve: A fold of mucous membrane that guards the connection between the ileum and the cecum.
ileum: The distal segment of the small intestine.
ilium: The largest of the three bones whose fusion creates a coxal bone.
immunoglobulins: Circulating antibodies.
implantation: The erosion of a blastocyst into the uterine wall.
incisors: Two pairs of flattened, bladelike teeth located at the front of the dental arches in both the upper and lower jaws.
inclusions: Aggregations of insoluble pigments, nutrients, or other materials in the cytoplasm.
incus: The central auditory ossicle, situated between the malleus and the stapes in the middle ear cavity.
infarct: An area of dead cells that results from an interruption of blood flow.
inferior vena cava: The vein that carries blood from the parts of the body inferior to the heart to the right atrium.
inferior: A directional reference meaning below, in reference to a particular structure, with the body in the anatomical position.
infundibulum: A tapering, funnel-shaped structure; in the nervous system, refers to the connection between the pituitary gland and the hypothalamus; the infundibulum of the uterine tube is the entrance bounded by fimbriae that receives the oocyte at ovulation.
inguinal canals: Passages through the abdominal wall that mark the path of testicular descent and that contain the testicular arteries, veins, and ductus deferens.
inhibin: A hormone, produced by nurse cells of the testes and follicular cells of the ovaries, that inhibits the secretion of follicle-stimulating hormone by the anterior lobe of the pituitary gland.
initial segment: The proximal portion of the axon, adjacent to the axon hillock, where an action potential first appears.
inner cell mass: Cells of the blastocyst that will form the body of the embryo.
innervation: The distribution of sensory and motor nerves to a specific region or organ.
insertion: A point of attachment of a muscle; the end that is easily movable.
insula: A region of the temporal lobe that is visible only after opening the lateral sulcus.
insulin: Hormone secreted by the beta cells of the pancreatic islets; causes a decrease in blood glucose concentrations.
intercalated discs: Regions where adjacent cardiocytes interlock and where gap junctions permit electrical coupling between the cells.
internal capsule: The collection of afferent and efferent fibers of the white matter of the cerebral hemispheres, visible on gross dissection of the brain.
internal ear: The membranous labyrinth that contains the organs of hearing and equilibrium; also called *inner ear*.
Interneurons: Groups of neurons between sensory and motor neurons that control a coordinated activity.
internodes: Sections of myelinated nerve fibers between two successive nodes.
interoceptors: Sensory receptors monitoring the functions and status of internal organs and systems.
interosseous membrane: Fibrous connective tissue membrane between the shafts of the tibia and fibula or the radius and ulna; an example of a fibrous amphiarthrosis.
interphase: Stage in the life of a cell during which the chromosomes are uncoiled and all normal cellular functions except mitosis are under way.
interstitial growth: Form of cartilage growth through the growth, mitosis, and secretion of chondrocytes inside the matrix.
interventricular foramen: The opening that permits fluid movement between the lateral and third ventricles.
intervertebral discs: Fibrous cartilage pads between the bodies of successive vertebrae that act as shock absorbers.
intervertebral foramina: Openings that permit fluid movement between the lateral and third ventricles of the brain.
intramembranous ossification: The formation of bone within a connective tissue without the prior development of a cartilaginous model.
intraperitoneal: Covered on all sides by the visceral peritoneum.
inversion: A turning inward.
iris: A contractile structure made up of smooth muscle that forms the colored portion of the eye.
ischium: One of the three bones whose fusion creates a coxal bone.
isthmus: A narrow band of tissue connecting two larger masses.

J

jejunum: The middle part of the small intestine.
joints: Areas where adjacent bones interact; also called *articulations*.
juxtaglomerular cells: Modified smooth muscle cells in the walls of the afferent and efferent arterioles adjacent to the glomerulus and the macula densa.
juxtaglomerular complex: The macula densa and the juxtaglomerular cells; a complex responsible for the release of renin and erythropoietin.
juxtamedullary nephrons: The 15 percent of nephrons whose nephron loops, or loops of Henle, extend into the medulla; these nephrons are responsible for creating the osmotic gradient within the medulla.

K

keratin: Tough, fibrous protein component of nails, hair, calluses, and the general integumentary surface.
keratinization: The production of keratin by epithelial cells.
keratinized: Containing large quantities of keratin.
keratohyalin: A protein within maturing keratinocytes.
kidneys: Components of the urinary system; organs that function in the regulation of blood composition, including the excretion of wastes and the maintenance of normal fluid and electrolyte balances.
kyphosis: Exaggerated thoracic curvature.

L

lacrimal gland: Tear gland on the posterior, lateral surface of the eye.
lactation: The production of milk by the mammary glands.

lacteal: A terminal lymphatic within an intestinal villus.
lactiferous duct: Duct draining one lobe of the mammary gland.
lactiferous sinus: An expanded portion of a lactiferous duct adjacent to the nipple of a breast.
lacuna/lacunae: A small pit or cavity.
lambdoid suture: Synarthrotic articulation between the parietal and occipital bones of the cranium.
lamellae: Concentric layers of bone within an osteon.
lamellar corpuscles: Receptors sensitive to vibration.
lamina propria: The reticular tissue that underlies a mucous epithelium and forms part of a mucous membrane.
laminae: Thin sheets or layers.
Langerhans cells: Cells in the epithelium of the skin and digestive tract that participate in the immune response by presenting antigens to T cells; also called *dendritic cells*.
large intestine: The terminal portions of the intestinal tract, consisting of the colon, the rectum, and the anal canal.
laryngopharynx: Division of the pharynx inferior to the epiglottis and superior to the esophagus.
larynx: A complex cartilaginous structure that surrounds and protects the glottis and vocal cords; the superior margin is bound to the hyoid bone and the inferior margin is bound to the trachea.
lateral apertures: Openings in the roof of the fourth ventricle that permit the circulation of cerebrospinal fluid into the subarachnoid space.
lens: The transparent refractive structure of the eye that is between the iris and the vitreous humor.
lesser omentum: A small pocket in the mesentery that connects the lesser curvature of the stomach to the liver.
leukocytes: White blood cells.
leukopoiesis: White blood cell formation.
ligaments: Dense bands of connective tissue fibers that attach one bone to another.
ligamentum arteriosum: The fibrous strand found in the adult that is the remnant of the ductus arteriosus of the fetus.
ligamentum nuchae: An elastic ligament that extends between the vertebra prominens and the external occipital crest.
limbic system: The nuclei and centers in the cerebrum and diencephalon that are involved with emotional states, memories, and behavioral drives.
linea alba: Tendinous band that runs along the midline of the rectus abdominis.
lipoproteins: Compounds containing relatively small lipids bound to a protein.
liver: An organ of the digestive system with varied and vital functions that include the production of plasma proteins, the excretion of bile, the storage of energy reserves, the detoxification of poisons, and the interconversion of nutrients.
lobules: Small lobes or subdivisions of a lobe.
loose connective tissue: A loosely organized, easily distorted connective tissue containing several different fiber types, a varied population of cells, and a viscous ground substance.
lordosis: An exaggeration of the lumbar curvature.
lumbricals: Four intrinsic plantar muscles.
luxation: Dislocation of a joint.
lymph node: Lymphoid organ that monitors the composition of lymph.
lymph: Fluid contents of lymphatic vessels, similar in composition to interstitial fluid.
lymphatic vessels: The vessels of the lymphatic system; also called *lymphatics*.
lymphocytes: Cells of the lymphatic system that play a role in the immune response.
lymphocytopoiesis: The formation of lymphocytes.
lymphopoiesis: The production of lymphocytes from lymphoid stem cells.
lysosomes: Intracellular vesicles containing digestive enzymes.
lysozyme: An enzyme present in some exocrine secretions that has antibiotic properties.

M

M line: The line in the center of the A band in striated muscle.
macula (of retina): The oval area in the retina whose center (the fovea centralis) is the region of sharpest vision; also called *macula lutea*.
macula densa: A group of specialized secretory cells that is located in a portion of the distal convoluted tubule, adjacent to the glomerulus and the juxtaglomerular cells; a component of the juxtaglomerular complex.
macula: A receptor complex, located in the saccule or utricle of the internal ear, that responds to linear acceleration or gravity.
malleus: An auditory ossicle, bound to the tympanic membrane and the incus.
mammary glands: Milk-producing glands of the female breast.
mammillary bodies: Nuclei in the hypothalamus that affect eating reflexes and behaviors; a component of the limbic system.
manubrium: The broad, roughly triangular, superior element of the sternum.
mast cells: Connective tissue cells that, when stimulated, release histamine, serotonin, and heparin, initiating the inflammatory response.
mastication: Chewing.
matrix: The ground substance of a connective tissue.
maxillary sinuses: Paranasal sinuses; air-filled chambers lined by a respiratory epithelium that are located in a maxilla and open into the nasal cavity.
mechanoreceptors: Receptors that can detect mechanical stimuli, such as touch, pressure, or vibration.
medial: Toward the midline of the body.
mediastinum: A septum between two parts of an organ or a cavity such as the central tissue mass that divides the thoracic cavity into two pleural cavities; includes the aorta and other great vessels, the esophagus, trachea, thymus, the pericardial cavity and heart, and a host of nerves, small vessels, and lymphatics.
medulla oblongata: The most caudal of the brain regions, also known as the *myelencephalon*.
medulla: Inner layer or core of an organ.
megakaryocytes: Red bone marrow cells responsible for the formation of platelets.
meiosis: Cell division that produces gametes with half of the normal somatic chromosome complement.
melanin: Yellow-brown pigment produced by the melanocytes of the skin.
melanocytes: Specialized cells in the deeper layers of the stratified squamous epithelium of the skin; responsible for the production of melanin.
melanocyte-stimulating hormone (MSH): A hormone, produced by the pars intermedia of the anterior lobe of the pituitary gland (adenohypophysis), that stimulates melanin production.
melanosomes: Pigment granule produced by melanocytes.
melatonin: Hormone secreted by the pineal gland; inhibits secretion of MSH and GnRH.
membrane flow: The movement of sections of membrane surface to and from the cell surface and components of the endoplasmic reticulum, the Golgi apparatus, and vesicles.
membrane potential: The potential difference, measured across a plasma membrane and expressed in millivolts, that results from the uneven distribution of positive and negative ions across the plasma membrane; also called *transmembrane potential*.
membrane: Any sheet or partition; a layer consisting of an epithelium and the underlying connective tissue.
membranous labyrinth: Endolymph-filled tubes that enclose the receptors of the internal ear.
memory T cells: T lymphocytes that provide immunologic memory, enabling an enhanced immune response when reexposed to a specific antigen.
menarche: The beginning of menstrual function.
menisci: Pads of fibrous cartilage between opposing surfaces in a joint.
menopause: The cessation of uterine cycles as a consequence of the aging process and exhaustion of viable follicles.
menstruation: The shedding of blood and endometrial tissue at menses.
mesencephalon: The midbrain.
mesenchyme: Embryonic or fetal connective tissue.
mesentery: A double layer of serous membrane that supports and stabilizes the position of an organ in the abdominopelvic cavity and provides a route for the associated blood vessels, nerves, and lymphatic vessels.
mesoderm: The middle germ layer that lies between the ectoderm and endoderm of the embryo.
metabolism: The sum of all of biochemical processes underway within the human body at a given moment; includes anabolism and catabolism.
metacarpals: The five bones of the palm of the hand.
metaphase: A stage of mitosis wherein the chromosomes line up along the equatorial plane of the cell.
metaphysis: The region of a long bone between the epiphysis and diaphysis, corresponding to the location of the epiphyseal cartilage of the developing bone.
metarterioles: Vessels that connect an arteriole to a venule and provide blood to a capillary plexus.
metastasis: The spread of cancer cells from one organ to another, leading to the establishment of secondary tumors.
metatarsasl bone: The five bones of the foot that articulate with the tarsal bones (proximally) and the phalanges (distally).
metencephalon: The pons and cerebellum of the brain.
microfilaments: Fine protein filaments visible with the electron microscope; components of the cytoskeleton.

microglia: Phagocytic neuroglia in the central nervous system, derived from the monocytes of the blood.
microtubules: Microscopic tubules that are part of the cytoskeleton and are found in cilia, flagella, the centrioles, and spindle fibers.
microvilli: Small, fingerlike extensions of the exposed plasma membrane of an epithelial cell.
micturition: Urination.
middle ear: Space between the external and internal ear that contains auditory ossicles.
midsagittal section: A plane passing through the midline of the body that divides it into left and right halves.
mineralocorticoids (MCs): Corticosteroids produced by the zona glomerulosa of the adrenal cortex; steroids such as aldosterone that affect mineral metabolism.
mitochondria: Intracellular organelles responsible for generating most of the ATP required for cellular operations.
mitosis: The division of a single cell nucleus that produces two identical daughter cell nuclei; an essential step in cell division.
mixed glands: Glands that contain exocrine and endocrine cells, or an exocrine gland that produces serous and mucous secretions.
modiolus: The bony central hub of the cochlea.
monocytes: Phagocytic agranulocytes (white blood cells) in the circulating blood.
monosynaptic reflex: A reflex where the sensory afferent synapses directly on the motor efferent.
morphology: The study of structure; the form of a living organism.
morula: A mulberry-shaped collection of cells produced through the mitotic divisions of a zygote.
motor neurons: Nerve cells forming a pathway from the brain or spinal cord to an effector (muscular or glandular) tissue.
motor unit: All of the muscle cells controlled by a single motor neuron.
mucins: Proteoglycans responsible for the lubricating properties of mucus.
mucosa: A mucous membrane; the epithelium plus the lamina propria; also called *mucous membrane*.
mucosa-associated lymphoid tissue (MALT): The extensive collection of lymphoid tissues linked with the epithelia of the digestive, respiratory, urinary, and reproductive tracts.
mucous cells: Goblet-shaped, mucus-producing, unicellular glands in certain epithelia of the digestive and respiratory tracts; they are called mucous cells when they are found in the stomach mucosa, respiratory mucosa, and salivary glands; they are called goblet cells in the mucosa of the small intestine, large intestine, terminal bronchioles, and conjunctiva.
mucous membrane: *See* **mucosa.**
multipennate muscle: A muscle whose internal fibers are organized around several different tendons.
muscarinic receptors: Membrane receptors sensitive to acetylcholine and to muscarine, a toxin produced by certain mushrooms; located at all parasympathetic neuromuscular and neuroglandular junctions and at a few sympathetic neuromuscular and neuroglandular junctions.
muscle tissue: A tissue characterized by the presence of cells capable of contraction; includes skeletal, cardiac, and smooth muscle tissue.
myelencephalon: The medulla oblongata.
myoblasts: Early muscle cells that differentiate into muscle fibers.
myocardial infarction (MI): Heart attack; damage to the heart muscle due to an interruption of regional coronary circulation.
myocardium: The cardiac muscle tissue of the heart.
myofibrils: Organized collections of myofilaments in skeletal and cardiac muscle cells.
myofilaments: Fine protein filaments, composed of the proteins actin (thin filaments) and myosin (thick filaments).
myoglobin: An oxygen-binding pigment especially common in slow skeletal muscle fibers and cardiac muscle fibers.
myometrium: The thick layer of smooth muscle in the wall of the uterus.
myosatellite cells: Cells that are the precursors to skeletal muscle cells (fibers).
myosin: Protein component of the thick myofilaments.

N

nails: Keratinous structures produced by epithelial cells of the nail root.
nasal cavity: The chamber on either side of the nasal septum.
nasolacrimal duct: Passageway that transports tears from the nasolacrimal sac to the nasal cavity.
nasopharynx: Region of the pharynx that lies superior to the soft palate.
natural killer (NK) cells: Type of lymphocyte that can kill target cells without previous sensitivity.
nebulin: Large protein making up skeletal muscle.
neonate: A newborn infant, or baby.
neoplasm: A tumor, or mass of abnormal tissue.
nephron loop: The segment of the nephron between the proximal and distal convoluted tubules that creates the concentration gradient; also called *loop of Henle*.
nephrons: Basic functional units of the kidneys.
neurofibrils: Microfibrils in the cytoplasm of a neuron.
neurofilaments: Microfilaments in the cytoplasm of a neuron.
neuroglia: Cells of the central nervous system and peripheral nervous system that support and protect neurons; also called *glial cells*.
neurohypophysis: The posterior lobe of the pituitary gland, or pars nervosa; contains the axons of hypothalamic neurons, which release OXT and ADH.
neurolemma: The outer surface of a Schwann cell that encircles an axon in the peripheral nervous system.
neuromuscular junction (NMJ): A synapse between a neuron and a muscle cell.
neuron: A cell in nervous tissue that is specialized for intercellular communication through changes in membrane potential and synaptic connections.
neurotubules: Microtubules in the cytoplasm of a neuron.
neutrophils: White blood cells that are very numerous and normally the first of the mobile phagocytic cells to arrive at an area of injury or infection.
nicotinic receptors: Acetylcholine receptors on the surfaces of sympathetic and parasympathetic ganglion cells; respond to the compound nicotine.
nipple: An elevated epithelial projection on the surface of the breast, containing the openings of the lactiferous sinuses.
Nissl bodies: The ribosomes, Golgi apparatus, rough endoplasmic reticulum, and mitochondria of the perikaryon of a typical neuron.
nociceptors: Pain receptors.
normovolemic: Having a normal blood volume.
nucleolus/nucleoli: The dense region in the nucleus that is the site of RNA synthesis.
nucleoplasm: Fluid content of the nucleus.
nucleus pulposus: The gelatinous core of an intervertebral disc.
nucleus/nuclei: Cellular organelle that contains DNA, RNA, and proteins; a mass of gray matter in the CNS.
nurse cells: Supporting cells of the seminiferous tubules of the testis; responsible for the differentiation of spermatids, the maintenance of the blood testis barrier, and the secretion of inhibin, androgen-binding protein, and Müllerian-inhibiting factor; also called *Sertoli cells*.
nystagmus: Involuntary, continual movement of the eyes as if to adjust to constant motion.

O

occlusal surface: The opposing surfaces of the teeth that come into contact when chewing food.
oculomotor nerves (N III): Cranial nerves that control the extra-ocular muscles other than the superior oblique and the lateral rectus.
olfaction: The sense of smell.
olfactory bulbs: The expanded ends of the olfactory tracts; the sites where the axons of the first cranial nerves (N I) synapse on central nervous system interneurons that lie inferior to the frontal lobes of the cerebrum.
oligodendrocyte: Central nervous system neuroglia that maintains cellular organization within gray matter and provides a myelin sheath in areas of white matter.
oncologists: Physicians specializing in the study and treatment of tumors.
oocyte: A cell whose meiotic divisions will produce a single ovum and three polar bodies.
oogenesis: Formation and development of an oocyte.
oogonia: Stem cells in the ovaries whose divisions give rise to oocytes.
ooplasm: The cytoplasm of the oocyte.
optic chiasm: The crossing point of the optic nerves.
optic nerves (N II): The second cranial nerves, which carry signals from the retina of the eye to the optic chiasm.
optic tracts: The tracts over which nerve impulses from the retinas are transmitted between the optic chiasm and the thalamus.
ora serrata: The anterior edge of the neural retina.
organ: Combinations of tissues that perform complex functions.
organelles: Intracellular structures with specific functions or group of functions.
organogenesis: The formation of organs during embryological and fetal development.
origin: In a skeletal muscle, the point of attachment that does not change position when the muscle contracts; usually defined in terms of movements from the anatomical position.
oropharynx: The middle portion of the pharynx, bounded superiorly by the nasopharynx, anteriorly by the oral cavity, and inferiorly by the laryngopharynx.

osmosis: The movement of water across a semipermeable membrane toward a solution containing a relatively high solute concentration.

osseous tissue: A strong connective tissue containing specialized cells and a mineralized matrix of crystalline calcium phosphate and calcium carbonate; also called *bone tissue*.

ossification: The formation of bone; osteogenesis.

osteoblasts: Cells that produce fibers and matrix of bone within connective tissue (intramembranous ossification) or cartilage (endochondral ossification); may differentiate into osteocytes.

osteoclast: A cell that dissolves the fibers and matrix of bone.

osteocytes: Bone cells responsible for the maintenance and turnover of the mineral content of the surrounding bone.

osteogenesis: Bone production.

osteogenic layer: The inner cellular layer of the periosteum that participates in bone growth and repair.

osteoid: The organic components of the bone matrix, produced by osteoblasts and osteocytes.

osteolysis: The breakdown of the mineral matrix of bone.

osteomalacia: Adult bone disease characterized by the gradual softening and bending of bones with accompanying pain; also called *adult rickets*.

osteon: The basic histological unit of compact bone, consisting of osteocytes organized around a central canal and separated by concentric lamellae.

osteopenia: The condition of inadequate bone production in the adult, leading to a loss in bone mass and strength.

osteoporosis: A reduction in bone mass and strength sufficient to compromise normal bone function.

osteoprogenitor cells: Stem cells that give rise to osteoblasts.

otoliths: Calcium carbonate crystals embedded in a gelatinous matrix; located on each macula of the vestibule.

oval window: Opening in the bony labyrinth where the stapes attaches to the membranous wall of the scala vestibuli (vestibular duct).

ovarian cycle: Monthly cycle of gamete development in the ovaries, associated with cyclical changes in the production of sex hormones (estrogens and progestins).

ovulation: The release of a secondary oocyte, surrounded by cells of the corona radiata, after the rupture of the wall of a tertiary follicle; in females, the periodic release of an oocyte from an ovary.

oxytocin: A hormone produced by hypothalamic cells and secreted into capillaries at the posterior lobe of the pituitary gland (neurohypophysis); stimulates smooth muscle contractions of the uterus or mammary glands in females and the prostate gland in males.

P

pacemaker cells: Cells of the sinoatrial (SA) node that set the pace of cardiac contraction.

pancreas: A digestive organ containing exocrine and endocrine tissues; the exocrine portion secretes pancreatic juice, and the endocrine portion secretes hormones, including insulin and glucagon.

pancreatic duct: A tubular duct that carries pancreatic juice from the pancreas to the duodenum.

pancreatic islets: Aggregations of endocrine cells in the pancreas; also called *islets of Langerhans*.

pancreatic juice: A mixture of buffers and digestive enzymes that is discharged into the duodenum under the stimulation of the enzymes secretin and cholecystokinin.

paranasal sinuses: Bony chambers lined by respiratory epithelium that open into the nasal cavity; include the frontal, ethmoidal, sphenoidal, and maxillary sinuses.

paraplegia: Paralysis of the upper limbs.

parasagittal section: A plane that parallels the midsagittal plane but that does not pass along the midline.

parasympathetic division: One of the two divisions of the autonomic nervous system; generally responsible for activities that conserve energy and lower the metabolic rate; the "rest and digest" division; also called *craniosacral division*.

parathyroid glands: Four small glands embedded in the posterior surface of the thyroid gland that secrete parathyroid hormone.

parathyroid hormone (PTH): A hormone secreted by the parathyroid glands when blood calcium levels decrease below the normal range; causes increased osteoclast activity, increased intestinal calcium uptake, and decreased calcium ion loss by the kidneys.

parietal cells: Cells of the gastric glands that secrete hydrochloric acid and intrinsic factor.

parietal layer of the serous pericardium: The outer part of the serous pericardium.

parietal peritoneum: The layer of the peritoneum (serous sac) that lines the abdominal walls.

parietal pleura: The serous membrane that lines the wall of the pulmonary cavity.

parotid glands: Large salivary glands that secrete saliva with high concentrations of salivary (alpha) amylase.

pars distalis: The large, anterior portion of the anterior lobe of the pituitary gland (adenohypophysis).

pars intermedia: The portion of the anterior lobe of the pituitary gland (adenohypophysis) that is immediately adjacent to the posterior lobe of the pituitary gland (neurohypophysis) and the infundibulum.

pars tuberalis: The portion of the anterior lobe of the pituitary gland (adenohypophysis) that wraps around the infundibulum superior to the posterior lobe (neurohypophysis).

parturition: Childbirth, delivery.

patella: The sesamoid bone of the knee; also called the *kneecap*.

pathology: The study of disease.

pedicles: Thick bony struts that connect the vertebral body with the articular and spinous processes.

pelvic cavity: Inferior subdivision of the abdominopelvic (peritoneal) cavity; encloses the urinary bladder, the sigmoid colon and rectum, and male or female reproductive organs.

pelvis: A bony complex created by the articulations among the coxal bones, the sacrum, and the coccyx.

penis: A component of the male external genitalia; a copulatory organ that surrounds the urethra and introduces semen into the female vagina; the developmental equivalent of the female clitoris.

pepsin: Proteolytic enzyme secreted by the chief cells of the gastric glands in the stomach.

pepsinogen: The inactive proenzyme that is secreted by chief cells of the gastric pits; after secretion it is converted to the proteolytic enzyme pepsin.

peptidases: Enzymes that split peptide bonds and release amino acids.

perforating canals: Passageways within compact bone that extend perpendicular to the surface.

pericardial cavity: The space between the parietal pericardium and the epicardium (visceral pericardium) that covers the outer surface of the heart.

pericarditis: Inflammation of the pericardium.

pericardium: The fibrous sac that surrounds the heart, and whose inner, serous lining is continuous with the epicardium.

perichondrium: Layer that surrounds a cartilage, consisting of an outer fibrous and an inner cellular region.

perikaryon: The cytoplasm that surrounds the nucleus in the cell body of a neuron.

perilymph: A fluid similar in composition to cerebrospinal fluid; found in the spaces between the bony labyrinth and the membranous labyrinth of the internal ear.

perimysium: Connective tissue partition that separates adjacent fasciculi in a skeletal muscle.

perineurium: Connective tissue partition that separates adjacent bundles of nerve fibers in a peripheral nerve.

periodontal ligament: Collagen fibers that anchor a tooth within its alveolus.

periosteum: Layer that surrounds a bone, consisting of an outer fibrous and inner cellular region.

peripheral nervous system (PNS): All nervous tissue outside the central nervous system.

peristalsis: A wave of smooth muscle contractions that propels materials along the lumen of a tube such as the digestive tract, the ureters, or the ductus deferens.

peritoneal dialysis: Removal of soluble substances and water by transfer across the peritoneum.

peritoneum: The serous membrane that lines the peritoneal (abdominopelvic) cavity.

permeability: Ease with which dissolved materials can cross a membrane; if freely permeable, any molecule can cross the membrane; if impermeable, nothing can cross; most biological membranes are selectively permeable.

peroxisomes: Membranous vesicles containing enzymes that break down hydrogen peroxide (H_2O_2).

phalanx/phalanges: Bone(s) of the finger(s) or toe(s).

pharynx: The throat; a muscular passageway shared by the digestive and respiratory tracts.

photoreceptors: Receptors that are sensitive to light.

physiology: The study of how living organisms perform their functions.

pia mater: The innermost layer of the meninges bound to the underlying nervous tissue.

pineal gland: Nervous tissue in the posterior portion of the roof of the diencephalon, responsible for secreting melatonin.

pinealocytes: Secretory cells of the pineal gland.

pituitary gland: An endocrine organ that is located in the sella turcica of the sphenoid bone and is connected to the hypothalamus by the infundibulum; includes the posterior lobe (neurohypophysis) and the anterior lobe (adenohypophysis); also called the *hypophysis*.

placenta: A temporary structure in the uterine wall that permits diffusion between the fetal and maternal circulatory systems.

placentation: Formation of a functional placenta following implantation of a blastocyst in the endometrium.

plantar flexion: Ankle extension; toe pointing.

plasma cells: Activated B cells that secrete antibodies.

plasma membrane: A cell membrane; also called a *plasmalemma* or *cell membrane*.

plasma: The fluid ground substance of whole blood; what remains after the cells have been removed from a sample of whole blood.

platelets: Small packets of cytoplasm that contain enzymes important in the clotting response; manufactured in the bone marrow by cells called megakaryocytes.

pleura: The serous membrane lining the pleural (lung) cavities.

pleural cavity: Body cavity of the thoracic region that surrounds the lung.

pneumothorax: The introduction of air into the pleural cavity.

polar body: A nonfunctional packet of cytoplasm containing chromosomes eliminated from an oocyte during meiosis.

pollex: The thumb.

polycythemia: An unusually high hematocrit due to the presence of excess numbers of formed elements, especially RBCs.

polysynaptic reflex: A reflex with interneurons interposed between the sensory fiber and the motor neuron(s).

pons: The portion of the metencephalon anterior to the cerebellum.

postcentral gyrus: The primary sensory cortex, where touch, vibration, pain, temperature, and taste sensations arrive and are consciously perceived.

posterior: Toward the back; dorsal.

precentral gyrus: The primary motor cortex of a cerebral hemisphere, located anterior to the central sulcus.

prefrontal cortex: The anterior portion of each cerebral hemisphere; thought to be involved with higher intellectual functions, predictions, and calculations.

premotor cortex: The motor association area between the precentral gyrus and the prefrontal area.

pre-optic area: The hypothalamic nucleus that coordinates thermoregulatory activities.

prepuce: Loose fold of skin that surrounds the glans penis (males) or the clitoris (females).

preputial glands: Glands on the inner surface of the prepuce that produce a viscous, odorous secretion called smegma.

prime mover: A muscle that performs a specific action; also called an *agonist*.

progesterone: The most important progestin secreted by the corpus luteum following ovulation.

progestins: Steroid hormones structurally related to cholesterol.

projection fibers: Axons carrying information from the thalamus to the cerebral cortex.

prolactin (PRL): The hormone that stimulates functional development of the mammary glands in females; a secretion of the anterior lobe of the pituitary gland (adenohypophysis).

pronation: Rotation of the forearm that makes the palm face posteriorly.

prone: Lying face down with the palms facing the floor.

prophase: The initial phase of mitosis, characterized by the appearance of chromosomes, breakdown of the nuclear membrane, and formation of the spindle apparatus.

proprioceptors: Sensory organs that monitor the position and movement of skeletal muscles and joints.

prostate: An accessory gland of the male reproductive tract, contributing about one-third of the volume of semen.

protraction: Movement anteriorly in the horizontal plane.

proximal convoluted tubule (PCT): The segment of the nephron between glomerular's capsule and the nephron loop; the major site of active reabsorption from the filtrate.

pubic symphysis: Fibrocartilaginous amphiarthrosis between the pubic bones of the hip.

pubis: The anterior, inferior component of the hip bone.

pulmonary circuit: Blood vessels between the pulmonary semilunar valve of the right ventricle and the entrance to the left atrium; the blood flow through the lungs.

pulmonary ventilation: Movement of air in and out of the lungs.

pulp cavity: Internal chamber in a tooth, containing blood vessels, lymphatics, nerves, and the cells that maintain the dentin.

pulvinar nuclei: The thalamic nuclei involved in the integration of sensory information prior to projection to the cerebral hemispheres.

pupil: The opening in the center of the iris through which light enters the eye.

Purkinje cell layer: A large, branching neuron of the cerebellar cortex.

putamen: Thalamic nucleus involved in the integration of sensory information prior to projection to the cerebral hemispheres.

pyelonephritis: Inflammation of the kidneys.

pyloric sphincter: Ring of smooth muscle that regulates the passage of chyme from the stomach to the duodenum.

pylorus: Gastric region between the body of the stomach and the duodenum; includes the pyloric sphincter.

Q

quadriceps femoris: Four anterior thigh muscles that include the rectus femoris, vastus lateralis, vastus intermedius, and vastus medialis.

quadriplegia: Paralysis of the upper and lower limbs.

R

radiological procedures: Methods for visualizing internal structures using various forms of radiation.

radius: The lateral and shorter of the two bones of the forearm.

ramus/rami: Branch/branches.

raphe: A seam.

receptive field: The area monitored by a single sensory receptor.

recto-uterine pouch: Peritoneal pocket between the anterior surface of the rectum and the posterior surface of the uterus.

rectum: The distal 15 cm (6 in.) of the digestive tract.

red blood cells (RBCs): *See* **erythrocytes.**

reflex arc: The receptor, sensory neuron, motor neuron, and effector involved in a particular reflex; interneurons may or may not be present, depending on the reflex considered.

reflex: A rapid, automatic response to a stimulus.

regulatory T cells: Population of T lymphocytes that suppress the immune response.

renal corpuscle: The initial portion of the nephron, consisting of an expanded chamber that encloses the glomerulus.

renin: The enzyme released by cells of the juxtaglomerular complex when renal blood flow decreases; converts angiotensinogen to angiotensin I.

rennin: Gastric enzyme that breaks down milk proteins.

respiration: Exchange of gases between living cells and the environment; includes pulmonary ventilation, external respiration, and internal respiration.

respiratory rhythmicity centers: The centers in the medulla oblongata that set the background pace of respiration; includes inspiratory and expiratory centers.

rete testis: An interwoven network of blood vessels or passageways.

reticular formation: Diffuse network of gray matter that extends the entire length of the brainstem.

retina: The innermost layer of the eye, lining the vitreous chamber; also known as the *inner layer*.

retraction: Movement posteriorly in the horizontal plane.

retroflexion: A posterior tilting of the uterus that has no clinical significance.

retroperitoneal: Situated behind or outside of the peritoneal cavity.

reverberation: Positive feedback along a chain of neurons such that they remain active once stimulated.

Rh factor: A surface antigen that may be present (Rh positive) or absent (Rh negative) from the surfaces of red blood cells.

rheumatism: A condition characterized by pain in muscles, tendons, bones, or joints.

ribosomes: Organelles that contain rRNA and proteins and are essential to mRNA translation and protein synthesis.

rickets: Bone disease in children caused by vitamin D deficiency characterized by inadequate calcification, skeletal deformities, and bowed legs.

rigor mortis: Extended muscular contraction and rigidity that occurs after death, as the result of calcium ion release from the SR and the exhaustion of cytoplasmic ATP reserves.

rods: Photoreceptors responsible for vision in dim lighting.

rotator cuff: Four muscles of the shoulder joint that include the supraspinatus, infraspinatus, teres minor, and Subscapularis.

rough endoplasmic reticulum (RER): A membranous organelle that is a site of protein synthesis and storage.

round window: An opening in the bony labyrinth of the internal ear that exposes the membranous wall of the tympanic duct to the air of the middle ear cavity.

rubrospinal tracts: Descending tracts that carry involuntary motor commands issued by the red nucleus of the mesencephalon.

rugae: Mucosal folds in the lining of the empty stomach that disappear as gastric distension occurs; folds in the urinary bladder.

S

saccule: A portion of the vestibular apparatus of the internal ear; contains a macula important for providing sensations of gravity and linear acceleration in a vertical dimension.

sacrum: Triangular bone that articulates with the last lumbar vertebra, coccyx, and the hip bone on either side.

sagittal plane: Sectional plane that divides the body into left and right portions.

sarcolemma: The plasma membrane of a muscle cell.

sarcomere: The smallest contractile unit of a striated muscle cell.

sarcoplasm: The cytoplasm of a muscle cell.

sarcoplasmic reticulum (SR): The endoplasmic reticulum of skeletal muscle and cardiac muscle.

scala tympani: The perilymph-filled chamber of the internal ear, adjacent to the basilar membrane; pressure changes there distort the round window; also called *tympanic duct*.

scala vestibuli: The perilymph-filled chamber of the internal ear, adjacent to the vestibular membrane; pressure waves are induced by movement of the stapes at the oval window; also called *vestibular duct*.

scapula: The shoulder blade.

scar tissue: Thick, collagenous tissue that forms at an injury site.

Schwann cells: Neuroglia responsible for the neurilemma that surrounds axons in the peripheral nervous system.

sciatica: Pain resulting from compression of the roots of the sciatic nerve.

sclera: The fibrous, outer layer of the eye forming the white area of the anterior surface; a portion of the fibrous layer of the eye.

scoliosis: An abnormal, exaggerated lateral curvature of the spine.

scrotum: Loose-fitting, fleshy pouch that encloses the testes of the male.

sebaceous glands: Glands that secrete sebum, usually associated with hair follicles.

sebum: A waxy secretion that coats the surfaces of hairs.

secretin: A hormone, secreted by the duodenum, that stimulates the production of buffers by the pancreas and inhibits gastric activity.

sectional planes: Views or slices along a two-dimensional flat surface.

semen: The fluid ejaculate that contains spermatozoa and the secretions of accessory glands of the male reproductive tract.

semicircular duct: The tubular component of the membranous labyrinth of the internal ear; responds to rotational movements of the head.

seminal glands: Glands of the male reproductive tract that produce roughly 60 percent of the volume of semen. Also known as *seminal vesicles*.

seminiferous tubules: Coiled tubules where sperm production occurs in the testis.

sensible perspiration: Water loss due to secretion by sweat glands.

serosa: *See* **serous membrane.**

serous membrane: A squamous epithelium and the underlying loose connective tissue; the lining of the pericardial, pleural, and peritoneal cavities.

serum: Blood plasma from which clotting agents have been removed.

sesamoid bones: Bone that form within tendons.

sex cells: Sperm or oocytes.

sigmoid colon: The S-shaped region of the colon between the descending colon and the rectum.

signs: Visible, objective evidence of the presence of a disease.

simple epithelium: An epithelium containing a single layer of cells above the basement membrane.

sinusoids: Exchange vessels that are similar in general structure to fenestrated capillaries. The two differ in size (sinusoids are larger and more irregular in cross section), continuity (sinusoids have gaps between endothelial cells), and support (sinusoids have thin basement membranes, if present at all).

skeletal muscle tissue: Contractile tissue dominated by skeletal muscle fibers; characterized as striated, voluntary muscle.

skeletal muscle: A contractile organ of the muscular system.

sliding filament theory: The concept that a sarcomere shortens as the thick and thin filaments slide past one another.

small intestine: The duodenum, jejunum, and ileum; the digestive tract between the stomach and large intestine.

smegma: Secretion of the preputial glands of the penis or clitoris.

smooth endoplasmic reticulum (SER): Membranous organelle in which lipid and carbohydrate synthesis and storage occur.

smooth muscle tissue: Muscle tissue in the walls of many visceral organs; characterized as nonstriated, involuntary muscle.

smooth muscle: Muscle tissue found in the walls of many visceral organs; characterized as nonstriated, involuntary muscle.

soft palate: Fleshy posterior extension of the hard palate, separating the nasopharynx from the oral cavity.

somatic cells: Cells of an organism other than the sex cells (sperm and oocytes).

somatic nervous system (SNS): The efferent division of the nervous system that innervates skeletal muscles.

somatic: Pertaining to the body.

somatostatin: GH–IH, a hypothalamic regulatory hormone that inhibits GH secretion by the anterior pituitary.

sperm: Male gametes; also called *spermatozoon/spermatozoa*.

spermatic cords: Spermatic vessels, nerves, lymphatics, and the ductus deferens, extending between the testes and the proximal end of the inguinal canal.

spermatids: The product of meiosis in the male, cells that differentiate into sperm.

spermatogenesis: Sperm production.

spermatogonia: Stem cells whose mitotic divisions give rise to other stem cells and spermatocytes.

spermiogenesis: The process of spermatid differentiation that leads to the formation of physically mature sperm.

sphincter: Muscular ring that contracts to close the entrance or exit of an internal passageway.

spinal meninges: Specialized membranes that line the vertebral canal and provide protection, stabilization, nutrition, and shock absorption to the spinal cord.

spinal nerve: One of 31 pairs of nerves that originate on the spinal cord from anterior and posterior roots.

spinocerebellar tracts: Ascending tracts carrying sensory information to the cerebellum.

spinothalamic tracts: Ascending tracts carrying poorly localized touch, pressure, pain, vibration, and temperature sensations to the thalamus.

spinous process: Prominent posterior projection of a vertebra, formed by the fusion of two laminae.

spiral organ: A receptor complex in the scala media of the cochlea that includes the inner and outer hair cells, supporting cells and structures, and the tectorial membrane; provides the sensation of hearing; also called the *organ of Corti*.

spleen: Lymphoid organ important for red blood cell phagocytosis, the immune response, and lymphocyte production.

spongy bone: Bone that consists of an open network of struts and plates that resembles a three-dimensional garden lattice.

sprains: Injuries to a ligament resulting from excessive forces at the joint but without joint dislocation or fracture.

squamous epithelium: An epithelium whose superficial cells are flattened and platelike.

stapedius: A muscle of the middle ear whose contraction tenses the auditory ossicles and reduces the forces transmitted to the oval window.

stapes: The auditory ossicle attached to the tympanic membrane.

stellate macrophages: Phagocytic cells of the liver sinusoids; also called *Kupffer cells*.

sternum: Long, flat bone that articulates with the cartilages of the first seven ribs and with the clavicle and forms the anterior wall of the thorax; has three parts: manubrium, body, and xiphoid process.

strains: Overuse injuries to a muscle; also called *muscle tear*.

stratum basale: The deepest epidermal layer.

stratum corneum: Layers of flattened, dead, keratinized cells covering the epidermis of the skin.

stratum granulosum: Layer of epidermis lying just superficial to the stratum spinosum and deep to the stratum lucidum.

stratum lucidum: Layer of epidermis found between the stratum corneum and stratum granulosum present only in thick skin.

stratum spinosum: Layer of the epidermis between the stratum basale and stratum granulosum.

stretch reflex: Tonic muscle contraction in response to stimulation of muscle proprioceptors.

stroma: The connective tissue framework of an organ; distinguished from the functional cells (parenchyma) of that organ.

subarachnoid space: Meningeal space containing cerebrospinal fluid; the area between the arachnoid mater and the pia mater.

subcutaneous layer: The layer of loose connective tissue below the dermis; also called the *superficial fascia*.

subluxation: A partial dislocation of a joint.

submucosa: Region between the muscularis mucosae and the muscularis externa.

subserous fascia: Loose connective tissue layer beneath the serous membrane lining the anterior body cavity.

sulcus/sulci: Groove(s) or furrow(s).

superficial fascia: *See* **subcutaneous layer.**

superficial: Located nearer the body surface.

superior vena cava (SVC): The vein that carries blood to the right atrium from parts of the body that are superior to the heart.

superior: Above, in reference to a portion of the body in the anatomical position.

supination: Rotation of the forearm so that the palm faces anteriorly.

supine: Lying face up, with palms facing anteriorly.

suprarenal gland: See **adrenal gland**.

surface anatomy: Referring to superficial anatomical structures; the study of the body surface in relation to its deeper parts.

surfactant: Lipid secretion that coats alveolar surfaces of the lungs and prevents their collapse.

sutural bones: Irregular bones that form in fibrous tissue between the flat bones of the developing cranium; also called *Wormian bones*.

suture: Fibrous joint between flat bones of the skull.

sympathetic division: The division of the autonomic nervous system that is responsible for "fight or flight" reactions; primarily concerned with the elevation of metabolic rate and increased alertness; also called the *thoracolumbar division*.

symphysis: A fibrous amphiarthrosis, such as those between adjacent vertebrae or between the pubic bones of the coxae.

symptoms: Abnormality in functions as a result of disease; subjective experiences of patient.

synapse: Site of communication between a nerve cell and some other cell; if the other cell is not a neuron, the term *neuromuscular junction* or *neuroglandular junction* is often used.

synaptic cleft: The space between the axolemma and the postsynaptic surface.

synaptic vesicles: Membrane-bound structures near the presynaptic membrane that contain neurotransmitters.

synarthrosis: A joint that does not permit relative movement between the articulating elements.

synchondrosis: A cartilaginous synarthrosis, such as the articulation between the epiphysis and diaphysis of a growing bone.

syndesmosis: A fibrous amphiarthrosis.

synergist: A muscle that assists a prime mover in performing its primary action.

synostosis: A synarthrosis formed through the fusion of the articulating elements.

synovial fluid: Substance secreted by synovial membranes that lubricates joints.

synovial joint: A freely movable joint where the opposing bone surfaces are separated by synovial fluid; also called a *diarthrosis*.

synovial membrane: An incomplete layer of fibroblasts confronting the synovial cavity, plus the underlying loose connective tissue.

systemic circuit: Vessels between the aortic valve and the entrance to the right atrium; the circulatory system other than vessels of the pulmonary circuit.

T

T cells: Lymphocytes responsible for cellular immunity, and for the coordination and regulation of the immune response; include regulatory T cells (helpers and suppressors) and cytotoxic (killer) T cells.

T tubules: Transverse, tubular extensions of the sarcolemma that extend deep into the sarcoplasm to contact cisternae of the sarcoplasmic reticulum.

tachycardia: Abnormally rapid heart rate, usually over 90 beats per minute.

tactile corpuscles: Touch receptors located within dermal papillae adjacent to the basement membrane of the epidermis; also called *Meissner's corpuscles*.

tactile discs: Sensory nerve endings that contact special receptors called Merkel cells, located within the deeper layers of the epidermis; also called *Merkel's discs*.

tarsal bones: The bones of the ankle (the talus, calcaneus, navicular, and cuneiform bones).

tarsus: The ankle.

tectorial membrane: Gelatinous membrane suspended over the hair cells of the spiral organ.

tectospinal tracts: Descending extrapyramidal tracts carrying involuntary motor commands issued by the colliculi.

tectum: The roof of the mesencephalon of the brain.

telencephalon: The forebrain or cerebrum, including the cerebral hemispheres, the internal capsule, and the cerebral nuclei.

telodendria: Terminal axonal branches that end in axon terminals.

telophase: The final stage of mitosis, characterized by the disappearance of the spindle apparatus, the reappearance of the nuclear membrane and the disappearance of the chromosomes, and the completion of cytokinesis.

temporal: Pertaining to time (temporal summation) or pertaining to the temples (temporal bone).

tendon: A collagenous band that connects a skeletal muscle to an element of the skeleton.

tentorium cerebelli: Dural partition that separates the cerebral hemispheres from the cerebellum.

teratogen: Any agent or factor that induces or increases the risk of abnormal prenatal development.

terminal cisternae: Pairs of transversely oriented tubules of the sarcoplasmic reticulum in skeletal muscle that together with an intermediate T tubule form a triad.

tertiary follicle: A mature ovarian follicle containing a large, fluid-filled chamber.

testes: The male gonads, sites of gamete production and hormone secretion; also called *testicles*.

testosterone: The main androgen produced by the interstitial endocrine cells of the testes.

thalamus: The walls of the diencephalon.

thermoreceptors: Receptors sensitive to temperature changes.

thermoregulation: Homeostatic maintenance of body temperature.

thick filament: A cytoskeletal filament in a skeletal or cardiac muscle cell; composed of myosin, with a core of titin.

thin filament: A cytoskeletal filament in a skeletal or cardiac muscle cell; consists of actin, troponin, nebulin, and tropomyosin.

thrombocytopenia: Abnormally low platelet count in the circulating blood.

thrombus: Blood clot.

thymosin: Thymic hormone essential to the development and differentiation of T cells.

thymus: Lymphoid organ, site of T cell development and maturation.

thyroglobulin: Circulating transport globulin that binds thyroid hormones.

thyroid gland: An endocrine gland whose lobes are lateral to the thyroid cartilage of the larynx.

thyroid hormones: Thyroxine (T_4) and triiodothyronine (T_3), hormones of the thyroid gland; hormones that stimulate tissue metabolism, energy utilization, and growth.

thyroid-stimulating hormone (TSH): Anterior pituitary hormone that triggers the secretion of thyroid hormones by the thyroid gland.

thyroxine: A thyroid hormone; also called T_4 or *tetraiodothyronine*.

tibia: The large, medial bone of the leg.

tissues: Collections of specialized cells and cell products that perform specific functions.

titin: Fibrous protein that connects think myosin filaments to Z discs in the sarcomere.

tonsillectomy: Removal of the tonsil(s).

tonsillitis: Inflammation of the tonsil(s).

tonsils: Lymphoid nodules in the wall of the pharynx; the palatine, pharyngeal, and lingual tonsils.

trabeculae carneae: Muscular ridges projecting from the walls of the ventricles of the heart.

trabeculae: Connective tissue partitions that subdivide an organ.

trachea: The windpipe, an airway extending from the larynx to the primary bronchi.

tracheostomy: Surgical opening of the anterior tracheal wall to permit airflow.

transfusion: Transfer of blood from a donor directly into the bloodstream of another person.

transient ischemic attack (TIA): Sudden loss of neurologic function with complete recovery.

transudate: Fluid that diffuses across a serous membrane and lubricates opposing surfaces.

transverse plane: A horizontal plane that divides the body into superior and inferior portions.

transverse tubules: The transverse, tubular extensions of the sarcolemma that extend deep into the sarcoplasm, contacting cisternae of the sarcoplasmic reticulum; also called *T tubules*.

triad: The transverse tubule (T tubule) and the terminal cisternae on each side of it in skeletal muscle fibers.

trigeminal nerves (N V): Cranial nerves that provide sensory information from the lower portions of the face (including the upper and lower jaws) and deliver motor commands to the muscles of mastication (chewing).

trigone: Triangular region of the urinary bladder bounded by the exits of the ureters and the entrance to the urethra.

triiodothyronine (T_3): One of the thyroid hormones.

trochlea: A pulley; the spool-shaped medial portion of the condyle of the humerus.

trochlear nerves (N IV): Cranial nerves controlling the superior oblique muscle of the eye.

trophoblast: Superficial layer of the blastocyst that will be involved with implantation, hormone production, and placenta formation.

tropomyosin: Fibrous muscle protein that covers active sites on G actin and prevents actin-myosin interaction.

troponin: Globular muscle protein that binds to tropomyosin.

tubulin: Protein subunit of microtubules.

tumor: A tissue mass formed by the abnormal growth and replication of cells.

tympanic membrane: Membrane that separates the external acoustic meatus from the middle ear; membrane whose vibrations are transferred to the auditory ossicles and ultimately to the oval window; the "eardrum."

U

ulna: The medial and larger of the two bones of the forearm.

umbilical cord: Connecting stalk between the fetus and the placenta; contains the allantois, the umbilical arteries, and the umbilical vein.

unicellular glands: Mucous cells.

ureters: Muscular tubes, lined by transitional epithelium, that carry urine from the renal pelvis to the urinary bladder.

urethra: A muscular tube that carries urine from the urinary bladder to the exterior.

urethritis: Inflammation of the urethra.

urinary bladder: Muscular, distensible sac that stores urine prior to urination.

urination: The voiding of urine; micturition.

uterine cycle: Cyclical changes in the uterine lining that occur in reproductive-age women. Each uterine cycle, which occurs in response to circulating hormones (*see* **ovarian cycle**), lasts 21–35 days.

uterus: Muscular organ of the female reproductive tract where implantation, placenta formation, and fetal development occur.

utricle: The largest chamber of the vestibular apparatus of the internal ear; contains a macula important for providing sensations of gravity and linear acceleration in a horizontal dimension.

uvula: A dangling, fleshy extension of the soft palate.

V

vagina: A muscular tube extending between the uterus and the vestibule.

vallate papillae: One of the large, dome-shaped papillae on the superior surface of the tongue that forms a V, separating the body of the tongue from the root.

valves: Flap-like structure that maintains unidirectional blood flow.

vas deferens: *See* **ductus deferens**.

vasa vasorum: Blood vessels that supply the walls of large arteries and veins.

vasoconstriction: A reduction in the diameter of arterioles due to the contraction of smooth muscles in the tunica media; increases peripheral resistance; may occur in response to local factors, through the action of hormones, or from the stimulation of the vasomotor center.

vasodilation: An increase in the diameter of arterioles due to the relaxation of smooth muscles in the tunica media; decreases peripheral resistance; may occur in response to local factors, through the action of hormones, or after decreased stimulation of the vasomotor center.

veins: Blood vessels carrying blood from capillary beds toward the heart.

ventricles: Fluid-filled chambers; in the heart, one of the large chambers discharging blood into the pulmonary or systemic circuits; in the brain, one of four fluid-filled interior chambers.

venules: Thin-walled veins that receive blood from capillaries.

vermis: Midsagittal band of neural cortex on the surface of the cerebellum.

vertebra prominens: The vertebra in the cervicothoracic region that has the most prominent spinous process, usually the seventh cervical vertebra, but sometimes the sixth cervical vertebra or the first thoracic vertebra.

vertebral canal: Passageway that encloses the spinal cord, a tunnel bounded by the neural arches of adjacent vertebrae.

vertebral column: The cervical, thoracic, and lumbar vertebrae, the sacrum, and the coccyx.

vestibular folds: Mucosal folds in the laryngeal walls that do not play a role in sound production; the false vocal cords.

vestibular nuclei: The processing centers for sensations that arrive from the vestibular apparatus of the internal ear, located near the border between the pons and the medulla oblongata.

vestibule: A chamber; in the internal ear, the term refers to the utricle, saccule, and semicircular ducts; also refers to (1) a region of the female external genitalia, (2) the space within the fleshy portion of the nose between the nostrils, and (3) the space between the ventricular folds and the vocal folds of the larynx.

vestibulospinal tracts: Descending tracts of the extrapyramidal system, carrying involuntary motor commands issued by the vestibular nucleus to stabilize the position of the head.

viscera: Internal organs of the thoracic and abdominopelvic cavities.

visceral layer of the serous pericardium: The inner part of the serous pericardium positioned directly on the heart.

visceral peritoneum: The layer of peritoneum on the abdominal organs.

visceral pleura: The serous membrane on the lungs and dipping into the fissures between the lung lobes.

visceral: Pertaining to viscera (internal organs) or their outer coverings.

viscosity: The resistance to flow that a fluid exhibits as a result of molecular interactions within the fluid.

vocal folds: Folds in the laryngeal wall containing elastic ligaments whose tension can be voluntarily adjusted; the true vocal cords, responsible for phonation.

vulva: The female pudendum (external genitalia).

W

wallerian degeneration: Disintegration of an axon and its myelin sheath distal to an injury site.

Wernicke's area: Region of the brain involved with language comprehension, located in the cortex of the dominant temporal lobe.

white blood cells (WBCs): The granulocytes and agranulocytes of whole blood; also called *leukocytes*.

white matter: Regions in the central nervous system (CNS) that are dominated by myelinated axons.

white ramus communicans: A nerve bundle containing the myelinated preganglionic axons of sympathetic motor neurons en route to the sympathetic chain or to a collateral ganglion.

X

xiphoid process: Slender, inferior extension of the sternum.

Y

yolk sac: One of the three extra-embryonic membranes, composed of an inner layer of endoderm and an outer layer of mesoderm.

Z

Z lines: Cross striations that bisect the I bands in striated muscle fibers; they anchor the actin filaments at either end of the sarcomere.

zona fasciculata: Region of the adrenal cortex that secretes glucocorticoids.

zona glomerulosa: Region of the adrenal cortex that secretes mineralocorticoids.

zona pellucida: Region between a developing oocyte and the surrounding follicular cells of the ovary.

zona reticularis: Region of the adrenal cortex that secretes androgens.

zygote: The fertilized ovum prior to the start of cleavage.

Credits

Visible Human Data courtesy of the Library of Medicine and the Visible Human Project.

Photo Credits

Chapter 1

Chapter Opener Sciepro/Science Photo Library/Getty Images
1.1a Rubberball/AGE Fotostock
1.1b Rangizzz/Fotolia
1.1c Loannis Ioannou/Shutterstock
1.1d Scott Bauer/U.S. Department of Agriculture/Science Source
1.1e DPA Picture Alliance Archive/Alamy Stock Photo
1.2c Anita Impagliazzo/Pearson Education, Inc.
1.7 The Cartoon Bank
1.9a, b, c Sciencefoto.De-Dr. Andre Kempe/Stockbyte/Getty Images
1.13a Anita Impagliazzo/Pearson Education, Inc.
1.13b Ralph T. Hutchings
1.13c Anita Impagliazzo/Pearson Education, Inc.
1.CN-01a, b Science Source
1.CN-01c Biophoto Associates/Science Source
1.CN-01d Frederic H. Martini
1.CN-01e Alexander Tsiaras/Science Source
1.CN-01f Fovia Inc
1.CN-01g Frederic H. Martini
1.CN-01h Timof/Shutterstock
1.CC-01 VStock/Alamy Stock Photo
1.CN-01i Zzcapture/Shutterstock

Chapter 2

Chapter Opener Stocktrek Images/Getty Images
2.CC-01 Bogdanhoda/Shutterstock
2.5b Don W. Fawcett/Science Source
2.5c Dr. Torsten Wittmann /Science Source
2.6 Don W. Fawcett/Science Source
2.CN-01a, b Ruth Anne O'Keefe
2.7 Frederic H. Martini
2.8 Bill Longcore/Science Source
2.9a,c Don W. Fawcett/Science Source
2.11 Don W. Fawcett/Science Source
2.12 Biophoto Associates/Science Source
2-13 Dr. Birgit H. Satir
2.17a, b Ed Reschke
2.17c, d, e Ed Reschke/Photolibrary/Getty Images
2.17g Centers for Disease Control and Prevention

Chapter 3

Chapter Opener Pasieka/Science Photo Library/Getty Images
3.1a Frederic H. Martini
3.1b Biophoto Associates/Science Source
3.1c Robert B. Tallitsch/Pearson Education, Inc.
3.1d Pearson Education, Inc.
3.2 P.M. Motta/Univ. La Sapienza, Rome/Science Source
3.4a, b Robert B. Tallitsch
3.5a, b Robert B. Tallitsch
3.6a Frederic H. Martini
3.6b Robert B.Tallitsch
3.7a, b, c Robert B. Tallitsch
3.8a, b Robert B. Tallitschc
3.8c Pearson Education, Inc.
3.8d William C. Ober
3.10 H. Jastrow from Dr. Jastrow's electron microscopic atlas on the www, http://www.drjastrow.de
3.12 Ward's Natural Science Establishment
3CN.1 LIU JIN/AFP/Stringer/Getty Images
3.13a, b Robert B. Tallitsch
3.14a, b Robert B.Tallitsch
3.14c Ward's Natural Science Establishment Robert B.Tallitsch
3.15a, b Robert B. Tallitsch
3.15c Frederic H. Martini
3.17 Robert B. Tallitsch
3.18b Biophoto Associates/Science Source
3.18c Robert B. Tallitsch
3.18d William C. Ober
3.19 Robert B. Tallitsch
3.21 William C. Ober
3.22a, b, c Robert B. Tallitsch
3.23 Robert B. Tallitsch
3.EOC-01a, b, c, d, e Robert B. Tallitsch
3.CC-01 Joeff Davis

Chapter 4

Chapter Opener Antoine Arraou/PhotoAlto/AGE Fotostock
4.1a, b, c Robert B.Tallitsch
4.3 Robert B. Tallitsch
4.4b, c Robert Tallitsch/Pearson Education
4.CN-01 Zoonar/Christine Lan/AGE Fotostock
4.CN-02 Tina Lorien/E+/Getty Images
4.5 Clouds Hill Imaging/Last Refuge Ltd.
4.6a Robert Tallitsch/Pearson Education
4.7a Eye of Science/Science Source
4.7b Steve Gschmeissner/Science Source
4.7c Science Photo Library/Science Source
4.CN-03a Biophoto Associates/Science Source
4.9b Kent Wood/Science Source
4.10c Michael Abbey/Science Source
4.CN-04 Family Business/Shutterstock
4.13 Frederic H. Martini
4.14a, b Frederic H. Martini
4.CN-05 William C. Ober
4.16 William C. Ober
4.CC-01 M.A. Ansary/Science Source

Chapter 5

Chapter Opener George Mattei/Science Source
5.1b Andrew Syred/Science Source
5.1c, d Robert Tallitsch/Pearson Education
5.3a1, a2 Ralph T. Hutchings
5.3b, c Ralph T. Hutchings
5.4 Frederic Martini
5.5a, b Frederic Martini
5.7 Frederic Martini
5.8a Matthew Borkoski Photography/Photolibrary/Getty Images
5.8b SMC Images/Oxford Scientific/Getty Images
5.CN-02a Alliance/Shutterstock
5.CN-02b Dr. Kathleen Welch
5.CN-02c James Cavallini/BSIP/AGE Fotostock
5.CN-02d Zephyr/Science Source
5.CN-02e Dr. Kathleen Welch
5.CN-02f Dan Zika
5.CN-02g Mark Aiken
5.CN-02h Frederic Martini
5.CN-02i Scott Camazine/Science Source
5.CN-02j Neil Borden/Science Source
5.CN-02k Kameel/Fotolia
5.12a, b, c, d Ralph T. Hutchings
5.12e Ralph T. Hutchings
5.CN-03a William C. Ober
5.CN-03b SPL/Science Source
5.CN-01a Bettmann/Getty Images
5.CN-01b Rick Wilking/Reuters
5.CN-01c John Radcliffe Hospital/Science Source
5.CN-01d Dr. LR/Science Source
5.CN-03c Lisa F. Young/Shutterstock
5.CN-03d Prof. P. Motta/Science Source
5.CN-03e David Effron, M.D.
5.CC-01 Southern Illinois University/Science Source
5.EOC-01 Ralph T. Hutchings

Chapter 6

Chapter Opener Sciepro/Science Photo Library/Getty Images
6.CC-01 Bob Harmeyer/Archive Photos/Getty Images
6.1b-top, bottom Ralph T. Hutchings
6.2a Ralph T. Hutchings
6.3a, b, c, d, e Ralph T. Hutchings
6.4 Ralph T. Hutchings
6.5 Ralph T. Hutchings
6.6a, b, c Ralph T. Hutchings
6.7a, b, c Ralph T. Hutchings
6.8a, b, c, d Ralph T. Hutchings
6.9a, b Ralph T. Hutchings
6.10a, b, c Ralph T. Hutchings
6.11b Karen Krabbenhoft/Pearson Education, Inc.
6.12a, b Ralph T. Hutchings
6.13b, c Ralph T. Hutchings
6.14a, b Ralph T. Hutchings
6.15 Ralph T. Hutchings
6.16a William C. Ober
6.16c, e Ralph T. Hutchings
6.17a Pearson Education, Inc.
6.17b Ralph T. Hutchings
6.18b Michael J. Timmons
6.18d Ralph T. Hutchings
6.19a Pearson Education, Inc.
6.19b Ralph T. Hutchings
6.19c Sovereign, ISM/Science Source
6.CN-01 Caters News Agency
6.CN-02 Phartisan/Dreamstime
6.CN-03 Princess Margaret Rose Orthopaedic Hospital/Science Source
6.21a, b, c Ralph T. Hutchings
6.22a, b, c, d, e Ralph T. Hutchings
6.CN-04 Biophoto Associates/Science Source
6.23a, b, c, d Ralph T. Hutchings
6.24a, b Ralph T. Hutchings
6.25a, b, c Ralph T. Hutchings
6.CN-05 Shaun Best/Thomson Reuters (Markets) LLC
6.26a, b, c, d Ralph T. Hutchings
6.CC-01 George Tiedemann/Getty Images
6.CN-06 Hong xia/Shutterstock

Chapter 7

Chapter Opener Sciepro/Science Photo Library/Getty Images
7.CC-01 Du Cane Medical Imaging Ltd./Science Source
7.CN-02 Ruth Anne O'Keefe
7.1a, b Ralph T. Hutchings
7.2a Ralph T. Hutchings
7.2b Digital Vision/Photodisc/Getty Images
7.3a, b Ralph T. Hutchings
7.4 Ralph T. Hutchings

7.5d, e, f Ralph T. Hutchings
7.6a, b, c, d Ralph T. Hutchings
7.7a, b, c, d, e, f Ralph T. Hutchings
7.8a, b, c Ralph T. Hutchings
7.CN-01 Ruth Anne O'Keefe
7.9a Ralph T. Hutchings
7.9b Sovereign, ISM/Science Source
7.10a, b Ralph T. Hutchings
7.11a, b Ralph T. Hutchings
7.13a, b, c, d, e, f Ralph T. Hutchings
7.14a, b Ralph T. Hutchings
7.15a, b, c, d Ralph T. Hutchings
7.16a, b, c, d Ralph T. Hutchings
7.EOC-02 Ralph T. Hutchings

Chapter 8

Chapter Opener Sebastian Kaulitzki/Science Photo Library/Sciepro/Getty Images
8.CN-01 Ruth Anne O'Keefe
8.CN-04 Ruth Anne O'Keefe
8.CN-05 Anagoria-Stuttgart-Landesmuseum Württemberg
8.CN-06 Living Art Enterprises/Science Source
8.CN-09 Zephyr/Science Source
8.CN-10 Praisaeng/Shutterstock
CO-08 Ugreen/Fotolia
8.3a-left, right Ralph T. Hutchings
8.3b-left, right Ralph T. Hutchings
8.3c-left, right Ralph T. Hutchings
8.3d Ralph T. Hutchings
8.4-left, middle, right Ralph T. Hutchings
8.5a, b, c, d, e, f Ralph T. Hutchings
8.CN-02 Ralph T. Hutchings
8.CN-07 James Cavallini/BSIP/AGE Fotostock
8.8 Ralph T. Hutchings
8.9d Ralph T. Hutchings
8.10b Ralph T. Hutchings
8.10c Frederic H. Martini
8.10d Ralph T. Hutchings
8.12d Michael J. Timmons
8.14b Frederic Martini
8.14c Ralph T. Hutchings
8.CN-08 CNRI/Science Source
8.CN-08, left Yellow Dog Productions/Stockbyte/Getty Images
8.15c Antoine Rosset/Science Source
8.15d Susan Leavines/Science Source
8.16b, c Karen Krabbenhoft/Pearson Education
8.17a Ralph T. Hutchings
8.17b Simon Fraser/RVI, Newcastle Upon Tyne/Science Source
8.18b, e Ralph T. Hutchings
8.18e Ralph T. Hutchings

Chapter 9

Chapter Opener Sciepro/Science Photo Library/Getty Images
9.CC-01 Rupak De Chowdhuri/Thomson Reuters (Markets) LLC
9.1 Pearson Education, Inc.
9.2a Ed Reschke/Oxford Scientific/Getty Images
9.2a, b Don W. Fawcett/Science Source
9.7 Graham Heywood/E+/Getty Images
9.11a, top, bottom Frederic H. Martini
9.11b Frederic H. Martini
9.CN-01 Philip Date/Shutterstock

Chapter 10

Chapter Opener Sebastian Kaulitzki/Science Photo Library/Getty Images
10.4 Karen Krabbenhoft/Pearson Education, Inc.
10.8a-1, 2 William C. Ober
10.9a William C. Ober
10.11a Sciencefoto.De-Dr. Andre Kempe/Stockbyte/Getty Images
10.11d Ralph T. Hutchings
10.12 Karen Krabbenhoft/Pearson Education, Inc.
10.CC-01 Dr P. Marazzi/Science Source

Chapter 11

Chapter Opener Sebastian Kaulitzki/Science Photo Library/Getty Images
11.2 Ralph T. Hutchings
11.4 Anita Impagliazzo/Pearson Education, Inc.
11.5 Anita Impagliazzo/Pearson Education, Inc.
11.6 Anita Impagliazzo/Pearson Education, Inc.
11.7 Anita Impagliazzo/Pearson Education, Inc.
11.8a, b William C. Ober
11.9 William C. Ober
11.10 Pearson Education, Inc.
11.11a-f William C. Ober
11.13b, c Ralph T. Hutchings
11.15a Karen Krabbenhoft/Pearson Education, Inc.
11.15d Ralph T. Hutchings
11.17a, b Ralph T. Hutchings
11.18a JRP Studio/Shutterstock
11.19a Pearson Education, Inc.
11.19a Pixologicstudio/Science Photo Library/Getty Images
11.20a, b Ralph T. Hutchings
11.20b Ralph T. Hutchings
11.20c Karen Krabbenhoft/Pearson Education, Inc.
11.22b Ralph T. Hutchings
11.23c Ralph T. Hutchings
11.24c Ralph T. Hutchings
11.26a Ralph T. Hutchings
11.CC-01 Ruth Anne O'Keefe
11.CN-01 Christian MartA-nez Kempin/E+/Getty Images
11.CN-03 Ruth Anne O'Keefe

Chapter 12

Chapter Opener Dyod/AGE Fotostock
12.1a, b, c Pearson Education, Inc.
12.2a, b Pearson Education, Inc.
12.3a, b Pearson Education, Inc.
12.4a, b Pearson Education, Inc.
12.5 Pearson Education, Inc.
12.CC-01 B. Boissonnet/BSIP/AGE Fotostock
12.6a, b, c Pearson Education, Inc.
12.7a, b, c, d Pearson Education, Inc.
12.8 left William C. Ober
12.8 right National Library of Medicine, Visible Human Project
12.9 left William C. Ober
12.9 right National Library of Medicine, Visible Human Project
12.10 left Pearson Education, Inc.
12.10 right National Library of Medicine, Visible Human Project
12.11 left Pearson Education, Inc.
12.11 right National Library of Medicine, Visible Human Project
12.12 left Pearson Education, Inc.
12.12 right National Library of Medicine, Visible Human Project
12-13 left Pearson Education, Inc.
12.13 right National Library of Medicine, Visible Human Project
12.14 left Pearson Education, Inc.
12.14 right National Library of Medicine, Visible Human Project

Chapter 13

Chapter Opener SCIEPRO/Science Photo Library/Getty Images
13.CC-01 Ruth Anne O'Keefe.
13.1 William C. Ober
13.3 William C. Ober
13.7a Robert B.Tallitsch
13.7b Biophoto Associates/Science Source
13.8 Robert B.Tallitsch
13.9 Biophoto Associates/Science Source
13.9 Don W. Fawcett/Science Source
13.10 PH College/Pearson Education, Inc.

Chapter 14

Chapter Opener Sebastian Kaulitzki/ Science Photo Library/Getty Images
14.CN-01 Ruth Anne O'Keefe
14-CN-02 Biophoto Associates/Science Source
14-CN-03 Ralph T. Hutchings
14.CC-01 Rich Carey/Shutterstock
14.2c Ralph T. Hutchings
14.2d Frederic H. Martini
14.3 Michael J. Timmons
14.4 Steve Gschmeissner/Science Source
14.7 Pearson Education, Inc.
14.8 Pearson Education, Inc.
14.9a, b William C. Ober
14.9c Ralph T. Hutchings
14.11d, e Karen Krabbenhoft/Pearson Education, Inc.

Chapter 15

15.CC-01 Dennis Van Tine/Future Image/WENN/Newscom

Chapter 16

Chapter Opener SCIEPRO/Science Photo Library/Getty Images
16.CC-01 Dr. Jill Bolte Taylor
16.CN-01 Ruth Anne O'Keefe
16.CN-04 Ueslei Marcelino/Thomson Reuters (Markets) LLC
16.CN-06 Józefowicz, M.D., Ralph F.
16.CN-03 Scott Camazine/Alamy Stock Photo
16.2b Ralph T. Hutchings
16.CN-01 Living Art Enterprises, LLC/Science Source
16.4 Ralph T. Hutchings
16.10a,c Pearson Education, Inc.
16.12 Ralph T. Hutchings
16.13 inset William C. Ober
16.13a Ralph T. Hutchings
16.13b Pearson Education, Inc.
16.14b Ralph T. Hutchings
16.14d Pearson Education, Inc.
16.15a Ralph T. Hutchings
16.15b-1 Ward's Natural Science Establishment
16.15b-2 Ralph T. Hutchings
16.16a, b Ralph T. Hutchings
16.16c Pearson Education, Inc.
16.CN-03 Southern Illinois University/Science Source
16.17 Ralph T. Hutchings
16.22a,c Pearson Education, Inc.
16.29 Pearson Education, Inc.
16.30 William C. Ober
16.31 William C. Ober
16.EOC-01 Ralph T. Hutchings

Chapter 17

Chapter Opener Sebastian Kaulitzki/Science Photo Library/Getty Images
17.CN-01 Mediscan/Alamy Stock Photo
17.1a Perkmeup/iStock/Getty Images
17.1b Nicolas hansen/E+/Getty Images
17.5a Anita Impagliazzo/Pearson Education, Inc.
17.5b Ward's Natural Science Establishment
17.CC-01 Jochem Wijnands/Horizons WWP/TRVL/Alamy
17.CC-02 Dr P. Marazzi/Science Source
17.9a, b Anita Impagliazzo/Pearson Education, Inc.

Credits

Chapter 18

Chapter Opener Tetra Images/Brand X Pictures/Getty Images
18.CC-01 Alexander Raths/Shutterstock
18.CN-01a Steve Snowden/Shutterstock
18.CN-01c Biophoto Associates/Science Source
18.CN-01d Marcio Jose Bastos Silva/Shutterstock
18.3d, f Frederic H. Martini
18.5 Frederic H. Martini
18.7c top, bottom Robert B.Tallitsch/Pearson Education, Inc.
18.10c Ralph T. Hutchings
18.10d Lennart Nilsson/TT News Agency
18.CN-01b Burger/Phanie/Alamy Stock Photo
18.15b Susumu Nishinaga/Science Source
18.15c left, right Anita Impagliazzo/Pearson Education, Inc.
18.17c Michael J.Timmons
18.17e Ward's Natural Science Establishment
18.17f P.M. Motta/Univ. La Sapienza, Rome/Science Source
18.19a, c Ralph T. Hutchings
18.19b Ralph T. Hutchings
18.20 Ralph T. Hutchings
18.21d Michael J.Timmons
18.21f Ralph T. Hutchings
18.23a Ed Reschke/Photolibrary/Getty Images
18.23c Thalerngsak Mongkolsin/Shutterstock
18.25 Ralph T. Hutchings

Chapter 19

Chapter Opener Science Photo Library-Steve Gschmeissner./Brand X Pictures/Getty Images
19.2 Robert B.Tallitsch
19.CC-01 Shannon Hibbs
19.CN-01 Kevin Landwer Johan/Alamy
19.CN-02 Elsevier Inc.-Health Sciences Division
19.CN-03 Mediscan/Alamy Stock Photo
19.CN-04 BSIP/Science Source
19.CN-05 Biophoto Associates/Science Source
19.4a William C. Ober
19.4b, c Robert B. Tallitsch/Pearson Education, Inc.
19.6b Frederic H. Martini
19.6c Robert B. Tallitsch/Pearson Education, Inc.
19.7a Anita Impagliazzo/Pearson Education, Inc.
19.7c Ward's Natural Science Establishment, Inc
19.8a Pearson Education, Inc.
19.8b Robert B. Tallitsch/Pearson Education, Inc.
19.8c-01, 02 Michael S. Ballot, Prof. Emeritus, ESB, University of the Pacific.
19.9 Robert B. Tallitsch/Pearson Education, Inc.

Chapter 20

Chapter Opener Sebastian Kaulitzki/Science Photo Library/Getty Images
20.CC-01b Jose Luis Calvo/Shutterstock
20.1 Shevelev Vladimir/Shutterstock
20.CN-01 Comstock/Stockbyte/Getty Images
20.2a Frederic H. Martini
20.2b Cheryl Power/Science Source
20.2d Ed Reschke
20.CN-03 Stockbyte/Getty Images
20.CN-04 Keith Morris/Alamy Stock Photo
20.5-01 Robert B. Tallitsch/Pearson Education, Inc.
20.5-02 Robert B. Tallitsch/Pearson Education, Inc.
20.5-03 Robert B. Tallitsch/Pearson Education, Inc.
20.5-04 Robert B.Tallitsch/Pearson Education, Inc.
20.5-05 Robert B. Tallitsch/Pearson Education, Inc.
20.6 Frederic H. Martini
20.7 Susumu Nishinaga/Science Source
20.CC-01a Jarun011/Fotolia
20.CN-02 Eye of Science/Science Source

Chapter 21

Chapter Opener PASIEKA/Science Photo Library/Getty Images
21.CC-01b Tara C Gangadhar, Elisabeth Von der Lohe, Stephen G Sawada and Paul R Helft
21.CN-01 SPL/Science Source
21.CN-02 Zephyr/Science Source
21.CN-03 Larry Mulvehill/Science Source
21.CN-04 Southern Illinois University/Science Source
21.CN-05 SPL/Science Source
21.2a Anita Impagliazzo/Pearson Education, Inc.
21.2d Ralph T. Hutchings
21.3c Robert B Tallitsch
21.4 William C. Ober
21.6a, b Karen Krabbenhoft/Pearson Education, Inc.
21.7a Philippe Plailly/Look at Sciences/Science Source
21.8a, b Karen Krabbenhoft/Pearson Education, Inc.
21.9a Biophoto Associates/Science Source
21.9b SPL/Science Source
21.10c Ralph T. Hutchings
21.10d Courtesy of TeraRecon, Inc.
21.CC-01a John A. Rizzo/Stockbyte/Getty Images
21.11 Sebastian Kaulitzki/Shutterstock

Chapter 22

22.CC-01 Living Art Enterprises, LLC/Science Source
22.CN-01 Richard Kirby, David Spears Ltd./Science Source
22.CN-02 Ed Reschke/Photolibrary/Getty Images
22.1 Biophoto Associates/Science Source
22.2d Don W. Fawcett/Science Source
22.2e Steve Gschmeissner/Science Source
22.3b Biophoto Associates/Science Source
22.7a Anita Impagliazzo/Pearson Education, Inc.
22.7b Gondelon/Science Source
22.9 Frederic H. Martini
22.10b, c Ralph T. Hutchings
22.11 Karen Krabbenhoft/Pearson Education, Inc.
22.11 Anita Impagliazzo/Pearson Education, Inc.
22.12 Courtesy of TeraRecon, Inc.
22.13b, c Ralph T. Hutchings
22.14 Anita Impagliazzo/Pearson Education, Inc.
22.15 Courtesy of TeraRecon, Inc.
22.16 Ralph T. Hutchings
22.20 Anita Impagliazzo/Pearson Education, Inc.
22.CN-04 Dmitry Meinkov/Shutterstock

Chapter 23

Chapter Opener Sciepro/Science Photo Library/Getty Images
23.3 Frederic H. Martini
23.4 Karen Krabbenhoft/Pearson Education, Inc.
23.5 Anita Impagliazzo/Pearson Education, Inc.
23.7a Robert B. Tallitsch
23.7c Biophoto Associates/Science Source
23.8 Ralph T. Hutchings
23.9 William C. Ober
23.11b Ralph T. Hutchings
23.11c Frederic H. Martini
23.12 Anita Impagliazzo/Pearson Education, Inc.
23.13a Anita Impagliazzo/Pearson Education, Inc.
23.13c, d Robert B. Tallitsch
23.14a Anita Impagliazzo/Pearson Education, Inc.
23.14c Robert B. Tallitsch
23.CC-01 Frederick A. Murphy/Centers for Disease Control and Prevention

Chapter 24

24.1 Anita Impagliazzo/Pearson Education, Inc.
24.2a Anita Impagliazzo/Pearson Education, Inc.
24.2b Frederic H. Martini
24.2c Photo Insolite Realite/Science Source
24.3 William C. Ober
24.4a, b William C. Ober
24.4b Ralph T. Hutchings
24.5 inset William C. Ober
24.5 Ralph T. Hutchings
24.7 CNRI/Science Source
24.9 Lester V. Bergman/Corbis Documentary/Getty Images
24.10a left Anita Impagliazzo/Pearson Education, Inc.
24.10a right Karen Krabbenhoft
24.12b, left, right Ralph T. Hutchings
24.12c Ralph T. Hutchings
24.12d Ralph T. Hutchings
24.13b, c Robert B. Tallitsch
24.15 left Anita Impagliazzo/Pearson Education, Inc.
24.15 Ralph T. Hutchings
24.EOC-10 Anita Impagliazzo/Pearson Education, Inc.
24.CC-01 Sunnysky69/Fotolia

Chapter 25

Chapter Opener Science Photo Library-SCIEPRO/Brand X Pictures/Getty Images
25.1 Anita Impagliazzo/Pearson Education, Inc.
25.2 Robert B. Tallitsch
25.4c Anita Impagliazzo/Pearson Education, Inc.
25.6a Anita Impagliazzo/Pearson Education, Inc.
25.6b-01, 02, 03 Robert B. Tallitsch
25.7c Ralph T. Hutchings
25.9 left Anita Impagliazzo/Pearson Education, Inc.
25.9a Alfred Pasieka/Photolibrary/Getty Images
25.9b Robert B. Tallitsch
25.10 Karen Krabbenhoft/Pearson Education, Inc.
25.10a Ralph T. Hutchings
25.11 Anita Impagliazzo/Pearson Education, Inc.
25.12b SPL/Science Source
25.12d, e, f Robert B. Tallitsch
25.14d Robert B. Tallitsch
25.14e-left, right Robert B. Tallitsch
25.15a, b, c Karen Krabbenhoft
25.16a Anita Impagliazzo/Pearson Education, Inc.
25.16b Karen Krabbenhoft
25.17 Frederic H. Martini
25.18 Robert B. Tallitsch
25.19a Anita Impagliazzo/Pearson Education, Inc.
25.19b, e Ralph T. Hutchings
25.20c, d Robert B. Tallitsch
25.21c Frederic H. Martini
25.22 Robert B. Tallitsch
25.23c Frederic H. Martini
25.CN-02 David M. Martin/Science Source
25.EOC-01 Anita Impagliazzo/Pearson Education, Inc.

Chapter 26

Chapter Opener Sebastian Kaulitzki/Science Photo Library/Getty Images
26.2 Ralph T. Hutchings
26.3 Ralph T. Hutchings
26.4a Ralph T. Hutchings
26.4c Philippe Garo/Science Source
26.6a Ralph T. Hutchings
26.6b Ralph T. Hutchings
26.8b, d Robert B. Tallitsch
26.8c Frederic H. Martini
26.9-01 Susumu Nishinaga/Science Source
26.9-02 Steve Gschmeissner/Science Source
26.10a Alain Pol, ISM/Science Source
26.10b Frederic H. Martini
26.10c CNRI/Science Source
26.11d Ralph T. Hutchings
26.12a, c Robert B. Tallitsch
26.12b Frederic H. Martini
26.CC-01 Ernesto R. Ageitos/Moment Open/Getty Images

Chapter 27

Chapter Opener SCIEPRO/Science Photo Library/Getty Images
27.1 Ralph T. Hutchings
27.4 Frederic H. Martini
27.5a, c Robert B. Tallitsch
27.6 Eye of Science/Science Source
27.7b Ralph T. Hutchings
27.7c, d Frederic H. Martini
27.8b Robert B. Tallitsch
27.8c, d, e Frederic H. Martini
27.9b Ed Reschke/Photolibrary/Getty Images
27.9d Ralph T. Hutchings
27.10 Ralph T. Hutchings
27.11c Karen Krabbenhoft
27.12-01 Frederic H. Martini
27.12-02 Frederic H. Martini
27.12-03 Frederic H. Martini
27.12-04 Frederic H. Martini
27.12-05 C. Edelmann/Science Source
27.12-06 BSIP SA/Alamy Stock Photo
27.12-07 Biophoto Associates/Science Source
27.14b Frederic H. Martini
27.14c Steve Gschmeissner/Science Source
27.15b Frederic H. Martini
27.16c Robert B. Tallitsch
27.17a, b Frederic H. Martini
27.17c Michael Timmons
27.17d Frederic H. Martini
27.19 Robert B. Tallitsch
27.21b Ralph T. Hutchings
27.21c, d Robert B. Tallitsch
27.CC-01 Brian Evans/Science Source

Chapter 28

Chapter Opener Science Picture Co/Collection Mix/Getty Images
28.1a Francis Leroy/Science Source
28.6b Frederic H. Martini
28.7b Biophoto Associates/Science Source
28.7c Pr. M. Forest, CNRI/Science Source
28.7d Claude Edelmann/Science Source
28.8a Nestle/Petit Format/Science Source
28.8b Dr. Najeeb Layyous/Science Source
28.CC-01 Barcroft Media/Getty Images

VW-01 Artjazz/Shutterstock
VW-02 Designsstock/Shutterstock
VW-03 Nik Merkulov/Shutterstock

Index

Index

Index

C

Index

D

Index

E

Index

F

G

Index

H

Index

Index

M

N

O

P

Index

Q

R

S

Index

T

Index

U

V

W

X

Y

Z